CONCEPTS in BIOLOGY

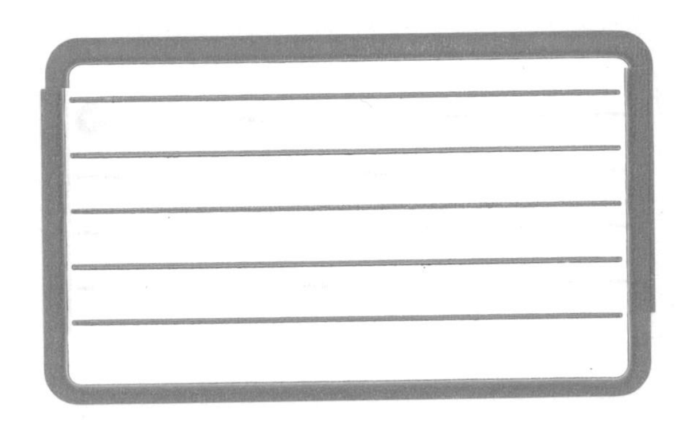

"A personal library is a lifelong source of enrichment and distinction. Consider this book an investment in your future and add it to your personal library."

CONCEPTS in BIOLOGY

seventh edition

Eldon D. Enger

J. Richard Kormelink

Frederick C. Ross

Rodney J. Smith

Delta College
University Center, Michigan

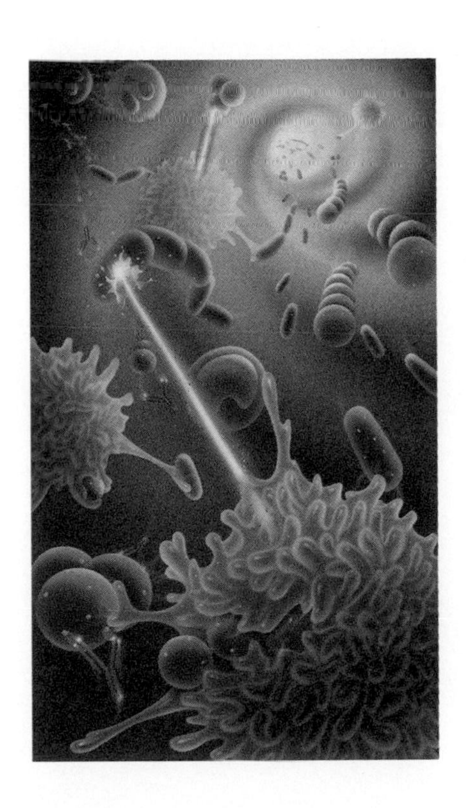

WCB **Wm. C. Brown Publishers**
Dubuque, Iowa • Melbourn, Australia • Oxford, England

Book Team

Developmental Editor *Margaret J. Kemp*
Production Editor *Catherine S. Di Pasquale*
Designer *Jeff Storm*
Art Editor *Rachel Imsland*
Photo Editor *Lori Gockel*
Permissions Coordinator *Gail Wheatley*
Art Processor *Brenda A. Ernzen*
Visuals/Design Developmental Consultant *Donna Slade*

Wm. C. Brown Publishers
A Division of Wm. C. Brown Communications, Inc.

Vice President and General Manager *Beverly Kolz*
Vice President, Publisher *Kevin Kane*
Vice President, Publisher *Earl McPeek*
Vice President, Director of Sales and Marketing *Virginia S. Moffat*
Marketing Manager *Carol J. Mills*
Advertising Manager *Janelle Keeffer*
Director of Production *Colleen A. Yonda*
Publishing Services Manager *Karen J. Slaght*

Wm. C. Brown Communications, Inc.

President and Chief Executive Officer *G. Franklin Lewis*
Corporate Senior Vice President, President of WCB Manufacturing *Roger Meyer*
Corporate Senior Vice President and Chief Financial Officer *Robert Chesterman*

Cover illustration by Todd Buck

Copyedited by Ann Mirels

The credits section for this book begins on page 442 and is considered an extension of the copyright page.

A Times Mirror Company

Library of Congress Catalog Card Number: 92-75380

ISBN 0-697-13645-0
0-697-13644-2

Printed in the United States of America by Wm. C. Brown Communications, Inc., 2460 Kerper Boulevard, Dubuque, IA 52001

10 9 8 7 6 5 4 3

BRIEF CONTENTS

C O N T E N T S

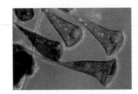

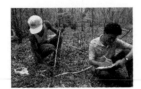

CHAPTER 5

CHAPTER 6

CHAPTER 7

PART III

Cell Division and Heredity

CHAPTER 8

CHAPTER 9

CHAPTER 10

PART V

Physiological Processes

CHAPTER 18

CHAPTER 19

CHAPTER 20

CHAPTER 21

PART VI

The Origin and Classification of Life

CHAPTER 22

CHAPTER 23

CHAPTER 24

CHAPTER 25

CHAPTER 26

Purpose
Sets the "goals" for the chapter and shows how the chapter topics play off the themes of the entire book.

For Your Information
Offers an introduction to practical information that will help the students gain a better perspective of chapter topics to be discussed.

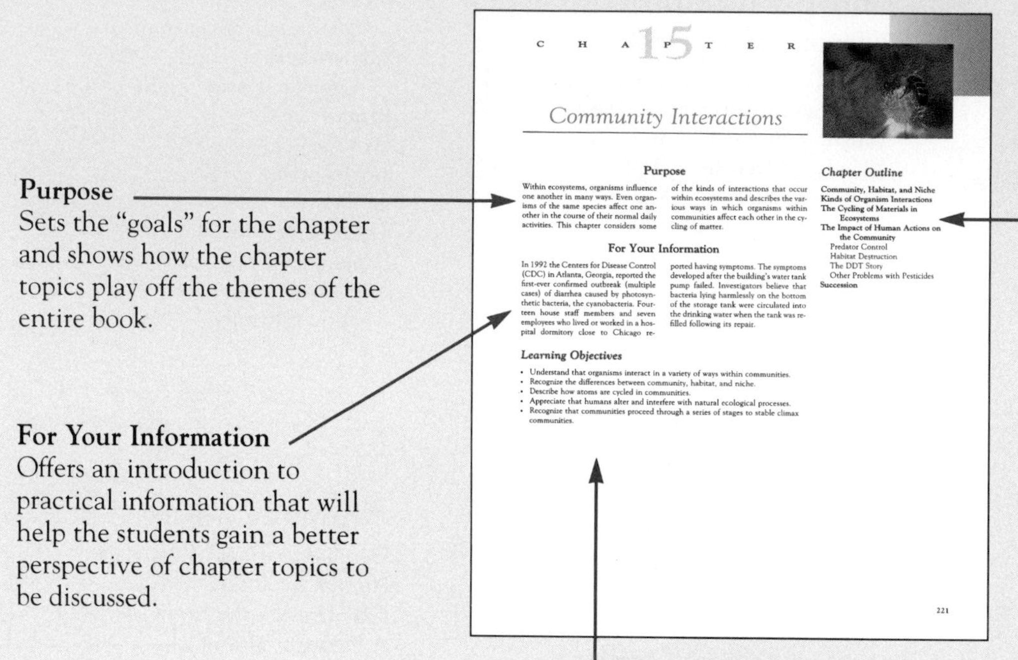

CHAPTER 15

Community Interactions

Purpose

Within ecosystems, organisms influence one another in many ways. Even organisms of the same species affect one another in the course of their normal daily activities. This chapter considers some of the kinds of interactions that occur within ecosystems and describes the various ways in which organisms within communities affect each other in the cycling of matter.

For Your Information

In 1992 the Centers for Disease Control (CDC) in Atlanta, Georgia, reported the first-ever confirmed outbreak (multiple cases) of diarrhea caused by photosynthetic bacteria, the cyanobacteria. Fourteen house staff members and seven employees who lived or worked in a hospital dormitory close to Chicago reported having symptoms. The symptoms developed after the building's water tank pump failed. Investigators believe that bacteria lying harmlessly on the bottom of the storage tank were circulated into the drinking water when the tank was refilled following its repair.

Learning Objectives

- Understand that organisms interact in a variety of ways within communities.
- Recognize the differences between community, habitat, and niche.
- Describe how atoms are cycled in communities.
- Appreciate that humans alter and interfere with natural ecological processes.
- Recognize that communities proceed through a series of stages to stable climax communities.

Chapter Outline

Community, Habitat, and Niche
Kinds of Organism Interactions
The Cycling of Materials in Ecosystems
The Impact of Human Actions on the Community
Predator Control
Habitat Destruction
The DDT Story
Other Problems with Pesticides
Succession

221

Chapter Outline
Appears at the beginning of each chapter as a quick guide to the chapter's organization. Optional readings of more in-depth topics, are delineated by a red arrow, and can be assigned or skipped at the instructor's discretion.

Learning Objectives
Presents a list of objectives that students can use to establish goals for understanding key ideas in the chapter.

Tables
Offer a clear and concise presentation of important data.

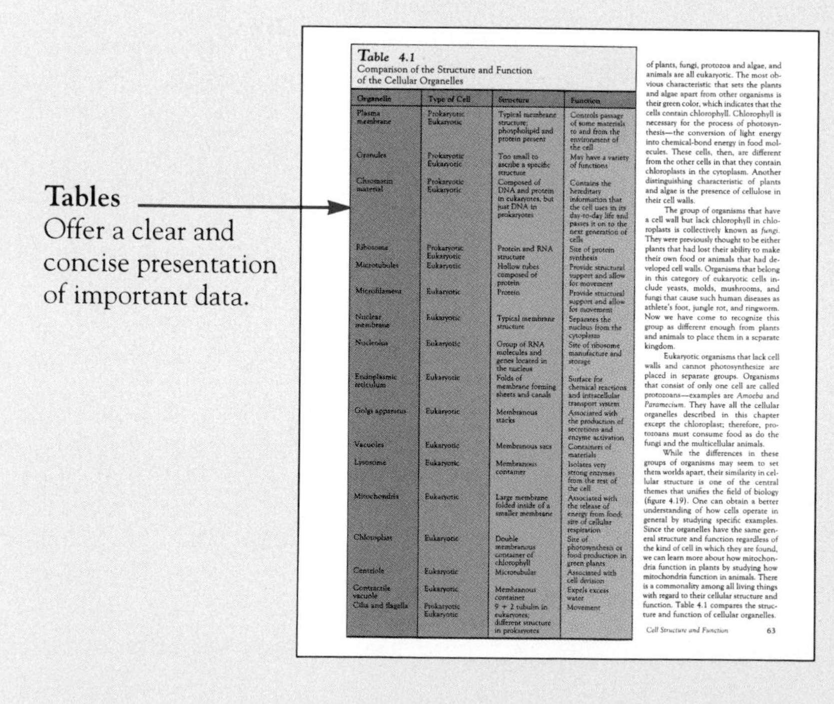

Table 4.1
Comparison of the Structure and Function of the Cellular Organelles

Organelle	Type of Cell	Structure	Function
Plasma membrane	Prokaryotic Eukaryotic	Typical membrane structure; phospholipid and protein present	Controls passage of some materials to and from the environment of the cell
Granules	Prokaryotic Eukaryotic	Too small to ascribe a specific structure	May have a variety of functions
Chromatin material	Prokaryotic Eukaryotic	Composed of DNA and protein in eukaryotes, but just DNA in prokaryotes	Contains the hereditary information that the cell uses in its day-to-day life and passes it on to the next generation of cells
Ribosome	Prokaryotic Eukaryotic	Protein and RNA structure	Site of protein synthesis
Microtubules	Eukaryotic	Hollow tubes composed of protein	Provide structural support and allow for movement
Microfilament	Eukaryotic	Protein	Provide structural support and allow for movement
Nuclear membrane	Eukaryotic	Typical membrane structure	Separates the nucleus from the cytoplasm
Nucleolus	Eukaryotic	Group of RNA molecules and genes located in the nucleus	Site of ribosome manufacture and storage
Endoplasmic reticulum	Eukaryotic	Folds of membrane forming sheets and canals	Surface for chemical reactions and intracellular transport system
Golgi apparatus	Eukaryotic	Membranous stacks	Associated with the production of secretions and enzyme activation
Vacuoles	Eukaryotic	Membranous sacs	Containers of materials
Lysosome	Eukaryotic	Membranous container	Isolates very strong enzymes from the rest of the cell
Mitochondria	Eukaryotic	Large membrane folded inside of a smaller membrane	Associated with the release of energy from food; site of cellular respiration
Chloroplast	Eukaryotic	Double membranous container of chlorophyll	Site of photosynthesis or food production in green plants
Centriole	Eukaryotic	Microtubular	Associated with cell division
Contractile vacuole	Eukaryotic	Membranous container	Expels excess water
Cilia and flagella	Prokaryotic Eukaryotic	9 + 2 tubulin in eukaryotes; different structure in prokaryotes	Movement

of plants, fungi, protozoa and algae, and animals are all eukaryotic. The most obvious characteristic that sets the plants and algae apart from other organisms is their green color, which indicates that the cells contain chlorophyll. Chlorophyll is necessary for the process of photosynthesis—the conversion of light energy into chemical-bond energy in food molecules. These cells, then, are different from the other cells in that they contain chloroplasts in the cytoplasm. Another distinguishing characteristic of plants and algae is the presence of cellulose in their cell walls.

The group of organisms that have a cell wall but lack chlorophyll in chloroplasts is collectively known as *fungi*. They were previously thought to be either plants that had lost their ability to make their own food or animals that had developed cell walls. Organisms that belong in this category of eukaryotic cells include yeasts, molds, mushrooms, and fungi that cause such human diseases as athlete's foot, jungle rot, and ringworm. Now we have come to recognize this group as different enough from plants and animals to place them in a separate kingdom.

Eukaryotic organisms that lack cell walls and cannot photosynthesize are placed in separate groups. Organisms that consist of only one cell are called protozoans—examples are *Amoeba* and *Paramecium*. They have all the cellular organelles described in this chapter except the chloroplast; therefore, protozoans must consume food as do the fungi and the multicellular animals.

While the differences in these groups of organisms may seem to set them worlds apart, their similarity in cellular structure is one of the central themes that unifies the field of biology (figure 4.19). One can obtain a better understanding of how cells operate in general by studying specific examples. Since the organelles have the same general structure and function regardless of the kind of cell in which they are found, we can learn more about how mitochondria function in plants by studying how mitochondria function in animals. There is a commonality among all living things with regard to their cellular structure and function. Table 4.1 compares the structure and function of cellular organelles.

Cell Structure and Function 63

Boxed Readings
Additional, timely topics are presented throughout the text to complement the themes of the chapters.

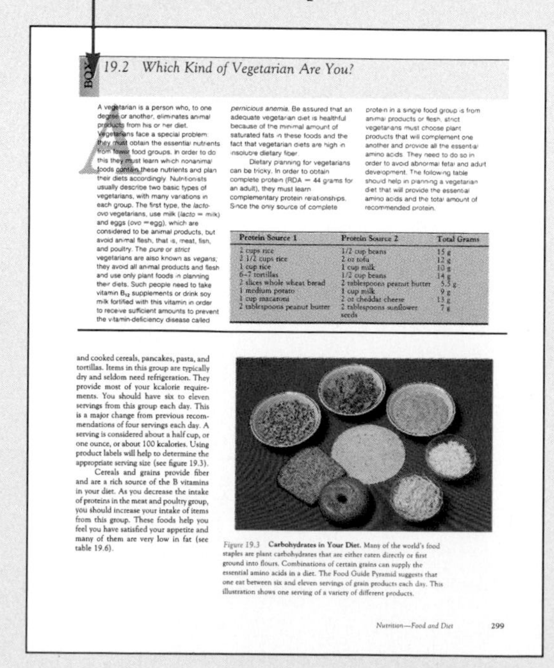

BOX 19.2 *Which Kind of Vegetarian Are You?*

A vegetarian is a person who, to one degree or another, eliminates animal products from his or her diet. Vegetarians face a special problem: they must obtain the essential nutrients from fewer food groups. In order to do this they must learn which nonanimal foods contain these nutrients and plan their diets accordingly. Nutritionists usually describe two basic types of vegetarians, with many variations in each group. The first type, the *lacto-ovo* vegetarians, use milk (*lacto* = milk) and eggs (*ovo* = egg), which are considered to be animal products, but avoid animal flesh, that is, meat, fish, and poultry. The *pure* or *strict* vegetarians are also known as *vegans;* they avoid all animal products and flesh and use only plant foods in planning their diets. Such people need to take vitamin B_{12} supplements or drink soy milk fortified with this vitamin in order to receive sufficient amounts to prevent the vitamin-deficiency disease called *pernicious anemia*. Be assured that an adequate vegetarian diet is healthful because of the minimal amount of saturated fats in these foods and the fact that vegetarian diets are high in insoluble dietary fiber.

Dietary planning for vegetarians can be tricky. In order to obtain complete protein (RDA = 44 grams for an adult), they must learn complementary protein relationships. Since the only source of complete protein in a single food group is from animal products or flesh, strict vegetarians must choose plant products that will complement one another and provide all the essential amino acids. They need to do so in order to avoid abnormal fetal and adult development. The following table should help in planning a vegetarian diet that will provide the essential amino acids and the total amount of recommended protein.

Protein Source 1	Protein Source 2	Total Grams
2 cups rice	1/2 cup beans	15 g
2 1/2 cups rice	2 oz tofu	12 g
1 cup rice	1 cup milk	10 g
6–7 tortillas	1/2 cup beans	14 g
2 slices whole wheat bread	2 tablespoons peanut butter	5.5 g
1 medium potato	1 cup milk	9 g
1 cup macaroni	2 oz cheddar cheese	13 g
2 tablespoons peanut butter	2 tablespoons sunflower seeds	7 g

and cooked cereals, pancakes, pasta, and tortillas. Items in this group are typically dry and seldom need refrigeration. They provide most of your kcalorie requirements. You should have six to eleven servings from this group each day. This is a major change from previous recommendations of four servings each day. A serving is considered about a half cup, or one ounce, or about 100 kcalories. Using product labels will help to determine the appropriate serving size (see figure 19.3).

Cereals and grains provide fiber and are a rich source of the B vitamins in your diet. As you decrease the intake of proteins in the meat and poultry group, you should increase your intake of items from this group. These foods help you feel you have satisfied your appetite and many of them are very low in fat (see table 19.6).

Figure 19.3 **Carbohydrates in Your Diet.** Many of the world's food staples are plant carbohydrates that are either eaten directly or first ground into flours. Combinations of certain grains can supply the essential amino acids in a diet. The Food Guide Pyramid suggests that one eat between six and eleven servings of grain products each day. This illustration shows one serving of a variety of different products.

Nutrition—Food and Diet 299

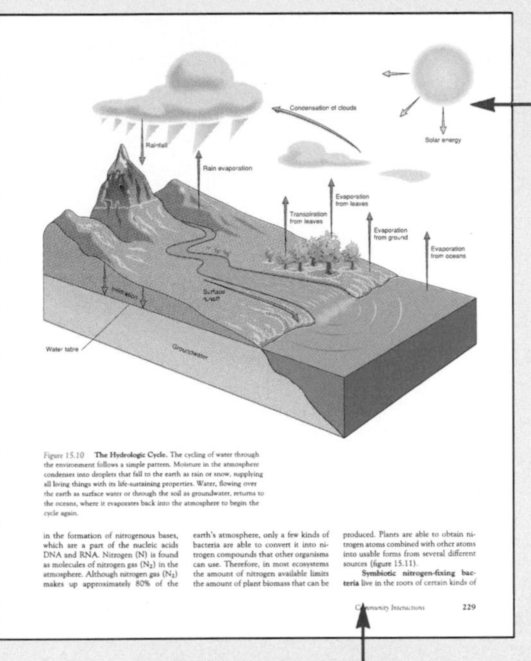

Figure 15.10 **The Hydrologic Cycle.** The cycling of water through the environment follows a simple pattern. Moisture in the atmosphere condenses into droplets that fall to the earth as rain or snow, supplying all living things with its life-sustaining properties. Water, flowing over the earth as surface water or through the soil as groundwater, returns to the oceans, where it evaporates back into the atmosphere to begin the cycle again.

in the formation of nitrogenous bases, which are a part of the nucleic acids DNA and RNA. Nitrogen (N) is found as molecules of nitrogen gas (N_2) in the atmosphere. Although nitrogen gas (N_2) makes up approximately 80% of the earth's atmosphere, only a few kinds of bacteria are able to convert it into nitrogen compounds that other organisms can use. Therefore, in most ecosystems the amount of nitrogen available limits the amount of plant biomass that can be produced. Plants are able to obtain nitrogen atoms combined with other atoms into usable forms from several different sources (figure 15.11).

Symbiotic nitrogen-fixing bacteria live in the roots of certain kinds of

Community Interactions 229

Visuals

Interesting, colorful line art and photos have been specifically created or chosen to enhance key ideas and stimulate learning.

Bold Face Terms

Key terms for each chapter appear in bold face type when first defined in the text.

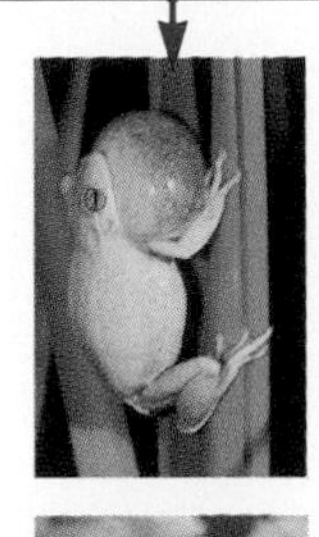

from that of the parent, meiosis would result in gametes that had different chromosome numbers from the original, and successful reproduction with the parent species would be difficult. In one step the polyploid could be isolated reproductively from its original species. However, since most plants can reproduce asexually, they can create an entire population of organisms that have the same polyploid chromosome number and are capable of sexual reproduction among themselves. In effect a new species has been created within a couple of generations. Although it is rare in animals, polyploidy is found in a few groups that typically use asexual reproduction in addition to sexual reproduction.

Some groups of plants, such as the grasses, may have 50% of their species produced as a result of polyploidy. Many economically important species are polyploids. Cotton, potatoes, sugarcane, wheat, and many of our garden flowers are examples (figure 13.7).

The Development of Evolutionary Thought

Although most scientists accept evolutionary processes as central to an understanding of how various life-forms arose and continue to change today, this was not always the case. For centuries people believed that the various species of plants and animals were fixed and unchanging—that is, they were thought to have remained unchanged from the time of their creation. This was a reasonable assumption because people knew nothing about DNA, meiosis, or population genetics. Furthermore, the process of evolution is so slow that people could not see the accumulation of changes we call evolution during their lifetime. It is even difficult for modern scientists to recognize this slow change in many kinds of organisms. In the mid-1700s, George Buffon, a French naturalist, expressed

Figure 13.6 **Animal Communication by Displays.** Most animals have specific behaviors that they use to communicate with others of the same species. The croaking of a male frog is specific to its species and is different from that of males of other species. The visual displays of peacocks and Siamese fighting fish are also used to communicate with others of the same species.

Figure 13.7 **Polyploidy.** Many species of plants have been created by increasing the chromosome number. Many large-flowered varieties of plants have been produced artificially by means of this technique. The smaller flower shown on the left is a plant with the normal diploid number; the larger one in the middle is a tetraploid variety; and on the right is a double-diploid variety.

194 *Chapter 13*

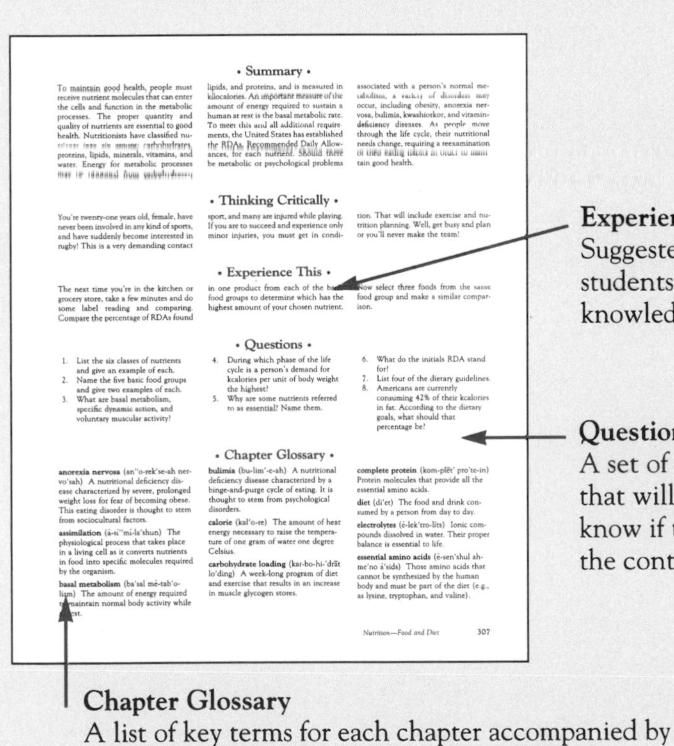

• Summary •

To maintain good health, people must receive nutrient molecules that can enter the cells and function in the metabolic processes. The proper quantity and quality of nutrients are essential to good health. Nutritionists have classified nutrients into six groups: carbohydrates, proteins, lipids, minerals, vitamins, and water. Energy for metabolic processes may be released from carbohydrates, lipids, and proteins, and is measured in kilocalories. An important measure of the amount of energy required to sustain a human at rest is the basal metabolic rate. To meet this and all additional requirements, the United States has established the RDAs, Recommended Daily Allowances, for each nutrient. Should there be metabolic or psychological problems associated with a person's normal metabolism, a variety of disorders may occur, including obesity, anorexia nervosa, bulimia, kwashiorkor, and vitamin-deficiency diseases. As people move through the life cycle, their nutritional needs change, requiring a reexamination of their eating habits in order to maintain good health.

• Thinking Critically •

You're twenty-one years old, female, have never been involved in any kind of sports, and have suddenly become interested in rugby! This is a very demanding contact sport, and many are injured while playing. If you are to succeed and experience only minor injuries, you must get in condition. That will include exercise and nutrition planning. Well, get busy and plan or you'll never make the team!

• Experience This •

The next time you're in the kitchen or grocery store, take a few minutes and do some label reading and comparing. Compare the percentage of RDAs found in one product from each of the basic food groups to determine which has the highest amount of your chosen nutrient. Now select three foods from the same food group and make a similar comparison.

• Questions •

1. List the six classes of nutrients and give an example of each.
2. Name the five basic food groups and give two examples of each.
3. What are basal metabolism, specific dynamic action, and voluntary muscular activity?
4. During which phase of the life cycle is a person's demand for kcalories per unit of body weight the highest?
5. Why are some nutrients referred to as essential? Name them.
6. What do the initials RDA stand for?
7. List four of the dietary guidelines.
8. Americans are currently consuming 42% of their kcalories in fat. According to the dietary goals, what should that percentage be?

• Chapter Glossary •

anorexia nervosa (an"o-rek'se-ah ner-vo'sah) A nutritional deficiency disease characterized by severe, prolonged weight loss for fear of becoming obese. This eating disorder is thought to stem from sociocultural factors.

assimilation (a-si"mi-la'shun) The physiological process that takes place in a living cell as it converts nutrients in food into specific molecules required by the organism.

basal metabolism (ba'sal me-tab'o-lizm) The amount of energy required to maintain normal body activity while at rest.

bulimia (bu-lim'-e-ah) A nutritional deficiency disease characterized by a binge-and-purge cycle of eating. It is thought to stem from psychological disorders.

calorie (kal'o-re) The amount of heat energy necessary to raise the temperature of one gram of water one degree Celsius.

carbohydrate loading (kar-bo-hi-'drāt lo'ding) A week-long program of diet and exercise that results in an increase in muscle glycogen stores.

complete protein (kom-plēt' pro'te-in) Protein molecules that provide all the essential amino acids.

diet (di'et) The food and drink consumed by a person from day to day.

electrolytes (e-lek'tro-līts) Ionic compounds dissolved in water. Their proper balance is essential to life.

essential amino acids (e-sen'shul ah-me'no a'sids) Those amino acids that cannot be synthesized by the human body and must be part of the diet (e.g., as lysine, tryptophan, and valine).

Nutrition—Food and Diet 307

Experience This Topics

Suggested projects that require students to apply their knowledge of the chapter.

Questions

A set of objective questions that will help the student know if they have mastered the contents of the chapter.

Chapter Glossary

A list of key terms for each chapter accompanied by a pronunciation guide and complete definition.

Summary

Key ideas are reviewed at the end of each chapter, to help students prepare for the next chapter or as a review for tests.

• Summary •

Throughout this chapter we have been comparing the functions of the nervous and endocrine systems, the kinds of effects they have, and their characteristics. Table 20.2 summarizes these differences.

A nerve impulse is caused by sodium ions entering the cell as a result of a change in the permeability of the cell membrane. Thus, a wave of depolarization passes down the length of a neuron to the synapse. The axon of a neuron secretes a neurotransmitter, such as acetylcholine, into the synapse, where these molecules bind to the dendrite of the next cell in the chain, resulting in an impulse in it as well. The acetylcholinesterase present in the synapse destroys acetylcholine so that it does not repeatedly stimulate the dendrite.

Several kinds of sensory inputs are possible. Many kinds of chemicals can bind to cell surfaces and be recognized. This is probably how the sense of taste and the sense of smell function. Light energy can be detected because light causes certain molecules in the retina of the eye to decompose and stimulate neurons. Sound can be detected because fluid in the cochlea of the ear is caused to vibrate, and special cells detect this movement and stimulate neurons. The sense of touch consists of a variety of receptors that respond to pressure, cell damage, and temperature.

Table 20.2
Comparison of the Nervous and Endocrine Systems

System	Method of Action	Effects
Nervous	1. Nerve impulse travels along established routes. 2. Neurotransmitters allow impulse to cross synapses. 3. Rapid action.	1. Causes skeletal-muscle contraction. 2. Modifies contraction of smooth and cardiac muscle. 3. Causes gland secretion.
Endocrine	1. Hormones released into bloodstream. 2. Receptors bind hormones to their target organs. 3. Often slow to act.	1. Stimulates smooth-muscle contraction. 2. Stimulates gland secretion. 3. Regulates growth.

Muscles shorten because of the ability of actin and myosin to bind to one another. A portion of the myosin molecule is caused to bend when ATP is used, resulting in the sliding of actin and myosin molecules past each other. Skeletal muscle responds to nervous stimulation to cause movements of the skeleton. Smooth muscle and cardiac muscle have internally generated contractions that may be modified by nervous stimulation or hormones.

Glands are of two types: exocrine glands, which secrete through ducts into the cavity of an organ or to the surface of the skin, and endocrine glands, which release their secretions into the circulatory system. Digestive glands and sweat glands are examples of exocrine glands. Endocrine glands such as the ovaries, testes, and pituitary gland change the activities of cells and often cause responses that result in growth over a period of time. It is becoming clear that the endocrine system and the nervous system are interrelated. Actions of the endocrine system can change how the nervous system functions, and the reverse is also true. Much of this interrelation takes place in the brain–pituitary gland association.

• Thinking Critically •

Humans are considered to have a poor sense of smell. However, when parents are presented with baby clothing, they are able to identify the clothing with which their own infant had been in contact with a high degree of accuracy. Specially trained individuals, such as wine and perfume testers, are able to identify large numbers of different kinds of molecules that the average person cannot identify. Birds rely primarily on sound and sight for information about their environment; they have a poor sense of smell. Most mammals are known to have a very well-developed sense of smell. Is it possible that we have evolved into sound-and-sight-dependent organisms like birds and have lost the keen sense of smell of our mammal ancestors? Or is it that we just don't use our sense of smell to its full potential? Can you devise an experiment that would help to shed light on this issue?

326 *Chapter 20*

Thinking Critically

Questions challenge students to think logically through problems based on chapter concepts, and arrive at their own conclusions.

Purpose

The origin of this book is deeply rooted in our concern for the education of college students in the field of biology. With each new edition, we've worked hard to maintain our original goal of writing a book that is both useful and interesting. But with this seventh edition, we've redoubled our efforts by significantly changing the text's format and design. After carefully surveying numerous adopters of the sixth edition, we decided to change from a one-column to a three-column format. Our reasons for doing this are threefold:

1. *Large, thick books intimidate introductory-level students who often are already anxious about taking science courses.* Further, thick books tend to put off those students who are simply uninterested in science. By converting to a three-column format, the length of the book was significantly reduced without reducing the extent of its coverage at all.
2. *Longer books are usually more expensive.* The reduction in length enabled our publisher to carry out our request to lower the price of this edition.
3. *Longer books expend more natural resources.* Consistent with the printing of the text on recycled paper, this shorter edition is more resource efficient and friendlier to the environment.

With the format and design changes, students will find this edition even easier to understand and enjoy than previous editions.

Organization

Concepts in Biology is arranged in a traditional manner. It begins with a discussion of the meaning, purpose, and future of biology as a scientific endeavor. It then turns to the coverage of biological concepts based on an expanding spiral of knowledge. Thus, chemistry is followed by cell biology, cell division, genetics, ecology, evolution, anatomy and physiology, and the diversity and classification of living things. The book progresses from the basic to the complex.

New to This Edition

As always, we greatly appreciate the suggestions of users of the text and reviewers of the current edition. We especially value the responses we received from our letter to adopters (over 50 replies). We carefully examined those features of the text that received criticism and responded by making appropriate changes.

The order of material has been changed in the seventh edition. Former Part 4, "Physiological Processes," and former Part 5, "Evolution and Ecology," have been interchanged. This allows for a better flow of content from population genetics to variation and selection.

The text material has been significantly rewritten to ensure a flow of ideas that will best enable students to link information in a logical way. "For Your Information" sections and the feature boxes and summaries were substantially revised to include topical information of interest to students. We also wrote several new boxes that will pique student interest. A new feature entitled "Thinking Critically" allows students to use material from the chapter to analyze information and make interpretations from their personal knowledge and experience.

The illustrations are a vital and integrated part of the text. They have been carefully chosen to clarify textual material and provide new insights. Fully one-quarter of all the illustrations have been revised. Captions have been critically examined for accuracy and appropriateness.

Chapter 1 has been substantially rewritten to include a new section that deals with fundamental attitudes in science, including discussion of the concept of pseudoscience.

Chapter 2 on introductory chemistry includes *both* the historically important Bohr atom and the quantum mechanics approach to the structure of the atom. The new organization allows instructors to select portions of the chapter that best suit their approaches to this material and meet the needs of their students.

Chapter 5 on enzymes has been completely rewritten to better explain the nature of chemical reactions and catalysis. A new "adjustable wrench" analogy has been employed to enhance students' understanding of how enzymes are thought to function, and a new section on cellular control process has been added. Many new illustrations have been included.

Chapter 6 has been organized to allow instructors to select portions of the chapter to provide either broad coverage or more detailed analysis of the specific biochemical pathways for their students.

Chapter 7, "DNA-RNA: The Molecular Basis of Heredity," has many revised illustrations to incorporate new information and to create a better picture of how these molecules function. The biotechnology applications and implications of DNA-RNA function have been strengthened by adding sections on DNA fingerprinting and the polymerase chain reaction.

Chapter 12, "Variation and Selection," has been significantly restructured to provide a better flow of ideas. It is structured around the role of natural selection in evolution, strengthening this important concept.

Chapter 18 on materials exchange in the body now includes material on the immune system.

Chapter 19 incorporates the new federal guidelines relating to Recommended Daily Allowances to provide a balanced diet.

Aids to the Reader

The text has a number of features intended to involve students in an active learning process. These features help them relate general concepts of biology to daily living.

Six Parts divide the chapters into sections of closely related material. Each part begins with an overview of the material contained in that section.

Each chapter contains these elements:

Chapter Outline As part of the chapter opening, the outline presents the material to be covered in that chapter.

Purpose This statement explains the value of each chapter to the understanding of a complete biology course.

For Your Information This introductory item of interest consists of timely information related to the chapter content.

Learning Objectives At the beginning of each chapter, a list of objectives focuses students' attention on specific goals.

Topical Headings Throughout the chapter, headings emphasize the essential concepts for understanding biology as a science.

Full-Color Graphics Numerous line drawings and photographs illustrate concepts or associate new concepts with previously mastered information. Every illustration is used to emphasize a point or to help teach a concept.

Chapter Summary At the end of each chapter, the summary clearly reviews the concepts presented.

Thinking Critically This feature focuses on issues that challenge the student to think logically through problems and arrive at conclusions based on the concepts of the chapter.

Experience This Through this feature, students can apply knowledge gained from the chapter.

Questions This review of the material helps students determine whether they have mastered the contents of the chapter.

Phonetic Pronunciations in Chapter Glossaries

You will notice that phonetic spellings follow each glossary entry. This can be helpful if you know how to read the symbols.

An unmarked vowel (a,e,i,o,u) at the end of a syllable has the long sound, as in the word "prey"—PRA. An unmarked vowel followed by a consonant has the short sound, as in the phonetic spelling of the word "cell"—SEL.

A vowel in the middle of a syllable may have a mark over it to indicate a short or long sound. A straight bar (ā) indicates the long sound and a small arc (ă), the short sound. The word "acetyl" = ă-sēt′l shows these two marks plus an accent (′) that tells us to stress the second syllable. Some phonetic spellings may also have a double accent (″). The double-accented syllable is stressed, too, but not as much as the single-accented syllable; for example, res″pĭ-ra′shun.

Writing Style and Readability

The Fry Readability Graph has been used to verify the appropriateness of the language level for an introductory biology course. The informal, easy-to-read writing has been praised by reviewers and adopters. A number of features enhance the readability:

Boldface focuses student attention on a key term when it is first defined in the text.

Italics emphasize important terms, phrases, names, and titles.

Graphics—often in the form of logical flow diagrams, analogy diagrams, and charts—clarify the textual material.

Chapter Glossaries immediately reinforce the terms necessary for student comprehension of concepts.

Master Glossary located at the end of the text serves as a single resource for essential terminology used throughout the text.

Useful Ancillaries

The following supplementary materials have been developed to provide instructors with a complete educational package:

Instructor's Manual/Test-Item File provides a rationale for the use of the text, with objectives, explanatory information, and an answer key for text questions. In addition, text sequences have been adapted to accommodate differing academic calendars.

Laboratory Manual features carefully designed, class-tested exercises with stated objectives.

Laboratory Resource Guide provides information on acquiring, organizing, and preparing laboratory equipment and supplies. The guide follows the arrangement of exercises in the

Laboratory Manual, enabling instructors to efficiently select those learning experiences most appropriate for their students. Estimates of the time required for students to complete individual laboratory experiences are also provided, along with answers to questions in the Laboratory Manual.

Student Study Guide features an overview, a chapter outline, and objectives for each chapter. Sections with multiple-choice, fill-in-the-blank, and label/diagram/explain questions are also included. Answers to the objective questions are provided in a unique foldover feature to allow for immediate feedback. There is a place entitled "Key Terms/Notes" for students to write review notes or explanations of specific material.

The Study Guide is available through your college bookstore under the title *Student Study Guide to Accompany Concepts in Biology* by Enger, Kormelink, Ross, and Smith. This Study Guide will help you master course material in preparing for your exams. If you don't see a copy of it on your bookstore's shelves, ask the bookstore manager to order a copy for you.

Transparencies Seventy color acetates are available free to adopters of *Concepts in Biology*. The transparencies are taken from the text and represent the important figures that merit extra visual review and discussion.

Testing Software A computerized testing service provides instructors with either a mail-in/call-in testing program or the complete test item file on diskette for use with the Apple™, MacIntosh and IBM PC computers.

How to Study Science A new workbook, prepared by Fred Drewes, Suffolk County Community College, offers students helpful suggestions for meeting the considerable challenges of a science course. It offers tips on how to take notes; how to get the most out of laboratories; and how to overcome science anxiety. The book's unique design helps to stir critical thinking skills, while facilitating careful note-taking on the part of the student. (ISBN 14474)

The Life Science Lexicon This inexpensive reference, by William Marchuk, Red Deer College, helps introductory-level students quickly master the vocabulary of the life sciences. Not a dictionary, it carefully explains the rules of word construction and derivation, while giving complete definitions of all important terms. (ISBN 12133)

Study Cards in Biology A boxed set of 300 cards, by Van De Graaff, Rhees, and Creek, is available to students as an additional study aid for understanding the terminology of biology. These double-sided cards contain pronunciation guides, word derivations, definitions, and illustrations depicting many biological terms and concepts. (ISBN 10977)

Physiological Concepts of Life Science Thirteen illustrations taken from WCB's library of Anatomy and Physiology texts are brought to life through animation. The animations have been placed on videotape (approximately 35 minutes in duration) and illustrate various physiological processes such as: conduction of an action potential; synaptic transmission; electron transport and oxidative phosphorylation; cyclic AMP action; and replication of the HIV virus. These full color animations provide invaluable support for understanding complex physiological processes. (ISBN 21512) Free to qualified adopters.

Videodisk with Linkway and Hypercard Software Bio Sci II from Videodiscovery is a general biology videodisk that features more than ten thousand still and moving images, including ones taken directly from this text. It comes with an Instructor's Guide that indexes each image on the disk. The disk can be used with the most elaborate lecture hall video equipment or with the simplest of home players, and it can be run manually or with an IBM or MacIntosh computer. For set-up with MacIntosh or IBM interface, the Bio Sci II stacks provide instructors with Linkway or Hypercard software that allow instructors to organize their lectures or seminars with a series of pre-programmed menus.

Acknowledgments

A large number of people have knowingly or unknowingly helped us write this text. Our families continued to give understanding and support as we worked on this revision. We acknowledge the thousands of students in our classes who have given us feedback over the years concerning the material and its relevancy. They were the best possible source of criticism.

We gratefully acknowledge the invaluable assistance of many reviewers throughout the development and preparation of the manuscript:

Market Survey Respondents

Dwight Meyer, *City University of New York–Queensborough Community College*

Richard A. Laddaga, *Bowling Green State University*

Michael Drake, *Sinclair Community College*

A. J. Russo, *Mount Saint Mary's College*

Ralph Morrison, *University of North Carolina–Greensboro*

Patricia M. Sheppard, *Shippensburg University of Pennsylvania*

Ralph F. Ungar, *Travecca Nazarene College*
A. Jay Fisher, *Indiana Vocational Technical College*
Betty Mosher, *East Kootenay Community College*
Kenneth R. Hille, *Firelands College of Bowling Green State University*
James G. Erickson, *Fort Lewis College*
Robert L. Henn, *Sinclair Community College*
Roland D. Gassler, *Cayuga Community College*
Kenneth L. Petersen, *Monmouth College*
Philip D. Nelson, *Barstow College*
Barbara L. Stewart, *J. Sargeant Reynolds Community College*
Daryl H. Johnson, *Mississippi County Community College*
Doris S. Powell, *University of Baltimore*
Andrew H. Barnum, *Dixie College*
Ligia Arango, *Stone Child College*
Ann Stemler, *De Anza College*
Peter Gauthier, *Mount Saint Mary's College*
Robert C. Romans, *Bowling Green State University*
Melvin B. Hathaway, *Muskingum Area Technical College*
Laurence A. Larson, *Ohio University*
Randall M. Brand, *Southern Union State Junior College*
Patricia McCarroll, *Fisk University*
L. Duane Thurman, *Oral Roberts University*
Herbert T. Hendrickson, *University of North Carolina–Greensboro*
Eileen Gregory, *Rollins College*
Gail F. Baker, *La Guardia Community College*
Joann Bowns, *Southern Utah University*
Dean L. Winward, *Southern Utah University*
Barbara North Beck, *Rochester Community College*
Robert D. Muckel, *Doane College*
R. Harvard Riches, *Pittsburg State University*
Marjorie McCann Collier, *Saint Peter's College*
Amanda L. Woods, *Florida Institute of Technology*
Harvey P. Friedman, *University of Missouri–St. Louis*
Stephen G. Lebsack, *Linn-Benton Community College*
Robert J. Swanson, *North Hennepin Community College*
Robert S. Turner, Sr., *Western Oregon State College*
Arthur L. Tyson, *Middle Georgia College*
Gerald E. Willey, *South Suburban College*
Mary Anne McMurray, *Henderson Community College*
James L. Botsford, *New Mexico State University*
Dean A. Adkins, *Marshall University*

Reviewers of the Seventh Edition

Pablo Mendoza, Jr., *El Paso Community College*
Carol W. Barker, *Fayetteville Technical Community College*
Diane Ryerson, *Humboldt State University*
Richard Meyer, *Humboldt State University*
Jay F. Davidson, *Shippensburg University of Pennsylvania*
Christopher L. Arnold, *York College*
Marion E. Rohlinger, *Lake Land College*
Gail F. Baker, *La Guardia Community College*
David W. Bushman, *Mount St. Mary's College*
Carl E. Estrella, *Merced College*

Reviewers of the Sixth Edition

Dawn Adrian Adams, *Baylor University*
Andrew N. Ash, *Pembroke State University*
Janice A. Blum, *Fayetteville Technical Community College*
Roy E. Cameron, *Aims Community College*
Richard Meyer, *Humboldt State University*
Doris S. Powell, *University of Baltimore*
Harold J. Grau, *Indiana University of Pennsylvania*
Margaret H. Peaslee, *Louisiana Tech.*
Charles Leavell, *Fullerton College*
Dan A. Alex, *Chabot College*
Virginia G. Latta, *Jefferson State Community College*
Charissa M. Urbano, *Delta College*
Gary Shields, *Kirkwood Community College*
Kenneth Allen, *University of Maine–Orono*
Herbert D. Papenfuss, *Boise State University*
James Nivison, *Mid Michigan Community College*
Jane Glasgow Vance, *Northeast Alabama State Junior College*
James B. Ebert, *Pembroke State University*
Perry V. Mack, *North Carolina A & T State University*
Diane Prather, *Colorado Northwestern Community College*
Rudy Locklear, *Robeson Community College*
William Montgomery, *Charles County Community College*

Reviewers of the Fifth Edition

Bonnie Amos, *Baylor University*
Ronald Basmajian, *Merced College*
Neal D. Buffaloe, *University of Central Arkansas*
Lynn Elkin, *California State University–Hayward*
Richard E. McKeeby, *Union College*
R. Harvard Riches, *Pittsburg State University*
E. Russell TePaske, *University of Northern Iowa*
Robert Wischmann, *Lakewood Community College*

Reviewers of the Fourth Edition

Harold G. Brotzman, *North Adams State College*
Terry A. Larson, *College of Lake County*
James Lipp, *Lewistown College Center*
Sr. Maureen Webb, *Holy Names College*
Stanley M. Wiatr, *Eastern Montana College*

Reviewers of the Third Edition

Frank Bonham, *San Diego Mesa College*
Richard Boutwell, *Missouri Western State College*
Clyde L. Britchett, *Brigham Young University*
Gil Desha, *Tarrant County Junior College*
Albert J. Grennan, *San Diego Mesa College*
Robert Kirkwood, *University of Central Arkansas*
Cathryn McDonald, *Jefferson State Junior College*
Don Misumi, *Los Angeles Trade-Technical College*
Robert Romans, *Bowling Green State University*

Jean Salter, *Valencia Community College*
Melvin Urschel, *Tacoma Community College*

Reviewers of the Second Edition

Ferron Andersen, *Brigham Young University*
Donald A. Denison, *Laney College*
Steven A. Fink, *West Los Angeles College*
James Hiser, *Normandale Community College*
Robert Kirkwood, *University of Central Arkansas*
Virginia Latta, *Jefferson State Junior College*
Cathryn McDonald, *Jefferson State Junior College*
Janice Roberts, Jefferson State Junior College
Robert Romans, *Bowling Green State University*
Melvin Urschel, *Tacoma Community College*

Reviewers of the First Edition

Gil Desha, *Tarrant County Junior College*
Albert J. Grennan, *San Diego Mesa College*
Rhoda Love, formerly of *Lane Community College*
Don Misumi, *Los Angeles Trade-Technical College*
Robert P. Quellett, Massasoit Community College
Ernest L. Rhamstine, *Valencia Community College*
John H. Standing, *Delaware Valley College*
Michael J. Timmons, *Moraine Valley Community College*

Market Research Respondents

We would like to thank the following adopters of the fifth edition for their help in preparing the current one. Each contributed greatly to our understanding of the relative strengths and weaknesses of the fourth edition by responding to a user's survey of the text.

Janice A. Blum, *Fayetteville Tech Community College*
William Montgomery, *Charles County Community College*
Rudy Locklear, *Robeson Community College*
John Dropp, *Mount St. Mary's College*
Jane Vance, *Northeast Alabama State Jr. College*
Debrah Kendall, *Fort Lewis College*
Diane Prather, *Colorado Northeast Community College*
Roy E. Cameron, *Aims Community College*
James Hunt, *Florida Keys Community College*
Doris S. Powell, *University of Baltimore*
Perry V. Mack, *North Carolina A & T State University*
James B. Ebert, *Pembroke State University*
James Nivison, *Mid Michigan Community College*
Herbert D. Papenfuss, *Boise State University*
Virginia G. Latta, *Jefferson State Community College*
Ann Escal, *National College*
Kenneth M. Allen, *University of Maine, Orono*
Gary Shields, *Kirkwood Community College*
Godfrey, *Ricks College*

PART I

Introduction

Science is a very important part of our life. Without scientific advances we would not enjoy the standard of living we have. Materials science developed the computer chip. Medical science developed vaccines against measles, smallpox, mumps, whooping cough, chicken pox, and many other "childhood diseases" that were once considered normal experiences of childhood. Engineering science has given us safer buildings, manufacturing processes, and automobiles. Yet it is not always easy to differentiate science from nonscience. Furthermore, there are those who use the image of science to sell products or ideas that are not based on scientific principles. This initial chapter examines the nature of a scientific approach and contrasts it with nonscientific approaches. It also begins to describe the condition known as life. The complexity of life will become more meaningful as you progress through this course. Let's now begin our exploration of the science of biology. ■

CHAPTER 1

What Is Biology?

Chapter Outline

The Significance of Biology in Your Life
Science and the Scientific Method
- Observation
- Questioning and Exploration
- The Formation and Testing of Hypotheses
- The Development of Theories and Laws

Science, Nonscience, and Pseudoscience
- Fundamental Attitudes in Science
- Applied and Theoretical Science
- Science and Nonscience
- Pseudoscience
- Limitations of Science

The Science of Biology
- Characteristics of Life
- The Value of Biology

Box 1.1 Edward Jenner and the Control of Smallpox
- Problems in the Field of Biology
- Future Directions in Biology

Purpose

This chapter is a general introduction to the nature of science and the significance of biological science in your everyday life. It presents a scientist's view of the world and describes what living things are and how they differ from nonliving things. This chapter lays the groundwork for helping you understand and answer questions about living things you encounter. You will be better able to understand and answer biological questions after you have an understanding of how science works.

For Your Information

As a result of recent, rapid advances in science, most newspapers have added a science page as a weekly feature. These sections are intended to keep the general public aware of the most significant advances in all areas of science. Subjects such as recombinant DNA theory, biological amplification, and punctuated evolution are no longer found exclusively in scientific journals that only the most well-informed, practicing scientist can understand.

Learning Objectives

- Understand the difference between science and nonscience.
- Know the steps in the scientific method.
- Recognize that science has limitations.
- Recognize that pseudoscience appears to be scientific but is really used to mislead.
- Differentiate between applied and theoretical science.
- Know the characteristics used to differentiate between living and nonliving things.
- Understand that many advances in the quality of life are the result of biological science.
- Be able to give examples of problems caused by unwise use of biological information.

The Significance of Biology in Your Life

Many college students question the need for biology and other science classes in their curriculum when they know they are not going to major in science. However, it is becoming increasingly important that all citizens understand the ways of science and how the actions of society affect living things. Consider how your future will be influenced by how the following questions are ultimately answered:

- Does electromagnetic radiation from electric power lines, computer monitors, or microwave ovens affect living things?
- Will the thinning of the ozone layer of the upper atmosphere result in increased incidence of skin cancer?
- Will a vaccine for AIDS be developed in the next ten years?
- Will new, inexpensive, socially acceptable methods of birth control be developed that can slow world population growth?

As an informed citizen in a democracy, you can have a great deal to say about the solutions to these problems. In a democracy it is assumed that the public has gathered enough information to make intelligent decisions (figure 1.1). This is why an understanding of biological concepts is so important for any person, regardless of his or her vocation. *Concepts in Biology* was written with this philosophy in mind. The concepts covered in this book were selected to help you become more aware of how biology influences nearly every aspect of your life.

Most of the important questions of today can be considered from philosophical, social, and scientific standpoints. None of these approaches individually presents a solution to most problems. For example, it is a fact that the human population of the world is growing very rapidly. Philosophically, we may all agree that the rate of population growth should be slowed. Killing infants or sterilizing women after they have had two children will work to control population. Both of these methods have been tried in some parts of the world within the last three centuries. However, most would agree that these "solutions" are philosophically or socially unacceptable. Science can provide information about the reproductive process and how it can be controlled, but society must answer the more fundamental social and philosophical questions about reproductive rights and the morality of certain controls. It is important to recognize that science has a role to play, but that it is not the answer to all our problems.

Figure 1.1 **Biology in Everyday Life.** It is very easy to identify biological problems and situations. These news headlines reflect a few of the biologically based issues that face us every day. While articles such as these seldom propose solutions, they do serve to make the general public aware of situations so that people can begin to explore the possibilities for making intelligent decisions leading to solutions.

Science and the Scientific Method

You probably have the idea that biology has something to do with plants and animals. Most textbooks would define **biology** as the science that deals with life. This definition seems tidy enough until you begin to think about what the words *science* and *life* mean.

The word *science* is a noun derived from a Latin term (*scientia*) meaning *to have knowledge* or *to know*. However, this is misleading because there are many

Table 1.1
The Scientific Method

Steps	Activity	Example
Observation	Recognize something has happened and that it occurs repeatedly. (Empirical evidence is gained from experience or observation.)	Students in a classroom are stricken with a disease that causes red rashes on their faces. This same situation has been described in several schools in your region. Skin cultures taken from the students indicate that there are some unusual bacteria present.
Question Formulation	Write many different kinds of questions about the observation and keep the ones that will be answerable.	Is the disease psychosomatic (i.e., is this a case of hysteria in which there is nothing organically wrong)? Is the rash caused by a bacterium? Is the disease caused by a virus?
Exploration of Alternative Resources	Go to the library to obtain information about this observation. Also, talk to others who are interested in the same problem.	A search of the medical literature reveals that physicians who used antibiotic X in similar circumstances reported cures, even though they never found a bacterium to be present. Attend scientific meetings where this disease outbreak will be discussed. Contact scientists who are reported to be interested in the same problem.
Hypothesis Formation	Pose a possible answer to your question. Be sure that it is testable and that it accounts for all the known information. Recognize that your hypothesis may be wrong.	Antibiotics do not usually affect viruses. Further, the disease has been reported elsewhere, which tends to rule out psychosomatic disease. Therefore, your hypothesis is that the disease is caused by a bacterium and that antibiotic X can cure the disease by controlling the rate of growth of the bacterial population.

fields that have accumulated great volumes of knowledge that are not "sciences."

Science is really distinguished by *how* knowledge is acquired, rather than by the act of accumulating facts. **Science** is actually a process or way of arriving at a solution to a problem or understanding an event in nature that involves testing possible solutions. The process has become known as the *scientific method.* The **scientific method** is a way of gaining information (facts) about the world by forming possible solutions to questions followed by rigorous testing to determine if the proposed solutions are valid. What scientists learn by using this approach may help them solve a particular practical problem, such as how to improve milk production in cows. It may also advance their understanding of an important concept, such as evolution, but have little immediate practical value. The scientific method requires a systematic search for information and a continual checking and rechecking to see if previous ideas are still supported by new information.

People who use the scientific method follow a sequence of thought processes and activities outlined in table 1.1. As with your own thought processes, the minds of scientists bounce from thought to thought and from category to category as they wrestle with the problem at hand. Keep in mind that scientists try not to "reinvent the wheel"; therefore, they do not always start at the beginning. The sequence presented in table 1.1 is idealistic and a simplification of a very complex series of events.

Observation

Scientific inquiry usually begins with an observation that an event has occurred repeatedly. An **observation** occurs when we use our senses (smell, sight, hearing, taste, touch) or an extension of our senses (microscope, tape recorder, X-ray

Table 1.1
The Scientific Method (*continued*)

Steps	Activity	Example
Experimentation	Set up an experiment that will allow you to test your hypothesis using a control group and an experimental group. Be sure to collect and analyze the data carefully.	To test the cause-and-effect relationship between administering antibiotic X and curing the illness, you set up two groups. A control group will be given a placebo (a pill with no active ingredient). The experimental group will receive pills containing antibiotic X. The pills will look identical and will be coded so that neither the person receiving the pill nor the person administering the pill will know which individuals receive the medication and which receive the placebo. After five days you collect the data and find that 90% of those receiving antibiotic X no longer have the rash. By contrast, only 10% of those receiving the placebo have recovered. You conclude that the disease is not psychosomatic and that a bacterium is probably the cause. You publish your results and others in the country report back that they have had similar results.
Theory Formation	Repeat the experiment and share the information with others over a long period of time. Should your information continue to be considered valid and consistent with other closely related research, the scientific community will recognize that a theory has been established or that your information is consistent with existing theories.	Your results support the generally held theory that many kinds of diseases are caused by microorganisms. This generalization is called the *germ theory of disease.*
Law Formation	If your findings are seen to fit with many other major blocks of information that tie together many different kinds of scientific information, it will be recognized by the scientific community as being consistent with current scientific laws. If it is a major new finding, a new law may be formulated.	Your experimental results are consistent with the *biogenetic law* that states that all living things come from previously living things. Your results strongly suggest that the disease was caused by the multiplication of certain bacteria, and that the antibiotic stopped their multiplication.

machine, pH meter, thermometer) to record an event. Observation is more than just taking note of something. You may hear a sound or see something without really observing. Do you know what music is being played in a shopping mall? You hear it but you don't observe. What color was the car that just drove past? You saw it but you didn't observe. When scientists talk about their observations, they are referring to careful, thoughtful, recognition of an event—not just a casual notice.

The information gained by direct observation of the event is called **empirical evidence.** Empirical evidence is capable of being verified or disproved by further observation. If the event only occurs once or cannot be repeated in an artificial situation, it is impossible to use the scientific method to gain further information about the event and explain it.

Questioning and Exploration

As scientists gain more empirical evidence about the event, they begin to develop *questions* about it. How does this happen? What causes it to occur? When will it take place again? Can I control the event to my benefit? The formation of the question is not as simple as it might seem because the way the question is asked will determine how you go about answering it. A question that is too broad or too complex may be impossible to answer; therefore, a great deal of effort is put into asking the question in the right way. In some situations, this can be the most time-consuming part of the scientific method; asking the right question is critical to how you look for answers.

Let's say, for example, that you observed a cat catch, kill, and eat a mouse. You could ask several kinds of questions:

1a. Does the cat like the taste of the mouse?
1b. If given a choice between mice and catfood, which would a cat choose?
2a. What motivates a cat to hunt?
2b. Do cats hunt only when they are hungry?

Obviously, the second of each pair of questions is much easier to answer than the first even though the two questions are attempting to obtain similar information.

Once the question is written, scientists *explore other sources of knowledge* to arrive at an acceptable answer. They turn first to the research of others to avoid wasting time and energy answering an already-answered question. This usually means a trip to the library or contact with fellow scientists interested in the same field of study. Even if their particular question has not already been answered, scientific literature and fellow scientists can provide insight into the problem that will help in its solution. It is during this time that a hypothesis is formed.

The Formation and Testing of Hypotheses

A **hypothesis** is a possible answer to a question or an explanation of an observation. It must account for all the observed facts and must be testable. Just as writing a good scientific question is important, hypothesis formation is critical and may be difficult. If the hypothesis does not account for all the observed facts in the situation, doubt will be cast on the work and may eventually invalidate the hypothesis. Doubt may also surround a hypothesis that is not testable. If a possible explanation cannot be proven true, it would only be hearsay and no more useful than mere speculation. Keep in mind that a hypothesis is an educated guess based on observations and information gained from other knowledgeable sources. To test the soundness of a hypothesis, scientists begin the next step in the scientific method—experimentation.

An **experiment** is a re-creation of an event or occurrence in a way that enables a scientist to support or disprove a hypothesis. This can be difficult since a particular event may involve a great many separate happenings. For example, the production of songs by birds involves many activities of the nervous system and the muscular system, and is stimulated by a wide variety of environmental factors. It might seem that developing an understanding of the factors involved in bird-song production would be an impossible task. To help unclutter such situations, scientists have devised what is known as a *controlled experiment*. A **controlled experiment** allows scientists to compare two situations that are identical in all but one respect. The situation used as the basis of comparison is called the **control group;** the other situation is called the **experimental group.** The single factor that is allowed to be different in the experimental group but is controlled in the other group is called the **variable.** The situation involving bird-song production would need to be broken down into a large number of simple questions, as previously mentioned. Each question would provide the basis on which experimentation would occur. Each experiment would provide information about a small part of the total process of bird-song production. For example, in order to test the hypothesis that male sex hormones are involved in stimulating male birds to sing, an experiment could be performed in which one group of male birds had their testes removed (the experimental group), while the control group was allowed to develop normally. After the experiment, the new data (facts) gathered would be analyzed. If there were no differences between the two groups, the variable evidently did not have a cause-and-effect relationship (i.e., was not responsible for the event). However, if there was a difference, it is likely that the variable was responsible for the difference between the control and experimental groups. In the case of songbirds, removal of the testes does change their singing behavior.

Any experiment needs to be repeated many times. Scientists encourage challenges to their work and admit that no experiment is perfect. In our bird-song example, this means that large numbers of birds would need to be involved in the experiments and that the experiments would need to be repeated several times by other experimenters. Furthermore, scientists will apply statistical tests to the results to help decide if the results obtained are **valid** (meaningful; fit with other knowledge) and **reliable** (give the same results repeatedly) or if they are the result of random events. Random events can give the impression that the hypothesis is valid when in fact it is not. For example, the operation necessary to remove the testes of male birds might cause illness in some birds, resulting in less singing. During experimentation, scientists learn new information and formulate new questions that can lead to even more experiments. Some scientists speculate that one good experiment can result in a hundred new questions and experiments. The discovery of the structure of the DNA molecule by Watson and Crick resulted in thousands of experiments and

(a)

(b)

Figure 1.2 **The Growth of Knowledge.** James D. Watson and Francis W. Crick are theoretical scientists who determined the structure of the DNA molecule, which contains the genetic information of a cell. (a) This photograph shows the model of DNA they constructed. The discovery of the structure of the DNA molecule was followed by much research into how the molecule codes information, how it makes copies of itself, and how the information is put into action. Ultimately, these lines of research have led to altering the DNA of bacteria so that they produce useful materials, such as vitamins, proteins, and antibiotics. (b) Genetically altered bacteria can be grown in special vats and the useful materials harvested.

stimulated the development of the entire field of molecular biology (figure 1.2). Similarly, the discovery of molecules that regulate the growth of plants resulted in much research about how the molecules work and which molecules might be used for agricultural purposes.

If the processes of questioning and experimentation continue, and evidence continually and consistently supports the original hypothesis and other closely related hypotheses, the scientific community will begin to see how these hypotheses and facts fit together into a broad pattern. When this happens, a theory has come into existence.

The Development of Theories and Laws

A **theory** is a plausible, scientifically acceptable generalization. An example of a biological theory is the germ theory of disease. This theory states that certain diseases, called *infectious* diseases, are caused by microorganisms that are capable of being transmitted from one individual to another. As you can see, this is a very broad statement. It is the result of years of questioning, experimentation, and pulling data together. While a hypothesis is an answer to a specific question, a theory encompasses the answers to many complex questions. As a result, there are fewer theories than hypotheses.

However, just because a theory exists does not mean that testing stops. In fact, many scientists see this as a challenge and exert even greater efforts to disprove the theory, and experimentation continues. Should the theory survive this skeptical approach and continue to be supported by experimental evidence, it will become a *scientific law*. A **scientific law** is a uniform or constant fact of nature. An example of a biological law is the biogenetic law, which states that all living things come from preexisting living things. You can see from this example that laws are even more general than theories and encompass the answers to even more complex questions. Therefore, there are relatively few scientific laws.

Science, Nonscience, and Pseudoscience

Fundamental Attitudes in Science

As you can see from this discussion of the scientific method, a scientific approach to the world requires a certain way of thinking. There is an insistence on ample proof supported by numerous studies rather than easy acceptance of strongly stated opinions. A scientist is a healthy skeptic.

Careful attention to detail is also important. Since scientists publish their findings and their colleagues examine their work, there is a strong desire to produce careful work that can be easily defended. This does not mean that scientists do not speculate and state opinions. When they do, however, they take great care to clearly distinguish fact from opinion.

There is also a strong ethic of honesty. Scientists are not saints, but the fact that science is conducted out in the open in front of one's peers tends to reduce the incidence of dishonesty. In addition, the scientific community strongly condemns and severely penalizes those who steal the ideas of others, perform shoddy science, or falsify data. Any of these infractions would lead to the loss of one's job.

Applied and Theoretical Science

The scientific method has helped us to understand and control many aspects of our natural world. Some information is extremely important in understanding the structure and functioning of things but is of little practical value. For example, an understanding of the life cycle of a star or how meteors travel through the universe may be very important for those who are trying to answer questions about how the universe was formed, but it is of little value to the average citizen. Thus, science can be divided into two categories, theoretical and applied, depending on whether experimentation is done to gain knowledge for its own sake or motivated primarily by a desire to better our daily lives.

Theoretical science is interested in obtaining new information regardless of its practical value. Little attention is paid to how the new information may be used in any specific or practical situation. Studies of evolution or particle physics, for example, have tremendous theoretical importance but few immediate, practical applications. On the other hand, a scientist who actively seeks solutions to practical problems is involved in **applied science.** The primary goal of an applied scientist is to solve a practical problem so that a job can be done more efficiently, people will have more to eat, or people will not have to put up with an inconvenience of some kind. Engineers and agricultural researchers are primarily motivated by specific problems that require solutions.

Although they seem different, theoretical and applied science are related. Applied science makes practical use of the theories provided by theoretical science by using the new knowledge to solve everyday problems. For example, applied scientists known as *genetic engineers,* have altered the chemical code system of small organisms (microorganisms) so that they may produce many new drugs such as antibiotics, hormones, and enzymes. The ease with which these complex chemicals are produced would not have been possible had it not been for the information gained from the theoretical sciences of microbiology, molecular biology, and genetics (figure 1.2).

As another example, Louis Pasteur was interested in the theoretical problem of whether life could be generated from nonliving material. Much of his theoretical work led to practical applications in disease control. His theory that there are microorganisms that cause diseases and decay led to the development of vaccinations against rabies and the development of pasteurization for the preservation of foods (figure 1.3).

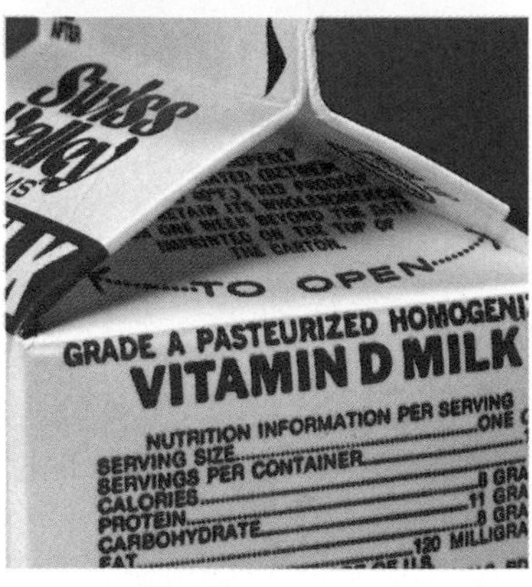

Figure 1.3 **Louis Pasteur and Pasteurized Milk.** Louis Pasteur (1822–1895) performed many experiments while he studied the question of the origin of life, one of which led directly to the food-preservation method now known as pasteurization.

Science and Nonscience

The differences between science and nonscience are often based on the assumptions and methods used to gather and organize information and, most importantly, the testing of these assumptions. The difference between a scientist and a nonscientist is that a scientist continually challenges and tests the principles and laws, whereas a nonscientist may not feel that this is important.

Once you understand the scientific method, you won't have any trouble identifying astronomy, chemistry, physics, and biology as sciences. But what about economics, sociology, anthropology, history, philosophy, and literature? All of these fields may make use of certain laws that are derived in a logical way, but they are also nonscientific in some ways. Some things are beyond science and cannot be approached using the scientific method. Art, literature, theology, and philosophy are rarely thought of as sciences. They are concerned with beauty, human emotion, and speculative thought rather than with facts and verifiable laws. On the other hand, physics, chemistry, geology, and biology are almost always considered sciences. Music is an area of study in a middle ground where scientific approaches may be used to some extent. "Good" music is certainly unrelated to science, but the study of how the human larynx generates the sound of a song is based on scientific principles. Any serious student of music will study the anatomy of the human voice box and how the vocal cords vibrate to generate sound waves. Similarly, economics makes use of mathematical models and established economic laws to help make predictions about future economic conditions. However, the regular occurrence of unpredicted economic changes indicates that economics is far from scientific, since the reliability of predictions is a central criterion of science. Anthropology and sociology also have many aspects that are scientific in nature, but they cannot be considered to be true sciences because many of the generalizations they have developed cannot be tested by repeated experimentation (table 1.2).

Pseudoscience

Pseudoscience (pseudo = false) takes on the flavor of science but is not supportable as valid or reliable. Often, the purpose of pseudoscience is to confuse or mislead. The area of nutrition is one that is flooded with pseudoscience (figure 1.4). We all know that we must obtain certain nutrients like amino acids, vitamins, and minerals from the food that we eat or we may become ill. Many scientific experiments have been performed that reliably demonstrate the validity of this information. However, in most cases, it has not been demonstrated that the nutritional supplements so vigorously advertised are useful or even desirable. Rather, selected bits of scientific information (amino acids, vitamins, and minerals are essential to good health) have been used to create the feeling that additional amounts of these nutritional supplements are necessary or that they can improve your health. In reality, the average person eating a varied diet will obtain all of these nutrients in adequate amounts, and nutritional supplements are not required.

Figure 1.4 **"Nine Out of Ten Doctors Surveyed Recommend Brand X."** It is obvious that there are many things wrong with this statement. First of all, is the person in the white coat really a doctor? Second, if only ten doctors were asked, the sample size is too small. Third, only selected doctors might have been asked to participate. Finally, the question could have been asked in such a way as to obtain the desired answer: "Would you recommend Brand X over Dr. Pete's snake oil?"

In addition, many of these products are labeled as organic or natural, with the implication that they have greater nutritive value because they are organically grown (grown without pesticides or synthetic fertilizers) or because they come from nature. The poisons curare, strychnine, and nicotine are all organic molecules that are produced in nature by plants that could be grown organically, but we wouldn't want to include them in our diet.

Table 1.2
Classification of Traditional Fields of Study

Definitely Science				Definitely Not Science
Geology		Psychology		Astrology
Chemistry			Music	Theology
Physics	Economics		History	Literature
Biology		Sociology	Political Science	
Astronomy	Anthropology			Philosophy
Geography				Art

Limitations of Science

By definition, science is a way of thinking and seeking information to solve problems. Therefore, the scientific method can be applied only to questions that have a factual basis. Questions concerning morals, value judgments, social issues, and attitudes cannot be answered using the scientific method. What makes a painting great? What is the best type of music? Which wine is best? What color should I paint my car? These questions are related to values, beliefs, and tastes; therefore, the scientific method cannot be used to answer them.

Just because scientists say something is true does not necessarily make it true. Everyone makes mistakes, and quite often, as new information is gathered, old laws must be changed or discarded. For example, at one time scientists were sure that the sun went around the earth. They observed that the sun rose in the east and traveled across the sky to set in the west. Since scientists could not feel the earth moving, it seemed perfectly logical that the sun traveled around the earth. Once they understood that the earth rotated on its axis, they began to understand that the rising and setting of the sun could be explained in other ways. A completely new concept of the relationship between the sun and the earth developed (figure 1.5).

Although this kind of study seems rather primitive to us today, this change in thinking about the sun and the earth was a very important step in understanding the universe and how the various parts are related to one another. This background information was built upon by many generations of astronomers and space scientists, and finally led to space exploration.

Many people view science as a powerful tool that will come up with answers to the major problems of our time. This is not necessarily true. Most of the problems we face are generated by the behavior and desires of people. Famine, drug abuse, and pollution are human-caused and must be resolved by humans. Science may provide some tools for the social planners, politicians, and ethical thinkers, but science does not have, nor

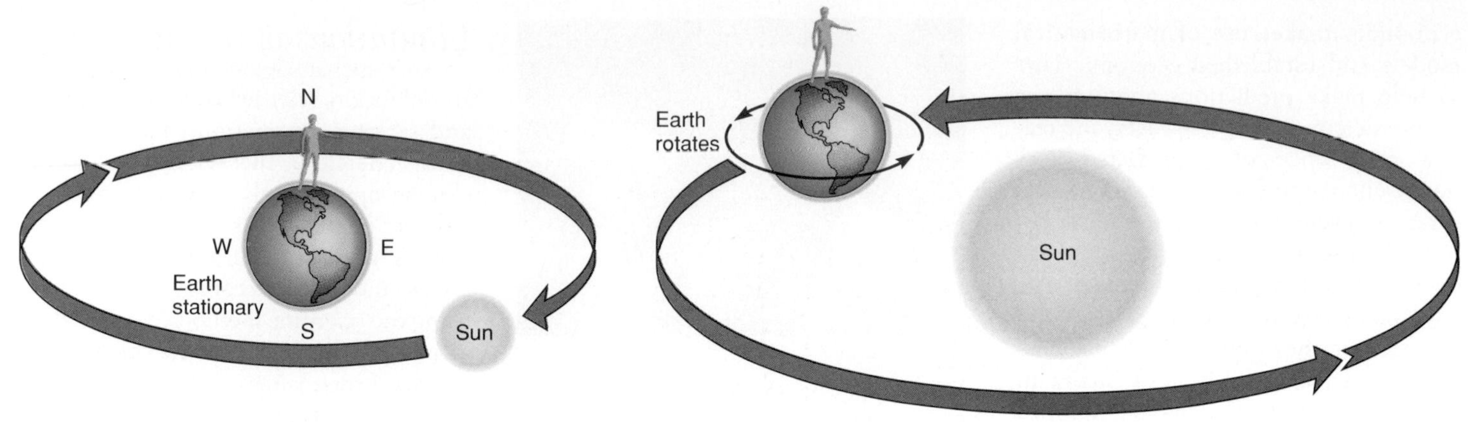
N
W
E
Earth
stationary
S
Sun
Earth
rotates
Sun
Scientists thought that the sun revolved around the earth.
We now know that the earth rotates on its axis and also revolves around the sun.

Figure 1.5 **Science Must Be Willing to Challenge Previous Beliefs.** Science always must be aware that new discoveries may force a reinterpretation of previously held beliefs. Early scientists thought that the sun revolved around the earth in a clockwise direction. This was certainly a reasonable theory at the time. Subsequently, we have learned that the earth revolves around the sun in a counterclockwise direction, at the same time rotating on its axis in a counterclockwise direction. It is this rotation of the earth on its axis that gives us the impression that the sun is moving.

does it attempt to provide, all the answers to the problems of the human race. Science is merely one of the tools at our disposal.

The Science of Biology

The science of biology is, broadly speaking, the study of living things. It draws on chemistry and physics for its foundation and applies these basic physical laws to living things. Because there are many kinds of living things, there are many special areas of study in biology. Practical biology—such as medicine, crop science, plant breeding, and wildlife management—is balanced by more theoretical biology—such as medical microbiological physiology, photosynthetic biochemistry, plant taxonomy, and animal behavior (ethology). There is also just plain fun biology like insect collecting and bird watching. Specifically, biology is a science that deals with living things and how they interact with all of the things around them.

At the beginning of the chapter, biology was defined as the science that deals with living things. But what does it mean to be alive? You would think that a biology textbook could answer this question very easily. However, it is more than just a theoretical question since it has been necessary in recent years to construct some legal definitions of what life is and especially when it begins and ends. The legal definition of death is important since it may determine whether or not a person will receive life insurance benefits or if body parts may be used in transplants. In the case of heart transplants, the person donating the heart may be legally "dead," but the heart certainly isn't since it can be removed while it still has "life." In other words, there are different kinds of death. There is the death of the whole living unit and the death of each cell within the living unit. A person actually "dies" before every cell has died. Death, then, is the absence of life, but that still doesn't tell us what life is. At this point, we won't try to define life but will describe some of the basic characteristics of living things.

Characteristics of Life

The ability to manipulate energy and matter is unique to living things. Just how this is accomplished is helpful in understanding how living things differ from nonliving objects. Living things show four characteristics that the nonliving do not display: (1) metabolic processes, (2) generative processes, (3) responsive processes, and (4) control processes.

Metabolic processes are the total of all chemical reactions taking place within an organism. There are three essential aspects of metabolism: (1) *nutrient uptake,* (2) *nutrient processing,* and (3) *waste elimination.* All living things expend energy to take in nutrients (raw materials) and energy from their environment. Many animals take in these materials by eating or swallowing other organisms. Microorganisms and plants absorb raw materials into their cells to maintain their lives. Once inside, nutrients enter a network of chemical reactions. These reactions process nutrients in order to manufacture new parts, make repairs, and reproduce. However, not all materials entering a living thing are valuable to it. There may be portions of nutrients that are valuable, but the rest may be useless or even harmful. In that situation, organisms eliminate waste. Heat energy may also be considered a waste product.

The second group of characteristics of life, **generative processes,** are reactions that result in an increase in the size of an individual organism—*growth*—or an increase in the number of individuals in a population of organisms—*reproduction.* During growth, living things add to their structure, repair parts, and store nutrients for later use. However, growth cannot go on indefinitely because as organisms get larger, they become inefficient. The result could be chemical chaos, with organisms wasting energy and nutrients. Living things respond to this problem by reproducing. Reproduction is one of the most important life functions because it is the only way that living things can perpetuate themselves. There are a number of different ways that kinds of organisms reproduce and guarantee their continued existence.

Survival also depends on an organism's ability to react to external and internal changes in its environment. The group of characteristics involved is called the **responsive processes.** Three types of responsive processes have been identified: *irritability, individual adaptation,* and *population adaptation,* or *evolution.* Irritability is an individual's rapid response to a stimulus, such as a knee-jerk reflex. This type of response occurs only in the individual receiving the stimulus and is rapid because the mechanism that allows the response to occur (i.e., muscles, bones, and nerves) is already in place. Individual adaptation is also an individual response but is slower since it requires a genetic action. For example, a weasel changes from its brown summer coat to its white winter coat by turning off the genes that are responsible for the production of brown pigment. Population adaptation is also known as *evolution,* which is a slow change in the genetic makeup of a *population* of organisms. This enables a group of organisms to adapt and better survive threatening changes in its environment over many generations.

The **control processes** of *coordination* and *regulation* constitute the fourth characteristic of life. Control processes are mechanisms that ensure that an organism will carry out all metabolic activities in the proper sequence (coordination) and at the proper rate (regulation). All the chemical reactions of an organism are coordinated and linked together in specific pathways. The orchestration of all the reactions ensures that there will be specific stepwise handling of the nutrients needed to maintain life. The molecules responsible for coordinating these reactions are known as *enzymes.* **Enzymes** are molecules produced by organisms that are able to increase and control the rate at which chemical reactions occur. Enzymes also regulate the amount of nutrients processed into other forms. This limits the chance that the organism will die from using nutrients too quickly or generating too much waste.

In addition to the four basic processes that are typical of living things, it is important to point out that living things have some basic structural similarities. All living things are organized into complex, structural units called *cells.* These cells have an outer limiting membrane consisting of complex molecular arrangements and internal structural units that have specific functions. Some living things, like you, consist of millions of cells with specialized abilities that interact to provide the independently functioning unit called an **organism.** Typically, in such large, multicellular organisms as humans, cells cooperate with one another in units called **tissues** (muscle). Groups of tissues are organized into larger units known as **organs** (heart), and in turn, into **organ systems** (circulatory system). Other organisms, such as bacteria or yeasts, carry out all four life processes within a single cell. Nonliving materials, such as rocks, water, or gases, do not share a structurally complex common subunit. Figure 1.6 summarizes the differences between living and nonliving things.

The Value of Biology

To a great extent, we owe our current high standard of living to biological advances in two areas: food production and disease control. Plant and animal breeders have developed organisms that provide better sources of food than the original varieties. One of the best examples of this is the changes that have occurred in corn (figure 1.7, top). Corn is a grass that produces its seed on a cob. The original corn plant had very small ears that were perhaps only three or four centimeters long. Through selective breeding, varieties of corn with much larger ears and more seeds per cob have been produced. This has increased the yield greatly. In addition, the corn plant has been adapted to produce other kinds of corn, such as sweet corn and popcorn.

Corn is not an isolated example. Improvements in yield have been brought about in wheat, rice, oats, and other cereal grains. The improvements in the plants, along with changed farming practices (also brought about through biological experimentation), have led to greatly increased production of food.

Animal breeders also have had great successes. The pig, chicken, and cow of today are much different animals from those available even one hundred years ago. Chickens lay more eggs, dairy cows give more milk, and beef cattle grow faster (figure 1.7, bottom). All of these improvements raise our standard of living. One interesting example is the change in the kinds of hogs that are raised. At one time, farmers wanted pigs that were fatty. The fat could be made into lard, soap, and a variety of other useful products. As the demand for the fat products of pigs began to decline, animal breeders began to develop pigs that gave a high yield of meat and relatively little fat. Today, plant and animal breeders can produce plants and animals almost to specifications.

Much of the improvement in food production has resulted from the control of plants and animals that compete with or eat the organisms we use as food. Control of insects and fungi that weaken plants and reduce yields is as important as the invention of new varieties of plants. Since these are "living" pests, biologists have been involved in the study of them also.

There has been fantastic progress in the area of health and disease control. Many diseases such as polio, whooping cough, measles, and mumps can be easily controlled by vaccinations or "shots" (box 1.1). Unfortunately, the vaccines have worked so well that some people no longer worry about getting their shots, and some of these diseases are reappearing (polio, diphtheria). These diseases have not been eliminated, and people who are not protected by vaccinations are still susceptible to them.

The understanding of how the human body works has led to treatments

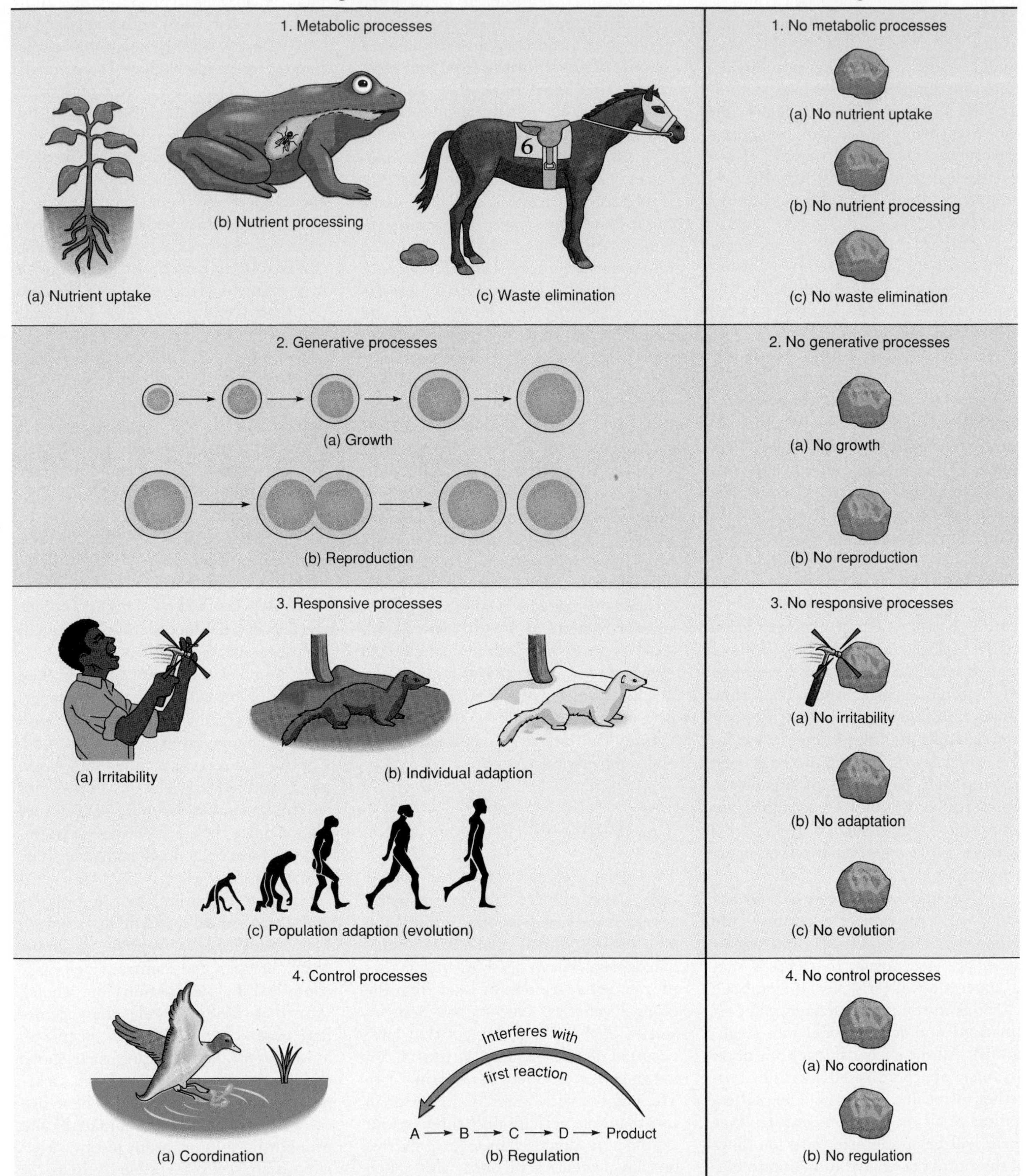

Figure 1.6 **Characteristics of Life.** Living and nonliving things differ in a number of ways. Some of these differences are shown here.

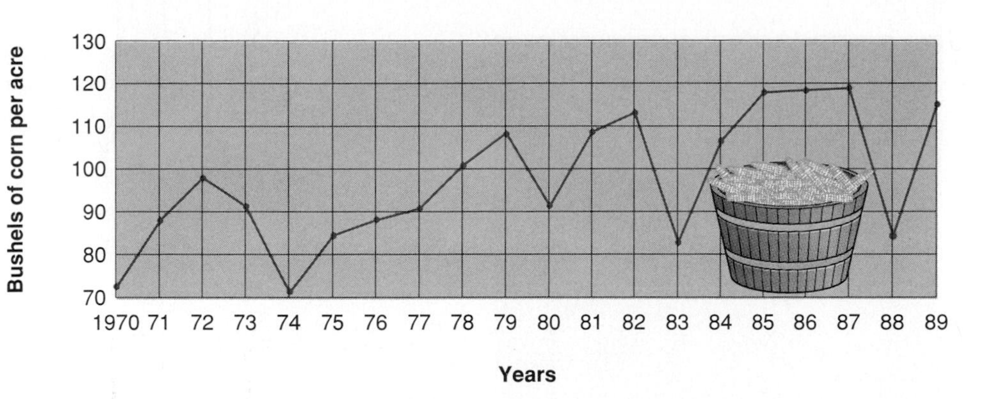

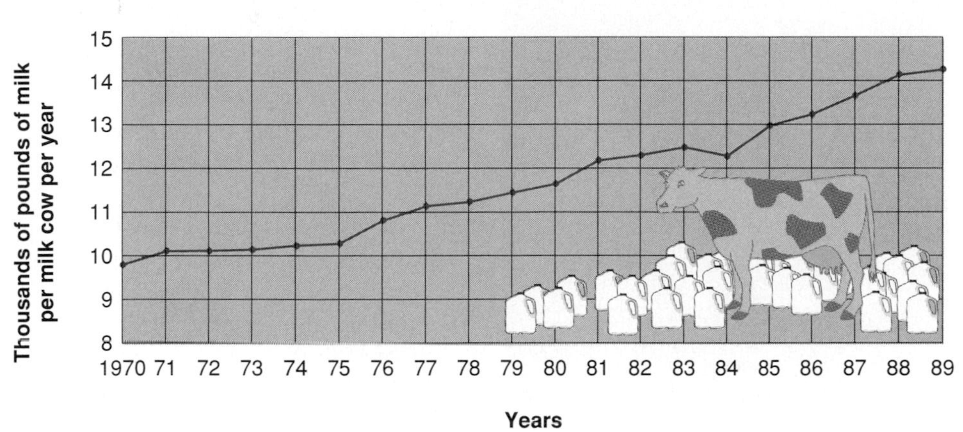

Figure 1.7 **Food Production in the United States.** Both of these graphs illustrate a steady increase in yield, largely because of changing farming practices and selective breeding programs.

Source: Data from the Department of Agriculture, *Agricultural Statistics*, 1990 and previous years.

that can control such diseases as diabetes, high blood pressure, and even some kinds of cancer. Paradoxically, these advances contribute to a major biological problem: the increasing size of the human population.

Problems in the Field of Biology

Now that you have seen some progress that can be credited to biologists, we will look at some of the problems that have been created by improperly applied biological principles. For example, the drive to preserve nature has in some cases resulted in conflicting goals. Many of our western forests have been preserved as parks. In order to preserve the trees, fire was not allowed in these forests. The lack of fire led to a dangerous buildup of debris, which could result in abnormally devastating fires. Before our involvement, natural fires periodically cleaned out the debris and helped to preserve the trees from more intense fires. These fires also eliminated some undesirable species of trees. As a result, the U.S. Park Service changed its policy on fires in parks and began to allow naturally caused fires

BOX 1.1 *Edward Jenner and the Control of Smallpox*

Edward Jenner first developed the technique of vaccination in 1795. This was the result of a twenty-six-year study of two diseases: cowpox and smallpox. Cowpox was known as *vaccinae.* From this word evolved the present terms *vaccination* and *vaccine.* Jenner observed that milkmaids developed pocklike sores after milking cows infected with cowpox, but they rarely became sick with smallpox. He asked the question, Why don't milkmaids get smallpox? He developed the hypothesis that the mild reaction milkmaids had to cowpox protected them from the often fatal smallpox. This led him to perform an experiment in which he transferred puslike material from the cowpox to human skin and discovered that *vaccinated* people were protected from smallpox. When these results became known, public reaction was mixed. Some people thought that vaccination was the work of the devil. Many European rulers supported Jenner by encouraging their subjects to be vaccinated. Napoleon and the Empress of Russia were very influential and, in the United States, Thomas Jefferson had some members of his family vaccinated. Many years later, following the development of the germ theory of disease, it was discovered that both cowpox and smallpox are caused by viruses that are very similar in structure. Exposure to the cowpox virus allows the body to develop immunity against the cowpox virus and the smallpox virus at the same time. In 1979, almost two hundred years after Jenner developed his vaccination, the Centers for Disease Control and Prevention in the United States and the World Health Organization of the United Nations declared that smallpox was extinct.

to burn themselves out. This caused considerable debate among park managers, forest interests, and the public. A whole generation of people had grown up with the idea that all forest fires were bad. During the extremely dry summer of 1988, several huge fires burned in Yellowstone National Park and other parks in the West (figure 1.8). As a result of these fires, the Park Service has modified its policy and will take measures to control fires if conditions are such that they may alter huge portions of the park ecosystem.

A second major biological mistake has been the introduction into some countries of exotic (foreign) species of plants and animals. In North America, this has had disastrous consequences in a number of cases. Both the American chestnut and the American elm have been nearly eliminated by diseases that were introduced by accident. Other organisms have been introduced on purpose through shortsightedness or a total lack of understanding about biology. The starling and the English (house) sparrow were both introduced into this country by people who thought that they were doing good. Both of these birds have multiplied greatly and displaced some native birds. The gypsy moth is also an introduced species; they were brought to the United States by silk manufacturers in hopes of interbreeding the gypsy moth with the silkworm moth to increase silk production. When the scheme fell short of its goal, and moths were accidentally set free, the moths quickly took advantage of their new environment by feeding on native forest trees. Even with these examples before us, there are still people who try to sneak exotic plants and animals into the country without thinking about the possible consequences.

Bioethics is another major area in which problems have been caused by biological advances. In many ways, technological advances have presented us with a series of ethical situations that we have not been able to resolve satisfactorily. Major advances in health care in this generation have prolonged the lives of people who would have died a generation earlier. Many of the techniques and machines that allow us to preserve and extend life are extremely expensive and are therefore unavailable to most citizens of the world. Furthermore, many people in the world are lacking even the most basic health care, while the rich nations of the world spend money on cosmetic surgery and keep comatose patients alive with the assistance of machines.

Figure 1.8 **Fire Hazard.** The control of fire in national parks became a major issue during the summer of 1988. Many biologists and park managers feel that periodic fires are part of the natural order of things in many kinds of living systems. This thinking was behind the policy that allowed naturally initiated fires to burn in national parks. The exceedingly dry weather of 1988 resulted in disastrous fires that have caused the Park Service to reevaluate the "let-burn policy" and to develop guidelines about which fires should be allowed to burn and which should be controlled.

Future Directions in Biology

Where do we go from here? Many problems remain to be solved. Major breakthroughs in the control of the human population are being sought. There is a continued interest in the development of more efficient methods of producing food.

One of the major areas that will receive attention in the next few years will be the relationship between genetic information and such diseases as Alzheimer's disease, stroke, arthritis, and cancer. Many kinds of diseases are caused by abnormal body chemistry. These changes are the result of hereditary characteristics. Curing certain hereditary diseases is a big job. It requires a thorough understanding of genetics and the introduction or subtraction of hereditary information from all of the trillions of cells of the organism.

Another area that will receive more attention in the next few years is ecology. Worldwide ecosystem damage is occurring, the human population is increasing rapidly, and pollution is still a problem. The majority of people still need to learn that some environmental changes may be acceptable, whereas other changes will ultimately lead to our destruction. We have two tasks. The first is improving technology and increasing our understanding about how things work in our biological world. The second, and probably the hardest, is educating, pressuring, and reminding people that their actions determine the kind of world in which the next generation will live.

• Summary •

The science of biology is the study of living things and how they interact with their surroundings. Science and nonscience can be distinguished by the kinds of laws and rules that are constructed to unify the body of knowledge. Science involves the continuous testing of rules and principles by the collection of new facts. In science these rules are usually arrived at by using the scientific method—observation, questioning, exploring resources, hypothesis formation, experimentation, theory formation, and law formation. If the rules are not testable, or if no rules are used, it is not science. Pseudoscience uses scientific appearances to mislead.

Living things show the characteristics of (1) metabolic processes, (2) generative processes, (3) responsive processes, and (4) control processes. In addition to these functional characteristics, living things have a structural unit called a cell. Biology has been responsible for major advances in the areas of food production and health. The incorrect application of biological principles has sometimes led to the elimination of useful organisms and to the destruction of organisms we wish to preserve. Many biological advances have led to ethical dilemmas that have not been resolved. In the future, biologists will study many things. Two areas that are sure to receive attention are the relationship between heredity and disease, and ecology.

• Thinking Critically •

The scientific method is central to all work that a scientist does. Can this process be used in the ordinary activities of life? How might a scientific approach to life change how you value such things as your choice of clothing, recreational activities, or the kind of car to buy. Can these things be analyzed scientifically? Should they be analyzed scientifically? Is there anything wrong with looking at these things from a scientific point of view?

• Experience This •

Take ten minutes to look through your local newspaper and count just how many advertised products have supposedly been "scienced."

• Questions •

1. What is biology?
2. What is the difference between science and nonscience? Give examples.
3. Why is testing so important in science?
4. What is the difference between theoretical science and applied science?
5. List four characteristics of living things.
6. List three advances that have occurred as a result of biology.
7. List three mistakes that could have been avoided had we known more about living things.
8. The scientific method cannot be used to deny or prove the existence of God. Why?
9. What are controlled experiments? Why are they necessary to support a hypothesis?
10. List the steps in the scientific method.
11. How could you identify pseudoscience?

• Chapter Glossary •

applied science (ap-plīd si′ens) Science that makes practical use of the theories provided by scientists to solve everyday problems

biology (bi-ol′o-je) The science that deals with life.

control group (con-trōl′ grūp) The situation used as the basis for comparison in a controlled experiment.

control processes (con-trōl′ pro′ses-es) Mechanisms that ensure that an organism will carry out all metabolic activities in the proper sequence (coordination) and at the proper rate (regulation).

controlled experiment (con-trōld′ ek-sper′ĭ-ment) An experiment that allows for a comparison of two events that are identical in all but one respect.

empirical evidence (em-pir′ĭ-cal ev′ĭ-dens) The information gained by observing an event.

enzymes (en′zīms) Molecules produced by organisms that are able to control the rate at which chemical reactions occur.

experiment (ek-sper'ĭ-ment) A re-creation of an event in a way that enables a scientist to gain valid and reliable empirical evidence.

experimental group (ek-sper-ĭ-men'tal grūp) The group in a controlled experiment that is identical to the control group in all respects but one.

generative processes (jen'uh-ra''tiv pros'es-es) Actions that increase the size of an individual organism (growth) or increase the number of individuals in a population (reproduction).

hypothesis (hi-poth'e-sis) A possible answer to or explanation of a question that accounts for all the observed facts and is testable.

metabolic processes (me-ta-bol'ik pros'es-es) The total of all chemical reactions within an organism; for example, nutrient uptake and processing, and waste elimination.

observation (ob-sir-vā'shun) The process of using the senses or extensions of the senses to record events.

organ (or'gun) A structure composed of two or more kinds of tissues.

organ system (or'gun sis'tem) A group of organs that performs a particular function.

organism (or'gun-izm) An independent living unit.

pseudoscience (su-dō-sī'ens) Use of the appearance of science to mislead. The assertions made are not valid or reliable.

reliable (re-li'a-bul) A term used to describe results that remain consistent over successive trials.

responsive processes (re-spon'siv pros'es-es) Those abilities to react to external and internal changes in the environment; for example, irritability, individual adaptation, and evolution.

science (si'ens) A process or way of arriving at a solution to a problem or understanding an event in nature using the scientific method.

scientific law (si-en-tif'ik law) A uniform or constant feature of nature supported by several theories.

scientific method (si-en-tif'ik meth'ud) A way of gaining information (facts) about the world around you that involves observation, hypothesis formation, testing of hypotheses, theory formation, and law formation.

theoretical science (the-o-ret'i-kul si'ens) Science interested in obtaining new information for its own sake.

theory (the'o-re) A plausible, scientifically acceptable generalization supported by several hypotheses and experimental trials.

tissue (tish'yu) A group of specialized cells that work together to perform a specific function.

valid (val'id) A term used to describe meaningful data that fit into the framework of scientific knowledge.

variable (var'e-ă-bul) The single factor that is allowed to be different.

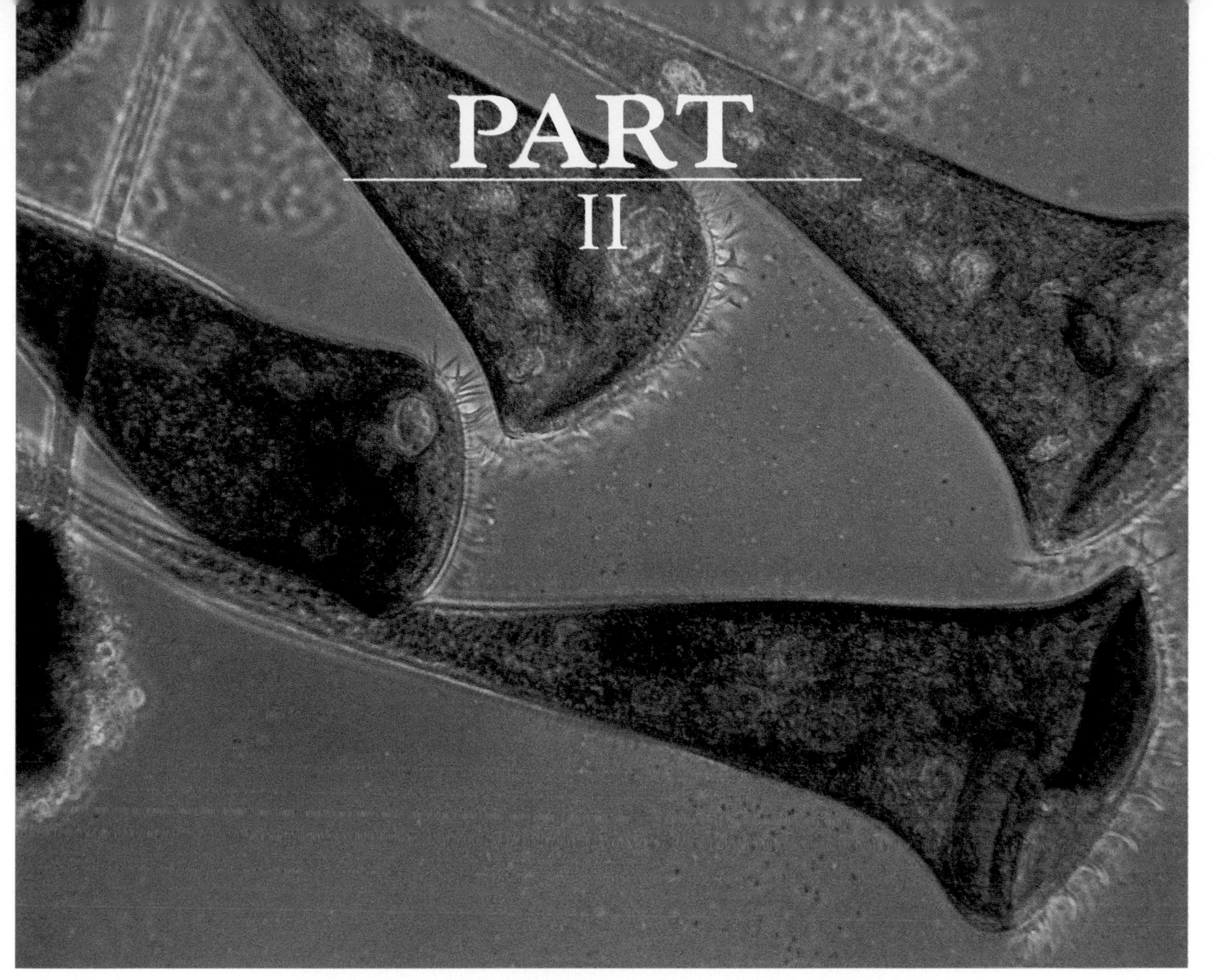

PART II

Cells: Anatomy and Action

The study of living things has revealed that there is a minimal unit that displays all the characteristics of life—the cell. It is necessary to understand the structure and workings of this most basic unit in order to understand a simple or complex organism. Analysis shows that a cell is composed of many basic chemicals found in nature. Those elements are bound together in special ways and arranged into complex molecules, which are in turn fashioned into more complex structures. Two fundamental cell types have been identified, differing in their chemical composition and structure. They also differ in how they perform the chemical reactions required to maintain their existence and reproduce. Biologically important chemical reactions are controlled by the cell and are responsible for the breakdown of nutrients and the manufacture of the cell's essential molecules. By understanding how cells are constructed and how they work, scientists can control them for the benefit of all. ■

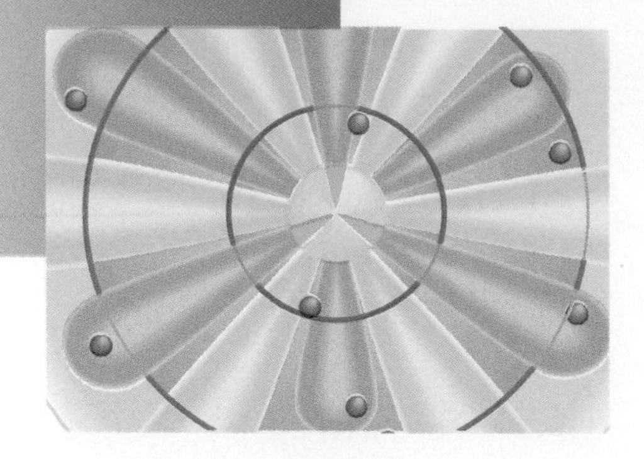

C H A P T E R 2

Simple Things of Life

Chapter Outline

Purpose

In order to understand the structure and activities of living organisms, you must know something about the materials from which they are made. In this chapter we will discuss the structure of matter and the energy it contains. As you read this chapter, you should consciously try to build a vocabulary that will help you describe matter.

For Your Information

Baking soda is sodium bicarbonate. It has several uses in the home. You brush your teeth with it to neutralize acids generated by decay-causing bacteria. Because it neutralizes acids, it is used in cooking to make foods less sour. It is also used in baking as a source of carbon dioxide gas to leaven breads and other baked goods.

Learning Objectives

- Understand that all matter is composed of atoms.
- Explain how molecular motion relates to diffusion and the laws of thermodynamics.
- Know the basic structure of an atom.
- Recognize how isotopes differ from one another.
- Differentiate between ionic, covalent, and hydrogen bonds.
- Describe how acids, bases, salts, and the pH scale interrelate.
- Describe how the three states of matter—solids, liquids, and gases—differ at the molecular level.
- Recognize that atoms may enter into different types of reactions as they become bonded and more stable.

Basic Structures

Everything on earth is part of what we call *matter.* **Matter** is anything that has weight (mass) and also takes up space (volume). Both of these characteristics depend on the amount of matter you are dealing with; the greater the amount, the greater its mass and volume.

Characteristics that are independent of the amount of matter include density and activity. **Density** is the weight of a certain volume of material; it is frequently expressed as grams per cubic centimeter. For example, a cubic centimeter of lead is very heavy and a cubic centimeter of aluminum is very light. Lead has a higher density than aluminum. The activity of matter depends almost entirely on its composition. All matter is composed of one or more types of substances called *elements.* **Elements** are the basic building blocks from which all things are made. You already know the names of some of these elements: oxygen, iron, aluminum, silver, carbon, and gold. The sidewalk, water, air, and your body are all composed of various types of elements.

The Atomic Nucleus

In order to understand the way elements act, we need to understand what they are composed of. The smallest part of an element that still acts like that element is called an **atom.** When we use a **chemical symbol** such as Al for aluminum or C for carbon, it represents one atom of that element. The atom is constructed of three major particles; two of them are in a central region called the **atomic nucleus.** The third type of particle is in the region surrounding the nucleus (figure 2.1). The weight, or mass, of the atom is concentrated in the nucleus, which is composed of neutrons and protons. One major group of particles located in the nucleus is the **neutrons;** they were named *neutrons* to reflect their lack of electrical charge. **Protons,** the second type of particle in the nucleus, have a positive electrical charge. **Electrons,** found in the area surrounding the nucleus, have a negative charge.

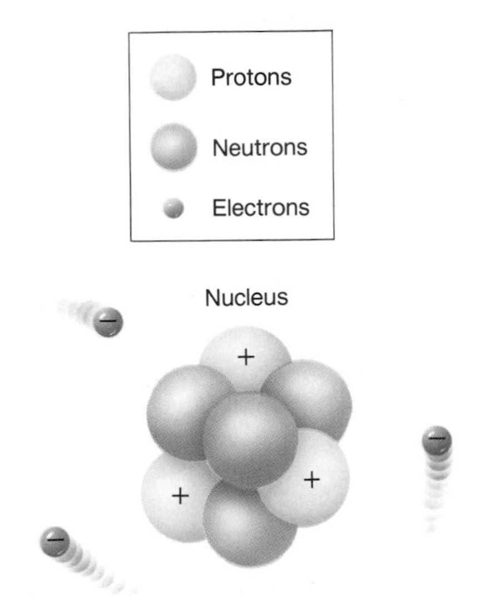

Figure 2.1 **Atomic Structure.** The nucleus of the atom contains the protons and the neutrons, which are the massive particles of the atom. The electrons, much less massive, are in constant motion about the nucleus.

An atom is neutral in charge because the number of positively charged protons is balanced by the number of negatively charged electrons. You can determine the number of either of these two particles in a balanced atom if you know the number of the other particle. For instance, carbon with six protons would have six electrons, oxygen with eight electrons would have eight protons, and hydrogen with one proton would have one electron.

The atoms of each kind of element have a specific number of protons. For example, oxygen always has eight protons and no other element has that number. Carbon always has six protons. The **atomic number** of an element is the number of protons in an atom of that element; therefore, each element has a unique atomic number. Since oxygen has eight protons, its atomic number is eight. The mass of a proton is 1.67×10^{-24} grams. Since this is an extremely small mass and is awkward to express, it is said to be equal to one **atomic mass unit,** abbreviated **AMU.** One AMU is actually 1/12 of the mass of a particular carbon atom, but is very close to the mass of each proton (table 2.1).

Although all atoms of the same element have the same number of protons, they do not always have the same number of neutrons. In the case of oxygen, over 99% of the atoms have eight neutrons, but there are others with more or fewer neutrons. Each atom of an element with a particular number of neutrons is called an **isotope.**

The most common isotope of oxygen has eight neutrons, but another isotope of oxygen has nine neutrons. We can determine the number of neutrons by comparing the masses of the isotopes. The **mass number** of an atom is the number of protons plus the number of neutrons in the nucleus. The mass number is customarily used to compare different isotopes of the same element. An oxygen isotope with a mass number of sixteen AMUs is composed of eight protons and eight neutrons and is identified as ^{16}O. Oxygen 17, or ^{17}O, has a

Table 2.1
Comparison of Atomic Particles

	Protons	Electrons	Neutrons
Location	Nucleus	Outside nucleus	Nucleus
Charge	Positive (+)	Negative (−)	None (neutral)
Number present	Identical to the atomic number	Equal to number of protons	Mass number minus atomic number
Mass	1 AMU	1/1,836 AMU	1 AMU

BOX 2.1 The Periodic Table of the Elements

Traditionally, elements are represented in a shorthand form by letters. For example, the symbol for water, H_2O, shows that a molecule of water consists of two atoms of hydrogen and one atom of oxygen. These chemical symbols can be found on any periodic table of elements. Using the periodic table, we can determine the number and position of the various parts of atoms. Notice that the atoms numbered 3, 11, 19, and so on, are in column one. The atoms in this column act in a similar way since they all have one electron in their outermost layer. In the next column, Be, Mg, Ca, and so on, act alike because these metals all have two electrons in their outermost electron layer. Similarly, atoms 9, 17, 35, and so on, all have seven electrons in their outer layer. Knowing how fluorine, chlorine, and bromine act, you can probably predict how iodine will act under similar conditions. At the far right in the last column, argon, neon, and so on, all act alike. They all have eight electrons in their outer electron layer. Atoms with eight electrons in their outer electron layer seldom form bonds with other atoms.

The Periodic Table of the Elements

IA	IIA	IIIB	IVB	VB	VIB	VIIB	VIII	VIII	VIII	IB	IIB	IIIA	IVA	VA	VIA	VIIA	O	
1 H 1.008																1 H 1.008	2 He 4.003	2
3 Li 6.939	4 Be 9.012											5 B 10.81	6 C 12.01	7 N 14.01	8 O 16.00	9 F 19.00	10 Ne 20.18	2·8
11 Na 22.99	12 Mg 24.31											13 Al 26.98	14 Si 28.09	15 P 30.97	16 S 32.06	17 Cl 35.45	18 Ar 39.95	2·8·8
19 K 39.10	20 Ca 40.08	21 Sc 44.96	22 Ti 47.90	23 V 50.94	24 Cr 52.00	25 Mn 54.94	26 Fe 55.85	27 Co 58.93	28 Ni 58.71	29 Cu 63.54	30 Zn 65.37	31 Ga 69.72	32 Ge 72.59	33 As 74.92	34 Se 78.96	35 Br 79.91	36 Kr 83.80	2·8·18·8
37 Rb 85.47	38 Sr 87.62	39 Y 88.91	40 Zr 91.22	41 Nb 92.91	42 Mo 95.94	43 Tc (97)	44 Ru 101.1	45 Rh 102.9	46 Pd 106.4	47 Ag 107.9	48 Cd 112.4	49 In 114.8	50 Sn 118.7	51 Sb 121.8	52 Te 127.6	53 I 126.9	54 Xe 131.3	2·8·18·18·8
55 Cs 132.9	56 Ba 137.3	57 La 138.9	72 Hf 178.5	73 Ta 180.9	74 W 183.9	75 Re 186.2	76 Os 190.2	77 Ir 192.2	78 Pt 195.1	79 Au 197.0	80 Hg 200.6	81 Tl 204.4	82 Pb 207.2	83 Bi 209.0	84 Po (210)	85 At (210)	86 Rn (222)	2·8·18·32·18·8
87 Fr (223)	88 Ra (226)	89 Ac (227)																

Lanthanum Series

58	59	60	61	62	63	64	65	66	67	68	69	70	71	
Ce 140.1	Pr 140.1	Nd 144.2	Pm (147)	Sm 150.4	Eu 152.0	Gd 157.3	Tb 158.9	Dy 162.5	Ho 164.9	Er 167.3	Tm 168.9	Yb 173.0	Lu 175.0	2·8·18·32·9·2

Actinium Series

90	91	92	93	94	95	96	97	98	99	100	101	102	103	
Th 232.0	Pa (231)	U 238.0	Np (237)	Pu (242)	Am (243)	Cm (247)	Bk (247)	Cf (249)	Es (254)	Fm (253)	Md (256)	No (253)	Lw (259)	2·8·18·32·?·9·2

Key: 6 — Atomic number (# of protons); C — Symbol; 12.01 — Atomic mass

1—Hydrogen (H)
2—Helium (He)
3—Lithium (Li)
4—Beryllium (Be)
5—Boron (B)
6—Carbon (C)
7—Nitrogen (N)
8—Oxygen (O)
9—Fluorine (F)
10—Neon (Ne)
11—Sodium (Na)
12—Magnesium (Mg)
13—Aluminum (Al)
14—Silicon (Si)
15—Phosphorus (P)
16—Sulfur (S)
17—Chlorine (Cl)
18—Argon (Ar)
19—Potassium (K)
20—Calcium (Ca)
26—Iron (Fe)
29—Copper (Cu)
30—Zinc (Zn)
35—Bromine (Br)
47—Silver (Ag)
53—Iodine (I)
79—Gold (Au)
80—Mercury (Hg)
82—Lead (Pb)
90—Thorium (Th)
92—Uranium (U)

mass of seventeen AMUs. Eight of these units are due to the eight protons that every oxygen atom has; the rest of the mass is due to nine neutrons ($17 - 8 = 9$). Figure 2.2 shows different isotopes of hydrogen.

The **periodic table of the elements** (see box 2.1) lists all the elements in order of increasing atomic number (number of protons). In addition, this table lists the mass number of each element. You can use these two numbers to determine the number of the three major particles in an atom—protons, neutrons, and electrons. Look at the periodic table and find helium in the upper right-hand corner (He). Two is its atomic number; thus, every helium atom will have two protons. Since the protons are positively charged, the nucleus will have two positive charges that must be balanced by two negatively charged electrons. The mass of helium is given as 4.003. This is the calculated

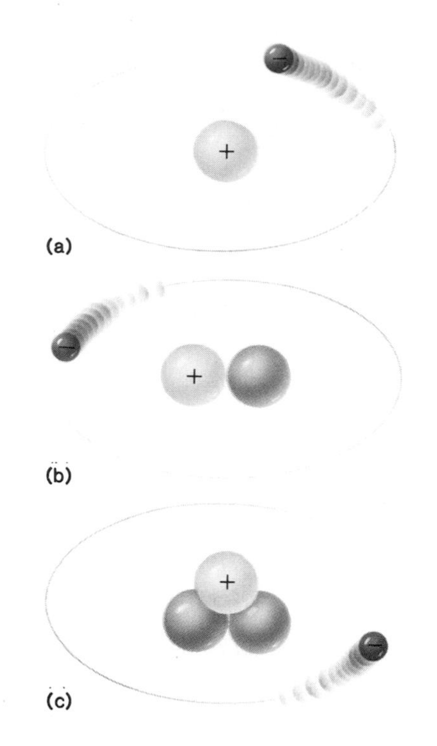

Figure 2.2 **Isotopes of Hydrogen.** (a) The most common form of hydrogen is the isotope that is 1 AMU. It is composed of one proton and no neutrons. (b) The isotope deuterium is 2 AMU and has one proton and one neutron. (c) Tritium, 3 AMU, has two neutrons and one proton. Each of these isotopes of hydrogen also has one electron, but since the mass of an electron is so small, they do not contribute significantly to the mass as measured in AMU.

average mass of a group of helium atoms. Most of them have a mass of four—two protons and two neutrons. Generally, you will need to work only with the most common isotope, so the mass number should be rounded to the nearest whole number. If it is a number like 4.003, use 4 as the most common mass. If the mass number is a number like 39.95, use 40 as the nearest whole number. Look at several atoms in the periodic table. You can easily determine the number of protons and the number of neutrons in the most common isotopes of almost all of these atoms.

Since isotopes differ in the number of neutrons they contain, it is logical to assume that some isotopes would have characteristics that were different from those of the most common form of the element. For example, there are many isotopes of iodine. The most common isotope of iodine is ^{127}I; it has a mass number of 127. A different isotope of iodine is ^{131}I; its mass number is 131 and it is **radioactive.** This means that it is not stable and that its nucleus disintegrates, releasing energy and particles from its nucleus. The energy can be detected by using photographic film or a Geiger counter. If a physician suspects that a patient has a thyroid gland that is functioning improperly, ^{131}I may be used to help confirm the diagnosis. The thyroid normally collects iodine atoms from the blood and uses them in the manufacture of the body-regulating chemical thyroxine. If the thyroid gland is working properly to form thyroxine, the radioactive iodine will collect in the gland, where its presence can be detected. If no iodine has collected there, the physician knows that the gland is not functioning correctly and can take steps to help the patient.

Electron Distribution

Electrons are the negatively charged particles of an atom that balance the positive charges of the protons in the atomic nucleus. Notice in table 2.1 that the mass of an electron is a tiny fraction of the mass of a proton. This mass is so slight that it usually does not influence the AMU of an element. But electrons are important even though they do not have a major effect on the mass of the element. The number and position of the electrons in an atom are responsible for the way atoms interact with each other.

Electrons are constantly moving at great speeds and tend to be found in specific regions some distance from the nucleus (figure 2.3). The position of an electron at any instant in time is determined by several factors. First, since protons and electrons are of opposite charge, electrons are attracted to the protons in the nucleus of the atom. Second, counterbalancing this is the force created by the movement of the electrons, which tends to cause them to move away from the nucleus. Third, the electrons repel one another because they have identical charges. The balance of these three forces creates a situation in which the electrons of an atom tend to remain in the neighborhood of the nucleus but are distributed apart from one another. Electron distribution is not random, but electrons are likely to be found in certain locations.

When chemists first described the atom, they tried to account for the fact that electrons seemed to be traveling at one of several different speeds about the

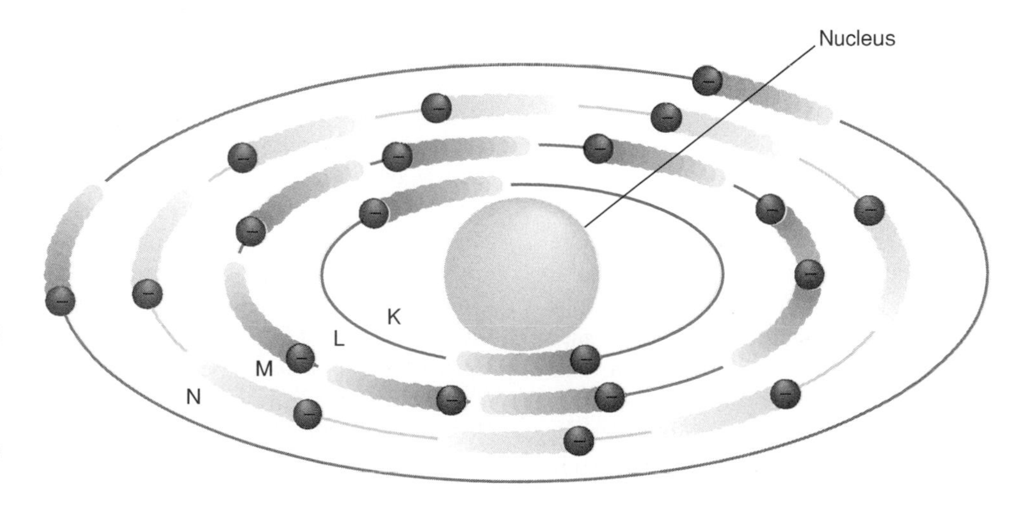

Figure 2.3 **The Bohr Atom.** Several decades ago we thought that electrons revolved around the nucleus of the atom in particular paths, or tracks. Each track was labeled with a letter: K, L, M, N, and so on. Each track was thought to be able to hold a specific number of electrons moving at a particular speed. These electron tracks were described as quanta of energy.

atomic nucleus. Electrons did not travel at intermediate speeds. Because of this, it was thought that electrons followed a particular path, or orbit, similar to the orbits of the planets about the sun.

The model of an atom shown in figure 2.3 is called the Bohr atom because Niels Bohr, a Danish physicist, advanced the theory that electrons move in discrete circular orbits about a nucleus. In the Bohr model, the electrons with the greatest amount of energy are farther from the nucleus. Think of swinging a weight on an elastic strap. As you swing the weight around your head, it makes a path a certain distance from you. If you swing it harder to make it go faster, to give it more energy, the path is a bigger circle a greater distance from your head. It was thought that there were only certain paths that electrons could follow. No electrons were thought to go at intermediate speeds between the particular paths. The speeds or paths were called *quanta* (singular *quantum*), meaning a certain amount of energy. From the collection of early experimental data, it was thought that only two electrons could exist in the first quantum, or *shell*, eight electrons could occupy the second shell, eight (or sometimes eighteen) the third shell, and so on. These shells, also known as *energy levels*, were labeled K, L, M, and so forth.

The Modern Model of the Atom

Several decades ago, as more experimental data were gathered and interpreted, we began to think of the K shell not as a particular pathway, but as a region, or space, within which electrons were likely to be. In this more modern model of the atom, each region or **orbital** is able to hold a maximum of two electrons. Each orbital is designated with a number that indicates the major energy level and a letter that indicates the kind of space the electrons occupy. The first orbital is lowest in energy and is designated as "1s." The "1" indicates it is the first energy level from the nucleus and the "s" is used to help us remember that the space is spherical in shape. (Originally the "s" indicated something entirely different but a lucky happenstance allows us to use it to remember the spherically shaped space.) Thus, the electrons in a helium atom would be located in the area described as an electron cloud in figure 2.4. The area is labeled the "*1s*" orbital, and that orbital is full with its two electrons.

If an atom has more than two electrons, not all of these have the same amount of energy. Neon, for example, has ten electrons. The first two we would say are located in the first energy level, just as the two in the helium atom. They are designated as being in the "*1s*" orbital. The rest of the electrons—the other eight—would be in a higher, second energy level. (This second energy level is similar to the Bohr model of atomic structure with eight electrons located on the second orbit or path.) All eight of these electrons, however, do not have exactly the same energy and they are not likely to occupy the same spacial area.

We now think that two of these eight electrons have an amount of energy that makes it likely that they will occupy a special area of the second energy level, designated as the "*2s*" orbital. The "2" indicates that it is the second energy level and the "*s*" helps us remember that the shape of the space the electrons occupy is spherical. The other six electrons have slightly more energy and they tend to occupy three areas as far away from each other as possible, but still on the second energy level. You might think of these three areas as propeller-shaped areas on the "*x*," "*y*," and "*z*" axes (figure 2.5). Each propeller-shaped area can hold a maximum of two electrons, so the eight electrons of the *L* shell of the Bohr atomic model can now be more accurately described as being located in one spherical area and three propeller-shaped areas at right angles to each other. By convention, we indicate these areas as the "*2s*," the "*2px*," the "*2py*," and the "*2pz*" areas.

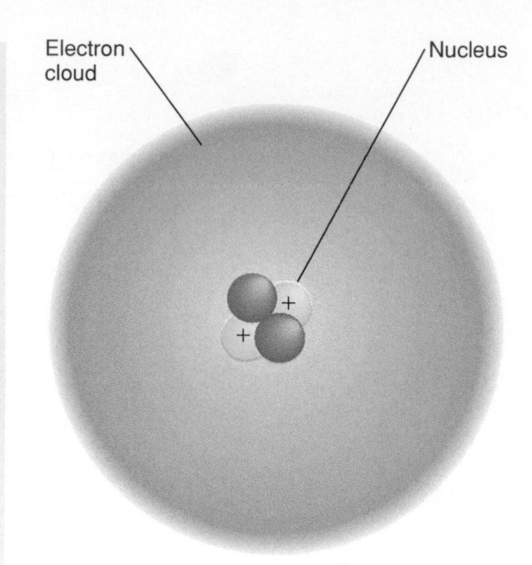

Figure 2.4 **The Electron Cloud.** So fast are the electrons moving around the nucleus that they can be thought of as forming a cloud around it, rather than an orbit or track. You might think of the electron cloud as hundreds of photographs of an atom. Each photograph shows where an electron was at the time the picture was taken. But when the next picture is taken, the electron is somewhere else. Although we are able to determine where an electron is at a given time, we do not know the path it uses to go from there to where it is the next time we determine its position.

The third energy level (formerly called the "M" shell) contains electrons that have a greater amount of energy than those in the second energy level. These electrons are distributed in four different orbitals, which are designated as the "*3s*," "*3px*," "*3py*," and "*3pz*" (figure 2.6). You can see how cluttered the graphic representation of the atom in figure 2.6 becomes when you try to account for the number and location of all its protons, neutrons, and electrons. This will become even more difficult as we deal with larger and larger atoms. A simpler

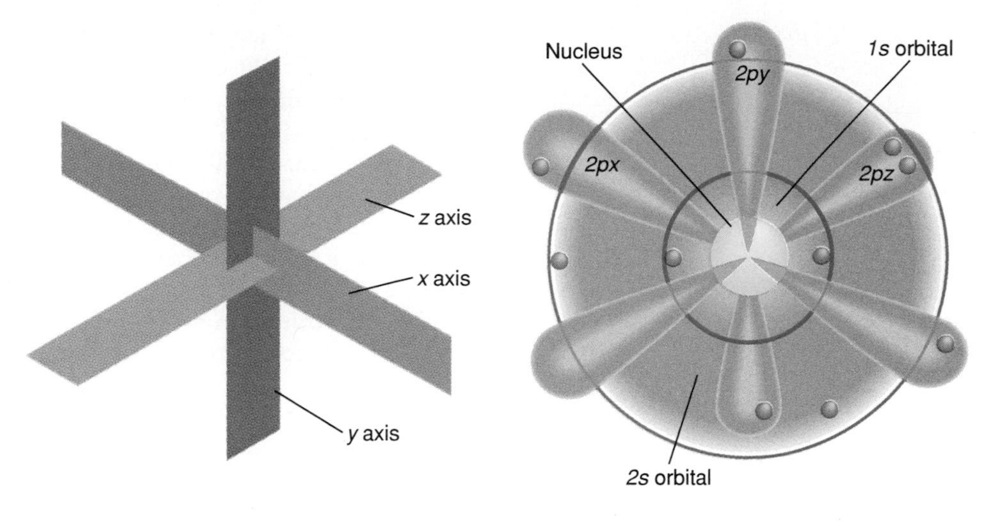

Figure 2.5 **The Second Energy Level of Electrons.** The electrons on the second energy level all have about the same amount of energy, but more energy than the electrons in the first energy level. The electrons are most likely to be located in the four regions labeled *2s*, *2px*, *2py*, and *2pz*.

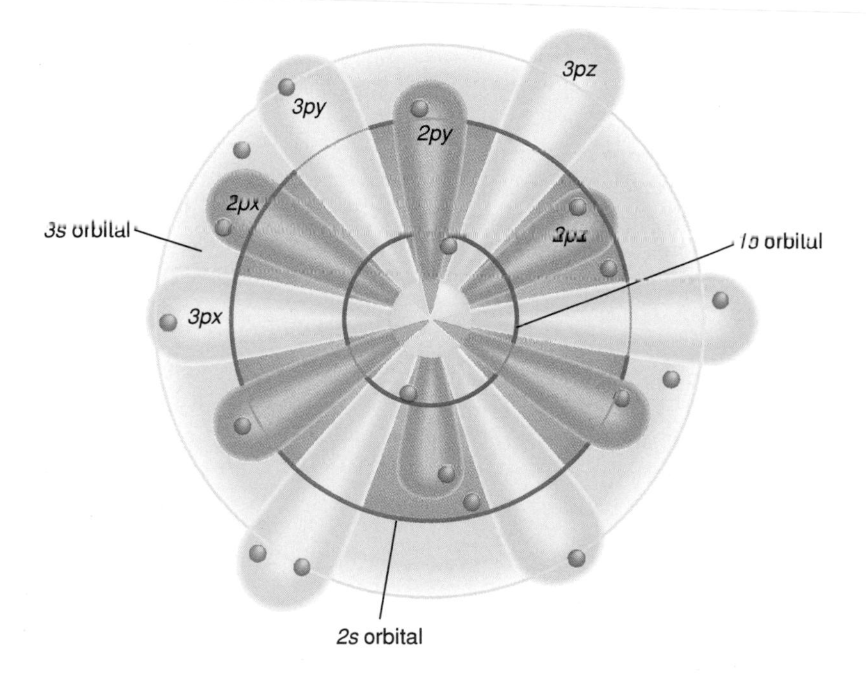

Figure 2.6 **The Third Energy Level of Electrons.** The electrons in these outer clouds all have about the same amount of energy. The areas where they are located are labeled *3s*, *3px*, *3py*, and *3pz*.

way to represent the atom is shown in figure 2.7. The arrows on the diagram represent the electrons. In order to diagram the structure of an atom and place the electrons in their proper orbitals, you must start filling the spaces at the "*1s*" level and move outward. Each orbital is filled with two electrons. If the atom contains more than two electrons, proceed to the second energy level. At the second energy level there are four different orbitals ("*2s*," "*2px*," "*2py*," "*2pz*"). The "*2s*" is filled with electrons first, before any additional electrons are placed in the "*p*" orbitals. After you have filled the "*2s*" orbital, begin adding electrons one at a time to each of the three "*p*" orbitals. An electron is added to the "*2px*," a second is added to the "*2py*," and a third to the "*2pz*." Additional electrons are then added in this same sequence until each orbital contains two electrons. Then you can continue to the third energy level and beyond using the same pattern.

An atom such as potassium, with nineteen protons and nineteen electrons, would have two electrons in the first energy level (*1s*). In the second energy level, there would be two electrons in the *2s* orbital; two electrons in each of the *2p* orbitals; two electrons in the *3s* orbital; two in each of the *3px*, *3py*, and *3pz* orbitals; and one electron in the *4s* orbital.

Ions

Now that you know the rules for positioning electrons in their proper orbitals, it would be convenient if all atoms always followed these rules. Remember that atoms are electrically neutral when they have equal numbers of protons and electrons. Certain atoms, however, are able to exist with an unbalanced charge. These unbalanced, or charged, atoms are called **ions.** The ion of sodium is formed when one of the eleven electrons of the sodium atom escapes. Let's look at the electron distribution to help explain how and why this happens. The sodium nucleus is composed of eleven positive charges (protons) insulated from each other by twelve neutrons. (The most common isotope of sodium is sodium 23, which has twelve neutrons.) The eleven electrons that balance the charge are most likely positioned as follows: two electrons in the first energy level, eight in the second energy level, and one in the third energy level. Focus your attention on the outermost electron. It has more energy than any of the other electrons. But because it is further from the nucleus than any other electron, it is not as strongly attracted to the positive charges

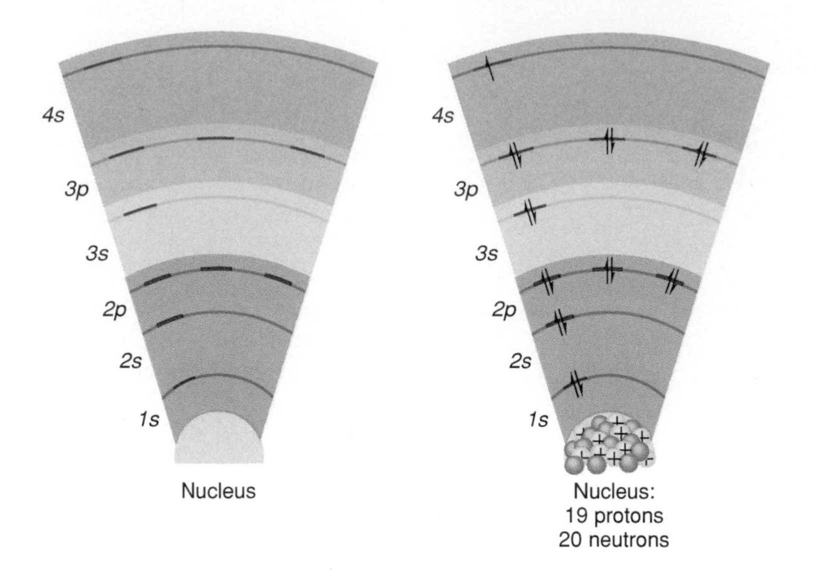

Figure 2.7 **Electron Configuration.** This chart is like a theater seating chart. The lines represent places where electrons can be. Each line can hold a maximum of two electrons. The electrons are most likely to be in the lower energy levels. They will go in an empty area of the same energy level before they will occupy an area that already has a negative electron in it. The filled-in chart on the right is of the atom potassium, number 19.

in the nucleus. This is similar to gravitational attraction—the closer to earth an object is, the greater the gravitational pull. Since this electron is the least attracted to the nucleus and has the most kinetic energy, when conditions are right it might escape from the sodium atom. What remains when the electron leaves the atom is called the *ion*. In this case, the sodium ion is composed of the eleven positively charged protons and the twelve neutral neutrons—but it has only ten electrons. The fact that there are eleven positive and only ten negative charges means that there is an excess of one positive charge. We still use the chemical symbol Na to represent the ion, but we add the $^{+}$ to indicate that it is no longer a neutral atom, but an electrically charged ion (i.e., Na^{+}). It is easy to remember that a positive ion is formed because it loses negative electrons.

The sodium ion is relatively stable because its outermost energy level is full. A sodium atom will lose one electron from its third major energy level so that the second energy level becomes outermost and is full of electrons. Similarly, magnesium loses two electrons from its third major energy level so that the second major energy level, which is full with eight electrons, becomes outermost. When a magnesium atom (Mg) loses two electrons, it becomes a magnesium ion (Mg^{++}). The periodic table of the elements is arranged so that all atoms in the first column become ions in a similar way. That is, when they form ions, they do so by losing one electron. Each becomes a $^{+}$ ion. Atoms in the second column of the periodic table become $^{++}$ ions when they lose two electrons. Those atoms at the extreme right of the periodic table of the elements do not become ions; they tend to be stable as atoms. These atoms are called *inert* because of their lack of activity. They seldom react because their protons and electrons are equal in number and they have a full outer energy level; therefore, they are not likely to lose electrons.

The column to the left of these gases contains atoms that lack a full outer energy level. They all require an additional electron. Fluorine with its nine electrons would have two in the *K* shell (*1s* orbital) and seven in the *L* shell (two in *2s*, two in *2px*, two in *2py*, and one in *2pz*). The second major energy level can hold a total of eight electrons. You can see that one additional electron could fit into the *2pz* orbital. Whenever the atom of fluorine can, it will accept an extra electron so that its outermost energy level is full. When it does so, it no longer has a balanced charge. When it accepts an extra electron, it has one more negative electron than positive protons; thus, it has become a negative ion (F^{-}) (figure 2.8).

Similarly, chlorine will form a $^{-}$ ion. Oxygen, in the next column, will accept two electrons and become a negative ion with two extra negative charges (O^{--}). If you know the number and position of the electrons, you are better able to hypothesize whether or not an atom will become an ion and, if it does, whether it will be a positive ion or a negative ion. You can use the periodic table of the elements to help you determine an atom's ability to form ions. This information is useful as we see how ions react to each other.

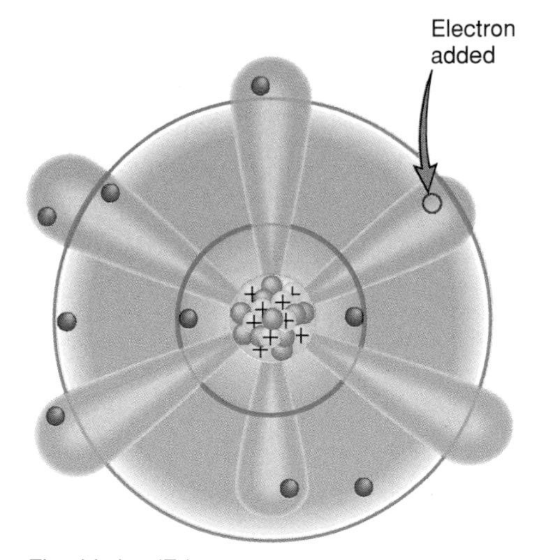

Figure 2.8 **Fluoride Ion.** When the fluorine atom accepts an additional electron, it becomes a negative ion. Negative ions are indicated with a minus sign and often end in *-ide*.

Chemical Bonds

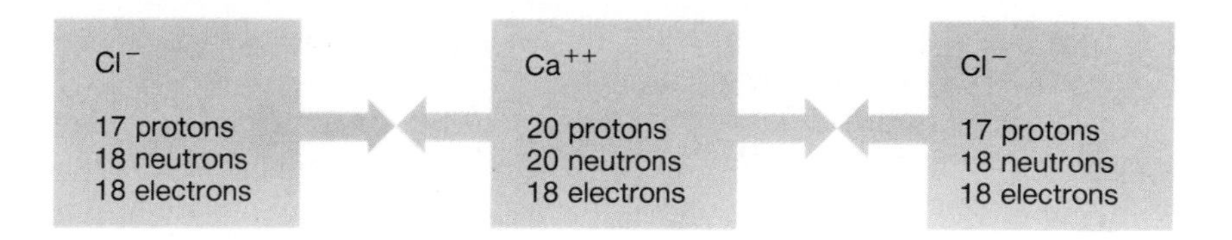

Figure 2.9 **Calcium Chloride.** This combination of a calcium ion and two chloride ions makes up the compound calcium chloride. The formula of the compound is $CaCl_2$. Notice that the overabundance of two positive charges on the calcium ion is offset by the two chloride ions, each of which has an overabundance of only one negative charge.

There are a variety of physical and chemical forces that act on atoms and make them attractive to each other. Each of these results in a particular arrangement of atoms or association of atoms. The forces that combine atoms and hold them together are called **chemical bonds.** There are several types of chemical bonds. They differ from each other with respect to the kinds of attractive forces holding the atoms together. The bonding together of atoms results in the formation of a *compound.* This **compound** is composed of a specific number of atoms (or ions) joined to each other in a particular way. We generally use the chemical symbols for each of the component atoms when we designate a compound. Sometimes there will be a small number behind the chemical symbol. This number indicates how many atoms of that particular element are used in the compound. The group of chemical symbols and numbers is termed a **formula;** it will tell you what elements are in a compound and also how many atoms of each element are required. For example, $CaCl_2$ tells us that the compound of calcium chloride is composed of one calcium atom and two chlorine atoms (figure 2.9).

The properties of a compound are very different from the properties of the atoms that make up the compound. Table salt is composed of the elements sodium and chlorine bound together. Both sodium and chlorine are very poisonous when they are by themselves. Yet, when they are combined as salt, the compound is a nontoxic substance, essential for living organisms.

Ionic Bonds

When positive and negative ions are near each other, they are mutually attracted because of their opposite charges. This attraction between ions of opposite charge results in the formation of a stable group of ions. This attraction is termed an **ionic bond.** Compounds that form as a result of attractions between ions are called *ionic compounds* (see figure 2.9) and are very important in living systems. We can categorize these ionic compounds into three different groups.

Acids, Bases, and Salts

Acids and bases are two classes of biologically important compounds. Their characteristics are determined by the nature of their chemical bonds. When acids are dissolved in water, hydrogen ions (H^+) are set free. The hydrogen ion is positive because it has lost its electron and now has only the positive charge of the proton. An **acid** is any ionic compound that releases a hydrogen ion in a solution. You can think of an acid, then, as a substance able to donate a proton to a solution. However, this is only part of the definition of an acid. We also think of acids as compounds that act like the hydrogen ion—they attract negatively charged particles. An example of a common acid with which you are probably familiar is the sulfuric acid (H_2SO_4) in your automobile battery.

A **base** is the opposite of an acid in that it is an ionic compound that releases a group known as a **hydroxyl ion,** or OH^- group. This group is composed of an oxygen atom and a hydrogen atom bonded together, but with an additional electron. The hydroxyl ion is negatively charged. It is a base because it is able to donate electrons to the solution. A base can also be thought of as any substance that is able to attract positively charged particles. A very strong base used in oven cleaners is NaOH, sodium hydroxide.

The degree to which a solution is acidic or basic is represented by a quantity known as **pH.** The pH scale is a measure of hydrogen ion concentration. A pH of seven indicates that the solution is neutral and has an equal number of H^+ ions and OH^- ions to balance each other. As the pH number gets smaller, the number of hydrogen ions in the solution increases. A number higher than seven indicates that the solution has more OH^- than H^+. As the pH number gets larger, the number of hydroxyl ions increases (figure 2.10).

An additional group of biologically important ionic compounds is called the *salts.* **Salts** are compounds that do not release either H^+ or OH^-; thus, they are neither acids nor bases. They are generally the result of the reaction between an acid and a base in a solution. For example, when an acid such as HCl is mixed with NaOH in water, the H^+ and the OH^- combine with each other to form water, H_2O. The remaining ions (Na^+ and Cl^-) join to form the salt NaCl.

$$HCl + NaOH \rightarrow (Na^+ + Cl^- + H^+ + OH^-) \rightarrow NaCl + H_2O$$

The chemical process that occurs when acids and bases react with each other is called **neutralization.** The acid no longer acts as an acid (it has been neutralized) and the base no longer acts like a base.

As you can see from figure 2.10, not all acids or bases produce the same pH. Some compounds release hydrogen ions very readily, cause low pHs, and are called *strong acids.* Hydrochloric acid (HCl) and sulfuric acid (H_2SO_4) are examples of strong acids. Many other compounds give up their hydrogen ions grudgingly and therefore do not change the pH very much. They are known as *weak acids*. Carbonic acid (H_2CO_3) and many organic acids found in living things are weak acids. Similarly, there are strong bases like sodium hydroxide ($NaOH$) and weak bases like sodium bicarbonate ($NaHCO_3$).

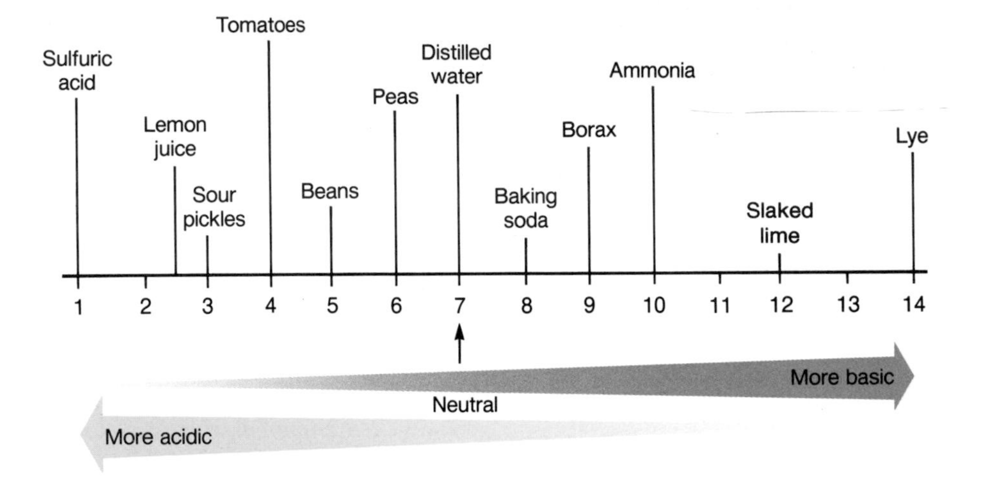

Figure 2.10 **The pH Scale.** The concentration of acid (proton donor or electron acceptor) is greatest when the pH number is lowest. As the pH number increases, the concentration of base (proton acceptor or electron donor) increases. At a pH of 7.0, the concentrations of H^+ and OH^- are equal.

Covalent Bonds

In addition to ionic bonds, there is a second strong chemical bond known as a *covalent bond*. A **covalent bond** is formed by two atoms that share a pair of electrons. This sharing can occur when two atoms have orbitals that overlap one another. A covalent bond should be thought of as belonging to each of the atoms involved. You can visualize the bond as people shaking hands: the people are the atoms, the hands are electrons to be shared, and the handshake is the combining force (figure 2.11). Generally, this sharing of a pair of electrons is represented by a single straight line between the atoms involved. The reason covalent bonds form relates to the arrangement of electrons within the atoms. There are many elements that do not tend to form ions. They will not lose electrons, nor will they gain electrons. Instead, these elements get close enough to other atoms that have unfilled outer orbitals and share electrons with each other. If the two elements have orbitals that overlap, the electrons can be shared. By sharing electrons, the unfilled outer energy levels of each atom will be filled. Both atoms become more stable as a result of the formation of this covalent bond.

Molecules are defined as the smallest particles of chemical compounds. They are composed of a specific

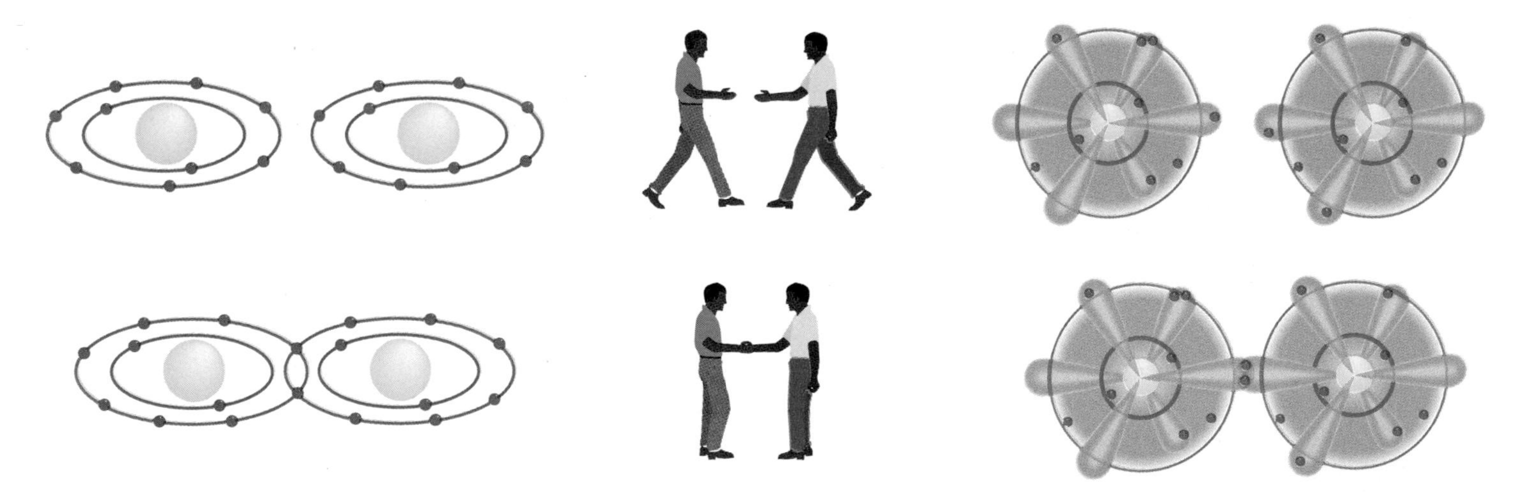

Figure 2.11 **Covalent Bonds.** When two atoms come sufficiently close to each other that the locations of the outermost electrons overlap, an electron from each one can be shared to "fill" that outermost energy-level area. Each person above has to get close enough to the other person so that their hands can overlap to form a handshake. At the left, using the Bohr model, the L-shells of the two atoms overlap, and so each shell appears to be full. Using the modern model at the right, the propeller-shaped orbitals of the second energy level of each atom overlap, so that each propeller appears to have a full orbital. Notice that just as it takes two hands to form a handclasp, it takes two electrons to form a covalent bond.

number of atoms arranged in a particular pattern. For example, a molecule of water is composed of one oxygen atom bonded covalently to two atoms of hydrogen. The shared electrons are in the second energy level of oxygen, and the bonds are almost at right angles to each other. Now that you realize how and why bonds are formed, it makes sense that only certain numbers of certain atoms will bond with each other to form molecules. Chemists also use the term *molecule* to mean the smallest naturally occurring part of an element or compound. Using this definition, one atom of iron is a molecule because one atom is the smallest natural piece of the element. Hydrogen, nitrogen, and oxygen tend to form into groups of two atoms. Molecules of these elements are composed of two atoms of hydrogen, two atoms of nitrogen, and two atoms of oxygen, respectively.

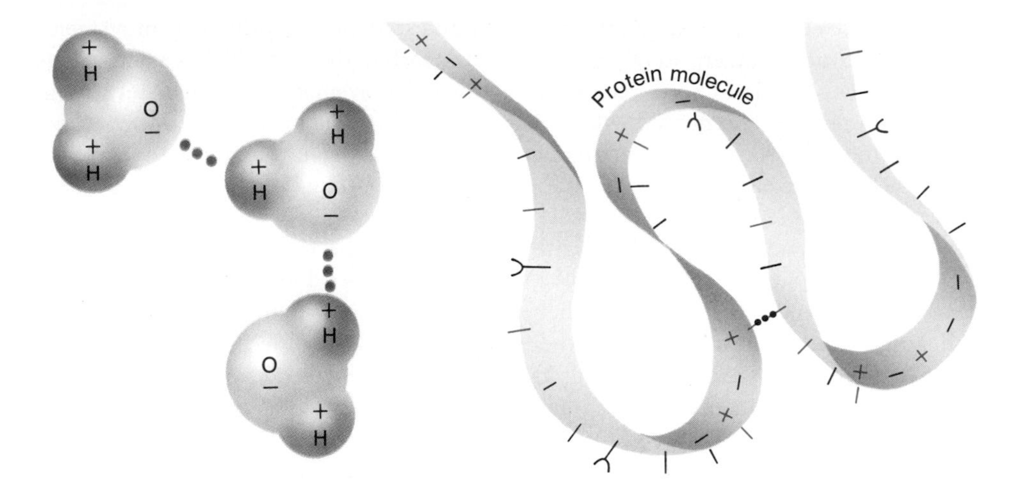

Figure 2.12 **Hydrogen Bonds.** Water molecules arrange themselve so their positive portions are near the negative portions of other water molecules. The attractions are indicated as three dots. The large protein molecule here also has polar areas. When the molecule is folded so that the partially positive areas are near the partially negative areas, a slight attraction forms that tends to keep it folded.

Hydrogen Bonds

Molecules that are composed of several atoms sometimes have an uneven distribution of charge. This may occur because the electrons involved in the formation of bonds may be located on one side of the molecule. This makes that side of the molecule slightly negative and the other side slightly positive. One side of the molecule has possession of the electrons more than the other side. When a molecule is composed of several atoms that have this uneven charge distribution, the whole molecule may show a positive side and a negative side. We sometimes think of such a molecule as a tiny magnet with a positive pole and a negative pole. This polarity of the molecule may influence how the molecule reacts with other molecules. When several of these polar molecules are together, they orient themselves so that the partially positive end of one is near the partially negative end of another. This attraction between two molecules is called a **hydrogen bond.** Since hydrogen has the least attractive force for electrons when it is combined with other elements, the hydrogen electron tends to spend more of its time encircling the other atom's nucleus than its own. The result is the formation of a polar molecule. When the negative pole of this molecule is attracted to the positive pole of another similar polar molecule, the hydrogen will usually be located between the two molecules. Since the hydrogen serves as a bridge between the two molecules, this weak bond has become known as a *hydrogen bond.*

We usually represent this attraction as three dots between the attracted regions. This weak bond is not responsible for forming molecules, but it is important in determining how groups of molecules are arranged. Water, for example, is composed of polar molecules that form hydrogen bonds (figure 2.12, left). Because of this, individual water molecules are less likely to separate from each other. They need a large input of energy to become separated. This is reflected in the relatively high boiling point of water. In addition, when a very large molecule, such as a protein or DNA (which is long and threadlike), has parts of its structure slightly positive and other parts slightly negative, these two areas will attract each other and result in coiling or folding of the chain of molecules in particular ways (figure 2.12, right).

Molecular Energy

Molecules have a certain amount of energy, and therefore are able to move. While we cannot see the movement of the individual molecules, we can deduce several things about their movement by measuring their activity and noting the results of their movement.

Molecular Motion

All molecules have a certain amount of **kinetic energy,** the energy of motion. The amount of energy that a molecule has is related to how fast it moves. **Temperature** is a measure of this velocity or energy of motion. The higher the temperature, the faster the molecules are moving. The three **states of matter**—solid, liquid, and gas—can be explained by thinking of the relative amounts of energy possessed by the molecules of each. A **solid** contains molecules packed tightly together. The molecules vibrate in place and are strongly attracted to each other. They are moving rapidly and constantly bump into each other. The amount of kinetic energy in a solid is less than that in a liquid of the same material. A **liquid** has molecules still strongly attracted to each other, but slightly farther apart. Since they are moving more rapidly, they sometimes slide past each other as they move. This gives the flowing property to a liquid. Still more energetic are the molecules of a **gas.** The attraction the gas molecules have for each other is overcome by the speed with which the individual molecules move. Since they are moving the fastest, their collisions tend to push them farther apart and so a gas expands to fill its container. A common example of a substance that displays the three states of matter is water. Ice, liquid water, and water vapor are all composed of the same chemical—H_2O. The molecules are moving at different speeds in each state because of the difference in kinetic energy. Considering the amount of energy in the molecules of each state of matter helps us explain changes such as freezing and melting. When a liquid becomes a solid, its molecules lose some of their energy; when it becomes a gas, its molecules gain energy.

Diffusion

There is a natural tendency for molecules of different types to completely mix with each other. This is because they are moving constantly. Their movement is random and is due to the energy found in the individual molecules. Consider two types of molecules. As the molecules of one type move about, they tend to disperse from a central location. The other type of molecule also tends to disperse. The result of this random motion is that the two types of molecules are eventually mixed with each other. Remember that the motion of the molecules is completely random. They do not move because of conscious thought—they move because of their kinetic energy. If you follow the paths of molecules from a sugar cube placed in a glass of water, you would find that some of the sugar molecules would move away from the cube, while others would move in the opposite direction. However, there would be more sugar molecules moving away from the original cube because there were more there to start with. We generally are not interested in the individual movement, but rather the overall movement. This overall movement is termed **net movement.** It is the movement in one direction minus the movement in the opposite direction. The direction of greatest movement (net movement) is determined by the relative concentration of the molecules. **Diffusion** is the resultant movement; it is defined as the net movement of a kind of molecule from a place where that molecule is in higher concentration to a place where that molecule is more scarce (figure 2.13). As we will see later, diffusion is one of the important ways that materials enter and leave cells.

When a kind of molecule is completely dispersed, and movement is equal in all directions, we say that the system has reached a state of **dynamic equilibrium.** There is no longer a net movement because movement in one direction equals movement in the other. It is dynamic, however, because the system still has energy, and the molecules are still moving. The kind of energy we have just dealt with is termed *kinetic energy* or *energy of motion. Energy in the universe remains constant: it can neither be created nor destroyed.* This concept is termed the **first law of thermodynamics.** An object that appears to be motionless does not necessarily lack energy. Its individual molecules will still be moving, but the object itself appears to be stationary. An object on top of a mountain may be motionless, but still may contain significant amounts of *potential energy*. **Potential energy** is defined as the energy an object has due to its position. If an object were to start rolling down a mountain, the potential energy it contained at the top of the mountain would be converted into kinetic energy. Kinetic and potential energy have many forms, such as light, heat, sound, X rays, radio waves, and electricity. All forms of energy can be interconverted. In biology, we are concerned with many types of energy conversions. Whenever such a conversion takes place, there is always a loss of some "usable" energy. For example, in an ordinary light bulb, electrical energy is converted to usable light energy; however, some heat energy is lost as unusable energy. *Whenever energy is converted from one form to another, some useful energy is lost.* This statement is the **second law of thermodynamics.**

Figure 2.13 **Diffusion.** Random molecular movement causes the molecules to move in all directions; however, more of them move from where they are in greater concentration to where they are in lesser concentration. This net movement is called diffusion.

Reactions

When molecules interact with each other and rearrange their chemical bonds, we say that they have undergone a **chemical reaction.** A chemical reaction usually involves a change in energy as well as some rearrangement in the molecular structure. We frequently use a chemical shorthand to express what is going on. An arrow ($\rightarrow$) indicates that a chemical reaction is occurring. The arrowhead points to the materials that are produced by the reaction; we call these the **products.** On the other side of the arrow, we generally show the materials that are going to react with each other; we call these the **reactants.** Some of the most fascinating information we have learned recently concerns the way in which living things manipulate chemical reactions to release or store chemical energy. This material is covered in detail in chapters 5 and 6. Figure 2.14 shows the chemical shorthand used to indicate several reactions. The chemical shorthand is called an *equation.* Look closely at the equations and identify the reactants and products in each.

$$HCl + NaOH \longrightarrow NaCl + H_2O$$

$$C_6H_{12}O_6 + 6\ O_2 \longrightarrow 6\ H_2O + 6\ CO_2 + \text{energy}$$

$$C_6H_{12}O_6 + C_6H_{12}O_6 \longrightarrow C_{12}H_{22}O_{11} + H_2O$$

Figure 2.14 **Chemical Equations.** The three equations here use chemical shorthand to indicate that there has been a rearrangement of the chemical bonds in the reactants to form the products. Along with the rearrangement of the chemical bonds, there has been a change in the energy content.

• Summary •

All matter is composed of atoms, which contain a nucleus of neutrons and protons. The nucleus is surrounded by moving electrons. There are many kinds of atoms, called elements. These differ from one another by the number of protons and electrons they contain. Each is given an atomic number, based on the number of protons in the nucleus and an atomic mass number, determined by the total number of protons and neutrons. Atoms of an element that have the same atomic number but differ in their atomic mass number are called isotopes. Some isotopes are radioactive, which means that they fall apart releasing energy and smaller, more stable particles. Atoms may be combined into larger units called molecules. Two kinds of chemical bonds allow molecules to form—ionic bonds and covalent bonds. A third bond, the hydrogen bond, is a weaker bond that holds molecules together and may also help large molecules maintain a specific shape.

Energy can neither be created nor destroyed, but it can be converted from one form to another. Potential energy and kinetic energy can be interconverted. When energy is converted from one form to another, some of the useful energy is lost. The amount of kinetic energy that the molecules of various substances contain determines whether they are solids, liquids, or gases. The random motion of molecules, which is due to their kinetic energy, results in their distribution throughout available space. This is called diffusion.

An ion is an atom that is electrically unbalanced. Ions interact to form ionic compounds, such as acids, bases, and salts. Those compounds that release hydrogen ions when dissolved in water are called acids; those that release hydroxyl ions are called bases. A measure of the hydrogen ions present in a solution is known as the pH of the solution. Molecules that interact and exchange parts are said to undergo chemical reactions. The changing of chemical bonds in a reaction may release energy or require the input of additional energy.

• Thinking Critically •

Hydrogen peroxide (H_2O_2) is an antiseptic commonly found in the medicine cabinet. This molecule breaks down into harmless water (H_2O) and reactive oxygen (O_2).

$$2H_2O_2 \rightarrow 2H_2O + O_2$$

The oxygen oxidizes many molecules found in living organisms; thus, disease-causing microorganisms can be destroyed in open wounds. This same reaction takes place in a bottle of hydrogen peroxide over time and results in the loss of antimicrobial properties. Can you describe what happens to the molecules of hydrogen peroxide, water, and oxygen? Can you describe why the bottle will finally contain nothing but water and oxygen? In your description, include diffusion, changes in chemical bonds, and kinetic energy.

• Experience This •

There are several chemicals available in your home that you can use to demonstrate some of the principles of matter. Dissolve an envelope of unflavored gelatin in a cup of very hot water. Allow it to cool and become solid. Then heat it up again and once more allow it to cool. Notice that it changes from solid to liquid and back to solid.

Mix an egg white with a cup of very hot water. Allow it to cool and then warm it up again. Does it also change from solid to liquid and back to solid?

• Questions •

1. How many protons, electrons, and neutrons are there in an atom of potassium having a mass number of thirty-nine?
2. Diagram an atom showing the positions of electrons, protons, and neutrons.
3. Name three kinds of chemical bonds that hold atoms or molecules together. How do these bonds differ from one another?
4. Diagram two isotopes of oxygen.
5. What does it mean if a solution has a pH number of three, twelve, two, seven, or nine?
6. What relationship does diffusion have to molecular motion?
7. Define the following terms: AMU, atomic number, orbital, second energy level, and covalent bond.
8. Define the term *chemical reaction* and give an example.
9. What is the difference between an atom and an element? between a molecule and a compound?
10. How do acids, bases, and salts differ from one another?

• Chapter Glossary •

acid (ăs'id) Any compound that releases a hydrogen ion (or other ion that acts like a hydrogen ion) in a solution.

atom (ă'tom) The smallest part of an element that still acts like that element.

atomic mass unit (ă-tom'ik mas yu'nit) A unit of measure used to describe the mass of atoms and is equal to 1.67×10^{-24} grams, approximately the mass of one proton.

atomic nucleus (ă-tom'ik nu'kle-us) The central region of the atom.

atomic number (ă-tom'ik num'bur) The number of protons in an atom.

base (bās) Any compound that releases a hydroxyl group (or other ion that acts like a hydroxyl group) in a solution.

chemical bonds (kem'ĭ-kal bonds) Forces that combine atoms or ions and hold them together.

chemical reaction (kem'ĭ-kal re-ak'shun) The formation or rearrangement of chemical bonds, usually indicated in an equation by an arrow from the reactants to the products.

chemical symbol (kem'ĭ-kal sim'bol) "Shorthand" used to represent one atom of an element, such as Al for aluminum or C for carbon.

compound (kom'pound) A kind of matter that consists of a specific number of atoms (or ions) joined to each other in a particular way and held together by chemical bonds.

covalent bond (ko-va'lent bond) The attractive force formed between two atoms that share a pair of electrons.

density (den'sĭ-te) The weight of a certain volume of a material.

diffusion (dĭ-fiu'zhun) Net movement of a kind of molecule from a place where that molecule is in higher concentration to a place where that molecule is more scarce.

dynamic equilibrium (di-nam'ik e-kwĭ-lib're-um) The condition in which molecules are equally dispersed, movement is equal in all directions.

electrons (e-lek'trons) The negatively charged particles moving at a distance from the nucleus of an atom that balance the positive charges of the protons.

elements (el'ĕ-ments) Matter consisting of only one kind of atom.

first law of thermodynamics (furst law uv thur"mo-di-nam'iks) Energy in the universe remains constant; it can neither be created nor destroyed.

formula (for'miu-lah) The group of chemical symbols that indicate what elements are in a compound and the number of each kind of atom present.

gas (gas) The state of matter in which the molecules are more energetic than the molecules of a liquid, resulting in only slight attraction for each other.

hydrogen bond (hi'dro-jen bond) Weak attractive forces between molecules. Important in determining how groups of molecules are arranged.

hydroxyl ion (hi-drok'sil i'on) A negatively charged particle (OH^-) composed of oxygen and hydrogen atoms released from a base when dissolved in water.

ionic bond (i-on'ik bond) The attractive force between ions of opposite charge.

ions (i'ons) Electrically unbalanced or charged atoms.

isotopes (i'so-tōps) Atoms of the same element that differ only in the number of neutrons.

kinetic energy (kĭ-net'ik en'er-je) Energy of motion.

liquid (lik'wid) The state of matter in which the molecules are strongly attracted to each other, but because they are farther apart than in a solid, they move past each other more freely.

mass number (mas num'ber) The weight of an atomic nucleus expressed in atomic mass units (the sum of the protons and neutrons).

matter (mat'er) Anything that has weight (mass) and also takes up space (volume).

molecule (mol'ĕ-kūl) The smallest particle of a chemical compound; also the smallest naturally occurring part of an element or compound.

net movement (net muv'ment) The movement in one direction minus the movement in the opposite direction.

neutralization (nu'tral-ĭ-za''shun) A chemical reaction involved in mixing an acid with a base; results in formation of a salt and water.

neutrons (nu'trons) Particles in the nucleus of an atom that have no electrical charge; they were named *neutrons* to reflect this lack of electrical charge.

orbital (or'bĭ-tal) The area of an atom able to hold a maximum of two electrons.

periodic table of the elements (pĭr-ē-od'ik tā'bul uv the el'ĕ-ments) A list of all of the elements in order of increasing atomic number (number of protons).

pH A scale used to indicate the strength of an acid or base.

potential energy (po-ten'shul en'er-je) The energy an object has because of its position.

products (prŏ'dukts) New molecules resulting from a chemical reaction.

protons (pro'tons) Particles in the nucleus of an atom that have a positive electrical charge.

radioactive (ra-de-o-ak'tiv) A term used to describe the property of releasing energy or particles from an unstable atom.

reactants (re-ak'tants) Materials that will be changed in a chemical reaction.

salts (salts) Ionic compounds formed from a reaction between an acid and a base.

second law of thermodynamics (sek'ond law uv ther''mo-di-nam'iks) Whenever energy is converted from one form to another, some useful energy is lost.

solid (sol'id) The state of matter in which the molecules are packed tightly together; they vibrate in place.

states of matter (stātes uv mat'er) Physical conditions of matter (solid, liquid, and gas) determined by the relative amounts of energy of the molecules.

temperature (tem'per-ă-chiur) A measure of molecular energy of motion.

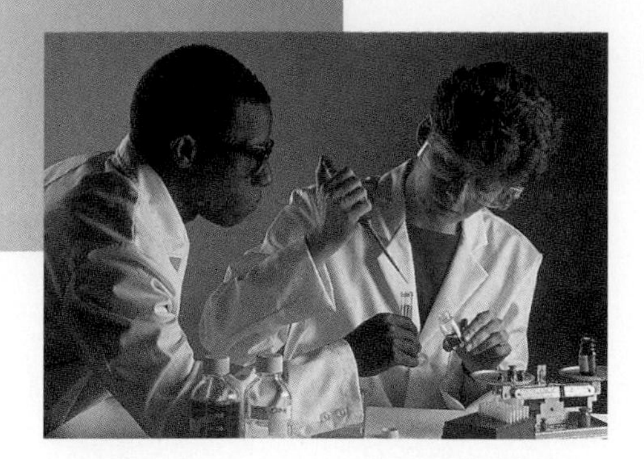

C H A P T E R 3

Organic Chemistry—The Chemistry of Life

Chapter Outline

Purpose

The chemistry of living things is really the chemistry of the carbon atom and a few other atoms that can combine with carbon. In order to understand some aspects of the structure and function of living things we will cover later, you should first learn some basic organic chemistry.

For Your Information

One subject that often comes up when planning a healthy diet relates to the so-called natural or organic foods. Remember that proteins, carbohydrates, fats, and vitamins are all organic molecules; only the minerals and water are inorganic elements or molecules. There is no legal, government-sanctioned definition of these terms. What most people mean when they use the term *natural* or *organic* foods is that the foods have not been supplemented with additives, or that they have not been sprayed with insecticides, chemical fertilizers, preservatives, or agents to enhance the marketability of the food.

Learning Objectives

- Recognize the difference between inorganic and organic molecules.
- Understand the importance of structure in organic molecules.
- Know the major functional groups found in organic molecules.
- Describe the structure and function of the four major groups of organic molecules.

Molecules Containing Carbon

The principles and concepts discussed in chapter 2 apply to all types of matter—nonliving as well as living matter. Living systems are composed of various types of molecules. The things we described in the previous chapter did not contain carbon atoms and so were classified as **inorganic molecules.** This chapter is mainly concerned with more complex structures, **organic molecules,** which contain carbon atoms arranged into rings or chains.

The original meaning of the terms *inorganic* and *organic* is related to the fact that organic materials were thought to be either alive or produced only by living things. Therefore, a very strong link exists between organic chemistry and the chemistry of living things, which is called **biochemistry,** or biological chemistry. Modern chemistry has considerably altered the original meaning of the terms *organic* and *inorganic,* since it is now possible to manufacture unique organic molecules that cannot be produced by living things. Many of the materials we use daily are the result of the organic chemist's art. Nylon, aspirin, polyurethane varnish, silicones, Plexiglas, food wrap, Teflon, and insecticides are just a few of the unique molecules that have been invented by organic chemists (figure 3.1).

In other instances, organic chemists have taken their lead from living organisms and have been able to produce organic molecules more efficiently, or in forms that are slightly different from the original natural molecule. Some examples of these are rubber, penicillin, some vitamins, insulin, and alcohol (figure 3.2).

Figure 3.1 **Some Common Synthetic Organic Materials.** These items are examples of useful organic compounds invented by chemists.

Figure 3.2 **Natural and Synthetic Organic Compounds.** Some organic materials, such as rubber, were originally produced by plants but are now synthesized in industry. The photograph on top shows the collection of latex from the rubber tree; the photograph on the bottom shows an organic chemist testing one of the steps in a manufacturing process.

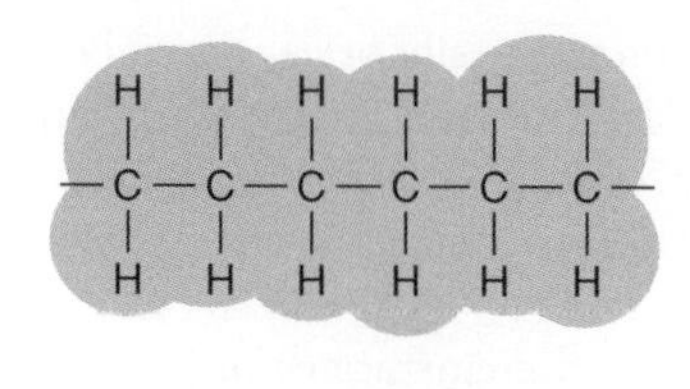

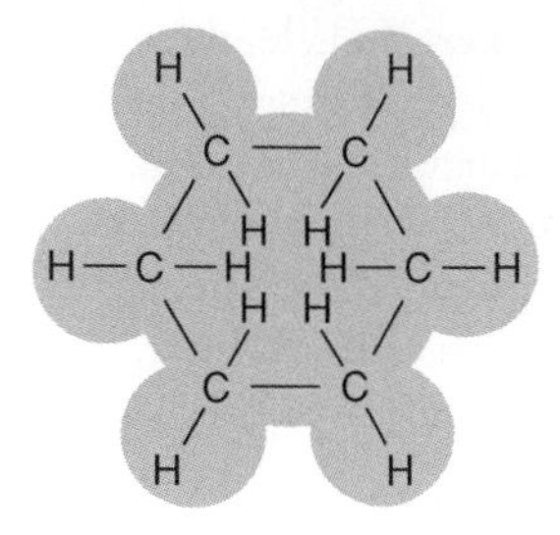

Figure 3.3 **A Ring or Chain Structure.** The ring structure shown on the bottom is formed by joining the two ends of a chain of carbon atoms.

Carbon—The Central Atom

All organic molecules, whether they are natural or synthetic, have certain common characteristics. The carbon atom, which is the central atom in all organic molecules, has some unusual properties. Carbon is unique in that it can combine with other carbon atoms to form long chains. In many cases, the ends of these chains may join together to form ring structures (figure 3.3). Only a few other atoms have this ability. What is really unusual is that these bonding sites are all located at equal distances from one another. If you were to take a rubber ball and stick four nails into it so that they were equally distributed around the ball, you would have a good idea of the geometry involved. These bonding sites are arranged this way because in the carbon atom there are four electrons in the second energy level. These four electrons in the *L* shell, or the *2s*, *2px*, *2py* and *2pz* orbitals, are all as far away from each other as possible (figure 3.4). Carbon atoms are usually involved in covalent bonds. Since carbon has four places it can bond, the carbon atom can combine with four other atoms. This is the case with the methane molecule, which has four hydrogen atoms attached

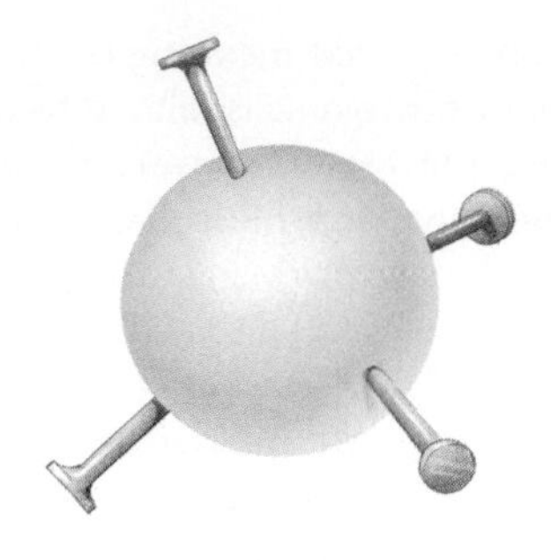

Figure 3.4 **Bonding Sites of a Carbon Atom.** The arrangement of bonding sites around the carbon is similar to a ball with four equally spaced nails in it. Each of the four bondable electrons inhabits an area as far away from the other three as possible.

to a single carbon atom. Methane is a colorless and odorless gas usually found in natural gas (figure 3.5).

Some atoms may be bonded to a single atom more than once. This results in a slightly different arrangement of bonds around the carbon atom. An example of this type of bonding occurs when oxygen is attracted to a carbon. Oxygen has two bondable electrons—if it shares one of these with a carbon and then shares the other with the same carbon, it forms a *double bond.* A **double bond** is two covalent bonds formed between two atoms that share two pairs of electrons. Oxygen is not the only atom that can form double bonds, but double bonds are common between it and carbon. The double bond is denoted by two lines between the two atoms:

$$-\underset{|}{C}=O$$

Two carbon atoms might form double bonds between each other and then bond to other atoms at the remaining bonding sites. Figure 3.6 shows several compounds that contain double bonds.

Although most atoms can be involved in the structure of an organic molecule, only a few are commonly found. Hydrogen (H) and oxygen (O) are almost always present. Nitrogen (N), sulfur (S), and phosphorus (P) are also very important in specific types of organic molecules.

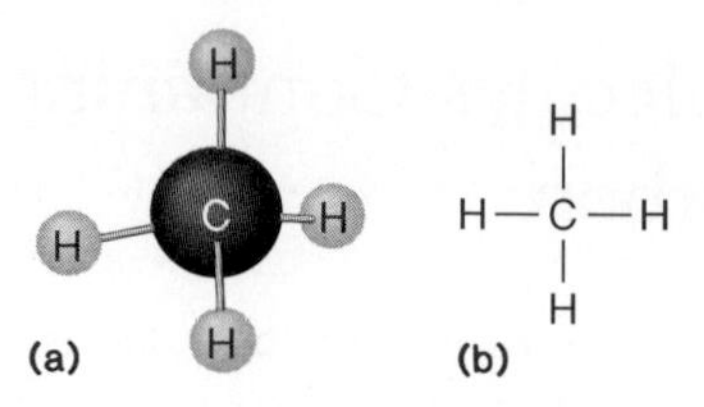

Figure 3.5 **A Methane Molecule.** A methane molecule is composed of one carbon atom bonded with four hydrogen atoms. These bonds are formed at the four bonding sites of the carbon. (a) For the sake of simplicity, all future diagrams of molecules will be two-dimensional drawings, although in reality they are three-dimensional molecules. (b) Each line in the diagram represents a covalent bond between the two atoms where a pair of electrons is being shared.

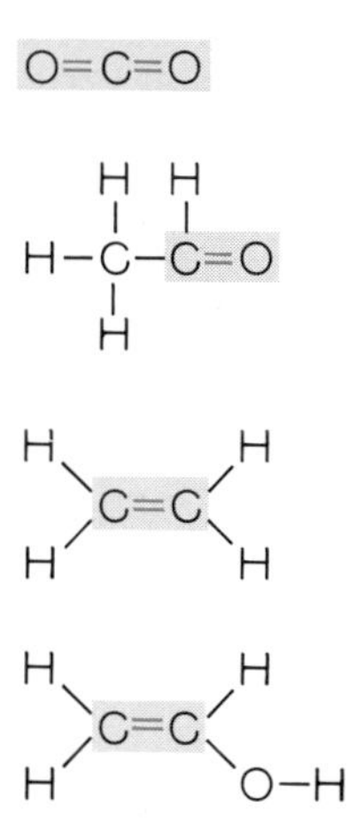

Figure 3.6 **Double Bonds.** These diagrams show several molecules that contain double bonds. A double bond is formed when two atoms share two pairs of electrons with each other.

An enormous variety of organic molecules is possible because carbon is able to bond at four different sites, form long chains, and combine with many other kinds of atoms. The types of atoms in the molecule are important in determining the properties of the molecule. The three-dimensional arrangement of the atoms within the molecule is also important. Since most inorganic molecules are small and involve few atoms, there is usually only one way in which a group of atoms can be arranged to form a molecule. There is only one arrangement for a single oxygen atom and two hydrogen atoms in a molecule of water. In a molecule of sulfuric acid, there is only one arrangement for the sulfur atom, the two hydrogen atoms, and the four oxygen atoms.

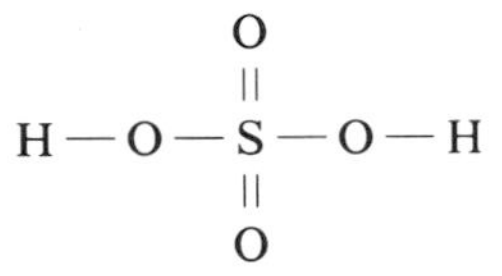

```
         O
         ||
H — O — S — O — H
         ||
         O
```

However, consider these two organic molecules:

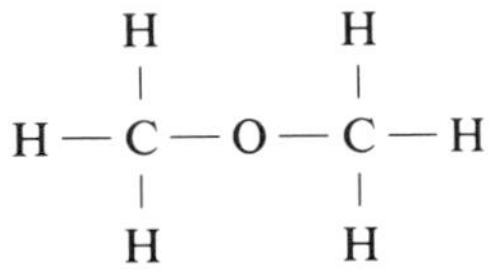

```
    H       H
    |       |
H — C — O — C — H
    |       |
    H       H
```

dimethyl ether

and

```
    H   H
    |   |
H — C — C — O — H
    |   |
    H   H
```

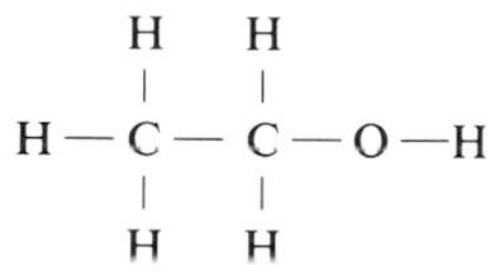

ethyl alcohol

Both the dimethyl ether and the ethyl alcohol contain two carbon atoms, six hydrogen atoms, and one oxygen atom, but they are quite different in their arrangement of atoms and in the chemical properties of the molecules. While the first is an ether, the second is an alcohol. Since the ether and the alcohol have the same number and kinds of atoms, they are said to have the same **empirical formula,** which in this case is written C_2H_6O. An empirical formula simply indicates the number of each kind of atom within the molecule. When the arrangement of the atoms and their bonding within the molecule is indicated, we call this a **structural formula.** Figure 3.7 shows several structural formulas for the empirical formula $C_6H_{12}O_6$. Molecules that have the same empirical formula but different structural formulas are called **isomers.**

The Carbon Skeleton and Functional Groups

To help us understand organic molecules a little better, let's consider some of their similarities. All organic molecules have a **carbon skeleton,** which is composed of rings or chains of carbons. It is this carbon skeleton in the organic molecule that determines the overall shape of the molecule. The differences between various organic molecules depend on the length and arrangement of the carbon skeleton. In addition, the kinds of atoms that are bonded to this carbon skeleton determine the way the organic compound acts. Attached to the carbon skeleton are specific combinations of atoms called **functional groups.** These functional groups determine specific chemical properties. By learning to recognize some of the functional groups, it is possible to identify an organic molecule and to predict something about its activity. Figure 3.8 shows some of the functional groups that are important in biological activity. Remember that a functional group does not exist by itself; it must be a part of an organic molecule (see box 3.1).

Common Organic Molecules

One way to make organic chemistry more manageable is to organize different kinds of compounds into groups on the basis of their similarity of structure or the chemical properties of the molecules. Frequently you will find that organic molecules are composed of subunits that are attached to each other. If you recognize the subunit, then the whole organic molecule is much easier to identify. It is similar to distinguishing between a passenger train and a freight train by recognizing the individual cars unique to each.

When there are several subunits (*monomers*) bonded together, the molecule is referred to as a macromolecule or a *polymer.* The word *monomer* means a single unit, while the term *polymer* means composed of many parts. The plastics industry has polymer chemistry as its foundation. The monomers in a

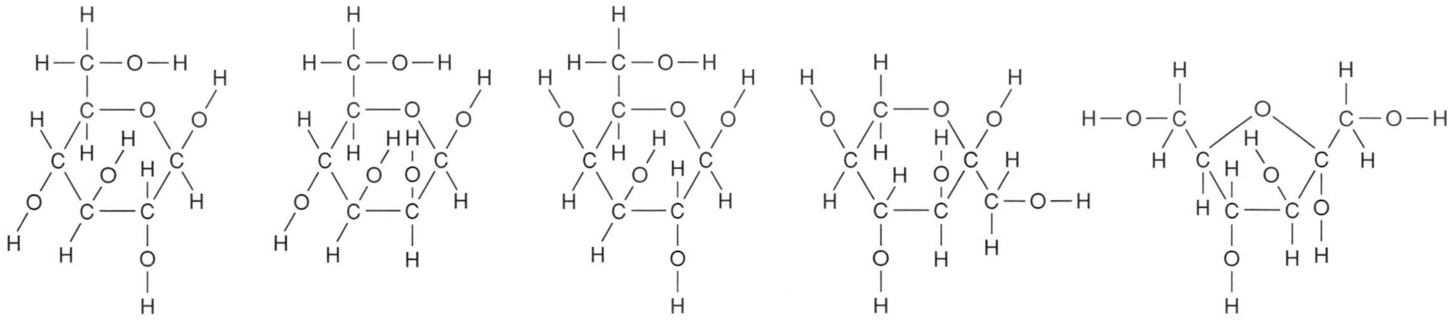

Glucose $C_6H_{12}O_6$ | Mannose $C_6H_{12}O_6$ | Galactose $C_6H_{12}O_6$ | Sorbose $C_6H_{12}O_6$ | Fructose $C_6H_{12}O_6$

Figure 3.7 **Structural Formulas for Several Hexoses.** Several six-carbon sugars are represented here. Each has the same empirical formula, but they each have a different structural formula. They will also act differently from each other.

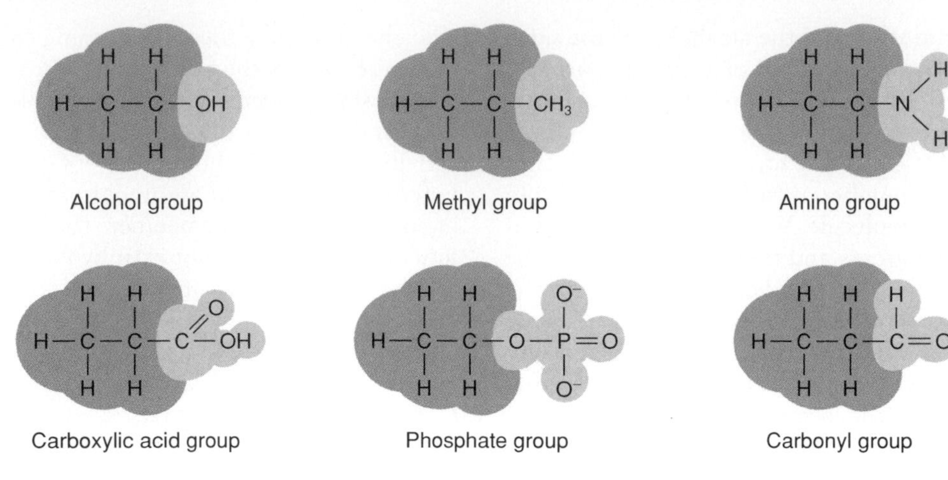

Figure 3.8 **Functional Groups.** These are some of the groups of atoms that frequently attach to a carbon skeleton. Notice that in each case the carbon skeleton is unchanged; just the group attached to it is changed. The functional group (in color) determines how the molecule will act.

polymer are usually combined by a **dehydration synthesis reaction.** This reaction results in the synthesis or formation of a macromolecule when water is removed from between the two smaller component parts. For example, when a monomer with an "—OH group" attached to its carbon skeleton approaches another monomer with an available hydrogen, dehydration synthesis can occur. Figure 3.9 shows the removal of water from between two such subunits. Notice that in this case, the structural formulas are used to help identify just what is occurring. However, the chemical equation also indicates the removal of the water. You can easily recognize a dehydration synthesis reaction because the reactant side of the equation shows numerous small molecules, while the product side lists fewer, larger products and water.

The reverse of a dehydration synthesis reaction is known as *hydrolysis.* **Hydrolysis** is the process of splitting a

BOX 3.1 *Chemical Shorthand*

You have probably noticed that sketching the entire structural formula of a large organic molecule takes a great deal of time. If you know the structure of the major functional groups, you can use several shortcuts to more quickly describe chemical structures.

When multiple carbons with two hydrogens are bonded to each other in a chain, we sometimes write it as follows:

```
  H   H   H   H   H   H   H   H   H   H   H   H
  |   |   |   |   |   |   |   |   |   |   |   |
— C — C — C — C — C — C — C — C — C — C — C — C —
  |   |   |   |   |   |   |   |   |   |   |   |
  H   H   H   H   H   H   H   H   H   H   H   H
```

or we might write it this way:

$— CH_2— CH_2— CH_2— CH_2— CH_2— CH_2— CH_2— CH_2— CH_2— CH_2— CH_2— CH_2—$

or more simply, we may write it as follows: $(—CH_2—)_{12}$. If the twelve carbons were in a pair of two rings, we probably would not label the carbons or hydrogens unless we wished to focus upon a particular group or point. We would probably draw the two 6-carbon rings with only hydrogen attached as follows:

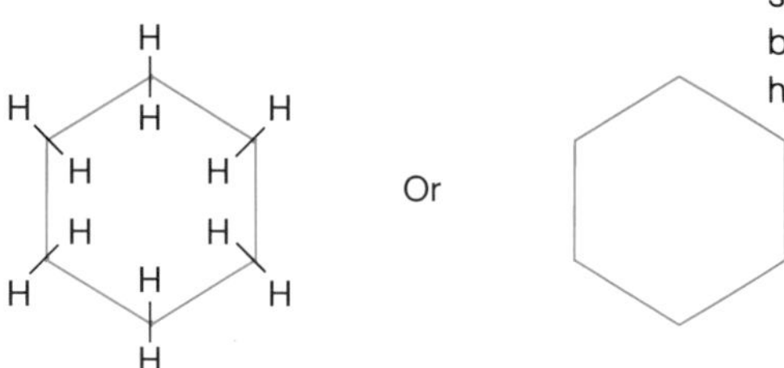

Don't let these shortcuts throw you. You will soon find that you will be putting an —OH group onto a carbon skeleton and neglecting to show the bond between the oxygen and hydrogen, just like a professional.

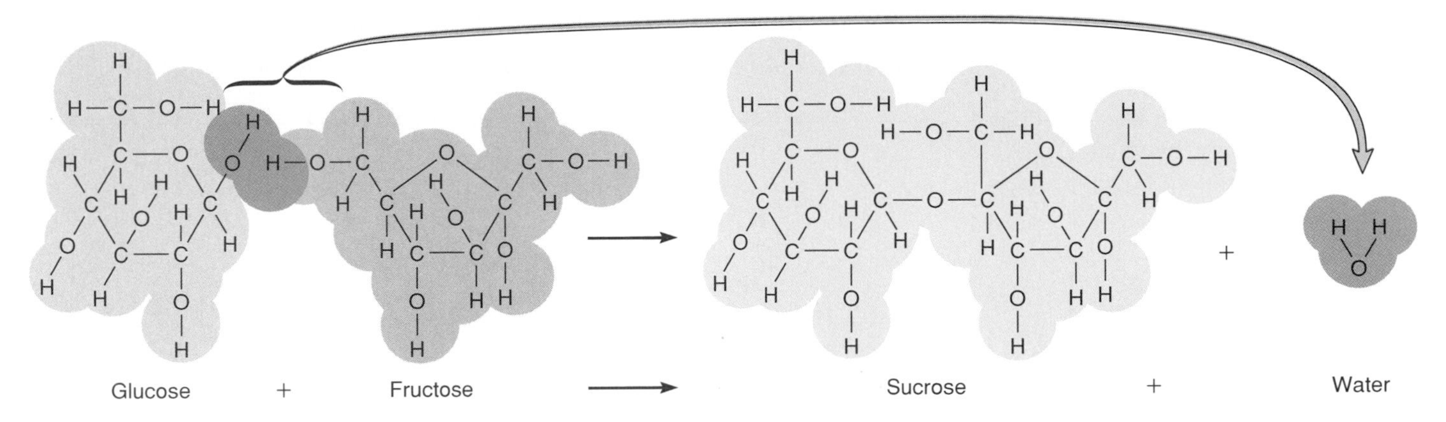

Figure 3.9 **The Dehydration Synthesis Reaction.** In the reaction illustrated here, the two —OH groups form water, and the oxygen that remains acts as an attachment site between the two larger sugar molecules. Many structural formulas appear to be complex at first glance, but if you look for the points where subunits are attached and dissect each subunit, they become much simpler to deal with.

larger organic molecule into two or more component parts by the addition of water. Digestion of food molecules in the stomach is an important example of hydrolysis.

Carbohydrates

One class of organic molecules, **carbohydrates,** is composed of carbon, hydrogen, and oxygen atoms linked together to form monomers called *simple sugars* or *monosaccharides.* The empirical formula for a simple sugar is easy to recognize because there are equal numbers of carbons and oxygens and twice as many hydrogens—for example, $C_3H_6O_3$ or $C_5H_{10}O_5$. We usually describe the kinds of simple sugars by the number of carbons in the molecule. The ending *-ose* is a clue that indicates you are dealing with a carbohydrate. A tri*ose* has three carbons, a pent*ose* has five, and a hex*ose* has six. If you remember that the number of carbons equals the number of oxygen atoms and that the number of hydrogens is double that number, these names tell you the empirical formula for the simple sugar. Simple sugars, such as glucose, fructose, and galactose, provide the chemical energy necessary to keep organisms alive. These simple sugars combine with each other by dehydration synthesis to form **complex carbohydrates** (figure 3.9). When two simple sugars bond to each other, a *disaccharide* is formed; when three bond together, a *trisaccharide* is formed (figure 3.10). Generally we call a complex carbohydrate that is larger than this a *polysaccharide* (many sugar units). In all cases, the complex carbohydrates are formed by the removal of water from between the sugars. Some common examples of polysaccharides are starch and glycogen. Cellulose is an important polysaccharide used in constructing the cell walls of plant cells. Humans cannot digest (*hydrolyze*) this complex carbohydrate, so we are not able to use it as an energy source. Plant cell walls add bulk or fiber to our diet, but no calories. Fiber is an important addition to your diet because it helps to control weight, reduce the risk of colon cancer, and control constipation and diarrhea.

Simple sugars can be used by the cell as components of other, more complex molecules, such as the molecule

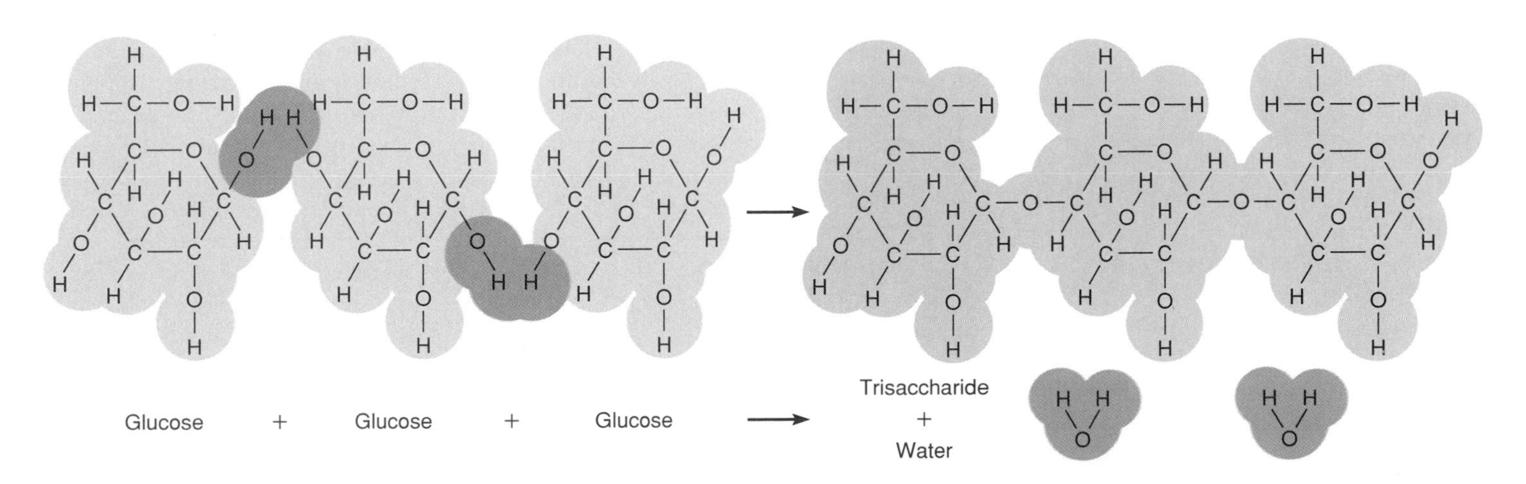

Figure 3.10 **A Trisaccharide.** Three simple sugars are attached to each other by the removal of two waters from between them. This is an example of a complex carbohydrate.

adenosine triphosphate (ATP). This molecule is important in energy transfer. It has a simple sugar (ribose) as part of its structural makeup. The building blocks of the genetic material (DNA) also have a sugar component (see figure 3.17).

Lipids

We generally describe **lipids** as large organic molecules that do not easily dissolve in water. Just like carbohydrates, the lipids are composed of carbon, hydrogen, and oxygen. They do not, however, have the same ratio of carbon, hydrogen, and oxygen in their empirical formulas. Lipids generally have very small amounts of oxygen in comparison to the amounts of carbon and hydrogen. Fats, phospholipids, and steroids are all examples of lipids, but they are all quite different from each other in their structure.

Fats are important organic molecules that are used to provide energy. The building blocks of a fat are a glycerol molecule and fatty acids. The **glycerol** is a carbon skeleton that has three alcohol groups attached to it. Its chemical formula is $C_3H_5(OH)_3$. A **fatty acid** is a long-chain carbon skeleton that has a carboxylic acid functional group. If the carbon skeleton has as much hydrogen bonded to it as possible, we call it **saturated.** The saturated fatty acid above is stearic acid, a component of solid meat fats such as mutton tallow. Notice that at every point in this structure the carbon has as much hydrogen as it can hold. Saturated fats are generally found in animal tissues—they tend to be solids at room temperatures. Some examples of saturated fats are butter, whale blubber, suet, lard, and fats associated with such meats as steak or pork chops.

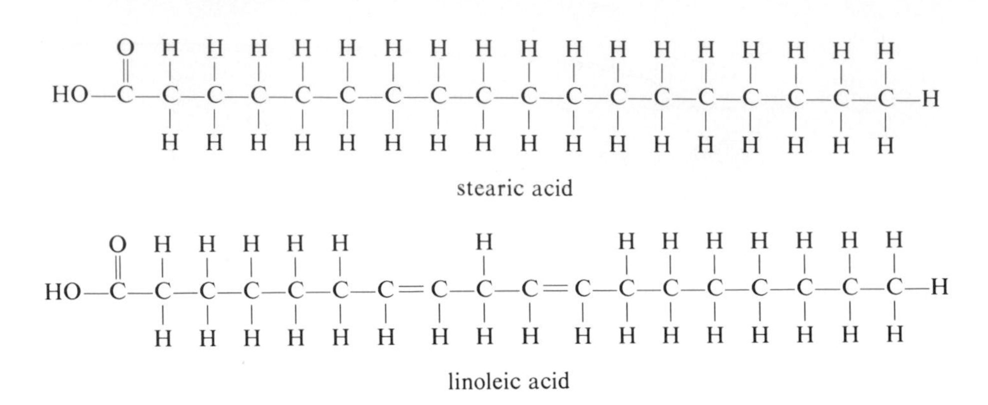

If the carbons are double-bonded to each other at one or more points, the fatty acid is said to be **unsaturated.** The unsaturated fatty acid above is linoleic acid, a component of sunflower and safflower oils.

Notice that there are several double bonds between the carbons and fewer hydrogens than in the saturated fatty acid. Unsaturated fats are frequently plant fats or oils—they are usually liquids at room temperature. Peanut oil, corn oil, and olive oil are considered unsaturated because they have double bonds between the carbons of the carbon skeleton. A polyunsaturated fatty acid is one that has a great number of double bonds in the carbon skeleton. When glycerol and three fatty acids are combined by three dehydration synthesis reactions, a fat is formed. Notice that dehydration synthesis is almost exactly the same as the reaction that causes simple sugars to bond together.

Fats are important molecules for storing energy. There is twice as much energy in a gram of fat as in a gram of sugar. This is important to an organism because fats can be stored in a relatively small space and still yield a high amount of energy. Fats in animals also provide protection from heat loss. Some animals have a layer of fat under the skin that serves as an insulating layer. The thick layer of blubber in whales, walruses, and seals prevents the loss of internal body heat to the cold, watery environment in which they live. This same layer of fat, together with the fat deposits around some internal organs—such as the kidneys and heart—serves as a cushion that protects these organs from physical damage. If a fat is formed from a glycerol molecule and three attached fatty acids, it is called a *triglyceride;* if two, a *diglyceride;* and if one, a *monoglyceride* (figure 3.11).

Phospholipids are a class of water-insoluble molecules that resemble fats but contain a phosphate group (PO_4) in their structure (figure 3.12). One of the reasons phospholipids are important is that they are a major component of membranes in cells. Without these lipids in our membranes, the cell contents would not be able to be separated from the exterior environment. Some of the phospholipids are better known as the *lecithins.* Lecithins are found in cell membranes and also help in the emulsification of fats. They help to separate large portions of fat into smaller units. This allows the fat to mix with other materials. Lecithins are added to many types of food for this purpose (chocolate bars, for example). Some people take lecithin as nutritional supplements because they believe it leads to healthier hair and better reasoning ability. But once inside your intestines, lecithins are destroyed by enzymes, just like any other phospholipid (see box 3.2).

Steroids, a third group of lipid molecules, are characterized by their arrangement of interlocking rings of carbon. They often serve as hormones that aid in regulating body processes. We have already mentioned one steroid molecule that you are probably familiar with: cholesterol. While serum cholesterol (the kind found in your blood associated with lipoproteins) has been implicated in many cases of atherosclerosis, this steroid is made by your body for use as a component of cell membranes. It is also used by your body to make bile acids. These products of your liver are channeled into your intestine to emulsify fats. Cholesterol is also necessary for the manufacture of vitamin D. Cholesterol molecules in the skin react with ultraviolet light to produce vitamin D, which assists in the proper development of bones and teeth.

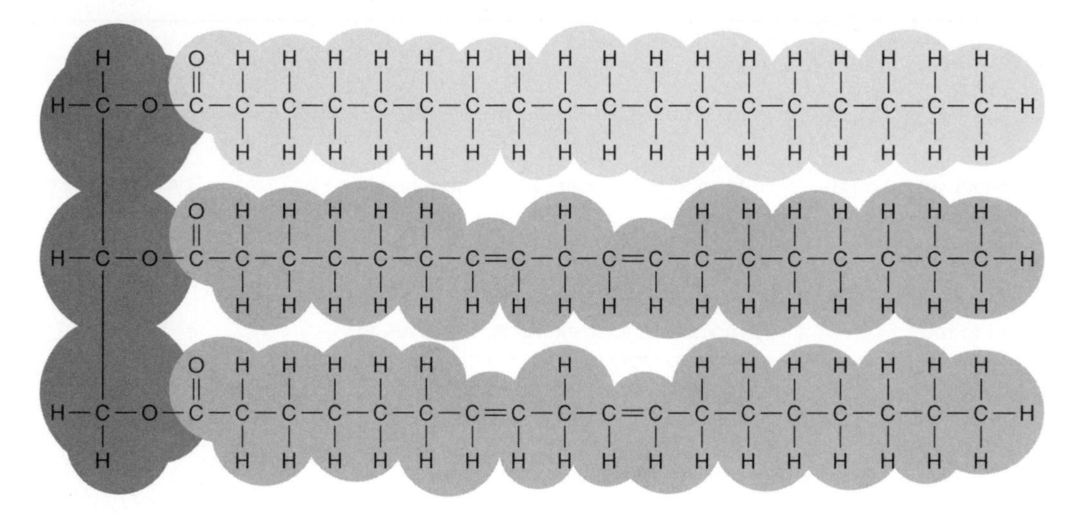

Figure 3.11 **A Fat Molecule.** The arrangement of the three fatty acids attached to a glycerol molecule is typical of the formation of a fat. The structural formula of the fat appears to be very cluttered until you dissect the fatty acids from the glycerol; then it becomes much more manageable. This example of a triglyceride contains a glycerol molecule, two unsaturated fatty acids (linoleic acid), and a third saturated fatty acid (stearic acid).

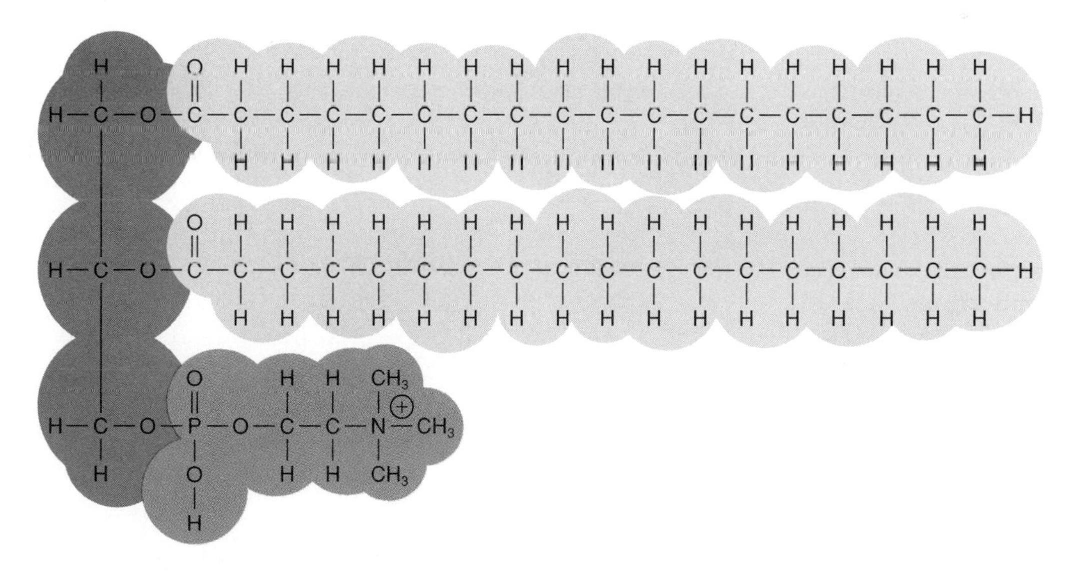

Figure 3.12 **A Phospholipid Molecule.** This molecule is similar to a fat but has a phosphate group in its structure. The phosphate group is bonded to the glycerol by a dehydration synthesis reaction. This phospholipid contains glycerol, two fatty acids, and the phosphate-containing portion. Molecules like this are known as the lecithins.

A large number of steroid molecules are hormones. Some of them regulate reproductive processes such as egg and sperm production (see chapter 21), while others regulate such things as salt concentration in the blood. Athletes have been known to use certain hormonelike steroids to increase their muscular bulk. The medical community is certain that use of these chemicals is potentially harmful, possibly resulting in liver disfunction, sex-characteristic changes, changes in blood chemistry, and even death.

Proteins

Proteins are polymers made up of monomers known as *amino acids.* An **amino acid** is a short carbon skeleton that contains an amino group (a nitrogen and two hydrogens) on one end

BOX 3.2 Fat and Your Diet

When triglycerides are eaten in fat-containing foods, digestive enzymes hydrolyze them into monoglycerides and fatty acids. These molecules are absorbed by the intestinal tract and coated with protein to form *lipoprotein,* as shown in the accompanying diagram.

There are four types of lipoproteins in the body: (1) chylomicrons, (2) very-low-density-lipoproteins (VLDL), (3) low-density-lipoproteins (LDL), and (4) high-density-lipoproteins (HDL). Chylomicrons are very large particles formed in the intestine and are between 80% and 95% triglycerides in composition. As the chylomicrons circulate through the body, the triglycerides are removed by cells in order to make hormones, store energy, and build new cell parts. When most of the triglycerides have been removed, the remaining portions of the chylomicrons are harmlessly destroyed. The VLDLs and LDLs are formed in the liver. VLDLs contain all types of lipid, protein, and 10%–15% cholesterol, while the LDLs are about 50% cholesterol. As with the chylomicrons, the body taps these molecules for their lipids. However, in some people, high levels of LDLs in the blood are associated with the disease *atherosclerosis* (hardening of the arteries). While in the blood, LDLs may stick to the insides of the vessels, forming hard deposits that restrict blood flow and contribute to high blood pressure, strokes, and heart attacks. Even though they are 30% cholesterol, a high level of HDLs (made in the intestine) in comparison to LDLs is associated with a lower risk of atherosclerosis. One way to reduce the risk of this disease is to lower your intake of LDLs. This can be done by reducing your consumption of saturated fats, since these are most easily converted by your body into LDLs and cholesterol.

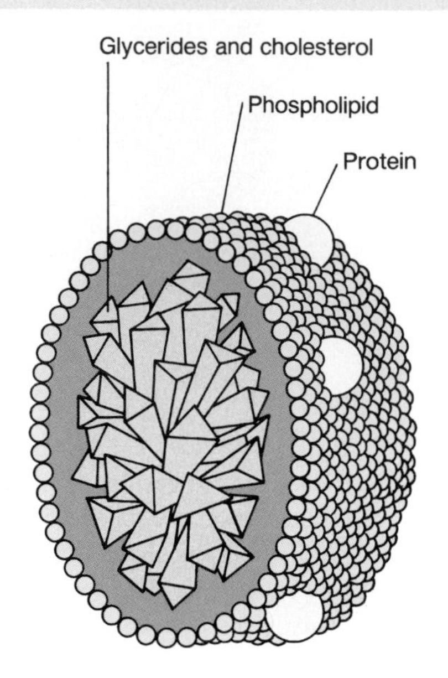

of the skeleton and a carboxylic acid group at the other end (figure 3.13). In addition, the carbon skeleton may have one of several different side chains on it. There are about twenty amino acids that are important to cells. All are identical except for their side chains.

The amino acids can bond together by dehydration synthesis reactions. When two amino acids form a bond by removal of water, the nitrogen of the amino group of one is bonded to the carbon of the acid group of another. This bond is termed a **peptide bond** (figure 3.14).

Any amino acid can form a peptide bond with any other amino acid. They fit together in a specific way, with the amino group of one bonding to the acid group

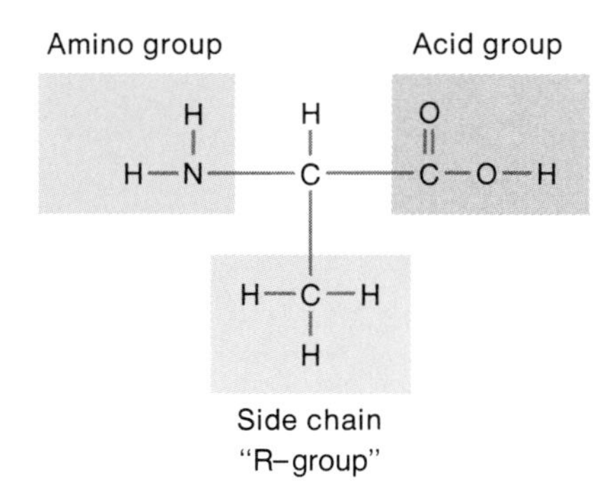

Figure 3.13 **The Structure of an Amino Acid.** An amino acid is composed of a short carbon skeleton with three functional groups attached: an amino group, a carboxylic acid group (acid group) and an additional variable group ("R-group"). It is the variable group that determines which specific amino acid is constructed.

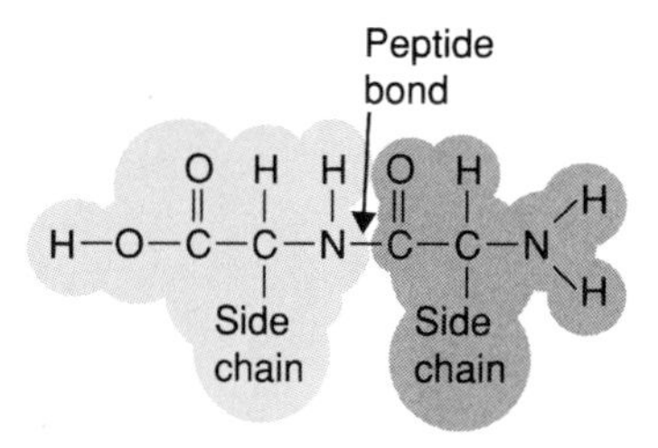

Figure 3.14 **A Peptide Bond.** The bond that results from a dehydration synthesis reaction between two amino acids is called a peptide bond. This bond forms as a result of the removal of the hydrogen and hydroxyl groups. In the formation of this bond, the nitrogen is bonded directly to the carbon.

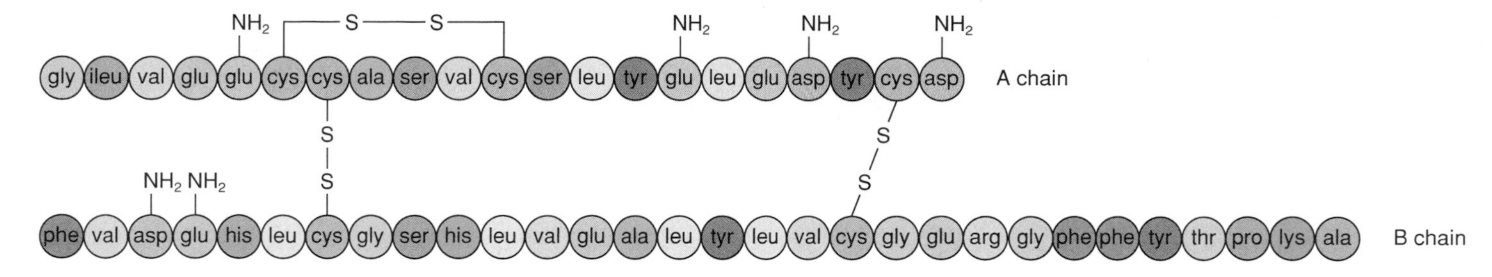

Figure 3.15 **An Insulin Molecule.** The protein insulin is composed of two polypeptide chains bonded together at specific points by reactions between the side chains of particular amino acids. The side chains of one interact with the side chains of the other and form a particular three-dimensional shape. The bonds that form between the polypeptide chains are called disulfide bonds.

of the next. You can imagine that by using twenty different amino acids as building blocks, you can construct millions of different combinations. Each of these combinations is termed a **polypeptide chain.** A specific polypeptide is composed of a specific sequence of amino acids bonded end to end. This is called its *primary structure.* A listing of the amino acids in their proper order within a particular polypeptide constitutes its primary structure. The specific sequence of amino acids in a polypeptide is controlled by the genetic information of an organism.

The string of amino acids in a polypeptide is likely to twist into a particular shape: a coil or a pleated sheet. These forms are referred to as the *secondary structure* of polypeptides. For example, some proteins (e.g., hair) take the form of a *helix:* a shape like that of a coiled telephone cord. The helical shape is maintained by hydrogen bonds formed between different amino acid side chains at different locations in the polypeptide. Remember from chapter 2 that hydrogen bonds do not form molecules but result in the orientation of one part of a molecule to another part of a molecule. Other polypeptides form hydrogen bonds that cause them to make several flat folds that resemble a pleated skirt.

It is also possible for a single polypeptide to contain one or more coils and pleated sheets along its length. As a result, these different portions of the molecule can interact to form an even more complex three-dimensional structure. This occurs when the coils and pleated sheets twist and combine with each other. The complex three-dimensional structure formed in this manner is the polypeptide's *tertiary* (third degree) *structure.* A good example of tertiary structure can be seen when a coiled phone cord becomes so twisted that it folds around and back on itself in several places. The oxygen-holding protein found in muscle cells, myoglobin, displays tertiary structure: it is composed of a single (153 amino acids), helical molecule folded back and bonded to itself in several places.

Frequently, several different polypeptides, each with its own tertiary structure, twist around each other and chemically combine. The larger, three-dimensional structure formed by these interacting polypeptides is referred to as the protein's *quaternary* (fourth degree) *structure.* The individual polypeptide chains are bonded to each other by the interactions of certain side chains, which can form disulfide bonds (figure 3.15). Quaternary structure is displayed by the protein molecules called *antibodies,* which are involved in fighting diseases such as mumps and chicken pox.

Individual polypeptide chains or groups of chains forming a particular configuration are proteins. The structure of a protein is closely related to its function. We will consider two aspects of the structure of proteins: the sequence of amino acids within the protein and the overall three-dimensional shape of the molecule. Any changes in the arrangement of amino acids within a protein can have far-reaching effects on its function. For example, normal hemoglobin found in red blood cells consists of two kinds of polypeptide chains called the alpha and beta chains. The beta chain is 146 amino acids long. If just one of these amino acids is replaced by a different one, the hemoglobin molecule may not function properly. A classic example of this results in a condition known as *sickle-cell anemia.* In this case, the sixth amino acid in the beta chain, which is normally glutamic acid, is replaced by valine. This minor change causes the hemoglobin to fold differently, and the red blood cells that contain this altered hemoglobin assume a sickle shape when the body is deprived of an adequate supply of oxygen.

When a particular sequence of amino acids forms a polypeptide, the stage is set for that particular arrangement to bond with another polypeptide in a certain way. Think of a telephone cord that has curled up and formed a helix (its secondary structure). Now imagine that at several irregular intervals along that cord, you have attached magnets. You can see that the magnets at the various points along the cord will attract each other, and the curled cord will form a particular three-dimensional shape. You can more closely approximate the complex structure of a protein (its tertiary structure) if you imagine several curled cords, each with magnets attached at several points. Now imagine these magnets as bonding the individual cords together. The globs or ropes of

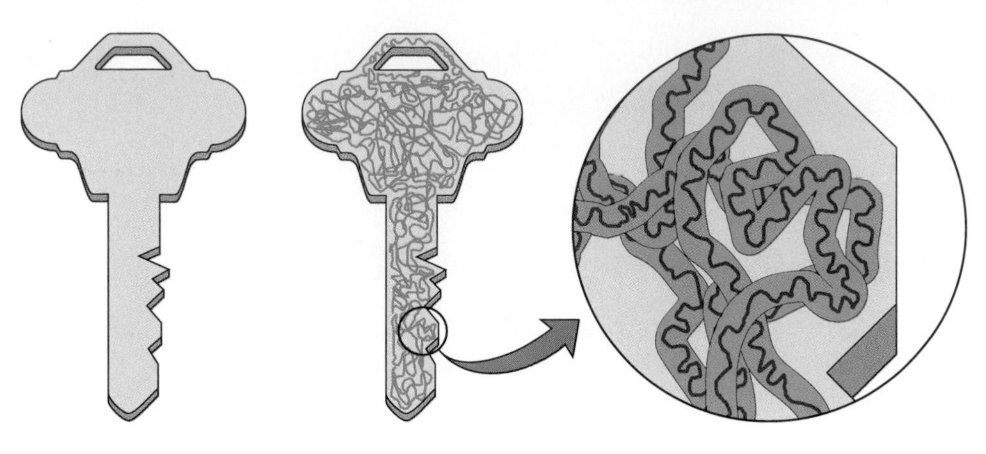

Figure 3.16 **The Three-Dimensional Shape of Proteins.** The specific arrangement of amino acids results in side chains that are available to bond with other side chains. The results are specific three-dimensional proteins that have a specific surface geometry. We frequently compare this three-dimensional shape to the three-dimensional shape of a specific key.

telephone cords approximate the quaternary structure of a protein. This shape can be compared to the shape of a key. In order for a key to do its job effectively, it has to have particular bumps and grooves on its surface. Similarly, if a particular protein is to do its job effectively, it must have a particular shape. The protein's shape can be altered by changing the order of the amino acids that causes different cross linkages to form. Changing environmental conditions also influences the shape of the protein. Figure 3.16 shows the importance of the three-dimensional shape of the protein.

Energy in the form of heat or light may break the hydrogen bonds within protein molecules. When this occurs, the chemical and physical properties of the protein are changed and the protein is said to be **denatured.** A common example of this occurs when the gelatinous, clear portion of an egg is cooked and the protein changes to a white solid. Some medications are proteins and must be protected from denaturation so as not to lose their effectiveness. Insulin is an example. For protection, such medications may be stored in brown-colored bottles or kept under refrigeration.

The thousands of kinds of proteins can be placed into two categories. Some proteins are important for maintaining the shape of cells and organisms—they are usually referred to as **structural proteins.** The proteins that make up the cell membrane, muscle cells, tendons, and blood cells are examples of structural proteins. The other kinds of proteins, **regulator proteins,** help determine what activities will occur in the organism. These regulator proteins include enzymes and some hormones. These molecules help control the chemical activities of cells and organisms. Enzymes are important, and they are dealt with in detail in chapter 5. Some examples of enzymes are the digestive enzymes in the stomach and the mouth. Two hormones that are regulator proteins are insulin and oxytocin. Insulin is produced by the pancreas and controls the amount of glucose in the blood. If insulin production is too low, or if the molecule is improperly constructed, glucose molecules are not removed from the bloodstream at a fast enough rate. The excess sugar is then eliminated in the urine. Other symptoms of excess sugar in the blood include excessive thirst and even loss of consciousness. The disease caused by improperly functional insulin is known as *diabetes.* Oxytocin, a second protein-type hormone, stimulates the contraction of the uterus during childbirth. It is also an example of an organic molecule that has been produced artificially and is used by physicians to induce labor.

Nucleic Acids

The last group of organic molecules that we will consider are the *nucleic acids.* **Nucleic acids** are complex molecules that store and transfer information within a cell. They are constructed of fundamental monomers known as **nucleotides.** Each nucleic acid is a polymer composed of nucleotides bonded together. There are eight different nucleotides, but each is constructed of a phosphate group, a sugar, and an organic nitrogenous base.

The two kinds of sugar that can be part of the nucleotide are ribose and deoxyribose. These are five-carbon simple sugars. The phosphate group is attached to the sugar and the nitrogenous base is attached to another part of the sugar. There are five common organic molecules containing nitrogen that are likely to be part of the nucleotide structure. The nitrogen-containing bases are adenine, guanine, cytosine, thymine, and uracil.

The nucleotides can then connect to form long chains by dehydration synthesis reactions. The long chains of nucleotides are of two types—RNA, ribonucleic acid, and DNA, deoxyribonucleic acid. The RNA forms a single polymer, whereas the DNA is generally composed of two matching polymers twisted together and held by hydrogen bonds (figure 3.17). These two types of molecules contain the information needed for the formation of particular sequences of amino acids; they determine what kinds of proteins an organism can manufacture. The mechanism of storing and using this information is the topic of chapter 8.

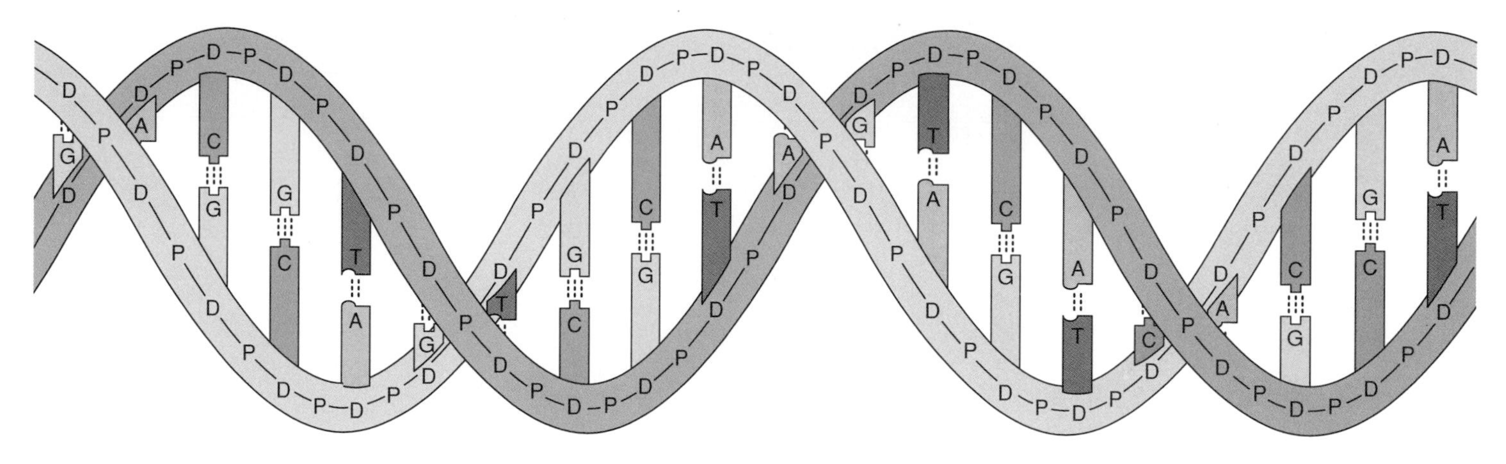

Figure 3.17 **Nucleic Acid.** Deoxyribonucleic acid (DNA) is an organic molecule composed of four nucleotides. Each nucleotide is composed of a sugar, deoxyribose (D); an inorganic acid, phosphoric acid (P); and one of four nitrogen-containing bases: adenine (A), cytosine (C), guanine (G), or thymine (T).

• Summary •

The chemistry of living things involves a variety of large and complex molecules. This chemistry is based on the carbon atom and the fact that carbon atoms can connect to form long chains or rings. This results in a vast array of molecules. The structure of each molecule is related to its function. Changes in the structure may result in abnormal functions, which we call disease. Some of the most common types of organic molecules found in living things are carbohydrates, lipids, proteins, and nucleic acids. Table 3.1 summarizes the major types of biologically important organic molecules and how they function in living things.

Table 3.1
A Summary of the Types of Organic Molecules Found in Living Things

Type of Organic Molecule	Basic Subunit	Function	Examples
Carbohydrates	Simple sugar	Provide energy Provide support	Glucose Cellulose
Lipids			
1. Fats	Glycerol and fatty acids	Provide energy Provide insulation Serve as shock absorber	Lard Olive oil Linseed oil Tallow
2. Steroids	A complex ring structure	Often serve as hormones that control the body processes	Testosterone Vitamin D Cholesterol
3. Phospholipids	Glycerol, fatty acids, and phosphorus compounds	Form a major component of the structure of the cell membrane	Cell membrane
Proteins	Amino acid	Maintain the shape of cells and parts of organisms	Cell membrane Hair Muscle
		As enzymes, regulate the rates of cell reactions	Ptyalin in the mouth
		As hormones, effect physiological activity, such as growth or metabolism	Insulin
Nucleic acids	Nucleotide	Store and transfer genetic information that controls the cell	DNA RNA

• Thinking Critically •

Amino acids and fatty acids are both organic acids. What property must they have in common with inorganic acids such as sulfuric acid? How do they differ? Consider such aspects as structure of molecules, size, bonding, and pH.

• Experience This •

Look at the labels on five items from your pantry. Make a list of ingredients that you think are organic chemicals. They may be additives to the food, preservatives, coloring materials, or flavor enhancers. They may also be the active ingredients in such things as pesticides or pharmaceuticals. As a result of reading these ingredients, can you identify the class of organic molecules to which each belongs?

• Questions •

1. Diagram an example of each of the following: amino acid, simple sugar, glycerol, fatty acid.
2. Give an example of each of the following classes of organic molecules: carbohydrate, protein, lipid, and nucleic acid.
3. What is the structural difference between a saturated fat and an unsaturated fat?
4. Describe three different kinds of lipids.
5. How do the primary, secondary, tertiary, and quaternary structures of proteins differ?
6. What two characteristics of the carbon atom make it unique?
7. What is the difference between inorganic and organic molecules?
8. What is meant by HDL, LDL, VLDL, and chylomicron? Where are they found? How do they relate to disease?
9. Describe five functional groups.
10. List three monomers and the polymers that can be constructed from them.

• Chapter Glossary •

amino acid (ah-mēn′o ă′sid) A basic subunit of protein consisting of a short carbon skeleton that contains an amino group, a carboxylic acid group, and one of various side groups.

biochemistry (bi-o-kem′iss-tre) The chemistry of living things, often called biological chemistry.

carbohydrate (kar-bo-hi′drāt) One class of organic molecules composed of carbon, hydrogen, and oxygen usually in a ratio of 1:2:1. The basic building block of a carbohydrate is a simple sugar (= monosaccharide).

carbon skeleton (kar′bon skel′uh-ton) The central portion of an organic molecule composed of rings or chains of carbon atoms.

complex carbohydrates (kom′pleks kar-bo-hi′drāts) Macromolecules composed of simple sugars combined by dehydration synthesis to form a polymer.

dehydration synthesis reaction (de-hi-dra′shun sin′thuh-sis re-ak′shun) A reaction that results in the formation of a macromolecule when water is removed from between the two smaller component parts.

denature (de-nā′chur) To alter the structure of a protein so that some of its original properties are diminished or eliminated.

double bond (dub′l bond) A pair of covalent bonds formed between two atoms when they share two pairs of electrons.

empirical formula (em-pir′ĭ-cal for′miu-lah) Chemical shorthand that indicates the number of each kind of atom within a molecule.

fat (fat) A class of water-insoluble macromolecules composed of a glycerol and fatty acids.

fatty acid (fat′ē ă′sid) One of the building blocks of a fat, composed of a long-chain carbon skeleton with a carboxylic acid functional group.

functional groups (fung′shun-al grūps) Specific combinations of atoms attached to the carbon skeleton that determine specific chemical properties.

glycerol (glis′er-ol) One of the building blocks of a fat, composed of a carbon skeleton that has three alcohol groups (—OH) attached to it.

hydrolysis (hi-drol′ĭ-sis) A process that occurs when a large molecule is broken down into smaller parts by the addition of water.

inorganic molecules (in-or-gan′ik mol′uh-kiuls) Molecules that do not contain carbon atoms in rings or chains.

isomers (i′so-meers) Molecules that have the same empirical formula but different structural formulas.

lipids (lī'pids) Large organic molecules that do not easily dissolve in water; classes include fats, phospholipids, and steroids.

nucleic acids (nu-kle'ik ă'sids) Complex molecules that store and transfer information within a cell. They are constructed of fundamental monomers known as nucleotides.

nucleotide (nu''kle-o-tid') A fundamental subunit of nucleic acid constructed of a phosphate group, a sugar, and an organic nitrogenous base.

organic molecules (or-gan'ik mol'uh-kiuls) Complex molecules whose basic building blocks are carbon atoms in chains or rings.

peptide bond (pep'tid bond) A covalent bond between amino acids in a protein.

phospholipid (fos''fo-lī'pid) A class of water-insoluble molecules that resemble fats but contain a phosphate group ($-PO_4$) in their structure.

polypeptide chain (pŏ''le-pep'tid chān) A macromolecule composed of a specific sequence of amino acids.

protein (pro'te-in) Macromolecules made up of amino acid subunits attached to each other by peptide bonds.

regulator proteins (reg'yu-la-tor pro'te-ins) Proteins that influence the activities that occur in an organism—for example, enzymes and some hormones.

saturated (sat'yu-ra-ted) A term used to describe the carbon skeleton of a fatty acid that contains no double bonds between carbons.

steroid (stēr'oid) One of the three kinds of lipid molecules characterized by their arrangement of interlocking rings of carbon.

structural formula (struk'chu-ral for'miu-lah) An illustration showing the arrangement of the atoms and their bonding within a molecule.

structural proteins (struk'chu-ral pro'te-ins) Proteins that are important for holding cells and organisms together, such as the proteins that make up the cell membrane, muscles, tendons, and blood.

unsaturated (un-sat'yu-ra-ted) A term used to describe the carbon skeleton of a fatty acid containing carbons that are double bonded to each other at one or more points.

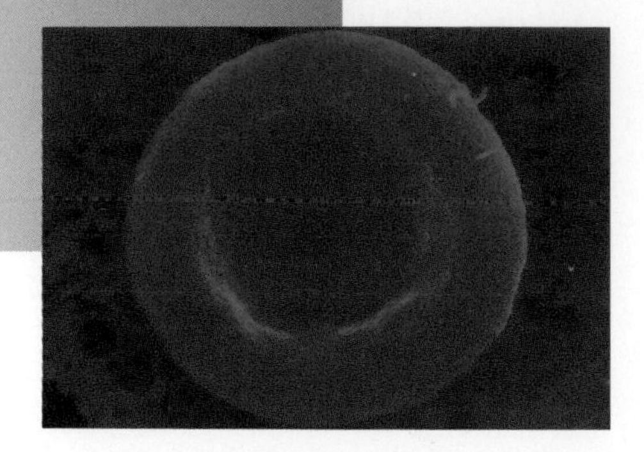

C H A P T E R 4

Cell Structure and Function

Chapter Outline

Purpose

The cell is the simplest structure capable of existing as an individual living unit. Within this unit, certain chemical reactions are required for maintaining life. These reactions do not occur at random, but are associated with specific parts of the many kinds of cells. This chapter deals with certain cellular structures found within most types of cells and discusses their functions.

For Your Information

Here are some tidbits of information about cells:

- The smallest cells are the bacteria, some of which are only 0.1 to 0.2 micrometers in diameter. *E. coli,* a very common bacterium in human intestines, is only 0.5 micrometers long.
- The largest cells are the egg cells of birds and mammals. The yolk of a chicken egg is a single cell which can be 50 millimeters in diameter. The nerve cell from the human spine to the big toe can be 1 meter or more in length.
- The animal cells that live the longest are the nerve or muscle cells of long-lived reptiles and mammals. The sea tortoise, which lives two hundred years, has the same nerve cells at birth as it does at death.
- Cells of certain bacteria can divide every fifteen minutes under optimum conditions. This means that if you start with only one bacterial cell and it divides as often as possible, in only twelve hours there could be as many as 281 trillion individual cells.

Learning Objectives

- Recognize the historical perspective of the development of the cell theory.
- Describe the molecular structure of a membrane and relate this structure to the processes whereby a cell accumulates and releases materials.
- Describe the processes whereby a cell accumulates some materials and releases others to the environment and the conditions controlling these various processes.
- Identify the cytoplasmic organelles in most eukaryotic cells.
- Associate the organelle structure with its major functions in eukaryotic cells.
- Identify the nuclear components of a cell and associate functions with the nuclear structures.

BOX 4.1 The Microscope

In order to view very small objects we use a magnifying glass. A magnifying glass is a lens that bends light in such a way that the object appears larger than it really is. Such a lens might magnify objects ten or even fifty times. Anton van Leeuwenhoek (1632–1723), a Dutch draper and haberdasher, was one of the first individuals to carefully study magnified cells. He made very detailed sketches of the things he viewed with his simple microscopes and communicated his findings to Robert Hooke and the Royal Society of London. His work stimulated further investigation of magnification techniques and description of cell structure. These first microscopes were developed in the 1600s.

Compound microscopes, developed soon after the simple microscopes, are able to increase magnification by bending light through a series of lenses. One lens, the *objective lens*, magnifies a specimen that is further magnified by the second lens, known as the *ocular lens*. With the modern technology of producing lenses, the use of specific light waves, and the immersion of the objective lens in oil to collect more of the available light, objects can be magnified one hundred to fifteen hundred times. Microscopes typically available for student use are compound light microscopes.

The major restriction of magnification with a light microscope is the limited ability of the viewer to distinguish two very close objects as two distinct things. The ability to separate two objects is termed *resolution* or *resolving power.* Some people have extremely good eyesight and are able to look at letters on a page and recognize that they are separate objects, while other persons see the individual letters as being "blurred together." Their eyes have different resolving powers. We can enhance the resolving power of the human eye by using lenses as in eye glasses or microscopes. All lens systems, whether in the eye or in microscopes, have a limited resolving power.

If two structures in a cell are very close to each other, you may not be able to determine that there are actually two structures close to each other rather than one structure. The limits of resolution of a light microscope are related to the wavelengths of the light being transmitted through the specimen. If you could see ultraviolet light waves, which have shorter wavelengths, it would be possible to resolve more individual structures.

An electron microscope makes use of this principle: the moving electrons have much shorter wavelengths than visible light. Thus, they are able to magnify 200,000 times and still resolve individual structures. The difficulty is, of course, that you are unable to see electrons with your eyes. Therefore, in order to use the electron microscope, the electrons strike a photographic film or television monitor, and this "picture" shows the individual structures. Heavy metals scattered on the structures to be viewed with the electron microscope increase the contrast between areas where there are structures that interfere with the transmission of the electrons and areas where the electrons are transmitted easily. The techniques for preparing the material to be viewed—slicing the specimen very thinly and focusing the electron beam on the specimen—make electron microscopy an art as well as a science.

The Cell Concept

The concept of a cell is one of the most important ideas in biology because it applies to all living things. It did not emerge all at once, but has been developed and modified over many years. It is still being modified today.

Several individuals made key contributions to the cell concept. Anton van Leeuwenhoek (1632–1723) was one of the first to make and use a *microscope* (box 4.1). A **microscope** is an instrument constructed of lenses that enlarge and focus an image of a small object. When van Leeuwenhoek discovered that he could see things moving in pond water using his microscope, his curiosity stimulated him to look at a variety of other things. He studied blood, semen, feces, pepper, and tartar, for example. He was the first to see individual cells and recognize them as living units, but he did not call them cells. The name he gave to these "little animals" that he saw moving around in the pond water was *animalcules*.

The first person to use the term *cell* was Robert Hooke (1635–1703) of England, who was also interested in how things looked when magnified. He chose to study thin slices of cork, the tissue from the bark of a cork oak tree. He saw a mass of cubicles fitting neatly together, which reminded him of the barren rooms in a monastery. Hence, he called them *cells*. As it is currently used, the term **cell** refers to the basic structural unit that makes up all living things. When Hooke looked at cork, the tiny boxes he saw were, in fact, only the cell walls that surrounded the living portions of plant cells. We now know that the cell wall is composed of the complex carbohydrate cellulose, which provides strength and

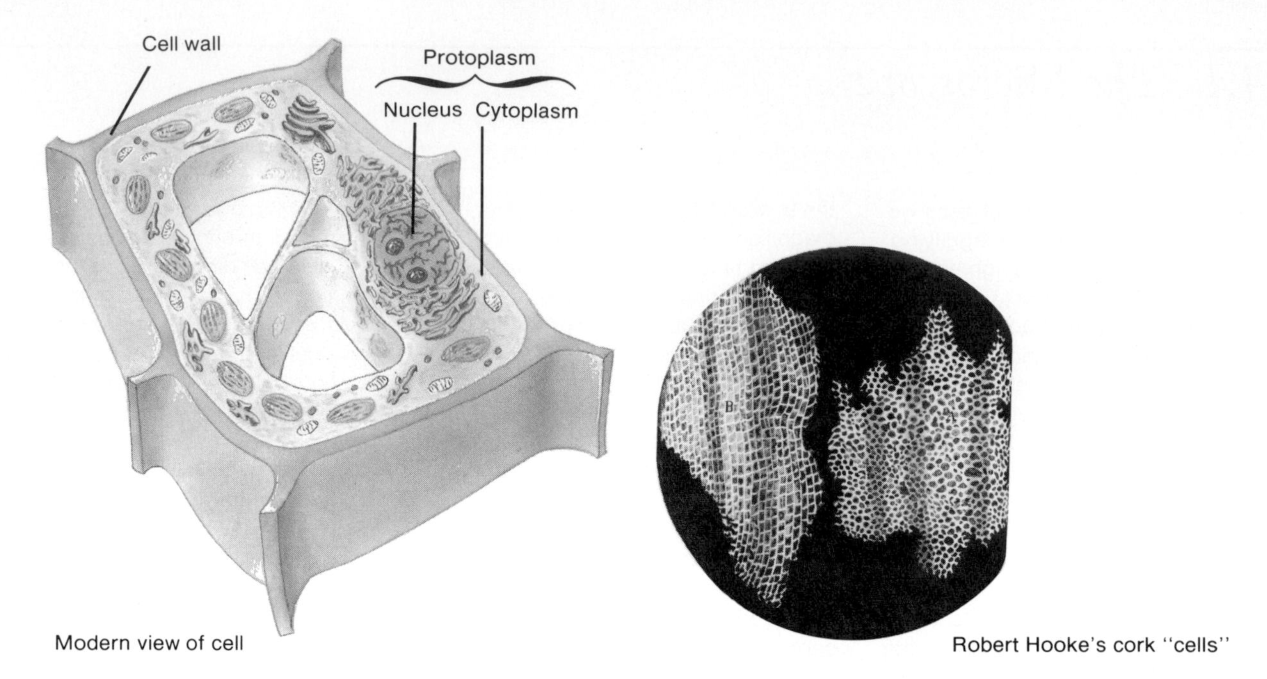

Figure 4.1 **Cells—Basic Structure of Life.** The cell concept has changed considerably over the last 300 years. Robert Hooke's idea of a cell was based on his observation of slices of cork. One of the first subcellular differentiations was to divide the protoplasm into cytoplasm and nucleus. We know now that cells are composed of many kinds of subcellular particles.

protection to the living contents of the cell. The cell wall appears to be a rigid, solid layer of material, but in reality it is composed of many interwoven strands of cellulose molecules. Its structure allows certain very large molecules to pass through it readily, but it acts as a screen to other molecules.

Hooke's use of the term *cell* in 1666 in his publication *Micrographia* was only the beginning, for it took nearly two hundred years before it was generally recognized that all living things were made of cells and that these cells could reproduce themselves. In 1838 Mathias Jakob Schleiden stated that all plants were made up of smaller cellular units. In 1839 Theodor Schwann published the idea that all animals were composed of cells.

Soon after the term *cell* caught on, it was recognized that its vitally important portion was inside the cell wall. This living material was termed **protoplasm,** which means *first-formed substance*. The use of the term *protoplasm* allowed the living portion of the cell to be distinguished from the nonliving cell wall. Very soon microscopists were able to distinguish two different regions of protoplasm. One type of protoplasm was more viscous and darker than the other. This region, called the **nucleus,** appeared as a central body within a more fluid material surrounding it. **Cytoplasm** is the name given to the more fluid portion of the protoplasm (figure 4.1).

The development of better light microscopes and, ultimately, the electron microscope revealed that protoplasm contains many structures called **organelles.** It has been determined that these organelles perform particular functions in each cell. The job that an organelle does is related to its structure. Each organelle is dynamic in its operation, changing shape and size as it works. Organelles move throughout the cell, and some even self-duplicate.

Cell Membranes

One feature common to all cells and many of the organelles they contain is a thin layer of material called *membrane*. Membrane can be folded and twisted into many different structures, shapes, and forms. The particular arrangement of membrane within an organelle is related to the functions it is capable of performing. This is similar to the way a piece of fabric can be fashioned into a pair of pants, a shirt, sheets, pillowcases, or a rag doll. All cellular membranes have a fundamental molecular structure that allows them to be fashioned into a variety of different organelles, each with a very specific function.

Cellular membranes are thin sheets composed primarily of phospholipids and proteins. The current concept of how membranes are constructed is known as the **fluid-mosaic model,** which proposes that the various molecules in the membrane are able to flow and move about, but that the membrane maintains its form because of the physical interaction of its molecules with surrounding water molecules. The phospholipid molecules of the membrane have one end that is soluble in water and is therefore called **hydrophilic** (water-loving) and

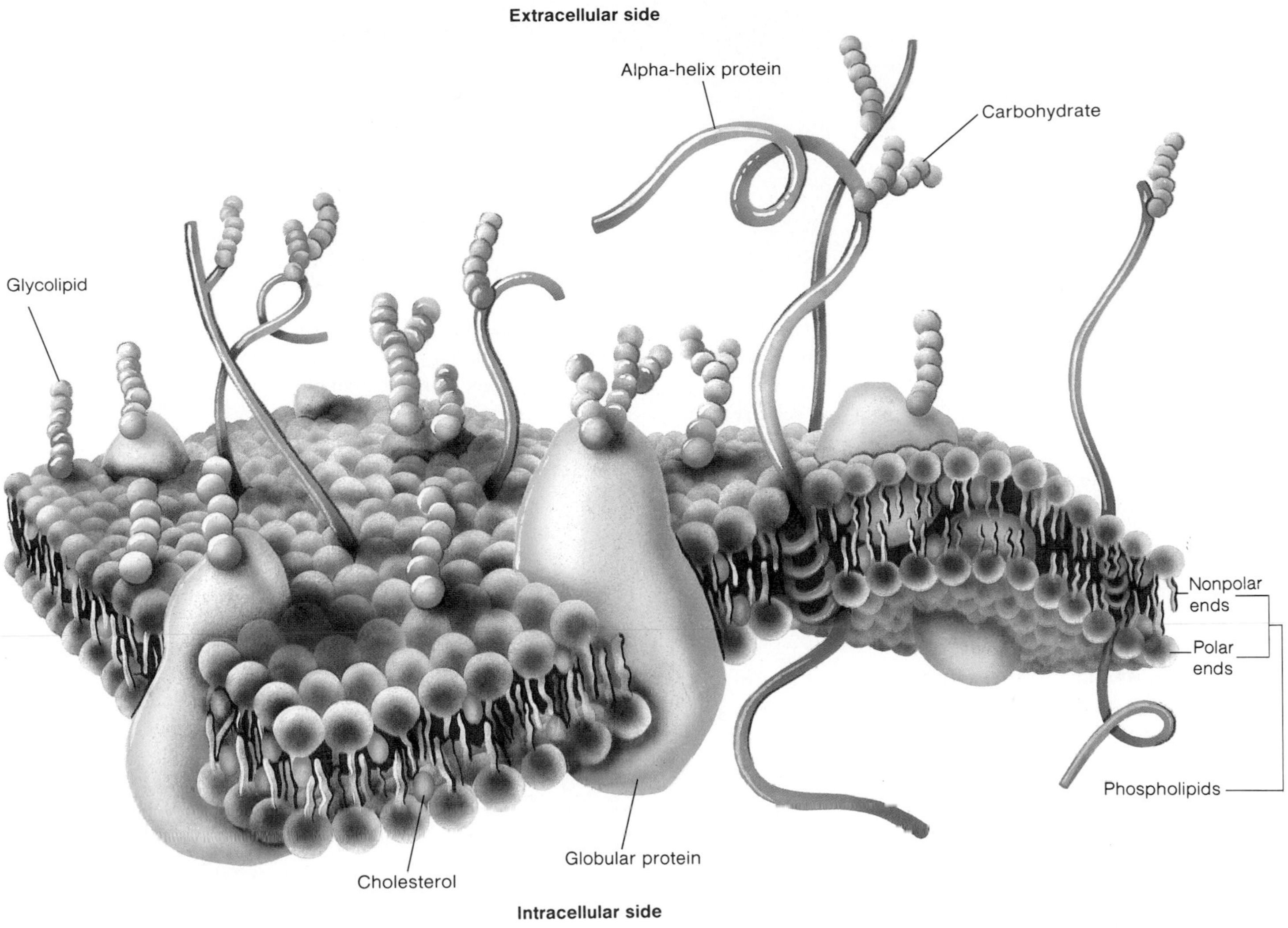

Figure 4.2 **Membrane Structure.** Membranes in cells are composed of protein and phospholipids. Two layers of phospholipid are oriented so that the hydrophobic fatty ends extend toward each other and the hydrophilic glycerol portions are on the outside. The phosphate-containing chain of the phospholipid is coiled near the glycerol portion. Buried within the phospholipid layer and/or floating on it are the globular proteins. Some of these proteins accumulate materials from outside the cell; others act as sites of chemical activity. Carbohydrates are often attached to one surface of the membrane.

another end that is not, called **hydrophobic** (water-hating). Consequently, when phospholipid molecules are placed in water, they form a double-layered sheet, with the water-soluble (hydrophilic) portions of the molecules facing away from each other. If phospholipid molecules are shaken in a glass of water, the molecules will automatically form double-layered membranes. It is important to understand that the membrane formed is not rigid or stiff. The component phospholipids are in constant motion as they move with the surrounding water molecules and slide past one another.

The protein component of cellular membranes can be found on either surface of the membrane or in the membrane, among the phospholipid molecules. Many of the protein molecules are capable of moving from one side to the other. These proteins can help with the chemical activities of the protoplasm. Others are involved in the uptake of materials from the environment. Still others aid in the movement of molecules across the membrane (figure 4.2). In addition to phospholipid and protein, some protein molecules found on the outside surfaces of cellular membranes have carbohydrates or fats attached to them. These combination molecules are important in determining the "sidedness" (inside-outside) of the membrane and help organisms recognize differences between types of cells. Your body can recognize disease-causing organisms because their surface proteins are different from those of its own cellular membranes.

Other molecules that are found in cell membranes are cholesterol and carbohydrates. Cholesterol is found in the middle of the membrane, in the hydrophobic region, because cholesterol is not water-soluble. It appears to play a role in stabilizing the membrane. Carbohydrates are usually found on the outside of the membrane, where they are bound to proteins or lipids. They appear to play a role in cell-to-cell interactions and are involved in binding with regulatory molecules.

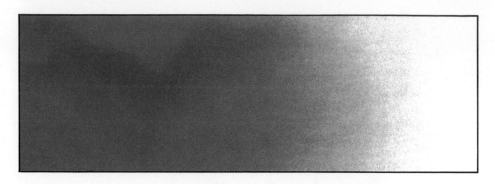

Figure 4.3 **The Concentration Gradient.** Gradual changes in concentrations of molecules over distance are called concentration gradients. This bar shows a color gradient with full color at one end and no color at the other end. A concentration gradient is necessary for diffusion to occur. Diffusion results in net movement of molecules from an area of higher concentration to an area of lower concentration.

Diffusion

Because the cell membrane is composed of phospholipid and protein molecules that are in constant motion, temporary openings are formed that allow small molecules to cross from one side of the membrane to the other. Molecules close to the membrane are in constant motion as well. They are able to move into and out of a cell by passing through these openings in the membrane. **Net movement** is the movement of molecules in one direction minus the movement of molecules in the opposite direction. The net movement of a particular kind of molecule from an area of higher concentration to an area of lower concentration is **diffusion.** If the concentration of a specific kind of molecule is higher on the outside of a cell membrane than on the inside, those molecules will diffuse through the membrane into the cell if the membrane is permeable to the molecules.

The rate of diffusion is related to the kinetic energy and size of the molecules. Since diffusion only occurs when molecules are unevenly distributed, the relative concentration of the molecules is important in determining how fast diffusion occurs. The difference in concentration of the molecules is known as a **concentration gradient** or **diffusion gradient.** When the molecules are equally distributed, no such gradient exists (figure 4.3).

Diffusion can take place only as long as there are no barriers to the free movement of molecules. In the case of a cell, the membrane permits some molecules to pass through, while others are not allowed to pass or only allowed to pass more slowly. This permeability is based on size, ionic charge, and solubility of the molecules involved. The membrane does not, however, distinguish direction of movement of molecules; therefore, the membrane does not influence the direction of diffusion. The direction of diffusion is determined by the relative concentration of specific molecules on the two sides of the membrane, and the energy that causes diffusion to occur is supplied by the kinetic energy of the molecules themselves (figure 4.4).

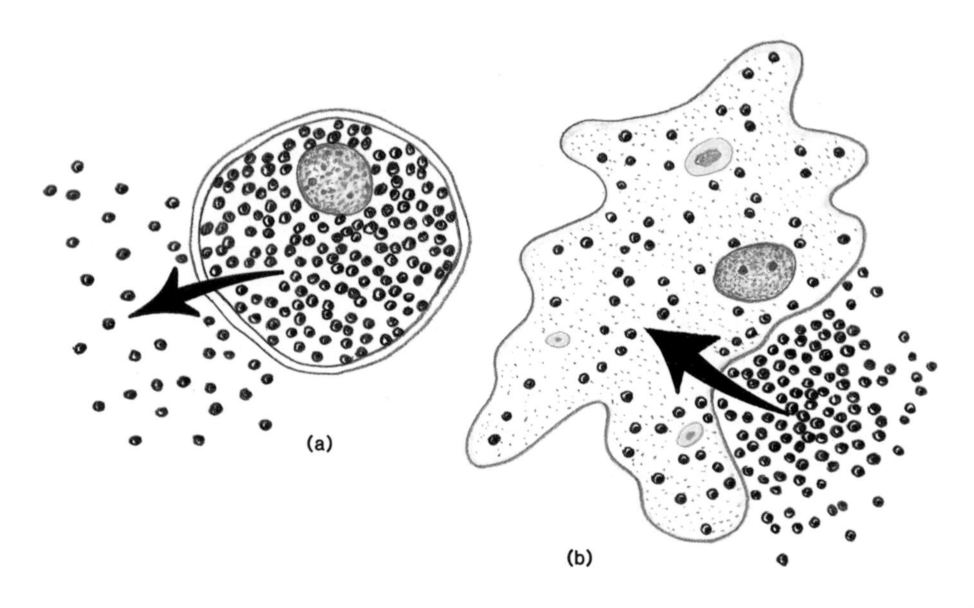

Figure 4.4 **Diffusion.** As a result of molecular motion, molecules move from areas where they are concentrated to areas where they are less concentrated. This figure shows (a) molecules leaving a cell by diffusion and (b) molecules entering a cell by diffusion. The direction is controlled by concentration, and the energy necessary is supplied by the kinetic energy of the molecules themselves.

Diffusion is an important means by which materials are exchanged between a cell and its environment. Since the movement of the molecules is random, the cell has little control over the process; thus, diffusion is considered a passive process. For example, animals are constantly using oxygen in various

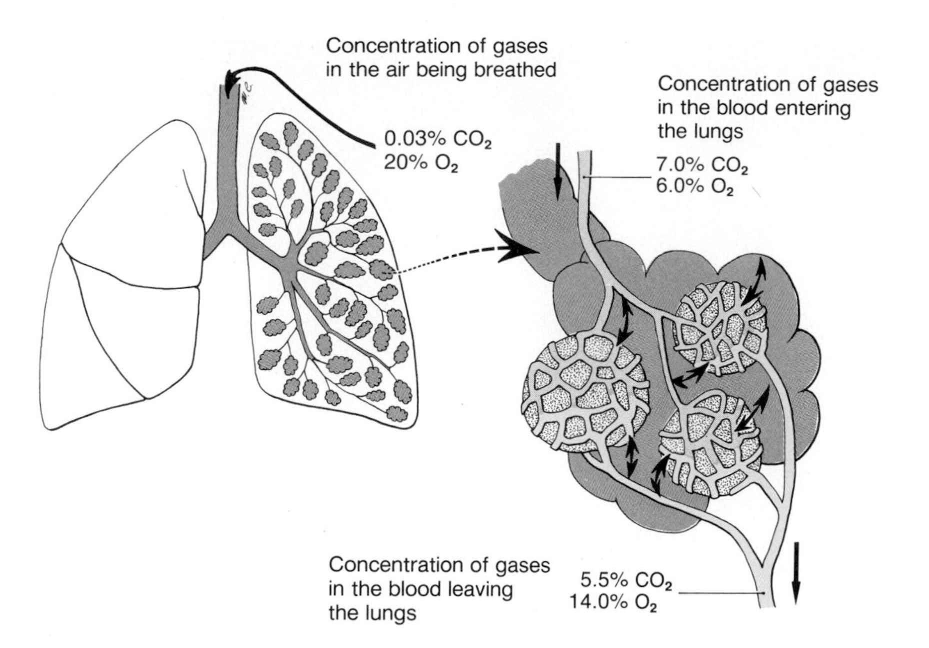

Figure 4.5 **Diffusion in the Lungs.** As blood enters the lungs, it has a higher concentration of carbon dioxide and a lower concentration of oxygen than the air in the lungs. The concentration gradient of oxygen is such that the oxygen diffuses from the lungs into the blood, and the concentration gradient of the carbon dioxide is such that it diffuses from the blood into the lungs. These two different diffusions happen simultaneously, and the direction of diffusion is controlled by the relative concentrations of each kind of molecule in the blood and in the lungs.

chemical reactions. Consequently, the oxygen concentration in cells always remains low. The cells then contain a lower concentration of oxygen in comparison to the oxygen level outside of the cell. This creates a diffusion gradient, and the oxygen molecules diffuse from the outside of the cell to the inside of the cell.

In large animals, many of the cells are buried deep within the body; if it were not for their circulatory systems, there would be little opportunity for cells to exchange gases directly with their surroundings. The circulatory system is a transportation system within a body composed of blood vessels of various sizes. These vessels carry many different molecules from one place to another. Oxygen may diffuse into blood through the membranes of the lungs, gills, or other moist surfaces of the animal's body. The circulatory system then transports the oxygen-rich blood throughout the body. The oxygen automatically diffuses into cells that are low in oxygen. The opposite is true of carbon dioxide. Animal cells constantly produce carbon dioxide, and so there is always a high concentration of it within the cells. These molecules diffuse from the cells into the blood, where the concentration of carbon dioxide is lower. The blood is pumped to the moist surface (gills, lungs, etc.), and the carbon dioxide again diffuses into the surrounding environment, which has a lower concentration of this gas. In a similar manner, many other types of molecules constantly enter and leave cells (figure 4.5).

Dialysis and Osmosis

Another characteristic of all membranes is that they are differentially permeable. (The terms *selectively permeable* and *semipermeable* are synonyms.) **Differential permeability** means that a membrane will allow certain molecules to pass across it and will prevent others from doing so. Molecules that are able to dissolve in phospholipids, such as vitamins A and D, can pass through the membrane rather easily; however, many molecules cannot pass through at all. In certain cases, the membrane differentiates on the basis of molecular size; that is, the membrane allows small molecules, such as water, to pass through and prevents the passage of larger molecules. The membrane may also regulate the passage of ions. If a particular portion of the membrane has a large number of positive ions on its surface, positively charged ions in the environment will be repelled and prevented from crossing the membrane.

We make use of diffusion across a differentially permeable membrane when we use a dialysis machine to remove wastes from the blood. If a kidney is unable to function normally, blood from a patient is diverted to a series of tubes composed of differentially permeable membrane. The toxins that have concentrated in the blood diffuse into the surrounding fluids in the dialysis machine, and the cleansed blood is returned to the patient. Thus, the machine functions in place of the kidney.

Water is a molecule that easily diffuses through cell membranes. The net movement of water molecules through a differentially permeable membrane is known as **osmosis.** In any osmotic situation, there must be a differentially permeable membrane separating two solutions. For example, a solution of 90% water and 10% sugar separated by a differentially permeable membrane from a different sugar solution, such as one of 80% water and 20% sugar, demonstrates osmosis. The membrane allows water molecules to pass freely but prevents the larger sugar molecules from crossing. There is a higher concentration of water molecules in one solution (compared to the concentration of water molecules in the other), so more of the water molecules move from the solution with 90% water to the other solution with 80% water. Be sure that you recognize that osmosis is really diffusion in which the diffusing substance is water, and that the regions of different concentrations are separated by a membrane that is more permeable to water.

A proper amount of water is required if a cell is to function efficiently. Too much water in a cell may dilute the cell contents and interfere with the chemical reactions necessary to keep the cell alive. Too little water in the cell may result in a buildup of poisonous waste products. As with the diffusion of other molecules, osmosis is a passive process because the cell has no control over the diffusion of water molecules. This means that the cell can remain in balance with an environment only if that environment does not cause the cell to lose or gain too much water.

Many organisms have a concentration of water and dissolved materials within their cells that is equal to that of their surroundings. When the concentration of water is equal on both sides of the cell membrane, the cell is said to be **isotonic** to its surroundings. This is particularly true of simple organisms that live in the ocean. The ocean has many kinds of salts dissolved in it, and such organisms as sponges, jellyfishes, and protozoa can be isotonic because the amount of material dissolved in their cellular water is equal to the amount of salt dissolved in the ocean's water.

However, if an organism is going to survive in an environment that has a different concentration of water than its cells, it must expend energy to maintain this difference. Organisms that live in fresh water have a lower concentration of water (higher concentration of dissolved materials) than their surroundings and tend to gain water by osmosis very rapidly. They are said to be **hypertonic** to their surroundings and the surroundings are **hypotonic.** These two terms are always used to compare two different solutions. The hypertonic solution is the one with more dissolved material and less water; the hypotonic solution has less dissolved material and more water. Organisms whose cells gain water by osmosis must expend energy to eliminate any excess if they are to keep from swelling and bursting (figure 4.6).

Under normal conditions, when we drink small amounts of water the cells of the brain will swell a little, and signals are sent to the kidneys to rid the body of excess water. By contrast, marathon runners may drink large quantities of water

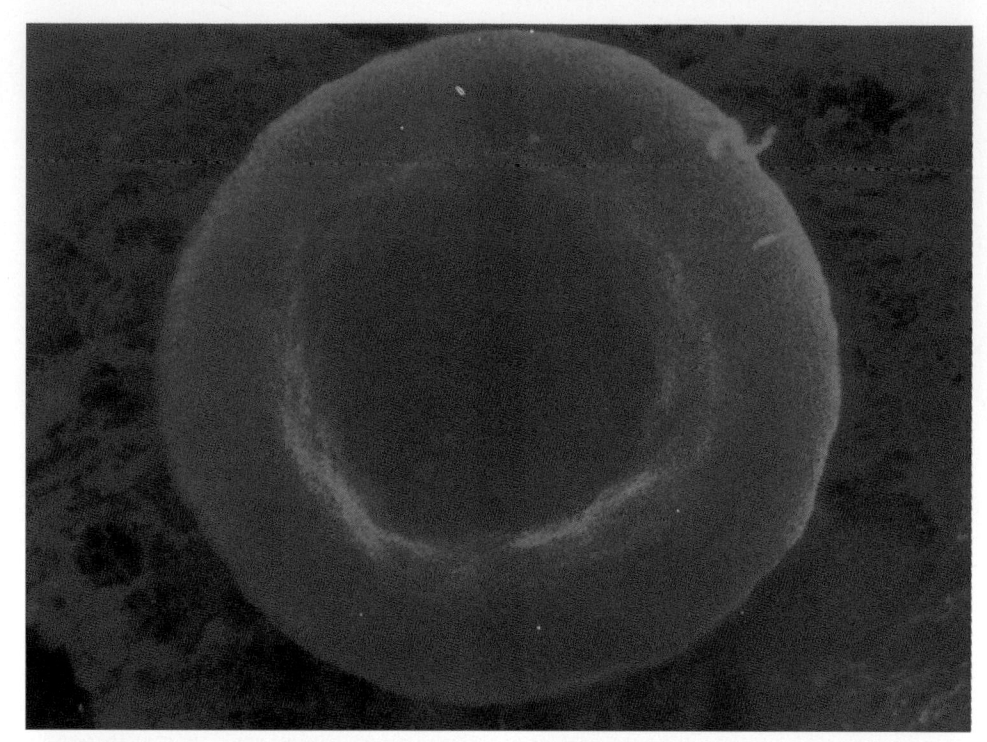

(a) Isotonic

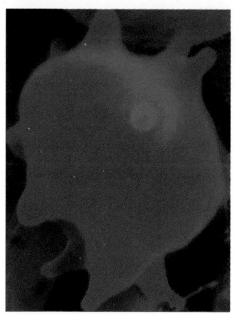

(b) Cell in hypertonic solution

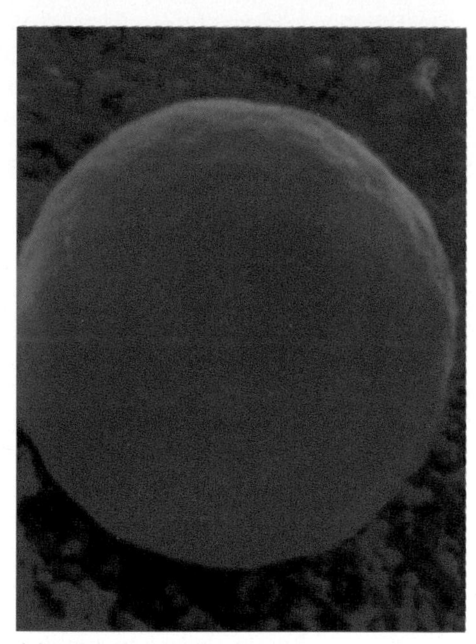

(c) Cell in hypotonic solution

Figure 4.6 **Osmotic Influences on Cells.** The cells in these three photographs were subjected to three different environments. In (a), the cell is isotonic to its surroundings. The water concentration inside the red blood cell and the water concentration in the environment are in balance with each other, so movement of water into the cell equals movement of water out of the cell, and the cell has its normal shape. In (b), the cell is in a hypertonic solution. Water has diffused from the cell to the environment because a higher concentration of water was in the cell, and the cell has shrunk. Part (c) shows a cell that has accumulated water from the environment because a higher concentration of water was outside the cell than in its protoplasm. The cell is in a hypotonic solution so it has swollen.

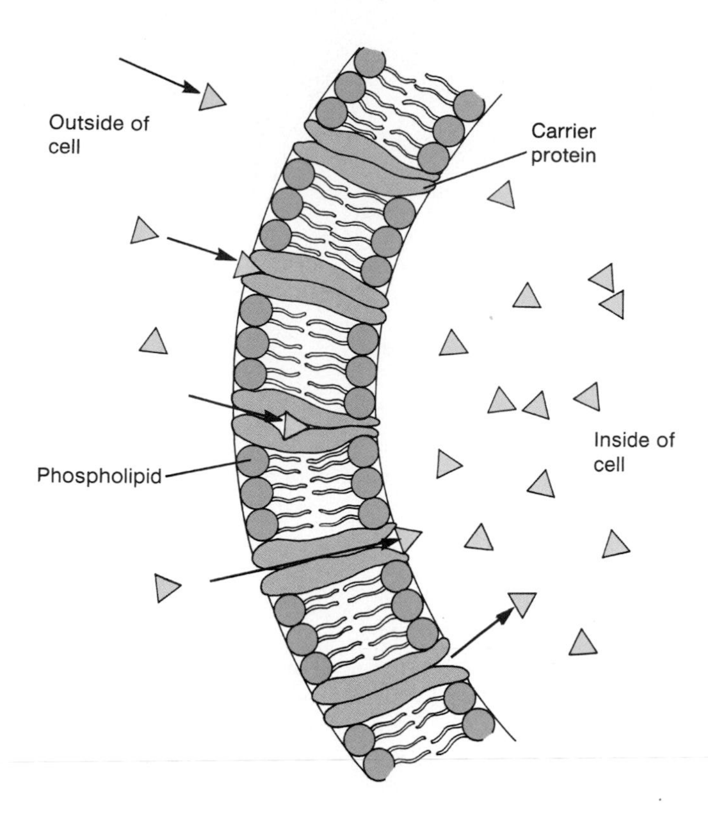

Figure 4.7 **Active Transport.** One possible method whereby active transport could cause materials to accumulate in a cell is illustrated here. Notice that the concentration gradient is such that if simple diffusion were operating, the molecules would leave the cell. The action of the carrier protein requires an active input of energy other than the kinetic energy of the molecules—therefore, this process is termed active transport.

in a very short time following a race. This rapid addition of water to the body may cause abnormal swelling of brain cells because the excess water cannot be gotten rid of rapidly enough. If this happens, the person may lose consciousness or even die because the brain cells have swollen too much.

Plant cells also experience osmosis. If the water concentration outside the plant cell is higher than the water concentration inside, more water molecules enter the cell than leave. This creates internal pressure within the cell. But plant cells do not burst because they are surrounded by a rigid cell wall. Lettuce cells that are crisp are ones that have gained water so that there is high internal pressure. Wilted lettuce has lost some of its water to its surroundings so that it has only slight internal cellular water pressure. Osmosis occurs when you put salad dressing on a salad. Because the dressing has a very low water concentration, water from the lettuce diffuses from the cells into the surroundings. Salad that has been "dressed" too long becomes limp and is unappetizing.

So far, we have considered only those situations in which the cell has no control over the movement of molecules. Cells cannot rely solely on diffusion and osmosis, because many of the molecules they require either cannot pass through the membrane or occur in relatively low concentrations in the cell's surroundings.

Controlled Methods of Transporting Molecules

Some molecules move across the membrane by combining with specific carrier proteins. When the rate of diffusion of a substance is increased in the presence of a carrier, we call this **facilitated diffusion.** Since this is diffusion, the net direction of movement is in accordance with the concentration gradient. Therefore, this is considered a passive transport method, although it can only occur in living organisms with the necessary carrier proteins. One example of facilitated diffusion is the movement of glucose molecules across the membranes of certain cells. In order for the glucose molecules to pass into these cells, specific proteins are required to carry them across the membrane. The action of the carrier does not require an input of energy other than the kinetic energy of the molecules.

When molecules are moved across the membrane from an area of *low* concentration to an area of *high* concentration, the cell must be expending energy. The process of using a carrier protein to move molecules up a concentration gradient is called **active transport** (figure 4.7). Active transport is very specific: only certain molecules or ions are able to be moved in this way, and they must be carried by specific proteins in the membrane. The action of the carrier requires an input of energy other than the kinetic energy of the molecules; therefore, this process is termed *active transport.* For example, some ions, such as sodium and potassium, are actively pumped across cell membranes. Sodium ions are pumped out of cells up a concentration gradient. Potassium ions are pumped into cells up a concentration gradient.

In addition to active transport, there are two other methods that are used to actively move materials into cells. **Phagocytosis** is the process that cells use to wrap membrane around a particle (usually food) and engulf it (figure 4.8). This is the process leucocytes (white blood cells) in your body use to surround invading bacteria, viruses, and other foreign materials. Because of this, these kinds of cells are called *phagocytes.* When phagocytosis occurs, the material to be engulfed touches the surface of the phagocyte and causes a portion of the outer cell membrane to be indented. The indented cell membrane is pinched off inside the cell to form a sac containing the engulfed material. This sac, composed of a single membrane, is called a **vacuole.** Once inside the cell, the membrane of the vacuole is broken down, releasing its contents inside the cell.

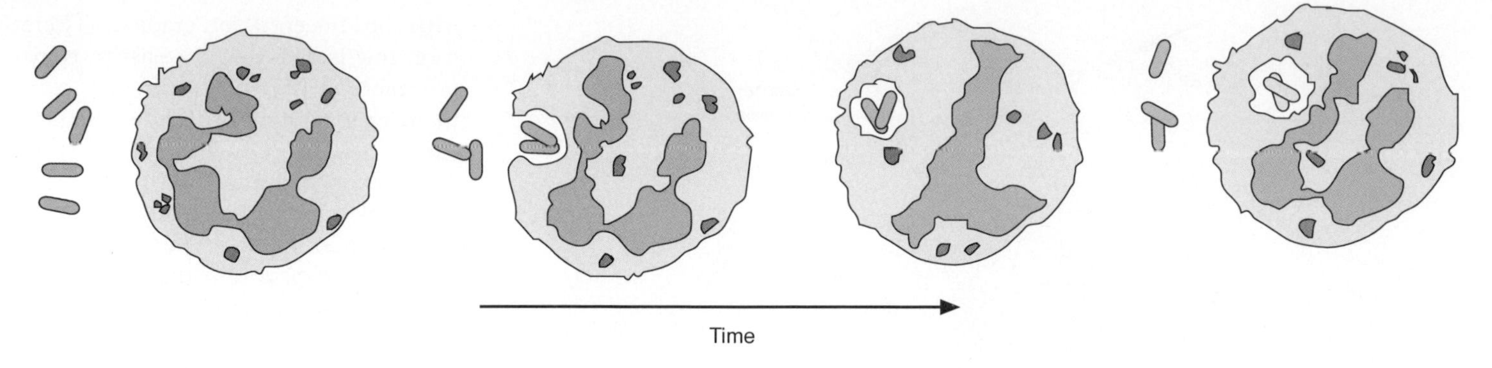

Figure 4.8 **Phagocytosis.** This cell is engulfing a large amount of material at one time and surrounding it with a membrane. A lysosome adds its digestive enzymes to aid in the breakdown. Finally, the digested material is able to be absorbed from the vacuole into the cytoplasm of the cell.

Phagocytosis is used by many types of cells to acquire large amounts of material from their environment. However, if the cell is not surrounding a large quantity of material, but is merely engulfing some molecules dissolved in water, the process is termed **pinocytosis.** In this process, the sacs that are formed are very small in comparison to those formed during phagocytosis. Because of this size difference, they are called **vesicles.** In fact, an electron microscope is needed in order to see them.

The processes of phagocytosis and pinocytosis differ from active transport in that the cell surrounds large amounts of material with a membrane rather than taking the material in through the membrane, molecule by molecule. The movement of materials into the cell by either phagocytosis or pinocytosis is referred to as *endocytosis*. The transport of material from the cell by the reverse processes is called *exocytosis*.

Cell Size

The size of a cell is directly related to its level of activity and the rates of movements of molecules across cell membranes. In order to stay alive, a cell must have a constant supply of nutrients, oxygen, and other molecules. It must also be able to get rid of carbon dioxide and other waste products that are harmful to it. The larger a cell becomes, the more difficult it is to satisfy these requirements; consequently most cells are very small. There are a few exceptions to this general rule, but they are easily explained. Egg cells, like the yolk of a hen's egg, are very large cells. However, the only part of an egg cell that is metabolically active is a small spot near its surface. The central portion of the egg is simply inactive stored food called *yolk*. Similarly, some plant cells are very large but consist of a large, centrally located region filled with water. Again, the metabolically active portion of the cell is at the surface, where exchange by diffusion or active transport is possible.

Organelles Composed of Membranes

Now that you have some background concerning the structure and the function of membranes, let's turn our attention to the way cells use the membranes to build specific structural components of their protoplasm. The outer boundary of the cell is termed the **cell membrane** or **plasma membrane.** It is associated with a great variety of metabolic activities including the uptake and release of molecules, the sensing of stimuli in the environment, the recognition of other cell types, and attachment to other cells and nonliving objects. In addition to the cell membrane, there are many other organelles composed of membranes. Each of these membranous organelles has a unique shape or structure that is associated with particular functions. One of the most common organelles found in cells is the *endoplasmic reticulum*.

The Endoplasmic Reticulum

The **endoplasmic reticulum, ER,** is a set of folded membranes and tubes throughout the cell. This system of membranes provides a large surface upon which chemical activities take place. Since the ER has an enormous surface area, many chemical reactions can be carried out in an extremely small space. Picture the vast surface area of a piece of newspaper crumpled into a tight little ball. The surface contains hundreds of thousands of tidbits of information in an orderly arrangement, yet it is packed into a very small volume. Proteins on the surface of the ER are actively involved in controlling and encouraging chemical activities—whether they are reactions involving growth and development of the cell or those resulting in the accumulation of molecules from the environment. The arrangement of the proteins allows them to control the sequences of metabolic activities so that chemical reactions can be carried out very rapidly and accurately.

Upon close examination using an electron microscope, it becomes apparent that there are two different types of ER—*rough* and *smooth*. The rough ER appears rough because it has ribosomes attached to its surface. *Ribosomes* are nonmembranous organelles that are associated with the synthesis of proteins from amino acids. Therefore, cells with an extensive amount of rough ER—for example, your pancreas cells—are capable of synthesizing large quantities of proteins. Smooth ER lacks attached ri-

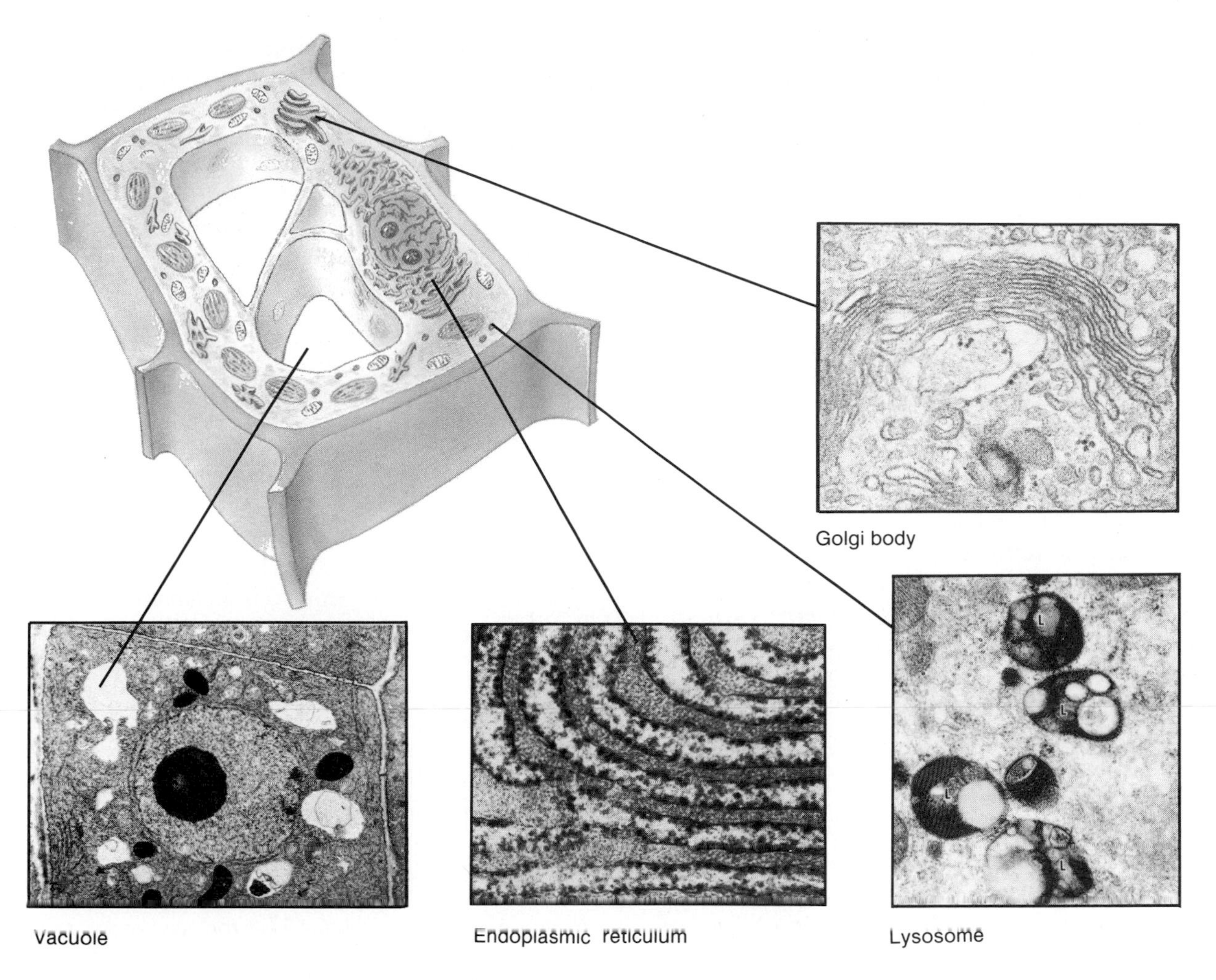

Figure 4.9 **Membranous Cytoplasmic Organelles.** Certain structures in the cytoplasm are constructed of membranes. Membranes are composed of protein and phospholipids. The four structures here—the endoplasmic reticulum, the Golgi body, vacuoles, and lysosomes—are constructed of simple membranes.

bosomes but is the site of many other important cellular chemical activities, including fat metabolism and detoxification reactions—for example, your liver cells contain extensive smooth ER.

In addition, the spaces between the folded membranes may serve as canals for the movement of molecules within the cell. Some researchers suggest that this system of membranes allows for rapid distribution of molecules within a cell (figure 4.9).

The Golgi Apparatus

Another organelle composed of membrane is the **Golgi apparatus.** Even though this organelle is also composed of membrane, the way in which it is structured enables it to perform jobs different from those performed by the ER. The typical Golgi is composed of from five to seven flattened, smooth, membranous sacs, which resemble a stack of pancakes. The Golgi apparatus is the site of the synthesis and packaging of certain molecules produced in the cell. It is also the place where particular chemicals are concentrated prior to their release from the cell. Some Golgi vesicles are used to transport such molecules as mucous, insulin, and digestive enzymes to the outside of the cell. The molecules are concentrated inside the Golgi, and tiny vesicles are budded off the outside surfaces of the Golgi sacs. These vesicles contain high concentrations of specific molecules. The vesicles move to the inside of the cell membrane and merge with it. In so doing, the contents are placed outside of the cell membrane.

Another important group of molecules that is necessary to the cell includes the hydrolytic enzymes. This group of enzymes is capable of destroying proteins and lipids. Since cells contain large amounts of proteins and lipids in their membranes and other structures, these enzymes must be controlled in order to prevent the destruction of the cell. The Golgi is the site where these enzymes are converted from their inactive to their active forms and packaged in membranous sacs. These vesicles are budded off from the outside surfaces of the Golgi sacs and given the special name **lysosomes.** The lysosomes are used by cells in four major ways:

1. When a cell is damaged, the membranes of the lysosomes break and the enzymes are released. These enzymes then begin to break down the contents of the damaged cell so that the component parts can be used by surrounding cells.
2. Lysosomes also play a part in the normal development of an organism. For example, as a tadpole slowly changes into a frog, the cells of the tail are destroyed by the action of lysosomes. In humans, the developing embryo has paddle-shaped hands. At a prescribed point in the development of the hand, the cells between the bones of the fingers release the enzymes that had been stored in the lysosomes. As these cells begin to disintegrate, the hand with individual fingers begins to take shape. Occasionally this process does not take place, and infants are born with "webbed fingers." This developmental defect, called *syndactylism,* may be surgically corrected soon after birth.
3. In many kinds of cells, the lysosomes are known to combine with food vacuoles. When this occurs, the enzymes of the lysosome break down the food particles into smaller and smaller molecular units. This process is common in one-celled organisms such as *Paramecium.*
4. Lysosomes are also used in the destruction of engulfed, disease-causing microorganisms such as bacteria, viruses, and fungi. As these invaders are taken into the cell by phagocytosis, lysosomes fuse with the phagocytic vacuole. When this occurs, the hydrolytic enzymes of the lysosome move into the vacuole to destroy the microorganisms.

The many kinds of vacuoles and vesicles contained in cells are frequently described by their function. Thus, food vacuoles hold food, and water vacuoles store water. Specialized water vacuoles called *contractile vacuoles* are able to forcefully expel excess water that has accumulated in the cytoplasm as a result of osmosis. The contractile vacuole is a necessary organelle in cells that live in fresh water. The water constantly diffuses into the cell and must be actively pumped out. The special containers that hold the contents resulting from pinocytosis are called *pinocytic vesicles.* In all cases, these simple containers are constructed of a surrounding membrane. In most plants, there is one huge, centrally located vacuole in which water, food, wastes, and minerals are stored.

The Nuclear Membrane

The nucleus is a place in the cell—not a solid mass. Just as a room is a place created by the walls, floor, and ceiling, the nucleus is a place in a cell created by the **nuclear membrane.** This membrane separates the *nucleoplasm,* liquid material in the nucleus, from the cytoplasm. If the membrane was not formed around the genetic material, the organelle we call the *nucleus* would not exist. The nuclear membrane is formed from many flattened sacs fashioned into a hollow sphere around the genetic material, DNA. It has large openings, called nuclear pores, that allow relatively large molecules to pass.

Energy Converters

All of the membranous organelles described above are able to be interconverted from one form to another (figure 4.10). For example, phagocytosis results in the formation of vacuolar membrane from cell membrane that fuses with lysosomal membrane, which in turn came from Golgi membrane. However, there are two other organelles composed of membranes that are chemically different and incapable of interconversion. Both types of organelles are associated with energy conversion reactions in the cell.

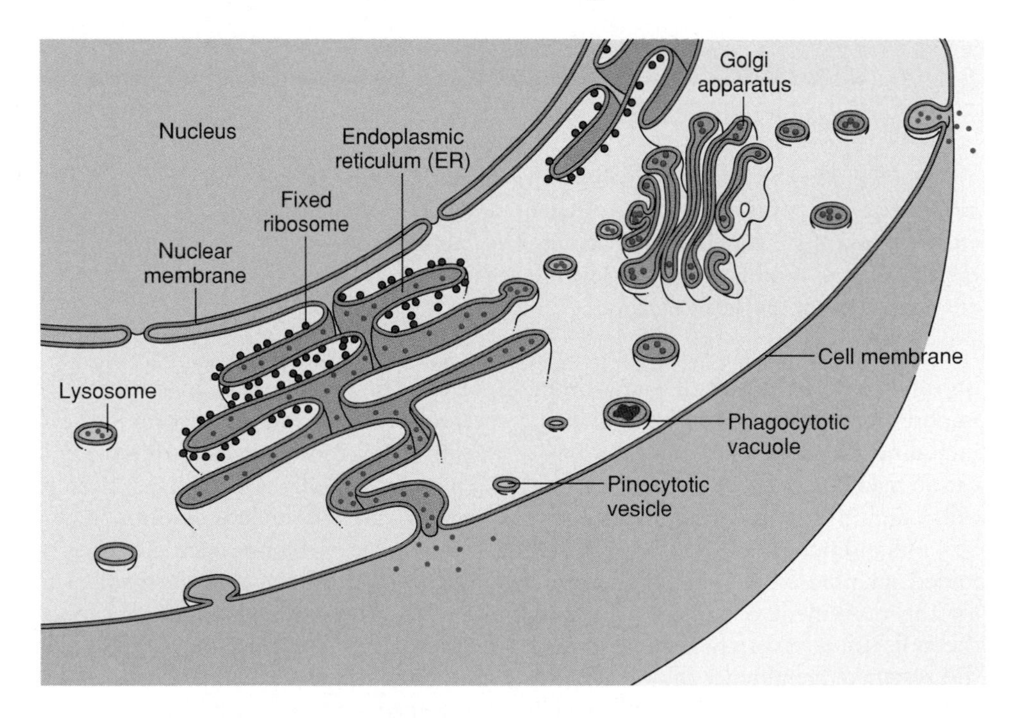

Figure 4.10 **Interconversion of Membranous Organelles.** Eukaryotic cells contain a variety of organelles composed of phospholipid and protein. Each has a unique shape and function. Many of these organelles are interconverted from one to another as they perform their essential functions. Cell membranes can become vacuolar membrane or endoplasmic reticulum, which can become vesicular membrane, which in turn can become Golgi or nuclear membrane. However, mitochondria and chloroplasts cannot exchange membrane parts.

These organelles are the *mitochondrion* and the *chloroplast* (figure 4.11).

The **mitochondrion** is an organelle resembling a small bag with a larger bag inside that is folded back on itself. These inner folded surfaces are known as the **cristae.** Located on the surface of the cristae are particular proteins and enzymes involved in *aerobic cellular respiration*. **Aerobic cellular respiration** is the series of reactions involved in the release of usable energy from food molecules, which requires the participation of oxygen molecules. Enzymes that speed the breakdown of food molecules are arranged in a sequence on the mitochondrial membrane. The average human cell contains upwards of ten thousand mitochondria. When properly stained, they can be seen with the compound light microscope. When cells are functioning aerobically, the mitochondria swell with activity. But when this activity diminishes, they shrink and appear as threadlike structures.

A second energy-converting organelle is the **chloroplast.** This membranous, saclike organelle contains the green pigment **chlorophyll.** Some cells contain only one large chloroplast, while others contain hundreds of smaller chloroplasts. It is in this organelle that light energy is converted to chemical-bond energy in the process known as **photosynthesis.** Chemical-bond energy is found in food molecules. A study of the ultrastructure—that is, the structures differentiated only by use of the electron microscope—of the chloroplasts shows that the entire organelle is enclosed by a membrane, while other membranes are folded and interwoven throughout. As shown in figure 4.11a, in some areas concentrations of these membranes are stacked up or folded back on themselves. Chlorophyll molecules are attached to these membranes. These areas of concentrated chlorophyll are called the **grana** of the chloroplast. The space between the grana, which has no chlorophyll, is known as the **stroma.**

Mitochondria and chloroplasts are different from other kinds of membranous structures in several ways. First, their membranes are chemically different from those of other membranous organelles; second, they are composed of double layers of membrane—an inner and an outer membrane; third, both of these structures have ribosomes and DNA that are similar to those of bacteria; and fourth, these two structures have a certain degree of independence from the rest of the cell—they have a limited ability to reproduce themselves but must rely on nuclear DNA for assistance. The functions of these two organelles are discussed in chapter 6.

All of the organelles just described are composed of membranes. Many of these membranes are modified for particular functions. Each membrane is composed of the double phospholipid layer with protein molecules associated with it.

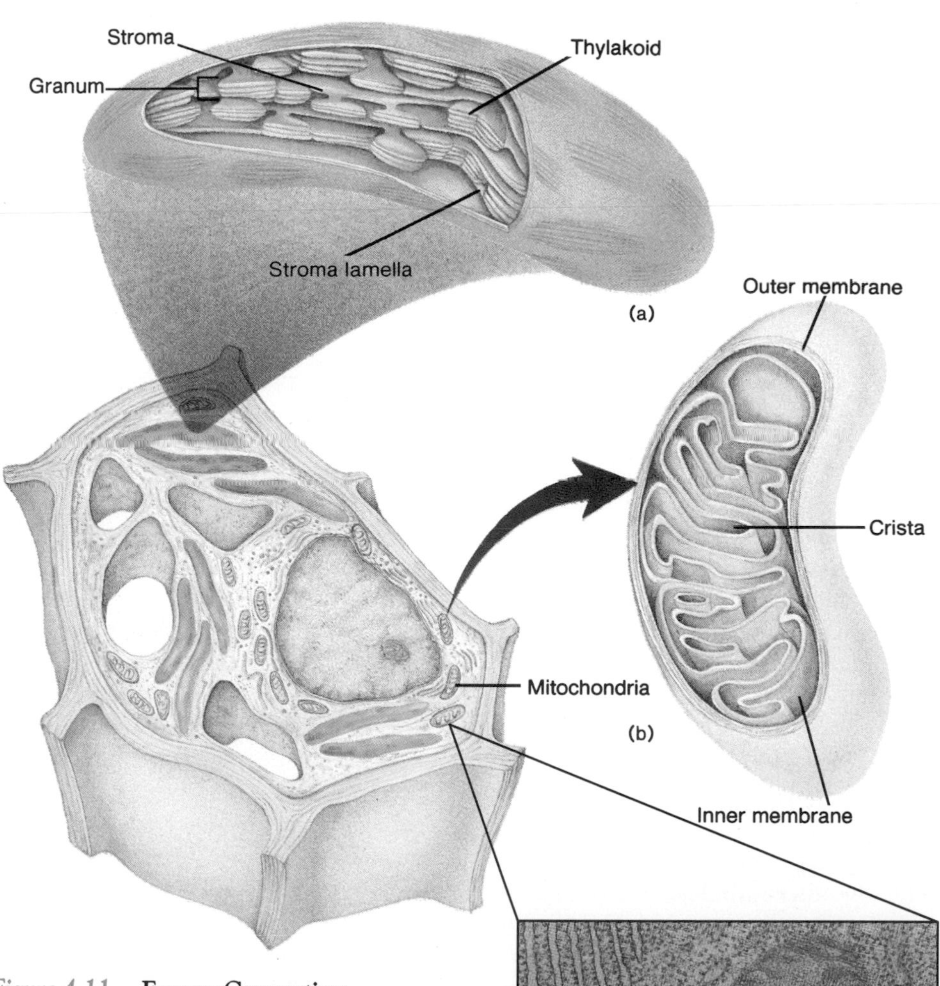

Figure 4.11 **Energy-Converting Organelles.** (a) The chloroplast, the container of the pigment chlorophyll, is the site of photosynthesis. The chlorophyll, located in the grana, captures light energy that is used to construct organic, sugarlike molecules in the stroma. (b) The mitochondria with their inner folds, called cristae, are the site of aerobic cellular respiration, where food energy is converted to usable cellular energy. Both of these organelles are constructed of protein and phospholipid membranes.

Nonmembranous Organelles

Suspended in the cytoplasm and associated with the membranous organelles are various kinds of structures that are not composed of phospholipids and proteins arranged in sheets.

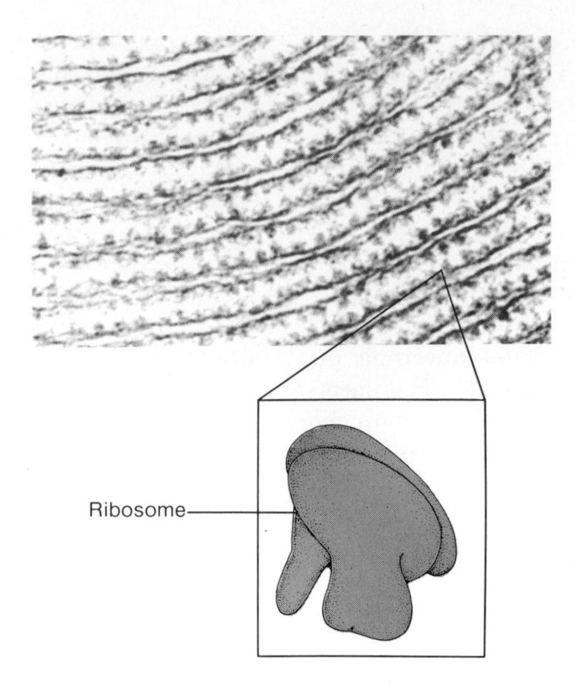

Figure 4.12 **Ribosomes.** Each ribosome is constructed of two subunits of protein and ribonucleic acid. These globular organelles are associated with the construction of protein molecules from individual amino acids. They are sometimes located individually in the cytoplasm where protein is being assembled, or they may be attached to endoplasmic reticulum (ER). They are so obvious on the ER when using electron micrograph techniques that when they are present, we label the ER as rough ER.

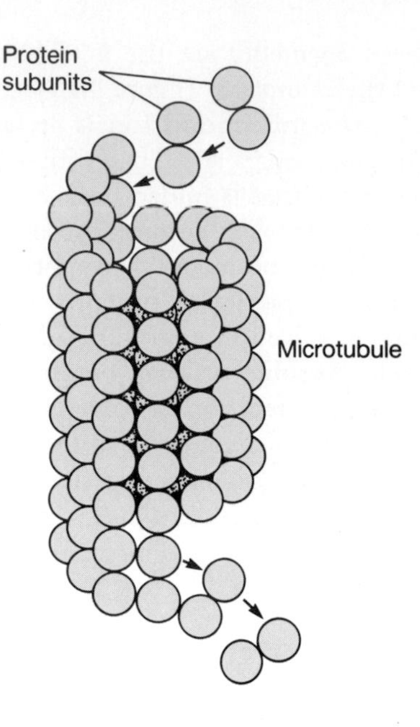

Figure 4.13 **Microtubules.** Microtubules are hollow tubes constructed of proteins. The dynamic nature of the microtubule is useful in the construction of certain organelles in a cell, such as centrioles and cilia or flagella.

Ribosomes

In the cytoplasm are many very small structures called **ribosomes** that are composed of ribonucleic acid (RNA) and protein. Ribosomes function in the manufacture of protein. Each ribosome is composed of two oddly shaped subunits—a large one and a small one. The larger of the two subunits is composed of a specific type of RNA associated with several kinds of protein molecules. The smaller is composed of RNA with fewer protein molecules than the large one. These globular organelles are involved in the assembly of proteins from amino acids—they are frequently associated with the endoplasmic reticulum to form rough ER. Areas of rough ER have been demonstrated to be active sites of protein production. Cells actively producing nonprotein materials, such as lipids, are likely to contain smooth ER rather than rough ER. Many ribosomes are also found floating freely in the cytoplasm (figure 4.12), wherever proteins are being assembled. Cells that are actively producing protein have great numbers of free and attached ribosomes. The details of how ribosomes function in protein synthesis are discussed in chapter 7.

Microtubules and Microfilaments

Another type of nonmembranous organelle, the **microtubules,** consist of small, hollow tubes composed of protein called *tubulin*. They function throughout the cytoplasm, where they provide structural support and enable movement. The microtubules are dynamic structures that are capable of being lengthened or shortened by the addition or subtraction of tubulin units (figure 4.13). As a result, there seems to be a constant shifting of microtubular material within a cell. Many of the structures with which microtubules are associated are able to move or grow. Microtubules participate in the movement of chromosomes during nuclear division, in the movement of flagella and cilia, and in the positioning of cellulose molecules during cell-wall synthesis. Important structures composed of microtubules are the *centrioles*.

Microfilaments are long, fiberlike structures made of protein found in cells. There are several kinds of microfilaments. Some are involved in the contraction of cells, while others serve as a cellular framework. It appears that microtubules and microfilaments are often interconnected to form a kind of cellular skeleton that allows the cell to have shape. However, this is not a rigid structural support because both microtubules and microfilaments are involved in movement and may be assembled and disassembled. Figure 4.14 shows how these different elements might be used to form a cellular skeleton.

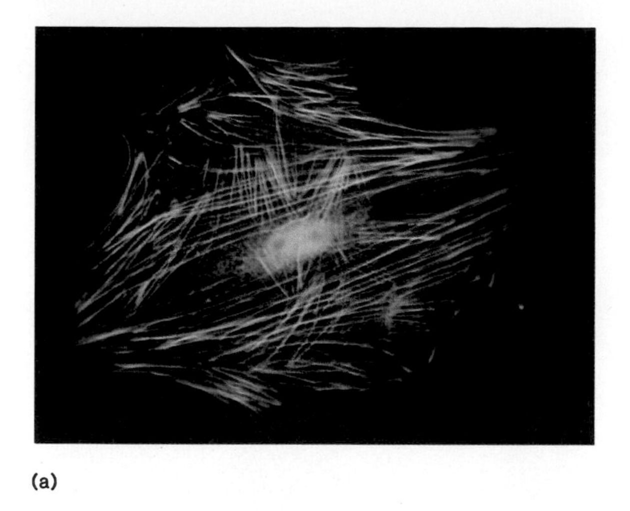

(a)

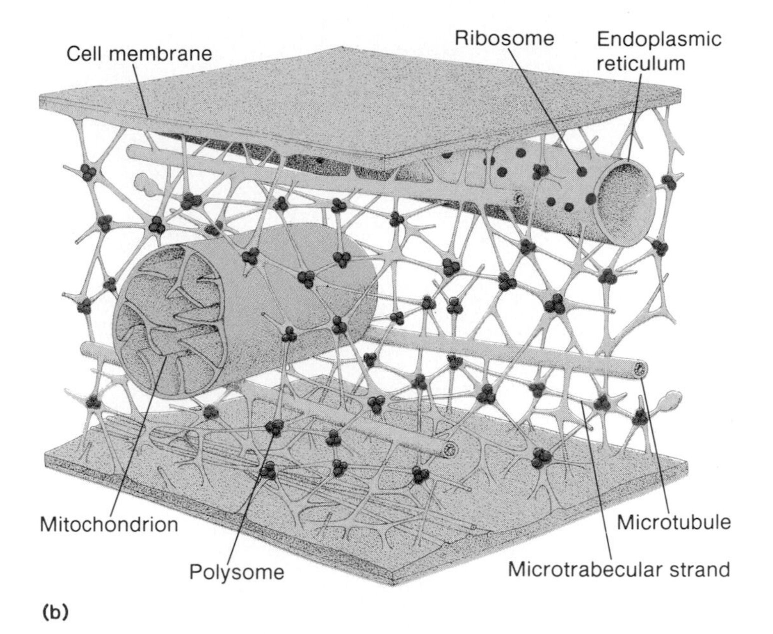

(b)

Figure 4.14 **The Cytoskeleton.** A complex array of microfilaments and microtubules provides a structure for the cell. The cellular skeleton is not a rigid structure but changes as the microfilaments and microtubules assemble and disassemble. Part (a) shows elements of the cytoskeleton that have been labeled with a fluorescent dye to make them visible. Part (b) shows how the various parts of the cytoskeleton are interconnected.

Centrioles

An arrangement of two sets of microtubules at right angles to each other makes up a structure known as the **centriole.** Each set is composed of nine groups of short microtubules arranged in a cylinder (figure 4.15). The centriole functions in cell division and is referred to again in chapter 9. One curious fact about centrioles is that they are present in most animal cells but not in many types of plant cells.

Cilia and Flagella

Many cells have microscopic, hairlike structures projecting from their surfaces; these are **cilia** or **flagella** (figure 4.16). In general, we call them *flagella* if they are long and few in number, and *cilia* if they are short and more numerous. They are similar in structure, and each functions to move the cell through its environment or to move the environment past the cell. They are constructed of a cylinder of nine sets of microtubules similar to those in the centriole, but they have an additional two microtubules in the center. These long strands of microtubules project from the cell surface and are covered by cell membrane. When cilia and flagella are sliced crosswise, their cut ends show what is referred to as the *9 + 2 arrangement* of microtubules. The cell has the ability to control the action of these microtubular structures, enabling them to be moved in a variety of different ways. Their coordinated actions either propel the cell through the environment or the environment past the cell surface. The protozoan *Paramecium* is covered with thousands of cilia that actively beat a rhythmic motion to move the cell through the water. The cilia on the cells that line your trachea move mucous-containing particles from deep within your lungs.

Inclusions

Inclusions are collections of materials that do not have as well-defined a structure as the organelles we have discussed so far. They might be concentrations of stored materials, such as starch grains, sulfur, or oil droplets, or they might be a collection of miscellaneous materials known as **granules.** Unlike organelles, which are essential to the survival of a cell, the inclusions are only temporary sites for the storage of nutrients and wastes.

Some inclusion materials may be harmful to other cells. For example, rhubarb leaf cells contain an inclusion composed of an organic acid—oxalic acid. Needle-shaped crystals of calcium oxalate can cause injury to the kidneys of an organism that eats rhubarb leaves. The sour taste of this particular compound aids in the survival of the rhubarb plant by discouraging animals from eating it. Similarly, certain bacteria store crystals of a substance in their inclusions that is known to be harmful to insects. Spraying plants with these bacteria is a biological method of controlling the population of the insect pests, while not interfering with the plant or with humans.

In the past, cell structures such as ribosomes, mitochondria, and chloroplasts were also called *granules,* since

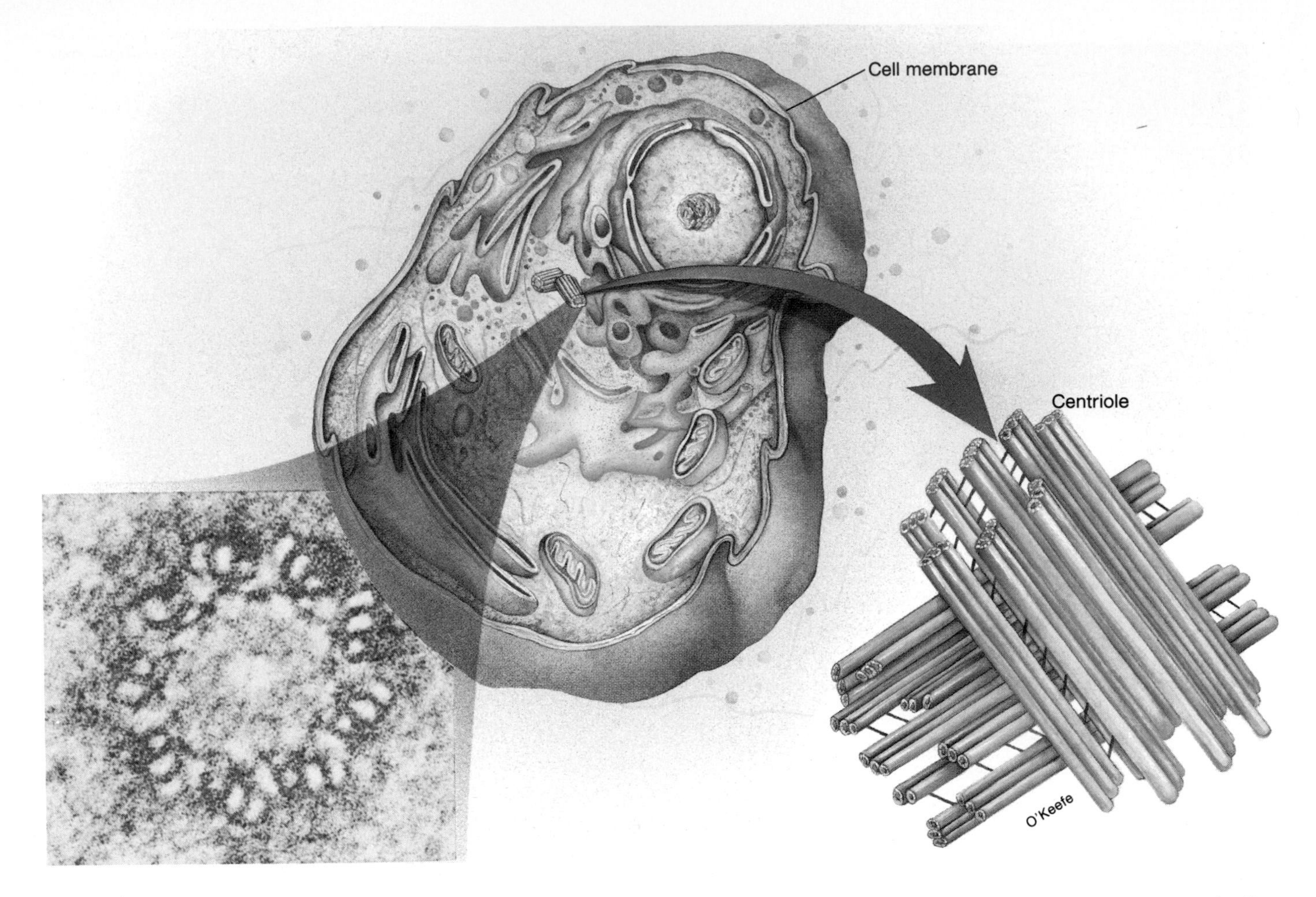

Figure 4.15 **The Centriole.** These two sets of short microtubules are located just outside the nuclear membrane in many types of cells. The micrograph shows an end view of one of these sets. Magnification is about 160,000 times.

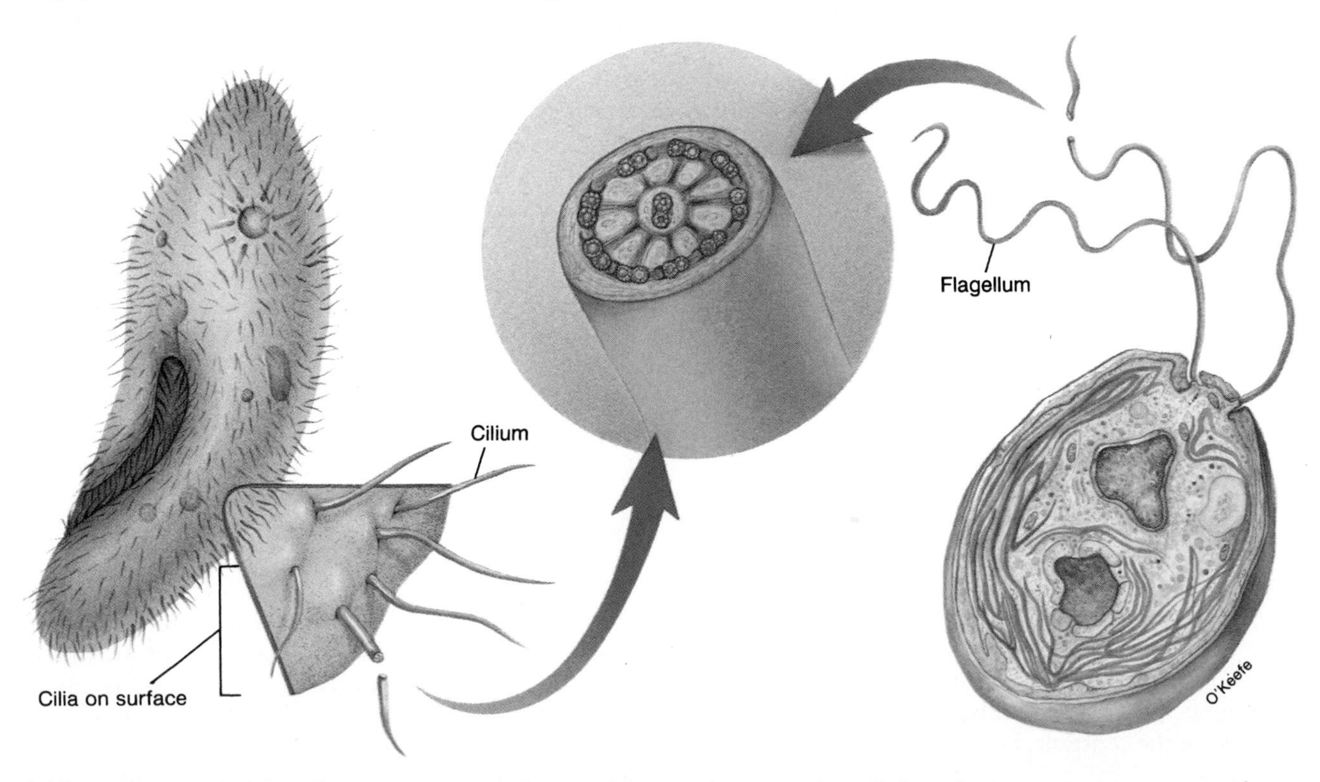

Figure 4.16 **Cilia and Flagella.** These two structures function like oars that move the cell through its environment or move the environment past the cell. Cilia and flagella are constructed of groups of microtubules. Flagella are longer than cilia.

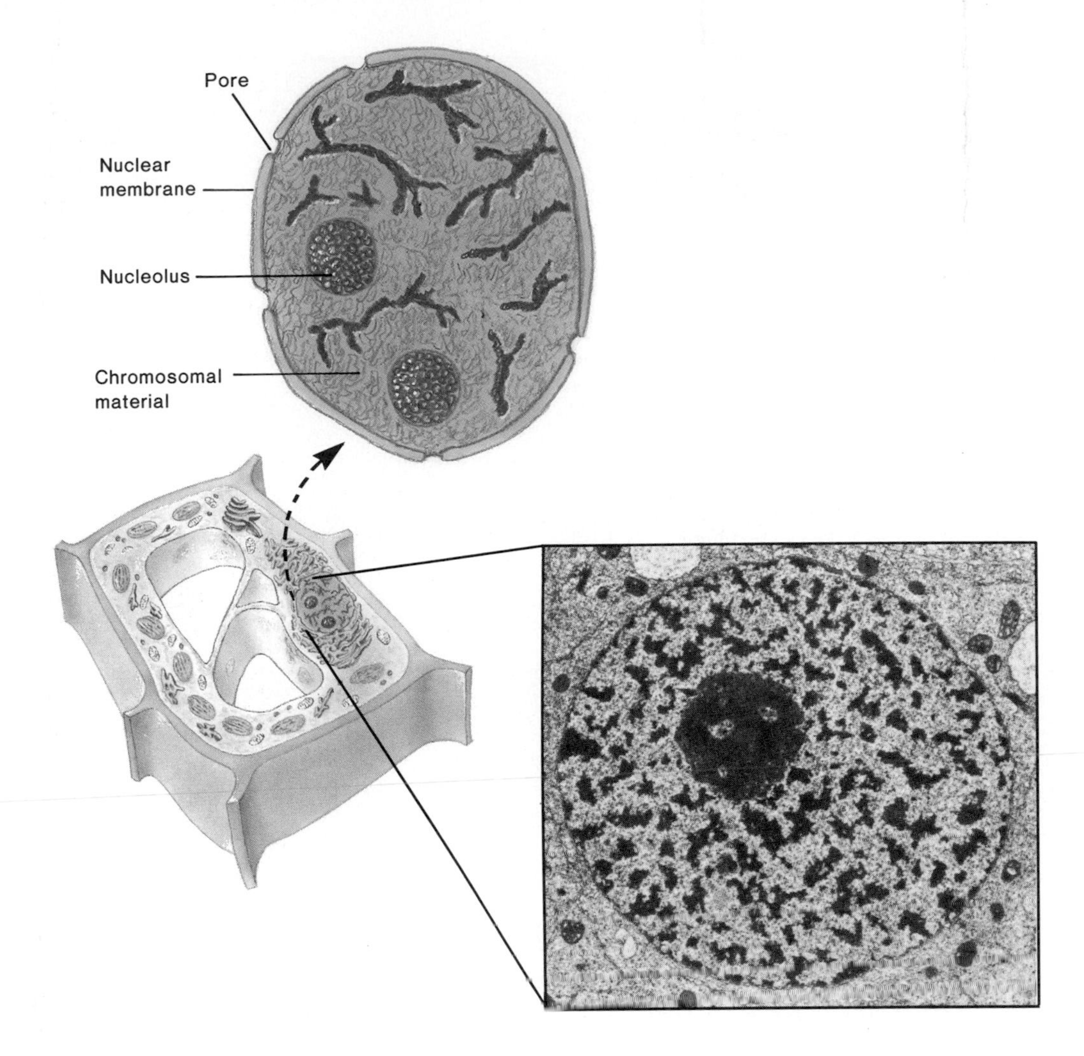

Figure 4.17 **The Nucleus.** One of the two major regions of protoplasm, the nucleus has its own complex structure. It is bounded by a double membrane that separates it from the cytoplasm. Inside the nucleus are the nucleoli, chromosomes or chromatin material composed of DNA, and the liquid matrix (nucleoplasm). Magnification is about 20,000 times.

their structure and function were not clearly known. As scientists learn more about unidentified particles in the cells, they too will be named and more fully described.

Nuclear Components

As stated at the beginning of this chapter, one of the first structures to be identified in cells was the nucleus. The nucleus was referred to as the cell center. If the nucleus is removed from a cell, the cell can only live a short time. For example, human red blood cells begin life in bone marrow, where they have nuclei. Before they are released into the bloodstream to serve as oxygen and carbon dioxide carriers, they lose their nuclei. As a consequence, red blood cells are only able to function for about 120 days before they disintegrate.

When nuclear structures were first identified, it was noted that certain dyes stained some parts more than others. The parts that stained more heavily were called **chromatin,** which means colored material. Chromatin is composed of long molecules of deoxyribonucleic acid (DNA) in association with proteins. These DNA molecules contain the genetic information for the cell, the blueprints for its construction and maintenance. Chromatin is loosely organized DNA in the nucleus. When the chromatin is tightly coiled into shorter, denser structures, we call them **chromosomes** (chromo = color; some = body). Chromatin and chromosomes are really the same molecules but differ in structural arrangement. In addition to chromosomes, the nucleus may also contain one or more *nucleoli.* A **nucleolus** is the site of ribosome manufacture. Nucleoli appear to be composed of specific parts of chromosomes that contain the information for the construction of ribosomes. These regions, together with the completed or partially completed ribosomes, are called nucleoli.

The final component of the nucleus is its liquid matrix called the **nucleoplasm.** It is a mixture composed of water and the molecules used in the construction of ribosomes, nucleic acids, and other nuclear material (figure 4.17).

Prokaryotic cells Characterized by few membranous organelles; nuclear material not separated from the cytoplasm by a membrane	Eukaryotic cells Cells larger than prokaryotic cells; nucleus with a membrane separating it from the cytoplasm; many complex organelles composed of many structures including membranes			
Kingdom Prokaryotae	Kingdom Protista	Kingdom Mycetae	Kingdom Plantae	Kingdom Animalia
Unicellular organisms	Unicellular organisms; some in colonies; both photosynthetic and heterotrophic nutrition	Multicellular organisms or loose colonial arrangement of cells; organism is a row or filament of cells; decay fungi and parasites	Multicellular organisms; cells supported by a rigid cell wall of cellulose; some cells have chloroplasts; complex arrangement into tissues	Multicellular organisms with division of labor into complex tissues; no cell wall present; acquire food from the environment
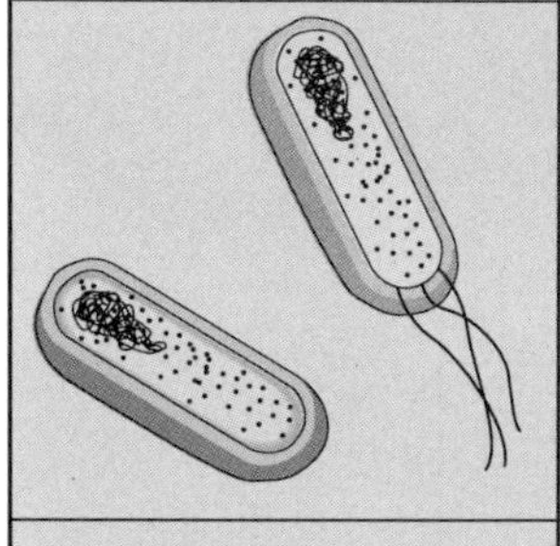	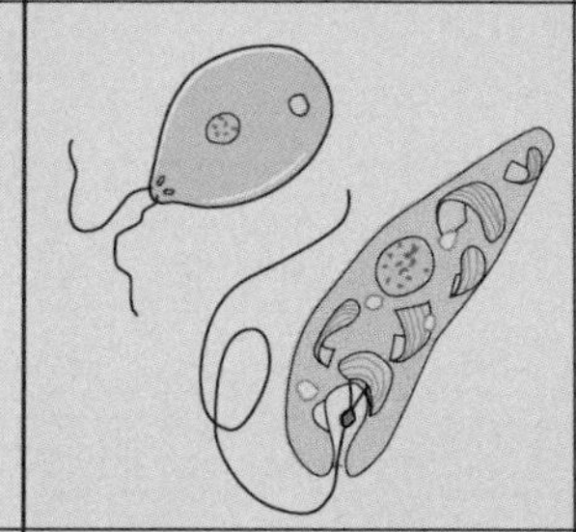	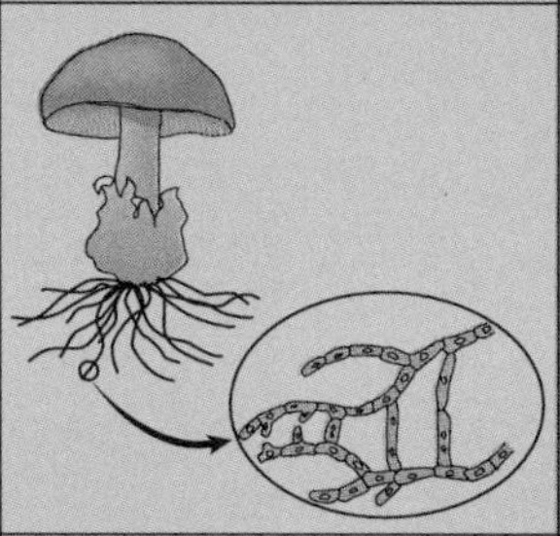	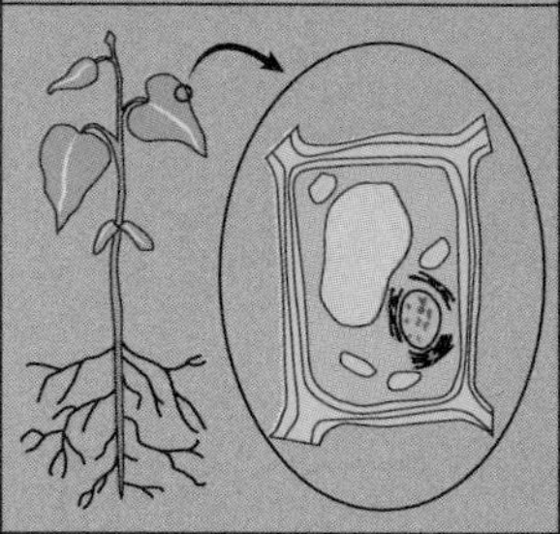	
Examples: bacteria and cyanobacteria	Examples: protozoans such as *Amoeba* and *Paramecium* and algae such as *Chlamydomonas* and *Euglena*	Examples: *Penicillium*, morels, button mushrooms, galls, and rusts	Examples: moss, ferns, cone-bearing trees, and flowering plants	Examples: worms, insects, starfish, frogs, reptiles, birds, and mammals

Figure 4.18 **Comparison of Cell Types.** The five types of cells illustrated here indicate the major patterns of construction found in all living things. Note the similarities of all five and the subtle differences among them.

Major Cell Types

Not all of the cellular organelles we have just described are located in every cell. Some cells typically have combinations of organelles that differ from others. For example, some cells have nuclear membrane, mitochondria, chloroplasts, ER, and Golgi, while others have mitochondria, centrioles, Golgi, ER, and nuclear membrane. Other cells are even more simple and lack most of the complex membranous organelles described in this chapter. Because of this fact, biologists have been able to classify cells into two major types: *prokaryotic* and *eukaryotic* (figure 4.18).

The Prokaryotic Cell Structure

Prokaryotic cells do not have a typical nucleus bound by a nuclear membrane, nor do they contain mitochondria, chloroplasts, Golgi, or extensive networks of ER. However, prokaryotic cells contain DNA and enzymes and are able to reproduce and engage in metabolism. They perform all of the basic functions of living things with fewer and more simple organelles. Examples of prokaryotic cells include bacteria that cause the diseases tuberculosis, strep throat, gonorrhea, and acne. Other prokaryotic cells are responsible for the breakdown of organic molecules. While some prokaryotic cells have a type of green photosynthetic pigment and carry on photosynthesis, they do so without chloroplasts and use different chemical reactions.

One significant difference between prokaryotic and eukaryotic cells is in the chemical makeup of their ribosomes. The ribosomes of prokaryotic cells contain different proteins than those found in eukaryotic cells. Prokaryotic ribosomes are also smaller. This discovery was important to medicine because many cells that cause common diseases are prokaryotic (bacteria). As soon as differences in the ribosomes were noted, researchers began to look for ways in which to interfere with the prokaryotic ribosome's function but *not* interfere with the ribosomes of eukaryotic cells. **Antibiotics,** such as streptomycin, are the result of this research. This drug combines with prokaryotic ribosomes and causes the death of the prokaryote by preventing the production of proteins essential to its survival. Since eukaryotic ribosomes differ from prokaryotic ribosomes, streptomycin does not interfere with the normal function of ribosomes in human cells.

The Eukaryotic Cell Structure

Eukaryotic cells contain a true nucleus and most of the membranous organelles described earlier. Eukaryotic organisms can be further divided into several categories based on the specific combination of organelles they contain. The cells

Table 4.1
Comparison of the Structure and Function of the Cellular Organelles

Organelle	Type of Cell	Structure	Function
Plasma membrane	Prokaryotic Eukaryotic	Typical membrane structure; phospholipid and protein present	Controls passage of some materials to and from the environment of the cell
Granules	Prokaryotic Eukaryotic	Too small to ascribe a specific structure	May have a variety of functions
Chromatin material	Prokaryotic Eukaryotic	Composed of DNA and protein in eukaryotes, but just DNA in prokaryotes	Contains the hereditary information that the cell uses in its day-to-day life and passes it on to the next generation of cells
Ribosome	Prokaryotic Eukaryotic	Protein and RNA structure	Site of protein synthesis
Microtubules	Eukaryotic	Hollow tubes composed of protein	Provide structural support and allow for movement
Microfilament	Eukaryotic	Protein	Provide structural support and allow for movement
Nuclear membrane	Eukaryotic	Typical membrane structure	Separates the nucleus from the cytoplasm
Nucleolus	Eukaryotic	Group of RNA molecules and genes located in the nucleus	Site of ribosome manufacture and storage
Endoplasmic reticulum	Eukaryotic	Folds of membrane forming sheets and canals	Surface for chemical reactions and intracellular transport system
Golgi apparatus	Eukaryotic	Membranous stacks	Associated with the production of secretions and enzyme activation
Vacuoles	Eukaryotic	Membranous sacs	Containers of materials
Lysosome	Eukaryotic	Membranous container	Isolates very strong enzymes from the rest of the cell
Mitochondria	Eukaryotic	Large membrane folded inside of a smaller membrane	Associated with the release of energy from food; site of cellular respiration
Chloroplast	Eukaryotic	Double membranous container of chlorophyll	Site of photosynthesis or food production in green plants
Centriole	Eukaryotic	Microtubular	Associated with cell division
Contractile vacuole	Eukaryotic	Membranous container	Expels excess water
Cilia and flagella	Prokaryotic Eukaryotic	9 + 2 tubulin in eukaryotes; different structure in prokaryotes	Movement

of plants, fungi, protozoa and algae, and animals are all eukaryotic. The most obvious characteristic that sets the plants and algae apart from other organisms is their green color, which indicates that the cells contain chlorophyll. Chlorophyll is necessary for the process of photosynthesis—the conversion of light energy into chemical-bond energy in food molecules. These cells, then, are different from the other cells in that they contain chloroplasts in the cytoplasm. Another distinguishing characteristic of plants and algae is the presence of cellulose in their cell walls.

The group of organisms that have a cell wall but lack chlorophyll in chloroplasts is collectively known as *fungi*. They were previously thought to be either plants that had lost their ability to make their own food or animals that had developed cell walls. Organisms that belong in this category of eukaryotic cells include yeasts, molds, mushrooms, and fungi that cause such human diseases as athlete's foot, jungle rot, and ringworm. Now we have come to recognize this group as different enough from plants and animals to place them in a separate kingdom.

Eukaryotic organisms that lack cell walls and cannot photosynthesize are placed in separate groups. Organisms that consist of only one cell are called protozoans—examples are *Amoeba* and *Paramecium*. They have all the cellular organelles described in this chapter except the chloroplast; therefore, protozoans must consume food as do the fungi and the multicellular animals.

While the differences in these groups of organisms may seem to set them worlds apart, their similarity in cellular structure is one of the central themes that unifies the field of biology (figure 4.19). One can obtain a better understanding of how cells operate in general by studying specific examples. Since the organelles have the same general structure and function regardless of the kind of cell in which they are found, we can learn more about how mitochondria function in plants by studying how mitochondria function in animals. There is a commonality among all living things with regard to their cellular structure and function. Table 4.1 compares the structure and function of cellular organelles.

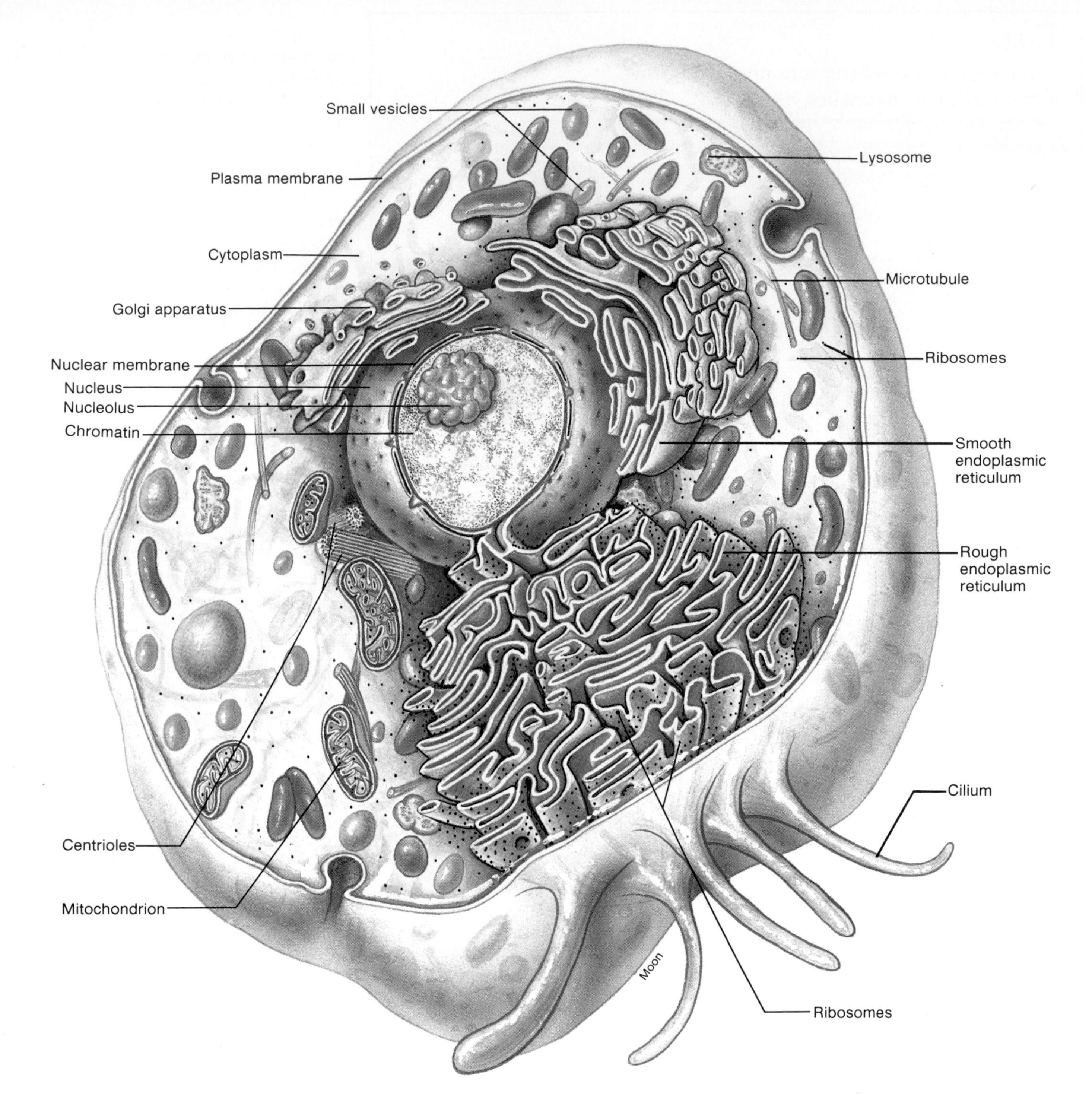

Figure 4.19 **Cellular Organelles.** The major organelles of a typical animal cell are shown here. Use this illustration to identify structures when you are reviewing table 4.1.

• Summary •

The concept of the cell has developed over a number of years. It passed through a stage where only two regions, the cytoplasm and the nucleus, could be identified. At present, numerous organelles are recognized as essential components of both prokaryotic and eukaryotic cell types. The structure and function of some of these organelles are compared in table 4.1. This table also indicates whether the organelle is unique to prokaryotic or eukaryotic cells or if it is found in both.

The cell is the common unit of life. We study individual cells and their structure to understand how they function as individual living organisms and as parts of many-celled beings. Knowing how prokaryotic and eukaryotic cell types resemble or differ from each other helps physicians control some organisms dangerous to humans.

• Thinking Critically •

A primitive type of cell consists of a membrane and a few other cell organelles. This protobiont lives in a sea that contains three major kinds of molecules with the following characteristics:

X	Y	Z
Inorganic	Organic	Organic
High concentration outside of cell	High concentration inside of cell	High concentration inside of cell
Essential to life of cell	Essential to life of cell	Poisonous to the cell
Small and can pass through the membrane	Large and cannot pass through the membrane	Small and can pass through the membrane

With this information and your background in cell structure and function, osmosis, diffusion, and active transport, decide whether or not this protobiont will continue to live in this sea, and explain why or why not.

• Experience This •

To demonstrate the movement of materials into and out of cells of living things you do not necessarily need sophisticated equipment. Obtain some dry beans. Count out 30 beans. Set 10 dry beans aside for later comparisons and place the remaining 20 beans in tap water over night. Compare the size of the dry beans and those that were soaked overnight. (If you have access to a sensitive household scale, you can weigh the beans as well as observe the change in size.) Remove 10 of the soaked beans and place them in a strong saltwater solution overnight. Compare the size of the beans after being soaked in saltwater to the original dry beans and those that had only been soaked in tap water. Can you explain why these changes occurred?

• Questions •

1. Make a list of the membranous organelles of a eukaryotic cell and describe the function of each.
2. Describe how the concept of the cell has changed over the past two hundred years.
3. What three methods allow the exchange of molecules between cells and their surroundings?
4. How do diffusion, facilitated diffusion, osmosis, and active transport differ?
5. What are the differences between the cell wall and the cell membrane?
6. Diagram a cell and show where proteins, nucleic acids, carbohydrates, and lipids are located.
7. Make a list of the nonmembranous organelles of the cell and describe their functions.
8. Define the following terms: cytoplasm, stroma, grana, cristae, chromatin, and chromosome.
9. Why does putting salt on meat preserve it from spoilage by bacteria?
10. In what ways do mitochondria and chloroplasts resemble one another?

• Chapter Glossary •

active transport (ak'tive trans'port) Use of a carrier molecule to move molecules across a cell membrane in a direction opposite that of the concentration gradient. The carrier requires an input of energy other than the kinetic energy of the molecules.

aerobic cellular respiration (a''ro'bik sel'yu-lar res''pĭ-ra'shun) A series of reactions in the mitochondria involved in the release of usable energy from food molecules by combining them with oxygen molecules.

antibiotics (an-te-bi-ot'iks) Drugs that selectively kill or inhibit the growth of a particular cell type.

cell (sel) The basic structural unit that makes up all living things.

cell membrane (sel mem'brān) The outer boundary membrane of the cell; also known as the plasma membrane.

cellular membranes (sel'yu-lar mem'brāns) Thin sheets of material composed of phospholipids and proteins. Some of the proteins have attached carbohydrates or fats.

centriole (sen'tre-ōl) Two sets of nine short microtubules, each arranged in a cylinder.

chlorophyll (klo'ro-fil) The green pigment located in the chloroplasts of plant cells associated with trapping light energy.

chloroplast (klo'ro-plast) An energy-converting, membranous, saclike organelle in plant cells containing the green pigment chlorophyll.

chromatin (kro'mah-tin) Areas or structures within the nucleus of a cell composed of long molecules of deoxyribonucleic acid (DNA) in association with proteins.

chromosomes (kro'mo-sōmz) Structures visible in the nucleus that consist of DNA and protein.

cilia (sil'e-ah) Numerous short, hairlike structures projecting from the cell surface that enable locomotion.

concentration gradient (kon''sen-tra'shun gra'de-ent) The gradual change in the number of molecules per unit of volume over distance.

cristae (kris̆'te) Folded surfaces of the inner membranes of mitochondria.

cytoplasm (si''to-plazm) The more fluid portion of the protoplasm that surrounds the nucleus.

differentially permeable (di''fur-ent'shu-le per'me-uh-bul) The property of a membrane that allows certain molecules to pass through it but interferes with the passage of others.

diffusion (dĭ-fiu'zhun) Net movement of a kind of molecule from an area of higher concentration to an area of lesser concentration.

diffusion gradient (di''fiu'zhun gra'de-ent) The difference in the concentration of diffusing molecules over distance.

endoplasmic reticulum (ER) (en''do-plaz'mik re-tĭk'yu-lum) Folded membranes and tubes throughout the eukaryotic cell that provide a large surface upon which chemical activities take place.

eukaryotic cells (yu'ka-re-ah''tik sels) One of the two major types of cells; characterized by cells that have a true nucleus, as in plants, fungi, protists, and animals.

facilitated diffusion (fah-sil'ĭ-ta''ted dĭ-fiu'zhun) Diffusion assisted by carrier molecules.

flagella (flah-jel'luh) Long, hairlike structures projecting from the cell surface that enable locomotion.

fluid-mosaic model (flu'id mo-za'ik mod'l) The concept that the cell membrane is composed primarily of protein and phospholipid molecules that are able to shift and flow past one another.

Golgi apparatus (gōl'je ap''pah-rat'us) A stack of flattened, smooth, membranous sacs; the site of synthesis and packaging of certain molecules in eukaryotic cells.

grana (gra'nuh) Areas of the chloroplast membrane where chlorophyll molecules are concentrated.

granules (gran'yūls) Materials whose structure is not as well defined as that of other organelles.

hydrophilic (hi'dro-fil'ik) Readily absorbing or dissolving in water.

hydrophobic (hi'dro-fo'bik) Tending not to combine with, or incapable of dissolving in water.

hypertonic (hi'pur-tŏn'ik) A comparative term describing one of two solutions. The hypertonic solution is the one with the higher amount of dissolved material.

hypotonic (hi'po-tŏn'ik) A comparative term describing one of two solutions. The hypotonic solution is the one with the lower amount of dissolved material.

inclusions (in-klu'zhuns) A general term referring to materials inside a cell that are usually not readily identifiable; stored materials.

isotonic (i'so-tŏn'ik) A term used to describe two solutions that have the same concentration of dissolved material.

lysosome (li'so-sōm) A specialized organelle that holds a mixture of hydrolytic enzymes.

microfilaments (mi''kro-fil'ah-ments) Long, fiberlike structures made of protein and found in cells, often in close association with the microtubules; provide structural support and enable movement.

microscope (mi'kro-skōp) An instrument used to produce an enlarged image of a small object.

microtubules (mi'kro-tū''byūls) Small, hollow tubes of protein that function throughout the cytoplasm to provide structural support and enable movement.

mitochondrion (mi-to-kahn'dre-on) A membranous organelle resembling a small bag with a larger bag inside that is folded back on itself; serves as the site of aerobic cellular respiration.

net movement (net muv'ment) Movement in one direction minus the movement in the other.

nuclear membrane (nu'kle-ar mem'brān) The structure surrounding the nucleus that separates the nucleoplasm from the cytoplasm.

nucleoli (singular, **nucleolus)** (nu-kle'o-li) Nuclear structures composed of completed or partially completed ribosomes and the specific parts of chromosomes that contain the information for their construction.

nucleoplasm (nu'kle-o-plazm) The liquid matrix of the nucleus composed of a mixture of water and the molecules used in the construction of the rest of the nuclear structures.

nucleus (nu'kle-us) The central body that contains the information system for the cell.

organelles (or-gan-elz') Cellular structures that perform specific functions in the cell. The function of an organelle is directly related to its structure.

osmosis (os-mo'sis) The net movement of water molecules through a differentially permeable membrane.

phagocytosis (fă''jo-si-to'sis) The process by which the cell wraps around a particle and engulfs it.

photosynthesis (fo-to-sin'thuh-sis) A series of reactions that take place in chloroplasts and result in the storage of sunlight energy in the form of chemical-bond energy.

pinocytosis (pĭ''no-si-to'sis) The process by which a cell engulfs some molecules dissolved in water.

plasma membrane (plaz'muh mem'brān) The outer-boundary membrane of the cell; also known as the cell membrane.

prokaryotic cells (pro'ka-re-ot''ik sels) One of the two major types of cells. They do not have a typical nucleus bound by a nuclear membrane and lack many of the other membranous cellular organelles; for example, bacteria.

protoplasm (pro'to-plazm) The living portion of a cell as distinguished from the nonliving cell wall.

ribosomes (ri'bo-sōmz) Small structures composed of two protein and ribonucleic acid subunits involved in the assembly of proteins from amino acids.

stroma (stro'muh) The region within a chloroplast that has no chlorophyll.

vacuole (vak'yu-ōl) A large sac within the cytoplasm of a cell, composed of a single membrane.

vesicles (vĕ'sĭ-kuls) Small, intracellular, membrane-bound sacs in which various substances are stored.

CHAPTER 5

Enzymes

Chapter Outline

Reactions, Catalysts, and Enzymes
How Enzymes Speed Chemical Reaction Rates
Environmental Effects on Enzyme Action
Cellular Controlling Processes and Enzymes

Purpose

Living cells require various chemical reactions to conduct their vital functions. To prevent the malfunction and death of the cell, these reactions must be rapid and controlled. The problem is not starting reactions but controlling their rate. Regulation of the rates of the many reactions in cells is the task of enzymes.

For Your Information

As a result of genetic engineering and other advanced research into enzymes, a great variety of products containing enzymes are now available. Meat tenderizer contains enzymes that break down tough protein fibers. Soft contact lenses are cleansed of mucoprotein with enzyme solutions. Pet urine odor is removed from soiled carpets with enzymes. Stained clothes are made bright and white again with enzyme-active presoaks and detergents. Hospitals use bacterial enzymes to remove dead skin from burn victims. The wide variety of enzyme-containing products currently in use is only a sample of what is to come.

Learning Objectives

- Explain the need for enzymes in the maintenance of living things.
- Describe what happens when an enzyme and substrate combine.
- Relate the three-dimensional structure of an enzyme to its ability to catalyze a reaction.
- Explain the role of coenzymes and vitamins in enzyme operation.
- Describe the influences of such environmental factors as temperature, pH, and concentration on turnover number.
- Define the mechanisms of enzyme competition and inhibition.
- Explain the methods used by cells to control and regulate their biochemical activities.

Reactions, Catalysts, and Enzymes

All living things require a source of energy and building materials in order to grow and reproduce. Energy may be in the form of visible light, or it may be derived from energy-containing covalent bonds found in nutrients. **Nutrients** are molecules required by organisms for growth, reproduction, and/or repair—they serve as a source of energy and molecular building materials. The formation, breakdown, and rearrangement of molecules to provide organisms with essential energy and building blocks are known as *biochemical reactions.* These reactions occur when atoms or molecules come together and form new, more stable relationships. This results in the formation of new molecules and a change in the energy distribution among the reactants and end products. Most chemical reactions require an input of energy to get them started. This is referred to as **activation energy.** This energy is used to make the reactants unstable and more likely to react (figure 5.1).

If organisms are to survive, they must gain sizable amounts of energy and building materials in a very short time. Experience tells us that the sucrose in candy bars contains the potential energy needed to keep us active, as well as building materials to help us grow (sometimes to excess!). Yet, it could be millions of years before a candy bar is broken down by random chemical processes, releasing its energy and building materials. Of course, living things cannot wait that long. To sustain life, biochemical reactions must occur at extremely rapid rates. One way to increase the rate of any chemical reaction and make its energy and component parts available to a cell is to increase the temperature of the reactants. In general, the hotter the reactants, the faster they will react. However, this method of increasing reaction rates has a major drawback when it comes to living things. The organism will die because cellular proteins are denatured before the temperature reaches the point required to sustain the biochemical reactions necessary for life. This is of practical concern to people who are experiencing a fever. Should the fever stay too high for too long, major disruptions of cellular biochemical processes can be fatal.

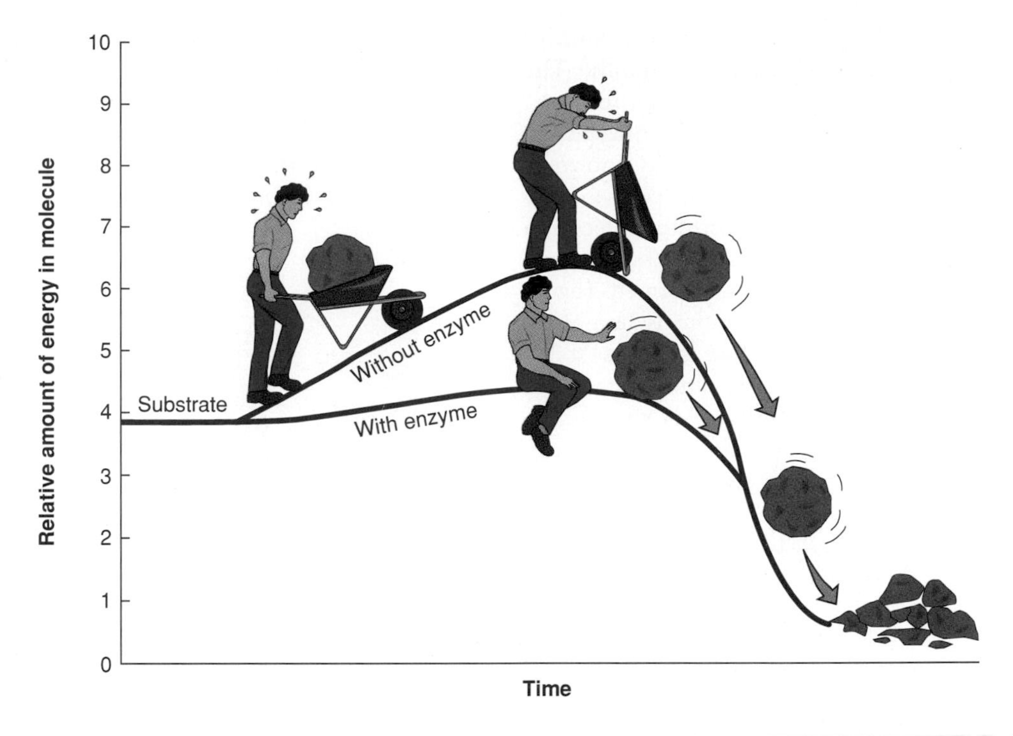

Figure 5.1 **The Lowering of Activation Energy.** Enzymes operate by lowering the amount of energy needed to get a reaction going—the activation energy. When this energy is lowered, the nature of the bonds is changed so they are more easily broken. While the cartoon shows the breakdown of a single reactant into many end products (as in a hydrolysis reaction), the lowering of activation energy can also result in bonds being broken so that new bonds may be formed in the construction of a single, larger end product from several reactants (as in a synthesis reaction).

However, there is a way of increasing the rate of chemical reactions without increasing the temperature. This involves using substances called *catalysts.* A **catalyst** is a chemical that speeds the reaction but is not used up in the reaction. It is able to be recovered unchanged when the reaction is complete. Catalysts function by lowering the amount of activation energy needed to start the reaction. A cell manufactures specific proteins that act as catalysts. A protein molecule that acts as a catalyst to change the rate of a reaction is called an **enzyme.** Enzymes can be used over and over again until they are worn out. The production of these protein catalysts is under the direct control of an organism's genetic material (DNA). The instructions for the manufacture of all enzymes are found in the genes of the cell. Organisms make their own enzymes. How the genetic information is used to direct the synthesis of these specific protein molecules is discussed in chapter 7.

How Enzymes Speed Chemical Reaction Rates

As the instructions for the production of an enzyme are read from the genetic material, a specific sequence of amino acids is linked together at the ribosomes. Once bonded, the chain of amino acids folds and twists to form a globular molecule. It is the nature of its three-dimensional shape that allows this enzyme to combine with a reactant and lower the activation energy. Each enzyme has a very specific three-dimensional shape which, in turn, is very specific to the kind of reactant with which it can combine. The enzyme physically fits with the reactant. The molecule to which the enzyme attaches itself (the reactant) is known as the **substrate.** When the enzyme attaches itself to the substrate molecule, a new, temporary molecule—the **enzyme-substrate complex**—is formed (figure 5.2). When the substrate is combined with the enzyme, its bonds are less stable and more likely to be altered and form new bonds. The enzyme is specific because it has a particular shape that can combine only with specific parts of certain substrate molecules.

You might think of an enzyme as a tool that makes a job easier and faster. For example, the use of an open-end crescent wrench can make the job of removing or attaching a nut and bolt go much faster than that same job done by hand. In order to accomplish this job, the proper wrench must be used. Just any old tool (screwdriver or hammer) won't work! The enzyme must also physically attach itself to the substrate; therefore, there is a specific **binding site** or **attachment site** on the enzyme surface. Figure 5.3 illustrates the specificity of both wrench and enzyme. Note that the wrench and enzyme are recovered unchanged after they have been used. This means that the enzyme and wrench can be used again. Eventually, like wrenches, enzymes wear out and need to be replaced by synthesizing new ones using the

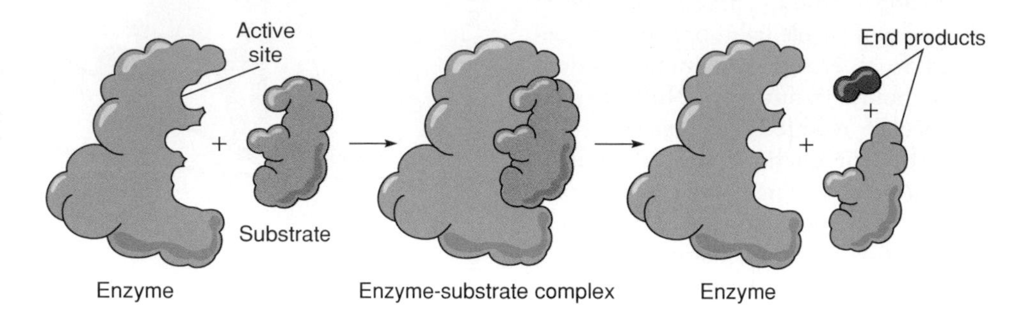

Figure 5.2 **Enzyme–Substrate Complex Formation.** During an enzyme-controlled reaction, the enzyme and substrate come together to form a new molecule—the enzyme–substrate complex molecule. This molecule exists for only a very short time. During that time, activation energy is lowered and bonds are changed. The result is the formation of a new molecule or molecules called the end products of the reaction. Notice that the enzyme comes out of the reaction intact and ready to be used again.

(a)

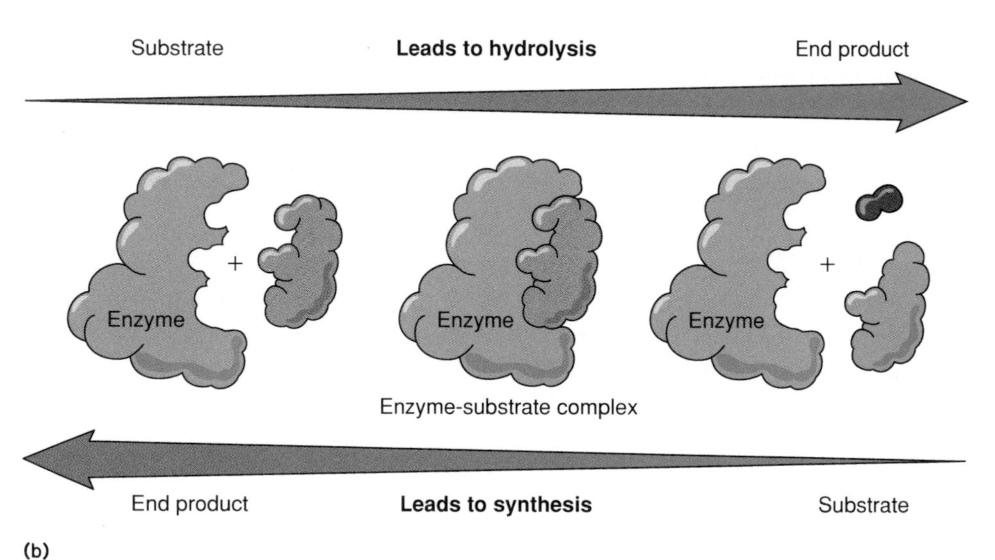

(b)

Figure 5.3 **It Fits, It's Fast, and It Works.** (a) While it could be done by hand, an open-end crescent wrench can be used to remove the wheel from this bicycle more efficiently. The wrench is adjusted and attached, temporarily forming a nut–bolt–wrench complex. Turning the wrench loosens the bonds holding the nut to the bolt and the two are separated. The use of the wrench makes the task much easier. Keep in mind that the same wrench that is used to disassemble the bicycle can be used to reassemble it. Enzymes function in the same way. (b) An enzyme will "adjust itself" as it attaches to its substrate, forming a temporary enzyme–substrate complex. The presence and position of the enzyme in relation to the substrate lowers the activation energy required to alter the bonds. Depending on the circumstances (what job needs to be done), the enzyme might be involved in synthesis (constructive) or hydrolysis (destructive) reactions.

instructions provided by the cell's genes. Generally, only very small quantities of enzymes are necessary because they work so fast and can be reused.

Both enzymes and wrenches are specific in that they have a particular surface geometry or shape that matches the geometry of their respective substrates. Note that both the enzyme and wrench are flexible. The enzyme can bend or fold to fit the substrate just as the wrench, to a limited extent, can be "adjusted" to fit the nut. This is called the *induced fit hypothesis*. The fit is induced because the presence of the substrate causes the enzyme to "mold" or "adjust" itself to the substrate as the two come together. The place on the enzyme that causes a specific part of the substrate to change is called the **active site** of the enzyme. (Note in the case illustrated in figure 5.3 that the "active site" is the same as the "binding site." This is typical of many enzymes.) This site is the place where the activation energy is lowered and the electrons are shifted to change the bonds. The active site may enable a positively charged surface to combine with the negative portion of a reactant. While the active site does mold itself to a substrate, enzymes do not have the ability to fit all substrates. Enzymes are specific to a certain substrate or group of very similar substrate molecules. One enzyme cannot speed the rate of all types of biochemical reactions. Rather, a special enzyme is required to control the rate of each type of chemical reaction occurring in an organism.

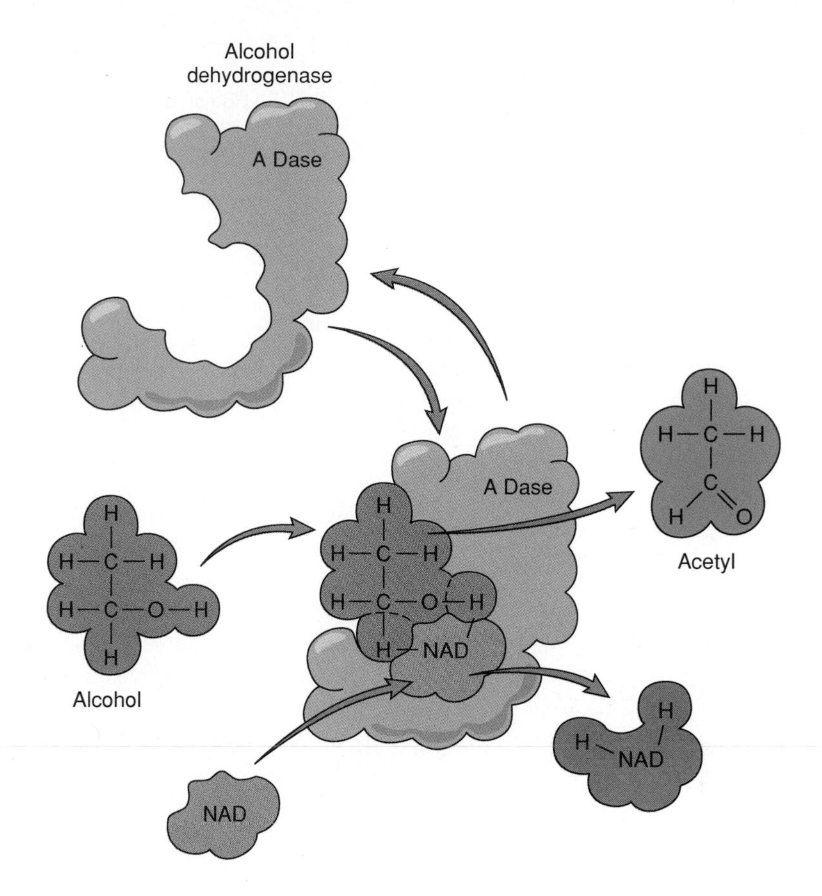

Figure 5.4 **The Role of Coenzymes.** NAD is a coenzyme that works with the enzyme alcohol dehydrogenase during the decomposition of alcohol. The coenzyme carries the hydrogen from the alcohol molecule after it is removed by the enzyme. Notice that the hydrogen on the alcohol is picked up by the NAD, converting it to $NADH_2$. The end products of this reaction are the $NADH_2$ and the molecule acetyl. The use of the coenzyme NAD makes the enzyme function more efficiently because one of the end products of this reaction (hydrogen) is removed from the reaction site. Because the hydrogen is no longer close to the reacting molecules, the overall direction of the reaction is toward the formation of acetyl. This encourages more alcohol to be broken down.

Because the enzyme is specific to both the substrate to which it can attach and the reaction that it can encourage, a unique name can be given to each enzyme. The first part of an enzyme's name is the name of the molecule to which it can become attached. The second part of the name indicates what type of reaction it facilitates. The third part of the name is "-ase," which is the ending that tells you it is an enzyme. For example, *DNA polymerase* is the name of the enzyme that attaches to the molecule DNA and is responsible for increasing its length through a polymerization reaction. The enzyme responsible for the dehydration synthesis reactions among several glucose molecules to form glycogen is known as *glycogen synthetase*. The enzyme responsible for breaking the bond that attaches the amino group to the amino acid arginine is known as *arginine aminase*. When an enzyme is very common, we often shorten its formal name. The salivary enzyme involved in the digestion of starch should be *amylose* (starch) *hydrolase*, but it is generally known as *amylase*.

Certain enzymes need an additional molecule to enable them to function. This additional molecule is not a protein, but attaches to the enzyme and works with the protein catalyst to speed up a reaction. This enabling molecule is called a *coenzyme*. A **coenzyme** aids a reaction by removing one of the end products or by bringing in part of the substrate. Many coenzymes cannot be manufactured by organisms and must be obtained from their foods. Coenzymes are frequently constructed from *vitamins*. You know that a constant small supply of vitamins in your diet is necessary for good health. The reason your cells require vitamins is to serve in the manufacture of certain coenzymes. A coenzyme can work with a variety of enzymes; therefore, you need extremely small quantities of vitamins. An example of enzyme–coenzyme cooperation is shown in figure 5.4. The metabolism of alcohol consists of a series of reactions resulting in its breakdown to carbon dioxide (CO_2), water (H_2O), and energy. During one of the reactions in this sequence, the enzyme alcohol dehydrogenase picks up hydrogen from alcohol and attaches it to NAD. In this reaction, NAD (*n*icotinamide *a*denine *d*inucleotide, manufactured from the vitamin niacin) acts as a coenzyme because NAD carries the hydrogen away from the reaction as the alcohol is broken down. Later, the NAD will acquire an additional hydrogen from another such reaction, causing it to become $NADH_2$. The presence of the coenzyme NAD is necessary for the enzyme to function properly.

Environmental Effects on Enzyme Action

An enzyme forms a complex with one substrate molecule, encourages a reaction to occur, detaches itself, and then forms a complex with another molecule of the same substrate. The number of molecules of substrate that a single enzyme molecule can react with in a given time (e.g., reactions per minute) under ideal conditions is called the **turnover number.** Sometimes the number of jobs an enzyme can perform during a particular time period is incredibly large—ranging between a thousand (10^3) and ten thousand trillion (10^{16}) times faster than uncatalyzed reactions per minute! Without the enzyme, perhaps only fifty or one hundred substrate molecules might be altered in the same time. With this in mind, let's identify the ideal conditions for an enzyme and consider how these conditions influence the turnover number.

An important environmental condition affecting enzyme-controlled reactions is temperature (figure 5.5), which has two effects on enzymes: (1) it can change the rate of molecular motion and (2) it can cause changes in the shape of an enzyme. As the temperature of an enzyme–substrate system increases, you would expect an increase in the amount of product molecules formed. This is true up to a point. The temperature at which the rate of formation of enzyme–substrate complex is fastest is termed the *optimum temperature. Optimum* means the best or most productive quantity or condition. In this case, the optimum temperature is the temperature at which the product is formed most rapidly.

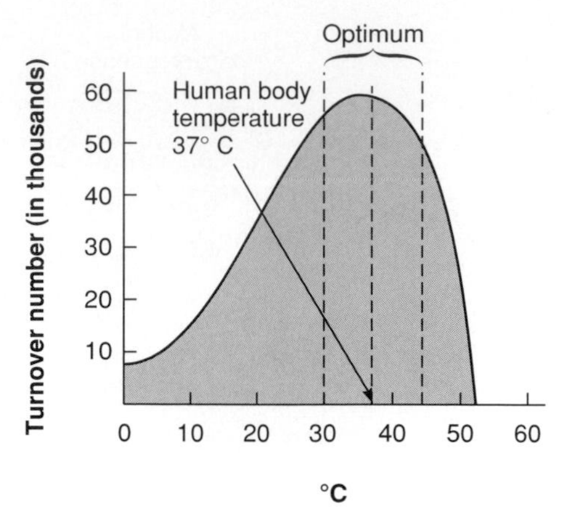

Figure 5.5 **The Effect of Temperature on the Turnover Number.** As the temperature increases, the rate of the enzymatic reaction increases. The increasing temperature may increase the number of times an enzyme contacts and combines with a substrate molecule. Temperature may also influence the shape of the enzyme molecule, making it fit better with the substrate. At certain temperatures, the enzyme molecule is irreversibly changed so that it can no longer function as an enzyme. At that point, it has been denatured. Notice that the enzyme represented in this graph has an optimum (best) temperature range of between 30° C and 45° C.

As one lowers the temperature below the optimum, molecular motion slows, and the rate at which the enzyme–substrate complexes form decreases. Even though the enzyme is still able to operate, it does so very slowly. Therefore, it is possible to preserve foods for long periods by storing them in freezers or refrigerators.

When the temperature is raised above the optimum, some of the molecules of enzyme are changed in such a way that they can no longer form the enzyme–substrate complex; thus, the reaction is not encouraged. If the temperature continues to increase, more and more of the enzyme molecules will become inactive. When heat is applied to an enzyme, it causes permanent changes in the three-dimensional shape of the molecule. The surface geometry of the enzyme molecule will not be recovered, even when the temperature is reduced. We can again use the wrench analogy. When a wrench is heated above a certain temperature, the metal begins to change shape. The shape of the wrench is changed permanently so that even if the temperature is reduced, the surface geometry of the end of the wrench is permanently lost. When this happens to an enzyme, we say that it has been *denatured.* A **denatured** enzyme is one whose protein structure has been permanently changed so that it has lost its original biochemical properties. Because enzymes are molecules and are not alive, they are not "killed," but denatured. Although egg white is not an enzyme, it is a protein and provides a common example of what happens when denaturation occurs as a result of heating. As heat is applied to the egg white, it is permanently changed (denatured).

Another environmental condition that influences enzyme action is pH. The three-dimensional structure of a protein leaves certain side chains exposed. These side chains may attract ions from the environment. Under the right conditions, a group of positively charged hydrogen ions may accumulate on certain parts of an enzyme. In an environment that lacked these hydrogen ions, this would not happen. Thus, variation in the most effective shape of the enzyme could be caused by a change in the number of hydrogen ions present in the solution. Because the environmental pH is so important in determining the shapes of protein molecules, there is an optimum pH for each specific enzyme. The enzyme will fit with the substrate only when it is

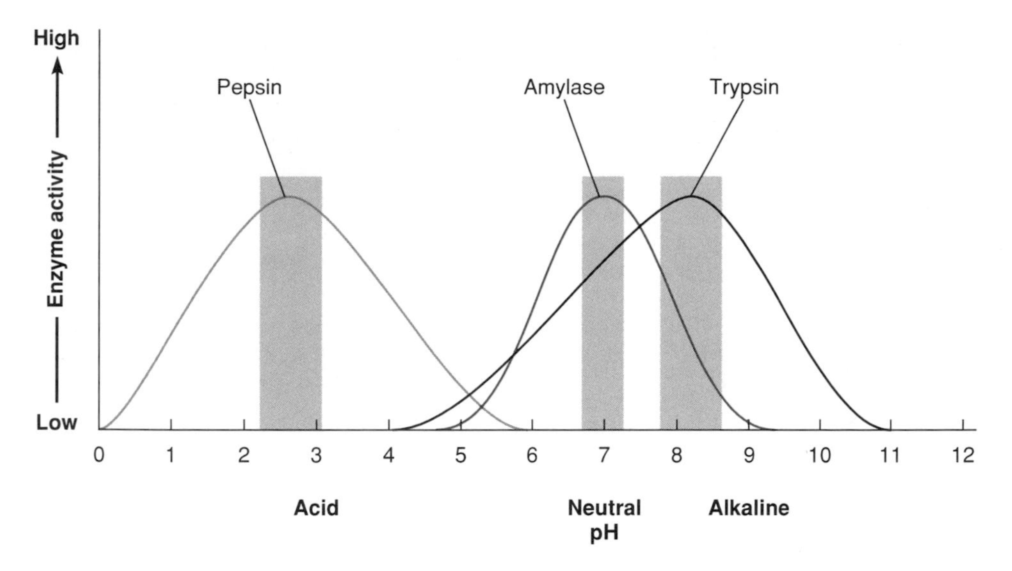

Figure 5.6 **The Effect of pH on the Turnover Number.** As the pH changes, the rate of the enzymatic reaction changes. The ions in solution alter the environment of the enzyme's active site and the overall shape of the enzyme. The enzymes illustrated here are amylase, pepsin, and trypsin. Amylase is found in saliva and is responsible for hydrolyzing starch to glucose. Pepsin is found in the stomach and hydrolyzes protein. Trypsin is produced in the pancreas and enters the small intestine where it also hydrolyzes protein. Notice that each enzyme has its own pH range of activity, the optimum (shown in the color bars) being different for each.

at the proper pH. Many enzymes function best at the pH close to neutral (7.0). However, a number of enzymes perform best at pHs quite different from seven. Pepsin, an enzyme found in the stomach, works well at an acid pH of 1.5 to 2.2, while arginase, an enzyme in the liver, works well at a basic pH of 9.5 to 9.9 (figure 5.6).

In addition to temperature and pH, the concentration of enzymes, substrates, and products influences the rates of enzymatic reactions. Although the enzyme and the substrate are in contact with one another for only a short period of time, when there are huge numbers of substrate molecules it may happen that all the enzymes present are always occupied by substrate molecules. When this occurs, the rate of product formation cannot be increased unless the number of enzymes is increased. Cells can actually do this by synthesizing more enzymes. However, just because there are more enzyme molecules does not mean that any one enzyme molecule will be working any faster. The turnover number for each enzyme stays the same. As the enzyme concentration increases, the amount of product formed increases in a specified time. A greater number of enzymes are turning over substrates; they are not turning over substrates faster. Similarly, if enzyme numbers are decreased, the amount of product formed declines.

We can also look at this from the point of view of the substrate. If substrate is in short supply, enzymes may have to wait for a substrate molecule to become available. Under these conditions as the amount of substrate increases, the amount of product formed increases. The increase in product is the result of more substrates available to be changed. If there is a very large amount of substrate, even a small amount of enzyme can eventually change all the substrate to product; it will just take longer. Decreasing the amount of substrate results in reduced product formation because some enzymes will go for long periods without coming in contact with a substrate molecule.

Cellular Controlling Processes and Enzymes

In any cell there are thousands of kinds of enzymes. Each controls specific chemical reactions and is sensitive to changing environmental conditions such as pH and temperature. In order for a cell to stay alive in an ever-changing environment, its innumerable biochemical reactions must be controlled. **Control processes** are mechanisms that ensure that an organism will carry out all metabolic activities in the proper sequence (coordination) and at the proper rate (regulation). *Coordination* of enzymatic activities in a cell results when specific reactions occur in a given sequence; for example, $A \rightarrow B \rightarrow C \rightarrow D \rightarrow E$. This ensures that a particular nutrient will be converted to a particular end product that is necessary to the survival of the cell. Should a cell not be able to coordinate its reactions, essential products might be produced at the wrong time or never be produced at all, and the cell would die. *Regulation* of biochemical reactions refers to how a cell controls the amount of chemical product produced. The old expression "having too much of a good thing" applies to this situation. For example, if a cell manufactures too much lipid, the presence of those molecules could interfere with other life-sustaining reactions, resulting in the death of the cell. On the other hand, if a cell does not produce enough of an essential molecule, such as a digestive enzyme, it might also die. The cellular control process involves both enzymes and genes.

Keep in mind that any one substrate may be acted upon by several different enzymes. While these different enzymes may all combine with the same substrate, they do not all have the same chemical effect on the substrate, each converting the substrate to different end products. For example, acetyl is a substrate that can be acted upon by three different enzymes: citrate synthetase, fatty acid synthetase, and malate synthetase (figure 5.7). Which enzyme has the

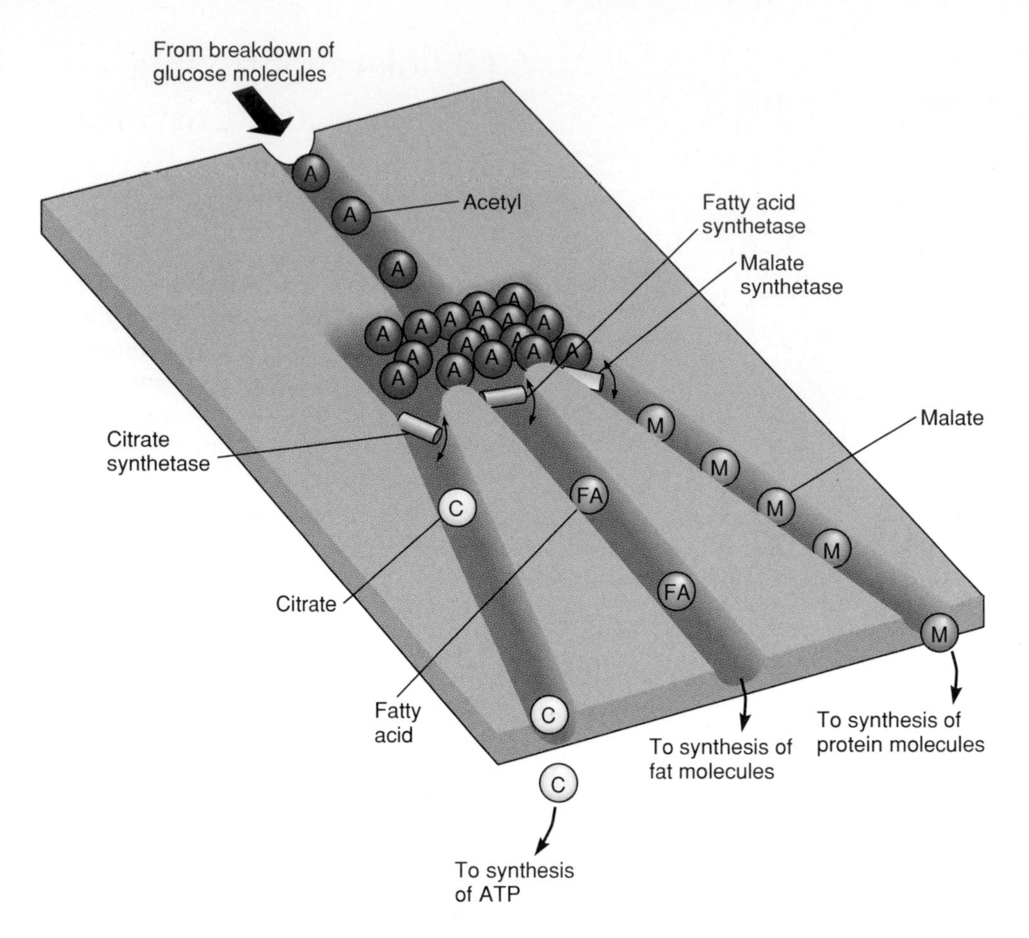

Figure 5.7 **Enzymatic Competition.** Acetyl can serve as a substrate for a number of different reactions. Whether it becomes a fatty acid, malate, or citrate is determined by the enzymes present. Each of the three enzymes may be thought of as being in competition for the same substrate—the acetyl molecule. The cell can partially control which end product will be produced in the greatest quantity by producing greater numbers of one kind of enzyme and fewer of the other kind. If citrate synthetase is present in the highest quantity, more of the acetyl substrate will be acted upon by that enzyme and converted to citrate rather than to the other two end products, malate and fatty acids. The illustration represents the action of each enzyme as an "enzyme gate."

greatest success depends on the number of each type of enzyme available and the suitability of the environment for the enzyme's operation. The enzyme that is present in the greatest number and/or is best suited to the job in the environment of the cell wins, and the amount of its end product becomes greatest. Whenever there are several different enzymes available to combine with a given substrate, **enzymatic competition** results. For example, the use a cell makes of the substrate molecule acetyl is directly controlled by the amount and kinds of enzymes it produces. The number and kind of enzymes produced are regulated by the cell's genes. It is the job of chemical messengers to inform the genes as to whether specific enzyme-producing genes should be turned on or off or whether they should have their protein-producing activities increased or decreased. Such chemical messengers are called **gene-regulator proteins.** Gene-regulator proteins that decrease protein production are called *gene-repressor proteins* while those that increase protein production are *gene-activator proteins.* Returning to our example, if the cell is in need of protein, the acetyl could be metabolized to provide one of the building blocks for the construction of protein by turning up the production of the enzyme malate synthetase. If the cell requires energy to move or grow, more acetyl can be metabolized to release this energy by producing more citrate synthetase. When the enzyme fatty acid synthetase outcompetes the other two, the acetyl is used in fat production and storage.

Another method of controlling the synthesis of many molecules within a cell is called **negative-feedback inhibition.** This control process occurs within an enzyme-controlled reaction sequence. As the number of end products increases, some product molecules *feed back* to one of the previous reactions and have a *negative* effect on the enzyme controlling that reaction; i.e., they *inhibit* or prevent that enzyme from performing at its best. Since the end product can no longer be produced at the same rapid rate, its concentration falls. When there are too few end-product molecules to feed back and cease their inhibition, the enzyme resumes its previous optimum rate of operation, and the end-product concentration begins to increase. This also helps regulate the number of end products formed but does not involve the genes.

In addition, the operation of enzymes can be influenced by the presence of other molecules. An **inhibitor** is a molecule that attaches itself to an enzyme and interferes with its ability to form an enzyme–substrate complex (figure 5.8). One of the early kinds of pesticides used to spray fruit trees contained arsenic. The arsenic attached itself to insect enzymes and inhibited the normal growth and reproduction of insects. Organophosphates are pesticides that inhibit several enzymes necessary for the operation of the nervous system. When they are incorporated into nerve

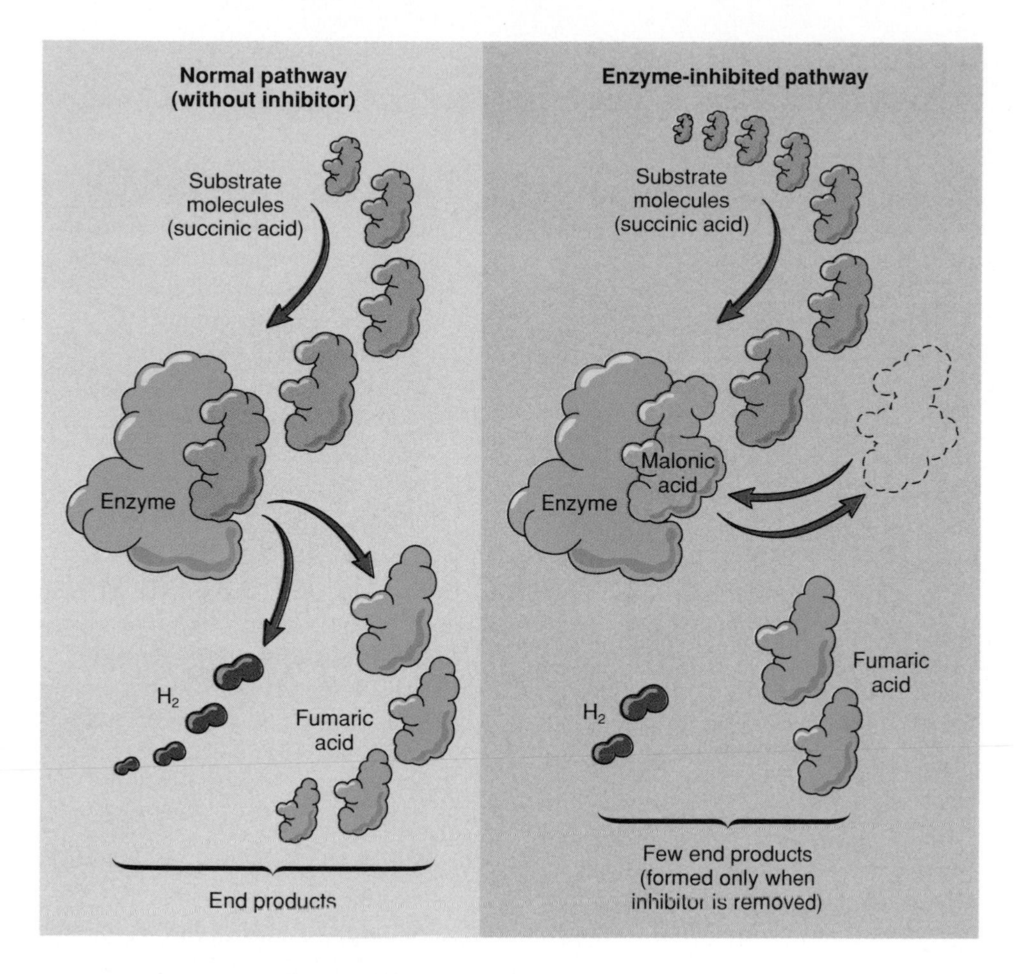

Figure 5.8 **Enzymatic Inhibition.** The left-hand side of the illustration shows the normal functioning of the enzyme. On the right-hand side, the enzyme is unable to function. This is because an inhibitor, malonic acid, is attached to the enzyme and prevents the enzyme from forming the normal complex with succinic acid. As long as malonic acid is present, the enzyme will be unable to function. If the malonic acid is removed, the enzyme will begin to function normally again. Its attachment to the inhibitor in this case is not permanent but has the effect of reducing the number of product molecules from per unit of time.

cells, they disrupt normal nerve transmission and cause the death of the affected organisms. In humans, death due to pesticides is usually caused by uncontrolled muscle contractions, resulting in breathing failure.

Some inhibitors have a shape that closely resembles the normal substrate of the enzyme. The enzyme is unable to distinguish the inhibitor from the normal substrate and so it combines with either or both. As long as the inhibitor is combined with an enzyme, the enzyme is ineffective in its normal role. Some of these enzyme-inhibitor complexes are permanent. An inhibitor removes a specific enzyme as a functioning part of the cell: the reaction that enzyme catalyzes no longer occurs, and none of the product is formed. This is termed *competitive inhibition* because the inhibitor molecule competes with the normal substrate for the active site of the enzyme.

We use enzyme inhibition to control disease. The sulfa drugs are used to control a variety of bacteria, such as the bacterium *Streptococcus pyogenes*, the cause of strep throat and scarlet fever. The drug resembles one of the bacterium's necessary substrates and so prevents some of the cell's enzymes from producing an essential cell component. As a result, the bacterial cell dies because its normal metabolism is not maintained.

• Summary •

Enzymes are protein catalysts that speed up the rate of chemical reactions without any significant increase in the temperature. They do this by lowering activation energy. Enzymes have a very specific structure that matches the structure of particular substrate molecules. Actually, the substrate molecule comes in contact with only a specific part of the enzyme molecule—the attachment site. The active site of the enzyme is the place where the substrate molecule is changed. The enzyme–substrate complex reacts to form the end product. The protein nature of enzymes makes them sensitive to environmental conditions, such as temperature and pH, that change the structure of proteins. The number and kinds of enzymes are ultimately controlled by the genetic information of the cell. Other kinds of molecules, such as coenzymes, inhibitors, or competing enzymes, can influence specific enzymes. Changing conditions within the cell shift the enzymatic priorities of the cell by influencing the turnover number.

• Thinking Critically •

The following data were obtained by a number of Nobel-prize-winning scientists from Lower Slobovia. As a member of the group, interpret the data with respect to the following:

1. Enzyme activities.
2. Movement of substrates into and out of the cell.
3. Competition among different enzymes for the same substrate.
4. Cell structure.

Data

a. A lowering of the atmospheric temperature from 22° C to 18° C causes organisms to form a thick protective coat.
b. Below 18° C, no additional coat material is produced.
c. If the cell is heated to 35° C and then cooled to 18° C, no coat is produced.
d. The coat consists of a complex carbohydrate.
e. The coat will form even if there is a low concentration of simple sugars in the surroundings.
f. If the cell needs energy for growth, no cell coats are produced at any temperature.

• Experience This •

Several cleaning products that contain enzymes are sold in supermarkets. Select one of these and another household product to test for enzyme activity and specificity in removing stains. Take two clean white pieces of cloth and stain them with a row of blood (from meat), chocolate, berry juice, grass stain, egg white, cooking oil, and grease. On one stained cloth, put a drop of the enzyme-active detergent in the middle of each stain. On the second cloth, put a drop of ordinary dishwashing detergent. Let the cloths sit for an hour. Wash these two cloths separately and compare results.

• Questions •

1. What is the difference between a catalyst and an enzyme?
2. Describe the sequence of events in an enzyme-controlled reaction.
3. How does changing temperature affect the rate of an enzyme-controlled reaction?
4. Would you expect a fat and a sugar molecule to be acted upon by the same enzyme? Why or why not?
5. What factors in the cell can speed up or slow down enzyme reactions?
6. What is the turnover number? Why is it important?
7. What is the relationship between vitamins and coenzymes?
8. What is enzyme competition, and why is it important to all cells?
9. What effect might a change in pH have on enzyme activity?
10. Where in a cell would you look for enzymes?

• Chapter Glossary •

activation energy (ak″ti-va′shun en′ur-je) Energy required to start a reaction.

active site (ak′tive sīt) The place on the enzyme that causes the substrate to change.

attachment site (uh-tatch′munt sīt) A specific point on the surface of the enzyme where it can physically attach itself to the substrate; also called **binding site.**

binding site (bin′ding sīt) See **attachment site.**

catalyst (cat′uh-list) A chemical that speeds up a reaction but is not used up in the reaction.

coenzyme (ko-en′zīm) A molecule that works with an enzyme to enable the enzyme to function as a catalyst.

competitive inhibition (kum-pet′ĭ-tiv in″hĭ-bĭ′shun) The formation of a temporary enzyme-inhibitor complex that interferes with the normal formation of enzyme–substrate complexes, resulting in a decreased turnover.

control processes (con-trōl′ pro′ses-es) Mechanisms that ensure that an organism will carry out all metabolic activities in the proper sequence (coordination) and at the proper rate (regulation).

denature (de-nā′chur) To permanently change the protein structure of an enzyme so that it loses its ability to function.

enzymatic competition (en-zi-mă′tik com-pĕ-tĭ′shun) Competition among several different available enzymes to combine with a given substrate material.

enzyme (en′zīm) A specific protein that acts as a catalyst to change the rate of a reaction.

enzyme–substrate complex (en′zīm–sub′strāt kom′pleks) A temporary molecule formed when an enzyme attaches itself to a substrate molecule.

gene-regulator proteins (jēn reg′yu-la-tor pro′te-ins) Chemical messengers within a cell that inform the genes as to whether protein-producing genes should be turned on or off or whether they should have their protein-producing activities increased or decreased, for example, gene-repressor proteins and gene-activator proteins.

inhibitor (in-hib′ĭ-tōr) A molecule that temporarily attaches itself to an enzyme, thereby interfering with the enzyme's ability to form an enzyme–substrate complex.

negative-feedback inhibition (neg′ă-tiv fēd′băk in-hib′ĭ-shun) A metabolic control process that operates at the surfaces of enzymes. This process occurs when one of the end products of the pathway alters the three-dimensional shape of an essential enzyme in the pathway and interferes with its operation long enough to slow its action.

nutrients (nu′tre-ents) Molecules required by organisms for growth, reproduction, and/or repair.

substrate (sub′strāt) A reactant molecule with which the enzyme combines.

turnover number (turn′o-ver num′ber) The number of molecules of substrate that a single molecule of enzyme can react with in a given time under ideal conditions.

CHAPTER 6

Biochemical Pathways

Chapter Outline

Purpose

This chapter deals with some of the major chemical reactions that occur in living things. Because these reactions are dependent on one another and occur in specific series, they are commonly referred to as *biochemical pathways.* An understanding of these biochemical pathways will help you understand how energy is utilized within an organism.

There are hundreds of such pathways, all of which interlink, but we will deal only with those that form the core of all chemical reactions in a living cell. The two major pathways are photosynthesis and cellular respiration.

For Your Information

An issue currently before the world concerns the interplay between photosynthesis and the amount of carbon dioxide in the atmosphere. Many human activities, such as burning gasoline, coal, and wood, release carbon dioxide. The increased amount of this gas in our atmosphere is thought to cause a global warming known as the *greenhouse effect.* The major method of removing carbon dioxide from the atmosphere is photosynthesis. Any significant decrease in the number of photosynthesizing organisms further complicates the situation. Many people have suggested that saving forests from destruction is an important step in slowing the greenhouse effect. Some countries, such as Australia and the United States, have plans for planting billions of trees to alleviate the problem. Another alternative is to stop the destruction of major forested areas, such as tropical rain forests.

Learning Objectives

- Associate the major parts of photosynthesis with the ultrastructure of the chloroplast.
- List the raw materials and products and describe the processes involved in the two major parts of photosynthesis.
- List the things a plant can do with the product of photosynthesis.
- Associate the major parts of aerobic respiration with the ultrastructure of the mitochondrion.
- List the raw materials and products and describe the processes involved in the three major parts of aerobic respiration.
- Describe the processes involved in the alternate pathway known as *anaerobic respiration* (fermentation).
- Follow a molecule through the steps of interconversion of fats, proteins, and carbohydrates.

Energy and Cells

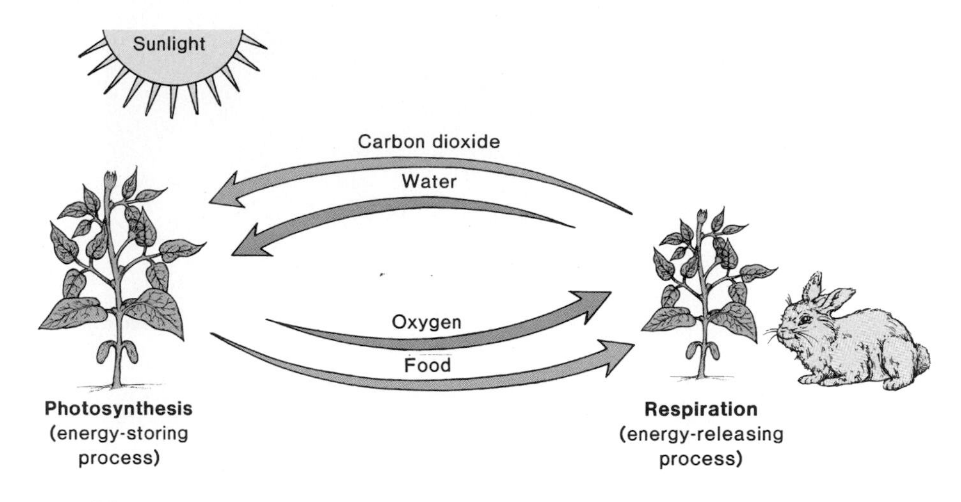

Figure 6.1 **The Relationship between Photosynthesis and Respiration.** The raw materials for photosynthesis—carbon dioxide and water—are the end products of respiration. The raw materials for respiration—food and oxygen—are the end products of photosynthesis. These materials are cycled between those organisms that are involved in the two major pathways, photosynthesis and respiration.

All living organisms require energy to conduct the many functions necessary to sustain life. The source of this energy for cells is the chemical bonds of food molecules. Cells can be thought of as chemical factories that conduct a variety of chemical reactions. These reactions are frequently linked to each other because the products of one reaction are used as the reactants for another. A series of enzyme-controlled reactions linked together is termed a **biochemical pathway.** Some organisms (green plants and algae) are capable of trapping sunlight energy and converting it to chemical bonds in food molecules. This important energy transformation is the source of all food and energy for all organisms, including animals. Green plants and algae do not have to eat because they are able to manufacture the food molecules necessary for their energy-requiring processes. Organisms that are able to make their food molecules from sunlight and inorganic raw materials are **autotrophs** (self-feeders). Organisms that eat food to obtain energy are termed **heterotrophs** (different feeders).

The process of converting sunlight energy to chemical-bond energy, **photosynthesis,** is one of the major biochemical pathways. In this series of biochemical reactions, plants produce food molecules, such as carbohydrates, for themselves as well as for all the other organisms on earth. **Cellular respiration,** a second major biochemical pathway, is a series of reactions during which cells release the chemical-bond energy from food and convert it into usable forms. All organisms must be able to accomplish this energy conversion regardless of the source of their food molecules. Whether organisms manufacture food or take it in from the environment, they use food molecules as a source of energy. The process of energy conversion is essentially the same in all organisms (figure 6.1).

Within cells, particular biochemical pathways are carried out in specific organelles. The organelles provide surfaces upon which chemical reactions occur and the enzymes necessary to control these reactions. Chloroplasts are the site of photosynthesis, and mitochondria are the site of most of the reactions of cellular respiration. As is the case with most things in life, there are exceptions. Prokaryotic cells lack mitochondria and chloroplasts, yet some are capable of processes very similar to the cellular respiration and photosynthesis processes that occur in eukaryotic cells.

A biochemical pathway may consist of several steps, each involving a molecular change and an energy change. Different chemical bonds have different amounts of chemical energy. If the products of a reaction do not have the same amount of energy as the reactants do, energy is either released or required. Some chemical reactions—like the burning of methane—may have a net release of energy, whereas others—like the synthesis of sugar by plants—require an input of energy. Cells can encourage reactions that require an input of energy by coupling energy-requiring reactions with others that yield net energy. These reactions are often called **coupled reactions.**

Generating Energy in a Useful Form: ATP

The second law of thermodynamics states that some usable energy is lost whenever energy is converted from one form to another. This wasted energy is generally released into the environment. If the conversion is to be useful, the amount of energy lost must not be too great. In a cell, the chemical bonds of a molecule of food are broken, releasing small amounts of energy. While some of the energy is lost as heat to the environment, some of it is used to form the energy-carrier molecule ATP. ATP is one of the major molecules involved in coupling energy-requiring reactions with energy-demanding reactions.

Adenosine triphosphate (ATP) is formed from adenine, ribose, and phosphates (figure 6.2). These three are chemically bonded to form AMP, adenosine monophosphate (one phosphate). When a second phosphate group is added to the AMP, a molecule of ADP (diphosphate) is formed. The covalent bond that attaches the second phosphate to the AMP molecule is easily broken to release energy for energy-requiring cell processes. Because this covalent bond is

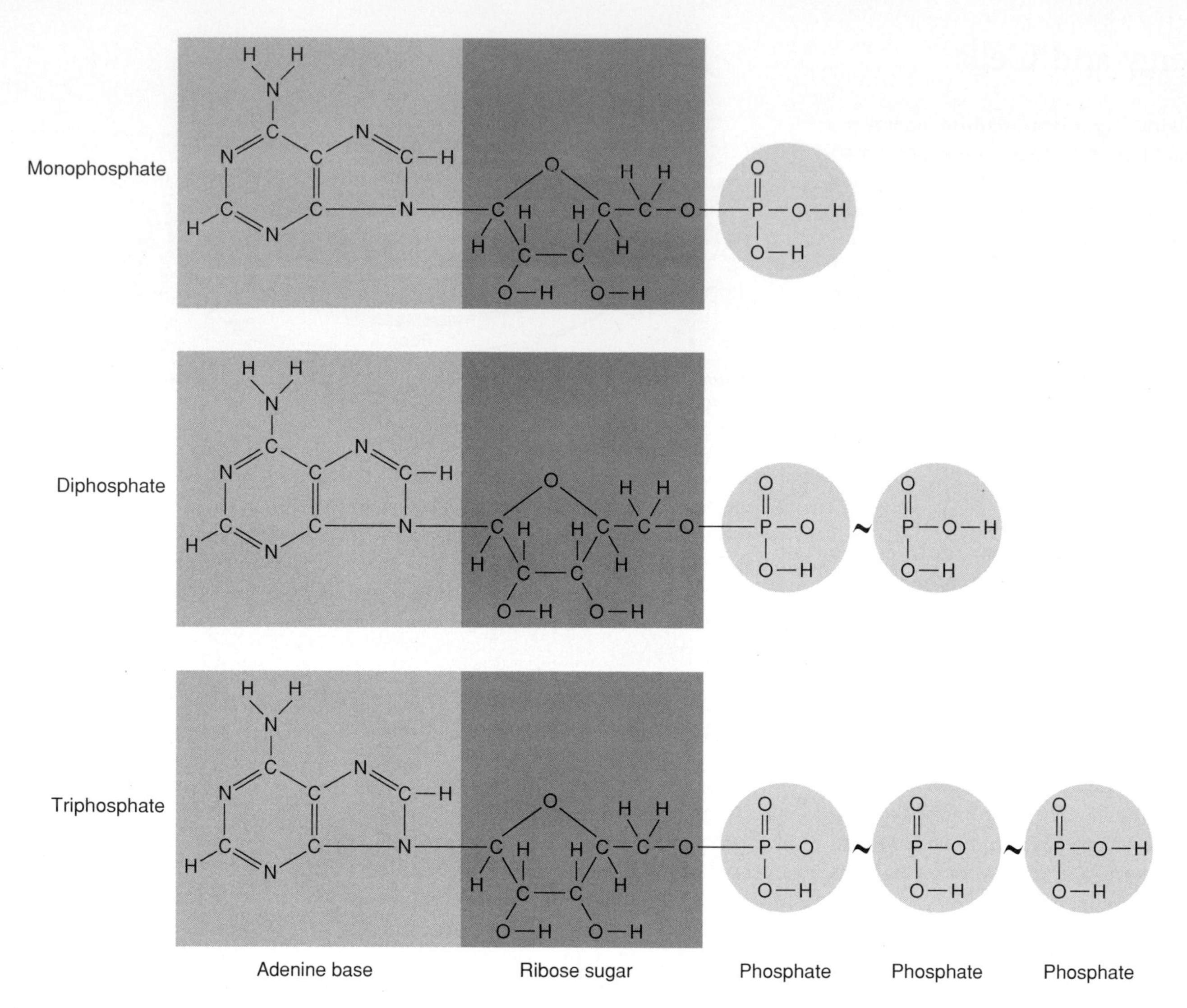

Figure 6.2 **Adenosine Triphosphate (ATP).** A macromolecule of ATP consists of a molecule of adenine, ribose, and three phosphate groups. The two end phosphate groups are bonded together by high-energy bonds. When these bonds are broken, they release an unusually great amount of energy; therefore, they are known as high-energy bonds. These bonds are represented by the curved lines. The ATP molecule is considered to be an energy carrier.

such a readily available source of energy, it is called a **high-energy phosphate bond.** Thus, the bond acts as an energy holder. The ADP, with the addition of more energy, is able to bond to a third phosphate group and form ATP. (The addition of phosphate to a molecule is called a *phosphorylation reaction.*) ATP has two high-energy phosphate bonds represented by curved solid lines. Both ADP and ATP, because they contain high-energy bonds, are very unstable molecules and readily lose their phosphates. When this occurs, the energy held in their high-energy bonds is released to the environment. Within a cell, enzymes direct this release of energy as ATP is broken down into its components. This channeling of energy enables the cell to better utilize this readily available source of energy.

Photosynthesis—An Overview

The generalized chemical equation for the total process of photosynthesis may be written as follows:

$$\underset{\text{Energy}}{\text{Sunlight}} + \underset{\text{carbon dioxide}}{6CO_2} + \underset{\text{water}}{6H_2O} \xrightarrow[\text{chlorophyll and enzymes in chloroplasts}]{\text{helped by}} \underset{\text{simple sugar}}{C_6H_{12}O_6} + \underset{\text{oxygen}}{6O_2}$$

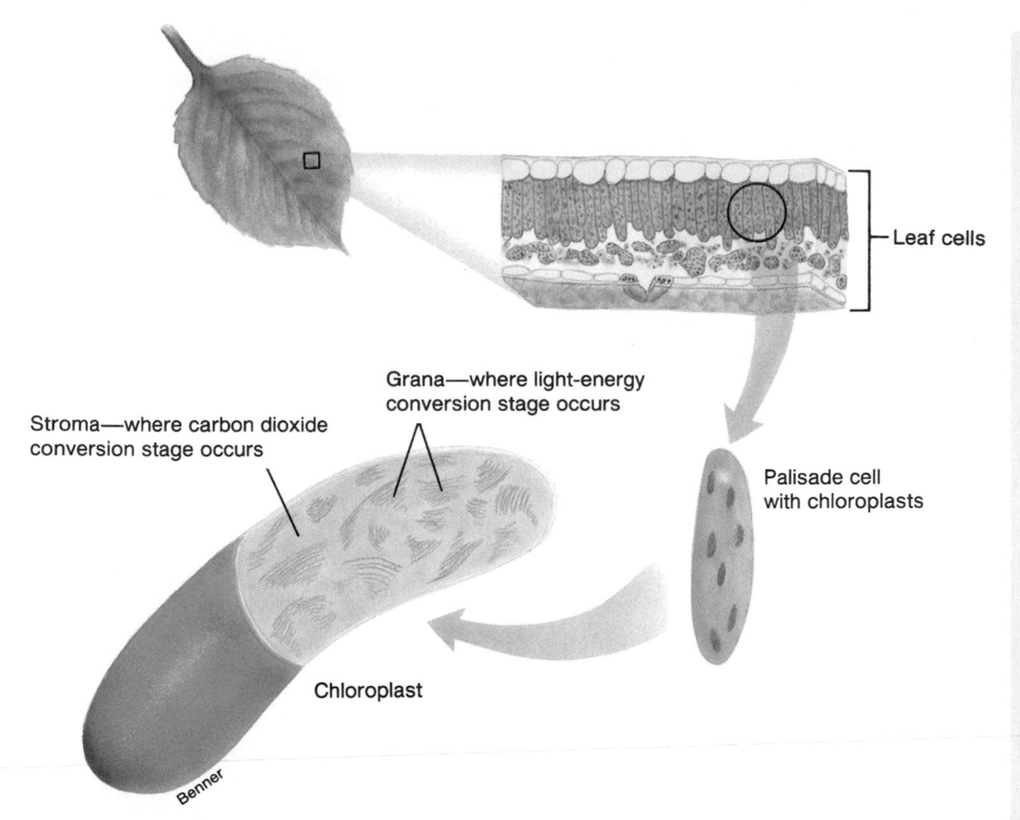

Figure 6.3 **Plant Structure and Photosynthesis.** The major structure associated with photosynthesis in plants is the leaf. Certain cells within the leaf contain large numbers of chloroplasts. It is within the chloroplasts that the light energy is converted to chemical-bond energy. It is here also that the carbohydrate is constructed from atmospheric carbon dioxide, water, and the energy from the sun.

The process of capturing light energy and converting it into the chemical bonds of an organic molecule occurs in the chloroplast. The structure of the chloroplast is directly related to both the light-trapping parts of the process and the energy-conversion steps. In the first parts of the process, light energy is captured by the green pigment **chlorophyll** and is converted to the kinetic energy of the moving electrons of chlorophyll. The kinetic energy of these electrons is then converted to the chemical-bond energy of an ATP molecule. The energy in ATP is used in later steps to combine hydrogen from water with carbon dioxide from the atmosphere to form simple sugar molecules (figure 6.3). The oxygen from the water molecules is released into the atmosphere. The plant can then use the simple sugar to construct other kinds of molecules, provided there are a few additional raw materials, such as minerals and nitrogen-containing molecules.

For most plants, the entire process of photosynthesis takes place in the leaf, in cells containing large numbers of chloroplasts. Recall from chapter 4 that chloroplasts are membranous, saclike organelles containing many thin flat disks. These disks, called *thylakoids,* contain the chlorophylls, some other pigments, electron-transport molecules, and enzymes. They are stacked in groups, called *grana* (singular *granum*). The regions between the grana are called the *stroma* of the chloroplast. The first of the two stages of photosynthesis, the **light-energy conversion stage,** takes place in the grana. It is here that the light energy is converted into kinetic energy and then into chemical-bond energy. The second stage of photosynthesis is known as the **carbon dioxide conversion stage.** This series of reactions occurs outside the grana, in the stroma. In this stage, carbon dioxide becomes incorporated into a simple sugar molecule (figure 6.4).

The Light-Energy Conversion Stage of Photosynthesis

The green pigment, chlorophyll, which is present in the chloroplasts of plants, is a complex molecule with many loosely attached electrons. Some of the electrons separate from the chlorophyll molecule when they are struck by certain wavelengths of light. The chlorophyll becomes oxidized in this process. The lost electron may then be picked up by electron-transfer molecules. This portion of photosynthesis is a series of oxidation–reduction reactions (figure 6.5). When the light energy is transferred to the electrons, the electrons move more rapidly and are called *excited electrons.* After excited electrons leave the chlorophyll molecule, they may follow one of two different pathways during the light-energy conversion stage of photosynthesis.

In the first pathway, the excited electrons release their newly acquired energy as they move through the cytochromes in an electron-transport system. The electron carriers are located on the membranes that make up the grana of the chloroplast. The series of cytochromes in the chloroplast takes the energy from the excited electrons by the chemiosmotic process and uses it to bind a phosphate to an ADP molecule to from ATP (see box 6.1). This energy-conversion process begins with sunlight energy exciting the electrons of chlorophyll to a higher energy level. The electrons, having lost their excess energy, return to their original positions in the chlorophyll molecule. Because the electrons that left the chlorophyll molecule eventually return to their original positions, this pathway is a complete cycle. It is called *cyclic photophosphorylation* because the electrons that leave the chlorophyll molecule eventually return to that same molecule (cyclic), the process is stimulated by light (photo), and phosphate is added to ADP to form ATP (phosphorylation). The shorthand chemical equation for this portion of photosynthesis may be written as follows:

$$\text{ADP} + \text{Ⓟ} + \text{sunlight energy} \longrightarrow \text{ATP}$$

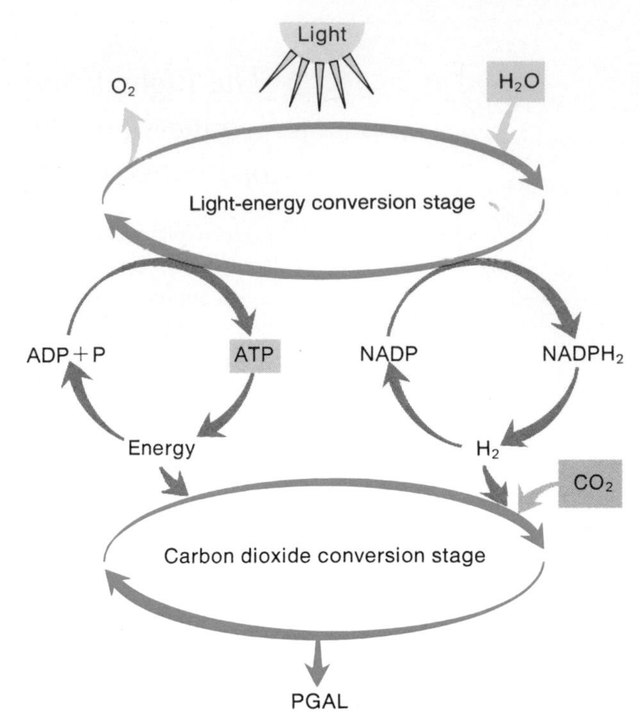

Figure 6.4 **Photosynthesis.** The process of photosynthesis is composed of the interrelated stages of light-energy conversion and carbon dioxide conversion. The carbon dioxide conversion stage requires the ATP and $NADPH_2$ produced in the light-energy conversion stage. The light-energy conversion stage in turn requires the ADP and NADP that are released from the carbon dioxide conversion stage. Therefore, each stage is dependent upon the other.

In the second pathway, the excited electrons are picked up by a different electron carrier known as **NADP** (**n**icotinamide **a**denine **d**inucleotide **p**hosphate), which becomes reduced. Each NADP molecule has the ability to capture two electrons. As a result, a molecule of NADP is converted to a molecule of $NADP^{--}$ that has two negative charges. At the same time, water in the cell is broken into H^+ and OH^- ions. Since the H^+ ions are positively charged, they are attracted to the negatively charged $NADP^{--}$. This results in the formation of $NADPH_2$, which carries two atoms of hydrogen.

The remaining hydroxyl ions (OH^-) accumulate and combine with each other to form water, oxygen (O_2), and free electrons (figure 6.6). The water becomes part of the cytoplasm, the oxygen is released into the atmosphere, and the free electrons attach to the chlorophyll molecule that had previously lost electrons to the NADP. Since the electrons that left the chlorophyll molecule in this pathway do not return to their original position, but are picked up by NADP, the pathway is

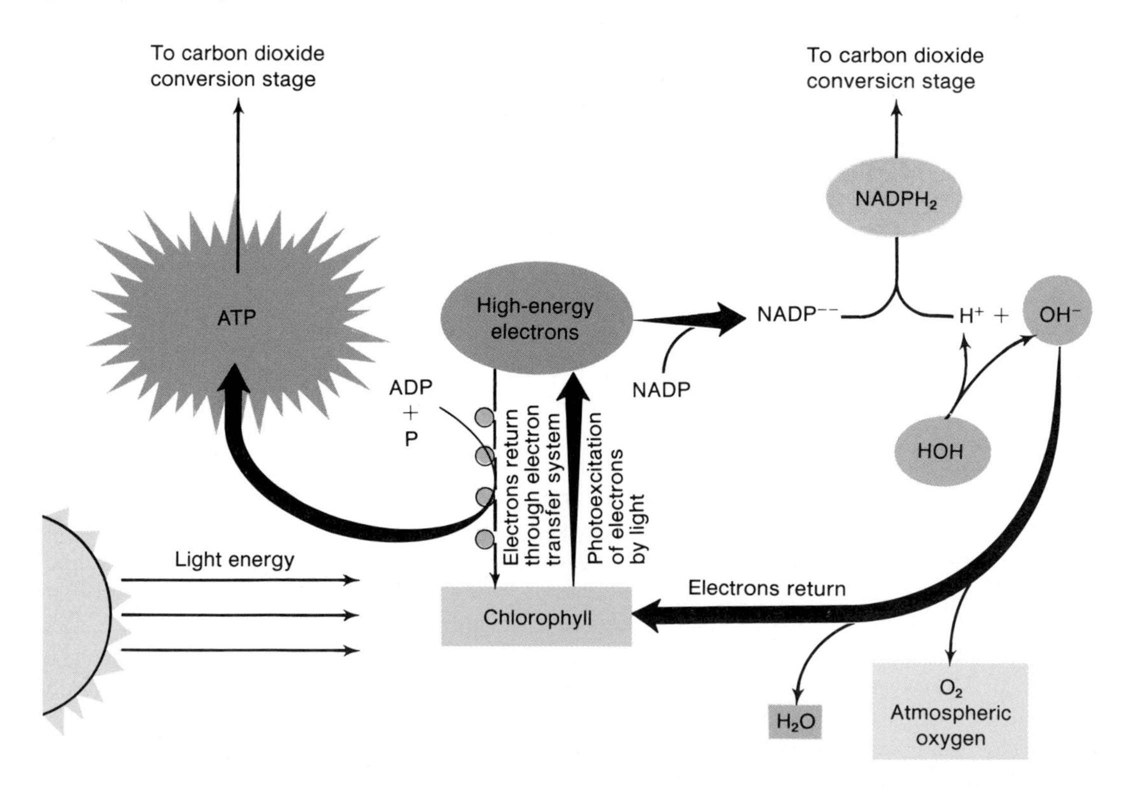

Figure 6.5 **The Light-Energy Conversion Stage of Photosynthesis.** In the light-energy conversion stage, sunlight is the source of energy. This stage converts the light energy to kinetic energy of excited electrons and then forms ATP molecules to carry the energy. $NADPH_2$ is produced to carry H_2, which is produced from water. Oxygen is a waste product formed from the water molecules. These processes take place inside the chloroplast and are associated with the structures known as the grana.

BOX 6.1 Chemiosmosis

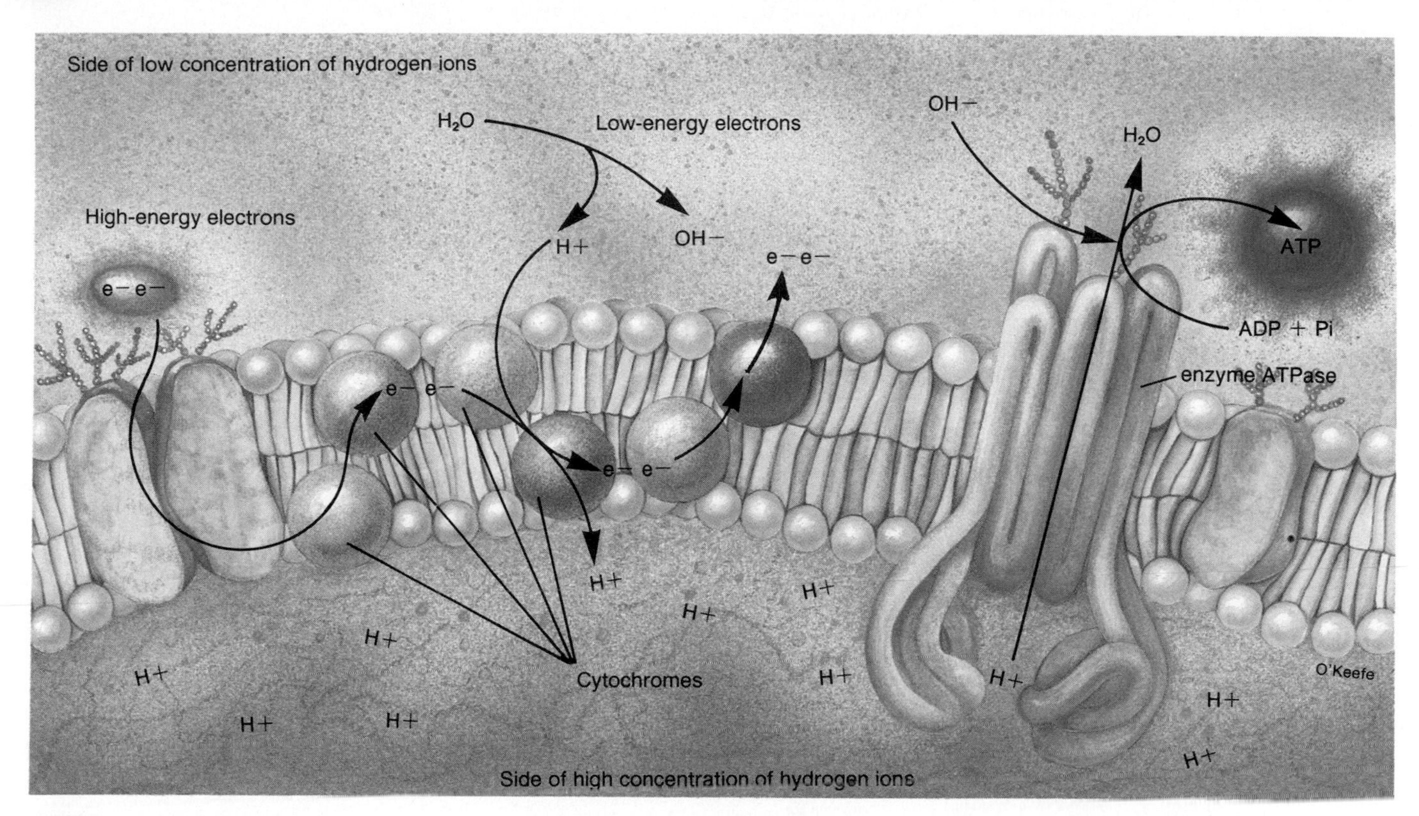

It has been known for many years that ATP and other phosphorylated compounds are very important in providing the energy needed for movement, synthesis, and other energy-demanding activities. The production of ATP is associated with two particular membranous organelles: mitochondria and chloroplasts.

The energy needed to form the high-energy phosphate bonds of ATP comes from electrons that are rich in kinetic energy. It appears that the processes that result in the formation of ATP occur on the membranes of mitochondria and chloroplasts. Research indicates that iron-containing cytochrome molecules (enzymes) are located on these membranes. The energy-rich electrons are passed from one cytochrome to another, and the energy is used to pump hydrogen ions from one side of the membrane to the other.

The result of this is a higher concentration of hydrogen ions on one side of the membrane. As the concentration of hydrogen ions increases on one side, a concentration gradient is established and a "pressure" develops. This causes these hydrogen ions to be forced back through pores in the membrane to the side from which they were pumped. As they stream through the pores an enzyme, ATPase, stimulates the formation of an ATP molecule by bonding a phosphate to an ADP molecule (phosphorylation).

This process of manufacturing ATP by first creating a high concentration of hydrogen ions on one side of a membrane, which leads to their flow back through the membrane, is called *chemiosmosis.* The word chemiosmosis was chosen because the process just described resembles osmosis. However, hydrogen ions move through a membrane from an area of high concentration to an area of lower concentration, not water.

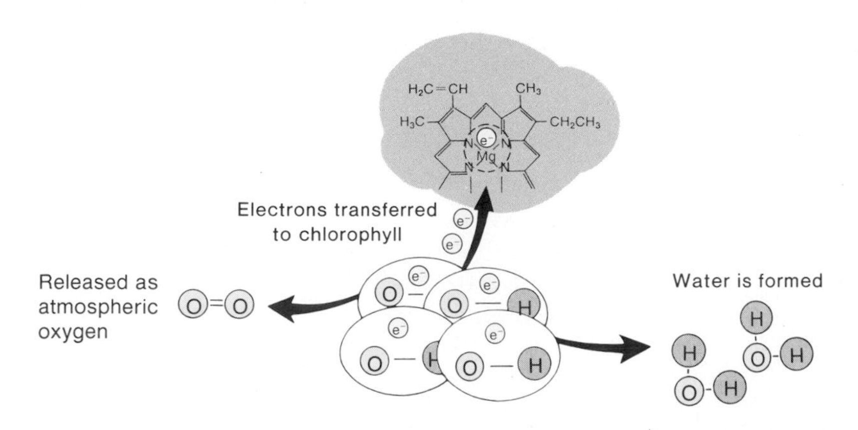

Figure 6.6 **The Fate of Hydroxyl Ions.** In the light-energy conversion stage of photosynthesis, the hydroxyl ions (OH^-) combine with one another to form water molecules and oxygen gas (O_2). At the same time, the electrons are returned to a chlorophyll molecule.

noncyclic. The electrons that eventually attach to the chlorophyll molecule come from water. At the completion of the light-energy conversion stage, the plant has acquired usable energy in the form of ATP and a source of hydrogen in the form of $NADPH_2$. The energy and hydrogen are necessary for the next stage of photosynthesis. During the light-energy conversion stage, water molecules are used up and oxygen is released to the environment. The shorthand chemical equation for this portion of photosynthesis may be written as follows:

$$2NADP + ADP + \textcircled{P} + 4H_2O + \underset{\text{energy}}{\text{sunlight}} \xrightarrow[\text{chlorophyll and enzymes}]{\text{helped by}} 2NADPH_2 + ATP + 2H_2O + O_2$$

The Carbon Dioxide Conversion Stage of Photosynthesis

The second major series of reactions involved in photosynthesis takes place within the stroma of the chloroplast (figure 6.7). The materials needed for the carbon dioxide conversion stage are ATP, $NADPH_2$, CO_2, and a 5-carbon starter molecule called *ribulose*. The first two ingredients (ATP and $NADPH_2$) are made available from the light-energy conversion reactions. The carbon dioxide molecules come from the atmosphere and the ribulose starter molecule is already present in the stroma of the chloroplast from previous reactions. The major event that occurs in the carbon dioxide conversion stage involves the use of energy from ATP to bond hydrogen from $NADPH_2$ and the carbon dioxide to ribulose in order to ultimately form **PGAL** (**p**hospho**g**lycer**al**dehyde).

The carbon dioxide molecule does not become PGAL directly; it is first attached to the 5-carbon starter molecule, ribulose, to form an unstable 6-carbon molecule. This 6-carbon molecule immediately breaks down into two 3-carbon molecules, which then undergo a series of reactions that involve a transfer of energy from ATP and a transfer of hydrogen from the $NADPH_2$. This series of reactions produces PGAL molecules. The general chemical equation for the CO_2 conversion stage is as follows:

$$CO_2 + ATP + NADPH_2 + \text{5-carbon starter (ribulose)} \longrightarrow PGAL + NADP + ADP + \textcircled{P}$$

PGAL: The Product of Photosynthesis

Phosphoglyceraldehyde (PGAL) is the actual product of the process of photosynthesis. If you look at the generalized equation, however, it would appear as if a 6-carbon sugar (hexose) should be the product. The reason a hexose ($C_6H_{12}O_6$) is usually listed as the end product is simply because, in the past, the simple sugars were easier to detect than PGAL. If a plant goes through photosynthesis and produces twelve PGALs, ten of the twelve are rearranged by a series of complex chemical reactions to regenerate the molecules needed to operate the carbon dioxide conversion stage. The other two PGALs can be considered profit from the process. As the PGAL profit accumulates, it is frequently changed into a hexose. So those who first examined photosynthesis chemically saw additional sugars as the product and did not realize that PGAL was the initial product.

There are a number of things the cell can do with the PGAL profit from photosynthesis in addition to manufacturing hexose (figure 6.8). Many other organic molecules can be constructed using PGAL as the basic construction unit. PGAL can be converted to glucose molecules, which can be combined to form complex carbohydrates, such as starch for energy storage or cellulose for cell-wall construction. In addition, other simple sugars can be used as building blocks for ATP, RNA, DNA or other carbohydrate-containing materials.

The cell may convert the PGAL into lipids, such as oils for storage, phospholipids for cell membranes, or steroids for vitamins. The PGAL can serve as the carbon skeleton for the construction of

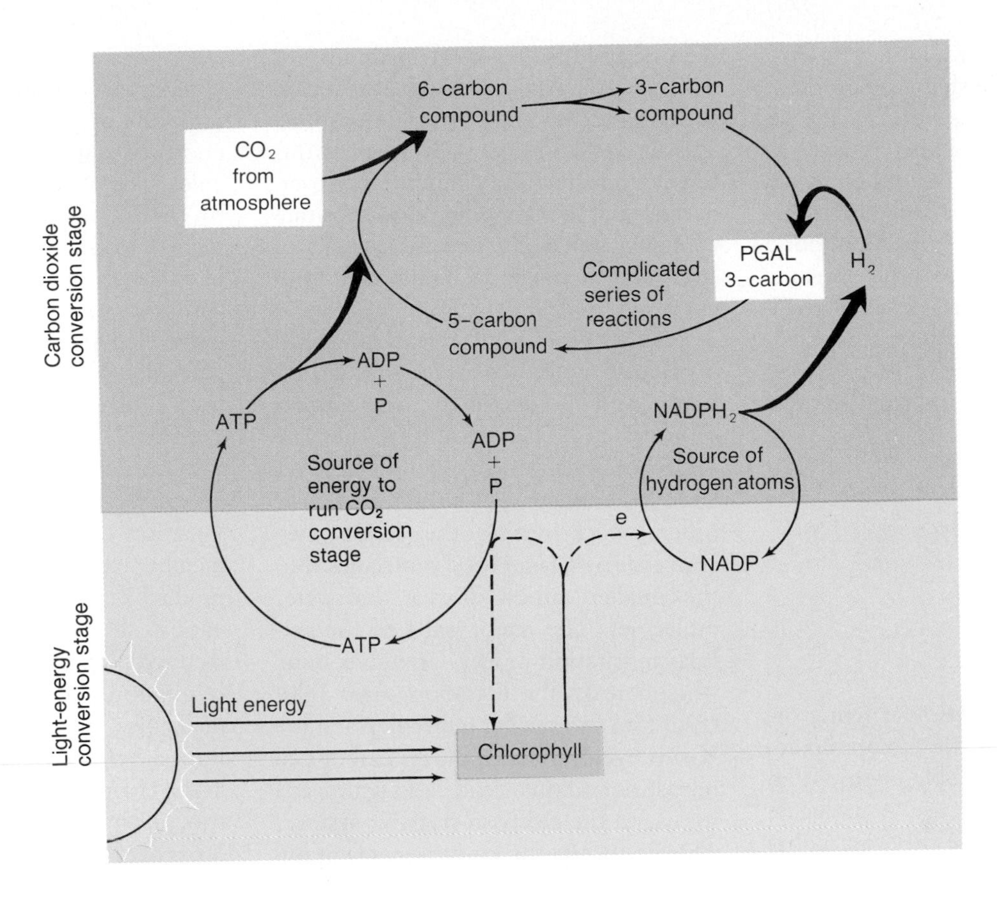

Figure 6.7 **The Carbon Dioxide Conversion Stage of Photosynthesis.** During this process, ATP molecules from the light-energy conversion stage are used to drive the reactions, which include the incorporation of carbon dioxide molecules and hydrogen atoms into an organic molecule. An important molecule produced is phosphoglyceraldehyde (PGAL). This process takes place in the stroma of the chloroplast.

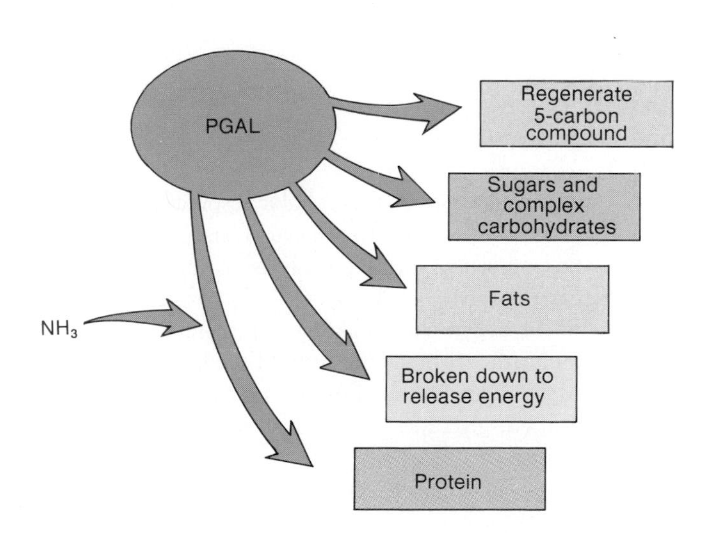

Figure 6.8 **Uses of PGAL.** The PGAL that is produced as the end product of photosynthesis is used for a variety of things. The plant cell can make simple sugars, complex carbohydrates, or even the original 5-carbon starter from it. It can also serve as an ingredient of lipids and amino acids (proteins). In addition, it provides a major source of metabolic energy when it is sent through the respiratory pathway.

amino acids needed to form proteins. Almost any molecule that a green plant can manufacture begins with this PGAL molecule. Finally (and this is easy to overlook), PGAL can be broken down during cellular respiration. Cellular respiration releases the chemical-bond energy from PGAL and other organic molecules and converts it into ATP energy. This conversion of chemical-bond energy enables the plant cell to do things that require energy, such as grow and move materials.

Aerobic Cellular Respiration—An Overview

One of the characteristics of living organisms is control of the metabolic activities that make available energy from food. Most organisms use a molecule such as PGAL or some other carbohydrate as a primary food source. To release energy, plants, animals, fungi, protozoa, algae, and many bacteria rely on the same basic chemical pathway—**aerobic cellular respiration.** The generalized equation for the total process is

$$\underset{\text{Glucose}}{C_6H_{12}O_6} + \underset{\text{oxygen}}{6O_2} \longrightarrow \underset{\text{carbon dioxide}}{6CO_2} + \underset{\text{water}}{6H_2O} + \underset{\text{energy}}{36ATP}$$

In aerobic cellular respiration, food and oxygen are chemically combined and rearranged to form carbon dioxide and water. Very importantly, this chemical rearrangement allows the cell to convert the chemical-bond energy from the food molecule into chemical-bond energy of ATP molecules. These ATP molecules with their high-energy phosphate bonds can then be used to power cellular activity that requires an input of energy.

Ths conversion of energy can be compared to a campfire that converts firewood, with its chemical-bond energy, into heat energy. The heat energy can then be used for a variety of things; for example, to warm your hands, to prepare your dinner, or to heat water for washing. The energy conversion in a cell is different from that in a campfire, however, because the cell carefully controls the rate of energy release to prevent damage to itself and to maximize its production of ATP.

The process of releasing energy from food molecules begins in the cytoplasm and is completed in the mitochondrion. The structure of the mitochondrion with its two membranes contains enzymes, placed in a particular order, which enable the step-by-step conversion of large food molecules into the smaller carbon dioxide and water molecules. The major parts of the cellular respiration process are: **glycolysis,** which breaks the 6-carbon sugar (glucose) into two smaller molecules, and the **Krebs cycle,** which removes carbons to release as carbon dioxide and hydrogens for use in the **electron-transfer system,** which converts the kinetic energy of hydrogen electrons to the high-energy phosphate bonds of ATP.

Glycolysis

The first stage of the cellular respiration process takes place in the cytoplasm. This first step, known as **glycolysis** (carbohydrate = glyco; splitting = lysis), consists of the enzymatic breakdown of a glucose molecule without the use of molecular oxygen (figure 6.9). Because no oxygen is rquired, glycolysis is called an **anaerobic** process.

Some energy must be put in to start glycolysis because glucose is a very stable molecule and will not spontaneously decompose to release energy. For each molecule of glucose entering glycolysis, energy is supplied by two ATP molecules. The phosphates are released from two ATP molecules and become attached to glucose to form phosphorylated sugar, ($P-C_6-P$). We term this reaction a *phosphorylation reaction*. It is controlled by an enzyme named *phosphorylase*. The phosphorylated glucose is then broken down through several enzymatically controlled reactions into two 3-carbon compounds, each with one attached phosphate, (C_3-P). These 3-carbon compounds are PGAL. (Remember that PGAL is also the end product of photosynthesis. In some situations, a plant will use PGAL manufactured during photosynthesis in cellular respiration.) Each of the two PGAL molecules acquires a second phosphate from a phosphate pool normally found in the cytoplasm. Each molecule now has two phosphates attached ($P-C_3-P$). A series of reactions follows in which energy is released by breaking chemical bonds, causing each of these 3-carbon compounds to lose their phosphates. These high-energy phosphates combine with ADP to form ATP. In addition, four hydrogen atoms detach from the carbon skeleton (oxidation) and become bonded to two hydrogen-carrier molecules (reduction) known as *NAD*. **NAD** (**n**icotinamide **a**denine **d**inucleotide) is very similar in structure and function to the NADP, which is the hydrogen carrier used in photosynthesis. The molecules of $NADH_2$ contain a large amount of potential energy that may be released in a usable form in later chemical reactions. The 3-carbon molecules that result from glycolysis are called **pyruvic acid.**

In summary, the process of glycolysis takes place in the cytoplasm of a cell. In this process, glucose undergoes reactions requiring the use of two ATPs, leading to the formation of four molecules of ATP, producing two molecules of $NADH_2$ and two 3-carbon molecules of pyruvic acid.

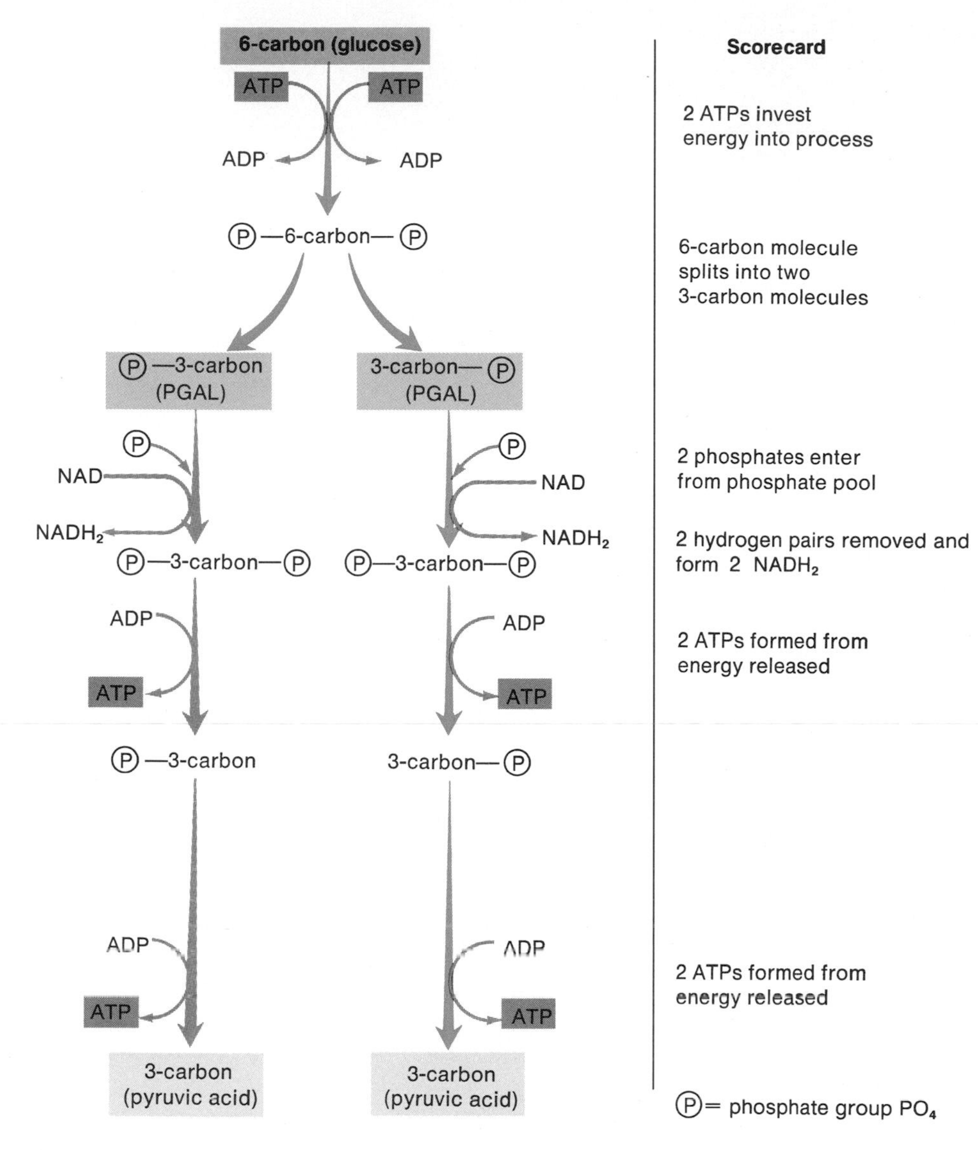

Figure 6.9 **Glycolysis.** The glycolytic pathway results in the breakdown of 6-carbon sugars under anaerobic conditions. Each molecule of sugar releases enough energy to produce a profit of two ATPs. In addition, two molecules of pyruvic acid and two molecules of hydrogen (carried as $NADH_2$) are produced.

The Krebs Cycle

The **Krebs cycle** is a series of oxidation–reduction reactions that complete the breakdown of pyruvic acid produced by glycolysis (figure 6.10). In order for pyruvic acid to be used as an energy source, it must enter the mitochondrion. Once inside, an enzyme converts the 3-carbon pyruvic acid molecule to a 2-carbon molecule called **acetyl.** When the acetyl is formed, the carbon removed is released as carbon dioxide. In addition to releasing carbon dioxide, each pyruvic acid molecule is oxidized, since it loses two hydrogens that become attached to NAD molecules (reduction) to form $NADH_2$. NAD is serving as a hydrogen carrier.

The carbon dioxide is a waste product that is eventually released by the cell into the atmosphere. The 2-carbon acetyl compound temporarily combines with a large molecule called *coenzyme A* (CoA) to form acetyl-CoA and transfers the acetyl to a 4-carbon compound called *oxaloacetic acid* to become part of a 6-carbon molecule. This new 6-carbon compound is broken down in a series of reactions to regenerate oxaloacetic acid in this cyclic pathway. In the process of breaking down pyruvic acid, three molecules of carbon dioxide are formed. In addition, five pairs of hydrogens are removed and become attached to hydrogen carriers. Four pairs become attached to NAD and one pair becomes attached to a different hydrogen carrier known as **FAD** (**f**lavin **a**denine **d**inucleotide). As the molecules move through the Krebs cycle, enough energy is released to allow the synthesis of one ATP molecule for each acetyl that enters the cycle. The ATP is formed from ADP and a phosphate already present in the mitochondria.

For each pyruvic acid molecule that enters the mitochrondrion and is processed through the Krebs cycle, three carbons are released as three carbon dioxide molecules, five pairs of hydrogens are removed and become attached to hydrogen carriers, and one ATP molecule is generated. When both pyruvic acid molecules have been processed through the Krebs cycle (1) all of the original carbons from the glucose have been released into the atmosphere as six carbon dioxide molecules, (2) all of the hydrogen originally found on the glucose has been transferred to either NAD or FAD to form $NADH_2$ or $FADH_2$, and (3) two ATPs have been formed from the addition of phosphates to ADPs.

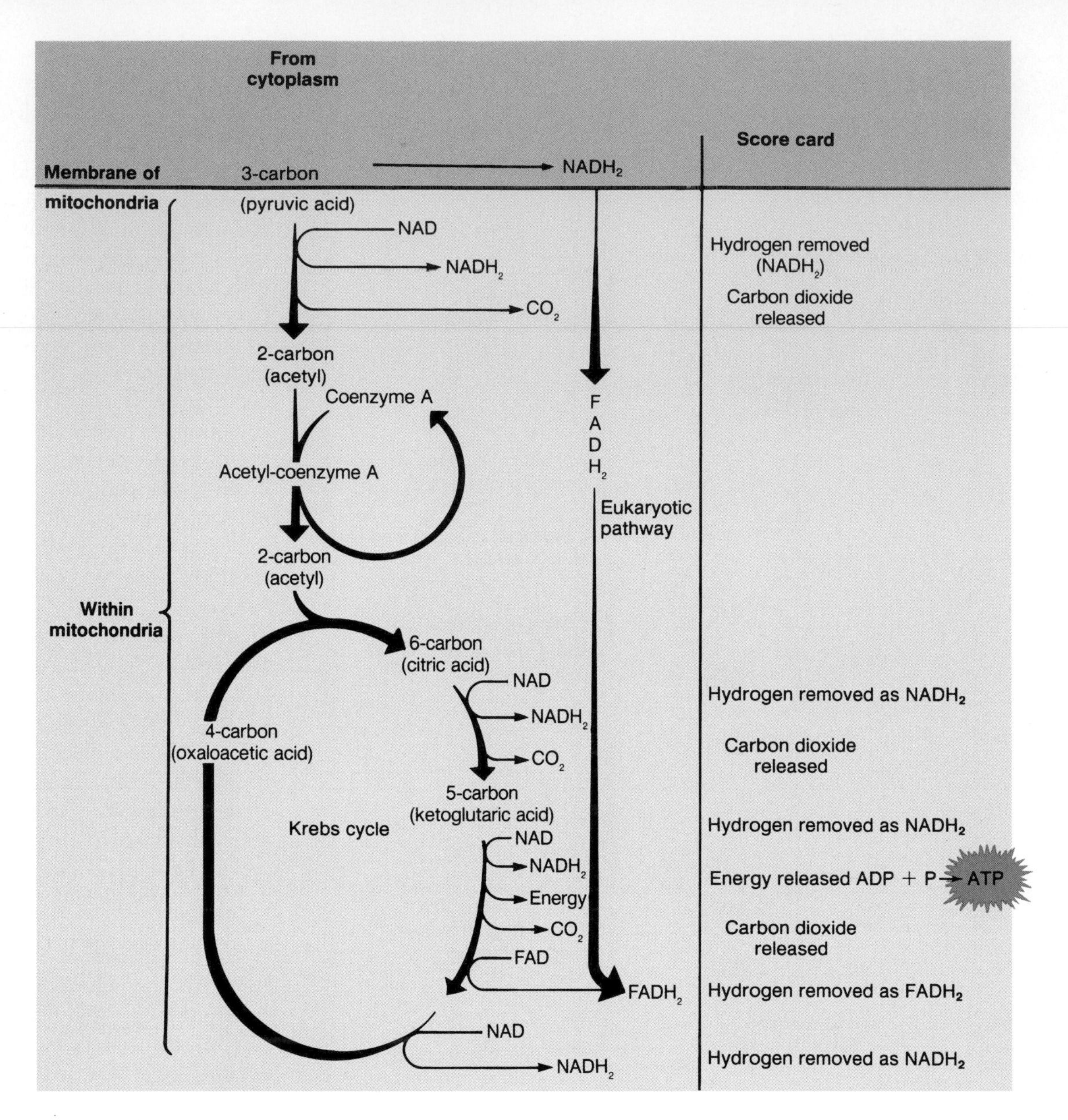

Figure 6.10 **The Krebs Cycle.** During the Krebs cycle, the pyruvic acid from glycolysis is broken down. The carbon ends up in carbon dioxide and the hydrogens are carried away to the electron-transfer system. Also, small amounts of energy are released to make ATP.

The Electron-Transfer System

The series of reactions in which energy is removed from the hydrogens carried by NAD and FAD is known as the **electron-transfer system (ETS)** (figure 6.11). This is the final stage of aerobic cellular respiration. The reactions that make up the electron-transfer system are a series of oxidation–reduction reactions in which the electrons from the hydrogen atoms are passed from one electron-carrier molecule to another until they ultimately are accepted by oxygen atoms. The oxygen combines with the hydrogens to form water. It is this step that makes the process aerobic. Water, a waste product of the process, is released into the cytoplasm.

Let's now look at the hydrogen and its carriers in just a bit more detail to account for all of the energy that becomes available to the cell. At three points in the series of oxidation–reductions in the ETS, sufficient energy is released from the $NADH_2$'s to produce an ATP molecule. Therefore, twenty-four ATPs are released from these eight pairs of hydrogen electrons carried on $NADH_2$. In eukaryotic cells, the two pairs of hydrogens released during glycolysis are carried as $NADH_2$ and converted to $FADH_2$ in order to shuttle them into the

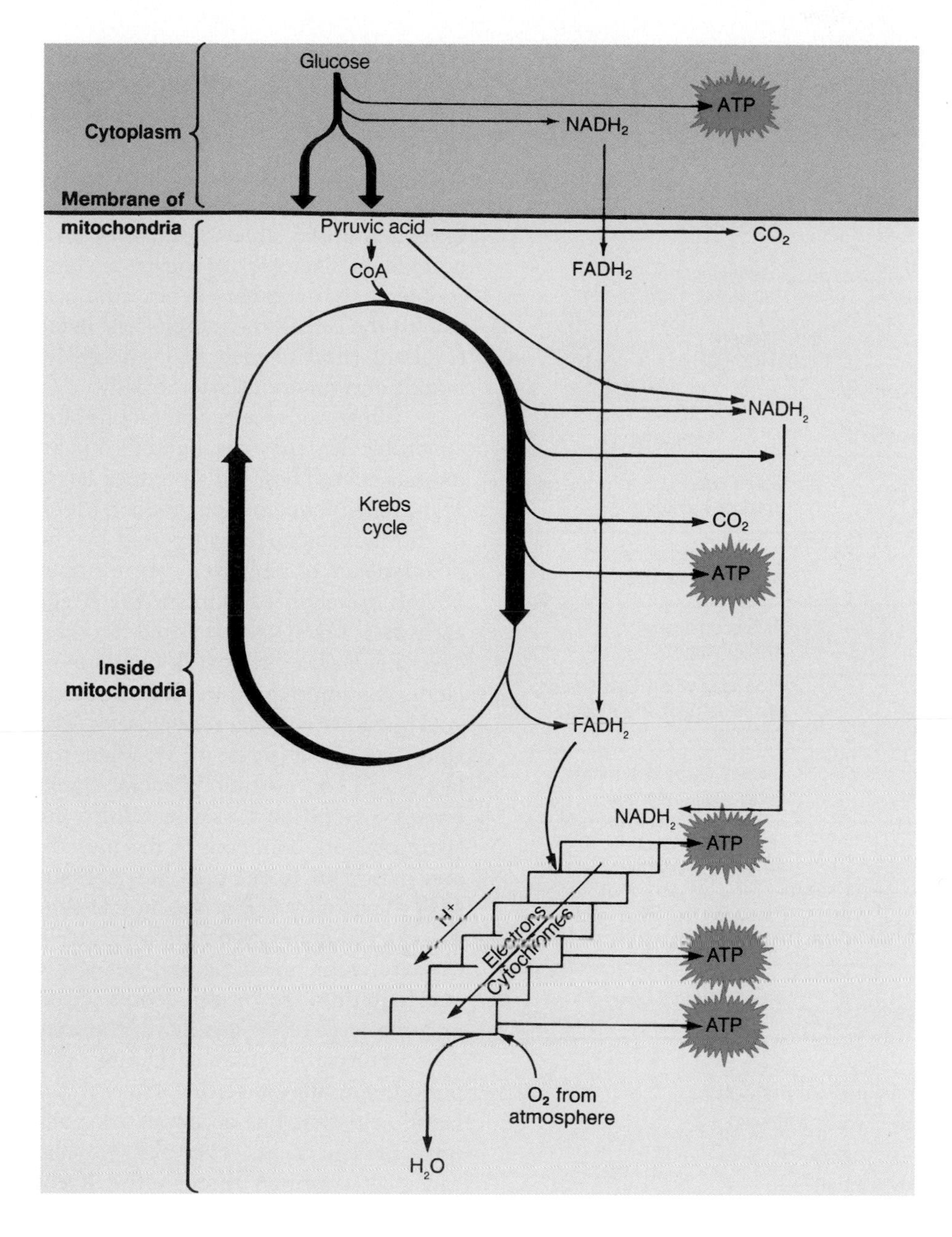

Figure 6.11 **Aerobic Respiration.** In this process, glucose, a 6-carbon sugar, is completely broken down into carbon dioxide and water and its energy is released in the form of ATP. The total process involves three major portions: glycolysis, the Krebs cycle, and the electron-transfer system. This final series of steps, known as the ETS, requires oxygen. It is the part of the pathway that releases the major quantities of energy.

mitochondria. Once they are inside the mitochondria, they follow the same pathway as the other $FADH_2$'s. The four pairs of hydrogen electrons carried by FAD are lower in energy. When these hydrogen electrons go through the series of oxidation–reduction reactions, they release enough energy to produce ATP at only two points. They produce a total of eight ATPs; therefore, we have a grand total of thirty-two ATPs produced from the hydrogens that enter the ETS.

Alternative Pathways

Certain cells do not have mitochondria or lack the necessary enzymes to use oxygen; thus, they are not able to go through the entire process of aerobic cellular respiration. These organisms must use an alternative biochemical pathway to generate ATP. Pathways that generate ATP energy in the absence of oxygen are called *fermentation*. Many fermentations include the first steps in the generalized process of respiration (glycolysis) but are followed by sequences of reactions that vary depending on the organism involved and its enzymes. Some organisms are capable of returning the hydrogens removed from the sugar to form the products ethyl alcohol and carbon dioxide. Other organisms produce enzymes that enable them to convert the pyruvic acid into lactic acid, acetone, or other organic molecules (figure 6.12).

Although many different products can be formed from pyruvic acid, we will look at only two pathways. **Alcoholic fermentation** is the anaerobic respiration pathway that yeast cells follow when oxygen is lacking in their environment. In this pathway, the pyruvic acid is converted to ethanol (a 2-carbon alcohol) and carbon dioxide. Yeast cells then are only able to generate four ATPs from glycolysis. The cost for glycolysis is still two ATPs; thus, for each glucose molecule a yeast cell oxidizes, it profits by two ATP. The waste products of carbon dioxide and ethanol are useful to humans. In making bread, the carbon dioxide is the important end product; it becomes trapped in the bread dough and makes it rise. The alcohol evaporates during the baking process. In the brewing industry, ethanol is the desirable product produced by yeast cells. Champagne, other sparkling wines, and beer are products that contain both carbon dioxide and alcohol. The alcohol accumulates, and the carbon dioxide in the bottle makes them sparkling (bubbly) beverages. In the manufacture of many wines, the carbon dioxide is allowed to escape so they are not sparkling but "still" wines.

Certain bacteria are unable to use oxygen even though it is available,

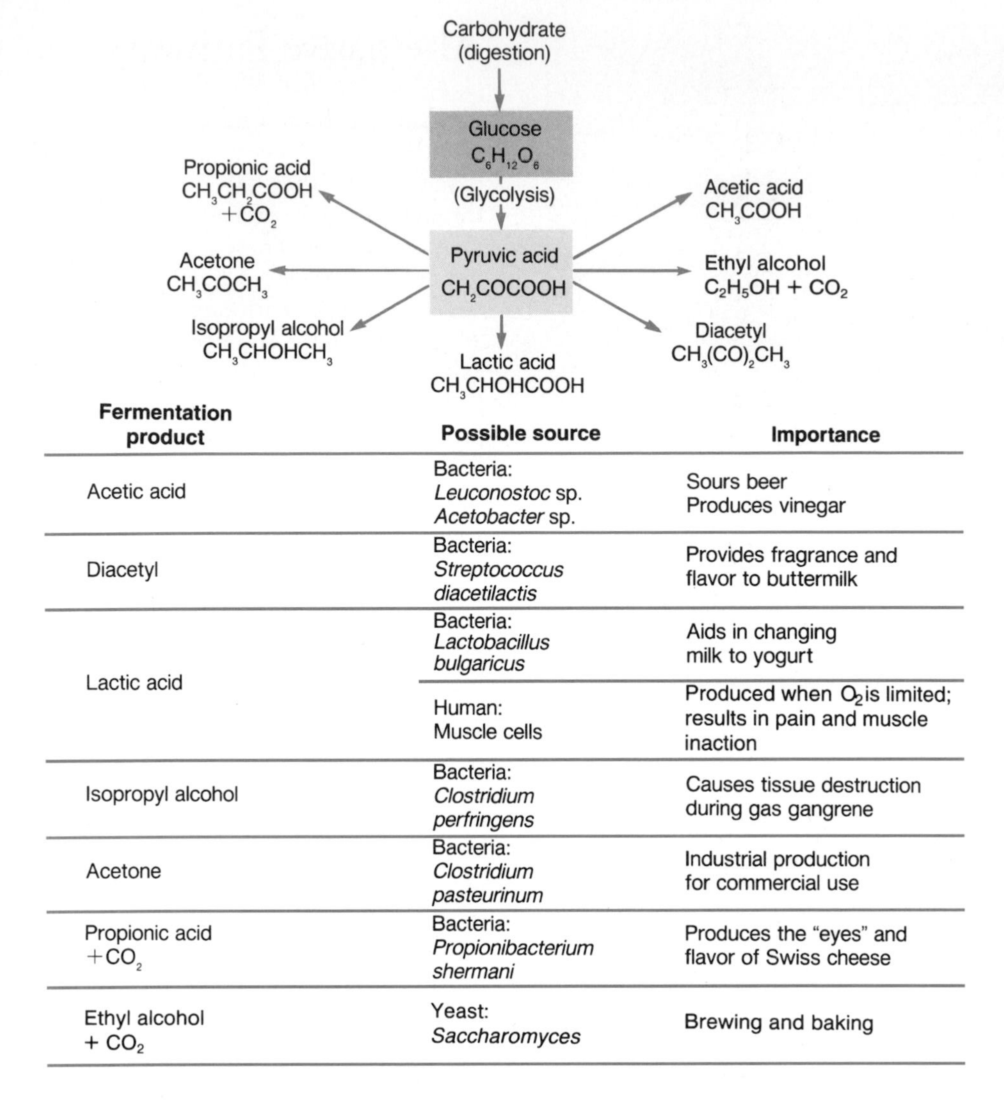

Fermentation product	Possible source	Importance
Acetic acid	Bacteria: *Leuconostoc* sp. *Acetobacter* sp.	Sours beer Produces vinegar
Diacetyl	Bacteria: *Streptococcus diacetilactis*	Provides fragrance and flavor to buttermilk
Lactic acid	Bacteria: *Lactobacillus bulgaricus*	Aids in changing milk to yogurt
	Human: Muscle cells	Produced when O_2 is limited; results in pain and muscle inaction
Isopropyl alcohol	Bacteria: *Clostridium perfringens*	Causes tissue destruction during gas gangrene
Acetone	Bacteria: *Clostridium pasteurinum*	Industrial production for commercial use
Propionic acid $+CO_2$	Bacteria: *Propionibacterium shermani*	Produces the "eyes" and flavor of Swiss cheese
Ethyl alcohol $+ CO_2$	Yeast: *Saccharomyces*	Brewing and baking

Figure 6.12 **A Variety of Fermentations.** This biochemical pathway illustrates the digestion of a complex carbohydrate to glucose followed by the glycolytic pathway forming pyruvic acid. Depending on the genetic makeup of the organisms and the enzymes they are able to produce, different end products may be synthesized from the pyruvic acid. The synthesis of these various molecules is the organism's particular way of oxidizing $NADH_2$ to NAD and reducing pyruvic acid to a new end product. Many bacteria use the fermentation process to generate their ATPs and in the process produce a variety of end products that are important in our lives.

making aerobic cellular respiration impossible. The pyruvic acid that results from glycolysis is converted to lactic acid by the addition of the hydrogens that had been removed from the original glucose. In this case, the net profit is again only two ATPs per glucose molecule. The lactic acid buildup eventually interferes with normal metabolic functions and the bacteria die. We use the lactic acid waste product from these types of anaerobic bacteria when we make yogurt, cultured sour cream, cheeses, and other fermented dairy products. The lactic acid makes the milk protein coagulate and become puddinglike or solid. It also gives the products their tart flavor, texture, and aroma.

In the human body, different cells have different metabolic capabilities. Red blood cells lack mitochondria and must rely on lactic acid fermentation to provide themselves with energy. Nerve cells can only use glucose aerobically. As long as oxygen is available to muscle cells, they function aerobically. However, when oxygen is unavailable—because of long periods of exercise, or heart or lung problems that prevent oxygen from getting to the cells—the muscle cells make a valiant effort to meet your energy demands and function anaerobically.

While your cells are functioning anaerobically, they are building up an oxygen debt. These cells produce lactic acid as their fermentation product. Much of the lactic acid is transported by the bloodstream to the liver, where about 20% is metabolized through the Krebs cycle and 80% is resynthesized into glucose. Even so, there is still a buildup of lactic acid in the muscles. It is the lactic acid buildup that makes the muscles tired when exercising (figure 6.13). When the lactic acid concentration becomes great enough, lactic acid fatigue results. Its symptoms are cramping of the muscles and pain. Due to the pain, we generally stop the activity before the muscle cells die. As you cool down after a period of exercise, your breathing and heart rate stay high until the oxygen debt is repaid and the level of oxygen in your muscle cells returns to normal. During this period, you are converting some of the lactic acid that has accumulated, back into pyruvic acid. The pyruvic acid can now continue through the Krebs cycle and the ETS as you make oxygen available.

Metabolism of Other Molecules

Up to this point we have described the methods and pathways that allow organisms to release the energy tied up in carbohydrates. Frequently, cells lack sufficient carbohydrates but have other materials from which energy can be removed. Fats and proteins, in addition to carbohydrates, make up the diet of

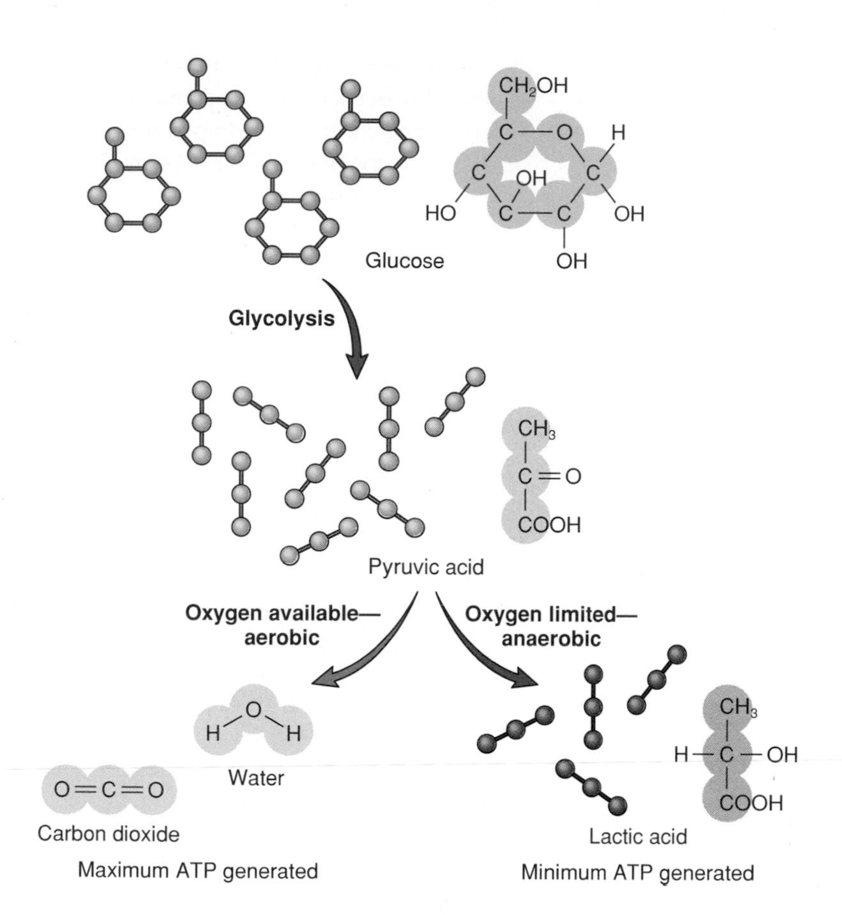

Figure 6.13 **Oxygen Starvation.** When oxygen is available to all cells, the pyruvic acid from glycolysis is converted into acetyl Co-A, which is sent to the Krebs cycle and the hydrogens pass through the electron-transfer system. When oxygen is not available in sufficient quantities (because of a lack of environmental oxygen or a temporary inability to circulate enough oxygen to cells needing it), some of the pyruvic acid from glycolysis is converted to lactic acid. The lactic acid builds up in cells when this oxygen starvation occurs.

many organisms. These three foods provide the building blocks for the cells, and all can provide energy. The pathways that organisms use to extract this chemical-bond energy are summarized here.

Fat Respiration

Fats consist of a molecule of glycerol with three fatty acids attached to it. Before fats can undergo oxidation and release energy, they must be broken down into glycerol and fatty acids. The 3-carbon glycerol molecule can be converted into PGAL, which can then enter the glycolytic pathway. However, each of the fatty acids must be processed before they can enter the pathway. Each long chain of carbons that makes up the carbon skeleton is hydrolyzed into 2-carbon fragments. Next, each of the 2-carbon fragments is converted into an acetyl molecule. The acetyl molecules are carried into the Krebs cycle by coenzyme A molecules. If you follow the glycerol and each 2-carbon fragment through the cycle, you can see that each molecule of fat has the potential to release several times as much ATP as a molecule of glucose. Each glucose molecule has six pairs of hydrogen, while a typical molecule of fat has up to ten times that number. This is why fat makes such a good long-term storage material. It is also why the removal of fat on a weight-reducing diet takes so long! It takes time to use all the energy contained in the hydrogen pairs of fatty acids. On a weight basis, there are twice as many calories in a gram of fat as there are in a gram of carbohydrate. Notice in figure 6.14 that both carbohydrates and fat can enter the Krebs cycle and release energy. Although you require both fat and carbohydrates in your diet, they need not be in precise ratios; your body can make some interconversions. This means that people who eat excessive amounts of carbohydrates will deposit body fat. It also means that people who starve can generate glucose by breaking down fats and using the glycerol to synthesize glucose.

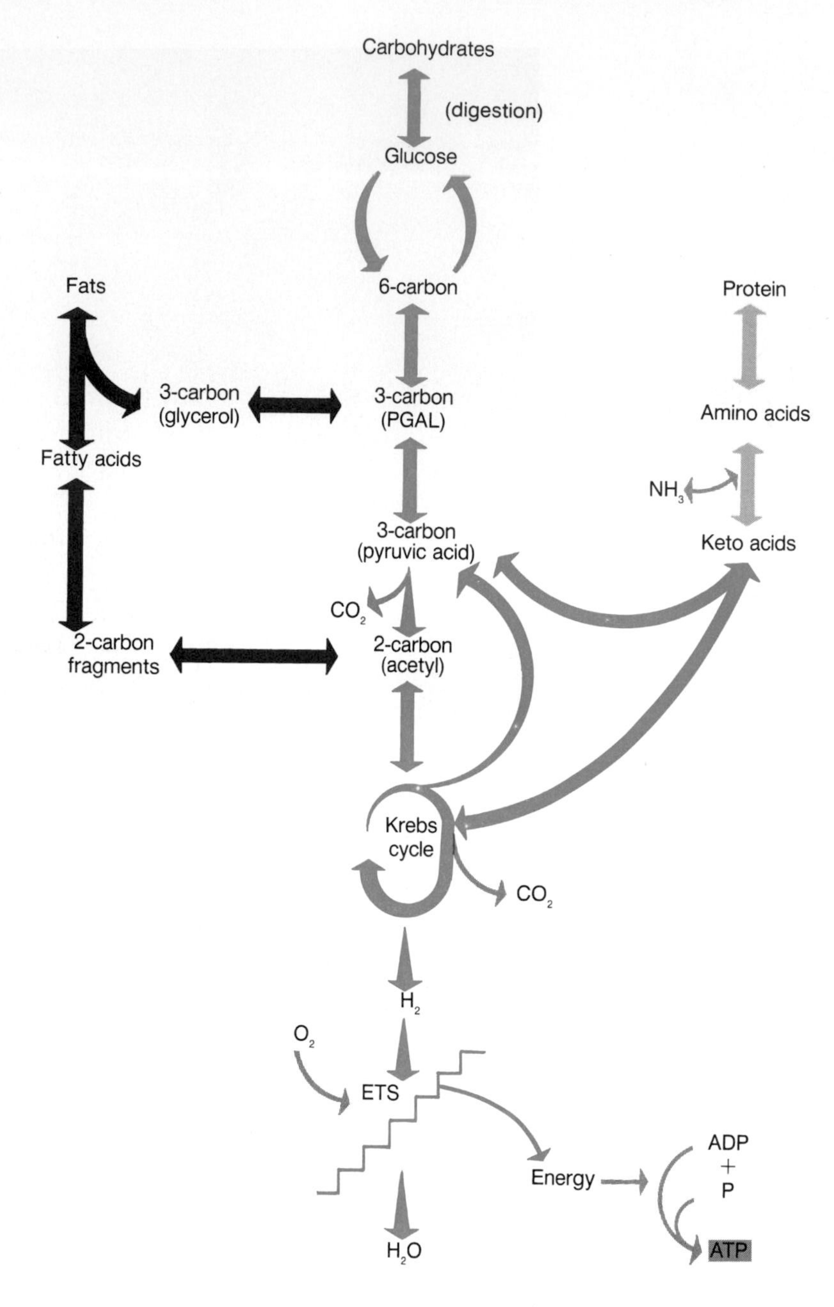

Figure 6.14 **The Interconversion of Fats, Carbohydrates, and Proteins.** Cells do not necessarily utilize all food as energy. One type of food can be changed into another type to be used as raw materials for construction of needed molecules or for storage.

Protein Respiration

Proteins can be interconverted just as fats and carbohydrates are. The first step in utilizing protein for energy is to digest the protein into individual amino acids. Each amino acid then needs to have the amino group ($-NH_2$) removed. The remaining carbon skeleton, a keto acid, is changed and enters the respiratory cycle as pyruvic acid or as one of the other types of molecules found in the Krebs cycle. These acids have hydrogens as part of their structure. As the acids progress through the Krebs cycle and the ETS, the hydrogens are removed and their energy is released. The amino group that was removed is converted into ammonia. Some organisms excrete ammonia directly, while others convert ammonia into other nitrogen-containing compounds, such as urea or uric acid. All of these molecules are toxic and must be eliminated. They are transported in the blood to the kidneys, where they are eliminated. In the case of a high-protein diet, you need to increase your fluid intake to allow the kidneys to efficiently remove the urea or uric acid.

When you eat any protein, you are able to digest it into its component amino acids. These amino acids are then available to be used to construct other proteins. If there is no need to construct protein, the amino acids are metabolized to provide energy, or they can be converted to fat for long-term storage. One of the most important concepts you need to recognize from this discussion is that carbohydrates, fats, and proteins can all be used to provide energy. The fate of any type of nutrient in a cell depends on the momentary needs of the cell.

An organism whose daily food-energy intake exceeds its daily energy expenditure will convert only the necessary amount of food into energy. The excess food will be interconverted according to the enzymes present and the needs of the organism at that time. In fact, glycolysis and the Krebs cycle allow molecules of the three major food types (carbohydrates, fats, and proteins) to be interconverted.

As long as a person's diet has a certain minimum of each of the three major types of molecules, the cell's metabolic machinery can interconvert molecules to satisfy its needs. If a person is on a starvation diet, the cells will use stored carbohydrates first. Once the carbohydrates are gone (about two days), cells will begin to metabolize stored fat. When the fat is gone (a few days to weeks), the proteins will be used. A person in this condition is likely to die.

If excess carbohydrates are eaten, they are often converted to other carbohydrates for storage or converted into fat. A diet that is excessive in fat results in the storage of fat. Proteins cannot be stored. If they or their component amino acids are not needed immediately, they will be converted into fat, carbohydrates, or energy. This presents a problem for those individuals who do not have ready access to a continuous source of amino acids (i.e., individuals on a low-protein diet). They must convert important cellular components into protein as they are needed. This is the reason why protein and amino acids are considered an important daily food requirement.

Plant Metabolism

At the beginning of this chapter we considered the conversion of carbon dioxide and water into PGAL through the process of photosynthesis. We described PGAL as a very important molecule because of its ability to be used as a source of energy. Plants and other autotrophs obtain energy from food molecules in the same manner as animals and other heterotrophs. They process the food through the respiratory pathways. This means that plants, like animals, require oxygen for the ETS portion of aerobic cellular respiration. Many people believe that plants only give off oxygen and never require it. This is incorrect! Plants do give off oxygen in the light-energy conversion stage of photosynthesis, but in aerobic cellular respiration they use oxygen just like any other organism. During their life span, green plants give off more oxygen to the atmosphere than they take in for use in respiration. The surplus they give off is the source of oxygen for aerobic cellular respiration in both plants and animals. Animals are not only dependent on plants for oxygen, but are ultimately dependent on plants for the organic molecules necessary to construct their bodies and maintain their metabolism (figure 6.15).

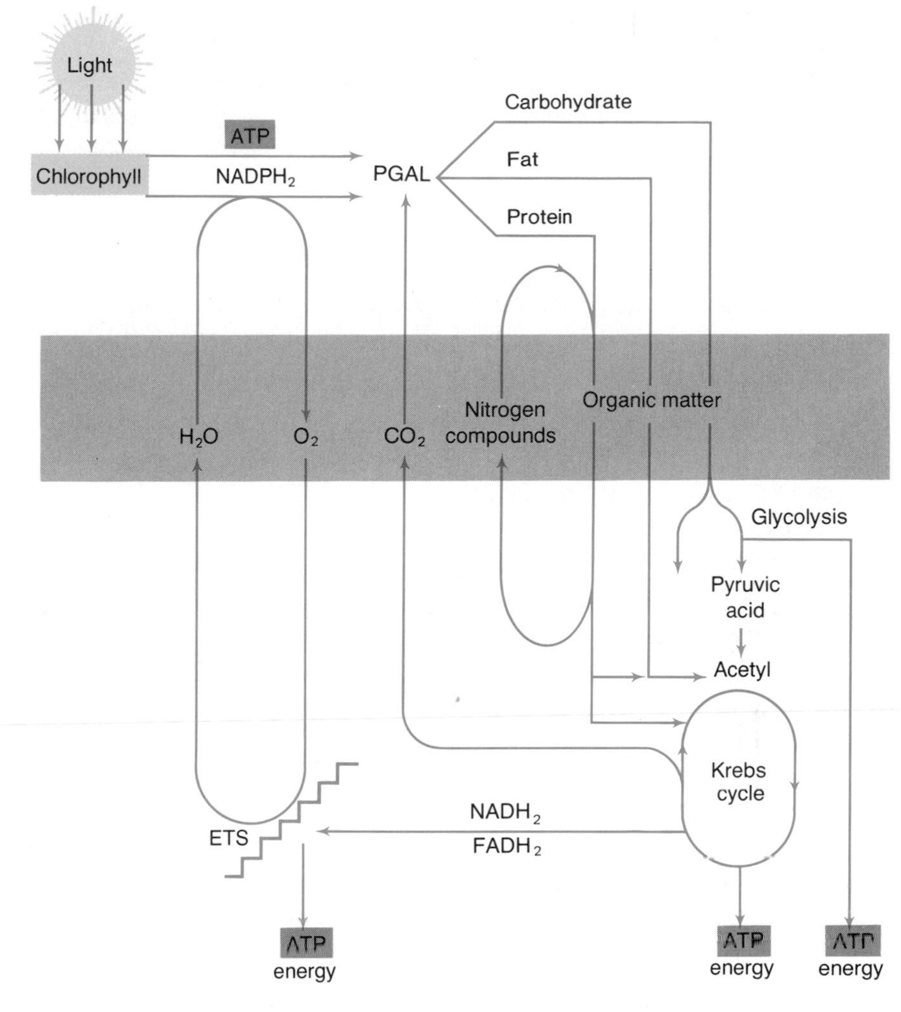

Figure 6.15 **The Interdependency of Photosynthesis and Respiration.** Plants use the end products of plant and animal respiration—carbon dioxide, water, and nitrogen compounds—to produce various foods. Plants and animals use the end products of plant photosynthesis—food and oxygen—as a source of energy. Therefore, plants are dependent upon animals and animals are dependent upon plants. Those materials that link the two processes together are seen in the colored bar.

By a series of reactions, plants produce the basic foods for animal life. To produce PGAL, which can be converted into carbohydrates, proteins, and fats, plants require carbn dioxide and water as raw materials. The carbon dioxide and water are available from the atmosphere, where they have been deposited as waste products of aerobic cellular respiration. To make the amino acids that are needed for proteins, plants require a source of nitrogen. This is available in the waste materials from animals.

Thus, animals supply raw materials—CO_2, H_2O, and nitrogen—needed by plants, while plants supply raw materials—sugar, oxygen, amino acids, fats, and vitamins—needed by animals. This constant cycling is essential to life on earth. As long as the sun shines and plants and animals remain in balance, the food cycles of all living organisms will continue to work properly.

• Summary •

In the light-energy conversion stage of photosynthesis, plants use chlorophyll to trap the energy of sunlight and manufacture a source of chemical energy, ATP, and a source of hydrogen, $NADPH_2$. Atmospheric oxygen is released in this stage. In the carbon dioxide conversion stage of photosynthesis, the ATP energy is used in a series of reactions to join the hydrogen from the $NADPH_2$ to a molecule of carbon dioxide and form a simple carbohydrate, PGAL.

In subsequent reactions, the plant uses the PGAL as a source of energy and raw materials to make complex carbohydrates, fats, and other organic molecules. With the addition of ammonia, the plant can form proteins.

In the process of respiration, organisms convert foods into energy (ATP) and waste materials (carbon dioxide, water, and nitrogen compounds). Organisms that have oxygen (O_2) available can employ the Krebs cycle and electron transfer system (ETS), which yield much more energy per sugar molecule than does fermentation; fermenters must rely entirely on glycolysis. Glycolysis and the Krebs cycle serve as a molecular interconversion system: fats, proteins, and carbohydrates are interconverted according to the needs of the cell.

The waste materials of respiration, in turn, are used by the plant. Therefore, there is a constant recycling of materials between plants and animals. Sunlight supplies the essential initial energy for making the large organic molecules necessary to maintain the forms of life we know.

• Thinking Critically •

Both plants and animals carry on metabolism. From a metabolic point of view, which are more complex? Include in your answer the following topics:

1. cell structure
2. biochemical pathways
3. enzymes
4. organic molecules
5. autotrophy and heterotrophy

• Experience This •

For this experience, you will need a large glass jar (with a lid) that you can fit your hand into. Place several cups of pond water with green "scum" into the jar. The green scum is a population of green plantlike organisms that are probably algae. Place the jar in a partially sunny spot. Watch it for the next several days. Note that inside the jar, bubbles develop when sunlight is available. You can collect the bubbles of gas in a test tube filled with water, inverted over the green algae. To determine what kind of gas has been released from the algae, remove the test tube from the water and immediately insert a "just-blown-out" wooden match into it. If the glowing match bursts into flame, the gas collected was oxygen. The release of oxygen by these organisms indicates that photosynthesis was taking place.

• Questions •

1. What is a biochemical pathway? Give two examples.
2. List four ways in which photosynthesis and aerobic respiration are similar.
3. Photosynthesis is a biochemical pathway that occurs in two stages. What are the two stages and how are they related to each other?
4. Why does aerobic respiration yield more energy than anaerobic respiration?
5. Even though animals do not photosynthesize, they rely on the sun for their energy. Why is this so?
6. Explain the importance of each of the following:
   NADP in photosynthesis
   PGAL in photosynthesis and in respiration
   oxygen in aerobic cellular respiration
   hydrogen acceptors in aerobic cellular respiration
7. In what way does ATP differ from other organic molecules?
8. Pyruvic acid can be converted into a variety of molecules. Name three.
9. Which cellular organelles are involved in the processes of photosynthesis and respiration?
10. Aerobic cellular respiration occurs in three stages. Name these and briefly describe what happens in each stage.

• Chapter Glossary •

acetyl (ă-sēt'l) The 2-carbon remainder of the carbon skeleton of pyruvic acid that is able to enter the mitochondrion.

adenosine triphosphate (ATP) (uh-den'o-sēn tri-fos'fāt) A molecule formed from the building blocks of adenine, ribose, and phosphates. It functions as the primary energy carrier in the cell.

aerobic cellular respiration (a-ro'bik sel'yu-lar res"pĭ-ra'shun) The biochemical pathway that requires oxygen and converts food, such as carbohydrates, to carbon dioxide and water. During this conversion, it releases the chemical-bond energy as ATP molecules.

alcoholic fermentation (al-ko-hol'ik fur"men-ta'shun) The anaerobic respiration pathway in yeast cells. During this process, pyruvic acid from glycolysis is converted to ethanol and carbon dioxide.

anaerobic respiration (an'uh-ro"bik res"pĭ-ra'shun) A biochemical pathway that does not require oxygen for the production of ATP.

autotrophs (aw'to-trōfs) Organisms that are able to make their food molecules from sunlight and inorganic raw materials.

biochemical pathway (bi'o-kem"ĭ-kal path'wa) A major series of enzyme-controlled reactions linked together.

carbon dioxide conversion stage (kar'bon di-ok'sīd kon-vur'zhun stāj) The second stage of photosynthesis, during which inorganic carbon from carbon dioxide becomes incorporated into a sugar molecule.

cellular respiration (sel'yu-lar res"pĭ-ra'shun) A major biochemical pathway along which cells release the chemical-bond energy from food and convert it into a usable form (ATP).

chemiosmosis (kem"e-os-mo'sis) The process of generating ATP as a result of creating a hydrogen-ion gradient across a membrane by using an electron-transport system.

chlorophyll (klo'ro-fil) A molecule directly involved in the light-trapping and energy-conversion process of photosynthesis.

coupled reactions (kup'ld re-ak'shuns) Reactions in which there is a linkage of a set of energy-requiring reactions with energy-releasing reactions.

electron-transfer system (ETS) (e-lek'tron trans'fur sis'tem) The series of oxidation–reduction reactions in aerobic cellular respiration in which the energy is removed from hydrogens and transferred to ATP.

FAD (**f**lavin **a**denine **d**inucleotide) A hydrogen carrier used in respiration.

glycolysis (gli-kol'ĭ-sis) The anaerobic first stage of cellular respiration, consisting of the enzymatic breakdown of a sugar into two molecules of pyruvic acid.

heterotrophs (hĕ'tur-o-trōfs) Organisms that require an external supply of food to provide a source of energy.

high-energy phosphate bond (hi en'ur-je fos'fāt bond) The bond between two phosphates in an ADP or ATP molecule that readily releases its energy for cellular processes.

Krebs cycle (krebs si'kl) The series of reactions in aerobic cellular respiration, resulting in the production of two carbon dioxides, the release of four pairs of hydrogens, and the formation of an ATP molecule.

light-energy conversion stage (līt en'ur-je kon-vur'zhun stāj) The first of the two stages of photosynthesis, during which light energy is converted to chemical-bond energy.

NAD (**n**icotinamide **a**denine **d**inucleotide) An electron acceptor and hydrogen carrier used in respiration.

NADP (**n**icotinamide **a**denine **d**inucleotide **p**hosphate) An electron acceptor and hydrogen carrier used in photosynthesis.

PGAL (**p**hospho**g**lycer**al**dehyde) The end product of the carbon dioxide conversion stage of photosynthesis produced when a molecule of carbon dioxide is incorporated into a larger organic molecule.

photosynthesis (fo-to-sin'thuh-sis) A major biochemical pathway in green plants, resulting in the manufacture of food molecules.

pyruvic acid (pi-ru'vik as'id) A 3-carbon carbohydrate that is the end product of the process of glycolysis.

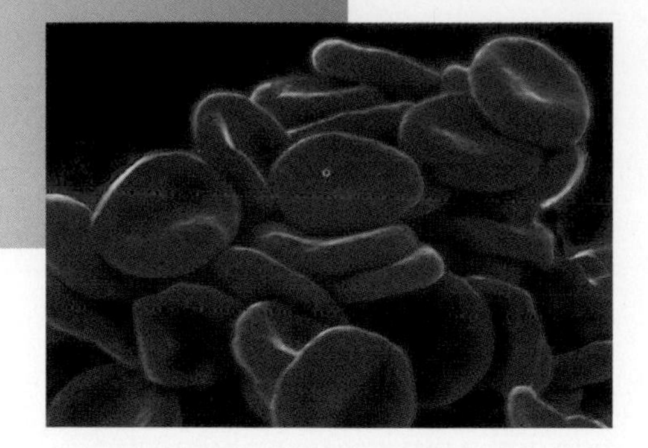

C H A P T E R 7

DNA-RNA: The Molecular Basis of Heredity

Chapter Outline

Purpose

In previous chapters we have considered a variety of biological structures and their functions. Organic molecules found in living cells are not haphazard arrangements of atoms; they are highly organized and can be classified into major groups. The group known as the *nucleic acids* has a unique structure and is the primary control molecule of the cell. This chapter considers how the structure of these complex molecules is converted into actions by living cells.

For Your Information

Molecular biologists are revolutionizing archaeology and paleontology with new ways of identifying the genetic material, DNA, found in specimens of plants, animals, and microorganisms from previous geological periods. These methods enable researchers to better understand the evolutionary relationships between long-extinct species and their rates of evolution. They also provide clues as to the racial makeup and the kinship, distribution, and migration patterns of our ancient ancestors. The old picture of archaeologists digging for dinosaur bones is giving way to a new image. We now see them performing DNA analysis of the genetic "remains" of preserved fossils found in the field and in museums. Ancient DNA has been successfully copied and analyzed by molecular paleontologists from cells taken from the fossil Quagga, an animal that resembles a cross between a horse and a zebra, extinct for more than 100 years; a 2,400-year-old Egyptian mummy; and magnolia leaves, 18 million years old.

Learning Objectives

- Recognize the structure of DNA and RNA.
- Distinguish between DNA, nucleoprotein, chromatin, and chromosome.
- Diagram the DNA replication process.
- Diagram the DNA transcription process.
- Diagram the process of translation.
- Give examples of mutagenic agents and how they might affect DNA.
- Describe the processes involved in recombinant DNA procedures.

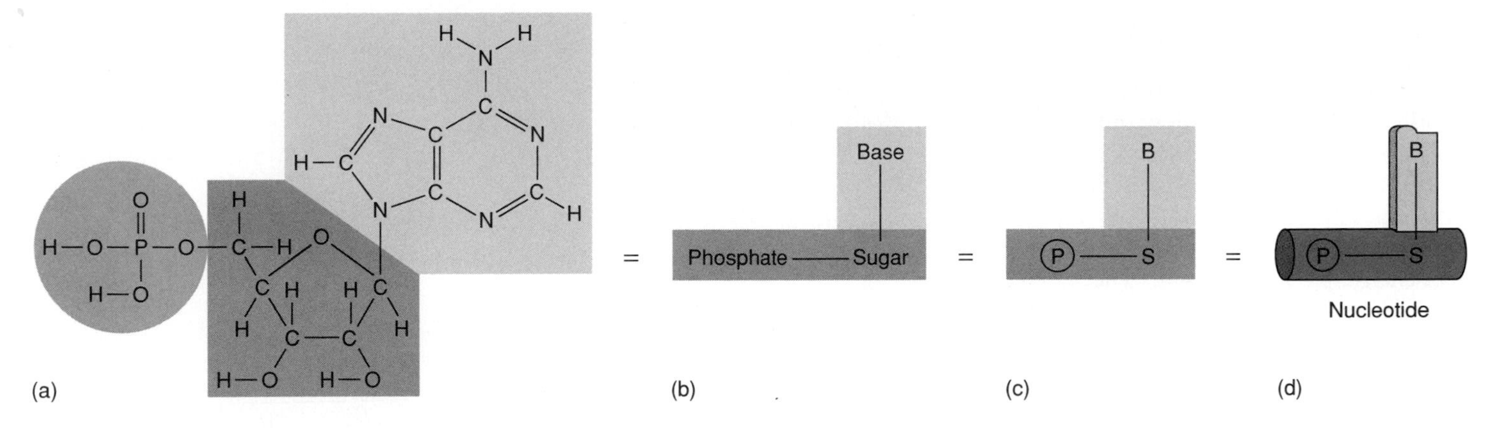

Figure 7.1 **Nucleotide Structure.** (a) All nucleotides are constructed in the basic way shown in part (a). The nucleotide is the basic structural unit of all nucleic acid molecules. Notice in part (c) that the phosphate group is written in "shorthand" form as a *P* inside a circle. Part (d) is a stylized version of a nucleotide and will be used throughout the chapter. Remember that this style is only representative of the kind of complex organic molecule shown in (a).

The Structure of DNA and RNA

In the nucleus of a eukaryotic cell is a very important library of molecular information. This library contains all the directions for making the structural and regulatory proteins required for life processes. It is like a library of how-to books that does not allow the books to circulate. It is a reference library only. You can copy information from the books and remove the copy, but the original always stays in the library (the nucleus). It may seem impossible that the directions for growth and development are stored in a place as small as the nucleus of a cell. The secret lies in the language these directions are written in. This language is deoxyribonucleic acid (DNA). DNA has four properties that enable it to function as genetic material. It is able to (1) *replicate* by directing the manufacture of copies of itself; (2) *mutate,* or chemically change, and transmit these changes to future generations; (3) *store* information that determines the characteristics of cells and organisms; and (4) use this information to *direct* the synthesis of structural and regulatory proteins essential to the operation of the cell or organism.

Like the other groups of organic macromolecules, the **nucleic acids** are made up of subunits. The subunits of nucleic acids are called **nucleotides.** Each nucleotide is composed of a *sugar* molecule (S) containing five carbon atoms, a **phosphate group** (P), and a kind of molecule called a **nitrogenous base** (B) (figure 7.1).

There are eight common types of nucleotides available in a cell for building nucleic acids. Nucleotides differ from one another in the kind of sugar and nitrogenous base they contain. Because of these differences, it is possible to classify nucleic acids into two main groups: ribonucleic acid (RNA) and deoxyribonucleic acid (DNA). The name of each tells you about the structure of the molecules. For example, the prefix *ribo-* in RNA tells you that the sugar part of this nucleic acid is **ribose.** Similarly, DNA contains a ribose sugar that has been deoxygenated (lost an oxygen atom) and is called **deoxyribose** (figure 7.2a). The nucleotide units contain nitrogenous bases of two sizes. The larger nitrogenous bases are **adenine** (A) and **guanine** (G), which differ in the kinds of atoms attached to their double-ring structure (figure 7.2b). The smaller nitrogenous bases are **cytosine** (C), **thymine** (T), and **uracil** (U). Each of these differs from the others in the kinds of atoms attached to its single-ring structure (figure 7.2c). These differences in size are important, as you will see later.

DNA is a nucleic acid that functions as the original blueprint for the synthesis of polypeptides. It contains deoxyribose sugar, phosphates, and the nitrogenous bases adenine, thymine, guanine, and cytosine. **RNA** is a type of nucleic acid that is directly involved in the synthesis of polypeptides at ribosomes. It contains ribose sugar, phosphates, and the four nitrogenous bases adenine, guanine, cytosine, and uracil (no thymine in RNA). The construction of a nucleotide involves the bonding of a nitrogenous base to a 5-carbon sugar. For example, when adenine is chemically bonded to ribose sugar, the result is a molecule called *adenosine.* When a phosphate is added to the ribose of the adenosine, a new molecule called *adenosine monophosphate* (AMP) is formed. This is a complete nucleotide. Monophosphate nucleotides are energetically stable, and they are the building blocks of RNA. However, it is possible to increase the potential energy of a monophosphate nucleotide by adding a second phosphate group to form a *diphosphate* nucleotide. For example, when a second phosphate is bonded to AMP, it becomes *adenosine diphosphate* (ADP).

This molecule contains potential energy in the bond formed between the two phosphate groups. Because this covalent bond contains chemical-bond energy, it is easily made available to power many types of chemical reactions in the cell. This special covalent bond is represented by a curved line and called a *high-energy phosphate bond.* The maximum amount of potential chemical-bond energy is packed into nucleotides when a third and final phosphate group is added to diphosphates to form *triphosphate* nucleotides. ADP becomes *adenosine triphosphate* (*ATP*), the most widely used source of chemical-bond energy in all cells.

Chapter 6 dealt with the function of ATP as a source of chemical-bond energy. Keep in mind that all nucleotides exist in mono-, di-, and triphosphate forms (figure 7.3). As in ATP, the other triphosphates contain high-energy phosphate bonds used in the dehydration synthesis reaction that binds nucleotides together as RNA or DNA. The result of this nucleic acid synthesis is a polymer that can be compared to a comb. The protruding "teeth" (different nitrogenous bases) are connected to a common "backbone" (sugar and monophosphate molecules). This is the basic structure of both RNA and DNA (figure 7.4).

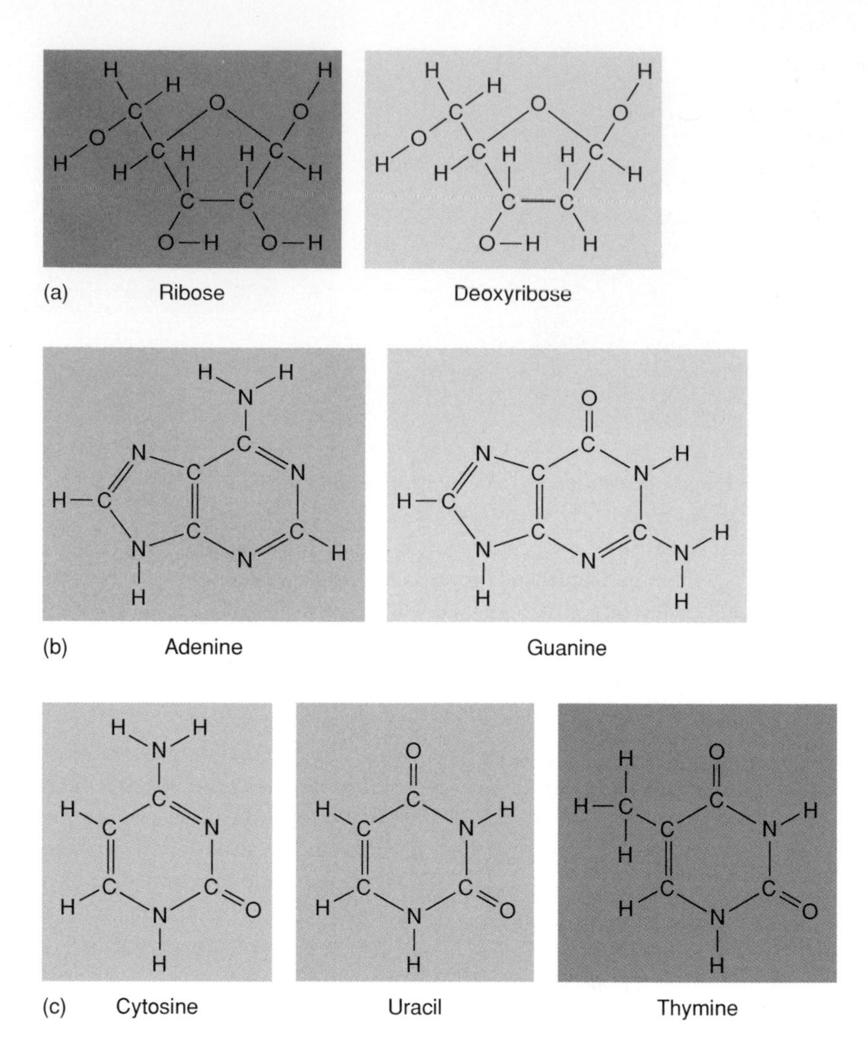

Figure 7.2 **The Building Blocks of Nucleic Acids.** All nucleic acids are composed of two organic components: a 5-carbon sugar molecule and a nitrogenous base. Notice the difference (highlighted in color) between the two sugar molecules in part (a). The nitrogenous bases are divided into two groups according to their size. The large purines—adenine and guanine molecules—differ from each other in their attached groups (in color), as do the three smaller pyrimidine nitrogenous bases—cytosine, thymine, and uracil—in (c). The two types of nucleic acids—DNA and RNA—are composed of these eight building blocks. Note that each building block has a sugar, base, and phosphate component. These nucleotides are color coded throughout the chapter so that you can recognize the difference between DNA and RNA.

The Molecular Structure of DNA

DNA is actually a double molecule. It consists of two comblike strands held together between their protruding bases by hydrogen bonds. The two strands are twisted about each other in a double helix known as **duplex DNA** (figure 7.5). This ladderlike molecule has as its "rungs" the hydrogen-bonded base pairs. The four kinds of bases always pair in a definite way: adenine (A) with thymine (T), and guanine (G) with cytosine (C). The bases that pair are said to be **complementary bases.** Notice that the large bases (A and G) pair with the small ones (T and C), thus keeping the backbones of two complementary strands parallel. Three hydrogen bonds are formed between guanine and cytosine:

G: : :C

and two between adenine and thymine:

A: : :T

You can "write" a message in the form of a stable DNA molecule by combining the four different DNA nucleotides (A, T, G, C) in particular sequences. Notice in figure 7.4 that it is possible to make sense out of the sequence of nitrogenous bases. If you "read" them from left to right in groups of three, you can read three words—CAT, ACT, and TAG. In this case, the four DNA nucleotides are being used as an alphabet to construct three-letter words. In order to make sense out of such a code, it is necessary to read in a consistent direction. Reading the sequence

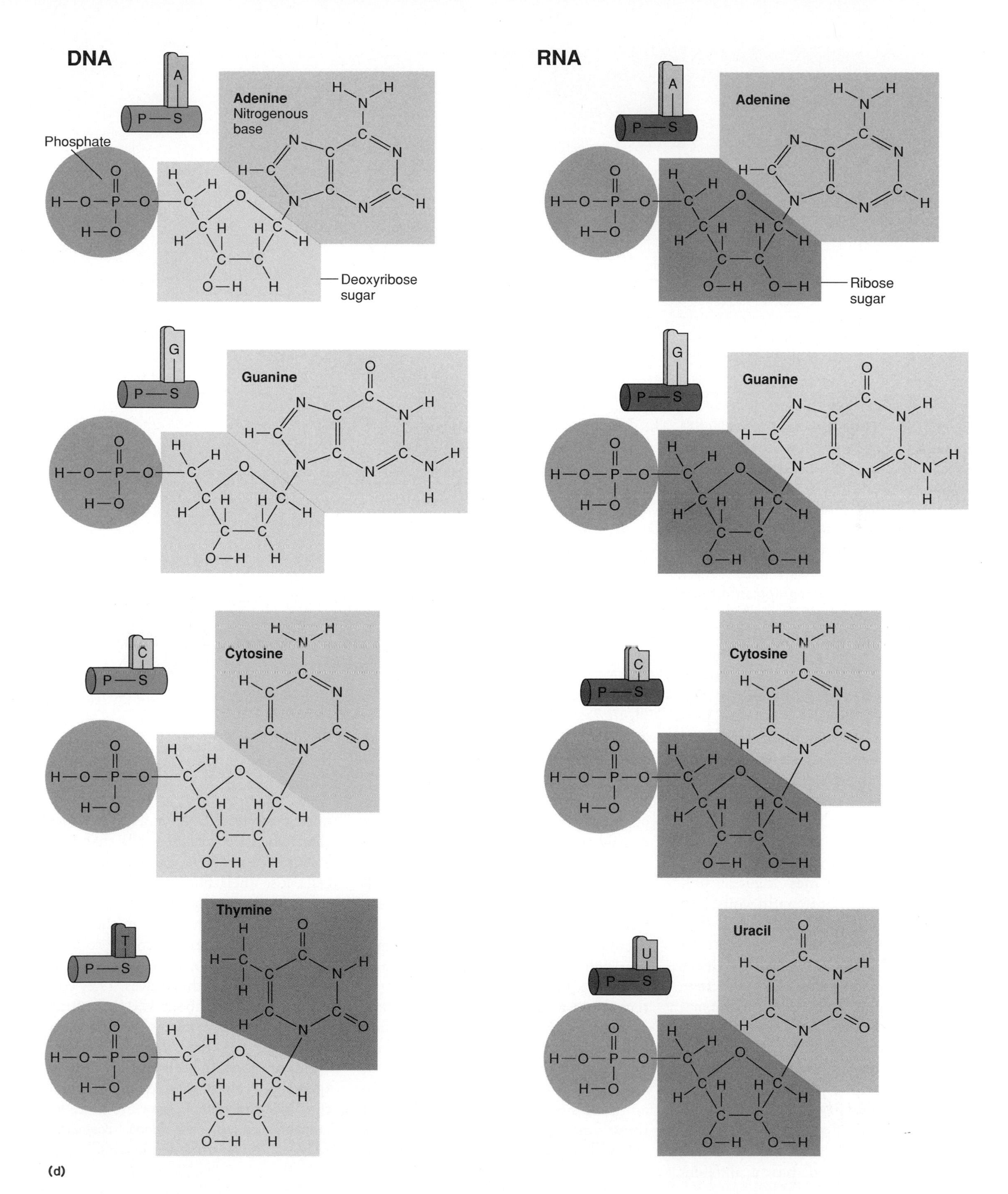

Figure 7.2 *(continued)*

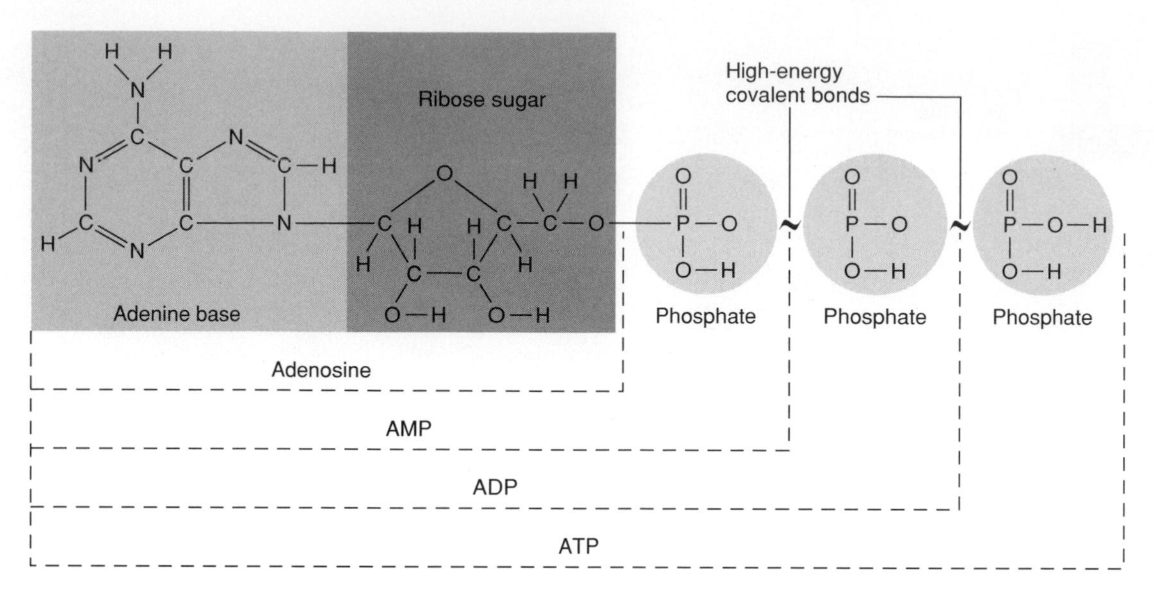

Figure 7.3 **Three Forms of a Nucleotide.** The three structural formulas pictured here illustrate the mono-, di-, and triphosphorylated forms of the adenosine nucleotide. Notice that the last two phosphate groups are attached by special high-energy phosphate bonds. It is this bond energy that is used to energize cellular reactions.

in reverse does not always make sense, just as reading this paragraph in reverse would not make sense.

There are two different forms of DNA. The genetic material of eukaryotes is duplex DNA, but it has histone proteins attached along its length. The duplex DNA strands with attached proteins are called **nucleoproteins,** or **chromatin fibers.** Histone and DNA are not arranged haphazardly, but come together in a highly organized pattern. The duplex DNA spirals around repeating clusters of eight histone spheres. Histone clusters with their encircling DNA are called **nucleosomes** (figure 7.6a). When eukaryotic chromatin coils into condensed, highly knotted bodies, they are seen easily through a microscope after staining with dye. Condensed like this, a chromatin fiber is referred to as a **chromosome** (figure 7.6b). The genetic material in prokaryotic cells is also duplex DNA, with the ends of the polymer connected to form a loop (figure 7.6c). While prokaryotic cells do not have histone protein, they do have an attached protein called *HU protein.*

Each chromatin strand is different because each strand has a different chemical code. Coded DNA serves as a central cell library. Tens of thousands of messages are in this storehouse of information. This information tells the cell such things as (1) how to produce enzymes required for the ingestion of nutrients, (2) how to manufacture enzymes that will metabolize the nutrients and eliminate harmful wastes, (3) how to repair and assemble cell parts, (4) how to reproduce healthy offspring, (5) when and how to react to favorable and unfavorable changes in the environment, and (6) how to coordinate and regulate all of life's essential functions. If any of these functions are not performed, the cell will die. The importance of maintaining essential DNA in a cell becomes clear when we consider cells that have lost it. For example, human red blood cells lose their nuclei as they become specialized for carrying oxygen and carbon dioxide throughout the body. Without DNA they are unable to manufacture the essential cell components needed to sustain themselves. They continue to exist for about 120 days, functioning only on enzymes manufactured earlier in their lives. When these enzymes are gone, the cells die. Since these specialized cells begin to die the moment they lose their DNA, they are more accurately called *red blood corpuscles* (*RBCs*): "little dying red bodies."

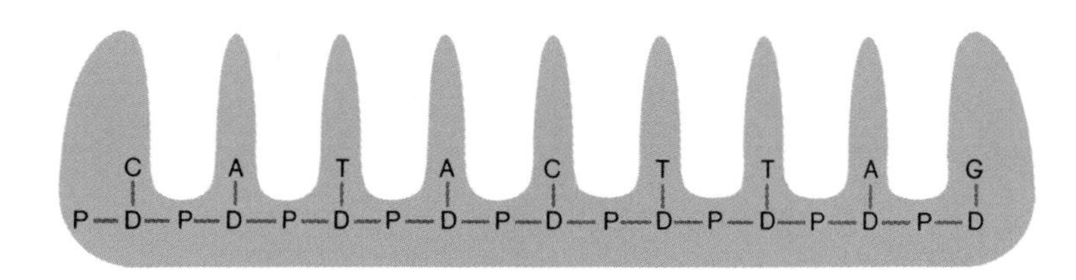

Figure 7.4 **A Single Strand of DNA.** A single strand of DNA resembles a comb. The molecule is much longer than pictured here and is composed of a sequence of linked nucleotides.

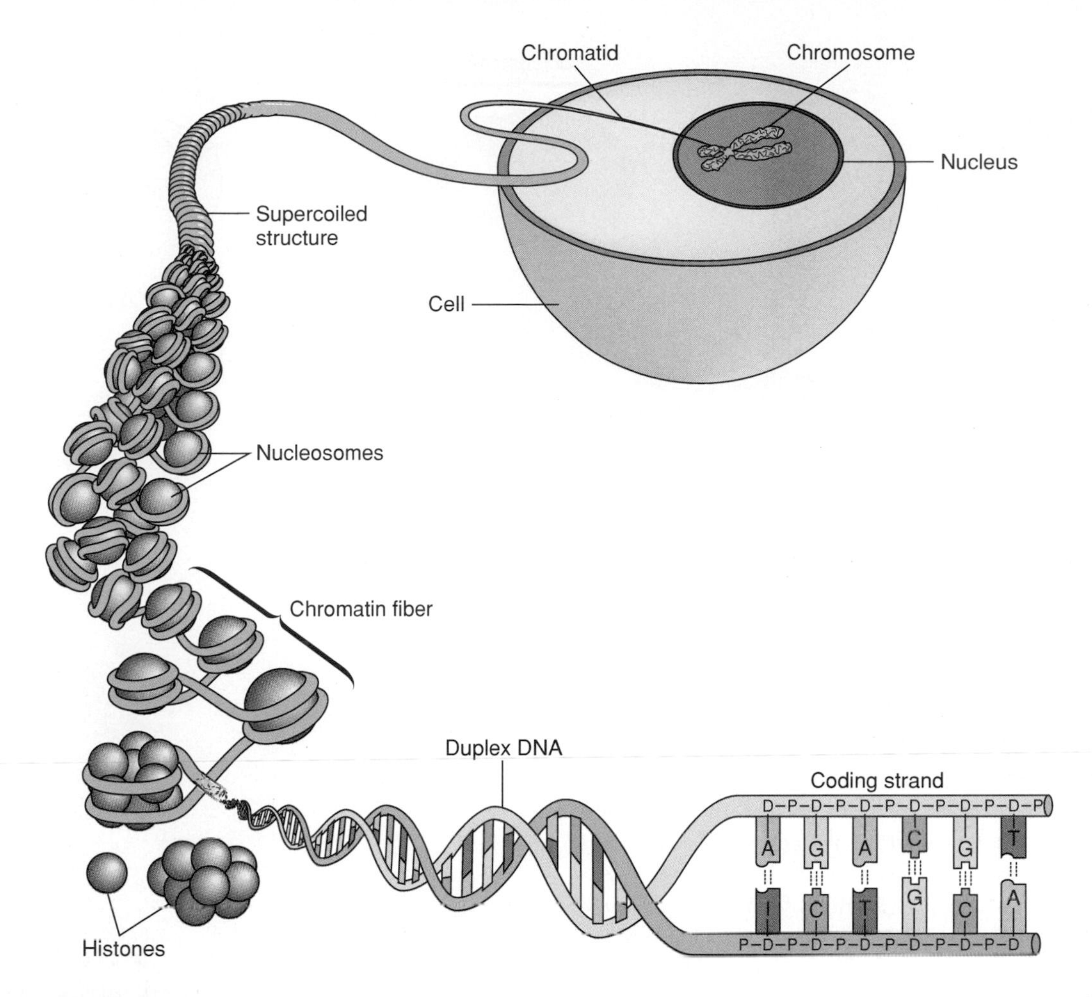

Figure 7.5 **Duplex DNA.** Eukaryotic cells contain duplex DNA in their nuclei, which takes the form of a three-dimensional helix. One strand is a chemical code (the coding strand) that contains the information necessary to control and coordinate the activities of the cell. The two strands are bound together by weak hydrogen bonds formed between the protruding nitrogenous bases according to the base-pairing rule: A pairs with T and C pairs with G. The length of a DNA molecule is measured in numbers of "base pairs"—the number of rungs on the ladder.

DNA Replication

Since all cells must maintain a complete set of genetic material, there must be a doubling of DNA in order to have enough to pass on to the offspring. **DNA replication** is the process of duplicating the genetic material prior to its distribution to daughter cells. When a cell divides into two daughter cells, each new cell must receive a complete copy of the parent cell's genetic information, or it will not function long. Accuracy of duplication is essential in order to guarantee the continued existence of that type of cell. Should the daughters not receive exact copies, they may be unable to manufacture the structural and regulatory proteins essential for their survival.

The DNA replication process requires many enzymes. It begins when an enzyme breaks the hydrogen bonds between the bases of the two strands of DNA. In eukaryotic cells, this occurs in hundreds of different "forks" along the length of the DNA. Moving along the DNA, the enzyme "unzips" the halves of the duplex DNA. Proceeding in opposite directions on each side, enzymes known as **DNA polymerase** proceed down the length of the DNA, bonding new DNA nucleotides into position. In addition, DNA polymerase, which speeds the addition of new nucleotides to the growing chain, works along with another enzyme called *exonuclease* to make sure that no mistakes are made. Mismatched nucleotides are either prevented from being bonded into the wrong position or misplaced nucleotides are cut out and replaced with the proper unit. Replication proceeds in both directions from every fork, appearing as replication "bubbles" (figure 7.7). The complementary bases (A: : :T, G: : :C) pair with the exposed nitrogenous bases of both DNA strands by forming new hydrogen bonds. Once properly aligned, a covalent bond is

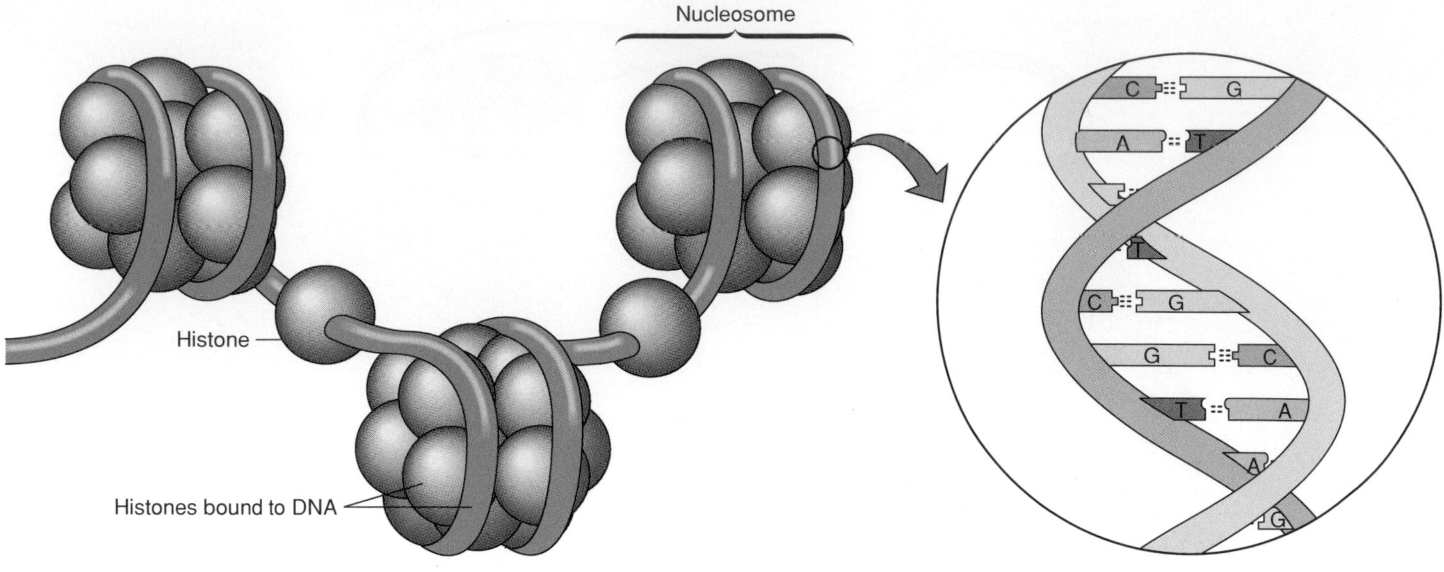

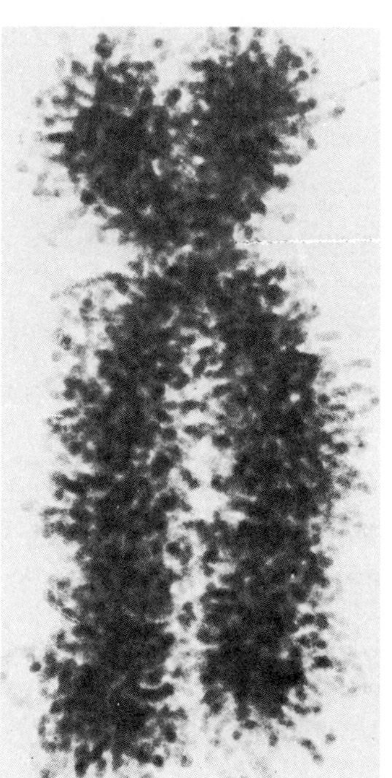

Figure 7.6 **The Many Forms of Deoxyribonucleic Acid, DNA.** The term *nucleoprotein* is used to describe the combination of DNA and its associated protein in eukaryotic cells. When these giant nucleoprotein molecules are found loose inside of a cell's nucleus, they are called chromatin. (a) Upon close examination of a portion of a eukaryotic nucleoprotein, it is possible to see how the DNA and protein are arranged. The protein component, histone, is found along the DNA in globular masses; they may be individual histones or arranged in groups, the nucleosomes. (b) During certain stages in the reproduction of a eukaryotic cell, the nucleoprotein coils and "supercoils," forming tightly bound masses. When stained, these are easily seen through the microscope. In their supercoiled form, they are called chromosomes, meaning colored bodies. (c) The nucleic acid of prokaryotic cells (the bacteria) does not have histone protein; rather, it has proteins called HU protein. In addition, the ends of the giant nucleoprotein molecule overlap and bind with one another to form a loop.

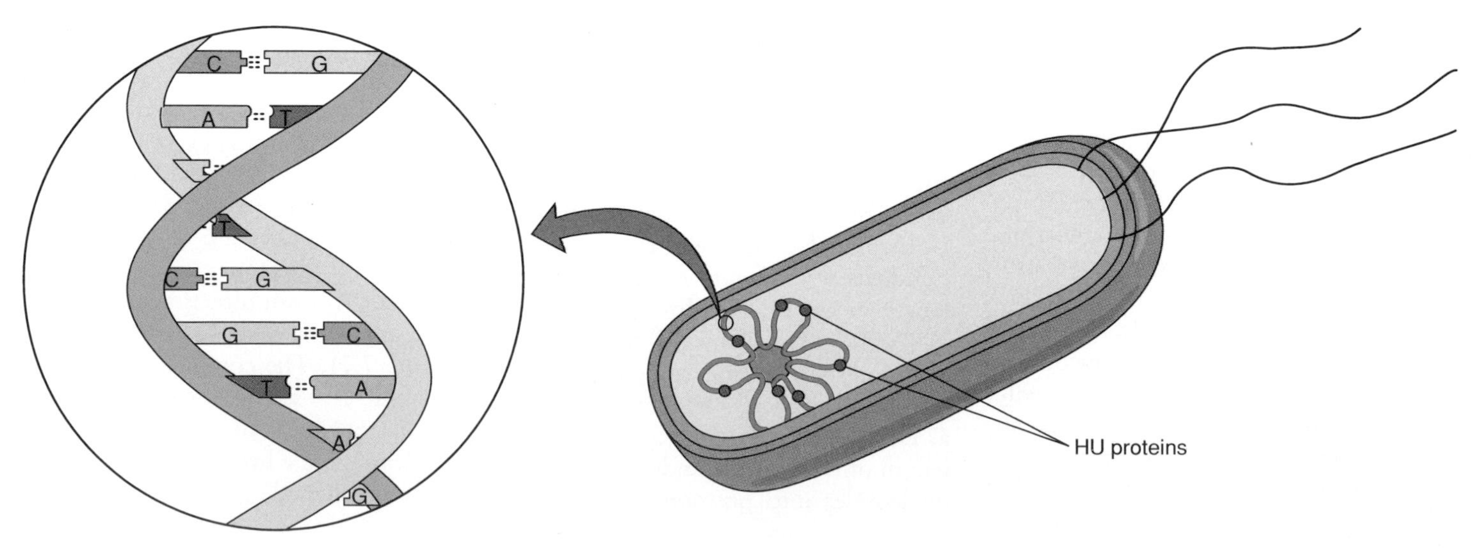

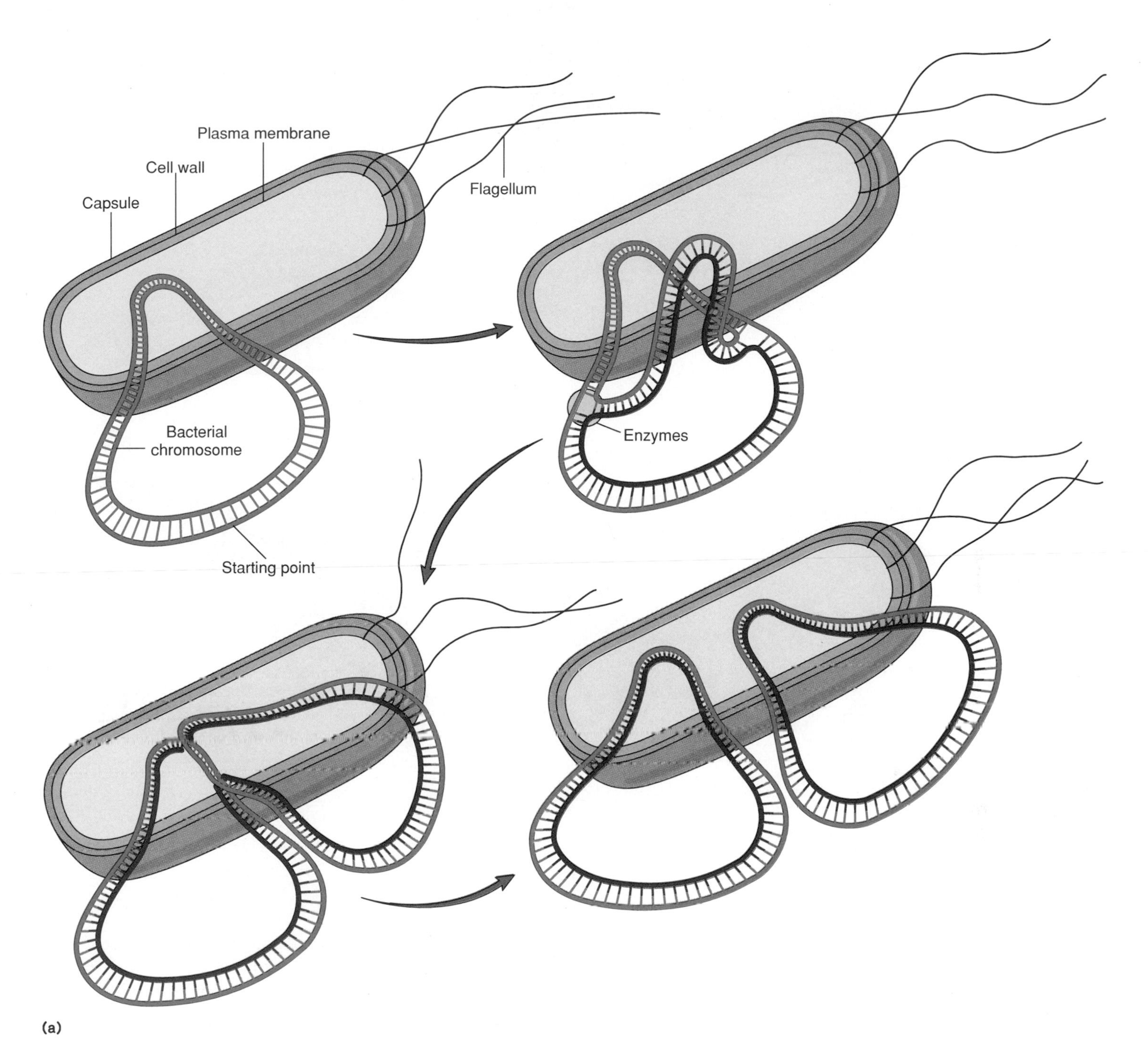

Figure 7.7 **DNA Replication.** (a) The illustration on this page and the following page summarize the basic events that occur during the replication of duplex DNA. In prokaryotic cells "unzipping" enzymes attach at only one place. As these enzymes move in both directions down the duplex DNA, new complementary DNA nucleotides are base-paired on the exposed strands and linked together by other enzymes (DNA polymerase and exonuclease), forming new strands that are identical to the originals and appear as ever-enlarging double loops.

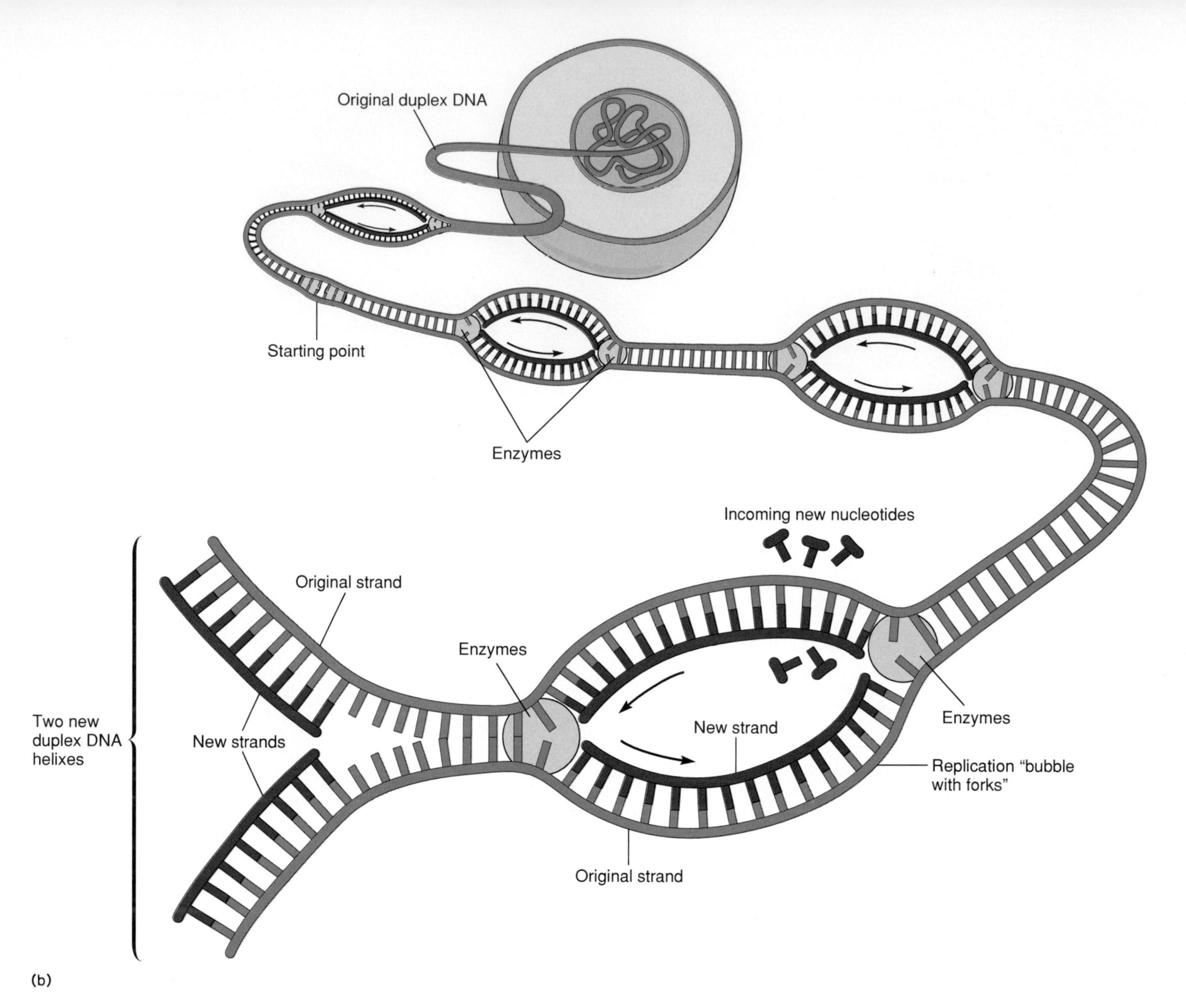

Figure 7.7 (continued) **DNA Replication.** (b) In eukaryotic cells the "unzipping" enzymes attach to the DNA at numerous points, breaking the hydrogen bonds that bind the complementary strands. As the DNA replicates, numerous replication "bubbles" or "forks" appear along the length of the duplex DNA. Eventually all the forks come together, completing the replication process.

formed between the sugars and phosphates of the newly positioned nucleotides using DNA polymerase. A strong sugar and phosphate backbone is formed in the process. This process continues until all the replication "bubbles" join. In prokaryotic cells, replication of the circular DNA molecule begins at one point and proceeds in opposite directions until the circle is completely replicated.

A new complementary strand of DNA forms on each of the old DNA strands, resulting in the formation of two double-stranded duplex DNA molecules. In this way, the exposed nitrogenous bases of the original DNA serve as a **template,** or pattern, for the formation of the new DNA. As the new DNA is completed, it twists into its double-helix shape.

The completion of the process yields two double helices that are identical in their nucleotide sequences. The DNA replication process is highly accurate. It has been estimated that there is only one error made for every 2×10^9 nucleotides. A human cell contains 46 chromosomes consisting of about 5,000,000,000 (5 billion) base pairs. This averages to about five errors per cell! Don't forget that this figure is an estimate, and while some cells may have five errors per replication, others may have more, and some may have no errors at all. It is also important to note that some errors may be major and deadly, while others are insignificant. Since this error rate is so small, DNA replication is considered by most to be essentially error-free. Following DNA replication, the cell now contains twice the amount of

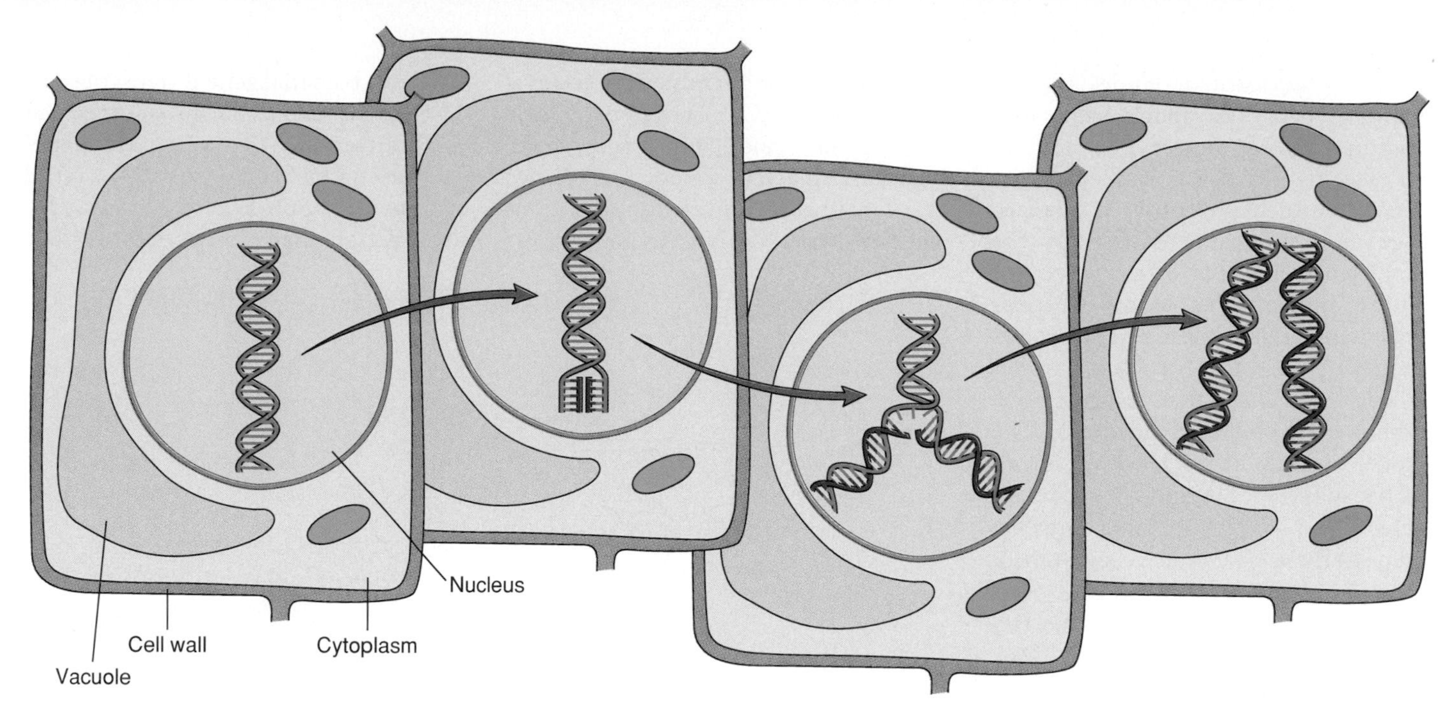

Figure 7.8 **The Process of DNA Replication.** These are the generalized events in the nucleus of a eukaryotic cell during the process of DNA replication. Notice that the final cell has two double helices; they are identical to each other and identical to the original duplex strands.

genetic information and is ready to begin the process of distributing one set of genetic information to each of its two daughter cells.

The distribution of DNA involves splitting the cell and distributing a set of genetic information to the two new daughter cells. In this way, each new cell has the necessary information to control its activities. The mother cell ceases to exist when it divides its contents between the two smaller daughter cells (figure 7.8).

A cell does not really die when it reproduces itself; it merely starts over again. This is called the *life cycle* of a cell. A cell may divide and redistribute its genetic information to the next generation in a number of ways. These processes will be dealt with in detail in chapters 8 and 9.

DNA Transcription

As noted earlier, DNA functions in the manner of a reference library that does not allow its books to circulate. Information from the originals must be copied. The second major function of DNA is to make a single-stranded, complementary RNA copy of DNA. This operation is called **transcription,** which means to transfer data from one form to another. In this case, the data is copied from DNA language to RNA language. The same base-pairing rules that control the accuracy of DNA replication apply to the process of transcription. Using this process, the genetic information stored as a DNA chemical code is carried in the form of an RNA copy to other parts of the cell. It is RNA that is used to guide the assembly of amino acids into structural and regulatory molecules, such as polypeptides and enzymes. Without the process of transcription, genetic information would be useless in directing cell functions. Although many types of RNA are synthesized from the genes, the three most important are messenger RNA (mRNA), transfer RNA (tRNA), and ribosomal RNA (rRNA).

Transcription begins in a way that is similar to DNA replication. The duplex DNA is separated by an enzyme, exposing the nitrogenous-base sequences of the two strands. However, unlike DNA replication, transcription only occurs on one of the two DNA strands, which serves as a template, or pattern, for the synthesis of RNA. But which strand is copied? Where does it start and when does it stop? Where along the sequence of thousands of nitrogenous bases does the chemical code for the manufacture of a particular enzyme begin and where does it end? If transcription begins randomly, the resulting RNA may not be an accurate copy of the code, and the enzyme product may be useless or deadly to the cell. To answer these questions, it is necessary to explore the nature of the genetic code itself.

We know that genetic information is in chemical-code form in the DNA molecule. When the coded information

DNA → { mRNA, tRNA, rRNA } → Protein → Structural protein / Regulatory proteins
DNA → Other types of RNA → Various biochemical reactions

is used or *expressed,* it guides the assembly of particular amino acids into structural and regulatory polypeptides and proteins. If DNA is molecular language, then each nucleotide in this language can be thought of as a letter within a four-letter alphabet. Each word, or code, is always three letters (nucleotides) long, and only three-letter words can be written. A **DNA code** is a triplet nucleotide sequence that codes for one of the twenty common amino acids. The number of codes in this language is limited because there are only four different nucleotides, which are used only in groups of three. The order of these three letters is just as important in DNA language as it is in our language. We recognize that CAT is not the same as TAC. If all the possible three-letter codes were written using only the four DNA nucleotides for letters, there would be a total of sixty-four combinations. When codes are found at a particular place along a coding strand of DNA, and the sequence has meaning, the sequence is called a **gene.** "Meaning" in this case refers to the fact that the gene can be transcribed into an RNA molecule, which in turn may control the assembly of individual amino acids into a polypeptide.

Prokaryotic Transcription

All the genes in prokaryotic cells are attached end-to-end forming the typical bacterial loop of DNA. Each bacterial gene is made of attached nucleotides that are transcribed in order into a single strand of RNA. This RNA molecule is used to direct the assembly of a specific sequence of amino acids to form a polypeptide. This system follows the pattern of:

one DNA → one RNA → one
gene polypeptide

The beginning of each gene on a DNA strand is identified by the presence of a region known as the **promoter,** just ahead of an **initiation code** that has the base sequence TAC. The gene ends with a terminator region, just in back of one of three possible **termination codes**—ATT, ATC, or ACT. These are the "start reading here" and "stop reading here" signals. The actual genetic information is located between initiation and termination codes:

promoter: :initiator code: : : : :gene: : : : :terminator code: :terminator region

When a bacterial gene is transcribed into RNA, the duplex DNA is "unzipped," and an enzyme known as **RNA polymerase** attaches to the DNA at the promoter region (figure 7.9). It is from this region that the enzymes will begin

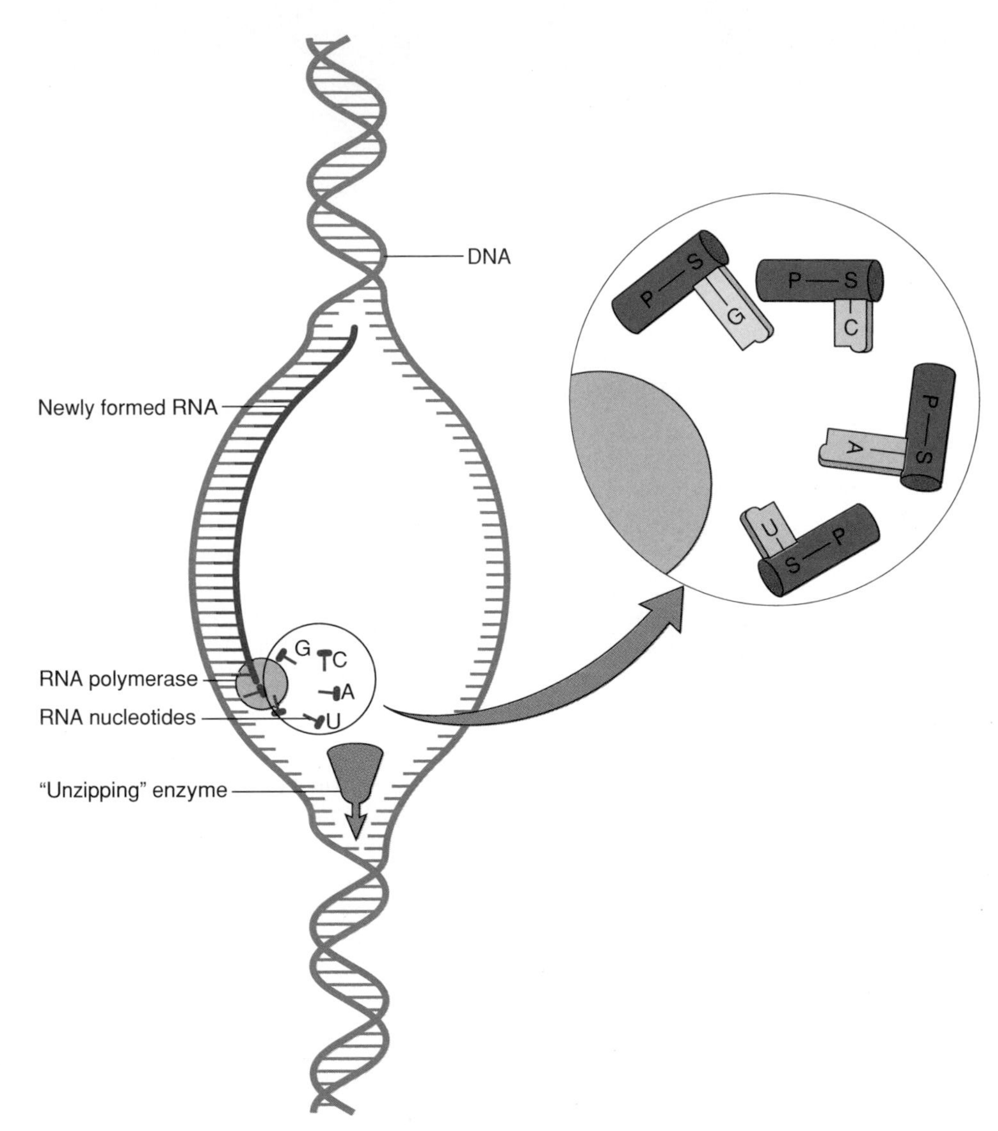

Figure 7.9 **Transcription of mRNA in Prokaryotic Cells.** This is a summary that illustrates the basic events that occur during the transcription of one side (the coding strand) of duplex DNA. The "unzipping" enzyme attaches to the DNA at a point that allows it to break the hydrogen bonds that bind the complementary strands. As this enzyme moves down the duplex DNA, new complementary RNA nucleotides are base-paired on one of the exposed strands and linked together by another enzyme (RNA polymerase), forming a new strand that is complementary to the nucleotide sequence of the DNA. The newly formed RNA is then separated from its DNA complement. Depending on the DNA segment that has been transcribed, this RNA molecule may be a messenger RNA (mRNA), a transfer RNA (tRNA), a ribosomal RNA (rRNA), or an RNA molecule used for other purposes within the cell.

to assemble RNA nucleotides into a complete, single-stranded copy of the gene, including initiation and termination codes. Triplet RNA nucleotide sequences complementary to DNA codes are called **codons.** Remember that there is no thymine in RNA molecules; it is replaced with uracil. Therefore, the initiation code in DNA (TAC) would be base-paired by RNA polymerase to form the RNA codon AUG. When transcription is complete, the newly assembled RNA is separated from its DNA template and made available for use in the cell; the DNA recoils into its original double-helix form.

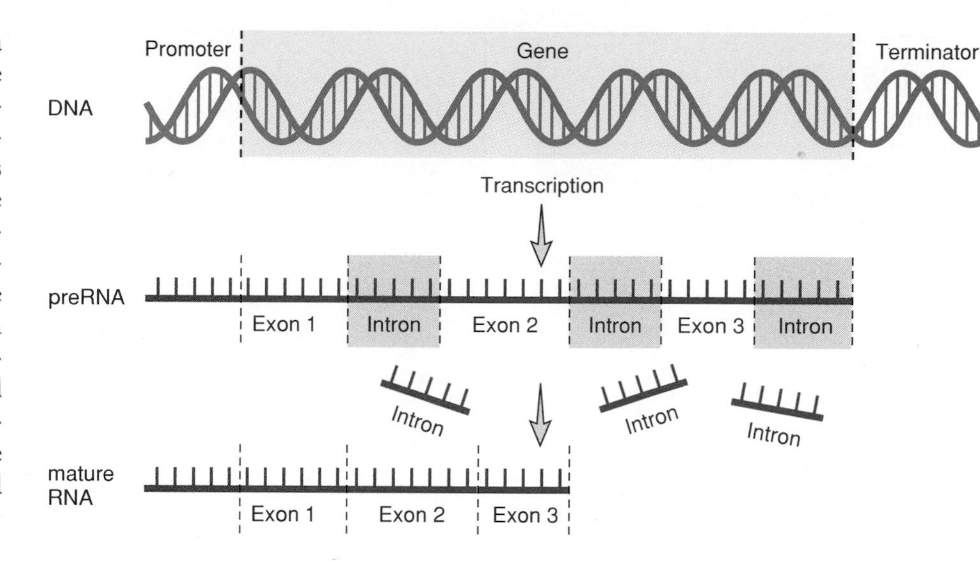

Figure 7.10 **Transcription of mRNA in Eukaryotic Cells.** This is a summary of the events that occur in the nucleus during the manufacturing of mRNA in a eukaryotic cell. Notice that the original nucleotide sequence is first transcribed into an RNA molecule that is later "clipped" with "snurp" molecules and then rebonded to form a shorter version of the original. It is during this time that the introns are removed.

Eukaryotic Transcription

The transcription system is different in eukaryotic cells. A eukaryotic gene begins with a promoter region and an initiation code and ends with a termination code and region. However, the intervening gene sequence contains patches of nucleotides that have no meaning. If they were used in protein synthesis, the resulting proteins would be worthless. To remedy this problem, eukaryotic cells prune these segments from the mRNA after transcription. When such *split genes* are transcribed, RNA polymerase synthesizes a strand of pre-mRNA that initially includes copies of both *exons* (meaningful mRNA coding sequences) and *introns* (meaningless mRNA coding sequences). Soon after its manufacture, this pre-mRNA molecule has the meaningless introns clipped out and the exons spliced together into the final version, or *mature = mRNA,* which is used by the cell (figure 7.10). The molecules that are responsible for cutting the pre-mRNA are RNA-protein complexes called *"snurps."*

As previously mentioned, three general types of RNA are produced by transcription: messenger RNA, transfer RNA, and ribosomal RNA. Each kind of RNA is made from a specific gene and performs a specific function in the synthesis of polypeptides from individual amino acids at ribosomes. **Messenger RNA (mRNA)** is a mature, straight-chain copy of a gene that describes the exact sequence in which amino acids should be bonded together to form a polypeptide.

Transfer RNA (tRNA) molecules are responsible for picking up particular amino acids and transferring them to the ribosome for assembly into the polypeptide. All tRNA molecules are shaped like a cloverleaf. This shape is formed when they fold and some of the bases form hydrogen bonds that hold the molecule together. One end of the tRNA is able to attach to a specific amino acid. Toward the midsection of the molecule, a triplet nucleotide sequence can base-pair with a codon on mRNA. This triplet nucleotide sequence on tRNA that is complementary to a codon of mRNA is called an **anticodon. Ribosomal RNA (rRNA)** is a highly coiled molecule and is used along with protein molecules in the manufacture of all ribosomes, the cytoplasmic organelles where tRNA, mRNA, and rRNA come together to help in the synthesis of proteins.

Translation, or Protein Synthesis

The mRNA molecule is a coded message written in the biological world's universal nucleic acid language. The code is read in one direction starting at the initiator. The information is used to assemble amino acids into protein by a process called **translation.** The word *translation* refers to the fact that nucleic acid language is being changed to protein language. To translate mRNA language into protein language, a dictionary is necessary. Remember, the four letters in the nucleic acid alphabet yield sixty-four possible three-letter words. The protein language has twenty words in the form of twenty common amino acids (table 7.1). Thus, there are more than enough nucleotide words for the twenty amino acid molecules because each nucleotide triplet codes for an amino acid.

Table 7.2 is an amino acid–nucleic acid dictionary. Notice that more than one codon may code for the same amino acid. Some would contend that this is

Table 7.1
Amino Acids

Amino Acid	Abbreviation
Alanine	Ala
Arginine	Arg
Asparagine	AspN
Aspartic acid	Asp
Cysteine	Cys
Glutamic acid	Glu
Glutamine	GluN
Glycine	Gly
Histidine	His
Isoleucine	Ileu
Leucine	Leu
Lysine	Lys
Methionine	Met
Phenylalanine	Phe
Proline	Pro
Serine	Ser
Threonine	Thr
Tyrptophan	Try
Tryosine	Tyr
Valine	Val

There are twenty common amino acids used in the protein synthesis operation of a cell. Each has a known chemical structure.

Table 7.2
The Amino Acid–Nucleic Acid Dictionary

Amino Acid	mRNA Codons
Alanine	GCU GCC GCA GCG
Arginine	CGU CGC CGA CGG AGA AGG
Asparagine	AAU AAC
Aspartic acid	GAU GAC
Cysteine	UGU UGC
Glutamic acid	GAA GAG
Glutamine	CAA CAG
Glycine	GGU GGC GGA GGG
Histidine	CAU CAC
Isoleucine	AUU AUC AUA
Leucine	UUA UUG CUU CUC CUA CUG
Lysine	AAA AAG
Methionine	AUG
Phenylalanine	UUU UUC
Proline	CCU CCC CCA CCG
Serine	UCU UCC UCA UCG AGU AGC
Threonine	ACU ACC ACA ACG
Tryptophan	UGG
Tyrosine	UAU UAC
Valine	GUU GUC GUA GUG
Terminator	UAA UAG UGA
Initiator	AUG

A dictionary can come in handy for learning any new language. This one is used to translate nucleic acid language into protein language.

needless repetition, but such "synonyms" can have survival value. If, for example, the gene or the mRNA becomes damaged in a way that causes a particular nucleotide base to change to another type, the chances are still good that the proper amino acid will be read into its proper position. But not all such changes can be compensated for by the codon system, and an altered protein may be produced (figure 7.11). Changes can occur that cause great harm. Some damage is so extensive that the entire strand of DNA is broken, resulting in improper **protein synthesis,** or a total lack of synthesis. Such a change in DNA is called a **mutation.**

The construction site of the protein molecules (i.e., the translation site) is on the ribosome, a cellular organelle that serves as the meeting place for mRNA and the tRNA that is carrying amino acid building blocks. Proteins destined to be part of the cell membrane or packaged for export from the cell are synthesized on ribosomes attached to the endoplasmic reticulum. Proteins that are to perform their function in the cytoplasm are synthesized on unattached or free ribosomes. The mRNA molecule is placed on the ribosome two codons (six nucleotides) at a time (figure 7.12). The tRNA, which is carrying an amino acid, forms hydrogen bonds with the mRNA (between the codon and anticodon) long enough only for certain reactions to occur. The ribosome tRNA–mRNA complex is formed. Both RNA molecules combine first with the smaller of the two ribosomal units, and then the larger

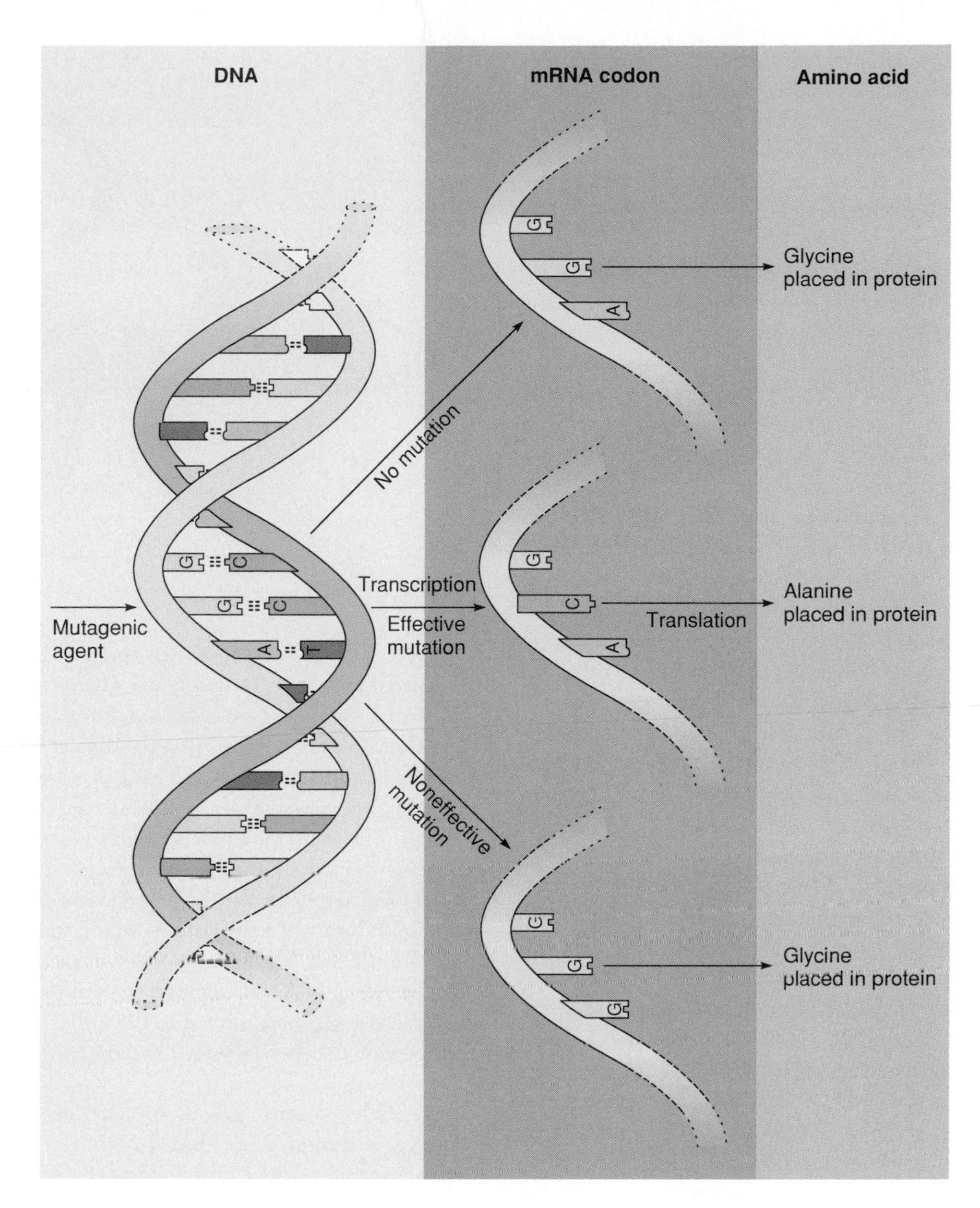

Figure 7.11 **Noneffective and Effective Mutation.** A nucleotide substitution changes the genetic information only if the changed codon results in a different amino acid being substituted into a protein chain. This feature of DNA serves to better ensure that the synthesized protein will be functional.

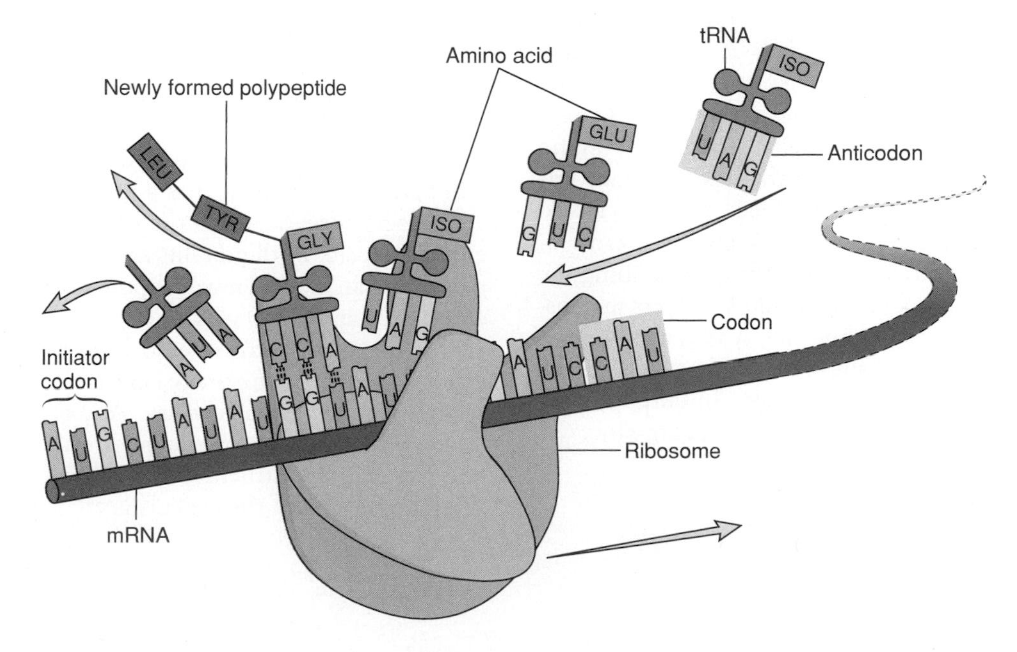

Figure 7.12 **Translation.** Translation results in the formation of a specific sequence of amino acids, each covalently bonded to another by a peptide bond. This amino acid aligning and bonding process takes place on the ribosomes. Each amino acid is brought into its proper position by a particular tRNA molecule. Each tRNA molecule has its own particular coded message that enables it to bind to a specific amino acid. As a result, a particular tRNA will only pick up one type of amino acid for transport to the translation reactions on the ribosomes.

ribosomal unit is added. The molecules of tRNA carrying amino acids move in to combine with the mRNA on the ribosome.

Amino acids are transferred to the ribosome by tRNA molecules that are so specific that they are only capable of transferring one particular amino acid. There are at least twenty different coding types of tRNA, and each tRNA transfers a specific amino acid to a ribosome for synthesis into a polypeptide. The tRNA properly aligns each amino acid so that it may be chemically bonded to another amino acid to form a long chain.

Each amino acid is bonded in sequence to form the new protein. As each is bonded in order, the ribosomal unit is moved along the mRNA to allow the next molecule of tRNA and its amino acid to fit into position. Once the final amino acid is bonded into position, all the molecules are released from the ribosome. The termination codons signal this action. The intact ribosome is again free to become engaged in another protein-synthesis operation. In most cells the mRNA travels through more than one ribosome at a time. When viewed with the electron microscope, this appears as a long thread (the mRNA) with several dark knots (the ribosomes) along its length. This sequence of several translating ribosomes attached to the same mRNA is known as a **polysome** (figure 7.13). The newly synthesized chains of amino acids leave the ribosomes and fold into their typical three-dimensional structure for use in the cell.

Thus, the mRNA moves through the ribosomes, its specific codon sequence allowing for the chemical bonding of a specific sequence of amino acids. Remember that the sequence was originally determined by the DNA. Figure 7.14 shows a possible result of protein synthesis. After transcribing the DNA code, the mRNA delivers its message to a ribosome, where a protein is made. In the case of figure 7.14, the amino acid sequence contains phenylalanine, serine, lysine, and arginine.

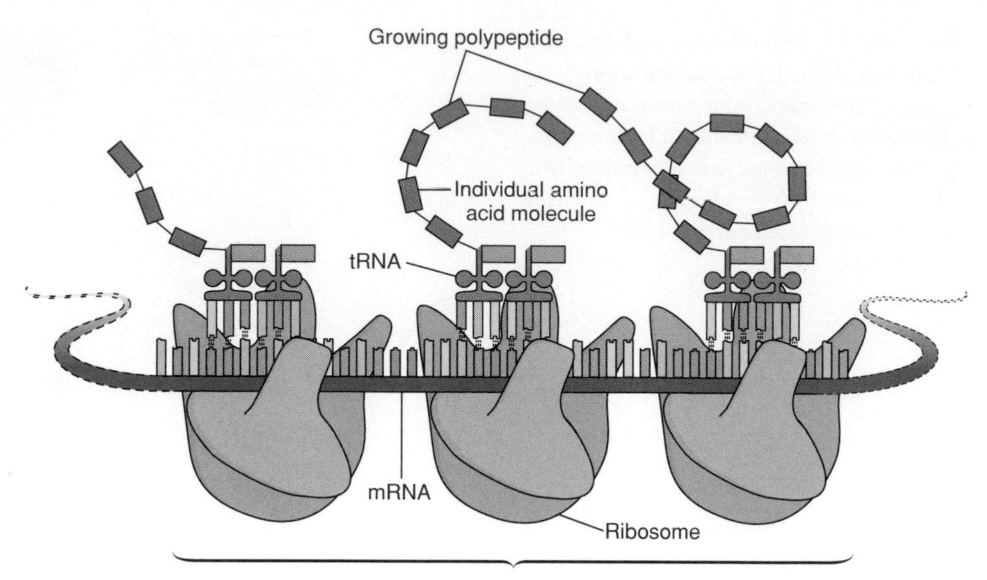

Figure 7.13 **A Polysome.** Any single mRNA molecule can be translated on a series of ribosomes. The mRNA and its several attached translating ribosomes are called a polysome. The mRNA moves along and through the ribosomes in a way similar to an audiocassette tape moving between the heads of a cassette player. As the tape passes through, the music signals on the tape are translated into audible sounds. The translation process used to assemble amino acids into protein would be comparable to feeding a single cassette tape through a series of cassette players. The first player would broadcast the music first and finish first, while the last one would start playing the music last and would finish last. Within the polysome complex, the first ribosome to bind to the mRNA will complete the translation of a polypeptide first, and the last ribosome to bind to the mRNA will synthesize the polypeptide last. Both the tape and the mRNA can be "played" again.

Each protein has a specific sequence of amino acids that determines its three-dimensional shape. This shape determines the activity of the protein molecule. The protein may be a structural component of a cell or a regulatory protein, such as an enzyme. Any changes in amino acids or their order changes the action of the protein molecule. The protein insulin, for example, has a different amino acid sequence than the digestive enzyme trypsin. Both proteins are essential to human life and must be produced constantly and accurately. The amino acid sequence of each is determined by a different gene. Each gene is a particular sequence of DNA nucleotides. Any alteration of that sequence can directly alter the protein structure and, therefore, the survival of the organism.

Alterations of DNA

Several kinds of changes to DNA may result in mutations. Phenomena that are either known or suspected causes of DNA damage are called **mutagenic agents.** Two such agents known to cause damage to DNA are x-radiation (X rays) and the chemical nicotine, found in tobacco. Both have been studied extensively, and there is little doubt that they cause mutations. *Jumping genes*, yet another cause of mutation, are segments of DNA capable of moving from one position in a strand of DNA to another. When the jumping gene is spliced into its new location, it alters the normal nucleotide sequence, causing normally stable genes to be misread during transcription. The result may be a mutant gene (figure 7.15).

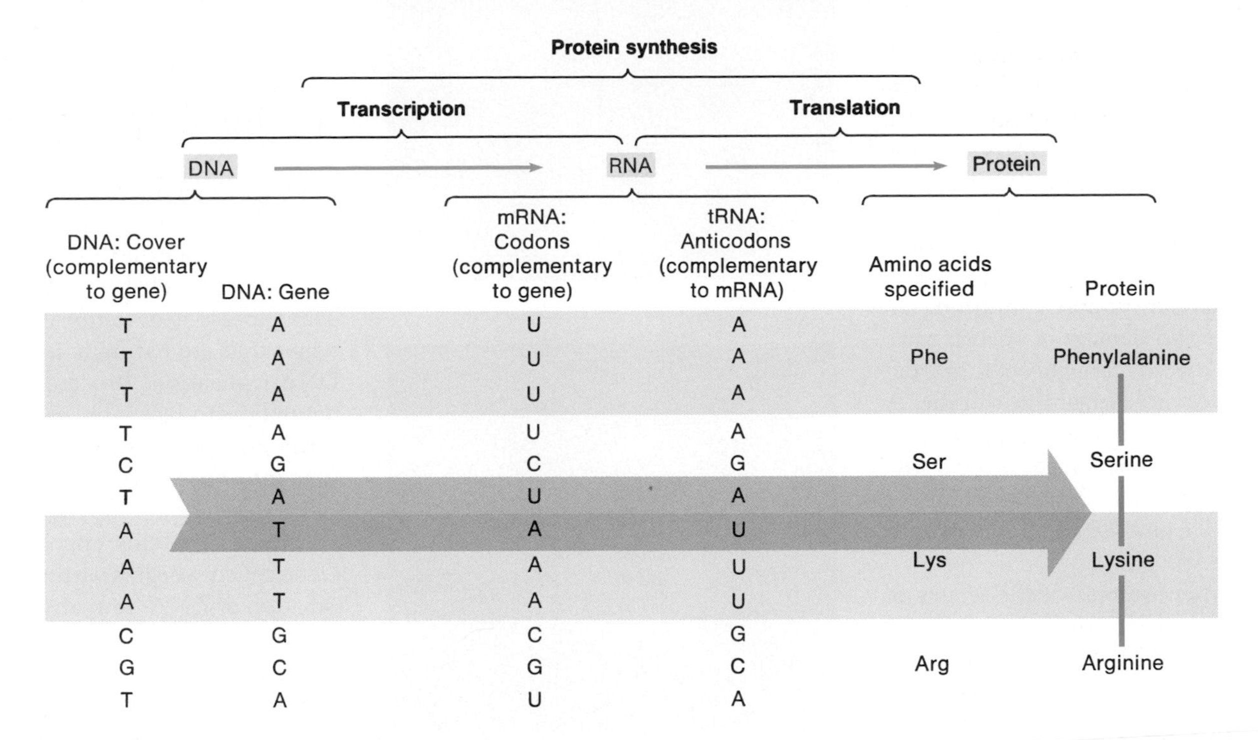

Protein synthesis
Transcription
Translation
DNA
RNA
Protein
DNA: Cover (complementary to gene)
DNA: Gene
mRNA: Codons (complementary to gene)
tRNA: Anticodons (complementary to mRNA)
Amino acids specified
Protein
T A U A
T A U A Phe Phenylalanine
T A U A
T A U A
C G C G Ser Serine
T A U A
A T A U
A T A U Lys Lysine
A T A U
C G C G
G C G C Arg Arginine
T A U A

Figure 7.14 **Protein Synthesis.** There are several steps involved in protein synthesis. (1) mRNA is manufactured from a DNA molecule in the transcription operation; (2) the mRNA enters the cytoplasm and attaches to ribosomes; (3) the tRNA carries amino acids to the ribosome and positions them in the order specified by the mRNA in the translation operation; (4) the amino acids are combined chemically to form a protein.

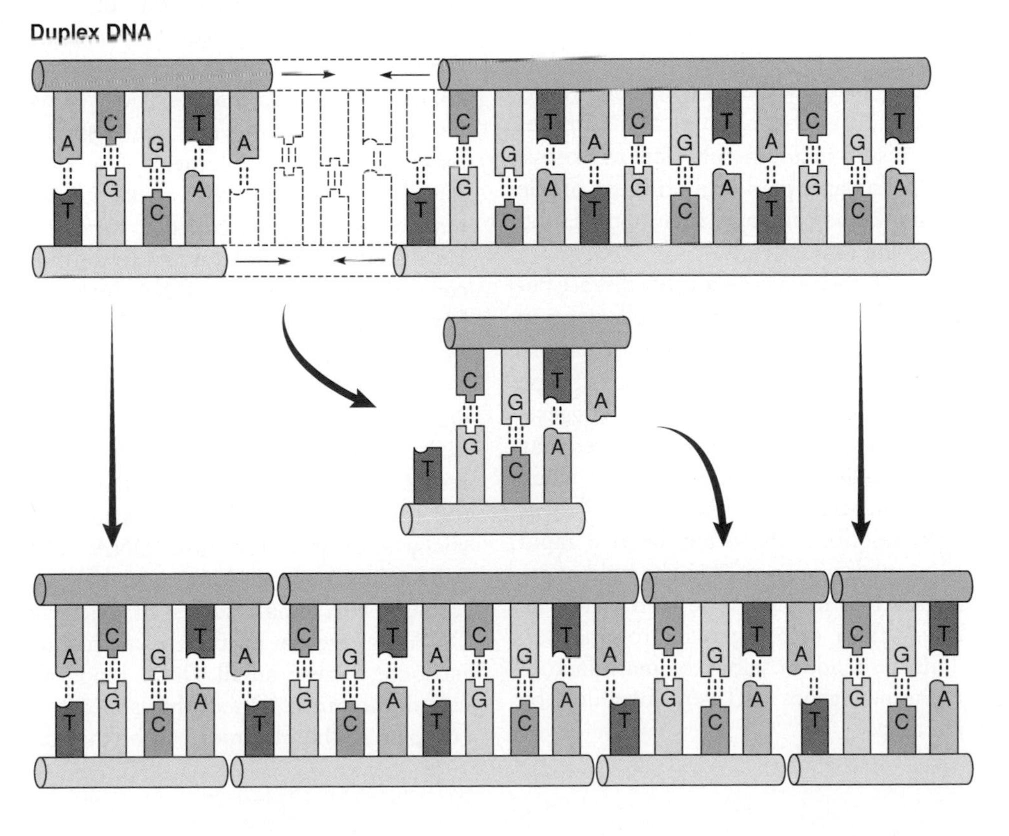

Figure 7.15 **Jumping Genes.** Alteration in the sequence of DNA nucleotides occurs when one segment of DNA "jumps" out of its original position in the duplex DNA and inserts itself into a different position in the strand. This changes the original nucleotide sequence and, therefore, the encoded message.

Changes in the structure of DNA have harmful effects on the next generation if they occur in the sex cells. Some damage to DNA is so extensive that the entire strand of DNA is broken, resulting in improper protein synthesis or a total lack of synthesis. This changes more than just a nucleotide base and is called a **chromosomal mutation.** In some cases the damage is so extensive that cells die. If enough cells are destroyed, the whole organism will die. A number of experiments indicate that many "street drugs," such as LSD (lysergic acid diethylamide), are mutagenic agents and cause DNA to break into smaller pieces.

Another example of the effects of altered DNA may be seen in human red blood cells. Red blood cells contain the oxygen-transport molecule, hemoglobin. Normal hemoglobin molecules are composed of 150 amino acid in four chains—two alpha and two beta chains. The nucleotide sequence of the gene for the beta chain is known, as is the amino acid sequence for this chain. In normal individuals, the sequence begins like this:

Val-His-Leu-Thr-Pro-***Glu***-Glu-Lys

. . . .

In some individuals, a single nucleotide of the gene that controls synthesis of the beta chain changes. This type of mutation is called a **point mutation.** The result is a new amino acid sequence in all the red blood cells:

Val-His-Leu-Thr-Pro-***Val***-Glu-Lys

. . . .

This single nucleotide change, which causes a single amino acid to change, may seem minor. However, it is the cause of **sickle-cell anemia,** a disease that affects the red blood cells by changing them from a circular to a sickle shape when oxygen levels are low (figure 7.16). When this sickling occurs, the red blood cells do not flow smoothly through capillaries. Their irregular shapes cause them to clump, clogging the blood vessels. This prevents them from delivering their oxygen load to the oxygen-demanding tissues. A number of physical disabilities may result, including physical weakness, brain damage, pain and stiffness of the joints, kidney damage, rheumatism, and in severe cases, death.

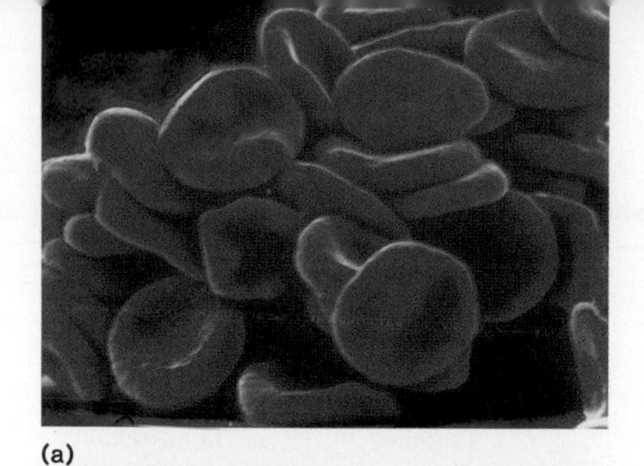

(a)

(b)

Figure 7.16 **Normal and Sickled Red Blood Cells.** Normal red blood cells (a) are shown in comparison with cells having the sickle shape (b). This sickling is the result of a single amino acid change in the hemoglobin molecule.

There is no cure for this disease because all of the individual's cells contain the same wrong genetic information. However, a powerful new science of gene manipulation, **biotechnology,** suggests that in the future, genetic diseases may be controlled or cured. Since 1953, when the structure of the DNA molecule was first described, there has been a rapid succession of advances in the field of genetics. It is now possible to transfer the DNA from one organism to another. This has made possible the manufacture of human genes and gene products by bacteria.

Manipulating DNA to Our Advantage

Biotechnology includes the use of a method of splicing genes from one organism into another, resulting in a new form of DNA called **recombinant DNA.** This process is accomplished using enzymes that are naturally involved in the DNA-replication process and others naturally produced by bacteria. When genes are spliced from different organisms into host cells, the host cell replicates these new, "foreign" genes and synthesizes proteins encoded by them. Gene splicing begins with the laboratory isolation of DNA from an organism that contains the desired gene; for example, from human cells that contain the gene for the manufacture of insulin. If the gene is short enough and its base sequence is known, it may be synthesized in the laboratory from separate nucleotides. If the gene is too long and complex, it is cut from the chromosome with enzymes called *restriction endonucleases.* They are given this name because these enzymes (-ases) only cut DNA (nucle-) at certain base sequences (restricted in their action) and work inside (endo-) the DNA. These particular enzymes do not cut the DNA straight across, but in a zigzag pattern that leaves one strand slightly longer than its complement. The short nucleotide sequence that sticks out and remains unpaired is called a *sticky end* because it can be reattached to another complementary strand. DNA segments have successfully been cut from rats, frogs, bacteria, and humans.

This isolated gene with its "sticky end" is spliced into microbial DNA. The host DNA is opened up with the proper restriction endonuclease, and ligase (tie together) enzymes are used to attach the sticky ends into the host DNA. This gene-splicing procedure may be performed with small loops of bacterial DNA that are not part of the main chromosome. These small DNA loops are called *plasmids.* Once the splicing is completed, the plasmids can be inserted

BOX 7.1 The PCR and Genetic Fingerprinting

In 1989 the American Association for the Advancement of Science named DNA polymerase "Molecule of the Year." The value of this enzyme in the polymerase chain reaction (PCR) is so great that it could not be ignored. Just what is the PCR, how does it work, and what can you do with it?

The PCR is a laboratory procedure for copying selected segments of DNA. A single cell can provide enough DNA for analysis and identification! Having a large number of copies of a "target sequence" of nucleotides enables biochemists to more easily work with DNA. This is like increasing the one "needle in the haystack" to such large numbers (*100 billion in only a matter of hours*) that they're not hard to find, recognize, and work with. The types of specimens that can be used include semen, hair, blood, bacteria, protozoa, viruses, mummified tissue, and frozen cells. The process requires the DNA specimen, free DNA nucleotides, synthetic "primer" DNA, DNA polymerase, and simple lab equipment, such as a test tube and source of heat.

Having decided which target sequence of nucleotides (which "needle") is to be replicated, scientists heat the specimen of DNA to separate the coding and noncoding strands. Molecules of synthetic "primer" DNA are added to the specimen. These primer molecules are specifically designed to attach to the ends of the target sequence. Next, a mixture of triphosphorylated nucleotides is added so that they can become the newly replicated DNA. The presence of the primer, attached to the DNA and added nucleotides, serves as the substrate for the DNA polymerase. Once added, the polymerase begins making its way down the length of the DNA from one attached primer end to the other. The enzyme bonds the new DNA nucleotides to the strand, replicating the molecule as it goes. It stops when it reaches the other end, having produced a new copy of the target sequence. Since the enzyme will continue to operate as long as enzyme and substrates are available, the process continues, and in a short time there are billions of small pieces of DNA, all replicas of the target sequence.

So what, you say? Well, consider the following. Using the PCR, scientists have been able to

1. more accurately diagnose such disease as sickle-cell anemia, cancer, Lyme disease, AIDS, and legionnaires disease;
2. perform highly accurate tissue typing for matching organ-transplant donors and recipients;
3. help resolve criminal cases of rape, murder, assault, and robbery by matching suspect DNA to that found at the crime scene;
4. detect specific bacteria in environmental samples;
5. monitor the spread of genetically engineered microorganisms in the environment;
6. check water quality by detecting bacterial contamination from feces;
7. identify viruses in water samples;
8. identify disease-causing protozoa in water;
9. determine specific metabolic pathways and activities occurring in microorganisms;
10. determine races, distribution patterns, kinships, migration patterns, evolutionary relationships, and rates of evolution of long-extinct species;
11. accurately settle paternity suits;
12. confirm identity in amnesia cases;
13. identify a person as a relative for immigration purposes;
14. provide the basis for making human antibodies in specific bacteria;
15. possibly, provide the basis for replicating genes that could be transplanted into individuals suffering from genetic diseases; and
16. identify nucleotide sequences peculiar to the human genome (an application currently underway as part of the Human Genome Project).

---

into the bacterial host by treating the cell with special chemicals that encourage it to take in these large chunks of DNA. A more efficient alternative is to splice the desired gene into the DNA of a bacterial virus so that it can carry the new gene into the bacterium as it infects the host cell. Once inside the host cell, the genes may be replicated along with the rest of the DNA to clone the "foreign" gene, or they may begin to synthesize the encoded protein.

As this highly sophisticated procedure has been refined, it has become possible to quickly and accurately splice genes from a variety of species into host bacteria, making possible the synthesis of large quantities of medically important products. For example, recombinant DNA procedures are responsible for the production of human insulin, used in the control of diabetes; interferon, used as an antiviral agent; human growth hormone, used to stimulate growth in children lacking this hormone; and somatostatin, a brain hormone also implicated in growth.

The possibilities that open up with the manipulation of DNA are revolutionary (box 7.1). These methods enable cells to produce molecules that they would not normally make. Some research laboratories have even spliced genes into laboratory cultured human cells. Should such a venture prove to be practical, genetic diseases such as sickle-cell anemia could be controlled. The process of recombinant DNA gene splicing also enables cells to be more efficient at producing molecules that they normally synthesize. Some of the likely rewards are (1) production of additional, medically useful proteins; (2) mapping of the locations of genes on human chromosomes; (3) more complete understanding of how genes are regulated; (4) production of crop plants with increased yields; and (5) development of new species of garden plants.

The discovery of the structure of DNA over forty years ago seemed very far removed from the practical world. The importance of this "pure" research is just now being realized. Many companies are involved in recombinant DNA research with the aim of alleviating or curing disease.

• Summary •

The successful operation of a living cell depends on its ability to accurately reproduce genes and control chemical reactions. DNA replication results in an exact doubling of the genetic material. The process virtually guarantees that identical strands of DNA will be passed on to the next generation of cells.

The enzymes are responsible for the efficient control of a cell's metabolism. However, the production of protein molecules is under the control of the nucleic acids, the primary control molecules of the cell. The structure of the nucleic acids DNA and RNA determines the structure of the proteins, while the structure of the proteins determines their function in the cell's life cycle. Protein synthesis involves the decoding of the DNA into specific protein molecules and the use of the intermediate molecules, mRNA and tRNA, at the ribosome. Errors in any of the codons of these molecules may produce observable changes in the cell's functioning and lead to the death of the cell.

Methods of manipulating DNA have led to the controlled transfer of genes from one kind of organism to another. This has made it possible for bacteria to produce a number of human gene products.

• Thinking Critically •

An 18-year-old college student reported that she had been raped by someone she identified as a "large, tanned, white man." A fellow student in her biology class fitting that description was said by eyewitnesses to have been, without a doubt, in the area at approximately the time of the crime. The suspect was apprehended and upon investigation was found to look very much like someone who lived in the area and who had a previous record of criminal sexual assaults. Samples of semen from the woman's vagina were taken during a routine rape-case follow-up physical exam. Cells were also taken from the suspect. He was brought to trial but found to be innocent of the crime based on evidence from the criminal investigations laboratory. His alibi that he had been working alone on a research project in the biology lab held up. Without PCR-genetic fingerprinting, the suspect would surely have been wrongly convicted, based solely on circumstantial evidence provided by the victim and the "eyewitnesses."

Place yourself in the position of the expert witness from the criminal laboratory who performed the PCR-genetic fingerprinting tests on the two specimens. The prosecuting attorney has just asked you to explain to the jury what led you to the conclusion that the suspect could not have been responsible for this crime. Remember, you must explain this to a jury of twelve men and women who in all likelihood have little or no background in the area of biological sciences. Please, tell the whole truth and nothing but the truth.

• Experience This •

The method of DNA replication described in this chapter is called *semiconservative* replication. It is called this because the "new" strands of DNA are formed from the "old" strands that remain intact during the entire process. The discovery of this method (since found to be the valid method) came after considerable thought and "playing around" with other hypotheses. One discarded method was called *conservative* replication because both original strands remained together during the process of replication. Diagram a short section of duplex DNA and use it to show the difference between semiconservative and conservative DNA replication.

• Questions •

1. What is the difference between a nucleotide, a nitrogenous base, and a codon?
2. What are the differences between DNA and RNA?
3. List the sequence of events that takes place when a DNA message is translated into protein.
4. Chromosomal and point mutations both occur in DNA. In what ways do they differ? How is this related to recombinant DNA?
5. Why is DNA replication necessary?
6. What is polymerase and how does it function?
7. How does DNA replication differ from the manufacture of an RNA molecule?
8. If a DNA nucleotide sequence is CATAAAGCA, what is the mRNA nucleotide sequence that would base-pair with it?
9. What amino acids would occur in the protein chemically coded by the sequence of nucleotides in question 8?
10. How do tRNA, rRNA, and mRNA differ in function?

• Chapter Glossary •

adenine (ad'ĕ-nēn) A double-ring nitrogenous-base molecule in DNA and RNA. It is the complementary base of thymine or uracil.

anticodon (an''te-ko'don) A sequence of three nitrogenous bases on a tRNA molecule capable of forming hydrogen bonds with three complementary bases on an mRNA codon during translation.

biotechnology (bi-o-tek-nol'uh-je) The science of gene manipulation.

chromatin fibers (kro'mah-tin fi'bers) See **nucleoproteins.**

chromosomal mutation (kro-mo-sōm'al miu-ta'shun) A change in the gene arrangement in a cell as a result of breaks in the DNA molecule.

chromosome (kro'mo-sōm) A duplex DNA molecule with attached protein (nucleoprotein) coiled into a short, compact unit.

codon (ko'don) A sequence of three nucleotides of an mRNA molecule that directs the placement of a particular amino acid during translation.

complementary base (kom''plĕ-men'tah-re bās) A base that can form hydrogen bonds with another base of a specific nucleotide.

cytosine (si'to-sēn) A single-ring nitrogenous-base molecule in DNA and RNA. It is complementary to guanine.

deoxyribonucleic acid (DNA) (de-ok''se-ri-bo-nu-kle'ik as'id) A polymer of nucleotides that serves as genetic information. In prokaryotic cells, it is a duplex DNA (double-stranded) loop and contains attached HU proteins. In eukaryotic cells, it is found in strands with attached histone proteins. When tightly coiled, it is known as a chromosome.

deoxyribose (de-ok''se-ri'bōs) A 5-carbon sugar molecule; a component of DNA.

DNA code (D-N-A cōd) A sequence of three nucleotides of a DNA molecule.

DNA polymerase (po-lim'er-ās) An enzyme that bonds DNA nucleotides together when they base pair with an existing DNA strand.

DNA replication (rep''lĭ-ka'shun) The process by which the genetic material (DNA) of the cell reproduces itself prior to its distribution to the next generation of cells.

duplex DNA (du'pleks) DNA in a double-helix shape.

gene (jēn) Any molecule, usually a segment of DNA, that is able to (1) replicate by directing the manufacture of copies of itself; (2) mutate, or chemically change and transmit these changes to future generations; (3) store information that determines the characteristics of cells and organisms; and (4) use this information to direct the synthesis of structural and regulatory proteins.

guanine (gwah'nēn) A double-ring nitrogenous-base molecule in DNA and RNA. It is the complementary base of cytosine.

initiation code (ĭ-nĭ'she-a''shun cōd) The code on DNA with the base sequence TAC that begins the process of transcription.

messenger RNA (mRNA) (mes'-en-jer) A molecule composed of ribonucleotides that functions as a copy of the gene and is used in the cytoplasm of the cell during protein synthesis.

mutagenic agent (miu-tah-jen'ik a-jent) Anything that causes permanent change in DNA.

mutation (miu-ta'shun) Any change in the genetic information of a cell.

nitrogenous base (ni-trah'jen-us bās) A category of organic molecules found as components of the nucleic acids. There are five common types: thymine, guanine, cytosine, adenine, and uracil.

nucleic acids (nu'kle-ik as'ids) Complex molecules that store and transfer information within a cell. They are constructed of fundamental monomers known as nucleotides.

nucleoproteins (nu-kle-o-pro'te-inz) The duplex DNA strands with attached proteins; also called **chromatin fibers.**

nucleosomes (nu'kle-o-sōmz) Histone clusters with their encircling DNA.

nucleotide (nu'kle-o-tīd) The building block of the nucleic acids. Each is composed of a 5-carbon sugar, a phosphate, and a nitrogenous base.

phosphate (fos'-fāt) Part of a nucleotide; composed of phosphorus and oxygen atoms.

point mutation (point miu-ta'shun) A change in the DNA of a cell as a result of a loss or change in a nitrogenous-base sequence.

polysome (pah'le-sōm) A sequence of several translating ribosomes attached to the same mRNA.

promoter (pro-mo'ter) A region of DNA at the beginning of each gene, just ahead of an initiator code.

protein synthesis (pro'te-in sin'thĕ-sis) The process whereby the tRNA utilizes the mRNA as a guide to arrange the amino acids in their proper sequence according to the genetic information in the chemical code of DNA.

recombinant DNA (re-kom'bĭ-nant) DNA that has been constructed by inserting new pieces of DNA into the DNA of another organism, such as a bacterium.

ribonucleic acid (RNA) (ri-bo-nu-kle'ik as'id) A polymer of nucleotides formed on the template surface of DNA by transcription. Three forms that have been identified are mRNA, rRNA, and tRNA.

ribose (ri'bōs) A 5-carbon sugar molecule that is a component of RNA.

ribosomal RNA (rRNA) (ri-bo-sōm'al) A globular form of RNA; a part of ribosomes.

RNA polymerase (po-lim'er-ās) An enzyme that attaches to the DNA at the promoter region of a gene when the genetic information is transcribed into RNA.

sickle-cell anemia (sĭ'kul sel ah-ne'me-ah) A disease caused by a point mutation. This malfunction produces sickle-shaped red blood cells.

template (tem'plet) A model from which a new structure can be made. This term has special reference to DNA as a model for both DNA replication and transcription.

termination code (ter-mĭ-na'shun cōd) The DNA nucleotide sequence just in back of a gene with the code ATT, ATC, or ACT that signals "stop here."

thymine (thi'mēn) A single-ring nitrogenous-base molecule in DNA but not in RNA. It is complementary to adenine.

transcription (tran-skrip'shun) The process of manufacturing RNA from the template surface of DNA. Three forms of RNA that may be produced are mRNA, rRNA, and tRNA.

transfer RNA (tRNA) (trans'fur) A molecule composed of ribonucleic acid. It is responsible for transporting a specific amino acid into a ribosome for assembly into a protein.

translation (trans-la'shun) The assembly of individual amino acids into a polypeptide.

uracil (yu'rah-sil) A single-ring nitrogenous-base molecule in RNA but not in DNA. It is complementary to adenine.

Cell Division and Heredity

All living things are capable of reproducing themselves so that they can perpetuate their own kind. Several methods of reproduction have been identified. One method ensures the exact replication of a cell. When a cell divides by this method, it ceases to exist and becomes its own offspring, identical to the parent cell. Another method follows a series of events that results in the formation of cells that display different gene combinations than are found in the parent cell. These differences may better enable the offspring to survive in a changing environment. When multicellular organisms reproduce, they go through a very complex series of cellular events that lead to the formation of the next generation. These events are controlled by chemicals produced in the body and by outside factors. Exactly which characteristics are displayed in the offspring depends on which genetically controlled traits have been transmitted by the parent cell. Characteristics are also affected by factors that influence how, when, or if these genes are expressed in the offspring. Should genetically controlled traits be found to enhance the survival and further reproduction of the organism, more individuals in the population may display these features as time goes by. An increase in the frequency of such genes in a population may change the population's appearance, behavior, or other characteristics. ■

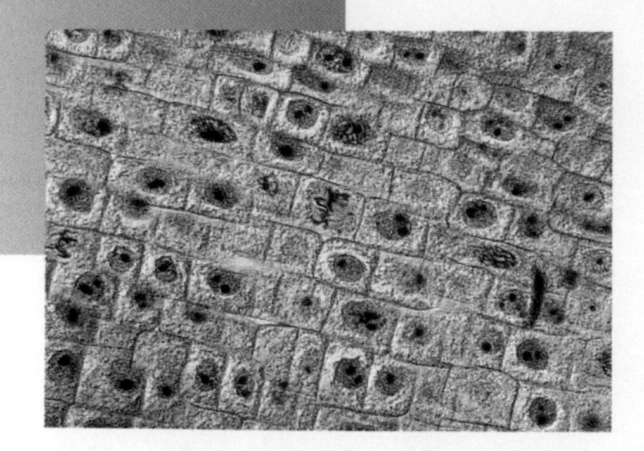

CHAPTER 8

Mitosis—The Cell-Copying Process

Chapter Outline

Purpose

In the previous chapter we saw how the molecule DNA replicates. Once this process is complete, doubled DNA is distributed to two new daughter cells. The way in which this cell-splitting process occurs ensures that the daughter cells will have the same genetic message as the original DNA molecule. However, cells with identical genetic messages may differ in how they are built and how they function. The relationships among DNA, genes, and chromosomes are concepts for later consideration of genetics, evolution, and sex-cell formation.

For Your Information

As you grow up, cell division results in an increase in the number of cells that make up your body. Many tissues, such as nerve and muscle tissue, stop dividing early in childhood. Other tissues, such as skin and blood, continue to divide throughout your life. As we pass middle age, we become smaller. This is the result of a decrease in the rate of cell division. Cells are dying faster than they can be replaced. This is made more obvious by the fact that tone is lost as the cells die and are replaced by connective tissues. Cell division is, therefore, one of the areas currently being researched in the area of gerontology.

Learning Objectives

- List the purposes of cell division.
- Explain the cell cycle.
- State the processes that occur during interphase.
- Name the stages of mitosis and explain what is happening during each stage.
- Define differentiation.
- Explain how cancer is caused and treated.

The Importance of Cell Division

Even though individual cells may die, the process of cell division replaces dead cells with new ones, repairs damaged tissues, and allows living organisms to grow. For example, you began as a single cell that resulted from the union of a sperm and an egg. One of the first activities of this single cell was to divide. As this process continued the number of cells in your body increased, so that as an adult your body consists of several trillion cells. The second function of cell division is to provide for the maintenance of the body. Certain cells in your body, such as red blood cells, gut lining, and skin, wear out. As they do, they must be replaced with new cells. Altogether, you lose about one million cells per second; this means that cell division is beginning a million times in your body at any given time. A third purpose of cell division is repair. When a bone is broken, the break heals because cells divide, increasing the number of cells available to knit the broken pieces together. If some skin cells are destroyed by a cut or abrasion, cell division produces new cells to repair the damage.

During cell division, two events occur. The replicated genetic information of a cell is equally distributed to two daughter nuclei in a process called **mitosis.** Following the division of the nucleus, there is a division of the cytoplasm into two new cells. This division of the cell's cytoplasm is called **cytokinesis**—cell splitting. Each new cell gets one of the two daughter nuclei so that both have a complete set of genetic information. These two processes usually happen in sequence.

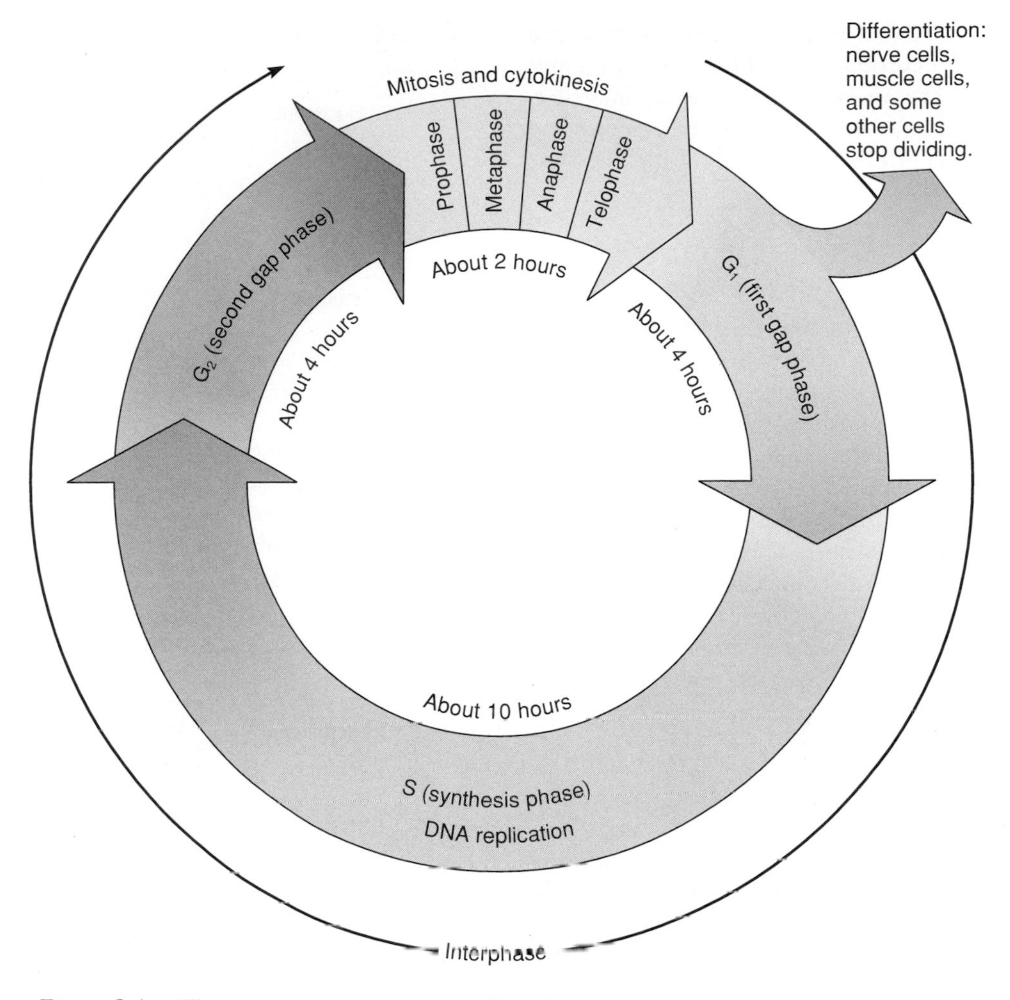

Figure 8.1 **The Cell Cycle.** During the cell cycle, tRNA, mRNA, ribosomes, and enzymes are produced in the G_1 stage. DNA replication occurs in the S stage. Proteins required for the spindles are synthesized in the G_2 stage. The nucleus is replicated in mitosis, and two cells are formed by cytokinesis. Once some organs, such as the brain, have completely developed, certain types of cells, such as nerve cells, remain in the G_2 stage. The time periods indicated are relative and vary depending on the type of cell and the age of the organism.

The Cell Cycle

All cells go through a basic life cycle, but they vary in the amount of time they spend in the different stages. A generalized picture of a cell's life cycle may help you to understand it better (figure 8.1). Once begun, cell division is a continuous process without a beginning or an end. It is a cycle in which cells continue to grow and divide. There are four stages to the life cycle of a eukaryotic cell: (1) G_1, gap-phase one; (2) S, synthesis; (3) G_2, gap-phase two; and (4) cell division (mitosis and cytokinesis).

The first three phases of the cell cycle—G_1, S, and G_2—occur during a period of time known as *interphase.* **Interphase** is the stage between cell divisions. During the G_1 stage, the cell grows in volume as it produces tRNA, mRNA, ribosomes, enzymes, and other cell components. During the S stage, DNA replication occurs in preparation for the distribution of genes to daughter cells. During the G_2 stage that follows, final preparations are made for mitosis with the synthesis of spindle-fiber proteins.

During interphase, the cell is not dividing but is engaged in metabolic activities, such as muscle-cell contractions, nerve-cell transmission, or glandular-cell secretion. During interphase, the nuclear membrane is intact and the individual chromosomes are not visible (figure 8.2). The individual chromatin strands are too thin and tangled to be seen. Remember that **chromosomes** are composed of various kinds of histone proteins and DNA that contain the cell's genetic information. The double helix of

DNA and the nucleosomes are arranged as a chromatid, and there are two chromatids in each replicated chromosome. It is these chromatids (chromosomes) that will be distributed during mitosis.

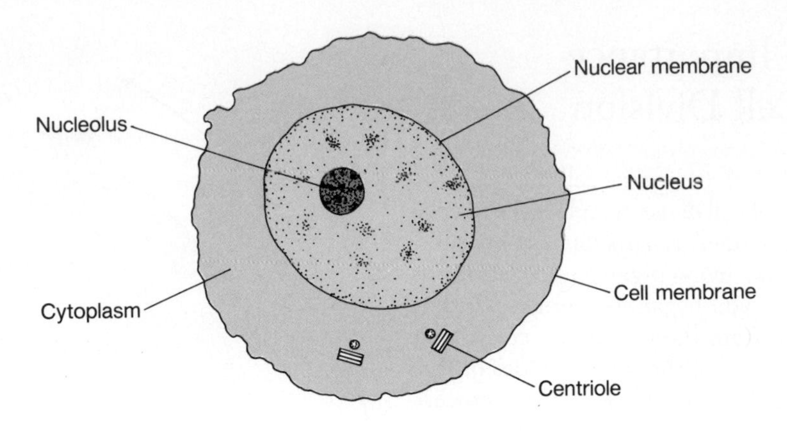

Figure 8.2 **Interphase.** Growth and the production of necessary organic compounds occur during this phase. If the cell is going to divide, DNA replication also occurs during interphase. The individual chromosomes are not visible, but a distinct nuclear membrane and nucleolus are present.

The Stages of Mitosis

All stages in the life cycle of a cell are continuous; there is no precise point when the G_1 stage ends and the S stage begins, or when the interphase period ends and mitosis begins. Likewise, in the individual stages of mitosis, there is a gradual transition from one stage to the next. However, to enable communication, scientists have subdivided the process into four recognizable stages based on recognizable events. These four phases are prophase, metaphase, anaphase, and telophase.

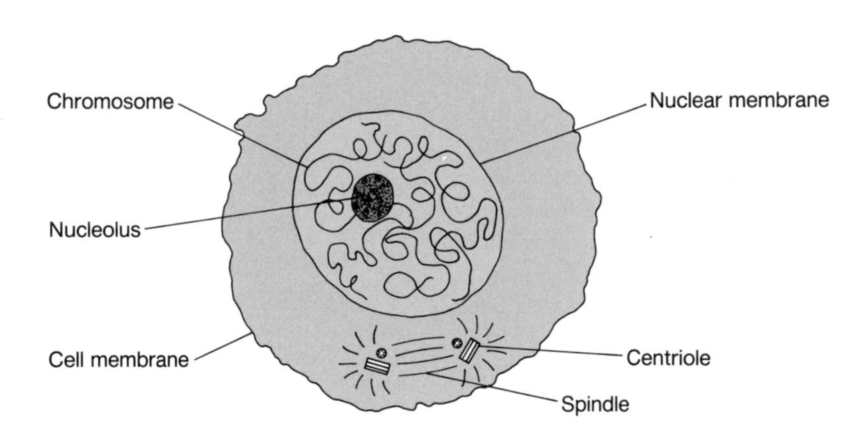

Figure 8.3 **Early Prophase.** Chromosomes begin to appear as thin tangled threads, and the nucleolus and nuclear membrane are present. The two sets of microtubules known as the centrioles begin to separate and move to opposite poles of the cell. A series of fibers known as the spindle will shortly begin to form.

Prophase

As the G_2 stage of interphase ends, mitosis begins. **Prophase** is the first stage of mitosis. One of the first noticeable changes is that the individual chromosomes become visible (figure 8.3). The thin, tangled chromatin present during interphase gradually coils and thickens, becoming visible as separate chromosomes. The DNA portion of the chromosome carries genes that are arranged in a specific order. Each chromosome carries its own set of genes that is different from the sets of genes on other chromosomes.

As prophase proceeds, and as the chromosomes become more visible, we recognize that each chromosome is made of two parallel, threadlike parts lying side by side. Each parallel thread is called a **chromatid** (figure 8.4). These chromatids were formed during the S stage of interphase, when DNA synthesis occurred. The two identical chromatids are attached at a region called the **centromere.**

In the diagrams in this text, a few genes are shown as they might occur on human chromosomes. The diagrams show fewer chromosomes and fewer genes on each chromosome than are actually present. Normal human cells have ten billion nucleotides arranged into forty-six chromosomes, each chromosome with thousands of genes. In this book, smaller numbers of genes and chromosomes are used to make it easier to follow the events that happen in mitosis.

Several other events occur as the cell proceeds to the late prophase stage (figure 8.5). One of these events is the duplication of the **centrioles.** Remember that some cells contain centrioles, which are microtubule-containing organelles located just outside the nucleus. As they duplicate, they move to the poles of the cells. As the centrioles move to the poles, the microtubules form the *spindle*. The **spindle** is an array of microtubules extending from pole to pole that is used in the movement of chromosomes.

As prophase is occurring, the nuclear membrane gradually disintegrates. Although it is present at the beginning of prophase, it disappears by the time this stage is completed. In addition to the nuclear membrane, the nucleoli within the nucleus disappear. Because of the

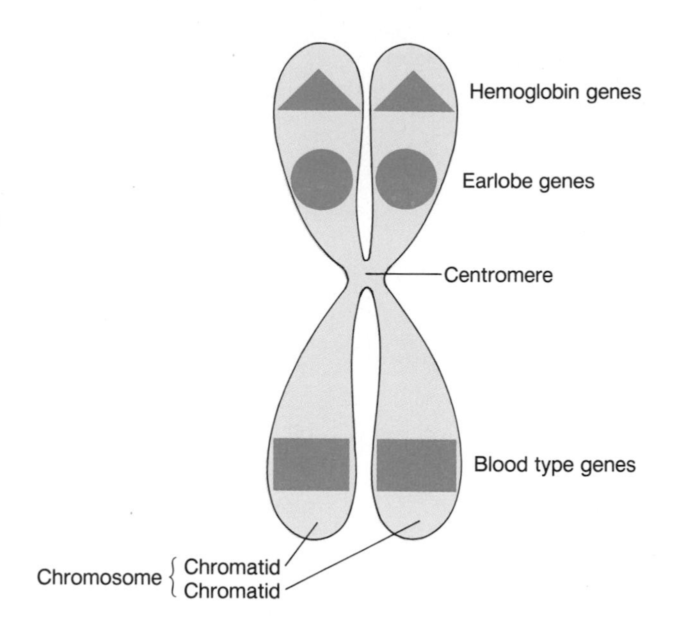

Figure 8.4 **Chromosomes.** During interphase, when chromosome replication occurs, the original duplex DNA unzips and forms two identical double strands that are attached at the centromere. Each of these double strands is a chromatid. The two identical chromatids of the chromosome are sometimes termed a dyad, to reflect that there are two duplex DNA molecules, each located in a chromatid. The DNA contains the genetic data. (We really don't know on which chromosomes most human genes are located. The examples presented here are for illustrative purposes only.)

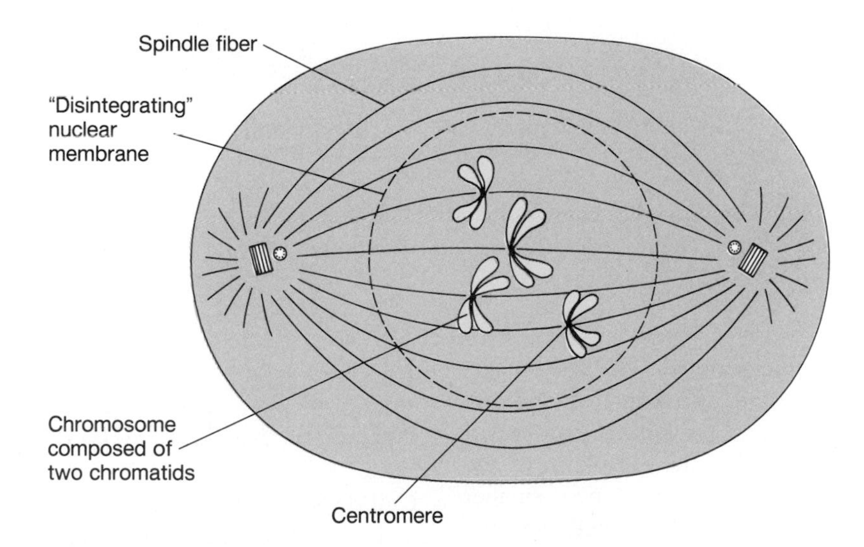

Figure 8.5 **Late Prophase.** In late prophase, the chromosomes appear as two chromatids (a dyad) connected at a centromere. The nucleolus and the nuclear membrane have disintegrated. The centrioles have moved farther apart, and the spindle is produced.

disintegration of the nuclear membrane, the chromosomes are free to move anywhere within the cytoplasm of the cell. As this movement occurs, the cell enters into the next stage of mitosis.

Metaphase

During **metaphase,** the second stage in mitosis, the chromosomes align at the equatorial plane. There is no nucleus present during metaphase, and the spindle, which started to form during prophase, is completed. The centrioles are at the poles, and the microtubules extend between them to form the spindle. At the beginning of metaphase, the chromosomes become attached to the spindle fibers at their centromeres. Initially they are distributed randomly throughout the cytoplasm. Then the chromosomes move until all their centromeres align themselves along the equatorial plane at the equator of the cell (figure 8.6). At this stage in mitosis, each chromosome still consists of two chromatids. In a human cell, there are forty-six chromosomes or ninety-two chromatids aligned at the cell's equatorial plane during metaphase.

If we view a cell in the metaphase stage from the side (figure 8.6), it is an equatorial view. In this view, the chromosomes appear as if they were in a line. If we view the cell from the pole, it is a polar view. The chromosomes are seen on the equatorial plane (figure 8.7). Chromosomes viewed from this direction look like hot dogs scattered on a plate.

Anaphase

Anaphase is the third stage of mitosis. The nuclear membrane is still absent and the spindle extends from pole to pole. In anaphase, each chromosome splits at the centromere, and the two chromatids within the chromosome separate as they move along the spindle fibers toward the

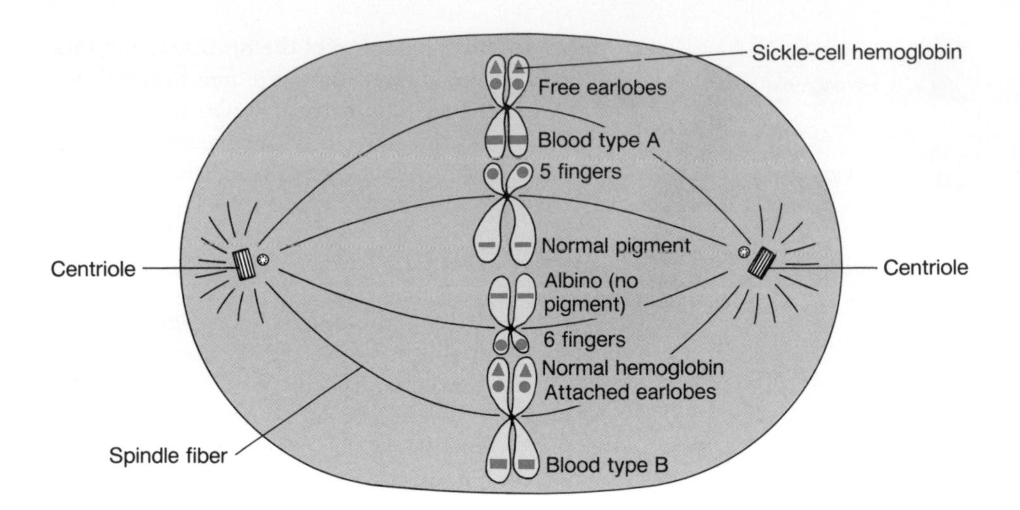

Figure 8.6 **Metaphase.** During metaphase the chromosomes travel along the spindle and align at the equatorial plane. Notice that each chromosome still consists of two chromatids.

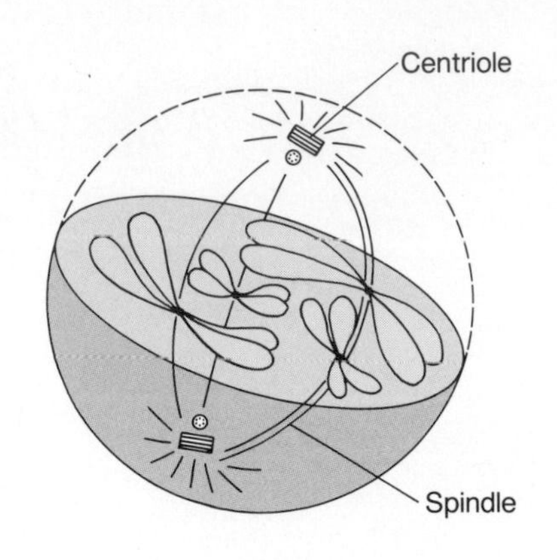

Figure 8.7 **A Polar View of Metaphase.** The polar view shows the chromosomes spread out on the equatorial plane.

poles (figure 8.8). Although this movement of chromosomes has been observed repeatedly, no one knows the exact mechanism of its action. After this separation of chromatids occurs, the chromatids are called **daughter chromosomes.** Daughter chromosomes contain identical genetic information.

Examine figure 8.8 closely and notice that the four chromosomes moving to one pole have exactly the same genetic information as the four moving to the opposite pole. It is the alignment of the chromosomes in metaphase and their separation in anaphase that causes this type of distribution. At the end of anaphase, there are two identical groups of chromosomes, one group at each pole. The next stage completes the mitosis process.

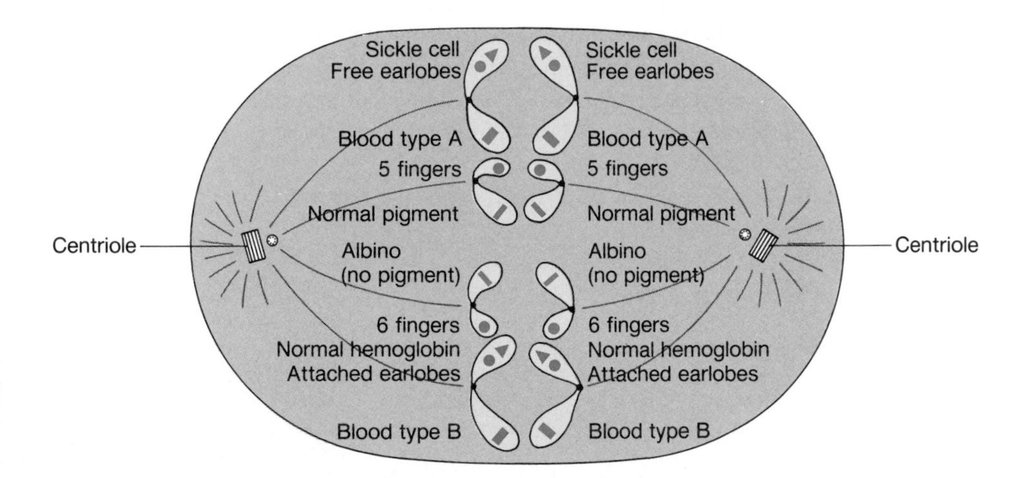

Figure 8.8 **Anaphase.** The pairs of chromatids separate as the centromeres divide. The chromatids, now called chromosomes, are separating and moving toward the poles.

Telophase

Telophase is the last stage in mitosis. It is during telophase that daughter nuclei are formed. Each set of chromosomes becomes enclosed by a nuclear membrane and the nucleoli reappear. Now the cell has two identical **daughter nuclei** (figure 8.9). In addition, the microtubules disintegrate, so the spindle disappears. With the formation of the daughter nuclei, mitosis, the first process in cell division, is completed and the second process, cytokinesis, can occur. Cytokinesis splits the cytoplasm of the original cell and forms two smaller daughter cells. **Daughter cells** are two cells formed by cell division that have identical genetic information. Each of the newly formed daughter cells then enters the G_1 stage of interphase. These cells can grow, replicate their DNA, and enter another round of mitosis and cytokinesis to complete the cell cycle.

Plant- and Animal-Cell Differences

Cell division is similar in plant and animal cells. However, there are some minor differences. One difference concerns the centrioles (figure 8.10). Centrioles are essential in animal cells, but they are not found in plant cells. However, by some process, plant cells do produce a spindle. There is also a difference in the process of cytokinesis (figure 8.11). In animal

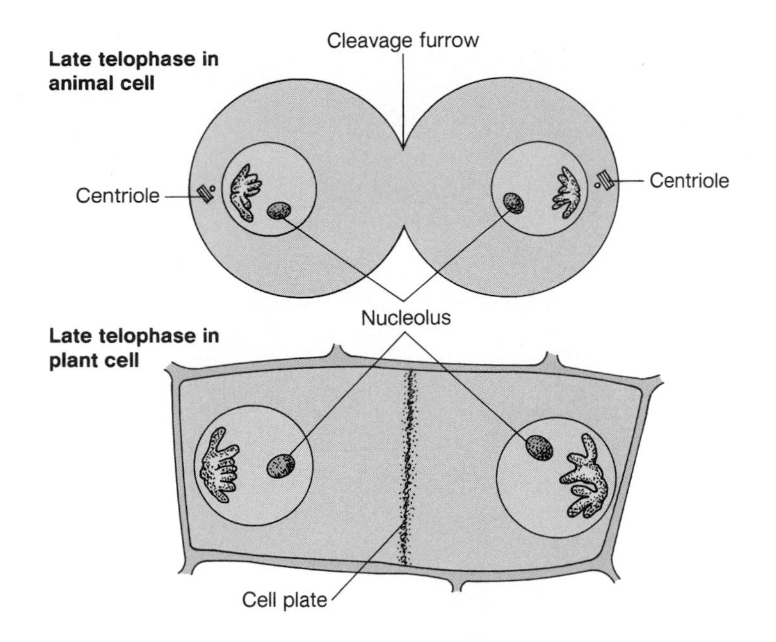

Figure 8.9 **Telophase.** During telophase the spindle disintegrates and the nucleolus and nuclear membrane form. Daughter cells are formed as a result of the division of the cytoplasm. This division, called cytokinesis, differs in plants and animals.

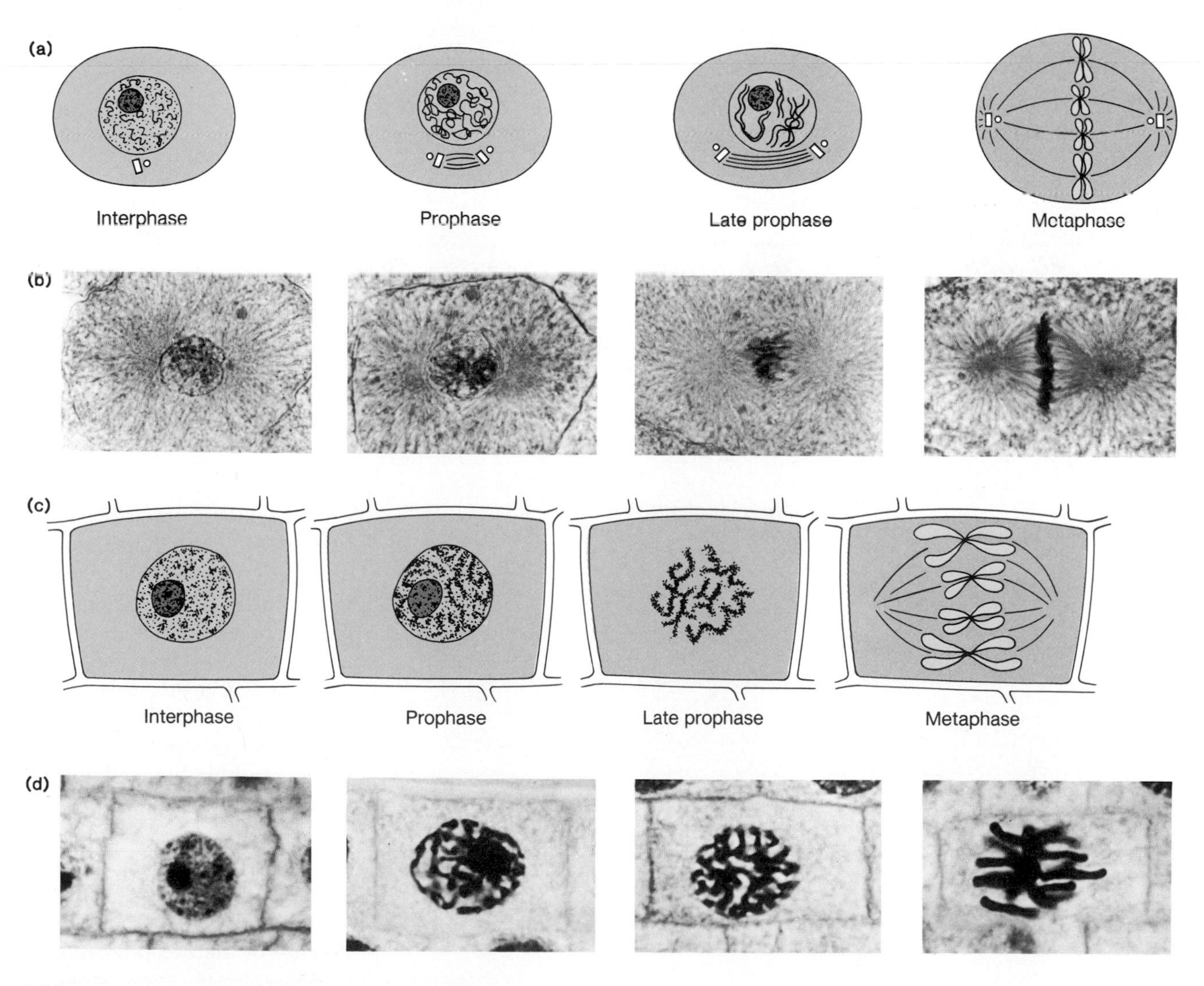

Figure 8.10 **A Comparison of Plant and Animal Mitosis.** (a) Illustration of mitosis in an animal cell. (b) Photographs of mitosis in a whitefish blastula. (c) Illustration of mitosis in a plant cell. (d) Photographs of mitosis in an onion-root tip.

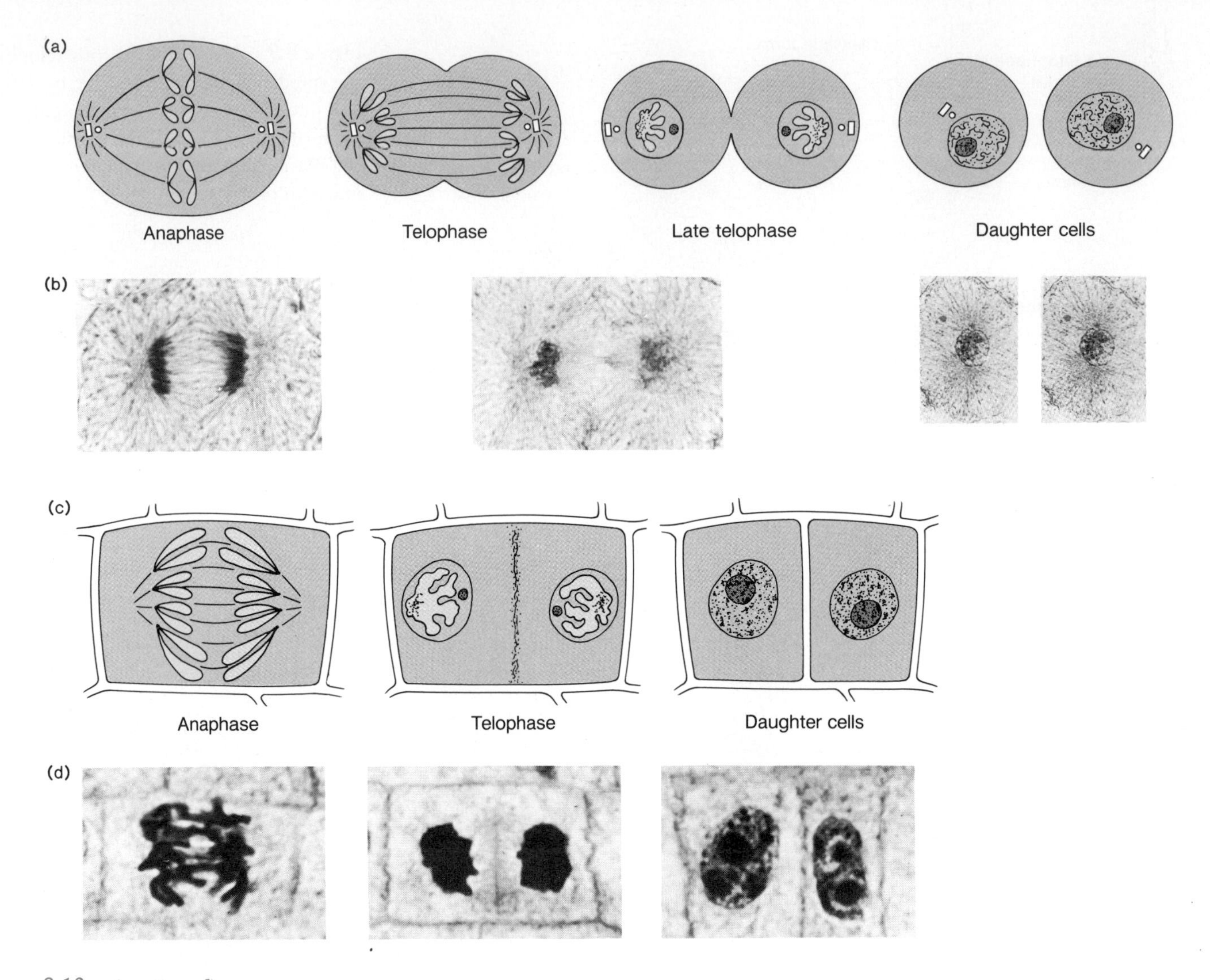

Figure 8.10 *(continued)*

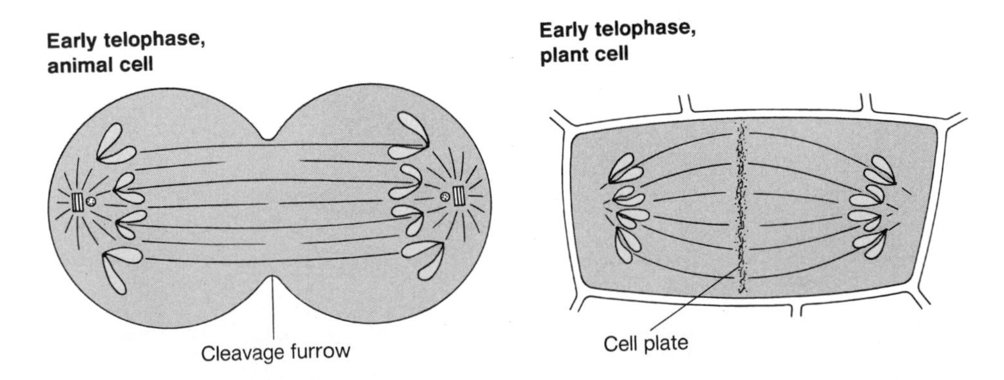

Figure 8.11 **Cytokinesis.** In animal cells there is a pinching in of the cytoplasm that eventually forms two daughter cells. Daughter cells in plants are formed when a cell plate separates the cell into two cells.

cells, cytokinesis results from a **cleavage furrow.** This is an indentation of the cell membrane of an animal cell that pinches the cytoplasm into two parts as if a string were tightened about its middle. In an animal cell, cytokinesis begins at the cell membrane and proceeds to the center. In plant cells, a **cell plate** begins at the center and proceeds to the cell membrane, resulting in a cell wall that separates the two daughter cells.

Differentiation

Because of the two processes in cell division, mitosis and cytokinesis, the daughter cells have the same genetic composition. You received a set of genes from your father in his sperm, and a set of genes from your mother in her egg. By cell division, this cell formed two daughter cells. This process was repeated, and there were four cells, all of which had the same genes. All the trillions of cells in your body were formed by the process of cell division. This means that, except for mutations, all the cells in your body have the same genes.

However, all the cells in your body are not the same. There are nerve cells, muscle cells, bone cells, skin cells, and many other types. How is it possible that cells with the same genes can produce different types of cells? Think of the genes in a cell as individual recipes in a cookbook. You could give a copy of the same cookbook to one hundred people and, although they all have the same book, each person could prepare a different dish. If you use the recipe to make a chocolate cake, you ignore the directions for making salads, fried chicken, and soups, although these recipes also are in the book.

It is the same with cells. Although some genes are used by all cells, some cells only activate certain genes. Muscle cells produce proteins capable of contraction. Most other cells do not use these genes. Pancreas cells use genes that result in the formation of digestive enzymes, but they never produce contractile proteins. **Differentiation** is the process of forming specialized cells within a multicellular organism. Some cells, such as muscle and nerve cells, lose their ability to divide; they remain permanently in the interphase condition. Other cells retain their ability to divide as their form of specialization. Cells that line the digestive tract or form the surface of your skin are examples of dividing cells.

In growing organisms, such as infants, seedlings, or embryos, most cells are capable of division and divide at a rapid rate. In older organisms, many cells lose their ability to divide as a result of differentiation, and the frequency of cell division decreases. As the organism ages, the lower frequency of cell division may affect many bodily processes, including healing. In some older people, there may be so few cells capable of dividing that a broken bone may never heal. It is also possible for a cell to undergo mitosis but not cytokinesis. In skeletal muscle cells, the cells undergo mitosis but not cytokinesis, which results in multinucleated cells (figure 8.12).

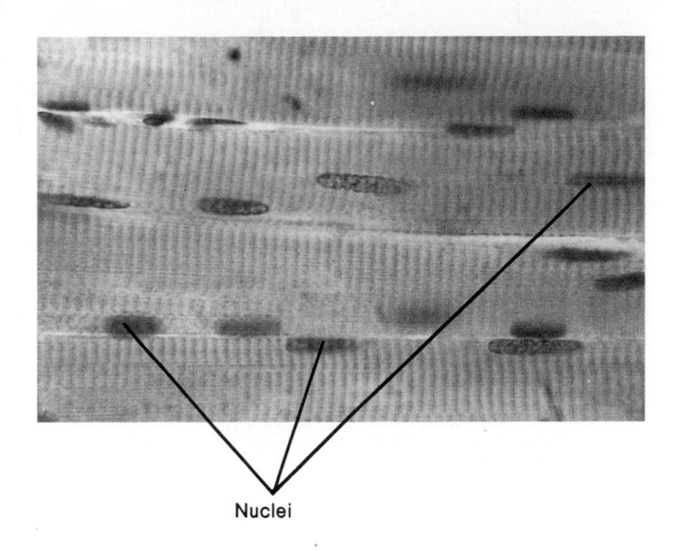

Figure 8.12 **Multinucleated Cells.** Skeletal muscles are used to move your arms, legs, and other parts of your body. As these muscles grow, they do not increase in number of cells. Only the size of the cells increases. Because the muscle cells shown here went through the process of mitosis but not cytokinesis, individual cells contain several nuclei.

Abnormal Cell Division

Understanding mitosis can help you understand certain biological problems and how to solve them. All cells do not divide at the same rate, but each kind of cell has a regulated division rhythm. Regulation of the cycle can come from inside or outside the cell. When human white blood cells are grown outside the body under special conditions, they develop a regular cell-division cycle. The cycle is determined by the DNA of the cells. However, white blood cells in the human body may increase their rate of mitosis as a result of outside influences. Disease organisms entering the body, tissue damage, and changes in cell DNA all may alter the rate at which white blood cells divide. An increase in white blood cells in response to the invasion of disease organisms is valuable because these white blood cells are capable of destroying the disease-causing organisms.

On the other hand, an uncontrolled increase in the rate of mitosis in white blood cells causes a kind of cancer known as *leukemia*. This condition causes a general weakening of the body because the excess number of white blood cells diverts necessary nutrients from other cells of the body and interferes with their normal activities.

Normally, cells become specialized for a particular function. Each cell type has its cell-division process regulated so that it does not interfere with the activities of other cells or the whole organism. However, some cells may revert to an embryonic state and begin to divide. Sometimes this division occurs in an uncontrolled fashion. When this happens, a group of cells forms what is known as a *tumor* (figure 8.13). A *benign* tumor is a cell mass that does not fragment and spread beyond its original area of growth. A benign tumor can become harmful by growing to a point at which it interferes with normal body functions. Some tumors are *malignant*. Cells of these tumors move from the original site (*metastasize*) and establish new colonies in other regions of the body. **Cancer** is an abnormal growth of cells that has a malignant potential.

Once cancer has been detected, treatment requires the elimination of the tumor. There are three kinds of treatment. If the cancer is confined to a few

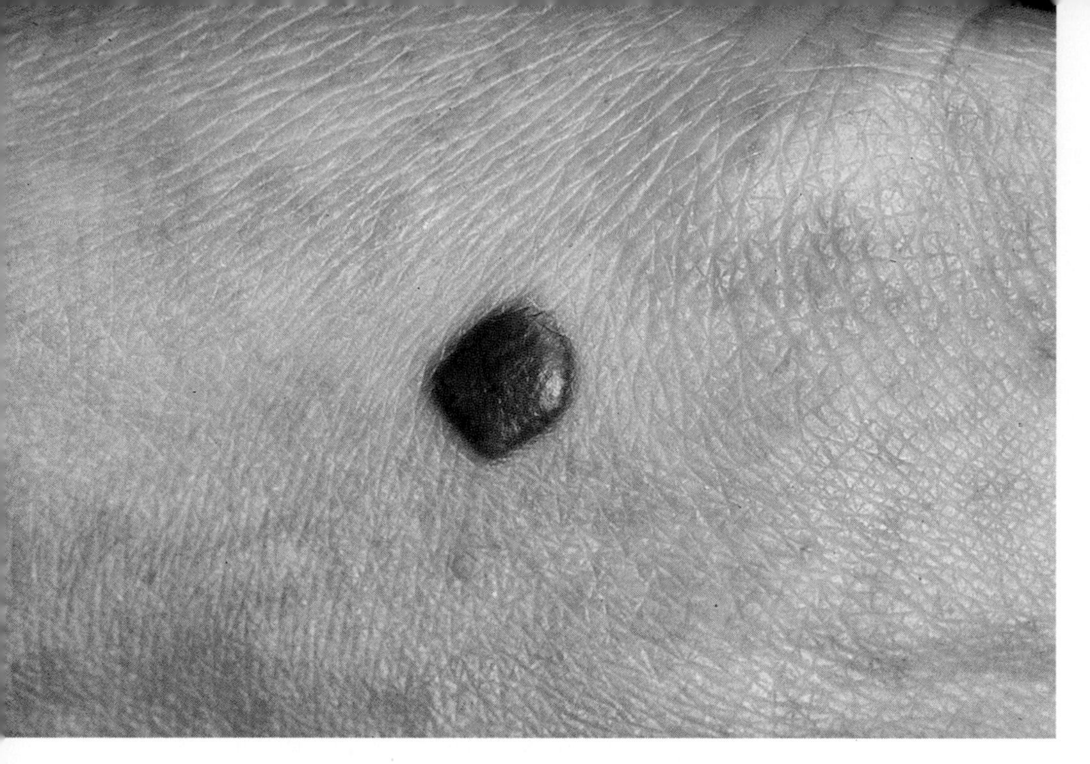

Figure 8.13 **Cancer.** Malignant melanoma is a type of skin cancer. It forms as a result of a mutation in a pigmented skin cell. Only the dark area in the photograph is the cancer; the surrounding cells have the genetic information to develop into normal, healthy skin.

specific locations, it may be possible to surgically remove it. However, there are cases where surgery is impractical. If the tumor is located where it can't be removed without destroying healthy vital tissue, surgery may not be used. For example, the removal of certain brain cancers could severely damage the brain. In these cases, two other methods may be used to treat cancer.

Chemotherapy uses various types of chemicals to destroy cancer cells. This treatment may be used without knowing exactly where the cancer is located. However, it has negative effects on normal cells. It lowers the body's immune reaction because it decreases its ability to produce white blood cells. Other side effects include intestinal disorders and the loss of hair, which are caused by damage to healthy dividing cells in the scalp and the intestinal tract.

Radiation therapy uses large amounts of X rays or gamma rays. Since this treatment damages surrounding healthy cells, it is used only very cautiously when surgery is impractical. Radiation therapy can be effective because cells undergoing division are usually not in the G_1 stage. Nondividing cells pause in the G_1 stage and are less likely to be damaged by treatment. Cancer cells are usually in the S or G_2 stage. In these stages, DNA replication or protein synthesis is occurring and the cell is more subject to damage. Therefore, while radiation therapy may destroy healthy cells, it usually destroys more cancer cells.

Chemotherapy and radiation treatment are often used to control leukemia by taking advantage of the fact that these cells are undergoing an unusual mitosis. Dividing cells are likely to be damaged by x-radiation because the radiation can more easily destroy essential molecules (DNA) of the cell. Since cancer cells spend more time dividing than normal cells, they have a greater chance of being killed by radiation. Physicians, therefore, often prescribe cobalt therapy for certain cancer patients with widely dispersed cancers such as leukemia. Cobalt is radioactive and releases radiation.

As a treatment for cancer, radiation is dangerous for the same reasons that it is beneficial. In cases of extreme exposure to radiation, people develop what is called *radiation sickness.* The symptoms of this disease include loss of hair, bloody vomiting and diarrhea, and a reduced white blood cell count. These symptoms occur in parts of the body where mitosis is common. The lining of the intestine is constantly being lost as food travels through and it must be replaced by the process of mitosis. Hair growth is the result of the continuous division of cells at the roots. White blood cells are also continuously reproduced in the bone marrow and lymph nodes. When radiation strikes these rapidly dividing cells and kills them, the lining of the intestine wears away and bleeds, hair falls out, and few new white blood cells are produced to defend the body against infection.

• Summary •

Cell division is necessary for growth, repair, and reproduction. Cells go through a cell cycle that includes cell division (mitosis and cytokinesis) and interphase. Interphase is the period of growth and preparation for division. Mitosis is divided into four stages: prophase, metaphase, anaphase, and telophase. During mitosis, two daughter nuclei are formed from one parent nucleus. These nuclei have identical sets of chromosomes and genes that are exact copies of those of the parent. Although the process of mitosis has been presented as a series of phases, you should realize that it is a continuous, flowing process from prophase through telophase. Following mitosis, cytokinesis divides the cytoplasm, and the cell returns to interphase.

The regulation of mitosis is important if organisms are to remain healthy. Regular divisions are necessary to replace lost cells and allow for growth. However, uncontrolled cell division may result in cancer and disruption of the total organism's well-being.

• Thinking Critically •

A chemical known as *colchicine* is extracted from the seeds of a small, crocuslike plant. This chemical is used in biological laboratories because it can prevent the formation of the spindle. Which parts of the cell cycle would proceed normally and which parts would be altered if this chemical were used on cells? If you know that the cells are not killed by colchicine and begin mitosis normally, what changes might occur in the number of chromosomes of the next cell generation, and how might this change the metabolism of the cell?

• Experience This •

What Is Cancer?

Cancer is a large group of diseases characterized by uncontrolled growth and the spread of abnormal cells. If the spread is not controlled or checked, it results in death. However, many cancers can be cured if detected and treated promptly.

Cancer's Seven Warning Signals

1. A change in bowel or bladder habits
2. A sore that does not heal
3. Unusual bleeding or discharge
4. A thickening or lump in the breast or elsewhere
5. Indigestion or difficulty in swallowing
6. An obvious change in a wart or mole
7. A nagging cough or hoarseness

Thirty percent of Americans will eventually have cancer. Cancer will strike three out of four families. If you notice one of the preceding symptoms, see your doctor.

To become better informed on cancer, contact your local Cancer Society office or write the American Cancer Society, Inc., 90 Park Ave., New York, N.Y. 10016. Request a copy of the latest publication *Cancer Facts and Figures*.

• Questions •

1. Name the four stages of mitosis and describe what occurs in each stage.
2. What is meant by cell cycle?
3. During which stage of a cell's cycle does DNA replication occur?
4. At what phase of mitosis does the DNA become most visible?
5. What are the differences between plant and animal mitosis?
6. Why can X-ray treatment be used to control cancer?
7. What is the purpose of mitosis?
8. What is the difference between a cell plate and a cell wall?
9. What types of activities occur during interphase?
10. List five differences between an interphase cell and a cell in mitosis.

• Chapter Glossary •

anaphase (an'ă-fāz) The third stage of mitosis, characterized by splitting of the centromeres and movement of the chromosomes to the poles.

cancer (kan'sur) A tumor that is malignant.

cell plate (sel plāt) A plant-cell structure that begins to form in the center of the cell and proceeds to the cell membrane, resulting in cytokinesis.

centrioles (sen'tre-ōls) Organelles containing microtubules located just outside the nucleus.

centromere (sen'tro-mēr) The region where two chromatids are joined.

chromatid (kro'mah-tid) One of two component parts of a chromosome formed by replication and attached at the centromere.

chromosomes (kro'mo-sōmz) Complex structures within the nucleus composed of various kinds of histone proteins and DNA that contains the cell's genetic information.

cleavage furrow (kle'vaj fuh'ro) An indentation of the cell membrane of an animal cell that pinches the cytoplasm into two parts.

cytokinesis (si-to-kĭ-ne'sis) Division of the cytoplasm of one cell into two new cells.

daughter cells (daw'tur sels) Two cells formed by cell division.

daughter chromosomes (daw'tur kro'mo-sōmz) Chromosomes produced by DNA replication that contain identical genetic information; formed after chromosome division in anaphase.

daughter nuclei (daw'tur nu'kle-i) Two nuclei formed by mitosis.

differentiation (dif''fur-ent-she-a'shun) The process of forming specialized cells within a multicellular organism.

interphase (in'tur-fāz) The stage between cell divisions in which the cell is engaged in metabolic activities.

metaphase (me'tah-fāz) The second stage in mitosis, characterized by alignment of the chromosomes at the equatorial plane.

mitosis (mi-to'sis) A process that results in equal and identical distribution of replicated chromosomes into two newly formed nuclei.

prophase (pro'fāz) The first phase of mitosis during which individual chromosomes become visible.

spindle (spin'dul) An array of microtubules extending from pole to pole; used in the movement of chromosomes.

telophase (tel'uh-fāz) The last phase in mitosis characterized by the formation of daughter nuclei.

C H A P T E R 9

Meiosis—Sex-Cell Formation

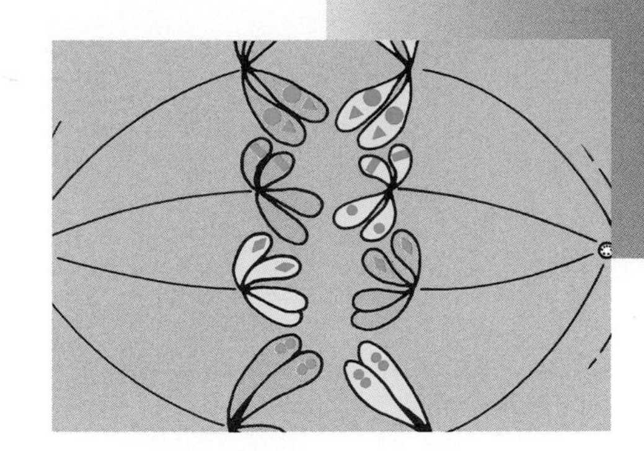

Purpose

How can the chromosome number in humans remain at forty-six generation after generation if both parents contribute equally to the genetic information of the child? In this chapter, we will discuss the mechanics of the process of meiosis. Meiosis is a specialized cell division that results in the formation of sex cells. Knowing the mechanics of this process is essential to understanding how genetic variety can occur in sex cells. This variety ultimately shows up as differences in offspring.

For Your Information

Decreased sperm counts have been noted in American males. For a species to survive and not be in danger of extinction, enough healthy sperm and eggs must be produced by individual mating members of the population. The reproductive success of any species is influenced by many factors. As levels of toxins increase in the environment and people use and abuse more medicines, street drugs, and other chemicals, the risk of sterility increases due to a decrease in the production of viable gametes. Numerous agents have been shown to interfere with sperm production, including radiation, lead, marijuana, alcohol, certain pesticides, and the antibiotic tetracycline. Only detailed research will indicate whether or not these agents are responsible for reduced sperm counts and if the trend toward lower sperm counts will result in less reproductive success.

Learning Objectives

- Explain why sexually reproducing organisms must form cells with the haploid number of chromosomes.
- Describe the stages in meiosis I.
- Describe the stages in meiosis II.
- Diagram how trisomy results from errors in the meiotic process.
- Understand how genetic variety in offspring is generated by mutation, crossing-over, segregation, independent assortment, and fertilization.
- Explain the similarities and differences between mitosis and meiosis.

Chapter Outline

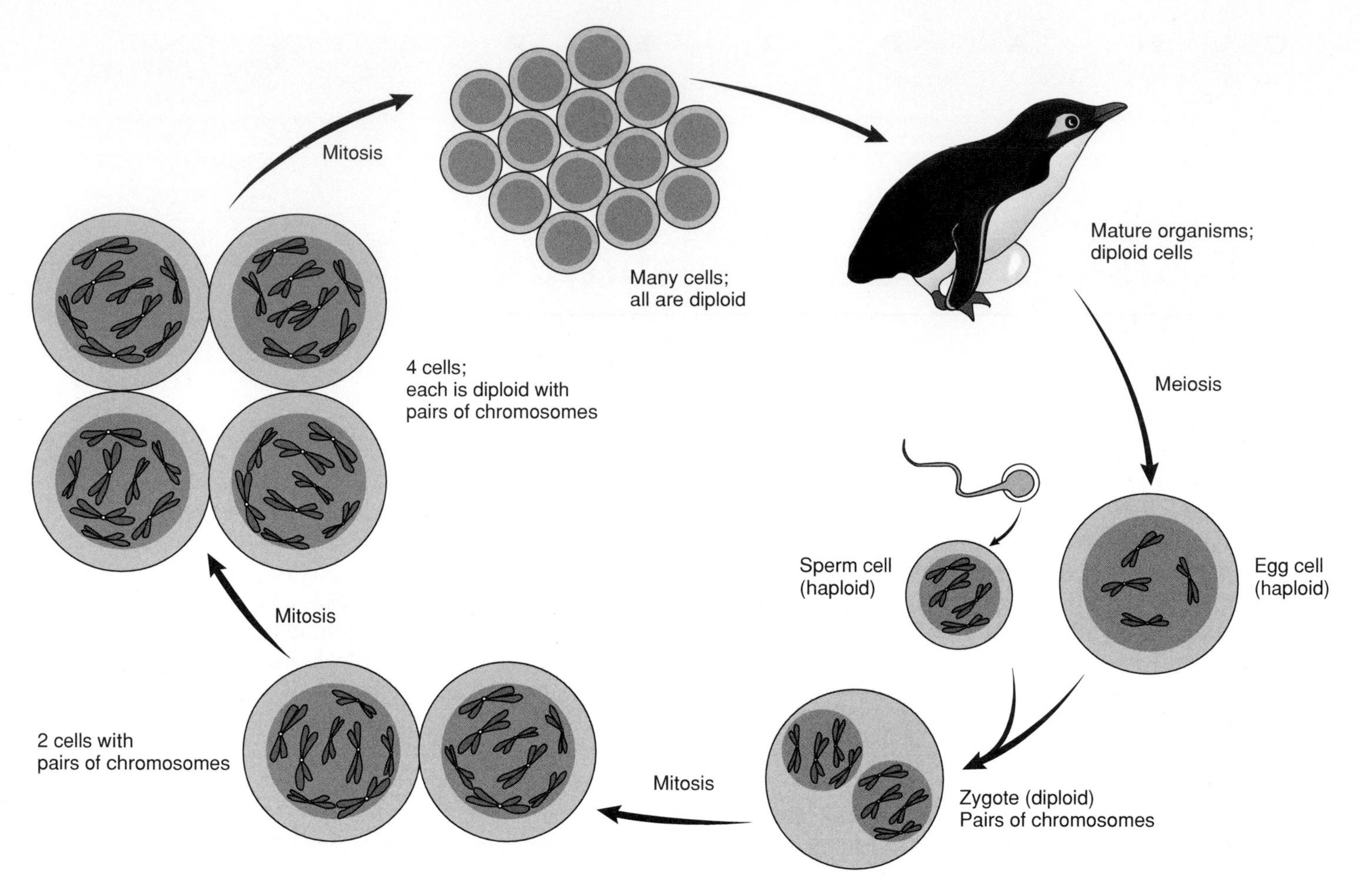

Figure 9.1 **Life Cycle.** The cells of this adult penguin have, for our purpose, eight chromosomes in their nuclei. In preparation for sexual reproduction, the number of chromosomes must be reduced by half so that fertilization will result in the original number of eight chromosomes in the new individual. The offspring will grow and produce new cells by mitosis. (The actual number of chromosomes has not been shown.)

Sexual Reproduction

The most successful kinds of plants and animals are those that have developed a method of shuffling and exchanging genetic information. This usually involves organisms that have two sets of genetic data, one inherited from each parent. **Sexual reproduction** is the formation of a new individual by the union of sex cells. Before sexual reproduction can occur, the two sets of genetic information must be reduced to one set. This is somewhat similar to shuffling a deck of cards and dealing out hands: the shuffling and dealing assure that each hand will be different. An organism with two sets of chromosomes can produce many combinations of chromosomes when it produces sex cells, just as many different hands can be dealt from one pack of cards. When one of these sex cells unites with another, a new organism containing two sets of genetic information is formed. This new organism's information might very well be superior to the information found in either parent; this is the value of sexual reproduction.

In chapter 8, we discussed the cell cycle and pointed out that it is a continuous process, without a beginning or an end. The process of mitosis followed by growth is important in the life cycle of any organism. Thus, the *cell cycle* is part of an organism's *life cycle* (figure 9.1).

The sex cells produced by male organisms are called **sperm,** and those produced by females are called **eggs.** A general term sometimes used to refer to either eggs or sperm is the term **gamete.** The cellular process that is responible for generating gametes is called **gametogenesis.** The uniting of an egg and sperm (gametes) is known as **fertilization.**

The **zygote,** which results from the union of an egg and a sperm, divides repeatedly by mitosis to form the complete organism. Notice in figure 9.1 that the zygote and its descendants have two sets of chromosomes. However, the male gamete and the female gamete each contain only one set of chromosomes. These sex cells are said to be **haploid.** The haploid number of chromosomes is noted as n. A zygote contains two sets and is said to be **diploid.** The diploid number of chromosomes is noted as $2n$ ($n + n = 2n$). Diploid cells have two sets of chromosomes, one set from each parent. Remember, a chromosome is composed of two chromatids, each containing duplex DNA. These two chromatids are attached to each other at a point called the *centromere.* In a diploid nucleus, the chromosomes occur as **homologous chromosomes**—a pair of chromosomes in a diploid cell that contain similar genes throughout their length. One of the chromosomes of a homologous pair was donated by the father, the other by the mother (figure 9.2). Different species of organisms vary in the number of chromosomes they contain. Table 9.1 lists several different organisms and their haploid and diploid chromosome numbers.

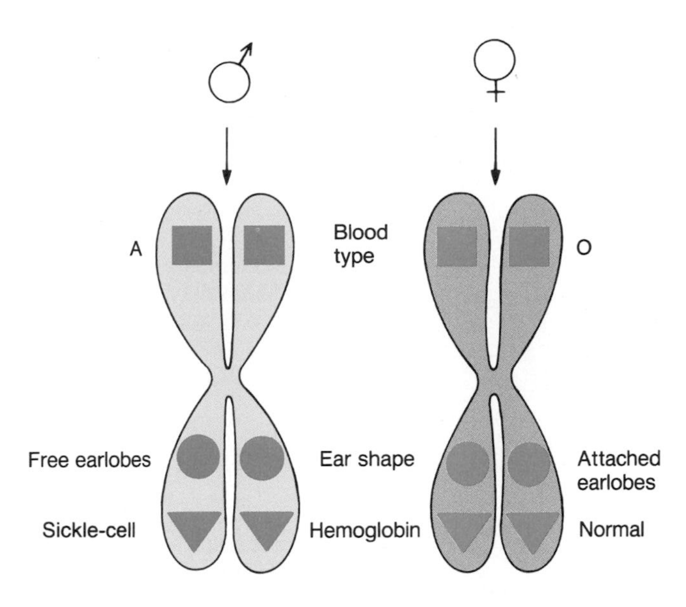

Figure 9.2 **A Pair of Homologous Chromosomes.** A pair of chromosomes of similar size and shape that have genes for the same traits are said to be homologous. Notice that the genes may not be identical but code for the same type of information.

It is necessary for organisms that reproduce sexually to form gametes having only one set of chromosomes. If gametes contained two sets of chromosomes, the zygote resulting from their union would have four sets of chromosomes. The number of chromosomes would continue to double with each new generation, which could result in death. However, this does not happen; the number of chromosomes remains constant generation after generation. Since cell division by mitosis and cytokinesis results in cells that have the same number of chromosomes as the parent cell, two questions arise: how are sperm and egg cells formed, and how do they get only one-half the chromosomes of the diploid cell? The answers lie in the process of **meiosis,** the specialized pair of cell divisions that reduce the chromosome number from diploid ($2n$) to haploid (n). The major function of meiosis is to produce cells that have one set of genetic information. Therefore, when fertilization occurs, the zygote will have two sets of chromosomes, as did each parent.

Not every cell goes through the process of meiosis. Only specialized organs are capable of producing haploid cells (figure 9.3). In animals, the organs in which meiosis occurs are called **gonads.** The female gonads that produce eggs are called **ovaries.** The male gonads that produce sperm are called **testes.** Organs that produce gametes are also found in flowering plants. In plants, the **pistil** produces eggs or ova, and the **anther** produces pollen, which contain sperm.

To illustrate meiosis in this chapter, for clarity we have chosen to

Table 9.1
Chromosome Numbers

Organism	Haploid Number	Diploid Number
Mosquito	3	6
Fruit fly	4	8
Housefly	6	12
Toad	18	36
Cat	19	38
Human	23	46
Hedgehog	23	46
Chimpanzee	24	48
Horse	32	64
Dog	39	78
Onion	8	16
Kidney bean	11	22
Rice	12	24
Tomato	12	24
Potato	24	48
Tobacco	24	48
Cotton	26	52

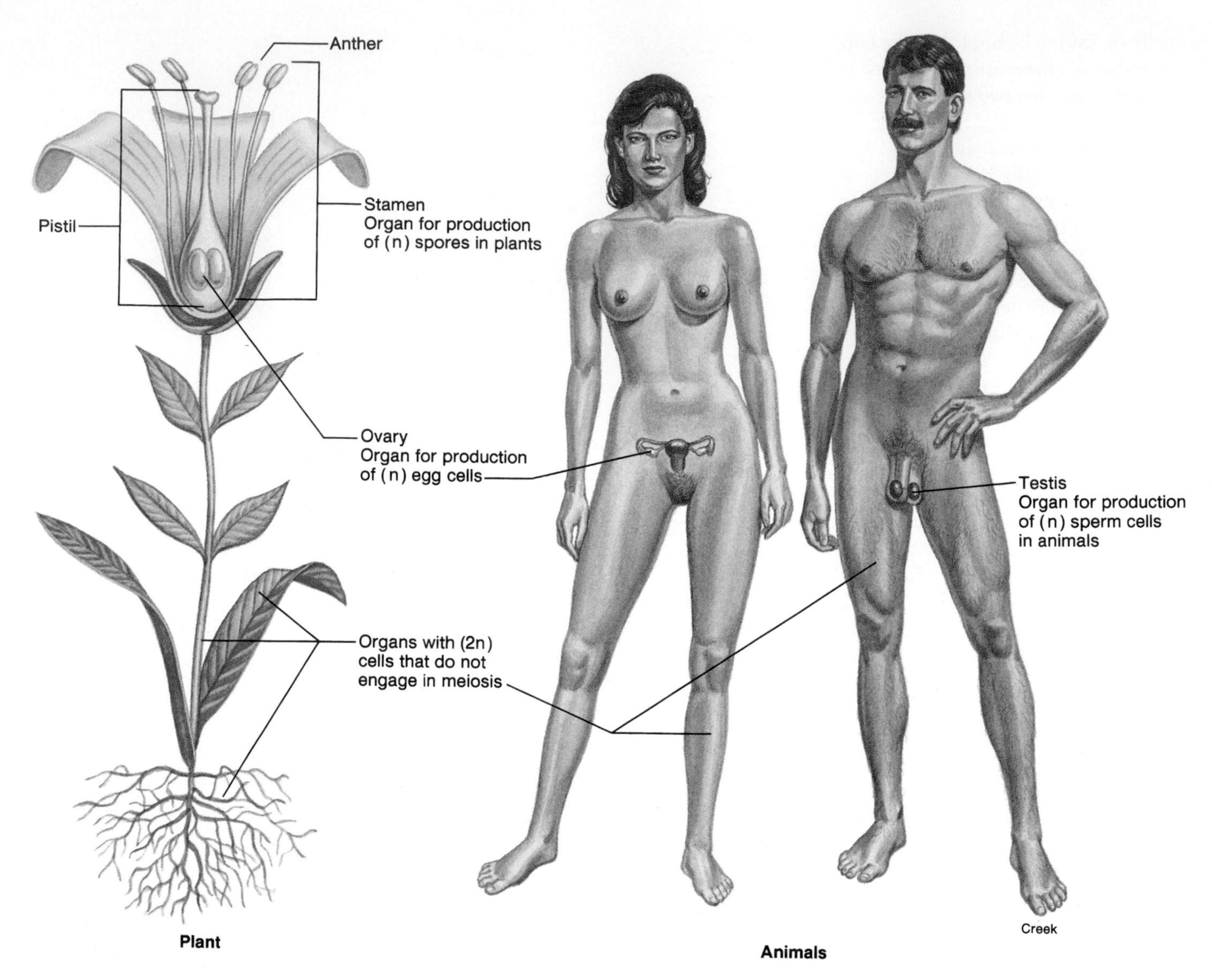

Figure 9.3 **Haploid and Diploid Cells.** Both plants and animals produce cells with a haploid number of chromosomes. The male anther in plants and the testes in animals produce haploid male cells, sperm. In both plants and animals, the ovaries produce haploid female cells, eggs.

show only 8 chromosomes (figure 9.4). In reality humans have 46 chromosomes (23 pairs). The haploid number of chromosomes in this cell is four, and these haploid cells contain only one complete set of four chromosomes. You can see, there are eight chromosomes in this cell—four from the mother and four from the father. A closer look at figure 9.4 shows you there are only four types of chromosomes, but two of each type:

1. long chromosomes consisting of chromatids attached near the center;
2. long chromosomes consisting of chromatids attached near one end;
3. short chromosomes consisting of chromatids attached near one end; and
4. short chromosomes consisting of chromatids attached near the center.

We can therefore talk about the number of chromosomes in two ways. We can say that our hypothetical diploid cell has eight chromosomes, or we can say that it has four pairs of homologous chromosomes.

Haploid cells, on the other hand, do not have homologous chromosomes. They have only one of each type of chromosome. The whole point of meiosis is to distribute the chromosomes and the genes they carry so that each daughter cell gets one member of each homologous pair. In this way, each daughter cell gets one complete set of genetic information.

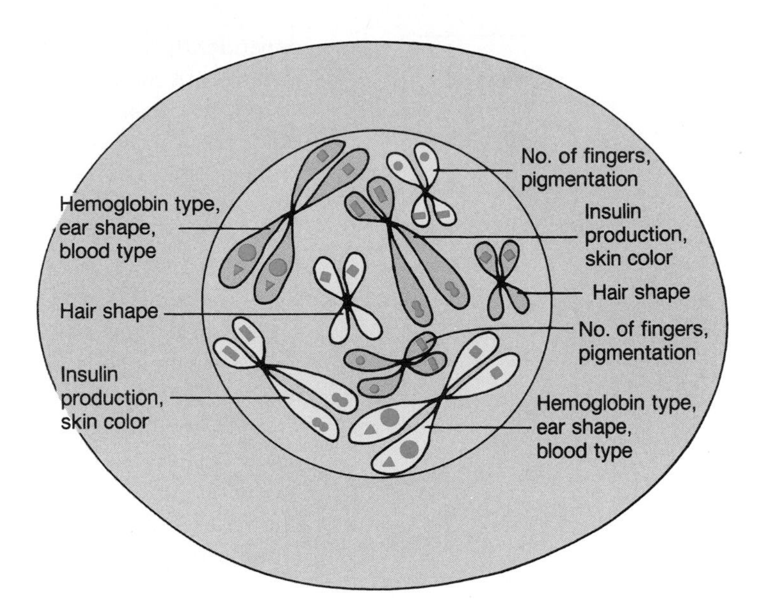

Figure 9.4 **Chromosomes in a Cell.** In this diagram of a cell, the eight chromosomes are scattered in the nucleus. Even though they are not arranged in pairs, note that there are four pairs of homologous chromosomes. Check to be sure you can pair them up using the list of characteristics.

The Mechanics of Meiosis: Meiosis I

Meiosis is preceded by an interphase stage when DNA replication occurs. In a sequence of events called *meiosis I,* members of homologous pairs of chromosomes divide into two complete sets. This is sometimes called a **reduction division,** a type of cell division in which daughter cells get only half the chromosomes from the parent cell. The division begins with chromosomes composed of two chromatids. The sequence of events in meiosis I is artificially divided into four phases: prophase I, metaphase I, anaphase I, and telophase I.

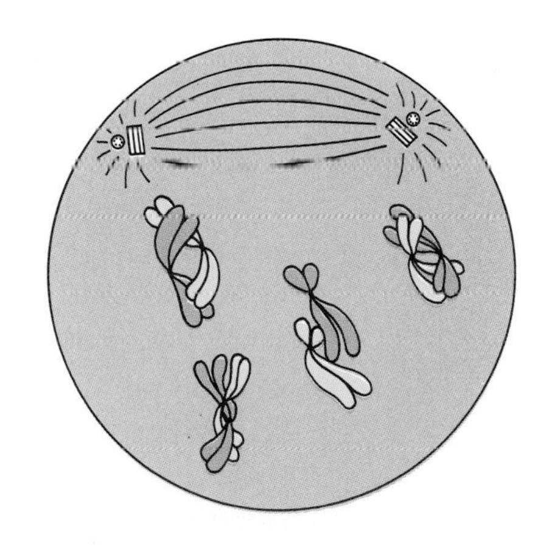

Figure 9.5 **Prophase I.** During prophase I, the cell is preparing for division. A unique event that occurs in prophase I is the synapsis of the chromosomes. Notice that the nuclear membrane is no longer apparent and that the paired homologues are free to move about the cell.

Prophase I

During prophase I, the cell is preparing itself for division (figure 9.5). The chromatin material coils and thickens into chromosomes, the nucleoli disappear, the nuclear membrane disintegrates, and the spindle begins to form. The spindle is formed in animals when the centrioles move to the poles. There are no centrioles in plant cells, but the spindle does form. However, there is an important difference between the prophase stage of mitosis and prophase I of meiosis. During prophase I, homologous chromosomes come to lie next to each other in a process called **synapsis.** While the chromosomes are synapsed, a unique event called *crossing-over* can occur. **Crossing-over** is the exchange of equivalent sections of DNA on homologous chromosomes. We will fit crossing-over into the whole picture of meiosis later.

Metaphase I

The synapsed pair of homologous chromosomes now move into position on the equatorial plane of the cell. In this stage, the centromere of each chromosome attaches to the spindle. The synapsed homologous chromosomes move to the equator of the cell as single units. How they are arranged on the equator (which one is on the left and which one is on the right) is determined by chance (figure 9.6). In the cell in figure 9.6, three green chromosomes from the father and one purple chromosome from the mother are lined up on the left. Similarly, one green chromosome from the father and three purple chromosomes from the mother are on the right. They could have aligned themselves in several other ways. For instance, they could have lined up as shown in the rectangular box at the right in figure 9.6.

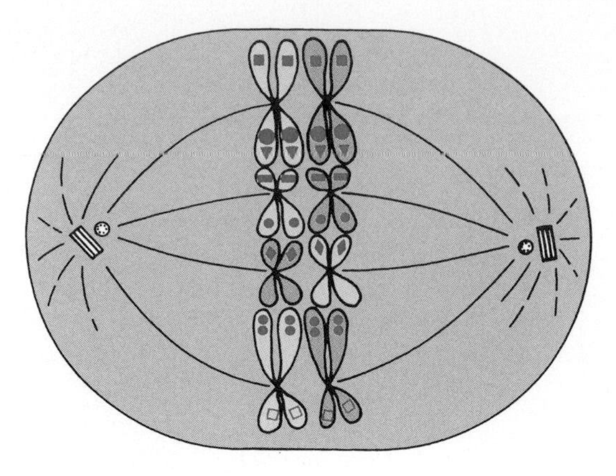

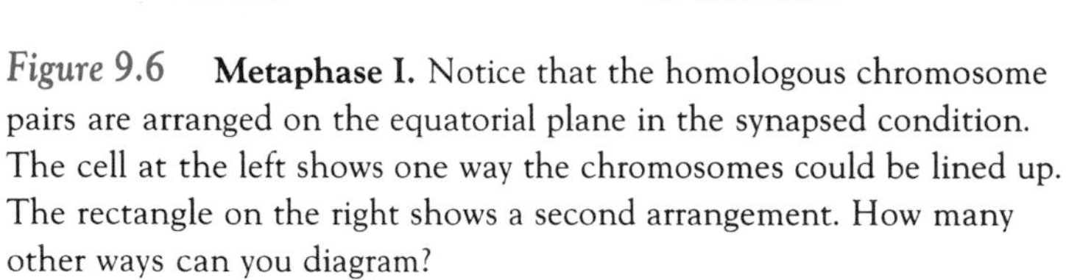

Figure 9.6 **Metaphase I.** Notice that the homologous chromosome pairs are arranged on the equatorial plane in the synapsed condition. The cell at the left shows one way the chromosomes could be lined up. The rectangle on the right shows a second arrangement. How many other ways can you diagram?

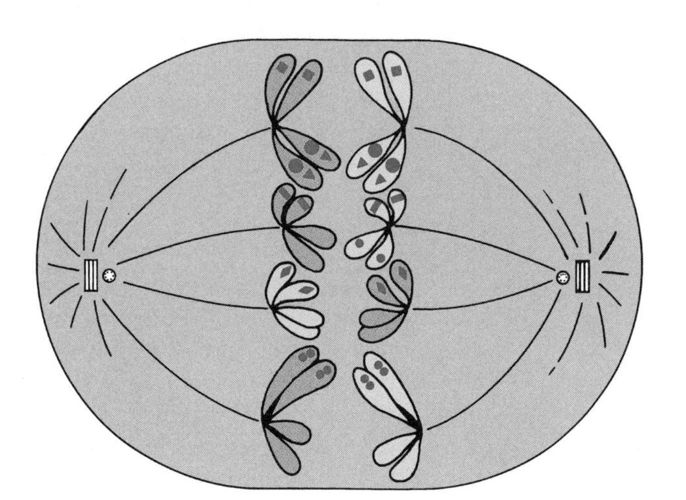

Figure 9.7 **Anaphase I.** During this phase, one member of each homologous pair is segregated from the other member of the pair. Notice that the centromeres of the chromosomes do not split.

Anaphase I

Anaphase I is the stage during which homologous chromosomes separate (figure 9.7). During this stage, the chromosome number is reduced from diploid to haploid. The two members of each pair of homologous chromosomes move away from each other toward opposite poles. The direction each takes is determined by how each pair was originally arranged on the spindle. Each chromosome is independently attached to a spindle fiber at its centromere. Unlike the anaphase stage of mitosis, the centromeres that hold the chromatids together *do not divide* during anaphase I of meiosis. Each chromosome still consists of two chromatids. Because the chromosomes and the genes they carry are being separated from one another, this process is called **segregation.** The way in which a single pair of homologous chromosomes segregates does not influence how other pairs of homologous chromosomes segregate. That is, each pair segregates independently of other pairs. This is known as **independent assortment** of chromosomes.

Telophase I

Telophase I consists of changes that return the cell to an interphase condition (figure 9.8). The chromosomes uncoil and become long, thin threads, the nuclear membrane re-forms around them, and nucleoli reappear. During this activity, cytokinesis divides the cytoplasm into two separate cells.

Because of meiosis I, the total number of chromosomes is divided equally, and each daughter cell has one member of each homologous chromosome pair. This means that the genetic data each cell receives is one half of the total, but each cell still has a complete set of the genetic information. Each individual chromosome is still composed of two chromatids joined at the centromere, and the chromosome number is reduced from diploid ($2n$) to haploid (n). In the cell we have been using as our example, the number of chromosomes is reduced from eight to four. The four pairs of chromosomes have been distributed to the two daughter cells.

Depending on the type of cell, there may be a time following telophase I when a cell engages in normal metabolic activity that corresponds to an interphase stage. However, the chromosomes do not replicate before the cell enters meiosis II. Figure 9.9 shows the events in meiosis I.

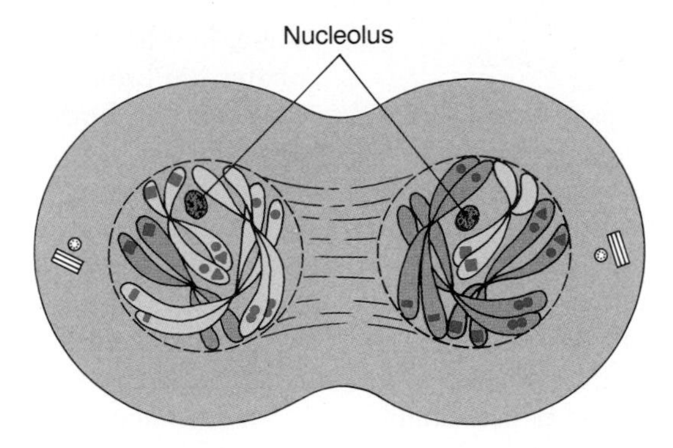

Figure 9.8 **Telophase I.** What activities would you expect during the telophase stage of cell division? What term is used to describe the fact that the cytoplasm is beginning to split the parent cell into two daughter cells?

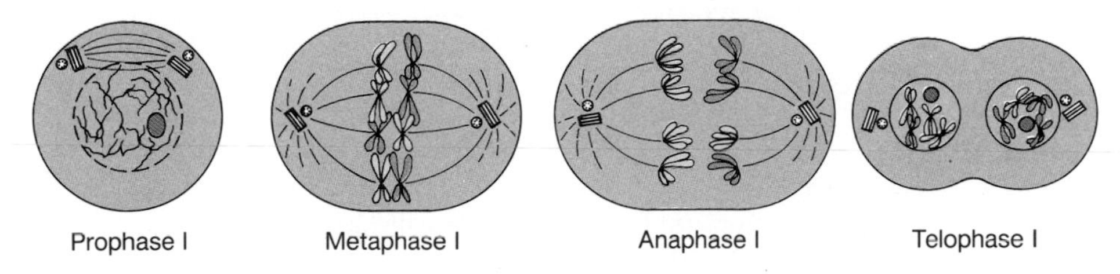

Figure 9.9 **Meiosis I.** The stages in meiosis I result in reduction division. This reduces the number of chromosomes in the parental cell from the diploid number to the haploid number in each of the two daughter cells.

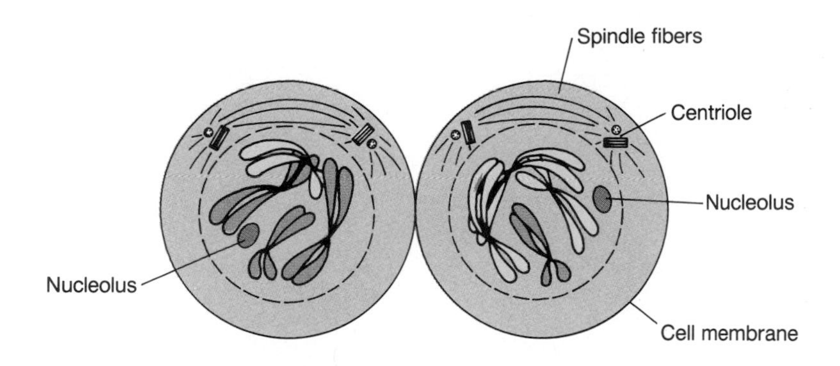

Figure 9.10 **Prophase II.** The two daughter cells are preparing for the second division of meiosis. Study this diagram carefully. Solely from your observations, can you list the events of this stage?

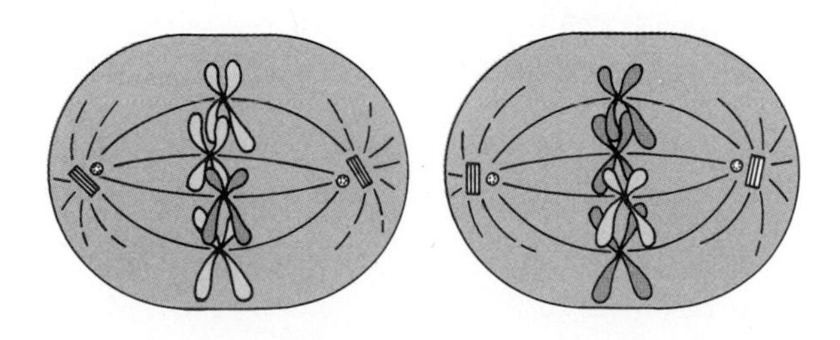

Figure 9.11 **Metaphase II.** During this metaphase, each chromosome lines up on the equatorial plane. Each chromosome is composed of two chromatids joined at a centromere. How does metaphase II of meiosis compare to metaphase of mitosis?

The Mechanics of Meiosis: Meiosis II

Meiosis II includes four phases: prophase II, metaphase II, anaphase II, and telophase II. The two daughter cells formed during meiosis I continue through meiosis II, so that usually four cells result from the two divisions.

Prophase II

Prophase II is similar to prophase in mitosis; the nuclear membrane disintegrates, nucleoli disappear, and the spindle apparatus begins to form. However, it differs from prophase I because these cells are haploid, not diploid (figure 9.10). Also, synapsis, crossing-over, segregation, and independent assortment do not occur during prophase II.

Metaphase II

The metaphase II stage is typical of any metaphase stage because the chromosomes attach by their centromeres to the spindle at the equatorial plane of the cell. Since pairs of chromosomes are no longer together in the same cell, each chromosome moves as a separate unit (figure 9.11).

Anaphase II

Anaphase II differs from anaphase I because the centromere of each chromosome splits in two, and the chromatids, now called *daughter chromosomes,* move to the poles (figure 9.12). Remember, there are no paired homologs in this stage; therefore, segregation and independent assortment cannot occur.

Telophase II

During telophase II, the cell returns to a nondividing condition. As cytokinesis occurs, new nuclear membranes form, chromosomes uncoil, nucleoli re-form, and the spindles disappear (figure 9.13).

This stage is followed by differentiation; the four cells mature into gametes—either sperm or eggs. The events of meiosis II are summarized in figure 9.14.

In many organisms, egg cells are produced in such a manner that three of the four cells resulting from meiosis in a female disintegrate. However, since the one that survives is randomly chosen, the likelihood of any one particular combination of genes being formed is not affected. The whole point of learning the mechanism of meiosis is to see how variation happens. Now we can look at variation and how it comes about.

Sources of Variation

The formation of a haploid cell by meiosis and the combination of two haploid cells to form a diploid cell by sexual reproduction results in variety in the offspring. There are five factors that influence genetic variation in offspring: mutations, crossing-over, segregation, independent assortment, and fertilization.

Two types of mutations were discussed in chapter 7: point mutations and chromosomal mutations. In point mutations, there is a change in a DNA nucleotide that results in the production of a different protein. In chromosomal mutations, genes are rearranged. By causing the production of different proteins, both types of mutations increase variation. The second source of variation is crossing-over.

Crossing-Over

Crossing-over is the exchange of a part of a chromatid from one homologous chromosome with an equivalent part of a chromatid from the other homologous chromosome. This exchange results in a new gene combination. Crossing-over occurs during meiosis I while homologous chromosomes are synapsed. Remember that a chromosome is a double strand of DNA. To break a chromosome, bonds between sugars and phosphates are broken. This is done at the same spot on both chromatids, and the two pieces switch places. After switching places, the two pieces of DNA are bonded together by re-forming the bonds between the sugar and the phosphate molecules. Examine figure 9.15 carefully to note precisely what occurs during crossing-over.

This figure shows a pair of homologous chromosomes close to each other.

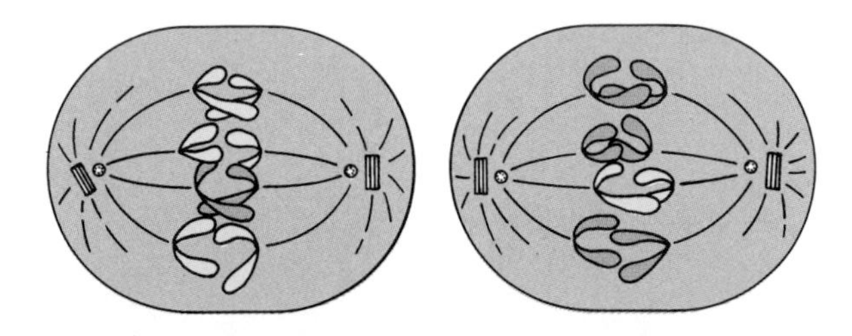

Figure 9.12 **Anaphase II.** This anaphase stage is very similar to the anaphase of mitosis. The centromere of each chromosome divides and one chromatid separates from the other. As soon as this happens, we no longer refer to them as chromatids; we now call each strand of nucleoprotein a chromosome.

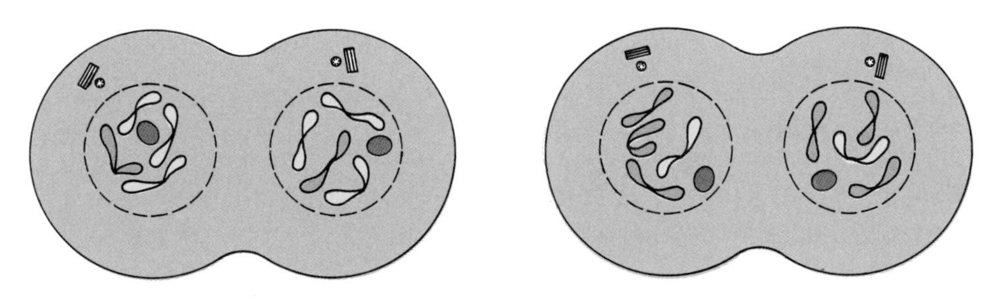

Figure 9.13 **Telophase II.** During the telophase stage, what events would you expect?

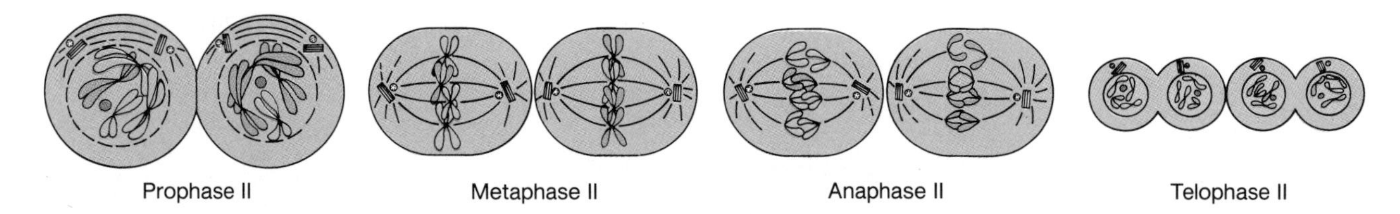

Figure 9.14 **Meiosis II.** During meiosis II, the centromere splits and each chromosome divides into separate chromatids. Four haploid cells are produced from one diploid parent cell.

Notice that each gene occupies a specific place on the chromosome. This is the *locus*, a place on a chromosome where a gene is located. Homologous chromosomes contain an identical order of genes. For the sake of simplicity, only a few loci are labeled on the chromosomes used as examples. Actually, the chromosomes contain hundreds or possibly thousands of genes.

What does crossing-over have to do with the possible kinds of cells that result from meiosis? Consider figure 9.16. Notice that without crossing-over, only two kinds of genetically different gametes result. Two of the four gametes have one type of chromosome, while the other two have the other type of chromosome. With crossing-over, four genetically different gametes are formed.

With just one crossover, we double the number of kinds of gametes possible from meiosis. Since crossing-over can occur at almost any point along the length of the chromosome, great variation is possible. In fact, crossing-over can occur at a number of different points on the same chromosome; that is, there can be more than one crossover per chromosome pair (figure 9.17).

Crossing-over helps to explain why a child can show a mixture of family characteristics (figure 9.18). If the violet chromosome was the chromosome that a mother received from her mother, the child could receive some genetic information not only from the mother's mother, but also from the mother's father. When crossing-over occurs during

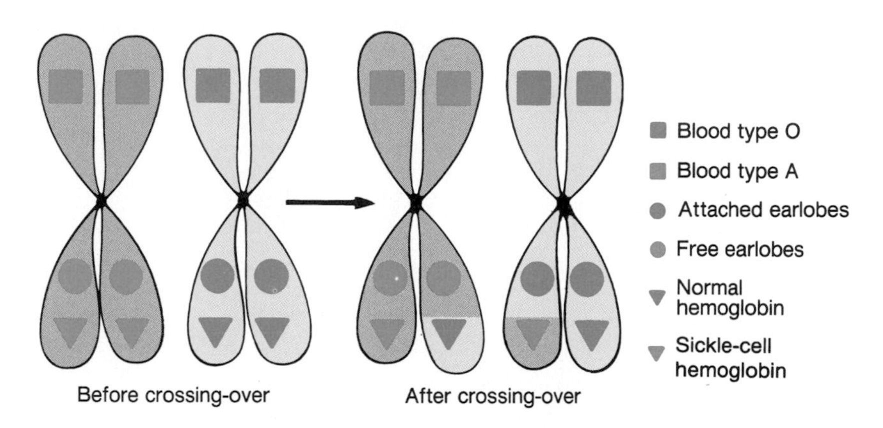

Figure 9.15 **Synapsis and Crossing-Over.** While pairs of homologous chromosomes are in synapsis, one part of one chromatid can break off and be exchanged for an equivalent part of its homologous chromatid. List the new combination of genes on each chromatid that has resulted from the crossing-over.

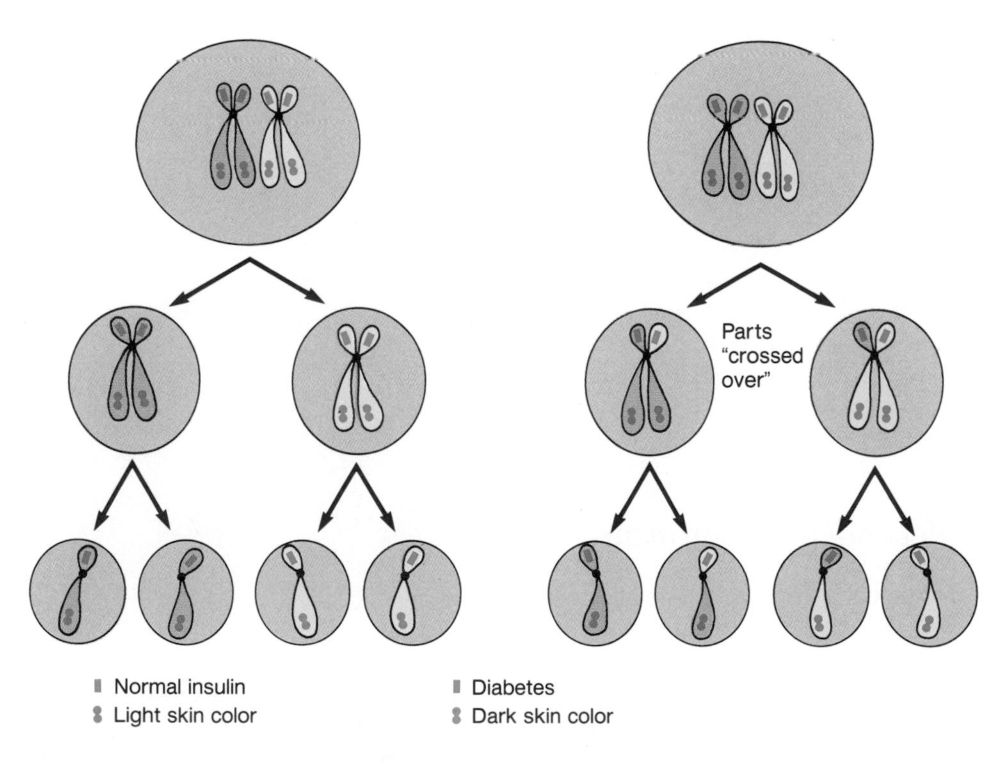

Figure 9.16 **Variations Resulting from Crossing-Over.** The cells on the left resulted from meiosis without crossing-over; those on the right had one crossover. Compare the results of meiosis in both cases.

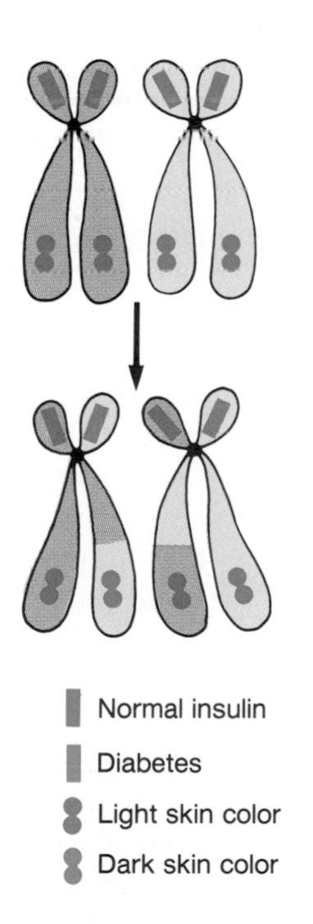

Figure 9.17 **Multiple Crossovers.** Crossing-over can occur several times between one pair of homologous chromosomes. List the new combination of genes on each chromatid that has resulted from the crossing-over.

the meiotic process, pieces of genetic material are exchanged between the chromosomes. This means that genes that were originally on the same chromosome become separated. They are moved to their synapsed homologue, and therefore into different gametes. The closer two genes are to each other on a chromosome (i.e., the more closely they are *linked*), the more likely they will stay together and not be separated during crossing-over. Thus, there is a high probability that they will be inherited together. The further apart two genes are, the more likely it is that they will be separated during crossing-over. This fact enables biologists to construct chromosome maps (see box 9.1).

Segregation

After crossing-over has taken place, segregation occurs. This involves the separation and movement of homologous chromosomes to the poles. Let's say a person has a gene for insulin production on one chromosome and a gene for diabetes on the homologous chromosome. Such a person would produce enough insulin to be healthy. When this pair of chromosomes segregates during anaphase I, one daughter cell receives a chromosome with an allele for insulin production and the second daughter cell receives a chromosome with an allele for diabetes. The process of segregation causes genes to be separated from one another so that they have an equal chance of being transmitted to the next generation. If the mate also has one allele for insulin production and one allele for diabetes, that person also produces two kinds of gametes.

Both of the parents have normal insulin production. If one or both of them contributed an allele for normal insulin production during fertilization, the offspring would produce enough insulin to be healthy. However, if, by chance, both parents contributed the gamete with the allele for diabetes, the child would be a diabetic. Thus, parents may produce offspring with traits different from their own. In this variation,

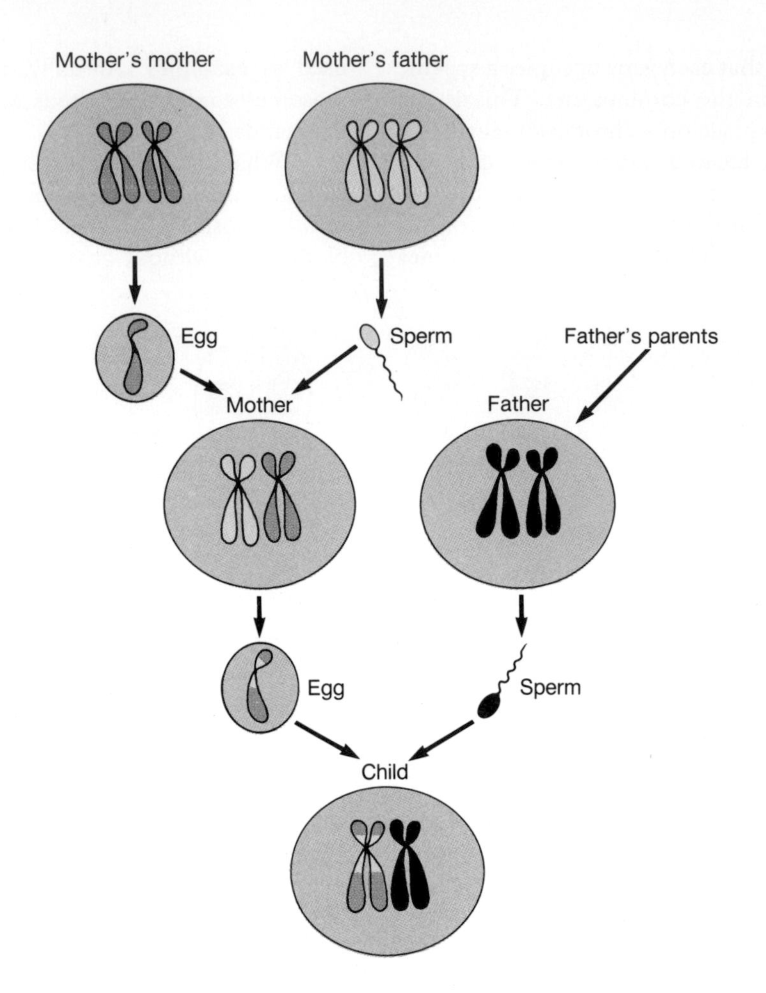

Figure 9.18 **Mixing of Genetic Information through Several Generations.** The mother of this child has information from both of her parents. The child receives a mixture of this information from the mother. Note that only the maternal line has been traced in this diagram. Can you imagine how many more combinations would result after including the paternal heritage?

no new genes are created; they are simply redistributed in a fashion that allows for the new combination of alleles in the offspring to be different from the parents' gene combinations. This will be explored in greater detail in chapter 10.

Independent Assortment

So far in discussing variety, we have only dealt with one pair of chromosomes, which allows two varieties of gametes. Now let's consider how variation increases when we add a second pair of chromosomes (figure 9.19).

In figure 9.19 chromosomes carrying insulin-production information

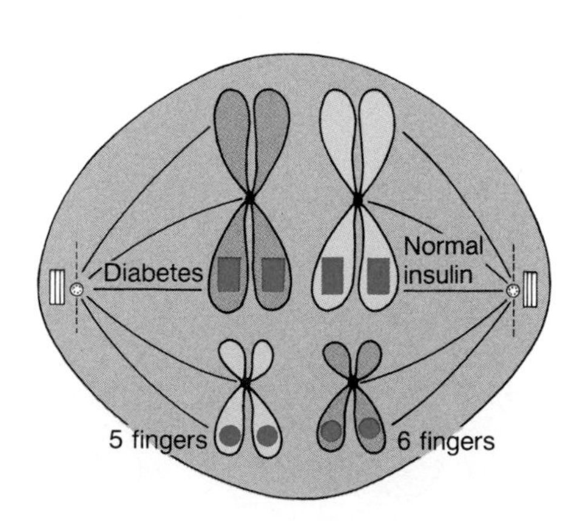

Figure 9.19 **The Independent Orientation of Homologous Chromosome Pairs.** The orientation of one pair of chromosomes on the equatorial plane does not affect the orientation of a second pair of chromosomes. This results in increased variety in the haploid cells.

BOX 9.1 The Human Genome Project

The human genome project was first proposed in 1986 and is one of the most ambitious projects ever undertaken in the biological sciences. The goal is nothing less than the complete characterization of the genetic makeup of humans. If the effort is successful, scientists will have produced a map of each of the twenty-three pairs of human chromosomes that will show the names and places of all of our genes. This international project involving about one hundred laboratories is expected to take fifteen years. Work began in many of these labs in 1990. Powerful computers are used to store and share the enormous amount of information derived from the analyses of human DNA. To get an idea of the size of this project, a human Y chromosome (one of the smallest of the human chromosomes) is estimated to be composed of 28 million nitrogenous bases. The larger X chromosome may be composed of 160 million nitrogenous bases!

Two kinds of work are progressing simultaneously. Physical maps are being constructed by determining the location of specific "markers" (known sequences of bases) and their closeness to genes. A kind of chromosome map already exists that pictures patterns of colored bands on chromosomes, a result of chromosome-staining procedures. Using these banded chromosomes, the markers can then be related to these colored bands on a specific region of a chromosome. It was expected that these correlations and this physical mapping could be nearly complete by 1995. Since progress has been more rapid than expected, the work may be accomplished even earlier.

The second goal to determine the exact order of nitrogenous bases of the DNA for each chromosome will take longer. Techniques exist for determining base sequences, but it is a time-consuming job to sort out the several billion bases that may be found in any one chromosome. It is estimated, for example, that there are over 100,000 genes yet to have their base sequences determined and their exact positions identified. At the end of 1992 certain chromosomes had some of their base sequences identified, and over 2,000 genes had been located at specific places.

When the maps are completed for all of the human chromosomes, it will be possible to examine a person's DNA and identify genetic abnormalities. This could be extremely useful in diagnosing diseases and providing genetic counseling to those considering having children. This kind of information would also create possibilities for new gene therapies. Once it is known where an abnormal gene is located and how it differs in base-sequence form from normal DNA sequence, steps could be taken to correct the abnormality. However, there is also a concern that as knowledge of our genetic makeup becomes easier to determine, some people may attempt to use this information for profit or political power. This is a real concern, since some insurance companies refuse to insure people at "genetic risk." Refusing to provide coverage would save these companies the expense of future medical bills incurred by "less-than-perfect people." Another fear is that attempts may be made to "breed out" certain genes and people from the human population in order to create a "perfect race."

always separate from each other. The second pair of chromosomes with the information for the number of fingers also separates. Since the pole to which a chromosome moves is a chance event, half of the time the chromosomes divide so that insulin production and six-fingeredness move in one direction, while diabetes and five-fingeredness move in the opposite direction. The other half of the time, insulin production and five-fingeredness go together, while diabetes and six-fingeredness go to the other pole. With four chromosomes (two pair), four kinds of gametes are possible (figure 9.20). With three pairs of homologous chromosomes, there are eight possible kinds of cells with respect to chromosome combinations resulting from

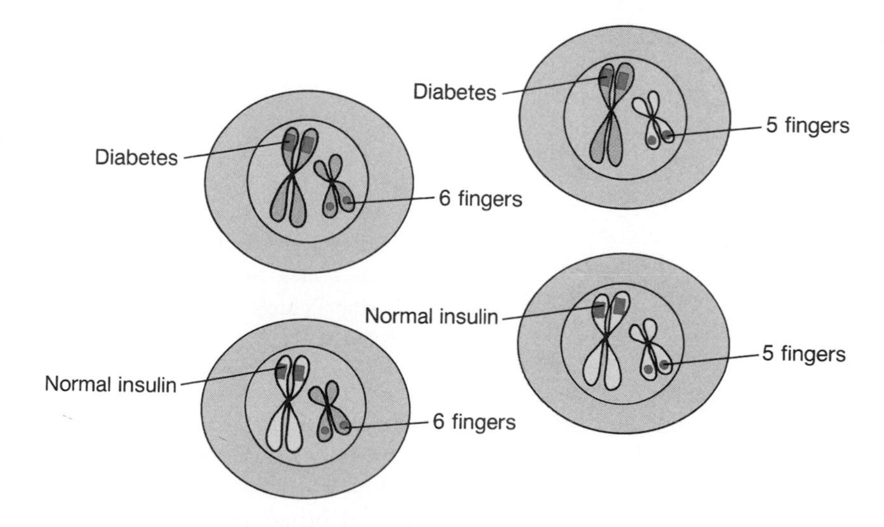

Figure 9.20 **Variation Resulting from Independent Assortment.** When a cell has two pairs of homologous chromosomes, four kinds of haploid cells can result from independent assortment. How many kinds of haploid cells could result if the parental cell had three pairs? Four pairs?

meiosis. See if you can list them. The number of possible chromosomal combinations of gametes is found by the expression 2^n, where n equals the number of pairs of chromosomes. With three pairs of chromosomes, n equals 3, and so $2^n = 2^3 = 2 \times 2 \times 2 = 8$. With twenty-three pairs of chromosomes, as in the human cell, $2^n = 2^{23} = 8,388,608$. More than eight million kinds of sperm cells or egg cells are possible from a single human parent organism. This huge variation is possible because each pair of homologous chromosomes assorts independently of the other pairs of homologous chromosomes (independent assortment). In addition to this variation, crossing-over creates new gene combinations, and mutation can cause the formation of new genes, thereby increasing this number greatly.

Fertilization

Because of the large number of possible gametes resulting from independent assortment, segregation, mutation, and crossing-over, an incredibly large number of types of offspring can result. Since human males can produce millions of genetically different sperm and females can produce millions of genetically different eggs, the number of kinds of offspring possible is infinite for all practical purposes. With the possible exception of identical twins, every human that has ever been born is genetically unique.

Nondisjunction

In the normal process of meiosis, diploid cells have their number of chromosomes reduced to haploid. This involves segregating homologous chromosomes into separate cells during the first meiotic division. Occasionally, a pair of homologous chromosomes does not segregate properly during gametogenesis and both chromosomes of a pair end up in the same gamete. This kind of division is known as **nondisjunction** (figure 9.21). As you can see in this figure, two cells are missing a chromosome and the genes that were carried on it. This usually results in the death of the cells. The other cells have a double dose of one chromosome. Apparently the genes of an organism are balanced against one another. A double dose of some genes and a single dose of others results in abnormalities that may lead to the death of the cell. Some of these abnormal cells, however, do live and develop into sperm or eggs. If one of these abnormal sperm or eggs unites with a normal gamete, the offspring will have an abnormal number of chromosomes. There will be three of one of the kinds of chromosomes instead of the normal two, a condition referred to as **trisomy.** All the cells that develop by mitosis from that zygote will also be trisomic.

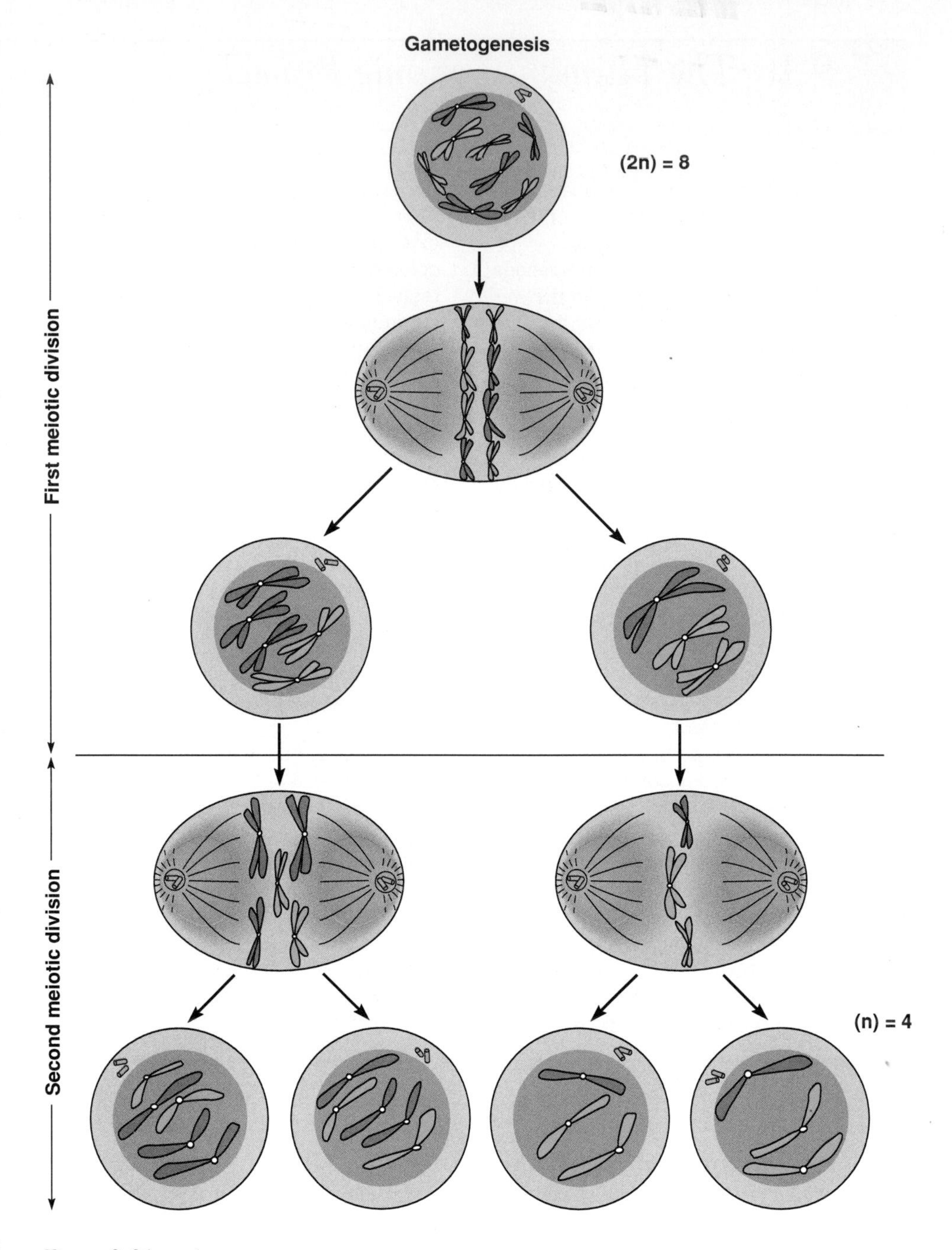

Figure 9.21 **Nondisjunction during Gametogenesis.** When a pair of homologous chromosomes fails to separate properly during meiosis I, gametogenesis results in gametes that have an abnormal number of chromosomes. Notice that two of the highlighted cells have an additional chromosome while the other two are deficient by that same chromosome.

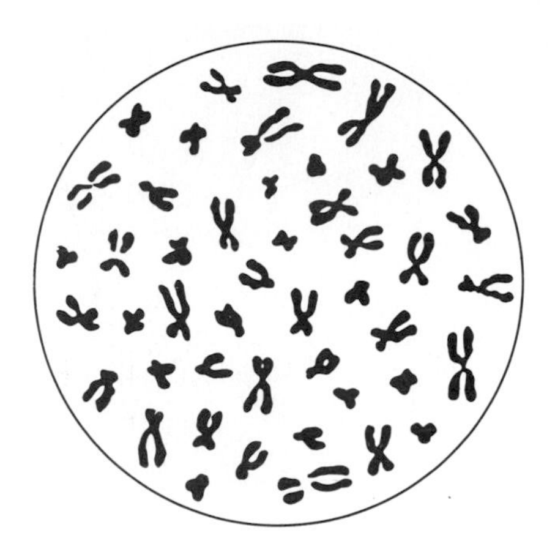

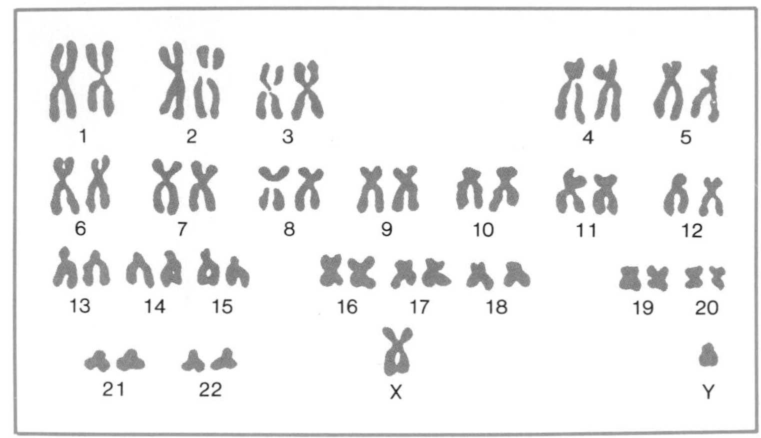

Figure 9.22 **Human Male Chromosomes.** The randomly arranged human male chromosomes shown on the left were photographed from metaphase cells spattered onto a microscope slide. Those on the right are the result of having cut out the chromosomes and arranging them into pairs of homologous chromosomes.

It is possible to examine cells and count chromosomes. Among the easiest cells to view are white blood cells. They are dropped onto a microscope slide so that the cells are broken open and the chromosomes separated. Photographs are taken of chromosomes from cells in the metaphase stage of mitosis. The chromosomes in the pictures can then be cut and arranged for comparison to known samples (figure 9.22).

One example of the effects of nondisjunction is the condition known as **Down syndrome** (mongolism). If an egg cell with two number 21 chromosomes has been fertilized by a sperm containing the typical one copy of chromosome number 21, the resulting zygote would have forty-seven chromosomes (twenty-four from the female plus twenty-three from the male parent) (figure 9.23). The child who developed from this fertilized egg would have forty-seven chromosomes in every cell of his or her body as a result of mitosis, and the symptoms characteristic of Down syndrome. These could include thickened eyelids, some mental impairment, and faulty speech (figure 9.24). Premature aging is probably the most significant impact of this genetic disease.

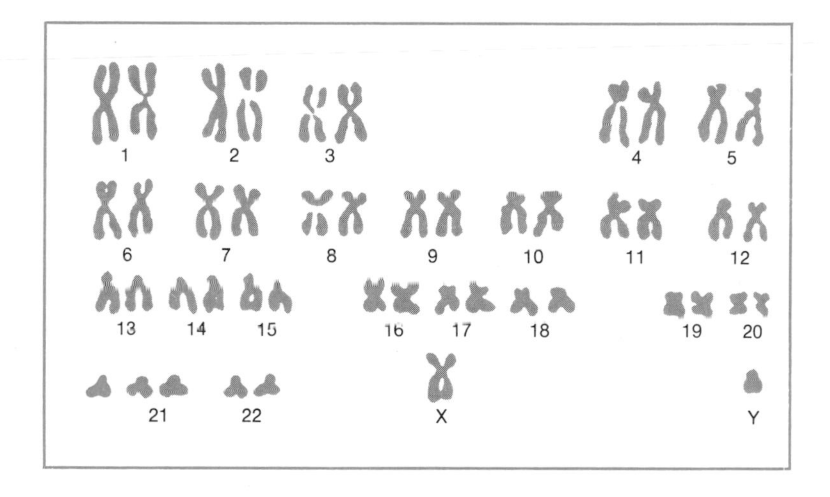

Figure 9.23 **Chromosomes from an Individual Displaying Down Syndrome.** Notice that each pair of chromosomes has been numbered and that the person from whom these chromosomes were taken has an extra chromosome number 21. The person with this trisomic condition could display a variety of physical characteristics, including slightly slanted eyes, flattened facial features, a large tongue, and a tendency toward short stature and fingers. Some individuals also display mental retardation.

The woman's age is another important consideration in the occurrence of trisomies such as Down syndrome. In women, gametogenesis begins early in life, but cells destined to become eggs are put on hold during meiosis I (see chapter 21). One of these cells completes the meiotic process monthly, beginning at puberty and ending at menopause. This means that eggs released for fertilization later in life are older than those released earlier in life. Therefore, the chances of abnormalities such as nondisjunction increase as the age of the mother increases. Figure 9.25 illustrates the frequency of occurrence of nondisjunction at different ages in women. Notice that the frequency of nondisjunction increases very rapidly after age thirty-seven. For this reason, many physicians encourage couples to have their children in their early to mid-twenties and not in their late thirties or early forties. Physicians normally encourage older women who are pregnant to have the cells of their fetus checked to see if they have the normal chromosome number.

Chromosomes and Sex Determination

You already know that there are several different kinds of chromosomes, that each chromosome carries genes unique to it, and that these genes are found at specific places. Furthermore, diploid organisms have homologous pairs of chromosomes. Sexual characteristics are determined by genes in the same manner as other types of characteristics. In many organisms, sex-determining genes are located on specific chromosomes known as **sex chromosomes.** All other chromosomes not involved in determining the sex of an individual are known as **autosomes.** In humans and all other mammals, and in some other organisms (e.g., fruit flies), the sex of an individual is determined by the presence of a certain chromosome combination. The genes that determine maleness are located on a small chromosome known as the *Y chromosome.* This Y chromosome behaves as if it and another larger chromosome, known as the *X chromosome,* were homologs. Males have one X and one Y chromosome. Females have two X chromosomes. Some animals have their sex determined in a completely different way. In bees, for example, the females are diploid and the males are haploid. Other plants and animals have still other chromosomal mechanisms for determining their sex.

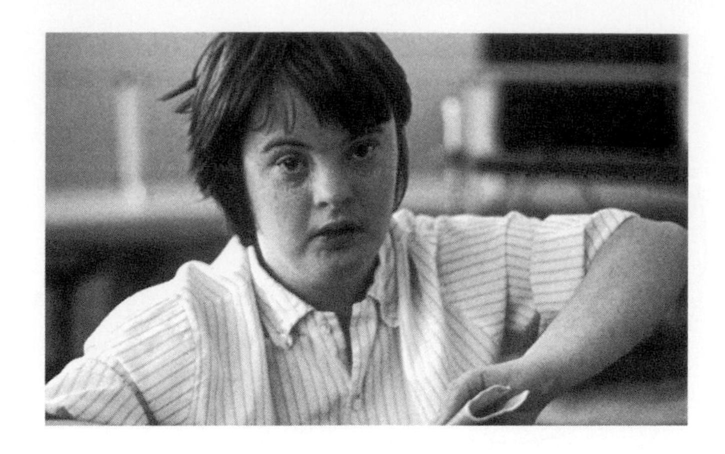

Figure 9.24 **Down Syndrome.** Every downic child's body has one extra chromosome. With special care, planning, and training, people with this syndrome can lead useful, productive lives.

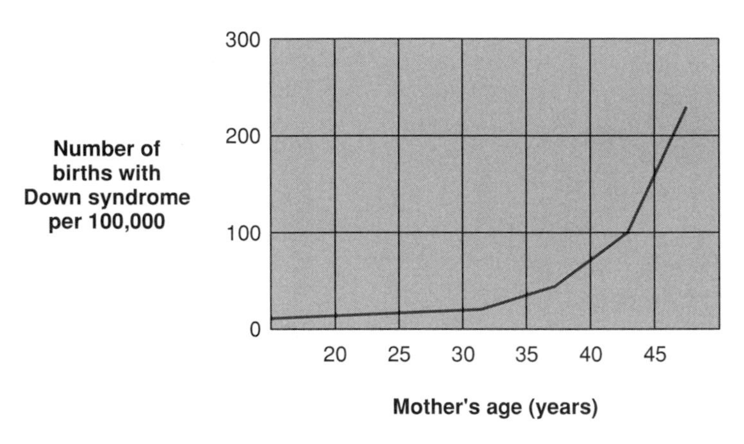

Figure 9.25 **Nondisjunction as a Function of a Mother's Age.** Notice that as the age of the female increases, the rate of nondisjunction increases only slightly until the age of approximately thirty-seven. From that point on, the rate increases drastically. How would a comparable graph for males be drawn? Why would it be so different?

A Comparison of Mitosis and Meiosis

Some of the similarities and differences between mitosis and meiosis were pointed out earlier in this chapter. Study table 9.2 to familiarize yourself with the differences between these two processes.

Table 9.2
A Comparison of Mitosis and Meiosis

Mitosis	Meiosis
1. One division completes the process.	1. Two divisions are required to complete the process.
2. Chromosomes do not synapse.	2. Homologous chromosomes synapse in prophase I.
3. Homologous chromosomes do not cross over.	3. Homologous chromosomes do cross over.
4. Centromeres divide in anaphase.	4. Centromeres divide in anaphase II, but not in anaphase I.
5. Daughter cells have the same number of chromosomes as the parent cell ($2n \rightarrow 2n$ or $n \rightarrow n$).	5. Daughter cells have half the number of chromosomes as the parent cell ($2n \rightarrow n$).
6. Daughter cells have the same genetic information as the parent cell.	6. Daughter cells are genetically different from the parent cell.
7. Results in growth, replacement of worn-out cells, and repair of damage.	7. Results in sex cells.

• Summary •

Meiosis is a specialized process of cell division resulting in the production of four cells, each of which has the haploid number of chromosomes. The total process involves two sequential divisions during which one diploid cell reduces to four haploid cells. Since the chromosomes act as carriers for genetic information, genes separate into different sets during meiosis. Crossing-over and segregation allow hidden characteristics to be displayed, while independent assortment allows characteristics donated by the mother and the father to be mixed in new combinations.

Together, crossing-over, segregation, and independent assortment ensure that all sex cells are unique. Therefore, when any two cells unite to form a zygote, the zygote will also be one of a kind. The sex of many kinds of organisms is determined by specific chromosome combinations. In humans, females have two X chromosomes, while males have an X and a Y chromosome.

• Thinking Critically •

Assume that corn plants have a diploid number of only 2. Each plant's chromosomes are diagrammed below.

Show sex-cell formation in the male and female plant. How many variations in sex cells can occur and what are they? What variations can occur in the production of chlorophyll and starch in the descendants of these parent plants?

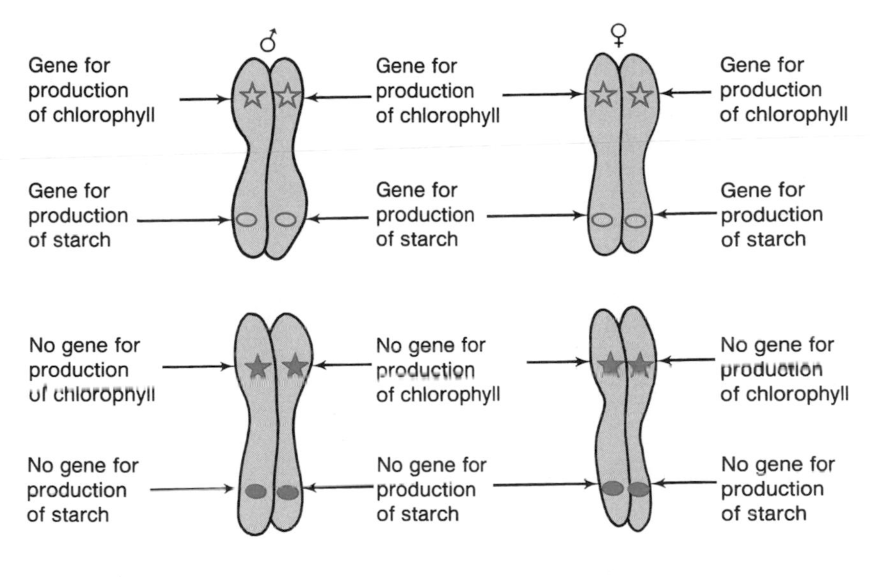

Note:

Gene for production of chlorophyll	= green plant
No gene for chlorophyll	= white, dead plant
Gene for production of starch	= regular corn
No gene for starch	= sweet corn

• Experience This •

Models can be very useful in helping us to understand complex biological events such as meiosis. You can create model chromosomes very easily by using various lengths of colored strings, threads, or yarns to simulate the twenty-three pairs of homologous chromosomes in a human cell. Each homologous pair should be different from the other pairs, either in color, length, or both. Begin your modeling with each chromosome in its replicated form (i.e., two chromatids per chromosome). Attach the two chromatids with a loose twist. Manipulate your twenty-three pairs of model chromosomes through the stages of meiosis I and II. If you have performed the actions properly, you should end up with four cells, each haploid ($n = 23$).

• Questions •

1. List three differences between mitosis and meiosis.
2. How do haploid cells differ from diploid cells?
3. What are the major sources of variation in the process of meiosis?
4. Can a haploid cell undergo meiosis?
5. What is unique about prophase I?
6. Why is meiosis necessary in organisms that reproduce sexually?
7. Define the terms *zygote*, *fertilization*, and *homologous chromosomes*.
8. How much variation as a result of independent assortment can occur in cells with the following diploid numbers: 2, 4, 6, 8, and 22?
9. Diagram the metaphase I stage of a cell with the diploid number of 8.
10. Diagram fertilization as it would occur between a sperm and an egg with the haploid number of 3.

• Chapter Glossary •

anther (an'ther) The sex organ in plants that produces the sperm.

autosomes (aw'to-sōmz) Chromosomes not involved in determining the sex of individuals.

crossing-over (kro'sing o'ver) The exchange of a part of a chromatid from one chromosome with an equivalent part of a chromatid from a homologous chromosome.

diploid (dip'loid) A cell that has two sets of chromosomes: one set from the maternal parent and one set from the paternal parent.

Down syndrome (down sin'drōm) A genetic disorder resulting from the presence of an extra chromosome number 21. Symptoms include slightly slanted eyes, flattened facial features, a large tongue, and a tendency toward short stature and fingers. Some individuals also display mental retardation.

egg cells (eg sels) The haploid sex cells produced by sexually mature females.

fertilization (fer"tĭ-lĭ-za'shun) The joining of haploid nuclei, usually from an egg and a sperm cell, resulting in a diploid cell called the zygote.

gamete (gam'ēt) A haploid sex cell.

gametogenesis (gă-me"to-jen'ĕ-sis) The generating of gametes; the meiotic cell-division process that produces sex cells.

gonad (go'nad) In animals, the organs in which meiosis occurs.

haploid (hap'loid) Having a single set of chromosomes resulting from the reduction division of meiosis.

homologous chromosomes (ho-mol'o-gus kro'mo-sōmz) A pair of chromosomes in a diploid cell that contain similar genes at corresponding loci throughout their length.

independent assortment (in"de-pen'dent ă-sort'ment) The segregation, or assortment, of one pair of homologous chromosomes, independently of the segregation, or assortment, of any other pair of chromosomes.

meiosis (mi-o'sis) The specialized pair of cell divisions that reduce the chromosome number from diploid ($2n$) to haploid (n).

nondisjunction (non"dis-junk'shun) An abnormal meiotic division that results in sex cells with too many or too few chromosomes.

ovaries (o'var-ēz) The female sex organs that produce haploid sex cells—the eggs or ova.

pistil (pis'til) The sex organ in plants that produces eggs or ova.

reduction division (re-duk'shun dĭ-vĭ'zhun) A type of cell division in which daughter cells get only half the chromosomes from the parent cell.

segregation (seg"rĕ-ga'shun) The separation and movement of homologous chromosomes to the poles of the cell.

sex chromosomes (seks kro'mo-sōmz) Chromosomes that carry genes that determine the sex of the individual.

sexual reproduction (sek'shu-al re"pro-duk'shun) The propagation of organisms involving the union of gametes from two parents.

sperm cells (spurm selz) The haploid sex cells produced by sexually mature males.

synapsis (sin-ap'sis) The condition in which the two members of a pair of homologous chromosomes come to lie close to one another.

testes (tes'tēz) The male sex organs that produce haploid cells—the sperm.

trisomy (tris'oh-me) An abnormal number of chromosomes resulting from the nondisjunction of homologous chromosomes during meiosis; for example, as in Down syndrome.

zygote (zi'gōt) A diploid cell that results from the union of an egg and a sperm.

CHAPTER 10

Mendelian Genetics

Purpose

This chapter considers the fundamentals of inheritance. In previous chapters we introduced the concept of DNA as a molecule for storing the genetic information used to manufacture proteins and to guide the processes of mitosis and meiosis. Here we will describe how characteristics are passed from one generation to the next, using many human characteristics to illustrate these patterns of inheritance.

For Your Information

The field of bioengineering is advancing as quickly as the electronics industry. The first bioengineering efforts focused on the development of genetically altered crops that displayed improvements over past varieties, such as increased resistance to infectious disease. The second wave of research involved manipulating DNA, which resulted in improved food handling and processing, such as the slowing of ripening in tomatoes. Currently, crops are being genetically manipulated to manufacture large quantities of specialty chemicals and biopolymers. While some of these products have been produced from genetically engineered microorganisms, crops of turnips, potatoes, and tobacco can generate tens or hundreds of kilograms of specialty product per year. Researchers have shown, for example, that turnips can produce interferon (an antiviral agent), tobacco can create antibodies to fight human disease, oilseed rape plants can serve as a source of human brain hormones, and potatoes can synthesize human serum albumin that is indistinguishable from the genuine human blood protein.

Chapter Outline

Learning Objectives

- Be able to work single-factor and double-factor genetic problems dealing with traits that show dominance, recessiveness, and lack of dominance.
- Be able to work genetic problems dealing with multiple alleles, polygenic inheritance, and X-linked characteristics.
- Explain how environmental conditions influence an organism's phenotype.

Genetics, Meiosis, and Cells

Why do you have a particular blood type or hair color? Why do some people have the same skin color as their parents, while others have a skin color different from that of their parents? These questions can be better answered if you understand how genes work. A **gene** is a portion of DNA that determines characteristics. Through meiosis and reproduction, these genes can be transmitted from one generation to another. The study of genes, how genes produce characteristics, and how the characteristics are inherited is the field of biology called **genetics.** The first person to systematically study inheritance and formulate laws about how characteristics are passed from one generation to the next was an Augustinian monk named Gregor Mendel (1822–1884). However, his work was not generally accepted until 1900, when three men, working independently, rediscovered some of the ideas that Mendel had formulated over thirty years earlier. Because of his early work, the study of the pattern of inheritance that follows the laws formulated by Gregor Mendel is often called **Mendelian genetics.**

To understand this chapter, you need to know some basic terminology. One term that you have already encountered is *gene.* Mendel thought of a gene as a particle that could be passed from the parents to the **offspring** (children, descendants, progeny). Today we know that genes are actually composed of specific sequences of DNA nucleotides. The particle concept is not entirely inaccurate because genes are located on specific portions of chromosomes; however, they are not like beads on a string.

Another important idea to remember is that most sexually reproducing organisms are diploid. Since gametes are haploid and most organisms are diploid, the conversion of diploid to haploid cells during meiosis is an important process. The diploid cells have two sets of chromosomes—one set inherited from each parent. Therefore, they have two chromosomes of each kind and have two genes for each characteristic. When sex cells are produced by meiosis, reduction division occurs, and the diploid number is reduced to haploid. Therefore, the sex cells produced by meiosis have one chromosome of each of the pairs that was in the diploid cell that began meiosis. Diploid organisms usually result from the fertilization of a haploid egg by a haploid sperm. Therefore, they inherit one gene of each type from each parent. For example, each of us has two genes for earlobe shape: one came with our father's sperm, the other with our mother's egg (figure 10.1).

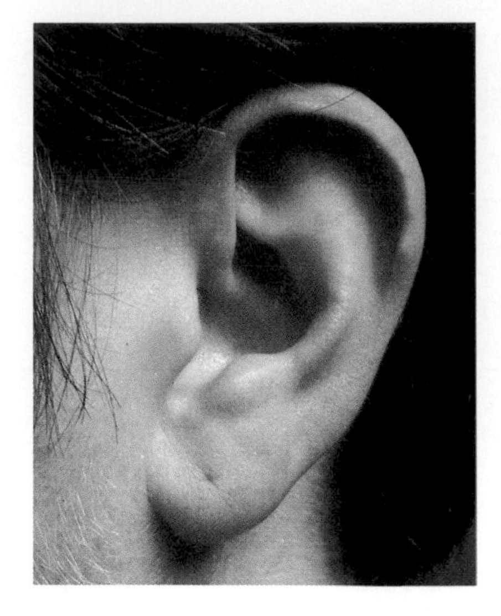

(a)

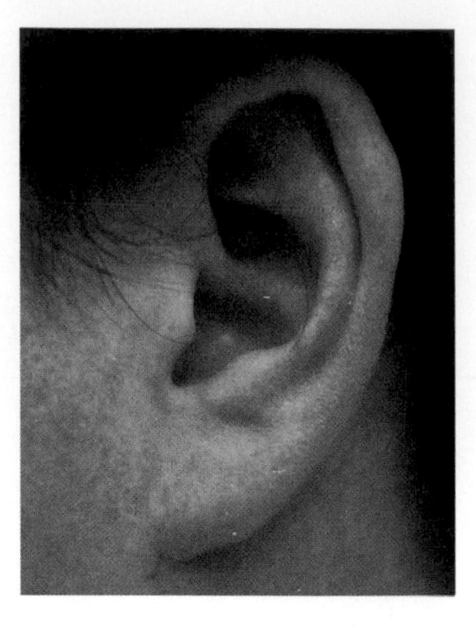

(b)

Figure 10.1 **Genes Control Structural Features.** Whether your earlobe is free (a), or attached (b) depends on the genes you have inherited. As genes express themselves, their actions affect the development of various tissues and organs. Some people's earlobes do not separate from the side of their head in the same fashion as do those of others. How genes control this complex growth pattern and why certain genes function differently than others is yet to be clarified.

Genes and Characteristics

Each diploid organism has two genes for each characteristic. These two genes for the same trait are termed **alleles.** There may be several alternative forms of each gene within the population. In people, there are two alleles for earlobe shape. One allele produces an earlobe that is fleshy and hangs free, while the other allele produces a lobe that is attached to the side of the face and does not hang free. The type of earlobe that is present is determined by the type of allele (gene) received from each parent and the way in which these alleles interact with one another. Alleles are always located on the pair of homologous chromosomes—one allele on each chromosome. These alleles are also always at the same specific location (**locus**) and they are always on the same type of chromosome in all individuals of a species (figure 10.2).

The **genome** is a set of all the genes necessary to specify an organism's complete list of characteristics. A diploid ($2n$) cell has two genomes and a haploid cell (n) has one genome. The **genotype** of an organism is a listing of the genes present in that organism. It consists of the cell's DNA code; therefore, you cannot see the genotype of an organism. It is impossible to know the complete genotype of most organisms, but it is often possible to figure out the genes present that determine a particular characteristic. For example, there are three possible genotypic combinations of the two alleles for earlobe shape. A person's genotype could be (1) two genes for attached earlobes, (2) one gene for attached and one gene for free earlobes, or (3) two genes for free earlobes.

How would individuals with each of these three genotypes appear? The way each combination of genes expresses itself is known as the **phenotype** of the organism. A person with two alleles for attached earlobes will have earlobes that do not hang free. A person with one allele for attached earlobes and one allele for free earlobes will have a phenotype that exhibits free earlobes. An individual with two alleles for free earlobes will also have free earlobes. Notice that there are three genotypes, but only two phenotypes. The individuals with the free-earlobe phenotype have different genotypes.

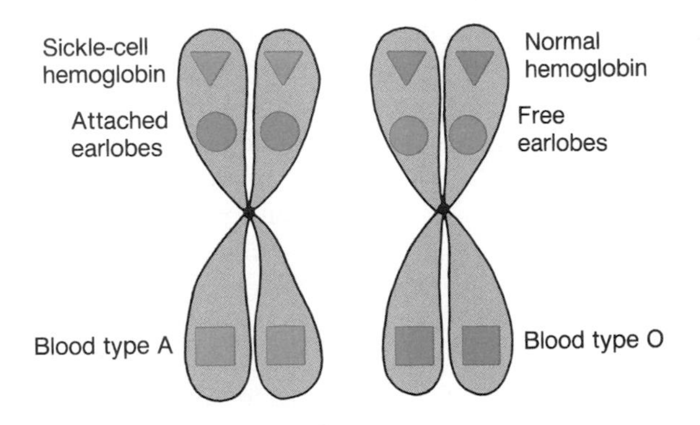

Figure 10.2 **A Pair of Homologous Chromosomes.** Homologous chromosomes contain genes for the same characteristics at the same place. Note that the attached-earlobe allele is located at the ear-shape locus on one chromosome, and the free-earlobe allele is located at the ear-shape locus on the other member of the homologous pair of chromosomes. The other two genes are for hemoglobin structure (alleles normal and sickled) and blood type (alleles type A and type O). We really don't know on which chromosomes most human genes are located. It is hoped that the Human Genome Project described in chapter 9 will resolve this problem. The examples presented here are for illustrative purposes only.

For various reasons, certain genes may not express themselves. Sometimes, the physical environment determines whether or not certain genes function. For example, some cats have coat-color genes that do not reveal themselves unless the temperature of the skin is below a certain point. Often, the only parts of a cat that become cool enough to allow the genes to express themselves are the tips of the ears and the feet. Consequently these areas will differ in color from the rest of the cat's body. Another example in humans is the presence of genes for freckles that do not show themselves fully unless the person's skin is exposed to sunlight (figure 10.3).

The expression of some genes is directly influenced by the presence of other alleles in the organism. For any particular pair of alleles in an individual organism, the two alleles from the two parents are either identical or not identical. The organism is **homozygous** for a trait when it has two identical alleles for that particular characteristic. A person with two alleles for freckles is said to be homozygous for that trait. A person with two alleles for no freckles is also homozygous. If an organism is homozygous, the characteristic expresses itself in a specific manner. A person homozygous for free earlobes has free earlobes, and a person homozygous for attached earlobes has attached earlobes.

Figure 10.3 **The Environment and Gene Expression.** The expression of many genes is influenced by the environment. The gene for dark hair in the cat is sensitive to temperature and expresses itself only in the parts of the body that stay cool. The gene for freckles expresses itself more fully when a person is exposed to sunlight.

An individual is designated as **heterozygous** when it has two different allelic forms of a particular gene. The heterozygous individual received one form of the gene from one parent and a different allele from the other parent. For instance, a person with one allele for freckles and one allele for no freckles is heterozygous. If an organism is heterozygous, these two different alleles interact to determine a characteristic.

Often, one gene expresses itself and the other does not. A **dominant allele** expresses itself and masks the effect of other alleles for the trait. For example, if a person has one allele for free earlobes and one allele for attached earlobes, that person has a phenotype of free earlobes. The allele for free earlobes is dominant. A **recessive allele** is one that, when present with another allele, does not express itself; it is masked by the effect of the other allele. Having attached earlobes is a recessive characteristic. A person with one allele for free earlobes and one allele for attached earlobes has a phenotype of free earlobes. Recessive traits only express themselves when the organism is homozygous for the recessive alleles. If you have attached earlobes, you have two alleles for that trait. Recessive genes are not necessarily bad. The term *recessive* has nothing to do with the significance of the gene—it simply describes how it can be expressed. Recessive genes are not less likely to be inherited but must be present in a homozygous condition to express themselves.

Mendelian genetics involves the study of the transfer of genes from one generation to another and the ways in which the genes received from the parents influence the traits of the offspring. Before you go on, be certain that you understand how meiosis works and how the gametes formed from this process are combined by fertilization. It is in meiosis that the two alleles in a pair of genes segregate. Although we will talk about the segregation of genes in this chapter, remember that it is the chromosomes and not the individual genes that actually segregate.

Notice that the chromosomes in figure 10.2 have two different alleles for hemoglobin: one for sickle-cell hemoglobin and one for normal hemoglobin. By meiosis, the parent may contribute a gamete containing a gene for sickle-cell hemoglobin or a gene for normal hemoglobin. By fertilization, the offspring will receive two genes for hemoglobin production, but only one from each parent. For most of the remainder of this chapter, we will deal with how to determine which genes are passed on by the parents and how to determine the genetic makeup of the offspring resulting from fertilization.

Mendel's Laws of Heredity

Heredity problems are concerned with determining which alleles are passed from the parents to the offspring and how likely it is that various types of offspring will be produced. The first person to develop a method of predicting the outcome of inheritance patterns was Mendel, who performed experiments concerning the inheritance of certain characteristics in sweet pea plants. From his work, Mendel concluded which traits were dominant and which were recessive. Some of his results follow below.

What made Mendel's work unique was that he studied only one trait at a time. Previous investigators had tried to follow numerous traits at the same time. When this was attempted, the total set of characteristics was so cumbersome to work with that no clear idea could be formed of how the offspring inherited traits. Mendel used traits with clear-cut alternatives, such as purple or white flower color, yellow- or green-colored seed pods, and tall or dwarf pea plants. He was very lucky to have chosen pea plants in his study because they naturally self-pollinate. When self-pollination occurs in pea plants over many generations, it is possible to develop a population of plants that is homozygous for a number of characteristics. Such a population is known as a *pure line*.

Mendel took a pure line of pea plants having purple flower color, removed the male parts (anthers) and discarded them so that they could not self-pollinate. He then took anthers from a pure-breeding white-flowered plant and pollinated the antherless purple flower. When the pollinated flowers produced seeds, Mendel collected, labeled, and planted them. When these seeds germinated and grew, they eventually produced flowers. You might be surprised to learn that all of the plants resulting from this cross had purple flowers. One of the prevailing hypotheses of Mendel's day would have predicted that the purple and white colors would have blended together, resulting in flowers that were lighter than the parental purple flowers. Another hypothesis would have predicted that the offspring would have had a mixture of white and purple flowers. The unexpected result—all of the offspring produced flowers like those of one parent and no flowers like those of the other—caused Mendel to examine other traits as well and form the basis for much of the rest of his work. He repeated his experiments using pure strains for other traits. Pure-breeding tall plants were crossed with pure-breeding dwarf plants. Pure-breeding plants with yellow pods were crossed with pure-breeding plants with green pods. The results were all the

Characteristic	*Alleles*	*Dominant*	*Recessive*
Plant height	Tall and dwarf	Tall	Dwarf
Pod shape	Full and constricted	Full	Constricted
Pod color	Green and yellow	Green	Yellow
Seed surface	Round and wrinkled	Round	Wrinkled
Seed color	Yellow and green	Yellow	Green
Flower color	Purple and white	Purple	White

same: the offspring showed the characteristic of one parent and not the other.

Next, Mendel crossed the offspring of the white-purple cross (all of which had purple flowers) with each other to see what the third generation would be like. Had the characteristic of the original white flowered-parent been lost completely? This second-generation cross was made by pollinating these purple flowers that had one white parent among themselves. The seeds produced from this cross were collected and grown. When these plants flowered, three-fourths of them produced purple flowers and one-fourth produced white flowers.

As a result of analyzing his data, Mendel formulated several genetic laws to describe how characteristics are passed from one generation to the next and how they are expressed in an individual.

Mendel's law of dominance When an organism has two different alleles for a trait, the allele that is expressed, overshadowing the expression of the other allele, is said to be *dominant*. The gene whose expression is overshadowed is said to be *recessive*.

Mendel's law of segregation When gametes are formed by a diploid organism, the alleles that control a trait separate from one another into different gametes, retaining their individuality.

Mendel's law of independent assortment Members of one gene pair separate from each other independently of the members of other gene pairs.

At the time of Mendel's research, biologists knew nothing of chromosomes or DNA or of the processes of mitosis and meiosis. Mendel assumed that each gene was separate from other genes. It was fortunate for him that each characteristic he picked to study was found on a separate chromosome. If two or more of these genes had been located on the same chromosome (*linked genes*), he probably would not have been able to formulate his laws. The discovery of chromosomes and DNA have led to modifications of Mendel's laws. However, it was Mendel's work that formed the foundation for the science of genetics.

Probability versus Possibility

In order to solve heredity problems, you must have an understanding of probability. **Probability** is the chance that an event will happen, and is often expressed as a percent or a fraction. *Probability* is not the same as *possibility*. It is possible to toss a coin and have it come up heads. But the probability of getting a head is more precise than just saying it is possible to get a head. The probability of getting a head is one out of two (1/2 or 0.5 or 50%) because there are two sides to the coin, only one of which is a head. Probability can be expressed as a fraction:

$$\textit{Probability} = \frac{\textit{the number of events that can produce a given outcome}}{\textit{the total number of possible outcomes}}$$

What is the probability of cutting a deck of cards and getting the ace of hearts? The number of times that the ace of hearts can occur is one. The total number of possible outcomes (number of cards in the deck) is fifty-two. Therefore, the probability of cutting an ace of hearts is 1/52.

What is the probability of cutting an ace? The total number of aces in the deck is four, and the total number of cards is fifty-two. Therefore, the probability of cutting an ace is 4/52 or 1/13.

It is also possible to determine the probability of two independent events occurring together. *The probability of two or more events occurring simultaneously is the product of their individual probabilities.* If you throw a pair of dice, it is possible that both will be fours. What is the probability that they both will be fours? The probability of one die being a four is 1/6. The probability of the other die being a four is also 1/6. Therefore, the probability of throwing two fours is

$$1/6 \times 1/6 = 1/36$$

Steps in Solving Heredity Problems—Single-Factor Crosses

The first type of problem we will work is the easiest type, a single-factor cross. A **single-factor cross** is a genetic cross or mating in which a single characteristic is followed from one generation to the next.

In humans, the allele for free earlobes is dominant and the allele for attached earlobes is recessive. If both parents are heterozygous (have one allele for free earlobes and one allele for attached earlobes), what is the probability that they can have a child with free earlobes? with attached earlobes?

In solving a heredity problem there are five basic steps:

Step 1: Assign a symbol for each allele. *Usually a capital letter is used for a dominant allele and a small letter for a recessive allele.* Use the symbol *E* for free earlobes and *e* for attached earlobes.

E = free earlobes *e* = attached earlob

Genotype		**Phenotype**
EE	=	free earlobes
Ee	=	free earlobes
ee	=	attached earlobes

Step 2: Determine the genotype of each parent and indicate a mating. Since both parents are heterozygous, the male genotype is *Ee*. The female genotype is also *Ee*. The × between them is used to indicate a mating.

$$Ee \times Ee$$

Step 3: Determine all the possible kinds of gametes each parent can produce. Remember that gametes are haploid; therefore, they can only have one allele instead of the two present in the diploid cell. Since the male has both the free-earlobe allele and the attached-earlobe allele, half of his gametes will contain the free-earlobe allele and the other half will contain the attached-earlobe allele. Since the female has the same genotype, the genotype of her gametes will be the same as his.

For genetic problems, a *Punnett square* is used. A **Punnett square** is a box figure that allows you to determine the probability of genotypes and phenotypes of the offspring of a particular cross. Remember, because of the process of meiosis, each gamete receives only one allele for each characteristic listed. Therefore, the male will give either an *E* or *e*; the female will also give either an *E* or *e*. The possible gametes produced by the male parent are listed on the left side of the square, while the female gametes are listed on the top. In our example, the Punnett square would show a single dominant allele and a single recessive allele from the male on the left side. The alleles from the female would appear on the top.

Female genotype *Ee*

Possible female gametes E & e

Male genotype *Ee*

Possible male gametes E & e

	E	e
E		
e		

Step 4: Determine all the gene combinations that can result when these gametes unite.
To determine the possible combinations of alleles that could occur as a result of this mating, simply fill in each of the empty squares with the alleles that can be donated from each parent. Determine all the gene combinations that can result when these gametes unite.

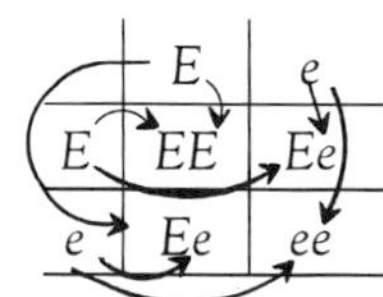

Step 5: Determine the phenotype of each possible gene combination.
In this problem, three of the offspring, *EE*, *Ee*, and *Ee*, have free earlobes. One offspring, *ee*, has attached earlobes. Therefore, the answer to the problem is that the probability of having offspring with free earlobes is 3/4; for attached earlobes, it is 1/4.

Take the time to learn these five steps. All single-factor problems can be solved using this method; the only variation in the problems will be the types of alleles and the number of possible types of gametes the parents can produce. Now let's work a problem with one parent heterozygous and the other homozygous for a trait.

Some people are unable to convert the amino acid phenylalanine into the amino acid tyrosine. Such individuals suffer from phenylketonuria (PKU) and may become mentally retarded. The normal condition is to convert phenylalanine to tyrosine. It is dominant over the condition for PKU. If one parent is heterozygous and the other parent is homozygous for PKU, what is the probability that they can have a child who is normal? with PKU?

Step 1:
Use the symbol *N* for normal and *n* for PKU.

N = normal   *n* = PKU

Genotype		***Phenotype***
NN	=	normal metabolism of phenylalanine
Nn	=	normal metabolism of phenylalanine
nn	=	PKU disorder

Step 2:

Nn × *nn*

Step 3:

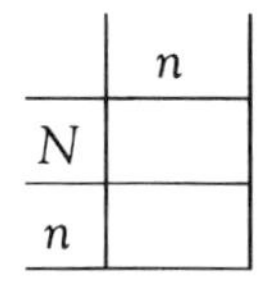

Step 4:

	n
N	*Nn*
n	*nn*

Step 5:
In this problem, one-half of the offspring will be normal, and one-half will have PKU.

The Double-Factor Cross

A **double-factor cross** is a genetic study in which two pairs of alleles are followed from the parental generation to the offspring. This problem is worked in basically the same way as a single-factor cross. The main difference is that in a double-factor cross you are working with two different characteristics from each parent.

It is necessary to use Mendel's law of independent assortment when working double-factor problems. Recall that according to this law, members of one allelic pair separate from each other independently of the members of other pairs of alleles. This happens during meiosis when the chromosomes segregate. (Mendel's law of independent assortment applies only if the two pairs of alleles are located on separate chromosomes. We will use this assumption in double-factor crosses.)

In humans, the allele for free earlobes dominates the allele for attached earlobes. The allele for dark hair dominates the allele for light hair. If both parents are heterozygous for earlobe shape and hair color, what types of offspring can they produce, and what is the probability for each type?

Step 1:
Use the symbol *E* for free earlobes and *e* for attached earlobes. Use the symbol *D* for dark hair and *d* for light hair.

E = free earlobes   *D* = dark hair
e = attached earlobes   *d* = light hair

Genotype		***Phenotype***
EE	=	free earlobes
Ee	=	free earlobes
ee	=	attached earlobes
DD	=	dark hair
Dd	=	dark hair
dd	=	light hair

Step 2:
Determine the genotype for each parent and show a mating. The male genotype is *EeDd*, the female genotype is *EeDd*, and the × between them indicates a mating.

EeDd × *EeDd*

Step 3:
Determine all the possible gametes each parent can produce and write the symbols for the alleles in a Punnett square. Since there are two pairs of alleles in a double-factor cross, each gamete must contain one allele from each pair—one from the earlobe pair (either *E* or *e*) and one from the hair color pair (either *D* or *d*). In this example, each parent can produce four different kinds of gametes. The four squares on the left indicate the gametes produced by the male; the four on the top indicate the gametes produced by the female.

To determine the possible gene combinations in the gametes, select one allele from one of the pairs of alleles and match it with one allele from the other pair of alleles. Then match the second allele from the first pair of alleles with each of the alleles from the second pair. This may be done as follows:

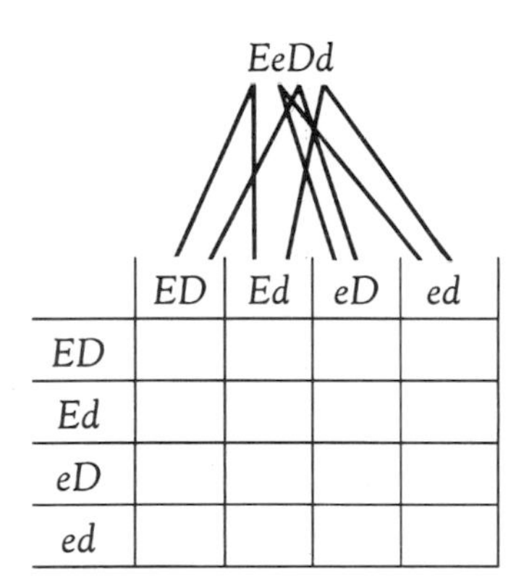

Step 4:
Determine all the gene combinations that can result when these gametes unite. Fill in the Punnett square.

	ED	Ed	eD	ed
ED	EEDD	EEDd	EeDD	EeDd
Ed	EEDd	EEdd	EeDd	Eedd
eD	EeDD	EeDd	eeDD	eeDd
ed	EeDd	Eedd	eeDd	eedd

Step 5:
Determine the phenotype of each possible gene combination. In this double-factor problem there are sixteen possible ways in which gametes could combine to produce offspring. There are four possible phenotypes in this cross. They are represented in the chart below.

The probability of having a given phenotype is

9/16 free earlobes, dark hair
3/16 free earlobes, light hair
3/16 attached earlobes, dark hair
1/16 attached earlobes, light hair

For our next problem, let's say a man with attached earlobes is heterozygous for hair color and his wife is homozygous for free earlobes and light hair. What can they expect their offspring to be like?

This problem has the same characteristics as the previous problem. Following the same steps, the symbols would be the same, but the parental genotypes would be as follows:

eeDd × *EEdd*

The next step is to determine the possible gametes that each parent could produce and place them in a Punnett square. The male parent can produce two different kinds of gametes, *eD* and *ed*. The female parent can only produce one kind of gamete, *Ed*.

	Ed
eD	
ed	

If you combine the gametes, only two kinds of offspring can be produced:

	Ed
eD	EeDd
ed	Eedd

They should expect either a child with free earlobes and dark hair or a child with free earlobes and light hair.

Real-World Situations

So far we have considered a few simple cases in which a characteristic is determined by simple dominance and recessiveness between two alleles. Other situations, however, do not fit these patterns. Some genetic characteristics are determined by more than two alleles; moreover, some traits are influenced by gene interactions and some traits are inherited differently, depending on the sex of the offspring.

Genotype		***Phenotype***	***Symbol***
EEDD or *EEDd* or *EeDD* or *EeDd*	=	free earlobes and dark hair	= *
EEdd or *Eedd*	=	free earlobes and light hair	= ∧
eeDD or *eeDd*	=	attached earlobes and dark hair	= ″
eedd	=	attached earlobes and light hair	= +

	ED	Ed	eD	ed
ED	EEDD *	EEDd *	EeDD *	EeDd *
Ed	EEDd *	EEdd ∧	EeDd *	Eedd ∧
eD	EeDD *	EeDd *	eeDD ″	eeDd ″
ed	EeDd *	Eedd ∧	eeDd ″	eedd +

Lack of Dominance

In the cases that we have considered so far, one allele of the pair was clearly dominant over the other. Although this is common, it is not always the case. In some combinations of alleles, there is a **lack of dominance.** This is a situation in which two unlike alleles both express themselves, neither being dominant. A classic example involves the color of the petals of snapdragons. There are two alleles for the color of these flowers. Because neither allele is recessive, we cannot use the traditional capital and small letters as symbols for these alleles. Instead, the allele for white petals is given the symbol F^W, and the one for red petals is given the symbol F^R. There are three possible combinations of these two alleles:

Genotype		*Phenotype*
F^WF^W	=	white flower
F^RF^R	=	red flower
F^RF^W	=	pink flower

Notice that there are only two different alleles, red and white, but there are three phenotypes, red, white, and pink. Both the red-flower allele and the white-flower allele partially express themselves when both are present, and this results in pink.

Heredity problems dealing with lack of dominance are worked according to the five steps used in other problems. In the following lack-of-dominance problem, only one trait is involved; therefore, it is worked as a single-factor problem.

If a pink snapdragon is crossed with a white snapdragon, what phenotypes can result, and what is the probability of each phenotype?

Step 1:

F^W = white flowers F^R = red flowers

Genotype		*Phenotype*
F^WF^W	=	white flower
F^WF^R	=	pink flower
F^RF^R	=	red flower

Step 2:

$F^RF^W \times F^WF^W$

Step 3:

	F^W
F^R	
F^W	

Step 4:

	F^W
F^R	F^WF^R pink
F^W	F^WF^W white

Step 5:
This cross results in two different phenotypes—pink and white. No red flowers can result because this would require that both parents be able to contribute at least one red allele. The white flowers are homozygous for white, and the pink flowers are heterozygous.

Multiple Alleles

So far we have discussed only those traits that are determined by two alleles. However, there can be more than two different alleles for a trait. The fact that some characteristics are determined by three or more different alleles is called **multiple alleles.** However, an individual can have only a maximum of two of the alleles for the characteristic. A good example of a characteristic that is determined by multiple alleles is the ABO blood type. There are three alleles for blood type:

I^A = blood type A
I^B = blood type B
i = blood type O

A and B show *codominance* when they are together in the same individual, but both are dominant to the O allele. These three alleles can be combined as pairs in six different ways, resulting in four different phenotypes:

Genotype		*Phenotype*
I^AI^A	=	blood type A
I^Ai	=	blood type A
I^BI^B	=	blood type B
I^Bi	=	blood type B
I^AI^B	=	blood type AB
ii	=	blood type O

Multiple-allele problems are worked as single-factor problems. Some examples are in the practice problems at the end of this chapter.

Polygenic Inheritance

Thus far we have considered phenotypic characteristics that are determined by alleles at a specific, single place on homologous chromosomes. However, some characteristics are determined by the interaction of genes at several different loci (on different chromosomes or at different places on a single chromosome). This is called **polygenic inheritance.** A number of different pairs of alleles may combine their efforts to determine a characteristic. Skin color in humans is a good example of this inheritance pattern. According to some experts, genes for skin color are located at a minimum of three different loci. At each of these loci, the allele for dark skin is dominant over the allele for light skin. Therefore, a wide variety of skin colors is possible depending on how many dark-skin alleles are present (figure 10.4). Polygenic inheritance is very common in determining characteristics that are quantitative in nature. In the skin-color example, and in many others as well, the characteristics cannot be categorized in terms of *either/or*, but the variation in phenotypes can be classified as *how much* or *what amount* (see box 10.1). For instance, people show great variations in height. There are not just tall and short people—there is a wide range. Some people are as short as one meter, and others are taller than two meters. This quantitative trait is probably determined by a number of different genes. Intelligence, also, varies significantly from those who are severely retarded to those who are geniuses. Many of these traits may be influenced by outside environmental factors, such as diet, disease, accidents, and social factors as well. These are just some examples of polygenic inheritance patterns.

Locus 1	d^1d^1	d^1D^1	d^1D^1	D^1D^1	D^1d^1	D^1d^1	D^1D^1
Locus 2	d^2d^2	d^2d^2	d^2D^2	D^2d^2	D^2d^2	D^2D^2	D^2D^2
Locus 3	d^3d^3	d^3d^3	d^3d^3	d^3d^3	D^3D^3	D^3D^3	D^3D^3
Total number of dark-skin genes	0	1	2	3	4	5	6
	Very light			Medium			Very dark

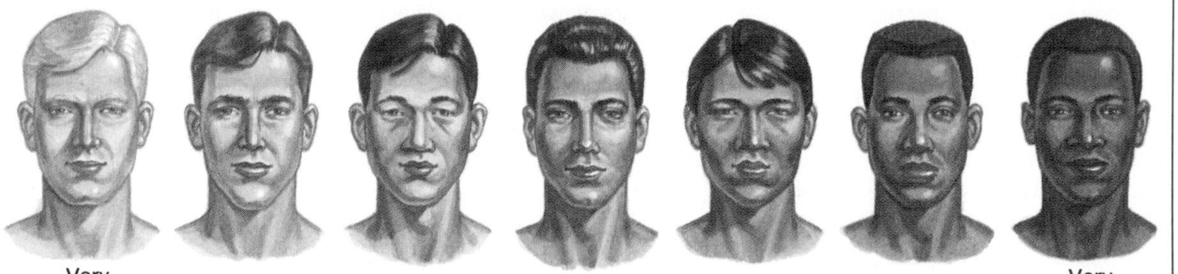

Creek

Figure 10.4 **Polygenic Inheritance.** Skin color in humans is an example of polygenic inheritance. The darkness of the skin is determined by the number of dark-skin genes a person inherits from his or her parents.

Pleiotropy

A gene often has a variety of effects on the phenotype of an organism. In fact every gene probably affects or modifies the expression of many different characteristics exhibited by an organism. This is called *pleiotropy*. **Pleiotropy** is a term used to describe the multiple effects that a gene may have on the phenotype of an organism. For example, the gene for sickle-cell hemoglobin has two major effects. One is good and one is bad. Having the allele for sickle-cell hemoglobin can result in abnormally shaped red blood cells. This occurs because the hemoglobin molecules are synthesized with the wrong amino acid sequence. These abnormal hemoglobin molecules tend to attach to one another in long, rodlike chains when oxygen is in short supply. These rodlike chains distort the shape of the red blood cells into a sickle shape. When these abnormal red blood cells change shape, they clog small blood vessels. The sickled red cells are also destroyed more rapidly than normal cells. This results in a shortage of red blood cells, causing anemia and an oxygen deficiency in the tissues that have become clogged. People with sickle-cell anemia may experience pain, swelling, and damage to organs such as the heart, lungs, brain, and kidneys.

Although sickle-cell anemia is usually lethal in the homozygous condition, it can be beneficial in the heterozygous state. A person with a single sickle-cell allele is more resistant to malaria than a person without this gene. A heterozygous person may not demonstrate any ill effects, but under laboratory conditions with low oxygen, there is a change in the red blood cells. Three genotypes can exist (Hb^A = normal hemoglobin, Hb^S = sickle-cell hemoglobin):

Genotype		*Phenotype*
$Hb^A\ Hb^A$	=	normal hemoglobin and nonresistance to malaria
$Hb^A\ Hb^S$	=	normal hemoglobin and resistance to malaria
$Hb^S\ Hb^S$	=	resistance to malaria but death from sickle-cell anemia

Sickle-cell anemia was originally found throughout the world in places where malaria was common. Today, however, this genetic disease can be found anywhere in the world. In the United States, it is most common among black populations whose ancestors came from equatorial Africa.

Let's look at another example of pleiotropy. In this example, a single gene affects many different chemical reactions that depend on the way a cell metabolizes the amino acid phenylalanine (figure 10.5). People normally have a gene for the production of an enzyme that converts the amino acid phenylalanine to tyrosine. If this gene is functioning properly, phenylalanine will be converted to tyrosine, which will be available to be converted into thyroxine and melanin by other enzymes. If the enzyme that normally converts phenylalanine to tyrosine is absent, toxic materials can accumulate and result in a loss of nerve cells, causing mental retardation. Because less tyrosine is produced, there is also less of the growth hormone, thyroxine, resulting in abnormal body

BOX 10.1 The Inheritance of Eye Color

It is commonly thought that eye color is inherited in a simple dominant/recessive manner. Brown eyes are considered to be dominant over blue eyes. The real pattern of inheritance, however, is considerably more complicated than this. Eye color is determined by the amount of a brown pigment, known as *melanin,* that is present in the iris of the eye. If there is a large quantity of melanin present on the anterior surface of the iris, the eyes are dark. Black eyes have a greater quantity of melanin than brown eyes.

If there is not a large amount of melanin present on the anterior surface of the iris, the eyes will appear to be blue, not because of a blue pigment but because blue light is returned from the iris. The iris appears blue for the same reason that deep bodies of water tend to appear blue. There is no blue pigment in the water, but blue wavelengths of light are returned to the eye from the water. People appear to have blue eyes because the blue wavelengths of light are reflected from the iris.

Just as black and brown eyes are determined by the amount of pigment present, colors such as green, gray, and hazel are produced by the various amounts of melanin in the iris. If a very small amount of brown melanin is present in the iris, the eye tends to appear green, whereas relatively large amounts of melanin produce hazel eyes.

Several different genes are probably involved in determining the quantity and placement of the melanin and, therefore, in determining eye color. These genes interact in such a way that a wide range of eye color is possible. Eye color is probably determined by polygenic inheritance, just as skin color and height are. (Some newborn babies have blue eyes that later become brown. This is because, at the time of birth, they have not yet begun to produce melanin in their irises.)

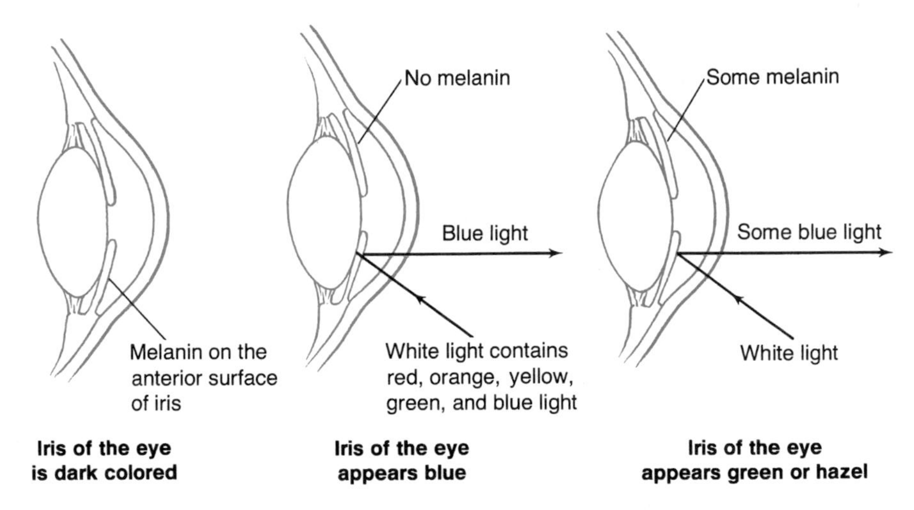

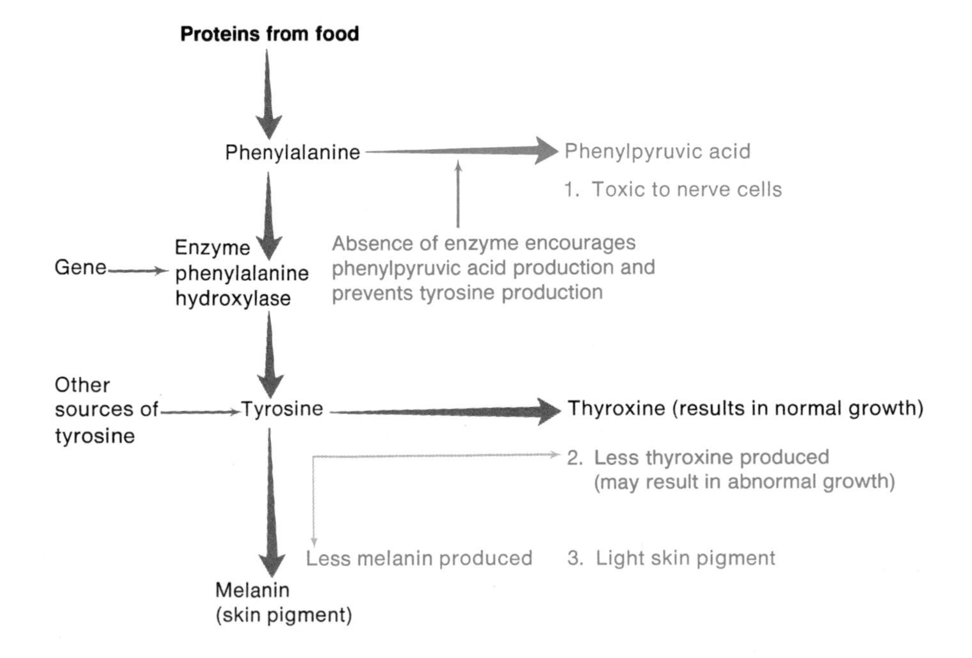

Figure 10.5 **Pleiotropy.** Pleiotropy is a condition in which a single gene has more than one effect on the phenotype. This diagram shows how the normal pathways work (these are shown in black). If the enzyme phenylalanine hydroxylase is not produced because of an abnormal gene, there are three major results: (1) mental retardation because phenylpyruvic acid kills nerve cells, (2) abnormal body growth because less of the growth hormone thyroxine is produced, and (3) pale skin pigmentation because less melanin is produced (abnormalities are shown in color). It should also be noted that if a woman who has PKU becomes pregnant, her baby is likely to be born retarded. While the embryo may not have the genetic disorder, the phenylpyruvic acid produced by the pregnant mother will damage the developing brain cells. This is called *maternal PKU.*

growth. Because tyrosine is necessary to form the pigment melanin, people who have this condition will have lighter skin color because of an absence of this pigment. The one abnormal allele produces three different phenotypic effects: mental retardation, abnormal growth, and light skin.

Linkage

Pairs of alleles located on nonhomologous chromosomes separate independently of one another during meiosis when the chromosomes separate into sex cells. Since each chromosome has many genes on it, these genes tend to be inherited as a group. Genes located on the same chromosome that tend to be inherited together are called a **linkage group.** The closer two genes are to each other on a chromosome, the more probable it is that they will be inherited together. The process of crossing-over, which occurs during prophase I of meiosis, may split up these linkage groups. Crossing-over happens between homologous chromosomes donated by the mother and the father, and results in a mixing of genes.

Sex-Linked Genes

Many organisms have two types of chromosomes. **Autosomes** are not involved in sex determination and have the same genes on both members of the homologous pair of chromosomes. **Sex chromosomes** are a pair of chromosomes in mammals and some other animals that determine the sex of an organism. In humans and some other animals there are two types of sex chromosomes—the X chromosome and the Y chromosome. The Y chromosomes are much shorter than the X chromosomes and probably have no genes for traits found on the X chromosome. One portion of the Y chromosome contains the male-determining genes. Females are produced when two X chromosomes are present. Males are produced when one X chromosome and one Y chromosome are present.

Genes found on the X chromosome are said to be **X-linked.** Because the Y chromosome is shorter than the X chromosome, it does not have many of the alleles that are found on the comparable portion of the X chromosome. Therefore, in men, the presence of a single allele on his only X chromosome will be expressed, regardless of whether it is dominant or recessive. The only confirmed Y-linked trait in humans is the SRY gene. This gene controls the differentiation of the embryonic gonad to a male testis. By contrast, there are over one hundred genes that are linked to the X chromosome. Some of these X-linked genes are color blindness, hemophilia, brown teeth, and a form of muscular dystrophy. To better understand an X-linked gene, let's now use the five steps of solving genetics problems to work an X-linked problem.

In humans the gene for normal color vision is dominant and the gene for color blindness is recessive. Both genes are X-linked. A male who has normal vision mates with a female who is heterozygous for normal color vision. What type of children can they have in terms of these traits, and what is the probability for each type?

Step 1:
Because this condition is linked to the X chromosome, it has become traditional to symbolize the allele as a superscript on the letter X. Since the Y chromosome does not contain a homologous allele, only the letter Y is used.

X^N = normal color vision
X^n = color-blind
Y = male (no gene present)

Genotype	***Phenotype***
X^NY =	male, normal color vision
X^nY =	male, color-blind
X^NX^N =	female, normal color vision
X^NX^n =	female, normal color vision
X^nX^n =	female, color-blind

Step 2:
Male's genotype = X^NY (normal color vision)
Female's genotype = X^NX^n (normal color vision)

$$X^NY \times X^NX^n$$

Step 3:
The genotype of the gametes are listed in the Punnett square:

	X^N	X^n
X^N		
Y		

Step 4:
The genotypes of the probable offspring are listed in the body of the Punnett square:

	X^N	X^n
X^N	X^NX^N	X^NX^n
Y	X^NY	X^nY

Step 5:
The phenotypes of the offspring are determined:

normal female	carrier female
normal male	color-blind male

1/4 normal female 1/4 normal male
1/4 carrier female 1/4 color-blind male

A **carrier** is any individual who is heterozygous for a trait. In this situation, the recessive allele is hidden. In X-linked situations, only the female can be heterozygous because the males lack one of the X chromosomes. Heterozygous females will exhibit the dominant trait (normal vision in this problem), but have the recessive allele hidden. If a male has a recessive allele on the X chromosome, it will be expressed because there is no other allele on the Y chromosome to dominate it. If a heterozygous carrier (must be female) has sons, she should expect half of them to be color-blind and half to have normal color vision. For these reasons, there are many more color-blind males than there are color-blind females.

Environmental Influences on Gene Expression

The specific phenotype an organism exhibits is determined by the interplay of the genotype of the individual and the conditions the organism encounters as it develops. Therefore, it is possible for two organisms with identical genotypes (identical twins) to differ in their phenotypes. All genes must express themselves through the manufacture of proteins. These proteins may be structural or enzymatic, and the enzymes may be more or less effective depending on the specific biochemical conditions when the enzyme is in operation. The expression of the genes will vary depending on the environmental conditions while the gene is operating.

Maybe you assumed that the dominant allele would always be expressed in a heterozygous individual. It is not so simple! Here, as in other areas of biology, there are exceptions. For example, the allele for six fingers is dominant over the allele for five fingers in humans. Some people who have received the allele for six fingers have a fairly complete sixth finger; in others, it may appear as a little stub. In another case, a dominant allele causes the formation of a little finger that cannot be bent as a normal little finger. However, not all people who are believed to have inherited that allele will have a stiff little finger. In some cases, this dominant characteristic is not expressed or perhaps only shows on one hand. Thus, there may be variation in the degree to which a dominant allele expresses itself, and in some cases it may not even be expressed. Other genes may be interacting with these dominant alleles, causing the variation in expression. It is important to recognize that the environment affects the expression of our genes in many ways.

Both internal and external environmental factors can influence the expression of genes. For example, at conception, a male receives genes that will eventually determine the pitch of his voice. However, these genes are expressed differently after puberty. At puberty, male sex hormones are released. This internal environmental change results in the deeper male voice. A male who does not produce these hormones retains a higher-pitched voice in later life. A comparable situation in females occurs when an abnormally functioning adrenal gland causes the release of large amounts of male hormones. This results in a female with a deeper voice.

Another influence on gene expression is called *gene imprinting*. This occurs when a gene donated by one of the parents has its expression altered by a gene donated by the other parent. Certain genes donated by the father affect those donated by the mother, and certain genes donated by the mother affect those donated by the father. It is also true that certain genes only function properly when donated by the father, and other genes only function properly when donated by the mother. Although the mechanics of the process are not yet well understood, it is known that paternal and maternal genes contribute in different ways to the developing embryo. Two forms of mental retardation illustrate gene imprinting in humans—Angelman and Prader-Willi syndromes. Prader-Willi is transmitted from the mother and Angelman from the father. Patients with Angelman syndrome display excessive laughter, jerky movements, and other symptoms of physical and mental retardation. Those with Prader-Willi syndrome show mental retardation, extreme obesity, short stature, and unusually small hands and feet. Both are disorders of chromosome 15.

Many external environmental factors can influence the phenotype of an individual. One such factor is diet. Diabetes mellitus, a metabolic disorder in which glucose is not properly metabolized and is passed out of the body in the urine, has a genetic basis. Some people who have a family history of diabetes are thought to have inherited the trait for this disease. Evidence indicates that they can delay the onset of the disease by reducing the amount of sugar in their diet. This change in the external environment influences gene expression in much the same way that temperature influences the expression of color production in cats or sunlight affects the expression of freckles in humans (figure 10.3).

• Summary •

Genes are units of heredity composed of specific lengths of DNA that determine the characteristics an organism displays. Specific genes are at specific loci on specific chromosomes. The phenotype displayed by an organism is the result of the effect of the environment on the ability of the genes to express themselves. Diploid organisms have two genes for each characteristic. The alternative genes for a characteristic are called alleles. There may be many different alleles for a particular characteristic. Those organisms with two identical alleles are homozygous for a characteristic; those with different alleles are heterozygous for a characteristic. Some alleles are dominant over other alleles that are said to be recessive.

Sometimes two alleles both will express themselves, and often a gene will have more than one recognizable effect on the phenotype of the organism. Some characteristics may be determined by several different pairs of alleles. In humans and some other animals, males have an X chromosome with a normal number of genes and a Y chromosome with fewer genes. Although they are not identical, they behave as a pair of homologous chromosomes. Since the Y chromosome is shorter than the X chromosome and has fewer genes, many of the recessive characteristics present on the X chromosome appear more frequently in males than in females, who have two X chromosomes.

• Experience This •

Some humans inherit the ability to taste the chemical phenylthiocarbamide (PTC), and others are unable to taste PTC. Ask your instructor to furnish you with a supply of paper impregnated with PTC. Place a piece of this paper on your tongue. If you experience a bitter taste, you are a taster. If you do not, you are a nontaster. Take enough strips to test those members of your family you can readily contact. These may include your grandparents, parents, siblings, children, aunts, and uncles. After you have tested the members of your family, construct a pedigree. Is it possible for you to determine the genotypes of your family members? Is tasting PTC a dominant or a recessive trait? Could it be a case of lack of dominance? Might it be linked to another trait?

In a pedigree, circles represent females and squares represent males. Symbols of parents are connected by a horizontal mating line, and the offspring are shown on a horizontal line below the parents. Individuals who are tasters are represented by a solid symbol, nontasters by an open symbol. A typical pedigree might look something like this one, which shows three generations. Fill in your pedigree. You may need to modify this pedigree to fit your particular family situation.

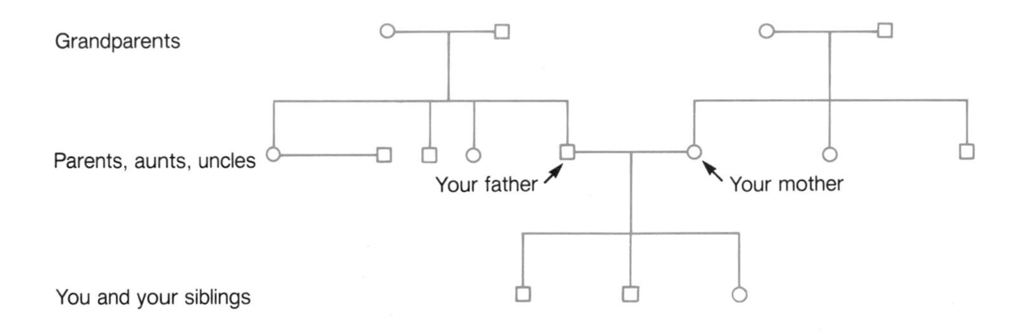

• Questions •

1. How many kinds of gametes are possible with each of the following genotypes?
   a. *Aa*
   b. *AaBB*
   c. *AaBb*
   d. *AaBbCc*
2. What is the probability of getting the gamete *ab* from each of the following genotypes?
   a. *aabb*
   b. Aabb
   c. *AaBb*
   d. *AABb*
3. What is the probability of each of the following sets of parents producing the given genotypes in their offspring?

	Parents	Offspring Genotype
a.	*AA* × *aa*	*Aa*
b.	*Aa* × *Aa*	*Aa*
c.	*Aa* × *Aa*	*aa*
d.	*AaBb* × *AaBB*	*AABB*
e.	*AaBb* × *AaBB*	*AaBb*
f.	*AaBb* × *AaBb*	*AABB*

4. If an offspring has the genotype *Aa*, what possible combinations of parental genotypes can exist?
5. In humans, the allele for albinism is recessive to the allele for normal skin pigmentation.
   a. What is the probability that a child of a heterozygous mother and father will be an albino?
   b. If a child is normal, what is the probability that it is a carrier of the recessive albino allele?
6. In certain pea plants, the allele *T* for tallness is dominant over *t* for shortness.
   a. If a homozygous tall and homozygous short plant are crossed, what will be the phenotype and genotype of the offspring?
   b. If both individuals are heterozygous, what will be the phenotypic and genotypic ratios of the offspring?
7. Smoos are strange animals with one of three shapes: round, cuboidal, or pyramidal. If two cuboidal smoos mate, they always have cuboidal offspring. If two pyramidal smoos mate, they always produce pyramidal offspring. If two round smoos mate, they produce all three kinds of offspring. Assuming only one locus is involved, answer the following questions.
   a. How is smoo shape determined?
   b. What would be the phenotypic ratio if a round and cuboidal smoo were to mate?
8. What is the probability of a child having type AB blood if one of the parents is heterozygous for A blood and the other is heterozygous for B? What other genotypes are possible in these children?
9. A color-blind woman marries a man with normal vision. They have ten children—six boys and four girls.
   a. How many are normal?
   b. How many are color-blind?
10. A light-haired man has blood type O. His wife has dark hair and blood type AB, but her father had light hair.
    a. What is the probability that this couple will have a child with dark hair and blood type A?

b. What is the probability that they will have a light-haired child with blood type B?
c. How many different phenotypes could their children show?

11. Certain kinds of cattle have two alleles for coat color: R = red, and r = white. When an individual cow is heterozygous, it is spotted with red and white (roan). When two red alleles are present, it is red. When two white alleles are present, it is white. The allele L, for lack of horns, is dominant over l, for the presence of horns.
    a. If a bull and a cow both have the genotype $RrLl$, how many possible phenotypes of offspring can they have?
    b. How probable is each phenotype?

12. Hemophilia is a disease that prevents the blood from clotting normally. It is caused by a recessive allele located on the X chromosome. A boy has the disease; neither his parents nor his grandparents have the disease. What are the genotypes of his parents and grandparents?

Answers

1. a. 2—*A*,*a*
   b. 2—*AB*, *aB*
   c. 4—*AB*, *Ab*, *aB*, *ab*
   d. 8—*ABC*, *ABc*, *Abc*, *AbC*, *aBC*, *aBc*, *abC*, *abc*
2. a. 100%—only, *ab* is possible
   b. 50%—*Ab* and *ab* are equally possible
   c. 25%—*AB*, *Ab*, *aB* and *ab* are equally possible
   d. 0%—*ab* not possible
3. a. 100%
   b. 1/2 or 50%
   c. 1/4 or 25%
   d. 1/8 or 12.5%
   e. 1/4 or 25%
   f. 1/16 or 6.25%
4. *AA* × *aa*
   AA × *Aa*
   Aa × *Aa*
   Aa × *aa*
5. a. 1/4 or 25%
   b. 2/3 or 67%
6. a. Tall, *Tt*
   b. Phenotypic ratio—3 tall to 1 short
   Genotypic ratio—1 homozygous tall, 2 heterozygous tall, 1 homozygous short
7. a. This is a case of lack of dominance
   b. Fifty % or ½ round, 50% or ½ cuboidal
8. 1/4 or 25%
   I^Ai, I^Bi, ii
9. a. (4) All the girls have the normal phenotype but are carriers.
   b. (6) All the boys are color-blind
10. a. 1/4 or 25%
    b. 1/4 or 25%
    c. 4
11. a. Six possible phenotypes:
    b. Red with horns—*RRll* = 1/16 or 6.25%
    Roan with horns—*Rrll* = 2/16 or 12.5%
    White with horns—*rrll* = 1/16 or 6.25%
    Red, hornless—*RrLL* or *RrLl* = 3/16 or 18.75%
    Roan, hornless—*RrLL* or *RrLl* = 6/16 or 37.5%
    White, hornless—*rrLL* or *rrLl* = 3/16 or 18.75%
12. | | |
    |---|---|
    | Father | X^NY |
    | Mother | X^HX^N |
    | Mother's father | X^NY |
    | Mother's mother | X^HX^N |
    | Father's father | X^NY |
    | Father's mother | $X^NX^?$ |

• Chapter Glossary •

alleles (a-lēlz') Alternative forms of a gene for a particular characteristic (e.g., attached-earlobe genes and free-earlobe genes are alternative alleles for ear shape).

autosomes (aw'to-sōmz) Chromosomes that are not involved in determining the sex of an organism.

carrier (ka're-er) Any individual having a hidden, recessive gene.

dominant allele (dom'in-ant a-lēl') An allele that expresses itself and masks the effect of other alleles for the trait.

double-factor cross (dub'l fak'tur kros) A genetic study in which two pairs of alleles are followed from the parental generation to the offspring.

gene (jēn) A unit of heredity located on a chromosome and composed of a sequence of DNA nucleotides.

genetics (jĕ-net'iks) The study of genes, how genes produce characteristics, and how the characteristics are inherited.

genome (je'nōm) A set of all the genes necessary to specify an organism's complete list of characteristics.

genotype (je'no-tīp) The catalog of genes of an organism, whether or not these genes are expressed.

heterozygous (hĕ"ter-o-zi'gus) A diploid organism that has two different allelic forms of a particular gene.

homozygous (ho"mo-zi'gus) A diploid organism that has two identical alleles for a particular characteristic.

lack of dominance (lak uv dom'in-ans) The condition of two unlike alleles both expressing themselves, neither being dominant.

law of dominance (law uv dom'in-ans) When an organism has two different alleles for a trait, the allele that is expressed and overshadows the expression of the other allele is said to be dominant. The allele whose expression is overshadowed is said to be recessive.

law of independent assortment (law uv in"de-pen'dent ă-sort'ment) Members of one allelic pair will separate from each other independently of the members of other allele pairs.

law of segregation (law uv seg"rĕ-ga'shun) When gametes are formed by a diploid organism, the alleles that control a trait separate from one another into different gametes, retaining their individuality.

linkage group (lingk'ij grūp) Genes located on the same chromosome that tend to be inherited together.

locus (loci) (lo'kus) (lo'si) The spot on a chromosome where an allele is located.

Mendelian genetics (Men-dĕ'le-an jĕ-net'iks) The pattern of inheriting characteristics that follows the laws formulated by Gregor Mendel.

multiple alleles (mul'tĭ-pul a-lēlz') A term used to refer to conditions in which there are several different alleles for a particular characteristic, not just two.

offspring (of'spring) Descendants of a set of parents.

phenotype (fēn'o-tīp) The physical, chemical, and psychological expression of the genes possessed by an organism.

pleiotropy (pli-ot'ro-pe) The multiple effects that a gene may have on the phenotype of an organism.

polygenic inheritance (pol"e-jen'ik in-her'ĭ-tans) The concept that a number of different pairs of alleles may combine their efforts to determine a characteristic.

probability (prob"a-bil'ĭ-te) The chance that an event will happen, expressed as a percent or fraction.

Punnett square (pun'net sqwār) A method used to determine the probabilities of allele combinations in a zygote.

recessive allele (re-sĕ'siv a-lēl') An allele that, when present with its homolog, does not express itself and is masked by the effect of the other allele.

sex chromosomes (seks kro'mo-sōmz) A pair of chromosomes that determine the sex of an organism.

single-factor cross (sing'ul fak'tur kros) A genetic study in which a single characteristic is followed from the parental generation to the offspring.

X-linked gene (eks-lingt jēn) A gene located on one of the sex-determining X chromosomes.

CHAPTER 11

Population Genetics

Chapter Outline

Purpose

This chapter is designed to help you understand why plants and animals of the same kind vary in different parts of the world, and how we artificially maintain certain groups of characteristics in domesticated species. The significance of population genetics to understanding how human genetic diseases are transmitted is also discussed. Later chapters on evolution will build on this information.

For Your Information

One of the forms that prejudice takes is a desire to "keep one's ancestral genes clean." Koreans living in Japan are considered "less able" and "unfit" for many higher level jobs in Japan. The South African government has attempted to rigidly segregate Indians, blacks, whites, and colored from one another. Many Native Central Americans are being eliminated by having their political and social freedoms denied. In the United States during World War II, Japanese-Americans were "detained" in relocation camps. Blacks and Native Americans also have been discriminated against for decades.

Learning Objectives

- Know the difference between the terms *species* and *population.*
- Be able to distinguish between the terms *gene pool* and *deme.*
- Describe the occurrence of a gene in a population in terms of gene frequency.
- Relate the concepts of cloning and hybridization to asexual and sexual reproduction.
- Recognize the role of mutation, sexual reproduction, population size, and migration in gene frequency.
- Describe the importance and potential danger of the practice of monoculture.
- Describe the role of a genetic counselor.
- Recognize that population-genetic principles can be valuable in understanding the occurrence of human genetic disease.

Figure 11.1 **Genetic Variety in Dogs.** Although these four breeds of dogs look quite different, they all have the same number of chromosomes and are capable of interbreeding. Therefore, they are members of the same species. The considerable difference in phenotypes is evidence of the genetic variety among breeds.

Genes in Populations

To understand the principles of genetics in chapter 10, we concerned ourselves with small numbers of organisms having specific genotypes. Plants and animals, however, don't usually exist as isolated individuals but as members of populations. Before we go any further, we need to define two terms that are used throughout this chapter, *species* and *population*.

A **species** is a group of organisms of the same kind that has the ability to interbreed to produce offspring that are also capable of reproducing. The members of a species usually look quite similar, although there are some exceptions. For example, all the dogs in the world are of the same species, but a Great Dane does not look very much like a Pekingese. However, mating can occur between these two quite different-appearing organisms (figure 11.1). If you examine the chromosomes of reproducing organisms, you find that they are identical in number and size and usually carry very similar groups of genes. In the final analysis, the species concept concerns the genetic similarity of organisms regardless of where or when they exist.

The concepts of population and species are interwoven: A **population** is considered to be all the organisms of the same species found within a specific geographic region. Population, however, is primarily concerned with numbers of organisms in a particular place at a particular time. The material in this chapter incorporates both concepts. It deals with populations and how they differ from each other genetically. For example, why are there more blue-eyed people in Scandinavia than in Spain, or why do some populations have a high frequency of blood type A?

The Gene Pool

Previously, we related the species concept to genetic similarity; however, you know that not all individuals of a species are genetically identical. Any one organism has a specific genotype consisting of all the genes that organism has in its DNA. It can have a maximum of two different alleles for a characteristic at one locus on homologous chromosomes. In a population, there may be many more than two kinds of alleles for a specific characteristic (multiple alleles). Since, theoretically, all organisms of a species are able to exchange genes, we can think of all the genes of all the individuals of the same species as a giant **gene pool.** Since each individual organism is like a container of a set of these genes, the gene pool contains many more kinds of genes than any one of the individuals. The gene pool is like a refrigerator full of cartons of different kinds of milk—chocolate, regular, skim, buttermilk, low-fat, and so on. If you were blindfolded and reached in with both hands and grabbed two cartons, you might end up with two chocolate, a skim and a regular, or one of the many other possible combinations. The cartons of milk represent different alleles, and the refrigerator (gene pool) contains a greater variety than could be determined by randomly selecting two cartons of milk at a time.

Individuals are usually not found evenly distributed but occur in clusters as a result of several different factors, such as geographic barriers that restrict movement or the local availability of resources. Gene clusters, called **demes,** may differ quite a bit from one place to another. There may be differences in the kinds of genes and the numbers of each kind of gene in different demes. Figure 11.2 indicates the relationship of genes to individuals, individuals to demes, and demes to the entire gene pool. Note, for example, that while all the demes contain the same kinds of genes, the relative number of alleles A and *a* differ from one deme to another.

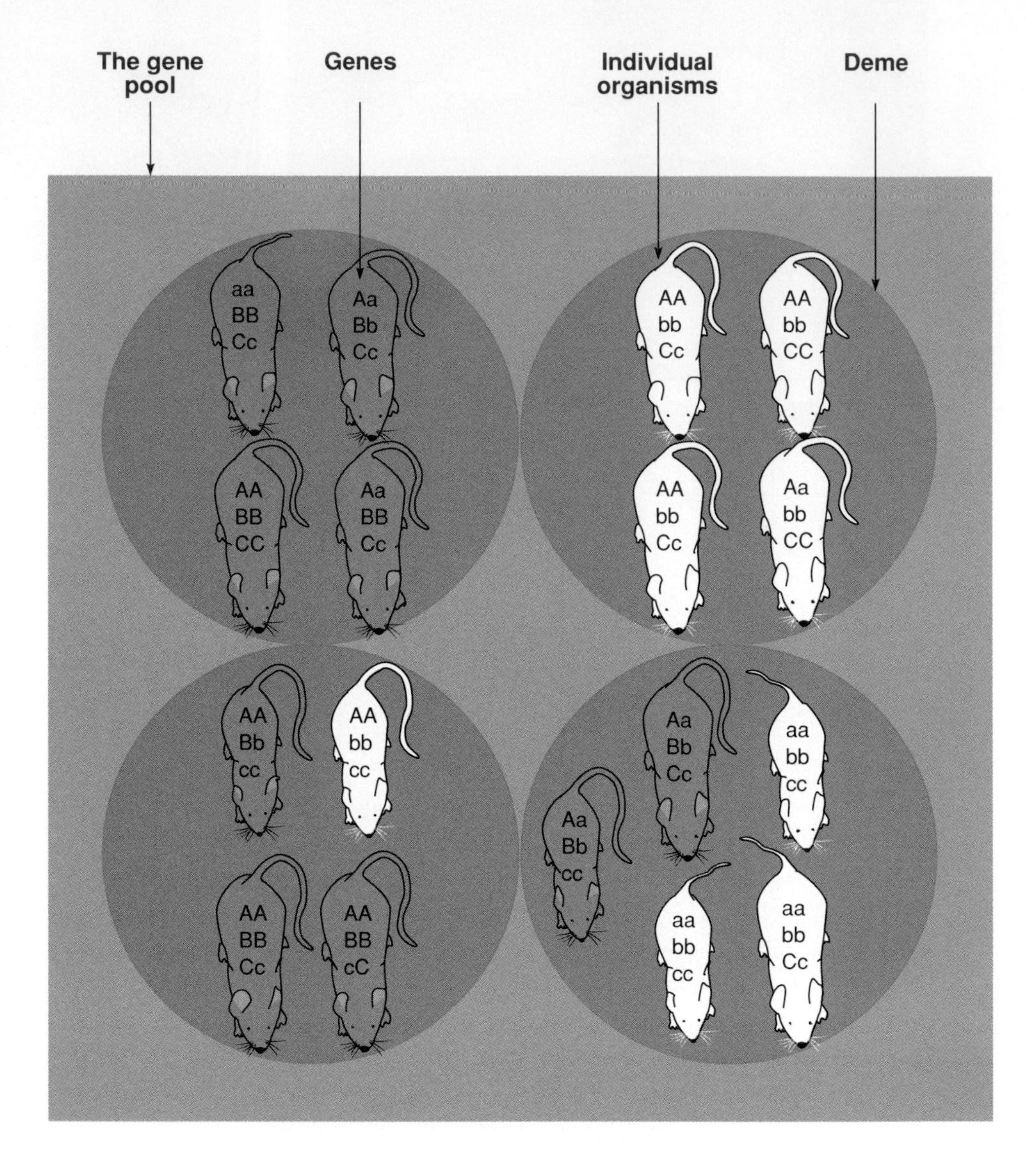

Figure 11.2 **Genes, Demes, and Gene Pools.** Each individual shown here has a specific genotype. Demes—local breeding populations—differ from one another in the frequency of each gene, but all demes have each of the different genes represented within the population. The gene pool includes all of the individuals present. Assume that A = long tail, a = short tail, B = brown color, b = white color, C = large size, and c = small size. Notice how the different frequencies of genes affect the appearance of the organisms in the different demes.

Since organisms tend to interbreed with other organisms located close by, local collections of genes tend to remain the same unless, in some way, genes are added to or subtracted from this local population (deme). The common water snake is found throughout the eastern portion of the United States (figure 11.3). The island water snake is one of the several demes within this species, which is confined to the islands in western Lake Erie. Most common water snakes of the mainland have light and dark bands. The island populations do not have this banded coloration; most individuals have genes for solid coloration with very few individuals having genes for banded coloration. The island snakes are geographically isolated from the main gene pool, and mate only with one another. Thus, the different color patterns shown by island snakes and mainland snakes result from a high incidence of solid-color genes in the island populations and a high incidence of banded-color genes in the mainland populations.

Within a deme genes are repackaged into new individuals from one generation to the next. Often there will be very little adding of new genes or subtracting of other genes from a deme, and a widely distributed species will consist of a number of more or less separate groups (demes) that are known as

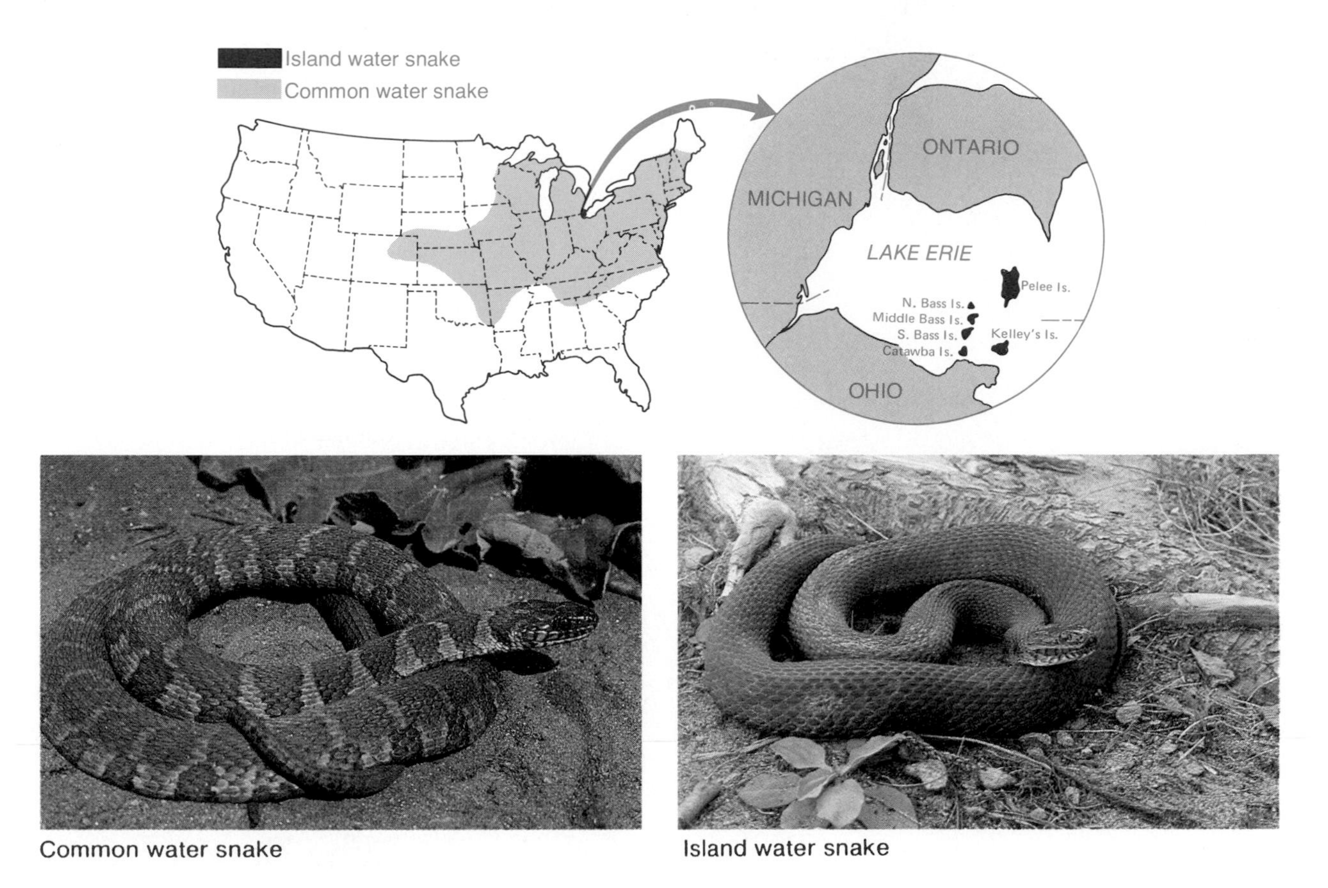

Figure 11.3 **The Range and Appearance of the Common Water Snake and the Island Water Snake.** The common water snake is found throughout the eastern part of the United States. The Island water snake is limited to the islands in the western section of Lake Erie.

subspecies, races, breeds, strains, or **varieties.** All these terms are used to describe different forms of organisms that are all members of the same species. However, certain terms are used more frequently than others, depending on one's field of interest. For example, dog breeders use the term *breed,* horticulturalists use the term *variety,* microbiologists use the term *strain,* and anthropologists use the term *race.* The most general and widely accepted term is *subspecies.*

Gene Frequency

How often a gene is found in a population is known as its **gene frequency.** Gene frequency is typically stated in terms of a percentage or decimal fraction and is a mathematical statement of how frequently a particular gene shows up in the sex cells of a population (for example, 10%, or 0.1; 50% or 0.5). It is possible for two demes to have all the same genes, but with very different frequencies.

As an example, all humans are of the same species and, therefore, constitute one large gene pool. There are, however, many distinct local populations scattered across the surface of the earth. These more localized populations (races, or demes) show many distinguishing characteristics that have been perpetuated from generation to generation. In Africa, genes for dark skin, tightly curled hair, and a flat nose have very high frequencies. In Europe, the frequencies of genes for light skin, straight hair, and a narrow nose are the highest. The Chinese tend to have moderately colored skin, straight hair, and broad noses (figure 11.4). All three of these populations have genes for dark skin and light skin, straight hair and curly hair, narrow noses and broad noses. The three differ, however, in the frequencies of these genes. Many other genes show differences in frequency from one race to another, but these three characteristics are easy to see. Once a particular mixture of genes is present in a population, that mixture tends to maintain itself unless something is operating to change the frequencies. In other words, gene frequencies are not going to change without reason. With the development of transportation, more people have moved from one geographic area to another and human gene frequencies have begun to change. Ultimately, as barriers to interracial marriage (both geographic and sociological) are leveled, the human gene pool will show fewer and fewer racial differences. However, it may be thousands of years before significant changes are seen.

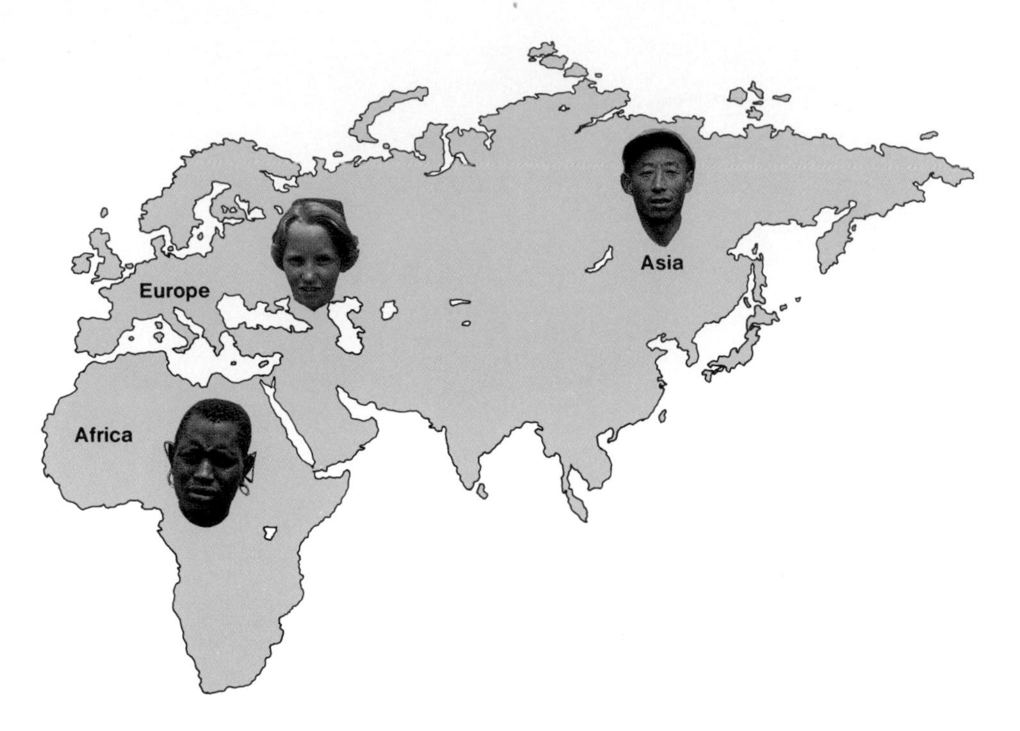

Figure 11.4 **Gene Frequency Differences among Humans.** Different physical characteristics displayed by people from different parts of the world are an indication that gene frequencies differ as well.

Table 11.1
Recessive Traits with a High Frequency of Expression. *Many recessive characteristics are extremely common in some human populations. The corresponding dominant characteristic is also shown here.*

Recessive	Dominant
Light skin color	Dark skin color
Straight hair	Curly hair
Five fingers	Six fingers
Type O blood	Type A or B blood
Normal hip joints	Dislocated hip birth defect
Blue eyes	Brown eyes
Normal eyelids	Drooping eyelids
No tumor of the retina	Tumor of the retina
Normal fingers	Short fingers
Normal thumb	Extra joint in the thumb
Normal fingers	Webbed fingers
Ability to smell	Inability to smell
Normal tooth number	Extra teeth
Presence of molars	Absence of molars
Normal palate	Cleft palate

For some reason, people tend to think that gene frequency has something to do with the dominance or recessiveness of genes. This is not true. Often in a population, recessive genes are more frequent than their dominant counterparts. Straight hair, blue eyes, and light skin are all recessive characteristics, yet they are quite common in the populations of certain European countries. See table 11.1 for other examples.

What really determines the frequency of a gene in a deme is the value that the gene has to the organisms possessing it. The dark-skin genes are very valuable to people living under the bright sun in tropical Africa. These genes are less valuable to those living in the less intense sunlight of the cooler European countries. This idea of gene value and what it can do to gene frequencies will be dealt with more fully when the process of natural selection is discussed.

Why Demes Exist

Since individual organisms within a population are not genetically identical, some individuals may possess combinations of genes that are particularly valuable for survival in the local environment. As a result, some individuals find the environment less hostile than others. The individuals with unfavorable combinations of genes leave the population more often, either by death or migration, and remove their genes from the deme. Therefore, local populations that occupy sites that differ greatly would be expected to consist of individuals having genes suited to local conditions. For example, a blind fish living in a lake is at a severe disadvantage. A blind fish living in a cave where there is no light, however, is not at the same disadvantage. Thus, these two environments might allow or encourage different genes to be present in the two populations at different frequencies (figure 11.5).

A second mechanism, which tends to create small demes with gene frequencies different from other demes, involves

Figure 11.5 **Blind Cave Fish.** This fish lives in caves where there is no light. Its eyes do not function and it has very little color in its skin. Because of its unusual habitat, the presence of genes for eyes and skin color are not important. If, at some time in the past, these genes were lost or mutated, it did not negatively affect the organism; hence, the present population has high frequencies of genes for the absence of color and eyes.

the founding of a new population. The collection of genes from a small founding population is likely to be different from the genes present in the larger parent population from which they came. After all, a few individuals leaving a population would be unlikely to carry copies of all the genes found within the original population. They may even carry an unrepresentative mixture of genes. Once a small founding population establishes itself, it tends to maintain its collection of genes because the organisms mate only among themselves. This results in a reshuffling of genes from generation to generation and discourages the introduction of new genes into the population. A third factor that tends to encourage the maintenance of demes is the presence of barriers to free movement. Animals and plants that live in lakes tend to be divided into small, separate populations by barriers of land. Whenever such barriers exist, there will very likely be differences in the gene frequencies from lake to lake because each lake was colonized separately and their environments are not identical. Other species of organisms like migratory birds (robins, mallard ducks) experience few barriers; therefore, subspecies are quite rare.

How Genetic Variety Comes About

A large gene pool with a great variety of genes is more likely to contain some genes that will better adapt the organisms to a new environment. A number of mechanisms introduce this necessary variety into a population.

Mutations

Mutations introduce new genes into a population. Sometimes a mutation is a first-time event, while at other times a mutation may have occurred before. All alleles for a particular trait originated as a result of mutations sometime in the past and have been maintained within the gene pool of the species as a result of sexual reproduction. If a mutation produces a bad gene, that gene will remain uncommon in the population. While many mutations are bad, very rarely one will occur that is valuable to the organism. For example, at some time in the past, mutations occurred in the DNA of certain insect species that made some individuals tolerant to the insecticide DDT, even though the chemical had not yet been invented. These alleles remained very rare in these insect populations until DDT was used. Then, these genes became very valuable to the insects that carried them. Since insects that lacked the gene for tolerance died when they came in contact with DDT, more of the DDT-tolerant individuals were left to reproduce the species and, therefore, the DDT-tolerant gene became much more common within these populations.

Sexual Reproduction

Although the process of *sexual reproduction* does not create new genes, it tends to generate new combinations of genes when the genes from two individuals mix during fertilization, generating a unique individual. This doesn't directly change gene frequency within the gene pool, but the new member may have a unique combination of characteristics so superior to those of other members of the population that the new member will be much more successful in producing offspring. In a corn population, there may be genes for resistance to corn blight (a fungal disease) and genes for resistance to attack by insects. Corn plants that possess both of these genes are going to be more successful than corn plants that have only one of these genes. They will probably produce more offspring (corn seeds) than the others because they will survive fungal and insect attacks; moreover, they will tend to pass on this same combination of resistant genes to their offspring (figure 11.6).

Migration

The *migration* of individuals from one deme to another is also an important way for genes to be added to or subtracted from a local population. Whenever an organism leaves one population and enters another, it subtracts its genes from the population it left and adds them to the population it joins. If it contains rare genes, it may significantly affect the gene frequency of both populations. The

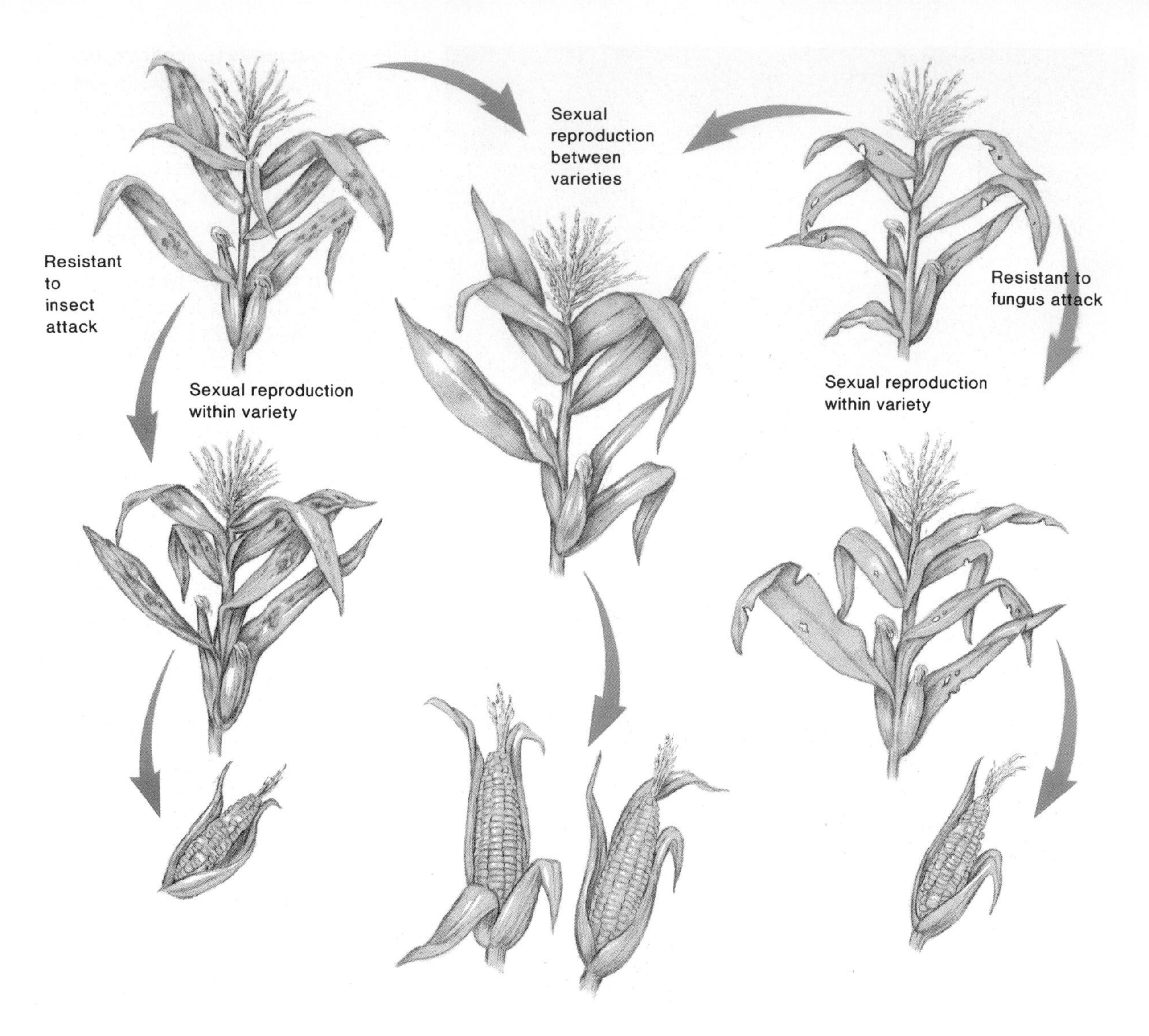

Figure 11.6 **New Combinations of Genes.** Sexual reproduction can bring about new combinations of genes that are extremely valuable. These valuable new gene combinations tend to be perpetuated.

extent of migration need not be great. As long as genes are entering or leaving a population, the gene pool will change.

Many captive populations of animals in zoos are in danger of dying out because of severe inbreeding (breeding with near relatives) and resulting reduced genetic variety. Most zoo managers have recognized the importance of increasing variety in their animals and have instituted programs of loaning breeding animals to distant zoos in an effort to increase genetic variety. In effect, they are simulating natural migration so that new genes can be introduced into distant populations.

Many domesticated plants and animals also have significantly reduced genetic variety. Corn, wheat, rice and other crops are in danger of losing their genetic variety. The establishment of gene banks in which wild or primitive relatives of domesticated plants are grown is one way that a source of genes can be kept for later introduction if domesticated varieties are threatened by new diseases or environmental changes.

The Importance of Population Size

The *size of the population* has a lot to do with how effective any of these mechanisms are at generating variety in a gene pool. The smaller the population, the less genetic variety it can contain. Therefore, migrations, mutations, and accidental death can have great effects on the genetic makeup of a small population. For example, if a town had a population of twenty people and only two had brown

Cuttings

A clone

Figure 11.7 **Clones.** All the plants in the right-hand photograph were produced asexually from cuttings and are identical genetically. The left-hand photograph shows how cuttings are made. The original plant is cut into pieces. Then, the cut ends are treated with a growth stimulant and placed in moist sand or other material. Eventually, they will root and become independent plants.

eyes and the rest had blues eyes, what happens to those two brown-eyed people would be more critical than if the town had twenty thousand people and two thousand had brown eyes. While the ratio of brown eyes to blue eyes is the same in both cases, even a small change in a population of twenty could significantly change the gene frequency of the brown-eye gene.

Genetic Variety in Domesticated Plants and Animals

Humans often work with small, select populations of plants and animals in order to artificially construct specific gene combinations that are useful or desirable. This is particularly true of the plants and animals used for food. If we can produce domesticated animals and plants with genes for rapid growth, high reproductive capacity, resistance to disease, and other desirable characteristics, we will be better able to supply ourselves with energy in the form of food. Plants are particularly easy to work with in this manner since we can often increase the numbers of specific organisms by asexual (without sex) reproduction. Potatoes, apple trees, strawberries, and many other plants can be reproduced by simply cutting the original plant into a number of parts and allowing these parts to sprout roots, stems, and leaves. If a single potato has certain desirable characteristics, it may be reproduced asexually. All of the individual plants reproduced asexually have exactly the same genes and are usually referred to as **clones.** Figure 11.7 shows how a clone is developed.

Humans can also bring together specific combinations of genes by selective breeding. This is not as easy as cloning. Sexual reproduction tends to mix up genes rather than preserve desirable combinations of genes. If, however, two different demes of the same species—each with particular desirable characteristics—are found, they can be crossed in hopes of producing a **hybrid** that has the desirable characteristics of both demes. For hybridization to be most effective, the desirable characteristics in each of the two demes should have homozygous genotypes. In small, controlled populations it is relatively easy to produce individuals that are homozygous for one specific trait. To make two characteristics homozygous in the same individual is more difficult. Therefore, hybrids are developed by crossing two different populations to collect several desirable characteristics in one organism.

The kinds of genetic manipulations we have just described result in reduced genetic variety within the population. Furthermore, most agriculture in the world is based on extensive plantings of the same varieties of a species over large expanses of land (figure 11.8). This agricultural practice is called **monoculture.** The plants have been extremely specialized through selective breeding to have just the qualities that people want. It is certainly easier to manage fields in which there is only one kind of plant growing. This is particularly true today when herbicides, pesticides, and fertilizers are tailored to meet the needs of specific crop species. However, with monoculture comes a significant risk.

Our primary food plants are derived from wild ancestors with combinations of genes that allowed them to compete successfully with other organisms in their environment. However, when humans use selective breeding within small populations to increase the frequency of certain desirable genes in our food plants, other valuable genes are lost from the gene pool. When we select specific good characteristics, we often get bad ones along with them. Therefore, these "special" plants and animals require constant attention. Insecticides, herbicides, cultivation, and irrigation practices are all used to aid the plants and animals that we need to maintain our dominant food-producing position in the world. In effect, these plants are able to live only under conditions that people carefully maintain. Furthermore, we plant vast expanses of the same plant, creating tremendous potential for extensive crop loss from diseases.

Whether we are talking about a clone or a hybrid population, there is the danger of the environment changing and affecting the population. Since these organisms are so similar, most of them will be affected in the same way. If the environmental change is a new variety of disease to which the organism is susceptible, the whole population may be killed or severely damaged. Since new diseases do come along, plant and animal breeders are constantly developing new clones, strains, or hybrids that are resistant to the new diseases.

Another related problem in plant and animal breeding is the tendency of heterozygous organisms to mate and reassemble new combinations of genes by chance from the original heterozygotes. Thus, hybrid organisms must be carefully managed to prevent the formation of gene combinations that would be unacceptable. Since most economically important animals cannot be propagated asexually, the development and maintenance of specific gene combinations in animals is a more difficult undertaking.

Figure 11.8 **Monoculture.** This wheat field is an example of monoculture, a kind of agriculture in which large areas are exclusively planted with a single crop. Monoculture makes it possible to use large farm machinery, but it also creates conditions that can encourage spread of disease.

Demes and Human Genetics

At the beginning of this chapter, we pointed out that the human gene pool consists of a number of groups called *races*. The particular characteristics that set one race apart from another originated many thousands of years ago before travel was as common as it is today, and we still associate certain racial types with certain geographic areas. Although there is much more movement of people and a mixing of racial types today, people still tend to have children with others who are of the same social, racial, and economic background who live in the same locality. This nonrandom mate selection can sometimes bring together two individuals, each of whom has genes that are relatively rare. Information about human gene frequencies within specific subpopulations can be very important to anyone who wishes to know the probability of having children with particular bad combinations of genes. This is particularly common if both individuals are descended from a common ancestral deme (tribe, religious group). Tay-Sachs disease causes degeneration of the nervous system and early death of children. Since it is caused by a recessive gene, both parents must pass the gene to their child in order for the child to have the disease. By knowing the frequency of the gene in the background of both parents, we can determine the probability of their having a child with this disease. Ashkenazic Jews have a higher frequency of this recessive gene than do people of any other group of racial or social origin (figure 11.9). Therefore, people of this particular background should be aware of the probability that they can have children who will develop Tay-Sachs disease. Likewise, sickle-cell anemia is more common in people of specific African ancestry than in any other human deme (figure 11.10). Since many black slaves came from regions where sickle-cell anemia is common, Afro-Americans should be aware that they might be carrying the gene for this type of defective hemoglobin. If they are, they should consider their chances of having children with this disease. These and other cases make it very important that trained **genetic counselors** have information about the frequencies of genes in specific human demes so they can help couples with genetic questions.

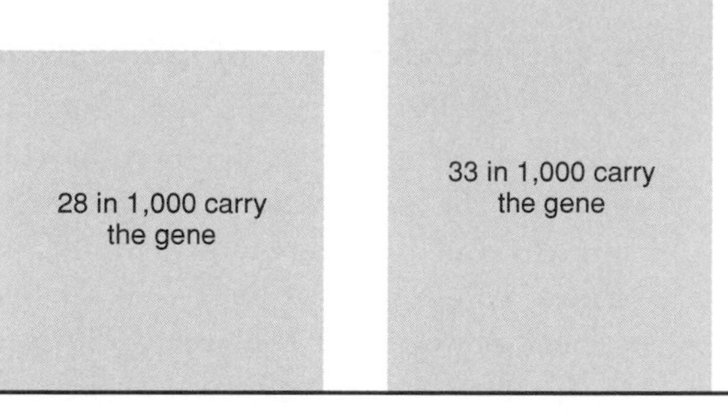

Figure 11.9 **The Frequency of Tay-Sachs Gene.** The frequency of a gene can vary from one population to another. Genetic counselors use this information to advise people of their chances of having specific genes and of passing them on to their children.

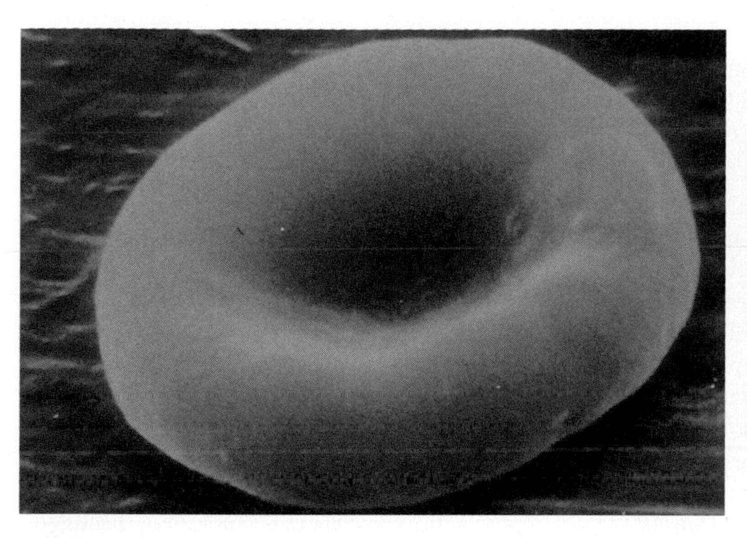

(a)

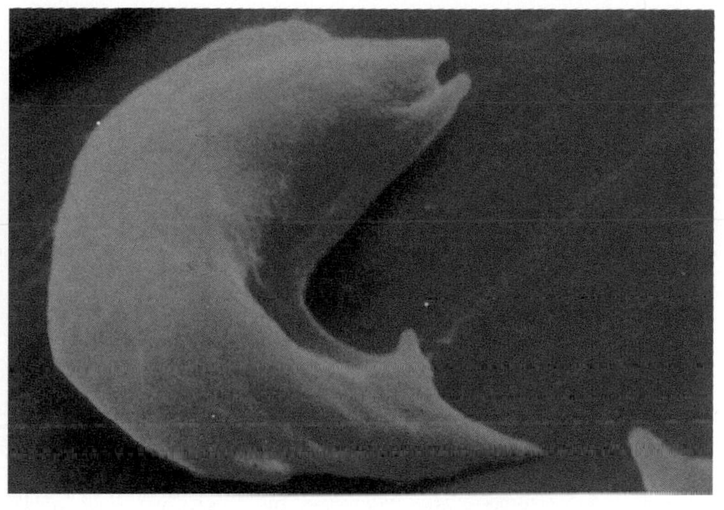

(b)

Figure 11.10 **Normal and Sickle-Shaped Cells.** Sickle-cell anemia is caused by a recessive allele that changes one amino acid in the structure of the oxygen carrying hemoglobin molecule within red blood cells. (a) Normal cells are disk-shaped. (b) The abnormal hemoglobin molecules tend to stick to one another and distort the shape of the cell when the cells are deprived of oxygen.

Ethics and Human Genetics

Misunderstanding of the principles of heredity has resulted in bad public policy. Often when there is misunderstanding there is mistrust. Even today, many prejudices against certain genetic conditions persist.

Modern genetics had its start in 1900 with the rediscovery of the fundamental laws of inheritance proposed by Mendel. For the next forty or fifty years, this rather simple understanding of genetics resulted in unreasonable expectations on the part of both scientists and laypeople. People generally assumed that much of what a person was in terms of structure, intelligence, and behavior was inherited. This led to the passage of **eugenics laws.** Their basic purpose was to eliminate "bad" genes from the human gene pool and encourage "good" gene combinations. These laws often prevented the marriage or permitted the sterilization of people who were "known" to have "bad" genes (figure 11.11).

Often these laws were thought to save money because sterilization would prevent the birth of future "defectives" and, therefore, reduce the need for expensive mental institutions or prisons. These laws were also used by people to legitimize racism and promote prejudice.

The writers of eugenics laws overestimated the importance of genes and underestimated the significance of such environmental factors as disease and poor nutrition. They also overlooked the fact that many genetic abnormalities are caused by recessive genes. In most cases, the negative effects of these "bad" genes can be recognized only in homozygous individuals. Removing only the homozygous individuals from the gene pool would have little influence on the frequency of the "bad" genes in the population. There would be many "bad" genes masked by dominant alleles in heterozygous individuals, and these genes would continue to show up in future generations. In addition, we now know that most characteristics are not inherited in a simple dominant/recessive fashion and that often many genes cooperate in the production of a phenotypic characteristic.

720.301 Sterilization of mental defectives; statement of policy

Sec. 1. It is hereby declared to be the policy of the state to prevent the procreation and increase in number of feebleminded and insane persons, idiots, imbeciles, moral degenerates and sexual perverts, likely to become a menace to society or wards of the state. The provisions of this act are to be liberally construed to accomplish this purpose. As amended 1962, No. 106, § 1, Eff. March 28, 1963.

Figure 11.11 **A Eugenics Law.** This particular state law was enacted in 1929 and is typical of many such laws passed during the 1920s and 1930s. A basic assumption of this law is that the conditions listed are inheritable; therefore, the sterilization of affected persons would decrease the frequency of these conditions. Prior to 1962, the law also included epileptics. The law was repealed in 1974.

Today, genetic diseases and the degree to which behavioral characteristics and intelligence are inherited are still important social and political issues. The emphasis, however, is on determining the specific method of inheritance or the specific biochemical pathways that result in what we currently label as insanity, lack of intelligence, or antisocial behavior. Although progress is rather slow, several genetic abnormalities have been "cured," or at least made tolerable, by medicines or control of the diet. For example, *Phenylketonuria* (*PKU*) is a genetic disease caused by an abnormal biochemical pathway. If children with this condition are allowed to eat foods containing the amino acid phenylalanine, they will become mentally retarded. However, if the amino acid phenylalanine is excluded from the diet, and certain other dietary adjustments are made, the person will develop normally. NutraSweet® is a phenylalanine-based sweetener, people with this genetic disorder must use caution when buying such products. This abnormality can be diagnosed very easily with a simple test of the urine of newborn infants.

Effective genetic counseling has become the preferred method of dealing with genetic abnormalities. A person known to be a carrier of a "bad" gene can be told the likelihood of passing that characteristic on to the next generation before deciding whether or not to have children. In addition, *amniocentesis* (a medical procedure that samples amniotic fluid) and other tests make it possible to diagnose some genetic abnormalities early in pregnancy. If an abnormality is diagnosed, an abortion can be performed. Since abortion is unacceptable to some people, the counseling process must include a discussion of the facts about an abortion and the alternatives. It would be inappropriate for counselors to be advocates; their role is to provide information that better allows individuals to make the best decisions possible for them.

• Summary •

All organisms with similar genetic information and the potential to reproduce are members of the same species. A species usually consists of several local groups of individuals known as populations. Groups of interbreeding organisms are members of a gene pool. Although individuals are limited in the number of alleles they can contain, within the population there may be many different kinds of alleles for a trait. Subpopulations may have different gene frequencies from one another and are called demes.

Demes exist because local conditions may demand certain characteristics, founding populations may have had unrepresentative gene frequencies, and barriers may prevent free flow of genes from one locality to another. Demes are often known as subspecies, varieties, strains, breeds, or races.

Genetic variety is generated by mutations, which can introduce new genes; sexual reproduction, which can generate new gene combinations; and migration, which can subtract genes from or add genes to a local population. The size of the population is also important because small populations typically have reduced genetic variety.

Knowledge of population genetics is useful for plant and animal breeders and for people who specialize in genetic counseling. Domesticated plants and animals have had their genetic variety reduced as a result of striving to produce high frequencies of particular valuable genes. Clones and hybrids are examples.

Understanding gene frequencies and how they differ in various demes sheds light on why certain genes are common in some human populations. Such understanding is also valuable in counseling members of populations with high frequencies of specific genes that are relatively rare.

• Thinking Critically •

Albinism is a condition caused by a recessive allele that prevents the development of pigment in the skin and other parts of the body. Albinos need to protect their skin and eyes from sunlight. The allele has a frequency of about 0.00005. What is the likelihood that a couple would both be carrying the gene? Why might two cousins or two members of a small tribe be more likely to both have the gene than two nonrelatives from a larger population? If an island population has its first albino baby in history, why might it have suddenly appeared? Would it be possible to eliminate this gene from the human population? Would it be desirable to do so?

• Experience This •

The next time you visit your local library, ask the reference librarian to refer you to the section that deals with the law. Take some time and see if you can locate laws or local ordinances that have a eugenic component to them. Look up topics that might deal with sterilization, conception control, fertility drugs, medicines that help select the genetically "unfit" to survive and therefore reproduce, and so on.

• Questions •

1. How does the size of a population affect the gene pool?
2. List three factors that change gene frequencies in a population.
3. Why do races or subspecies develop?
4. Give an example of a gene pool containing a number of separate demes.
5. How is a clone developed? What are its benefits?
6. How is a hybrid formed? What are its benefits?
7. What forces maintain racial differences in the human gene pool?
8. How do the concepts of species, deme, and population differ?
9. What is meant by the term gene frequency?
10. How do the gene frequencies in clones and normally reproducing populations differ?

• Chapter Glossary •

clones (klōnz) All of the individuals reproduced asexually that have exactly the same genes.

deme (dēm) A local, recognizable population that differs in gene frequencies from other local populations of the same species. (See also **subspecies.**)

eugenics laws (yu-jen'iks laws) Laws designed to eliminate "bad" genes from the human gene pool and encourage "good" gene combinations.

gene frequency (jēn fre'kwen-se) A measure of the number of times that a gene occurs in a population. The percentage of sex cells that contain a particular gene.

gene pool (jēn pool) All the genes of all the individuals of the same species.

genetic counselor (jĕ-ne'tik kown'sel-or) A professional with specific training in human genetics.

hybrid (hy'brid) The offspring of two different genetic lines produced by sexual reproduction.

monoculture (mon''o-kul'chur) The agricultural practice of planting the same varieties of a species over large expanses of land.

population (pop''u-la'shun) All the organisms of the same species within a specified geographic region.

species (spe'shēz) A group of organisms of the same kind that have the ability to interbreed to produce offspring that are also capable of reproducing.

subspecies (races, breeds, strains, or **varieties)** (sub'spe-shēz) A number of more or less separate groups (demes) within the same gene pool. These groups differ from one another in gene frequency.

Evolution and Ecology

Organisms do not live alone. All organisms are members of populations that interact with other populations and their nonliving environment. Exactly when, how, and to what extent they interact influences their ability to survive. Organisms channel energy and materials through the environment and interact with members of their own and other populations. A major factor in maintaining the variety of species in a changing world is that all populations experience genetic change over generations of time. These changes may be small, or they may be extensive enough to result in the production of organisms that are so different from their ancestors that they form a new species. Understanding what factors play roles in this type of long-term population change enables biologists to predict and control them. Such efforts, we hope, will enhance the survival of organisms that are on the verge of extinction and possibly have value for our own species. ■

CHAPTER 12

Variation and Selection

Chapter Outline

Purpose

Previous chapters have presented background material on chemistry, information systems (DNA), sexual reproduction, heredity, and population genetics. These topics were preparation for understanding how the environment influences the transmission of genes from one generation to the next. Since the surroundings are always changing, the survival of living things depends on their ability to withstand these changes. The characteristics that ensure survival might occur within any individual in the population, but unless they are genetically (DNA) determined and transmitted to the next generation, they are of little value to the survival of the species. It is the purpose of this chapter to identify how genetic differences come about and how they may change a sexually reproducing species over thousands of generations.

For Your Information

Old Disease Becomes New Problem

Tuberculosis is caused by the bacterium *Mycobacterium tuberculosis*. It is a disease that many thought had been eliminated from the United States and the rest of the economically developed world. But it is rebounding to the point that some U.S. cities now have higher rates of tuberculosis than those of developing African nations. The incidence of tuberculosis has steadily become more frequent since 1985. What caused this recurrence of an old enemy? Several factors are involved. Poverty is important in that it creates the conditions for rapid spread of this airborne disease. People with lack of access to medical care often live in crowded conditions, allowing for easy spread. AIDS sufferers have an impaired immune system; therefore, they are susceptible to the disease if they encounter it. However, a major contributor to the problem is the evolution of drug-resistant strains of the *Mycobacterium tuberculosis*. After years of being subjected to the same kinds of antibiotics, selection has taken place and certain populations of the bacterium are now able to live and reproduce in the presence of antibiotics that once controlled their spread. When the environment of *Mycobacterium tuberculosis* was changed by the addition of antibiotics, some individual tuberculosis organisms had genes that allowed them to resist certain drugs. Years of selection have given rise to strains that are resistant to many different kinds of drugs, resulting in disease outbreaks that are very difficult to control.

Learning Objectives

- Recognize that evolutionary change is a result of natural selection.
- Understand that genetic variety is essential for natural selection to occur.
- Understand that genes must be expressed to be subjected to selection.
- Recognize that reproduction provides excess individuals from which selection can occur.
- Know that differential survival, reproduction, or mate selection can lead to changes in gene frequency.
- Recognize the conditions under which the Hardy–Weinberg law applies.

Evolution and the Theory of Natural Selection

In many cultural contexts, the word *evolution* means progressive change. We talk about the evolution of economies, fashion, or musical tastes. From a biological perspective, the word has a more specific meaning. **Evolution** is the genetic adaptation of a population of organisms to its environment. Individual organisms are not able to evolve—only populations can. And the mechanism by which evolution occurs involves the biased passage of genes from one generation to the next through sexual reproduction. The various mechanisms that encourage the passage of beneficial genes to future generations and discourage harmful or less valuable genes are collectively known as **natural selection.** Therefore, evolution occurs as a result of natural selection.

The idea that some individuals whose gene combinations favor life in their surroundings will be most likely to survive, reproduce, and pass their genes on to the next generation is known as the **theory of natural selection.** Recall that a theory is a well-established generalization supported by many different kinds of evidence. The theory of natural selection was first proposed by Charles Darwin and Alfred Wallace and was clearly set forth in 1859 by Darwin in his book *On the Origin of Species by Means of Natural Selection, or the Preservation of Favored Races in the Struggle for Life* (see box 12.1).

There are two common misconceptions associated with the process of natural selection. The phrase "survival of the fittest" is commonly associated with the theory of natural selection. Survival is important because those that do not survive will not reproduce. But the more important factor is the number of descendants that an organism leaves. An organism that has survived for hundreds of years but has not reproduced has not contributed any of its genes to the next generation and has been selected against. The key, therefore, is not survival alone but survival and reproduction of the more fit organisms.

Second, the phrase, "struggle for life" does not necessarily refer to open conflict and fighting. It is usually much more subtle than that. When a resource such as nesting material, water, sunlight, or food is in short supply, some individuals survive and reproduce more effectively than others. For example, many kinds of birds require holes in trees as nesting places. If these are in short supply, some birds will be fortunate and find a top-quality nesting site, others will occupy less suitable holes, and some may not find any. There may or may not be fighting for possession of a site. If a site is already occupied, a bird may not necessarily try to dislodge its occupant but may just continue to search for suitable but less valuable sites. Those that successfully occupy good nesting sites will be much more successful in raising young than will those that must occupy poor sites or those that do not find any.

Similarly, with low light levels at the forest floor, some small plants may grow faster and obtain sunlight while shading out those that grow more slowly. The struggle for life in this instance involves a subtle difference in the rate at which the plants grow. But the plants are indeed engaged in a struggle, and a superior growth rate is the weapon for survival.

What Influences Natural Selection?

Now that we have a basic understanding of how natural selection works, we can look in more detail at those factors that influence it. Genetic variety within a species, genetic recombination as a result of sexual reproduction, the degree to which genes are expressed, and the ability of most species to reproduce in excess all exert an influence on the process of natural selection.

The Role of Genetic Variety

In order for natural selection to occur, there must be genetic differences among the many individuals of an interbreeding population of organisms. If all individuals are identical genetically, it does not matter which ones reproduce—the same genes will be passed to the next generation and natural selection cannot occur. Genetic variety is generated in two different ways. First of all, mutations may introduce entirely new genes into a species' gene pool. **Spontaneous mutations** are changes in DNA that cannot be tied to a particular causative agent. It is suspected that cosmic radiation or naturally occurring mutagenic chemicals might be the cause of many of these mutations. It is known that subjecting organisms to high levels of radiation or to certain chemicals increases the rate at which mutations occur. It is for this reason that people who work with radioactive materials or other mutagenic agents take special safety precautions (figure 12.1).

Naturally occurring mutation rates are low (perhaps one chance in one hundred thousand that a gene will be altered), and mutations usually result in a gene that is not valuable. However, in populations of millions of individuals, each of whom has thousands of genes, over thousands of generations it is quite possible that a new beneficial gene could come about as a result of mutation. When we look at the various alleles that exist in humans or in any other organism, we should remember that every allele originated as a modification of a previously existing gene. For example the allele for blue eyes may be a mutated brown-eye allele or blond hair may have originated as a mutated brown-hair gene. Thus, mutations have been very important for introducing new genetic material into species over time.

In order for mutations to be important in the evolution of organisms, they must influence the gamete-producing cells. Mutations to the cells of the skin or liver will only affect those specific cells and will not be passed to the next generation.

12.1 *The Voyage of HMS* Beagle, *1831–1836*

Probably the most significant event in Charles Darwin's life was his appointment in 1831 as naturalist on the British survey ship HMS *Beagle.* Surveys were common at this time, helping to refine maps and chart hazards to shipping. Darwin was just twenty-two years old at the time and probably would not have gotten the appointment had his uncle not persuaded Darwin's father to allow him to go. The post of naturalist was not a paid position.

The voyage of the *Beagle* lasted nearly five years. During the trip, the ship visited South America, the Galápagos Islands, Australia, and many Pacific Islands. Darwin suffered greatly from seasickness, and perhaps because of it he made extensive journeys by mule and on foot some distance inland from wherever the *Beagle* happened to be at anchor. His experience was unique for a man so young and very difficult to duplicate because of the slow methods of travel.

Although many people had seen the places that Darwin visited, never before had a student of nature collected volumes of information on these places. Also, most other people who had visited these faraway places were not trained to recognize the significance of what they saw. Darwin's notebooks included information on plants, animals, rocks, geography, climate, and the native peoples he encountered. The natural history notes he took during the voyage served as a vast storehouse of information which he used in his writings for the rest of his life.

Since Darwin was wealthy, he did not need to work to earn a living and could devote a good deal of his time to the further study of natural history and the analysis of his notes. He was a semi-invalid during much of his later life. Many people feel this was caused by a tropical disease he contracted during the voyage of the *Beagle.* As a result of his experiences, he wrote books on the formation of coral reefs, how volcanos might have been involved in their formation, and, finally, the *Origin of Species.* This last book, written twenty-three years after his return from the voyage, changed biological thinking for all time.

The Voyage of HMS *Beagle,* 1831–1836

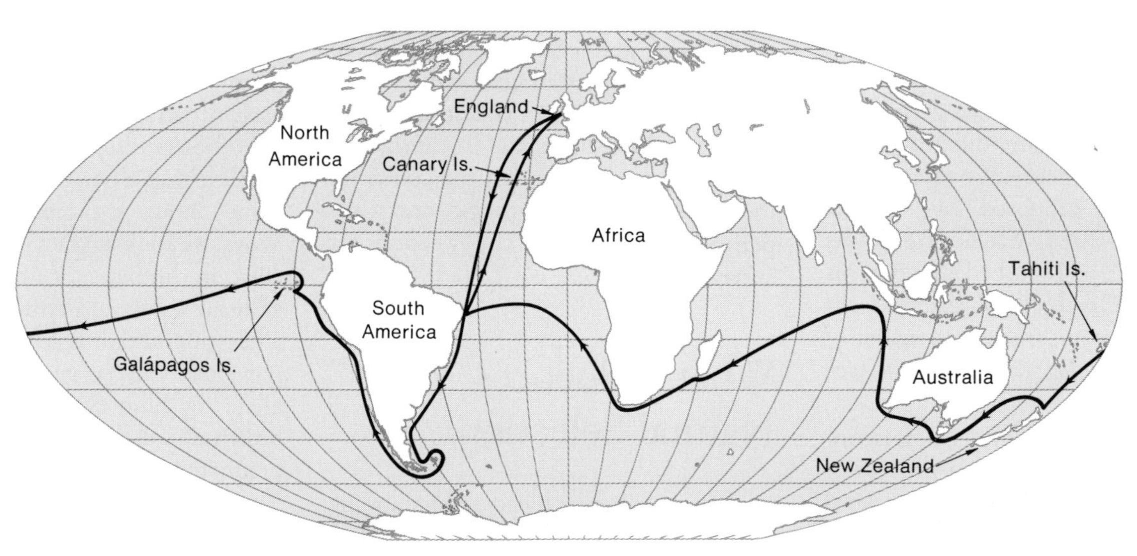

0 1000 2000 3000
Equatorial scale of miles

Figure 12.1 **Protection from Mutagenic Agents.** Because radiation and certain chemicals increase the likelihood of mutations, those people who work in hazardous environments receive special training and use protective measures to reduce their exposure to mutagenic agents.

The Role of Sexual Reproduction

A second very important process involved in generating genetic variety is sexual reproduction. While sexual reproduction does not generate new genes the way mutation does, it allows for the recombination of genes into combinations that did not occur previously. Each individual entering a population by sexual reproduction carries a unique combination of genes. During meiosis, variety is generated in the gametes through crossing-over between homologous chromosomes and independent assortment of nonhomologous chromosomes. This results in millions of possible combinations of genes in the gametes of any individual. When fertilization occurs, one of the millions of possible sperm unites with one of the millions of possible eggs, resulting in a genetically unique individual. The gene mixing that occurs during sexual reproduction is known as **genetic recombination.** The new individual has a **genome**—a complete set of genes—that is different from that of any other organism that ever existed.

There are many kinds of organisms that reproduce primarily asexually and, therefore, do not benefit from genetic recombination. In most cases, however, when their life history is studied closely, it is apparent that they also have the ability to reproduce sexually at certain times. Organisms that reproduce exclusively by asexual methods are not able to generate new gene combinations but still experience mutations and generate variety through mutations.

The Role of Gene Expression

In order for genes to be selected for or against, they must be expressed in the phenotype of the individuals possessing them. Genes may not express themselves for a number of different reasons. Some genes express themselves only during specific periods in the life of an organism. If the organism dies before the gene has had a chance to express itself, the gene never had the opportunity to contribute to the fitness of the organism. Say, for example, a tree has genes for producing very attractive fruit. If the tree dies before it can reproduce, the characteristic may never be expressed.

In addition, many genes require an environmental trigger to initiate their expression. If the trigger is not encountered, the gene never expresses itself. It is becoming clear that many kinds of human cancers are caused by the presence of genes that require an environmental trigger. Therefore, we seek to identify the triggers and prevent these negative genes from being turned on and causing disease.

Many genes are recessive and must be present in a homozygous condition before they have an opportunity to express themselves. For example, the gene for albinism is recessive. There are people who carry this recessive allele but never express it because it is masked by the dominant gene for normal pigmentation (figure 12.2).

Some genes may have their expression hidden because of the negative effects of a completely different gene. Just because an individual organism has a "good" gene does not guarantee that that gene will be passed on. The organism may also have "bad" genes in combination with the good. All individuals produced by sexual reproduction probably have certain genes that are extremely valuable for survival and others that are less valuable or harmful. However, natural selection evaluates the total phenotype of the organism. Therefore, it is the combination of characteristics that is

Figure 12.2 **Gene Expression.** Genes must be expressed to allow the environment to select for or against them. The recessive gene *c* for albinism only shows itself in individuals who are homozygous for the recessive characteristic. The man in this photo is an albino who has the genotype *cc*. The characteristic is absent in those who are homozygous dominant and is hidden in those who are heterozygous. The dark-skinned individuals could be either *Cc* or *CC*.

Figure 12.3 **Acquired Characteristics.** The ability to play an outstanding game of tennis is learned through long hours of practice. The tennis skills this person acquires by practice cannot be passed on to her offspring.

evaluated—not each characteristic individually. For example, fruit flies may show resistance to insecticides or lack of it, well-formed or shriveled wings, and normal vision or blindness. An individual with insecticide resistance, shriveled wings, and normal vision has two good characteristics and one negative one, but it would not be as successful as an individual with insecticide resistance, normal wings, and normal vision.

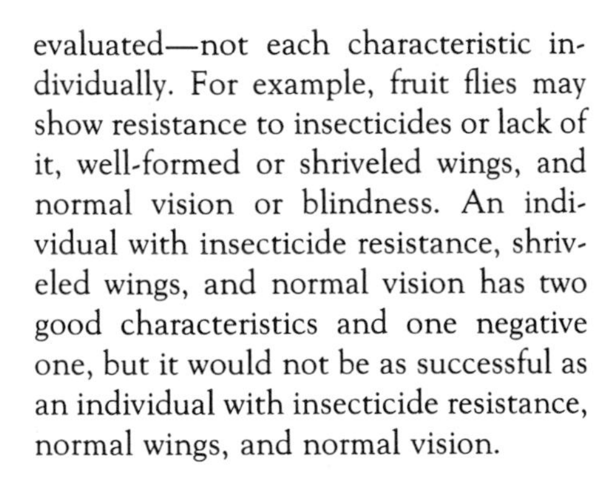

Acquired Characteristics Do Not Influence Natural Selection

Many organisms survive because they have characteristics that are not genetically determined. These **acquired characteristics** are gained during the life of the organism; they are not genetically determined and, therefore, cannot be passed on to future generations through sexual reproduction. Consider an excellent tennis player's skill. This ability is acquired through practice, not through genes. An excellent tennis player's offspring will not automatically be excellent tennis players. They may inherit some of the genetically determined physical characteristics necessary to become excellent tennis players, but the skills are still acquired through practice (figure 12.3).

We often desire a specific set of characteristics in our domesticated animals. For example, the breed of dog known as boxers are "supposed" to have short tails. However, the genes for short tails are rare in this breed (deme). Consequently, the tails of these dogs are amputated—a procedure called docking. Similarly, the tails of lambs are also usually amputated. These acquired characteristics are not passed on to the next generation. Removing the tails of these animals does not remove the genes for tail production from their genomes.

The Role of Excess Reproduction

Whenever a successful organism is examined, it can be shown that it reproduces at a rate in excess of that necessary to merely replace the parents when they die (figure 12.4). For example, geese have a life span of about ten years and, on the average, a single pair can raise a brood of about eight young each year. If these two parent birds and all their offspring were to survive and reproduce at this same rate for a ten-year period, there would be a total of 19,531,250 birds in the family.

However, the size of goose populations and most other populations remains relatively constant over time. Minor changes in number may occur, but if the species is living in harmony with its environment, it does not experience dramatic increases in population size. A high death rate tends to offset the high reproductive rate and population size remains stable. But don't think of this as

Figure 12.4 **Reproductive Potential.** The ability of a population to reproduce greatly exceeds the number necessary to replace those who die. Here are some examples of the prodigious reproductive abilities of some species.

a "static population." Although the total number of organisms in the species may remain constant, the individuals that make up the population change. It is this extravagant reproduction that provides the large surplus of genetically different individuals that allows natural selection to take place.

In fact, to maintain itself in an ever-changing environment, each species must change in ways that enhance its ability to adapt to its new environment. For this to occur, members of the population must be eliminated in a nonrandom manner. Those individuals that survive are those that are, for the most part, better suited to the environment than other individuals. They reproduce more of their kind and transmit more of their genes to the next generation than do individuals with genes that do not allow them to be well adapted to the environment in which they live.

How Natural Selection Works

There are several different mechanisms that allow for selection of certain individuals for successful reproduction. The specific environmental factors that favor certain characteristics are called **selecting agents.** All selecting agents influence the likelihood that certain characteristics will be passed to subsequent generations.

Differential Survival

As stated previously, the phrase "survival of the fittest" is often associated with the theory of natural selection. Although this is recognized as an oversimplification of the concept, survival is an important factor in influencing the flow of genes to subsequent generations. If a population consists of a large number of genetically and phenotypically different individuals and some of them possess characteristics that make their survival difficult, they are likely to die early in life and not have an opportunity to pass their genes on to the next generation. The English peppered moth provides a classic example. Two color types are found in the species: One form is light-colored and one is dark-colored. These moths normally rest on the bark of trees during the day, where they may be spotted and eaten by birds. About 150 years ago, the light-colored moths were most common. However, with the growth of the Industrial Revolution in England, which involved an increase in the use of coal, air pollution increased. The fly ash in the air settled on the trees, changing the bark to a darker color. Because the light moths were more easily seen against a dark background, the birds ate them (figure 12.5). The darker ones were less conspicuous; therefore, they were less frequently eaten and more likely to reproduce successfully. The light-colored moth, which was originally the more common type, became much less common. This change in gene frequency occurred within the short span of fifty years. Scientists who have studied this situation have estimated that the dark-colored moths had a 20% better chance of reproducing than did the light-colored moths. This study is continuing today. As England solves some of its air-pollution problems and the tree barks become lighter in color, the light-colored form of the moth is increasing in frequency.

As another example of how differential survival can lead to changed gene frequencies, consider what has happened to many insect populations as we have subjected them to a variety of insecticides. Since there is genetic variety within all species of insects, an insecticide that is used for the first time on a particular species kills all those that are genetically susceptible. However, individuals with slightly different genetic compositions may not be killed by the insecticide.

Figure 12.5 **The Peppered Moth.** This photo of the two variations of the peppered moth shows that the light-colored moth is much more conspicuous against the dark tree trunk. (The two dark moths are indicated by arrows.) The trees are dark because of an accumulation of pollutants from the burning of coal. The more conspicuous light-colored moths are more likely to be eaten by bird predators, and the genes for light color should become more rare in the population.

Suppose that in a population of a particular species of insect, 5% of the individuals have genes that make them resistant to a specific insecticide. The first application of the insecticide could, therefore, kill 95% of the population. However, tolerant individuals would then constitute the majority of the breeding population that survived. This would mean that many insects in the second generation would be tolerant. The second use of the insecticide on this population would not be as effective as the first. Many species of insects produce a new generation each month. With continued use of the same insecticide, each generation would become more tolerant. In organisms with a short generation time, 99% of the population could become resistant to the insecticide in just five years. As a result, the insecticide would no longer be useful in controlling the species. As a new factor (the insecticide) was introduced into the environment of the insect, natural selection resulted in a population that was tolerant of the insecticide. Figure 12.6 indicates that over four hundred species of insects have populations that are resistant to many kinds of insecticides.

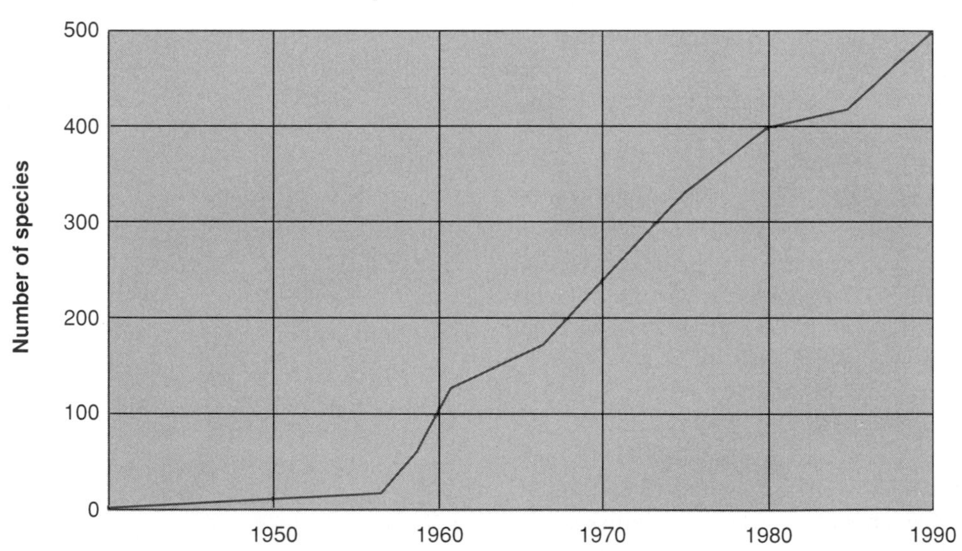

Figure 12.6 **Resistance to Insecticides.** The continued use of insecticides has constantly selected for genes that give resistance to a particular insecticide. As a result, many species of insects and other arthropods are now resistant to many kinds of insecticides, and the number continues to increase.

Source: Data from George Georghiou, University of California at Riverside.

Figure 12.7 **Selection for Shortness in Clover.** The clover field to the left of the fence is undergoing natural selection; the grazing cattle are eating the tall plants and causing them to reproduce less than the short plants. The other field is not subjected to this selection pressure, so its clover population has more genes for tallness.

Differential Reproductive Rates

Survival does not always ensure reproductive success. For a variety of reasons, some organisms may be better able to utilize available resources to produce offspring. If one individual leaves 100 offspring and another leaves only 2, the first organism has passed more copies of its genetic information to the next generation than has the second. If we assume that all 102 individual offspring have similar survival rates, the first organism has been selected for.

Among plants, there have been studies of the frequencies of genes for the height of clover plants (figure 12.7). Two identical fields of clover were planted and cows were allowed to graze in only one of them. Cows acted as a selecting agent by eating the taller plants first. These tall plants rarely got a chance to reproduce. Only the shorter plants flowered and produced seeds. After some time, seeds were collected from both the grazed and ungrazed fields and grown in a greenhouse under identical conditions. The average height of the plants from the ungrazed field was compared to that of the plants from the grazed field. The seeds from the ungrazed field produced some tall, some short, but mostly medium-sized plants. However, the seeds from the grazed field produced many more shorter plants than medium or tall ones. The cows had selectively eaten the plants that had the genes for tallness. Since the flowers are at the tip of the plant, tall plants were less likely to successfully reproduce, even though they might have been able to survive grazing by cows.

Differential Mate Selection

Within animal populations, some individuals may be chosen as mates more frequently than others. Obviously, those that are frequently chosen have an opportunity to pass on more copies of their genes than those that are rarely chosen. Characteristics of the more frequently chosen may involve general characteristics, such as body size or aggressiveness, or specific conspicuous characteristics attractive to the opposite sex.

For example, male redwing blackbirds establish territories in cattail marshes where female redwing blackbirds will build their nests. A male redwing blackbird will chase out all other males but not females. Some males have large territories, some have small territories, and some are unable to establish territories. Although it is possible for any male to mate, it has been demonstrated that those who have no territory are least likely to mate. Those who defend large territories may have two or more females nesting in their territories and are very likely to mate with those females. It is unclear exactly why females will choose one male's territory over another, but the fact is that some males are chosen as mates and others are not.

In other cases, it appears that the females select males that display specific conspicuous characteristics. Male peacocks have very conspicuous tail-feathers. Those with spectacular tails are more likely to mate and have offspring (figure 12.8).

Figure 12.8 **Mate Selection.** In many animal species the males display very conspicuous characteristics that are attractive to females. Since the females choose the males they will mate with, those males with the most attractive characteristics will have more offspring and, in future generations, there will be a tendency to enhance the characteristic. With peacocks, those individuals with large colorful displays are more likely to mate.

Gene-Frequency Studies and Hardy–Weinberg Equilibrium

Throughout this chapter there have been frequent references to changing gene frequencies. (Mutations introduce new genes into a species causing gene frequencies to change. Successful organisms pass on more of their genes to the next generation, and the frequency of certain genes changes.) In the early 1900s an English mathematician, G. H. Hardy, and a German physician, Wilhelm Weinberg, recognized that it was possible to apply a simple mathematical relationship to the study of gene frequencies. Their basic idea was that if certain conditions existed, gene frequencies would remain constant and that the distribution of genotypes could be described by the relationship $A^2 + 2Aa + a^2 = 1$, where A^2 represents the frequency of the homozygous dominant genotype, $2Aa$ represents the frequency of the heterozygous genotype, and a^2 represents the frequency of the homozygous recessive genotype. Constant gene frequencies would imply that evolution was not taking place.

The conditions necessary for gene frequencies to remain constant are

1. Mating must be completely random.
2. No mutations can occur.
3. Migration of individual organisms into and out of the population must not occur.
4. The population must be very large.
5. Genes must not be selected for or against.

The concept that gene frequencies will remain constant if these five conditions are met has become known as the **Hardy–Weinberg law.**

Determining Genotype Frequencies

It is possible to apply the Punnett square method from chapter 10 to an entire gene pool to illustrate how the Hardy–Weinberg law works. Consider a gene pool composed of only two alleles, *A* and *a*. Sixty percent (0.6) of the alleles in the population are *A* and 40% (0.4) are *a*. In this hypothetical gene pool, we do not know which individuals are male or female, nor do we know their genotypes. With these gene frequencies how many of the individuals would be homozygous dominant (*AA*), homozygous recessive (*aa*), and heterozygous (*Aa*)? To find the answer, we treat these genes and their frequencies as if they were individual genes being distributed into sperm and eggs. The sperm produced by the males of the population will be 60% (0.6) *A* and 40% (0.4) a. The females will produce eggs with the same relative frequencies. We can now set up a Punnett square as follows:

		Possible female gametes	
		A = 0.60	***a* = 0.40**
Possible male gametes	**A = 0.60**	Genotype of offspring $AA = 0.60 \times 0.60 = 0.36 = 36\%$	Genotype of offspring $Aa = 0.60 \times 0.40 = 0.24 = 24\%$
	***a* = 0.40**	Genotype of offspring $Aa = 0.40 \times 0.60 = 0.24 = 24\%$	Genotype of offspring $aa = 0.40 \times 0.40 = 0.16 = 16\%$

The Punnett square gives the frequency of occurrence of the three possible genotypes in this population: $AA = 36\%$, $Aa = 48\%$, and $aa = 16\%$.

If we use the relationship $A^2 + 2Aa + a^2 = 1$, you can see that if $A^2 = 0.36$ then A would be the square root of 0.36, which is equal to 0.6—our original frequency for the A allele. Similarly, $a^2 = 0.16$ and a would be the square root of 0.16, which is equal to 0.4. In addition, $2Aa$ would equal $2 \times 0.6 \times 0.4 = 0.48$. If this population were to reproduce randomly, it would maintain a gene frequency of 60% A and 40% a alleles.

It is important to understand that Hardy–Weinberg equilibrium conditions rarely exist; therefore, we usually see changes in gene frequency over time. Let's now examine why this is the case.

Why Hardy–Weinberg Conditions Rarely Exist

First of all, random mating does not occur. Many segments of a gene pool are isolated, so that no mating with other segments occurs during the lifetime of the individuals. In human populations, these isolations may be geographic, political, or social. Therefore, the Hardy–Weinberg law is not valid because nonrandom mating is a factor that leads to differential reproduction and natural selection.

Second, you will recall that DNA is constantly being changed (mutated) spontaneously. Totally new kinds of genes are being introduced into a population. Mutations automatically change the frequency of genes in the gene pool.

Third, environmental conditions may encourage immigration or emigration of individual organisms, thus changing the frequency of genes within a particular deme. In many parts of the world, severe weather disturbances have lifted animals and plants and moved them over great distances, isolating them from their original gene pool. The island of Krakatoa, for example, was blown to bits in 1883 by a volcanic explosion. For two months it remained so hot that rain falling on what remained of the island turned into steam. Essentially all life was eliminated from the island. The nearest possible source of new organisms was the island of Java forty kilometers away. Yet only one year after this disaster plants were found growing on Krakatoa, and by 1908 two hundred species of animals had established new populations on the island as they migrated from neighboring islands.

The fourth assumption of the Hardy–Weinberg law is that the population is infinitely large. If numbers are small, random events might give results that are quite different from the expected statistical results. Take coin flipping as an analogy. If you flip a coin once, there is a fifty-fifty (50:50) chance that the coin will turn up heads. If you flip two coins, you may come up with two heads, two tails, or one head and one tail. To come closer to the statistical probability of flipping 50% heads and 50% tails, you would need to flip many coins at the same time. The more coins you flip, the more likely it is that you will end up with 50% of all coins showing heads and the other 50% showing tails. The same is true of gene frequencies.

Gene-frequency differences that result from chance are more likely to occur in small populations than in large populations. A population of ten organisms of which 20% have curly hair and 80% have straight hair is significantly changed by the death of one curly-haired individual. An example of this kind of frequency difference occurs in a Pennsylvania settlement of the German Baptist Brethren, or Dunkers. This group originated from twenty-seven families who migrated to America in 1719.

Because Dunkers and other similar groups originated from a small founding population and are socially and reproductively isolated from the rest of the American population, many of their gene frequencies can be maintained differently from those of the American population as a whole. For example, "hitchhiker's thumb" is the ability to bend the thumb backwards so that it points toward the elbow (figure 12.9). The frequency of this gene in the general population is 0.496, while the gene frequency in the Dunker population is 0.410. This means that about one-half of the individuals in the population at large will carry the gene and about one-quarter will exhibit this recessive trait, whereas only about 40% of the Dunker population carry the gene, and only about 17% show it.

Finally, it is important to understand that genes differ in their value to the species. Since not all genes have equal value, natural selection will be operating and some genes will be more likely to be passed on to the next generation than will others.

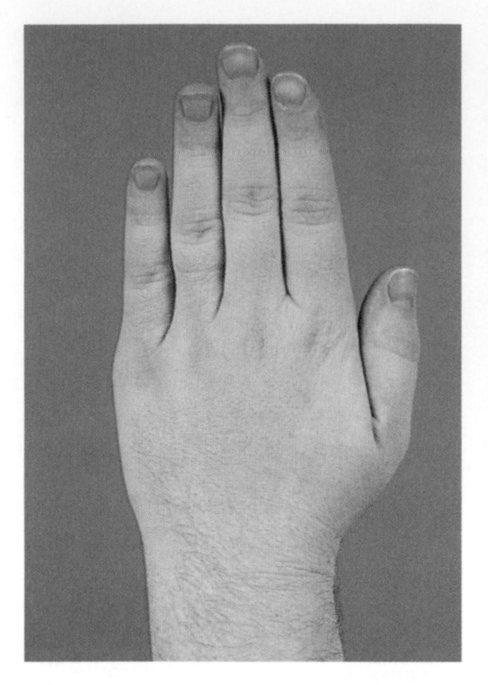

(a)

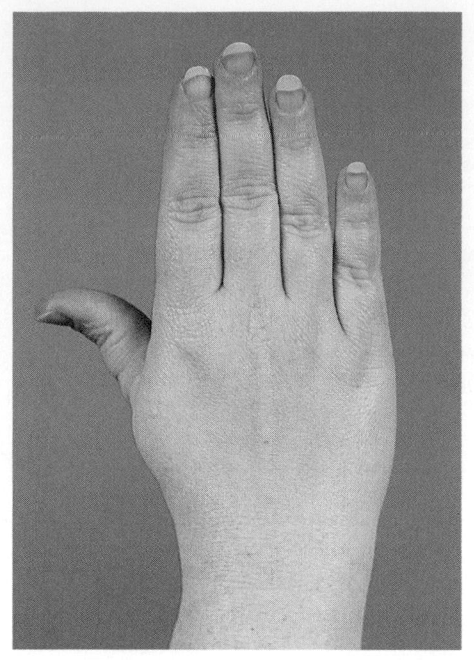

(b)

Figure 12.9 **Hitchhiker's Thumb.** Individuals with the dominant allele (a) cannot bend the thumb as far back as those homozygous for the recessive allele for hitchhiker's thumb (b). This allele is much less frequent in the Dunker sect than in the United States' population in general.

Table 12.1

Differential Mortality *The percentage of each genotype in the offspring differs from the percentage of each genotype in the original population as a result of differential reproduction.*

Original Gene Frequencies and Genotypes	Total Number of Individuals within Population of 100,000	Number Lost Due to 50% Death	Total of Each Genotype in Reproducing Population of 58,000	New Percentage of Each Genotype in Population
AA = 36%	36,000	36,000 −18,000 = 18,000	18,000	18,000/58,000 = 32%AA
Aa = 48%	48,000	48,000 −24,000 = 24,000	24,000	24,000/58,000 = 41%Aa
aa = 16%	16,000	16,000 − 0 = 16,000	16,000	16,000/58,000 = 27%aa

An Example of Gene-Frequency Change

Now we can return to our original example of genes A and *a* to show how natural selection based on nonrandom mating can result in gene-frequency changes in only one generation. Again, assume that the parent generation has the following genotype frequencies: AA = 36%, A*a* = 48%, and *aa* = 16%, with a total population of 100,000 individuals. Suppose that 50% of all the individuals having at least one A gene do not reproduce because they are more susceptible to disease. The parent population of 100,000 would have 36,000 individuals with the AA genotype, 48,000 with the A*a* genotype, and 16,000 with the *aa* genotype. Because of the 50% loss, only 18,000 AA individuals and 24,000 A*a* individuals will reproduce. All 16,000 of the *aa* individuals will reproduce, however. Thus, there is a total reproducing population of only 58,000 individuals out of the entire original population of 100,000. What percentage of A and *a* will go into the gametes produced by these 58,000 individuals?

The percentage of A-containing gametes produced by the reproducing population will be 31% from the AA parents and 20.5% from the A*a* parents (table 12.1). The frequency of the A gene in the gametes is 51.5%. The percentage of *a*-containing gametes is 48.5%—20.5% from the A*a* parents and 28% from the *aa* parents. The original parental gene frequencies were A = 60% and *a* = 40%. These have changed to A = 51.5% and *a* = 47.5%. More individuals in the population will have the *aa* genotype, and fewer will have the AA and A*a* genotypes.

If this process continued for several generations, the gene frequency would continue to shift until the A gene became rare in the population (figure 12.10). This is natural selection in action. Differential reproduction has changed the frequency of characteristics in this population.

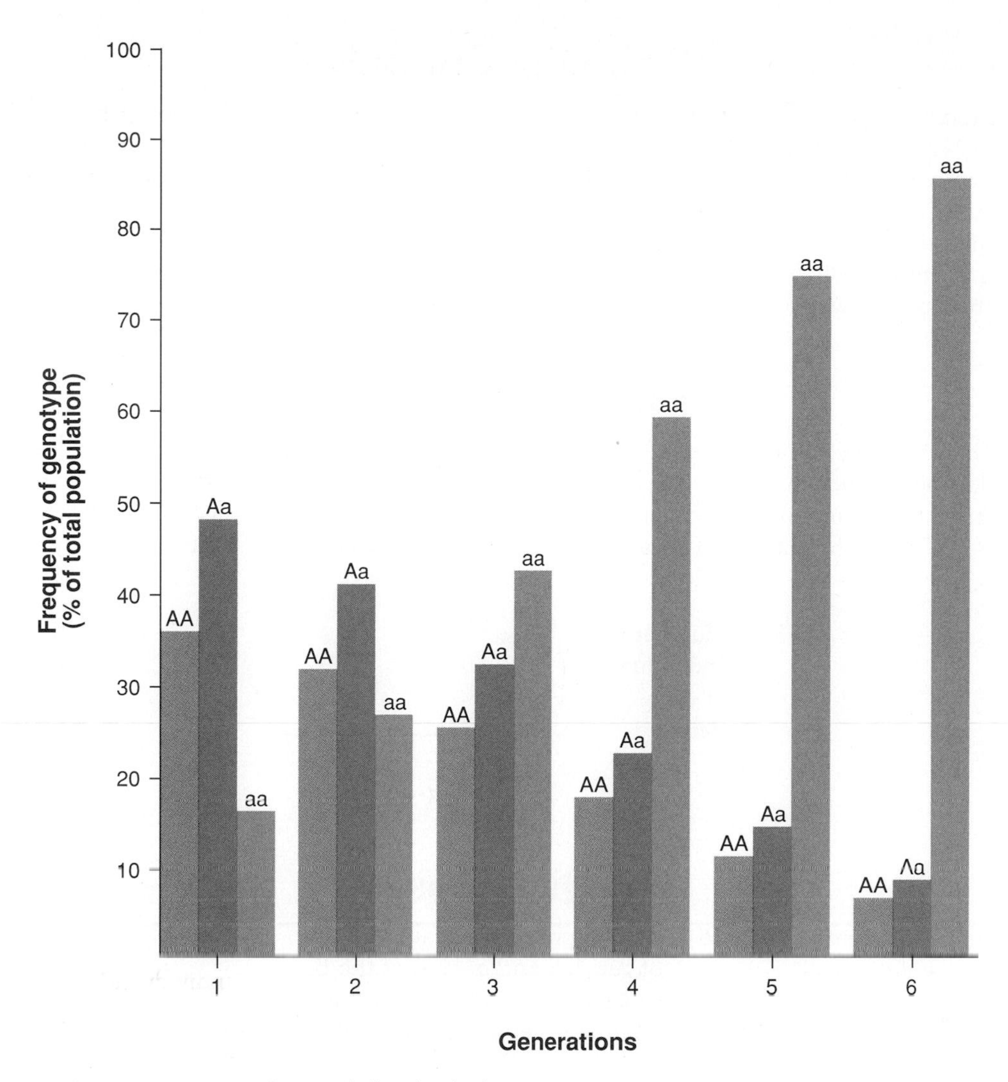

Figure 12.10 **Changing Gene Frequency.** If 50% of all individuals with the genotypes *AA* and *Aa* die in each generation, the frequency of the *aa* genotype will increase as the other two genotypes decrease in frequency.

• Summary •

All sexually reproducing organisms naturally exhibit genetic variety among the individuals in the population as a result of mutations and the genetic recombination resulting from meiosis and fertilization. These genetic differences are important for the survival of the species because natural selection must have genetic variety to select from. Natural selection by the environment results in better-suited individual organisms that have greater numbers of offspring than those that are less well off genetically. Not all genes are equally expressed. Some express themselves only during specific periods in the life of an organism and some may be recessive genes that only show themselves when in the homozygous state. Characteristics that are acquired during the life of the individual and not determined by genes cannot be raw material for natural selection.

Selecting agents act to change the gene frequencies of the population if the conditions of the Hardy–Weinberg law are violated. The conditions of the Hardy–Weinberg law are random mating, no mutations, no migration, large population size, and no selection for genes. These conditions are met only rarely, however, so that typically, after generations of time, the genes of the more favored individuals will make up a greater proportion of the gene pool. The process of natural selection allows the maintenance of a species in its environment, even as the environment changes.

• Thinking Critically •

Penicillin was first introduced as an antibiotic in the early 1940s. Since that time, it has been found to be effective against the bacteria that cause gonorrhea, a sexually transmitted disease. The drug acts on dividing bacterial cells by preventing the formation of a new protective cell wall. Without the wall, the bacteria can be killed by normal body defenses. Recently, a new strain of this disease-causing bacterium has been found. This particular bacterium produces an enzyme that metabolizes penicillin. How can gonorrhea be controlled now that this organism is resistant to penicillin? How did a resistant strain develop? Include the following in your consideration: DNA, enzymes, selecting agents, and gene-frequency changes.

• Experience This •

You can demonstrate variability in a population in the following manner. As members of the class enter the room, ask them how tall they are. Plot height against the number of people at each height. Plot men and women separately. Perhaps you could do this on the blackboard and use pink and blue chalk to distinguish males from females. Calculate the average height for males and females. What are the extremes for both sexes?

Number of Individuals at each height

Height

• Questions •

1. Why are acquired characteristics of little interest to evolutionary biologists?
2. What factors can contribute to variety in the gene pool?
3. Why is overreproduction necessary for evolution?
4. What is natural selection? How does it work?
5. The Hardy–Weinberg law is only theoretical. What factors do not allow it to operate in a natural gene pool?
6. A gene pool has equal numbers of genes *B* and *b*. Half of the *B* genes mutate to *b* genes in the original generation. What will the gene frequencies be in the next generation?
7. Why is sexual reproduction important to the process of natural selection?
8. How might a bad gene remain in a gene pool for generations without being eliminated by natural selection?
9. Give two examples of selecting agents and explain how they operate.
10. The smaller the population, the more likely it is that random changes will influence gene frequencies. Why is this true?
11. List three factors that can lead to changed gene frequencies from one generation to the next.

• Chapter Glossary •

acquired characteristics (ă-kwīrd' kar''ak-ter-iss'tiks) A characteristic of an organism gained during its lifetime, not determined genetically, and therefore not transmitted to the offspring.

evolution (ĕv-o-lu'shun) The genetic adaptation of a population of organisms to its environment.

genetic recombination (jĕ-net'ik re-kom-bĭ-na'shun) The gene mixing that occurs during sexual reproduction.

genome (je-nōm) The complete set of genes of an individual.

Hardy–Weinberg law (har'de wīn'burg law) Populations of organisms will maintain constant gene frequencies from generation to generation as long as mating is random, the population is large, mutation does not occur, migration does not occur, and all genes have equal value.

natural selection (nat'chu-ral se-lek'shun) A broad term used in reference to the various mechanisms that encourage the passage of beneficial genes to future generations and discourage harmful or less valuable genes from being passed to future generations.

selecting agent (se-lek'ting a'jent) Any factor that affects the probability that a gene will be passed to the next generation.

spontaneous mutation (spon-ta'ne-us miu-ta'shun) Natural changes in the DNA caused by unidentified environmental factors.

theory of natural selection (the'o-re uv nat'chu-ral se-lek'shun) In a species of genetically differing organisms, the organisms with the genes that enable them to survive better in the environment and thus reproduce more offspring than others will transmit more of their genes to the next generation.

CHAPTER 13

Speciation and Evolutionary Change

Chapter Outline

Purpose

The process of natural selection under certain circumstances can lead to the development of new species. Since species are reproductively isolated from other kinds of organisms, this event is irreversible. Once we understand how new species are produced, we can think about how it would be possible for major lines of evolution to develop. Evolution is a progressive process that continually allows for the adaptation of all kinds of organisms to their continually changing environment. This chapter explores these concepts.

For Your Information

Humans have invented several kinds of plants and animals for their own purposes. Most cereal grains are special plants that rely on human activity for their survival: most would not live without fertilizer, cultivation, and other helps. These grains are the descendants of wild plants, but in many cases, the wild ancestors have been lost and may be extinct. Thousands of generations of selection have, in effect, caused the development of new species.

Learning Objectives

- Understand the importance of interrupting gene flow to the process of speciation.
- Recognize the role of reproductive isolation in maintaining species distinct from one another.
- Know that many plant species originate as a result of polyploidy.
- Appreciate that the concept of evolution has changed as more information has been gained.
- Recognize the steps necessary for speciation to occur.
- Understand that the basic pattern of evolution is divergence, but that several evolutionary patterns can occur above the level of speciation.
- Know that the rate of evolutionary change differs with different organisms and at different times.

Figure 13.1 **Hybrid Sterility.** Even though they don't do so in nature, donkeys and horses can be mated. The offspring is called a mule and is sterile. Because the mule is sterile, the donkey and the horse are considered to be of different species.

Species—A Working Definition

Before we consider how new species are produced, let's consider how one species is distinguished from another. A **species** is commonly defined as a population of organisms that have the potential to interbreed naturally to produce fertile offspring but do not interbreed with other groups. This is a working definition that applies in most cases but must be interpreted to take care of some exceptions. There are two key ideas within this definition. First, a species is a population of organisms. An individual—you, for example—is not a species. You can only be a member of the group that is recognized as a species. The human species, *Homo sapiens*, consists of over 5 billion individuals, whereas the endangered California condor species, *Gymnogyps californianus*, consists of about twenty individuals.

Second, the definition involves the ability of individuals within the group to produce fertile offspring. Obviously, we cannot check every individual to see if it is capable of mating with any other individual that is similar to it, so we must make some judgement calls. Do most individuals within the group potentially have the capability of interbreeding to produce fertile offspring? In the case of humans we know that some individuals are sterile and cannot reproduce, but we don't exclude them from the human species because of this. If they were not sterile, they would have the potential to interbreed. We recognize that, although humans and other organisms normally choose mating partners from the local area that are of the same deme, humans from all parts of the world are potentially capable of interbreeding because of the large number of instances of interbreeding between people of different ethnic and racial backgrounds.

Another way to look at this question is to think about gene flow. **Gene flow** is the movement of genes from one generation to the next or from one region to another. Two or more populations that demonstrate gene flow between them constitute a single species. Conversely, two or more populations that do not demonstrate gene flow between them are generally considered to be different species. Some examples will clarify this working definition.

The mating of a male donkey and a female horse produces young that grow to be adult mules, incapable of reproduction (figure 13.1). Since mules are sterile, there can be no gene flow between horses and donkeys and they are considered to be separate species. Similarly, lions and tigers can be mated in zoos to produce offspring. However, this does not happen in nature and the offspring are not likely to be fertile. Gene flow does not occur naturally and they are considered to be two separate species. As a third example, coyotes have been known to mate with dogs and produce fertile pups. However, although the young are fertile, this mating does not happen often in nature. In this case, gene flow is possible, but because of the different behavior of dogs and coyotes, gene flow is greatly restricted and they are considered to be separate species. This example is not nearly as clear-cut as the previous ones.

The spread of the African honeybee from Brazil to the southern United States is a good example of how genes can flow from one region to another as a result of migration and sexual reproduction between different populations of the same species. The chain of events began in 1957 when a researcher in São Paulo, Brazil, accidentally released an African strain of honeybees into the local population. The African type of honeybee is of the same species as the honeybees that are most commonly used for honey production, the Italian honeybee, which has a relatively gentle disposition and is a better honey-producer than the African strain. The African type of honeybee has two characteristics that make it much more difficult to work with; it is much more aggressive, tending to attack in large numbers when anything disturbs the hive, and it often abandons old hives and establishes new ones. Both of these characteristics are negative from the point of view of the beekeeper. This aggressive type of honeybee steadily moved north through Central America to the United States (figure 13.2). Some of these honeybees were found in southern California in June 1985; they had been accidentally brought into the state with oil-drilling equipment from Central America. These colonies were eliminated, but another colony was discovered in Texas in 1992.

Since the African honeybees are the same species as the more commonly kept honeybees, it is possible that the genes for aggressiveness could be introduced into the gene pool of the commercially valuable types. This has caused much concern on the part of the public and some concern among commercial beekeepers in the southern United States. It was thought that the unfavorable genes for aggression would be diluted in the much larger, nonaggressive honeybee population and not lead to major disruptions in honey production. However, experience in Central America shows that the aggressive bees have become very common, and many beekeepers have gone out of business. Fortunately, the African type of honeybee cannot stand cold temperatures and therefore would not be as successful in most of North America.

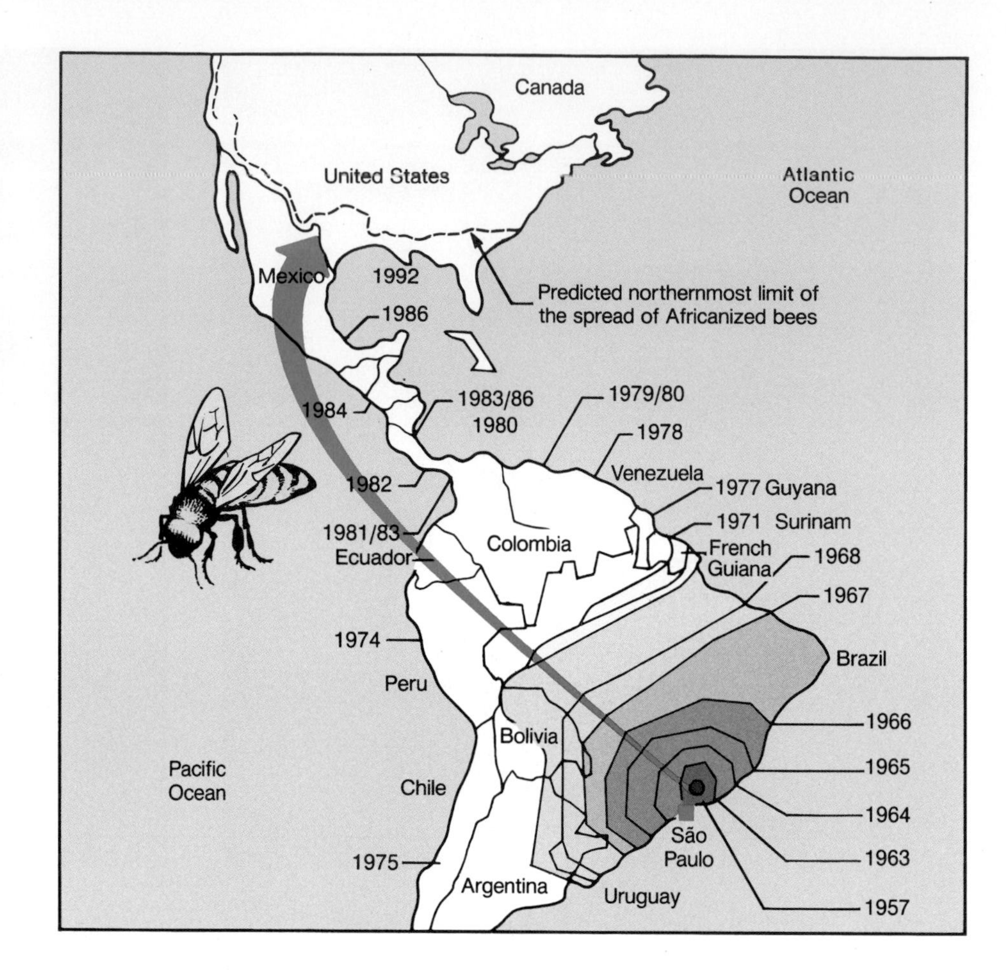

Figure 13.2 **Gene Flow in the Honeybee Species.** The honeybee species consists of several subspecies. The African type was accidentally released in Brazil in 1957. This map shows the spread of this type of honeybee. As indicated, a colony was discovered in Texas in 1992.

How New Species Originate

The geographical distribution of a species is known as its **range.** As a species expands its range, portions of the population can become separated from the rest. Thus, many species consist of partially isolated populations that display characteristics that differ significantly from other local populations. Many of the differences observed may be directly related to adaptations to local environmental conditions. This means that new colonies may have infrequent gene exchange with their geographically distant relatives. As you will recall from chapter 11, these genetically distinct populations are known as subspecies or demes.

When a portion of a species becomes totally isolated from the rest of the gene pool by some geographic change, such as the formation of a mountain range, river valley, desert, or ocean, we say that the local population is in **geographic isolation** from the rest of the species. The geographic features that keep the different portions of the species

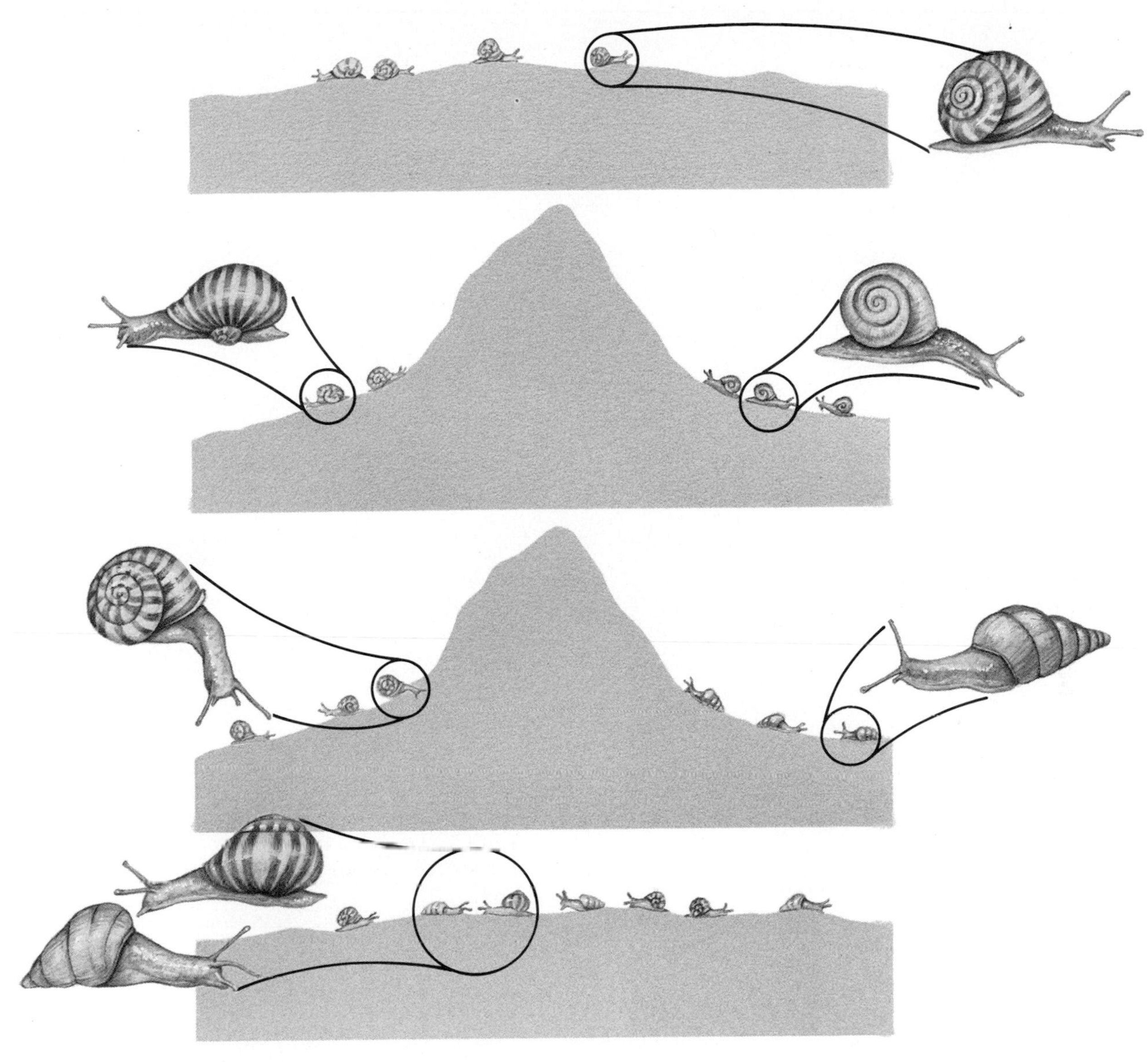

Figure 13.3 **The Effect of Geographic Isolation.** If a single species of snail was to be divided into two different populations by the development of a ridge between them, the two populations could be subjected to different environmental conditions. This could result in a slow accumulation of changes that could ultimately result in two populations that would not be able to interbreed even if the ridge between them were to erode away. They would be different species.

from exchanging genes are called **geographic barriers.** The uplifting of mountains, the rerouting of rivers, and the formation of deserts all may separate one portion of a gene pool from another. For example, two kinds of squirrels are found on opposite sides of the Grand Canyon. Some people consider them to be separate species, while others consider them to be different isolated demes of the same species. Even small changes may cause geographic isolation in species that have little ability to move. A fallen tree, a plowed field, or even a new freeway may effectively isolate populations within such species. Snails in two valleys separated by a high ridge have been found to be closely related but different species. The snails cannot get from one valley to the next because of the height and climatic differences presented by the ridge (figure 13.3).

The separation of a species into two or more isolated demes is not enough to generate new species. Even after many generations of geograpic isolation, these separate groups may still be able to exchange genes (mate and produce fertile offspring) if they overcome the geographic barrier because they have not accumulated enough genetic differences to prevent reproductive success. Natural selection and differences in environments play very important roles in the

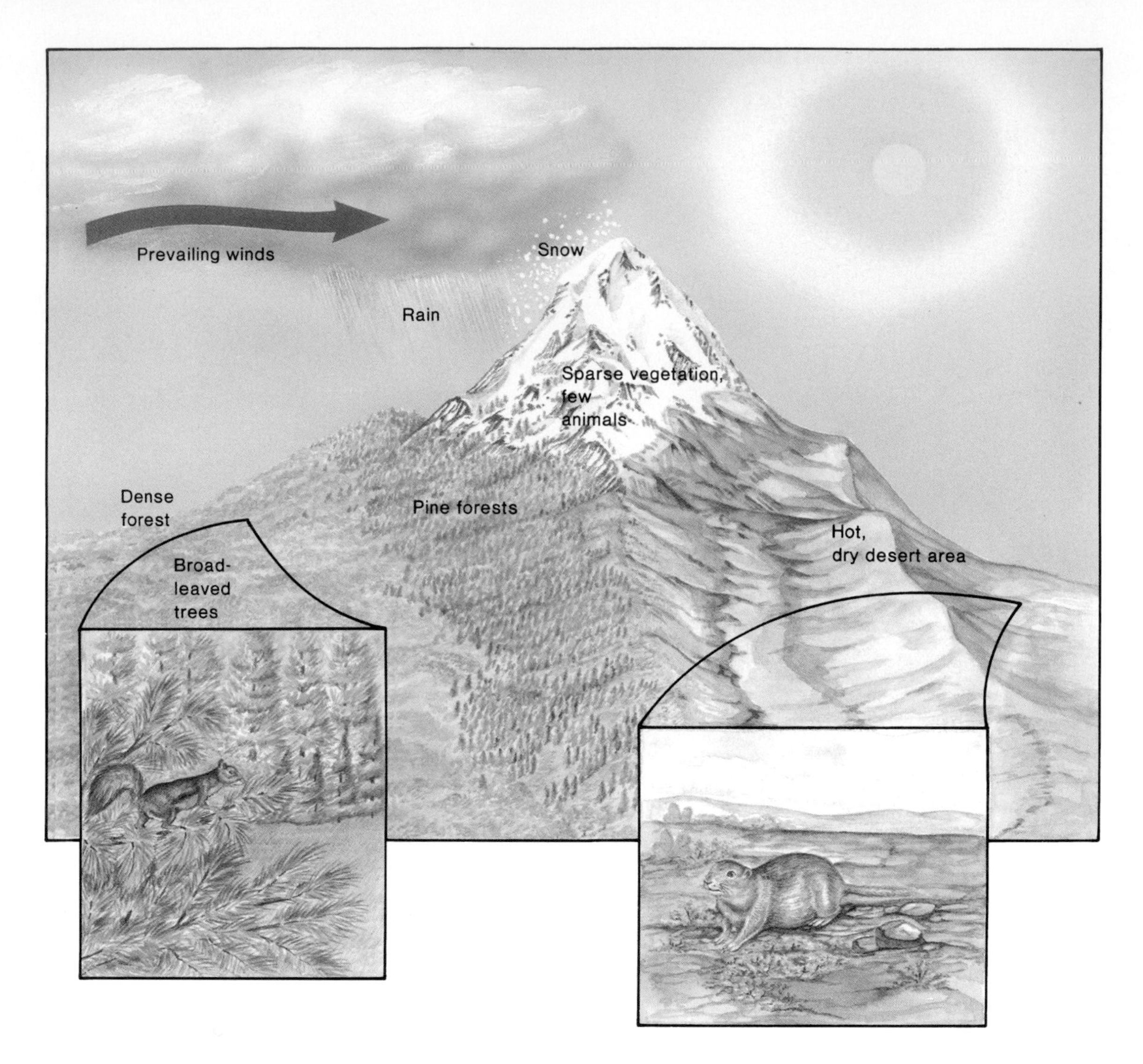

Figure 13.4 **Environmental Differences Caused by Mountain Ranges.** Most mountain ranges affect the local environment. Because of the prevailing winds, most rain falls on the windward side of the mountain. This supports abundant vegetation. The other side of the mountain receives much less rain and is dryer. Often a desert may exist. Both plants and animals must be adapted to the kind of climate typical for their region. Cactus and ground squirrels would be typical of the desert and pine trees and tree squirrels would be typical of the windward side of the mountain.

process of forming new species. Following separation from the main portion of the gene pool by geographic isolation, the organisms within the small, local population are likely to experience different environmental conditions. If, for example, a mountain range has separated a species into two populations, one of them may receive more rain or more sunlight than the other (figure 13.4). These environmental differences act as natural selecting agents on the two gene pools and account for different genetic combinations in the two places. Furthermore, different mutations may occur in the two populations and there may be different random combinations of genes as a result of sexual reproduction. This would be particularly true if one of the demes was small. As a result, the two populations may show differences in color, height, enzyme production, time of seed germination, or many other characteristics.

Over a long period of time, the genetic differences that accumulate may result in regional populations called **subspecies** that are significantly modified structurally, physiologically, or behaviorally. The differences among some subspecies may be so great that they reduce reproductive success when the subspecies mate. **Speciation** is the process of generating new species. This process occurs only if gene flow between isolated populations does not occur even after barriers are removed. In other words, the process of speciation can begin with the geographic isolation of a portion of the species, but new species are generated only when isolated populations become separate from one another *genetically*. Speciation is really a three-step process. It begins with geographic isolation, is followed by the action of selective agents that choose specific genetic combinations as being valuable, and ends with the genetic differences becoming so great that reproduction between the two groups is impossible.

Figure 13.5 **Courtship Behavior (Behavioral Isolation).** The dancing of a male prairie chicken attracts female prairie chickens, but not females of other species. This behavior tends to keep prairie chickens reproductively isolated from other species.

Maintaining Genetic Isolation

Organisms that allow mating across species lines will not be very successful because most cross-species matings result in no offspring or offspring that are sterile. Part of the speciation process typically involves the development of **reproductive isolating mechanisms** or **genetic isolating mechanisms.** These mechanisms prevent cross-species matings. A great many different types of genetic isolating mechanisms can be recognized.

In central Mexico, two species of robin-sized birds called *towhees* live in different environmental settings. The collared towhee lives on the mountainsides in the pine forests, while the spotted towhee is found at lower elevations in oak forests. Geography presents no barriers to these birds. They are perfectly capable of flying to each other's habitats, but they do not. Because of their **habitat preference** or **ecological isolation,** mating between these two similar species does not occur. Similarly, areas with wet soil have different species of plants than nearby areas with drier soils.

Some plants flower only in the spring of the year, while other species which are closely related, flower in midsummer or in the fall; therefore, the two species are not very likely to pollinate one another. Among many species of insects there is a similar spacing of the reproductive periods of closely related species so that they do not overlap. Thus, **seasonal isolation** (differences in the time of the year at which reproduction takes place) is an effective genetic isolating mechanism.

Inborn behavior patterns that prevent breeding between species result in **behavioral isolation.** The mating calls of frogs and crickets are highly specific. The sound pattern produced by the males is species-specific and invites only females of the same species to come. The females have a built-in response to the particular species-specific call.

The courtship behavior of birds involves both sound and visual signals that are species-specific. For example, groups of male prairie chickens gather on meadows shortly before dawn in the early summer and begin their dances. The air sacs on either side of the neck are inflated so that the bright red skin is exposed. Their feet move up and down very rapidly, while their wings are spread out and quiver slightly (figure 13.5). This combination of sight and sound attracts females. When they arrive, the males compete for the opportunity to mate with them. Other related species of birds conduct their own similar, but distinct, courtship displays. The differences among the dances are great enough so that a female can recognize the dance of a male of her own species.

Behavioral isolating mechanisms such as these occur among other types of animals as well. The strutting of a peacock, the fin display of Siamese fighting fish, and the croaking of male frogs of different species are all examples of behaviors that help to prevent different species from interbreeding and producing hybrids (figure 13.6).

Polyploidy—Instant Speciation

So far, we have considered only those hereditary changes that take a long time to add up to enough differences to generate a new species. In plants, there are many examples of *polyploidy*. **Polyploidy** is the increase in number of chromosomes as a result of abnormal mitosis or meiosis in which the chromosomes do not separate properly. Polyploids have multiple sets of chromosomes. For example, if a cell had the normal diploid chromosome number of six ($2n = 6$), and the cell went through mitosis but did not divide into two cells, it would then consist of twelve chromosomes. It is also possible to have crosses between two species followed by a doubling of the chromosome number that result in polyploid species. Since the number of chromosomes of the polyploid is different

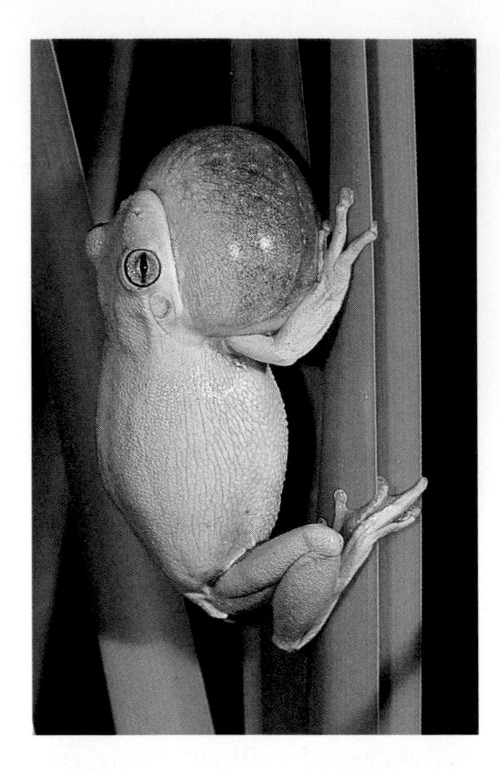

Figure 13.6 **Animal Communication by Displays.** Most animals have specific behaviors that they use to communicate with others of the same species. The croaking of a male frog is specific to its species and is different from that of males of other species. The visual displays of peacocks and Siamese fighting fish are also used to communicate with others of the same species.

from that of the parent, meiosis would result in gametes that had different chromosome numbers from the original, and successful reproduction with the parent species would be difficult. In one step the polyploid could be isolated reproductively from its original species. However, since most plants can reproduce asexually, they can create an entire population of organisms that have the same polyploid chromosome number and are capable of sexual reproduction among themselves. In effect a new species has been created within a couple of generations. Although it is rare in animals, polyploidy is found in a few groups that typically use asexual reproduction in addition to sexual reproduction.

Some groups of plants, such as the grasses, may have 50% of their species produced as a result of polyploidy. Many economically important species are polyploids. Cotton, potatoes, sugarcane, wheat, and many of our garden flowers are examples (figure 13.7).

The Development of Evolutionary Thought

Although most scientists accept evolutionary processes as central to an understanding of how various life-forms arose and continue to change today, this was not always the case. For centuries people believed that the various species of plants and animals were fixed and unchanging—that is, they were thought to have remained unchanged from the time of their creation. This was a reasonable assumption because people knew nothing about DNA, meiosis, or population genetics. Furthermore, the process of evolution is so slow that people could not see the accumulation of changes we call evolution during their lifetime. It is even difficult for modern scientists to recognize this slow change in many kinds of organisms. In the mid-1700s, George Buffon, a French naturalist, expressed

Figure 13.7 **Polyploidy.** Many species of plants have been created by increasing the chromosome number. Many large-flowered varieties of plants have been produced artificially by means of this technique. The smaller flower shown on the left is a plant with the normal diploid number; the larger one in the middle is a tetraploid variety; and on the right is a double-diploid variety.

some concerns about the possibilities of change (evolution) in animals, but he did not suggest any mechanism that would result in evolution.

In 1809, Jean Baptiste de Lamarck, a student of Buffon's, suggested a process by which evolution could occur. He proposed that acquired characteristics could be transmitted to offspring. For example, he postulated that giraffes originally had short necks. Since giraffes constantly stretched their necks to obtain food, their necks got slightly longer. This slightly longer neck could be passed to the offspring, who were themselves stretching their necks, and over time, the necks of giraffes would get longer and longer. Although we now know Lamarck's theory was wrong (because acquired characteristics are not inherited), it stimulated further thought as to how evolution could occur. All during this period, from the mid-1700s to the mid-1800s, lively arguments continued about the possibility of evolutionary change. Some, like Lamarck and others, thought that change did take place, while many others said that it was not possible. It was the thinking of two English scientists that finally provided a mechanism that explained how evolution could occur.

In 1858, Charles Darwin and Alfred Wallace suggested the theory of natural selection as a mechanism for evolution. They based their theory on the following assumptions about the nature of living things:

1. All organisms produce more offspring than can survive.
2. No two organisms are exactly alike.
3. Among organisms, there is a constant struggle for survival.
4. Those individuals that possess favorable characteristics have a higher rate of survival and produce more offspring.
5. Favorable characteristics become more common in the species, and unfavorable characteristics are lost.

With these assumptions, the Darwin–Wallace theory of evolution by natural selection would have a different explanation for the development of long necks in giraffes (figure 13.8):

1. In each generation, more giraffes would be born than the food supply could support.
2. In each generation, some giraffes would inherit longer necks, and some would inherit shorter necks.
3. All giraffes would compete for the same food sources.
4. Those giraffes with longer necks would obtain more food, have a higher survival rate, and produce more offspring.
5. As a result, succeeding generations would show an increase in the neck length of the giraffe species.

This logic seems simple and obvious today, but remember that at the time Darwin and Wallace proposed their theory, the processes of meiosis and fertilization were poorly understood and the concept of the gene was only beginning to be discussed. Nearly fifty years after Darwin and Wallace suggested their theory, the work of Gregor Mendel (chapter 10) provided an explanation for how characteristics could be transmitted from one generation to the next. Not only did Mendel's idea of the gene provide a means of passing traits from one generation to the next, it also provided the first step in understanding mutations, gene flow, and the significance of reproductive isolation. All of these ideas are interwoven into the modern concept of evolution. If we look at the same five ideas and update them with modern information, they might look something like this:

1. An organism's capacity to overreproduce results in surplus organisms.
2. Because of mutation new genes enter the gene pool. Because of sexual reproduction involving meiosis and fertilization, new combinations of genes are present in every generation. These processes are so powerful that each individual in a sexually reproducing population is genetically unique. The genes present are expressed as the phenotype of the organism.
3. Resources such as food, water, mates, and nest materials are in short supply, so some individuals will need to do without. Other environmental factors, such as disease organisms, predators, and defense mechanisms, affect survival. These are called selecting agents.
4. Individuals with the best combination of genes will be more likely to survive and reproduce, passing more of their genes on to the next generation. An organism can be selected against if it has fewer offspring than better adapted species members. It does not need to die to be selected against.
5. Therefore, genes or gene combinations that produce characteristics favorable to survival will become more common, and the species will become better adapted to its environment.

Evolution above the Species Level

Speciation is generally considered to be an irreversible event. Species do not evolve back into an earlier stage in their development. The development of new species is, therefore, the smallest unit of evolution. Since species are reproductively isolated from one another, they can only diverge further—they do not combine with other species to make something new. Higher levels of evolutionary change are the result of differences accumulated from a long series of speciation events leading to greater and greater

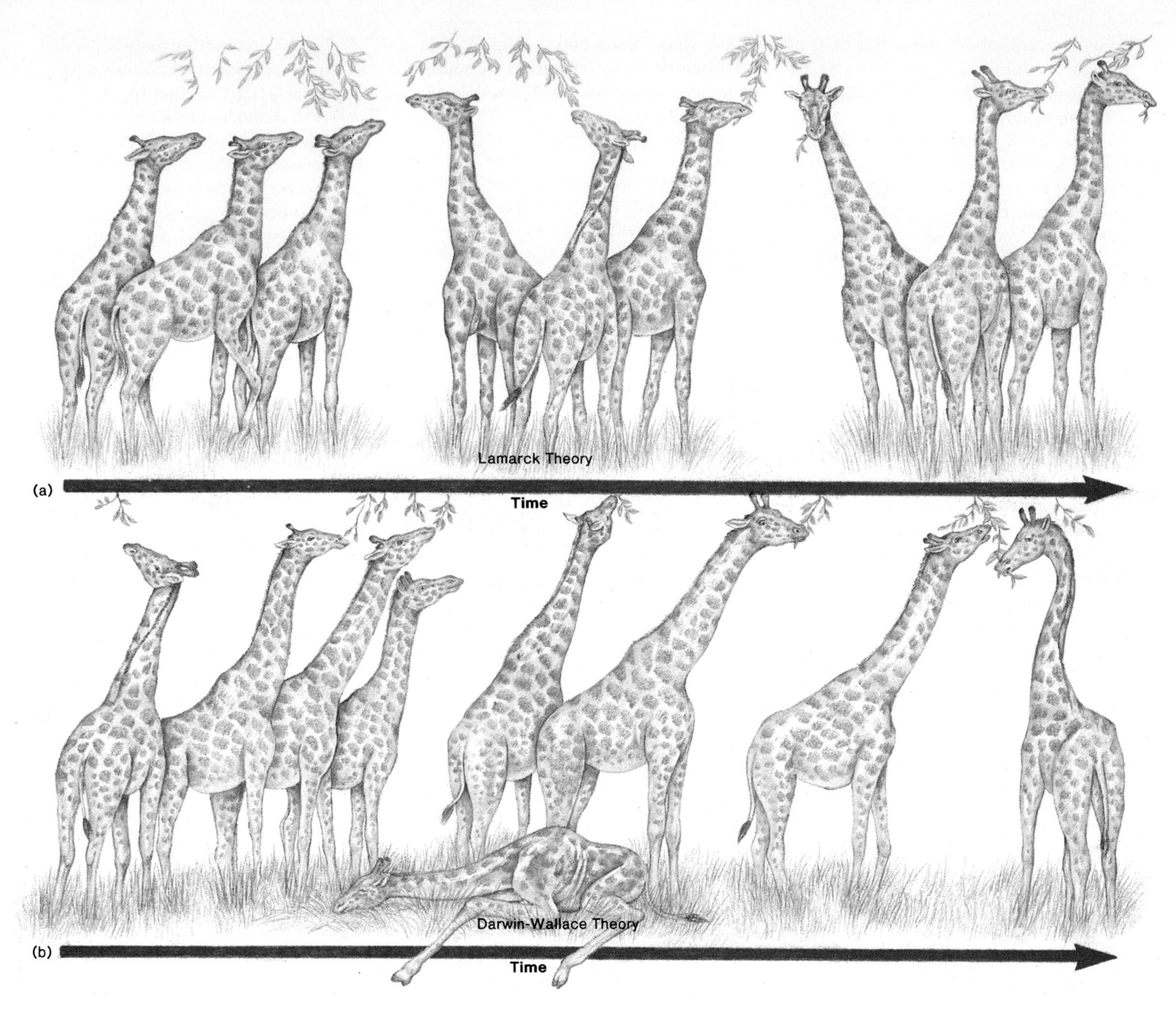

Figure 13.8 **Two Theories of How Evolution Occurs.** (a) Lamarck thought that acquired characteristics could be passed on to the next generation. Therefore, he postulated that as giraffes stretched their necks to get food, their necks got slightly longer. This characteristic was passed on to the next generation, which would have longer necks. (b) The Darwin–Wallace theory states that there is variation within the population and that those with longer necks would be more likely to survive and reproduce and pass their genes for long necks on to the next generation.

diversity. The basic pattern in evolution is one of **divergent evolution** in which individual speciation events cause successive branches in the evolution of a group of organisms. This basic pattern is well illustrated by the evolution of the horse shown in figure 13.9. The modern horse, with its large size, single toe on each foot, and teeth designed for grinding grasses, is thought to be the result of accumulated changes beginning from a small, dog-sized animal with four toes on its front feet, three toes on its hind feet, and teeth designed for chewing leaves and small twigs. Even though we know much about the evolution of the horse, there are still many gaps that need to be filled before we have a complete evolutionary history.

Another basic pattern in the evolution of organisms is extinction. Notice

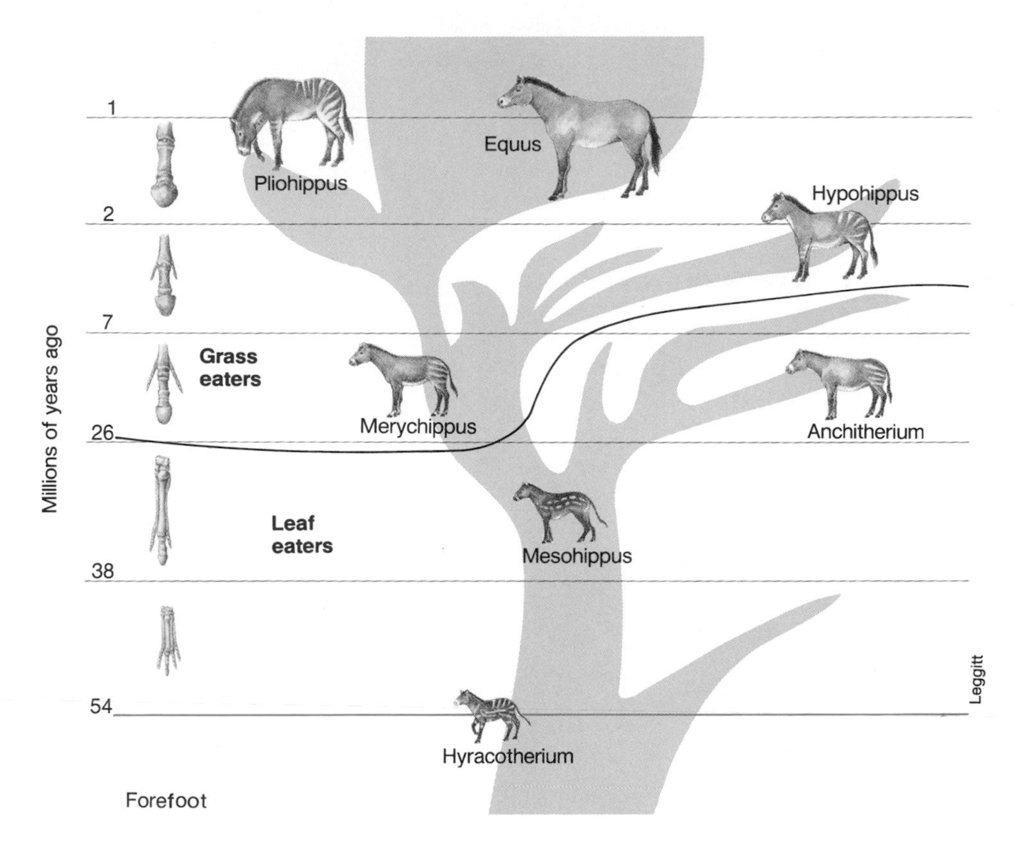

Figure 13.9 **Divergent Evolution.** In the evolution of the horse, there have been many speciation events that have followed one after another. What began as a small, leaf-eating, four-toed animal of the forest has evolved into a large, grass-eating, single-toed animal of the plains. There are many related animals alive today, but early ancestral types are extinct.

in figure 13.9 that most of the species that developed during the evolution of the horse are extinct. This is typical. Most of the species of organisms that have ever existed are extinct. Estimates of extinction are around 99%; that is, 99% or more of all the species of organisms that ever existed are extinct. Given this high rate of extinction, we can picture evolutionary change as the result of many "experiments," most of which result in failure. This is not the complete picture though. From chapter 12 we recognize that organisms are continually being subjected to selection pressures that lead to a high degree of adaptation to a particular set of environmental conditions. Organisms become more and more specialized. However, the environment does not remain constant and often changes in such a way that the species that were originally present are unable to adapt to the new set of conditions. The early ancestors of the modern horse were well adapted to a moist tropical environment, but when the climate became drier, most were no longer able to survive. Only some kinds had the genes necessary to lead to the development of modern horses.

Furthermore, it is important to recognize that many extinct species were very successful organisms for millions of years. They were not failures for their time but simply did not survive to the present. It is also important to realize that many currently existing organisms will eventually become extinct.

Tracing the evolutionary history of an organism back to its origins is a very difficult task because many of its ancestors no longer exist. Keep in mind that the fossil record is incomplete and only provides limited information about the biology of the organism represented in that record. We may know a lot about the structure of the bones and teeth of an extinct ancestor, but know almost nothing about its behavior, physiology, and natural history. Biologists must use a great deal of indirect evidence to piece together the series of evolutionary steps that lead to a current species. Figure 13.10 is typical of evolutionary trees that help us understand how time and structural changes are related in the evolution of birds, mammals, and reptiles.

Although divergence is the basic pattern in evolution, it is possible to superimpose several other patterns on it. One special evolutionary pattern, characterized by a rapid increase in the number of kinds of closely related species, is known as **adaptive radiation.** Adaptive radiation results in an evolutionary explosion of new species from a common ancestor. There are basically two kinds of situations that are thought to favor adaptive radiation. One is a condition in which an organism invades a previously unexploited environment. For example, at one time there were no animals on the land masses of the earth. The amphibians were the first vertebrate animals able to spend part of their lives on land. A variety of different kinds of amphibians evolved rapidly and exploited several different kinds of lifestyles.

Another good example of adaptive radiation is found among the finches of the Galápagos Islands, one thousand kilometers west of Ecuador in the Pacific Ocean. These birds were first studied by Charles Darwin. Since these islands are volcanic, the assumption is that they have always been isolated from South America and originally lacked finches and other land-based birds. It is thought that one kind of finch arrived from South America to colonize the islands and that adaptive radiation from the common ancestor resulted in the many different kinds of finches found on the islands today (figure 13.11).

Some of these finches took roles normally filled by other kinds of birds elsewhere in the world. Some even became warblerlike, and one uses a cactus spine as a tool to probe for insects as woodpeckers do.

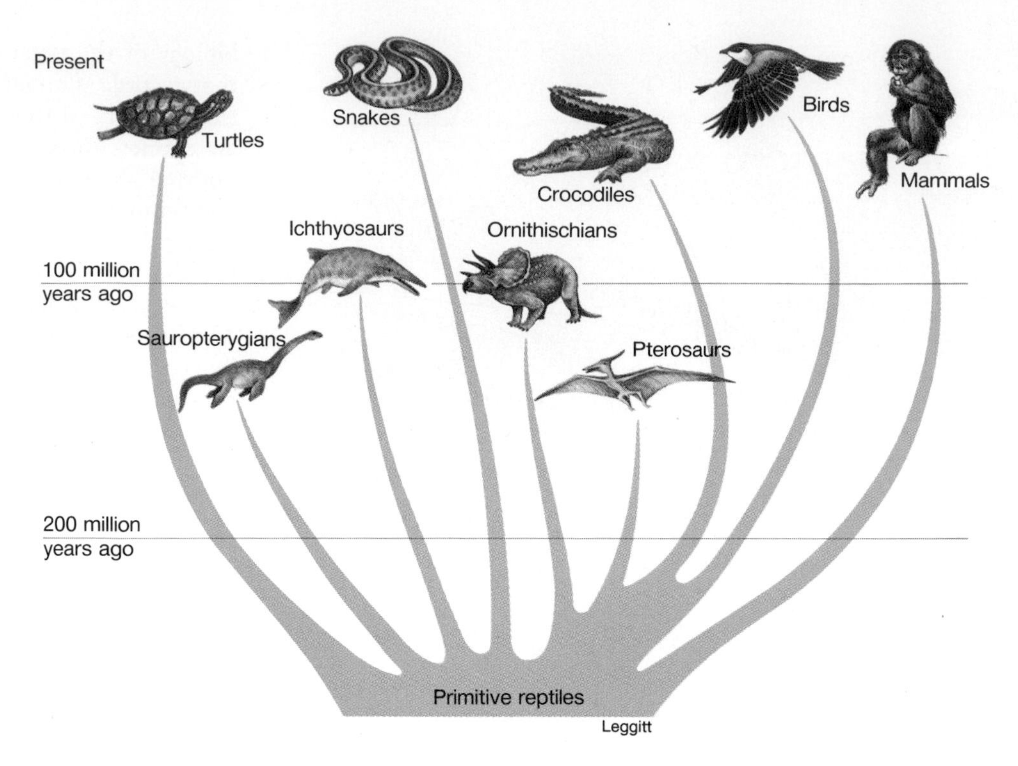

Figure 13.10 **An Evolutionary Tree.** This is a typical evolutionary tree that shows how present-day animals evolved from a primitive reptilian ancestor. Notice that an extremely long period of time is involved (220 million years) and that many of the species illustrated are extinct.

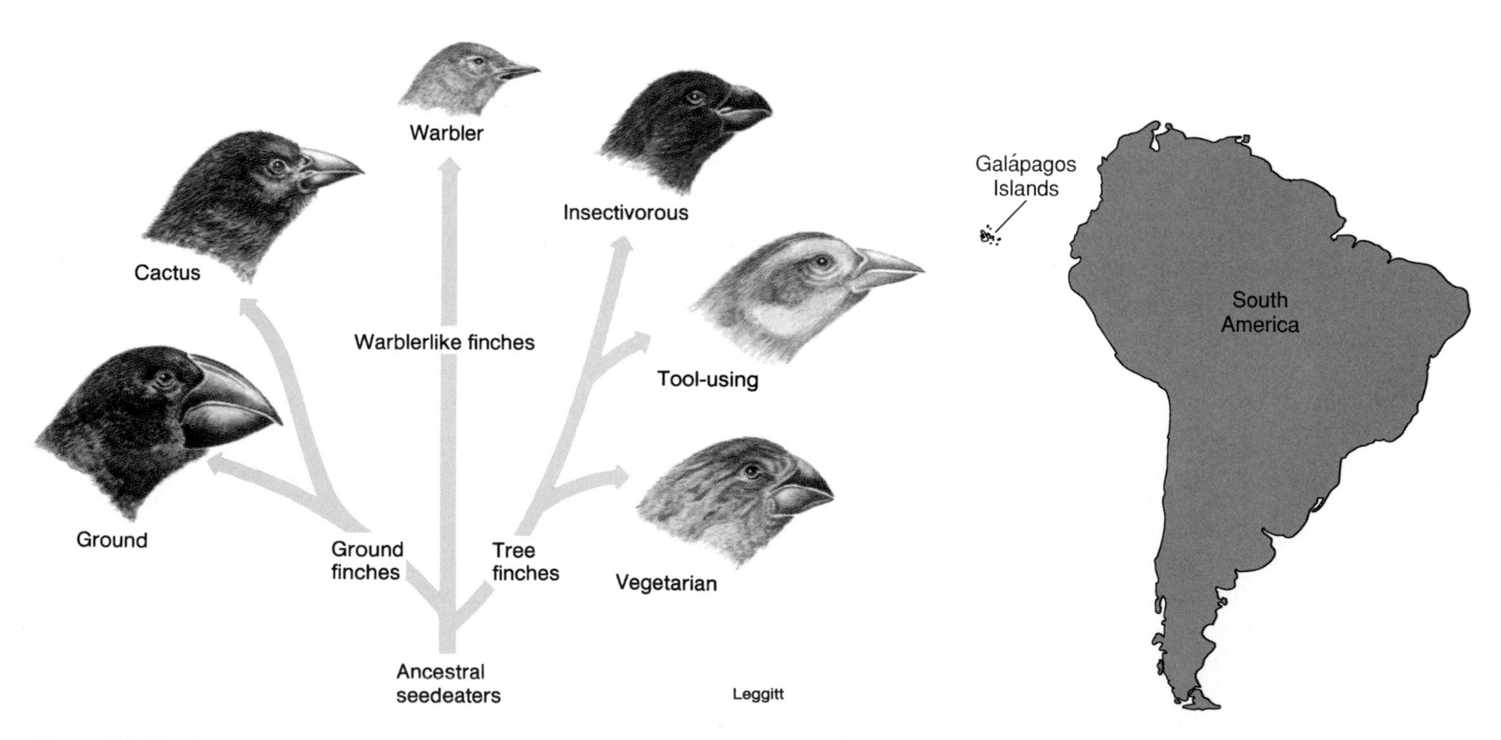

Figure 13.11 **Adaptive Radiation.** When Darwin discovered the finches of the Galápagos Islands, he thought they might all have derived from one ancestor that arrived on these relatively isolated islands. If they were the only birds to inhabit the islands, they could have evolved very rapidly into the many different types shown here. The drawings show the specializations of beaks for different kinds of food.

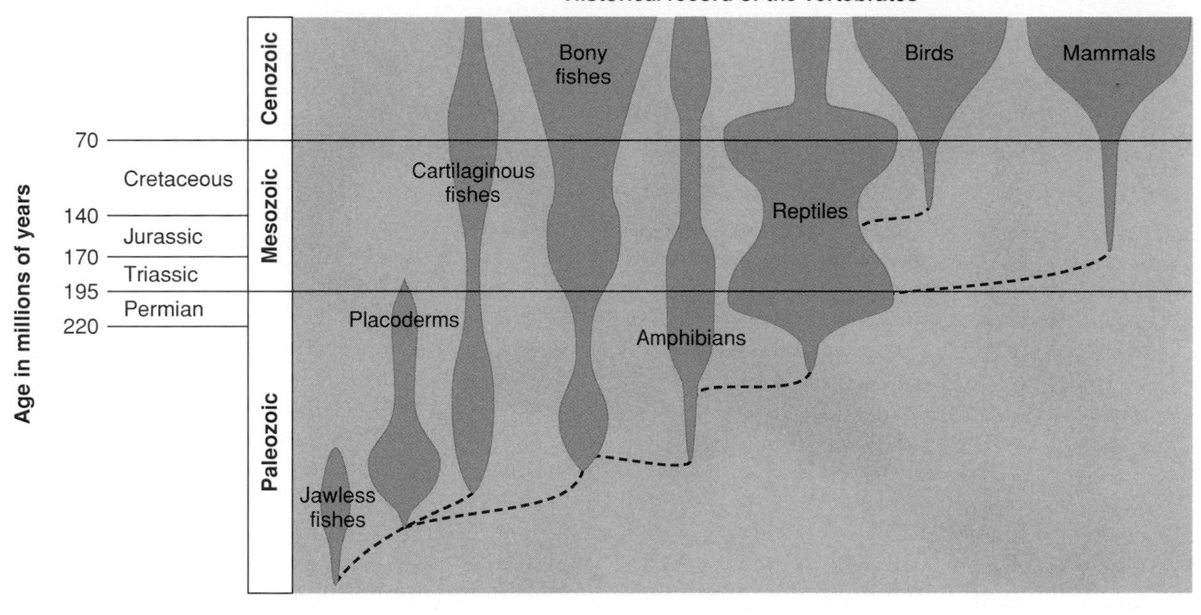

Figure 13.12 **Adaptive Radiation in Terrestrial Vertebrates.** The amphibians were the first vertebrates to live on land. They were replaced by the reptiles, that were better adapted to land. The reptiles, in turn, were replaced by the adaptive radiation of birds and mammals. (Note: the width of the colored bars indicates the number of species present.)

A second set of conditions that can favor adaptive radiation is one in which a type of organism evolves a new set of characteristics that enable it to displace organisms that previously filled roles in the environment. For example, although amphibians were the first vertebrates to occupy land, they were replaced by reptiles because reptiles had characteristics, such as dry skin and an egg that could develop on land. These characteristics allowed them to replace most of the amphibians, which could only live in relatively moist surroundings where they would not dry out and where their aquatic eggs could develop. The adaptive radiation of reptiles was extensive. They invaded most terrestrial settings and even evolved forms that flew and lived in the sea. Subsequently, the reptiles were replaced by the mammals, who went through a similar radiation. Figure 13.12 shows the sequence of radiations that occurred within the vertebrate group. The number of species of amphibians and reptiles has declined, while the number of species of birds and mammals has increased.

When organisms of widely different backgrounds develop similar characteristics, we see an evolutionary pattern known as **convergent evolution.** This particular pattern often leads people to misinterpret the evolutionary history of organisms. For example, many kinds of plants that live in desert situations have thorns and lack leaves during much of the year. Superficially they appear similar, but are often quite different from one another. They have not become one species, although they may resemble one another to a remarkable degree. The presence of thorns and the absence of leaves are adaptations to a desert type of environment—the thorns discourage herbivores, and the absence of leaves reduces water loss. Another example involves animals that survive by catching insects while flying. Bats, swallows, and dragonflies all obtain food in this manner. They all have wings, but they are derived from the modification of different structures (figure 13.13). At first glance, they may appear to be very similar and perhaps closely related, but detailed study of their wings and other structures shows that they are quite different kinds of animals. They have simply converged in structure, type of food eaten, and method of obtaining food. Likewise, whales, sharks, and tuna appear to be similar, but are different kinds of animals that all happen to live in the open ocean.

Rates of Evolution

Although it is commonly thought that evolutionary change takes long periods of time, you should understand that rates of evolution can vary greatly. Remember that natural selection is driven by the environment. If the environment is changing rapidly, one would expect rapid changes in the organisms that are present. Periods of rapid change also result in extensive episodes of extinction. During some periods in the history of the earth when little change was taking place, the rate of evolutionary change was probably slow. Nevertheless, when we talk about evolutionary time, we are generally thinking in thousands or millions

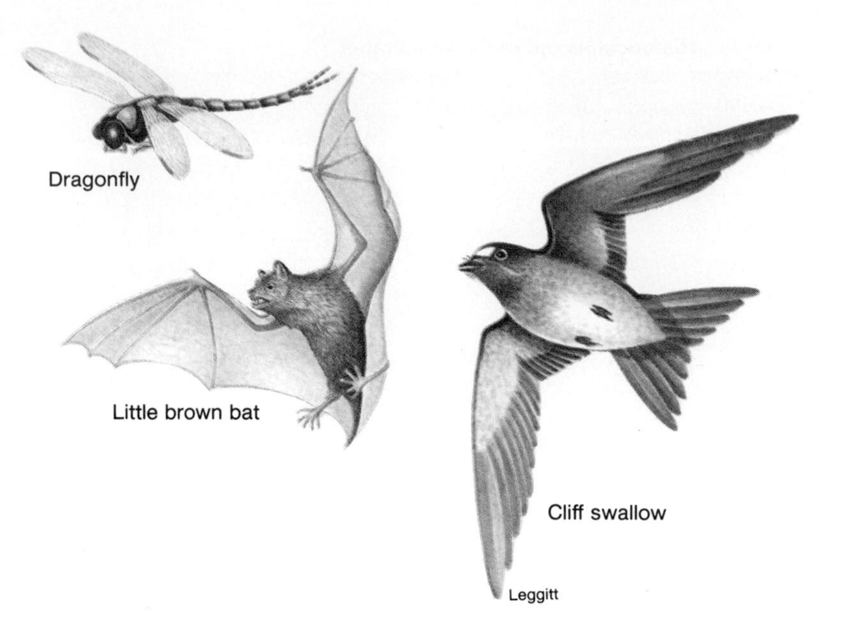

Figure 13.13 **Convergent Evolution.** All of these animals have evolved wings as a method of movement, and capture insects for food as they fly. However, they have completely different evolutionary origins.

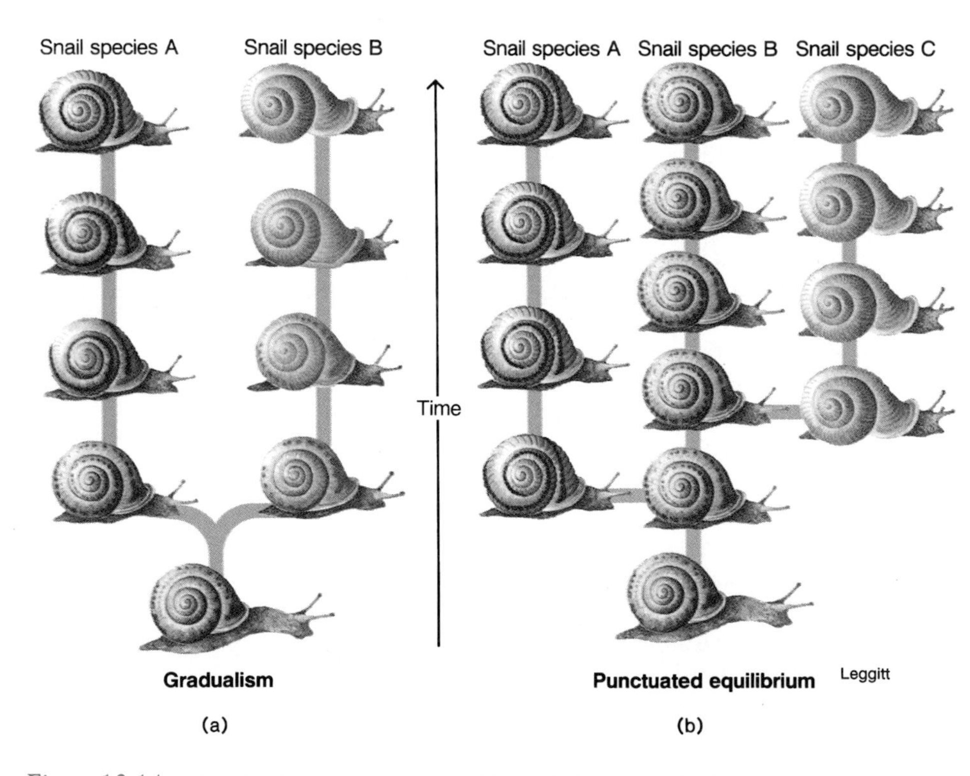

Figure 13.14 **Gradualism vs. Punctuated Equilibrium.** Gradualism (a) is the evolution of new species from the accumulation of a series of small changes over a long period of time. Punctuated equilibrium (b) is the evolution of new species from a large number of changes in a short period of time.

of years. Although these time periods are both long compared to the human life span, the difference between thousands or millions of years in the evolutionary time scale is still very significant.

When we examine the fossil record, we can see gradual changes in physical features of organisms over time. The accumulation of these changes could result in such extensive change from the original species that we would consider the current organism to be a different species from its ancestor. This is such a common feature of the evolutionary record that biologists refer to this kind of evolutionary change as **gradualism** (figure 13.14a).

Charles Darwin's view of evolution was based on the gradual change in species he perceived from his studies of geology and natural history. However, as early as the 1940s, some biologists challenged this idea. They pointed out that the fossils of some species were virtually unchanged over millions of years. If gradualism were the only explanation of species evolution, then gradual changes in the fossil record of a species would always be found. However, some organisms seem to have appeared from nowhere in the fossil record. They appeared suddenly and showed rapid change from the time they first appeared. In 1972, two biologists, Niles Eldredge of the American Museum of Natural History and Stephen Jay Gould of Harvard University, proposed the idea of **punctuated equilibrium.** This hypothesis suggests that evolution occurs in spurts of rapid change followed by long periods with little evolutionary change (figure 13.14b).

At the present time, the scientific community has no uniform concept about how such rapid bursts of evolution take place or what conditions favor such rapid change. The gradualists point to the fossil record as proof that evolution is a slow, steady process. Those who support punctuated equilibrium point to the gaps in the fossil record as evidence that rapid change occurs. As with most controversies of this nature, more information is required to resolve the question. It will take decades to collect all the information and, even then, the differences of opinion may not be reconciled.

• Summary •

Populations are usually genetically diverse. Mutations, meiosis, and sexual reproduction tend to introduce genetic variety into a population. Organisms with wide geographic distribution often show different gene frequencies in different parts of their range. A species is a group of organisms that can interbreed to produce fertile offspring. The process of speciation usually involves the geographic separation of the species into two or more isolated populations. While they are separated, natural selection operates to adapt each population to its environment. If this generates enough change, the two populations may become so different that they cannot interbreed. Similar organisms that have recently evolved into separate species normally have mechanisms to prevent interbreeding. Some of these are habitat preference, seasonal isolation, and behavioral isolation. Plants have a special way of generating new species by increasing their chromosome number as a result of abnormal mitosis or meiosis.

At one time people thought that all organisms had remained unchanged from the time of their creation. Lamarck suggested that change did occur and thought that acquired characteristics could be passed from generation to generation. Darwin and Wallace proposed the theory of natural selection as the mechanism that drives evolution. Evolution is basically a divergent process upon which other patterns can be superimposed. Adaptive radiation is a very rapid divergent evolution, while convergent evolution involves the development of superficial similarities among widely different organisms. The rate at which evolution has occurred probably varies. The fossil record shows periods of rapid change interspersed with periods of little change. This has caused some to look for mechanisms that could cause the sudden appearance of large numbers of new species in the fossil record and to challenge the traditional idea of slow, steady change accumulating enough differences to cause a new species to be formed.

• Thinking Critically •

Explain how all of the following are related to the process of speciation: mutation, natural selection, meiosis, the Hardy–Weinberg law, geographic isolation, changes in the earth, gene pool, and competition.

• Experience This •

Spend a few minutes in a greenhouse, a park, or your back yard. How many examples of reproductive isolation can you discover?

• Questions •

1. Why is geographic isolation important in the process of speciation?
2. How does speciation differ from the formation of subspecies or races?
3. Why aren't mules considered a species?
4. Describe three kinds of genetic isolating mechanisms that prevent interbreeding between different species.
5. How does a polyploid organism differ from a haploid or diploid organism?
6. Can you always tell by looking at two organisms whether or not they belong to the same species?
7. Why has Lamarck's theory been rejected?
8. Describe two differences between convergent evolution and adaptive radiation.
9. Give an example of seasonal isolation, ecological isolation, and behavioral isolation.
10. List the series of events necessary for speciation to occur.
11. What is the difference between gradualism and punctuated equilibrium?

• Chapter Glossary •

adaptive radiation (uh-dap'tiv ra-de-a'shun) A specific evolutionary pattern in which there is a rapid increase in the number of kinds of closely related species.

behavioral isolation (be-hāv'yu-ral i-so-la'shun) A genetic isolating mechanism that prevents interbreeding between species because of differences in behavior.

convergent evolution (kon-vur'jent ev-o-lu'shun) An evolutionary pattern in which widely different organisms show similar characteristics.

divergent evolution (di-vur'jent ev-o-lu'shun) A basic evolutionary pattern in which individual speciation events cause many branches in the evolution of a group of organisms.

ecological isolation (e-kō-loj'ĭ-kal i-so-la'shun) A genetic isolating mechanism that prevents interbreeding between species because they live in different areas; also called **habitat preference.**

gene flow (jēn flo) The movement of genes from one generation to another or from one place to another.

genetic isolating mechanism (jĕ-net'ic i-so-la'ting mek'an-izm) See **reproductive isolating mechanism.**

geographic barriers (je-o-graf'ik băr'yurz) Geographic features that keep different portions of a species from exchanging genes.

geographic isolation (je-o-graf'ik i-so-la'shun) A condition in which part of the gene pool is separated by geographic barriers from the rest of the population.

gradualism (grad'u-al-izm) The theory stating that evolution occurred gradually with an accumulated series of changes over a long period of time.

habitat preference (hab'i-tat pref'urents) See **ecological isolation.**

polyploidy (pah''lĭ-ploy'de) A condition in which cells contain multiple sets of chromosomes.

punctuated equilibrium (pung'chu-a-ted e-kwĭ-lib're-um) The theory stating that evolution occurs in spurts, between which there are long periods with little evolutionary change.

range (rānj) The geographical distribution of a species.

reproductive isolating mechanism (re-pro-duk'tiv i-so-la'ting me'kan-izm) A mechanism that prevents interbreeding between species; also called **genetic isolating mechanism.**

seasonal isolation (se'zun-al i-so-la'shun) A genetic isolating mechanism that prevents interbreeding between species because their reproductive periods differ.

speciation (spe-she-a'shun) The process of generating new species.

species (spe'shēz) A group of organisms that can interbreed naturally to produce fertile offspring.

subspecies (sub'spe-shēz) Regional groups within a species that are significantly different structurally, physiologically, or behaviorally, yet are capable of exchanging genes by interbreeding.

C H A P T E R 14

Ecosystem Organization and Energy Flow

Purpose

All living things require a continuous source of energy in order to grow, move about, reproduce, and perform many other functions. Certain physical laws describe how energy changes occur. The second law of thermodynamics states that during the process of converting energy from one form to another, some useful energy is lost as useless heat. Many of the world's problems result from our failure to recognize the limits imposed by the laws of thermodynamics. The purpose of this chapter is to show how energy is used and converted within groups of interacting organisms, and how the laws of thermodynamics apply to living systems. The major types of ecosystems are also described.

For Your Information

About 1985 a clamlike organism called the zebra mussel, *Dressenia polymorpha*, was introduced into the waters of the Great Lakes. It probably arrived when ballast water was discharged by a ship from the former Soviet Union. This organism has since spread to many areas of the Great Lakes, where it attaches to any hard surface and greatly increases in number. Densities of more than twenty thousand individuals per square meter have been documented in Lake Erie. These invaders are of concern for two reasons. First, they coat the intake pipes of municipal water plants and other facilities and, second, they introduce a new organism into the food chain. Zebra mussels filter small aquatic organisms from the water very efficiently and may remove food organisms required by native species. There is concern that they will significantly change the ecological organization of the Great Lakes.

Chapter Outline

Learning Objectives

- Recognize the relationships that organisms have to each other in an ecosystem.
- Understand that useful energy is lost as energy passes from one trophic level to the next.
- Appreciate that it is difficult to quantify the energy flow through an ecosystem.
- List characteristics of several different biomes.
- Understand that humans have converted natural ecosystems to human use.

Ecology and Environment

Today we hear people from all walks of life using the terms *ecology* and *environment*. Students, homemakers, politicians, planners, and union leaders speak of "environmental issues" and "ecological concerns." Often these terms are interpreted in different ways, so we need to establish some basic definitions.

Ecology is the branch of biology that studies the relationships between organisms and their environment. This is a very simple definition for a very complex branch of science. Most ecologists define the word **environment** very broadly as anything that affects an organism during its lifetime. These environmental influences can be divided into two categories. Other living things that affect an organism are called **biotic factors,** and nonliving influences are called **abiotic factors** (figure 14.1). If we consider a fish in a stream, we can identify many environmental factors that are important to its life. The temperature of the water is extremely important as an abiotic factor, but it may be influenced by the presence of trees (biotic factor) along the stream bank that shade the stream and prevent the sun from heating it. Obviously, the kind and number of food organisms in the stream are important biotic factors as well. The type of material that makes up the stream bottom and the amount of oxygen dissolved in the water are other important abiotic factors, both of which are related to how rapidly the water is flowing.

As you can see, characterizing the environment of an organism is a complex and interrelated process; everything seems to be influenced or modified by other factors. A plant is influenced by many different factors during its lifetime: the types and amounts of minerals in the soil, the amount of sunlight hitting the plant, the animals that eat the plants, and the wind, water, and temperature. Each item on this list can be further subdivided into other areas of study. For instance, water is important in the life of plants, so rainfall is studied in plant ecology. But even the study of rainfall is not simple. The rain could come during one part of the year, or it could be evenly distributed throughout the year. The rainfall could be hard and driving, or it could come as gentle, misty showers of long duration. The water could soak into the soil for later use, or it could run off into streams and be carried away.

(a)

(b)

Figure 14.1 **Biotic and Abiotic Environmental Factors.** (a) The woodpecker feeding its young in the hole in this tree is influenced by several biotic factors. The tree itself is a biotic factor as is the disease that weakened it causing conditions that allowed the woodpecker to make a hole in the rotting wood. (b) The irregular shape of the trees is the result of wind and snow, both abiotic factors. Snow driven by the prevailing winds tends to "sandblast" one side of the tree and prevent limb growth.

Temperature is also very important to the life of a plant. For example, two areas of the world can have the same average daily temperature of 10° C* but not have the same plants because of different temperature extremes. In one area, the temperature may be 13° C during the day and 7° C at night, for a 10° C average. In another area, the temperature may be 20° C in the day and only 0° C at night, for a 10° C average. Plants react to extremes in temperature as well as to the daily average. Furthermore, different parts of a plant may respond differently to temperature. Tomato plants will grow at temperatures below 13° C but will not begin to develop fruit below 13° C.

*See the metric conversion chart inside the back cover for conversion to Fahrenheit.

The animals in an area are influenced as much by abiotic factors as are the plants. If nonliving factors do not favor the growth of plants, there will be little food and few hiding places for animal life. Two types of areas that support only small numbers of living animals are deserts and polar regions. Near the polar regions of the earth, the low temperature and short growing season inhibits plant growth; therefore, there are relatively few species of animals with relatively small numbers of individuals. Deserts receive little rainfall and therefore have poor plant growth and low concentrations of animals. On the other hand, tropical rain forests have high rates of plant growth and large numbers of animals of many kinds.

As you can see, living things are themselves part of the environment of other living things. If there are too many

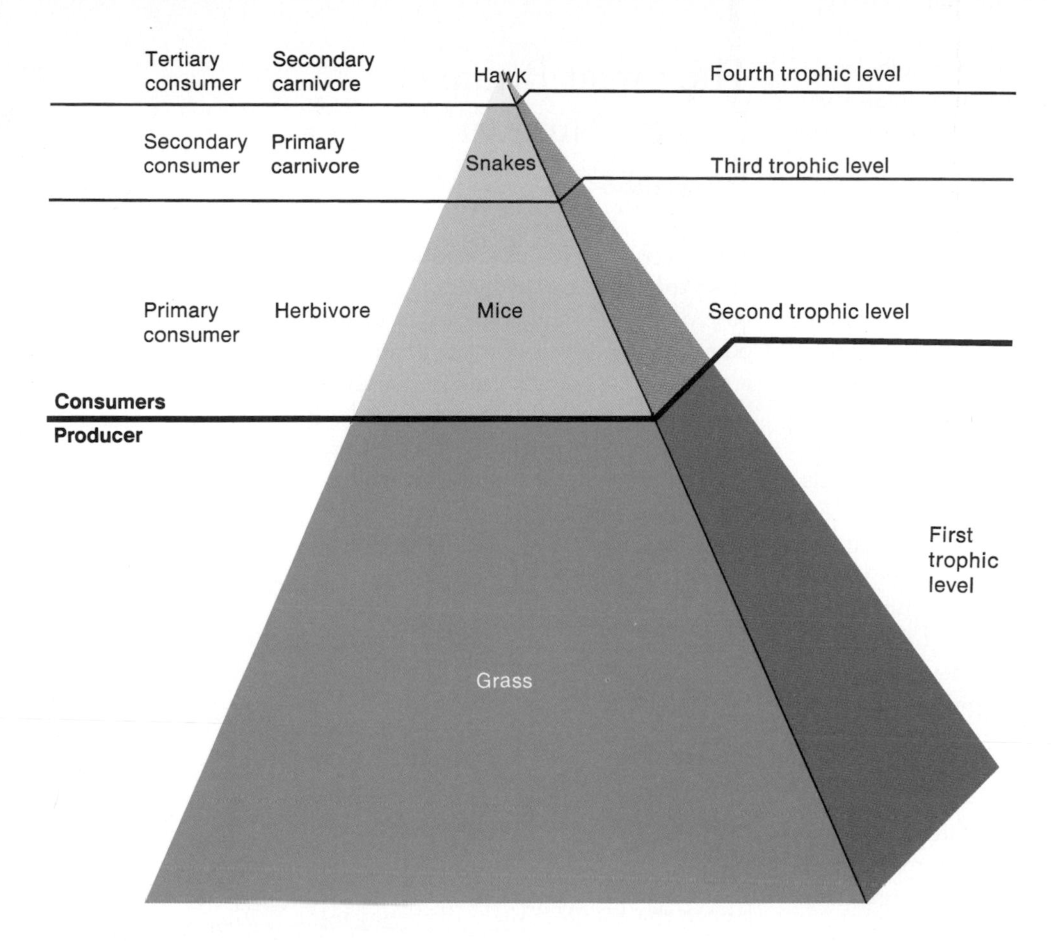

Figure 14.2 **The Organization of an Ecosystem.** Organisms within ecosystems can be divided into several different trophic levels based on how they obtain energy. Several different sets of terminology are used to identify these different roles. This illustration shows how the different sets of terminology are related to one another.

animals in an area, they could demand such large amounts of food that they would destroy the plant life, and the animals themselves would die.

So far we have discussed how organisms interact with their environment in rather general terms. Ecologists have developed several concepts that help us understand how biotic and abiotic factors interrelate to form a complex system.

The Organization of Living Systems

All living things require a continuous supply of energy to maintain life. Therefore, many people like to organize living systems by the energy relationships that exist among the different kinds of organisms present. An **ecosystem** is an interacting collection of organisms and the abiotic factors that affect them. An ecosystem contains several different kinds of organisms. Those that trap sunlight for photosynthesis, resulting in the production of organic material from inorganic material, are called **producers.** Green plants and other photosynthetic organisms are, in effect, converting sunlight energy into the energy contained within the chemical bonds of organic compounds. There is a flow of energy from the sun into the living matter of plants.

The energy that plants trap can be transferred through a number of other organisms in the ecosystem. Since all of these organisms must obtain energy in the form of organic matter, they are called **consumers.** Consumers cannot capture energy from the sun as plants do. They eat plants directly or feed on plants indirectly by eating organic matter of other living things. Each time the energy enters a different consumer organism, it is said to enter a different **trophic level,** which is a step, or stage, in the flow of energy through an ecosystem (figure 14.2). The plants (producers) receive their energy directly from the sun and are said to occupy the *first trophic level.* Various kinds of consumers can be divided into several categories, depending on how they fit into the flow of energy through an ecosystem. Animals that feed directly on plants are called **herbivores,** or **primary consumers,** and occupy the *second tropic level.* Animals that eat other animals are called **carnivores,** or **secondary consumers,** and can be subdivided into different tropic levels depending on what specific animals they eat. Animals that feed on herbivores occupy the *third trophic level* and are known as **primary carnivores.** Animals that feed on the primary carnivores are known as **secondary carnivores** and occupy the *fourth trophic level.* For example, a human may eat a fish that ate a frog that ate a spider that ate an insect that consumed plants for food. This sequence of organisms feeding on one another is known as a **food chain.** Figure 14.3 shows the six different trophic levels in the food chain. Obviously, there can be higher categories, and some organisms don't fit neatly into this theoretical scheme. Some animals are carnivores at some times and herbivores at others; they are called **omnivores.** They are classified into different trophic levels depending on what they happen to be eating at the moment.

If an organism dies, the energy contained within the organic compounds of its body is finally released to the environment as heat by organisms that decompose the dead body into carbon dioxide, water, ammonia, and other simple inorganic molecules. Organisms of decay, called **decomposers,** are things such as bacteria, fungi, and other organisms that use dead organisms as sources of energy (box 14.1).

This group of organisms efficiently converts nonliving organic matter into simple inorganic molecules that can be used by producers in the process of trapping energy. Decomposers are thus very important components of ecosystems that cause materials to be recycled. As long as the sun supplies the energy, elements are cycled through ecosystems repeatedly. Table 14.1 summarizes the various categories of organisms within an ecosystem.

Figure 14.4 illustrates a forest ecosystem. Can you identify producers, herbivores, carnivores, scavengers, and decomposers? Now that we have a better idea of how ecosystems are organized, we can look more closely at energy flow through ecosystems.

The Great Pyramids: Energy, Number, Biomass

The ancient Egyptians constructed elaborate tombs we call *pyramids.* The broad base of the pyramid is necessary to support the upper levels of the structure, and it narrows to a point at the top. This same kind of relationship exists when we look at how the various trophic levels of ecosystems are related to one another.

The Pyramid of Energy

At the base of the pyramid is the producer trophic level, which contains the largest amount of energy of any of the trophic levels within an ecosystem. In an ecosystem, the total energy can be measured in several ways. The total producer trophic level can be harvested and burned. The number of calories of heat energy produced by burning is equivalent to the energy content of the organic material of the plants. Another way of determining the energy present is to measure the rate of photosynthesis and respiration and calculate the amount of energy being trapped in the living material of the plants.

Since only the plants in the producer trophic level are capable of capturing energy from the sun, all other organisms are directly or indirectly dependent on the producer trophic level. The second trophic level consists of herbivores that eat plants. This trophic level has significantly less energy in it for several reasons. *In general, there is about a 90% loss of energy as we proceed from one trophic level to the next higher level.* Actual measurements will vary from one ecosystem to another, but 90% is a good rule of thumb. This loss in energy content at the second and subsequent trophic levels is primarily due to the second law of thermodynamics. This law states that whenever energy is converted from one form to another, some energy is converted to useless heat. Think of any energy-converting machine; it probably releases a great deal of heat energy. For example, an automobile engine must have a cooling system to get rid of the heat energy produced. An incandescent light bulb also produces large amounts of heat. Living systems are somewhat different, but they must follow the same energy rules.

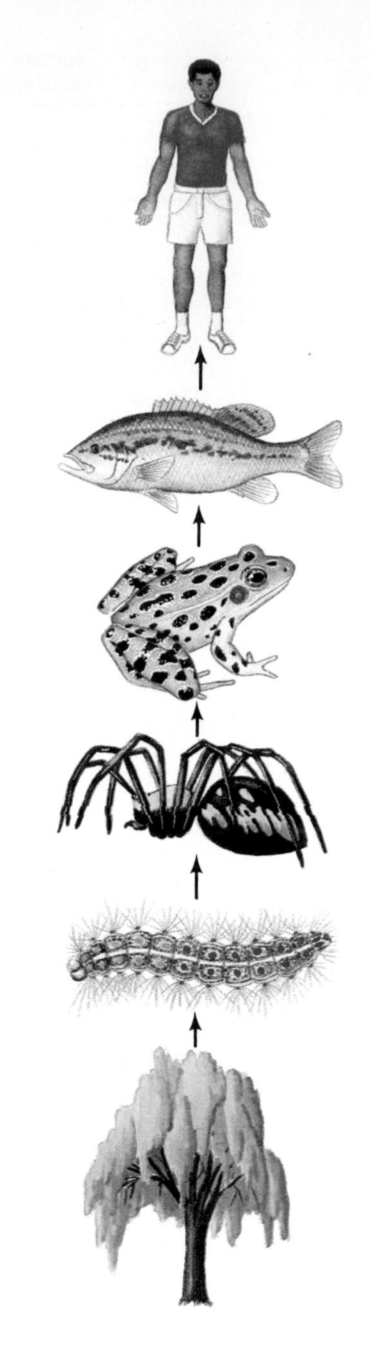

Figure 14.3 **Trophic Levels in a Food Chain.** As one organism feeds on another organism, there is a flow of energy from one trophic level to the next. This illustration shows six trophic levels.

BOX 14.1 Detritus Food Chains

Although most ecosystems receive energy directly from the sun through the process of photosynthesis, some ecosystems obtain most of their energy from a constant supply of dead organic matter. For example, forest floors and small streams receive a rain of leaves and other bits of material that small animals use as a food source. The small pieces of organic matter, such as broken leaves, feces, and body parts, are known as *detritus.* The insects, slugs, snails, earthworms, and other small animals that use detritus as food are often called *detritivores.* In the process of consuming leaves, detritivores break the leaves and other organic material into smaller particles that may be used by other organisms for food. The smaller size also allows bacteria and fungi to more effectively colonize the dead organic matter, further decomposing the organic material and making it available to still other organisms as a food source. The bacteria and fungi are in turn eaten by other detritus feeders. Some biologists believe that we greatly underestimate the energy flow through detritus food chains.

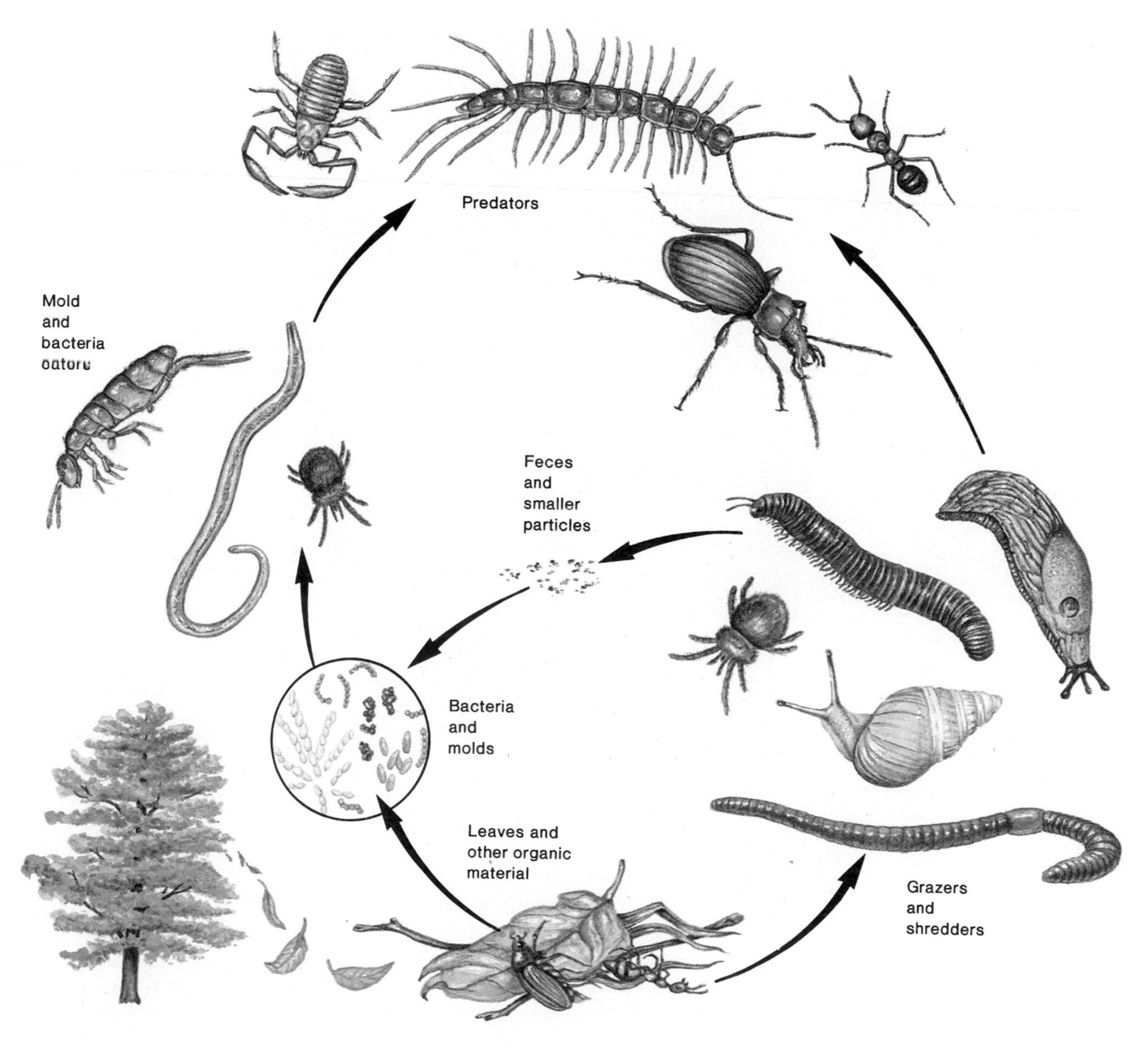

Table 14.1
Roles in an Ecosystem

Classification	Description	Examples
Producers	Plants that convert simple inorganic compounds into complex organic compounds by photosynthesis.	Trees, flowers, grasses, ferns, mosses, algae
Consumers	Organisms that rely on other organisms as food. Animals that eat plants or other animals.	
Herbivore	Eats plants directly.	Deer, goose, cricket, vegetarian human, many snails
Carnivore	Eats meat.	Wolf, pike, dragonfly
Omnivore	Eats plants and meat.	Rat, most humans
Scavenger	Eats food left by others.	Coyote, skunk, vulture, crayfish
Parasite	Lives in or on another organism, using it for food.	Tick, tapeworm, many insects
Decomposers	Organisms that return organic compounds to inorganic compounds. Important components in recycling.	Bacteria, fungi

Figure 14.4 **A Forest Ecosystem.** This illustration shows many of the organisms in a forest ecosystem. Can you identify the trophic level of each organism within the ecosystem?

In addition to the loss of energy as a result of the second law of thermodynamics, there is an additional loss involved in the capture and processing of food material by herbivores. Although herbivores don't need to chase their food, they do need to travel to where food is available, then gather, chew, digest, and metabolize it. All these processes require energy.

Just as the herbivore trophic level experiences a 90% loss in energy content, the higher trophic levels of primary carnivores, secondary carnivores, and tertiary carnivores also experience a reduction in the energy available to them. Figure 14.5 shows an energy pyramid in which the energy content decreases by 90% as we pass from one trophic level to the next.

The Pyramid of Numbers

Since it may be difficult to measure the amount of energy in any one trophic level of an ecosystem, people often use other methods to quantify the different trophic levels. One method is to simply count the number of organisms at each trophic level. This generally gives the same pyramid relationship, called a *pyramid of numbers* (figure 14.6). Obviously this is

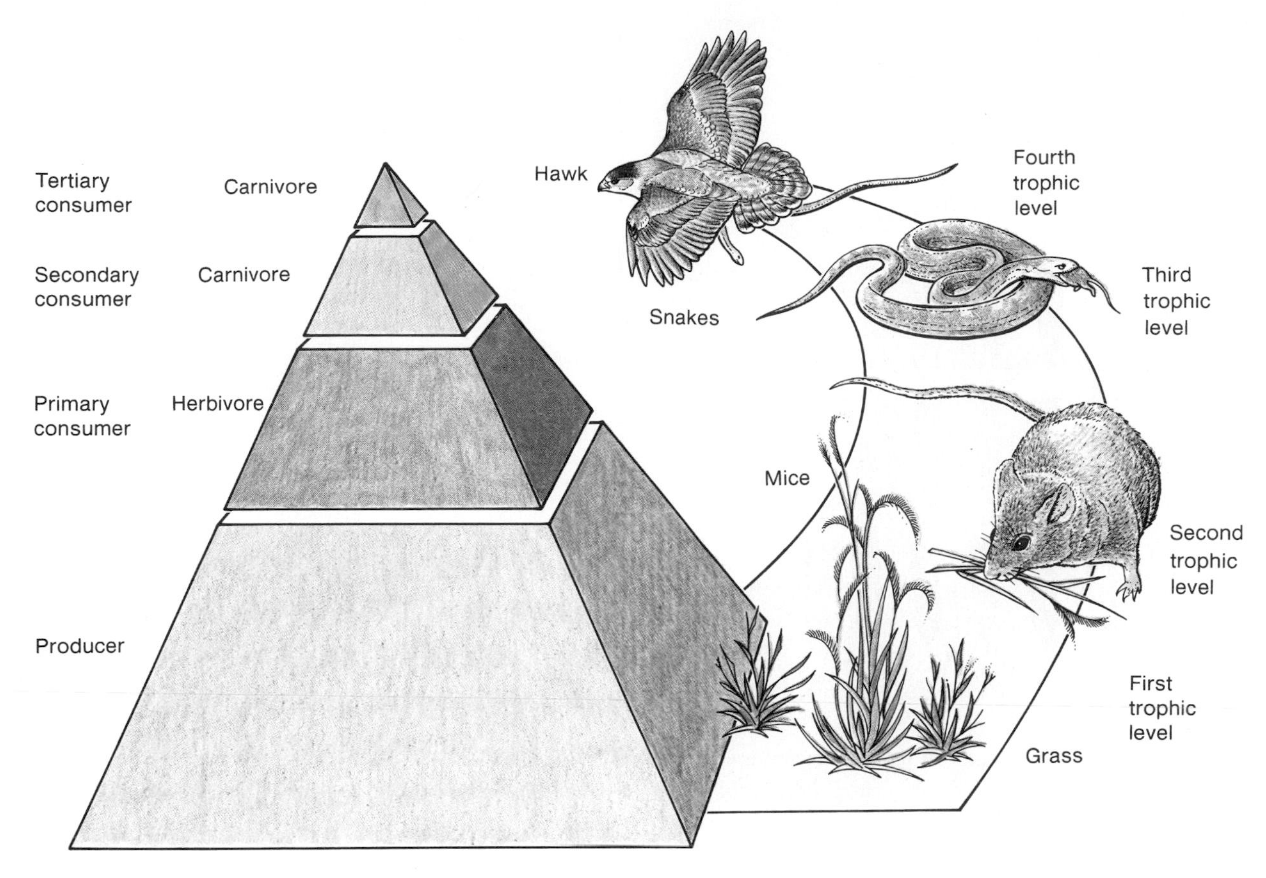

Figure 14.5 **Energy Flow through an Ecosystem.** As energy flows from one trophic level to the next, approximately 90% of it is lost. This means that the amount of energy at the producer level must be ten times larger than the amount of energy at the herbivore level.

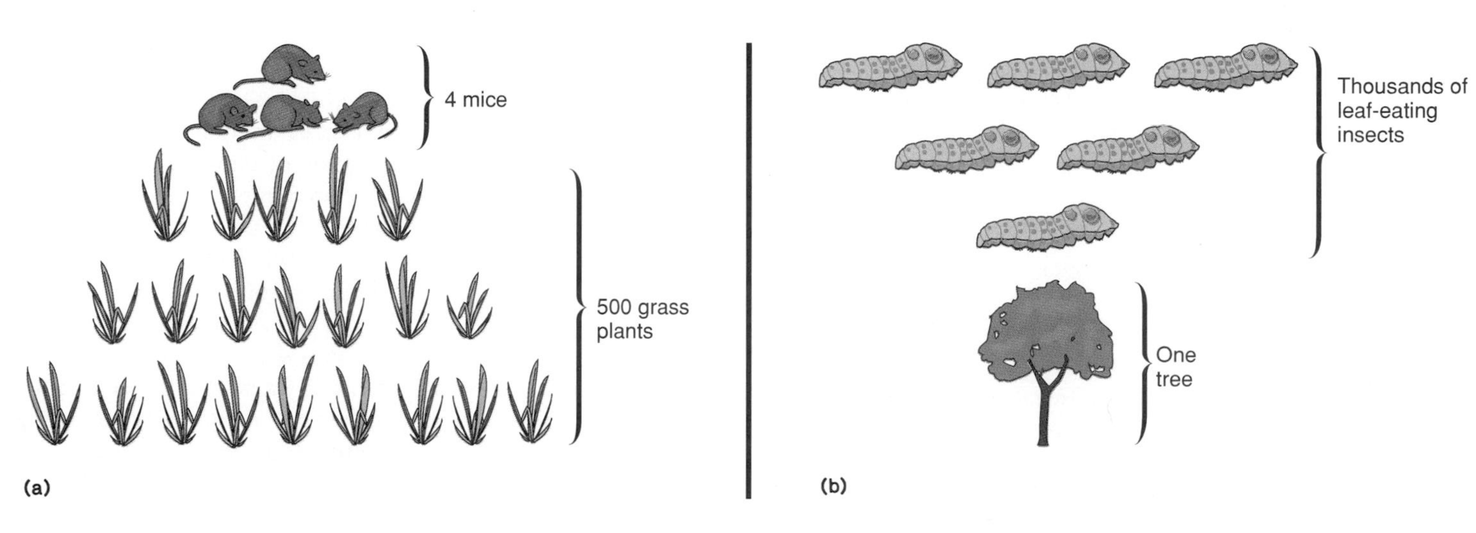

Figure 14.6 **A Pyramid of Numbers.** One of the easiest ways to quantify the various trophic levels in an ecosystem is to count the number of individuals in a small portion of the ecosystem. As long as all the organisms are of similar size and live about the same length of time, this method gives a good picture of how different trophic levels are related. (a) The relationship between grass and mice is a good example. However, if the organisms at one trophic level are much larger or live much longer than those at other levels, our picture of the relationship may be distorted. (b) This is what happens when we look at the relationship between forest trees and the insects that feed on them. A pyramid of numbers becomes inverted in this instance.

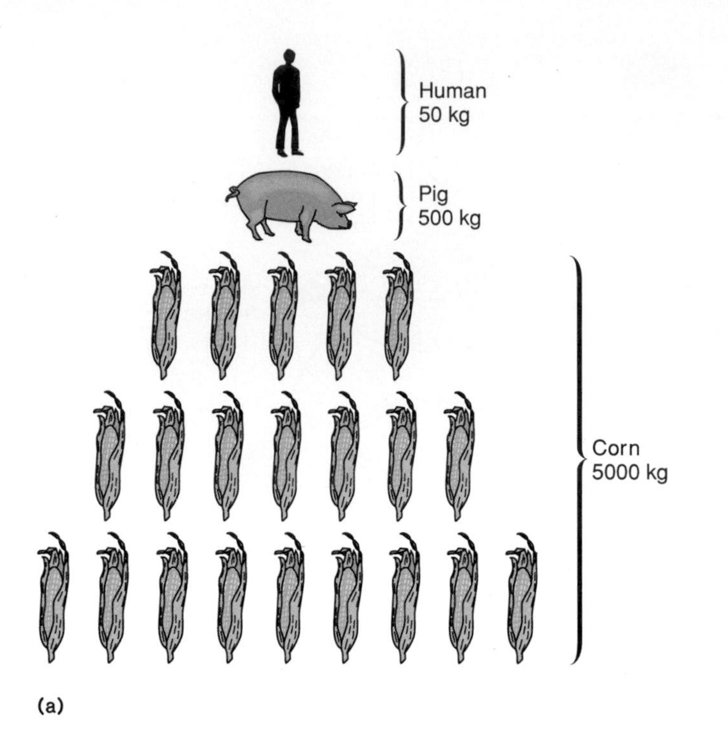

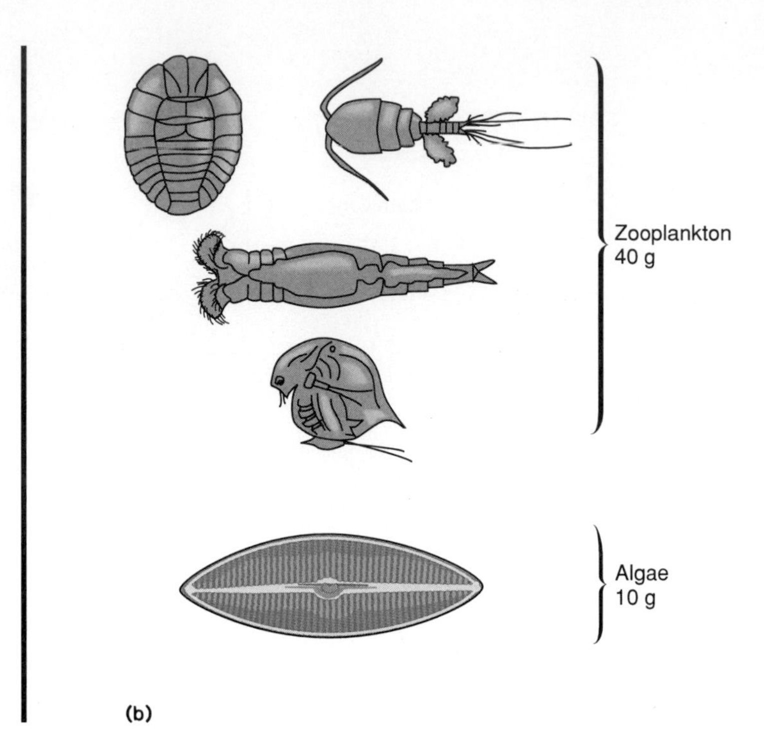

Figure 14.7 **A Pyramid of Biomass.** Biomass is determined by collecting and weighing all the organisms in a small portion of an ecosystem. (a) This method of quantifying trophic levels eliminates the problem of different-sized organisms at different trophic levels. However, it does not always give a clear picture of the relationship between trophic levels if the organisms have widely different lengths of life. (b) For example, in aquatic ecosystems, many of the small producers may divide several times per day. The zooplankton that feed on them live much longer and tend to accumulate biomass over time. The single-celled algae produce much more living material, but it is eaten as fast as it is produced, so it is not allowed to accumulate.

not a very good method to use if the organisms at the different trophic levels are of greatly differing size. For example, if you count all the small insects, feeding on the leaves of one large tree, you would actually get an inverted pyramid.

The Pyramid of Biomass

Because of the size-difference problem, many people like to use biomass as a way of measuring ecosystems. **Biomass** is usually determined by collecting all the organisms at one trophic level and measuring their dry weight. This eliminates the size-difference problem because all the organisms at each trophic level are weighed. This *pyramid of biomass* also shows the typical 90% loss at each trophic level. Although a biomass pyramid is better than a pyramid of numbers in measuring some ecosystems, it has some shortcomings.

Some organisms tend to accumulate biomass over long periods of time, while others do not. Many trees live for hundreds of years, while their primary consumers, insects, generally live only one year. Likewise, a whale is a long-lived animal, while its food organisms are relatively short-lived. Figure 14.7 shows two biomass pyramids.

Ecological Communities

The way energy flows through an ecosystem involves many interchanges between organisms. We can distinguish between the ecosystem and the interacting organisms that are a part of it. An ecosystem is a unit that consists of the physical environment and all the interacting organisms within that area. The collection of interacting organisms within an ecosystem is called a **community** and consists of many kinds of organisms. The number of individuals of a particular species in an area is called a **population.** Therefore, we can look at the same organism from several points of view. We can look at it as an individual, as a part of a population of similar individuals, as a part of a community that includes other populations, and as a part of an ecosystem, which includes abiotic factors as well as living organisms.

As you know from the discussion in the previous section, one of the ways that organisms interact is by feeding on one another. A community includes many different food chains. Some organisms may be involved in several of the food chains at the same time, so the food chains become interwoven into a **food web** (figure 14.8). In a community, the interacting food chains usually result in a relatively stable combination of populations. If a particular kind of organism

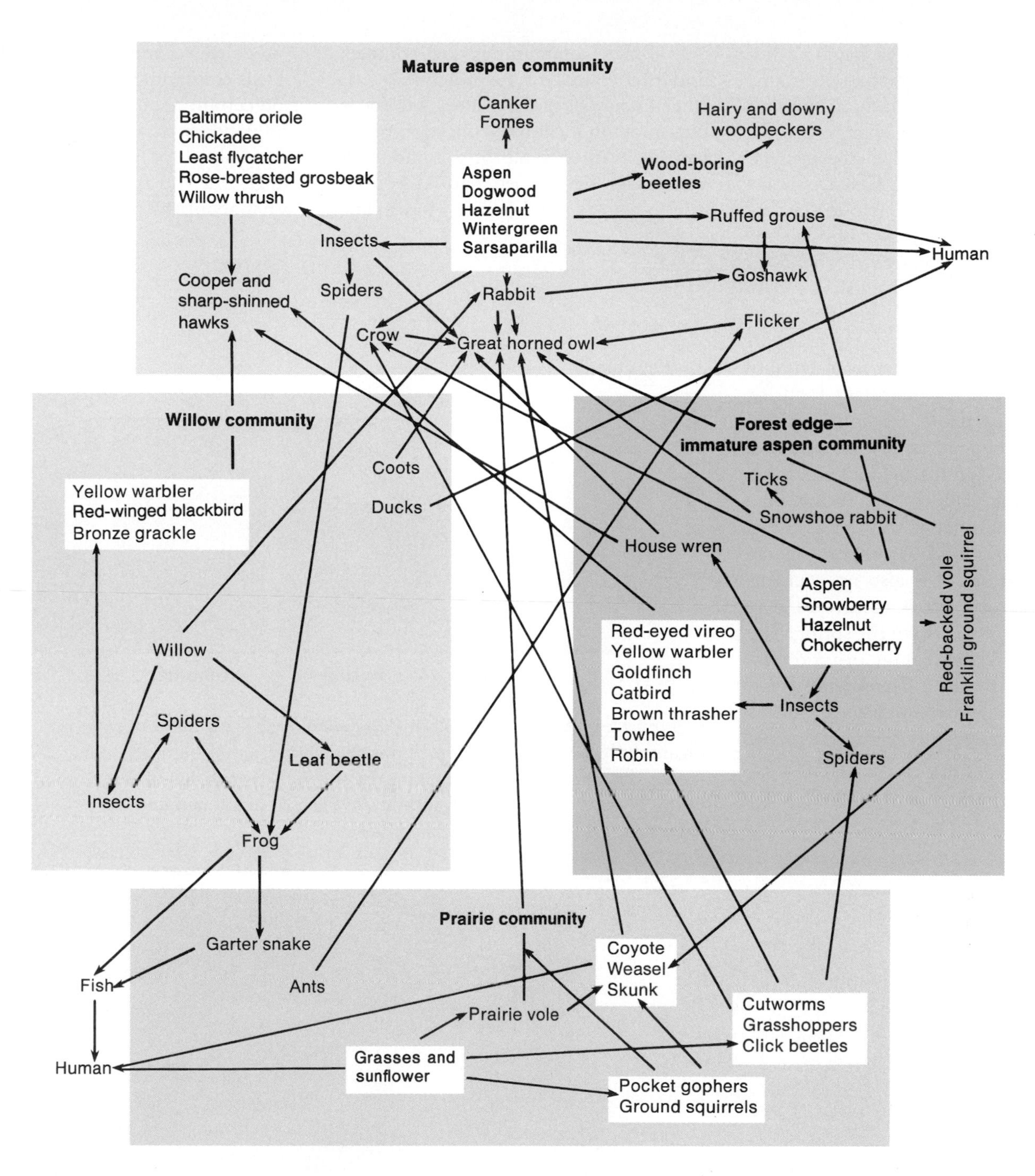

Figure 14.8 **A Food Web.** When many different food chains are interlocked with one another, a food web results. Notice that some organisms are a part of several food chains—the great horned owl in particular. Because of the interlocking nature of the food web, changing conditions may shift the way in which food flows through this system.

is removed from a community, some adjustment usually occurs in the populations of other organisms within the community. For example, humans have used insecticides to control the populations of many kinds of insects. Reduced insect populations may result in lower numbers of insect-eating birds. Often the indiscriminate use of insecticides actually increases the insect problem because insecticides kill many beneficial predator insects rather than just the one or two target pest species. Herbivorous insects may even develop increased populations following insecticide use because there are fewer carnivorous insects to eat them.

Communities are dynamic collections of organisms: as one population increases, another decreases. This might occur over several years, or even in the period of one year. This happens because most ecosystems are not constant. There may be differences in rainfall throughout the year or changes in the amount of sunlight and in the average temperature. We should expect populations to fluctuate as abiotic factors change. A change in the size of one population will trigger changes in other populations as well. Figure 14.9 shows what happens to the size of a population of deer as the seasons change. The area can support one hundred deer from January through February, when plant food for deer is least available. As spring arrives, plant growth increases. It is no accident that deer breed in the fall and give birth in the spring. During the spring producers are increasing, and the area has more available food to support a large deer population. It is also no accident that wolves and other carnivores that feed on deer give birth in the spring. The increased available energy of producers means more food for deer (herbivores), which, in turn, means more energy for the wolves (carnivores) at the next trophic level.

Since communities are complex and interrelated, it is helpful if we set artificial boundaries that allow us to focus our study on a definite collection of organisms. An example of a community with easily determined natural boundaries is a small pond (figure 14.10). The water's edge naturally defines the limits of this community. You would expect to find certain animals and plants living in the pond, such as fish, frogs, snails, insects, algae, pondweeds, bacteria, and fungi. But you might ask at this point, What about the plants and animals that

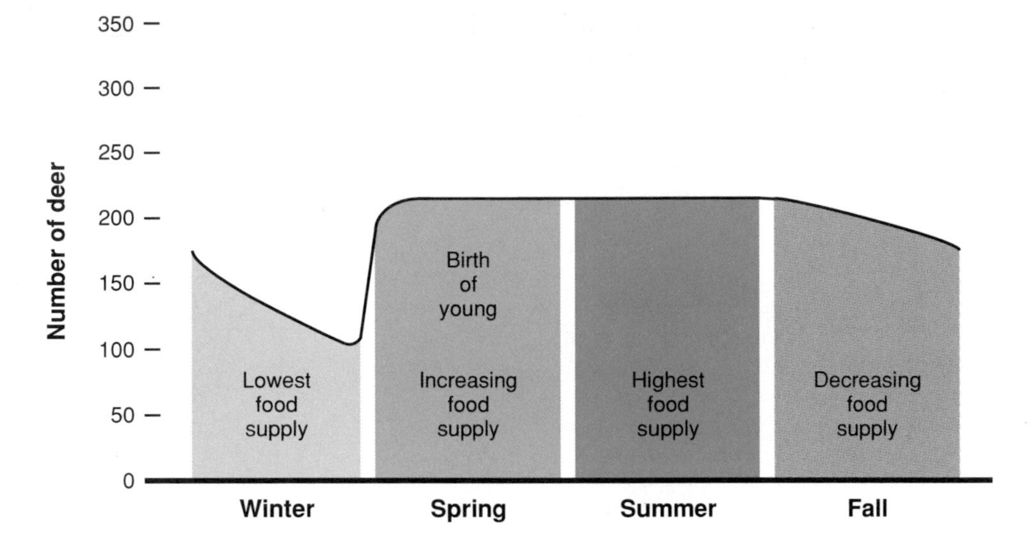

Figure 14.9 **Annual Changes in Population Size.** The number of organisms living in an area varies during the year. The availability of food is the primary factor determining the size of the population of deer in this illustration, but water availability, availability of soil nutrients, and other factors could also be important.

Figure 14.10 **A Pond Community.** Although a pond would seem to be an easy community to characterize, it interacts extensively with the surrounding land-based communities. Some of the organisms associated with a pond community are always present in the water (fish, pond weeds, clams); others occasionally venture from the water to the surrounding land (frogs, dragonflies, turtles, muskrats); still others are occasional or rare visitors (minks, heron, ducks).

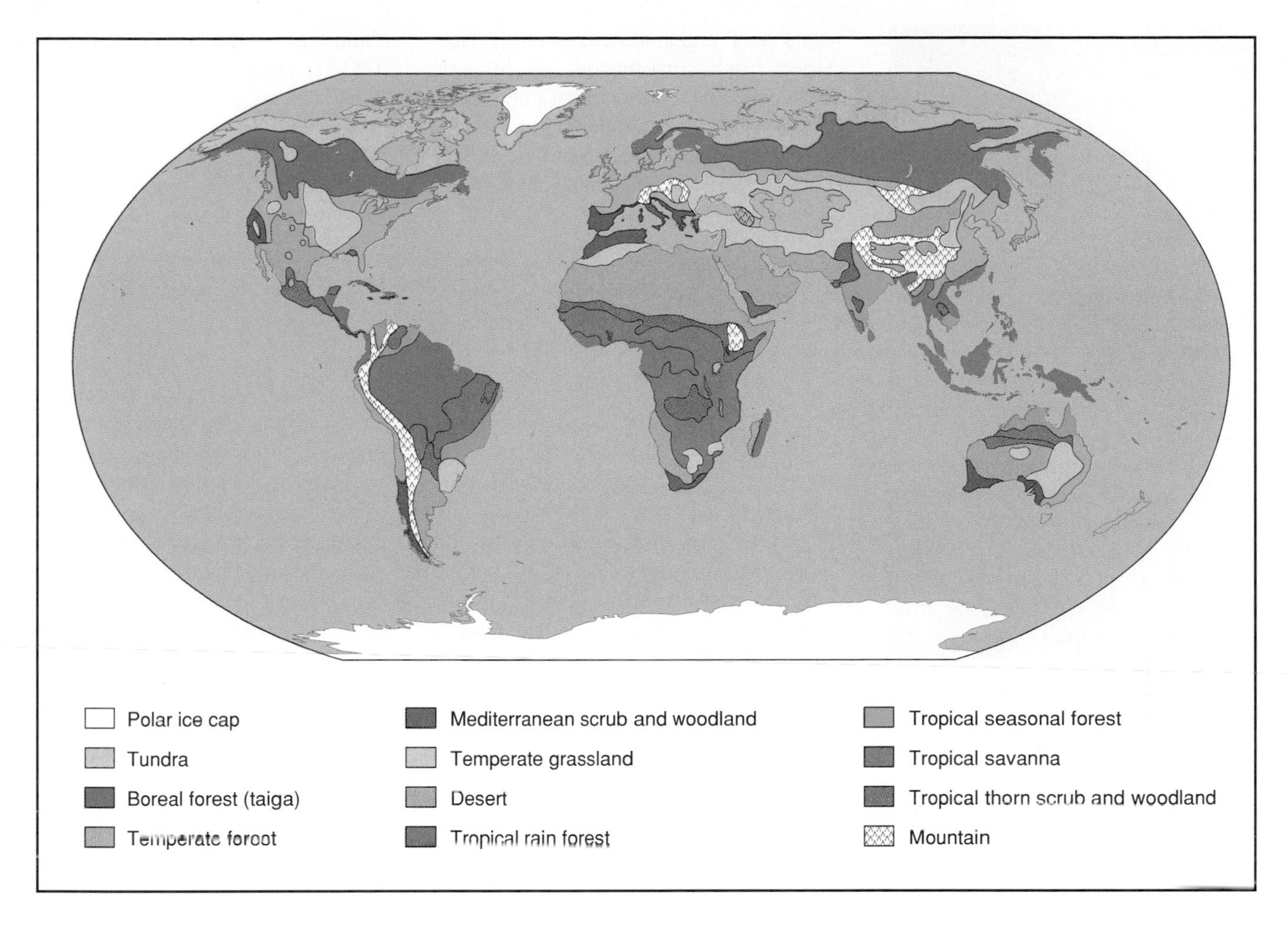

Figure 14.11 **Biomes of the World.** Major climate differences determine the kind of vegetation that can live in a region of the world. Associated with specialized groups of plants are particular communities of animals. These regional ecosystems are called biomes.

live right at the water's edge? That leads us to think about the animals that only spend part of their lives in the water. That awkward looking, long-legged bird wading in the shallows and darting its long beak down to spear a fish has its nest atop some tall trees away from the water. Should it be considered part of the pond community? Should we also include the deer that comes to drink at dusk and then wanders away? Small parasites could enter the body of the deer as it drinks. The immature parasite would develop into an adult within the deer's body. That same parasite must spend part of its life cycle in the body of a certain snail. Are these parasites part of the pond community? Several animals are members of more than one community. What originally seemed to be a clear example of a community has become less clear-cut. Although the general outlines of a community can be arbitrarily set for the purposes of a study, we must realize that the boundaries of a community, or any ecosystem for that matter, must be considered somewhat artificial.

Types of Communities

Ponds and the other small communities all can be gathered into large regional communities called **biomes.** Refer to the map of the various biomes in figure 14.11. One of the large, land-based biomes in the eastern part of North America is the *temperate deciduous forest.* This biome, like other land-based biomes, is named for the major feature of the ecosystem, which in this case happens to be the dominant vegetation. The predominant plants are large trees that lose their leaves more or less completely during the fall of the year and are therefore called *deciduous* (figure 14.12). Most of the trees require a considerable amount of rainfall.

This naming system works fairly well, since the major type of plant determines the other kinds of plants and animals that can occur. Of course, since the region is so large and has different climatic conditions in different areas, we can find some differences in the particular species of trees (and other organisms) in this biome. For instance, in Maryland the tulip tree is one of the common large trees, while in Michigan it is so unusual that people plant it in

Figure 14.12 **A Temperate Deciduous Forest Biome.** This kind of biome is found in parts of the world that have significant rainfall (100 centimeters or more) and cold weather for a significant part of the year when the trees are without leaves.

lawns and parks as a decorative tree. Aspen, birch, cottonwood, oak, hickory, beech, and maple are typical trees found in this geographic region. Typical animals of this biome are skunks, porcupines, deer, frogs, opossums, owls, mosquitoes, and beetles.

The temperate deciduous forest covers a large area from the Mississippi River to the Atlantic Coast, and from Florida to southern Canada. This type of biome is also found in parts of Europe and Asia. Many local spots within this biome are quite different from one another. Many of them have no trees at all. For example, the tops of some of the mountains along the Appalachian Trail, the sand dunes of Lake Michigan, and the scattered grassy areas in Illinois are natural areas within this biome that lack trees. In much of this region, the natural vegetation has been removed to allow for agriculture, so the original character of the biome is gone except where farming is not practical or the original forest has been preserved.

The biome located to the west of the temperate deciduous forest in North America is the *prairie biome* (figure 14.13). This kind of biome is also common in parts of Eurasia, Africa, Australia, and South America. The dominant vegetation in this region is made up of various species of grasses. The rainfall in this grassland is not adequate to support the growth of trees, which are common in this biome only along streams where they can obtain sufficient water. The common plants in this area are those that can grow in drier conditions. Animals found in this area include the prairie dog, pronghorn antelope, prairie chicken, grasshopper, rattlesnake, and meadowlark. Most of the original grasslands, like the temperate deciduous forest, have been converted to agricultural uses. Breaking the sod (the thick layer of grass roots) so that wheat, corn, and other grains can be grown exposes the soil to the wind, which may cause excess drying and result in soil erosion that depletes the fertility of the soil.

A biome that is similar to a prairie is a *savanna* (figure 14.14). Savannas are typical biomes of central Africa and parts of South America and typically consist

Figure 14.13 **A Prairie Biome.** This typical short-grass prairie of the western United States is associated with an annual rainfall of 25–50 centimeters. This community contains a unique grouping of plant and animal species.

of grasses with scattered trees. Such areas generally have wet and dry seasons and experience fires during the dry part of the year.

Very dry areas are known as *deserts* and are found throughout the world wherever rainfall is low and irregular. Some deserts are extremely hot, while others can be quite cool during much of the year. The distinguishing characteristic of desert biomes is low rainfall, not high temperature. Furthermore, deserts show large daily fluctuations in air temperature. When the sun goes down at night, the land cools off very rapidly. There is no insulating blanket of clouds to keep the heat from radiating into space. A desert biome is characterized by scattered, thorny plants that lack leaves or have reduced leaves (figure 14.15). Many of the plants, like cacti, are capable of storing water in their fleshy stems. Although this is a very harsh environment, many kinds of flowering plants, insects, reptiles, and mammals can live in this biome. The animals usually avoid the hottest part of the day by staying in burrows or other shaded, cool areas.

Through parts of southern Canada, extending southward along the mountains of the United States, and in much of Northern Asia we find communities that are dominated by evergreen trees. This is the *boreal,* or *coniferous, forest biome* (figure 14.16). The evergreen trees are especially adapted to withstand long, cold winters with abundant snowfall. Most of the trees in the wetter, colder areas are spruces and firs, but some drier, warmer areas have pines. The wetter areas generally have dense stands of small trees intermingled with many other kinds of vegetation and broken up by many small lakes and bogs. In the mountains of the western United States, the pines are often widely scattered and very large, with few branches near the ground. The area has a parklike appearance because there is very little vegetation on the forest floor. Typical animals in this biome are mice, wolves, squirrels, moose, midges, and flies.

Figure 14.14 **A Savanna Biome.** A savanna is likely to develop in areas that have a rainy season and a dry season. During the dry season, fires are frequent. The fires kill tree seedlings and prevent the establishment of forests.

Figure 14.15 **A Desert Biome.** The desert gets less than 25 centimeters of precipitation per year, but it teems with life. Cacti, sagebrush, lichens, snakes, small mammals, birds, and insects inhabit the desert. Because daytime temperatures are high, most animals are only active at night, when the air temperature drops significantly.

Figure 14.16 **A Boreal Forest Biome.** Conifers are the dominant vegetation in most of Canada, in a major part of the former Soviet Union, and at high altitudes in sections of western North America. The boreal forest biome is characterized by cold winters with abundant snowfall.

North of the coniferous forest biome is an area known as the *tundra* (figure 14.17). It is characterized by extremely long, severe winters and short, cool summers. The deeper layers of the soil remain permanently frozen and are known as the *permafrost.* Under these conditions, very few kinds of animals and plants can survive. No trees can live in this region. Typical plants and animals of the area are dwarf willow and some other shrubs, reindeer moss, some flowering plants, caribou, wolves, musk oxen, fox, snowy owls, mice, and many kinds of insects. Many kinds of birds are summer residents only. The tundra community is relatively simple, so any changes may have drastic and long-lasting effects. The tundra is easy to injure and slow to heal; therefore, we must treat it gently. The construction of the Alaskan pipeline has left scars that could still be there a hundred years from now.

The *tropical rain forest* is at the other end of the climate spectrum from the tundra. Tropical rain forests are found primarily near the equator in Central and South America, Africa, parts of southern Asia, and some Pacific Islands (figure 14.18). The temperature is high, rain falls nearly every day, and there are thousands of species of plants in a small area. Balsa (a very light wood), teak (used in furniture), and ferns the size of trees are examples of plants from the tropical rain forest. Typically, every plant has other plants growing on it. Tree trunks are likely to be covered with orchids, many kinds of vines, and mosses. Tree frogs, bats, lizards, birds, monkeys, and an almost infinite variety of insects inhabit the rain forest. These forests are very dense, and little sunlight reaches the forest floor. When the forest is opened up (by a hurricane or the death of a large tree) and sunlight reaches the forest floor, the opened area is rapidly overgrown with vegetation. Since plants grow so quickly in these forests, many attempts have been made to bring this land under cultivation. North American agricultural methods require the clearing of large areas and the planting of a single species of crop, such as corn. The constant rain falling on these fields quickly removes the soil's nutrients so that heavy applications of fertilizer are required. Often these soils become hardened when exposed in this way. Although most of these forests are not suitable for agriculture, large expanses of tropical rain forest are being cleared yearly because of the pressure for more farmland in the highly populated tropical countries and the desire for high-quality lumber from many of the forest trees.

Figure 14.17 **A Tundra Biome.** The tundra biome is located in northern parts of North America and Eurasia. It is characterized by short, cool summers and long, extremely cold winters. There is a layer of soil below the surface that remains permanently frozen; consequently, there are no large trees in this biome. Relatively few kinds of plants and animals can survive this harsh environment.

Figure 14.18 **A Tropical Rain Forest Biome.** The tropical rain forest is a moist, warm region of the world located near the equator. The growth of vegetation is extremely rapid. There are more kinds of plants and animals in this biome than in any other.

Human Use of Ecosystems

The extent to which humans use an ecosystem is often tied to its productivity. **Productivity** is the rate at which an ecosystem can accumulate new organic matter. Since the plants are the producers, it is their activities that are most important. Ecosystems in which conditions are most favorable for plant growth are the most productive. Warm, moist, sunny areas with high levels of nutrients in the soil are ideal. Some areas will have low productivity because one of the essential factors is missing. Deserts have low productivity because water is scarce, arctic areas because temperature is low, and the open ocean because nutrients are in short supply. Some communities, such as coral reefs and tropical rain forests, have high productivity. Marshes and estuaries are especially productive because the waters running into them are rich in the nutrients that aquatic photosynthesizers need. Furthermore, these aquatic systems are usually shallow so that light can penetrate through most of the water column.

Humans have been able to make use of naturally productive ecosystems by harvesting the food from them. However, in most cases, we have altered certain ecosystems substantially to increase productivity for our own purposes. In so doing, we have destroyed the original ecosystem and replaced it with an agricultural ecosystem. For example, the Native Americans living in the Great Plains area of the United States used buffalo as a source of food. There was much grass, many buffalo, and few humans. Therefore, in the Native Americans' pyramid of energy, the base was more than ample. However, with the exploitation and settling of America, the caucasion population in North America increased at a rapid rate. The top of the pyramid became larger. The food chain (prairie grass—buffalo—human) could no longer supply the food needs of the growing population. As the top of the pyramid grew, it became necessary for the producer base to grow larger. Since wheat and corn yield more biomass for humans than the original prairie grasses could, the settlers' domestic grain and cattle replaced the prairie grass and buffalo. This was fine for the settlers, but unfortunate for the buffalo and Native Americans.

In our pursuit of more productivity for our own purposes, we often overlook the alterations we have been making in the worldwide ecosystem known as the **biosphere.** In many parts of the world, the human demand for food is so large that it can only be met if humans occupy the herbivore trophic level rather than the carnivore trophic level. Humans are omnivores that can eat both plants and animals as food, so they have a choice. However, as the size of the human population increases, it cannot afford the 90% loss that occurs when plants are fed to animals that are in turn eaten by humans. In much of the less developed world, the primary food is grain; therefore, the people are already at the herbivore level. It is only in the developed countries that people can afford to eat meat. This is true from both an energy point of view and a monetary point of view. Figure 14.19a shows a pyramid of biomass having a producer base of 100 kilograms of grain. The second trophic level only has 10 kilograms of cattle because of the 90% loss typical when energy is transferred from one trophic level to the next. The consumers at the third trophic level, humans in this case, experience a similar 90% loss. Therefore, only 1 kilogram of humans could be sustained by the two-step energy transfer. There has been a 99% loss in energy: 100 kilograms of grain are necessary to sustain 1 kilogram of humans.

Humans do not need to be carnivores at the third trophic level; they can switch most of their food consumption to the second trophic level. There would then only be a 90% loss rather than a 99% loss, and the 100 kilograms of grain could support 10 kilograms of humans (figure 14.19b). By eliminating cattle from the human food chain, ten times as much human life can be supported by the same amount of plant material. In parts of the world where food is scarce, people cannot afford the energy loss involved in passing food through the herbivore trophic level. Consequently, most of the people of the world are consumers at the second trophic level and rely on corn, wheat, rice, and other organisms of the first trophic level as food. Because much of the world's population is already feeding at the second trophic level, we cannot expect food production to increase to the extent that we could feed ten times more people than exist today.

It is unlikely that most people will be able to fulfill all of their nutritional needs by just eating grains. In addition to calories, people need a certain amount of protein in their diet—and one of the best sources of protein is meat. Although protein is available from plants, the concentration is greater from animal sources. Major parts of Africa, Asia, and Latin America have diets that are deficient in both calories and protein. These people have very little food, and what food they do have is mainly from plant sources. These are also the parts of the world where human population growth is most rapid. In other words, these people are poorly nourished, and as the population increases, they will probably experience greater calorie and protein deficiency. This example reveals that even when people live as consumers at the second trophic level, they may still not get enough food, and if they do, it may not have the protein necessary for good health.

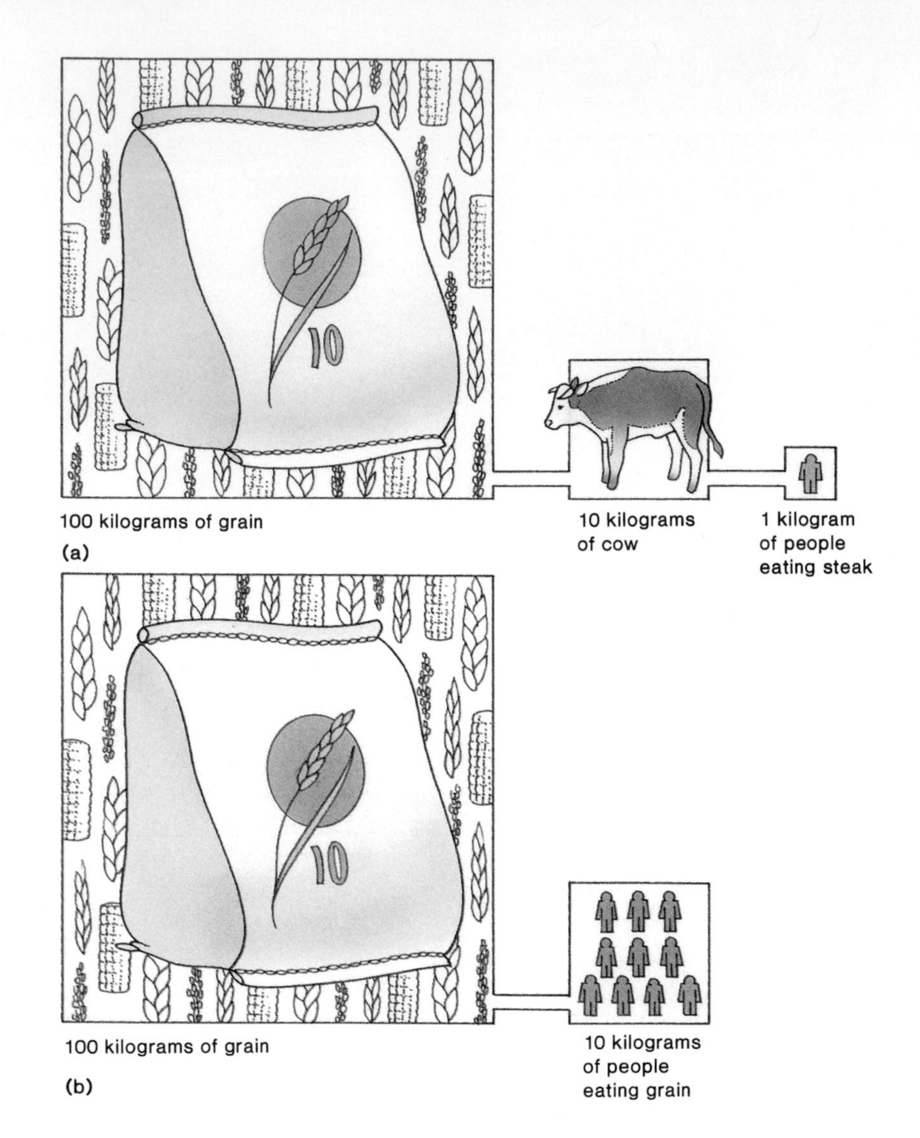

Figure 14.19 **Human Biomass Pyramids.** Since approximately 90% of the energy is lost as energy passes from one trophic level to the next, more people can be supported if they eat producers directly than if they feed on herbivores. Much of the less developed world is in this position today. Rice, corn, wheat, and other producers provide the majority of food for the world's people.

• Summary •

Ecology is the study of how organisms interact with their environment. The environment consists of biotic and abiotic components that are interrelated in an ecosystem. All ecosystems must have a constant input of energy from the sun. Producer organisms are capable of trapping the sun's energy and converting it into biomass. Herbivores feed on producers and are in turn eaten by carnivores, which may be eaten by other carnivores. Each level in the food chain is known as a trophic level. Other kinds of organisms involved in food chains are omnivores, which eat both plant and animal food, and decomposers, which break down dead organic matter and waste products. All ecosystems have a large producer base with successively smaller amounts of energy at the herbivore, primary carnivore, and secondary carnivore trophic levels. This is because each time energy passes from one trophic level to the next, about 90% of the energy is lost from the ecosystem. A community consists of the interacting populations of organisms in an area. The organisms are interrelated in many ways in food chains that interlock to create food webs. Because of this interlocking, changes in one part of the community can have effects elsewhere.

Major land-based regional ecosystems are known as biomes. The temperate deciduous forest, coniferous forest, tropical rain forest, desert, savanna, and tundra are examples of biomes.

Humans use ecosystems to provide themselves with necessary food and raw materials. As the human population increases, most people will be living as herbivores at the second trophic level because we cannot afford to lose 90% of the energy by first feeding it to a herbivore, which we then eat. Humans have converted most productive ecosystems to agricultural production and continue to seek more agricultural land as population increases.

• Thinking Critically •

Farmers are managers of ecosystems. Consider a cornfield in Iowa. Describe five ways in which the cornfield ecosystem differs from the original prairie it replaced. What trophic level does the farmer fill?

• Experience This •

The next time you are in the grocery store, determine the price per gram of the following kinds of items:

Dry beans
Whole wheat bread
Flour
Sugar
Rice
Hamburger
Frozen fish
Ham
Chicken

Why are there differences in prices?

• Questions •

1. Why are rainfall and temperature important in an ecosystem?
2. Describe the flow of energy through an ecosystem.
3. What is the difference between the terms *ecosystem* and *environment?*
4. What role does each of the following play in an ecosystem: sunlight, plants, the second law of thermodynamics, consumers, decomposers, herbivores, carnivores, and omnivores?
5. Give an example of a food chain.
6. What is meant by the term *trophic level?*
7. Why is there usually a larger herbivore biomass than a carnivore biomass?
8. List a predominant abiotic factor in each of the following biomes: temperate deciduous forest, coniferous forest, desert, tundra, tropical rain forest, and savanna.
9. Can energy be recycled through an ecosystem?
10. What is the difference between an ecosystem and a community?

• Chapter Glossary •

abiotic factors (a-bi-ot'ik fak'tōrz) Nonliving parts of an organism's environment.

biomass (bi'o-mas) The dry weight of a collection of designated organisms.

biomes (bi'ōmz) Large regional communities.

biosphere (bi'o-sfēr) The worldwide ecosystem.

biotic factors (bi-ot'ik fak'tōrz) Living parts of an organism's environment.

carnivores (kar'nĭ-vōrz) Those animals that eat other animals.

community (ko-miu'nĭ-te) A collection of interacting organisms within an ecosystem.

consumers (kon-soom'urs) Organisms that must obtain energy in the form of organic matter.

decomposers (de-kom-po'zurs) Organisms that use dead organic matter as a source of energy.

ecology (e-kol'o-je) The branch of biology that studies the relationships between organisms and their environment.

ecosystem (e''ko-sis-tum'') An interacting collection of organisms and the abiotic factors that affect them.

environment (en-vi'ron-ment) Anything that affects an organism during its lifetime.

food chain (food chān) A sequence of organisms that feed on one another, resulting in a flow of energy from a producer through a series of consumers.

food web (food web) A system of interlocking food chains.

herbivores (her'bĭ-vōrz) Those animals that feed directly on plants.

omnivores (om'nĭ-vōrz) Those animals that are carnivores at some times and herbivores at others.

population (pop''u-la'shun) The number of individuals of a specific species in an area.

primary carnivores (pri'mar-e kar'nĭ-vōrz) Those carnivores that eat herbivores and are therefore on the third trophic level.

primary consumers (pri'mar-e kon-su'merz) Those organisms that feed directly on plants—herbivores.

producers (pro-du'surz) Organisms that produce new organic material from inorganic material with the aid of sunlight.

productivity (pro-duk-tiv'ĭ-te) The rate at which an ecosystem can accumulate new organic matter.

secondary carnivores (sěk'on-děr-e kar'nĭ-vorz) Those carnivores that feed on primary carnivores and are therefore on the fourth trophic level.

secondary consumers (sěk'on-děr-e kon-su'merz) Those animals that eat other animals—carnivores.

trophic level (tro'fik lě'vel) A step in the flow of energy through an ecosystem.

C H A P T E R 15

Community Interactions

Purpose

Within ecosystems, organisms influence one another in many ways. Even organisms of the same species affect one another in the course of their normal daily activities. This chapter considers some of the kinds of interactions that occur within ecosystems and describes the various ways in which organisms within communities affect each other in the cycling of matter.

For Your Information

In 1992 the Centers for Disease Control (CDC) in Atlanta, Georgia, reported the first-ever confirmed outbreak (multiple cases) of diarrhea caused by photosynthetic bacteria, the cyanobacteria. Fourteen house staff members and seven employees who lived or worked in a hospital dormitory close to Chicago reported having symptoms. The symptoms developed after the building's water tank pump failed. Investigators believe that bacteria lying harmlessly on the bottom of the storage tank were circulated into the drinking water when the tank was refilled following its repair.

Learning Objectives

- Understand that organisms interact in a variety of ways within communities.
- Recognize the differences between community, habitat, and niche.
- Describe how atoms are cycled in communities.
- Appreciate that humans alter and interfere with natural ecological processes.
- Recognize that communities proceed through a series of stages to stable climax communities.

Chapter Outline

Community, Habitat, and Niche

People approach the study of organism interactions in two major ways. Many people look at interrelationships from the broad ecosystem point of view, while others focus on individual organisms and the specific things that affect them in their daily lives. The first approach involves the study of all the organisms that interact with one another—the community—and usually looks at general relationships among them. Chapter 14 described categories of organisms—producers, consumers, and decomposers—that perform different kinds of functions in a community.

Another way of looking at interrelationships is to study in detail the ecological relationships of particular species of organisms. Each organism has particular requirements for life and lives where the environment provides what is needed. The environmental requirements of a whale include large amounts of salt water—you would not expect to find a whale in the desert or an elephant in the ocean. The kind of place, or part of an ecosystem, occupied by an organism is known as its **habitat.** Habitats are usually described in terms of conspicuous or particularly significant features in the area where the organism lives. For example, the habitat of a prairie dog is usually described as a grassland, while the habitat of a tuna is described as the open ocean. The habitat of the fiddler crab is sandy ocean shores, and the habitat of various kinds of cacti is the desert. The key thing to keep in mind when you think of habitat is the *place* in which a particular kind of organism lives. When describing the habitats of organisms, we sometimes use the terminology of the major biomes of the world. However, it is also possible to describe the habitat of the bacterium *Escherichia coli* as the human gut, or the habitat of a fungus as a rotting log. Organisms that have very specific places in which they live simply have more restricted habitats.

Each species has particular requirements for life and places specific demands on the habitat in which it lives. The specific functional role of an organism is its **niche.** Its niche is the way it goes about living its life. Just as the word *place* is the key to understanding the concept of habitat, the word *function* is the key to understanding the concept of a niche. Understanding the niche of an organism involves a detailed understanding of the impacts an organism has on its biotic and abiotic surroundings and all of the factors that affect the organism. For example, the niche of an earthworm includes abiotic items such as soil particle size, soil texture, and the moisture, pH, and temperature of the soil. The same niche also includes biotic impacts such as serving as food for birds, moles, and shrews; as bait for anglers; or as a consumer of dead plant organic matter (figure 15.1). In addition, an earthworm serves as a host for a variety of parasites, transports minerals and nutrients from deeper soil layers to the surface, and creates burrows that allow air and water to penetrate the soil more easily.

Some organisms have rather broad niches; others, with very specialized requirements and limited roles to play, have niches that are quite narrow. The opossum (figure 15.2a) is an animal with a very broad niche. It eats a wide variety of plant and animal foods, can adjust to a wide variety of different climates, is used as food by many kinds of carnivores (including humans), and produces large numbers of offspring. By contrast, the koala of Australia (figure 15.2b) has a very narrow niche. It can live only in areas of Australia with specific species of *Eucalyptus* trees, because it eats the leaves of only a few kinds of these trees. Furthermore, it cannot tolerate low temperatures and does not produce large numbers of offspring. As you might guess, the opossum is expanding its range, and the koala is endangered in much of its range.

The complete description of an organism's niche is a very detailed inventory of influences, activities, and impacts. It involves what the organism does and what is done to the organism. Some of the impacts are abiotic, others are biotic. Since the niche of an organism is a complex set of items, it is often easy to overlook important roles played by some organisms.

For example, when Europeans introduced cattle into Australia—a continent where there had previously been no large, hoofed mammals—they did not think about the impact of cow manure or the significance of a group of beetles called *dung beetles.* These beetles rapidly colonize fresh dung and cause it to be broken down. No such beetles existed in Australia; therefore, a significant amount of land was covered with accumulated cow dung. The problem was eventually solved by the importation of several species of dung beetles from Africa, where large, hoofed mammals are common. The cattle were consumers of plant material, and the dung beetles made use of what the cattle did not digest, returning it to a form that plants could more easily recycle into plant biomass.

Kinds of Organism Interactions

When organisms encounter one another in their habitats, they can influence one another in numerous ways. One kind of organism interaction is predation. **Predation** occurs when one animal captures, kills, and eats another animal. The organism that is killed is called the **prey,** and the one that does the killing is called the **predator.** The predator obviously benefits from the relationship, while the prey organism is harmed. Most predators are relatively large compared to their prey and have specific adaptations that aid them in catching prey. Many spiders build webs that serve as nets to catch flying insects. The prey are quickly paralyzed by the spider's bite and wrapped in a tangle of silk threads. Other rapidly moving spiders, like wolf spiders and

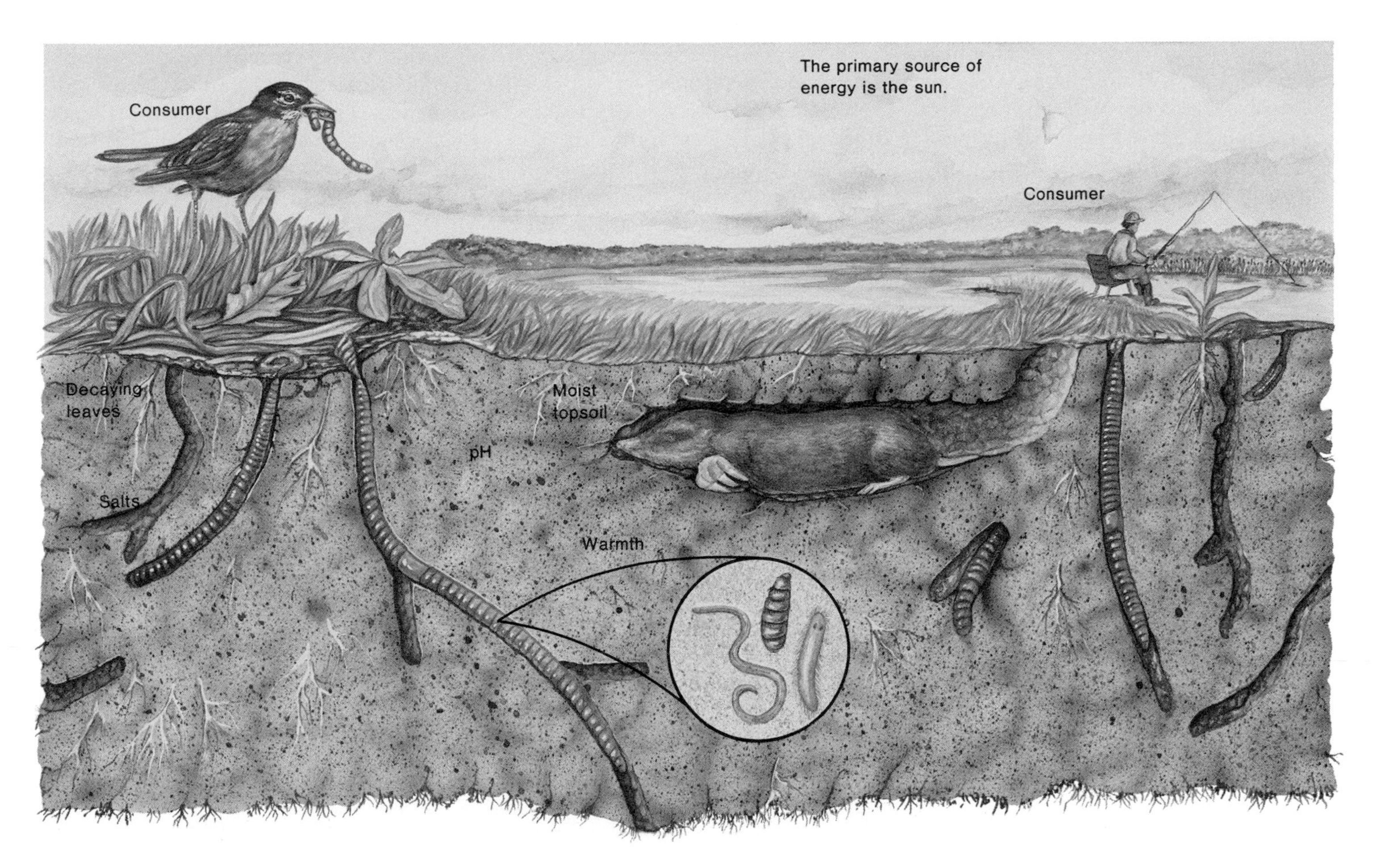

Figure 15.1 **The Niche of an Earthworm.** The niche of an earthworm involves a great many factors. It includes the fact that the earthworm is a consumer of dead organic matter, a source of food for other animals, a host to parasites, and bait for an angler. Furthermore, it includes the fact that the earthworm loosens the soil by its burrowing and "plows" the soil when it deposits materials on the surface. Additionally, the pH, texture, and moisture content of the soil have an impact on the earthworm. Keep in mind that this is but a small part of what the niche of the earthworm includes.

(a)

(b)

Figure 15.2 **Broad and Narrow Niches.** (a) The opossum has a very broad niche. It eats a variety of foods, is able to live in a variety of habitats, and has a large reproductive capacity. It is generally extending its range in the United States. (b) The koala has a narrow niche. It feeds only on the leaves of the eucalyptus tree, is restricted to relatively warm, forested areas, and is generally endangered in much of its habitat.

jumping spiders, have large eyes that help them find prey without webs. Dragonflies patrol areas where they can capture flying insects. Hawks and owls have excellent eyesight that allows them to find their prey. Many predators, like leopards, lions, and cheetahs, use speed to run down their prey (figure 15.3).

Many kinds of predators are useful to us because they control the populations of organisms that do us harm. For example, snakes eat many kinds of rodents that eat stored grain and other agricultural products. Many birds eat insects that are agricultural pests. It is even possible to think of a predator as having a beneficial effect on the prey species. Certainly the *individual* organism that is killed is harmed, but the *population* can benefit. Predators can prevent starvation by preventing overpopulation in prey species or reduce the likelihood of epidemic disease by eating sick or diseased individuals. Furthermore, predators act as selecting agents. The individuals who fall to them as prey are likely to be those that are less well adapted than the ones that escape predation. Predators usually kill the slow, stupid, sick, or injured individuals. Thus, the genes that may have contributed to slowness, stupidity, illness, or the likelihood of being injured are removed from the gene pool and a better adapted population remains. Because predators eliminate poorly adapted *individuals,* the *species* benefits. What is bad for the individual can be good for the species.

Another kind of interaction in which one organism is harmed and the other aided is the relationship of parasitism. In fact, there are more species of parasites in the world than there are nonparasites, making this a very common kind of relationship. **Parasitism** involves one organism living in or on another living organism from which it derives nourishment. The **parasite** derives the benefit and harms the **host,** the organism it lives in or on (figure 15.4). Many kinds of fungi live on trees and other kinds of plants, including those that are commercially valuable. Dutch elm disease is caused by a fungus that infects the living, sap-carrying parts of the tree. Many kinds of worms, protozoa, bacteria, and viruses are important parasites. Parasites that live on the outside

Figure 15.3 **The Predator–Prey Relationship.** Many predators capture prey by making use of speed. Since strength is needed to kill the prey, the predator is generally larger than the prey. Obviously, predators benefit from the food they obtain to the detriment of the prey organism. The cheetah can reach speeds of 112 kilometers per hour (70 miles per hour) during the sprint to capture its prey.

(a)

(b)

Figure 15.4 **The Parasite–Host Relationship.** Parasites benefit from the relationship because they obtain nourishment from the host. Tapeworms (a) are internal parasites and lamprey (b) are external parasites. The host may not be killed directly by the relationship, but it is often weakened, thus becoming more vulnerable to predators or diseases. There are more parasites in the world than organisms that are not parasites.

of their hosts are called **external parasites.** For example, fleas live on the outside of rats' bodies, where they suck blood and do harm to the rats. At the same time, a rat could have a tapeworm in its intestine. Since the tapeworm lives inside the host, it is called an **internal parasite.** Another kind of parasite found in the blood of the rat may be the bacterium *Yersinia pestis.* It does little harm to the rat but causes a disease known as *plague* or *black death* if it is transmitted to humans. The flea can serve as a carrier of the bacterium between rats and humans. An organism that can carry a disease from one individual to another is called a **vector.** If a flea sucks blood from an infected rat and then bites a human, the bacterium may enter the human bloodstream and cause plague. During the mid-1300s, when living conditions were poor and rats and fleas were common, epidemics of plague killed millions of people. In some countries in Western Europe, 50% of the population was killed by this disease. *Lyme's disease,* also a vector-borne disease, is currently spreading through the United States (figure 15.5); however, 97% of the cases are centered in three regions: the Northeast (e.g., Connecticut, Rhode Island, New Jersey); the Middle West (e.g., Michigan, Indiana, Ohio); and the West (e.g., California, Oregon, Utah).

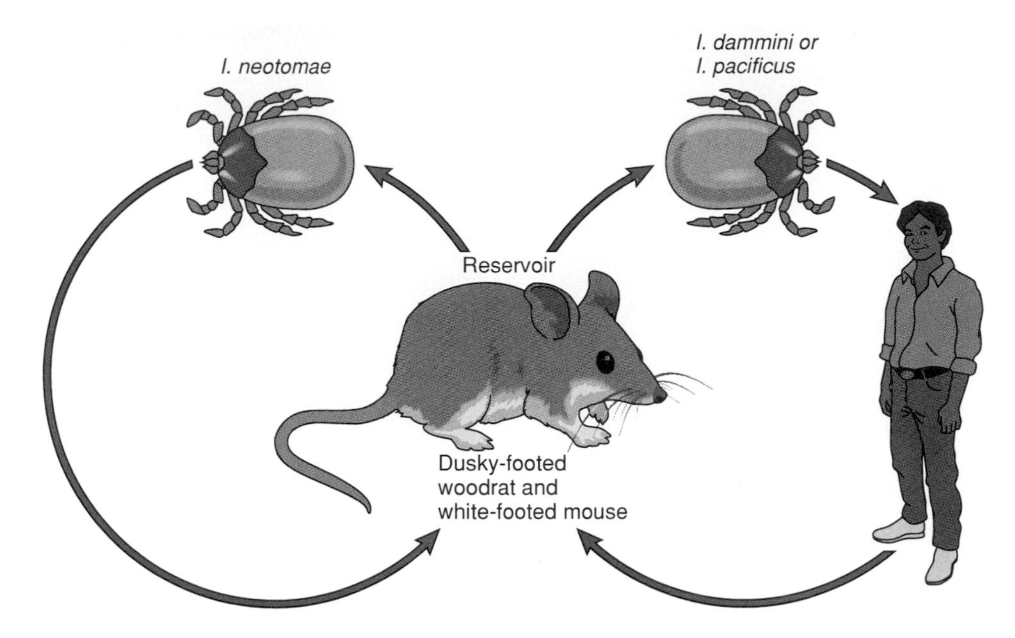

Figure 15.5 **Lyme's Disease—Hosts, Parasites, and Vectors.** Lyme's disease is a bacterial disease originally identified in a small number of individuals in the Old Lyme, Connecticut area. Once the parasite, *Borrelia burgdorferi,* has been transferred into a suitable susceptible host (e.g., humans, mice, horses, domestic cats, and dogs), it causes symptoms that have been categorized into three stages. The first-stage symptoms may appear three to thirty-two days after an individual is bitten by an infected tick (*Ixodes dammini, I. neotomae,* or *I. pacificus*) and include a spreading red rash, headache, nausea, fever, aching joints and muscles, and fatigue. Stage two may not appear for weeks or months after the infection and may affect the heart and nervous system. The third stage may appear months or years later and typically appear as severe arthritis attacks. The main reservoir of the disease appears to be the white-footed mouse and Dusky-footed woodrat.

Predation and parasitism are both relationships in which one member of the pair is helped and the other is harmed. There are several related kinds of interactions that are similar but don't fit our definitions very well. For example, when a cow eats grass, it is certainly harming the grass while deriving benefit from it. We could call cows *grass predators,* but we usually refer to them as *herbivores.* Likewise, such animals as mosquitoes, biting flies, vampire bats, and ticks take blood meals but don't usually live permanently on the host, nor do they kill it. Are they temporary parasites, or specialized predators? Finally, birds like cowbirds and cuckoos lay their eggs in the nests of other species of birds, who raise these foster young rather than their own. The adult cowbird and cuckoo or their offspring remove the eggs or the young of the host-bird species, so that it is usually only the cowbird or cuckoo that is raised by the foster parents. This kind of relationship has been called *nest parasitism.* Many kinds of interactions between organisms don't fit neatly into the classification scheme dreamed up by scientists.

In another specialized kind of interrelationship called **amensalism,** one member of the pair of interacting organisms is harmed, but the other is not affected either positively or negatively. The classic example of an amensal relationship is the inhibitory influence of the chemical penicillin on the growth of certain bacteria. Penicillin is a product of the mold *Penicillium.* It is obvious that the bacteria are harmed by the relationship, since they cannot grow and reproduce, but if there is a benefit for the mold, it must be a very subtle benefit.

There is also a kind of relationship in which one organism is benefited, while the other is not affected. This is known as **commensalism.** For example, sharks often have another fish, the remora, attached to them. The remora has a sucker on the top side of its head that allows it to attach to the shark and get a free ride (figure 15.6). While the remora benefits from the free ride and by eating leftovers from the shark's meals, the shark does not appear to be troubled by this uninvited guest, nor does it benefit from the presence of the remora. Another example of commensalism is the relationship between trees and epiphytic plants. **Epiphytes** are plants that live on the surface of other plants but do not derive nourishment from them. Many kinds of plants (e.g., orchids, ferns, and mosses) use the surface of trees as a place to live.

Figure 15.6 **Commensalism.** In the relationship called commensalism, one organism benefits and the other is not affected. The remora fish shown here hitchhike a ride on the shark. They eat scraps of food left over by the messy eating habits of the shark. The shark does not seem to be hindered in any way.

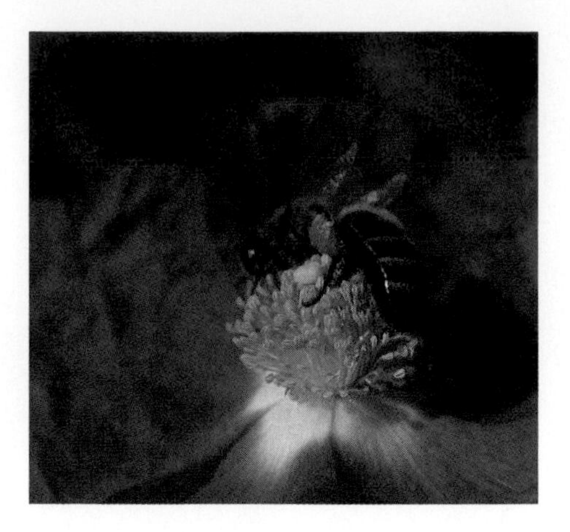

Figure 15.7 **Mutualism.** Mutualism is an interaction between two organisms in which both benefit. The plant benefits because cross-fertilization (exchange of gametes from a different plant) is more probable, the bee benefits by acquiring food—nectar.

These kinds of organisms are particularly common in tropical rain forests. Many epiphytes derive benefit from the relationship because they are able to be located in the top of the tree, where they receive more sunlight and moisture. The trees derive no benefit from the relationship, nor are they harmed; they simply serve as a support surface for epiphytes.

So far in our examples, only one species has benefited from the association of two species. There are also many situations in which two species live in close association with one another, and both benefit. This is called **mutualism.** One interesting example of mutualism involves digestion in rabbits. Rabbits eat plant material that is high in cellulose, even though they do not produce the enzymes capable of breaking down cellulose molecules into simple sugars. They manage to get energy out of these cellulose molecules with the help of special bacteria living in their digestive tracts. The bacteria produce cellulose-digesting enzymes, called *cellulases,* that break down cellulose into smaller carbohydrate molecules that the rabbit's digestive enzymes can break down into smaller glucose molecules. The bacteria benefit because the gut of the rabbit provides them with a moist, warm, nourishing environment in which to live. The rabbit benefits because the bacteria provide them with a source of food. Termites, cattle, buffalo, and antelope also have collections of bacteria and protozoa that live in their digestive tracts and help them to digest cellulose.

Another kind of mutualistic relationship exists between flowering plants and bees. Undoubtedly you have observed bees and other insects visiting flowers to obtain nectar from the blossoms. Usually the flowers are constructed in such a manner that the bees pick up pollen (sperm-containing packages) on their hairy bodies and transfer it to the female part of the next flower they visit (figure 15.7). Because bees normally visit the same species of flower for several minutes and ignore other species, they can serve as pollen carriers between two flowers of the same species. Plants pollinated in this manner produce less pollen than do plants that rely on the wind to transfer pollen. This saves the plant energy because it doesn't need to produce huge quantities of pollen. It does, however, need to transfer some of its energy savings into the production of showy flowers and nectar to attract the bees. The bees benefit from both the nectar and pollen; they use both for food.

One additional term that relates to parasitism, commensalism, and mutualism is *symbiosis.* **Symbiosis** literally means "living together." Unfortunately, this word is used in several ways, none of which are very precise. It is often used as a synonym for mutualism, but it is also often used to refer to commensalistic relationships and parasitism. The emphasis, however, is on interactions that involve a close physical relationship between the two kinds of organisms.

So far in our discussion of organism interactions we have left out the most common one. **Competition** is a kind of interaction between organisms in which both organisms are harmed to some extent. Competition occurs whenever two organisms both need a vital resource that is in short supply (figure 15.8). The vital resource could be food, shelter, nesting sites, water, mates, or space. It can be a snarling tug-of-war between two dogs over a scrap of food, or it can be a silent struggle between plants for access to available light. If you have ever started tomato seeds (or other garden plants) in a garden and failed to eliminate the weeds, you have witnessed competition. If the weeds are not removed, they compete with the garden plants for available sunlight, water, and nutrients, resulting in poor growth of both the garden plants and the weeds.

The more similar the requirements of two species of organisms, the more intense the competition. According to the **competitive exclusion principle,** no two species of organisms can occupy the same niche at the same time. If two species of organisms do occupy the same niche, the competition will be so intense that one will become extinct, one may be forced to migrate to a different area, or the two species may evolve into slightly different niches. Thus, even though both kinds of organisms are harmed during competition, there can still be winners and losers: one may be harmed more than the other.

Figure 15.8 **Competition.** Whenever a needed resource is in limited supply, organisms compete for it. This competition may be between members of the same species (*intraspecific*), illustrated by the vultures shown in the photograph, or may involve different species (*interspecific*).

Competition provides a major mechanism for natural selection. Because of the necessity to reduce interspecies competition, organisms often develop slight differences in their niches that prevent direct competition. For example, many birds catch flying insects as food. However, they do not compete directly with each other because some feed at night, some feed high in the air, some feed only near the ground, and still others perch on branches and wait for insects to fly past.

Any community of organisms offers many similar examples of specialization to very specific niches. These niches overlap and interrelate into communities that are capable of using their resources very efficiently through a process of cycling materials.

The Cycling of Materials in Ecosystems

The earth is a closed ecosystem in that no significant amount of new matter comes to the earth from space. Only sunlight energy comes to the earth in a continuous stream, and even this is ultimately returned to space as heat energy. However, it is this flow of energy through the biosphere that drives all biological processes. Living systems have evolved ways of using this energy to continue life through growth and reproduction. Although some new atoms are being added to the earth from cosmic dust and meteorites, this amount is not significant in relation to the entire biomass of the earth. Therefore, living things must reuse the existing atoms again and again. In this recycling process, inorganic molecules are combined to form the organic compounds of living things. If there were no way of recycling this organic matter back into its inorganic forms, organic material would build up as the bodies of dead organisms. Decomposers play a vital role in this recycling process, if conditions allow them to operate. If they are kept from destroying organic matter, it builds up as deposits of organic matter. This is thought to have occurred millions of years ago when the present deposits of coal, oil, and natural gas were formed.

Living systems contain many kinds of atoms, but some are more common than others. Carbon, nitrogen, oxygen, hydrogen, and phosphorus are found in all living things and must be recycled when an organism dies. Let's look at some examples of this recycling process. Carbon and oxygen combine to form the molecule carbon dioxide (CO_2), which is a gas found in small quantities in the atmosphere. During photosynthesis, carbon dioxide (CO_2) combines with water (H_2O) to form complex organic molecules ($C_6H_{12}O_6$). At the same time, oxygen molecules (O_2) are released into the atmosphere.

The organic matter in the bodies of plants may be used by herbivores as food. When an herbivore eats a plant, it breaks down the complex organic molecules into more simple molecules, like simple sugars, amino acids, glycerol, and fatty acids. These can be used as building blocks in the construction of its own body. Thus, the atoms in the body of the herbivore can be traced back to the plants that were eaten. Similarly, when herbivores are eaten by carnivores, these same atoms are transferred to them. The waste products of plants and animals and the remains of dead organisms are also used by decomposer organisms as sources of carbon and oxygen atoms. Finally, all the organisms in this cycle—plants, herbivores, carnivores, and decomposers—are involved in the process of respiration, in which oxygen (O_2) is used to break down organic compounds into carbon dioxide (CO_2) and water (H_2O). Thus, the carbon atoms that started out as components of carbon dioxide (CO_2) molecules have passed through the bodies of living organisms as parts of organic molecules in their bodies and returned to the atmosphere as carbon dioxide, ready to be cycled again. Similarly, the oxygen atoms (O) released as oxygen molecules (O_2) during photosynthesis have been used during the process of respiration (figure 15.9).

Water molecules are essential for life. They are involved in the process of photosynthesis as raw material. The hydrogen atoms (H) from water (H_2O) molecules are added to carbon atoms to make carbohydrates and other organic molecules. Furthermore, the oxygen in

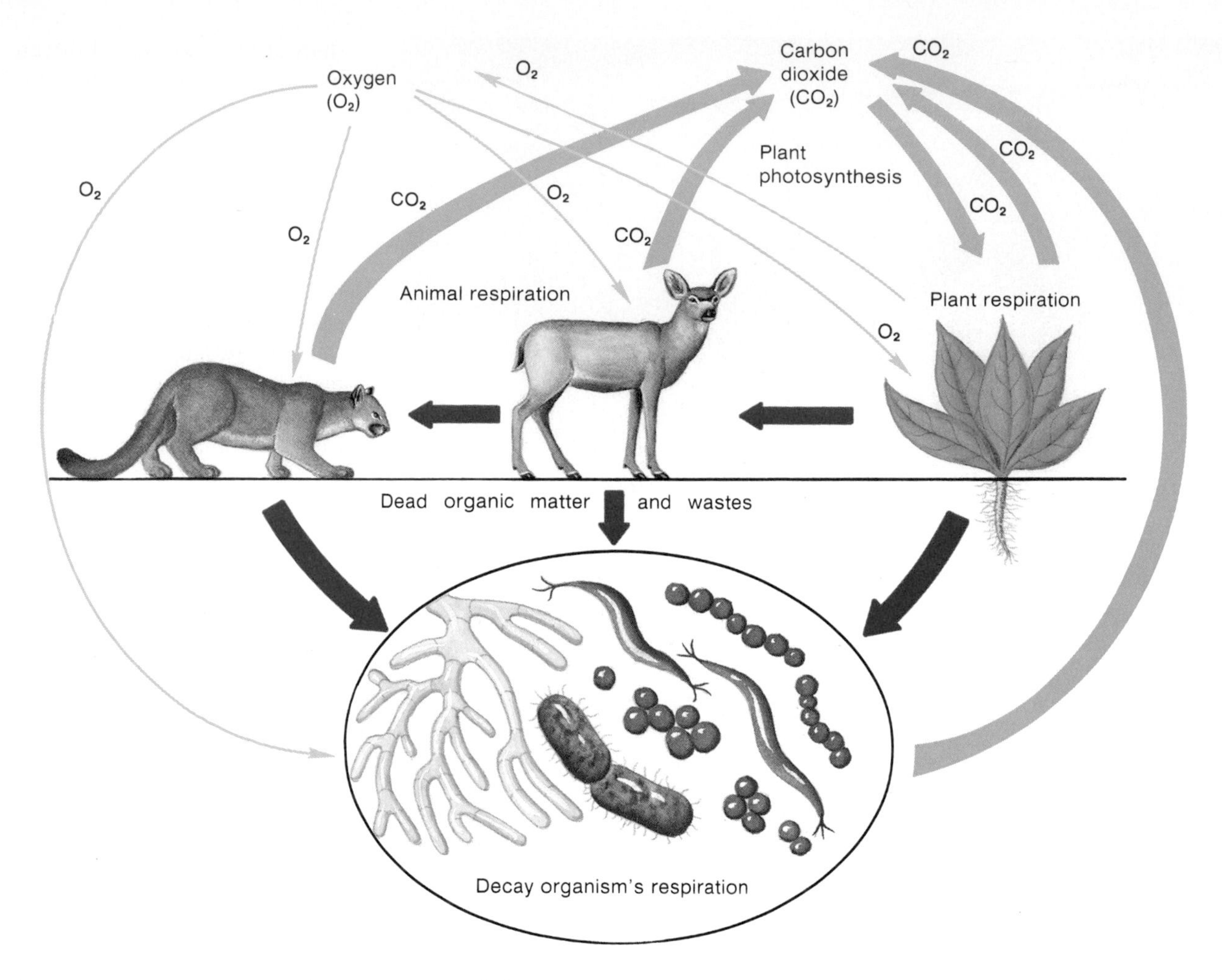

Figure 15.9 **The Carbon Cycle.** Carbon atoms are cycled through ecosystems. Carbon dioxide (green arrows) produced by respiration is the source of carbon that plants incorporate into organic molecules when they carry on photosynthesis. These carbon-containing organic molecules (pink arrows) are passed to animals and decomposers when they eat plants and animals. Organic molecules in waste or dead organisms are consumed by decay organisms. All organisms (plants, animals, and decomposers) return carbon atoms to the atmosphere as carbon dioxide when they carry on cellular respiration. Oxygen (blue arrows) is being cycled at the same time that carbon is. The oxygen is released to the atmosphere and into the water during photosynthesis and taken up during cellular respiration.

water molecules is released during photosynthesis as oxygen molecules (O_2). All of the metabolic reactions that occur in organisms take place in a watery environment. The most common molecule in the body of any organism is water. This important molecule is circulated in a hydrologic cycle (figure 15.10).

Most of the forces that cause water to be cycled do not involve organisms, but are the result of normal physical processes. Because of the motion of molecules, liquid water evaporates into the atmosphere. This can occur wherever water is present; it evaporates from lakes, rivers, soil, or the surface of organisms. Since the oceans contain most of the world's water, an extremely large amount of water enters the atmosphere from the oceans. Plants also transport water from the soil to leaves, where it evaporates in a process called **transpiration.** Once the water molecules are in the atmosphere, they are moved by prevailing wind patterns. If warm, moist air encounters cooler temperatures, which often happens over land masses, the water vapor condenses into droplets and falls as rain or snow. When the precipitation falls on land, some of it runs off the surface, some of it evaporates, and some penetrates into the soil. The water in the soil may be taken up by plants and transpired into the atmosphere, or it may become groundwater. Much of the groundwater also eventually makes its way into lakes and streams and ultimately arrives at the ocean from which it originated.

Another important element for living things is nitrogen (N). Nitrogen is essential in the formation of amino acids, which are needed to form proteins, and

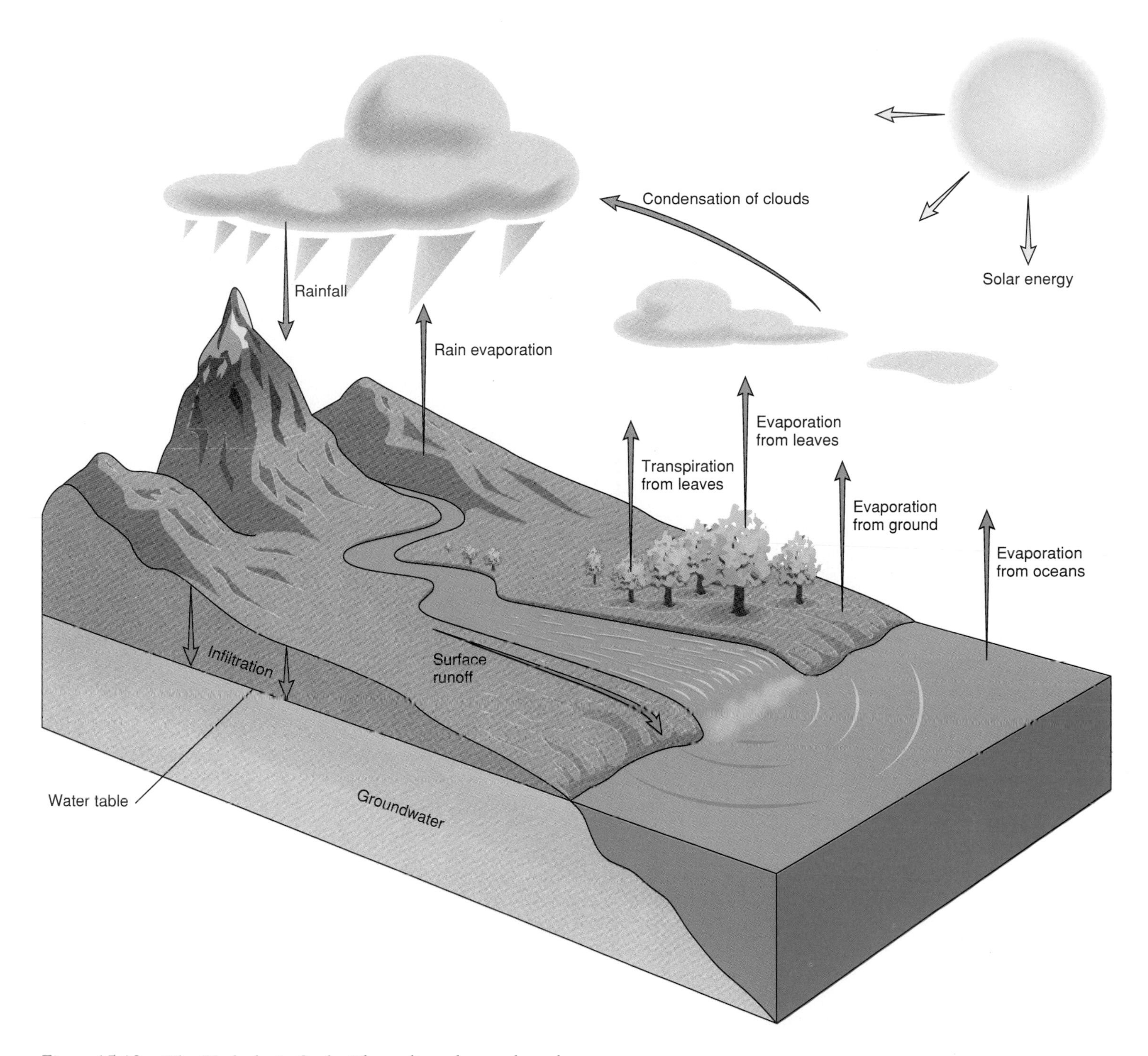

Figure 15.10 **The Hydrologic Cycle.** The cycling of water through the environment follows a simple pattern. Moisture in the atmosphere condenses into droplets that fall to the earth as rain or snow, supplying all living things with its life-sustaining properties. Water, flowing over the earth as surface water or through the soil as groundwater, returns to the oceans, where it evaporates back into the atmosphere to begin the cycle again.

in the formation of nitrogenous bases, which are a part of the nucleic acids DNA and RNA. Nitrogen (N) is found as molecules of nitrogen gas (N_2) in the atmosphere. Although nitrogen gas (N_2) makes up approximately 80% of the earth's atmosphere, only a few kinds of bacteria are able to convert it into nitrogen compounds that other organisms can use. Therefore, in most ecosystems the amount of nitrogen available limits the amount of plant biomass that can be produced. Plants are able to obtain nitrogen atoms combined with other atoms into usable forms from several different sources (figure 15.11).

Symbiotic nitrogen-fixing bacteria live in the roots of certain kinds of

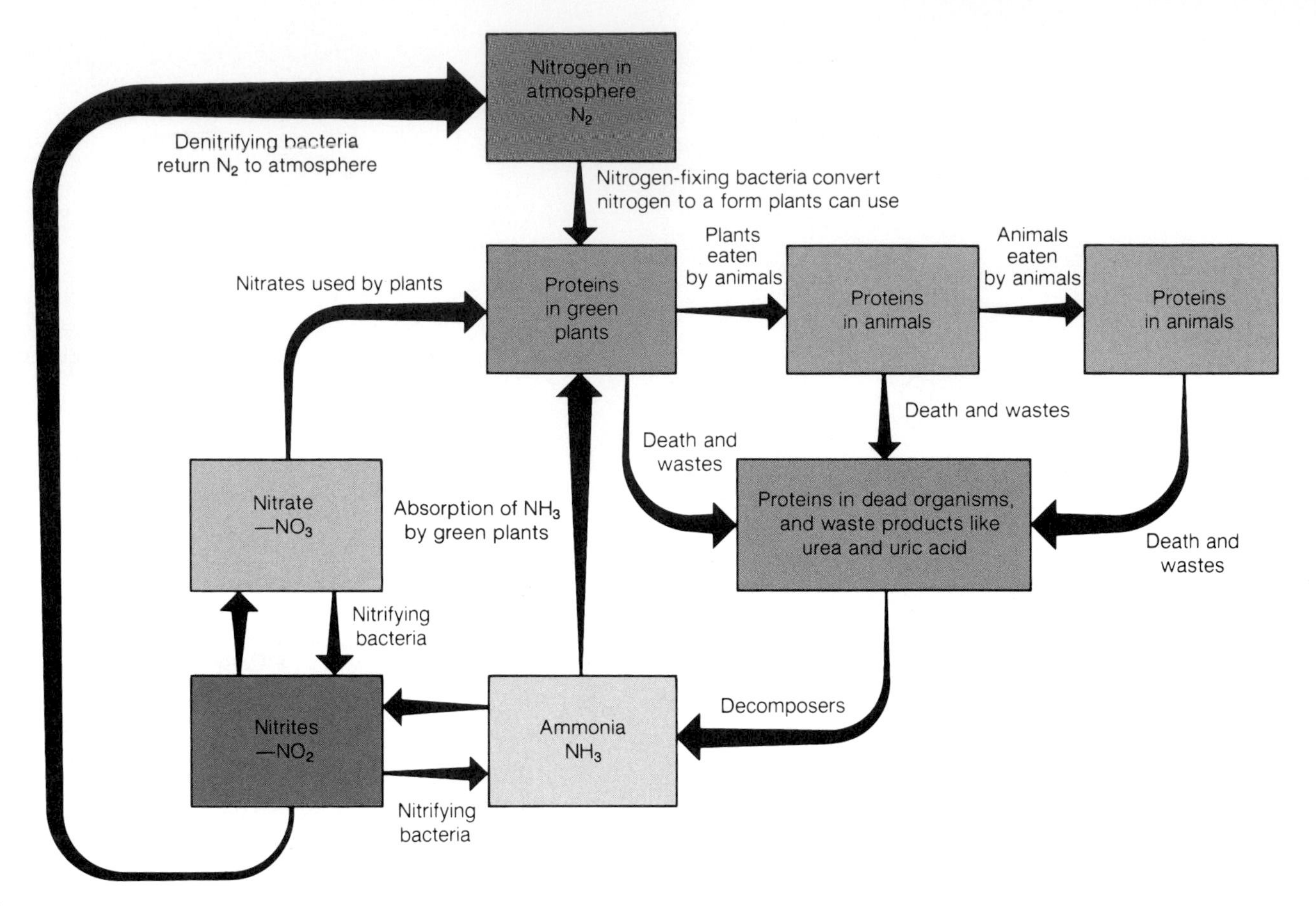

Figure 15.11 **The Nitrogen Cycle.** Nitrogen atoms are cycled through ecosystems. Atmospheric nitrogen is converted by nitrogen-fixing bacteria to nitrogen-containing compounds that plants can use to make proteins and other compounds. Proteins are passed to other organisms when one organism is eaten by another. Dead organisms and their waste products are acted upon by decay organisms to form ammonia, which may be reused by plants and converted to other nitrogen compounds by nitrifying bacteria. Denitrifying bacteria return nitrogen as a gas to the atmosphere.

plants, where they convert nitrogen gas molecules into compounds that the plants can use to make amino acids and nucleic acids. The most common plants that enter into this mutualistic relationship with bacteria are the legumes, such as beans, clover, peas, alfalfa, and locust trees. Some other organisms, such as alder trees, can also participate in this relationship. There are also **free-living nitrogen-fixing bacteria** in the soil that provide nitrogen compounds that can be taken up through the roots, but the bacteria do not live in a close physical union with plants.

Another way plants get usable nitrogen compounds involves a series of different bacteria. Decomposer bacteria convert organic nitrogen-containing compounds into ammonia (NH_3). **Nitrifying bacteria** can convert ammonia (NH_3) into nitrite- (NO_2^-) containing compounds, which in turn can be converted into nitrate- (NO_3^-) containing compounds. Many kinds of plants can use either ammonia (NH_3) or nitrate (NO_3^-) from the soil as building blocks for amino acids and nucleic acids.

All animals obtain their nitrogen from the food they eat. The ingested proteins are digested to amino acids, which can be assembled into new proteins. All dead organic matter and waste products of plants and animals are acted upon by decomposer organisms, and the nitrogen is released as ammonia (NH_3), which is acted upon by nitrifying bacteria.

Finally, other kinds of bacteria called **denitrifying bacteria** are capable of converting nitrite (NO_2^-) into nitrogen gas (N_2), which is released into the atmosphere. Thus, there is a nitrogen cycle in which nitrogen from the atmosphere is passed through a series of organisms, many of which are bacteria, and ultimately returned to the atmosphere to be cycled again. It is a much more complicated cycle than the carbon cycle and just as important.

Since nitrogen is in short supply in most ecosystems, farmers usually find it necessary to supplement the natural nitrogen sources in the soil to obtain maximum plant growth. This can be done in a number of ways. Alternating nitrogen-producing crops with nitrogen-demanding crops helps to maintain high levels of usable nitrogen in the soil. One year, a crop can be planted that has symbiotic nitrogen-fixing bacteria associated with its roots, such as beans or clover. The following year, the farmer can plant a nitrogen-demanding crop, such as corn. The use of manure is another way of improving nitrogen levels. The waste products of animals are broken down by decomposer bacteria and nitrifying bacteria, resulting in enhanced levels of ammonia and nitrate. Finally, the farmer can use industrially produced fertilizers containing ammonia or nitrate. These compounds can be used directly by plants or converted into other useful forms by nitrifying bacteria.

Fertilizers usually contain more than just nitrogen compounds. The numbers on a fertilizer bag tell you the percentages of nitrogen, phosphorus, and potassium in the fertilizer. For example, a 6–24–24 fertilizer would have 6% nitrogen compounds, 24% phosphorus compounds, and 24% potassium-containing compounds. These other elements (phosphorus and potassium) are also cycled through ecosystems. In natural, nonagricultural ecosystems these elements would be released by decomposers and enter the soil, where they would be available for plant uptake through the roots. However, when crops are removed from fields, these elements are removed with them and must be replaced by adding more fertilizer.

The Impact of Human Actions on the Community

As you can see from this discussion and from the discussion of food webs in chapter 14, all organisms are associated in a complex network of relationships. A community consists of all these sets of interrelations. Therefore, before one decides to change a community, it is wise to analyze how the organisms are interrelated.

Predator Control

During the formative years of wildlife management, it was thought that populations of game species could be increased if the populations of their predators were reduced. Consequently, many states passed laws that encouraged the killing of foxes, eagles, hawks, owls, coyotes, cougars, and other predators that use game animals as a source of food. Often bounties were paid to people who killed these predators. In South Dakota it was decided to increase the pheasant population by reducing the numbers of foxes and coyotes. However, when the supposed predator populations were significantly reduced, there was no increase in the pheasant population. There was rapid increase in the rabbit and mouse populations, however, and they became serious pests. Evidently the foxes and coyotes were major factors in keeping rabbit and mouse populations under control but had only a minor impact on pheasants.

Habitat Destruction

Some communities are quite fragile, whereas others seem to be able to resist major human interference. Communities that have a wide variety of organisms and a high level of interaction are more resistant than those with few organisms and little interaction. In general, the more complex an ecosystem is, the more likely it is to recover after being disturbed. The tundra biome is an example of a community with relatively few organisms and interactions. It is not very resistant to change, and because of its very slow rate of repair, damage caused by human activity may persist for hundreds of years.

Some species are much more resistant to human activity than others. Rabbits, starlings, skunks, and many kinds of insects and plants are able to maintain high populations despite human activity. Indeed, some may even be encouraged by human activity. By contrast, whales, condors, eagles, and many plant and insect species are not able to resist human interference very well. For most of these endangered species, direct action of humans is not the problem; very few organisms have been driven to extinction by hunting or direct exploitation. Usually, the human influence is indirect. Habitat destruction is the main cause of extinction and the endangering of species. As humans convert land to farming, grazing, commercial forestry, and special wildlife management areas, the natural ecosystems are disrupted, and those plants and animals with narrow niches tend to be eliminated because they lose critical resources in their environment. Table 15.1 lists several endangered species and the probable causes of their difficulties.

The DDT Story

Humans have developed a variety of chemicals to control specific pest organisms. One of these was the insecticide DDT. DDT is an abbreviation for the chemical name dichloro-diphenyl-trichloro-ethane. DDT is one of a group of organic compounds called *chlorinated hydrocarbons*. Because DDT is a poison that was used to kill a variety of insects, it was called an **insecticide.** Another term that is sometimes used is **pesticide,** which implies that the poison is effective against pests. Although it is no longer used in the United States (its use was banned in the early 1970s), DDT is still manufactured and used in many parts of the world, including Mexico.

DDT was a very valuable insecticide for the U.S. Armed Forces during World War II. It was sprayed on clothing and dusted on the bodies of soldiers, refugees, and prisoners to kill body lice and other insects. Lice, besides being a nuisance, carry the bacteria that can cause a disease known as *typhus fever.* When bitten by a louse, a person can develop typhus fever. Because body lice could be transferred from one person to another by contact or by wearing infested clothing, DDT was important in maintaining the health of millions of people. Since DDT was so useful in controlling these insects, people could see

Table 15.1
Endangered and Threatened Species

Species	Reason for Endangerment
Hawaiian crow *Corvis hawaiinsis*	Predation by cat and mongoose; disease; habitat destruction
Sonora Chub *Gila ditaenia*	Competition with introduced species
Black-footed ferret *Mustela nigripes*	Poisoning of prairie dogs (their primary food)
Snail kite *Rostrhamus sociabilis*	Specialized eating habits (only eat apple snails); draining of marshes
Grizzly bear *Ursus arctos*	Loss of wilderness areas
California condor *Gymnogyps californianus*	Slow breeding; lead poisoning
Ringed sawback turtle *Graptemys oculifera*	Modification of habitat by construction of reservoir that reduced their primary food source
Scrub mint *Dicerandra frutescens*	Conversion of habitat to citrus groves and housing

the end of pesky mosquitoes and flies, as well as the elimination of many disease-carrying insects.

Although DDT was originally very effective, many species of insects developed a resistance to it. The genetic variety present in all species relates to how sensitive an organism is to many environmental factors, including manufactured ones such as DDT. When DDT or any pesticide is applied to a population of insects, susceptible individuals die, and those with some degree of resistance have a greater chance of living. Now the reproducing population consists of many individuals that have DDT-resistant genes, which are passed on to the offspring. When this happens repeatedly over a long period of time, a DDT-resistant population develops, and the insecticide is no longer useful. DDT acts as a selecting agent, killing the normal insects but allowing the resistant individuals to live. This happened in the orange groves of California, where many populations of pests became DDT-resistant. Similarly, throughout the world, DDT was used (and in many areas is still used) to control malaria-carrying mosquitoes. Many of these populations have become resistant to DDT and other kinds of insecticides. The people who had foreseen the elimination of insect pests had not reckoned with the genetic diversity of the gene pools of these insects.

Another problem that was not anticipated was the effect of pesticide use on the food chain. DDT was a very effective insecticide because it was extremely toxic to insects but not very toxic to birds and mammals. It was also a very stable compound, which meant that once it was applied it would remain present and effective for a long period of time. It sounds like an ideal insecticide. What went wrong? Why was its use banned?

When DDT was applied to an area to get rid of insect pests, it was usually dissolved in an oil or fatty compound. It was then sprayed over an area and fell on the insects and on the plants that the insects used for food. Eventually the DDT entered the insect either directly through the body wall or through the food it was eating. When ingested with food, DDT interferes with the normal metabolism of the insect. If small quantities were taken in, the insect could digest and break down the DDT just like any other large organic molecule. Since DDT is soluble in fat or oil, the DDT or its breakdown products were stored in the fat deposits of the insect. Some insects could break down and store all of the DDT they encountered and, therefore, they survived. If an area had been lightly sprayed with DDT, some insects died, some were able to tolerate the DDT, and others broke down and stored nonlethal quantities of DDT. As much as one part DDT per one billion parts of insect tissue could be stored in this manner. This is not much DDT! It is equivalent to one drop of DDT in one hundred railroad tank cars. However, when an aquatic area is sprayed with a small concentration of DDT, many kinds of organisms in the area can accumulate such tiny quantities in their bodies. Even algae and protozoa found in aquatic ecosystems accumulate pesticides. They may accumulate concentrations in their cells that are 250 times more concentrated than the amount sprayed on the ecosystem. The algae and protozoa are eaten by insects, which in turn are eaten by frogs, fish, or other carnivores.

The concentration in frogs may be two thousand times what was sprayed. The birds that feed on the frogs and fish may accumulate concentrations that are as much as eighty thousand times the original amount. What was originally a dilute concentration became more concentrated as it accumulated in the food chain because DDT is relatively stable and is stored in the fat deposits of the organisms that take it in.

Many animals at higher trophic levels died as a result of this accumulation of pesticide in food chains. This process is called **biological amplification** (figure 15.12). Even if they were not killed directly by DDT, many birds such as eagles, pelicans, and the osprey at higher trophic levels suffered reduced populations because the DDT interfered with the female bird's ability to produce eggshells. Thin eggshells are easily broken, and there were no live young hatched.

What was originally used as an insecticide to control insect pests has been shown to have many harmful community consequences. Instead of controlling the pest, it selects for resistance and creates populations that can tolerate the

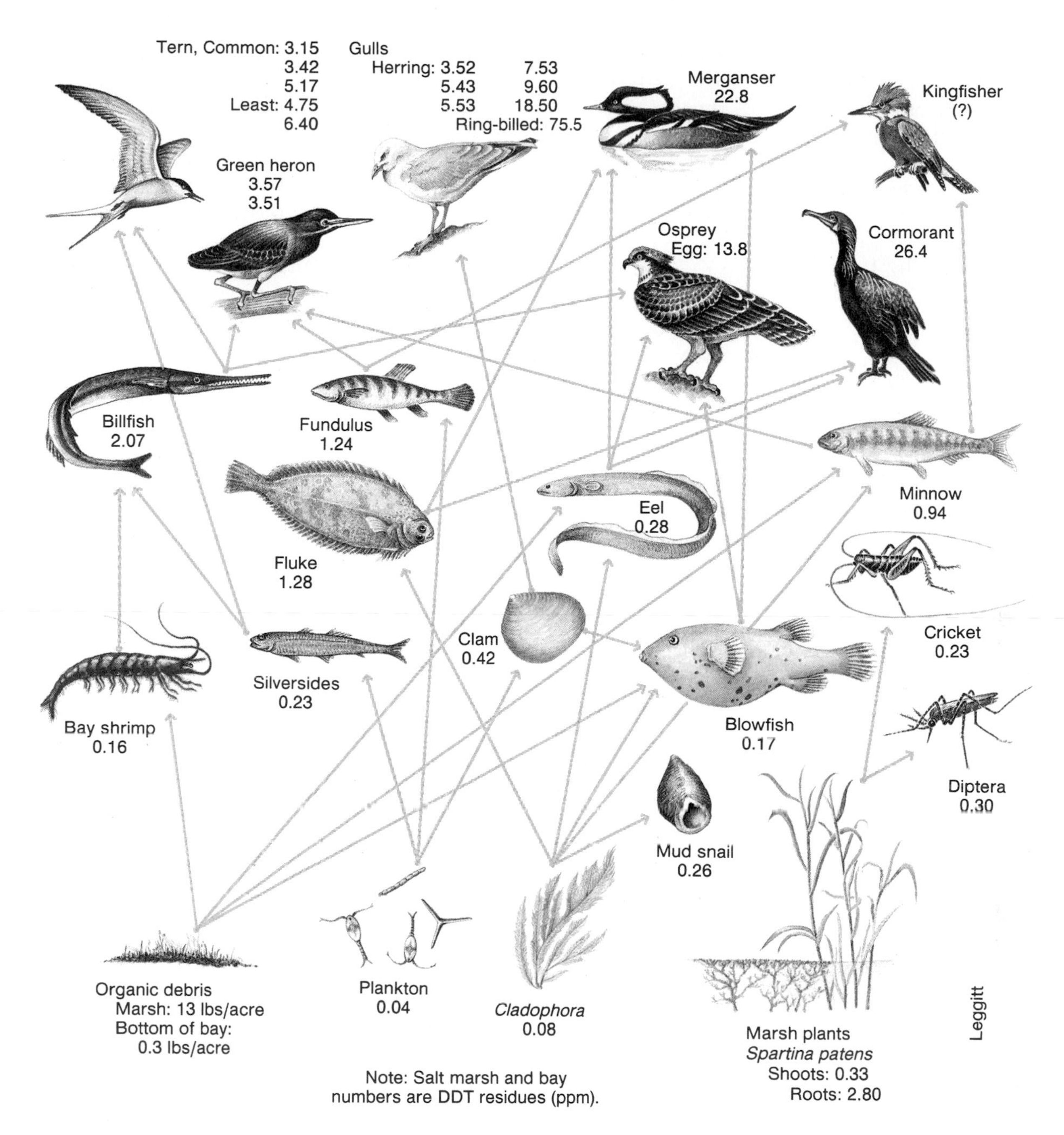

Figure 15.12 **The Biological Amplification of DDT.** All of the numbers shown are in parts per million (ppm). A concentration of one part per million means that in a million equal parts of the organism, one of the parts would be DDT. Notice how the amount of DDT in the bodies of the organisms increases as we go from producers to herbivores to carnivores. Since DDT is persistent, it builds up in the top trophic levels of the food chain.

poison. Instead of harming only the pest species, it accumulates in food chains to kill nontarget species as well. Furthermore, it is not specific to pest species only, but kills many beneficial species of insects. DDT has not been the gift to humankind that we originally thought it would be, and its sale and use have been banned in the United States and many other countries.

Another widely used group of synthetic compounds of environmental concern are polychlorinated biphenyls (PCBs). PCBs are highly stable compounds that resist changes from heat, acids, bases, and oxidation. These characteristics make PCBs desirable for industrial use but also make them persistent pollutants when released into the environment. About half of the PCBs are used in transformers and electrical capacitors. Other uses include inks, plastics, tapes, paints, glues, waxes, and polishes. PCBs are harmful to fish and other aquatic forms of life because they interfere with reproduction. In humans, PCBs produce liver ailments and skin lesions. In high concentration, they can damage the nervous system and they are suspected carcinogens. In 1970, PCB production was limited to those cases where satisfactory substitutes were not available.

Other Problems with Pesticides

A number of factors determine how successful you will be in controlling a pest with a pesticide. You must choose a pesticide that will cause the least amount of damage to the harmless or beneficial organisms in the community. The ideal pesticide or insecticide would only affect the target pest. Because many of the insects we consider pests are herbivores, you would expect that carnivores in the community would use the pest species as prey, and parasites would use the pest as a host. These predators and parasites would have important roles in controlling the numbers of a pest species. Generally, predators and parasites reproduce more slowly than their prey or host species. Because of this, the use of a nonspecific pesticide may actually make matters worse; if such a pesticide is applied to an area, the pest is killed but so are its predators and parasites. Since the herbivore pest reproduces faster than its predators and parasites, the pest population rebounds quickly, unchecked by natural predation and parasitism. This may necessitate more frequent and more concentrated applications of pesticides. This has actually happened in many cases of pesticide use; the pesticides made the problem worse, and it became increasingly costly to apply them.

We have just looked at how human actions can cause changes in ecosystems. However, not all changes are abnormal. Communities of organisms experience normal fluctuations in numbers and sometimes slow progressive changes that convert one kind of community into another.

Succession

Many communities like the biomes we discussed in chapter 14 are relatively stable over long periods of time. A relatively stable, long-lasting community is called a **climax community.** The word *climax* implies the final step in a series of events. That is just what the word means in this context because communities can go through a series of predictable, temporary stages that eventually result in a long-lasting stable community. The process of changing from one type of community to another is called **succession,** and each intermediate stage leading to the climax community is known as a **successional stage, successional community,** or **sere.**

Two different kinds of succession are recognized: **primary succession,** in which a community of plants and animals develops where none existed previously, and **secondary succession,** in which a community of organisms is disturbed by a natural or human-related event (e.g., hurricane, fire, forest harvest) and returned to a previous stage in succession. Primary succession is much more difficult to observe than secondary succession because there are relatively few places on earth that lack communities of organisms. The tops of mountains, newly formed volcanic rock, and rock newly exposed by erosion or glaciers can be said to lack life. However, bacteria, algae, fungi, and lichens quickly begin to grow on the bare rock surface, and the process of succession has begun. The first organisms to colonize an area are often referred to as **pioneer organisms,** and the community is called a **pioneer community.** Lichens are frequently important in pioneer communities. They are unusual organisms that consist of a combination of algae cells and fungi cells—a combination that is very hardy and is able to grow on the surface of bare rock (figure 15.13). Since algae cells are present, the lichen is capable of photosynthesis and can form new organic matter. Furthermore, many tiny consumer organisms can make use of the lichens as a source of food and a sheltered place to live. The action of the lichens also tends to break down the rock surface upon which they grow. This fragmentation of rock by lichens is aided by the physical weathering processes of freezing and thawing, dissolution by water, and wind erosion. It is the first step in the development of soil. Lichens trap dust particles, small rock particles, and the dead remains of lichens and other organisms that live in and on them, resulting in a thin layer of soil.

As the soil layer becomes thicker, small plants such as mosses may become established, increasing the rate at which energy is trapped and adding more organic matter to the soil. Eventually, the soil may be able to support larger plants that are even more efficient at trapping sunlight, and the soil-building process continues at a more rapid pace. Associated with each of the producers in each successional stage is a variety of small animals, fungi, and bacteria. Each change in the community makes it more difficult for the previous group of organisms to maintain themselves. Tall plants shade smaller producers; consequently, the smaller organisms become less common, and some may disappear entirely. Only shade-tolerant species are able to compete successfully. One stage has succeeded the other.

Depending on the physical environment and the availability of new colonizing species, succession from this point can lead to different kinds of climax communities. If the area is dry, it might stop at a grassland stage. If it is cold and wet, a coniferous forest might be the climax community. If it is warm and wet, it may be a tropical rain forest. The rate at which this successional process takes place is variable. In some warm, moist, fertile areas the entire process might take place in less than one hundred years. In

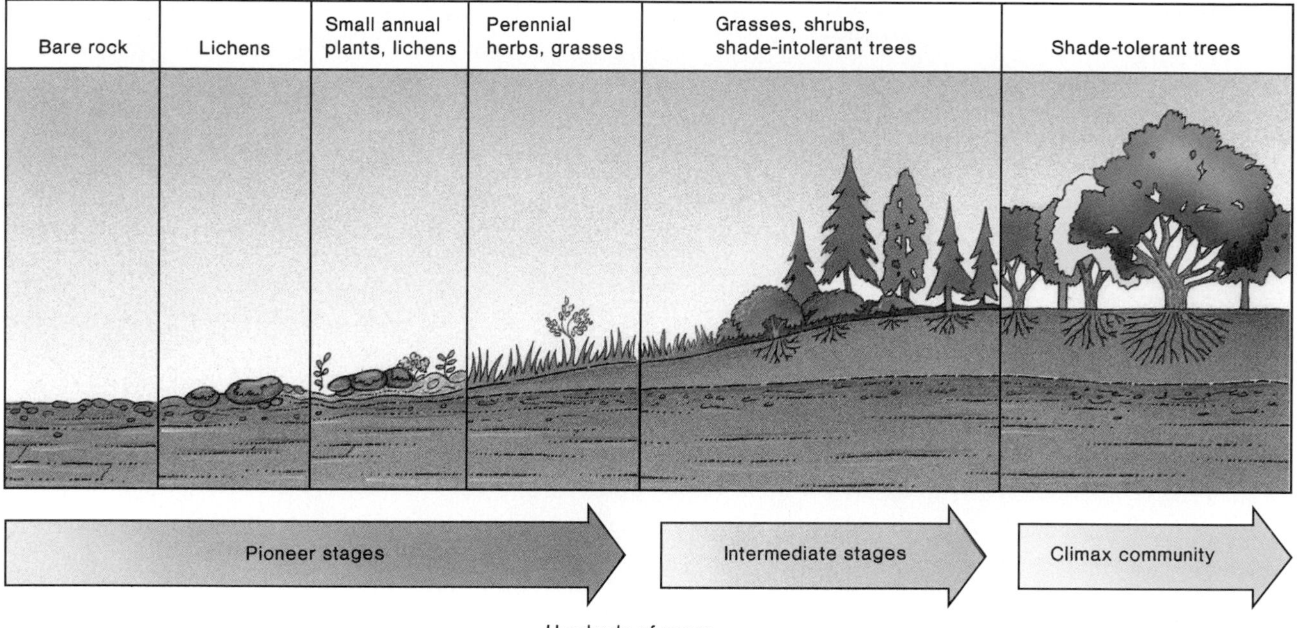

Figure 15.13 **Primary Succession.** The formation of soil is a major step in primary succession. Until soil is formed, the area is unable to support large amounts of vegetation. The vegetation modifies the harsh environment and increases the amount of organic matter that can build up in the area. The presence of plants eliminates the earlier pioneer stages of succession. If given enough time, a climax community may develop.

harsh environments, like mountain tops or very dry areas, it may take thousands of years.

Another situation that is often called primary succession is the progression from an aquatic community to a terrestrial community. Lakes, ponds, and slow-moving parts of rivers accumulate organic matter. Where the water is shallow, this organic matter supports the development of rooted plants. In deeper water, we find only floating plants like water lilies that send their roots down to the mucky bottom. In shallower water, upright rooted plants like cattails and rushes develop. The cattail community contributes more organic matter, and the water level becomes more shallow. Eventually, a mat of mosses, grasses, and even small trees may develop on the surface along the edge of the water. If this continues for perhaps one hundred to two hundred years, an entire pond or lake will become filled in. More organic matter accumulates because of the large number of producers and because the depression that was originally filled with water becomes drier. This will usually result in a wet grassland, which in many areas will be replaced by the climax forest community typical of the area (figure 15.14).

Secondary succession occurs when a climax community or one of the successional stages leading to it is changed to an earlier stage. For example, this is what happens when agricultural land is abandoned. One obvious difference between primary succession and secondary succession is that in the latter there is no need to develop a soil layer. If we begin with bare soil the first year, it is likely to be invaded by a pioneer community of weed species that are annual plants. Within a year or two, perennial plants like grasses become established. Since most of the weed species need bare soil for seed germination, they are replaced by the perennial grasses and other plants that live in association with grasses. The more permanent grassland community is able to support more insects, small mammals, and birds than the weed community could. If rainfall is adequate, several species of shrubs and fast-growing trees that require lots of sunlight (e.g., birch, aspen, juniper, hawthorn, sumac, pine, spruce, and dogwood) will become common. As the trees become larger, the grasses fail to get sufficient sunlight and die out. Eventually, shade-tolerant species of trees (e.g., beech, maple, hickory, oak, hemlock, and cedar) will replace the shade-intolerant species, and a climax community results (figure 15.15).

Most human use of ecosystems involves replacing the natural climax community with an artificial early successional stage. Agriculture involves replacing natural forest or prairie communities with specialized grasses such as

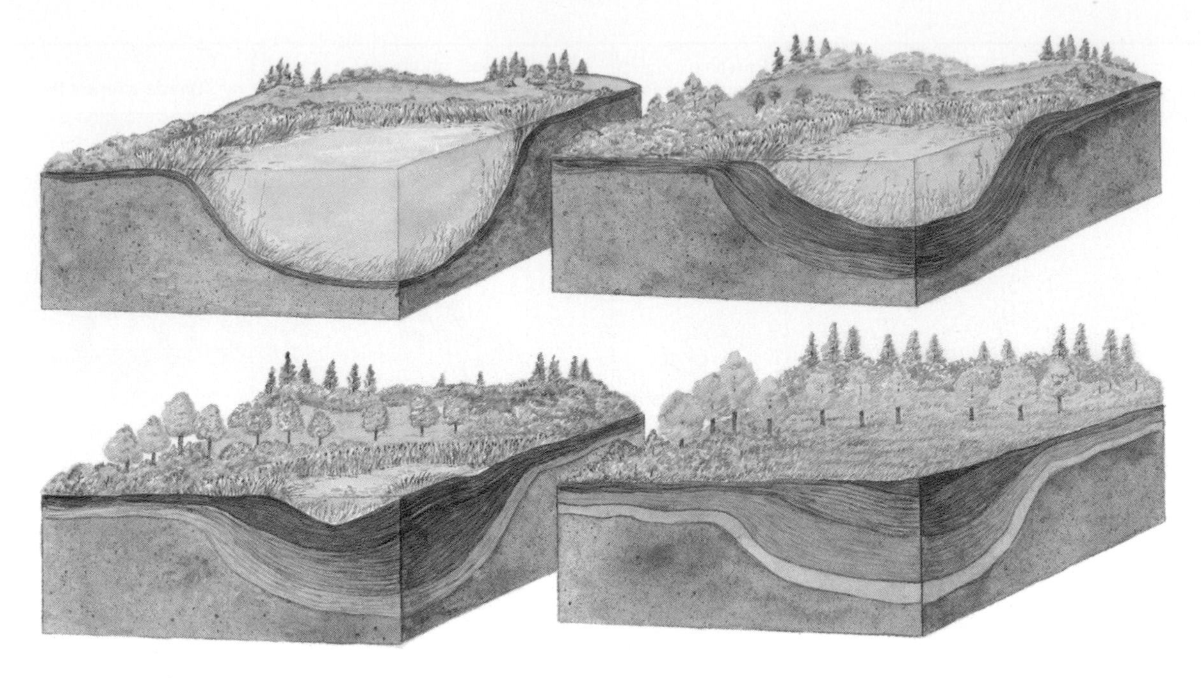

Figure 15.14 **Succession from a Pond to a Wet Meadow.** A shallow pond will slowly fill with organic matter from producers in the pond. Eventually, a floating mat will form over the pond and grasses will become established. In many areas this will be succeeded by a climax forest.

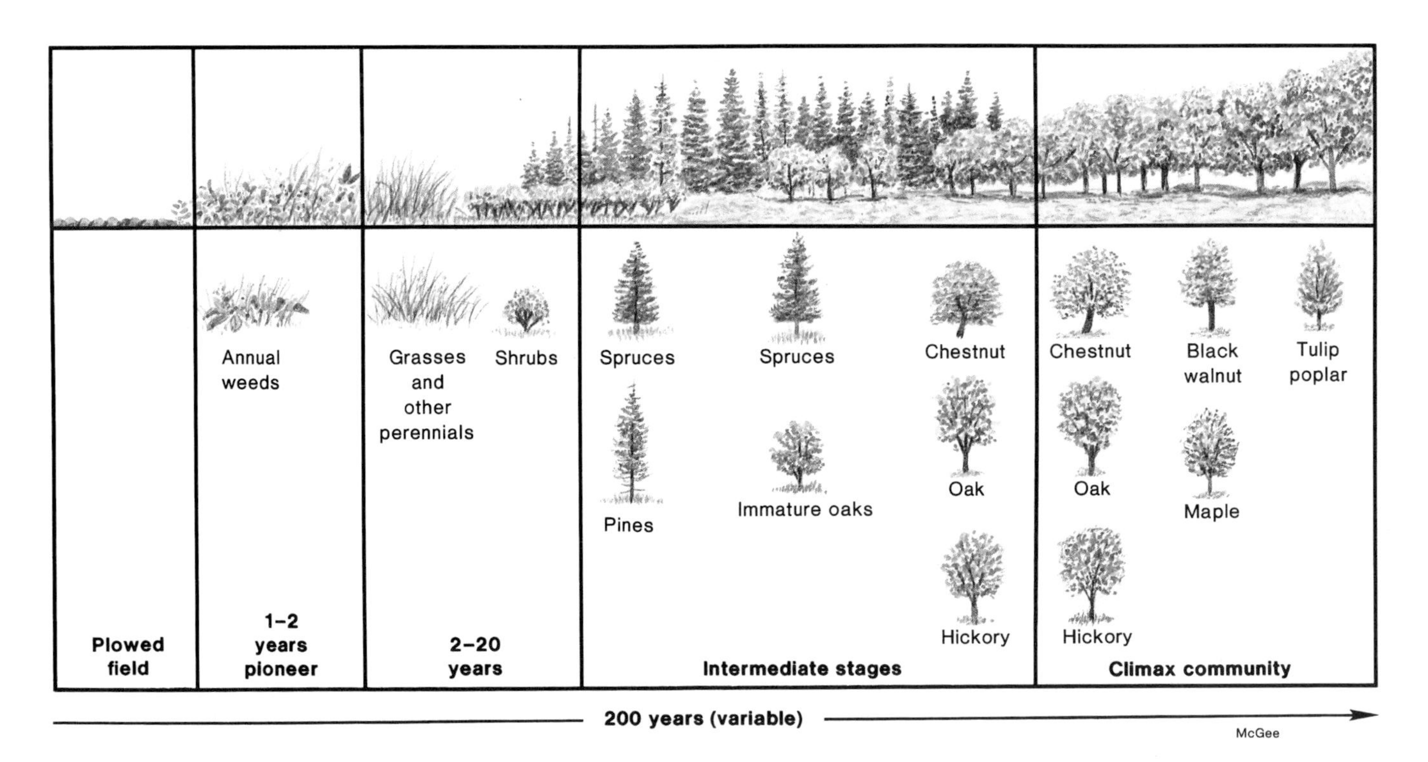

Figure 15.15 **Secondary Succession of Land.** A plowed field in the southeastern United States shows a parade of changes over time involving plant and animal associations. The general pattern is for annual weeds to be replaced by grasses and other perennial herbs, which are replaced by shrubs, which are replaced by trees. As the plant species change, so do the animal species.

wheat, corn, rice, and sorghum. This requires considerable effort on our part because the natural process of succession tends toward the original climax community. This is certainly true if refuges of the original natural community are still locally available to colonize agricultural land. Small woodlots in agricultural areas of the eastern United States serve this purpose. Much of the work and expense of farming is necessary to prevent succession to the natural climax community. It takes a lot of energy to fight nature.

Forestry practices often seek to simplify the forest by planting single-species forests of the same age. This certainly makes management and harvest practices easier and more efficient, but these kinds of communities do not contain the variety of plants, animals, fungi, and other organisms typically found in natural ecosystems.

Human-constructed lakes or farm ponds often have weed problems because they are shallow and provide ideal conditions for the normal successional processes that lead to their being filled in. Often we do not recognize what a powerful force succession is.

• Summary •

Each organism in a community occupies a specific space known as its habitat and has a specific functional role to play known as its niche. An organism's habitat is usually described in terms of some conspicuous element of its surroundings. The niche is very difficult to describe because it involves so many different interactions with the physical environment and other living things.

Interactions between organisms fit into several categories. Predation involves one organism benefiting (predator) at the expense of the organism killed and eaten (prey). Parasitism involves one organism benefiting (parasite) by living in or on another organism (host) and deriving nourishment from it. Organisms that carry parasites from one host to another are called vectors. Amensal relationships exist when one organism is harmed but the other is not affected. Commensal relationships exist when one organism is helped but the other is not affected. Mutualistic relationships benefit both organisms. Symbiosis is any interaction in which two organisms live together in a close physical relationship. Competition causes harm to both of the organisms involved, although one may be harmed more than the other and may become extinct, evolve into a different niche, or be forced to migrate.

Many atoms are cycled through ecosystems. The carbon atoms of living things are trapped by photosynthesis, passed from organism to organism as food, and released to the atmosphere by respiration. Water is necessary as a raw material for photosynthesis and as the medium in which all metabolic reactions take place. Water is cycled by the physical processes of evaporation and condensation. Nitrogen originates in the atmosphere, is trapped by nitrogen-fixing bacteria, passes through a series of organisms, and is ultimately released to the atmosphere by denitrifying bacteria.

Organisms within a community are interrelated with one another in very sensitive ways; thus, changing one part of a community can lead to unexpected consequences. Predator-control practices, habitat destruction, and pesticide use have all caused changes in parts of communities not directly associated with the part changed.

Succession occurs when a series of communities replace one another as each community changes the environment to make conditions favorable for a subsequent community and unfavorable for itself. Most successional processes result in a relatively stable stage called the climax community. The stages leading to the climax community are called successional stages. If the process begins with bare rock or water, it is called primary succession. If it begins as a disturbed portion of a community, it is called secondary succession.

• Thinking Critically •

This is a thought puzzle—put it together! Here are the pieces:

People are starving.
Commercial fertilizer production requires temperatures of 900° C.
Geneticists have developed plants that grow very rapidly and require high amounts of nitrogen to germinate during the normal growing season. Fossil fuels are stored organic matter.
The rate of the nitrogen cycle depends on the activity of bacteria.
The sun is expected to last for several million years.
Crop rotation is becoming a thing of the past.
The clearing of forests for agriculture changes weather in the area.

• Experience This •

Take a walk in a park, zoo, or natural area. Look for examples of predation, parasitism, mutualism, and competition. Write a short description of the niche of one of the animals you observe.

• Questions •

1. Describe your niche.
2. What is the difference between a habitat and a niche?
3. What do parasites, commensal organisms, and mutualistic organisms have in common? How are they different?
4. Describe two situations in which competition may involve combat and two that do not involve combat.
5. Trace the flow of carbon atoms through a community that contains plants, herbivores, decomposers, and parasites.
6. Describe four different roles played by bacteria in the nitrogen cycle.
7. Describe the flow of water through the hydrologic cycle.
8. How does primary succession differ from secondary succession?
9. Describe the impact of DDT on communities.
10. How does a climax community differ from a successional community?

• Chapter Glossary •

amensalism (a-men'sal-izm) A relationship between two organisms in which one organism is harmed and the other is not affected.

biological amplification (bi-o-loj'ĭ-cal am''plĭ-fĭ-ka'shun) The accumulation of a compound in increasing concentrations in organisms at successively higher trophic levels.

climax community (klī'maks ko-miu'nĭ-te) A relatively stable, long-lasting community.

commensalism (ko-men'sal-izm) A relationship between two organisms in which one organism is helped and the other is not affected.

competition (com-pe-tĭ'shun) A relationship between two organisms in which both organisms are harmed.

competitive exclusion principle (com-pĕ'tĭ-tiv eks-klu'zhun prin'sĭ-pul) No two species can occupy the same niche at the same time.

denitrifying bacteria (de-ni'trĭ-fi-ing bak-te're-ah) Several kinds of bacteria capable of converting nitrite to nitrogen gas.

epiphyte (ep'e-fīt) A plant that lives on the surface of another plant.

external parasite (eks-tur'nal pĕr'uh-sīt) A parasite that lives on the outside of its host.

free-living nitrogen-fixing bacteria (ni'tro-jen fik'sing bak-te're-ah) Soil bacteria that convert nitrogen gas molecules into nitrogen compounds that plants can use.

habitat (hab'ĭ-tat) The place or part of an ecosystem occupied by an organism.

host (hōst) An organism that a parasite lives in or on.

insecticide (in-sek'tĭ-sīd) A poison used to kill insects.

internal parasite (in-tur'nal pĕr'uh-sīt) A parasite that lives inside its host.

mutualism (miu'chu-al-izm) A relationship between two organisms in which both organisms benefit.

niche (nitch) The functional role of an organism.

nitrifying bacteria (ni'trĭ-fi-ing bak-te're-ah) Several kinds of bacteria capable of converting ammonia to nitrite or nitrite to nitrate.

parasite (pĕr'uh-sīt) An organism that lives in or on another organism and derives nourishment from it.

parasitism (pĕr'uh-sit-izm) A relationship between two organisms that involves one organism living in or on another organism and deriving nourishment from it.

pesticide (pes'tĭ-sīd) A poison used to kill pests. This term is often used interchangeably with insecticide.

pioneer community (pi''o-nēr' ko-miu'nĭ-te) The first community of organisms in the successional process established in a previously uninhabited area.

pioneer organisms (pi''o-nēr' or'gun-izms) The first organisms in the successional process.

predation (prē-da'shun) A relationship between two organisms that involves the capturing, killing, and eating of one by the other.

predator (pred'uh-tōr) An organism that captures, kills, and eats another animal.

prey (prā) An organism captured, killed, and eaten by a predator.

primary succession (pri'mar-e suk-sē'shun) The orderly series of changes that begins in a previously uninhabited area and leads to a climax community.

secondary succession (sĕk'on-dĕr-e suk-sē'shun) The orderly series of changes that begins with the disturbance of an existing community and leads to a climax community.

sere (sēr) See **successional community (stage).**

succession (suk-sē'shun) The process of changing one type of community to another.

successional community (stage) (suk-sē'shun-al ko-miu'nĭ-te) An intermediate stage in succession; also called **sere.**

symbiosis (sim-be-o'sis) A close physical relationship between two kinds of organisms. It usually includes parasitism, commensalism, and mutualism.

symbiotic nitrogen-fixing bacteria (sim-be-ah'tik ni-tro-jen fik'sing bak-te're-ah) Bacteria that live in the roots of certain kinds of plants, where they convert nitrogen gas molecules into compounds that plants can use.

transpiration (trans''pĭ-ra'shun) The process of water evaporation from the leaves of a plant.

vector (vek'tōr) An organism that carries a disease or parasite from one host to the next.

C H A P T E R 16

Population Ecology

Chapter Outline

Purpose

Populations of organisms exhibit many kinds of characteristics that can vary significantly from one population to another. Recognizing that populations differ from one another is necessary to understanding differences in population growth rates. While not specifically about growth of the human population, much of the material in this chapter relates to the problems associated with the human population explosion.

For Your Information

China is the most populous country in the world, with over one billion people—approximately one-fourth of the world's population. The government has tried to limit the size of families by encouraging later marriages, providing conception-control information, making small family size a patriotic duty, and giving financial incentives to couples who have no children or limit themselves to one child. Most families desire a male child. Consequently, many baby girls are left to die so the parents can have another opportunity to have a male child. The government must now deal with the problems of infanticide and an aging work force.

Learning Objectives

- Recognize that populations vary in gene frequency, age distribution, sex ratio, size, and density.
- Describe the characteristics of a typical population growth curve.
- Understand why populations grow.
- Recognize the pressures that ultimately limit population size.
- Understand that human populations obey the same rules of growth as populations of other kinds of organisms.

Population Characteristics

A **population** is a group of organisms of the same species located in the same place at the same time. Examples are the number of dandelions in your front yard, the rat population in the sewers of your city, or the number of people in your biology class. On a larger scale, all the people of the world constitute the human population. The terms *species* and *population* are interrelated because a species is a population—the largest possible population of a particular kind of organism. Population, however, is often used to refer to portions of a species by specifying a space and time. For example, the size of the human population in a city changes from hour to hour during the day and varies according to where you set the boundaries of the city.

Since most populations are small portions of a species, we should expect different populations of the same species to show differences. One of the ways in which they can differ is in gene frequency. Chapter 11 on population genetics introduced you to the concept of **gene frequency,** which is a measure of how often a specific gene shows up in the gametes of a population. Two populations of the same species often have quite different gene frequencies. For example, many populations of mosquitoes have high frequencies of insecticide-resistant genes, whereas others do not. The frequency of the genes for tallness in humans is greater in certain African tribes than in any other human population. Figure 16.1 shows that the frequency of the allele for type B blood differs significantly from one human population to another.

Since members of a population are of the same species, sexual reproduction can occur, and genes can flow from one generation to the next. Genes can also flow from one place to another as organisms migrate or are carried from one geographic location to another. **Gene flow** is used to refer to both the movement of genes within a population due to migration and the movement from one generation to the next as a result of gene replication and sexual reproduction.

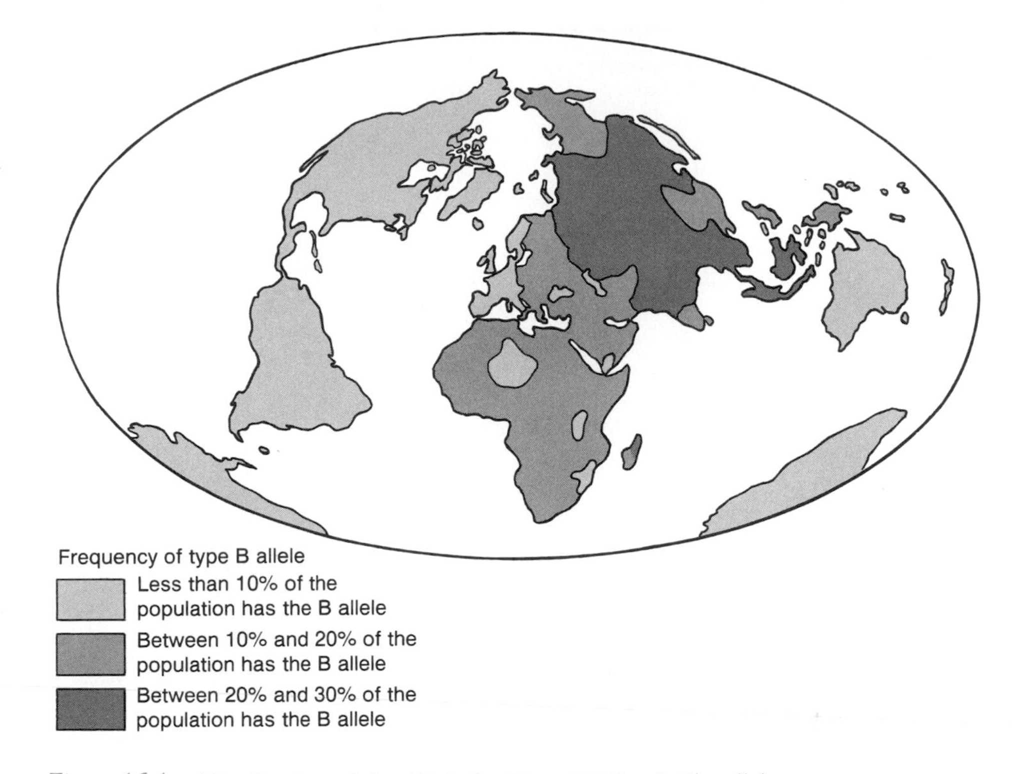

Figure 16.1 **Distribution of the Allele for Type B Blood.** The allele for type B blood is not evenly distributed in the world. This map shows that the type B allele is most common in parts of Asia and has been dispersed to the Middle East and parts of Europe and Africa. There has been very little flow of the allele to the Americas.

Another feature of a population is its **age distribution,** which is the number of organisms of each age in the population. Organisms are often grouped into the following categories: (1) prereprodutive juveniles—insect larvae, plant seedlings, or babies; (2) reproductive adults—mature insects, plants producing seeds, or humans in early adulthood; or (3) postreproductive adults no longer capable of reproduction—annual plants that have shed their seeds, salmon that have spawned, and many elderly humans. A population is not necessarily divided into equal thirds (figure 16.2). In some situations, a population may be made up of a majority of one age group. If the majority of the population is prereproductive, then a "baby boom" should be anticipated in the future. If a majority of the population is reproductive, the population should be growing rapidly. If the majority of the population is postreproductive, a population decline should be anticipated.

Populations can also differ in their sex ratios. The **sex ratio** is the number of males in a population compared to the number of females. In bird and mammal species where strong pair-bonding occurs, the sex ratio may be nearly one to one (1:1). Among mammals and birds that do not have strong pair-bonding, sex ratios may show a larger number of females than males. This is particularly true among game species, where more males are shot than females. Since one male can fertilize several females, the population can remain large. However, if the population of these managed game species becomes large enough to cause a problem, it becomes necessary to harvest some of the females as well, since their number determines how much reproduction can take place. In addition to these examples, many species of animals like bison, horses, and elk have mating systems in which one male maintains a harem of females. The sex ratio in these small groups is quite different from a 1:1 ratio (figure 16.3). There are very few

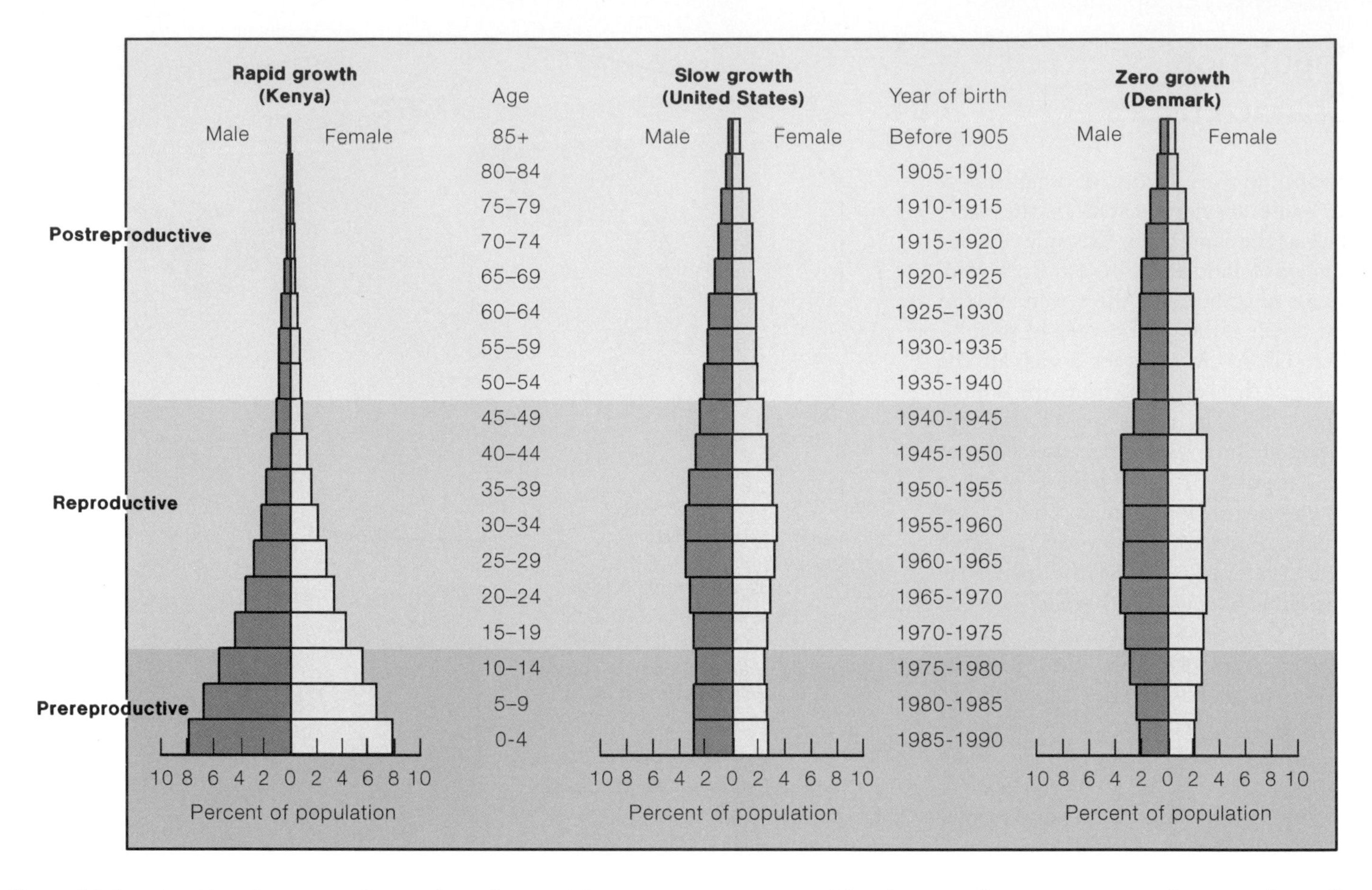

Figure 16.2 **Age Distribution in Human Populations.** The relative number of individuals in each of the three categories (prereproductive, reproductive, and postreproductive) can give a good clue to the future of the population. Kenya has a large number of young individuals who will become reproducing adults. Therefore, this population will grow rapidly and will double in about nineteen years. The United States has a declining proportion of prereproductive individuals, but a relatively large reproductive population. Therefore, it will continue to grow for a time, but will probably stabilize in the future. Denmark's population has a large proportion of postreproductive individuals and a small proportion of prereproductive individuals. Its population is stable.

Source: U.S. Bureau of the Census and the United Nations, as reported in Joseph A. McFalls, Jr., "Population: A Lively Introduction," *Population Bulletin,* Vol. 46, No. 2 (Washington, D.C.: Population Reference Bureau, Inc., October 1991).

Figure 16.3 **Sex Ratio in Elk.** Some male animals defend a harem of females; therefore, the sex ratio in these groups is several females per male.

situations in which the number of males exceeds the number of females. In some human and other populations, there may be sex ratios in which the males dominate if female mortality is unusually high or if some special mechanism separates most of one sex from the other.

Regardless of the specific sex ratio in a population, most species can generate large numbers of offspring, producing a concentration of organisms in an area. **Population density** is the number of organisms of a species per unit area. Some populations are extremely concentrated into a limited space, while others are well dispersed. As the population density increases, competition among members of the population for the necessities of life increases. This increases the likelihood that some individuals will explore new habitats and migrate to new areas. Increases in the intensity of competition that cause changes in the environment and lead to dispersal are often referred to as **population pressure.** The dispersal of individuals to new areas can relieve the pressure on the home area and lead to the establishment of new populations. Among animals, it is often the juveniles who participate in this dispersal process. If dispersal cannot relieve population pressure, there is usually an increase in the rate at which individuals die due to predation, parasitism, starvation, and accidents. In plant populations, dispersal is not very useful for relieving population density; instead, the death of weaker individuals usually results in reduced population density.

Reproductive Capacity

Sex ratios and age distributions within a population have a direct bearing on the rate of reproduction. Each species has an inherent **reproductive capacity** or **biotic potential,** which is the theoretical maximum rate of reproduction. Generally this biotic potential is many times larger than the number of offspring needed simply to maintain the population. For example, a female carp may produce one to three million eggs in her lifetime. This is her reproductive capacity. However, only two or three of these offspring would ever develop into sexually mature adults. Therefore, her reproductive rate is much smaller than her reproductive potential.

A high reproductive capacity is valuable to a species because it provides many opportunities for survival. It also provides many slightly different individuals for the environment to select among. With most plants and animals, many of the potential gametes are never fertilized. An oyster may produce a million eggs a year, but not all of them are fertilized, and most that are fertilized die. An apple tree may have thousands of flowers but only produce a few apples because the pollen that contains the sperm cells may not be transferred to the female part of the flower in the process of pollination. Even after the new individuals are formed, mortality is usually high among the young. Most seeds that fall to the earth do not grow, and most young animals die as well. But usually enough survive to ensure continuance of the species. Organisms that reproduce in this way spend large amounts of energy on the production of potential young, with the probability that a small number of them will reach reproductive age.

A second way of approaching reproduction is to produce relatively fewer individuals but provide care and protection that ensure a higher probability that the young will become reproductive adults. Humans generally produce a single offspring per pregnancy, but nearly all of them live. In effect, the energy has been channeled into the care and protection of the young produced rather than into the production of incredibly large numbers of potential young. Even though fewer young are produced by animals like birds and mammals, their reproductive capacity still greatly exceeds the number required to replace the parents when they die.

The Population Growth Curve

Because most species of organisms have a high reproductive capacity, there is a tendency for populations to grow if environmental conditions permit. For example, if the usual litter size for a pair of mice is four, the four would produce eight, which in turn would produce sixteen and so forth. Figure 16.4 shows a graph of change in population size over time known as a **population growth curve.** This kind of curve is typical for situations where a species is introduced into a previously unutilized area.

The change in the size of a population depends on the rate at which new organisms enter the population compared to the rate at which they leave. The number of new individuals added to the population by reproduction per thousand individuals is called **natality.** The number of individuals leaving the population by death per thousand individuals is called **mortality.** When a small number of organisms (two mice) first invades an area, there is a period of time before reproduction takes place when the population remains small and relatively constant. This part of the population growth curve is known as the **lag phase.** Mortality and natality are similar during this period of time. In organisms that take a long time to mature and produce young, such as elephants, deer, and many kinds of plants, the lag phase may be measured in years. With the mice in our example, it will be measured in weeks. The first litter of young will reproduce in a matter of weeks. Furthermore, the original parents will probably produce an additional litter or two during this time period. Now we have several pairs of mice reproducing more than just once. With several pairs of mice reproducing, natality increases and mortality remains low; therefore, the population begins to grow at an ever-increasing (accelerating) rate. This portion of the population growth curve is known as the **exponential growth phase.** The number of mice (or any other organism) cannot continue to increase at a faster and faster rate because, eventually, something in the environment will cause an increase in the number of deaths. Eventually, the number of individuals entering the population will equal the number of individuals leaving it by death or migration, and the population size becomes stable. Often there is both a decrease in natality and an increase in mortality at this point. This portion of the population growth

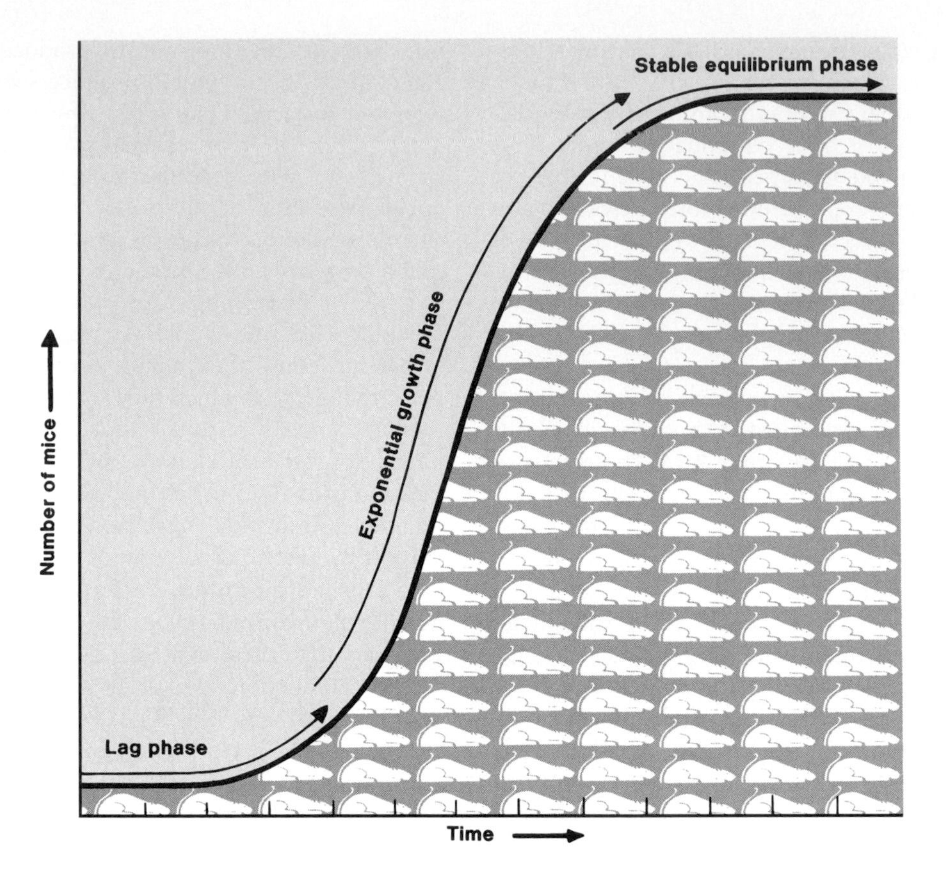

Figure 16.4 **A Typical Population Growth Curve.** In this mouse population, the period of time in which there is little growth is known as the lag phase. This is followed by a rapid increase in population as the offspring of the originating population begin to reproduce themselves; this is known as the exponential growth phase. Eventually the population reaches a stable equilibrium phase, during which the birth rate equals the death rate.

curve is known as the **stable equilibrium phase.** Reproduction continues and the birthrate is still high, but the death rate increases and larger numbers of individuals migrate from the area.

Population-Size Limitations

Populations cannot continue to increase indefinitely; eventually, some factor or set of factors acts to limit the size of a population leading to the development of a stable equilibrium phase or even to a reduction in population size. The specific identifiable factors that prevent unlimited population growth are know as **limiting factors.** All of the different limiting factors that act on a population are collectively known as **environmental resistance.** In general, those organisms that are small and have short life spans tend to have fluctuating populations, while large organisms that live a long time tend to reach an optimum population size that can be sustained over an extended period known as the **carrying capacity** (figure 16.5). For example, a forest ecosystem contains populations of many insect species that fluctuate widely, but the number of specific tree species or large animals such as owls or deer is relatively constant.

Carrying capacity, however, is not an inflexible rule. Often such environmental changes as successional changes, climate changes, disease epidemics, forest fires, or floods can change the capacity of an area to support life. In addition, a change that negatively affects the carrying capacity for one species may increase the carrying capacity for another. For example, the cutting down of mature forests followed by the growth of young trees increases the carrying capacity for deer and rabbits, which use the new growth for food, but decreases the carrying capacity for squirrels, which need matured, fruit-producing trees as a source of food and hollow trees for shelter.

The size of the organisms in a population also affects the carrying capacity. For example, an aquarium of a certain size can support only a limited number of fish, but the size of the fish makes a difference. If all the fish are tiny, a large number can be supported, and the carrying capacity is high; however, the same aquarium may be able to support only one large fish. In other words, the biomass of the population makes a difference (figure 16.6). Similarly, when an area is planted with small trees, the population size is high. But as the trees get larger, competition for nutrients and sunlight becomes more intense, and the number of trees present declines, while the biomass increases.

Limiting Factors

Limiting factors can be placed in four broad categories: (1) availability of raw materials, (2) availability of energy, (3) production and disposal of waste products, and (4) interaction with other organisms.

The first category of limiting factors is the availability of raw materials. For plants, magnesium is necessary for the manufacture of chlorophyll, nitrogen is necessary for protein production, and water is necessary for the transport of materials and as a raw material for photosynthesis. If these are not present in the soil, the growth and reproduction of plants is inhibited. However, if fertilizer supplies these nutrients, or irrigation is used to supply water, the effects of these limiting factors can be removed, and some other factor becomes limiting. For animals, the amount of

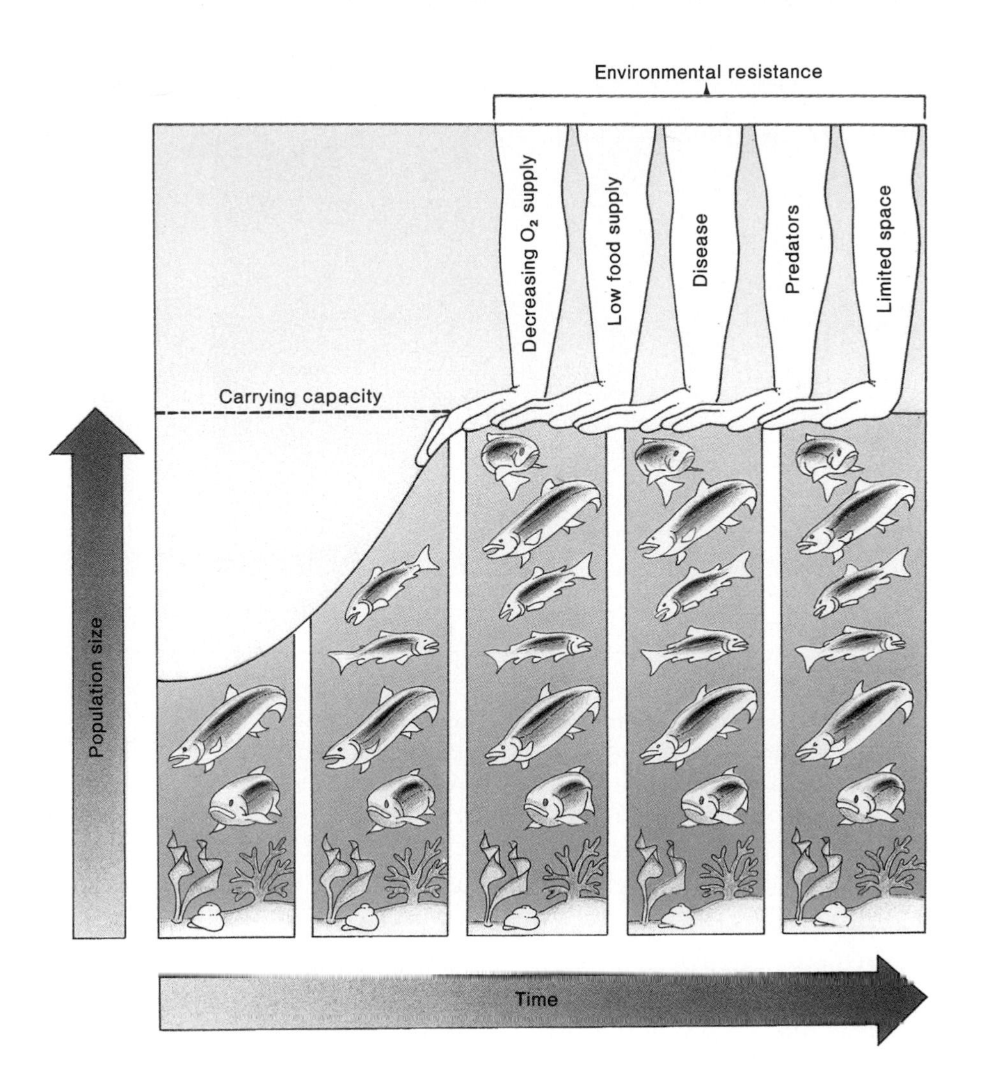

Figure 16.5 **Carrying Capacity.** A number of factors in the environment, such as food, oxygen supply, diseases, predators, and space, determine the number of organisms that can survive in a given area—the carrying capacity of that area. The environmental factors that limit populations are collectively known as environmental resistance.

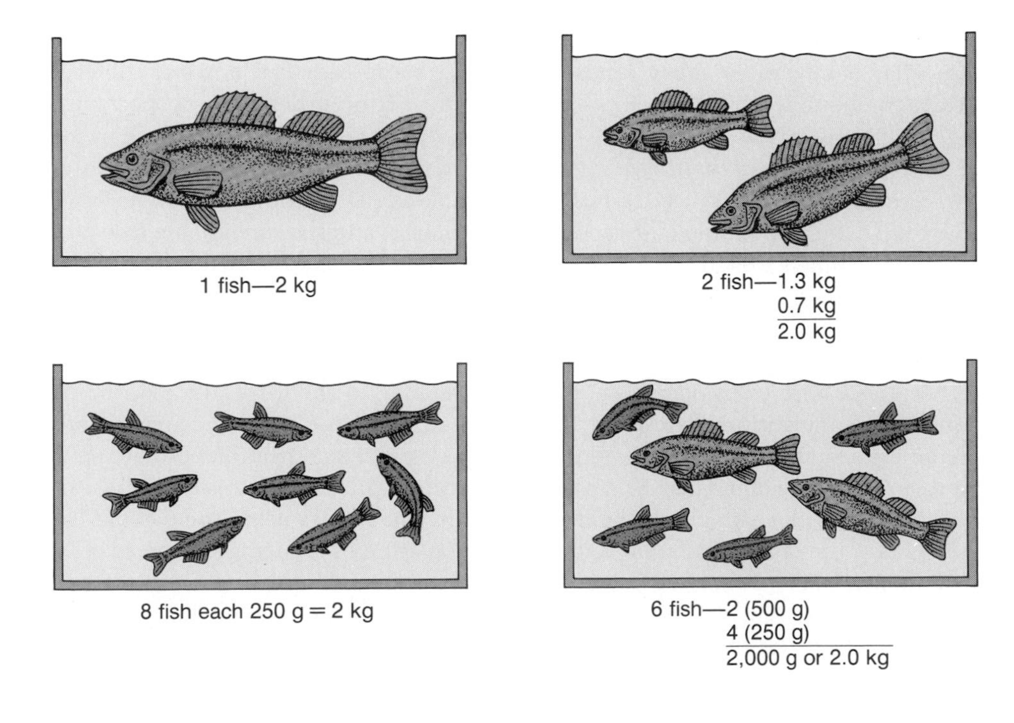

Figure 16.6 **The Effect of Biomass on Carrying Capacity.** Each aquarium can support a biomass of 2 kilograms of fish. The size of the population is influenced by the body size of the fish in the population.

water, minerals, materials for nesting, suitable burrow sites, or food may be limiting factors. Food for animals really fits into both this category and the next because it supplies both raw materials and energy.

The second major type of limiting factor is the availability of energy. The amount of light available is often a limiting factor for plants, which require light as an energy source for photosynthesis. Since all animals use other living things as sources of energy and raw materials, a major limiting factor for any animal is its food source.

The accumulation of waste products is the third general category of limiting factors. It does not usually limit plant populations because they produce relatively few wastes. However, the buildup of high levels of self-generated waste products is a problem for bacterial populations and populations of tiny aquatic organisms. As wastes build up, they become more and more toxic, and eventually reproduction stops, or the population may even die out. When a few bacteria are introduced into a solution containing a source of food, they go through the kind of population growth curve typical of all organisms. As expected, the number of bacteria begins to increase following a lag phase, increases rapidly during the exponential growth phase, and eventually reaches stability in the stable equilibrium phase. But as waste products accumulate, the bacteria literally drown in their own wastes. When space for disposal is limited, and no other organisms are present that can convert the harmful wastes to less harmful products, a population decline known as the **death phase** follows (figure 16.7).

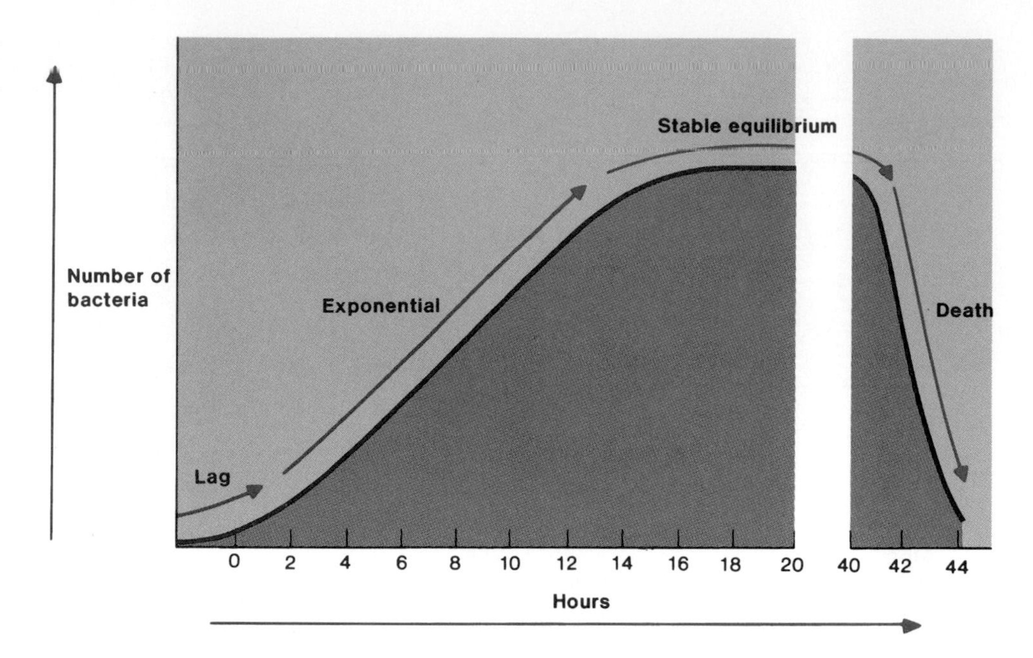

Figure 16.7 **Bacterial Population Growth Curve.** The rate of increase in the population of these bacteria is typical of population growth in a favorable environment. As the environmental conditions change as a result of an increase in the amount of waste products, the population first levels off, then begins to decrease. This period of decreasing population size is known as the death phase.

Wine makers deal with this same situation. When yeasts ferment the sugar in grape juice, they produce ethyl alcohol. When the alcohol concentration reaches a certain level, the yeast population stops growing and eventually declines. Therefore, wine can naturally reach an alcohol concentration of only 12% to 15%. To make any drink stronger than that (of a higher alcohol content), water must be removed (to distill) or alcohol must be added (to fortify). In small aquatic pools like aquariums, it is often difficult to keep populations of organisms healthy because of the buildup of ammonia in the water from the waste products of the animals. This is the primary reason that activated charcoal filters are commonly used in aquariums. The charcoal removes many kinds of toxic compounds and prevents the buildup of waste products.

The fourth set of limiting factors is organism interaction. As we learned in chapter 15 on community interaction, organisms influence each other in many different ways. Some organisms are harmed and others are benefited. The population size of any organism would be negatively affected by parasitism, predation, or competition. Parasitism and predation usually involve interactions between two different species. Cannibalism is rare, but competition among members of the same population is often very intense. Many kinds of organisms perform services for others that have beneficial effects on the population. For example, decomposer organisms destroy toxic waste products, thus benefiting populations of animals. They also recycle materials needed for the growth and development of all organisms. Mutualistic relationships benefit both of the populations involved.

Often, the population sizes of two kinds of organisms will be interdependent because each is a primary limiting factor of the other. This is most often seen in parasite–host relationships and predator–prey relationships. A good example is the relationship of the *lynx* (a predator) and the *varying hare* (the prey) as it was studied in Canada. The varying hare has a high reproductive capacity that the lynx helps to control by using the varying hare as food. The lynx can capture and kill the weak, the old, the diseased, and the unwary varying hares, leaving stronger, healthier ones to reproduce. Because the lynx is removing unfit individuals, it benefits the varying hare population by reducing the spread of disease and reducing the amount of competition among varying hares. At the same time, the varying hare gene pool benefits because individuals that are less fit have their genes removed from the gene pool. While the lynx is helping to

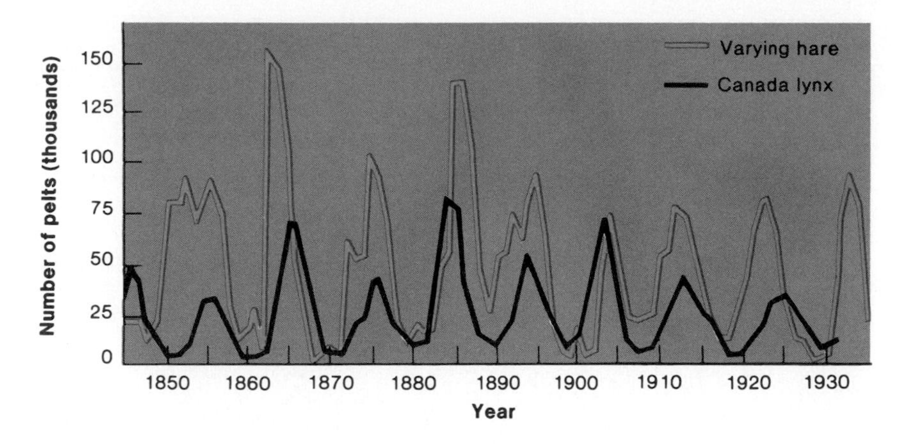

Figure 16.8 **Organism Interaction.** The interaction between predator and prey species is complex and often difficult to interpret. These data were collected from the records of the number of pelts purchased by the Hudson Bay Company. It shows that the two populations fluctuate, with changes in the lynx population usually following changes in varying hare population.

Source: Data from D. A. MacLulich, *Fluctuations in the Numbers of the Varying Hare (Lepus americanus)*. University of Toronto Press, 1937, reprinted 1974.

limit the varying hare population, the size of the varying hare population determines how many lynx can live in the area, since varying hares are their primary food source. If such events as disease epidemics or unusual weather conditions cause a decline in the varying hare population, the population of the lynx also falls (figure 16.8).

Extrinsic and Intrinsic Limiting Factors

Some of the factors that help control populations come from outside the population and are known as **extrinsic factors.** Predators, loss of a food source, lack of sunlight, or accidents of nature are all extrinsic factors. However, many kinds of organisms self-regulate their population size. The mechanisms that allow them to do this are called **intrinsic factors.** For example, a study of rats under crowded living conditions showed that as conditions became more crowded, abnormal social behavior became common. There was a decrease in litter size, fewer litters per year were produced, mothers were more likely to ignore their young, and many young were killed by adults. Thus, changes in the behavior of the members of the rat population itself resulted in lower birth rates and higher death rates, leading to a reduction in the population growth rate. As another example, trees that are stressed by physical injury or disease often produce extremely large numbers of seeds (offspring) the following year. The trees themselves alter their reproductive rate. The opposite situation is found among populations of white-tailed deer. It is well known that reproductive success is reduced when the deer experience a series of severe winters. When times are bad, the female deer are more likely to have single offspring rather than twins.

Density-Dependent and Density-Independent Limiting Factors

Many populations are controlled by limiting factors that become more effective as the size of the population increases. Such factors are referred to as **density-dependent factors.** Many of the factors we have already discussed are density-dependent. For example, the larger a population becomes, the more likely it is that predators will have a chance to catch some of the individuals. Furthermore, a prolonged period of increasing population allows the size of the predator population to increase as well. Large populations with high population density are more likely to be affected by epidemics of parasites than are small populations of widely dispersed individuals, since dense populations allow for the easy spread of parasites from one individual to another. The rat example discussed previously is another good example of a density-dependent factor operating, since the amount of abnormal behavior increased as the size of the population increased. In general, whenever there is competition among members of a population, its intensity increases as the population increases. Large organisms that tend to live a long time and have relatively few young are most likely to be controlled by density-dependent factors.

A second category, made up of population-controlling influences that are not related to the size of the population, is known as **density-independent factors.** Density-independent factors are usually accidental or occasional extrinsic factors in nature that happen regardless of the size or density of a population. A sudden rainstorm may drown many small plant seedlings and soil organisms. Many plants and animals are killed by frosts that come late in spring or early in the fall. A small pond may dry up, resulting in the death of many organisms. The organisms most likely to be controlled by density-independent factors are small, short-lived organisms that can reproduce very rapidly.

So far we have looked at populations primarily from a nonhuman point of view. Now it is time to focus on the human species and the current problem of the world population.

Human Population Growth

It is important to realize that human populations follow the same patterns of growth and are acted upon by the same kinds of limiting factors as populations of other organisms. When we look at the curve of population growth over the past several thousand years, estimates are that the human population remained low and constant for thousands of years but has increased rapidly in the last few hundred years (figure 16.9). For example, it has been estimated that when Columbus discovered America, the Native American population was about one million. Today, the population of North America is over 280 million people. Does this mean that humans are different from other animal species? Can the human population continue to grow forever?

The human species is no different from other animals. It has a carrying capacity but has been able to continuously shift the carrying capacity upward through technology and the displacement of other species. Much of the exponential growth phase of the human population can be attributed to the removal of diseases, improvement in agricultural methods, and destruction of natural ecosystems in favor of artificial agricultural ecosystems. But even this has its limits. There must be some limiting factors that will eventually cause a leveling off of our population growth curve. We cannot increase beyond our ability to get raw materials and energy, nor can we ignore the waste products we produce or the other organisms with which we interact.

To many of us, raw materials consist simply of the amount of food available, but we should not forget that in a technological society, iron ore, lumber, irrigation water, and silicon chips are also raw materials. However, most people of the world have only more basic needs. For the past several decades, large portions of the world's population have not had enough food. Although it is biologically accurate to say that the world can currently produce enough food to feed all the people of the world, there are many reasons why people can't get food or won't eat it. Many cultures have food taboos or traditions that prevent the use of some available food sources. For example, pork is forbidden in some cultures. Certain groups of people find it almost impossible to digest milk. Some African cultures use a mixture of cow's milk and cow's blood as food, which people of other cultures might be unable to eat.

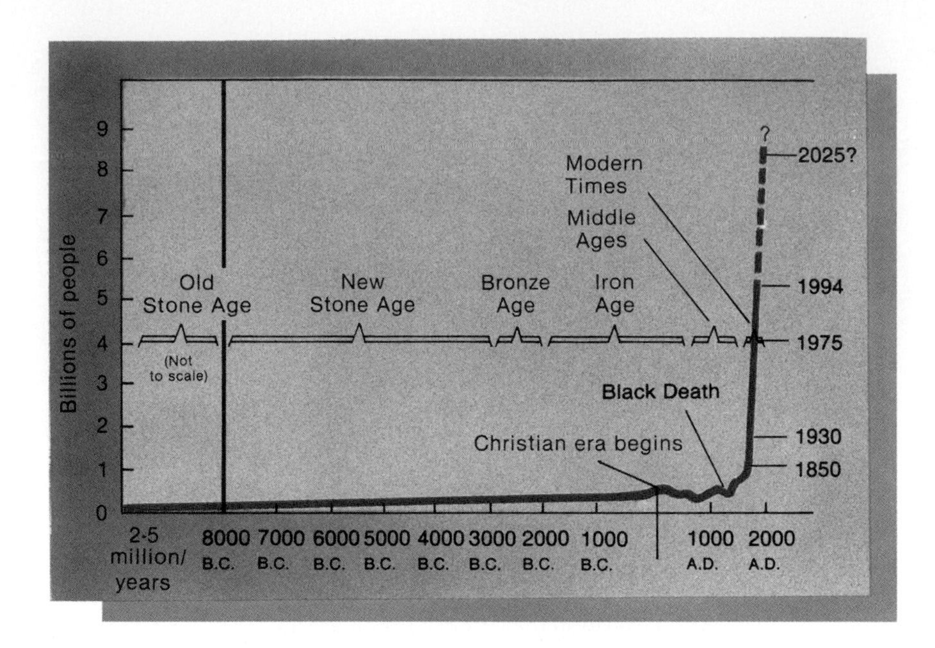

Figure 16.9 **Human Population Growth.** The number of humans doubled from A.D. 1850 to 1930 (from one billion to two billion), then doubled again by 1975 (four billion), and could double again (eight billion) by the year 2025. How long can the human population continue to double before the earth's ultimate carrying capacity is reached?

In addition, complex political, economic, and social problems are related to the production and distribution of food. In some cultures, farming is a low-status job, which means that people would rather buy their food from someone else than grow it themselves. This can result in underutilization of agricultural resources. Food is sometimes used as a political weapon when governments want to control certain groups of people. But probably most important is the fact that transportation of food from centers of excess to centers of need is often very difficult and expensive.

A more fundamental question is whether the world can continue to produce enough food. In 1992 the world population was growing at a rate of 1.7% per year. This amounts to nearly three new people being added to the world population every second, which will result in a doubling of the world population in forty years. With a continuing increase in the number of mouths to feed, it is unlikely that food production will be able to keep pace with the growth in human population (see box 16.1). A primary indicator of the status of the world food situation is the amount of grain produced for each person in the world (per capita gain production). World per capita grain production peaked in 1984. The less-developed nations of the world have a disproportionately large increase in population and a decline in grain production because they are less able to afford costly fertilizer, machinery, and the energy necessary to run the machines and irrigate the land to produce their own grain.

The availability of energy is the second broad limiting factor that affects human populations as well as other kinds of organisms. All species on earth ultimately depend on sunlight for energy—including the human species. Whether one produces electrical power from a hydroelectric dam, burns fossil fuels, or uses a solar cell, the energy is derived from the sun. Energy is needed for transportation, building and maintaining homes, and food production. It is very difficult to develop unbiased, reasonably accurate estimates of global energy "reserves" in the

BOX 16.1 *Thomas Malthus and His Essay on Population*

In 1798 Thomas Robert Malthus, an Englishman, published an essay on human population. It presented an idea that was contrary to popular opinion. His basic thesis was that human population increased in a geometric or exponential manner (2, 4, 8, 16, 32, 64, etc.), while the ability to produce food increased only in an arithmetic manner (1, 2, 3, 4, 5, 6, etc.). The ultimate outcome of these different rates would be that population would outgrow the ability of the land to produce food. He concluded that wars, famines, plagues, and natural disasters would be the means of controlling the size of the human population. His predictions were hotly debated by the intellectual community of his day. His assumptions and conclusions were attacked as being erroneous and against the best interest of society. At the time he wrote the essay, the popular opinion was that human knowledge and "moral constraint" would be able to create a world that would supply all human needs in abundance. One of Malthus's basic postulates was that "commerce between the sexes" (sexual intercourse) would continue unchanged, while other philosophers of the day believed that sexual behavior would take less procreative forms and human population would be limited. Only within the last fifty years, however, have really effective conception-control mechanisms become widely accepted and used, and they are used primarily in developed countries.

Malthus did not foresee the use of contraception, major changes in agricultural production techniques, or the exporting of excess people to colonies in the Americas. These factors, as well as high death rates, prevented the most devastating of his predictions from coming true. However, in many parts of the world today, people are experiencing the forms of population control (famine, epidemic disease, wars, and natural disasters) predicted by Malthus in 1798. Many people feel that his original predictions were valid—only his time scale was not correct—and that we are seeing his predictions come true today.

Another important impact of Malthus's essay was the effect it had on the young Charles Darwin. When Darwin read it, he saw that what was true for the human population could be applied to the whole of the plant and animal kingdoms. As overreproduction took place, there would be increased competition for food, resulting in the death of the less fit organisms. This theory he called *natural selection*.

---

form of petroleum, natural gas, and coal. Therefore, it is difficult to predict how long these "reserves" might last. We do know, however, that the quantities are limited and that the rate of use has been increasing, particularly in the developed and developing countries.

If the less developed countries were to attain a standard of living equal to that of the developed nations, the global energy "reserves" would disappear overnight. Since the United States constitutes 4.72% of the world's population and consumes approximately 25% of the world's energy resources, raising the standard of living of the entire world population to that of the United States would result in a 500% increase in the rate of consumption of energy and reduce theoretical reserves by an equivalent 500%. Humans should realize there is a limit to our energy resources; we are living on solar energy that was stored over millions of years, and we are using it at a rate that could deplete it in hundreds of years. Will energy availability be the limiting factor that determines the ultimate carrying capacity for humans, or will problems of waste disposal predominate?

One of the most talked-about aspects of human activity is the problem of waste disposal. Not only do we have normal biological wastes, which can be dealt with by decomposer organisms, but we generate a variety of technological wastes and by-products that cannot be efficiently degraded by decomposers. Most of what we call pollution results from the waste products of technology. The biological wastes can usually be dealt with fairly efficiently by the building of waste-water treatment plants and other sewage facilities. Certainly these facilities take energy to run, but they rely on decomposers to degrade unwanted organic matter to carbon dioxide and water. Earlier in this chapter we discussed the problem that bacteria and yeasts face when their metabolic waste products accumulate. In this situation, the organisms so "befoul their nest" that their wastes poison them. Are humans in a similar situation on a much larger scale? Are we dumping so much technological waste, much of which is toxic, into the environment that we are being poisoned? Some people believe that disregard for the quality of our environment will be a major factor in decreasing our population growth rate. In any case, it makes good sense to do everything possible to stop pollution and work toward cleaning our nest.

The fourth category of limiting factors that determine carrying capacity is interaction among organisms. Humans interact with other organisms in as many ways as other animals do. We have parasites and occasionally predators. We are predators in relation to a variety of animals, both domesticated and wild. We have mutualistic relationships with many of our domesticated plants and animals, since they could not survive without our agricultural practices and we would not survive without the food they provide. Competition is also very important. Insects and rodents compete for the food we raise, and we compete directly with many kinds of animals for the use of ecosystems.

As humans convert more and more land to agricultural and other purposes, many other organisms are displaced. Many of these displaced organisms are not able to compete successfully and must leave the area, have their populations reduced, or become extinct. The American bison (buffalo), African and Asian elephants, the panda, and the grizzly bear are a few species that are much reduced in number because they were not able to compete successfully with the human species. The passenger pigeon, Carolina parakeet, and great auk are a few that have become extinct. Our parks and natural areas have become tiny refuges for plants and animals that once occupied vast expanses of the world. If these refuges are lost, many organisms will become extinct. What today might seem to be an insignificant organism that we can easily do without may tomorrow be seen as a link to our very survival. We humans have been extremely successful in our efforts to convert ecosystems to our own uses at the expense of other species.

Competition with one another (intraspecific competition), however, is a different matter. Since competition is negative to both organisms, humans must be harmed. We are not displacing another species, we are displacing some of our own kind. Certainly when resources are in short supply, there is competition. Unfortunately, it is usually the young that are least able to compete, and high infant mortality is the result.

Humans are different from most other organisms in a fundamental way: we are able to predict the outcome of a specific course of action. Current technology and medical knowledge are available to control human population and improve the health and well-being of the people of the world. Why then does the human population continue to grow, resulting in human suffering and stressing the environment in which we live? Since we are social animals that have freedom of choice, we frequently do not do what is considered "best" from an unemotional, unselfish point of view. People make decisions based on historical, social, cultural, ethical, and personal considerations. What is best for the population as a whole may be bad for you as an individual. The biggest problems associated with control of the human population are not biological problems, but require the efforts of philosophers, theologians, politicians, sociologists, and others. As population increases, so will political, social, and biological problems; there will be less individual freedom, and herd politics will prevail. The knowledge and technology necessary to control the human population are available, but the will is not. What will eventually limit the size of our population? Will it be lack of resources, lack of energy, accumulated waste products, competition among ourselves, or rational planning of family size?

• Summary •

A population is a group of organisms of the same species in a particular place at a particular time. Populations differ from one another in gene frequency, age distribution, sex ratio, and population density. Organisms typically have a reproductive capacity that exceeds what is necessary to replace the parent organisms when they die. This inherent capacity to overreproduce causes a rapid increase in population size when a new area is colonized. A typical population growth curve consists of a lag phase in which population rises very slowly, followed by an exponential growth phase in which the population increases at an accelerating rate, followed by a leveling-off of the population in a stable equilibrium phase. In some populations, a fourth phase may occur, known as the death phase. This is typical of bacterial and yeast populations.

The carrying capacity is the number of organisms that can be sustained in an area over a long period of time. It is set by a variety of limiting factors. Availability of energy, availability of raw materials, accumulation of wastes, and interactions with other organisms are all categories of limiting factors. Because organisms are interrelated, population changes in one species sometimes affect the size of other populations. This is particularly true when one organism uses another as a source of food. Some limiting factors operate from outside the population and are known as extrinsic factors; other are properties of the species itself and are called intrinsic factors. Some limiting factors become more intense as the size of the population increases; these are known as density-dependent factors. Other limiting factors that are more accidental and not related to population size are called density-independent factors.

Humans as a species have the same limits and influences that other organisms do. Our current problems of food production, energy needs, pollution, and habitat destruction are outcomes of uncontrolled population growth. However, humans can reason and predict, thus providing the possibility of population control through conscious population limitation.

• Thinking Critically •

If you return to figure 16.9, you will note that it has very little in common with the population growth curve shown in figure 16.4. What factors have allowed the human population to grow so rapidly? What natural limiting factors will eventually bring this population under control?

What is the ultimate carrying capacity of the world? What alternatives to the natural processes of population limitation could bring human population under control?

Consider the following in your answer: reproduction, death, diseases, food supply, energy, farming practices, food distribution, cultural biases, and anything else you consider to be appropriate.

• Experience This •

Place a male and a female fruit fly in a bottle with half a banana. (You may use prepared fruit fly medium if you have it available.) The fruit flies can be wild or from cultures. Males have solid black tail ends, while females have striped tail ends. Count the fruit flies each day for four weeks and plot them on a graph.

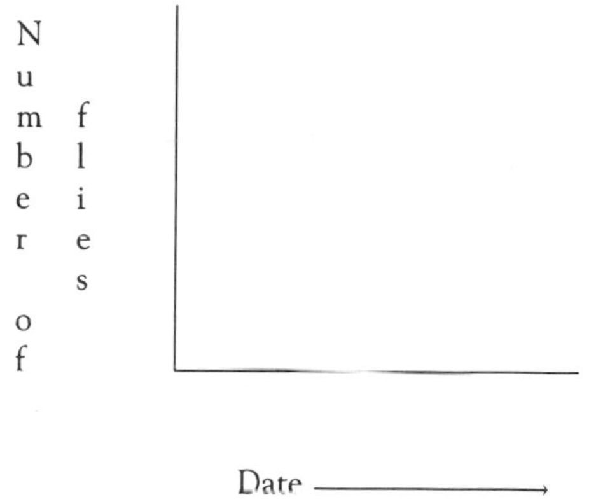

Number of flies

Date →

• Questions •

1. Draw the population growth curve of a yeast culture during the wine-making process. Label the lag, exponential growth, stable equilibrium, and death phases.
2. List four ways in which two populations of the same species could be different.
3. Why do populations grow?
4. List four kinds of limiting factors that help to set the carrying capacity for a species.
5. How do the concepts of biomass and population size differ?
6. Differentiate between density-dependent and density-independent limiting factors. Give an example of each.
7. Differentiate between intrinsic and extrinsic limiting factors. Give an example of each.
8. As the human population continues to grow, what should we expect to happen to other species?
9. How does the population growth curve of humans compare with that of other kinds of animals?
10. All organisms overreproduce. What advantage does this give to the species? What disadvantages?

• Chapter Glossary •

age distribution (āj dis″tri-biu′shun) The number of organisms of each age in the population.

biotic potential (bi-ah′tik po-ten′shul) See **reproductive capacity.**

carrying capacity (ka′re-ing kuh-pas′ĭ-te) The optimum population size an area can support over an extended period of time.

death phase (deth fāz) The portion of some population growth curves in which the size of the population declines.

density-dependent factors (den′sĭ-te de-pen′dent fak′tōrz) Population-limiting factors that become more effective as the size of the population increases.

density-independent factors (den′sĭ-te in″de-pen′dent fak′tōrz) Population-controlling factors that are not related to the size of the population.

environmental resistance (en-vi-ron-men′tal re-zis′tants) The collective set of factors that limit population growth.

exponential growth phase (eks-po-nen′shul grōth fāz) A period of time during population growth when the population increases at an accelerating rate.

extrinsic factors (eks-trin′sik fak′tōrz) Population-controlling factors that arise outside the population.

gene flow (jēn flo) The movement of genes within a population due to migration or the movement of genes from one generation to the next by gene replication and reproduction.

gene frequency (jēn fre′kwen-se) A measure of how often a specific gene shows up in the gametes of a population.

intrinsic factors (in-trin′sik fak′tōrz) Population-controlling factors that arise from within the population.

lag phase (lag fāz) A period of time following colonization when the population remains small or increases slowly.

limiting factors (lim′ĭ-ting fak′tōrz) Environmental influences that limit population growth.

mortality (mor-tal′ĭ-te) The number of individuals leaving the population by death per thousand individuals in the population.

natality (na-tal′ĭ-te) The number of individuals entering the population by reproduction per thousand individuals in the population.

population (pop″u-la′shun) A group of organisms of the same species located in the same place at the same time.

population density (pop″u-la′shun den′sĭ-te) The number of organisms of a species per unit area.

population growth curve (pop″u-la′shun grōth kurv) A graph of the change in population size over time.

population pressure (pop″u-la′shun presh-yur) Intense competition that leads to changes in the environment and dispersal of organisms.

reproductive capacity (re-pro-duk′tiv kuh-pas′ĭ-te) The theoretical maximum rate of reproduction; also called **biotic potential.**

sex ratio (seks ra′sho) The number of males in a population compared to the number of females.

stable equilibrium phase (stā′bul e-kwi-lib′re-um fāz) A period of time during population growth when the number of individuals entering the population and the number leaving the population are equal, resulting in a stable population.

C H A P T E R 17

Behavioral Ecology

Purpose

This chapter focuses on the activities of individual organisms. We look at several examples of behavior and the significance of that behavior for the welfare of the species. The dangers of misinterpreting observed behavior are discussed, and the ecological importance of behavior is a central theme.

For Your Information

The two most popular exhibits at the zoo are the primates and the reptiles. Our fascination with reptiles is generally based on fear and hate, while interest in primates is based on their similarity to humans. Many of the bizarre behaviors we see in captive animals are not normal. The pacing of many zoo animals does not occur when they are in the wild. Similarly, the begging behavior of bears and elephants is learned.

Chapter Outline

Learning Objectives

- Understand that behavior has evolutionary and ecological significance.
- Distinguish between instinctive and learned behaviors.
- Recognize that there are several kinds of learning.
- Know that animals use sight, sound, and chemicals to communicate for reproductive purposes.
- Appreciate that territoriality and dominance hierarchies allocate resources.
- Know several methods used by animals in navigation.
- Describe why the evolution of social animals is different from that of nonsocial animals.

Understanding Behavior

Behavior is how an organism acts, what it does, and how it does it. When we think about the behavior of an animal, we should keep in mind that behavior is like any other characteristic displayed by an animal. It has a value or significance to the animal as it goes about exploiting resources and reproducing more of its own species. Many behaviors are inherited; consequently, they have evolved just as structures have. In this respect, behavioral characteristics are no different from structural characteristics. The evolution of behavior is much more difficult to study, however, since behavior is transient and does not leave the fossils that structures do.

Behavior is a very important part of the ecological role of any animal. It allows animals to escape predators, seek out mates, gain dominance over others of the same species, and respond to changes in the environment. Plants, for the most part, must rely on structures, physiological changes, or chance to accomplish the same ends. For example, a rabbit can run away from a predator—a plant cannot. But the plant may have developed thorns or toxic compounds within its leaves that discourage animals from eating it. Mate selection in animals often involves elaborate behaviors that assist them in identifying the species and sex of the potential mate. Most plants rely on a much more random method for transferring male gametes to the female plant. Dominance in plants is often achieved by depriving competitors of essential nutrients or by inhibiting the development of the seeds of other plants. Animals have a variety of behaviors that allow them to exert dominance over members of the same species.

It is not always easy to identify the significance of a behavior without careful study of the behavior pattern and the impact it has on other organisms. For example, a hungry baby herring gull pecks at a red spot on its parent's bill. What possible value can this behavior have for either the chick or the parent? If we watch, we see that when the chick pecks at the spot, the parent regurgitates food onto the ground, and the chick feeds (figure 17.1). This looks like a simple behavior, but there is more to it than meets the eye. Why did the chick peck to begin with? How did it know to peck at that particular spot? Why did the pecking cause the parent to regurgitate food? These questions are not easy to answer, and many people assume that the actions have the same motivation and direction as similar human behaviors. For example, when a human child points to a piece of candy and makes appropriate noises, it is indicating to its parent that it wants some candy. Is that what the herring-gull chick is doing?

Figure 17.1 **Animal Behavior.** Baby herring gulls cause their parent to regurgitate food onto the ground by pecking at the red spot on their parent's bill. The parent then picks up the food and feeds it to the baby.

Some people believe that a bird singing on a warm, sunny spring day is making that beautiful sound because it is so happy. Students of animal behavior do not accept this idea and have demonstrated that a bird sings to tell other birds to keep out of its territory.

The barbed stinger of a honeybee remains in your skin after you are stung, and the bee tears the stinger out of its body when it flies away. The damage to its body is so great that it dies. Has the bee performed a noble deed of heroism and self-sacrifice? Was it defending its hive from you? We need to know a great deal more about the behavior of bees to understand the value of such behavior to the success of the bee species. The fact that bees are social animals like us makes it particularly tempting to think that they are doing things for the same reasons we are.

The idea that we can ascribe human feelings, meanings, and emotions to the behavior of animals is called **anthropomorphism.** The fable of the grasshopper and the ant is another example of crediting animals with human qualities. The ant is pictured as an animal that, despite temptations, works hard from morning until night, storing away food for the winter (figure 17.2). The grasshopper, on the other hand, is represented as a lazy good-for-nothing that fools away the summer when it really ought to be saving up for the tough times ahead. If one is looking for parallels to human behavior, these are pretty good illustrations. But they really are not accurate statements about the lives of the animals from an ecological point of view. Both the ant and the grasshopper are very successful organisms, but each has a different way of satisfying its needs and ensuring that some of its offspring will

Figure 17.2 **The Fable of the Ant and the Grasshopper.** In many ways we give human meaning to the actions of animals. The ant is portrayed as an industrious individual who prepares for the future, and the grasshopper as a lazy fellow who sits in the sun and sings all day. This view of animal behavior is anthropomorphic.

be able to provide another generation of organisms. One method of survival is not necessarily better than another, as long as both animals are successful. This is what the study of behavior is all about—looking at the activities of an organism during its life cycle and determining the value of the behavior in the ecological niche of the organism. The scientific study of the nature of behavior and its ecological and evolutionary significance in its natural setting is known as **ethology.**

Before we go much further, we need to discuss how animals generate specific behaviors. Both instinct and learning are involved in the behavior patterns of most organisms.

Instinct

Many animal behaviors are automatic, preprogrammed, and genetically determined. Such behaviors are called **instinctive behavior,** and they are found in a wide range of organisms from simple one-celled protozoans to complex vertebrates. These behaviors are performed correctly the first time without previous experience when the proper stimulus is given. A **stimulus** is some change in the internal or external environment of the organism that causes it to react. The reaction of the organism to the stimulus is called a **response.**

In our example of the herring-gull chick, the red spot on the bill of the adult bird serves as a stimulus to the chick. The chick responds to this spot in a very specific, genetically programmed way. The behavior is innate—it is done correctly the first time without prior experience. The pecking behavior of the chick is in turn the stimulus for the adult bird to regurgitate food. It is obvious that these behaviors have adaptive value for the gull species, since they leave little to chance. Instinctive behavior has great value because it allows correct, perfect, and necessary behavior to occur without prior experience.

Instinctive behavior cannot be modified when a new situation presents itself, but it can be very effective for the survival of a species if it is involved in fundamental, essential activities that rarely require modification. Instinctive behavior is most common in animals that have short life cycles, simple nervous systems, and little contact with parents. Over long periods of evolutionary time, these genetically determined behaviors have been selected for and have been useful to most of the individuals of the species. However, some instances of inappropriate behavior may be generated by unusual stimuli or circumstances in which the stimulus is given. For example, many insects use the sun, moon, or stars as aids to navigation. Over the millions of years of insect evolution, this has been a valuable and useful tool for the animal. However, the human species has invented a variety of artificial lights that generate totally inappropriate behavior, such as when insects collect at a light and batter themselves to death. This mindless, mechanical behavior seems incredibly stupid to us, but it is still valuable for the species since the majority do not encounter artificial lights and complete their life cycles normally.

Certain species of geese will go through a behavior pattern that involves rolling eggs back into the nest. If the egg is taken from the goose when it is in the middle of egg-rolling behavior, it will continue its egg rolling until it gets back to the nest, even though there is no egg to roll (figure 17.3). It was also discovered that many other somewhat egg-shaped structures would generate the same behavior. For example, beer cans and baseballs were good triggers for egg-rolling behavior. So not only was the bird unable to stop the egg-rolling behavior in midstride, but several non-egg objects generated inappropriate behavior because they had approximately the correct shape.

Some activities are so complex that it seems impossible for an organism to be born with such abilities. For example, you may have seen a caterpillar spin its cocoon. This is not just a careless jumble of silk threads. A cocoon is so precisely made that you can recognize what species of caterpillar made it. But cocoon spinning is not a learned ability. A caterpillar has on opportunity to learn how to spin a cocoon, since it never observes others doing it. Furthermore, caterpillars do not practice several times before they get a proper, workable cocoon. It is as if a "program" for making a cocoon is in the caterpillar's "computer." If you interrupt its cocoon-making effort in the middle and remove most of the finished part, the caterpillar will go right on making the last half of the cocoon. The caterpillar's "program" does not allow for this kind of incident (figure 17.4). This inability to adapt as circumstances change is a prominent characteristic of instinctive behavior.

Could these behavior patterns be the result of natural selection? It is well established that many kinds of behaviors are controlled by genes. The "computer" in our example is really the DNA of the organism, and the "program" consists of a specific package of genes. Through the millions of years that insects have been in existence, natural selection has modified the cocoon-making program to refine the process. Certain genes of the program have undergone mutation, resulting in changes in behavior. Imagine various ancestral caterpillars, each with a slightly different program. The inherited program that gave the caterpillar the best chance of living long enough to produce a new generation is the program selected for and most likely to be passed to the next generation.

Figure 17.3 **Egg-Rolling Behavior in Geese.** Geese will use a specific set of head movements to roll any reasonably round object back to the nest. There are several components to this instinctive behavior, including recognition of the object and head-tucking movements. If the egg is removed during the head-tucking movements, the behavior continues as if the egg were still there.

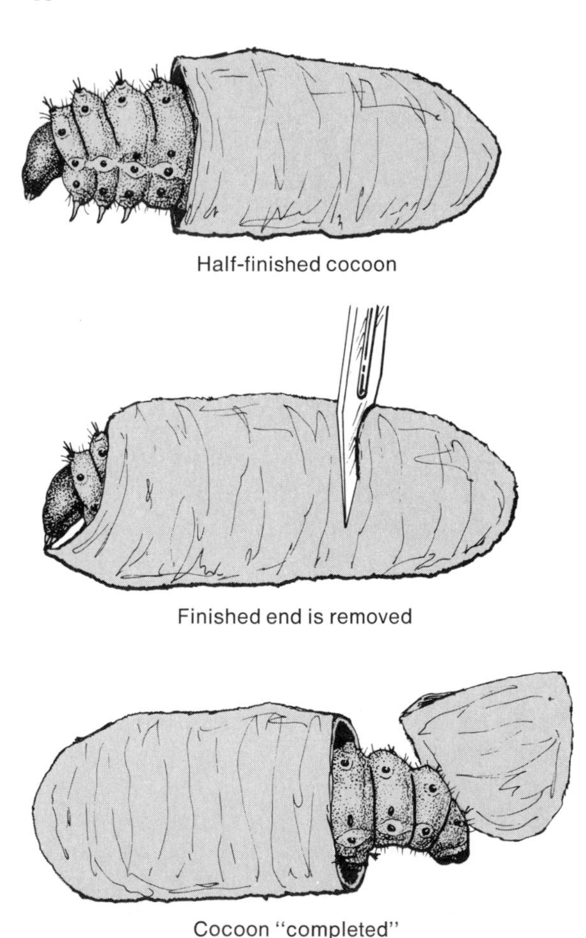

Figure 17.4 **Inflexible Instinctive Behavior.** If part of the caterpillar's half-complete cocoon is removed while it is still spinning, the animal will finish the job but never repair the damaged end.

Learned Behavior

The alternative to preprogrammed, instinctive behavior is learned behavior. **Learning** is a change in behavior as a result of experience. (Your behavior will be different in some way as a result of reading this chapter.) Many kinds of birds must learn parts of their songs. Experimenters have raised young song sparrows in the absence of any adult birds so there was no song for the young birds to imitate. These isolated birds would sing a series of notes that was like the normal song of the species, but not exactly correct. Birds from the same nest that were raised with their parents developed a song nearly identical to that of their parents. If bird songs were totally instinctive, there would be no difference between these two groups. It appears that the basic melody of the song was inherited by the birds and that the refinements of the song were the result of experience. Therefore, the characteristic song of that species was partly learned behavior (a change in behavior as a result of experience) and partly unlearned (instinctive). This is probably true of the behavior of many organisms; they show complex behaviors that are a mixture of instinct and learning. It is important to note that many kinds of birds learn most

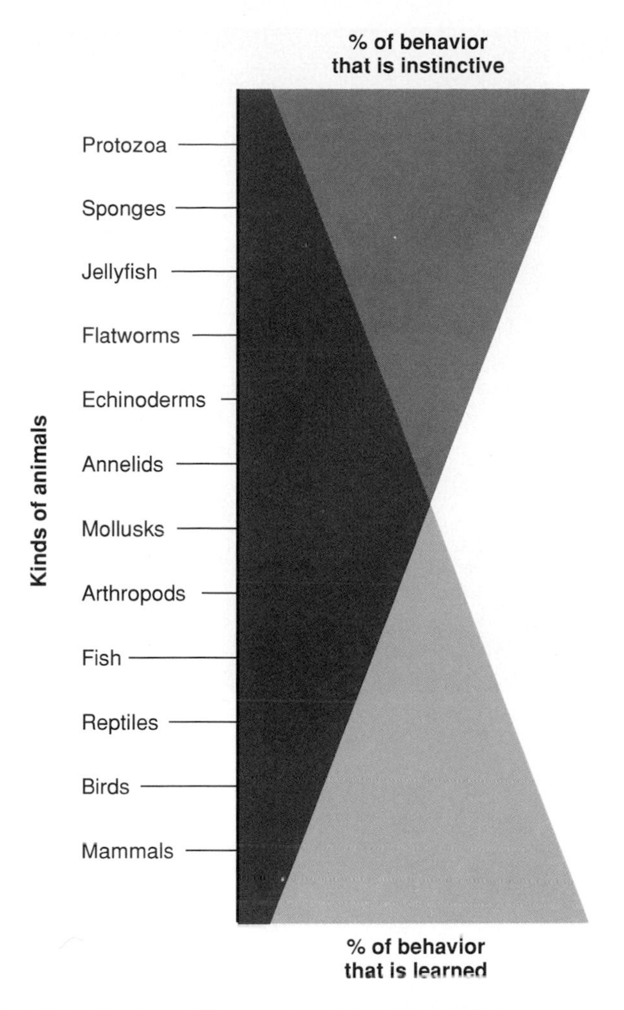

Figure 17.5 **The Distribution of Learned and Instinctive Behavior.** Different groups of animals show different proportions of instinctive and learned behavior in their behavior patterns.

of their song with very few innate components. A mockingbird is very good at imitating the songs of a wide variety of bird species.

This mixture of learned and instinctive behavior is not the same for all species. Many invertebrate animals rely on instinct for the majority of their behavior patterns, whereas many of the vertebrates (particularly birds and mammals) make use of a great deal of learning (figure 17.5).

Learning becomes more significant in long-lived animals that care for their young because the young can imitate their parents and develop behaviors that are appropriate to local conditions. These behaviors take time to develop but have the advantage of adaptability. In order for learning to become dominant in an animal's life, the animal must also have a large brain to store the new information it is learning. This is probably why learning is a major part of life for only a few kinds of animals like the vertebrates.

Learning is not just one kind of activity, but is subdivided into several categories: conditioning, imprinting, and insight learning.

Conditioning

A Russian physiologist, Ivan Pavlov (1849–1936), was investigating the physiology of digestion when he discovered that dogs can associate an unusual stimulus with a natural stimulus. He was studying the production of saliva by dogs, and he knew that a natural stimulus, such as the presence or smell of food, would cause the dogs to start salivating. Then he rang a bell just prior to the presentation of the food. After a training period, the dogs would begin to salivate when the bell was rung, even though no food was presented. This kind of learning, in which a "neutral" stimulus (the sound of a bell) is associated with a "natural" stimulus (the taste of food), is called **classical conditioning** or **associative learning.** The response produced by the neutral stimulus is called a **conditioned response.**

In this case the dogs were receiving positive reinforcement, but it is also possible to apply a negative reinforcement procedure, in which a response eliminates or prevents the occurrence of a painful or distressing stimulus. For example, if a dog receives an electrical stimulus on the right foot at the same time that a bell is rung, it soon learns to associate the bell with the painful stimulus and lifts its foot even without an electrical stimulus upon hearing the sound of the bell.

Associative learning takes place in the natural environment of animals as well. If certain kinds of fruits or insects have unpleasant tastes, animals will learn to associate the bad tastes with the colors and shapes of the offending objects and avoid them in the future (figure 17.6).

Imprinting

Imprinting is a special kind of learning in which a very young animal is genetically primed to learn a specific behavior in a very short period. This type of learning was originally recognized by Konrad Lorenz (1903–1989) in his experiments with geese and ducks. He determined that shortly after hatching, a duckling would follow an object if the object was fairly large, moved, and made noise. In one of his books, Lorenz described himself squatting on the lawn one day, waddling and quacking, followed by

Figure 17.6 **Associative Learning.** Many animals learn to associate unpleasant experiences with the color or shape of offensive objects and thus avoid them in the future. The blue jay is eating a monarch butterfly. These butterflies contain a chemical that makes the blue jay sick. After one or two such experiences, blue jays learn not to eat the monarch or any other butterfly having a similar coloration.

newly hatched ducklings. He was busy being a "mother duck." He was surprised to see a group of tourists on the other side of the fence watching him in amazement. They couldn't see the ducklings hidden by the tall grass. All they could see was this strange performance by a big man with a beard!

Ducklings will follow only the object on which they were originally imprinted. Under normal conditions, the first large, noisy, moving object newly hatched ducklings see is their mother. Imprinting ensures that the immature birds will follow her and learn appropriate feeding, defensive tactics, and other behaviors by example. Since they are always near their mother, she can also protect them from enemies or bad weather. If animals imprint on the wrong objects, they are not likely to survive. Since these experiments by Lorenz in the early 1930s, we have discovered that many young animals can be imprinted on several types of stimuli and that there are responses other than following (figure 17.7).

Figure 17.7 **Imprinting.** Imprinting is a special kind of irreversible learning that occurs during a very specific part of the life of an animal. These geese have been imprinted on Konrad Lorenz and exhibit the "following response" that is typical of this type of behavior.

For song sparrows, the learning of their song appears to be a kind of imprinting. It has been discovered that the young birds must hear the correct song during a specific part of their youth or they will never be able to perform the song correctly as adults. This is true even if they are surrounded by other adult song sparrows that are singing the correct song. Furthermore, the period of time when they learn the song is prior to the time that they begin singing. Recognizing and performing the correct song is important because it has particular meaning to other song sparrows. For males it conveys the information that a male song sparrow has a space reserved for himself. For females, the male song is an advertisement of the location of a male of the correct species that could be a possible mate.

Insight Learning

Insight learning is a special kind of learning in which past experiences are reorganized to solve new problems. When you are faced with a new problem, whether it is a crossword puzzle, a math problem, or any one of a hundred other everyday problems, you sort through your past experiences and locate those that apply. You may not even realize that you are doing it, but you put these past experiences together in a new way that may give the solution to your problem. Because this process is internal and can be demonstrated only through some response, it is very difficult to understand exactly what goes on during insight learning. Behavioral scientists have explored this area for many years, but the study of insight learning is still in its infancy.

Insight learning in animals is particularly difficult to study since it is impossible to know for sure whether a novel solution to a problem is the result of "thinking it through" or an accidental occurrence. For example, a small group of Japanese macaques (monkeys) was studied on an island. They were fed by simply dumping food, such as sweet potatoes or wheat, onto the beach. Eventually, one of the macaques discovered that she could get the sand off the sweet potato by washing it in a nearby stream. She also discovered that she could sort the wheat from the sand by putting the mixture into water because the wheat would float. Are these examples of insight learning? We will probably never know, but it is tempting to think so.

Of the examples used so far, some were laboratory studies, some where field studies, and some included aspects of both. Often they overlap with the field of psychology. This is particularly true for many of the laboratory studies. You can see that the science of animal behavior is a broad one that draws on information from several fields of study and can be used to explore many different kinds of questions. The topics that follow avoid the field of psychology and concentrate on the significance of behavior from an ecological and an evolutionary point of view.

Now that we have some understanding of how organisms generate behavior, we can look at a variety of behaviors in several different kinds of animals and see how they are useful to the animals in their ecological niches.

Reproductive Behavior

Obviously, if a species is to survive, it must reproduce. There are many stages in any successful reproductive strategy. First of all, animals must be able to recognize individuals of the same species that are of the opposite sex. Several different techniques are used for this purpose. For instance, frogs of different species produce sounds that are just as distinct as the calls of different species of birds. The call is a code system that delivers a very private message, since it is only meant for one species. It is, however, meant for any member of that species near enough to hear. The call produced by male frogs, which both male and female frogs can receive by hearing, results in frogs of both sexes congregating in a small area. Once they gather in a small pond, it is much easier to have the further communication necessary for mating to take place.

Chemicals can also serve to attract animals. **Pheromones** are chemicals produced by animals and released into the environment that trigger behavioral or developmental changes in other animals of the same species. They have the same effect as sound through a different code system. The classic example of a pheromone is the chemical that female moths release into the air. The large, fuzzy antennae of the male moths can receive the chemical in unbelievably tiny amounts. The male then changes its direction of flight and flies upwind to the source of the pheromone, which is the female (figure 17.8). Some of these sex-attractant pheromones have been synthesized in the laboratory. One of these, called Disparlure®, is widely used to attract and trap male gypsy moths. Since gypsy moths cause considerable damage to trees by feeding on the leaves, the sex attractant is used to estimate population size so that control measures can be taken to prevent large population outbreaks.

The firefly is probably the most familiar organism that uses light signals to bring together males and females. Several different species may live in the same area, but each species flashes its own code. The code is based on the length of the flashes, their frequency, and their overall pattern (figure 17.9). There is also a difference between the signals given by males and females. For the most part, males are attracted to and mate with females of their own species. Once male and female animals have attracted one another's attention, the second stage in successful reproduction takes place. However, in one species of firefly, the female has the remarkable ability to signal the correct answering code to species other than her own. After she has mated, she will continue to signal to passing males of other species. She is not hungry for sex, she is just hungry. The luckless male who responds to her "come-on" is going to be her dinner.

The second important activity in reproduction is fertilizing eggs. Many marine organisms simply release their gametes into the sea simultaneously and allow fertilization and further development to take place without any input

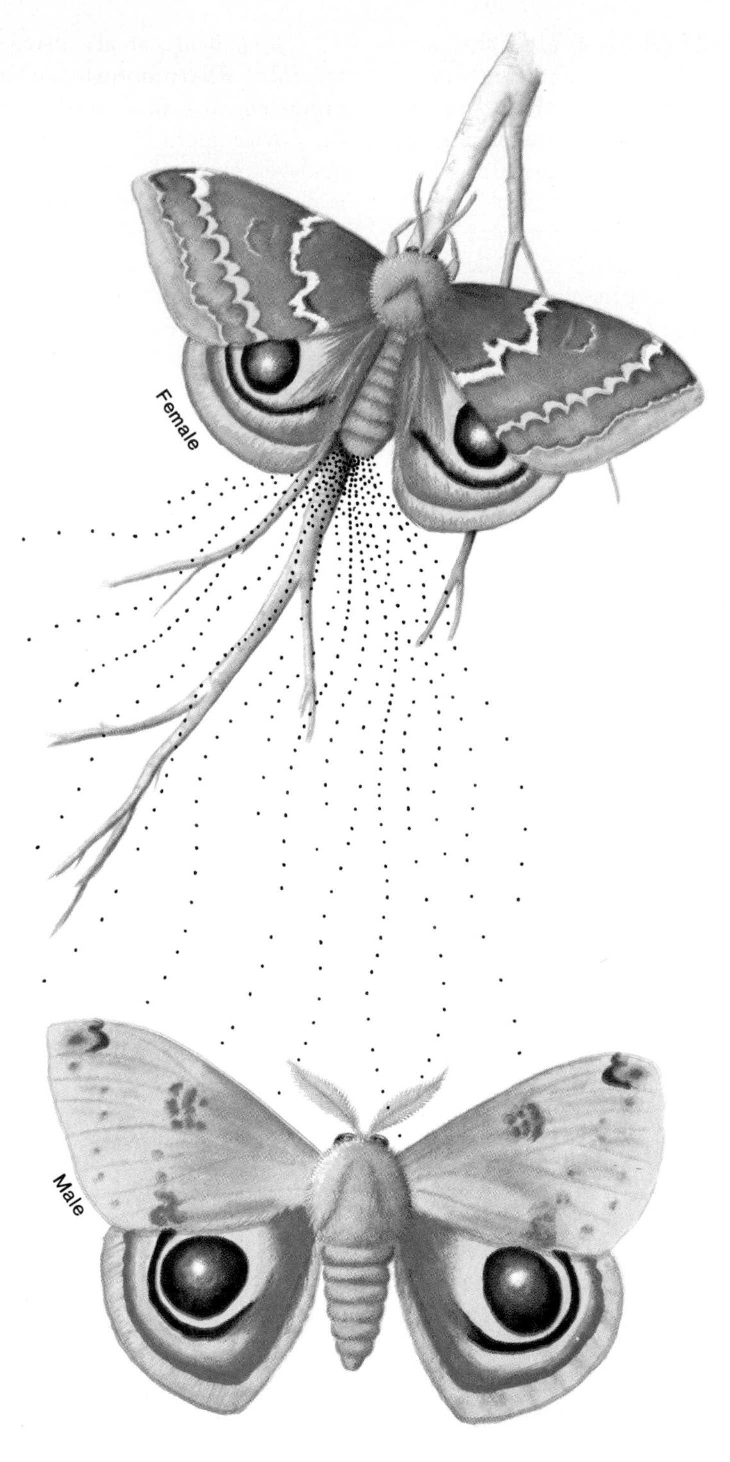

Figure 17.8 **Communication.** The female moth signals her readiness to mate and attracts males by releasing a pheromone that attracts males from long distances downwind.

from the parents. Sponges, jellyfishes, and many other marine animals fit into this category. Other aquatic animals congregate so that the chances of fertilization are enhanced by the male and female being near one another as the gametes are shed. This is typical of many fish and some amphibians, such as frogs. In most terrestrial organisms, internal fertilization occurs, in which the sperm are introduced into the reproductive tract of the female. Some spiders and other terrestrial animals produce packages of sperm that the female picks up with her reproductive structures. Many of these mating behaviors require elaborate, species-specific communication prior to the mating act. Several examples were given in the previous paragraphs.

A third element in successful reproduction is providing the young with the resources they need to live to adulthood. Many invertebrate animals spend little energy on the care of the young, leaving them to develop on their own. Usually the young become free-living larvae that eat and grow rapidly. In some species, females make preparations for the young by laying their eggs in particularly suitable sites. Many insects lay their eggs on the particular species of plant that the larva will use as food as it develops. Many parasitic species seek out the required host in which to lay their eggs. The eggs of others may be placed in spots that provide safety until the young hatch from the egg. Turtles, many fish, and some insects fit into this category. In most of these cases, however, the female lays large numbers of eggs, and most of the young die before reaching adulthood. This is an enormously expensive process: the female invests considerable energy in the production of the eggs but has a low success rate.

An alternative to this "wasteful" loss of potential young is to produce fewer young but invest large amounts of energy in their care. This is typical of birds and mammals. They build nests, parents share in the feeding and protection of the young, and parents often assist the young in learning appropriate behavior. Many insects, such as bees, ants, and termites, have elaborate social organizations in which one or a few females produce large numbers of young that are cared for by sterile offspring of the fertile females. Some of the female's offspring will be fertile, reproducing individuals. The activity of caring for young involves many complex behavior patterns. It appears that most animals that feed and raise young are able to recognize their own young from those of other nearby families and may even kill the young of another family unit. Elaborate greeting ceremonies are usually performed when animals return to the nest or the den. Perhaps this has something

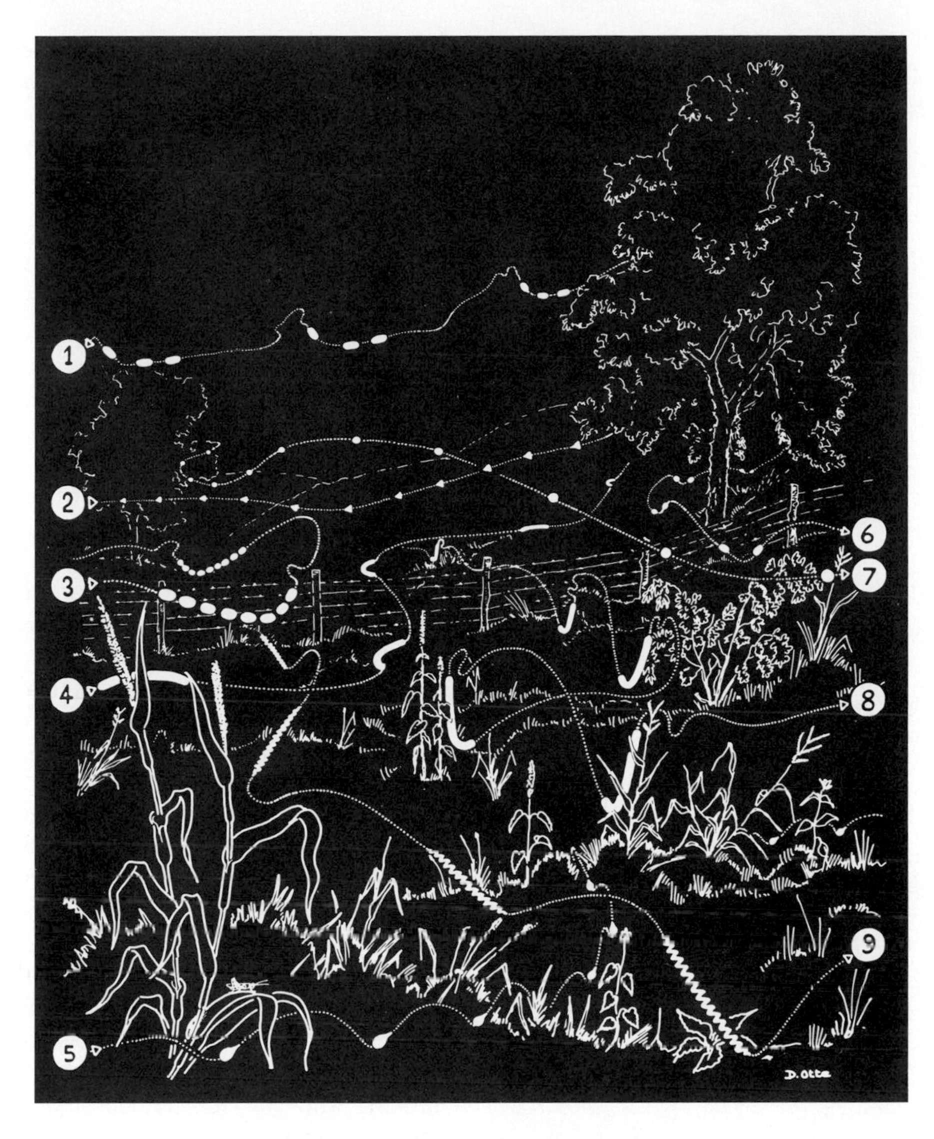

Figure 17.9 **Firefly Communication.** The pattern of light flashes, the location of light flashes, and the duration of light flashes all help fireflies identify members of the opposite sex who are of the appropriate species.

to do with being able to identify individual young. Often this behavior is shared among adults as well. This is true for many colonial nesting birds, such as gulls and penguins, and for many carnivorous mammals, such as wolves, dogs, and hyenas.

Allocating Resources

For an animal to be successful, it must receive sufficient resources to live and reproduce. Therefore, we find many kinds of behaviors that divide the available resources so that the species as a whole is benefited, even though some individuals may be harmed. One kind of behavior pattern that is often tied to successful reproduction is territoriality. **Territoriality** consists of the setting aside of space for the exclusive use of an animal for food, mating, or other purposes. A **territory** is the space an animal defends against others of the same species. This territory has great importance because it reserves exclusive rights to the use of a specific piece of space. When territories are first being established, there is much conflict between individuals. This eventually gives way to the use of a series of signals that define the territory and communicate to others that the territory is occupied. The male redwing blackbird has red shoulder patches, but the female does not. The male will perch on a high spot, flash his red shoulder patches, and sing to other male redwing blackbirds that happen to venture into his territory. Most other males get the message and leave his territory; those that do not leave, he attacks. He will also attack a stuffed, dead male redwing blackbird in his territory, or even a small piece of red cloth. Clearly, the spot of red is the characteristic that stimulates the male to defend his territory. Such key characteristics that trigger specific behavior patterns are called **sign stimuli.** (Refer back to figure 17.1 for another example of a sign stimulus.)

During the mating period, many animals are highly territorial. Within a gull colony, each nest is in a territory of about one square meter (figure 17.10). When one gull walks or lands on the territory of another, the defender walks toward the other in the upright threat posture. The head is pointed down with the neck stretched outward and upward. The folded wings are raised slightly as if to be used as clubs. The upright threat posture is one of a number of **intention movements** that signal what an animal is likely to do in the near future. The bird is communicating an intention to do something, to fight in this case, but it may not follow through. If the invader shows no sign of retreating, then one or both gulls may start pulling up the grass very vigorously with their beaks. This seems to make no sense. The gulls were ready to fight one moment; the next moment they apparently have forgotten about the conflict and are pulling grass. But the struggle has not been forgotten: pulling grass is an example of redirected aggression. In **redirected aggression,** the animal attacks something other than the natural opponent. If the intruding gull doesn't leave at this point, there will be an actual battle. (A person who starts

Figure 17.10 **Territoriality.** Colonial nesting seabirds typically have very small nest territories. Each territory is just out of pecking range of the neighbors'.

Figure 17.11 **A Dominance Hierarchy.** Many animals maintain order within their groups by establishing a dominance hierarchy. For example, whenever you see a group of cows or sheep walking in single file, it is likely that the dominant animal is at the head of the line, while the lowest ranking individual is at the end.

pounding the desk during an argument is showing redirected aggression. Look for examples of this behavior in your neighborhood cats and dogs—maybe even in yourself!)

The possession of a territory is often a requirement for reproductive success. In a way, then, territorial behavior has the effect of allocating breeding space and limiting population size to that which the ecosystem can support. This kind of behavior is widespread in the animal kingdom and can be seen in such diverse groups as insects, spiders, fish, reptiles, birds, and mammals.

Another way of allocating resources is by the establishment of a **dominance hierarchy,** in which a relatively stable, mutually understood order of priority within the group is maintained. A dominance hierarchy is often established in animals that form social groups. One individual in the group dominates all others. A second-ranking individual dominates all but the highest-ranking individual, and so forth, until the lowest-ranking individual must give way to all others within the group. This kind of behavior is seen in barnyard chickens, where it is known as a *pecking order*. Figure 17.11 shows a dominance hierarchy; the lead animal has the highest ranking and the last animal has the lowest ranking.

A dominance hierarchy allows certain individuals to get preferential treatment when resources are scarce. The dominant individual will have first choice of food, mates, shelter, water, and other resources because of the position occupied. Animals low in the hierarchy may fail to mate or be malnourished in times of scarcity. In many social animals, like wolves, only the dominant males and females reproduce. This ensures that the most favorable genes will be passed to the next generation. Poorly adapted animals with low rank may never reproduce. Once a dominance hierarchy is established it results in a more stable social unit with little conflict, except perhaps for an occasional altercation that reinforces the knowledge of which position an animal occupies in the hierarchy. Such a hierarchy frequently results in low-ranking individuals emigrating from the area. Migrating individuals are often subject to heavy predation. Thus, the dominance hierarchy serves as a population-control mechanism and a way of allocating resources. Resource allocation becomes most critical during periods of scarcity. In some areas, the dry part of the year is most stressful. In temperate areas, winter reduces many sources of food and forces organisms to adjust. Animals have several ways of coping with seasonal stress.

Some animals simply avoid the stress by hibernating. Hibernation is a physiological slowing of all body processes that allows an animal to survive on food it has stored within its body. Hibernation is typical of many insects, bats, marmots, and some squirrels. Other animals have built-in behavior patterns that cause them to store food during seasons of plenty for periods of scarcity. These behaviors are instinctive and are seen in a variety of animals. Squirrels bury nuts, acorns, and other seeds. (They also plant trees because they never find all the seeds they bury.) Chickadees stash seeds in cracks and crevices when seeds are plentiful and spend many hours during the winter exploring similar places for food. Some of the food they find is food they stored. Honeybees store honey, which allows them to live through the winter when nectar is not available. This requires a rather complicated set of behaviors that coordinates the activities of thousands of bees in the hive.

Navigation

The activities of honeybees involve communication among the various individuals that are foraging for nectar. The bees are able to communicate information about the direction and distance of the nectar source from the hive. If the source of nectar is some distance from the hive, the scout bee performs a "wagging dance" in the hive. The bee walks in a straight line for a short distance, wagging its rear end from side to side. It then circles around back to its starting position and walks the same path as before (figure 17.12). This dance is repeated many times. The direction of the straight-path portion of the dance indicates the direction of the nectar relative to the position of the sun. For instance, if the bee walks straight upward on a vertical surface in the hive, that tells the other bees to fly directly toward the sun. If the path is thirty degrees to the right of vertical, the source of the nectar is thirty degrees to the right of the sun's position.

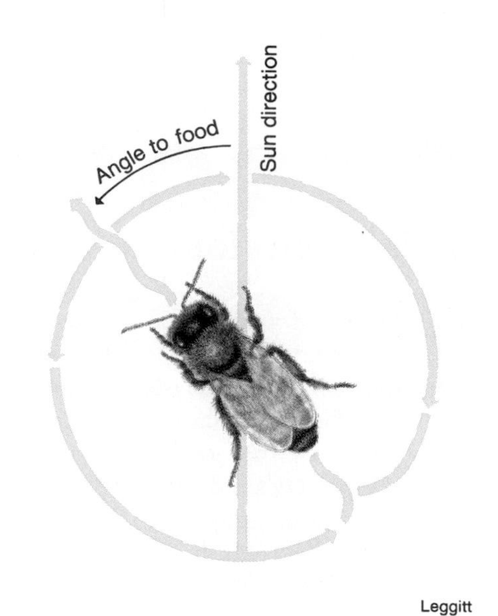

Figure 17.12 **Honeybee Communication and Navigation.** The direction of the straight, tail-wagging part of the dance indicates the direction to a source of food. The angle that this straight run makes with the vertical is the same angle at which the bee must fly in relation to the sun to find the food source. The length of the straight run and the duration of each dance cycle indicate the flying time necessary to reach the food source.

The duration of the entire dance and the number of waggles in the straight-path portion of the dance are positively correlated with the time the bee must fly to get to the nectar source. So the dance is able to communicate the duration of flight as well as the direction. Since the recruited bees have picked up the scent of the nectar source from the dancer, they also have information about the kind of flower to visit when they arrive at the correct spot. Since the sun is not stationary in the sky, the bee must constantly adjust its angle to the sun. It appears that they do this with some kind of internal clock. Bees that are prevented from going to the source of nectar or from seeing the sun will still fly in the proper direction sometime later, even though the position of the sun is different.

The ability to sense changes in time is often used by animals to prepare for seasonal changes. In areas away from the equator, the length of the day changes as the seasons change. The length of the day is called the **photoperiod.** Many birds prepare for migration and have their migration direction determined by the changing photoperiod. For example, in the fall of the year many birds instinctively change their behavior, store up fat, and begin to migrate from northern areas to areas closer to the equator. This seasonal migration allows them to avoid the harsh winter conditions signaled by the shortening of days. The return migration in the spring is triggered by the lengthening photoperiod. This migration certainly requires a lot of energy, but it allows many birds to exploit temporary food resources in the north during the summer months.

Like honeybees, some daytime-migrating birds use the sun to guide them. We need two instruments to navigate by the sun—an accurate clock and a sextant for measuring the angle between the sun and the horizon. Can a bird perform such measurements without instruments when we, with our much bigger brains, need these instruments to help us? It is unquestionably true! For nighttime migration, some birds use the stars to help them find their way. In one interesting experiment, warblers, which migrate at night, were placed in a planetarium. The pattern of stars as they appear at any season could be projected onto a large domed ceiling. During autumn, when these birds would normally migrate southward, the stars of the autumn sky were shown on the ceiling.

The birds responded with much fluttering activity at the south side of the cage, as if they were trying to migrate southward. Then the experimenters tried projecting the stars of the spring sky, even though it was autumn. Now the birds tended to try to fly northward, although there was less unity in their efforts to head north; the birds seemed somewhat confused. Nevertheless, the experiment showed that the birds recognized star patterns and were influenced by them.

There is evidence that some birds navigate by compass direction—that is, they fly as if they had a compass in their heads. They seem to be able to sense magnetic north. Their ability to sense magnetic fields has been proven at the U. S. Navy's test facility in Wisconsin. The weak magnetism radiated from this test site has changed the flight pattern of migrating birds, but it is yet to be proven that birds use the magnetism of the earth to guide their migration.

Many animals besides birds and bees have a time sense built into their bodies. For instance, you have one. Travelers who fly part way around the world by nonstop jet plane need some time to recover from "jet lag." Their digestion, sleep, or both, may be upset. Their discomfort is not caused by altitude, water, or food, but by having rapidly crossed several time zones. There is a great difference in the time as measured by the sun or local clocks and that measured by the body; the body's clock adjusts more slowly.

In the animal world, mating is the most obviously timed event. In the Pacific Ocean, off some of the tropical islands, lives a marine worm known as the *palolo worm.* Its habit of making a well-timed brief appearance in enormous swarms is a striking example of a biological-clock phenomenon. At mating time, these worms swarm into the shallows of the islands and discharge sperm and eggs. There are so many worms that the sea looks like noodle soup. The people of the islands find this an excellent time to change their diet. They dip up the worms much as North Americans dip up smelt or other small fish that are making a spawning run. The worms appear around the third quarter of the moon in October or November, the time varying somewhat according to local environmental conditions.

Social Behavior

Many species of animals are characterized by interacting groups called **societies,** in which there is division of labor. Societies differ from simple collections of organisms by the greater specialization of the society's individuals. The individuals performing one function cooperate with others having different special abilities. As a result of specialization and cooperation, the society has characteristics not found in any one member of the group: the whole is more than the sum of its parts. But if cooperation and division of labor are to occur, there must be communication among individuals and coordination of effort.

Honeybees, for example, have an elaborate communication system and are specialized for specific functions. A few individuals known as *queens* and *drones* specialize in reproduction, while large numbers of *worker* honeybees are involved in collecting food, defending the hive, and caring for the larvae. These roles are quite rigidly determined by inherited behavior patterns. Each worker honeybee has a specific task, and all tasks must be fulfilled for the group to survive and prosper. As they age, the worker honeybees move through a series of tasks over a period of weeks. When they first emerge from their wax cells, they clean the cells. Several days later, their job is to feed the larvae. Next they build cells. Later they become guards that challenge all insects that land near the entrance to the hive. Finally they become foragers who find and bring back nectar and pollen to feed the other bees in the hive. Foraging is usually the last job before the worker honeybee dies. Although this progression of tasks is the usual order, workers can shift from their main task to others if there is a need. Both the tasks performed and the progression of tasks are instinctively (genetically) determined.

A hive of bees may contain thousands of individuals, but under normal conditions only the queen bee and the male drones are capable of reproduction. None of the thousands of workers who are also females will reproduce. This does not seem to make sense because they appear to be giving up their chance to reproduce and pass their genes on to the next generation. Is this some kind of self-sacrifice on the part of the workers, or is there another explanation? In general, the workers in the hive are the daughters or sisters of the queen and therefore share a large number of her genes. This means they are really helping a portion of their genes get to the next generation by assisting in the raising of their own sisters, some of whom will become new queens. This argument has been used to partially explain behaviors in societies that might be bad for the individual but advantageous for the society as a whole.

Animal societies exhibit many levels of complexity, and types of social organization differ from species to species. Some societies show little specialization of individuals other than that determined by sexual differences or differences in physical size and endurance. The African wild dog illustrates such a flexible social organization. These animals are nomadic and hunt in packs. Although an individual wild dog can kill prey about its own size, groups are able to kill fairly large animals if they cooperate in the chase and the kill, which often involves a chase of several kilometers. When the dogs are young, they do not follow the pack. When adults return from a successful hunt, they regurgitate food if the proper begging signal is presented to them. Therefore, the young and adults that remained behind to guard the young are fed by the hunters. The young are the responsibility of the entire pack, which cooperates in their feeding and protection. During the time that the young are at the den site, the pack must give up its nomadic way of life. Therefore, the young are born during the time of year when prey are most abundant. Only one or two of the females in the pack have young each year. If every female had young, the pack couldn't feed them all. At about two months of age, the young begin traveling with the pack, and the pack can return to its nomadic way of life.

In many ways the honeybee and African wild dog societies are similar. Not all females reproduce, the raising of young is a shared responsibility, and there is some specialization of roles. The analysis and comparison of animal societies has led to the thought that there may be fundamental processes that shape all societies. The systematic study of all forms of social behavior, both animal and human, is a newly emerging area of study called **sociobiology.**

How did various types of societies develop? What selective advantage does a member of a social group have? In what ways are social groups better adapted to their environment than nonsocial organisms? How does social organization affect the way populations grow and change? These are difficult questions because, although evolution occurs at the population level, it is individual organisms that are selected. Thus, we need new ways of looking at evolutionary processes when describing the evolution of social structures.

The ultimate step in this new science is to analyze human societies according to sociobiological principles. Such an analysis is difficult and controversial, however, since humans have a much greater ability to modify behavior than other animals. Human social structure changes very rapidly compared to that of other animals. Sociobiology will continue to explore the basis of social organization and behavior and will continue to be an interesting and controversial area of study.

• Summary •

Behavior is how an organism acts, what it does, and how it does it. The kinds of responses that organisms make to environmental changes (stimuli) may be simple reflexes, very complex instinctive behavior patterns, or learned responses.

From an evolutionary viewpoint, behaviors represent adaptations to the environment. They increase in complexity and variety the more highly specialized and developed the organism is. All organisms have inborn or instinctive behavior, while higher animals also have one or more ways of learning. These include conditioning, imprinting, and insight. Communication for purposes of courtship and mating is accomplished by sounds, visual displays, and chemicals called pheromones. Many animals have special behavior patterns that are useful in the care and raising of young.

Territoriality is communicated through a series of behaviors involving aggressive displays, redirected aggression, intention movements, and sign stimuli. Dominance hierarchies and territorial behavior are both involved in the allocation of scarce resources. To escape from seasonal stress, some animals hibernate, others store food, and others migrate. Migration to avoid seasonal extremes involves a timing sense and some way of determining direction. Animals navigate by means of sound, celestial light cues, and magnetic fields.

Societies consist of groups of animals that specialize and cooperate. Sociobiology attempts to analyze all social behavior in terms of evolutionary principles, ecological principles, and population dynamics.

• Thinking Critically •

If you were going to teach an animal to communicate a message new to that animal, what message would you select? How would you teach the animal to communicate the message at the appropriate time?

• Experience This •

Spend ten minutes watching a specific individual animal. Birds, squirrels, and most insects are good subjects because many are active during the daytime. See if you can identify territorial behavior, sign stimuli, aggression, dominance, or any other behavior that was discussed in this chapter.

• Questions •

1. Why do students of animal behavior reject the idea that a singing bird is a happy bird?
2. Briefly describe the behavior of some animals as an example of unlearned behavior. Name the animal.
3. Briefly describe the behavior of some animals as an example of unlearned behavior. Name the animal.
4. Give an example of a conditioned response. Can you describe one that is not mentioned in this chapter?
5. Name three behaviors typically associated with reproduction.
6. How do territorial behavior and dominance hierarchies help to allocate scarce resources?
7. How do animals use chemicals, light, and sound to communicate?
8. What is sociobiology? Ethology? Anthropomorphism?
9. What is imprinting, and what value does it have to the organism?
10. Describe how honeybees communicate the location of a nectar source.

• Chapter Glossary •

anthropomorphism (an-thro-po-mōr'fizm) The ascribing of human feelings, emotions, or meanings to the behavior of animals.

associative learning (ă-so'shuh-tiv lur'ning) See **classical conditioning.**

behavior (be-hav'yur) How an organism acts, what it does, and how it does it.

classical conditioning (klas'ĭ-kul kon-dĭ-shun-ing) A kind of learning in which a neutral stimulus is associated with a natural stimulus to produce a particular response; also called **associative learning.**

conditioned response (kon-dĭ'shund re-spons') The behavior displayed when the neutral stimulus is given after association has occurred.

dominance hierarchy (dom'in-ants hi'ur-ar-ke) A relatively stable, mutually understood order of priority within a group.

ethology (e-thol'uh-je) The scientific study of the nature of behavior and its ecological and evolutionary significance in its natural setting.

imprinting (im'prin-ting) Learning in which a very young animal is genetically primed to learn a specific behavior in a very short period.

insight learning (in-sīt lur'ning) Learning in which past experiences are reorganized to solve new problems.

instinctive behavior (in-stink'tiv be-hāv'yur) Automatic, preprogrammed, or genetically determined behavior.

intention movements (in-ten'shun moov'ments) Behavior that signals what the animal is likely to do in the near future.

learning (lur'ning) A change in behavior as a result of experience.

pheromone (fĕr-uh-mōn) A chemical produced by an animal and released into the environment to trigger behavioral or developmental processes in some other animal of the same species.

photoperiod (fō''tō-pir'ē-ud) The length of the light part of the day.

redirected aggression (re-di-rek'ted ă-grĕ'shun) A behavior in which the aggression of an animal is directed away from an opponent and to some other animal or object.

response (re-spons') The reaction of an organism to a stimulus.

sign stimulus (sīn stim'yu-lus) A specific object or behavior that triggers a specific behavioral response.

society (so-si'uh-te) Interacting groups of animals of the same species that show division of labor.

sociobiology (so-sho-bi-ol'o-je) The systematic study of all forms of social behavior, both human and nonhuman.

stimulus (stim'yu-lus) Some change in the internal or external environment of an organism that causes it to react.

territoriality (tĕr''ĭ-tor'e-al'ĭ-te) A behavioral process in which an animal protects space for its exclusive use for food, mating, or other purposes.

territory (tĕr'ĭ-tor-e) A space that an animal defends against others of the same species.

Physiological Processes

All animals must solve several kinds of problems to survive. Many of these revolve around a central problem of the exchange of molecules between the organism and its environment. Since all animals require a constant input of organic matter and oxygen, you will find that many different systems are involved in providing these necessary molecules. The digestive, respiratory, and circulatory systems are all involved in meeting these needs. The circulatory system is also instrumental in transporting waste materials to the lungs and kidneys. These systems are discussed in chapter 18. Nutrition involves an understanding of the body's needs for different kinds of molecules for energy, building blocks, and the regulation of body activities. Chapter 19 discusses these basic concepts and also provides background that allows for evaluation of the many conflicting claims made for various foods and nutrient supplements.

The endocrine and nervous systems are involved in the integration of various systems of the body. Sense organs, muscles, and the nervous system are designed to respond to short-term changes; the endocrine system is better suited to monitoring the long-term growth responses involved in development, growth, and reproduction. Chapter 20 discusses the functions of the nervous and endocrine systems, and chapter 21 deals with a variety of topics related to reproduction. Because human reproductive activity is more than just a biological process, this chapter also includes consideration of psychological and sociological aspects of sexual behavior. ■

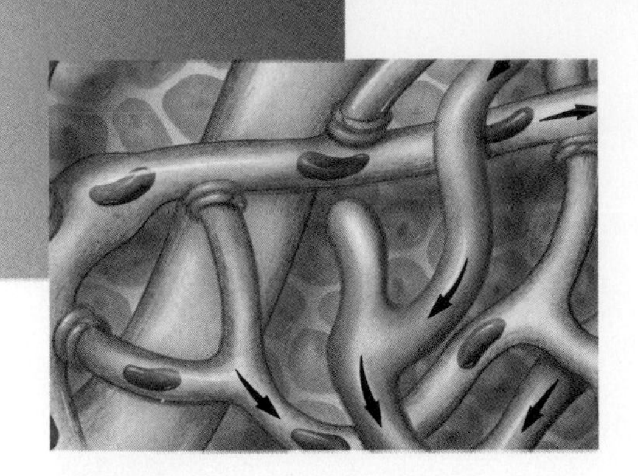

C H A P T E R 18

Materials Exchange in the Body

Chapter Outline

Purpose

The significance of cell surface area and cell volume must be appreciated in order to understand the movement of materials across cell surfaces. The transport of materials in and out of cells is an important concept, whether the organism is single-celled or multicellular. In addition, multicellular animals must have special organs with large surface areas to accomplish materials exchange. The first part of this chapter discusses the concept of the surface-area-to-volume ratio and is followed by a description of how this principle applies to the respiratory, circulatory, digestive, and excretory systems.

For Your Information

At one time it was fashionable for obese people to have a section of their small intestine removed to help them control weight. This had the effect of reducing the surface area available for nutrient absorption and allowed the individual to enjoy eating food without the negative side effect of gaining weight. This surgery seemed to be a simple answer to a problem shared by many. However, it was later discovered that, for some reason, people who had this operation tended to have a greater incidence of intestinal cancer later in life. Therefore, this surgical procedure is rarely performed today.

Learning Objectives

- Recognize that the size of cells is limited by surface area and volume considerations.
- Recognize that the circulatory system transports molecules, cells, and heat.
- Understand the important role of capillaries in allowing molecular exchanges between the circulatory system and cells.
- Understand that hemoglobin in red blood cells increases the oxygen-carrying capacity of the blood.
- Describe the ways in which carbon dioxide is carried in the blood.
- Understand how carbon dioxide level, blood pH, and breathing rate are interrelated.
- Describe the two mechanisms involved in reducing the size of food particles.
- List the different kinds of molecules absorbed by the gut.
- List three functions performed by the liver.
- Describe how the kidneys actively filter the blood.

Basic Principles

Cells are highly organized units that require a constant flow of energy in order to maintain themselves. The energy they require is provided in the form of nutrient molecules that enter the cell. Oxygen is required for the efficient release of energy from the large organic molecules that serve as fuel. Inevitably, as oxidation takes place, waste products are formed that are useless or toxic. These must be removed from the cell. All these exchanges of food, oxygen, and waste products must take place through the cell surface.

As a cell grows its volume increases, and the amount of metabolic activity required to maintain it rises. There is an increase in the quantity of materials that must be exchanged between the cell and its surroundings. Growth cannot increase indefinitely. Size becomes limiting because of two interrelated factors:

1. The ability to transport materials across cell membranes is limited by the surface area of the cell.
2. Diffusion is only effective over short distances, and as a cell increases in size, its surface area and its volume do not increase at the same rate. This relationship between surface area and volume is often expressed as the **surface-area-to-volume ratio** (SA/V).

Assume that we have a cube 1 centimeter on a side, as shown in figure 18.1. This cube will have a volume of 1 cubic centimeter (1 cm^3). Each side of the cube will have a surface area of 1 square centimeter (1 cm^2), and since there are six surfaces on a cube, it will have a total surface area of 6 square centimeters (6 cm^2). It has a surface-area-to-volume ratio of 6:1 (6 cm^2 of surface to 1 cm^3 of volume). If we increase the size of the cube so that each side has an area of 4 cm^2, the total surface area of the cube will be 24 cm^2 (six surfaces times 4 cm^2 per square equals 24 cm^2). However, the volume now becomes 8 cm^3, since each side of this new, larger cube is 2 cm (2 cm $\times$ 2 cm $\times$ 2 cm = 8 cm^3). Therefore, the surface-area-to-volume ratio is 24:8, which reduces to 3:1. So you can see that as an object increases in size its volume increases faster than its surface area.

The ability to transport materials into or out of a cell is determined by its surface area, while its metabolic demands are determined by its volume. So, the larger a cell becomes, the more difficult it is to satisfy its needs. Some cells overcome this handicap by having highly folded cell membranes that substantially increase their surface area. This is particularly true of cells that line the intestine or are involved in the transport of large numbers of nutrient molecules. These cells have many tiny, folded extensions of the cell membrane called **microvilli** (see figure 18.2).

In a similar way, the structure of an automobile radiator increases the efficiency of heat exchange between the engine and the air. The radiator has many fins attached to tubes through which a coolant fluid is pumped. Because of the large surface area provided by the fins, heat from the engine can be efficiently radiated away.

In addition to the limitation that surface area presents to the transport of materials, large cells also have a problem with diffusion. The molecular process of diffusion is quite rapid over short distances but becomes very slow over longer distances. Diffusion is generally insufficient to handle the needs of cells if it must take place over a distance of more

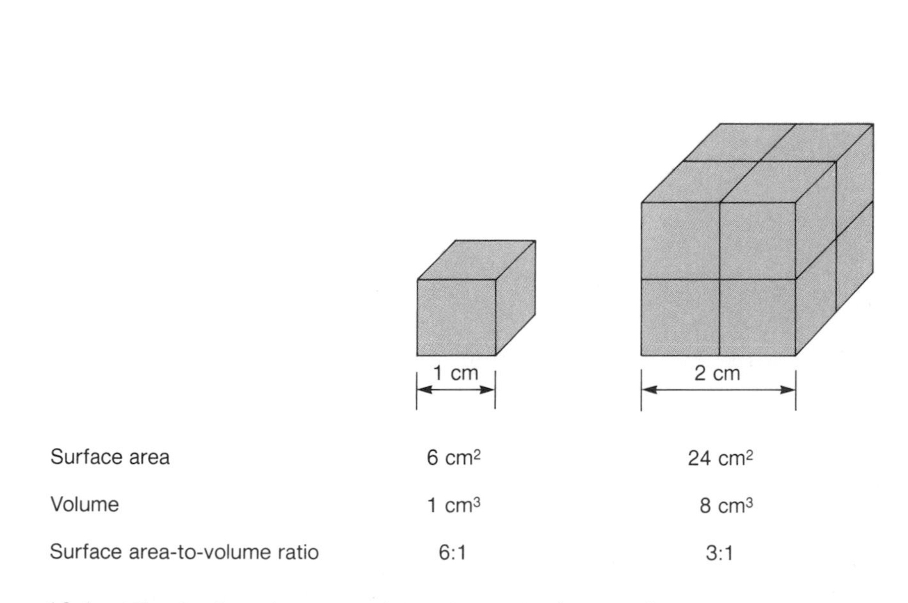

1 cm
2 cm
Surface area
6 cm2
24 cm2
Volume
1 cm3
8 cm3
Surface area-to-volume ratio
6:1
3:1

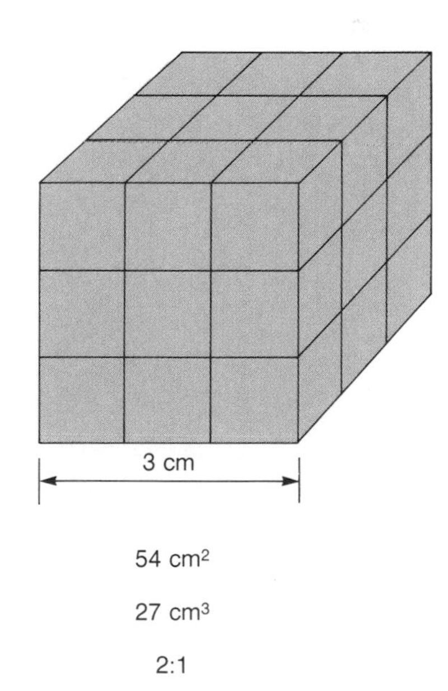

Figure 18.1 **The Surface-Area-to-Volume Ratio.** As the size of an object increases, its volume increases faster than its surface area. Therefore, the surface-area-to-volume ratio decreases.

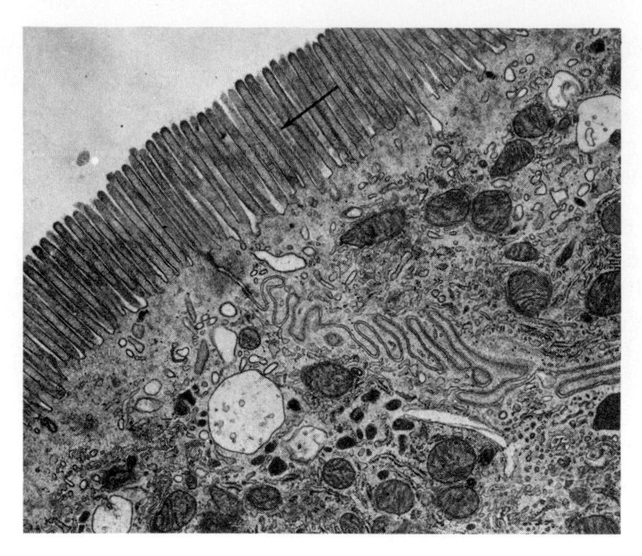

(a)

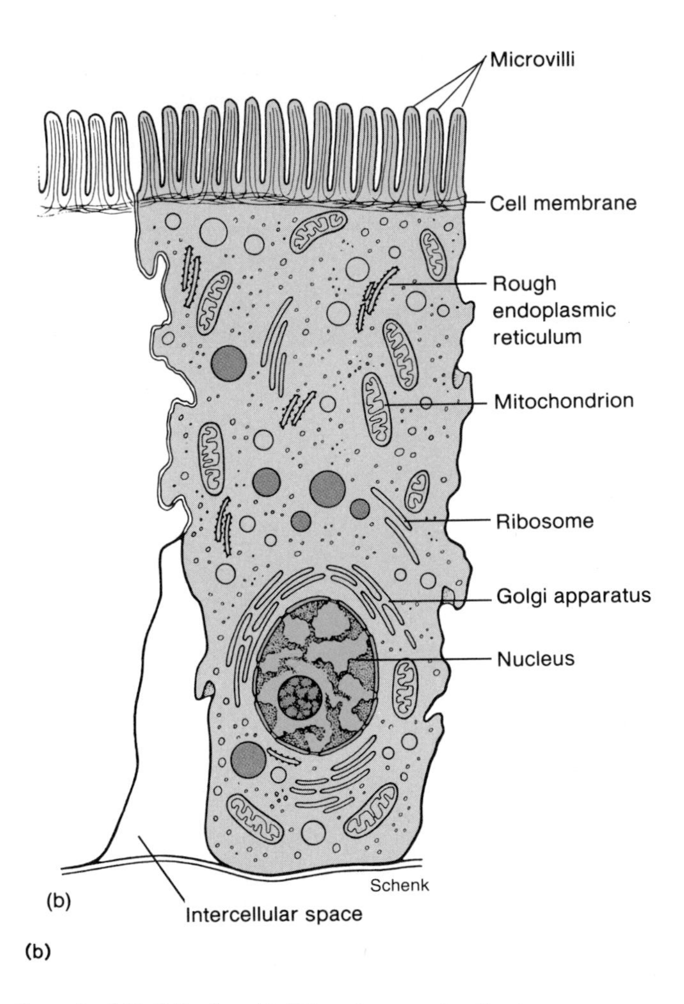

Figure 18.2 **Intestinal Cell Surface Folding.** Intestinal cells that are in contact with the food in the gut have highly folded surfaces. The tiny projections of these cells are called microvilli. These can be clearly seen in the photomicrograph in (a). The drawing in (b) shows that only one surface has these projections.

than 1 millimeter. The center of the cell would die before it received the molecules it needed if the distance were greater. Because of this and the problems presented by the surface-area-to-volume ratio, it is understandable that the basic unit of life, the cell, must remain small.

All single-celled organisms are limited to a small body size because they handle the exchange of molecules through their cell membranes. Large, multicellular organisms consist of a multitude of cells, many of which are located far from the surface of the organism. Each cell within a multicellular organism must solve the same materials-exchange problems as single-celled organisms. Large organisms have several interrelated systems that are involved in the exchange and transport of materials so that each cell can meet its metabolic needs.

Circulation

Large, multicellular organisms like humans consist of trillions of cells. Since many of these cells are buried within the organism far from the body surface, there must be some sort of distribution system to assist them in solving their exchange problems. The primary mechanism used is the circulatory system.

The circulatory system consists of several fundamental parts. **Blood** is the fluid medium that assists in the transport of materials and heat. The **heart** is a pump that forces the fluid blood from one part of the body to another. The **arteries** and **veins** are, respectively, the vessels that distribute blood to organs and return blood from organs. **Capillaries** are tiny vessels through which exchange takes place between cells and the blood.

The Nature of Blood

Blood is a fluid that consists of a watery matrix called **plasma.** This fluid plasma contains many kinds of dissolved molecules and larger suspended cellular components. The primary function of the

BOX 18.1 Buffers

Since many kinds of chemical activities, such as enzyme-controlled reactions, are sensitive to changes in the pH of the surroundings, it is important to regulate the pH of the blood and other body fluids within very narrow ranges. The normal blood pH is about 7.4. Although the respiratory system and kidneys are involved in regulating the pH of the blood, there are several buffer systems in the blood that prevent wide fluctuations in pH. **Buffers** are mixtures of weak acids and the salts of weak acids. A buffer system tends to maintain constant pH because it can either accept or release hydrogen ions (H^+). The weak acid can release hydrogen ions (H^+) if a base is added to the solution, and the negatively charged ion of the salt can accept hydrogen ions (H^+) if an acid is added to the solution.

One example of a buffer system frequently encountered in the body is the carbonic acid/bicarbonate ion buffer system. In this system carbonic acid (H_2CO_3) is a weak acid and bicarbonate ion (HCO_3^-) is the weak base.

$$H_2CO_3 \rightleftharpoons HCO_3^- + H^+$$

Addition of an acid to the mixture causes the equilibrium to shift to the left. The additional hydrogen ions attach to HCO_3^- to form H_2CO_3. This removes the additional hydrogen ions from solution and ties them up in the H_2CO_3, so that the amount of free hydrogen ions remains constant.

$$H_2CO_3 \rightleftharpoons HCO_3^- + H^+ + \text{added } H^+$$

If a base is added to the mixture, the equilibrium shifts to the right and additional hydrogen ions are released to tie up the hydroxyl ions; the pH remains unchanged.

$$H_2CO_3 + \text{added } OH^- \rightleftharpoons H^+ + HCO_3^- + HOH$$

blood is to transport molecules, cells, and heat from one place to another. The major kinds of molecules that are distributed by the blood are respiratory gases (oxygen and carbon dioxide), nutrients of various kinds, waste products, and chemical messengers (hormones). Blood has special characteristics that allow it to distribute respiratory gases very efficiently. Although little oxygen is carried as free, dissolved oxygen in the plasma, *red blood cells* (*RBCs*) contain **hemoglobin,** an iron-containing molecule, to which oxygen molecules readily bind.

Because hemoglobin is inside red blood cells, it is possible to assess certain kinds of health problems by counting the number of red blood cells. If the number is low, the person will not be able to carry oxygen efficiently and will tire easily. This condition, in which a person has reduced oxygen-carrying capacity, is called **anemia.** Anemia can also result when a person does not get enough iron. Since iron is a central atom in hemoglobin molecules, people with an iron deficiency will not be able to manufacture sufficient hemoglobin. They would therefore be anemic even though their number of red blood cells may be normal.

Red blood cells are also important in the transport of carbon dioxide. Carbon dioxide is produced as a result of normal aerobic respiration of food materials in the cells of the body. If it is not eliminated, it causes the blood to become more acidic (lowers its pH), eventually resulting in death (see box 18.1).

Table 18.1 lists the variety of cells found in blood. While the red, hemoglobin-containing erythrocytes serve in the transport of oxygen, the *white blood cells* (*WBCs*) carried in the blood are involved in defending against harmful agents. These cells help the body resist many diseases. They constitute the core of the **immune system.** *Neutrophils, eosinophils, basophils,* and *monocytes* are capable of phagocytosis. When harmful microorganisms (e.g., bacteria, viruses, fungi); cancer cells; or toxic molecules enter the body, WBCs (1) recognize, (2) boost their abilities to engulf, (3) move toward, (4) engulf, and (5) destroy the problem-causers. While most can move from the bloodstream into the surrounding tissue, monocytes undergo such a striking change that they are given a different name—*macrophages.* Macrophages can be found throughout the body and are the most active of the phagocytes.

The other white cells, *lymphocytes,* work with phagocytes to provide protection. The two major types are *T-lymphocytes* (*T-cells*) and *B-lymphocytes* (*B-cells*). T-cells are the heart of cell-mediated immune response. This highly complex response involves the release of chemical messengers that coordinate the response, an increase in the population of T- and B-cells, and stimulation of B-cell and macrophage activities. Some T-cells are capable of killing dangerous cells by destroying their cell membranes. B-cells are the source of protein molecules known as *antibodies* or *immunoglobulins.* They are the heart of antibody-mediated immunity. Immunoglobulins (globular blood proteins that aid in resisting disease) are released into the body when B-cells are stimulated by the presence of dangerous agents. The antibodies are specifically constructed so that they will bind to particular agents. Agents that stimulate the production of antibodies and then combine with the antibodies are called *antigens* or *immunogens.* Examples include viruses, bacteria, fungi, poisons, transplanted tissues, and cancer cells. The combination of antibody–antigen renders the antigen harmless by destroying its poisonous properties, preventing it from infecting human cells, or stimulating phagocytosis.

Table 18.1
The Composition of Blood

Component	Quantity Present
Plasma	55%
Water	91.5%
Protein	7.0%
Other materials	1.5%
Cellular material	45%
Red blood cells (erythrocytes)	4.3–5.8 million/mm³
White blood cells (leukocytes)	5–9 thousand/mm³
Lymphocytes	25%–30% of white cells present
Monocytes	3%–7% of white cells present
Neutrophils	57%–67% of white cells present
Eosinophils	1%–3% of white cells present
Basophils	less than 1% of white cells present
Platelets	130–360 thousand/mm³

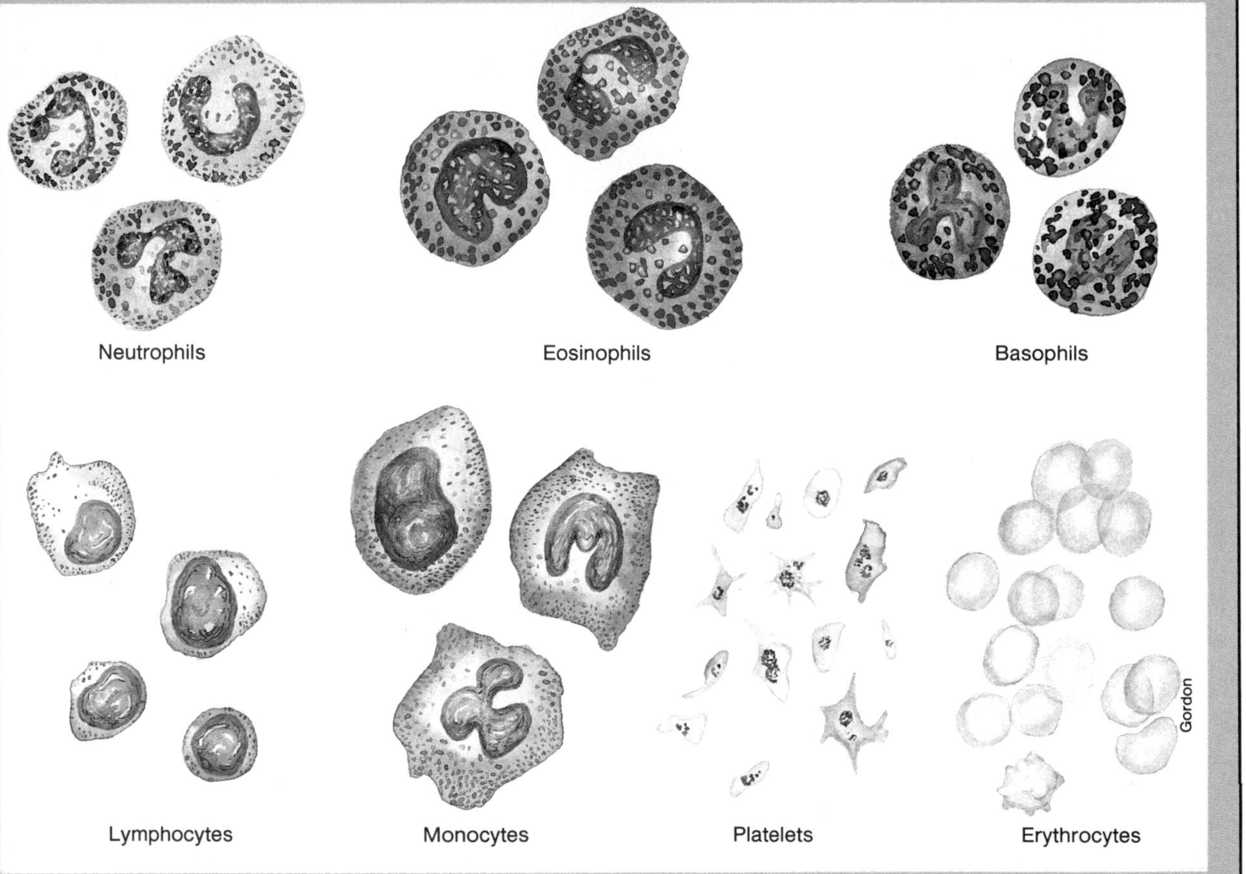

Heat is also transported by the blood. Heat is generated by metabolic activities and must be lost from the body. To handle excess body heat, blood is shunted to the surface of the body, where heat can be radiated away. In addition, humans and some other animals have the ability to sweat. The evaporation of sweat from the body surface also gets rid of excess heat. If the body is losing heat too rapidly, blood flow is shunted away from the skin, and metabolic heat is conserved. Vigorous exercise produces an excess of heat so that, even in cold weather, blood is shunted to the skin and the skin feels hot.

Carbon dioxide can be carried in the blood in three forms: about 10% is

$$CO_2 + H_2O \underset{\text{anhydrase}}{\overset{\text{carbonic}}{\rightleftharpoons}} H_2CO_3 \rightleftharpoons H^+ + HCO_3^-$$

dissolved in the plasma of the blood, about 20% is carried attached to hemoglobin molecules, and 70% is carried as bicarbonate ions. An enzyme in red blood cells known as **carbonic anhydrase** assists in converting carbon dioxide into bicarbonate ions (HCO_3^-), which can be carried as dissolved ions in the plasma of the blood. The reversible chemical equation above shows the changes that occur.

When the blood reaches the lungs, dissolved carbon dioxide is lost from the plasma, and carbon dioxide is released from the hemoglobin molecules as well. In addition, the bicarbonate ions reenter the red blood cells and can be converted back into molecular carbon dioxide by the same enzyme-assisted process that converts carbon dioxide to bicarbonate ions. The importance of this mechanism will be discussed later when the exchange of gases at the lung surface is described.

The plasma also carries nutrient molecules from the gut to other locations where they are modified, metabolized, or incorporated into cell structures. Amino acids and simple sugars are carried dissolved in the blood. Lipids, which are not water-soluble, are carried as suspended particles, the lipoproteins. Most lipids do not enter the bloodstream directly from the gut but are carried to the bloodstream by the lymphatic system. Other organs, like the liver, manufacture or modify molecules for use elsewhere; therefore, they must constantly receive raw materials and distribute their products to the cells that need them.

Many different kinds of hormones are produced by the brain, reproductive organs, digestive organs, and glands of the body. These are secreted into the bloodstream and transported throughout the body. Tissues with appropriate receptors take up these molecules and respond to these chemical messengers.

Blood can perform its transportation function only if it moves. The organ responsible for providing the energy to pump the blood is the heart.

The Heart

In order for a fluid to flow through a tube, there must be a pressure difference between the two ends of the tube. Water flows through pipes because it is under pressure. Because the pressure is higher behind a faucet than at the spout, water flows from the spout when the faucet is opened. The circulatory system can be analyzed from the same point of view. The heart is a muscular pump that provides the pressure necessary to propel the blood throughout the body. It must continue its cycle of contraction and relaxation, or blood stops flowing and body cells are unable to satisfy their immediate needs. Some cells, such as brain cells, are extremely sensitive to having their flow of blood interupted because they require a constant supply of glucose and oxygen. Others, such as muscle cells or skin cells, are much more able to withstand temporary interruptions of blood flow.

The heart consists of four chambers and four sets of valves that work together to ensure that blood flows in one direction only. Two of these chambers, the **right and left atria,** are relatively thin-walled, weak structures that collect blood from the major veins and empty it into the larger, more muscular ventricles (figure 18.3). Most of the flow of blood from the atria to the ventricles is caused by the lowered pressure produced within the ventricles as they relax. The contraction of the thin-walled atria assists in emptying them more completely.

The **right and left ventricles** are powerful muscles whose contraction forces blood to flow through the arteries to all parts of the body. The valves between the atria and ventricles, known as **atrioventricular valves,** are important one-way valves that allow the blood to flow from the atria to the ventricles but prevent flow in the opposite direction. Similarly, there are valves in the aorta and pulmonary artery, known as **semilunar valves.** The **aorta** is the large artery that carries blood from the left ventricle to the body, and the **pulmonary artery** carries blood from the right ventricle to the lungs. The semilunar valves prevent blood from flowing back into the ventricles. If the atrioventricular or semilunar valves are damaged or function improperly, the efficiency of the heart as a pump is diminished, and the person may develop an enlarged heart or other symptoms. Malfunctioning heart valves are often diagnosed because they cause abnormal sounds as the blood passes through them. These sounds are referred to as heart murmurs. Similarly, if the ventricles are weakened because of infection, damage from a heart attack, or lack of exercise, the pumping efficiency of the heart is reduced and the person develops symptoms which may include chest pain, shortness of breath, or fatigue. The pain is caused by the heart muscle not getting sufficient blood to satisfy its needs. If this persists, the portion of the heart muscle not receiving blood will die. Shortness of breath and fatigue have the same cause, since the heart is not able to pump blood efficiently to the lungs, muscles, and other parts of the body.

The right and left sides of the heart have slightly different jobs, since they pump blood to different parts of the body. The right side of the heart receives blood from the general body and pumps it through the pulmonary arteries to the lungs, where exchange of oxygen and carbon dioxide takes place. This is called **pulmonary circulation.** The larger, more powerful left side of the heart receives blood from the lungs and delivers it through the aorta to all parts of the body. This is known as **systemic circulation.** Both circulatory pathways are shown in figure 18.4. The systemic circulation is responsible for gas, nutrient, and waste exchange in all parts of the body except the lungs.

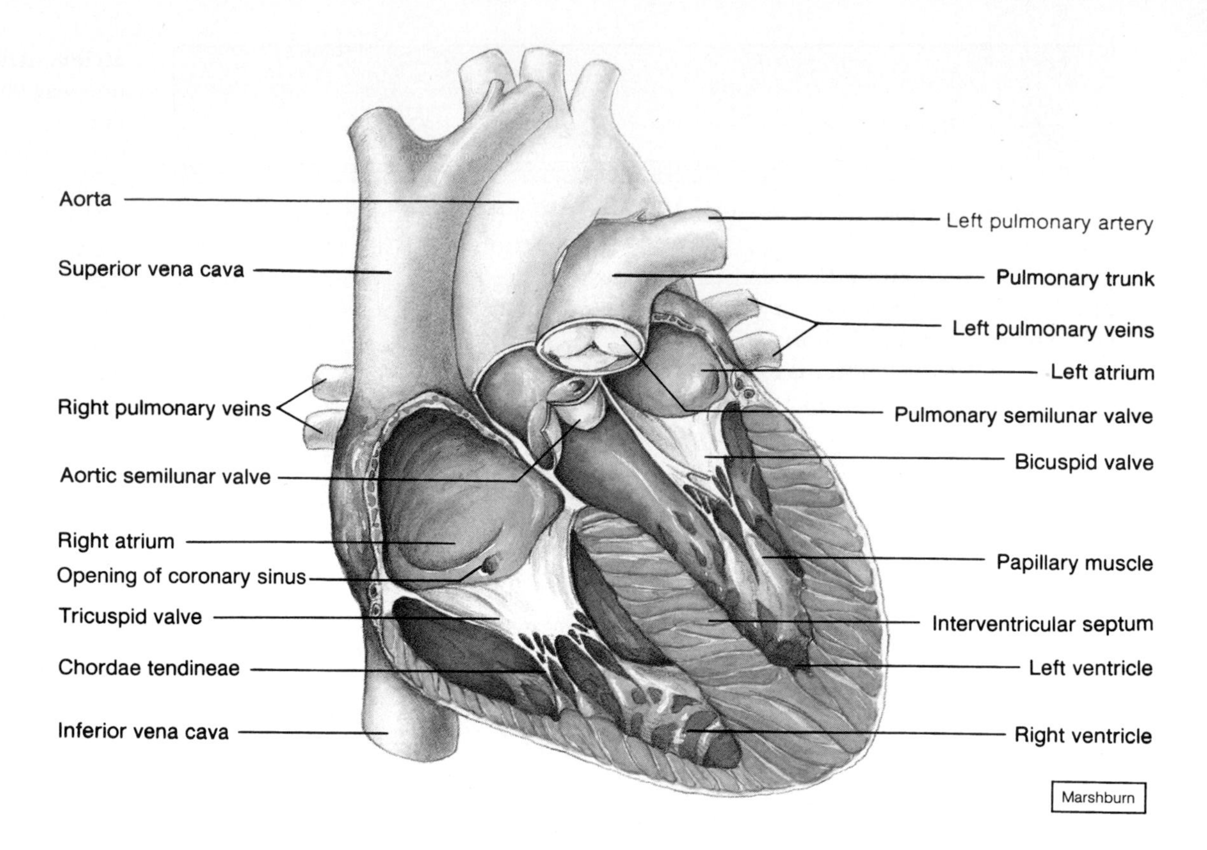

Figure 18.3 **The Anatomy of the Heart.** The heart consists of two thin-walled chambers called atria that contract to force blood into the two ventricles. When the ventricles contract, the atrioventricular valves (bicuspid and tricuspid) close, and blood is forced into the aorta and pulmonary artery. Semilunar valves in the aorta and pulmonary artery prevent the blood from flowing back into the ventricles when they relax.

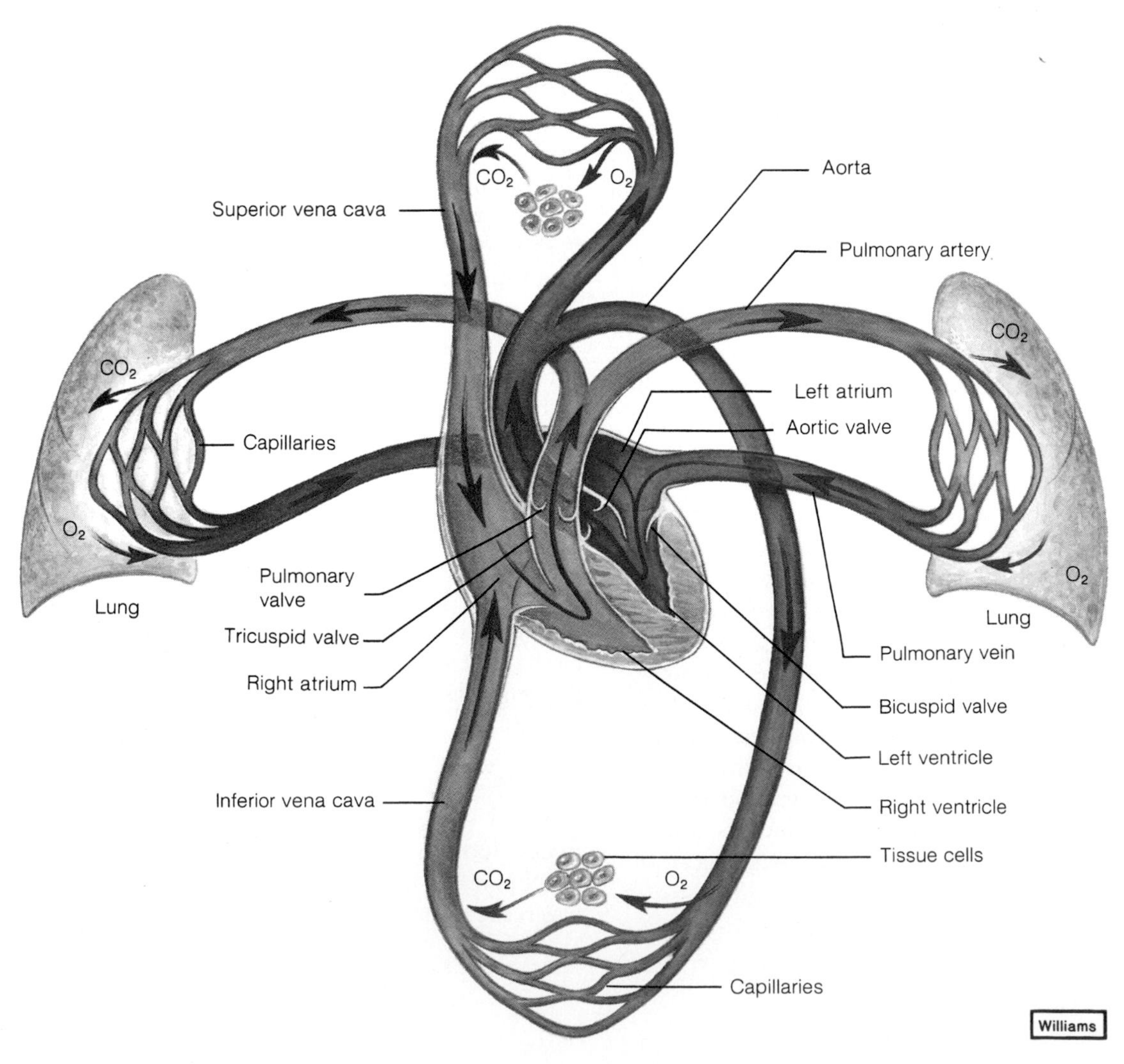

Figure 18.4 **Pulmonary and Systemic Circulation.** The right ventricle pumps blood that is poor in oxygen to the two lungs, where it receives oxygen and turns bright red. The blood is then returned to the left atrium by way of four pulmonary veins. This part of the circulatory system is known as the pulmonary circulation. The left ventricle pumps oxygen-rich blood to all parts of the body except the lungs. This blood returns to the right atrium, depleted of its oxygen, by way of the superior vena cava from the head region and the inferior vena cava from the rest of the body. This portion of the circulatory system is known as the systemic circulation.

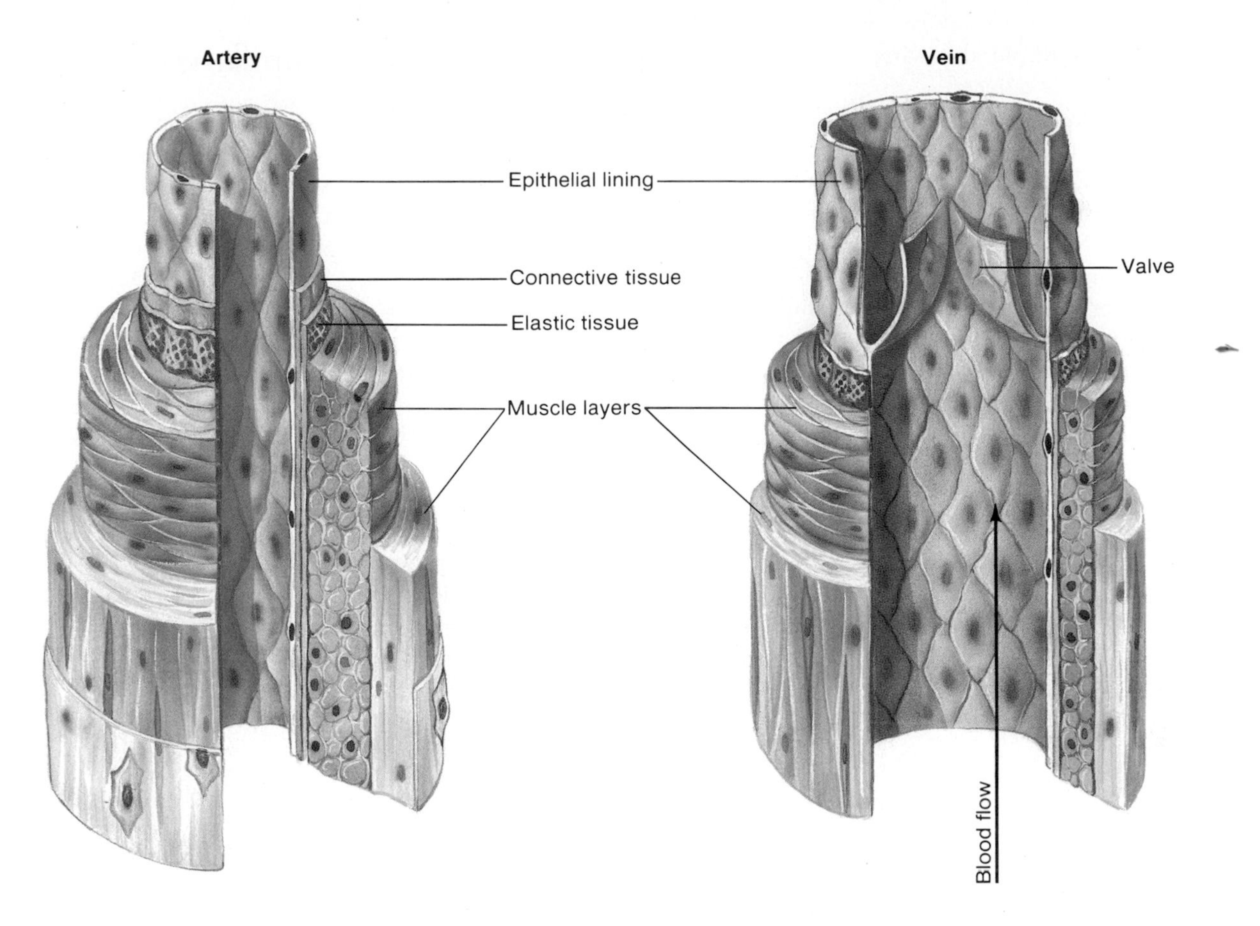

Figure 18.5 **The Structure of Arteries and Veins.** The walls of arteries are much thicker than the walls of veins because the pressure in arteries is much higher than the pressure in veins. The pressure generated by the ventricles of the heart forces blood through the arteries. Veins often have very low pressure. The valves in the veins prevent the blood from flowing backward, away from the heart.

Arteries and Veins

Arteries and veins are the tubes that transport blood from one place to another within the body. Figure 18.5 compares the structure and function of arteries and veins. Arteries carry blood away from the heart, because it is under considerable pressure from the contraction of the ventricles. A typical pressure that would be recorded in a large artery while the heart is contracting would be about 120 millimeters of mercury. This is known as the **systolic blood pressure.** The pressure that would be recorded while the heart is not contracting would be about 80 millimeters of mercury. This is known as the **diastolic blood pressure.** A blood pressure reading includes both of these numbers and would be recorded as 120/80. (Originally blood pressure was measured by how high the pressure of the blood would cause a column of mercury [Hg] to rise in a tube. Although many of the devices used today have dials or digital readouts and contain no mercury, they are still calibrated in mmHg.)

The walls of arteries are relatively thick and muscular. Healthy arteries have the ability to expand as blood is pumped into them and return to normal as the pressure drops. This dampens the peak pressure within the arteries and reduces the likelihood that they will burst. If arteries become hardened and less resilient, the peak blood pressure rises and they are more likely to rupture. The elastic nature of the arteries is also responsible for assisting the flow of blood. When they return to normal from their stretched condition they give a little push to the blood that is flowing through them.

Arteries branch into smaller and smaller blood vessels as the blood is distributed from the large aorta to millions of tiny capillaries. Some of the smaller arteries, called **arterioles,** may contract or relax to regulate the flow of blood to specific parts of the body. Major parts of the body that receive differing amounts of blood, depending on need, are the digestive system, muscles, and skin. When light-skinned people *blush,* it is because many arterioles in the skin have expanded, allowing a large volume of blood to flow to the capillaries of the skin. Since the blood is red, their skin reddens. Similarly, when you exercise, there is an increased blood flow to muscles to accommodate their increased metabolic needs for oxygen and glucose and to get rid of wastes. Exercise also results in an increased flow of blood to the skin, which allows for heat loss. At the same time, the amount of blood flowing to the digestive system is reduced.

Veins collect blood from the capillaries and return it to the heart. The pressure in these blood vessels is very low. Some of the largest veins may have a blood pressure of 0.0 mmHg. The walls of veins are not as muscular as those of arteries. Because of the low pressure, veins must have valves that prevent the blood from flowing backwards, away from the heart. Veins are often found at the surface of the body and are seen as blue lines. *Varicose veins* result when veins contain faulty valves that do not allow efficient return of blood to the heart. Therefore, blood pools in these veins, and they become swollen bluish networks.

Since pressure in veins is so low, muscular movements of the body are important in helping to return blood to the heart. When muscles of the body contract, they compress nearby veins, and this pressure pushes blood along in the veins. Because the valves allow blood to flow only toward the heart, this activity acts as an additional pump to help return blood to the heart. People who sit or stand for long periods without using their leg muscles tend to have a considerable amount of blood pool in the veins of their legs and lower body. Thus, less blood may be available to go to the brain and the person may faint.

Although the arteries are responsible for distributing blood to various parts of the body and arterioles regulate where blood goes, it is the function of capillaries to assist in the exchange of materials between the blood and cells.

Capillaries

Capillaries are tiny thin-walled tubes that receive blood from arterioles. They are so small that red blood cells must go through them in single file. They are so numerous that each cell in the body has a capillary located near it. It is estimated that there are about 1,000 square meters of surface area represented by the capillary surface in a typical human. Each capillary wall consists of a single layer of cells and therefore presents only a thin barrier to the diffusion of materials between the blood and cells. It is also possible for liquid to flow through tiny spaces between the individual cells of most capillaries (figure 18.6). The flow of blood through these smallest blood vessels is relatively slow. This allows time for the

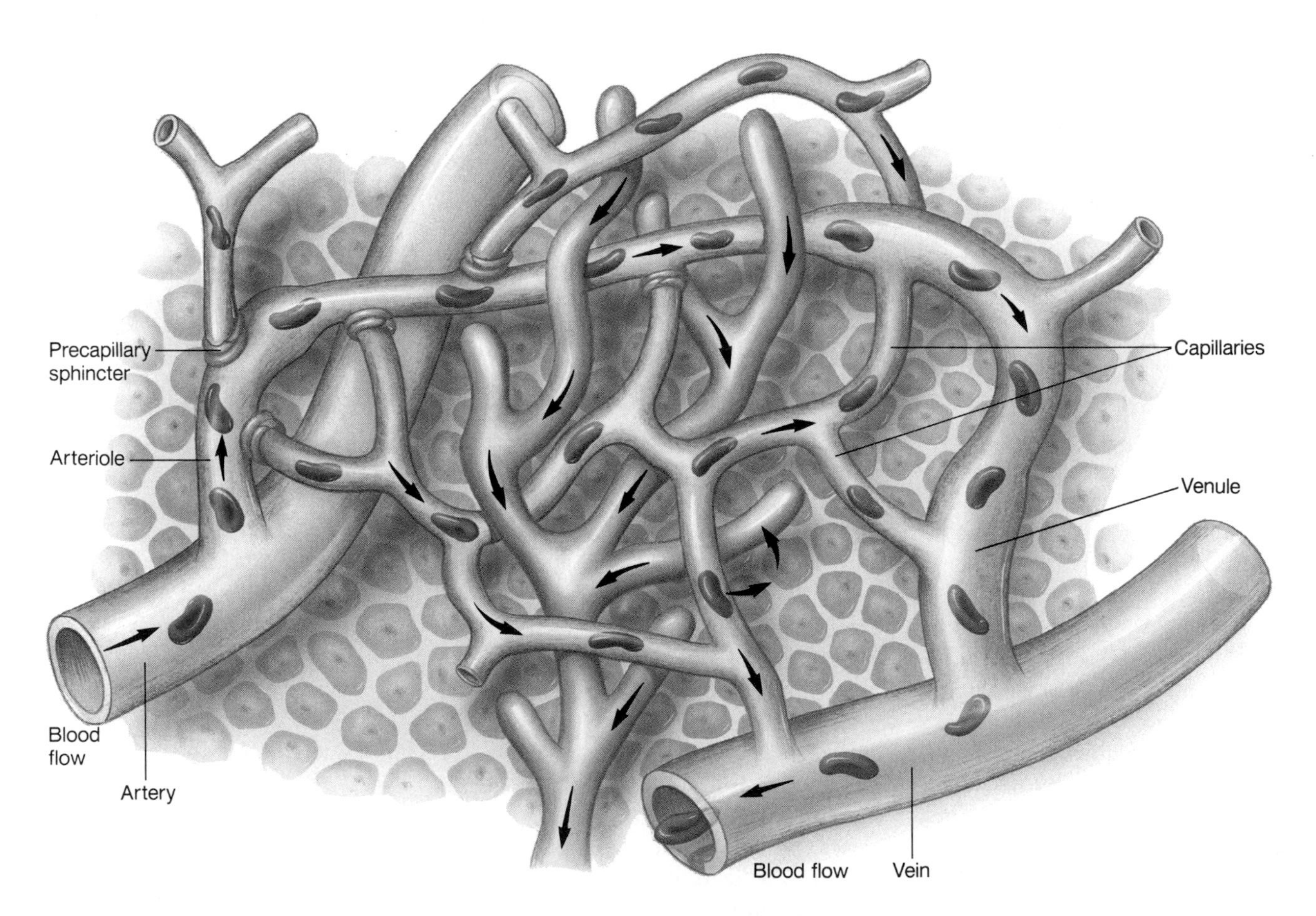

Figure 18.6 **Capillaries.** Capillaries are tiny blood vessels. Exchange of cells and molecules can occur between blood and tissues through their thin walls. Molecules diffuse in and out of the blood, and cells such as monocytes can move from the blood through the thin walls into the surrounding tissue. There is also a flow of liquid through holes in the capillary walls. This liquid, called lymph, bathes the cells and eventually enters small lymph vessels that return lymph to the circulatory system near the heart.

diffusion of such materials as oxygen, glucose, and water from the blood to surrounding cells, and for the movement of such materials as carbon dioxide, lactic acid, and ammonia from the cells into the blood.

In addition to molecular exchange, considerable amounts of water and dissolved materials leak through the small holes in the capillaries. This liquid is known as **lymph.** Lymph is produced when the blood pressure forces water and some small dissolved molecules through the walls of the capillaries. Lymph bathes the cells but must eventually be returned to the circulatory system by lymph vessels or swelling will occur. Return is accomplished by the **lymphatic system,** a collection of thin-walled tubes that branch throughout the body. These tubes collect lymph that is filtered from the circulatory system and ultimately empty it into major blood vessels near the heart. Figure 18.7 shows the structure of the lymphatic system.

Some of this leakage through the capillary walls is normal, but the flow is subject to changes in pressure inside the capillaries, pressure in the tissues, and changes in the permeability of the capillary wall. If pressure inside the capillary increases, more fluid may leak from the capillaries into the tissues and cause swelling. This swelling is called *edema,* and it is common in circulatory disorders. Another cause of edema is an increase in the permeability of the capillaries. This is commonly associated with injury to a part of the body: a sprained ankle or smashed thumb are examples you have probably experienced.

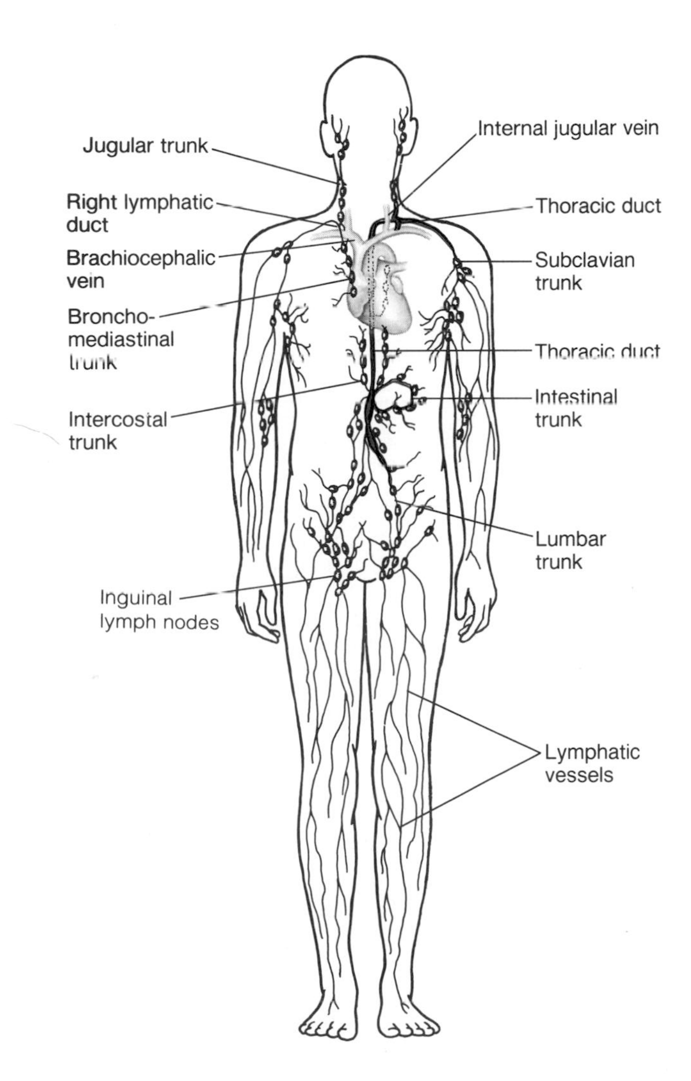

Figure 18.7 **The Lymphatic System.** The lymphatic system consists of many ducts that transport lymph fluid back toward the heart. Along the way the lymph is filtered in the lymph nodes, and bacteria and other foreign materials are removed before the lymph is returned to the circulatory system near the heart.

Gas Exchange

One of the places in the body where this interplay between blood flow, capillary exchange, and surface area can be readily appreciated is in the lungs.

Respiratory Anatomy

The **lungs** are organs of the body that allow gas exchange to take place between the air and blood. Associated with the lungs is a set of tubes that conducts air from outside the body to the lungs. The single large-diameter **trachea** is supported by rings of cartilage that prevent its collapse. It branches into two major **bronchi** that deliver air to smaller and smaller branches. Bronchi are also supported by cartilage. The smallest tubes, known as **bronchioles,** contain smooth muscle and are therefore capable of constricting. Finally, the bronchioles deliver air to clusters of tiny sacs, known as **alveoli,** where the exchange of gases takes place between the air and blood.

The nose, mouth, and throat are also important parts of the air-transport pathway because they modify the humidity and temperature of the air and clean the air as it passes. Figure 18.8 illustrates the various parts of the respiratory system.

Breathing-System Regulation

Breathing is the process of pumping air in and out of the lungs. It is accomplished by the movement of a muscular organ known as the **diaphragm,** which separates the lung cavity from the abdominal cavity. In addition, muscles

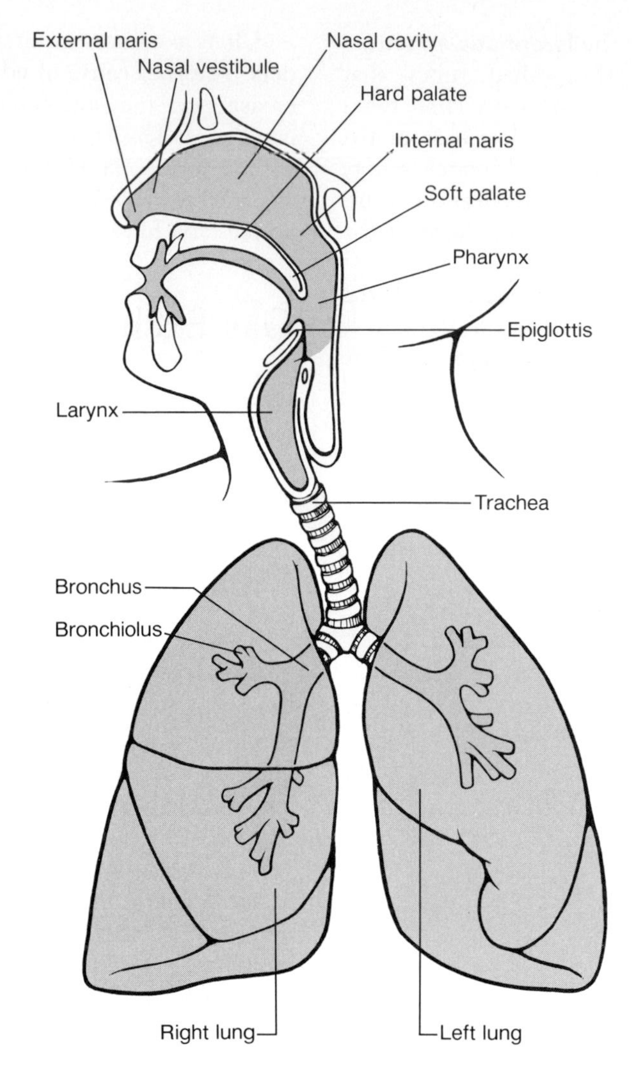

Figure 18.8 **Respiratory Anatomy.** Although the alveoli of the lungs are the places where gas exchange takes place, there are many other important parts of the respiratory system. The nasal cavity cleans, warms, and humidifies the air entering the lungs. The trachea is also important in cleaning the air going to the lungs.

located between the ribs (*intercostal* muscles) are attached to the ribs in such a way that their contraction causes the chest wall to move outward and upward, which increases the size of the chest cavity. During inhalation, the diaphragm moves downward and the external intercostal muscles of the chest wall contract, causing the volume of the chest cavity to increase. This results in a lower pressure in the chest cavity compared to the outside air pressure. Consequently, air flows from the outside high-pressure area through the trachea, bronchi, and bronchioles to the alveoli. During normal relaxed breathing, exhalation is accomplished by the chest wall and diaphragm simply returning to their normal position. Muscular contraction is not involved (figure 18.9).

However, when the body's demand for oxygen increases during exercise, the only way that the breathing system can respond is by exchanging the gases in the lungs more rapidly. This can be accomplished both by increasing the breathing rate and by increasing the volume of air exchanged with each breath. Increase in volume exchanged per breath is accomplished in two ways. First, the muscles of inhalation can contract more forcefully, resulting in a greater change in the volume of the chest cavity. In addition, the lungs can be emptied more completely by contracting the muscles of the abdomen, which force the abdominal

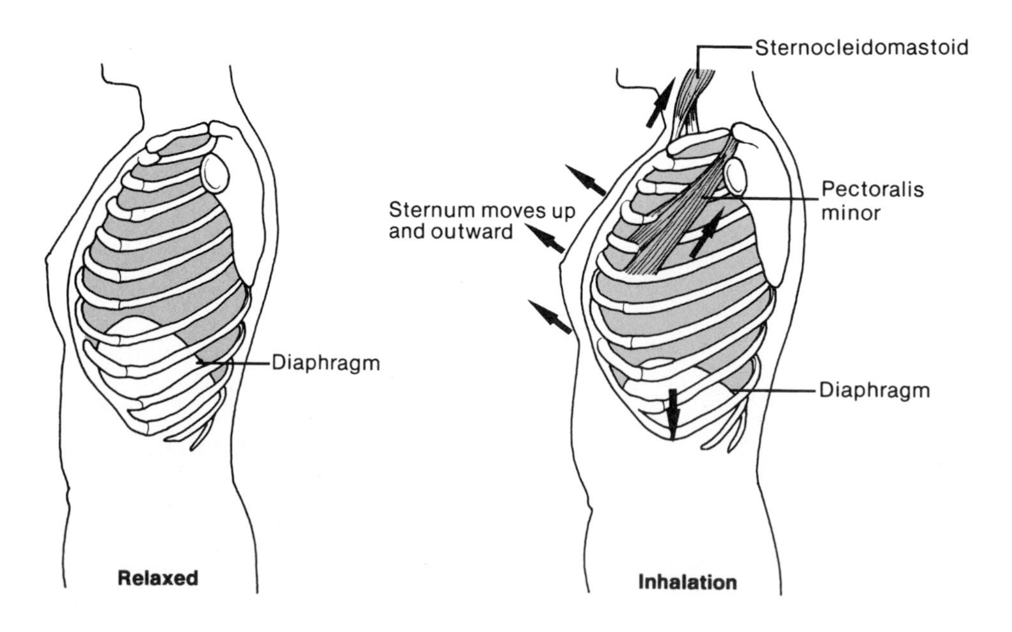

Figure 18.9 **Breathing Movements.** During inhalation, the diaphragm and external intercostal muscles between the ribs contract, causing the volume of the chest cavity to increase. During a normal exhalation, these muscles relax, and the chest volume returns to normal.

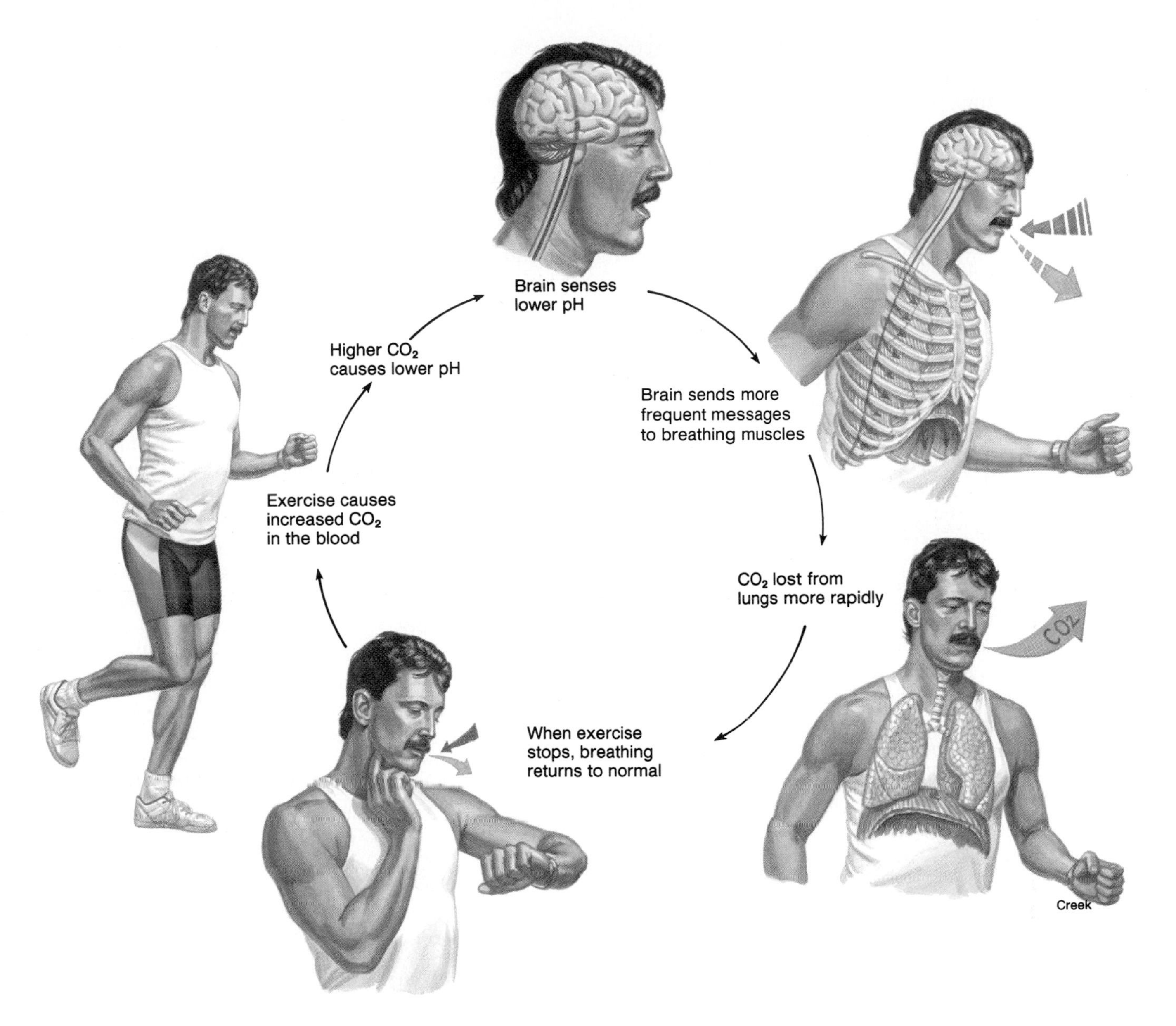
Brain senses
lower pH
Higher CO_2
causes lower pH
Brain sends more
frequent messages
to breathing muscles
Exercise causes
increased CO_2
in the blood
CO_2 lost from
lungs more rapidly
CO_2
When exercise
stops, breathing
returns to normal
Creek

Figure 18.10 **The Control of Breathing Rate.** The rate of breathing is controlled by specific cells in the brain that sense the pH of the blood. When the amount of CO_2 increases, the pH drops (becomes more acid) and the brain sends more frequent messages to the diaphragm and intercostal muscles, causing the breathing rate to increase. More rapid breathing increases the rate at which CO_2 is lost from the blood; thus, the blood pH rises (becomes less acid) and the breathing rate decreases.

contents upward and compress the lungs. A set of internal intercostal muscles also helps to compress the chest. You are familiar with both of these mechanisms. When you exercise you breathe more deeply and more rapidly.

Several mechanisms can cause changes in the rate and depth of breathing, but the primary mechanism involves the amount of carbon dioxide present in the blood. Carbon dioxide is a waste product of aerobic cellular respiration and becomes toxic in high quantities because it combines with water to form carbonic acid:

$$CO_2 + H_2O \rightarrow H_2CO_3$$

As mentioned previously, if carbon dioxide cannot be eliminated, the pH of the blood is lowered. Eventually, this may result in death.

Exercising causes an increase in the amount of carbon dioxide in the blood because muscles are oxidizing glucose more rapidly. This lowers the pH of the blood. Certain brain cells are sensitive to changes in blood pH. When they sense a lower blood pH, nerve impulses are sent more frequently to the diaphragm and intercostal muscles. These muscles contract more rapidly and more forcefully, resulting in more rapid, deeper breathing. Since more air is being exchanged per minute, carbon dioxide is lost from the lungs more rapidly. When exercise stops, blood pH rises, and breathing eventually returns to normal (figure 18.10). Bear in mind, however, that moving air in and out of the lungs is of no value unless oxygen is diffusing into the blood and carbon dioxide is diffusing out.

Lung Function

The lungs are organs that allow blood and air to come in close contact with each other. Air is pumped in and out of the lungs during breathing. The blood flows through capillaries in the lungs and is in close contact with the air in the cavities of the lungs. Since oxygen that is going to enter the body or carbon dioxide that is going to exit the body must pass through a surface, the efficiency of exchange is limited by the surface area available. This problem is solved in the lungs by the large number of tiny sacs, the alveoli. Each alveolus is about 0.25–0.5 millimeters across. However, alveoli are so numerous that the total surface area of all these sacs is about 70 square meters—comparable to the floor space of many standard-sized classrooms. Associated with these alveoli are large numbers of capillaries (figure 18.11). The walls of both the capillaries and alveoli are very thin, and the close association of alveoli and capillaries in the lungs facilitates the diffusion of oxygen and carbon dioxide across these membranes.

Another factor that increases the efficiency of gas exchange is that both the blood and air are moving. Because blood is flowing through capillaries in the lungs, the capillaries continually receive new blood that is poor in oxygen and high in carbon dioxide. As blood passes by the alveoli, it is briefly exposed to the gases in the alveoli, where it gains oxygen and loses carbon dioxide. Thus, blood that leaves the lungs is high in oxygen and low in carbon dioxide. Although the movement of air in the lungs is not in one direction, as is the case with blood, the cycle of inhalation and exhalation allows air that is high in carbon dioxide and low in oxygen to exit the body and brings in new air that is rich in oxygen and low in carbon dioxide.

Any factor that interferes with the flow of blood or air or alters the effectiveness of gas exchange in the lungs reduces the efficiency of the organism. A poorly pumping heart sends less blood to the lungs, and the person will experience shortness of breath as a symptom. Similarly, diseases like *asthma,* which causes constriction of the bronchioles, reduce the flow of air into the lungs and inhibit gas exchange.

Any process that reduces the number of alveoli will also reduce the efficiency of gas exchange in the lungs. *Emphysema* is a progressive disease in which some of the alveoli are lost and replaced with connective tissue. As the disease progresses, those afflicted have less and less respiratory surface area and experience greater and greater difficulty in getting adequate oxygen, even though they may be breathing more rapidly.

The breathing mechanism is designed to get oxygen into the bloodstream so that it can be distributed to the cells that are carrying on the oxidation of food molecules, such as glucose and fat. Obtaining food molecules involves a variety of organs and activities associated with the digestive system.

Obtaining Nutrients

All cells must have a continuous supply of nutrients to provide the energy they require and the building blocks needed to construct the macromolecules typical of living things. The specific functions of various kinds of nutrients are discussed in chapter 19. This section will deal with the processing and distribution of different kinds of nutrients. The digestive system consists of a muscular tube with several specialized segments. In addition, there are glands that secrete digestive juices into the tube. Four different kinds of activities are involved in getting nutrients to the cells that need them: mechanical processing, chemical processing, nutrient uptake, and chemical alteration.

Mechanical Processing

The digestive system is designed as a disassembly system. Its purpose is to take large chunks of food and ultimately break them down to smaller molecules that can be taken up by the circulatory system and distributed to cells. The first step in this process is mechanical processing.

It is important to grind large particles into small pieces in order to increase their surface area and allow for more efficient chemical reactions. It is also important to add water to the food, which further disperses the particles and provides the watery environment needed for these chemical reactions. Materials must also be mixed, so that all the molecules that need to interact with one another have a good chance of doing so. The mouth and the stomach are the major body regions involved in reducing the size of food particles. The teeth are involved in cutting and grinding food to increase its surface area. This is another example of the surface-area-to-volume concept presented at the beginning of this chapter. The watery mixture that is added to the food in the mouth is known as *saliva,* and the three pairs of glands that produce saliva are known as **salivary glands.** Saliva contains the enzyme **salivary amylase,** which initiates the breakdown of starch. Saliva also lubricates the mouth and helps to bind food before swallowing.

In addition to having taste buds that help to identify foods, the tongue performs the important service of helping to position the food between the teeth and pushes it to the back of the throat for swallowing. The mouth is very much like a food processor in which mixing and grinding take place. Figure 18.12 describes and summarizes the functions of these structures.

Once the food has been chewed, it is swallowed and passes down the esophagus to the stomach. The process of swallowing involves a complex series of events. First, a ball of food, known as a *bolus,* is formed by the tongue and moved to the back of the mouth cavity. Here it

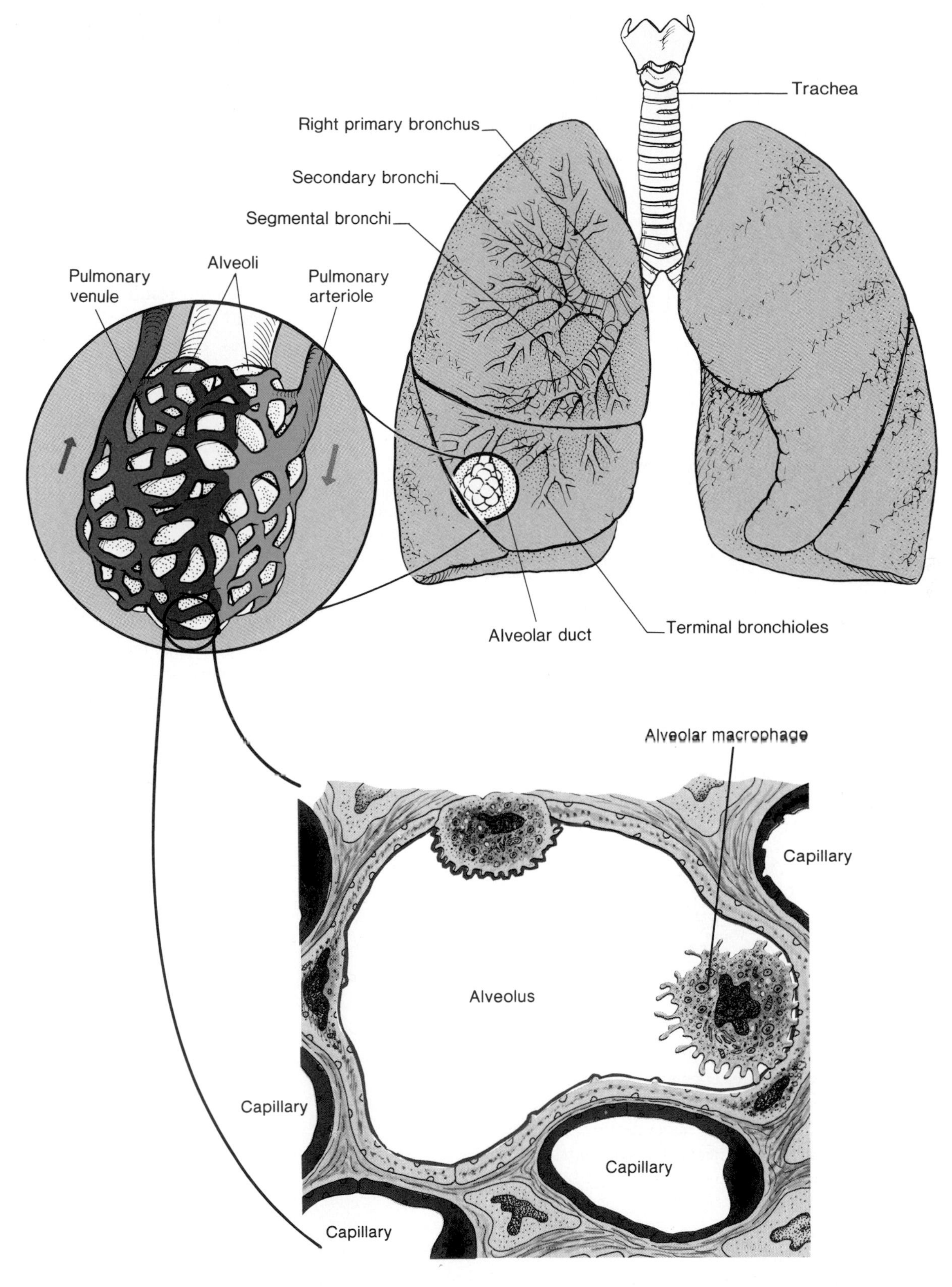

Figure 18.11 **The Association of Capillaries with Alveoli.** The exchange of gases takes place between the air-filled alveolus and the blood-filled capillary. The capillaries form a network around the saclike alveoli. The thin walls of the alveolus and capillary are in direct contact with one another; their combined thickness is usually less than one micrometer (a thousandth of a millimeter).

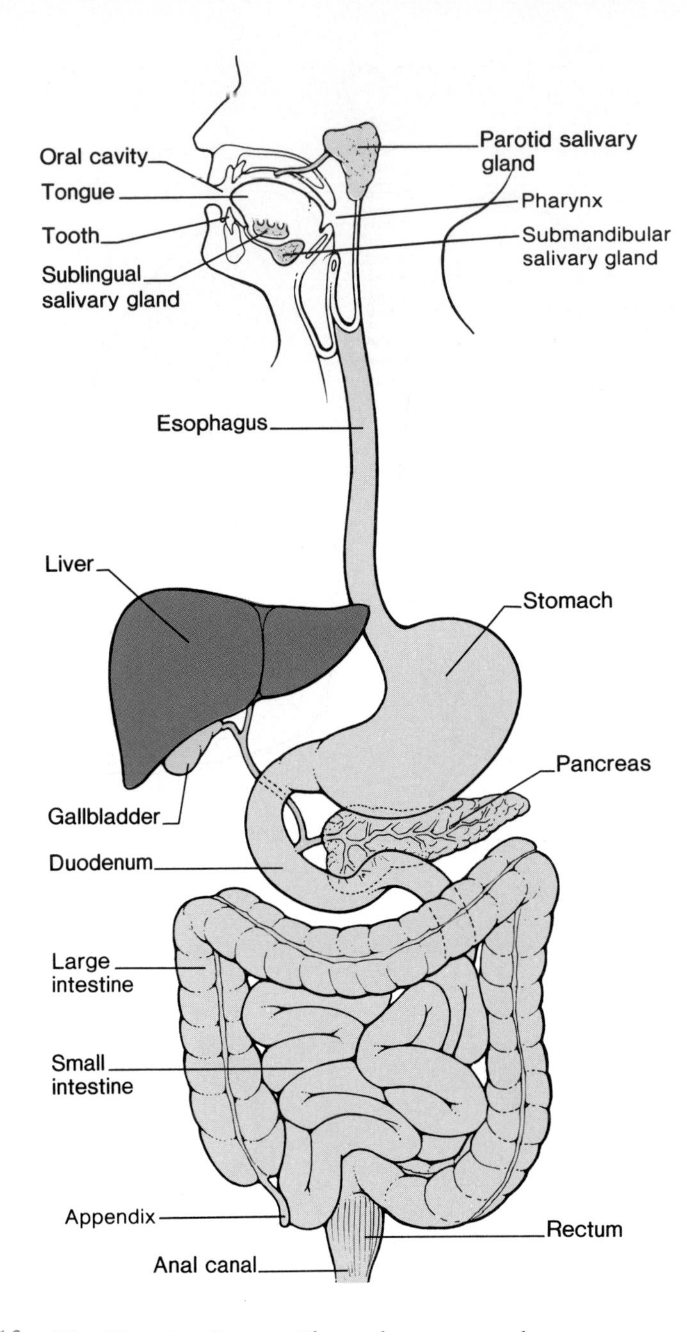

Figure 18.12 **The Digestive System.** The teeth, tongue, and enzymes from the salivary glands modify the food before it is swallowed. The stomach adds acid and enzymes and further changes the texture of the food. The food is eventually emptied into the duodenum, where the liver and pancreas add their secretions. The small intestine also adds enzymes and is involved in absorbing nutrients. The large intestine is primarily involved in removing water.

stimulates the walls of the throat, which is also known as the **pharynx.** Nerve endings in the lining of the pharynx are stimulated, causing a reflex contraction of the walls of the esophagus, which transports the bolus to the stomach. In the stomach, additional liquid, called **gastric juice,** is added to the food. Gastric juice contains enzymes and hydrochloric acid. The major enzyme of the stomach is **pepsin,** which initiates the breakdown of protein. The pH of gastric juice is very low, generally around pH–2. Consequently, very few bacteria or protozoa emerge from the stomach alive. Those that do have special protective features that allow them to survive as they pass through the stomach. The entire mixture is churned by the contractions of the three layers of muscle in the stomach wall. The combined activities of enzymatic breakdown, chemical breakdown by hydrochloric acid, and mechanical processing by muscular movement results in a thoroughly mixed liquid called *chyme.* Chyme eventually leaves the stomach through a valve known as the **pyloric sphincter** and enters the small intestine.

The first part of the small intestine is known as the **duodenum.** In addition to producing enzymes, the duodenum secretes several kinds of hormones that regulate the release of food from the stomach and the release of secretions from the pancreas and liver. The **pancreas** produces a number of different digestive enzymes and also secretes large amounts of bicarbonate ions, which neutralize stomach acid. The **liver** is a large organ in the upper abdomen that performs several functions. One of its functions is the secretion of **bile.** When bile leaves the liver, it is stored in the **gallbladder** prior to being released into the duodenum. When bile is released from the gallbladder, it assists mechanical mixing by breaking large fat globules into smaller particles. This process is called *emulsification.*

Emulsification is important because fats are not soluble in water, yet the reactions of digestion must take place in a water solution. Bile causes large globules of fat to be broken into much smaller units (increasing the surface-area-to-volume ratio) much as soap breaks up fat particles into smaller units that are suspended in water and washed away. The activity of bile is important for the further digestion of fats in the intestine.

Along the length of the intestine, additional watery juices are added until the mixture reaches the **large intestine.** The large intestine is primarily involved in reabsorbing the water that has been added to the food tube along with saliva, gastric juice, bile, pancreatic secretions, and intestinal juices.

Table 18.2
Digestive Enzymes and Their Functions

Enzyme	Site of Production	Molecules Altered	Molecules Produced
Salivary amylase	Salivary glands	Starch	Smaller polysaccharides (many sugar molecules attached together)
Pepsin	Stomach lining	Proteins	Peptides (several amino acids)
Gastric lipase	Stomach lining	Fats	Fatty acids and glycerol
Chymotrypsin	Pancreas	Polypeptides (long chains of amino acids)	Peptides
Trypsin	Pancreas	Polypeptides	Peptides
Carboxypeptidase	Pancreas	Peptides	Smaller peptides and amino acids
Pancreatic amylase	Pancreas	Polysaccharides	Disaccharides
Pancreatic lipase	Pancreas	Fats	Fatty acids and glycerol
Nuclease	Pancreas	Nucleic acids	Nucleotides
Aminopeptidase	Intestinal lining	Peptides	Smaller peptides and amino acids
Dipeptidase	Intestinal lining	Dipeptides	Amino acids
Lactase	Intestinal lining	Lactose	Glucose and galactose
Maltase	Intestinal lining	Maltose	Glucose
Sucrase	Intestinal lining	Sucrose	Glucose and fructose
Nuclease	Intestinal lining	Nucleic acids	Nucleotides

Although the physical processing of food by mechanical forces, pH changes, and emulsification is important in reducing the size of the food particles, food cannot enter the circulatory system unless it is broken down into fundamental chemical units by enzymes and other chemical processes.

Chemical Processing

Chapter 5 introduced the topic of enzymes and how they work. Some enzymes, such as those involved in glycolysis, the Krebs cycle, and protein synthesis, are produced and used inside cells; others, such as the digestive enzymes, are produced by cells and secreted into the digestive tract. Digestive enzymes are simply a special class of enzymes and have the same characteristics as the enzymes you studied previously. They are protein molecules that speed up specific chemical reactions and are sensitive to changes in temperature or pH. The various digestive enzymes, the sites of their production, and their functions are listed in table 18.2.

Nutrient Uptake

As we move simple sugars, amino acids, glycerol, and fatty acids into the circulatory system, we encounter another situation where surface area is important. The amount of material that can be taken up is limited by the surface area available. This problem is solved by increasing the surface area of the intestinal tract in several ways. First, the intestine is a very long tube; the longer the tube, the greater the internal surface area. In a typical adult human it is about 6 meters long. In addition to being long, the lining of the intestine consists of millions of fingerlike projections called **villi,** which increase the surface area. When we examine the cells that make up the villi, we find that they also have folds in their surface membranes. All of these characteristics increase the surface area available for the transport of materials from the gut into the circulatory system (figure 18.13). We estimate that the cumulative effect of all of these features produces a total intestinal surface area of about 250 square meters. This is equivalent to about half the area of a football field.

The surface area by itself would be of little value if it were not for the intimate contact of the circulatory system with this lining. Each villus contains several capillaries and a branch of the lymphatic system called a **lacteal.** The close association between the intestinal surface and the circulatory and lymphatic systems allows for the efficient uptake of nutrients from the cavity of the gut into the circulatory system.

Several different kinds of processes are involved in the transport of materials from the intestine to the circulatory system. Some molecules, such as water and many ions, simply diffuse through the wall of the intestine into the circulatory system. Other materials, such as amino acids and simple sugars, are assisted across the membrane by carrier molecules. Fatty acids and glycerol are absorbed into the intestinal lining cells where they are resynthesized into fats and enter lacteals in the villi. Since the lacteals are part of the lymphatic system, which eventually empties its contents into the circulatory system, fats are also transported by the blood. They just reach the blood by a different route.

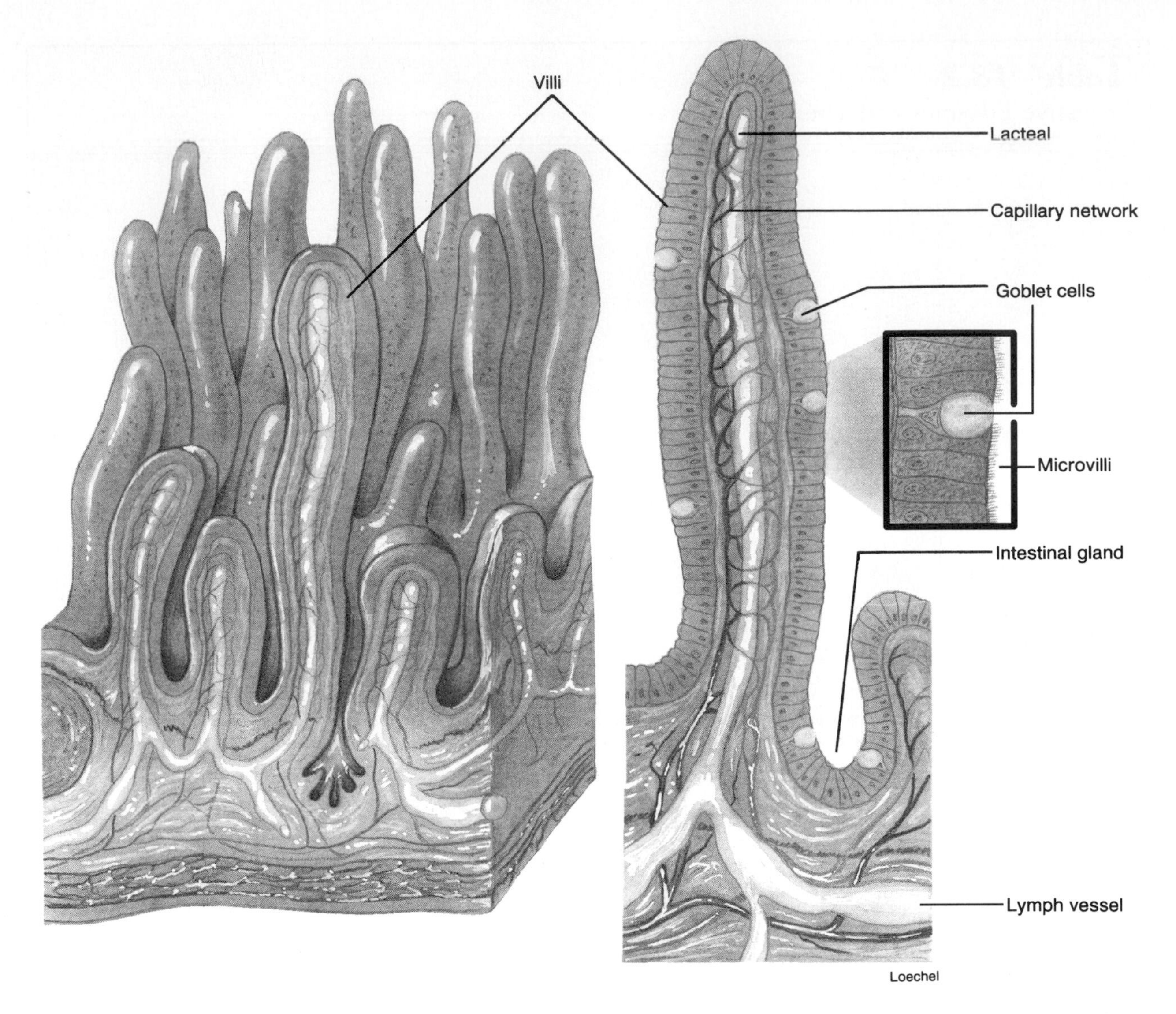

Figure 18.13 **The Exchange Surface of the Intestine.** The surface area of the intestinal lining is increased by the many fingerlike projections known as villi. Within each villus are capillaries and lacteals. Most kinds of materials enter the capillaries, but most fat-soluble substances enter the lacteals, giving them a milky appearance. Lacteals are part of the lymphatic system. Since the lymphatic system empties into the circulatory system, fat-soluble materials also eventually enter the circulatory system. The close relationship between these vessels and the epithelial lining of the villus allows for efficient exchange of materials from the intestinal cavity to the circulatory system.

Chemical Alteration—The Role of the Liver

When the blood leaves the intestine, it flows directly to the liver through the **hepatic portal vein.** Portal veins are blood vessels that collect blood from capillaries in one part of the body and deliver it to a second set of capillaries in another part of the body without passing through the heart. Thus, the hepatic portal vein collects nutrient-rich blood from the intestine and delivers it directly to the liver. As the blood flows through the liver, enzymes in the liver cells modify many of the molecules and particles that enter them. One of the functions of the liver is to filter any foreign organisms from the blood that might have entered through the intestinal cells. It also detoxifies many dangerous molecules that might have entered with the food.

Many foods contain toxic substances that could be harmful if not destroyed by the liver. Ethyl alcohol is one obvious example, but many plants contain various kinds of toxic molecules that are present in small quantities and could accumulate to dangerous levels if the liver did not perform its role of detoxification.

In addition, the liver is responsible for modifying nutrient molecules. The liver collects glucose molecules and synthesizes glycogen, which can be stored in the liver for later use. When glucose is in short supply the liver can convert some of its stored glycogen back into glucose. Although amino acids are not stored, the liver can change the relative numbers of different amino acids circulating in the blood. It can remove the amino group from one kind of amino acid and attach it to a different carbon skeleton, generating a different amino acid. The liver is also able to take the amino group off

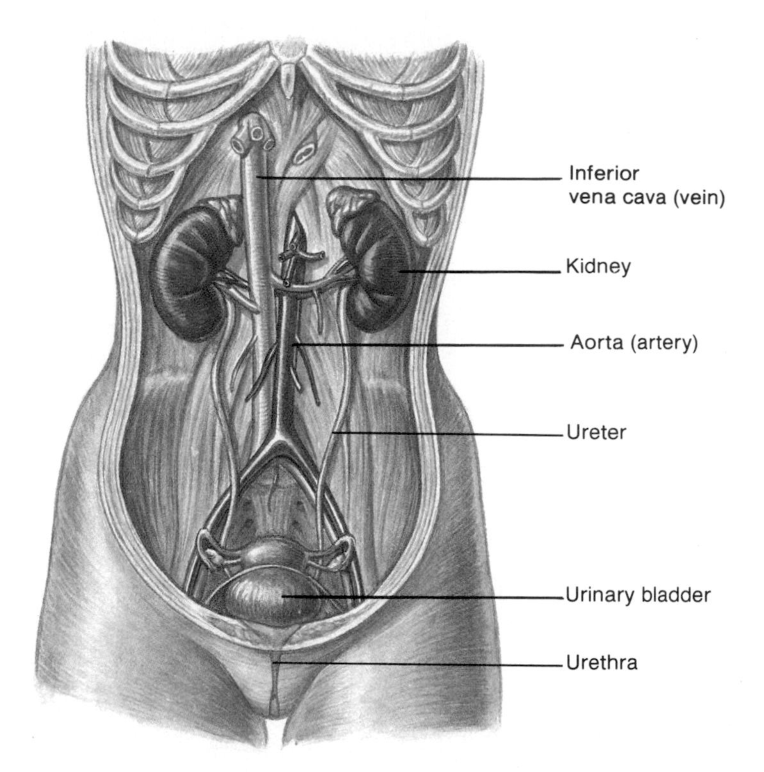

Figure 18.14 **The Urinary System.** The primary organs involved in removing materials from the blood are the kidneys. The urine produced by the kidneys is transported by the ureters to the urinary bladder. From the bladder, the urine is emptied to the outside of the body by way of the urethra.

amino acids so that what remains of the amino acid can be used in aerobic respiration. The toxic amino groups are then converted to urea by the liver. Urea is secreted back into the bloodstream and is carried to the kidneys for disposal in the urine.

Waste Disposal

Because cells are modifying molecules during metabolic processes, harmful waste products are constantly being formed. Urea is a common waste, but many other toxic materials must be eliminated as well. Among these are large numbers of hydrogen ions produced by metabolism. This excess of hydrogen ions must be removed from the bloodstream. Other molecules, such as water and salts, may be consumed in excessive amounts and must be removed. The primary organs involved in regulating the level of toxic or unnecessary molecules are the **kidneys** (figure 18.14).

Kidney Structure

The kidneys consist of about 2.4 million tiny tubules called **nephrons.** At one end of a nephron is a cup-shaped structure called **Bowman's capsule,** which surrounds a knot of capillaries known as a **glomerulus** (figure 18.15). In addition to Bowman's capsule, a nephron consists of three distinctly different regions: the **proximal convoluted tubule,** the **loop of Henle,** and the **distal convoluted tubule.** The distal convoluted tubule of a nephron is connected to a collecting duct that transports fluid to the ureters, and ultimately to the urinary bladder, where it is stored until it can be eliminated.

Kidney Function

As in the other systems discussed in this chapter, the excretory system involves a close connection between the circulatory system and a surface. In this case the large surface is provided by the walls of the millions of nephrons, which are surrounded by capillaries. Three major activities occur at these surfaces; filtration, reabsorption, and secretion. The glomerulus presents a large surface for the filtering of material from the blood to Bowman's capsule. Blood that enters the glomerulus is under pressure from the muscular contraction of the heart. The capillaries of the glomerulus are quite porous and provide a large surface area for the movement of water and small dissolved molecules from the blood into Bowman's capsule. Normally, only the smaller molecules, such as glucose, amino acids, and ions, are able to pass through the glomerulus into the nephron. The various kinds of blood cells and larger molecules like proteins do not pass into the nephron. This physical filtration process allows many kinds of molecules to leave the blood and enter the nephron. The volume of material filtered in this way is about 7.5 liters per hour. Since your entire blood supply is about 5–6 liters, there must be some method of recovering much of this fluid.

The walls of the nephron are made of cells that actively assist in the transport of materials. Some molecules are reabsorbed from the nephron and picked up by the capillaries that surround them, while other molecules are actively secreted into the nephron. Each portion of the nephron has cells with specific secretory abilities. Surrounding the various portions of the nephron are capillaries that passively accept or release molecules on the basis of diffusion gradients.

The proximal convoluted tubule is primarily responsible for reabsorbing valuable materials from the fluid in it. Molecules like glucose, amino acids, and sodium ions are actively transported across the membrane of the proximal convoluted tubule and returned to the circulatory system. In addition, water moves across the membrane, since it follows the absorbed molecules and diffuses to the area where water molecules are less common. By the time the fluid has reached the end of the proximal convoluted tubule, about 65% of the fluid has been reabsorbed into the capillaries surrounding this region.

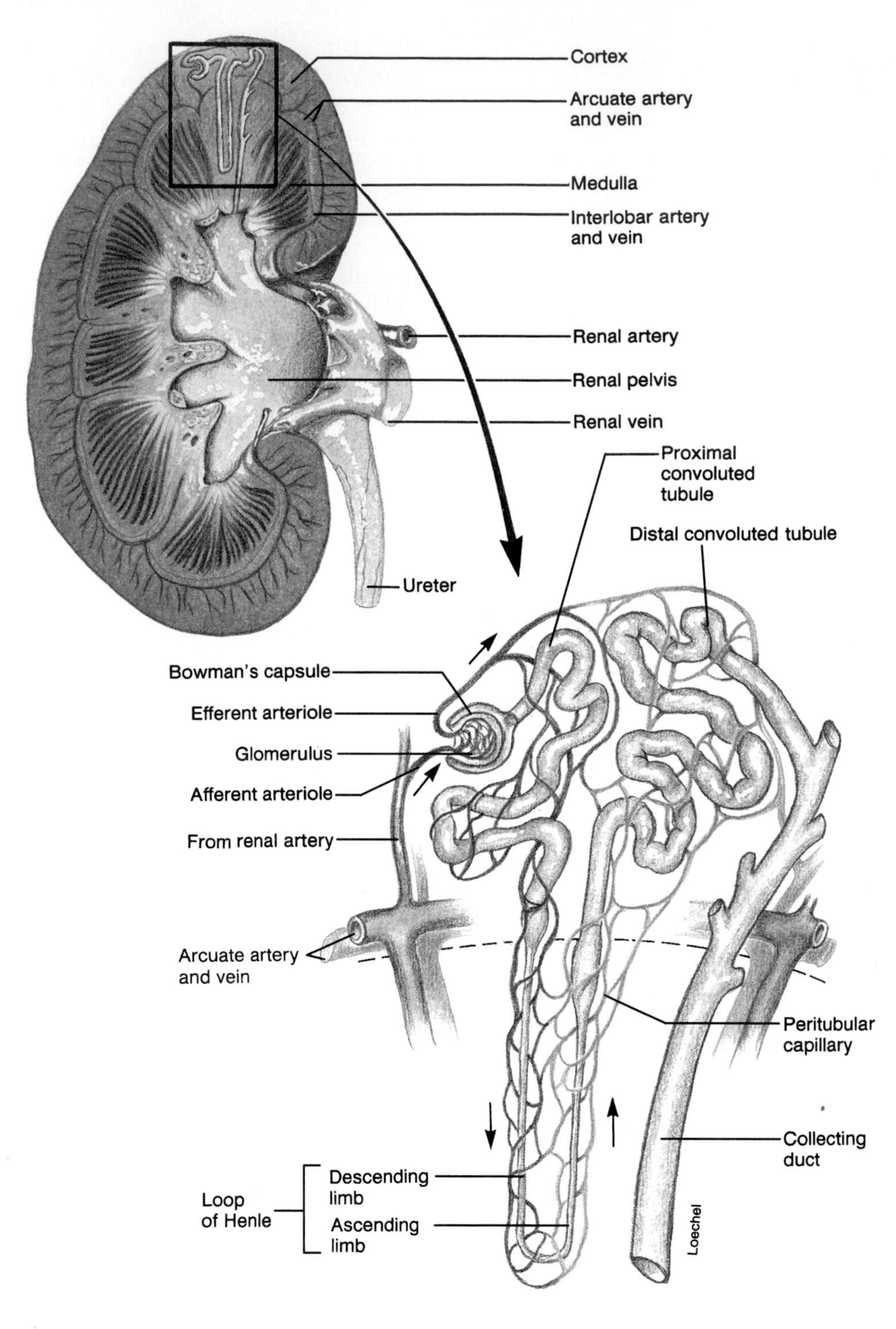

Figure 18.15 **The Structure of the Nephron.** The nephron and the closely associated blood vessels create a system that allows for the passage of materials from the circulatory system to the nephron by way of the glomerulus and Bowman's capsule. Materials are added to and removed from the fluid in the nephron via the tubular portions of the nephron.

The next portion of the tubule, the loop of Henle, is primarily involved in removing additional water from the nephron. Although the details of the mechanism are complicated, the principles are rather simple. The cells of the ascending loop of Henle actively transport sodium ions from the nephron into the space between nephrons where sodium ions accumulate in the fluid that surrounds the loop of Henle. The collecting ducts pass through this region as they carry urine to the ureters. Since the area these collecting ducts pass through is high in sodium ions, water within the collecting ducts diffuses from the ducts and is picked up by surrounding capillaries. However, the ability of water to pass through the wall of the collecting duct is regulated by hormones. Thus, it is possible to control water loss from the body by regulating the amount of water lost from the collecting ducts. For example, if you drank a liter of water or some other liquid, the excess water would not be allowed to leave the collecting duct and would exit the body as part of the urine. However, if you were dehydrated, most of the water passing through the collecting ducts would be reabsorbed, and very little urine would be produced.

The distal convoluted tubule is primarily involved in fine-tuning the amounts of various kinds of molecules that are lost in the urine. Hydrogen ions (H^+), sodium ions (Na^+), chloride ions (Cl^-), potassium ions (K^+), and ammonium ions (NH_4^+) are regulated in this way.

Some molecules that pass through the nephron are relatively unaffected by the various activities going on in the kidney. One of these is urea, which is filtered through the glomerulus into Bowman's capsule. As it passes through the nephron, much of it stays in the tubule and is eliminated in the urine. Many other kinds of molecules, such as minor metabolic waste products and some drugs, are also treated in this manner. Figure 18.16 summarizes the major functions of the various portions of the kidney tubule system.

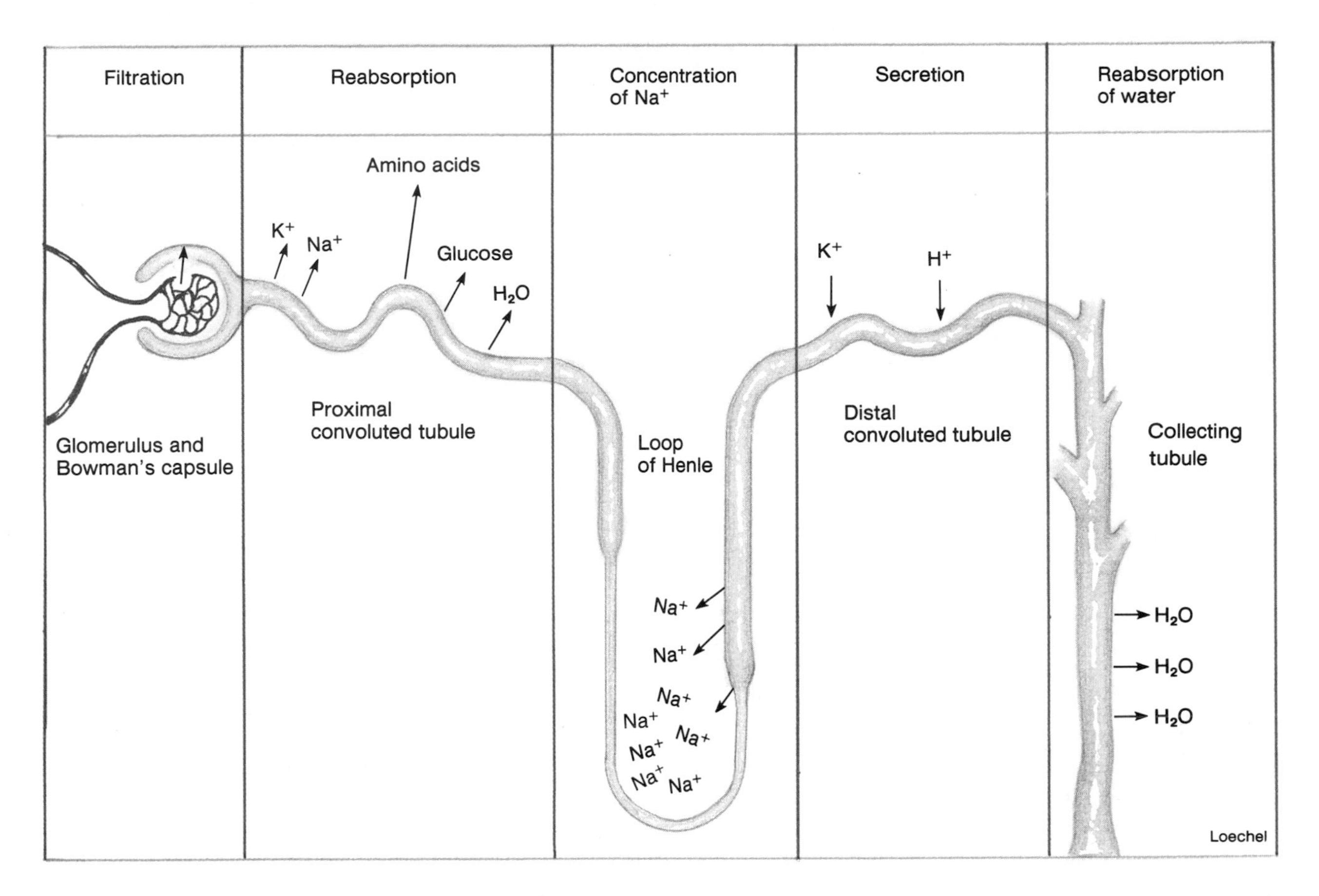

Figure 18.16. **Specific Functions of the Nephron.** Each portion of the nephron has specific functions. The glomerulus and Bowman's capsule accomplish the filtration of fluid from the bloodstream into the nephron. The proximal convoluted tubule reabsorbs a majority of the material filtered. The loop of Henle concentrates Na^+ so that water will move from the collecting tubule. The distal convoluted tubule regulates pH and ion concentration by differential secretion of K^+ and H^+.

• Summary •

This chapter surveys four systems of the body—the circulatory, respiratory, digestive, and excretory systems—and describes how they are integrated. All of these systems are involved in the exchange of materials across cell membranes. Because of problems of exchange, cells must be small. Exchange is limited by the amount of surface area present, so all of these systems have special features that provide large surface areas to allow for necessary exchanges.

The circulatory system consists of a pump, the heart, and blood vessels that distribute the blood to all parts of the body. The blood is a carrier fluid that transports molecules and heat. The exchange of materials between the blood and body cells takes place through the walls of the capillaries. Since the flow of blood can be regulated by the contraction of arterioles, blood can be sent to different parts of the body at different times. Hemoglobin in red blood cells is very important in the transport of oxygen. Carbonic anhydrase is an enzyme in red blood cells that converts carbon dioxide into bicarbonate ions that can be easily carried by the blood.

The respiratory system consists of the lungs and associated tubes that allow air to enter and leave the lungs. The diaphragm and muscles of the chest wall are important in the process of breathing. In the lungs, tiny sacs called alveoli provide a large surface area in association with capillaries, which allows for rapid exchange of oxygen and carbon dioxide.

The digestive system is involved in disassembling food molecules. This involves several processes: grinding by the teeth and stomach, emulsification of fats by bile from the liver, addition of water to dissolve molecules, and enzymatic action to break complex molecules into simpler molecules for absorption. The intestine provides a large surface area for the absorption of nutrients because it is long and its wall contains many tiny projections that increase surface area. Once absorbed, the materials are carried to the liver, where molecules can be modified.

The excretory system is a filtering system of the body. The kidneys consist of nephrons into which the circulatory system filters fluid. Most of this fluid is useful and is reclaimed by the cells that make up the walls of these tubules. Materials that are present in excess or those that are harmful are allowed to escape. Some molecules may also be secreted into the tubules before being eliminated from the body.

• Thinking Critically •

It is possible to keep a human being alive even if the heart, lungs, kidneys, and digestive tract are not functioning by using heart–lung machines in conjunction with kidney dialysis and intravenous feeding. This implies that the basic physical principles involved in the functioning of these systems is well understood because the natural functions can be duplicated with mechanical devices. However, these machines are expensive and require considerable maintenance. Should society be spending money to develop smaller, more efficient mechanisms that could be used to replace diseased or damaged hearts, lungs, and kidneys? Debate this question.

• Experience This •

Obtain three glasses of water. Place a sugar cube in the first, a teaspoon of granulated sugar into the second, and a teaspoon of powdered sugar into the third. Stir each of the glasses gently. Record the time it takes to dissolve the sugar. Which one totally dissolved first? Which one dissolved last? Why did this happen?

• Questions •

1. List three reasons that cells must be small.
2. What are the functions of the heart, arteries, veins, arterioles, the blood, and capillaries?
3. Describe three ways in which the digestive system increases its ability to absorb nutrients.
4. How do red blood cells assist in the transportation of oxygen and carbon dioxide?
5. List three functions of the liver.
6. Name five digestive enzymes and their functions.
7. What is the role of bile in digestion?
8. What is the function of the glomerulus, proximal convoluted tubule, loop of Henle, and distal convoluted tubule?
9. Describe the mechanics of breathing.
10. How are blood pH and breathing interrelated?
11. How is fat absorption different from the absorption of carbohydrate and protein?

• Chapter Glossary •

alveoli (al-vē'o-lī'') Tiny sacs found in the lungs where gas exchange takes place.

anemia (uh-nēm'e-ah) A disease condition in which the oxygen-carrying capacity of the blood is reduced.

aorta (ā-or'tah) The large blood vessel that carries blood from the left ventricle to the majority of the body.

arteries (ar'tē-rēz) The blood vessels that carry blood away from the heart.

arterioles (ar-tēr'e-ōlz) Small arteries located just before capillaries that can expand and contract to regulate the flow of blood to parts of the body.

atria (ā'trē-ah) Thin-walled sacs of the heart that receive blood from the veins of the body and empty into the ventricles.

atrioventricular valves (ā''trē-o-ven-trik'u-lar valvz) Valves located between the atria and ventricles of the heart that prevent the blood from flowing backwards from the ventricles into the atria.

bile (bīl) A product of the liver, stored in the gallbladder, which is responsible for the emulsification of fats.

blood (blud) The fluid medium consisting of cells and plasma that assists in the transport of materials and heat.

Bowman's capsule (bo'manz kap'sl) A saclike structure at the end of a nephron that surrounds the glomerulus.

breathing (bre'thing) The process of pumping air in and out of the lungs.

bronchi (brŏng'ki) Major branches of the trachea that ultimately deliver air to bronchioles in the lungs.

bronchioles (brŏng'kē-ōlz) Small tubes capable of contracting and delivering air to the alveoli in the lung.

buffer (bŭ'fer) A mixture of a weak acid and the salt of a weak acid that operates to maintain a constant pH.

capillaries (cap'ĭ-lair-ēz) Tiny blood vessels through which exchange between cells and the blood takes place.

carbonic anhydrase (car-bon'ik an-hi'drās) An enzyme present in red blood cells that assists in converting carbon dioxide to bicarbonate ions.

diaphragm (di'uh-fram) A muscle separating the lung cavity from the abdominal cavity that is involved in exchanging the air in the lungs.

diastolic blood pressure (di''uh-stol'ik blud presh'yur) The pressure present in a large artery when the heart is not contracting.

distal convoluted tubule (dis'tul kon'vo-lu-ted tūb'yūl) The downstream end of the nephron of the kidney, which is primarily responsible for regulating the amount of hydrogen and potassium ions in the blood.

duodenum (dōō"ŏ-dē'num) The first part of the small intestine, which receives food from the stomach and secretions from the liver and pancreas.

gallbladder (gol'blad"er) An organ attached to the liver that stores bile.

gastric juice (gas'trik jūs) The secretions of the stomach that contain enzymes and hydrochloric acid.

glomerulus (glō-mer'u-lus) A cluster of blood vessels in the kidney, surrounded by Bowman's capsule.

heart (hart) The muscular pump that forces the blood through the blood vessels of the body.

hemoglobin (he'mo-glo-bin) An iron-containing molecule found in red blood cells, to which oxygen molecules bind.

hepatic portal vein (hĕ-pat'ik pōr'tul vān) A blood vessel that collects blood from capillaries in the intestine and delivers it to a second set of capillaries in the liver.

immune system (i-mūn' sis'tem) A system of white blood cells specialized to provide the body with resistance to disease. There are two types: antibody-mediated immunity and cell-mediated immunity.

kidneys (kid'nēz) The primary organs involved in regulating blood levels of water, hydrogen ions, salts, and urea.

lacteals (lak'tēlz) Tiny lymphatic vessels located in the villi.

large intestine (larj in-tes'tin) The last portion of the food tube. It is primarily involved in reabsorbing water.

liver (lĭ'vur) An organ of the body responsible for secreting bile, filtering the blood, detoxifying molecules, and modifying molecules absorbed from the gut.

loop of Henle (loop uv hen'le) The middle portion of the nephron, which is primarily involved in regulating the amount of water lost from the kidney.

lung (lung) A respiratory organ in which air and blood are brought close to one another and gas exchange occurs.

lymph (limf) Liquid material that leaves the circulatory system to surround cells.

lymphatic system (lim-fă'tik sis'tem) A collection of thin-walled tubes that collects, filters, and returns lymph from the body to the circulatory system.

microvilli (mi-kro-vil'ē) Tiny projections from the surfaces of cells that line the intestine.

nephrons (nef'ronz) Tiny tubules that are the functional units of kidneys.

pancreas (pan'kre-as) An organ of the body that secretes many kinds of digestive enzymes into the duodenum.

pepsin (pep'sin) An enzyme produced by the stomach that is responsible for beginning the digestion of proteins.

pharynx (far'inks) The region at the back of the mouth cavity; the throat.

plasma (plaz'mah) The watery matrix that contains the molecules and cells of the blood.

proximal convoluted tubule (prok'sĭ-mal kon'vo-lu-ted tūb'ūl) The upstream end of the nephron of the kidney, which is responsible for reabsorbing most of the valuable molecules filtered from the glomerulus into Bowman's capsule.

pulmonary artery (pul'muh-nă-rē ar'tuh-rē) The major blood vessel that carries blood from the right ventricle to the lungs.

pulmonary circulation (pul'muh-nă-rē ser-ku-la'shun) The flow of blood through certain chambers of the heart and blood vessels to the lungs and back to the heart.

pyloric sphincter (pi-lor'ik sfingk'ter) A valve located at the end of the stomach that regulates the flow of food from the stomach to the duodenum.

salivary amylase (să'lĭ-vĕ-rē ă'mĭ-lās) An enzyme present in saliva that breaks starch molecules into smaller molecules.

salivary glands (să'li-vĕ-rē glanz) Glands that produce saliva.

semilunar valves (sĕ-me-lu'nar valvz) Valves located in the pulmonary artery and aorta that prevent the flow of blood backwards into the ventricles.

surface-area-to-volume ratio (ser'fas a're-ah to vol'ūm ra'sho) The relationship between the surface area of an object and its volume. As objects increase in size, their volume increases more rapidly than their surface area.

systemic circulation (sis-tē'mik ser-ku-la'shun) The flow of blood through certain chambers of the heart and blood vessels to the general body and back to the heart.

systolic blood pressure (sis-tah'lik blud presh'yur) The pressure generated in a large artery when the ventricles of the heart are in the process of contracting.

trachea (trā'ke-ah) A major tube supported by cartilage that carries air to the bronchi; also known as the windpipe.

veins (vānz) The blood vessels that return blood to the heart.

ventricles (ven'trĭ-klz) The powerful muscular chambers of the heart whose contractions force blood to flow through the arteries to all parts of the body.

villi (vil'ē) Tiny fingerlike projections in the lining of the intestine that increase the surface area for absorption.

CHAPTER 19

Nutrition—Food and Diet

Chapter Outline

Purpose

The biochemical pathways and physiological processes described earlier in the text occur in humans as in all other living organisms. You truly are what you eat. The chemicals we call food undergo reactions that convert them into components of human cells, tissues, and organs. A better understanding of foods, dietary requirements, and how foods are processed can be helpful when selecting healthful combinations of foods.

For Your Information

It has been estimated that 75% of U.S. households currently have at least one microwave oven and that percentage is expected to rise. Some manufacturers advertise "improved nutritional value" as a result of microwave cooking. Within the past few years, nutritionists have embarked on numerous research projects to confirm or refute this claim. Some have studied the loss of heat-sensitive thiamine from microwave-cooked meats such as beef and pork as an indicator of nutritional value. The loss of other more stable vitamins, such as riboflavin and niacin, has also been investigated. Others have studied the loss of ascorbic acid, carotene, and folic acid from microwave-cooked vegetables. The results of these investigations indicate that nutrient retention in microwave-cooked or reheated foods is equal to or better than that in the same products prepared in a more conventional manner.

Learning Objectives

- List the six classes of nutrients.
- Recognize the functions of the six types of nutrients.
- Recognize the difference between a complete and incomplete protein.
- Be familiar with the concept of Recommended Dietary Allowances (RDAs).
- Be familiar with the five basic food groups; know examples of each, their sources, and their benefits.
- Understand the concepts of basal metabolic rate, specific dynamic action, and voluntary activity.
- Distinguish between calorie and kcalorie.
- Give examples of psychological eating disorders and deficiency diseases.
- Be familiar with the unique nutritional requirements of each stage of the human life cycle.

Living Things as Chemical Factories—Matter and Energy Manipulators

Organisms are chemical processors capable of sustaining themselves in a changing environment. All molecules required to support these marvelous reactions are obtained from food and are called **nutrients.** Many of the nutrients that enter living cells undergo chemical changes before they are incorporated into the body. These changes are referred to as *assimilation*. **Assimilation** converts nutrients into specific molecules required by the organism to support its cells and cell activities. These interconversion processes are ultimately under the control of the genetic material DNA. It is DNA that codes the information necessary to manufacture the enzymes required to extract energy from chemical bonds and to convert raw materials (nutrients) into the structure (anatomy) of the organism.

The food and drink consumed from day to day constitutes a person's **diet.** It must contain the minimal nutrients necessary to manufacture and maintain the body's structure (bones and muscle) and regulatory molecules (enzymes) and to supply the energy (ATP) needed to run the body's machinery. If the diet is deficient in nutrients, or if a person's body cannot process nutrients efficiently, a dietary deficiency and ill-health may result. The processes involved in assimilating and utilizing nutrients are collectively known as **nutrition.** A good understanding of nutrition can promote good health and help people to avoid disease. An important concept in nutrition is the amount of energy and nutrients contained in various foods.

Kilocalories, Basal Metabolism, and Weight Control

The unit used to measure the amount of energy in foods is the **kilocalorie (kcalorie).** One kilocalorie is the amount of energy needed to raise the temperature of one *kilo*-gram of water one degree Celsius. Remember that the prefix *kilo-* means "one thousand times" the value listed. Therefore, a kilocalorie is one thousand times more heat energy than a **calorie,** which is the amount of heat energy needed to raise the temperature of one *gram* of water one degree Celsius. The energy requirements for different types of activities are listed in table 19.1.

All the activities listed in table 19.1 are physical exercises; the majority of energy expenditure occurs through muscular activities. However, everyone requires a certain amount of energy to maintain basic body functions while at rest. Much of the energy required is used to keep your body temperature constant. The **basal metabolic rate (BMR)** is an estimate of this amount of energy and is measured in kcalories. To determine your basal metabolism, calculate your skin-surface area from table 19.2. To use the table, locate your height in the left column and your weight in the right column. Place a straightedge between these two points. The straightedge will cross the middle column and show your skin-surface area in square meters. For example, a person 160 centimeters tall (5′3″) who weighs 50 kilograms (110 pounds) has 1.5 square meters of skin-surface area. The heat production in kcalories released per square meter of skin varies with a person's age and sex (see table 19.3). A twenty-year-old, 160-centimeter, 50-kilogram female uses 886 kcalories per day for each square meter of skin. Therefore, her basal metabolism is 1,329 kcalories per day (1.5 × 886 = 1,329 kcalories). To calculate your basal metabolism, determine your skin-surface area using table 19.2 and multiply this figure by the kcalorie figure determined from table 19.3.

Remember, basal metabolism represents the amount of energy your body requires at rest. Since few of us rest for twenty-four hours a day, we normally require more than the energy needed for basal metabolism. Besides age, sex, weight, and height, basal metabolism depends upon a number of other factors, such as climate, altitude, physical condition, race, previous diet, and time of the year. A good general indicator of the number of kcalories needed above basal metabolism is the type of occupation a person has (table 19.4). If a twenty-year-old, 160-centimeter, 50-kilogram female were a bank teller, she would need between 750 and 1,200 kcalories per day above her basal metabolism of 1,329 kcalories. Therefore, her total daily need would be somewhere between 2,079 and 2,529 kcalories per day. Using tables 19.2, 19.3, and 19.4, calculate your daily caloric requirements.

Table 19.1
Energy Requirements

Kinds of Activity	Kilocalories (per Hour)
Walking up stairs	1,100
Running (a jog)	570
Swimming	500
Vigorous exercise	450
Slow walking	200
Dressing and undressing	118
Sitting at rest	100

This list of activities shows the amount of energy expended (measured in kilocalories) if the activity is performed for an hour.

Table 19.2

Table for Determining Body-Surface Area

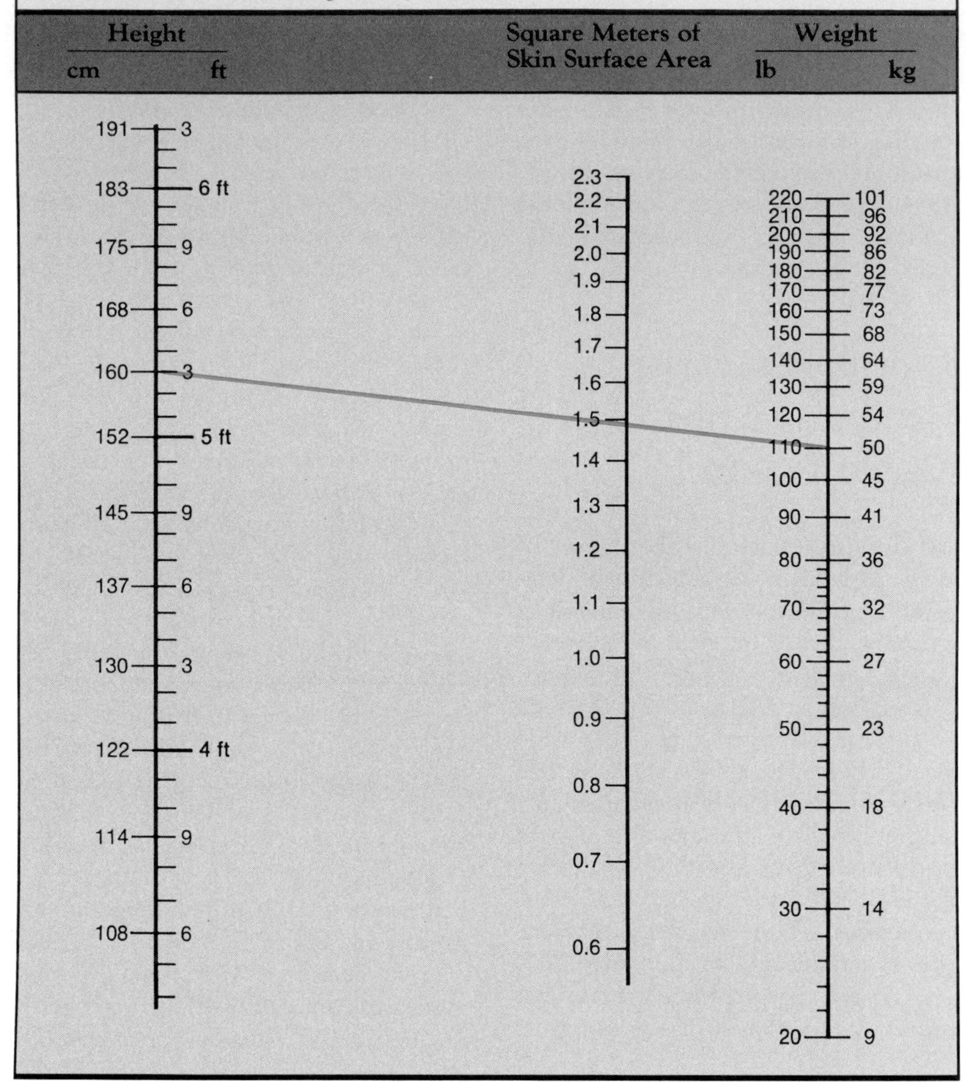

From *Dynamic Anatomy and Physiology*, 4th ed., by Langley et al. Copyright © 1974 by McGraw-Hill Inc. Used with permission of McGraw-Hill Book Company.

Table 19.3

Kilocalories per Day per Square Meter of Skin

Age	Male	Female	Age	Male	Female
6	1,265	1,217	20–24	984	886
8	1,229	1,154	25–29	967	878
10	1,188	1,099	30–40	948	876
12	1,147	1,042	40–50	912	847
14	1,109	984	50–60	886	826
16	1,073	924	60–70	859	806
18	1,030	895	70–80	828	782

The body's metabolism is designed to convert carbohydrates (glucose) or proteins to fat, and under certain circumstances, fat is converted to glucose; only a limited amount of protein is ever converted to glucose. Although energy doesn't weigh anything, the nutrients that contain the energy do. Weight control is a matter of balancing dietary intake with output measured in kcalories. There is a limit to the rate at which a moderately active human body can use fat as an energy source. At the most, one or two pounds of fat per week are lost by an average person when dieting. These pounds represent true weight loss and not a decrease in weight from water loss. Many diets promise large and rapid weight loss but in fact only result in temporary water loss. Decreasing your kcalorie intake by 500–1,000 kcalories per day while maintaining a balanced diet, including proteins, carbohydrates, and fats, will result in a meaningful loss. For those who need to gain weight, increasing kcalorie intake by 500–1,000 kcalories per day will result in an increase of one or two pounds per week, provided the low weight is not the result of a health problem.

How do you go about determining where you fit into this weight picture so that you can adjust your diet? After you have determined your basal metabolism, add to that kcalorie value the kcalorie energy you expend on a daily basis. That includes the energy you use in voluntary muscular activities such as work and sports plus the amount of energy required to digest and assimilate your food. This latter kcalorie amount is called your **specific dynamic action (SDA)** and is equal to approximately 10% of your total daily kcalorie intake.

You must also keep track of your kcalorie input. That means keeping an accurate diet record for at least a week. Record everything you eat and drink and determine the number of kcalories in those nutrients. This can be done by estimating the amounts of protein, fat, and carbohydrate (including alcohol) in your foods. Roughly speaking, 1 gram of carbohydrate is the equivalent of 4 kcalories, 1 gram of fat is the equivalent of 9

Table 19.4
Additional Kilocalories as Determined by Occupation

Occupation	Kilocalories Needed above Basal Metabolism
Sedentary (student)	500–700
Light work (business person)	750–1,200
Moderate work (laborer)	1,250–1,500
Heavy work (professional athlete)	1,550–5,000 and up

These are general figures and will vary from person to person depending on the specific activities performed in the job.

kcalories, 1 gram of protein is the equivalent of 4 kcalories, and 1 gram of alcohol, 7 kcalories. Most nutrition books have food-composition tables that tell you how much protein, fat, and carbohydrate is in a particular food. Do the arithmetic and determine your total kcalorie intake for the week. If your intake (from your diet) in kcalories equals your output (from basal metabolism plus voluntary activity plus SDA), you should not have gained any weight! You can double-check this by weighing yourself before and after your week of record keeping. To lose two pounds each week, reduce your kcalorie intake by 1,000 kcalories per day. Be careful not to reduce your total daily intake below 1,200 kcalories unless you are under the care of a physician. Below that level, you may not be getting all the vitamins required for efficient metabolism and you could cause yourself harm. To gain two pounds, increase your intake by 1,000 kcalories per day.

If, like millions of others, you feel that you are overweight, you have probably tried numerous diet plans. Not all of these plans are the same, and not all are suitable to your particular situation. If a diet plan is to be valuable in promoting good health, it must satisfy your needs in several ways. It must provide you with needed kcalories, protein, fat, and carbohydrates. It should also contain readily available foods from all the basic food groups, and it should provide enough variety to prevent you from becoming bored with the plan and going off the diet too soon. A diet should not be something you follow only to abandon it and regain the lost weight.

The Chemical Composition of Your Diet

Nutritionists have divided nutrients into six major classes: carbohydrates, lipids, proteins, vitamins, minerals, and water. Chapters 2 and 3 presented the fundamental structures and examples of these types of molecules. A look at each of these classes from a nutritionist's point of view should help you to better understand how your body works and how you might best meet its nutritional needs.

Carbohydrates

When the word *carbohydrate* is mentioned, many people think of things like table sugar, pasta, and potatoes. The term *sugar* is usually used to refer to mono- or disaccharides, but the carbohydrate group also includes more complex polysaccharides, such as starch, glycogen, and cellulose. Each of these has a different structural formula, different chemical properties, and plays a different role in the body (figure 19.1). Many carbohydrates taste sweet and stimulate the appetite. When complex carbohydrates like starch or glycogen are hydrolyzed to monosaccharides, these may then be utilized in aerobic cellular respiration, and the energy of carbohydrates is used to manufacture ATP. Carbohydrates are also converted by the body into molecules that can be used to manufacture necessary components of molecules such as nucleic acids. Carbohydrates can also be a source of fibers that slow the absorption of nutrients and stimulate peristalsis in the intestinal tract (box 19.1).

A diet deficient in carbohydrates results in a condition in which fats are primarily oxidized and converted to ATP. In this situation, fats are metabolized to keto acids, resulting in a potentially dangerous change in the body's pH. A carbohydrate deficiency may also result in the body's use of proteins as a source of energy. In extreme cases this can be fatal, since the oxidation of protein results in an increase in toxic, nitrogen-containing compounds. And keep this in mind: As with other nutrients, if there is an excess of carbohydrates in the diet, they are converted to lipids and stored by the body in fat cells—and you gain weight.

Lipids

An important class of nutrients is technically known as *lipids*, but many people use the term *fats*. This is unfortunate and may lead to some confusion because fat is only one of three subclasses of lipids. The subclasses—phospholipids, steroids, and true fats—play important roles in human nutrition. Phospholipids are essential components of all cell membranes. Many steroids are hormones that help to regulate a variety of body processes. The lipids sometimes referred to as *true fats* (also called *triglycerides*) are an excellent source of energy. They are able to release 9 kcalories of energy per gram compared to 4 kcalories per gram of carbohydrate or protein. When hydrolyzed to glycerol and fatty acids, some true fats provide the body with an **essential fatty acid,** linoleic acid. Linoleic acid cannot be synthesized by the human body and, therefore, must be a part of the diet. This essential fatty acid is required by the body for such things as normal growth, blood clotting, and healthy hair and skin. A diet high in linoleic acid has also been shown to help in reducing the amount of the steroid cholesterol in the blood. Some dietary fats also contain dissolved vitamins, such as vitamins A, D, E, and K. Therefore, dietary fats can serve as a source of these fat-soluble vitamins.

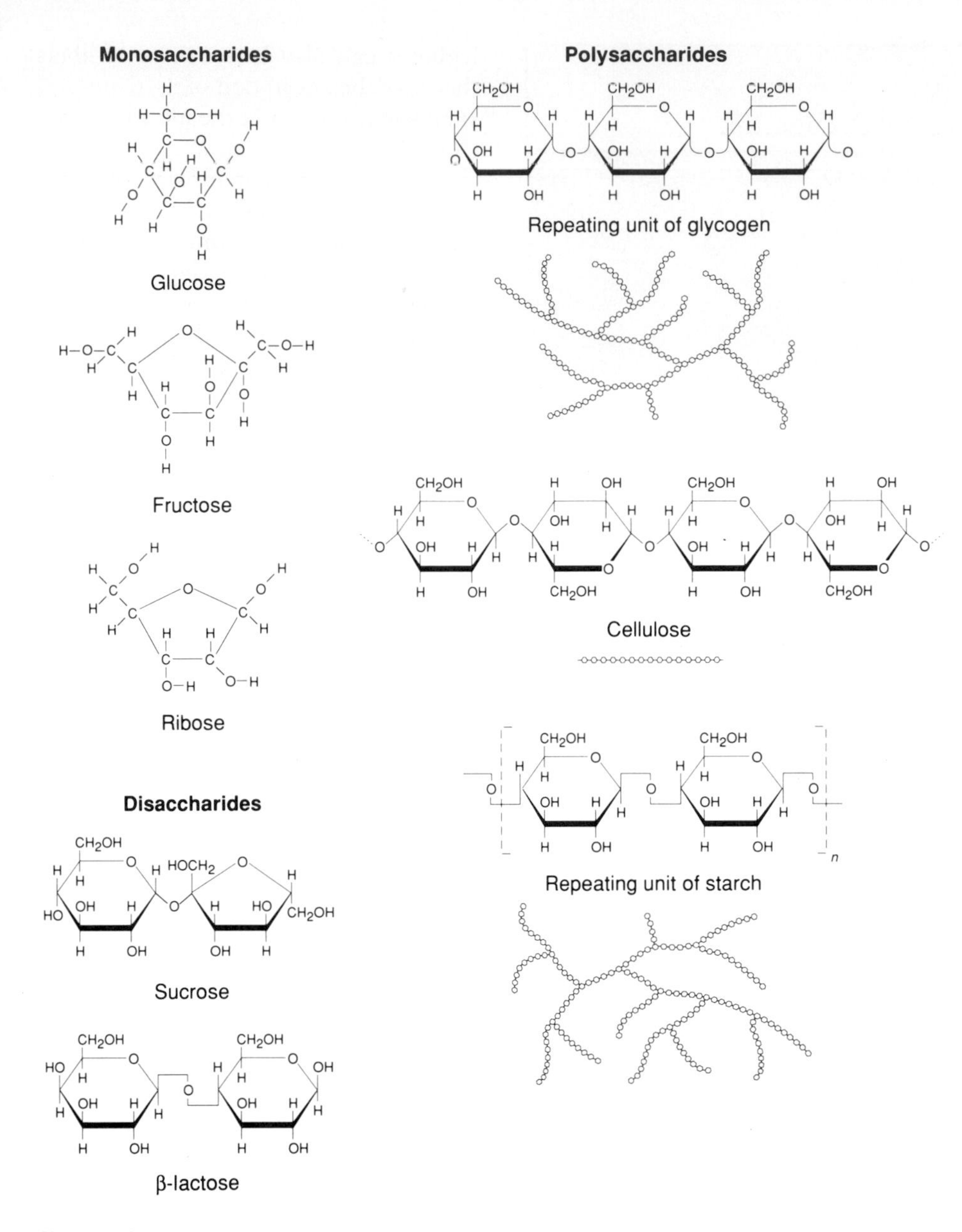

Figure 19.1 **The Structure and Role of Various Carbohydrates.** The diet includes a wide variety of carbohydrates. Some are monosaccharides (simple sugars) while others are more complex disaccharides, trisaccharides, and polysaccharides. The complex carbohydrates differ from one another depending upon the type of monosaccharides that are linked together by dehydration synthesis. Notice that the complex carbohydrates shown are primarily from plants. With the exception of milk, animal products are not a good nutritional source of carbohydrates because animals do not store them in great quantities or use large amounts of them as structural materials.

Fat is an insulator against outside cold and internal heat loss and is an excellent shock absorber. Deposits in back of the eyes serve as cushions when the head suffers a severe blow. During starvation, these deposits are lost, and the eyes become deep-set in the eye sockets, giving the person a ghostly appearance. The pleasant taste and "mouth feel" of many foods is the result of fats. Their ingestion provides that full feeling after a meal because they leave the stomach more slowly than other nutrients. You may have heard people say, "When you eat Chinese food, you're hungry a half hour later." Since Chinese foods contain very little animal fat, it's understandable that after such a meal, the stomach will empty soon, and people don't have that full feeling very long.

Proteins

Proteins are composed of amino acids linked together by peptide bonds; however, not all proteins contain the same amino acids. Proteins can be divided into two main groups, the **complete proteins** and the **incomplete proteins.** Complete proteins contain all the amino acids necessary for good health, while incomplete proteins lack certain amino acids that the body must have to function efficiently. Table 19.5 lists the **essential amino acids,** those that cannot be synthesized by the human body. Without adequate amounts of these amino acids in the diet, a person may develop protein-deficiency disease. Proteins are essential components of hemoglobin and cell membranes, as well as antibodies, enzymes, some hormones, hair, muscle, and the connective tissue fiber, collagen. Plasma proteins are important because they can serve as buffers. Proteins also provide a last-ditch source of energy when carbohydrate and fat consumption falls below protective levels.

The body's need for a mixture of proteins and carbohydrates is vitally important. Proteins are present in the structures of the human body, but they cannot be stored during times of protein deficiency. If you are on a high-protein diet, the amino acids are not stored; they are either used in protein synthesis or oxidized. A unique relationship exists between carbohydrates and proteins called **protein-sparing.** When adequate amounts of carbohydrates are present in the diet, the body's proteins do not have to be tapped as a source of energy. They are spared from being oxidized. Only when carbohydrate consumption falls below an adequate level will the body use proteins as an energy source. Most people in developed countries have a misconception with regard to the amount of protein necessary in their diets. The total amount necessary is actually quite small and can be easily met.

BOX

19.1 *Fiber in Your Diet*

Fiber has also been called *bulk* and *roughage*. Most fibers are a variety of different kinds of indigestible polysaccharides. The five types of fiber are described in table 19.A.

Most fiber is indigestible by human enzymes, but a small amount can be hydrolyzed by beneficial bacteria normally found in the intestinal tract. Since only a negligible amount of kcalories are obtained through dietary fiber, they are not considered to be a source of energy; however, they do serve other important functions, as shown in table 19.B.

While dietary fiber is of great benefit in many situations, increasing your fiber intake or increasing it too rapidly may result in diarrhea, loss of vitamins and minerals, and damage to the lining of the intestinal tract.

Table 19.A
The Five Types of Fiber

Type	Dietary Source	Nature
Cellulose and hemicellulose	Fruits, vegetables, beans, oat and wheat bran, nuts, seeds, whole-grain flour	Wood fiber; plant-cell glue; polysaccharide of glucose
Gums, mucilages, and pectins	Vegetables, fruits, seeds, beans, oats, barley	Plant secretions; galactose-containing polysaccharide
Lignin	Seeds, vegetables, whole grains	Stiffens plant-cell walls; complex of alcohols and organic acids
Algal polysaccharides	Colloids used in chocolate milk, puddings, pie fillings	Carrageenan; polysaccharide extracted from marine algae and seaweeds
Methyl cellulose	Synthetic products	Polysaccharides

Table 19.B
Important Functions of Fiber

Benefit	Positive Effect of Dietary Fiber
1. Relieves constipation/diarrhea	Fiber holds water; softens stool to prevent constipation; forms gels to thicken stool to prevent diarrhea.
2. Hemorrhoid control	Softer stools ease elimination to prevent weakening of rectal muscles and protruding of swollen veins.
3. Weight control	Creates feeling of fullness that promotes weight loss; can be used as replacement for fats and sweets.
4. Reduces colon cancer	Speeds movement through intestinal tract reducing exposure time of cancer-causing agents; does same for bile (associated with cancer risk).
5. Reduces blood lipids/ cardiovascular disease	Binds bile, cholesterol, and other lipids to carry them out of the body.
6. Benefits blood glucose/insulin controls diabetes	Mildly stimulates insulin production and causes a gradual increase in blood glucose.
7. Controls appendicitis	Loosens stool to prevent packing in the appendix and possible infection.
8. Controls diverticulosis	Exercises digestive-tract muscles so that they retain their tone; resists bulging of the wall of the intestinal tract into pouches, called diverticula, that could become infected.

Table 19.5
Sources of Essential Amino Acids

Essential Amino Acids	Food Sources
Threonine	Dairy products, nuts, soybeans, turkey
Lysine	Dairy products, nuts, soybeans, green peas, beef, turkey
Methionine	Dairy products, fish, oatmeal, wheat
Arginine (essential to infants only)	Dairy products, beef, peanuts, ham, shredded wheat, poultry
Valine	Dairy products, liverwurst, peanuts, oats
Phenylalanine	Dairy products, peanuts, calves' liver
Histidine (essential to infants only)	Human and cow's milk and standard infant formulas
Leucine	Dairy products, beef, poultry, fish, soybeans, peanuts
Tryptophan	Dairy products, sesame seeds, sunflower seeds, lamb, poultry, peanuts
Isoleucine	Dairy products, fish, peanuts, oats, macaroni, lima beans

The essential amino acids are required in the diet for protein building and, along with the nonessential amino acids, allow the body to metabolize all nutrients at an optimum rate. Combinations of different plant foods can provide essential amino acids even if complete protein foods (*e.g., meat, fish, and milk*) are not in the diet.

Vitamins

Vitamins are the fourth class of nutrients. Like essential amino acids and linoleic acid, vitamins cannot be manufactured by the body but are essential in minute amounts to the body's metabolism. Vitamins do not serve as a source of energy, but they help in many enzymatically controlled reactions. They function with specific enzymes to speed the rate of certain chemical reactions. Some enzymes do not function alone but acquire the attachment of a vitamin to complete their structure. For this reason such vitamins are called *coenzymes.* For example, a B-complex vitamin (niacin) helps enzymes in the respiration of carbohydrates. Most vitamins are acquired from food; however, vitamin D may be formed when ultraviolet light strikes a molecule already in your skin, coverting this molecule to vitamin D. This means that vitamin D is not really a vitamin at all. It came to be known as a vitamin because of the mistaken idea that it is only acquired through food rather than being formed in the skin on exposure to sunshine. It would be more correct to call vitamin D a hormone, but most people do not.

Minerals

All **minerals** are inorganic elements found throughout nature, and they cannot be synthesized by the body. Because they are elements, they cannot be broken down or changed by metabolism or cooking. They commonly occur in many foods and in water. Minerals retain their characteristics whether they are in foods or in the body, and each plays a different role in metabolism. Minerals can function as regulators, activators, transmitters, and controllers of various enzymatic reactions. For example, sodium ions (Na^+) and potassium ions (K^+) are important in the transmission of nerve impulses, while magnesium ions (Mg^{++}) facilitate energy release during reactions involving ATP. Without iron, not enough hemoglobin would be formed to transport oxygen, a condition called *anemia,* and a lack of calcium may result in *osteoporosis.* **Osteoporosis** is a condition that results from calcium loss leading to painful, weakened bones. There are many minerals that are important in your diet. In addition to those just mentioned, you need chlorine, cobalt, copper, iodine, phosphorus, potassium, sulfur, and zinc to remain healthy.

Water

This last nutrient is crucial to all life and plays many essential roles. You may be able to survive weeks without food, but you would die in a matter of days without water. It is known as the universal solvent because so many types of molecules are soluble in it. The human body is about 65% water. Even dense bone tissue consists of 33% water. All the chemical reactions in living things take place in water. It is the primary component of blood, lymph, and body-tissue fluids. Inorganic and organic nutrients and waste molecules are also dissolved in water. Dissolved inorganic ions, such as sodium (Na^+), potassium (K^+), and chloride (Cl^-), are called **electrolytes** because they form a solution capable of conducting electricity. The concentration of these ions in the body's water must be regulated in order to prevent electrolyte imbalances.

Excesses of many types of wastes are eliminated from the body dissolved in water; that is, they are excreted from the kidneys as urine or in small amounts from the lungs or skin through evaporation. In a similar manner, water acts as a conveyor of heat. Water molecules are also essential reactants in all the various hydrolytic reactions of metabolism. Without it, the breakdown of molecules such as starch, proteins, and lipids would be impossible. With all these important roles played by water, it's no wonder that nutritionists recommend that you drink the equivalent of at least eight glasses each day. This amount of water can be obtained from tap water, soft drinks, juices, and numerous food items, such as lettuce, cucumbers, tomatoes, and applesauce.

Amounts and Sources of Nutrients

In order to give people some guidelines for planning a diet that provides adequate amounts of the six classes of nutrients, nutritional scientists in the United States and many other countries have developed nutrient standards. In the

United States, these guidelines are known as the **Recommended Dietary Allowances,** or **RDAs.** RDAs are dietary recommendations, not requirements or minimum standards. They are based on the needs of a healthy person already eating an adequate diet. RDAs do not apply to a person with medical problems who is under stress or suffering from malnutrition. The amount of each nutrient specified by the RDAs has been set relatively high so that most of the population eating those quantities will be meeting their nutritional needs. Keep in mind that since everybody is different, eating the RDA amounts may not meet your personal needs. You may have a special need for additional amounts of a particular nutrient if you have an unusual metabolic condition.

General sets of RDAs have been developed for four groups of people: infants, children, adults, and pregnant and lactating women. The U.S. RDAs are used when preparing product labels. The federal government requires by law that labels list ingredients from the greatest to the least in quantity. The volume in the package must be stated along with the weight, and the name of the manufacturer or distributor. If any nutritional claim is made, it must be supported by factual information. Many manufacturers voluntarily provide nutritional labeling for the increasing number of consumers concerned with their diets.

A product label that proclaims, for example, that a serving* of cereal provides 25% of the RDA for vitamin A means that you are getting at least one-fourth of your RDA of vitamin A from a single serving of that cereal. To figure your total RDA of vitamin A, consult a published RDA table for adults. It tells you that an adult male requires 1,000 and a female 800 international units of vitamin A per day. Twenty-five percent of this is 250 and 200 international units, respectively—the amount you are getting in a serving of that cereal. You will need to get the additional amounts (750 for men and 600 for women) by having more of that cereal or eating other foods that contain vitamin A. If a product claims to have 100% of the RDA of a particular nutrient, that amount must be present in the product. However, restricting yourself to that one product will surely deprive you of many of the other nutrients necessary for good health. Ideally, you should eat a variety of complex foods containing a variety of nutrients to ensure that all your health requirements are met.

*The measurement of ingredients during food preparation varies throughout the world. Some people measure by weight (e.g., grams), others by volume (e.g., cups); some use the metric system (grams), others the English system (pounds). Still others use units of measure that are even less uniform (e.g., "pinches"). Because of this variability, this chapter describes quantities of nutrients using the units of measure most familiar to people in the United States.

The Food Guide Pyramid with Five Food Groups

Using RDAs and product labels is a pretty complicated way for a person to plan a diet. Planning a diet around basic food groups is generally easier. The four basic food groups first developed and introduced in 1953 have been modified and updated several times to serve as guidelines in maintaining a balanced diet (figure 19.2). In May 1992, the U.S. Department of Agriculture released the results of its most recent study on how best to educate the public about daily nutrition. The federal government adopted the **Food Guide Pyramid** of the Department of Agriculture as one of its primary tools to help the general public plan for good nutrition. The Food Guide Pyramid contains five basic groups of foods with guidelines for the amounts one needs daily from each group for ideal nutritional planning. Some of the important aspects of the Food Guide Pyramid that differ from previous information provided by the federal government include decreasing our emphasis on fats and sugars, while increasing our daily servings of fruits and vegetables. In addition, the new guidelines suggest significantly increasing the amount of grain products we eat each day.

Group 1—Meat, Poultry, Fish, and Dry Beans

This group contains most of the things we eat as a source of protein; for example, nuts, peas, tofu, and eggs are considered a part of this group. It is recommended that we include 4–6 ounces of these items in our daily diet. In the past, people in the economically developed world ate at least this quantity and frequently much more. With the recent emphasis on decreasing fats in our diets, more attention is being paid to the quantity of the protein-rich foods, and to the foods themselves. Not only have we decreased our intake of items from this group, but we have also shifted from the high-fat-content foods, such as beef and eggs, to foods that are high in protein but lower in fat content, such as fish and poultry. Modern food preparers tend to use smaller portions and cook foods in ways that decrease fat rather than increase it. Broiled fish, rather than fried, and baked, skinless portions of chicken or turkey are seen more and more often on restaurant menus and on dining-room tables at home.

Remember that 4–6 ounces is the recommended daily portion from this group. This means that one double cheeseburger greatly exceeds this recommendation for the daily intake. Eating excessive amounts of protein can stress the kidneys by causing higher concentrations of calcium in the urine, increase the demand for water to remove toxic keto acids, and lead to weight gain. It should be noted, however, that vegetarians must pay particular attention to acquiring adequate sources of protein because they have eliminated a major source from their diet (box 19.2).

Group 2—Dairy Products

All of the cheeses, ice cream, yogurt, and milk are in this group. Two servings from this group are recommended each day. Each of these servings should be about 1 cup or 100 calories. Using product labels will help you determine the appropriate serving size of individual items. This group provides minerals, such as calcium, in your diet, but also provides

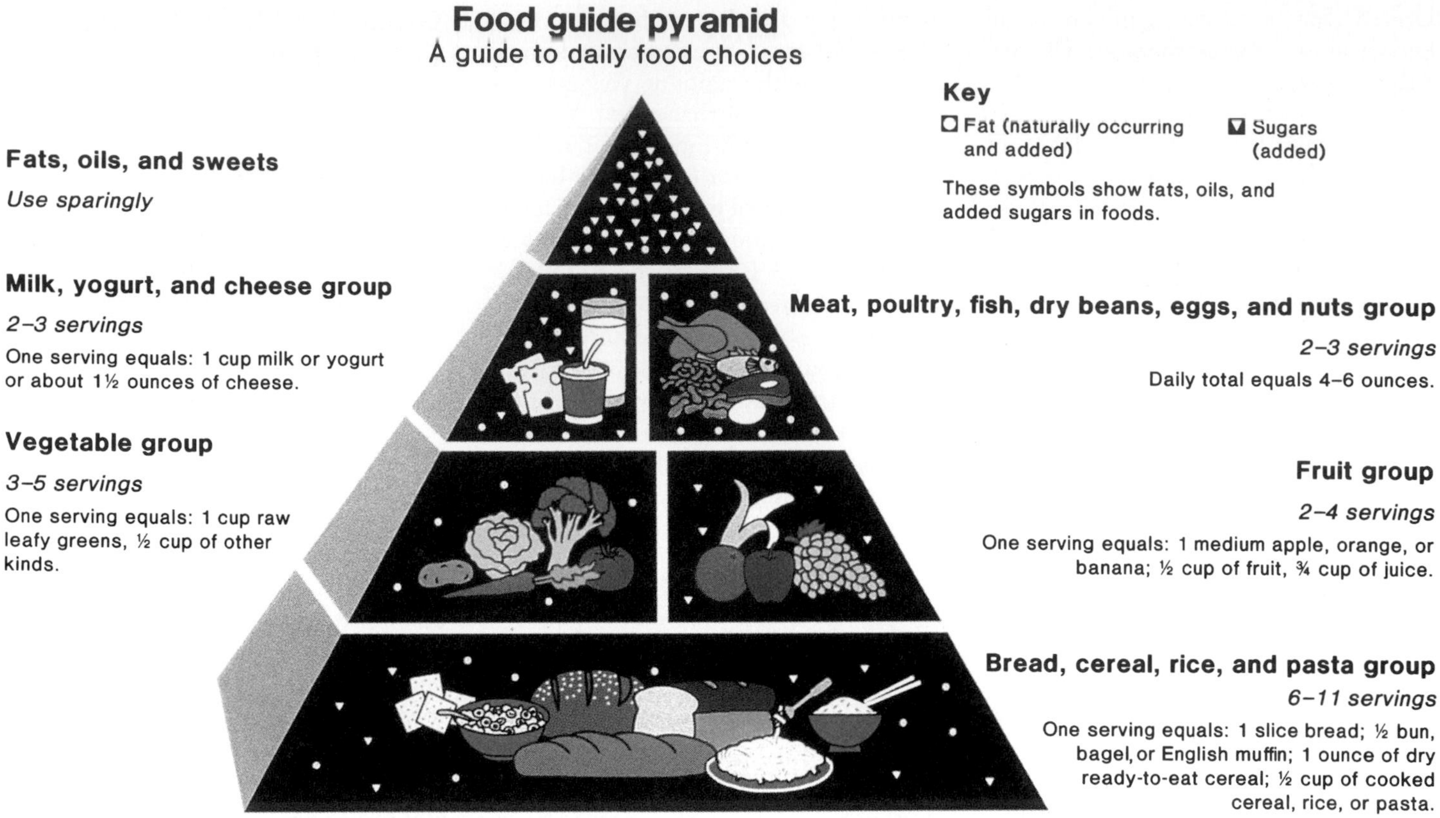

Food guide pyramid
A guide to daily food choices

Key
Fat (naturally occurring and added)
Sugars (added)
These symbols show fats, oils, and added sugars in foods.

Fats, oils, and sweets
Use sparingly

Milk, yogurt, and cheese group
2–3 servings
One serving equals: 1 cup milk or yogurt or about 1½ ounces of cheese.

Meat, poultry, fish, dry beans, eggs, and nuts group
2–3 servings
Daily total equals 4–6 ounces.

Vegetable group
3–5 servings
One serving equals: 1 cup raw leafy greens, ½ cup of other kinds.

Fruit group
2–4 servings
One serving equals: 1 medium apple, orange, or banana; ½ cup of fruit, ¾ cup of juice.

Bread, cereal, rice, and pasta group
6–11 servings
One serving equals: 1 slice bread; ½ bun, bagel, or English muffin; 1 ounce of dry ready-to-eat cereal; ½ cup of cooked cereal, rice, or pasta.

Figure 19.2 **The Food Guide Pyramid.** In May 1992, the Department of Agriculture released a new guide to good eating. This Food Guide Pyramid suggests that we eat particular amounts of five different food groups while decreasing our intake of fats and sugars. This guide should simplify our menu planning while helping to ensure that we get all of the recommended amounts of basic nutrients.

Source: U.S. Department of Agriculture, 1992.

water, vitamins, carbohydrates, and protein. You must also remember that cheese contains large amounts of cholesterol and fat for each serving.

Group 3—Vegetables

The Food Guide Pyramid suggests from three to five servings from this group each day. Items in this group include non-sweet plant materials, such as broccoli, carrots, cabbage, corn, green beans, potatoes, lettuce, and spinach. A serving is considered to be one cup of raw leafy vegetables or a half cup of other types. It is wise to include as much variety as possible in this group. If you eat only carrots, several cups each day can become very boring. There is increasing evidence indicating that cabbage, broccoli, and cauliflower can provide some protection from certain types of cancers. This is a good reason to include these foods in your diet.

Foods in this group provide vitamins A and C as well as water and minerals. They also provide fiber, which assists in the proper functioning of the digestive tract.

Group 4—Fruits

You probably all can remember discussions of whether a tomato is a vegetable or a fruit. This controversy arises from the fact that the term *vegetable* is not a scientifically precise term but means a plant material eaten during the main part of the meal. A *fruit,* on the other hand, is a botanical term for the structure that is produced from the female part of the flower to protect and nourish the ripening seeds. Although green beans, peas, and corn are all fruits, botanically speaking, they are placed in the vegetable category because they are generally eaten during the main part of the meal. In this group we include such sweet plant products as melons, berries, apples, oranges, and bananas. Since these foods tend to be high in natural sugars, you should carefully regulate your intake to two to four servings daily. A small apple, half a grapefruit, a half cup of grapes, or six ounces of fruit juice is considered a serving.

Group 5—Grain Products

Group 5 includes vitamin-enriched or whole-grain cereals and grain products such as breads, bagels, buns, crackers, dry

BOX 19.2 *Which Kind of Vegetarian Are You?*

A vegetarian is a person who, to one degree or another, eliminates animal products from his or her diet. Vegetarians face a special problem: they must obtain the essential nutrients from fewer food groups. In order to do this they must learn which nonanimal foods contain these nutrients and plan their diets accordingly. Nutritionists usually describe two basic types of vegetarians, with many variations in each group. The first type, the *lacto-ovo* vegetarians, use milk (*lacto* = milk) and eggs (*ovo* =egg), which are considered to be animal products, but avoid animal flesh; that is, meat, fish, and poultry. The *pure* or *strict* vegetarians are also known as *vegans;* they avoid all animal products and flesh and use only plant foods in planning their diets. Such people need to take vitamin B_{12} supplements or drink soy milk fortified with this vitamin in order to receive sufficient amounts to prevent the vitamin-deficiency disease called *pernicious anemia.* Be assured that an adequate vegetarian diet is healthful because of the minimal amount of saturated fats in these foods and the fact that vegetarian diets are high in insoluble dietary fiber.

Dietary planning for vegetarians can be tricky. In order to obtain complete protein (RDA = 44 grams for an adult), they must learn complementary protein relationships. Since the only source of complete protein in a single food group is from animal products or flesh, strict vegetarians must choose plant products that will complement one another and provide all the essential amino acids. They need to do so in order to avoid abnormal fetal and adult development. The following table should help in planning a vegetarian diet that will provide the essential amino acids and the total amount of recommended protein.

Protein Source 1	Protein Source 2	Total Grams
2 cups rice	1/2 cup beans	15 g
2 1/2 cups rice	2 oz tofu	12 g
1 cup rice	1 cup milk	10 g
6–7 tortillas	1/2 cup beans	14 g
2 slices whole wheat bread	2 tablespoons peanut butter	5.5 g
1 medium potato	1 cup milk	9 g
1 cup macaroni	2 oz cheddar cheese	13 g
2 tablespoons peanut butter	2 tablespoons sunflower seeds	7 g

and cooked cereals, pancakes, pasta, and tortillas. Items in this group are typically dry and seldom need refrigeration. They provide most of your kcalorie requirements. You should have six to eleven servings from this group each day. This is a major change from previous recommendations of four servings each day. A serving is considered about a half cup, or one ounce, or about 100 kcalories. Using product labels will help to determine the appropriate serving size (see figure 19.3).

Cereals and grains provide fiber and are a rich source of the B vitamins in your diet. As you decrease the intake of proteins in the meat and poultry group, you should increase your intake of items from this group. These foods help you feel you have satisfied your appetite and many of them are very low in fat (see table 19.6).

Figure 19.3 **Carbohydrates in Your Diet.** Many of the world's food staples are plant carbohydrates that are either eaten directly or first ground into flours. Combinations of certain grains can supply the essential amino acids in a diet. The Food Guide Pyramid suggests that one eat between six and eleven servings of grain products each day. This illustration shows one serving of a variety of different products.

Table 19.6
Dietary Goals and Guidelines for the United States

Dietary Goals for the United States

The Select Committee on Nutrition and Human Needs of the U.S. Senate (1977) has established seven dietary goals for United States citizens.

1. To avoid overweight, consume only as much energy (kcalories) as is expended; if overweight, decrease energy intake and increase energy expenditure.
2. Increase the consumption of complex carbohydrates and "naturally occurring" sugars from about 22% of energy intake to about 48% of energy intake.
3. Reduce the consumption of refined and other processed sugars by about 45% to account for about 10% of the total energy intake.
4. Reduce overall fat consumption from approximately 42% to about 30% of energy intake.
5. Reduce saturated fat consumption to account for about 10% of total energy intake; balance that with polyunsaturated and monounsaturated fats, which should account for about 10% of the energy intake each.
6. Reduce cholesterol consumption to about 300 milligrams a day.
7. Limit the intake of sodium by reducing the intake of salt (sodium chloride) to about 5 grams a day.

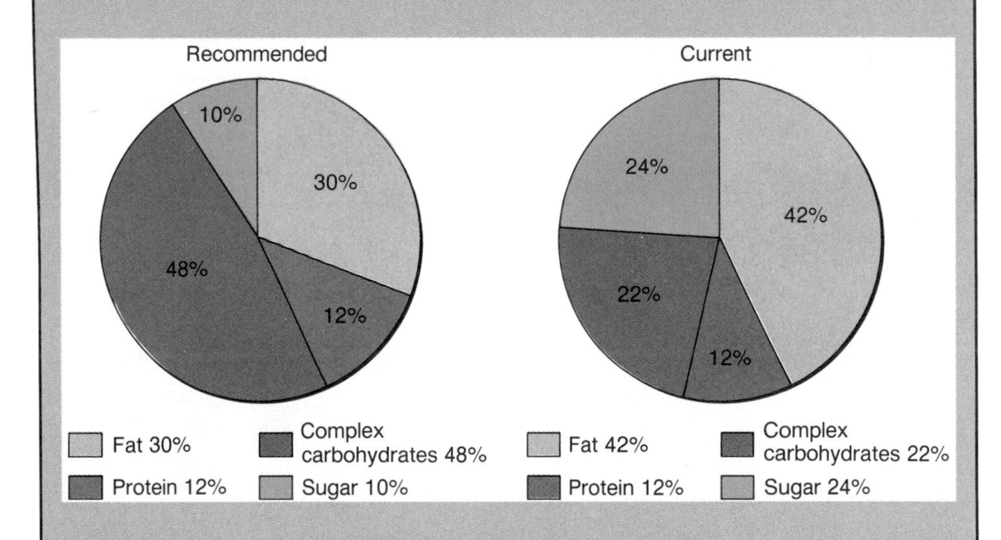

Dietary Guidelines for the United States

The U.S. Department of Health and Human Services recommends the following guidelines.

1. Eat a variety of foods.
2. Maintain desirable weight.
3. Avoid too much fat, saturated fat, and cholesterol.
4. Eat foods with adequate starch and fiber.
5. Avoid too much sugar.
6. Avoid too much sodium.
7. If you drink alcoholic beverages, do so in moderation.

The dietary goals differ from the dietary guidelines; the goals state nutritional objectives in terms of nutrients, while the guidelines translate them into types of food.

Psychological Eating Disorders

Eating disorders are grouped into three categories—obesity, bulimia, and anorexia nervosa. All three disorders are founded in psychological problems of one kind or another and are strongly influenced by the culture in which we live. People who gain a great deal of unnecessary weight and are 15%–20% above their ideal weight are **obese.** As a person engages in the activities of everyday life, the body constantly uses energy acquired from food. At no time are the food molecules entering the body permanently locked into specific parts of cells. They are constantly being changed and exchanged; fat molecules, for example, are completely exchanged about every four weeks. This exchange is easily seen when a person either gains or loses weight. In each case, the molecules of the body are rearranged. The nature of the rearrangement depends on the amount of incoming food and the amount of activity. If a person's activity level increases and eating habits remain the same, there will be a loss of weight. On the other hand, if the activity level drops and eating habits remain the same, the person will gain weight. Of course, there are extremes of both cases. Obesity occurs when people consistently take in more food energy than is necessary to meet their daily requirements.

Obesity

Obesity is probably the most familiar eating disorder and the one that is most publicized. Obesity is the condition of being overweight to the extent that a person's health and life span are adversely affected. The majority of people suffering from this condition use food as a psychological crutch. These people attempt to cope with the problems they face by overeating. Overeating to solve problems is encouraged by our culture. Americans celebrate almost all occasions with food. Social gatherings of almost every type are considered incomplete without some sort of food and drink. If snacks (usually high-calorie foods) are not made available by the host, many people feel uneasy or even unwelcome. It is also true that Americans and people of other cultures show love and friendship by sharing a meal. Many photographs in family albums have been taken at mealtime. Controlling obesity can be very difficult, since it requires basic changes in a person's eating habits, lifestyle, and value system. For some, these changes may require professional help.

While the majority of cases of obesity are psychologically based, some have been demonstrated to have a strong biological component. It appears that some obese individuals have a chemical imbalance of the nervous system that prevents them from feeling "full" until they have eaten an excessive amount of food. This imbalance prevents the brain from "turning off" the desire to eat after a reasonable amount of food has been eaten. Research into the nature and action of this brain chemical indicates that if obese people lacking this chemical receive it in pill form, they can feel "full" even when their food intake is decreased by 25%.

Bulimia

Bulimia ("Hunger of an ox" in Greek) is sometimes called the *silent killer* because it is difficult to detect. Bulimics are usually of normal size or overweight. This disorder involves a cycle of eating binges and purges. The cause is thought to be psychological, stemming from depression, low self-esteem, displaced anger, a need to be in control of one's body, or a personality disorder. The cycle usually begins with an episode of overeating followed by elimination of the food by induced vomiting and/or excessive use of laxatives and diuretics. Vomiting may be induced physically or by the use of some nonprescription drugs. Case studies have shown that bulimics may take forty to sixty laxatives a day to rid themselves of food. For some, the laxative becomes addictive. Diuretics are also used by bulimics to increase water loss through urination. The binge-purge cycle results in a variety of symptoms that can be deadly. The following is a list of the major symptoms observed in many bulimics:

- Excessive water loss
- Diminished blood volume
- Extreme potassium, calcium, and sodium deficiencies
- Kidney malfunction
- Increase in heart rate
- Loss of rhythmic heartbeat
- Lethargy
- Diarrhea
- Severe stomach cramps
- Damage to teeth and gums
- Loss of body proteins
- Migraine headaches
- Fainting spells
- Increased susceptibility to infections

Anorexia Nervosa

Anorexia nervosa (figure 19.4) is a nutritional deficiency disease characterized by severe, prolonged weight loss. An anorexic person's fear of becoming overweight is so intense that even though weight loss occurs, it does not lessen the fear of obesity, and the person continues to diet, often even refusing to maintain the optimum body weight for his or her age, sex, and height. This nutritional deficiency disease is thought to stem from sociocultural factors. Our society's preoccupation with weight loss and the desirability of being thin strongly influences this disorder. Individuals with anorexia are mostly adolescent and preadolescent females, although the disease does occur in males. Anorexic individuals starve themselves to death. Just turn on your television or radio, or look at newspapers, magazines, or billboards, and you can see how our culture encourages people to be thin. Male and female models are thin. Muscle protein is considered to be healthy and fat to be unhealthy. Unless you are thin, so the ads imply, you will never be popular, get a date, or even marry. In fact, you may die early. Are these prophecies self-fulfilling? Our culture's constant emphasis on being thin has influenced many people to lose too much weight and become anorexic. Here are some of the symptoms of anorexia nervosa.

- Thin, dry, brittle hair
- Degradation of fingernails
- Constipation
- Amenorrhea (lack of menstrual periods)
- Decreased heart rate
- Loss of body proteins
- Weaker-than-normal heartbeat
- Calcium deficiency
- Osteoporosis
- Hypothermia (low body temperature)
- Hypotension (low blood pressure)
- Increased skin pigmentation
- Reduction in size of uterus
- Inflammatory bowel disease
- Slowed reflexes
- Fainting
- Weakened muscles

Deficiency Diseases

Without minimal levels of the essential amino acids in the diet, a person may develop health problems that could ultimately lead to death. In many parts of the world, large populations of people live on diets that are very high in carbohydrates and fats but low in complete protein. This is easy to understand, since carbohydrates and fats are inexpensive to grow and process in comparison to proteins. For example, corn, rice, wheat, and barley are all high-carbohydrate foods. Corn and its products (meal, flour) contain protein, but it is an incomplete protein that lacks the amino acid tryptophan. Without this amino acid, many necessary enzymes cannot be made in sufficient amounts to keep a person healthy. One protein-deficiency disease is called **kwashiorkor,** and the symptoms are easily seen (figure 19.5). A person with this deficiency has a distended belly, slow growth, slow movement, and is emotionally depressed. If the disease is caught in time, brain damage may be prevented and death averted. This requires a change in diet that includes expensive protein, such as poultry, fish, beef, shrimp, or milk. As the world

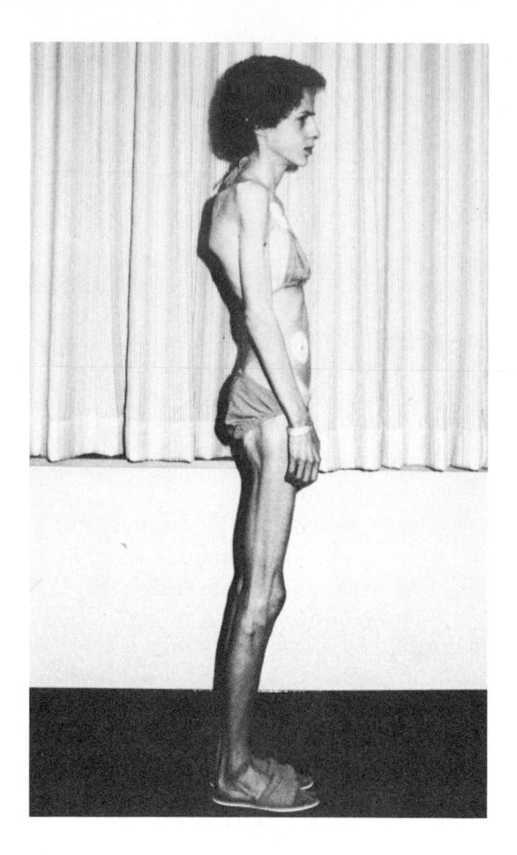
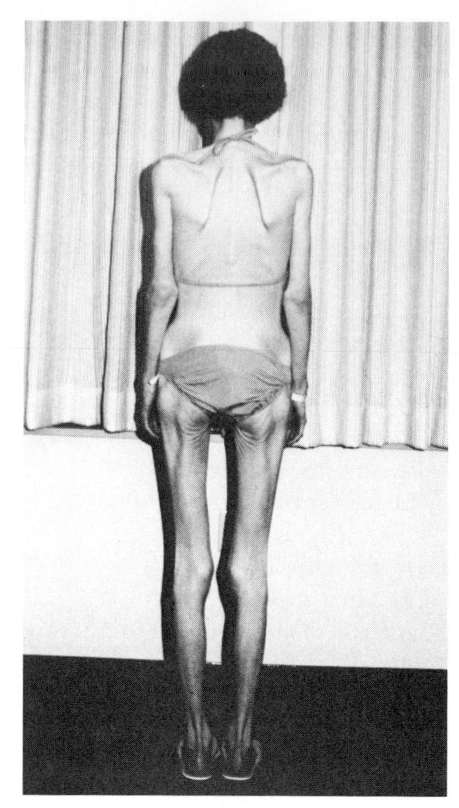
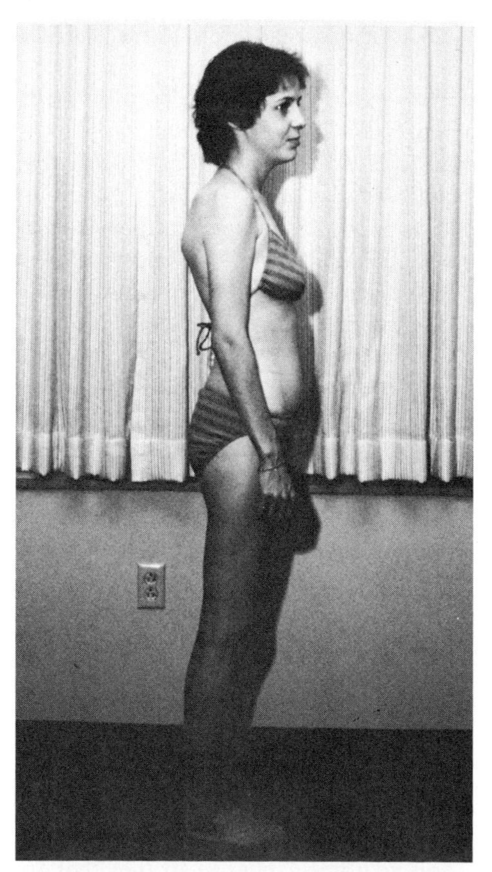
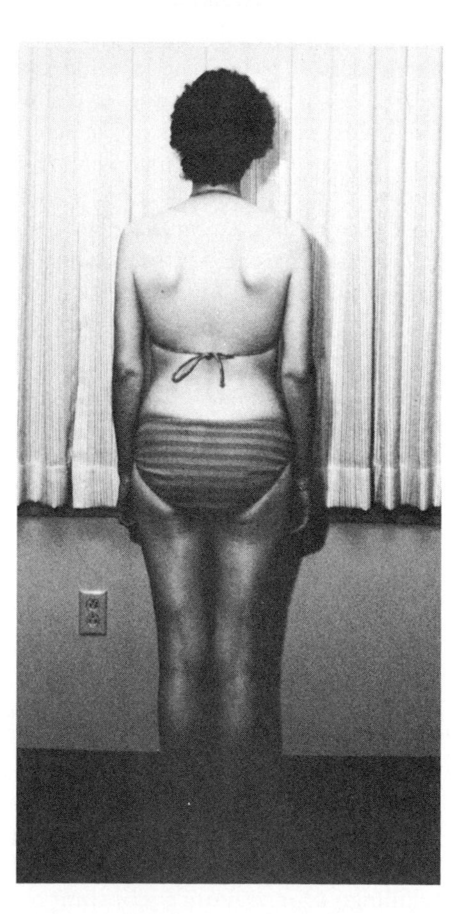

Figure 19.4 **Anorexia nervosa.** Anorexia nervosa is a psychological eating disorder afflicting many Americans. These photographs were taken of an individual before and after treatment. Restoring a person with this disorder requires both medical and psychological efforts.

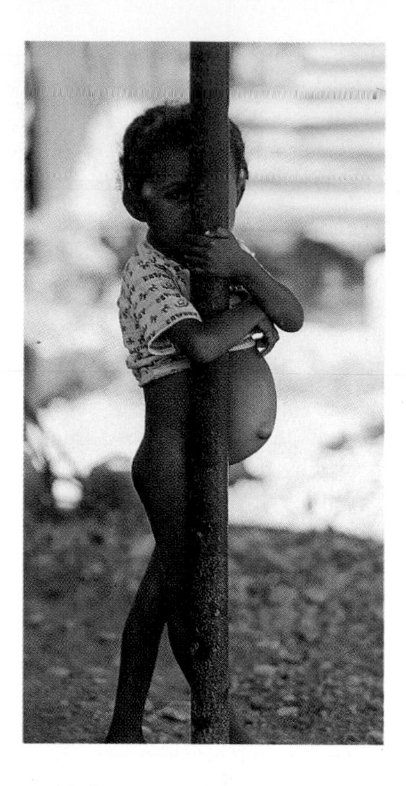

Figure 19.5 **Kwashiorkor.** This starving child shows the symptoms of kwashiorkor, a protein-deficiency disease. If treated with a proper diet containing all amino acids, the disease can be cured.

food problem increases, these expensive foods will be in even shorter supply and will become more and more costly.

Very little carbohydrate is stored in the body. If you starve yourself, this small amount will last as a stored form of energy only for about two days. After the stored carbohydrate has been used, your body begins to use its stored fat deposits as a source of energy; the proteins will be used last. During the early stages of starvation, the amount of fat in the body will steadily decrease, but the amount of protein will drop only slightly (figure 19.6). This can continue only up to a certain point. For about the first six weeks of this starvation period, the fat acts as a protein protector. You can see the value of this kind of protection when you remember the vital roles that protein-sparing plays in cellular metabolism. After about six weeks, however, so much fat has been lost from the body that proteins are no longer protected, and cells begin to use them as a source of energy. This results in a loss of proteins from the cells that prevents them from carrying

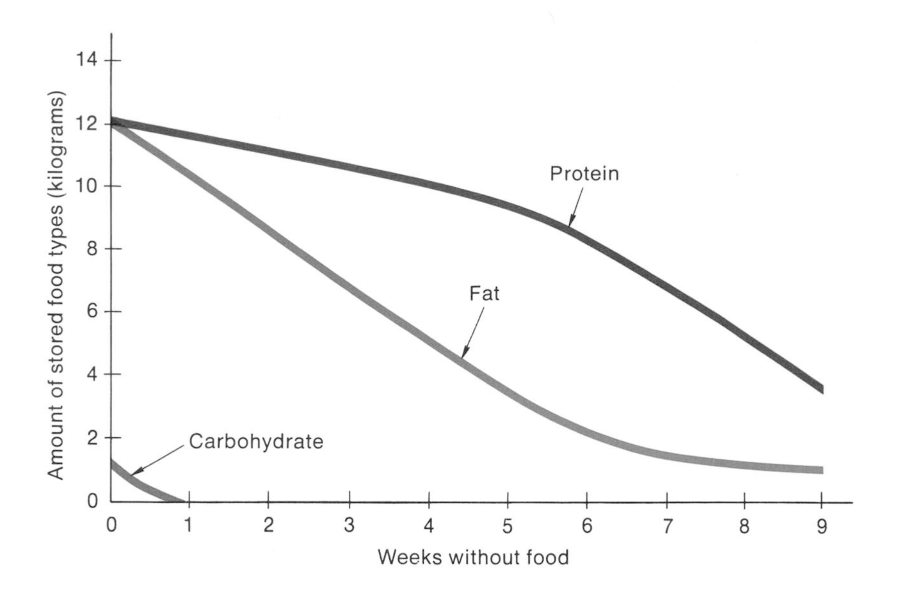

Figure 19.6 **Starvation and Stored Foods.** Starving yourself results in a very selective loss of the kinds of nutrients stored in the body. Notice how the protein level in the body has the slowest decrease of the three nutrients. This protein-conservation mechanism enables the body to preserve essential amounts of enzymes and other vital proteins.

Table 19.7
A Comparison of Human Breast Milk and Cow's Milk

Nutrient	Human Milk	Cow's Milk
Energy (kcalories/1,000g)	690	660 (whole milk)
Protein (grams per liter)	9	35
Fat (grams per liter)	40	38
Lactose (grams per liter)	68	49
Vitamins		
A (International Units)	1,898	1,025
D (activity units)	40	14
E (International Units)	3.2	0.4
K (micrograms)	34	170
Thiamine (B_1) (micrograms)	150	370
Riboflavin (B_2) (micrograms)	380	1,700
Niacin (B_3) (milligrams)	1.7	0.9
Pyridoxine (B_6) (micrograms)	130	460
Cyanocobalamine (B_{12}) (micrograms)	0.5	4
Folic acid (micrograms)	41–84.6	2.9–68
Ascorbic acid (C) (micrograms)	44	17
Minerals (all in milligrams)		
Calcium	241–340	1,200
Phosphorus	150	920
Sodium (depends on distributor)	160	560
Potassium	530	1,570
Iron	0.3–0.56	0.5
Iodine	200	80

All milks are not alike. Each milk is unique to the species that produces it for its young, and each infant has its own special growth rate. Humans have one of the slowest infant growth rates, and human milk contains the least amount of protein. Since cow's milk is so different, many pediatricians recommend that human infants be fed either human breast milk or formulas developed to be comparable to breast milk during the first twelve months of life. The use of cow's milk is discouraged. This table lists the relative amounts of different nutrients in human breast milk and cow's milk.

out their normal functions. When not enough enzymes are available to do the necessary cellular jobs, the cells die.

The lack of a particular vitamin in the diet can result in a **vitamin-deficiency disease.** A great deal has been said about the need for vitamin and mineral supplements in diets. Some people claim that supplements are essential, while others claim that a well-balanced diet provides adequate amounts of vitamins and minerals. Supporters of vitamin supplements have even claimed that extremely high doses of certain vitamins can prevent ill-health or even create "supermen." It is very difficult to evaluate many of these claims, however, since the functioning of vitamins and minerals and their regulation in the body is not completely understood. In fact, the minimum daily requirement of a number of vitamins has not been determined.

Nutrition through the Life Cycle

Nutritional needs vary throughout life and involve many factors, including age, sex, reproductive status, and level of physical activity. Infants, children, adolescents, adults, and the elderly all require slightly different amounts and kinds of nutrients.

Infancy

A person's total energy requirements are highest during the first twelve months of life: 100 kcalories per kilogram of body weight per day. Fifty percent of this energy is required for an infant's BMR. Infants (birth to twelve months) triple their weight and increase their length by 50% during that first year; this is their so-called first growth spurt. They require nutrients that are high in easy-to-use kcalories, proteins, vitamins, minerals, and water because of their need to grow and produce new cells and tissues. For many reasons, the food that most easily meets these needs is human breast milk (table 19.7). Even with breast milk's many nutrients, many physicians strongly recommend multivitamin supplements as a part of an infant's diet.

Childhood

As infants reach childhood, their dietary needs change. The rate of growth generally slows between one year of age and puberty, and girls increase in height and weight slightly faster than boys. During childhood, the body becomes more lean, bones elongate, and the brain reaches 100% of its adult size between the ages of six and ten. To adequately meet growth and energy needs during childhood, protein intake should be high enough to take care of the development of new tissues. Minerals, such as calcium, zinc, iron, and vitamins, are also necessary to support growth and prevent anemia. While many parents continue to provide their children with multivitamin supplements, such supplements should be given only after a careful evaluation of their children's diets. There are four groups of children who are at particular risk and should receive such supplements:

1. children from deprived families and those suffering from neglect or abuse;
2. children who have anorexia or poor eating habits, or who are obese;
3. pregnant teens; and
4. children who are strict vegetarians.

During childhood, eating habits are very erratic and often cause parental concern. Children often limit their intake of milk, meat, and vegetables while increasing their intake of sweets. To get around these problems, parents can provide calcium by serving cheeses, yogurt, and soups as alternatives to milk. Meats can be made more acceptable if they are in easy-to-chew, bite-size pieces, and vegetables may be more readily accepted if smaller portions are offered on a more frequent basis. Steering children away from sucrose by offering sweets in the form of fruits can help reduce dental caries. You can better meet the dietary needs of children by making food available on a more frequent basis, such as every three to four hours.

Adolescence

The nutrition of an adolescent is extremely important because during this period the body changes from nonreproductive to reproductive. Puberty is usually considered to last between five and seven years. Before puberty, males and females tend to have similar proportions of body fat and muscle. Both body fat and muscle make up between 15% and 19% of the body's weight. Lean body mass, primarily muscle, is about equal in males and females. However, this soon changes. Female body fat increases to about 23%, and in males it decreases to about 12%. Males double their muscle mass in comparison to females.

The changes in body form that take place during puberty constitute the second growth spurt. Because of their more rapid rate of growth and unique growth patterns, males require more protein, iron, zinc, and calcium than females. During adolescence, youngsters will gain as much as 20% of their adult height and 50% of their adult weight, and many body organs will double in size. Nutritionists have taken these growth patterns and spurts into account by establishing RDAs for males ten to twenty years old and for females ten to twenty years old, including requirements at the peak of their growth spurt. RDAs at the peak of the growth spurt are about twice what they are for adults and children.

Adulthood

People who have completed the changes associated with adolescence are considered to have entered adulthood. Most of the information available to the public through the press, television, and radio focuses on this stage in the life cycle. During adulthood, the body has entered a plateau phase, and diet and nutrition focus on maintenance and disease prevention. Nutrients are used primarily for tissue replacement and repair, and changes such as weight loss occur slowly. Since the BMR slows, as does physical activity, the need for food-energy decreases from about 2,700 kcalories in average young adult males (ages twenty to forty) to about 2,050 for elderly men. For women, the corresponding numbers reduce from 2,000 to 1,600 kcalories. Protein intake for most U.S. citizens is usually in excess of the recommended amount. The RDA standard for protein is about 56 grams for men and 44 grams for women each day. About 25%–50% should come from animal foods to ensure intake of the essential amino acids. The rest should be from plant-protein foods, such as whole grains, legumes, nuts, and vegetables.

An adult who follows a well-balanced diet should have no need for vitamin supplements; however, improper diet, disease, or other conditions might require that supplements be added. The two minerals that demand special attention are calcium and iron, especially for women. A daily intake of 1,200 milligrams of each should prevent calcium loss from bones (osteoporosis [see figure 19.7]) and allow adequate amounts of hemoglobin to be manufactured to prevent anemia in women over fifty and men over sixty. In order to reduce the risk of chronic diseases such as heart attack and stroke, adults should be sure to eat a balanced diet, participate in regular exercise programs, control their weight, avoid cigarettes and alcohol, and practice stress management.

Nutritional Needs Associated with Pregnancy and Lactation

Risk-management practices that help in avoiding chronic adult diseases become even more important when planning pregnancy. Studies have shown that an inadequate supply of the essential nutrients can result in infertility, spontaneous abortion, and abnormal fetal development. The period of pregnancy and milk production (lactation) requires that special attention be paid to the diet to ensure proper fetal development, a safe delivery, and a healthy milk supply.

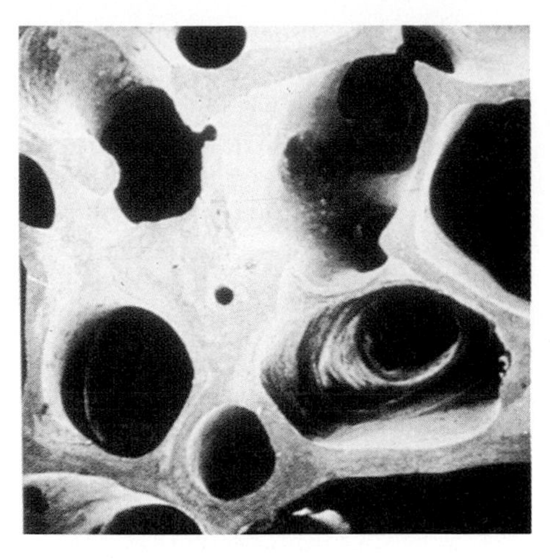

(a)

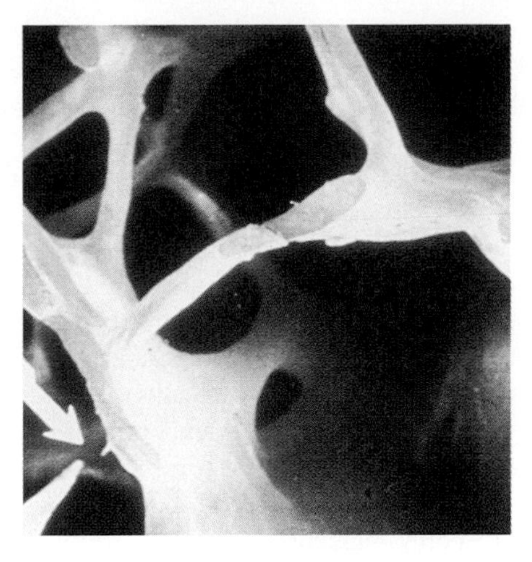

(b)

Figure 19.7 **Osteoporosis.** These photographs are of a healthy bone (a) and a section of bone from a person with osteoporosis (b). This nutritional deficiency disease results in a change in the density of the bones as a result of the loss of bone mass. Bones that have undergone this change look "lacy" or like Swiss cheese, with larger than normal holes. A few risk factors found to be associated with this disease are being female and fair skinned; having a sedentary life-style; using alcohol, caffeine, and tobacco; and having reached menopause.

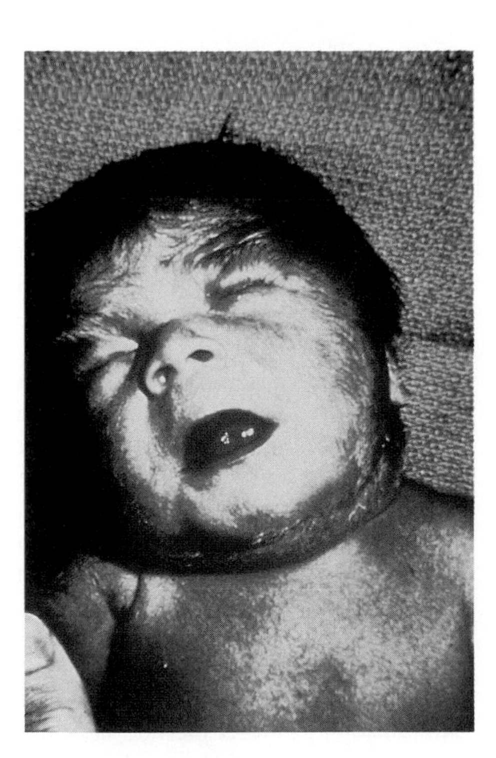

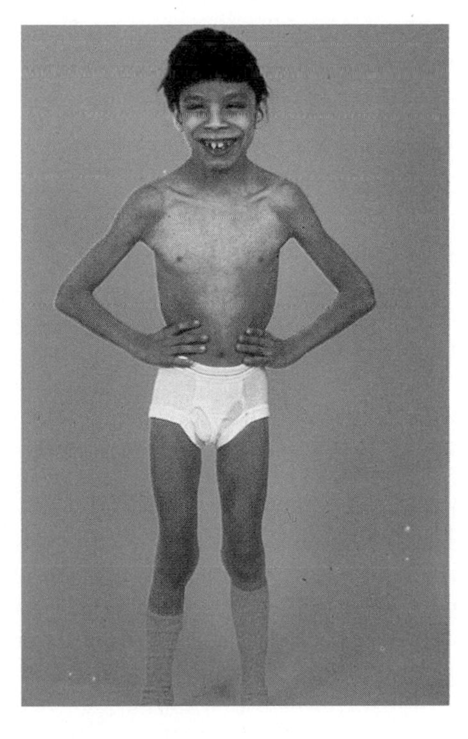

Figure 19.8 **Fetal Alcohol Syndrome (FAS).** The one-day-old infant on the left displays fetal alcohol syndrome while the child on the right with the syndrome is eight years old. The eight-year-old was diagnosed at birth and has spent all of his life in a foster home where the quality of care has been excellent. His IQ has remained stable at 40 to 45.

The daily amount of essential nutrients must be increased, as should the kcaloric intake. Kilocalories must be increased by 300 per day to meet the needs of an increased BMR, the development of the uterus, breasts, and placenta, and the work required for fetal growth. Some of these kcalories can be obtained by drinking milk, which simultaneously supplies calcium needed for fetal bone development. In addition, the daily intake of protein should be increased by 75%. Two essential nutrients, folic acid and iron, should be obtained through prenatal supplements since they are so essential to cell division and development of the fetal blood supply.

The mother's nutritional status affects the developing baby in several ways. If she is under fifteen years of age or has had three or more pregnancies in a two-year period, her nutritional stores are inadequate to support a successful pregnancy. The use of drugs such as alcohol, caffeine, nicotine, and "hard" drugs (e.g., heroin) can result in decreased nutrient exchange between the mother and fetus. In particular, heavy smoking can result in low birth weights, while alcohol abuse is responsible for fetal alcohol syndrome (figure 19.8).

Old Age

As people move into their sixties and seventies, digestion and absorption of all nutrients through the intestinal tract is not impaired but does slow. The number of cells undergoing mitosis is reduced, resulting in an overall loss in the number of body cells. With age, complex organs such as the kidneys and brain function less efficiently, and protein synthesis becomes inefficient. With regard to nutrition, energy requirements for the elderly decrease as the BMR slows, physical activity decreases, and eating habits also change.

The change in eating habits is particularly significant, since it can result in dietary deficiencies. For example, linoleic acid, the essential fatty acid, may fall below required levels as an older person reduces the amount of food eaten. The same is true for some vitamins and minerals. Therefore, it may be necessary to supplement the diet daily with one tablespoon of vegetable oil. Vitamin E, multiple vitamins, or a mineral supplement may also be necessary. The loss of body protein means that people must be sure to meet their daily RDA for protein. As with all stages of the life cycle, regular exercise is important in maintaining a healthy, efficiently functioning body.

Nutrition for Fitness and Sports

In the past few years there has been a heightened interest in fitness and sports. Along with this, an interest has developed in the role nutrition plays in fueling activities, controlling weight, and building muscle. The cell-respiration process described in chapter 7 is the source of the energy needed to take a leisurely walk or run a marathon. However, just which molecules are respired and when depends on whether or not you warm up before you exercise and how much effort you exert during exercise. The molecules respired by muscle cells to produce ATP may be glucose or glycogen, triglycerides or fatty acids. During a long, brisk walk, the heart and lungs of most people should be able to keep up with the muscle cells' requirement for oxygen. The oxygen is used by mitochondria to run the aerobic Krebs cycle and electron-transport system. Studies have shown that under these moderate exercise conditions, triglycerides stored in cells and fatty acids from the circulatory system are used as the prime sources of energy. Glycogen stored in muscle cells and glucose in the blood are not utilized. This is why such moderate, short-term exercise is most beneficial in weight loss.

But what about vigorous exercise for longer periods? Without a warm-up period, your muscle cells will have to begin respiring muscle-stored glycogen. At first the heart and lungs are not able to supply all the oxygen needed by the mitochondria. Glycolysis will provide ATP for muscle contraction, but there will be an increase in the amount of pyruvic and lactic acids. About five minutes into vigorous exercise, 20% of muscle glycogen is gone and little fat has been respired. By this time, the heart and lungs are "warmed up" and are able to provide the oxygen necessary to respire the triglycerides and fatty acids. ATP output increases dramatically. A person who has not warmed up experiences this metabolic shift as a "second wind."

From this point on, all four sources of energy—glucose, glycogen, triglycerides, and fatty acids—are utilized. If the exercise continues, stored lipids will be hydrolyzed into fatty acids and blood levels of these molecules will increase sixfold. At approximately fifty minutes into exercise, cells shift to using glycogen. If exercise is suddenly stopped, the high concentration of fatty acids will be converted to keto acids and the exerciser could experience kidney problems; a loss of sodium, calcium, and other minerals; and a change in the pH of the blood. For this reason, many exercise physiologists suggest a cool-down period of light exercises that allow the body to slowly shift back to normal metabolism. If the exerciser continues full-out, the next metabolic shift takes place at about eighty minutes. It is known as "hitting the wall." This is the sudden onset of disabling fatigue that occurs when the limited amount of stored glycogen falls below a certain level.

To avoid or postpone hitting the wall, many marathon athletes practice **carbohydrate loading.** This should only be done by those engaged in periods of hard exercise or competition lasting for ninety minutes or more. It requires following a week-long diet and exercise program. On the first day, the muscles needed for the event are exercised for ninety minutes and carbohydrates, such as fruits, vegetables, or pasta, provide a source of dietary kcalories. On the second and third days of carbohydrate loading, the person continues the 50% carbohydrate diet, but the period of exercise is reduced to forty minutes. On the fourth and fifth days, the workout period is reduced to twenty minutes and the carbohydrates are increased to 70% of total kcalorie intake. On the sixth day, the day before competition, the person rests and continues the 70% carbohydrate diet. Following this program increases muscle glycogen levels and makes it possible to postpone "hitting the wall."

As the body is conditioned, there is an increase in the number of mitochondria per cell, the Krebs cycle and the ETS run more efficiently, the number of capillaries increases, fats are respired more efficiently and for longer periods, and weight control becomes easier.

The amount of protein in an athlete's diet has also been investigated. Understand that increasing dietary protein does not automatically increase strength, endurance, or speed. In fact, most Americans eat the 10% additional protein that athletes require as a part of their normal diets. The additional percentage is used by the body for many things, including muscle growth. But increasing protein intake will not automatically increase muscle size. Only when there is a need will the protein be used to increase muscle mass. That means exercise. Your body will build the muscle it needs in order to meet the demands you place on it. Vitamins and minerals operate in much the same way. No supplements should be required as long as your diet is balanced and complex. Your meals should provide the vitamins and minerals needed to sustain your effort.

Athletes must monitor their water intake because dehydration can cripple an athlete very quickly. A water loss of only 5% of the body weight can decrease muscular activity by as much as 30%. One way to replace water is to drink 1 to 1½ cups of water fifteen minutes before exercising and a half cup during exercise. In addition, drinking 16 ounces of cool tap water for each pound of body weight lost during exercise is enough to prevent dehydration. Another method is to use diluted orange juice (1 part juice to 5 parts water). This is an excellent way to replace water and resupply a small amount of lost glucose and salt. Salt pills (so-called electrolyte pills) are not recommended.

• Summary •

To maintain good health, people must receive nutrient molecules that can enter the cells and function in the metabolic processes. The proper quantity and quality of nutrients are essential to good health. Nutritionists have classified nutrients into six groups: carbohydrates, proteins, lipids, minerals, vitamins, and water. Energy for metabolic processes may be obtained from carbohydrates, lipids, and proteins, and is measured in kilocalories. An important measure of the amount of energy required to sustain a human at rest is the basal metabolic rate. To meet this and all additional requirements, the United States has established the RDAs, Recommended Daily Allowances, for each nutrient. Should there be metabolic or psychological problems associated with a person's normal metabolism, a variety of disorders may occur, including obesity, anorexia nervosa, bulimia, kwashiorkor, and vitamin-deficiency diseases. As people move through the life cycle, their nutritional needs change, requiring a reexamination of their eating habits in order to maintain good health.

• Thinking Critically •

You're twenty-one years old, female, have never been involved in any kind of sports, and have suddenly become interested in rugby! This is a very demanding contact sport, and many are injured while playing. If you are to succeed and experience only minor injuries, you must get in condition. That will include exercise and nutrition planning. Well, get busy and plan or you'll never make the team!

• Experience This •

The next time you're in the kitchen or grocery store, take a few minutes and do some label reading and comparing. Compare the percentage of RDAs found in one product from each of the basic food groups to determine which has the highest amount of your chosen nutrient. Now select three foods from the same food group and make a similar comparison.

• Questions •

1. List the six classes of nutrients and give an example of each.
2. Name the five basic food groups and give two examples of each.
3. What are basal metabolism, specific dynamic action, and voluntary muscular activity?
4. During which phase of the life cycle is a person's demand for kcalories per unit of body weight the highest?
5. Why are some nutrients referred to as essential? Name them.
6. What do the initials RDA stand for?
7. List four of the dietary guidelines.
8. Americans are currently consuming 42% of their kcalories in fat. According to the dietary goals, what should that percentage be?

• Chapter Glossary •

anorexia nervosa (an″o-rek′se-ah ner-vo′sah) A nutritional deficiency disease characterized by severe, prolonged weight loss for fear of becoming obese. This eating disorder is thought to stem from sociocultural factors.

assimilation (ă-sĭ″mĭ-la′shun) The physiological process that takes place in a living cell as it converts nutrients in food into specific molecules required by the organism.

basal metabolism (ba′sal mĕ-tab′o-lizm) The amount of energy required to maintain normal body activity while at rest.

bulimia (bu-lim′-e-ah) A nutritional deficiency disease characterized by a binge-and-purge cycle of eating. It is thought to stem from psychological disorders.

calorie (kal′o-re) The amount of heat energy necessary to raise the temperature of one gram of water one degree Celsius.

carbohydrate loading (kar-bo-hi-′drāt lo′ding) A week-long program of diet and exercise that results in an increase in muscle glycogen stores.

complete protein (kom-plēt′ pro′te-in) Protein molecules that provide all the essential amino acids.

diet (di′et) The food and drink consumed by a person from day to day.

electrolytes (ĕ-lek′tro-līts) Ionic compounds dissolved in water. Their proper balance is essential to life.

essential amino acids (ĕ-sen′shul ah-me′no ă′sids) Those amino acids that cannot be synthesized by the human body and must be part of the diet (e.g., as lysine, tryptophan, and valine).

essential fatty acid (ĕ-sen'shul fă'tē ă'sid) The fatty acid linoleic acid. It cannot be synthesized by the human body and must be part of the diet.

fiber (fi'ber) Natural (plant) or industrially produced polysaccharides that are resistant to hydrolysis by human digestive enzymes.

Food Guide Pyramid (food gīd pĭ'ra-mĭd) A tool developed by the U.S. Department of Agriculture to help the general public plan for good nutrition. It contains guidelines for required daily intake from each of the five food groups.

incomplete protein (in-kom-plēt' pro'te-in) Protein molecules that do not provide all the essential amino acids.

kilocalorie (kcalorie) (kil''o-kal'o-re) A measure of heat energy one thousand times larger than a calorie.

kwashiorkor (kwa''she-or'kor) A protein-deficiency disease.

minerals (mĭn'er-alz) Inorganic elements that cannot be manufactured by the body but are required in low concentrations; essential to metabolism.

nutrients (nu'tre-ents) Molecules required by the body for growth, reproduction, and/or repair.

nutrition (nu-trĭ'shun) Collectively, the processes involved in taking in, assimilating, and utilizing nutrients.

obese (o-bēs) A term describing a person who gains a great deal of unnecessary weight and is 15%–20% above his or her ideal weight.

osteoporosis (os''te-o-po-ro'sis) A disease condition resulting from the demineralization of the bone, resulting in pain, deformities, and fractures; related to a loss of calcium.

protein-sparing (pro'te-in spĕ'ring) The conservation of proteins by first oxidizing carbohydrates and fats as a source of ATP energy.

Recommended Dietary Allowances (RDA) (rĕ-ko-men'ded di'ĕ-tĕ-rē ă-lao'an-ses) U.S. dietary guidelines for a healthy person that focus on the six classes of nutrients.

specific dynamic action (SDA) (spĕ-sĭ'fik di-nă'mik ak'shun) The amount of energy required to digest and assimilate food. SDA is equal to approximately 10% of your total daily kcalorie intake.

vitamin-deficiency disease (vi'tah-min de-fish'en-se dĭ-zēz) Poor health caused by the lack of a certain vitamin in the diet; for example, scurvy for lack of vitamin C.

vitamins (vi'tah-minz) Organic molecules that cannot be manufactured by the body but are required in very low concentrations.

C H A P T E R 20

The Body's Control Mechanisms

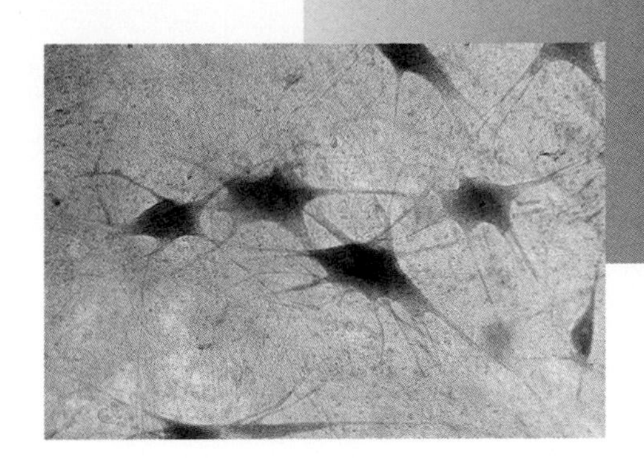

Purpose

The control of various body functions is important to the survival of an organism. Because many different systems each perform a specific set of tasks in the human body, it is necessary to coordinate these various activities with one another. This chapter describes the characteristics of the two major control systems: the nervous system and endocrine system. It also discusses various sensory inputs and response mechanisms that are involved in the control of the body.

For Your Information

In the spring of 1989 a major trade war almost erupted between the European Economic Community (EEC) and the United States. Some members of the EEC banned the importation of beef from the United States if the animal's growth had been enhanced by the use of synthetic growth hormones. It was claimed that the countries involved feared that beef reared in this manner presented a health hazard to their citizens. Farmers in the United States argued that the use of hormones was not a health hazard and that the EEC was simply using this as an excuse to stop U.S. beef from being imported. The use of hormones is extremely cost-effective. A tiny container of hormone is inserted under the skin of an animal, where it slowly releases the hormone. The hormone results in faster growth and reduces the time needed before the animal can be marketed, thus reducing the cost of production. The political problem was eventually resolved by negotiations between officials of the EEC and the United States. There is now no import ban on U.S. beef, even if it has been raised using synthetic hormones.

Chapter Outline

Learning Objectives

- Describe the ionic events of a nerve impulse.
- Describe the molecular events at the synapse.
- List the structural differences between the endocrine system and the nervous system.
- Describe how information is sent by both the nervous system and the endocrine system.
- Recognize that the endocrine system is under negative-feedback control.
- Understand that the endocrine system is able to regulate growth.
- Describe how chemicals, light, and sound are detected.
- Describe how a muscle cell contracts from a molecular point of view.
- Compare skeletal, smooth, and cardiac muscle.

Integration of Input

A large, multicellular organism, which consists of many different kinds of systems, must have some way of integrating various functions so that it can survive. If the organism does not respond appropriately to stimuli, it may die. A **stimulus** is any change in the environment that the organism can detect. Some stimuli, like light or sound, are typically external to the organism; others, like the pain generated by an infection, are internal. The reaction of the organism to a stimulus is known as a **response** (figure 20.1).

The nervous and endocrine systems are the major systems of the body that integrate stimuli and generate appropriate responses. The **nervous system** consists of a network of cells with fibrous extensions that carry information throughout the body. The **endocrine system** consists of a number of glands that communicate with one another and with other tissues through chemicals distributed throughout the organism. **Glands** are organs that manufacture molecules that are secreted either through ducts or into surrounding tissue, where they are picked up by the circulatory system. **Endocrine glands** have no ducts and secrete their products into the circulatory system; other glands, such as the digestive glands, empty their contents through ducts. Glands that have ducts are called **exocrine glands.**

Although the functions of the nervous and endocrine systems can overlap and be interrelated, these two systems have quite different methods of action. The nervous system functions very much like a telephone system. A message is sent along established pathways from a specific initiating point to a specific end point, and the transmission is very rapid. The endocrine system functions in a manner analogous to a radio broadcast system. Messenger molecules are distributed throughout the body by the circulatory system so that all cells receive them. However, only those cells that have the proper receptor sites can receive and respond to the molecules. In the same way, a radio signal will go to all the radios in a particular area, but only those that are tuned to the correct frequency can receive the message.

The Structure of the Nervous System

The basic unit of the nervous system is a specialized cell called a **neuron,** or **nerve cell.** A typical neuron consists of a central body that contains the nucleus and is called the **soma,** or **cell body,** and several long, protoplasmic extensions called **fibers.** There are two kinds of fibers: **axons,** which carry information away from the cell body, and **dendrites,** which carry information toward the cell body (figure 20.2). Typically, a cell has one axon and several dendrites.

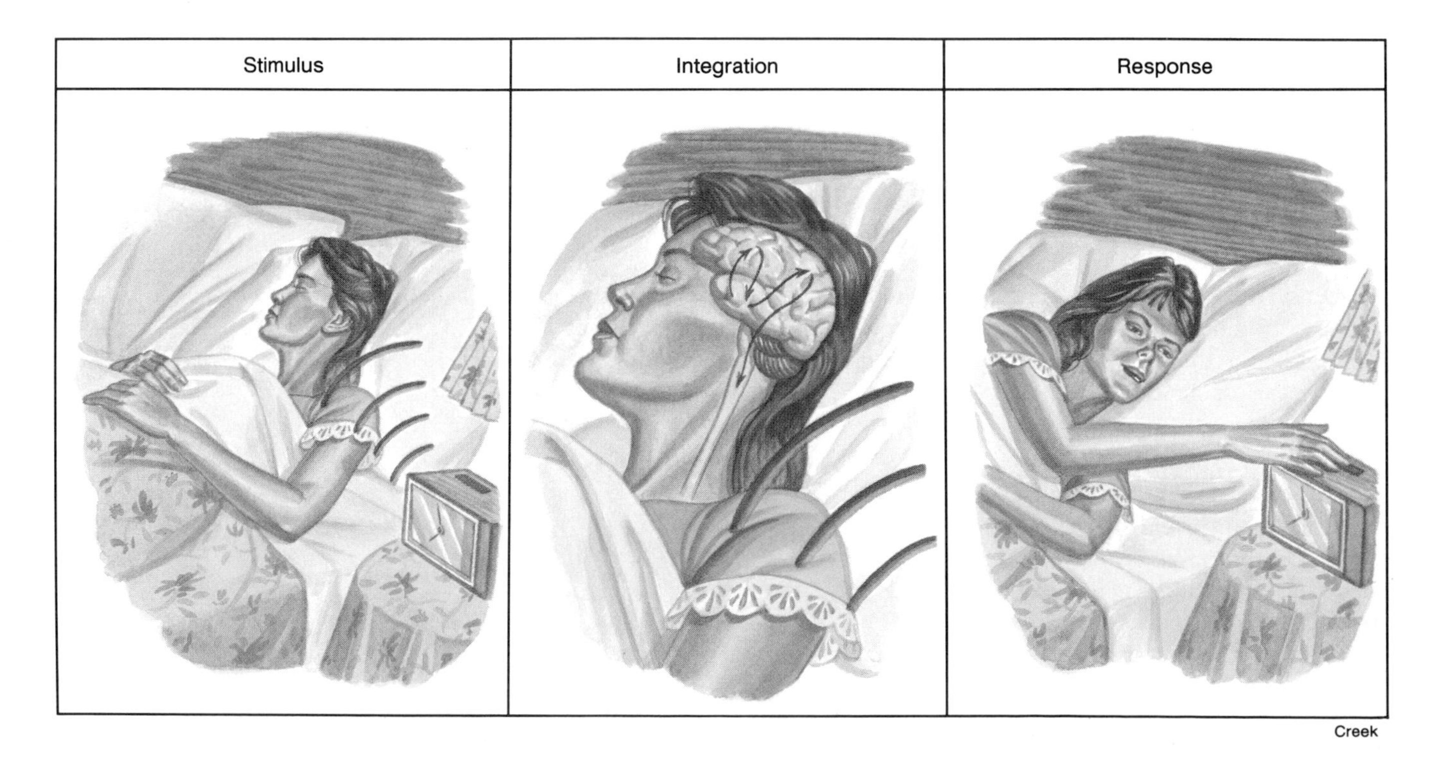

Figure 20.1 **Stimulus–Response.** A stimulus is any detectable change in the surroundings of an organism. When an organism receives a stimulus, it processes the information and may ignore the stimulus or generate a response to it.

Neurons are arranged into two major systems. The **central nervous system,** which consists of the brain and spinal cord, is surrounded by the skull and the vertebrae of the spinal column. It receives input from sense organs, interprets information, and generates responses. The **peripheral nervous system** is located outside the skull and spinal column and consists of bundles of long fibers called **nerves.** There are two different sets of neurons in the peripheral nervous system. **Motor neurons** carry information from the central nervous system to muscles and glands, and **sensory neurons** carry input from sense organs to the central nervous system. Motor neurons typically have one long axon that runs from the spinal cord to a muscle or gland, while sensory neurons have long dendrites that carry input from the sense organs to the central nervous system.

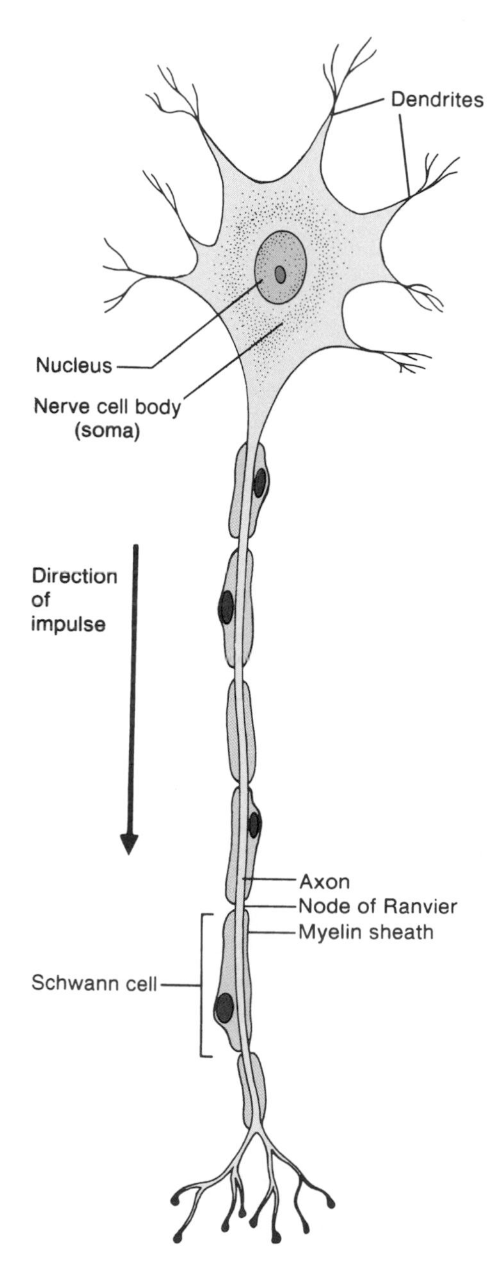

Figure 20.2 **The Structure of a Nerve Cell.** Nerve cells consist of a nerve-cell body that contains the nucleus and several fibrous extensions. The shorter, more numerous fibers that carry impulses to the nerve-cell body are dendrites. The long fiber that carries the impulse away from the cell body is the axon.

The Nature of the Nerve Impulse

The message that travels along a neuron is known as a **nerve impulse.** This transmission of information involves a series of chemical events. An impulse occurs because of several characteristics of the cell membrane. The cell membrane is differentially permeable; that is, it only allows certain kinds of ions to diffuse through it. There are also proteins in the membrane that can actively transport specific ions from one side of the membrane to the other. One of the ions that is actively transported from cells is the sodium ion (Na^+). At the same time as sodium ions (Na^+) are being transported out of cells, potassium ions (K^+) are being transported into the normal resting cells. However, there are more sodium ions (Na^+) transported out than potassium ions (K^+) transported in.

Because a normal resting cell has more positively charged Na^+ ions on the outside of the cell than on the inside, a small but measurable voltage exists across the membrane of the cell. (**Voltage** is a measure of the electrical charge difference that exists between two points or objects.) The voltage difference between the inside and outside of a cell membrane is about 70 millivolts (.007 volt). The cell membrane is therefore polarized in the same sense that a battery is polarized, with a positive and negative pole. A resting neuron has its positive pole on the outside of the cell membrane and its negative pole on the inside of the membrane (figure 20.3).

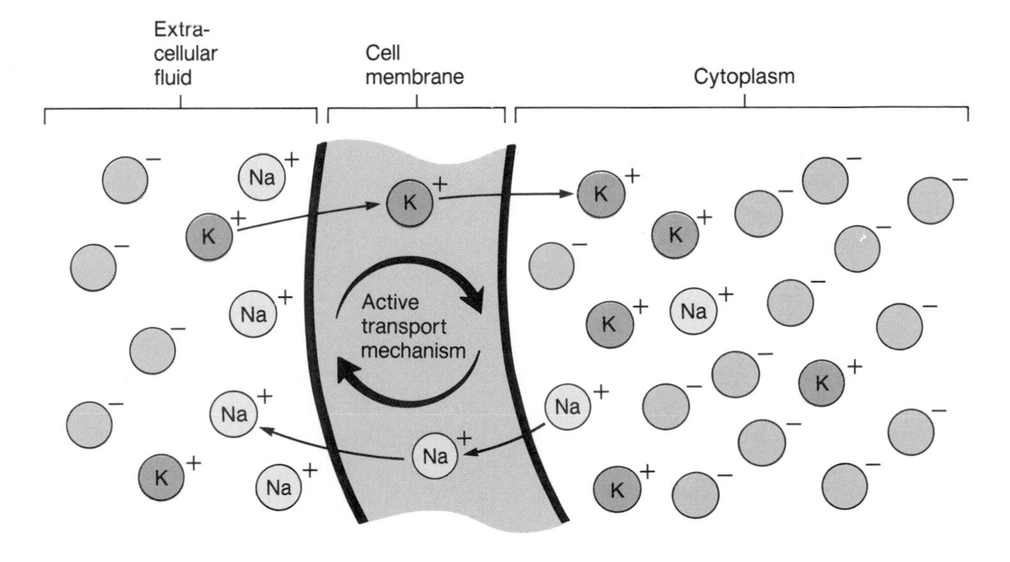

Figure 20.3 **The Polarization of Cell Membranes.** All cells, including nerve cells, have an active transport mechanism that pumps Na^+ out of cells and simultaneously pumps K^+ into them. The end result is that there are more Na^+ ions outside the cell and more K^+ ions inside of the cell. In addition, negative ions such as Cl^- are more numerous inside the cell. Consequently, the outside of the cell is positive (+) compared to the inside, which is negative (−).

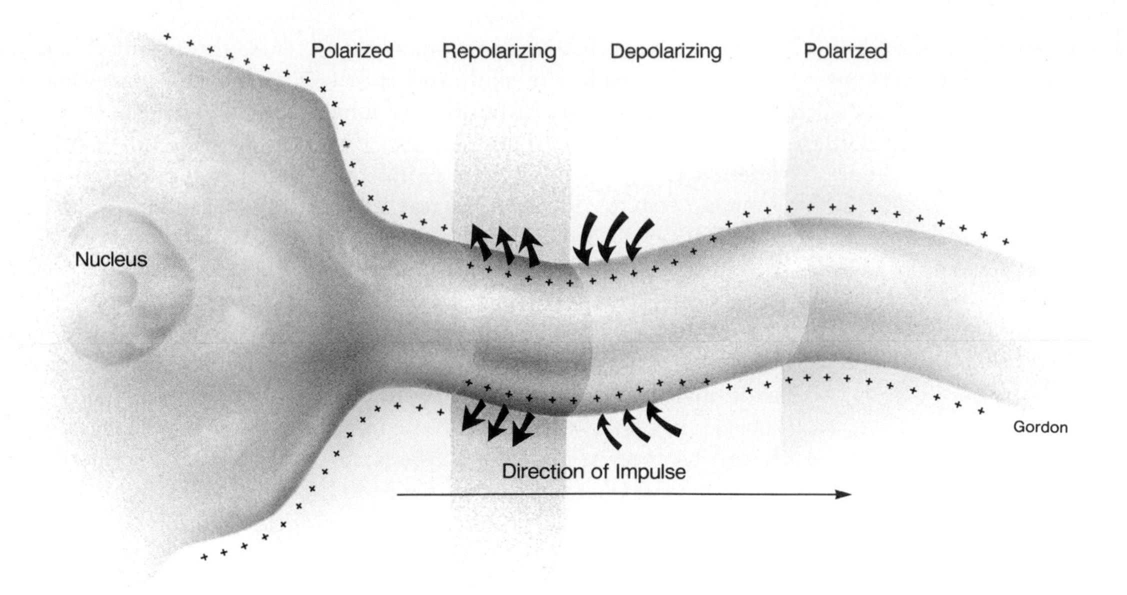

Figure 20.4 **A Nerve Impulse.** When a nerve cell is stimulated, a small portion of the cell membrane depolarizes as Na^+ flows into the cell through the membrane. This encourages the depolarization of an adjacent portion of the membrane, and it depolarizes a short time later. In this way a wave of depolarization passes down the length of the nerve cell. Shortly after a portion of the membrane is depolarized, the ionic balance is reestablished. It is repolarized and ready to be stimulated again.

When a cell is stimulated, the cell membrane changes its permeability and lets sodium ions (Na^+) pass through it. The membrane is thus **depolarized;** it loses its difference in charge as sodium ions (Na^+) diffuse into the cell from the outside. They do so because they are in greater concentration outside the cell than inside. When the membrane becomes more permeable, they are able to diffuse into the cell, toward the area of lower concentration. The depolarization of one point on the cell membrane causes the adjacent portion of the cell membrane to change its permeability, and it also depolarizes. Thus, a wave of depolarization passes along the length of the neuron from one end to the other (figure 20.4). The passage of an impulse along any portion of the neuron is a momentary event, since sodium ions (Na^+) begin to be actively pumped out of the cell just as soon as they enter. This reestablishes the original polarized state, and the membrane is said to be *repolarized.* When the nerve impulse reaches the end of the axon, it stimulates the release of a chemical that stimulates depolarization of the next neuron in the chain.

Activities at the Synapse

Between the fibers of adjacent neurons in a chain is a space called the **synapse.** When a neuron is stimulated, an impulse passes along its length from one end to the other. When the impulse reaches a synapse, a molecule called a **neurotransmitter** is released into the synapse from the axon. It diffuses across the synapse and binds to specific receptor sites on the dendrite of the next neuron. When enough neurotransmitter molecules have bound to the second neuron, an impulse is initiated in it as well. Several kinds of neurotransmitters are produced by specific neurons. These include dopamine, epinephrine, acetylcholine, and several other molecules. The first neurotransmitter identified was **acetylcholine.** Acetylcholine molecules are manufactured in the soma and migrate down the axon where they are stored until needed (figure 20.5).

However, if acetylcholine continues to occupy receptors, the neuron continues to be stimulated again and again. An enzyme called **acetylcholinesterase** destroys acetylcholine and prevents this from happening. The destruction of acetylcholine allows the second neuron in the chain to return to normal and be ready to accept another burst of acetylcholine from the first neuron when it arrives. Neurons must also constantly manufacture new acetylcholine molecules or they will exhaust their supply and be unable to conduct an impulse across a synapse.

Certain drugs, such as curare and strychnine, interfere with activities at the synapse and may cause paralysis or overstimulation. Many of the modern insecticides are also nerve poisons and are therefore quite hazardous.

Because of the way the synapse works, impulses can go in only one direction: only axons secrete acetylcholine, and only dendrites have receptors. This explains why there are sensory and motor neurons to carry messages to and from the central nervous system.

The nervous system is organized in a fashion similar to a computer. Information enters the computer by way of wires and is interpreted in the computer, and messages can be sent by way of cables

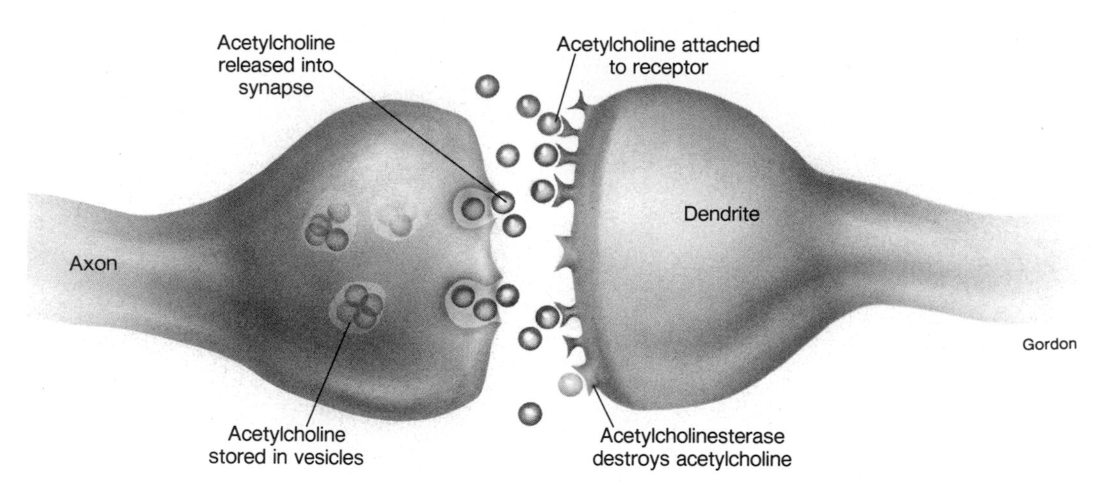

Figure 20.5 **Events at the Synapse.** When a nerve impulse reaches the end of an axon, it releases a neurotransmitter into the synapse. In this illustration, the neurotransmitter is acetylcholine. When acetylcholine is released into the synapse, acetylcholine molecules diffuse across the synapse and bind to receptors on the dendrite, initiating an impulse in the next neuron. Acetylcholinesterase is an enzyme that destroys acetylcholine, preventing continuous stimulation of the dendrite.

to drive external machinery. This concept allows us to understand how the functions of various portions of the nervous system have been identified. It is possible to electrically stimulate specific portions of the nervous system or to damage certain parts of the nervous system in experimental animals and determine the functions of different parts of the brain and other parts of the nervous system. For example, since peripheral nerves carry bundles of both sensory and motor fibers, damage to a nerve may result in both a lack of feeling and an inability to move.

The functions of specific portions of the brain have also been identified. Certain parts of the brain are involved in controlling fundamental functions such as breathing and heart rate. Others are involved in decoding sensory input, and still others are involved in coordinating motor activity. The human brain also has considerable capacity to store information and create new responses to environmental stimuli. Figure 20.6 shows a diagram of the brain and some of the major locations of specific functions. The brain has many specialized regions where neurons produce specific neurotransmitter molecules that are used only to stimulate specific sensitive cells that have the proper receptor sites. As we learn more about the functioning of the brain, we are finding more kinds of specialized neurotransmitter molecules.

Endocrine-System Function

As mentioned previously, the nervous system functions much like a telephone system. By contrast, the endocrine system is basically a broadcasting system in which glands secrete messenger molecules, called **hormones,** that are distributed throughout the body by the circulatory system. However, each kind of hormone attaches only to appropriate receptor molecules on the surfaces of certain cells. The cells that receive the message typically respond in one of three ways. Some cells release products that have been previously manufactured, other cells are stimulated to synthesize molecules or to begin metabolic activities, and some are stimulated to divide and grow.

These different kinds of responses mean that some endocrine responses are relatively rapid, while others are very slow. For example, the release of the hormones **epinephrine** and **norepinephrine*** from the adrenal medulla causes a rapid change in the behavior of an organism. The heart rate increases, blood pressure rises, blood is shunted to muscles, and the breathing rate increases. You have certainly experienced this reaction many times in your lifetime.

*Epinephrine and norepinephrine were formerly called adrenalin and noradrenalin.

Another hormone, called **antidiuretic hormone** acts more slowly. It is released from the posterior pituitary gland and regulates the rate at which the body loses water through the kidneys by encouraging the reabsorption of water from their collecting ducts (see chapter 18). The effects of this hormone can be noticed in a matter of minutes to hours. Insulin is another hormone whose effects are quite rapid. **Insulin** is produced by the pancreas and stimulates cells—particularly muscle, liver, and fat cells—to take up glucose from the blood. After a meal that is high in carbohydrates, the level of glucose in the blood begins to rise, stimulating the pancreas to release insulin. The increased insulin causes glucose levels to fall as the sugar is taken up by cells. People with diabetes have insufficient or improperly acting insulin and therefore have difficulty regulating glucose levels in their blood.

The responses that result from the growth of cells may take weeks or years to occur. For example **growth-stimulating hormone** is produced by the anterior pituitary gland over a period of years and results in typical human growth. After sexual maturity, the amount of this hormone generally drops, and body growth stops. Sexual development is also largely the result of the growth of specific tissues and organs. The male sex hormone **testosterone,** produced by the testes, causes the growth of

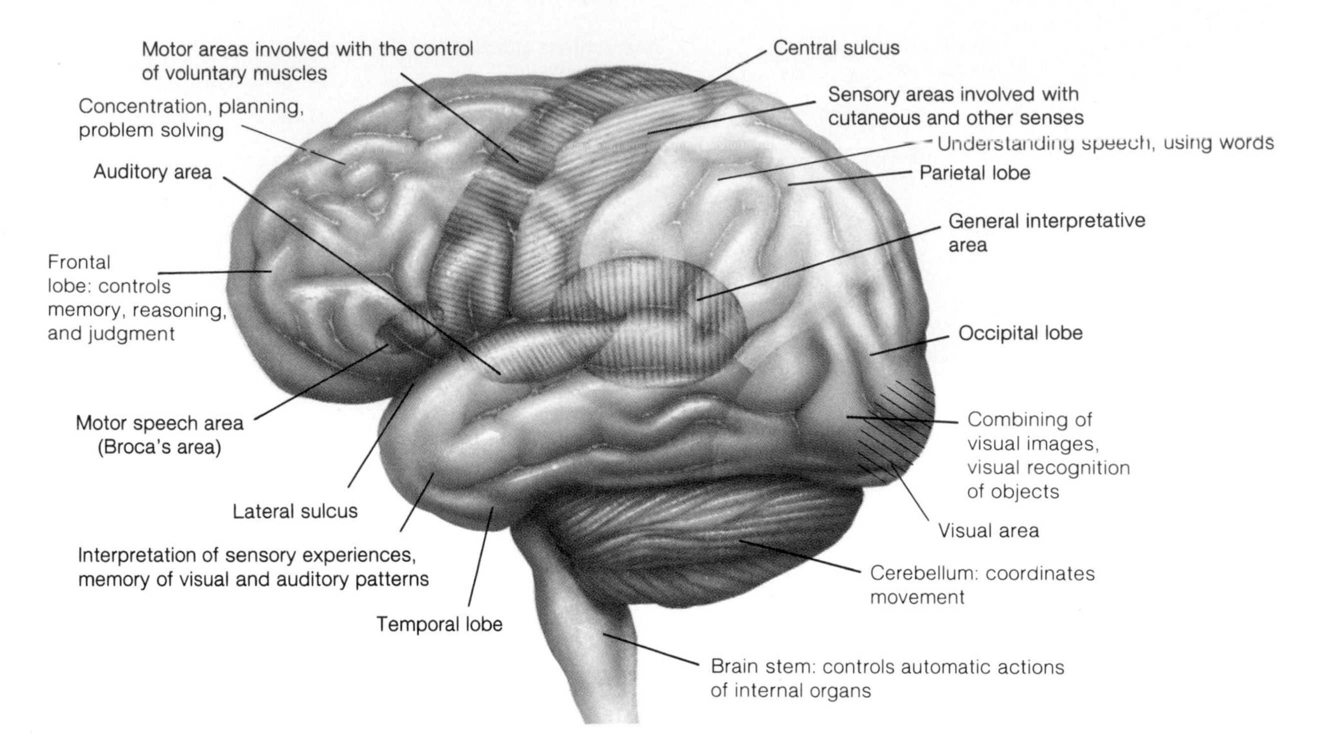

Figure 20.6 **Specialized Areas of the Brain.** Each portion of the brain has particular functions. Although we do not know all of the brain's functions and where they are located, general regions and their functions have been identified.

male sex organs and a change to the adult body form. The female counterpart, **estrogen,** results in the development of female sex organs and body form. In all of these cases, it is the release of hormones over long periods, continually stimulating the growth of sensitive tissues, that results in the normal developmental pattern. The absence or inhibition of any of these hormones early in life changes the normal growth process.

Glands within the endocrine system typically interact with one another and control production of hormones. One common control mechanism is called *negative-feedback control.* In **negative-feedback control** the increased amount of one hormone interferes with the production of a different hormone in the chain of events. The production of **thyroxine** and **triiodothyronine** by the thyroid gland exemplifies this kind of control. The production of these two hormones is stimulated by increased production of a hormone from the anterior pituitary called **thyroid-stimulating hormone.** The control lies in the quantity of the hormone produced. When the anterior pituitary produces high levels of thyroid-stimulating hormone, the thyroid is indeed stimulated. But when increased amounts of thyroxine and triiodothyronine are produced, these hormones have a negative effect on the pituitary so that it decreases its production of thyroid-stimulating hormone, leading to reduced production of thyroxine and triiodothyronine. As a result of the interaction of these hormones, their concentrations are maintained within certain limits (figure 20.7).

It is possible for the nervous and endocrine systems to interact. The pituitary gland is located at the base of the brain and is divided into two parts. The posterior pituitary is directly connected to the brain and develops from nerve tissue. The other part, the anterior pituitary, is produced from the lining of the roof of the mouth in early fetal development. Certain pituitary hormones are produced in the brain and transported down axons to the posterior pituitary where they are stored before being released. The anterior pituitary also receives a continuous input of messenger molecules from the brain, but these are delivered by way of a special set of blood vessels that pick up hormones produced by the hypothalamus and deliver them to the anterior pituitary.

The pituitary gland produces a variety of hormones that are responsible for causing other endocrine glands, such as the thyroid, ovaries and testes, and adrenals, to secrete their hormones. Pituitary hormones also influence milk production, skin color, body growth, mineral regulation, and blood glucose levels (figure 20.8).

Because the pituitary is constantly receiving information from the brain, many kinds of sensory stimuli to the body can affect the functioning of the endocrine system. One example is the way in which the nervous system and endocrine system interact to influence the menstrual cycle. At least three different hormones are involved in the cycle of changes that affect the ovary and the

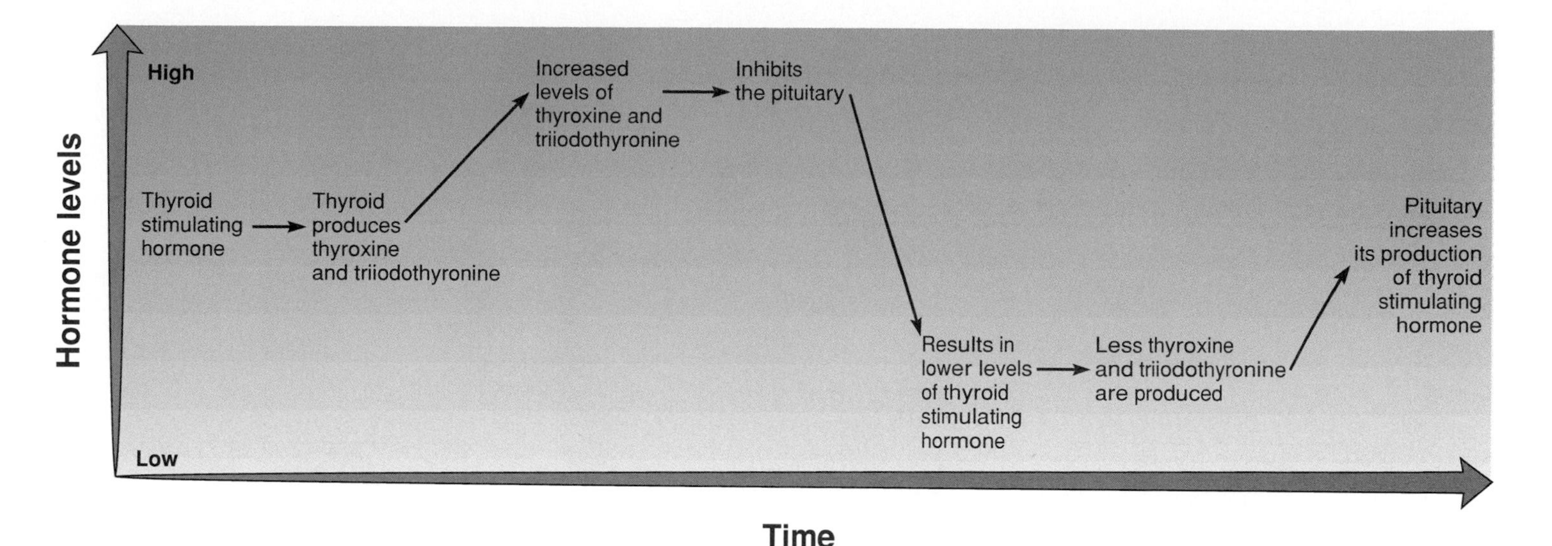

Figure 20.7 **Control of Thyroid Hormone Levels.** The levels of thyroxine and triiodothyronine increase and decrease in response to the amount of thyroid-stimulating hormone that is present. Increased levels of the thyroid hormones cause the pituitary to stop the production of thyroid-stimulating hormone so that eventually the thyroid hormone levels decrease. This is an example of negative-feedback control.

lining of the uterus (see chapter 21 for details). It is well documented that stress caused by tension or worry can interfere with the normal cycle of hormones and delay or stop menstrual cycles. In addition, young women living in groups, such as in college dormitories, often have their menstrual cycles synchronized. Although the exact mechanism involved in this phenomenon is unknown, it is suspected that input from the nervous system causes this synchronization. (Odors and sympathetic feelings also have been suggested as causes.)

In many animals, the changing length of the day causes hormonal changes related to reproduction. In the spring, birds respond to lengthening days and begin to produce hormones that gear up their reproductive systems for the summer breeding season. The pineal body, a portion of the brain, serves as the receiver of light stimuli and changes the amounts of hormones secreted by the pituitary, resulting in changes in the levels of reproductive hormones. These hormonal changes modify the behavior of birds. Courtship, mating, and nest-building behaviors increase in intensity. Therefore, it appears that a change in hormone level is affecting the behavior of the animal; the endocrine system is influencing the nervous system (figure 20.9).

It has been known for centuries that changes in the levels of sex hormones cause changes in the behavior of animals. Castration (removal of the testes) of male domesticated animals, such as cattle, horses, and pigs, is sometimes done in part to reduce their aggressive behavior and make them easier to control. In humans, the use of anabolic steroids to increase muscle mass is known to cause behavioral changes and "moodiness."

Although we still tend to think about the nervous and endocrine systems as being separate and different, it is becoming clear that they are interconnected. These two systems cooperate to bring about appropriate responses to environmental challenges. The nervous system is specialized for receiving and sending short-term messages, while activities that require long-term, growth-related actions are handled by the endocrine system.

Sensory Input

The activities of the nervous and endocrine systems are often responses to some kind of input received from the sense organs. Sense organs of various types are located throughout the body. Many of them are located on the surface, where environmental changes can be easily detected. Hearing, sight, and touch are good examples of such senses. Other sense organs are located within the body and indicate to the organism how its various parts are changing. For example, pain and pressure are often used to monitor internal conditions. The sense organs detect changes, but the brain is responsible for **perception**—the recognition that a stimulus has been received. Sensory abilities involve many different kinds of mechanisms, including chemical recognition, the detection of energy changes, and the monitoring of forces.

Chemical Detection

All cells have receptors on their surfaces that can bind selectively to molecules they encounter. This binding process can

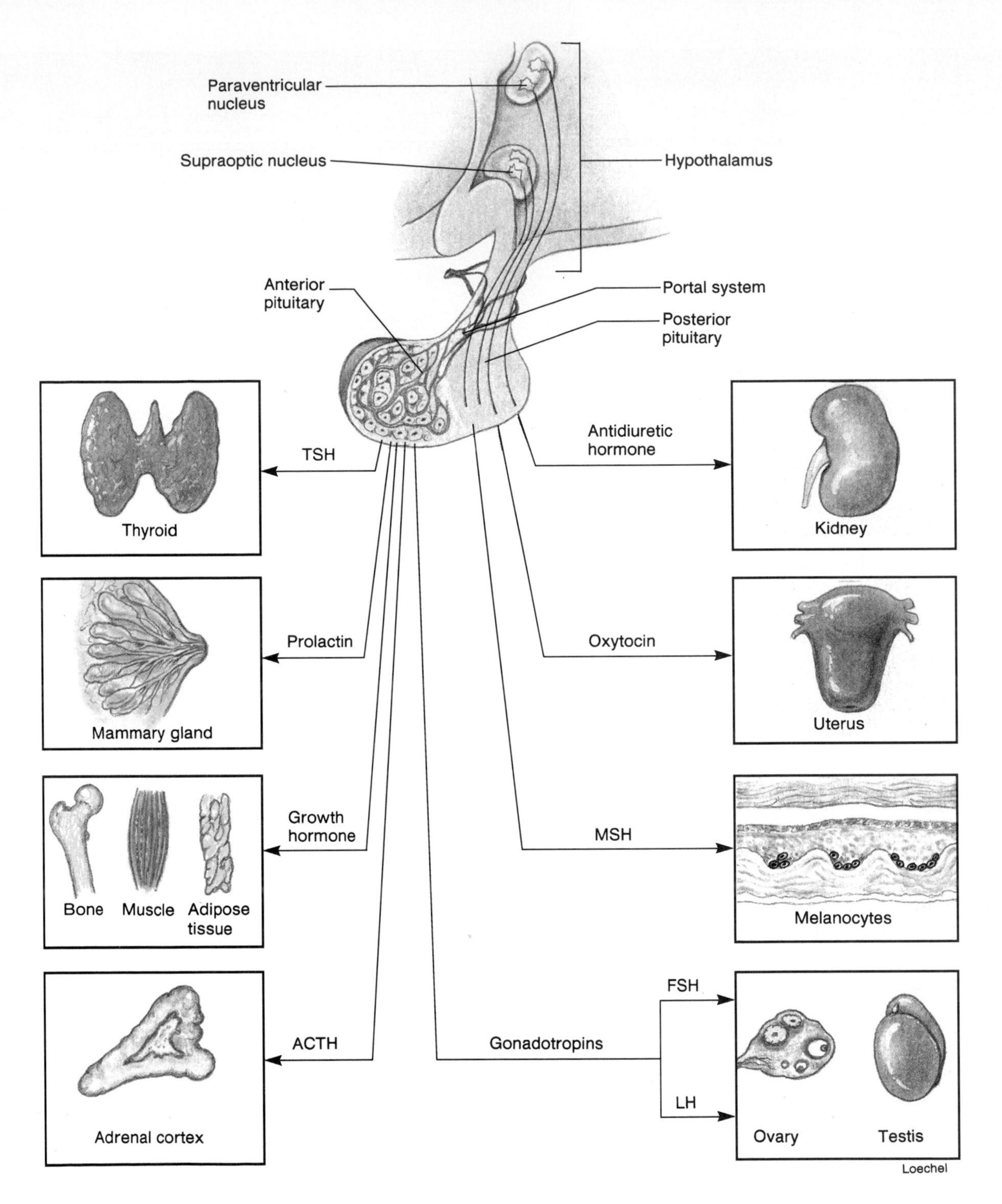

Figure 20.8 **Hormones of the Pituitary.** The anterior pituitary gland produces several hormones that regulate growth and the secretions of target tissues. The posterior pituitary produces hormones that change the behavior of the kidney and uterus but do not influence the growth of these organs.

Light
Changing length of day stimulates growth of reproductive organs
Reproductive organs secrete hormones that cause changed behavior
O'Keefe

Figure 20.9 **Interaction between the Nervous and Endocrine Systems.** In birds and many other animals, the brain receives information about the changing length of day that causes the pituitary to produce hormones that stimulate sexual development. The testes or ovaries grow and secrete their hormones in increased amounts. Increased levels of testosterone or estrogen result in changed behavior, with increased mating, aggression, and nest-building activity.

cause changes in the cell in several ways. In some cells it causes depolarization. When this happens, they can stimulate neurons and cause messages to be sent to the central nervous system, informing it of some change in the surroundings. In other cases, a molecule binding to the cell surface may cause certain genes to be expressed, and the cell responds by changing the molecules it produces. This is typical of the way the endocrine system receives and delivers messages.

Many kinds of cells have specific binding sites for particular molecules that are limited to a few kinds of detectable molecules. Others, such as the taste buds on the tongue, appear to respond to classes of molecules. They are able to distinguish four kinds of tastes: sweet, sour, salt, and bitter. These different kinds of taste buds are found at specific

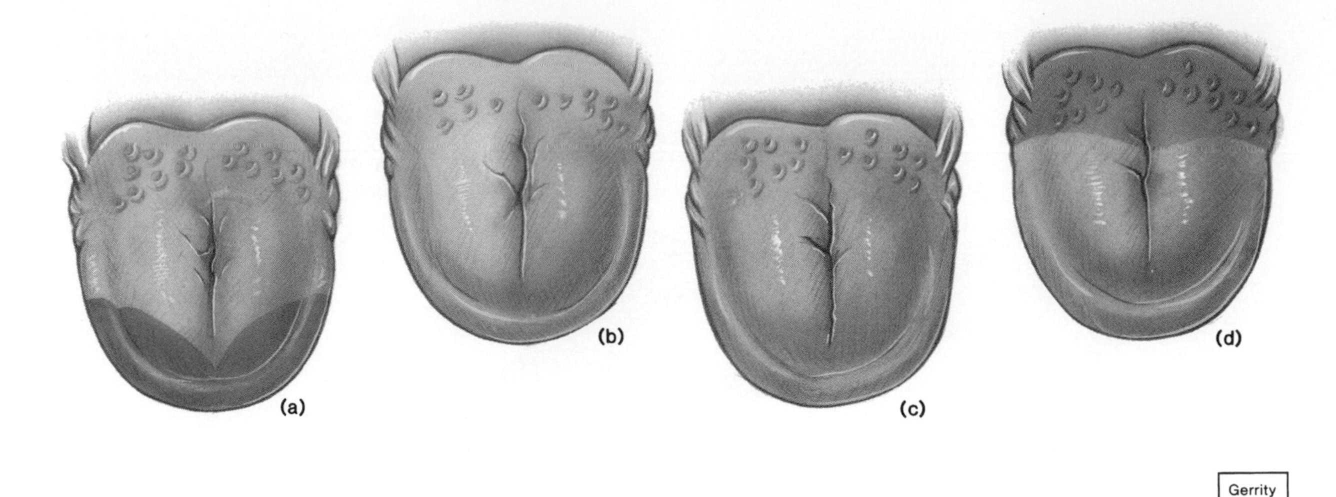

Figure 20.10 **The Location of Different Taste Sensors.** The four primary tastes are sweet, sour, salt, and bitter. These different kinds of taste buds are located on specific regions of the tongue. (a) Sweet receptors are located at the tip of the tongue; (b) sour receptors are located on the sides of the tongue; (c) salt receptors are located at the tip and on the sides of the tongue; (d) bitter receptors are located at the back of the tongue.

locations on the surface of the tongue (figure 20.10). The taste buds that give us the sensation of sour appear to respond to the presence of hydrogen ions (H^+), since acids taste sour. Many kinds of ionic compounds, including sodium chloride, can stimulate the taste buds that give us the sensation of a salty taste. However, the sensation of sweetness can be stimulated by many kinds of organic molecules, including sugars, artificial sweeteners and lead salts. The sweet taste of lead salts in old paints partly explains why children may eat paint chips. Since the lead interferes with normal brain development, this can have disastrous results. Many other kinds of compounds of diverse structures give the bitter sensation.

It is also important to understand that much of what we often refer to as *taste* involves such inputs as temperature, texture, and smell. Cold coffee has a different taste than hot coffee even though they are chemically the same. Lumpy, cooked cereal and smooth cereal have different tastes. If you are unable to smell food, it doesn't taste as it should, which is why you sometimes lose your appetite when you have a stuffy nose. We still have much to learn about how the tongue detects chemicals and the role that other associated senses play in modifying taste.

The other major chemical sense, the sense of smell, is much more versatile; it can detect thousands of different molecules at very low concentrations. The cells that make up the **olfactory epithelium,** the part of the nasal cavity that responds to smells, apparently bind molecules to receptors on their surface. Exactly how this can account for the large number of recognizably different odors is unknown, but the receptor cells are extremely sensitive. In some cases a single molecule of a substance is sufficient to cause a receptor cell to send a message to the brain, where the sensation of odor is perceived. These sensory cells also fatigue rapidly. You have probably noticed that when you first walk into a room, specific odors are readily detected, but after a few minutes you are unable to detect them. Most perfumes and aftershaves are undetectable after 15 minutes of continuous stimulation.

Many internal sense organs also respond to specific molecules. For example, the brain and aorta contain cells that respond to concentrations of hydrogen ions, carbon dioxide, and oxygen in the blood. Remember, too, that the endocrine system relies on the detection of specific messenger molecules to trigger its activities.

Light Detection

The eyes primarily respond to changes in the flow of light energy. The structure of the eye is designed to focus light on a light-sensitive layer of the back of the eye known as the **retina** (figure 20.11). There are two kinds of receptors in the retina of the eye. The cells called **rods** respond to a broad range of wavelengths of light and are responsible for black-and-white vision. Since rods are very sensitive to light, they are particularly useful in dim light. Rods are located over most of the retinal surface except for the area of most acute vision known as the **fovea centralis.** The other receptor cells, called **cones,** are found throughout the retina but are particularly concentrated in the fovea centralis. Cones are not as sensitive to light, but they can detect different wavelengths of light. This combination of receptors gives us the ability to detect color when light levels are high, but we rely on black-and-white vision at night.

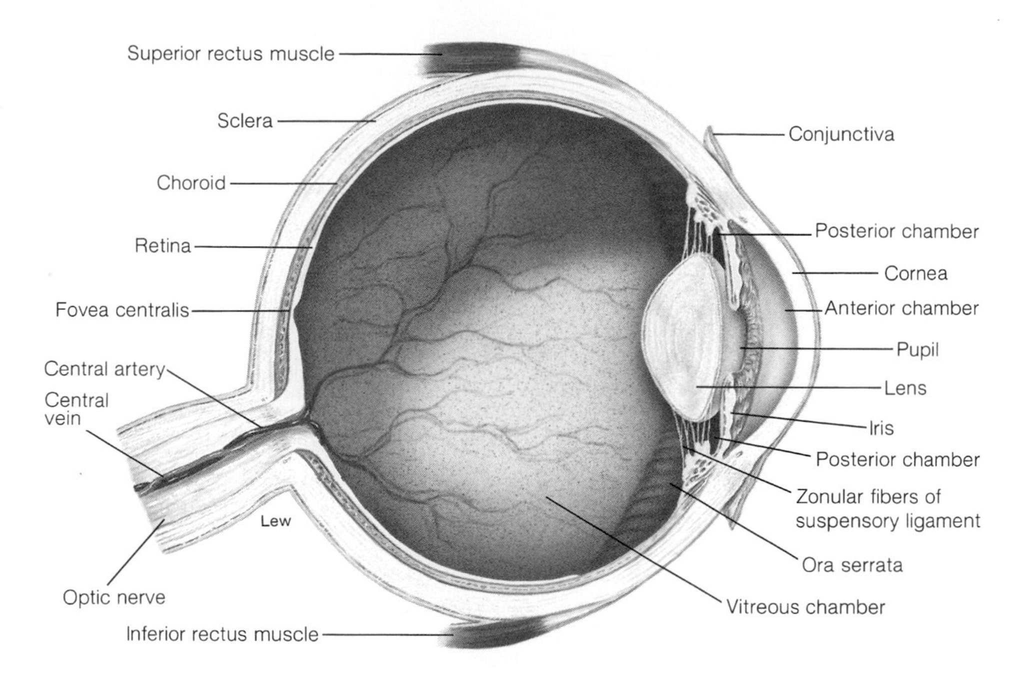

Figure 20.11 **The Structure of the Eye.** The eye contains a cornea and lens that focus the light on the retina of the eye. The light causes pigments in the rods and cones of the retina to decompose. This leads to the depolarization of these cells and the stimulation of neurons that send messages to the brain.

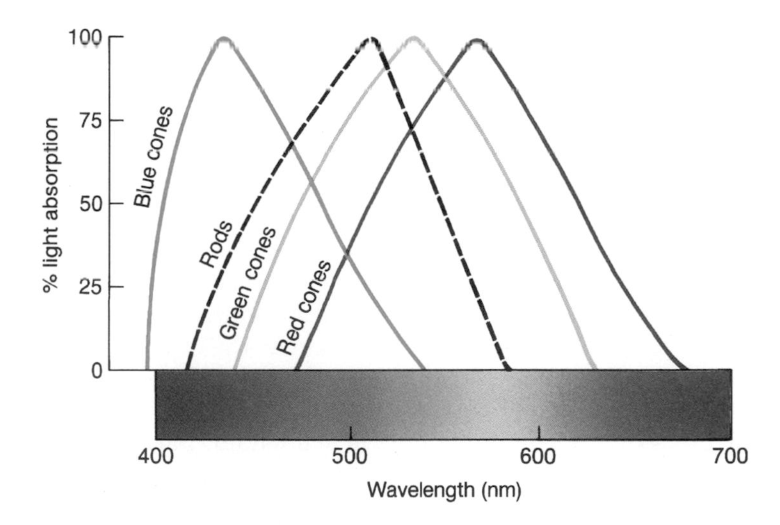

Figure 20.12 **Light Reception by Cones.** There are three different kinds of cones that respond differently to red, green, and blue wavelengths of light. Stimulation of combinations of these three kinds of cones gives us the ability to detect many different shades of color.

Sources: Data from W. B. Marks, W. H. Dobelle, and E. F. MacNichol, "Visual Pigments of Single Primate Cones," *Science* 143:45–52 (1964), and P. K. Brown and G. Wald, "Visual Pigments in Single Rods and Cones of the Human Retina," *Science* 144:45–52 (1964).

There are three different varieties of cones: one type responds best to red light, another responds best to green light, and the third responds best to blue light. Stimulation of various combinations of these three kinds of cones allows us to detect different shades of color (figure 20.12).

Rods and the three different kinds of cones each contain a pigment that decomposes when struck by light of the proper wavelength and sufficient strength. The pigment found in rods is called **rhodopsin.** This change in the structure of rhodopsin causes the rod to depolarize. Cone cells have a similar mechanism of action, and each of the three kinds of cones has a different pigment. Since rods and cones synapse with neurons, they stimulate a neuron when depolarized and cause a message to be sent to the brain. Thus, the pattern of color and light intensity recorded on the retina is detected by rods and cones and converted into a series of nerve impulses that are received and interpreted by the brain.

Sound Detection

The ears respond to changes in sound waves. Sound is produced by the vibration of molecules. Consequently, the ears are detecting changes in the quantity of energy and the quality of sound waves. Sound has several characteristics. Loudness, or volume, is a measure of the intensity of sound energy that arrives at the ear. Very loud sounds will literally vibrate your body, and can cause hearing loss if they are too intense. Pitch is a quality of sound that is determined by the frequency of the sound vibrations. High-pitched sounds have short wavelengths; low-pitched sounds have long wavelengths.

Figure 20.13 shows the anatomy of the ear. The sound that arrives at the ear is first funneled by the external ear to the **tympanum,** also known as the *eardrum.* The cone-shaped nature of the external ear focuses sound on the tympanum and causes it to vibrate at the same frequency as the soundwaves reaching it. Attached

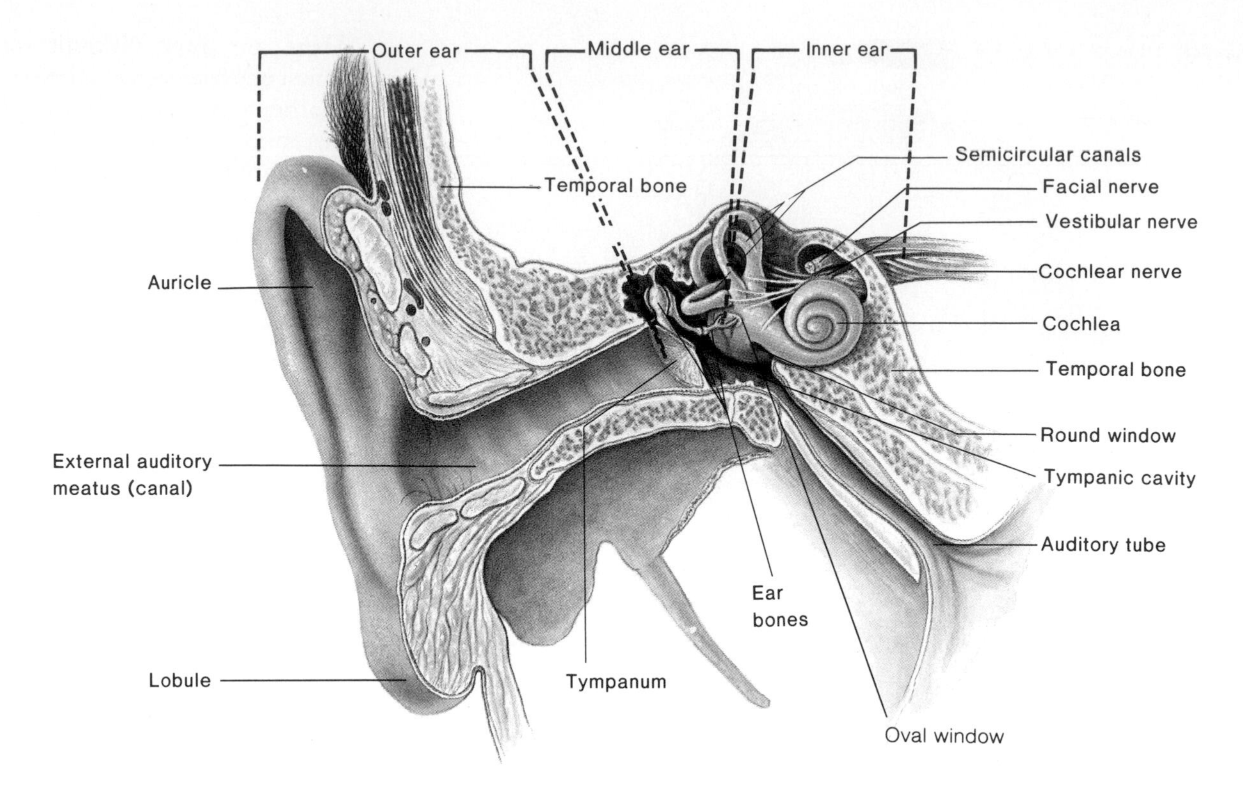

Figure 20.13 **The Anatomy of the Ear.** The ear consists of an external cone that directs sound waves to the tympanum. Vibrations of the tympanum move the ear bones and vibrate the oval window of the cochlea, where the sound is detected. The semicircular canals monitor changes in the position of the head, helping us to maintain balance.

to the tympanum are three tiny bones known as the **malleus** (hammer), **incus** (anvil), and **stapes** (stirrup). The malleus is attached to the tympanum, the incus is attached to the malleus and stapes, and the stapes is attached to a small, membrane-covered opening called the **oval window** in a snail-shaped structure known as the **cochlea.** The vibration of the tympanum causes the tiny bones (malleus, incus, and stapes) to vibrate, and they in turn cause a corresponding vibration in the membrane of the oval window.

The cochlea of the ear is the structure that detects sound and consists of a snail-shaped set of fluid-filled tubes. When the oval window vibrates, sound waves are transferred to the fluid in the cochlea, causing a membrane in the cochlea, called the **basilar membrane,** to vibrate. High-pitched, short-wavelength sounds cause the basilar membrane to vibrate at the base of the cochlea near the oval window. Low-pitched, long-wavelength sounds vibrate the basilar membrane far from the oval window. Loud sounds cause the basilar membrane to vibrate more vigorously than do faint sounds. Cells on this membrane depolarize when they are stimulated by its vibrations. Since they synapse with neurons, messages can be sent to the brain (figure 20.14).

Because sounds of different wavelengths stimulate different portions of the cochlea, the brain is able to determine the pitch of a sound. Most sounds consist of a mixture of pitches that are heard. Louder sounds stimulate the membrane more forcefully, causing the sensory cells in the cochlea to send more nerve impulses per second. Thus, the brain is able to perceive the loudness of various sounds, as well as the pitch.

Associated with the cochlea is a set of fluid-filled tubes called the **semicircular canals.** In the walls of these canals and chambers are cells similar to those found on the basilar membrane. These cells are stimulated by movements of the head and by the position of the head with respect to the force of gravity. The constantly changing position of the head results in sensory input that is important in maintaining balance.

Touch

What we normally call the sense of *touch* consists of a variety of different kinds of input. Some receptors respond to pressure, others to temperature, and others, which we call *pain receptors,* usually respond to cell damage. When these receptors are appropriately stimulated, they send a message to the brain. Since receptors are stimulated in particular parts

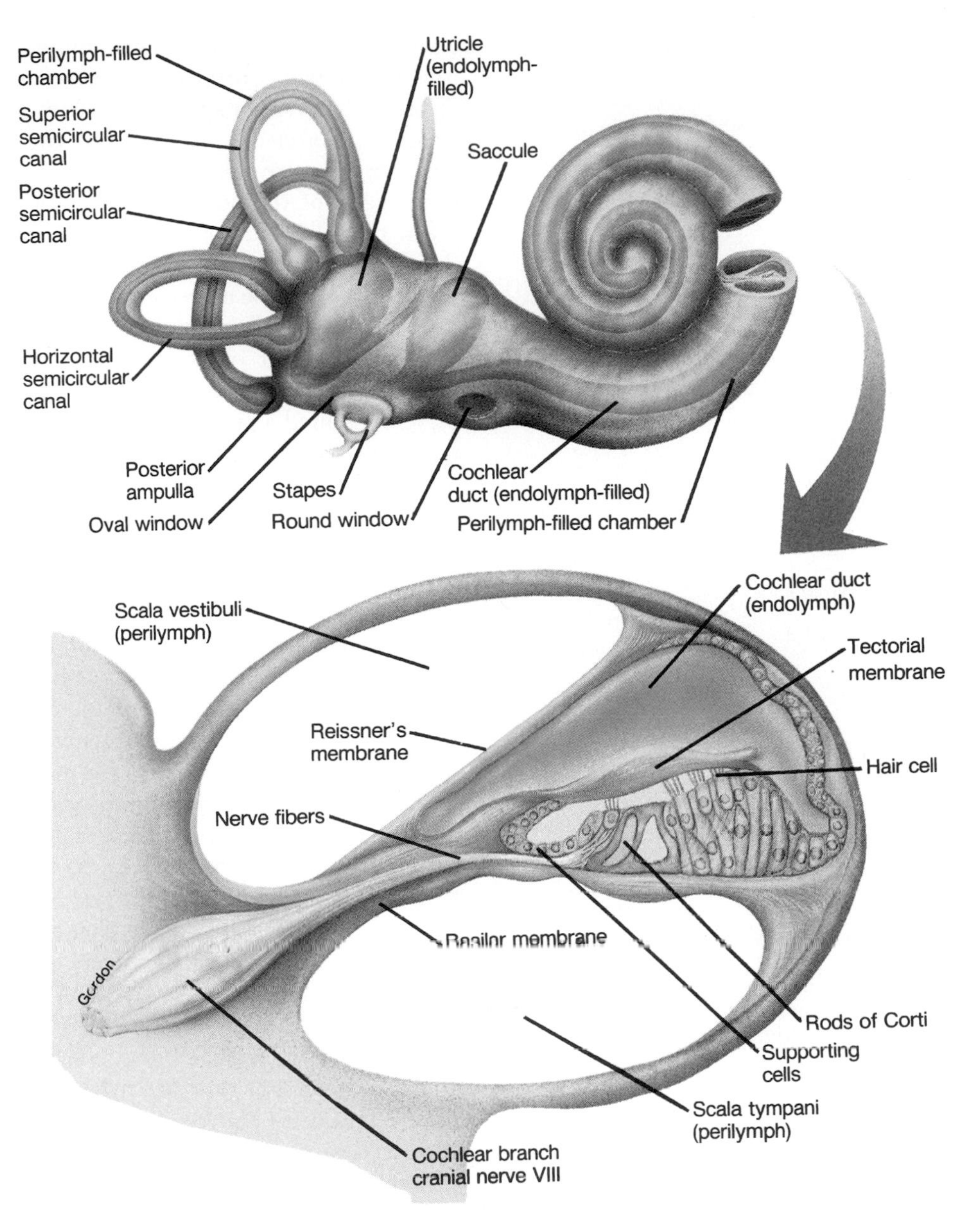

Figure 20.14 **The Basilar Membrane.** The cells that respond to vibrations and stimulate neurons are located in the cochlea. Vibrations of the oval window cause the fluid in the cochlea to vibrate, and the basilar membrane moves also. This movement causes the receptor cells to depolarize and send a message to the brain.

of the body, the brain is able to localize the sensation. Not all parts of the body are equally supplied with these receptors. The tips of the fingers, lips, and external genitals have the highest density of these nerve endings, while the back, legs, and arms have far fewer receptors.

Some internal receptors, such as pain and pressure receptors, are important in allowing us to monitor our internal activities. Many pains generated by the internal organs are often perceived as if they were somewhere else. For example, the pain associated with heart attack is often perceived to be in the left arm. Pressure receptors in joints and muscles are important in providing information about the degree of stress being placed on a portion of the body. This is also important information to feed back to the brain so that adjustments can be made in movements to maintain posture. If you have ever had your foot "go to sleep" because the nerve stopped functioning, you have experienced what it is like to lose this constant input of nerve messages from the pressure sensors to assist in guiding the movements you make. Your movements become uncoordinated until the nerve function returns to normal.

Output Coordination

There are several ways in which the nervous system and endocrine system cause changes. Both systems can stimulate muscles to contract and glands to secrete. The endocrine system is also able to change metabolism of cells and regulate the growth of tissues. The nervous system acts upon two kinds of organs: muscles and glands. The actions of muscles and glands are simple and direct: muscles contract and glands secrete.

Muscles

The ability to move is one of the fundamental characteristics of animals. Through the coordinated contraction of many muscles, the intricate, precise movements of a dancer, basketball player, or writer are accomplished. It is important to recognize that muscles can pull only by contracting; they are unable to push by lengthening. The work of any muscle is done during its contraction. Relaxation is the passive state of the muscle. There must always be some force available that will stretch a muscle after it has stopped contracting and relaxes. Therefore, the muscles that control the movements of the skeleton are present in antagonistic sets. For every muscle's action there is another muscle that has the opposite action. For example, the biceps muscle causes the arm to flex (bend) as the muscle shortens. The contraction of its antagonist, the triceps muscle, causes the arm to extend (straighten) and at the same time stretches the relaxed biceps muscle (figure 20.15).

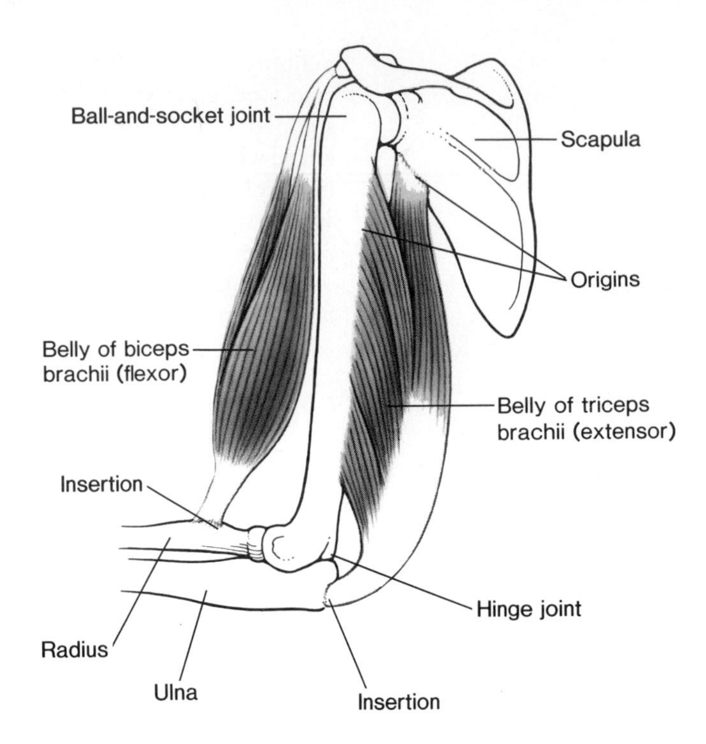

Figure 20.15 **Antagonistic Muscles.** Since muscles cannot actively lengthen, it is necessary to have sets of muscles that oppose one another. The contraction and shortening of one muscle causes the stretching of a relaxed muscle.

What we recognize as a muscle is composed of many muscle cells, which are in turn made up of *myofibrils* that are composed of two kinds of *myofilaments* (figure 20.16). The mechanism by which muscle contracts is well understood and involves the movement of protein filaments past one another as ATP is utilized. The filaments in muscle cells are of two types, arranged in a particular pattern. Thin filaments composed of the proteins **actin, tropomyosin,** and **troponin** alternate with thick filaments composed primarily of a protein known as **myosin** (figure 20.17).

The myosin molecules have a shape similar to a golf club. The head of the club-shaped molecule sticks out from the thick filament and can combine with the actin of the thin filament. The troponin and tropomyosin proteins associated with the actin cover the actin in such a way that myosin cannot bind with it. When actin is uncovered and myosin can bind to it, contraction of a muscle can occur when ATP is utilized.

The process of muscle-cell contraction involves several steps. When a nerve impulse arrives at a muscle cell, it causes the cell to depolarize. When muscle cells depolarize, calcium ions (Ca^{++}) contained within membranes are released among the actin and myosin filaments. The calcium ions (Ca^{++}) combine with the troponin molecules, causing the troponin-tropomyosin complex to expose actin so that it can bind with myosin. While the actin and myosin molecules are attached, the head of the myosin molecule can flex as ATP is used and the actin molecule is pulled past the myosin molecule. Thus, a tiny section of the muscle cell shortens (figure 20.18). When one of our muscles contracts, thousands of such interactions take place within a tiny portion of a muscle cell, and many cells within a muscle all contract at the same time.

There are three major types of muscle: skeletal, smooth, and cardiac. These differ from one another in several ways. *Skeletal muscle* is voluntary muscle: it is under the control of the nervous system. The brain or spinal cord sends a message to skeletal muscles, and they contract to move the legs, fingers, and other parts of the body. This does not mean that you must make a conscious decision every time you want to move a muscle. Many of the movements we make are learned initially, but become automatic as a result of practice. For example, walking, swimming, or riding a bicycle required a great amount of practice originally, but now you probably perform these movements without thinking about them. They are, however, still considered to be voluntary actions.

Skeletal muscles are constantly bombarded with nerve impulses that result in repeated contractions of differing strength. Many neurons end in each muscle, and each one stimulates a specific set of muscle cells called a **motor unit.** Since each muscle consists of many motor units, it is possible to have a wide variety of intensities of contraction within one muscle organ. This allows a single set of muscles to serve a wide variety of functions. For example, the same muscles of the arms and shoulders that are used to play a piano can be used together in other combinations to tightly grip and throw a baseball. If the nerves going to a muscle are destroyed, the muscle becomes paralyzed and begins to shrink. Regular nervous stimulation of skeletal muscle is necessary for muscle to maintain size and strength. Any kind of prolonged inactivity leads to the degeneration of muscles. Muscle maintenance is one of the primary functions of physical therapy and a benefit of regular exercise.

Skeletal muscles are able to contract quickly, but they cannot remain contracted for long periods. Even when we contract a muscle for a minute or so, the muscle is constantly shifting the individual motor units within it that are in a state of contraction. A single skeletal muscle cell cannot stay in a contracted state.

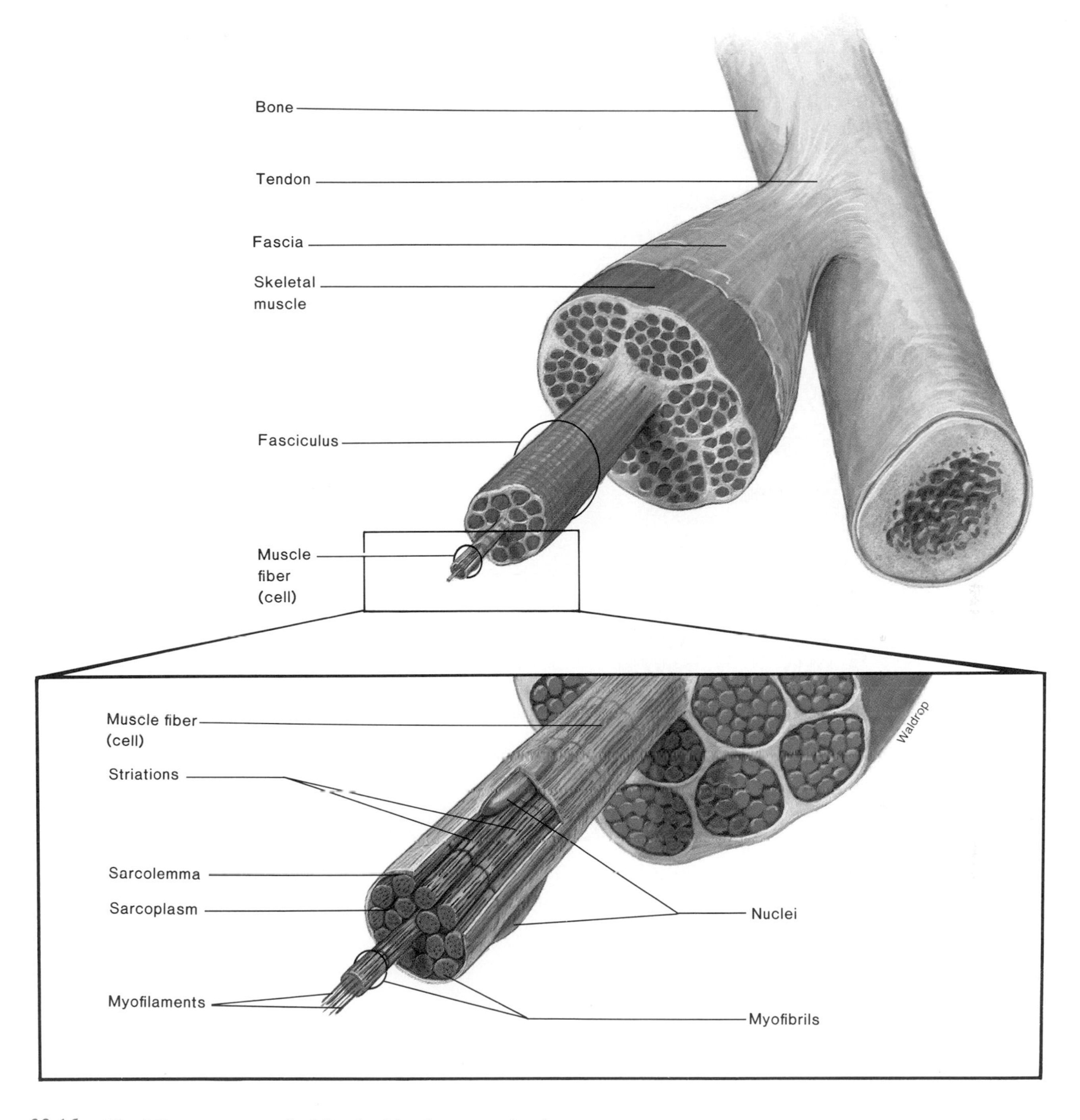

Figure 20.16 **The Microanatomy of a Muscle.** Muscles are made of cells that contain bundles known as myofibrils. The myofibrils are composed of myofilaments of two different kinds: thick myofilaments composed of myosin, and thin myofilaments containing actin, troponin, and tropomyosin.

Smooth muscles make up the walls of muscular internal organs, such as the gut, blood vessels, and reproductive organs. They have the property of contracting as a response to being stretched. Since much of the digestive system is being stretched constantly, the responsive contractions contribute to the normal rhythmic movements associated with the digestive system. These are involuntary muscles; they can contract on their own without receiving direct messages from the nervous system. This can be demonstrated by removing portions of the gut or uterus from experimental animals. When these muscular organs are kept moist with special solutions, they go through cycles of contraction without any possible stimulation from neurons. However, they do receive nervous stimulation, which can modify the rate and

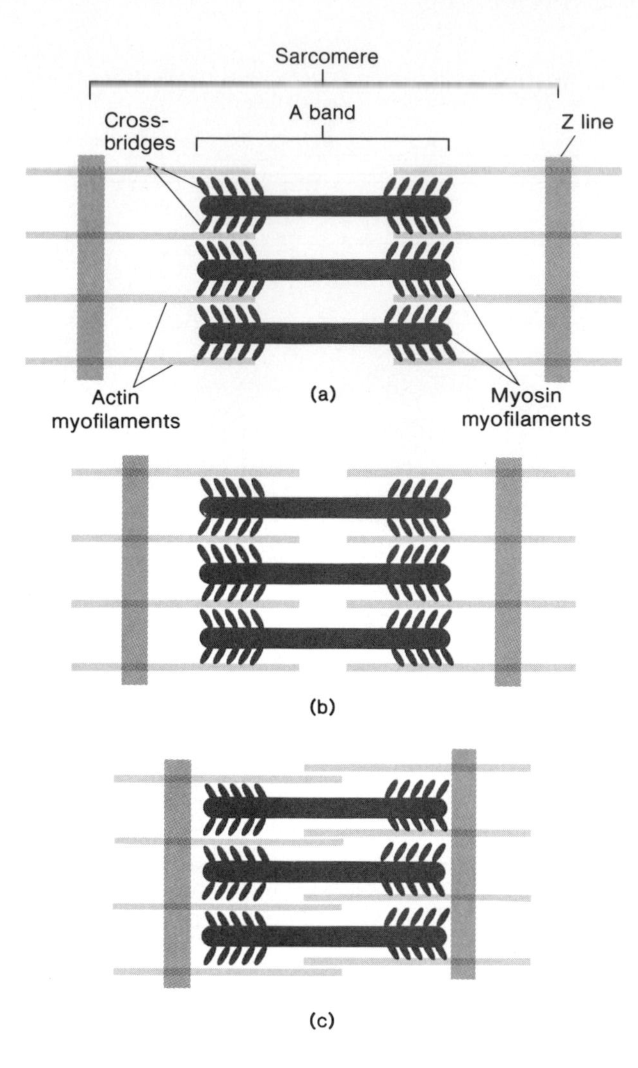

Figure 20.17 **The Subcellular Structure of Muscle.** The actin- and myosin-containing myofilaments are arranged in a regular fashion into units called sarcomeres (a). Each sarcomere consists of two sets of actin-containing myofilaments inserted into either end of bundles of myosin-containing myofilaments (b). The actin-containing myofilaments slide by the myosin-containing myofilaments, shortening the sarcomere (c).

strength of their contraction. This kind of muscle also has the ability to stay contracted for long periods without becoming fatigued. Many kinds of smooth muscle, such as the muscle of the uterus, also respond to the presence of hormones. Specifically, the hormone **oxytocin,** which is released from the posterior pituitary, causes strong contractions of the uterus during labor and birth. Similarly, several hormones produced by the duodenum influence certain muscles of the digestive system to either contract or relax.

Cardiac muscle is the muscle that makes up the heart. It has the ability to contract rapidly like skeletal muscle, but does not require nervous stimulation to do so. Nervous stimulation can, however, cause the heart to speed or slow its rate of contraction. Hormones, such as epinephrine and norepinephrine, also influence the heart by increasing its rate and strength of contraction. Cardiac muscle also has the characteristic of being unable to stay contracted. It will contract quickly but must have a short period of relaxation before it will be able to contract a second time. This makes sense in light of its continuous, rhythmic, pumping function. Table 20.1 summarizes the differences between skeletal, smooth, and cardiac muscle.

Glands

The glands of the body are of two different kinds. Those that secrete into the blood stream are called **endocrine glands.** We have already talked about several of these: the pituitary, thyroid, ovary, and testis are examples. The **exocrine glands** are those that secrete to the surface of the body or into one of the tubular organs of the body, such as the gut or reproductive tract. Examples are the salivary glands, intestinal mucous glands, and sweat glands. Some of these glands, such as salivary glands and sweat glands, are under nervous control. When stimulated by the nervous system, they secrete their contents.

The Russian physiologist Ivan Petrovich Pavlov showed that salivary glands were under the control of the nervous system when he trained dogs to salivate in response to hearing a bell. You may recall from chapter 17 that initially the animals were presented with food at the same time the bell was rung. Eventually they would salivate when the bell was rung even if food was not present. This demonstrated that saliva production was under the control of the central nervous system.

However, many other exocrine glands are under hormonal control. The secretion of many of the digestive enzymes of the stomach and intestine are secreted in response to local hormones produced in the gut. These are circulated through the blood to the digestive glands, which respond by secreting the appropriate digestive juice.

Growth Responses

The hormones produced by the endocrine system can have a variety of effects. As mentioned earlier, hormones can stimulate smooth muscle to contract and

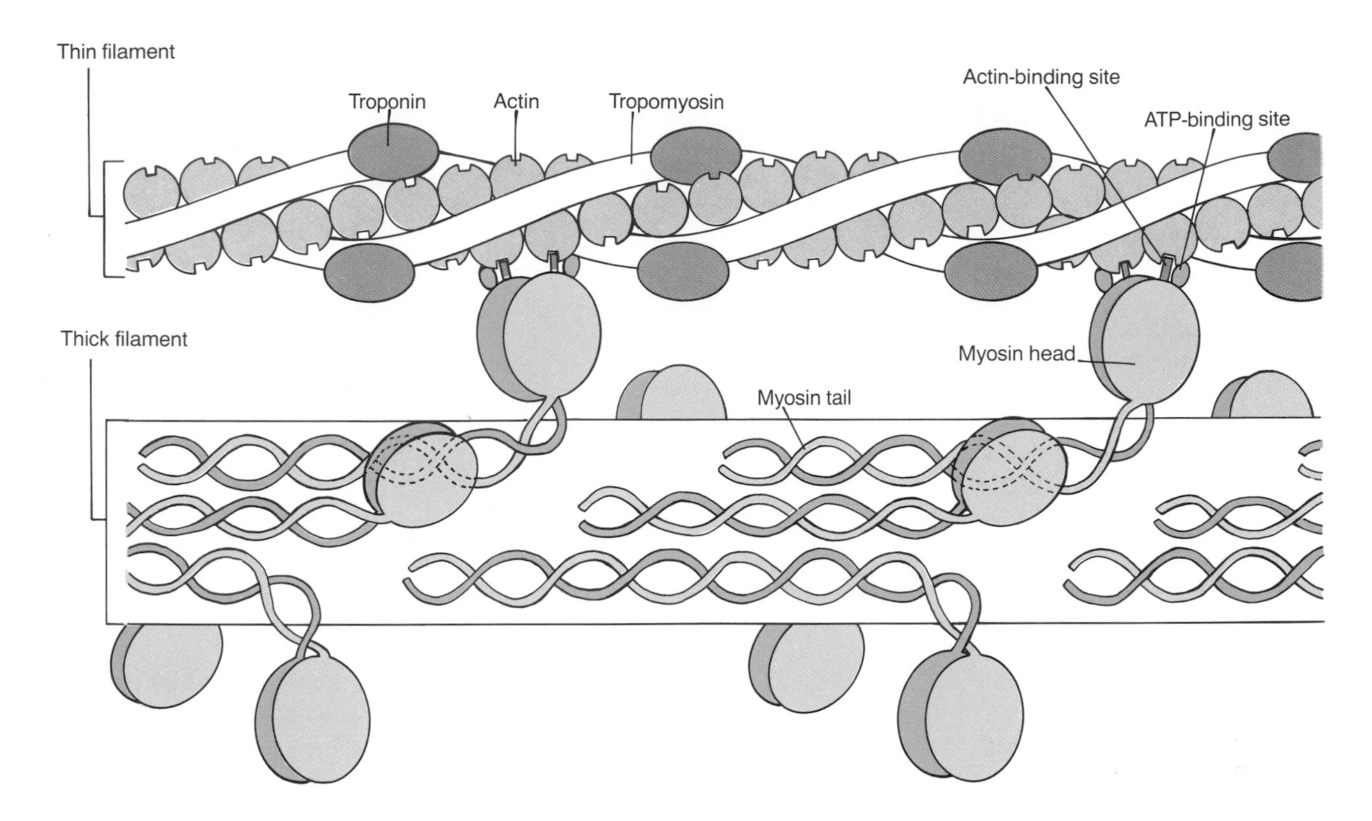

Figure 20.18 **Interaction between Actin and Myosin.** When calcium ions (Ca^{++}) enter the region of the muscle cell containing actin and myosin, they allow the actin and myosin to bind to each other. The club-shaped head of the myosin flexes and moves the actin along, causing the two molecules to slide past each other.

Table 20.1
Characteristics of Different Kinds of Muscle

Kind of Muscle	Stimulus	Length of Contraction	Rapidity of Response
Skeletal	Nervous	Short, tires quickly	Most rapid
Smooth	1. Self-stimulated 2. Also responds to nervous and endocrine systems	Long, doesn't tire quickly	Slow
Cardiac	1. Self-stimulated 2. Also responds to nervous and endocrine systems	Short, cannot stay contracted	Rapid

can influence the contraction of cardiac muscle as well. Many kinds of glands, both endocrine and exocrine, are caused to secrete as a result of a hormonal stimulus. However, the endocrine system has one major effect that is not equaled by the nervous system: hormones regulate growth. Several examples of the many kinds of long-term growth changes that are caused by the endocrine system were given earlier in the chapter. Growth-stimulating hormone is produced over a period of years to bring about the increase in size of most of the structures of the body. The absence of this hormone results in a person with small body size. It is important to recognize that the amount of growth-stimulating hormone present varies from time to time. It is present in fairly high amounts throughout childhood and results in steady growth. It also appears to be present at higher levels at certain times, resulting in growth spurts. Finally, as adulthood is reached, the level of this hormone falls, and growth stops.

Similarly, testosterone produced during adolescence influences the growth of bone and muscle to provide men with larger, more muscular bodies. In addition, there is growth of the penis, growth of the larynx, and increased growth of hair on the face and body. The primary female hormone, estrogen, causes growth of reproductive organs and development of breast tissue. It is also involved, along with other hormones, in the cyclic growth and sloughing of the wall of the uterus.

• Summary •

Throughout this chapter we have been comparing the functions of the nervous and endocrine systems, the kinds of effects they have, and their characteristics. Table 20.2 summarizes these differences.

Table 20.2
Comparison of the Nervous and Endocrine Systems

System	Method of Action	Effects
Nervous	1. Nerve impulse travels along established routes. 2. Neurotransmitters allow impulse to cross synapses. 3. Rapid action.	1. Causes skeletal-muscle contraction. 2. Modifies contraction of smooth and cardiac muscle. 3. Causes gland secretion.
Endocrine	1. Hormones released into bloodstream. 2. Receptors bind hormones to their target organs. 3. Often slow to act.	1. Stimulates smooth-muscle contraction. 2. Stimulates gland secretion. 3. Regulates growth.

A nerve impulse is caused by sodium ions entering the cell as a result of a change in the permeability of the cell membrane. Thus, a wave of depolarization passes down the length of a neuron to the synapse. The axon of a neuron secretes a neurotransmitter, such as acetylcholine, into the synapse, where these molecules bind to the dendrite of the next cell in the chain, resulting in an impulse in it as well. The acetylcholinesterase present in the synapse destroys acetylcholine so that it does not repeatedly stimulate the dendrite.

Several kinds of sensory inputs are possible. Many kinds of chemicals can bind to cell surfaces and be recognized. This is probably how the sense of taste and the sense of smell function. Light energy can be detected because light causes certain molecules in the retina of the eye to decompose and stimulate neurons. Sound can be detected because fluid in the cochlea of the ear is caused to vibrate, and special cells detect this movement and stimulate neurons. The sense of touch consists of a variety of receptors that respond to pressure, cell damage, and temperature.

Muscles shorten because of the ability of actin and myosin to bind to one another. A portion of the myosin molecule is caused to bend when ATP is used, resulting in the sliding of actin and myosin molecules past each other. Skeletal muscle responds to nervous stimulation to cause movements of the skeleton. Smooth muscle and cardiac muscle have internally generated contractions that may be modified by nervous stimulation or hormones.

Glands are of two types: exocrine glands, which secrete through ducts into the cavity of an organ or to the surface of the skin, and endocrine glands, which release their secretions into the circulatory system. Digestive glands and sweat glands are examples of exocrine glands. Endocrine glands such as the ovaries, testes, and pituitary gland change the activities of cells and often cause responses that result in growth over a period of time. It is becoming clear that the endocrine system and the nervous system are interrelated. Actions of the endocrine system can change how the nervous system functions, and the reverse is also true. Much of this interrelation takes place in the brain–pituitary gland association.

• Thinking Critically •

Humans are considered to have a poor sense of smell. However, when parents are presented with baby clothing, they are able to identify the clothing with which their own infant had been in contact with a high degree of accuracy. Specially trained individuals, such as wine and perfume testers, are able to identify large numbers of different kinds of molecules that the average person cannot identify. Birds rely primarily on sound and sight for information about their environment; they have a poor sense of smell. Most mammals are known to have a very well-developed sense of smell. Is it possible that we have evolved into sound-and-sight-dependent organisms like birds and have lost the keen sense of smell of our mammal ancestors? Or is it that we just don't use our sense of smell to its full potential? Can you devise an experiment that would help to shed light on this issue?

• Experience This •

Grasp your textbook in your hand and hold it out in front of you with your arm straight. How long can you hold it there before your arm starts to drop? Why does this happen? What is happening to the motor units in the muscles? What characteristics of skeletal muscle are illustrated by this experience? Explain all of the events responsible for the contraction and the eventual fatigue you experienced.

• Questions •

1. Describe how changing permeability of the cell membrane and the movement of sodium ions cause a nerve impulse.
2. What is the role of acetylcholine in a synapse? What is the role of acetylcholinesterase?
3. List three ways in which the nervous system differs from the endocrine system.
4. Give an example of the interaction between the endocrine system and the nervous system.
5. Give an example of negative-feedback control in the endocrine system.
6. How do skeletal, cardiac, and smooth muscle differ in (1) speed of contraction, (2) ability to stay contracted, and (3) cause of contraction.
7. What is actually detected by the nasal epithelium, taste buds, cochlea of the ear, and retina of the eye?
8. What is the role of each of the following in muscle contraction: actin, myosin, ATP, troponin, and tropomyosin?
9. List three hormones and give their functions.
10. List the differences between the following:
    a. central and peripheral nervous systems;
    b. motor and sensory nervous systems; and
    c. anterior and posterior pituitary.

• Chapter Glossary •

acetylcholine (ă-sēt″l-kō′lēn) A neurotransmitter secreted into the synapse by many axons and received by dendrites.

acetylcholinesterase (a-sēt″l-kō″li-nes′tĕ-rās) An enzyme present in the synapse that destroys acetylcholine.

actin (ak′tin) A protein found in the thin filaments of muscle fibers that binds to myosin.

antidiuretic hormone (an-ti-di″u-rē′tik hōr′mōn) A hormone produced by the pituitary gland that stimulates the kidney to reabsorb water.

axon (ak′sahn) A neuronal fiber that carries information away from the nerve cell body.

basilar membrane (ba′si-lar mem′brān) A membrane in the cochlea containing sensory cells that are stimulated by the vibrations caused by sound waves.

central nervous system (sen′trul ner′vus sis′tem) The portion of the nervous system consisting of the brain and spinal cord.

cochlea (kŏk′lē-ah) The part of the ear that converts sound into nerve impulses.

cones (kōnz) Light-sensitive cells in the retina of the eye that respond to different colors of light.

dendrites (den′drīts) Neuronal fibers that receive information from axons and carry it toward the nerve-cell body.

depolarized (de-po′lă-rīzd) Having lost the electrical difference existing between two points or objects.

endocrine glands (en′do-krin glandz) Glands that secrete into the circulatory system.

endocrine system (en′do-krin sis′tem) A number of glands that communicate with one another and other tissues through chemical messengers transported throughout the body by the circulatory system.

epinephrine (e′pi-nef′rin) A hormone produced by the adrenal medulla that increases heart rate, blood pressure, and breathing rate.

estrogen (es′tro-jen) One of the female sex hormones.

exocrine glands (ek′sa-krin glandz) Glands that secrete through ducts to the surface of the body or into hollow organs of the body.

fovea centralis (fo′ve-ah sen-tral′is) The area of sharpest vision on the retina, where light is normally focused.

gland (gland) An organ that manufactures and secretes a material either through ducts or directly into the circulatory system.

growth-stimulating hormone (grōth sti′mu-la-ting hōr′mōn) A hormone produced by the anterior pituitary gland that stimulates tissues to grow.

hormones (hōr′mōnz) Chemical messengers secreted by endocrine glands.

incus (in'kus) The ear bone that is located between the malleus and the stapes.

malleus (mă'le-us) The earbone that is attached to the tympanum.

motor neurons (mo'tur noor'onz'') Those neurons that carry information from the central nervous system to muscles or glands.

motor unit (mo'tur yōō'nit) All of the muscle cells stimulated by a single neuron.

myosin (mi'o-sin) A protein molecule found in the thick filaments of muscle fibers that bends and moves along actin molecules.

negative-feedback control (nĕ'gă-tiv feed'bak con-trōl') A kind of control mechanism in which the product of one activity inhibits an earlier step in the chain of events.

nerves (nervz) Bundles of neuronal fibers.

nerve impulse (nerv im'puls) A series of changes that take place in the neuron, resulting in a wave of depolarization that passes from one end of the neuron to the other.

nervous system (ner'vus sis'tem) A network of neurons that carry information from sense organs to the central nervous system and from the central nervous system to muscles and glands.

neuron (noor'on'') The cellular unit consisting of a cell body and fibers that makes up the nervous system; also called nerve cell.

neurotransmitter (noor''o-trans'mĭt-er) A molecule released by the axons of neurons that stimulates other cells.

norepinephrine (nor-ĕ''pĭ-nef'rin) A hormone produced by the adrenal medulla that increases heart rate, blood pressure, and breathing rate.

olfactory epithelium (ōl-fak'to-re ĕ''pĭ-thē'le-um) The cells of the nasal cavity that respond to chemicals.

oval window (o'val win'do) The membrane-covered opening of the cochlea, to which the stapes is attached.

oxytocin (ok''sĭ-to'sin) A hormone released from the posterior pituitary that causes contraction of the uterus.

perception (per-sep'shun) Recognition by the brain that a stimulus has been received.

peripheral nervous system (pŭ-rĭ'fĕ-ral ner'vus sis'tem) The fibers that communicate between the central nervous system and other parts of the body.

response (re-spons') The reaction of an organism to a stimulus.

retina (rĕ'tĭ-nah) The light-sensitive region of the eye.

rhodopsin (ro-dop'sin) A light-sensitive, purple-red pigment found in the retinal rods that is important for vision in dim light.

rods (rahdz) Light-sensitive cells in the retina of the eye that respond to low-intensity light but do not respond to different colors of light.

semicircular canals (sĕ-mi-ser'ku-lar că-nalz') A set of tubular organs associated with the cochlea that sense changes in the movement or position of the head.

sensory neurons (sen'so-re noor'onz'') Those neurons that send information from sense organs to the central nervous system.

soma (so'mah) The cell body of a neuron, which contains the nucleus.

stapes (sta'pēz) The ear bone that is attached to the oval window.

stimulus (stim'yu-lus) Any change in the environment of an organism that it can detect.

synapse (si'naps) The space between the axon of one neuron and the dendrite of the next, where chemicals are secreted to cause an impulse to be initiated in the second neuron.

testosterone (tes-tah'stĕ-rōn) The male sex hormone.

thyroid-stimulating hormone (thi'roid stĭ'mu-la-ting hōr'mōn) A hormone secreted by the pituitary gland that stimulates the thyroid to secrete thyroxine.

thyroxine (thi-rok'sin) A hormone produced by the thyroid gland that speeds up the metabolic rate.

triiodothyronine (trī''ī-ō''dō-thī'rownen) A hormone produced by the thyroid gland that speeds up the metabolic rate; similar to thyroxine but more potent.

tropomyosin (tro''po-mi'o-sin) A molecule found in thin filaments of muscle that helps to regulate when muscle cells contract.

troponin (tro'po-nin) A molecule found in thin filaments of muscle that helps to regulate when muscle cells contract.

tympanum (tim'pă-num) The eardrum.

voltage (vōl'tij) A measure of the electrical difference that exists between two different points or objects.

CHAPTER 21

Human Reproduction, Sex, and Sexuality

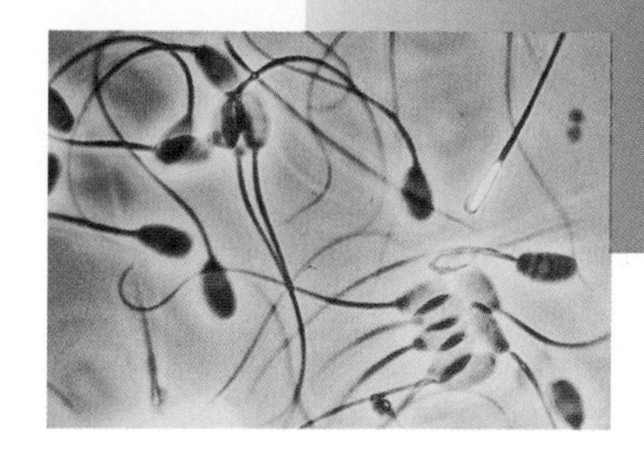

Purpose

Sex and our sexuality influences us in many different ways throughout life. Before birth, sex-determining chromosomes direct the formation of hormones that control the development of sex organs, after which the effects of these hormones diminish. With the start of puberty, renewed hormonal activity causes major structural and behavioral changes that influence us for the remainder of our lives. The power of our sexuality and the reproductive urge demand that the production of offspring be considered an important aspect of our nature. This chapter examines the influences of sex and sexuality on human life stages.

For Your Information

New evidence suggests that homosexuality is at least in part a biological phenomenon. Should this be confirmed, it could explain the occurrence of male homosexuality in human populations despite cultural pressures. It has been found that in homosexual men, the INAH-3 "nuclei" of the anterior hypothalamus—a section of the brain that governs sexual behavior—have the anatomical form usually found in females rather than the form typical of heterosexual males. The INAH-3 "nuclei" of heterosexual males are twice as large as those of homosexual males or heterosexual females. This difference in size may be dependent on levels of testosterone just before birth and immediately after birth. Unusual levels of sex hormones at any given time may switch the development of susceptible brain areas such as the hypothalamus from one sex to the other. While studies clearly show that size differences do exist between homo- and heterosexual male brains, it is important to note that this does not rule out the possibility that childhood or adolescent experiences also alter INAH-3 "nuclei."

Chapter Outline

Learning Objectives

- Describe the sex-determining chromosomes and how they function in determining gender.
- Know the different hormones that regulate the maturation of the female and male human reproductive systems at puberty and describe their involvement in sexual function in an adult.
- List the structures of the male and female human reproductive systems.
- Describe how hormones function in the process of ovulation and pregnancy.
- Recognize the role of the placenta in the development of the human fetus.
- Understand the differences between spermatogenesis and oogenesis.
- List the events necessary for conception and pregnancy to occur in humans.
- Describe the maturing processes involved in the development of a fetus.

Sexuality from Different Points of View

Probably nothing interests people more than sex and sexuality. By **sexuality,** we mean all the factors that contribute to one's female or male nature. These include the structure and function of the sex organs and the behaviors that involve these structures. We have an intense interest in the facts about our own sexual nature and the sexual behavior of members of the opposite sex and that of peoples of other cultures.

Several areas of study view sex and sexuality in slightly different ways. Sex is considered a strong drive by psychologists. They describe the sex drive as a basic impulse to satisfy a biological, social, or psychological need. The sex drive does not have life-sustaining implications for the individual, as do the drives for food or water. Other social scientists describe sexuality differently. They classify it as an appetite, strong desire, or urge that is not as strong as a drive. Whether we call this interest in sexual matters a drive, an appetite, or an urge, it still provides a great deal of motivation for many activities from birth to death.

Biologists have long considered the function of sex and sexuality in light of its value to the population or species. Without sexual exchange of genetic material, variation within a population is reduced, which makes improvement of the gene pool less likely. When we analyze this statement, we must place human behavior in evolutionary perspective. Certainly the behaviors of courtship, mating, child-rearing, and the division of labor within groups are more complex in social animals. These are demonstrated in the elaborate social behaviors surrounding mate selection and the establishment of families. It is difficult to draw the line between the biological development of sexuality and the social establishment of customs related to the sexual aspects of human life. However, the biological mechanism that determines female or male gender has been well documented.

Chromosomal Determinaton of Sex

When a human egg or sperm cell is produced, it contains twenty-three chromosomes. Twenty-two of these are **autosomes** that carry most of the genetic information used by the organism. The other chromosome is a **sex-determining chromosome.** There are two kinds of sex-determining chromosomes: the **X chromosome** and the **Y chromosome** (refer back to figure 9.16). The two sex-determining chromosomes, X and Y, do not carry equivalent amounts of information, nor do they have equal functions. X chromosomes carry typical information about the production of specific proteins in addition to their function in determining sex.

When a human sperm cell is produced, it carries twenty-two autosomes and a sex-determining chromosome. Unlike eggs, which always carry an X chromosome, half of the sperm cells carry an X chromosome, and the other half carry a Y chromosome. If an X-carrying sperm cell fertilizes the X-containing egg cell, the resultant embryo will develop into a female. A typical human female has an X chromosome from each parent. If a Y-carrying sperm cell fertilizes the egg, a male embryo develops. It is the presence or absence of the Y chromosome that determines the sex of the developing individual.

Evidence that the Y chromosome controls male development comes as a result of studying individuals who have an abnormal number of chromosomes. An abnormal meiotic division that results in sex cells with too many or too few chromosomes is called *nondisjunction* (nondisjunction is explained in chapter 9). If nondisjunction affects the X and Y chromosomes, a gamete might be produced that has only twenty-two chromosomes and lacks a sex-determining chromosome, or it might have twenty-four, with two sex-determining chromosomes. If a cell with too few or too many sex chromosomes is fertilized, an abnormal embryo develops. If a normal egg cell is fertilized by a sperm cell with no sex chromosome, the offspring will have only one X chromosome. These people are designated as XO. They develop a collection of characteristics known as *Turner's syndrome.* An individual with this condition is female but is generally sterile and matures sexually later than normal. In addition, she may have a thickened neck (termed webbing), hearing impairment, and some abnormalities in the cardiovascular system.

An individual who has XXY chromosomes is basically male. This genetic anomaly is termed *Klinefelter's syndrome,* and the symptoms include mental deficiencies and sexual dysfunction due to small testes that do not usually produce viable sperm. Since both of these conditions involve abnormal numbers of X or Y chromosomes, this is strong evidence that these chromosomes are involved in determining sexual development.

The embryo resulting from fertilization and cell division becomes female or male based on the sex-determining chromosomes that control the specialization of the cells of the gonads into female sex organs, the **ovaries,** or male sex organs, the **testes.** This specialization of embryonic cells is termed **differentiation.** The embryonic gonads begin to differentiate into testes about seven weeks after conception if the Y chromosome is present. The Y chromosome seems to control this differentiation process in males, since the gonads do not differentiate into female sex organs until later, and then only if two X chromosomes are present (box 21.1). It is the absence of the Y chromosome that determines femaleness. Femaleness appears to be the "default" condition.

Researchers were interested in how females, with two X chromosomes, handled the double dose of genetic material in comparison to males, who have only one X chromosome. M. L. Barr discovered that a darkly staining body was generally present in female cells but was not present in male cells. It was postulated, and has since been confirmed, that this

Box 21.1 Gender Anomalies

Sometimes the hormone levels are out of balance at critical times in the development of an embryo. This hormonal imbalance may be due to an abnormal number of sex-determining chromosomes, or it may be the result of abnormal functioning of the endocrine glands. In some instances partial development of the internal and external genitalia of both sexes may occur. The individual is then classified as a hermaphrodite. Corrective surgery and appropriate counseling are sometimes helpful in allowing the individual to live a more normal life. Some hermaphrodites are primarily female with a slightly enlarged clitoris, while others are primarily male with an underdeveloped penis and normal labia and vagina. Such people must be assessed by a physician to determine which sexual structures should be retained or surgically reconstructed. The physician may also decide that hormone therapy might be a more successful treatment. These decisions are not easily made because they involve children who have not fully developed their sexual nature.

More and more frequently, we are becoming aware of individuals whose physical gender does not match their psychological gender. These people are referred to as being *gender-confused.* A male with normal external male genitals may "feel" like a female to one degree or another. The same situation may occur with structurally female individuals. This may mean that, in private, such people might dress as the other sex. However, in social situations, they would behave normally for their sex. Others have completely changed their public and private behavior to reflect their inner desire to function as the other sex. A male may dress as a female, work in a traditionally female occupation, and make social contacts as a female. Tremendous psychological and emotional pressures develop from this condition. Frequently, these extreme individuals would like to have gender reassignment surgery: a sex-change operation. This surgery and the follow-up hormonal treatment can cost tens of thousands of dollars and take several years.

Because the most frequent behavior of gender-confused individuals is dressing as a member of the opposite sex, we label them *cross-dressers,* or *transvestites.* (Other psychological conditions can cause this same symptom of cross-dressing. Therefore, it is not accurate to say that all transvestites are gender-confused individuals.)

The sexual orientation of a person seems to be unrelated to whether they are transvestites. Some males who are attracted to males (are "gay") can also be gender-confused, while others are not. The reverse is also true. Some females who are attracted to females are gender-confused, while some are not. Homosexuality is a complex behavioral pattern that appears to be separate and distinct from gender-confusion.

---

structure is an X chromosome that is largely nonfunctional. Therefore, female cells have only one dose of information that is functional; the other X chromosome coils up tightly and does not direct the manufacture of proteins. The one X chromosome of the male functions as expected, and the Y chromosome directs only male-determining activities. The tightly coiled structure in the cells of female mammals is called a **Barr body** after its discoverer.

Planning the Sex of Your Child

Frequently, due to cultural or personal desires, a couple may want to choose the sex of their child. In the past, the sex of offspring was determined strictly by chance. If the father produced typical sperm cells, half would be X-containing and therefore produce female children, and half would be Y-containing and would produce male children. Recently, it has become possible to change the probability in order to favor production of the desired sex. Reproductive specialists claim a success rate of from 65% to 85% in obtaining the desired sex of a child. This is done by taking advantage of what we know about the mobility of sperm cells, their viability and pH requirements, and the relationship between the age of the egg cell and its ability to be fertilized.

Generally speaking, a sperm cell carrying an X chromosome is stronger than the Y-carrying sperm cell. It is larger and has an oval head and a shorter tail. Anything that favors the stronger sperm cell favors the production of a female child. An acid douche (vinegar and water) creates a slightly acidic environment, which favors sperm cells carrying an X chromosome. Any condition that makes it difficult for sperm to reach the egg works in favor of the stronger X-carrying sperm. Shallow penetration during intercourse, lack of female orgasm, and refraining from intercourse several days before ovulation all achieve this purpose. The opposite conditions favor the production of a male offspring. An alkaline douche (baking soda and water), deeper penetration, female orgasm, and frequent intercourse centered around the time of ovulation will favor fertilization by a Y sperm cell.

Male and Female Fetal Development

Development of embryonic gonads begins very early during fetal growth. First, a group of cells begins to differentiate into primitive gonads. Within a matter of weeks, these gonads will become testes if a Y chromosome is present; they will develop later into ovaries if two X chromosomes are present. If the embryo has an X and a Y chromosome, the central portion of the gonad (medulla) will begin to develop structures that are characteristic of a testis. If not influenced by the presence of the Y chromosome, the embryo will begin to develop ovaries at about the twelfth week after **conception** (fertilization), when the outer portion (cortex) of the embryonic gonad develops into an ovary.

As soon as the gonad has differentiated into an embryonic testis or ovary it begins to produce **hormones** (figure 21.1). These chemical control molecules produced by the ovaries and testes influence the further development of the embryo, causing it to complete its sexual differentiation.

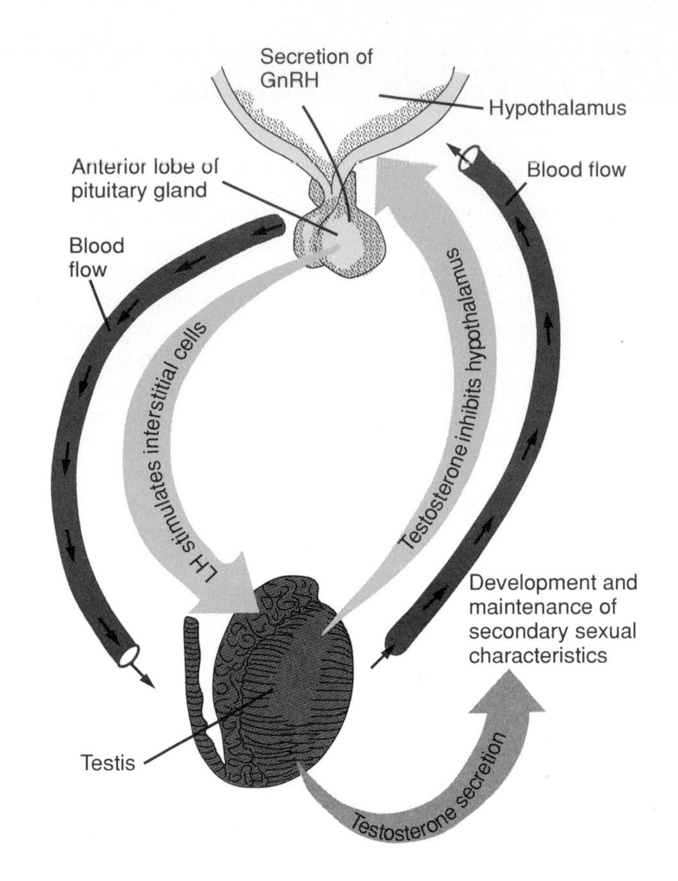

Figure 21.1 **Interaction of Hormones.** Here we see the interaction of the pituitary, hypothalamus, and testes, and how the secretion of one influences the production of hormones in another.

In normal males, the testes move from a position in the abdominal cavity to the external sac, called the scrotum, via an opening called the **inguinal canal.** This descent occurs at about the seventh month of development; afterward, the inguinal canal is closed off. This canal continues to be a weakened area in the abdominal wall and may rupture later in life. This can happen when strain (e.g., from improperly lifting heavy objects) causes a portion of the intestine to push through the inguinal canal into the scrotum. This condition is known as an **inguinal hernia.**

When the testes do not descend, a condition known as **cryptorchidism** (hidden testes) develops. Normal sperm-cell development cannot occur while the testes are in the abdomen because the temperature is higher than the temperature in the scrotum. The temperature of the testes is very carefully regulated by muscles that control their distance from the body. Physicians have diagnosed cases of male infertility as being caused by tight-fitting pants that hold the testes so close to the body that the temperature increase interferes with normal sperm development.

If the testes are retained in the abdominal cavity after birth, the slightly higher temperature causes sterility. Sometimes the descent occurs later, during puberty; if not, there is an increased incidence of testicular cancer. Because of this increased risk, undescended testes can be surgically moved to their normal position in the scrotum.

Sexual Maturation of Young Adults

Following birth, sexuality plays only a small part in physical development for several years. Culture and environment shape the responses that the individual will come to recognize as normal behavior. During **puberty,** increasing production of sex hormones causes major changes as the individual reaches sexual maturity. After puberty, humans are sexually mature and have the capacity to produce offspring. While sexual intercourse frequently does not result in the fertilization and development of an embryo, the capacity has been developed.

The Maturation of Females

Female children typically begin to produce quantities of sex hormones from the hypothalamus, pituitary gland, ovaries, and adrenal glands at eight to twelve years of age. This marks the onset of puberty. The **hypothalamus** controls the functioning of many other glands

Table 21.1
Human Reproductive Hormones

Hormone	Production Site	Target Organ	Function
1. Prolactin (lactogenic or luteotropic hormone)	Pituitary gland	Breasts, ovary	Stimulates milk production; also helps maintain normal ovarian cycle
2. Follicle-stimulating hormone	Pituitary gland	Ovary, testis	Stimulates ovary and testis development; stimulates egg production in females and sperm production in males
3. Luteinizing hormone (interstitial cell stimulating hormone)	Pituitary gland	Ovary, testis	Stimulates ovulation in females and sex-hormone (estrogen and testosterone) production in both males and females
4. Estrogen	Follicle of the ovary	Entire body	Stimulates development of female reproductive tract and secondary sexual characteristics
5. Testosterone	Testes	Entire body	Stimulates development of male reproductive tract and secondary sexual characteristics
6. Progesterone	Ovaries	Uterus, breasts	Causes uterine thickening and maturation; maintains pregnancy
7. Oxytocin	Pituitary gland	Breasts, uterus	Causes uterus to contract and breasts to release milk

throughout the body, including the pituitary gland. At puberty, the **pituitary gland** in girls begins to produce **follicle-stimulating hormone (FSH).** The increasing amount of this hormone circulating in the blood causes the ovaries to begin producing larger quantities of **estrogen.** The increasing supply of estrogen is responsible for the many changes in sexual development that can be noted at this time. These changes include breast growth, changes in the walls of the uterus and vagina, increased blood supply to the clitoris, and changes in the pelvic bone structure.

Estrogen also stimulates the female adrenal gland to produce **androgens,** male sex hormones. The androgens are responsible for the production of pubic hair and they seem to have an influence on the female sex drive. The adrenal gland secretions may also be involved in the development of acne. Those features that are not primarily involved in sexual reproduction but are characteristic of a sex are called **secondary sex characteristics.** In women, the distribution of body hair, patterns of fat deposits, and a higher voice are examples.

Curiosity about the changing female body form and new feelings leads to self-investigation. Studies have shown that sexual activity such as manipulation of the clitoris, which causes pleasurable sensations, is performed by a large percentage of young women. Self-stimulation, frequently to orgasm, is a common result. **Orgasm** is a complex response to mental and physical stimulation that causes rhythmic contractions of the muscles of the reproductive organs and an intense frenzy of excitement. This stimulation is termed **masturbation,** and it should be stressed that it is considered to be a normal part of sexual development.

A major development during this time is the establishment of the **menstrual cycle.** This involves the periodic growth and shedding of the lining of the uterus. These changes are under the control of a number of hormones produced by the ovaries. The ovaries are stimulated to release their hormones by the pituitary gland, which is in turn influenced by the ovarian hormones. Follicle-stimulating hormone (FSH), and luteinizing hormone (LH) are both produced by the pituitary gland. FSH causes the maturation and development of the ovaries, and LH is important in causing ovulation and in maintaining the menstrual cycle. In addition, secretions of the hypothalamus at puberty bring about the development of a menstrual cycle (table 21.1). Associated with the menstrual cycle is the periodic release of sex cells from the surface of the ovary, called **ovulation** (figure 21.2). Initially, these two cycles, menstruation and ovulation, may be irregular, which is normal during puberty. Eventually, hormone production becomes regulated so that ovulation and menstruation take place on a regular monthly basis in most women.

The Maturation of Males

Males also typically begin puberty with a change in hormone levels. Testosterone causes the differentiation of internal and external genital anatomy in a male embryo. One hormone involved in the development of the male sex organs is follicle-stimulating hormone (FSH), the

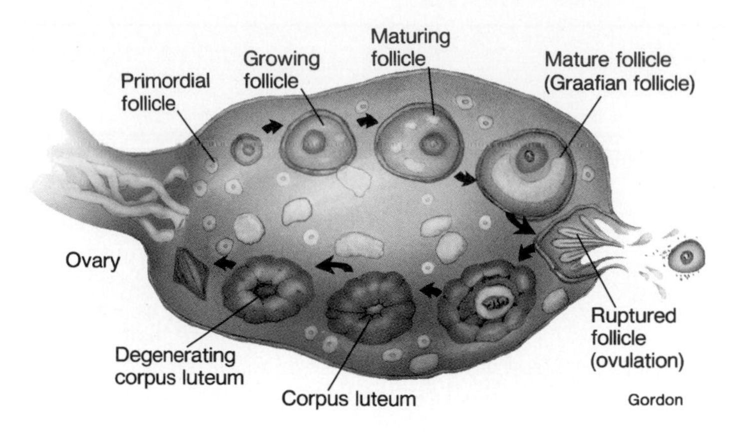

Figure 21.2 **Ovulation.** In the ovary, the egg begins development inside a sac of cells known as a follicle. Each month, one of these follicles develops and releases its product. This release through the wall of the ovary is known as ovulation.

same pituitary-gland secretion produced by females. This is the primary growth stimulator of the testes and ovaries. FSH produced by the male is responsible for the embryonic development of testes and, later, their production of sperm cells. **Interstitial cell-stimulating hormone (ICSH)** stimulates the testes to produce testosterone, which is also important in the maturation and production of sperm.

Male children reach puberty about two years later than female children. At about age ten or eleven, the pituitary gland begins to produce increased quantities of FSH. It also increases the production of ICSH, which in turn stimulates increased production of **testosterone,** the primary male sex hormone.

The major changes during puberty include growth of the testes and scrotum, pubic-hair development, and increased size of the penis. Secondary sex characteristics begin to become apparent at age thirteen or fourteen. Facial hair, underarm hair, and chest hair are some of the most obvious. The male voice changes as the larynx (voice box) begins to change shape. Body contours also change, and a growth spurt increases height. In addition, the proportion of the body that is muscle increases, while the proportion of body fat decreases. At this time, a boy's body begins to take on the characteristic adult male shape, with broader shoulders and heavier muscles.

In addition to these external changes, the FSH released from the pituitary causes the production of seminal fluid by the **seminal vesicles,** prostate gland, and the bulbo-urethral glands. Later, FSH stimulates the production of sperm cells. The release of sperm cells and seminal fluid also begins during puberty and is termed **ejaculation.** This release is generally accompanied by the pleasurable sensations of orgasm. The sensations associated with ejaculation may lead to self-stimulation, or masturbation. Studies of sexual behavior have shown that nearly all men masturbate at some time during their lives.

Spermatogenesis

One of the biological reasons for sexual activity is the production of offspring. The process of producing gametes include meiosis and is called **gametogenesis** (gamete formation) (figure 21.3).

The term **spermatogenesis** is used to describe gametogenesis in males and takes place in the testes. The two bean-shaped testes are composed of many small sperm-producing tubes, or **seminiferous tubules,** and collecting ducts that store sperm. These are held together by a thin covering membrane (figure 21.4). The seminiferous tubules join together and eventually become the epididymis, a long, narrow convoluted tube in which sperm cells are stored and mature before ejaculation.

Leading from the epididymis is the vas deferens, or sperm duct; this empties into the urethra, which conducts the sperm out of the body through the **penis.** Before puberty, the seminiferous tubules are packed solid with diploid cells called spermatogonia. These cells, which are found just inside the tubule wall, undergo *mitosis* and produce more spermatogonia. Beginning about age eleven, some of the spermatogonia specialize and begin the process of *meiosis,* while others continue to divide by mitosis, assuring a constant and continuous supply of spermatogonia. Once spermatogenesis begins, the seminiferous tubules become hollow and can transport the mature sperm.

Spermatogenesis involves several steps. Some of the spermatogonia in the walls of the seminiferous tubules differentiate and enlarge to become **primary spermatocytes.** These diploid cells undergo the first meiotic division, which produces two haploid **secondary spermatocytes.** The secondary spermatocytes go through the second meiotic division, resulting in four haploid **spermatids,** which lose much of their cytoplasm and develop long tails. These cells are then known as **sperm** (figure 21.5). The sperm have only a small amount of food reserves. Therefore, once they are released and become active swimmers, they live no more than seventy-two hours. However, the life of a sperm may be increased in some cases. If the temperature is lowered drastically, the sperm freeze, become deactivated, and can live for years outside the testes. This has led to the development of sperm banks.

Spermatogenesis in human males takes place continuously throughout a male's reproductive life, although the number of sperm produced decreases as a man ages. Sperm counts can be taken and used to determine the probability of successful fertilization. For reasons not totally understood, a man must be able to release at least 100 million sperm at one insemination to be fertile. A healthy male probably releases about 300 million sperm during each act of **sexual intercourse,** also known as **coitus** or **copulation.**

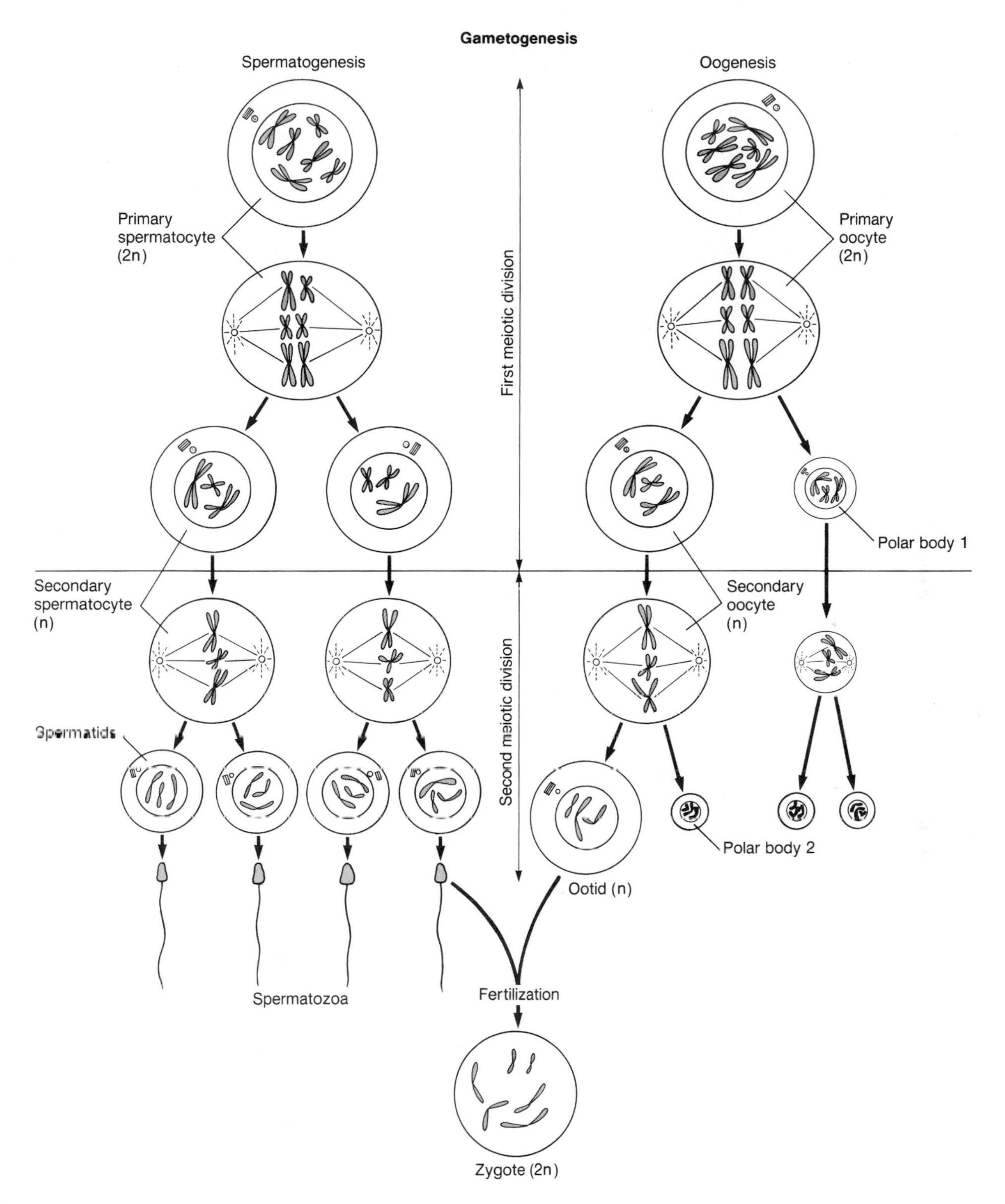

Figure 21.3 **Gametogenesis.** This diagram illustrates the process of gametogenesis in human males and females. Not all of the forty-six chromosomes are shown. Carefully follow the chromosomes as they segregate, recalling the details of the process of meiosis explained previously. Note the sex-determining X and Y chromosomes.

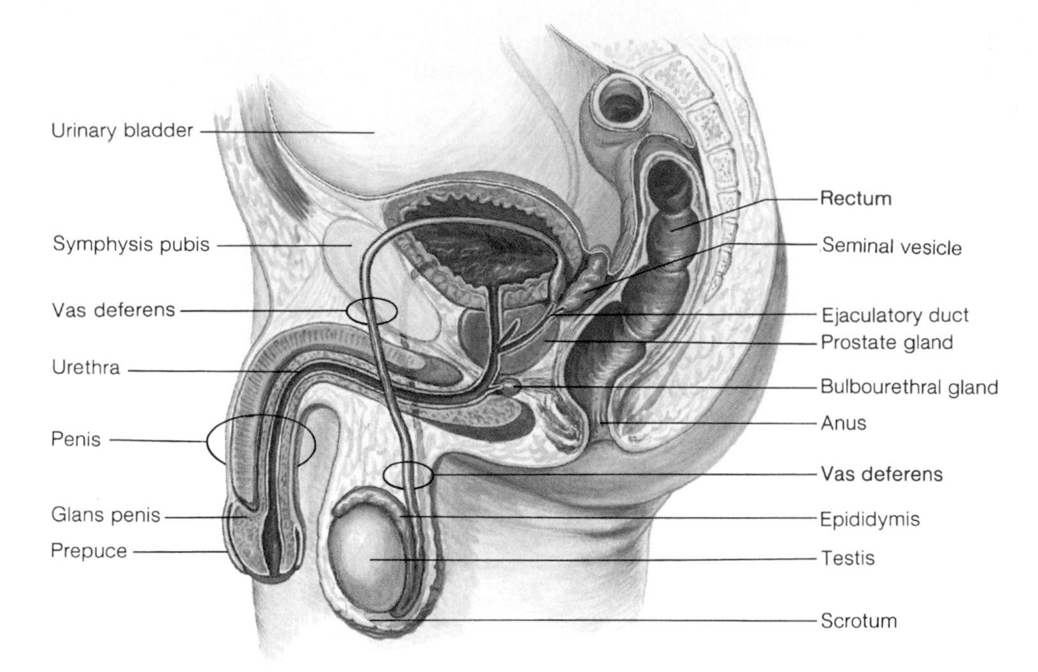

Figure 21.4 **The Human Male Reproductive System.** The male reproductive system consists of two testes that produce sperm, ducts that carry the sperm, and various glands. Muscular contractions propel the sperm through the vas deferens past the seminal vesicles, prostate gland, and bulbo-urethral gland, where most of the liquid of the semen is added. The semen passes through the urethra of the penis to the outside of the body.

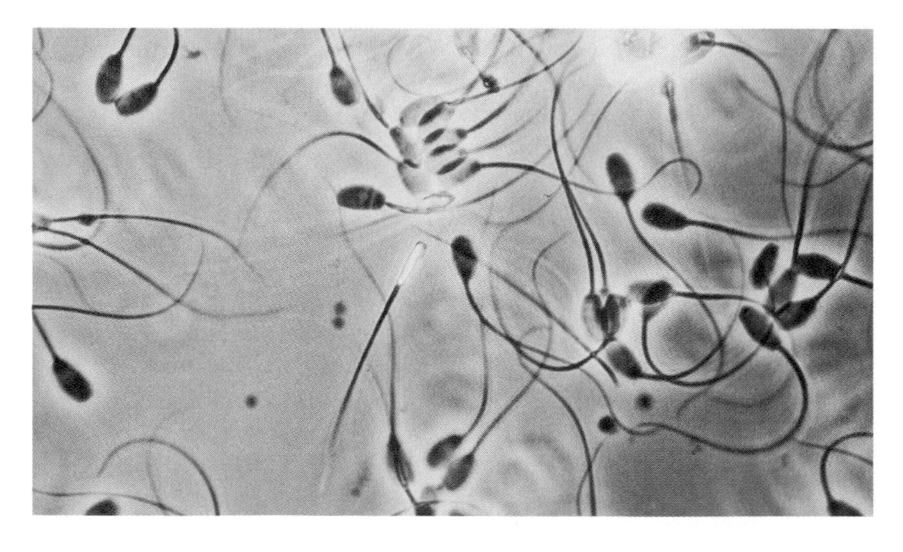

Figure 21.5 **Human Sperm Cells.** These cells are primarily DNA-containing packages produced by the male.

Oogenesis

The term **oogenesis** refers to the production of egg cells. This process starts during prenatal development of the ovary, when diploid oogonia cease dividing by *mitosis* and enlarge to become **primary oocytes.** Note that all of the primary oocytes that a woman will ever have are already formed prior to her birth. This is approximately 2 million, but that number is reduced by cell death to between 300,000 to 400,000 cells by the time of puberty. Oogenesis halts at this point, and all the primary oocytes remain just under the surface of the ovary.

Primary oocytes begin to undergo *meiosis* in the normal manner at puberty. At puberty and on a regular basis thereafter, the sex hormones stimulate a primary oocyte to continue its maturation process, and it goes through the first meiotic division. But in telophase I, the two cells that form receive unequal portions of cytoplasm. You might think of it as a lopsided division (figure 21.3). The smaller of the two cells is called a **polar body,** and the larger haploid cell is the **secondary oocyte.** The other primary oocytes remain in the ovary. Ovulation begins when the soon-to-be released secondary oocyte, encased in a saclike structure known as a **follicle,** grows and moves near the surface of the ovary. When this maturation is complete, the follicle erupts and the secondary oocyte is released. It is swept into the **oviduct** (fallopian tube) by ciliated cells and travels toward the **uterus** (figure 21.6). Because of the action of the luteinizing hormone, the rest of the follicle develops into a glandlike structure, the **corpus luteum,** which produces hormones (progesterone and estrogen) that prevent the release of other secondary oocytes.

If the secondary oocyte is fertilized, it completes meiosis by proceeding through meiosis II with the sperm DNA inside. During the second meiotic division, the secondary oocyte again divides unevenly, so that a second polar body forms. None of the polar bodies survive; therefore, only one large secondary oocyte is produced from each primary oocyte that begins oogenesis. If the cell is not fertilized, the secondary oocyte passes through the **vagina** to the outside during menstruation. During her lifetime, a female releases about three hundred to five hundred secondary oocytes. Obviously, few of these cells are fertilized.

One of the characteristics to note here is the relative age of the sex cells. In males, sperm production is continuous throughout life. Sperm do not remain in the tubes of the male reproductive system for very long. They are either released shortly after they form or die and are harmlessly absorbed. In females, meiosis begins before birth, but the oogenesis process is not completed, and the cell is

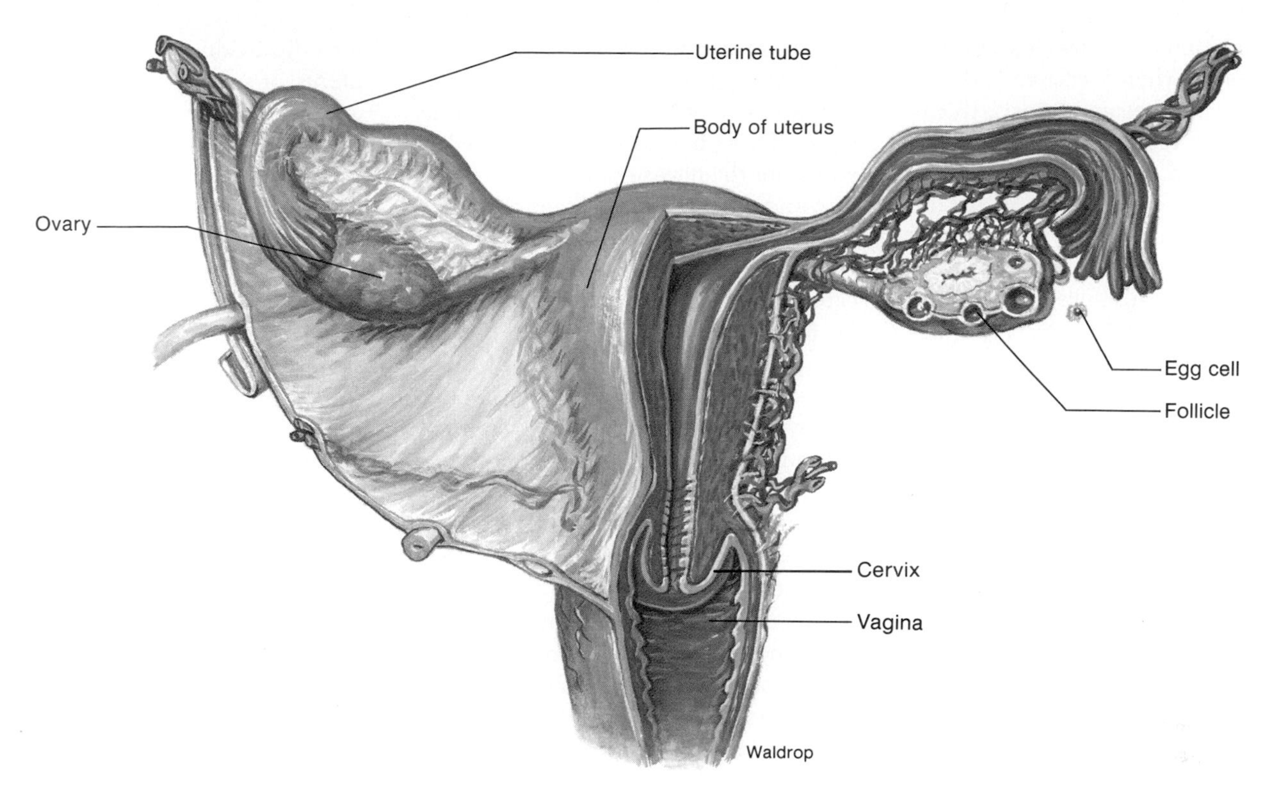

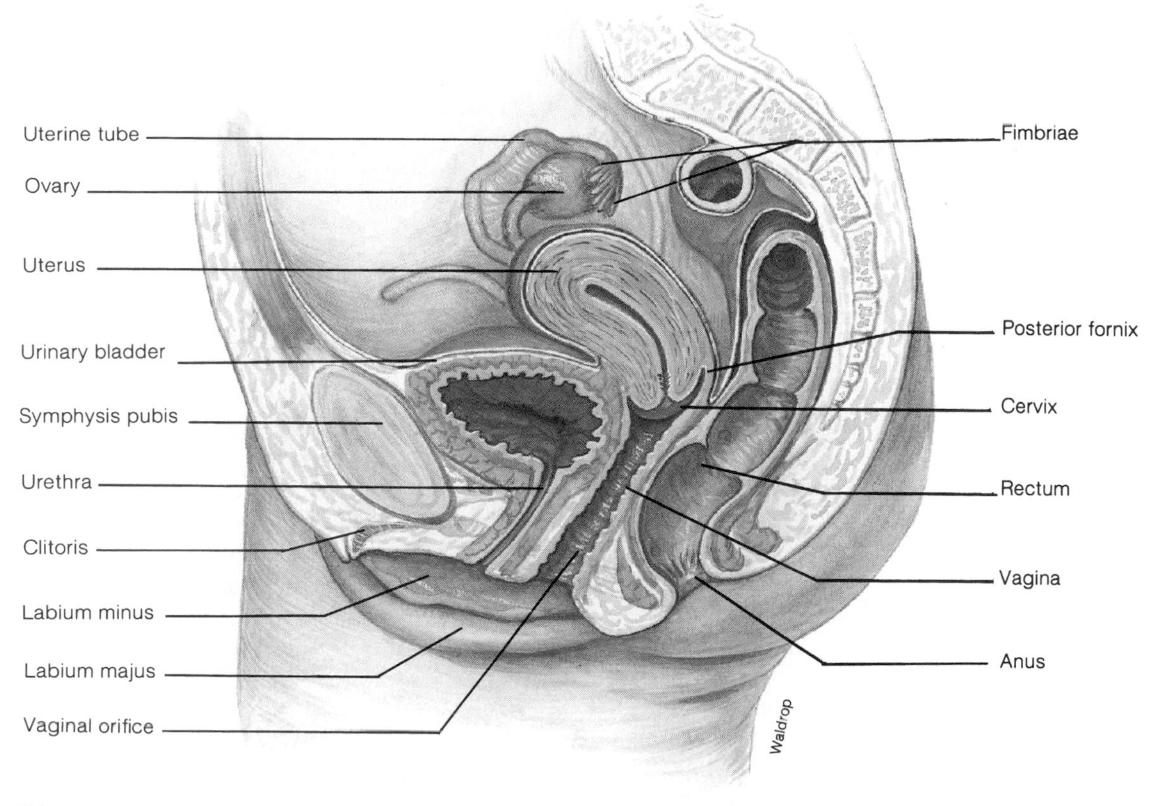

Figure 21.6 **The Human Female Reproductive System.** (a) After ovulation, the cell travels down the oviduct to the uterus. If it is not fertilized, it is shed when the uterine lining is lost during menstruation. (b) The human female reproductive system, side view.

not released for many years. A secondary oocyte released when a woman is thirty-seven years old began meiosis thirty-seven years before! During that time, the cell was exposed to many influences, a number of which may have damaged the DNA or interfered with the meiotic process. Such alterations are less likely to occur in males because new gametes are being produced continuously. Also, defective sperm are much less likely to be involved in fertilization.

Hormones control the cycle of changes in breast tissue, in the ovaries, and in the uterus. In particular, estrogen and progesterone stimulate milk production by the breasts and cause the lining of the uterus to become thicker and more vascularized prior to the release of the secondary oocyte. This ensures that if the secondary oocyte becomes fertilized, the resultant embryo will be able to attach itself to the wall of the uterus and receive nourishment. If the cell is not fertilized, the lining of the uterus is shed. This is known as *menstruation, menstrual flow, the menses,* or *a period.* Once the wall of the uterus has been shed, it begins to build up again. As noted previously, this continual building up and shedding of the wall of the uterus is known as the menstrual cycle.

At the same time that hormones are regulating the release of the secondary oocyte and the menstrual cycle, changes are taking place in the breasts. The same hormones that prepare the uterus to receive the embryo also prepare the breasts to produce milk. These changes in the breasts, however, are relatively minor unless pregnancy occurs.

Hormonal Control of Fertility

An understanding of how various hormones regulate the menstrual cycle, ovulation, milk production, and sexual behavior has led to the medical use of certain hormones. Some women are unable to have children because they do not release oocytes from their ovaries or they release them at the wrong time. Physicians can now regulate the release of oocytes from the ovary using certain hormones, commonly called *fertility drugs.* These hormones can be used to stimulate the release of oocytes for capture and use in what is called *in vitro* fertilization (*test-tube* fertilization) or to increase the probability of natural conception; that is, *in vivo* fertilization (*in-life* fertilization).

Unfortunately, the use of these drugs often results in multiple implantations, since they may cause too many secondary oocytes to be released at one time. The implantation of multiple embryos makes it difficult for one embryo to develop properly and be carried through the entire gestation period. The gestation period is the nine-month period of pregnancy. When we better understand the action of hormones, we may be able to control the effects of fertility drugs and eliminate the problem of multiple implantations.

A second medical use of hormones is in the control of conception by the use of birth-control pills—oral contraceptives. Birth-control pills have the opposite effect of fertility drugs. They raise the levels of estrogen and progesterone, which suppresses the production of FSH and LH, preventing the release of secondary oocytes from the ovary. Hormonal control of fertility is not as easy to achieve in men because there is no comparable cycle of gamete release.

Fertilization and Pregnancy

In most women, a secondary oocyte is released from the ovary at about the middle of the menstrual cycle. The menstrual cycle is usually said to begin on the first day of menstruation. Therefore, if a woman has a regular twenty-eight day cycle, the cell is released approximately on day fourteen (figure 21.7). Some women, however, have very irregular menstrual cycles, and it is difficult to determine just when the oocyte will be released to become available for fertilization. Once the cell is released, it is swept into the oviduct and moved toward the uterus. If sperm are present, they swarm around the secondary oocyte as it passes down the oviduct, but only one sperm penetrates the outer layer to fertilize it and cause it to complete meiosis II.

During this division, the second polar body is pinched off and the *ovum* (egg) is formed. Since chromosomes from the sperm are already inside, they simply intermingle with those of the ovum, forming a **zygote** or fertilized egg. As the zygote continues to travel down the oviduct, it begins dividing by mitosis into smaller and smaller cells (figure 21.8). This division process is called cleavage. Eventually, a solid ball of cells is produced, known as the morula stage of embryological development. Following the morula stage, the solid ball of cells becomes hollow and is then known as the blastula stage. During this stage, when the embryo is about six days old, it becomes embedded, or implanted, in the lining of the uterus. In mammals, the blastula has a region of cells, called the *inner-cell mass,* that develops into the embryo proper. The outer cells become membranes associated with the embryo.

The next stage in the development is known as the gastrula stage, since the gut is formed during this time. In many kinds of animals, the gastrula is formed by an infolding of one side of the blastula, a process similar to poking a finger into a balloon. Gastrula formation in mammals is a much more complicated process, but the result is the same. The embryo develops a tube that eventually becomes the gut. The formation of the primitive gut is just one of a series of changes that eventually results in an embryo that is recognizable as a miniature human being (figure 21.8). Most of the time during its development, the embryo is enclosed in a water-filled membrane, the amnion, which protects it from blows and keeps it moist. Two other membranes, the chorion and allantois, fuse with the lining of the uterus to form the **placenta** (figure 21.9). A fourth sac, the yolk sac, is well developed in birds, fish, and reptiles, but poorly developed in mammals. The placenta produces hormones that prevent men-

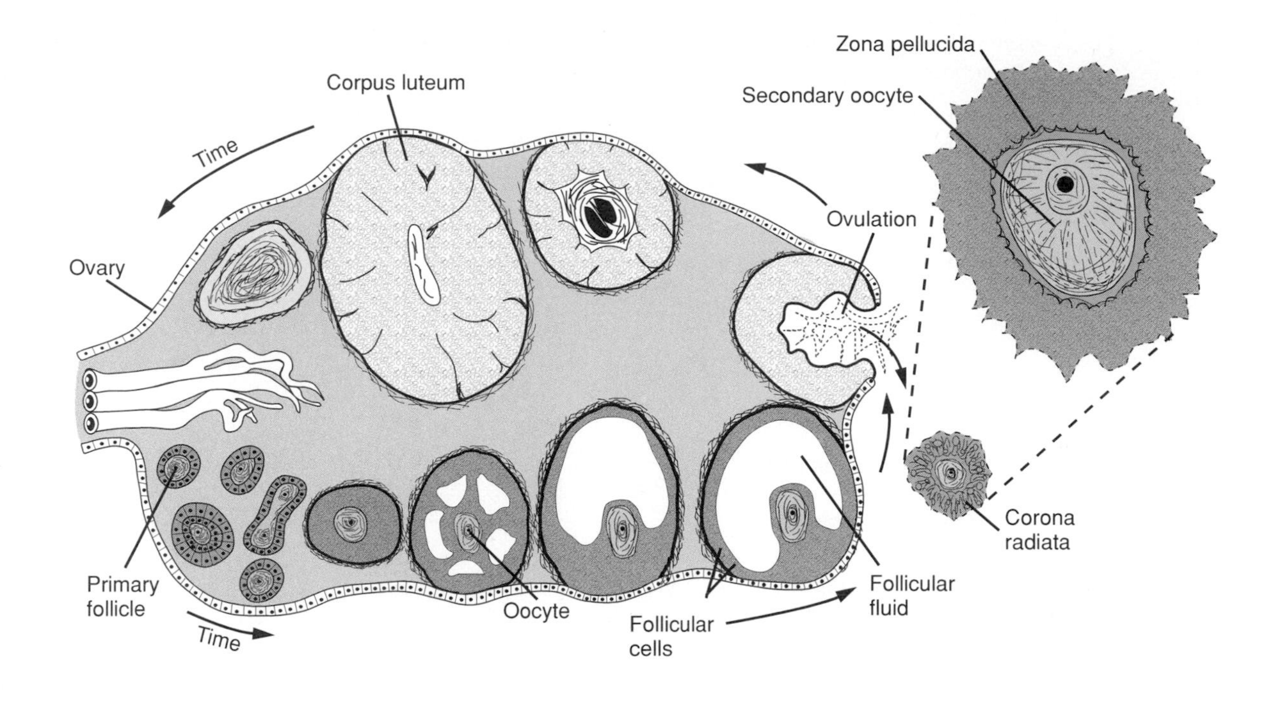

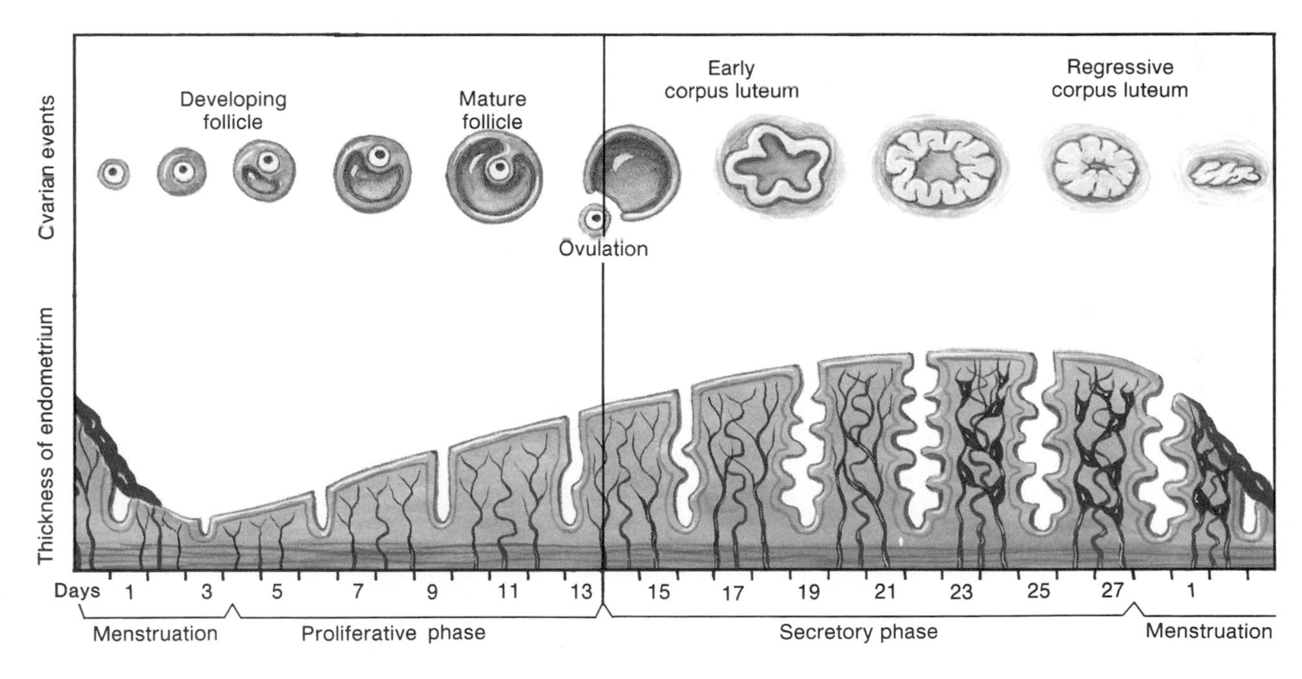

Figure 21.7 **The Ovarian Cycle in Human Females.** The release of a secondary oocyte (ovulation) is timed to coincide with the thickening of the lining of the uterus. The uterine cycle in humans involves the preparation of the uterine wall to receive the embryo if fertilization occurs. Knowing how these two cycles compare, it is possible to determine when pregnancy is most likely to occur.

struation and ovulation during gestation. It also provides for the metabolic needs of the embryo.

As the embryo's cells divide and grow, some of them become differentiated into nerve cells, bone cells, blood cells, or other specialized cells. In order to divide, grow and differentiate, cells must receive nourishment. This is provided by the mother through the placenta, a specialized organ in which both fetal and maternal blood vessels are abundant allowing for the exchange of substances between the mother and embryo. The materials diffusing across the placenta include oxygen, carbon dioxide, nutrients, and a variety of waste products. The materials entering the embryo travel through blood vessels in the umbilical cord. The embryo is contained within the mother's body and

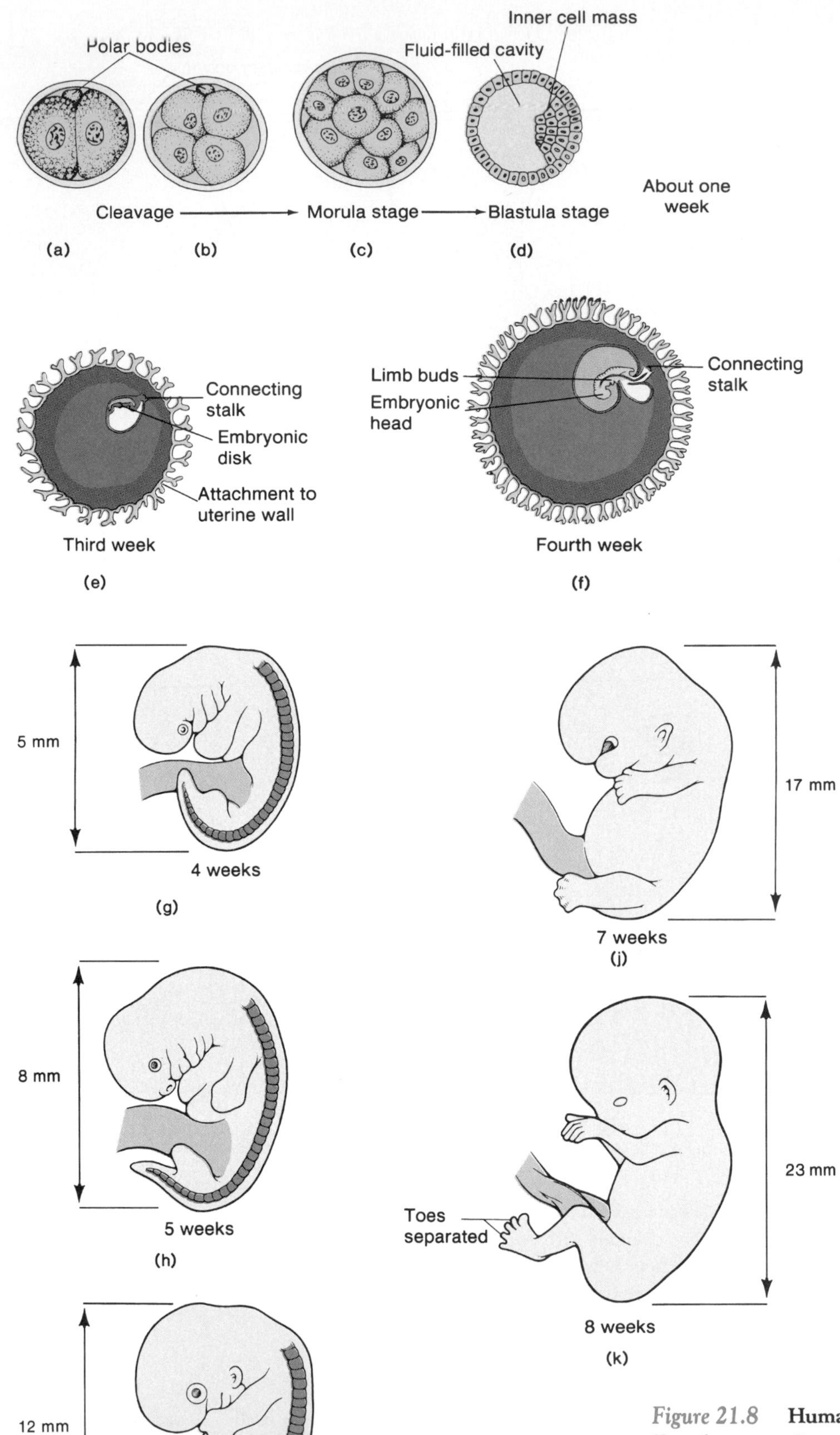

Figure 21.8 **Human Embryonic Development.** During the period of time between fertilization and birth, many changes are taking place in the embryo. Here we see some of the changes that take place during the first eight weeks.

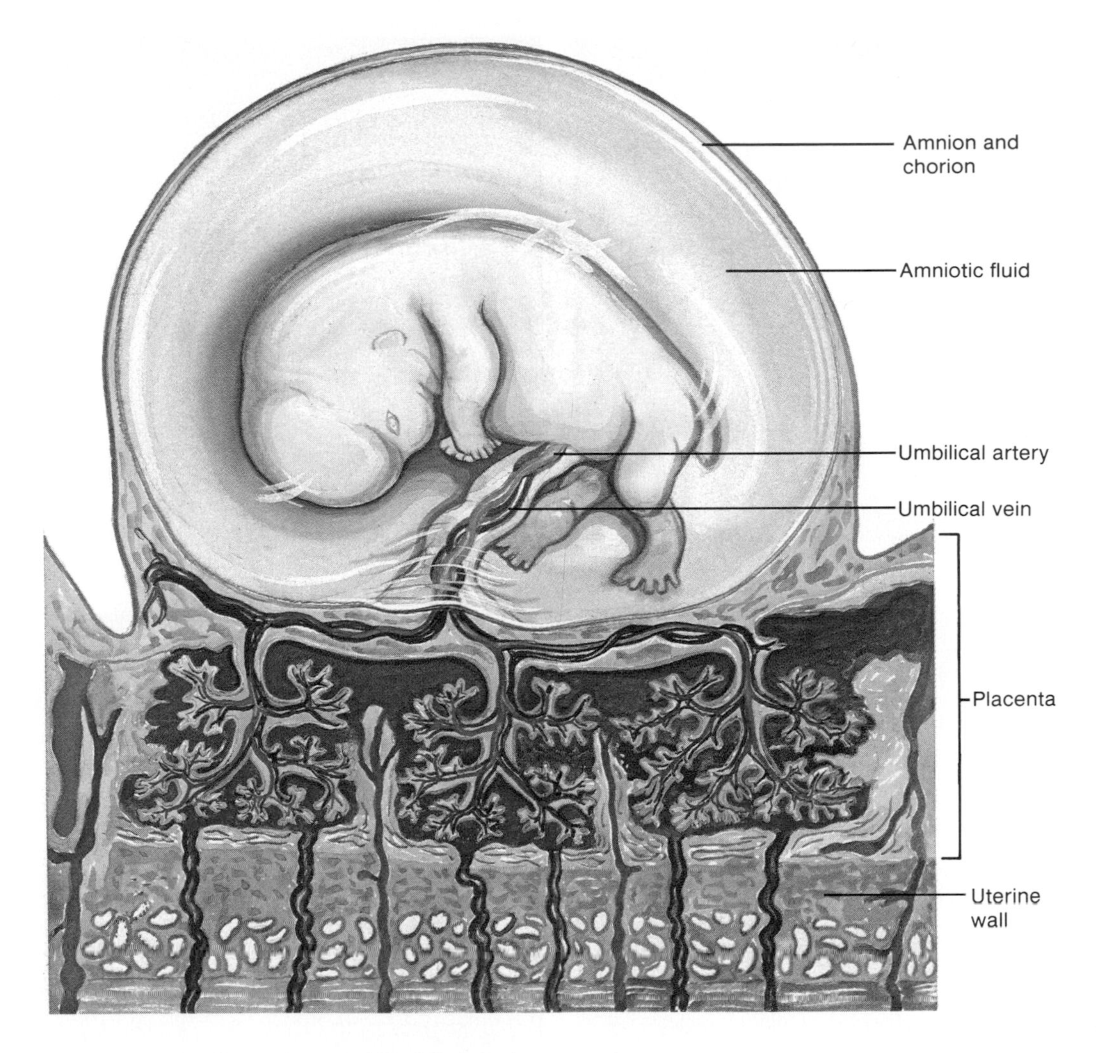

Figure 21.9 **Placental Structure.** The embryonic blood vessels that supply the developing child with nutrients and remove the metabolic wastes are separate from the blood vessels of the mother. Because of this separation, the placenta can selectively filter many types of incoming materials and microorganisms.

relies upon her body to provide all its needs. The major parts of the body develop by the tenth week of pregnancy. After this time, the embryo increases in size, and the structure of the body is refined.

Twins

The occasional production of twins happens in two ways. In the case of identical twins, when the zygote divides by cleavage it splits into two separate groups of cells. Each develops into an independent embryo. Since they came from the same single fertilized ovum, they have the same genes and are of the same sex.

Fraternal twins do not contain the same genetic information and may be of different sexes. They result from the fertilization of two separate oocytes by different sperm. Therefore, they no more resemble each other than ordinary brothers and sisters.

Birth

At the end of about nine months, hormone changes in the mother's body stimulate contractions of the muscles of the uterus during a period prior to birth called labor. These contractions are stimulated by the hormone oxytocin, which is released from the posterior pituitary. The contractions normally move the baby headfirst through the vagina, or birth canal. One of the first effects of these contractions may be bursting of the amnion (bag of water) surrounding the baby. Following this, the uterine contractions become stronger, and shortly thereafter the baby is born. In some cases, the baby becomes turned in the uterus before labor. If this occurs, the feet or buttocks appear first. Such a birth is called a *breech birth*. This can be a dangerous situation since the baby's source of oxygen is being cut off as the placenta begins to separate from the mother's body.

If for any reason the baby does not begin breathing on its own, it will not be receiving enough oxygen to prevent the

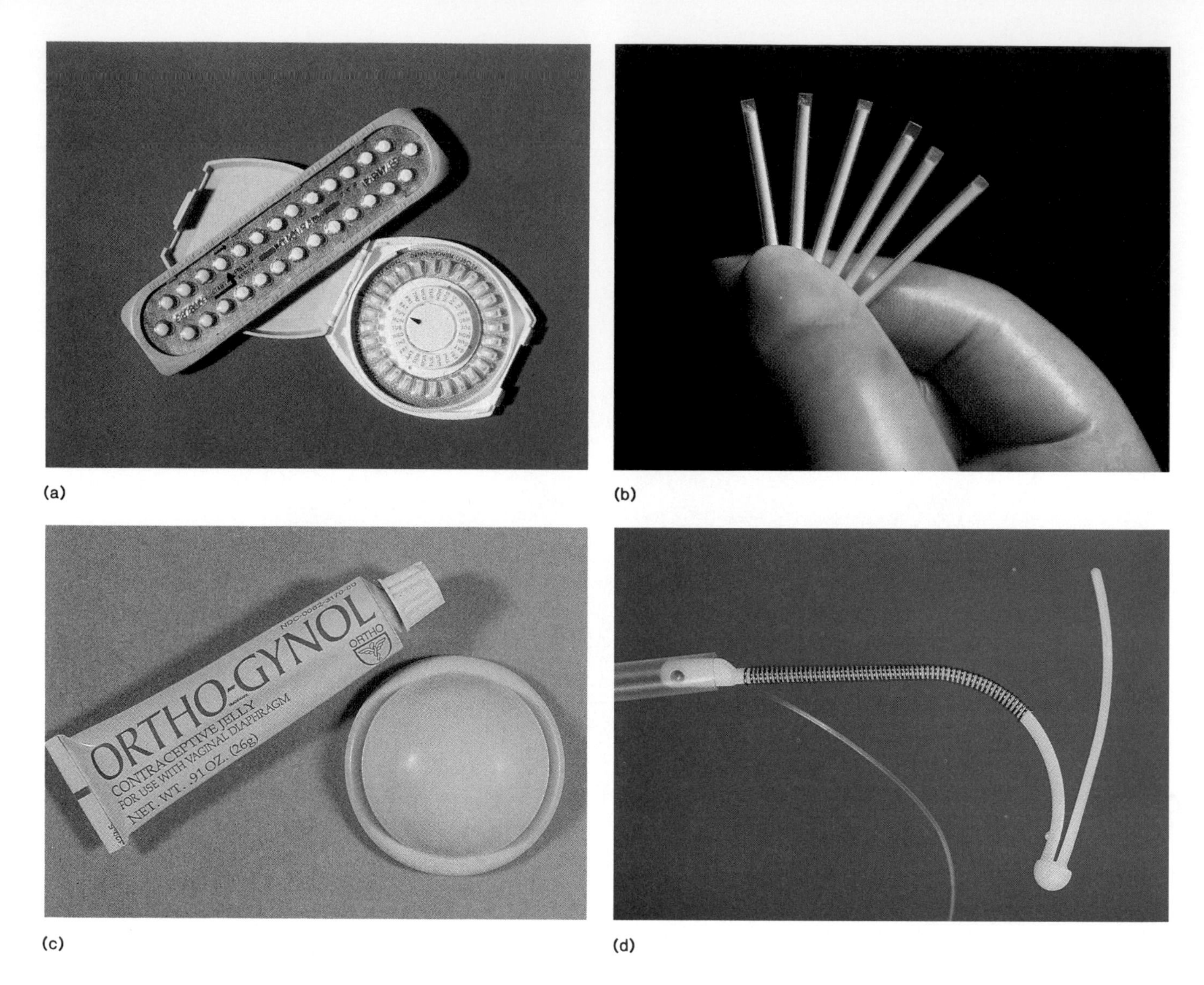

Figure 21.10 **Contraceptive Methods.** These are the primary methods of conception control used today: (a) oral contraception (pills), (b) contraceptive implants, (c) diaphragm and spermicidal jelly, (d) intrauterine device, (e) spermicidal vaginal foam, (f) contraceptive sponge, (g) male condom, and (h) female condom.

death of nerve cells; thus, brain damage or death can result. A common procedure to resolve this problem is the surgical removal of the baby through the mother's abdomen. This procedure is known as a cesarean, or C-section.

Following the birth of the baby, the placenta, also called the *afterbirth,* is expelled. Once born, the baby begins to function on its own. The umbilical cord collapses and the baby's lungs, kidneys, and digestive system must now support all bodily needs. This change is quite a shock, but the baby's loud protests fill the lungs with air and stimulate breathing.

Over the next few weeks, the mother's body returns to normal, with one major exception. The breasts, which have undergone changes during the period of pregnancy, are ready to produce milk to feed the baby. Following birth, progesterone stimulates the production of milk, and oxytocin stimulates its release. If the baby is breast-fed, the stimulus of the baby's sucking will prolong the time during which milk is produced.

In some cultures, breast-feeding continues for two to three years and the continued production of milk often delays the reestablishment of the normal cycles of ovulation and menstruation. Many people believe that a woman cannot become pregnant while she is nursing a baby. However, because there is so much variation among women, relying on this as a natural conception-control method is not a good choice. Many women have been surprised to find themselves pregnant again a few months after delivery.

Contraception

Throughout history people have tried various methods of conception control (figure 21.10). In ancient times, conception control was encouraged during times of food shortage or when tribes were on the move from one area to another in

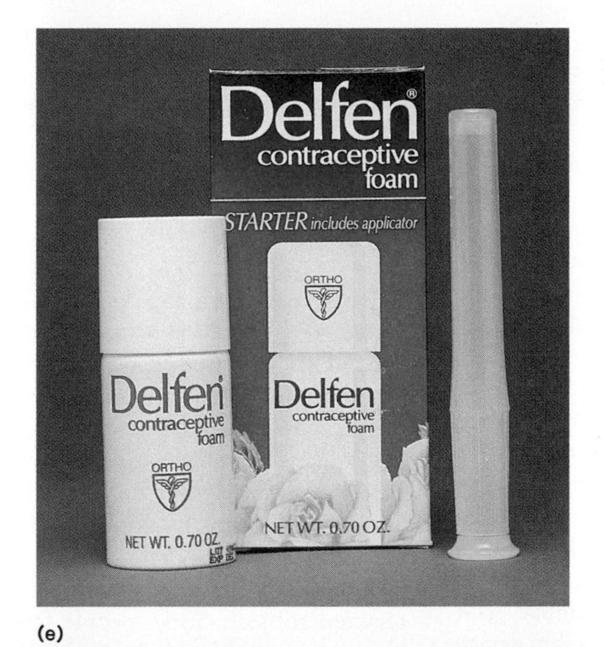

(e)

(f)

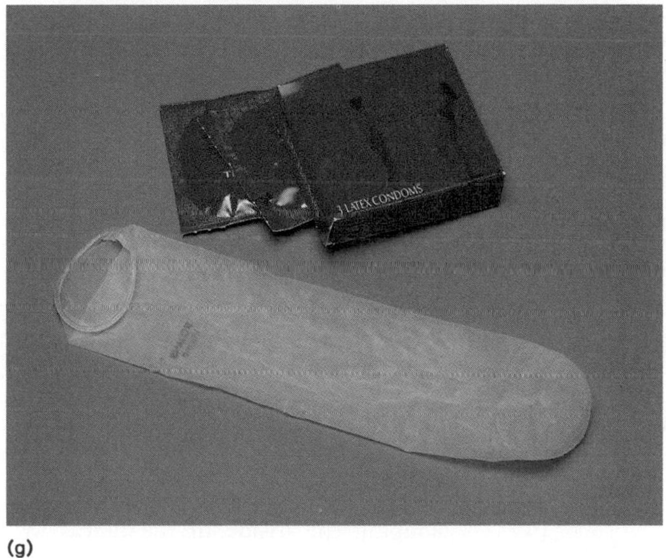

(g)

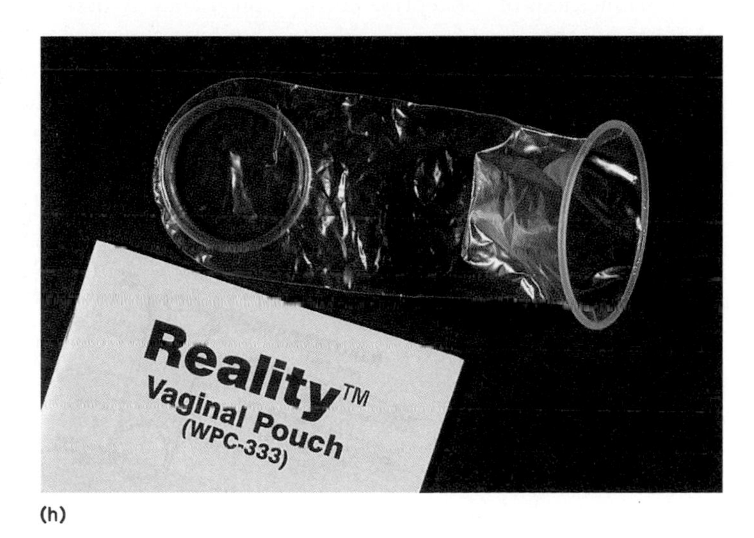

(h)

Figure 21.10 **Contraceptive Methods.** (continued)

search of a new home. Writings as early as 1500 B.C. indicate that the Egyptians used a form of tampon medicated with the ground powder of a shrub to prevent fertilization. This may sound primitive, but we use the same basic principle today to destroy sperm in the vagina.

Contraceptive jellies and foams make the environment of the vagina more acidic, which diminishes the sperm's chances of survival. The spermicidal (sperm killing) foam or jelly is placed in the vagina before intercourse. When the sperm make contact with the acidic environment, they stop and soon die. Aerosol foams are an effective method of conception control, but interfering with the hormonal regulation of ovulation is more effective. The first successful method of hormonal control was "the pill." One of the newest methods of conception control also involves hormones. The hormones are contained within small rods or capsules, which are placed under a woman's skin. These rods, when properly implanted, slowly release hormones and prevent the maturation and release of oocytes from the follicle. The major advantage of the implant is its convenience. Once the implant has been inserted, the woman can forget about contraceptive protection for several years.

Killing sperm or preventing ovulation are not the only methods of preventing conception. Any method that prevents the sperm from reaching the oocyte prevents conception. One method is to avoid intercourse during those times of the month when a secondary oocyte may be present. This is known as the *rhythm method* of conception control. While at first glance it appears to be the simplest and least expensive, determining just when a secondary oocyte is likely to be present can be very difficult. If a woman has an irregular menstrual cycle, there may be only a few days each month for intercourse without fear of pregnancy. In addition to calculating safe days based on the length of the menstrual cycle, a woman can better estimate the time of ovulation by keeping a record of changes in her body temperature and vaginal pH. Both of these changes are tied to the menstrual cycle and can therefore help a woman predict ovulation. In particular, at about the time of ovulation, a woman has a slight rise in body temperature (less than 1° C). Thus, one should use an extremely sensitive thermometer. (There is even a digital-readout thermometer on the market that spells out the word yes or no.)

Other methods of conception control that prevent the sperm from reaching the secondary oocyte include the diaphragm, cap, sponge, and condom. The diaphragm is a specially fitted membranous shield that is inserted into the vagina before intercourse and positioned so that it covers the opening of the uterus. Because of anatomical differences among females, diaphragms must be fitted by a physician. The effectiveness of the diaphragm is increased if spermicidal foam or jelly is also used. The vaginal cap functions in a similar way. The contraceptive sponge, as the name indicates, is a small amount of absorbent material that is soaked in a spermicide. The sponge is placed within the vagina, and chemically and physically prevents the sperm cells from reaching the oocyte.

The male condom is probably the most popular contraceptive device. It is a thin sheath that is placed over the erect penis before intercourse. In addition to preventing sperm from reaching the secondary oocyte, this physical barrier also helps to prevent sexually transmitted diseases (STDs), such as syphilis, gonorrhea, and AIDS, from being passed from one person to another during sexual intercourse (box 21.2). The most desirable condoms are made of a thin layer of latex that does not reduce the sensitivity of the penis. The condom is most effective if it is prelubricated with a spermicidal material such as nonoxynol-9. This lubricant also has the advantage of providing some protection against the spread of the HIV virus. Recently developed condoms for women are now available for use. One called the Femidom is a polyurethane sheath that, once inserted, lines the contours of the woman's vagina. It has an inner ring that sits over the cervix and an outer ring that lies flat against the labia. Research shows that this device protects against STDs and is as effective a contraceptive as the condom used by men.

The intrauterine device (IUD) is not a physical barrier that prevents the gametes from uniting. How this device works is not completely known. It may in some way interfere with the implantation of the embryo. The IUD must be fitted and inserted into the uterus by a physician, who can also remove it if pregnancy is desired. One such device has been shown to be dangerous, and injured women have collected damages from the company that developed it. As a result of the legal action, many American physicians are less willing to suggest these devices for their patients. However, IUDs continue to be used successfully in many countries. Current research with new and different intrauterine implants indicates that they are able to prevent pregnancy, and one is currently available in the United States.

Two contraceptive methods that require surgery are tubal ligation and vasectomy (figure 21.11). Tubal ligation involves the cutting and tying off of the oviducts and can be done on an outpatient basis in most cases. Ovulation continues as usual, but the sperm and egg cannot unite. Vasectomy can be performed in a physician's office and does not require hospitalization. A small opening is made above the scrotum, and the spermatic cord (vas deferens) is cut and tied. This prevents sperm from moving through the ducts to the outside. Because most of the sperm-carrying fluid, **(semen),** is produced by the seminal vesicles, prostate gland, and bulbourethral glands, a vasectomy does not interfere with normal ejaculation. The sperm that are still being produced die and are reabsorbed in the testes. Neither tubal ligation nor vasectomy interferes with normal sex drives. However, these medical procedures are generally not reversible and should not be considered by those who may want to have children at a future date. The effectiveness of various contraceptive methods is summarized in table 21.2.

Abortion

Another medical procedure often associated with birth control is abortion, which has been used throughout history. Abortion involves various medical procedures that cause the death and removal of the developing embryo. Abortion is obviously not a method of conception control; rather, it prevents the normal development of the embryo and causes its death. Abortion is a highly charged subject. Some people feel that abortion should be prohibited by law in all cases. Others feel that abortion should be allowed in certain situations, such as in pregnancies that endanger the mother's life or in pregnancies that are the result of rape or incest. Still others feel that abortion should be available to any woman under any circumstances. Regardless of the moral and ethical issues that surround abortion, it is still a common method of terminating unwanted pregnancies.

The abortion techniques used in the United States today all involve the possibility of infections, particularly if done by poorly trained personnel. The three most common techniques are scraping the inside of the uterus with special instruments (called a *D and C* or *dilation and curettage*), injecting a saline solution into the uterine cavity, or using a suction device to remove the embryo from the uterus. In the future, abortion may be accomplished by a medication prescribed by a physician. One drug, RU-486, is currently used in about 15% or more of the elective abortions in France. It has received approval for use in the United States. The medication is administered orally under the direction of a physician, and several days later, a hormone is administered. This usually results in the onset of contractions that expel the fetus. A follow-up examination of the woman is made after several weeks to ensure that there are no serious side effects of the medication.

BOX 21.2 *Sexually Transmitted Diseases*

Diseases currently referred to as *sexually transmitted diseases* (*STDs*) were formerly called *venereal diseases* (*VDs*). The term *venereal* is derived from the name of the Roman goddess for love, Venus. Although these kinds of illnesses are most frequently transmitted by sexual activity, many can also be spread by other methods of direct contact, by hypodermic needles, by blood transfusions, and by blood-contaminated materials. Currently the Centers for Disease Control and Prevention in Atlanta, Georgia, recognizes eighteen diseases as being sexually transmitted and a nineteenth, gay bowel syndrome, which is actually caused by a great variety of different microorganisms.

Some of the most important STDs are described here because of their high incidence in the population and our inability to bring some of them under control. For example, there is no known cure for the HIV virus that is responsible for AIDS. There has also been a sharp rise in the number of gonorrhea cases in the United States caused by a form of the bacterium *Neisseria gonorrhoeae* that has become resistant to the drug penicillin by producing an enzyme that actually destroys the antibiotic. However, most of the infectious agents can be controlled if diagnosis occurs early and treatment programs are carefully followed by the patient. The spread of STDs during sexual intercourse is significantly diminished by the use of condoms. Other types of sexual contact (i.e., hand, oral, anal) and congenital transmission (i.e., from the mother to the fetus during pregnancy) help to maintain some of these diseases in the population at high enough levels to warrant attention by public health officials, the United States Public Health Service, the Centers for Disease Control and Prevention, and state and local public health agencies. All of these agencies are involved in attempts to raise the general public health to a higher level. Their investigations have resulted in the successful control of many diseases and the identification of special problems, such as those associated with the STDs.

Members of all public health agencies are responsible for warning the public about things that may be dangerous to them. In order to meet these obligations when dealing with sexually transmitted diseases, such as AIDS and syphilis, they encourage the use of one of their most potent weapons, sex education. Individuals must know about their own sexuality if they are to understand the transmission and nature of STDs. Then it will be possible to alter their behavior in ways that will prevent the spread of these diseases. The intent is to present people with biological facts, not to scare them. Public health officials do not have the luxury of advancing their personal opinions when it comes to their jobs. The biological nature of sexual behavior is not a moral issue, but biological facts are needed if people are to make intelligent decisions relating to their sexual behavior. It is hoped that through education, people will alter their high-risk sexual behaviors and avoid situations where they could become infected with one of the STDs. As one health official stated, we should be knowledgeable enough about our own sexuality and the STDs to answer the question, Is what I'm about to do worth dying for?

Disease	Agent
Genital herpes	Virus
Gonorrhea	Bacterium
Syphilis	Bacterium
Acquired Immune Deficiency Syndrome (AIDS)	Virus
Candidiasis	Yeast
Chancroid	Bacterium
Condyloma acuminatum (venereal warts)	Virus
Gardnerella vaginalis	Bacterium
Genital *Chlamydia* infection	Bacterium
Genital Cytomegalovirus infection	Virus
Genital *Mycoplasma* infection	Bacterium
Group B *Streptococcus* infection	Bacterium
Nongonococcal urethritis	Bacterium
Pelvic inflammatory disease (PID)	Bacterium
Reiter's syndrome	Bacterium
Crabs	Body lice
Scabies	Mite
Trichomoniasis	Protozoan
Viral hepatitis (HBV)	Virus
Gay bowel syndrome	Variety of agents

Sexual Function in the Elderly

At about the age of fifty, a woman's hormonal balance begins to change due to changes in the production of hormones by the ovaries. At this time, the menstrual cycle becomes less regular and ovulation is often unpredictable. The changes in hormone levels cause many women to experience mood swings and physical symptoms, including cramps and hot flashes. This period when the ovaries stop producing viable secondary oocytes and the body becomes nonreproductive is known as the **menopause.** Occasionally the physical impairment becomes so severe that it interferes with the normal life and the enjoyment of sexual activity, and a physician might recommend hormonal treatment to augment the natural production of hormones. Normally the sexual enjoyment of a healthy woman continues during the time of menopause and for many years thereafter.

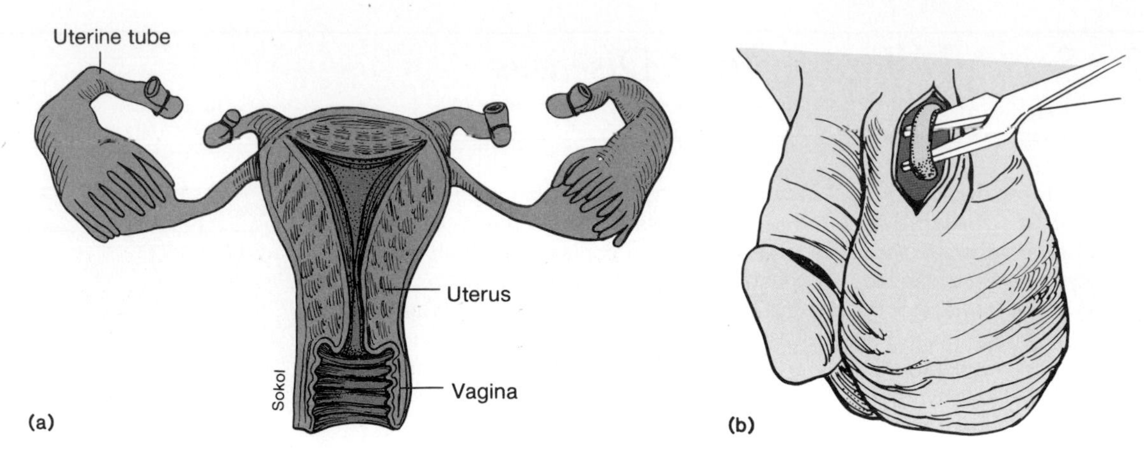

Figure 21.11 **Tubal Ligation and Vasectomy.** Two very effective contraceptive methods require surgery. Tubal ligation (a) involves severing the oviducts and suturing or sealing the cut ends. This prevents the sperm cell and the secondary oocyte from meeting. This procedure is generally considered ambulatory surgery, or at most requires a short hospitalization period. Vasectomy (b) requires minor surgery, usually in a clinic under local anesthesia. Following the procedure, minor discomfort may be experienced for several days. The severing and sealing of the vas deferens prevents the release of sperm cells from the body by ejaculation.

Human males do not have a cyclic release of sex cells; therefore, they do not experience a finale to their reproductive or sexual lives. Rather, their sexual desires tend to wane slowly as they age. They produce fewer sperm cells and less seminal fluid. Healthy individuals can experience a satisfying sex life during aging. The sex lives of older persons show the same kind of variation as that of younger people. The whole range of responses to sexual partners continues but generally in a diminished form. People who were very active sexually when young continue to be active but are less active as they reach middle age. Those who were less active tend to decrease their sexual activity also. It is reasonable to state that one's sexuality continues from before birth until death.

Table 21.2
The Effectiveness of Contraceptive Methods

Method	Pregnancies per 100 Women per Year[a]	
	High[b]	Low
No contraceptive[c]	80	40
Coitus interruptus	23	15
Condom[d]	17	8
Douche	61	34
Chemicals (spermicides)[e]	40	9
Diaphragm and jelly	28	11
Rhythm[f]	58	14
Pill	2	0.03
IUD	8	3
Sterilization	0.003	0

a Data describe the number of women per 100 who will become pregnant in a one-year period while using a given method.
b High and low values represent best and worst estimates from various demographic and clinical studies.
c In the complete absence of contraceptive practice, 8 out of 10 women can expect to become pregnant within one year.
d Effectiveness increases if spermicidal jelly or cream is used in addition.
e Aerosol foam is considered to be the best of the chemical barriers.
f Use of a clinical thermometer to record daily temperatures increases effectiveness.

• Summary •

The human sex drive is a powerful motivator for many activities in our lives. While it provides for reproduction and improvement of the gene pool, it also has a nonbiological dimension. Sexuality begins before birth, as gender is determined by the sex-determining chromosome complement that we receive at fertilization. Females receive two X sex-determining chromosomes. Only one of these is functional; the other remains tightly coiled as a Barr body. A male receives one X and one Y sex-determining chromosome. It is the presence of the Y chromosome that causes male development.

At puberty, hormones influence the development of secondary sex characteristics and the functioning of gonads. At this time fertilization is possible.

Sexual reproduction involves the production of gametes by meiosis in the ovaries and testes. The production and release of these gametes is controlled by the interaction of hormones. In males, each cell that undergoes spermatogenesis results in four sperm; in females, each cell that undergoes oogenesis results in one oocyte and two polar bodies. Humans have specialized structures for the support of the developing embryo, and many factors influence its development in the uterus. Successful sexual reproduction depends on proper hormone balance, proper meiotic division, fertilization, placenta formation, proper diet of the mother, and birth. Hormones regulate ovulation and menstruation and may also be used to encourage or discourage ovulation. Fertility drugs and birth-control pills, for example, involve hormonal control. In additon to the pill, a number of contraceptive methods have been developed, including the diaphragm, condom, IUD, spermicidal jellies and foams, the sponge, tubal ligation, and vasectomy.

Hormones continue to direct our sexuality throughout our lives. Even after menopause, when fertilization and pregnancy are no longer possible for a female, normal sexual activity can continue in both men and women.

• Thinking Critically •

A great world adventurer discovered a tribe of women in the jungles of Brazil. After many years of very close study and experimentation, he found that sexual reproduction was not possible. He also noticed that the female children resembled their mothers to a great degree and found that all the women had a gene that prevented meiosis. Ovulation occurred as usual, and pregnancy lasted nine months. The mothers nursed their children for three months after birth and became pregnant the next month. This cycle was repeated in all the women of the tribe.

Consider the topics of meiosis, mitosis, sexual reproduction, and regular hormonal cycles in women, and explain in detail what may be happening in this tribe.

• Experience This •

The continuation of the human species is based on sexual reproduction. Yet countless millions know more about how an automobile engine works than how their own bodies function. Because of this, many school systems have either expressed a desire to introduce sex-education programs into their curricula or have made commitments to do so. Where does the school system in your community stand with regard to sex education? What do they see as the pros and cons of such a program? At what age is the topic introduced to students? What do you think the content of such a course should include?

• Questions •

1. Describe the processes that cause about 50% of babies to be born male and 50% female.
2. What are the effects of the secretions of the pituitary, the gonads, and the adrenal glands at puberty?
3. What structures are associated with the human female reproductive system? What are their functions?
4. What structures are associated with the human male reproductive system? What are their functions?
5. What are the differences between oogenesis and spermatogenesis in humans?
6. How are ovulation and menses related to each other?
7. What changes occur in ovulation and menstruation during pregnancy?
8. What are the functions of the placenta?
9. Describe the methods of conception control.
10. List the events that occur as an embryo matures.

• Chapter Glossary •

androgens (an'dro-jenz) Male sex hormones produced by the testes that cause the differentiation of the internal and external genital anatomy along male lines.

autosomes (aw'to-sōmz) Chromosomes that typically carry genetic information used by the organism for characteristics other than the primary determination of sex.

Barr bodies (bar bod'ēz) Tightly coiled X chromosomes in the cells of female mammals, described by Dr. M. L. Barr.

coitus (ko'ĕ-tus) See **sexual intercourse.**

conception (kon-sep'shun) Fertilization.

copulation (kop-yu-la'shun) See **sexual intercourse.**

corpus luteum (kōr'pus lu'te-um) Remainder of the follicle after the release of the secondary oocyte. It develops into a glandlike structure that produces hormones (progesterone and estrogen) that prevent the release of other eggs.

cryptorchidism (krip-tōr'kĭ-dizm) A developmental condition in which the testes do not migrate through the inguinal canal.

differentiation (dif-ĕ-ren"she-a'shun) Specialization of embryonic cells.

ejaculation (e-jak"u-lā'shun) The release of sperm cells and seminal fluid through the penis of a male.

estrogens (es'tro-jens) Female sex hormones that cause the differentiation in the female embryo of the internal and external genital anatomy along female lines; responsible for the changes in breasts, vagina, uterus, clitoris, and pelvic bone structure at puberty.

follicle (fol'ĭ-kul) The saclike structure near the surface of the ovary that encases the soon-to-be-released secondary oocyte.

follicle-stimulating hormone (FSH) (fol'ĭ-kul stim'yu-lā-ting hōr'mōn) The pituitary secretion that causes the ovaries to begin to produce larger quantities of estrogen and to develop the follicle and prepare the egg for ovulation.

gametogenesis (gă-me"to-jen'ĕ-sis) The generating of gametes; the meiotic cell-division process that produces sex cells; oogenesis and spermatogenesis.

gonads (go'nadz) Sex organs, such as ovaries or testes.

hormone (hōr'mōn) A chemical substance that is released from glands in the body to regulate other parts of the body.

hypothalamus (hi"po-thal'ă-mus) The region of the brain that causes the cyclic production of hormones in females and the constant production of hormones in males.

inguinal canal (ing'gwĭ-nal că-nal') An opening in the abdominal cavity through which the testes in a human male embryo move into the scrotum.

inguinal hernia (ing'gwĭ-nal her'ne-ah) A rupture in the abdominal wall that allows a portion of the intestine to push through the abdominal wall in the area of the inguinal canal.

interstitial cell-stimulating hormone (ICSH) (in"ter-stĭ'shal sel stim'yu-lā-ting hōr'mōn) The chemical messenger molecule released from the pituitary that causes the testes to produce testosterone, the primary male sex hormone.

masturbation (măs"tur-ba'shun) Stimulation of one's own sex organs.

menopause (mĕn'o-pawz) The period beginning at about age fifty when the ovaries stop producing viable secondary oocytes.

menstrual cycle (men'stru-al si'kul) **(menses, menstrual flow, period)** The repeated building up and shedding of the lining of the uterus.

oogenesis (oh"ō-jen'ĕ-sis) The specific name given to the gametogenesis process that leads to the formation of eggs.

orgasm (or'gaz-um) A complex series of responses to sexual stimulation that result in intense frenzy of sexual excitement.

ovary (o'vah-re) The female sex organ responsible for the production of the haploid egg cells.

oviduct (o'vĭ-dukt) The tube (*fallopian tube*) that carries the oocyte to the uterus.

ovulation (ov-yu-la'shun) The cyclic release of a secondary oocyte from the surface of the ovary every twenty-eight days.

penis (pe'nis) The portion of the male reproductive system that deposits sperm in the female reproductive tract.

pituitary gland (pĭ-tu'ĭ-tĕ-re gland) The structure in the brain that controls the functioning of other glands throughout the organism.

placenta (plah-sen'tah) An organ made up of tissues from the embryo and the uterus of the mother that allows for the exchange of materials between the mother's bloodstream and the embryo's bloodstream. It also produces hormones.

polar body (po'lar bod'ē) The smaller cell formed by the unequal meiotic division during oogenesis.

primary oocyte (pri'mar-e o'o-sīt) The cell of the ovary that begins to undergo the first meiotic division in the process of oogenesis.

primary spermatocyte (pri'mar-e spur-mat'o-sīt) The diploid cell in the testes that undergoes the first meiotic division in the process of spermatogenesis.

puberty (pu'ber-te) A time in the life of a developing individual characterized by the increasing production of sex hormones, which cause it to reach sexual maturity.

secondary oocyte (sĕk'on-dĕr-e o'o-sīt) The larger of the two cells resulting from the lopsided cytoplasmic division of a primary oocyte in meiosis I of oogenesis.

secondary sex characteristics (sĕk'on-dĕr-e seks kăr-ak-tē-ris'tiks) Characteristics of the adult male or female, including the typical shape that develops at puberty: broader shoulders, heavier long-bone muscles, development of facial hair, axillary hair, and chest hair, and changes in the shape of the larynx in the male; rounding of the pelvis and breasts and changes in deposition of fat in the female.

secondary spermatocyte (sĕk'on-dĕr-e spur-mat'o-sīt) Cells in the seminiferous tubules that go through the second meiotic division, resulting in four haploid spermatids.

semen (se'men) The sperm-carrying fluid produced by the seminal vesicles, prostate glands, and bulbo-urethral glands of males.

seminal vesicle (sĕm'ĭ-nal ves'ĭ-kul) A part of the male reproductive system that produces a portion of the semen.

seminiferous tubules (sem''ĭ-nif'ur-us tūb-yūlz) Sperm-producing tubes in the testes.

sex-determining chromosome (seks de-ter'mĭ-ning kro'mo-sōm) The chromosome that is primarily responsible for the development of male or female sexual structures.

sexual intercourse (sek'shoo-al in'ter-kors) The mating of male and female; the deposition of the male sex cells, or sperm cells, in the reproductive tract of the female; also known as **coitus** or **copulation.**

sexuality (sek''shoo-al'ĭ-te) A term used in reference to the totality of the aspects—physical, psychological, and cultural—of our sexual nature.

sperm (spurm) Cells that develop from the spermatids by losing much of their cytoplasm and developing long tails; the male gamete.

spermatids (spurm'ah-tids) Haploid cells produced by spermatogenesis that change into sperm.

spermatogenesis (spur-mat-o-jen'uh-sis) The specific name given to the gametogenesis process that leads to the formation of sperm.

testes (tes'tēz) The male organs that produce sperm.

testosterone (tes-tos'tur-ōn) The male sex hormone produced in the testes that controls the secondary sex characteristics.

uterus (yu'tur-us) The organ in female mammals in which the embryo develops.

vagina (vuh-ji'nah) The passageway between the uterus and outside of the body; the birth canal.

X chromosome (eks kro'mo-sōm) The chromosome in a human female egg (and in one-half of sperm cells) that is associated with the determination of sexual characteristics.

Y chromosome (wi kro'mo-sōm) The sex-determining chromosome in one-half of the sperm cells of human males.

zygote (zi'gōt) The fertilized egg.

The Origin and Classification of Life

The origin of life on Earth has long puzzled scientists. Evidence from such fields as chemistry, physics, astrophysics, biochemistry, and biology have provided support for some hypotheses; however, none can ever be validated. We know that since the first living things appeared, there has been an explosion of numbers and kinds of organisms on Earth. Biologists have attempted to make their study of organisms easier by classifying them into five major groups. The five kingdoms of life include Prokaryotae, Protista, Mycetae, Plantae, and Animalia. While this system reflects our current understanding of life on Earth, it continues to be modified to account for new information discovered every day. ■

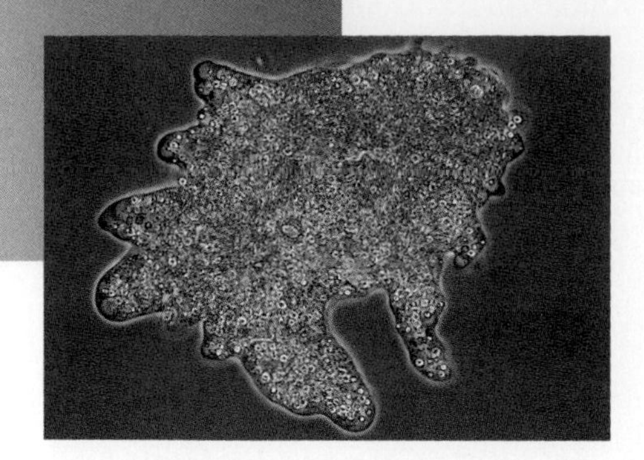

CHAPTER 22

The Origin of Life and Evolution of Cells

Chapter Outline

Purpose

This chapter deals with theories of the origin of cellular life, which began with the formation of our solar system. Early in Earth's history, inorganic materials were converted to organic materials. These organic materials have become combined into living units called cells.

For Your Information

In 1976 an unmanned spacecraft, *Viking I*, landed on the planet Mars. One of its missions was to determine if life existed on the planet. Soil samples generated carbon dioxide, which suggests that some living organisms might have been present. It is also possible that the carbon dioxide came from inorganic sources. Scientists have recently postulated that life could have developed on Mars if liquid water was available. Iron-rich clay particles would have allowed organic molecules to be concentrated on their surfaces as a prelude to the formation of primitive cell types. However, water does not exist on Mars in the liquid form, but it is found at the poles as ice caps.

Learning Objectives

- State the significance of Pasteur's experiment.
- Describe the formation of our solar system, including planet Earth.
- Describe the physical conditions on early Earth and the changes thought to have happened before life could exist.
- Beginning with the gases in the early atmosphere, trace the events that led to the first living cells.
- Explain why spontaneous generation can only occur in a reducing atmosphere.
- Explain why the evolution of autotrophs assured a continuation of life on Earth.
- Explain the evolution of eukaryotic cells.

Spontaneous Generation versus Biogenesis

For centuries curiosity has spurred humans to study the basic nature of their environment. The vast amount of chemical and biological information presented in previous chapters is evidence of our ability to gather and analyze information. These efforts have resulted in solutions to many problems and have simultaneously revealed new and more challenging areas of concern. Despite these efforts, two questions have continued to be subjects of speculation: What is the nature of life, and how did it originate?

In earlier times, no one ever doubted that life originated from nonliving things. The Greeks, Romans, Chinese, and many other ancient peoples believed that maggots arose from decaying meat; mice developed from wheat stored in dark, damp places; lice formed from sweat; and frogs originated from damp mud. The concept of **spontaneous generation**—the theory that living organisms arise from nonliving material—was widely believed until the seventeenth century (figure 22.1). However, there were some who doubted this theory. These people subscribed to an opposing theory, called *biogenesis*. **Biogenesis** is the concept that life originates only from preexisting life. One of the earliest challenges to the theory of spontaneous generation came in 1668.

Figure 22.1 **Life from Nonlife.** Many works of art explore the idea that living things could originate from very different types of organisms or even from nonliving matter. M. C. Escher's work entitled "The Reptiles, 1943" shows the life cycle of a little alligator. Amid all kinds of objects, a drawing book lies open at a drawing of a mosaic of reptilian figures in three contrasting shades. Evidently, one of them is tired of lying flat and rigid among its fellows, so it puts one plastic-looking leg over the edge of the book, wrenches itself free, and launches out into "real" life. It climbs up the back of the zoology book and works its way laboriously up the slippery slope of the set-square to the highest point of its existence. Then after a quick snort, tired but fulfilled, it goes downhill again, via an ashtray, to the level surface, to that flat drawing paper, and meekly rejoins its erstwhile friends, taking up once more its function as one element of surface division.

Francesco Redi, an Italian physician, set up a controlled experiment designed to disprove the theory of spontaneous generation (figure 22.2). He used two sets of jars that were identical except for one aspect. Both sets of jars contained decaying meat, and both were exposed to the atmosphere; however, one set of jars was covered by gauze, and the other was uncovered. Redi observed that flies settled on the meat in the open jar, but the gauze blocked their access to the covered jars. When maggots appeared on the meat in the uncovered jars but not on the meat in the covered ones, Redi concluded that the maggots arose from the eggs of the flies and not from spontaneous generation in the meat.

Even after Redi's experiment, there were still some who supported the theory of spontaneous generation. After all, a belief that has been prevalent for over two thousand years does not die a quick death. In 1748 John T. Needham, an English priest, placed a solution of boiled mutton broth in containers that he sealed with corks. Within several days, the broth became cloudy and contained a large population of microorganisms. Needham reasoned that boiling killed all the organisms and that the corks prevented any microorganisms from entering the broth. He concluded that life in the broth was the result of spontaneous generation.

In 1767 another Italian scientist, Abbe Lazzaro Spallanzani, challenged Needham's findings. Spallanzani boiled a meat and vegetable broth, placed this medium in clean glass containers, and sealed the openings by melting the glass over a flame. He placed the sealed containers in boiling water to make certain all microorganisms were destroyed. As a control, he set up the same conditions but did not seal the necks, allowing air to enter the flasks (figure 22.3). Two days later, the open containers had a large population of microorganisms, but there were none in the sealed containers.

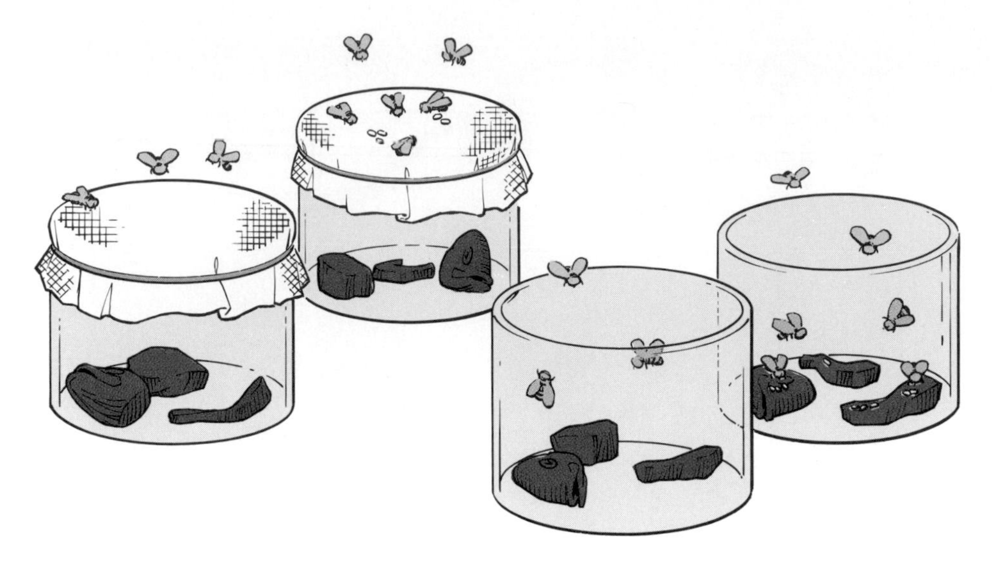

Figure 22.2 **Redi's Experiment.** The two sets of jars here are identical in every way except one—the gauze covering. The set on the left is called the control group; the set on the right is the experimental group. Any differences seen between the control and the experimental groups are the result of a single variable. In this manner, Redi concluded that the presence of maggots in meat was due to flies laying their eggs on the meat and not spontaneous generation.

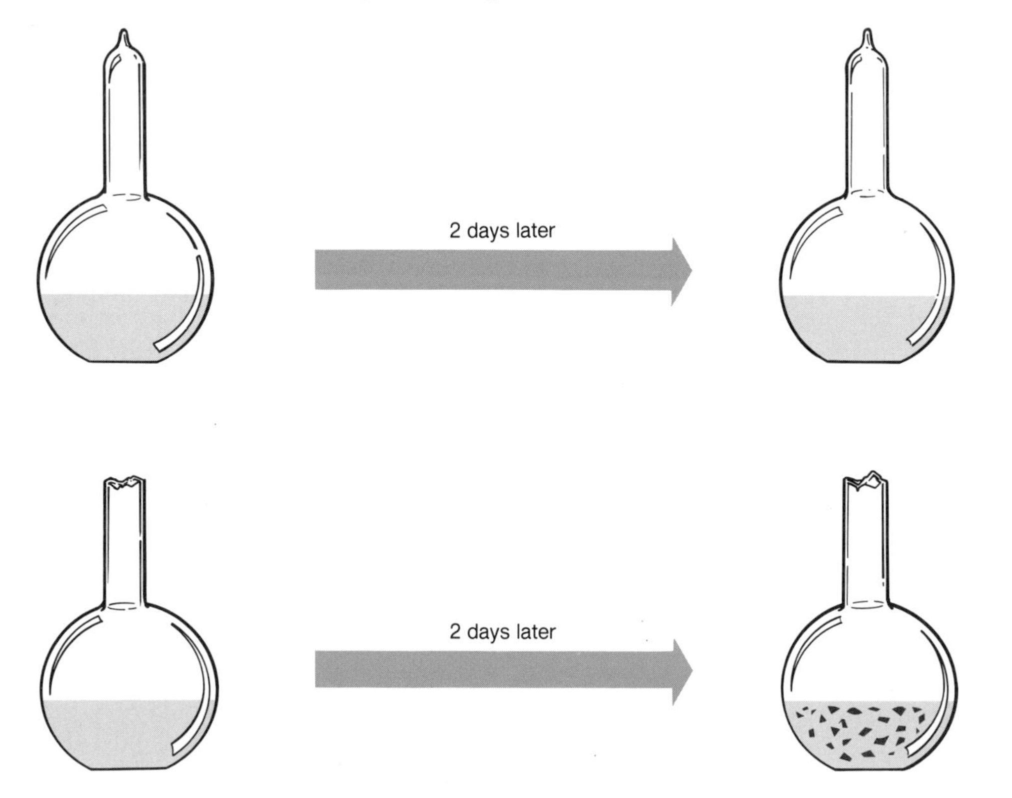

Figure 22.3 **Spallanzani's Experiment.** Spallanzani carried the experimental method of Redi one step further. He boiled a meat and vegetable broth and placed this medium into clean flasks. He sealed one and put it in boiling water. As a control, he subjected another flask to the same conditions, except he left it open. Within two days, the open flask had a population of microorganisms. Spallanzani demonstrated that spontaneous generation could not occur unless the broth was exposed to the "germs" in the air.

Spallanzani's experiment did not completely disprove the theory of spontaneous generation to everyone's satisfaction. The supporters of the theory attacked Spallanzani by stating that he excluded air, a factor believed necessary for spontaneous generation. Supporters also argued that boiling had destroyed a "vital element." When Joseph Priestly discovered oxygen in 1774, the proponents of spontaneous generation claimed that oxygen was the "vital element" that Spallanzani had excluded in his sealed containers.

In 1861 the French chemist Louis Pasteur convinced most scientists that spontaneous generation could not occur. He placed a fermentable sugar solution and yeast mixture in a flask that had a long swan neck. The mixture and the flask were boiled for a long time. The flask was left open to allow oxygen, the "vital element," to enter, but no organisms developed in the mixture. The organisms that did enter the flask settled on the bottom of the curved portion of the neck and could not reach the sugar-water mixture. As a control, he cut off the swan neck (figure 22.4). This allowed microorganisms from the air to fall into the flask, and within two days the fermentable solution was supporting a population of microorganisms. In his address to the French Academy, Pasteur stated, "Never will the doctrine of spontaneous generation arise from this mortal blow."

The Modern Theory of the Origin of Life

Pasteur's prediction regarding the theory of spontaneous generation was questioned some sixty years later. In the 1920s, a Russian biochemist, Alexander I. Oparin, and a British biologist, J. B. S. Haldane, working independently, proposed the idea of spontaneous generation in a new form.

As the name implies, spontaneous generation proposes that nonliving material is naturally converted into living material. The first advocates of the theory

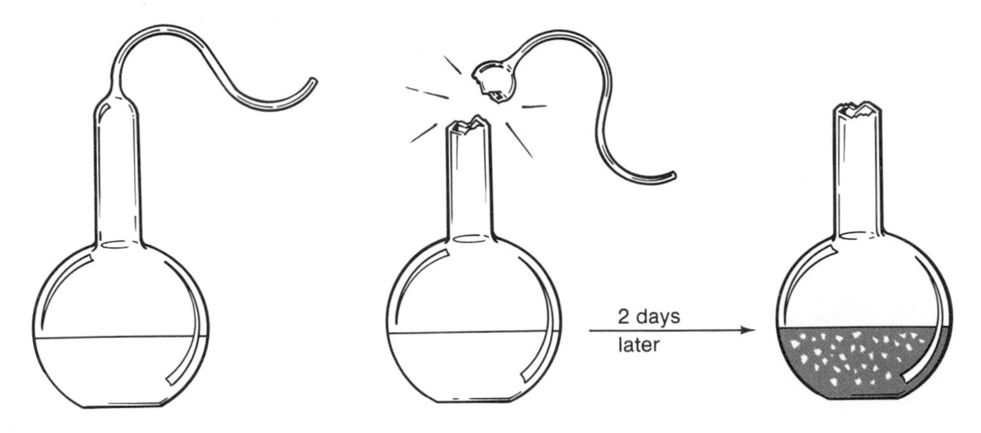

Figure 22.4 **Pasteur's Experiment.** Pasteur used the swan-neck flask that allowed oxygen, but not the airborne organisms, to enter the flask. He broke the neck off of another flask. Within two days, there was growth in this second flask. Pasteur demonstrated that germfree air with its oxygen does not cause spontaneous generation.

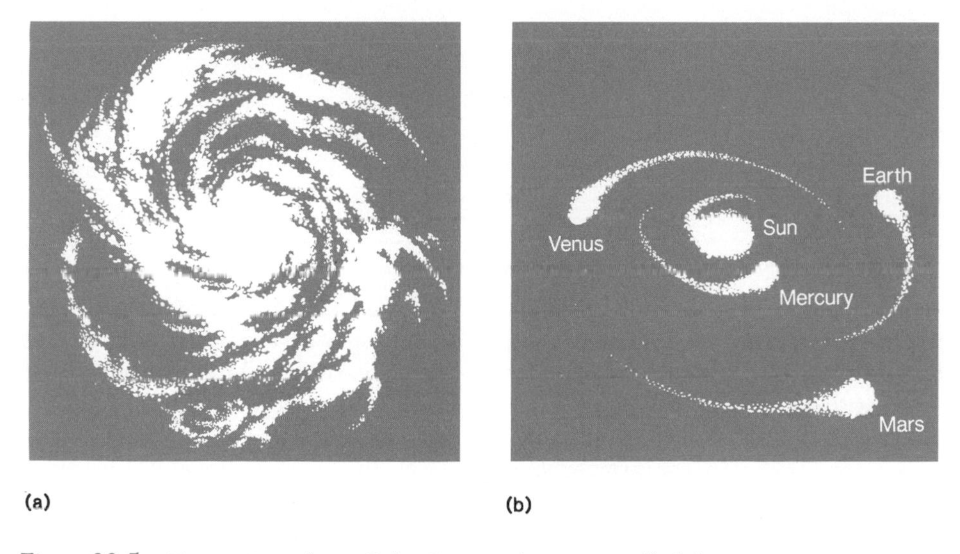

Figure 22.5 **Formation of our Solar System.** As gravity pulled the gas particles into the center, the sun developed (a). In other regions, smaller gravitational forces caused the formation of the sun's planets (b).

of spontaneous generation believed that the creation of life was happening during their lifetime. They also believed that it happened in a matter of days. However, Oparin and Haldane proposed that it required about a billion years for nonliving matter to be organized into units that could be called living. Modern supporters view this process as something that could have happened only very early in Earth's history, when physical conditions were quite different from those that exist today. In fact, Oparin and Haldane were endeavoring to answer the fundamental questions posed at the beginning of this chapter: What is the nature of life, and how did it originate?

Oparin first presented his ideas in a paper published in Moscow in 1924. However, his ideas did not come to the attention of the general scientific community because the work was never translated. A later work by Oparin, a book entitled *Origin of Life,* published in 1931 and translated into other languages, led to a renewed interest in developing theories that could account for the origin of life.

Early Earth

To understand Oparin's theory and variations of it, it is necessary to understand how our solar system and Earth are thought to have formed. The *solar nebula theory* proposes that the solar system was formed from a large cloud of gases that developed some 10 to 20 billion years ago (figure 22.5). A gravitational force was created by the collection of particles within this cloud that caused other particles to be pulled from the outer edges to the center. As gravity caused these particles to collect, they formed a very large disk with our sun at its center. The condensation of atoms into the sun resulted in the release of large amounts of thermonuclear energy. Atoms were forced to fuse with one another because of their self-generated gravitational force.

As the particles within this solar nebula were pulled into the center to form the sun, other gravitational centers developed. The material collected in these areas formed the sun's planets, including Earth. Many scientists believe that Earth was formed at least 4½ billion years ago. A large amount of heat was generated as the particles became concentrated to form Earth. While not as hot as the sun, the material of Earth formed a molten core that became encased by a thin outer crust as it cooled. In its early stages of formation, about 4 billion years ago, there may have been a considerable amount of volcanic activity on Earth (figure 22.6).

Physically, Earth was probably much different from what it is today. Because the surface was hot, there was no water on the surface or in the atmosphere. In fact, the tremendous amount of heat probably prevented any atmosphere from forming around early Earth. The gases associated with our present atmosphere (nitrogen, oxygen, carbon dioxide, and water vapor) were contained in the planet's molten core. These hostile conditions on early Earth could not have supported any form of life.

Over hundreds of millions of years, Earth slowly changed. Volcanic activity caused the release of water vapor (H_2O), carbon dioxide (CO_2), methane (CH_4), ammonia (NH_3), and

Figure 22.6 **Formation of Organic Molecules in the Atmosphere.** The environment of the primitive Earth was harsh and lifeless. But many scientists believe that it contained the necessary molecules to fashion the first living cell by the process of spontaneous generation. The energy furnished by volcanos, lightning, and ultraviolet light broke the bonds in the simple inorganic molecules in the atmosphere. New bonds formed as the atoms from the smaller molecules were rearranged and bonded to form simple organic compounds in the atmosphere. The rain carried these chemicals into the oceans. Here they reacted with each other to form more complex organic molecules.

hydrogen (H_2), and the early atmosphere was formed. These gases formed a **reducing atmosphere**—an atmosphere that does not contain molecules of oxygen (O_2). Any oxygen would have quickly combined with other atoms to form compounds, so it would have been highly unlikely that a significant quantity of molecular oxygen was present. Further cooling enabled the water vapor in the atmosphere to condense into droplets of rain. The water ran over the land and collected to form the oceans we see today.

The First Organic Molecules

According to Oparin and Haldane, the first organic molecules were formed in this early reducing atmosphere. The molecules of water vapor, ammonia, methane, carbon dioxide, and hydrogen supplied the atoms of carbon, hydrogen, oxygen, and nitrogen necessary to form simple organic molecules. Lightning, heat from volcanoes, and ultraviolet radiation furnished the energy needed for these synthetic reactions.

After these simple organic molecules were formed in the atmosphere, they were washed from the air and carried into the newly formed oceans by the rain. Here, the molecules could have reacted with each other to form the more complex molecules of simple sugars, amino acids, and nucleic acids. This accumulation is thought to have occurred over half a billion years, resulting in oceans that were a dilute organic soup. These simple organic molecules in the ocean served as the building materials for more complex organic macromolecules, such as complex carbohydrates, proteins, lipids, and nucleic acids. But what scientific evidence is there to support these ideas?

In the early 1950s, support was growing for the idea that organic material was synthesized from inorganic material. One person who advanced this idea was Harold Urey of the University of Chicago. In 1953, one of Urey's students, Stanley L. Miller, conducted an experiment to test this idea. Miller constructed a model of early Earth (figure 22.7). In this glass apparatus he placed distilled water to represent the early oceans. The reducing atmosphere was duplicated by adding hydrogen, methane, and ammonia to the water. Electrical sparks provided the energy needed to produce organic compounds. By heating parts of the apparatus and cooling others, he simulated the rains that are thought to have fallen into the early oceans. After a week of operation, he removed some of the water from the apparatus. When this water was analyzed, it was found to contain many simple organic compounds. Although Miller demonstrated nonbiologic synthesis of simple organic molecules like amino acids and simple sugars, his results did not account for complex organic molecules like proteins and DNA.

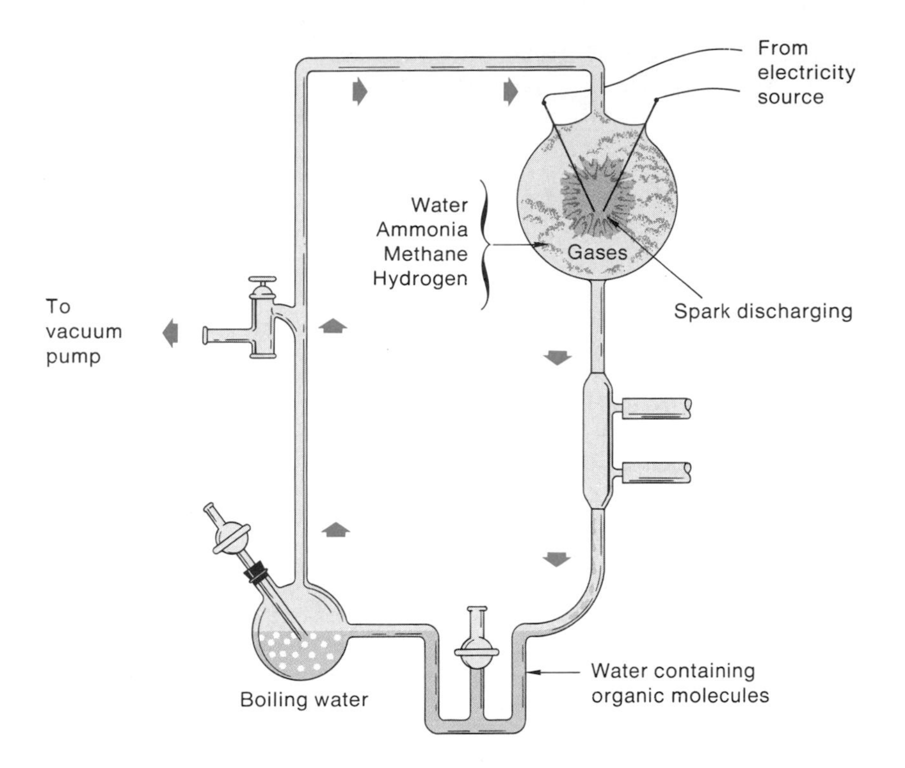

Figure 22.7 **Miller's Apparatus.** Stanley Miller developed this apparatus to demonstrate that the spontaneous formation of complex organic molecules could take place in a reducing atmosphere.

Several ideas have been proposed for the concentration of simple organic molecules and their combination into macromolecules. A portion of the early ocean could have been separated from the main ocean by geological changes. The evaportion of water from this pool could have concentrated the molecules, which might have led to the manufacture of macromolecules by dehydration synthesis. It has also been proposed that freezing may have been the means of concentration. When a mixture of alcohol and water is placed in a freezer, the water freezes solid and the alcohol becomes concentrated into a small portion of liquid. A similar process could have occurred on Earth's early surface, resulting in the concentration of simple organic molecules. In this concentrated solution, dehydration synthesis in a reducing atmosphere could have occurred, resulting in the formation of macromolecules. A third theory proposes that clay particles may have been a factor in concentrating simple organic molecules. Small particles of clay have electrical charges that can attract and concentrate organic molecules like protein from a watery solution. Once the molecules became concentrated, it would have been easier for them to interact to form larger macromolecules.

Coacervates and Microspheres

Geologists and biologists typically measure the history of life by looking back from the present. Therefore, time scales are given in "years ago." It has been estimated that the formation of simple organic molecules in the atmosphere began about 4 billion years ago and lasted approximately 1½ billion years. The oldest known fossils of living cells are thought to have formed 3½ billion years ago. The question is, How do you get from the spontaneous formation of macromolecules to primitive cells in half a billion years?

There are two theories proposed for the formation of **prebionts,** nonliving structures that led to the formation of the first living cells. Oparin speculated that a prebiont consisted of carbohydrates, proteins, lipids, and nucleic acids that accumulated to form a **coacervate.** Such a structure could have consisted of a collection of organic macromolecules surrounded by a film of water molecules. This arrangement of water molecules, while not a membrane, could have functioned as a physical barrier between the organic molecules and their surroundings.

Coacervates have been synthesized in the laboratory. They can selectively absorb chemicals from the surrounding water and incorporate them into their structure. Also, the chemicals within coacervates have a specific arrangement—they are not random collections of molecules. Some coacervates contain enzymes that direct a specific type of chemical reaction. No one claims coacervates are alive, since they lack a definite membrane, but they do exhibit some lifelike traits: they are able to grow and divide if the environment is favorable.

More recently, it has been suggested that a possible prebiotic structure could have been a *microsphere.* A **microsphere** is a nonliving collection of organic macromolecules with a double-layered outer boundary. **Proteinoids** are proteinlike structures consisting of branched chains of amino acids. Proteinoids are formed by the dehydration synthesis of amino acids at a temperature of 180° C. Sidney Fox, from the University of Miami, showed that it was feasible to combine single amino acids into polymers of proteinoids. He also demonstrated the ability to build microspheres from these proteinoids.

Microspheres can be formed when proteinoids are placed in boiling water and slowly allowed to cool. Some of the proteinoid material produces a double-boundary structure that encloses the microsphere. Although these walls do not contain lipids, they do exhibit some membranelike characteristics and suggest the structure of a cellular membrane. Microspheres swell or shrink depending upon the osmotic potential in the surrounding solution. They also display a type of internal movement (streaming) similar to that exhibited by cells and contain some proteinoids that function as enzymes. Using ATP as a source of energy, microspheres can direct the formation of polypeptides and nucleic acids. They can absorb material from the surrounding medium and form buds, which results in a second generation of microspheres. Given these characteristics, some investigators believe that microspheres can be considered **protocells,** the first living cells.

Other scientists throughout the world have replicated Miller and Fox's experiments and have built upon their findings. Many different combinations of gases have been shown to furnish the raw materials necessary to produce organic compounds when supplied with a suitable source of energy. But these experiments have one thing in common—they only work in a reducing atmosphere. It is widely believed that spontaneous generation did happen, but that it could only occur under the past conditions of a reducing atmosphere and not in today's oxygen-rich atmosphere. The laboratory synthesis of coacervates and microspheres helps us understand how the first primitive living cells might have developed. However, it leaves a large gap in our understanding because it does not explain how these first cells might have become the highly complex living cells we see today.

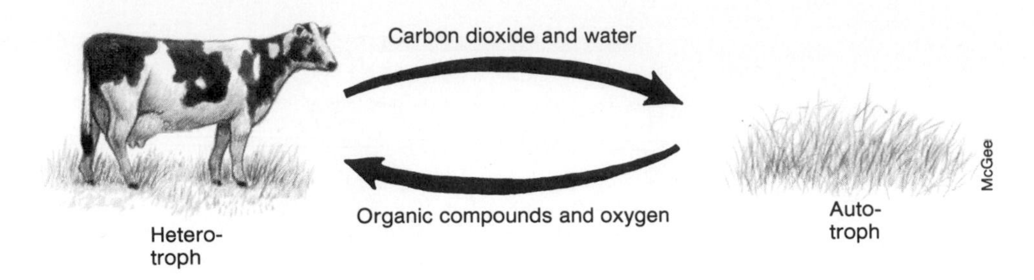

Figure 22.8 **Autotroph–Heterotroph Interrelations.** Autotrophs and heterotrophs have developed an interdependence. Today, the autotrophs depend on the inorganic waste products of the heterotrophs—water and carbon dioxide—for the raw material with which they produce the organic compounds and oxygen essential for life. The heterotrophs also depend on the organic material produced by the autotrophs as material for energy and structural growth.

Heterotrophs to Autotrophs

Regardless of how it developed, the first living thing was probably the result of spontaneous generation. Fossil evidence indicates that there were primitive forms of life on Earth about 3½ billion years ago. These first primitive cells were thought to be **heterotrophs,** which require a source of organic material from their environment. Since the early heterotrophs are thought to have evolved in a reducing atmosphere that lacked oxygen, they were of necessity anaerobic organisms; therefore, they did not obtain the maximum amount of energy from the organic molecules in the environment. At first, this would not have been a problem. The organic molecules that had been accumulating in the ocean for millions of years served as an ample source of organic material for the heterotrophs. However, as the population of heterotrophs increased through reproduction, the supply of organic material would have been consumed faster than it was being spontaneously produced in the atmosphere. If there was no other source of organic compounds, the heterotrophs would have eventually exhausted their nutrient supply, and they would have become extinct.

Even though the early heterotrophs probably contained nucleic acids and were capable of producing enzymes that could regulate chemical reactions, they probably carried out a minimum of biochemical activity. There is evidence to suggest that a wide variety of compounds were present in the early oceans, some of which could have been used unchanged by the heterotrophs. There was no need for the heterotrophs to modify the compounds to meet their needs.

Those compounds that could be easily used by heterotrophs would have been the first to become depleted from the early environment. However, some of the heterotrophs may have contained a mutated form of nucleic acid, which allowed them to convert material that was not directly usable into a compound that could be used. Heterotrophs with this mutation could have survived, while those without it would have become extinct as the compounds they used for food became scarce. It has been suggested that through a series of mutations in the early heterotrophs, a more complex series of biochemical reactions originated within some of the cells. Such cells could use chemical reactions to convert ingestible chemicals into usable organic compounds. Possibly because of these kinds of mutations, new metabolic pathways evolved that led to the evolution of **autotrophs.**

Although it required over a billion years after the formation of Earth, once the heterotrophs had been formed by spontaneous generation and the autotrophs had evolved as a result of mutations, the pattern of life on Earth as it exists today was established (figure 22.8). Table 22.1 presents a summary of events.

An Oxidizing Atmosphere

Ever since its formation, Earth has undergone constant change. In the beginning, it was too hot to support an atmosphere. Later, as it cooled and gases issued forth from volcanoes, a reducing atmosphere was formed. The emergence of heterotrophic cells led to the development of autotrophic cells. The metabolic activities of cells resulted in the release of waste products that changed Earth's environment. One of the most significant changes was the development of an **oxidizing atmosphere,** which contains molecular oxygen. The develop-

Table 22.1
Spontaneous Generation and the Evolution of Autotrophs

1. The conversion of inorganic compounds in the atmosphere into organic compounds.
2. The accumulation of these organic compounds in the oceans.
3. The concentration of some of these organic compounds into prebionts.
4. The formation of the heterotrophic protocells.
5. The evolution of the autotrophic cells.

ment of an oxidizing atmosphere created an environment unsuitable for spontaneous generation of organic molecules and life: organic molecules tend to break down (oxidize) when oxygen is available to accept electrons. However, the presence of molecular oxygen opened the door for the evolution of aerobic organisms.

As the result of autotrophic activity, an oxidizing atmosphere began to develop about 2 billion years ago. Although various chemical reactions released small amounts of molecular oxygen into the atmosphere, it was photosynthesis that generated most of the oxygen. Since oxygen is a highly reactive atom, the environment of Earth was now unfit for spontaneous generation of life. The oxygen molecules also reacted with one another to form ozone (O_3). Ozone collected in the upper atmosphere and acted as a screen to prevent most of the ultraviolet light from reaching Earth's surface. The lack of ultraviolet light diminished the spontaneous formation of complex organic molecules. It also reduced the number of mutations in primitive cells that had produced a great variety of cellular life forms. In an oxidizing atmosphere, it was no longer possible for organic molecules to accumulate over millions of years to be later incorporated into living material.

The appearance of oxygen in the atmosphere also allowed for the evolution of aerobic respiration. Since the first heterotrophs were of necessity anaerobic organisms, they did not derive large amounts of energy from the organic materials available as food. With the evolution of aerobic heterotrophs, there could be a much more efficient conversion of food into usable energy. Aerobic organisms would have a significant advantage over anaerobic organisms: they could use the newly generated oxygen as a final hydrogen acceptor and, therefore, generate many more ATPs from the food molecules they consumed.

The Origin of Eukaryotic Cells

The early heterotrophs and autotrophs were probably simple one-celled organisms like bacteria. They were **prokaryotes** that lacked nuclear membranes and other membranous organelles, such as mitochondria, an endoplasmic reticulum, chloroplasts, and a Golgi apparatus. Present-day bacteria and blue-green bacteria are prokaryotes. The types of cells found in all other forms of life are **eukaryotes,** possessing a nuclear membrane and other membranous organelles. Biologists generally believe that the eukaryotes evolved from the prokaryotes.

The **endosymbiotic theory** attempts to explain this evolution. This theory suggests that present-day eukaryotic cells evolved from the combining of several different types of primitive cells. It is thought that some organelles found in eukaryotic cells may have originated as free-living prokaryotes. Since mitochondria and chloroplasts contain bacterialike DNA and ribosomes, control their own reproduction, and synthesize their own enzymes, it has been suggested that they originated as free-living prokaryotic bacteria. These bacterial cells could have established a symbiotic relationship with another primitive nuclear-membrane-containing cell type (figure 22.9).

If these cells adapted to one another and were able to survive and reproduce better as a team, it is possible that a relationship may have evolved into present-day eukaryotic cells. If this relationship had included only a nuclear-membrane-containing cell and aerobic bacteria, the newly evolved cell would have been similar to present-day heterotrophic protozoa, fungi, and animal cells. If this relationship had included both aerobic bacteria and photosynthetic bacteria, the newly formed cell would have been similar to present-day autotrophic algae and plant cells.

Regardless of the type of cell (prokaryotic or eukaryotic), or whether the organisms are heterotrophic or autotrophic, all organsims have a common basis. DNA is the universal genetic material; protein serves as structural material and enzymes, and ATP is the source of energy. Although there is a wide variety of organisms, they all are built from the same basic molecular building blocks. Therefore, it is probable that all life derived from a single origin and that the variety of living things seen today evolved from the first protocells. In this chapter we have studied how this protocell originated. It is thought that from these cells evolved the great diversity we see in living organisms today (table 22.2). The remaining chapters of this book are concerned with the study of this diversity.

Table 22.2
Timetable of Events

20 billion years ago	*Solar nebula*
4–5 billion years ago	*Sun and planets*
3.5 billion years ago	*Prokaryotes*
2.3 billion years ago	*Oxidizing atmosphere*
1.5 billion years ago	*Eukaryotes*
700 million years ago	*Multicellular organisms*

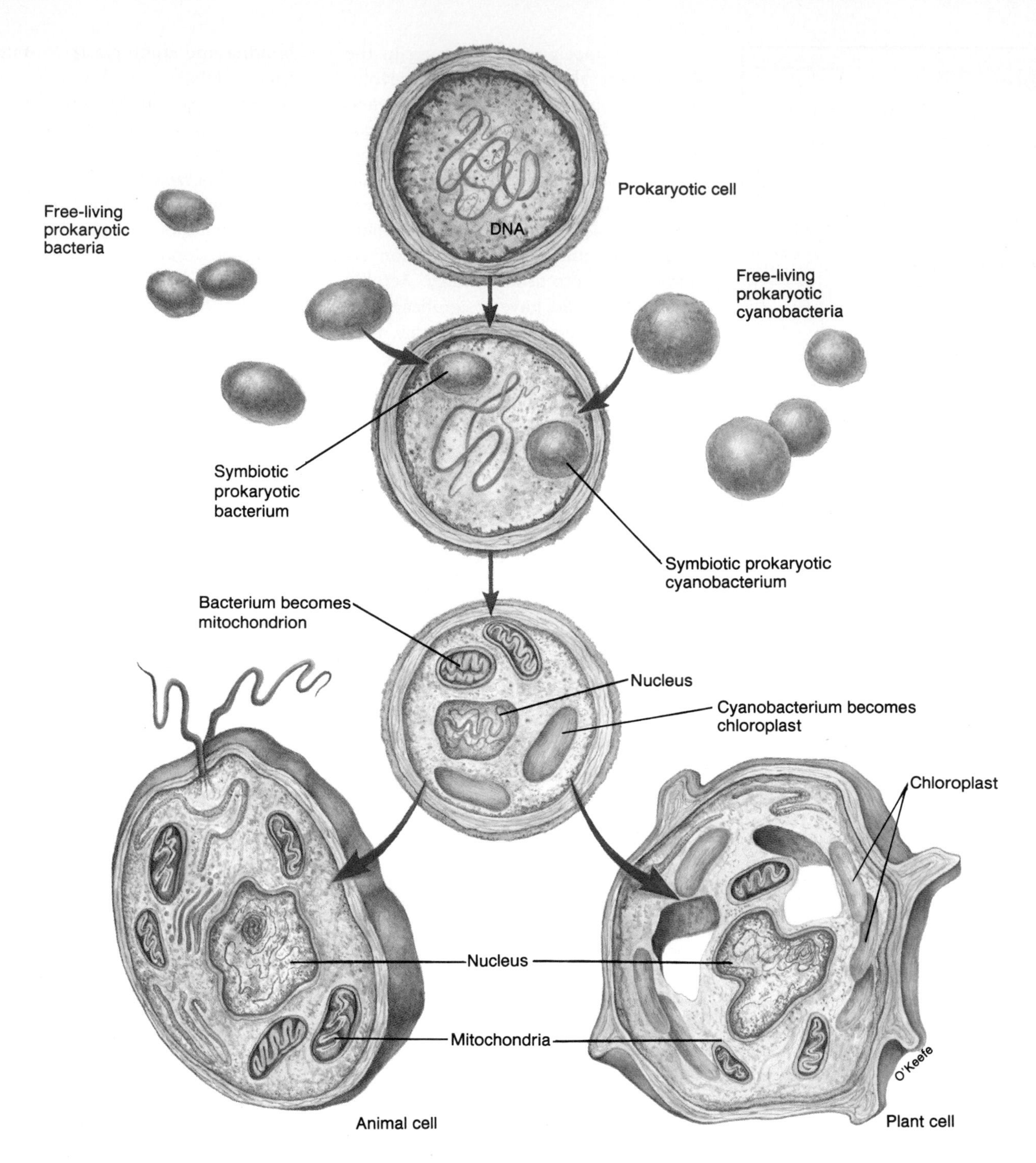

Figure 22.9 **The Endosymbiotic Theory.** This theory proposes that some free-living prokaryotic bacteria and cyanobacteria (blue-green algae) developed symbiotic relationships with a host cell. When the bacteria developed into mitochondria and the cyanobacteria developed into chloroplasts, a eukaryotic cell evolved. These cells evolved into eukaryotic plant and animal cells.

• Summary •

The centuries of research outlined in this chapter illustrate the development of our attempts to understand the origin of life. The current theory of the origin of life speculates that primitive Earth's environment led to the spontaneous organization of chemicals that became organized into primitive cells. These basic units of life then changed through time as a result of mutation and in response to a changing environment. The likelihood of these occurrences is supported by experiments that have simulated primitive Earth environments. Similarities between blue-green bacteria and chloroplasts and between bacteria and mitochondria suggest that eukaryotic cells may really be a combination of ancient cell ancestors that live together symbiotically. Despite volumes of information, the question of how life began remains unanswered. Although no one can prove how life began, a generally accepted scheme is presented in figure 22.10.

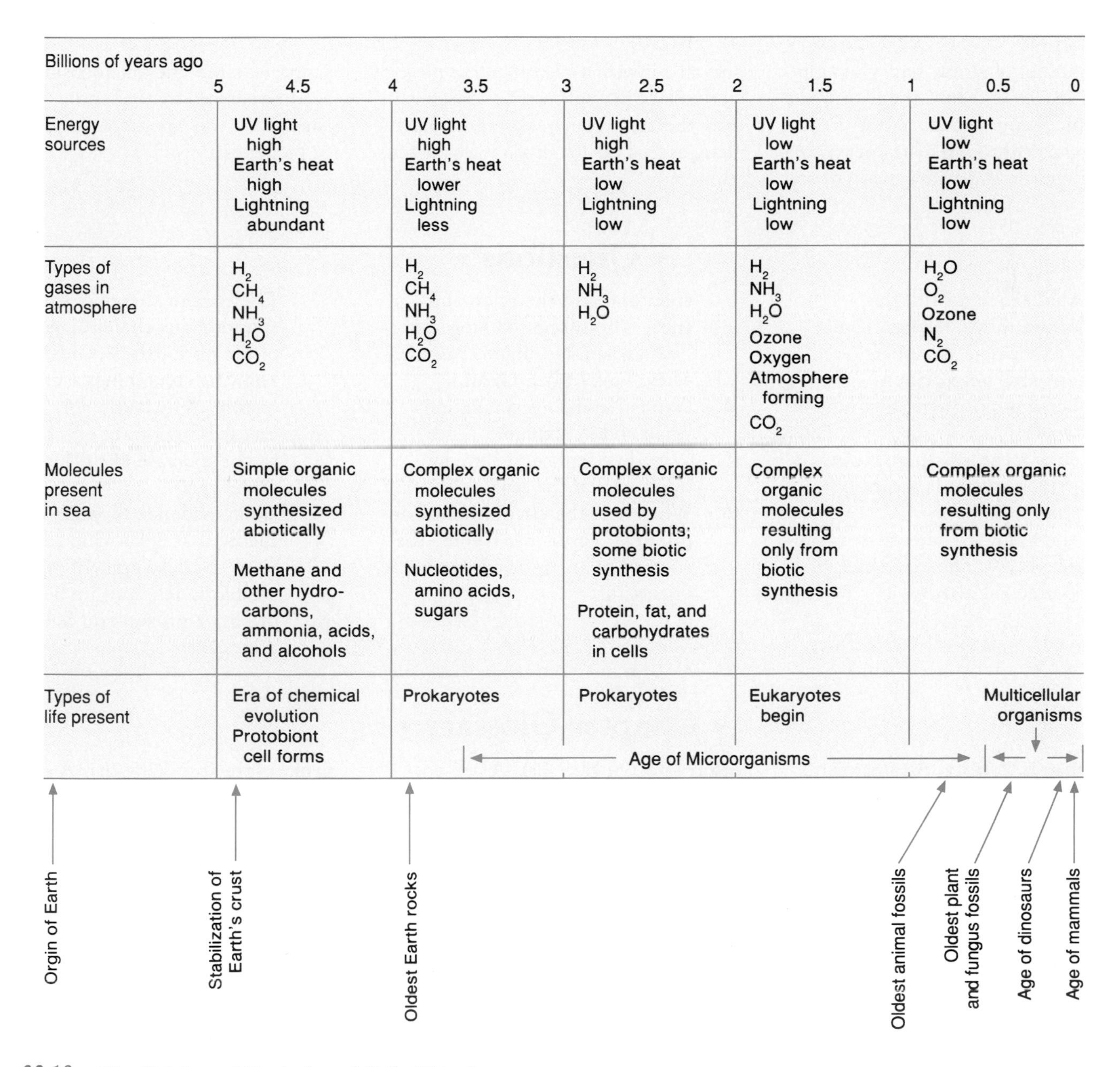

Billions of years ago
5
4.5
4
3.5
3
2.5
2
1.5
1
0.5
0
Energy sources
UV light high Earth's heat high Lightning abundant
UV light high Earth's heat lower Lightning less
UV light high Earth's heat low Lightning low
UV light low Earth's heat low Lightning low
UV light low Earth's heat low Lightning low
Types of gases in atmosphere
H_2 CH_4 NH_3 H_2O CO_2
H_2 CH_4 NH_3 H_2O CO_2
H_2 NH_3 H_2O
H_2 NH_3 H_2O Ozone Oxygen Atmosphere forming CO_2
H_2O O_2 Ozone N_2 CO_2
Molecules present in sea
Simple organic molecules synthesized abiotically
Methane and other hydro-carbons, ammonia, acids, and alcohols
Complex organic molecules synthesized abiotically
Nucleotides, amino acids, sugars
Complex organic molecules used by protobionts; some biotic synthesis
Protein, fat, and carbohydrates in cells
Complex organic molecules resulting only from biotic synthesis
Complex organic molecules resulting only from biotic synthesis
Types of life present
Era of chemical evolution
Protobiont cell forms
Prokaryotes
Prokaryotes
Eukaryotes begin
Multicellular organisms
Age of Microorganisms
Orgin of Earth
Stabilization of Earth's crust
Oldest Earth rocks
Oldest animal fossils
Oldest plant and fungus fossils
Age of dinosaurs
Age of mammals

Figure 22.10 **The Origin and Evolution of Cells.** This chart summarizes the events that are thought to have led to the origin and evolution of cells.

• Thinking Critically •

It has been postulated that there is "life" on another planet in our galaxy. The following data concerning the nature of this life have been obtained from "reliable" sources. Using these data, what additional information is necessary, and how would you go about verifying these data in developing a theory of the origin of life on planet X? Data:

1. The age of the planet is 10 billion years.
2. Water is present in the atmosphere.
3. The planet is farther from its sun than our Earth is from our sun.
4. The molecules of various gases in the atmosphere are constantly being removed.
5. Chemical reactions on this planet occur at approximately half the rate at which they occur on Earth.

• Experience This •

Take a handful of straw, hay, grass clippings, or leaves and boil them to make an organic soup broth. After it has cooled, pour it through a strainer into a clear glass bottle. Add a teaspoon of soil and plug it with a cotton ball or piece of cloth. Place in a sunny spot. Observe this over the next two weeks and note the changes that occur. If you have a microscope, sample the microcosm with an eyedropper and try to identify the organisms. Was this spontaneous generation or biogenesis?

• Questions •

1. In what sequence did the following things happen: living cell, oxidizing atmosphere, autotrophy, heterotrophy, reducing atmosphere, first organic molecule?
2. What is meant by *spontaneous generation*? What is meant by *biogenesis*?
3. Of the following scientists, name those who supplied evidence that supported the theory of spontaneous generation and those who supported biogenesis: Spallanzani, Needham, Pasteur, Urey, Fox, Miller, Oparin.
4. Can spontaneous generation occur today? Explain.
5. Why do scientists believe life originated in the seas?
6. What were the circumstances on primitive Earth that favored the survival of an autotrophic type of organism?
7. The current theory of the spontaneous chemical generation of life on Earth depends on our knowing something of Earth's history. Why is this so?
8. List two important effects caused by the increase of oxygen in the atmosphere.
9. What evidence supports the theory that eukaryotic cells arose from the development of a symbiotic relationship between primitive prokaryotic cells and protocells?

• Chapter Glossary •

autotroph (aw'to-trōf) An organism able to produce organic nutrients from inorganic materials; a self-feeder.

biogenesis (bi-o-jen'uh-sis) The concept that life originates only from preexisting life.

coacervate (ko-as'ur-vāt) A collection of organic macromolecules surrounded by water molecules, aligned to form a sphere.

endosymbiotic theory (en"do-sim-be-ot'ik the'o-re) A theory suggesting that some organelles found in eukaryotic cells may have originated as free-living prokaryotes.

eukaryote (yu-kār'e-ōt) A cell possessing a nuclear membrane and other membranous organelles.

heterotroph (het'ur-o-trōf) An organism that requires a source of organic material from its environment. It cannot produce its own food.

microsphere (mi'kro-sfēr) A collection of organic macromolecules in a structure with a double-layered outer boundary.

oxidizing atmosphere (ok'sĭ-di-zing at'mos-fēr) An atmosphere that contains molecular oxygen.

prebionts (pre"bi'onts) Nonliving structures that led to the formation of the first living cells.

prokaryote (pro-kār'e-ōt) A cell that lacks a nuclear membrane and other membranous organelles.

proteinoid (pro'te-in-oid) The proteinlike structure of branched amino acid chains that is the basic structure of a microsphere.

protocell (pro'to-sel) The first living cell.

reducing atmosphere (re-du'sing at'mos-fēr) An atmosphere that does not contain molecular oxygen (O_2).

spontaneous generation (spon-ta'ne-us jen-uh-ra'shun) The theory that living organisms arose from nonliving material.

CHAPTER 23

The Classification and Evolution of Organisms

Purpose

It has been estimated that there are over 30 million different kinds of organisms. The scientific community can communicate effectively only if each form of life has its own universally recognized name. In this chapter we investigate how biologists catalog closely related forms of life into larger groups to help see the order in life forms. Primitive cells are thought to have given rise to present-day species. We introduce the five kingdoms into which biologists currently classify life forms and also consider viruses, which don't fit conveniently into any of the major kingdoms.

Chapter Outline

For Your Information

WARNING TO HIKERS
DO NOT DRINK THE WATER

Giardia lamblia is a protozoan found in streams and lakes throughout the world, including "pure" water in wilderness areas. Over forty species of animals harbor this organism in their small intestines. Its presence may cause diarrhea, vomiting, cramps, or nausea. *Giardia* may be found even if good human sanitation is practiced. No matter how inviting it may be to drink directly from that cold mountain stream, DON'T. Deer, beaver, or other animals could have contaminated the water with *Giardia*. Treat the water before drinking. The most effective way to eliminate the spores formed by this protozoan is to use special filters that can filter out particles as small as 1 micrometer; otherwise, boil the water for at least five minutes before drinking.

Learning Objectives

- Know the importance of Linnaeus's work.
- Know the categories used in the science of classification.
- Explain the benefits of a system of classification.
- Know the five kingdoms.
- List several characteristics for each kingdom.
- Explain how eukaryotic single-celled organisms could have evolved into multicellular forms.
- Diagram a typical plant life cycle.
- List three theories of the origin of viruses.
- Describe the structure and life cycle of a virus.

(a)

(b)

Figure 23.1 **Fish Identification.** Using the scientific name *Micropterus salmoides* for largemouth black bass (a) and *Salmo trutta* for brown trout (b) correctly indicates which of these two species of fish a biologist is talking about.

The Classification of Organisms

Every day you see a great variety of living things. Just think of how many different species of plants and animals you have observed. What names do you assign to each? Is the name you use the same as that used in other sections of the country or regions of the world? In much of the United States and Canada, the fish pictured in figure 23.1a is known as a *largemouth black bass;* but in sections of the southern United States it is called a *trout.* This use of local arbitrary names can lead to confusion. If a student in Mississippi writes to a friend in Wisconsin that he has caught a six-pound trout, the person in Wisconsin thinks that his friend has caught the kind of fish pictured in figure 23.1b.

In the scientific community, accuracy is essential; local names cannot be used. When a biologist is writing about a species, all biologists in the world who read that article must know exactly what that species is. **Taxonomy** is the science of naming and grouping organisms into logical categories. Biologists at the Smithsonian Institution estimated that there are over 30 million species in the world; approximately 1.5 million of these have been named.

Various approaches have been used to classify organisms. The Greek philosopher Aristotle (384–322 B.C.) had an interest in nature and was the first person to attempt a logical classification system. The root word for *taxonomy* is the Greek word *taxis,* which means *arrangement.* Aristotle used the size of plants to divide them into the categories of trees, shrubs, and herbs.

During the Middle Ages, Latin was widely used as the scientific language. As new species were identified, they were given Latin names, often using as many as fifteen words. Although using Latin meant that most biologists, regardless of their native language, could understand a species name, it did not completely do away with duplicate names. Because many of the organisms could be found over wide geographic areas and communication was slow, there could be two or more Latin names for a species. To make the situation even more confusing, ordinary people still called organisms by their common local names.

The modern system of classification began in 1758 when Carolus Linnaeus (1707–1778), a Swedish doctor and botanist, published his tenth edition of *Systema Naturae* (figure 23.2). In the previous editions, Linnaeus had used a polynomial (many-names) Latin system. However, in the tenth edition he introduced the **binomial** (two-name) **system of nomenclature.** This system used two Latin names, genus and species, for each type of organism. A **genus** (plural, *genera*) is a group of closely related organisms; a **species** is the unique name given to a particular type of organism. In order to clearly identify the scientific name, binomial names are

Figure 23.2 **Carolus Linnaeus (1707–1778).** Linneaus, a Swedish doctor and botanist, originated the modern system of taxonomy.

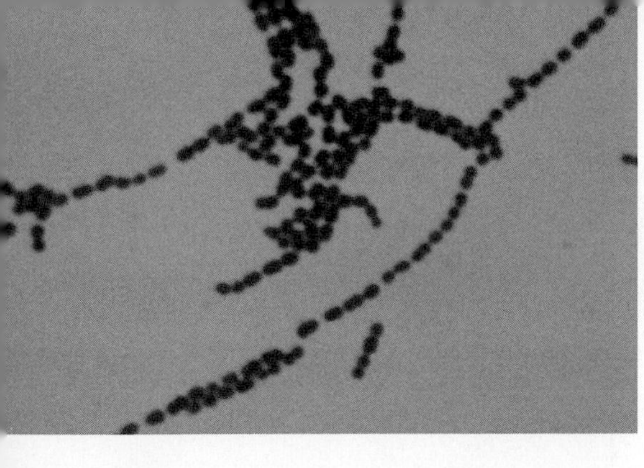

(a)

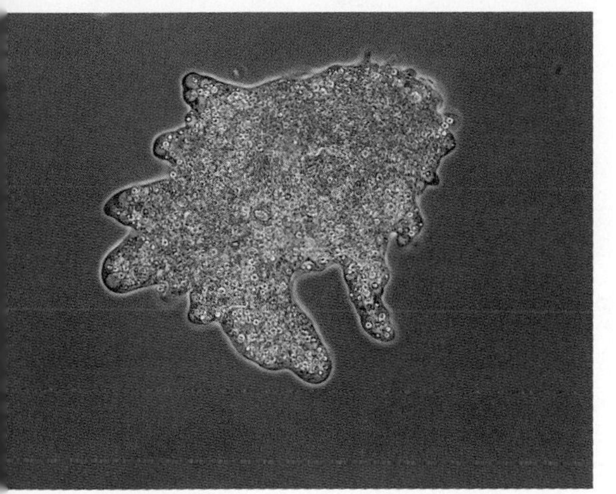

(c)

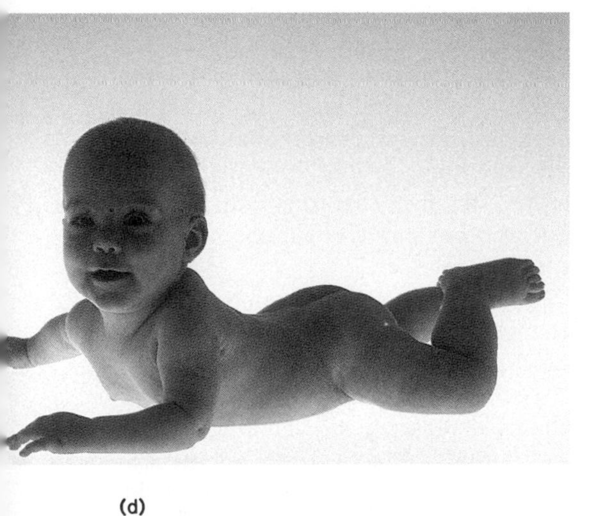

(d)

(b)

(e)

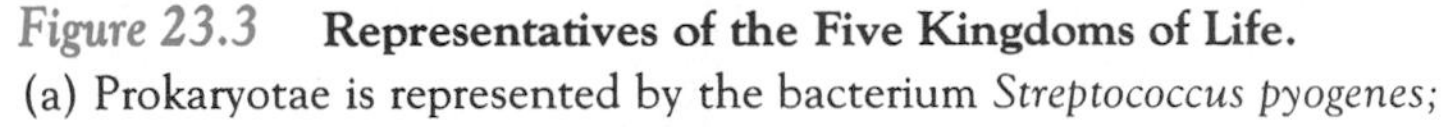

Figure 23.3 **Representatives of the Five Kingdoms of Life.**
(a) Prokaryotae is represented by the bacterium *Streptococcus pyogenes;* (b) Mycetae, by the mushroom *Morchella esculenta;* (c) Protista, by one-celled *Amoeba proteus;* (d) Animalia, by the animal *Homo sapiens;* and (e) Plantae, by the tree *Acer saccharum.*

either *italicized* or underlined. The first letter of the genus name is capitalized. The species name is always written in lowercase. *Micropterus salmoides* is the binomial name for the largemouth black bass. When biologists adopted Linnaeus's binomial method, they eliminated the confusion that was the result of using common local names. For example, with the binomial system the white water lily is known as *Nymphaea odorata.* Regardless of which of the 245 common names is used in a botanist's local area, when botanists read *Nymphaea odorata,* they know exactly which plant is being referred to. The binomial name cannot be changed unless there is compelling evidence to justify doing so. The rules that govern the worldwide classification of species are expressed in the International Rules for Botanical Nomenclature, the International Rules for Zoological Nomenclature, and the International Bacteriological Code of Nomenclature.

In addition to assigning a specific name to each species, Linnaeus recognized a need for placing organisms into groups (figure 23.3). This system divides all forms of life into **kingdoms,** the largest grouping used in the classification of organisms. Originally there were two kingdoms, Plantae and Animalia. Today most biologists recognize five kingdoms of life (figure 23.3). It is convenient to divide each of these kingdoms into smaller units and give specific names to each. The taxonomic subdivision under each kingdom is usually called a **phylum,** although microbiologists and botanists replace this term with the word *division.* All kingdoms have more than one phylum. For example, the kingdom Prokaryotae contains four phyla. In order to determine an organism's place in one of these phyla (divisions), each organism is investigated to determine the specific nature of its structure, metabolism, and biochemistry.

A **class** is a grouping within a phylum. For example, within the phylum Chordata there are seven classes: mammals, birds, reptiles, amphibians, and three classes of fishes. An **order** is a grouping within a class. Carnivora is an order of meat-eating animals within the class Mammalia. A **family** consists of a group of closely related species within an order. The cat family is a subgrouping of the order Carnivora. Thus, in the present-day science of taxonomy each organism that has been classified has its own unique binomial name, the genus and species. In turn, it is assigned to larger groupings that have a common evolutionary history.

Phylogeny is the science that explores the evolutionary relationships among organisms and seeks to reconstruct evolutionary history. Taxonomists and phylogenists work together so that the products of their work are compatible. A taxonomic ranking should reflect the evolutionary relationships among the organisms being classified. Although taxonomy and phylogeny are sciences, there is no complete agreement as to how organisms are classified or how they are related. Just as there was dissension two hundred years ago when biologists disagreed on the theories of spontaneous generation and biogenesis, there are differences in opinion about the evolutionary relationships of organisms. Different people arrive at different conclusions because they use different kinds of evidence or interpret this evidence differently. Phylogenists use several lines of evidence to develop evolutionary histories: fossils, comparative anatomy, life-cycle information, and biochemical/molecular evidence.

Fossils are physical evidence of previously existing life. They may be preserved whole in an unaltered form. For example, mammoths have been found frozen in glaciers, and bacteria and insects have been preserved after becoming embedded in plant resins. Other fossils are only parts of once-living organisms. The outlines or shapes of extinct plant leaves are often found in coal deposits, and individual animal bones that have been chemically altered over time are often dug up (figure 23.4). Animal tracks have also been discovered in the dried ancient mud of river beds. It is important to understand that some organisms are more easily fossilized than others. Those that have hard parts like

(a)

(b)

Figure 23.4 **Fossil Evidence.** Fossils are either the remains of prehistoric organisms or evidence of their existence. (a) The remains of an ancient fly preserved in amber. (b) A trilobite specimen. Trilobites, an extinct class of arthropods, make good fossils because of their hard exoskeleton.

cell walls, skeletons, and shells are more likely to be preserved than are tiny, soft-bodied organisms. Aquatic organisms are much more likely to be buried in the sediments at the bottom of the oceans or lakes than are their terrestrial counterparts. Later, when they are pushed up by geologic forces, aquatic fossils are found in their layers of sediments, which have often been uplifted to become dry land.

Evidence obtained from the discovery and study of fossils allows biologists to place organisms in a time sequence. This can be accomplished by comparing one type of fossil with another. As geological time passes and new layers of sediment are laid down, the older organisms should be in deeper layers providing the sequence of layers has not been disturbed. In addition, it is possible to age-date rocks by comparing the amounts of certain radioactive isotopes that they contain. The older sediment layers have less of these specific radioactive isotopes than do younger layers (figure 23.5). A comparison of the layers gives an indication of the relative age of the fossils found in the rocks. Therefore, fossils found in the same layer must have been alive during the same geological period. Even though two fossils may have obvious structural differences, the fact that they are next to one another provides support for their coexistence.

It is also possible to compare subtle changes in particular kinds of fossils over time. For example, the size of the leaf of a specific fossil plant has been found to change extensively through long geological periods. A comparison of the extremes, the oldest with the newest, would lead to their classification into different categories. However, the fossil links between the extremes clearly show that the younger plant is a descendent of the older.

The comparative anatomy of fossil or currently living organisms can be very useful in developing a phylogeny. Since the structures of an organism are determined by its genes and developmental processes, those organisms having similar structures are thought to be related.

Figure 23.5 **Determining the Age of Fossils.** Since new layers of sedimentary rock are formed on top of older layers of sedimentary rock, it is possible to determine the relative ages of fossils found in various layers. The layers of rock shown here represent on the order of hundreds of millions of years of formation. The fossils of the lower layers are millions of years older than the fossils in the upper layers.

Plants can be divided into several categories: all plants that have flowers are thought to be more closely related to one another than to plants like ferns, which do not have flowers. In the animal kingdom, all organisms that nurse their young from mammary glands are grouped together, and all animals in the bird category have feathers and beaks and lay eggs with shells. Reptiles also have shelled eggs but differ from birds in that reptiles lack feathers and have scales covering their bodies. The fact that these two groups share this fundamental eggshell characteristic implies that they are more closely related to each other than they are to other groups.

Another line of evidence useful to phylogenists and taxonomists comes from the field of developmental biology. Many organisms have complex life cycles that include many completely different stages. After fertilization, some organisms grow into free-living developmental stages that do not resemble the adults of their species. These are called *larvae* (singular, *larva*). Larval stages often provide clues to the relatedness of organisms. For example, barnacles live attached to rocks and other solid marine objects and look like small, hard cones. Their outward appearance does not suggest that they are related to shrimp; however, the larval stages of barnacles and shrimp are very similar. Detailed anatomical studies of barnacles confirm that they share many structures with shrimp; their outward appearance tends to be misleading (figure 23.6). This same kind of evidence is available in the plant kingdom. Many kinds of plants, such as peas, peanuts, and lima beans, produce large, two-parted seeds in pods (you can easily split the seeds into two parts). Even though peas grow as vines, lima beans grow as bushes, and peanuts have their seeds underground, all these plants are considered to be related.

Like all aspects of biology, the science of taxonomy is constantly changing as new techniques develop. Recent advances in DNA analysis are being used to determine genetic similarities among species. In the field of ornithology, which deals with the study of birds, there are those who believe that storks and flamingos are closely related; others believe that flamingos are more closely related to geese. An analysis of the DNA points to a higher degree of compatibility between flamingos and storks than between flamingos and geese. This is interpreted to mean that the closest relationship is between flamingos and storks. Algae and plants have several different kinds of chlorophyll: chlorophyll *a*, *b*, *c*, *d*, and *e*. Most photosynthetic organisms contain a combination of two of

(a)

(b)

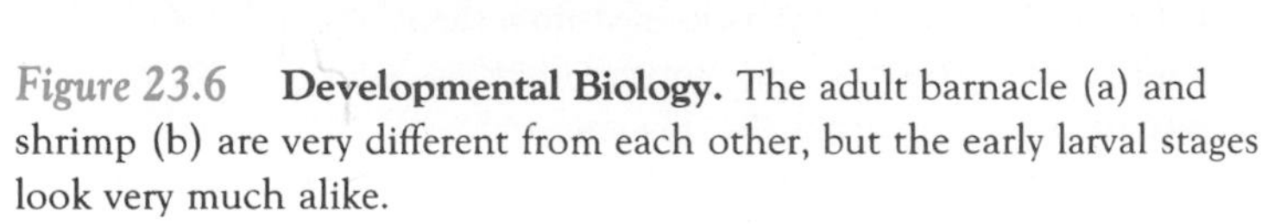

Figure 23.6 **Developmental Biology.** The adult barnacle (a) and shrimp (b) are very different from each other, but the early larval stages look very much alike.

these chlorophyll molecules. Members of the kingdom Plantae have chlorophyll *a* and *b*. The large seaweeds, like kelp, superficially resemble terrestrial plants like trees and shrubs. However, a comparison of the chlorophylls present shows that kelp has chlorophyll *a* and *d*, while plants have chlorophyll *a* and *b*. When another group of algae, called the *green algae,* are examined, they are found to have chlorophyll *a* and *b*. Along with other anatomical and developmental evidence, this biochemical information has helped to establish an evolutionary link between the green algae and plants. All of these kinds of evidence (fossils, comparative anatomy, developmental stages, and biochemical evidence) have been used to develop the various taxonomic categories, including kingdoms.

Given all these sources of evidence, biologists have developed a hypothetical picture of how all organisms are related (figure 23.7). At the base of this evolutionary scheme is the biochemical evolution of cells first postulated by Oparin (see chapter 22). These first cells are thought to be the origin of the five kingdoms. While protocells no longer exist, their descendants have diversified over millions of years. Of these groups, the Prokaryotae have the simplest structure and are probably most similar to some of the first cellular organisms on Earth.

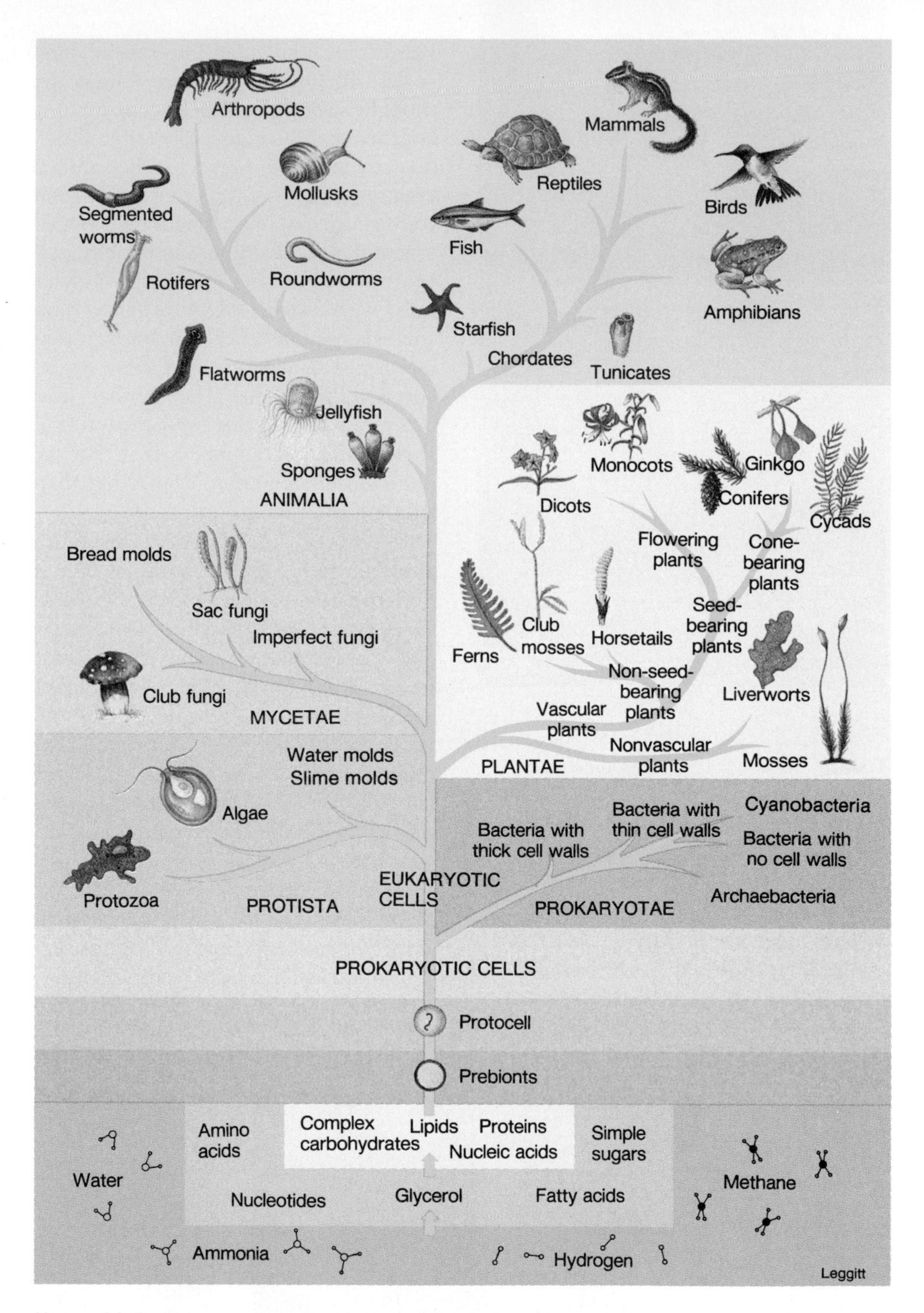

Figure 23.7 **Molecules to Organisms.** The theory of spontaneous generation proposes that the molecules in the early atmosphere and early oceans accumulated to form prebionts—nonliving structures composed of carbohydrates, proteins, lipids, and nucleic acids. The prebionts are believed to be the forerunners of the protocells—the first living cells. These protocells probably evolved into prokaryotic cells, on which the kingdom Prokaryotae is based. Some prokaryotic cells probably gave rise to eukaryotic cells. The organisms formed from these early eukaryotic cells were members of the kingdom Protista. Members of this kingdom evolved into the kingdoms Animalia, Plantae, and Mycetae. Thus, all present-day organisms evolved from the protocells.

Kingdom Prokaryotae

Members of the kingdom Prokaryotae are grouped together because they all have the same cellular structure. They are small, single-celled organisms ranging from 1 to 10 micrometers (μm). Their cell walls typically contain complex organic molecules not found in other kinds of organisms. Some of these are peptidoglycan, polymers of unique sugars, and unusual amino acids. These organisms are commonly known as *bacteria*. Some are disease-causing, such as *Streptococcus pneumoniae,* but most are not disease-causing. In addition, many are able to photosynthesize.

Prokaryotes have no nucleus and the genome is a single loop of DNA.

Some genomes have as few as 5,000 genes. The DNA in the Prokaryotae does not undergo the process of mitosis or meiosis. Rather, the cells reproduce by more primitive methods, such as binary fission. This is a type of asexual cell division that does not involve the more complex structures used by eukaryotes in mitosis or meiosis. As a result, daughter cells are produced that have a single copy of the parental DNA loop (figure 23.8). Some motile cells move by secreting a slime that glides over the cell's surface, causing it to move through the environment. Others move by means of flagella, the structure of which is unique to prokaryotes.

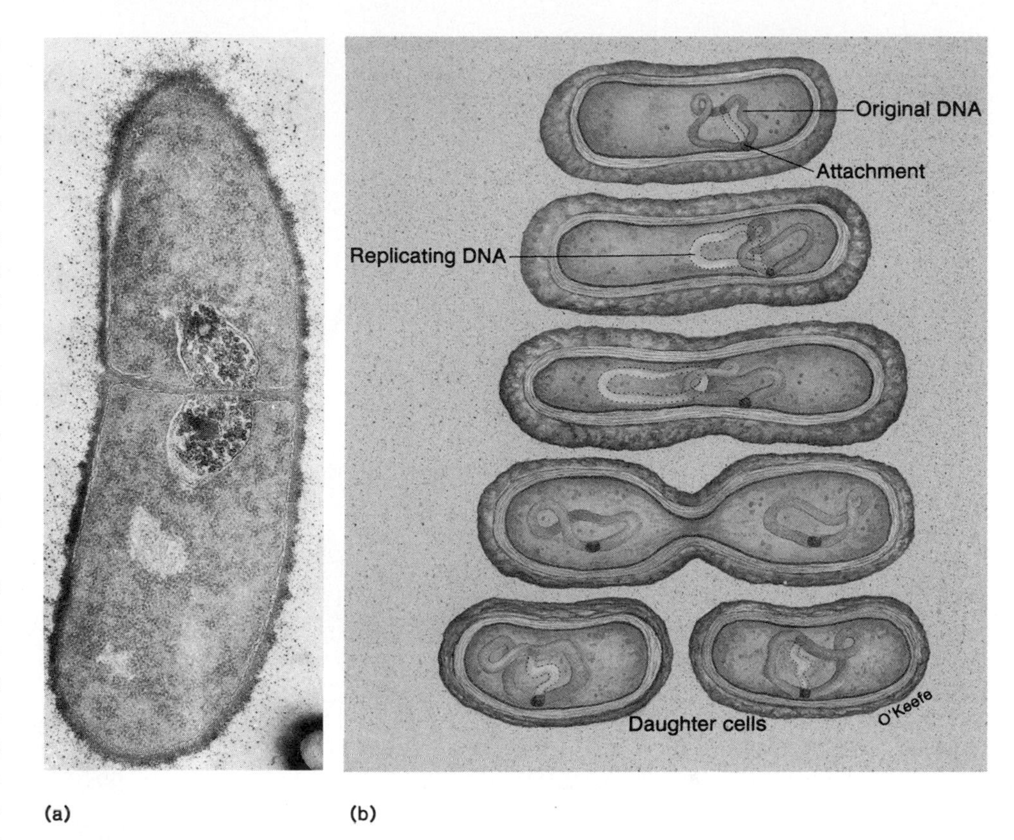

Figure 23.8 **Binary Fission.** Two cells of bacterium *Bacillus megaterium* formed by binary fission (a). (b) Binary fission consists of DNA replication and cytoplasmic division.

Since the early atmosphere is thought to have been a reducing atmosphere, the first Prokaryotae were probably anaerobic organisms. Today there are both anaerobic and aerobic Prokaryotae. There are some autotrophic Prokaryotae, but the majority are heterotrophs. Some heterotrophs are **saprophytes,** organisms that obtain energy by the decomposition of dead organic material; others are parasites that obtain energy and nutrients from living hosts; still others are mutualistic or commensalistic with host organisms.

Some biologists hypothesize that eukaryotic cells evolved from prokaryotic cells by a process of endosymbiosis. This hypothesis proposes that structures like mitochondria, chloroplasts, and other membranous organelles originated from separate cells that were ingested by larger, more primitive cells. Once inside, these structures and their functions became integrated with the host cell and ultimately became essential to its survival. This new type of cell was the forerunner of present-day eukaryotic cells. (See "The Origin of Eukaryotic Cells," page 359 and figure 22.10.) Single-celled eukaryotic organisms are members of the kingdom Protista (figure 23.7).

Kingdom Protista

The changes in cell structure that led to eukaryotic organisms most probably gave rise to single-celled organisms similar to those currently grouped in the kingdom Protista. Most members of this kingdom are one-celled organisms, although there are some colonial forms. Eukaryotic cells are usually much larger than the prokaryotes, typically having more than a thousand times the volume of prokaryotic cells. Their larger size was made possible by the presence of specialized membranous organelles, such as mitochondria, the endoplasmic reticulum, chloroplasts, and nuclei.

There is a great deal of diversity within the sixty thousand known species of Protista. Many species live in fresh water; others are found in marine or terrestrial habitats, and some are parasitic, commensalistic, or mutualistic. All species can undergo mitosis, resulting in asexual reproduction. Some species can also undergo meiosis and reproduce sexually. Many contain chlorophyll in chloroplasts and are autotrophic; others require organic molecules as a source of energy and are heterotrophic. Both autotrophs and heterotrophs have mitochondria and respire aerobically.

Because members of this kingdom are so diverse with respect to form, metabolism, and reproductive methods, most biologists do not feel that the Protista form a valid phylogenetic unit. However, it is still a convenient taxonomic grouping. By placing these organisms together in this group it is possible to gain a useful perspective on how they relate to other kinds of organisms. After the origin of eukaryotic organisms, evolution proceeded along several different pathways. Three major lines of evolution can be seen today in the plantlike autotrophs (algae), animal-like heterotrophs (protozoa), and the funguslike heterotrophs (slime molds). *Amoeba* and *Paramecium* are commonly encountered examples of protozoa. Many seaweeds and pond scums are collections of large numbers of algal cells. Slime molds are less frequently seen because they live in and on the soil in moist habitats; they are most often encountered as slimy masses on decaying logs.

Through the process of evolution, the plantlike autotrophs probably gave rise to the kingdom Plantae, the animal-like heterotrophs probably gave rise to the kingdom Animalia, and the funguslike heterotrophs were probably the forerunners of the kingdom Mycetae (figure 23.7).

Kingdom Mycetae

Fungus is the common name for members of the kingdom Mycetae. The majority of fungi are nonmotile. They have a rigid cell wall, which in most species is composed of chitin. Members of the kingdom Mycetae are nonphotosynthetic eukaryotic organisms. The majority are multicellular, but a few, like yeasts, are single-celled. In the multicellular fungi the basic structural unit is made up of multicellular filaments. Because all of these organisms are heterotrophs, they must obtain nutrients from organic sources. Most secrete enzymes that digest large molecules into smaller units that are absorbed.

Fungi can either be free-living or parasitic. Those that are free-living feed on a variety of nutrients ranging from dead organisms to such products as shoes, foodstuffs, and clothing. Most synthetic organic molecules are not attacked as readily by fungi; this is why plastic bags, foam cups, and organic pesticides are slow to decompose. The fungi that are parasitic are responsible for athlete's foot, vaginal yeast infections, valley fever, "ringworm," and other diseases.

Figure 23.9 **Plant Evolution.** Two lines of plants are thought to have evolved from the plantlike Protista, the algae. The nonvascular mosses evolved as one type of plant. The second type, the vascular plants, evolved into the seed and nonseed plants.

Kingdom Plantae

Another major group that has its roots in the kingdom Protista are the green, photosynthetic plants. The ancestors of plants were most likely specific kinds of algae commonly called *green algae.* Members of the kingdom Plantae are nonmotile, multicellular organisms that contain chlorophyll and produce their own organic compounds. All plant cells have a cellulose cell wall. Over 300,000 species of plants have been classified; 80% are flowering plants, 5% are mosses and ferns, and the remainder are cone-bearing.

A wide variety of plants exist on earth today. Members of the Plantae range from simple mosses to vascular plants with stems, roots, leaves, and flowers. Most biologists feel that the evolution of this kingdom began about 400 million years ago when the green algae of the kingdom Protista evolved along two lines: the nonvascular plants like the mosses evolved as one type of plant and the vascular plants like the ferns evolved as a second type (figure 23.9). Some of the vascular plants evolved into seed-producing plants, which today are the cone-bearing and flowering plants. The development of vascular plants was a major step in the evolution of plants from an aquatic to a terrestrial environment.

Plants have a unique life cycle. There is a haploid **gametophyte stage** that produces a haploid sex cell by mitosis. There is also a diploid **sporophyte stage** that produces haploid spores by meiosis (figure 23.10). This **alternation of generations,** which is a unifying theme that ties together all members of this kingdom, is fully explained in chapter 25. In addition to sexual reproduction, plants are able to reproduce asexually.

The evolution of the plants was closely paralleled by the evolution of

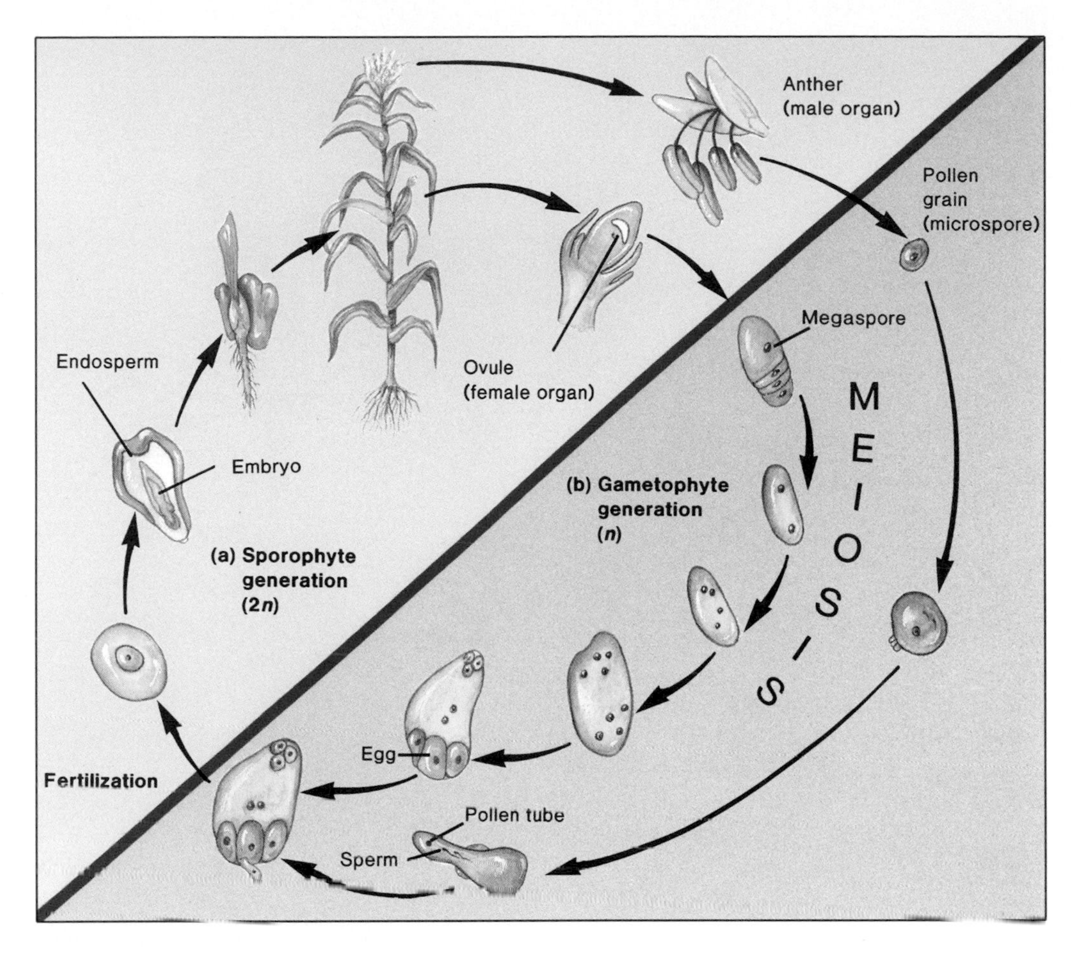

Figure 23.10 **Alternation of Generations.** (a) Plants have a multicellular structure (sporophyte generation) that undergoes meiosis to form spores—haploid cells. (b) These spores give rise to haploid organisms (gametophyte generation) that form the gametes—sperm cells and egg cells. The union of these gametes forms the diploid stage, which develops into the sporophyte.

animals. Ever since the early interdependency between autotrophic and heterotrophic species, there has been a close relationship between plants and animals. Like the plants, the animals are thought to have evolved from the Protista (figure 23.7).

Kingdom Animalia

Over a million species of animals have been classified. These range from microscopic types, like mites or aquatic plankton, to huge animals like elephants or whales. Regardless of their type, all animals have some common traits. They all are composed of eukaryotic cells, and all species are heterotrophic and multicellular. Most animals are motile; however, some, like the sponges, are sessile (not able to move). All animals are capable of sexual reproduction, but most can reproduce asexually.

It is thought that animals evolved from the flagellated Protista (figure 23.7). This idea proposes that colonies of flagellated Protista gave rise to simple multicellular forms of animals like sponges. These first animals lacked specialized tissues and organs. Division of labor eventually gave rise to the more complex forms of animal life.

Although taxonomists have grouped organisms into five kingdoms, some organisms do not easily fit into these categories. Viruses, which lack all cellular structures, still show some characteristics of life. In fact some people consider them to be nonliving. For this reason they are considered separately from the five kingdoms.

Viruses

A **virus** is a nucleic acid particle coated with protein (figure 23.11). Viruses are **obligate intracellular parasites,** which means they are infectious particles that can function only when inside a living cell. Due to their unusual characteristics, viruses are not a member of any kingdom. Biologists do not consider them to be living because they are not capable of living by themselves and show the characteristics of life only when inside living cells.

Soon after viruses were discovered in the late part of the nineteenth century, biologists began speculating on how they originated. One early hypothesis on the origin of viruses was that they were either prebionts or parts of prebionts that did not evolve into cells. This idea was

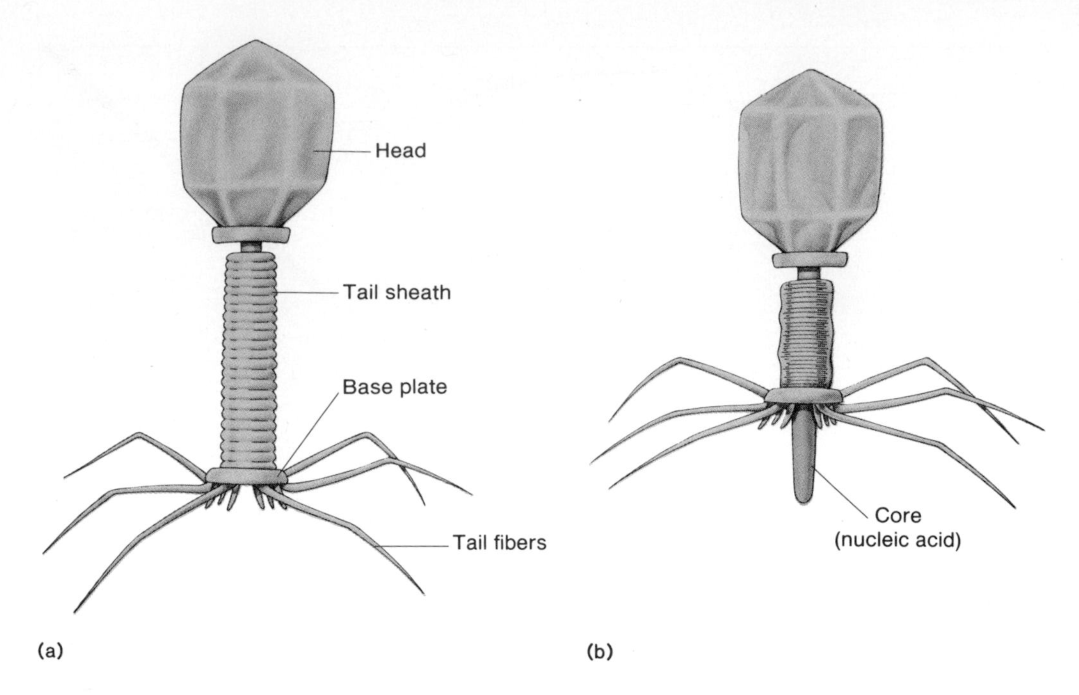

Figure 23.11 **A Typical Virus.** (a) The head and tail sheath are made of protein, which encloses the viral nucleic acid (RNA or DNA). (b) When the virus comes into contact with the proper host cell, the tail sheath contracts and the nucleic acid is injected into the host cell.

discarded as biologists learned more about the complex relationship between viruses and host cells. A second hypothesis was that viruses developed from intracellular parasites that became so efficient that they needed only the nucleic acid to continue their existence. Once inside, this nucleic acid can take over a host cell that provides for all of the virus's needs. A third hypothesis is that viruses are runaway genes that have escaped from cells and must return to a host cell to replicate. Regardless of how the viruses came into being, today they are important as parasites in all forms of life.

Viruses are host-specific. The **host** is a specific kind of cell that provides what the virus needs to function. Viruses can infect only those cells that have the proper receptor sites to which the virus can attach. This site is usually a glycoprotein molecule on the cell membrane. For example, the virus responsible for measles attaches to skin cells, hepatitis viruses attach to liver cells, and mumps viruses attach to cells in the salivary glands. Host cells for the HIV virus include some types of human brain cells and several types belonging to the immune system (box 23.1, see page 374).

Upon entering a host cell, the virus loses its protein coat. Once free in the cell, the nucleic acid portion of the virus may remain free in the cell or it may link with the host's genetic material. Some viruses contain as few as three genes, others contain as many as five hundred. A typical eukaryotic cell contains tens of thousands of genes.

Viral genes are able to take command of the host's metabolic pathways and direct it to carry out the work of making new copies of the original virus. The virus makes use of the host's available enzymes and ATP for this purpose. When enough new viral nucleic acid and protein coat are produced, complete virus particles are assembled and released from the host (figure 23.12). The number of viruses released ranges from ten to thousands. The virus that causes polio releases about ten thousand new virus particles after it has invaded its human host cell.

Viruses vary in size and shape, which helps in classifying them. Some are rod-shaped, others are round, and still others are in the shape of a coil or helix. Viruses are some of the smallest infecting agents known to humans. Only a few can be seen with a standard laboratory microscope; most require an electron microscope to make them visible. A great deal of work is necessary to isolate viruses from the environment and prepare them for observation with an electron microscope. For this reason, most viruses are more quickly identified by their activities in host cells. Almost all of the species in the five kingdoms serve as hosts to some form of virus (table 23.1).

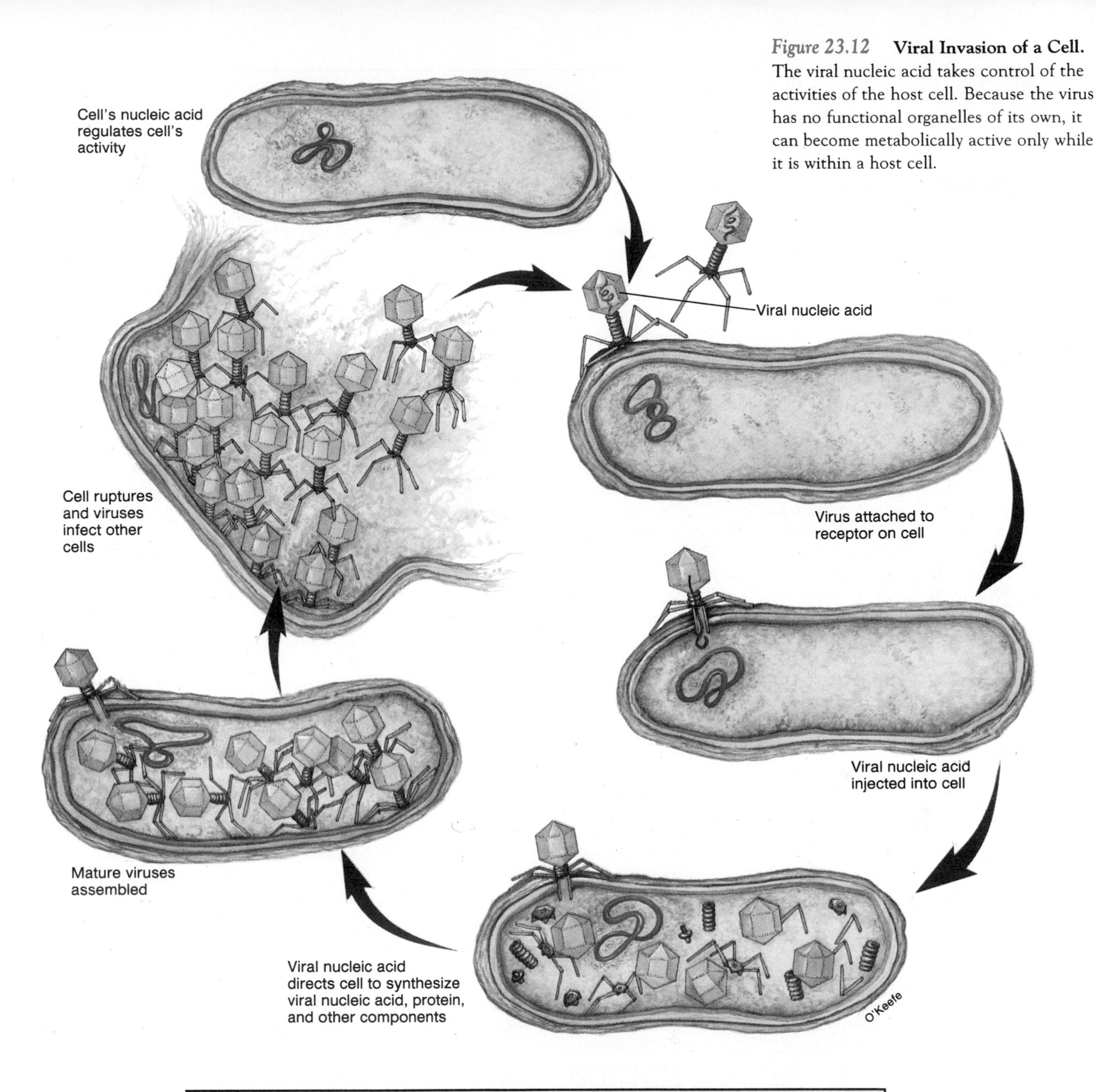
Cell's nucleic acid
regulates cell's
activity
Viral nucleic acid
Virus attached to
receptor on cell
Viral nucleic acid
injected into cell
Viral nucleic acid
directs cell to synthesize
viral nucleic acid, protein,
and other components
Mature viruses
assembled
Cell ruptures
and viruses
infect other
cells
O'Keefe

Figure 23.12 **Viral Invasion of a Cell.** The viral nucleic acid takes control of the activities of the host cell. Because the virus has no functional organelles of its own, it can become metabolically active only while it is within a host cell.

Table 23.1
Viral Diseases

Type of Virus	Disease
Papovaviruses	Warts in humans
Paramyxoviruses	Mumps and measles in humans; distemper in dogs
Adenoviruses	Respiratory infections in most mammals
Poxviruses	Smallpox
Wound-tumor viruses	Diseases in corn and rice
Potexviruses	Potato diseases
Bacteriophage	Infections in many types of bacteria

BOX 23.1 The AIDS Pandemic

Epidemiology is the study of the transmission of diseases through a population. Diseases that occur throughout the world population at extremely high rates are called *pandemics*. Influenza, the first great pandemic of the first part of the twentieth century, killed hundreds of thousands of people. AIDS has become the greatest pandemic of the second half. This viral disease has been reported in all countries around the world and is estimated to have infected between 10 and 12 million people.

AIDS is an acronym for *a*cquired *i*mmuno*d*eficiency *s*yndrome and is caused by human immunodeficiency viruses (HIV-I & II), members of the retrovirus family (see the figure on page 375). Evidence strongly supports the belief that this RNA-containing virus originated through many mutations of an African monkey virus sometime during the late 1950s or early 1960s. The virus probably moved from its original monkey host to humans as a result of an accidental scratch or bite. It wasn't until the late 1970s that the virus was identified in human populations. It has since spread to all corners of the globe. The first reported case of AIDS was diagnosed in the United States in 1981 at the UCLA Medical Center.

While the virus first entered the United States through the homosexual population, it is not a disease unique to that group; no virus is known that shows a sexual preference. Transmission of HIV can occur in homosexual and heterosexual individuals.

Data from the Centers for Disease Control and Prevention, Atlanta, Georgia, indicate that the mortality rate among those with the disease is over 60%! Today, over 200,000 cases have been diagnosed in the United States. AIDS is believed by many to be the greatest health-care concern we have ever had in the world. They point out that the crisis has not reached its peak because large numbers of people who test HIV+ for the virus infection have not yet developed symptoms, but all are expected to die as a result of their infection. The numbers of people suspected of being HIV+ is not exactly known; however, there are an estimated million HIV+ people in the United States alone.

HIV is a spherical virus containing an RNA genome, including a gene for an enzyme called *reverse transcriptase,* a protein shell, and a lipid-protein envelope. The virus gains entry into a suitable host cell through a very complex series of events involving the virus envelope and the host-cell membrane. Certain types of human cells can serve as hosts since they have a specific viral receptor site in their surface identified as CD-4. CD-4-containing cells include some types of brain cells and several types of cells belonging to the immune system; namely, monocytes, macrophages, and T4-helper/inducer lymphocytes. Once inside the host cell, the RNA of the HIV virus is used to make a DNA copy with the help of reverse transcriptase. This is the reverse of the normal transcription process, in which a DNA template is used to manufacture an RNA molecule. When reverse transcriptase has completed its job, the DNA genome is spliced into the host cell's DNA. In this integrated form, the virus is called a *provirus.* As a provirus, it may remain inside some host cells for an extended period without causing any harm. Some estimate this time to be more than thirty years. Eventually, the virus replicates, and new viruses are released into surrounding body fluids where they can be transmitted to other cells in the body or to other individuals.

The virus is probably transferred to CD-4-containing brain cells from infected macrophages (normally protective phagocytes) that move from other parts of the body into the spinal column and to the brain. Once inside the brain cells, they go to work. The result of such an infection is progressive memory loss; mimicry of other neurological diseases, such as multiple sclerosis; loss of coordination; dementia (senility); and ultimately death.

Because HIV has a unique gene, it is able to replicate much more rapidly than many other types of viruses. This transactivator gene controls the way the whole virus reproduces. When this occurs, the host cell's metabolism is disrupted to the point that it dies. AIDS patients, therefore, have a decrease in the number of CD-4 cell types. A decrease in one type of CD-4 cell—the T4 lymphocytes—is an important diagnostic indicator of HIV infection and an indicator of the onset of AIDS symptoms.

Another unique feature of HIV is its rapid mutation rate. Studies have indicated over one hundred mutant strains of the virus developing from a single parental strain over the course of the infection in one individual. Such an astronomical mutation rate makes a vaccine against HIV very difficult to develop, since the vaccine would have to stimulate an immune response that would protect against all possible mutant forms.

Since lymphocyte host cells are found in the blood and other body fluids, it is logical that these fluids serve as carriers for the transmission of HIV. The virus is transmitted through contact with contaminated blood, semen, mucous secretions, serum, or blood-contaminated hypodermic needles. If these body fluids contain the free viruses or infected cells (monocytes, macrophages, T4-helper/inducer lymphocytes) in sufficient quantity, they can be a source of infection. There is no evidence indicating transmission through the air; by toilet seats; by mosquitoes; by casual contact, such as shaking hands, hugging, touching, or closed-mouth kissing; by utensils, such as silverware or glasses; or by caring for AIDS patients. The virus is just too fragile to survive transmission by these routes.

T4-helper/inducer lymphocytes are only one of several types of

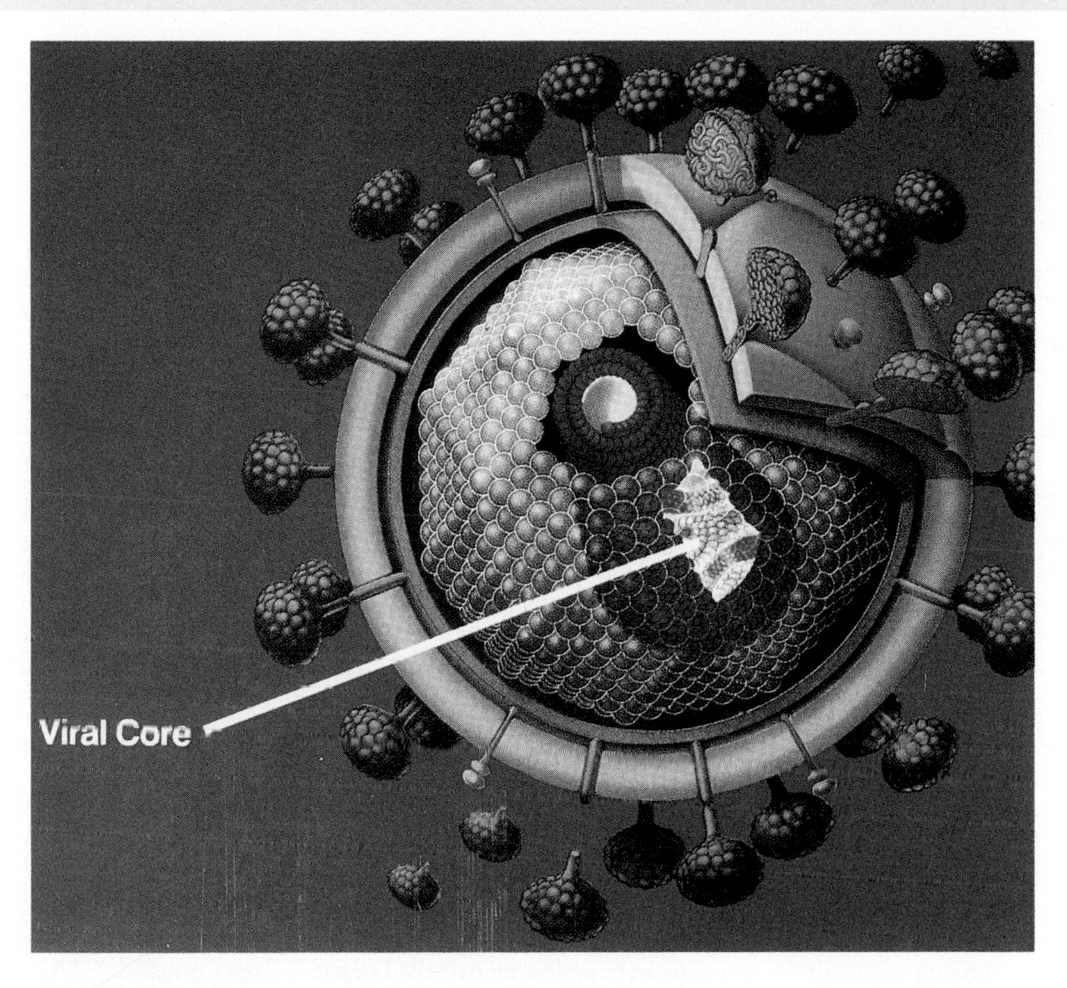

lymphocytes that are responsible for defending the body against a great variety of dangerous chemicals, tumor cells, and infectious organisms. These cells are responsible for five essential functions:

1. They help B lymphocytes change into plasma cells and encourage them to produce protective antibody molecules (immunoglobulins) against specific dangerous agents.
2. They induce T8 lymphocytes to become "killer" cells that attack and kill invading microorganisms, body cells invaded by microorganisms, and tumor cells.
3. They induce T8 lymphocytes to stop operating once the infection is brought under control.
4. They induce T4 lymphocytes to reproduce (clone).
5. They help T4 cells to stop functioning once infection is brought under control.

When T4 lymphocytes are destroyed by HIV, all these defensive efforts of the immune system are depressed. This leaves the body vulnerable to invasion by many types of infecting microbes or to being overtaken by body cells that have changed into tumor cells. This means that the HIV virus does not directly cause the death of the infected individual unless it destroys sufficient brain cells. AIDS is a progressive disease that can occur over many years or even decades. It is a series of bodily changes that begins with destruction of brain cells and ends in death as a result of infections caused by otherwise harmless organisms or rare forms of cancer. The initial symptoms of the disease have been referred to as *ARC*, or *AIDS Related Complex*, or *pre-AIDS*. Some of the more common microbial infections include (1) a rare lung infection, *Pneumocysitis carinii* pneumonia (PCP), caused by a fungus; (2) gastroenteritis (severe diarrhea) caused by the protozoan *Isopora;* (3) cytomegalovirus infections of the retina of the eye. One of the most common forms of cancer found among AIDS patients is Kaposi's sarcoma, a form of skin cancer that shows up as purple-red bruises.

At the present time the infection is controlled (but not cured) by using drugs that can kill infected cells, improve the body's immune system, or selectively interfere with the life cycle of the virus. The life cycle may be disrupted when the virus enters the cell and the reverse transcriptase converts the RNA to DNA. If this enzyme does not operate, the virus is unable to function. The drugs AZT (azidothymidine or zodovudine) and DDC (dideoxycytosine) are to date the most effective of the antiviral agents. They disrupt the operation of reverse transcriptase. However, there are two

BOX 23.1 continued

reasons why these drugs should be used with caution. AZT and DDC can have severe side effects, including anemia. In addition, HIV has been shown to become resistant to AZT. To deal with the resistance problem, AZT and DDC may be used alternatively over the course of an AIDS infection. In addition, the drug dideoxyinosine (ddl) was approved for the treatment of HIV infection in 1991. This product is used in conjunction with AZT or by itself for people with advanced HIV infection who have not responded to AZT treatment.

What about the development of a vaccine to prevent the virus from infecting the body? Experimental vaccines have been developed based on the body's ability to produce antibodies against the virus. Vaccines, however, have been shown to be effective only in monkeys. In addition, it will be necessary to deal with the problem of genetic differences among the many kinds of HIV viruses. The greater the variety of viruses, the greater the variety of vaccine types needed to prevent infection.

To control the spread of the virus, there must be wide public awareness of the nature of the disease and how it is transmitted. People must be able to recognize high-risk behavior and take action to change it. The most important risk factor is promiscuous sexual behavior (i.e., sex with large numbers of partners). This increases the probability that one of the partners may be a carrier. Other high-risk behaviors include intravenous drug use with shared needles, contact with blood-contaminated articles, and intercourse (vaginal, anal, oral) without the use of a condom. Babies born to women known to be HIV+ are at high risk.

Blood tests (the ELISA and Western Blot) can be performed that indicate exposure to the virus. The tests should be taken on a voluntary basis, absolutely anonymously, and with intensive counseling before and after. People who test positively (HIV+) should not expose anyone else or place themselves in a situation where they might be reinfected. They should do everything to maintain good health—exercise regularly, eat a balanced diet, get plenty of rest, and reduce stress. We cannot stop this pandemic in its tracks, but it can be slowed.

• Summary •

To facilitate accurate communication, biologists assign a specific name to each species that is cataloged. The various species are cataloged into larger groups on the basis of similar traits.

The taxonomic ranking of organisms reflects their evolutionary relationships. In addition, fossil evidence, comparative anatomy, developmental stages, and biochemical evidence are employed in the science of taxonomy.

The first organisms to evolve were single-celled organisms of the kingdom Prokaryotae. From this simple beginning, more complex, many-celled organisms evolved, creating members of the kingdoms Protista, Mycetae, Plantae, and Animalia.

Although viruses are not considered to be living organisms, they are able to invade many types of cells. Because of their pathogenic effects, the viruses are an important factor in the world of living organisms.

• Thinking Critically •

A minimum estimate of the number of species of insects in the world is 750,000. Perhaps then it would not surprise you to see a fly with eyes on stalks as long as its wings, a dragonfly with a wingspread over 100 centimeters, an insect that can revive after being frozen at −35° C, and a wasp that can push its long, hairlike, egg-laying tool directly into a tree. Only the dragonfly is not presently living, but it once was!

What other curious features of this fascinating group can you discover? Have you tried to look at a common beetle under magnification? It will hold still if you chill it.

• Experience This •

Using a Dichotomous Key

A valuable tool for students interested in taxonomy is a dichotomous key. Dichotomous keys help to correctly identify organisms. There are a great variety of dichotomous keys, including keys for fish, plants, reptiles, insects, and all other forms of life.

All dichotomous keys are constructed on the same basis. To identify a species properly, the user is given a series of choices, usually two. Most libraries have copies of these keys. Pick a subject that interests you, such as plants, insects, or fish. Then obtain a specimen that is new to you, follow the directions in the key, and key the organism out to its genus and species.

Use the following key to identify this object.

Key to Common Methods of Travel

1a. No motor 2
1b. Motor 8
2a. Two wheels or fewer 3
2b. Three wheels or more 5
3a. One wheel . . . . unicycle
3b. Two wheels 4
4a. Pedals . . . . bicycle
4b. No pedals . . . . scooter
5a. Three wheels . . . . tricycle
5b. Four wheels 6
6a. Usually 30 cm in length or under . . . . roller skate
6b. Usually longer than 30 cm 7
7a. No front handle . . . . skate board
7b. Front handle . . . . wagon
8a. Two wheels . . . . motorcycle
8b. Four wheels . . . . automobile

• Questions •

1. What are the five kingdoms of living things?
2. What is the difference between the kingdom Prokaryotae and the kingdom Plantae?
3. What is the value of taxonomy?
4. An order is a collection of what similar groupings?
5. How are viruses thought to have originated?
6. Eukaryotic cells are found in which kingdoms?
7. How do viruses reproduce?
8. What are the components of a viral particle?
9. What characteristics are there in common between the members of the kingdoms Mycetae and Plantae?
10. Why are Latin names used for genus and species?
11. Who designed the present-day system of classification? How does this system differ from previous systems?
12. Why do viruses invade only specific types of cells?

• Chapter Glossary •

alternation of generations (awl″tur-na′shun uv jen″uh-ra′shunz) A term used to describe that aspect of the life cycle in which there are two distinctly different forms of an organism. Each form is involved in the production of the other and only one form is involved in producing gametes.

binomial system of nomenclature (bi-no′mi-al sis′tem ov no′men-kla-ture) A naming system that uses two Latin names, genus and species, for each type of organism.

class (class) A group of closely related families found within a phylum.

family (fam′ĭ-ly) A group of closely related species within an order.

fungus (fun′gus) The common name for the kingdom Mycetae.

gametophyte stage (gă-me′to-fīt stāj) A life-cycle stage in plants in which a haploid sex cell is produced by mitosis.

genus (je′nus) (plural, genera) A group of closely related species within a family.

host (host) A specific cell that provides what a virus or other parasitic organism needs to function.

kingdom (king′dom) The largest grouping used in the classification of organisms.

obligate intracellular parasites (ob′lĭ-gat in″trah-sel′yu-lar pĕr′uh-sīt) Infectious particles (viruses) that can function only when inside a living cell.

order (or′der) A group of closely related classes within a phylum.

phylogeny (fi-laj′uh-ne) The science that explores the evolutionary relationships among organisms and seeks to reconstruct evolutionary history.

phylum (fi′lum) A subdivision of a kingdom.

saprophyte (sap′ruh-fīt) An organism that obtains energy by the decomposition of dead organic material.

species (spe′shēz) The scientific name given to a group of organisms that can potentially interbreed to produce fertile offspring.

sporophyte stage (spor′o-fīt stāj) A life-cycle stage in plants in which a haploid spore is produced by meiosis.

taxonomy (tak-son′uh-me) The science of classifying and naming organisms.

virus (vi′rus) A nucleic acid particle coated with protein that functions as an obligate intracellular parasite.

CHAPTER 24

Prokaryotae, Protista, and Mycetae

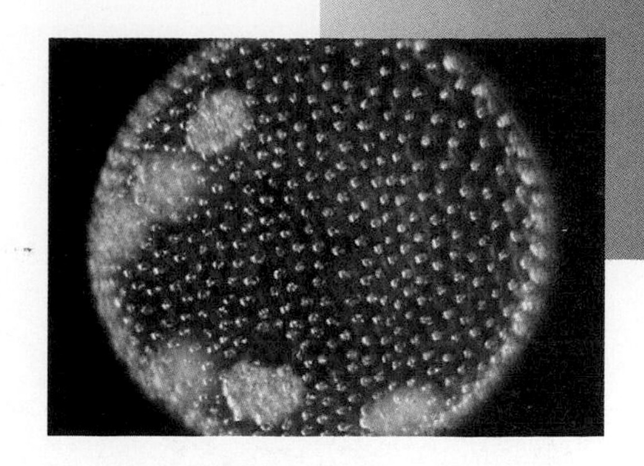

Purpose

Of the five kingdoms described in the previous chapter, three are called *microorganisms* (*microbes*), because most are so small that microscopes are needed to see them. In addition, they are relatively uncomplicated in comparison to members of the plant and animal kingdoms. This chapter gives you a glimpse into the nature of these three groups: bacteria, protists, and fungi.

For Your Information

Throughout history, fungi and humans have had a close association. For humans it has been sometimes happy and sometimes not. The pharaohs of Egypt considered mushrooms too delicious for anyone but the ruling class to eat. Roquefort cheese, mold-ripened in the famous French caves of the same name, was considered a delicacy in the time of the Roman Empire. Fungi gave a boost to education for women in 1861 when Vassar Female College was established from a fortune earned from beer.

Do you like to feed pigeons in the park? How do you feel about starlings? Did you know that humans can become infected with the disease histoplasmosis by breathing in spores from a fungus found in bird droppings? A fungus disease of grain, ergot, has been used to aid in childbirth. It has also been used for abortions and has accidentally caused death. This is only a very small sampling of the ways in which fungi have affected humans.

Chapter Outline

Learning Objectives

- List the beneficial and harmful effects of microorganisms.
- Explain how a high reproductive rate and the ability to form spores have enabled microorganisms to survive in changing environments.
- Recognize the importance of algae in aquatic environments.
- Understand the major roles played by microorganisms in ecosystems.
- List some commercial uses of microorganisms.
- Explain the life history of the organism that causes malaria.
- List the three groups of Protista and describe their differences.

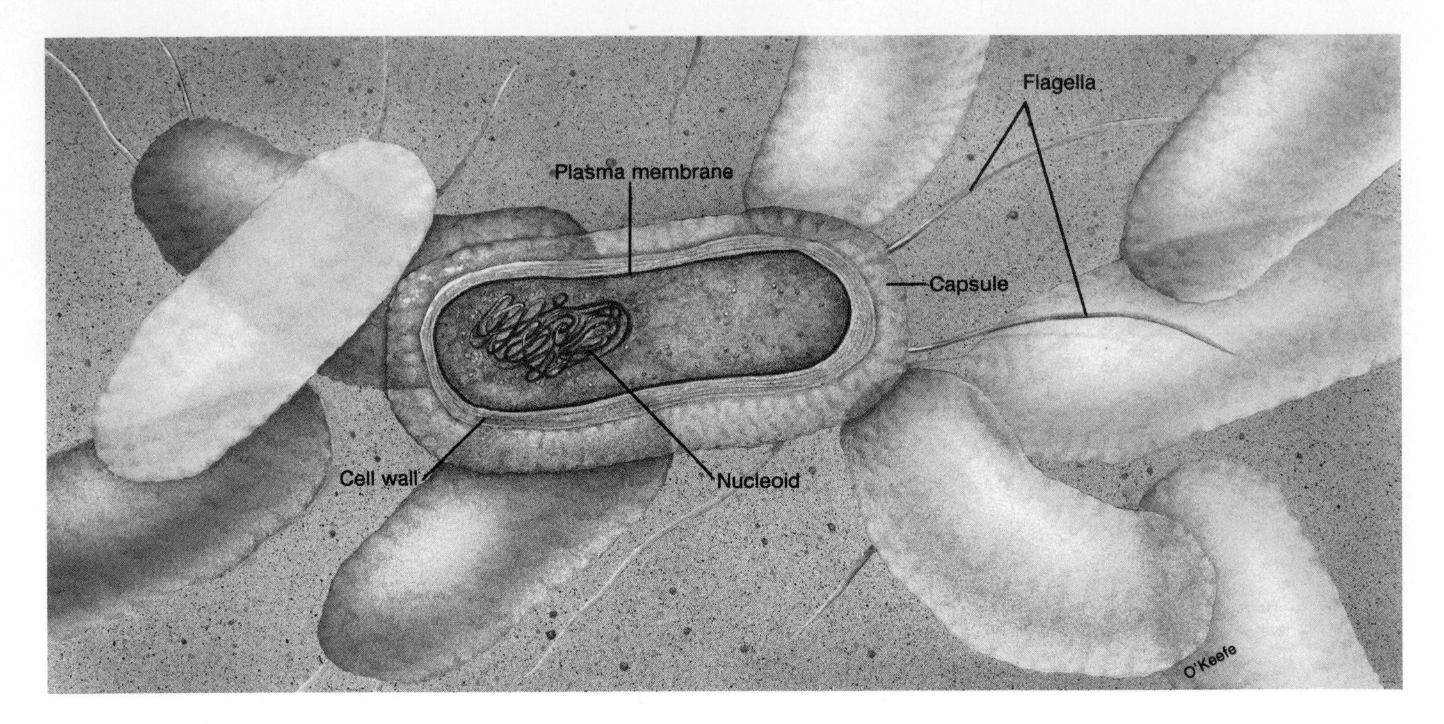

Figure 24.1 **Bacteria Cell.** The plasma membrane regulates the movement of material between the cell and its environment. A rigid cell wall protects the cell and determines its shape. Some bacteria, usually pathogens, have a capsule to protect them from the host's immune system. The genetic material is a single loop of DNA.

Microorganisms

The kingdoms Prokaryotae, Protista, and Mycetae share several characteristics that set them apart from the kingdoms Plantae and Animalia. These three kingdoms include organisms that rely primarily on asexual reproduction and have cells that routinely satisfy all of their own nutritional needs. Because the majority of organisms in these kingdoms are small and cannot be seen without some type of magnification, they are called **microorganisms,** or **microbes.**

There is minimal cooperation among the different cells of microorganisms. Some microbes are free-living, single-celled organisms, while others are collections of cells that cooperate to a limited extent. The latter type are called **colonial.** The limited cooperation of individual cells within a colony may take several forms. Some cells within a colony may specialize for reproduction, while others do not. Some colonial microbes coordinate their activities so that the colony moves as a unit. Some cells are specialized to produce chemicals that are nutritionally valuable to other cells in the colony.

Microbes are typically found in aquatic or very moist environments; most lack the specialization required to withstand drying. Since they are small, the moist habitat does not need to be large. Microbes can maintain huge populations in very small moist places like the skin of your armpits, temporary puddles, and tiled bathroom walls. Others have the special ability to become dormant and survive long periods without water. When moistened, they become actively growing cells again. The simplest of microbes are the bacteria of the kingdom Prokaryotae.

Kingdom Prokaryotae

The kingdom Prokaryotae contains organisms that are commonly referred to as **bacteria.** Other common names for them are *germs* or *microorganisms.* Bacteria of various kinds have the genetic ability to function in various environments. They are single-celled prokaryotes that lack an organized nucleus and other complex organelles (figure 24.1). *Bergey's Manual of Determinative Bacteriology* lists over seventeen hundred species of bacteria and describes the subtle differences among them. For general purposes, bacteria are divided into the three groups shown in table 24.1, although these are not true taxonomic divisions.

Many forms of bacteria are beneficial to humans. Some forms of saprophytic bacteria decompose dead material, sewage, and other forms of waste into simpler molecules that can be recycled. The food industry uses bacteria to produce cheeses, yogurt, sauerkraut, and many other forms of food. Alcohols, acetones, acids, and other chemicals are produced by bacterial cultures. The pharmaceutical industry employs bacteria to produce antibiotics and vitamins. Some bacteria can even metabolize oil and are use to clean up oil spills.

Table 24.1
Types of Bacteria

Type	Characteristics
Eubacteria (true bacteria)	Nitrogen-fixing Major pathogens Decomposers
Cyanobacteria (blue-green bacteria)	Photosynthetic; release oxygen
Archaebacteria	Tolerate extreme environments Anaerobic Oxidize inorganic molecules as a source of energy

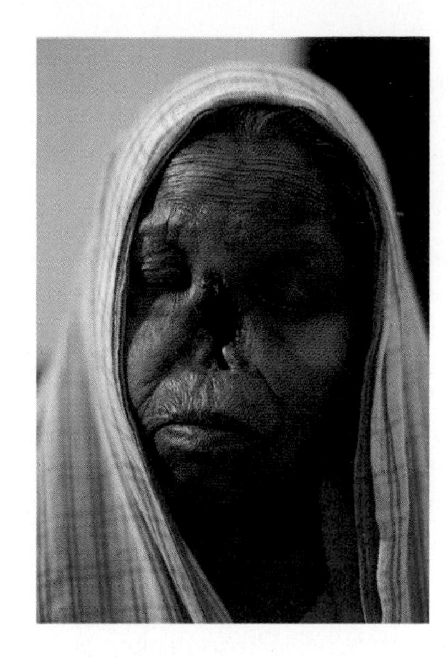

Figure 24.2 **Leprosy.** Over twenty million people worldwide are infected with *Mycobacterium leprae* and have leprosy. This disease alters the host's physiology, resulting in these open sores.

There are also mutualistic relationships between bacteria and other organisms. Some intestinal bacteria benefit humans by producing antibiotics that inhibit the development of pathogenic bacteria. They also compete with disease-causing bacteria for nutrients, thereby helping to keep the pathogens in check. They aid digestion by releasing various nutrients. They produce and release vitamin K. Mutualistic bacteria establish this symbiotic relationship when they are ingested along with food or drink. When people travel, they consume local bacteria along with their food and drink and may have problems in establishing a new symbiotic relationship with these bacteria. Both the host and the symbionts have to make adjustments to their new environment, which can result in a very uncomfortable situation for both. Some people develop traveler's diarrhea as a result.

Animals do not produce the enzymes needed for the digestion of cellulose. Methanogens, bacteria that obtain metabolic energy by reducing carbon dioxide (CO_2) to methane (CH_4), digest the cellulose consumed by herbivorous animals, such as cows. There is a mutualistic relationship between the cow and the methanogens. Some methanogens are also found in the human gut and are one of the organisms responsible for the production of gas. In some regions of the world methanogens are used to digest organic waste, and the methane is used as a source of fuel.

The Romans knew that bean plants somehow enriched the soil, but it was not until the 1800s that bacteria were recognized as being the enriching agents. Certain types of bacteria have a symbiotic relationship with the roots of bean plants and other legumes. These bacteria are capable of converting atmospheric nitrogen into a form that is usable to the plants.

Early forms of life consisted of prokaryotic cells living in a reducing atmosphere. Photosynthetic bacteria released oxygen, and the Earth's atmosphere began to change to an oxidizing atmosphere. Photosynthetic, colonial blue-green bacteria are still present in large numbers on Earth and continue to release significant quantities of oxygen. Colonies of blue-green bacteria are found in aquatic environments, where they form long, filamentous strands commonly called *pond scum*. Some of the larger cells in the colony are capable of nitrogen fixation and convert atmospheric nitrogen, N_2, to ammonia, NH_3. This provides a form of nitrogen usable to other cells in the colony—an example of division of labor.

The word *bacteria* usually brings to mind visions of tiny things that cause diseases; however, the majority are free-living and not harmful. Their roles in the ecosystem include those of decomposers, nitrogen-fixers, and other symbionts. It is true that some diseases are caused by bacteria, but only a minority of bacteria are **pathogens:** microbes that cause infectious diseases. It is normal for all organisms to have symbiotic relationships with bacteria. Most organisms are lined and covered by populations of bacteria called *normal flora*. In fact, if an organism lacks bacteria it is considered abnormal. Some pathogenic bacteria may be associated with an organism yet do not cause disease. For example, *Streptococcus pneumoniae* may grow in the throats of healthy people without any pathogenic effects. But if a person's resistance is lowered, as after a bout with viral flu, *Streptococcus pneumoniae* may reproduce rapidly in the lungs and cause pneumonia; the relationship has changed from commensalistic to parasitic.

Bacteria may invade the healthy tissue of the host and cause disease by altering the tissue's normal physiology. Bacteria living in the host release a variety of enzymes that cause the destruction of tissue. The disease ends when the pathogens are killed by the body's defenses or some outside agent, such as an antibiotic. Examples are the infectious diseases strep throat, syphilis, pneumonia, tuberculosis, and leprosy (figure 24.2).

Many other bacterial illnesses are caused by toxins or poisons produced by bacteria, which may be consumed with food or drink. In this case, disease can be caused even though the pathogens may never enter the host. For example, botulism is an extremely deadly disease that is caused by the presence of bacterial toxins in food or drink. Some other bacterial diseases are caused by toxins released from bacteria growing inside the host tissue; tetanus and diphtheria are examples. In general, toxins may cause tissue damage, fever, and aches and pains.

Bacterial pathogens are also important factors in certain plant diseases. Bacteria are the causative agents in many types of plant blights, wilts, and soft rots. Apples and other fruit trees are susceptible to fire blight, a disease that lowers the fruit yield because it kills the tree's branches. Citrus canker, a disease of citrus fruits that causes cancerlike growth, can cause widespread damage. In a three-year period, Florida citrus growers lost $2.5 billion because of this disease (figure 24.3).

Figure 24.3 **Plant Disease.** Citrus canker growth on an orange tree. This growth promotes rotting of the infected part of the tree.

Despite large investments of time and money, scientists have found it difficult to control bacterial populations. Two factors operate in favor of the bacteria: their reproductive rate and their ability to form spores. Under ideal conditions some bacteria can grow and divide every twenty minutes. If one bacterial cell and all of its offspring were to reproduce at this ideal rate, in forty-eight hours there would be 2.2×10^{43} cells. In reality, bacteria cannot achieve such incredibly large populations because they would eventually run out of food and be unable to dispose of their wastes.

Because bacteria reproduce so rapidly, a few antibiotic-resistant cells in a population can increase to dangerous levels in a very short time. This requires the use of stronger doses of antibiotics or of new types in order to bring the bacteria under control. Furthermore, these resistant strains can be transferred from one host to another. For example, sulfa drugs and penicillin, once widely used to fight infections, are now ineffective against many strains of pathogenic bacteria. As new antibiotics are developed, natural selection encourages the development of resistant bacterial strains. Therefore, humans are constantly waging battles against new strains of resistant bacteria.

Another factor that enables some bacteria to survive a hostile environment is their ability to form *endospores*. An **endospore** is a unique bacterial structure with a low metabolic rate that germinates under favorable conditions to form a new, actively growing cell (figure 24.4). For example, people who preserve food by canning often boil it to kill the bacteria. But not all bacteria are killed by boiling; some of them form endospores. For example, botulism poison is usually found in foods that are improperly canned. The endospores of *Clostridium botulinum,* the bacterium that causes botulism, can withstand boiling and remain for years in the endospore state. However, endospores do not germinate and produce botulism toxin if the pH of the canned goods is in the acid range; in that case, the food remains preserved and edible. If conditions become favorable for endospores to germinate, they become actively growing cells and produce toxin. Home canning is the major source of botulism. Using a pressure cooker and heating the food to temperatures higher than 121° C for fifteen to twenty minutes destroys both botulism toxin and the endospores.

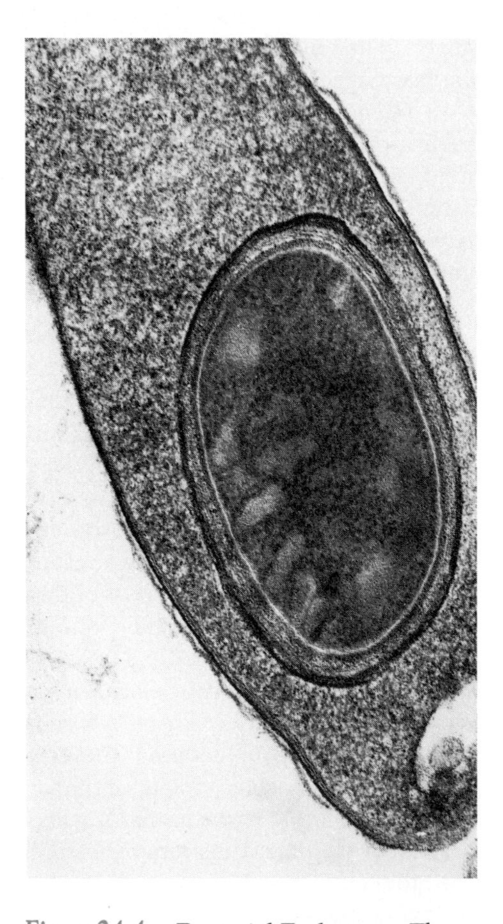

Figure 24.4 **Bacterial Endospore.** The darker area in the cell is the endospore. It contains the bacterial DNA as well as a concentration of cytoplasmic material that is surrounded and protected by a thick wall (magnification 63,000×).

Table 24.2
Types of Protista

Phylum	Algae Characteristics
Dinoflagellophyta (Dinoflagellates)	Marine; some produce toxins; colonial forms; major photosynthesizers in ocean
Chrysophyta (Golden algae)	Freshwater; form large colonies; no sexual reproduction
Bacillariophyta (Diatoms)	Aquatic; single cells or colonial; two-part silica shell; major aquatic photosynthesizers
Euglenophyta (Euglena)	Freshwater; some both photosynthetic and heterotrophic
Chlorophyta (Green algae)	Marine and freshwater; ancestors of green plants
Phaeophyta (Brown algae)	Marine; large colonial forms; form large seaweed beds
Rhodophyta (Red algae)	Mainly marine, some freshwater; multicellular
Phylum	**Protozoa Characteristics**
Zoomastigina (Flagellates)	Freshwater and marine; some parasitic; one or more flagella
Rhizopoda, Actinopoda, Foraminifera (Amoebae)	Marine and freshwater; some terrestrial; some parasitic
Sporozoa (Nonmotile)	Parasitic; complex life cycle; spore-forming
Ciliophora (Ciliates)	Marine and freshwater; cilia
Phylum	**Funguslike Protist Characteristics**
Myxomycota (Plasmodial slime mold)	Feed by phagocytosis; both haploid and diploid stages
Acrasiomycota (Cellular slime molds)	Freshwater, damp soil; feed by phagocytosis; spore-forming
Oomycota (Water mold)	Parasites or decomposers, filamentous forms, spore-forming

Kingdom Protista

The first protists evolved about 1.5 billion years ago. Like the prokaryotes, most of the protists are one-celled organisms. However, there is a significant difference between the two kingdoms: all of the protists are eukaryotic cells and all of the prokaryotes are prokaryotic cells. Prokaryotic cells usually have a volume of 1–5 cubic micrometers. Most eukaryotic cells have a volume greater than 5,000 cubic micrometers. This means that eukaryotic cells usually have a volume at least one thousand times greater than prokaryotic cells. The presence of membranous organelles such as the nucleus, ribosomes, mitochondria, and chloroplasts allows protists to be larger than prokaryotes. These organelles provide a much greater surface area within the cell upon which specialized reactions may occur. This allows for more efficient cell metabolism than is found in prokaryotic cells.

Because of the great diversity within the more than 60,000 species, it is difficult to separate the kingdom Protista into subgroupings. Usually the species are divided into three groups: **algae,** autotrophic unicellular organisms; **protozoa,** heterotrophic unicellular organisms; and funguslike protists (table 24.2).

Plantlike Protists

Algae are protists that have a cellulose cell wall. They contain chlorophyll and can therefore carry on photosynthesis. Unicellular and colonial types occur in a variety of habitats. There are two major forms of algae in a variety of marine and freshwater habitats: planktonic and benthic. **Plankton** consists of small floating or weakly swimming organisms. **Benthic** organisms live attached to the bottom or to objects in the water. **Phytoplankton** consists of photosynthetic plankton that forms the basis for most aquatic food chains (figure 24.5). The large number of benthic and planktonic

(a)

(b)

Figure 24.5 **Algae.** Algae may be found in a variety of types and colors: (a) is a single-celled green alga, *Micrasterias;* (b) is a colonial red alga, *Antithamnium.*

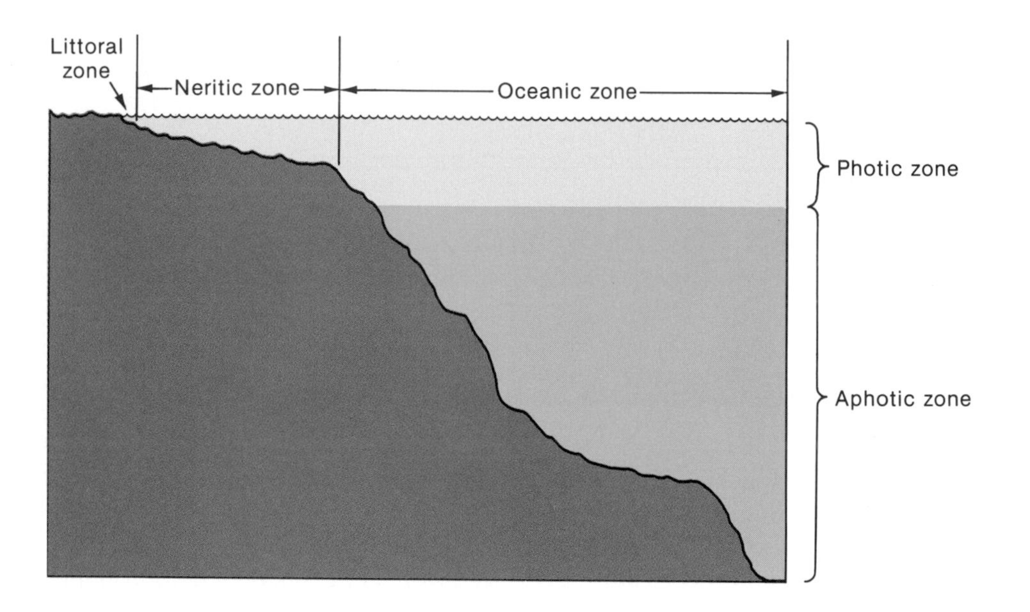

Figure 24.6 **Zones of the Ocean.** Photosynthesis can occur only in the photic zone, the top 100 meters of the oceans. This photosynthetic area contains most of the marine life. The prime source of food in the aphotic zone is the material that filters down from the photic zone. For this reason there is relatively little life in the deep ocean.

algae makes them an important source of atmospheric oxygen (O_2). It is estimated that 30%–50% of atmospheric oxygen is produced by algae.

Since algae require light, phytoplankton is found only near the surface of the water. Even in the clearest water, photosynthesis does not usually occur any deeper than 100 meters. To remain near the surface, some of the phytoplankton are capable of locomotion. Others maintain their position by storing food as oil, which is less dense than water and enables the cells to float near the surface.

Three common forms of single-celled algae typically found as phytoplankton are *Euglena* (Euglenophyta), diatoms (Bacillariophyta), and dinoflagellates (Dinoflagellata). *Euglena* are found mainly in fresh water. They are widely studied because they are easy to culture. Under certain conditions, these photosynthetic species can ingest food. *Euglena* can be either autotrophic or heterotrophic.

There are over 10,000 species of diatoms. These algae are unique because their cell walls contain silicon dioxide (silica). The algal walls fit together like the lid and bottom of a shoebox; the lid overlaps the bottom. Because their cell walls contain silicon dioxide, they readily form fossils. The fossil cell walls have large, abrasive surface areas with many tiny holes and can be used in a number of commercial processes. They are used as filters for liquids and as abrasives in specialty soaps, toothpastes, and scouring powders.

Diatoms are commonly found in freshwater and marine environments. They can reproduce both sexually and asexually. When conditions are favorable, asexual reproduction can result in what is called an algal **bloom**—a rapid increase in the population of microorganisms in a body of water. The population can become so large that the water looks murky. Along with diatoms, dinoflagellates are the most important producers in the ocean's ecosystem (figure 24.6). All members of this group of algae have two flagella, which is the reason for their name: *dino* = two. Many marine forms are bioluminescent; they are responsible for the twinkling lights seen at night in ocean waves or in a boat's wake.

Some species of dinoflagellates have a symbiotic relationship with marine animals, such as the reef corals; the dinoflagellates provide a source of nutrients for the reef-building coral. Corals that live in the light and contain dinoflagellates grow ten times faster than corals without this symbiont. Thus, in coral-reef ecosystems, dinoflagellates form the foundation of the food chain. Some forms of dinoflagellates produce toxins that can be accumulated by such filter-feeding marine animals as clams and oysters. Filter-feeding shellfish ingest large amounts of the toxin, which has no effect on the shellfish but can cause sickness or death in animals that feed on them, such as fish, birds, and mammals. Many of the toxin-producing dinoflagellates contain red pigment. Blooms of this kind are responsible for *red tides.* Red tides usually occur in the warm months, during which people should refrain from collecting and eating oysters. The expression "Oysters 'R' in season" comes from the fact that most of the months with an *R* in their spelling are cold weather months, during which oysters are safer to eat. Commercially available shellfish are tested for toxin content; if they are toxic, they are not marketed.

Multicellular algae, commonly known as *seaweed,* are large colonial forms usually found attached to objects in shallow water. Two types, red algae (Rhodophyta) and brown algae (Phaeophyta), are mainly marine forms. The green algae (Chlorophyta) are a third kind of seaweed; they are primarily freshwater species.

Red algae live in warm oceans and attach to the ocean floor by means of a holdfast structure. They may be found from the splash zone, the area where waves are breaking, to depths of 100 meters. Some red algae become encrusted with calcium carbonate and are important in reef building; other species are of commercial importance because they produce agar and carrageenin. *Agar* is widely used as a jelling agent for growth media in microbiology. *Carrageenin* is a gelatinous material used in paints, cosmetics, and baking. It is also used to make gelatin desserts harden faster and to make ice cream smoother. In Asia and Europe some red algae are harvested and used as food.

Brown algae are found in cooler marine environments than the red algae. Most species of brown algae have a holdfast organ. Colonies of these algae can reach 100 meters in length (figure 24.7). Brown algae produce *alginates,* which are widely used as stabilizers in frozen desserts, emulsifiers in salad dressings, and as thickeners that give body to foods such as chocolate milk and cream cheeses; they are also used to form gels in such products as fruit jellies.

Figure 24.7 **A Kelp Grove.** These multicellular brown algae are attached to the ocean floor by holdfasts. Their blades may reach a length of 100 meters and float upward because of a bladderlike container of air.

The Sargasso Sea is a large mat of free-floating brown algae between the Bahamas and the Azores. It is thought that this huge mass (as large as the European continent) is the result of brown algae that have become detached from the ocean bottom, have been carried by ocean currents, and accumulate in this calm region of the Atlantic Ocean. This large mass of floating algae provides a habitat for a large number of marine animals, such as marine turtles, eels, jellyfish, and innumerable crustaceans.

Green algae are found primarily in freshwater ecosystems, where they may attach to a variety of objects. Members of this group can also be found growing on trees, in the soil, and even on snow fields in the mountains. Like land plants, green algae have cellulose cell walls and store food as starch. Green algae also have the same types of chlorophyll as the plants. Biologists believe that land plants evolved from the green algae. A second major group of organisms in the kingdom Protista, the protozoa, lack all types of chlorophyll.

Animal-like Protists

The word *Protozoa* literally means "first animal." It is a descriptive term that includes all eukaryotic, heterotrophic, unicellular organisms. The protozoa are classified according to their method of locomotion. Most members of the Zoomastigina have flagella and live in fresh water. They have no cell walls and no chloroplasts, and they can be parasitic or free-living. There is a mutualistic relationship between some flagellates and their termite hosts. Certain protozoa live in termite guts and are capable of digesting cellulose into simple sugars that serve as food for the termite. Of the parasitic protozoa, two different species produce sleeping sickness in humans and domestic cattle. In both cases the protozoan enters the host as the result of an insect bite. The parasite develops in the circulatory system and moves to the cerebrospinal fluid surrounding the brain. When this occurs, the infected person develops the "sleeping" condition, which, if untreated, is eventually fatal.

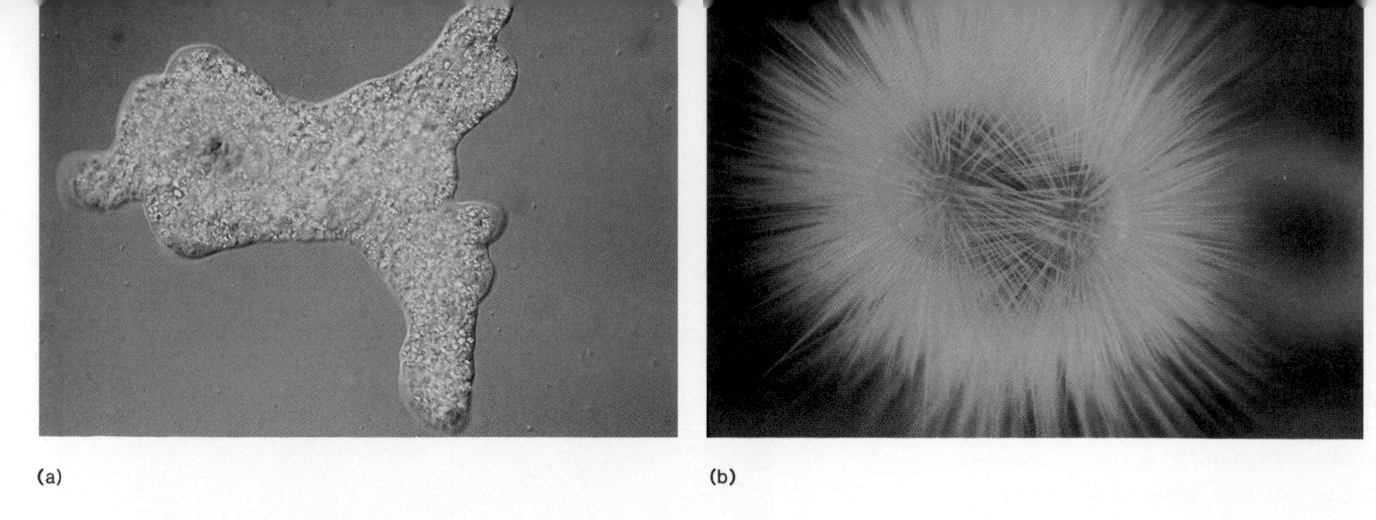

Figure 24.8 **Sarcodina.** These protozoa range from (a) the *Amoeba,* which changes shape to move and feed, to (b) organisms that are enclosed in a shell.

Many biologists believe that all other types of protozoa, and even the multicellular animals, evolved from primitive flagellated microorganisms similar to the Zoomastigina.

Members of the phyla Rhizopoda, Actinopoda, and Foraminifera range from the most well known *Amoeba,* with its constantly changing shape, to species having a rigid outer cover (figure 24.8). *Amoeba* uses pseudopods to move about and to engulf food. A pseudopod is a protoplasmic extension of the cell that contains moving cytoplasm. Many pseudopods are temporary extensions that form and disappear as the cell moves. Most amoeboid protists are free-living and feed on bacteria, algae, or even small multicellular organisms. Some forms are parasitic, such as the one that causes amoebic dysentery in humans.

Foraminiferans live in warm oceans and are enclosed in a shell. As these cells die, the shells collect on the ocean floor, and their remains form limestone. The cliffs of Dover, England, were formed from such shells. Oil companies have a vested interest in foraminiferans because they are often found where oil deposits are located.

All members of the Sporozoa are nonmotile parasites that have a sporelike stage in their life cycle. Malaria, one of the leading causes of disability and death in the world, is caused by a type of sporozoan. Two billion people live in malaria-prone regions of the world. There are an estimated 150 to 300 million new cases of malaria each year, and the disease kills 2 to 4 million people annually.

Like most sporozoans the one that causes malaria has a complex life cycle involving a mosquito vector for transmission (figure 24.9). Recall from chapter 15 that a *vector* is an organism capable of transmitting a parasite from one organism to another. While in the mosquito vector, the parasite goes through the sexual stages of its life cycle. One of the best ways to control this disease is to eliminate the vector, which usually involves using some sort of pesticide. Many of us are concerned about the harmful effects of pesticides in the environment. However, in parts of the world where malaria is common, the harmful effects of pesticides are of less concern than the harm caused by the disease. Many diseases of domestic and wild animals are also caused by members of this phylum.

The Ciliophora are the most structurally complex protozoans. They are commonly known as *ciliates* and derive their name from the fact that they have numerous short, flexible filaments called *cilia* (figure 24.10). These move in an organized, rhythmic manner and propel the cell through the water. Some types of ciliates, such as *Paramecium,* have nearly 15,000 cilia per cell and move at a rapid speed of one millimeter per second. Most ciliates are free-living cells found in fresh and salt water, where they feed on bacteria and other small organisms.

Funguslike Protists

Funguslike protists have a motile amoeboid reproductive stage, which differentiates them from true fungi. There are two kinds of funguslike protists: the slime molds and water molds. Some slime molds, members of Myxomycota, can be found growing on rotting damp logs, leaves, and soil. They look like giant amoebae whose nucleus and other organelles have divided repeatedly within a single large cell (figure 24.11). No cell membranes partition this mass into separate segments. They vary in color from white to bright red or yellow, and may reach relatively large sizes (45 centimeters in length) when in an optimum environment.

Other kinds of slime molds, members of Acrasiomycota, exist as large numbers of individual, amoebalike cells. These haploid cells get food by engulfing microorganisms. They reproduce by mitosis. When their environment becomes dry or otherwise unfavorable, the cells come together into an irregular mass. This mass glides along rather like an ordinary garden slug and is labeled the sluglike stage. This sluglike form may flow about for hours before it forms spores. When the mass gets ready to form spores, it forms a stalk with cells that have cell walls. At the top of this specialized structure, cells are modified to become haploid spores. When released, these spores may be carried by the wind and, if they

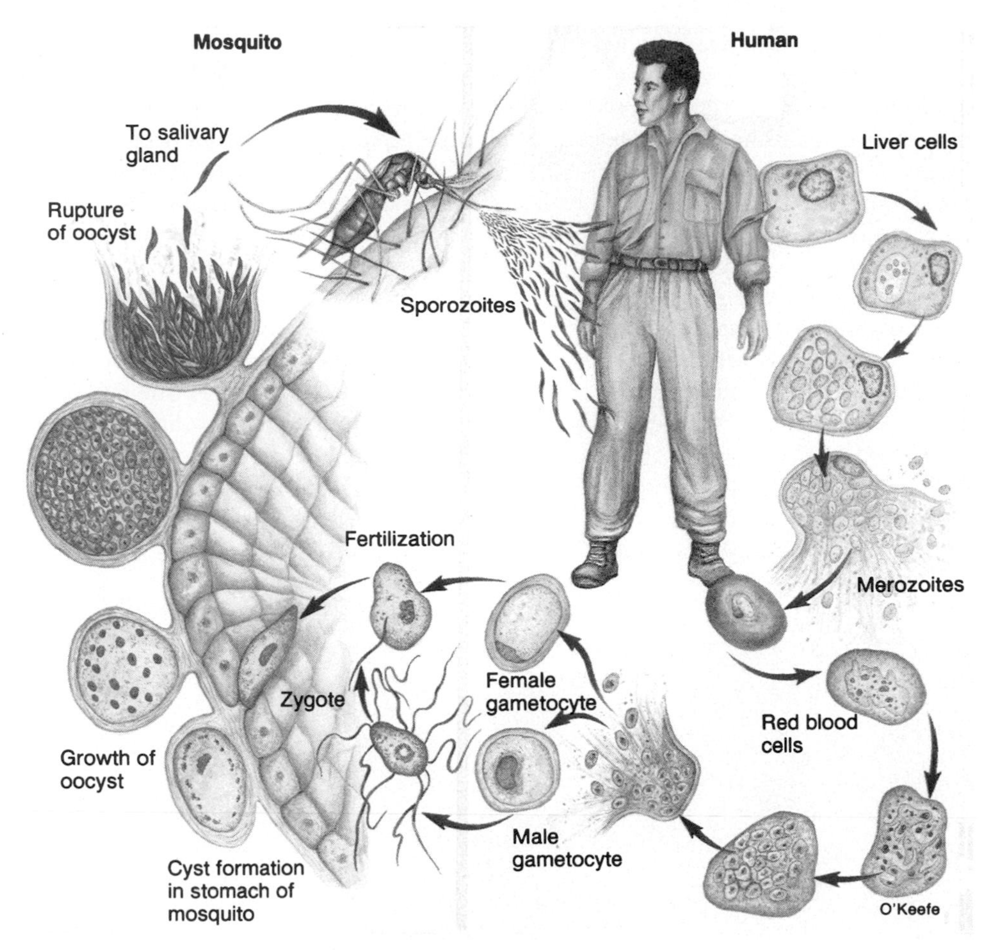

Figure 24.9 **The Life Cycle of *Plasmodium vivax*.** The complex life cycle of the member of the Protista that causes malaria requires two hosts, the *Anopheles* mosquito and the human. Humans get malaria when they are bitten by a mosquito carrying the larval stage of *Plasmodium*. The larva undergoes asexual reproduction and releases thousands of individuals that invade the red blood cells. Their release causes the chills, fever, and headache associated with malaria. Inside the red blood cell, more reproduction occurs to form male gametocytes and female gametocytes.

When the mosquito bites a person with malaria, it ingests some gametocytes. Fertilization occurs and zygotes develop. The resulting larvae are housed in the mosquito's salivary gland. Then, when the mosquito bites someone, some saliva containing the larvae is released into the person's blood and the cycle begins again.

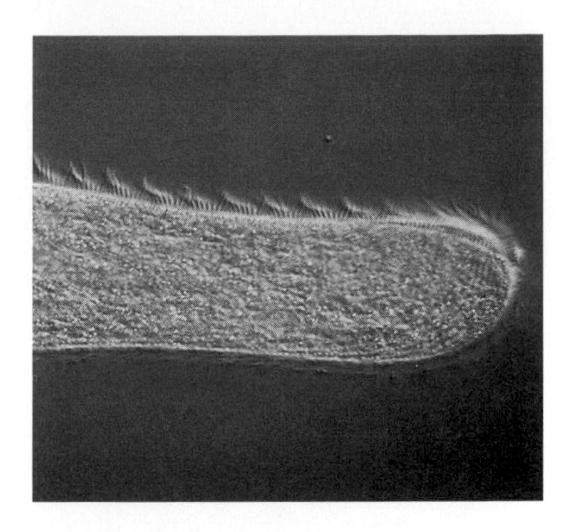

Figure 24.10 **Ciliated Protozoa.** The many hairlike cilia on the surface of this cell are used to propel the protozoan through the water.

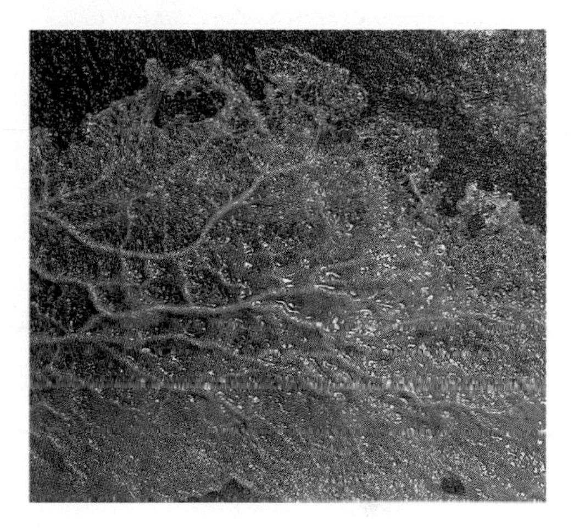

Figure 24.11 **Slime Mold.** Slime molds grow in moist conditions and are important decomposers. As the slime mold grows, additional nuclei are produced by mitosis, but there is no cytoplasmic division. Thus, at this stage, it is a single mass of cytoplasm with many nuclei.

land in a favorable place, develop into new amoebalike cells.

Another group of funguslike protists includes the water molds. This group, the Oomycota, has reproductive cells with two flagella. A wide variety of water molds are saprophytes, which are usually found growing in a moist environment. They differ in structure from the true fungi in that some filaments have no cross walls, thus allowing the cell contents to flow from cell to cell.

Water molds are important saprophytes and parasites in aquatic ecosystems. They are often seen as fluffy growths on dead fish or other organic matter floating in water. A parasitic form of this fungus is well known to people who rear tropical fish; it causes a cottonlike growth on the fish. Although these organisms are usually found in aquatic habitats, they are not limited to this environment. Some species cause downy mildew on plants such as grapes. In the 1880s this mildew almost ruined the French wine industry when it spread throughout the vineyards. A copper-based fungicide called *Bordeaux mixture*—the first chemical used against plant diseases—was used to save the vineyards. A water mold was also responsible for the Irish potato blight. In the nineteenth century, potatoes were the staple of the Irish diet. Cool, wet weather in 1845 and 1847 damaged much of the potato crop, and over a million people died of starvation. Nearly one-third of the survivors left Ireland and moved to Canada or the United States.

Multicellularity in the Protista

The three major types of the kingdom Protista (algae, protozoa, and funguslike protists) include both single-celled and multicellular forms. Biologists believe that there has been a similar type of evolution in all three of these groups. The most primitive organisms in each group are thought to have been single-celled, and we believe these gave rise to the more advanced multicellular forms. Most protozoan organisms are single-celled; however, there is a group that contains numerous colonial forms. The multicellular forms of funguslike protists are the slime molds, which have both single-celled and multicellular stages. Perhaps the most widely known example of this trend from a single-celled to a multicellular condition is found in the green algae. A very common single-celled green alga is *Chlamydomonas*, which has a cell wall and two flagella. It looks just like the individual cells of the colonial green algae *Volvox*. *Volvox* can be composed of more than half a million cells (figure 24.12). All the flagella of each cell in the colony move in unison, allowing the colony to move in a given direction. Many of the cells cannot reproduce sexually; other cells assume this function for the colony. In some *Volvox* species, certain cells have even specialized to produce sperm or eggs. Biologists believe that the division of labor seen in colonial protists represents the beginning of specialization that led to the development of true multicellular organisms with many different kinds of specialized cells. Three types of multicellular organisms—fungi, plants, and animals—eventually developed.

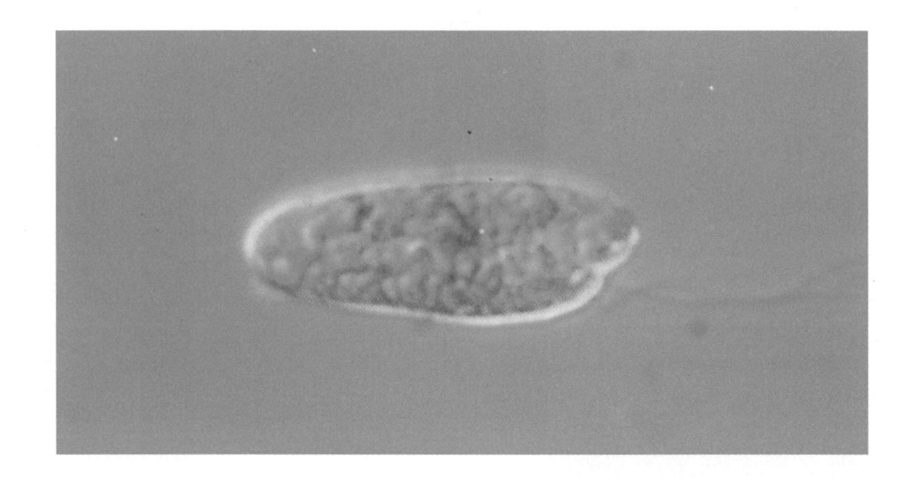

(a)

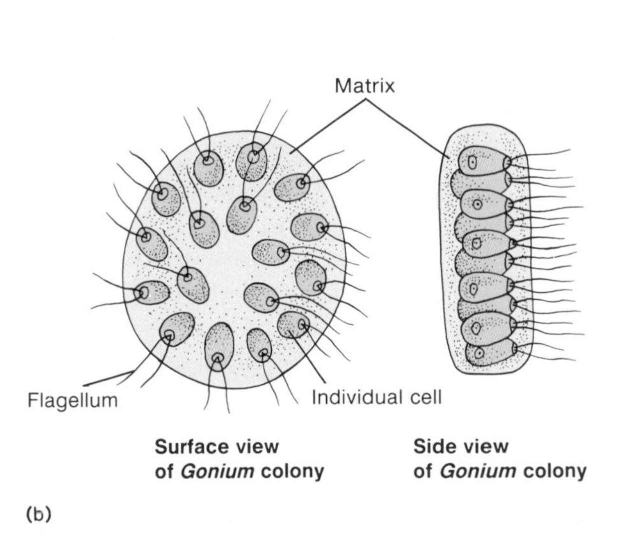

(b)

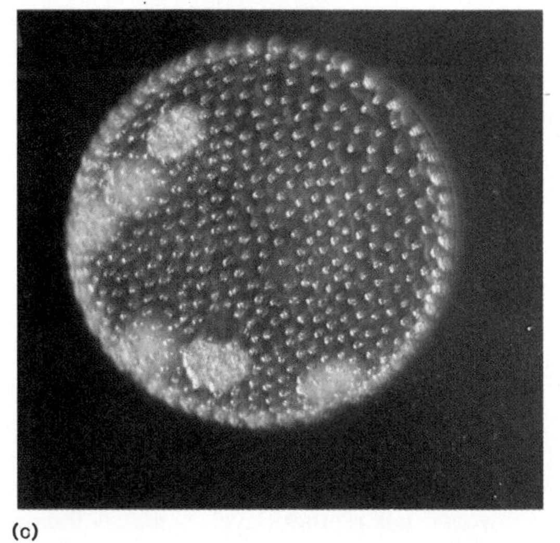

(c)

Figure 24.12 **Algae.** (a) *Chlamydomonas*, a green, single-celled alga containing the same type of chlorophyll as that found in green plants, is thought to be the ancestor of green plants. (b) *Gonium*, a green alga similar to *Chlamydomonas*, forms colonies composed of four to thirty-two cells. (c) Volvox, another green alga, is a more complex form in the evolution of colonial green algae. (Magnifications: (a) 800×, (c) 100×.)

Kingdom Mycetae

Members of the kingdom Mycetae are nonphotosynthetic, eukaryotic organisms, and the majority are multicellular. There is no unanimity regarding the divisions within the kingdom Mycetae. Originally, fungi were thought to be members of the Plantae kingdom. In fact the term *division* is used with this kingdom because this is the term used by the botanists in place of *phylum* (table 24.3).

Even though fungi are nonmotile, they are very successful in surviving and dispersing because of their ability to form spores, which some produce sexually and others produce asexually. Spores may be produced internally or externally (figure 24.13). An averge-sized mushroom can produce over 20 billion spores; a good sized puffball can produce as many as 8 trillion spores. When released, the spores can be transported by wind or water. Because of their small size, spores can remain in the atmosphere for long periods of time and travel thousands of kilometers. Fungal spores have been collected as high as 50 kilometers above the earth.

In a favorable environment, a fungus produces dispersal spores, which are short-lived and germinate quickly under suitable conditions. If the environment becomes unfavorable—too cold or hot, or too dry—the fungus produces survival spores. These may live for years before germinating. Fungi are so prolific that their spores are almost always present in the air; as soon as something dies, fungal spores settle on it, and decomposition usually begins (figure 24.14).

Fungi play a variety of roles. They are used in the processing of food and are vital in the recycling processes within ecosystems. As decomposers, they destroy billions of dollars worth of material each year; as pathogens, they are responsible for certain diseases. They are beneficial in the production of antibiotics and other chemicals used in the treatment of diseases. *Penicillium chrysogenum* is a mold that produces the

Table 24.3
Fungi Characteristics

Division (phylum)	Characteristics
Zygomycota (bread mold)	No cross wall in hypha; mainly saprophytic; form spores asexually; some sexual reproduction
Ascomycota (sac fungi)	Form visible fruiting bodies and spores; sexual reproduction; some one-celled species (yeasts)
Basidiomycota (club fungi)	Spore-forming; sexual reproduction; decomposers; some parasitic on plants; mushrooms
Deuteromycota (imperfect fungi)	Sexual reproduction has never been seen; spore-forming
Mycophycophyta (lichens)	Not a single species, but a symbiotic relationship between alga and fungus

Figure 24.13 **Spore Production.** Some fungi, like the puffball (a), produce spores on the inside. The puffball must be broken (b) to release the spores. Other forms, like the club fungus (c), have exposed gills with spore-producing basidia (d).

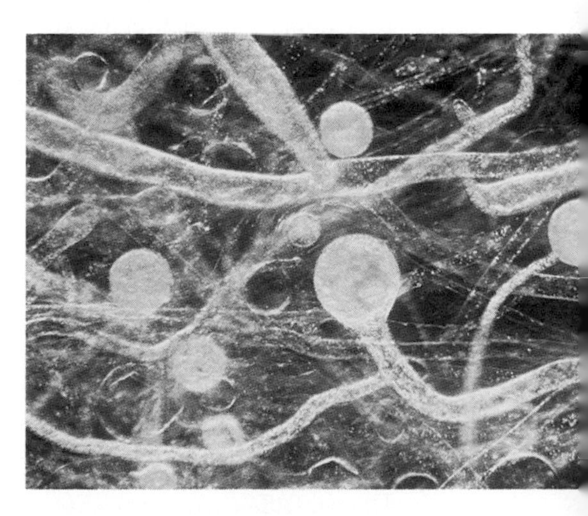

Figure 24.14 **Water Mold.** Rapidly reproducing water molds are fungi that quickly produce a large mass of filamentous hyphae. These hyphae are the cause of fuzzy growth often seen on dead fish or other dead material in the water.

BOX 24.1 Penicillin

The discovery of the antibiotic penicillin is an interesting story. In 1928, Dr. Alexander Fleming was working at St. Mary's Hospital in London. As he sorted through some old petri dishes on his bench, he noticed something unusual. The mold *Penicillium notatum* was growing on some of the petri dishes. Apparently, the mold had found its way through an open window and onto a bacterial culture of *Staphylococcus aureus.* The bacterial colonies that were growing at a distance from the fungus were typical, but there was no growth close to the mold (see figure). Fleming isolated the agent responsible for this destruction of the bacteria and named it *penicillin.*

Through Fleming's research efforts and those of several colleagues, the chemical was identified and used for about ten years in microbiological work in the laboratory. Many suspected that penicillin might be used as a drug, but the fungus could not produce enough of the chemical to make it worthwhile. When World War II began, and England was being firebombed, there was an urgent need for a drug that would control bacterial infections in burn wounds. Two scientists from England were sent to the United States to begin research into the mass production of penicillin.

Their research in isolating new forms of *Penicillium* and purifying the drug were so successful that cultures of the mold now produce over one hundred times more of the drug than the original mold discovered by Fleming. In addition, the price of the drug dropped considerably—from a 1944 price of $20,000 per kilogram to a current price of less than $250.00. The species of *Penicillium* used to produce penicillin today is *P. chrysogenum,* which was first isolated in Peoria, Illinois, from a mixture of molds found growing on a cantaloupe. The species name, *chrysogenum,* means "golden" and refers to the golden-yellow droplets of antibiotic that the mold produces on the surface of its hyphae. The spores of this mold were isolated and irradiated with high dosages of ultraviolet light, which caused mutations to occur in the genes. When some of these mutant spores were germinated, the new hyphae were found to produce much greater amounts of the antibiotic.

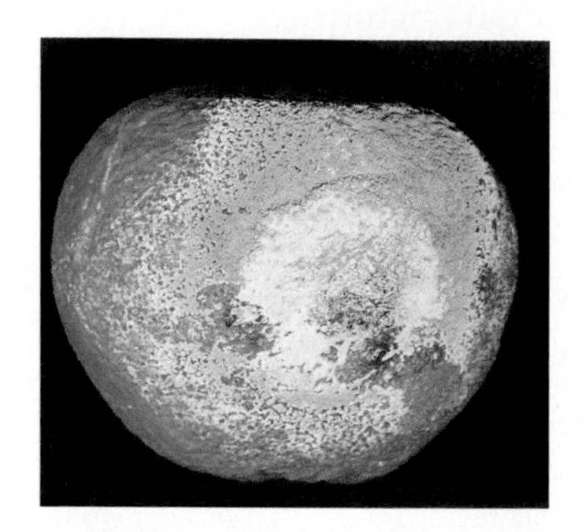

antibiotic penicillin, which was the first commercially available antibiotic and is still widely used (see box 24.1).

There are over one hundred species of *Penicillium,* and each characteristically produces spores in a brushlike border; the word *penicillus* means *little brush.* Members of this group do more than just produce antibiotics; they are also widely used in processing food. Many people are familiar with the blue, cottony growth that sometimes occurs on citrus fruits. The *P. italicum* growing on the fruit appears to be blue because of the pigment produced in the spores. The blue cheeses, such as Danish, American, and the original Roquefort cheese, all have this color. Each has been aged with *P. roquefortii* to produce the color, texture, and flavor. Differences in the cheese are determined by the kind of milk used and the conditions under which the aging occurs. Roquefort cheese is made from sheep's milk and aged in Roquefort, France, in particular caves. American blue cheese is made from cow's milk and aged in many places around the United States. The blue color has become a very important feature of these cheeses. The same research laboratory that first isolated *P. chrysogenum* also found a mutant species of *P. roquefortii* that would produce spores having no blue color. The cheese made from this mold is "white" blue cheese. The flavor is exactly the same as "blue" blue cheese, but commercially it is worthless: people want the blue color.

Fungi and their by-products have been used as sources of food for centuries. When we think of fungi and food, mushrooms usually come to mind. The common mushroom found in the grocer's vegetable section is grown in seventy countries and has an annual market value in the billions of dollars. But there are other uses for fungi as food. *Shoyu* (soy sauce) was originally made by fermenting a mixture of wheat, soybeans, and an ascomycote fungus for a period of a year. Most of the soy sauce used today, however, is made by a cheaper method of processing soybeans with hydrochloric acid. However, true connoisseurs still prefer soy sauce made the original way. Another mold is important to the soft-drink industry. The citric acid that gives a soft drink its sharp taste was originally produced by squeezing juice from lemons and purifying the acid. Today, however, a mold is grown on a nutrient medium with table sugar (sucrose) to produce great quantities of citric acid at a low cost.

All fungi are capable of breaking down organic matter to provide themselves with the energy and building materials they need. This may be either beneficial or harmful, depending on what is being broken down. In order for any ecosystem to survive, it must have a source of carbon, nitrogen, phosphorus, and other elements that can be incorporated into new carbohydrates, fats, proteins, and other molecules necessary for growth. The fungi, along with bacteria, are the primary recycling agents for these elements in ecosystems.

Spores are an efficient method of dispersal, and when they land in a favorable environment with moist conditions, they germinate and begin the process of decomposition. As decomposers, fungi cause billions of dollars worth of damage each year. Clothing, wood, leather, and all types of food are susceptible to damage by fungi. One of the best ways to protect against such damage is to keep the material dry because fungi grow best in a moist environment. Millions of dollars are also spent each year on fungicides to limit damage due to fungi.

Some fungi have a symbiotic relationship with plant roots; **mycorrhizae** usually grow inside a plant's root-hair cells—the cells through which plants absorb water and nutrients. The hyphae from the fungus grow out of the root-hair cells and greatly increase the amount of absorptive area (figure 24.15). Plants with mycorrhizal fungi can absorb as much as ten times more minerals than those without the fungi. Some types of fungi also supply plants with growth hormones, while the plants supply carbohydrates and other organic compounds to the fungi. Mycorrhizal fungi are found in 80%–90% of all plants.

Figure 24.15 **Mycorrhiza.** The symbiotic relationship between a fungi and the roots of the two plants on the right increases the intake of water and nutrients into the plant. As a result these plants have more growth than the control plant on the left.

Figure 24.16 **Fairy Ring.** Legend tells us that fairies danced in a circle in the moonlight and rested on the mushrooms. Mycologists tell us that the mushrooms began to grow in the center; as the organic material was consumed, the mushrooms grew in an ever-widening circle and formed this "fairy ring."

In some situations, mycorrhizae may be essential to the life of a plant. Botanists are investigating a correlation between mycorrhizae and acid-rain damage to trees. Acid-rain conditions can leach certain necessary plant minerals from the soil, making them less accessible to plants. The increased soil acidity also makes certain toxic chemicals, such as copper, more accessible to plants. When the roots of trees suspected of being killed by acid rain are examined, there is often no evidence of the presence of mycorrhizal fungi, while a healthy tree growing next to a dead one has the root fungus.

One of the most interesting formations caused by mushroom growth can be seen in soil that is rich in mushroom hyphae, such as in lawns, fields, and the forest. These formations, known as *fairy rings* (figure 24.16), result from the expanding growth of the mushrooms. The inner circle is normal grass and vegetation. The mushroom population originally began to grow at the center, but grew out from there because it exhausted the soil nutrients necessary for fungal growth. As the microscopic hyphae grow outward from the center, they stunt the growth of grass, forming a ring of short, inhibited grass. Just to the outside of this growth ring, the grass is luxuriant because the hyphae excrete enzymes that decompose soil material into rich nutrients for growth. The name *fairy ring* comes from an old superstition that such rings were formed by fairies tramping down the grass while dancing in a circle.

Figure 24.17 **Corn Smut.** Most people who raise corn have seen corn smut. Besides being unsightly, it decreases the corn yield.

(a)

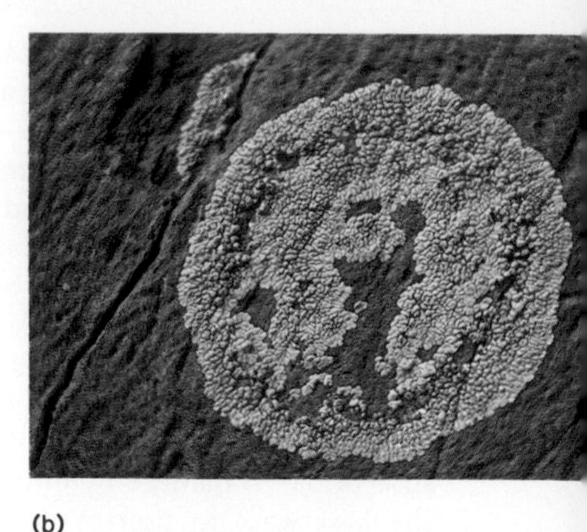

(b)

Figure 24.18 **Lichens.** Lichens grow in a variety of habitats: (a) the shrubby lichen is growing on soil; (b) the crustlike lichen is growing on rock. The different coloring is due to the different species of algae or cyanobacteria in the lichens.

There are also pathogenic fungi that feed on living organisms; those that cause ringworm and athlete's foot are two examples. A number of diseases are caused by fungi that grow on human mucous membranes, such as those of the vagina, lungs, and mouth.

Plants are also susceptible to fungal attacks. The chestnut blight and later the Dutch elm disease almost caused these two species of trees to become extinct. The fungus that causes Dutch elm disease is a parasite that kills the tree; then it functions as a saprophyte and feeds on the dead tree. Fungi also damage certain domestic crops. Wheat rust gets its common name because infected plants look as if they are covered with rust. Corn smut is also due to a fungal pathogen of plants (figure 24.17).

A number of fungi produce deadly poisons called **mycotoxins.** There is no easy way to distinguish those that are poisonous from those that are safe to eat. The poisonous forms are sometimes called *toadstools* and the nonpoisonous ones, *mushrooms.* However, they are all properly called mushrooms. The origin of the name toadstools is unclear. One idea is that toadstools are mushrooms on which toads sit; another is that the word is derived from the German *todstuhl* "seat of death." The most deadly of these, *Aminita verna,* is known as "the destroying angel" and can be found in woodlands during the summer. Mushroom hunters must learn to recognize this deadly, pure white species. This mushroom is believed to be so dangerous that food accidentally contaminated by its spores can cause illness and possible death. Another mushroom, *Psilocybe mexicana,* has been used for centuries in religious ceremonies by certain Mexican tribes because of the hallucinogenic chemical that it produces. These mushrooms have been grown in culture, and the drug psilocybin has been isolated. In the past, it was used experimentally to study schizophrenia. *Claviceps purpurea,* a sac fungus, is a parasite on rye and other grains. The metabolic activity of *C. purpurea* produces a toxin that can cause hallucinations, muscle spasms, insanity, or even death. However, it is also used to treat high blood pressure, to stop bleeding after childbirth, and to treat migraine headaches.

Lichens

Lichens are usually classified with the Mycetae, but they actually represent a very close mutualistic relationship between a fungus and an algal protist or a cyanobacterium. Algae and cyanobacteria require a moist environment. Certain species of these photosynthetic organisms grow surrounded by fungus. The fungal covering maintains a moist area, and the photosynthesizers in turn provide nourishment for the fungus. These two species growing together are what we call a *lichen* (figure 24.18). Lichens grow slowly; a patch of lichen may only grow 1 centimeter per year in diameter.

Since the fungus provides a damp environment and the algae produce the food, lichens require no soil for growth. For this reason, they are commonly found growing on bare rock, and are the pioneer plants in the process of succession. Lichens are important in the process of soil formation. They secrete an acid that weathers the rock and makes minerals available for use by plants. When lichens die, they provide a source of humus—dead organic material—that mixes with the rock particles to form soil.

Lichens are found in a wide variety of environments, ranging from the frigid arctic to the scorching desert. One reason for this success is their ability to withstand drought conditions. Some lichens can survive with only 2% water by weight. In this condition they stop photosynthesis and go into a dormant stage, remaining so until water becomes available and photosynthesis begins again.

Another factor in the success of lichens is their ability to absorb minerals. However, because air pollution has increased the amounts of minerals they absorb many lichens are damaged. Some forms of lichens absorb concentrations of sulfur one thousand times greater than those found in the atmosphere. This increases the amount of sulfuric acid in the lichen, resulting in damage or death. For this reason, areas with heavy air pollution are "lichen deserts." Because they can absorb minerals, certain forms of lichens have been used to monitor the amount of various pollutants in the atmosphere, including radioactivity. The absorption of radioactive fallout from Chernobyl by arctic lichens made the meat of the reindeer that fed on them unsafe for human consumption.

• Summary •

The kingdoms Prokaryotae, Protista, and Mycetae rely mainly on asexual reproduction and each cell usually satisfies its own nutritional needs. In some species, there is minimal cooperation between cells. The members of the kingdom Prokaryotae are bacteria, which have the genetic ability to function in various environments. Most species of bacteria are beneficial, although some are pathogenic.

Members of the kingdom Protista are one-celled organisms. They differ from the prokaryotes in that they are eukaryotic cells, whereas the prokaryotes are prokaryotic cells. Protists include algae, autotrophic cells that have a cell wall and carry on photosynthesis; protozoa, which lack cell walls and cannot carry on photosynthesis; and funguslike protists, whose motile, amoeboid reproductive stage distinguishes them from true fungi. Some species of Protista developed a primitive type of specialization, and from these evolved the multicellular fungi, plants, and animals.

The kingdom Mycetae consists of nonphotosynthetic, eukaryotic organisms. Most species are multicellular. Fungi are nonmotile organisms that disperse by producing spores. Lichens are a combination of organisms involving a mutualistic relationship between a fungus and an algal protist or cyanobacterium.

• Thinking Critically •

Throughout much of Europe there has been a severe decline in the mushroom population. On study plots in Holland, data collected since 1912 indicate that the number of different mushroom species has dropped from thirty-seven to twelve per plot in recent years. Along with the reduction in the number of species there is a parallel decline in the number of individual plants; moreover, the surviving plants are smaller.

The phenomenon of the disappearing mushrooms is also evident in England. One study noted that in sixty fungus species twenty exhibited declining populations. Mycologists are also concerned about a decline in the United States; however, there are no long-term studies, such as those in Europe, to provide evidence for such a decline.

Consider the niche of fungi in the ecosystem. How would an ecosystem be affected by a decline in their numbers?

• Experience This •

Practical Application of Microbes, Yeasts, and Bacteria:

Yeast Raised Sourdough Bread

This bread depends on a bacterial sourdough starter culture for its flavor and a commercial yeast to raise (leaven) the dough.

To prepare sourdough starter:

In a nonmetallic container, and using a nonmetallic spoon, mix the following:

- 3 cups milk
- 3 cups flour
- ¼ cup yogurt (unflavored, with active culture)

Cover and place mixture on top of the refrigerator for twenty-four hours to incubate. This will result in an active growing culture of acid-producing bacteria.

To make sourdough bread:

- 2 cups sourdough starter (see above)
- 1 cup scalded milk
- 3 tablespoons butter or oleo (softened)
- 3 tablespoons sugar
- 2 teaspoons salt
- 1 package yeast
- ¼ cup warm water
- 6½ cups flour
- 1 teaspoon baking soda

Mix scalded milk with butter to melt. Add sugar and salt. Stir yeast into warm water and let stand five minutes. Then add the yeast and sourdough starter to the cooled milk mixture. Beat in 2 cups flour to make a smooth batter and over this sprinkle the baking soda. Mix well, cover with a cloth, and set aside to rise for about thirty minutes.

Gradually mix in the rest of the flour until dough is stiff. Knead the dough on a floured surface until smooth and elastic, adding more flour to prevent excessive stickiness. Let stand about ten minutes.

Oil two loaf pans, divide the dough into two halves and place them into the prepared pans. Turn each loaf to coat lightly with the oil. Let rise about one hour. Bake at 375° F for fifty minutes.

To maintain starter:
To the leftover starter add 1 cup milk and 1 cup flour. Place on top of the refrigerator for twenty-four hours, after which it can be refrigerated until further use.

Before using the starter in another recipe, be sure to build it up (create additional volume) the night before by adding milk and flour in sufficient amounts to make enough for your recipe and have some left over for future use.

• Questions •

1. Why are the protozoa and the algae in different subgroups of the kingdom Protista?
2. What is meant by the term *bloom?*
3. What is a pathogen? Give two examples.
4. What are the two types of poison produced by bacteria?
5. Name two beneficial results of fungal growth and activity.
6. Define the term *saprophytic.*
7. Give an example of a symbiotic relationship.
8. What is a bacterial endospore?
9. What is phytoplankton?
10. Name three commercial uses of algae.
11. What is the best method to prevent the spread of malaria?
12. What types of spores do fungi produce?

• Chapter Glossary •

algae (al′je) Protists that have cell walls and chlorophyll and can therefore carry on photosynthesis.

bacteria (bak-tĭr′ē-ah) Unicellular organisms of the kingdom Prokaryotae that have the genetic ability to function in various environments.

benthic (ben′thik) A term used to describe organisms that live in bodies of water, attached to the bottom or to objects in the water.

bloom (bloom) A rapid increase in the number of microorganisms in a body of water.

colonial (ko-lo′ne-al) A term used to describe a collection of cells that cooperate to a small extent.

endospore (en′do-spōr″) A unique bacterial structure with a low metabolic rate that germinates under favorable conditions to grow into a new cell.

lichen (li′kĕn) A mutualistic relation between fungi and algal protists or cyanobacteria.

microorganisms (microbes) (mi″kro-or′guh-niz″mz) Small organisms that cannot be seen without some type of magnification.

mycorrhiza (my″ko-rye′zah) A symbiotic relation between fungi and plant roots.

mycotoxin (mi″ko-tok′sin) A deadly poison produced by fungi.

pathogen (path′uh-jen) An agent that causes a specific disease.

phytoplankton (fye-tuh-plank′tun) Photosynthetic species that form the basis for most aquatic food chains.

plankton (plank′tun) Small floating or weakly swimming organisms.

protozoa (pro″to-zo′ah) Heterotrophic, unicellular organisms.

C H A 25 P T E R

Plantae

Purpose

You know a plant when you see it—it's the green organism that doesn't run away from you. Their green pigment and nonmobility are just two of the many interesting and important characteristics of plants. In this chapter you will become familiar with some of these characteristics and begin to understand the basis for classifying and naming plants. Plant biologists, *botanists*, have devised a scheme that they believe expresses the evolutionary advances that have occurred over millions of years. This chapter is organized around that scheme.

Chapter Outline

For Your Information

Myths of the Mandrake

Plants have been the subject of some fascinating myths, among which are those concerning the mandrake. This plant is of the genus *Mandragora,* which is native to the Mediterranean area. We are not concerned here with the American mandrake, or mayapple, which belongs to an entirely different family.

The mandrake has a taproot much like a carrot, but it sometimes branches to form two leglike structures. With a little imagination, a human shape can be visualized in some of the roots, especially when someone has skillfully carved the root a little. Some of the ancient books on herbs contain sketches of mandrake plants with roots that look like male and female nude figures. All of this gave rise to a belief in the curative powers of the plant for all sorts of ailments, including lovesickness. The belief that it counteracted sterility seems to be implied in the biblical account of Leah and Jacob in Genesis 30:14–16. Shakespeare reflected another myth in *Romeo and Juliet* in Juliet's soliloquy in Act IV. It was believed that when one dug up a mandrake plant, the plant gave a terrible shriek and released such a foul odor that anyone in the area would have difficulty withstanding it.

Learning Objectives

- Describe alternation of generations in the mosses.
- Discuss two distinctive ways in which mosses depend on water.
- State the advantage of vascular tissue to the ferns.
- State the advantage of the diploid condition of the fern sporophyte.
- Describe the life cycle of gymnosperms.
- State the differences between the two major groups of angiosperms.

Mosses

The first classification of plants was based on their methods of reproduction. Carolus Linnaeus later categorized plants according to the number and position of male parts in flowers. Although almost everyone has looked closely at a **flower,** few people recognize it as a structure associated with sexual reproduction in plants. To understand how and why the flower evolved, we need to go back in time and examine the primitive structures that were modified over and over again until the flower came into existence.

One group of plants that shows many primitive characteristics consists of the **mosses.** Mosses grow as a carpet composed of many parts. Each individual moss plant is composed of a central stalk under 5 centimeters tall with short, leaflike structures that are the sites of photosynthesis. If you look at the individual cells in the leafy portion of a moss, you can distinguish the cytoplasm, cell wall, and chloroplasts (figure 25.1). You can also distinguish the nucleus of the cell. This nucleus is haploid, meaning that it has only one set of chromosomes. In fact, every cell in the moss plant is haploid. Although all the cells have the haploid number of chromosomes (the same as gametes), not all of them function as gametes. Because this plant produces cells that are capable of acting as gametes, it is called the **gametophyte generation,** or gamete-producing, plant. Special structures in the moss, called **antheridia,** produce mobile sperm cells capable of swimming to a female egg cell (figure 25.2). The sperm cells are enclosed within a jacket of cells that opens when the sperm are mature. The sperm swim by the undulating motion of flagella through a film of dew or rainwater, carrying their packages of genetic information. Their destination is the egg cell of another moss plant with a different package of genetic information. The egg is produced within a jacket called the **archegonium** (figure 25.2). There is usually only one egg cell in each archegonium. The sperm and egg nuclei fuse, resulting in a diploid cell. The diploid zygote grows, divides, and differentiates into an embryo.

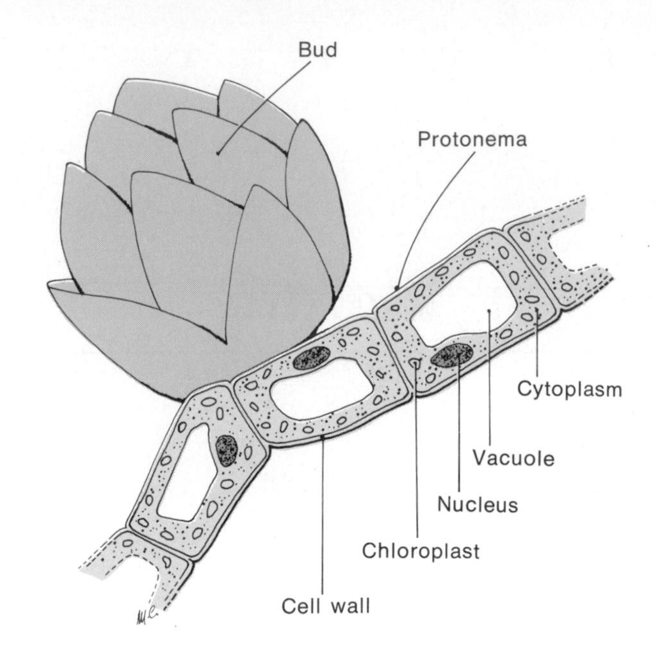

Figure 25.1 **The Cells of a Moss Plant.** These are typical plant cells. Note the large central vacuole and cell wall. One characteristic that separates mosses from the photosynthetic protists is the presence of the individual chloroplasts.

Alternation of Generations

The embryo grows into a structure that is called the **sporophyte** plant because it is the spore-producing part of the **life-cycle.** The sporophyte plant grows a stalk with a swollen tip, the **capsule** (figure 25.2). Inside the capsule, special cells undergo meiosis and form haploid **spores.** These spores are released and carried by air currents until they reach the ground. Some land in areas that are too wet, too dry, or too sunny; these will not **germinate** (begin to grow) and will not survive. Others land in a suitable environment and grow into gametophyte plants. Thus, the haploid plant produces the diploid plant, which in turn produces the haploid plant. This aspect of the life cycle is called **alternation of generations** (figure 25.2). The gametophyte generation is dominant over the sporophyte generation in mosses. This means that the gametophyte generation is independent of the sporophyte generation and is more likely to be seen.

Adjustments to Land

Why do botanists consider mosses the lowest step of the evolutionary ladder in the plant kingdom? First of all, they are considered primitive because they have not developed an efficient way of transporting water throughout their bodies; they must rely on the physical processes of diffusion and osmosis to move materials. The fact that mosses do not have a complex method of moving water limits their size to a few centimeters and their location to moist environments. Another characteristic of mosses points out how closely related they are to their aquatic ancestors: they require water for fertilization. The sperm cells "swim" from the antheridia to the archegonia. Small size, moist habitat, and swimming sperm are considered characteristics of a primitive organism. In a primitive way, mosses have adapted to a terrestrial niche.

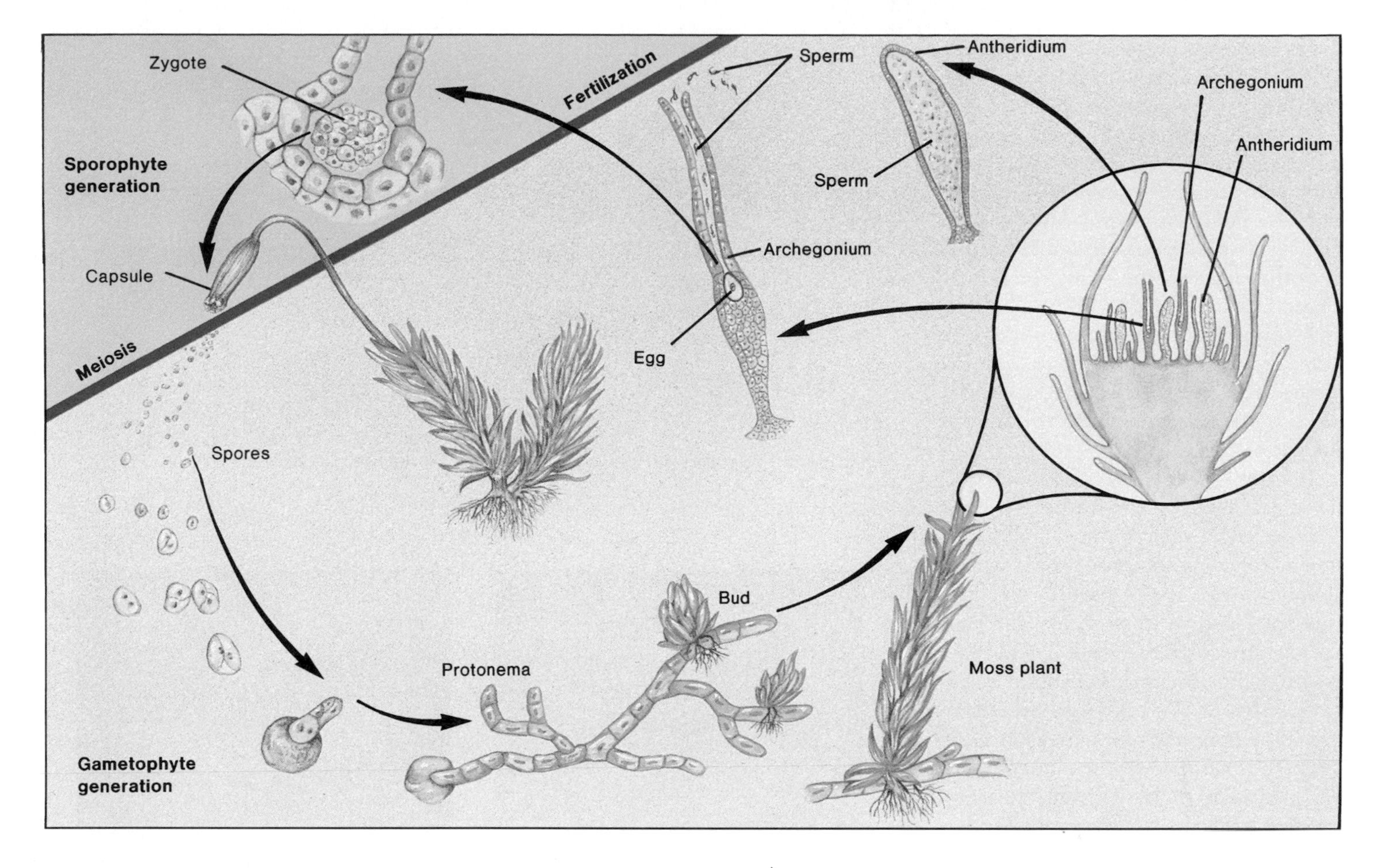

Figure 25.2 **The Life Cycle of a Moss.** In this illustration the portion with the lightly colored background represents cells that have the haploid nuclei (gametophyte generation). The portion with a darker colored background represent cells that have the diploid nuclei (sporophyte generation). Notice that the haploid and diploid portions of the life cycle alternate.

Vascular Tissue

A small number of currently existing organisms show some of the unsuccessful directions of evolution. You might think of these evolutionary groups as experimental models that couldn't quite "cut it" but were important steps in the evolution of more successful plants. The advances all concern cell specialization, which enables a plant to do a better job of acquiring, moving, and keeping water. Cells that are specialized to perform a particular function are called **tissues,** and the tissue important in moving water is called **vascular tissue.** The vascular tissues in plants are of two types: *xylem tissue* and *phloem tissue*. **Xylem** is a series of hollow cells arranged end to end so that they form a tube. These cells carry water absorbed from the soil into the roots to the upper parts of the plant. Associated with these tubelike cells are cells with thickened cell walls that provide strength and support for the plant.

The second type of vascular tissue, **phloem,** carries the organic molecules produced in one part of the plant to storage areas in other parts. The specialization of cells into vascular tissues has allowed for the development of the specialized parts of plants. **Roots, stems,** and **leaves** are examples of specialized parts that contain vascular tissue (figure 25.3). The root is specialized for picking up water. It has special outgrowths called **root hairs** that increase its efficiency in absorbing water from the soil. The stem has well-developed vascular tissue and transports water from the roots to the

Figure 25.3 **Vascular Tissue in Roots, Stems, and Leaves.** In this diagram, the black lines represent the series of tubes in the roots, stems, and leaves that distribute fluids throughout the plant.

leaves and organic molecules from the leaves to storage areas in the roots. The leaf is the site of photosynthesis.

The experimental evolutionary groups have some vascular tissue and are links between the nonvascular mosses and the more successful land plants. They also show some specialized structures that suggest true roots, stems, and leaves. But they tend to be small plants that require moist environments because they still have swimming sperm. The liverworts and the club mosses represent two of these experimental evolutionary groups.

Liverworts (figure 25.4) are usually overlooked by the casual observer because they are rather small, low-growing plants composed of a green ribbon of cells. While they do not have well-developed roots or stems, the leaflike ribbon of tissue is well suited to absorb light for photosynthesis. Club mosses (figure 25.5) are a group of low-growing plants that are somewhat more successful than liverworts in adapting to life on land. They have a stemlike structure that holds the leafy parts above other low-growing plants, enabling them to compete better for the available sunlight. Thus, they are larger than mosses and not as closely tied to wet areas. While not as efficient as the stems in higher plants, the stem of the club moss, with its vascular tissue, is a hint of what's to come.

Figure 25.4 **Liverworts.** These ribbon-shaped plants are related to the mosses.

Figure 25.5 **Club Mosses.** These plants are sometimes called ground pines because of their slight resemblance to the evergreen trees.

Ferns

With fully developed vascular tissue, plants are no longer limited to wet areas. They can absorb water and distribute it to leaves many meters above the surface of the soil. The ferns are the most primitive vascular plants that are truly successful at terrestrial living. Not only do they have a wider range and greater size than mosses and club mosses, they have an additional advantage: the sporophyte generation has assumed more importance and the gametophyte generation has decreased in size and complexity. Figure 25.6 illustrates the life cycle of a fern. The diploid condition of the sporophyte is an advantage because a recessive gene can be masked until it is combined with another identical recessive gene. In other words, the plant does not suffer because it has one bad allele. On the other hand, a mutation may be a good change, but time is lost by having it hidden in the heterozygous condition. In a haploid plant, any change, whether recessive or not, shows up. Not only is a diploid condition beneficial to an individual, but the population benefits when many alleles are available for selection.

Ferns take many forms, including the delicate, cloverlike maidenhair fern of northern wooded areas; the bushy bracken fern (figure 25.7); and the tree fern, known primarily from the fossil record and seen today in tropical areas. In spite of all this variety, however, they still lack one tiny but very important structure—the seed. Without seeds, the ferns must rely on spores to spread the species from place to place.

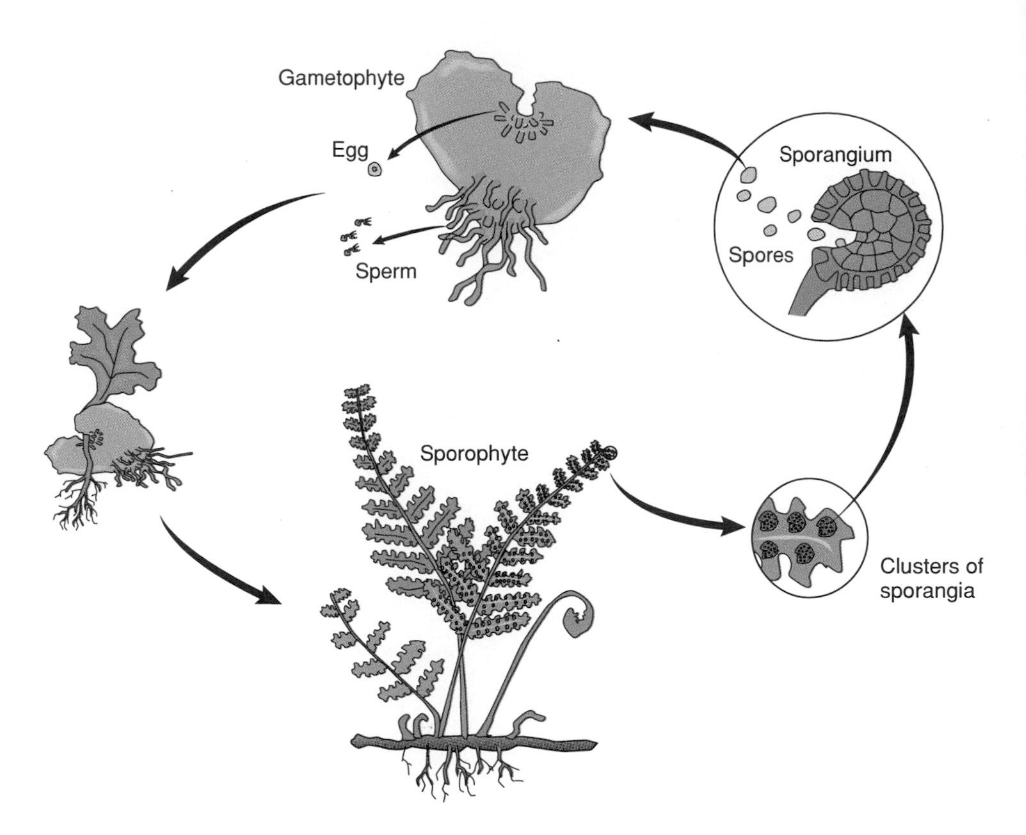

Figure 25.6 **The Life Cycle of a Fern.** In this illustration the dark green color represents cells that have diploid nuclei (sporophyte generation). On the back of some fern leaves there are small dots—clusters of sporangia. The sporangia produce spores. These spores develop into cells that have haploid nuclei (gametophyte generation). This stage is shown in a light green color. Notice that the gametophyte and sporophyte generations alternate. Compare this life cycle with that of the moss (figure 25.2). In the moss the gametophyte generation is considered dominant, whereas in the fern the sporophyte is dominant.

Figure 25.7 **A Typical Fern.** Most ferns, such as this one live in shaded areas of the forest.

Gymnosperms

The next advance made in the plant kingdom was the evolution of the **seed.** A seed has an embryo enclosed in a protective covering called the **seed coat.** It also has some stored food for the embryo. The first attempt at seed production is exhibited in the conifers, which are cone-bearing plants such as pine trees. **Cones** are reproductive structures that develop either pollen-bearing sacs or naked ovules or seeds. The male cone produces **pollen.** Pollen grains are miniaturized male gametophyte plants. Each of these small dustlike particles contains a sperm nucleus. The female cone is larger than the male cone and produces the female gametophyte plants. The archegonia in the female gametophyte contain eggs. The pollen is carried by wind to the female cone, which holds the archegonium in a position to gather the airborne pollen. The process of getting the pollen from the male cone to the female cone is called **pollination.** The production of seeds and pollination by wind are features of conifers that place them higher on the evolutionary ladder than ferns.

Because seeds with their embryos are produced on the surface of the woody, leaflike structure (the female cone), they are said to be *naked,* or out in the open (figure 25.8).The cone-producing plants are sometimes called **gymnosperms,** which means "naked seed" plants. Producing seeds out in the open makes this very important part of the life cycle vulnerable to adverse environmental influences, such as attack by insects, birds, and other organisms.

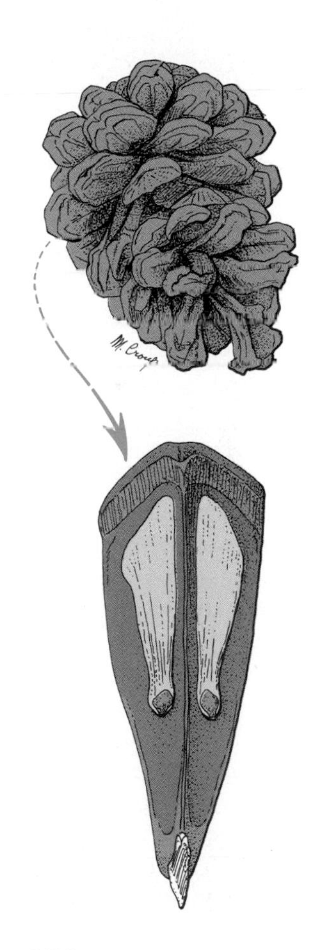

Figure 25.8 **A Pine Cone with Seeds.** On the leaflike portions of the cone are the seeds. Since these cones produce seeds out in the open, they are aptly named the naked seed plants—gymnosperms.

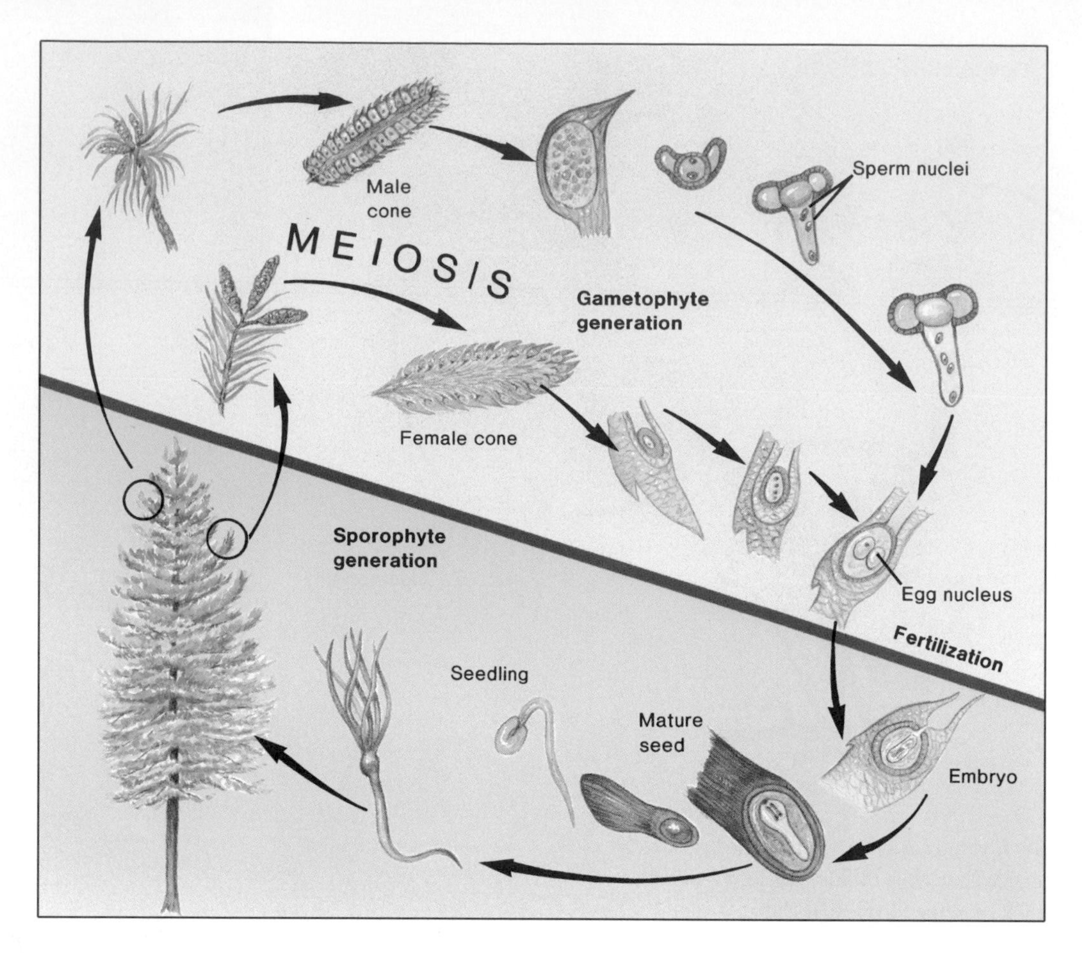

Figure 25.9 **The Life Cycle of a Pine.** In this illustration, the portion with the lightly colored background represents cells that have haploid nuclei (gametophyte generation). The darker colored background represents cells that have diploid nuclei (sporophyte generation). Notice that the gametophyte and sporophyte generations alternate. Compare this life cycle of the pine with the life cycle of the moss (figure 25.2) and that of the fern (figure 25.6). Notice the ever-increasing dominance in the sporophyte generation.

Gymnosperms generally produce needle-shaped leaves, or needles, which do not all fall off at once. Such trees are said to be **nondeciduous.** This term is misleading because it suggests that the needles do not fall off at all. Actually, they are constantly being shed a few at a time. Perhaps you have seen the mat of needles under a conifer. The tree retains some leaves year-round, and therefore is called an *evergreen*. The portion of the evergreen with which you are familiar is the sporophyte generation; the gametophyte, or haploid, stages have been reduced to only a few cells. Look closely at figure 25.9 which shows the life cycle of a pine with its alternation of haploid and diploid generations.

Gymnosperms are **perennials;** that is, they live year after year. Unlike **annuals,** which complete their life cycle in one year, gymnosperms take many years to grow from seeds to reproducing adults. The trees get higher and larger in diameter each year, continually adding layers of strenghtening cells and vascular tissue. As a tree becomes larger, the strengthening tissue in the stem becomes more and more important. A layer of cells in the stem, called the **cambium,** is responsible for this increase in size. Xylem tissue is the innermost part of the tree trunk or limb, and phloem is outside of the cambium. The cambium layer of cells is positioned between the xylem and the phloem. These cambium cells go through a mitotic cell division, and two cells form. One cell remains cambium tissue, and the other specializes to form vascular tissue. If the cell is on the inside of the cambium ring, it becomes xylem; if it is on the outside of the cambium ring, it becomes phloem. As cambium cells divide again and again, one cell always remains cambium, and the other becomes vascular tissue. Thus, the tree constantly increases in diameter (figure 25.10).

The accumulation of the xylem in the trunk of gymnosperms is called **wood.** Wood is one of the most valuable biological resources of the world. We get lumber, paper products, turpentine, and many other valuable materials from gymnosperms. You are already familiar with many examples of gymnosperms, three of which are pictured in figure 25.11.

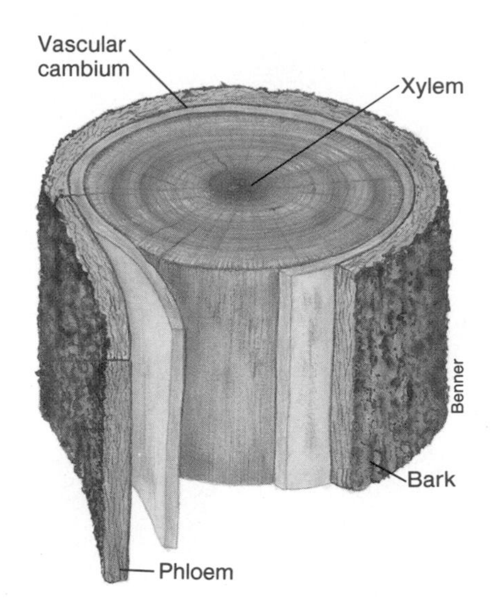

Figure 25.10 **A Cross Section of Woody Stem.** Notice that the xylem makes up most of what we call wood. Can you picture the relative position of the labeled structures twenty years from now?

Angiosperms

The group of plants considered most highly evolved are known as **angiosperms.** This name means that the seeds, rather than being produced naked, are contained within the surrounding tissues of the **ovary.** The ovary and other tissues mature into a protective structure known as the **fruit.** Many of the foods we eat are the seed-containing fruits of angiosperms: green beans, melons, tomatoes, and apples are only a few of the many edible fruits (figure 25.12).

The flower is the structure that produces the sex cells and enables the sperm cells to get to the egg cells. The important parts of the flower are the **pistil** (female part) and the **stamen** (male part). In figure 25.13, notice that the egg cell is located inside the ovary. Any flower that has both male and female parts is called **perfect;** a flower containing just female or male parts is **imperfect.** Any additional parts of the flower are called **accessory structures** because fertilization can occur without them. **Sepals,** which form the outermost whorl of the flower, are accessory structures that serve a protective function. **Petals,** also accessory structures, increase the probability of fertilization. Before the sperm cell (contained in the pollen) can join with the egg cell, it must somehow get to the egg. This is the process called *pollination.* Some flowers with showy petals are adapted to attracting insects, who unintentionally carry the pollen to the pistil. Others have become adapted for wind pollination. The important thing is to get the genetic information from one parent to the other.

(a)

(b)

(c)

Figure 25.11 **Several Gymnosperms.** How many of these trees do you recognize? They are (a) redwood, (b) Torrey pine, and (c) cedar.

All of the flowering plants have retained the evolutionary advances of previous groups. That is, they have well-developed vascular tissue with true roots, stems, and leaves. They have pollen and produce seeds within the protective structure of the ovary.

There are thousands of kinds of plants that produce flowers, fruits, and seeds. Almost any plant you can think of is an angiosperm. If you made a list of these familiar plants, you would quickly see that they vary a great deal in structure and habitat. The mighty oak, the delicate rose, the pesky dandelion, and the expensive orchid are all flowering plants. How do we organize this diversity into some sensible and useful arrangement? Botanists classify all angiosperms into one of two groups: **dicots** or **monocots.** The names *dicot* and *monocot* refer to a structure (called a *cotyledon*) in the seeds of these plants. If the embryo has two **seed leaves,** the plant is a dicot; those with only one seed leaf are the monocots (figure 25.14). A peanut is a dicot; lima beans and apples are also dicots; grass, lilies, and orchids are all monocots. Even with this separation, the diversity is staggering. The characteristics used to classify and name plants are listed in figure 25.15, which includes a comparison of the extremes of these characteristics.

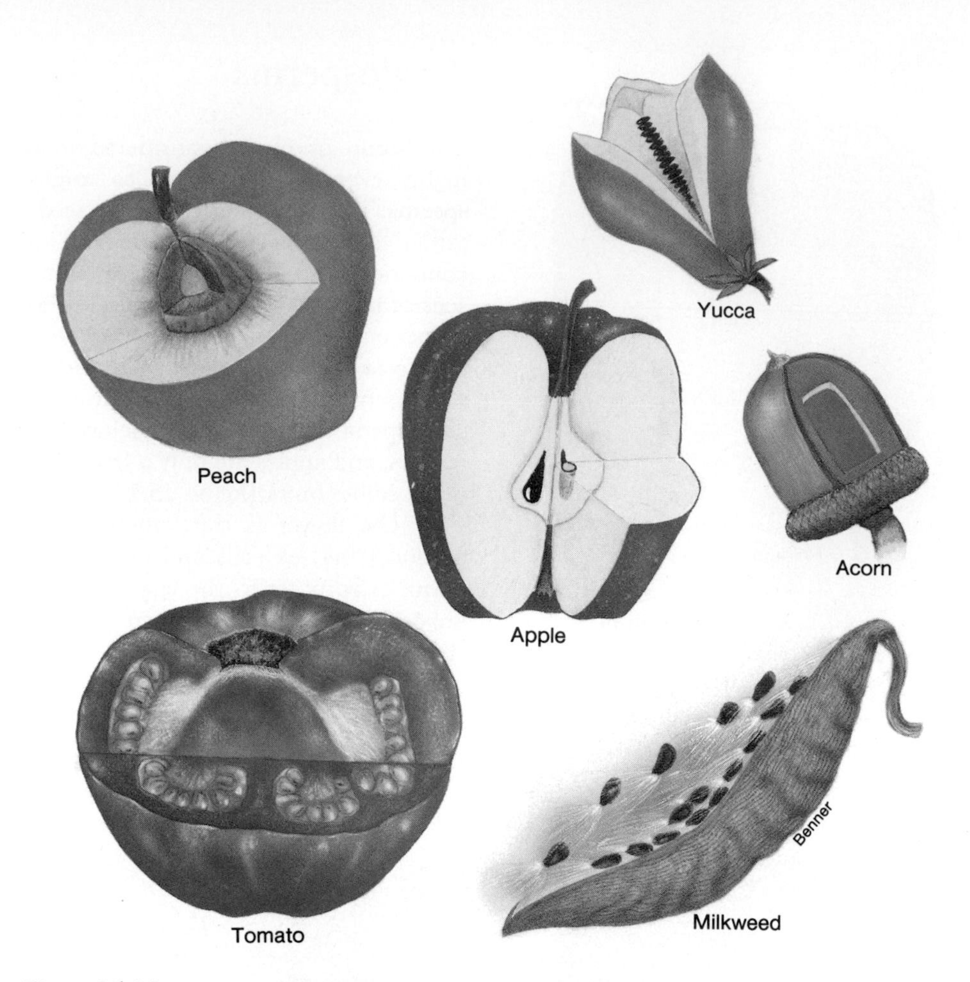

Figure 25.12 **Types of Edible and Inedible Fruits.** Fruits are the structures that contain seeds. The seed containers of the peach, apple, and tomato are used by humans as food. The other fruits are not usually used by humans as food.

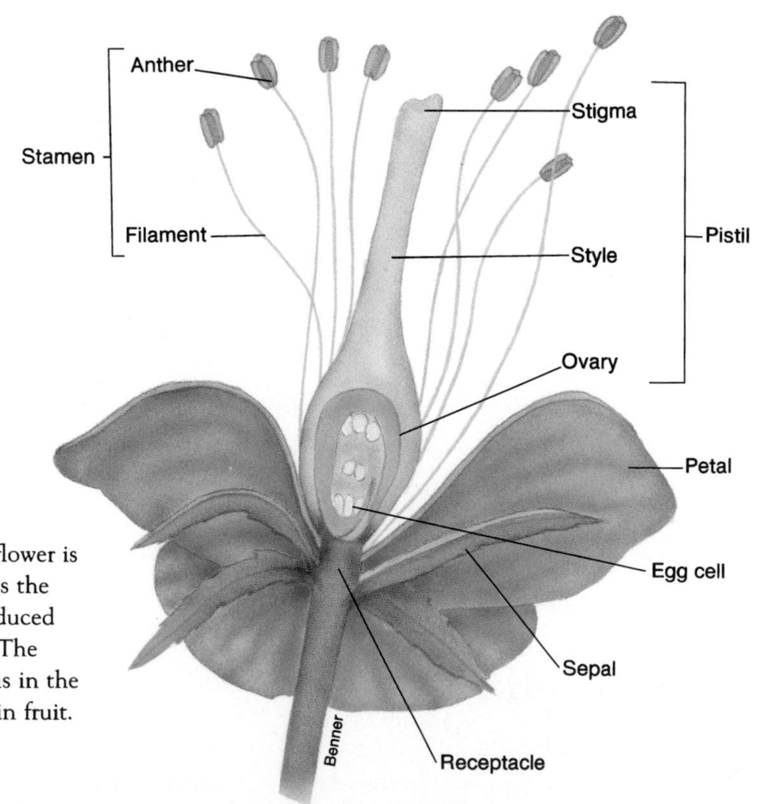

Figure 25.13 **The Flower.** The flower is the structure in plants that produces the sex cells. Notice that the egg is produced within a structure called the ovary. The seeds, therefore, will not be naked as in the gymnosperms, but will be enclosed in fruit.

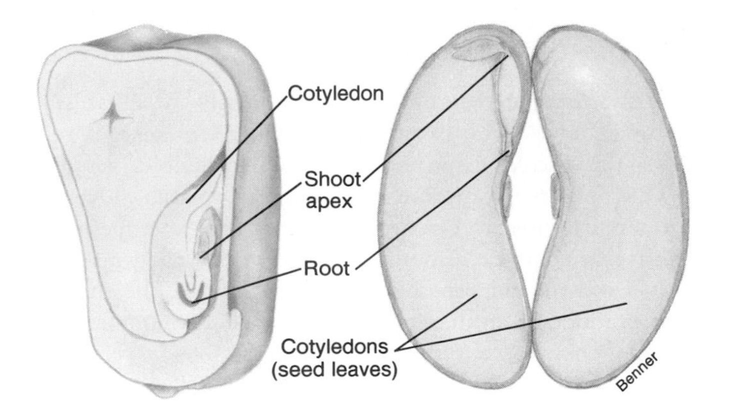

Figure 25.14 **Embryos in Dicots and Monocots.** The number of seed leaves attached to an embryo is one of the characteristics botanists use to classify flowering plants.

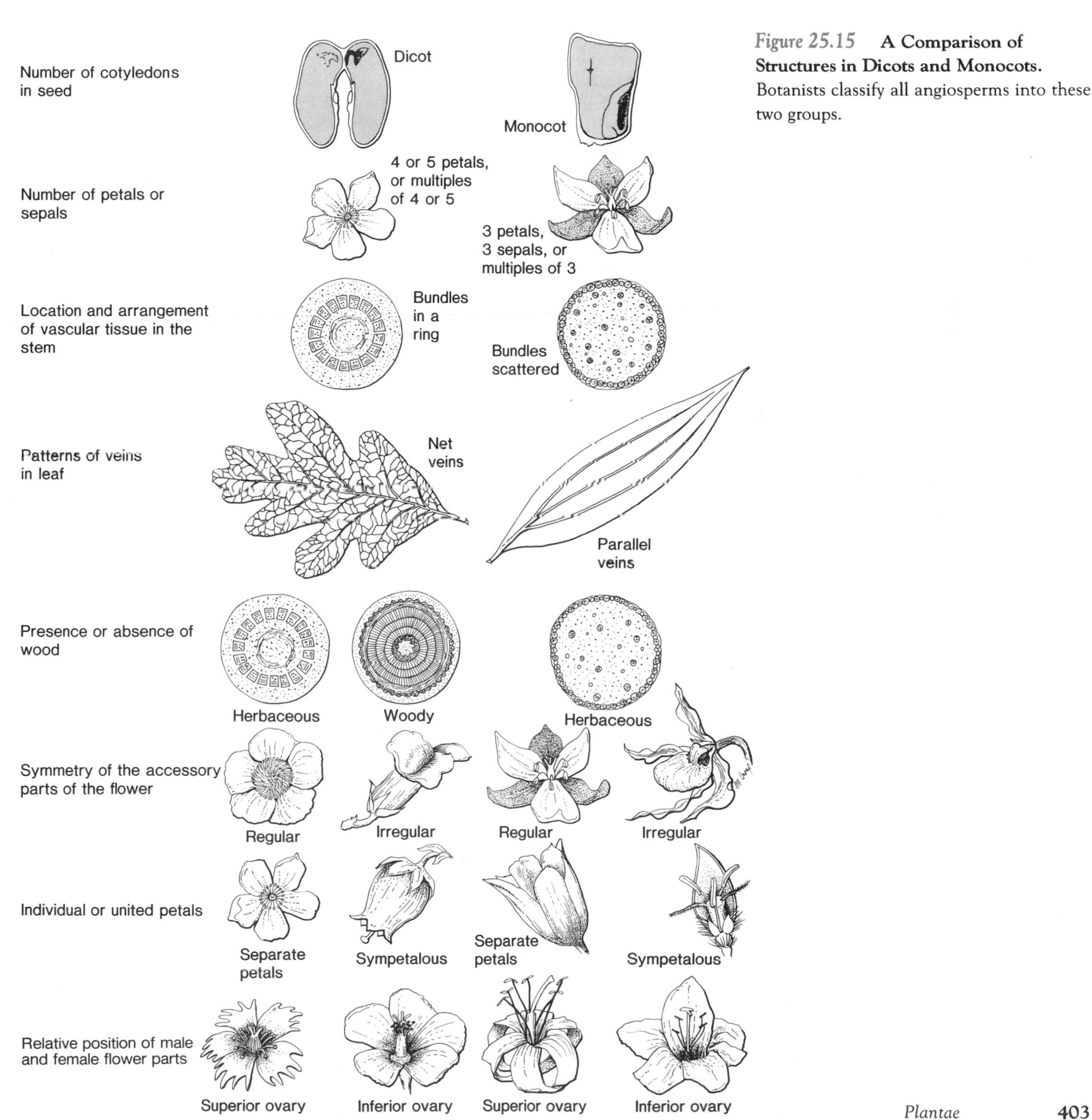

Figure 25.15 **A Comparison of Structures in Dicots and Monocots.** Botanists classify all angiosperms into these two groups.

• Summary •

The plant kingdom is composed of organisms that are able to manufacture their own food by the process of photosynthesis. They have specialized structures for producing the male sex cell (the sperm) and the female sex cell (the egg). The relative importance of the haploid gametophyte and the diploid sporophyte that alternate in plant life cycles is a major characteristic used to determine an evolutionary sequence. The extent and complexity of the vascular tissue and the degree to which plants rely on water for fertilization are also used to classify plants as primitive or complex. Among the gymnosperms and the angiosperms, the methods of production, protection, and dispersal of pollen are used to name and classify the organisms into an evolutionary sequence. Based on the information available, mosses are the most primitive plants. Liverworts and club mosses are experimental models. Ferns, seed-producing gymnosperms, and angiosperms are the most advanced and show the development of roots, stems, and leaves.

• Thinking Critically •

Some people say the ordinary "Irish" potato is poisonous when the skin is green, and they are at least partly correct. A potato develops a green skin if the potato tuber grows so close to the surface of the soil that it is exposed to light. An alkaloid called *solanine* develops under this condition and may be present in toxic amounts. Eating such a potato raw may be dangerous. However, cooking breaks down the solanine molecules and makes the potato as edible and tasty as any other.

The so-called Irish potato is of interest to us historically. Its real country of origin is only part of the story. Check your local library to find out about this potato and its relatives. Are all related organisms edible? Where did this group of plants develop? Why is it called the "Irish" potato?

• Experience This •

New Plants from Old

Anyone can have a variety of flowering plants at little expense with just a bit of effort. One need only take a sharp knife and cut a piece of the stem of a desirable plant (with your neighbor's permission!). Woody stems are apt to be more difficult to root than plants with soft stems. Place some soil in a container, make a hole in the soil with a pencil, and insert the cut stem. The stem piece should include the tip and about five leaves. The container can be anything, as long as there are a few holes in the bottom for drainage. Cover the container tightly with a transparent plastic bag to prevent water loss through the leaves before the roots develop. Also trim the lower leaves for the same purpose. Water the plant once and keep it out of direct sun. Roots may take ten days or longer to develop. If the soil should feel dry to the touch, water the plant again.

If potting soil or vermiculite is available at a hardware or garden store, it might be better than the soil you could dig locally. Another aid is to dip the cut end of the stem into some rooting hormone before planting. However, people have had success for years without these refinements.

• Questions •

1. What characteristics distinguish algae in the kingdom Protista from the organisms of the kingdom Plantae?
2. List three characteristics shared by mosses, ferns, gymnosperms and angiosperms.
3. What were the major advances that led to the development of angiosperms?
4. How is a seed different from pollen, and how do both of these differ from a spore?
5. How are cones and flowers different?
6. How are cones and flowers similar?
7. What are the dominant generations in mosses, ferns, gymnosperms, and angiosperms?
8. What is the difference between the xylem and the phloem?
9. Ferns have not been as successful as gymnosperms and angiosperms. Why?
10. What is the significance of the cambium tissue in perennials?

• Chapter Glossary •

accessory structures (ak-ses'o-re struk'churs) The parts of some flowers that are not directly involved in gamete production.

alternation of generations (awl"tur-na'shun uv jen"uh-ra'shunz) The cycling of a diploid sporophyte generation and a haploid gametophyte generation in plants.

angiosperms (an'je-o"spurmz) Plants that produce flowers and fruits.

annual (an'yu-uhl) A plant that completes its life cycle in one year.

antheridia (singular, **antheridium**) (an"thur-id'e-ah) The structures in lower plants that produce sperm.

archegonium (ar"ke-go'ne-um) The structure in lower plants that produces eggs.

cambium (kam'be-um) A tissue in higher plants that produces new xylem and phloem.

capsule (kap'sūl) The part of the sporophyte generation of mosses that contains spores.

cone (kōn) A reproductive structure of gymnosperms that produces pollen in males or eggs in females.

dicot (di'kot) An angiosperm whose embryo has two seed leaves.

flower (flow'er) A complex structure made from modified stems and leaves. It produces pollen in the males and eggs in the females.

fruit (froot) The structure in angiosperms that contains seeds.

gametophyte generation (gă-me'to-fīt jen"uh-ra'shun) The haploid generation in plant life cycles, which produces gametes.

germinate (jur'min-āt) To begin to grow (as from a seed).

gymnosperms (jim'no-spurmz) Plants that produce their seeds in cones.

imperfect flowers (im"pur'fekt flow'erz) Flowers that contain either male or female reproductive structures, but not both.

leaves (lēvz) Specialized portions of higher plants that are the sites of photosynthesis.

life cycle (līf sī'kl) The series of stages in the life of any organism.

monocot (mon'o-kot) An angiosperm whose embryo has one seed leaf.

mosses (mŏ'sez) Lower plants that have a dominant gametophyte generation, spores, and swimming sperm. They lack vascular tissue.

nondeciduous (non"de-sid'yu-us) A term used to describe trees that do not lose their leaves all at once.

ovary (o'vah-re) The female structure that produces eggs.

perennial (pur-en'e-uhl) A plant that requires many years to complete its life cycle.

perfect flowers (pur'fekt flow'erz) Flowers that contain both male and female reproductive structures.

petals (pĕ'tuls) Modified leaves of angiosperms; accessory structures of a flower.

phloem (flo'em) One kind of vascular tissue found in higher plants. It transports food materials from the leaves to other parts of the plant.

pistil (pis'til) The female reproductive structure in flowers.

pollen (pol'en) The male gametophyte in gymnosperms and angiosperms.

pollination (pol"ĭ-na'shun) The transfer of pollen in gymnosperms and angiosperms.

root (root) A specialized structure for the absorption of water and minerals in higher plants.

root hairs (root hārs) Tiny root outgrowths that improve the ability of plants to absorb water and minerals.

seed (sēd) A specialized structure produced by gymnosperms and angiosperms that contains the embryonic sporophyte.

seed coat (sēd kōt) A protective layer around a seed.

seed leaves (sēd lēvz) Embryonic leaves in seeds.

sepals (se'pals) Accessory structures of flowers.

spores (spōrz) Haploid structures produced by sporophytes.

sporophyte (spōr'o-fīt) The diploid generation in the life cycle of plants, which produces spores.

stamen (sta'men) The male reproductive structure of a flower.

stem (stem) The upright portion of a higher plant.

tissue (tish'yu) A group of specialized cells that work together to perform a particular function.

vascular tissue (vas'kyu-lar tish'yu) Specialized tissue that transports fluids in higher plants.

wood (wo͞od) The xylem of gymnosperms and angiosperms.

xylem (zi'lem) A kind of vascular tissue that transports water from the roots to other parts of the plant.

CHAPTER 26

Animalia

Chapter Outline

Purpose

Processes such as respiration, cell division, reproduction, and heredity are common to all forms of life. This was true for the first living organisms and remains true for the many types of organisms that have evolved since. The evolution of the great variety of species was presented previously in chapter 23. At that point, only a passing reference was made to the various types of animals. In this chapter, we take a closer look at the variety of animals found on Earth today and also provide insights into the ecological niches that animals occupy. Particular attention is paid to the differences between terrestrial and aquatic animals.

For Your Information

Several years ago, as a result of an administrative decision, a large number of black bears were systematically killed by park rangers in Yosemite National Park. The park administration believed that the national park was set aside for people, not bears.

The Forest Service in Wyoming schedules clear cutting in one of the few remaining grizzly bear habitats. This forestry practice forces the grizzlies into areas where their activities are more likely to get them killed as dangerous nuisances.

Wolves introduced into the Upper Peninsula of Michigan by the Michigan Department of Natural Resources were killed by some outraged citizens. Those in favor of this introduction of wolves believe that stories of unprovoked attacks on humans were fabricated.

Learning Objectives

- Name and describe the three forms of animal bodies.
- Explain the basic body structure of animals with bilateral symmetry.
- Explain why animal life probably originated in the ocean.
- Describe the animal life cycle that exhibits alternation of generations.
- Explain the role of the host and the parasite in a parasitic life-style.
- List some benthic marine organisms.
- Describe the characteristics of a pelagic marine animal.
- Describe the problems that animals needed to overcome in order to adapt to a terrestrial environment.
- Describe the amniotic egg and discuss its importance.

General Features of Animals

Today there are at least 4 million known species of animals, ranging in size from microscopic rotifers, 40 micrometers in length, to giant blue whales, 30 meters long. Animals not only vary in size, they inhabit widely diverse habitats—from the frigid arctic to the scorching desert, from the dry African savanna to the South American tropical rain forest, and from violent tidal zones to the ocean's depths.

All animals are multicellular and heterotrophic. Multicellular organisms have some advantages over one-celled organisms. In multicellular organisms, individual cells are able to specialize. Thus, multicellular organisms can perform some tasks better than single-celled organisms. For example, some unicellular organisms can move by using cilia, flagella, or amoeboid motion. However, such movement is relatively slow, weak, and allows the organism to move only a short distance. Many animals have muscle cells that are specialized for movement and allow for a greater variety of more efficient kinds of movement. For example, some animals use muscle power to migrate thousands of kilometers—a feat beyond the capability of unicellular organisms.

Functions such as ingesting food, exchanging gases, and removing waste are more complicated in animals than in unicellular organisms. One-celled organisms are in direct contact with the environment; any exchange between the organism and the external environment occurs through the plasma membrane. However, in animals the majority of cells are not on the body surface and are therefore not in direct contact with the external environment. Animals must have a specialized means of exchange between the internal environment and the external environment (figure 26.1). They must also have developed a method of transporting materials between the body surface and internal cells. As the size of an animal increases, the amount of body surface increases more slowly than the volume, and the body systems for exchanging and transporting material become more complex (see chapter 18).

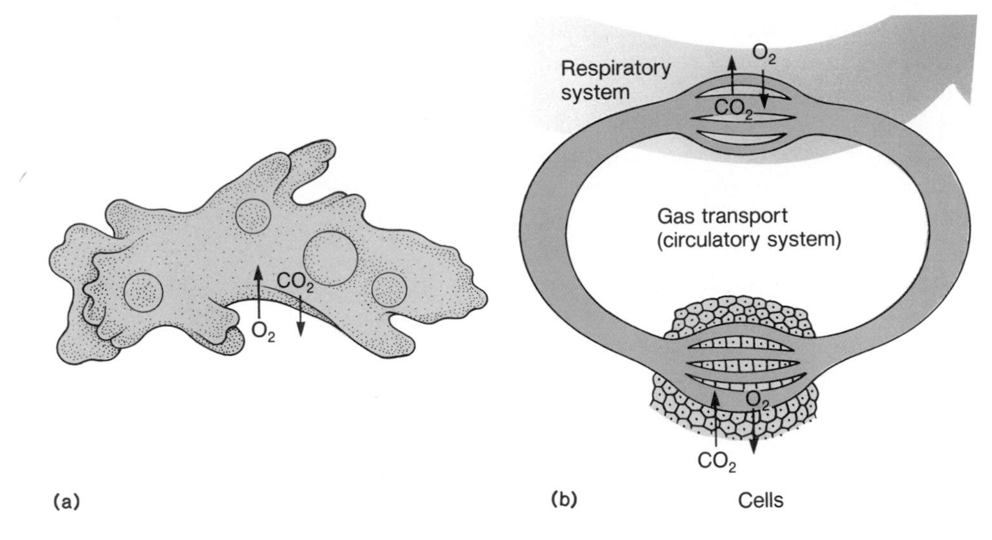

Figure 26.1 **Simple Systems.** (a) In the unicellular organism the exchange of oxygen and carbon dioxide between the cell and the environment occurs directly by diffusion. (b) In the multicellular organism a respiratory system—a network of tubes—brings the oxygen into the animal. The oxygen then diffuses from the respiratory system to the circulatory system—a network of blood vessels—which transports the oxygen to the inner cells. At this point the oxygen diffuses from the blood into the cells. Also, carbon dioxide diffuses from the cell to the blood. The blood that brings the oxygen to the cells from the respiratory system then transports the carbon dioxide from the cells to the respiratory system.

The majority of cells in an animal are internal and do not directly perceive changes in the external environment, such as changes in light and temperature. Some of the surface cells have developed specialized sensory sites that do perceive environmental changes. These external changes can then be communicated to the internal regions of the body through a network of sensory neurons. Again, as an animal gets larger, the body systems become more complex (see chapter 20).

Chapters 18, 19, 20, and 21 describe the functions of some of the systems in the human body. Most animals have systems that function in basically the same ways. Naturally there are some modifications—fish have gills, not lungs—but the purpose of the respiratory system is the same in fish and in animals with lungs.

Unicellular organisms are all **poikilotherms**—organisms whose body temperature varies. Their body temperature changes as the external temperature changes, which means that at cooler temperatures, poikilotherms also have lower metabolic rates. Most animals, including insects, worms, and reptiles, are poikilotherms. Some, however, are **homeotherms,** which means that they maintain a constant body temperature that is generally higher than the environmental temperature, regardless of the external temperature. These animals, birds and mammals, all have high metabolic rates.

The form of animal bodies is diverse. Some animals have no particular body shape, a condition called **asymmetry.** Asymmetrical body forms are rare and occur only in certain species of sponges, which are the simplest kinds of animals. **Radial symmetry** (figure 26.2a) occurs when a body is constructed around a central axis. Any division of the body along this axis results in two similar halves. Although many animals with radial symmetry are capable of movement, they do not always lead with the same portion of the body; that is, there is no anterior, or head, end.

Animals with **bilateral symmetry** (figure 26.2b) are constructed along a plane running from a head to a tail region. There is only one way to divide bilateral animals into two mirror halves. Animals with bilateral symmetry move head first, and the head typically has sense organs and a mouth. The vast majority of animals display bilateral symmetry.

All animals with bilateral symmetry have a basic body structure composed of three layers. In most of them, this resembles a tube within a tube (figure 26.3). The outer layer contains muscles

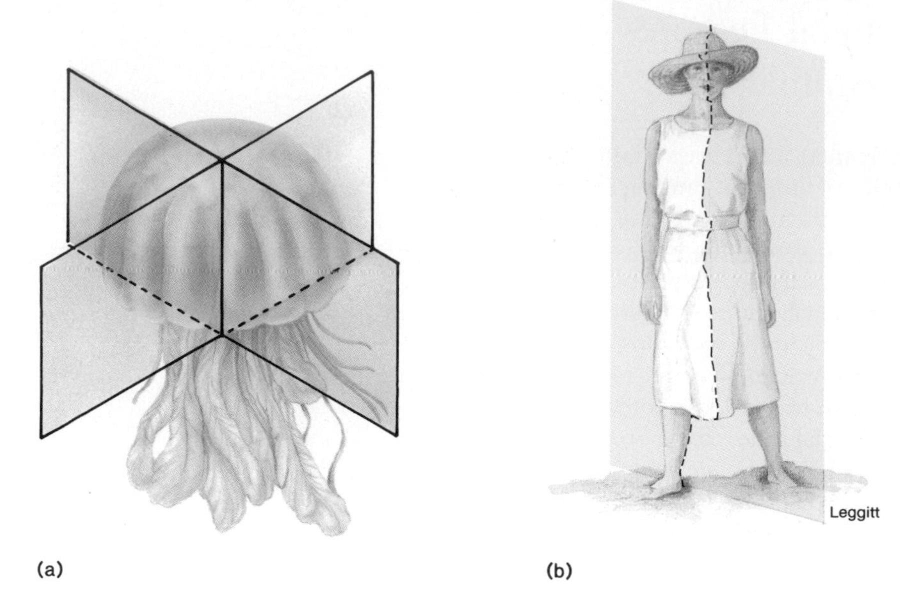

Figure 26.2 **Radial and Bilateral Symmetry.** (a) In animals with radial symmetry, any cut along the central body axis results in similar halves. (b) In animals with bilateral symmetry, only one cut along one plane results in similar halves.

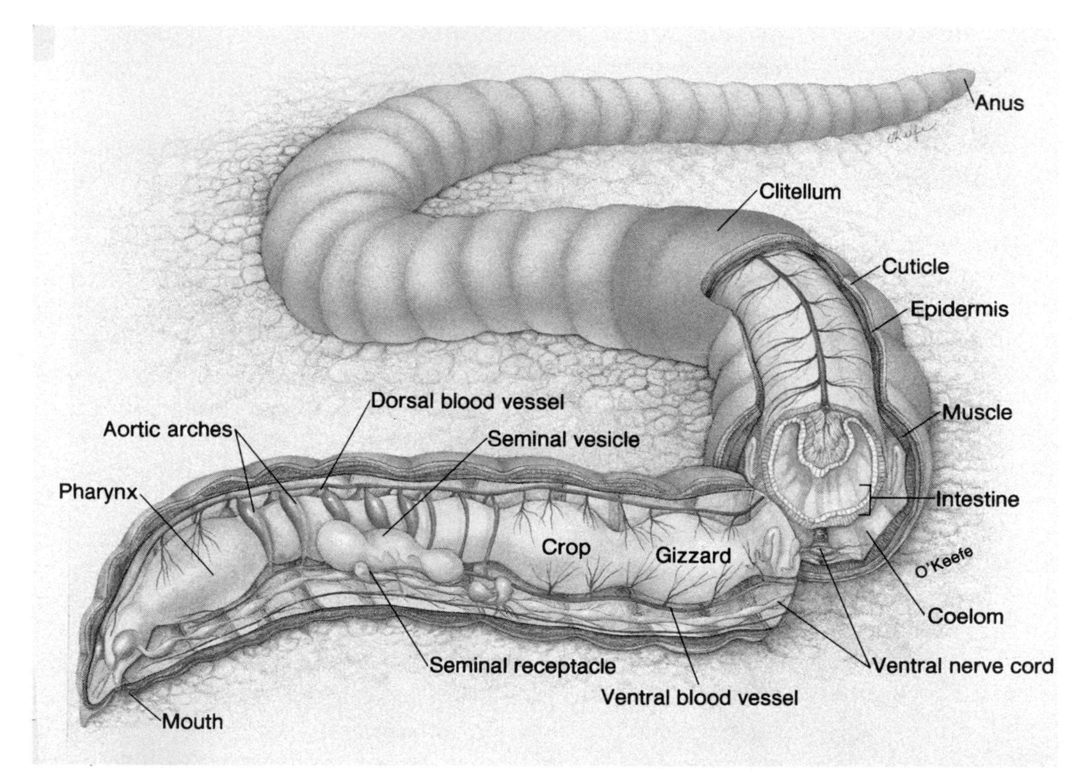

Figure 26.3 **Body Structure.** All animals with bilateral symmetry have a body consisting of a digestive tube running through an outer tube. This is often called a tube-within-a-tube body plan. The outer tube consists of two layers of different origin. The outermost thin layer forms the surface of the skin and its derivatives, such as a cuticle, hair, scales, or feathers. This is shown in blue in the drawing. Attached to the outer layer of the skin is a layer of muscle and connective tissue (shown in red). The gut is lined with a thin layer of cells called the epithelial lining. This is shown in yellow in the drawing. Surrounding the gut lining is a layer of muscle and connective tissue (shown in red). Most of the internal organs of the body have the same origin as the muscle and connective tissue.

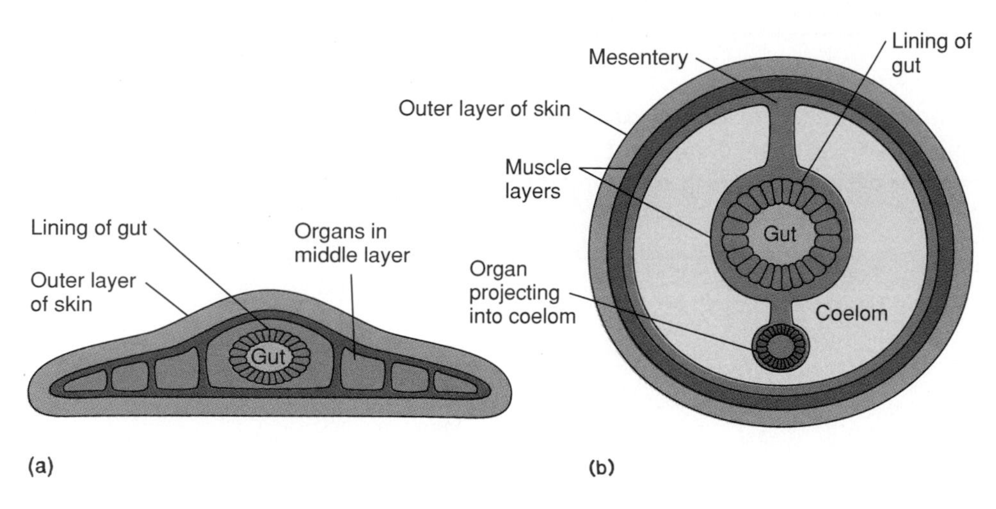

Figure 26.4 **The Coelom.** (a) Some animals, like the flatworms, have no open space between the gut and outer body layer. (b) Other animals, including all vertebrates, have a coelom, an open area within the middle layer. Organs form from this middle layer projecting into the coelom. The mesenteries are thin sheets of connective tissue that hold the organs in place within the coelom.

and nerves, and is exposed to the environment. In many animals the outer layer is not protected by specialized structures, but other animals have such things as shells, scales, feathers, or hair protecting the outer layer of skin.

The inner tube layer constitutes the digestive system, with a mouth at one end and anus at the other. Many portions of this food tube are specialized for the digestion, absorption, and reabsorption of nutrients. Other organs associated with the food tube secrete digestive enzymes into it, and are located between the digestive tube and the outside body wall. Also located in this area are other organs that are involved in excretion of waste, circulation of material, exchange of gases, and body support.

Simple animals, such as jellyfish and flatworms, have no space between the inner and outer tubes (figure 26.4a). More advanced animals, such as earthworms, insects, reptiles, birds, and mammals, have a **coelom,** or body cavity, between these two tubes (figure 26.4b). The coelom in a turkey is the cavity where you stuff the dressing. In the living bird this cavity contained a number of organs, including those of the digestive, excretory, and circulatory systems. The development of the coelom was significant in the evolution of animals. In **acoelomates,** animals without a coelom, the internal organs are packed closely together. In coelomates there is less crowding of organs and less interference among them. Organs such as the heart, lungs, stomach, liver, and intestines have ample room to grow, move, and function. The coelom allows for separation of the inner tube and the body-wall musculature; thus, the inner tube functions freely—independent of the outer wall. This results in organ systems that are more highly specialized than acoelomate systems. Organs are not loose in the coelom but are held in place by sheets of connective tissue called **mesenteries.** Mesenteries also serve as support for blood vessels connecting the various organs (see figure 26.4b).

Animal Evolution

Scientists estimate that the Earth is at least 4.5 billion years old (table 26.1) and that life originated in the ocean about 3.8 billion years ago (chapter 22). For approximately 2 billion years, one-celled organisms were the only forms of life present in the ocean, and there were no life forms on land. These early life forms probably evolved into unicellular plant-like and animal-like organisms that were the forerunners of present-day plants and animals (refer back to figure 23.7).

The early forms of life remained in the ocean presumably because it provided a more hospitable environment than land. In the ocean, animals did not have a problem with dehydration. Also, the ion content of the early ocean approximated that of the animals' cells, so little energy was required to keep the cell in osmotic balance. Finally, the temperature range in the ocean is not as great as that on land and the rate of temperature change is lower. Therefore, animals in the ocean did not require mechanisms to deal with rapid or extreme changes in the environment.

For the first billion years of their existence, animals were limited to the oceans. Most of them were plankton: small floating or weakly swimming organisms. Even though the ocean provided a hospitable environment, the amount of light was a limitation. As mentioned in chapter 24 (see figure 24.6), the *photic zone*— the region suitable for photosynthesis—is limited to the top 100 meters. This limited most food chains to the upper layer of the open ocean or the shallow water along shore lines.

Animals that live in shallow coastal areas must withstand tidal changes and the forces of wave action. Some are free-moving and migrate with the tidal changes. Others are firmly attached to objects; these are said to be **sessile.** Most sessile animals are **filter feeders** that use cilia or other appendages to create water currents to filter food out of the water. Mussels, oysters, and barnacles are sessile marine animals.

Reproduction presents special problems for sessile organisms because they cannot move to find mates. However, since they are in an aquatic environment, it is possible for the sperm to swim to the egg and fertilize it. The fertilized egg develops into a larval stage—the juvenile stage of the organism (figure 26.5). The larvae are usually ciliated or have appendages that enable them to

Table 26.1

The Geological Time Scale. *Major Divisions of Geological Time with Some of the Major Evolutionary Events of Each Geological Period.*

Era	Period	Millions of Years Ago	Major Biological Events	
			Plants	*Animals*
Cenozoic	Quaternary	2.5	Rise of herbaceous plants	Age of humans
	Tertiary	65	Dominance of the angiosperms	First hominids Rise of modern forms of animals Mammals and insects dominate the land
Mesozoic	Cretaceous	130	Spread of angiosperms Decline of gymnosperms	Extinction of the dinosaurs
	Jurassic	180	First flowering plants (angiosperms)	Age of dinosaurs First mammals and first birds
	Triassic	230	Land plants dominated by gymnosperms	First appearance of the dinosaurs
Paleozoic	Permian	280	Land covered by forests of primitive vascular plants	Expansion of the reptiles Decline of the amphibians Extinction of the trilobites
	Carboniferous	350	Land covered by forests of coal-forming plants	Age of amphibians First appearance of the reptiles
	Devonian	400	Expansion of primitive vascular plants over land	Fishes dominate the seas First insects First amphibians move onto land
	Silurian	435	First appearance of primitive vascular plants on land	Expansion of the fishes
	Ordovician	500	Marine algae	Invertebrates dominate the seas First fishes (jawless)
	Cambrian	600	Primitive marine algae	Age of invertebrates Trilobites abundant
Precambrian		4,600		*Aquatic Life Only* Origin of the invertebrates Origin of complex (eukaryotic) cells Origin of photosynthetic organisms Origin of primitive (prokaryotic) cells Origin of life

move, even though the adults are sessile. The free-swimming, ciliated larval stages allow the animal to disperse through its environment.

The larva differs from the sessile adult not only because it is free-swimming, but because it usually uses a different source of food and often becomes part of the plankton community. The larval stages of most organisms are subjected to predation, and the mortality rate is high. The larvae move to new locations, settle down, and develop into adults. Even marine animals that do not have sessile stages typically produce free-swimming larvae. For example, crabs, starfish, and eels move freely about and produce free-swimming larvae.

The backbone is a recent development in the evolution of animals. Animals with backbones made of vertebrae are called **vertebrates;** those without backbones are called **invertebrates.** All of the early animals lacked backbones and still constitute 99.9% of all animal species in existence today.

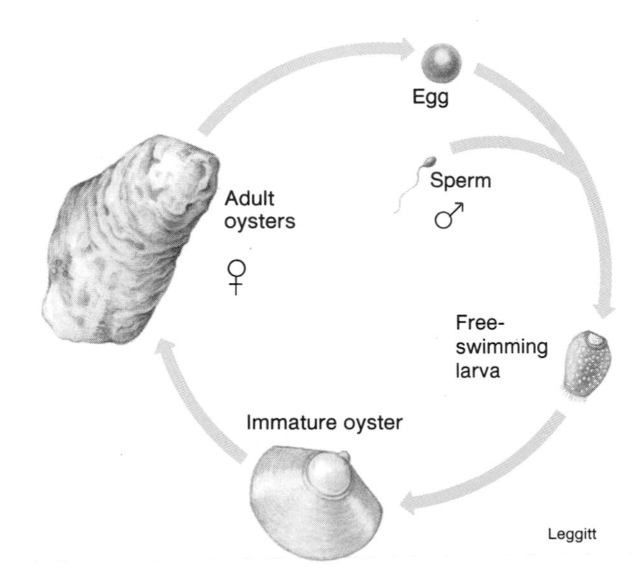

Figure 26.5 **The Life Cycle of an Oyster.** Each individual oyster can be either male or female at different times in its life. Sperm are released and swim to the egg. Fertilization results in the formation of a free-swimming larva. This larva undergoes several changes during the first twelve to fourteen days and eventually develops into an immature oyster that becomes attached and develops into the adult oyster.

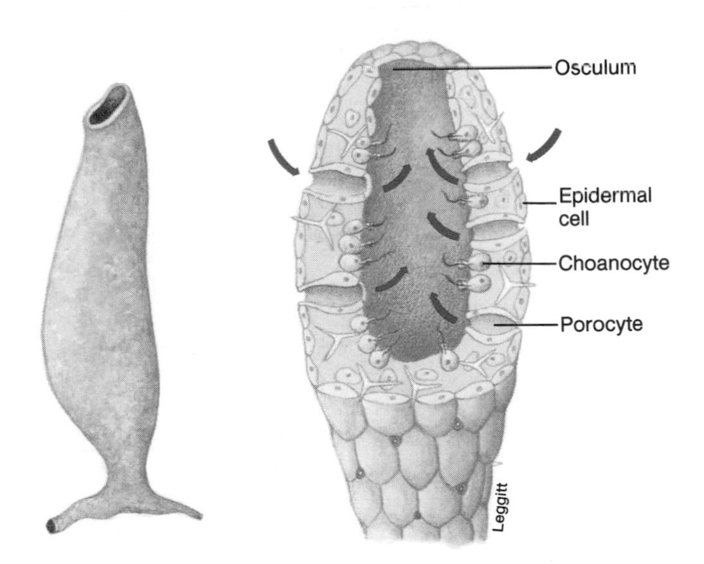

Figure 26.6 **A Sponge.** *Verongia* is an example of a tube sponge. The cells are in direct contact with the environment. The choanocytes form an inner layer of flagellated cells. These flagella create a current that brings water in through the openings formed by the porocyte cells, and it flows out through the osculum. The current brings food and oxygen to the inner layer of cells. The food is filtered from the water as it passes through the animal.

Primitive Marine Animals

Sponges, jellyfish, and corals are the simplest multicellular animals. They evolved about 600 million years ago and are usually found in saltwater environments. Even though sponges are classified as multicellular, in many ways they are colonial. Most cells are in direct contact with the environment. All adult sponges are sessile filter feeders with ciliated cells that cause a current of water to circulate within the organism. The individual cells obtain their nutrients directly from the water (figure 26.6).

Reproduction in sponges can occur by fragmentation. Wave action may tear off a part of a sponge, which eventually settles down, attaches itself, and begins to grow. Sponges also reproduce by **budding,** a type of asexual reproduction in which the new organism is an outgrowth of the parent. They also reproduce sexually, and external fertilization results in a free-swimming, ciliated larval stage. The larva swims in the plankton and eventually settles to the bottom, attaches, and grows into an adult sponge.

Cnidarians include the jellyfish, corals, and sea anemones. Like many sponges, they have radial symmetry. Many species of Cnidaria exhibit alternation of generations and have both sexual and asexual stages of reproduction. The **medusa** is a free-swimming adult stage that reproduces sexually. The **polyp** is a sessile larval stage that reproduces asexually (figure 26.7). All species have a single opening that leads into a saclike interior. Surrounding the opening

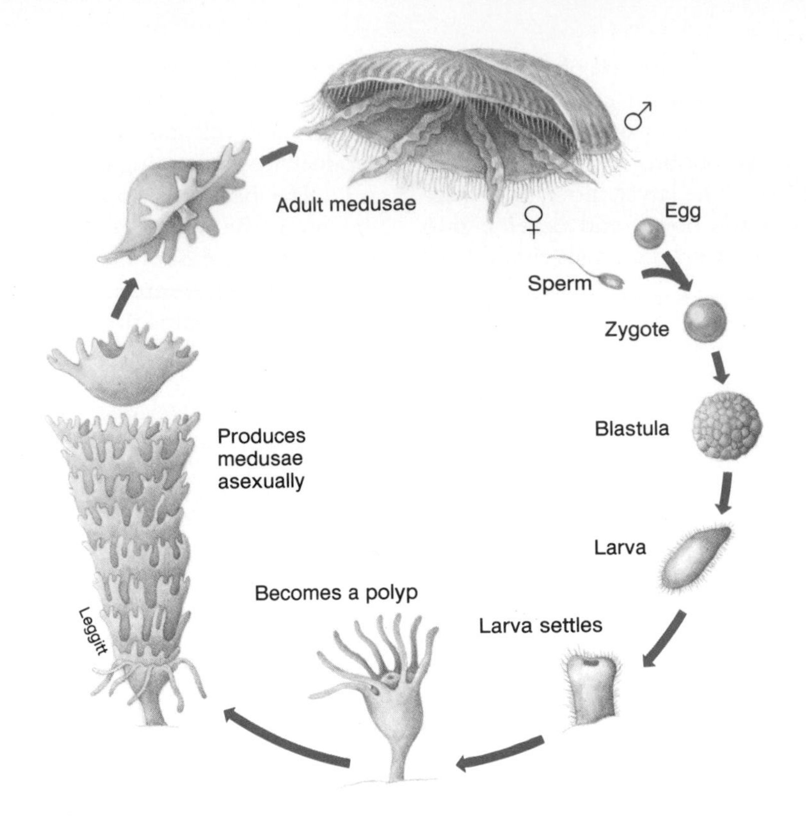

Figure 26.7 **The Life Cycle of Cnidaria.** The life cycle of *Aurelia* is typical of the alternation of generations seen in most species of cnidaria. The free-swimming adult medusae (jellyfish) reproduce sexually, and the resulting larva develops into a polyp. The polyp undergoes asexual reproduction, which produces the free-swimming medusa stage.

is a series of tentacles (figure 26.8). These long, flexible, armlike tentacles have specialized cells that can sting and paralyze small organisms. Even though they are primitive organisms, cnidarians are carnivorous.

A Parasitic Way of Life

There are three basic types of flatworms (the Platyhelminthes): free-living flatworms (often called *planarians*), flukes, and tapeworms (figure 26.9). The majority of free-living flatworms are nonparasitic bottom-dwellers in marine or fresh water. A few species are found in moist terrestrial habitats. Free-living flatworms have muscular, nervous, and excretory systems.

All of the flukes and tapeworms are parasites. Some of the flukes are external parasites on the gills and scales of fish, but the majority are internal parasites. Most flukes have a complex life cycle involving more than one host. Usually, the larval stage infects an invertebrate host, whereas the adult parasite infects a vertebrate host.

(a)

(b)

Figure 26.8 **Phylum Cnidaria.** The Portuguese man-of-war (a) and the sessile sea anemone (b) are typical cnidarians. Each has a saclike body structure with a single opening into the gut.

Schistosomiasis, which causes dysentery, anemia, and a lowering of the body's resistance, is caused by adult *Schistosoma mansoni* flukes that live in the blood vessels of the human digestive system. Fertilized eggs pass out with the feces. Eggs released into the water hatch into free-swimming larvae. If a larva infects a snail, it undergoes additional reproduction and produces a second larval stage. A single infected snail may be the source of thousands of larvae. These new larvae swim freely in the water. Should they encounter a human, the larvae bore through the skin and enter the circulatory system, which carries them to the blood vessels of the intestine (figure 26.10).

Two hosts are also involved in the tapeworm's life cycle, but both hosts are usually vertebrate animals. An herbivore eats tapeworm eggs that have been passed from another infected host through its feces. The eggs are eaten along with the vegetation the herbivore uses for food. An egg develops into a larval stage that encysts in the muscle of the herbivore. When the herbivore is eaten by a carnivore, the tapeworm cyst develops into the adult form in the intestine of the carnivore. When the worms reproduce, the eggs can be easily dispersed in the feces (figure 26.11).

Few animals are found in as many diverse habitats or in such numbers as

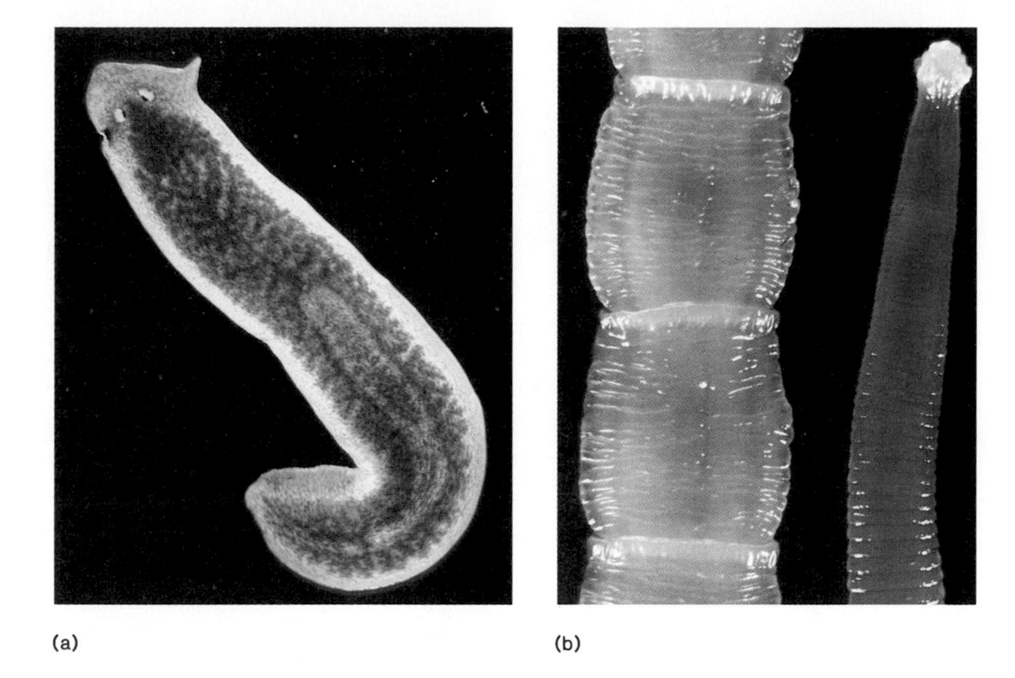

Figure 26.9 **Flatworms.** (a) Planarians are free-swimming, nonparasitic flatworms that inhabit fresh water. (b) Adult tapeworms are parasites found in the intestines of many carnivores.

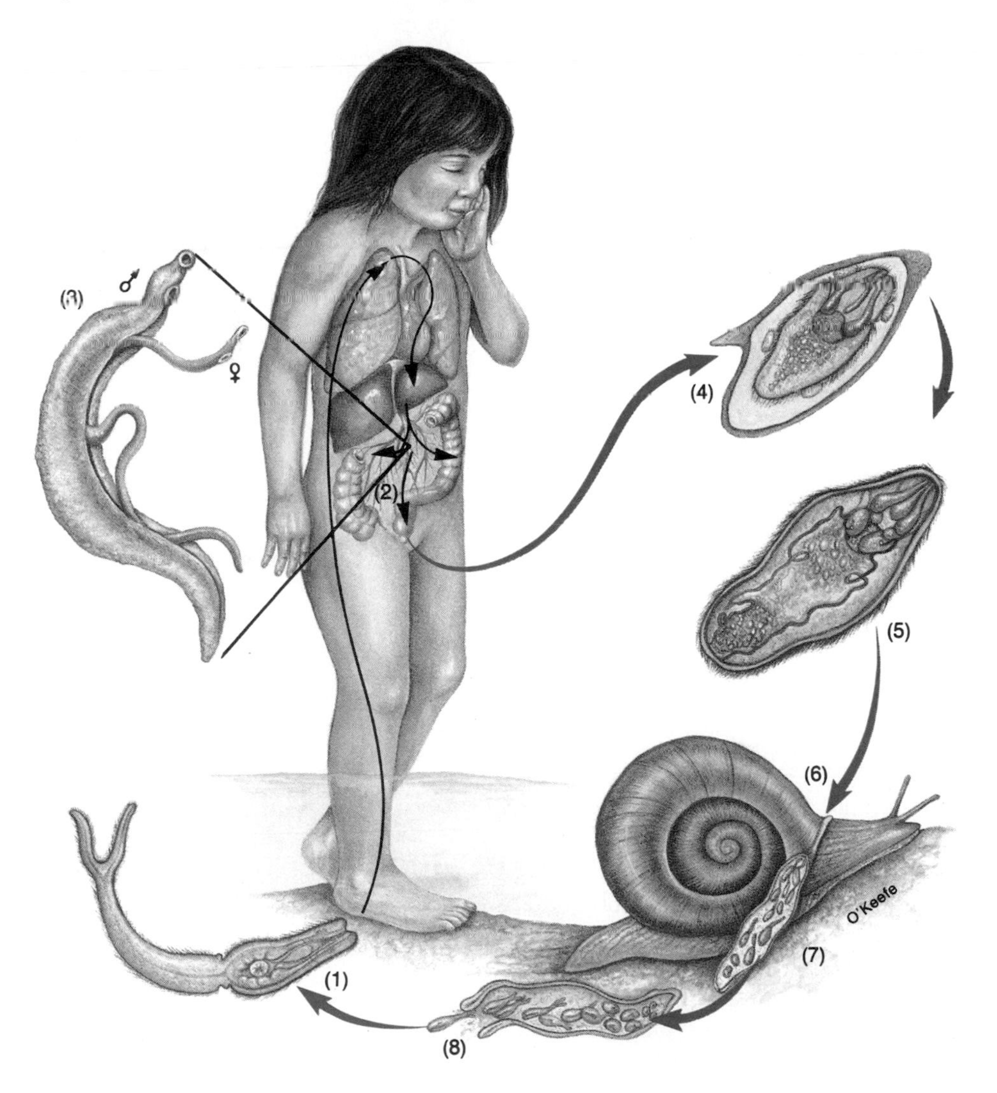

Figure 26.10 **The Life History of *Schistosoma mansoni*.** (1) Cercaria larvae in water penetrate human skin and are carried through the circulatory system to the veins of the intestine. They develop into (2) adult worms, which live in the blood vessels of the intestine. In (3) copulating worms are shown. The female produces eggs, which enter the intestine and leave with the feces. In (4) a miracidium larva within an eggshell is shown. (5) The miracidium larva hatches in water and burrows into a snail (6) where it develops into a mother sporocyst (7). The mother sporocyst produces many daughter sporocysts (8), each of which produces many cercaria larvae. These leave the snail's body and enter the water, thus completing the life cycle.

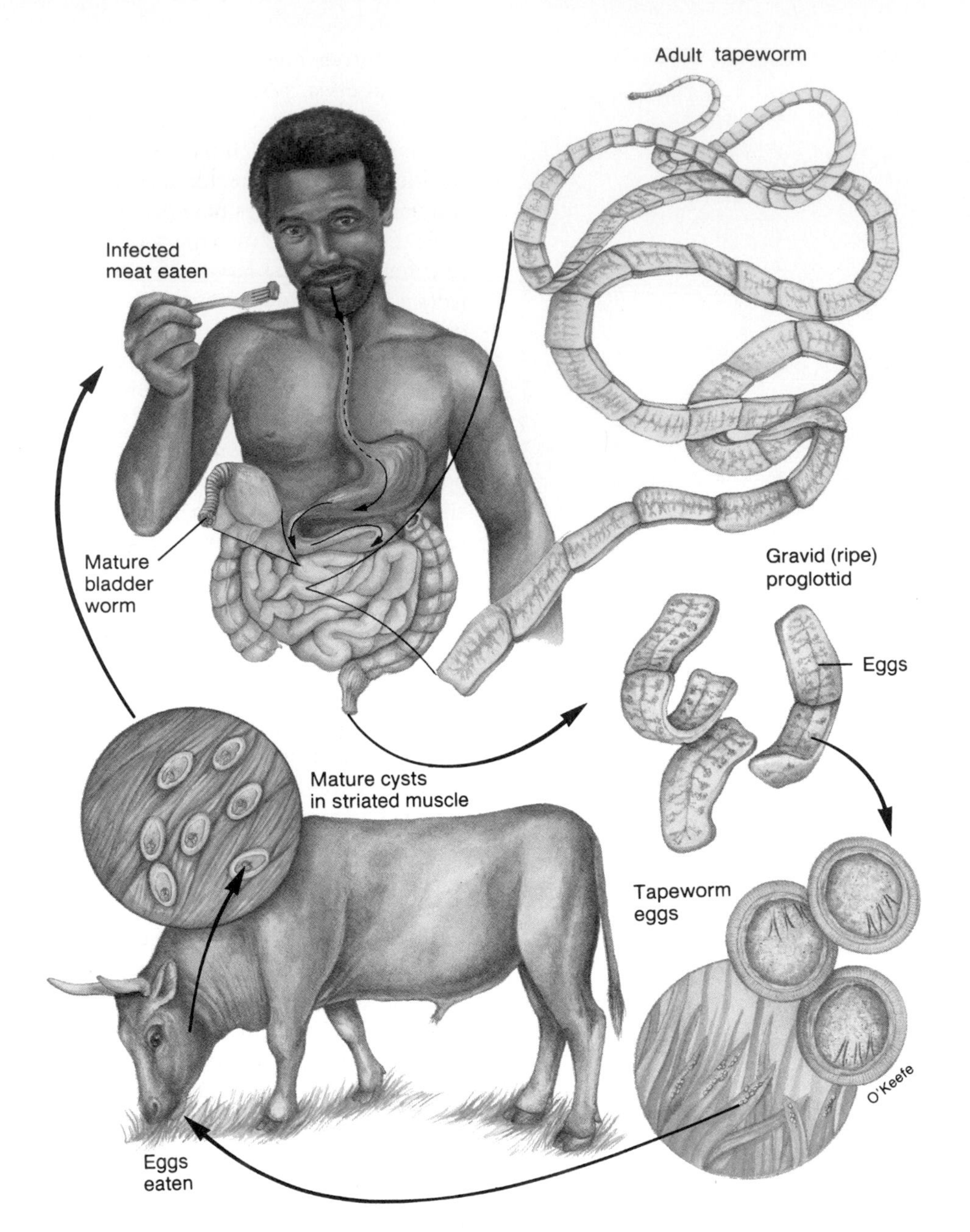

Figure 26.11 **The Life Cycle of a Tapeworm.** The adult beef tapeworm lives in the human small intestine. Proglottids, individual segments containing male and female sex organs, are the site of egg production. When the eggs are ripe, the proglottids drop off the tapeworm and pass out in the feces. If a cow eats an egg, it will develop into a cyst in the cow's muscles. When humans eat the cyst in the meat, the cyst develops into an adult tapeworm.

the roundworms, the Nematoda. Most are free-living, but many are economically important parasites. Some are parasitic on plants, while others infect animals, and collectively they do untold billions of dollars worth of damage to our crops and livestock (figure 26.12).

Roundworms parasites range from the relatively harmless human intestinal pinworms *Enterobius,* which may cause irritation but no serious harm, to *Dirofilaria,* which can cause heartworm disease in dogs. If untreated, this infection may be fatal. Often the amount of damage inflicted by roundworms is directly proportional to their number. For example, hookworms (figure 26.13) feed on the host's blood. A slight infestation often results in anemia, but a heavy infestation of hookworms may result in mental or physical retardation.

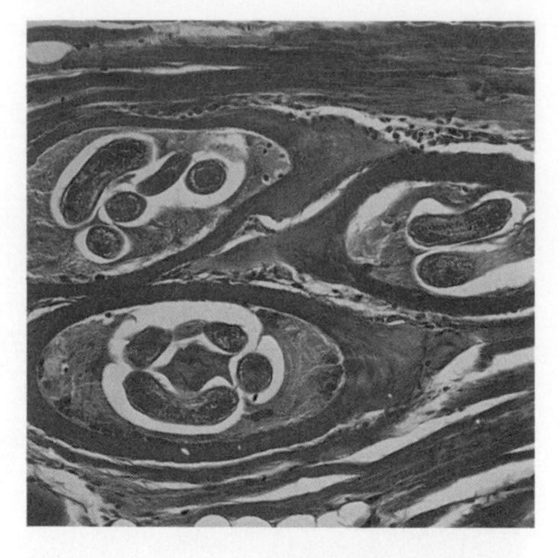

(a)

(b)

Figure 26.12 **Roundworms.** When meat infected with trichina cysts is eaten by a host, the cysts develop into adults. The adult worms reproduce, and the resulting larva encyst in the muscles of the host. A heavy infection can result in the death of the host. (a) A cross section of an infected muscle. Within each circle is an encysted larva. (b) Some roundworms are parasitic on plants and may cause extensive damage.

Advanced Benthic Marine Organisms

A major ecological niche in the oceans includes large numbers of organisms that live on the bottom, called **benthic** organisms. Among benthic organisms are

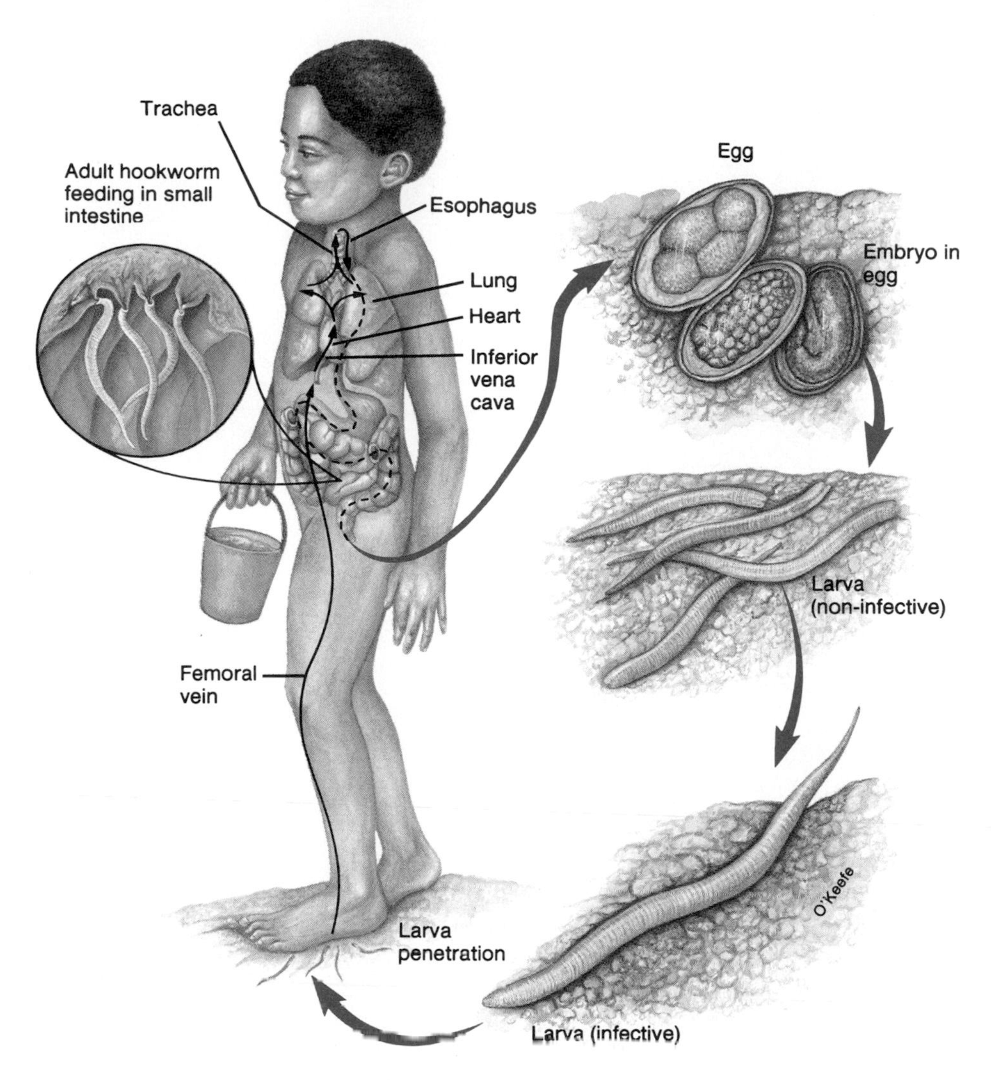

Figure 26.13 **The Life Cycle of a Hookworm.** Fertilized hookworm eggs pass out of the body in the feces and develop into larvae. In the soil, the larvae feed on bacteria. The larvae bore through the skin of humans and develop into adults. The adult hookworms live in the digestive system, where they suck blood from the host. The loss of blood can result in anemia, mental and physical retardation, and a loss of energy.

such animals as the segmented worms, the Annelida; mollusks, such as clams and snails; and many kinds of arthropods, such as lobsters, crabs, and shrimp. When people think of annelids, they commonly think of the terrestrial earthworm. However, most annelids are not terrestrial but are mainly found in marine benthic habitats, where most burrow into the ocean floor (figure 26.14).

The bilaterally symmetrical annelids have a well-developed musculature and circulatory, digestive, excretory, and nervous systems that are organized into repeating segments. For this reason the annelids are called *segmented worms*. Depending upon the species, the individual may be male, female, or hermaphroditic (contain both male and female reproductive organs). Since most marine annelids live on the bottom and do not travel great distances, a free-swimming larval stage is important in their distribution. Like many other marine animals, many annelids are filter feeders, straining small organic materials from their surroundings. Others are primarily scavengers, and a few are predators of other small animals.

Another major group of benthic animals are the mollusks. Like most other forms of animal life, the mollusks originated in the ocean, and even though some forms have made the move to fresh water and terrestrial environments, the majority still live in the oceans. They range from microscopic organisms to the giant squid, which is up to 18 meters long (figure 26.15).

A primary characteristic of mollusks is the presence of a soft body enclosed by a hard shell. Clams and oysters have two shells, whereas snails have a single shell. Some forms, such as the slugs, have no shell; they are unprotected. In the squid and octopus, the shell is located internally and serves as a form of support structure. Reproduction is generally sexual; some species have separate sexes and others are hermaphroditic.

Except for the squids and octopi, mollusks are slow-moving benthic animals. Some are herbivores and feed on marine algae; others are scavengers and feed on dead organic matter. A few are even predators of other slow-moving or sessile neighbors. As with most other marine animals, the mollusks produce free-swimming larval stages that aid in dispersal.

Echinoderms are strictly marine benthic animals and are found in all regions, from the shoreline to the deep portions of the ocean. Echinoderms are often the most common type of animal on much of the ocean floor. Most species are free-moving and are either carnivores or feed on detritus. They are unique among more advanced invertebrates in that they display radial symmetry. However, the larval stage has bilateral symmetry, leading many biologists to believe that the echinoderm ancestors were bilaterally symmetrical. Another unique characteristic of this group is the water vascular system (figure 26.16). In this system, water is taken in through a structure on the top side of the animal and then moves through a series of canals. The passage of water through this water vascular system is involved in the organism's locomotion.

(a)

(b)

Figure 26.14 **Annelids.** Annelids include (a) the sandworm, which is common in marine environments, and (b) sessile forms that are filter feeders.

(a)

(b)

Figure 26.15 **Mollusks.** Mollusks may range in complexity from (a) a small slow-moving, grazing animal like a chiton to (b) intelligent, rapidly moving carnivores like an octopus.

Pelagic Marine Animals: Fish

Animals that swim freely as adults are called **pelagic.** Many kinds of animals belong in this ecological niche, including squid, swimming crabs, sea snakes, and whales. However, the major kinds of pelagic animals are commonly called *fish.* There are several different kinds of fish that are as different from one another as reptiles are from birds, or birds from mammals.

Hagfish and lampreys lack jaws and are the most primitive of the fish. Hagfish are strictly marine forms and are scavengers; lampreys are mainly marine but may also be found in fresh water (figure 26.17). Adult lampreys suck blood from their larger fish hosts. Lampreys reproduce in freshwater streams, where the eggs develop into filter-feeding larvae. After several years, the larvae change to adults and migrate to open water.

Sharks and rays are marine animals that have an internal skeleton made entirely of cartilage (figure 26.18). These animals have no swim bladder to adjust their body density in order to maintain their position in the water; therefore, they must constantly swim or they will sink. Many of us have developed some misconceptions about sharks and rays as a result of movies or TV. Rays feed by gliding along the bottom and dredging up food, usually invertebrates. Sharks are predatory and feed primarily on other fish. They travel great distances in search of food. Of the forty species of sharks, only seven are known to attack people.

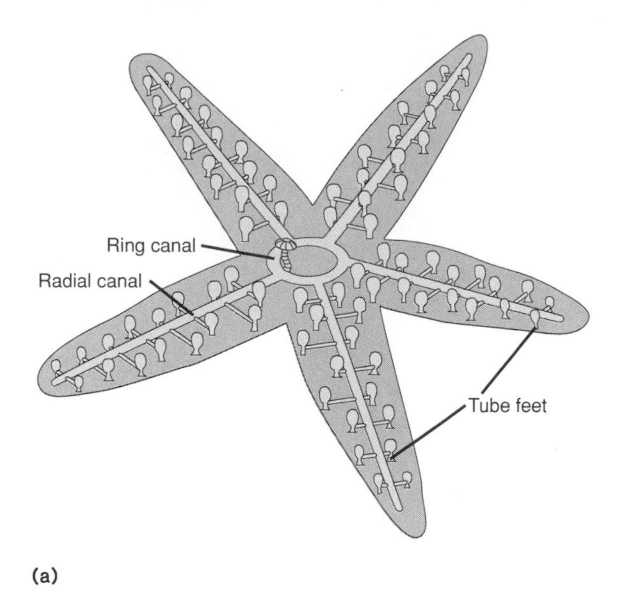

Figure 26.16 **Echinoderms.** (a) Starfish move by means of a water vascular system. Water enters the system through an opening, travels to the radial canals, and is forced into the tube feet. Other echinoderms include (b) the sea urchin and (c) the sea cucumber.

Figure 26.17 **The Lamprey.** The lamprey uses its round mouth (a) to attach to a fish (b) and then sucks blood from the fish.

Most sharks grow no longer than a meter. The whale shark, the largest shark, grows to 16 meters, but it is strictly a filter feeder.

The bony fish are the class that is most familiar to us (figure 26.19). The skeleton is composed of bone. Most species have a swim bladder and can regulate the amount of gas in the bladder to control their density. Thus, the fish can remain at a given level in the water without expending large amounts of energy. Bony fish are found in marine and freshwater habitats, and some, like the salmon, can live in both. Bony fish feed on a wide variety of materials, including algae, detritus, and other animals. Like the sharks, many range widely in search of food. However, many fish are highly territorial and remain in a small area their entire lives.

(a)

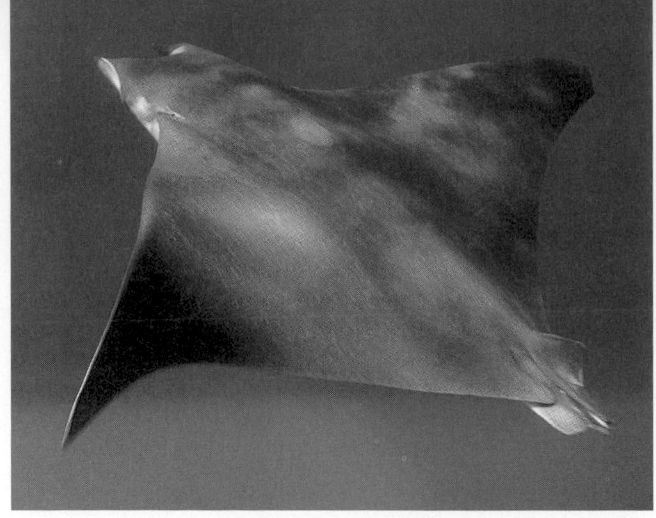

(b)

Figure 26.18 **Cartilaginous Fish.** Although sharks (a) and rays (b) are large animals, they do not have bones; their skeletal system is made entirely of cartilage.

(a) (b) (c)

Figure 26.19 **Bony Fish.** Fish have a variety of body shapes. The animals pictured here have skeletons made of bones and, like most bony fish, they have a swim bladder. (a) Sea perch, (b) moray eel, and (c) sea horse.

The Movement to Land

Plants were the first organisms to adapt to a terrestrial environment. They began to colonize land about 410 million years ago, during the Silurian period, and they were well established on land before the animals. Thus, they served as a source of food and shelter for the animals. When the first terrestrial animals evolved, there were many unfilled niches; therefore, much adaptive radiation occurred, resulting in a large number of different animal species. Of all the many phyla of animals in the ocean, only a few made the transition from the ocean to the extremely variable environments found on the land. The annelids and the mollusks evolved onto the land but were confined to moist habitats. Many of the arthropods (insects and spiders) and vertebrates (reptiles, birds, and mammals) adapted to a wide variety of dryer terrestrial habitats.

Regardless of their type, all animals that live on land must overcome certain common problems. Terrestrial animals must have (1) a moist membrane that allows for adequate gas exchange between the atmosphere and the organism, (2) a means of support and locomotion suitable for land travel, (3) methods to conserve internal water, (4) a means of reproduction and early embryonic development in which large amounts of water are not required, and

(a)

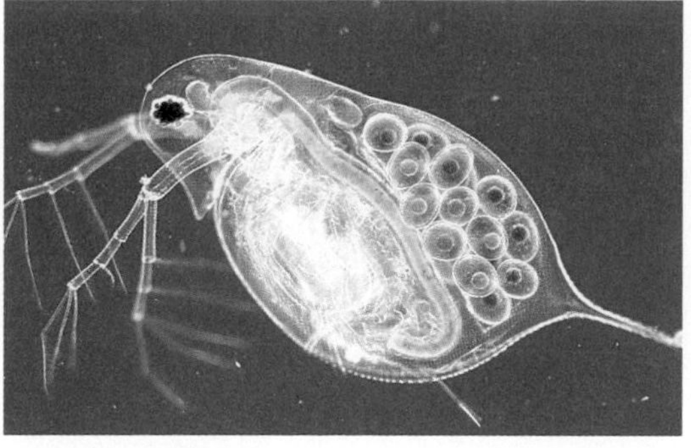

(b)

(c)

Figure 26.20 **Crustaceans.** Crustaceans include marine forms such as the king crab shown in (a). The microscopic water flea (b) is a freshwater organism, and the pill (sow) bug (c) is terrestrial.

(5) methods to survive the rapid and extreme climatic changes that characterize many terrestrial habitats. When we consider the transition of animals from an aquatic to a terrestrial environment, it is important to understand that this process required millions of years. There had to be countless mutations resulting in altered structures, functions, and behavioral characteristics that enabled animals to successfully adapt.

One large group of animals, the arthropods, have been incredibly successful in all kinds of habitats. They can be found in the plankton, as benthic inhabitants, and as pelagic organisms. This phylum includes nearly three quarters of all known animal species. No other phylum lives in such a wide range of habitats. Although they include carnivores and omnivores, the majority of arthropods are herbivores.

The crustaceans are the best-known class of aquatic arthropods (figure 26.20). Copopods are common in the plankton of the oceans, crabs and their relatives are found as benthic organisms, and shrimp and krill are pelagic. However, the major success of this group is seen in the huge variety of terrestrial insects. Other terrestrial anthropod groups include the millipedes, centipedes, spiders, and scorpions.

Insects and other arthropods probably adapted to the land at about the same time as the plants. They developed an internal tracheal system of thin-walled tubes extending into all regions of the body, thus providing a large surface area for gas exchange (figure 26.21a). These tubes have small openings to the outside, which reduces the amount of water lost to the environment. They also developed a rigid outer layer that provides body support and an area for muscle attachment that permits rapid muscular movement. Since it is waterproof, this outer layer also reduces water loss. Another important method of conserving water in insects and spiders is the presence of malphigian tubules, thin-walled tubes that surround the gut (figure 26.21b). If the insect is living in a dry

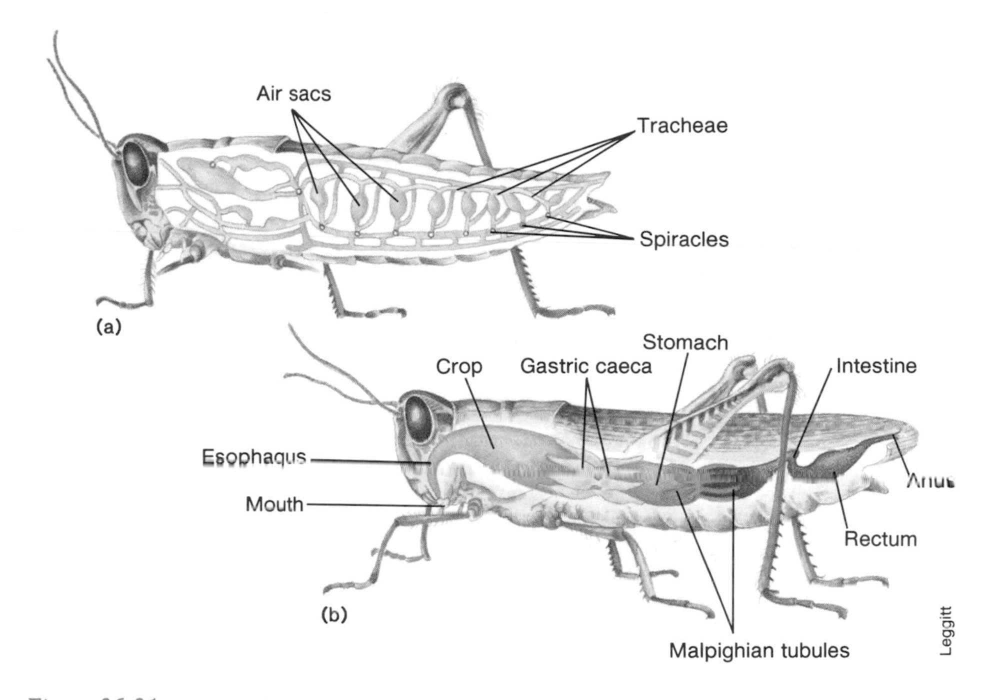

Figure 26.21 **Insect Respiratory and Waste-Removal Systems.** (a) Spiracles are openings in the exoskeleton of an insect. These openings connect to a series of tubes (tracheae) that allow for the transportation of gases in the insect's body. (b) Malpighian tubules are used in the elimination of waste materials and the reabsorption of water into the insect's body. Both systems are means of conserving body water.

(a) (b) (c) (d)

Figure 26.22 **The Life History of a Moth.** The fertilized egg (a) of the moth hatches into the larval stage (b). This wormlike stage then covers itself with a case and becomes a pupa (c). After emerging from the pupa (d), the adult moths wings expand to their full size.

environment, most of the water in waste materials is reabsorbed into the body by the malphigian tubules and conserved.

Insects have separate male and female individuals and fertilization is internal, which means that the insects do not require water to reproduce. Insects have evolved a number of means of survival under hostile environmental conditions. Their rapid reproductive rate is one means. Most of a population may be lost because of an unsuitable environmental change, but when favorable conditions return, the remaining individuals can quickly increase in number. Other insects survive unfavorable conditions in the egg or larval form and develop into adults when conditions become suitable (figure 26.22). Some insects survive because of a lower metabolic rate during unfavorable conditions.

The terrestrial arthropods occupy an incredible variety of niches. Many are herbivores that compete directly with humans for food. They are capable of completely decimating plant populations that serve as food for human consumption. Many farming practices, including the use of pesticides, are directed at controlling insect populations. Other kinds of insects are carnivores that feed primarily on herbivorous insects. These insects are beneficial in controlling herbivore populations. Wasps and ladybird beetles have been used to reduce the devastating effects of insects that feed on agricultural crops. Insects have evolved in concert with the flowering plants; their role in pollination is well understood. Bees, butterflies, and beetles transfer pollen from one flower to another as they visit the flowers in search of food. Many kinds of crops rely on bees for pollination, and farmers even rent beehives to ensure adequate seed or fruit production.

The first vertebrate animals to evolve onto land were probably the ancestors of present-day amphibians (frogs, toads, and salamanders). The first amphibians made the transition to land some 360 million years ago during the Devonian period. This was 50 million years after plants and arthropods had become established on land. Thus, when the first vertebrates developed the ability to live on land, there was shelter and food for herbivorous as well as carnivorous animals. But the same five problems that the insects and spiders faced in their transition ashore also faced the vertebrates.

(a)

(b)

Figure 26.23 **Amphibians.** Amphibian larvae (a) are aquatic organisms that have external gills and feed on vegetation. The adults (b), such as the salamander, are terrestrial and feed on insects, worms, and other small animals.

In amphibians (figure 26.23), the development of lungs was an adaptation that provided a means for land animals to exchange oxygen and carbon dioxide with the atmosphere. However, amphibians do not have an efficient method of breathing; they swallow air to fill the lungs, and most gas exchange between amphibians and the atmosphere must occur through the skin. In addition to needing water to keep their skin moist, amphibians must reproduce in water. When they mate, the female releases eggs into the water, and the male releases sperm amid the eggs. External fertilization occurs in the water, and the fertilized eggs must remain in water or they will dehydrate. Thus, with the appearance of amphibians, vertebrate animals moved onto "dry" land, but the processes of gas exchange and reproduction still limit the range of movement of amphibians from water.

Their buoyancy in water helps support the bodies of aquatic animals. This form of support is lost when animals move ashore; thus, the amphibians developed a skeletal structure that prevented the collapse of their bodies on land. Even though they have an appropriate skeletal structure, amphibians must always be near water because they dehydrate and require water for reproduction. The extreme climatic changes were a minor problem: when conditions on land became too hostile, the amphibians retreated to an aquatic environment.

(a)

(b)

Figure 26.24. **Reptiles.** Present-day reptiles include (a) turtles and (b) lizards.

For 40 million years amphibians were the only vertebrate animals on land. During this time, mutations continued to occur, and valuable modifications were passed on to future generations. One change allowed the male to deposit sperm directly within the female. Because the sperm could directly enter the female and remain within a moist interior, it was no longer necessary for the animals to return to water to reproduce. The reptiles had evolved (figure 26.24).

Internal fertilization was not enough to completely free the reptiles from returning to water, however. The developing young still required a moist

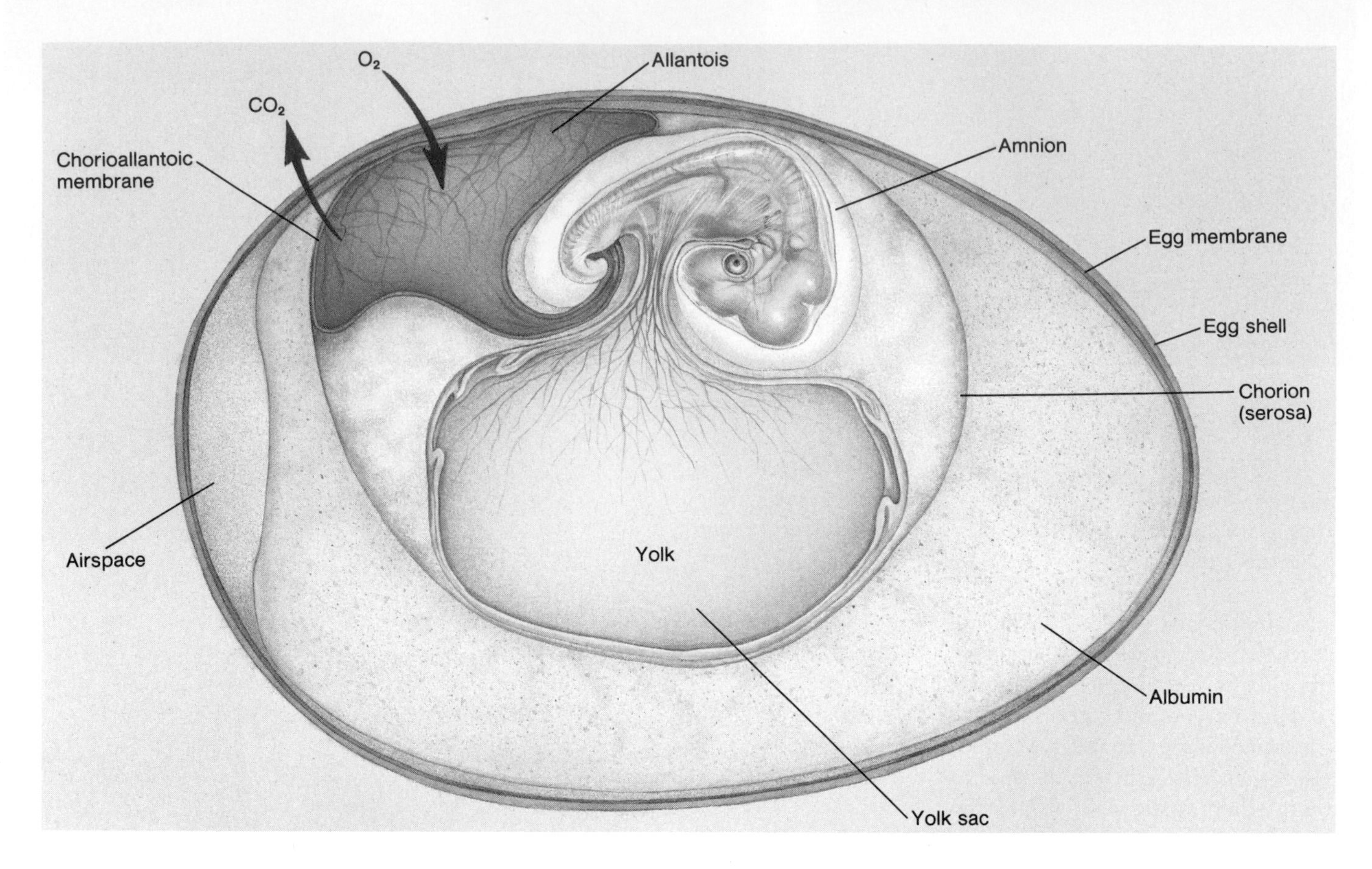

Figure 26.25 **The Amniotic Egg.** An amniotic egg has a shell and a membrane that prevent the egg from dehydrating and allow for the exchange of gases between the egg and the environment. The egg yolk provides a source of nourishment for the developing young. The embryo grows three extraembryonic membranes: the amnion is a fluid-filled sac that allows the embryo to develop in a liquid medium, the allantois collects the embryo's metabolic waste material and exchanges gases, and the chorion is a membrane that encloses the embryo and the other two membranes.

environment for their early growth. Reptiles became completely independent of an aquatic environment with the development of the amniotic egg, which protects the developing young from injury and dehydration (figure 26.25). The covering on the egg seals in the moisture and protects the developing young from dehydration while allowing for the exchange of gases. The reptiles were the first animals to develop such an egg.

The development of a means of internal fertilization and the amniotic egg allowed the reptiles to spread over much of the Earth and occupy a large number of previously unfilled niches. For about 200 million years they were the only large vertebrate animals on land. The evolution of reptiles increased competition with the amphibians for food and space. The amphibians generally lost in this competition; consequently, most became extinct. Some, however, were able to evolve into the present-day frogs, toads, and salamanders.

There have been several periods of mass extinction on the earth. One such period occurred about 65 million years ago, when many kinds of reptiles became extinct. Before that period of mass extinction, about 150 million years ago, birds evolved (figure 26.26). Although the amniotic egg remained the method of protecting the young, a series of changes in the reptiles produced animals with a more rapid metabolism, feathers, and other adaptations for flight. These were the first birds. They also possessed behavioral instincts, such as nest building, defense of their young, and feeding of the young. Because of these adaptations and their invasion of the air, a previously unoccupied niche, birds became one of the very successful groups of animals.

Even though the reptiles and birds had mastered the problems of coming ashore, mutations and natural selection continued, and so did evolution. As good as the amniotic egg is, it does have drawbacks: it lacks sufficient protection from sudden environmental changes and from predators that use eggs as food. Other mutations in the reptile line of evolution resulted in animals that overcame the disadvantages of the external egg by providing for internal development of the young. Such development allowed for a higher survival rate. The internal development of the young, along with milk-gland development, a constant body temperature, a body covered with hair, and care of young by parents marked the emergence of mammals.

The first mammals to evolve were egg-laying mammals (figure 26.27), whose young still developed in an external egg. The marsupials (pouched mammals) have internal development of the young. However, the young are all born prematurely and must be reared in a pouch (figure 26.28). In the pouch, the young attach to a nipple and remain until they are able to forage for themselves. The young of placental mammals remain within the female much longer, and they are born in a more advanced stage of development than is typical for marsupials (figure 26.29).

(a)

(b)

Figure 26.26 **Birds.** Birds range in size from (a) the small hummingbird to (b) the large ostrich.

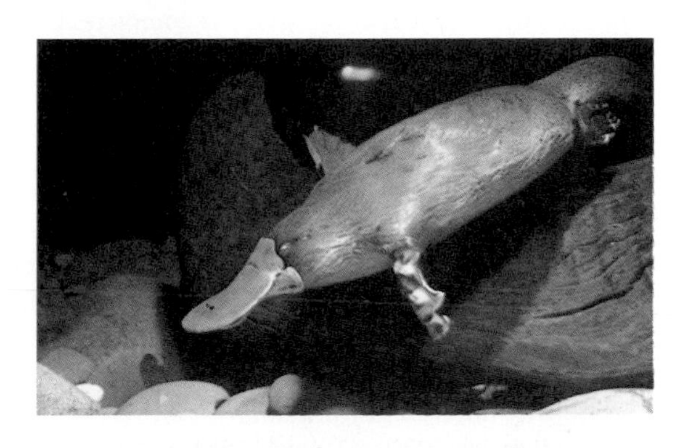

Figure 26.27 **The Duck-billed Platypus.** The duck-billed platypus is a primitive type of mammal. It has the mammalian characteristics of fur and milk production, but the young are hatched from eggs.

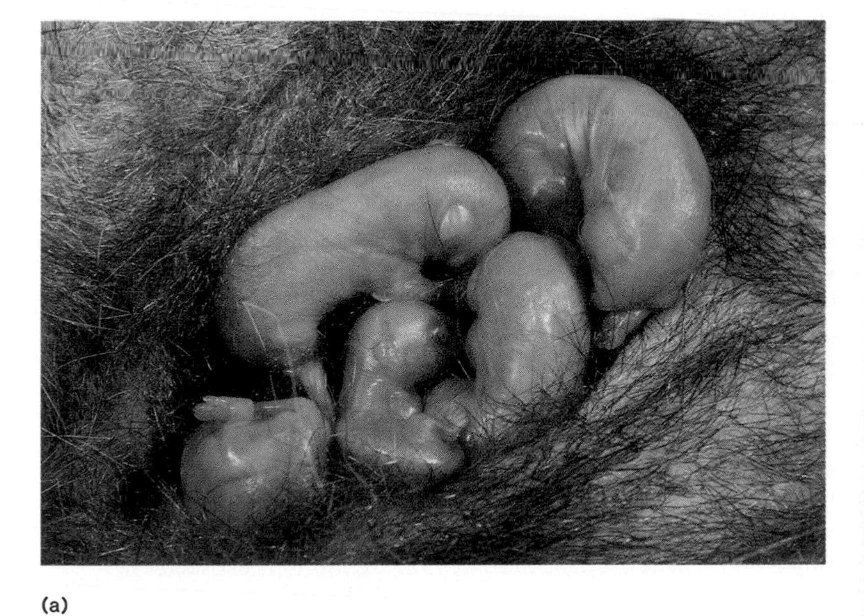

(a)

(b)

Figure 26.28 **Marsupials.** All marsupials are born prematurely. (a) The young of the opossum crawl into a pouch and complete development there. (b) Even after the young are fully developed, some marsupials still carry the young in a pouch.

(a)

(b)

Figure 26.29 **Placental Mammals.** Placental mammals are not born until the development of the young is complete. Included in this group are animals ranging from (a) small terrestrial animals like the lease shrew (5 centimeters) to (b) large marine animals, like the sperm whale (20 meters).

• Summary •

The 4 million known species of animals, which inhabit widely diverse habitats, are all multicellular and heterotrophic. Animal body shape is asymmetrical, radial, or bilateral. All animals with bilateral symmetry have a body structure composed of three layers.

Animal life originated in the ocean about 3.8 billion years ago, and for the first 2 billion years, because of the favorable environment, all animal life remained in the ocean. Many simple marine animals have life cycles that involve alternation of generations.

For many animals, a parasitic lifestyle is appropriate. A major ecological niche for many marine animals is the ocean bottom—the benthic zone. Large, free-swimming marine animals dominate the pelagic ocean zone.

Animals that adapted to a terrestrial environment had to have (1) a moist membrane for gas exchange, (2) support and locomotion suitable for land, (3) a means of conserving body water, (4) a means of reproducing and providing for early embryonic development out of water, and (5) a means of surviving in rapid and extreme climatic changes.

Major Groups of Animals

Sponges (Phylum Porifera)
1. Body has many pores and canals
2. Attached to an object
3. Two cell layers
4. No movable appendages
5. Intracellular digestion
6. Aquatic organisms

Cnidaria (Phylum Cnidaria)
1. Two life stages: attached polyps and free-swimming medusae
2. Radial symmetry
3. Digestive cavity
4. Tentacles and stinging cells
5. Nerve net
6. Aquatic organisms

Flatworms (Phylum Platyhelminthes)
1. Flattened body
2. Bilateral symmetry
3. Three germ layers
4. Incomplete digestive system
5. Nervous system
6. Usually hermaphroditic

Roundworms (Phylum Nematoda)
1. Cylindral body with cuticle
2. Unsegmented body
3. Unlined body cavity
4. Complete digestive tract
5. Sexes usually separate
6. Many parasitic forms

Segmented Worms (Phylum Annelida)
1. Segmented body
2. Circular and longitudinal muscle layers
3. Organ systems present
4. Closed circulatory system
5. Respiration by skin or gills
6. Usually have a larval form

Arthropods (Phylum Arthropoda)
1. Jointed legs
2. Chitinous exoskeleton
3. Three body regions: head, thorax (may be combined into a cephalothorax), and abdomen
4. Compound eyes
5. Striated muscle fibers
6. Fertilization usually internal

Mollusks (Phylum Mollusca)
1. Body has anterior head region, dorsal visceral hump, and ventral foot
2. Bilateral symmetry
3. Many forms have a shell
4. Usually has an open circulatory system
5. Chambered heart
6. Sexes usually separate

Echinoderms (Phylum Echinodermata)
1. Spiny-skinned
2. Radial symmetry, usually five-parted
3. Only found in saltwater environment
4. No head
5. Larval stages
6. Water-vascular system

Jawless Fish (Class Agnatha)
1. Sucking mouth
2. Unpaired fins
3. Two-chambered heart
4. Poikilotherms
5. Single gonad, no duct

Cartilaginous Fish (Class Chondrichthyes)
1. Cartilaginous skeleton
2. No operculum
3. No swim bladder
4. Lateral line
5. Paired fins
6. Internal fertilization

Bony Fish (Class Osteichthyes)
1. Bony skeleton, scales present
2. Operculum present
3. Swim bladder usually present
4. Well-developed jaws with teeth
5. Two-chambered heart
6. Usually external fertilization

Amphibians (Class Amphibia)
1. Usually moist skin, no scales
2. Usually two pairs of legs
3. Lungs present in adults
4. Three-chambered heart
5. Aquatic larvae and terrestrial adults
6. Usually external fertilization

Reptiles (Class Reptilia)
1. Scales present
2. Internal fertilization
3. Amniotic eggs
4. Imperfectly formed four-chambered heart
5. Respiration by lungs
6. Poikilotherms

Birds (Class Aves)
1. Body covered with feathers
2. Amniotic eggs
3. Forelimbs usually adapted for flight
4. Complete four-chambered heart
5. Beak present, no teeth
6. Homeotherms

Mammals (Class Mammalia)
1. Body covered with hair
2. Limbs usually have five digits
3. Male copulatory organ, internal fertilization
4. Mammary glands present
5. Placenta present
6. Complete four-chambered heart

• Thinking Critically •

Animals have been used routinely as models for the development of medical techniques and strategies. They have also been used in the development of pharmaceuticals and other biomedical products such as heart valves, artificial joints, and monitors. The techniques necessary to perform heart, kidney, and other organ transplants were first refined using chimpanzees, rats, and calves. Antibiotics, hormones, and chemotherapeutic drugs have been tested for their effectiveness and for possible side-effects using laboratory animals that are very sensitive and responsive to such agents. Biologists throughout the world have bred research animals that readily produce certain types of cancers that resemble cancers found in humans. By using these animals to screen potential drugs instead of humans, the risk to humans is greatly reduced. The emerging field of biotechnology is producing techniques that enable researchers to manipulate the genetic makeup of organisms. Research animals are used to perfect these techniques and highlight possible problems.

Animal-rights activists are very concerned about using animals for these purposes. They are concerned about research that seems to have little value in relation to the suffering these animals are forced to endure. Members of the American Liberation Front (ALF), an animal-rights organization, vandalized a laboratory at Michigan State University where mink were used in research to assess the toxicity of certain chemicals. Members of this group poured acid on tables and in drawers containing data, smashed equipment, and set fires in the laboratory. This attack destroyed thirty-two years of research records, including data used for developing water-quality standards. In one year, eighty similar actions were carried out by groups advocating animal rights.

What type of restrictions or controls should be put on such research? Where do you draw the line between "essential" and "nonessential" studies? Do you support the use of live animals in experiments that may alleviate human suffering?

• Experience This •

Tie a slip knot with a thread around the thorax of a living grasshopper, being careful not to interfere with its wings.

Lift the grasshopper into the air and blow gently at its head. What do you observe?

Suspend the grasshopper by the thread so that its feet touch the table top and again blow gently at the grasshopper's head. In what way does the grasshopper's behavior differ from that of the first trial?

Suspend the grasshopper above the table, but allow its feet to grasp a piece of tissue paper. Blow gently at the grasshopper's head and observe its behavior once more.

What is your conclusion concerning grasshopper behavior as a result of these three experiments?

• Questions •

1. Describe asymmetrical, radial, and bilaterally symmetrical body forms.
2. What is a sessile filter feeder?
3. How does the medusa stage of an animal differ from the polyp stage?
4. Explain the tapeworm's life cycle.
5. Describe a benthic environment.
6. How does a shark differ from most freshwater fish?
7. List the problems animals had to overcome to adapt to a terrestrial environment.
8. Why can't amphibians live in all types of terrestrial habitats?
9. What is the importance of the amniotic egg?
10. How does a marsupial differ from a placental mammal?

• Chapter Glossary •

acoelomates (a-se'lĕ-māts) Animals without a coelom. The internal organs are closely packed together.

asymmetry (ā-sim'-ĭ-tre) The characteristic of animals with no particular body shape.

benthic (ben'thik) A term used to describe organisms that live on the ocean bottom.

bilateral symmetry (bi-lat'er-al sĭm'ĭ-tre) The characteristic of animals that are constructed along a plane running from a head to a tail region, so that only a cut along one plane of this axis results in two mirror halves.

budding (bud'ing) A type of asexual reproduction in which the new organism is an outgrowth of the parent.

coelom (se'lem) A body cavity in which internal organs are suspended.

filter feeders (fil'ter fēd'er) Animals that use cilia or other appendages to create water currents and filter food out of the water.

homeotherms (ho'me-o-thurmz) Animals that maintain a constant body temperature.

invertebrates (in-vur'tuh-brāts) Animals without backbones.

medusa (muh-du'sah) A free-swimming adult stage in the phylum Cnidaria that reproduces sexually.

mesenteries (mes'en-ter''ēz) Connective tissues that hold the organs in place and also serve as support for blood vessels connecting the various organs.

pelagic (pĕ-lăj'ĭk) A term used to describe animals that swim freely as adults.

poikilotherms (poy-ki'luh-thermz) Animals with a variable body temperature that changes with the external environment.

polyp (pol'ip) A sessile larval stage in the phylum Cnidaria that reproduces asexually.

radial symmetry (ra'de-ul sĭm'ĭ-tre) The characteristic of an animal with a body constructed around a central axis. Any division of the body along this axis results in two similar halves.

sessile (ses'il) Firmly attached.

vertebrates (vur'tuh-brāts) Animals with backbones.

GLOSSARY

abiotic factors (a-bi-ot'ik fak'tōrz) Nonliving parts of an organism's environment.

accessory structures (ak-ses'o-re struk'churs) The parts of some flowers that are not directly involved in gamete production.

acetyl (ă-sēt'l) The 2-carbon remainder of the carbon skeleton of pyruvic acid that is able to enter the mitochondrion.

acetylcholine (ă-sēt''l-kō'lēn) A neurotransmitter secreted into the synapse by many axons and received by dendrites.

acetylcholinesterase (ă-sēt''l-kō''li-nes'tĕ-rās) An enzyme present in the synapse that destroys acetylcholine.

acid (ăs'id) Any compound that releases a hydrogen ion (or other ion that acts like a hydrogen ion) in a solution.

acoelomates (a-se'lă-māts) Animals without a coelom. The internal organs are closely packed together.

acquired characteristic (å-kwīrd' kar''ak-ter-iss'tik) A characteristic of an organism gained during its lifetime, not determined genetically, and therefore not transmitted to the offspring.

actin (ak'tin) A protein found in the thin filaments of muscle fibers that binds to myosin.

activation energy (ak''tī-va'shun en'ur-je) Energy required to start a reaction. It may be used to increase the number of effective collisions, or it may be required to form the transitional molecule in the progress of the reaction pathway.

active site (ak'tive sīt) The place on the enzyme that causes the substrate to change.

active transport (ak'tive trans'port) Use of a carrier molecule to move molecules across a cell membrane in a direction opposite that of the concentration gradient. The carrier requires an input of energy other than the kinetic energy of the molecules.

adaptive radiation (uh-dap'tiv ra-de-a'shun) A specific evolutionary pattern in which there is a rapid increase in the number of kinds of closely related species.

adenine (ad'ĕ-nēn) A double-ring nitrogenous-base molecule in DNA and RNA. It is the complementary base of thymine or uracil.

adenosine triphosphate (ATP) (uh-den'o-sēn tri-fos'fāt) A molecule formed from the building blocks of adenine, ribose, and phosphates. It functions as the primary energy carrier in the cell.

aerobic cellular respiration (a-ro'bik sel'yu-lar res''pī-ra'shun) The biochemical pathway that requires oxygen and converts food, such as carbohydrates, to carbon dioxide and water. During this conversion, it releases the chemical-bond energy as ATP molecules.

age distribution (āj dis''tri-biu'shun) The number of organisms of each age in the population.

alcoholic fermentation (al-ko-hol'ik fur''men-ta'shun) The anaerobic respiration pathway in yeast cells. During this process, pyruvic acid from glycolysis is converted to ethanol and carbon dioxide.

algae (al'je) Protists that have cell walls and chlorophyll and can therefore carry on photosynthesis.

alleles (a-lēlz') Alternative forms of a gene for a particular characteristic (e.g., attached earlobe genes and free earlobe genes are alternative alleles for ear shape).

alternation of generations (awl''tur-na'shun uv gen''uh-ra'shunz) A term used to describe that aspect of the life cycle in which there are two distinctly different forms of the organism. Each form is involved in the production of the other and only one form is involved in producing gametes. The cycling of a diploid sporophyte generation and a haploid gametophyte generation in plants.

alveoli (al-vē'o-lī'') Tiny sacs found in the lungs where gas exchange takes place.

amensalism (a-men'sal-izm) A relationship between two organisms in which one organism is harmed and the other is not affected.

amino acid (ah-mēn'o ă'sid) A basic subunit of proteins consisting of a short carbon skeleton that contains an amino group, a carboxylic acid group, and one of various side groups.

anaerobic respiration (an'uh-ro''bik res''pī-ra'shun) A biochemical pathway that does not require oxygen for the production of ATP.

anaphase (an'ă-fāz) The third stage of mitosis, characterized by splitting of the centromeres and movement of the chromosomes to the poles.

androgens (an'dro-jenz) Male sex hormones produced by the testes that cause the differentiation of the internal and external genital anatomy along male lines.

anemia (uh-nēm'e-ah) A disease condition in which the oxygen-carrying capacity of the blood is reduced.

angiosperms (an'je-o''spurmz) Plants that produce flowers and fruits.

annual (an'yu-uhl) A plant that completes its life cycle in one year.

anorexia nervosa (an''o-rek'se-ah ner-vo'sah) A nutritional deficiency disease characterized by severe, prolonged weight loss for fear of becoming obese. This eating disorder is thought to stem from sociocultural factors.

anther (an-ther) A sex organ in plants that produces the male gametophyte.

antheridia (singular, **antheridium)** (an''thur-id'e-ah) The structures in lower plants that produce sperm.

anthropomorphism (an-thro-po-mōr'fizm) The ascribing of human feelings, emotions, or meanings to the behavior of animals.

antibiotics (an-te-bi-ot'iks) Drugs that selectively kill or inhibit the growth of a particular cell type.

anticodon (an''te-ko'don) A sequence of three nitrogenous bases on a tRNA molecule capable of forming hydrogen bonds with three complementary bases on an mRNA codon during translation.

antidiuretic hormone (an-tĭ-di″u-rē′tik hōr′mōn) A hormone produced by the pituitary gland that stimulates the kidney to reabsorb water.

aorta (ā-or′tah) The large blood vessel that carries blood from the left ventricle to the majority of the body.

applied science (ap-plīd si′ens) Science that makes practical use of the theories provided by scientists to solve everyday problems.

archegonium (ar″ke-go′ne-um) The structure in lower plants that produces eggs.

arteries (ar′tē-rēz) The blood vessels that carry blood away from the heart.

arterioles (ar-tēr′e-ōlz) Small arteries located just before capillaries that can expand and contract to regulate the flow of blood to parts of the body.

asexual reproduction (ā-sek-shoo-al re″pro-duk′shun) Reproduction in which no gametes are formed; that is, reproduction without sex.

assimilation (ă-sĭ″mĭ-la′shun) The physiological process that takes place in a living cell as it converts nutrients in food into specific molecules required by the organism.

associative learning (ă-so′shuh-tiv lur′ning) See **classical conditioning.**

asymmetry (ā-sim′ĭ-tre) The characteristic of animals with no particular body shape.

atom (ă′tom) The smallest part of an element that still acts like that element.

atomic mass unit (ă-tom′ik mas yu′nit) A unit of measure used to describe the mass of atoms and is equal to 1.67×10^{-24} grams, approximately the mass of one proton.

atomic nucleus (ă-tom′ik nu′kle-us) The central region of the atom.

atomic number (ă-tom′ik num′bur) The number of protons in an atom.

atria (ā′trē-ah) Thin-walled sacs of the heart that receive blood from the veins of the body and empty into the ventricles.

atrioventricular valves (ā″trē-o-ven-trik′u-lar valvz) Valves located between the atria and ventricles of the heart that prevent the blood from flowing backwards from the ventricles into the atria.

attachment site (uh-tatch′munt sīt) A specific point on the surface of the enzyme where it can physically attach itself to the substrate; also called **binding site.**

autosomes (aw′to-sōmz) Chromosomes that typically carry genetic information used by the organism for characteristics other than the primary determination of sex.

autotroph (aw′to-trōf) An organism able to produce organic nutrients from inorganic materials; a self-feeder.

axon (ak′sahn) A neuronal fiber that carries information away from the nerve cell body.

---

bacteria (bak-tir′ē-ah) Unicellular organisms of the kingdom Prokaryotae that have the genetic ability to function in various environments.

Barr bodies (bar bod′ēz) Tightly coiled X chromosomes in the cells of female mammals, described by Dr. M. L. Barr.

basal metabolism (ba′sal mē-tab′o-lizm) The amount of energy required to maintain normal body activity while at rest.

base (bās) Any compound that releases a hydroxyl group (or other ion that acts like a hydroxyl group) in a solution.

basilar membrane (ba′sĭ-lar mem′brān) A membrane in the cochlea containing sensory cells that are stimulated by the vibrations caused by sound waves.

behavior (be-hav′yur) How an organism acts, what it does, and how it does it.

behavioral isolation (be-hāv′yu-ral i-so-la′shun) A genetic isolating mechanism that prevents interbreeding between species because of differences in behavior.

benthic (ben′thik) A term used to describe organisms that live on the ocean bottom.

bilateral symmetry (bi-lat′er-al sim′ĭ-tre) The characteristic of animals that are constructed along a plane running from a head to a tail region, so that only a cut along one plane of this axis results in two mirror halves.

bile (bīl) A product of the liver, stored in the gallbladder, which is responsible for the emulsification of fats.

binding site (bīn′ding sīt) See **attachment site.**

binomial system of nomenclature (bi-no′mi-al sis′tem ov no′men-kla-ture) A naming system that uses two Latin names, genus and species, for each type of organism.

biochemical pathway (bi′o-kem″ĭ-kal path′wa) A major series of enzyme-controlled reactions linked together.

biochemistry (bi-o-kem′iss-tre) The chemistry of living things, often called biological chemistry.

biogenesis (bi-o-jen′uh-sis) The concept that life originates only from preexisting life.

biological amplification (bi-o-loj′ĭ-cal am″plĭ-fi-ka′shun) The accumulation of a compound in increasing concentrations in organisms at successively higher trophic levels.

biology (bi-ol′o-je) The science that deals with life.

biomass (bi′o-mas) The dry weight of a collection of designated organisms.

biomes (bi′ōmz) Large regional communities.

biosphere (bi′o-sfēr) The worldwide ecosystem.

biotechnology (bi-o-tek-nol′uh-je) The science of gene manipulation.

biotic factors (bi-ot′ik fak′tōrz) Living parts of an organism's environment.

biotic potential (bi-ah′tik po-ten′shul) See **reproductive capacity.**

blood (blud) The fluid medium consisting of cells and plasma that assists in the transport of materials and heat.

bloom (bloom) A rapid increase in the number of microorganisms in a body of water.

Bowman's capsule (bo′manz kap′sl) A saclike structure at the end of a nephron that surrounds the glomerulus.

breathing (bre′thing) The process of pumping air in and out of the lungs.

bronchi (brŏng′ki) Major branches of the trachea that ultimately deliver air to bronchioles in the lungs.

bronchioles (brŏng′kē-ōlz) Small tubes that are capable of contracting and deliver air to the alveoli in the lung.

budding (bud′ing) A type of asexual reproduction in which the new organism is an outgrowth of the parent.

buffer (bŭ′fer) A mixture of a weak acid and the salt of a weak acid that operates to maintain a constant pH.

bulimia (bu-lim′e-ah) A nutritional deficiency disease characterized by a binge-and-purge cycle of eating. It is thought to stem from psychological disorders.

calorie (kal'o-re) The amount of heat energy necessary to raise the temperature of one gram of water one degree Celsius.

cambium (kam'be-um) A tissue in higher plants that produces new xylem and phloem.

cancer (kan'sur) A tumor that is malignant.

capillaries (cap'ĭ-lair-ēz) Tiny blood vessels through which exchange between cells and the blood takes place.

capsule (kap'sūl) The part of the sporophyte generation of mosses that contains spores.

carbohydrate (kar-bo-hi'drāt) One class of organic molecules composed of carbon, hydrogen, and oxygen in a ratio of usually 1:2:1. The basic building block of a carbohydrate is a simple sugar (= monosaccharide).

carbohydrate loading (kar-bo-hi-'drāt lo'ding) A week-long program of diet and exercise that results in an increase in muscle glycogen stores.

carbon dioxide conversion stage (kar'bon di-ok'sīd kon-vur'zhun stāj) The second stage of photosynthesis, during which inorganic carbon from carbon dioxide becomes incorporated into a sugar molecule.

carbon skeleton (kar'bon skel'uh-ton) The central portion of an organic molecule composed of rings or chains of carbon atoms.

carbonic anhydrase (car-bon'ik an-hi'drās) An enzyme present in red blood cells that assists in converting carbon dioxide to bicarbonate ions.

carnivores (kar'nĭ-vōrz) Those animals that eat other animals.

carrier (ka're-er) Any individual having a hidden, recessive gene.

carrying capacity (ka're-ing kuh-pas'ĭ-te) The optimum population size an area can support over an extended period of time.

catalyst (cat'uh-list) A chemical that speeds up a reaction but is not used up in the reaction.

cell (sel) The basic structural unit that makes up all living things.

cell membrane (sel mem'brān) The outer boundary membrane of the cell; also called **plasma membrane.**

cell plate (sel plāt) A plant-cell structure that begins to form in the center of the cell and proceeds to the cell membrane, resulting in cytokinesis.

cellular membranes (sel'yu-lar mem'brāns) Thin sheets of material composed of phospholipids and proteins. Some of the proteins have attached carbohydrates or fats.

cellular respiration (sel'yu-lar res"pĭ-ra'shun) A major biochemical pathway along which cells release the chemical-bond energy from food and convert it into a usable form (ATP).

central nervous system (sen'trul ner'vus sis'tem) The portion of the nervous system consisting of the brain and spinal cord.

centrioles (sen'tre-ōls) Microtubule-containing organelles located just outside the nucleus. The centriole divides and organizes spindle fibers during mitosis and meiosis.

centromere (sen'tro-mēr) The region where two chromatids are joined.

chemical bonds (kem'ĭ-kal bonds) Forces that combine atoms or ions and hold them together.

chemical reaction (kem'ĭ-kal re-ak'shun) The formation or rearrangement of chemical bonds, usually indicated in an equation by an arrow from the reactants to the products.

chemical symbol (kem'ĭ-kal sim'bol) "Shorthand" used to represent one atom of an element, such as Al for aluminum or C for carbon.

chemiosmosis (kem"e-os-mo'sis) The process of generating ATP as a result of creating a hydrogen-ion gradient across a membrane by using an electron-transport system.

chlorophyll (klo'ro-fil) The green pigment located in the chloroplasts of plant cells associated with trapping light energy.

chloroplast (klo'ro-plast) An energy-converting, membranous, saclike organelle in plant cells containing the green pigment chlorophyll.

chromatid (kro'mah-tid) One of two component parts of a chromosome formed by replication and attached at the centromere.

chromatin (kro'mah-tin) Areas or structures within the nucleus of a cell composed of long molecules of deoxyribonucleic acid (DNA) in association with proteins.

chromatin fibers (kro'mah-tin fi'bers) See **nucleoproteins.**

chromosomal mutation (kro-mo-sōm'al miu-ta'shun) A change in the gene arrangement in a cell as a result of breaks in the DNA molecule.

chromosome (kro'mo-sōm) A duplex DNA molecule with attached protein (nucleoprotein) coiled into a short, compact unit.

cilia (sil'e-ah) Numerous short, hairlike structures projecting from the cell surface that enable locomotion.

class (class) A group of closely related families found within a phylum.

classical conditioning (klas'ĭ-kul kon-dĭ'shun-ing) A kind of learning in which a neutral stimulus is associated with a natural stimulus to produce a particular response; also called **associative learning.**

cleavage furrow (kle'vaj fuh'ro) An indentation of the cell membrane of an animal cell that pinches the cytoplasm into two parts.

climax community (klī'maks ko-miu'nĭ-te) A relatively stable, long-lasting community.

clones (klōnz) All of the individuals reproduced asexually that have exactly the same genes.

coacervate (ko-as'ur-vāt) A collection of organic macromolecules surrounded by water molecules, aligned to form a sphere.

cochlea (kŏk'lē-ah) The part of the ear that converts sound into nerve impulses.

codon (ko'don) A sequence of three nucleotides of an mRNA molecule that directs the placement of a particular amino acid during translation.

coelom (se'lem) A body cavity in which internal organs are suspended.

coenzyme (ko-en'zīm) A molecule that works with an enzyme.

coitus (ko'ĕ-tus) See **sexual intercourse.**

colonial (ko-lo'ne-al) A term used to describe a collection of cells that cooperate to a small extent.

commensalism (ko-men'sal-izm) A relationship between two organisms in which one organism is helped and the other is not affected.

community (ko-miu'nĭ-te) A collection of interacting organisms within an ecosystem.

competition (com-pe-tĭ'shun) A relationship between two organisms in which both organisms are harmed.

competitive exclusion principle (com-pē'ti-tiv eks-klu'zhun prin'sĭ-pul) No two species can occupy the same niche at the same time.

competitive inhibition (kum-pet'ĭ-tiv in''hĭ-bĭ'shun) The formation of a temporary enzyme-inhibitor complex that interferes with the normal formation of enzyme–substrate complexes, resulting in a decreased turnover.

complementary base (kom''plĕ-men'tah-re bās) A base that can form hydrogen bonds with another base of a specific nucleotide.

complete protein (kom-plēt' pro'te-in) Protein molecules that provide all the essential amino acids.

complex carbohydrates (kom'pleks kar-bo-hi'drāts) Macromolecules composed of simple sugars combined to form a polymer.

compound (kom'pound) A kind of matter that consists of a specific number of atoms (or ions) joined to each other in a particular way and held together by chemical bonds.

concentration gradient (kon''sen-tra'shun gra'de-ent) The gradual change in the number of molecules per unit of volume over distance.

conception (kon-sep'shun) Fertilization.

conditioned response (kon-dĭ'shund re-spons') The behavior displayed when the neutral stimulus is given after association has occurred.

cone (kōn) A reproductive structure of gymnosperms that produces pollen in males or eggs in females.

cones (kōnz) Light-sensitive cells in the retina of the eye that respond to different colors of light.

consumers (kon-soom'urs) Organisms that must obtain energy in the form of organic matter.

control group (con-trōl' grūp) The situation used as the basis for comparison in a controlled experiment.

control processes (con-trōl' pro'ses-es) Mechanisms that ensure that an organism will carry out all metabolic activities in the proper sequence (coordination) and at the proper rate (regulation).

controlled experiment (con-trōld' ek-sper'ĭ-ment) An experiment that allows for a comparison of two events that are identical in all but one respect.

convergent evolution (kon-vur'jent ev-o-lu'shun) An evolutionary pattern in which widely different organisms show similar characteristics.

copulation (kop-yu-la'shun) See **sexual intercourse.**

corpus luteum (kōr'pus lu'te-um) Remainder of the follicle after the release of the secondary oocyte. It develops into a glandlike structure that produces hormones (progesterone and estrogen) that prevent the release of other eggs.

coupled reactions (kup'ld re-ak'shuns) Reactions in which there is a linkage of a set of energy-requiring reactions with energy-releasing reactions.

covalent bond (ko-va'lent bond) The attractive force formed between two atoms that share a pair of electrons.

cristae (krĭs'te) Folded surfaces of the inner membranes of mitochondria.

crossing-over (kro'sing o'ver) The exchange of a part of a chromatid from one chromosome with an equivalent part of a chromatid from a homologous chromosome.

cryptorchidism (krip-tōr'kĭ-dizm) A developmental condition in which the testes do not migrate through the inguinal canal.

cytokinesis (si-to-kĭ-ne'sis) Division of the cytoplasm of one cell into two new cells.

cytoplasm (si''to-plazm) The more fluid portion of the protoplasm that surrounds the nucleus.

cytosine (si'to-sēn) A single-ring nitrogenous-base molecule in DNA and RNA. It is complementary to guanine.

daughter cells (daw'tur sels) Two cells formed by cell division.

daughter chromosomes (daw'tur kro'mo-sōmz) Chromosomes produced by DNA replication that contain identical genetic information; formed after chromosome division in anaphase.

daughter nuclei (daw'tur nu'kle-i) Two nuclei formed by mitosis.

death phase (deth fāz) The portion of some population growth curves in which the size of the population declines.

decomposers (de-kom-po'zurs) Organisms that use dead organic matter as a source of energy.

dehydration synthesis (de''hi-dra'shun sin'thĕ-sis) A reaction that results in the joining of smaller molecules to make larger molecules by the removal of water.

dehydration synthesis reaction (de-hi-dra'shun sin'thuh-sis re-ak'shun) A reaction that results in the formation of a macromolecule when water is removed from between the two smaller component parts.

deme (dēm) A local, recognizable population that differs in gene frequencies from other local populations of the same species. (See also **subspecies.**)

denature (de-nā'chur) To permanently change the protein structure of an enzyme so that it loses its ability to function.

dendrites (den'drīts) Neuronal fibers that receive information from axons and carry it toward the nerve-cell body.

denitrifying bacteria (de-ni'trĭ-fi-ing bak-te're-ah) Several kinds of bacteria capable of converting nitrite to nitrogen gas.

density (den'sĭ-te) The weight of a certain volume of a material.

density-dependent factors (den'sĭ-te de-pen'dent fak'tōrz) Population-limiting factors that become more effective as the size of the population increases.

density-independent factors (den'sĭ-te in''de-pen'dent fak'tōrz) Population-controlling factors that are not related to the size of the population.

deoxyribonucleic acid (DNA) (de-ok''se-ri-bo-nu-kle'ik ă'sid) A polymer of nucleotides that serves as genetic information. In prokaryotic cells, it is a duplex DNA (double-stranded) loop and contains attached HU proteins. In eukaryotic cells, it is found in strands with attached histone proteins. When tightly coiled, it is known as a chromosome.

deoxyribose (de-ok''se-ri'bōs) A 5-carbon sugar molecule; a component of DNA.

depolarized (de-po'lă-rīzd) Having lost the electrical difference existing between two points or objects.

diaphragm (di'uh-fram) A muscle separating the lung cavity from the abdominal cavity that is involved in exchanging the air in the lungs.

diastolic blood pressure (di''uh-stol'ik blud presh'yur) The pressure present in a large artery when the heart is not contracting.

dicot (di'kot) An angiosperm whose embryo has two seed leaves.

diet (di'et) The food and drink consumed by a person from day to day.

differentially permeable (di''fur-ent'shu-le per'me-uh-bul) The property of a membrane that allows certain molecules to pass through it but interferes with the passage of others.

differentiation (dif-ĕ-ren''she-a'shun) The process of forming specialized cells within a multicellular organism.

diffusion (dĭ-fiu'zhun) Net movement of a kind of molecule from a place where that molecule is in higher concentration to a place where that molecule is more scarce.

diffusion gradient (di''fiu'zhun gra'de-ent) The difference in the concentration of diffusing molecules over distance.

diploid (dip'loid) Having two sets of chromosomes: one set from the maternal parent and one set from the paternal parent.

distal convoluted tubule (dis'tul kon'vo-luted tūb'yūl) The downstream end of the nehpron of the kidney, which is primarily responsible for regulating the amount of hydrogen and potassium ions in the blood.

divergent evolution (di-vur'jent ĕv-o-lu'shun) A basic evolutionary pattern in which individual speciation events cause many branches in the evolution of a group of organisms.

DNA code (D-N-A cōd) A sequence of three nucleotides of a DNA molecule.

DNA polymerase (pō-lim'er-as) An enzyme that bonds new DNA nucleotides together when they base pair with an existing DNA strand.

DNA replication (rep''lĭ-ka'shun) The process by which the genetic material (DNA) of the cell reproduces itself prior to its distribution to the next generation of cells.

dominance hierarchy (dom'in-ants hi'ur-ar-ke) A relatively stable, mutually understood order of priority within a group.

dominant allele (dom'in-ant a-lēl') An allele that expresses itself and masks the effect of other alleles for the trait.

double bond (dub'l bond) A pair of covalent bonds formed between two atoms when they share two pairs of electrons.

double-factor cross (dub'l fak'tur kros) A genetic study in which two pairs of alleles are followed from the parental generation to the offspring.

Down syndrome (down sin'drom) A genetic disorder resulting from the presence of an extra chromosome number 21. Symptoms include slanted eyes, flattened facial features, a large tongue, and a tendency toward short stature and fingers. Some individuals also display mental retardation.

duodenum (dōō''ō-dē'num) The first part of the small intestine, which receives food from the stomach and secretions from the liver and pancreas.

duplex DNA (du'pleks) DNA in a double-helix shape.

dynamic equilibrium (di-nam'ik e-kwĭ-lib're-um) The condition in which molecules are equally dispersed, therefore movement is equal in all directions.

ecological isolation (e-kō-loj'ĭ-kal i-so-la'shun) A genetic isolation mechanism that prevents interbreeding between species because they live in different areas; also called **habitat preference.**

ecology (e-kol'o-je) The branch of biology that studies the relationships between organisms and their environment.

ecosystem (ĕk'o-sis''tem) An interacting collection of organisms and the abiotic factors that affect them.

egg cells (eg sels) The haploid sex cells produced by sexually mature females.

ejaculation (e-jak''u-lā'shun) The release of sperm cells and seminal fluid through the penis of a male.

electrolytes (ē-lek'tro-līts) Ionic compounds dissolved in water. Their proper balance is essential to life.

electrons (e-lek'trons) The negatively charged particles moving at a distance from the nucleus of an atom that balance the positive charges of the protons.

electron-transfer system (ETS) (e-lek'tron trans'fur sis'tem) The series of oxidation–reduction reactions in aerobic cellular respiration in which the energy is removed from hydrogens and transferred to ATP.

elements (el'ĕ-ments) Matter consisting of only one kind of atom.

empirical evidence (em-pir'ĭ-cal ev'ĭ-dens) The information gained by observing an event.

empirical formula (em-pir'ĭ-cal for'miu-lah) Chemical shorthand that indicates the number of each kind of atom within a molecule.

endocrine glands (en'do-krin glandz) Glands that secrete into the circulatory system.

endocrine system (en'do-krin sis'tem) A number of glands that communicate with one another and other tissues through chemical messengers transported throughout the body by the circulatory system.

endoplasmic reticulum (ER) (en''do-plaz'mik re-tik'yu-lum) Folded membranes and tubes throughout the eukaryotic cell that provide a large surface upon which chemical activities take place.

endospore (en'do-spōr'') A unique bacterial structure with a low metabolic rate that germinates under favorable conditions to grow into a new cell.

endosymbiotic theory (en''do-sim-be-ot'ik the'o-re) A theory suggesting that some organelles found in eukaryotic cells may have originated as free-living prokaryotes.

environment (en-vi-ron-ment) Anything that affects an organism during its lifetime.

environmental resistance (en-vi-ron-men'tal re-zis'tants) The collective set of factors that limit population growth.

enzymatic competition (en-zi-mă'tik com-pĕ-ti'shun) Competition among several different available enzymes to combine with a given substrate material.

enzyme (en'zīm) A specific protein molecule that acts as a catalyst to control the rate at which a chemical reaction occurs.

enzyme-substrate complex (en'zīm-sub'strāt kom'pleks) A temporary molecule formed when an enzyme attaches itself to a substrate molecule.

epinephrine (ĕ''pĭ-nef'rin) A hormone produced by the adrenal medulla that increases heart rate, blood pressure, and breathing rate.

epiphyte (ep'e-fīt) A plant that lives on the surface of another plant.

essential amino acids (ĕ-sen'shul ah-me'no ă'sids) Those amino acids that cannot be synthesized by the human body and must be part of the diet (e.g., lysine, tryptophan, and valine).

essential fatty acid (ĕ-sen'shul fă'tē ă'sid) The fatty acid linoleic acid. It cannot be synthesized by the human body and must be part of the diet.

estrogens (es'tro-jens) A group of female sex hormones that cause the differentiation in the female embryo of the internal and external genital anatomy along female lines; responsible for the changes in breasts, vagina, uterus, clitoris, and pelvic bone structure at puberty.

ethology (e-thol'uh-je) The scientific study of the nature of behavior and its ecological and evolutionary significance in its natural setting.

eugenics law (yu-jen'iks laws) Laws designed to eliminate "bad" genes from the human gene pool and encourage "good" gene combinations.

eukaryote (yu-kār'e-ōt) A cell possessing a nuclear membrane and other membranous organelles.

eukaryotic cells (yu'ka-re-ah"tik sels) One of the two major types of cells; characterized by cells that have a true nucleus, as in plants, fungi, protists, and animals.

evolution (ĕv-o-lu'shun) The genetic adaptation of a population of organisms to its environment.

exocrine glands (ek'sa-krīn glandz) Glands that secrete through ducts to the surface of the body or into hollow organs of the body.

experiment (ek-sper'ĭ-ment) A recreation of an event in a way that enables a scientist to gain valid and reliable empirical evidence.

experimental group (ek-sper-ĭ-men'tal grūp) The group in a controlled experiment that is identical to the control group in all respects but one.

exponential growth phase (eks-po-nen'shul grōth fāz) A period of time during population growth when the population increases at an accelerating rate.

external parasite (eks-tur'nal pĕr'uh-sīt) A parasite that lives on the outside of its host.

extrinsic factors (eks-trin'sik fak'tōrz) Population-controlling factors that arise outside the population.

facilitated diffusion (fah-sil'ĭ-ta"ted dĭ-fiu'zhun) Diffusion assisted by carrier molecules.

FAD (**f**lavin **a**denine **d**inucleotide) A hydrogen carrier used in respiration.

family (fam'ĭ-ly) A group of closely related species within an order.

fat (fat) A class of water-insoluble macromolecules composed of a glycerol and fatty acids.

fatty acid (fat'ē ă'sid) One of the building blocks of a fat, composed of a long-chain carbon skeleton with a carboxylic acid functional group.

fertilization (fer"tĭ-lĭ-za'shun) The joining of haploid nuclei, usually from an egg and a sperm cell, resulting in a diploid cell called the zygote.

fiber (fi'ber) Natural (plant) or industrially produced polysaccharides that are resistant to hydrolysis by human digestive enzymes.

filter feeders (fil'ter fēd'ers) Animals that use cilia or other appendages to create water currents and filter food out of the water.

first law of thermodynamics (furst law uv thur"mo-di-nam'iks) Energy in the universe remains constant; it can neither be created nor destroyed.

flagella (flah-jel'luh) Long, hairlike structures projecting from the cell surface that enable locomotion.

flower (flow'er) A complex structure made from modified stems and leaves. It produces pollen in the males and eggs in the females.

fluid-mosaic model (flu'id mo-za'ik mod'l) The concept that the cell membrane is composed primarily of protein and phospholipid molecules that are able to shift and flow past one another.

follicle (fol'ĭ-kul) The saclike structure near the surface of the ovary that encases the soon-to-be-released secondary oocyte.

follicle-stimulating hormone (FSH) (fol'ĭ-kul stim'yu-lā-ting hōr'mōn) The pituitary secretion that causes the ovaries to begin to produce larger quantities of estrogen and to develop the follicle and prepare the egg for ovulation.

food chain (food chān) A sequence of organisms that feed on one another, resulting in a flow of energy from a producer through a series of consumers.

Food Guide Pyramid (food gīd pĭ'ra-mĭd) A tool developed by the U.S. Department of Agriculture to help the general public plan for good nutrition. It contains guidelines for required daily intake from each of five food groups.

food web (food web) A system of interlocking food chains.

formula (for'miu-lah) The group of chemical symbols that indicate what elements are in a compound and the number of each kind of atom present.

fovea centralis (fo've-ah sen-tral'is) The area of sharpest vision on the retina, where light is normally focused.

free-living nitrogen-fixing bacteria (ni'tro-jen fik'sing bak-te're-ah) Soil bacteria that convert nitrogen gas molecules into nitrogen compounds that plants can use.

fruit (froot) The structure in angiosperms that contains seeds.

functional groups (fung'shun-al grūps) Specific combinations of atoms attached to the carbon skeleton that determine specific chemical properties.

fungus (fun'gus) The common name for the kingdom Mycetae.

gallbladder (gol'blad"er) An organ attached to the liver that stores bile.

gamete (gam'ēt) A haploid sex cell.

gametogenesis (gă-me"to-jen'ĕ-sis) The generating of gametes; the meiotic cell-division process that produces sex cells; oogenesis and spermatogenesis.

gametophyte generation (gă-me'to-fīt jen"uh-ra'shun) The haploid generation in plant life cycles, which produces gametes.

gametophyte stage (gă-me'to-fīt stāj) A life-cycle stage in plants in which a haploid sex cell is produced by mitosis.

gas (gas) The state of matter in which the molecules are more energetic than the molecules of a liquid, resulting in only slight attraction for each other.

gastric juice (gas'trik jūs) The secretions of the stomach that contain enzymes and hydrochloric acid.

gene (jēn) Any molecule, usually a segment of DNA, that is able to (1) replicate by directing the manufacture of copies of itself; (2) mutate, or chemically change and transmit these changes to future generations; (3) store information that determines the characteristics of cells and organisms; and (4) use this information to direct the synthesis of structural and regulatory proteins.

gene flow (jēn flo) The movement of genes from one generation to another or from one place to another.

gene frequency (jēn fre'kwen-se) A measure of the number of times that a gene occurs in a population. The percentage of sex cells that contain a particular gene.

gene pool (jēn pool) All the genes of all the individuals of the same species.

gene-regulator proteins (jēn reg'yu-la-tor pro'te-ins) Chemical messengers within a cell that inform the genes as to whether protein-producing genes should be turned on or off, or whether they should have their protein-producing activities increased or decreased; for example, gene-repressor proteins and gene-activator proteins.

generative processes (jen'uh-ra''tiv pros'es-es) Actions that increase the size of an individual organism (growth) or increase the number of individuals in a population (reproduction).

genetic counselor (jĕ-net'ik kown'sel-or) A professional with specific training in human genetics.

genetic isolating mechanism (jĕ-net'ik i-so-la'ting mek'an-izm) See **reproductive isolating mechanism**.

genetic recombination (jĕ-net'ik re-kom-bī-na-shun) The gene mixing that occurs during sexual reproduction.

genetics (jĕ-net'iks) The study of genes, how genes produce characteristics, and how the characteristics are inherited.

genome (je'nōm) A set of all the genes necessary to specify an organism's complete list of characteristics.

genotype (je'no-tīp) The catalog of genes of an organism, whether or not these genes are expressed.

genus (je'nus) (plural, **genera**) A group of closely related species within a family.

geographic barriers (je-o-graf'ik băr'yurz) Geographic features that keep different portions of a species from exchanging genes.

geographic isolation (je-o-graf'ik i-so-la'shun) A condition in which part of the gene pool is separated by geographic barriers from the rest of the population.

germinate (jur'min-āt) To begin to grow (as from a seed).

gland (gland) An organ that manufactures and secretes a material either through ducts or directly into the circulatory system.

glomerulus (glō-mer'u-lus) A cluster of blood vessels in the kidney, surrounded by Bowman's capsule.

glycerol (glis'er-ol) One of the building blocks of a fat, composed of a carbon skeleton that has three alcohol groups (OH) attached to it.

glycolysis (gli-kol'ĭ-sis) The anaerobic first stage of cellular respiration, consisting of the enzymatic breakdown of a sugar into two molecules of pyruvic acid.

Golgi apparatus (gōl 'je ap''pah-rat'us) A stack of flattened, smooth, membranous sacs; the site of synthesis and packaging of certain molecules in eukaryotic cells.

gonad (go'nad) In animals, the organ in which meiosis occurs.

gradualism (grad'u-al-izm) The theory stating that evolution occurred gradually with an accumulated series of changes over a long period of time.

grana (gra'nuh) Areas of the chloroplast membrane where chlorophyll molecules are concentrated.

granules (gran'yūls) Materials whose structure is not as well defined as that of other organelles.

growth-stimulating hormone (grōth stī'mu-la-ting hōr'mōn) A hormone produced by the anterior pituitary gland that stimulates tissues to grow.

guanine (gwah'nēn) A double-ring nitrogenous-base molecule in DNA and RNA. It is the complementary base of cytosine.

gymnosperms (jim'no-spurmz) Plants that produce their seeds in cones.

habitat (hab'ĭ-tat) The place or part of an ecosystem occupied by an organism.

habitat preference (hab'i-tat pref'ur-ents) See **ecological isolation.**

haploid (hap'loid) Having a single set of chromosomes resulting from the reduction division of meiosis.

Hardy–Weinberg law (har'de wīn'burg law) Populations of organisms will maintain constant gene frequencies from generation to generation as long as mating is random, the population is large, mutation does not occur, migration does not occur, and all genes have equal value.

heart (hart) The muscular pump that forces the blood through the blood vessels of the body.

hemoglobin (he'mo-glo-bin) An iron-containing molecule found in red blood cells, to which oxygen molecules bind.

hepatic portal vein (hĕ-pat'ik pōr'tul vān) A blood vessel that collects blood from capillaries in the intestine and delivers it to a second set of capillaries in the liver.

herbivores (her'bĭ-vōrz) Those animals that feed directly on plants.

heterotroph (het'ur-o-trōf) An organism that requires a source of organic material from its environment. It cannot produce its own food.

heterozygous (hĕ''ter-o-zi'gus) A term used to describe a diploid organism that has two different allelic forms of a particular gene.

high-energy phosphate bond (hi en'ur-je fos'fāt bond) The bond between two phosphates in an ADP or ATP molecule that readily releases its energy for cellular processes.

homeotherms (ho'me-o-thurmz) Animals that maintain a constant body temperature.

homologous chromosomes (ho-mol'o-gus kro'mo-sōmz) A pair of chromosomes in a diploid cell that contain similar genes at corresponding loci throughout their length.

homozygous (ho''mo-zi'gus) A term used to describe a diploid organism that has two identical alleles for a particular characteristic.

hormones (hōr'mōnz) Chemical messengers secreted by endocrine glands to regulate other parts of the body.

host (hōst) A specific cell or organism that provides a virus or other parasitic organism what is needed to function.

hybrid (hi'brid) The offspring of two different genetic lines produced by sexual reproduction.

hydrogen bond (hi'dro-jen bond) Weak attractive forces between molecules. Important in determining how groups of molecules are arranged.

hydrolysis (hi-drol'ĭ-sis) A chemical reaction that occurs when a large molecule is broken down into smaller parts by reacting with a molecule of water.

hydrophilic (hi'dro-fil'ik) Readily absorbing or dissolving in water.

hydrophobic (hi'dro-fo'bik) Tending not to combine with, or incapable of dissolving in water.

hydroxyl ion (hi-drok'sil i'on) A negatively charged particle (OH^-) composed of oxygen and hydrogen atoms released from a base when dissolved in water.

hypertonic (hi'pur-tŏn'ik) A comparative term describing one of two solutions. The hypertonic solution is the one with the higher amount of dissolved material.

hypothalamus (hi''po-thal'ă-mus) The region of the brain that causes the cyclic production of hormones in females and the constant production of hormones in males.

hypothesis (hi-poth'e-sis) A possible answer to or explanation of a question that accounts for all the observed facts and is testable.

hypotonic (hi'po-tŏn'ik) A comparative term describing one of two solutions. The hypotonic solution is the one with the lower amount of dissolved material.

immune system (i-mun'sis'tem) A system of white blood cells specialized to provide the body with resistance to disease. There are two types: antibody-mediated immunity and cell-mediated immunity.

imperfect flowers (im''pur'fekt flow'erz) Flowers that contain either male or female reproductive structures, but not both.

imprinting (im'prin-ting) Learning in which a very young animal is genetically primed to learn a specific behavior in a very short period.

inclusions (in-klu'zhuns) A general term referring to materials inside a cell that are usually not readily identifiable; stored materials.

incomplete protein (in-kom-plēt' pro'te-in) Protein molecules that do not provide all the essential amino acids.

incus (in'kus) The ear bone that is located between the malleus and the stapes.

independent assortment (in''de-pen'dent ă-sort'ment) The segregation, or assortment, of one pair of homologous chromosomes independently of the segregation, or assortment, of any other pair of chromosomes.

inguinal canal (ing'gwi-nal că-nal') An opening in the abdominal cavity through which the testes in a human male embryo move into the scrotum.

inguinal hernia (ing'gwi-nal her'ne-ah) A rupture in the abdominal wall that allows a portion of the intestine to push through the abdominal wall in the area of the inguinal canal.

inhibitor (in-hib'ĭ-tōr) A molecule that temporarily attaches itself to an enzyme, thereby interfering with the enzyme's ability to form an enzyme-substrate complex.

initiation code (ĭ-ni'she-a''shun cōd) The code of DNA with the base sequence TAC that begins the process of transcription.

inorganic molecules (in-or-gan'ik mol'uh-kiuls) Molecules that do not contain carbon atoms in rings or chains.

insecticide (in-sek'tĭ-sīd) A poison used to kill insects.

insight learning (in-sīt lur'ning) Learning in which past experiences are reorganized to solve new problems.

instinctive behavior (in-stink'tiv be-hāv'yur) Automatic, preprogrammed, or genetically determined behavior.

intention movements (in-ten'shun moov'ments) Behavior that signals what the animal is likely to do in the near future.

internal parasite (in-tur'nal pĕr'uh-sīt) A parasite that lives inside its host.

interphase (in'tur-fāz) The stage between cell divisions in which the cell is engaged in metabolic activities.

interstitial cell-stimulating hormone (ICSH) (in''ter-stĭ'shal sel stim'yu-lā-ting hōr'mōn) The chemical messenger molecule released from the pituitary that causes the testes to produce testosterone, the primary male sex hormone.

intrinsic factors (in-trin'sik fak'tōrz) Population-controlling factors that arise from within the population.

invertebrates (in-vur'tuh-brāts) Animals without backbones.

ionic bond (i-on'ik bond) The attractive force between ions of opposite charge.

ions (i'ons) Electrically unbalanced or charged atoms.

isomers (i'so-meers) Molecules that have the same empirical formula but different structural formulas.

isotonic (i'so-tŏn'ik) A term used to describe two solutions that have the same concentration of dissolved material.

isotopes (i'so-tōps) Atoms of the same element that differ only in the number of neutrons.

kidneys (kid'nēz) The primary organs involved in regulating blood levels of water, hydrogen ions, salts, and urea.

kilocalorie (kcalorie) (kil''o-kal'o-re) A measure of heat energy one thousand times larger than a calorie.

kinetic energy (k-net'ik en'er-je) Energy of motion.

kingdom (king'dom) The largest grouping used in the classification of organisms.

Krebs cycle (krebs si'kl) The series of reactions in aerobic cellular respiration, resulting in the production of two carbon dioxides, the release of four pairs of hydrogens, and the formation of an ATP molecule.

kwashiorkor (kwa''she-or'kor) A protein-deficiency disease.

lack of dominance (lak uv dom'in-ans) The condition of two unlike alleles both expressing themselves, neither being dominant.

lacteals (lak'tēlz) Tiny lymphatic vessels located in the villi.

lag phase (lag fāz) A period of time following colonization when the population remains small or increases slowly.

large intestine (larj in-tes'tin) The last portion of the food tube. It is primarily involved in reabsorbing water.

law of dominance (law uv dom'in-ans) When an organism has two different alleles for a trait, the allele that is expressed and overshadows the expression of the other allele is said to be dominant. The allele whose expression is overshadowed is said to be recessive.

law of independent assortment (law uv in''de-pen'dent ă-sort'ment) Members of one allelic pair will separate from each other independently of the members of other allele pairs.

law of segregation (law uv seg''rē-ga'shun) When gametes are formed by a diploid organism, the alleles that control a trait separate from one another into different gametes, retaining their individuality.

learning (lur'ning) A change in behavior as a result of experience.

leaves (lēvz) Specialized portions of higher plants that are the sites of photosynthesis.

lichen (li'kĕn) A mutualistic relation between fungi and algal protists or cyanobacteria.

life cycle (līf sī'kl) The series of stages in the life of any organism.

light-energy conversion stage (līt en'ur-je kon-vur'zhun stāj) The first of the two stages of photosynthesis, during which light energy is converted to chemical-bond energy.

limiting factors (lim'ĭ-ting fak'tōrz) Environmental influences that limit population growth.

linkage group (lingk'ij grūp) Genes located on the same chromosome that tend to be inherited together.

lipids (lĭ'pids) Large organic molecules that do not easily dissolve in water; classes include fats, phospholipids, and steroids.

liquid (lik'wid) The state of matter in which the molecules are strongly attracted to each other, but because they are farther apart than in a solid, they move past each other more freely.

liver (li'vur) An organ of the body responsible for secreting bile, filtering the blood, detoxifying molecules, and modifying molecules absorbed from the gut.

locus (loci) (lo'kus) (lo'si) The spot on a chromosome where an allele is located.

loop of Henle (loop uv hen'le) The middle portion of the nephron, which is primarily involved in regulating the amount of water lost from the kidney.

lung (lung) The respiratory organ in which air and blood are brought close to one another and gas exchange occurs.

lymph (limf) Liquid material that leaves the circulatory system to surround cells.

lymphatic system (lim-fă'tik sis'tem) A collection of thin-walled tubes that collects, filters, and returns lymph from the body to the circulatory system.

lysosome (li'so-sōm) A specialized organelle that holds a mixture of hydrolytic enzymes.

malleus (mă'le-us) The ear bone that is attached to the tympanum.

mass number (mas num'ber) The weight of an atomic nucleus expressed in atomic mass units. (The sum of the protons and neutrons.)

masturbation (mas''tur-ba'shun) Stimulation of one's own sex organs.

matter (mat'er) Anything that has weight (mass) and also takes up space (volume).

medusa (muh-du'sah) A free-swimming adult stage in the phylum Cnidaria that reproduces sexually.

meiosis (mi-o'sis) The specialized pair of cell divisions that reduces the chromosome number from diploid ($2n$) to haploid (n).

Mendelian genetics (men-dē-le-an jĕ-net'iks) The pattern of inheriting characteristics that follows the laws formulated by Gregor Mendel.

menopause (mĕn'o-pawz) The period beginning at about age fifty when the ovaries stop producing viable secondary oocytes.

menstrual cycle (men'stru-al si'kul) **(menses, menstrual flow, period)** The repeated building up and shedding of the lining of the uterus.

mesenteries (mes'en-ter''ēz) Connective tissues that hold the organs in place and also serve as support for blood vessels connecting the various organs.

messenger RNA (mRNA) (mes'en-jer) A molecule composed of ribonucleotides that functions as a copy of the gene and is used in the cytoplasm of the cell during protein synthesis.

metabolic processes (me-ta-bol'ik pros'es-es) The total of all chemical reactions within an organism; for example, nutrient uptake and processing, and waste elimination.

metaphase (me'tah-fāz) The second stage in mitosis, characterized by alignment of the chromosomes at the equatorial plane.

microfilaments (mi''kro-fil'ah-ments) Long, fiberlike structures made of protein and found in cells, often in close association with microtubules; provide structural support and enable movement.

microorganisms (microbes) (mi''kro-or'guh-niz''mz) Small organisms that cannot be seen without some type of magnification.

microscope (mi'kro-skōp) An instrument used to produce an enlarged image of a small object.

microsphere (mi'kro-sfēr) A collection of organic macromolecules in a structure with a double-layered outer boundary.

microtubules (mi'kro-tū''byūls) Small, hollow tubes of protein that function throughout the cytoplasm to provide structural support and enable movement.

microvilli (mi-kro-vil'ē) Tiny projections from the surfaces of cells that line the intestine.

minerals (min'er-alz) Inorganic elements that cannot be manufactured by the body but are required in low concentrations; essential to metabolism.

mitochondrion (mi-to-kahn'dre-on) A membranous organelle resembling a small bag with a larger bag inside that is folded back on itself; serves as the site of aerobic cellular respiration.

mitosis (mi-to'sis) A process that results in equal and identical distribution of replicated chromosomes into two newly formed nuclei.

molecule (mol'ĕ-kūl) The smallest particle of a chemical compound; also the smallest naturally occurring part of an element or compound.

monocot (mon'o-kot) An angiosperm whose embryo has one seed leaf.

monoculture (mon''o-kul'chur) The agricultural practice of planting the same varieties of a species over large expanses of land.

mortality (mor-tal'i-te) The number of individuals leaving the population by death per thousand individuals in the population.

mosses (mŏ'sez) Lower plants that have a dominant gametophyte generation, spores, and swimming sperm. They lack vascular tissue.

motor neurons (mo'tur noor'onz'') Those neurons that carry information from the central nervous system to muscles or glands.

motor unit (mo'tur yoo'nit) All of the muscle cells stimulated by a single neuron.

multiple alleles (mul'ti-pul a-lēlz) A term used to refer to conditions in which there are several different alleles for a characteristic, not just two.

mutagenic agent (miu-tah-jen'ik a-jent) Anything that causes permanent change in DNA.

mutation (miu-ta'shun) Any change in the genetic information of a cell.

mutualism (miu'chu-al-izm) A relationship between two organisms in which both organisms benefit.

mycorrhiza (my''ko-rye'zah) A symbiotic relation between fungi and plant roots.

mycotoxin (mi''ko-tok'sin) A deadly poison produced by fungi.

myosin (mi'o-sin) A protein molecule found in the thick filaments of muscle fibers that bends and moves along actin molecules

NAD (**n**icotinamide **a**denine **d**inucleotide) An electron acceptor and hydrogen carrier used in respiration.

NADP (**n**icotinamide **a**denine **d**inucleotide **p**hosphate) An electron acceptor and hydrogen carrier used in photosynthesis.

natality (na-tal'i-te) The number of individuals entering the population by reproduction per thousand individuals in the population.

natural selection (nat'chu-ral se-lek'shun) A broad term used in reference to the various mechanisms that encourage the passage of beneficial genes to future generations and discourage harmful or less valuable genes from being passed on to future generations.

negative-feedback control (neg'a-tiv fēd'bak con-trol') A kind of control mechanism in which the product of one activity inhibits an earlier step in the chain of events.

negative-feedback inhibition (neg'a-tiv fēd'bak in-hib'ĭ-shun) A metabolic control process that operates at the surfaces of enzymes. This process occurs when one of the end products of the pathway alters the three-dimensional shape of an essential enzyme in the pathway and interferes with its operation long enough to slow its action.

nephrons (nef'ronz) Tiny tubules that are the functional units of kidneys.

nerve impulse (nerv im'puls) A series of changes that take place in the neuron, resulting in a wave of depolarization that passes from one end of the neuron to the other.

nerves (nervz) Bundles of neuronal fibers.

nervous system (ner'vus sis'tem) A network of neurons that carry information from sense organs to the central nervous system and from the central nervous system to muscles and glands.

net movement (net muv'ment) The movement in one direction minus the movement in the opposite direction.

neuron (noor'on'') The cellular unit consisting of a cell body and fibers that makes up the nervous system; also called nerve cell.

neurotransmitter (noor''o-trans'mĭt-er) A molecule released by the axons of neurons that stimulates other cells.

neutralization (nu'tral-ĭ-za''shun) A chemical reaction involved in mixing an acid with a base; results in formation of a salt and water.

neutrons (nu'trons) Particles in the nucleus of an atom that have no electrical charge; they were named *neutrons* to reflect this lack of electrical charge.

niche (nitch) The functional role of an organism.

nitrifying bacteria (ni'trĭ-fi-ing bak-te're-ah) Several kinds of bacteria capable of converting ammonia to nitrite or nitrite to nitrate.

nitrogenous base (ni-trah'jen-us bās) A category of organic molecules found as components of the nucleic acids. There are five common types: thymine, guanine, cytosine, adenine, and uracil.

nondeciduous (non''de-sid'yu-us) A term used to describe trees that do not lose their leaves all at once.

nondisjunction (non''dis-junk'shun) An abnormal meiotic division that results in sex cells with too many or too few chromosomes.

norepinephrine (nor-ĕ''pĭ-nef'rin) A hormone produced by the adrenal medulla that increases heart rate, blood pressure, and breathing rate.

nuclear membrane (nu'kle-ar mem'brān) The structure surrounding the nucleus that separates the nucleoplasm from the cytoplasm.

nucleic acids (nu-kle'ik ă'sids) Complex molecules that store and transfer information within a cell. They are constructed of fundamental monomers known as nucleotides.

nucleoli (singular, **nucleolus**) (nu-kle'o-li) Nuclear structures composed of completed or partially completed ribosomes and the specific parts of chromosomes that contain the information for their construction.

nucleoplasm (nu'kle-o-plazm) The liquid matrix of the nucleus composed of a mixture of water and the molecules used in the construction of the rest of the nuclear structures.

nucleoproteins (nu-kle-o-pro'te-inz) The duplex DNA strands with attached proteins; also called **chromatin fibers.**

nucleosomes (nu'kle-o-sōmz) Histone clusters with their encircling DNA.

nucleotide (nu'kle-o-tīd) The building block of the nucleic acids. Each is composed of a 5-carbon sugar, a phosphate, and a nitrogenous base.

nucleus (nu'kle-us) The central body that contains the information system for the cell. (See also **atomic nucleus.**)

nutrients (nu'tre-ents) Molecules required by the body for growth, reproduction, and/or repair.

nutrition (nu-trĭ'shun) Collectively, the processes involved in taking in, assimilating, and utilizing nutrients.

obese (o-bēs) A term describing a person who gains a great deal of unnecessary weight and is 15%–20% above his or her ideal weight.

obligate intracellular parasites (ob'lĭ-gat in''trah-sel'yu-lar pĕr'uh-sīts) Infectious particles (viruses) that can function only when inside a living cell.

observation (ob-sir-vā'shun) The process of using the senses or extensions of the senses to record events.

offspring (of'spring) Descendants of a set of parents.

olfactory epithelium (ōl-fak'to-re ĕ''pĭ-thē'le-um) The cells of the nasal cavity that respond to chemicals.

omnivores (om'nĭ-vōrz) Those animals that are carnivores at some times and herbivores at others.

oogenesis (oh''ŏ-jen'ĕ-sis) The specific name given to the gametogenesis process that leads to the formation of eggs.

orbital (or'bĭ-tal) The area of an atom able to hold a maximum of two electrons.

order (or'der) A group of closely related classes within a phylum.

organ (or'gun) A structure composed of two or more kinds of tissues.

organelles (or-gan-elz') Cellular structures that perform specific functions in the cell. The function of an organelle is directly related to its structure.

organic molecules (or-gan'ik mol'uh-kiuls) Complex molecules whose basic building blocks are carbon atoms in chains or rings.

organism (or'gun-izm) An independent living unit.

organ system (or'gun sis'tem) A group of organs that performs a particular function.

orgasm (or'gaz-um) A complex series of responses to sexual stimulation that result in intense frenzy of sexual excitement.

osmosis (os-mo'sis) The net movement of water molecules through a differentially permeable membrane.

osteoporosis (os''te-o-po-ro'sis) A disease condition resulting from the demineralization of the bone, resulting in pain, deformities, and fractures; related to a loss of calcium.

oval window (o'val win'do) The membrane-covered opening of the cochlea, to which the stapes is attached.

ovaries (o'var-ēz) The female sex organs that produce haploid sex cells—the eggs or ova.

oviduct (o'vĭ-dukt) The tube (*fallopian tube*) that carries the oocyte to the uterus.

ovulation (ov-yu-la'shun) The cyclic release of a secondary oocyte from the surface of the ovary every twenty-eight days.

oxidizing atmosphere (ok'sĭ-di-zing at'mos-fēr) An atmosphere that contains molecular oxygen.

oxytocin (ok''sĭ-to'sin) A hormone released from the posterior pituitary that causes contraction of the uterus.

pancreas (pan'kre-as) An organ of the body that secretes many kinds of digestive enzymes into the duodenum.

parasite (pĕr'uh-sīt) An organism that lives in or on another organism and derives nourishment from it.

parasitism (pĕr'uh-sit-izm) A relationship between two organisms that involves one organism living in or on another organism and deriving nourishment from it.

pathogen (path'uh-jen) An agent that causes a specific disease.

pelagic (pē-lăj'ĭk) A term used to describe animals that swim freely as adults.

penis (pe'nis) The portion of the male reproductive system that deposits sperm in the female reproductive tract.

pepsin (pep'sin) An enzyme produced by the stomach that is responsible for beginning the digestion of proteins.

peptide bond (pep'tīd bond) A covalent bond between amino acids in a protein.

perception (per-sep'shun) Recognition by the brain that a stimulus has been received.

perennial (pur-en'e-uhl) A plant that requires many years to complete its life cycle.

perfect flowers (pur'fekt flow'erz) Flowers that contain both male and female reproductive structures.

periodic table of the elements (pĭr-ē-od'ik tā'bul uv the el'e-ments) A list of all of the elements in order of increasing atomic number (number of protons).

peripheral nervous system (pŭ-rĭ'fĕ-ral ner'vus sis'tem) The fibers that communicate between the central nervous system and other parts of the body.

pesticide (pes'tĭ-sīd) A poison used to kill pests. This term is often used interchangeably with insecticide.

petals (pĕ'tuls) Modified leaves of angiosperms; accessory structures of a flower.

PGAL (**p**hospho**gl**ycer**al**dehyde) The end product of the carbon dioxide conversion stage of photosynthesis produced when a molecule of carbon dioxide is incorporated into a larger organic molecule.

pH A scale used to indicate the strength of an acid or base.

phagocytosis (fă''jo-si-to'sis) The process by which the cell wraps around a particle and engulfs it.

pharynx (far'inks) The region at the back of the mouth cavity; the throat.

phenotype (fēn'o-tīp) The physical, chemical, and psychological expression of the genes possessed by an organism.

pheromone (fĕr-uh-mōn) A chemical produced by an animal and released into the environment to trigger behavioral or developmental processes in some other animal of the same species.

phloem (flo'em) One kind of vascular tissue found in higher plants. It transports food materials from the leaves to other parts of the plant.

phosphate (fos'fāt) Part of a nucleotide; composed of phosphorus and oxygen atoms.

phospholipid (fos''fo-lĭ'pid) A class of water-insoluble molecules that resemble fats but contain a phosphate group (PO_4) in their structure.

photoperiod (fō''tō-pĭr'ē-ud) The length of the light part of the day.

photosynthesis (fo-to-sin'thuh-sis) A series of reactions that take place in chloroplasts and result in the storage of sunlight energy in the form of chemical-bond energy.

phylogeny (fi-laj'uh-ne) The science that explores the evolutionary relationships among organisms and seeks to reconstruct evolutionary history.

phylum (fi'lum) A subdivision of a kingdom.

phytoplankton (fye-tuh-plank'tun) Photosynthetic species that form the basis for most aquatic food chains.

pinocytosis (pĭ''no-si-to'sis) The process by which a cell engulfs some molecules dissolved in water.

pioneer community (pi''o-nēr' ko-miu'ni-te) The first community of organisms in the successional process established in a previously uninhabited area.

pioneer organisms (pi''o-nēr' or'gu-nizms) The first organisms in the successional process.

pistil (pis'til) The sex organ in plants that produces eggs or ova.

pituitary gland (pĭ-tu'ĭ-tĕ-re gland) The structure in the brain that controls the functioning of other glands throughout the organism.

placenta (plah-sen'tah) An organ made up of tissues from the embryo and the uterus of the mother that allows for the exchange of materials between the mother's bloodstream and the embryo's bloodstream. It also produces hormones.

plankton (plank'tun) Small floating or weakly swimming organisms.

plasma (plaz'mah) The watery matrix that contains the molecules and cells of the blood.

plasma membrane (plaz'muh mem'brān) See **cell membrane.**

pleiotropy (pli-ot'ro-pe) The multiple effects that a gene may have on the phenotype of an organism.

poikilotherms (poy-ki'luh-thermz) Animals with a variable body temperature that changes with the external environment.

point mutation (point miu-ta'shun) A change in the DNA of a cell as a result of a loss or change in a nitrogenous base sequence.

polar body (po'lar bod'ē) The smaller cell formed by the unequal meiotic division during oogenesis.

pollen (pol'en) The male gametophyte in gymnosperms and angiosperms.

pollination (pol''ĭ-na'shun) The transfer of pollen in gymnosperms and angiosperms.

polygenic inheritance (pol''e-jen'ik in-her'ĭ-tans) The concept that a number of different pairs of alleles may combine their efforts to determine a characteristic.

polyp (pol'ip) A sessile larval stage in the phylum Cnidaria that reproduces asexually.

polypeptide chain (pŏ''le-pep'tīd chān) A macromolecule composed of a specific sequence of amino acids.

polyploidy (pah''lĭ-ploy'de) A condition in which cells contain multiple sets of chromosomes.

polysome (pah'le-sōm) A sequence of several translating ribosomes attached to the same mRNA.

population (pop''u-la'shun) A group of organisms of the same species located in the same place at the same time.

population density (pop''u-la'shun den'sĭ-te) The number of organisms of a species per unit area.

population growth curve (pop''u-la'shun grōth kurv) A graph of the change in population size over time.

population pressure (pop''u-la'shun presh'yur) Intense competition that leads to changes in the environment and dispersal of organisms.

potential energy (po-ten'shul en'er-je) The energy an object has because of its position.

prebionts (pre″bi′onts) Nonliving structures that led to the formation of the first living cells.

predation (prē-da′shun) A relationship between two organisms that involves the capturing, killing, and eating of one by the other.

predator (pred′uh-tōr) An organism that captures, kills, and eats another animal.

prey (prā) An organism captured, killed, and eaten by a predator.

primary carnivores (pri′mar-e kar′nĭ-vōrz) Those carnivores that eat herbivores and are therefore on the third trophic level.

primary consumers (pri′mar-e konsu′merz) Those organisms that feed directly on plants—herbivores.

primary oocyte (pri′mar-e o′o-sīt) The diploid cell of the ovary that begins to undergo the first meiotic division in the process of oogenesis.

primary spermatocyte (pri′mar-e spurmat′o-sīt) The diploid cell in the testes that undergoes the first meiotic division in the process of spermatogenesis.

primary succession (pri′mar-e suk-sĕ′shun) The orderly series of changes that begins in a previously uninhabited area and leads to a climax community.

probability (prob″a-bil′ĭ-te) The chance that an event will happen, expressed as a percent or fraction.

producers (pro-du′surz) Organisms that produce new organic material from inorganic material with the aid of sunlight.

productivity (pro-duk-tiv′ĭ-te) The rate at which an ecosystem can accumulate new organic matter.

products (prŏ′dukts) New molecules resulting from a chemical reaction.

prokaryote (pro-kăr′e-ōt) A cell that lacks a nuclear membrane and other membranous organelles.

prokaryotic cells (pro′ka-re-ot″ik sels) One of the two major types of cells. They do not have a typical nucleus bound by a nuclear membrane and lack many of the other membranous cellular organelles; bacteria.

promoter (pro-mo′ter) A region of DNA at the beginning of each gene, just ahead of an initiator code.

prophase (pro′fāz) The first phase of mitosis during which individual chromosomes become visible.

protein (pro′te-in) Macromolecules made up of amino acid subunits attached to each other by peptide bonds; groups of polypeptides.

proteinoid (pro′te-in-oid) The proteinlike structure of branched amino acid chains that is the basic structure of a microsphere.

protein-sparing (pro′te-in spĕ′ring) The conservation of proteins by first oxidizing carbohydrates and fats as a source of ATP energy.

protein synthesis (pro′te-in sin′thĕ-sis) The process whereby the tRNA utilizes the mRNA as a guide to arrange the amino acids in their proper sequence according to the genetic information in the chemical code of DNA.

protocell (pro′to-sel) The first living cell.

protons (pro′tons) Particles in the nucleus of an atom that have a positive electrical charge.

protoplasm (pro′to-plazm) The living portion of a cell as distinguished from the nonliving cell wall.

protozoa (pro″to-zo′ah) Heterotrophic, unicellular organisms.

proximal convoluted tubule (prok′sĭ-mal kon′vo-lu-ted tūb′yūl) The upstream end of the nephron of the kidney, which is responsible for reabsorbing most of the valuable molecules filtered from the glomerulus into Bowman's capsule.

pseudoscience (su-dō-si′ens) The use of the appearance of science to mislead. The assertions made are not valid or reliable.

puberty (pu′ber-te) A time in the life of a developing individual characterized by the increasing production of sex hormones, which cause it to reach sexual maturity.

pulmonary artery (pul′muh-nă-rē ar′tuh-rē) The major blood vessel that carries blood from the right ventricle to the lungs.

pulmonary circulation (pul′muh-nă-rē ser-ku-la′shun) The flow of blood through certain chambers of the heart and blood vessels to the lungs and back to the heart.

punctuated equilibrium (pung′chu-a-ted e-kwĭ-lib′re-um) The theory stating that evolution occurs in spurts, between which there are long periods with little evolutionary change.

Punnett square (pun′net sqwār) A method used to determine the probabilities of allele combinations in a zygote.

pyloric sphincter (pi-lor′ik sfingk′ter) A valve located at the end of the stomach that regulates the flow of food from the stomach to the duodenum.

pyruvic acid (pi-ru′vik as′id) A 3-carbon carbohydrate that is the end product of the process of glycolysis.

radial symmetry (ra′de-ul sĭm′ĭ-tre) The characteristic of an animal with a body constructed around a central axis. Any division of the body along this axis results in two similar halves.

radioactive (ra-de-o-ak′tiv) A term used to describe the property of releasing energy or particles from an unstable atom.

range (rānj) The geographical distribution of a species.

reactants (re-ak′tants) Materials that will be changed in a chemical reaction.

recessive allele (re-sĕ′siv a-lēl′) An allele that, when present with its homolog, does not express itself and is masked by the effect of the other allele.

recombinant DNA (re-kom′bĭ-nant) DNA that has been constructed by inserting new pieces of DNA into the DNA of another organism, such as a bacterium.

Recommended Dietary Allowances (RDA) (rĕ-ko-men′ded dī′ĕ-tĕ-rē ă-lao′an-ses) U.S. dietary guidelines for a healthy person that focus on the six classes of nutrients.

redirected aggression (re-di-rek′ted ă-grĕ′shun) A behavior in which the aggression of an animal is directed away from an opponent and to some other animal or object.

reducing atmosphere (re-du′sing at′mos-fēr) An atmosphere that does not contain molecular oxygen (O_2).

reduction division (re-duk′shun dĭ-vĭ′zhun) A type of cell division in which daughter cells get only half the chromosomes from the parent cell.

regulator proteins (reg′yu-la-tor pro′te-ins) Proteins that influence the activities that occur in an organism—for example, enzymes and some hormones.

reliable (re-li′a-bul) A term used to describe results that remain consistent over successive trials.

reproductive capacity (re-pro-duk′tiv kuh-pas′ĭ-te) The theoretical maximum rate of reproduction; also called **biotic potential.**

reproductive isolating mechanism (re-pro-duk'tiv i-so-la'ting me'kan-izm) A mechanism that prevents interbreeding between species; also called **genetic isolating mechanism.**

response (re-spons') The reaction of an organism to a stimulus.

responsive processes (re-spon'siv pros'es-es) Those abilities to react to external and internal changes in the environment; for example, irritability, individual adaptation, and evolution.

retina (rĕ'tĭ-nah) The light-sensitive region of the eye.

rhodopsin (ro-dop'sin) A light-sensitive, purple-red pigment found in the retinal rods that is important for vision in dim light.

ribonucleic acid (RNA) (ri-bo-nu-kle'ik ă'sid) A polymer of nucleotides formed on the template surface of DNA by transcription. Three forms that have been identified are mRNA, rRNA, and tRNA.

ribose (ri'bōs) A 5-carbon sugar molecule that is a component of RNA.

ribosomal RNA (rRNA) (ri-bo-sōm'al) A globular form of RNA; a part of ribosomes.

ribosomes (ri'bo-sōmz) Small structures composed of two protein and ribonucleic acid subunits involved in the assembly of proteins from amino acids.

RNA polymerase (po-lim'er-ās) An enzyme that attaches to the DNA at the promoter region of a gene when the genetic information is transcribed into RNA.

rods (rahdz) Light-sensitive cells in the retina of the eye that respond to low-intensity light but do not respond to different colors of light.

root (root) A specialized structure for the absorption of water and minerals in higher plants.

root hairs (root hārs) Tiny root outgrowths that improve the ability of plants to absorb water and minerals.

salivary amylase (să'lĭ-vĕ-rē ă'mĭ-lās) An enzyme present in saliva that breaks starch molecules into smaller molecules.

salivary glands (să'lĭ-vĕ-rē glanz) Glands that produce saliva.

salts (salts) Ionic compounds formed from a reaction between an acid and a base.

saprophyte (sap'ruh-fīt) An organism that obtains energy by the decomposition of dead organic material.

saturated (sat'yu-ra-ted) A term used to describe the carbon skeleton of a fatty acid that has as much hydrogen bonded to it as possible.

science (si'ens) A process or way of arriving at a solution to a problem or understanding an event in nature using the scientific method.

scientific law (si-en-tif'ik law) A uniform or constant feature of nature supported by several theories.

scientific method (si-en-tif'ik meth'ud) A way of gaining information (facts) about the world around you that involves observation, hypothesis formation, testing of hypotheses, theory formation, and law formation.

seasonal isolation (se'zun-al i-so-la'shun) A genetic isolating mechanism that prevents interbreeding between species because their reproductive periods differ.

second law of thermodynamics (sek'ond law uv ther''mo-di-nam'iks) Whenever energy is converted from one form to another, some useful energy is lost.

secondary carnivores (sĕk'on-dĕr-e kar'nĭ-vorz) Those carnivores that feed on primary carnivores and are therefore on the fourth trophic level.

secondary consumers (sĕk-on-dĕr-e kon-su'merz) Those animals that eat other animals—carnivores.

secondary oocyte (sek'on-dĕr-e o'o-sīt) The larger of the two cells resulting from the lopsided cytoplasmic division of a primary oocyte in meiosis I of oogenesis.

secondary sex characteristics (sĕk'on-dĕr-e seks kăr-ak-tĕ-ris'tiks) Characteristics of the adult male or female, including the typical shape that develops at puberty: broader shoulders, heavier long-bone muscles, development of facial hair, axillary hair, and chest hair, and changes in the shape of the larynx in the male; rounding of the pelvis and breasts and changes in deposition of fat in the female.

secondary spermatocyte (sek'on-dĕr-e spur-mat'o-sīt) Cells in the seminiferous tubules that go through the second meiotic division, resulting in four haploid spermatids.

secondary succession (sek'on-dĕr-e suk-sĕ'shun) The orderly series of changes that begins with the disturbance of an existing community and leads to a climax community.

seed (sēd) A specialized structure produced by gymnosperms and angiosperms that contains the embryonic sporophyte.

seed coat (sēd kōt) A protective layer around a seed.

seed leaves (sēd lēvz) Embryonic leaves in seeds.

segregation (seg''rĕ-ga'shun) The separation and movement of homologous chromosomes to the poles of the cell.

selecting agent (se-lek'ting a'jent) Any factor that affects the probability that a gene will be passed to the next generation.

semen (se'men) The sperm-carrying fluid produced by the seminal vesicles, prostate glands, and bulbo-urethral glands of males.

semicircular canals (sĕ-mi-ser'ku-lar că-nalz') A set of tubular organs associated with the cochlea that sense changes in the movement or position of the head.

semilunar valves (sĕ-me-lu'ner valvz) Valves located in the pulmonary artery and aorta that prevent the flow of blood backwards into the ventricles.

seminal vesicle (sĕm'ĭ-nal ves'ĭ-kul) A part of the male reproductive system that produces a portion of the semen.

seminiferous tubules (sem''ĭ-nif'ur-us tūb'yūlz) Sperm-producing tubes in the testes.

sensory neurons (sen'so-re noor'onz'') Those neurons that send information from sense organs to the central nervous system.

sepals (se'pals) Accessory structures of flowers.

sere (ser) See **successional community (stage).**

sessile (ses'il) Firmly attached.

sex chromosomes (seks kro'mo-sōmz) A pair of chromosomes that determine the sex of an organism.

sex-determining chromosome (seks de-ter'mĭ-ning kro'mo-sōm) The chromosome that is primarily responsible for the development of male or female sexual structures.

sex ratio (seks ra'sho) The number of males in a population compared to the number of females.

sexual intercourse (sek'shool-al in'ter-kors) The mating of male and female. The deposition of the male sex cells, or sperm cells, in the reproductive tract of the female; also known as **coitus** or **copulation.**

sexual reproduction (sek'shu-al re''pro-duk'shun) The propagation of organisms involving the union of gametes from two parents.

sexuality (sek''shu-al'ĭ-te) A term used in reference to the totality of the aspects—physical, psychological, and cultural—of our sexual nature.

sickle-cell anemia (sĭ'kul sel ah-ne'me-ah) A disease caused by a point mutation. This malfunction produces sickle-shaped red blood cells.

sign stimulus (sīn stim'yu-lus) A specific object or behavior that triggers a specific behavioral response.

single-factor cross (sing'ul fak'tur kros) A genetic study in which a single characteristic is followed from the parental generation to the offspring.

society (so-si'uh-te) Interacting groups of animals of the same species that show division of labor.

sociobiology (so-sho-bi-ol'o-je) The systematic study of all forms of social behavior, both human and nonhuman.

solid (sol'id) The state of matter in which the molecules are packed tightly together; they vibrate in place.

soma (so'mah) The cell body of a neuron, which contains the nucleus.

speciation (spe-she-a'shun) The process of generating new species.

species (spe'shēz) The scientific name given to a group of organisms that can potentially interbreed to produce fertile offspring.

specific dynamic action (SDA) (spĕ-sĭ'fik di-nă'mik ak'shun) The amount of energy required to digest and assimilate food. SDA is equal to approximately 10% of your total daily kcalorie intake.

sperm (spurm) Cells that develop from the spermatids by losing much of their cytoplasm and developing long tails; the male gamete.

sperm cells (spurm selz) The haploid sex cells produced by sexually mature males.

spermatids (spurm'ah-tids) Haploid cells produced by spermatogenesis that change into sperm.

spermatogenesis (spur-mat-o-jen'uh-sis) The specific name given to the gametogenesis process that leads to the formation of sperm.

spindle (spin'dul) An array of microtubules extending from pole to pole; used in the movement of chromosomes.

spontaneous generation (spon-ta'ne-us jen-uh-ra'shun) The theory that living organisms arose from nonliving material.

spontaneous mutation (spon-ta'ne-us miu-ta'shun) Natural changes in the DNA caused by unidentified environmental factors.

spores (spōrz) Haploid structures produced by sporophytes.

sporophyte stage (spor'o-fit stāj) A life-cycle stage in plants in which a haploid spore is produced by meiosis.

stable equilibrium phase (stā'bul e-kwi-lib're-um fāz) A period of time during population growth when the number of individuals entering the population and the number leaving the population are equal, resulting in a stable population.

stamen (sta'men) The male reproductive structure of a flower.

stapes (sta'pēz) The ear bone that is attached to the oval window.

states of matter (stātes uv mat'er) Physical conditions of matter (solid, liquid, and gas) determined by the relative amounts of energy of the molecules.

stem (stem) The upright portion of a higher plant.

steroid (stēr'oid) One of the three kinds of lipid molecules characterized by their arrangement of interlocking rings of carbon.

stimulus (stim'yu-lus) Some change in the internal or external environment of an organism that causes, it to react.

stroma (stro'muh) The region within a chloroplast that has no chlorophyll.

structural formula (struk'chu-ral for'miu-lah) An illustration showing the arrangement of the atoms and their bonding within a molecule.

structural proteins (struk'chu-ral pro'te-ins) Proteins that are important for holding cells and organisms together, such as the proteins that make up the cell membrane, muscles, tendons, and blood.

subspecies (sub'spe-shēz) Regional groups within a species that are significantly different structurally, physiologically, or behaviorally, yet are capable of exchanging genes by interbreeding.

substrate (sub'strāt) A reactant molecule with which the enzyme combines.

succession (suk-sĕ'shun) The process of changing one type of community to another.

successional community (stage) (suk-sĕ'shun-al ko-miu'nĭ-te) An intermediate stage in succession; also called **sere.**

surface-area-to-volume ratio (ser'fas a're-ah to vol'ūm ra'sho) The relationship between the surface area of an object and its volume. As objects increase in size, their volume increases more rapidly than their surface area.

symbiosis (sim-be-o'sis) A close physical relationship between two kinds of organisms. It usually includes parasitism, commensalism, and mutualism.

symbiotic nitrogen-fixing bacteria (sim-be-ah'tik ni-tro-jen fik'sing bak-te're-ah) Bacteria that live in the roots of certain kinds of plants, where they convert nitrogen gas molecules into compounds that plants can use.

synapse (si'naps) The space between the axon of one neuron and the dendrite of the next, where chemicals are secreted to cause an impulse to be initiated in the second neuron.

synapsis (sin-ap'sis) The condition in which the two members of a pair of homologous chromosomes come to lie close to one another.

systemic circulation (sis-tĕ'mik ser-ku-la'shun) The flow of blood through certain chambers of the heart and blood vessels to the general body and back to the heart.

systolic blood pressure (sis-tah'lik blud presh'yur) The pressure generated in a large artery when the ventricles of the heart are in the process of contracting.

taxonomy (tak-son'uh-me) The science of classifying and naming organisms.

telophase (tel'uh-fāz) The last phase in mitosis, characterized by the formation of daughter nuclei.

temperature (tem'per-ă-chiur) A measure of molecular energy of motion.

template (tem'plet) A model from which a new structure can be made. This term has special reference to DNA as a model for both DNA replication and transcription.

termination codes (ter-mĭ-na'shun cōdz) The DNA nucleotide sequence just in back of a gene with the code ATT, ATC, or ACT that signals "stop here."

territoriality (tĕr''ĭ-tor'e-al'ĭ-te) A behavioral process in which an animal protects space for its exclusive use for food, mating, or other purposes.

territory (tĕr'ĭ-tor-e) A space that an animal defends against others of the same species.

testes (tes'tēz) The male sex organs that produce haploid cells—the sperm.

testosterone (tes-tos'tur-ōn) The male sex hormone produced in the testes that controls the secondary sex characteristics.

theoretical science (the-o-ret'i-kul si'ens) Science interested in obtaining new information for its own sake.

theory (the'o-re) A plausible, scientifically acceptable generalization supported by several hypotheses and experimental trials.

theory of natural selection (the'o-re uv nat'chu-ral se-lek'shun) In a species of genetically differing organisms, the organisms with the genes that enable them to survive better in the environment and thus reproduce more offspring than others will transmit more of their genes to the next generation.

thymine (thi'mēn) A single-ring nitrogenous-base molecule in DNA but not in RNA. It is complementary to adenine.

thyroid-stimulating hormone (thi'roid sti'mu-la-ting hōr'mōn) A hormone secreted by the pituitary gland that stimulates the thyroid to secrete thyroxine.

thyroxine (thi-rok'sin) A hormone produced by the thyroid gland that speeds up the metabolic rate.

tissue (tish'yu) A group of specialized cells that work together to perform a particular function.

trachea (trā'ke-ah) A major tube supported by cartilage that carries air to the bronchi; also known as the windpipe.

transcription (tran-skrip'shun) The process of manufacturing RNA from the template surface of DNA. Three forms of RNA that may be produced are mRNA, rRNA, and tRNA.

transfer RNA (tRNA) (trans'fur) A molecule composed of ribonucleic acid. It is responsible for transporting a specific amino acid into a ribosome for assembly into a protein.

translation (trans-la'shun) The assembly of individual amino acids into a polypeptide.

transpiration (trans''pi-ra'shun) The process of water evaporation from the leaves of a plant.

triiodothyronine (tri''ī-ō''dō-thī'row-nēn) A hormone produced by the thyroid gland that speeds up the metabolic rate; similar to thyroxine but more potent.

trisomy (tris'oh-me) An abnormal number of chromosomes resulting from the nondisjunction of homologous chromosomes during meiosis; for example, as in Down syndrome in which the individual has three of one of the kinds of chromosomes instead of the normal two.

trophic level (tro'fik lē'vel) A step in the flow of energy through an ecosystem.

tropomyosin (tro''po-mi'o-sin) A molecule found in thin filaments of muscle that helps to regulate when muscle cells contract.

troponin (tro'po-nin) A molecule found in thin filaments of muscle that helps to regulate when muscle cells contract.

turnover number (turn'o-ver num'ber) The number of molecules of substrate that a single molecule of enzyme can react with in a given time under ideal conditions.

tympanum (tim'pā-num) The eardrum.

unsaturated (un-sat'yu-ra-ted) A term used to describe the carbon skeleton of a fatty acid containing carbons that are double bonded to each other at one or more points.

uracil (yu'rah-sil) A single-ring nitrogenous-base molecule in RNA but not in DNA. It is complementary to adenine.

uterus (yu'tur-us) The organ in female mammals in which the embryo develops.

vacuole (vak'yu-ōl) A large sac within the cytoplasm of a cell, composed of a single membrane.

vagina (vuh-ji'nah) The passageway between the uterus and outside of the body; the birth canal.

valid (val'id) A term used to describe meaningful data that fit into the framework of scientific knowledge.

variable (var'e-ā-bul) The single factor that is allowed to be different.

vascular tissue (vas'kyu-lar tish'yu) Specialized tissue that transports fluids in higher plants.

vector (vek'tōr) An organism that carries a disease or parasite from one host to the next.

veins (vānz) The blood vessels that return blood to the heart.

ventricles (ven'tri-klz) The powerful muscular chambers of the heart whose contractions force blood to flow through the arteries to all parts of the body.

vertebrates (vur'tuh-brāts) Animals with backbones.

vesicles (vĕ'si-kuls) Small, intracellular, membrane-bound sacs in which various substances are stored.

villi (vil'ē) Tiny fingerlike projections in the lining of the intestine that increase the surface area for absorption.

virus (vi'rus) A nucleic acid particle coated with protein that functions as an obligate intracellular parasite.

vitamin-deficiency disease (vi'tah-min de-fish'en-se di-zēz) Poor health caused by the lack of a certain vitamin in the diet; for example, scurvy for lack of vitamin C.

vitamins (vi'tah-minz) Organic molecules that cannot be manufactured by the body but are required in very low concentrations.

voltage (vōl'tij) A measure of the electrical difference that exists between two different points or objects.

wood (wŏŏd) The xylem of gymnosperms and angiosperms.

X chromosome (eks kro'mo-sōm) The chromosome in a human female egg (and in one-half of sperm cells) that is associated with the determination of sexual characteristics.

X-linked gene (eks-lingt jēn) A gene located on one of the sex-determining X chromosomes.

xylem (zi'lem) A kind of vascular tissue that transports water from the roots to other parts of the plant.

Y chromosome (wi kro'mo-sōm) The sex-determining chromosome in one-half of the sperm cells of human males.

zygote (zi'gōt) A diploid cell that results from the union of an egg and a sperm.

C R E D I T S

Photographs

Part Openers and Table of Contents

Part One: © Chip & Jill Isenhart/Tom Stack & Associates; **Part Two:** © Eric Grave/Science Source/Photo Researchers, Inc.; **Part Three:** © Hans Pfletschinger/Peter Arnold, Inc.; **Part Four:** © Ed Reschke/Peter Arnold, Inc.; **Part Five:** © John Carter/Photo Researchers, Inc.; **Part Six:** © David Dennis/Tom Stack & Associates

Chapter 1

Opener, 1.1: © Wm. C. Brown Communications, Inc., Bob Coyle, photographer; **1.2a:** From J. D. Watson, *The Double Helix,* p. 215, Atheneum, New York, 1968, by J. D. Watson, Cold Spring Harbor Laboratory. Barrington Brown, photographer; **1.2b:** © Hank Morgan/Photo Researchers, Inc.; **1.3(top):** The Bettmann Archives; **(bottom):** © Bob Coyle; **1.4:** © Wm. C. Brown Communications/Jim Shaffer, photographer; **1.8:** © Tom Ulrich/ Visuals Unlimited

Chapter 3

Opener: © Will & Deni McIntyre/Photo Researchers, Inc.; 3.1**(top left):** © SIU/Visuals Unlimited; **(top right):** © Bob Coyle; **3.1(bottom left):** © SIU/Visuals Unlimited; **3.1(bottom right):** © Bob Coyle; **3.2(top):** © Phil A. Dotson/Photo Researchers, Inc.; **(bottom):** © Will & Deni McIntyre/Photo Researchers, Inc.

Chapter 4

Opener: © David M. Phillips/Visuals Unlimited; **4.1:** Historical Picture Service/Stock Montage; **4.6a–c:** © David M. Phillips/Visuals Unlimited; **4.9(top right):** © R. Rodewald, Univ. of Virginia/BPS; **(bottom left):** © David M. Phillips/Visuals Unlimited; **(bottom middle):** © Warren Rosenberg, Iona College/BPS; **(bottom right):** © K. G. Murti/Visuals Unlimited; **4.11c:** Courtesy of Dr. Keith Porter; **4.12:** © Don W. Fawcett/Photo Researchers, Inc.; **4.14a:** Sandra L. Wolin; **4.15:** © W. Rosenberg, Iona College/BPS; **4.17:** William Jensen and Roderick B. Park, *Cell Ultrastructure,* 1967 by Wadsworth Publishing Co., Inc., p. 57

Chapter 5

Opener, 5.3(both): © Wm. C. Brown Communications, Inc./Jim Shaffer, photographer

Chapter 6

6.13b: © Brian Parker/Tom Stack & Associates

Chapter 7

Opener: © CNRI/Photo Researchers, Inc.; **7.6b:** E. J. Dupraw; **7.16a:** © CNRI/Photo Researchers, Inc.; **7.16b:** © Jackie Lewin/Royal Free Hospital/Photo Researchers, Inc.

Chapter 8

Opener: © J. F. Gennaro, Jr. and L. R. Grillome/ Photo Researchers, Inc.; **8.10b,d:** Carolina Biological Supply Co.; **8.12:** © Edwin A. Reschke; **8.13:** © James Stevenson/Photo Researchers, Inc.

Chapter 9

Figure 9.24: © M. Coleman/Visuals Unlimited

Chapter 10

Opener: © Renee Lynn/Photo Researchers, Inc.; **10.1a,b:** © Tom Ballard/EKM Nepenthe; **10.3a:** © Renee Lynn/Photo Researchers, Inc.

Chapter 11

Opener: © George Jones/Photo Researchers, Inc.; **11.1(top left):** © Reynolds Photography; **(top right):** © Jacana/Photo Researchers, Inc.; **(bottom left):** © Reynolds Photography; **(bottom right):** © Jeanne White/Photo Researchers, Inc.; **11.3(left):** © John Cancalosi/Tom Stack & Associates; **(right):** © John Serrao/Visuals Unlimited; **11.4 (Europe):** © Henry E. Bradshaw/Photo Researchers, Inc.; **(Africa):** © Linda Bartlett/Photo Researchers, Inc.; **(Asia):** © Lawrence Migdale/Photo Researchers, Inc.; **11.5:** © Gary Milburn/Tom Stack & Associates; **11.7a:** Courtesy of Ball Seed Company; **11.7b:** © George Jones/Photo Researchers, Inc.; **11.8:** © Earl Roberge/Photo Researchers, Inc.; **11.10a,b:** © Stanley L. Flegler/Visuals Unlimited

Chapter 12

Opener: © John Colwell/Grant Heilman Photography; **12.1:** © Robert Houser/Comstock; **12.2:** © Joe McDonald/Visuals Unlimited; **12.3:** © Comstock; **12.4a:** © Betty Derig/Photo Researchers, Inc.; **12.4b:** © John Colwell/Grant Heilman Photography; **12.4c:** © Kees Van Den Berg/Photo Researchers, Inc.; **12.5:** © John Cunningham/Visuals Unlimited; **12.8:** © T. Kitchin/Tom Stack & Associates; **12.9a,b:** © Bob Coyle

Chapter 13

Opener: © Mero/Jacana/Photo Researchers, Inc.; **13.1a:** © Walt Anderson/Visuals Unlimited; **13.1b:** © John D. Cunningham/ Visuals Unlimited; **13.1c:** © William J. Weber/ Visuals Unlimited; **13.5:** © Mike Blair; **13.6a:** © Joel Arrington/Visuals Unlimited; **13.6b:** © Mero/Jacana/Photo Researchers, Inc.; **13.7:** © Courtesy of W. Altee Burpee & Co.

Chapter 14

Opener: © Willard Clay; **14.1a:** © Gregory K. Scott/Photo Researchers, Inc.; **14.1b:** © Fred Marsik/ Visuals Unlimited.; **14.12:** © Willard Clay; **14.14:** © C. P. Hickman/Visuals Unlimited; **14.15:** © G. R. Roberts/Nelson, New Zealand; **14.17:** © Michael Giannechini/Photo Researchers, Inc.; **14.18:** © S. L. Pimm/Visuals Unlimited

Chapter 15

Opener, 15.2a: © John D. Cunningham/Visuals Unlimited; **15.2b:** © G. R. Roberts/Nelson, New Zealand; **15.3:** © David Waters/Envision; **15.4a:** © J. H. Robinson/Photo Researchers, Inc.; **15.4b:** © Gary Milburn/Tom Stack & Associates; **15.6:** © Douglas Faulkner/Sally Faulkner Collection; **15.7:** © John D. Cunningham/Visuals Unlimited; **15.8:** © Susanna Pashko/Envision

Chapter 16

Opener: © Harry Rogers/Photo Researchers, Inc.; **16.3:** © G. R. Higbee/Photo Researchers, Inc.

Chapter 17

Opener: © G. R. Roberts/Nelson/New Zealand; **17.6:** © Lincoln P. Brower, Univ. of Florida, Gainesville; **17.7:** Sybille Kalas; **17.10:** © G. R. Roberts/Nelson/New Zealand; **17.11:** © Harry Rogers/Photo Researchers, Inc.

Chapter 18

Figure 18.2a: Courtesy of Dr. Keith Porter

Chapter 19

Opener, 19.3: © Wm. C. Brown Communications/Bob Coyle, photographer; **19.4:** Randy A. Sansone; **19.5:** © Paul A Souders; **19.7:** National Osteoporosis Foundation; **19.8:** Dr. Ann Pytkowicz Streissguth, Univ. of Washington, Seattle

Chapter 20

Opener: © Dwight Kuhn

Chapter 21
Opener, 21.5: © SIU/Peter Arnold, Inc.; **21.10a:** © Jim Shaffer; **21.10b:** © H. Morgan/ Photo Researchers, Inc.; **21.10c:** © Bob Coyle; **21.10d:** © Ray Ellis/Photo Researchers, Inc.; **21.10e:** © Bob Coyle; **21.10f:** © SIU/Photo Researchers, Inc.; **21.10g,h:** © Bob Coyle

Chapter 22
Opener: © Paul W. Johson/BPS; **22.1:** Courtesy Cordon Art, Holland

Chapter 23
Opener: © Don Valenti/Tom Stack & Associates; **23.1a:** © Tom McHugh/Photo Researchers, Inc.; **23.1b:** © Russ Kinne/ Comstock; **23.2:** Historical Pictures Service/ Stock Montage; **23.3a:** © A. M. Siegelman/ Visuals Unlimited; **23.3b:** © John Gerlach/Tom Stack & Associates; **23.3c:** © Paul W. Johnson/ BPS; **23.3d:** © Comstock; **23.3e:** © Glenn Oliver/Visuals Unlimited; **23.4a:** © W. B. Saunders, Bryn Mawr College/BPS; **23.4b:** © Edward S. Ross; **23.5:** © Frank T. Aubrey/ Visuals Unlimited; **23.6a:** © Tom Stack/Tom Stack & Associates; **23.6b:** © Don Valenti/Tom Stack & Associates; **23.8a:** © Donald F. Lundgren; **23.9(algae):** © P. Nuridsany/Photo Researchers, Inc.; **(ferns):** © John Cunningham/ Visuals Unlimited; **(gymnosperm):** © Fritz Polking/Peter Arnold, Inc.; **(angiosperm):** © Rod Planck/Tom Stack & Associates; **(mosses):** © Claudia E. Mills/Univ. of Washington/BPS; **page 375:** Coulter Corporation; **page 377:** © James L. Shaffer

Chapter 24
Opener: © Dwight Kuhn; **24.2:** © Y. Arthus-Bertrand/Peter Arnold; **24.3:** USDA; **24.4:** © T. J. Beveridge, Univ. of Guelph/BPS; **24.5a:** © Leland Johnson; **24.5b:** Carolina Biological Supply Co.; **24.7:** © David J. Wrobel/BPS; **24.8a:** © Michael Abbey/Photo Researchers, Inc.; **24.8b:** © Science VV/Visuals Unlimited; **24.10:** © M. Abbey/Visuals Unlimited; **24.11:** © E. S. Ross; **24.12a:** © Cabisco/Visuals Unlimited; **24.12c:** © Dwight Kuhn; **24.13a:** © Barbara J. Miller/BPS; **24.13b:** © Bill Keogh/ Visuals Unlimited; **24.13c:** © Bruce Iverson/ Visuals Unlimited; **24.13d:** © S. Flegler/Visuals Unlimited; **24.14:** © Cabisco/Visuals Unlimited; **page 390:** © Richard Humbert/BPS; **24.15:** © R. Roncardori/Visuals Unlimited; **24.16:** © Nancy M. Wells/Visuals Unlimited; **24.17:** © David M. Dennis/Tom Stack & Associates; **24.18a,b:** © E. S. Ross

Chapter 25
Opener: © John Shaw/Tom Stack & Associates; **25.5(left):** © John Sohlden/Visuals Unlimited; **25.5(right), 25.7:** © John Shaw/Tom Stack & Associates; **25.11a:** © John Gerlach/Tom Stack & Associates; **25.11b:** © Bruce Wilcox/BPS; **25.11c:** © L. West/Photo Researchers, Inc.

Chapter 26
Opener: © Don & Esther Phillips/Tom Stack & Associates; **26.8a:** © Peter Parks/Oxford Scientific Films; **26.8b:** © Robert Evans/Peter Arnold, Inc.; **26.9a:** © Bruce Russell/Bio Media Associates; **26.9b:** © CBS/Visuals Unlimited; **26.12a:** © Ed Reschke/Peter Arnold, Inc.; **26.12b:** © Norm Thomas/Photo Researchers, Inc.; **26.14a:** © Michael DiSpezio; **26.14b:** © Paul L. Janosi/Valan Photos; **26.15a:** © Robert A. Ross; **26.15b:** © David D. Fleetham/Tom Stack & Associates; **26.16b,c:** © Michael Dispezio; **26.17a:** © Russ Kinne/ Photo Researchers, Inc.; **26.17b:** © Tom Stack/ Tom Stack & Associates; **26.18a:** © Marty Snyderman; **26.18b:** © Ed Robinson/Tom Stack & Associates; **26.19a:** © Tom McHugh/Photo Researchers, Inc.; **26.19b:** © David Doubilet; **26.19c:** © Michael DiSpezio; **26.20a:** © Kent Dannen/Photo Researchers, Inc.; **26.20b:** © Bio Media Associates; **26.20c:** © Michael DiSpezio; **26.22a–d:** © Peter J. Bryant/BPS; **26.23a,b:** © Dwight Kuhn; **26.24a:** © E. S. Ross; **26.24b:** © Don & Esther Phillips/Tom Stack & Associates; **26.26a:** © Tom Ulrich/Visuals Unlimited; **26.26b:** © Joe McDonald/Visuals Unlimited; **26.27:** © Tom McHugh/Photo Researchers, Inc.; **26.28a:** © T. J. Cawley/Tom Stack & Associates; **26.28b:** © Tom McHugh/ Photo Researchers, Inc.; **26.29a:** © John Gerlach/Tom Stack & Associates; **26.29b:** © C. Allan Morgan/Peter Arnold, Inc.

Illustrators

Beverly Benner
Figures 6.3, 25.3, 25.10, 25.12, 25.13, 25.14

Chris Creek
Figures 9.3, 10.4, 18.10, 20.1

Dennis
Figures 17.1, 17.8

Peg Gerrity
Figure 20.10a–d

Rob Gordon
Figures 20.4, 20.5, 20.14, 21.2

Illustrious, Inc.
Figures 1.5, 1.6, 1.7, 2.3, 2.4, 2.5, 2.6, 2.7, 2.8, 2.11, 2.14, 3.3, 3.5, 3.7, 3.8, 3.9, 3.10, 3.11, 3.12, 3.15, 3.16, 3.17, 4.10, 4.18, 5.1, 5.2, 5.3b, 5.4, 5.5, 5.6, 5.7, 5.8a, 5.8b, 6.2, 6.13a, 7.1, 7.2a–c, 7.2d, 7.3, 7.5, 7.6a,c, 7.7, 7.8, 7.9, 7.10, 7.11, 7.12, 7.13, 7.15, 8.1, 9.1, 9.21, 11.2, 11.9, 12.6, 12.10, 13.11b, 14.6, 14.7, 14.9, 15.10, 17.5, 20.7, 25.6, 26.4

Carlyn Iverson
Figure 22.6

Marjorie Leggitt
Figures 13.9, 13.10, 13.11a, 13.13, 13.14a–b, 15.12, 17.12, 23.7, 26.2b, 26.5, 26.6, 26.7, 26.21

Bill Loechel
Figures 18.13, 18.15, 18.16, 20.8

Nancy Marshburn
Figure 18.3

John McGee
Figures 15.15, 22.8

McLean
Figure 21.7a

Steve Moon
Figure 4.19

Laurie O'Keefe
Figures 4.11a–b, 4.15a, 4.16, Box 6.1 (page), 10.3b, 11.6, 12.7, 13.3, 13.4, 13.8, Box 14.1 (page), 15.1, 15.14, 20.9, 22.9, 23.8b, 23.10, 23.12, 24.1, 24.9, 25.2, 25.9, 26.2a, 26.3, 26.10, 26.11, 26.13, 26.25

Rolin Graphics, Inc.
Figures 2.1, 2.2, 2.9, 2.12, 3.4, 3.14, 4.13, 8.2, 8.3, 8.4, 8.5, 8.6, 8.7, 8.8, 8.9, 8.10a,c, 8.11, 9.2, 9.4, 9.5, 9.6, 9.7, 9.8, 9.9, 9.10, 9.11, 9.12, 9.13, 9.14, 9.15, 9.16, 9.17, 9.18, 9.19, 9.20, 10.2, 11.4, 15.11, 16.4, 16.6, 18.1, 20.3, 20.6, 20.12, 21.3, 26.1a–b

Mike Schenk
Figure 18.2b

Nadine Sokol
Figure 21.11a–b

Alice Thiede
Figure 24.6

Tom Waldrop
Figures 20.16, 21.4, 21.6a, 21.6b, 21.7b

Williams
Figure 18.4

Line Art

Chapter 3
Page 40 top: Reprinted with permission, Best Foods, a division of CPC International Inc.

Chapter 4
Figure 4.2: From Stuart Ira Fox, *Human Physiology*, 4th ed. Copyright © 1993 Wm. C. Brown Communications, Inc., Dubuque, Iowa. All Rights Reserved. Reprinted by permission. **Figure 4.19:** From John W. Hole, Jr., *Human Anatomy and Physiology*, 5th ed. Copyright © 1990 Wm. C. Brown Communications, Dubuque, Iowa. All Rights Reserved. Reprinted by permission.

Chapter 13
Figure 13.2: Reprinted with permission of the *Los Angeles Herald Examiner*.

Chapter 14
Figure 14.8: From Ralph D. Bird, "Biotic Communities of the Aspen Packland of Central Canada," *Ecology* 11(2):410. Used by permission.

Chapter 16
Figure 16.9: From Jean Van der Tak, et al., "Our Population Predicament: A New Look," *Population Bulletin*, 34(5), Dec. 1979. Reprinted with permission from Population Reference Bureau, Inc.

Chapter 17
Figure 17.9: Courtesy of James E. Lloyd.

Chapter 18
Figure 18.2b: From Kent M. Van De Graaff and Stuart Ira Fox, *Concepts of Human Anatomy and Physiology*, 2d ed. Copyright © 1989 Wm. C. Brown Communications, Inc., Dubuque, Iowa. All Rights Reserved. Reprinted by permission.

Page 272: From Kent M. Van De Graaff and Stuart Ira Fox, *Concepts of Human Anatomy and Physiology*, 3d ed. Copyright © 1992 Wm. C. Brown Communications, Inc., Dubuque, Iowa. All Rights Reserved. Reprinted by permission. **Figure 18.3:** From John W. Hole, Jr., *Human Anatomy and Physiology*, 6th ed. Copyright © 1992 Wm. C. Brown Communications, Inc., Dubuque, Iowa. All Rights Reserved. Reprinted by permission. **Figures 18.4, 18.7, 18.9:** From John W. Hole, Jr., *Human Anatomy and Physiology*, 5th ed. Copyright © 1990 Wm. C. Brown Communications, Inc., Dubuque, Iowa. All Rights Reserved. Reprinted by permission. **Figures 18.5, 18.8, 18.11 top, 18.12:** From Kent M. Van De Graaff, *Human Anatomy*, 3d ed. Copyright © 1992 Wm. C. Brown Communications, Inc., Dubuque, Iowa. All Rights Reserved. Reprinted by permission. **Figure 18.11 bottom:** From Stuart Ira Fox, *Human Physiology*, 3d ed. Copyright © 1990 Wm. C. Brown Communications, Inc., Dubuque, Iowa. All Rights Reserved. Reprinted by permission. **Figure 18.14:** From Kent M. Van De Graaff and Stuart Ira Fox, *Concepts of Human Anatomy and Physiology*, 3d ed. Copyright © 1992 Wm. C. Brown Communications, Inc., Dubuque, Iowa. All Rights Reserved. Reprinted by permission.

Chapter 20

Figure 20.2: From Stuart Ira Fox, *Human Physiology*, 3d ed. Copyright © 1990 Wm. C. Brown Communications, Inc., Dubuque, Iowa. All Rights Reserved. Reprinted by permission. **Figures 20.6, 20.13, 20.16, 20.17:** From John W. Hole, Jr., *Human Anatomy and Physiology*, 5th ed. Copyright © 1990 Wm. C. Brown Communications, Dubuque, Iowa. All Rights Reserved. Reprinted by permission. **Figure 20.10:** From John W. Hole, Jr., *Human Anatomy and Physiology*, 6th ed. Copyright © 1993 Wm. C. Brown Communications, Inc., Dubuque, Iowa. All Rights Reserved. Reprinted by permission. **Figure 20.11:** From Kent M. Van De Graaff and Stuart Ira Fox, *Concepts of Human Anatomy and Physiology*, 3d ed. Copyright © 1992 Wm. C. Brown Communications, Inc., Dubuque, Iowa. All Rights Reserved. Reprinted by permission. **Figure 20.15:** From Kent M. Van De Graaff and Stuart Ira Fox, *Concepts of Human Anatomy and Physiology*, 2d ed. Copyright © 1989 Wm. C. Brown Communications, Inc., Dubuque, Iowa. All Rights Reserved. Reprinted by permission. **Figure 20.18:** Reproduced with permission. Eugene Morkin, "Contractile Proteins of the Heart," *Hospital Practice*, Volume 18, issue 6, page 98. Illustration by Bunji Tagawa.

Chapter 21

Figures 21.1, 21.4, 21.6b, 21.7, 21.8: From John W. Hole, Jr., *Human Anatomy and Physiology*, 5th ed. Copyright © 1990 Wm. C. Brown Communications, Dubuque, Iowa. All Rights Reserved. Reprinted by permission. **Figure 21.6a:** From Kent M. Van De Graaff and Stuart Ira Fox, *Concepts of Human Anatomy and Physiology*, 3d ed. Copyright © 1992 Wm. C. Brown Communications, Inc., Dubuque, Iowa. All Rights Reserved. Reprinted by permission. **Figure 21.8k:** From Kent M. Van De Graaff, *Human Anatomy*, 3d ed. Copyright © 1992 Wm. C. Brown Communications, Inc., Dubuque, Iowa. All Rights Reserved. Reprinted by permission. **Figure 21.9:** From Stuart Ira Fox, *Human Physiology*, 4th ed. Copyright © 1993 Wm. C. Brown Communications, Inc., Dubuque, Iowa. All Rights Reserved. Reprinted by permission. **Figure 21.11:** From Ross M. Durham, *Human Physiology*. Copyright © 1989 Wm. C. Brown Communications, Inc., Dubuque, Iowa. All Rights Reserved. Reprinted by permission.

Chapter 23

Figure 23.11: From Leland G. Johnson, *Biology*, 2d ed. Copyright © 1987 Wm. C. Brown Communications, Inc., Dubuque, Iowa. All Rights Reserved. Reprinted by permission.

Chapter 24

Figures 24.6, 24.12b: From Leland G. Johnson, *Biology*, 2d ed. Copyright © 1987 Wm. C. Brown Communications, Inc., Dubuque, Iowa. All Rights Reserved. Reprinted by permission.

Chapter 25

Figure 25.4: Courtesy of Carolina Biological Supply Company.

Chapter 26

Figure 26.1b: From Leland G. Johnson, *Biology*, 2d ed. Copyright © 1987 Wm. C. Brown Communications, Inc., Dubuque, Iowa. All Rights Reserved. Reprinted by permission.

I N D E X

An *italicized* page number indicates where the term is defined in the chapter glossary. A **boldface** page number indicates where the term appeared boldfaced in the text. Reference to figures, tables, and boxes are designated after the page number.

Author and Consultant

ALAN CACKETT is a freelance journalist best known for his contributions to *Country Music People*. In the late '60s he edited and published Britain's first monthly country music magazine – *Country Record Exchange* and *Country Music Monthly*. He has contributed a regular country music column for the *Kent Evening Post/Kent Today* and has also written for *Record Mirror* and *Sounds*, as well as providing pop music and country music columns for the Kent Messenger Group of publications. For the past seven years Alan has been promoting country music shows and festivals, featuring both UK and US talent, through his company Good 'N' Country. He contributed to the last edition of this encyclopedia and continues to be one of Britain's most informed music journalists, with strong connections to the US country scene.

Special Consultant

ALEC FOEGE is a contributing editor at *SPIN* Magazine, for which he has covered music. A former senior editor at *SPIN*, he has also written for many other publications, including *Rolling Stone*, *Variety*, *Vogue* and *New York Woman*. He lives in New York, where he recently wrote his first book.

Contributors

FRED DELLAR is one of the busiest journalists on the British music scene. He began his career providing rock and country articles for *Audio Record Review* (now *Hi-Fi News*) in 1968 and has since contributed to a huge number of publications, though he is probably best known for his Fred Fact column in *New Musical Express*.

DOUGLAS B. GREEN is one of America's leading country music historians. Once head of the Oral History Project at the Country Music Foundation in Nashville, he has written several well-received books and has contributed to some two score music publications, including liner notes for several country albums. He heads a band of his own, Riders In The Sky, and is a member of the Grand Ole Opry.

ROY THOMPSON is a country music fanatic who met Fred Dellar while working for a publishing company. Roy, who owns one of the largest collections of country records in England, has worked with Fred on a number of musical projects.

Foreword

Country music is the music of the past and the present. It has a lot to say to people of all ages and from all walks of life. At one time, country's core fans were much older than today's audiences, which more closely resembles a crowd of young rock and roll fans. They, of course, represent the future of country music. Yet, today's country music is really not all that different from the country music of yesteryear. You can still hear the influences of the singing cowboys from western movies – the lonesome yodel of Jimmie Rodgers, the heartache of Hank Williams and the western swing of Bob Wills. All of them are there in our country music of today.

Country music is all about communication, that ability to move the listener, whether it be to tears with the sad songs of real life or to whoop it up with the kickin' beat of a Texas dance hall. And that's the way it's always been with country music. Our music breaks down all kinds of barriers as it crosses international boundaries, bringing the people of the world together in an appreciation of the music.

In the pages of this reference work, you will come across singers, musicians and performers from all around the world who have helped to make country music truly international. It's a music that has absorbed the sounds and traditions of such far flung countries as Ireland, Scotland, Czechoslovakia, Germany, Mexico, England, Australia and many others.

For more than 200 years America has been home to immigrants from all corners of the world who come in search of the American Dream. Nowhere is this dream more apparent than within country music. For years, aspiring writers, singers and musicians have flocked to Nashville in the hope of gaining fame and fortune. I'm living proof that it can happen, that those dreams really can come true. This book has stories of others who not only sought that pot of gold, but eventually found it. They have all enriched our musical heritage with their own styles, which over the years has evolved into the whole of country music.

I believe that this book brings that rich heritage into perspective, historically and musically, and is essential reading for anyone with an appreciation of country music.

Introduction

It has been 18 years since the first edition of this encyclopedia was published, and since that time the fortunes of country music have undergone many changes. At the time of writing this the music is enjoying a boom in America with country music record sales at an all-time high, and several of the major stars registering high placings on the American pop charts.

I guess the biggest phenomenon of the early '90s has to be Garth Brooks, who has amassed record sales exceeding 30 million albums, and set new attendance records virtually everywhere he has appeared. Garth's success in both the American pop and country charts has opened the door for others, and so we have witnessed many other country names, including Reba McEntire, Wynonna, Travis Tritt, Billy Ray Cyrus, George Strait, Brooks & Dunn, etc, storming the American pop charts.

All of these relatively new stars owe a great debt to the pioneers of country music. Reba will be the first to pay tribute to such stars of the past as Patsy Cline, Loretta Lynn and Ray Price. Garth openly admits it was George Strait who opened his ears to country music, and that George Jones is the king.

Going back further, both George Jones and George Strait owe their start to such legends as Roy Acuff and Merle Haggard respectively. So the country music story is on-going, and within the pages of this book I have attempted to document the history of country music; from its early beginnings of the very first commercial country music recordings of the '20s through the early days of the Nashville Sound of the '50s to the revitalized New Traditional Sound of the '80s and then on to the future stars of the '90s.

It is a rich heritage of sounds and styles; bluegrass to comedy, smooth pop-country balladry to hard-core honky-tonk, Texas swing to Louisiana Cajun, West Coast country to cowboy ballads. These and many other facets of country music are covered within the pages of this book.

Major artists are listed in alphabetical order, classified under their surnames, while groups are listed under their full title – The Kentucky HeadHunters, for instance, being logged under 'K'. The past few years has seen the emergence of many new names into country music, so it has been necessary to enlarge the all-embracing appendix, which lists performers who, though making important contributions to the history of country music, were deemed not of the stature to be included in the main body of the book. Also included in the appendix is information on such items as the Country Music Association, important centres such as Austin, musical styles and other relevant information to the history of country music.

The emphasis of this book is placed heavily on the musicians, writers and singers who have made country music a worldwide phenomenon of cultural and commercial standing. Major hit singles and albums are noted, plus details of awards, whether it be for sales or from the various governing bodies such as the CMA or the Academy of Country Music.

Some choose to claim a huge array of hit singles in their press releases; a 'nationwide No.1' in one singer's handout probably relates to a record that merely went Top 20! To keep accuracy and consistency to the fore, unless otherwise stated, the chart references all relate to *Billboard* magazine's country music charts, which I feel is the most authentic and accurate of its type. I have also used the *Billboard* pop and rock charts for any references to country music artists and records that have 'crossed over'.

Since the last edition of the encyclopedia in 1986 much has changed. Many of the singers and musicians whose lives were documented have sadly gone, while others have taken their places. This has resulted in a completely revised and newly written book. I thank Sandie Paine for her exhaustive research work, and I am also indebted to the invaluable input of Fred Dellar, Roy Thompson, Douglas B. Green and David Redshaw to the original edition, which has formed the basis for much of this new edition.

I am extremely proud of the results. We have not achieved perfection, because I do not think any such thing is possible on such an all-embracing project. For instance, it is difficult to even pin down definite dates of birth because some people will not reveal such details, while others just downright lie about their time of arrival. This is very much the case with many of the new young stars currently emerging on the scene. Image is all important to their career standing, and sometimes it is felt that truth about age might damage that image.

Despite providing the most comprehensive country music encyclopedia it is possible to devise, perfection is not claimed. In fact I look to you to supply any details that you feel should be revised, and information about any artist you feel may have been neglected.

Alan Cackett

Roy Acuff

Son of a Baptist minister, Roy Claxton Acuff was born September 15, 1903, in a three-room shack in Maynardsville, Tennessee. As a child, Roy learnt jew's-harp and harmonica. However, it seemed that he was destined to become an athlete. Following a move to Fountain City, near Knoxville, Acuff (at that time nicknamed 'Rabbit' because he weighed only 130 pounds) gained 13 letters at high school, eventually playing minor league ball and being considered for the New York Yankees. Severe sunstroke put an end to this career, confining Acuff to bed for much of 1929 and 1930.

Following this illness, Acuff, whose jobs had included that of callboy on the L&N Railroad, hung around the house, learning fiddle and listening to records by old-time players – also becoming adept with a yoyo.

In spring 1932, he joined a travelling medicine show, led by a Dr Haver, playing small towns in Virginia and Tennessee. By 1933 he formed a group, the Tennessee Crackerjacks, in which Clell Summey played dobro, thus providing the distinctive sound that came to be associated with Acuff (Pete 'Bashful Brother Oswald' Kirby providing the dobro chords in later Acuff aggregations). Soon he obtained a programme on Knoxville radio station WROL, moving on to the rival KNOX for the Mid-day Merry-Go-Round show. On being refused a raise (each musician received fifty cents per show) the band returned to WROL once more, adopting the name of the Crazy Tennesseans.

Acuff married Mildred Douglas in 1936, that same year recording two sessions for ARC (a company controlling a host of labels, later merged with Columbia). Tracks from these sessions included **Great Speckled Bird** and **Wabash Cannonball**, two classic items, the latter having a vocal by Dynamite Hatcher.

Making his first appearance on the Grand Ole Opry in 1938, Acuff soon became a regular on the show, changing the name of the band once more to the Smoky Mountain Boys. He won many friends with his sincere, mountain-boy vocal style and his dobro-flavoured band sound, and eventually became as popular as Uncle Dave Macon, who was the Opry main attraction at the time.

In 1942, together with songwriter Fred Rose, Acuff organized Acuff-Rose, a music publishing company destined to become one of the most important in country music. During that same period, Acuff's recordings became so popular that he headed Frank Sinatra in some major music polls and reportedly caused Japanese troops to yell 'To hell with Roosevelt, to hell with Babe Ruth, to hell with Roy Acuff' as they banzai-charged at Okinawa. The war years also saw some of his biggest hits, including **Wreck On The Highway** (1942), **Fireball Mail** (1942), **Night Train to Memphis** (1943), **Pins And Needles** and **Low And Lonely** (1944).

Nominated to run as governor of Tennessee in 1944 and 1946, Acuff failed to get past the primaries. But in 1948 he won the Republican primary, although failed to win the ensuing election. Nevertheless he gained tremendous support, earning a larger slice of the vote than any previous Republican candidate had ever earned in that particular political confrontation. Also in 1948, Acuff opened his Dunbar Cave resort, a popular folk music park, which he owned for several years.

Four years later, after being requested to change his style by Columbia, Acuff left the label, switching in turn to MGM, Decca and Capitol. And though his live performances still continued to go well and his publishing empire seemed ever-expanding, Acuff's record sales failed to maintain their previous high – **So Many Times** (1959), **Come And Knock** (1959) and **Freight Train Blues** (1965), all on his own Hickory label, being the only releases to create any real interest during the '50s and '60s.

However, his tremendous contribution to country music was recognized in November 1962, when Acuff became the first living musician to be honoured as a member of the Country Music Hall Of Fame. Known as the 'King of Country Music', Roy Acuff has sold more than 30 million records throughout the years – his most successful disc being his Columbia version of **Wabash Cannonball**, which went gold in 1942. His film appearances include 'Grand Ole Opry' (1940), 'Hi Neighbor' (1942), 'My Darling Clementine' (1943), 'Sing, Neighbor, Sing' (1944), 'Cowboy Canteen' (1944) and 'Night Train To Memphis' (1946).

Severely injured in a road accident during 1965, Acuff was back and touring within a few months, at one stage making several visits to the Vietnam War front. On May 24, 1973, he entertained returned POWs at the White House and, on March 16, 1974, was chosen to provide the President with yoyo lessons at the opening of the new Nashville Opryhouse, an incident which Acuff considered to be one of the high points in his career.

He guested on the Nitty Gritty Dirt Band's triple album set **Will The Circle Be Unbroken?** in 1972, lending credence to contemporary and country-rock music. He continued to appear regularly on the Grand Ole Opry throughout the '70s and '80s, but cut down on his previously extensive touring schedule, until by the early '90s his only appearances were infrequent guest spots at Opryland. He died on November 23, 1992, following a short illness, and was buried just four hours later. He had requested a swift service and burial because he did not want his funeral turned into a circus.

Above: The King of Country Music, Roy Acuff, died in 1992, after a career spanning over half a century. He was the first living member of the Country Music Hall Of Fame.

Recommended:

Roy Acuff And His Smoky Mountain Boys (Capitol/–)
How Beautiful Heaven Must Be (Pickwick/–)
King Of Country (Hickory/–)
Smoky Mountain Memories (Hickory/DJM)
That's Country (Hickory/–)
Columbia Historic Edition (Columbia/–)
Steamboat Whistle Blues (Rounder/–)
The Great Roy Acuff (–/Stetson)
Two Different Worlds (–/Sundown)

Time, Roy Acuff, a 1971 offering. Courtesy Hickory Records.

Country Music Hall Of Fame, Roy Acuff. Courtesy Hickory Records.

Alabama

Jeff Cook (born August 27, 1949, Fort Wayne, Alabama), lead guitar, keyboards, fiddle; Teddy Gentry (born January 22, 1952, Alabama), bass; Randy Owen (born December 13, 1949), guitar; Mark Herndon, drums.

This American country-rock group has been one of the most successful country acts of recent years, with the majority of their singles hitting No.1 on the country charts, and all albums having reached gold or platinum status. Initially formed in 1969 at Fort Wayne, Alabama, as Wildcountry, the group was a semi-professional outfit with the nucleus of cousins Jeff Cook and Randy Owen, plus Teddy Gentry. They turned fully professional in 1973 when they landed a club residency in Myrtle Beach, South Carolina.

Southern Star, Alabama. Courtesy RCA Records.

By this time they had started writing songs and Teddy Gentry's **I May Never Be Your Love, But I'll Always Be Your Friend**, was recorded by Bobby G. Rice and made the country charts in 1975. The band had recorded for small labels as Wildcountry in the mid-'70s and made the name change to Alabama when they signed to GRT Records at the beginning of 1977, making their first mark on the country charts with **I Want To Be With You**.

At this time Alabama were undertaking tours all across the southern states, but without a major record deal or hit singles they were struggling. In 1976 original drummer John Vartanian decided to quit, and the group spent several months as a three-piece until they found Mark Herndon, the fourth member of Alabama. Larry McBride, a Dallas businessman, took an interest in the group and signed them to a management deal. He set up MDJ Records and the group's first record, **I Wanna Come Over**, made the country charts in the autumn of 1979. Under the production of Harold Shedd they came up with another hit, **My Home's In Alabama**.

The group signed with RCA Records at the beginning of 1980 and hit the top of the country charts with **Tennessee River**, following up with **Why Lady Why**, **Feels So Right** (also Pop Top 20), **Mountain Music**, **Love In The First Degree** (also Pop Top 20), **The Closer You Get**, **40 Hour Week**, **Take Me Down** (also Pop Top 20), **Can't Keep A Good Man Down**, **Song Of The South** and **Jukebox In My Mind**. All albums have gone gold and many have gone multi-platinum.

Alabama have been named CMA Group Of The Year consistently since 1981, and also won Entertainer Of The Year in 1982, 1983 and 1984. In 1986 they joined pop-soul singer Lionel Ritchie in the studio and sang harmonies on **Deep River Woman**, a major pop-country hit. Alabama's commercial success changed the thinking in Nashville away from the solo performer. They have created the group sound rather than a singer accompanied by a group of musicians and set things in motion for other outfits such as Atlanta, Exile and Bandana, and, later, Restless Heart, Confederate Railroad, Desert Rose Band and Kentucky HeadHunters.

Though they could have turned their back on country music, Alabama are keen to retain their country connection. This is reflected in their song content, instrumental arrangements and overall musical presentation.

Below: Alabama created the group sound in country music during the '70s, leading the way for such '90s bands as Confederate Railroad and the Desert Rose Band.

Recommended:
The Closer You Get (RCA/–)
Feels So Right (RCA/–)
Mountain Music (RCA/–)
Southern Star (RCA/–)
Pass It On Down (RCA/–)
American Pride (RCA/–)

Deborah Allen

One of Nashville's leading songwriters, Deborah Allen was born Deborah Lynn Thurmond on September 30, 1953 in Memphis, Tennessee, and always dreamt of being a country singer. The black-haired beauty first made the move to Nashville in 1972, worked as a waitress, then landed a job at Opryland, where she sang and danced in the chorus line. Two years later she was in Los Angeles, working on Jim Stafford's TV Variety Show.

She made a move back to Nashville in 1977 and signed a writer's contract with MCA Music, providing hit songs for such stars as: Janie Fricke, who scored her first No.1 country hit with Deborah's **Don't Worry 'Bout Me Baby**; John Conlee with **I'm Only In It For The Love**; Tanya Tucker with **Can I See You Tonight**; and Patty Loveless with **Hurt Me Bad (In a Real Good Way)**.

Deborah first made an impression on the record-buying public during 1979 when her vocals were added to already-recorded tracks by the late Jim Reeves, resulting in a trio of Top 10 singles. This led to a recording contract with Capitol for whom she recorded the album, **Trouble In Mind**, a highly acclaimed mix of country, folk and gospel, and scored Top 20 hits with **Nobody's Fool** and **You (Make Me Wonder Why)**.

In 1983 she joined RCA and immediately came up with the crossover hit, **Baby I Lied**, which she co-wrote with Rafe Van Hoy, whom she married in 1982. Van Hoy was also Deborah's producer and her first RCA album, **Cheat The Night**, contained the further Top 10 hits **I've Been Wrong Before** and **I Hurt For You**. Her next album, **Let Me Be The First**, was the first all-digitally produced Nashville album in history. It received overwhelming critical acclaim, and Allen was touted as a major star of the future. Instead RCA went through traumatic executive changes, and when her contract was up for renewal in 1988, Deborah declined and found herself in an artistic wilderness.

She continued with her writing and singing on demo sessions, then late in 1992 she landed a new record deal with Giant, and produced **Delta Dreamland**, a superb album of self-composed songs with a distinctive nod to her rich Memphis musical heritage blending in country, R&B, pop and gospel influences.

Recommended:
Trouble In Mind (Capitol/–)
Cheat The Night (RCA/–)
Delta Dreamland (Giant/–)

Rex Allen

Known as the 'Arizona Cowboy', Allen was born Willcox, Arizona, December 31, 1924. A rodeo rider in his teens, he learnt to play guitar and fiddle at an early age. He took an electronics course at University College of Los Angeles but opted instead for a singing career, finding his first job with radio station WTTM, Trenton, New Jersey, during the mid '40s. Like Gene Autry before him, he was a popular singer on the NBD before entering films. In 1951 he was awarded his own Hollywood radio show by CBS, subsequently getting high ratings.

He has since recorded for Decca, Mercury, Buena Vista and others, and has made films for Fox, Republic and Universal, having the distinction of being the last of the singing cowboys on screen. He had a Top 20 hit with **Don't Go Near The Indians** in September 1962, having previously won a gold disc for his version of **Crying In The Chapel** in 1953.

He is perhaps best known nowadays for his singing and narration chores in various Disney productions.

Recommended:
Golden Songs Of The West (Vocalion/–)
Under Western Skies (–/Stetson)
Boney-Kneed, Hairy-Legged Cowboy Songs (–/Bear Family)
Mister Cowboy (–/Stetson)
Voice Of The West (–/Bear Family)

Rex Allen Jr

The son of singing cowboy Rex Allen was born on August 23, 1947 in Chicago and made quite an impression on the country scene with such hit singles as **Two Less Lonely People**, **Lonely Street** and **Me And My Broken Heart**. He started while still a youngster, playing rodeos and state fairs with his father.

A move to Nashville in the early '70s set him up for a promising career in country music, signing with Shelby Singleton's SSS International label. A change to Warner Brothers Records in 1973 started a long run of hit singles and distinctive albums, such as **Oklahoma Rose** and **The Singing Cowboys**, featuring guest appearances by his father and Roy Rogers. He moved to the smaller Moonshine Records in 1984 and produced the excellent album **On The Move** which, like many of his previous recordings, he co-produced. By the early '90s he had become a regular on the Statler Brothers Show on American cable television.

Recommended:
Ridin' High (Warner Bros/–)
Brand New (Warner Bros/Warner Bros)
Oklahoma Rose (Warner Bros/–)
Cats In The Cradle (Warner Bros/–)
On The Move (Moonshine/–)

Rosalie Allen

Known as 'The Prairie Star' in her heyday, Rosalie Allen was a great cowgirl yodeller in an era full of them. Actually, she was born Julie Marlene Bedra, the daughter of a Polish-born chiropractor, in Old Forge, Pennsylvania, on June 6, 1924.

Entranced by the cowboy image and music, she gained a radio spot with long-time New York City favourite Denver Darling in the late '30s, and through the '40s and '50s she was a fixture of the northeast states. She signed with RCA, her biggest solo efforts being yodelling spectaculars **I Want To Be A Cowboy's Sweetheart** and **He Taught Me How To Yodel**. She was frequently paired with

Elton Britt, another legendary yodeller, for a number of records.

She turned to a career as a disc jockey over WOV in the '50s, preferring not to travel, and gradually left her performing and recording career behind.

For several years she owned a record shop in New Jersey, specializing in country music. She currently lives in rural Alabama.

Allman Brothers Band

Although they are not a country outfit, this southern-rock band, formed by brothers Duane and Gregg Allman in 1969, has been influential with many country acts, especially the bands that came to the forefront in Nashville in the '80s and '90s. Originally an R&B based rock band, they started leaning more towards a melodic country-rock style following Duane's fatal motorcycle crash on October 29, 1971, which was typified in such numbers as the instrumental hit, **Jessica**, and **Ramblin' Man**. The latter became a small country hit for Gary Stewart in 1973. The band split in 1977, but have reunited on several occasions since, making it back into the rock charts with the **Seven Turns** album in 1990. The Allmans were the most important and influential of all the southern boogie outfits.

Bill Anderson

Nicknamed 'Whispering Bill' because of his lack of any real voice, Anderson was born Columbia, South Carolina, November 1, 1937. Training initially to be a journalist, he obtained a BA degree at the University of Georgia, singing and acting as a disc jockey in his spare time. Along the way, he worked as a sports writer for the weekly 'DeKalb New Era' and as correspondent for the 'Atlanta Journal', opting for a full-time musical career in 1958 after Ray Price heard him singing his self-penned **City Lights** on a car radio and promptly covered it, thus earning a gold disc.

Many other Anderson songs were then recorded by Hank Locklin, Jim Reeves, Porter Wagoner, Faron Young, Jean Shepard, and others, while his own discs (for Decca) also sold well. However, his real breakthrough as a recording artist came in 1962, with a cross-over hit **Mama Sang A Song**, this being followed by **Still** and **8 × 10**, both covered in Britain by singing comedian Ken Dodd. Since that time, Anderson has waxed a stream of hits including **I Get The Fever** (1966), **For Loving You** with Jan Howard (1967), **Wild Weekend** (1968), **Happy State Of Life** (1968). **My Life** (1969), **But You Know I Love You** (1969), **Quits** (1971), **The Corner Of My Life** (1973), and many others.

The Bill Anderson Story. Courtesy MCA Records.

Live From London, Bill Anderson. Courtesy MCA Records.

Anderson discovered Mary Lou Turner, who became another successful duet partner in the '70s (**Sometimes** and **I Can't Sleep With You**). When his records failed to make an impression, he turned to disco songs and came up with disco-country numbers such as **I Can't Wait Any Longer** and **Three Times A Lady**.

In recent years he has become a TV personality, starting with his own syndicated Bill Anderson Show in the '60s and leading up to game shows like Mister and Mrs. Bill is now an elder statesman of the Grand Ole Opry, where he regularly reprises his many hits for country fans, old and new.

Recommended:
The Bill Anderson Story (MCA/MCA)
Don't She Look Good? (MCA/–)
Always Remember (MCA/–)
Whispering Bill (MCA/–)
Live From London (–/MCA)
Ladies Choice (MCA/Bulldog)
Bright Lights And Country Music (Decca/Stetson)
Golden Greats (–/MCA)
Country Music Heaven (Curb/–)

John Anderson

This honky-tonk country singer with a rich bluesy vocal style was born December 12, 1955 in Apopka, Florida. He played in a rock'n'roll band called Living End while still in high school in Florida. One week after graduation he moved to Nashville and for two years he was singing in lounge bars with his older sister Donna. Anderson signed a recording contract with the small Ace Of Hearts label, releasing his first single **Swoop Down Sweet Jesus** in 1974 and landed a writer's contract with Al Gallico Publishing. For the next three years he tried his hand at various jobs during the day while working the Nashville clubs and bars. A move to Warner Bros Records in 1977 started a consistent run of chart successes with hard country songs like **The Girl At The End Of The Bar**, **Your Lying Blue Eyes** and the reflective **1959** in 1980.

Hailed as the new George Jones and Lefty Frizzell, John scored No.1 hits with **Wild And Blue**, **Swingin'**, a hardcore country blues number which crossed into the American pop charts in 1983, and **Black Sheep**. A further Top 3 hit with **She Sure Got Away With My Heart** in 1984 saw John's records start to miss the higher reaches, though he did make a return to the Top 10 with **Honky Tonk Crowd** in 1986. The following year he joined MCA Records and made a couple of good albums, but his only notable success came with **Somewhere Between Ragged And Right**, which featured Waylon Jennings on harmony vocals in 1988. Next he signed with Universal, and, although he was working with Bernie Taupin and Paul Kennerley and touring regularly, he failed to make an impact on the charts.

Above: Lynn Anderson has spent a lot of time collecting funds for charity.

In 1991 he signed with BNA Records, and he was back at the top of the charts with **Straight Tequila Night**. Taken from his **Seminole Wind** album, this went platinum in 1992 and Anderson was back as one of the finest country vocalists and writers on the contemporary scene. He worked with Mark Knopfler, who provided him with the song **When It Comes To You**; other guests on the album included Carl Jackson, Buddy Emmons and Dann Huff. Another track from the album, **Let Go Of The Stone** also hit No.1 in the charts.

Below: John Anderson made an impact in 1992 with Seminole Wind.

Recommended:
Wild And Blue (Warner Bros/Warner Bros)
I Just Came Home To Count The Memories (Warner Bros/–)
Best Of (Warner Bros/Warner Bros)
Eye Of A Hurricane (Warner Bros/–)
Tokyo, Oklahoma (Warner Bros/–)
Seminole Wind (BNA/–)
Solid Ground (BNA/–)
Blue Skies Again (MCA/–)

Lynn Anderson

Singer Lynn Rene Anderson, daughter of Casey and Liz Anderson, country performers of the '50s and '60s, was born in Grand Forks, North Dakota, September 26, 1947. Later, the family moved to Sacramento, where Lynn became an equestrian success.

During 1966 she joined Chart Records, recording around 100 songs for the label and producing 17 hits during the late '60s, the biggest of these being a cover version of Ben Peters' **That's A No No** (1969). After marrying producer-songwriter Glen Sutton, Lynn signed for Columbia Records, her first single for the label being **Stay There Till I Get There**, a hit written by her husband. In the same year (1970), Lynn's recording of **Rose Garden** became a monster hit winning her a Grammy and the CMA Female Vocalist Of The Year awards.

Other hits include **You're My Man**, **Keep Me In Mind**, **Top Of The World**, **I've Never Loved Anyone More** and **Isn't It Always Love**. She and Sutton were divorced in the mid-'70s and for several years Lynn concentrated on her horse-riding skills. In 1983 she signed to Permain Records and made a return to the Top 10 with **You're Welcome Tonight**, a duet with Gary Morris. This was followed by a short stay with Mercury, for whom she recorded a few minor hits, including a revival of the Drifters' **Under The Boardwalk** in 1988.

Below: Lynn Anderson comes from a country music family.

Recommended:
Rose Garden (Columbia/CBS)
What A Man My Man Is (Columbia/CBS)
I've Never Loved Anyone More (Columbia/CBS)
Outlaw Is Just A State Of Mind (Columbia/CBS)
All The King's Horses (Columbia/–)
Rose Garden/How Can I Unlove You? (Columbia/–)
Country Girl (–/Embassy)
Lynn Anderson Is Back (Permain/–)
Country Store (–/Starblend)

Eddy Arnold

A country crooner with a smooth, very commercial voice, Arnold has probably sold more records than any other C&W artist. Born on a farm near Henderson, Tennessee, on May 15, 1918, Richard Edward Arnold first became interested in music while at elementary school, his father – an old-timer fiddler – teaching him guitar at the age of ten.

Arnold left high school during the early '30s to help his family run their farm. During this period he played at local barn dances, sometimes travelling to such dates on the back of a mule. He made his radio debut in Jackson, Tennessee during 1936, six years later gaining a regular spot on Jackson station WTJS. His big break came as singer/guitarist with Pee Wee King's Golden West Cowboys providing exposure on Grand Ole Opry.

As a solo act he signed for RCA in 1944, sparking off an amazing tally of hit records with **It's A Sin** and **I'll Hold You In My Heart** in 1947, the latter becoming a million-seller. This achievement was matched by later Arnold recordings: **Bouquet Of Roses, Anytime, Just A Little Lovin' Will Go A Long Long Way** (1948); **I Wanna Play House With You** (1951) and **Cattle Call** (1955), while many others sold nearly as many.

Arnold's records sold to people who normally bought straight pop, so his TV appearances were not confined to just Grand Ole Opry and country shows; he guested on programmes hosted by Perry Como, Milton Berle, Arthur Godfrey, Dinah Shore, Bob Hope, Spike Jones and other showbiz personalities. Arnold also had his own syndicated TV series, Eddy Arnold Time, plus other shows on NBC and ABC networks.

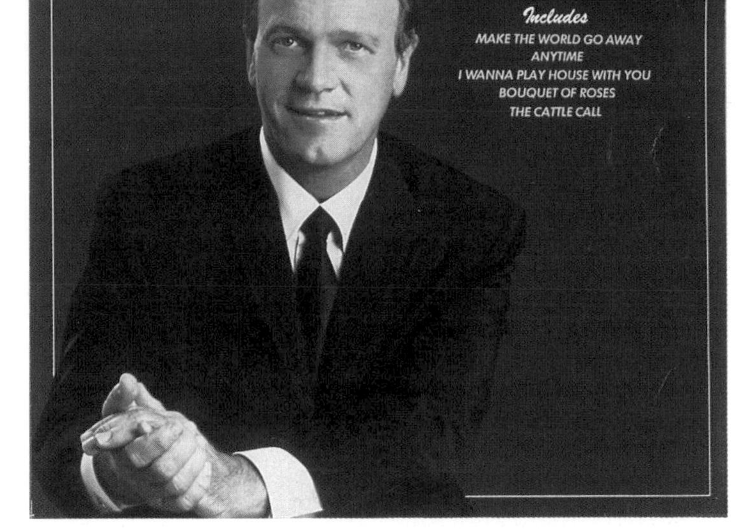

Above: 20 Of The Best, Eddy Arnold. Courtesy RCA Records. The album contains some of his finest – including Cattle Call, Make The World Go Away and Bouquet Of Roses.

Eddy. Courtesy RCA Records. A fine Owens Bradley-produced album.

Though his record sales dropped slightly by the end of the '50s. **What's He Doing In My World? Make The World Go Away**, **I Want To Go With You**, **Somebody Like Me**, **Lonely Again**, **Turn The World Around** and **Then You Can Tell Me Goodbye**, all topped the country charts during the '60s. Arnold, nicknamed 'The Tennessee Plowboy', had sold well in excess of 60 million discs.

In 1966, Arnold was elected to the Country Music Hall Of Fame.

The Best Of Eddy Arnold. Courtesy RCA Records.

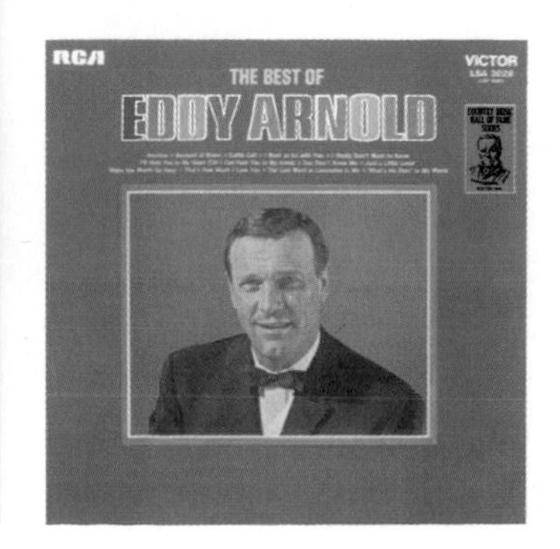

Recommended:
Cattle Call (RCA/Bear Family)
So Many Ways/If The World Stopped Turnin' (MGM/–)
Country Gold (RCA/–)
Pure Gold (RCA/–)
Eddy (RCA/RCA)
A Legend And His Lady (RCA/–)
Famous Country Music Makers (–/RCA)
Hand Holding Songs (RCA/–)
Last Of The Love Song Singers (RCA/–)

Ernie Ashworth

Born in Huntsville, Alabama on December 15, 1928, Ernie Ashworth played guitar and sang on local radio stations in his teens, later moving to Nashville and working for station WSIX. Carl Smith and Little Jimmy Dickens were among those to record Ashworth's songs, prompting MGM to sign him to a recording contract in 1955, a number of the discs being cut for the label under the name of Billy Worth. Wesley Rose, who had initiated the MGM contract, won Ashworth another deal, this time with Decca, his first release **Each Moment (Spent With You)** becoming a major hit in May 1960. After further successes with **You Can't Pick A Rose In December** and **Forever Gone** came a move to Hickory Records, the hits continuing via **Everybody But Me** (1962), his first Hickory single, and **Talk Back Trembling Lips**, a No.1 in the 1963 country charts.

A consistent hit-maker throughout the '60s, Ashworth's **A Week In The Country**, **I Love To Dance With Annie** and **The DJ Cried**, all figured in the country charts.

Asleep At The Wheel

Wheel, a western swing unit, began life on vocalist-guitarist Ray Benson's rent-free 1,500-acre farm, near Paw Paw, West Virginia, where he, Leroy Preston (vocals, guitar) and Reuben Gosfield often called 'Lucky Oceans' (pedal steel guitar) formed a small country band. With various changes, the band gained further shape when female vocalist Chris O'Connell, just out of high school, became the fourth permanent member. Following a move to San Francisco, pianist Floyd Domino joined, bringing jazz influence to the band.

Above: Asleep At The Wheel's fiddle sound recreated '40s western swing.

The first UA album by the Wheel was near country in character but subsequent albums for Epic, Capitol, MCA have seen the band employing more diverse material, linking country, R&B and jazz in best western swing tradition, often employing contributions by guesting ex-Texas Playboy Johnny Gimble. They won a Grammy for their version of Count Basie's **One O'Clock Jump** in 1978, and in 1980 appeared on the soundtrack of 'Roadie'.

They do what they do and it's often full of surprises. In fact the only thing about the band that can be guaranteed is that the Wheel will feature a changed personnel next time you hear them – at the last count some 60 musicians claim to have worked with the band at one point or another!

Recommended:
Ten (Epic/Epic)
Alive And Kicking (Arista/Arista)
Western Standard Time (Epic/Epic)
Comin' Right At Ya (UA/–)
Texas Gold (Capitol/Capitol)
Wheelin' And Dealin' (Capitol/Capitol)
Collision Course (Capitol/Capitol)
Pasture Prime (MCA Dot/Demon)

Collision Course, Asleep At The Wheel. Courtesy Capitol Records.

Bob Atcher

Robert Owen Atcher, born in Hardin County, Kentucky on May 11, 1914, grew up in North Dakota in a family of folk singers and championship fiddlers, and the duality of these locations gave him a broad knowledge of both traditional folk songs and cowboy songs, although he proved to be a very commercial country singer during his long association with Columbia Records (1937–58).

Bob's clear tenor voice was a fixture and a cornerstone of the once-thriving country music scene in Chicago; he appeared there on WJJD and WBBM and on a host of network programmes from 1931–34 and 1937–48, then joined the National Barn Dance as its top star from 1948 right through to its demise in 1970.

He recorded for Kapp and Capitol as well as Columbia Records (his biggest hit for them was a comedy version of **Thinking Tonight Of My Blue Eyes**), wrote songs, pioneered television in Chicago, and appeared as a singing cowboy in several Columbia pictures.

Although he continued to perform, Atcher's interest gradually turned to the civic, and he spent nearly two decades as mayor of the Chicago suburb of Schaumburg.

His past albums have included **Early American Folk Songs**, **Songs Of The Saddle**, **Dean Of The Cowboy Singers** (all Columbia); **Bob Atcher's Best** (Harmony) and **Saturday Night At The Old Barn Dance** (Kapp).

Chet Atkins

Chester Burton Atkins was born June 20, 1924, on a 50-acre farm near Luttrell, Tennessee. His elder half-brother, Jim, was a proficient guitarist who later went on to play with Les Paul. At 18, Chet became fiddler on station WNOX in Knoxville, Tennessee, then toured with Archie Campbell and Bill Carlisle, playing fiddle and guitar. After marrying Leona Johnson in 1945, Atkins joined Red Foley in 1946, then moved to Nashville in 1950 with the Carter Sisters and Mother Maybelle, signing a record contract with RCA and initially cutting more vocals than instrumentals – the latter got the airplay.

He became top session guitarist for RCA's Nashville sessions at the end of the '40s, moving up to the post of A&R assistant under Steve Sholes in 1952. After

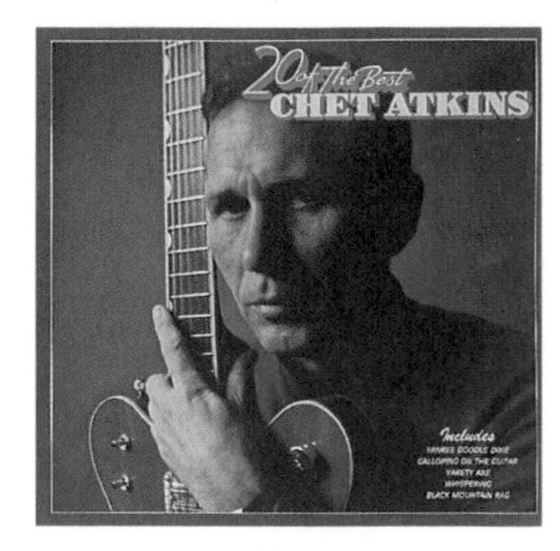

20 Of The Best, Chet Atkins. Courtesy RCA Records.

assisting Sholes on Presley's **Heartbreak Hotel** sessions in 1955. Atkins was placed in charge of the new RCA studio, becoming Nashville A&R manager in 1960 and vice-president of RCA Records just eight years later, holding the post until 1982, when he left the label.

A guitarist able to tackle many styles, Atkins has appeared at the Newport Jazz Festival and was featured soloist with the Atlanta Symphony Orchestra. He has produced hit records for Hank Snow, Waylon Jennings, Perry Como, Al Hirt and many, many others, and also had his own major hits with **Poor People Of Paris** (1956), **Boo Boo Stick Beat** (1959), **One Mint Julep** (1960), **Teensville** (1960) and **Yakety Axe** (1965), among others. Along with Floyd Cramer, Hank Garland and others, he is credited with creating the highly commercial Nashville Sound that brought pop acts scurrying to record in the area. For 14 consecutive years he won Best Instrumentalist award in the 'Cashbox' poll.

Above: Chet Atkins teamed with Mark Knopfler on his Neck And Neck album.

He signed with Columbia Records in 1983 and at that time added the initials C.G.P. (Country Guitar Picker) after his name. Though he had now returned to his first love of guitar picking, his recordings tended to lean towards Jazz/New Age. He made a big impact with his album **Stay Tuned**, which featured guest guitarists Larry Carlton, George Benson, Mark Knopfler and Earl Klugh. Further albums for Columbia, such as **Street Dreams**, **Sails** and **Chet Atkins C.G.P.**, all maintained a modern jazz sound. He achieved notable commercial success in 1991 with **Neck And Neck**, which found him teaming up once again with Mark Knopfler, who produced the album and also sang.

With more than 100 albums to his credit covering all styles of music and having guided the careers of Jerry Reed, Jim Reeves, Hank Locklin, etc, he is one of Nashville's leading country figures. He was elected to the Country Music Hall Of Fame in 1973 as the youngest inductee.

Recommended:

Country Pickin' (Camden/–)
Finger Pickin' Good (Camden/–)
For The Good Times (RCA/–)
Me And Jerry Reed (RCA/–)
Superpickers (RCA/RCA)
Chester And Lester (RCA/RCA)
Me And Chet Atkins with Jerry Reed (RCA/RCA)
Famous Country Music Makers (–/RCA)
Atkins-Travis Traveling Show (RCA/RCA)
Stay Tuned (Columbia/CBS)
C.G.P. (Columbia/Columbia)
Sneaking Around with Jerry Reed (Columbia/–)
Neck And Neck with Mark Knopfler (Columbia/Columbia)

Gene Autry

Orvon Gene Autry, the most successful of all singing cowboys to break into movies, was born in Tioga, Texas, September 29, 1907. Taught to play guitar by mother Elnora Ozment Autry, Gene joined the Fields Brothers Marvelous Medicine Show while still at high school, but after graduation in 1925 became a railroad telegrapher with the Frisco Railway at Sapulpar, Oklahoma. Encouraged by Will Rogers following a chance meeting, Autry took a job on radio KVOO, Tulsa, in 1930, billing himself as 'Oklahoma's Singing Cowboy', and singing much in the style of Jimmie Rodgers.

In 1929 he had visited New York and began recording with such labels as Victor, Okeh, Columbia, Grey Gull and Gennett (often under a pseudonym), sometimes working with Jimmy Long, a singer-songwriter-guitarist, once Autry's boss on the Frisco line. On many he was assisted by Frank and Johnny Marvin. Shortly after, Autry began broadcasting regularly on the WLS Barn Dance programme for Chicago, his popularity gaining further momentum with the 1931 release of **Silver Haired Daddy Of Mine** (penned by Autry and Long), a recording that eventually sold over five million copies.

Next came a move to Hollywood where, following a performance in a Ken Maynard western 'In Old Santa Fe', he was asked to star in a serial 'The Phantom Empire'. Thereafter, Autry appeared in innumerable B movies, usually with the horse, Champion. His list of hit records during the '30s and '40s – he was easily the most popular singer of the time – is awesome, including **Yellow Rose Of Texas** (1933), **The Last Roundup** (1934), **Tumbling Tumbleweeds** (1935), **Mexicali Rose** (1936), **Back In The Saddle Again** (1939), **South Of The Border** (1940), **You Are My Sunshine** (1941), **It Makes No Difference Now** (1941), **Be Honest With Me** (1941), **Tweedle-O-Twill** (1942) and **At Mail Call Today** (1945).

Autry enlisted in the Army Air Corps in July 1942 and became a pilot flying in the Far East and North Africa with Air Transport Command. Discharged on June 17, 1945, he formed a film company, continuing to star in such movies as 'Sioux City Sue' (1947), 'Guns And Saddles' (1949) and 'Last Of The Pony Riders' (1953). He also appeared in the long-running 'Melody Ranch' radio programme from 1939 until 1956.

His other activities included opening a chain of radio and TV stations and running a record company, a hotel chain and a music publishing firm, plus a major league baseball club, the California Angels. Since **Silver Haired Daddy**, Autry has had three other million-selling discs in **Here Comes Santa Claus** (1947), **Peter Cottontail** (1949) and the other nine million-seller **Rudolph The Red Nosed Reindeer** (1948).

Writor of scores of hit songs, Gene Autry has also starred at a series of annual rodeos held in Madison Square Garden and even had an Oklahoma town named after him. In 1969 he was elected to the Country Music Hall Of Fame.

Recommended:

Country Music Hall Of Fame (Columbia/–)
All American Cowboy (Republic/–)
Cowboy Hall Of Fame (Republic/–)
Favorites (Republic/Ember)
South Of The Border (Republic/Ember)
Live From Madison Square Garden (Republic/Ember)

Below: Singing cowboy Gene Autry appears here in a still from the movie 'The Last Roundup'.

Hoyt Axton

Born on March 25, 1938 in Comanche, Oklahoma, the son of teachers John Thomas Axton and his wife Mae – a lady who worked for Grand Ole Opry and wrote many fine songs including **Heartbreak Hotel**, Axton began playing guitar and singing in West Coast clubs during 1958.

After a brief spell in the Navy, he cut his first record **Follow The Drinking Gourd** in Nashville, along with sessionmen Jimmy Riddle (harmonica) and Grady Martin (guitar). Axton's **Greenback Dollar** became a hit when recorded by the Kingston Trio in 1963, while **The Pusher**, another Axton song, was heard by John Kay in 1964 and subsequently became an enormous success for Kay's rock group, Steppenwolf. Other Axton compositions, **Joy To The World** and **Never Been To Spain**, proved winners for Three Dog Night and **No No Song** scored for Ringo Starr in 1976.

While Axton was once thought of primarily as a folk singer, he is now considered as one of the finest writers in country music, his songs being recorded by Waylon Jennings, Tanya Tucker, John Denver, Glenn Yarborough, Lynn Anderson, Glen Campbell, Commander Cody and many others. He featured in the country charts during the late '70s with **Flash Of Fire**, **A Rusty Old Halo** and **Della And The Dealer**, the latter making an impression in Britain. Axton has also made his mark in films, appearing in 'The Black Stallion', 'Gremlins', and many more. He also has his own record company, Jeremiah, named after the bullfrog in the song **Joy To The World**.

Above: Hoyt Axton is probably best known as the songwriter of Joy To The World.

Recommended:
Life Machine (A&M/A&M)
Southbound (A&M/A&M)
A Rusty Old Halo (Jeremiah/Youngblood)
Spin Of The Wheel (DPI/–)
Road Songs (A&M/A&M)

Free Sailin', Hoyt Axton. Courtesy MCA Records.

The Bailes Brothers

Homer Bailes, fiddle, vocals; Johnny Bailes, guitar, vocals; Kyle Bailes, string bass, vocals; Walter Bailes, guitar, vocals.

Despite the number of Bailes Brothers members, the act of that name rarely consisted of all four brothers. The heart of the act, however, comprised Johnny and Walter, whose song writing and singing made them one of the most popular groups of the 1940s.

Johnny was actually the first to work professionally, teaming up with Red Sovine in 1937. By the time he had gone to Beckley, West Virginia (1939), he had not only acquired the services of Skeets Williamson but also those of his sister Laverne, who became known as Molly O'Day. Yet another band member was Little Jimmy Dickens. During the same period, Kyle and Walter were billing themselves as the Bailes Brothers; before long it was Walter and Johnny.

In 1942, Roy Acuff heard them and arranged for an audition – which was successful – for the Grand Ole Opry, a tenure which lasted through to 1948 when they joined the Louisiana Hayride. It was during the Opry years that they recorded many of the songs that made them famous: **Dust On The Bible**, **I Want To Be Loved**, **Remember Me**, **As Long As I Live** and others, mainly for the Columbia Records label.

The act eventually broke up in 1949, Homer and Walter entering the ministry, although both Johnny and Walter, and Homer and Kyle, worked as gospel duets intermittently in the '50s. By the late '70s, Kyle was in the air-conditioning business and Johnny was managing one of Webb Pierce's radio stations, leaving Walter, though actively involved in church work, as the other brother still performing on a semi-regular basis, doing gospel numbers (largely self-written) like his classic **Whiskey Is The Devil In Liquid Form**.

Recommended:
The Bailes Brothers: Johnny And Homer (Old Homestead/–)
I've Got My One Way Ticket (Old Homestead/–)

Razzy Bailey

A rough-voiced singer with a penchant for country blues, Rasie (later Razzy) Bailey was born on February 14, 1939 on a Five Points Alabama farm which had no running water or electricity. A back porch picker, he learned guitar and joined a string band sponsored by the Future Farmers Of America. He then became, in turn, a truck driver, an insurance salesman, a furniture company representative and a meat cutter for a butcher, trying to keep both a band and a family together, the former wanting to quit and his wife first seeking a divorce and then psychiatric help. In 1972, he recorded just one release for MGM and in 1975 he cut **Peanut Butter** for Capicorn, but nothing really jelled, though his marriage miraculously held together.

But in 1976, amid a contract hassle, he was told by a psychic that not only his contractual problems would be cleared up but another artist would record a Bailey song and that it would change his whole way of life. Soon after, Dickey Lee recorded Razzy's **9,999,999 Tears** and had a Top 5 hit. Another of the psychic's predictions came true when RCA signed Bailey in 1978 and his first single, **What Time Do You Have To Be Back In Heaven** went Top 20, followed by **Tonight She's Gonna Love Me (Like There Was No Tomorrow)** (1978), **If Love Had A Face**, **I Ain't Got No Business Doin' Business Today**, **I Can't Get Enough Of That** (all 1979) and **Too Old To Play Cowboys** (1980), his first chart-toppers coming with **Loving Up A Storm** and **I Keep Coming Back**, in 1980.

By 1981 he was 'Billboard' magazine's Country Singles Artist Of The Year, and played many major venues.

Many artists wait until their record sales plummet before they jump labels but Razzy switched from RCA to MCA while still a happy hit-maker in 1984. And at last his feel for things raunchy and bluesy seemed to be paying off. His last Top 20 hit for RCA was a version of Wilson Pickett's **In The Midnight Hour**, while his first album with his new label, **Cut From A Different Stone**, saw Bailey linking with soul guitar ace Steve Cropper to write several songs, also reprising Eddie Floyd's R&B classic **Knock On Wood**.

He continued his association with country-blues with the excellent **Blues Juice** album for King Snake Records in 1989, which featured a whole host of country and blues pickers lending their support to his raspy vocal work.

Makin' Friends, Razzy Bailey. Courtesy RCA Records.

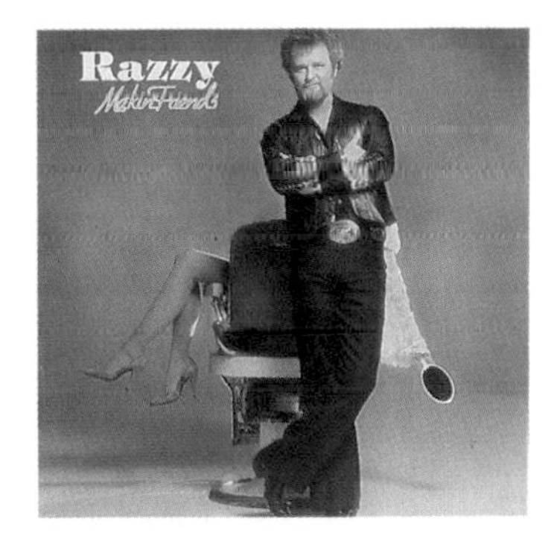

Above: People used to ask Razzy if he was a pop singer or an R&B artist. But he is really into the country blues.

Recommended:
Makin' Friends (RCA/RCA)
Razzy Bailey's Greatest Hits (RCA/–)
The Midnight Hour (RCA/–)
Cut From A Different Stone (MCA/–)
Blues Juice (King Snake/–)

Baillie And The Boys

This East Coast trio hit the big-time in Nashville in the late '80s after a twenty year wait in the wings. Michael Bonagura and Alan LeBoeuf teamed up in New Jersey in 1968 to form a band called London Fog, playing a mixture of originals and Top 40 covers. They met Kathie Baillie, a talented session singer, in 1973, and the trio provided back-up vocals for such diverse pop acts as Gladys Knight, Talking Heads and the Ramones. The trio split in 1977 when LeBoeuf landed a starring role in the Broadway show 'Beatlemania'. In the meantime, Kathie and Michael married, and, in 1981, with Alan back with them, the re-formed trio moved to Nashville, where they landed a job as back-up vocalists for Ed Bruce.

Their talents as writers, singers and musicians led to steady work in the studios and in 1984 they were signed as writers to Picalic Music. Two years later they landed a record deal with RCA, and the blend of three-part harmony, strong musicianship and impressive writing skills created such country Top 10 entries as **Oh Heart** (1987), **Wilder Days**, **Longshot** (1988), **She Deserves You**, **(I Wish I Had A) Heart Of Stone**, **Can't Turn The Tide** (1989) and **Fool Such As I** (1990).

LeBoeuf left again in 1989 and Kathy and Michael continued as a duo, utilizing back-up musicians for live shows. They put out a cover version of Fairground Attraction's **Perfect**, which became a minor country hit in 1990, and made it back into the Top 20 with **Treat Me Like A Stranger** (1991), but have done nothing chart-wise, since 1991.

The Lights Of Home, Baillie And The Boys. Courtesy RCA Records.

Recommended:
Turn The Tide (RCA/–)
The Lights Of Home (RCA/–)

Moe Bandy

A provider of undiluted honky-tonk, Bandy was born in Meridan, Mississippi, February 12, 1944, one of six children born to a piano-playing mother and a guitar-picking father. The family settled near San Antonio, Texas, when Moe was six, his father there organizing a band called the Mission City Playboys. A guitar player from an early age, Bandy initially preferred bronco-busting and opted for a rodeo career until a surfeit of broken bones convinced him to return to a less hazardous occupation. At the age of 19, he formed a group, Moe Bandy and the Mavericks, eventually gaining his first record contract with Satin, a local label. Then came stints with Shannon (during which Bandy first met producer Ray Baker) and GRC, the latter releasing **I Just Started Hating Cheatin' Songs Today** (1974), Bandy's first hit. The label released three more Top 20 hits – **It Was Always So Easy** (1974), **Don't Anybody Make Love At Home Anymore?** and **Bandy The Rodeo Clown** (1975), after which GRC folded and Moe moved on to Columbia and renewed success with **Hank Williams You Wrote My Life** (1975).

With Baker masterminding his career (acting as producer and also as song-finder, sometimes finding well over a thousand songs which were whittled down to a final 20), Bandy became a hit-making machine, logging two Top 20 songs in 1976, three in 1977, and three in 1978, peaking in 1979 when he provided five records including two chart-toppers, **Just Good Ol' Boys** (with Joe Stampley) and **I Cheated Me Right Out Of You**.

Still singing songs about seeing life through a bottle or of love gone sour, Moe has refused to cross over in search of pop success and throughout the '80s he continued to hit the charts with such honky-tonkers as **She's Not Really Cheatin' (She's Just Gettin' Even)**, **It Took A Lot Of Drinkin' (To Get That Woman Over Me)**, **Till I'm Too Old To Die Young**, **You Haven't Heard The Last Of Me** and **Daddy's Honky Tonk**. The latter was another duet with Joe Stampley, Bandy's partner on the 1984 Top 10 **Where's The Dress?**, a hilarious send-up of the pop transvestite scene.

A frequent award winner, Moe Bandy was also linked with Stampley at the 1980 CMA Awards ceremony; on this occasion the twosome won the Vocal Duo Of The Year category. **Americana**, a Top 10 success for Bandy in 1988, painted a patriotic look at how small town America has hardly changed over the years. Sadly, things have changed for Moe Bandy. He is no longer on a major label, but has established a nice niche in Branson, Missouri, where he has his own theatre, packing the country fans in with a healthy diet of good ole honky-tonk music.

Recommended:
Bandy The Rodeo Clown (GRC/–)
Hank Williams You Wrote My Life (Columbia/–)
Soft Lights And Hard Country Music (Columbia/CBS)
It's A Cheatin' Situation (Columbia/CBS)
Just Good Ol' Boys – with Joe Stampley (Columbia/CBS)
Many Mansions (MCA/MCA)
You Haven't Heard The Last Of Me (MCA/–)
No Regrets (Curb/–)
Live In Branson (Laserlight/–)

R. C. Bannon

From Dallas, Texas, Bannon initially moved into Nashville as a DJ, holding down a five year residency at the city's Smuggler's Inn. Prior to this, he had spent many years on the road, performing at clubs throughout the Southwest, at one point acting as warm-up act for Marty Robbins.

In Nashville he was introduced to songwriter Harlan Sanders who nudged Warner Brothers Music into signing Bannon as a writer. By 1977 he had gained a recording contract with Columbia, logging a trio of minor hits that year and a couple in 1978. But the big breakthrough came in 1979 when he married Louise Mandrell, the two of them also getting together on record for Epic and recording a number of chart duets, one of which, **Reunited**, went Top 20. However, his own solo career refused to move up a gear and 1980 brought only three further mid-chart singles, leaving Bannon searching for the right song to really establish himself as a headline act. He has, however, worked on arrangements for television and wrote **One Of A Kind Pair Of Fools**, a No.1 hit for Barbara Mandrell in 1983.

Recommended:
R. C. Bannon Arrives (Columbia/–)
Inseparable – with Louise Mandrell (Epic/–)

Bobby Bare

The provider of such million-selling singles as **Detroit City** and **500 Miles Away From Home** and one of country's ranconteurs supreme, Bare was born in Ironton, Ohio, April 7, 1935. Motherless at the age of five, his sister being sent for adoption because of the father's inability to feed the whole family, Bare became a farm worker at 15, later obtaining a job in a clothing factory. He built his own guitar and learned to play, eventually winning a job with a country band in the Springfield-Portsmouth area, for which he received no pay. He recorded his own song **All American Boy** in 1958, then joined the army, the tapes of his song being offered to various record companies and taken up by Fraternity, who released the disc as by 'Bill Parsons'. But although the single became the second biggest selling record in the USA in December, Bare hardly benefited financially, having sold the song rights for $50.

Upon service discharge, Bare began performing and writing once more, contributing three songs for the Jimmy Clanton-Chubby Checker movie 'Teenage Millionaire', a year later having his own hit record with an RCA release, **Shame On Me**. Richard Anthony then recorded a French version of **500 Miles Away From Home** (a folk song adapted and arranged by Bare, Hedy West and Charles Williams) and obtained a gold disc; Bare also achieved hit status with the same song. A year later, Bare appeared in an acting role in the cavalry western 'A Distant Trumpet' and also provided RCA with further hits via his versions of Hank Snow's **Miller's Cave** and Ian and Sylvia's **Four Strong Winds**.

After numerous other hit singles including **The Streets Of Baltimore** (1966) and **(Margie's At) The Lincoln Park Inn** (1966), Bare left RCA to join Mercury in 1970 supplying his new label with such Top 10 country hits as **How I Got To Memphis** (1970), **Come Sundown** (1970) and actually charting with every release. But he rejoined RCA (1972) and began working on a series of fine albums, commencing with one of his most successful, **Lullabies, Legends and Lies** (1973), an album penned almost entirely by Shel Silverstein and one that spawned such Top 10 hits as **Daddy What If** (1973) and **Marie Laveau**, a 1974 No.1. After an album with his whole family, **Singing In The Kitchen** (1974), he moved on to produce **Hard Time Hungries** (1975), another critically acclaimed LP. Other hit singles followed but, unhappy with RCA, Bare signed for Columbia in 1978, first cutting **Bare** (1979), a well-received, self-produced album. He then moved on to make **Sleeper Wherever I Fall**, reputed to have cost nearly $100,000, making it one of the most expensive albums ever to have been made in Nashville at that time.

Managed by rock mogul Bill Graham at this period, it seemed that Bare would once more cross over into the pop market. But, despite a number of other exemplary albums and a potentially chart-busting single in **Numbers** (1980), a Shel Silverstein-written parody of the movie '10', Bare's recording career gradually lost impetus and, despite the kudos gained by hosting the highly acclaimed Bobby Bare And Friends TV show on the Nashville Network. By the mid-'80s, Bare – still without any well-deserved CMA award – eventually signed to EMI America in 1985, but failed to score any major chart successes, and is now content to make regular tours of Europe.

Recommended:
Famous Country Music Makers (RCA/RCA)
Lullabies, Legends And Lies (RCA/RCA)
Hard Time Hungries (RCA/RCA)
Bare (Columbia/CBS)
Sleeper Wherever I Fall (Columbia/CBS)
Drinkin' From The Bottle, Singin' From The Heart (Columbia/CBS)
The Mercury Years (–/Bear Family)
Ain't Got Nothin' To Lose (Columbia/CBS)

Below: Bobby Bare is still commanding a loyal following in Europe, though less of one at home.

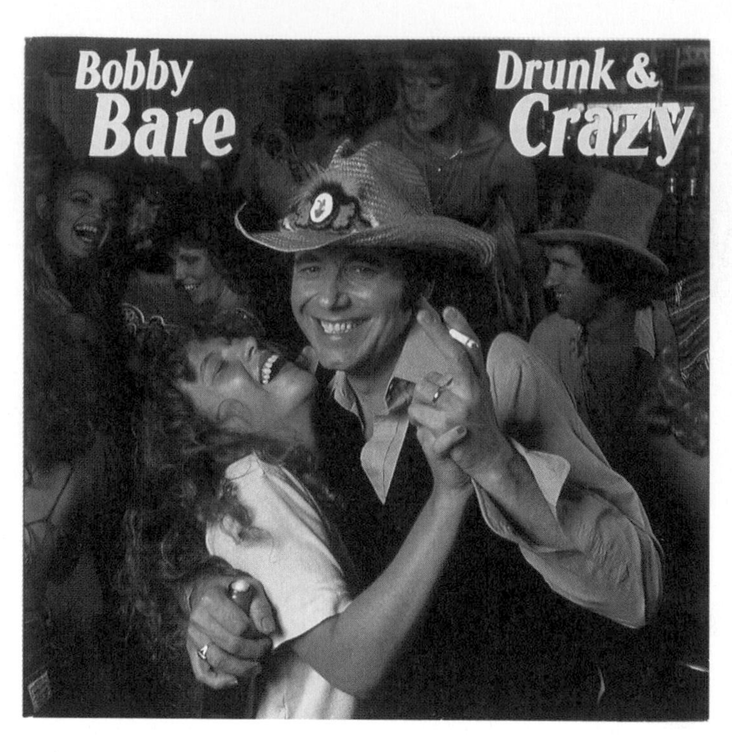

Drunk & Crazy, Bobby Bare. Courtesy CBS Records.

Dr Humphrey Bate (And The Possum Hunters)

Oscar Albright, bass; Alcyone Bate, vocals, ukelele, piano; Buster Bate, guitar, tipple, harmonica, jew's-harp; Dr Humphrey Bate, harmonica; Burt Hutcherson, guitar; Walter Leggett, banjo; Oscar Stone, fiddle; Staley Walton, guitar.

The harmonica-playing leader of the most popular string band on the Grand Ole Opry, Dr Bate – a graduate of the Vanderbilt Medical School who earned his living as a physician – was born in Summer County, Tennessee, in 1875. He had fronted a great many popular local string bands before he began to play on Nashville radio in 1925, joining the forerunner of the Opry late in that year. Not long after, they recorded a number of tunes and songs for Brunswick Records. After Dr Bate's death in 1936, Oscar Stone headed the band until 1949, when Staley Walton and Alcyone Bate took over the leadership of the Possum Hunters, whose remaining members were absorbed when four old-time Opry bands were amalgamated into two during the '60s. Bate's daughter, Alcyone Beasley, joined her father's band at the age of 13 as a vocalist, eventually becoming pianist with the Crook Brothers and logging well over half a century as an Opry member.

Above: The first string band to play the Opry – Dr Bate And The Possum Hunters.

Beausoleil

Tommy Alesi, drums; Jimmy Breaux, accordion; Tommy Comeaux, bass, mandolin; David Doucet, guitar, vocals; Michael Doucet, guitar, vocals; Billy Ware, percussion.

One of the most popular American Cajun bands, Beausoleil have probably found as much fame as it is possible to find by staying true to the tradition of their music. Formed by Michael Doucet, a superb fiddle player, excellent singer and knowledgeable scholar on Cajun music, Beausoleil have successfully combined the essence of traditional Cajun music with a contemporary edge.

Doucet came from a musical family in Lafayette, Louisiana, but it was a mish-mash of jazz, classical, French and big band that his relatives played. Though he grew up with Cajun music, it was not until he went to college at the Louisiana State University in the late '60s and studied American folklore, that he started taking Cajun music seriously. By the mid-'70s he was playing traditional music with Marc and Ann Savoy as the Savoy-Doucet Cajun Band. He later formed Beausoleil and has been successful in promoting Cajun music to an increasingly enthusiastic and ever-widening audience. Their music was featured in the 1987 movie 'The Big Easy', while Doucet has found time to be involved in other projects, including producing radio and TV programmes. The nucleus of the band backed up Mary-Chapin Carpenter on her 1991 hit recording **Down At The Twist And Shout**. They also appeared on that year's CMA Awards Show, backing up Carpenter. Beausoleil have recorded for various labels since the early '80s.

Bayou Cadillac, Beausoleil. Courtesy Rounder Records.

Recommended:
Dance De La Vie (Rhino/–)
Hot Chili Mama (Arhoolie/–)
Bayou Deluxe (Rhino/–)
Bayou Boogie (Rounder/–)
Bayou Cadillac (Rounder/–)

Carl Belew

Singer-songwriter and guitarist Carl Robert Belew was born in Salina, Oklahoma, April 21, 1931. For many years he played minor venues and country fairs, eventually gaining a spot on Shreveport KWKH's Louisiana Hayride and obtaining a recording contract with Decca. His first major hit for the label was **Am I That Easy To Forget?**, a self-penned song that became a Top 10 country hit in 1959, the success of this disc being matched by **Hello Out There**, an RCA release waxed by Belew in 1962. Other hits followed throughout the mid-'60s, namely **In The Middle Of A Memory**, **Crystal Chandelier**, **Boston Jail**, **Walking Shadow**, **Talking Memory**, **Girl Crazy** and **Mary's Little Lamb**.

After 1968 Belew's name was absent from the charts for a long period, returning in 1971 when **All I Need Is You**, a duet recorded with Betty Jean Robinson, became a mini hit. A further success with **Welcome Back To My World**, an MCA release in 1974, augered well for furthering his career, but chart success was not to be his. He continued to tour regularly until cancer claimed his life in November 1990.

Recommended:
Carl Belew (Vocalion/–)
12 Shades (Victor/–)
Songs (Vocalion/–)

Bellamy Brothers

David, who plays guitar and keyboards, was born September 16, 1950, and Howard, who plays guitar, was born February 2, 1946 in Darby, Florida. The sons of a farmer who played dobro and fiddle in a bluegrass band, they were raised in a musical environment and made their first professional appearance in 1958. David joined soul band the Accidents in 1965 and played organ behind Percy Sledge, Little Anthony and the Imperials,.

During 1968 the brothers formed a pop band, Jericho, playing the club circuit in Georgia, Mississippi and South Carolina, but disbanded in 1971 and returned to the family farm. There the duo began songwriting, also providing jingles for the local radio and TV stations.

After one of David's songs, **Spiders And Snakes**, proved to be a two-million-seller for Jim Stafford in 1973, they headed for California, and embarked on a recording career, making an initial impact with **Nothin' Heavy**, a regional hit (1975). They moved on to gain an international reputation with their version of **Let Your Love Flow**, a song written by Neil Diamond's roadie Larry Williams, thus providing the brothers with a US pop No.1 in 1976. Fashioning a rock-styled, yet often acoustic-based, kind of country music, the Bellamy Brothers achieved their second major success with **If I Said You Had A Beautiful Body Would You Hold It Against Me** (1979), a song which was voted Single Of The Year by the CMA of Great Britain.

Since that time there have been no further pop hits for the Bellamys, but plenty of success on the country circuit with such No.1s as **Sugar Daddy** (1980), **Dancin' Cowboys** (1980), **Do You Love As Good As You Look** (1981), **For All The Wrong Reasons** (1982), **I Need More Of You** (1985) and **Too Much Is Not Enough** (1986), the latter in partnership with the Forester Sisters.

In 1987, now signed to MCA/Curb, the brothers started a series of nostalgic songs, such as **Kids Of The Baby Boom** and **Rebels Without A Clue**, which kept them high in the country charts. Another label change in 1991 saw them on Atlantic, but struggling. They continue as a popular touring act, and such songs as **Let Your Love Flow** and **If I Said You Had A Beautiful Body** have become standards.

Recommended:
Rebels Without A Clue (MCA-Curb/–)
Crazy From The Heart (MCA-Curb/–)
Country Rap (MCA-Curb/–)
Reality Check (MCA-Curb/–)

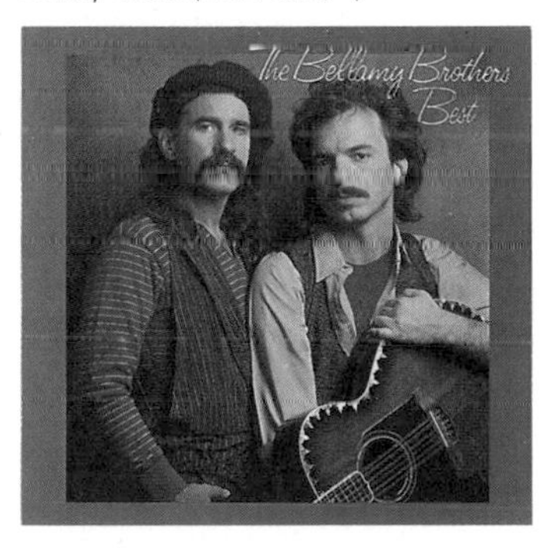

The Bellamy Brothers Best. Courtesy MCA/Curb Records.

Matraca Berg

A dynamic singer-writer, Matraca Berg looked set to take country music by storm in 1990, but has so far failed to fulfil the commercial success her talent promises. She was born in 1954 into a musical family. Her mother, Icie Callaway Kirby, was a member of the Callaway Sisters, a quartet that performed on country shows throughout the '50s and '60s. The family moved to Nashville in the early '60s and Icie soon became an in-demand session singer and songwriter. Matraca was constantly around the music business and naturally followed in her mother's footsteps. Her first breakthrough came as a writer, penning **Fakin' Love**, a country No.1 for T. G. Sheppard and Karen Brooks in 1982 and **The Last One To Know**, this time a No. 1 for Reba McEntire in 1987, which was also nominated for a Grammy. This led to a record deal with RCA, and her acclaimed **Lying To The Moon** album, which spawned such country hits as **Baby, Walk On** and **The Things You Left Undone** (both 1990), and **I Got It Bad** and

Above: **The Bellamy Brothers have become country chart regulars.**

I Must Have Been Crazy (both 1991). Matraca made a big impact in Europe with **I Got It Bad**, which became a hit in Holland and gained extensive radio plays throughout Britain. Although she has continued to write and sing on sessions, Matraca Berg has not yet built upon the solid base of her excellent first album.

Recommended:
Lying To The Moon (RCA/–)

Clint Black

Black is one of the most visible country stars on the American tour circuit, which has led to him gaining multi-platinum awards for all three of his albums and an impressive run of chart-topping singles.

Though born in Long Beach, New Jersey on February 4, 1962, Black and his family are all Texans. The family home was in Houston, and Clint grew up to the strains of classic George Jones and Merle Haggard. He was playing harmonica and guitar by the age of 15. An older brother had a band and Clint sang harmony and played bass with them. A chance meeting with local musician Hayden Nicholas in 1980 led to a promising songwriting partnership. The pair formed a band and for the next seven years played the bars and honky-tonks right across Texas.

Demos of their songs reached Bill Ham, the manager of ZZ Top, who immediately struck up a management agreement with Black and negotiated a record contract with RCA Nashville at the end of 1987. Armed with a supply of original songs, Black produced a classic album of stirring honky-tonk ballads that endeared him to the working man. His first single, **Better Man**, climbed to the top of the charts in early 1989, followed by the album's excellent title song, **Killin' Time**. Two more singles from the album, **Nobody's Home** and **Walkin' Away** also topped the charts, while a fifth single, **Nothing's New**, reached No.3. No performer in any style of music had achieved that level of success with their debut album. **Killin' Time** not only topped the country album charts, but also crossed over into the pop listings, and within a year of release had reached platinum status.

Black's dance style of honky-tonk had evolved from his Texas upbringing and years of playing the clubs. His quivering, traditional-sounding vocals and common-man songs gained him a huge following and his second album, **Put Yourself In My Shoes**, went platinum just two months after release in November 1990. The title song went to No.5, then he was back at No.1 with **Loving Blind** and **Where Are You Now** both 1991 hits. That same year he married soap-star Lisa Hartman.

A management disagreement between Black and Bill Ham kept Clint out of the studios for almost two years, though he did record a duet with legendary cowboy-star Roy Rogers – **Hold On Partner**, a minor country hit at the end of 1991. Meanwhile Clint maintained a busy touring schedule, dispensed with Ham's services, and returned to the studio to co-produce his third album, **The Hard Way**. This was released in the summer of 1992, but saw him in dispute with RCA, because the album was completed behind schedule. The wait was well worthwhile. This was Black's finest album, containing such hit singles as **We Tell Ourselves**, **Burn One Down** (1992) and **When My Ship Comes In** (1993).

In 1992 Black put together the most ambitious concert tour ever undertaken by a mainstream country performer. He played 150 dates before 1.5 million people from July 1992 through to March 1993. He added four additional band members to his five-piece backing outfit, and had a quarter-of-a-million dollars stage set specially built with dual video screens.

Recommended:
Killing' Time (RCA/RCA)
Put Yourself In My Shoes (RCA/RCA)
The Hard Way (RCA/RCA)

The Blue Sky Boys

Bill Bolick, vocals, mandolin; Earl Bolick, vocals, guitar.

Both born in Hickory, North Carolina: Bill on October 28, 1917, Earl on December 16, 1919, they were sons of Garland Bolick, who grew tobacco and worked in a textile mill. The Bolicks began playing traditional material, working in and around the Hickory area. In 1935, Bill sang for the East Hickory String Band – a name later changed to the Crazy Hickory Nuts, after J. W. Fincher of the Crazy Water Crystal Company who had offered them a job in Nashville – and the same year, the brothers began singing duets on the local radio station. In 1936, they recorded for Victor, whose A&R man believed them to be copies of the Monroe Brothers.

Their first release – on Bluebird – was **The Sunny Side Of Life**, written by Bill. During the late '30s and early '40s, the Bolicks' mixture of old-time and religious music became very popular but, following World War II, country fans began seeking something more commercial.

In 1951, the Blue Sky Boys broke up. Persuaded by Starday Records to come out of retirement in 1963 and cut an album titled **Together Again**, they also decided to play the odd date or two at folk festivals and colleges. They also recorded an album for Capitol in 1965 and another for Rounder as late as 1976.

But these albums and the occasional live dates were but gestures. After 1951 the Boys – soured by their experiences in the music business – never returned to full-time music careers.

Recommended:
The Blue Sky Boys (Rounder/–)
The Sunny Side Of Life (Rounder/–)
Presenting The Blue Sky Boys (JEMF/–)

Dock Boggs

Possessor of a unique banjo style, Moran Lee 'Dock' Boggs was born Norton, Virginia, February 7, 1898. A miner for 41 years, Boggs played five string banjo merely as a hobby.

Acquiring his unusual playing style (using two fingers and a thumb instead of the normal one finger and thumb claw hammer method) from a black musician he met in Norton, Boggs gradually developed his technique and recorded a number of sides for Brunswick in 1927. But upon his retirement from mining, Boggs turned more of his attention to music and found himself to be an in-demand performer at various folk festivals where his playing, his eerie, haunting songs like **Oh Death** and his nasal vocal delivery, attracted much attention, encouraging Mike Seeger to record Boggs for the Folkways label. He died in 1971.

Recommended:
Dock Boggs Volumes 1 & 2 (Folkways/–)
Dock Boggs Volumes 3 & 4 (Folkways/–)

The Blue Sky Boys (Bill and Earl Bolick). Courtesy RCA Records.

Suzy Bogguss

A real country girl, Suzy Bogguss was born on December 30, 1956 in Aledo, Illinois; growing up in a farming family near a small town in rural America. The youngest in a family of four, she recalls listening to a diverse mix of music from country to big band and '60s pop. She sang in the Aledo Presbyterian Church choir, then later branched out to accompanying herself on guitar and playing coffee houses at Peoria, Illinois. At college she graduated as a metalsmith, gaining an art degree and planning to open a studio for jewellery design. But the pull of music became too strong, so she took to booking dates into clubs and coffee houses. Driving a camper truck and accompanied by her pet dog and cat, she worked all across the Midwest.

A determined lady, Suzy recorded her own cassette in Nashville in 1985 and sold it at her show dates. The following year she was signed as the headlining female performer at Dollywood for the summer

season. A contract with Capitol Records followed, and she made her first impact on the charts with **I Don't Want To Set The World On Fire** (1987). Further minor hits followed and Suzy finally cracked the Top 20 with **Cross My Broken Heart** (1989). Her first album, **Somewhere Between**, proved influential with British female performers, with several of the songs gaining cover versions.

Although she was touring regularly, Suzy was struggling to make a big impact with her records, but she did return to the Top 20 with a duet, **Hopelessly Yours** (1991), with Lee Greenwood, and a solo revival of **Someday Soon** (1991). Since then she has hardly looked back. Her third album, **Aces**, was the big breakthrough she needed with such Top 10 singles as **Outbound Plane**, **Letting Go** and the title song. The album went gold in the summer of 1992 and that September Suzy Bogguss was presented with the Country Music Association Horizon award. She was back in the Top 10 in the spring of 1993 with **Drive South**, a song taken from her second gold album, **Voices In The Wind**.

Recommended:
Somewhere Between (Capitol/–)
Moment Of Truth (Capitol/Capitol)
Aces (Capitol/Liberty)
Voices In The Wind (Liberty/Liberty)

Johnny Bond

Singing cowboy film star, Johnny Bond, was born Cyrus Whitfield Bond, at Enville, Oklahoma, on June 1, 1915.

After his family moved to a Marietta, Oklahoma farm in the '20s, he bought a 98 cent ukelele through a Montgomery Ward catalogue and began playing it, moving on to guitar and becoming proficient on that instrument by the time he'd entered high school. In 1934, Bond made his debut on an Oklahoma City radio station. Three years later, he joined the Jimmy Wakely Trio (known as the Bell Boys) – a group heard by Gene Autry – and signed for Autry's Melody Ranch show in 1940.

That same year, Bond began recording for Columbia Records, with whom he recorded **Cimarron**, **Divorce Me, C.O.D.**,

Legendary Singer And Banjo Player, Dock Boggs. Courtesy Folkways.

Tennessee Saturday Night, **Cherokee Waltz**, **A Petal From A Faded Rose** and many others. His strong association with Autry also led to minor parts and sidekick roles in scores of films, while during the '50s and '60s he guested on TV shows hosted by Autry, Spade Cooley and Jimmy Wakely, becoming co-host of Compton's Town Hall Party Show and partner with Tex Ritter in music publishing.

A fine songwriter, Bond is responsible for such standards as **Cimarron**, **I Wonder Where You Are Tonight**, **Gone And Left Me Blues**, **Your Old Love Letters**, **Tomorrow Never Comes** plus around 500 other compositions. His contract with Autry's Republic Records provided that label with **Hot Rod Lincoln**, a Top 30 pop hit in 1960; a 1965 recording of **Ten Little Bottles** giving another company, Starday, a top-selling single. In the late '60s he signed for Capitol, cutting an album with Merle Travis, and then, after seeing one of his Starday tracks **Here Come The Elephants** become a mild hit (1971), moved on to Lamb And Lion who released one album, **How I Love Them Old Songs** (1975).

Bond toured the UK in 1976, the same year that his book 'The Tex Ritter Story' was published by Chappell. A writer of some talent he also wrote a book on Gene Autry plus 'Reflections', an autobiography.

His death occurred in Burbank, California, on June 12, 1978, after an illness which lasted for almost a year.

Recommended:
The Best Of Johnny Bond (Starday-Gusto/–)

Debby Boone

Daughter of singer Pat Boone and granddaughter of Red Foley, Debby was born Hackensack, New Jersey, September 22, 1956.

An onstage performer at an early age, Debby, her sisters and her mother (singer-actress Shirley Jones) provided the backing for Pat's act as the Boone Girls. Debby and her sisters later formed a gospel quartet, the Boones, before she opted to go solo, gaining a massive hit with **You Light Up My Life** (1977), claimed to be the biggest-selling single in over 20 years, the record providing her with a Grammy as Best New Artist. It also gave Debby her first major country hit and won her the Top New Vocalist category at the Academy Of Country Music presentations in 1978. That same year Debby achieved another country chart hit **God Knows**, following this with such climbers as **My Heart Has A Mind Of Its Own** and **Breakin' In A Brand New Broken Heart** in 1979, the year that she was nominated in the Top Female Vocalist category at the CMA Awards. By 1980 she had her first country No. 1 in **Are You On The Road To Lovin' Me Again**.

An autobiography 'Debby Boone – So Far' was published in 1981.

Recommended:
Debby Boone (Warner Bros/–)
Love Has No Reason (Warner Bros/–)

Larry Boone

One of the very best honky-tonk singers and country songwriters to emerge in the '80s, Larry Boone, from Cooper City, Florida, has probably enjoyed more success

The Best Of Debby Boone. Courtesy Warner Bros Records.

as a writer than recording star, although he has consistently hit the country charts and produced several acclaimed albums.

Before taking the route to Nashville and a career in country, Boone dabbled with the idea of becoming an athlete. For a time he was a sports writer, and with his academic credentials he became a substitute teacher. He eventually reached Nashville in 1980, but it was to be several years before he was to make his mark. Initially he played at the Wax Museum, a Nashville tourist attraction. His warm, mellow country voice came to the attention of Gene Ferguson, who signed him to a management contract. He started to make an impact as a writer, then he landed a record deal with Mercury in 1986 and proceeded to chalk up several minor hits such as **Stranger Things Have Happened** (1986), **Roses In December** (1987) and making the Top 10 with **Don't Give Candy To A Stranger** (1988). By this time he was an accomplished stage performer and was regularly providing fellow country stars Lacy J. Dalton, Ed Bruce, John Conlee and Don Williams with hit songs.

Boone continued to make the Top 20 with **I Just Called To Say Goodbye Again** (1988) and **Wine Me Up** (1989), and looked set for success when he joined Columbia Records in 1990. He co-wrote the majority of the songs on **One Way To Go**, his first album for the new label, but instead of hitting the big-time, his career faltered. He only managed a couple of minor hits with **I Need A Miracle** and **To Be With You** (both 1991). Boone was still with Columbia in 1993, and went back in the charts with **Get In Line**.

Recommended:
Swinging Doors, Sawdust Floors (Mercury/Mercury)
Down That River Road (Mercury/–)
One Way To Go (Columbia/–)

Don Bowman

Top country comedian of the '60s, Bowman was born in Lubbock, Texas, August 26, 1937. While a child he sang in church, later learning to play guitar – though part of his stock in trade is that he professes to play it badly! During Bowman's school years he became a DJ though he was forced to sell hub caps and pick cotton.

Becoming more established as a DJ, at one time working with Waylon Jennings, he began working more of his own routines into shows, eventually opting to become a full-time fun maker, appearing at clubs in the South and Southwest. In the mid-'60s, Chet Atkins signed Bowman to RCA, the result being **Chet Atkins Made Me A Star**. In 1966 came smaller successes with **Giddy Up Do-Nut** and **Surely Not**, which helped Bowman win the 'Billboard' award as favourite C&W Comedian Of The Year.

His others hits have included **Folsom Prison Blues No.2** (1968) and **Poor Old Ugly Gladys Jones** (1970).

Recommended:
Support Your Local Prison (RCA/–)
All New (Mega/–)

Boxcar Willie

A singer whose train whistle impressions helped him become a UK favourite before he had established any real reputation in his homeland, Boxcar was born Lecil Martin, Sterret, Texas, September 1, 1931. The son of a railroad worker, he worked in many jobs while pursuing a part-time profession as an entertainer for 37 years. Determined to become a top-line country artist, he assumed a hobo-like persona during the mid-'70s and moved to Nashville, giving himself three years in which to make the grade. Seen by Scottish agent Drew Taylor at Nashville's Possum Hollow Club in 1977, he was signed to play his first British tour the following year, playing to small clubs. Gradually, after further tours, he built something of a reputation and gained a spot on the 1979 Wembley Festival bill where he proved a sensation. Also in 1979 he made his first appearance on the Grand Ole Opry winning a standing ovation.

Boxcar Willie, a 1979 release by the sage of the train who has a strong following in the US and Europe. Courtesy Big R Records.

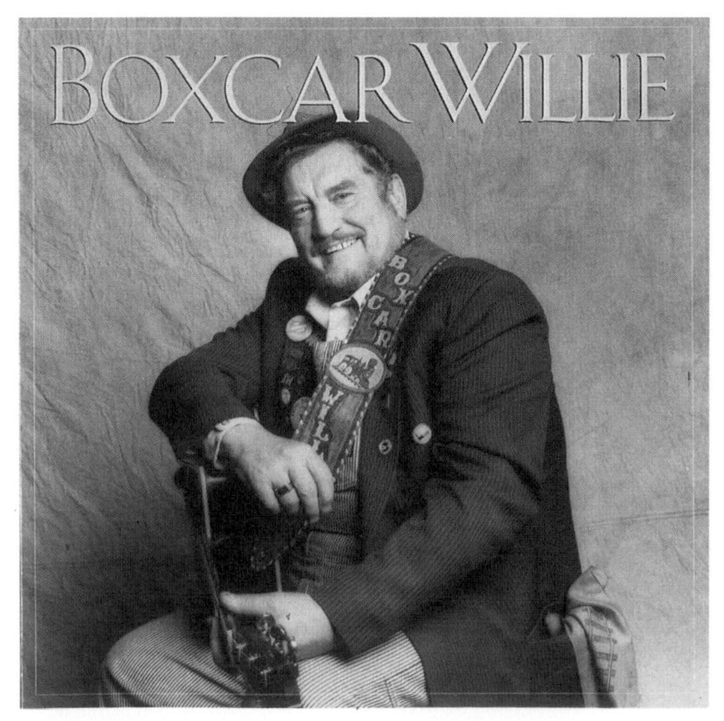

He returned to Wembley for a further triumph in 1980, reaching the Top 5 in the UK album charts that year with a TV-advertised album, **King Of The Road**. In the US he logged his first single in the country charts, his **Train Medley** nudging into the Top 100.

A fine entertainer who deals mainly in material of a traditional nature, Boxcar has remained a major crowd-puller in Europe, but has failed to make much of an impact in the USA. He now owns a theatre in Branson, Missouri, where his larger-than-life personality delights the inquisitive country fans.

Recommended:
King Of The Road (–/Prism)
Daddy Was A Railway Man (–/Big R)
Live In Concert (–/Pickwick)

Bill Boyd

Singer, guitarist, bandleader and film actor, William Boyd was born in Fannin County, Texas, on September 29, 1910.

A true cowboy raised on a ranch, Boyd, together with his brother Jim (born on September 28, 1914), formed Alexander's Daybreakers in the late '20s, this group eventually becoming the Cowboy Ramblers, a Greenville, Texas, band. They began recording for Bluebird in 1934, initially cutting versions of traditional numbers plus cowboy songs like **Strawberry Roan** – of these, **Under The Double Eagle** (1935), a fiddle and guitar version of the popular march, proved most successful. Boyd, whose band grew from a four-man string band in 1934 to a ten-man western swing outfit in 1938, recorded over 300 sides for RCA, including such best-sellers as **New Spanish Two Step** and **Spanish Fandango** (1939) plus **Lone Star Rag** (1949), and had his own radio show on station WRR, Dallas for many years. His last recording session was February 7, 1950 and he retired soon after. He died on December 7, 1977.

After his brother's death, Jim Boyd, who had worked with the Light Crust Doughboys in 1938–39 and the Men Of The West in 1949–51, continued the family tradition, performing with the Light Crust Doughboys before modern-day audiences in the Dallas-Forth Worth area.

Recommended:
Bill Boyd's Cowboy Ramblers (Bluebird/–)
Bill Boyd And His Cowboy Ramblers 1934–47 (Texas Rose/–)

Owen Bradley

A Nashville producer, musician and executive, Owen Bradley had as much, if not more, to do with the creation of the Nashville Sound and the growth of the studios than any other individual.

Born on October 10, 1915, in Westmoreland, Tennessee, Bradley was a band-leading pianist who had one of the most popular dance bands in the Nashville area for some time and served as director of WSM radio from 1940–58, leading their staff orchestra from 1947. That same year, Decca's Paul Cohen asked Bradley to do some producing for him in Nashville. Bradley accepted and by 1952 he and his younger brother Harold (later a leading Nashville session musician) had built their first studio.

By 1956 they had built their legendary Quonset hut on 16th Avenue South, which was the beginning of large scale Nashville studio activity and of the area which has come to be known as Music Row. In the Studio, which became known as Bradley's Barn, Bradley became instrumental in pioneering the so-called Nashville Sound during the '50s, smoothing out country music and providing it with a more pop, uptown treatment, often featuring lush strings and background voices.

Bradley, whose own recording career had started with the Bullet label, made many successful records of his own during the '50s. But as MCA's chief staff producer he achieved more fame, producing just about every one of their major artists.

He was quite open about dispensing with steel guitars, fiddles etc, and pursuing a sophisticated recording sound which many felt was castrating country as an individual form. But Bradley, with the Nashville Sound, formulated a safe, broadly appealing sound and can justifiably be said to have brought country music in from the cold when it needed help most. Had the music not gone through this stage of its development, it might well not enjoy its current acceptability.

In 1958 he moved up the corporate ladder to become country A&R director, a position he held for over a decade until his promotion to vice-president of MCA's Nashville operation. In 1974 he was elected to the Country Music Hall Of Fame and, after retirement from MCA, continued to produce independently until he retired in 1982. Canadian new-wave country singer k. d. lang coaxed Bradley out of retirement in 1988 to produce her **Shadowlands** album, which was as much a tribute to Bradley, the producer, as it was to lang's distinctive vocal work. The project turned out to be one of the biggest successes of Bradley's long career and helped rekindle interest in the recordings of Patsy Cline, which he had produced more than 25 years earlier.

Recommended:
Big Guitar (–/Charly)

Elton Britt

Born in Marshall, Arkansas, June 27, 1917, Elton Britt's real name was James Britt Baker. He was the son of a champion fiddle player. While still at school he learnt guitar on a model purchased from Sears Roebuck for $5.

When only 14, he was discovered by talent scouts who signed him to a year's contract with station KMPC, Los Angeles, where he appeared with the Beverly Hillbillies. In 1937 he signed for RCA Records, staying with the label for over 20 years, during which time he recorded 672 singles and over 50 albums. Following this, he moved on to record for Decca, Ampar and ABC-Paramount.

Britt, who was considered to be one of the world's greatest yodellers, obtained the first gold disc awarded to any country star when, in 1944, his version of **There's A Star Spangled Banner Waving Somewhere**, originally released in May 1942 as a B-side, reached the million sales mark. His other successful singles include **Chime Bells** (1948), **Jimmie Rodgers Blues** (1968) and **Quicksilver** (1949), the latter being one of the many duets he recorded with Rosalie Allen.

Britt, who died June 23, 1972, appeared in several films and made many TV appearances during the '50s and '60s, although he was a semi-retired gentleman farmer from around 1954 to 1968.

Recommended:
16 Great Country Performances (ABC/–)
Elton Britt Yodel Songs (RCA/Stetson)

Brooks & Dunn

This pair of rowdy, good-natured country rockers teamed up in 1991 and delivered **Brand New Man**, the most successful debut album ever released by a country duo or group. The album not only went multi-platinum, but also produced four No.1 singles and a fifth that hit Top 5. In 1993, the album won ACM award for Album Of The Year, and single **Boot Scootin' Boogie** won Single Of The Year.

Kix Brooks was born Leon Eric Brooks, May 12, 1956, Shreveport, Louisiana, just down the street from Johnny Horton. His first paying gig was performing with Horton's daughter, and later he became a veteran of the road, settling for a time in both Fairbanks, Alaska and New Orleans. He moved to Nashville in the early '80s and started penning hits for John Conlee (**I'm Only In It For The Love**), Nitty Gritty Dirt Band (**Modern Day Romance**) and Highway 101 (**Who's Lonely Now**). He recorded as a solo for Avion Records (1983), then later joined Capitol Records, releasing an excellent rock-styled album in 1989 and scoring a minor hit with the self-penned **Sacred Ground**.

Ronnie Dunn, like Brooks, spent years working the road. He was born June 1, 1954, in Coleman, Texas, and attended Abilene Christian College, but was thrown out for performing in honky-tonks. He moved to Tulsa where he formed the house band at Duke's night club (his bass player

Hard Workin' Man, Brooks & Dunn. Courtesy Arista Records.

was Garth Brooks' sister Betsy Smittle). In Oklahoma he recorded for Churchill Records, making his chart debut with **It's Written All Over Your Face** (1983). After winning the 1989 Marlboro National Talent Contest he relocated to Nashville.

Both Brooks and Dunn were trying to land solo deals in 1990 when they were introduced to each other by Arista Records' Tim DuBois, who suggested they write together. The ploy paid off, and DuBois signed them as a duo to Arista, and their first single, **Brand New Man** (1991), soared to the top of the charts and paved the way for the album of the same title, which crossed over to the pop charts. Further No.1s came with **Next Broken Heart** (1991), **Neon Moon** and **Boot Scootin' Boogie** (both 1992), the latter gaining a special dance mix, which took Brooks & Dunn's music to a whole new audience outside mainstream country.

Above: Brooks & Dunn keep up their boot scootin' stage antics at the 1993 Nashville Fan Fair.

The pair make a formidable combination in concert; Dunn is the less flamboyant of the two, but his vocal intensity somehow matches Brooks' manic leaps, duckwalks and near-violent guitar work and his more rock-edged voice. They picked up the Country Music Association Vocal Duo Of The Year Award in 1992. Their second album, **Hard Workin' Man** (1993), also gained platinum status and presented the duo with another No.1 in the title song.

Recommended:
Brand New Man (Arista/–)
Hard Workin' Man (Arista/–)
Kix Brooks (Capitol/–)

Garth Brooks

The Country Music Superstar of the '90s, Brooks was born on February 7, 1962, in Tulsa, but raised in Yukon, Oklahoma, where country music played a role in the Brooks' household, but not a dominant one. His mother, Colleen Carroll Brooks, had been a country performer in the mid-'50s, recording for Capitol Records and she was a regular on Red Foley's Ozark Jubilee TV show. By the time Garth was born she had retired from a professional career and the Brooks' house reverberated with as much rock and pop music as country.

A keen sportsman, Garth attended Oklahoma State University, Stillwater, on a track scholarship (javelin) with no set plans for a music career, although he had sung occasionally at Yukon High School. It was not until 1984 when he realized a sports career was not for him, so he went into advertising and started to take music seriously. He had been performing around Stillwater for some months, and in the summer of 1985 he left for Nashville and a career in country music, only to return home four days later, dejected by rejection.

He joined a local band, Santa Fe, as lead singer, and for the next year played the Southwest circuit performing a mix of nostalgic pop hits and modern country. Garth had become a big fan of singer-writers James Taylor and Dan Folgelberg, rock bands Boston and Kansas, and the modern country of George Strait.

In early 1987 Garth, his new wife Sandy (whom he married in May 1986) and Santa Fe moved to Nashville. It was Garth who made the impact, singing jingles and on demos for publishers, and holding down a day-time job in a boot store at the same time. He signed a writer's contract in November 1987 and by the end of the next year had joined Capitol Records, working in the studio with Allen Reynolds. His first single, **Much Too Young (To Feel This Damn Old)**, slowly climbed into the Top 10 in the summer of 1989, and his first album, **Garth Brooks**, gave no indication of what was to come. It was a straight country set with plenty of fiddle and steel guitar and good songs, gaining favourable reviews in all the country music press. His second single, **If Tomorrow Comes**, hit the top of the charts, but it was his fourth release, **The Dance**, accompanied by an evocative video, which opened the floodgates. **The Dance** soared to the top of the charts and pushed the album into the pop charts, and from there was no stopping Garth Brooks.

During the next few years he was everywhere. He had further country No.1s with **Friends In Low Places** and **Unanswered Prayers** (1990), **Two Of A Kind, Workin' On A Full House**, **The Thunder Rolls** and **Shameless** (1992). With his third album, **Ropin' The Wind**, he set a new precedent, when on day of release in September 1991, advance orders stood at 4 million and the record entered the pop charts at No.1. This had never happened to a singer wearing a cowboy hat who phrased his words with an unmistakable southern drawl. At the same time, his two previous albums, **Garth Brooks** and **No Fences**, which had both gained platinum status, were in the pop Top 10, and for some weeks Brooks held both the No.1 and No.2 position in the pop charts. This kind of success helped lead to a new awareness of country music in America and opened the doors for many other country stars to taste pop chart success.

Ropin' The Wind, Garth Brooks. Courtesy Capitol Records.

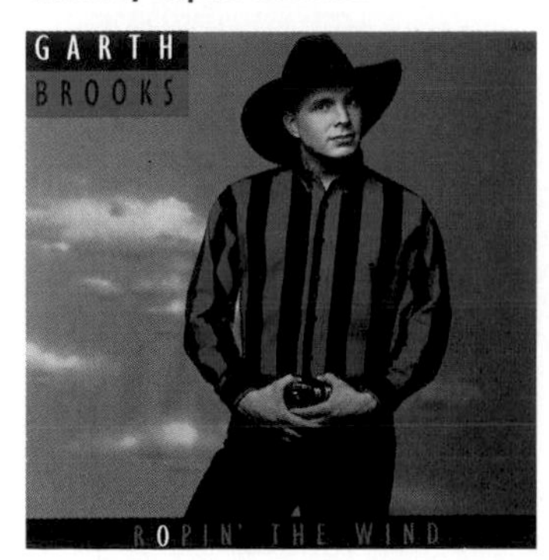

As successful as his recordings have been, it is as an onstage performer that Brooks has taken country music to a new high and reached out to a younger audience. He is very visual, and rides a continual wave of energy in true rock rather than country, as he runs around the stage, involves his fellow musicians in his crazy antics and really communicates with his fans. In 1991 and 1992 he outsold every pop and rock act at the major American stadiums, setting new records for selling thousands of show tickets in mere minutes.

The Brooks phenomenon continued with his fourth album, **The Chase**, which also had high advance orders prior to release in September 1992, gaining instant platinum status, entering the pop charts at No.1 and providing such country chart-toppers as **We Shall Be Free**, **Somewhere Other Than The Night** and **Learning To Live Again**. Simultaneously, he released **Beyond The Season**, a Christmas album which hit No.2 on the pop and country charts and gained yet another multi-platinum award.

He has won more than 50 major industry awards, three times winning CMA's Entertainer Of The Year. His album sales (for six releases) have topped 35 million in a career of less than five years, which makes him possibly the most successful country music entertainer of all time. But the man maintains his close affinity to country music. His Stillwater Band utilizes fiddle and steel guitar, his songs cover a wide range from cowboy ballads through raucous honky-tonkers to country swing and rock. His family are closely involved with his career, with his sister, Betsy Smittle, playing bass in the band, and his brother, Kelly, acting as road manager.

Although Garth talked about retiring from touring to spend more time with his family in 1992, all the indications are that he will be around for a good many years to come. He is a natural writer, a versatile singer and an exciting performer, who has made country music the music of the early 1990s, showing that it can be acceptable right across the USA and as an equal to rock, pop, soul, jazz or any other musical style.

Recommended:
Garth Brooks (Capitol/Capitol)
No Fences (Capitol/Capitol)
Ropin' The Wind (Capitol/Capitol)
The Chase (Capitol/Capitol)
Beyond The Season (Capitol/Capitol)

The Browns

The Browns – Ella Maxine Brown, born in Sampti, Louisiana, April 27, 1932; her brother Jim Edward, born in Sparkman, Arkansas, March 1, 1934; and younger sister Bonnie, born in Sparkman, Arkansas, July 31, 1937 – began as a duo in the early '50s when Jim and Maxine won a talent contest on Little Rock's KLRA radio station. In 1955, following an extensive concert tour, Bonnie came in to form a trio, the threesome becoming a headline act on the Ozark Jubilee show. Initially signed to Abbott Records, the Browns moved to RCA after being brought to the label's attention by Jim Reeves. They scored an initial hit with a version of the Louvin Brothers' **I Take A Chance** in 1956 and after Jim Ed had completed his army service, they recorded **The Three Bells**, an adaptation of Edith Piaf's Euro-hit **Les Trois Cloches**, which proved to be a 1959 million-seller.

Hits such as **Scarlet Ribbons** (1959), **The Old Lamplighter**, **Teen-Ex**, **Send Me The Pillow You Dream On**, **Blue Christmas** (all 1960) and **Ground Hog** (1961) followed, plus a number of well-received overseas tours.

In 1963, the Browns became Opry members, but Bonnie and Maxine, both married, wanted to spend more time with their families. As a result, the group disbanded in 1967, despite having chart success that same year with **I Hear It Now** and **Big Daddy**. Maxine later returned to record as a solo artist for Chart, scoring with **Sugar Cane Country** (1968), while Jim Ed began a solo career on RCA.

Recommended:
20 Of The Best Of The Browns (–/RCA)
Looking Back To See (–/Bear Family)
Rockin' Rollin' Browns (–/Bear Family)

Above: The Browns' biggest record was a cover of an Edith Piaf song.

Jim Ed Brown

In 1965, Bonnie and Maxine of the Browns persuaded Chet Atkins to record Jim Ed as a solo artist, his first single, **I Heard From A Memory Last Night**, being a chart entry. Following this, Jim Ed began to record more frequently as a soloist, enjoying hits with **I'm Just A Country Boy**, **A Taste Of Heaven** (1966), **Pop A Top**, **Bottle, Bottle** (1967) and others. When the Browns disbanded in 1967, Jim Ed was easily able to reshape his stage act and continue as a star attraction.

Throughout the '70s and early '80s, Jim Ed's Midas touch continued to function. In 1970 he charted with **Morning**, a song successfully covered by singer Val Doonican in Britain, while Top 10 entries also came with **Southern Loving**, **Sometime Sunshine** (1973), and **It's That Time Of Night** (1974). During 1976 he teamed with Helen Cornelius and grabbed a country No.1 with **I Don't Want To Marry You**, followed this with another duet, **Saying Hello, Saying I Love You, Saying Goodbye**, which only just missed the top spot. Since that time Brown and Cornelius have also had many major records, such as **If The World Ran Out Of Love Tonight** (1978), **Lying In Love With You** (1979), and **Morning Comes Too Early** (1980), the twosome's final chart record for RCA being **Don't Bother To Knock** (1981). An award winner way back in 1967, when the Browns won the Cash Box poll as the best country vocal group,

Angel's Sunday, Jim Ed Brown. Courtesy RCA Records.

Jim Ed also shared the CMA vocal duo award with Helen Cornelius ten years later. These days you'll find Jim Ed regularly hosting the Grand Ole Opry in Nashville, where he is rightly regarded as one of the great country stars of yesteryear.

Recommended:
Morning (RCA/RCA)
Barrooms And Pop A Tops (RCA/–)
I Don't Want To Have To Marry You – with Helen Cornelius (RCA/RCA)
I'll Never Be Free – with Helen Cornelius (RCA/–)

I Don't Want To Have To Marry You, Jim Ed Brown. Courtesy RCA Records.

Marty Brown

Marty Brown, the kid from the small town, burst on the country scene in 1992 with the catchy single, **Every Now And Then**, which, with the help of an excellent video, rapidly climbed the charts. Born and raised in Maceo, Kentucky, Brown served his dues working clubs and honky-tonks. With a yodel and a hint of sadness in his voice, Brown evokes comparisons to the legendary Hank Williams Sr.

After years of knocking on doors in Nashville, he finally landed a contract with MCA Records in 1991, and his first album, **High And Dry**, introduced a sound that was so country it made the New Traditionalists sound slick and ultra-modern. In support of the album, Marty and his band, The Maceo Misfit, undertook a 12-state High And Dry Cadillac Tour, when he visited 45 Wal-Mart stores in the south-west in 42 days in his '69 red Cadillac DeVille Convertible. His second album, **Wild Kentucky Skies**, further showcased his solid writing and undiluted country styling.

Recommended:
High And Dry (MCA/–)
Wild Kentucky Skies (MCA/–)

Milton Brown

One of the founders of western swing, Brown was born in Stephenville, Texas, September 8, 1903. In 1918 his family moved to Fort Worth. Following a meeting with Bob Wills in 1931, he began singing professionally as part of Wills' Fiddle Band. Then followed a number of sponsored radio shows, the band name-changing to the Aladdin Laddies when promoting Aladdin Mantle Lamps, and the Light Crust Doughboys as pluggers of Light Crust Flour.

Following some personnel changes, the band cut some sides for Victor in the guise of the Fort Worth Doughboys. But soon after, Brown formed his own unit, the Musical Brownies, to play on radio KTAT, Fort Worth, the lineup eventually stabilizing at Milton Brown (vocals), Durwood Brown (guitar and vocals), Jesse Ashlock and Cecil Brower (fiddles), Wanna Coffman (bass) and Fred 'Papa' Calhoun (piano). This line-up – minus Ashlock – recorded a number of sides for Bluebird in April 1934, the fiddler returning for the band's second Bluebird session in August of that year.

With the addition of jazz-playing guitarist Bob Dunn – the first in country to electrify his instrument – a complete western swing sound was achieved. At the same time, Brown concluded a contract with the new Decca and with this label the Brownies began cutting a miscellany of titles ranging from jazz items like **St Louis Blues**, **Memphis Blues** and **Mama Don't Allow** through to such western songs as **Carry Me Back To The Lone Prairie** and **The Wheel Of The Wagon Is Broken**. But, just when the band began moving into top gear, Brown died following a car accident. Though rushed to hospital, Brown died on April 13, 1935. A passenger, vocalist Katherine Prehoditch, was instantly killed.

For a while the Brownies continued to fulfil contractual obligations but the band finally folded in early 1938. Brown is considered, along with Wills, as the co-founder of western swing.

Recommended:
Taking Off (String/–)
Dance-O-Rama (Rambler/–)
Easy Ridin' Papa (–/Charly)

Above: Ed Bruce was once a Memphis rockabilly who sold used cars.

T. Graham Brown

T. Graham Brown turned out to be the real revelation of the New Country Movement that started gaining momentum in Nashville in the late '80s. A throwback to '50s rockabilly and '60s R&B, he brought a black, soulful edge to country, resulting in a string of Top 10 country hits including **Hell And High Water** (1986), **Don't Go To Strangers** (1987) and **Darlene** (1988).

Anthony Graham Brown was born October 30, 1954, in Atlanta, Georgia. Following graduation, he moved to Athens to attend the University of Georgia. Brown and college buddy Dirk Howell, played rock, pop and beach music as Dirk And Tony at the Athens Holiday Inn, but when Dirk opted for the straight life, Brown decided to form a band.

Modelling himself on David Allan Coe, he fronted a band called Reo Diamond. With the advent of Urban Cowboy at the end of the '70s, Brown's raunchy country was no longer in vogue, so he formed T. Graham Brown's Rack Of Spam, and played soul and R&B music. By this time he was married, and with his wife Sheila's support he moved to Nashville in 1982, hoping to make a career in country music.

He signed a writer's contract with CBS Songs and was taken under the wings of veteran tunesmith Harlan Howard. He gained work singing on demos and also did jingles for Kraft, McDonald's and Coca Cola, among others. In 1984 he was signed to a recording contract by Terry Choate at Capitol and made the country charts with **Drowning In Memories** the following year. His first Top 10 record was **I Tell It Like It Used To Be** (1985), and for the next few years he was hardly absent from the charts. His mix of blues, rock, R&B, soul and country blended with his big, gravelly voice, and gave Brown one of the most distinctive sounds in country. He has continued to make jingles, a lucrative sideline to his hits, which have included **Brilliant Conversationalist** (1987), **The Last Resort** (1988), **Come As You Were** (1989) and **Don't Go Out** (1990), the latter a duet with Tanya Tucker.

Recommended:
Brilliant Conversationalist (Capitol/Capitol)
Bumper To Bumper (Capitol/Capitol)
I Tell It Like It Used To Be (Capitol/Capitol)
You Can't Take It With You (Capitol/–)

I Write It Down, Ed Bruce. Courtesy MCA Records.

Ed Bruce

Recording for Phillips under the name Edwin Bruce (he was born William Edwin Bruce Jr on December 29, 1940 in Keiser, Arkansas), this singer had a brief career as a '50s rocker. By the mid-'60s he'd moved to Nashville, there recording over a dozen singles for RCA with the aid of producer Bob Ferguson, having mild chart reaction with **Walker's Woods** (1967), **Last Train To Clarksville** (1967), and **Painted Girls And Wine** (1968), his rich voice also being heard on an album, **If I Could Just Go Home**.

In 1968, Bruce joined Monument Records cutting a fine album, **Shades Of Ed Bruce**, also having minor singles success with **Song For Jenny** and **Everybody Wants To Go Home** (1969). However, despite his strong commercial appeal, it wasn't until the '70s and an association with UA Records that Bruce finally established himself as a record seller of any consequence, his **Mamas Don't Let Your Babies Grow Up To Be Cowboys** moving high into the charts during early 1976. The song, written by Bruce and his wife Patsy, was later covered by Waylon Jennings and Willie Nelson whose joint version climbed to No.1 in 1978, gaining Bruce a nomination both for a Grammy and for the CMA's Song Of The Year.

In the wake of his last single for UA, a version of Alex Harvey's **Sleep All Mornin'**, Bruce signed for Epic (1977), only mustering mid-chart singles from 1977 to 1979.

Throughout 1979 Bruce spent his time writing and recording, fashioning new material for his new label, MCA, for whom he signed in 1980. That year he logged three major singles, **Diane**, **The Last Cowboy Song** and **Girls, Women And Ladies** plus **Ed Bruce**, which re-established him as a major country music name. Well-known on TV ads where he had dressed in Daniel Boone manner to advertise Tennessee – gaining Bruce the title of The Tennessean – his acting career flourished initially when he was signed to appear in The Chisholms mini-series.

Since that time there have been such other excellent albums as **One To One** (1981) and **I Write It Down** (1982), along with hit singles that include **Everything's A Waltz**, **You're The Best Break This Heart Ever Had** (1981), **Love's Found You And Me, Ever Never Lovin' You** (1982), **My First Taste Of Texas** and **After All** (1984). Over the years Ed Bruce has written hits for such artists as Kitty Wells, Tanya Tucker, Crystal Gayle, Tommy Roe, Kenny Price and Charlie Louvin. 1984 found Ed back on RCA gaining Top 5 hits with **You Turn Me On (Like A Radio)** and **Nights**. Now divorced from wife Patsy, Ed's career in recent years has gone into rapid decline and he no longer has a major record label deal.

Recommended:
Shades Of Ed Bruce (Monument/Monument)
Ed Bruce (UA/–)
The Tennessean (Epic/–)
I Write It Down (MCA/MCA)
The Best Of Ed Bruce (–/MCA)
Night Things (RCA/RCA)
Rock Boppin' Baby (–/Bear Family)

Cliff Bruner

The leader of an early swing/honky-tonk band, Cliff Bruner was born April 25, 1915, in Houston, Texas, and got his start as a fiddler in Milton Brown's Musical Brownies, cutting 48 sides with Brown before the bandleader's death in 1935.

At that point Bruner formed his own band, the Texas Wanderers and began a long recording association with Decca in 1937. His biggest hit on the label was the first version released of Floyd Tillman's **It Makes No Difference Now** (1938), although one of the band's most historic milestones was cutting the first truck driving song on record: Ted Daffan's **Truck Driver's Blues** (1939).

Bruner – who also recorded for the Ayo label in the '40s – became increasingly inactive as the years passed and by the early '50s had become an insurance salesman and executive, performing only occasionally. By the late '70s he was living in the Houston suburb, League City.

Boudleaux And Felice Bryant

Boudleaux Bryant, born in Shellman, Georgia on February 13, 1920, originally aimed at a career in classical music, studying to become a concert violinist and, in 1938, playing a season with the Atlanta Philharmonic. Then came a switch to more popular music forms, Bryant joining a jazz group for a period. It was during this time he met his wife-to-be Felice Scudato (born Milwaukee, Wisconsin, August 7, 1925), then an elevator attendant at Milwaukee's Shrader Hotel. After marriage, the duo began writing songs together, in 1949 sending one composition, **Country Boy**, to Fred Rose who published the song, thus providing Little Jimmy Dickens with a Top 10 hit.

In 1950 the Bryants moved to Nashville and began writing hit after hit, supplying Carl Smith with a constant supply of chart-busters, one of these songs, **Hey Joe** (1953), becoming a million-seller when covered by Frankie Laine. During 1955, Eddy Arnold charted with the Bryants' **I've Been Thinkin'** and **The Richest Man**, but it was the duo's association with the Everly Brothers that brought the songwriting team their biggest string of successes with **Bye Bye Love**, **Wake Up Little Susie**, **Problems**, **Bird Dog**, **All I Have To Do Is Dream**, **Poor Jenny** and **Take A Message To Mary**, all ending up on million-selling discs.

Other Bryant hits have included: **Raining In My Heart** (Buddy Holly, 1959); **Let's Think About Living** (Bob Luman, 1960); **Mexico** (Bob Moore, 1960); **Baltimore** (Sonny James, 1964); **Come Live With Me** (Roy Clark, 1973); and the oft-recorded **Rocky Top**, which Felice and Boudleaux wrote in just ten minutes. In 1974 'Billboard' was able to publish a list of well over 400 artists who had recorded songs by the Bryants, and in 1986 the duo were honoured by being elected to the Songwriters' Hall Of Fame. The following year Boudleaux died on June 30, following a short illness. Deservedly this talented husband and wife team was inducted into the Country Music Hall Of Fame in 1991, when Felice was able to make a tearful speech of acceptance.

Jimmy Buffett

Singer-songwriter Jimmy Buffett brought a good-time feel to country music in the mid-'70s with his songs about boats, beaches, bars and beautiful women. Born December 25, Pascagoula, Mississippi, Buffett was raised in Mobile, Alabama. He gained a BS degree in history and journalism from the University of Southern Mississippi and started his singing career in the bars of New Orleans. In 1969 he moved to Nashville, landed a job as the Nashville correspondent for 'Billboard' magazine, and started building a reputation for his lyrical songs and solo performances.

Above: Jimmy Buffett is also a Florida resident and popular novelist.

He signed a recording contract with Barnaby Records and, working in the studio with Buzz Cason, made the **Down To Earth** album (1970), which flopped. A second album, **High Cumberland Jubilee**, was never released, and a disillusioned Buffett moved to Florida. In the meantime, Tompall & The Glaser Brothers recorded some of his songs, and, in 1972, he received an offer from ABC Dunhill Records to return to Nashville.

A White Sport Coat And A Pink Crustacean received rave reviews and presented Buffett with his first minor country hit in **The Great Filling Station Holdup** (1973). He made a bigger impact the following year with **Come Monday**, which made the pop charts. During the next few years he emerged as one of the biggest concert attractions and album-sellers on the American music scene. He made the country Top 20 with **Margaritaville** (1977) and also crossed over into the pop Top 10. Soon albums like **Changes In Latitudes, Changes In Attitudes** (1977), **Son Of A Son Of A Sailor** (1978), **You Had To Be There** (1978) and **Coconut Telegraph** (1980), all gained platinum awards.

Many of his songs reflect his lifestyle in the Florida Keys where he spends much of his time on his boat or in his Margaritaville store. He has maintained a large and loyal following over the years with a glossy fan magazine. He is also now a popular novelist. His albums have continued to sell in vast quantities and he was one of the first artists to have all his back catalogue (16 albums) made available on CD. He was back in the country Top 20 with **If The Phone Doesn't Ring, It's Me** (1985), and though he has had songs recorded by Lefty Frizzell, Merle Haggard, Crystal Gayle and Waylon Jennings, Buffett has remained outside the country music mainstream.

Recommended:
Boats, Beaches, Bars & Ballads (MCA/–)
Floridays (MCA/–)
Son Of A Son Of A Sailor/Coconut Telegraph (MCA/–)
Feedin' Frenzy (MCA/–)

The Burch Sisters

Cathy, Charlene and Cindy Burch came to prominence in Nashville in the late '80s when they landed a recording contract with Mercury Records. This followed several years of building up a healthy reputation around their home town of Screven, Georgia.

The three sisters were born in Jacksonville, Florida, and moved to Georgia in their teens. Coming from a musical family, the girls perfected their pure-country harmonies on a mixture of country standards and gospel songs. In 1986 they won a State talent contest and started to take their music more seriously. Financing a recording session in Atlanta, they hawked the finished tape around the Nashville studios and were lucky enough to be signed to Mercury in 1987. The Burch Sisters scored a handful of chart singles during the next three years, including

Everytime You Go Outside I Hope It Pours (1988) and **Old Flame, New Fire** (1989). The girls have always remained in the shadow of more famous girl vocal groups, and by 1992 the act had split and Cathy Burch signed a solo recording contract with Giant Records in March 1993.

Recommended:
New Five (Mercury/–)

Billy Burnette

William Burnette III was born May 8, 1953, in Memphis, Tennessee. His father was rockabilly guitarman Dorsey Burnette, and his uncle was rockabilly legend Johnny Burnette, and they comprised two-thirds of the Johnny Burnette Rock'n'Roll Trio, a red-hot rockin' '50s bop act that hiccupped its way to fame and fortune. Billy grew up surrounded by music, and when he was only seven he had Ricky Nelson's band back him up on his very first single, **Hey Daddy** on Dot Records. At this time the Burnette brothers were based in Los Angeles, making their mark as writers with hits for Nelson (**Believe What You Say**, **Waitin' In School**), and scoring their own solo pop hits.

Young Billy learned to play guitar at an early age and performed worldwide as part of Brenda Lee's shows. In the early '70s he led the band behind his father, later moving back to Memphis where he worked in the studios with Chips Moman. A contract with Polydor Records followed in 1979 and he made his debut on the country charts with **What's A Little Love Between Friends** (1979). He then moved to Nashville where he began writing hit songs for such diverse acts as Charlie Rich, Eddy Raven, Ray Charles, Everly Brothers and Charley Pride. He also spent two years as a singer and guitarist with Roger Miller.

In 1985 Billy signed with Curb Records and was nominated by the Academy of Country Music as the Top New Male Vocalist the following year and seemed poised for major success, but was sidetracked when Fleetwood Mac asked him to join the band when Lindsey Buckingham left in 1987. He spent six years with Mac, touring around the world and recording two multi-platinum albums with them. He brought a country influence to the group and his song, **When The Sun Goes Down**, received a smattering of country airplay. Billy left Fleetwood Mac in 1992 and headed back to Nashville to resume his solo career. He signed with Capricorn Records, and with a batch of songs he had co-written with Paul Kennerley, Dennis Morgan, Ronnie Rogers, Deborah Allen and Rafe VanHoy, he produced the **Coming Home** album (1993) which aptly displayed his energetic brand of country music with an edge.

Recommended:
Coming Home (Capricorn/–)

Johnny Bush

Born in Houston, Texas, on February 17, 1935, Bush – voted Most Promising Male Vocalist Of The Year by Record World in 1968, an accolade duplicated by Music City News 12 months later – moved to San Antonio, Texas in 1952, obtaining his first musical job at the Texas Star Inn, where he played rhythm guitar and sang.

At a later stage, he opted to become a drummer, eventually joining a band organized by his friend Willie Nelson during the early '60s. After a year's stay with this outfit, he then became a member of Ray Price's Cherokee Cowboys, with whom he played for three years before returning to Nelson's side once more. With Nelson he became front man for the band, the Record Men, also branching out as a solo artist on Stop Records, his first release for the label, a Nelson original titled **You Ought To Hear Me Cry**, being a mild hit in 1967. His next release, **What A Way To Live**, yet another Nelson song, climbed even further up the charts.

Primarily a honky-tonk singer in the Ray Price tradition, Bush switched from Stop – with whom he had major hits with **Undo The Right** (1968) and **You Gave Me A Mountain** (1969) – to RCA in 1972, enjoying his greatest-ever disc success with Willie Nelson's **Whiskey River**.

Throughout 1973 the hits kept on coming but began to peter out during the following year, Bush's name then being absent from the chart until 1977 when he had low-level success with **You'll Never Leave Me Completely**, a Gusto-Starday release.

Recommended:
You Gave Me A Mountain (Stop/Stop)
Undo The Right (Gusto/–)
Whiskey River (RCA/RCA)

Carl And Pearl Butler

A highly popular duo during the '60s, honky-tonk vocalist Carl Roberts Butler (born Knoxville, Tennessee, June 2, 1927) and his wife Pearl (born Pearl Dee Jones, Nashville, Tennessee, September 20, 1930) first performed as a team in 1962. Prior to this, Carl had been a highly successful recording artist with Capitol and Columbia, having hits for Columbia with **Honky Tonkitis** (1961) and **Don't Let Me Cross Over** (1962), the latter being released as a Carl Butler solo item but featuring Pearl on harmony vocals.

When **Don't Let Me Cross Over** became a country No.1, the Butlers realized they had hit on a winning formula and began recording as a duo, logging a fair number of chart entries during the '60s, including: **Loving Arms** (1963), **Too Late To Try Again** (1964), **I'm Hanging Up The Phone** (1964) and **Just Thought I'd Let You Know** (1965).

Granted Opry status in 1962, the Butlers were influential, with Gram Parsons borrowing their country-gospel style when he recorded their **We'll Sweep Out The Ashes In The Morning**. Carl was also a gifted songwriter, penning **If Teardrops Were Pennies**, a hit for Carl Smith (1951). Later Ricky Skaggs had a No.1 hit with a revival of **Crying My Heart Out Over You** (1982). Members of the Salvation Army, the Butlers were involved in various rehabilitation programmes leading to a Meritorious Service Award in 1970. The couple continued performing and recording well into the '80s. Pearl died of thyroid complications on March 3, 1988, and Carl suffered a fatal heart attack on September 4, 1992.

Recommended:
Greatest Hits (Columbia/–)

Jerry Byrd

Born in Lima, Ohio, March 9, 1920, Jerry Byrd went on to become one of the genuine giants of the electric steel guitar. Unequalled for purity of tone and taste, he was in demand for record sessions for years, although he was never comfortable with the increasingly popular pedal steel style, preferring Hawaiian stylings. Finally, growing weary of Nashville and the music business, he chucked it all and caught a plane to Honolulu where he became revered as practically a national monument for his advancement of the Hawaiian guitar.

Recommended:
Master Of Touch And Tone (Midland/–)

Below: After starting with the family rockabilly band, then two years with Roger Miller and a six-year-stint with Fleetwood Mac, Billy Burnette went solo with a back-up band.

Above: Sweetheart Of The Rodeo, the Byrds. Courtesy CBS Records. One of the first and finest of all country rock albums. The design was taken from a catalogue of western clothes.

The Byrds

Initially an LA folk-rock group formed in 1964, the Byrds recorded a country-influenced album **The Notorious Byrd Brothers** in 1968, using such guest musicians as Lloyd Green, John Hartford, Earl Ball and Byrd-to-be, Clarence White. That same year, the group, heavily influenced by newcomer Gram Parsons, produced **Sweetheart Of The Rodeo**, arguably the first real country-rock album, containing songs penned by the Louvin Brothers, Woody Guthrie, Merle Haggard and others. Thereafter, most Byrds albums featured some country-style tracks, despite the fact that Gram Parsons quit the group later in 1968.

The group, which folded in 1972, featured several musicians who have made some contribution to the expansion of country music. Gram Parsons and Chris Hillman formed the Flying Burrito Brothers in 1969. Parsons was also a major influence on '70s country-rock and he discovered Emmylou Harris. Hillman has played in several country-styled outfits, the most recent being the highly successful Desert Rose Band.

Recommended:
Sweetheart Of The Rodeo (Columbia/Edsel)

The Callahan Brothers

Homer C. ('Bill') Callahan, guitar, mandolin, bass and vocals; Walter T. ('Joe') Callahan, guitar and vocals.

Natives of Laurel, North Carolina (Walter born on January 27, 1910 and Homer born on March 27, 1912), the Callahan Brothers became a popular duet team of the south-eastern style in the 1930s, and by 1933 were already busy on radio and were recording for the ARC labels. They spent some time at WHAS in Louisville and WWVA in Wheeling before serving other stretches at WLW (1937–39) and KVOO in Tulsa, before settling down in the North Texas area, basing their operations from either Dallas or Wichita Falls for over a period of ten years.

It was here, for reasons best known to themselves, that they changed their names from Walter and Homer to Bill and Joe, and changed their music as well, performing more and more western and swing material, the highlight probably being their double-yodel version of **St Louis Blues**. Except for a single session with Decca (1941) and one Bill Callahan session with Cowboy Records, their 91 recorded sides were with ARC or Columbia, over a period stretching from 1934 to 1951. They became increasingly inactive in the '50s and '60s. Bill Callahan died on September 10, 1971.

Recommended:
The Callahan Brothers (Old Homestead/–)

Archie Campbell

Honoured as Comedian Of The Year in 1969 by the CMA, Campbell has been writer and star of the Hee Haw TV show.

Born in Bulls Gap, Tennessee, on November 17, 1914, his career really rocketed through stints on WNOX, Knoxville in 1949, an eventual TV show on WATE, Knoxville (1952–58), coming his way. He joined the Prince Albert portion of the Grand Ole Opry in 1958, also signing a recording contract with RCA, for whom he cut a number of comedy routines including **Beeping Sleauty** and **Rindercella**.

On the serious side,he scored with the narration **The Men In My Little Girl's Life** (1965), as well as duets with Lorene Mann, the most popular being **At The Dark End Of The Street** (1968). Away from showbusiness, Campbell was a talented sculptor, poet and painter and also a keen golfer. He died of a heart attack in Knoxville on August 29, 1987.

Recommended:
Live At Tupelo (Elektra/–)
Bedtime Stories For Adults (RCA/–)

Glen Campbell

The seventh son of a seventh son, singer-songwriter-guitarist-banjoist Glen Campbell was born in Delight, Arkansas, on April 22, 1936. A reasonable guitarist at the age of six, he joined Dick Bills' (his uncle) western band while a teenager, later forming his own outfit in New

Below: Archie Campbell – a multi-talented comedian and country singer.

Glen Campbell, a 1962 album. Courtesy Capitol Records.

Mexico, where he met Billie Nunley and married her.

Armed with his 12-string, in 1960 he opted to become one of Hollywood's busiest session musicians but found time to cut sides as a solo performer, one, **Turn Around, Look At Me**, becoming a 1961 pop hit on the local Crest label. Signed immediately to Capitol, Campbell graced the 1962 charts with Al Dexter's **Too Late To Worry, Too Blue To Cry** and Merle Travis' **Kentucky Means Paradise**, and in the mid-'60s supplied a couple of other minor hits while continuing his work as a sideman with the Beach Boys, Jan and Dean, Association, Rick Nelson, Elvis Presley and many others.

Then in 1967 he recorded John Hartford's **Gentle On My Mind**, following this monster hit with an even bigger one in **By The Time I Get To Phoenix**, a song written by Jimmy Webb. From then on came a succession of high-selling albums and singles in the '70s. The most successful of the latter were **Wichita Lineman**, **Galveston**, **Where's The Playground, Susie?**, **Try A Little Kindness**, **Honey Come Back**, **It's Only Make Believe**, **All I Have To Do Is Dream** – with Bobbie Gentry, **Dream Baby**, **Country Boy**, **Rhinestone Cowboy** and **Southern Nights**.

Above: Glen Campbell and a friend of Colonel Sanders on the Louisiana Hayride in a scene from 'Norwood'.

Campbell made the headlines over his on-off affair with singer Tanya Tucker while they toured and recorded together, and, in a barrage of publicity, they finally split up in 1981. He left Capitol Records that same year, and has since recorded with Atlantic, MCA, Universal and a return to Capitol. Although he has never regained the commercial success he enjoyed in the late '60s and early '70s, Campbell has been a regular in the country Top 10 with such songs as **Faithless Love** (1984), **Still Within The Sound Of My Voice** (1987), **The Hand That Rocks The Cradle** (a duet with Steve Wariner in 1987) and **She's Gone, Gone, Gone** (1989).

Possessor of 12 gold records, featured artist on countless TV shows, including his own Glen Campbell Show, co-star with John Wayne in the film 'True Grit' (1969) and star of 'Norwood' (1969), Campbell is also a golf fanatic, hosting the Glen Campbell Los Angeles Open.

Recommended:
Glen Travis Campbell (Capitol/Capitol)
I Remember Hank Williams (Capitol/Capitol)
Rhinestone Cowboy (Capitol/Capitol)
Ernie Sings And Glen Picks (Capitol/Capitol)
Old Home Town (Atlantic-America/Atlantic-America)
It's Just A Matter Of Time (Atlantic-America/Atlantic-America)
Unconditional Love (Capitol/Capitol)
Still Within The Sound Of My Voice (MCA/MCA)
Walkin' In The Sun (Capitol/Capitol)
Somebody Like That (Liberty/Liberty)

Left: Glen duetted with Steve Wariner on The Hand That Rocks The Cradle.

Henson Cargill

From a family of political and legal background, Henson Cargill was born February 5, 1941 in Oklahoma City. He was brought up on a ranch. He studied at Colorado State, worked for a time as a deputy sheriff in Oklahoma City, then headed for Nashville.

When he first arrived in Nashville he started out performing with the Kimberleys, then, with the aid of guitar-playing producer Fred Carter Jr, he made his recording debut with **Skip A Rope**, which earned him a contract with Monument Records and a 1967 million-seller. He followed with such major hits as **Row, Row, Row** (1968), **None Of My Business** (1969) and **The Most Uncomplicated Goodbye** (1970). Then Cargill went through such labels as Mega, Atlantic and Copper Mountain without exactly setting the charts on fire. In the mid-'70s he moved to Stillwater, Oklahoma where he operated a large cattle ranch. After a ten-year lay-off, Henson returned to the studios in 1990 to produce the **All-American Cowboy** album for Amethyst.

Recommended:
Coming On Strong (Monument/Monument)
Henson Cargill Country (Atlantic/–)
All-American Cowboy (Amethyst/–)

Bill And Cliff Carlisle

Cliff Carlisle, born in Taylorsville, Kentucky, on May 6, 1904, was among the first top-line dobro players. As a boy he toured as a vaudeville act, first recording for Gennett in 1930 with guitarist Wilbert Ball. An excellent yodeller, Cliff – who backed Jimmie Rodgers on some of his recordings – eventually formed a duo, the Carlisle Brothers, with his younger brother, Bill (born in Wakefield, Kentucky, December 19, 1908), playing dates in the Louisville-Cincinnati area. The brothers, who spiced their vocal and instrumental act with a fair degree of comedy, obtained regular radio exposure on station WLAP, Lexington, Kentucky during 1931, and six years later had their own show, The Carlisle Family Barn Dance, on radio station WLAP, Louisville.

In 1947 Cliff retired and for many years lived in Lexington, Kentucky. He died, following a heart attack on April 2, 1983. Bill went on to form a new group, the Carlisles. In 1954, following hits with **Rainbow At Midnight**, **No Help Wanted** and **Too Old To Cut The Mustard**, the Carlisles joined the Grand Ole Opry, remaining cast members to this day, scoring a further hit with **What Kinda Deal Is This?** (1966).

During his career, Bill Carlisle has won over 60 various country awards, his past albums include **Fresh From The Country** (King) and **The Best Of Bill Carlisle** (Hickory).

Recommended:
The Carlisles – Busy Body Boogie (–/Bear Family)

Paulette Carlson

Paulette Tenae Carlson from Northfield, Minnesota, made a big impression in the late '80s as lead singer with Highway 101, a country-rock outfit that scored No.1 hits **Somewhere Tonight** (1987) and **(Do You Love Me) Just Say Yes** (1988).

This talented singer, writer and guitarist moved to Nashville in 1981 and signed a writer's contract with the Oak Ridge Boys' publishing company. The following year she demoed some songs for an RCA singles deal and enjoyed a handful of minor hits including **You Gotta Get To My Heart (Before You Lay A Hand On Me)** (1983) and **I'd Say Yes** (1984). At this time she also had songs recorded by Tammy Wynette and Gail Davies and also sang back-up for the latter.

In 1986 she was brought in as lead singer for Highway 101, her crisp, rocky and clear vocals sitting perfectly in the blend of traditional country and rock'n'roll backbeat. Paulette used the group as a stepping stone for a solo career, and, at the end of the 1990, branched out with a solo deal with Capitol/Liberty Records that resulted in a Top 20 entry with **I'll Start With You** (1991). Her first album, **Love Goes On**, turned out to be a patchy affair. Paulette took almost a year off to have daughter Cali, so her fortunes as a solo star have yet to be realized, though she was back in the studios in the spring of 1993 working on her second solo album.

Recommended:
Love Goes On (Capitol/–)

Below: Paulette Carlson no longer travels Highway 101 as she steps out on her own.

Mary-Chapin Carpenter

Mary-Chapin Carpenter, third of four daughters, was born on February 21, 1958, in Princeton, New Jersey. Her father was an executive for 'Life' magazine and for some years the family lived in Japan. In 1974 they moved to Washington DC, by which time Mary-Chapin had learned guitar. She thrived on the local music scene and, after gaining a degree from Brown University, she regularly appeared at the Birchmere, an acoustic-based night-club. She met guitar player John Jennings in 1982, and he has since played a major role in her career development. Four years later she collected five awards at Washington DC's Wammie Awards and then started work on a demo tape. The original plan was to produce a gig tape, but the tape reached Columbia Records and in 1987 that demo became the basis of her debut album, **Hometown Girl**.

Come On Come On, Mary-Chapin Carpenter. Courtesy Columbia.

A throwback to singer-songwriters of the '70s, Mary-Chapin's music was literate, personal and beautifully melodic. That first album, co-produced by Jennings, was dismissed by country as being too 'folksy'. Gradually she has opened her music to more commercial possibilities, stretching country music's parameters along the way. A second album, **State Of The Heart** (1989), made both the pop and country charts and produced such Top 20 country hits as **How Do** and **Never Had It So Good** (1989), as well as **Quittin' Time** and **Something Of A Dreamer** (1990).

By this time she was working with a band, and, although she had initially written for herself to perform as a solo act, she was now creating songs that could utilize fuller instrumentation. This was evident in her breakthrough album, **Shooting Straight In The Dark** (1990), which gained gold status. The album included a revival of an old Gene Vincent hit, **Right Now** (1991), which was another Top 20 country hit, and **Down At The Twist And Shout**, a rousing Cajun number featuring three members of Beausoleil. This song hit No.2 on the country charts, won a Grammy in 1992, and led to Mary-Chapin Carpenter being named CMA's 1992 Female Vocalist Of The Year.

Carpenter is rightly hailed as a musical phenomenon, always crossing borders and breaking new ground with her blending of blues, folk, pop, rock, modern country and much more. There have been further major country hits with **Going Out Tonight** (top 20 in 1991), the raunchy **I Feel Lucky** (1992), a duet with Joe Diffie on **Not Too Much To Ask** (1992) and Top 5 again with Lucinda Williams' **Passionate Kisses** (1993). The **Come On, Come On** album was named one of the Top 50 albums by Britain's Q Magazine. The album soared high into the country and pop charts and gained Carpenter her first platinum.

The CMA nominated the album Album Of The Year in 1993, and Mary-Chapin was named for Female Vocalist again.

Recommended:
State Of The Heart (Columbia/CBS)
Hometown Girl (Columbia/–)
Shooting Straight In The Dark (Columbia/Columbia)
Come On, Come On (Columbia/Columbia)

Fiddling John Carson

An old-timer fiddler, he became (on June 14, 1923, in Atlanta, Georgia) the first country musician to be recorded by field recordist Ralph Peer. The tracks cut, **Little Old Log Cabin In The Lane** and **That Old Hen Cackled And The Rooster's Goin' To Crow**, were initially released on 500 unlabelled discs, all immediately sold at a local old-time fiddlers' convention. Following a re-release on Okeh, Carson was awarded a contract with the label.

Born on March 23, 1868, in Fannin County, Georgia, Carson was the fiddler champion of Georgia seven times, and he made his radio debut on September 9, 1922. Often working with a string band, the Virginia Reelers (which included his daughter Rosa Lee Carson, also known as Moonshine Kate), he cut around 150 discs for Okeh between 1923 and 1931. Following the Depression, Carson moved to RCA, mainly re-cutting earlier successes.

In later life an elevator operator, Carson died December 11, 1949.

Recommended:
That Old Hen Cackled (Rounder/–)

Martha Carson

A country gospel singer whose repertoire appealed not only to Opry fans but also the audiences at such ritzy venues as New York's Waldorf Astoria, Martha Carson was one of the most popular vocalists in her genre during the 50s.

Born Irene Ambergay, in Neon, Kentucky, on March 19, 1921, her first broadcasts were relayed over station WHIS, Bluefield, West Virginia in 1939. During the '40s she toured as one half of Martha and James Carson, the other half of the duo being her husband, the singing and mandolin-playing son of Fiddlin' Doc Roberts. They were longtime fixtures of the WSB Barn Dance in Atlanta becoming known as the Barn Dance Sweethearts. Together they recorded many magnificent sides, including **The Sweetest Gift**, **Man Of Galilee** and **Budded On Earth**.

Divorced in 1951, Martha began gracing the Opry with her fervent style during the following year. Martha is the writer of well over 100 songs, including **I'm Gonna Walk And Talk With My Lord**, **I Can't Stand Up Alone** and **Satisfied**.

Recommended:
Explodes (–/Bear Family)
Satisfied (–/Stetson)

Left: Anita, June, Helen and Carlene Carter at the 1986 Wembley Festival in London.

Carlene Carter

Keeping the famous Carter Family name alive, Carlene Carter, the daughter of June Carter and her husband Carl Smith, was born in Nashville, Tennessee, September 26, 1955. Married at 15, she toured as part of the Carter Family while still at school, but later rebelled, working as a model, then moving to London where she signed with Warner Brothers Records and made her first album, **Carlene Carter** (1977) with rock singer Nick Lowe, who became her third husband in 1979; the couple later separated. She made her first appearance on the country charts with **Do It In A Heartbeat** that same year, and also recorded with rocker Dave Edmunds, Graham Parker's the Rumour and members of the Squeeze, as she directed her music towards a young rock audience.

In 1984 she was in the London production of the show 'Pump Boys And Dinettes' and made a short film, 'Too Drunk (To Remember)'. A return to Nashville found her recording with country-rock band Southern Pacific (1989) and finally making a big impact in country the following year when she signed with Reprise Records and scored two Top 5 hits with **I Fell In Love** and **Come On Back**. Both were taken from her acclaimed **I Fell In Love** album, comprised mainly of self-composed songs backed up by some of music's finest pickers and singers.

Recommended:
I Fell In Love (Reprise/–)
Musical Shapes/Blue Nun (–/Demon)

The Carter Family

One of the most influential groups in country music, the original line-up was headed by Alvin Pleasant (A.P.) Delaney Carter, born in Maces Spring, Virginia, on April 15, 1891. One of nine children, A.P. initially sang in a church quartet alongside two uncles and an elder sister. Later he met Sara Dougherty (born in Wise County, Virginia, on July 21, 1898), a singer, guitarist, autoharp and banjo player, and they married on June 18, 1915. The third member of the group, Maybelle Addington (born in Nickelsville, Virginia, on May 10, 1909), joined after marrying A.P.'s brother, Ezra Carter, in 1926. She too played guitar, autoharp and banjo.

The Carter Family were first recorded by Ralph Peer for Victor on August 1, 1927 (at the same session that Jimmie Rodgers cut his first sides) completing six titles, including **Single Girl, Married Girl**, at a makeshift studio in Bristol, Tennessee. After some success the family began a whole series of sessions for Victor, often recording A.P.'s own songs, though **Wildwood Flower**, a traditional item cut on May 9, 1928, proved to be the group's biggest seller, registering over a million sales for 78 rpm discs alone. Another important recording date occurred during June 1931, when Peer cut sides featuring the collective talents of the Carters and Jimmie Rodgers.

After recording around 20 songs at one Victor session on December 11, 1934, the family moved on to ARC, waxing some 40 titles for that label during May 5–10, 1935.

Sara and A.P. obtained a divorce during the following year but continued working together in the group, next recording for Decca before moving to Texas to appear on various radio stations. During this three-year period, other members of the Carter Family joined the group – Anita, June and Helen (Maybelle and Ezra's three daughters) and Janette and Joe (Sara and A.P.'s children). Sara then remarried, to Coy Bayes, and though the Family cut further sides for Columbia and Victor (the last session by the original Carter Family taking place on October 14, 1941) they disbanded in 1943, having waxed over 250 of their songs including **Wabash Cannonball**, **Lonesome Valley** and **I'm Thinking Tonight Of My Blue Eyes**.

Maybelle then formed a group with her three daughters and began a five-year stint on station WRVA, Richmond, Virginia, after which came a switch to Nashville, where the quartet became regulars on Grand Ole Opry. A.P. began his career again too, working on some sides for the Acme label during 1952–6. But this version of the Carter Family – employing Sara, Joe and Janette – made little impact. A.P. died on November 7, 1960, ten years before the Carter Family's election to the Country Music Hall Of Fame. Sara Carter died on January 8, 1979, in Lodi, California following a long illness.

Following A.P.'s death, Maybelle and her daughters began working as the Carter Family (previously they'd appeared as the Carter Sisters And Mother Maybelle). June Carter eventually went solo and became part of the Johnny Cash Road Show in 1961. Maybelle Carter, plus Helen and Anita, also joined Johnny Cash, and became regulars on his TV show in 1966. After Mother Maybelle passed away on October 23, 1978, the Carter Family carried on with Anita's daughter Lori, Helen's son David and June's daughter Carlene, along with their mother, all appearing as part of the Carter Family Road Show.

Recommended:
Famous Country Music Makers (–/RCA)
Carter Sisters – Maybelle, Anita, June & Helen (–/Bear Family)
Musical Shapes – Carlene Carter (–/F-Beat)
Favorite Family Songs (Liberty/–)
Traveling Minstrel Band (Columbia/–)
Clinch Mountain Treasure (County/–)

Maybelle Carter

During the '60s Maybelle became a mother figure to the New Generation folkies (it was at a time when singers like Joan Baez had begun to discover and re-record Carter Family songs) and appeared on many folk bills throughout the country, winning much acclaim at the Newport Folk Festival of 1963. At a later Newport Festival, in 1967, she reunited with Sara Carter to record the live album **An Historic Reunion**, their first recording together for 25 years. In 1971, she also appeared on Nitty Gritty Dirt Band's **Will The Circle Be Unbroken** album. She died on October 23, 1978 in Nashville.

Recommended:
Mother Maybelle Carter (Columbia/–)
Sara And Maybelle Carter (Columbia/–)

Lionel Cartwright

Cartwright, a hillbilly from West Virginia, has progressed steadily with his own songs, going against the trend of the twangy, cowboy-hatted country of the early '90s and hitting paydirt with his chart-topping **Leap Of Faith** (1991).

He had piano lessons as a child and later picked up chords on his elder brother's guitar. In his early teens he landed a spot on a radio show in Milton, West Virginia and later featured as singer/musician on Columbus, Ohio's WMNI Country Cavalcade. The next step was the WWVA Jamboree in Wheeling, West Virginia.

Lionel moved to Nashville in the early '80s and soon landed a position as musical director for the Nashville Network and worked on the TNN sit-com 1–40 Paradise as musical director and had a small acting role in the series. A recording contract with MCA in 1988 led to a chart debut with **You're Gonna Make Her Mine** (1988), and the following year he made the Top 5 with **Give Me His Last Chance**. Further Top 10 hits came with **I Watched It All (On My Radio)** and **My Heart Is Set On You** (both 1990), as well as the chart-topping **Leap Of Faith** and **Family Tree** (both 1991). His albums have been high quality, with his self-penned songs about real-life situations.

Heroes, Johnny Cash and Waylon Jennings. Courtesy CBS Records.

A multi-talented instrumentalist, brilliant songwriter and heart-tugging singer, with rugged, boyish good looks, Lionel Cartwright has built up a sizeable following among younger female country music fans.

Recommended:
Lionel Cartwright (MCA/MCA)
I Watched It On The Radio (MCA/–)
Chasin' The Sun (MCA/–)

Johnny Cash

Winner of six CMA awards in 1969 (Best Male Vocalist, Entertainer Of The Year, Best Single, Best Album, Outstanding Service and even one award with June Carter for Best Vocal Group), John R. Cash was born in Kingsland, Arkansas, February 26, 1932, son of a poverty-stricken cotton farmer, Ray Cash and his wife Carrie. In 1935 the Cash family moved to the government resettlement Dyess Colony, surviving the Mississippi river flood of 1937, an event documented in a 1959 Cash song, **Five Feet High And Rising**. One son, Roy, put together a country band, the Delta Rhythm Ramblers, which broadcast on KCLN, Blytheville. However, tragedy struck the Cash family in 1944 when son Jack died after an accident with a circular saw, which had a lasting effect on the young John.

After high school graduation in 1950, J.R. took various menial jobs, but by July of that year he had enlisted in the air force for a four-year term. It was while serving in Germany that Cash learned guitar and wrote his first songs. Upon discharge in July 1954, he married Vivian Liberto and headed for Memphis where he became an electric appliance salesman.

In Memphis he met guitarist Luther Perkins and bassist Marshall Grant and began performing with them – for no pay – on station KWEM. Eventually they gained an audition with Sam Phillips at Sun Records and their first single, **Hey Porter/Cry, Cry, Cry** – listed as by Johnny Cash and The Tennessee Two – became a hit, selling more than 100,000 copies. The follow-up, **Folsom Prison Blues**, another Cash original, was also a success and led to Cash joining KWKH's Louisiana Hayride in December 1955. He also began touring extensively, and after **I Walk The Line**, a cross-over hit that sold a million, and **There You Go**, another 1956 winner, he joined the Grand Ole Opry.

He played a maniacal killer in a 'B' movie entitled 'Five Minutes To Live' (1958), then came tours in Canada and Australia and a lucrative new record deal with Columbia. Drummer W. S. Holland joined the Tennessee Two in 1960 (at which point they became the Tennessee Three), Cash and his enlarged band being much in demand and playing nearly 300 gigs a year; he was pill-popping to provide enough energy.

Cash first began working with June Carter in December 1961. The following year saw a heavier work schedule that included a 30-hour tour of Korea and a disastrous Carnegie Hall date. At the Newport Folk Festival of 1964 he sang with Bob Dylan and started widening his audience outside of the usual mainstream country. But the pill-popping worsened and, in October 1965, Cash was arrested by the narcotics squad in El Paso and received a 30-day suspended sentence and $1,000 fine. The following year he was jailed once more – for embarking on a 2am flower-picking spree! Also in 1966, Carl Perkins became a regular part of the Johnny Cash touring show.

Though in poor health and with his weight down to 140 pounds, his recorded work maintained a high standard, but his personal life continued heading ever downwards, with Vivian Cash suing him for divorce in mid-1966. The pill addiction continued, but with help from June Carter he undertook treatment, gradually fighting back until by the close of the '60s he was restored to complete health once more.

Since signing to Columbia in 1958, Cash had cut a tremendous quota of hit singles – the biggest of these being **All Over Again** (1958), **Don't Take Your Guns To Town** (1959), **I Got Stripes** (1959), **Ring Of Fire** (1963), **Understand Your Man** (1964), **It Ain't Me Babe** (1964), **Orange Blossom Special** (1965), **The One On Your Right Is On Your Left** (1966), **Folsom Prison Blues** (1968), **Daddy Sang Bass** (1968) and **A Boy Named Sue** (1969). More importantly, he had also cut some of the most striking albums to emerge from country music – including **Ride This Train** (1960), a kind of musical hobo-ride through America; **Blood, Sweat And Tears** (1964), a tribute to the working man; **Bitter Tears** (1964), regarding the mistreatment of the American Indian; **Ballads Of The True West** (1965), a double-album glance at western folklore; and **At Folsom Prison** (1968), an award-winning affair recorded in front of perhaps the world's toughest, but most appreciative audience.

Look At Them Beans, Johnny Cash. Courtesy CBS Records.

His partnership with June Carter, whom he married in March, 1968 (Merle Kilgore was best man), proved successful both on and off stage, the duo gaining a Grammy for **Jackson**, judged the Best Country Performance by a group during 1967. And as the '60s rolled away, the CMA showered a whole flood of awards on the craggy-faced man in black, while the film critics applauded his portrayal of an ageing gun-fighter in 'The Gunfight', a 1970 release in which he pitted his acting ability against Kirk Douglas. During the '70s, Johnny Cash turned increasingly towards religion, visiting Israel to make a film about life in Holy Land, and appearing on shows headed by evangelist Billy Graham.

Though he continued to make superb albums, his run of hit singles slowed down to a mere trickle. Since **One Piece At A Time** in 1976, he has only hit the Top 5 in

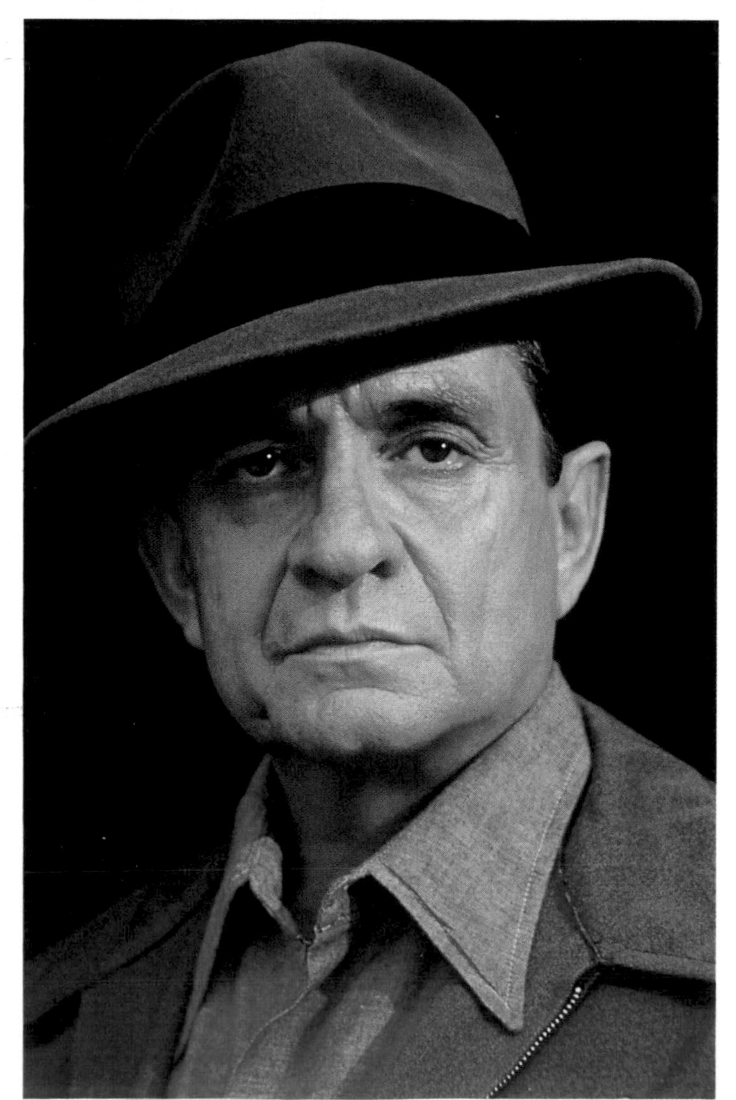

Right: Johnny Cash has cut albums in San Quentin and Folsom prisons and shot a documentary film in the Holy Land. But his album with Bob Dylan still remains unissued.

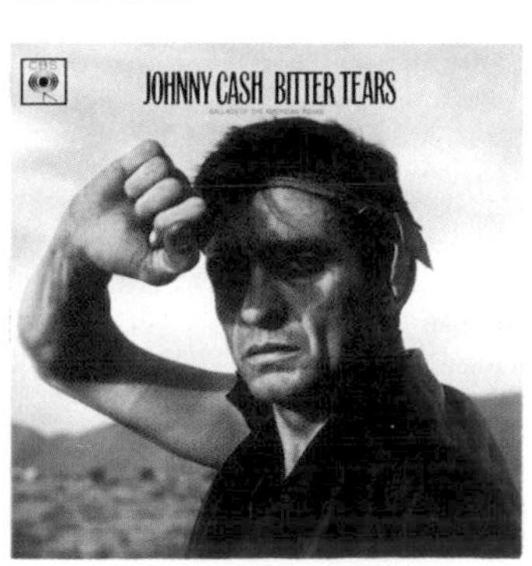

Bitter Tears, Johnny Cash. Courtesy CBS Records.

the country charts with **There Ain't No Good Chain Gang** (a duet with Waylon Jennings, 1978), **(Ghost) Riders In The Sky** (1979) and **Highwayman** (with Waylon Jennings, Willie Nelson and Kris Kristofferson, 1985). He was elected to the Country Music Hall Of Fame in 1980, but in 1983 Cash nearly died before undergoing abdominal treatment at Nashville's Baptist Hospital. "It was abuse", he claimed, admitting that he had once more begun dabbling with pills, "Many, many long years of abuse."

But he has since pulled back, resuming his touring career, signing with Mercury Records in 1987 and tackling songs of various dimensions. But thanks to his obvious sincerity and his crumbling rock of a voice (often placed out front over the most meagre of rhythm patterns) everything he sings about, no matter how sentimental, always comes out sounding completely believable.

Recommended:
At Folsom Prison (Columbia/CBS)
At San Quentin (Columbia/CBS)
Blood, Sweat And Tears (Columbia/–)
Man In Black (Columbia/CBS)
Ride This Train (Columbia/Bear Family)
Ballads Of The True West (Columbia/Embassy)
America (Columbia/CBS)
Silver (Columbia/CBS)
Rockabilly Blues (Columbia/CBS)
The Adventures Of Johnny Cash (Columbia/CBS)
Water From The Wells Of Home (Mercury/Mercury)
Bitter Tears (Columbia/Bear Family)
Heroes – with Waylon Jennings (Columbia/CBS)
Johnny 99 (Columbia/CBS)

Rosanne Cash

The eldest daughter of Johnny Cash and his first wife, Vivian Liberto, Rosanne was born on May 24, 1955, in Memphis, Tennessee at a time when her father's career was just starting to take off. Her parents were divorced when she was eleven, and Rosanne and her three sisters lived with their mother in California. Inevitably, she moved to Nashville following graduation, working with the Johnny Cash Roadshow, singing back-up vocals to her famous father. In 1975 she left to pursue her interest in acting, moving to London with the intention of enrolling in drama school. Instead she landed a job at CBS Records, the same label her father recorded for.

A year later she returned to America and studied drama at Nashville's Vanderbilt University, then moved back to California to study method acting at the Lee Strasberg's Theatre Institute. A vacation in Germany during the Christmas break led to a change in career direction. She was offered a recording contract by Ariola Records and after completing her studies, returned to Munich in the summer of 1978 to record her first album. The album completed, she moved back to Nashville to make music her career.

Teaming up with Rodney Crowell, whom she married in 1979, she signed with Columbia Records. With Crowell handling production, she recorded the acclaimed **Right Or Wrong** album. Though not a huge success, there were three country hits, including a duet with Bobby Bare, **No Memories Hangin' 'Round** (Top 20 in

Seven Year Ache, Rosanne Cash. Courtesy CBS Records.

1979). The next album, **Seven Year Ache**, continued with progressive country meshed with a definite rock attitude. Surprisingly, it provided Rosanne with her first country No.1s in the self-penned title song, **My Baby Thinks He's A Train** and **Blue Moon With A Heartache** (both 1981). The album gained gold status.

Uncompromising in her attitude to music and her life, Rosanne has refused to take the accepted 'star route' in promoting her career. **Somewhere In The Stars**, released towards the end of 1982, failed to match the commercial success of its predecessor, mainly because Rosanne decided it was more important to organize her life around her family. She spent the next two years raising her two daughters, Caitlin and Chelsea. The album still produced two Top 10 hits with **Ain't No Money** and **I Wonder** (both 1982).

This period of inactivity gave her the time to write material for **Rhythm And Romance** (1985), which not only went gold, but included such No.1 hits as **I Don't Know Why You Don't Want Me** and **Never Be You**. Rosanne won a Grammy for Best Female Vocalist Performance in 1986, and the next album, **King's Record Shop**, followed much quicker. Titled after a record store in Louisville, Kentucky, this album had No.1s with **The Way We Make A Broken Heart** (1987), a revival of her father's **Tennessee Flat Top Box**, **If You Change Your Mind** and **Runaway Train** (all 1988). Sandwiched between these was yet another No.1, **It's Such A Small World**, a duet with Rodney Crowell that was included on his **Diamonds & Dirt** album in 1988.

Maintaining an integrity rare in country music, Rosanne has been a pioneer for

Below: Rosanne cut her newest album, The Wheel, in 1993, without Rodney Crowell.

Six White Horses, Tommy Cash. Courtesy Epic Records.

female rights. This came through strongly on **Rosie Strikes Back**, a song which encourages a woman who suffers physical abuse to stand up for herself. A remarkable decade was captured in the compilation **Hits 1979–1989**, which featured ten of her past hits, plus yet another No.1, a revival of the Beatles' **I Don't Want To Spoil The Party** (1989).

The following year rumours were rife that her marriage to Crowell was in trouble, and were heightened by the doomy, introspective songs of **Interiors** (1990). There were no major hit singles, though an excellent video of the album was also released and, artistically, it was a superb set of meaningful songs. The pair did separate and filed for divorce in November 1991, but they made it clear they remained close friends. It was the start of a new beginning when Rosanne entered the studio without Crowell to record **The Wheel** (1993). There was an optimistic air to many of the songs, which were all self-written, and a commercial edge which should keep Rosanne at the forefront of modern-day country music.

Recommended:
Right Or Wrong (CBS/–)
Seven Year Ache (CBS/–)
Somewhere In The Stars (CBS/CBS)
Rhythm And Romance (CBS/CBS)

Tommy Cash

Younger brother of superstar Johnny Cash, Tommy was born in Mississippi County, Arkansas, on April 5, 1940.

He learnt the guitar at 16 after watching his brother play chords and listening to other guitarists. His first public appearance was as a performer at Treadwell High School in 1957; he then joined the army at 18, becoming a disc jockey for AFN, Frankfurt, Germany and having his own show, Stickbuddy Jamboree. He met his wife Barbara during this period.

Cash began appearing with a band at various service clubs and eventually launched his professional career by performing alongside Hank Williams Jr at Montreal in January, 1965. He recorded for Musicor and United Artists, but it was not until he signed to Epic in 1969 that he made a commercial breakthrough with **Six White Horses**, a hit in 1969, and **Rise And Shine** and **One Song Away**, hits in 1970. During the '70s, Cash continued to tour with his band, the Tom Cats.

A regular visitor to Britain and Europe, Cash recorded a **25th Anniversary Album** with guest appearances from George Jones, Tom T. Hall, Connie Smith and brother Johnny Cash, in 1990.

Recommended:
Six Horses/Lovin' Takes Leavin' (Epic/–)
Only A Stone (Elektra/–)
25th Anniversary Album (Playback/Cottage)

Ray Charles

Although best known as an R&B performer, Ray Charles has made a telling contribution to the cause of country music. Born Ray Charles Robinson, in Albany, Georgia, on September 30, 1930, he could play piano before he was five. At six he contracted glaucoma, which eventually left him blind. Later he learned composition and became proficient on several instruments, leaving school to work with dance bands around Florida.

He went on to work with a Nat Cole R&B-styled group, then switched style and utilized a more fervent, gospel-styled approach, signing with Atlantic Records in 1953 and piling up an imposing array of pop, rock and R&B hits. Though he covered country songs, it was not until 1962, when he made the remarkable **Modern Sounds In Country Music** for ABC Records, that he really edged into country. The album sold a million copies and spawned some massive pop hit singles. Charles' success paved the way for many other black artists to cut Nashville-oriented albums. These include Esther Phillips, Dobie Gray, Joe Tex, Bobby Womack and Millie Jackson.

After fashioning a sequel to **Modern Sounds**, he moved on yet again, touching all bases, but settling for none, seemingly losing direction. However, in 1982, he returned to country music once more, recording in Nashville and releasing the excellent **Wish You Were Here Tonight** album for Columbia. This time the Nashville connection lasted for three years and included the subsequent album, **Friendship** (1984), a collection of duets that found Ray swopping vocal licks with Johnny Cash, Mickey Gilley, Ricky Skaggs, Hank Williams Jr and others. The partnerships with George Jones (**We Didn't See A Thing**), B. J. Thomas (**Rock And Roll Shoes**) and Willie Nelson (**Seven Spanish Angels**) all furnished hit singles. Ray Charles was inducted into the Rock'n'Roll Hall Of Fame in 1986, and still continues to record and perform all styles of music in his own inimitable fashion.

Recommended:
Modern Sounds In Country & Western Music (ABC/HMV)
Wish You Were Here Tonight (Columbia/CBS)
Do I Ever Cross Your Mind (Columbia/CBS)
Friendship (Columbia/CBS)

Below: Ray Charles adds a soulful edge to traditional country music.

Mark Chesnutt

Born September 6, 1963, in Beaumont, Texas, Mark Chesnutt has established himself as one of the most solid new honky-tonk traditionalists with a series of Top 10 hits. His first two albums achieved gold status. Known as 'The Human Jukebox', Mark is said to have the words to as many as 1,000 country songs all ready to sing when required. This has come from years of hanging around honky-tonks and listening to nothing but classic country albums.

His father, Bob Chesnutt, was a regional Texas star who had several singles on an independent Nashville label in the late '60s and early '70s. When Mark was 16 he started going to clubs, initially to watch his father perform, but then started to sing himself. He quit school in the 11th grade to develop a career in country music and made his recording debut for the AXBAR label in San Antonio. The next few years saw Mark and his band working the night-clubs around Beaumont. With his father's guidance he made some demo tapes which were sent to Nashville. Another independent deal was struck up with Cherry Records in Houston. One of his singles, **Too Cold At Home**, came to the attention of MCA Records, who signed Mark and re-released the song, pushing it up to No.3 (1990) in the country charts.

The first album, titled after that hit, was endorsed by George Jones in the sleeve notes, and contained Mark's No.1, **Brother Jukebox** (1990) and two more Top 5 hits with **Blame It On Texas** and **Your Love Is A Miracle** (both 1991). The album made a big impact on both the pop and country charts and became the biggest selling debut LP in MCA Nashville's long history. To capitalize on that success, Mark and his band undertook more than 300 show dates in 1991 and he still found time to record a second album, **Longnecks And Short Stories**. He enjoyed another No.1 with **I'll Think Of Something** (1992), and maintained his Top 10 consistency with the wild **Bubba Shot The Jukebox** (1992) and **Ol' Country** (1993). Chesnutt's only setback was the sudden death of his father in 1991 from a massive heart attack.

Recommended:
Longnecks And Short Stories (MCA/–)
Too Cold At Home (MCA/–)
Almost Goodbye (MCA/–)

Above: Texas honky-tonker Mark Chesnutt plays one of his many hits.

Guy Clark

One of country music's finest songwriters, Clark, who was born on November 6, 1941, in Rockport, Texas, spent most of his early life in the town of Monahans, living with his grandmother in a run-down hotel, which provided the inspiration for many of his later songs.

During the '60s he moved to Houston, there working as an art director on a local TV station, and meeting Jerry Jeff Walker and Townes Van Zandt. He briefly performed in a folk trio with K. T. Oslin, playing the coffee house circuits of Houston, Dallas and Austin. Clark moved to Los Angeles, where he utilized his talent for constructing classical guitars and dobros and landed a writer's contract with Sunbury Music.

He headed back home via Oklahoma City, where he met and married Susanna, a gifted artist and songwriter in her own right. Finally the creative trail brought Clark to Nashville, where he recorded his first album, but scrapped it due to personal dissatisfaction. By this time his songs, such as **Desperados Waiting For A Train**, **L.A. Freeway** and **Texas 1947**, were being recorded by Jerry Jeff Walker, Johnny Cash and others. A return to the studio in late 1974 resulted in the **Old No.1** album, which critics acclaimed as the finest album for many years.

Clark's second RCA album, **Texas Cookin'**, was equally impressive, but failed to provide any hit singles, even though there were guest appearances by Waylon Jennings, Emmylou Harris, Rodney Crowell and Hoyt Axton. Clark writes almost poetically about losers and low-life ladies. He joined Warner Bros Records in 1978 and during the next five years recorded three more superb albums. He made brief appearances on the charts with singles **The Partner Nobody Chose** (1981) and **Homegrown Tomatoes** (1983).

A father figure to the new breed of singer-songwriters, Clark has never enjoyed the commercial success his talent would suggest, but has become an in-demand performer and well-respected songwriter. He has provided hit songs for Emmylou Harris, Steve Wariner, Gary Stewart, Ricky Skaggs and others. In 1988 he was back in the studios again, recording **Old Friends**, an album for Sugar Hill, which was picked up for European release on U2's Mother Records. This led to several European trips, where his intimate solo shows gained him a cult following among young music fans. He's now based mainly in Nashville, where his wife Susanna is busy as a visual artist, and he spends time co-writing with modern country writers, such as Richard Leigh, Roger Murrah, Jim McBride, Verlon Thompson and Lee Roy Parnell. Another brilliant album, **Boats To Build**, surfaced on Asylum in 1992, but lack of promotion

Right: The well-respected songwriter Guy Clark provides hit songs for many country stars.

from the record company meant that not too many know of its existence.

Recommended:
Old No. 1 (RCA/Edsel)
Texas Cookin' (RCA/Edsel)
The South Coast Of Texas (Warner Bros/–)
Better Days (Warner Bros/–)
Boats To Build (Elektra/–)

Roy Clark

Multi-talented star of the Hee Haw TV series, Roy Linwood Clark was born in Meherrin, Virginia, on April 15, 1933. Son of a guitar-playing tobacco farmer, Clark soon picked up the rudiments of guitar techniques but at an early age became even more proficient on banjo. Following a move to Washington DC, Roy played as part of the Clark family group, performing at local square dances and eventually winning a solo spot on the Jimmy Dean TV Show – but getting fired due to perpetual lateness. A subsequent job found him getting fired by Marvin Rainwater – this time for earning more applause than the star himself. Better luck followed when, as a cast member of a George Hamilton IV TV series, he gained wide recognition.

Recording initially for Four Star as Roy Clark And The Wranglers, he later moved on to Debbie, Coral and Capitol, enjoying his first sizeable hit with a Capitol single, **Tips Of My Fingers**, in 1963. Following a contract with Dot, the hits really began to proliferate, via titles like **Yesterday When I Was Young** (1969), **I Never Picked Cotton**, **Thank God And Greyhound** (both 1970), **A Simple Thing Called Love** (1971), **Come Live With Me** (1973), **Honeymoon Feeling** (1974) and **If I Had To Do It All Over Again** (1976).

A jovial international ambassador for country music, Clark was presented with the CMA Entertainer Of The Year award in 1973. Since the early '80s he has recorded for Churchill Records (a label formed by his manager Jim Halsey), Silver D and Hallmark, providing each with minor hits. He joined the Grand Ole Opry in 1987.

Recommended:
My Music And Me (Dot/–)
Roy Clark Sings Gospel (Word/Word)
A Pair Of Fives – with Buck Trent (Dot/ABC)
Back To The Country (MCA/MCA)
Live From Austin City Limits (Churchill/–)
The Entertainer (Dot/Ember)
I'll Paint You A Song (–/Ember)
20 Golden Pieces (–/Bulldog)
Makin' Music (MCA/MCA)

Jack Clement

Few people have mastered more phases of the music business than Jack 'Cowboy' Clement, who was born on April 5, 1931, in Memphis, Tennessee. As a youth he mastered a variety of instruments and played all kinds of music from big band to bluegrass as he worked his way into the production end of the business, where he fell in with Sam Phillips at Sun Records as a session musician, engineer and producer. During that era, Clement also wrote a great many songs including **Guess Things Happen That Way** and **Ballad Of A Teenage Queen** for Johnny Cash.

Since the mid-'60s he has concentrated his energies in Nashville, where he built an extremely successful publishing firm – Jack Music – ran Jack Clement Studio, assisted in the careers of Charley Pride, Tompall and the Glasers and continued to write as well. He even found time to produce a horror film, 'Dear Dead Delilah'. In 1971 he formed JMI Records, working with Allen Reynolds, Bob McDill and Don Williams, but the company folded in 1974.

Clement made a comeback in the late '70s as a producer with Johnny Cash and Waylon Jennings and started recording for the first time in almost 20 years for Elektra. Amongst his most notable songs are **Miller's Cave, I Know One, A Girl I Used to Know** and **The One On The Right Is On The Left**.

Recommended:
All I Want To Do In Life (Elektra/Elektra)

Vassar Clements

A much respected fiddle player who worked for some time as a sessioneer in Nashville before finding a degree of solo fame. Born on April 25, 1928, at Kinard, South Carolina, Vassar has been in the bands of Bill Monroe, Jim and Jesse, Faron Young and the contemporary-slanted Earl Scruggs Revue.

He appeared on the Nitty Gritty Dirt Band's **Will The Circle Be Unbroken** and his consequent familiarity to the general public after years of being a name among recording sessions credits has meant that he has been able to get his own show together and travel the road.

Recommended:
Superbow (Mercury/–)
Hillbilly Jazz (Flying Fish/Sonet)
Bluegrass Session (Flying Fish/Sonet)
Grass Routes (Rounder/–)

Zeke Clements

'The Alabama Cowboy' was born near Empire, Alabama, on September 9, 1911, and has had one of the longest careers in country music, having appeared on radio shows in all sections of the country, as well as having the unusual distinction of having been a member of all three of the major barn dances at one time or another.

Zeke began on the National Barn Dance in 1928, toured for some years with Otto Gray's Oklahoma Cowboys, then joined the Grand Ole Opry in 1933 as a member of their first cowboy group, the Bronco Busters. After spending some time on the West Coast on radio and in films, he returned to the Opry in 1939, where he became one of the Opry's major stars throughout the '40s. He also became known as a songwriter during this era especially for **Blue Mexico Skies, There's Poison In Your Heart**, and as co-writer of **Smoke On The Water**, the top country record of 1945.

Clements later appeared on the Louisiana Hayride and on many other Deep South stations. He pursued a business career in Nashville in the late '50s and '60s, then moved to Miami, Florida, where he spent nearly a decade playing tenor banjo in a dixieland band before returning to the Nashville area.

Right: Grand Ole Opry and Hayride star Zeke Clements appeared on radio throughout his early career.

Nashville Jam, Vassar Clements. Courtesy Flying Fish Records.

Bill Clifton

Perennial globetrotter Clifton was born at Riderwood, Maryland, in 1931. A vocalist, guitarist and autoharp player, he became interested in the music of the Carter Family during the 1940s, later forming his own group, the Dixie Mountain Boys, which established a considerable reputation among bluegrass aficionados. In 1961, he recorded 22 Carter Family songs for a Starday album but some months later turned up in Britain where he became instrumental in setting up tours by Bill Monroe, the Stanley Brothers, the New Lost City Ramblers and others during the '60s. He also enhanced his own reputation as a bluegrass musician, embarking on tours of Europe and recording a programme of old-time music for transmission on Radio Moscow in 1966.

By 1967, Clifton was on the move once more, he, his wife and seven children sailing for the Philippines, where he became an active member of the Peace Corps. Still in the Pacific area at the commencement of the '70s, he arrived in New Zealand, playing at a banjo players' convention and cutting an album with

A Bluegrass Jam Session 1952, Bill Clifton. Courtesy Bear Family.

Hamilton County Bluegrass Band, a local outfit. Since that time, Clifton has been active in Europe, the States and Japan.

Recommended:
Happy Days (–/Golden Guinea)
Blue Ridge Mountain Blues (County/ Westwood)
Going Back To Dixie (–/Bear Family)
Mountain Folk Songs (Starday/–)

Patsy Cline

Patsy Cline (real name Virginia Patterson Hensley) was born in Winchester, Virginia, on September 8, 1932. Winner of an amateur tap-dancing contest at the age of four, she began learning piano at eight, and in her early teens became a singer at local clubs. In 1948, an audition won her a trip to Nashville, where she appeared in a few clubs before returning home – but her big break came in 1957 when she won an Arthur Godfrey Talent Scout show, singing **Walking After Midnight**. Her Decca single of the contest-winning song then entered the charts, both pop and country. In 1961 came **I Fall To Pieces**, one of Patsy's biggest hits, followed in quick succession by **Crazy**, **Who Can I Count On?**, **She's Got You**, **Strange** and **When I Get Through With You**, most of them being massive sellers.

During the same period she became a featured singer on the Opry, soon attaining the rank of top female country singer. Such hits as **Imagine That**, **So Wrong** and **Leavin' On Your Mind** continued to proliferate until, on March 5, 1963, Patsy died in an air disaster at Camden, Tennessee. She had been returning home from a Kansas City benefit concert with Hawkshaw Hawkins and Cowboy Copas, both of whom were also killed in the crash.

But even after her death, Patsy's records continued to sell, **Sweet Dreams (Of You)** and **Faded Love** being top hits during '63, **When You Need A Laugh** and **He Called Me Baby** entering the country charts in 1964, and **Anytime** appearing in the latter as late as 1969.

Patsy has continued to be a major influence on singers like Loretta Lynn, who recorded a tribute album in 1977, Reba McEntire and Sylvia. In 1973 she was elected to the Country Music Hall Of Fame and her recordings and those of Jim Reeves were spliced together to produce a duet effect resulting in hits with **Have You Ever Been Lonely** and **I Fall To Pieces** during 1981. A film tracing her career, 'Sweet Dreams', was made in 1985 and led to Patsy's recordings making yet another comeback on the country charts.

This renewed interest and success has continued with two big-selling video programmes tracing her life and career through archive film and interviews. In 1992 a **Greatest Hits** compilation gained multi-platinum status with sales reaching 4 million copies, and all of her original albums were made available on CD in a massive promotional campaign. In Europe Patsy's success has been even more startling. A re-release of **Crazy** made the British pop charts at the end of 1990, and her albums also made the charts.

Recommended:
The Patsy Cline Story (MCA/–)
Always (MCA/MCA)
Sweet Dreams (MCA/MCA)
Live At The Opry (MCA/MCA)
Songwriters' Tribute (MCA/MCA)
Sentimentally Yours (MCA/MCA)

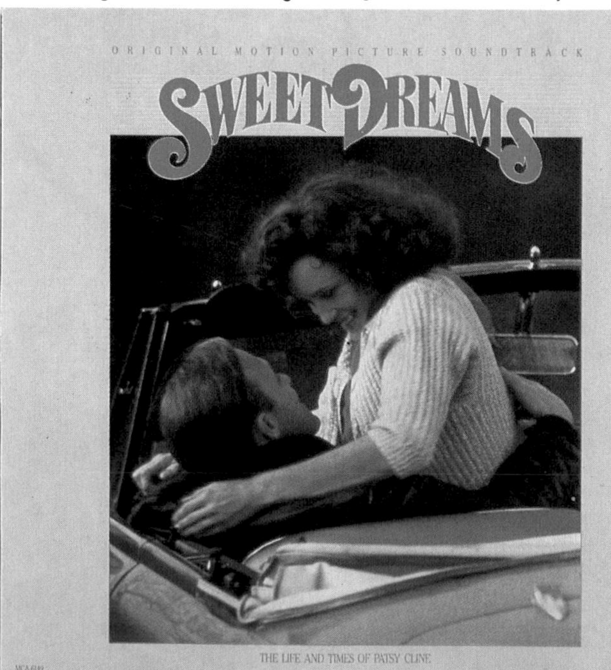

Sweet Dreams – the soundtrack to the Patsy Cline biopic. Courtesy MCA.

Jerry Clower

Country comedian Jerry Clower was born at Liberty, Mississippi, on September 28, 1926, and grew up in Amite County. After graduation in 1944 he joined the Navy. Upon discharge came a football scholarship at Southwest Junior College, Summit, where he won a further scholarship to Mississippi State University.

After becoming a field representative with the Mississippi Chemical Company, Clower rose to become Director of Field Services with the company. Sales talks became part of his stock in trade and so his **Coon Hunt Story** and other routines were introduced to make such speeches more acceptable. A friend then suggested that Clower should record an album of his routines and, it was agreed that an album be cut on the Lemon label. Named **Jerry Clower From Yazoo City Talkin'** and advertised only by word of mouth, the album sold over 8,000 copies in a short period, gaining Clower a contract with MCA in 1971. His album later went into the 'Billboard' charts for a lengthy stay.

Following the release of other high-selling LPs, he became accepted as country music's funniest man, winning many spots on TV shows. The father of four children, Clower is a Yazoo City Baptist deacon, an active member of the Gideon Bible Society and author of the best-selling book, 'Ain't God Good'. For several years he was co-host of the syndicated TV show, Nashville On The Road.

Ambassador Of Goodwill, Jerry Clower. Courtesy MCA Records.

Recommended:
Clower Power (MCA/–)
Country Ham (MCA/–)
Live in Picayune (MCA/–)
The Ambassador Of Goodwill (MCA/–)
Runaway Truck (MCA/–)

The Heart Of Hank Cochran. Courtesy Monument.

Hank Cochran

Singer-songwriter Henry 'Hank' Cochran was born in Greenville, Mississippi, on August 2, 1935. After completing school he moved to New Mexico, working in the oil-fields during the mid-'50s, and eventually made his way to California, where he began entertaining in small clubs. In 1954, he and Eddie Cochran (no relation) formed a duo, the Cochran Brothers, initially recording country material but later switching to rock after watching Elvis Presley perform in Dallas.

After two years, the Cochrans went their separate ways, Hank going on to join the California Hayride TV show in Stockton, then in 1960 moving to Nashville in order to sell his songs. One such composition, entitled **I Fall To Pieces**, written with the aid of Harlan Howard, became a 1961 winner for Patsy Cline, after which came **Make The World Go Away** (a hit for both Ray Price and Eddy Arnold), **A Little Bitty Tear** and **Funny Way Of Laughing** (Burl Ives), **I Want To Go With You** (Eddy Arnold), and others. Signed by Liberty as a recording artist in 1961, Cochran's name appeared in the charts during '62 with **Sally Was A Good Old Girl** and **I'd Fight The World**. The singer later scored with **A Good Country Song** on Gaylord in 1963, and **All Of Me Belongs To You** on Monument in 1967.

Married for many years to Jeannie Seely, Hank re-emerged in the mid-'70s writing hit songs for Willie Nelson and Mel Tillis. He once again took up performing and recording, signing to Capitol Records in 1978 and making the album **With A Little Help From His Friends**, featuring Merle Haggard, Willie Nelson, Jack Greene and Jeannie Seely. Two years later he recorded an album for Elektra, **Make The World Go Away**.

Recommended:
Make The World Go Away (Elektra/–)
With A Little Help From His Friends (Capitol/–)
Going In Training (RCA/–)
Heart (Monument/–)

Commander Cody And His Lost Planet Airmen

This zany, modern western-swing outfit was named after space movie characters. Formed in 1967 by leader George 'Commander Cody' Frayne (born Boise, Idaho), previously Farfisa organ player with various small-time rock bands, it acquired a sizeable reputation in California. The Airmen signed to ABC-Paramount in 1969, recording the fine **Lost In The Ozone** album and subsequently gaining a Top 20 hit with a version of Johnny Bond's **Hot Rod Lincoln** (1972). In 1974 they moved to Warner Bros, releasing two studio albums and a live double, **We've Got A Live One Here**, after which the band split.

Line-up changes have always affected Cody's unwieldy aggregation – among those who have worked with the band are Bobby Black (pedal steel), Lance Dickerson (drums), Bruce Barlow (bass), Norton Buffalo (harmonica, trombone) and vocalist Nicolette Larson – but somehow he has always managed to put something together. In the early '80s he logged a hit in Europe with **Two Triple Cheeseburgers (Side Order Of Fries)**, on MCA release.

Recommended:
Lost In The Ozone (Paramount/ABC)
Hot Licks, Cold Steel (Paramount/ABC)
Live – Deep In The Heart Of Texas (Paramount/ABC)
Tales From The Ozone (Warner Bros/Warner Bros)
We've Got A Live One Here (Warner Bros/Warner Bros)
Aces High (Relix/–)

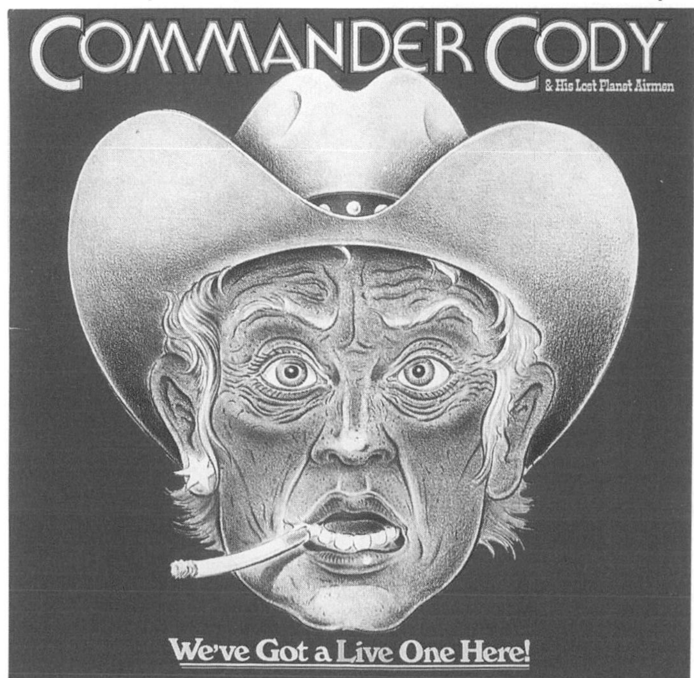

We've Got A Live One Here, Commander Cody. Courtesy Warner.

David Allan Coe

This flamboyant and outrageous country star, born in Akron, Ohio, on September 6, 1939, made his initial impression in country music writing the hit songs, **Would You Lay With Me In A Field Of Stone** (Tanya Tucker) and **Take This Job And Shove It** (Johnny Paycheck). He spent most of his youth in various reform schools and prisons. Released from jail in 1967, Coe attempted a career as a singer and eventually went to Nashville. At that time, however, Coe was primarily a blues artist. He signed a recording contract with Shelby Singleton in Nashville and came up with **Penitentiary Blues**. He followed this with **Requiem For A Harlequin**.

Next came a switch to country music. He signed with Columbia Records in 1972 and with the album **The Mysterious Rhinestone Cowboy** he launched a successful career that saw him turning out top-selling albums such as **Once Upon A Rhyme**, **Longhaired Redneck**, **David Allan Coe Rides Again**, **Human Emotions** and **Family Album**. He finally made a big impression on the country singles chart with **The Ride**, the tale of a ghostly meeting with Hank Williams, which topped the charts in 1983. He had further success with the reflective **Mona Lisa's Lost Her Smile** in 1984, but has remained very much an album artist.

Recommended:
Once Upon A Rhyme (Columbia/CBS)
Family Album (Columbia/–)
Human Emotions (Columbia/–)
Rough Rider (Columbia/–)
Tennessee Whiskey (Columbia/–)
Castles In The Sand (Columbia/CBS)
A Matter Of Life And Death (Columbia/CBS)

Mark Collie

A rock'n'roll kid, Mark Collie was born on January 18, 1956, in Waynesboro, Tennessee, and spent the best part of 20 years perfecting his craft before making the commercial breakthrough in 1992 with **Even The Man In The Moon Is Cryin'**.

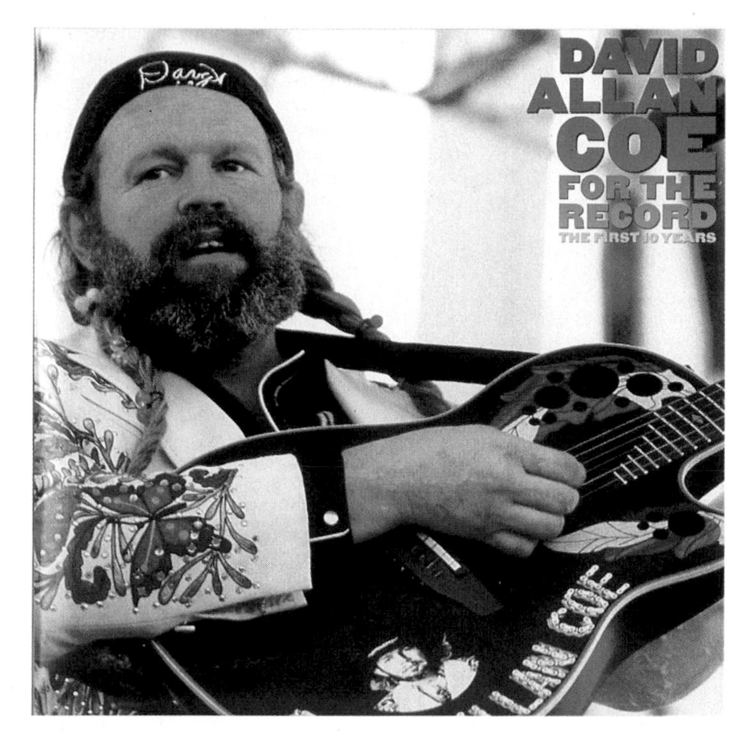

For The Record, David Allan Coe. Courtesy CBS Records.

Waynesboro, a small town near the Tennessee River, is where Mark cut his musical teeth. Playing piano and guitar as a child, music was his life, and during his high school days he was a DJ on local radio station WAAM. In his teens he played clubs around his hometown, then headed west to Memphis, where he formed a band and played local clubs. A few years later he took a vacation to Hawaii and spent the next 18 months singing country and rock'n'roll at a beach club.

When he returned home he decided to present his talent in Nashville. Moving to Music City in 1982, he signed as a writer with a large publisher, but, though he had a few album cuts, nothing much happened during the next five years. Looking for ways to gain recognition, he started doing showcases at Douglas Corner (a Nashville writers' venue) and was spotted by Tony Brown and Bruce Hinton, who offered him a contract with MCA Records. His first album, **Hardin County Line**, was an

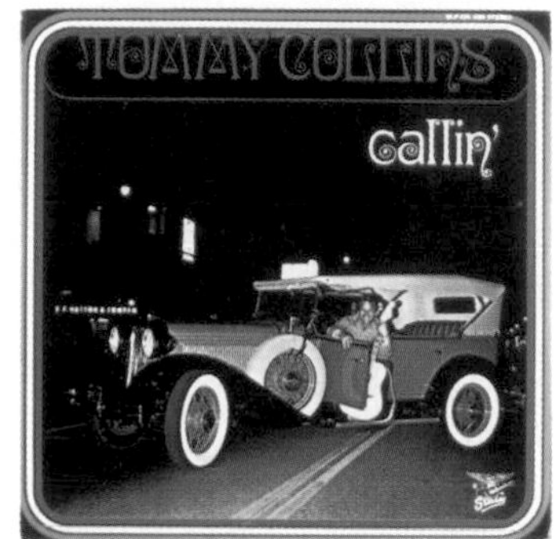

Callin', Tommy Collins. Courtesy Gusto-Starday Records.

impressive mix of modern country and rockabilly, and Mark made his chart debut with **Something With A Ring To It** (1990). Gradually he built up a following, making Top 20 with **Let Her Go** (1991) and then into the Top 5 with **Even The Man In The Moon Is Cryin'** (1992).

Recommended:
Hardin County Line (MCA/–)
Born And Raised In Black & White (MCA/–)

Tommy Collins

Tommy Collins (real name Leonard Raymond Sipes) was born in Oklahoma City, Oklahoma, on September 28, 1930. A boyhood guitarist, he began performing at local clubs while at Oklahoma State College, also making appearances on radio stations. During the early '50s, he became a resident of Bakersfield and a Capitol recording artist, his halcyon days being in 1954–55 when Collins followed **You Better Not Do That**, a half-million seller, with a quartet of other Top 20 discs. In 1966 he presented Columbia with a Top 10 item in **If You Can't Bite, Don't Growl**, but his later discs only achieved moderate chart placings.

As a songwriter, he provided Merle Haggard with some of his best numbers, including the chart-topping **Roots Of My Raising** in 1976.

Collins' past albums include **This Is Tommy Collins** (Capitol), **Dynamic** (Columbia), **On Tour** (Columbia) and two albums recorded in Britain during 1980, **Country Souvenir** and **Cowboys Get Lucky Some Of The Time**.

Jessi Colter

Writer and singer of the 1975 No.1, **I'm Not Lisa**, Jessi Colter (real name Miriam Johnson) was born in Phoenix, Arizona, on May 25, 1947, the sixth of seven children.

Jessi, Jessi Colter. Courtesy Capitol Records.

A church pianist at the age of 11, Jessi married guitarist Duane Eddy just five years later, touring England, Germany, South Africa and other countries as part of the Eddy show. A singer-songwriter, she was recorded by Eddy and Lee Hazelwood for the Jamie and RCA labels, meeting Waylon Jennings during some Phoenix-based sessions.

She split from Eddy in 1965 and, changing her name to Jessi Colter because her great, great uncle was a member of the Jesse James Gang, she moved to Nashville, signed a new contract with RCA and recorded the critically acclaimed album, **A Country Star Is Born**, in 1966.

By this time she had made an impression writing songs recorded by Eddy Arnold, Dottie West, Anita Carter, Patsy Sledd and Don Gibson. She teamed up with Waylon Jennings and the pair married in 1969, recording a series of duets including **Suspicious Minds** and **Under Your Spell Again**.

Jessi signed a solo contract with Capitol Records in 1974 and scored hits with **I'm Not Lisa**, which she wrote in five minutes, **What Happened To Blue Eyes** (1975), **It's Morning** (1976) and **Maybe You Should've Been Listening** (1978). She appeared on the best-selling RCA album titled **Outlaws** (1976) and has continued touring and recording with Waylon.

Recommended:
I'm Jessi Colter (Capitol/–)
Jessi (Capitol/Capitol)
Diamond In The Rough (Capitol/–)
That's The Way A Cowboy Rocks 'n' Rolls (Capitol/–)
Ridin' Shotgun (Capitol/–)
Leather & Lace – with Waylon Jennings (RCA/RCA)

Confederate Railroad

Mark DuFrene, drums; Michael Lamb, guitar; Chris McDaniel, keyboards; Gates Nichol, steel guitar; Wayne Secrest, bass; Danny Shirley, lead vocals and guitar.

Every now and then there is an act that draws a whole generation of rock'n'rollers to country music. Such an act is Confederate Railroad, a Georgia-based outfit that took a long, hard twelve-year ride to Nashville and a major record label deal in 1992. Their road-house mix of honky-tonk, rock, boogie and modern country has made them one of the fastest-rising new bands of the '90s.

Front-man Danny Shirley, born in Soddy Daisy, Tennessee, but raised in Chattanooga, Georgia, worked the club circuit with his own band and was in the back-up band for David Allan Coe for several years. The Danny Shirley Band had a regular stint at Miss Kitty's, Marietta, Georgia for six years, and Shirley also landed a deal with independent Amor Records, resulting in three albums and several minor country hits between 1984 and 1988.

The current band has been together the best part of eight years. Thousands of miles on the road have given these guys their rollicking approach on good-time songs, which conjures up vivid images of Southern manhood – aggressively independent, boisterous, self-opinionated, yet with a strong sense of decency shining through. It's all good ole whisky-drenched honky-tonk music.

Right: Earl Conley added a middle name, Thomas, to avoid confusion with John Conlee and Con Hunley.

The Danny Shirley Band changed its name to Confederate Railroad when they signed a record deal with Atlantic Nashville in the summer of 1991. The name refers to a Civil War locomotive, long associated with Chattanooga and Kennesaw, Georgia, where Shirley has lived since 1985. Their first self-titled album featured Nashville session players, and the initial single, **She Took It Like A Man**, helped by a tongue-in-cheek video, climbed into the Top 10. That was followed by the out-of-character ballad **Jesus And Mama** which reached No.3 (1992) and **The Queen Of Memphis** (1993), a raucous bar-room ditty.

Recommended:
Confederate Railroad (Atlantic/–)

John Conlee

The singer with the saddest voice in modern country music, Conlee was born on August 11, 1946 in Versailles, Kentucky and raised on a tobacco farm. By his early teens he was a member of a folk trio and later joined a rock group.

Following graduation he worked for six years as a mortician but finally landed himself a DJ stint on a Fort Knox Station. A move to Nashville in 1971, playing rock music, enabled him to make important music contacts, which led to signing with ABC Records.

His initial releases failed to make much impression but his fourth release, **Rose Colored Glasses**, a song he co-wrote with a newsreader at the radio station, made the country Top 5 in May 1978. That same year ABC Records were absorbed by MCA, for whom John scored more than a dozen Top 10 hits, including such chart-toppers as **Lady Lay Down** (1978), **Backside Of Thirty** (1979), **Common Man** (1983), **I'm Only In It For The Love** (1983), **In My Eyes** (1984), **As Long As I'm Rockin' With You** (1984). At the beginning of 1986, John signed with Columbia Records and continued to dominate the Top 10 with songs **Got My Heart Set On You** (1986), **The Carpenter** (1987) and **Domestic Life** (1987). This contract only lasted three years, then he joined 16th Avenue Records, but by 1990 was unable to make an impact.

Throughout his career Conlee has championed the ordinary working man. This has been typified in such songs as **Busted**, **Common Man**, **Working Man** and **American Faces**, and for several years he was chairman of the Family Farm Defense Fund. He still continues to tour regularly and make TV appearances. When he joined the Grand Ole Opry in 1979, John was the first new member in five years.

Recommended:
Rose Colored Glasses (MCA/MCA)
Busted (MCA/–)
In My Eyes (MCA/–)
Blue Highway (MCA/–)
Harmony (Columbia/–)
Fellow Travelers (16th Avenue/–)
American Faces (Columbia/–)
Songs For The Working Man (MCA/–)

Confederate Railroad's 1992 debut album. Courtesy Atlantic Records.

Earl Thomas Conley

Country music singer, songwriter and philosopher, Conley, who was born October 17, 1941, in Portsmouth, Ohio, made a major breakthrough in the 1980s with a string of chart-topping singles and acclaimed albums. His father was a railroad man, but with the advent of diesel locomotives was put out of work in the early '50s, which led to the large Conley family being on the breadline for many years. The young Earl took to carving, painting and drawing; in his teens he drifted around living off money earned from his artistic skills. During Army service in Germany he became interested in country music, and, following demob, he started singing in clubs around Huntsville, Alabama.

A move to Nashville led to songwriting. He provided Mel Street with **Smokey Mountain Memories**, a Top 10 hit in 1975, and Conway Twitty with the chart-topping **This Time I've Hurt Her More Than She Loves Me** the following year. Earl was then signed to GRT Records and had a few minor hits including **I Have Loved You Girl** (1975) and **High And Wild** (1976).

He added the Thomas to the middle of his name in 1978 to save confusion with John Conlee and Con Hunley and signed with Warner Brothers, scoring his biggest hit with **Dreamin's All I Do** (1979). A move to the small Sunbird label the following year and the release of his debut album, **Blue Pearl**, led to a Top 10 hit with **Silent Treatment** (1980) and a No.1 chart-placer with **Fire And Smoke** the following August.

Conley's contract with Sunbird was acquired by RCA and the next few years saw him emerge as the most successful country singer of the '80s. Using mainly self-written tunes, he's been hailed for his sensitive, introspective writing talents, blending traditional strains with elements of modern country and soul, bringing a fresh spirit to the country forms he celebrates.

He hit the top of the charts with almost every release, the most notable being **Somewhere Between Right And Wrong** (1982), **Holding Her And Loving You** (1983), **Nobody Falls Like A Fool** (1985), **Once In A Blue Moon** (1986), **Right From The Start** (1987), **What She Is (Is A Woman In Love)** (1988), **We Believe In Happy Endings** (a duet with Emmylou Harris, 1988) and **Love Out Loud** (1989). Following the 1988 release of **The Heart Of It All** (Conley's most successful album, providing four No.1 singles), he had problems with his voice due to heavy smoking, and also a series of differences with RCA which held up further releases until the **Yours Truly** album in 1991. This album gave him the Top 10 hits **Shadow Of A Doubt** and **Brotherly Love**, the latter a duet with Keith Whitley. At the beginning of 1993, Conley, one of the most consistent country hit-makers of the previous twelve years, found himself without a record deal.

Recommended:
Blue Pearl (Sunbird/–)
Fire And Smoke (RCA/–)
Don't Make It Easy For Me (RCA/–)
Treadin' Water (RCA/–)
Somewhere Between Right And Wrong (RCA/–)
Greatest Hits (RCA/–)
The Heart Of It All (RCA/RCA)
Yours Truly (RCA/–)

Swinging The Devil's Dream, Spade Cooley. Courtesy Charly Records.

Above: Ry Cooder, the master of the guitar and American film music.

Ry Cooder

One-time Los Angeles session man, Cooder is of interest to country fans in that his music is soaked generally in southern folk styles. Born in Los Angeles, California, on March 15, 1947, he first came to public notice with his tasteful and well-timed mandolin excursions on bluesman Taj Mahal's first CBS album and he played with Taj's band, the Rising Sons.

His solo career began in 1970 with the first in a series of highly acclaimed albums for Warner Brothers, which featured long forgotten or little known gospel, old-timey country, Cajun and blues tunes given a very distinctive treatment. He also became proficient on slide guitar and is now a rated exponent of that style. One of his finest albums is **Chicken Skin Music**, which introduced Tex-Mex music to a wide audience and featured accordionist Flaco Jiminez and his Conjuncto musicians from San Antonio.

Ry has also been in demand for writing and playing on film soundtracks such as 'The Long Riders', for which he created western music of a style played in saloons during the gunfighter era of Jesse James and Cole Younger, and the country blues used as the backcloth for the 1984 movie, 'Paris, Texas'.

Recommended:
Chicken Skin Music (Reprise/Reprise)
The Long Riders (Reprise/Reprise)
Into The Purple Valley (Reprise/Reprise)
Paradise And Lunch (Reprise/Reprise)

Spade Cooley

One-time King Of Western Swing, Donell C. Cooley had literally one of the biggest bands in country music, sometimes numbering round two dozen musicians.

Born in Grand, Oklahoma, on February 22, 1910, Cooley, of Scottish-Irish descent plus Cherokee Indian, acquired the nickname of Spade from an exceptional run of spades he once held during a poker game.

Both his father and his grandfather were fine fiddle players and Cooley, who was first taught cello, also became adept on fiddle, playing at square dances while still a boy. During the late '30s he became a Hollywood extra.

By 1942 Cooley was leading his own very successful band, in '46 leasing the Santa Monica ballroom for long-term use as the band's headquarters. A radio performer during the early '40s, he was signed for a Hollywood country TV show in 1948, shortly after returning from a lengthy tour with his band. Playing a mixture of jazz, country and pure dance music, the Cooley outfit gained further fame, but the fiddle-playing leader's career came to a dramatic end when he was jailed for wife-slaying. His death occurred shortly after his release from prison: Cooley suffered a heart attack on November 23, 1969 while playing a sheriff's benefit concert in Oakland, California.

During his career, his records were released on such labels as Okeh, Columbia, RCA Victor and Decca, Cooley's biggest hit being with the self-penned **Shame On You** in 1945.

Recommended:
The King Of Western Swing (Club Of Spade/–)
Swinging The Devil's Dream (–/Charly)
Rompin', Stompin', Singin', Swinging' (–/Bear Family)

Rita Coolidge

Nashville-born, on May 1, 1944, Rita Coolidge and her two sisters first began singing in a church choir. Rita later worked with a rock band, the Moonpies, playing at fraternity parties at the University of Florida and at Florida State. In college, she also sang with a folk rock unit, then, after leaving, went to Memphis and began working for Pepper Records, cutting some sides with the label. Her sister Priscilla began working with and then married the popular R&B organist-bandleader Booker T. Jones, which brought Rita into further contact with the rock scene.

Following a tour with Joe Cocker's Mad Dogs And Englishmen in 1970, on which she was featured as a vocalist and pianist, Rita was signed to A&M as a solo artist, her first album, **Rita Coolidge**, being released in 1971. Forming a band known as the Dixie Flyers, she then began touring, playing engagements in Britain and Canada during 1971. Some time later she met Kris Kristofferson, whom she married in 1973.

The duo appeared on several tours and many TV shows, also recording together and scoring such hits as **A Song I'd Like To Sing** (1973) and **Loving Arms** (1974). Known as the Delta Lady (Leon Russell named the song in her honour), Rita is considered to be one of the leading performers in country rock with such solo hits as **Higher And Higher** and **We're All Alone** (1977), **The Way You Do The Things You Do** (1978) and **One Fine Day** (1980). Kris and Rita went their separate ways and were divorced in 1980.

Recommended:
Rita Coolidge (A&M/–)
Fall Into Spring (A&M/–)
The Lady's Not For Sale (A&M/A&M)
Breakaway – with Kris Kristofferson (Monument/Monument)
Natural Act – with Kris Kristofferson (A&M/A&M)
Very Best (A&M/A&M)

The Lady's Not For Sale, Rita Coolidge. Courtesy A&M Records. Rita is known as the 'Delta Lady'.

The Coon Creek Girls

Violet Koehler, vocals, guitar, mandolin; Daisy Lange, vocals, bass; Black Eyed Susan Ledford, bass, Lily Mae Ledford, vocals, banjo, fiddle; Rosie Ledford, vocals, guitar.

The Coon Creek Girls were an extremely popular all-girl string band of the '30s and '40s, led by the singer and multi-instrumentalist Lily Mae Ledford of Pinch-em-tight Holler, Kentucky.

Lily Mae began her career as a fiddler on the National Barn Dance in 1936, but it was John Lair who conceived the idea of an all-girl band built around Lily Mae. It first consisted of Daisy Lange and Violet Koehler as well as younger sister Rosie Ledford, but in later years the band consisted of the three Ledford sisters only. They spent their entire career on the Renfro Valley Barn Dance (1938–58), first over WLW and then over WHAS, and regrouped sporadically for events like the Newport Folk Festival.

Their most popular record was their theme song, **You're A Flower That Is Blooming There For Me** (Vocalion).

Wilma Lee And Stoney Cooper

Born at Harman, West Virginia, on October 16, 1918, singer-songwriter-fiddler Dale T. 'Stoney' Cooper came from a farming family of some considerable musical ability. By the time he was 12, he had become an accomplished musician and upon leaving school joined the Leary Family, a religious singing group, who were in need of a fiddle player. One member of the group was Wilma Lee Leary (born at Valley Head, West Virginia, on February 7, 1921), a singer-songwriter, guitarist and organist, whom Stoney courted and eventually married in 1939.

For several years the duo remained members of the Leary Family, appearing on radio shows and performing in churches and other venues. then, in the mid-'40s, the Coopers decided to go their own way and began playing dates on many radio stations throughout the country, in 1947 joining the WWVA Jamboree at Wheeling West Virginia, on a regular basis.

After a ten-year stay with WWVA, during which time they switched record labels from Columbia to Hickory (1955), the Coopers headed for Nashville, becoming members of the Opry in 1957. Soon after they scored hits with **Come Walk With Me, Big Midnight Special** and **There's A Big Wheel**, all three discs being Top 5 country hits during 1959. Further successes followed, including **Johnny My Love** (1960), **This Old House** (1960), and **Wreck On The Highway** (1961), then, though the duo left Hickory for Decca, the hit supply seemed to dry up. However, the Coopers continued to be a tremendous onstage attraction and in the mid '70s were still rated as one of the Opry's most popular acts. Stoney Cooper's death, caused by a heart attack, on March 22, 1977 put the future of the act in doubt, although Wilma Lee continued with daughter Carol Lee Snow (married to Hank Snow's son, Jimmie Rodgers Snow) and the Clinch Mountain Clan.

Recommended:
Satisfied (–/DJM)
Wilma Lee And Stoney Cooper (Rounder/–)
A Daisy A Day (Leather/–)
Early Recordings (County/–)
Wilma Lee Cooper (Rounder/–)

Cowboy Copas

A victim of the 1963 plane crash that also claimed the lives of Patsy Cline and Hawkshaw Hawkins, Copas' 1960 recording of **Alabam** had just placed him

Wilma Lee And Stoney Cooper. Courtesy Rounder Records.

back on top of the heap again, following his virtual disappearance from the country charts during the late '50s. Born in Muskogee, Oklahoma, on July 15, 1913, Lloyd 'Cowboy' Copas was brought up on a ranch, where his grandfather taught him Western folklore and songs. Learning guitar at the age of 16, Copas then formed a duo with an Indian fiddler, Natchee. After winning a talent contest, the twosome then played a series of dates throughout the country. Copas himself performed on 204 radio stations in North America between 1939 and 1950. When Natchee went his own way in 1940, Copas obtained a regular spot on a Knoxville station, eventually being featured on WLM's Boone County Jamboree and signing for King Records, for whom he made a number of hits including **Filipino Baby, Tragic Romance, Gone And Left Me Blues** and **Signed, Sealed And Delivered**.

Becoming a featured vocalist with Pee Wee King's Golden West Cowboys on the Opry in 1946, he furthered his reputation with **Kentucky Waltz** and **Tennessee Waltz**. for a while he became a performer with an SRO reputation, but after the advent of **Strange Little Girl**, a 1951 chart-climber, Copas dropped from sight. Signed to Starday in 1959, his hit recording of **Alabam** seemed set to launch him on a new career. Then came the plane disaster.

Recommended:
The Best Of Cowboy Copas (Starday/Starday-Midland)

Helen Cornelius

Finding fame in 1976 as Jim Ed Brown's singing partner and as a member of the Nashville On The Road TV Show, Helen Cornelius has also made a considerable impression as a songwriter, her compositions being recorded by Dottsy, LaCosta, Liz Anderson, Bonnie Guitar, Barbara Fairchild and many other artists.

Below: The married duo Wilma Lee and Stoney Cooper.

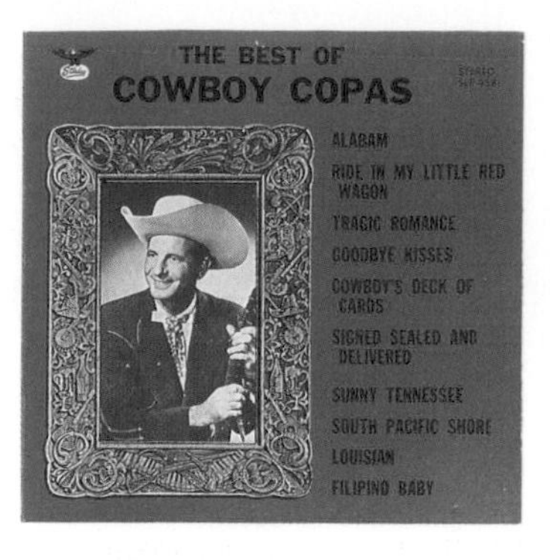

The Best Of Cowboy Copas. Courtesy Gusto Records.

Brought up on a farm in Hannibal, Missouri, where she was born on December 6, 1941, Helen was one of an eight-strong musical family, with her sisters Judy and Sharon combining with Helen to form a vocal trio, their father chauffeuring them from town to town in order to provide the threesome with opportunities to perform. As a solo act, Helen began obtaining work at various local gigs, playing some radio and TV dates, and won the Ted Mack Amateur Hour. But by 1970 she was songwriting as a profession, signing a contract with Columbia-Screen Gems and cutting some discs with Columbia Records.

Signing for RCA Records in September, 1975, her first release was **We Still Love**

Songs In Missouri, a well received single. However, being cast as the ideal vocal foil for Jim Ed Brown proved to be the real breakthrough, the duo's version of **I Don't Want To Have To Marry You** becoming both a controversial release – many radio stations banning the disc – and a country No.1. In early 1977, **Saying Hello, Saying I Love You, Saying Goodbye** found the Cornelius-Brown partnership chart-topping yet again. Further hits followed with **Born Believer** (1977), **If The World Ran Out Of Love Tonight** (1978) and **Lying in Love With You** (1979). Helen has been concentrating on a solo career since 1982, but has so far failed to score a major hit.

Recommended:
I Don't Want To Have To Marry You – with Jim Ed Brown (RCA/RCA)
Born Believer – with Jim Ed Brown (RCA/–)
Helen Cornelius (MCA-Dot/–)

Above: Country Gazette in 1974 – Alan Munde, Byron Berline, Roger Bush and Kenny Wertz.

Country Gazette

With the famous Flying Burrito Brothers now in abeyance, it was left to Country Gazette, the remains of the Burrito aggregation, to perpetuate the line. Byron Berline had originally been a championship-winning fiddle player and in-demand session man and he teamed up with Roger Bush (string bass), Kenny Wertz (guitar, vocals), Roland White (guitar, mandolin, vocals) and Alan Munde (banjo, vocals), to bring a souped-up, '70s-style bluegrass sound to the nation.

Since bluegrass was commonplace in America, their appeal there was rather limited, but in Britain and Europe they found themselves preaching to a new audience.

They have perhaps never bettered their first album for UA, **Traitor In Our Midst**, but in spite of personnel changes they have always maintained a standard of instrumental slickness and attractive vocal harmonies. On more than one occasion they have been voted Top Country Group in British award polls. Berline eventually left the band to form Sundance and was replaced for a time by Dave Ferguson. Far more popular abroad than at home, the band has continued regardless of numerous line-up changes.

Recommended:
All This And Money Too (Ridge Runner/–)
American And Clean (Flying Fish/–)
Traitor In Our Midst (UA/–)
Don't Give Up Your Day Job (UA/–)
From The Beginning (–/Sunset)

Country Gentlemen

One of the first progressive bluegrass groups, the Country Gentlemen played their first date on July 4, 1957.

During the group's embryonic days the personnel consisted of Charlie Waller, John Duffy, Bill Emerson and Eddy Adcock, though by the mid-'70s, lead singer and flat top guitarist Waller (born in Jointerville, Texas, on January 19, 1935) was the only remaining member of the original line-up.

Initially from the Washington DC area, the group played two nights a week for ten years at a Georgetown niterie known as the Shamrock Club – but during the '60s the Gentlemen gained a nationwide reputation, being featured on many networked TV shows, the group's popularity gaining ground along with the bluegrass explosion. Voted Best Band by 'Muleskinner News' in 1972 and 1973, they became the first Washington bluegrass band to play in Japan.

Eclectic in their choice of material – drawing equally from Bob Dylan, Charlie Poole, Letty Frizzell and even Hollywood film composers – the Country Gentlemen have recorded for such labels as Folkways, Rebel, Mercury, Starday and Vanguard.

Recommended:
Country Songs Old And New (Folkways/–)
Folksongs and Bluegrass (Folkways/–)
Bringing Mary Home (Rebel/–)
Play It Like It Is (Rebel/–)
New Look, New Sound (Rebel/–)
The Award Winning Gentlemen (Rebel/–)
Live At Roanoke (Zap/–)
Sit Down Young Stranger (Sugar Hill/–)

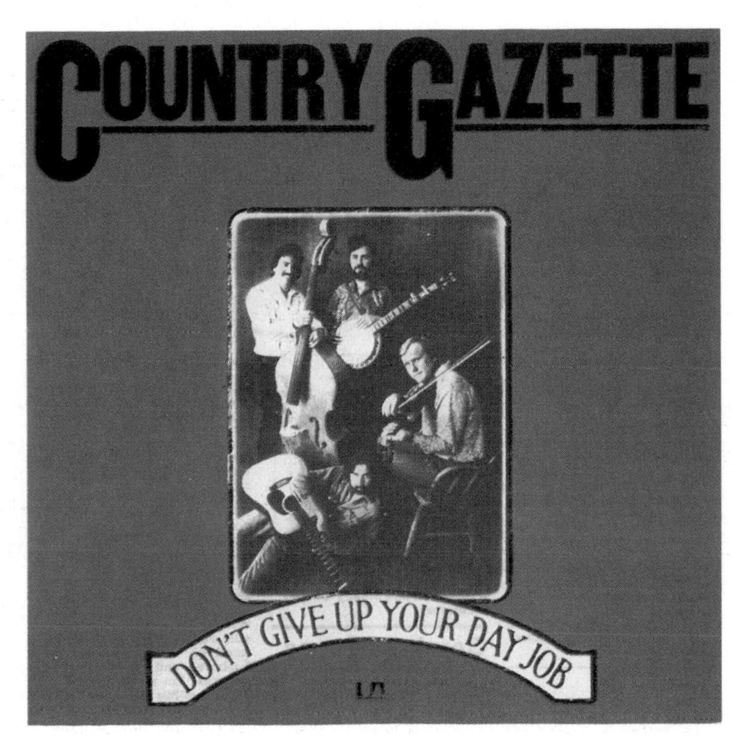

Don't Give Up Your Day Job, Country Gazette. Courtesy UA Records.

Cousin Jody

James Clell Summey began his career as a straight musician and was one of the fine early practitioners of the dobro. In fact, it was as a member of Roy Acuff's Smoky Mountain Boys that he recorded several of the early Acuff hits, including **Wabash Cannonball** and **The Great Speckled Bird**. It was he, as well, who first joined the Opry with Acuff in 1938.

He branched out as a musician, playing electric steel (one of the first to do so on the Opry stage) with Pee Wee King's Golden West Cowboys before delving into comedy with Oral Rhodes as Odie and Jody, then joining Lonzo and Oscar for a number of years prior to gaining his own solo Opry spot for over a decade.

Summey was born near Sevierville, Tennessee, on December 14, 1914. Plagued by ill health in his late years, he died of cancer in 1976.

Billy 'Crash' Craddock

Known as 'Mr Country Rock', Craddock was born on June 16, 1939 in Greensboro, North Carolina. Initially part of a group called the Four Rebels, along with his brother Clarence, Billy later signed to Columbia as a solo artist, scoring heavily with a 1959 single, **Don't Destroy Me**, which made a big impression in Australia. He claims he then quit the business because Columbia insisted on casting him in a pop role, while he saw his future in country music.

After taking many menial jobs, he eventually signed for Ron Chancey's Cartwheel label, coming up with a whole string of rock-oriented country hits, including **Ain't Nothin' Shakin'**, **Dream Lover** and **Knock Three Times**, a revamped version of Dawn's pop hit. Cartwheel was then absorbed into ABC-Dot and Craddock enjoyed further hits with **Rub It In** (1974), **Easy As Pie** (1976) and **Broken Down In Tiny Pieces** (1977). He joined Capitol Records in 1978 and continued with such hits as **I Cheated On A Good Woman's Love** (1978) and **I Just Had You On My Mind** (1980). Highly respected for his stunning in-concert antics, which recall the early days of Elvis Presley, he is regarded as one of the most exciting performers in country music.

Recommended:
Two Sides Of Crash (ABC/–)
The Country Side Of (–/MFP)
Greatest Hits Volume 1 (ABC/–)
The New Will Never Wear Off Of You (Capitol/–)
Laughing And Crying (Capitol/–)

Floyd Cramer

Pianist on a large proportion of Nashville hits during the late '50s – including Elvis Presley's **Heartbreak Hotel** – Cramer was born at Shreveport, Louisiana, on October 27, 1933, and grew up in Huttig, Arkansas, playing his first dates at local dances.

On completing high school in 1951 he returned to Shreveport where he appeared on KWKH's Louisiana Hayride show, played on sessions at the Abbott Record

20 Of The Best, Floyd Cramer. Courtesy RCA Records.

company and fitted in tours with Presley and other major acts.

Following various Nashville session dates, Chet Atkins advised Cramer to become a regular Music City sideman. This he did in 1955, quickly establishing himself as one of the city's most active musicians, helping to create the new Nashville Sound with his distinctive 'slipnote' piano style. He also toured and performed on radio and TV shows including Grand Ole Opry.

Signed to RCA, his first hit record was **Flip, Flop And Bop** in 1958, following this with two self-penned million sellers, **Last Date** (1960) and **On The Rebound** (1961).

Winner of countless polls and awards, Cramer has enjoyed many other high selling discs including **San Antonio Rose** (1961), **Chattanooga Choo Choo, Hot Pepper** (1962) and **Stood Up** (1967).

In 1977 Floyd emerged with **Keyboard Kick Band**, an album on which he played no less than eight keyboard instruments including various ARP synthesizers.

Recommended:
Super Hits (RCA/–)
Best Of Class Of (RCA/–)
Last Date (RCA/–)
Plays The Big Hits (Camden/–)
In Concert (RCA/–)
Piano Masterpieces (RCA/–)
This is Floyd (RCA/–)
Keyboard Kick Band (RCA/–)
Forever Floyd Cramer (Step One/–)

Rob Crosby

Blond singer-songwriter Rob Crosby, born on March 25, 1954, in Sumter, South Carolina, is one of the '90s most promising and formidable new voices. He received his first guitar when he was nine years old, and instead of hitting the college books he hit the college club circuit with his own band. Eventually he ended up in Nashville, making his mark as a writer, with Lee Greenwood, taking **Holding A Good Hand** to No.2. Crosby signed with Arista Records and made Top 20 with **Love Will Bring Her Around** (1990). His strong voice with its expressive, gravelly edge helps him put his strong and vivid lyrics across. There have been further Top 20 successes with **She's A Natural** (1991) and **Still Burnin' For You** (1992).

Another Time And Place, Rob Crosby. Courtesy Arista Records.

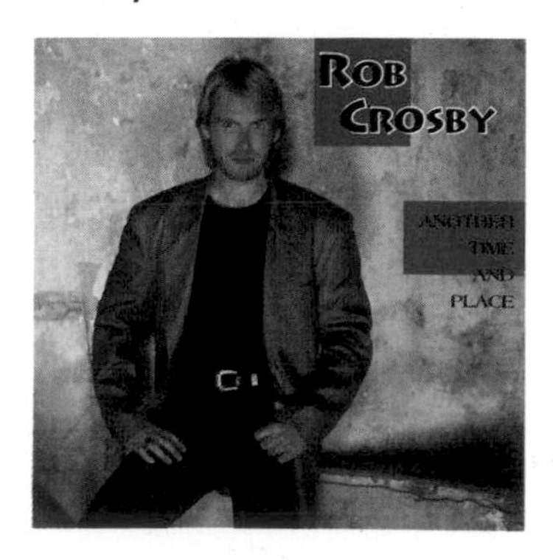

Recommended:
Solid Ground (Arista/–)

Rodney Crowell

One of Nashville's most successful contemporary songwriters and producers, Rodney was born on August 7, 1950, in Houston, Texas and started his musical career in his teens playing in a high school band, the Arbitrators, in 1965. A few years later he was playing drums in his Kentucky-born father's band in Houston clubs. He dropped out of college and moved to Nashville in the early '70s, determined to become a songwriter. He was signed to Jerry Reed's publishing firm, but struggled to make a living. His luck changed when a chance meeting with Brian Ahern, husband of Emmylou Harris, led to a move to California where he joined Emmylou's Hot Band in 1975.

Rodney worked with Emmylou for just over two years, during which time he wrote several songs, such as **Bluebird Wine**, **'Til I Gain Control Again** and **Leaving Louisiana In The Broad Daylight**, which Emmylou recorded. He embarked on a solo career towards the end of 1977, coming up with the critically-acclaimed **Ain't Living Long Like This** album. In 1979 he married Rosanne Cash.

Two further solo albums followed on Warners, though at the time he was having more success with rock fans than country. He scored a pop Top 40 hit with **Ashes By Now** (1980). A fourth album was recorded in 1984, but Warners rejected it, so Crowell changed four tracks and signed with Columbia, Rosanne Cash's label. **Street Language** (1986) was a superbly produced album, but it failed to produce any major country hits, and Rodney was now living in the shadow of his more famous wife. He decided to change direction, and moved away from his artistic, meaningful songs towards a more relaxed, country sound for **Diamonds & Dirt** (1988), an album that produced five No.1 country singles: **It's Such A Small World** (a duet with Rosanne Cash, 1988), **I Couldn't Leave You If I Tried**, **She's Crazy For Leavin'** (1988), **After All This Time** and **Above And Beyond** (both 1989). He received five nominations at the 1989 CMA Awards, but was still not fully accepted within country music.

His commercial success was not to last. The impressive **Keys To The Highway** (1989) album produced just two Top 10 hits, the excellent **Many A Long & Lonesome Highway** (1989) and **If Looks Could Kill** (1990). Another single, the superb **Things I Wish I'd Said**, a moving tribute to his late father, just managed to scrape to No.72 in 1991. By this time Rodney's marriage to Rosanne was gong through rough times, leading to the pair filing for divorce in November 1991. Crowell has always tended to write from personal experiences and **Life Is Messy**, a 1992 album, reflected many of the pair's marriage problems in one of his strongest sets of songs. **What Kind Of Love** (a song he had written in partnership with Will Jennings and the late Roy Orbison) made the country Top 20 (1992), but this album was not to be so readily accepted by the country fans or the critics as the more commercially slanted **Diamonds & Dirt**.

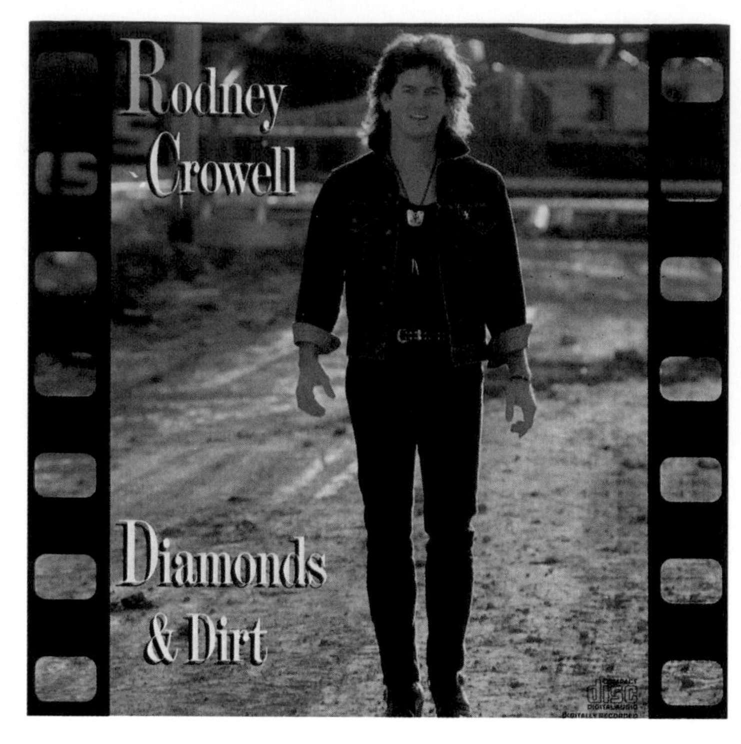

Diamonds & Dirt, Rodney Crowell. Courtesy CBS Records.

As a songwriter, Crowell has penned songs covered by Waylon Jennings, Crystal Gayle, the Oak Ridge Boys, Johnny Cash, Bob Seger, Nitty Gritty Dirt Band and others. He has produced albums for Guy Clark, Bobby Bare and Sissy Spacek. He seems to put others' musical endeavours in front of his own, and has not given his own career the commitment it has needed.

Recommended:
Ain't Living Long Like This (Warner Bros/ Warner Bros)
Rodney Crowell (Warner Bros/–)
Keys To The Highway (Columbia/CBS)
Life Is Messy (Columbia/Columbia)
Diamonds And Dirt (Columbia/CBS)

Ramblin' Country, Dick Curless. Courtesy Capitol Records.

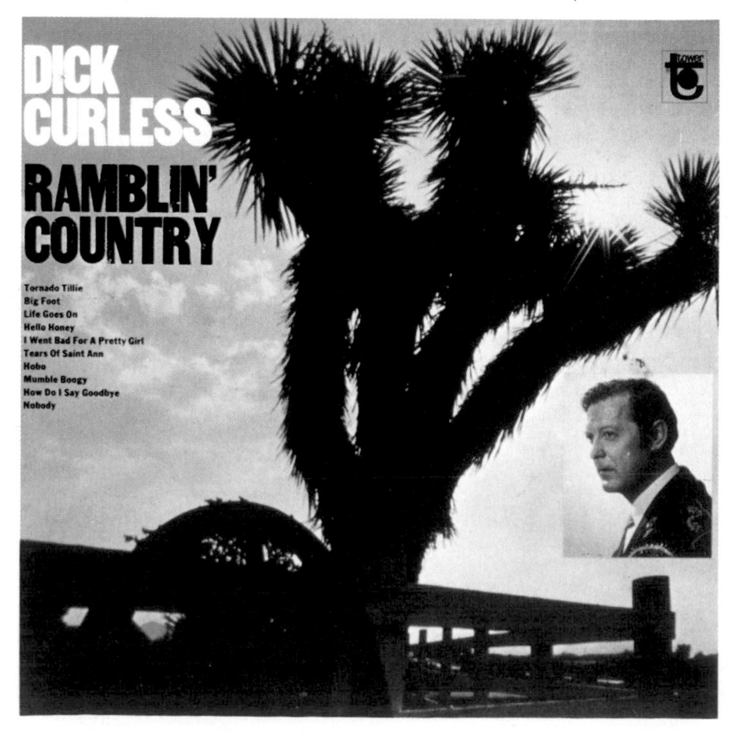

The Cumberland Ridge Runners

Hugh Cross, banjo; Karl Davis, mandolin; Red Foley, bass; Doc Hopkins, guitar, vocals; John Lair, leader, announcer, harmonica; Slim Miller, fiddle; Linda Parker, vocals, guitar, banjo, dulcimer; Harty Taylor, guitar.

The Cumberland Ridge Runners were a popular string band brought to the National Barn Dance by John Lair in 1930. It had more or less dissolved in 1935, mainly because Karl and Harty, Doc Hopkins, and especially Red Foley had gone on to stardom in their own right. Hugh Cross had left WLS, and Linda Parker, 'The Sunbonnet Girl', who was the real star of the act, had met an early death.

As popular as they were for a time, their place in history is assured far more by their talented individual members than by the entire band itself.

Dick Curless

Born in Fort Fairfield, Maine, on March 17, 1932, Richard Curless joined the Trail Blazers soon after leaving high school. He had his own radio show as the 'Tumbleweed Kid' in Ware, Massachusetts in 1948, but, shortly after his marriage to wife Pauline in 1951, he was drafted into the army and sent to Korea. There he had his own programme on the Armed Forces Network, and achieved great popularity as the 'Rice Paddy Ranger'.

Discharged in 1954, Curless returned to his home in Bangor and worked as a vocalist in a local club. By 1957 he was playing Las Vegas and Hollywood clubs, having won first place on Arthur Godfrey's TV Talent Show. Then he faded from the scene, first involving himself in a logging business, then later playing small clubs in the Maine area.

His luck changed for the better in 1965 when he recorded **A Tombstone Every Mile** for Allagesh Records. It picked up national airplay, then Tower Records, a subsidiary of Capitol, bought the master, signed Curless to a contract and **Tombstone** became a Top 5 hit. Curless provided Tower with more high-selling singles before moving on to the parent Capitol label in 1970, scoring with **Big Wheel Cannonball** and **Drag 'Em Off The Interstate Sock It To 'Em J.P. Blues**.

Although he has always been – and still is – a true country singer, Curless, who wears a pirate-like patch over his right eye, has never been a Nashvillite, remaining a resident of New England to this day.

Recommended:
Hard Traveling Man (Capitol/–)
Last Blues Song (Capitol/–)
Keep On Truckin' (Capitol/–)
The Great Race (Rocade/–)

Billy Ray Cyrus

The overnight success of country hunk Billy Ray Cyrus in the summer of 1992 became known as the 'Cyrus Virus' as his popularity spread across America. The 30-year-old sex symbol turned country music upside-down with his debut single, the million-selling **Achy Breaky Heart**, the first country single in more than six years to make the American pop charts. It started a national dance craze with Melanie Greenwood, former wife of Lee Greenwood, choreographing the Achy-Breaky Dance, which became a massive hit on video and in dance clubs.

Cyrus was born August 25, 1962, in Flatwoods, Kentucky. His grandfather was a Pentecostal preacher, which is at odds with his sexy, sensual gyrations on stage, more blatantly sexual than Presley's were in the mid-'50s. Music didn't really feature in Billy Ray's plans as a child. His parents were divorced when he was six and he led an aimless type of life. He drifted through colleges without graduating. It wasn't until he turned twenty that he picked up a guitar and started to think about a music career.

His first band, Sly Dog, played its debut gig in Ironton, Ohio and played the club and bar circuit around the Kentucky-Ohio-West Virginia border for the next few years. By 1986 Cyrus was headlining five nights a week at the Ragtime Lounge in Huntington, West Virginia. A year later he was making the Nashville connection. Knocking on doors finally paid off when veteran Opry star Del Reeves listened to the young hopeful and produced a few demo tapes.

Reeves was impressed with Billy Ray's stage show, and after months of pitching Cyrus to different labels with no luck, he connected the singer with his present manager, Jack McFadden, who signed him in July 1989. It was to be another two years before McFadden gained Cyrus a record deal with Mercury. In the meantime Billy Ray and his band continued with the club dates and the occasional larger country package show. Del Reeves recorded Billy Ray's song **It Ain't Over 'Til It's Over**, and the singer had six of his own songs included on his debut album, **Some Gave All** (1992).

Released simultaneously with the **Achy Breaky Heart** single, **Some Gave All** was the fastest-rising debut album in any musical category in the history of the Billboard charts. Within weeks it went multi-platinum as it sat comfortably at the top of both the country and pop charts. In six months the album amassed sales in excess of five million and Cyrus was seriously challenging Garth Brooks as the biggest name in country music.

This sudden success led to the media and fellow country stars taking swipes at Billy Ray, calling him the 'Fabian of Country Music' and Nashville's quickest flash-in-the-pan. Travis Tritt, in a much publicized outburst, dismissed him as a one-hit wonder, but Billy Ray has enjoyed more tabloid press than any other country star.

Billy Ray did consolidate his position on the country charts with **Could've Been Me** (No.3, 1992) and **She's Not Cryin' Anymore** (No.6, 1993), though his self-penned **Where'm I Gonna Live** (1992), written after his ex-wife Cindy kicked him out of the house, failed to make the Top 20. His band, Sly Dog, is still the same outfit since those early club dates, but now they are more likely to be playing large stadiums. Billy Ray's stage presence has the female fans in ecstasy with his combination of romping energy, sex appeal and innocent country-boy good fun, all wrapped up into a muscle-rippling body. In a music not renowned for sex appeal, Billy Ray Cyrus has certainly made a major impact in a very short period. Only time will tell if he really does possess the talent and staying power to build a long-lasting country music career.

Recommended:
Some Gave All (Mercury/Mercury)
It Won't Be The Last (Mercury/Mercury)

Ted Daffan

Born in Beauregard County, Louisiana, on September 21, 1912, singer-songwriter-guitarist Theron Eugene Daffan spent his childhood in Texas, graduating from high school in 1930. During the early '30s he led

Left: Billy Ray Cyrus, 1992 Nashville Fan Fair's biggest success story. Billy Ray's Achy Breaky Heart catapulted him to fame and started a national dance craze.

the Blue Islanders, a Hawaiian band. In 1934 he moved on to become steel guitarist with the Blue Ridge Playboys, a unit featuring Floyd Tillman on lead guitar. He later worked with the Bar X Cowboys, a Houston band, and, after a long stay, formed a band of his own. In 1939 he wrote **Truck Driver's Blues**, reputed to be the first trucking song. This became such a high-selling disc that Daffan and his band, the Texans, were signed by Columbia, providing that label with a 1940 hit in **Worried Mind**, which sold about 350,000 copies.

The future seemed assured for Daffan, but, after cutting some two dozen of his own songs for Columbia (sometimes using the nom-de-plume of Frankie Brown), World War II intervened and the Texans were forced to disband. Within two years Daffan was recording once more, cutting **No Letter Today** and **Born To Lose**, a double-sided hit that won him a gold disc for a million sale. He also formed a new band to play at the Venice Ballroom, Los Angeles, California, during the mid-'40s heading back to Texas to organize bands in the Fort Worth-Dallas area. Throughout the '50s, many singers recorded Daffan's songs – including Faron Young and Hank Snow, the latter becoming a partner in a Daffan music publishing enterprise.

By 1961, Daffan was once more a resident of Houston, this time as general manager of a publishing house. And that same year yet another of his compositions ended up on a million-selling disc: Joe Barry won an award for his recording of Daffan's **I'm A Fool To Care**, which had been a hit nearly a decade earlier for Les Paul and Mary Ford. Other hits from the prolific Daffan pen include **Blue Steel Blues**, **Heading Down The Wrong Highway**, **I've Got Five Dollars And It's Saturday Night**, and **A Tangled Mind**.

Vernon Dalhart

A seminal figure in country music development, Vernon Dalhart (real name Marion Try Slaughter) was born in Jefferson, Texas, on April 6, 1883, the son of a ranch owner. While a teenager, he and his mother moved to Dallas. He began attending Dallas Conservatory of Music.

Next came a series of jobs in New York, during which time Dalhart sang in churches and vaudeville, auditioning for light opera. In 1912, he obtained a part in Puccini's 'Girl Of The Golden West' and later worked in other similar productions, though his first traceable recording, Edison cylinder **Can't Yo' Heah Me Callin' Caroline**, released June 1917, featured a 'coon' song. Following this came a deluge of Dalhart recordings on various labels, the singer tackling operatic arias, popular songs and patriotic World War I ditties. He recorded mountain musician Henry Whitter's **The Wreck Of The Old '97** for Edison in May 1924. He then cut the same song, backed with **The Prisoner's Song** (an adaptation of a traditional folk tune) for Victor, the disc having a November 1924 release. It promptly became a massive hit, encouraging Dalhart to record more hillbilly material, though more often than not he was content to re-record his hits. **The Wreck Of The Old '97**, sung by Dalhart in various guises, appears on more than 50 labels. Meanwhile, Victor sold over six million copies of the original version, making the disc their biggest seller of the pre-electric period.

Above: Lacy J. Dalton continues to output albums and singles in the '90s.

Lacy J. Dalton

"A voice so unique it rises above the rest. Lacy J. Dalton possesses that exciting style and quality that make her special." So said Billy Sherrill about Lacy J. Dalton shortly before the release of her CBS debut album in 1980.

Born Jill Byrem, October 13, 1948, near Bloomsburg, Pennsylvania, Lacy J. has enjoyed mixed fortunes since Sherrill produced her debut album. She comes from a musical family, her father being a guitarist and mandolinist. She began her own musical career as a folk singer in her teens, performing protest songs in local clubs in Minnesota. In 1967 she moved to Los Angeles and, soon after, settled in Santa Cruz where she played the local club circuit for 12 years. At one point she became lead singer of a psychedelic rock group, Office, and accrued a hefty catalogue of self-penned songs. After a series of demo discs was recorded, one, **Crazy Blue Eyes**, found its way to CBS, who signed Lacy to a contract in May 1979. CBS immediately released a Sherrill-embellished version of **Crazy Blue Eyes** which went Top 20. Lacy, with her band the Dalton Gang, then began a heavy schedule of gigs, sometimes headlining, sometimes opening, for such country artists as Willie Nelson, Emmylou Harris, Christopher Cross and the Oak Ridge Boys. Her first two albums, **Lacy J. Dalton** and **Hard Times**, became Top 20 chart items in 1980.

Lacy J. enjoyed notable chart success with her singles, making Top 10 with **Hard Times** (1980), **Hillbilly Girl With The Blues**, **Whisper**, **Takin' It Easy** (all 1981), **Everybody Makes Mistakes** and **16th Avenue** (1982), and a revival of Roy Orbison's **Dream Baby** (1983). She's retained a solid country feel, yet also appeals to a rock audience with her gutsy blues/rock, passion-filled vocal style. Her 1986 album, **Highway Diner**, moved her into Bruce Springsteen territory, but there was still a definite country edge to such Top 20 hits as **If That Ain't Love** (1984), **You Can't Run Away From Your Heart** (1985), **Size 7 Round** (a duet with George Jones in 1985) and **Working Class Man** (1986).

In 1989 Dalton joined the new and ill-fated Universal Records, releasing the acclaimed **Survivor** album which contained Kristofferson's **The Heart** – another Top 20 hit. She moved to Capitol/Liberty when Jimmy Bowen closed Universal and took up an executive position there. She has maintained her high quality of output with the album **Crazy Love** (1991) and **Chains On The Wind** (1992), and with her most recent Top 20 single, **Black Coffee** (1990).

Hard Times, Lacy J. Dalton. Courtesy CBS Records. This album became a Top 20 chart item in 1980.

Recommended:
Hard Times (Columbia/CBS)
Takin' It Easy (Columbia /CBS)
Highway Diner (Columbia/CBS)
Blue Eyed Blues (Columbia/CBS)
Survivor (Universal/–)
Lacy J. (Capitol/–)
Crazy Love (Capitol/–)
Chains On The Wind (Capitol/–)

Charlie Daniels

Once categorized as a southern boogie band, the Charlie Daniels Band has moved further and further into the pure country fold over the years. This was hardly surprising, considering Daniels' history as a bluegrass player and regular Nashville sessionman.

Born the son of a lumberjack in Wilmington, North Carolina, on October 28, 1937, the guitar- and fiddle-playing Daniels spent the years 1958–67 (except for a short period during which he found employment in a Denver junkyard) playing with a band known as the Jaguars. He claims that the band played every honky-tonk, dive and low-life joint from Raleigh to Texas. It was in Texas that Daniels met producer Bob Johnston, who guided him to Nashville where he became a sessioneer with Flatt and Scruggs, Marty Robbins, Claude King, and Pete Seeger, also playing on Ringo Starr's country album and Bob Dylan's **Nashville Skyline**.

High Lonesome, Charlie Daniels Band. Courtesy Epic Records.

A classy songwriter, Daniels started pitching his songs in the late '50s and his first taste of success came with **It Hurts Me**, a tender ballad that became an Elvis Presley B-side in 1964. He's since had songs recorded by Tammy Wynette, Gary Stewart and many others. In the late '60s he produced two albums on Youngblood, **Elephant Mountain** and **Ride The Wind**. In 1970, **Charlie Daniels**, a self-titled album, was released on Capitol.

The Charlie Daniels Band was formed in 1970 and for the next four years recorded for Kama Sutra, gaining a gold album with **Fire On The Mountain** and hit singles in **Uneasy Rider** (1973) and **The South's Gonna Do It Again** (1975). During 1974, Daniels began his annual Volunteer Jam concerts, featuring some of the biggest names in country and rock. During the following year he signed the band to Epic Records in a contract worth a reported three million dollars. His first album for the new label, **Saddle Tramp**, was released in 1976 and went gold within a year. The CDB were blazing new territory as their records crossed boundaries with music that strained at country's fences.

The quality of the albums kept moving up and up – and so did their public acceptance. In 1979 the boundaries were forever re-defined when the multi-million-selling **Million Mile Reflections** provided a huge spin-off single in **The Devil Went Down To Georgia**, a Top 5 pop chart single that was full of country fiddling. Charlie took home a Grammy for Best Vocal Performance, wowing a primarily rock-oriented audience with his blazing performance. At the CMA Awards that year, Charlie not only picked up Single Of The Year, but the CDB was named Band Of The Year. The following year the band appeared in the film 'Urban Cowboy' and recorded another hit single, **In America**, which went Top 20 on both pop and country charts.

Since then Daniels has kept up a heavy road schedule and charted such singles as **The Legend Of Wooley Swamp** (1981), **Drinkin' My Baby Goodbye** (1985), **Boogie Woogie Fiddle Country Blues** (1986) and **Simple Man** (1989). There have also been several gold and platinum albums, such as **Full Moon** (1980), **Me And The Boys** (1985) and **Simple Man** (1989). In 1992 they signed a new contract with Liberty Records; the CDB's music still remaining raw, abandoned and frenetic with Daniels himself at the fore.

Recommended:
Nightrider (Kama Sutra/Kama Sutra)

Full Moon, Charlie Daniels Band. Courtesy Epic Records.

Dave And Sugar – the group's debut album. Courtesy RCA Records.

Volunteer Jam (Capricorn/Capricorn)
Saddle Tramp (Epic/Epic)
Fire On The Mountain (Epic/Epic)
High And Lonesome (Epic/Epic)
Full Moon (Epic/Epic)
Simple Man (Epic/Epic)

Dave And Sugar

Dave And Sugar were a one-man, two-girl vocal group that was momentarily hailed as Nashville's answer to Abba in the late '70s. Original members were Dave Rowland (born on January 26, 1942, in Los Angeles), a beefy, soulful singer, Vicki Hackman from Louisville, Kentucky and Jackie Franc (Frantz) from Sidney, Ohio.

Rowland started out as a dance band vocalist, becoming a trumpeter with the 75th Army Band after being drafted. Graduating from the Stamps School of Music in Texas he became a member of the Stamps Quartet, toured with Elvis Presley and eventually became a member of the Four Guys. While still a member of the Guys he backed Charley Pride and heard that Pride required a new backup group for vocal harmony work. So Rowland held auditions, signed ex-trumpeter Franc and Hackman (who later married Pride's guitarist) and had the group signed to Chardon, the company that so astutely masterminded Pride's own meteoric rise.

The trio had an immediate No.1 single with **The Door Is Always Open**, in mid-1976, following this with a second chart-topper, **I'm Gonna Love You**, just a few months later. Rowland then experienced some changes in partners but the hits – all Top 10s – continued to flow profusely. 1977 brought **Don't Throw It All Away**, **That's The Way Love Should Be** and **I'm Knee Deep In Loving You**; 1978 providing **Gotta Quit Lookin' At You Baby** and **Tear Time** (a No.1); and 1979 supplying **Golden Tears** (yet another No.1), **Stay With Me** and **My World Begins And Ends With You**, all RCA releases.

By 1980, the group, now labelled Dave Rowland And Sugar, began to slide in terms of record sales with only one single, **New York Wine And Tennessee Shine**, going Top 20. Then followed a label switch, to Elektra, where **Fool By Your Side** (1981) saw the group momentarily back in the Top 10 again, after which they joined MCA Records in 1985.

Below: Gail Davies became lead singer with Wild Choir in 1986, then went solo again.

By this time Dave Rowland had embarked on a solo career. Although he had label deals with MCA and Elektra, it seems that the glory days of the late '70s were just a faded memory.

Recommended:
Dave And Sugar (RCA/RCA)
New York Wine And Tennessee Shine (RCA/RCA)
Stay With Me (RCA/RCA)
Greatest Hits (RCA/–)

Gail Davies

A talented singer-songwriter, Davies was born in Broken Bow, Oklahoma on April 4, 1948, the daughter of a country musician who kept a home juke-box stacked with classic country singles. When her parents split up, Gail, her mother and her brother Ron relocated to Seattle. A one-time keypunch operator for Westinghouse, quitting after only two weeks, she opted for a singing career. Davies and her brother formed a folk-rock duo and recorded an unreleased album for A&M.

Gail, who had lost her voice and had been advised to stop singing for a while, turned to songwriting and headed for Nashville, scoring immediately as the writer of **Bucket To The South**, a 1978 Top 20 hit for Ava Barber. That same year Gail herself began logging country hits, signing for the Lifesong label and notching two Top 20 singles in **No Love Have I** and **Poison Love**. In the wake of one further Lifesong hit, **Someone Is Looking For Someone Like You** (1979), came a

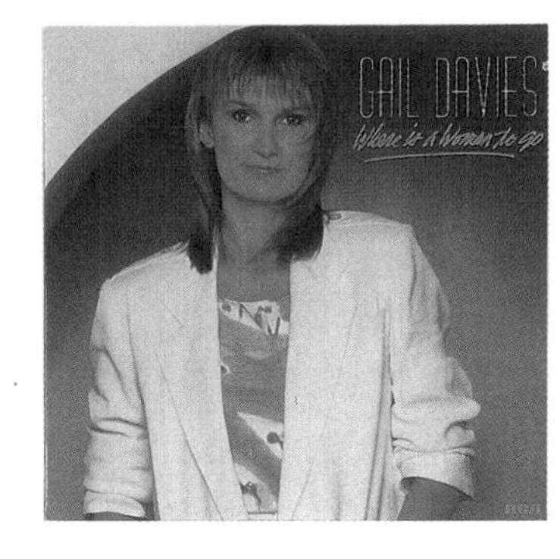

Where Is A Woman To Go?, Gail Davies. Courtesy RCA Records. She recorded a Best Of compilation album in 1991.

Warner contract and Top 10 success with **Blue Heartache** (1979), **I'll Be There (If You Ever Want Me)** (1980), **It's A Lovely, Lovely World** (1981), **Grandma's Song** (1981) and **'Round The Clock Lovin'** (1982).

Two years later Gail switched labels again, signing with RCA, and immediately charted with **Jagged Edge Of A Broken Heart**, a single that had all the hallmarks of a high-grade pop cross-over, though one that ultimately did not gain the attention it deserved. RCA seemed unsure how to handle the headstrong Davies, who was the first woman in Nashville to produce her own recordings. Further singles like **Unwed Fathers** and **Break Away** in 1985 failed to match their potential in sales. Gail formed a band, Wild Choir, the following year, notching up two minor hits for RCA, then opting once again for a solo career. She signed with MCA in 1988, but, even with production by Jimmy Bowen, was unable to recapture her past glories. She followed Bowen to Capitol Records, and in 1991 re-recorded some of her past hits for **The Best Of Gail Davies**, artistically a great album, but, commercially, yet another flop. Gail continues to sing harmony on Nashville sessions, where she is held in high esteem by fellow singers and musicians but sadly not by the record-buying public.

Recommended:
What Can I Say? (Warner Bros/–)
Where Is A Woman To Go? (RCA/RCA)
The Game (Warner Bros/–)
The Other Side Of Love (Capitol/–)
Pretty Words (MCA/MCA)

Danny Davis

Davis, real name George Nowlan, was born in Randolph, Massachusetts, on April 29, 1925. He first played trumpet with high school bands during the '30s. At 17 he became a sideman with some of the best bands of the swing era, including those of Gene Krupa and Bob Crosby. Later, he recorded under his own name, having a hit with **Trumpet Cha Cha Cha**. In 1958 he became a record producer, first with Joy Records, then with MGM, where he helped Connie Francis on her way to several No.1 singles.

During a trip to Nashville, Davis – who terms himself 'a Yankee Irishman' – met publisher Wesley Rose and Chet Atkins and became the latter's production assistant at RCA in 1965. He then conceived the idea of adding a brass sound to a pop-oriented country rhythm section, recording under the name Nashville Brass.

The band proved a success right from the start with the Nashville-cum-Alpert sound of the Davis Outfit appealing to such a cross-section of the public that the band's first 14 albums all sold in excess of 100,000 copies. Such singles as **Wabash Cannonball** and **Columbus Stockade Blues** also made their way into the charts.

By the early '70s, Davis was living in a fabulous ranch house near Nashville, had a stake in several oil wells and a seaside motel, and was also the proud owner of a private airliner. For six straight years, from 1969 through to 1974, Davis and the Nashville Brass were voted Instrumental Band Of The Year at the CMA Awards. Davis also picked up a Grammy in 1969 for his **More Nashville Sounds** album.

Recommended:
Bluegrass (RCA/–)
The Best Of Danny Davis (RCA/RCA)
Moving On (RCA/–)
Danny Davis & Willie Nelson With The Nashville Brass (RCA/–)

Jimmie Davis

Elected Governor of Louisiana in both 1944 and 1960, James Houston Davis (born in Quitman, Louisiana, on September 11, 1902) is also an eminently successful recording artist and songwriter. His writing credits include such standards as **You Are My Sunshine**, **It Makes No Difference Now**, **Sweethearts Or Strangers** and **Nobody's Darlin' But Mine**.

Gaining a BA at Louisiana College and an MA at Louisiana State University, Davis became a professor of history at Dodd College in the later '20s. During the next decade he forwarded his musical career – recording for RCA, Victor and Decca – at the same time still managing to hold down various positions of public office. A popular performer by the end of the '30s, in 1944 he appeared in the movie 'Louisiana'. When his first term of office as State Governor came to an end in 1948, Davis returned to the entertainment industry once more. He concentrated much of his activity within the sphere of gospel music where he won an award as Best Male Sacred Singer in 1957.

In 1960, Davis was asked to run in the primary, proving successful again. He also won against the might of the Huey Long machine in the election. During this second stint as Governor he had a hit in 1962, **Where The Old Red River Flows**.

Returning to active duty on the recording front in 1964, he provided Decca with an album, **Jimmie Davis Sings**, his other late '60s and early '70s releases including **At The Crossing**, **Still I Believe**, **Amazing Grace** and **Christ Is Sunshine**. In 1972, Davis was elected to the Country Music Hall Of Fame.

Recommended:
Greatest Hits (MCA/–)
You Are My Sunshine (MCA/–)
Christ Is Sunshine (Canaan/Canaan)
Country Side Of Jimmie Davis (MCA/–)
Barnyard Stomp (–/Bear Family)

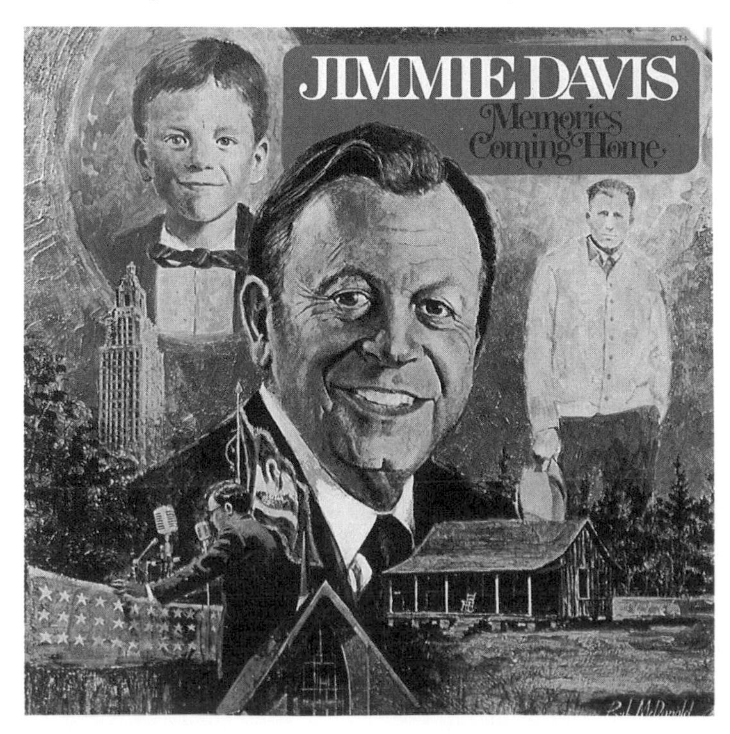

Memories Coming Home, Jimmie Davis. Courtesy MCA Records.

Linda Davis

Davis, born on November 26, 1956, in Dotson, Texas, has been around the country music scene for a number of years without making a major breakthrough. She grew up in the small Texas town of Gary, and, prior to moving to Nashville in 1982, she was a regular on both the Louisiana Hayride in Shreveport and the Grapevine Opry. Linda teamed up with Skip Eaton, and, working as Skip & Linda, the duo signed to MDJ Records. They enjoyed chart success with **If You Could See Through My Eyes** (1982).

The next few years found Linda singing on demos and jingles. In 1988 she signed a recording contract with Epic Records and made her chart debut as a solo star with **All The Good Ones Are Taken** (1988). Two years later she joined Capitol Records and was back in the charts with **In A Different Light** (1991). In 1993 Linda duetted with Reba McEntire on **Does He Love You**.

Recommended:
In A Different Light (Capitol/–)
Linda Davis (Capitol/–)

In A Different Light, Linda Davis. Courtesy Capitol Records.

Mac Davis

Born at Lubbock, Texas, on January 21, 1941, Mac Davis's career has taken in rock'n'roll, songwriting, performing and record company work. He lived much of his early life in Atlanta, Georgia, and after attending high school he worked for the Georgia State Board of Probation. In his spare time he formed a band and toured in the south, playing mainly rock'n'roll. Later he took a job as a regional manager (in Atlanta) for Vee-Jay Records, then going to Liberty Records.

In 1968 Lou Rawls recorded one of Davis's songs, **You're Good For Me**, and Elvis Presley recorded his **In The Ghetto**. The latter was a funky departure for Presley at the time and it really brought Davis to prominence. At this point he was writing under noms-de-plume to avoid confusion with lyricist Mack David, although he finally switched back to using his own name.

He wrote some material for Presley's first TV spectacular and also provided some material for such films as 'Norwood', starring Glen Campbell. Additionally, he wrote hits for Kenny Rogers And The First Edition (**Something's Burning**) and Bobby Goldsboro (**Watching Scotty Grow**).

In 1970 he guested on TV shows with Johnny Cash and Glen Campbell. The following year he cut a debut album for CBS called **I Believe In Music**, the title track being covered by over 50 artists, the most successful version being that released by Gallery in 1972. That same year Davis came up with the big one, **Baby Don't Get Hooked On Me**, his first US pop chart-topper, following this with such hits as **One Hell Of A Woman**, **Stop And Smell The Roses** and **Rock'n'Roll (I Gave You The Best Years Of My Life)**, all in 1974. In 1974, too, he began hosting his own TV show.

After this period, the flood of pop chart hits began to ebb, but Davis's country audience stayed faithful and he continued to log country hits including **Forever Lovers** (1976). In 1979 he signed a new deal with Casablanca, scoring five Top 10 singles – **It's Hard To Be Humble**, **Let's Keep It That Way** and **Texas In My Rear View Mirror** (all 1980) as well as **Hooked On Music** and **You're My Bestest Friend** (both 1981). His next Top 10 success came with **I Never Made Love ('Til I Made Love With You)** (1985), his first single for MCA Records. In the late '80s he worked closely with Dolly Parton and the pair co-wrote several songs including **White Limozeen** and duetted on the playful **Wait 'Til I Get You Home** (1989).

A brilliant live performer, always being able to ring the changes – which is why he has been able to command high fees at all the best Las Vegas night-spots – Davis has also developed a career as an actor, appearing in movies such as 'North Dallas Forty', 'Cheaper To Keep Her' and 'The Sting II'. Glen Campbell once said of him: "Mac Davis don't write songs, he paints them". Hence the album title, **Song Painter**.

Recommended:
Song Painter (Columbia/CBS)
Its Hard To Be Humble (Casablanca/Casablanca)

Skeeter Davis

Suspended by the Opry in December 1973 for criticizing the Nashville Police Department on a WSM broadcast – following a week in which two Opry performers had been murdered and Tom T. Hall's house burned down – Skeeter Davis has always had an eventful career.

She began life (in Dry Ridge, Kentucky, on December 30, 1931) as Mary Frances Penick, the eldest of seven children. At high school, she and her friend Betty Jack Davis (born in Corbin, Kentucky, on March 3, 1932) formed a harmony vocal team, the Davis Sisters, that led to a regular programme on radio station WLEX, Lexington, Kentucky. This, in turn, led to other radio shows in Detroit and Cincinnati and, eventually, to a recording contract first with Fortune, then with RCA Victor. In 1953, their first RCA effort, **I Forgot More Than You'll Ever Know**, became a No.1 record, claiming a chart position for 26 weeks – and it appeared that the Davis Sisters were set for a long career. But while travelling to Cincinnati on August 2, 1953, the girls became involved in a car accident which killed Betty Jack and critically injured Skeeter.

It was some considerable time before she resumed work, but, after a brief spell working as a duo with Betty Jack's sister Georgia, Skeeter began a solo career. Her first real breakthrough occurred in 1959 when her recording of **Set Him Free** established her as a chart name. A 1962 release, **The End Of The World**, proved the real clincher, the disc earning Skeeter a gold record and worldwide reputation. Though she asked her agency not to book her into clubs where liquor was being

Song Painter, Mac Davis. Courtesy Columbia Records.

served (Skeeter claimed that she did not want her non-drinking fans drawn into a situation where they might be tempted to imbibe) her bookings became more and more prestigious. And her records continued to sell. **I'm Saving My Love** (1963), **Gonna Get Along Without You Now** (1964), **What Does It Take?** (1967) and **I'm A Lover, Not A Fighter** (1969) proved to be her biggest sellers. Along the way she also recorded two best-selling duets with Bobby Bare – **A Dear John Letter** (1965) and **Your Husband, Your Wife** (1971) – and appeared on disc with such artists as Porter Wagoner and George Hamilton IV.

Ever pop-connected – during the 1960s she toured with the Rolling Stones and in 1985 she did an album with nutty rock outfit NRBQ, one cut proving to be a version of **Someday My Prince Will Come** done in 4/4 time! – Skeeter has often been accused of betraying her country heritage. But she claims: "I've been with the Opry since joining in 1959 – which proves that my heart's in country."

Recommended:
The Hillbilly Singer (RCA/RCA)
She Sings They Play – with NRBQ (Rounder)
Tunes For Two – with Bobby Bare (RCA/RCA)

Billy Dean

In less than three years, Billy Dean, born on April 2, 1962, in Quincy, Florida, has become one of the top writer-performers in country music. The Academy Of Country Music named him Top New Male Vocalist and his self-titled No.1 hit, **Somewhere In My Broken Heart**, Song Of The Year for 1992. He also gained a Grammy nomination, and his first two albums, **Young Man** and **Billy Dean**, netted six Top 5 hits, including the No.1s **Billy The Kid** and **If There Hadn't Been You** (both 1992).

Billy was attracted to music at an early age. His father, an auto mechanic, moonlighted as a bandleader and encouraged his son's musical talents. Billy began playing guitar in grade school, making his first appearance with his father's band, the Country Rock, at the age of eight. By the time he was fifteen he was writing songs, and he spent much of his teenage years performing along the Gulf Coast of Florida. Following a year at college in Decatur, Mississippi, on a basketball scholarship, Billy became a national finalist in the Wrangler Star Search Competition, which resulted in an appearance at Nashville's Grand Ole Opry.

He then made a move to Nashville, where he found himself in demand as a jingle singer and back-up vocalist. He had songs recorded by the Oak Ridge Boys, Les Taylor, Ronnie Milsap, Shelly West and Randy Travis. He signed a publishing deal with EMI Music and a recording contract with SBK/Liberty Records in 1990. His first single, **Only Here For A Little While**, soared to No.3 (1991) and Dean enjoyed further Top 10 hits with **You Don't Count The Cost** (1991), **Only The Wind** (1992) and **Tryin' To Hide A Fire In The Dark** (1993). His second album, **Billy Dean**, gained a gold disc, and, along with Verlon Thompson, Dean has written and recorded the theme music for the ABC-TV series – Wild West C.O.W. Boys Of Moo Mesa.

Below: Newcomer to the '90s Billy Dean is making a big impact with Top 10 hits and his gold album.

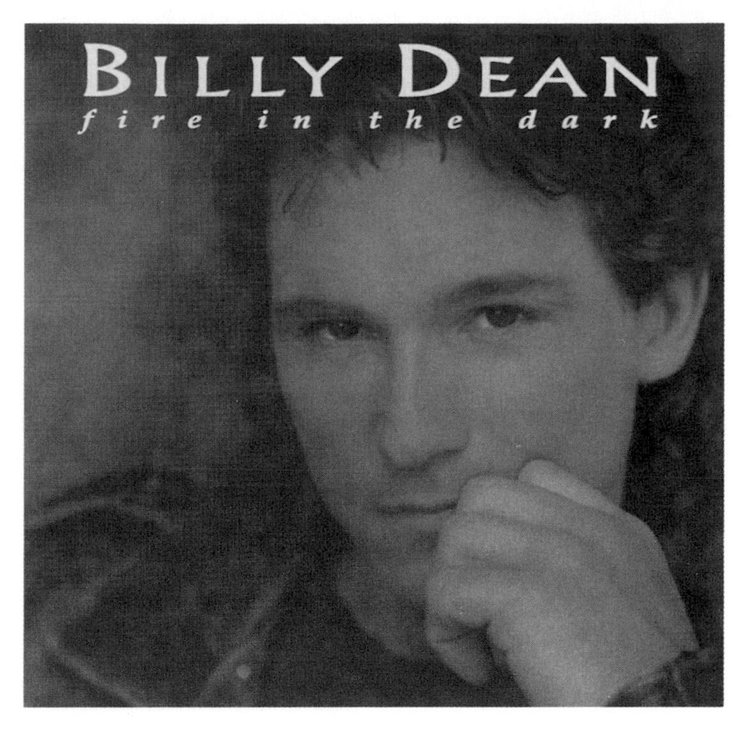

Fire In The Dark, Billy Dean. Courtesy Liberty Records. This 1993 album contains Billy's Top 10 hit, Tryin' To Hide A Fire In The Dark.

Recommended:
Young Man (Capitol SBK/–)
Billy Dean (Capitol-SBK/Capitol-SBK)
Fire In The Dark (Liberty/Liberty)

Jimmy Dean

The writer and performer of **Big Bad John**, a five-million-selling disc of semi-recitative nature, Jimmy Dean was born on a farm near Plainview, Texas, on August 10, 1928. He began his musical career at the age of ten, first learning piano and then mastering accordion, guitar and mouth harp. While in the Air Force during the '40s, he joined the Tennessee Haymakers, a country band comprised of service personnel who played off-duty gigs around Washington DC. Dean continued to play in that area after discharge in 1948.

Impresario Connie B. Gay hired him to perform for US Forces in the Caribbean during 1952 and, following this tour, Dean and his band, the Texas Wildcats, began playing on radio station WARL, Arlington, Virginia, obtaining a hit record on the Four Star label with **Bummin' Around** (1953). By 1957 he had his own CBS-TV show. Dean gained a CBS recording contract that same year. Initially he found hits hard to come by, but in 1961 he wrote **Big Bad John**, a somewhat dramatic tale of mineshaft heroism. He then supplied CBS with a run of Top 20 discs that included **Dear Ivan**, **Cajun Queen**, **PT109** (all 1962) and **The First Thing Every Morning**, the latter becoming a country No.1 in 1965.

A star of ABC-TV during the mid-1960s, Dean switched his record allegiance to RCA in 1966. But though this relationship began encouragingly with **Stand Beside Me**, a Top 10 record, Dean's country pop approach seemed to lose much of its appeal as the '70s drew near – though such releases as **I'm A Swinger** (1967), **A Thing Called Love** (1968), **A Hammer And Nails** (1968) and **Slowly**, a duet with Dottie West (1971), all fared well.

After something of a lull, Dean was back in the Top 10 once more in 1976, with **I.O.U.**, a single which was re-released by the Churchill label in 1983. And though such records proved all too few at the start of the 1980s, Dean was hardly forgotten. On TV he had a syndicated music show, Jimmy Dean's Country Beat, while honours-wise he was still gaining nominations, becoming the 11th inductee into the Texas Hall Of Fame, his presentation being made by Roy Orbison and Buddy Holly's widow.

Recommended:
Greatest Hits (Columbia/–)
I.O.U. (GRT/–)
American Originals (Columbia/–)

Penny De Haven

From Winchester, Virginia (born on May 17, 1948), Penny De Haven was the first female country star to entertain the armed forces in Vietnam. A frequent Opry guest in the '60s and early '70s, she appeared in three films, 'Valley Of Blood', 'Traveling Light' and 'The Country Music Story' and had a number of minor hits on the Imperial and UA labels. Her highest chart placing came with **Land Mark Tavern**, a 1970 duet with Del Reeves.

Since that time her name has continued to pop up in the lower reaches of the country charts. De Haven achieved some success with **(The Great American) Classic Cowboy** on the Starcrest label in 1976, and, more recently, with such Main St releases as **Only The Names Have Been Changed** (1983) and **Friendly Game Of Hearts** (1984).

Recommended:
Penny De Haven (UA/–)

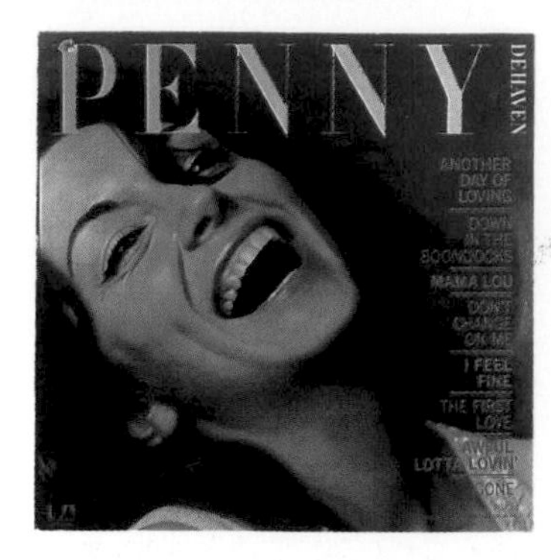

Penny De Haven. An excellent 1972 offering. Courtesy UA Records.

Delmore Brothers

Longtime Opry favourites and writers of a huge number of songs, including the oft-recorded **Blues Stay Away From Me**, the Delmore Brothers – Alton (born on December 25, 1908) and Rabon (born on December 3, 1910) – both hailed from Elkmont, Alabama. Farm raised, they were taught fiddle by their mother, Aunt Mollie Delmore, in 1930 winning an old-time fiddle contest in Athens, Alabama. Equally adept on guitar, the brothers soon won a contract with Columbia Records.

During the '40s came appearances on scores of radio stations, plus record dates for King. The duo's single of **Blues Stay Away From Me** became a Top 5 hit in 1949 and enjoyed a chart stay of no less than 23 weeks. Soon after, the Delmores became based in Houston, where Alton began drinking heavily due to the death of his daughter Sharon. Rabon became seriously ill with lung cancer, returning to Athens, where he died on December 4, 1952. Alton later moved to Huntsville, where his death occurred on June 9, 1964, the cause being diagnosed as a haemorrhage brought on by a liver disorder. The Delmores were posthumously honoured in October 1971, when they became elected to the Songwriters Hall Of Fame.

John Denver

Born John Henry Deutschendorf on December 31, 1943, John Denver grew up in an Air Force family. His father was a pilot and John also had flying ambitions until the music bug caught him. He took guitar lessons early as a boy on an old 1910 Gibson but it was during his time at Texas Tech that he felt he should try for a showbiz career, subsequently playing in West Coast clubs and eventually replacing Chad Mitchell in the trio of that name. Four years later, his own solo talents were sufficiently developed for RCA to sign him.

John has never been purely country. His albums tend to contain elements of country, folk, rock and ballads. However he has written one all-time country standard, **Take Me Home Country Roads**. He has also logged many country chart hits. He was back on the charts with **Some Days Are Diamonds**, **Wild Montana Skies** and **Dreamland Express** in the '80s.

Recommended:
Back Home Again (RCA/RCA)
Rocky Mountain High (RCA/RCA)
Some Days Are Diamonds (RCA/RCA)

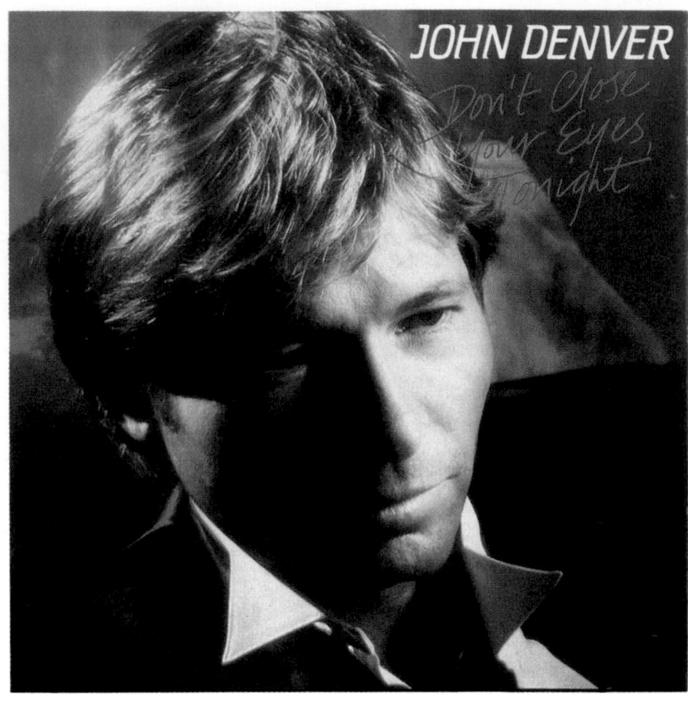

Don't Close Your Eyes Tonight, a 1985 RCA single from John Denver.

Desert Rose Band

Reviving glorious memories of the West Coast country-rock movement, the Desert Rose Band burst upon the country scene in the late '80s with a series of excellent albums and Top 10 country hits. Lead singer and main songwriter Chris Hillman made a couple of solo albums for Sugar Hill in the early '80s that laid the foundation for the formation of the Desert Rose Band in 1985.

Hillman brought in long-time music ally Herb Pedersen (lead and harmony vocals, and guitar), as well as John Jorgenson (background vocals, acoustic and electric guitars, mandolin), Jay Dee Maness (steel guitar), Bill Bryson (background vocals, bass) and Steve Duncan (drums). Hillman and the band signed with Curb Records. Their first self-titled album, produced by Paul Worley, included such Top 10 hits as **Love Reunited** and **One Step Forward** (1987) as well as **He's Back And I'm Blue** (No.1, 1988).

A second album, **Running**, produced another No.1 single, **I Still Believe In You** (1988). The Top 10 hits continued, but the heavy touring schedule began to take its toll on musicians who had spent years on the road. In 1991 Maness left because he didn't want to be away from his family. A fourth album, **True Love**, featured session steel guitarist Paul Franklin. Later that year Tom Brumley was recruited, but then Jorgenson left in March 1992. An ideal replacement turned out to be Jeff Ross, but by the end of the year drummer Steve Duncan had also departed, with Tim Grogan becoming the newest DRB member.

Rocky Mountain High, a country-flavoured offering from John Denver. Courtesy RCA Records.

Back Home Again, John Denver. Courtesy RCA Records.

Recommended:
Desert Rose Band (MCA-Curb/RCA)
True Love (MCA-Curb/–)
Pages Of Life (MCA-Curb/–)
Running (MCA-Curb/–)

Al Dexter

Born in Jacksonville, Texas, on May 4, 1902, Dexter was the leader of a country outfit known as the Texas Troopers. The singer (real name Albert Poindexter) is best remembered for his self-penned **Pistol Packin' Mama**, a song that provided him and Bing Crosby with million-selling discs.

One of the first artists to use the term 'honky tonk' in a song (**Honky Tonk Blues**), Dexter died in Lewisville, Texas, on January 28, 1984. He was a prolific hitmaker in the '40s – he will also be remembered for **Too Late To Worry, Too Blue To Cry**, **Rosalita**, **Guitar Polka** and **So Long Pal**, all No.1 songs in that era.

Recommended:
Pistol Packin' Mama (Harmony/–)
Sings And Plays (–/Stetson)

De Zurich Sisters

Mary and Caroline DeZurich were a popular yodelling team on the WLS National Barn Dance for years, specializing in sky-high Alpine yodels. During their long career they were also known as the Cackle Sisters. Their career lasted from the mid-'30s to the early '50s.

Below: Al Dexter was best known for Pistol Packin' Mama.

Above: The yodelling DeZurich Sisters, once stars of WLS.

Little Jimmy Dickens

The provider of **May The Bird Of Paradise Fly Up Your Nose**, a monster cross-over hit in 1965, the four-foot-eleven-inch Dickens had, at that time, already been hitmaking for some 16 years.

Born in Bolt, West Virginia, on December 19, 1925, Dickens was the youngest of 13 children. At 17, he won a spot on a Beckley, West Virginia, early morning radio show, moving on to appear on WIBC Indianapolis, WLW Cincinnati and WKNX Saginaw, Michigan, there meeting Roy Acuff, who invited him to appear on a duet spot on the Grand Ole Opry.

Signed to Columbia Records in the late '40s, Dickens' first Top 10 disc was **Take An Old Cold Tater And Wait** (1949). This the diminutive showman followed with hits such as **Country Boy**, **Pennies For Papa** (1949), **A-Sleeping At The Foot Of The Bed**, **Hillbilly Fever** (1950), **The Violet And The Rose** (1962) and **May The Bird Of Paradise Fly Up Your Nose** (1965).

His TV credits are impressive, but the biggest night of his life came in 1982, when he climbed on stage at the Opry and made a short acceptance speech after being elected to the CMA Country Music Hall Of Fame. Dickens can be seen regularly at the Opryland Theme Park and on the Grand Ole Opry.

Roots And Branches, The Dillards. Courtesy UA Records.

Recommended:
May The Bird Of Paradise (Columbia/–)
Little Jimmy Dickens Sings (Decca/–)
Straight From The Heart (Rounder/–)

Diamond Rio

This self-contained sextet is the first country band in history to reach No.1 with its debut single, **Meet Me In The Middle** (1991). The single paved the way for them to collect Vocal Group Of The Year from both the ACM (1991) and the CMA (1992), and see their self-titled debut album gain platinum status.

Diamond Rio won over country music fans with their superb harmonies and knack for picking commercial, but appealing songs, anchored by Marty Roe's assertive lead vocals. The band's biggest strength is its ingenious hybrid of style. Marty Roe leans towards traditional country; lead guitarist and banjoist Jimmy Olander had played with Rodney Crowell and the Nitty Gritty Dirt Band; Gene Johnson, mandolin and vocals, spent many years on the bluegrass circuit; and keyboard player Dan Truman is classically trained and played in jazz groups. Drummer Brian Prout played in rock bands before joining country group Heartbreak Mountain with Shenandoah vocalist Marty Raybon and he is married to Wild Rose drummer Nancy Given-Prout. Dana Williams, bass and vocals, is a nephew of bluegrass legends Bobby and Sonny Osborne, and has played with the likes of Vassar Clements and Jimmy C. Newman.

It was the chemistry and diverse musical history of this unlikely combination that jelled. The group originated as an entertainment attraction at the Opryland Theme Park. In the late '80s they started touring and changed their name to Diamond Rio (a trucking company based in Harrisburg, Pennsylvania), when they signed to Arista Records in 1990.

Though Diamond Rio are foremost a live band, they have chalked up impressive chart success with Top 10 hits such as: **Mirror, Mirror** (1991); **Mama Don't Forget To Pray For Me**, **Norma Jean Riley** and **Nowhere Bound** (all 1992); and **In A Week Or Two** and **Oh Me, Oh My Sweet Baby** (both 1993). Their second album, **Close To The Edge**, gained gold status within a few weeks of release, and there is little doubt that Diamond Rio are set to establish themselves as one of the most successful country bands of the '90s.

Below: Little Jimmy Dickens was elected to the Country Music Hall Of Fame in 1982.

Recommended:
Close To The Edge (Arista/–)
Diamond Rio (Arista/–)

Joe Diffie

A regular Joe, Diffie was born on December 28, 1958, in Tulsa, Oklahoma, and grew up in nearby Velma. He was in a high-school rock group called Blitz, then later joined Genesis II, a gospel outfit. After attending college, he married and moved to Texas. In 1977 he moved his family to Oklahoma where he worked in a foundry. In his spare time he continued with his music, joining another gospel group, Higher Purpose, and also a bluegrass group, Special Edition.

In 1986 Joe moved to Nashville. He had already had one of his songs, **Love On The Rocks**, recorded by Hank Thompson, but soon found it was quite difficult to break into the Nashville music scene. Eventually he signed a writer's contract with Forest Hills Publishing and for the next three years was busy co-writing and singing on demos while holding down a day-time job with Gibson Guitars. A recording contract with Epic in 1989 coincided with his first major success as a writer, with **There Goes My Heart**, a No.4 hit for Holly Dunn, on which he sang back-up harmonies.

Diffie became the only debut artist with his first single, **Home**, to go No.1 on all three major country charts and stay there for two weeks (1990). The follow-up, **If You Want Me To**, made No.2, then he was back at the top with **If The Devil Danced (In Empty Pockets)** (1991). Further Top 5 hits came in quick succession with **New Way (To Light Up An Old Flame)** (1991), **Is It Cold In Here** (1992) and **Ships That Don't Come In** (1992).

Destined for big things, Diffie has produced albums such as **A Thousand Winding Roads** and **Honky Tonk Attitude** with a mixture of down-home tear-jerkers and snappy up-tempo honky-tonk songs. He joined Mary-Chapin Carpenter on the Top 10 duet **Not Too Much To Ask** (1992), and has toured extensively during the past few years with his Heartbreak Highway band.

Recommended:
Honky Tonk Attitude (Epic/–)
A Thousand Winding Roads (Epic/–)
A Regular Joe (Epic/–)

The Dillards

Initially an ethnically rated bluegrass band who scored on cross-over appeal to a rock audience, Rodney (born on May 18, 1942) and Doug Dillard (born on March 6, 1937) from Salem, Missouri were the nucleus. They joined up with Mitch Jayne, a local radio announcer, and Dean Webb (from Independence, Missouri) and travelled to California where they were signed by Elektra Records.

The Dillards came from a strong bluegrass tradition but their novel, lighthearted approach, coupled with the fact that they themselves were a younger group, won them the plaudits of a wide public. They cut their **Back Porch Bluegrass** and **Live! Almost!** albums before meeting fiddler Byron Berline, with whom they made **Pickin' And Fiddlin'**, an album now much rated and sought after by fans of old-time music. Doug Dillard left to be replaced by Herb Pedersen and the Dillards then pursued a more commercial, rock direction. **Wheatstraw Suite** and **Copperfields** are albums from this period.

After recording albums for the Anthem and Poppy labels, the Dillards – then Rodney (vocals, guitar), Dean Webb (mandolin), Jeff Gilkinson (vocals, bass, cello), Paul York (drums) and Billy Ray Latham (banjo, guitar) – moved on to Flying Fish in 1977, releasing an album titled **The Dillards Vs The Incredible L.A. Time Machine**.

Since that time Rodney has continued to head the band, cutting such albums as **Decade Waltz** (1979), **Mountain Rock** (on Crystal Clear, 1980) and **Homecoming And Family Reunion** (1980), while Doug, who recorded solo in the early '70s, opted for a career as a studio musician.

Recommended:
Back Porch Bluegrass (Elektra/Elektra)
Pickin' And Fiddlin' (Elektra/Elektra)
Wheatstraw Suite (Elektra/Elektra)
I'll Fly Away (–/Edsel)

Dean Dillon

With his ponytailed hair, handlebar moustache and quiet drawl, Dillon (born Dean Rutherford on March 26, 1955, in Lake City, near Knoxville, Tennessee) looks and sounds more like a real cowboy on a time travel from the 1880s than one of Nashville's most successful songwriters. He first came to Music City in 1972, having already served a four-year apprenticeship on the Kathy Hill Show in Knoxville, and played in high school bands.

He landed a job at the Opryland Theme Park, impersonating Hank Williams, as a member of the Mac McGahey Quartet. Young and immature, Dillon tried to live out the role he was playing, becoming hooked on drugs and booze. He stayed at Opryland for four years, then turned to writing. He was signed to Pi-Gem Music, working a disciplined shift, perfecting his craft. The first song he had recorded was **She Called It Love** by Johnny Rodriguez in 1977. Two years later he provided Jim Ed Brown and Helen Cornelius with the chart-topping **Lying In Love With You**. While still in his early twenties, he became one of the hottest young writers around, providing George Strait with approximately 30 songs, several of them making No.1, such as **Ocean Front Property**, **The Chair** and **Famous Last Words Of A Fool**. He also wrote major hits for George Jones, Hank Williams Jr, Con Hunley, Vern Gosdin and Steve Wariner.

A recording contract with RCA in 1979 saw Dillon hit the country charts with several discs, the biggest two being **What Good Is A Heart** and **Nobody In His Right Mind (Would've Left Her)** (1980). RCA teamed him with fellow honky-tonker Gary Stewart. The pair made two albums and scored minor hits with **Brotherly Love** (1982) and **Those Were The Days** (1983). In retrospect, it was a big mistake, and Dillon started concentrating on his writing. By this time he was straightening out his hell-raising life; settling down to married life and raising kids. In November 1987 he was signed to Capitol Records and looked set to establish himself as a performer. Randy Scruggs produced two excellent albums, but the biggest chart-placing he achieved was with a revival of Peter & Gordon's **I Go To Pieces**.

Three years later he was on Atlantic Records. He produced a New-Traditionalist sound, laced with heartfelt singing and tough, melodic songs, as on **Friday's Night Woman** and **A Country Boy (Who Rolled The Rock Away)**, a tribute to Hank Williams, Elvis Presley and Buddy Holly. But Dillon was still unable to gain radio plays which could lead to public acceptance. He continues as a prolific tunesmith; his songs have a straight, no-nonsense approach to country music, and the lyrics are chock-full of wry humour, satire and lyrical sensitivity.

Recommended:
Hot, Country And Single (Atlantic/–)
Slick Nickel (Capitol/Capitol)
I've Learned To Live (Capitol/–)
Out Of Your Ever Lovin' Mind (Atlantic/–)

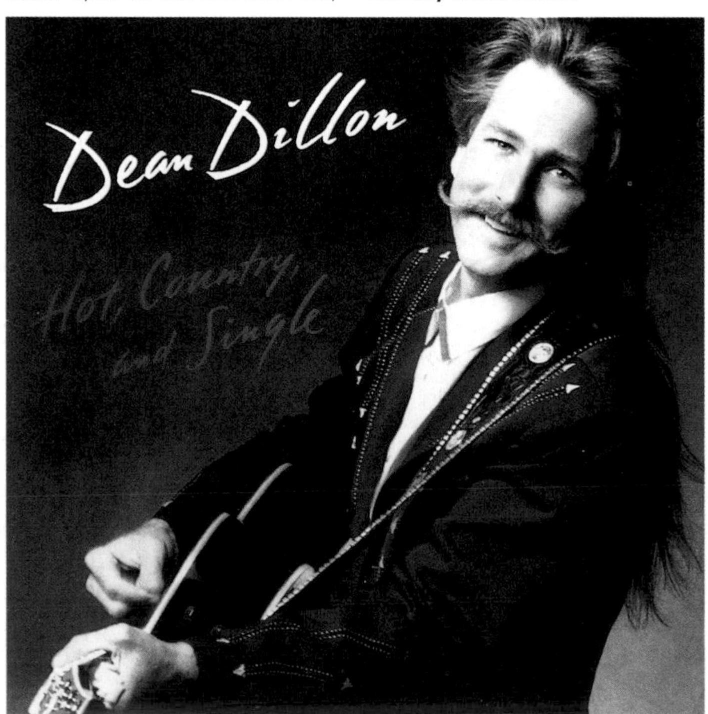

Hot, Country And Single, Dean Dillon. Courtesy WEA Records.

Peter Drake

Session man supreme, producer, owner of a studio and part-owner of a record company and music publishing firm, Pete Drake has played with everyone, from Jim Reeves to Bob Dylan, George Harrison and Ringo Starr. Born in Atlanta, Georgia, on October 8, 1932, Drake did not take up guitar until 1951 but rapidly became so proficient on the instrument that, within a year, he was leading his own band, the Sons Of The South.

After playing on radio station WLWA, Atlanta and WTJH, East Point, Georgia, he worked with Wilma Lee and Stoney Cooper, moving to Nashville with the duo in 1959.

He claims that he "starved for a year and a half" before catching the ears of Roy Drusky and George Hamilton IV, who both asked him to play on their sessions – after which he became one of the most sought-after sessioneers in Nashville. Drake also began cutting solo discs, his 1964 **Forever** becoming a Top 30 pop hit.

Recordings bearing Drake's name as an artist have also appeared on such labels as Starday, Stop, Hillside, Cumberland and Canaan, while he can be heard playing on the soundtracks of several Elvis movies.

As a producer, he is particularly proud of his **Amazing Grace** album for B. J. Thomas, a Grammy and Dove Award winner in 1982; as a music publisher he is equally proud of backing Linda Hargrove, who was named Writer Of The Year for providing Olivia Newton-John with **Let It Shine**. Pete Drake died in Nashville on July 29, 1988 from lung disease.

Recommended:
Forever (Smash/–)
Talking Steel Guitar (Smash/–)

Jimmie Driftwood

Singer-guitarist-fiddler-banjoist Driftwood (real name James Morris) was born in Mountain View, Arkansas, on June 20, 1917, and grew up in the Ozark Mountains, where he learned the songs and traditions of the early settlers. During high school he played at local dances, continuing his role as a part-time musician even after qualifying to become a teacher.

During the '50s he performed at various festivals and concerts and in 1958 signed for RCA producing **Newly Discovered Early Folk Songs**.

One song from the album, a revamped version of an old fiddle tune, **The 8th Of January**, recorded under the title of **The Battle Of New Orleans**, became a hit single for Johnny Horton during 1959, and was covered in Britain by skiffle star Lonnie Donegan. Other Driftwood songs, including **Sal's Got A Sugar Lip** and **Soldier's Joy**, also came into popular use. Another song, entitled **Tennessee Stud**, provided Eddy Arnold with a big hit.

Driftwood ceased recording in 1966 but still managed to devote much of his time to the cause of folk music, helping to run the Rackensack Folklore Society and assisting with some folk festivals. His **The Battle Of New Orleans** continues to be a much recorded number, Harpers Bizarre charting with the song yet again in 1968. The song proved popular again in 1975 with the Nitty Gritty Dirt Band and Buck Owens both cutting versions.

As well as being heavily involved in American folklore, Driftwood has gained a great deal of inspiration from the American Civil War.

Recommended:
Famous Country Music Makers (–/RCA)
Songs Of Billie Yank And Johnny Reb (–/RCA)
Americana (–/Bear Family)

Roy Drusky

Born in Atlanta, Georgia, on June 22, 1930, Roy Frank Drusky did not acquire an interest in country music until the late '40s, when he signed for a two-year term in the navy. While on ship, he met some C&W fans who had organized their own band – at which point Drusky bought a guitar and taught himself to play. Upon return to civilian life, he initially tried for a degree in veterinary medicine at Emory University. Then in 1951 he formed a band, the Southern Ranch Boys, who played a daily show on a Decatur radio station where Drusky subsequently became a DJ. Next came a three-year residency as a vocalist at a local venue, during which time Drusky made his TV debut and signed for Starday Records.

Following a later DJ stint on KEVE, Minneapolis, he took over another residency, this time at the city's Flame Club, where he began writing more of his own songs. These he began recording with Decca, one – **Alone with You** – was covered by Faron Young. However, after a move to Nashville, the Drusky hitmaking machine really went into action, his first solo successes coming with **Another** and **Anymore** in 1960, followed by five more winners before the singer opted for a switch to the Mercury label in 1963. With Mercury he maintained his supply of chart records right through to the early '70s, **Peel Me A Nanner** (1963), **From Now On All My Friends Are Going To Be Strangers** (1965), **Yes, Mr Peters** (a duet with Priscilla Mitchell that went to No.1 in 1965), **The World Is Round**, **If The Whole World Stopped Lovin'** (both 1966), **Where The Blue And The Lonely Go**, **Such A Fool** (both 1969), and **I'll Make Amends**, **Long Texas Road** and **All The Hard Times** (all 1970), being Top 10 entries.

Also during this period, Drusky appeared in two movies – 'The Golden Guitar' and 'Forty Acres Feud' – enjoying success with **White Lightning Express**.

In 1974, Drusky, still a consistent, if lower level hitmaker, joined Capitol. But after a brace of minor successes, he decided to label-hop to Scorpion, supplying two more mini-hits in **Night Flying** and **Betty's Song** (both 1977). Since that time, however, Drusky's name has been sadly absent from the charts, though, as the Como-like, smoothed-voiced singer proved during his Wembley Festival visits to the UK of the early 1980s, he remains very much a crowd-pleaser.

Recommended:
All My Hard Times (Mercury/–)
Country Special (Mercury/–)
Anymore (Decca/Stetson)
Night Flying (Scorpion/Big R)

Above: Roy Drusky, the Perry Como of country music.

Dave Dudley

One of the several country stars who could have made the grade in baseball, Dudley (born in Spencer, Wisconsin, on May 3, 1928) gained his first radio date after receiving an arm injury playing for the Gainsville Owls, Texas. While recuperating, he stopped by radio station WTWT and began playing along with the DJ's choice of discs. It was then suggested he should sing live, which he did, obtaining a positive reaction from listeners. Following stints on stations in the Idaho region during 1951–52, Dudley moved on to lead a couple of small groups, his career taking a setback in the early '60s when he was hit by a car. However, this period proved kind to the near-rockabilly performer, hits coming on Vee, Jubilee and Golden Wing, the most important being the song **Six Days On The Road**, a release reputed to have commenced the whole truck song cycle.

In 1964 came a contract with Mercury Records and four Top 10 discs **in Last Day In The Mines** (1963), **Mad** (1964), **What We're Fighting For** and **Truck Drivin' Son-Of-A-Gun** (both 1965). The Nashville local branch of the truckers' union provided Dudley with a solid gold security card in appreciation of his musical efforts on behalf of their chosen profession. With his band, the Roadrunners, he has since continued on his truckstop way, switching labels and moving from Mercury to Rice, UA and Sun, providing each of them with hits of various sizes, including **The Pool Shark** (a No.1 in 1970), **Fireball Rolled A Seven** and **Me And Ole CB** (both 1975), **One A.M. Alone** (1978) and **Rolaids, Doan's Pills And Preparation H** (1980).

Recommended:
Truck Songs (–/Mercury)

Dave Dudley Sings. Courtesy Mercury Records.

Duke Of Paducah

A homespun comedian, always ready to close his routine with a rousing banjo solo, Benjamin Francis 'Whitey' Ford was born in DeSoto, Missouri, on May 12, 1901.

Brought up by his grandmother in Little Rock, Arkansas, he joined the navy in 1918, learning banjo while at sea. During the '20s he toured with his own Dixieland band. Then, after some time in vaudeville and a stay with Otto Gray's Oklahoma Cowboys, he became MC on Gene Autry's WLS Chicago show, acquiring the title 'Duke Of Paducah'.

In 1937 he, John Lair, Red Foley and Red's brother Cotton Foley, founded the Renfro Valley Barn Dance. Ford gained his first date on the Grand Ole Opry in 1942, creating such an impression that he remained a member until 1959, thereafter becoming a constant visitor to the show.

Ford was a regular member of the Hank Williams Jr Road Show and a performer on countless TV shows originating out of Nashville. With his famous wind-up line – "I'm going to the wagon, these shoes are killin' me" – he toured for many years, often providing a serious lecture, 'You Can Lead A Happy Life', at colleges, sales functions and various conventions. Known as 'Mr Talent', Whitey Ford was elected to the Country Music Hall Of Fame in 1986, four months after his death on June 20, 1986, following a long fight against cancer.

Right: Whitey Ford, the Duke Of Paducah, died in Nashville in 1986.

Johnny Duncan

Born on a farm near Dublin, Texas, on October 5, 1938, Johnny Duncan came from a music-loving family. At first, Duncan thought of himself as an instrumentalist and Chet Atkins, Les Paul and Merle Travis were his idols. Then during his mid-teens, he realized that his singing ability was an equal asset. After attending high school in Dublin, he went on for a stay at Texas Christian University, meeting and marrying his wife Betty during this period. Shortly after their marriage, in 1959, the Duncans moved to Clovis, New Mexico, where John joined forces with Buddy Holly and producer Norman Petty with whom he worked for the next three years. Then in 1964, following a spell as a DJ in the Southwest, he headed for Nashville, where he applied his talents to a number of menial jobs while waiting to break into the music industry. After an appearance on WSM-TV Nashville, Don Law of Columbia Records signed Duncan.

His first hit came in 1967 with the release of **Hard Luck Joe**, after which Duncan, who for a while became a regular part of the Charley Pride band, began slotting two or three records into the country charts, going Top 10 with **Sweet**

Country Woman (1973) and achieving No.1s with **Thinking Of A Rendezvous** (1976) and **It Couldn't Have Been Any Better** (1978). Several of his most successful records featured vocal assistance from Janie Fricke. It seemed that Duncan would have no difficulty extending his run of hits well into the '80s, but in the wake of an ill-titled and ill-received album, **The Best Is Yet To Come**, the Texan's luck ran out and the hits inexplicably began to dry up. In 1986 he turned up on the indie Pharoah label.

Recommended:
There's Something About A Lady (Columbia/–)
The Best Of Johnny Duncan (Columbia/–)
Johnny Duncan (–/CBS)
Nice 'n' Easy (Columbia/CBS)

Tommy Duncan

Although best known as Bob Wills' longtime lead singer, Tommy Duncan also had a long and significant solo career. Born January 11, 1911, in Hillsboro, Texas, he was steeped in the music of Jimmie Rodgers and loved both 'hillbilly' and the blues when he joined the Light Crust Doughboys in 1932. When their fiddler, Wills, split from the Doughboys, Duncan joined him in the newly formed Playboys, soon known as the Texas Playboys.

Duncan's years with Wills were great for both, his mellow, bluesy, high baritone appearing on hundreds of records, including **New San Antonio Rose**, the biggest of them all. He struck out on a career of his own in 1948 (taking a number of Texas Playboys with him) and, though many of his Capitol recordings were worthy enough, it was generally considered that neither Wills nor Duncan were as great apart as they had been together. He spent many years touring with his new band, the Western All Stars, but it was to everyone's delight when he and Bob Wills joined forces once more for a series of albums for Liberty in the '60s.

Although not to the inspired greatness of the '30s and '40s, these albums are tributes to the genius of two of country music's greats and now form part of the legacy of Duncan, who died from a heart attack on July 25, 1967.

Recommended:
Hall Of Fame – Bob Wills And Tommy Duncan (UA/–)

Greatest Hits, Johnny Duncan. Courtesy CBS Records.

Holly Dunn

A multi-talented, radiant, singer-songwriter and a down-to-earth country girl, Holly Suzette Dunn was born on August 22, 1957, in San Antonio, Texas. Her father was a Church of Christ minister. Holly learned guitar while still a young girl and naturally sang in church. She was lead singer with the Freedom Folk Singers, representing Texas at the White House Bicentennial Celebrations in 1976. At university she took a degree in advertising and public relations and also became a member of the Hilltop Singers. Once studies were completed, Holly moved to Nashville to join her elder brother, Chris Waters Dunn, who had made his mark writing hit songs for Dr Hook and others.

She sang lead and harmony vocals on demo tapes and finally signed as a writer with CBS Songs. She co-wrote songs for Louise Mandrell, Cristy Lane, Terri Gibbs, Marie Osmond and others. In 1985 she gained a recording contract with MTM Records, making her chart debut with **Playing For Keeps** (1985). The next year she made Top 10 with the self-penned **Daddy's Hands**, and had further Top 10 hits with **A Face In The Crowd** (a duet with Michael Martin Murphey, 1987), **Love Someone Like Me** and **Only When I Love** (1987), **Strangers Again** and **That's What Your Love Does To Me** (both 1988). That year MTM went into liquidation, but Holly was able to secure a contract with Warner Brothers, scoring her first No.1 with **Are You Ever Gonna Love Me** (1989). She recorded a duet with Kenny Rogers, **Maybe**, that made No.25 in 1990. However, she was having more success with solo efforts such as **There Goes My Heart** (No.4) and **You Really Had Me Going** (No.1).

Holly still writes much of her own material, often working with her brother Chris. The pair also act as co-producers on her recordings. She caused controversy in 1991 with the song, **Maybe I Mean Yes**, which was lifted off her **Milestone** album. The lyrics put the onus on the woman not to give too many come-on signs to a man on a date, or face the consequences; it gained few radio plays. Since then Holly has failed to score a major hit, but still has excellent songs, as she showed with the album, **Getting It Dunn** (1992).

Recommended:
The Blue Rose Of Texas (Warner Bros/Warner Bros)
Getting It Dunn (Warner Bros/–)
Across The Rio Grande (MTM/–)
Milestones (Warner Bros/–)
Heart Full Of Love (Warner Bros/–)

Eagles

The Eagles is a now defunct West Coast country and soft rock band which featured lucid, flowing instrumental work and exquisite rough harmonies. They followed on from the Byrds and the Flying Burrito Brothers in bringing country to rock fans.

Founded by Randy Meisner (ex-Poco and Rick Nelson's Stone Canyon Band), Bernie Leadon (ex-Linda Ronstadt, Dillard & Clark and Flying Burrito Brothers), Glenn Frey (ex-Linda Ronstadt and John David Souther) and Don Henley (ex-Shiloh), they emanated from the ethically loose musical scene in Los Angeles but cut their first album in England under Rolling Stones' producer Glyn Johns. Later personnel changes saw guitarist and slide guitarist Don Felder boosting the line-up and Joe Walsh replacing Bernie Leadon.

The Eagles scored a No.1 single in the US with the memorable **The Best Of My Love**, and subsequent singles, **Lyin' Eyes** and **One Of These Nights**, notched up hit status in Britain.

In 1980 they disbanded to pursue solo careers, but many of their best-known songs have since been recorded by a number of established country acts. These include Conway Twitty, Tanya Tucker and Nat Stuckey.

Recommended:
Eagles (Asylum/Asylum)
Desperado (Asylum/Asylum)
On The Border (Asylum/Asylum)
One Of These Nights (Asylum/Asylum)
Hotel California (Asylum/Asylum)
Their Greatest Hits (Asylum/Asylum)

Steve Earle

The mid-'80s found Steve Earle, a young Texas country-rockabilly singer, poised to take country music by storm, but after making a big impact with **Guitar Town** in 1986, he rapidly turned his back on

Right: One Of These Nights, the Eagles. Courtesy WEA Records. The title song of this album went to No. 1, although the group disbanded in 1980.

country, heading towards modern rock and alienating himself from country music fans.

Earle was born on January 17, 1955, in Fort Monroe, Virginia, but grew up in San Antonio, Texas into a musical family. By the time he was eleven he had mastered guitar and was a wild tearaway, running away from home to make music when he was fourteen. He went to Austin, Dallas and Houston, working bars and coffee houses. In 1974 he followed other Texas writers Guy Clark and Townes Van Zandt to Nashville, and almost immediately landed a job as an 'extra' in the Robert Altman film 'Nashville'. For the next few years he performed in local Nashville bars, linking up with other singer-writers, such as Guy and Susanna Clark, Rodney Crowell and Townes Van Zandt.

In 1980, Steve relocated to Mexico for a while, then returned to San Antonio where he picked up a couple of musicians and took to the road as Steve Earle And The Dukes. A move back to Nashville in 1981 led to him being signed as a writer by Pat Carter and Roy Dea, and also being asked to run their music publishing company in Music Row. Carl Perkins cut his **Mustang Wine**, then Johnny Lee took **When You Fall In Love** into the Top 20 (1982). He cut demos of his songs, and, using his two-piece back-up group, he recorded a four-song EP which perfectly captured the excitement of 1950s rockabilly with a definite 1980s contemporary edge.

Consequently, Steve was offered a contract by Epic Records in early 1983, and his first single, **Nothin' But You**, was the first of a handful of minor country hits. Subsequently, other country stars such as Johnny Cash, Waylon Jennings and Janie Fricke started cutting his songs. A move to MCA Records at the beginning of 1986 resulted in the acclaimed **Guitar Town** album, one of the most exciting records by a new Nashville-based artist in years. The title song became a Top country hit in 1986. Earle enjoyed further chart success with **Goodbye's All We've Got Left** (1987) and **Six Days On The Road** (1988), the latter featured in the film 'Planes, Trains and Automobiles'. At this time his style was fast, rockin' country, guitar-dominated and laden with heavy southern rock drumming. He became a major star in Europe, but with the album, **Copperhead Road**, he moved into heavy rock, and the Pogues featured on **Johnny Come Lately**. Struggling to become a country Bruce Springsteen, Earle has never fully recaptured the freshness and vitality of his first two albums and now works outside of mainstream country music.

Above: The Eagles in 1974 – Bernie Leadon, Glenn Frey, Don Henley, Randy Meisner and Don Felder. The group brought a taste of country music to rock and pop fans throughout the '70s.

Right: Connie Eaton had her biggest success with Lonely Men, Lonely Women.

Recommended:

Guitar Town (MCA/–)
Copperhead Road (MCA/MCA)
Early Tracks (–/Epic)
Exit-O (MCA/MCA)

Connie Eaton

A recording artist with Chart, GRC, Stax and ABC, Connie is the daughter of Bob Eaton, a one-time artist best known for his 1950 hit **Second Hand Heart**, and was born that same year, on March 1, in Nashville, Tennessee. Following an acting award in 1968, she met Chart Records' A&R man Cliff Williamson (whom she later married) who signed her to the label.

In 1970 Connie was voted Most Promising Female Artist by both Cashbox and Record World magazines, also having a fair-sized hit that year with her version of **Angel Of The Morning**. After providing several other minor hits (including two duets with Dave Peel) for Chart, Connie moved on briefly to GRC and Stax before signing to ABC in 1974 and having her most successful release in **Lonely Men, Lonely Women**, a Top 20 single in the spring of 1975. An album, **Connie Eaton**, was released later that year, but has since been deleted.

Stoney Edwards

Stoney, a soulful black country singer of Afro-Indian parents, was born Frenchy Edwards, on December 24, 1937 in Seminole, Oklahoma. He grew up listening to the country sounds of Hank Williams and Lefty Frizzell, as depicted in his 1973 hit, **Hank And Lefty Raised My Country Soul**.

Life was hard for Stoney who worked as a farm hand, truck driver, janitor, crane operator and dock worker before becoming a singer in small Texas clubs and finally in Oakland, California, where he made a minor breakthrough in 1970 when he signed with Capitol Records.

For the next decade or so, Stoney built up a cult following with his traditionally-styled honky-tonk sound that was firmly rooted in the 1950s. He scored minor country hits with **Poor Folks Stick Together** (1971), **She's My Rock** (1972), **Mississippi You're On My Mind** (1975) and **Blackbird (Hold Your Head High)** (1976). In his own modest way, he paved the way for the acceptance in the '80s of similarly styled singers, such as Con Hunley, John Anderson and John Conlee, but sadly Stoney has never really achieved the success he deserved. He was dropped by Capitol in 1977 and has since recorded for JMI Records and Music America. He lost a leg in a shooting accident and virtually retired from music in 1982, then unexpectedly returned in 1991 with a new album, **Just For Old Times' Sake**, which found him once again in top-notch form.

Below: Joe Ely brought Texas music to a wider audience.

Recommended:
Mississippi You're On My Mind (Capitol/–)
Down Home In The Country (Capitol/–)
Just For Old Times' Sake (–/Ragged But Right)

Joe Ely

One of the most promising new singers to emerge from Texas in the late 1970s, Ely, born February 9, 1947, has never matched his cult following and rave reviews with commercial success.

Raised in Lubbock, a stone's throw away from where Buddy Holly lived, Joe never finished school, but instead hit the road playing bars and cafés, jamming with the likes of Johnny Winter. For a time he joined a theatre company and ended up on a European tour playing the arts circuit including shows in London and Edinburgh.

On his return to Texas in the early '70s, Joe teamed up with Butch Hancock and Jimmie Dale Gilmore to form the Flatlanders, an acoustic country group. They played throughout the South, and in 1972 the trio recorded an album for Shelby Singleton in Nashville. The record was not released at the time, though it did surface in the late '70s.

A move back to Lubbock in 1974 led to Joe forming a new band, an electric country-rock outfit designed to play up-tempo honky-tonk music for dancing audiences. A regular gig at the Cotton Club, a large honky-tonk on the outskirts of Lubbock, led to the Joe Ely Band gaining record company interest. An album cut at Chip Young's studio in Murfreesboro, Tennessee was released by MCA in 1977, and Joe Ely gained widespread acclaim but minimal sales.

Further albums like **Honky Tonk Masquerade** and **Down On The Drag** showed that the first album was no fluke. Each one was a mind-blowing celebration of Texas honky-tonk music, mixing rock sounds with the traditional to bring it slap-bang up-to-date. Ely and band proved to be a very popular live act in Texas and also Europe, where they toured several times in the late '70s and early '80s.

In 1984 he recorded **Hi-Res**, which was no more successful than his previous albums and he now found himself without a major label deal. Three years later, with a new band and a batch of new original songs, he was back in the studio recording **Lord Of The Highway** (1987) for Hightone Records, a strong mix of Texas honky-tonk. This gave Ely and his band a new lease of life; they were kept busy on the road building up a reputation with younger music fans. A dynamic **Live At Liberty Lunch**, recorded in 1990, gained Ely and the band a new contract with MCA, who immediately remastered four of his earlier albums – **Joe Ely**, **Honky Tonk Masquerade**, **Down On The Drag** and **Musta Notta Gotta Lotta** – and released them on CD. He was back in the Nashville studios in early 1992 working with producer Tony Brown on **Love And Danger**, his first studio album in four years. He also teamed up with musical colleagues John Mellancamp, John Prine, Dwight Yoakam and James McMurtry as the Buzzin' Cousins on the soundtrack to the Mellencamp movie, 'Falling From Grace'. Joe Ely continues to command a very large cult following, but sometimes his high-energy presentation falls between country and rock music, making radio programming in America very difficult.

Recommended:
Joe Ely (MCA/MCA)
Honky Tonk Masquerade (MCA/MCA)
Hi-Res (MCA/MCA)
Flatlanders – One Road More (–/Charly)
Down On The Drag (MCA/MCA)
Live At Liberty Lunch (MCA/MCA)
Lord Of The Highway (–/Demon)
Love And Danger (MCA/MCA)

Buddy Emmons

A multi-instrumentalist of exceptional ability – he is a first class pianist, an able bass player and not a bad singer, either – Emmons is generally considered in terms of his brilliantly inventive steel guitar work.

Born in Mishawaka, Indiana, on January 27, 1937, he was given a six-string 'lap' guitar at the age of 11 and subsequently studied at the Hawaiian Conservatory of Music, South Bend, Indiana. At 16 he appeared in Calumet City, Illinois, playing around the local clubs for most nights of the week and jamming in Chicago at weekends. By 1955 he had moved on to Detroit where, after deputizing for steelman Walter Haynes at a Little Jimmy Dickens gig, Emmons was awarded a permanent position with Dickens' band, which, in turn, led to dates on the Opry.

Steel Guitar, Buddy Emmons. Courtesy Flying Fish Records.

During the years that followed came lengthy stints with Ernest Tubb and Ray Price, also the founding of the Sho-Bud company. He and his then partner Shot Jackson marketed the first steel guitar with push-rod pedals. In 1969 Emmons left Nashville, joined Roger Miller as a bassist, and became based in LA, playing West Coast sessions with Ray Charles, Linda Ronstadt, Henry Mancini and many others between tours with Miller.

When he and the King Of The Road parted company in December '73, Emmons returned to Nashville once more and since that time has involved himself in an incredible amount of session work, also spending some time in promoting his own Emmons Guitar Company, his association with Sho-Bud having terminated some years before.

Recommended:
Sings Bob Wills (Flying Fish/Sonet)
Steel Guitar (Flying Fish/Sonet)
Buddies – with Buddy Spicher (Flying Fish/Sonet)
Minors Aloud – with Lenny Breau (Flying Fish/Sonet)

Dale Evans

Wife of singing cowboy superstar Roy Rogers, Dale was born Frances Smith, at Ulvalda, Texas, on October 31, 1912.

Brought up in Texas and Arkansas, Dale married Thomas Fox in 1928, she and her husband parting two years later – at which time she began concentrating upon a career as a popular vocalist.

During the '30s she became a band singer with the Anson Weeks Orchestra, then became resident vocalist on the CBS News And Rhythm Show. Following many appearances on major radio shows, Dale moved into films, appearing in such productions as 'Orchestra Wives' (1942), 'Swing Your Partner' (1943), 'Casanova In Burlesque' (1944), 'Utah' (1945), 'Bells of Rosarita' (1945), 'My Pal Trigger' (1946), 'Apache Pass' (1947), 'Slippy McGee' (1948), 'Susanna Pass' (1949), 'Twilight In The Sierras' (1950), 'Pals Of The Golden West' (1951), many of these movies starring Roy Rogers whom she married in 1947.

With Rogers, she has recorded a number of albums for such labels as RCA, Capitol and Word, some of the these being in the gospel vein, and is also the author of an armful of small books of an inspirational/religious nature.

Recommended (with Roy Rogers):
In The Sweet Bye And Bye (Word/Word)
The Good Life (Word/Word)
The Bible Tells Me So (Capitol/–)

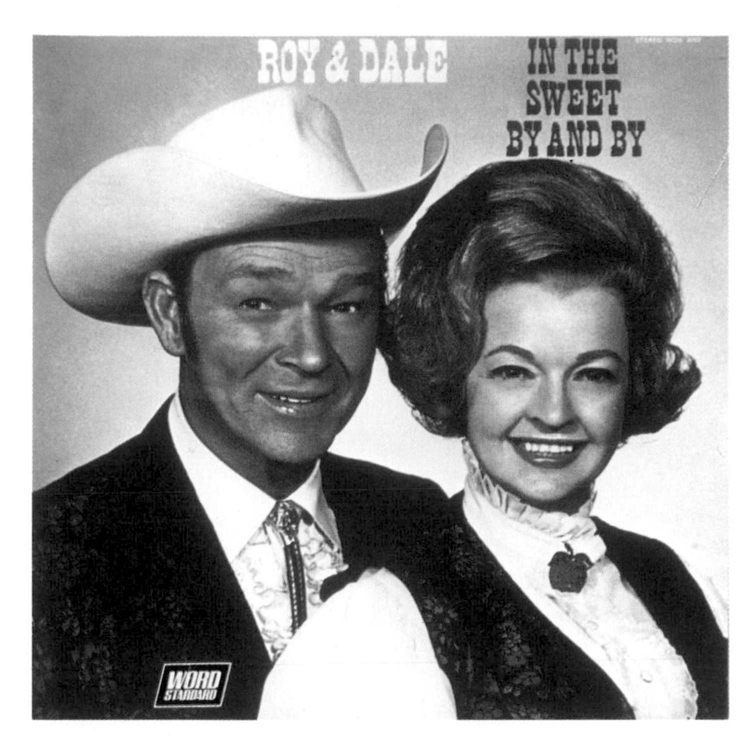

In The Sweet By And By, a 1973 release, Roy Rogers and Dale Evans. Courtesy Word.

Leon Everette

In the space of a few short years, Leon Everette (born Leon Everette Baughman, June 21, 1948 in Aiken, South Carolina, but raised in Queens, New York), emerged as one of the most exciting and successful names on the country music scene. However, the transition from obscurity to overnight fame was not easy for the rugged, good-looking performer.

Rejection by the major labels in Nashville led to some frustrating years with labels like Doral Records and True Records, a company organized with the sole intention of establishing Leon as a major name in country music.

Carroll Fulmer, a Florida businessman, financed Orlando Records in the spring of 1978 and subsequent extensive recording sessions in Nashville were produced jointly by Jerry Foster, Bill Rice, Ronnie Dean and Leon himself. The first two singles, **We Let Love Fade Away** and **Giving Up Easy**, failed to make much impression, but the third release, **Don't Feel Like The Lone Ranger**, made the country Top 40 in the summer of 1979.

Further hits on Orlando, including **I Love That Woman (Like The Devil Loves Sin)**, **I Don't Want To Lose** and **Over You**, the latter his first entry into the Top 10, led to interest from major labels. In October 1980 Everette signed with RCA Records and was given creative freedom and control of his own recordings.

He soon proved that RCA's faith in him was justified with such Top 5 country hits as **Giving Up Easy**, **If I Keep On Going Crazy**, **Hurricane** and **Don't Be Angry**. With his own five-piece band, initially called Tender Loving Care but changed to Hurricane in 1981, Leon has also made an impact as a dynamic stage performer with a fast-moving act that has gained rave reviews and praise from fellow performers Hank Williams Jr and Waylon Jennings.

Though he continued to chalk up Top 10 hits for RCA with **Soul Searchin'**, **Give Me What You Think Is Fair** and **Midnight Rodeo**, Leon was not entirely happy with the promotion he was receiving from the label, and at the end of 1984 he signed with Mercury Records. Bill Rice was engaged as producer and gave Everette a contemporary country styling that resulted in such hit singles as **Feels Like Forever**, **Too Good To Say No To** and **'Til A Tear Becomes A Rose.**

However, by 1986, Leon was back on his own Orlando Records, scoring minor hits with **Danger List (Give Me Someone I Can Love)** and **Still In The Picture**.

Recommended:
This Is Leon Everette (–/RCA)
Where's The Fire? (Mercury/–)

This Is Leon Everette – a British compilation. Courtesy RCA Records.

Everly Brothers

A mid-'50s teen heart-throb duo, the Everly Brothers came from a solid country music background, made rock'n'roll history, and then returned to country.

Born in Brownie, Kentucky, Don in 1937 and Phil in 1939, they were the sons of Ike and Margaret Everly, well-known local country stars (Ike was an influential guitar player). Early on the boys were joining their parents on tour and their first radio appearance with the show was on KMA, Shenandoah, Iowa.

After high school, Don and Phil procured a record contract with Cadence in 1957. They had previously made one unsuccessful Columbia single. At Acuff-Rose music publishers they were introduced to Felice and Boudleaux Bryant, the result of this liaison being their first hit, **Bye Bye Love**. The Everlys' sound was characterized by harmony singing which contrived to be velvety and whining at the same time, and which was matched with pounding, open-chord guitars and songs which perfectly caught the era's mood of teenage frustration. In no time at all they were causing riots at theatres, yet they were also acceptable to adults because of their clean-cut looks.

Bye Bye Love was a huge international hit for them, topping pop and country charts in America and becoming the most hummed international hit that year. With rock'n'roll losing some of its early steam the time was right for this distinctive and poignant teen sound. Subsequent hits included **Wake Up Little Susie**, **Bird Dog**, **Claudette** and **All I Have To Do Is Dream**, all huge sellers and the last-named making No. 1 in the American and British pop charts.

In 1960 they left Cadence for Warner Bros, thus losing the Nashville production team and the Bryant songwriting team. However, the self-composed **Cathy's Clown** was an immediate hit for them and they followed up with **Ebony Eyes** and **Walk Right Back**. After a spell in the Marines they managed an excellent 1965 hit with **Price Of Love** but it was to be their last really big pop winner. Don had suffered a nervous breakdown on their 1963 tour of England and gradually the brothers began to go their separate ways.

The Everlys returned to their country roots with the albums **Great Country Hits**, **Roots** and **Pass The Chicken And Listen**, the latter album recorded in Nashville in 1972, reuniting them with producer Chet Atkins. Shortly after this Don and Phil split to pursue solo careers; Don based himself in Nashville and Phil in Los Angeles. Apart from a few spasmodic successes, they never really made a big impression as solo artists.

Pass The Chicken, Everly Brothers. Courtesy RCA Records.

In 1983, they reunited for a concert at London's Royal Albert Hall. It was such a huge success that they undertook a world wide tour the following year, signed a new recording contract with Phonogram, and made a return to the charts with the Paul McCartney-penned **On The Wings Of A Nightingale** and **Born Yesterday** (1984), the latter became their first country Top 20 chart entry since 1959. Deservedly, the Everly Brothers were inducted into the Rock 'n' Roll Hall Of Fame in 1986.

Below: Phil and Don, the Everly Brothers, reunited for their famous Royal Albert Hall comeback concert in 1983.

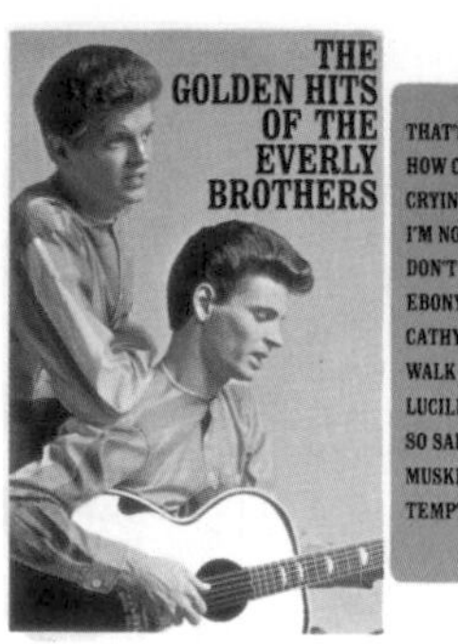

Golden Hits, Everly Brothers. Courtesy WEA Records.

Recommended:
Born Yesterday (Mercury/Mercury)
1984 (Mercury/Mercury)
Very Best Of The Everly Brothers (Warner Bros/Warner Bros)
Great Country Hits (Warner Bros/Warner Bros)
Stories We Could Tell (RCA/RCA)
Pass The Chicken And Listen (RCA/RCA)

Don Everly solo:
Brother Jukebox (Hickory/DJM)
Sunset Towers (Ode/A&M)

Phil Everly solo:
Phil's Diner (Pye/Pye)
Star Spangled Springer (RCA/RCA)
Louise (–/Magnum Force)

Skip Ewing

Singer-songwriter Skip Ewing, born on March 6, 1965, in Redlands, California, has brought a fresh and dynamic voice to a traditional sound, and found himself in demand as a writer for such acts as George Jones, Charley Pride, Reba McEntire, George Strait, Sawyer Brown and many others.

Skip was hooked on country music since he first heard Merle Haggard as a youngster. It is reputed that he could play guitar before he could read. Having no funds for college, he headed out on the road with his music as soon as he graduated from high school. In 1984 he moved to Nashville and started out as a performer at Opryland; already a competent musician and budding songwriter, he landed a writer's contract. Jimmy Bowen heard Skip's demos and signed him to MCA Records in 1987. His first record, **Your Memory Wins Again**, made Top 20 in 1988. Other Top 10 hits followed, such as **I Didn't Have Far To Fall** and **Burning A Hole In My Heart** (both 1988), and **The Gospel According To Luke** and **It's You Again** (both1989).

Possessing one of the most distinctive voices on the Nashville scene, Ewing is a multi-instrumentalist who has absorbed his musical influences – James Taylor, Dan Fogelberg and Jim Croce – into a modern-day country singer-songwriter with a soulful and energetic edge to his work. When Bowen moved to Liberty/Capitol, Skip followed in 1991 and produced two excellent albums for his new label.

Recommended:
The Will To Love (MCA/MCA)
Home Grown Love (Liberty/–)
A Healin' Fire (MCA/–)
Naturally (MCA/–)

Exile

Steve Goetzman, drums; Marlon Hargis, keyboards; Sonny Le Maire, bass, vocals; J. P. Pennington, guitar, vocals; Les Taylor, guitar, vocals.

This spirited five-piece band silenced any arguments about its rock'n'roll origins with a string of No.1 country singles throughout 1985. The band rivalled Alabama as the top country band of the mid-'80s.

Originally a pop group who scored with the chart-topping **Kiss You All Over** in 1978, Exile made their initial impact as a country group with **High Cost Of Leaving**, a country Top 10 entry in August 1983. Previously J. P. Pennington and Sonny Le Maire had made an impression as songwriters when Alabama, Kenny Rogers, Dave And Sugar, Janie Fricke and Bill Anderson all recorded their songs.

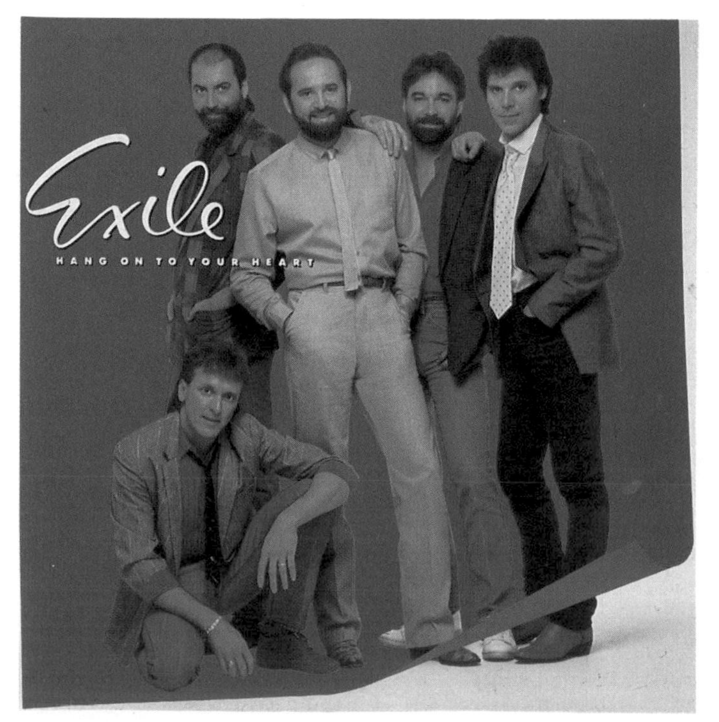

Hang On To Your Heart, Exile. Courtesy Epic Records.

Above: Barbara Fairchild, who signed to Capitol Records in 1986.

After their demise from the pop charts, they moved naturally towards country. The group had often played on country sessions long before they made it as a pop attraction. J. P. Pennington had a further connection with country music in his aunt, Lila May Ledford, who was one of the Coon Creek Girls.

Signed to Epic Records in Nashville, Exile enjoyed No. 1 hits with: **Woke Up In Love**, **I Don't Want To Be A Memory**, **Give Me One More Chance** (all 1984); **Crazy For Your Love**, **She's A Miracle** and **Hang On To Your Heart** (all 1985); **I Could Get Used To You**, **It'll Be Me** (both 1986); **She's Too Good To Be True** and **I Can't Get Close Enough** (both 1987), In 1985 keyboard player Marlon Hargis left to be replaced by Lee Carroll. Problems within the band started to boil to the surface, and in 1988 the hit singles suddenly dried up. Founding members J. P. Pennington and Les Taylor left to pursue solo careers. With newcomers Mark Jones and Paul Martin, the new-look Exile signed with Arista Records in 1989. They were back in the Top 20 with **Nobody's Talking**, **Yet** (1980) and **Even Now** (1991) – all up-mood numbers with infectious country arrangements.

Recommended:
Kentucky Hearts (Epic/Epic)
Hang On To Your Heart (Epic/Epic)
I Love Country (–/Epic)

Love's Old Song, Barbara Fairchild. Courtesy CBS Records.

On The Move, Donna Fargo. Courtesy WEA Records.

Barbara Fairchild

A husky-voiced blonde, Barbara Fairchild was born in Knobel, Arkansas on November 12, 1950. Spending her high school years in St Louis, Missouri, at 15 she had recorded for a local station and had a regular spot on a weekly TV show. Later came the inevitable move to Nashville and a meeting with MCA staffman Jerry Crutchfield, who signed her to the company as a songwriter.

Still working primarily as a songwriter, she made some demos that came to the attention of Columbia's Billy Sherrill, the result was a new recording contract and a subsequent flow of minor hits with **Love Is A Gentle Thing** (1969), **A Girl Who'll Satisfy Her Man** (1970), **(Loving You Is) Sunshine** (1971), **Love's Old Song** (1971), **Thanks For The Memories** (1972) and others. Her real breakthrough came with **Teddy Bear Song** (1973), **Kid Stuff** (1973) and **Baby Doll** (1974), all three becoming Top 5 singles.

For a couple of years Barbara faltered on the charts, but made a minor comeback with the Top 20 hit **Cheatin' Is** towards the end of 1976 and **Let Me Love You Once Before You Go** the following year. Though she has faded from the scene in recent years, Barbara made an impression with British country fans following tours in 1978 and 1979 and recorded a successful duet with Billy Walker on **The Answer Game** in 1982.

Recommended:

Free And Easy (Columbia/CBS)
This Is (Columbia/CBS)
Mississippi (Columbia/CBS)
Greatest Hits (Columbia/CBS)
The Answer Game – with Billy Walker (–/RCA)

Donna Fargo

Born Yvonne Vaughn on November 10, 1949, the daughter of a Mount Airey, North Carolina tobacco farmer, Donna attended High Point, North Carolina, Teachers' College and also spent some time at the University of Southern California, her musical education consisting of just four piano lessons taken at the age of ten. Nevertheless, she found herself torn between two careers, teaching, in the Corvin, California area, and singing, which she did under a stage name, in LA clubs.

After meeting record producer Stan Silver, whom she married in 1969, Donna set out for Phoenix where she cut some sides for Ramco Records. Her initial releases flopped so Donna continued with her teaching chores, later switching her recording activities to the Challenge label, also with little success.

Taught guitar by Silver, who also encouraged her to songwrite, Donna finally won a contract with a major company, ABC-Dot, repaying her belief in her talent via a self-penned No. 1 in **Happiest Girl In The Whole USA**, the CMA Single Of The Year for 1972. All possibilities of her being a one-hit wonder were soon dispelled when **Funny Face**, another 1972 Fargo original, climbed the charts, to be followed by **Super Man**, **You Were Always There**, **Little Girl Gone** (all 1973), **You Can't Be A Beacon** (1974), **It Do Feel Good** (1975) and **Don't Be Angry** (1976), all Top 10 entries that benefited from the distinctive, dry-throated, Fargo vocal style.

In 1977 she became part of the growing roster of Warner Bros Records' country artists, scoring such hits as **That Was Yesterday** (1977), **Do I Love You (Yes In Every Way)** (1978) and **Somebody Special** (1979). In 1979 she was stricken with Multiple Sclerosis, but has fought against the crippling disease and continued with her career, though on a lesser scale than previously. She has recorded for RCA, Cleveland, International and MCA-Songbird, as well as duetted with Billy Joe Royal on the **Members Only** single in 1988, and she hit the country charts with **Soldier Boy**.

Recommended:

The Happiest Girl In The Whole World (Dot/–)
All About A Feeling (Dot/–)
Country Sounds Of (–/MFP)
On The Move (Warner Bros/Warner Bros)
Dark Eyed Lady (Warner Bros/–)
Brotherly Love (MCA-Songbird/–)
Just For You (Warner Bros/–)
Shame On Me (Warner Bros/Warner Bros)

Narvel The Marvel, Narvel Felts. Courtesy MCA/Dot Records.

Left: Narvel Felts, who started out as a rocker in the 1950s.

Narvel Felts

A native of Missouri, born on November 11, 1938, Felts grew up with country music, became known as a rock'n'roller in the mid-'50s and returned to country later in his career. The first singer he remembers hearing was Ernest Tubb: "I used to wonder what his girlfriend was doing on the floor and him walking over her."

In 1956 he won a high school talent contest by singing **Blue Suede Shoes** and in an effort to trace him and have him sing on the station, KDEX in Bernie, Missouri, put out a message for him. Narvel and his father drove the eight miles into Bernie in their pickup truck to find a telephone.

Performing on the station's Saturday show led to Narvel landing the bass guitar spot in Jerry Mercer's band. When Mercer left, Narvel became band leader. Felts worked with both Conway Twitty (then plain Harold Jenkins) and Charlie Rich at Sun Records in 1957.

He then signed first with Mercury and then with Pink Records where **Honey Love** and **3000 Miles** made the charts. In 1970, while contracted to Hi Records, he came to Nashville hoping to find a solid country label with good distribution. Then, discussing his problems with friend and DJ Johnny Morris, they decided to evolve a new label, Cinnamon Records. In 1973, **Drift Away** (written by Mentor Williams and a pop hit for Dobie Gray) was an impressive country hit for Narvel. Later that year he scored again with **All In The Name Of Love** and in 1974 with **When Your Good Love Was Mine**.

When Cinnamon folded in 1975, Narvel joined ABC-Dot Records, coming up with **Reconsider Baby**, which was a No. 1 hit and was chosen by 'Billboard' as No. 1 Song Of The Year and 'Cashbox' as No. 1 Country Song Of The Year. It also crossed over to the pop charts. **Somebody Hold Me (Until She Passes By)** (1975), **Lonely Teardrops** (1976) and **The Feeling's Right** (1977) have all been Top 20 country hits for Narvel. Since 1979 he has recorded for a variety of small labels, including College, GMC, Lobo, Compleat and Evergreen.

Recommended:
Drift Away (Cinnamon/–)
Inside Love (ABC-Dot/–)
Memphis Days (–/Bear Family)
Pink And Golden Days (Fox/–)

Below: Freddy Fender joined the Texas Tornados in 1990, and now has a solo contract with Warner Bros.

Freddy Fender

A maverick country artist, Fender (real name Baldermar Huerta) waited 20 years for record success. He was born on June 4, 1937, in South Benito, a South Texas border town. His family worked as casual farm labourers throughout the year, but made it back to the San Benito valley each Christmas. Fender remembers that music helped to make a hard life happy and that he always managed to persuade his mother to buy him a new guitar when the old one wore out.

He dropped out of high school at 16 and joined the Marines for three years. In the late '50s he was back in San Benito playing bars and Chicano dances. By 1958 his records, in which he utilized all-Spanish lyrics, were doing well in Texas and Mexico. Gradually he turned to the more commercial fields of R&B and country for inspiration. Local club owner Wayne Duncan formed Duncan Records, and the established Imperial Records took an interest in Fender's records. Imperial released **Wasted Days And Wasted Nights**, written by Fender and Duncan, and made it into a big pop hit in 1960.

In May 1960, Fender was arrested for possession of drugs, betrayed by a paroled informer. Fender served three years of his five-year sentence in the Angola State Penitentiary, Louisiana, cutting several titles for the Goldband and Powerpack labels while inside. The Governor of Louisiana, Jimmie Davis, himself something of a country singer, helped secure Fender's release. However, a condition of his parole was that Fender should leave the entertainment business.

Fender managed to pick up his career again though, forgoing record hits, but gigging steadily. He also worked as a mechanic and even went to college for two years. In 1974 he was introduced to noted Louisiana R&B producer Huey Meaux, who put Fender's distinctive voice in a country setting. In Houston, Texas, they put down many tracks, among them a re-recording of **Wasted Days And Wasted Nights** and an update of **Before The Next Teardrop Falls**, performed partly in English and partly in Spanish. The latter was picked up by ABC-Dot and became a big pop and country hit, which led to the record being named CMA Single Of The Year in 1975, and Fender being named Top Male Vocalist by the ACM.

Further country and pop successes followed with **Secret Love** (1975), **You'll Lose A Good Thing** and **Vaya Con Dios** (both 1976), **The Day That The Rains Came** (1977) and **Talk To Me** (1978). He Joined Starflite Records in 1979, then signed with Warner Brothers in 1982, but was unable to regain the enormous success he enjoyed in the late '70s. Fender was still indulging in drugs and hitting the bottle. In 1985 he entered a clinic, and recovered. He appeared in the 1987 film 'The Milagro Beanfield War', directed by Robert Redford. Three years later he became a member of the all-star Texas Tornados, with long-time friends Doug Sahm, Augie Meyers and Flaco Jimenez. This rekindled interest in Fender's older recordings and led to a new solo contract with Warner Brothers in 1991.

Recommended:
Before The Next Teardrop Falls (Dot/ABC)
Are You Ready For Freddy? (Dot/ABC)
If You're Ever In Texas (Dot/ABC)
Rock 'n' Country (Dot/ABC)
Swamp Gold (ABC-Dot/–)
The Texas Balladeer (Starflite/–)
Best Of (–/MCA)
Freddy Fender Collection (Reprise/Reprise)

Flatt And Scruggs

Lester Raymond Flatt, born on June 19, 1914 in Overton County, Tennessee, and Earl Eugene Scruggs, born on January 6, 1924, in Cleveland County, North Carolina, pioneered a particular type of bluegrass under Bill Monroe's leadership – especially Scruggs' 'three-finger banjo' technique – and thus helped to popularize bluegrass immensely.

Both came from highly musical families. Lester's parents both played the banjo (in the old 'frailing' style) and Lester practised on both guitar and banjo. He also sang in the church choir. Earl came from an area east of the Appalachians which was already using a three-finger style on the five-string banjo. The style was not new anyway (although the strict universal style then involved either two-finger picking or simply brushing or frailing the strings): a three-finger style had been used by Uncle Dave Macon and Charlie Poole, and Earl himself had heard such banjoists as Snuffy Jenkins use it locally. But Scruggs evolved a newer style, syncopated and rhythmic, blending in his three-finger banjo to make the bluegrass style sound fresh and alive.

Lester became a textile worker but still listened to a lot of 'hillbilly' music, while also continuing to play instruments. His wife could also play guitar and sing. He was a fan of Bill and Charlie Monroe who were heard frequently on Carolina radio stations in the years before World War II. Lester was living in Covington, Virginia, and he got together with some old friends from Tennessee to play. By 1939 they had become pretty proficient and were to be hard on Radio WDBJ, Roanoke, as the Harmonizers. In 1943 Lester and his wife Gladys were hired by Charlie Monroe. Lester sang tenor harmony and played mandolin. He tired of the travelling and quit, then procured a position with a North Carolina radio station. It was there that he received a telegram from Bill Monroe asking Lester to come and play with him on the Grand Ole Opry.

Earl had played with his brothers from the age of six and by 15 he was playing on a North Carolina radio station with the Carolina Wildcats. He became a textile worker too (during the war years) but the end of the war saw him playing with 'Lost' John Miller in Knoxville. Shortly after, he began to be heard widely when Miller started broadcasting on Radio WSM from Nashville. Miller then stopped touring and Earl, out of work, was hired by Bill Monroe.

At this time Monroe was known mainly around the Southeast but the Opry was becoming more and more popular, and Bill's show rapidly gained a broader appeal. Monroe switched musical duties around and Lester Flatt's high tenor voice easily adapted to singing lead when necessary. Less easy was the pace Lester was required to keep on guitar. He sometimes shortcut his guitar part by developing a characteristic run to catch up and finish the lines. This became known as the 'Lester Flatt G Run' since it was usually played in that chord position.

Scruggs was given full rein by Monroe to develop his fluid banjo technique and helped popularize songs such as **Blue Grass Breakdown**, numbers which would remain associated with him after his departure from Monroe. Monroe even put the names of Flatt and Scruggs on some of his records. This precision and teamwork, which characterized Monroe's sound, was attracting many new listeners to the music and, by association, to the Opry itself.

In 1948, within weeks of each other, Earl and Lester resigned from Monroe to escape the constant travelling (Monroe has always been a dedicated touring man). Almost inevitably the two then decided to team up and do some radio work. They recruited ex-Monroe men Jim Shumate, on fiddle, and Howard Watts (stage name Cedric Rainwater), on bass, and then moved to Hickory, North Carolina, where they were joined by Mac Wiseman. That year, 1948, they made their first recordings for Mercury Records.

The band took its name from an old Carter Family tune, **Foggy Mountain Top**, calling themselves Foggy Mountain Boys. Wiseman left and was replaced by mandolin player Curly Seckler. Many of the fiddle players they used had previously worked with Bill Monroe. Earl's banjo was now more to the fore and the mandolin was used less often. In most other respects they promoted a harder, more driving music than that of Monroe, but Lester's vocals provided the mellowness.

They moved to Bristol on the Tennessee-Virginia border at the end of 1948 and while broadcasting on radio WCYB met the Stanley Brothers and Don Reno, both of whom developed this new sound into what we now know as 'bluegrass' and 'Scruggs-style banjo'.

In 1949 they recorded **Foggy Mountain Breakdown** and it was released the

following year. It has remained one of their most consistently popular numbers and was included in the film 'Bonnie And Clyde' as background to the famous car chase. In 1950 they were offered a lucrative contract by Columbia Records, a recording association that was to last for 20 years.

Earl introduced the 'Scruggs' peg', a device which allowed him to change the tuning of his banjo strings easily, for the number **Earl's Breakdown** (1951). That year a boost was given to their career when they appeared on a show headlined by the then fashionable Ernest Tubb and Lefty Frizzell. In 1953 the band began broadcasting 'Martha White Biscuit Time' on Nashville's Radio WSM, a show which not only ran for years, but which saw them coming well and truly into country music prominence. In 1955 their position was consolidated with an equivalent syndicated TV show and at this time they also became Grand Ole Opry members.They were travelling more than they had ever done with Bill Monroe, and they were also winning fan polls and industry awards.

The '60s folk revival also helped them, since by this time 'Scruggs picking' was already in instrument tutor terminology. Folkways released an album, compiled by Mike Seeger, titled **American Banjo Scruggs' Style** and both Mercury and Columbia released similar albums with the artist himself featured. Further recognition came in the shape of the CBS-TV series, The Beverly Hillbillies. The theme tune, **The Ballad Of Jed Clampett**, played by Lester and Earl was No. 1 on the country charts for three months from December 1962. They became a household name and a symbol of this exciting, syncopated musical styling.

They consolidated their position as leaders of the bluegrass movement and sold a vast amount of records. Towards the end of the '60s (mainly pushed by Earl), they began experimenting with new folk songs, with drums and gospel-style harmonies in an effort to build on a younger audience. Some of their older fans were unhappy about these changes and in 1969 they split up. Lester returned to more traditional sounds and made reunion albums with his old buddy Mac Wiseman. He also formed the Nashville Grass, composed mainly of the Foggy Mountain Boys. Earl defiantly went off in new directions with his Earl Scruggs Revue, utilizing his own sons and later dobro player Josh Graves in a unit which could also appeal to young, rock audiences. Earl also played a big part in getting together the old stars for the 1971 Nitty Gritty Dirt Band album, **Will The Circle Be Unbroken**.

Lester Flatt died on May 11, 1979, and in recent years Earl Scruggs has cut back his activities, while his sons have made their mark as songwriters, producers and multi-instrumentalists in country music.

Recommended:
Foggy Mountain Breakdown (Hillside/–)
Carnegie Hall (Columbia/–)
Changin' Times (Columbia/–)
Blue Ridge Cabin Home (County/–)
The Golden Era (Rounder/–)
Mercury Sessions (Rounder/–)
Flatt & Scruggs Volume 1 & 2 (–/Bear Family)

Earl Scruggs:
Nashville's Rock (Columbia/–)
Duelling Banjos (Columbia/–)
Kansas State (Columbia/–)
I Saw The Light (Columbia/–)
Earl Scruggs Revue (Columbia/–)
Scruggs Revue Volume 2 (Columbia/–)
Rockin' Cross (Columbia/–)
Family Portrait (Columbia/–)
Live From Austin City Limits (Columbia/–)

Lester Flatt:
Before You Go (RCA/–)
Foggy Mountain Breakdown (RCA/RCA)
Over The Hills To The Poorhouse – with Mac Wiseman (RCA/–)
Best Of (RCA/RCA)
Flatt Gospel (Canaan/–)
Lester Raymond Flatt (Flying Fish/–)
Live Bluegrass Festival – with Bill Monroe (RCA/–)
Living Legend (CMH/–)
The One And Only (Nugget/–)

Rosie Flores

Rosie, another Texas-born singer-songwriter, came to the fore in the mid-'80s as part of the West Coast country movement. Born in San Antonio, Texas,

Above: Flatt And Scruggs with their band, the Foggy Mountain Boys, on the Grand Ole Opry in the mid-'50s.

Rosie's earliest musical influences were a mixture of Mexican, Tex-Mex and country music. At 12 her family moved to San Diego, California, where the guitar-playing youngster started to take music seriously. She joined Penelope's Children, an all-girl band playing country, rock and psychedelic music. Starting to write songs, Rosie also did some solo gigs, then teamed up with the Screamers, an all-male band that had made quite an impact in Los Angeles. A few years later she was in the Screamin' Sirens, an all-girl band that appeared in the movie 'The Running Kind'. By this time Rosie wanted to play country, so she put a four-track demo of her own songs together and eventually gained a record contract with Reprise (1986).

Dwight Yoakam produced her first album, and Rosie scored minor country hits with **Crying Over You** (1987) and **Somebody Loves, Somebody Wins** (1988). With her rock-tinged country music and raunchy vocal style, Rosie won over a younger audience for her music, but has been unable to break through to mainstream country acceptance. She continues to work the Californian club scene and at one time looked set to make an impact in Europe.

Recommended:
Rosie Flores (Warner Bros/Warner Bros)
After The Farm (Hightone/–)

Flying Again, Flying Burrito Brothers. Courtesy Columbia/CBS Records.

Flying Burrito Brothers

The band was formed in 1968 by ex-Byrds Gram Parsons and Chris Hillman, to bring country music to the rock fans of California. A&M Records felt that the charismatic Parsons might help to generate some big sales with this new concept, and they subsequently put much promotional money behind the first album, **Gilded Palace Of Sin**. Bizarre photo sessions in the desert resulted in an album sleeve depicting the Burritos in extravagant Nudie suits. The marijuana leaves embroidered on the suits emphasized a new approach. **Gilded Palace** featured some of Parsons' best-ever songs and beefed up his sensitive but non-too-strong voice with a rock production, and Chris Hillman was also prominent on vocals. The line-up for this album was: Parsons, guitar, vocals; Chris Hillman, guitar, mandolin, vocals; Chris Ethridge, bass; Sneaky Pete Kleinow, pedal steel guitar; and Jon Corneal, drums.

In 1969, Corneal and Ethridge dropped out, and Bernie Leadon (guitar, vocals) and Mike Clarke (drums) joined. Hillman switched to bass. The next two albums, **Burrito De Luxe** and **Flying Burrito Brothers**, were straighter productions but still with a good dash of country included. Parsons left between the two albums (in 1970) and in 1971 Bernie Leadon also left, to subsequently form the highly successful Eagles. Sneaky Pete also left to undertake production and session work.

1971 saw a vastly expanded line-up in which Byron Berline (a top fiddle player who had recorded with the Rolling Stones), Al Perkins (pedal steel guitar), Kenny Wertz (guitar, banjo, vocals) and Roger Bush (string bass), all joined. This line-up saw the release of a good live album, **Last Of The Red Hot Burritos**.

The addition of Alan Munde (banjo) in 1971 completed a floating aggregation which rejoiced under the title, Hot Burrito Revue. The bluegrass-oriented members of this loose set-up who finally decided to stay became known as Country Gazette.

However, the Burritos were to re-form again in 1974. With a line-up of Sneaky Pete, Gib Guilbeau (a Cajun fiddle player), Gene Parsons (drums), Chris Ethridge and Joel Scott-Hill, they toured America and Europe, cutting albums for CBS.

By 1979, the Burritos were down to a two-piece, consisting of Guilbeau and John Beland. Known as the Burrito Brothers they moved from the West Coast to Nashville in 1981, determined to make an impression as a country act. They signed a recording contract with Curb Records, who were licensed through Columbia, and scored Top 20 country hits with **Does She Wish She Was Single Again** and **She Belongs To Everyone But Me** (1983). By 1985 the pair had split. Beland concentrated on his songwriting skills and Guilbeau teamed up with Sneaky Pete Kleinow for yet another re-formed Flying Burrito Brothers line-up.

Recommended:
Gilded Palace Of Sin (A&M/Demon)
Last Of The Red Hot Burritos (A&M/–)
Flying Burrito Brothers (A&M/A&M)
Close Up The Honky Tonks (A&M/A&M)
Flying Again (Columbia/CBS)
Sleepless Nights – with Gram Parsons (A&M/A&M)
The Flying Burrito Brothers – Live From Tokyo (–/Sundown)
Cabin Fever (Relix Records/–)
Dim Lights, Thick Smoke And Loud Music (–/Edsel)
Hollywood Nights, 1979–1981 (–/Sundown)

Dan Fogelberg

Fogelberg grew up in the little town of Preoria, Illinois (where he was born on August 13, 1951), and attended the University of Illinois as an art student before settling on a musical career. He established himself in the mid-'70s as a pop-country singer-songwriter with a series of albums on Full Moon Records.

Although he has lived in Colorado for a number of years, most of his songs have a midwestern setting. His writings on freedom, lost love and women have proved to be decidedly effective and timely pieces. Dan showcased his country roots on the highly acclaimed 1985 album, **High Country Snows**.

Utilizing the talents of Ricky Skaggs, Herb Pedersen, Chris Hillman, Doc Watson, Al Perkins and Vince Gill, Dan produced a first-rate bluegrass-flavoured album, blending his own self-penned songs with those of Carter Stanley and Flatt And Scruggs. He made inroads on the country charts with **Go Down Easy** and **Down The Road (Mountain Pass)** during1985.

Recommended:
HIgh Country Snows (Full Moon/Full Moon)
The Wild Places (Full Moon/–)
Dan Fogelberg Live (Full Moon/–)

Red Foley

Elected to the CMA Hall Of Fame in 1967, Clyde Julian 'Red' Foley was born in Bluelick, Kentucky, on June 17, 1910.

A star athlete at high school and college, at the age of 17 he won the Atwater-Kent contest in Louisville. In 1930 he moved to Chicago to become a member of John Lair's Cumberland Ridge Runners on the WLS National Barn Dance Show. Seven years later he helped to originate the Renfro Valley Show with Lair, by 1939 appearing on Avalon Time, a programme in which he co-starred with Red Skelton, thus becoming the first country star to have a network radio show.

His Decca records soon proved eminently popular. Foley's versions of **Tennessee Saturday Night**, **Candy Kisses**, **Tennessee Polka**, and **Sunday Down In Tennessee** all became Top 10 discs during 1949. In 1950 sales escalated even further, with no less than three Foley titles – **Chattanoogie Shoe Shine Boy** and the spirituals **Steal Away** and **Just A Closer Walk With Thee** – becoming million-sellers.

The following year, his success with religious material continued. Foley's recording of Thomas A. Dorsey's **Peace In The Valley** sold well enough to become an eventual gold disc winner. Meanwhile, his more commercial songs also accrued huge sales, **Birmingham Bounce** becoming a 1950 No.1, and **Mississippi** (1950), **Cincinatti Dancing Pig** (1950), **Hot Rod Race** (1951), **Alabama Jubilee** (1951), **Midnight** (1952), **Don't Let The Stars Get In Your Eyes** (1953), **Hot Toddy** (1953), **Shake A Hand** (1953), **Jilted** (1954), **Hearts of Stone** (1954) and **A Satisfied Mind** (with Betty Foley, 1955), all providing him with Top 10 placings.

Right: Red Foley on the Grand Ole Opry in the early '50s.

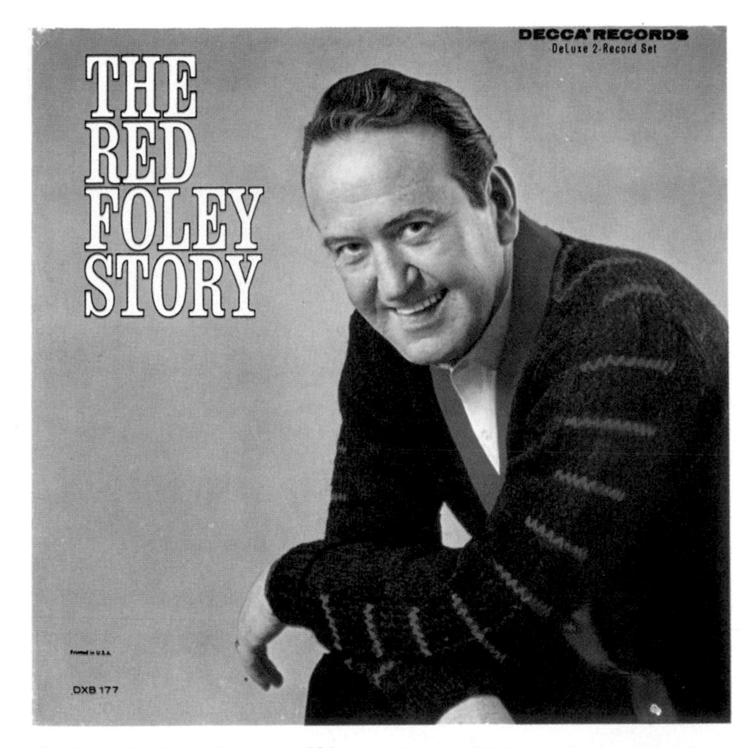

The Red Foley Story. Courtesy MCA Records.

An Opry star during the '40s, in 1954 he moved to Springfield, Missouri, where he hosted the Ozark Jubilee –one of the first successful country TV series. During the early '60s, Foley co-starred with Fess Parker on an ABC-TV series Mr Smith Goes To Washington. He continued appearing on radio and TV, and making many personal appearances, right up to the time of his death on September 19, 1968, in Fort Wayne, Indiana.

Recommended:
Beyond The Sunset (MCA/–)
Red Foley Story (MCA/–)
Red & Ernie – with Ernest Tubb (–/Stetson)
Tennessee Saturday Night (–/Charly)
Company's Coming (Decca/Stetson)

Tennessee Ernie Ford

The singer who recorded **Sixteen Tons**, a mining song which sold over four million copies during the mid-'50s, Ernest Jennings Ford was born in Bristol, Tennessee, on February 13, 1919. At school he sang in the choir and played trombone in the school band – but he also spent much time at the local radio station, WOAI, where he began working as an announcer in 1937.

Four years later, following a period of study at the Cincinnati Conservatory of Music and further announcing stints with various radio stations, Ford enlisted in the Air Force, becoming first a bombadier on heavy bombers, then an instructor. After discharge, he returned to announcing, working on C&W station KXLA, Pasadena, where he met Cliffie Stone and appeared as a singer with Stone's quartet on the Hometown Jamboree show.

In 1948, Ford signed with Capitol Records, for whom his warm bass voice provided immediate hits in **Mule Train** and **Smokey Mountain Boogie** (1949). He subsequently scored with **Anticipation Blues** (1949), **I'll Never Be Free** (with Kay Starr, 1950), **The Cry Of The Wild Goose** (1950) and **Shotgun Boogie**, a Ford original that became a million-seller in 1950.

His own radio shows over the CBS and ABC networks gained Ford further popularity, then, in 1955, following another handful of hits, he recorded **Sixteen Tons**, a superb Merle Travis song that was decked out in a fine Jack Marshall arrangement. This became a massive hit, winning Ford his own NBC-TV show – which the singer hosted until 1961.

Since the early '60s Ford increasingly concentrated on religious music, the biggest seller being **Hymns**, reputed to be the first million-selling album in country music, and one which gained him a platinum-plated master in 1963. Gradually he cut down on his work, although he appeared on many TV shows and enjoyed some chart success with **Hicktown** (1965), **Honey-Eyed Girl** (1969) and **Happy Songs Of Love** (1971). Inducted into the Country Music Hall Of Fame in 1990, Tennessee Ernie died of liver disease, October 17, 1991 at HCA Hospital, Reston, Virginia, where he had been recuperating after being taken ill at a White House dinner on September 28.

Ernie Sings & Glen Picks, Tennessee Ernie Ford. Courtesy Capitol Records.

Recommended:
Hymns (Capitol/–)
America The Beautiful (Capitol/–)
Civil War Songs (Capitol/–)
Country Hits Feelin' Blue (Capitol/–)
Ernie Sings And Glen Picks – with Glen Campbell (Capitol/–)
Precious Memories (Capitol/–)
Sixteen Tons (–/Bear Family)
Farmyard Boogie (–/See For Miles)

The Forester Sisters

Kathy (born January 4, 1955, Lookout Mountain), June (born September 25, 1956, Chattanooga), Kim (born December 4, 1960, Chattanooga) and Christy (born October 10, 1962, Chattanooga), four sisters from Georgia, burst upon the country scene at the beginning of 1985 when their very first single, **(That's What You Do) When You're In Love**, gave them a country Top 10 entry.

The girls, who count Bonnie Raitt, Linda Ronstadt and Emmylou Harris as being early influences and main inspiration, began by singing at weddings, funerals and community events around their Lookout Mountain home. Specializing in sweet harmonies, with Kim and eldest sister Kathy handling most of the lead vocals, they cut a demo tape in Muscle Shoals, Alabama, which eventually found its way to Warner Brothers Records In Nashville.

Jim Ed Norman signed the girls to the label at the end of 1984, and, under the guidance of Jerry Wallace and Terry Skinner, they recorded their debut album in Muscle Shoals. Their first single hit No.1 in the country charts, and two other No.1s followed, with **I Fell In Love Again Last Night** and **Just In Case** (both 1985). Further chart-toppers came with **Mama's Never Seen Those Eyes** (1986), a duet with the Bellamy Brothers on **Too Much Is Not Enough** (1986), and **You Again** (1987).

In addition to Kathy's keyboards and Kim's guitar, the girls make use of a four-piece band, including Kathy's husband on bass, who also acts as the Foresters' road manager. The four sisters have five children between them, and they all travel on the road together. Unlike most other country stars, they have not moved to Nashville, preferring to live near their relatives in Lookout Mountain. With their roots in old-time country, they perfectly bridge the gap between traditional and contemporary country without edging too close to pop. The Forester Sisters have continued a consistent run of Top 10 hits, including **Lying In His Arms Again** (1987), **Letter Home** (1988), **Love Will** (1989) and **Men** (1991), plus several acclaimed albums that show they can be gutsy as well as sweet, serious and humorous, and, most importantly, musically entertaining.

Recommended:
The Forester Sisters (Warner Bros/–)
Talkin' 'Bout Men (Warner Bros/–)
Sincerely (Warner Bros/Warner Bros)
Come Hold Me (Warner Bros/–)
I Got A Date (Warner Bros/–)
Perfumes, Ribbons And Pearls (Warner Bros/Warner Bros)

Foster And Lloyd

This unlikely country duo first joined as songwriters in 1985 and went on to score several Top 10 country hits in the late '80s. They then went their separate ways with Radney Foster emerging as a promising '90s solo performer.

Kentucky-born Bill Lloyd worked in New York, then put a band together in Kentucky before moving to Nashville in 1982, where he was signed as a writer with the MTM Music Group. Radney Foster, born on July 10, 1960, in Del Rio, Texas, played local clubs before moving to Nashville in 1981, where he waited on tables and pitched his songs. He signed with MTM, and soon after the pair started writing together. Their first success came in 1986 with **Since I Found You**, in a Top 10 hit for Sweethearts Of The Rodeo. Foster, the main writing force, co-wrote **Someone Like Me** with Holly Dunn, which gave the Texas lady a No.2 hit in 1987, the same year that Foster And Lloyd secured a record contract with RCA.

Their dynamic blend of '60s pop, country and R&B soon found them riding the charts with such Top 10 hits as **Crazy Over You** and **Sure Thing** (1987), **What Do You Want From Me This Time** (1988) and **Fair Shake** (1989), plus three superb albums, one of which, **Faster and Llouder**, crossed into the pop charts, gaining a gold disc. Both wanted to pull in different directions, so in 1991 they split.

Radney Foster signed with Arista Records and his first single, **Just Call Me Lonesome** (1992), went Top 20. His album, **Del Rio Tx 1959** contained all self-penned songs, his obvious strong point. The songs evoked memories of late '50s rock, pop and country dressed up in contemporary country arrangements. It proved to be a successful formula, with his second single, **Nobody Wins** (1993), climbing to No.2 on the country charts, and opening the doors for a highly successful solo career.

Recommended:
Version Of The Truth (RCA/RCA)
Faster And Llouder (RCA/–)

Radney Foster Solo:
Del Rio Tx, 1959 (Arista/–)

Curly Fox And Texas Ruby

Curly Fox, the fiddling son of a fiddler, was born Arnum LeRoy Fox in Graysville, Tennessee, on November 9, 1910. He joined a medicine show at the age of 13, and recorded as early as 1929, with the Roane County Ramblers. He also played with the Carolina Tarheels and headed his own band, the Tennessee Firecrackers, over WSB in Atlanta in 1932.

Texas Ruby was Tex Owen's sister, born on June 4, 1910 in Wise Country, Texas, who had come to the Grand Ole Opry as early as 1934 with Zeke Clements and his Bronco Buster. She and Clements worked for a while on WHO in Des Moines, before she and Curly Fox teamed up on WSM in 1937. Not long after this, they became one of the most popular acts on the Opry and in country music, with their winning combination of Ruby's deep, strong, sultry voice and Curly's masterful trick fiddling.

After marrying in 1939, their biggest years were on the Grand Ole Opry in the '40s, where they were stars of the Purina portion, and they recorded for Columbia (1945–46) and King (1947). In 1948 they journeyed to New York and then Houston, where they were to spend seven years on KPRC-TV before returning to the Grand Ole Opry.

They recorded an album for Starday during this second Nashville period, just prior to Texas Ruby's death in March 1963, when a fire raged through their house trailer. Curly went into virtual retirement afterwards, living in rural Illinois, but in the mid-'70s he began to appear at occasional bluegrass festivals and other gatherings.

My Baby Packed Up My Mind And Left Me, Dallas Frazier. Courtesy RCA Records.

Dallas Frazier

A respected singer-songwriter, Frazier is capable of penning country songs that 'cross-over' without losing too much individuality in the process (although it must be said that one big success in this area, **Alley Oop**, moulded into a novelty hit by Kim Fowley for the Hollywood Argyles, was pure gimmickry).

Born in Spiro, Oklahoma, in 1939, Frazier was a featured stage performer before he reached his teens and a best-selling songwriter by 21. Early on, his family moved to the country centre at Bakersfield, California. In a talent contest sponsored by Ferlin Husky he won first prize and Husky offered him a place on his show.

He was signed by Capitol Records and moved to Nashville to pitch his songs, and also starred on radio and TV there. Ferlin Husky had one of his most famous hits with Frazier's **Timber, I'm Falling** and there were other cross-overs with **There Goes My Everything** (Engelbert Humperdinck) and **Son Of Hickory Holler's Tramp** (O. C. Smith). But he could also pen a convincing dues-paying country song, as he showed with **California Cotton Fields**, a title recorded by Merle Haggard.

Frazier made a comeback as a writer in the last ten years with the Oak Ridge Boys' updated version of **Elvira**, a million seller in 1981, and several other singers raiding his vast catalogue of songs. He has been helped in his writing career by having a stronger voice than many composers and this has enabled him to succeed as a performer also. His albums have included **My Baby Packed Up My Mind And Left Me** and **Singin' My Song** (RCA).

Janie Fricke

Regarded as the most versatile female vocalist in country music, Janie, who was born December 19, 1947 on a farm near South Whitney, Indiana, came from a musical family. Her father was a guitarist and her mother taught piano and played organ at the local church.

Between study lessons at Indiana University, Bloomington, she broke into the thriving Memphis jingle scene. After completing her studies she tried her luck in Los Angeles as a background vocalist but failed to make any impression. Consequently, she returned to Memphis in the summer of 1972 and started building a first-rate reputation as a jingle singer and performed in the group Phase II.

In 1975 she moved to Nashville where she joined the Lea Jane Singers and Janie's natural vocal ability allowed her to stand out in the cut-throat Nashville session business. On a Johnny Duncan session she was asked to sing some solo lines on **Jo And The Cowboy** and on subsequent Johnny Duncan recordings of **Stranger** and **Thinkin' Of A Rendezvous** (No.1s in 1976), **It Couldn't Have Been Any Better** and **Come A Little Bit Closer** (No.1s in 1977). During this period she contributed to more than 20 Top 10 country singles by Ronnie Milsap, Eddie Rabbitt, Crystal Gayle, Mel Tillis and Vern Gosdin.

Above: Janie Fricke changed her name to Frickie in the '80s, then back again.

Singer Of Songs, Janie Fricke. Courtesy CBS Records.

After being coaxed into a solo recording contract in 1977 with Columbia Records, where she was initially produced by Billy Sherrill, Janie enjoyed Top 20 hits with **What're You Doing Tonight**, **Baby It's You** and **Please Help Me I'm Falling** from her first album. Due to her reluctance to give up her session work to form a band and do show dates, she was dubbed 'the reluctant superstar'.

A duet with Charlie Rich, **On My Knees**, hit the top of the charts in 1978, then finally she made the breakthrough as a solo singer with **Down To My Last Broken Heart** (1980), which was followed by other Top 10 hits, such as **I'll Need Someone To Hold Me (When I Cry)** (1981), **Don't Worry 'Bout Me Baby** and **It Ain't Easy Bein' Easy** (both 1982).

By this time Janie had made the transition from secure anonymity as one of Nashville's most successful jingle and session singers to the forefront as a country star. She was named CMA Female Vocalist in both 1982 and 1983, and enjoyed further chart-toppers with **He's A Heartache (Looking For A Place To Happen)** (1983), **Let's Stop Talking About It** (1984) and **She's Single Again** (1985). Her recordings were now being produced by Bob Montgomery and her career was managed by her husband Randy Jackson, with her own Heart City Band, which she used on the acclaimed **It Ain't Easy** LP, Janie gained a reputation as one of country music's most dynamic female entertainers.

In 1986 she started spelling her name 'Frickie', and though she topped the charts with **Always Have, Always Will** that same year, suddenly her records stopped making such a big impact, though the quality of her albums were consistently high. By 1989 her contract with Columbia expired and the next two years found her without a record deal. Janie signed with the independent Intersound in 1991, reverting back to the original spelling of her name. She has continued to make good albums and tours regularly, but has failed to make a return to the charts – the standard by which all country music acts are gauged as far as star status is concerned.

Recommended:

Sleeping With Your Memory (Columbia/CBS)
Love Notes (Columbia/CBS)
The First Word In Memory (Columbia/CBS)
Janie Fricke (Intersound/–)
After Midnight (Columbia/CBS)
Black And White (Columbia/CBS)

Kinky Friedman

Leader of outlandish country rock band, the Texas Jewboys, Richard Friedman was born in Palestine, Texas, October 31, 1944, the son of a Texas University professor.

Brought up on a ranch, he later attended university in nearby Austin, in which town he formed his first band, King Arthur And The Carrots. Then came some time spent in Borneo, where he was a member of the Peace Corps.

In 1971 he headed for LA with his band, the Texas Jewboys, establishing a reputation as the Frank Zappa of country music in **Sold American**, for Vanguard.

He was signed to ABC Records during

Sold American, Kinky Friedman. Courtesy Vanguard Records.

1974, but by 1976 Friedman had moved on once more, to Epic, cutting **Lasso From El Paso**, an all-star album featuring such dignitaries as Bob Dylan and Eric Clapton. Although he has appeared on the Opry, Friedman's country is generally far-out and not really meant for mainstream fans. He has become a country music critic for 'Rolling Stone' magazine and something of a detective novelist with 'Greenwich Killin' Time', a story about a country singer turned detective. He still occasionally returns to the stage, but his live sets are mainly re-runs of his old songs.

Recommended:

Sold American (Vanguard/Vanguard)
Kinky Friedman (ABC/ABC)
Lasso From El Paso (Epic/Epic)
Old Testaments And New Revelations (Fruit Of The Tune/–)

David Frizzell

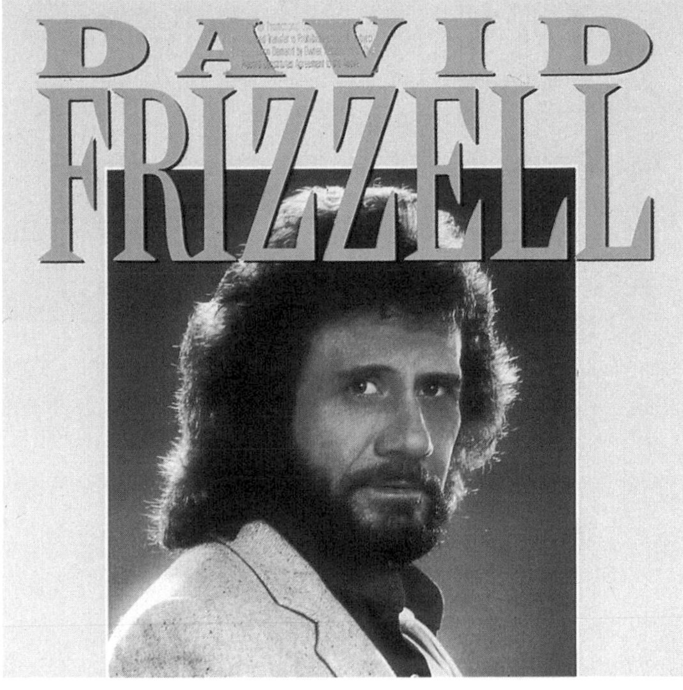

Solo, David Frizzell. Courtesy WEA/Viva Records.

A younger brother of legendary Lefty Frizzell, David was born on September 26, 1941, in El Dorado, Texas. He spent 23 years recording for seven different labels under eight separate contracts before he made a major breakthrough in 1981, when he teamed up with Shelly West (daughter of singer Dottie West) on the chart-topping duet, **You're The Reason God Made Oklahoma**.

David hitch-hiked from his Texas home to California to be with Lefty in the late '50s, and his first recordings were made under the guidance of Don Law for Columbia Records in 1958. A handful of country-rockabilly singles were released, but all sank without trace. Following a stint in the US Army, he re-signed with Columbia in the late '60s, scoring a minor hit with **I Just Can't Help Believing** (1970). David spent a few years making regular appearances on Buck Owens' Ranch Show TV programme. A few recordings for Capitol were released, David scoring minor hits with **Words Don't Come Easy** (1973) and **She Loved Me Away From You** (1974).

Further recording stints followed without too much success. David invested in his own club in Concord, California, in 1977, and it was the following year that Shelly West, along with David's younger brother Allen Frizzell, joined him at the club. Shelly and David toyed with a few duets and did a demo tape of **We're Lovin' On Borrowed Time**. Producer Snuffy Garrett heard the tape and recorded an album. He set up a deal with Casablanca West, but it fell through, so he began shopping around for another one. He found no takers in Nashville. However, he played the tape to actor Clint Eastwood, his partner in Viva Records, who decided to use **You're The Reason God Made Oklahoma** on the soundtrack of his upcoming film, 'Any Which Way You Can'.

The rest was like a dream for David and Shelly. The song was put out as a single at the beginning of 1981, made it to the top of the country charts, and the pair walked off with a CMA Award for Top Country Duo for 1981, a feat they repeated the following year. More duet hits followed

Left: Friedman is an outlandish character who is a novelist, critic and singer.

with **A Texas State Of Mind** (1981), **Honky-Tonk Night On Broadway** (1982), **I Just Came Here To Dance** (1983) and **It's A Be Together Night** (1984).

Both were keen to follow solo careers, and David was first to record a solo album, **The Family's Fine, But This One's All Mine**. The album produced a No.1 country hit, **I'm Gonna Hire A Wino To Decorate Our Home**. He has continued to score solo hits with **Lost My Baby Blues** (1982), **A Million Light Beers Ago** (1983), **When We Get Back To The Farm** (1984) and **Country Music Love Affair** (1985). With changing trends in country, Frizzell's hits became harder to find, though he has continued to record for Viva, Nashville America and Compleat. He still tours regularly, preferring to play the club and cabaret circuit.

Recommended:
On My Own Again (Warner-Viva/–)
The David Frizzell & Shelly West Album (Warner-Viva/–)
Our Best To You – with Shelly West (Warner-Viva/–)
My Life Is Just A Bridge (BFE/–)

Lefty Frizzell

Acquiring the nickname 'Lefty' after disposing of several opponents with his left hand during an unsuccessful attempt to become a Golden Gloves boxing champion, the Texas-born (Corsicana, March 31, 1928) singer-songwriter-guitarist began life as William Orville Frizzell.

The Legendary Lefty Frizzell. Courtesy MCA Records.

A childhood performer, at 17 he could be found playing the honky-tonks and dives of Dallas and Waco, moulding his early, Jimmie Rodgers-stylings to his environment, thus formulating a sound that was very much his own.

In 1950, Frizzell's Columbia recording of **If You've Got The Money, I've Got The Time** became a massive hit, claiming a chart position for some 20 weeks. The ex-pugilist followed this with two 1951 No.1s in **I Want To Be With You Always** and **Always Late**.

He became an Opry star, and throughout the rest of the decade he continued to supply a series of chart high-flyers, many of these in honky-tonk tradition. The '60s too found Frizzell obtaining more than a dozen hits, though only **Saginaw, Michigan** – a 1964 No.1 – and **She's Gone, Gone, Gone** (1965) proved of any consequence. His last hit for Columbia was **Watermelon Time In Georgia** (1970).

He joined ABC Records in 1973 and was beginning to make a comeback with **I Never Go Around Mirrors** and **Lucky Arms** (both 1974) and **Falling** (1975), when he died on July 19, 1975 after suffering a stroke. Elected to the Country Music Hall Of Fame in 1982, Frizzell's influence has played a major role in much of the country music of the '90s. You can hear strains of his work in the style of Merle Haggard, which has been continued through George Strait, Keith Whitley and lately in the music of Clint Black and Doug Stone, among others.

Recommended:
The Classic Style (ABC/–)
The Legend Lives On (Columbia/–)
Songs Of Jimmie Rodgers (Columbia/–)
Lefty Goes To Nashville (Rounder/–)
American Originals (Columbia/–)
Treasures Untold (Rounder/Bear Family)
Life's Like Poetry (–/Bear Family)

Treasures Untold, Lefty Frizzell. Courtesy Bear Family Records.

Steve Fromholz

Purveyor of what he terms 'free-form, country-folk, science-fiction, gospel, cum existential bluegrass-opera music', the hirsute Fromholz (born June 8, 1945, Temple, Texas) once looked likely to become the most talented has-been in Austin, a situation later reflected in the title of his first solo album – **A Rumor In My Own Time**.

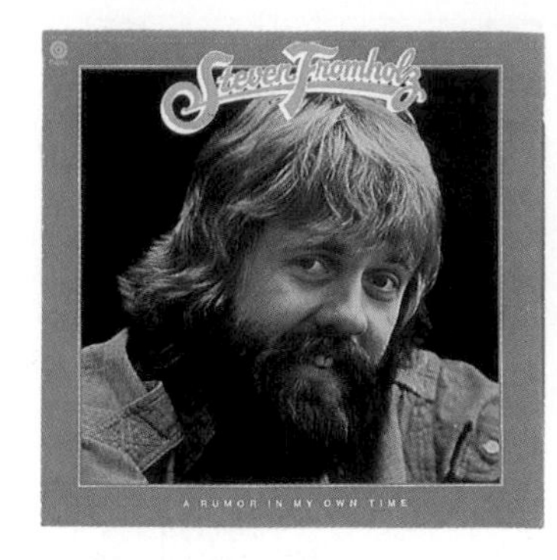

A Rumor In My Own Time, Steve Fromholz. Courtesy Capitol Records.

At 18, he attended North Texas State University, meeting singer-songwriter Michael Murphey, the duo becoming part of the Dallas County Jug Band. After an abbreviated stay in the navy, Fromholz befriended another singer-songwriter, Dan McCrimmon. The twosome formed Frummox and recorded an album **From Here To There** for Probe (1969), the disc featuring Fromholz's ambitious **Texas Trilogy**.

He moved to Austin (1974), and there became an accepted part of the outlaw community, providing material and singing on Willie Nelson's **Sound In Your Mind** LP. The first real Fromholz solo album, **A Rumour In My Own Time**, an all-star soirée featuring Red Rhodes, Willie Nelson, Doug Dillard, John Sebastian, B. W. Stevenson and the Lost Gonzo Band, found a Capitol release in 1976 and fulfilled all the hopes of his cult following. However, a later album, **Frolicking In The Myth**, proved him to be moving on in search of new frontiers to breach.

Recommended:
A Rumor In My Own Time (Capitol/–)
Frolicking In The Myth (Capitol/–)
Jus' Playin' Along (Lone Star/–)

Larry Gatlin

Born in Seminole, Texas, on May 28, 1948, but raised in nearby Odessa, clear-voiced Larry Wayne Gatlin is a singer-songwriter whose roots are in gospel music. When only five he could be found watching the Blackwood Brothers. At the same age he appeared in a talent contest as part of the Gatlins (along with his two brothers and one of his sisters), a gospel group that toured throughout the southern states. But his breakthrough came while he was working with the Imperials in Vegas, as part of the Jimmy Dean Show.

There Gatlin met Dottie West, one of Dean's guests, who offered to help him. In May 1971, he sent her eight songs – from which she selected and recorded two, **Once You Were Mine** and **You're The Other Half Of Me**. A few months later, when Dottie formed her own First Generation Music Company, Gatlin was the first writer to gain a contract. Following further songs for Dottie, including **My Mind's Gone Away**, Gatlin sang the verse plus harmony vocals on Kristofferson's **Why Me?** hit. Johnny Cash employed several Gatlin compositions for his 'Gospel Road' movie. Due to Kristofferson's insistence, Monument signed Gatlin in 1972, releasing singles by Gatlin and his band that year. Larry's first album, **The Pilgrim**, appeared in 1974. Also in 1974 he made his Top 20 debut

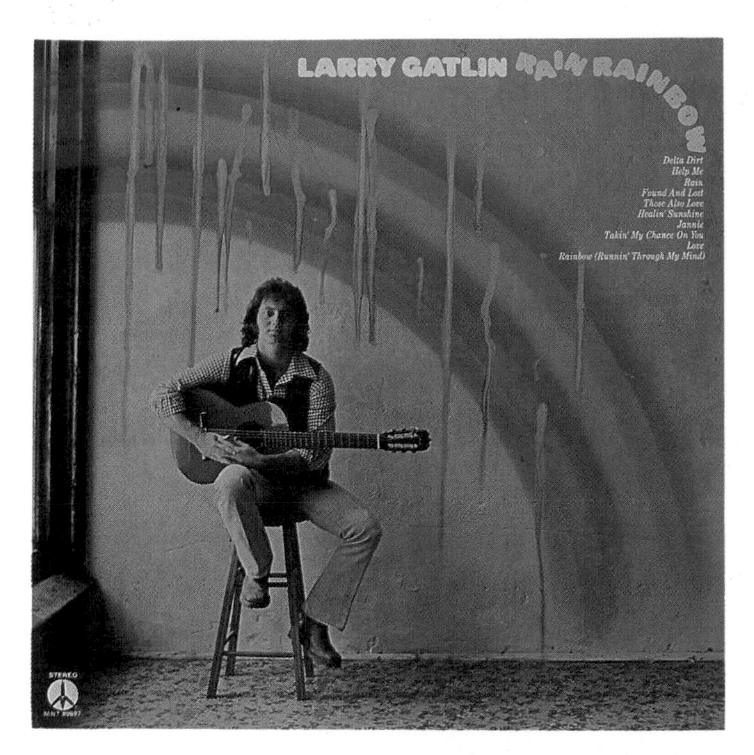

Rain Rainbow, Larry Gatlin. Courtesy Monument Records.

with **Delta Dirt**, going Top 5 with a late 1975 release, **Broken Lady**, a song that won a Grammy in 1976.

During 1975, too, Gatlin produced some sides for Johnny Duncan, one of which, **Jo And The Cowboy**, provided Janie Fricke with her debut as a backup singer.

He also reunited with his brothers Rudy and Steve, forming a band and reshaping his whole musical approach once more. The next major hit, **Statues Without Hearts** (1976), was released under Larry Gatlin, With Family And Friends. During 1977, in the wake of a flood of hits, he scored his first No.1 with **I Wish You Were Someone I Love**. There were three more Top 10 singles in 1978 along with two hit albums, **Oh Brother** and **Greatest Hits Volume 1**, but the Monument label was slowly folding. In early 1979 Gatlin switched to Columbia, his act now billed as Larry Gatlin And The Gatlin Brothers Band. The result was another No.1, **All The Gold In California** (1979) as well as Top 10 singles **Take Me To Your Lovin' Place** (1980), **What Are We Doin' Lonesome** (1981), **Sure Feels Like Love** (1982), **Houston** (a No.1 in 1983), **Denver** and **The Lady Takes The Cowboy Everytime** (both 1984). But the strain of constant writing, recording and touring took its toll. On December 10, 1984 he voluntarily checked himself into a California drug and alcohol abuse centre. Happily, he was soon back to full health once more, he and his brothers cutting **Smile**, an album produced by jazz-funk guitarist Larry Carlton, in 1985.

Recommended:

The Pilgrim (Monument/–)
Rain, Rainbow (Monument/Monument)
Larry Gatlin With Family And Friends (Monument/Monument)
Straight Ahead (Columbia/CBS)
Live At 8pm (Capitol/–)
Alive And Well (Columbia/–)
Pure 'n' Simple (Universal/–)

Crystal Gayle

Loretta Lynn's younger sister (real name Brenda Gail Webb), Crystal was the last of eight children born to the Webbs and the only one who arrived in a hospital. Born in Paintsville, Kentucky, in 1951, she toured with Conway Twitty and Loretta at the age of 16. Her name change was inspired by the Krystal hamburger chain. In 1970, she placed her first Decca release, **I Cried (The Blue Right Out Of My Eyes)** on the country charts, but there was little other chart action during this period.

After the Decca deal ended, Crystal moved to UA Records, refusing to record any material associated with Loretta. There began an association with producer Allen Reynolds. Reynolds' own composition, **Wrong Road Again**, brought Crystal back into the country charts, this being followed by two other singles, all of them from her successful debut album **Crystal Gayle**. This 1975 comeback was the start of a consistent run of hitmaking. Crystal's **Somebody Loves You** single went Top 10 that year. The following year

Below: Larry Gatlin (centre) formed a band with his brothers in the late '70s.

Crystal, Crystal Gayle. Courtesy UA Records.

saw the singer having two country chart-toppers, **I'll Get Over You** and **You Never Miss A Real Good Thing**, plus a best-selling third album, **Crystal**, all bearing the Reynolds' hallmark of quality. 1977 saw the advent of two further monster hits in **I'd Do It All Over Again** and the bluesy **Don't It Make My Brown Eyes Blue**, Crystal's first major cross-over single. 1978 saw her logging three No.1s in a row with **Ready For The Times To Get Better**, **Talking In Your Sleep** and **Why Have You Left The One You Left Me For**.

The flood of hits continued in 1979, the year that saw Crystal become the first country artist to tour China. Now with Columbia Records, she provided her new label with further No.1s: **It's Like We Never Said Goodbye** and **If You Ever Change Your Mind** (1980). There was another No.1, **Too Many Lovers**, for Columbia in 1981 before she signed for Elektra and carried on where she left off – logging yet another chart-topper with **'Til I Gain Control Again** (1982). The same year she duetted with Eddie Rabbitt on **You And I**, a country No.1 and pop Top 10 entry. Switching to Warner Brothers, further No.1s in the mid-'80s came with

These Days, Crystal Gayle. Courtesy CBS Records.

Our Love Is On The Faultline, **Baby, What About You**, **The Sound Of Goodbye**, **Turning Away** and a duet with Gary Morris on **Makin' Up For Lost Time**, the theme from the Dallas TV series.

Crystal has been devoted to charitable causes and received the Waterford 'Celebration Of Light' Award in 1988, in recognition of her involvement in such activities. Away from showbiz, she has her own fine gifts and jewellery business, 'Crystals', which is located in Nashville. But this did not interfere with her music and there were two more chart-topping hits for Warners, a revival of **Cry** (1986) and **Straight From The Heart** (1987). The late '80s saw a movement away from Crystal's smooth pop/country to more traditional-sounding music in Nashville, and she found it more difficult to gain radio plays. In 1990 she moved over to Capitol Records and was reunited with Allen Reynolds, who produced the **Ain't Gonna Worry** LP, but this only resulted in one minor hit with a revival of **Never Ending Song Of Love** (1990). Two years later she was without a major label deal. Although there had been much talk of recording with her sisters Loretta Lynn and Peggy Sue Wright, this has not come to fruition, and Crystal has ended up re-recording several of her old hits for **Best Always** (1993).

Recommended:
Somebody Loves You (UA/UA)
Crystal (UA/UA)
When I Dream (UA/UA)
Ain't Gonna Worry (Capitol/Capitol)
Three Good Reasons (Capitol/Capitol)
Cage The Songbird (Warner Bros/Warner Bros)
Country Girl (–/Music For Pleasure)
True Love (Elektra/Elektra)

Bobbie Gentry

In July 1967, **Ode to Billy Joe**, a song about the suicide of a certain Billy Joe Macallister, was released. Bedecked in an imaginative, swamp-flavoured, Jimmy Haskell arrangement, it was one of the year's finest singles. Thus was the public introduced to the singing and songwriting talent of Bobbie Gentry.

Of Portuguese descent, Bobbie was born Roberta Streets, in Chickasaw County, Mississippi, on July 27, 1944, later changing her name to Gentry after seeing 'Ruby Gentry', a movie about swampland passion. A childhood singer and guitarist, Bobbie spent her early years in Greenwood, Mississippi, and then moved with her family to California, there attending high school in Palm Springs and UCLA, where she majored in philosophy. Already proficient on several instruments (she plays guitar, banjo, bass, piano and vibes) Bobbie attended the LA Conservatory of Music, studying theory.

Signed to Capitol Records in 1967, she cut **Ode To Billy Joe** at one half-hour session and became a star virtually overnight. Then followed a number of other cross-over hits including **Okolona River Bottom Band** (1967), **Fancy** (1969), **Let It Be Me** (1969) and **All I Have To Do Is Dream** (1970), the last two being duets with Glen Campbell.

Extremely popular in Britain where she had her own BBC-TV series and gained a No.1 with **I'll Never Fall In Love Again** (1970), Bobbie's sales waned as the '70s moved on. But in 1976 a film 'Ode To Billie Joe', based on the events documented in the Gentry song, reactivated some interest in her Delta ditties. Her appeal was boosted on the Vegas-Reno circuit and at the Hughes hotel chain, her contract with the latter touted to be in the multi-million-dollar bracket.

Recommended:
Ode To Billy Joe (Capitol/Capitol)
Bobbie Gentry's Greatest (Capitol/Capitol)

Don't Stop Loving Me, Don Gibson. Courtesy Hickory Records.

Above: Talented singer and multi-instrumentalist Vince Gill has had a steady climb to the top – from early years as a popular sessionman and singer to superstardom in the '90s.

Don Gibson

A rich-voiced singer-songwriter whose wares enabled him to move into the pop market of the 1960s, Don Gibson was born in Shelby, North Carolina, on April 3, 1928.

A competent guitarist before he left school, Gibson built up a regional following via live gigs and radio broadcasts. After finishing his education, he moved to Knoxville, where he was heard on the WNOX Tennessee Barn Dance. His first big writing success came from **Sweet Dreams**, a song which was a hit for Faron Young. In 1958 **I Can't Stop Loving You**, his best-known composition, became a hit for Kitty Wells. Later, Ray Charles was to have international success with the same song. Gibson himself recorded the number as a B-side of **Oh Lonesome Me**, but it broke through for him and gave him a name-making pop hit in the process. Total sales of **I Can't Stop Loving You** were not long in reaching the one million mark. Other songs that provided hit records for Don included: **Give Myself A Party**, **Blue Blue Day**, **Sea Of Heartbreak** and **Lonesome Number One**. The last three, and particularly **Sea Of Heartbreak**, showed that Gibson's deep voice and neatly novel songs could cross over into the pop charts.

Consistently successful in the country charts, Gibson has always kept his country image and has often been critical of the inroads rock has made into his chosen style of music. Though regarded mainly as a 1960s singer (because of his cross-over hits during that period), he was a fairly prolific hitmaker in the 1970s when, signed to Hickory and ABC-Hickory, he logged nearly 40 chart records, including a number of duets with Sue Thompson. One of his solo cuts, **Woman (Sensuous Woman)**, reached No.1 in 1972.

He recorded for MCA (1979) and Warner (1980), providing both with some chart action, but is now virtually retired form the music business. A purveyor of classic songs dealing with heartbreak and loneliness, Gibson once wrote: "If loneliness meant world acclaim, then everyone would know my name – I'd be a legend in my time." And he really is.

Recommended:
Don't Stop Loving Me (Hickory/DJM)
Rockin' Rollin' Gibson (–/Bear Family)
A Legend In My Time (–/Bear Family)
Sings Country Favorites (–/Pickwick)

Vince Gill

The son of an attorney who is now a federal appellate judge, Vincent Grant Gill (born on April 4, 1957, in Norman, Oklahoma) learned his craft in the bluegrass world before becoming a mainstream country star of the '80s and walking off with the CMA Male Vocalist award in 1991, 1992 and 1993.

Vince made his radio debut on a local radio station when he was eight. Two years later he had not only mastered guitar, but was leading his own band. While still in high school he was in a bluegrass outfit called Mountain Smoke. At the time he had to decide between being a musician or a professional golfer. A call from Sam Bush and Dan Crary asking him to join the Bluegrass Alliance solved the dilemma, and in 1974 he moved to Kentucky. A year later he was in Los Angeles, where he joined Byron Berline's Sundance for a two-year stint. It was at this time he met his future wife, Janis Oliver, who, with sister Kristine, was working the West Coast clubs as Sweethearts Of The Rodeo. Three years later they were married, and Vince was beginning to make his mark as lead singer with Pure Prairie League. He recorded three albums with the band and sang on their Top 10 pop hit, **Let Me Love You Tonight** (1980).

In 1982 Vince Gill lined up alongside Emory Gordy Jr and Tony Brown in Rodney Crowell's Cherry Bombs. This meant a move to Nashville, where Brown was also working at RCA. He signed Vince to the label in 1983, but before any sessions took

Pocket Full Of Gold, Vince Gill. Courtesy MCA Records.

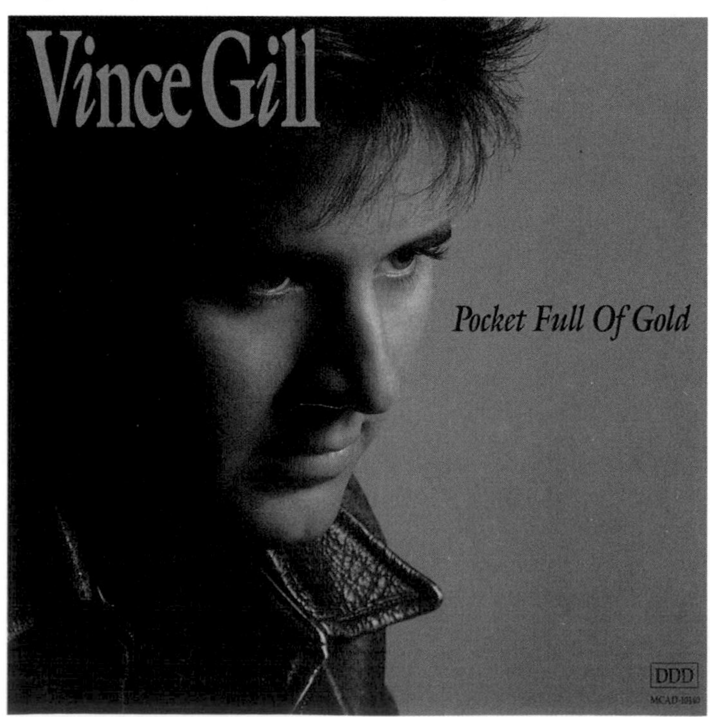

Above: Night-club owner Gilley became a star during the 'Urban Cowboy' craze.

place Brown left for greener pastures, joining MCA. Ironically, Gill's first solo sessions were produced by Gordy, and he made his country chart debut with **Victim Of Life's Circumstances** (1984). During the next four years Vince recorded three albums for RCA. Although he scored such Top 10 hits as **If It Weren't For Him** (1985), **Oklahoma Borderline** (1986), **Cinderella** (1987) and **Everybody's Sweetheart** (1988), and the Academy Of Country Music named him Best New Male Vocalist of 1984, it was hardly the successful solo career he had expected.

All this time Vince has been in demand as a session musician and singer, making guest appearances on recordings by Rosanne Cash, Rodney Crowell, Emmylou Harris, Dan Fogelberg, Vern Gosdin, Reba McEntire, Conway Twitty and many others. When his contract with RCA had run its course, Tony Brown signed Vince to MCA and produced the career-breaking album, **When I Call Your Name**. The title song, co-written by Gill and Tim DuBois, and featuring Patty Loveless on harmonies, became his first No.1. It was named Single Of The Year and Song Of The Year at the 1990 CMA Awards. The album crossed over onto the pop charts and gained a platinum award. The next album, **Pocket Full Of Gold**, contained Top 10 hits in the title song, the upbeat **Liza Jane** and **Look At Us** (Song Of The Year at the 1991 CMA Awards), plus another No.1 with **Take Your Memory With You**. It was another platinum album, and Gill made it a hat-trick for MCA with the album, **I Still Believe In You**. The title song, **Don't Let Our Love Start Slippin' Away**, and a duet with Reba McEntire on **The Heart Won't Lie** (1993) became chart-toppers.

One of Nashville's most talented performers, Gill can play guitar, banjo, mandolin and dobro. He teamed up with Steve Wariner, Ricky Skaggs, Mark O'Connor and the New Nashville Cats for a revival of Carl Perkins' **Restless** (1991), which won the CMA Instrumental Award. He also added harmonies to Dire Straits' album, **On Every Street**. Mark Knopfler invited him to join Dire Straits for their 1992 World Tour, but he turned it down, preferring to concentrate on his own solo career. In 1993 Gill won five CMA awards, including Entertainer Of The Year. Now firmly established as one of country music's major stars, Vince Gill possesses the right temperament to handle the stardom and adulation now showered upon him.

Recommended:
Pocket Full Of Gold (MCA/MCA)
When I Call Your Name (MCA/MCA)
I Still Believe In You (MCA/MCA)
I Never Knew Lonely (RCA/–)
The Way Back Home (RCA/–)

Mickey Gilley

The piano-playing cousin of Jerry Lee Lewis and a performer in a similar vein, Mickey Gilley was born on March 9, 1936 in Ferriday, Louisiana. He moved to Houston at the age of 17 and began playing at local clubs, cutting the rock'n'roll songs **Tell Me Why** and **Oo-ee-baby** for the Minor label. This brought no rewards, so Gilley travelled on, recording for Dot, Rex and Khoury. In 1960 he recorded a Warner Mack song, **Is It Wrong**, for Potomac, and gained a regional best-seller – but the label folded and Gilley continued label-hopping.

In 1964, he formed his own record company, Astro. His second release, **Lonely Wine**, proved another regional hit. An album of the same name (later retitled **Down The Line** when reissued by Paula) was also released. However, 1965 found Gilley on 20th Century Fox. From there he moved to Paula, where he enjoyed a mid-1968 hit in **Now I Can Live Again**. It was not until 1974 and some reaction to his Astro version of George Morgan's old **Roomful Of Roses**, that things started coming together for Gilley. The single was picked up by Hugh Hefner's Playboy label and immediately went to No.1 in the charts. Chart-toppers followed in **I Overlooked An Orchid**, **City Lights** and **Window Up Above**. Two more No.1s came in 1976 – **Don't The Girls All Get Prettier At Closing Time** and **Bring It On Home To Me** – at which point he garnered the Entertainer Of The Year, Top Male Vocalist, Song Of The Year, Single Of The Year and Album Of The Year awards from the ACM.

After further hits for Playboy, including **She's Pulling Me Back Again**, a 1977 No.1, Gilley signed with Epic. His career gained an added fillip in 1980 when the movie 'Urban Cowboy', shot at Gilley's – the Houston club the singer bought in 1971 – provided him with international exposure and a pop hit in **Stand By Me**. The song was one of three country No.1s that year, the others being **True Love Ways** and **That's All That Matters To Me**. For the next few years the flow of hits continued, including **A Headache Tomorrow (Or A Heartache Tonight)**, **You Don't Know Me**, **Lonely Nights**, **Put Your Dreams Away**, **Paradise Tonight** (a duet with Charly McClain), **Fool For Your Love** and **You've Really Got A Hold On Me**. Based mainly in his thriving club, it appeared that Gilley was set up for life. It was not to be, though, and by 1986 he had scored has last Top 10 hit with **Doo-Wah Days** for Epic. The following year he signed with Airborne and only managed to place a few minor hits on the charts.

Recommended:
At His Best (Paula/–)
Welcome To Gilley's (Playboy/Pye)
Down The Line (–/Charly)
That's All That Matters To Me (Epic/Epic)
Gilley's Smokin' (Epic/–)
It Takes Believers – with Charly McClain (Epic/–)
From Pasadena With Love (–/Sundown)

Jimmie Dale Gilmore

Texas singer-songwriter Jimmie Dale Gilmore's music is steeped in traditionalism, while maintaining a contemporary edge. This may help explain why it took him more than twenty years to gain any kind of following within mainstream country music. Jimmie was born in 1945 in the small Texas town of Tulia, into a family where country music was almost like a religion. Jimmie's father played electric guitar in an Ernest Tubb-inspired West Texas honky-tonk band. The family moved to Buddy Holly's nearby hometown of Lubbock in 1950, and his father stopped playing in the band to enrol in college. However, his father still maintained a keen interest in country music, and Jimmy can recall accompanying him to shows during the '50s.

Jimmie didn't get the music bug until the early '60s when the folk revival started. He picked up an acoustic guitar and hung out with Butch Hancock and Joe Ely. By the early '70s the three had teamed up as the Flatlanders, an acoustic band that also featured Steve Wesson and Tony Pearson. They played around Lubbock and Austin, and in 1971 travelled to Nashville to record an album. A single was released, Jimmie's lovely **Dallas**, and the other tracks were put out by Plantation Records.

Shortly after this the Flatlanders split. From 1974 to 1980 Jimmie lived in Denver at a spiritual community that followed teenage guru Maharaj Ji. In 1981 he moved back to Austin and started picking up the threads of his music career. Due to the success of Joe Ely, interest in the Flatlanders started to build, and Gilmore found he had a cult following. A contract with Hightone Records led to **The Fair And Square** album, which produced minor country hits **White Freight Liner Blues** (1988) and **Honky Tonk Song** (1989). A second album surfaced at this time and Gilmore teamed up with Butch Hancock for successful tours in England and Australia. His grainy, weather-beaten voice and timeless West Texas country songs gained a following among country music fans in these countries. Word of mouth praise for Gilmore and his music spread, and in 1991 he was signed to Elektra/Nonesuch's American Explorer series. He then released the acclaimed **After Awhile** album. For the first time Gilmore's music received recognition in the mainstream of country music. In early 1993 he was preparing a new album, to be released and marketed on Elektra itself, unlike its predecessor, which had been aimed squarely at the rootsy pop audience.

Recommended:
Fair And Square (–/Demon)
After Awhile (Elektra/Elektra)

Johnny Gimble

A brilliant session fiddle player and mandolinist, also the writer of **Fiddlin' Around**, a tune nominated for a 1974 Grammy Award, Johnny Gimble was born on May 30, 1926, in Tyler, Texas, and grew up on a farm nearby.

With his brothers Gene, Jerry, Jack and Bill, the 12-year-old Johnny began playing at local gigs. Gene, Jerry and Johnny combined with James Ivie during their high school days to form the Rose City Swingsters, a group that played on radio station KGKB. Leaving home in 1943, Gimble played fiddle and banjo with Bob and Joe Shelton at KWKH, Shreveport, Louisiana, also working as part of the Jimmie Davis band.

Gimble spent two or three stints with Bob Wills and his Texas Playboys. When western swing's popularity sagged, he left the music business and settled down.

A resident of Dallas for a while, he began working as a studio musician with Lefty Frizzell, Ray Price, Marty Robbins and

After Awhile, Jimmie Dale Gilmore. Courtesy Elektra Records.

others. He eventually moved to Nashville, where he gained more work, both in the studio and as a touring back-up man. He also played on such TV shows as Hee Haw and Austin City Limits. He won the CMA Instrumentalist Of The Year award in 1975, while the Academy Of Country Music adjudged him Top Fiddle Player in 1978 and 1979.

In the late '70s Johnny moved from Nashville to Austin, but he still turns up on countless Nashville sessions and tours with Merle Haggard, Asleep At The Wheel, Willie Nelson, and various all-star bands, playing both jazz and pure country with equal fluency.

Recommended:
Texas Dance Party (Columbia/–)
Still Swingin' – with the Texas Swing Pioneers (CMH/–)
The Texas Fiddle Collection (CMH/–)

Texas Dance Party, Johnny Gimble. Courtesy CBS Records.

Girls Next Door

Stretching the boundaries of country music with their close harmony singing, the Girls Next Door were specifically formed to fulfil a musical role in the early '80s, in much the same way that the Monkees had been created to perform pop music in the '60s.

Formed as Belle in 1982, they were the brainchild of record producer Tommy West and session singer Doris King (born February 13, 1957, Nashville, Tennessee), who thought it would be great to have an all-girl harmony group that could blend country with soul and big band swing music. King recruited the members: Cindy Nixon (born August 3, 1958, Nashville, Tennessee), whose father and uncle had worked in country music as the Nixon Brothers; Tammy Stephens (born April 13, 1961, Arlington, Texas), who had been singing with the Wills Family Gospel Group since the age of six, and was married to Hee Haw regular Jeff Smith; and Diane Williams (born August 9, 1959, Hahn AFB, Germany). All four girls had performed on different shows at Opryland and worked sessions as back-up vocalists. They started playing small clubs, perfecting their repertoire and harmonies, then West took them to the studio to produce a demo tape. Eventually he gained the girls a contract with MTM Records in 1985 when Belle became the Girls Next Door. They made their chart debut in 1986 with **Love Will Get You Through Times With No Money**. The same year they made the Top 10 with a revival of **Slow Boat To China**. Another major hit came with **Walk Me In The Rain** (1987) before MTM Records folded. The Girls Next Door are still a popular touring act and signed with Atlantic Records, returning to the charts with **He's Gotta Have Me** (1989).

Recommended:
How About Us? (Atlantic/–)

Girls Of The Golden West

Authentic westerners, both from Muleshoe, Texas, Dorothy Laverne 'Dolly' Good (born December 11, 1915) and Mildred Fern 'Millie' Good (born April 11, 1913) were one of the most popular acts in early country music, and helped pave the way for other women singers. They were among the earliest to exploit the cowboy image in dress and song.

They began their career on WIL and KMOX in St Louis in 1930, then spent three years in Milford, Kansas, and on XER in Mexico, before coming to nationwide attention on the National Barn Dance from 1933 to 1937. The Goods, both of whom sang and played guitar, were even more popular on the Boone County Jamboree and the Midwestern Hayride (both over WLW) in Cincinnati, where they were voted the most popular act on WLW in 1945. Their appearances and performance tailed off in the '50s and they did not perform after about 1963. Dolly died on November 12, 1967, but Millie was still living in Cincinnati, Ohio at the start of the 1980s.

They never had any great success on record, but were among the most popular groups of their era – and one of the most influential.

Recommended:
The Girls Of The Golden West (Old Homestead/–)

Below: Johnny Gimble is a fine fiddle-player and superb mandolinist.

Still Swingin', Johnny Gimble. Courtesy CMH Country Classics.

Vern Gosdin

A singer whose success was a long time in coming, Vern Gosdin has played bluegrass, West Coast country-rock and rock'n'roll, but he is most at home with his impeccably pure honky-tonk country style, which finally gained him full recognition as a 50-year-old 'New Traditional' country star of the late '80s.

Born in Woodland, Alabama on August 5, 1936, Gosdin joined the Gosdin Family Gospel Radio Show on WKOK, Birmingham, Alabama, broadcast six days a week during the early '50s. In 1953 he moved to Atlanta where he worked as a singer in the evening. Three years later found him in Chicago, initially working as a welder, then later managing a country music night-club. He moved to California in 1960, where he formed the bluegrass outfit, Golden State Boys, with his younger brother, Rex. Two years later the Gosdin brothers were playing with Chris Hillman in the Hillmen, yet another bluegrass group. He was invited to join the Byrds in 1964, but declined and teamed up with Rex to work as the Gosdin Brothers. The pair recorded an album with Gene Clark in 1966, then a year later had a fair-sized hit with **Hangin' On** on the Bakersfield International label. This led to signing with Capitol Records, for whom they cut the album **Sounds Of Goodbye**, the title song an early Eddie Rabbitt composition.

By the early '70s they were finding it hard to make a good living from music. In 1972 Vern, his wife Cathy and two children moved back to Atlanta. He started selling glassware door-to-door, eventually operating his own successful glass and mirror business. Gary S. Paxton, who had produced the Capitol recordings, encouraged Vern to make a fresh start in music, this time in Nashville. Vern cut new versions of **Hangin' On** (1976), with Emmylou Harris singing harmony, and **Till The End** (1977), both these Elektra releases going Top 10, along with **Yesterday's Gone,** another Gosdin Top 10 entry in 1977. Vern was asked to make an entry on his revered Opry, following this with such major hits as **Never My Love**, **Break My Mind** and **You've Got Somebody, I've Got Somebody**. He switched to the Ovation label in 1981 and had a couple of hits, including **Dream Of Me**.

A consistent label-hopper, he was back in the Top 10 with **Today My World Slipped Away** in 1982, this time on AMI. The following year he signed with Compleat, scoring Top 10 hits with: **If You're Gonna Do Me Wrong (Do It Right)**, **Way Down Deep**, **I Wonder Where We'd Be Tonight**, **I Can Tell By The Way You Dance (You're Gonna Love Me Tonight)** (his first No.1), **What Would Your Memories Do** and **Slow Burning Memories**. In 1987 he landed back on his feet on a major label, this time Columbia, who were looking for a 'New Traditional' country artist to cash in on the success of the back-to-basics country artists. 51-year-old Gosdin didn't let them down, going Top 10 with **Do You Believe Me Now** (1987) and right to the top with **Set 'Em Up Joe** (1988).

His first album for Columbia, **Chiseled in Stone** (the title song another Top 10 single), is one of the all-time classic honky-tonk albums. Further Top 10 entries followed with **Who You Gonna Blame It On This Time**, another No.1 with **I'm Still Crazy, That Just About Does It, Right In The Wrong Direction** and **Is It Raining At Your House**. Though held in high esteem by many of his contemporaries, and commanding a large and loyal fan following, Gosdin's age, looks and stature were alien to the country image of the early '90s. He has continued to produce classic albums for Columbia, but found it increasingly difficult to place his singles in the Top 10. He has invested in his future with his own music showplace, the Country Music Amphitheater in Ardmore, Alabama.

Recommended:
Till The End (Elektra/Elektra)
The Best Of (Elektra/–)
There Is A Season (Compleat/PRT Compleat)
Nickles & Dimes & Love (Columbia/–)
Chiseled In Stone (Columbia/–)
Alone (Columbia/–)

Above: Billy Grammer hit paydirt with his million-selling hit Gotta Travel On.

Billy Grammer

Originator of the Grammer guitar, a fine flat-top instrument, Billy Grammer's first guitar was installed in the Country Music Hall Of Fame in March, 1969.

One of 13 children fathered by an Illinois coalminer, Grammer (born Benton, Illinois, August 28, 1925) became a major star during the late '50s and early '60s. He performed on WRAL, Arlington in 1947 and by 1955 had earned a regular spot on the Washington-based Jimmy Dean TV Show, moving with Dean on to a CBS network programme later.

A popular bandleader, Grammer signed with Monument Records in 1958, having his first hit – a million seller – with **Gotta Travel On**, a song adapted by the Weavers from a traditional melody. Becoming an Opry regular in 1959 and obtaining a double-sided hit that year with **Bonaparte's Retreat/The Kissing Tree**, Grammer recorded for numerous labels throughout the '60s, with minor hits on most of them, via such titles as **I Wanna Go Home** (Decca, 1963), **The Real Thing** (Epic 1966), **Mabel** (Rice, 1967), **Ballad Of John Dillinger** (Mercury, 1968) and **Jesus Is A Soul Man** (Stop, 1969).

Recommended:
Country Guitar (Decca/–)
Favorites (Vocalion/–)

Dobie Gray

Dobie Gray (born Leonard Victor Ainsworth in Brookshire, Texas on July 26, 1942) is a much respected singer-songwriter in Nashville, who has never quite made that breakthrough to commercial acceptance that his talent has deserved.

A son of a Protestant minister, Dobie listened to country and blues music as a small boy. In 1960 he left his Texas home and headed for California to become a film star in Hollywood. Instead he landed a role as singer and was soon recording pop and soul records for a variety of West Coast labels, scoring a minor pop hit with **Look At Me** (1963). Two years later he made a bigger impact with **The In Crowd**, which made the Top 20 and led to work on the lucrative club circuit. In 1971 he became a member of rock band Pollution, then started recording demos for Paul Williams. Through Williams he met Mentor Williams (his brother), and established a partnership that saw Gray signed to MCA Records in Nashville. Mentor wrote Dobie's biggest hit, **Drift Away** (1973). With MCA he recorded three albums that perfectly fused country simplicity and pop complexity. Utilizing the talents of David Briggs, Weldon Myrick and Buddy Spicher, Gray cut the original versions of such songs as **We Had It All**, **Lovin' Arms**, **I Never Had It So Good** and **There's A Honky Tonk Angel (Who'll Take Me Back In)**.

Three years later Dobie joined Capricorn Records, for whom he cut a couple of excellent southern-styled country-rock albums, achieving pop success with **If Love Must Go** and **Find 'Em, Fool 'Em And Forget 'Em** (both 1976). Shortly afterwards Capricorn folded, but Dobie continued writing and was also doing sessions in Nashville. He joined Infinity Records and had his last pop hit with **You Can Do It** (1979). His next recordings came six years later when he signed with Capitol and made his debut on the country charts with **That's One To Grow On** (1986). His Capitol album, **From Where I Stand**, was a classic modern country album with Nashville pickers, singers and writers involved .

Recommended:
Drift Away (–/Cottage)
From Where I Stand (Capitol/–)
Love's Talkin' (Capitol/–)

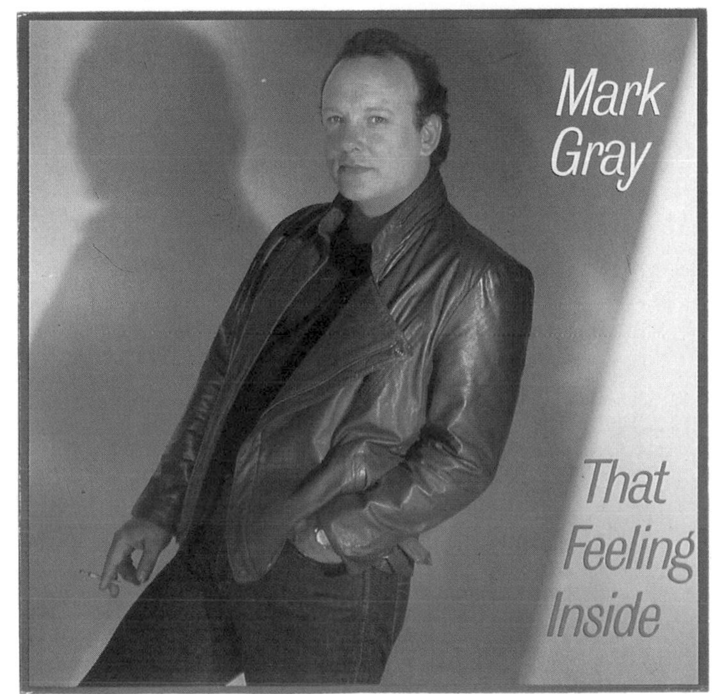

That Feeling Inside, Mark Gray. Courtesy CBS Records.

Mark Gray

A one-time gospel singer, a composer of jingles for 17 years and a writer of pop hits, Mark Gray was born in Vicksburg, Mississippi, in 1952. His aunt was part of a gospel group and Gray toured with them for several years. At 19 he moved on to work with the Oak Ridge Boys' publishing company and appeared with them onstage. After seven years Gray returned to Mississippi and began concentrating on his songwriting career.

Eventually he made his way back to Nashville and became lead singer with Exile. But he continued to write songs, including **Take Me Down** and **The Closer You Get**, both No.1s for Alabama, and **It Ain't Easy Bein' Easy**, a chart-topper for Janie Fricke. One of his demo tapes for Fricke was heard by a Columbia Records executive and led to a contract as a solo act with the label in 1983. Gray made an immediate impact, notching a Top 20 single that year with **Wounded Hearts**. By 1984 he was up among the frontrunners, logging three Top 10 records. In 1985 Gray teamed with Tammy Wynette in a Top 10 duet, **Sometimes When We Touch**, returning to solo mode for **Please Be Love** (1986). His last Top 20 hit came with **Back When Love Was Enough** in 1987, after which he signed with the small 615 label, scoring two minor duet hits with Bobbi Lace in 1988.

Recommended:
Magic (Columbia/CBS)
This Ol' Piano (Columbia/CBS)
That Feeling Inside (Columbia/CBS)

Lloyd Green And His Steel Guitar. Courtesy M&M Records.

Lloyd Green

One of Nashville's top sessionmen, Mensa member Green was born in Mississippi. on October 4, 1937, and grew up in Mobile, Alabama. He began taking lessons on steel guitar at the age of seven, playing professionally three years later. During his high school days he played weekends at clubs and bars where 'real rough fights, shootings and stabbings were common', using material drawn mainly from the Eddy Arnold and Hank Williams songbooks. He attended the University of Southern Mississippi as a psychology major but left after two years to play in Nashville. He initially worked with Hawkshaw Hawkins and Jean Shepard, then toured with Faron Young and George Jones, his first recording session in Music City being on Jones' **Too Much Water Runs Under The Bridge** single in 1957. Though he has had hard times since, Green is now an in-demand steelie and plays on some 500 sessions a year. A recording artist in his own right, he has also had a few hit singles, the biggest of these being **I Can See Clearly Now**, on Monument in 1973.

Recommended:
Steel Rides (Monument/Monument)
Cool Steel Man (Chart/Chart)
Green Velvet (Little Darlin'/President)

Greenbriar Boys

A New York-based bluegrass group, Greenbriar Boys was formed in 1958 by Bob Yellin, John Herald and Eric Weissberg. Extremely popular at folk festivals in the 1960s, they recorded for Elektra and Vanguard and produced several well-regarded albums. The group has included among its personnel legendary mandolinist Frank Wakefield, fiddler Buddy Spicher and Ralph Rinzler, one-time manager of Bill Monroe and a leading authority on old-time country music. Original member Weissberg teamed with Steve Mandell in 1972 to provide **Dueling Banjos** – an instrumental from the film 'Deliverance' – which turned a million-selling single the following year.

Recommended:
Best Of (Vanguard/–)
Ragged But Right (Vanguard/–)
Better Late Than Never (Vanguard/–)

Jack Greene

Yet another of the long list of country entertainers who could play guitar at an early age, Jack Henry Greene (born Maryville, Tennessee, January 7, 1930), also a fine drummer, first became a full-time musician with the Cherokee Trio, an Atlanta GA group, moving on to become stocksman with the Rhythm Ranch Boys in 1950. Then came two years of Army service, followed by a stint with another Atlanta band, the Peachtree Cowboys.

Below: Talented Lee Greenwood was once a Las Vegas croupier.

Joining Ernest Tubb's Texas Troubadours in 1962, the amiable six-footer, dubbed 'the Jolly Giant', soon became a favourite. While still a member of Tubb's band, he began having solo discs released by Decca. One single, **Ever Since My Baby Went Away**, charted in mild fashion during 1965, this being followed by two No. 1s in **There Goes My Everything** (1966) and **All The Time** (1967). During 1967, Greene gained four CMA awards – Best Male Vocalist, Best Album, Best Song and Single Of The Year (for **There Goes My Everything**, a Dallas–Frazier composition). Thereafter he continued on his hit-making way, providing Decca with five more top singles during the late '60s.

In 1969, Greene and Jeannie Seely, his co-vocalist on the Ernest Tubb TV show, put together a roadshow and began touring with a band called the Green Giants. The twosome enjoyed immediate success with **Wish I Didn't Have To Miss You**. The first country act to play the Rooftop Lounge, King Of The Road, Nashville, in 1972, Greene and Seely received considerable acclaim for their 1974 Madison Square Garden concert. The duo played host and hostess at the Wembley Country Festival two years later. But at the beginning of 1981, they went their separate ways. Greene claimed the split gave him the opportunity to provide a less Vegas-styled presentation and a back-to-basics approach. Also, during the start of the '80s he began having hits again, for the first time since 1975. But his deal with the Frontline label fell through. However, his happy relationship with the country charts was resumed yet again in 1983 when he signed with EMH Records.

Recommended:
Greatest Hits (MCA/–)
Best Of (–/MCA)
Greene Country (MCA/–)
Two For The Show – with Jeannie Seely (MCA/–)
Jack Greene And Jeannie Seely (MCA/–)

Lee Greenwood

The son of half-Cherokee parents who split-up when he was just a year old, Lee Greenwood (born on October 27, 1942, in Los Angeles) was left in the care of his grandparents, who had a chicken farm near Sacramento. A sax-player and a pianist, he became a schoolboy member of a local band known as My Moonbeams. Later, reunited with his mother in Los Angeles, he played for various jazz and rock bands in the LA area. After returning to Sacramento in 1958, he moved into country music, joining a band headed by Capitol recording artist Chester Smith, and appearing on TV at the age of 15. Hired by Del Reeves for his sax expertise (he can also play guitar, bass and banjo), he learned the art of showmanship from Reeves. He then formed his own band, Apollo, which became based in Las Vegas in 1962. By 1965 Apollo had evolved into the Lee Greenwood Affair, a pop band signed to Paramount Records.

The band moved to the West Coast in an attempt to break into the pop market, but the Paramount label folded and so did the Affair. Greenwood returned to Vegas where he took up various jobs including bandleader, back-up singer, musical arranger, bar-room singer and casino card-dealer. In 1979 Mel Tillis' bandleader, Larry McFadden, heard Greenwood singing in a bar and arranged for him to fly to Nashville to record some demo discs. Once completed, Greenwood returned to jobs in Vegas and Reno, but started concentrating on his songwriting, switching from showbiz material to more country-oriented fare at McFadden's insistence. The latter took Greenwood's demos to various Nashville-based labels and eventually got MCA to sign a deal in June 1981.

A distinctive singer with a voice loaded with finely sifted gravel, Greenwood went Top 20 with his first MCA single in mid-'81, **It Turns Me Inside Out**. He followed this with three Top 10 hits during the following year – **Ring On Her Finger, Time On Her Hands, She's Lying, Ain't No Trick (It Takes Magic)** – all of which appeared on his debut album **Inside And Out**. But 1983 was really Greenwood's year, providing him with three hit singles – **IOU, Somebody's Gonna Love You** and **Going, Going, Going**, the last two being No.1s and all three crossing over into the pop charts. He also won the Male Vocalist Of The Year title at the 1983 CMA Awards.

Since that time he has had further hits with: **God Bless The USA, Fool's Gold, You've Got A Good Love Comin', To Me** (a duet with Barbara Mandrell) (all 1984); **Dixie Road, I Don't Mind The Thorns (If You're The Rose)** and **Don't Underestimate My Love For You** (1985); **Hearts Aren't Made To Break (They're Made To Love)** (1965) and **Mornin' Ride** (1987). During the Gulf War the patriotic **God Bless The USA** made a comeback, while Greenwood continued a consistent run of Top 10 hits with **Someone, If There's Any Justice** and **Touch And Go Crazy**. In 1989 he joined Capitol Records and immediately scored a Top 5 hit in 1990 with **Holdin' A Good Hand**. In 1991 he teamed up with Suzy Bogguss for the duet hit, **Hopelessly Yours**.

Greenwood is a talented songwriter who has had his material recorded by Kenny Rogers, Mel Tillis, Brenda Lee and others. He is also the voice on many commercials. In 1988 he starred in the CBS-TV series, High Mountain Rangers, in which he also performed the theme song. In Britain, however, Lee is best known for **The Wind Beneath My Wings**, a single which entered the UK pop charts in 1984.

Recommended:
Inside And Out (MCA/MCA)
Somebody's Gonna Love You (MCA/MCA)
Meant For Each Other – with Barbara Mandrell (MCA/–)
American Patriot (Liberty/–)
Streamline (MCA/MCA)
The Wind Beneath My Wings (–/MCA)
If Only For One Night (Capitol/–)

Ray Griff

The writer of many hundreds of songs, Ray Griff was born in Vancouver, British Columbia, Canada, on April 22, 1940, moving with his family to Calgary, Alberta, shortly before reaching his teens. A drummer in a band at the age of eight, Griff also mastered guitar and piano, becoming a bandleader on the night-club circuit at 18. His reputation as a songwriter was enhanced when Johnny Horton recorded **Mr Moonlight**, a Griff composition, during the late '50s. Jim Reeves cut **Where Do I Go?**, another Griff original, in 1962.

Encouraged by Reeves, he became Nashville-based in 1964, initially involving

Above: Lyle Lovett became an extra in Nanci Griffith's Last Of The True Believers 1986 album sleeve. Courtesy MCA.

himself in songwriting and music publishing, but later recording some sides for RCA's Groove label. An MGM release, **Your Lily White Hands**, provided him with his first hit (1967). Griff followed this with **Sugar From My Candy**, on Dot, a few months later. Label-switching again, he recorded Clarence Carter's **Patches** for Royal American, gaining a 1970 success. He climbed even higher with **The Morning After Baby Let Me Down**, in 1971, also enjoying Top 10 discs with **You Ring My Bell** (1975) and **If I Let Her Come In** (1976), both on Capitol.

Since that time, Griff has supplied a few mini-hits during the '80s for Vision and RCA but it is as a songwriter that he has staked a claim to fame. His compositions include **Canadian Pacific** (recorded by George Hamilton IV), **Baby** (Wilma Burgess), **Better Move It On Home** (Porter Wagoner and Dolly Parton), **Step Aside** (Faron Young), **Who's Gonna Play This Old Piano** (Jerry Lee Lewis) and many others, the majority published by Griff's own Blue Echo company.

Below: Ray Griff, one of several Canadians who have enhanced country music.

Recommended:
Ray Griff (Capitol/–)
The Last Of The Winfield Amateurs (Capitol/–)
Canada (Boot/–)

Nanci Griffith

Story-teller and 'folkabilly-poet' Nanci Griffith (born on July 6, 1953 in Seguin, Texas) is not, strictly speaking, a country artist, although country music does echo in her songs. Several of her songs have become mainstream country hits when recorded by singers Kathy Mattea, Suzy Bogguss and Willie Nelson.

Griffith, the youngest of three children, had a middle-class upbringing; her mother was an amateur actress and her father sang in a barbershop quartet. They separated when Nanci was only six, and she absorbed herself in books and music. By her mid-teens she had already started performing as a solo singer in bars and honky-tonks in Austin. She attended the University of Texas where she majored in education. Although she became a kindergarten teacher, it was short-lived, and in 1977 she decided to pursue a career in music. By this time Nanci had married Texas singer-songwriter Eric Taylor. She made her first recordings in 1976 and they appeared on a folk sampler on B.F. Deal Records the following year. Her own debut album, **There's A Light Beyond These Woods**, was released by B.F. Deal in 1978. Shortly after this her marriage fell apart. Nanci took to handling her own career, driving up to 1,000 miles in her Toyota station wagon to play concerts in Minneapolis, New Mexico or San Francisco.

A second album, **Poet In My Window**, recorded in Kerrville, Texas, was released on Featherbed Records in 1982 and featured mainly self-penned songs. By this time Nanci's reputation was spreading, and in 1984 she met Jim Rooney, who produced her third album, **Once In A Very Blue Moon** for Philo/Rounder. Rooney also helped introduce Nanci and her music to a much wider audience. Recorded in Nashville, the title song from the album became a minor hit in 1986. The same year a fourth album was released, **Last Of The True Believers,** which started a European cult following for her music. Two songs, her own **Love At The Five And Dime** and Pat Alger's **Goin' Gone** became massive hits when later recorded by Kathy Mattea. Tony Brown, ex-keyboard player for the Oak Ridge Boys and Rodney Crowell's Cherry Bombs, was a young and dynamic A&R man for MCA Nashville, and he not only liked Nanci's music, but signed her to MCA in 1987. He produced her breakthrough album, **Lone Star State Of Mind**. The title song became her biggest country hit, but only peaked at No.37. In Europe her music was more readily accepted, with several songs from the album becoming standard fare among country music performers. This album also included the first recording of Julie Gold's **From A Distance**, a song which Griffith published and which was set to become a standard in the early '90s when covered by Cliff Richard, Bette Midler, Kathy Mattea and others.

Since then Nanci has maintained a high integrity on her recordings without in any way compromising her music for commercial gain. **Little Love Affairs** (1988) was a concept album filled with delicate little vignettes and tales of love. But, despite a high media profile, country radio failed to play her records, so MCA moved her promotion and publicity to Los Angeles in an effort to market her recordings more towards the rock and college market. This was borne out with **Storms** (1989), an album with a definite bias towards American radio, produced by Glyn Johns. It made both the American pop and country charts, but failed to attract American country fans towards Nanci and her music. By this time she had become a virtual superstar in Ireland, and in 1991 contributed to the Chieftains' Christmas album, **The Bells of Dublin**.

Above: Nanci Griffith broke through with Lone Star State Of Mind in 1987.

Though she has still not made a major impact on Nashville's mainstream country music, Nanci bought a 100-year-old farmhouse in Franklin on the outskirts of Music City in 1990, vowing, "I'm staying in Nashville and the music isn't going to change." With her band, the Blue Moon Orchestra, which sounds pure country, she has been a regular visitor to Britain where she is regarded as at the forefront of 'New Country', selling out major theatres, including London's Royal Albert Hall for four consecutive nights – a far cry from the Texas bars and honky-tonks where she used to play from 10pm until 2am. More of an album artist than a singles star, Nanci worked with the British team of Rod Argent and Peter Van Hook on **Late Night Grande Hotel** (1991). She reunited with Jim Rooney on **Other Voices, Other Rooms** (1993), a collection which took the Texas lady back to her roots as she performed songs by Bob Dylan, Tom Paxton, Gordon Lightfoot, John Prine and Woody Guthrie.

Recommended:
Other Voices, Other Rooms (MCA/MCA)
Last Of The True Believers (Philo/MCA)
Little Love Affairs (MCA/MCA)
Once In A Very Blue Moon (Philo/MCA)
There's A Light Beyond These Woods (Philo/MCA)

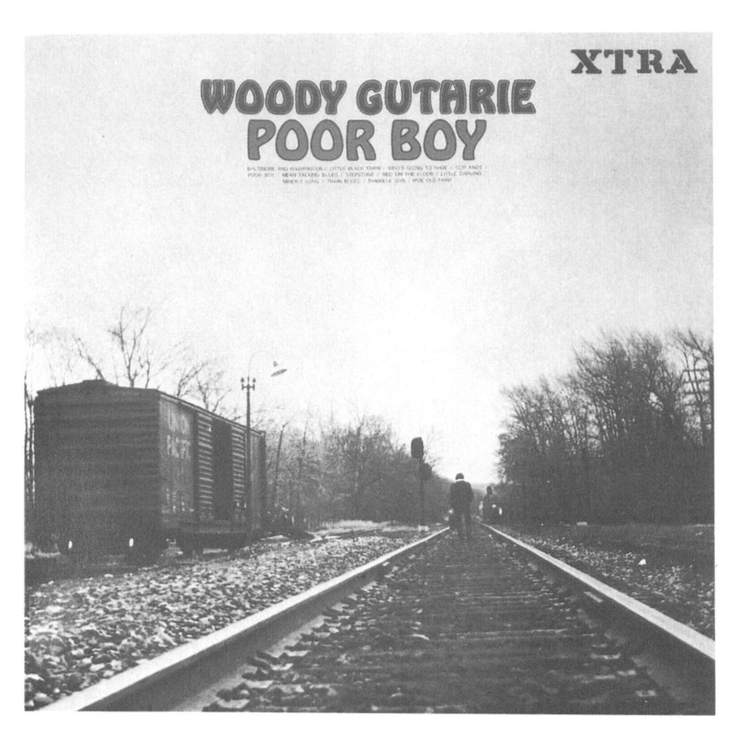

Poor Boy, Woody Guthrie. Courtesy Xtra Records. Guthrie found much of the material for his songs on the road.

The Gully Jumpers

Charlie Arrington, fiddle; Roy Hardison, banjo; Burt Hutcherson, guitar; Paul Warmack, mandolin and guitar.

The Gully Jumpers were one of the early popular Opry string bands and participated in that early Nashville recording session for Victor in October 1928. Led by Paul Warmack, an auto mechanic by trade, they remained with basically the same personnel for well over two decades (they had joined the Opry about 1927) and in fact were one of the most popular and most used bands of the Opry's early years. The group was dissolved in the mid-'60s when four of the old-time Opry bands were accordioned into two.

Though few recordings of the Gully Jumpers are available on vinyl, the band can be heard on **Nashville – The Early String Bands Volume 1** (County) playing **Robertson County** and **Stone Rag**.

Woody Guthrie

An influential country folk singer, Guthrie's visual attitude and thin, fragmented vocal style have been copied by many, most notably Bob Dylan.

Born Woodrow Wilson Guthrie, in Okema, Oklahoma, in 1912, a hard rural upbringing amid a background of natural disasters set the tone for many of his songs. He championed the rural poor and the loser (as on **Dustbowl Ballads**) yet he was also capable of joyous hymns to the country itself (**This Land Is Your Land**).

He roamed the land extensively, incorporating what he found into songs. He later wrote that he saw things happen to oil people, cattle people and wheat folks, and detailed these happenings in songs which he broadcast over the LA station KFVD. He gave rise to what is sometimes called 'The Dustbowl Tradition', other exponents of which include Cisco Houston and Rambling Jack Elliot. The slogan on his guitar read 'this machine kills fascists' and as he roamed America during the Depression, singing in union halls and for picket lines, it was hardly surprising when the authorities, already scared by the crisis, tried to tag the 'red' label on him. In a parallel with the '60s decade, those who pointed the need for social change could find themselves ostracized or even in danger.

Guthrie's mother had died of Huntingdon's Chorea (a hereditary nerve disease) and Woody himself succumbed to it in 1967, having been in hospital since 1954. He was a country singer in the very widest sense, a drifting son of the earth, crafting his simple songs out of experience and his own perception. During 1976, the singer's autobiography 'Bound For Glory' became the subject of a film directed by Hal Ashby with David Carradine in the role of Guthrie.

Recommended:
Dust Bowl Ballads (Folkways/–)
Bound For Glory (Folkways/–)
This Land Is Your Land (Folkways/–)
Columbia River Collection (Rounder/Topic)

Merle Haggard

Country's most charismatic living legend, Merle Haggard is proof that you do not have to forsake your musical roots to achieve fame. The Haggard family had been driven from their farm in dustbowl East Oklahoma and were living in a converted boxcar in Bakersfield, California, when Merle was born on April 6, 1937. Merle was nine when his father, a competent fiddle player, died, and without his father's influence he began to run wild. He embarked on a series of petty thefts and frauds and was in and out of local prisons. Then, in 1957, he was charged with attempted burglary and sentenced to six to fifteen years in San Quentin.

While in prison Merle did some picking and songwriting, and was in San Quentin when Johnny Cash performed one of his prison concerts in 1958. This convinced him that music could help him to straighten out his life, and when he left jail in 1960 he was determined to try and make a go of performing. He moved to Bakersfield, which was then growing into a respectable little country music centre. Helped initially by Bakersfield eminence gris, Buck Owens and his former wife Bonnie Owens, whom Merle eventually married, he started playing the local club scene. At this time Merle ran into Fuzzy Owen, an Arkansas musician who was also playing the Bakersfield clubs. Fuzzy, who is Merle's manager to this day, encouraged him and helped get Merle work locally.

In 1962 Fuzzy organized some recording sessions in a converted 'garage' studio and produced some singles, which were released on Tally, a label Owen had purchased from his cousin Lewis Tally. The next year Merle made his debut on the country charts with **Sing A Sad Song** which reached No.19. In 1964 **Sam Hill** made No.45. In 1965 they put out (**My Friends Are Gonna Be**) **Strangers**, which gave them a Top 10 hit. This led to Capitol acquiring Merle's contract, plus all the recordings made for Tally.

Merle's second Capitol single, the self-penned classic honky-tonker **Swinging Doors**, spent six months on the charts, reaching a Top 5 placing. Equally as impressive was **The Bottle Let Me Down**, which made No.3. This was followed by Haggard's first No.1, **I'm A Lonesome Fugitive**, which made 1966 a highly successful year for him. Merle had been trying to suppress the news of his prison record, but as the story emerged the hard-core country music public were fascinated by this man who had lived the songs he wrote. It appeared that his own life story was unfolding in such country No. 1s as **Branded Man** and **Sing Me Back Home** (1967); **Mama Tried**, which referred to his wild childhood and prison record (1968); and **Hungry Eyes** and **Workin' Man Blues** (1969). In reality it was closer to country romanticism, but Haggard was using his own background for the inspiration for many of his best songs. Few composers have had as much impact on country music as Merle Haggard, who has been referred to as 'the poet of the common man.'

Two other apparently innocent songs were committed to record in 1969: **Okie From Muskogee** and **The Fightin' Side Of Me**. **Okie** re-stated redneck values in the face of then current campus disturbances and Vietnam marches, yet Merle had written it as a joke, picking up a remark one of his band members had made about the conservative habits of Oklahoma natives as they rolled through Muskogee one day. **Fightin' Side Of Me** was another apparent putdown of those who were so bold as to disparage America's image. When Haggard premiered **Okie** for a crowd of NCOs at the Fort Bragg, North Carolina camp, they went wilder than he had expected, and from then on the song became a silent majority legend.

Haggard had been gaining a reputation as the new Woody Guthrie before **Okie** and his hippy following was stunned yet intrigued by this new turn of events. Even President Nixon was said to have written to congratulate Haggard on the song. In 1972, the singer received a pardon for his prison sentence from Ronald Reagan, then governor of California.

Merle himself has admitted to feeling scared at the reaction the song provoked, and he backed away from further right-wing involvement, refusing a proposal to endorse George Wallace politically. Indeed, for his next single, he wanted to record a song about an inter-racial love affair (**Irma Jackson**), but Capitol advised against it.

After the **Okie** controversy had died down, Merle was able to settle into the straightforward country career with which he felt most comfortable. He has not appeared often on television, as he lacks the easy, flip manner which TV companies seem to want from a host, and he has not bothered to cultivate the medium. He once walked out on an Ed Sullivan show when they tried to tell him which songs to sing and how to sing them. However, this principled non-conforming attitude, which probably lost him lucrative work, only strengthened the bond between Merle and country fans.

Since **Okie**, hits have come consistently: **Daddy Frank (The Guitar Man)** (1971); **Carolyn**, **Grandma Harp** and **It's Not Love (But It's Not Bad)** (all

Kern River, Merle Haggard. Courtesy Epic Records.

Above: The legendary Merle Haggard opened shows for Clint Black in the early '90s, but still releases albums on Epic.

1972); **Everybody's Had The Blues** and **If We Make It Through December** (1973); and **Old Man From The Mountain** and **Kentucky Gambler** (1974); all have been No.1s. Albums have provided an area for experimentation, often paying tribute to his influences, as in **Same Train, A Different Time**, in which he performs the songs of Jimmie Rodgers, or one of his most auspicious projects, **A Tribute To The Best Damn Fiddle Player In The World**, in which he teamed original members of Bob Wills' Band with his own.

Haggard had grown up with western swing, and Bob Wills returned the compliment by inviting him to appear on Wills' own album **For The Last Time**. This was a fateful occasion since Wills suffered a stroke during these sessions, from which he never recovered. Haggard has also made other concept albums including **The Land Of Many Churches**, recorded at various churches and featuring the Carter Family, Tommy Collins and other guest singers and musicians. On **Let Me Tell You About A Song**, a 1972 album, he offers a selection of strong story songs, some self-penned, others written by country greats such as Red Foley, Tommy Collins and Red Simpson. Fascinated by the old American railroads, he recorded **My Love Affair With Trains** in 1976. Another concept album, **I Love Dixie Blues (So I Recorded Live In New Orleans)**, had his band the Strangers augmented by a Dixieland jazz band.

In 1977, he joined MCA Records. Although he enjoyed major hits with **If We're Not Back In Love By Monday** (1977), **I'm Always On A Mountain When I Fall** (1978) and **The Way I Am** (1980), it was not altogether a successful association. His tribute album, **My Farewell To Elvis** (1977), was slated by the critics, though it came across in typical Haggard fashion. In no way did he try to mimic the Elvis style, but his identifiable delivery captured the soul of Presley's music, creating an enjoyable encounter with past Presley hits.

By this time Merle and Bonnie Owens had divorced, though she did continue to run his business affairs. Leona Williams, a country singer in her own right, joined the Haggard group as a backing vocalist and soon a stormy relationship developed. The pair were married on October 7, 1978 and recorded several duets which failed to make much of an impression. Five years later they were separated, obtaining a divorce in 1984.

A move to Epic Records towards the end of 1981 led to duet recordings with George Jones (**A Taste Of Yesterday's Wine** album in 1982) and Willie Nelson (**Poncho & Lefty**, which was named CMA Album Of The Year in 1983). His solo hits continued with **Big City** (1982). **That's The Way Love Goes** (1983) and **Natural High** (1984), but as the '80s drew to a close, not so many Haggard singles hit the Top 10. The rather sentimental **Twinkle Twinkle Lucky Star** (1987) has been his last chart-topper, and **A Better Love Next Time** (1989) his last Top 10 entry. Merle Haggard has continued to release albums regularly for Epic, but by 1991 he was opening shows for Clint Black, one of the new, young country traditionalists, who had used the Haggard style as the basis for his own; it was a case of carrying on the country music of yesterday to make the country music of today and tomorrow.

Usually, legendary figures are larger than life, but somehow Merle Haggard has managed to become a legend in his own time without losing the reality of being a down-to-earth human being. Perhaps this is because his songs deal so closely with the reality of being human. A classic, uncompromising country artist, his voice is hurting, yet subtle, with no showbiz nuances. He gives the impression, with his sparsely instrumented band the Strangers, of being more comfortable before audiences of working men than in Las Vegas hotel lounges.

Recommended:

Same Train, A Different Time (Capitol/Capitol)
Okie From Muskogee (Capitol/–)
The Fightin' Side Of Me (Capitol/Capitol)
I Love Dixie Blues (Capitol/Capitol)
It's All In The Movies (Capitol/Capitol)
My Love Affair With Trains (Capitol/Capitol)
The Roots Of My Raising (Capitol/Capitol)
My Farewell To Elvis (MCA/MCA)
Back To The Barrooms (MCA/MCA)
Going Where The Lonely Go (Epic/Epic)
Amber Waves (Epic/Epic)
Out Among The Stars (Epic/–)
Chill Factor (Epic/Epic)
Blue Jungle (Curb/–)
Land Of Many Churches (Capitol/Stetson)

Bill Haley

Bill Haley was born William John Clifton Haley, in Highland Park, Michigan, on July 6, 1925. The leader of a series of good local country bands in the late '40s and '50s, Haley was undoubtedly more surprised than anyone when his creative mixture of R&B, boogie and country music took off like a rocket in 1955, with the success of **Rock Around The Clock** and later **Shake, Rattle And Roll**, turning him into an international superstar overnight.

Haley had led bands which pretty much describe their musical approach – Bill Haley And The Four Acres Of Western Swing, Bill Haley And The Saddle Pals – before attempting to fuse the then all-black sound of R&B with that of swing, western and country music. The result met such a phenomenal reaction that it vaulted him out of the ranks of country into the ranks of rock, never to return.

It is more than significant, however, that until that turning point his roots and approach had been firmly – if experimentally – country, a trait he shared with many of rock's originators. Having become the first real star of rock'n'roll, Bill died in his sleep on February 9, 1981 at home in Harlingen, Texas.

Recommended:
Greatest Hits (MCA/MCA)
Rock The Joint (–/Roller Coaster)
Golden Country Origins (–/Australian Grass Roots)
Hillbilly Haley (–/Rollercoaster)

Tom T. Hall

The 'Mark Twain' of country music – even his band is called the Storytellers – Tom T. Hall's songs are full of colourful characters and intriguing or humorous situations. Born in Olive Hill, Kentucky, on May 25, 1936, the son of a preacher, he first learned to play on a broken Martin guitar, which his father, the Reverend Virgil L. Hall, restored to working order.

At the age of 14, Tom T. quit school and went to work in a clothing factory, two years later forming his first band, the Kentucky Travelers, playing local dates and appearing on radio station WMOR, Morehead, Kentucky. After the band broke up, Hall continued with WMOR as a DJ for a period of five years.

After enlistment in the US Army in 1957, Hall was posted to Germany, where he worked on the AFN radio network, taking the opportunity to try out a number of his own compositions – with some success. Discharged in 1961, he returned to WMOR, also working with the Technicians, another local band.

More stints as a DJ followed, during which time Hall penned **DJ For A Day**, a major hit for Jimmy Newman in 1963. Next, Dave Dudley scored with **Mad** (1964), another Hall composition, and Hall promptly moved to Nashville to begin supplying songs to such acts as Roy Drusky, Stonewall Jackson and Flatt And Scruggs, eventually having his own hit disc with **I Washed My Face In The Morning Dew**, a release on the Mercury label in 1967.

A year later, Jeannie C. Riley recorded **Harper Valley PTA** – a brilliant and highly commercial song about a fast-living woman and a band of small-town hypocrites – and Hall became the writer of a million-seller.

During the early '70s he became something of a star performer, sending the audience and press into raptures at his 1973 Carnegie Hall concert. Record buyers readily snapped up such Hall releases as: **The Ballad Of Forty Dollars** (1968); **A Week In A County Jail** (1969); **The Year That Clayton Delaney Died** (1971); **Old Dogs, Children And Watermelon Wine**, **Ravishing Ruby** (both 1973); **I Love, That Song Is Driving Me Crazy**, **Country Is** (all 1974); **I Care** (1975); and **Faster Horses** (1976), all chart-toppers from Nashville's prime yarn spinner.

Above: Tom T. Hall: "People who write songs are often as equally amazed by them as those who listen to them."

His albums include **Songs Of Fox Hollow**, which Hall described as 'an LP of songs for children of all ages', and **The Magnificent Music Machine**, a bluegrass collection that spawned a popular single in **Fox On The Run** (a Tony Hazzard song which had been a 1969 pop chartbusters for Manfred Mann).

In 1977, Hall signed with RCA Records. Although he recorded some fine singles such as **What Have You Got To Lose** (1978), **The Old Side Of Town** (1979) and **Soldier Of Fortune** (1980), his record sales slumped quite dramatically. At this point he took time off from performing to write books, resulting in the best-seller, 'The Storyteller's Nashville', and became host of the syndicated TV show 'Pop Goes The Country'.

On returning to the studio, a link-up with bluegrass musician Earl Scruggs led to the acclaimed album, **The Storyteller And The Banjoman** (1982). This stunning set mixed traditional country songs with contemporary tunes. Hall's laconic vocal style worked perfectly with Scruggs' fluid banjo work. the success of this album led to Hall rejoining Mercury Records, and his career took off again with the top-selling singles **Famous In Missouri** (1984) and **P.S. I Love You** (1985). Hall has also built a reputation as a novelist with the publication of 'The Laughing Man Of Woodmont Cove' and 'Spring Hill', while 'The Songwriter's Handbook' is of more interest to keen country music fans.

Recommended:
Homecoming (Mercury/Mercury)
I Witness Life (Mercury/Bear Family)
The Storyteller (Mercury/Mercury)
Songs Of Fox Hollow (Mercury/–)
Country Classics (–/Phillips)
The Magnificent Music Machine (Mercury/–)
Ol' T's In Town (RCA/RCA)
Places I've Done Time (RCA/RCA)
Everything From Jesus To Jack Daniels (Mercury/Mercury)
Music Man's Dreams (–/Range)

Wendell Hall

Although he was by no means a true country entertainer, it was Hall's hillbilly-like recording of **It Ain't Gonna Rain No Mo'**, a 1923 million-seller, that

We All Got Together And . . . , Tom T. Hall. Courtesy Mercury Records.

encouraged Victor to embark on a search for possible country hitmakers.

Born in St George, Kansas, on August 23, 1896, Hall attended the University of Chicago and, after military service during World War I, began touring in vaudeville, singing and playing ukelele. Known as the 'Red-Headed Music-Maker', Hall was a friend of Carson Robison. It was with Robison that he went to New York, where the pair recorded for Victor during the early '20s.

Director of many shows during the '30s, Hall was still active in the music business up to the time of his death in Alabama, on April 2, 1969.

The writer of such songs as **My Carolina Rose** and **My Dream Sweetheart**, he frequently guested on the WLS National Barn Dance show.

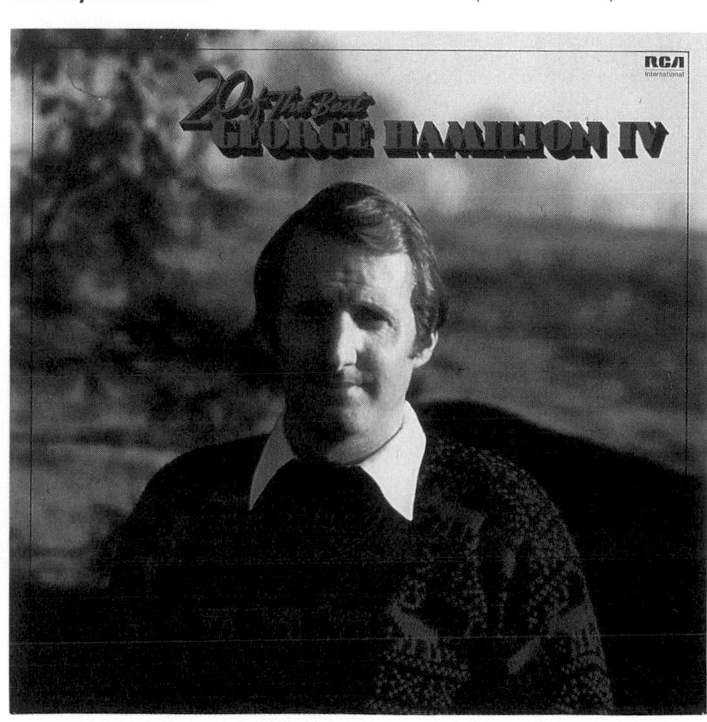

20 Of The Best, George Hamilton IV. Courtesy RCA Records.

Stuart Hamblen

Born in Kellyville, Texas, on October 20, 1908, singer and bandleader Stuart Hamblen achieved considerable fame during the '50s as a songwriter. He attended the McMurray State Teachers College, Abilene, Texas in the '20s but later switched to a musical career, working and broadcasting in the California area, sometimes appearing in minor roles in western films.

In 1949, Hamblen had a Top 10 hit with a Columbia release, **But I'll Go Chasin' Women**, following this with **(Remember Me) I'm The One Who Loves You** a few months later.

An attempt to run for the Presidency of the United States, on a Prohibition Party ticket, proved a predictable failure in 1952, but in '54 he had more luck when his self-penned **This Ole House** (a song written after Hamblen had discovered a dead man inside a dilapidated hut many miles from the nearest habitation) became a country hit, prompting a million-selling cover version by Rosemary Clooney. This same song was later successfully revived by Shakin' Stevens, who topped the British charts with his updated rendition during the summer of 1981.

Hamblen, who was responsible for many other popular songs of the '50s, later turned increasingly to religious material, including the gospel standard **It Is No Secret (What God Can Do)**. Other Hamblen-penned classics include **My Mary** and **Texas Plains**, both of which first became popular in the early '30s. Stuart Hamblen died on March 8, 1989, following surgery on a brain tumor.

Canadian Pacific, George Hamilton IV. Courtesy RCA Records.

Recommended:
Cowboy Church (Word/Word)
A Man And His Music (Lamb & Lion/–)

George Hamilton IV

A pleasant-voiced vocalist who has gained tremendous popularity in Canada and England as well as his native country, George Hamilton IV was born on July 19, 1937 and raised in Winston Salem, North Carolina. Becoming a country music fan after watching Gene Autry and Tex Ritter films at Saturday matinees, he bought his first guitar at the age of 12, earning the necessary cash on a newspaper round.

He then began buying Hank Williams discs, and frequently caught the Greyhound bus out to Nashville, where he saw the Grand Ole Opry and met people like Chet Atkins, Eddy Arnold, Hank Snow and others. Later he began a high school band at Reynolds High, Winston Salem. In his senior year, he made a demo recording of Little Jimmy Dickens' **Out Behind The Barn**, and sent the results to talent scout Orville Campbell. Through Campbell, Hamilton met John D. Loudermilk and recorded his **A Rose And A Baby Ruth**, which sold over a million in 1956–57 when released by ABC-Paramount. A teen pop star, Hamilton found himself booked on Alan Freed's show during the autumn of 1956, also gaining a place on various package shows featuring Buddy Holly, Gene Vincent and the Everly Brothers.

Frustrated with his teeny-booper image, Hamilton moved to Nashville in 1959, scoring his first country Top 10 hit with **Before This Day Ends** (1960). He joined the Grand Ole Opry and signed to RCA Records in 1961, making it back into the Top 10 with the self-penned **If You Don't Know I Ain't Gonna Tell You** (1962). His first No.1 came with **Abilene** (1963), a John D. Loudermilk song, which also crossed over to pop. After scoring another Top 10 winner in **Fort Worth, Dallas Or Houston** (1964), Hamilton became influenced by the folk revival of the early '60s. Becoming friendly with Gordon Lightfoot in 1965, he began recording the Canadian's songs, making a Top 10 entry with **Early Morning Rain** (1966), and eventually recording more Lightfoot compositions than any other artist. Through this Canuck connection, he began to work more and more with Canadian writers and later signed with RCA's Canadian division. Prior to this he had more Nashville-produced Top 10 successes with **Urge For Going** (1967), **Break My Mind** (1968) and **She's A Little Bit Country** (1970), his last Top 20 entry coming with **Anyway** (1971).

He first visited England in 1967, en route to Nashville, following a tour of US bases in Germany. He did a guest spot on the BBC's Country Meets Folk programme, and later became a regular on many British programmes. He was booked several times for Mervyn Conn's Country Music Festival at Wembley. In Canada, Hamilton hosted his own TV show, North Country, for five years, while in 1977 he became signed to Anchor, a British record label, rejoining ABC-Dot for American releases only.

For several years he has been managed by Mervyn Conn in Britain, where George spends much of his time, and his recordings have been geared very much to the British market. However, he has remained a member of the Grand Ole Opry in Nashville. Known as the 'International Ambassador of Country Music', he was the first American country singer to perform in Russia and Czechoslovakia, where he recorded an album with Czech country group, Jiri Brabeck and Country Beat. He has also hosted successful country music festivals in Sweden, Finland, Norway, Holland and Germany.

A deeply religious person, in recent years George has undertaken several gospel tours throughout the British Isles. He has played mainly churches with 'themed' performances based on Thanksgiving, Easter or Christmas, all portrayed in an entertaining mix of music and bible readings. He has maintained a close association with country, touring with Slim Whitman and other country stars. He has participated in country music shows with his son, George Hege Hamilton V, who had a country hit with **She Says** (1988) and is popular around the British club circuit.

Bluegrass Gospel, George Hamilton IV. Courtesy Lamb & Lion Records.

Recommended:
Bluegrass Gospel (Lamb & Lion/Lamb & Lion)
Canadian Pacific (RCA/RCA)
Country Music In My Soul (–/RCA)

Left: George Hamilton IV is probably the best country music ambassador.

Famous Country Music Makers (–/RCA)
Travelin' Light (–/RCA)
Fine Lace And Homespun Cloth (Dot/Anchor)
Feel Like A Million (Dot/Anchor)
Songs For A Winter's Night (–/Ronco)
Music Man's Dream (–/Range)
George Hamilton IV (MCA-Dot/MCA)
American Country Gothic – with the Moody Brothers (Lamont/Conifer)
Easter In The Country (–/Word)

Butch Hancock

Born and raised in and around Lubbock, Texas, Butch Hancock is best known for his songwriting. Joe Ely, Emmylou Harris, Jerry Jeff Walker, the Texas Tornados, Alvin Crow and Jimmie Dale Gilmore are just a few who have recorded his songs. He has played a major role in the Texas music scene for more than twenty years. Although he was held in high esteem by other performers, gained write-ups in such prestige publications as 'The New York Times', and released loads of recordings on various minor labels, no major record label has ever offered him a contract.

Butch started writing songs back in the heyday of the '60s folk movement. At the time he was driving a tractor for his father, an earth-moving contractor. He has favoured the Dylan/Guthrie guitar-and-harmonica set-up from the beginning, but it took a couple of false starts before he got serious about music. Even today music has to take its place alongside his many other projects. He attended architecture school at Texas Tech, studied arts and physics, and also pursued a photography course in San Francisco. In 1971 he joined up with Joe Ely, Jimmie Dale Gilmore, Steve Wesson and Tony Pearson in the Flatlanders, a short-lived folk-country outfit that recorded in Nashville and won a standing ovation at the 1972 Kerrville Folk Festival, then split up. A short stint doing construction work in Clarendon, Texas was followed by a move to Austin where he picked up on music and also undertook other projects, including carpentry, photography and redesigning a train station in Seguin.

He formed his own Rainlight Records in 1978. Over the next ten years he released six albums and two cassettes, sometimes featuring his own simple guitar and harmonica, other times his band the Sunspots, and occasionally a host of distinguished Texas players. Some of these recordings have subsequently been reissued by Demon in England and Sugar Hill in America. An extensive traveller, Hancock has made a habit of touring Europe and Australia, playing his music and also indulging in his photography. He is an accomplished, published photographer, and, in addition to exhibiting shows of photographs and other visual arts, he spent much of his time as a video cameraman and producer of the regional show, 'Dixie Bar And Bus Stop' in the mid-80s. For four years he and associates taped some 150 episodes featuring over 100 artists and bands from Texas. In 1990 he recorded over 140 of his own original songs during a five-night run at Austin's Cactus Café. Then, during the following year, he started marketing the recordings on a 'No 2 Alike' tape on a tape-a-month mail-order basis. That year he also established his own gallery/retail outlet 'Lubbock Or Leave It', in Austin's downtown Brazos Street.

Music does play a major role in Hancock's life, but in the end it's all of a piece – his finely crafted songs, photos of faraway places, videos of honky-tonkers, displays of art. Like other idiosyncratic artists who don't fit into today's mass-marketing systems, Butch Hancock has figured out that he can do just as well by doing it all himself.

Recommended:
Own The Way Over Here (Sugar Hill/–)
Own And Own (Sugar Hill/Demon)
West Texas Waltzes And Dust Bowl Tractor Blues (Rainbow/Rainlight)

Great Country Hits, the Harden Trio. Courtesy Harmony.

Arleen Harden

One-time secretary for an insurance company, Arleen Harden (born England, Arkansas, March 1, 1945) was part of the Harden Trio, a family group, whose **Tippy Toeing** charted for 21 weeks in 1966, gaining the trio Opry membership from 1966–68. During this time they supplied Columbia with other hits in **Seven Days Of Crying** (1966), **Sneakin' Across The Border** (1967) and **Everybody Wants To Be Somebody Else** (1968).

Arleen also became signed to the label as a solo artist, having a first hit with **Fairweather Lover** in 1967. Following the break up of the Harden Trio in '68 she enjoyed minor successes, **Lovin' Man** (1970) proving the most potent.

Following a stay with UA, Arleen later signed for Capitol, cutting a warm, easy-listening, Cam Mullins-arranged album, **I Could Almost Say Goodbye**, in 1975. She has since recorded for Elektra without too much success and spends much of her time working as a background vocalist on Nashville sessions.

Recommended:
Sings Roy Orbison (Columbia/–)
I Could Almost Say Goodbye (Capitol/–)

Linda Hargrove

A superior singer-songwriter and an outstanding guitarist (at least, Pete Drake and Mike Nesmith have said so), Linda

was born on February 3, 1950. She was raised in Tallahassee, Florida, where she took piano lessons at the age of five and moved on to become a French horn player in a high school band before getting bitten by the rock bug.

Influenced by Dylan's **Nashville Skyline**, she packed her bags and headed for Nashville in 1970. She hit hard times until Sandy Posey recorded one of her songs. Pete Drake, who sat in on the Posey session, then offered Linda a songwriting contract plus some session chores as a guitarist. Some time later, he taught her to handle the console at Drake's own studio.

An album featuring Linda was cut by Mike Nesmith for his ill-fated Countryside label but was never released. However, her songs met a better fate, Leon Russell employing two on his **Hank Wilson's Back** LP, Jan Howard, Billie Jo Spears, Melba Montgomery, David Rogers and many others also utilizing Linda's compositions on various recordings.

Since the abortive Nesmith dates, Linda has recorded for Elektra, cutting such albums as **Music Is Your Mistress** and **Blue Jean Country Queen**. After joining Capitol Records in 1975, she made a breakthrough to the singles chart with **Love Was (Once Around The Dance Floor)** (1975), and came up with the acclaimed album, **Love You're The Teacher**. Following a change of labels to RCA in 1978 and the release of two singles, Linda became a born-again Christian and no longer sings her secular material, devoting her life instead to her religious teachings.

Recommended:
Music Is Your Mistress (Elektra/–)
Impressions (Capitol/–)
Love You're The Teacher (Capitol/–)

Kelly Harrell

A country music pioneer who recorded as early as 1924 for Ralph Peer, then with Okeh Records, Crockett Kelly Harrell was born in Drapers Valley, Virginia, on September 13, 1899. A one-time rambler, he became a loom fixer in a mill around 1927, but also continued with a musical

Gliding Bird, Emmylou Harris. A pirate version of Emmylou's debut album.

career, making a number of important early records for Victor, including **Cuckoo**, **She's A Pretty Bird**, **New River Train**, **Rovin' Gambler**, **I Wish I Was Single Again**, **Charles Guiteau** and **The Butcher Boy**. Often accompanied by banjo, fiddle and guitar, Harrell himself did not play an instrument.

Despite the success of his early records, and his songwriting efforts in two popular early songs, **Away Out On The Mountain** (as recorded by Jimmie Rodgers) and **The Story Of The Mighty Mississippi** (as recorded by Ernest Stoneman), his musical career was a brief one, and he ended his short life working in a Virginia factory. He died of a heart attack on July 9, 1942.

Recommended:
Kelly Harrell And The Virginia String Band (County/–)

Emmylou Harris

The 'First Lady' of contemporary country music, Emmylou was born in Birmingham, Alabama, on April 2, 1949. She turned country upside-down when she took the music back to its roots while still retaining a contemporary country-rock edge to her work. With the influence of Gram Parsons and his songs, she has been among those responsible for making country music acceptable to a wider audience.

Emmylou developed an early interest in country music. When her family moved to Washington DC in the mid-'60s she performed in local folk clubs and also in

Below: Through her Hot Band, Emmylou discovered and nurtured the careers of such stars as Rodney Crowell and Ricky Scaggs.

Blue Kentucky Girl, Emmylou Harris. Courtesy WEA Records.

Greenwich Village, New York. Building up a reputation, she recorded a folksy album, **Gliding Bird**, for the Jubilee label in 1970, which featured some of her own songs plus others by Bob Dylan, Fred Neil and Hank Williams. The title song was penned by her first husband Tom Slocum. She was playing small clubs in Washington DC when Gram Parsons, who was looking for a female harmony singer, was urged to go and see her. Obviously impressed, he asked Emmylou to move to Los Angeles to work on his first solo album for Warner Brothers in 1972. After completing the recording of **GP**, Gram, Emmylou and the Fallen Angel Band embarked on a short tour. They then returned to the studio to start work on a second album. Parson's death occurred shortly after **Grievous Angel** was completed in 1973, the album becoming a

posthumous success and, in a way, paving the way for an Emmylou Harris solo career.

Emmylou had been close to Gram and was stunned by his death, and when she picked up the threads of her own career, it was to Gram's material that she turned. In 1975 she recorded **Pieces Of The Sky** for Reprise Records, an album which mixed country songs with some light rock'n'roll, and, although none of Gram's songs were included on this album, she was using his material in her act.

Pieces Of The Sky did not make a great impact, although her version of the Louvin Brothers' **If I Could Only Win Your Love** from the album did make the country Top 10 in 1975. But as Emmylou toured America and Europe her pure voice and impeccable backing band were to enchant listeners. She also has a fragile, Californian sort of beauty, and those who had admired her recorded work found that her stage act was everything they had hoped for. Over the years Emmylou's famous Hot Band, in all its various amalgamations has included such music greats as Ricky Skaggs, Tony Brown, Emory Gordy Jr, Glen D. Hardin, Albert Lee, Rodney Crowell and James Burton.

Her voice had a pure, innocent, classic quality and it lacked the nasal sound which so many non-country fans find hard to take. The second album, **Elite Hotel**, was released in 1976 and it featured three of Gram Parson's better compositions: **Wheels**, **Sin City** and **Ooh Las Vegas**. As usual, it was a well-balanced mix of country, ballads and rock. Emmylou's first No.1 came with **Together Again** in 1976, and further Top 10 entries followed with **One Of These Days** (1976), **Sweet Dreams** (a No.1 in 1976), **(You Never Can Tell) C'Est La Vie** (1977), **To Daddy** (1978) and **Two More Bottles Of Wine** (a No.1 in 1978).

Rodney Crowell proved himself a capable songwriter for her, having had a hand in **'Til I Gain Control Again**, **Leaving Louisiana In The Broad Daylight** and **Amarillo**. Ricky Skaggs, a fiddle player and mandolinist, brought a bluegrass influence to Emmylou's music and was largely responsible for the more traditional arrangements used on **Roses In The Snow**. This 1980 release finally brought recognition from a country audience and was to lead to Skaggs becoming a major country artist in his own right, and was also notable for being the first introduction many listeners had to the group the Whites, who, like Skaggs, became a top country act during the early '80s.

Always striving to vary her music, Emmylou had more Top 10 entries with **Save The Last Dance For Me** (1979) and **Beneath Still Waters** (No.1, 1980), as well as duet hits with Roy Orbison (**That Lovin' You Feelin' Again** in 1980) and Don Williams (**If I Needed You** in 1981). She almost certainly caught a lot of her fans unawares with her inventive re-workings of old pop songs like **Mister Sandman**, a pop and country hit in 1981. Throughout Emmylou's recordings, the greater the challenge the song provided the more inspired her performance became. She recorded such diverse material as Donna Summer's **On The Radio**, Jule Styne's **Diamonds Are A Girl's Best Friend**, Bruce Springsteen's **Racing In The Streets** and the early Presley classic, **Mystery Train**.

Up until the early '80s most of Emmylou's recordings were produced by Canadian Brian Ahern, her second husband. The final project they worked on was 1983's **White Shoes**, at which point Harris and Ahern separated, both personally and professionally. Emmylou had been working in the studio with British singer-songwriter Paul Kennerley on **The Legend Of Jesse James** concept album in 1982, and also recorded several of his songs. A relationship built, and, in 1985, Kennerley became Harris' third husband. Together they wrote and produced **The Ballad Of Sally Rose**, a superb, but commercially underrated concept album. Kennerley continued to produce her recordings and also provided some excellent song material, including **Born To Run** (Top 10 – 1982), **In My Dreams** (Top 10 – 1984) and **Heartbreak Hill**, which is, at the time of writing, Emmylou's last Top 10 entry, in 1989.

Emmylou and Kennerley separated in 1991, and around the same time the Hot Band ceased to exist. Emmylou formed a new band, the Nash Ramblers, an acoustic group comprising Sam Bush (fiddle, mandolin), Roy Huskey Jr (upright bass), Al Perkins (dobro, banjo, guitar), Jon Randall Stewart (mandolin, lead acoustic guitar) and Larry Mamanuik (drums). She recorded a live album, **Emmylou Harris At The Ryman**, in 1991, which was also filmed for a TNN special in concert at the Ryman Auditorium in Nashville, the former home of the Grand Ole Opry. The album was not a commercial success and at the end of 1992 Emmylou was dropped by Reprise. The following year a new contract was signed with Asylum, which, like Reprise, is part of the large WEA group.

Emmylou Harris has discovered and nurtured some of the finest writers, singers and musicians of the past twenty years, and she exposed new audiences to standard songwriters. Emmylou Harris has always performed great music and continues singing and playing.

Recommended:

Pieces Of The Sky (Reprise/Reprise)
Elite Hotel (Reprise/Reprise)
Luxury Liner (Reprise/Reprise)
Quarter Moon In A Ten Cent Town (Warner Bros/Warner Bros)
Roses In The Snow (Warner Bros/Warner Bros)
Cimarron (Warner Bros/Warner Bros)
White Shoes (Warner Bros/Warner Bros)
The Ballad Of Sally Rose (Warner Bros/Warner Bros)
Thirteen (Warner Bros/–)
Blue Bird (Reprise/Reprise)
Brand New Dance (Reprise/Reprise)
Duets (Reprise/Reprise)
At The Ryman (Reprise/Reprise)

Freddie Hart

Born in Lockapoka, Alabama, on December 21, 1926, Hart is said to have run away from home at seven, becoming – amongst other things – a cotton picker, a sawmill worker, a pipeline layer in Texas and a dishwasher in New York.

By the age of 14, he had become a Marine, three years later helping to take Guam, having already been to Iwo Jima and Okinawa. A physical fitness expert and the possessor of a black belt in karate, he taught this form of self defence at the LA Police Academy in the '50s, eventually moving into the music business with the aid of Lefty Frizzell with whom he worked until 1953 when Hart signed a recording contract with Capitol.

Please Don't Tell Her, Freddie Hart. Courtesy Pickwick Records.

He subsequently recorded for Columbia (having his first hit in 1959 with **The Wall**), Monument and Kapp throughout the mid-'60s, logging around a dozen chart entries, becoming a major artist after re-signing for Capitol in 1969 and having a million-selling single, **Easy Loving** (1971), which won CMA Song Of The Year.

For the next few years Hart enjoyed a run of success, most of his singles claiming Top 5 status: **My Hang Up Is You**, **Bless Your Heart**, **Got The All Overs For You** (all 1972), **Super Kind Of Woman**, **Trip To Heaven** (1973), **If You Can't Feel It**, **Hang On In There Girl**, **The Want To's** (1974), **The First Time** (1975) and **Why Lovers Turn To Strangers** (1977) were just a few of his major hits.

In 1980 he joined the small Sunbird label and scored minor hits with **Roses Are Red** (1980) and **You Were There** (1981). However, in recent years he has slipped in popularity. Now extremely wealthy, Hart owns many acres of plum trees, a trucking company and over 200 breeding bulls, and runs a school for handicapped children.

Recommended:

Easy Loving (Capitol/–)
That Look In Her Eyes (Capitol/–)
The Pleasure's Been All Mine (Capitol/Capitol)
Only You (Capitol/–)
The First Time (Capitol/–)
Greatest Hits (Capitol/–)
My Lady (Capitol/Capitol)

John Hartford

This banjoist, fiddler, guitarist, singer-songwriter is one of the most exciting solo entertainers in country music today. Born in New York, on December 30, 1937, but

Mark Twang, John Hartford. Courtesy Sonet Records.

Iron Mountain Depot, John Hartford. Courtesy RCA Records.

raised in St Louis by his doctor father and painter mother, he first learnt to play on a banjo which he claims was beaten-up and had no head. By the time he was 13 he had also mastered fiddle and played at local square dances; next he graduated to the dobro, then on to guitar.

Upon leaving school, he worked as a sign painter, a commercial artist, a deck-hand on a Mississippi riverboat and as a disc jockey. After marriage and the birth of a son, Hartford headed for Nashville, becoming a session musician. His work on these sessions gained him a recording contract with RCA, for whom he cut eight albums and several singles, the first of which was **Tall Tall Grass**, a single released in 1966.

Soon many acts began recording Hartford's songs, and one, **Gentle On My Mind**, from his 1967 **Earthwords And Music** album, became a million-seller when covered by Glen Campbell. It entered the charts in both July 1967 and September 1969, winning three Grammies in the process and becoming the most recorded song of the period.

After appearances on the Smothers Brothers Comedy Hour and a regular spot on the Glen Campbell Goodtime Hour, Hartford toured with his own band for a while but eventually opted to become a solo performer. His 1976 **Mark Twang** album presents him in this role, unaccompanied by any rhythm section, Hartford providing all the percussive sounds with his mouth and feet!

An entertaining performer who utilizes his own songs and those drawn from the traditions of country music, he often performs at bluegrass festivals and plays and records with bluegrass musicians. Due to his keen sense of humour and natural entertaining skills he has been a regular guest on TV shows.

Recommended:
The Love Album (RCA/–)
Aero Plain (Warner Bros/Warner Bros)
Mark Twang (Flying Fish/Sonet)
Nobody Knows What You Do (Flying Fish/Sonet)
All In The Name Of Love (Flying Fish/Sonet)
Slumbering On The Cumberland (Flying Fish/–)
Down On The River (Flying Fish/–)

Hawkshaw Hawkins

Another victim of the plane crash that killed Patsy Cline and Cowboy Copas, Harold Hawkins was born in Huntingdon, West Virginia, on December 22, 1921. A guitarist at the age of 15, he then won a local amateur talent show, the prize being a $15 a week spot on radio station WSAZ.

By the time of Pearl Harbor, Hawkins had established himself as a radio personality, but he then enlisted. By 1946 he was home again and singing on WWVA, Wheeling, West Virginia. Then came a recording contract with King and hit records in **I Wanted A Nickel** (1949) and **Slow Poke** (1951), plus a country classic in **Sunny Side Of The Mountain**.

Despite some recordings for RCA and a 1955 contract with the Grand Ole Opry, Hawkins enjoyed no further chart success until 1959, when Columbia single **Soldier's Joy** climbed high in the country list. Four years later – he was at this time married to Jean Shepard – his first country No.1 came with the release of **Lonesome 7-7203**. But on March 5, 1963, just two days after the disc had entered the charts, Hawkins was lying dead among the aircraft wreckage near Kansas.

Recommended:
16 Greatest Hits (Gusto/–)
Hawk 1953–1961 (–/Bear Family)

The All New Hawkshaw Hawkins. Courtesy London Records.

Ronnie Hawkins

Country rock'n'roller Ronnie was born on January 10, 1935 in Huntsville, Madison County, Arkansas and came from a country music background. He formed his first band while still in his teens, playing 'hopped-up hillbilly' music.

During the '50s he developed into a fully fledged rockabilly performer with his group the Hawks, who later went on to play with Bob Dylan and became the Band, the successful rock band of the late '60s.

Ronnie landed a contract with Roulette Records in 1958 and the following year enjoyed success on the American pop charts with **Mary Lou, 40 Days** and **Who Do You Love?**. Unlike most of the '50s rock'n'rollers, Ronnie has never changed his style to easy-listening pop or country music but has continued to perform genuine rock'n'roll, mainly in Canada, where he has lived since the early '60s.

Recommended:
Sings Songs Of Hank Williams (–/PRT)

George D. Hay

Founder of the Grand Ole Opry, George Dewey Hay (born Attica, Indiana, November 9, 1895) was once a reporter for the Memphis Commercial Appeal. Shortly after World War 1, while on an assignment in the Ozarks, he attended a mountain cabin hoedown, and conceived the idea of country music's most famous showcase.

When the Appeal moved into radio, setting up station WMC, Hay became radio editor. Later, in 1924, he took up an appointment on Chicago station WLS. With WLS he helped begin the National Barn Dance. This success led to the position of director with the new WSM, Nashville, in 1925.

Below: Ronnie Hawkins' band the Hawks became Bob Dylan's backing band, and then evolved into the Band, but without Ronnie.

Again he instigated a similar Barn Dance programme, the first broadcast taking place on November 28, 1925, although it did not become a regularly scheduled programme until December of that year. The show rapidly grew in quality and popularity.

It was on December 10, 1927, that the WSM Barn Dance became officially retitled Grand Ole Opry. The show had been preceded by a programme featuring the NBC Symphony Orchestra and, after an introductory number by DeFord Bailey, Hay, who announced the show, declared: "For the past hour we have been listening to music taken from Grand Opera – but from now on we will present the Grand Ole Opry." And so the Opry it became.

Hay, known as the Solemn Old Judge, continued to expand and develop the Opry throughout the rest of his career, extending the range of WSM's broadcasts, encouraging the best country entertainers in the country to appear in Nashville, and recruiting new talent to keep the show both vital and fresh. However, he began to show some signs of mental instability, and in 1951 he retired to live with his daughter in Virginia and died at Virginia Beach, Virginia on May 9, 1968, having been elected to the Country Music Hall Of Fame in 1966.

Above: The man who named the Grand Ole Opry – George D. Hay.

Roy Head

A rock'n'roll-based country singer, Roy was born on January 9, 1943, in Three Rivers, Texas. He started out in the early '60s with his own band, the Traits, reworking old rock'n'roll songs in clubs across Texas.

A recording contract with New York's Scepter Records in 1964 proved to be a failure, so Roy returned to Texas and joined the small Back Beat label, making a breakthrough with the R&B-styled **Treat Her Right**, which reached the Top 3 on the American pop charts towards the end of 1965.

Further pop success followed with **Apple Of My Eye** and **To Make A Big Man Cry** (both 1966), which led to Scepter re-releasing his singles from 1964 resulting in **Just A Little Bit** making the Top 40.

By the end of the '60s Roy was very much a pop has-been and was finding his music more closely aligned with country. He signed to Mega Records in Nashville and scored a minor hit with Mickey Newbury's **Baby's Not Home** (1974). A move to Shannon Records led to his biggest hit, **The Most Wanted Woman In Town** (1975).

This resulted in a move to the major labels, beginning with ABC-Dot, for whom he had further hits with **The Door I Used To Close** (1976) and **Come To Me** (1977). Next came Elektra and success with **In Our Room** (1979) and **The Fire Of Two Old Flames** (1980). Since then Roy has recorded with a number of smaller labels including Churchill, NSD and Avion without exactly setting the charts on fire.

Recommended:
Ahead Of His Time (ABC-Dot/–)
The Many Sides Of Roy Head (Elektra/–)

Bobby Helms

A cross-over performer, Helms had a Top 10 pop hit with **Jingle Bell Rock** in 1957, the same year that he was adjudged the nation's leading country singer by Cashbox magazine.

Born in Bloomington, Indiana, on August 15, 1933, guitarist, singer-songwriter Helms appeared on radio at the age of 13, making his debut on the Grand Ole Opry four years later. In 1957, he achieved a No.1 country hit with **Fraulein**, the disc remaining in the charts for a whole year. His version of Jimmy Duncan's **My Special Angel** became both a country and pop hit, selling over a million copies.

The impetus was maintained throughout 1958, with **Jacqueline** (from the film 'A Case Against Brooklyn') and **Just A Little Lonesome** providing him with best-sellers.

But, despite constant seasonal reappearances by **Jingle Bell Rock** (which took five years to become a million-seller), Helms' recording career faded rapidly. Between 1960 and 1967 his name was absent from the charts, but later he achieved a series of mini hits on such labels as Little Darlin' and Certon, as he drifted in and out of the business.

Recommended:
My Special Angel (–/President)
Sings His Greatest Hits (Power Pak/–)
My Special Angel (Vocalion/–)
Fraulein – The Decca Years, 1956–1962 (Harmony/Bear Family)

John Hiatt

Vocalist, guitarist and composer Hiatt was born in Indianapolis in 1952. Early albums **Hangin' Around The Observatory** and **Overcoats** were cut in Nashville where the artist maintains a steady workload with his writing and performing.

An unsuccessful stint in East Coast folk clubs was followed by employment with Ry Cooder's back-up band, where Hiatt played guitar and featured on a trio of Cooder albums, including **Borderline**, **Slide Area** and **The Border**.

His own recorded work was not commercially successful until Rosanne Cash cut Hiatt's **The Way We Make A Broken Heart** in 1987, which achieved prominence in the US country chart. Hiatt then formed a permanent band, the Goners, who cut **Slow Turning** and **Stolen Moments** for A&M Records.

Recommended:
Overcoats (Epic/–)
Two Bit Monsters (MCA/–)
All Of A Sudden (Geffen/–)
Warming Up The Ice Age (Geffen/–)
Bring The Family (A&M/Demon)
Slow Turning (A&M/–)
Y'All Caught (Geffen/–)
Stolen Moments (A&M/–)

Highway 101

A 'manufactured' group, Highway 101 gave country music a definite harder edge in the late '80s as they rose above the hype to produce some of the most refreshing

All New Just For You, Bobby Helms. Courtesy Little Darlin' Records.

Clear Sailin', Chris Hillman. Courtesy WEA Records.

country-rock. They picked up a 1988 CMA award for Best Group.

The brainchild of respected band manager Chuck Morris, Highway 101 had been 'assembled' to fulfil a need for a group that could play 'traditional country with a rock'n'roll backbeat'. Two years in the planning stage, Morris initially recruited session-player Scott 'Cactus' Moser to play drums, who brought in bassist Curtis Stone (the son of the legendary Cliffie Stone) and lead guitarist Jack Daniels. The final piece of the jigsaw fell in place when Minnesota-born singer-songwriter Paulette Carlson was brought into the picture. Morris heard some demos that Paulette had made in Nashville, and knew he had found the right front-person for his fledgling outfit. He negotiated a singles deal with Warner Brothers, but without all the publicity in place, the initial single **Some Find Love** (released in late 1986) failed to make any impact.

A second single, **The Bed You Made For Me**, a Carlson original, hit the country Top 5, and Highway 101 were signed to a full contract by Warners. Then the hits really flowed, with: **Whiskey**, **If You Were A Woman**, **Somewhere Tonight** (No.1, 1988), **All The Reasons Why**, **Setting Me Up**, **Honky Tonk Heart** and **Who's Lonely Now** (No.1, 1989). The band definitely discovered a winning formula, initially creating their own unique sound, which other acts started to 'duplicate'. Acclaim and praise was heaped on them; they undertook extensive roadwork and even played the New York Ritz to a wildly enthusiastic sell-out audience! Discontent started to creep into the group in 1990 when the three guys realized Paulette was attracting most of the attention and calling the shots. Paulette then left to pursue a solo career.

After listening to hundreds of demo tapes and sitting through extensive auditions, Nashville-based Nikki Nelson was chosen as new lead singer. She had performed in her father's band since the age of 12, and at the time was working as a waitress at the Nashville Palace. With Nikki upfront, the Highway 101 sound changed. There were more upbeat songs and fewer ballads devoted to the pain and heartache of lost love. They scored a Top 20 entry with **Bing, Bang, Boom**, the title song from the first album to feature Nikki in 1991, but have since failed to crack the Top 10. Obviously country music fans have been slow to accept change, and there is a strong possibility that Highway 101 may never regain the enthusiastic following and chart success they enjoyed for that three-year period at the end of the '80s.

Recommended:

Highway 101 (Warner Bros/–)
Highway 101 2 (Warner Bros/–)
Bing Bang Boom (Warner Bros/–)
Paint The Town (Warner Bros/–)

Chris Hillman

For most of his musical career, Chris Hillman, who was born on December 4, 1944, in San Diego, California, has chosen to take a back seat, leaving the spotlight on others. However, he has been recognized as the musical backbone of each band he's participated in.

He played mandolin in his first group, the bluegrass-oriented Scotsville Squirrel Barkers, while still in high school. This group also included Bernie Leadon and Kenny Wertz. Then came the Hillmen, which included Vern and Rex Gosdin. Later he had careers with several well-known bands where he was often underrated.

During a stint with the Byrds, Hillman (with Gram Parsons' support) urged the folk-rock group to make a country LP, **Sweetheart Of The Rodeo**, the pioneering country-rock album of 1968. Chris and Gram left the band soon afterwards, developing their country-rock ideas into the Flying Burrito Brothers.

Throughout the '70s, Chris worked with various country-rock outfits such as Manassas (with Steve Stills), the Souther-Hillman-Furay Band (with J.D. Souther and Richie Furay), Firefall (with Rick Roberts) and McGuinn, Clark And Hillman (with Roger McGuinn and Gene Clark).

In the early '80s Chris started to make his mark as a solo performer with a Sugar Hill Records contract. Two albums resulted, **Morning Sky** and **Desert Rose**, which neatly blended his country and bluegrass roots with rock. Scoring a minor country hit with **Somebody's Back In Town** (1984), he started building a bigger following and also worked regularly as a background vocalist and instrumentalist on recordings by Linda Ronstadt, Dan Fogelberg, John Denver and many other contemporary country acts. In 1989 he teamed up for a short period with former Byrds' members Dave Crosby and Roger McGuinn, and, with McGuinn, enjoyed a Top 10 country hit with a revival of the Byrds' **You Ain't Going Nowhere** (1989). However, Hillman remains totally committed to his highly successful Desert Rose Band, which he formed in 1986. This has finally given him the recognition he has worked towards so steadily for the past 30 years.

Recommended:

Desert Rose (Sugar Hill/Sundown)
Morning Star (Sugar Hill/Sundown)
Clear Sailin' (Asylum/Asylum)

Buddy Holly

One of rock's prime movers in its early years, Buddy Holly actually began his career as a country singer, and the sound

Nashville, Tennessee, Buddy Holly. Part of MCA's wonderful boxed set.

was never to leave him during his short but brilliant life; nor has the power of his songwriting seemed to diminish, as **That'll Be The Day**, **Everyday** and **It's So Easy** have all been pop and country hits in recent years.

Charles Hardin Holley (the 'e' in his last name was dropped only after he signed his first record contract) was born on September 7, 1936, in Lubbock, Texas, and grew up listening to the blues and Tex-Mex music as well as to Hank Williams and Bill Monroe. His first band, with longtime friend Bob Montgomery, tells the story of their musical approach; they were called Buddy and Bob: Western and Bop.

Holly's first professional session was, in fact, a country session for Decca, produced in Nashville by Owen Bradley early in 1956, and featured not Holly's own band, the Crickets, but a group of Nashville sidemen. However, the combination of slick Nashville sound and raw Texas rockabilly did not mix well and the records were not successful. It is ironic that Holly's great success came on Coral Records, a Decca subsidiary, after the parent label had dropped him.

His career as a rock star – although many country stations continued to play his records and many country fans continued to buy them – was brief and hectic, filled with hit records **Oh Boy!**, **Peggy Sue**, **Rave On**, **Fool's Paradise** and **Raining In My Heart**. It was on one of his hectic tours that he died in a plane crash on February 3, 1959.

His songs, his style, and his sidemen – Waylon Jennings, Tommy Allsup, Bob Montgomery and Sonny Curtis – have all left great marks on country music, and Holly was a genuine influence on country at this pivotal point in its history.

Recommended:
The Complete Buddy Holly (6 LP Box Set) (–/MCA)

Homer And Jethro

Both from Knoxville, Tennessee, Henry D. Haynes (Homer) was born July 29, 1917 and Kenneth C. Burns (Jethro) was born March 10, 1923. They formed a duo in 1932, the two boys winning a regular stop on station WNOX, Knoxville.

Discovering that their parodies gained more attention than their 'straight' material, they opted to become country comics, holding down a residency at Renfro Valley, Kentucky, until war service caused a temporary halt to their career. With Japan defeated, the duo re-formed, for a decade appearing as cast members of the National Barn Dance on Chicago WLS, also guesting on the Opry and many networked radio and TV shows.

Signed to RCA Records in the late '40s, they cut **Baby It's Cold Outside** with June Carter in 1948, obtaining later hits with **That Hound Dog In The Window** (1953), **Hernando's Hideaway** (1954), **The Battle Of Kookamonga** (1959) and **I Want To Hold Your Hand** (1964). The duo also recorded a number of instrumental albums (Haynes on guitar, Burns on mandolin), at one time teaming with Chet Atkins to form a recording group known as the Nashville String Band.

The 39-year-old partnership terminated on August 7, 1971, with the death of Henry Haynes, but Burns continued with his musical career. A brilliant mandolinist, he became involved with country-jazz, playing almost in Django Reinhardt fashion. Following a short illness, Burns died on February 4, 1989.

Recommended:
Country Comedy (–/RCA)
The Far Out World Of Homer & Jethro (RCA/–)
Assault The Rock 'n' Roll Era (–/Bear Family)

Jethro Burns solo:
Jethro Burns (Flying Fish/–)
Back To Back – with Tiny Moore (Kaleidoscope/–)

The Hoosier Hot Shots

Frank Kettering, banjo, guitar, flute, piccolo, bass, piano; Hezzie Triesch, song whistle, washboard, drums, alto horn; Kenny Triesch, banjo, tenor guitar, bass horn; Gabe Ward, clarinet.

"Are you ready, Hezzie?" always signalled the arrival of the Hoosier Hot Shots on the National Barn Dance, a first-rate group of comedians and musicians who had one of the most popular novelty acts in the country before Spike Jones came along.

They started out as a small dance band but their flair for comedy and unusual instruments got the better of them. When they joined WLS in 1935 it was as a novelty group, and their success was immediate. They appeared in many films both with and without other Barn Dance cast members, and retired to California.

Their records (for the ARC complex of labels and Vocalion) did well, but they were primarily a visual comedy act.

Doc Hopkins

Doctor Howard Hopkins – yes, that is his real name – was born on January 26, 1899 in Harlan County, Kentucky. He was associated for a long time (1930–49) with station WLS and the National Barn Dance. During that period he became well known as one of the best and most authentic of American folk singers.

Although he spent a great deal of time on WLS and has recorded for many labels (including Paramount, Decca and others), he has somehow never received the recognition as a a country music pioneer that he richly deserves.

Johnny Horton

With **Battle Of New Orleans**, a Jimmie Driftwood song said to be based on an old fiddle tune known as **The 8th Of January**, Horton achieved one of the biggest selling discs of 1959. A cover version by skiffle king Lonnie Donegan became a Top 5 record in Britain.

Born in Tyler, Texas, on April 3, 1925, Horton went to college in Jacksonville and Kilgore, Texas, later attending the University of Seattle in Washington. Spending some time in the fishing industry in Alaska and California, he then became a performer under the title of the Singing Fisherman, when he began starring on Shreveport's Louisiana Hayride during the mid-'50s.

Completing recording stints with both Mercury and Dot, he moved on to Columbia, his first hit being with **Honky Tonk Man** (1956) and his first country No.1 with **When It's Springtime In Alaska** (1959).

Following the runaway success of the million-selling **Battle Of New Orleans**, Horton became a nationwide star, having hits with **Johnny Reb/Sal's Got A Sugar Lip** (1959) and **Sink The Bismarck** (1960). He was also asked to sing Mike Phillips' **North To Alaska** in the John Wayne film of that title, the resulting record providing the Texan with yet another million-seller in 1960.

On November 5, 1960, Horton was killed in a car accident while travelling to Nashville, but his records continued to sell throughout the '60s and his songs have been recorded by Claude King, Dwight Yoakam and many other country stars over the years. In fact, Claude King recorded a tribute album to Johnny in the late '60s, and in 1983 a biography entitled 'Your Singing Fisherman' was published.

The Spectacular Johnny Horton. Courtesy Columbia Records.

Recommended:
Honky Tonk Man (Columbia/–)
Makes History (Columbia/–)
On Stage (Columbia/–)
Spectacular (Columbia/–)
Greatest Hits (–/CBS)
Rockin' Rollin' (–/Bear Family)
America Remembers (CSP-Gusto/–)

David Houston

A direct descendant of Sam Houston and Robert E. Lee, Houston was born in Shreveport, Louisiana, on December 9, 1938.

Brought up in Bossier City, where he was taught guitar by his aunt, Houston was aided in his career by his godfather Gene Austin (a pop singer who had 1920s million-sellers with **My Blue Heaven** and **Ramona**).

By the age of 12, Houston had won a guest spot on Shreveport's famed Louisiana Hayride radio show, later joining the cast as a regular member. During his teens he completed college, then, in the late '50s, began touring avidly, appearing on many TV and radio shows.

Signed to Epic in 1963, he gained an instant hit with **Mountain Of Love** which stayed in the charts for 18 weeks, winning Houston Most Promising Country Newcomer plaudits from magazines.

An accomplished yodeller and a talented guitarist-pianist, Houston went from strength to strength throughout the '60s, having No.1 hits with **Almost Persuaded** (1966), **With One Exception** (1967), **You Mean The World To Me** (1967), **My Elusive Dreams** (1968) and **Baby, Baby (I Know You're A Lady)** (1969), winning two Grammy awards for **Almost Persuaded** and earning a part in a 1967 film, 'Cottonpickin' Chickenpickers'.

During the early '70s, Houston's discs continued to chart regularly. Top 10 contenders were: **I Do My Swinging At Home** (1970), **After Closing Time** (with Barbara Mandrell, 1970), **Wonders Of The Wine** (1970), **A Woman Always Knows** (1971), **Soft Sweet And Warm** (1972), **Good Things** (1973) and **She's All Woman** (1973).

Since the mid-'70s David's decline has been rapid, as he has moved through a succession of labels, looking for the one song that might take him back into the Top 10. However, he has continued to work steadily, usually touring with his manager Tillman Franks, who also doubles as David's guitarist when he appears with a pick-up band.

Recommended:
Best Of Houston And Mandrell (Epic/–)
Day Love Walked In (Epic/–)
A Perfect Match – with Barbara Mandrell (Epic/–)
A Man Needs Love (Epic/–)
From The Heart Of Houston (Derrick/–)
From Houston To You (Excelsior/–)
American Originals (Columbia/–)

Harlan Howard

An outstanding performer, Howard (born Lexington, Kentucky, September 8, 1929) has generally preferred to remain a songwriter, picking up numerous awards, and running his Wilderness Music Publishing Company. Raised in Detroit, he began songwriting at the age of 12. Spending four years in the paratroops following high school, he became based in Fort Benning, Georgia, spending his weekends in Nashville.

Later, in Los Angeles, he met Tex Ritter and Johnny Bond who began publishing his songs, hits emerging with **Pick Me Up On Your Way Down** (Charlie Walker, 1958), **Mommy For A Day** (Kitty Wells, 1959) and **Heartaches By The Number** (Ray Price and Guy Mitchell, 1959). In 1960, Howard moved to Nashville with his wife, singer Jan Howard, where he became known as the 'king' of country songwriters, a title only challenged perhaps by Dallas Frazier and Bill Anderson.

His many songs have included: **I've Got A Tiger By The Tail**, **Under The Influence Of Love**, **A Guy Named Joe**, **Streets Of Baltimore**, **Heartbreak USA**, **Busted**, **No Charge**, **I Fall To Pieces** and **Three Steps To The Phone**.

As a recording artist, he has cut albums for Monument, RCA and Nugget.

Jan Howard

The daughter of a Cherokee maid and an Irish immigrant, Jan was born in West Plains, Missouri on March 13, 1932, acquiring her present surname after marriage to songwriter Harlan Howard.

An avid country music record collector early on, her first public performance came as a result of a meeting with Johnny Cash, a tour with Johnny Horton and Archie Campbell ensuing.

At the close of the 1950s, she began recording for the Challenge label, her first release being **Yankee Go Home**, a duet with Wynn Stewart. This followed with **The One You Slip Around With**, a 1960 hit that won her several awards in the Most Promising Newcomer category.

In the wake of recordings for such labels as Capitol and Wrangler, she moved to Nashville during the mid-'60s, there signing for Decca Records, and teaming with Bill Anderson as a featured part of his road and TV shows. With Anderson she cut a number of hit duets that included **I Know You're Married** (1966), **For Loving You** (a 1967 No.1), **If It's All The Same To You** (1969), **Someday We'll Be Together** (1960) and **Dissatisfied** (1971).

Proving similarly successful as a solo act, a score or so of her releases attained chart status, the most prominent of these being **Evil On Your Mind** (1966), **Bad Seeds** (1966), **Count Your Blessings, Woman** (1968) and **My Son** (1968), the last a self-penned tribute to her son Jim, who died in Vietnam just two weeks after the song had been recorded.

Following the tragic death of the second of her three sons, Jan opted for retirement during the early '70s. However she later joined the Carter Family appearing on Johnny Cash's road show, and in 1985 she recorded her first album in many years when she signed with the reactivated MCA-Dot.

Recommended:
Rock Me Back To Little Rock (Decca/–)
Sincerely (GRT/–)
Jan Howard (MCA-Dot/–)

Rock Me Back To Little Rock, Jan Howard. Courtesy MCA Records.

Above: Paul Howard and his Arkansas Cotton Pickers – a '40s western swing band.

Paul Howard

Although hot western swing on the stage of the staid Grand Ole Opry sounds a little far-fetched, that was exactly Paul Howard's role in the 1940s, the height of western swing's popularity. Born July 10, 1908 in Midland, Arkansas, Howard drifted in and out of music until 1940 when he joined the Opry as a solo singer.

Always entranced by western swing, he began to build a bigger and bigger band, which grew to some nine or ten pieces, sometimes with multiple basses to make up for the lack of drums which were then still taboo on the Opry stage. His band, the Arkansas Cotton Pickers, was one of the hottest of the era, and he recorded for Columbia and King.

Frustrated by the lack of attention western swing got in the Southeast, Howard left the Opry in 1949 for a circuit of radio programmes and dances in Louisiana, Arkansas and Texas.

For many years, Paul lived in Shreveport where he led a band playing dances. He died on June 18, 1984 in Little Rock, Arkansas.

Con Hunley

Conrad Logan Hunley, born on April 9, 1946, Knoxville, Tennessee, the eldest of six children, found success in country music towards the end of the '70s with a run of Top 20 successes that started with **Week-end Friend**.

Country music's blue-eyed soul man grew up listening to the music of Lefty Frizzell and George Jones, but he switched to a soul-country sound modelled on Ray Charles' country-pop successes of the early '60s. He played with various local bands for a dozen years. In 1976 he put together his own group and landed a regular gig at the Village Barn in Knoxville.

Businessman Sam Kirkpatrick took an interest in Hunley's career, setting up a new record label, Prairie Dust, and paying for the singer to travel to Nashville and record. During 1976 and '77, Con enjoyed several minor hits on Prairie Dust including **Breaking Up Is Hard To Do** and **I'll Always Remember That Song**.

This led to the major record companies showing an interest and in 1978 he signed to Warner Bros with **Week-end Friend** (1978), beginning a run of Top 20 hits which included: **You've Still Got A Place In My Heart** (1978), **I've Been Waiting For You All My Life** (1979), **They Never Lost You** (1980), **She's Stepping Out** (1981) and **Oh Girl** (1982).

A short stint recording for MCA Records (1983–84) was followed by a contract with Capitol, which resulted in minor hits with **I'd Rather Be Crazy**, **All American Country Boy** (both 1985) and **What Am I Gonna Do About You** (1986).

Recommended:
As Any Woman (Warner Bros/–)
Oh Girl (Warner Bros/–)

Ferlin Husky

Born in Flat River, Missouri, on December 3, 1927, comedian-singer-songwriter-guitarist Husky grew up on a farm. It is claimed that his first attempt to own a guitar was foiled when the hen that he swopped it for failed to lay eggs, causing neighbours to cancel the deal.

He did, however, obtain a guitar at a later date. Following stints in the Merchant Marines and as a DJ, he began performing in the Bakersfield, California area using the name Terry Preston and eventually being discovered by Tennessee Ernie Ford's manager, Cliffie Stone, who asked Husky to deputize for Ford during a vacation period. About this time, Husky also created a character called Simon Crum, a kind of hick philosopher who became so popular that Capitol signed the singer to cut several sides as his alter ego.

Later, recording as Terry Preston, Husky had his first hit with **A Dear John Letter**, a duet recorded with Jean Shepard in 1953. Husky eventually obtained a minor hit under his own name with **I Feel Better all Over**. During 1957, he appeared on a Kraft TV theatre show playing a dramatic role. Also, in the same year, he recorded **Gone**, a remake of a Smokey Rogers' song originally cut by Husky in his Terry Preston era, this new version becoming a million-seller. By 1958 it was Crum's turn to become a chartbuster, a comedy song **Country Music Is Here To Stay** hitting the No.2 spot in the country listings. 1958 also saw Husky obtain a film role in

'Country Music Holiday', alongside Zsa Zsa Gabor and Rocky Graziano.

Since 1957 his long list of record hits has included **A Fallen Star**, **Wings Of A Dove** and **The Waltz You Saved For Me**, all cross-over successes, and **Once** and **Just For You**, both country Top 10 items. Father of seven children – the youngest being named Terry Preston in memory of Husky's earlier identity – the singer has made many radio, TV and film appearances and recorded for ABC during the early '70s. He tours with his group the Hush Puppies.

Recommended:
Ferlin Husky (MCA-Dot/–)
True True Lovin' (ABC/–)
Sings The Foster-Rice Songbook (ABC/–)
Country Sounds Of Ferlin Husky (–/Music For Pleasure)
Freckles And Polliwog Days (ABC/–)
Audiograph Live (Audiograph/–)

Alan Jackson

The most creatively consistent honky-tonker of the new breed of country performers, Alan Eugene Jackson was born October 17, 1959 in Newnan, Georgia.

Married by the time he was twenty, Jackson did everything from driving a forklift at K-Mart to waiting tables to fuel his addiction to cars and his love of country music. It was a chance meeting between his wife Denise (an airline stewardess) and

Above: Ferlin Husky's alter ego – Simon Crum, hick philosopher

Glen Campbell, who was waiting for a change of flights at Atlanta, that opened the first doors in Nashville. Campbell gave Jackson's wife a card and introduction to his music publishing company in Nashville. That encounter provided the impetus for Alan to move to Music City, where he

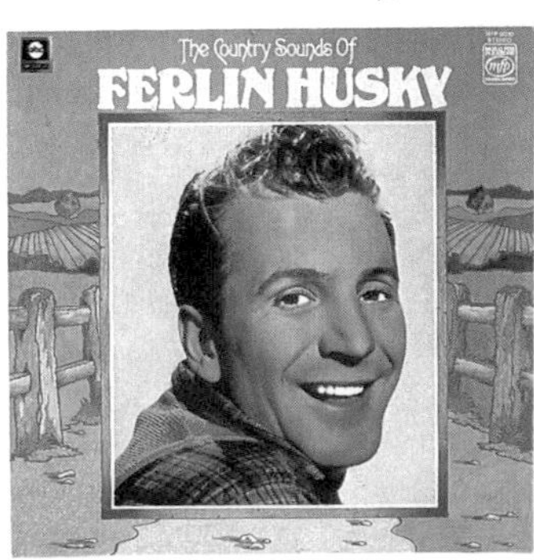

The Country Sounds Of Ferlin Husky. Courtesy MFP Records.

started making the right connections. He spent four years based in Nashville, writing songs and building a fan base by solid roadwork with his band, before the big break finally came.

New York-based rock label Arista Records decided to open a Nashville office and develop a country music roster. At the same time Jackson, along with songwriter/producer Keith Stegall, completed some new demos, and, in 1989, Alan Jackson became the first country artist signed to Arista. His debut single, **Blue Blooded Woman**, made No.45 on the charts at the end of 1989. Jackson's career really took off the following year. **Here In The Real World**, a classic country love ballad, spent six months on the charts, reaching a peak of No.4. Further Top 10 entries came with **Wanted** and **Chasin' That Neon Rainbow**. His debut

Banjo Player, Carl Jackson. Courtesy Capitol Records.

album, **Here In The Real World**, raced up both the country and pop charts, eventually reaching platinum status.

The hits have continued with amazing consistency, making No.1 with **I'd Love You All Over Again**, **Don't Rock The Jukebox** and **Someday** (all 1991), **Dallas**, **Love's Got A Hold On You** and **Midnight In Montgomery** (all 1992). The latter, about a visit to Hank Williams' grave, is probably the best song he's written. His second album, **Don't Rock The Jukebox**, was more successful than his debut, rapidly going double platinum and leading to TV appearances.

A third album, **A Lot About Livin' (And A Little 'Bout Lovin')**, reached platinum status within seven weeks of release. It is now a certified double platinum, and total sales of Jackson's first three albums have exceeded six million in little more than three years. Another chart-topper came with **She's Got The Rhythm (And I Got The Blues)** and a Top 10 entry with **I Climbed The Wall** in 1993. Amazingly, Jackson has failed to pick up any of the major CMA awards. Writing and performing some of the most authentic-sounding country music around, Jackson's image is of a good-looking, straight-singing man who remains modest and unassuming. He has built his career gradually and sensibly, and is assured of being around for a long time, utilizing solid country values of the past, but with a fresh style and sound.

Recommended
Here In The Real World (Arista/Arista)
Don't Rock The Jukebox (Arista/–)
A Lot About Livin' And A Little 'Bout Love (Arista/–)

Carl Jackson

A fast-pickin' banjo and guitar player who was an integral part of the Glen Campbell Show, Jackson was born in Louisville, Kentucky, in 1953. He learned banjo at the age of five and at 13 began playing with a family bluegrass outfit. During his schooldays he toured as part of Jim And Jesse's band, cutting an album, **Bluegrass Festival**, for Prize Records during this period. In 1972, he learned that Larry McNeely, Campbell's banjoist, was leaving the group. McNeely then set up a meeting between Jackson and Campbell, the latter being so impressed by the 19-year-old's playing that he signed him as part of his touring show, producing Jackson's first solo album for Capitol, **Banjo Player** (1973).

Jackson stayed 12 years with Campbell, during this time cutting another album for Capitol and three for Sugar Hill. Then he split, opted for a true solo career, signed with Columbia, and in 1984 gained his first ever hit single with **She's Gone, Gone, Gone**, a Lefty Frizzell standard. He followed this with another mid-chart entry, **Dixie Train**, in 1985. In the late '80s, he teamed up with John Starling And The Nash Ramblers to produce the acclaimed **Spring Training** album for Sugar Hill.

Recommended:
Banjo Player (Capitol/Capitol)
Banjo Man – A Tribute To Earl Scruggs (Sugar Hill/–)
Banjo Hits (Sugar Hill/–)
Songs Of The South (Sugar Hill/-)

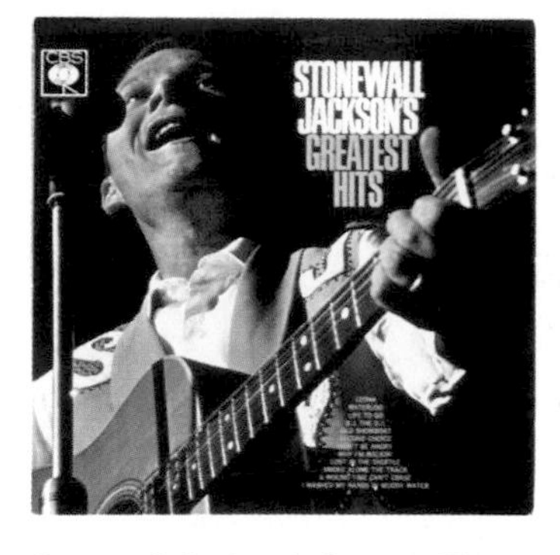

Stonewall Jackson's Greatest Hits. Courtesy CBS Records.

Stonewall Jackson

This is his real name; he was named after the Confederate general. Born in Tabor City, North Carolina, on November 6, 1932, Stonewall had an impoverished childhood and obtained his first guitar at the age of ten by trading an old bike. He figured out the chords by watching others and would listen to the radio, using what he heard as the basis for constructing his own songs.

By 1956, he had saved enough money to go to Nashville. Wesley Rose, head of Acuff Rose Publishing, heard him and signed him to a long-term contract. He was successful across the country via TV.

Up Against The Wall, Stonewall Jackson. Courtesy PRT Records.

Capitol Country Classics, Wanda Jackson. Courtesy Capitol Records.

In 1958 he had a big country hit with **Life To Go**. 1959 saw his monster cross-over smash **Waterloo**, an international pop hit in which a neat country backing was combined with novelty lyrics drawing military analogies to a love affair. As a result of the hit, he starred on Dick Clark's American Bandstand. Other Jackson-composed standards are **Don't Be Angry, Mary Don't You Weep** and **I Washed My Hands In Muddy Water**. In 1967 he came back strongly with **Stamp Out Loneliness** and also had a successful album based around the title.

Recommended:
At The Opry (Columbia/–)
Greatest Hits (Columbia/–)
American Originals (Columbia/–)

Wanda Jackson

Child prodigy Wanda Jackson was born Maud, in Oklahoma, on October 20, 1937, the daughter of a piano-playing barber. By the age of ten she could play both guitar and piano, three years later obtaining her own radio show. By 1954 she was cutting discs for Decca – charting with a Billy Gray-aided duet **You Can't Have My Love** – and she began touring with Hank Thompson's band. In 1955–56, Wanda toured with Elvis Presley, then became Capitol Records' leading female rocker, scoring heavily in the 1960 pop charts with **Let's Have A Party**. However, 1961 saw her return to more country-oriented fare – and her hits such as **Right Or Wrong** and **In The Middle Of A Heartache** went into both pop and country charts.

Throughout the '60s, Wanda racked up over a score of hits – even having a major success in Japan with the rocking **Fujiyama Mama** – and though her chart-

Make Me Like A Child, Wanda Jackson. Courtesy Myrrh Records.

busting continued into the '70s, in 1971 she asked for her release from Capitol. She switched to pure gospel music, cutting sides for the religious Word and Myrrh labels. Even so, Wanda can still easily be persuaded to revive her old rockabilly and country hits onstage.

Recommended:
The Best Of (Capitol/Capitol)
Capitol Country Classics (–/Capitol)
Early (–/Bear Family)

Sonny James

Sonny James is still thought of by many in terms of his pace-setting 1950s teen hit **Young Love** and there is no denying that his music has often been easily accessible to the MOR market.

Born in Hackleburg, Alabama, on May 1, 1929 (real name Jimmy Loden), into a showbiz family, he made his stage debut at the age of four, touring with his sisters after making his radio debut. He learned to play violin at seven, later becoming signed to a full-time contract with a Birmingham, Alabama, radio station. Following 15 months in Korea he returned home and pacted with Capitol Records, obtaining his first hit with **For Rent** in 1956, that same year recording **Young Love**, an eventual million-seller. After one more Top 10 entry, with the pop-slanted **First Date, First Kiss, First Love** (1957), he saw little chart action until 1963 when he scored with **The Minute You're Gone**. The song was successfully covered in Britain by Cliff Richard, which proves how suited James was to pop-oriented material.

During 1964, **Baltimore, Ask Marie** and **You're The Only World I Know** all charted for James, the last commencing an incredible run of Top 5 singles (most of them reaching the No.1 spot!) that extended well into the mid-'70s. After supplying one final No.1 hit, **That's Why I Love You Like I Do**, for Capitol in mid-1972, James switched to the rival Columbia label, claiming an immediate chart-topper with **When The Snow Is On The Roses**. He turned producer to fashion a hit record, **Paper Roses**, for Marie Osmond in 1973, and continued with his own flow of vinyl winners for a while, going to No.1 yet again with **Is It Wrong (For Loving You)** in 1974 and adding to his tally of Top 10 entries with **A Mi Esposa Con Amor** (1974), **A Little Bit South Of Saskatoon**, **Little Band Of Gold**, **What In The World's Come Over You** (1975), **When Something's Wrong With My Baby**, **Come On In** (1976), and **You're Free To Go** (1977). During this period he also made a couple of out-of-the-rut albums in **200 Years Of Country Music** (1976) and **In Prison, In Person** (1977), the latter produced by Tammy Wynette's husband George Richey inside Tennessee State Prison.

Since then, the career of the man known as the 'Southern Gentleman' has slowed right down, but he has been a remarkably consistent performer. He is capable of producing great licks on a number of instruments, and has even nudged his way into a number of minor films along the way, including 'Hillbilly In A Haunted House', a movie that also featured Lon Chaney and Basil Rathbone. In fact, the only thing he has missed out on is a CMA Award – which seems something of an oversight.

Recommended:
The Best Of (Capitol/Capitol)
Country Artist Of The Decade (Columbia/–)
200 Years Of Country Music (Columbia/CBS)
In Prison, In Person (Columbia/–)
American Originals (Columbia/–)

In Prison, In Person, Sonny James. Courtesy CBS Records.

Waylon Jennings

A strong voice and a strong personality have enabled this charismatic man to aspire to country music's heights. From a modestly successful career as a mainstream country and folk-country artist, he became the definitive 'outlaw' figure, a man who, with Willie Nelson, spear-headed the movement away from orchestral blandness in country towards exciting, gritty, more personalized music.

Born in Littlefield, Texas, on June 15, 1937, the son of a truck-driver, Waylon could play guitar by this teens. He gained a DJ job on a Littlefield radio station at 12 and, although interested in pop in his teens, he had developed an interest in country by 21.

In 1958 he moved to Lubbock, working as a DJ there, and meeting Buddy Holly. In 1958–59 he toured as Holly's bassist. When Holly's plane crashed in 1959, killing the singer and two others, it was Waylon who, at the last moment, had given up his seat to J. P. Richardson, the 'Big Bopper' of **Chantilly Lace** fame.

In the early '60s, Waylon settled in Phoenix, Arizona, forming the Waylors to back him and becoming locally known at Phoenix's famous JD's club. Chet Atkins signed him to RCA in 1965 and the following year he moved to Nashville. He appeared on the Grand Ole Opry, made several TV appearances and also starred in the film 'Nashville Rebel'. But Waylon was to become more than just a celluloid rebel. Nashville was tightly business-minded, and artists had to work with certain staff producers. The label's own Nashville studios had to be used and so did an elite band of session musicians. Artists were not encouraged to record with their own bands and, consequently, much of the Nashville product sounded similar. Waylon, initially, conformed to this, chalking up an impressive list of Top 10 hits including **(That's What You Get) For Lovin' Me** (1966), **The Chokin' Kind** (1967), **Only Daddy That'll Walk The Line** (1968), **Yours Love** (1969), and **Brown Eyed Handsome Man** and **The Taker** (both 1970).

Waylon wanted to break out from this tight regime in order to direct his own material, musicians and production. He upset the RCA Nashville hierarchy by negotiating directly with the New York bosses.. He was guaranteed an independent production package in which he would provide RCA with a number of sides each year for them to promote and sell. The big musical change for Waylon had already become apparent on the **Ladies Love Outlaws** album, where he at last succeeded in blending his own band, the Waylors, in with the regular session musicians, and picking some distinctive and evocative current song material. The Top 10 entries were now much more consistent, starting with **Good Hearted Woman** (1972) and continuing with **Sweet Dream Woman** and **Pretend I Never Happened** (also 1972), as well as **You Can Have Her** and **You Ask Me To** (both 1973).

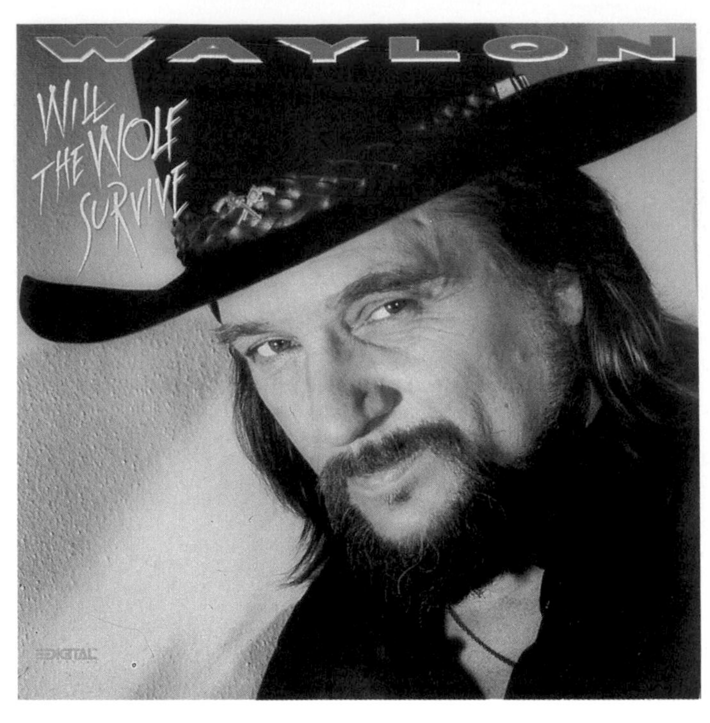

Will The Wolf Survive, Waylon Jennings. Courtesy MCA Records.

Honky Tonk Heroes from 1973 proved an even bigger watershed. Waylon extensively plundered the repertoire of Billy Joe Shaver to come up with an album variously produced by himself, Tompall Glaser, Ronnie Light and Ken Mansfield, and featuring music sounding as hard and 'outlaw' as Waylon's reputation. His first No.1 came with **This Time** (1974), followed by **I'm A Ramblin' Man** (also 1974). He reached Top 10 with **Rainy Day Woman** and **Dreamin' My Dreams With You** in 1975. He made the pop charts with a double-sided single, **Are You Sure Hank Done It This Way/Bob Wills Is Still The King** in 1975. That same year he teamed up with Willie Nelson on a live version of **Good Hearted Woman**, which also hit No.1 in the country charts and crossed over to pop.

In 1976 he scored with **Suspicious Minds**, an evocative duet with his wife Jessi Colter. The previous year had seen him making an inroad into the CMA Awards by winning Male Vocalist Of The Year. But in 1976 **Suspicious Minds** involved Waylon in two awards, Duo Of The Year and Single Of The Year. He was also involved in Album Of The Year (**The Outlaws**), and it was evident that the new contemporary strain of country had finally gained official acceptance.

It was a period during which Jennings could do little wrong. In 1977, he had two No.1 singles in **Luckenbach, Texas (Back To The Basics Of Love)** and **The Wurlitzer Prize.** In 1978 he came up with another chart-topping solo single, **I've Always Been Crazy**, and gained yet another by teaming with Willie Nelson for **Mammas, Don't Let Your Babies Grow Up To Be Cowboys**. Additionally the albums of identical titles also went to No.1 in their division, while a duet single with Johnny Cash, **There Ain't No Good Chain Gang** only just missed the top spot.

By 1979, grabbing No.1 singles had become almost routine. **Amanda** and **Come With Me** raised the total further, as did **I Ain't Living Long Like This** and **Good Ol' Boys**, Waylon's self-penned theme to the TV series The Dukes Of Hazzard in 1980. A **Greatest Hits** compilation also proved a popular seller. When the sales were added up they amounted to three million. There were no No.1s in 1981, although **Shine** climbed into the Top 5 and a couple of duets with Jessi Colter also sold well. Nevertheless, with the help of duet partner Willie Nelson, he returned to his chart-topping ways with **Just To Satisfy You** during 1982, following this with a solo effort, **Lucille (You Won't Do Your Daddy's Will)** in 1983.

Waylon provided further No.1 hits for RCA with **Never Could Toe The Mark** and **America** (both 1984), **Waltz Me To Heaven** and the superb **Drinkin' And Dreamin'** (1985), which came from the album called **Turn The Page**. The album also saw him quitting his label of 20 years. His next album, **Will The Wolf Survive,** came out on MCA and the hits continued with the title song and **Working Without A Net** (both 1986), the latter a reference to him kicking drugs. By this time he had formed a liaison with Willie Nelson, Johnny Cash and Kris Kristofferson, resulting in the best-selling album **Highwayman** in 1985, plus a No.1 single of that same title. Another chart-topper came with his fourth MCA single, **Rose In Paradise** (1987). He enjoyed further Top 10 entries with **Fallin' Out** and **My Rough And Rowdy Days** (both 1987).

In 1990 Waylon signed with Epic Records, scoring a Top 10 hit with **Wrong** (1990). Since then major chart success has eluded him, but he has continued to speak out through his music, as he showed with the album, **Too Dumb For New York**

City, Too Ugly For L.A. (1992). The title song is a tongue-in-cheek, self-penned ditty that pokes fun at the way those in high places think of country people and, more specifically, country music. Throughout his long career Jennings has always been patently country, but his use of a heavier instrumentation and his rock-star approach has tended to mislead people.

Recommended:
Ladies Love Outlaws (RCA/RCA)
Honky Tonk Heroes (RCA/RCA)
Leather And Lace – with Jessi Colter (RCA/RCA)
Waylon And Willie (RCA/RCA)
I've Always Been Crazy (RCA/RCA)
Wanted! The Outlaws – with Jessi Colter, Willie Nelson and Tompall Glaser (RCA/RCA)
Full Circle (MCA/–)
A Man Called Hoss (MCA/MCA)
Will The Wolf Survive (MCA/MCA)
The Eagle (Epic/Epic)
Files (–/Bear Family)
Too Dumb For New York City, Too Ugly For LA (Epic/–)

Jim And Jesse

Bluegrass-playing brothers Jim and Jesse McReynolds were both born in Coeburn, Virginia, Jim on February 13, 1927, Jesse on July 9, 1929. From a musical family – their grandfather was an old-time fiddler who recorded with Victor – the duo began playing at local get-togethers.

With Jim on guitar and Jesse on mandolin, they made their radio debut in 1947 and cut some records for the Kentucky label during the early '50s, later signing for Capitol. However, the duo's progress was terminated for a while during the Korean War when Jesse was called up for service. They reformed on Knoxville's WNOX Tennessee Barn Dance in 1954.

During the '60s, Jim and Jesse, with their band the Virginia Boys (which included such musicians as Bobby Thompson and Vassar Clements) signed for Epic Records. They began notching up a number of fair-sized chart entries with **Cotton Mill Man** (1964), **Diesel On My Tail, Ballad Of Thunder Road** (both 1967), **The Golden Rocket** (1968) and other titles before switching to Capitol for a 1971 success in **Freight Train**.

Regular on the Opry since 1964, Jim And Jesse are a fine, no-frills, bluegrass outfit, specializing in smooth, haunting, sky-high harmony vocals. They have played the Newport Folk Festival, have appeared at Britain's Wembley Festival many times and, in addition to their own TV show, have played on scores of TV shows throughout Europe and America. They have recorded well over 40 albums and in 1982 even made a return to the singles charts, having a mild hit with **North Wind**, a single that saw them teaming with Charlie Louvin.

Recommended:
We Like Trains/Diesel On My Tail (Epic/–)
The Jim And Jesse Show (Old Dominion/DJM)
All-Time Greatest Country Instrumentals (Columbia/–)
The Epic Bluegrass Years (Rounder/–)

Michael Johnson

A gifted songwriter, singer and acoustic guitarist, Michael Johnson was born in Alamosa, Colorado, on August 8, 1944. He first became involved in music while in his early teens in Denver and over the years he has covered the musical spectrum from rock'n'roll to the classics. Though he made his first recordings in Nashville in the early '70s, these were geared towards the pop market. It was not until 1985, when he teamed up with Sylvia for the duet country hit, **I Love You By Heart**, that he made a commitment to country that resulted in a highly successful chart run in the late '80s.

While attending Colorado State College in 1964, Johnson won a national amateur talent contest which earned him a recording contract with Epic Records. He took to the road as a folk singer touring the college and club circuit. Two years later he travelled to Spain where he spent a year in Barcelona studying classical guitar under Graciano Tarrego. On his return to Colorado he joined David Boise and John Denver in the 'new' Chad Mitchell Trio. This was a short-lived association, as Denver embarked on a solo career and Johnson became an actor in a touring production of 'Jacques Brel Is Alive And Well And Living In Paris'.

The Jim And Jesse Show. Courtesy Old Dominion/DJM Records. Jim And Jesse provided classic bluegrass.

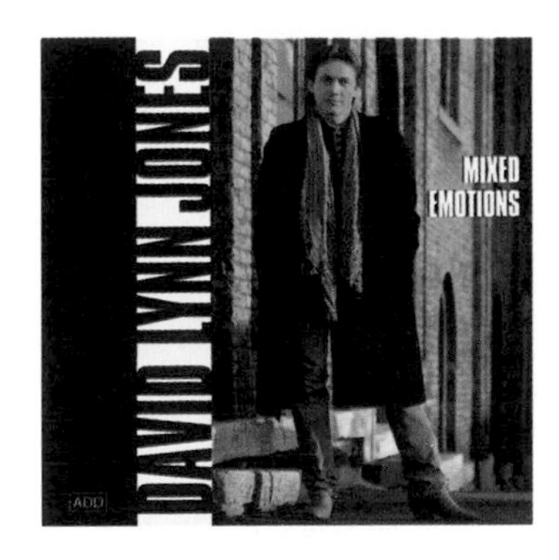

Mixed Emotions, David Lynn Jones. Courtesy Liberty Records.

In 1971 he recorded his first solo album, **There Is A Breeze**, for Atlantic. He formed his own Sanskrit label and continued working as a folk singer, mainly recording and performing his own self-penned songs. Two of his songs attracted the attention of Gene Cotton, resulting in a move to Nashville, where he worked sessions and teamed up with Brent Maher. This led to a contract with EMI-America, for whom he scored significant pop chart success with **Bluer Than Blue** and **Almost Like Being In Love** in 1978. Dropped by EMI in 1981, Johnson started all over again, writing songs and making demos. In 1984, his old friend Brent Maher (by this time producing the Judds and Sylvia), signed him to RCA, resulting in his duet with Sylvia on **I Love You By Heart** (1985), and solo success with **Give Me Wings** (No.1, 1986), **The Moon Is Still Over My Shoulder** (No.1, 1987), **Crying Shame** (1987), **I Will Whisper Your Name** and **That's That** (both 1988).

By 1990 Johnson's record sales had dropped off. He also had a lengthy period of ill-health and was subsequently dropped by RCA. By 1992 he had gained a new contract with Atlantic Records and was beginning to build a new career with his delicate, folk-country style of music.

Recommended:
That's That (RCA/RCA)
Life's A Bitch (RCA/RCA)
Michael Johnson (Atlantic/–)

David Lynn Jones

A performer who leans more towards the rock and blues side of country music, David Lynn Jones, from Bexar, Arkansas, is a singer-songwriter who has never slotted into the Nashville mainstream. He has been performing since the early '70s, where his uncompromising style of country with its Springsteen and Dylan overtones won him a cult following. He made his initial impact in Nashville as a writer, but his reputation as a dynamic performer led to a recording contract with Mercury in 1986 and a Top 10 entry with **Bonnie Jean (Little Sister)** the following year.

His debut album, **Hard Times On Easy Street**, received rave reviews, grabbing him some attention from outside of country music. Waylon Jennings added harmonies to **High Ridin' Heroes** (Top 20, 1988). Becoming disillusioned with Nashville, Jones maintained a base in Bexar where he had built his own studio and continued working with musicians who had played with him since the beginning. A second album for Mercury failed to gain any commercial acceptance, so he returned to his musical beginnings of playing honky-tonks and clubs. Another shot at the big time was offered when Jimmy Bowen signed him to Liberty Records in 1992. This time the performer was determined to do it his way, and his third album, **Mixed Emotions**, was recorded in his own studios. The project was a joint production between the singer and Richie Albright, who had produced some of Waylon Jennings' edgier work. For once Jones' highly charged blend of country, rock, gospel and soul had finally been allowed to fly free. From spicy rock'n'roll to blues-tingling numbers, this was a roots-oriented collage of every emotion that encapsulates the essence of country.

Recommended:
Hard Times On Easy Street (Mercury/Mercury)

George Jones

Known as the 'Rolls Royce Of Country Singers', George Glenn Jones' nasal, blues-filled, vocal styling has influenced a host of country performers.

Born in Saratoga, Texas, on September 12, 1931, Jones grew up against a musical background. His mother played piano at the local church and his father was an amateur guitarist. He got his first guitar at the age of nine and was soon performing at local events. In his late teens he served with the Marines in Korea, and upon discharge began working as a house painter while also playing evening gigs. By 1954 he had gained a sufficiently good reputation to attract the interest of industry executive H.W. 'Pappy' Dailey at Starday Records in Houston.

Jones saw his first big country hit, **Why Baby Why**, in 1955. He stayed with Starday until 1957, cutting both country and rockabilly sides, before moving on to Mercury. His first No.1 for his new label was **White Lightning**, an uptempo song with a novelty chorus, in 1959.

After 18 hits including another No.1, **Tender Years** (1961), he landed at UA Records, where he hit a fertile period, turning out songs that have become country standards. He proved particularly strong with the anguished, two-timing women type of song, singing in a manner that suggested he had lived the lyric.

So he hit big with **Window Up Above** (1961), **She Thinks I Still Care**, an archetypal Jones song and one covered by countless other artists (1962), **We Must Have Been Out Of Our Minds**, performed with his travelling co-star Melba Montgomery (1963), and the classic **Race Is On** (1964).

In 1965 he renewed acquaintance with Pappy Dailey. Dailey had left Starday to form Musicor, and Jones was an obvious target for him. This proved a successful liaison and Jones' tally of hits continued to grow via such singles as **Things Have Gone To Pieces**, **Love Bug**, **Take Me** (1965), **I'm A People, 4033** (1966), **Walk Through This World With Me**, **I Can't Get There From Here, If My Heart Had**

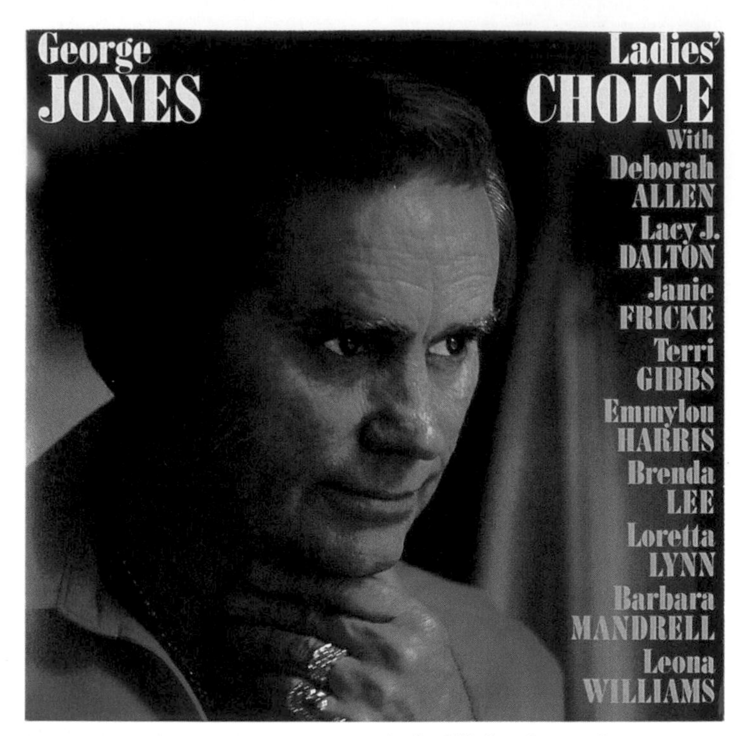

Windows and **Say It's Not You** (1967). But the singer became increasingly unhappy with the way he was being recorded and began a fight to shake off the Musicor contract. At this period of his life he went through a rough patch, having become divorced from his wife. He began drinking heavily as he undertook endless overseas tours. He seemed to be increasingly living out his honky-tonk songs and became associated with stories of wild and destructive living, sometimes having to be helped onstage. During 1967 he met Tammy Wynette, when the two played the same package tour. Tammy was having problems with her marriage and she and George became romantically entwined. When, in the wake of his 1968 hits, **Say It's Not You**, **As Long As I Live** and **When The Grass Grows Over Me,** he released **I'll Share My World With You** (1969), the fans knew it was Tammy he was singing about. The record went to

Below: George Jones shows up for a rare photo call.

Ladies' Choice, George Jones. Courtesy Epic Records.

No.2 – kept out of the No.1 spot by Tammy's **Stand By Your Man**. That year, both George and Tammy joined the Opry. They also got married, being hailed as 'Mr And Mrs Country Music'.

Though Jones continued having hits with Musicor, he still wished to quit the label and eventually did so in 1971, joining Tammy and producer Billy Sherrill on Epic. Tammy was a huge star during this period (she had five chart-toppers during 1972–73) and though George had his share of hits, it was only when the twosome worked together that he really hit the heights. The duo's **We're Gonna Hold On** (1973) provided him with a half-share in his first No.1 since 1967. In 1974, things began to swing around. Tammy has only one hit, while George had two No.1s with **The Grand Tour** and **The Door**. Also that year, the stars swung apart, Tammy filing for a legal separation. And in 1976 she remarried – in the same year that her and George's duets, **Golden Ring** and **Near You,** went top of the charts.

Though Jones was named Country Singer Of The Year by 'Rolling Stone' in 1976, the hits stopped flowing as regularly as they once did. He had one solo Top 10 record in 1978 with **Bartender's Blues** and duetted to good effect with Johnny Paycheck on **Mabellene**.

During 1980, he released three chart albums (one with Tammy, another with Johnny Paycheck) and five Top 20s. One, **He Stopped Loving Her Today**, reached No.1 and helped George win a Grammy Award for the Best Male Country Vocal Performance. Also in 1980 he was adjudged Male Vocalist Of The Year by the CMA, a title he claimed again in 1981.

Jones continued to create his own problems. He frequently failed to turn up at events and sometimes appeared to be living through a permanent nightmare. Nevertheless, he kept on supplying hits, such as: **If Drinkin' Don't Kill Me (Her Memory Will)**; **Still Doin' Time** (1981); **Same Ole Me** (1982); **Yesterday's Wine** and **C.C. Waterback** – both with Merle Haggard (1982); **Shine On**; **I Always Get Lucky With You**; **Tennessee Whiskey** (1983); **We Didn't See a Thing** – with Ray Charles (1983); **You've Still Got A Place In My Heart**; **She's My Rock** (1984); **Who's Gonna Fill Her Shoes** (1985); **The One I Loved Back Then** (1986) and **The Right Left Hand** (1987).

By the mid-'80s Jones had settled down, with his career being guided by divorcee Nancy Sepulveda, whom he married in 1983. Idolized by many of the young New Traditionalists, he had settled into a comfortable career as a legendary country veteran, but still maintained that high quality in his recordings. He teamed up with beautiful newcomer Shelby Lynne, the pair scoring a duet hit with **If I Could Bottle This Up** (1988). He made the Top 10 with his own distinctive revival of Johnny Horton's **I'm A One-Woman Man** (1989) and duetted with Randy Travis on **A Few Ole Country Boys** (1990). In 1991 Jones joined MCA Records, gaining minor hits with **You Couldn't Get The Picture**, **I Don't Need Your Rockin' Chair** and **Wrong's What I Do Best**.

George Jones is more than a classic country singer. He is *the* classic country singer – a product of honky-tonks and heartaches, hard living and hard loving. He finally received the recognition he deserved in 1992 when he was inducted into the Country Music Hall Of Fame.

Recommended:
White Lightning (–/Ace)
The King Of Country Music (–/Liberty)
My Very Special Guests – with various guests (Epic/Epic)
Burn The Honky Tonk Down (Rounder/–)
Alone Again (Epic/Epic)
I Am What I Am (Epic/Epic)
George Jones And Tammy Wynette – Greatest Hits (Epic/Epic)
George Jones Meets Hank Williams and Bob Wills (–/EMI)
Too Wild Too Long (Epic/Epic)
Don't Stop The Music (–/Ace)
You Ought To Be Here With Me (Epic/Epic)
Walls Can Fall (MCA/–)
A Good Year For The Roses (–/Castle)
Who's Gonna Fill Their Shoes (Epic/–)
Friends In High Places (Epic/–)

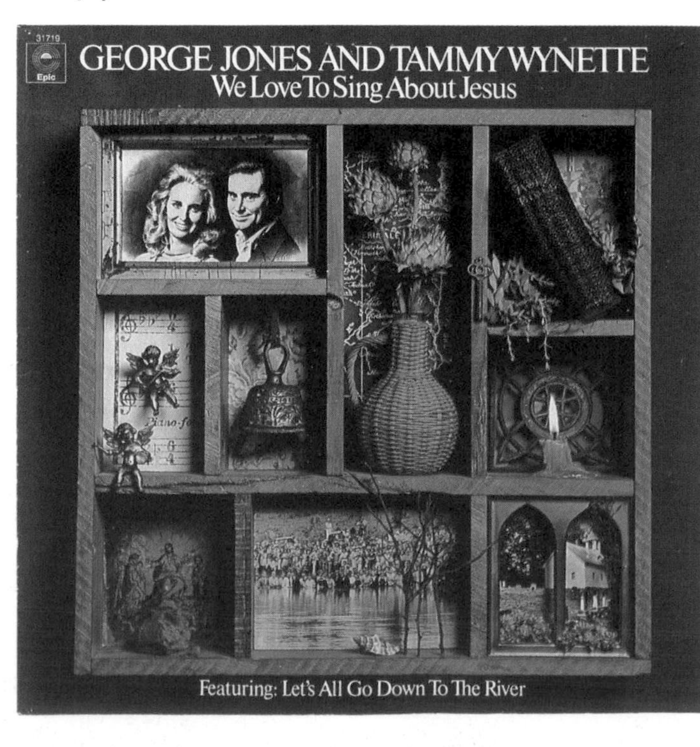

Epic
GEORGE JONES AND TAMMY WYNETTE
We Love To Sing About Jesus
Featuring: Let's All Go Down To The River

We Love To Sing About Jesus, George Jones and Tammy Wynette. Courtesy Epic Records.

Grandpa Jones

A long-time regular on both the Opry and Hee Haw, high-kicking, joke-cracking, story-telling, foot-stomping vaudevillian Grandpa Jones is one of the most colourful figures in country music.

Born Louis Marshall Jones, in Nigeria, Kentucky, on October 20, 1913, he began playing guitar on an instrument costing only 75 cents. At 16 he had become so proficient that he won a talent contest promoted by Wendell Hall, while 1935 found him with Bradley Kincaid's band.

Jones began disguising himself as an old-timer while only in his twenties. By 1937 he was leading an outfit known as Grandpa Jones And His Grandchildren, this unit becoming regulars on WWVA's Wheeling Jamboree, and then Cincinnati WLW, during the late '30s and early '40s. It was here that he began recording for King Records, by himself, with Merle Travis, and with the Delmore Brothers as Brown's Ferry Four. In 1944 Jones joined the US Army and was posted to Germany, where he played on AFN Radio until his discharge in 1946. Almost immediately he became a member of the Opry, remaining so for many years. Elected to the Country Music Hall Of Fame in 1984, he published an autobiography 'Grandpa: 50 Years Behind The Mike'.

Recommended:
The Grandpa Jones Story (CMH/–)
Everybody's Grandpa – Hits From Hee Haw (Monument/–)
20 Of The Best (–/RCA)

Jordanaires

A vocal group that has appeared on hundreds of Nashville recordings, the Jordanaires were formed in Springfield, Missouri during 1948. Initially a male barbershop quartet, performing mainly gospel material, they gained their first Opry appearance in 1949.

A year later they were featured on Red Foley's million-selling version of **Just A Closer Walk With Thee**, and in 1956

sprang to even wider fame by commencing a long and successful association with Elvis Presley. Grammy award winners in 1965 (for best religious album), the Jordanaires have appeared on many TV and radio shows, also lending their talents to an impressive number of movie scores.

The Man From Kentucky, Grandpa Jones. Courtesy Bulldog Records.

Recommended:
Sing Elvis' Gospel Favorites (–/Magnum Force)

Sarah Jory

The 'Queen of British Country Music', Sarah Elizabeth Jory was born on November 20, 1969 in Exeter, Devon, England. A multi-instrumentalist, Sarah can play banjo, mandolin, acoustic, rhythm and steel guitar and keyboards. Her ability to hold a tune is natural. It all started with a toy drum, then a small guitar, and most amazingly a steel guitar with a local band, Colorado Country. Within three years she was travelling all over southern England, playing country clubs, social evenings, fêtes and barbecues.

By this time Sarah had started making an impression in America. At only eleven she appeared at the annual Steel Guitar Convention in St Louis, Missouri. Recognized steel guitarists like Lloyd Green, Weldon Myrick and Pete Drake accepted this young girl as their equal. She returned to the States a dozen times, on a number of occasions as a guest at the Steel Guitar Convention and, at other times, to Nashville where she sat in on sessions, jammed at nightspots and guested on television, including Ralph Emery's Nashville Now. At 15 Sarah had become possibly Britain's finest steel guitarist. Sarah brings a feminine touch to her work that has made her much more than just another steel guitarist. There is a passion in her playing that, added to a considerable technique, makes her country themes an especially moving experience.

A precocious child prodigy, Sarah began developing an interest in expanding her talents to include singing and entertaining. At 18 Sarah formed her own band and started making inroads into Europe. In 1989 she gained a standing ovation at Holland's famed Floralia Festival, and less than two years later was the invited guest of Eric Clapton, when the 'king of blues-rock' appeared at The Point, Dublin. Since then she has toured the UK theatre circuit as an opening act for Glen Campbell and Charley Pride, played all the major festivals throughout Britain and Europe and undertaken her own headlining concert tours.

Sarah is a remarkably mature and sophisticated entertainer. With a powerful and versatile voice, and instrumental skills that would leave most musicians in awe, this lady really has little to worry about. Since the early '80s Sarah has recorded regularly, mainly with steel guitar instrumental albums, then later adding vocals, In 1992 she signed with the London-based Irish Ritz Records, and her first album for them, **New Horizons**, was named Country Album Of the Year by the British Country Music Association. On the threshold of a highly successful international career, Sarah Jory, a pure, natural and complete musician, a brilliant singer and all-round entertainer, was named European Country Music Artist Of The Year in April 1993 for the second consecutive year.

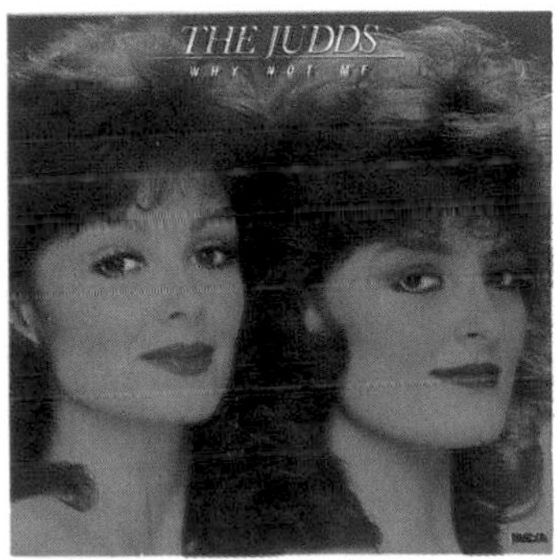

Why Not Me?, The Judds. Courtesy RCA Records.

Recommended:
New Horizons (–/Ritz)

The Judds

This mother (Naomi, born Diana Ellen Judd on January 11, 1946, Ashland, Kentucky) and daughter (Wynonna, born Christina Criminella on May 30, 1964, Ashland, Kentucky) vocal duo managed to go 'New Country' while retaining links with traditional roots. Naomi moved her one-parent family to Hollywood in 1968. During the next seven years Wynonna and her younger sister, Ashley, attended public school, while Naomi worked at various jobs, among them a model, a secretary for the pop group the Fifth Dimension, and Girl Friday to an Oriental millionaire.

In 1976 the family moved back to Morrill, Kentucky, close to their hometown, where Naomi pursued a career in nursing. Because they had no TV, they began singing duets at home and realized they

Left: The Jordanaires team up with their most famous client – Elvis.

were good enough to embark on a musical career. They began singing at local functions, bought a 30-dollar tape deck and began making demo tapes, all the time honing their harmonies to perfection. Following a move back to California, Naomi completed her nursing training in San Francisco, then decided to push for a career in music. She and her daughters headed for Nashville in 1979, where Wynonna finished high school, winning a talent contest in the 10th grade

As a duo they worked on WSM's Ralph Emery Show, but their big break came when Naomi met producer Brent Maher while nursing his daughter, taking the opportunity to pass on one of her demo tapes. Eventually with the help of Maher, Woody Bowles (a veteran Nashville manager and publicist) and manager Ken Stilts, they landed an RCA recording contract after an unprecedented live audition before the company's executives.

Immediate Top 20 chart-jumpers with **Had A Dream (For The Heart)**, a late 1983 release, the twosome's career really took off in 1984 when they gained two No.1s with **Mama He's Crazy** and **Why Not Me**, also netting a Grammy for Best Country Performance for the former title and Single Of The Year for the latter at the CMA Awards, a ceremony that also saw them win Best Vocal Group. Since that time, the Judds have logged further massive hits with **Love Is Alive, Girls Night Out, Have Mercy** (1985), **Grandpa (Tell Me About The Good Old Days)** and **Rockin' With The Rhythm Of The Rain** (1986).

The most glamorous duo in country music, Naomi and Wynonna won the CMA Duo Of The Year for six consecutive years and helped bring a new youth audience to country music. All of their albums have attained multi-platinum status, and they achieved further single hits with: **Cry Myself To Sleep**, **I Know Where I'm Going** and **Maybe Your Baby's Got The Blues** (all No.1s in 1987); **Turn It Loose**, **Give A Little Love** and **Change Of Heart** (1988); **Young Love, Let Me Tell You About Love** (1990); **Born To Be Blue** and **Love Can Build A Bridge** (1990). In 1988 the pair became the first female country act to form their own booking agency (Pro-Tours), but there was speculation that Wynonna (definitely the lead singer) would embark on a solo career. In the end, unforseen circumstances forced a decision. On October 17, 1990 it was announced that Naomi Judd was to retire at the end of the duo's 'Love Can Build A Bridge' 1990–91 concert tour. Having been diagnosed with hepatitis ten months earlier, the 44-year-old entertainer admitted her health condition had prompted her retirement.

The tour became one of the most successful in country music history. The show was simultaneously televised across America and set new records as the most watched musical event in pay-per-view cable history. The concert's gross revenues were estimated in excess of 4.4 million dollars. A tearful tribute to Naomi and Wynonna's career, the Judds' farewell concert can be considered one of the most memorable country shows of all time.

In 1992, Wynonna, having dropped the surname Judd, set out on a solo career, which is likely to see her emerge as one of country music's most successful ground-breaking female entertainers of the '90s.

Recommended:
The Judds (RCA mini-LP/–)
Why Not Me? (RCA/RCA)
Rockin' With The Rhythm (RCA/RCA)
Give A Little Love (–/RCA)
Love Can Build A Bridge (RCA/RCA)
River Of Time (RCA/RCA)

Karl And Harty

Karl Victor Davis (born December 17, 1905) and Hartford Connecticut Taylor (born April 11, 1905, died October, 1963), both of Mt Vernon, Kentucky, composed one of the earliest and most influential of the mandolin-guitar duets. They were brought to the WLS National Barn Dance by John Lair in 1930 as members of the Cumberland Ridge Runners and remained on the show for some 20 years.

Their records for the ARC complex of labels (and, later, Capitol) were popular in their era, especially **I'm Just Here To Get My Baby Out Of Jail** (1934), **The Prisoner's Dream** (1936) and **Kentucky** (1938), all written by Karl. They both left the recording and performing field in the '50s, Karl continuing to work at WLS as a record turner for many years.

Best known as a songwriter (all the above have been recorded and have been hits by several groups at different times), Davis was also able to write a hit song as late as the '60s – Hank Locklin's **Country Music Hall Of Fame** (1967) being one of Karl's compositions.

Wayne Kemp

Born in Greenwood, Arkansas on June 1, 1941, honky-tonk-style vocalist Kemp, the son of a motor mechanic, naturally enough became interested in automobile racing in his early teens.

He grew up in Oklahoma and, after forming his own band and touring throughout the Southwest, he met Buddy Killen, who signed him to a writer's contract with Free Music and a recording deal with the Dial label in 1963. He's made the biggest impact as a writer, with George Jones taking **Love Bug** into the Top 10 in 1965. Conway Twitty was next to pick up on Wayne's songs, recording **The Image Of Me**, **Next In Line**, **Darling, You Know I Wouldn't Lie** and **That's When She Started To Stop Loving You**, all No.1s between 1968 and 1970.

The association with Twitty enabled Kemp to gain a contract with Decca Records in 1969, for whom he supplied several minor hits, leading up to **Honky Tonk Wine**, which went Top 20 in 1973. He has consistently charted the lower reaches for MCA, UA, Mercury and Door Knob, with his **I'll Leave This World Loving You** hit of 1980 being revived by Ricky Van Shelton in 1988.

Recommended:
Wayne Kemp (MCA/–)
Kentucky Sunshine (MCA/–)

The Kendalls

A father and daughter team, Royce Kendall was born in St Louis, Missouri, on 25 September, 1934, while Jeannie hails from the same town, born on 30 November, 1954. Royce began guitar-picking at five and by the age of eight was on radio with his brother Floyce, the twosome forming an act called the Austin Brothers and touring. After serving in the US Army, Royce worked in various jobs, eventually settling in St Louis.

When the Kendall's only child, Jeannie, began singing duets with her father for fun, Royce realized that he and his daughter had an outstanding harmony sound. In 1969 Royce and Jeannie took a trip to Nashville, cutting a demo that caught the ear of Peter Drake, who signed the couple to his Stop label. The Kendalls' debut single **Leavin' On A Jet Plane**, climbed to 52 in the country charts in 1970. Encouraged, Royce moved the family to Nashville, where the Kendalls began recording for Dot. During this three-year period they made numerous appearances on TV shows and played the Opry, while Jeannie notched a Beatle connection, singing harmony on Ringo Starr's **Beaucoup Of Blues** album.

After Dot, they signed to UA for a year, but still could not make the desired breakthrough, at which point the duo took a six-month hiatus from recording to reassess. In 1977 they returned to the studio once more, this time working for the Ovation label. Their first release, a version of an old country standard **Making Believe**, nudged into the lower regions of the charts. The follow-up **Live And Let Live** was well received but looked no chart-buster until some country stations reported interest in the single's B-side and Ovation flipped the release and began plugging **Heaven's Just A Sin Away**, which soared to No.1. At the next CMA Awards ceremony in 1978 the Kendalls romped away with the Single Of The Year plaudit after being nominated in three categories. They were also judged Best Country Group at the Grammy Awards.

The Kendalls stuck with Ovation until 1982, having hits with **It Don't Feel Like Sinnin' To Me**, **Sweet Desire** (No.1, 1978), **I Had A Lovely Time**, **Put It Off Until Tomorrow** (1980) and **Heart Of The Matter** (1981), among others. The duo then began charting with Mercury and 1983 saw them go to the Top 20 with **Precious Love** and **Movin' Train**, and chart-top with **Thank God For The Radio**.

During the late '80s, the Kendalls did some label-hopping from MCA/Curb to Step One and Epic, scoring minor hits with each one, but failing to return to the Top 10. An unlikely pair in some ways – the idea of father and daughter swopping choruses on cheatin' songs has often led to the duo being asked if they are married to each other – they have stayed country and paid the price by never seeing their records cross over into the more lucrative pop market. It could be argued that Jeannie Kendall should have embarked upon a solo career in the early '80s. She has possibly left it too late in life, especially now that country music is dominated by new, younger stars, making it very difficult for veteran stars to make an impact.

Recommended:
Heaven's Just A Sin Away (Ovation/Polydor)
Hearts Of The Matter (Ovation/Ovation)
Movin' Train (Mercury/–)
Fire At First Sight (MCA/MCA)
Break The Routine (Step One/–)

Ray Kennedy

Known as the 'High-Tech Hillbilly', singer-guitarist Ray Kennedy (born in Buffalo, New York) is one of country music's more versatile and creative personalities. Prior to becoming a 'naturalized' Nashvillian in the early '80s, Kennedy rode the drifter's trail, living in Massachusetts, on the Virginia shore, in the Oregon mountains and in Vermont. He rode motorcycle escort for funerals by day, while working the club circuit by night. His father, Ray Sr, was a vice-president with Sears, and was responsible for creating the Discover Card.

In 1980 Ray Jr arrived in Nashville with plans to become a top songwriter. Within six months he had a song, **The Same Old Girl**, cut by John Anderson. He signed a

Below: New York-born Ray Kennedy debuted for Atlantic in 1990 with his Top 10 hit What A Way To Go.

writer's contract with Tree Music and had more cuts by Charley Pride, David Allan Coe and T. Graham Brown. This enabled him to build his own studio, where he churned out radio and TV jingles as well as demos of his own songs, while also working as a session musician and singer. His studio has also been used by many diverse artists, including Kevin Welch, T. Graham Brown, Stevie Nicks, Diamond Rio, Dude Mowery and Michelle Wright.

With his black leather jackets and rockabilly look, Ray Kennedy doesn't quite fit the standard image of a country singer. But, in 1989 he signed a recording contract with Atlantic, his debut album, **What A Way To Go**, being an almost totally solo affair with Kennedy arranging, producing and playing all instruments, except steel guitar and dobro. He also wrote, or co-wrote, all the songs. The title track became a Top 10 single in 1990. Since then Kennedy has scored minor hits with **Scars** (1991) and **No Way Jose** (1992).

Recommended:
Guitar Man (Atlantic/–)
What A Way To Go (Atlantic/–)

The Kentucky HeadHunters

A real modern 'hillbilly' band, the Kentucky HeadHunters, a rock quintet from Edmonton, Kentucky, brought a heavy metal and blues influence to country music in the late '80s. Their first album, **Pickin' On Nashville**, went multi-platinum as it raced up both the country and pop charts. After 20 years of struggling, the band had found 'instant success' with major awards, and hit following hit, then within three years the band was torn apart. Both their lead singer and bass player decided to leave and pursue their own careers.

The history of the Kentucky HeadHunters is intertwined with another group, Itchy Brother, formed in the late '60s when two brothers, Richard and Fred Young, and two of their cousins, Anthony Kenney and Greg Martin, got together to play a raunchy brand of blues, rock'n'roll and country. The name Itchy Brother came from a cartoon character and the foursome worked the club circuit for the next decade. In 1980 they added lead vocalist Mark Orr and pursued a recording contract with Led Zeppelin's Swan Song label. When it didn't materialize, the five split up.

In 1986 the Young brothers decided to re-form Itchy Brother. Kenney opted out, so Greg Martin suggested recruiting Doug Phelp, who was playing with him in Ronnie McDowell's band. They also needed a lead vocalist, so Doug's brother, Ricky Lee Phelps came into the line-up, alongside the three original members. The five-piece opted for a new name, taking their cue from Muddy Waters' band, the Headchoppers. In 1987 the Kentucky HeadHunters were born.

The following year they made an eight-song demo tape that set the Nashville community buzzing and led to a contract with Mercury Records. Several of the tracks from the demo were used on the debut album, **Pickin' On Nashville**, including their first country hit, a revival of Bill Monroe's **Walk Softly On This Heart Of Mine** (1989). The HeadHunters enjoyed further hits with **Dumas Walker** and **Oh Lonesome Me** (1990), but it's as album artists and a live act that they have really scored. They have a boisterous, fun-packed, rockin' country show; a southern outfit that specializes in head-banging party music. In 1991 their second album, **Electric Barnyard**, again featured revivals of old country classics, and soon went gold. Then, as they prepared to start a third album, the brothers Doug and Ricky Lee Phelps announced their departure. It was on June 2, 1992, when they sent a fax to the others, notifying them of their intention. On the very same day a filmed segment of their announcement was shown on the TNN Crook & Chase Show.

The remaining members re-grouped, bringing in 'new' members Mark Orr (lead vocals) and Anthony Kenney (bass), who had played with them in the old days of Itchy Brother. The change has brought about a more bluesy edge to the HeadHunters' music, which might well alienate them from mainstream country. Meanwhile the Phelpses have claimed that their music will be a return to the country roots and geared more towards mainstream country radio. Both sides emphasize that the split was due to different musical directions.

Above: The Kentucky HeadHunters rockin' it up in their inimitable way.

Recommended:
Rave On! (Mercury/Mercury)
Pickin' On Nashville (Mercury/Mercury)
Electric Barnyard (Mercury/Mercury)

Doug Kershaw

Cajun fiddler Douglas James Kershaw, born at Tiel Ridge, Louisiana, on January 24, 1936, first appeared onstage as a child, accompanying his mother (singer-guitarist-fiddler Mama Rita) at the Bucket Of Blood, Lake Arthur. In 1948, together with his brothers Russell Lee ('Rusty') and Nelson ('Pee Wee') Kershaw, he formed the Continental Playboys, gaining a spot on Lake Charles KPLC-TV in 1953. Rusty and Doug then began recording as a duo for the Feature label, later obtaining a contract with Hickory.

With an Everly-like treatment of a Boudleaux Bryant song, **Hey Sheriff**, the Kershaws made an indent on the country charts in October 1958 and even briefly joined the Opry. After Doug completed his military service, the duo resumed their joint career, scoring with country classics **Louisiana Man** and **Diggy Diggy Lo**.

After cutting sides for Victor and Princess, the twosome parted in 1964, Doug moving on to record for Mercury, MGM, Warner Bros, Starflyte and Scotti Bros. He has also guested on scores of sessions, appearing on albums with Longbranch Pennywhistle, Bob Dylan, Johnny Cash, John Stewart and even Grand Funk Railroad. After playing a cameo role in the film 'Zachariah' (1971), he also appeared in 'Medicine Ball Caravan' (1971) and 'Days Of Heaven' (1978).

Recommended
The Cajun Country Rockers – Rusty And Doug (–/Bear Family)
The Cajun Way (Warner Bros/–)
Devil's Elbow (Warner Bros/Warner Bros)
Douglas James Kershaw (Warner Bros/ Warner Bros)
Hot Diggity Doug (–/Sundown)

Sammy Kershaw

A self-taught Cajun chef, Sammy Kershaw was born in 1958, in Kaplan, Louisiana, a third cousin of Cajun fiddler Doug Kershaw. Before making a breakthrough with **Cadillac Style** in 1991, Sammy paid his dues, starting out in music when he was only 12, and coming through two divorces, a brace of bankruptcies and numerous jobs, including a stint as a professional baseball player. A man of many talents, he's a terrific story-teller (having worked as a stand-up comedian) and a soulful singer, with more than a hint of George Jones in his voice.

He played the Louisiana club scene for 20 years, and at one point quit music. Years on the road playing clubs six nights a week had taken its toll with heavy drinking and cocaine, and Sammy headed down a dead-end street. In 1988 he married for a third time and settled down. Two years later he got a call to head for Nashville for an audition with Mercury Records. Once safely in Nashville, Sammy recorded the album, **Don't Go Near The Water**, a collection of hard-country ballads and saloon songs which garnered a gold disc and spawned such big hits as **Cadillac Style** (1991), **Don't Go Near The Water** and **Anywhere But Here** (1992).

A self-named 'ballad singing fool', Sammy's warm, emotional vocal style with its country-to-the-core phrasing frequently elicits comparisons to George Jones. "I've been singing like this since I was a kid", Sammy's quick to point out. "So many times I was compared to George Jones, but it wasn't my fault. It's just natural harmonics." Sammy's management decided to cash in on the sweet smell of success with Starclone, a new fragrance for women made from Kershaw's natural 'body essences'. The sweat of his back and chest is distilled with flowers, herbs and oils. Slickly packaged with cassettes and photos of the Cajun star, you have 'Spray On Sammy'. This is a long way from the old days of country music when stars would sweat it out all day behind a plough, then drive into town in the evening to sing at the local school gym or church.

Recommended:
Haunted Heart (Mercury/–)
Don't Go Near The Water (Mercury/–)

Hal Ketchum

One of country music's fastest-rising stars, success for Hal Ketchum (born Michael Ketchum, April 12, 1953 in Greenwich, New York) came later in life than most.

His father was a linotype operator for Gannett Press, and also a keen country music fan. His mother, who suffered from multiple sclerosis, enjoyed crooners. Because of his mother's illness, he spent much of his childhood at his grandparents' home where his grandfather listened to jazz and classical music. A carpenter by trade, Hal absorbed all these musical influences, and was encouraged by his father who played banjo and can also boast to being a member of the Buck Owens Fan Club. Hal started out as a teenage drummer in a New York R&B outfit before relocating south. He first moved to Florida, where he played in a blues band, then to Texas, where he became a singer-songwriter on the Austin music scene.

By this time, Ketchum was married with two young children, and really had no idea in which direction to take his music. He

Past the Point Of Rescue, Hal Ketchum. Courtesy Curb.

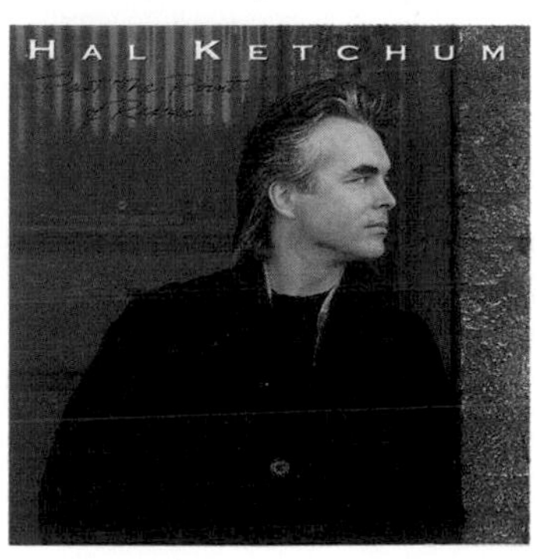

maintained a day-time job as a carpenter, working the bars night after night searching for some kind of musical recognition. This lifestyle put a heavy strain on his marriage and led to divorce. In 1987 he produced his own album, **Threadbare Alibis**, which was no more than a gig cassette for Watermelon Records. Two years later it was picked up by Line Records in Germany and put out on CD. It was an appearance at the Kerrville Folk Festival in 1990 that led to him signing a major record deal with Curb Records the following year. Noted producer Jim Rooney brought him to Nashville where he started writing and doing demos for Forerunner Music.

His first single, **Small Town Saturday Night**, helped by a video which was directed by photographer Jim McGuire, hit No.2 in the charts in 1991. His debut album, **Past The Point Of Rescue**, raced to gold status. That same year he married Terrell Tye, president of Forerunner Music, and also put together his four-piece band, the Alibis, consisting of musicians who had worked with him in Texas. Ketchum consolidated this initial success with such chart singles as **I Know Where Love Lives** (1991), **Past The Point Of Rescue**, **Five O'Clock World** and **Sure Love** (1992), and **Hearts Are Gonna Roll** (1993).

An unabashed fan of country music, Hal is more than just a songwriter. He's also a writer of short stories and plans to write children's books some day. At this time, though, it is his music that is bringing him fame and fortune. All his musical influences are united in a style that is totally his own, yet represents the eclectic traditions of the Texas song poets.

Recommended:
Sure Love (Curb/–)
Past The Point Of Rescue (Curb/–)
Threadbare Alibis (–/Sawdust)

Merle Kilgore

Born Wyatt Merle Kilgore, in Chickasha, Oklahoma, on September 8, 1934, Merle's family moved to Shreveport, Louisiana, when he was still young. He learned guitar as a boy and got a job as a DJ on Shreveport's KENT when he was just 16. At this time he was already attracting attention as a performer and writer, and by the age of 18 Kilgore had his first hit composition, **More, More, More**. Invited to join the Louisiana Hayride, he became the principal guitarist on the show. In 1952 he appeared on the Opry and in that same year attended Louisiana Tech. The following year saw him working at the American Optical Company but performing at night. He appeared on the Hayride throughout the '50s, initially recording for the Imperial and D labels, but his first big hit came in 1959 with **Dear Mama**, a Starday release. Johnny Horton scored with **Johnny Reb**, a Kilgore composition, around the same time.

Another of Kilgore's Starday releases, **Love Has Made You Beautiful**, charted in 1960, as did **Gettin' Old Before My Time**, while in 1962 he co-wrote **Wolverton Mountain** with Claude King and **Ring Of Fire** with June Carter, the later composition providing Johnny Cash with a million-selling single in 1963.

Though his record sales fared less successfully as the '60s wore on, Kilgore became established as an impressive Western actor, appearing in such movies as 'Nevada Smith' (1966), 'Five Card Stud' (1968) and a number of others. For many years Kilgore has acted as Hank Williams Jr's manager and masterminded his transition into a major country music superstar.

Above: Pee Wee King, co-writer of Slowpoke and You Belong To Me.

Recommended:
Merle Kilgore (Mercury/–)
Teenager's Holiday (–/Bear Family)

Bradley Kincaid

A pioneer broadcaster of traditional Kentucky mountain music, Kincaid (born Point Leavell, Kentucky, July 13, 1895) began singing folk songs on WLS, Chicago, in August, 1925, while still attending that city's George Williams College. By 1926 he had become a regular on the WLS Chicago Barn Dance, remaining with the show (later to be known as the National Barn Dance) until 1930. Following graduation in June, 1928, Kincaid began touring, at the same time collecting folk songs from a variety of sources, publishing these in a series of songbooks.

His recording career also began in 1928, when he made a number of sides for Gennett, these discs appearing on myriad labels, sometimes under a pseudonym. Throughout the '30s and '40s, Kincaid continued to record for many different labels.

A banjo picker at the age of five, Kincaid, who became known as the 'Kentucky Mountain Boy', purveyed such materials as **I Gave My Love A Cherry**, **The Letter Edged In Black** and **Barbara Allen**, singing the latter over WLS every Saturday night for four successive years. An ever-active radio performer, he bought his own station (WWSO, Springfield, Ohio) in 1949 but sold it again in 1953. His more recent recordings include albums for Bluebonnet and McMonigle.

Besides appearing on nearly every major barn dance, Kincaid played for years in the Northeast, introducing folk and country music to a whole new area before retiring in Springfield.

Recommended:
Mountain Ballads And Old Time Songs (Old Homestead/–)

Claude King

Wolverton Mountain – a distinctive and menacing country-styled 'Jack And The Beanstalk' saga – was a 1962 million-seller for King and co-writer Merle Kilgore.

Born in Shreveport, Louisiana, on February 5, 1933, King attended the University of Idaho and then returned to Shreveport, to business college. He had been interested in music, having bought his first guitar from a farmer for 50 cents when he was 12.

During the '50s he began writing and performing, playing clubs, radio and TV. He was signed to Columbia in 1961, gaining country hits with, among others: **Big River, Big Man** (1961), **The Comancheros** (1961), **Burning Of Atlanta** (1962), **I've Got The World By The Tail** (1962), **Building A Bridge** (1963), **Hey Lucille** (1963), **Sam Hill** (1963), **All For The Love Of A Girl** (1969), **Friend, Lover, Woman, Wife** (1969). **Big River, The Comancheros, Wolverton Mountain** and **Burning Of Atlanta** also crossed over into the pop charts.

During the early '60s, Nashville expected King to become a superstar, but it was not to be. Public taste seemed to move away from butch sagas of the great outdoors and, as a result, the '70s failed to prove kind to the singer, though he did see some chart action with **Montgomery Mabel** (1974) and **Cotton Dan** (1977).

Recommended:
Claude King's Best (Gusto/Gusto)
Meet Claude King (Columbia/–)

Pee Wee King

The leader of what was claimed to be the first band to use an electric guitar and drums on the Opry, King is also a noted songwriter, being writer or co-writer of such hits as **Slow Poke**, **Bonaparte's Retreat**, **You Belong To Me** and **Tennessee Waltz**, the last named being declared the State song of Tennessee in February 1965.

Born in Abrams, Wisconsin, on February 18, 1914, of Polish descent, Frank 'Pee Wee' King learned a harmonica, accordion and fiddle while still a boy, broadcasting over radio stations in Racine and Green Bay at the age of 14.

After stints with the WLS (Chicago) Barn Dance during the early '30s, he joined the

Don't Say Aloha, Bashful Brother Oswald (Pete Kirby). Courtesy Rounder Records.

Gene Autry Show, taking over the band in 1934 when Autry headed for Hollywood. Renamed the Golden West Cowboys, the band – which featured such stars as Eddy Arnold, Redd Stewart, Cowboy Copas, Ernest Tubb and guitarist Clem Sumney at various points in its history – first graced the Opry in the mid-'30s. In 1938 the band followed Autry to Hollywood to make 'Gold Mine In The Sky' for Republic Pictures, the first in a series of cowboy movies.

From 1947 to 1957 King hosted his own radio and TV show on Louisville WAVE, also in 1947 signing a record deal with RCA-Victor. Additionally, he did a weekly television circuit.

Success as a composer came when **Tennessee Waltz**, penned by King and Stewart, became a hit record for Cowboy Copas in 1948. Around the same time, King himself began logging a tally of hits. **Tennessee Tears** (1949), **Slow Poke** (a 1951 million-seller), **Silver And Gold** (1952) and **Bimbo** (1954) all became Top 10 entries.

The Golden West Cowboys were hit by the rise of rock'n'roll during the late '50s, King adding horns in an effort to compete with the all-conquering rockers. However, by 1959 the financial struggle had become uneven and King disbanded the Cowboys, forming another unit several months later when Minnie Pearl asked him to accompany her on a roadshow. When Minnie ceased touring in 1963, King kept the unit – which included Redd Stewart and the Collins Sisters – together until 1968. Then he disbanded once again, relying on local musicians to support him.

Though King's records have failed to sell in any tremendous quantities since he terminated his contract with RCA in 1959, he remains a popular and highly respected member of the country music profession, worthily being elected to the Country Music Hall Of Fame in 1974.

Recommended:
Ballroom King (–/Detour)
The Legendary Pee Wee King (Longhorn/–)
The Best Of Pee Wee King And Redd Stewart (Starday/–)
Rompin, Stompin', Singin', Swingin' (–/Bear Family)
Hog Wild Too (–/Zu Zazz)

Pete Kirby (Bashful Brother Oswald)

Real name Beecher Kirby, born in Sevier County, Tennessee, this guitarist, banjoist and dobro player was one of eight brothers and two sisters, all of whom played instruments, their father being proficient on fiddle, banjo and guitar. As a young man, Kirby worked in a sawmill, a cotton mill and on a farm before becoming a guitarist in an Illinois club.

During the World's Fair in Chicago, he played in local beer joints, passing the hat around, also working part-time in a restaurant in order to survive. Then came a move to Knoxville, Tennessee, where he joined Ray Acuff's Crazy Tennesseans on radio station WRL. As Bashful Brother Oswald, the bib-overall-clad Kirby sang and duetted with Acuff, playing the banjo for most solo work, and reverting to dobro whenever Acuff's distinctive band sound was required. A member of the Smoky Mountain Boys for many years, Kirby

Above: Kristofferson as Billy The Kid in 'Pat Garrett And Billy The Kid'.

looked after Acuff's Nashville museum in the '70s, often indulging in good-time pickin' in order to attract extra customers.

One of the stars to appear on the Nitty Gritty Dirt Band's **Will The Circle Be Unbroken** (1971), Kirby cut a fine series of albums for the Rounder label during the '70s, some of the sessions lining him up alongside fellow Smoky Mountain Boy Charlie Collins.

Recommended:
Brother Oswald (Rounder/–)
That's Country – with Charlie Collins (Rounder/–)

Alison Krauss

A fiddle prodigy for most of her young life, Alison Krauss (born in 1971 in Champaign, Illinois) has taken bluegrass into the country charts and country radio.

Alison began to play fiddle at five, formed her first band at 12, and signed with Rounder Records to make her first album at 14. She began entering fiddle contests while still a young child, and was seven times a fiddle champion in five Midwestern states by the time she was 16. By this time she had made a big impact at the Newport Folk Festival and was recording with some of the top session players in Nashville, including Jerry Douglas and Sam Bush for the album **Too Late To Cry** (1987). Putting together her own bluegrass band, Union Station, she began touring, playing coffee houses, festivals and concerts in a gruelling schedule. A third album, **I've Got That Old Feeling**, released in 1991, won a Grammy. Her baby-fine voice is a real surprise when you hear it with her authorative fiddle style and the equally impressive skills of her band. That combination has made Alison an in-demand Nashville studio player and singer on recordings by Dolly Parton, Vince Gill, Mark Chesnutt, Michelle Shocked and the Desert Rose Band. She plays her fiddle with the ease of a veteran, liberally dashed with youthful exuberance. Her music is rooted in traditional bluegrass, to which she remains deeply committed, while her willingness to draw in other influences has

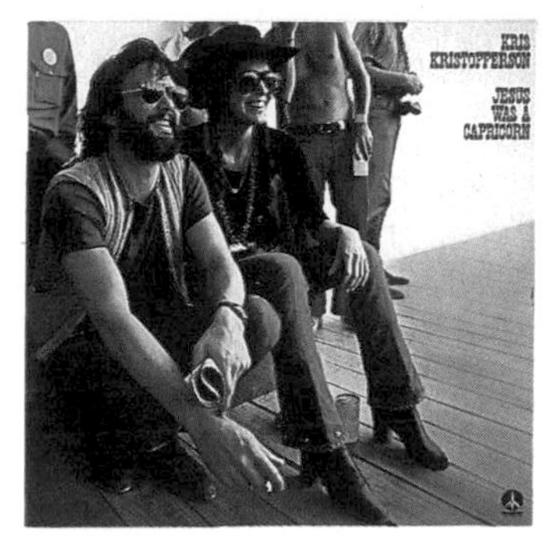

Jesus Was A Capricorn, Kris Kristofferson. Courtesy Monument Records.

made her music accessible to fans of country and other types of music.

Recommended:
I've Got That Old Feeling (Rounder/–)
Two Highways – with Union Station (Rounder/–)
Too Late To Cry (Rounder/–)

Kris Kristofferson

Born in Brownsville, Texas, on June 22, 1936, the son of a retired Air Force Major-General, Kristofferson's family moved to California during his high school days. Living in San Mateo, he went to Pomona College. In 1958, he won a Rhodes Scholarship to Oxford University, England, where he began writing his second novel, becoming a songwriter as a sideline, using the name of Kris Carson.

His novels rejected by publishers, he became disenchanted with a literary career and left Oxford after a year, first getting married, then joining the Army and becoming a helicopter pilot in Germany. He began singing at service clubs in Germany, also sending his songs to a Nashville publisher. Upon discharge in 1965, Kristofferson headed for Nashville, initially becoming a janitor in Columbia Records' studio. Broke and with his marriage in tatters, he was about to take a construction job when Roger Miller recorded one of his songs, **Me And Bobby**

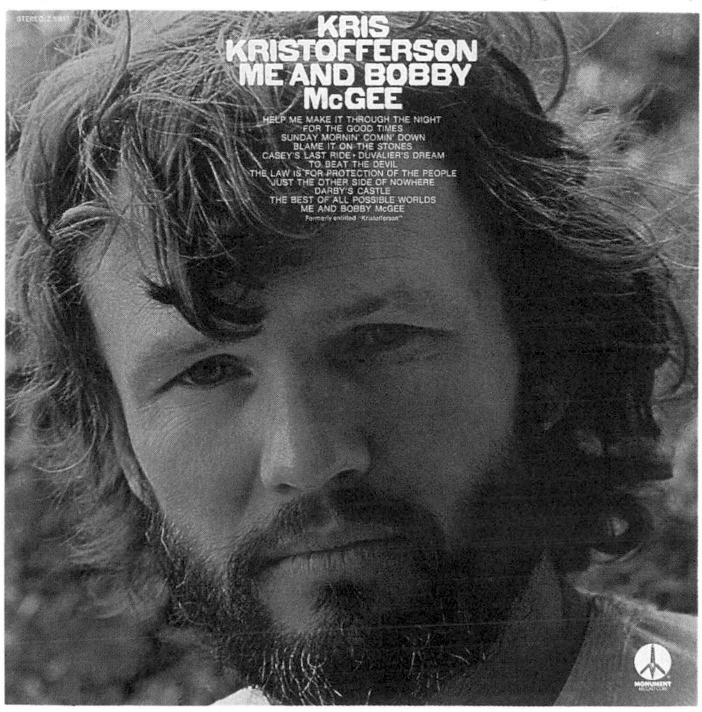

Me And Bobby McGee, Kris Kristofferson. Courtesy Monument Records.

McGee, the composition also being covered by Janis Joplin, whose version became a million-seller in 1971. During 1970, Johnny Cash waxed **Sunday Morning Coming Down**, another Kristofferson original, and the Texan cut his first album for Monument, Cash writing a poem documenting the singer-songwriter's lean years, for use as a sleeve note.

Appearances on Cash's TV show and other triumphs followed, including a debut engagement at a name club (The Troubadour, L.A.) and another hit via Sammi Smith's version of his **Help Me Make It Through The Night** – a million-seller in 1971. During the following year, Kristofferson's **Silver Tongued Devil And I** single went gold, while in November, 1973, another single, **Why Me?** also qualified as a gold disc. Additionally, in 1973, the year that he married singer Rita Coolidge, two albums, **The Silver Tongued Devil And I** and **Jesus Was A Capricorn**, provided the Texan with further gold awards.

He and Rita merged bands and began recording together, though still continuing with their solo careers. But Kris's recording career was burning out and sales started to dip. However, he had made his debut as an actor in 'Cisco Pike' (1972) and from there on gained role after role. He and Rita appeared in 'Pat Garrett And Billy The Kid' (1973), the real breakthrough coming with 'Alice Doesn't Live Here Anymore' (1974), after which came major roles in movies such as 'The Sailor Who Fell From Grace With The Sea' (1976), 'A Star Is Born' (1976), 'Convoy' (1978), 'Heaven's Gate' (1980), 'Rollover' (1981), etc.

He kicked a 20-year drinking problem at the end of the '70s (too late to save his marriage to Rita Coolidge, which ended in 1979), and was singing better than at any time in his life. He even got his name on a No.1 album in 1985 when he joined with Willie Nelson, Johnny Cash and Waylon Jennings to create the **Highwayman** LP and hit single. Shortly after this he signed with Mercury Records. His songwriting returned to the biting edge of his earlier work, with a hard-hitting political stance to the fore on **Third World Warrior**, a 1990 album. He also undertook more concert work, both as a member of the Highwaymen and with his own band, or with just a couple of back-up musicians.

Recommended:
Kristofferson (Monument/Monument)
Me And Bobby McGee (Monument/Monument)
Jesus Was A Capricorn (Monument/Monument)
Full Moon – with Rita Coolidge (A&M/A&M)
The Legendary Years (–/Connoisseur)
Third World Warrior (Mercury/Mercury)

k.d. lang

Farmer's daughter k.d. lang (born Kathy Dawn Lang in 1962 in Consort, Alberta, Canada) requests that her name is always printed in lowercase type, because she believes that you have to be different to stand out, and this young lady certainly is different. She dresses differently, and her whole lifestyle is different from any other country music performer. She has her own uncompromising musical style. As a child she heard a lot of country music, and by the time she reached her teens she had mastered piano and guitar, and had started writing songs. In the early '80s she began working with local bands, singing and writing country-influenced songs. By 1983 she had formed her own band, the Reclines (a nod towards her idol, the late Patsy Cline). In 1984 the group produced **A Truly Western Experience**, an independent Canadian release on Bumstead Records. This eventually led to a contract with Sire Records who brought in Dave Edmunds to produce **Angel With A Lariat** (1987), a rock-influenced country album that gained rave reviews in the rock press, but little mainstream country radio play. That same year lang duetted with Roy Orbison on **Cryin'**, which was used in the movie soundtrack for the comedy 'Hiding Out'. The duet introduced her to the country charts for the first time.

In October 1987, k.d. coaxed veteran producer Owen Bradley out of semi-retirement to produce **Shadowlands**, a kind of quirky tribute to Patsy Cline, but also an album that had a little bit of everything, from '50s country to pop and big band. One track, **The Honky Tonk Angels Medley**, featured Kitty Wells, Loretta Lynn and Brenda Lee, and the whole project brought k.d. closer to country. She scored on the country charts with **I'm Down To My Last Cigarette** and **Lock, Stock And Teardrops** in 1988, and was also included in the CMA's Route 88 European country music promotion. **Absolute Torch And Twang**, which she co-produced with band members Ben Pink and Greg Penny, hit both the pop and country charts in 1989 on its way to platinum status. k.d. won several major music awards in Canada, and was also nominated by the ACM in the New Female Vocalist category.

It was to be another three years before k.d. produced her next album, **Ingenue** (1992), a brilliant artistic and commercially successful record that gained platinum status in America and gold in Britain, where she made a big impact on the pop charts. This took her a long way from country, a style of music that k.d. has probably left behind her.

Recommended:
Ingenue (Sire/Sire)
Absolute Torch And Twang (Sire/Sire)
Angel With A Lariat (Sire/Sire)
Shadowland (Sire/Sire)

Above: Canadian k.d. lang, who brought a fresh approach to modern country music.

Jim Lauderdale

Singer-songwriter Jim Lauderdale (born April 11, 1957 in Statesville, North Carolina) draws from a wealth of American musical styles and fashions to produce a quirky combination of progressive country-rock, bar-room ballads and soul-searching blues. He has built a cult following in the early '90s that could break through to the mainstream of country music.

He grew up in various towns in the Carolinas, where his father was an Associate Reformed Presbyterian Minister and his mother served as a music teacher and choir director. Lauderdale began his own musical pursuits playing drums in the school band. He then delved into country, bluegrass and folk, music, learning banjo and guitar. Throughout his college years at the North Carolina School of the Arts, he was active as a solo performer. Once studies were completed and he had gained a degree in theatre, he moved to New York, where he played the country/folk scene with his own band, mixing country, swing and blues. Later landing a part in the off-Broadway show 'Cotton Patch Gospel', he joined several touring productions. One, 'Diamond Studs', found him playing Jesse James. He was in Los Angeles with the 'Pump Boys And Dinettes' show when he came to the attention of Pete Anderson, the well-known producer of Dwight Yoakam. Impressed by Lauderdale's songs and singing, Anderson produced the singer-songwriter for the second volume of **A Town South Of Bakersfield** compilation. This led to Lauderdale singing back-up on recordings by Carlene Carter, Dwight Yoakam, Darden Smith and Jann Browne, and also having his songs cut by Vince Gill, Shelby Lynne, Kelly Willis, Jann Browne and George Strait.

His own record deal came in 1991 when he signed with Reprise and teamed up with John Leventhal and Rodney Crowell. They co-produced his debut album, **Planet Of Love**, which combined a healthy roots reverence with some flat-out blues, honky-tonk and even a yodel.

Recommended:
Planet Of Love (Reprise/–)

Absolute Torch And Twang, k.d. lang. Courtesy Sire Records.

Tracy Lawrence

One of the young 'hunks' of country music, Tracy Lawrence (born January 27, 1968 in Atlanta, Texas) is determined to make his career in country music a long and successful one. Lawrence is still young enough to look boyish – he has long, curling, blondish-brown hair, a starter moustache, and, at six feet tall, he makes a big impact with the female fans. His blossoming career was almost cut short before it had a chance to get started. In May 1991, he was celebrating the completion of his first album with an old friend in Nashville when they were confronted by three men with two guns. The trio robbed the couple, and when Tracy put up a fight to protect his female friend, he was shot four times, resulting in several weeks in hospital and his album release put on hold until he had recovered.

Tracy had dreamt of being a singing star from quite a young age. In 1972 his family moved to Foreman, Arkansas. His stepfather was a banker, while his mother raised six kids, with Tracy being more trouble than the other five put together. Even so, he faithfully attended the Methodist church and sang in the choir. He learned to play guitar and was performing at jamborees by the time he was 15. Two years later he was gigging in honky-tonks and night-clubs. As soon as he graduated from high school, he moved around Louisiana, Arkansas, Texas and Oklahoma with bands. He studied at Southern Arkansas University, then moved to Louisiana to join a band as lead singer. When the group's routine of three-day weekend gigs seemed to be leading nowhere, he packed up and moved to Nashville, arriving in September of 1990. Seven months later he signed a record deal. Prior to that he had earned his living entering and winning singing contests at open-mike nights in Tennessee and Kentucky. The youngster also landed a regular gig on the Nashville radio show, 'Live At Libbys', and one night was spotted at the club in Kentucky by the man who would become his manager.

Definitely one of the hottest acts on the country music scene, his debut album, **Sticks And Stones**, has sold in excess of 800,000 copies, gaining a gold disc. Lawrence became the first country artist on Atlantic to score a No.1 single, with **Sticks And Stones** in 1992. Since then he has had more No.1s with **Today's Lonely Fool** and **Runnin' Behind** (also 1992). A fourth single from the album, **Somebody Paints The Wall**, was a Top 5 entry. His follow-up album, **Alibis**, is proving just as successful and hit-laden, with the title song making No.1 in 1993. Tracy and his band undertook more than 280 show dates during 1992, and he picked up ACM's New Male Vocalist award in 1993. He is now regarded as one of the finest and youngest new breed honky-tonkers to hit country music in the '90s.

Recommended:
Sticks And Stones (Atlantic/–)
Alibis (Atlantic/–)

Above: Chris LeDoux is known for a rodeo as well as for a music career.

Chris LeDoux

While many modern country stars like to think of themselves as cowboys, only Chris LeDoux (born on October 2, 1948 in Biloxi, Mississippi) can lay claim to being the real McCoy. Amazingly adept at bringing the rodeo life into vivid colour, he's not just blowing smoke. The singer is a world champion rodeo star, and since the early '80s has run his own 500-acre ranch in Wyoming.

His family settled in Texas after his father retired from the Air Force. Chris began his musical odyssey at 14 when he started playing guitar and writing songs. He was also heavily involved in youth rodeo, and his songs reflected his love of the sport. After the family relocated to Wyoming, LeDoux began to actively pursue a rodeo career. He had twice won the state's bareback title while still attending high school in Cheyenne. After graduation he won a rodeo scholarship and received a national title in his third year. Chris started singing his songs to fellow rodeo contestants, and they reacted favourably to early works such as **Bareback Jack**, **Rodeo Life** and **Hometown Cowboy**. Shortly after getting married in 1972, LeDoux made his first independent recordings in Sheridan, Wyoming, resulting in **Rodeo Songs Old And New** and **Songs Of Rodeo And Country** (on cassette). He made his first Nashville recordings after his parents moved to Tennessee, and over the next few years recorded 15 albums in Nashville for his own American Cowboy label. In those days, LeDoux supposedly regarded the music as just a sideline to being a cowboy. But he apparently took the music seriously enough to sell 14 million dollars' worth of cassettes, most of them manufactured by his parents in their own home tape-duplicating room.

By 1976, Chris was becoming known as a singer-songwriter of note and his rodeo career was riding high. He won the Bareback Bronc World Title, and also picked up awards in Wyoming and Nevada for his bronze sculptures of a Bull Rider and Bronc Rider.

Something of a cult figure, LeDoux continued to rodeo until 1984, when accumulated injuries made him hang up his spurs. Once he had his Wyoming ranch operational, he began working on his musical career. He was now booking himself and his Saddle Boogie Band. Finally his recording career bolted out of the chute when a mention from Garth Brooks, who sang about listening to 'a worn-out tape of Chris LeDoux' in his 1989 hit **Much Too Young (To Feel This Damn Old)** was enough to kick up new interest in Wyoming's singing cowboy.

In early 1991 Chris LeDoux was signed to Capitol Records. With his recordings released on Liberty, his first album, **Western Underground**, co-produced by Jimmy Bowen and Jerry Crutchfield, sold in excess of 100,000 units and included a minor country hit in **This Cowboy's Hat** (1991). Impressed by these sales, Capitol took over all of the singer's entire 22 independent cassettes, and re-released them on CD during 1991 and 1992. Chris teamed up with Garth Brooks for the title song of his 1992 album **Whatcha Gonna Do With A Cowboy**, and saw the single race up the country charts with the album attaining gold status. Another single, **Cadillac Ranch** (1992), went Top 20 and the cowboy singer co-starred with Suzy Bogguss in a TNN special, 'Ropin' And Rockin' in early 1993.

LeDoux describes his music as 'a combination of western soul, sagebrush blues, cowboy folk and rodeo rock 'n' roll; and captures a piece of modern-day Americana in his songs. An 18-year overnight sensation, he is now playing to packed audiences and attracting younger fans with his Western Underground Band.

Planet Of Love, Jim Lauderdale. Courtesy Reprise Records.

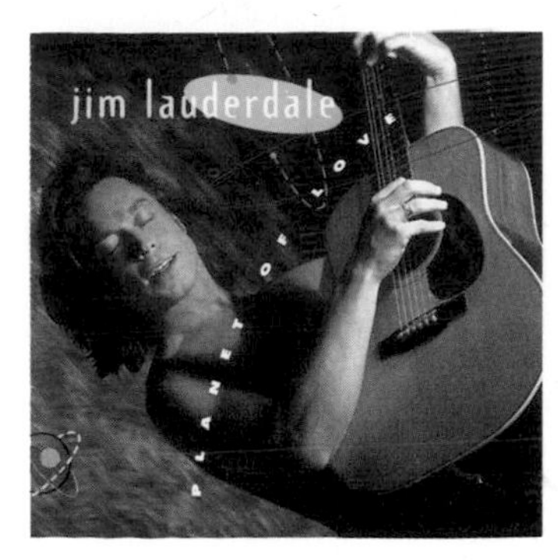

Recommended:
Whatcha Gonna Do With A Cowboy (Capitol/Capitol)
Life As A Rodeo Man (Capitol/Capitol)
Rodeo's Singing Bronco Rider (–/Westwood)
Western Underground (Capitol/–)
Under This Old Hat (Liberty/–)

Brenda Lee

Once known as 'Little Miss Dynamite' – she is only 4 feet 11 inches tall – Brenda Lee was born Brenda Mae Tarpley, in Lithonia, Georgia, on December 11, 1944. She won a talent contest at the age of six and in 1956 she was heard by Red Foley who asked her to appear on his Ozark Jubilee Show. Her success on that show led to further TV stints and a Decca record contract in May, 1956. After her initial record release – a version of Hank Williams' **Jambalaya** –she began chalking up an impressive tally of chart entries, commencing with **One Step At A Time**, which became both a country and pop hit in 1957. Later came a series of hard-headed rockers like **Dynamite** (1957) and **Sweet Nuthin's** (1959), Brenda proving equally adept at scoring with such ballads as **I'm Sorry** (1960) and **As Usual** (1963). For a while she was undoubtedly one of the world's most popular singers, but after a major hit with **Coming On Strong** in 1966, her record sales tapered off considerably.

After a few years in an easy-listening wilderness, Brenda made a return to the country Top 10 in 1973 with a version of Kris Kristofferson's **Nobody Wins**. She followed this with other major records such as **Sunday Sunrise** (1973), **Wrong Ideas**, **Big Four Poster Bed**, **Rock On Baby** (1974), **He's My Rock** (1975), **Tell Me What It's Like** (1979), **The Cowgirl And The Dandy**, **Broken Trust** (1980), and a joyous version of **Hallelujah I Love You So**, a 1984 duet with George Jones. At this time, Brenda had disagreements with MCA, her label for 30 years, and for a time she made no new recordings. In 1990 she signed with Warner Brothers, but has so far failed to make a return to the best-sellers' lists.

Still an energetic onstage performer, Brenda has appeared in the movie 'Smokey And The Bandit 2' and has her own syndicated radio show. She has also contributed her time on behalf of country music as a director of the CMA.

Recommended:
The Brenda Lee Story (MCA/MCA)
The Golden Decade (–/Charly)
L.A. Sessions (MCA/MCA)
Brenda Lee (Warner Bros/–)
Even Better (MCA/MCA)

Johnny Lee

East-Texan John Lee Ham was born in Texas City, Texas on July 3, 1946. Raised on a farm at Alta Loma, he formed a high school band called Johnny Lee And Roadrunners, winning various talent contests. After high school he joined the US Navy, becoming a bosun's mate. Upon discharge he lived for a while in California, but later returned to Texas, working in honky-tonks and gaining a job as a singer and trumpet-player with Mickey Gilley's band.

By 1971 he was heading Mickey's band at Gilley's Houston club. He also signed a number of short-lived record deals, grabbing a place in the country charts with **Sometimes**, a 1975 ABC-Dot release.

National fame really began heading his way in 1979. When the TV movie 'The Girls In The Office', starring Barbara Eden and Susan St James, was shot in Houston, Lee and his band being given a spot in the film. The following year, the John Travolta movie 'Urban Cowboy' used Gilley's band as a focal point. Travolta heard Johnny Lee sing **Looking For Love**, a pop song penned by two Gulfport, Mississippi, schoolteachers, and insisted it went on to the film soundtrack. Released as a single by Full Moon/Asylum Records, it proved a massive cross-over hit, reaching No.1 in the country charts, and Top 5 in the US pop listings, He recorded a best-selling album of the same title and three singles from this, **One In A Million** (No.1, 1980), **Pickin' Up Strangers** and **Prisoner Of Hope** (1981), all charted high, along with **Bet Your Heart On Me** (No.1, 1981) the title track of Lee's next album.

He became an in-demand act, working with the Urban Cowboy Band, either headlining or co-headlining with former boss Mickey Gilley. He also gained much publicity when he began dating Charlene Tilton (Lucy Ewing of TV's Dallas series), the couple getting married on St Valentine's Day, 1982. That same year he logged two further Top 10 singles. **Be There For Me Baby** and **Cherokee Fiddle.** Other notable hits were **Sounds Like Love**, **Hey Bartender** (1983), **The Yellow Rose,** with Lane Brody, **You Could Have A Heart Break** (both 1984 No.1s) and **Rollin' Lonely** (1985). During 1985 he also contributed a track called **Lucy's Eyes** to an album of songs by the stars of Dallas. He and Charlene were divorced in 1984 and Lee's recording career started a rapid downward spiral. He only managed a few minor hits for Warner Brothers and Curb, making his last chart entry in 1989 with **You Can't Fly Like An Eagle**.

Recommended:
Lookin' For Love (Asylum/–)
Bet Your Heart On Me (Full Moon/–)
Hey Bartender (Warner Bros/–)

Bobby Lewis

Born in Hodgerville, Kentucky, Bobby Lewis appeared on the Hi-Varieties TV show at the age of 13, later working on the Old Kentucky Barn Dance radio show, CBS's Saturday Night Country Style and the highly rated Hayloft Hoedown.

Only 5-feet 4-inches tall, Lewis had problems handling his heavy and bulky Gibson J-200 guitar and eventually bought a 'funny-shaped small guitar' in a Kentucky music shop. This 'guitar' proved to be a lute, which he fitted with steel strings and adopted as his main instrument.

Signed to UA Records during the mid-'60s, Lewis' first Top 10 hit came in 1964 with **How Long Has It Been**, which he followed with such other sellers as **Love Me And Make It Better** (1967), **From Heaven To Heartache** (1968), **Things For You and I** (1969), and **Hello Mary Lou** (1970) **From Heaven To Heartache**, a cross-over hit, earned him a Grammy nomination as Best Male Performer in 1969. In 1973, after supplying UA with 14 hits, he moved on to Ace Of Hearts and added three more, the biggest of these being **Too Many Memories**. Since then Lewis had had minor hits on GRT Records (1974), Ace Of Hearts (1975) and the ill-fated Capricorn label (1979). In 1985 he charted with **Love Is An Overload**.

Recommended:
The Best Of (UCA/–)
Tossin' And Turnin' (–/Line)

Below: Bobby Lewis, who turned lute into loot, was a frequent Opry guest. He last made the charts in 1985 with Love Is An Overload.

Jerry Lee Lewis

Yet another example of the 1950s' interplay between country and rock 'n' roll, this pianist-singer with the wild stage manner was originally influenced by the pumping, honky-tonk piano style of Moon Mullican. Vocally he incorporated the delivery of black singers into his routine and, finally, Jerry Lee returned to this country roots splitting his programmes into half rock, half country affairs.

Born in Ferriday, Louisiana, on September 29, 1935, Lewis was exposed to a wide range of music and particularly church music – his parents sang and played at the Assembly of God church. Jerry learned piano and played his first public gig in 1948 at Ferriday's Ford car agency, where to introduce a new model he sang **Drinkin' Wine Spo Dee O Dee**. Like many of his generation, Lewis spent time in the local black clubs and when he cut his first sides for Sam Phillips' Sun label in Memphis, they proved to be some of the wildest rock'n'roll sounds of their time. But amid the frantic rockers Jerry Lee also snuck in an array of pure country cuts, his first Sun release being a cover of the Ray Price hit **Crazy Arms** that made the country charts.

His career went through a traumatic period after he married his 13-year-old cousin. He was booed off the stage in Britain and the UK press crucified him. However, by the late '60s he was making a comeback. Jerry Lee was back making more country records this time. He scored big on the US country charts of the period with out and out honky-tonkers like **Another Place, Another Time** and **What Made Millwaukee Famous** (both 1968) and other hits for the Smash label, such as **To Make Love Sweeter For You** (No.1, 1968), **One Has My Name, She Even Woke Me Up To Say Goodbye** (1969) and **Once More With Feeling** (1970).

Ever controversial – in 1976, 'The Killer' was picked up by the police for waving a gun around and demanding entry to Elvis Presley's Memphis mansion – he continued accruing both publicity and hits for the Mercury label throughout the '70s, the latter including **Touching Home, Would You Take Another Chance On Me** (1971), **Chantilly Lace** (1972), **Sometimes A Memory Ain't Enough** (1973), **He Can't Fill My Shoes** (1974), **Let's Put It Back Together Again** (1976), **Middle Aged Crazy** (1977), **Come On In** and **I'll Find It Where I Can** (1978).

By 1979 Jerry Lee had moved over to Elektra, his first Top 10 record for his new label being a bluesy version of the Yip Harburg standard **Over The Rainbow** (1980). During 1981 he played London's Wembley Festival and jammed with Carl Perkins, the duo becoming a Sun rock trio when they were joined onstage by Johnny Cash in Germany later that same tour. Also in 1981 Jerry Lee went Top 10 yet again with **Thirty Nine and Holding**. Then he quit Elektra and joined the MCA roster. Unfortunately his health has been poor – a couple of times he has reportedly been on the point of death. He fought the law (or rather the income tax authorities) and happily managed to win through but he also lost his fifth wife, Shawn, who died from an overdose of methadone in 1983. Sometimes Jerry Lee's biography seems no less than a horror story – two of his sons have died in accidents, two of his wives have met untimely deaths, – but Lewis himself carried on, often making magnificent music, earthy, gutsy, the sort that is the very roots of both country and rock music.

A strange personality, combining arrogance and boastfulness with an apparent respect for religion and traditional Southern values, Jerry Lee is the ultimate country music enigma. He was one of the first performers to be inducted into the Rock'n'Roll Hall Of Fame, in 1986. Three years later, a bio-pic of his early career, 'Great Balls Of Fire', starring Dennis Quaid, brought him back into the pubic eye, with the soundtrack album making an appearance on the US pop charts.

Odd Man In, Jerry Lee Lewis. Courtesy Mercury Records. The album includes the classics Shake, Rattle And Roll and Your Cheatin' Heart.

Recommended:
The Sun Years (–/Charly)
Odd Man In (Mercury/–)
Country Class (Mercury/–)
The Best Of (Mercury/Mercury)
Southern Roots (Mercury/Mercury)
Killer Country (Elektra/Elektra)
I Am What I Am (MCA/–)
Classic Jerry Lee Lewis (–/Bear Family)
Pretty Much Country (–/Ace)
When Two Worlds Collide (Elektra/Elektra)
The Country Sound Of (–/Pickwick)

Country Class, Jerry Lee Lewis. Courtesy Mercury Records.

Light Crust Doughboys

The Doughboys, basically a western swing outfit, first came to life when Bob Wills and Herman Arnspiger began playing as Wills' Fiddle Band. With the addition of vocalist Milton Brown they became the Aladdin Laddies in 1931 and later gained a job advertising Light Crust Flour on Fort Worth radio station KFJZ, at which point they became the Fort Worth Doughboys, then in 1932 the Light Crust Doughboys.

The personnel of the Doughboys changed frequently during the band's career and even by 1933 – the year that the band switched to another Fort Worth station, WBAP – all the original members had departed. O'Daniel restocked the outfit with new members (including his sons).

But despite the changes, the band continued their long association with Burrus Mills until 1942, when they became the Coffee Grinders for a while under the sponsorship of the Duncan Coffee Co. Later, the Doughboys – who in various forms had recorded for Victor and Vocation – reverted to their former and better-known title, but never again achieved the fame that was theirs during the '30s.

Recommended:
The Light Crust Doughboys (Texas Rose/–)
String Band Swing Volume 2 (Longhorn/–)
Live 1936 (–/Flyright)

Little Texas

One of Nashville's 'long hair' acts, Little Texas have brought a fresh, high-energy approach to traditional country, highlighted by soaring vocal harmonies and versatile instrumental work. The band's beginnings can be traced back to 1984, when lead vocalist Tim Rushlow (son of Tom Rushlow, lead singer with '60s band Moby Dick And The Whales) and guitarist Dwayne O'Brien, both originally from Oklahoma, got together in Arlington, Texas to play local clubs. A year later the pair teamed up with two native Texans, lead guitarist Porter Powell and bass player Duane Propes. The foursome took to the road, honing their skills and learning their craft. At a fair in Massachusetts they met up with keyboard player Brady Seals (nephew of songwriter Troy Seals and cousin to Dan Seals) and drummer Del Gray.

At this time the six-man band didn't have a name, but, having cut some demos as Band X, they signed a development and record deal with Warner Brothers in 1988. After passing on the name Possum Flat, they came up with Little Texas, a hollow 35 miles south of Nashville where the band had their first rehearsals. The area was named for its tough characters; in the 1920s it was a lawless place where people on the run would hide out. Warners sent the band out on the road for two years, then took them in the studio at the end of 1990 to cut their first single, **Some Guys Have All The Love**, which started a steady ascent up the country charts the following summer and peaked at No.8.

Little Texas were then packed off to Memphis' legendary Ardent Studios to record an album, **First Time For Everything**, on which they sing all the vocals, play all basic instruments and have written all the songs. The title song became their second Top 10 single in early 1992, with the album also providing three more major hits in **You And Forever And Me** and **What Were You Thinkin'** (both 1992) and **I'd Rather Miss You** (1933). The album went on to reach gold status as

The Era Of Hank Locklin. Courtesy Ember Records, a British release.

Little Texas picked up nominations for both Top New Vocal Group and Top Vocal Group Of The Year at the 1993 ACM Awards. What makes their sound so unique is the combination of five voices representing a whole new generation of country bands. "I feel like we're the first country band that was influenced by young country," says Tim Rushlow. "Sure, we love bands like the Eagles and Poco, but our real influences were Alabama and Restless Heart – country's new sound."

Recommended:
First Time For Everything (Warner Bros/–)
Big Time (Warner Bros/–)

Hank Locklin

Elected mayor of his hometown during the 1960s, Locklin is the possessor of a vocal style that somehow endears him to audiences of Irish extraction.

Born in McLellan, Florida, on February 15, 1918, Lawrence Hankins Locklin played guitar in amateur talent shows at the age of ten. During the Depression years he did almost any job that came his way, gaining his first radio exposure on station WCOA, Pensacola.

At the age of 20 he made his first professional appearance in Whistler, Alabama, and then embarked on a series

Below: An odd individual, Jerry Lee Lewis makes fine country music.

of tours and broadcasts throughout the South, becoming a member of Shreveport's Louisiana Hayride during the late '40s. Record contracts with Decca and Four Star were proffered and duly signed, Locklin gaining two hits with Four Star in **The Same Sweet Girl** (1949) and **Let Me Be The One** (1953). The success of the latter helped him obtain Opry bookings and become an RCA recording artist.

With RCA he began to accrue a number of best-sellers – **Geisha Girl** (1957), **Send Me The Pillow You Dream On** and **It's A Little More Like Heaven** (both 1959), all being Top 10 items. but he surpassed these in sales with the self-penned **Please Help Me I'm Falling**, a 1960 No.1 that provided Locklin with a gold disc – the composer again recording the song in 1970, with Danny Davis' Nashville Brass.

A habitual tourer, Locklin, whose many other hits have included **Happy Birthday To Me** (1961), **Happy Journey** (1962) and **Country Hall Of Fame** (1967), was among the artists who, as part of the 'Concert In Country Music' made the first country music tour of Europe, in 1957. During the '70s, the singer based himself in Houston, appearing on KTR-TV and also on Dallas' KRID Big D Jamboree and by 1975 had become signed to MGM Records.

Recommended:
The First 15 Years (RCA/RCA)
Famous Country Music Makers (–/RCA)
Hank Locklin (MGM/MGM)
Irish Songs Country Style (RCA/RCA)

Josh Logan

A native of Kentucky, Josh Logan worked in an auto wrecking yard in Richmond, Kentucky during the week, and on weekends he performed at local clubs and private parties. That was in the mid-'70s when Nashville was not ready for the pure honky-tonk country style that has always been the Logan trademark. In the early '80s Josh decided to make music a full-time occupation. He recorded for several independent labels. One release, **I Made You A Woman For Somebody Else** (NSD, 1981), started to gain extensive radio plays, only for Conway Twitty to put the song on the B-side to his **Tight-Fittin' Jeans** hit.

Logan signed with Charley Pride's booking agency Chardon and undertook several major tours as Earl Thomas Conley's opening act. Following this, he worked several club dates with Sandy Powell (sister of Sue Powell – a member of Dave And Sugar). In 1987 he came to the attention of producer Nelson Larkin, who recorded some sides in the hope of setting up a deal with RCA. They liked what they heard, but were only signing girl acts, so Josh signed with Curb and made his debut on the country charts with **Everytime I Get To Dreamin'** (1988). The following year he gained another chart entry with **Somebody Paints The Wall**, the title song to what is, so far, his only album.

Recommended:
Somebody Paints The Wall (Curb/–)

Lone Justice

A West Coast rock band formed in 1984, Lone Justice was always a vehicle for lead singer Maria McKee, a talented songwriter, vocalist and guitarist, who was born on August 17, 1964 in Los Angeles. She started performing in local L.A. clubs from the age of 16 with her half-brother Bryan MacLean, who had cut his musical teeth playing in the rock band Love. Lone Justice landed supporting roles on tours with both U2 and the Alarm during 1984. Bob Dylan caught the band doing a couple of his songs, and, suitably impressed, he paved the way for them to sign with Geffen Records in 1985. Their debut album, **Lone Justice**, was aimed at country-rock fans with McKee's strident vocals having that edge to appeal to both markets. Most songs were penned by McKee, with **Sweet, Sweet Baby (I'm Falling)** and Tom Petty's **Ways To Be Wicked**, both hitting the pop singles chart, while the album also climbed the pop charts, earning the band a gold disc. Developing as a writer, McKee wrote **A Good Heart**, a No.1 British pop hit for Feargal Sharkey. She also duetted with Dwight Yoakam on **Bury Me**, a track included on **Guitars, Cadillacs**, Yoakam's debut album in 1985. Lone Justice topped the British pop charts in 1988 with **I Found Love**, while they continued to straddle a fine line between country and rock stylings. Maria McKee embarked on a solo career in 1991, but has remained close to that hard-edged country rock sound.

Below: Maria McKee of Lone Justice flavours her music with country.

Recommended:
Lone Justice (Geffen/–)

Lonzo And Oscar

Really the Sullivan brothers, John (Lonzo) was born in Edmonton, Kentucky, on July 7, 1917, and Rollin (Oscar) was born in Edmonton, Kentucky, on January 19, 1919. Lonzo And Oscar were the top comedy act on the Opry, their 20-year stint being terminated by the death of Johnny Sullivan on June 5, 1967. Originally there was another Lonzo, a performer named Ken Marvin (real name Lloyd George) teaming with Rollin in pre-World War II days and recording a nationwide comedy hit, **I'm My Own Grandpa**, a song penned by the Sullivans. The act went into store while the brothers became part of the armed forces. Shortly after their return to civilian life, Ken Marvin retired. John assumed the guise of Oscar, the duo touring with Eddy Arnold until 1947 – in which year the Sullivans became Opry regulars.

Some time after the death of John, Rollin Sullivan again resurrected Lonzo And Oscar using a new partner Dave Hooten. In its various permutations over the years, the act recorded for RCA, Decca, Starday, Nugget, Columbia and GRC.

Recommended:
Traces Of Life (GRC/–)

John D. Loudermilk

The writer of such hits as **Talk Back Trembling Lips**, **Tobacco Road**, **Abilene**, **Ebony Eyes**, **Indian Reservation**, **Language Of Love**, **Norman**, **Angela Jones**, **Sad Movies** and **A Rose And A Baby Ruth**, also co-writer (with Marijohn Wilkin) of **Waterloo**, Loudermilk was once a Salvation Army bandsman. Born in Durham, North Carolina, on March 31, 1934, he learned to play trumpet, saxophone, trombone and bass drum at Salvationist meetings, later learning to play a homemade ukelele which he took to square dances. Although he made his TV debut at the age of 12 – with Tex Ritter, no less – his big break came in the mid-'50s, when he set a poem to music and performed on TV. George Hamilton heard Loudermilk's composition and recorded it, the result – **A Rose And Baby Ruth** – released in 1956, selling more than a million copies.

After penning **Sittin' In The Balcony**, a 1958 smash hit for Eddie Cochran, Loudermilk married Gwen Cooke, a university student and headed for Nashville, there meeting Jim Denny and Chet Atkins. His own recording of **Language Of Love** became a huge hit on both sides of the Atlantic during the winter of 1961–62 but all his subsequent releases have made but slight chart indentations. He is, nevertheless, an onstage performer of considerable talent and charm and has always been a great favourite in Britain.

Recommended:
A Bizarre Collection Of The Most Unusual Songs (RCA/–)
Country Love Songs (RCA/–)
Encores (RCA/RCA)
Blue Train (–/Bear Family)
It's My Time (–/Bear Family)
Elloree (Warner Bros/–)

Louvin Brothers

The Louvin Brothers, Ira and Charlie, formed one of the finest duos in country music, offering superb close harmony vocals that often displayed their gospel roots.

Born in Rainesville, Alabama (Ira on April 21, 1924; Charlie on July 7, 1927), the Louvins (real name Loudermilk) were raised on a farm where they first learned to play guitar. Drafted into the forces during World War II, they returned to music at the cessation of their active service, gaining dates on Knoxville's KNOX Mid-Day Merry-Go-Round. However, just when the brothers seemed to be making the grade, Charlie was recalled for duty during the Korean crisis.

Once more, the Louvins had to re-establish themselves. Following some appearances on a radio show in Memphis, they signed a recording contract with MGM Records, followed by a signing with Capitol Records. By 1955 they had become Opry regulars, also having a hit record with their self-penned **When I Stop Dreaming**, a disc which sparked off a run of similarly successful singles by the Louvins during the period 1955–62. Then, after one last duo hit via **Must You Throw Dirt In My Face?**, the Louvins decided to go their separate ways. Charlie proved the more popular of the two, with three 1964 chart records, **I Don't Love You Anymore**, **See The Big Man Cry** and **Less And Less**. Just a few months later, Ira was dead, the victim of a head-on car accident near Jefferson City, Missouri (June 20, 1965). His wife Florence, who sang under the name of Anne Young, was also killed in the crash.

Since that time, Charlie Louvin has continued as a top flight country entertainer. He also provided Capitol with a number of chart entries before leaving the label to join United Artists in the fall of 1973 and immediately scoring with **You're My Wife, She's My Woman** (1974). He has still notched the odd hit or two in recent times – he and Emmylou Harris charted with **Love Don't Care**, a Little D release, in 1979, while in 1982 he teamed with Jim And Jesse for **North Wind**, a mid-chart entry for the Soundwaves label – and at the beginning of the '80s claimed to be averaging 100,000 miles a year playing concert dates.

Recommended:
Great Gospel Singing Of The Louvin Brothers (Capitol/–)
Tragic Songs Of Life (Rounder/–)
The Louvin Brothers (Rounder/–)
Running Wild (–/Sundown)
Close Harmony (–/Bear Family)
Live At New River Ranch (Copper Creek/–)
Sing Their Hearts Out (–/See For Miles)

Charlie Louvin:
Somethin' To Brag About – with Melba Montgomery (Capitol/–)
50 Years Of Making Music (Playback/Cottage)
I Forgot To Cry (Capitol/Stetson)

Patty Loveless

A genuine coal miner's daughter, Patty Loveless (born Patricia Ramey on January 4, 1957 in Pikeville, Kentucky) spent her younger years in the tiny eastern Kentucky community of Beecher Holler. She was one of eight children who watched their father's health ebb away to black lung disease. When Patty was ten the family moved north to Louisville so her father could receive medical treatment. Her brother, Roger Ramey, introduced her to local country music shows, and by 1971 they were working together as a duo. Patty had already started writing songs and the pair drove to Nashville armed with around 30 songs and high ambitions. The Wilburn Brothers heard her sing and invited her on their show to replace Loretta Lynn. They signed Patty to their publishing company, and featured her in their weekly TV series and as a member of their road show from 1973 to 1975.

In 1976 she married Terry Lovelace, the Wilburns' drummer, and moved to Charlotte, North Carolina, where she continued her singing, mainly with rock and pop bands. When the marriage didn't work out, she decided to pursue a country career again. With help from her brother, Roger, who became her manager, and using the name Patty Loveless, she worked local clubs and opened shows for such major acts as Jerry Reed, Pure Prairie League and Hank Williams Jr. In early 1985 Patty again travelled to Nashville and recorded some of her own tunes as demo tapes. She became a staff writer with Acuff-Rose and Tony Brown signed her to MCA Records later that year. After scoring several minor hits, Patty finally made a breakthrough with a Top 10 entry of **If My Heart Had Windows** (1988). Her soulful voice with its high lonesome mountain edge evoked real emotion, and for the next few years she seldom scored outside the Top 10. Her hits included **A Little Bit In Love** and **Blue Side Of Town** (1988), **Don't Toss Us Away**, **The Lonely Side Of Love**, and her first No.1 – **Timber I'm Falling In Love** (all 1989). Patty scored another chart-topper with **Chains** in 1990, plus further Top 10 entries with **On Down The Line** (1990), **I'm Not That Kind Of Girl** and **Hurt Me Bad (In A Real Good Way)** (1991).

By this time, Patty had married again New husband Emory Gordy Jr, one-time member of Emmylou Harris' Hot Band, had co-produced her first two albums with Tony Brown. The pair were living in Georgia, though they also had an apartment in Nashville, and by early 1993 they had bought a townhouse in Music City, obviously anticipating a move. Alongside her own recordings, Patty had built a reputation for the soulful, ethereal harmonies she provided for Vince Gill's hits, such as **When I Call Your Name**. She gained a gold disc for her impeccable **Honky Tonk Angel** album, but has yet to break through to superstar status. Feeling that there were too many female singers with MCA (Trisha Yearwood, Reba McEntire and Wynonna), Patty joined Epic Records in 1992.

In the midst of recording her first album for the new label (produced by Emory Gordy Jr), she had to undergo voice-saving laser surgery on one of her vocal chords, which proved a successful operation. With her winning combination of wholesome beauty, shy, sweet nature and powerful, emotional voice, Patty Loveless is a singer yet to reach her peak of achievements and accolades.

Above: This honky-tonk angel with an edge is Patty Loveless.

Recommended:
Only What I Feel (Epic/–)
If My Heart Had Windows (MCA/–)
Honky Tonk Angel (MCA/MCA)
On Down The Line (MCA/MCA)

Lyle Lovett

Lyle Lovett isn't your typical country singer – his music is as different as his distinctive high hair, exploring not only country, but jazz, gospel and blues. He was born on November 1, 1957 in Houston, Texas, but grew up in the rural Klein community, an area populated by farmers of German extraction. After graduating from Klein high school, Lovett attended Texas A&M University, where he received a degree in journalism in 1980 and a year later one in German. It was at this time that he first started writing songs and performing in the songwriter showcase clubs of Houston, Dallas and Austin. He visited Europe in 1979, to improve his German, and became friendly with a local country musician named Buffalo Wayne. When he completed his university studies, Lovett started playing further afield, from New Mexico to New York. In 1983 Wayne was involved in Luxembourg's annual fair, the Schueberfouer, and invited Lovett to appear. Also on the bill was a Texan outfit, J. David Sloan And The Rogues, whose members included Matt Rollings and Ray Herndon; both were later involved with Lovett's first recordings, made in Phoenix in the summer of 1984.

The tape was taken to Nashville by Guy Clark, who passed it to Tony Brown, who in turn wasted no time in signing Lyle to MCA/Curb Records in 1985. His songs started to be picked up by other performers including Nanci Griffith and Lacy J. Dalton.

Lyle made an impact on the country charts with **Cowboy Man** (1986), **Why I Don't Know** and **Give Me Back My Heart** (1987) and **She's No Lady** (1988). His albums, which combined elements of western swing, '70s singer-songwriter and contemporary country, gained rave reviews outside mainstream country music, and it was obvious that Lovett could never be constrained by the confines of country.

He writes most of his music, which leans towards dark humour and satire. His first albums featured such guests as Vince Gill, Rosanne Cash, Emmylou Harris and members of the Phoenix band that played on the original demo tape. In 1989 he won a Grammy for Best Vocal Performance By A Male Artist, but by this time was far removed from country music. He moved from Nashville and started recording in Los Angeles. In 1992 Lyle was chosen as the opening act for Dire Straits' first world tour of the '90s. In June of 1993 Lyle married Hollywood actress Julia Roberts in a highly secretive, surprise wedding.

Below: Cult country singer and film extra, Lyle Lovett.

Recommended:
Joshua Judges Ruth (MCA-Curb/MCA-Curb)
Cowboy Man (MCA/MCA)
Pontiac (MCA/MCA)
Lyle Lovett And His Large Band (MCA/MCA)

Lulu Belle And Scotty

Husband-wife teams have long been a staple of country music performance but one of the earliest and most popular was Lulu Belle And Scotty, mainstays of the National Barn Dance from 1933 to 1958.

Lulu Belle was born Myrtle Eleanor Cooper in Boone, North Carolina, on December 24, 1913. Active musically as a teenager, she auditioned for the National Barn Dance in 1932 and was hired, immediately becoming one of the stars of the show. She often teamed with Red Foley (then bass player with the Cumberland Ridge Runners) in duets.

In 1933 another cast member was added to the National Barn Dance: a guitarist, banjoist, singer and songwriter

Above: Husband and wife team Lulu Belle and Scotty were regulars on the National Barn Dance for 25 years.

named Scott Wiseman, known professionally as Skyland Scotty. Born near Spruce Pine, North Carolina, on November 8, 1909, Scotty had appeared on radio over WVRA in Richmond as early as 1927, and on WMMN, Fairport, West Virginia, while attending Fairport Teachers College.

The two hit it off and became a very popular team (although some listeners wrote angry letters to WLS, thinking Scotty had 'stolen' Red Foley's girl), largely on the basis of their smooth duet sound and on Scotty's prolific songwriting, which produced such country standards as **Mountain Dew** (co-written with Bascomb Lamar Lunsford), **Remember Me**, the folk favourite **Brown Mountain Light**, and their biggest hit, **Have I Told You Lately That I Love You?**

They also appeared as stars of several films based around the National Barn Dance cast, including 'Village Barn Dance', 'Hi Neighbor', 'Country Fair', 'Sing, Neighbor, Sing' and 'National Barn Dance'. They spent a brief time away from the Barn Dance at the Boone County Jamboree in Cincinnati (1938–41), but were closely associated with Chicago, where they had a daily TV show over WNBQ from 1949–57.

Scotty began working towards a masters degree in education during the '50s and when the act bowed out of the performing limelight in 1958 he and Lulu Belle retired to their native North Carolina, where Scotty finally fulfilled his early ambition to teach, Lulu Belle spending a couple of terms in state legislature. Scotty died of a heart attack in 1981, in Florida.

Recommended:

Have I Told You Lately That I Love You? (Old Homestead/–)
Lulu Belle And Scotty (Starday/–)
Sweethearts Still (Starday/–)

Bob Luman

Once a teenage rockabilly, Luman became an Opry regular in August 1969. Born in Nachogdoches, Texas, on April 15, 1937, Robert Glynn Luman spent much of his boyhood listening to various country and R&B shows on the radio. Encouraged to pursue a musical career by this father, he learned guitar but was torn between continuing as a musician or as a baseball player. But after being offered a trial with the Pittsburgh Pirates during the mid-'50s, he flunked it and from then on concentrated his energies on becoming a rock star in the Presley mould.

In 1955, Luman recorded tracks for a small Dallas company. A short time later he won an amateur talent contest that resulted in a spot on Shreveport's Louisiana Hayride. Record dates with Imperial and Capitol followed, plus Las Vegas bookings and a part in the rock movie 'Carnival' (1957) but it was not until 1960 and the Warner release of a Luman single called **Let's Think About Livin'** that the real breakthrough came. The Boudleaux Bryant song provided the singer with a residency in both the pop and

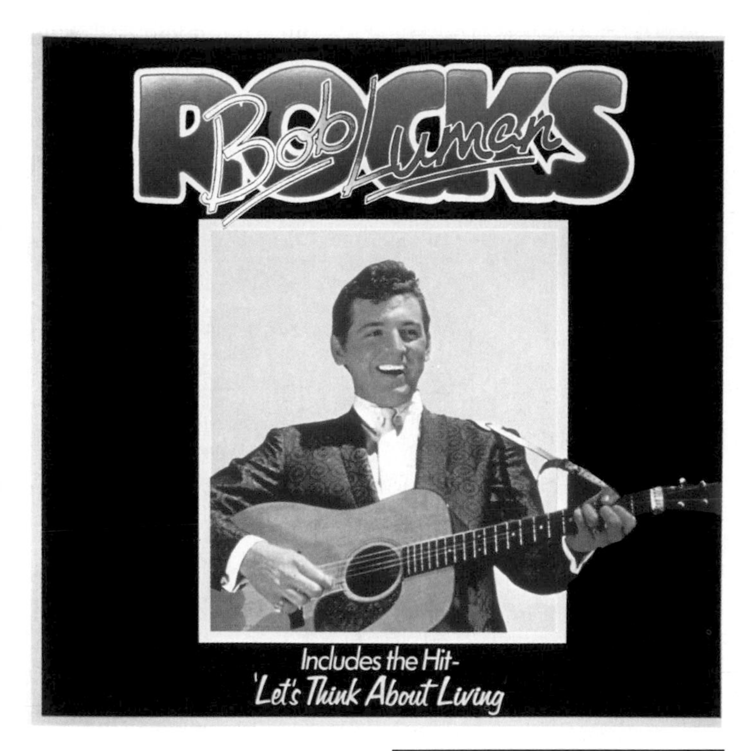

Bob Luman Rocks. A British compilation. Courtesy DJM Records.

country Top 10. He was, however, unable to capitalize on this position. A reservist, Luman was called up for active duty, Jim Reeves taking over his band during Luman's army stay.

With the advent of the Beatles, Luman once more turned to country, signing for Hickory and Epic and logging over three dozen hits, the biggest being: **The File** (1964), **Ain't Got The Time To Be Unhappy** (1968), **When You Say Love, Lonely Women Make Good Lovers** (1972), **Neither One Of Us, Still Loving You** (1973) and **The Pay Phone** (1977). From 1976 he suffered health problems which affected his career but in 1977 he made an album, produced by Johnny Cash, that proclaimed he was **Alive And Well**. But well he was not. He died in December 1978.

Recommended:
Alive And Well (Epic/–)
The Rocker (–/Bear Family)
Let's Think About Livin' (–/Sundown)
Loretta (–/Sundown)

Robert Lunn

As the 'Talking Blues Boy' (or the 'Talking Blues Man'), Robert Lunn brought an unusual form of both comedy and blues to the Grand Ole Opry for two decades and was long the country's foremost exponent of the talking blues, a style which was to become a staple of folk song revival.

Lunn was born in Franklin, Tennessee, on November 28, 1912. He apprenticed in vaudeville before joining Opry in 1938, and he stayed with the show, except for service in World War II, until 1958. A left-handed guitar player, he rarely sang, relying instead on his droll, dry, talking blues recitations, most of which he wrote himself. His only recording was a long out-of-print Starday album called **The Original Talking Blues Man**. Lunn died of a heart attack on March 8, 1966.

Right: Robert Lunn's forte, the talking blues, was popularized on the Opry. This style was brought to a greater prominence by the great Woody Guthrie.

Frank Luther

Often remembered best for his children's records, Frank Luther actually had a long career in country music as well as some success in the pop field. Born Frank Crow in Kansas, on August 5, 1905, he grew up in Bakersfield, California, and his early musical experience was as a singer and pianist with gospel quartets.

He moved to New York in the late '20s, where he teamed up with Carson J. Robison as a recording act (frequently called on record Bud and Joe Billings) and as songwriters, collaborating on **Barnacle Bill The Sailor** and **What Good Will It Do?** He and his wife Zora Layman also did extensive recording, some of it with Ray Whitley.

He recorded a wide variety of country material in the '20s and '30s, for such labels as Victor, Conqueror and Decca. He moved into the field of children's recording in the late '30s and '40s, recording stories, ballads and cowboy songs, largely for the Decca label.

In addition, he lectured on American music and even wrote a book on the subject: 'Americans And Their Songs'.

Other achievements include early country music films (c.1933), authorship of some 500 songs and a good bit of popular recording as well as country. In the '50s he moved into an executive role before finally retiring in the New York area.

Few of his country recordings are available, but he can be heard on Carson J. Robison's **Just A Melody**, an Old Homestead album.

Loretta Lynn

CMA Female Vocalist Of The Year in 1967, 1972, 1973 and, with Conway Twitty, three times winner of the Association's Vocal Duo Of The Year section, Loretta, the daughter of Melvin Webb, a worker in the Van Lear coal mines, was born in Butcher's Hollow, Kentucky, on April 14, 1935.

Part of a musical family, she sang at local functions in her early years, marrying Oliver 'Moonshine' Lynn (known as Mooney) immediately prior to her fourteenth birthday. In the '50s, the Lynns moved to Custer, Washington, where Loretta formed a band that included her brother Jay Lee Webb on guitar. Later, signed to Zero Records, the diminutive vocalist hit the charts with **Honky Tonk Girl**, a 1960 best-seller. She and Mooney toured in a 1955 Ford in order to promote the record.

The Wilburn Brothers were impressed enough to ask Loretta to come to Nashville, where Mooney took a job in a garage to support his four daughters, while Loretta and the Wilburns tried to negotiate a record deal – the singer eventually signing for Decca.

With a song appropriately titled **Success**, she broke into the charts in 1962, at the same time winning the first of her numerous awards. And for the rest of the '60s and all of the '70s, Loretta became the most prolific female country hit-maker

in Nashville. Virtually every one of her releases made the Top 10 during this period, many of them, including **Don't Come Home A-Drinkin'** (1966), **Fist City** (1968), **Woman Of The World** (1969), **Coal Miner's Daughter** (1970), **One's On The Way** (1971), **Rated X** (1972), **Love Is The Foundation** (1973), **Trouble In Paradise** (1974), **Somebody Somewhere (Don't Know What He's Missin' Tonight)** (1976), **She's Got You** (1977) and **Out Of My Head And Back In My Bed** (1977), reaching the premier position. And her series of duets with Conway Twitty claimed impressive sales figures, the duo enjoying joint No.1s with **After The Fire Is Gone**, **Lead Me On** (1971), **Louisiana Woman, Mississippi Man** (1973), **As Soon As I Hang Up The Phone** (1974), and **Feelin's** (1975).

A grandmother at 32, Loretta has six children. The owner of various business interests, she also owns the whole town of Hurricane Mills, Tennessee, where she resides. Ever-popular – even in the '80s she continued to notch up more than a dozen chart records including such Top 10 singles as **It's True Love** (1980) and **I Still Believe In Waltzes** (1981), both duets with Conway Twitty, and a solo effort with **I Lie** (1982) – the singer suddenly found herself with a whole new host of fans in 1980 when her autobiography, 'Coal Miner's Daughter', was turned into a much-hailed movie, with Sissy Spacek portraying Loretta and Tommy Jones playing Mooney. The first female artist to win the CMA's coveted Entertainer Of The Year award (1972), Loretta has never been above a bit of controversy when acting as a spokesperson on behalf of downtrodden womanhood. Her songs included such feminist banner-wavers as **The Pill**, a 1975 hit that endorsed birth control.

A very wealthy woman, Loretta owns property in Mexico and Hawaii, and is a popular guest on American chat shows. Still active on the touring circuit, her magnificent soprano is the very voice of rural heart and heartache. In 1988 she became one of the most popular stars to be inducted into the Country Music Hall Of Fame.

Recommended:
I Remember Patsy (MCA/MCA)
Greatest Hits Volumes 1 & 2 (MCA/–)
The Very Best Of Conway And Loretta (MCA/MCA)
Coal Miner's Daughter (MCA/MCA)
Just A Woman (MCA/MCA)
Don't Come Home A-Drinkin' (MCA/–)
Loretta Lynn Story (–/Music For Pleasure)
Sings Country (–/Music For Pleasure)
Country Partners – with Conway Twitty (MCA/–)

Shelby Lynne

A young lady with a distinctive blues-flavoured vocal style, Shelby Lynne (born on October 22, 1969 in Quantico, Virginia) made her initial impact in Nashville in 1987, when she appeared on TNN's Nashville Now series. The very next day she had four record label executives all eager to sign her. A tough young lady with a determined attitude, she decided to wait so that she could sign with Billy Sherrill at Epic Records. For this lady, no other label would do, she wanted what she considered to be the best.

Shelby had been around music for a number of years, previously working clubs in Alabama. She had grown up in Jackson, Alabama, where her father was a local bandleader. Along with younger sister, Alison, she would often add vocal harmony support at her father's night-club and Holiday Inn gigs. Music played a major role in the Lynne household. As a child she grew up on a rich diet that ranged from the Mills Brothers to Barbra Streisand, Bob Wills to Les Paul and Mary Ford. In 1986 the Lynne family was torn apart when their heavy-drinking father killed himself and their mother in a shooting tragedy in the driveway of their home. To overcome her grief, Shelby immersed herself in music, and entered singing contests. Following an unsuccessful Opryland audition, she was asked to demo some songs for a local songwriter. It was this demo tape that led to the appearance on Nashville Now.

Her first single, a duet with George Jones on **If I Could Bottle This Up**, made the country charts in 1988, and Shelby has since charted on a regular basis with such songs as **The Hurtin' Side** (1989), **I'll Lie Myself To Sleep** and **Things Are Tough All Over** (1990) and **What About The Love We Made** (1991), without making the breakthrough to the Top 10. In 1991 she gained the ACM's Horizon Award and has produced three albums for Epic.

Above: The diminutive, Virginia-born Shelby Lynne is a powder keg vocalist ready to explode.

Recommended:
Temptation (Morgan Creek/–)
Soft Talk (Columbia/–)
Tough All Over (Columbia/–)

Below: Loretta Lynn's biography was made into the Oscar-winning film, 'Coal Miner's Daughter', in 1980.

Mac And Bob

Lester McFarland (mandolin and vocals), born on February 2, 1902, in Gray, Kentucky, and Robert Alexander Gardner (guitar and vocals) born on March 16, 1897, in Olive Springs, Tennessee, met each other at the Kentucky School for the Blind in their middle teens and became one of the first of the mandolin-guitar duet teams that became so popular in America in the middle 1930s.

They spent several long stints with the National Barn Dance (1931–34, 1939–50) as well as on KNOX, Knoxville (1925–31) and KDKA, Pittsburgh and KMA, Shenandoah, Iowa. They began recording with Brunswick in 1926 and also recorded for the American Record Company complex of labels, Conqueror, Columbia, Dixie, Irene and others. They were best known for **When The Roses Bloom Again**, but introduced many old-time songs and ballads to the repertoires of the duet teams that followed. Although they were long favourites, their sound was rather stiff and they were superseded by later duos.

Bob retired in 1951, while Mac went on as a solo until 1953. A talented musician, he also played piano, trumpet, cornet and trombone.

Mac McAnally

Singer-songwriter Mac McAnally (born on July 1, 1959 in Belmont, Mississippi) first tasted success while still a teenager, when his self-penned song, **It's A Crazy World**, became a Top 40 pop hit in 1977. Musically, McAnally cut his teeth on his mother's gospel piano-playing. His first gig, when he was only 13, was as a piano player at 'state-line' clubs north of the Tennessee–Mississippi border. Alcohol sales were, and still are, illegal in Belmont. In search of another place to play, Mac took his tasteful finger-pickin guitar style to Muscle Shoals, Alabama, where he still lives. Working in the famous studios as a guitar player, he played some of his own songs when a session was cancelled. Before he knew it, his first album had been completed. **It's A Crazy World** went to No.2 on the Adult-Contemporary charts and broke into the pop Top 40. The song was later revived by country star Steve Wariner.

The song established McAnally as a singer-songwriter with a unique artistic voice, but he maintained his session work and blossomed as a songwriter, penning hits for Jimmy Buffett (**It's My Job**), Alabama (**Old Flame**), Shenandoah (**Two Dozen Roses**), Steve Wariner (**Precious Thing**) and Ricky Van Shelton (**Crime Of Passion**). He also became involved in production, co-producing Ricky Skaggs' album **My Father's Son**. All this behind-the-scenes work not only keeps the bills paid and his record-making skills intact, but allows Mac to spend time with friends, who are otherwise on the road.

In 1989, he signed with Warner Brothers Records, producing the album, **Simple Life**, a brilliant collection of his own songs that featured contributions from Ricky Skaggs, Vince Gill, Tammy Wynette, Mark Gray and Mark O'Connor. He made a Top 20 entry on the country charts with the autobiographical **Back Where I Came From** (1990). Two years later he landed on MCA, with Tony Brown producing yet another superb collection of vignettes from real life, under the title **Live And Learn**. Mac specializes in compelling stories told with sensitive lyrics, seasoned with a dash of wry humour, a style that is perhaps at odds with '90s honky-tonk.

Recommended:
Live And Learn (MCA/–)
Simple Life (Warner Bros/Warner Bros)

Leon McAuliffe

"Take it away, Leon", was Bob Wills' famous cry which made Leon McAuliffe's name a household word in the Southwest during the heyday of western swing. Although he rose to prominence with the Texas Playboys, he actually had a long career of his own as well.

Born William Leon McAuliffe on January 3, 1917, in Houston, Texas, he joined the Light Crust Doughboys in 1933, at the age of 16, and began his famous association with Wills in 1935. One of the first to electrify his steel, he popularized the sound for many years with Wills.

After his return from World War II, Leon set up his own band, the Cimarron Boys, in Tulsa. He recorded for Columbia through to 1955, then with Dot, ABC, Starday, Capitol, his own label Cimarron, and also Stoneway, before leading the newly revived Texas Playboys on Capitol. His biggest hits were **Blacksmith Blues** and **Cozy Inn**, though he wrote and performed many western swing and steel guitar classics while a Texas Playboy. These included the following notables: **Steel Guitar Rag**, **Panhandle Rag**, **Bluebonnet Rag**.

As western swing faded in popularity in the late '50s, Leon developed other business interests including two Arkansas radio stations. But by the late '70s Leon was back in the band business once more, leading a revitalized version of the Original

Texas Playboys, and recording for Capitol and Delta. He died on August 20, 1988, in Tulsa, Oklahoma.

Recommended:
Bob Wills' Original Texas Playboys Today (Capitol/–)
Cozy Inn (ABC/–)
Everybody Dance, Everybody Swing (–/Stetson)

McBride And The Ride

McBride And The Ride's electrifying three-part harmony vocal blend saw Billy Thomas, Terry McBride and Ray Herndon riding the crest of a wave of success. The trio came together in 1989, but the members brought to the band three lifetimes of country music experience.

Lead singer and bassist Terry McBride was born in Austin, Texas, and grew up 60 miles down the road in the small ranching community of Lampasas. He started playing guitar at the age of nine and by his early teens had already played with his father's band. After high school he spent three years on the road with his father, Dale McBride, who had eleven country chart hits on independent labels. Terry followed with a two-year stint in Delbert McClinton's band, then moved to Austin. Guitarist Ray Herndon also grew up with music. In his early teens he played in his father's band in Scottsdale, Arizona. Later, as a member of J. David Sloan And The Rogues, he played on some demos for the then unknown Lyle Lovett. The demos became the basis of Lovett's first MCA album and the band became the core ensemble of Lovett's Large Band. Drummer Billy Thomas started playing the kit in sixth grade and grew up working in rock bands around Fort Myers, Florida. In 1973 he moved to Los Angeles, where he toured with Rick Nelson, Mac Davis and the Hudson Brothers. After moving to Nashville in 1987, he toured and recorded with Vince Gill and Emmylou Harris.

MCA Nashville President Tony Brown brought the three together in 1989 and produced the debut album, **Burnin' Up The Road**, which didn't even get off the starting grid. The first two singles, **Every Step Of The Way** and **Felicia**, failed to chart, so the three veteran musicians loaded up a van and took to the road for a heavy schedule of live appearances. Slowly building a fan base, they made the Top 20 with **Can I Count On You** (1991). Another, **Same Old Star** (1991), also charted, but it was **Sacred Ground**, their first No.1 in 1992, which really established the strong, modern driving sound of McBride And The Ride. The album of the same title gained them a gold disc and provided further Top 10 hits in **Going Out Of My Mind** and **Just One Night** (1992).

Recommended:
Sacred Ground (MCA/–)
Hurry Sundown (MCA/–)
Burnin' Up The Road (MCA/–)

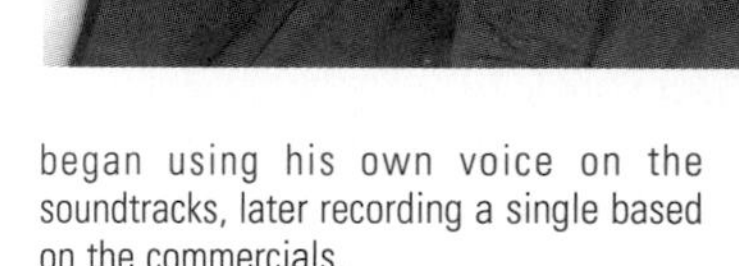

Above: McBride And The Ride are possessors of a trio of smooth voices.

Below: Martina McBride went from t-shirt saleswoman to major star in a year.

Martina McBride

One of the many young female singers to hit Nashville in the early '90s, Martina McBride (born Martina Schiff on her family's Kansas farm in 1969) was raised on the traditional country music of Merle Haggard, Buck Owens and Hank Williams. Growing up in a small Texas town, she sang at various VFW and barn dances in her father's band, the Schiffters, in which she also played keyboards. Following graduation from high school, Martina toured Kansas with several bands and ended up marrying soundman John McBride. In 1990 the couple moved to Nashville. John handled sound for Charlie Daniels, Ricky Van Shelton and other stars, while Martina waitressed and sang songwriter demos. Eventually John produced a five-song demo of his wife and they took it door-to-door around the Nashville record labels.

Eventually the foot-slogging paid off and Martina was offered a recording contract with RCA. By this time John had also moved up the ladder and was Garth Brooks' production manager, planning and organizing equipment for the superstar's tours. Martina took to the road with the Brooks' entourage, selling T-shirts and other merchandise throughout 1991. Her love for country music's traditionalism was an obvious accent on her first album, **The Time Has Come**. She had a minor hit single with the title song in 1992, and with the controversial **Cheap Whiskey** (with Garth Brooks adding harmonies, 1992) and **That's Me** (1993).

The association with Brooks paid dividends when she was selected as the opening act for his 1992 tour dates. Standing up for modern women, Martina says, "I try to portray women with dignity and respect. I don't want to do any songs about women getting walked on, treated badly and putting up with it all."

Recommended:
The Time Has Come (RCA/–)

C. W. McCall

McCall (real name William Fries, born Audubon, Iowa, November 15, 1928) worked his way up the ladder in the advertising industry, and won a 1973 Cleo award for a TV campaign he masterminded on behalf of the Metz Bread Company. Creating a fictional Old Home Bread truckdriver called C. W. McCall as lead character in this series of commercials, he began using his own voice on the soundtracks, later recording a single based on the commercials.

Adopting his McCall guise he cut **The Old Home Filler-Up and Keep On A-Truckin' Café** (1974), the result eventually becoming a national hit. With truckers fast becoming the folk heroes of the '70s, Fries aimed further narrative-type singles at this market, scoring with such releases as **Wolf Creek Pass**, **Classified**, **Black Bear Road** and **Convoy**, the last named becoming a worldwide multi-million seller in early 1976.

That same year, Fries/McCall left MGM Records and signed for Polydor, entering the country music charts with **There Won't Be No Country Music**. During 1977 he went Top 10 once more with **Roses For Mama**, but when his Polydor contract ended a few months later he decided to quit music and return to advertising, gaining some added income when Sam Pekinpah decided to make a 1978 film based on Fries' **Convoy** hit..

In 1982 Fries and his family moved to Ouray, Colorado, where the former performer became involved in local politics.

Recommended:
Black Bear Road (MGM/MGM)
Wilderness (Polydor/–)

The McCarters

A traditional-sounding trio from Sevierville in the Smoky Mountains of Tennessee, the McCarters, comprising Jennifer McCarter and her younger twin sisters, Lisa and Teresa, brought refreshing mountain harmonies to mainstream country music for an all-too-brief period in the late '80s.

The sisters grew up surrounded by country music. Their father, Gerald McCarter, a factory foreman, played banjo in local bands, while his wife was a popular gospel singer at the local Baptist church. Watching clog dancing on TV, the three girls learned the routines and appeared with their father's band. Soon their skills landed them a spot on local TV. Three years later, when Jennifer was 14, she made a move into music, learning guitar and singing, with the twins adding harmonies. During the next three years, they played local shows, gaining experience working with Opry performer Stu Phillips and comedian Archie Campbell. In 1986, the girls made a big push for the big time. Jennifer started contacting Nashville record companies, and eventually gained the attention of producer Kyle Lehning, who arranged an impromptu audition in his office. Suitably impressed, he signed the McCarters to Warner Brothers. With the help of top Nashville pickers, they produced **The Gift**, a beautiful debut album, which contained their two Top 10 hits **Timeless And True Love** and **The Gift** in 1988.

The McCarters toured extensively, initially as a opening act for Randy Travis. Their mountain harmonies were laced with energy and power, and expressed with infectious enthusiasm. By the time they released their second album, **Better Be Home Soon**, in 1990, they had become known as Jennifer McCarter And The McCarter Sisters. Failing to score any more major hits, the sisters were dropped by Warner Brothers in 1992, a prime example of a major label unable to promote and build a traditional-flavoured country act.

The Gift, the McCarters. Courtesy Warner Bros Records.

Recommended:
The Gift (Warner Bros/Warner Bros)
Better Be Home Soon (Warner Bros/–)

Charly McClain

Born Charlotte Denise McClain in Memphis, Tennessee, on March 26, 1956, Charly sang and played bass in her brother's band at the age of nine. At 17 she became a regular on Memphis' Mid-South Jamboree. Urged by Ray Pillow, she went onstage with Shylo at a fair, after which Shylo's producer Larry Rogers cut a demo, which he handed on to Billy Sherrill. By 1976 she had been signed to Epic, her first single being **Lay Down**. In 1978 Charly went Top 20 with **Let Me Be Your Baby**, climbing even higher with **That's What You Do To Me**.

Live At Randy's Rodeo, O.B. McClinton. Courtesy Enterprise.

She kept up the flow with four more major singles in 1979 – including **I Hate The Way I Love It**, a duet with Johnny Rodriguez – and an additional quartet during 1980, the year that she gained her first chart-topper with **Who's Cheatin' Who?** Awards were showered on her. She responded by placing three singles in the Top 10 during 1981 (**Surround Me With Love**, **Sleepin' With The Radio On**, **The Very Best Is You**) and also going Top 10 in the album charts.

The girl George Jones dubbed the 'Princess of Country Music' has added to her toll of Top 10 singles with: **Dancing With Your Memory**, **With You** (1982), **Sentimental Ol' You** (1983), **Paradise Tonight** (a duet with Mickey Gilley that went to No.1 in 1983), **Candy Man** (another Gilley–McClain duet, 1984) and **Radio Heart** (No.1 in 1985). Charly married singer/actor Wayne Massey in 1984 and the pair enjoyed two Top 10 duet hits, **With Just One Look In Your Eyes** (1985) and **You Are My Music, You Are My Song** (1986). Her most recent solo success came with **Don't Touch Me There**, which spent six months on the charts in 1986. In 1988 Charly signed with Mercury and scored a few minor hits.

Recommended:
Paradise (Epic/–)
Women Get Lonely (Epic/Epic)
Surround Me With Love (Epic/–)
Who's Cheatin' Who? (Epic/–)
I Love Country (–/Epic)
Still I Stay (Epic/–)
Ten Years Anniversary (Epic/–)

Obie McClinton

McClinton was one of the few black country stars who began carving a fairly impressive career in country music during the early '70s.

Born May 25, 1940 in Senatobia, Mississippi, Obie Burnett McClinton was raised on country music, though in his early days he tried to move into R&B as a singer. After completing high school, he headed for Memphis, and ended up as a dishwasher. Next came a choir scholarship

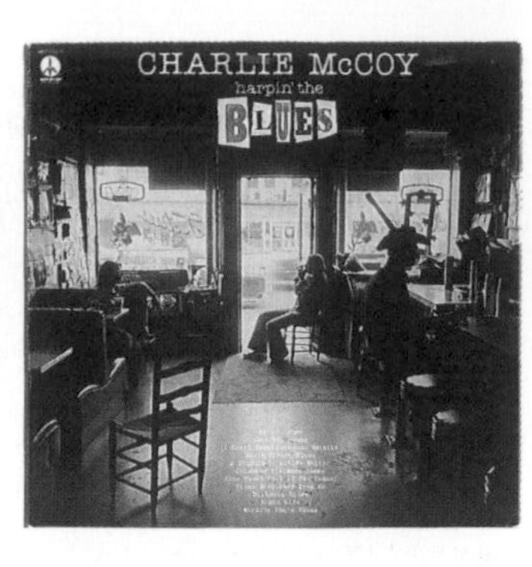

Harpin' The Blues, Charlie McCoy. Courtesy Monument.

to Rust College, Holly Springs, Mississippi, where he graduated in 1966. At the end of the year he volunteered for the Air Force and began singing on service talent shows. During this period he formed a songwriting relationship with the Stax label, contributing songs waxed by Otis Redding, Clarence Carter and others.

In 1971 he became a Stax-Enterprise artist himself, cutting several hits singles including **Don't Let The Green Grass Fool You** (1972) and **My Whole World Is Falling Down** (1973) before the label folded. Afterwards he moved on to Mercury, Epic and Sunbird, supplying all three with chart records before surfacing on Moonshine Records during 1984 with a wonderfully titled hit, **Honky Tonk Tan**. Following a lengthy illness, Obie died of abdominal cancer on September 23, 1987.

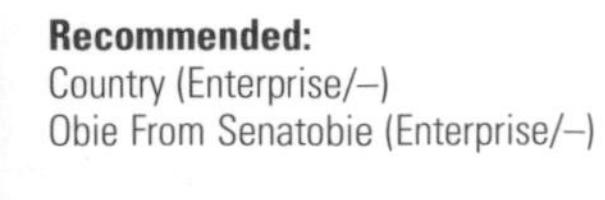
Recommended:
Country (Enterprise/–)
Obie From Senatobie (Enterprise/–)

Charlie McCoy

One of the finest harmonica players ever to grace the Nashville scene, McCoy was born at Oak Ridge, West Virginia, on March 28, 1941. Once a member of Stonewall Jackson's touring band, he opted for session work during the '60s, gaining a wide audience through his appearances on various Bob Dylan albums. He became a member of Area Code 615 in 1969 and was featured on the Code's **Stone Fox Chase**, a theme to BBC-TV's Whistle Test rock show.

Signed to Monument since 1963, McCoy had a minor pop hit with **Cherry Berry Wine** while recording for Cadence in 1961. Since then he has had a fairly active chart career, his biggest solo records being **I Started Loving You Again**, **I'm So Lonesome I Could Cry**, **I Really Don't Want To Know**, **Orange Blossom Special** (all 1972), **Boogie Woogie** (with Barefoot Jerry, 1974), and **Fair And Tender Ladies** (1978). He formed a music-making partnership with Laney Hicks and charted with **Until The Nights** (1981) and **The State Of Our Union** (1983). A brilliant all-round musician, McCoy was adjudged CMA Instrumentalist Of The Year in both 1972 and 1973.

Right: Hip harmonica man Charlie McCoy has played sessions and solo in his career.

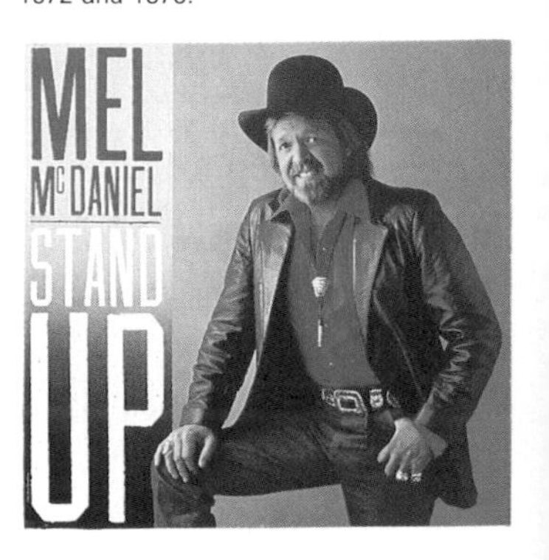

Stand Up, Mel McDaniel. Courtesy Capitol Records.

Recommended:
Goodtime Charlie's Got The Blues (Monument/–)
Nashville Hit Man (Monument/Monument)
Harpin' The Blues (Monument/–)
Stone Fox Chase (–/Monument)
Appalachian Fever (Monument/Monument)
Beam Me Up Charlie (Step One/–)

Neal McCoy

A long-haired, half-Filipino, half-Irish country singer, McCoy was born Neal McGaughey on July 30, 1963 in Jacksonville, Texas. By his mid-teens Neal had started playing clubs and singing at private functions. His parents divorced and his mother, Virginia, remarried, moving to Houston with her second husband, Don McCoy. When Neal took to the clubs, he adopted the name Neal McGoy, and in the early '80s entered a country music talent contest in Dallas, walking away with the first prize. He gained the attention of Charley Pride, who signed him to a management contract.

He joined the Pride road show and toured the world with Pride for seven years. In 1987 he landed a record deal with 16th Avenue Records in Nashville, making his country chart debut with **That's How Much I Love You** (1988). Two years later he changed his name to Neal McCoy, gained a new contract with Atlantic Records and has gradually been building a healthy following for his modern honky-tonk music. With his band, Justice, he has toured regularly, opening shows for Tracy Lawrence, Alan Jackson and Clint Black. He made an impression on the charts with **If I Built You A Fire** (1991), **There Ain't Nothin' I Don't Like About You** (1992) and **Now I Pray For Rain** (1993).

Recommended:
Where Forever Begins (Atlantic/–)

Mel McDaniel

A rugged but amiable singer, McDaniel was born on September 6, 1942 in the Creek Indian area of Checotah, Oklahoma. During his high school days he became a trumpet player but, influenced by Elvis Presley, he switched to guitar. He joined a band that played local gigs and cut some singles. One, **Lazy Me**, on the Galway label, was produced by J. J. Cale. After graduation from high school, he headed for Alaska where he found he could make a living as an entertainer. Two more years on and he was in Nashville, working at the Holiday Inn. He began selling his songs and also became a demo singer, recording songs for writers who lacked the vocal equipment to do the jobs themselves.

One of this own songs, **Roll Your Own** was recorded by Hoyt Axton, Commander Cody and Arlo Guthrie but it was not until 1976 that McDaniel gained a record contract of his own. His debut single for Capitol was a version of a Bob Morrison song, **Have A Dream On Me**, which went Top 50. Throughout the rest of the '70s, his singles charted regularly but modestly. Among the most successful were **Gentle To Your Senses** (1977), **God Made Love** (1977) and **Play Her Back To Yesterday** (1979). McDaniel was proficient in songwriting, providing hits for Bobby Goldsboro and Johnny Rodriguez.

As the '80s moved in, Mel McDaniel suddenly became fashionable. In 1981, two of his singles, **Louisiana Saturday Night** and **Right In The Palm Of Your Hand**, went Top 10, while **Preachin' Up A Storm** also sold well. Since then he has gone Top 20 in almost routine fashion, notching hit after hit.

In 1986 he joined the Grand Ole Opry. The same year he charted with Bruce Springsteen's **Stand On It**. Mel's most recent Top 10 entry came with **Real Good Feel Good Song** in 1988. Three years later, he was recording for the independent DPI Records, and like so many of the veteran country stars, made the Branson connection, signing a recording contract with Branson Records in 1993.

Recommended:
I'm Countryfied (Capitol/–)
Stand Up (Capitol/–)
Naturally Country (Capitol/–)
Mel McDaniel With Oklahoma Wind (Capitol/–)
Take Me To The Country (Capitol/–)
Country Pride (DPI/–)

Ronnie McDowell

A singer who favours the dramatic, sensual style of country crooning, Ronnie McDowell (born in Fountain Head, Tennessee) initially found fame through his ability to imitate the voice of Elvis Presley. He first started singing publicly while in the US Navy in the late '60s. After completing his service, he moved to Nashville working as a sign painter while building his reputation as a songwriter. His self-penned tribute to Elvis Presley, **The King Is Gone** (1977), became a major pop/country hit peaking at No.13 on both charts. Later that year he made the country Top 10 with **I Love You, I Love You, I Love You**, setting the tone for McDowell's later sexy love-song hits.

He continued to score minor hits for Scorpion during the next two years. Then, in 1979, he signed with Epic Records, making a Top 20 entry with **World's Most Perfect Woman** (1979), but his next few releases were only minor successes. In early 1981 he made it to No.2 with **Wandering Eyes**. This was followed by the chart-topping **Older Women** (1981) and **You're Gonna Ruin My Bad Reputation** (1983). He had provided the voice of Presley for the the soundtrack of the film 'Elvis' in 1979, otherwise all traces of Presley had been erased from his repertoire. He built up a reputation as a country sex symbol, concentrating on 'women' songs. Ronnie scored two more massive hits for Epic with **In A New York Minute** and **Love Talks** (1985). He then switched labels to MCA/Curb and had Top 10 winners in **All Tied Up** (1986) and **It's Only Make Believe** (1987), the latter featuring a guest vocal by Conway Twitty. By this time he was using his own band, the Rhythm Kings, on his recordings, including future Kentucky HeadHunters Doug Phelps and Greg Martin. Phelps also played a major role in the vocal arrangements, which evoked memories of late '50s teen-beat vocal groups. McDowell teamed up with Jerry Lee Lewis for **Never Too Old To Rock'n'Roll**, a minor chart entry in 1989. A revival of **Unchained Melody** in 1990 has been his last major country hit, though with his slick, choreographed show, he is still a popular live entertainer.

Recommended:
All Tied Up In Love (MCA/MCA)
Unchained Melody (Curb/–)
I'm Still Missing You (Curb/–)

Red River Dave McEnery

Although saga songs have long been a tradition in country music – dating back to broadside sheets – the foremost exponent of the style has been a tall, blue-eyed Texan named Red River Dave McEnery.

Born in San Antonio, on December 15, 1914, he began a professional career in 1935, playing a host of radio stations all

Below: Red River Dave McEnery onstage with Bill Fenner and Roy Huxton.

across the country but finding success in New York from 1938 through to 1941. Dave returned to Texas in the early '40s, playing the Mexican border stations and then basing himself in San Antonio. Though he found time to record for Decca, Sanora, MGM, Confidential and a whole host of smaller labels, he also appeared in a film for Columbia ('Swing In The Saddle', 1948) and a couple for Universal ('Hidden Valley' and 'Echo Ranch', both 1949).

He really found his niche when he wrote **Amelia Earhart's Last Flight**. Although he was long popular as a singer of country songs, it was these modern-day event songs which were to become his forte. Using the course of current events, he has written **The Ballad Of Francis Gary Powers**, **The Flight Of Apollo Eleven**, and **The Ballad Of Patty Hearst**.

In the early '70s Red River Dave moved to Nashville, where he became a well-known sight, with his gold boots, lariat strapped to his side, big hat and leonine white hair and goatee. He later moved back to Texas and makes occasional appearances at folk festivals.

Reba McEntire

Adjudged Female Vocalist Of The Year at the CMA awards in 1984 (and for the next three years), Reba McEntire was born in Chockie, Oklahoma, on March 28, 1954. Daughter of a world champion steer-roper, she became part of a country band while still in the ninth grade, sometimes playing at clubs until the early hours of the morning. A promising rodeo performer, she intended to become a school teacher and enrolled at Southeastern Oklahoma State University as an elementary education major. But in 1974 she was offered the opportunity to sing the national anthem at the National Rodeo Finals, where she met

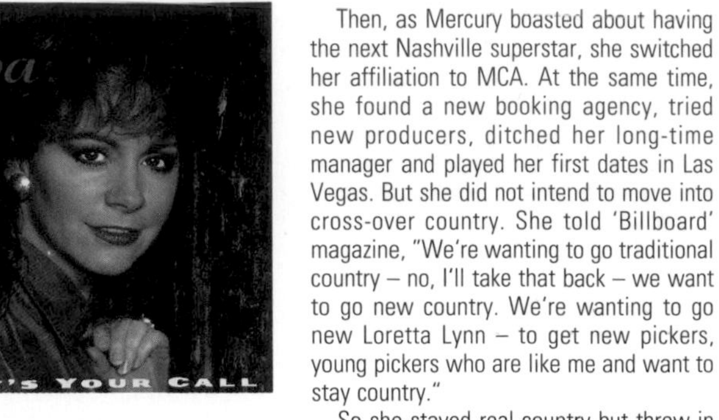

It's Your Call, Reba McEntire. Courtesy MCA Records.

Red Steagall. Impressed by her voice, Steagall arranged for her to cut a demo tape in Nashville. The results gained her a contract with Mercury Records.

Her first chart record, **I Don't Want To Be A One Night Stand**, came in mid-1976, coinciding with her marriage to rodeo champion Charlie Battles on June 21. For the first couple of years, her records sold well enough, but only a duet with Jacky Ward (**Three Sheets In The Wind**, 1978) pierced the barrier into the Top 20. By 1979 she had a Top 20 single of her own with **Sweet Dreams**, while 1980 saw her in the Top 10 with **(You Lift Me) Up To Heaven**. The next few years saw her chart with such singles as: **I Don't Think Love Ought To Be That Way**, **Today All Over Again** (1981), **I'm Not That Lonely Yet** (1982), **Can't Even Get The Blues** (No.1, 1982), **You're The First Time I've Thought About Leaving** (No.1, 1983), **Why Do We Want (What We Know We Can't Have)** and **There's No Future In This** (1983).

Below: Reba is a determined soprano who wraps her songs around woman-to-woman themes.

Then, as Mercury boasted about having the next Nashville superstar, she switched her affiliation to MCA. At the same time, she found a new booking agency, tried new producers, ditched her long-time manager and played her first dates in Las Vegas. But she did not intend to move into cross-over country. She told 'Billboard' magazine, "We're wanting to go traditional country – no, I'll take that back – we want to go new country. We're wanting to go new Loretta Lynn – to get new pickers, young pickers who are like me and want to stay country."

So she stayed real country but threw in little bits of business dreamed up by choreographers, lighting directors and others who could help her act stay imaginative. And the results paid off in record sales. Her first three singles for MCA, **Just A Little Love**, **He Broke Your Mem'ry Last Night** and **How Blue** all went Top 20 in 1984, the last named reaching the top of the charts.

During 1985 she logged yet another No.1 with **Somebody Should Leave** and clambered twice more into the Top 10 via **Have I Got A Deal For You** and **Only In My Mind**. The following year she had a trio of No.1s with **Whoever's In New England**, **Little Rock** and **What Am I Gonna Do About You**. She also walked off with the CMA's Entertainer Of The Year award, as well as being a third-time Female Vocalist Of The Year winner. **Whoever's In New England**, a tear-stained ballad, became her first music video. It led to her first million-selling album and in 1987 gained her a Grammy. The No.1s continued in 1987 with **One Promise Too Late** and **The Last One To Know** as her album sales took off. **My Kind Of Country**, **Whoever's In New England**, **What Am I Gonna Do About You** and **The Last One To Know**, all became gold albums. Turning the emphasis of her music to female themes, Reba succeeded in reaching the women record-buyers like no other female country singer. This has been reflected in her huge album sales and sell-out concert appearances. She is one of the few country music entertainers who can draw huge crowds just as easily in New England and New York as she can in the southern states.

Top-of-the-chart singles continued with **Love Will Find Its Way To You**, **I Know How He Feels** (1988), **New Fool At An Old Game** and **Cathy's Clown** (1989). In 1988 Reba was listed in a Gallup Youth Survey as one of the Top 10 Female Vocalists of any kind of music, the same year People magazine named her in the Top 3 female vocalists. Four years later she did even better in the People's Choice when she was named favourite female vocalist. These awards mirrored Reba's skyrocketing record sales. Her **Greatest Hits** album went platinum in 1990, as did **Rumor Has It**. The following year, **For My Broken Heart** climbed high on the pop charts and chalked up sales in excess of two million copies. Reba's success continued on the country charts with such Top 10 high climbers as: **Walk On**, **You Lie** (1990), **Fallin' Out Of Love**, **For My Broken Heart** (1991), **The Greatest Man I Never Knew** (1992), **Take It Back** and **The Heart Won't Lie** (1993), the last a chart-topping duet with Vince Gill.

In 1987, Reba separated from Charlie Battles, her husband of eleven years, and immersed herself in her career. On June 3, 1989, she took everyone by surprise when she married Narvel Blackstock, her one-time steel guitarist and, at the time, her road manager. With all the success, there was also tragedy, and in March 1991 Reba was devastated when eight members of her band were killed in a plane crash in California. One week after the accident, she braved her fears and faced the world at the Academy Of Country Music awards.

Since making her first video in 1986 for **Whoever's In New England**, Reba has used the format to explore the ambiguities of songs. No other artist in country music remotely matches Reba when it comes to chronicling the ups and downs of life for modern women.

Recommended:

Unlimited (Mercury/–)
Behind The Scene (Mercury/–)
My Kind Of Country (MCA/MCA)
Just A Little Love (MCA/MCA)
Whoever's In New England (MCA/MCA)
It's Your Call (MCA/MCA)
For My Broken Heart (MCA/MCA)
Live (MCA/–)
The Last One To Know (MCA/–)
Reba Nell McEntire (Mercury/–)
Reba (MCA/MCA)

Reba Live, Reba McEntire. Courtesy MCA Records.

Sam And Kirk McGee

The McGee brothers were both born in Franklin, Tennessee, Sam on May 1, 1894, Kirk on November 4, 1899. They were influenced by their father, an old-time street fiddle player, and the black street musicians of Perry, Tennessee, where the McGees spent part of their boyhood. With Sam on banjo and Kirk on guitar and fiddle, they joined Uncle Dave Macon's band in 1924, joining him on the Opry two years later. In 1930 they worked with fiddler Arthur Smith, forming the Dixieliners. Smith left in the early '40s, at which time the brothers occasionally joined a popular Opry act, Sara And Sally.

Occasional members of several of the Opry's old-time bands in ensuing years, the duo eventually opted to go their own way again, becoming favourites at many folk festivals in the '60s.

Sam, who claimed to be the first musician to play electric guitar on the Opry (a claim disputed by others), was killed on his farm on August 21, 1975 when his tractor fell on him. But Kirk soldiered on and was still appearing on the Opry up to his death on October 24, 1983.

Recommended:

Sam McGee – Grand Dad Of Country Guitar Pickers (Arhoolie/–)
The McGee Brothers With Arthur Smith (Folkways/–)

Warner Mack

Nashville-born, on April 2, 1938, singer-songwriter Warner McPherson was raised in Vicksburg, Mississippi. While at Vicksburg's Jett School he played guitar at various events, from there moving on to perform in local clubs. During the '50s, McPherson became a regular on KWHH's Louisiana Hayride and was also featured on Red Foley's Ozark Jubilee. His record career made progress in 1957 when his recording of **Is It Wrong?** charted in fairly spectacular manner. In the wake of this initial success, McPherson – who had become Warner Mack after his nickname had inadvertently been placed on a record label – decided to return to Music City. But there was a lull in his record-selling fortunes until 1964 when, following a moderate hit with **Surely**, Mack suddenly hit top gear, providing Decca with 14 successive Top 20 entries between 1964 and 1970. One of these, **The Bridge Washed Out**, was the best-selling country disc for a lengthy period of 23 weeks during 1965.

An ever-busy songwriter, Mack has written over 250 songs. Several of his other chart climbers include **Talkin' To The Wall** (1966) and **How Long Will It Take?** (1967).

During the early '70s, Mack continued on his chart-filling way with such singles as **Draggin The River**, **You're Burnin' My House Down** (1972), **Some Roads Have No Ending**, **Goodbyes** and **Don't Come Easy** (1973), all for Decca/MCA. Mack's last chart appearance came on Pageboy in late 1977 with **These Crazy Thoughts**, but still he continues playing successful tours. In recent times, Mack has completed two, well received jaunts around Europe.

Uncle Dave Macon

Known variously as the 'Dixie Dewdrop', the 'King Of The Hillbillies' and the 'King Of Banjo Players', David Harrison Macon was the first real star of the Grand Ole Opry.

Born in Smart Station, Tennessee, on October 7, 1870, he grew up in a theatrical environment, his parents running a Nashville boarding house catering for travelling showbiz folk.

Following his marriage to Mathilda Richardson, Macon moved to a farm near Readyville, Tennessee, there establishing a mule and wagon transport company which operated for around 20 years. A natural entertainer and a fine five-string banjoist, David Macon played at local functions for many years but remained unpaid until 1918 when, wishing to decline an offer to play at a pompous farmer's party, he asked what he thought was the exorbitant fee of 15 dollars, expecting to be turned down. But the fee was paid and Uncle Dave played his first paid function, being spotted there by a Loew's talent scout who offered him a spot at a Birmingham, Alabama, theatre.

In 1923, while playing in a Nashville barber's shop, he met fiddler and guitarist Sid Harkreader, the two of them, teaming to perform at the local Loew's Theatre, then moving on to tour the South as part of a vaudeville show. A year later, while

The Gayest Old Dude In Town, Uncle Dave Macon. Courtesy Bear Family Records. Uncle Dave was the first real star of the Grand Ole Opry.

playing at a furniture convention, the duo were approached by C. C. Rutherford of the Sterchi Brothers Furniture Company, who offered to finance a New York recording date with Vocalion. Macon and Harkreader accepted, cutting 14 sides at the initial sessions, returning in 1925 to produce another 28 titles. Macon's next New York sessions (1926) found him playing alongside guitarist Sam McGee, cutting such sides as **The Death Of John Henry** and **Whoop 'Em Up Cindy** – and that same year he first appeared on the Opry, where the jovial, exuberant Macon, clad in his waistcoat, winged collar and plug hat, soon became a firm favourite.

An Opry performer almost up to the time of his death, the fun-loving banjoist cut many records during his lifetime, sometimes recording solo, sometimes as part of the Fruit Jar Drinkers Band (not the same as the Opry band of the same name) or, on more religious sessions, as a member of the Dixie Sacred Singers, usually employing fiddler Mazy Todd and the McGee Brothers as supporting musicians. He appeared in the 1940 film 'Grand Ole Opry'.

He died aged 82 on March 22, 1952, in Readyville, just three weeks after his final appearance on the Opry, his burial taking place in Coleman County, Murfreesboro, Tennessee. Uncle Dave had never learned to drive a car, and even said in one song, "I'd rather ride a wagon to heaven than to hell in an automobile."

In October 1966, he was elected to the Country Music Hall Of Fame, his plaque recalling that the man known as the 'Dixie Dewdrop' was "a proficient banjoist and singer of old-time ballads who was, during his time, the most popular country music artist in America."

Recommended:

Gayest Old Dude in Town (–/Bear Family)
Uncle Dave Macon 1926–1939 (Historical/–)
Early Recordings (County/–)
Laugh Your Blues Away (Rounder/–)
At Home In 1950 (–/Bear Family)

Rose Maddox

Born in Boaz, Alabama, on December 15, 1926, Rose Maddox began her show business as part of a family band, an outfit justifiably known as 'the most colourful hillbilly band in the land'. With Cal on guitar and harmonica, Henry on mandolin, Fred on bass, Don providing the comedy and Rose handling the lead vocals in her full-throated, emotional style, the Maddox Brothers and Rose established a reputation first in California then on to the Louisiana Hayride in Shreveport.

During the '50s, the Maddoxes produced several fine records, also putting in appearances on the Grand Ole Opry and other leading country music shows, moving back to California as the decade came to a close. Shortly after the group disbanded and Rose became a solo act, recording for Capitol and supplying the label with such successful records as **Gambler's Love** (1959), **Kissing My Pillow** (1961), **Sing A Little Song Of Heartache** (1962), **Lonely Teardrops** (1963), **Somebody Told Somebody** (1963) and **Bluebird, Let Me Tag Along** (1964).

During this period she also recorded a number of duets with Buck Owens – one of which **Loose Talk/Mental Cruelty** proved a double-sided hit in 1961. She cut an album called **Bluegrass**, a disc made at the suggestion of Bill Monroe, who played mandolin on the sessions.

After being in semi-retirement for a while, she began working again, often in partnership with Vern Williams and his band. This unit played many benefits for Rose in order to pay her hospital bills when she became gravely ill in the late '70s. During 1983 Rose and Vern Williams again joined forces to record **A Beautiful Bouquet**, an album of gospel music, recorded in honour of Rose's son Donnie, who died in August 1982.

Recommended:

The Maddox Brothers And Rose – On The Air (Arhoolie/–)
Rockin' Rollin' Maddox Brothers And Rose (–/Bear Family)
Maddox Brothers And Rose 1946–1951 Volumes 1 and 2 (Arhoolie/–)
This Is Rose Maddox (Arhoolie/–)
A Beautiful Bouquet (Arhoolie/–)
Family Folks (–/Bear Family)

J. E. Mainer

The leader of the Mountaineers, one of the first string bands on record, banjoist and fiddler J. E. Mainer was born in Weaversville, North Carolina, on July 20, 1898.

A banjo player at the age of nine, by his early teens he had become a cotton mill hand, working alongside his father at a Glendale, South Carolina mill. In 1913, he moved to Knoxville, Tennessee, there witnessing an accident in which a fiddler was killed by a railroad train. He then claimed the musician's broken instrument as his own, had it repaired and learned to play it, soon becoming one of the finest fiddlers in his area.

By 1922, Mainer had hoboed his way to Concord, North Carolina, there marrying Sarah Gertrude McDaniel, and, later, forming a band with his brother Wade (banjo), Papa John Love (guitar) and Zeke Morris (mandolin/guitar). The band gained a fair degree of fame locally. But in the early '30s came a change of fortune, when Mainer's unit, adopting the title of the Crazy Mountaineers, began a series of broadcasts from Charlottesville, sponsored by the Crazy Water Crystal Company. Record dates for RCA Bluebird followed, the Mountaineers cutting such tracks as **John Henry**, **Lights In The Valley**, **Ol' Number 9** and **Maple On The Hill**.

The band's popularity continued throughout the '30s and '40s. Mainer and the Mountaineers recorded over 200 sides for RCA and broadcasted over WPTF, Raleigh, North Carolina. And by the late '60s J.E. was still performing and recording – one of his releases being a single featuring unaccompanied jew's-harp – but the grim reaper eventually caught up with him in 1971.

Recommended:

J. E. Mainer's Mountaineers Volumes 1 and 2 (Old Timey/–)
Good Old Mountain Music (King/–)

Wade Mainer

The banjo-playing younger brother of J. E. Mainer, Wade had a long and influential career of his own. Born on April 21, 1907, near Weaverville, North Carolina, Wade developed an advanced two-finger banjo picking style which led to a distinctive sound on his records as well as those he made with his brother.

After splitting from Mainer's Mountaineers, Wade formed his own group (which at times included Clyde Moody and Wade Morris of the Morris Brothers) called Wade Mainer And The Sons Of The Mountaineers.

Possessed of a strong clear voice, his **Sparkling Blue Eyes** was a major hit in 1939, one of the last commercial releases by a string band. Wade recorded for Bluebird until 1941, and after the war spent years with King but without much commercial success. After a brief retirement in North Carolina, he moved to Flint, Michigan, where he worked for Chevrolet until his retirement. In later years he recorded for Old Homestead and proved that his strong, pure country voice

and unique banjo style had not been affected by his advanced years.

Recommended:
Sacred Songs Of Mother And Home (Old Homestead/–)
Wade Mainer And The Mainer's Mountaineers (Old Homestead/–)

Barbara Mandrell

Barbara is the singer who, probably more than anyone else, has managed to weld Las Vegas glitz to the relatively more down-homey sounds of Nashville. She picks but she is slick, she sings songs like **I Was Country When Country Wasn't Cool** yet still manages to make them sound as though they had originally been penned for some Hollywood disco-flick. She is beautiful and glossy, but has brains, business-sense and enough talent to win the CMA Entertainer Of The Year award in 1980 and 1981, along with several other major plaudits.

Born in Houston, Texas, in 1948, but raised in L.A., she became part of the family band, headed by her father. At the age of 11 she could play pedal-steel, demonstrating her prowess at Las Vegas' Showboat Hotel that same year. Also before hitting her teens she could find her way around a piano, bass, guitar, banjo and saxophone. At 13 she toured with Johnny Cash and in 1966–67 played at military bases in Korea and Vietnam.

During the late '60s she began recording for the minor Mosrite label, cutting titles that included **Queen For A Day**. But following a family move to Tennessee, she set her sights on a Nashville contract, eventually signing for Columbia in March 1969, and gaining her first hit that year with her version of Otis Redding's **I've Been Loving You Too Long**, establishing a successful country-meets-soul format that saw her achieving further chart status with such R&B material as **Do Right Woman – Do Right Man** (1971), **Treat Him Right** (1971) and Joe Tex's **Show Me** (1972). During 1971 she had her first Top 10 record with **Tonight My Baby's Coming Home**, following this with **Midnight Oil** (1973). In 1975 she signed to ABC/Dot and immediately went Top 5 with **Standing Room Only**, providing the label with many other major hits, including the No.1s **Sleeping Single In A Double Bed** (1978) and **(If Loving You Is Wrong) I Don't Want To Be Right** (1979).

When ABC/Dot sold out to MCA, Barbara went on hitmaking. **Fooled By A Feeling** (1979) was succeeded by such hits as **Years** (No.1, 1979), **The Best Of Strangers** (1980), **I Was Country When Country Wasn't Cool** (No.1 1981), **Till You're Gone** (No.1, 1982), **In Times Like These** (1983), **One Of A Kind Pair Of Fools** (No.1, 1983), **Only A Lonely Heart Knows** (1984) and **To Me**, a duet with Lee Greenwood (1984).

In 1984 the Mandrell luck momentarily changed and Barbara narrowly escaped death when her car was hit head-on. Barbara and her two children were badly injured. Hospitalized for 19 days, she was unable to work for nearly a year. The singer lost much of her credibility when she sued,

Left: Mainer's Mountaineers, circa 1936. J.E. is top left, and brother Wade is bottom right.

Left: Barbara Mandrell wrote her best-selling biography and now has her own museum on Nashville's Division Street.

on her insurance company's advice, the late driver's family for 10 million dollars. Back in the studios, she continued her hit-making mode with **There's No Love in Tennessee**, **Angel In Your Arms** (1985), **Fast Lanes And Country Roads** and **No One Mends A Broken Heart Like You** (both 1986). In 1987 she signed with EMI–America/Capitol, and returned to a more solid country style with a Top 10 revival of **I Wish That I Could Fall In Love Today** (1988).

Barbara was a guest on many of the top chat shows, and her autobiography, 'Get To The Heart', became a best-seller. Following her accident, she was seen in many TV public service announcements, telling viewers to buckle-up. By the end of the '80s, the hit singles started tapering off, and a last Top 20 entry came in 1989, with **My Train Of Thought**.

Clean Cut, Barbara Mandrell. Courtesy MCA Records.

Recommended:
The Best Of (Columbia/–)
Live (MCA/MCA)
The Key's In The Mailbox (Capitol/–)
Sure Feels Good (EMI-America/EMI-America)
Get To The Heart (MCA/MCA)
I'll Be Your Jukebox (Capitol/–)

Louise Mandrell

Born in Corpus Christi, Texas, on July 13, 1954, Louise is the fiddle and bass playing member of the Mandrell girls. A member of Barbara's Do-Rites at 15, mainly playing bass, she toured with the group for several years, but during the '70s became part of the troupe headed by Opry star Stu Phillips. Next came a musical liaison with Merle Haggard, Louise singing both lead and back-ups on Haggard live dates and recordings, after which she decided to go it alone, signing a deal as a solo act with Epic. The label also signed R. C. Bannon, whom Louise married in 1979.

The first hit, a minor one, was **Put It On Me**, in 1978. But the following year her career took off. Louise garnered no less than five successful singles, including a Top 20 shot with **Reunited**, a duet with Bannon. By 1981, the couple had signed to RCA, immediately making some impression with **Where There's Smoke, There's Fire.** Louise was given considerable exposure by her weekly appearances on her sister's TV show, and during 1983 she increased her solo fortunes. Two singles, **Save Me** and **Too Hot To Sleep,** both reached Top 10, while a third, **Runaway Heart**, reached Top 20, During 1984 she again had a Top 10 record with **I'm Not Through Loving You Yet**, and the following year two more with **Maybe My Baby** and **I Wanna Say Yes**. By this time Louise had her own TV series and continued charting minor hits for RCA through 1988 and **As Long As We Got Each Other**, a duet with Eric Carmen.

Recommended:
Inseparable – with R. C. Bannon (Epic/–)
Too Hot To Sleep (RCA/–)
I'm Not Through Loving You Yet (RCA/–)
Louise Mandrell (Epic/Epic)

Zeke Manners

An accordion player, singer and songwriter, Zeke Manners is best known as a co-founder of the Beverly Hillbillies and for his association with Elton Britt.

Manners befriended Britt when he joined the Hillbillies in 1932. In 1935 the two of them headed for New York, where they proved popular for many years – sometimes playing together, sometimes as solo acts.

Aside from his Brunswick recordings with the Beverly Hillbillies, Manners recorded on his own for Variety, Bluebird and RCA, collaborating with Elton Britt in the late '50s for the **Wandering Cowboy** album on ABC-Paramount.

Joe And Rose Lee Maphis

A husband and wife team whose popularity peaked during the '50s and '60s, Otis W. 'Joe' Maphis (born Suffolk, Virginia, May 12, 1921) and Rose Lee (born Baltimore, Maryland, December 29, 1922) met on Richmond, Virginia's WRVA Old Dominion Barn Dance in 1948. Shortly after meeting, they married and moved out to the West Coast, there playing on Cliffie Stone's Hometown Jamboree and Crompton's Town Hall Party for several years, also performing vocally and instrumentally. Joe played fiddle, guitar, banjo, mandolin and bass, while Rose Lee played guitar on various recording sessions. Joe, who is Barbara Mandrell's

uncle, is possibly best known for the time be spent on the road and in the studio with Rick Nelson. He died in June 1986.

Recommended:
Honky Tonk Cowboy (CMH/–)
Joe And Rose Maphis With The Blue Ridge Mountain Boys (–/Stetson)

Jimmy Martin

Born in Sneedville, Tennessee, in 1927, James Henry Martin rose to prominence as a member of Bill Monroe's Bluegrass Boys. He was lead vocalist, guitarist and frontman for the group for much of 1949–53, his last session with the band taking place in January 1954. Next came some fine sides for Victor, which found Martin in the company of the Osborne Brothers, fiddler Red Taylor and bassist Howard Watts, a luminary of both Monroe's band and Hank Williams' Drifting Cowboys.

Some time later, Martin formed his own regular outfit, the Sunny Mountain Boys, and began recording for Decca. He sometimes employed material of a novelty nature but always used musicians of quality. J. D. Crowe, Vic Jordan (later with Lester Flatt), Bill Emerson (with Country Gentlemen) and Allan Munde (of Country Gazette) were among those who worked with the band.

Maligned for his use of drums – thus horrifying some purists – Martin is perhaps one of the more unsung heroes of the bluegrass scene, though his contribution, both vocally and instrumentally, has been of inestimable value.

Recommended:
Big Instrumentals (MCA/–)
Jimmy Martin (MCA/–)
Good'n'Country (MCA/–)
Jimmy Martin And The Sunny Mountain Boys (MCA/–)

Fly Me To Frisco, Jimmy Martin And The Sunny Mountain Boys. Courtesy MCA Records.

Frankie Marvin

Following in the footsteps of his elder brother Johnny, Frankie Marvin (born Butler, Oklahoma, 1905) journeyed to New York and joined him at the peak of his popularity, doing comedy as well as playing steel guitar and ukelele. He also worked with the Duke Of Paducah in a comedy team known as Ralph And Elmer, and cut numerous records for a host of companies, being among the earliest country sides recorded in New York.

He joined Gene Autry – whom he had befriended in New York in 1929 – in Chicago in the early '30s, as Autry's popularity began to grow. Frank toured, broadcast and recorded with Autry for the next two decades, adding the distinctive steel guitar styling that was so much part of the Autry sound. He left the Autry show in 1955 and later retired in the mountains near Frazier Park, California.

Johnny Marvin

Born in Butler, Oklahoma, in 1898, young Johnny ran away from home at the age of 12 to pursue a career as a musician, singer and entertainer. His quest eventually took him to New York City where he became popular on Broadway, also logging a couple of hit records on Victor: **Just Another Day Wasted Away** and **Wait For Me At the Close Of A Long, Long Day**, which was his theme song.

He also became popular in the country field as the 'Lonesome Singer Of The Air' and used his steel guitar playing alongside brother Frankie both as a comedian and as a musician. The two befriended Gene Autry in 1929, and, after the Depression, he went to work for Gene as songwriter and producer on the Melody Ranch radio show. He contracted an illness while entertaining GIs in the South Seas during World War II and died in 1945 at the age of 47.

Above: Jimmy Martin and his Sunny Mountain Boys. The girl is Lois Johnson.

Louise Massey And The Westerners

Milt Mable, vocals and various instruments; Allen Massey, vocals and various instruments; Curt Massey, fiddle, trumpet, piano, vocals; Dad Massey, vocals and various instruments; Louise Massey, vocals; Larry Wellington, vocals and various instruments.

One of the earliest, most popular and most professional western bands in country music was known as the Musical Massey Family, then as the Westerners and, finally, as Louise Massey And The Westerners. Natives of Texas, they were the first to dress in flashy cowboy outfits and exploit a western image. They appeared for several years on Plantation Party and other popular radio shows.

The main vocals were by Louise Massey, whose growth in popularity is reflected in the changing band name, with many other solos by her brother Curt. Curt had an active and successful career in popular music both before and after his association with the Westerners, ultimately working as musical director and writer of the theme songs for the Beverly Hillbillies and Petticoat Junction TV shows.

Supporting vocals were done by all the band members, most notably the third sibling Allen, and Louise's husband Milt Mable. They recorded for Vocalion, Okeh and Conqueror. Their biggest hit was probably **The Honey Song**, although they are best known for Louise's composition **My Adobe Hacienda**.

Above: Kathy Mattea hit the charts in 1993 with Standing Knee Deep In A River.

Kathy Mattea

A country song stylist who leans more towards the traditional strains of the music, Kathleen Alice Mattea was born on June 21, 1959 in South Charleston, West Virginia, and grew up in nearby rural Cross Lanes. Her musical career began when she was given her first guitar at the age of ten. She sang in the local choir and community theatre throughout high school. At West Virginia University she joined a bluegrass group, Pennsboro, but dropped out of college in 1978 to go to Nashville with the leader of the band. Initially she was waitressing at TGI Friday's, then came a stint as a tour guide in the Country Music Hall Of Fame Museum. By this time she was recording demos, jingles and commercials, and for a time worked as a secretary at a music publishers. In 1982, she became part of Bobby Goldsboro's road show. The following year she signed a recording contract with Mercury Records.

Kathy's first single, **Street Talk** (1983). made the country charts. Over the next three years she had several more minor hits, eventually cracking the Top 10 with **Love At The Five And Dime** (1986). This was followed by **Walk The Way The Wind Blows** (1986), **You're The Power**, **Train Of Memories**, and her first No.1, **Goin' Gone** (all 1987). When Kathy first arrived in Nashville, she wrote the hauntingly beautiful **Leaving West Virginia**, which has since been used by the West Virginia Department of Tourism as the centrepiece of a million-dollar advertising campaign encouraging tourism in her home state. Kathy's talent as a song interpreter was recognized early in her Nashville career and she gained nominations as Best Female Vocalist before she had made the Top 20.

A major breakthrough came with **Eighteen Wheels And A Dozen Roses**, her second No.1 in 1988, which also won the CMA Single Of The Year. The hit singles continued with **Untold Stories** (1988), **Life As We Knew It**, **Come From The Heart** and **Burnin' Old Memories** (all 1989). On February 12, 1988 Kathy married Nashville songwriter Jon Vezner, who was to provide his wife with one of her finest songs a year later. **Where've You Been**, written about his grandparents, was only an album track when it was named 1989 Song Of The Year by the Academy Of Country Music. Released as a single, it made the Top 10, and was also named the 1990 Song Of The Year by the CMA. Kathy was now at the peak of her career. She was named CMA Vocalist Of The Year in 1989 and 1990, and continued with amazing consistency placing her records in the Top 10 – **She Came From Fort Worth** (1990), a duet with Tim O'Brien on **The Battle Hymn Of Love** (1990), **A Few Good Things Remain** and **Time Passes By** (1991).

In 1990, she completed work on her seventh album, **Time Passes By**, a radical departure from her past albums and her most ambitious project so far. The album, an eclectic blend of folk, country, Scottish and a touch of classical, could hardly be termed radio-friendly, with one cut five minutes long, and another five about four minutes long. In country, three minutes is the absolute maximum length for a country hit. The following year, Kathy suffered serious throat problems and it was discovered she had ruptured a blood vessel on her vocal chords. This meant several months off the road. But she still succeeded in hitting the charts with **Lonesome Standard Time** (1992) and **Standing Knee Deep In A River (Dying Of Thirst)** (1993).

Recommended:

Lonesome Standard Time (Mercury/Mercury)
Time Passes By (Mercury/Mercury)
Untasted Honey (Mercury/Mercury)
Walk The Way The Wind Blows (Mercury/Mercury)
Willow In The Wind (Mercury/Mercury)

Ken Maynard

Ken Maynard (born July 21, 1895, Vevay, Indiana) was a major cowboy film star in both the silent and sound era, making and losing several screen fortunes in his many years on the screen. His place in country music was assured firstly by his being the first cowboy to sing on film, in 'The Wagon Master' in 1930; secondly by being the first to use a western song as a film title (**The Strawberry Roan**, 1933); thirdly by introducing Gene Autry to films, as a singer in 'In Old Sante Fe' in 1934; and lastly by his 1930 recording session for Columbia at which he cut eight cowboy songs.

A fiddler, banjoist, guitarist and rough but appealing singer, as well as a stunt man, rodeo star and film hero of over two decades (1922–45), Ken Maynard died in California on March 23, 1973.

Below: Ken Maynard fiddles his way through the movie 'Strawberry Roan'.

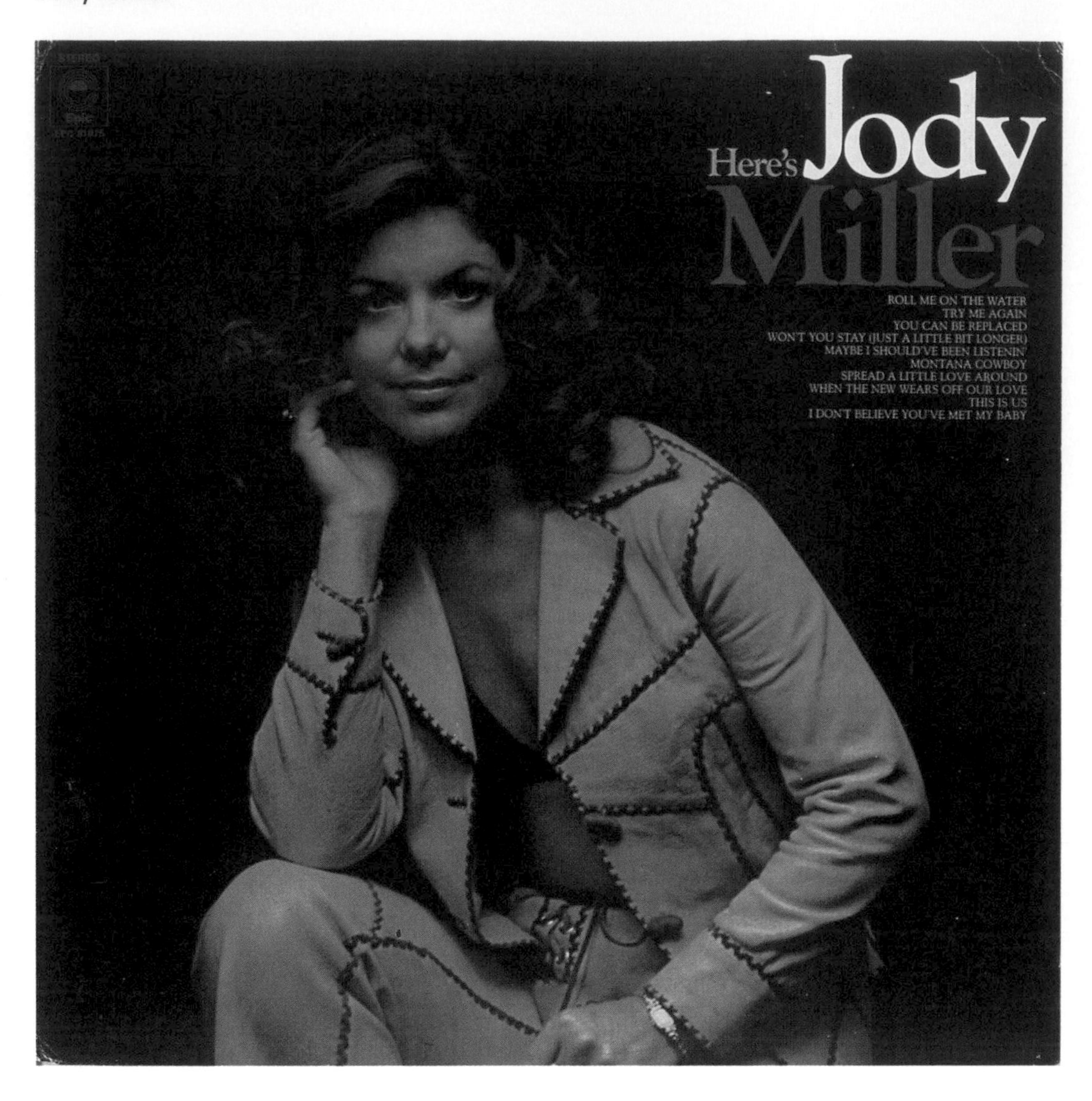

Here's Jody Miller. Courtesy Epic Records.

Jody Miller

The complete cross-over singer, Jody Miller is equally at home in pop, country or folk, a trait that leads to most articles on her talents commencing with the words: "Jody Miller is difficult to classify".

Born in Phoenix, Arizona, on November 29, 1941, the daughter of a country fiddle player, Jody grew up in Oklahoma, where she and school friends formed a trio known as the Melodies. Upon graduation she decided on a solo singing career and began establishing herself locally, joining the Tom Paxton TV Show, and gained a reputation as a folk singer. Through actor Dale Robertson she became signed to Capitol Records in 1963, making a fairly commercial folk album in **Wednesday's Child**, later cutting a hit single, **He Walks Like A Man** (1964), and obtaining a place in the Italian San Remo Song Festival. A year later, Jody's version of **Queen Of The House** – Mary Taylor's sequel to Roger Miller's **King Of The Road** – became a monster country and pop hit. But, despite some enjoyable pop–country albums that included **Jody Miller Sings The Hits Of Buck Owens** and **The Nashville Sound Of Jody Miller**, there were no further hit singles for Capitol.

Following a short retirement, during which Jody spent her time on her Oklahoma ranch raising her daughter Robin, she returned to performing once more, cutting sides for Epic. An association with producer Billy Sherrill provided her with a 1970 chart entry called **Look At Mine**, a Tony Hatch song. Her other Top 10 hits have included **He's So Fine, Baby I'm Yours** (1971), **There's A Party Goin' On, Darling, You Can Always Come Back, Good News** (1972), while another 1972 single, **Let's All Go Down The River**, a duet with Johnny Paycheck, also sold well.

A popular act at such diverse venues as the Wembley Country Music Festival and the Riviera and Frontier in Las Vegas, Jody spends most of her spare time on the family farm in Oklahoma, where she breeds and raises quarter horses.

Recommended:

There's A Party Going On (Epic/Epic)
He's So Fine (Epic/Epic)
Here's Jody Miller (Epic/Epic)

Roger Miller

The voice and songs of Roger Miller have been heard in films ranging from 'Waterhole 3' through to the Disney version of 'Robin Hood'. He has also paid some dues as a TV actor while his records made him one of the most original and successful pop stars during the '60s.

Born in Fort Worth, Texas, on January 2, 1936, Roger Dean Miller was raised by his uncle in Erick, Oklahoma. Influenced by the singing of Hank Williams, Miller saved enough money to buy a guitar, later also acquiring a fiddle.

Following a period spent as a ranch hand, Miller spent three years in the US Army in Korea. Upon discharge, he made his way to Nashville. In the sleeve notes to his **Trip In The Country** album, Miller describes himself at this period as being "a young ambitious songwriter, walking the streets of Nashville, trying to get anybody and everybody to record my songs. All in all, I wrote about 150 songs for George Jones, Ray Price, Ernest Tubb and others. Some were hits and some were not. In the beginning I created heavenly, earthy songs."

Ray Price was one of the first to benefit from the Miller songwriting skill, having a 1958 hit with **Invitation To The Blues**. And signed to RCA in 1960, Miller began accruing his own country winners with **You Don't Want My Love** (1960) and **When Two Worlds Collide** (1961). In 1962 he joined Faron Young's band as drummer, also guesting on the Tennessee Ernie TV Show. After one last hit for RCA in **Lock, Stock And Teardrops**, Miller moved to Smash, having an immediate million-seller with **Dang Me**, following this with other gold disc winners in the infectious **Chug-A-Lug** (1964) and the lightly swinging **King Of The Road** (1965). His endearing mixture of humour, musicianship and pure corn continued to pay dividends throughout the '60s, with many of his songs becoming Top 40 US pop hits.

The '70s, which saw Miller switch labels first to Mercury then to Columbia, brought less success to the then California-based singer-songwriter and hotel chain owner, though he kept up a steady flow of moderately sized hits. Contracts with Mercury Records (1979) and Elektra (1981) did not change the situation and it was not until 1982 and a liaison with Willie Nelson and Ray Price on **Old Friends**, a Columbia release, that Miller climbed back into the Top 10 for the first time since 1973.

But by 1985 he had moved on to a new stage in his career as 'Big River', a Miller musical based on the writings of Mark Twain, opened on Broadway. He won a coveted Tony Award for scoring the hit show. At the same time, MCA recorded an original cast album, the first album of its type to use a Nashville-based producer (Jimmy Bowen). He also signed an album deal with MCA, not that Miller really worried too much about records at the time. He had accumulated a haul of gold records and numerous industry awards, including an astonishing feat of 11 Grammies in two years, an achievement which has not yet been repeated by

Supersongs, Roger Miller. Courtesy CBS Records.

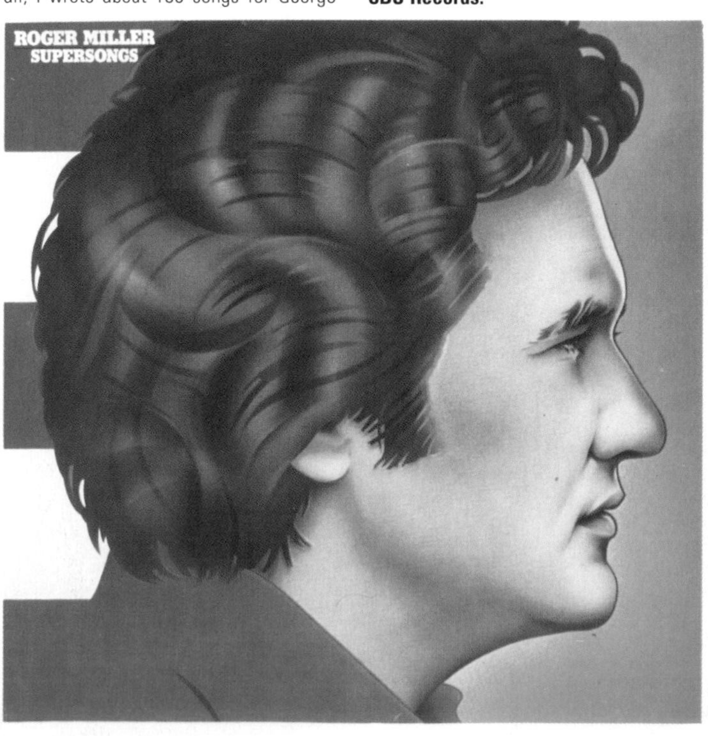

anyone. Roger Miller died on October 25, 1992 at Los Angeles' Century Hill Hospital. He had been diagnosed with cancer less than a year earlier.

Recommended:
Spotlight On (–/Philips)
Supersongs (Columbia/CBS)
King Of The Road (–/Bear Family)
Country Tunesmith (Mercury/–)
Roger Miller (MCA/MCA)

Ronnie Milsap

The winner of the CMA Male Vocalist Of The Year award in 1974 and Entertainer Of The Year 1977, Milsap is equally at home with the blues as with country ballads.

Born blind in Robbinsville, North Carolina (1944), he learned the violin at the age of seven and could play piano just a year later. By 12 he had mastered the guitar. Attending the State School for the Blind in Raleigh, he became interested in classical music but formed a rock group, the Apparitions. Upon completing high school, Milsap attended Young Harris Junior College, Atlanta, studying pre-law and planning to go on to law school at Emory College, where he had been granted a scholarship.

However, he quit studies to play with J. J. Cale and in 1965 formed his own band, playing blues, country and jazz, also signing with Scepter and cutting **Never Had It So Good/Let's Go Get Stoned**, two R&B tracks for his first single release.

By 1969, he and his band had moved to Memphis, becoming resident group at a club called TJ's, Milsap recording for the Chips label and coming up with a hit disc in **Loving You Is A Natural Thing** (1970). After a stint with Warner Brothers Records, Milsap, who had always featured some country material in his act, decided to devote his career to becoming a country entertainer. At that point he moved to Nashville, there gaining a residency at Roger Miller's King Of The Road motel and signing a management deal with Jack D. Johnson, the Svengali behind the rise of Charley Pride. In April 1973 he became an RCA recording artist, his first release on the label being **I Hate You**, a Top 10 single.

His hits since then are too numerous to list but some of his chart-toppers were: **Pure Love**, **(I'd Be) A Legend In My Time** (1974), **Daydreams About Night Things** (1975), **(I'm A) Stand By My Woman Man** (1976), **It Was Almost Like A Song** (1977), **Only One Love In My Life** (1978), **Nobody Likes Sad Songs** (1979), **Why Don't You Spend The Night**, **My Heart**, **Cowboys And Clowns**, **Smokey Mountain Rain** (1980), **Am I Losing You**, **(There's) No Getting Over Me**, **I Wouldn't Have Missed It For The World** (1981), **Any Day Now**, **He Got You**, **Inside** (1982), **Don't You Know How Much I Love You**, **Show Her** (1983) and **Still Losing You** (1984).

By this time Milsap had his own highly successful 'state of the art' GroundStar recording studios in Nashville and had also become heavily involved in video production. His 1985 album, **Lost In The Fifties Tonight**, had a doo-wop flavour captured in the title song and **Happy, Happy Birthday Baby**. Milsap kept edging further towards pop than he had previously, and though he ditched his honky-tonk approach, he remains one of the most potent performers on the scene.

Some of his more notable recent successes include: **In Love**, **How Do I Turn You On** (1986), **Snap Your Fingers** (1987), **Make No Mistakes She's Mine** (a duet with Kenny Rogers, 1987), **Don't You Ever Get Tired (Of Hurtin' Me)** (1989) and **A Woman In Love** (1990). In 1992, still regularly scoring Top 10 hits, after nearly 20 years of amazing pop and country success, a disgruntled Milsap announced that he had left RCA for Liberty Records: "You got to have a reason to do this sort of thing, so I gotta say it will be a revitalization of my career. For one thing, we'll have a record company behind us for a change."

Recommended:
A Legend In My Time (RCA/RCA)
Images (RCA/RCA)
Pure Love (RCA/RCA)
Live (RCA/RCA)
Night Things (RCA/RCA)
Vocalist Of The Year (Crazy Cajun/)
Milsap Magic (RCA/RCA)
Lost In The Fifties Tonight (RCA/RCA)
Stranger Things Have Happened (RCA/–)
True Believer (Liberty/–)
There's No Getting Over Me (RCA/RCA)

Lost In The Fifties Tonight, Ronnie Milsap. Courtesy RCA Records.

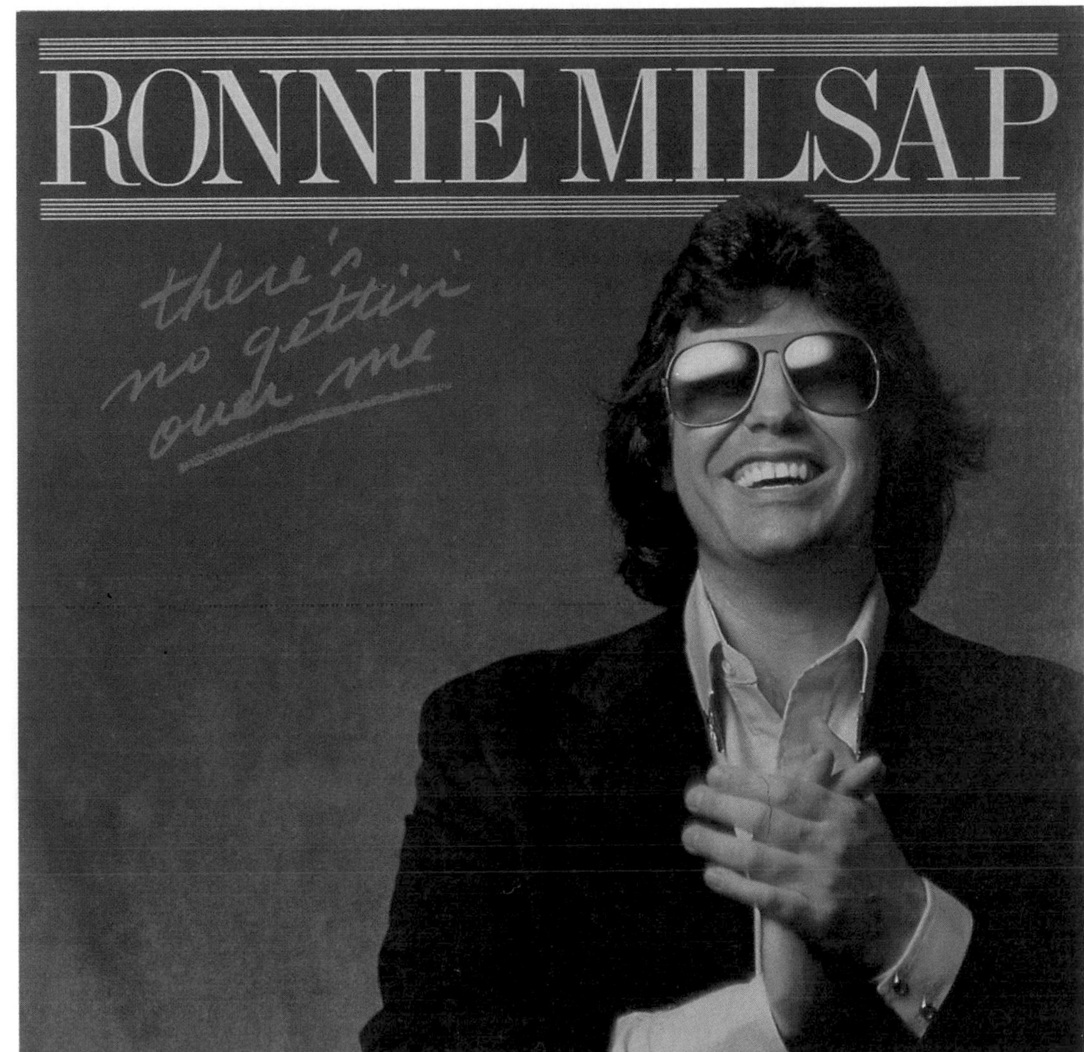

There's No Getting Over Me, Ronnie Milsap. Courtesy RCA Records.

Hugh Moffatt

Long-haired, folk baritone Hugh Moffatt (born on November 10, 1948 in Fort Worth, Texas) has been providing Nashville stars such as Ronnie Milsap, Dolly Parton, Bobby Bare, Lacy J. Dalton, Rex Allen Jr, Johnny Rodriguez and Alabama with hit songs since 1974. With a keen interest in music from a young age, he played trumpet in his high school band and developed a liking for big band jazz. He learned guitar and started performing at local clubs while obtaining a degree in English at Houston's Rice University. After a stint on the Austin music scene, Moffatt headed for Washington DC, but stopped off en-route in Nashville. He decided that Music City would be a better environment for his delicate song-poetry.

In 1974 Ronnie Milsap recorded Moffatt's **Just In Case**. The following year it became a Top 5 country hit. By this time the singer-writer had married Pebe Sebert, and the couple wrote **Old Flames (Can't Hold A Candle To You)**, which Joe Sun took into the Top 20 in 1978. The most recorded of Moffatt's songs, it became a chart-topper for Dolly Parton two years later. Irish duo Foster And Allen converted the song into a British easy-listening pop hit. Also in 1978, Hugh signed a recording contract with Mercury Records, scoring a minor hit with a cover version of Don Schlitz's **The Gambler** (1978). In the early '80s he formed the band Ratz, which included his wife, and produced a self-marketed five-track EP, **Putting On The Ratz**. Throughout the '80s Moffatt continued writing and performing at Nashville's songwriting bars and showcase stages. In 1987 he signed with Philco Records and produced the exquisite **Loving You** album. Very much a modern troubadour, Moffatt packs his guitar and takes his songs and music on the road, when the fancy takes him. He plays small clubs in America and Europe, where he has built a cult following for his invigorating and deeply enriching musical style.

Recommended:

Hugh And Katy – Dance Me Outside (Philo/–)
Live And Alone (Brambus/–)
Loving You (Philo/–)
Troubadour (Philo/–)

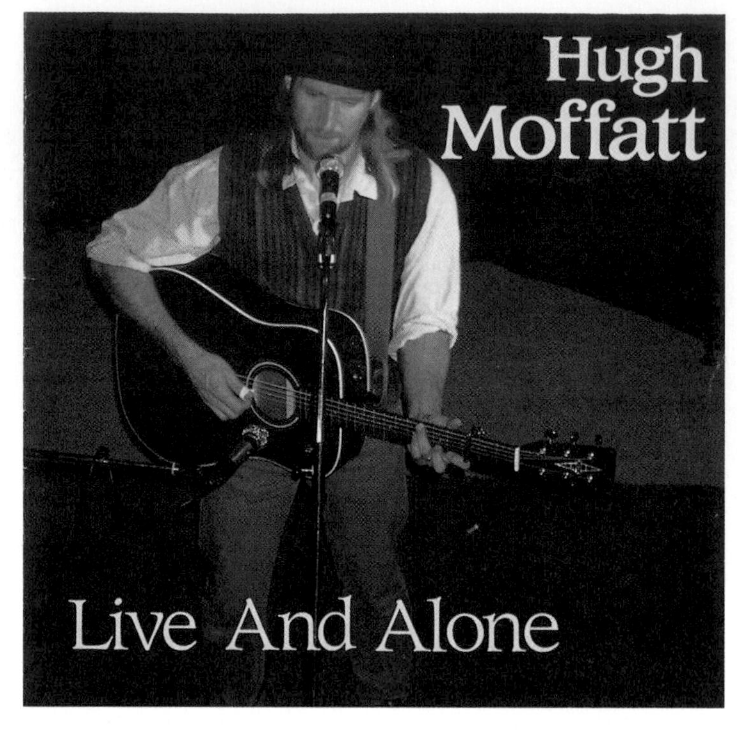

Live And Alone, Hugh Moffatt.
Courtesy Brambus Records.

Katy Moffatt

Often referred to as a country-rocker, Katy, sister of singer-songwriter Hugh Moffatt, was born on November 19, 1950 in Fort Worth, Texas. Her musical style is as varied and large as the Lone Star State. Although her parents were not musical, Katy started piano lessons at five and continued regularly until she was 11. Like most teenagers in the '60s she was influenced by the Beatles. She bought her first guitar at 14 and two years later was performing songs from songwriters as diverse as Lennon and McCartney, and Leonard Cohen. Blues and folk music played a major role in her musical development and she began writing her own songs and singing in small coffee houses in and around Fort Worth.

During the '70s she recorded a couple of country-rock albums for Columbia. Since then she has gradually edged closer to country, with all of the elements of her past musical loves filtering through to her current style. Many years on the club circuit, sometimes performing solo, others working with her own band, has made Katy equally at ease on acoustic and hard-rocking material. In the late '80s she started a song partnership with Tom Russell and has recorded a number of independent albums. Now living in Los Angeles, she has built up a sizeable following in Europe, mainly with the younger 'New Country' fans who enjoy her diverse, contemporary-slanted approach to country music.

Recommended:

The Greatest Show On Earth (–/Round Tower)
Child Bride (Philo/Heartland)
Walking On The Moon (Philo/Heartland)

Bill Monroe

The virtual base on which the whole of bluegrass music rests, William Smith (Bill) Monroe was born at Rosine, Kentucky, on September 13, 1911, the youngest of eight children. Brother Charlie was next youngest, having been born eight years previously on July 4, 1903. This gap, coupled with Bill's poor eyesight, inhibited the youngest son from many of the usual play activities and gave him an introverted nature which carried though into later life. However, this sense of isolation did allow him to develop his musical talents with great alacrity. Church was a major formative influence. Bill could not see to read the shape note hymnals too well and so learned the music by ear. The shaped notes were to appear later in Bill's bluegrass music and the ear training was also important since it has obviously contributed to the man's fine sense of harmony in high vocal ranges.

The Monroe family was musical on both sides. Brother Charlie could play guitar by 11 and Birch could play fiddle. Bill's mother's side, the Vandivers, were the more musical and Uncle Pendleton Vandiver (later immortailized in Bill's most famous composition, **Uncle Pen**) would often stay overnight at the household, occasions of great musical festivity. Uncle Pen was a rated fiddler locally and Bill was playing publicly with him by 13, backing Pen's fiddle with guitar.

Another influence was a black musician from Rosine, Arnold Shultz. Bill would gig with him and rated him a fine musician with an unrivalled feel for the blues. At this time he also started to hear gramophone records featuring such performers as Charlie Poole and the North Carolina Ramblers.

Birch and Charlie left to seek work in Indiana and Bill joined them in 1929, when he was 18. Until 1934, in East Chicago, Indiana, they worked manual jobs by day and played dances and parties at night. For a while they went on tour with the Chicago WLS station Barn Dance, doing exhibition dancing. In 1934, Radio WLS, for whom the three brothers (Birch on fiddle, Charlie on guitar and Bill on mandolin) had been working on a semi-professional basis, offered them full-time employment. Birch decided to give up music but Charlie and Bill reformed as a duet, the Monroe Brothers. In 1935 they were sponsored on Carolina radio by Texas Crystals. In that same year Bill married Caroline Brown – their children, Melissa (born 1936) and James (1941) have both performed with Bill. James eventually formed his own band, the Midnight Ramblers.

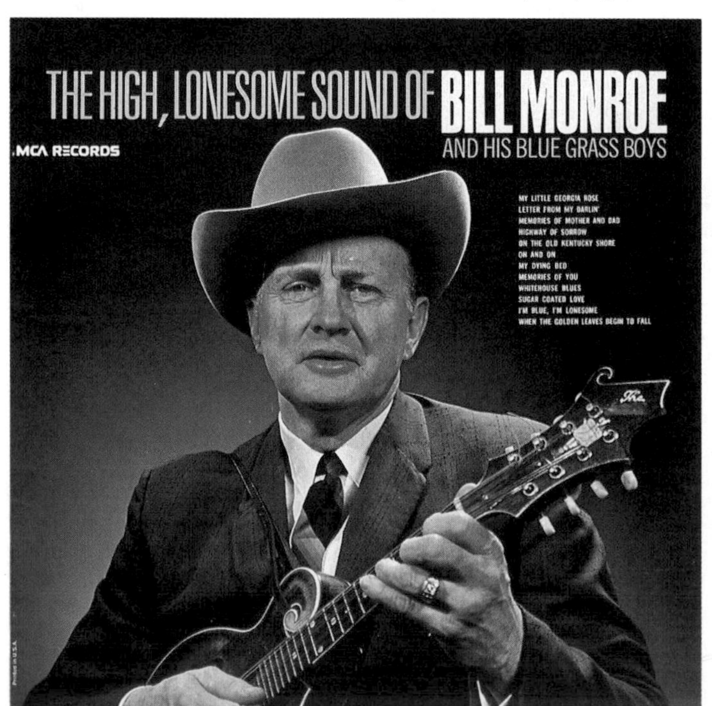

THE HIGH, LONESOME SOUND OF BILL MONROE
AND HIS BLUE GRASS BOYS
MCA RECORDS
MY LITTLE GEORGIA ROSE
LETTER FROM MY DARLIN'
MEMORIES OF MOTHER AND DAD
HIGHWAY OF SORROW
ON THE OLD KENTUCKY SHORE
ON AND ON
MY DYING BED
MEMORIES OF YOU
WHITEHOUSE BLUES
SUGAR COATED LOVE
I'M BLUE, I'M LONESOME
WHEN THE GOLDEN LEAVES BEGIN TO FALL

The High, Lonesome Sound Of Bill Monroe And His Blue Grass Boys.
Courtesy MCA Records.

The Monroe Brothers were engaged on radio work in Greenville, South Carolina, and Charlotte, North Carolina. They were then persuaded forcefully (by Eli Oberstein of Victor Records) to cut some records in 1936. That year, in the Radio Building at Charlotte, they cut ten sides during February, including an early best-seller **What Would You Give (In Exchange For Your Soul)**. At first they did not feel interested in the idea of recording since they were already doing well via radio and live broadcasts, but sales of the early sides were impressive enough to warrant five more such sessions during the next 12 months. Besides, featuring popular traditional material (the Monroes did not write their own songs at the time), they had pioneered a distinctive style in which then-advanced mandolin and guitar techniques were coupled with a high clear, recognizable vocal sound.

In 1938 they went their separate ways. Bill formed the Kentuckians in Little Rock, Arkansas and then moved on to Radio KARK, Atlanta, Georgia, where the first of the Blue Grass Boys line-ups was evolved. At this time, Bill began to sing lead and to take mandolin solos rather than just remaining part of the general sound. In 1939 he auditioned for the Opry and George D. Hay was impressed enough to sign him. The following Saturday Bill played his first Opry number, the famous **Mule Skinner Blues**.

Bill Monroe's music then started to undergo subtle changes. He added accordion and banjo (played by Sally Ann Forester and Stringbean respectively) in 1945 when he joined Columbia Records and this period saw the evolution of a fuller sound with the musicians taking more solos. But in 1945 the accordion was dispensed with never to return, and in that year the addition of Earl Scruggs, with a banjo style that was more driving and syncopated than anything heard previously, put the final, distinctive seal on Monroe's bluegrass sound. Flatt and Scruggs remained with Bill until 1948.

Songs from this period include **Blue Moon Of Kentucky**, **I Hear A Sweet Voice Calling**, **I'm Going Back To Old Kentucky** and **Will You Be Loving Another Man**? Other musicians with Bill at this time were Chubby Wise (fiddle) and Howard Watts – also known as Cedric Rainwater – (bass).

Bill left Columbia in 1949 because he objected to them signing the Stanley Brothers, a rival bluegrass group. With his next label, Decca, his main man was Jimmy Martin, a musician with a strong thin, highish voice, whose talents enabled Monroe to fill in more subtle vocal harmonies alongside. This was Monroe's golden age for compositions. He wrote **Uncle Pen**, **Roanoke, Scotland** (a nod towards the original source of string band jigs and reels) **My Little Georgia Rose**, **Walking In Jerusalem** and **I'm Working On A Building**, the last two being religious 'message' songs, always part of the Monroe tradition from the earlier days.

By the end of the decade, bluegrass, though it had added a new dimension to country, was in decline due to the onset of rock'n'roll (it is interesting to remember that when Elvis Presley released his **Blue**

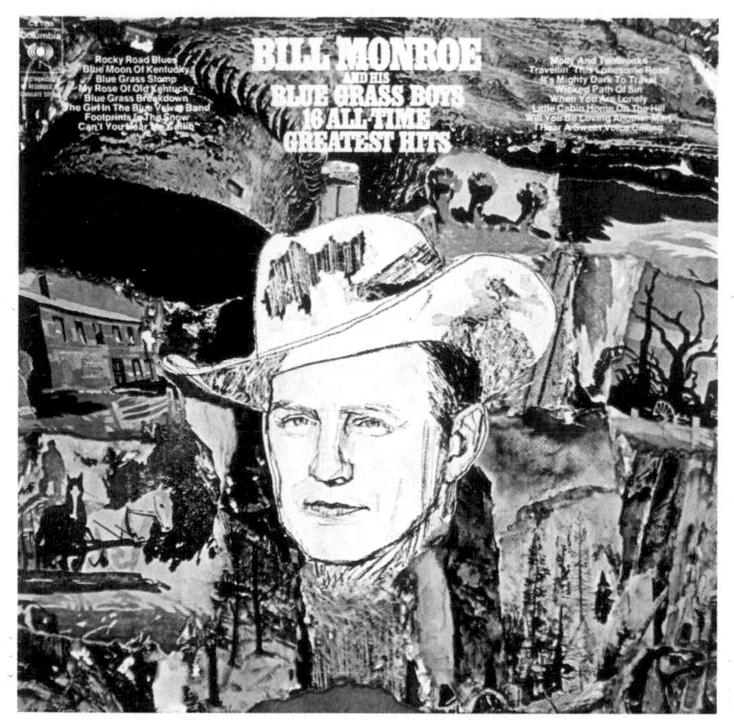

Moon Of Kentucky in 1954, Bill's record company rushed out a re-release of Monroe's (very different) original). But the '60s saw a folk revival and Bill now found hordes of students eager to embrace indigenous rural white folk music. In 1963 Bill made his first college appearance at the University of Chicago; later that same year he played to 15,000 people at the Newport Folk Festival.

Bill Monroe was elected to the Country Music Hall Of Fame in 1970. In 1985 his record company, MCA, went through several major changes, culminating in a new regime. Emory Gordy Jr, former bass player with Emmylou Harris and Rodney Crowell, was brought in as Bill's producer, which resulted in three albums that are among the most important recordings he has ever made. **Stars Of The Bluegrass Hall Of Fame** (1986) brought Bill together with Jim and Jesse McReynolds, Ralph Stanley, Mac Wiseman, the Osborne Brothers, Del McCoury, Carl Story, and members of Country Gentlemen and Seldom Scene. **Bluegrass '87** added greatly to the bluegrass repertoire with the inclusion of six newly written songs, four of them instrumentals. **Southern Flavor** included five new tunes and featured Blue Grass Boys' guitarist Tom Ewing on all but one of the lead vocal parts. The album won the first Grammy ever awarded for bluegrass in 1989.

Monroe has always trodden his own musical path, never bowing to commercial pressure, and his contribution to country music is inestimable. On August 13, 1986, one month to the day before his 75th birthday, the US Senate passed a resolution recognizing "his many contributions to American culture and his many ways of helping American people enjoy themselves". It also said, "As a musician, showman, composer and teacher, Mr Monroe has been a cultural figure and force of signal importance in our time."

As for Charlie, he had a long and successful career – though not nearly as spectacular as Bill's – with his own band, the Kentucky Pardners, well into the '50s. He returned from retirement in the early '70s to appear on the bluegrass circuit – displaying great grace and charm – before dying of cancer in 1975.

16 All-Time Greatest Hits, Bill Monroe. Courtesy CBS Records. The album contains the classic Blue Moon Of Kentucky.

Recommended:
The Original Bluegrass Band (Rounder/–)
Best Of Bill Monroe And The Bluegrass Boys (–/MCA)
Bluegrass Instrumentals (MCA/–)
Bluegrass Ramble (MCA/–)
Blue Moon Of Kentucky, 1950–1958 (–/Bear Family)
Bluegrass, 1959–1969 (–/Bear Family)
'87 (MCA/–)
Country Music Hall Of Fame (MCA/–)

Monroe Brothers:
Early Blue Grass Music (Camden/–)

Charlie Monroe:
Who's Calling You Sweetheart Tonight (Camden/–)
Charlie Monroe On The Noonday Jamboree (County/County)

Patsy Montana

Patsy Montana, born Rubye Blevins, on October 30, 1914, in Hot Springs, Arkansas, became the first woman in country music to have a million-selling record when **I Want To Be A Cowboy's Sweetheart** was released in 1935. She began her early career with silent film cowboy star Monte Montana (no relation), but was long associated with the Prairie Ramblers on the National Barn Dance radio show (1934–52). The Ramblers backed her on **Cowboy's Sweetheart** and most of her other hits. She recorded with them on the ARC complex of labels (1935–42) and then Decca (1942–49) and RCA (1949–51) before leaving them and Chicago for the West Coast in 1952.

She has been in and out of retirement ever since, occasionally appearing with her daughter Judy Rose, and recording in later years on Surf and Starday.

Recommended:
Early Country Favorites (Old Homestead/–)

John Michael Montgomery

Country's latest hunk, John Michael Montogemery, was born on January 20, 1965, in Nicholasville, Kentucky. His parents, Harold and Carol Montgomery, raised their three children to make music. Harold had his own country band, playing clubs in central Kentucky, with his wife on drums. All three children, Eddie, also on drums, John Michael on guitar and vocals, and daughter Becky on vocals, performed with the band.

When John was 17 his parents divorced. This marked the end of the band, so the youngster started solo stints. In 1984 he started a rock band called Erly Tymz (Early Times), playing the music of Bob Seger, Jimmy Buffett and Lynyrd Skynyrd. A couple of years later he returned to country, and by 1988 was performing at Lexington's Austin Saloon, where he played five nights a week over a four-year period. The gig paid off. Executives from Atlantic Records, tipped off by songwriter Steve Clark, saw him, and signed him to a contract in 1991. Not happy with the way his first album was shaping up, Montogemery took his career in his hands and phoned Atlantic–Nashville executive Rick Blackburn. Blackburn brought in Doug Johnson to complete the production, and the result was an immediate Top 5 country single in late 1992 with the album's title song, **Life's A Dance**. A second single, **I Love The Way You Love Me**, soared to No.1 in 1993.

Montgomery's powerful voice and compelling charisma jump out from an album that runs the gamut from laid-back country balladry to rowdy honky-tonk tunes. Instead of just working the honky-tonk bars of central Kentucky, John Michael now takes his music to a much wider audience, including opening shows for Reba McEntire and Alabama.

Recommended:
Life's A Dance (Atlantic/–)

Below: Melba Montgomery's eponymous album. Courtesy Elektra Records.

Melba Montgomery

Born in Iron City, Tennessee, on October 14, 1938, Melba was raised in Florence, Alabama. She began her singing at the local Methodist church where her fiddle- and guitar-playing father taught singing.

Melba moved to Nashville, there catching the attention of Roy Acuff, whose show she stayed with for four years. In 1962 she went solo and released her first singles, her initial Top 10 entry coming with **We Must Have Been Out Of Our Minds**, a duet with George Jones released by UA in 1963. Further duet hits with Jones followed, and she also made duets with pop singer Gene Pitney, placing **Baby, Ain't That Fine** into the Top 20 in 1966.

She continued recording for UA and Musicor in moderately successful fashion through to 1967, when her musical partnership with George Jones ended. Melba moved on to Capitol and a series of duets with Charlie Louvin, the most successful being **Something To Brag About** (1970).

Her time at Capitol saw the start of a liaison with Pete Drake that carried over fruitfully into her stint at Elektra. She began at Elektra in 1973 with **Wrap Your Love Around Me** (1973).

By 1974 she had a No.1 record with **No Charge**, a sentimental 'talkover' song penned by Harlan Howard that provided J. J. Barrie with a British pop hit in 1976. She also went Top 20 with **Don't Let The Good Times Fool You** (1975), but, by 1977, was back on UA once more, doing pretty well with a version of **Angel Of The Morning**. After this her name became absent from the annual list of chart contenders, although she has since recorded for Kari and Compass.

Recommended:
Baby You've Got What It Takes – with Charlie Louvin (Capitol/–)
Don't Let The Good Times Fool You (Elektra/–)
No Charge (Elektra/–)
We Must Have Been Out Of Our Minds – with George Jones (RCA/RCA)

Above: Clyde Moody – North Carolina's Hillbilly Waltz King.

Above: George Morgan's Candy Kisses was the biggest country song of 1949.

Clyde Moody

Born on a Cherokee Reservation in North Carolina on September 19, 1915, Clyde Moody rose to prominence in the '40s as the 'Hillbilly Waltz King', largely on the strength of his gold record for **Shenandoah Waltz** for King Records, which sold some three million copies.

Moody apprenticed with Mainer's Mountaineers, spent several successful years with Bill Monroe, with whom he recorded the classic **Six White Horses**, and spent a bit of time with Roy Acuff before joining the Opry on his own in the mid-'40s. He left the Opry in the late '40s to pioneer television in the Washington DC area, then returned to his native North Carolina, where he had a long-running TV show and several business interests. But he resumed his musical career in Nashville, annually appearing at bluegrass festivals (those fans did not forget his years as a Blue Grass Boy) and touring over 300 days a year with Ramblin' Tommy Scott's Show. He passed away on April 7, 1989.

Recommended:
Moody's Blue (Old Homestead/–)

George Morgan

The writer and singer of **Candy Kisses**, the biggest country song of 1949, Morgan, for a brief period, looked capable of usurping Eddy Arnold's position as the 'king of country pop'.

Born in Waverly, Tennessee, on June 28, 1925, he spent his teen years in Barberton, Ohio. On completing high school, Morgan worked as a part-time performer. Obtaining a regular singing spot on WWVA Jamboree, Wheeling, West Virginia, he established something of a reputation there and became signed to Columbia Records. His first release was **Candy Kisses**, which reached No.1 in the country charts (subsequently selling a million) while Elton Britt's cover version reached No.3. Invited to join WSM in 1948 in the dual role of DJ and vocalist, he soon became an Opry regular, scoring with further 1949 hits in **Rainbow In My Heart**, **Room Full Of Roses** and **Cry-Baby Heart**.

Establishing a smooth, easy style that came replete with fiddle and steel guitar, Morgan became a popular radio and concert artist. but, despite major successes with **Almost** (1952), **I'm In Love Again** (1959) and **You're The Only Good Thing** (1960), his record sales remained little more than steady.

In 1966 he joined Starday, who provided him with elaborate orchestral trappings, a ploy which gained Morgan a quintet of mini-hits during 1967–68. But soon he was on the move again, recording with Stop, Decca and Four Star, but only twice attaining Top 20 status, with **Lilacs And Rain**, a 1970 Stop release, and **Red Rose From The Blue Side Of Town**, a 1973 MCA item.

Morgan's death occurred in July 1975, following a heart attack sustained while on the roof of his house, fixing a TV aerial.

Sounds Of Goodbye, George Morgan. Courtesy Starday Records.

Recommended:
Remembering (Greatest Hits) (Columbia/–)
The Best Of (Starday/–)
American Originals (Columbia/–)

Lorrie Morgan

Lorrie Morgan is one of country music's most attractive women, and her life story is as dramatic and tragic as the words to many country songs. The youngest daughter of country star George Morgan, Loretta Lynn Morgan was born on June 27, 1959 in Nashville, Tennessee. A consummate professional reared in a country-star family, she has performed publicly since she was five. Lorrie made her debut on the Grand Ole Opry at 13, and for some years toured with her father. Having more than her fair share of heartache, her father died when she was 16, then came pregnancy and marriage at 20, followed by divorce at 21. She struggled for acceptance as a country singer for 15 years, and just when she began to taste success, that happiness was overshadowed by the death of her second husband, singer Keith Whitley, of alcohol poisoning on May 9, 1989.

She had married Whitley in November 1986, and the couple had a son, Jesse (Lorrie had a daughter, Morgan, from her previous marriage). The pair were ecstatically in love, and the marriage had given a boost to both their careers. Lorrie had previously recorded for Hickory Records in 1979 and MCA in the early '80s without too much success. She signed with RCA in 1986 and immediately made the Top 20 with her first single, **Trainwreck Of Emotion** (1988). This was followed by her Top 10 debut with **Dear Me**, a song she demoed ten years earlier.

"I always swore that when I got to do a record I was going to record **Dear Me**," Lorrie explained, "But every producer I had said, 'Naw, that doesn't sound like you, we're not doing that,' but when I played it for Barry Beckett [her producer at RCA], he just flipped over it. I knew he was the right producer then, because I knew it was a great song."

Astute at choosing the right material, Lorrie portrays a masterful combination of vocal emotion and vocal beauty. Her voice has a contagious feeling and sincerity. A third RCA single, **Out Of Your Shoes**, spent three weeks at No.2 on the country charts in 1989, while her first album, **Leave The Light On**, went gold. A chart-topper with **Five Minutes** (1990) lay the foundation for further Top 10 entries with **He Talks To Me** (1990), **We Both Walk** and **A Picture Of Me (Without You)** (1991). A posthumously released duet with Keith Whitley, **'Til A Tear Becomes A Rose**, went Top 20 in 1990. A second album, **Something In Red**, also gained a gold disc, and the title song and **Except For Monday** presented her with Top 10 hits in 1992. That same year she moved to the affiliated BNA label, scoring with **Watch Me** (1992) and a chart-topper with **What Part Of No** (1993).

Above: Gary Morris has rebuilt his career during the '90s via minor country hits.

Life was still no bed of roses for the determined Lorrie Morgan. In January 1992 she was forced to file Chapter 11 bankruptcy, though by the end of the year, she had paid all her creditors in full. A third marriage in October 1991 to Brad Johnson, a former bus driver for Clint Black, was reportedly on the rocks, and a routine operation in the summer of 1992 turned out to be major surgery. But despite all her setbacks, Lorrie has not stood still licking wounds. She has pushed her career forward, and, following acclaimed music videos, she is beginning to make her mark as an actress, with a role in a TV movie in production during 1993.

Recommended:
Watch Me (BNA/–)
Leave The Light On (RCA/RCA)
Something In Red (RCA/RCA)

Gary Morris

Like country pioneer Vernon Dalhart, Gary Morris is able to sing both country and opera. He was born on December 7, 1948 in Fort Worth, Texas. A junior high school guitarist who sang in the church choir, Morris formed a trio during the late '60s. After the trio played a Hank Williams' medley at a live, audience-attended audition, they were booked into a Denver night-club.

After several years at the club, Morris went solo and headed back to Texas, there meeting Lawton Williams, writer of **Fraulein**, a massive hit for Bobby Helms. Williams introduced Morris to various music-biz people and in 1978 he was invited to play at a White House party hosted by President Carter, after which he was asked to record a number of sides for MCA Nashville. For a while Morris went back to Colorado, forming a band called Breakaway. But in 1980 he flew to Nashville once more, meeting producer Norro Wilson, who had seen him at the White House gig. He cut **Sweet Red Wine** and **Fire In Your Eyes**, both going Top 40 on Warner.

Late in 1981 he had his first Top 10 single in **Headed For A Heartache**. The following year he had three major singles, the biggest of these being **Velvet Chains**. By 1983 he had become a fully fledged Top 10 act, logging three solo hit records – **The Love She Found In Me**, **The Wind Beneath My Wings**, **Why, Lady, Why?** – and **You're Welcome Tonight**, a high-selling duet with Lynn Anderson, recorded for the Permian label. The next year saw Gary netting three further Top 10 records, **Between Two Fires**, **Second Hand Heart** and **Baby Bye Bye** (the latter becoming his first No.1) and being acclaimed as country music's premier male sex symbol.

After that, everything headed his way. He was chosen to star opposite Linda Ronstadt in the New York Shakespeare Festival production of the opera 'La Bohème'. This led to regular guest appearances on the TV soap Dynasty II – The Colbys, in which he played Wayne Masterson, a blind musician. In 1987 Gary headed back to New York to star in Broadway's 'Les Misérables', and reprised the role in London for the complete symphonic recording in 1988.

Alongside all this activity, his country hit-making continued with **Lasso The Moon**, **Making Up For Lost Time** (a duet with Crystal Gayle), **I'll Never Stop Loving You** (all 1985), **100% Chance Of Rain** (1986), and a No.1 with **Leave Me Lonely**, as well as another duet with Crystal Gayle on **Another World** (both 1987). In 1988 he signed with the short-lived Universal Records, finally landing on Capitol in 1990 and slowly rebuilding his career via minor hits, such as **Miles Across The Bedroom** (1991) and **Love Hurts** (1992).

Recommended:
Faded Blue (Warner Bros/–)
Second Hand Love (Warner Bros/Warner Bros)
Full Moon, Empty Heart (Capitol/–)
Stones (Universal/–)
These Days (Capitol/–)

Michael Martin Murphey

A pioneer of the Texas 'cosmic' cowboy scene of the early '70s, Michael Martin Murphey was born in 1946 in Dallas, Texas, and was raised on the music of Hank Williams, Bob Wills and Woody Guthrie. He had sang in a folksy outfit, the Texas Twosome, while still at high school in Dallas. After graduation he attended the University of California in Los Angeles, where he started to take his songwriting more seriously. He signed a writer's contract with Screen Gems Music in 1965, and wrote mainly theme tunes and soundtrack material for television, though he did provide the Monkees with one of their better songs – **What Am I Doin' Hangin' Around**. In 1967 he toured as Travis Lewis of the pop outfit the Lewis And Clarke Expedition. He then embarked on a solo career, basing himself in Austin where he joined the thriving Texas music scene. Murphey worked with producer Bob Johnston, and signed to A&M Records, achieving a Top 40 pop placing for the self-penned **Geronimo's Cadillac**.

Building a career as a major singer-songwriter, Murphey moved to Colorado in 1974 and signed to Epic Records. He made his debut on the country charts with **A Mansion On The Hill** in 1976. Three years later, he moved again, this time to Taos, New Mexico. Gradually enlarging his country following, he appeared in the films 'Hard Country', 'Take This Job And Shove It' and 'Urban Cowboy', and made it to No.1 on the country charts for Liberty with **What's Forever For** (1982), a record that also made the pop Top 20. Further country success came with: **Still Taking Chances**, **Don't Count The Rainy Days** (1983), **Will It Be Loving By Morning** (1984), and **What She Wants** and **Carolina In The Pines** (1985). By this time his records were appearing on EMI-America.

In 1986 he signed with Warner Brothers Records, that year going Top 20 with **Rolling Nowhere**. The following year he was back in the Top 10 with **A Face In The Crowd** (a duet with Holly Dunn), and at No.1 with **A Long Line Of Love**. The Top 10 hits continued with **Talkin' To The Wrong Man** (1988) featuring his son, Ryan Murphey. Getting closer to his musical roots, in 1990 Murphey released **Cowboy Songs**, a collection of traditional campfire and trail songs that made an impact on the charts and helped to re-kindle interest in cowboys and their music.

Recommended:
Land Of Enchantment (Warner Bros/–)
Cowboy Songs (Warner Bros/–)
River Of Time (Warner Bros/–)
Michael Martin Murphey (Liberty/Liberty)

Anne Murray

A deceptively light-voiced Canadian singer, Anne Murray packed enough punch on her **Snowbird** hit in the '60s to score a major international pop hit.

Born in Spring Hill, Nova Scotia, on June 20, 1946, Anne was the only girl in a

Danny's Song, Anne Murray. Courtesy Capitol Records.

family of five brothers. She obtained a bachelor's degree at the University of New Brunswick and taught physical education. Eventually, finding that singing was taking more of her time, she quit and moved entirely into show-biz when offered a contract by Capitol Records.

Snowbird was one of her first releases and its light airy melody was soon on everyone's lips. It scored in both the pop and country charts in 1970 and provided Anne with two potential markets which she proceeded to tightrope-walk.

She made some high-selling singles during 1972, including **Cotton Jenny**, **Danny's Song** and **Love Song** but had to wait until 1974 and **He Thinks I Still Care** before gaining her second No.1. After this she had Top 10 singles with **Son Of A Rotten Gambler** (1974), **Walk Right Back, You Needed Me** (1978), hitting a winning streak in 1979 when **I Just Fall In Love Again**, **Shadows In The Moonlight** and **Broken Hearted Me** all went to No.1. Then came such hits as **Daydream Believer**, **Could I Have This Dance?** (1980), **Blessed Are The Believers**, **It's All I Can Do** (1981), **Another Sleepless Night**, **Hey! Baby!**, **Somebody's Always Saying Goodbye** (1982) and **A Little Good News** (No.1, 1983). The last named is, lyrically at least, the finest song that Murray has ever

Above: Feed This Fire proved a Top 10 entry for Anne Murray in 1990, getting her back in the limelight after a turbulent few years in the late '80s.

recorded and one that won her a Grammy award as Best Country Female Singer, helping her accrue a tally of major awards that includes four Grammys, three CMA awards and no less than 22 Canadian Juno plaudits.

During 1984 she logged two further country No.1s with **Just Another Woman In Love** and **Nobody Loves Me Like You Do**, a duet with Dave Loggins. The following year had a further hit **Time Don't Run Out On Me**. Another No.1 came with **Now And Forever (You And Me)** in 1986. Anne, completely confused by her continuing country success, claimed her album, **Something To Talk About**, was her first real pop album in six years. The late '80s saw her records struggling to make a major impact on the country charts. Then in 1990 she was back in the Top 10 with **Feed This Fire**. A regular presenter and host on the country music award shows, retirement is not in the cards: "This business of retirement and getting old and all that is garbage," says Anne. "There is no such thing. People work 'til they drop now. That's the way it should be, heading for new challenges when you're 70–75."

Recommended:
Anne Murray's Greatest Hits (Capitol/Capitol)
A Little Good News (Capitol/Capitol)
Something To Talk About (Capitol/Capitol)
Heart Over Mind (Capitol/Capitol)
Songs Of The Heart (Capitol/Capitol)

Jerry Naylor

Once front-man with the Crickets, Naylor won 'Billboard' awards in 1973 and 1974 for providing the best syndicated country radio show.

Born in Stephenville, Texas, on March 6, 1939, he formed his own group at the age of 14 and soon proved able enough to perform on the Louisiana Hayride show, touring alongside such acts as Johnny Cash, Elvis Presley and Johnny Horton. A DJ during his high school days, he later enrolled at the Elkins Electronic Institute, employing his radio know-how for AFRS in Germany during 1957.

Following discharge from the army after a spinal injury, he returned home and recorded for Sklya Records, also befriending Glen Campbell. The duo moved to L.A. where Naylor worked for KRLA and KDAY. In 1961 he became a member of the Crickets, replacing bassist Joe B. Maudlin (although both musicians are depicted on the sleeve of the **Bobby Vee Meets The Crickets** album!). But he suffered a heart attack in 1964 and left the group, taking up a solo career in country music, recording first for Tower, then for Columbia and MGM before signing for Melodyland – for which label he provided a hit in **Is That All There Is To A Honky Tonk?** in 1975.

Along the way, Naylor, who once recorded for Raystar under the name of Jackie Garrard, has notched hits for Hitsville, MC and Oak, also working for Hoyt Axton's Jeremiah without much success. He now concentrates on DJ work and lives in Angoura, California with his wife Pamela and three children.

Recommended:
Love Away Her Memory (MCA/–)

Rick Nelson

The child of a showbusiness family (his parents Ozzie and Harriet had a radio and later a TV show), Nelson made the transition from teenage idol to modern country artist.

Below: The late Rick Nelson, who crashed in a plane once owned by Jerry Lee Lewis.

Above: Tax problems gave Willie Nelson a major headache in 1991, but his Across The Borderline in 1993 put him back in pocket.

Born on May 8, 1940, in Teaneck, New Jersey, he signed as Ricky Nelson first for Verve and then Imperial; his lonesome teenthrob voice was allied to light, country-influenced backings on songs such as **Poor Little Fool**, **It's Late** and **Lonesome Town**, sagas of jilted love and dating frustrations, and perfectly in tune with the softening tone of rock'n'roll.

This was the major record companies' answer to the more raw, sharper music of Memphis. Since Nelson was also blessed with archetypal boy-next-door looks, his appeal was further propagated via concerts and TV.

The coming of the British Invasion swept away many of these clean-cut American teen rockers, although Nelson was by then signed to MCA and had shortened his name to Rick. Although he had James Burton in his band, he failed to make a real impression. However, two country albums, **Country Fever** and **Bright Lights And Country Music**, found favour with some people.

The formation of the Stone Canyon Band in the '70s saw Nelson being accepted as a viable country rocker. This band included, at one point, Randy Meisner, also known from Poco and the Eagles, and Tom Brumley, Buck Owens' rated steel player for five years). **Garden Party**, a poignant auto-biographical piece of country-rock, made the American pop Top 10 in 1972 and won Nelson yet another gold record. Although he recorded some outstanding country-rock albums during the '70s for MCA, Epic and Capitol, he failed to make any further inroads into the charts.

During the early '80s, Rick and his excellent Stone Canyon Band were mainly working rock'n'roll revival shows, but his musical integrity was always maintained by his insistence on giving many of his old songs fresh, new arrangements. He successfully toured Britain during November 1985 with Bo Diddley, Bobby Vee and Frankie Ford. Two months later, on December 31, 1985, Rick, along with his fiancée, Helen Blair, and members of his band were in a fatal air crash while en route to a show in Dallas, Texas.

Recommended:

Country (MCA/–)
Garden Party (MCA/MCA)
Rudy The 5th (MCA/MCA)
Sings Rick Nelson (MCA/MCA)
String Along with Rick (–/Charly)
Playing To Win (Capitol/Capitol)
Country Fever/Bright Lights And Country Music (–/See For Miles)

Willie Nelson

Willie Nelson has run a long, hard race in country music but has won through as a premier stylist. Born in Abbott, Texas, on April 30, 1933, Nelson was raised by his grandparents after his own parents separated.

His grandparents taught him some chords and by his teens he was becoming proficient on guitar. In 1950, he joined the Air Force and on his subsequent discharge married a Cherokee Indian girl by whom he had a daughter, Lana. Living in Waco, Texas, Nelson took various salesman jobs. But, anxious to gain a proper intro into music, he talked his way into an announcing job on a local station.

Soon after, he was hosting country shows on a Fort Worth station, doubling at night as a musician in some rough local honky-tonks and, whenever he could, he was jotting down songs. It was during this period that he wrote **Family Bible** and **Night Life**, songs which have become standards.

When he finally made his way to Nashville and found a job in Ray Price's band as a bass player, he found that he was finally placing his songs. Price, a huge name of that era, made **Night Life** his theme tune. Faron Young cut **Hello Walls**, Patsy Cline **Crazy** and Willie himself recorded **The Party's Over**. They were sombre but haunting melodies, true 'white man's blues', and Willie has since incorporated them tellingly into his sparse, bluesy act. More than 70 artists have recorded **Night Life**.

Help Me Make It Through The Night, Willie Nelson. Courtesy RCA Records.

After poaching most of Ray Price's band from him, Nelson went on the road. At this time his first marriage broke up and he went off with the wife of a DJ Association president and married again, settling in Fort Worth, Los Angeles and Nashville.

Besides recording 18 albums in three years, he also helped the career of Charley Pride, featuring him on his show in the deepest South during the racially sensitive years of civil rights.

During the '60s the smooth Nashville Sound was in its ascendant and Willie found himself becoming increasingly disillusioned with big business methods, hankering to make his mark as a singer rather than as a songwriter and preferably on his own terms. By the early '70s he was determined to get out of his RCA contract, and with the help of Neil Reshen (afterwards Nelson's manager) he landed a contract from the new Nashville offices of Atlantic. However country music was new to Atlantic, who had built their reputation on black music. But with Atlantic's Jerry Wexler producing in New York, Willie came up with **Shotgun Willie** and **Phases And Stages**, the second had a new, intense country sound and made no concessions to modern Nashville.

Nelson had by now settled in Austin with his third wife, Connie. By this time Atlantic's Nashville operation had folded and Willie signed with Columbia, for whom he made **Red Headed Stranger** and **The Sound In Your Mind.** Like **Phases And Stages**, **Red Headed Stranger** was a concept album, but even more personal. It threw up the national cross-over hit single **Blue Eyes Cryin' in The Rain** and established Willie Nelson as a nationally known figure.

Willie, recognized as the unofficial Mayor of Austin, reconciled hip and redneck musical interests and helped lead a new explosion of interest in country music. Teaming up with Waylon Jennings, they topped the country charts in 1976 with **Good Hearted Woman** and were both featured on the compilation album, **Wanted: The Outlaws**, the first certified platinum album in country music, and so started the Outlaw Movement. The two were voted into top positions in the annual Country Music Association awards, a sure sign of industry acceptance.

Refusing to be tied down to commercial considerations, Nelson has recorded such diverse album projects as **Stardust** (popular standards), **The Troublemaker** (a gospel set), **To Lefty From Willie** (a tribute to Lefty Frizzell) and **Angel Eyes** (featuring jazz guitarist Jackie King).

Willie also instigated the now legendary Fourth Of July Picnics, massive outdoor festivals in Texas which have featured stars such as Leon Russell, Kris

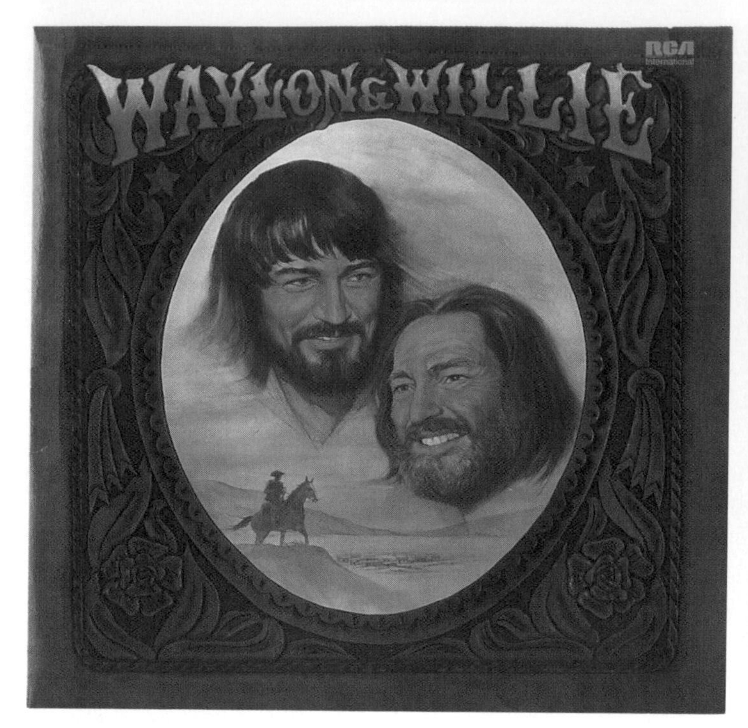

Waylon And Willie, Willie Nelson with Waylon Jennings. Courtesy RCA.

Kristofferson, Roy Acuff, Waylon Jennings, Tex Ritter and Asleep At The Wheel.

During the early '80s, Willie became acknowledged as the 'king' of the country duets. Alongside his highly successful duets with Waylon Jennings, which resulted in the chart-topping **Mammas Don't Let Your Babies Grow Up To Be Cowboys** winning a Grammy in 1978, he has recorded duet albums with Roger Miller, Ray Price, Faron Young, Webb Pierce and Merle Haggard, as well as guesting on albums and singles by Emmylou Harris, Pam Rose, Rattlesnake Annie and many others.

Poncho And Lefty, a duet album recorded with Merle Haggard, was named CMA Album Of The Year (1983) with the title song topping the country charts. Nelson achieved even more success the following year when he teamed up with Julio Iglesias for the pop and country hit, **To All The Girls I've Loved Before**, which also won CMA and Grammy awards.

Nelson continued to chalk up hits throughout the '80s and '90s, with such songs as **Always On My Mind** (1982), **Highwayman** (1986), **Ain't Necessarily So** (1990). He has also organized several Farm Aid benefits. In early 1991 his tax shelter was exposed as a scam and the singer was left owing 17 million dollars in unpaid taxes. In co-operation with the IRS Nelson released a two-record set, **Who'll Buy My Memories aka The IRS Tapes**, with the proceeds paying his tax bill.

A return to the commercial mainstream came with **Across The Borderline**, a 1993 album produced by Don Was and featuring collaborations with Paul Simon, Bob Dylan, Sinead O'Connor, Mark O'Connor, David Crosby, Mose Allison and others. A rich mixture of musical styles, it took Nelson a long way from his well-established country sounds, but once again put him in touch with a mass audience.

Recommended:

Phases And Stages (Atlantic/–)
Shotgun Willie (Atlantic/–)
Live (RCA/RCA)
Red Headed Stranger (Columbia/CBS)
Famous Country Music Makers (–/RCA)
The Sound In Your Mind (Columbia/CBS)
Tougher Than Leather (Columbia/CBS)
Me And Paul (Columbia/–)
Always On My Mind (Columbia/CBS)
Across The Borderline (Columbia/Columbia)
Beautiful Texas, 1936–1986 (–/Bear Family)
City Of New Orleans (Columbia/CBS)
I Love Country (–/CBS)

Michael Nesmith

One of the fabulously successful Monkees, a teenybopper quartet whose main claim to artistic fame was that they were usually given good commercial song material, Nesmith left in 1969 to carve a modestly notable career as a sort of freewheeling cosmic cowboy.

Although he claimed to be only nominally into country, his songs have provided good country fodder, and Nesmith's records with the First and Second National Bands have their own cult audience.

Born in Houston, Texas, on December 30, 1942, Nesmith only learned to play guitar after his Air Force discharge in 1962. However, he was writing songs and his **Different Drum** was covered by Linda Ronstadt.

Becoming increasingly disenchanted by the big business surrounding the Monkees,

The Troublemaker, Willie Nelson. Courtesy CBS Records.

Above: Willie Nelson set up Farm Aid concerts in 1985 and 1986.

he had, by 1968, produced an album of self-composed instrumentals, **The Wichita Train Whistle Sings**.

Nesmith was the creator of the First National Band, which included pedal steel player Red Rhodes, and was signed to RCA. In 1970 they put out the album **Magnetic South**, a Nesmith composition from his album, **Joanne**, being covered by Andy Williams among others. Another album that year was **Loose Salute**.

The First National Band split in 1971 and James Burton and Glen D. Hardin were brought in to help complete the album then being recorded, **Nevada Fighter**. Another line-up (again including Red Rhodes) recorded **Tantamount To Treason Volume 1**, and in 1972 only Nesmith and Rhodes made **And The Hits Just Keep On Comin'**.

Nesmith founded his own label, Countryside, a subsidiary of Elektra, with the intention of milking some of the country music talent that was going to waste in Los Angeles. But a change of leadership at Elektra, where David Geffen replaced Jac Holzman, saw Countryside closed down.

However, Nesmith had formed another band during this time, the Countryside Band (again including Rhodes) and 1973 saw them releasing **Pretty Much Your Standard Ranch Stash**. He then moved from RCA and formed Pacific Arts, which has seen him involved in mixed media projects, notably 'The Prison', a book with a soundtrack.

One of the first to recognize the importance of video in music promotion, Nesmith became an innovative director performer in the video field. His 1977 album, **From A Radio Engine To The Photon Wing**, was possibly the first record to utilize video images for effect, and resulted in a British hit single for the self-penned **Rio**.

It is generally agreed that Nesmith has written some very good country, or country-influenced songs, one of the most famous being **Some Of Shelley's Blues**.

Recommended:

And The Hits Just Keep On Comin' (RCA/Island)
Pretty Much Your Standard Ranch Stash (–/Island)
The Prison (Pacific Arts/Island)
From A Radio Engine To The Photon Wing (Pacific Arts/Island)
The Newer Stuff (–/Awareness)
The Older Stuff (Rhino/–)

New Grass Revival

A young, electric, bluegrass band, New Grass Revival evolved a distinctive hard-edged picking sound during the early '70s and helped to spread the appeal of bluegrass to a youth audience.

Based around the jazz-tinged fiddle of Sam Bush (who also plays guitar and sings, and is now regarded as one of the world's foremost mandolin players) the four-piece

group had some success with the single **Prince Of Peace**, an evocative reworking of a Leon Russell song. The number subsequently appeared on **New Grass Revival** (Starday), the band's debut album. Later, the Revival moved on to Flying Fish, cutting several albums before touring with Leon Russell. One of those gigs, at Pasadena's Perkin's Palace, was recorded, the results appearing on the Paradise Records album, titled **Leon Russell And The New Grass Revival**.

As the '80s moved on, the Revival signed to the Sugar Hill label, in 1984 cutting **On The Boulevard**, an album that featured the foursome playing material that ranged from Bob Marley's **One Love** and Curtis Mayfield's **People Get Ready**, through to **County Clare**, an original on which bluegrass renewed acquaintance with Irish folk music. In 1986 interest in New Grass Revival soared as they signed a major label deal with EMI-America. At this

Too Late To Turn Back Now, New Grass Revival. Courtesy Flying Fish.

Nevada Fighter, Mike Nesmith. Courtesy RCA Records.

time the line-up comprised Bush and long-time member John Cowan (bass, vocals), plus Pat Flynn (guitars, vocals) and banjoist Bela Fleck, who has been credited with virtually re-inventing the banjo, playing every conceivable type of music on an instrument traditionally scorned by sophisticates. The Revival scored minor country hits with **Ain't That Peculiar** (1986), **Unconditional Love** (1987) and **Can't Stop Now** (1988), making it into the country Top 40 with **Callin' Baton Rouge** (1989). Regarded as one of the finest live bands on the circuit, with their richly textured harmonies and dazzling instrumental work, New Grass Revival succeeded in taking acoustic music into the mainstream of country music. In 1991, Emmylou Harris formed the Nash Ramblers, an acoustic band led by Sam Bush, which could signal the end of one of country music's finest acoustic outfits of recent times.

Recommended:
New Grass Revival (Starday/–)
Fly Through The Country (Flying Fish/–)
Barren Country (Flying Fish/–)
Live Album – with Leon Russell (Paradise/Paradise)
Too Late To Turn Back Now (Flying Fish/Sonet)
On The Boulevard (Sugar Hill/–)
Friday Night In America (Capitol/Capitol)
Hold On To A Dream (Capitol/Capitol)

New Riders Of The Purple Sage

Originally formed as a splinter group from San Francisco's leading acid rock band the Grateful Dead, the New Riders eventually became a name in their own right.

The album **Workingman's Dead** saw the Grateful Dead moving from rock towards a more earthy, sometimes country sound, and the New Riders were the natural off-shoot of this movement.

The New Riders at first played gigs with the Dead and were able to utilize Dead guitarist Jerry Garcia's latent talents on pedal steel guitar. Garcia was eventually replaced by Buddy Cage. Famed West Coast rock artists who have been in NRPS include Mickey Hart and Phil Lesh (Grateful Dead), Spencer Dryden (Jefferson Airplane) and Skip Battin (Byrds).

The band originally recorded for Columbia, but in the mid-'70s they changed labels and recorded for MCA. However, they appeared to lose direction. To their credit, though, in their hey-day the NRPS were a fine country-rock band.

Recommended:
New Riders Of The Purple Sage (Columbia/CBS)
Powerglide (Columbia/–)
Gypsy Cowboy (Columbia/CBS)
The Adventures Of Panama Red (Columbia/CBS)
Home, Home On The Road (Columbia/CBS
New Riders (MCA/MCA)

New Riders Of The Purple Sage. Courtesy CBS Records.

Mickey Newbury

Called 'a poet' by Johnny Cash, Newbury's often ultra sad songs have been recorded by Elvis Presley, Jerry Lee Lewis, Ray Charles, Lynn Anderson, Andy Williams, Kenny Rogers and countless others.

Born in Houston, Texas, on May 19, 1940, Newbury travelled around in earlier years, eventually joining the Air Force for four years, during which time he was based in England.

"After that", he says, "I worked on the shrimp boats in the Gulf, diddled around, did a little writing and lots of other things. I started playing guitar when I was a kid, just enough to be able to go through three or four chords and sing something with it. But when I went into the Air Force, I ditched it all. One day I wound up at a place where they served snacks and had a piano. I began playing it because I just had to get my hands on something that made music. Later I borrowed a guitar from a guy because I didn't have enough money to buy one of my own – but I didn't really start trying to write until I was 24."

Moving to Nashville in the mid-'60s, Newbury began writing songs of many different styles, at one time having four songs simultaneously in the R&B, country, easy-listening and pop charts.

As a recording artist he began cutting albums for RCA and Mercury, without making much impact, though his Mercury release **It Looks Like Rain** became a collectors' item hauling in high bids before it became repackaged as part of a double album set following Newbury's signing with Elektra in 1971.

His biggest hit to date has been **American Trilogy** (1972), a composition formed from three Civil War era songs, which also became an international hit for Elvis Presley.

Throughout the '70s Newbury recorded a series of albums, carefully crafted works that won him high critical praise, but no large sales. He has continued to provide country acts with hit songs, including Tompall And The Glaser Brothers (**I Still Love You, After All These Years**), Marie

Osmond (**Blue Sky Shinin'**), Johnny Rodriguez (**Makes Me Wonder If I Ever Said Goodbye**) and Don Gibson (**When Do We Stop Starting Over**). Since that time Newbury has released few new songs, although still performs in the USA.

Recommended:
Frisco Mabel Joy (Elektra/Elektra)
Heaven Help The Child (Elektra/Elektra)
Live At Montezuma/It Looks Like Rain (Elektra/Elektra)
I Came To Hear The Music (Elektra/Elektra)
Rusty Tracks (ABC/ABC)
After All These Years (Mercury/–)
His Eye Is On The Sparrow (ABC/–)
Sweet Memories (MCA/–)
In A New Age (Airborne/–)

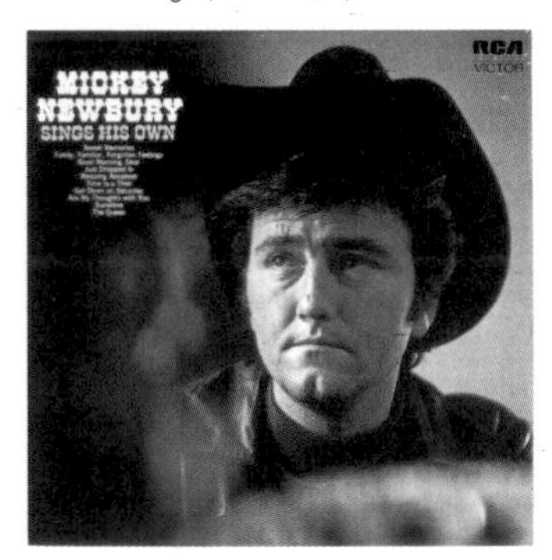

Mickey Newbury Sings His Own. Courtesy RCA Records.

Jimmy C. Newman

Born of part-French ancestry, in Big Mamou, Louisiana, on August 27, 1927, Jimmy began singing in the Lake Charles area, his style employing many Cajun characteristics. During the early '50s, he became a regular on Shreveport's Louisiana Hayride and signed with the major Dot label, obtaining a Top 10 disc with **Cry, Cry, Darling** in 1954. An Opry regular by 1956, he celebrated by cutting **A Fallen Star**, his most successful record, during the following year.

Next came an MGM contract and such winners as **You're Making A Fool Out Of Me** (1958), **Grin And Bear It** (1959), and **Lovely Work Of Art** (1960), before Newman became a long-term Decca artist. His run of hits continued with the chat-filled **Bayou Talk** (1962), **Artificial Rose** (1965), **Back Pocket Money** (1966), **Blue Lonely Winter** (1967), **Born To Love You** (1968) and others.

Above: Mickey Newbury formed his American Trilogy from Civil War songs.

Proficient on virtually any type of country material, it was with the formation of his band, Cajun Country, in the mid-'70s and a return to his Cajun roots that Jimmy made a big impact. With a musical style that mixed traditional Cajun with contemporary country, he built up a sizeable following in Europe, especially Britain, and recorded some fine modern Cajun albums.

Recommended:
Cajun Country (–/RCA)
Alligator Man (Rounder/Charly)
Progressive C.C. (–/Charly)
The Happy Cajun (–/Charly)
Bop-A-Hula (–/Bear Family)
Wild'n'Cajun (–/RCA)

Juice Newton

Country-rock singer Juice was born Judy Kaye Cohen on February 18, 1952 in Lakehurst, New Jersey. The daughter of a Navy man, she grew up in Virginia Beach, Virginia. The only musical person in a family of five, she started to play guitar and sing folk songs in her early teens. At college in North Carolina, she started working in local bars, some nights waiting on tables, and others taking to the stage with her guitar and entertaining.

In the late '60s, Juice moved to northern California where she attended Foothill College and first met her longtime boyfriend and partner Otha Young. At first she performed in local folk clubs then, combining her folk interest with rock'n'roll, formed an electric band with Young called Dixie Peach.

A move to Los Angeles in 1975 led to Juice and Otha forming a new band called Juice Newton And Silver Spur. They signed a recording deal with RCA and released their debut self-titled album, a mixture of country, rock and pop which spawned a minor country hit in **Love Is A Word** in 1976.

Two years later Juice and the band moved on to a new deal with Capitol Records. That same year she provided the Carpenters with **Sweet, Sweet Smile**, a Top 10 country hit. After **Come To Me**, the first album for the new label, was completed, Juice disbanded Silver Spur, opting to work as a solo artist. By the beginning of 1980 she was making inroads into the country charts with **Sunshine** and **Let's Keep It That Way**. A year later her major breakthrough came with the album simply titled **Juice**, which produced two big pop hits in **Angel Of The Morning** and **Queen Of Hearts**, both singles also making it into the country Top 20. Juice started picking up several awards, including a Grammy for **Angel Of The Morning** and several gold discs.

Her next single, **The Sweetest Thing (I've Ever Known)**, written by Otha Young back in the mid-'70s, also made the pop Top 10 and shot to No.1 on the country charts. Juice enjoyed another major pop and country hit with Capitol in her dramatic revival of Brenda Lee's **Break It To Me Gently** (1982). She then moved back to RCA, and hit No.1 in the country charts with her third single for the label, **You Make Me Want To Make You Mine** (1985). An accomplished equestrian, Juice married polo star Tom Goodspeed in 1986. That same year she dominated the country charts with: a revival of **Hurt** that went to No.1; **Old Flame**, a Top 10 entry; **Both To**

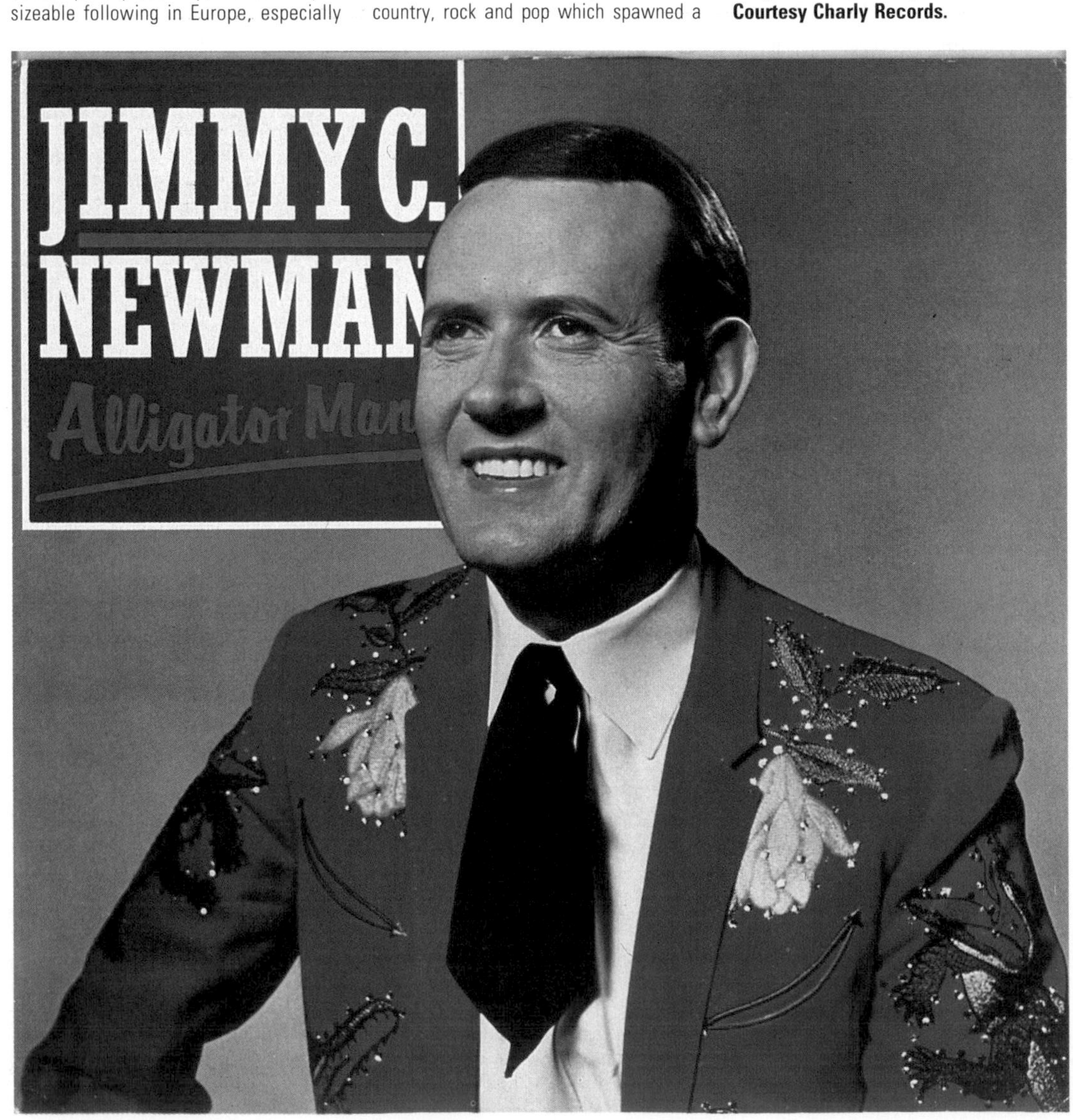

Alligator Man, Jimmy C. Newman. Courtesy Charly Records.

Left: Juice Newton's most recent hits have included What Can I Do With My Heart and Tell Me True.

Each Other (Friends And Lovers), a No.1 with Eddie Rabbit; and **Cheap Love**, a Top 10 entry.

After all that activity, Juice took time off the following year to give birth to a daughter, Jessica Ann. She still scored two more Top 10 hits with **What Can I Do With My Heart** and **Tell Me True**. These have been her most recent major hits, though she has recorded some quality albums and still makes regular TV and concert appearances.

Recommended:
Juice (Capitol/Capitol)
Quiet Lies (Capitol/Capitol)
Can't Wait All Night (RCA/RCA)
Ain't Gonna Cry (RCA/–)

The Nitty Gritty Dirt Band

A Californian country-rock band, the Nitty Gritty Dirt Band finally gained country acceptance in the early '80s with country chart-toppers **Sharecroppers' Dream (A Long Hard Road)**, **I Love Only You** and **High Horse**, following 17 years of releasing critically acclaimed albums covering a sort of all-American eclecticism with strands of a whole musical range: blues, hillbilly, Cajun, folk, boogie, traditional and modern country.

Formed in Long Beach in 1966 by Bruce Kunkel and Jeff Hanna as the Illegitimate Jug Band (Jackson Browne was a one-time member), they engaged other like-minded local students: Jimmie Fadden, Leslie Thompson, Ralph Barr and John McEuen. Then, changing their name to the Nitty Gritty Dirt Band, John's elder brother, Bill, became their manager and producer.

They signed with Liberty Records in 1967 and made the American pop charts with **Buy For Me The Rain** (1967), **Mr Bojangles**, **House At Pooh Corner** and **Some Of Shelly's Blues** (all 1971). Good musicians and fine songwriters, they made an impression with their albums. However, it was the historic recording sessions they undertook at Woodland Sound Studio, Nashville in 1971 which really put the NGDB on the musical map.

A three-record set entitled **Will The Circle Be Unbroken**, conceived by the NGDB but in no way dominated by them, was an ambitious project. It was the first of its kind and required total co-operation between contemporary and old country artists as the NGDB shared the studio with Doc Watson, Mother Maybelle Carter, Roy Acuff, Merle Travis, Jimmy Martin and the Scruggs Family. Even more astounding is the fact that the whole thing was mixed live on a two-track tape machine.

Despite the presence of all these greats of traditional country music, the Nitty Gritties kept their cool splendidly, with John McEuen's banjo and Jimmie Fadden's mouth harp proving particularly outstanding. It is probably the highest-energy acoustic music ever recorded (not a single electric instrument on the whole thing) and a testimonial to both the musicians involved and the music they play. When finally released in 1973 in a lavish booklet sleeve, it became one of the most discussed albums of the time and gave old-time country music a big boost.

Throughout the mid-'70s the NGDB continued to release albums which showed they were not to be tied down by musical labels, as they mixed old rock'n'roll songs, country tunes, folk numbers, self-penned songs and material by contemporary writers Michael Murphey, J. D. Souther and Jackson Browne. Sadly they failed to achieve any kind of commercial success on record, though they were in continuous demand as an exciting stage act.

In 1976 they shortened their name to the Dirt Band and began to formulate a more straightforward country-rock sound, making a brief return to the American pop charts in 1978 with **In For The Night**. After reverting to the name the Nitty Gritty Dirt Band at the end of 1982, the group had their first taste of success on the country charts with **Shot Full Of Love** and **Dance Little Jean**, which rose to Nos.19 and 9 respectively in 1983. The following year the band changed labels for the first time in 17 years, joining Warner Brothers Records and so beginning a series of country No.1s.

Their initial album for Warner Bros, **Plain Dirt Fashion**, established a fine modern country sound and set the pattern for the next few years, with hits: **Long Hard Road** (No.1, 1984), **I Love Only You** (1984), **High Horse**, **Modern Day Romance**, **Home Again In My Heart** (1985) and **Partners, Brothers And Friends** (1986), which paid tribute to their illustrious, sometimes disappointing, but always fun career. The following year, John McEuen left to pursue a solo career, and for a time ex-Eagle Bernie Leadon joined them. Due to various other commitments, and the heavy road schedule of the band, Leadon eventually declined full membership, opting to play on their recordings and undertake the occasional live show. Now down to a leaner, meaner four-piece, **Workin' Band**, their 1988 album, perfectly summed up how the band saw themselves. They had another No.1 with **Fishin' In The Dark** (1987) and Top 10 entries with **Baby's Got A Hold On Me**, **Oh What A Love** (1987), **Workin' Man (Nowhere To Go)**, **I've Been Lookin'** (1988) and **Down That Road Tonight** (1989).

In late 1988 the NGDB signed with the new Universal Records, and set out on a major recording project to produce **Will The Circle Be Unbroken, Volume 2**. The collective band members drew on the finest talent contemporary American music had to offer. Johnny Cash, the Carter Family, Emmylou Harris, Roy Acuff, Bruce Hornsby, John Prine, Jimmy Martin, John Denver, Michael Martin Murphey, John Hiatt, Rosanne Cash, and members of

Below: After 25 years together, the Nitty Gritty Dirt Band recorded their Not Fade Away album.

Above: In 1986, the Nitty Gritty Dirt Band performed at London's Wembley Festival.

Highway 101 and New Grass Revival, were among the artists who joined in the sessions. The album became a big seller and produced Top 20 country hits with **And So It Goes** (with John Denver) and **When It's Gone** in 1989.

With the demise of Universal Records in 1990, the NGDB found themselves on MCA for their next album, **The Rest Of The Dream**. Though in the same style as previous band albums, it failed to produce any major country hits, but this was due to record company changes and politics, rather than the quality of the music. That same year, Jimmy Ibbotson took some time off to work some dates with Jim Salestom and Jim Ratts as the Wild Jimbos, and produce a neat little off-the-wall country album for MCA.

The NGDB celebrated their 25th year together by signing with Capitol/Liberty (their first label, in 1967), releasing **Live Two Five**, a live album recorded in Red Deer, Alberta, Canada. Still true to their original music, the first studio album for their new label was appropriately titled **Not Fade Away**; fading away is certainly unlikely for a band that has become an American institution.

Recommended:

Will The Circle Be Unbroken (UA/UA)
Plain Dirt Fashion (Warner Bros/–)
Partners, Brothers and Friends (Warner Bros/–)
Let's Go (UA/UA)
Dirt, Silver And Gold (UA/–)
Workin' Band (Warner Bros/Warner Bros)
The Rest Of The Dream (MCA/MCA)
Live Two Five (Capitol/–)
Not Fade Away (Liberty/–)
All The Good Times (UA/Beat Goes On)
Uncle Charlie And His Dog Teddy (UA/Beat Goes On)

The Notting Hillbillies

It was only to be expected that one day Mark Knopfler, leader of rock band Dire Straits, would turn his hand to country music. From the beginning, Dire Straits' music and, more especially, Knopfler's writing leaned heavily towards country at times. Several Dire Straits' songs had been covered by American country stars Waylon Jennings, Gail Davies and Highway 101, and by the late '80s Knopfler was no stranger to the Nashville studios, having played alongside Chet Atkins, and guested on recordings by Wynonna Judd, Vince Gill and John Anderson.

The Notting Hillbillies came together in 1989 when two of Knopfler's old musical buddies, Steve Phillips and Brendan Croker, decided to make an album. They called in Knopfler to assist. He in turn recruited Dire Straits' keyboardist Guy Fletcher. The recordings took place in Knopfler's small home studio in Notting Hill, London, hence the name. Always planned as a one-off project, the sessions (that also featured Nashville steel guitarist Paul Franklin) resulted in the album, **Missing ... Presumed Having A Good Time**. The album was a mixture of country-blues, gospel and traditional country that harked back more to the old-timey country sounds of the '30s and '40s than the electrified music of the '50s and '60s.

Due to Knopfler's involvement, the album sold in excess of one million copies. The four musicians took to the road, playing a spring tour in 1990 that saw them performing sell-out shows at civic halls and university campuses. For Knopfler, it was a unique opportunity to get back to his roots: a tour of small, accessible venues. Knopfler had previously played with Steve Phillips in the early '60s in Leeds, Yorkshire as a duo known as the Duolian String Pickers, a name taken from a brand of National guitar. Crocker, also from Leeds, is a singer, songwriter and guitarist with his own band, the Five O'Clock Shadows. In recent years Crocker has had his songs recorded in Nashville by Wynonna Judd and others.

Pure Dirt, the Nitty Gritty Dirt Band's first album. Courtesy UA Records.

Oak Ridge Boys

Duane Allen, lead; Joe Bonsall, tenor; Steve Sanders, baritone; Richard Sterban, bass.

For many years, the Oaks (as they now prefer to be called) were one of the top groups in Gospel music, winning 14 Dove Awards from the Gospel Music Association and four Grammies. Then in 1975 they decided to make a move towards country music and within three years they won the first of many Country Music Association awards, as Best Vocal Group of 1978.

The Oaks were originally known as the Country Cut-Ups and were formed shortly after World War II. They performed at the atomic energy plant in Oak Ridge, Tennessee. The group was reformed in 1957 by Smitty Gatlin as the Oak Ridge Quartet and became fully professional in April 1961.

Their initial appeal lay with their high quality sacred material in an infectious, foot-stomping fashion. In 1971 they received a Grammy for their recording of **Talk About The Good Times** and in 1976 did the same thing for **Where The Soul Never Dies**.

William Lee Golden (born January 12, 1939, Brewton, Alabama) was the longest serving member of the group, having joined in 1964. Lead singer Duane Allen (born April 29, 1944, Taylortown, Texas), formerly a member of the Southernaires Quartet, joined in 1966 and is generally regarded as the Oaks' spokesman. Newer members are Richard Sterban (born April 24, 1944, Camden, New Jersey), who joined in 1972, and Joe Bonsall (born May 18, 1944, Philadelphia, Pennsylvania), who joined in 1973.

The foursome made a move towards country acceptance in 1975, scoring a minor hit with their secular recording of **Family Reunion** for Columbia the following year. Soon after, they provided the back-up on Paul Simon's **Slip Sliding Away**, which became a million-seller.

Encouraged by Johnny Cash, who asked them to open for him in Las Vegas, and signed by booking agent Jim Halsey, the Oaks began recording for ABC-Dot, and shortly had their first Top 5 country hit with **Y'all Come Back Saloon**.

Since then the Oaks have been red-hot with their gospel-flavoured country-pop songs, scoring hits with **I'll Be True To You** (1978), **Sail Away** (1979), **Leaving Louisiana In The Broad Daylight** (1979), **Trying To Love Two Women** (1980) and **Beautiful You** (1980).

The inevitable pop chart breakthrough came with the multi-million selling **Elvira** (1981) and **Bobbie Sue** (1982), each album achieving platinum status within weeks of release. They have continued to dominate the country charts with **American Made** (1983), **Love Song** (1983), **I Guess It Never Hurts To Hurt Sometimes** (1984), **Make My Life With You** (1985) and **Come On In (You Did The Best You Could)** (1986).

In March, 1987, Golden was dismissed from the group due to 'continual musical and personal differences', and set out on a solo career. He filed a $40 million suit against the three remaining members, which was eventually settled out of court. His replacement was the band's rhythm guitarist Steve Sanders (born September 17, 1952, Richland, Georgia), who had a long musical career in gospel music starting when he was five. By the time he was 12 Steve had appeared on Broadway, and two years later starred as Faye Dunaway's son in the film 'Hurry Sundown'. He had joined the Oaks' band in 1982, and took over the baritone vocal chores on their 1987 **Heartbeat** album and their '87 Fast Lane concert tour. The change didn't seem to make any considerable difference to the Oaks' ability as hitmakers, and they scored such No.1s as **This Crazy Love** (1987), **Gonna Take A Lot Of River** (1988) and **No Matter How High** (1990). The following year the Oak Ridge Boys left MCA to sign with RCA, immediately scoring yet another Top 10 hit with **Lucky Moon** (1991).

Above: Bill, Joe, Richard and Duane – the Oak Ridge Boys.

With their own six-piece Oak Ridge Boys Band, the Oaks have gained a reputation for their exciting live performances, a fast-paced showcase with dynamic presentations.

Recommended:
Room Service (MCA/–)
Deliver (MCA/MCA)
American Made (MCA/MCA)
Step On Out (MCA/MCA)
Heartbeat (MCA/–)
Where The Fast Lane Ends (MCA/–)
The Long Haul (RCA/–)
American Dreams (MCA/MCA)

Mark O'Connor

For many years Nashville's most in-demand session musician, Mark O'Connor was born on August 5, 1961 in Seattle, Washington. A virtuoso on the guitar and mandolin, and a varied and accomplished composer, it is as a violinist that he is recognized as a true master. He has done remarkable work in jazz, country, pop, fusion and the classics, combining them in ways that have re-defined both the instrument and its audience.

At age six, Mark began guitar lessons, winning his first contest at ten in a classical/flamenco competition at the University of Washington. By the time he was 11, he could play mandolin, banjo, steel-string guitar and dobro. Seven months after picking up a violin, he entered the National Old-Time Fiddle Championship and won second place in the 12-and-under division. By the time he was a senior in high school, he had won every major fiddle contest in the country. He had been introduced to Roy Acuff when he was 12, and he so impressed the father of country music that he was given a spot in Acuff's Grand Ole Opry set that same night. At this time he was signed to Rounder Records and the label put out six releases covering Mark's astonishing teenage career development.

After graduation in 1979, he played the bluegrass festival circuit, then toured with David Grisman and Stephane Grappelli as guitarist, and can be heard on several of Grisman's albums. In 1981 he joined country-fusion band, the Dregs, appearing on the album **Industry Standard**. Moving to Nashville in 1983, he soon became a popular session player. Devoting his time to studio work, he appeared on nearly 500 recordings in just over six years. He also continued with his own recordings, with Michael Brecker and James Taylor

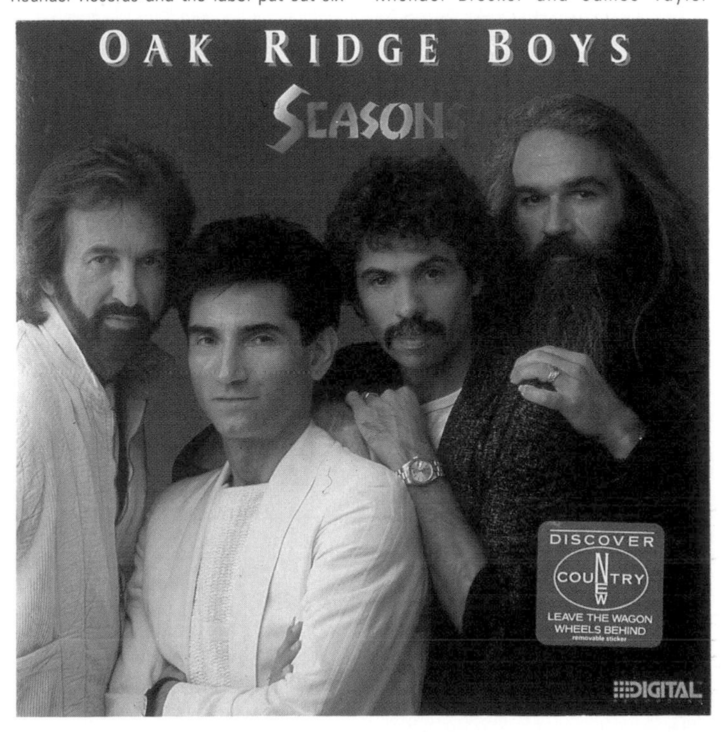

Seasons, the Oak Ridge Boys. Courtesy MCA Records.

The New Nashville Cats, Mark O'Connor. Courtesy Warner Brothers Records.

guesting on his 1989 album, **On The Mark**. By this time he was beginning to follow a more independent career and writing much of his own material. In 1990, with an assemblage of 53 prominent Nashville studio musicians, he recorded **The New Nashville Cats** album, which won a Grammy in 1991 for Best Country Instrumental Performance. A revival of Carl Perkins' **Restless** from the album, with vocals by country stars Vince Gill, Steve Wariner and Ricky Skaggs, was a Top 30 country hit and won the CMA Vocal Event. O'Connor picked up the Musician Of The Year in 1992.

An accomplished composer of contemporary classical music, he has composed for the Sante Fe Music Festival, appeared as a guest soloist with the Boston Pops Orchestra and is musical director of TNN's American Music Shop. His latest project is the production of **Violin Heroes**, which will pair him with Stephane Grappelli, Vassar Clements, Charlie Daniels, Jean-Luc Ponty, Buddy Spicher, Byron Berline, Johnny Gimble and Doug Kershaw.

Recommended:
Markology (Rounder/–)
Pickin' In The Wind (Rounder/–)
Soppin' The Gravy (Rounder/–)

Molly O'Day

Molly O'Day had an earnest exhortative style – not unlike that of Roy Acuff and Wilma Lee Cooper – which was the epitome of a style and brand of old-time music not found today.

She was born LaVerne Williamson in Pike County, Kentucky, on July 9, 1923, and embarked on a professional career in the summer of 1939, when she joined her fiddling brother Skeets in a band which also included Johnny Bailes. Here she went by the first of her many stage names, Mountain Fern, which she changed to Dixie Lee in the autumn of 1940. She married her longtime husband – and, at the time, fellow bandmember – Lynn Davis on April 5, 1941.

Molly and Lynn made the rounds of a number of Southeastern radio stations for the next few years – Beckley, West Virginia; Birmingham, Alabama; Louisville (where she finally chose the name Molly O'Day), Beckley again and then Dallas and finally Knoxville, where they were heard by Fred Rose who interested Satherley in recording them.

Their first Columbia session (December 16, 1946) produced many of her classics: **Tramp On The Street** (written by Hank Williams), **Six More Miles, Black Sheep Returned To The Fold** and others. The session also marked the recording debut of Mac Wiseman, who played bass on the recordings.

The records were minor hits, and they resumed the circuit of radio stations, also doing more Columbia recordings: **Poor Ellen Smith** (with Molly on the banjo), **The First Fall Of Snow**, **Matthew Twenty-four** and others.

In 1950 they began recording only sacred material for Columbia, and when Molly contracted tuberculosis in 1952, she and Lynn both left musical careers to become ministers in the Church of God, careers which they followed through to the mid-'80s. Molly occasionally broadcast over a Christian station based in Huntingdon, West Virginia. After several months of illness, Molly died of cancer on December 5, 1987.

There is no question that Molly O'Day quit performing and recording well before her prime; she certainly had all the talent and appeal to become country music's first really great and popular woman singer.

Recommended:
Molly O'Day And The Cumberland Mountain Folks (Old Homestead/–)
The Heart And Soul Of Molly O'Day (Queen City/–)
Living Legend Of Country Music (Starday/–)

Kenny O'Dell

One of Nashville's most successful songwriters, Kenny was born in Oklahoma, raised in California, and began writing songs in his early teens. After graduating from Santa Maria High School in California, he formed his own record label, Mar-Kay Records, and recorded his own song, **Old-Time Love**, which received only minimal attention from the public.

Success finally came his way with the self-penned **Beautiful People**, a release on the small Vegas Records which made Top 40 in the American pop charts in 1967, only to be overtaken by a cover version by Bobby Vee. Another of Kenny's songs, **Next Plane To London**, as recorded by Rose Garden, made the Top 20 at the end of that year.

Kenny made a move to Nashville in 1971 to write songs and take over the running of House of Gold, Bobby Goldsboro's publishing firm. With **Why Don't We Go Somewhere And Love** by Sandy Posey and **I Take It On Home** by Charlie Rich, his songs began appearing on the country charts.

In 1973 Kenny came up with **Behind Closed Doors**, a song that made Charlie Rich a major country-pop cross-over star and was named Song Of The Year by the CMA and ACM. This success led to Kenny signing a recording contract with Capricorn Records and scoring minor country hits with **You Bet Your Sweet Love** (1974) and **Soulful Woman** (1975), before scoring Top 10 hits with **Let's Shake Hands And Come Out Lovin'** and **As Long As I Can Wake Up In Your Arms** (both 1978).

Kenny, who has served on the Board of Directors of the CMA and the Board of the Nashville Songwriters Association, has continued to pen hits for such artists as Billie Jo Spears (**Never Did Like Whiskey**), Tanya Tucker (**Lizzie And The Railman**) and the Judds (**Mama He's Crazy**), though his own recording career seems to have come to a halt with no further releases since 1979.

Recommended:
Kenny O'Dell (Capricorn/Capricorn)
Let's Shake Hands And Come Out Lovin' (Capricorn/–)

Daniel O'Donnell

The most popular 'Country'n'Irish' entertainer, Daniel O'Donnell (born on December 12, 1961 in Kincasslagh, Co. Donegal, Eire) was out-selling all American country stars and most MOR acts in the British Isles during the late '80s and early '90s. Adding to this huge record success, he also undertook sell-out concert tours.

He came from a musical family and his elder sister, Margo, was a popular Irish country singer in the late '60s and throughout the '70s. During school days Daniel sang in local variety concerts staged in the Kincasslagh parish hall, and would also occasionally perform with his sister. After completion of his school education, he undertook several jobs, then attended Galway Regional College on a business studies course in 1980. He dropped out in December and joined Margo's band on a full-time basis, making his first professional appearance at The Rag, Thurles, Co. Tipperaray, on January 28, 1981. Daniel stayed with Margo's band for two years, then opted for a solo career. He made his first record, **My Donegal Shore**, in early 1983, and in July formed his first band, Daniel O'Donnell And Country Fever. Selling the record at his gigs, he gradually built up a following for his easy-listening, Irish-flavoured country music. The summer of 1984 found him organizing a new band, Grassroots, at the same time completing more recordings, which would later surface on **The Boy From Donegal**.

It was at his appearance at the Irish Festival in London in 1985 that O'Donnell received his first breakthrough. Michael Clerkin, the owner of the independent Ritz Records, an Irish label based in London, saw the young Irish singer's performance and immediately offered him a recording contract. His first album for Ritz, **The Two Sides Of Daniel O'Donnell**, was released at the end of 1985. In early 1986 he signed a management contract with Clerkin. A second album, **I Need You**, released in late 1986, was the first to reach the UK country charts. Like the first, it featured a blend of Irish favourites and country standards, all performed in an effortless and undemanding manner. With his youthful good looks, O'Donnell was gaining a female following, and a third album, **Don't Forget To Remember**, entered the UK country chart at No.1, as did all his subsequent albums.

By now, O'Donnell's concert tours were selling out. He was undertaking two tours each year, usually of three months' duration. In 1988 Ritz licensed his next album, **From The Heart**, to TV marketing company Telstar Records. As well as entering the UK country chart at No.1, the album also crossed over to the UK pop chart, gaining the singer his first gold disc. The following year brought **Thoughts Of Home**, an album and video. The album made the UK pop album chart and the video reached No.1 on the UK music video chart. Slowly broadening his scope, in 1990 O'Donnell went to Nashville to record with producer Allen Reynolds, and the subsequent album, **The Last Waltz**, was his most country offering so far.

O'Donnell was inadvertently involved in controversy when the UK CMA and the Chart Committee decided to remove most of his albums (plus those of contemporary acts like Steve Earle) from the UK country chart as 'not being country'. This produced an uproar in British country music, not just from O'Donnell fans, but also from noted country music experts, who could see British country music was being dismissed in favour of 'the real thing' from the USA. After a few months there was a change of heart. All of his albums were reinstated, and O'Donnell continued as the most successful country act in Britain. He has also started building a following in Australia and New Zealand, and has played selected concerts in America, including Carnegie Hall, but performing mainly in cities with a large Irish immigrant population. In 1992 he made further strides in the UK when his single, **I Just Want To Dance With You**, made the UK pop Top

The O'Kanes, their debut album. Courtesy CBS Records.

20, the first British country act to achieve this, leading to appearances on the BBC-TV show Top Of The Pops.

Recommended:
Follow Your Dream (–/Ritz)
From The Heart (–/Ritz)
The Last Waltz (–/Ritz)
Thoughts Of Home (–/Telstar)
Don't Forget To Remember (–/Telstar)

The O'Kanes

A pair of Nashville-based songwriters, Jamie O'Hara (born, Toledo, Ohio) and Kieran Kane (born Queens, New York), teamed up in 1986 to produce an exciting fusion of traditional country harmony singing, with a rock'n'roll backbeat and bluegrass-flavoured instrumentation. Several big-selling singles resulted, along with a Grammy nomination and write-ups in 'Time', 'Newsweek', 'USA Today' and 'Rolling Stone'.

O'Hara seemed destined for a career in professional sport until a knee injury changed his plans. A birthday gift of a guitar created an interest in songs and songwriting, and, after working clubs and honky-tonks in the Midwest, he arrived in Nashville in 1979 and signed a songwriter's contract with Cross-Keys. He gained his first cut with **Old Fashioned Love**, a John Conlee album track, in 1980.

Kieran Kane was raised in Mount Vernon, just outside New York City, and started playing drums in his brother's rock'n'roll band when he was nine. Later he turned to folk and bluegrass music. In the early '70s he was living in Los Angeles, where he worked as a lead guitarist, playing the club scene, writing songs and trying to land a record deal. Befriended by songwriter Rafe Van Hoy, who urged him to try Nashville, Kieran moved to Music City in 1979 and signed with Tree Music. Two years later he landed a recording contract with Elektra Records, making the country Top 20 with **You're The Best** (1981) and **It's Who You Love** (1982).

Throughout the early '80s both O'Hara and Kane built up a reputation for their writing, penning hits, individually, for Janie Fricke, T.G. Sheppard, the Judds and many others. In 1984 they started writing together and recorded a demo in Kane's attic studio. Producer Bob Montgomery thought the demos were good enough for a recording contract, so, adopting the name the O'Kanes, they signed to Columbia Records in 1986. The acoustic demo recordings (two guitars, bass, fiddle, banjo, accordion, drums) made up most of their debut album. The first single, **Oh Darlin'**, made the Top 10 in 1986, followed by **Can't Stop My Heart From Loving You**, a No.1 the following year. Further Top 10 entries came with **Daddies Need To Grow Up Too**, **Just Lovin' You** (1987), **One True Love** and **Blue Love** (1988). They took their music on the road with a fast-paced and exciting stage show. A second album, **Tired Of The Runnin'**, was released hot on the heels of their debut, but appealed more to a cult audience.

The O'Kanes had become part of the 'New Traditionalists'. In 1989 they recorded a third album, but it was scrapped because both they and Columbia were unhappy with it. Jamie and Kieran returned to the studios, this time with ace producer Allen Reynolds, and emerged with **Imagine That**, a masterpiece album that successfully blended all the best

Tired Of The Runnin', the O'Kanes. Courtesy CBS Records.

elements from their first two albums to create the definitive 'O'Kanes sound'. Unfortunately by the time the album hit the stores in 1990, the O'Kanes' career had lost momentum. Though the album made a brief appearance on the country charts, the singles failed to gain radio plays and Columbia didn't renew their contract in 1991. The pair returned to songwriting and playing sessions, with country music losing one of the finest acts that it had produced during the '80s.

Recommended:
The O'Kanes (Columbia/CBS)
Imagine That (Columbia/CBS)
Tired Of The Runnin' (Columbia/CBS)

Roy Orbison

Born in Vernon, Texas, on April 23, 1936, Orbison's roots were deep country. His first band, the Wink Westerners, were named after the town of Wink, where he was raised. He was one of the many youngsters who fell under the spell of rockabilly, and under the guidance of Sam Phillips, had his first hit on Sun Records, **Ooby Dooby**, in 1956. However, his totally unique voice owes nothing to any particular genre and he moved into popular music, where he became an international star.

A victim of tragedy since leaving Monument Records in 1965 – his career immediately plummeted, his wife Claudette was killed in a motorcycle accident and two of his children died in a fire during 1968 – Orbison resumed his relationship with Monument in 1977. A subsequent album, **Regeneration**, proved to be his strongest since the early '60s.

In 1980 he made his debut on the country charts with **That Lovin' You Feelin' Again**, a stylish duet with Emmylou Harris that was featured in the film 'Urban Cowboy' and won a Grammy award. During 1985, Orbison took part in the Sun reunion – with Johnny Cash, Jerry Lee Lewis and Carl Perkins – that produced the **Class Of '55** album. That same year he re-recorded several of his old hits and started to rebuild his career. Two years later he recorded **A Black And White Night**, a live concert at Coconut Grove that featured guest spots by Bruce Springsteen, Elvis Costello and Bonnie Raitt. Also in 1987 he became a member of the Travelling Wilburys. Orbison also joined k.d. lang on a new version of his pop classic **Cryin'**, which became a minor country hit. A new solo contract was signed with Virgin Records and he had just completed the **Mystery Girl** album, when he died of a heart attack on December 6, 1988 in Madison, Tennessee. Released the following year, the album made the country chart, with a single, **You Got It**, becoming a Top 10 entry in 1989.

Recommended:
All Time Greatest Hits (Monument/Monument)
Regeneration (Monument/Monument)
The Sun Years (–/Charly)
Big O Country (–/Decca)
In Dreams (Virgin/Virgin)
Mystery Girl (Virgin/Virgin)
A Black And White Night (Virgin/Virgin)
Laminar Flow (Asylum/Asylum)

Osborne Brothers

Among the first of the so-called progressive bluegrass outfits, the Osborne Brothers (both born at Hyden, Kentucky, Bob on December 7, 1931, Sonny on October 29, 1937) made their radio debut on station WROL, Knoxville, Tennessee, in the early '50s.

After teaming for a time with Jimmy Martin, they signed for MGM Records in 1956. They became regulars on WWVA's Wheeling Jamboree show, where they specialized in precise, sky-high three-part harmony with guitarist Benny Birchfield.

Their gig at Antioch College in 1959 was a milestone, sparking off a series of campus dates, while they also found themselves accepted on the Opry.

Constantly dismaying purists with their electric sounds and their use of steel guitar, drums and piano, the Osbornes became Decca artists in 1963, terminating their seven-year relationship with MGM. As electric bluegrass began to prosper, the group (later featuring Ronnie Reno or Dale Sledd as the third vocalist) began accumulating a number of low chart singles: **The Kind Of Woman I Got** (1966), **Rocky Top** (1968), **Tennessee Hound Dog** (1969) and **Georgia Pinewoods** (1971).

In 1976 they cut their first album for the new CMH label, featuring accompaniment more sparse than in past years, focusing as always on Bob's awesome tenor voice and impressive harmony singing. They have since recorded more albums for CMH.

Recommended:
Voices In Bluegrass (MCA/Stetson)
Ru-bee (MCA/–)
Pickin' Grass And Singin' Country (MCA/–)
Number One (CMH/–)
From Rocky Top To Muddy Bottom (CMH/–)

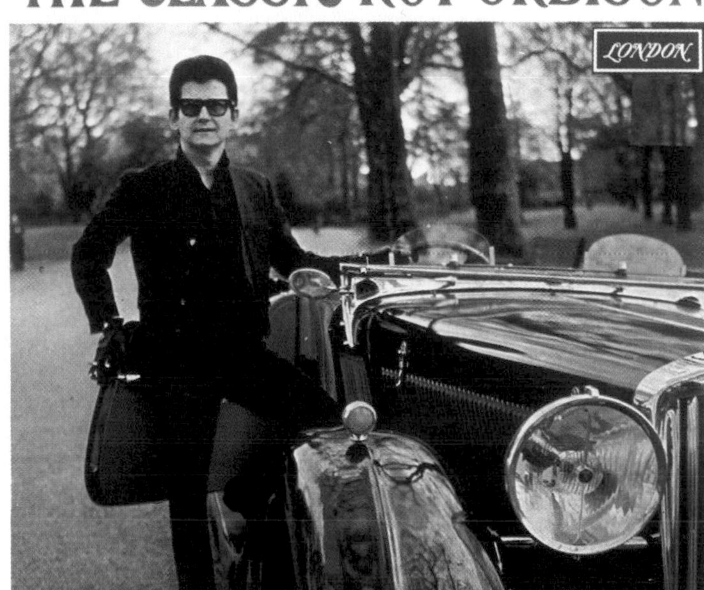

The Classic Roy Orbison. Courtesy London Records.

Some Things I Want To Sing About, Bob and Sonny – The Osborne Brothers. Courtesy Sugar Hill.

K.T. Oslin

Kay Toinette Oslin was born in Crossitt, Arkansas in 1943. She emerged as a major country star in the late '80s; well over 40, a little overweight, yet a hip dresser. It was just the right image needed to give a voice to older women. Musically a diverse artist, K.T. provided a view of love in the '80s from the perspective of a woman in her 40s, through her mainly self-penned songs. Her **'80s Ladies** album and single became best-sellers in 1987, and the lady won a Grammy and was named both CMA and ACM Female Vocalist in 1988.

Raised mainly in Mobile, Alabama, her Southern upbringing exposed her to honky-tonk music, blues, rock'n'roll and soul. By the early '60s she was living in Houston, Texas, where she made her professional debut as a folk singer. She teamed up with school friend David Jones and songwriter Guy Clark in a folk trio.

She moved to Nashville, sang harmony on Clark's 1978 album, and landed both a writer's contract and a record deal, signing with Elektra Records. She made her debut on the country charts as Kay T. Oslin with Chip Taylor's **Clean Your Own Tables** in 1981. She also cut an early version of **Younger Men**, but lack of radio plays led to her being dropped. Gail Davies recorded her **'Round The Clock Lovin'** and **Where Is A Woman To Go**, which gave Oslin the incentive to continue. She borrowed $7,000 from an aunt and hired a band for a Nashville showcase. RCA producer Harold Shedd was suitably impressed, and signed her to a new recording contract in 1986. The first single, **Walls Of Tears**, made the country Top 40 in 1987, but it was the title song of her debut album, **'80s Ladies**, that made K.T. Oslin into a major star. The single went into the country Top 10, while the album gained a platinum award.

K.T. made No.1 with **Do Ya** (1987), **I'll Always Come Back** and **Hold Me** (1988), while a second album, **This Woman**, gained her another platinum award. Her songs were felt to be too feminist for Nashville, but with her wry good humour and solid sense she broke down the barriers. Touring with Alabama in 1989, she guested on their chart-topping single, **Face To Face**, meanwhile enjoying further solo success with **Hey Bobby** and **This Woman** (1989), before starting work on a third album, **Love In A Small Town**. A single, **Come Next Monday**, was a No.1 in 1990, and the album gained a gold disc. There has not been a new studio album, but in 1993 she did release a **Greatest Hits** set.

Recommended:
Love In A Small Town (RCA/RCA)
'80s Ladies (RCA/RCA)
This Woman (RCA/RCA)

Below: Tommy Overstreet began work on TV, as Tommy Dean from Abilene.

Paul Overstreet

A multi-award-winning songwriter, Paul Overstreet (born on March 17, 1955, in Newton, Mississippi) is a country tunesmith who prefers to write with an optimistic outlook on love and happiness. It is obviously a successful ploy as he has been named BMI's Songwriter Of The Year for five straight years, 1987–1992. In addition, Paul has also managed to carve out his own comfortable niche as a performer, with several self-penned country hits to his credit.

Just Between The Two Of Us, Bonnie Owens with Merle Haggard. Courtesy Capitol Records.

The son of a Baptist minister, Paul came to Nashville after finishing high school in Newton. He had his years of struggle, and played the clubs. He started songwriting because he wanted his own material, but, recording briefly for RCA in 1982, he soon found it would be more profitable to let others cut his songs. In 1983 George Jones turned **Same Ol' Me** into a Top 10 hit and doors started opening. But Paul had already fallen prey to self-destructive tendencies. Briefly married to Dolly Parton's younger sister Frieda Parton, Overstreet came to rely on drugs and alcohol to fuel his writing inspiration. Finally, in 1984, he stopped drinking, kicked the drug habit, and in early 1985 married Julie, a make-up artist for a TV show in which he was appearing.

A steadying influence, Julie helped her husband find his feet. In 1986 he teamed up with fellow songwriters Thom Schuyler and Fred Knobloch as the country trio SKO. They signed to MTM Records, and charted Top 10 with **You Can't Stop Love** (1986), then a No.1 with **Baby's Got A New Baby** (1987). However, Paul soon left the group, intent on pursuing his own solo career. His first – and last – single for MTM was **Love Helps Those**, which rose to No.3 in 1988. Shortly after MTM closed down, so Overstreet took his material to his old label RCA. He scored two Top 10 hits with **Sowin' Love** and **All The Fun** in 1989. His RCA debut album, also titled **Sowin' Love**, was full of catchy melodies

Best Of Buck Owens – a compilation of hits. Courtesy Capitol Records.

and traditional country swing. Further hits followed with **Seeing My Father In Me**, **Richest Man On Earth** and **Daddy's Come Around** (all 1990), the latter making No.1. He joined Tanya Tucker and Paul Davis on the chart-topping **I Won't Take Less Than Your Love** in 1987, and picked up two Grammies for Best Country Song with **Forever And Ever Amen** (1988) and **Love Can Build A Bridge** (1991).

His optimistic lyrics brought Overstreet close to gospel music, though he is not a gospel performer. This came through strongly on his second album, **Heroes**, with the title song going Top 10 in 1991. More hits have come his way as a singer with **Ball And Chain** (1991), **Still Out There Swinging** (1992) and the infectious **Take Another Run** (1993).

Recommended:
Heroes (RCA/RCA)
Love Is Strong (RCA/–)
Sowin' Love (RCA/RCA)

Tommy Overstreet

Born in Oklahoma, on September 10, 1937, Overstreet began his career in Houston, Texas, working on a Saturday morning TV show in the guise of Tommy Dean from Abilene.

In 1956–57 he studied radio and TV production at the University of Texas. After a short stint as a touring performer and a spell in the Army, he claims to have just 'coasted' for a few years. His fortunes changed in 1967 when, following a move to Nashville, he became manager of Dot Records' Nashville office, at the same time becoming a Dot recording artist.

His initial singles failed to make much impression, but following the Top 5 success of **Gwen (Congratulations)** (1971), virtually every release made the Top 20. These hits include **Ann (Don't Go Running)**, **Heaven Is My Woman's Love** (1972), **Jeannie Marie (You Were A Lady)** (1974) and **Don't Go City Girl On Me** (1977).

Recommended:
Heaven Is My Woman's Love (Dot/–)
This Is Tommy Overstreet (Dot/–)
Welcome To My World Of Love (–/Ember)
I'll Never Let You Down (Elektra/–)

Bonnie Owens

The ex-wife of Merle Haggard and mother of Buddy Allan from a marriage to Buck Owens, Bonnie Campbell Owens was born in Blanchard, Oklahoma, on October 1, 1932.

In her early years she sang at clubs throughout Arizona, working with Buck Owens as part of the Buck And Britt Show on a Mesa radio station, later joining him in Mac's Skillet Lickers. During the '60s, Bonnie moved to Bakersfield, California, there meeting Merle Haggard, whom she married in 1965.

The twosome became signed to Capitol Records, cutting an album of duets entitled **Just Between The Two Of Us**, which led to them being voted Best Vocal Group of 1966 by the ACM. Prior to this, the duo had cut a single of the same name for Tally, which had become a 1964 hit. Bonnie also achieved two solo successes with Tally via **Daddy Don't Live Here Anymore** and **Don't Take Advantage Of Me**.

Following a small clutch of other minor chart entries, including **Number One Heel** (1965) and **Lead Me On** (1969), Bonnie officially retired from performing in 1975.

Recommended:
That Makes Two Of Us – with Merle Haggard (Hilltop/–)
Lead Me On (Capitol/–)
Just Between The Two Of Us – with Merle Haggard (Capitol/Stetson)

Buck Owens

Main man behind the 'California Sound' and the establishment of Bakersfield, Alvis Edgar 'Buck' Owens was born in Sherman, Texas, on August 12, 1929.

While Buck was still young, the Owens family moved to Arizona in search of a better standard of living but they failed to find prosperity. Buck left school while in his ninth grade to work in farm labouring. A fine guitarist and mandolin player, he began playing with a band over radio station KTYL, Mesa, Arizona when he was barely 17. At the same age he got married – by 18 he was a father.

In 1951 he moved to Bakersfield, forming a band, the Schoolhouse Playboys, with whom he played sax and trumpet. He also established himself as a first-class session man on guitar. Following a stint as a lead guitarist with Tommy Collins' band, he became signed as a Capitol recording artist on March 1, 1957.

His first chart entry was with **Second Fiddle** (1959), then followed a long sojourn in the Top 5 via such releases as **Under Your Spell Again** (1959), **Excuse Me, It Think I've Got A Heartache** (1960), **Fooling Around** and **Under The Influence Of Love** (1961).

With his band the Buckaroos – an outfit that has featured Don Rich, Doyle Holly and Tom Brumley – Owens began playing to sell-out crowds. Owens registered no less than 17 No.1 hits between 1963–69.

During the early '60s, the band leader had recorded a series of extremely successful duets with Rose Maddox. At the onset of the '70s Owens revived this practice, employing Susan Raye as a partner and logging up hits with **We're Gonna Get Together, Togetherness** and **The Great White Horse**. Although some had begun to suggest that Owens had become **Too Old To Cut The Mustard** – a 1971 single featuring his son Buddy Alan – the Baron of Bakersfield continued to supply Capitol with major discs.

However, in 1976 he terminated his long association with Capitol, signing with Warner Brothers and releasing **Buck 'Em!**. This marked the end of his days as a Top 10 regular, though he did make Top 10 with **Play Together Again, Again**, a duet with Emmylou Harris in 1979.

He continued as a regular on the TV show Hee-Haw until the early '80s, but had retired to run his many business interests. In 1987 Dwight Yoakam prised Owens out of retirement to join him on a remake of an Owens' oldie, **Streets Of Bakersfield**. It hit No.1 on the charts in 1988 and gave Owens a new lease of life. He signed a new contract with Capitol and scored two minor hits with **Hot Dog** (1988) and **Act Naturally** (1989), the latter featuring Ringo Starr. Owens also took to the road but he announced his retirement from performing in 1992.

Recommended:
Buck 'Em (Warner Bros/–)
Our Old Mansion (Warner Bros/–)
Kickin' In (Capitol/–)
Blue Love (–/Sundown)

Below: Vernon Oxford now makes religion a part of his shows.

Vernon Oxford

Highly regarded in Europe, Oxford was born on June 8, 1941, in Benton County, Arkansas, one of seven children of one of the area's leading fiddle players. Like his father, he became a fiddler, entering the well-known Cowtown contest and also the Kansas State championship.

After forming his own band and touring throughout the Midwest, Oxford then moved on to Nashville, where he was turned down by several record companies as being 'too country'. Eventually, in 1965, he became signed to RCA, who released seven singles and an album before dropping him. He also signed for Stop, with continued lack of real recognition.

In 1971, fans in Britain and Sweden organized a petition urging RCA to release Oxford's discs once more and, two years later, British RCA duly obliged with a double album in their **Famous Country Music Makers** series, a release which accrued impressive sales figures.

This encouraged RCA Nashville to re-sign Oxford in 1974, the singer immediately responding by providing a minor hit in **Shadows Of My Mind** (1975), then a major hit with **Redneck** (1975). Remaining a major star in Europe, Oxford turned increasingly to religion, often preaching about hell and damnation during his country performances. In 1989 he was filmed for a BBC documentary, Power In The Blood, in which he took his cowboy hat, guitar and his Hank Williams' voice to the sinners of Northern Ireland, performing in drinking dens, mission halls and the Maze Prison.

Recommended:
Famous Country Music Makers (–/RCA)
I Just Want To Be A Country Singer (RCA/RCA)
Keepin' It Country (Rounder/Sundown)
Tribute To Hank Williams (–/Meteor)
A Better Way Of Life (–/Sundown)
His And Hers (Rounder/–)

Ozark Mountain Daredevils

A country-rock band from Springfield, Missouri, the Daredevils sprang from an outfit known as Cosmic Corncob And His Amazing Mountain Daredevils.

Their first album for A&M Records came out in 1973 and was well received. By 1974 they had a pop hit single with **If You Want To Get To Heaven**. Their next album, **It'll Shine When It Shines** – generally considered to be their best – spawned yet another hit in **Jackie Blue**, penned by band members Steve Cash (vocals, harmonica) and Larry Lee (drums). Mixing pure rock'n'roll with slick country picking, the Daredevils pushed ahead for a while, heading for Nashville to cut their third LP, **The Car Over The Lake Album**.

But interest began to wane and by 1977 members Randle Chowning (guitar, vocals) and Buddy Brayfield (piano) had quit. Original members Steve Cash and John Dillon (vocals, guitar, keyboard, fiddle) appeared on Paul Kennerley's all-star **White Mansions** album in 1978, but they were not destined to stay with A&M much longer. In the wake of a live double, they moved on, signing with Columbia, for whom they cut an eponymously titled album in 1980. But by that time only Dillon, Cash and Mike Granda (bass) remained from the line-up that cut their first sides at a ranch in Missouri.

The band were back in the studios in 1988 working with producer Wendy Waldman on the **Modern History** album, an exciting blend of American heartland roots and a powerful contemporary sound, which is still popular at their live dates throughout North America.

The Car Over The Lake Album, the Ozark Mountain Daredevils' third album. Courtesy A&M Records.

Recommended:
Ozark Mountain Daredevils (A&M/A&M)
It'll Shine When It Shines (A&M/A&M)
The Car Over The Lake Album (A&M/A&M)
Modern History (–/Conifer)
Heart Of The County (–/Dixie Frog)

Lee Roy Parnell

With his musical and lyrical mix of blues, country, rock, heartache and hope, Lee Roy Parnell (born on December 21, 1957, in Stevenville, Texas) brought a harder, rocking edge to country in the early '90s. His music flaunted the gritty vocals and jagged rhythm licks more akin to R&B than country.

He started performing after high school, spending more than a decade plying his trade in Texas clubs before descending on Nashville in 1987. Life on the road had been hard, leading to a divorce in 1986 and drinking. The most difficult part of moving to Nashville was leaving his children behind, but he found the means to visit them at regular intervals, due to his increasing success as a songwriter.

Parnell's demos, which featured his own electric lead and slide guitar work, led to a recording contract with Arista Records in 1989. This coincided with a second marriage to a fellow Texan, Kim, who worked for Polygram Music in Nashville. Parnell scored minor hits with **Crocodile Tears** and **Oughta Be A Law** in 1990. However, his music, with elements of blues, old rock'n'roll and western swing, was at odds with the trend of traditional country. His next few singles all failed to chart, but Arista had faith in him, and **What Kind Of Fool Do You Think I Am** soared to No.2 on the charts in the summer of 1992. The title track of his second album, **Love Without Mercy,** reached the Top 10 and **Tender Moment** reached No.2 in 1993.

Recommended:
Love Without Mercy (Arista/–)
Lee Roy Parnell (Arista/–)

Gram Parsons

A seminal figure in the country-rock movement of the late '60s, Gram was born Cecil Connor in Winterhaven, Florida, on November 5, 1946. His father 'Coon Dog' Connor owned a packing plant in Waycross, Georgia. Coon Dog shot himself when Gram was 13 and Gram's mother married again, to Robert Parsons, a rich New Orleans businessman, Parsons formally adopted Cecil and changed the boy's name to Gram Parsons.

With much drinking in both real and adopted families, and a hitherto uprooted life, Gram ran away at 14 and two years later was in New York's Greenwich Village singing folk songs. At one point he formed a folk band with Jim Stafford, and with a later band, Shiloh, Gram specialized in a commercial brand of folk.

Studying theology at Harvard in 1965, Gram formed the International Submarine Band. After he dropped out of university, the band reformed and an album, on Lee Hazelwood's label, showed them to be following a fairly purist country path.

By the time that album came out, Gram had joined the Byrds. Meeting Chris Hillman of that band in L.A. in 1968, he convinced Hillman that the hitherto rock-oriented Byrds should experiment with country. The result was **Sweetheart Of The Rodeo**, the first real country-rock album. Although much of Gram's contribution was mixed out, the album set a new style among rock groups and reminded many a southern-bred rocker just where his roots lay. The Byrds appeared on the Grand Ole Opry and sang Gram's own composition **Hickory Wind**.

Gram's fantasy about marrying country with rock was nurtured by his association with the Rolling Stones. He quit the Byrds on the eve of a South African tour, causing a welter of ill feeling. However, he reunited with Chris Hillman in 1969 when his country aspirations were more fully realized in the Flying Burrito Brothers, that band's **Gilded Palace Of Sin** album being hailed as a country-rock classic, showcasing compositions that have since become standards via the talents of Emmylou Harris.

But Parsons was getting into the West Coast drug lifestyle and by the recording of the band's second album, **Burrito Deluxe**, he seemed more interested in hanging out with Jagger and Richard in Europe. With his trust fund providing him extensive monies to indulge his lifestyle and a fantasy about getting to rock superstar level, Gram's preoccupations were tending more towards drugs and drink than productive musical output. However, Warner Reprise came up with a contract and, better still, was the possibility of Merle Haggard producing his next album. Parsons went to visit Haggard but Merle appeared to have a change of heart at the last moment. Even so, the session went ahead using Haggard's engineer Hugh

Love Without Mercy, Lee Roy Parnell. Courtesy Arista Records.

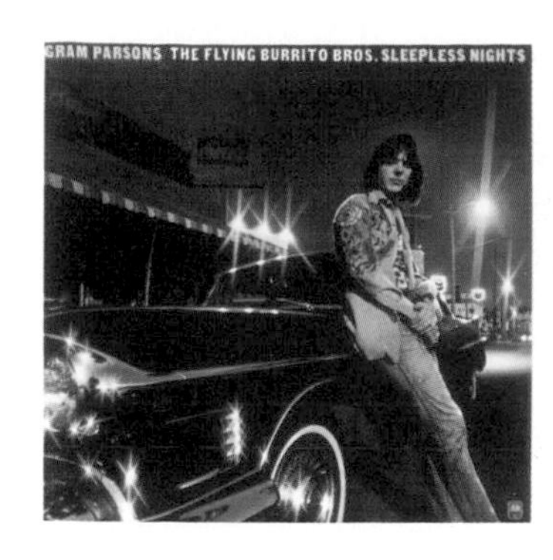

Sleepless Nights, Gram Parsons. Courtesy A&M Records.

Davis. Also booked were Glen D. Hardin, James Burton and a new girl singer from Baltimore, Emmylou Harris.

Gram was evidently drunk to the point of falling down for the first sessions. Nevertheless, the album showed that his writing ability was still there. **GP** was not a big commercial success on its release in 1972, and it was followed by **Grevious Angel**, which featured a similar line-up and more classic Parsons' songs. A tour around that time, with Emmylou Harris and the Fallen Angels Band gave hints of what might have been for Gram had he lived. It has been left to Emmylou to perpetuate the songs and the legend.

Parsons probably did not expect to live long and on September 19, 1973, he died of a heart attack at Joshua Tree, in the California desert. The causes were apparently a heavy mix of drink and drugs, followed by (according to Byrd Roger McGuinn) a lovemaking bout with his wife. Later, at L.A.'s International Airport, his body was snatched and instead of ending up in New Orleans for a family funeral, was driven to the desert at Joshua Tree and unofficially cremated, the result of a pact between Gram and manager Phil Kaufman that whoever died first would be taken to the desert and cremated.

Because of this bizarre incident no autopsy was possible and no official cause of death established. Parsons has since become a cult figure, following one of his country idols, Hank Williams, to an early and mysterious death. But, unlike Williams, Parsons had to wait until after his demise for recognition.

Recommended:

Sleepless Nights – with the Flying Burrito Brothers (A&M/A&M)
Live 1973 – with the Fallen Angels (Sierra/Sundown)
The International Submarine Band (LHI or Shiloh/Edsel)
GP/Grievous Angel (Warner Bros/Warner Bros)

Dolly Parton

Born on a farm in Locust Ridge, Seiver County, Tennessee, on January 19, 1946, Dolly Rebecca Parton was the fourth of 12 children born to a mountain family. At the age of ten, she was already an accomplished performer, her first regular radio and TV dates being on the shows of Cas Walker, in Knoxville. At 13 she was cutting sides for a small Louisiana record company, the same year making an appearance on Grand Ole Opry.

Graduating from Seiver County High School in June 1964, she immediately left for Nashville, where she, at first, scraped by as part of a songwriting team (with her uncle, Bill Owens). Her first success came in 1967, with two hit records on Monument (**Dumb Blonde** and **Something Fishy**) and a contract to join the Porter Wagoner TV and road show. Also that year, Dolly began recording for RCA, her duet with Wagoner on Tom Paxton's **Last Thing On My Mind** entering the charts in December 1967, and rapidly climbing into the Top 10.

For the next six years, the Parton–Wagoner partnership continued to flourish, over a dozen of their duets becoming RCA best-sellers. However, it seemed that Dolly was gradually becoming the major attraction on disc, obtaining a No.1 with her recording of **Joshua** in 1970. By 1974, she had branched out as a true solo act , though continuing to record duets with Wagoner and utilize his talents on some of her sessions. The result was a move away from mainstream country in an attempt to gain a wider audience. And the bid proved profitable. Dolly's recording of **Jolene** (1974) became a world-wide hit, this being followed by such other '70s winners as **The Bargain Store**, **We Used To** (1975), **Hey Lucky Lady**, **All I Can Do** (1976), **Light Of A Clear Blue Morning** (1977) and then five No.1s in a row with **Here You Come Again** (1977), **It's All Wrong But It's All Right**, **Heartbreaker**, **I Really Got The Feeling**, **Baby I'm Burnin'** (1978) and **You're The Only One** (1979).

During this period Dolly began exploring every pop avenue, from bluegrass through to disco. Her albums ranged from **Great Balls Of Fire**, on which she provided her own versions of earlier pop and rock hits, to **New Harvest, First Gathering**, a release which featured half of Nashville providing back-up assistance on **Applejack**, one of Dolly's own songs.

By this time Dolly had become a pop personality and the best-known country performer in the world. Unfortunately much of this was down to her cheesecake poses rather than her unquestioned singing ability, and her outstanding talent as a songwriter. A major film career was sparked by a starring role in the comedy '9 to 5', and this was followed by 'The Best Little Whorehouse In Texas' (1982), 'Rhinestone' (1984) and 'Steel Magnolias' (1990), by which time she had invested in her own film production company.

She continued to accrue an imposing array of awards, including CMA Female Vocalist Of The Year (1975 and 1976) and the coveted CMA Entertainer Of The Year (1978). There were further country No.1s with **Starting Over Again**, **Old Flames Can't Hold A Candle To You**, **9 To 5** – also one of her pop hits (1980), **But You Know I Love You** (1981), **I Will Always Love You** (1982), **Islands In The Stream** (a duet with Kenny Rogers that went to No.1 in many countries, 1983), **Tennessee Homesick Blues** (1984), **Real Love** (another Parton–Rogers duet, 1985) and **Think About Love** (1986).

Dolly saw another dream come true in 1986 with the opening of her Dollywood Theme Park in Pigeon Forge, Tennessee. The following year she hosted her own TV variety series, left RCA Records to sign with Columbia, and also completed the long-awaited **Trio** album with Emmylou Harris and Linda Ronstadt. It produced a No.1 country hit with a revival of **To Know Him Is To Love Him** (1987), won a Grammy for Best Country Album and also went on to gain a platinum award. Dolly's first Columbia album, **Rainbow**, was a pure pop record. Superbly produced, Dolly was in great voice, but it was slated by the critics and failed to sell. Unperturbed, she teamed up with Ricky Skaggs who produced **White Limozeen**, which saw her score No.1s with **Why'd You Come In Here Lookin' Like That** and **Yellow Roses** (1989). Further major country hits came with **Rockin' Years** (a No.1 duet with Ricky Van Shelton, 1991), **Silver And Gold** (1991), **Country Road** (1992) and **Romeo** (1993).

Her phenomenal success continued as both writer and performer. Soul singer Whitney Houston had the biggest-selling single in 1992 with Parton's composition, **I Will Always Love You**, a No.1 in Britain and America, which was featured in the film, 'The Bodyguard'. Meanwhile Dolly's own **Eagle When She Flies** album gained a platinum award. The following year she released **Slow Dancing With The Moon**, a subtle mix of country and pop themes that featured an all-star guest list. Unable to stand still for too long, and always moving on to new projects, Dolly Parton teamed with Loretta Lynn and Tammy Wynette for the 1993 album **Honky Tonk Angels**.

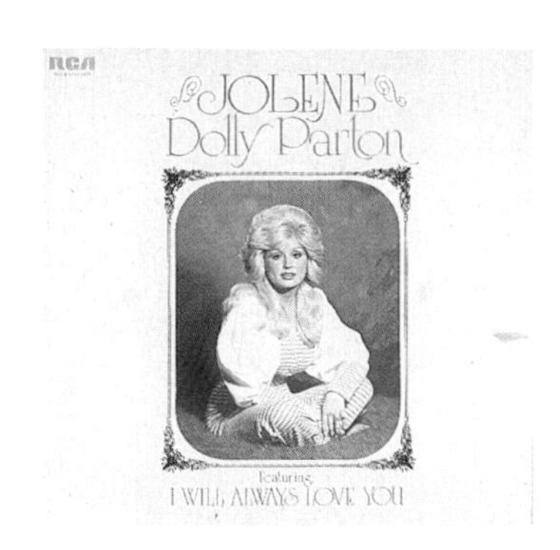

Jolene, Dolly Parton. Courtesy RCA Records.

Recommended:

Coat Of Many Colors (RCA/RCA)
My Tennessee Mountain Home (RCA/RCA)
Jolene (RCA/RCA)
New Harvest, First Gathering (RCA/RCA)
Burlap And Satin (RCA/RCA)
White Limozeen (Columbia/CBS)
Eagle When She Flies (Columbia/CBS)
Slow Dancing With The Moon (Columbia/Columbia)
Golden Streets Of Glory (RCA/–)

Stella Parton

The sister of Dolly Parton and the sixth of 12 children, Stella Parton (born Seiver County, Tennessee, May 4, 1949) married while still at high school and was pregnant by the time she graduated. A TV performer in Knoxville at the age of nine, she recorded early on for some minor Nashville labels and formed a gospel group the Stella Parton Singers, working as performer, manager and booking agent.

Below: Dolly – she once claimed she lost a Dolly Parton lookalike contest.

Stella Parton, Dolly's sister's eponymous album. Courtesy WEA Records.

She formed her own label Soul, Country And Blues Records, in the early '70s and in 1975 released **Ode To Olivia**, supporting Olivia Newton-John's right to be acclaimed Top Female Vocalist by the CMA. Also that same year she recorded a song of her own called **I Want To Hold You In My Dreams**, which provided her with her first Top 10 single.

By 1976 she was with Elektra-Asylum, her records being produced by Jim Malloy, The partnership proved successful and resulted in twin major records in 1977, **The Danger Of A Stranger** and **Standard Lie Number One**, the former becoming a pop hit in Britain, where she made a promotional tour.

1978 was another good year for Stella, bringing three sizable hits including **Four Little Letters**, a Top 20 entry. Two more chart entries followed in 1979 but since then it has been all quiet on the vinyl front for a lady who has valiantly fought a battle to be accepted for her own ability, generally steering clear of any situation that might make it seem that she owed anything to Dolly's influence.

Recommended:
Stella Parton (Elektra/Elektra)
Country Sweet (Elektra/Elektra)

Les Paul

One of the most influential guitarists in popular music history. Les Paul (born Lester William Polfus, Waukseha, Wisconsin, June 9, 1915) began his career as a country musician-comedian named Hot Rod, later becoming Rhubarb Red. He toured for some time with a popular Chicago group, Rube Tronson And His Texas Cowboys, but his growing interest in jazz led him to play with big bands and small combos for a time. At one point, from 1934 to 1935, he had a country show as Rhubarb Red over WJJD, Chicago, in the morning and an afternoon jazz show as Les Paul on WIND.

In 1936, he, plus vocalist and rhythm guitarist Jim Atkins (Chet's half-brother) and bassist Ernie Newton (a longtime Opry and Nashville session bassist in the '40s and '50s) auditioned in New York as the Les Paul Trio and spent the next five years with Fred Waring.

Les moved to L.A. late in 1941, where he spent much of his time in the studios as well as a stint in the services. In 1947 he teamed up with an ex-Gene Autry band member named Colleen Summers, who was such a fine guitarist she played lead guitar in Jimmy Wakely's band, singing harmony with him on **One Has My Name, The Other Has My Heart**.

Colleen, whom Paul later married, became Mary Ford, and the combination of her singing, his extraordinary playing and his then unique use of multiple track recording for guitar and voice made the team immensely popular in the '40s and early '50s, producing eleven No.1 pop hits for Capitol, including **Nola**, **Lover**, **How High The Moon** and **The World Is Waiting For The Sunrise**.

Les had experimented with electric guitars as early as the late '30s – in fact he had never stopped being fascinated by electronics since his first crystal set in 1927 – and in 1952 Gibson began putting out their fabulously successful series of Les Paul guitars, designed by Les himself.

After he and Mary Ford divorced in 1962, Les retired from performing, turning to inventing in his New Jersey home. The boom in interest in the Les Paul guitar, however, made his name a household word among musicians, and in 1973 he began performing again on a limited scale. Using a guitar called a Les Pulveriser which was light years ahead of the standard Gibson production model, he recorded a number of albums with Chet Atkins. One – **Chester And Lester** – won a Grammy for the Best Country Instrumental Performance of 1976.

Recommended:
Chester And Lester – with Chet Atkins (RCA/RCA)
Guitar Monsters – with Chet Atkins RCA/RCA)

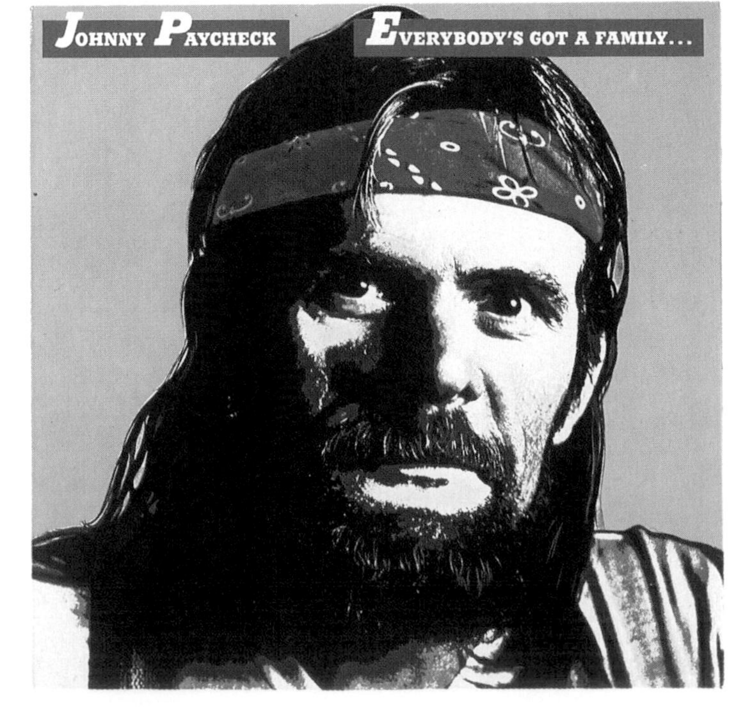

Everybody's Got A Family, Johnny Paycheck. Courtesy Epic Records.

Johnny Paycheck

A one-time Nashville renegade who, during the mid-'70s, temporarily adopted the name of John 'Austin' Paycheck in honour of the Music City's greatest rivals. This singer-songwriter (born Don Lytle, Greenfield, Ohio, May 31, 1941) began his career as a Nashville sideman, enjoying a brief stay as bass-guitarist with Porter Wagoner's Wagonmasters. Later he became a member of Faron Young's Deputies before moving on to play with both George Jones and Ray Price. During this period he switched to steel guitar, rejoining Jones' band as a guitarist during 1959–60.

As a rockabilly he cut some sides for Decca using the pseudonym Donny Young – then came sessions for Mercury and Hilltop, Paycheck having two fair-sized hits on the latter label with **A-11** (1965) and **Heartbreak, Tennessee** (1966). He then helped to form Little Darlin' Records, providing the label with several good sellers during the '60s, the biggest of these being **The Lovin' Machine**, a Top 10 entry during 1966. It was around this time that Paycheck also made the grade as a writer, his **Apartment No. 9** affording Tammy Wynette her first hit. **Touch My Heart**, another of his compositions, went Top 10 via a version cut by Ray Price.

Little Darlin' folded at the end of the '60s, a period during which Paycheck virtually hit rock bottom, becoming a self-confessed alcoholic. However, he proved not to be a quitter. He took the cure and fought back with considerable determination, teaming up with producer Billy Sherrill to cut some sides for Epic. His first release on the label was **She's All I Got,** which reached the top of the country charts in 1971.

For the next couple of years, he was rarely out of the Top 20, thanks to singles such as **Someone To Give My Love To**, **Love Is A Good Thing** (1972), **Something About You I Love**, **Mr Lovemaker** and **Song And Dance Man** (1973). And though the hits continued, for a while they came at lower-order level. Even so, it was surprising when in 1976 Paycheck was reported to be bankrupt. Once again he bounced back and in 1977 notched three major hits, **Slide Off Your Satin Sheets**, **I'm The Only Hell (Mama Ever Raised)** and **Take This Job And Shove It.** The latter, one of David Allan Coe's anti-establishment anthems, became one of the year's biggest records. He teamed up with George Jones for a session that produced hits in **Maybellene** (1978), **You Can Have Her** (1979), and **You Better Move On** (1980), the duo adopting a gruesome twosome guise for the sleeve of **Double Trouble**, a rock'n'fun album that came out in 1980.

By 1981 he was working with Merle Haggard, Hag turning up on Paycheck's album **Mr Hag Told My Story**, which provided a spin-off single in **I Can't Hold Myself In Line**. But, despite fine records, Paycheck's popularity seemed to be waning again. By 1983 he had slipped his Epic/Columbia connection and was on AMI, charting with **I Never Got Over You**. Three years later he joined Mercury Records, returning to the Top 20 with **Old Violin** (1986). Paycheck was making the headlines, with stories of his drunken behaviour and bankruptcy doing nothing for his already rock-bottom reputation. In 1988 he started a nine-year prison sentence for aggravated assault relating to a bar-room shooting incident in 1985. While in jail he studied and received his General Education Diploma and, having served less than three years, was released for model behaviour. Known for years as a renegade and a victim of drugs and booze, Paycheck came

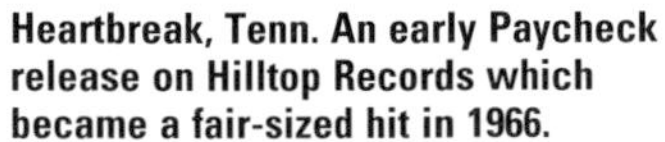

Heartbreak, Tenn. An early Paycheck release on Hilltop Records which became a fair-sized hit in 1966.

out of jail 'clean', and by 1993 was headlining in Branson, Missouri.

As Epic once claimed in a press handout: "With a life story that fits impeccably into the 'rags to riches to rags to riches' stereotype, the only truly amazing thing about Johnny Paycheck is that no-one has yet seen fit to put his biography on the silver screen. Change a couple of names to protect the guilty and avoid lawsuits and you'd have an instant smash."

Recommended:
Double Trouble – with George Jones (Epic/Epic)
Take This Job And Shove It (Epic/Epic)
Armed And Crazy (Epic/Epic)
Everybody's Got A Family (Epic/Epic)

Leon Payne

A smooth-voiced singer and multi-instrumentalist, Payne was blind from childhood. Born in Alba, Texas, on June 15, 1917, he attended the Texas School for the Blind between 1924 and 1935 learning to play guitar, piano, organ, drums, trombone and other instruments.

During the mid-'30s he began playing with various Texas bands, occasionally joining Bob Wills and his Texas Playboys. In the late '40s, he became a member of Jack Rhodes' Rhythm Boys, in 1949 forming his own outfit, The Lone Star Buddies, and playing on Grand Ole Opry. A prolific songwriter, Payne penned a great number of much-covered songs including

Above: Opry favourite from Grinder's Switch, Minnie Pearl – "Howdy. I'm jest so proud to be here!"

Lost Highway, **Blue Side Of Lonesome**, **They'll Never Take Her Love From Me** and **I Love You Because**, the latter providing him with his own hit in late 1949.

Payne, who recorded with such labels as MGM, Bullet, Decca, Capitol and Starday, suffered a heart attack in 1965 and had to curtail many of his performing activities. He died on September 11, 1969.

Minnie Pearl

Born Sarah Ophelia Colley in Centerville, Tennessee, on October 25, 1912, she majored in stage technique at Nashville's Ward-Belmont College during the '20s, then taught dancing for a while before joining an Atlanta production company as a drama coach in 1934. By 1940, Sarah had become Minnie Pearl from Grinder's Switch (Grinder's Switch is a railroad switching station just outside Centerville) and made her debut on the Grand Ole Opry in this guise. She became an instant Opry favourite and appeared on numerous tours, radio and TV shows, also appearing on the first country music show ever to play New York's Carnegie Hall (1947).

Much honoured by the music industry, Minnie – Nashville's Woman Of The Year in 1965 – was elected to the Country Music Hall Of Fame in 1975. Though she has recorded for such labels as Everest, Starday and RCA, her only chart success came with the Top 10 hit, **Giddyup Go – Answer** (1966), the woman's reply to Red Sovine's country No.1. Minnie continued with her stage and TV career right up until 1991 when she suffered a severe stroke and was forced into semi-retirement.

Recommended:
Stars Of The Grand Ole Opry – one track only (RCA/RCA)

Hank Penny

Hank Penny was one of the few East Coast musicians who tried to bring Western swing sound to country music in the '40s. He was born in Birmingham, Alabama, on August 18, 1918.

He began his career as early as 1933, and later founded the Radio Cowboys with stints on WWL in New Orleans and the Midwestern Hayride WLW in Cincinnati, before departing for the West Coast in the mid-'40s. An influential country comedian, he was a regular on Spade Cooley's LA TV show in the late '40s. In 1949 he co-founded the legendary Palomino night-club, then in the late '50s sold the club and took up residency in Vegas. He performed regularly in Vegas until 1972, when he moved back to Los Angeles and semi-retirement. Hank died of a heart attack on April 17, 1992, in California.

Recommended:
Rompin', Stompin', Singin', Swingin' (–/Bear Family)

Carl Perkins

Although Perkins is known as one of the seminal figures of Memphis rock'n'roll, he came from a solid country background and his albums have always been dotted with pure country songs.

Born in Lake City, Tennessee, on April 9, 1932, in a poor farming community, he started his career performing at local country dances and honky-tonks, along with brothers Jay and Clayton. Independently of Elvis Presley, Carl realized that country was moving in new directions. But, when he first approached Sam Phillips at Sun Studios in Memphis, Phillips insisted that he cut country music only since Elvis had the other scene tied up. The result was three country singles, **Turn Around**, **Let The Jukebox Keep On Playing** and **Gone, Gone, Gone**. But Perkins persuaded Phillips to let him do

Stars Of The Grand Ole Opry (RCA). Minnie performs Jealous Hearted Me.

some faster material. This plea resulted in **Blue Suede Shoes**, a Perkins original which topped the pop, country and R&B charts simultaneously in 1956. Perkins looked set to be the next superstar from the Sun stable, but late that year he was on his way to do the Perry Como and Ed Sullivan shows in New York when a car crash left Perkins with multiple injuries and a broken career. His brother Jay later died as a result of that same crash. However, Elvis had recorded **Blue Suede Shoes** and Perkins was assured of a place in the rock'n'roll honours list.

Nevertheless, Carl's solo career seemed to be at a standstill. Elvis had overtaken him as a rock star and Perkins began drinking heavily. But a tour of Britain in 1964, and another as a headliner in 1965, convinced him that he was a star in some countries. Also the Beatles had recorded his **Honey Don't, Matchbox** and **Everybody's Trying To Be My Baby**.

After this, Perkins was approached by Johnny Cash to become part of his road show and for many years remained a mainstay of the Cash package. He began cutting country records and also recorded

Above: Carl Perkins is a country singer who is also one of rock's seminal figures.

an album with the eclectic NRBQ in 1970. In 1978 he released a UK album titled **Ol' Blue Suede's Back**, which saw him rockin' once more. He also began touring with his sons on drums and bass.

A frequent visitor to Britain, where he has appeared on several Wembley Festival bills, he turned up on Paul McCartney's **Tug Of War** album in 1981. He also recorded live in Germany with Johnny Cash and Jerry Lee Lewis (forming three-quarters of Sun's legendary million-dollar quartet) and was the subject of a magnificent Charly Records boxed set, **The Sun Years**. In 1985 he signed with MCA Records, the same year that he starred in a television special, 'Blue Suede Shoes: A Rockabilly Session' to mark the 30th anniversary of **Blue Suede Shoes**. He also appeared with Johnny Cash, Jerry Lee Lewis and Roy Orbison on the **Class Of '55** album in 1986, the same year that his **Birth Of Rock And Roll** became his biggest country hit since **Restless** in 1969.

In 1987 Carl Perkins was inducted into the Rock'n'Roll Hall Of Fame, and two years later he released a new album, **Born To Rock**, a collection of inspired performances blending rockabilly, lazy low-down blues and soulful country ballads. Perkins was still writing, providing the Judds with **Let Me Tell You About Love**, a country No.1 in 1989, while his **Restless** was revived by the Nashville Cats in 1991, gaining a CMA award.

Recommended:
Long Tall Sally (–/CBS Embassy)
The Sun Years – boxed set (–/Charly)
Classics (–/Bear Family)
Country Boy's Dream (–/Bear Family)
Friends, Family And Friends (–/Magnum Force)
Carl Perkins (MCA/Dot/MCA)

Webb Pierce

Born in West Monroe, Louisiana, on August 8, 1926, Webb Pierce became a distinctive stylist in the heavily electric country of the '50s. Early in his youth he learned to play good guitar and was soon gaining notice playing at local events. After regular stints on Radio KMLB, Monroe, Pierce moved to Shreveport, home of the Louisiana Hayride. He was noticed by Horace Logan, programme director of KWKH, the sponsoring station of the Hayride, and subsequently joined the Hayride. During this period, the early '50s, his band included many who were themselves to find fame: Faron Young, Jimmy Day, Floyd Cramer.

Early '50s hits with Decca included **Wondering**, **That Heart Belongs To Me** and **Back Street Affair**, while Pierce also co-wrote his **The Last Waltz.** By 1953 he had become popular enough to win the No.1 singer award given by the American Juke Box Operators. Soon after he moved to Nashville and joined the Grand Ole Opry. In 1954 he recorded **Slowly**, a No.1 single that featured a ground-breaking pedal steel solo (by Bud Isaacs), while during 1955 he had three No.1 hits: **In The Jailhouse Now**, **Love, Love, Love** and **I**

Carl Perkins – The Man Behind Johnny Cash. Courtesy CBS Records.

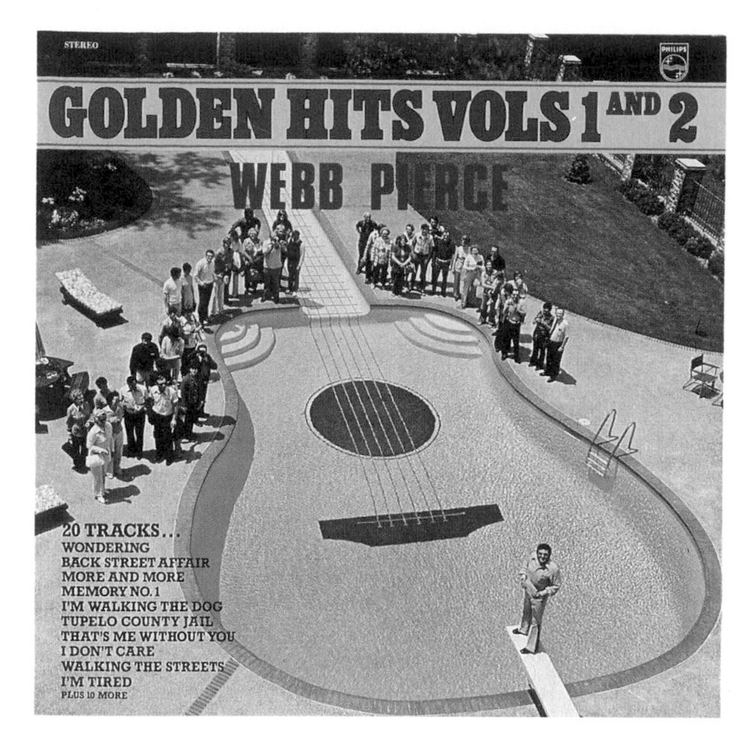

Golden Hits, Volumes 1 & 2, Webb Pierce. Courtesy Phonogram Records.

Don't Care. 1956 saw him scoring duet hits with Red Sovine – **Why, Baby, Why?** and **Little Rosa.** He recorded **Teenage Boogie**, an item now much sought after by rockabilly collectors. He followed up with more Top 10 hits including **Bye Bye Love**, **Honky Tonk Song** and **Tupelo County Jail**.

Below: A distinctive stylist of '50s country, Webb Pierce died in 1991.

His voice had the authentic, nasal, modern country ring and his songs had a bar-room edge, uncluttered by the excessive orchestration later to dominate Nashville recording.

He became extensively involved in the business side of music – a record company, radio stations and a publishing company or two. He was also the first star actually to own the Nashville cliché, the guitar-shaped swimming pool. However, though Webb had the occasional mini-hit during the '70s and even charted as recently as 1982 (when he teamed up with Willie Nelson on a revival of **In The Jailhouse Now**), he semi-retired from music in 1975. In later years his health caused many hospital visits, initially for heart problems, but in 1990 he was diagnosed with cancer. He died on February 24, 1991, in Nashville.

Recommended:
I Ain't Never (–/Charly)
The Wondering Boy, 1951–1958 (–/Bear Family)
Cross Country (Decca/Stetson)

Ray Pillow

Voted Most Promising Male Country Artist Of The Year by 'Billboard' and 'Cashbox' in 1966, singer-guitarist Pillow was born in Lynchburg, Virginia, on July 4, 1937. Following four years in the forces during the late '50s, he completed a stay at college, then opted for a singing career.

Later signed to Capitol Records, he gained his first sizeable hit with **Take Your Hands Off My Heart** in 1965 and obtained a Top 20 entry that same year with **Thank You Ma'am**. During 1966 came four more chart-fillers, including **I'll Take The Dog**, a duet with Jean Shepherd that was to become Pillow's most successful record. On April 30, 1966, he became a member of the Grand Ole Opry.

Still an Opry regular, Pillow has not had the best of luck with his records since the '60s; he has moved through such labels as Dot, Mega, Hilltop, MCA and First Generation.

Recommended:
Slippin' Around (Mega/–)
Countryfied (Dot/–)
One Too Many Memories (–/Allegianco)

Poco

A country-slanted rock band, Poco formed in 1969 when ex-Buffalo Springfield cohorts Richie Furay (guitar and vocals) and Jim Messina (guitar and vocals) linked with Rusty Young (pedal steel, banjo, guitar, vocals), George Grantham (drums) and Randy Meisner (bass).

Three desertions took place during the band's nine-album stay with the Epic label – Messina moving out to join Kenny Loggins as half of a hit-making duo; Meisner becoming part of Rick Nelson's Stone Canyon Band before joining the Eagles; and Furay eventually helping to form the Souther-Hillman-Furay Band. In the wake of some commercially unsuccessful but musically interesting attempts to create an 'orchestral country' style – as exemplified by the title track of their **Crazy Eyes** album (1973) – Poco completed their obligations to Epic and signed a deal with ABC Records. The immediate result was **Head Over Heels** (1975), one of their best-received albums. Though it had been rumoured the band would fold in 1973, a version of Poco still existed in 1982, cutting an album called **Ghost Town** for Atlantic, after cutting three albums for MCA.

In 1984 the band disbanded, but five years later reformed with the original line-up from 1968. They signed to RCA Records and released **Legacy**, an album that produced the Top 20 single **Call It Love**, while the album went gold. This marked the first time that the original line-up had ever recorded; so, 21 years after they were formed, Poco finally made what was, in a sense, its first album.

Recommended:
Poco (Epic/CBS)
Deliverin' (Epic/Epic)
Crazy Eyes (Epic/Epic)
Cantamos (Epic/Epic)
Rose Of Cimarron (ABC/MCA)
Legacy (RCA/RCA)
The Forgotten Trail – 1969–1974 (Epic/–)

Poco. Courtesy CBS Records. The band was originally named Pogo.

Charlie Poole

Leader of the North Carolina Ramblers, one of the most popular bands to emerge from that area in the '20s, five-string banjo player, singer and hard drinker Charlie Poole was born in Alamance County, North Carolina, on March 22, 1892.

Charlie Poole 1926–1930. Courtesy Historical Records.

A textile worker for most of his life, in 1917 he met Posey Rorer, a crippled miner who played fiddle. They teamed up and played together in the West Virginia–North Carolina area, eventually adding guitarist Norman Woodlieff and recording for Columbia on July 27, 1925. Among their first sides was **Don't Let Your Deal Go Down**, one of the band's most requested numbers.

Personnel changes followed throughout the ensuing years, Roy Harvey replacing Woodlieff on the band's Columbia sessions of September 1926 and Posey Rorer leaving in 1928 to be succeeded first by Lonnie Austin, and later by Odell Smith.

Invited to provide background music for a Hollywood movie in 1931, Poole readied himself for a move to California. But later that same year, on May 21, he suffered a heart attack and died having reached the age of 39.

Recommended:
Charlie Poole And The North Carolina Ramblers Volumes 1–3 (County/–)
Charlie Poole (Historical/–)

The Prairie Ramblers

The Prairie Ramblers, long associated with WLS and the National Barn Dance, were one of the most influential of the early string bands, although their style progressed through the years from Southeast string band to western swing in their National Barn Dance tenure of 1932–56.

There were numerous personnel changes in the band throughout its life but the nucleus of the Ramblers was formed by Chick Hurt (mandolin, tenor, banjo) and three of his neighbours in western Kentucky, Jack Taylor (string bass), Tex Atchison (fiddle) and Salty Holmes (guitar, harmonica, jug).

Originally called the Kentucky Ramblers, they began on radio on WOC, Davenport, Iowa, but within a few months were members of the National Barn Dance, where they teamed with Patsy Montana, backing her on her records and live. They introduced many important songs like **Feast Here Tonight**, **Shady Grove** and **Rolling On**.

As time went on, however, their style became increasingly swingy. An interesting aside – they also recorded a number of risqué songs under the name Sweet Violet Boys.

Atchison left the band in 1937, heading for California, where he appeared in many films and with the bands of Jimmy Wakely, Ray Whitley, Merle Travis and others. Holmes left and returned, then went on to a career that took him to the Opry with his wife Maddie (Martha Carson's sister) as Salty and Maddie. Atchison was replaced by Alan Crockett, who shot himself in 1947, and he was replaced in turn by Wade Ray, later to lead his own swing band and record for RCA. Wally Moore was his replacement, and when the band finally ground to a halt in 1956, Hurt and Taylor (the original members) were playing with a polka band, Stan Wallowick And His Polka Chips, which they continued to do for nearly another decade.

They recorded for the ARC complex of labels, Conqueror, Vocalion, Okeh, Mercury, Victor and Bluebird.

Elvis Presley

When Elvis Presley took country music into undreamed-of realms in 1955, many thought that he had killed it for all time. For there was no doubt that Presley was a country singer up until then. In fact, until the term rock'n'roll was coined, the new music was believed to be just country with an extra hard backbeat.

Elvis Aaron Presley was born in Tupelo, Mississippi, on January 8, 1935, and brought up in a religious family atmosphere. He sang with his parents at revival meetings, at concerts and in church, also learning to play some guitar. The family moved to Memphis when he was 13 and he began to sing at local dances. After graduating from high school, he was employed as a truck driver, playing with local groups at night.

That year he cut his first record, a private recording of **My Happiness**, to give his mother as a birthday present. The people at Sam Phillips' Memphis studio became interested in the country boy with the strange inflections in his voice and, later that year, they set up some experiments in the studio with Presley, Scotty Moore and Bill Black, to find a sound that suited Elvis. Country songs were not quite suitable, but when Elvis started a wild version of blues singer Arthur Crudup's **That's All Right, Mama**, they knew it was the missing piece of the puzzle, a piece that linked fast, almost breathless, country backups with Presley's frantic, uninhibited vocals. However, he was to keep his country connections, backing each single he released with a country title. Indeed, Charlie Feathers has described Presley's version of **Blue Moon Of Kentucky** as classic rockabilly – bluegrass, speeded up and with a black music feel. Presley was particularly influenced by the blues and by black musicians generally. He tended to dress in the extravagantly coloured suits of the black street hipsters.

That's All Right Mama was doing well locally, where DJ Dewey Phillips had plugged it. A local country agent, Bob Neal, Presley's manager for a while, got

Roustabout, Elvis Presley. Courtesy RCA Records.

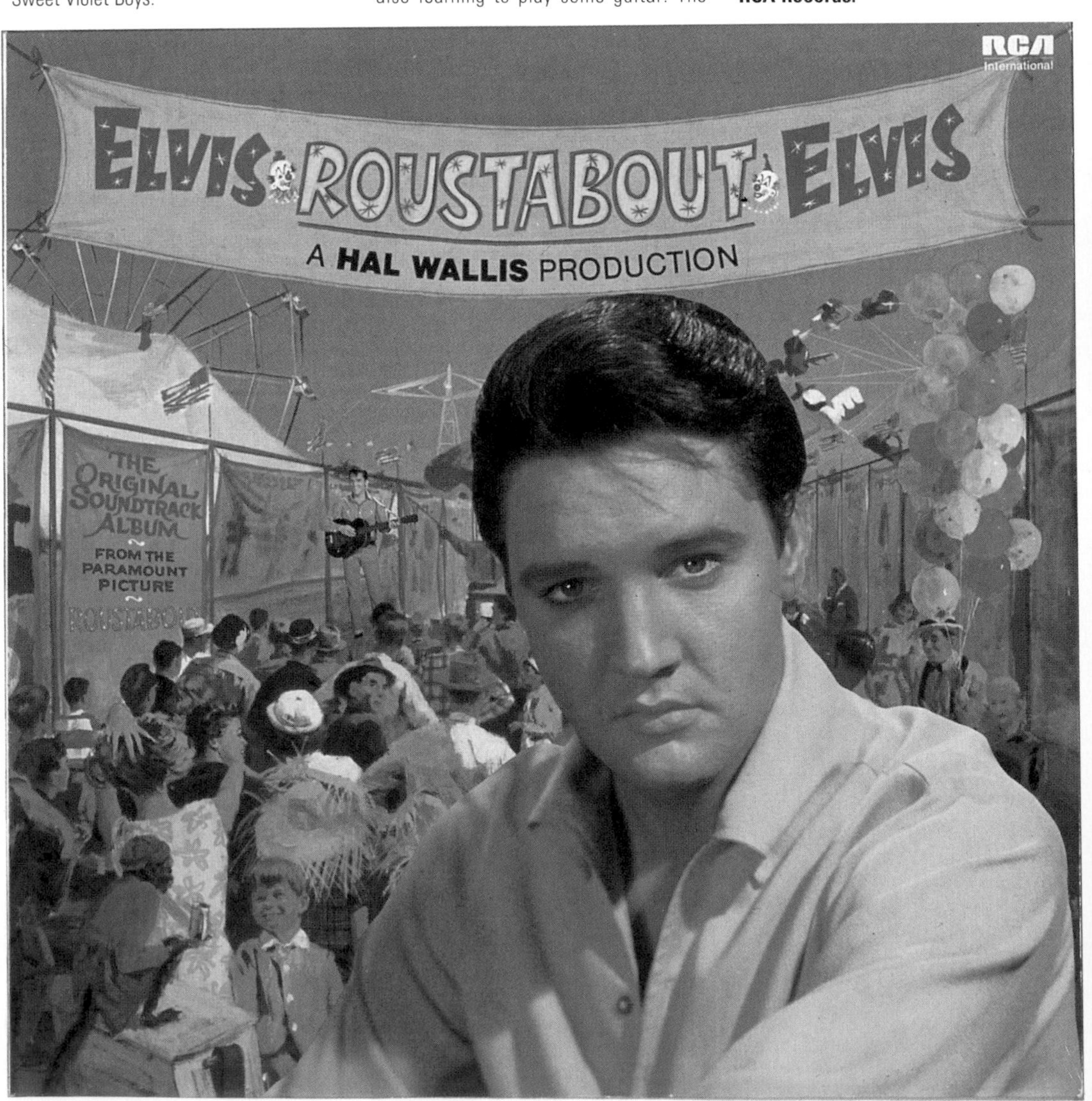

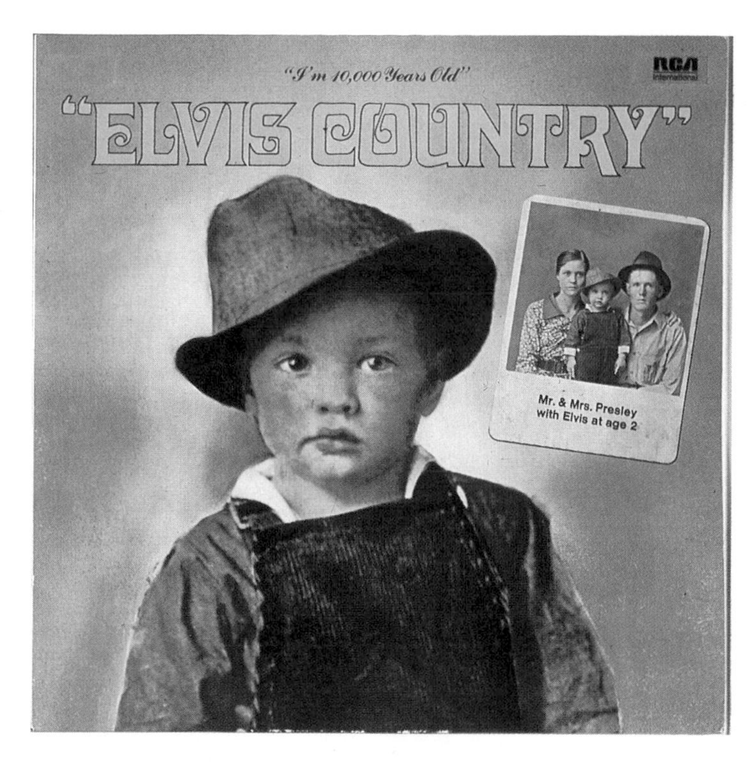

"Elvis Country", Elvis Presley. Courtesy RCA Records.

him some bookings on local country shows. But it was his next manager, 'Colonel' Tom Parker, a hustling wheeler-dealer who undoubtedly would have been running a medicine show in earlier days, who procured him vital exposure on the prestigious Louisiana Hayride on March 3, 1955. The Hayride was the next most important radio show to the Saturday night Grand Ole Opry from Nashville, which the then 'Hillbilly Cat' was also to play.

He toured on country bills with people like Hank Snow and Johnny Cash and the receptions got wilder, girls trying to get at him and tear his clothes. Parker eventually negotiated a deal for him to join the major RCA label and Presley's days as a country-styled rock'n'roller were over as RCA smoothed him gradually into a singer acceptable to both kids and parents.

Although the biggest of superstars, Presley would never again reach those primitive but exciting heights as he settled into a career of Las Vegas concerts and second-rate films. He had taken country music to the limit, making even honky-tonk country seem tame by comparison. But in so doing, he badly bruised country music for many years. Conversely, however, it is as well to remember country's contribution to rock'n'roll. When archivists rediscovered the roots of rock years later, they were led to a whole wealth of half-forgotten '50s music and were able to bring it from the shadow of the then prevailing 'Nashville Sound'. Elvis Presley may have utilized black music to launch rock'n'roll but country enthusiasts might argue that he sounded like the fashionable bar-room wailers of the day, taken to their wild, bopping conclusion.

Elvis was found dead at his Gracelands home in Memphis on August 16, 1977, his death attributed to 'acute respiratory distress' (though later investigation revealed that drugs may have been a factor). He amassed well over 100 US Top 40 hits during his lifetime, though he was outsold on the country chart.

Gracelands, where Elvis laid in state prior to being buried in a mausoleum at Forest Hill Cemetery, in Memphis, was opened to the public in late 1982. The singer's ex-wife claimed that funds were needed to maintain the property.

Elvis At Madison Square Garden. Courtesy RCA Records.

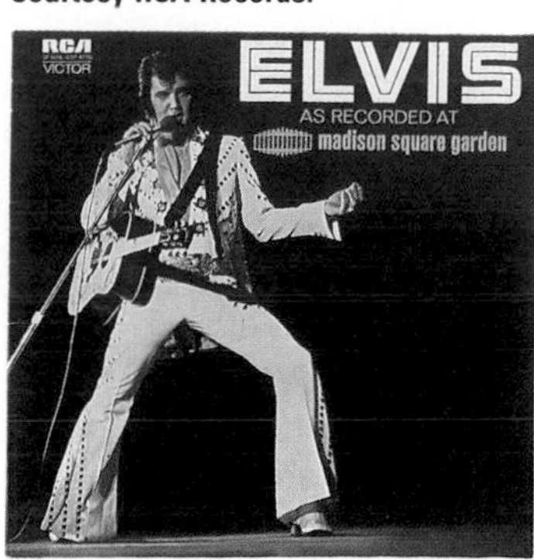

Recommended:

The Sun Sessions (–/RCA)
I'm 10,000 Years Old – Elvis Country (RCA/RCA)
World Wide 50 Gold Award Hits – four album set (RCA/RCA)
The Complete '50s Masters (–/RCA)
Promised Land (RCA/RCA)
Welcome To My World (RCA/RCA)

Kenny Price

Known as the 'Round Mound Of Sound', Price, a chunky singer-multi-instrumentalist and Hee Haw regular, was born in Florence, Kentucky on May 27, 1931. Raised on a farm in Boone County, Kentucky, he learned to play a guitar bought from Sears Roebuck and began playing country music at local functions.

During his service career in Korea he entertained the troops, and, upon discharge, he enrolled at the Cincinnati Conservatory of Music. He became a regular on Cincinnati's WLW Midwestern Hayride in 1954 as lead singer with the Hometowners. He then moved to Nashville and signed with the new Boone Records as a solo act, registering an immediate hit with **Walking On New Grass** in 1966. That same year he obtained a second Top 10 record with **Happy Tracks**. Further hits for Boone flowed, the biggest of these being **My Goal For Today** (1967). In 1969 Price took his happy sound to RCA – two of his 1970 releases, **Biloxi** and **Sheriff Of Boone County** becoming Top 10 hits. Price became a regular member of Hee Haw and continued to chalk up minor hits throughout the '70s for RCA, MRC and Dimension, his last chart entry being with **She's Leavin' (And I'm Almost Gone)** in 1980. He died of a heart attack on August 4, 1987.

Recommended:

Turn On Your Love Light (RCA/–)
Supersideman (RCA/–)
The Red Foley Songbook (RCA/–)
North East Arkansas Mississippi County Bootlegger (RCA/–)

Ray Price

Born on a farm in Perrybille, East Texas, on January 12, 1926, Ray Noble Price, the 'Cherokee Cowboy', was brought up in Dallas. After high school, he spent several years in the forces, returning to civilian life in 1946 and attending college to study veterinary surgery. However, an able singer-songwriter-guitarist, he began performing at college events and local clubs, eventually making his radio debut as an entertainer in 1948, on station KRBC, Abilene. Later came further exposure on Big D Jamboree, a Dallas show.

Below: Hee Haw regular and 'Round Mound Of Sound' Kenny Price.

Price began recording for Bullet during the early '50s, his first release being a song called **Jealous Lies**. Then in 1951 came a contract with Columbia. Many of his early records for the label reflected the influence of Hank Williams – Price's band, the Cherokee Cowboys, was formed from the remnants of Williams' outfit, the Drifting Cowboys.

By the end of 1952 Price was an Opry regular with two hit singles to his credit – **Talk To Your Heart** and **Don't Let The Stars Get In Your Eyes**, both charting during the year. And though his name was absent from the charts for the next 14 months, in February 1954 he began a run of major hits that continued to 1973. The most prominent of these were **Crazy Arms** (a 1956 million-seller), **I've Got A New Heartache** (1956), **My Shoes Keep Walking Back To You** (1957), **City Lights** (another million-seller, 1958), **Heartaches By The Number**, **The Same Old Me** (1959), **One More Time** (1960), **Soft Rain** (1961), **Make The World Go Away** (1963), **Burnin' Memories** (1964), **The Other Woman** (1965), **Touch My Heart** (1966), **For The Good Times**, **I Won't Mention It Again** (1970), **I'd Rather Be Sorry** (1971), **Lonesomest Lonesome**, **She's Got To Be A Saint** (1972), and **You're The Best Thing That Ever Happened To Me** (1973), all Columbia releases.

An astute judge of current trends and possessing an ear for up and coming songwriters – he was one of the first to recognize Kris Kristofferson's potential – Price realized that country had to appeal to a wider audience in order to forge ahead. Accordingly he began using large back-up units, often employing full string sections, in his plan to take country to a non-country audience. He ditched any pretensions to a cowboy image and began appearing onstage in a dress suit. Yet the ballads still sounded as though they came out of Texas and the ploy worked.

In 1974 Ray began recording for Myrrh, obtaining a Top 10 single for this mainly gospel label with **Like Old Times Again** and following it with another in **Roses And Love Songs** (1975). He then signed with ABC/Dot, and during 1975 had hits on three different labels, Columbia still cashing in on his back catalogue material. His records still sold better than most and the ABC/Dot hits continued, albeit at a lower level, through to 1978, when Ray decided to make a comeback. He signed to Monument Records and gained an immediate Top 20 single with **Feet**. During 1979 he provided three more hits, reaching the Top 20 again with **That's The Only Way To Say Goodbye**, while in 1980 he teamed with Willie Nelson to cut an excellent duet album that spawned the hit single **Faded Love**. By 1981 Ray had switched to the Dimension label, cutting three hits, including such Top 10 entries as **It Don't Hurt Me Half So Bad** and **Diamonds In The Stars**. His 1982 haul contained **Old Friends**, a single made with Roger Miller and Willie Nelson. The following year he had label-jumped yet again, this time to Warner, scoring with two tracks recorded for the soundtrack of the movie 'Honkytonk Man'. He continued label-hopping, scoring minor hits through to 1989 and recording albums, mainly in his old '50s country style. In 1992 he cut an album for Columbia titled **Sometimes A Rose**. But, despite such a long and distinguished career, Ray has only landed one CMA award: Album Of The Year in 1971 for **I Won't Mention It Again**.

Above: The first black country superstar, Charley Pride.

Recommended:
For The Good Times (Columbia/CBS)
Willie Nelson And Ray Price (Columbia/CBS)
Like Old Times Again (Myrrh/–)
Hank 'n' Me (ABC/–)
The Heart Of Country Music (Step One/–)
Sometimes A Rose (Columbia/–)
Essential (Columbia/–)

Charley Pride

Easily the most successful black entertainer to emerge from the country music scene, Charley Pride was born on a Delta cotton farm in Sledge, Mississippi, March 18, 1938. One of 11 children, Pride picked cotton alongside his parents during his boyhood days, eventually saving enough cash to purchase a $10 silvertone guitar from Sears Roebuck. Despite being born in blues territory, Pride preferred playing country music. However, he initially turned to baseball, not show business – playing for the Memphis Red Sox as a pitcher and outfield player in 1954. Two years later, he was drafted into the forces, during this period marrying Rozene, a girl he met in Memphis. Returning to civilian life in 1959, he quit baseball following a wage disagreement. Several jobs later he settled down in Helena, Montana, working at a zinc smelting plant, also playing semi-pro ball in the Pioneer League. And though he continued his efforts to break into major league ball, he was turned down by the California Angels in 1961 and by the New York Mets a year later.

However, his secondary career reaped rewards when, in 1963, Pride sang a song on a local show headed by Red Sovine and Red Foley, earning praise from Sovine, who urged him to try his luck in Nashville. Taking this advice, Pride made the move. Chet Atkins eventually heard some of his demo tapes and signed him to RCA.

Pride's first RCA single, **Snakes Crawl At Night**, was released with little in the way of accompanying publicity in December 1965. Few of the DJs who played the disc realized that the singer was black. But, by the following spring, the Mississippian was nationally known after coming up with **Just Between You And Me**, his performance gaining him a Grammy nomination. A couple of hits later, in January 1967, and he was on the Opry. His introduction, by Ernest Tubb, received a warm reception.

As the '70s rolled in, Pride reached superstar status. Having logged five No.1s in a row with **All I Have To Offer Is Me**, **I'm So Afraid Of Losing You** (1969), **Is Anybody Goin' To San Antone?**, **Wonder Could I Live There Anymore** and **I Can't Believe That You've Stopped Loving Me** (1970), he was adjudged Entertainer Of The Year in 1971, by the CMA, also winning the association's Male Vocalist Of The Year award, a title he retained in 1972.

Basically a honky-tonk singer who, like many others, moved towards a smoother, more pop-oriented country sound, Pride proved one of the most consistent record sellers on the RCA roster. Some of his No.1s have been: **I'd Rather Love You**, **Kiss An Angel Good Mornin'** (1971); **It's Gonna Take A Little Bit Longer**, **She's Too Good To Be True** (1972); **A Shoulder To Cry On**, **Amazing Love** (1973); **Then Who Am I?** (1974); **Hope You're Feelin' Me** (1975); **My Eyes Can Only See As Far As You** (1976); **She's Just An Old Love Turned Memory**, **I'll Be Leaving Alone**, **More To Me** (1977); **Someone Loves You Honey** (1978), **Where Do I Put Her Memory?**, **You're My Jamaica** (1979); **Honky Tonk Blues**, **You Win Again** (1980); **Never Been So Loved (In All My Life)**, **Mountain Of Love** (1981); **You're So Good When You're Bad**, **Why Baby Why?** (1982); and **Night Games** (1983).

The possessor of a warm baritone voice, Charley Pride sounds as pure country as Hank Williams, a hero whom he honoured in 1980 via **There's A Little Bit Of Hank In Me**, a No.1 album. Paradoxically, he had become RCA's biggest-selling country act since Elvis Presley – a country boy who tried his hardest to sound black.

In 1984, he scored his last Top 10 hit on RCA with **The Power Of Love**. Disillusioned with RCA's inability to market his recordings, he signed with 16th Avenue Records in 1987. After a three-year absence he was back in the Top 20 with **Have I Got Some Blues For You** (1987), soaring into the Top 10 with **Shouldn't It Be Better Than This** (1988). Having built a European fan base in the early '70s, Pride was able to capitalize on this career investment with annual sell-out tours, which have seen him more popular than the new superstars of the '80s and '90s.

Everybody's Choice, Charley Pride. Courtesy RCA Records.

Recommended:
(Country) Charley Pride (RCA/RCA)
There's A Little Bit Of Hank In Me (RCA/RCA)
Amy's Eyes (16th Avenue/Ritz)
In Person (RCA/RCA)
Classics With Pride (16th Avenue/Ritz)

John Prine

One of the most acclaimed singer-songwriters to come out of the '70s, John Prine was born on October 10, 1946 in Maywood, Illinois. He spent much of his childhood in Kentucky, where his grandparents lived, but he had his musical beginnings in Chicago, where he started writing songs as a teenager. Following a stint in the army, he worked as a postman, but, discovering he could make more money playing at local folk clubs, he took up music as a full-time profession. Prine's big break came through his close friend, singer-songwriter Steve Goodman, who introduced him to Paul Anka and Kris Kristofferson, who in turn introduced him to Jerry Wexler. Wexler signed Prine to an Atlantic record contract in 1971.

Prine made a series of critically acclaimed albums throughout the '70s, along the way writing such classic songs as **Hello In There**, **Paradise**, **Sam Stone** and **Angel From Montgomery**. In 1979 he recorded **Pink Cadillac**, a rockabilly-flavoured album at Sam Phillips' Sun Studios in Memphis. By the early '80s he was based in Nashville, writing country hits for everyone from Don Williams to Tammy Wynette, Johnny Cash to Gail Davies. Without a major label deal, Prine decided to form his own label, Oh Boy Records, with his long-time manager, Al Bunetta. He released three low-key albums. Two of them, the acoustic-flavoured **German Afternoons** and the two-record set, **John Prine Live**, gained Grammy nominations. Turning his hand to various projects, Prine spent time working with John Mellencamp on the soundtrack of his movie 'Falling From Grace', also making his screen debut in the film. He recorded a duet, **If You Were The Woman And I Was The Man**, with Margo Timmins and the Cowboy Junkies, and visited Britain to perform and act as compère for the Channel 4 TV series Town & Country.

A major turnaround came in Prine's career with **The Missing Years** album, which was produced by Howie Epstein, bassist for Tom Petty And The Heartbreakers. Whereas his previous studio albums had been completed in about a week with a limited budget, this one took almost ten months. Alongside fellow Heartbreakers Mike Campbell and Benmont Tench, noted Prine fans Tom Petty, Bruce Springsteen and Bonnie Raitt all dropped by to add background vocals. Again, released on Prine's own Oh Boy Records, **The Missing Years** was something of a sleeper. It eventually sold over 500,000 copies and won John a Grammy at the 1992 awards. This led to tours with Bonnie Raitt and the Cowboy Junkies, and recognition, at long last, for performing, rather than just songwriting.

Recommended:
Aimless Love (Oh Boy/This Way Up)
German Afternoons (Oh Boy/This Way Up)
John Prine Live (Oh Boy/This Way Up)
The Missing Years (Oh Boy/This Way Up)
Diamonds In The Rough (Atlantic/Atlantic)
Pink Cadillac (Asylum/Asylum)

Left: A talented writer, John Prine is finally making an impact onstage.

Ronnie Prophet. The all-rounder's debut RCA album.

Ronnie Prophet

Described by Chet Atkins as 'the greatest one-man show I've seen', singer-guitarist Prophet was born near Montreal, Canada, on December 26, 1937, and was raised on a farm at Calumet. He began playing at square dances in his early teens and soon moved to Montreal, later playing club dates in Fort Lauderdale. He first appeared in Nashville in 1969.

He obtained a residency at Nashville's Carousel Club. His drawing power proved such that the venue became renamed Ronnie Prophet's Carousel Club. His first American album was released on RCA in 1976, and the LP contained such tracks as **Shine On**, **Sanctuary** and **It's Enough**, all minor country hits. He played the British Wembley Festival of 1978 and broke the place up, looning around with compère George Hamilton IV and regaling the audience with his 'Harold The Horny Toad'. This led to further appearances at Wembley, regular tours and his own TV series, but over-exposure of his rather 'samey' act, without any major hits to back him up, led to fans tiring of Ronnie Prophet.

Recommended:
Ronnie Prophet Country (RCA/RCA)

Jeanne Pruett

Jeanne, a singer-songwriter, was born Norma Jean Bowman on January 30, 1937 in Pell City, Alabama. One of 10 children, she used to listen to the Grand Ole Opry as a youngster. She first headed for Nashville in 1956 along with her husband Jack Pruett, a guitarist who played lead guitar with Marty Robbins for almost 14 years.

It was Robbins who was responsible for her signing to RCA Records in 1963, at which time she cut six titles. And though there was little reaction, the following year found her making her debut on the Grand Ole Opry. Jeanne continued writing and performing, eventually securing a new record contract with Decca in 1969 and enjoying a minor hit in 1971 with **Hold On To My Unchanging Love**.

The real breakthrough came in 1973 when her recording of **Satin Sheets** became a phenomenal seller, crossing over to enter the pop charts. Throughout the

Welcome To The Sunshine, Jeanne Pruett. Courtesy MCA Records.

rest of the '70s, Jeanne graced the country charts with such MCA singles as **I'm Your Woman** (1973), **You Don't Need To Move A Mountain** (1974), **A Poor Man's Woman** (1975) and **I'm Living A Lie** (1977). She switched to Mercury for a while, then made something of a comeback with **Back To Back** (1979), **Temporarily Yours** and **It's Too Late** (both 1980), three IBC releases that all went Top 10. Since then there's been no major hits, but Jeanne still appears on the Opry, though she tends to spend much of her spare time cooking or gardening.

Recommended:
Encore (IBC/RCA)
Introducing Jeanne Pruett (–/MCA)

Riley Puckett

One of the pioneers of recorded old-time music, George Riley Puckett was born in Alpharetta, Georgia, on May 7, 1884. When only three months old he suffered an eye infection, and following incorrect treatment of the ailment lost his sight. Educated at a school for the blind in Macon, Georgia, he learned to play five-string banjo, later moving on to guitar. During the early '20s, Puckett worked with a band led by fiddle player Clayton McMichen. He then joined Gid Tanner's Skillet Lickers in 1924, remaining featured vocalist with the outfit until its disbandment some 10 years later. Puckett's many solo recordings include **Rock All Our Babies to Sleep**, reputed to be one of the first discs to feature a country yodeller, cut three years prior to Jimmie Rodgers' initial session.

From 1934 to 1941, Puckett recorded for RCA Victor – also cutting a few sides for Decca in 1937 – and worked on radio stations in Georgia, West Virginia, Kentucky and Tennessee up to his death in East Point Georgia, on July 13, 1946. At the time of his death, caused by blood poisoning from an infected boil on his neck, Puckett was working with a band called the Stone Mountain Cowboys, on radio station WACA, Atlanta.

His exuberant – sometimes even wild – bass-run guitar style was very influential on country guitarists of his day, the forerunner of the pulsing style which characterizes bluegrass.

Recommended:
The Skillet Lickers Volumes 1 & 2 (Country/–)
Riley Puckett (Old Homestead/–)
Red Sails In The Sunset (–/Bear Family)

Pure Prairie League

Formed in Cincinnati in 1971, this country-rock band took their name from a Women's Temperance Society that appeared in an Errol Flynn movie. Their trademark, which decorated all of their album covers, was 'Luke', an old-timer originally created by

Pure Prairie Collection. Courtesy RCA Records.

'Saturday Evening Post' artist Norman Rockwell.

League signed for RCA, the line-up then being Craig Fuller (guitars, vocals), George Ed Powell (guitars, vocals), Jim Lanham (bass, vocals), Jim Caughlan (drums) and John David Call (steel guitar). By the time their second album, **Bustin' Out**, appeared, only Fuller and Powell remained from the original band.

In 1973, after further personnel problems, PPL ceased recording and their record company thought they had broken

Pure Prairie League. Courtesy RCA Records. Norman Rockwell provided the sleeve.

up. But the group continued playing live dates and caused such a response by these appearances that RCA were forced to reissue **Bustin' Out**, plus **Amie**, a single taken from the album, the latter going into the US pop charts during 1975.

That year, the group, then comprising original members Fuller and Call plus Larry Goshorn (lead guitar), Michael Connor (keyboards), Mike Reilly (bass) and William Frank Hinds (drums), cut **Two Lane Highway**, an album that featured such guests as Chet Atkins, Emmylou Harris, Johnny Gimble and Don Felder. But, despite further albums and the odd hit like **That'll Be The Way** (1976), the line-up continued to fluctuate. Powell quit in 1977 along with Larry Goshorn and brother Timmy Goshorn, who had earlier replaced Call. The band, by then West Coast-based, regrouped yet again, one of the new members being Vince Gill, who had earlier worked with Boone Creek and Byron Berline's band.

By 1980, still unable to make the breakthrough that would take them up a notch on the concert circuits, PPL moved to the Casablanca label. The switch brought

Right: Pure Prairie League – John David Call, Mike Reilly, William Frank Hinds, Larry Goshorn, George Ed Powell and Michael O'Connor.

the band immediate benefits. By the middle of the year they had gone Top 10 in the pop charts with **Let Me Love You Tonight**, following this with other major pop hits in **I'm Almost Ready** (1980) and **Still Right Here In My Heart** (1981). But Gill left, first to work with Rodney Crowell and Rosanne Cash and then to pursue a solo career with RCA, and soon PPL were having problems once more and searching for a label deal.

Recommended:
Pure Prairie League (RCA/RCA)
Bustin' Out (RCA/RCA)
Two Lane Highway (RCA/RCA)
Dance (RCA/RCA)
Just Fly (RCA/RCA)

Eddie Rabbitt

Considered by many pundits to be pure MOR, singer Eddie Rabbitt (real name Edward Thomas, born in Brooklyn, New York on November 27, 1944) has, nevertheless, won many friends among country music buyers.

A one-time truck driver, soda jerk, boat helper and fruit picker, Rabbitt recorded for 20th Century Fox and Columbia during the 1960s. He achieved little until, as a staff writer with music publishers Hill and Range, he wrote **Kentucky Rain**, which Elvis Presley recorded and turned into a hit (1970). In 1973, Ronnie Milsap waxed Rabbitt's **Pure Love** and grabbed a No.1, this leading to the New Yorker gaining an Elektra recording contract of his own and getting his chart career underway with **You Get To Me**, a middle-order hit.

By 1975, his records were hitting the Top 20, the first No.1 being with **Drinkin' My Baby (Off My Mind)** in 1976, a year in which Rabbitt also went Top 10 with **Rocky Mountain Music** and **Two Dollars In The Jukebox**. But the best was yet to come for, in the wake of three more Top 10 singles, he provided Elektra with three chart-toppers in a row during 1978 **You Don't Love Me Anymore**, **I Just Want To Love You** and **Every Which Way But Loose**, the latter being the title track of a Clint Eastwood movie. There was another No.1 in 1979, **Suspicions**, then three in 1980: **Gone Too Far**, **Drivin' My Life Away** and **I Love A Rainy Day**. Two more in 1981 were **Step By Step** and **Someone Could Lose A Heart Tonight**. Rabbitt's **Step By Step** album became the best-selling album in the country market.

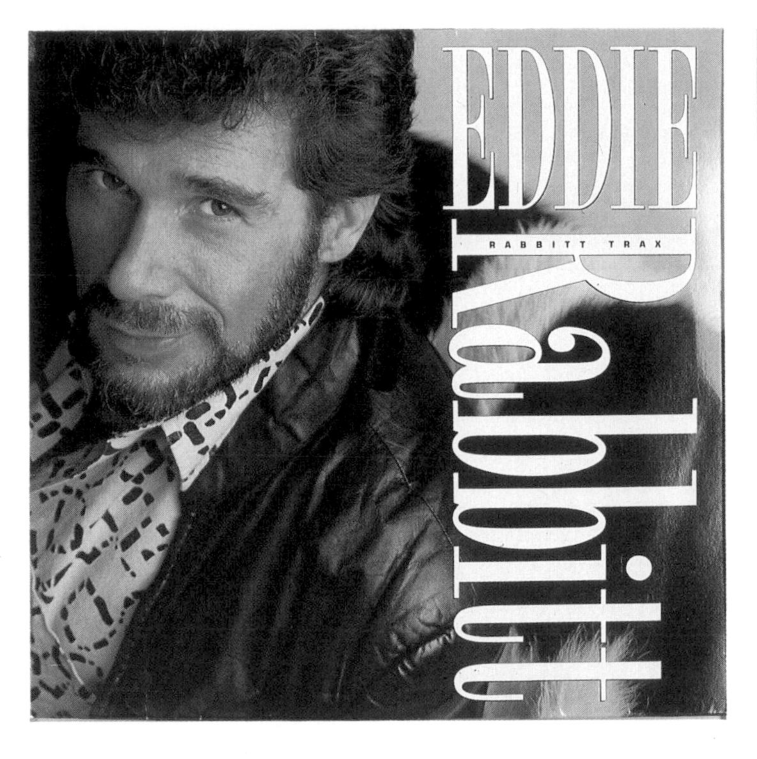

Rabbitt Trax, Eddie Rabbitt. Courtesy RCA Records.

He became guest host on the late TV show The Midnight Special and even had a TV special of his own. But CMA award nominations never came his way.

Not that the lack of acclaim in Music City meant much to Rabbitt. He just kept on doing the thing he was good at – notching hits. And in 1982 he notched a couple more; one, a duet with Crystal Gayle (**You And I**), provided the singer with yet another chart-topper that crossed over into the pop listings.

Throughout the rest of the '80s he has hardly faltered. He supplied another No.1 in 1983, **You Can't Run From Love**, and yet another in 1984, **The Best Year Of My Life**. During 1985 Rabbitt notched two Top 10 records with **Warning Sign** and **She's Coming Back To Say Goodbye**. That same year he joined RCA Records, making a Top 10 entry with **Repetitive Regret** early in 1986, and going all the way with Juice Newton to No.1 with **Both To Each Other (Friends And Lovers)**. Two more No.1s came his way in 1988 with the self-penned **I Wanna Dance With You** and a revival of Dion's rock'n'roll classic, **The Wanderer**. The following year he moved to Universal Records, immediately hitting No.1 with **On Second Thought**, before the label was absorbed into Capitol/Liberty with Rabbitt back in the Top 10 with **Runnin' With The Wind** (1990). But even still (gold records apart), his awards cupboard stays somewhat bare.

Below: Eddie Rabbitt once wrote hits for Elvis Presley and Ronnie Milsap.

Recommended:
Rabbitt (Elektra/Elektra)
Loveline (Elektra/Elektra)
Radio Romance (Elektra/Mercury)
Step By Step (Elektra/Elektra)
Horizon (Elektra/Elektra)
Jersey Boy (Capitol/–)
Rocky Mountain Music (Elektra/Elektra)
Ten Rounds (Capitol/–)

Gonna Find Me A Bluebird, Marvin Rainwater. Courtesy MGM Records.

Marvin Rainwater

Of Indian ancestry, Rainwater – a singer and prolific songwriter, who also plays guitar and piano – became a star during the 1950s when several of his records sold well over a million each. Born in Wichita, Kansas on July 2, 1925, his real name was Marvin Percy. Rainwater was his mother's maiden name, which he later adopted for his stage work. He trained as a veterinary surgeon, becoming a pharmacist's mate in the Navy during World War II. Upon discharge, he opted for a career in the music business. His first breakthrough came in 1946 when he debuted on Red Foley's Ozark Jubilee Radio Show, the station receiving so many enquiries about 'that singer with the Indian name' that Rainwater was signed to a regular spot on the programme.

During the early '50s he began touring and cutting records for Four Star and Coral. In 1955 he entered the Arthur Godfrey CBS Talent Scout TV Show, and was brought back for four consecutive wins. The following January he signed with MGM Records. With his second release, the self-penned **Gonna Find Me A Bluebird**, he won his first gold disc. Promoted by MGM as a full-blooded Cherokee brave, Rainwater usually appeared bedecked in Indian head-dress and similar paraphernalia. The idea seemed to work, for in 1958 his rocking **Whole Lotta Woman** became a world-wide hit, reaching the top of the British charts and ensuring Rainwater a season at the London Palladium. However, the USA reaction to the disc was mixed and some radio stations banned the song as being 'too suggestive'.

Soon after – following another million-seller with his version of John D. Loudermilk's **Half-Breed** (1959) – Rainwater and MGM parted company, the singer moving on to record for such labels as Warner Brothers, UA, Warwick and his own Brave Records. In the mid-'60s came a serious throat ailment that resulted in an operation, after which Rainwater was not

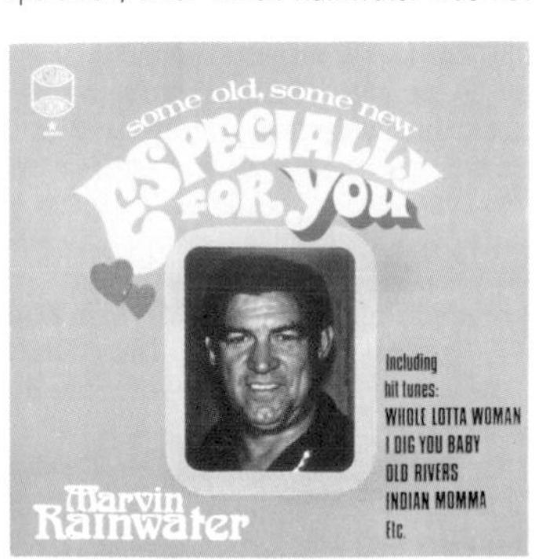

Especially For You, Marvin Rainwater. Courtesy Westwood Records.

to record for four years. Throughout the '70s he toured Britain, and, like George Hamilton IV and Vernon Oxford, he has enjoyed a higher reputation in Europe than in his homeland.

Recommended:
Rockin' Rolling Rainwater (–/Bear Family)
With A Heart With A Beat (–/Bear Family)

Bonnie Raitt

A white blues singer, songwriter and guitarist with a country-rock feel in her work, Bonnie Raitt was born on November 8, 1949 in Burbank, California. She grew up in a musical family; her father, John Raitt, starred in musicals during the '40s and '50s. Bonnie attended college in Boston in 1967, then two years later moved to the East Coast where she played blues in an acoustic duo with Freebo. Later she joined Paul Barrere in the Bluebusters.

Her gravelly, emotionally mature voice and dynamic slide guitar work resulted in a recording contract with Warner Bros in 1971. Throughout the '70s she produced excellent albums, many of which sold well, while her **Don't It Make Ya Wanna Dance**, featured in 'Urban Cowboy', made the US country charts in 1980.

Raitt finally broke through to superstar status with her **Nick Of Time** album on Capitol in 1989, which sold in excess of two million and gained a Grammy award as Album Of The Year. She has worked in Nashville on many occasions, playing sessions and concerts with Emmylou Harris and Wynonna Judd. The Judd youngster cites Bonnie Raitt as a major influence on her own music.

Recommended:
Sweet Forgiveness (Warner Bros/–)
The Glow (Warner Bros/–)
Nine Lives (Warner Bros/–)
Nick Of Time (Capitol/–)
Luck Of The Draw (Capitol/Capitol)

Boots Randolph

A premier Nashville session man, Homer Louis 'Boots' Randolph III was born in Paducah, Kentucky. He first learned to play trombone but later switched to saxophone, playing first in a high school unit then in bands throughout the Midwest.

Eventually he was spotted by Homer And Jethro, who saw him at a Decatur, Illinois club and promptly relayed their enthusiasm to Chet Atkins. Later, Atkins was to hear a tape of Randolph playing **Chicken Reel** in his good-timey, slap-tongued style, and was similarly impressed. He invited Randolph to Nashville and enthused about the sax-man's talents to his friends. Within a few days, Owen Bradley hired him for a Brenda Lee session. From then on he became a much-sought-after sessioner and an established part of the 'Nashville Sound', recording as a solo act for both RCA and Monument.

In 1963 he obtained a major hit with **Yakety Sax**, following this with such other pop successes as **Mr Sax Man** (1964), **The Shadow Of Your Smile** (1966) and **Temptation** (1967). But perhaps the most interesting Randolph disc, at least to country fans, is **Country Boots,** a 1974 album made in the company of Maybelle Carter, Chet Atkins, Uncle Josh Graves and a host of other pickers. The majority of his other album releases fall outside the scope of this book.

Recommended:
Yakety Sax (–/Bear Family)

Rattlesnake Annie

A singer whose stage name comes from the habit of wearing a rattlesnake's rattler on her ear, her real name is Annie McGowan and she was born Ann Gallimore on December 26, 1941 in Puryear, Tennessee, of Cherokee Indian heritage. From a poor family who worked in the tobacco and cotton fields near Puryear, she grew up in a music-making home environment, Annie singing and playing piano at the local church. At the age of 12, she and two cousins formed the Gallimore Sisters and appeared as regulars on the Junior Grand Ole Opry in 1954, but when she married Max McGowan her musical ambitions had to take a back seat.

Later, Annie headed for Memphis where she sang the blues on Beale Street, travelling through the South and ending up in Austin, where she befriended Willie Nelson, Billy Joe Shaver and David Allan Coe. Coe co-wrote **Texas Lullaby** with Annie and recorded it on his 1976 **Long Haired Redneck** album.

The first woman from the West to record a country album in Czechoslovakia, she has a cult following in Europe and has made a brace of highly acclaimed appearances at the UK's Wembley Festival. She released a brilliantly conceived album, **Rattlesnakes And Rusty Water**, on her own Rattlesnake label during 1980. Featuring such guests as John Hartford, Josh Graves, Vassar Clements and Charlie McCoy, the album merged blues and traditional country in a fashion Jimmie Rodgers would have approved of. Signed to Columbia Records in 1987, Annie scored some minor country hits, then returned to doing things her own way, marketing her own recordings, which include the albums **Indian Dream** and **Rattlesnake Annie Sings Hank Williams**. For several years she and her husband have lived in Spain, and more recently she has taken her traditional country-blues music to Japan.

Recommended:
Rattlesnakes And Rusty Water (Rattlesnake/–)
Country Livin' (Rattlesnake/–)
Sings Hank Williams (Montana/–)
Rattlesnake Annie (Columbia/CBS)

Below: During the '60s, Eddy Raven played in blues bands headed by Edgar and Johnny Winter.

Eddy Raven

Eddy was born Edward Garvin Futch on August 19, 1944 in the bayou country of Lafayette, Louisiana, one of nine children whose father travelled all over the South as a musician and trucker. At seven Eddy had his first guitar and at 13 his first band. Although his father geared him towards country music, Eddy followed the prevailing rock'n'roll path, even though he was concerned about lyrics.

He lived in Georgia and had his own radio slot on WHAB, even gaining a local hit on the Cosmo label. Then his father moved the family back to Lafayette, Louisiana, where Eddy met Lake Charles record entrepreneur, Bobby Charles, and wrote a country-blues hit for him which sold 60,000 on Charles' label.

After an experimental period in the 1960s when Eddy performed all around the Gulf Coast (at one time playing with albino brothers Edgar and Johnny Winter), he tired of travelling and, in 1970, went to Nashville where fellow Cajun Jimmy 'C' Newman put him on to Acuff-Rose.

Eddy's first writing success was **Country Green**, which provided a hit for Don Gibson. He did likewise for Jeannie C. Riley (**Good Morning Country Rain**) and Don Gibson again (**Touch The Morning**). A one-time lounge performer at Nashville's King Of The Road Inn, he became a headliner and grabbed the attention of ABC Records, a label for whom he began providing hits in 1974 (starting with **The Last Of The Sunshine Cowboys**), continuing in a moderately successful way through to 1976.

In an effort to get his own recording career into higher gear, Eddy signed for Monument in 1978, gaining just one mini-hit with **You're A Dancer**. He switched again, this time to Dimension, and at last began really chart-climbing. One 1980 single, **Dealin' With The Devil**, gained him access to the Top 25 for the first time. A year later he was on Elektra and reaching even higher through releases such as **I Should Have Called** and **Who Do You Know in California?**.

By 1982, 12 years after making the trip to Nashville, he made his Top 10 debut with **She's Playing Hard To Forget**. But the long hard climb to the very top was not achieved until 1984, when Raven, now with yet another record company, RCA, got to No.1 with **I Got Mexico**.

Raven continued his hit-making formula throughout the '80s and into the '90s with many Top 10 hits and a trio of No.1s with **Shine, Shine, Shine** (1987), **I'm Gonna Get You** and **Joe Knows How To Love** (1988). He then joined Universal Records, chalking up two more No.1s with **In A Letter To You** and **Bayou Boys** (1989). Universal was then absorbed into Capitol/Liberty, and Raven continued with **Sooner Or Later** and **Island**, two more Top 10 entries in 1990.

Recommended:
I Could Use Another You (RCA/–)
That Cajun Country Sound (La Louisianne/–)
This Is Eddie Raven (Dot/–)
Eyes (Dimension/–)
Desperate Dreams (Elektra/–)
Right For The Flight (Capitol/–)
Temporary Sanity (Universal/–)
Right Hand Man (RCA/–)

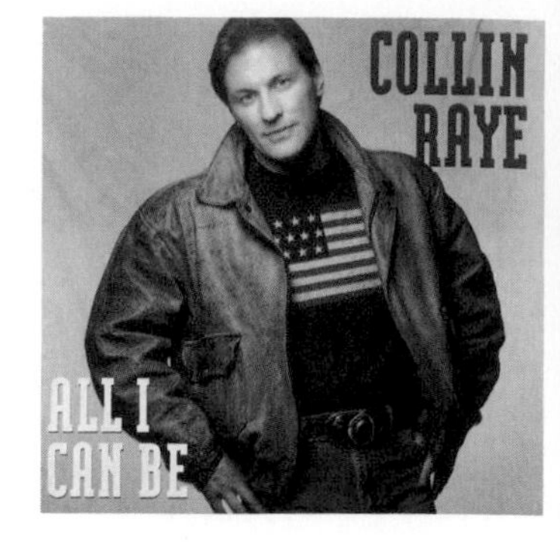

All I Can Be, Collin Raye. Courtesy Epic Records.

Collin Raye

The sweetest high-flying tenor to hit country music in years, Collin Raye (born in 1961 in Texarkana, Arkansas) has country music in the blood. Raised in Texas, his mother Lois was a regional star in the '50s, sharing bills on the back of flat-bed trucks with artists like Elvis Presley, Johnny Cash, Jerry Lee Lewis and Carl Perkins. His father played guitar and an uncle was a professional musician. Collin started singing harmony on stage when he was seven. He spent his teenage years in a band with brother Scott, playing in the clubs in Oregon in the Pacific Northwest. When he was 20 he married and briefly moved back to Texas. Then he put together a show band, playing a wide mixture of music in the casinos in Las Vegas and Reno. His band landed a recording contract with Mercury Records in 1985, released several singles, and completed an unreleased album. At this time his personal life was turned upside-down when in 1985 his wife Connie developed complications during pregnancy with the couple's second child and she lapsed into a two-month coma. The child, Jacob, was born with cerebral palsy, and Connie had to re-learn many basic skills. Collin was trying to hold down a seven-night-a-week club date in Reno. The couple later divorced, but have remained friends and the singer is very close to his two children, Britanny and Jacob, who live with their mother in Greenville, Texas.

A move to Nashville in 1989 eventually led to a recording contract with Epic Records. His first single, **All I Can Be (Is A Sweet Memory)**, made No.29 on the country charts in 1991. His debut album, also titled **All I Can Be**, was released virtually simultaneously with the single, rare for a new name in country music. Initially, the album was a sleeper, but the single's video spent weeks at the top of CMT's Chart, and the album was named one of the ten best of 1991 by 'USA Today'. Early the next year a second single, **Love Me**, a heart-rending tale of enduring romance, raced up the charts, spending three weeks at No.1. The song, again accompanied by a superb video, struck a chord with music fans. Collin Raye's 'career song', **Love Me** boosted the sales of his album, eventually gaining the singer a gold disc. A dynamic stage performer, Raye tears up his audience with a powerful mix of soaring rockers, dramatic ballads and, harking back to his Las Vegas days, a few Eagles' songs. He has continued with the country hits with **Every Second** (1992), **In This Life** (No.1 1992), and further Top 10 entries in **I Want You Bad (And That Ain't Good)** and **Somebody Else's Moon** (1993).

Recommended:
In This Life (Epic/Epic)
All I Can Be (Epic/Epic)

Susan Raye

A long-time regular on Hee Haw and the Buck Owens Show, singer Susan Raye had a number of hit singles during the late 1960s and early 1970s. Once a member of a rock group, Raye (born in Eugene, Oregon in 1944) switched to country after winning a regular spot with a radio station that was looking for a country singer. Later, after stints as a DJ and night-club singer, she was invited to Bakersfield to meet Buck Owens, whose show she joined. She also gained a contract with Capitol. Her first release, **Maybe If I Closed My Eyes** (1969), was an Owens original.

Her other hits have included a series of duets with Owens, plus such Top 20 solo efforts as: **L.A. International Airport**, **Pitty, Pitty, Patter** (1971), **My Heart Has A Mind Of Its Own**, **Wheel Of Fortune**, **Love Sure Feels Good In My Heart** (1972), **Cheating Game** (1973), **Stop The World (And Let Me Off)** and **Whatcha Gonna Do With A Dog Like That** (1974), all on Capitol. Later she signed with UA, without much success. But by 1984, her name still appeared in the charts, aboard **Put Another Notch In Your Belt**, a Westexas release.

Below: Collin Raye and his second wife Tammie. Raye burst on the scene in the early '90s with such hits as Love Me.

Above: Susan Raye had a number of hit singles in the '70s.

Recommended:
Best Of (Capitol/–)
Whatcha Gonna Do With A Dog Like That (Capitol/–)

Jerry Reed

One of Nashville's most remarkable guitarmen, Reed was born Jerry Hubbard in Atlanta, Georgia on March 20, 1937. A cotton mill worker in his early days, he began playing at local Atlanta clubs, obtaining a record contract with Capitol in 1955 and cutting some rockabilly tracks. However, these made little impact, Reed's first claim to fame arising through his songwriting ability, predominantly with **Crazy Legs**, which Gene Vincent waxed in 1956. Following a two-year stint in the forces, Reed then settled in Nashville, there providing Columbia with two minor 1962 hits in **Goodnight Irene** and **Hully Gully Guitars**.

Establishing himself as a superior session man, Reed was signed to RCA as a solo act in 1965, his first hit for the label being the rocking **Guitar Man** (1967), which Elvis Presley covered in 1968. That same year, Presley also scored with **US Male**, another Reed composition. The Georgian's own records all sold increasingly well, **Tupelo Mississippi Flash** (1967), **Remembering** (1968), **Are You From Dixie?** (1969), **Talk About The Good Times** (1970) and **Georgia Sunshine** (1970) all reaching high chart positions. Then, late in 1970, came **Amos Moses**, one of Reed's swamp-rock specials. The song proved to be a Top 10 pop hit and resulted in Reed being nominated CMA Instrumentalist Of The Year. He also won a Grammy for Best Country Male Vocal Performance of 1970.

A maker of somewhat erratic albums, his guitar duet LPs with Chet Atkins, cut during the mid-1970s, resulted in some fine music. Additionally, he has become much acclaimed as an actor, an appearance in 'WW And The Dixie Dance Kings' (1974) resulting in an association with Burt Reynolds that has since seen the duo pairing in 'Gator' (1976), 'Smokey And The Bandit' (1977) and 'Smokey And The Bandit II' (1980). A performer with a larger-than-life personality, Reed has also starred alongside Claude Akins in a TV series 'Nashville 99' and has continued his hit single way with such Top 10 records as **When You're Hot, You're Hot** (No.1, 1971), **Lord Mr Ford** (No.1, 1973), **(I Love You) What Can I Say?** (1978), **She Got The Goldmine (I Got The Shaft)** (No.1, 1982) and **The Birds** (1982). 1984 proved to be the first year since 1967 that failed to provide Reed with chart action of any kind.

Reed concentrated more on his acting than music during the mid-'80s, making several film appearances and starring in his own TV series, Concrete Cowboys. After a long absence he returned to the studios in 1992 to cut **Sneakin' Around**, a guitar duet album with Chet Atkins.

Recommended:
Me And Chet – with Chet Atkins (RCA/RCA)
20 Of The Best (–/RCA)
Sneakin' Around – with Chet Atkins (Columbia/–)

Ko-Ko Joe, Jerry Reed. Courtesy RCA Records.

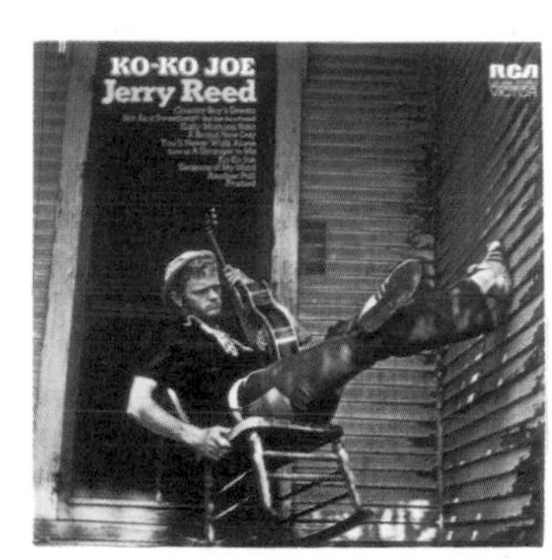

Live At The Palomino Club, Del Reeves' 1974 album. Courtesy UA Records. Del was with United Artists from his first No.1 in 1965 until 1978.

Del Reeves

Singer-songwriter, multi-instrumentalist Franklin Delano Reeves was born in Sparta, North Carolina on July 14, 1934. At the age of 12 he had his own radio show in North Carolina. Then, after attending Appalachian State College and spending four years in the Air Force, he became a regular on the Chester Smith TV show in California.

By the late 1950s, Reeves had his own TV show which he fronted for four years before moving to Nashville, signing for Frank Sinatra's Reprise label and writing songs with his wife Ellen Schiell Reeves – these being recorded by Carl Smith, Sheb Wooley, Roy Drusky and others. Reeves' own initial hit single came with **Be Quiet Mind**, a 1961 Decca release. But, despite label changes and a couple of minor chart entries, it was not until 1965 and a contract with UA that he obtained his first No.1 with **Girl On The Billboard**.

In October 1966, just four hits later, Reeves became a member of the Grand Ole Opry. His reputation as a hit artist was maintained by a stream of chart singles that included such Top 10 records as **Looking At The World Through a Windshield**, **Good Time Charlies** (1968), **Be Glad** (1969) and **The Philadelphia Fillies** (1971). Del continued supplying such hits for UA right through to 1978, cutting duets with Penny DeHaven and Billie Jo Spears.

By 1980 he was on the Koala label and back in the charts once more, logging minor hits – the biggest of these being **Slow Hand** (1981) – through to 1982.

Del Reeves has often been tagged the 'Dean Martin of Country Music' because of his laid-back stage manner. The multi-talented Reeves has appeared in such movies as 'Second Fiddle To A Steel Guitar', 'Sam Whiskey', 'Cottonpickin' Chickenpickers' and 'Forty Acre Feud'. He also has managed Billy Ray Cyrus.

Recommended:
Live At The Palamino (UA/–)
10th Anniversary (UA/–)
By Request – with Billie Joe Spears (UA/–)
Baby I Love You (–/Bear Family)

Goebel Reeves

An early-century Woody Guthrie figure, Reeves, known as the 'Texas Drifter', specialized in songs of hobos and hard times, writing from his own experience on the road.

Born in Sherman, Texas, on October 9, 1899, he came from a solid middle-class background but chose to live a rough travelling life. After a stint in the US Army, during which time he saw front-line action in World War I, he wandered around the USA, noting the things he saw and eventually turning his thoughts into songs, some of which he recorded for the Okeh and Brunswick labels. He played in vaudeville, the main outlet for music at that time, and also did some radio work. He claimed to have taught Jimmie Rodgers his yodelling style. His songs include **Hobo's Lullaby**, **Hobo And The Cop**, **Railroad Boomer**, **Bright Sherman Valley** and **Cowboy's Prayer**. A one-time member of the Workers Of The World Organization, Goebel Reeves died in California in 1959.

The Abbott Recordings, Jim Reeves. Courtesy RCA Records.

Jim Reeves

Originally a stone country singer, smooth-toned Jim Reeves from Texas reached amazing heights as a pop ballad singer and since his death in an air crash his fame has burgeoned into cult proportions.

James Travis Reeves was born on August 20, 1923, in Galloway, Panola County, Texas. His father died when he was young and his mother supported a large family by working in the fields. Early in his life he heard the sound of Jimmie Rodgers. He acquired a guitar at the age of

Below: Jim Reeves' posthumous hits outnumbered those during his life. A duet with Patsy Cline was recorded in 1981, long after they had both died.

RCA
International
DISTANT DRUMS
JIM REEVES
I Missed Me / Is It Really Over / This Is It / Good Morning Self / Losing Your Love
Not Until the Next Time / Where Does a Broken Heart Go / Snow Flake / Distant Drums
A Letter to My Heart / Overnight*/ The Gods Were Angry with Me*
*Stereo Electronically Reprocessed

Distant Drums, Jim Reeves. Courtesy RCA Records.

five; it had strings missing but an oil construction worker fitted it up for him and taught him some basic chords. At nine he made his first radio broadcast, a 15-minute programme on a Shreveport station.

At high school in Cathage, Texas, he was just as interested in sport as in music and became star of the school baseball team, although he still performed at local events. He entered the University of Texas in Austin, and his baseball prowess as a pitcher soon attracted the attention of the St Louis Cardinals scouts who signed him up. But an unlucky slip gave him an ankle injury that was to halt his career.

In 1947 he met and married schoolteacher Mary White, who encouraged his musical interest. Jim had studied phonetics and pronunciation and now sought a job in radio, becoming a DJ and newsreader at station KGRI in Henderson, Texas. (He later bought the station.)

Jim then made a momentous journey. He and Mary, having decided to make a determined effort to further Jim's career, drove to the crossroads of Highway 80 in Texas. They tossed a coin to determine whether to proceed to Dallas or Shreveport. Shreveport won and Jim moved there, ending up with a job as announcer on KWKH, the station that owned the Louisiana Hayride.

It was one of Reeves' jobs to announce the Saturday night Hayride show and he was even allowed to sing occasionally. One night in 1952, Hank Williams failed to arrive and Jim was asked to fill in. In the audience was Fabor Robinson, owner of Abbott Records, who immediately signed Reeves to a contract. Jim had already made four obscure sides for the Macy label – these records only coming to light in 1966. However, his Abbott deal soon began bearing fruit, via **Mexican Joe**, his second release, which went to No.1 in the country charts during 1953. That same year he also released **Bimbo**, one of the 36 tracks he recorded for Abbott. This too sold well and attracted the attention of RCA who signed him in 1955 amid considerable competition. That same year, he joined the Grand Ole Opry at the recommendation of Ernest Tubb and Hank Snow.

A string of country hits followed and from the release of **Yonder Comes A Sucker** in 1955 through to 1969, Reeves' name was never absent from the country chart. The February, 1957 release of **Four Walls** proved the real turning point. That year, Reeves had undertaken a European tour in the company of the Browns, Del Wood and Hank Locklin and was unaware the song had hit both pop and country fields, earning him his third gold disc. He returned to find radio and TV offers in abundance, including a spot on NBC-TV's prestigious Bandstand Show. He also gained his own daily show on ABC-TV.

In the wake of **Billy Bayou**, a 1958 No.1, Reeves recorded his all-time greatest hit, **He'll Have To Go**, a 1959 chart-buster. The theme was familiar enough. Some years earlier it might have been called a honky-tonk song. But the treatment, with Reeves' dark, intimate, velvet tones gliding over a muted backing, was something different again. The result brought him international stardom.

Over the next few years, Jim travelled to every state in America and to most parts of the world. He toured South Africa with Chet Atkins and Floyd Cramer.

During 1963 he returned to South Africa to star in his only film, 'Kimberley Jim', the story of a con man in South Africa's diamond strike era. He had not toured British venues in these years because of Musicians' Union restrictions, but in 1964 he arrived in Britain for some TV dates and to promote his current single, **I Love You Because**. During the early 1960s, he also continued to dominate the US country charts; some of his many hits during this period include: **I'm Getting Better** (1960), **Losing Your Love** (1961), **Adios Amigo** (1962), **I'm Gonna Change Everything** (1962), **Is This Me?** (1963), **Guilty** (1963) and **Welcome To My World** (1964). But on a flight back to Nashville from Arkansas on July 31, 1964, following the negotiation of a property deal, Jim and his manager Dean Manuel reported that their single-engine plane had run into heavy rain while crossing remote hills just a few miles from Nashville's Beery Field airport. The plane was making its approach to land when it disappeared from the airport radar screen. A search was instigated involving 12 planes, two helicopters and a ground party of 400. But it was not until two days later that the wreckage and the bodies were discovered amid thick foliage.

On August 2, 1964, services were held for Reeves and Manuel, most of the country music fraternity being present. Jim's body was flown back to Cathage where hundreds filed past the coffin. Honorary pall bearers included Chet Atkins and Steve Sholes, the man who had signed him to RCA.

But the legend lived on and Reeves' records continued to hit the charts, his after-death No.1s out-numbering those made while he was alive. Voted into the Country Music Hall Of Fame in 1967, Reeves continued to log hits posthumously as 1970s moved in. And even in the 1980s, Reeves' name has cropped up in the Top 10 via electronically created duets with Deborah Allen **(Take Me In Your Arms And Hold Me** – 1980) and Patsy Cline **(Have You Ever Been Lonely?** – 1981).

20 Of The Best, Jim Reeves. Courtesy RCA Records.

Recommended:

The Abbott Recordings Volumes 1 & 2 (–/RCA)
50 All-Time Worldwide Favourites – four album set (RCA/RCA)
Gentleman Jim, 1955–1959 (–/Bear Family)
Good 'n' Country (Camden/Camden)
Jim Reeves On Stage (RCA/RCA)
Songs From The Heart (RCA/Pickwick)

Mike Reid

A former All-American football player for Penn State University, Mike Reid (born 1948, Altoona, Pennsylvania) initially made an impact in country music as a songwriter. He was the recipient of ASCAP's Songwriter Of The Year award in 1985, and two of his songs Ronnie Milsap recorded – **Stranger In My House** and **Lost In The Fifties Tonight** – earned Grammy awards. Other successes as a writer include **One Good Well** (Don Williams), **Love Without Mercy** (Lee Roy Parnell), **Born To Be Blue** (the Judds), **There You Are** (Willie Nelson) and **He Talks To Me** (Lorrie Morgan).

Reid received his degree in music in 1970, but he was a star football player, winning the Outland Trophy as outstanding college football player in 1969, and he was a first-round draft pick for the Cincinnati Bengals, being named Rookie Of The Year in 1971. In between playing pro ball, he had made some off-season appearances with symphony orchestras in Cincinnati, Dallas and San Antonio. This whetted his appetite, and in 1975 he quit the Bengals and began touring as a keyboard player for the Apple Butter Band. A few months later he formed his own group, but after a year on the road, Reid split away and continued touring as a solo act. He worked the listening-room circuit up and down the East Coast.

In concentrating on his performing, Reid also worked more diligently on his songwriting. One of his demos caught the ear of a Nashville publisher, who offered him a job as a staff writer. So, in 1980, Reid and his family moved to Music City. Initially, he played the clubs, but as more of his songs were recorded the financial pressure was lifted and he settled in as a professional writer, working closely with Ronnie Milsap, providing him with one hit song after another. Reid sang a duet with Milsap on **Old Folks**, which went Top 5 on the country charts in 1988, but, ultimately, he and Milsap parted company. After taking stock, Reid decided to return once more to performing, and in 1990 signed a recording contract with Columbia, producing the album, **Turning For Home**, with a first single, **Walk On Faith**, hitting No.1 on the country charts in 1991.

With his songs concerning love and romance, many fueled by his love for his wife Susan, Reid scored further country hits with **Till You Were Gone**, **As Simple As That** (1991), **I'll Stop Loving You**, **Keep On Walkin'** and **Call Home** (1992), while providing songs for Bonnie Raitt, Gene Watson, Glen Campbell and Barbara Mandrell. Reid also found time to make his mark in the theatrical world with 'A House Divided', a musical saga he co-wrote, which debuted at Nashville's Performing Arts Center in January 1991.

Recommended:
Turning For Home (Columbia/–)
Twilight Town (Columbia/–)

The Remingtons

With their mature vocals and slick instrumentation, the Remingtons have made a big impression on the country scene of the early '90s with their carefully crafted albums and hit singles. Both individually and in various artistic configurations, the three members, Jimmy Griffin, Richard Mainegra and Rick Yancey, have weather the ups and downs of the music business for more than two decades.

Griffin was a founder member of '70s pop group Bread, which scored such hits as **Everything I Own**, **Make It With You**, **If**, plus six gold albums. He also won an Academy Award for co-writing **For All We Know**, from the 1971 film 'Lovers And Strangers'. When Bread split he embarked on a solo career, then in the mid-'80s he linked up with Randy Meisner and Billy Swan to form Black Tie, who produced a sleeper album **When The Night Falls** for Bench Records in 1988. The album's lushly romantic **Learning The Game** became a country hit in 1990.

At the time Bread were knocking out the hits, Mainegra and Yancey were in Memphis putting together a band called Cymarron. While it would prove to be a short-lived act, their first single, **Rings**, went to the pop Top 20 in 1971, while Tompall And The Glaser Brothers turned the song into a Top 10 country hit that same year. Mainegra relocated to Nashville to write songs and sing jingles, penning hits for Elvis Presley, Tanya Tucker, Reba McEntire and others. Yancey also maintained a Nashville connection, working as a session musician.

As the story goes, producer Josh Leo went out to a friend's home one Sunday afternoon, to listen to three songwriters who were in the process of working up some new material. The three were sitting in a circle, each picking acoustic guitars and harmonizing. Leo phoned RCA Nashville President Joe Galante and called him in to see the trio. After hearing the first song, Galante declared, "You got a deal – play another one."

The Remingtons' authentic blend, with its multiple chimes of lead, alto and high tenor wrapped snugly around alluring self-written material, was immediately radio-friendly, with the debut single, **Long Time Ago**, making the country Top 10 in 1991. Further hits have followed with **I Could Love You (With My Eyes Closed)**, **Two-Timin' Me** (1992) and **Nobody Loves You When You're Free** (1993).

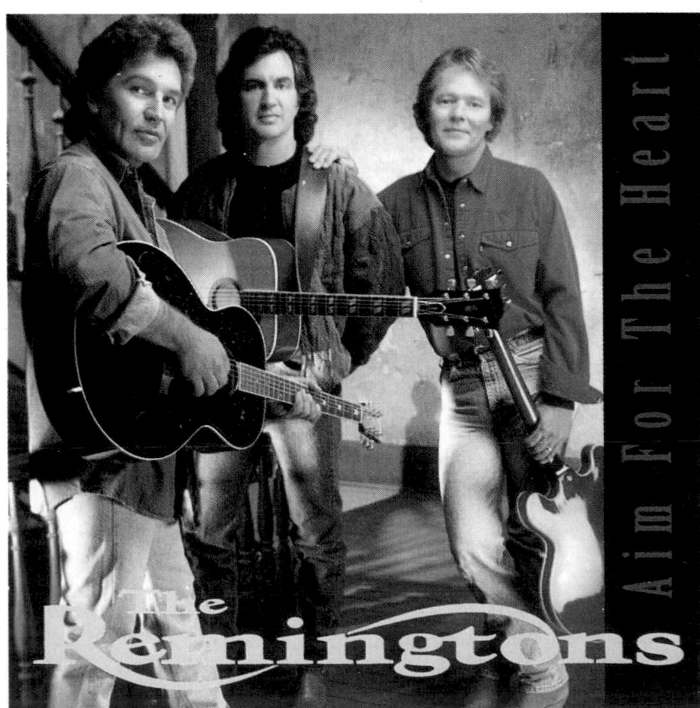

Aim For The Heart, the Remingtons. Courtesy RCA Records.

Recommended:
Blue Frontier (RCA/–)
Aim For The Heart (RCA/–)

Above: Restless Heart was reduced to a trio in 1993.

Restless Heart

Creating a sound which is a cross between contemporary music and the traditional country, Restless Heart's blending of lush country harmonies, folk-pop melodies and rock guitar resulted in a string of chart-topping country singles, five gold albums and being named the ACM's Vocal Group Of The Year in 1990.

Formed in 1984 when producer Tim DuBois brought together five studio musicians, John Dittrich, Paul Gregg, Dave Innis, Greg Jennings and Larry Stewart, to form a new country-rock band. Restless Heart's name was chosen after each member submitted 50 possible names, which were all discarded in favour of the title of a song they had previously recorded. Signed to RCA Records, the first single, **Let The Heartache Ride**, went to No.23 on the country charts in 1985. All subsequent releases have gone Top 10 with seven hitting No.1, including: **That Rock Won't Roll** (1986); **I'll Still Be Loving You** (which also crossed into the pop charts), **Why Does It Have To Be (Wrong Or Right)**, and **Wheels** (all 1987); **The Bluest Eyes In Texas**, **A Tender Lie** (1988); and **Fast Moving Train** (1990). Alongside these successes, Restless Heart have also been a major album act with **Wheels**, **Big Dreams In A Small Town** and **Fast Moving Train** going gold, while **The Best Of Restless Heart** gained them a platinum award.

From the very beginning the distinctive sound of Restless Heart has revolved around the forceful, sad-tinged lead vocals of Larry Stewart. The other four members always provided two, three and four-part harmonies behind Stewart, but never took the lead. At the beginning of 1992 Stewart announced that he would be leaving the band to pursue a solo career. Initially the remaining members started looking for a new lead singer, then decided to share the vocal honours among themselves. They produced a new studio album, **Big Iron Horses**, immediately scoring a Top 10 country hit with **When She Cries**, which also gained a Grammy nomination. The band were then dealt a double blow when keyboardist Dave Innis departed at the beginning of 1993. Now down to a three-piece, with only drummer John Dittrich, lead guitarist Greg Jennings and bassist Paul Gregg remaining from the original five members, Restless Heart scored another Top 20 country hit with **Mending Fences**, while **Big Iron Horses** crossed into the pop charts, earning them another gold disc. The three partners hired keyboardist Dwain Rowe and guitarist Chris Hicks to flesh out the Restless Heart sound, and they have maintained a gruelling road schedule.

Classic Rich, Charlie Rich. Courtesy UA Records.

Recommended:
Big Iron Horses (RCA/RCA)
Fast Movin' Train (RCA/RCA)
Big Dreams In A Small Town (RCA/RCA)

Charlie Rich

Born in Forrest City, Arkansas on December 14, 1932, the 'Silver Fox' (his hair turned prematurely white at 23) was the son of a hard-drinking father and a Bible-thumping mother. Rich's high school and University of Arkansas era saw him heavily influenced by jazz and blues. He studied music formally at college and when the USAF posted him to Oklahoma in the early 1950s, one of his first groups, the Velvetones, secured a spot on local TV. Upon his discharge, Rich moved to West Memphis, Arkansas, to work on his father's cotton farm. After sitting in one night with Bill Justis' band, Rich was invited to Sam Phillips' Memphis studio to lay down some trial tracks, but was told he was too jazzy.

After playing sessions with Warren Smith, Ray Smith and Billy Lee Riley, Rich landed his own rockabilly hit in 1959 – **Lonely Weekends**. The demise of Sun saw Rich without a record label and Bill Justis persuaded him to sign for Groove, an RCA subsidiary, in 1963. From this period came a foot-stomping hit single, **Big Boss Man**.

In 1965 Rich moved to Smash, where Shelby Singleton encouraged Rich to utilize both rock and country influences. That year he scored another hit with **Mohair Sam**.

After an unproductive stint with Hi, Epic signed him in 1968 and made him part of their modern country push under producer Billy Sherrill. However, even Sherrill had trouble stimulating more than average sales with Rich, though local critical acclaim was voiced for **Raggedy Ann** and **I Almost Lost My Mind**.

But in 1972, the smoothly soulful country **I Take It On Home**, backed with **Peace On You**, proved a country and pop hit and was nominated for a Grammy award. The album **The Best Of Charlie Rich** swept up many of the earlier Epic titles, re-presented them to the public and suddenly Rich was a big country name. But success really came with the next album, **Behind Closed Doors**. By this time, Sherrill and Rich had become a winning combination and the new urbane country sound known as 'countrypolitan' became the talk of the 1973 CMA awards. Rich was topping the country charts with such singles as **Behind Closed Doors** and **The Most Beautiful Girl** (both 1973 No 1s). Rich was so big that, when RCA released his old **There Won't Be Anymore** single, it also went to No.1 (1973).

In 1974, Charlie had four chart-toppers in a row, two (**A Very Special Love Song**, **I Love My Friend**) stemming from Epic, and two (**I Don't See Me In Your Eyes Anymore**, **She Called Me Baby**) from RCA. He also won CMA's Entertainer Of The Year in 1974.

There were further Top 10 records in 1975, but gradually interest waned. It looked as though Rich might be running out of time, but in 1977, following a Top 20 record in **Easy Look**, he returned to the top of the charts with **Rollin' With The Flow**. By 1978 he was as massive a seller as ever, with four hits on two different labels (Epic and UA) – his No.1 that year, **On My Knees**, achieved with the vocal aid of Janice Fricke. By 1979 Charlie had signed with Elektra and had a Top 10 record in **I'll Wake You Up When I Get Home**. But it was difficult to ascertain exactly who he was working for. He had five other hits that same year, four for UA and one for Epic. Perhaps there were just too many Rich records around, but by the mid-'80s the fans had tired of buying his records, so he just stopped making them.

In 1992 a new Charlie Rich album, **Pictures And Paintings**, a mixture of jazzy original and new versions of songs from his past, suddenly surfaced.

Recommended:
Behind Closed Doors (Epic/Epic)
Boss Man (Epic/Epic)
Classic Rich (Epic/–)
Nobody But You (UA/UA)
Silver Linings (Epic/–)
Pictures And Paintings (Warner Bros/Warner Bros)
American Originals (Columbia/–)
Original Hits And Midnight Demos (–/Charly)

Jeannie C. Riley

An international star on the strength of just one record – a multi-million-selling version of Tom T. Hall's **Harper Valley PTA** (1968) – Jeannie (born in Anson, Texas, on October 19, 1945) had only minimal experience in the entertainment industry prior to her arrival in Nashville.

In Music City she worked as a secretary for some time, cutting a few demo discs, but generally had little success until Shelby Singleton signed her to launch his new Plantation label with **Harper Valley PTA**, an ably constructed song dealing with small town hypocrisy. An immediate hit which sold four million copies in the US alone, the single sparked off an album which qualified for a gold disc, while Jeannie C. was awarded a Grammy as Best Female Country Vocalist of 1968.

The Girl Most Likely (1968), **There Never Was A Time** (1969), **Country Girl** (1970), **Oh Singer** (1971) and **Good Enough To Be Your Wife** (1971) were other Plantation releases that achieved Top 10 status. But, following a switch to MGM in 1971, sales of her discs began to taper off and it's now been a long time since she had a major record, although her name appeared on the charts in 1974 with **Plain Vanilla** and in 1976 with **The Best I've Ever Had**. A born-again Christian, these days she exclusively records gospel songs and only plays dates where the sale of alcohol is banned.

Recommended:
Harper Valley PTA (Plantation/Polydor)
Jeannie (Plantation/Polydor)
Total Woman (–/Sundown)
Here's Jeannie C. (Playback/Cottage)

Tex Ritter

Born near Murval, in Panola County, Texas on January 12, 1905, Woodward Maurice Ritter aspired to a career in law at the University of Texas and at Northwestern before heading for a career on the Broadway stage, where he appeared in five plays in the early 1930s, including 'Green Grow The Lilacs' in 1930. During his New York years, he also appeared as a dramatic actor on radio's popular Cowboy Tom's Round-Up and co-hosted the WHN Barn Dance with Ray Whitley, making his first records for ARC in 1934. One of the first to follow Gene Autry into films as a singing cowboy, Tex moved to Hollywood in 1936, where he was to star in some 60 films for Grand National, Monogram, Columbia, Universal and PRC up to 1945. And, after several unsuccessful years as a Decca recording artist, he was the first singer to sign with the new Capitol label in 1942, providing a long string of hits for them, thus becoming one of country music's biggest sellers of the 1940s.

When his film career declined, he turned to touring. In addition, he and Johnny Bond co-hosted 'Town Hall Party' from 1953 to 1960, during which time his rendition of the theme-song for the film 'High Noon' won an Academy Award (1953).

Ritter moved to Nashville in 1965, where he joined the Grand Ole Opry and took over a late night radio programme on WSM. He acted on a long-standing desire to run for political office, when he stood (unsuccessfully) for the US Senate in 1970.

A lifelong student of western history, he was instrumental in setting up the Country Music Foundation and the Country Music Hall Of Fame, to which he was elected in 1964. He did not have a great voice, but his unusual accent, odd slurs and phrasing, allied to a strong feeling of genuine honesty, made his voice one of the most appealing in country music history. His long string of hits included: **Jingle Jangle Jingle** (1942), **Jealous Heart** (1944), **There's A New Moon Over My Shoulder** (1944), **I'm Wasting My Tears On You** (1945), **You Two Timed Me Once Too Often** (1945), **Rye Whiskey** (1945), **Green Grow The Lilacs** (1945), **High Noon** (1952), **The Wayward Mind** (1956) and **I Dreamed Of A Hillbilly Heaven** (1961). Tex died on January 2,

Fall Away, Tex Ritter. Courtesy Capitol Records.

Above: **In 1982, after releasing Some Memories Just Won't Die, Marty Robbins died of a heart attack.**

1973, after a heart attack at the Metro Jail, Nashville (where he was arranging bail for one of his band) and was dead on arrival at Baptist Hospital.

Recommended:
An American Legend (Capitol/–)
Blood On The Saddle (Capitol/Capitol)
Songs Of The Golden West (Capitol/MFP)
High Noon (–/Bear Family)
Capitol Collectors Series (Capitol/Capitol)
Songs From The Western Screen (Capitol/Stetson)
Country Music Hall Of Fame (MCA/–)

Marty Robbins

In a music where 'western' has often been considered a misnomer, Marty Robbins emphasized the Western of C&W through a series of memorable cowboy/Mexican-style ballads, many of them crossing over to the pop fields.

Born in Glendale, Arizona, on September 26, 1925, he grew up in a desert area. His earliest musical recollections involved his harmonica-playing father and the songs and stories of his grandfather, Texas Bob Heckle, a travelling medicine man. Many of Robbins' own songs, such as **Big Iron**, owed much to his grandfather's tales.

Influenced by the films of Gene Autry, Robbins developed an ambition to become a singing cowboy. Following a three-year term of service in the Navy, he began playing clubs in the Phoenix area, also appearing on radio station KPHO. Soon he gained his own TV show, Western Caravan, at one time having Little Jimmy Dickens as guest. Dickens was so impressed by Robbins' performance that he contacted Columbia Records who immediately signed him, releasing **Love Me Or Leave Me Alone**, in 1952.

With his third Columbia release, **I'll Go On Alone**, Robbins hit the country music Top 10 in 1953, followed by another, **I Couldn't Keep From Crying**. He had already guested on the Grand Ole Opry and in 1953 became a regular on the show, celebrating his signing with two 1954 hits in **Pretty Words** and **That's All Right**, the latter being the same Arthur Crudup up-tempo blues with which Elvis Presley made his name. Robbins thus established a rock'n'roll connection that haunted him for many years, following up by recording the hummable **Singing The Blues**, **Knee Deep In The Blues**, **The Story Of My Life**, **White Sport Coat** and **Teenage Dream**. Robbins was a convincing rock'n'roller, a fact not lost on the British Teddy Boy fraternity, who tended to shout "You gotta rock, Marty" at Wembley Festivals when it became apparent that the singer was bent on providing an exclusively country set.

The cross-over hits continued, at least in America, where Robbins had success in 1958 with **She Was Only Seventeen** and **Stairway Of Love**. In 1959 came his biggest ever, **El Paso**, the lyrics of which would have made a convincing Western film and the rhythm of which slipped insistently along, tinged with Mexican nuances. This was the archetypal Robbins, purveying a mixture of macho western feel and melodic sentiment in a dramatically powerful but held-back voice.

Following **Don't Worry**, a 1961 country No.1, **Devil Woman** brought the same response as **El Paso** among pop fans and again they brought Robbins into the pop charts on both sides of the Atlantic. That same year (1962) he again climbed high on the pop charts with **Ruby Ann**. From then on, his wares were to be found in the country charts only, where he logged hits from 1956 through to his death, topping the chart with **Begging To You** (1963), **Ribbon Of Darkness** (1965), **Tonight Carmen** (1967), **I Walk Alone** (1968), **My Woman, My Woman, My Wife** (1970) during his first stay with Columbia.

He signed with Decca/MCA in 1972 and stayed with the label for three years, having minor hits. In 1976 he returned to Columbia and moved back into the area of former glories with **El Paso City**. A neatly crooned version of the old standard **Among My Souvenirs** provided him with his second No.1 of the year.

An actor of some substance – Robbins appeared in such films as 'The Gun And The Gavel', 'The Badge Of Marshal Brennan' and 'Buffalo Gun' – he was also a successful album artist and maintained a fairly prolific presence in this area, showing an ability in later years to appeal to the MOR market with his releases. He also appeared on most major American TV shows and toured heavily. An Opry favourite, the one-time desert rat held the distinction of being the last person to perform at the Ryman Auditorium. He survived major heart surgery in 1970 and made a return to the Opry where he was forced to remain on stage for 45 minutes by his appreciative fans.

With the irony that so often besets singers (Hank Williams releasing **You'll Never Get Out Of This World Alive** just prior to his death), Marty went Top 10 in 1982 with **Some Memories Just Won't Die**. But on December 8, 1982, just after the song had drifted out of the chart and two months after his induction to the Country Music Hall Of Fame, Marty himself died, the victim of a heart attack. The name Robbins, however, still crops up in the charts these days, Marty's son Ronnie taking over where his father left off and logging hits for such labels as Artic, Columbia and Epic, though his singles are often outsold by Marty's re-releases.

Recommended:
Gunfighter Ballads And Trail Songs (Columbia/CBS)
Rock'n'Rolling Robbins (–/Bear Family)
A Lifetime Of Song 1951–1982 (Columbia/CBS)
El Paso City (Columbia/CBS)
All Around Cowboy (Columbia/Pickwick)
Marty Robbins, 1951–1958 (–/Bear Family)
Just Me And My Guitar (–/Bear Family)

Eck Robertson

Born in Delaney, Madison County, Arkansas on November 20, 1887, old-time fiddler Alexander 'Eck' Robertson grew up in Texas and was probably the first country musician to make records.

Following a Confederate reunion held in Virginia in 1922, he and fiddler Henry Gilliland dressed as western plainsmen and travelled to New York where they persuaded Victor to let them record. On June 30 and July 1, 1922, they cut six titles, the first of these – including Robertson's

Devil Woman, Marty Robbins. Courtesy CBS Records.

version of **Sally Goodin** – being released in April 1923. A month earlier, Robertson had played **Sally Goodin** and **Arkansas Traveler** – the latter also being among the recorded tracks – over radio station WBAP, thus becoming the first country performer to promote his own discs over the air. The duo's **Arkansas Traveler** can be heard on the RCA release, **60 Years Of Country Music**.

Carson J. Robison

Composer of such songs as **Barnacle Bill The Sailor**, **Open Up Them Pearly Gates**, **Carry Me Back To the Lone Prairie**, **Little Green Valley**, **Blue Ridge Mountain Home**, **Left My Gal In the Mountains** and the hit monologue, **Life Gets Teejus Don't It?**, Robison's formula of vaudeville, new songs and pure hillbilly made him one of the most popular country songwriters of his era.

Born in Oswego, Kansas on August 4, 1890, he first sang at local functions in the Oswego area, moving to Kansas City in 1920, where he became one of the first country singers ever to appear on a radio show. In New York during 1924 he recorded as a whistler for Victor, teaming (as guitarist and co-vocalist) with ex-opera singer Vernon Dalhart to form a formidable hit-making partnership that lasted for four years.

After the termination of this association, Robison formed another duo – with Frank Luther (Francis Luther Crowe) whose voice resembled Dalhart's, also moving on to lead such bands as the Pioneers, the Buckaroos, the Carson Robison Trio and the Pleasant Valley Boys.

Based in Pleasant Valley, New York, during the 1940s and 1950s, Robison remained an active performer and writer right up to his death on March 24, 1957, one of his final MGM recordings being a rockabilly track, **Rockin' And Rollin' With Grandmaw!** Known as the Kansas Jayhawk, he recorded many hundreds of songs during his career, working for RCA, Conqueror, Supertone and a host of others.

Recommended:
Just A Melody (Homestead/–)

Jimmie Rodgers

Known as the 'Father Of Country Music', James Charles Rodgers was born in Meridan, Mississippi on September 8, 1897, the son of a section foreman on the Mobile and Ohio Railroad. Always in ill health (his mother died of TB when Rodgers was four), he left school in 1911, becoming a water carrier on the M&O.

He later moved on to perform other tasks on the railroad until ill health caught up with him once more and he was forced to seek a less strenuous occupation. An amateur entertainer for many years, he became a serious performer in 1925. In 1926 he appeared in Johnson City, Tennessee, as a yodeller, assisted by guitarist Ernest Helton. Also in 1926, Rodgers and his wife Carrie – whom he married in 1920 – moved to Ashville, North Carolina, there organizing the Jimmie Rodgers' Entertainers, a hillbilly band comprising Jack Pierce (guitar), Jack Grant (mandolin/banjo), Claude Grant (banjo) and Rodgers himself (banjo).

Together they broadcast on station WWNC, Ashville, in 1927 for a period of six weeks, then set out to play a series of dates throughout the Southeast. Upon hearing that Ralph Peer of Victor Records was setting up a portable recording studio in Bristol, on the Virginia–Tennessee border, the entertainers headed in that direction. But due to a dispute within their ranks, Rodgers eventually recorded as a solo artist, selecting a sentimental ballad, **The Soldier's Sweetheart**, and a lullaby, **Sleep, Baby, Sleep**, as his first offerings. These tracks were released in October 1927, alongside the first release by the Carter Family. The record met with instant acclaim, thus causing Victor to record further Rodgers' sides throughout 1927. These included **Ben Dewberry's Run**, **Mother Was A Lady** and **T For Texas**, the latter originally issued as **Blue Yodel** and becoming a million-seller.

By the middle of 1928, two more **Blue Yodels** (the series was eventually to include 13 such titles) were in the catalogue and Rodgers had become America's 'Blue Yodeller'. His amalgam of blues, country, folk music and down-to-earth songs had worldwide appeal, and made him the first true country superstar. Victor began to provide Rodgers with various backing groups, jazzmen accompanying him on some tracks, Hawaiian musicians being drafted in to assist on others, even a whistler named Bob McGimsey being signed to accompany on **Tuck Away My Lonesome Blues**.

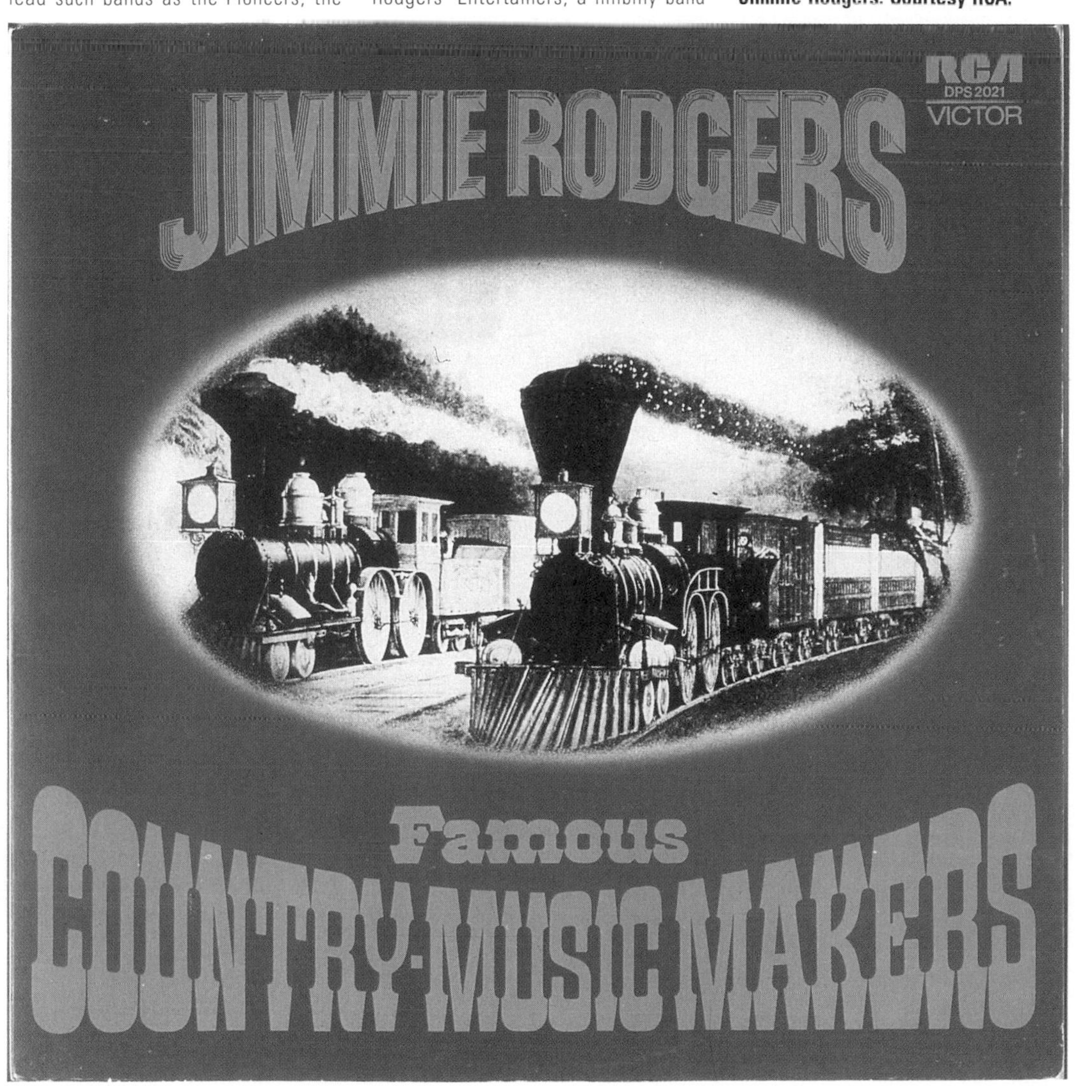

Famous Country Music Makers, Jimmie Rodgers. Courtesy RCA.

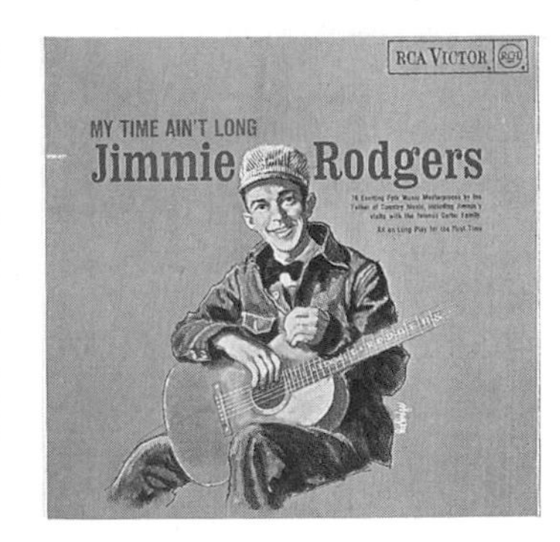

My Time Ain't Long, Jimmie Rodgers. Courtesy RCA Records.

Rodgers' popularity grew daily. **Brakeman's Blues** became his third million-selling disc. And even though the Depression had hit America, his fans still bought his records by the tens of thousands. But Rodgers was gradually wasting away. The illness that he sang about in songs such as **TB Blues** often forced him to cancel performances. In 1931 he joined humourist Will Rodgers in a series of concerts in aid of the drought sufferers in the southeastern regions, while in 1932 he began a twice-weekly radio show from station KMAC, San Antonio, Texas, but quit when he became hospitalized in early 1933. However, later that year, though critically ill, he returned to Victor's New York recording studios on 24th Street, there cutting 12 sides ove a period of eight days, a special cot being erected in which Rodgers rested between takes. The final song, **Fifteen Years Ago Today**, was completed on May 24, but during the following day Rodgers began to haemorrhage and then lapsed into a coma from which he never regained consciousness. He died on May 26, 1933.

Rodgers never appeared on any major radio show or even played the Grand Ole Opry during his lifetime. But he, Fred Rose and Hank Williams were the first persons to be elected to the Country Music Hall Of Fame in 1961, which is indicative of his importance in the history of country music.

Recommended:
Famous Country Music Makers Volumes 1 & 2 (–/RCA)
My Rough And Rowdy Ways (RCA/–)
Train Whistle Blues (RCA/–)
You And My Old Guitar (–/Conifer)
The Singing Brakeman (–/Bear Family)
Never No Mo' Blues (RCA/RCA)
My Old Pal (–/Living Era)

Judy Rodman

A singer who enjoyed brief success in the mid-'80s, Judy Rodman was born on May 23, 1951, in Riverside, California, into a musical family. She travelled extensively as a child, living in England, Mississippi, Tennessee and Alaska, but was raised mainly in Miami and Jacksonville, Florida, where she completed her college education. In the late '60s she moved to Memphis, becoming a roommate to Janie Fricke. The pair were both trying to make it as professional singers and they found work singing jingles. Later they teamed up with Karen Taylor to form Phase II and started working local clubs. This came to an end in 1975 when Judy married John Rodman and became a mother.

Above: Rodriguez is braced for a comeback career after an absence of several years.

In 1980 the Rodmans moved to Nashville where Judy worked on jingles and also landed studio work as a back-up vocalist for Crystal Gayle, Dolly Parton, George Jones and many others. When the new MTM Records was set up in 1985, Judy Rodman was the first act signed. After scoring a trio of Top 40 country hits, she shot to No.1 with **Until I Met You** (1986). That was followed by further Top 10 successes in **She Thinks That She'll Marry** (1986), **Girls Ride Horses Too** and **I'll Be Your Baby Tonight** (1987). When MTM closed down the following year, Judy was unable to find her way back into the Top 10, though she still keeps busy as a session singer and performs the odd concert and club date.

Recommended:
Judy Rodman (MTM/–)

Johnny Rodriguez

Chicano country star Juan Raul Davis Rodriguez was born in Sabinal, Texas, on December 10, 1952, the second youngest of nine children born to André and Isabel Rodriguez. He was given a guitar by his brother Andres at the age of seven, and, during his high school days he became vocalist and lead guitarist with a rock outfit. At 17 he recorded a demo disc in San Antonio but a possible record deal fell through. Then, following a couple of minor offences (including the barbecuing of a goat he and some friends had stolen), Rodriguez was taken by a friendly Texas Ranger to see Happy Shahan, the owner of the Alamo Village resort in Bracketville – the belief being that if he obtained a regular job in music he would be more likely to stay out of trouble. Employed by Shahan (who became Rodriguez's co-manager) he spent the summers of 1970 and 1971 at the village driving a stagecoach, riding horses and singing for the tourists. It was during 1971 that Tom T. Hall and Bobby Bare heard Rodriguez in Bracketville and urged him to come to Nashville.

This he did a few months later when, following the deaths of his father and his brother Andres, he headed for Music City to become a guitarist with Tom T. Hall's band, the Storytellers. Hall, signed to Mercury Records, obtained Johnny an audition with the label, Rodriguez gaining an immediate contract.

His first release, **Pass Me By**, became a Top 10 hit before the end of the 1972. His next three singles, **You Always Come Back (To Hurting Me)**, **Riding My Thumb To Mexico** and **That's The Way Love Goes** all hit the No.1 spot. In 1975 he logged three more No.1s with **I Just Can't Get Her Out Of My Mind**, **Just Get Up And Close The Door** and **Love Put A Song In My Heart**. His flood of winners (mostly Top 20) for Mercury continued through to 1979, when he signed to Epic Records. That year he had four further Top 20 records, the biggest of which was **Down On The Rio Grande**. One, **I Hate the Way I Love It**, was a duet with Charly McClain.

One of the better-looking of the younger singers on the scene in the early '80s, Rodriguez possessed both teen-appeal and the ability to sell to older audiences. But during 1981 his appeal started to ebb and his releases charted in lowly positions. Later, he admitted that a long stint with drugs had led to a careless attitude to his career. In late 1983 he sacked his band, quit drugs, formed a new band and took a more serious approach to his work. His record sales showed an almost immediate response. Johnny climbed back into the Top 10 with **Foolin'** and **How Could I Love Her So Much?** during 1984, also notching four more hits, including a Top 20 entry in **Too Late To Go Home**.

Reflecting, Johnny Rodriguez. Courtesy Mercury Records.

There followed another quiet period and in 1986 he was without a recording contract. He signed with Capitol Records in December 1987, and returned to the Top 20 with **I Didn't (Every Chance I Had)** the next year. Not heeding that song title's advice, Rodriguez backslid again with stories of cocaine use, a failing marriage, financial hardship and loss of his voice. In early 1993 he was back in Nashville. He had, supposedly, yet again kicked his drug habit and was looking for a new record deal. In his early 40s, Rodriguez still has many good opportunities to resurrect his career.

Recommended:
Rodriguez (Epic/Epic)
For Every Rose (Epic/–)
All I Ever Meant To Do Was Sing (Mercury/Mercury)
My Third Album (Mercury/Mercury)
Country Classics (–/Phillips)
Gracias (Capitol/–)

David Rogers

Rogers grew up in a country music environment, listening to the Grand Ole Opry every Saturday night of his boyhood. Born in Atlanta, Georgia, on March 27, 1936, the singer set his sights on becoming an entertainer at an early stage. In 1956 he was auditioned by Roger Miller, then part of the Third Army Special Services Division, but he failed to become a forces entertainer, merely being drafted in the normal manner. On his return to civilian life in 1958, he began working at various venues in the Atlanta area, in 1962 singing for the city's Egyptian Ballroom, where he continued to work for nearly six years.

In October 1967, he became a full member of Wheeling WWVA's Jamboree show, his first strong disc coming in 1968 via **I'd Be Your Fool Again**, a Columbia release.

Farewell To The Ryman, David Rogers. Courtesy WEA Records.

After further success that year with **I'm In Love With My Wife**, Rogers decided on a move to Nashville. It was here that he began appearing on various package shows, also receiving bookings for top-flight clubs and a number of syndicated TV programmes.

Following a run of money-spinning discs including **A World Called You** (1969), **I Wake Up In Heaven** (1970), **She Don't Make Me Cry**, **Ruby You're Warm** (both 1971) and **Need You** (1972), Rogers signed for Atlantic, becoming the label's first country act. Immediately he provided a Top 20 hit in **Just Thank Me** (1973), following this with **Loving You Has Changed My Life**, a Top 10 hit during 1974. After a hit album with **Farewell To**

The Ryman, Rogers switched to the Republic label and continued with his flow of chart contenders through to 1979. The biggest of these was **Darlin'** (1979). Since that time his name has appeared on releases from Kari (1981), Music Master (1982), Mr Music (1983) and Hal Kat (1984). David Rogers died on August 10, 1993, after a long illness.

Recommended:
Farewell To The Ryman (Atlantic)

Kenny Rogers

One of the real superstars of country music, Kenneth Donald Rogers was born in Houston, Texas on August 21, 1938, the son of a dock worker who played fiddle. A member of a high school band, the Scholars, who had a pop hit with **Crazy Feeling** in 1958, 'Rogers later attended the University of Houston where he studied music and commercial art. After working with the Bobby Doyle Trio, a harmony quartet called the Lively Ones, plus the New Christy Minstrels, he and other ex-Minstrels formed the First Edition in 1967. The group signed for Reprise Records and had a Top 10 pop hit in 1968 with a version of Mickey Newbury's **Just Dropped In (To See What Condition My Condition Was In)**. Other monster discs followed including **But You Know I Love You** (1969), **Ruby, Don't Take Your Love To Town**, an outstanding Mel Tillis song (1969), **Reuben James** (1969), **Something's Burning**, **Tell It All, Brother** and **Heed The Call** (1970). The group's popularity gradually tapered off following the success of **Someone Who Cares** in 1971.

Later becoming a solo artist – it had always been his grainy voice that had helped sell the Edition's mixture of folk, country and pop – Rogers gained a recording contract with UA, making some headway on the country scene with **Love Lifted Me**, a chart single in early 1976. By 1977 he was really on his way. UA released his version of **Lucille**, a pure country saga regarding an unfaithful wife, which reached No.1 in the country charts. It also crossed over to become the first of many of his pop Top 10 singles, and gained the bearded six-footer a Grammy award for Best Country Vocal Performance. Additionally, **Lucille** provided Rogers with Single Of The Year and Song Of The Year at the CMA awards ceremony.

From there on, it became difficult deciding just what CMA award he was not going to win. In 1978 and 1979, he and Dottie West claimed the Vocal Duo title. During 1979, too, Kenny was declared Male Vocalist Of The Year, while **The Gambler**, another Rogers' million-seller, was Song Of The Year. In the interim Rogers notched further No.1s with **Daytime Friends** (1977), **Love Or Something Like It** (1978) and **Every Time Two Fools Collide** (with Dottie West – 1978). 1979 provided four further chart-toppers in **All I Ever Need Is You** (with Dottie West), **She Believes In Me**, **You Decorated My Life** and **Coward Of The County**.

The 1980s saw little let-up in Rogers' career, though purists may knock him for being too pop, and rock fans may consider him pure MOR. He has also made further inroads into TV, in 1980 starring in 'The Gambler', a TV movie based on Don Schlitz's outstanding song, since when he has also appeared in 'Coward Of The County' (1981) and 'The Gambler II' (1983).

Above: The ever-popular Kenny Rogers has recently duetted with Travis Tritt on his 1993 album, If Only My Heart Had A Voice.

But it is his records that continue to provide Rogers with an entry into most homes. His 1980 album **Kenny** was reputed to have sold five million copies worldwide, while **Lady**, a song penned for him by Lionel Richie, not only topped the country charts but also provided Rogers with his first US pop No.1, hanging onto that position for six straight weeks in 1981. In the wake of further No.1s with **I Don't Need You** (1981), **Love Will Turn You Around** (1982) and **We've Got Tonight** (with Sheena Easton, 1983), after five years with UA/Liberty – during which time he sold some 35 million albums – Kenny signed to RCA for a reported 20 million dollars. He immediately provided them with one of his biggest ever hits, **Islands In The Stream**, a duet with Dolly Parton that again went to No.1 in both the US country and pop charts.

Since then there have been further chart-toppers with **Crazy** (1984), **Real Love** (another duet with Dolly Parton, 1985), **Morning Desire** (1985) and **Tombstone Of The Unknown Love** (1986). Kenny and Dolly also lit up the end of 1984 with a TV Christmas Special. In 1987 he teamed up with Ronnie Milsap for another country No.1 with **Make No Mistake She's Mine**, while his solo offerings of **Twenty Years Ago** and **I Prefer The Moonlight** made No.2. He joined Reprise Records in 1989, immediately scoring with **The Vows Go Unbroken (Always True to You)**, another Top 10 entry. The following year he duetted with Holly Dunn on **Maybe** and Dolly Parton on **Love Is Strange**, but was now finding it difficult to gain country radio plays with his pop-flavoured records. In a concerted effort to win his mainstream country support back, he cut **Back Home Again** in 1991, his first 'real country' album in nearly ten years.

In 1993 he signed with Giant Records. His first album for his new label, **If Only My Heart Had A Voice**, took him closer to the 'New Traditional' country of the '90s, including a duet with Travis Tritt. Not that Rogers really needed to compete with the younger stars. One of the richest men in country music, he has homes in Beverly Hills, Malibu, Bel Air and Georgia, and in 1991 started his own restaurant business.

Recommended:
Daytime Friends (UA/UA)
The Gambler (UA/UA)
Eyes That See In The Dark (RCA/RCA)
Duets (EMI-America/EMI-America)
Back Home Again (Reprise/–)
I Prefer The Moonlight (RCA/RCA)
The Heart Of The Matter (RCA/RCA)
If Only My Heart Had A Voice (Giant/–)

Roy Rogers

A major western movie star between 1938 and 1953 and known as the 'King Of The Cowboys', Rogers started out as Leonard Slye, born in Cincinnati, Ohio on November 5, 1911. His biggest early musical influence was his father, who played mandolin and guitar. He grew up on a farm in the Portsmouth, Ohio, area and, following high school, became employed in a Cincinnati shoe factory. During the 1920s he began playing and singing at local functions, and in 1930 hitched a ride to California, initially becoming a peach picker then a truck driver.

After stints with such groups as the Rocky Mountaineers and the Hollywood Hillbillies, he formed his own band, the International Cowboys. Later – with the aid of Tim Spencer and Bob Nolan – he formed the Sons Of The Pioneers.

Though this outfit established a considerable reputation, Slye set his sights higher and began playing bit parts in films, first under the name of Dick Weston and then assuming his guise as Roy Rogers, eventually winning a starring role in 'Under Western Skies', a 1938 production. With his horse Trigger and frequent female partner, Dale Evans (whom he married in 1947), and occasional help from such people as the Sons Of The Pioneers and Spade Cooley, Rogers became Gene Autry's only real rival, starring in over 100 movies and heading his own TV show in the mid-1950s. His films include 'Carson City Kid' (1940), 'Robin Hood Of The Pecos' (1942), 'The Man From Music Mountain' (1944), 'Along The Navajo Trail' (1946), 'Son Of Paleface' (1952). 'Pals Of The Golden West' (1953) and 'Mackintosh And TJ' (1975).

Seen on TV guesting on series such as The Fall Guy, Rogers was a recording artist with RCA-Victor for many years. He later recorded for Capitol, Word and 20th Century, gaining a Top 20 single **Hoppy, Gene And Me** (1974) with 20th Century. Even in 1980, then signed to MCA, Rogers was still charting. He and the Sons Of The Pioneers teamed up once more for **Ride Concrete Cowboy, Ride**, a song stemming from the movie 'Smokey And The Bandit II'. Inducted into the Country Music Hall Of Fame in 1988, three years later he was back in the country charts with **Hold On Partner**, a duet with Clint Black from Rogers' **Tribute** album. This album had the 80-year-old cowboy duetting with such youngsters as Lorrie Morgan, Kathy Mattea, Ricky Van Shelton, Randy Travis, Restless Heart and the Kentucky HeadHunters. The owner of a chain of restaurants, Rogers is estimated to be worth something over 100 million dollars.

Recommended:
Best Of Roy Rogers (Camden/–)
Happy Trails To You (20th Century/–)
The King Of The Cowboys (–/Bear Family)

Above: Roy Rogers with Bob Hope in the 1952 movie 'Son Of Paleface'.

Linda Ronstadt

These days, Linda is a rocker who has notched some of her biggest albums with Sinatra-styled standards and made an onstage impact by singing Gilbert And Sullivan on Broadway. But once she sang plenty of country and employed some of the best country-rockers in the business.

Born on July 16, 1946, in Tucson, Arizona, she arrived in L.A. in 1964 and formed the folksy Stone Poneys. After kicking around for a year, they signed to Capitol. 1967 saw the release of two albums, **Stone Poneys** and **Evergreen**, **Stone Poneys** yielding a hit single in Linda's version of Mike Nesmith's **Different Drum**.

Linda next went out as a solo act and made two more albums, **Hand Sown, Home Grown** (1969) and **Silk Purse** (1970), the first featuring the talents of such musicians as Clarence White, Red Rhodes and Doug Dillard, while the latter was made with the aid of Nashville pickers. By 1971, however, she made the first move in a new direction, forming a backing band consisting of future Eagles members Glenn Frey, Don Henley and Randy Meisner, and releasing an album called **Linda Ronstadt**.

Since that time Linda has moved further and further away from her original country-oriented sound, though every now and then she has cut a track to remind country fans of days gone by. Her hit singles include **Silver Threads And Golden Needles**, **I Can't Help It (If I'm Still In Love With You)** (both 1974), **When Will I Be Loved** (No.1, 1975), **Crazy** (1976) and **Blue Bayou** (1977). Ringing the changes in her musical output, the finishing touches were put to the **Trio** album in 1987. The project saw Ronstadt teamed with Dolly Parton and Emmylou Harris on traditional country music. The album also won a Grammy for Best Country Album and produced four Top 10 country singles – **To Know Him Is To Love Him** hitting No.1. That same year she also returned to her childhood memories in Arizona when she recorded **Canciones De Mi Padre**, a Mexican album that was so successful that she released another one four years later, **Más Canciones**. She made a return to pop success with the superlative album **Cry Like A Rainstorm – Howl Like The Wind**; her haunting duet with Aaron Neville on **Don't Know Much** became a Top 10 pop hit in Britain and America.

Recommended:
Linda Ronstadt – A Retrospective (Capitol/Capitol)
Don't Cry Now (Asylum/Asylum)
Prisoner In Disguise (Asylum/Asylum)
Más Canciones (Asylum/Asylum)

Hasten Down The Wind, Linda Ronstadt. Courtesy WEA Records.

Billy Joe Royal

A singer who first found fame as a pop idol in the '60s before moving over to country music, Royal was born on April 3, 1942 in Valdosta, Georgia, and spent most of his childhood in Marietta, Georgia. His father owned a trucking company and in 1952 moved his family and business to Atlanta. Music played a role in the Royal household, with country at the top of the agenda. Billy Joe got his first taste of entertaining in school concerts, and during high school formed his own band, the Corvettes. After graduation he worked in a Savannah night-club, already a talented guitarist, pianist and drummer. In the early '60s he recorded several unsuccessful singles, and in 1964 teamed up with Joe South, a local singer-songwriter. A year later they produced **Down In The Boondocks**, which made the pop Top 20 when released by Columbia. Billy Joe had several more pop hits in the '60s before ending up on the cabaret circuit.

Eventually he returned, musically, to his Georgia country roots. While recording in Nashville and Memphis, he landed a record deal with Atlantic America in 1985. Mixing up an R&B beat, a country attitude, heart-felt vocals and a touch of rock'n'roll, he started to make an impact on the country charts with: **Burned Like A Rocket** (No.10, 1985), **I Miss You Already** (No.14, 1986), **Old Bridges Burn Slow** (No.11, 1987) and **I'll Pin A Note On Your Pillow** (No.5, 1987). With his dynamic stage show and his varied mixture of material and styles, Royal built up a big following in country. Several more Top 10 hits resulted, including: **Out Of Sight And On My Mind** (1988), **Tell It Like It Is**, **Love Has No Right** and **Till I Can't Take It Anymore** (all 1989). Then he mysteriously stopped hitting the charts, though he still continues to tour regularly, mixing pop and country into a slick act.

Recommended:
Royal Treatment (Atlantic/Atlantic)
Out Of The Shadows (Warner Bros/–)

Mr & Mrs Untrue, Johnny Russell. Courtesy RCA Records.

Johnny Russell

John Bright Russell was born in Sunflower City, Mississippi on January 23, 1940, his family moving to California when he was 12. A singer-songwriter and guitarist, he got a job plugging for the Wilburn Brothers' music publishing company and also pushed his own songs. One, **Act Naturally**, became a chart-topper for Buck Owens in 1963, the song providing a favourite for the Beatles, who also recorded it.

A Burl Ives-type character himself, he wrote songs for Burl plus Loretta Lynn, Del Reeves, Patti Page, Dolly Parton and Porter Wagoner. Eventually, he got his own recording career underway with RCA in 1971. By 1972 he had Top 20 hits with **Catfish John**, **Rednecks**, **White Socks And Blue Ribbon Beer** and **Baptism Of Jesse Tylor**. But despite such singles as **She's In Love With A Rodeo Man** and **Obscene Phone Call**, Russell's only other major hit for the label came with **Hello I Love** (1975) and in 1978 he turned up on Mercury.

His onstage act simply gets better and better, while his songs continue to bring in an abundant supply of royalties. Russell's **You'll Be Back (Every Night In My Dreams)** went Top 5 for the Statlers in 1982, while **Let's Fall To Pieces Together** made it to No.1 in the charts for George Strait in 1984.

Recommended:
Rednecks, White Socks And Blue Ribbon Beer (RCA/RCA)
She's In Love With A Rodeo Man (RCA/RCA)
Mr Entertainer (–/RCA)

Tom Russell

Singer-songwriter Tom Russell, who since the late '60s has played the clubs and bars in North America and Europe, and had his songs recorded by Johnny Cash, Jerry Jeff Walker, Nanci Griffith, Ian Tyson, Suzy Bogguss and many others, has remained an under-appreciated artist.

He was born in California and grew up in Los Angeles, but has spent much of his life travelling. While living in Canada, he wrote **End Of The Trail**, a song that secured him a great deal of acclaim. His own version of the song was included on a compilation album for Buddha Records. Russell moved on through Boston to Austin, Texas. At this period he recorded two albums with pianist Patricia Hardin, but was soon on the move again, working as a performer in Puerto Rico.

He teamed up with guitarist Andrew Hardin, the pair touring Europe together, eventually travelling to Norway, where they settled. Also, during the mid-'80s he developed his songwriting, often co-writing with Steve Young, Tom Pacheco and Katy Moffatt. One of his songs, **Navajo Rug**, was recorded by Canadian Ian Tyson.The song became the Canadian Country Music Association's Single Of The Year in 1987, and was recorded as the title track for Jerry Jeff Walker's 1992 album.

In the meantime, Russell assembled a band to play his unique mixture of country, rock, Tex-Mex, blues and folk. The Tom Russell Band is capable of playing in wide-ranging styles, but with the freedom to really cook, which provides the ideal backdrop for Russell's grainy baritone.

Recommended:
Heart On A Sleeve (–/Bear Family)
Beyond St Olav's Gate (Rounder/Round Tower)
Hurricane Season (Philo/–)
Poor Man's Dreams (–/Sonet)

Doug Sahm

Although Doug Sahm became known originally through his teeny-bop hits and then moved on to utilize blues, Mexican music, rock and country in his recordings, it is his involvement in the so-called 'Outlaw' community of Austin, Texas that won him his large and loyal cult following.

Born on November 6, 1941, and raised in San Antonio, he was subject to the varied root musical influences of that area, but nevertheless found himself part of the mid-'60s garage band movement. **She's About A Mover** featured a pumping 4/4 beat, Sahm's strange whining vocal and the amateurish sounding organ dabs of Augie Meyers. It is a sound that has since become Sahm's trademark, with various refinements. The Sir Douglas Quintet, as his band was known, then moved to the San Francisco scene.

But Sahm was still a native Texan (a sentiment he has expressed in the song **I'm Just A Country Boy In This Great Big Freaky City**). His country-orientated albums, **Doug Sahm And Band** and **Texas Rock For Country Rollers**, have shown his natural flair for the music. Sahm recorded and toured throughout the '80s and in 1990 joined forces with long-time musical buddies Freddy Fender, Flaco Jimenez and Augie Meyers to form the Texas Tornadoes.

Recommended:
Texas Rock For The Country Rollers (ABC/ABC)
Wanted – Very Much Alive (Texas Records/Sonet)
Doug Sahm And Band (–/Edsel)
Back To The Dillo (–/Sonet)

Buffy Sainte-Marie

Hardly a pure country singer – she began as a folk singer and has since headed every which way – Buffy has nevertheless made several recordings of interest to country music enthusiasts.

Below: Buffy Sainte-Marie's I'm Gonna Be A Country Girl Again album was pure country.

Her birthplace is clouded in mystery though Sebago Lake, Maine (February 20, 1941), is the location most generally accepted. Born to Cree Indian parents, she was adopted at an early age and raised mainly in Massachusetts.

She broke into the folk scene in Greenwich Village during the early '60s, learning to play Indian mouth-bow from singer-songwriter and fellow Cree, Patrick Sky. A codeine addict at one point in her career, she wrote a classic song, **Cod'ine**, about her experiences, though it proved to be Donovan's version of her **Universal Soldier** that brought her songwriting into perspective.

Achieving considerable kudos through her appearances at the Newport Folk Festivals during the '60s, Buffy signed to Vanguard Records, cutting her first album, **It's My Way** (containing both **Cod'ine** and **Universal Soldier**) for the label in 1964.

By 1968, Buffy's intense vibrato was to be heard in a Nashville studio where she cut a pure country album **I'm Gonna Be A Country Girl Again**, achieving a mild pop hit with the title track, but having even more success with **Soldier Blue** (1971), the theme song from a film dealing with the massacre of the Indians during the last century.

Leaving Vanguard in 1973, she signed for MCA but after only two albums moved on to ABC, making a strong album, **Sweet America**, in 1976.

One Sainte-Marie composition, **Until It's Time For You To Go**, recorded by Buffy in 1965, later became a 1972 million-seller for Elvis Presley and she was also co-writer of the 1982 Grammy award-winning **Up Where We Belong**. Much of Buffy's recorded work lies outside the scope of this book but **I'm Gonna Be A Country Girl Again** and **A Native North-American Child** (both Vanguard), the latter album being a plea on behalf of the North-American Indians, should be heard.

Recommended:
I'm Gonna Be A Country Girl Again (Vanguard/Vanguard)
Sweet America (ABC/ABC)

Sawyer Brown

Duncan Cameron, lead guitar; Gregg (Hobie) Hubbard, keyboards; Mark Miller, lead guitar; Jim Scholten, bass; Joe Smyth, drums.

Country-rock band Sawyer Brown made their breakthrough in 1984 when they won Star Search, an American syndicated television talent contest. This led to a recording contract with Capitol and tours with such stars as Kenny Rogers, Dolly Parton, Crystal Gayle, the Oak Ridge Boys and Eddie Rabbitt.

Prior to their long run on Star Search, Sawyer Brown had been together for two years, working the American club circuit and building a reputation as one of the most musically proficient acts to have emerged in country music.

Their first single, **Leona**, made the country Top 20 at the end of 1984, and the self-contained five-piece outfit hit No.1 with **Step That Step** in 1985, followed by such Top 10 entries as **Used To Blue** and **Betty's Bein' Bad**, also in 1985. Flamboyant performers, Sawyer Brown appeal to both country and rock fans with their exciting live shows, but have not always been consistent on the charts. They made Top 20 entries in 1986 with **Heart Don't Fall Now** and **Shakin'**, then seemed to slip in popularity. They made a comeback with **This Missin' Heart Of Mine** in 1988 and a revival of George Jones' **The Race Is On** in 1989. The following year there were a couple of minor hits, coinciding with the first line-up change in ten years. Lead guitarist Bobby Randall left, to be replaced by Duncan Cameron, who had previously been in the Amazing Rhythm Aces.

This change seemed to breathe new life into the band. They signed with Curb (their records still being released through Capitol), and made a big impact with their 8th album, **Buick**, and a single, **The Walk**, which took them back into the Top 10 in 1991. The following year they had a trio of big hits with **The Dirt Road**, **Some Girls Do** and **Cafe On The Corner**, continuing into 1993 with **All These Years**. And after all these years, Sawyer Brown show no signs of slowing down, playing over 225 shows annually. 'Amusement Business', in its 1991 year-end issue, lists the band as the 9th ranked top-grossing country act.

Recommended:
Shakin' (Capitol/Capitol)
Dirt Road (Capitol-Curb/–)
Buick (Capitol-Curb/–)

John Schneider

Actor turned singer, John was born on April 8, 1954 in New York, and came to prominence as Bo Duke in the TV series The Dukes Of Hazzard. He then found success in country music, scoring his first hit in 1981 with **It's Now Or Never**.

John first demonstrated an inclination towards a career in showbusiness during his early schooldays, landing a part in 'L'il Abner', which was being staged by a community theatre group. Following his parents' divorce in 1968, John moved to Atlanta, Georgia with his mother.

While in high school he became active in the drama club and played the lead roles in many musicals and dramas. Following graduation he appeared in several community theatre productions, sang in Atlanta's clubs and co-wrote the musical score for a play called 'Under Odin's Eye'.

His big break came in 1978 when he won the coveted role of Bo Duke, the youngest Duke of Hazzard. John signed a recording contract with Scotti Brothers Records in 1981 and scored a Top 5 country hit with his update of the Elvis Presley classic **It's Now Or Never**.

In 1984 Schneider signed with MCA Records, and his first single, **I've Been Around Enough To Know**, was sent out to radio stations as an unmarked promo copy. Because it was on the same label as George Strait, many thought it was the Texan. Slowly the record climbed the country charts, eventually becoming Schneider's first No.1. The following year came another No.1 with **Country Girls**, while **It's A Short Walk From Heaven To Hell** and **I'm Going To Leave You Tomorrow** were both Top 10 entries. Alongside his singing career, Schneider continued with his acting, appearing in films 'Eddie Macon's Run', 'Dream House', 'Happy Endings' and 'Gus Brown And Midnight Brewster'. In 1986 he placed two more singles, **What's A Memory Like You (Doing In A Love Like This)** and **You're The Last Thing I Needed Tonight**, at No.1 on the charts. More Top 10 entries followed in: **At The Sound Of The Tone** (1986), **Take The Long Way Home** and **Love, You Ain't Seen The Last Of Me** (1987). Dividing his time between music and acting was not really working out, so finally he decided to concentrate on acting. He appeared in a remake of 'Stagecoach', also took up scriptwriting and directing, and in 1990 starred in the TV series The Grand Slam.

Recommended:
Tryin' To Outrun The Wind (MCA/–)
Now Or Never (Scotti Bros/–)
Take The Long Way Home (MCA/MCA)
You Ain't Seen The Last Of Me (MCA/MCA)

Dan Seals

Born on February 8, 1948 in McCamey, Texas, and raised in Dallas, singer-songwriter Dan Seals' background is steeped in country music. His family were originally settlers in mid-Tennessee, who moved to Texas in the 1920s. Here, Dan's father gained a reputation as an accomplished guitar player, and with his son Jimmy (Dan's older brother, who later became part of Seals And Crofts) on fiddle, he played backup for many of the country music stars who toured in the Midland-Odessa area of Texas.

Jimmy eventually joined the Champs (along with Glen Campbell) and played on such pop hits as **Tequila** and **Limbo Rock** in the late '50s and early '60. Dan started playing in bands while at high school and joined up with John Ford Coley and Shane Keister to form Southwest F.O.B. (Freight On Board), scoring a minor pop hit in 1968 with **Smell Of Incense**.

Moving to Los Angeles, Coley and Dan landed a recording contract with Atlantic, by this time working as a duo called England Dan And John Ford Coley. During the '70s they released a string of pop hits,

Below: Sawyer Brown, who took their name from a Nashville street.

including **I'd Really Love To See You Tonight**, **We'll Never Have To Say Goodbye Again** and **Nights Are Forever Without You**.

The act split in 1979, with Dan retaining the band's name and a mountain of unpaid tax bills. He recorded a couple of solo albums, **Stones** and **Harbinger**, for Atlantic without any success. Kyle Lehning, who had produced the duo's hits, suggested he make a move to Nashville. In 1982 he uprooted his family from L.A., determined to make his mark in country music, but arriving in Music City bankrupt. Separated from his wife and children, who were living with friends, he started his career again. With Lehning's help, Seals recorded a demo and landed a contract with Capitol, scoring a Top 20 hit with **Everybody's Dream Girl** in 1983. A couple of minor hits followed, then he hit the Top 10 with **God Must Be A Cowboy**, **(You Bring Out) The Wild Side Of Me** (1984), **My Baby's Got Good Timing** and **My Old Yellow Car** (1985).

A big breakthrough came when he teamed up with Marie Osmond for **Meet Me In Montana** in 1985, the first of an incredible run of nine consecutive country No.1s, many of them self-penned. The infectious **Bop** crossed over to the pop charts in 1986, but all the others were straight country – **Everything That Glitters (Is Not Gold)**, **You Still Move Me** (1986), **I Will Be There**, **Three Time Loser**, **One Friend** (1987), **Addicted** and **Big Wheels In The Moonlight** (1988). With his meticulous choice of song material and musical presentation, Seals became a big star. His albums **Won't Be Blue Anymore** and **On The Front Line** went gold. He scored a Top 5 hit with the superb **They Rage On** in 1989, and was back at No.1 the following year with **Love On Arrival** and **Good Times**.

In 1991 he signed with Warner Brothers, and though he has continued to produce impeccable recordings, he has only managed a few minor hits for his new label, the best being **Sweet Little Shoe** (1991), **When Love Comes Around The Bend** and **We Are One** (1992).

Recommended:
Won't Be Blue Anymore (EMI-America/–)
San Antone (EMI-America/–)
On Arrival (Capitol/Capitol)
Rage On (Capitol/Capitol)
On The Front Line (Capitol/Capitol)
Walking The Wire (Warner Bros/–)

Jeannie Seely

An Opry member since 1966, Jeannie was born in Pennsylvania on July 16, 1940, and grew up in the Titusville area. She began singing on local radio in Meadville, Pennsylvania at the age of 11, and during her high school era appeared on the prestigious Midwest Hayride show.

Although she studied banking and associated subjects at the American Institute of Banking, Jeannie preferred the trappings of showbiz and moved to L.A., signing with Four Star Music as a writer and cutting a couple of unsuccessful discs for Challenge Records.

With encouragement from Hank Cochran, who later became her husband, she moved to Nashville in 1965, becoming a writer for Tree International Music. She also signed for Monument Records and had an instant hit with Cochran's **Don't Touch Me** (1966), a release which won her a Grammy award for the Best Female C&W Vocal Performance.

In the wake of eight chart singles for Monument – these including **It's Only Love** (1966), **A Wanderin' Man** (1966) and **I'll Love You More** (1967) – Jeannie became a part of the Jack Greene Show in 1969. Joining Greene on the Decca label, the duo impressed record buyers via **I Wish I Didn't Have To Miss You**, a Top 5 disc in 1969. Jeannie's biggest solo hits arrived with **Can I Sleep In Your Arms?** in 1973 and **Lucky Ladies** (1974).

Recommended:
Greatest Hits On Monument (Sony/–)

Ronnie Sessions

A singer who looked set for the big time in the late '70s, Ronnie (born on December 7, 1948, in Henrietta, Oklahoma) had faded from the limelight by the mid-'80s, though he continues to work the club and honky-tonk circuit where his driving country music is best appreciated.

Something of a child prodigy, he made his first records for the small Pike Records in Bakersfield, California when he was nine, and was a regular on the Herb Henson Trading Post TV show. He continued his television career on the West Coast with an array of popular performers including Billy Mize, Wes Sanders and the Melody Ranch Show. In the mid-'60s he started recording for such small labels as Starview and Mosrite, then signed to Gene Autry's Republic Records in 1968. He notched up regional hits with **Life Of Riley** and **More Than Satisfied**.

A move to Nashville in 1972 saw him signed as a writer to Tree Publishing and he started recording at MGM, making his debut on the country charts with **Never Been To Spain** (1972). Then he joined MCA Records, achieving success with **Makin' Love** (1975) and his first Top 20 entry **Wiggle Wiggle** (1976). Finally establishing himself with his rocking country music, more hits came with **Me And Millie**, **Ambush** (both 1977) and **Juliet And Romeo** (1978). Then, with his records only just making the country Top 100, he was dropped by MCA in 1980 and has hardly recorded since.

Recommended:
Ronnie Sessions (MCA/MCA)

Above: Successful singer Ricky Van Shelton is a farmer who still ploughs and cultivates his own land.

Billy Joe Shaver

A 'New Wave' singer-songwriter who rose to prominence while in his mid-'30s, Shaver was a part-time poet who set his sights on Nashville after hearing Waylon Jennings sing.

Born in Corsicana, Texas, on September 15, 1941, he moved to Waco at the age of 12, spending his early life employed at a sawmill, punching cattle, working as a carpenter and performing various menial tasks as a farmhand. His ambitions as a songwriter later led him to Nashville where he was rejected by every publisher. Shaver, on the brink of starvation, was forced to return to Texas. However, on a later trip to the Music City he sold a song to Bobby Bare that became the B-side of a hit and also signed as a writer to Bare's own Return Music Publishing Company, achieving something of a breakthrough when Kristofferson recorded his **Good Christian Soldier** (1971).

Next, Tom T. Hall latched on to Shaver's songs, as did Dottie West, Jan Howard, Jerry Reed, Tex Ritter and Jim Ed Brown. His reputation as one of the most potent new writers in Nashville was established when Waylon Jennings cut **Honky Tonk Heroes** (1973), an album of songs nearly all penned by Shaver.

An often controversial writer – one of his songs **Black Rose** deals with the subjects of inter-racial marriage – he became a fully fledged recording artist when Kris Kristofferson produced **Old Five And Dimers Like Me**, Shaver's first album for the Monument label. Further recordings for Capricorn and Columbia received critical acclaim but were commercial failures, though he has continued to score as a writer, providing John Anderson with his first chart-topper, **I'm Just An Old Chunk Of Coal** (1981), as well as providing hits for Johnny Cash, George Jones and Conway Twitty.

Recommended:
Gypsy Boy (Capricorn/–)
I'm Just an Old Chunk Of Coal ... But I'm Gonna Be A Diamond Someday (Columbia/–)
Salt Of The Earth (Columbia/–)

Ricky Van Shelton

With his silky-smooth baritone, Ricky Van Shelton (born on January 12, 1952, in Danville, Virginia) has displayed an innate ability to pick great songs. He has amassed 13 No.1 singles since hitting the country charts in 1987, from the rollicking **Crime Of Passion** to **Keep It Between The Lines** four years later. He grew up in the small town of Grit, Virginia, with southern gospel all around him. His father played guitar and sang in a gospel quartet, but as a teenager Ricky was more interested in the music of the Beatles and the Rolling Stones. Displaying a vocal ability at a young age, country music was not an interest of young Shelton, until the intervention of an older brother in need of a lead singer for his country band. Ricky slotted in neatly, so he was working as a gas jockey and pipe fitter by day and playing honky-tonks by night in Virginia.

In the mid-'80s Ricky's wife Bettye landed a job in Nashville, and this enabled him to start working on a country music career. He continued working club dates and a demo tape led to a showcase for CBS executives. A contract with Columbia Records resulted in 1986. His first single, **Wild-Eyed Dream**, made the country Top 30 in 1987, followed by a Top 10 entry with **Crime Of Passion**. This started a consistent run of No.1s with: **Somebody Lied** (1987), **Life Turned Her That Way**, **Don't We All Have The Right**, **I'll Leave This World Loving You** (1988), **From A Jack To A King**, **Living Proof** (1989), **Statue Of A Fool**, **I've Cried My Last Tear For You** (1990), **Rockin' Years** (a duet with Dolly Parton, 1991), **I Am A Simple Man** and **Keep It Between The Lines** (1991).

Shelton made a success out of finding little-known songs from the past and making them into modern hits. It was a winning formula, because by the end of 1991 his first four albums, **Wild-Eyed Dream**, **Living Proof**, **RVS III** and **Backroads**, had all earned platinum awards. He won the CMA Horizon award in 1988, while winning as CMA Male Vocalist in 1989. Fans also voted him both Entertainer and Male Artist Of The Year in the MCN 1991 awards. The following year he remembered the gospel music he heard as a child, and, specially for his parents, he recorded **Don't Overlook Salvation**, a gospel album, which became a commercial success. Demonstrating that he is multi-talented, he wrote a children's book, 'Tales From A Duck Named Quacker'. Unable to get a publisher interested in it, he set up his own RVS Book Publishers, and the first of a proposed six-book series had sold 60,000 copies by the end of 1992. Meanwhile, hit singles continued with **Backroads** and **Wild Man** (1992), and Shelton participated in the 'Honeymoon In Las Vegas' soundtrack. His version of the

Presley classic, **Wear My Ring Around Your Neck**, provided him with a Top 20 country hit in 1992.

Recommended:
RVS (Columbia/–)
Don't Overlook Salvation (Columbia/–)
Loving Proof (Columbia/Columbia)
Wild-Eyed Dream (Columbia/Columbia)
Backroads (Columbia/–)

Shenandoah

Ralph Ezell, bass; Mike McGuire, drums; Marty Raybon, lead vocals; Jimmy Seales, lead guitar; Stan Thorn, keyboards.

A quintet formed in Muscle Shoals, Alabama, Shenandoah started out in the early '80s as the MGM Band, so named because they were the house band at the MGM Club in Muscle Shoals. The name change to Shenandoah came in 1986, when some demos they made at the legendary Fame studios with Rick Hall landed them a recording contract with Columbia Records.

All southern boys, lead singer Marty Raybon was raised in Florida. Heavily into bluegrass music, he joined his father and brothers in the American Bluegrass Band for nine years. He then moved to Nashville, where he worked as a songwriter, having material recorded by George Jones and Johnny Duncan. Marty was invited to Muscle Shoals when the MGM Band had a vacancy for a lead singer. At the same time they were looking for a bass player, so in stepped Ralph Ezell from Jackson, Mississippi, who had played recording sessions for David Allan Coe, Mac Davis and others. The three founder members of the MGM Band were Jimmy Seales, a session-man and songwriter from Illinois, Mike McGuire, from Alabama and who penned T. Graham Brown's 1987 Top 10 hit **She Couldn't Love Me Anymore**, and Stan Thorn, who started out in the family gospel group, then joined '70s group Funkadelic.

This line-up has stayed together as Shenandoah, emerging as one of the top country bands of the late '80s and early '90s. They scored minor country hits during 1987, then the following year hit the Top 10 with **She Doesn't Cry Anymore** and **Mama Knows**. Their gritty sound with a soulful edge behind Raybon's raspy, sawdust and splinters vocal style is just right for the mainly southern themes of their songs. It resulted in a trio of No.1s in 1989 – **The Church On Cumberland Road**, **Sunday In The South** and **Two Dozen Roses**. There was another chart-topper with **Next To You, Next To Me** the following year, while **See If I Care** and **Ghost In This House** were Top 10 entries. Two of their albums, **The Road Not Taken** and **Next To You**, went gold. At this time a dispute arose over the group name Shenandoah, and a tough legal battle ensued. The band had a couple more Top 10 hits with **I Got You** and **The Moon Over Georgia** in 1991, but for almost a year all their earnings were frozen while lawyers battled over the ownership of their name. Eventually it was decided in the band's favour. They left Columbia Records and signed with RCA Records, immediately returning to the charts with **Hey Mister (I Need This Job)** (1992), and **Leavin's Been A Long Time Comin'** (1993).

Recommended:
Long Time Comin' (RCA/–)
The Road Not Taken (Columbia/–)
Extra Mile (Columbia/–)

Right: T.G. Sheppard was most successful in the '80s for his country-pop music.

Jean Shepard

Born in Pauls Valley, Oklahoma, on November 21, 1933, Jean originally sang and played bass with an all-girl western swing outfit known as the Melody Ranch Girls.

After impressing Hank Thompson via a Melody Ranch Girls–Brazos Valley Boys joint gig, Jean became signed to Capitol Records in 1953, that same year gaining her first No.1 with **Dear John Letter**, a duet recorded with Ferlin Husky. Next came a sequel, **Forgive Me, John** (1953), then a brace of 1955 solo winners with **Satisfied Mind** and **Beautiful Lies**.

Jean became a regular on the Red Foley Show over KWTO, two years later moving to Nashville and attaining Opry status. But her hits remained infrequent until 1964 (shortly after the death of husband Hawkshaw Hawkins) when her recording of **Second Fiddle (To An Old Guitar)** sparked off a long flow of successes.

During the mid-'70s, Jean made a big impact with British country audiences through her outspoken insistence on keeping country free from pop contamination. Her repertoire of pure '50s-type country, full of hard-sob ballads and full-throttle honky-tonk won her standing ovations at the Wembley Festivals in 1977, 1978, and subsequent concert tours.

Recommended:
I'm A Believer (UA/Music For Pleasure)
Mercy Ain't Love Good (UA/UA)
I'll Do Anything It Takes (–/Sunset)
Lonesome Love (Capitol/Stetson)

T. G. Sheppard

T. G. (real name Bill Browder) was born on July 20, 1944, in Humboldt, Tennessee, and moved to Memphis in 1960 where he became a guitarist and backup singer in Travis Wammack's band.

Securing a contract with Atlantic Records and using the name Brian Stacy, he had a number of pop-styled singles released in the early '60s, including **High School Days**. In 1965 he got married and took a hard look at his modest singing

I'm A Believer, Jean Shepard. Courtesy UA Records.

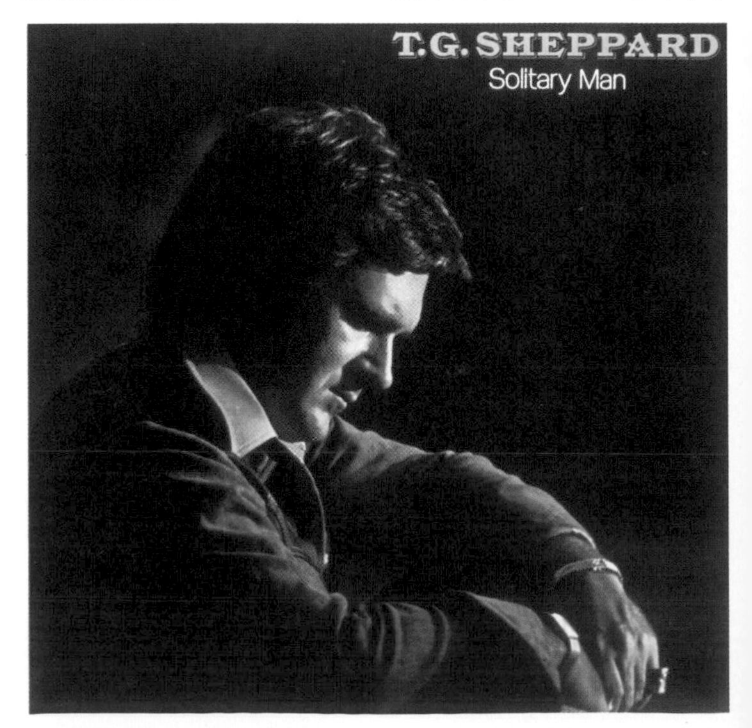

Solitary Man, T.G. Sheppard. Courtesy Hitsville Records.

career, but he wanted to stay in records so became a promotion man.

After a stint as RCA's Memphis promotion man, he formed his own independent company, Umbrella Productions. He picked up **Devil In The Bottle** from writer Bobby David and demoed it himself, eventually finding an unlikely outlet in Tamla Motown.

Since the name Bill Browder would coincide with his promotional activities, he called himself T. G. Sheppard, after spotting a bunch of German shepherd dogs through an office window. The record was released on Motown's Melodyland label in 1974 and rapidly climbed to the top of the country charts, crossing over to the pop charts the following spring.

Further hits followed with **Tryin' To Beat The Morning Home**, **Motel And Memories** (both 1975), **Solitary Man** and **Show Me A Man** (both 1976), by which time the record label Melodyland had become Hitsville.

Finally, Motown closed down its country division and T. G. signed with Warner Brothers, making a return to the Top 20 with **Mr DJ** in 1977. Following Top 10 hits **When Can We Do This Again** and **Daylight** (both 1978), he changed producers and, under the guidance of Buddy Killen, scored No.1s with **Last Cheater's Waltz**, **I'll Be Coming Back For More** (both 1979), **Smooth Sailin'** and **Do You Wanna Go To Heaven** (both 1980), and crossed over to the pop charts in 1981 with the million-selling **I Loved 'Em Every One**.

With his good looks, slightly sexy song lyrics and dynamic stage act, T.G. has appealed mainly to women, creating mob scenes that are usually associated with pop stars rather than mature country singers. His success on record has been maintained with such songs as **Finally**, **War Is Hell (On The Homefront Too)** (1982), **Faking Love** (a duet with Karen Brooks, 1982), **Slow Burn** (1983), **One Owner Heart** (1984) and **You're Going Out Of My Mind** (1985).

He changed labels in 1985, moving over to Columbia. His run of hits continued with **Doncha?** (1985), **Strong Heart**, **Half Past Forever (Till I'm Blue In The Heart)** (1986), **You're My First Lady** and **One For The Money** (1987). He only managed a few minor hits in the late '80s, though his concert tours have still proved highly successful. In 1991, he signed with Curb Records, but his **Born In A High Wind** never really took off.

Recommended:
3/4 Lonely (Warner Bros/–)
I Love 'Em All (Warner Bros/–)
Livin' On The Edge (Columbia/–)
Crossroads (Columbia/–)

Billy Sherrill

An ex-R&B producer who developed a smooth line in modern country production during the 1960s, Sherrill sparked off success for Tammy Wynette and Tanya Tucker. In early '70s he helped revitalize the career of ex-rock'n'roller Charlie Rich, the seductive, soporific sound being dubbed 'countrypolitan' by the critics.

Working for Columbia and Epic, Sherrill evolved a masterly way of balancing steel guitars and orthodox country instruments against orchestras. Such was the success of this fine balance, with all the rough edges knocked off, that records made by

Above: A mid-1970's shot of Billy Sherrill, Nashville's leading producer.

Sherrill sold in spectacular quantities and helped identify the 'Nashville Sound' as mainstream country to most people, when in fact it was just part of a larger whole.

He also worked in the studio with such singers as Johnny Paycheck, Elvis Costello, George Jones, Marty Robbins, Lacy J. Dalton and David Allan Coe, illustrating the scope of his musical styles. Over the years he has co-written dozens of country hits, including **Too Far Gone**, **Stand By Your Man** and **Almost Persuaded**.

The Shooters

Walt Aldridge, lead vocals; Gary Baker, lead vocals, bass; Barry Billings, guitar; Chalmers Davis, keyboards; Michael Dillon, drums.

A contemporary studio band, the Shooters were based around the multi-talented Walt Aldridge, a prolific sessionman, songwriter and producer at the Muscle Shoals' Fame studios. Aldridge had already started recording the band's first album in 1986 before the band was actually formed. Once he had a few ideas down on tape with studio musicians, he then brought in the other musicians. That first album, simply titled **The Shooters** and released by Epic, produced such country hits as **They Only Come Out At Night** and **Tell It To Your Teddy Bear** (1987) as well as **I Taught Her Everything She Knows About Love** (1988).

Aldridge was raised in the Muscle Shoals area and graduated from the commercial music programme at the University of North Alabama. He started working at the Fame studios and made a big impact in country music penning hits such as **There's No Getting Over Me** (Ronnie Milsap), **Holding Her And Loving You** (Earl Thomas Conley) and **Crime Of Passion** (Ricky Van Shelton). Gary Baker, who shares lead vocals with Aldridge, is a former member of pop groups Boatz, LeBlanc and Carr, and has penned country hits for Steve Wariner, Gary Morris and Alabama. Chalmers Davis began his session work in Jackson, Mississippi for Malaco Records and worked on the road and in the studios with the Gatlins, Tammy Wynette, T.G. Sheppard and Mac Davis. Barry Billings has played guitar since fourth grade. Based in Huntsville, Alabama, he went on the road with the Cornelius Brothers And Sister Rose, before moving to Muscle Shoals and joining the Shooters. Michael Dillon, from Florida, came to Muscle Shoals in the early '80s as a member of a revue band. He was part of the line-up of the Allman Brothers reunion tour before joining the Shooters.

A second album, **Solid As A Rock**, produced their first Top 20 records in **Borderline** (1988) and **If I Ever Go Crazy** (1989). Strictly into contemporary country, the Shooters had a hard-cutting edge in a style that was a rich mix of Eagles-sounding West Coast country and the more rootsy southern country rock. Not willing to tour, their career was short-lived.

Recommended:
Solid As A Rock (Columbia/–)

Shel Silverstein

Cartoonist with Playboy magazine for over 20 years, Silverstein is the most eccentric figure in popular music. A bald hipster from Chicago, who writes poetry and children's

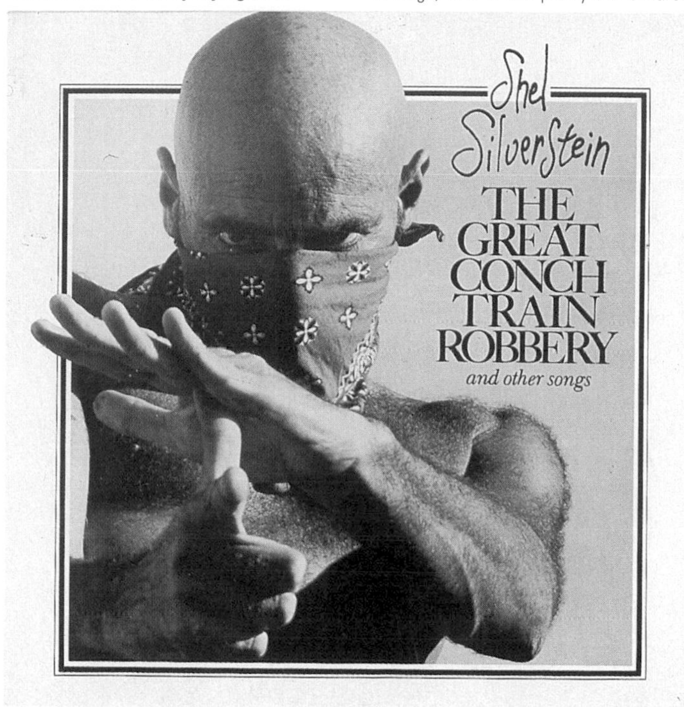

The Great Conch Train Robbery, Shel Silverstein. Courtesy Flying Fish.

books, he has made a number of bizarre solo albums, and is a highly professional country songwriter.

Silverstein's musical past does not suggest much affinity with country music. He grew up with Chicago jazz, blues and folk music. On his own records his humour comes from hipster tradition – he uses black language like Leiber and Stoller did with the Coasters. A highly sophisticated lyricist, it is hard to believe that he could approach country without cynicism, derision or calculation, but he does! His best songs use a conventional form to tell pointed stories, and he has kept faith with the popularity and reality that still underlie country sentiment.

Songs such as **One More On The Way** (a hit for Loretta Lynn) make their humorous point through self-commentary. He also wrote **A Boy Named Sue**, a Grammy award winner for Johnny Cash in 1969, and is best known in the wider pop world for his work with Dr Hook, for whom he wrote **Syvia's Mother** and such comic songs as **The Cover Of Rolling Stone** and **Freaker's Ball**.

In country circles, he is best known for his work with Bobby Bare, who enjoyed hits with **Daddy What If**, **Marie Laveau** and **Alimony**, and a series of concept albums including **Hard Times Hungrys**, **Lullabys, Legends And Lies** and **Drinkin' From The Bottle, Singin' From The Heart**, which were all penned by Silverstein.

Although as a songwriter he specializes in the catchy and clever, occasionally he proves himself capable of the sincerity and sensitivity inherent in country song-writings, as with his **Here I Am Again** (recorded by Loretta Lynn).

Recommended:
The Great Conch Train Robbery (Flying Fish/–)

Asher And Little Jimmy Sizemore

One of the early professional bands on the Grand Ole Opry consisted of Asher Sizemore (born on June 6, 1906, in Manchester, Kentucky) and his young son Jimmy (born on January 29, 1928), who specialized in sentimental hearth-and-home-type ballads and songs, and were one of the first Opry acts to put out a very successful songbook.

Asher put little Jimmy on the radio as early as the age of three, and the Opry's Harry Stone, having heard them on WHAS in Louisville, brought them to the Opry where they stayed for some ten years (1932–1942), recording for Bluebird Records as well, although no big hits emerged.

After World War II they appeared on KXEL, Waterloo, Iowa, SMOX, St. Louis, WHO in Des Moines, and WSB in Atlanta, often with Asher's younger son Buddy Boy, who was killed in Korea late in 1950. Jimmy Sizemore worked into the '80s as a radio executive, while Asher died some years ago, in the '70s.

Ricky Skaggs

Ricky Skaggs' roots and sound are pure bluegrass, western swing and traditional country, and his vocals mountain-flavoured, yet he dominated the country charts in the early 1980s with No.1 singles and gold albums. He won CMA awards as Male Vocalist Of The Year (1982) and Entertainer Of The Year (1985).

Born on July 18, 1954, near Cordell, Kentucky, Ricky had an old-fashioned, mountain upbringing, traditional music and religion being a vital part of family life. A child prodigy, in bluegrass terms, he appeared on TV with Flatt And Scruggs when he was seven and joined Ralph Stanley's band when he was just 15.

Below: Ricky Skaggs' traditional country didn't gain sales in the '90s.

His reputation as a tenor vocalist adept at high harmonies, and his skill on mandolin, fiddle, acoustic guitar and banjo, landed him a job with the bluegrass group Country Gentlemen in Washington D.C. Ricky later formed his own band, Boone Creek, and recorded two very good albums.

When Rodney Crowell left Emmylou Harris' Hot Band in 1977, Ricky filled the vacancy as acoustic guitarist, fiddle player, mandolinist and, above all, second voice to Emmylou. In his three years with the band, he played a big part in influencing her musical direction and had a key role in her **Roses In The Snow**, an LP widely acclaimed as a bluegrass masterpiece.

In 1980, Ricky decided to branch out on a solo career and recorded for the small North Carolina label Sugar Hill. With the help of Emmylou, Albert Lee, Jerry Douglas and Bobby Hicks, he came up with the acclaimed **Sweet Temptation** album and hit the country charts with **I'll Take The Blame**. This led to signing to Epic in Nashville in 1981, where he made the Top 20 with a fine update of Flatt And Scruggs' **Don't Get Above Your Raising**.

The album that followed, **Waitin' For The Sun To Shine**, resulted in the chart-topping singles **Crying My Heart Out Over You** and **I Don't Care** and his CMA award as Male Vocalist Of The Year (1982). Ricky's pure country styling and the sharp instrumentation of his band brought a breath of fresh air to the country charts and also won him a following by pop and rock audiences.

Ricky has cut an enviable number of country chart-toppers, including: **Heartbroke** (1982), **Don't Cheat In Our Hometown** (1983), **Honey (Open That Door)** (1984), **Country Boy** (1985), **Cajun Moon** (1986) and **Lovin' Only Me** (1989). In 1981 he married Sharon White of the Whites family group. The pair have recorded several duets, one of which, **Love Can't Ever Get Better Than This**, made the country Top 10 in 1987, the same year they won the CMA Vocal Duo award. Ricky also gained a Grammy for Best Country Instrumental for **Wheel Hoss**, using it later as the theme for his BBC Radio 2 series Hit It Boys. Ricky has shown that traditional country music can be commercially successful, having achieved his own worldwide success simply by being himself and playing and singing the music he loves. His hit singles continued into the '90s with such Top 20 entries as **Hummingbird** (1990) and **Same Ol' Love** (1992), but generally Skaggs' record sales had started to dip quite considerably and Epic did not renew his contract at the end of 1992.

Recommended:
Sweet Temptation (Sugar Hill/Ritz)
Country Boy (Epic/Epic)

Live In London, Ricky Skaggs. Courtesy Epic Records.

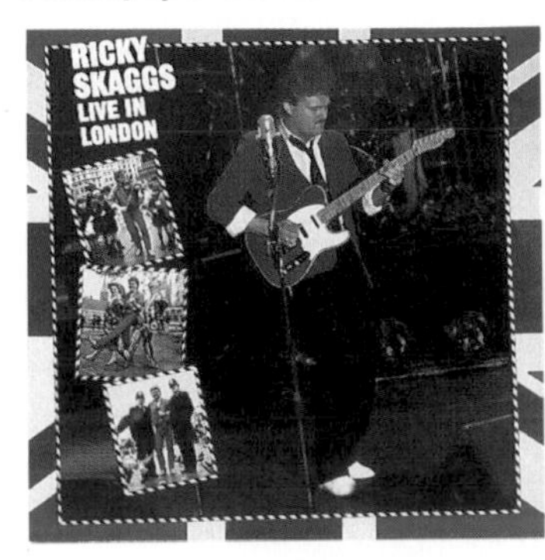

Highways And Heartaches (Epic/Epic)
Favorite Country Songs (Epic/Epic)
Kentucky Thunder (Epic/Epic)
Comin' Home To Stay (Epic/Epic)
My Father's Son (Epic/Epic)

Above: Songwriter Paul Overstreet left SKO in the late '80s to begin a successful solo career.

The Skillet Lickers

Ted Hawkins, mandolin, fiddle; Bert Layne, fiddle; Clayton McMichen, fiddle; Fate Norris, banjo, harmonica; Riley Puckett, guitar; Hoke Rice, guitar; Lowe Stokes, fiddle; Arthur Tanner, banjo, guitar; Gid Tanner, fiddle; Gordon Tanner, fiddle; Mike Whitten, guitar.

An extremely popular and influential Atlanta-based string band of the 1920s and 1930s, the Skillet Lickers was led by Gid Tanner. The band included two other country music figures of great importance in their own right: Clayton McMichen and Riley Puckett.

Tanner (1885–1960), a Georgian like all the band-members through the years, first recorded with Puckett (1894–1945), the blind guitarist, for Columbia in 1924. He continued to record with various permutations of the Skillet Lickers for Columbia and Victor for the next decade, although the band name did not actually exist until McMichen (1900–1970) joined Tanner, Puckett and Norris in 1926.

Their material, for the most part, was composed of fiddle breakdowns, minstrel songs and a bizarre and hilarious series of eighteen spoken comedy records called **A Corn Likker Still In Georgia**, although the fiddle breakdown **Down Yonder** is most closely associated with them.

Their sound together was rough and wild, typically featuring the fine twin fiddling of McMichen and Layne, falsetto shouts and snatches of verses by Tanner, and often Puckett's bluesy singing. All three mainstays went their separate ways after 1934, Tanner dying on May 13, 1960, but the legacy of the Skillet Lickers is one of humorous, extremely good-natured, old-time music played in the most spirited of styles. They were unique, but their sound was in many ways the sound of an old-time sub-style of country music already on the way out as they were recording it.

Recommended:
The Skillet Lickers Volumes 1 and 2 (County/–)
Gid Tanner And His Skillet Lickers (Rounder/–)

SKO/SKB

The group with the unusual name, Schuyler, Knobloch and Overstreet (SKO) was formed in 1986 and, less than a year later, had changed its name to Schuyler, Knobloch and Bickhardt (SKB), throwing country fans and DJs into total confusion. the original SKO came together when three of Nashville's most successful songwriters, Thom Schuyler, Fred Knobloch and Paul Overstreet, decided to work as a recording unit and touring band. Prior to that, all three individual songwriters had penned dozens of country hits and also recorded as solo acts.

Thom Schuyler, born in 1952 in Bethlehem, Pennsylvania, was a carpenter by trade, who moved to Nashville in 1978. He was doing carpentry work on Eddie Rabbitt's recording studio, and this enabled him to demonstrate some of his songs. This opened doors, and he initially penned hits for Lacy J. Dalton, Leon Everette and Eddie Rabbitt, then in 1983 Schuyler signed with Capitol Records. An album, **Brave Heart**, spawned the country hit, **A Little At A Time** (1983). Fred Knobloch was born in Jackson, Mississippi and started out with a rock band, Let's Eat, in the late '70s in Atlanta, Georgia. In 1980 a move to Los Angeles led to him signing with Scotti Bros Records, scoring on both the pop and country charts with **Why Not Me** (1980), a duet with Susan Anton on **Killin' Time** and a revival of Chuck Berry's **Memphis** (1981). Due to the country success of these records, he moved to Nashville in 1983. Paul Overstreet, from Newton, Mississippi, is a more traditional country songwriter, who moved to Nashville straight from high school. Originally working the clubs, he made his mark penning hits for George Jones, Michael Martin Murphey, the Judds and Randy Travis.

Signing to the new MTM Records, SKO immediately hit the Top 10 with **You Can't Stop Love** (1986), followed by the chart-topping **Baby's Got A New Baby** (1987) and **American Me**, which was a Top 20 entry. The trio cut an album, then Overstreet opted for a solo career, which has seen him achieving notable success on RCA. Rather than disband, Schuyler and Knobloch brought in another writer, Craig Bickhardt, the group now becoming SKB. Bickhardt was from Pennsylvania, where he was leader of Wire And Wood in 1972, a band that had opened for Bruce Springsteen, Stephen Stills and Harry Chapin. A few years later he formed the Craig Bickhardt Band and gained a sizeable following. He signed a writer's contract with Screen Gems in New York and had songs recorded by Art Garfunkel, B.B. King and Randy Meisner, who had a Top 40 pop hit with **Never Been In Love**. He moved to Nashville in 1984. One of his first projects was writing songs for the film 'Tender Mercies', which saw Bickhardt make the country charts with **You Are What Love Means To Me** (1984).

Like SKO, SKB recorded an album, **No Easy Horses**. the title song became a Top 20 hit in 1987, followed by **Givers And Takers**, which made the Top 10 the following year. MTM Records closed its doors at the end of 1988, and that, more or less, marked the end of SKB.

Recommended:
S-K-O (MTM/–)
SKB – No Easy Horses (MTM/–)

Arthur 'Guitar Boogie' Smith

Leader of an outfit known as the Crackerjacks, guitarist-banjoist-mandolin player Arthur Smith was born on April 1, 1921, in Clinton, South Carolina. One of the state's most popular performers, he played on radio station WBT, Charlotte, for over 20 years, achieving national fame when his recording of **Guitar Boogie**, originally issued by the Superdisc label, was subsequently released on MGM, providing Smith with a 1947 million-seller.

After several years of hits for MGM, many of which were of the eight-to-a-bar genre, Smith moved on to other labels. He signed to Starday and Dot during the '60s and has since recorded for Monument and CMH. His **Feudin' Banjos** (frequently called **Duellin' Banjos**) became internationally well known as a result of its prominent part in the film 'Deliverance'. He has copywrited over 500 songs and written a book, 'Apply It To Life', in 1991.

Recommended:
Battling Banjos (Monument/–)
Feudin' Again – with Don Reno (CMH/–)

Fiddlin' Arthur Smith

Born in 1898 in Dixon County, Tennessee, Smith was a railroad worker who, during the early '30s, joined Sam and Kirk McGee in a trio known as the Dixieliners. Becoming an Opry favourite, Smith toured under the auspices of WSM for several years, sometimes playing with the McGees, at other times working with his own trio. His trio acquired a national reputation with their recording of **There's More Pretty Girls Than One** in 1936.

Later, Smith moved on to play with the Delmore Brothers, and throughout the '40s and '50s worked variously as a sideman or with units of his own, much of it on the West Coast, where he appeared in western films.

The 1960s brought him a new lease of life, thanks to the advent of the folk festivals and the rediscovery of many folk heroes by a new and youthful audience. This period was marked by **Fiddlin' Arthur Smith**, an album for Starday in 1963, and a Mike Seeger-masterminded Smith–McGee Brothers set for Folkways. Smith died in 1973.

Recommended:
The McGee Brothers With Arthur Smith (Folkways/–)

Cal Smith

Born Calvin Grant Shofner on April 7, 1932, in Gans, Oklahoma, Smith was raised in Oakland, California. He became a regular on the California Hayride TV show, gaining his first regular club job in San Jose, California, during the early 1950s.

Later, after engagements that included a spell as a DJ, Smith became MC and vocalist with Ernest Tubb's Texas Troubadours. Through Tubb, Smith became signed to Kapp Records, initially charting with **The Only Thing I Want** in 1967, then continuing to keep the label supplied with a number of lower-level hits (**Drinking Champagne**, **It Takes All Night Long**, **Heaven Is Just A Touch Away**, etc.) through to 1971. From this time, his discs began appearing on Decca.

Around the same time, his chart placings began to rise. **I've Found Someone Of My Own** became a Top 5 hit in 1972. In 1973, he joined the MCA label and earned yet another top placing with **The Lord Knows I'm Drinking**, capping even this success with that of **Country Bumpkin**, a release that won him his first CMA award – for Single Of The Year in 1974.

Further hits followed with **Jason's Farm** (1975), **MacArthur's Hand** (1976) and **I Just Came Home To Count The Memories** (1977). Then, mysteriously, Cal faded from the country charts, though he continued to record for MCA until early 1980 and has since had recordings released on the small Soundwaves label.

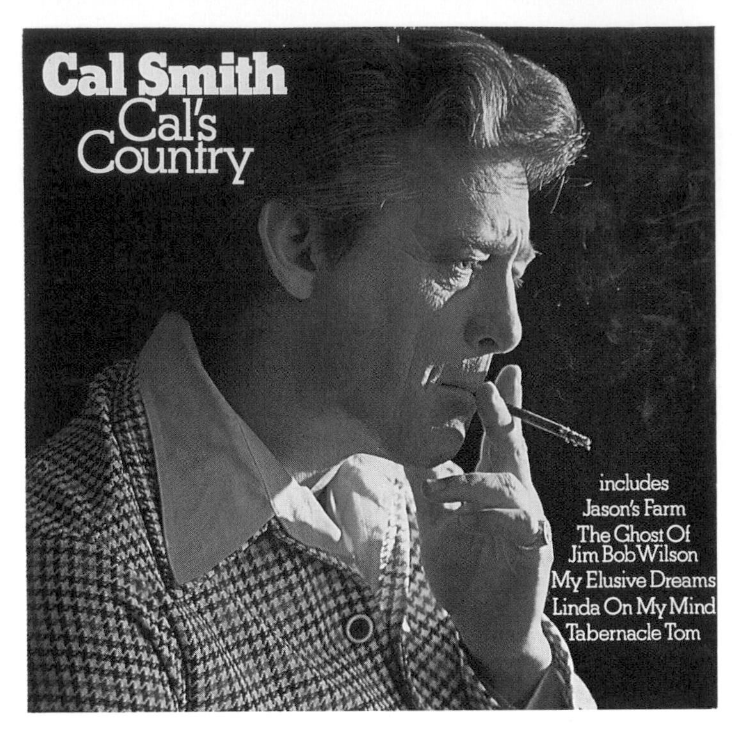

Cal's Country, Cal Smith. Courtesy MCA Records.

Famous Country Music Makers, Connie Smith. Courtesy RCA Records.

Recommended:
Cal's Country (MCA/MCA)
Introducing (MCA/MCA)
I Just Came Home To Count The Memories (MCA/–)

Carl Smith

Born in Maynardsville, Tennessee, on March 15, 1927, Carl Smith sold flower seeds to pay for his first guitar, then cut grass to pay for lessons.

His first break in show business came a few years later, with radio station WROL, Knoxville, Tennessee. Then in 1950, Jack Stapp, at that time programme director of WSM, asked Smith to come to Nashville and work on the WSM morning show.

Soon after, he won a place on the Opry, signed a contract with Columbia Records and, with his second release, **Let's Live A Little**, had a smash hit. That same year (1951), he was voted No.1 country singer by several polls and accrued three further chart-busters via **If Teardrops Were Pennies**, **Mr Moon** and **Let Old Mother Nature Have Her Way**.

During the '50s and '60s he was rarely out of the charts, averaging around three hits per year, the biggest of these being **Don't Just Stand There**, **Are You Teasing Me?** (1952); **Trademark**, **Hey, Joe** (1953); **Loose Talk**, **Go, Boy, Go** (1954), **Kisses Don't Lie** (1955) and **Ten Thousand Drums** (1959).

Following his success on radio, Smith moved on to TV, working for such shows as Four Star Jubilee (ABC/TV) and Carl Smith's Country Music Hall, a weekly networked show in Canada, syndicated to several US stations. He has also been featured in two films – 'The Badge Of Marshall Brennan' and 'Buffalo Guns'.

In 1957, Smith married country singer Goldie Hill (his second wife; the first having been June Carter) and moved to a ranch near Franklin, Tennessee.

Early in 1974, the singer left Columbia Records after a 24-year stay with the label and moved on to Hickory Records, for whom he had some minor successes. In recent years, however, Carl has preferred to spend most of his time on his ranch, though he has re-recorded his older hits for a successful TV-advertised album.

Recommended:
Sings Bluegrass (Columbia/–)
Legendary (Gusto-Lakeshore/–)
This Lady Loving Me (Hickory/DJM)
The Essential (Columbia/–)
Old Lonesome Times, 1951–1956 (Rounder/–)

Connie Smith

In 1963, Connie Smith won an amateur talent contest in Ohio, and was heard by Bill Anderson. The result was an offer to sing with Anderson and an opportunity to embark on a recording career which allowed her to accrue a tally of nearly 30 Top 10 hits.

Born in Elkhart, Indiana, on August 14, 1941, one of a family of 16, Connie learned to play guitar while hospitalized with a serious leg injury. A performer at local events, she later married and settled down. But following her discovery by Anderson, she became signed to RCA Records, creating a tremendous impact with her initial release **Once A Day**, an Anderson-penned song that became a country No.1 in 1964.

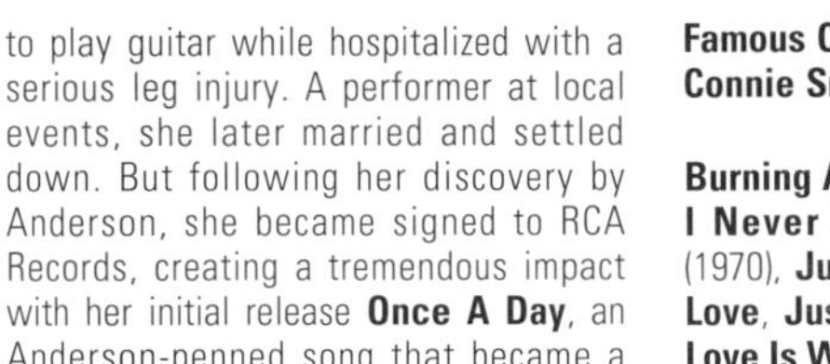

An overnight success, Connie was signed to make her TV debut on the Jimmy Dean Show, won the Most Promising Singer of 1964 from 'Billboard' and – in the wake of three early '65 hits with **Then And Only Then**, **Tiny Blue Transistor Radio** and **I Can't Remember** – she became an Opry star on June 13, 1965.

Since that time, Connie has appeared on scores of TV and radio shows and has featured in country-oriented films. Also, recording mainly Bill Anderson and Dallas Frazier numbers, she has had hits with **Ain't Had No Lovin'**, **The Hurtin's All Over** (both 1966), **Cincinnati, Ohio**, **Burning A Hole In My Mind** (both 1967), **I Never Once Stopped Loving You** (1970), **Just One Time** (1971), **If It Ain't Love**, **Just For What I Am** (both 1972), **Love Is What You're Looking For** (1973), **Till (I Kissed You)** and **I Don't Wanna Talk About It Anymore** (1976).

One of the many country artists with deep religious convictions, the diminutive Connie toured Australia, New Zealand and Japan during 1972, raising funds for the children of Bangladesh. She often plays on radio and TV shows, and appeared in Britain in 1990.

Recommended:
Famous Country Music Makers (–/RCA)
Songs We Fell In Love To (Columbia/CBS)
If It Ain't Love (RCA/RCA)
Live in Branson (Branson/–)

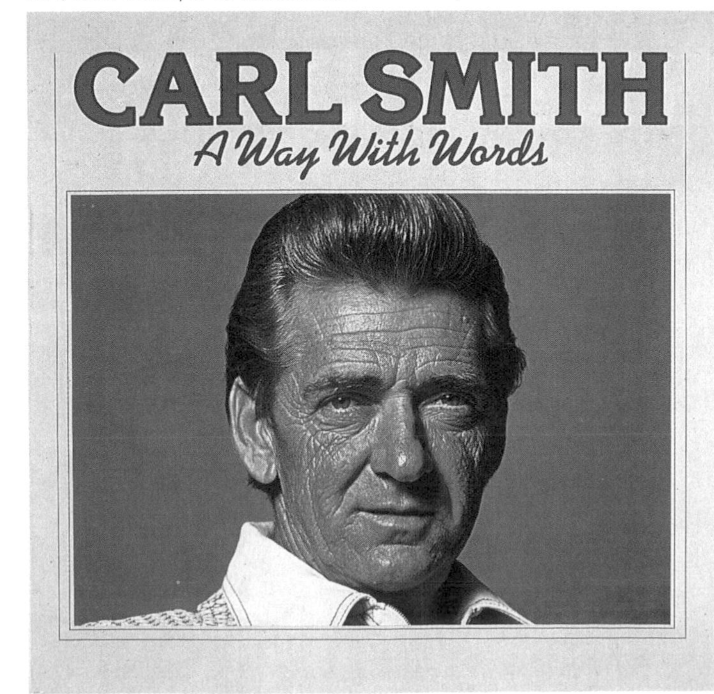

A Way With Words, Carl Smith. Courtesy DJM Records.

Darden Smith

Singer-songwriter Darden Smith was born on March 11, 1962 in Blenham, Texas. A modern troubadour, Smith's music has rich traces of rockabilly, blues, country and rock'n'roll, yet he neatly fits into the Texas singer-songwriter genre with his poetic lyrics and heart-tugging melodies. He grew up on a farm between Austin and Houston and was writing his first embryonic songs before he was ten. His family moved to Houston in 1975 and Darden played in his older brother's band during his high school days while pursuing his songwriting. When he started college in San Marcos, he put playing aside for some months, but, discovering the thriving Austin club scene, Smith soon strapped his guitar on and played three nights a week to earn a bit of money. He graduated from the University of Texas and immediately set out to put his own songs on record.

The self-produced **Native Soil**, put out on his own RediMix label in 1986, was a low-key collection of captivating vignettes of real people in real life situations. Working regularly on the Texas music scene with his own 'band', the Big Guns, comprised of himself on guitar, Roland

Denney on stand-up bass and Paul Pearcy on drums, Smith's reputation spread throughout country music circles. He signed with Epic Records in Nashville, and his debut, simply titled **Darden Smith**, was produced by Ray Benson of Asleep At The Wheel and featured guest appearances from Lyle Lovett, Nanci Griffith, C.J. Chenier, Sonny Landreth and various Asleep At The Wheelers.

Not fitting into the accepted confines of mainstream country, Darden only made a brief appearance on the country charts with **Little Maggie**, a track from his debut Epic album in 1988. Realizing that he was unlikely to gain country radio plays, Darden Smith's contract was moved away from Nashville and his career was handled from New York which gave him a better springboard for both an international breakthrough, and building a bigger following in America. He has succeeded in developing a cult following, and, now on the main Columbia label, has produced some very fine albums, which typify his wide-ranging, country-tinged music.

Recommended:
Little Victories (Columbia/Columbia)
Trouble No More (Columbia/Columbia)
Native Soil (Redimix/–)
Darden Smith (Columbia/CBS)

Margo Smith

A petite, ex-school teacher, born Betty Lou Miller on April 9, 1942 in Dayton, Ohio, Margo initially used her vocal talents singing country and folk songs as a teaching aid in her kindergarten classes.

Margo started writing songs in her spare time and cut a demo tape which she took to Nashville and promptly signed a recording contract with Chart Records.

A few singles were released, but they made little impression. However, in 1975 she signed a new contract with 20th Century Records and made her debut on the country charts with the self-penned **There I Said It**, which slowly headed towards the Top 10. At this point, Margo gave up teaching.

Shortly after this, 20th Century closed their Nashville offices and Margo was without a label. Eventually she signed with Warner Brothers and was soon back in the Top 10 with **Save Your Kisses For Me**, **Take My Breath Away** (both 1976) and **Love's Explosion** (1977), before making the coveted top spot with **Don't Break The Heart That Loves You** (1978).

For the next couple of years Margo was rarely out of the charts, scoring another No.1 with **It Only Hurts A Little While** (1978), a Top 5 hit with her update of **Little Things Mean A Lot** (1978) and Top 10 entries with **Still A Woman** and **If I Give My Heart To You**. She also recorded duets with Rex Allen Jr of **Cup O' Tea** and **While The Feeling's Good** (1980 and 1981 respectively), with plans for an album to follow.

Changes at Warner Brothers during 1981 meant that Margo was dropped by the label, but she has since recorded for AMI, Moonshine and Bermuda Dunes, achieving minor hits but failing to make a return to the Top 10. In 1986 she began her climb back to the top once more by signing with Dot/MCA and cutting an album, simply titled **Margo Smith**. A tremendously talented lady and exciting stage performer, Margo is one of the few country ladies who yodel on stage.

Recommended:
Margo Smith (MCA–Dot/–)
Diamonds And Chills (Warner Bros/–)
Just Margo (Warner Bros/–)
Song Bird (Warner Bros/–)

Below: Connie Smith had her heyday in the late '60s and the '70s, and now works for charity and makes guest appearances on television and radio.

As Long As There's A Sunday, Sammi Smith. Courtesy WEA Records.

Sammi Smith

Considered a country rebel, to be mentioned alongside Jennings and Nelson, Sammi originally made her mark as a cross-over artist of some potential. Her version of Kristofferson's **Help Me Make It Through The Night** (which gained CMA Single Of The Year in 1971) sold over two million copies.

Born in Orange, California on August 5, 1943, she grew up in Oklahoma, playing clubs in that area at the age of 12. Encouraged in her career by Oklahoma City songwriter/recording studio owner Gene Sullivan, she was heard by Tennessee Three bassist Marshall Grant, who persuaded her to make the inevitable trip to Nashville where she became a Columbia recording act.

Some moderate hits followed including **So Long, Charlie Brown** (1968) and **Brownsville Lumberyard** (1969), after which Sammi moved on to the Mega label in 1970, recording **He's Everywhere**, the title track becoming a chart single. A second track from the same album was then selected as a follow-up. The amazingly successful **Help Me Make It Through The Night** prompted Mega to title the parent album in line with the single and to embark on a re-promotion campaign.

However, Sammi found it hard to hit the charts consistently, though she has enjoyed Top 20 hits with **I've Got To Have You** (1972), **Today I Started Loving You Again** (1975), **Sunday School To Broadway** (1976), **Loving Arms** (1977) and **What A Lie** (1979) as she has moved through such labels as Elektra, Zodiac, Cyclone and Sound Factory.

One of the most soulful singers in country music, Sammi has the ability to transform even the most banal lyrics into a charged, emotional experience.

Recommended:
Girl Hero (Cyclone/–)
New Winds, All Quadrants (Elektra/–)
Mixed Emotions (Elektra/–)

Hank Snow

The most revered of all Canadian country performers, Clarence Eugene Snow was born in Liverpool, Nova Scotia, on May 9, 1914. Leaving home at 12, he became a cabin boy on a freighter for four years, but began singing in Nova Scotia clubs, obtaining his first radio show in 1934, on CHNS, Halifax, Nova Scotia.

Known initially as the 'Yodelling Ranger' and later as the 'Singing Ranger', he

Country Music Hall Of Fame, Hank Snow. Courtesy RCA Records.

became signed to RCA-Victor in 1934, his first sides being **Lonesome Blue Yodel** and **Prisoned Cowboy**.

Despite a move to the US during the mid-'40s and appearances on such shows as the WWVA Jamboree, Snow remained virtually unknown south of the Canadian border until 1949, when, following a disappointing Opry performance on January 7, his recording of **Marriage Vows** became a Top 10 hit.

In 1950 he became an Opry regular, that same year seeing the release of **I'm Moving On**, a self-penned hit that became a US No.1. Throughout the '50s and early '60s, Snow continued to provide RCA with a huge number of Top 10 singles – several of them being train songs, revealing his debt to Jimmie Rodgers.

Too numerous to list, Snow's hits include **Golden Rocket** (1950), **Rhumba Boogie** (1951), **I Don't Hurt Anymore** (1954) and the tongue-twisting **I've Been Everywhere** (1962), all country No.1s. **I'm Moving On** and **I Don't Hurt Anymore** both became million-sellers.

Snow, who has consistently fought against what he believes to be over-commercialization of country music, has proved to be one of country's most travelled ambassadors, appearing in his somewhat stagey cowboy attire at venues all over the world.

His eldest son, Jimmie Rodgers Snow, is both a country performer and a travelling evangelist. Hank still appears on the Opry, though he rarely tours nowadays, and he has had no inclination to record since being dropped by RCA after a record-breaking 45 years with the same company. He was elected to the Country Music Hall Of Fame in 1979.

Recommended:
Famous Country Music Makers (–/RCA)
Award Winners (RCA/RCA)
Grand Ole Opry Favorites (RCA/RCA)
Best Of (RCA/RCA)
The Yodeling Ranger, 1936–1947 (–/Bear Family)
The Singing Ranger – I'm Movin' On (–/Bear Family)
The Singing Ranger – Volume 2 (–/Bear Family)
The Singing Ranger – Volume 3 (–/Bear Family)
Country Music Hall Of Fame (–/RCA)
Hits Covered By Snow (RCA/RCA)

Jo-El Sonnier

Acclaimed French-Cajun accordionist, Jo-El Sonnier was born on October 2, 1946, in Rayne, a rural area of Louisiana. Something of a child prodigy, at only six he was singing and playing accordion on his own 15-minute radio spot in nearby Crowley, and recorded his first single, **Tes Yeux Bleus (Your Blue Eyes)** when he was 13. Working the Louisiana club circuit, he recorded prolifically for the local Swallow and Goldband records and won first prize in the Mamou Mardi Gras competition in 1968, but found that his strong French accent was proving something of a handicap. In 1972 he moved to Los Angeles, where he built up a reputation as accordion player in the bars and honky-tonks. Two years later he relocated to Nashville, where, as well as making an impression as a songwriter and studio musician, he landed a contract with Mercury Records, scoring a few minor country hits during 1975–76 under the name Joel Sonnier. In 1980 he returned to Louisiana and recorded **Cajun Life**, performed entirely in French and released by Rounder Records. Renewing friendships he'd built up in Nashville, Sonnier moved back to Los Angeles in 1982 and formed an all-star band, Jo-El Sonnier And Friends, with Sneaky Pete Kleinow, Albert Lee, Garth Hudson and David Lindley. Elvis Costello saw his show and recruited Sonnier for his **King Of America** album.

Commuting between the West Coast, Nashville and Louisiana, Sonnier eventually landed a recording contract with RCA in Nashville in 1987. His first single, **Come On Joe**, a song written for and about him by a New York songwriter in 1974, made the country Top 40 in 1987. The following year he made the Top 10 with **No More One More Time** and **Tear-Stained Letter**. With his exciting mixture of Louisiana swamp-rock, French-Cajun tunes, infectious country-rock and blues, and a tight-knit powerhouse band, Sonnier built up a reputation as one of the best live acts around. He has never really successfully duplicated the raw magic of his live shows on to record. He had further minor hits with **Rainin' In My Heart** (1988) and **If Your Heart Should Ever Roll This Way Again** (1989), and in 1991 took his rootsy Cajun music to Capitol.

Recommended:
Come On Joe (RCA/RCA)
Have A Little Faith (RCA/RCA)
Tears Of Joy (Capitol/–)
Cajun Life (Rounder/–)
The Complete Mercury Sessions (Mercury/–)

Sons Of The Pioneers

Originally a guitarist/vocals trio when formed by Roy Rogers, Bob Nolan and Tim Spencer in 1934 as the Pioneer Trio, the name changed to Sons Of The Pioneers in deference to the American Indian heritage of members Karl and Hugh Farr.

They did much radio work during the '30s and recorded variously for Decca, Columbia and RCA. Films also figured large for them and they appeared in many of those featuring Rogers.

Other members of the group have included Lloyd Perryman, Ken Carson, Ken Curtis, Pat Brady, Doye O'Dell, Dale Warren, Deuce Spriggins, Tommy Doss, Shug Fisher and Rusty Richards.

Bob Nolan composed the group's biggest hits, **Tumbling Tumbleweeds** and **Cool Water**, and Spencer, who left in 1950 but managed the Sons until 1955, composed **Cigarettes, Whiskey And Wild Wild Women**, **Careless Kisses** and **Roomful Of Roses**. Spencer died on April 26, 1974 aged 65, in California. However, the Sons were still performing at that time, led by Perryman who had originally joined the group back in 1936 and passed away on May 3, 1977, aged 60.

A word should be said about Bob Nolan, whose songwriting may be the finest ever to appear in country music. A brilliant poet with an inventive ear for melody and harmony, he virtually invented the sound and style of western harmony singing single-handedly. He supplied the once thriving field with the great majority of its classic songs, which, in addition to the above, include **Trail Herding Cowboy**, **A Cowboy Has To Sing**, **One More Ride**, **Way Out There** and **Song Of The Bandit**.

Nolan himself recorded an acclaimed solo album, **Sound Of A Pioneer**, for Elektra in 1979. The following year he died at his home in California on June 15, aged 72, and the Sons Of The Pioneers were elected to the Country Music Hall Of Fame the following October.

Recommended:
Cowboy Country (–/Bear Family)
Riders In The Sky (Camden/–)
Sons Of The Pioneers (Columbia/CBS)
Country Music Hall Of Fame (MCA/–)

Southern Pacific

Stu Cook, bass; Kurt Howell, keyboards; David Jenkins, lead singer; Keith Knudsen, drums; John McKee, guitar, fiddle.

A modern country-rock band, Southern Pacific are a 'supergroup' formed in Los Angeles in 1985, comprising ex-Doobie Brothers Keith Knudsen and John McKee, plus Stu Cook (formerly of Creedence Clearwater Revival), Kurt Howell (who had previously played with Waylon Jennings and Crystal Gayle) and Tim Goodman. They signed to Warner Brothers and made their debut on the charts with **Someone's Gonna Love Me Tonight** (1985).

Their self-titled album showed a strong resemblance to early Eagles, with Southern Pacific playing with all the energy of a blaring rock band but also free gliding and as poetic and graceful as classic country-rock outfits. Emmylou Harris added vocals to **Thing About You**, which became the band's first Top 20 country hit. This was followed by **Perfect Stranger**, but, just as the band was really beginning to take off, lead singer Goodman departed. He was replaced by David

Jo-El Sonnier's 1987 eponymous album. Courtesy RCA Records.

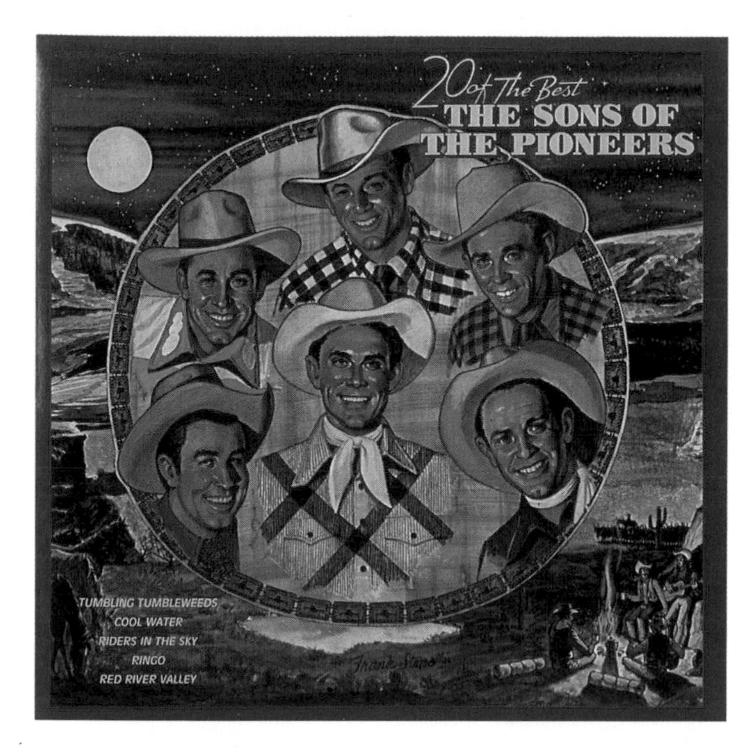

Jenkins, a former member of rock band Pablo Cruise, who also sang back-up vocals on the Huey Lewis album, **Fore**. The change didn't slow the band down as they scored a series of Top 10 hits with **Reno Bound** (1986), **New Shade Of Blue** (1988), **Honey, I Dare You** and **Anyway The Wind Blows** (1989), the latter from the film 'Pink Cadillac'.

Recommended:
Southern Pacific (Warner Bros/Warner Bros)
Zuma (Warner Bros/Warner Bros)
Country Line (Warner Bros/–)

Red Sovine

King of the truck-driving songs and narrations, Woodrow Wilson Sovine was born in Charleston, West Virginia on July 17, 1918. He learned guitar at an early age and tuned in to C&W radio stations, obtaining his own first radio job with Jim Pike's Carolina Tar Heels on WCHS, Charleston, West Virginia, in 1935. Later the unit moved on to play the WWVA Jamboree, Wheeling, West Virginia.

During the late 1940s, Sovine formed his own band, the Echo Valley Boys. He and the group gained their own show on WCHS. Then on June 3, 1949, Hank Williams left the Louisiana Hayride to become an Opry regular and Sovine's band was drafted in as a replacement, the Echo Valley Boys also taking over Williams' daily Johnny Fair Syrup Show stint.

Teddy Bear, Red Sovine. Courtesy RCA Records.

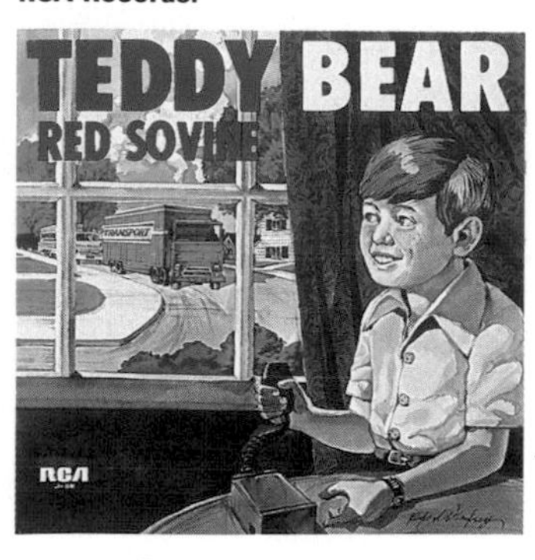

20 Of The Best, Sons Of The Pioneers. Courtesy RCA Records.

From 1949 until 1954, Sovine remained a star attraction on the Hayride, during that period striking up a friendship with Webb Pierce. The twosome performed duets on the show, combining to write songs and, in turn, both became Opry regulars. On disc they joined up for **Why, Baby Why?**, a country No.1 in 1956, following this with **Little Rosa**, a Top 10 hit that same year.

Though Sovine remained a top rated performer throughout the late '50s and early '60s, his name disappeared from the charts until 1964, when a Starday release, **Dream House For Sale**, climbed into the listings. This was followed a few months later by **Giddyup Go**, which provided the singer with yet another No.1.

From that time on, Sovine continued adding to his list of chart honours, making a major impression in 1967 with **Phantom 309**. Then, in the mid-'70s, following a flood of recordings based on tales of CB radio, Sovine came into his own, achieving one of his biggest-ever successes with **Teddy Bear**, a highly sentimental tale regarding a crippled boy, his CB radio and a number of friendly truckers. He had, at the age of 58, finally earned a million-selling record.

Red was killed in a motor accident in Nashville on April 4, 1980, but had a posthumous British pop hit with **Teddy Bear** when it was re-released and reached No.2 in the charts in 1981, selling more than half-a-million copies.

Recommended:
Little Rosa (–/Release)
Teddy Bear (Starday/RCA)
Woodrow Wilson Sovine (Starday/–)
Classic Narrations (Starday/–)

Billie Jo Spears

Though she has a voice that is hardly in Opry tradition – take away the country backings and you are left with a bluesy sound befitting an uppercrust torch singer – Billie Jo was country-raised (born on January 14, 1937, Beaumont, Texas) and appeared on the Louisiana Hayride in her early teens. She performed **Too Old For Toys, Too Young For Boys**, a ditty which she recorded on the reverse of a Mel Blanc Bugs Bunny-type disc at the age of 13.

In 1964, country songwriter Jack Rhodes heard her sing and talked her into a trip to Nashville where she became signed to UA Records. However, her first country hit came with **He's Got More Love in His Little Finger**, a Capitol release of 1968. Billie Jo reached the Top 5 during the following year with **Mr Walker It's All Over**. Apart from an elongated chart stay with **Marty Gray** (1970), her other Capitol sides, though fair sellers, failed to emulate the success of **Mr Walker**.

A change of fortune occurred following a switch to UA in 1974. **Blanket On The Ground** (1975), a Roger Bowling song dealing with the delights of alfresco lovemaking, established Billie Jo with an international reputation.

Since then she has scored with **What I've Got In Mind**, **Misty Blue** (both 1976), **If You Want Me** (1977), **Lonely Hearts Club**, **'57 Chevrolet** (both 1978) and **I Will Survive** (1979). She has maintained her successes with **What I've Got In Mind** and **Sing Me An Old-Fashioned Song**, though in America she has failed to gain a contract with a major label for a number of years.

Recommended:
We Just Came Apart At The Dreams (–/Premier)
For The Good Times (–/Music For Pleasure)
Special Songs (Liberty/Liberty)
50 Original Tracks (–/EMI)
Ode To Billie Jo (–/Capitol)

Carl T. Sprague

Known as the 'Original Singing Cowboy', Sprague was born near Houston, Texas, in 1895. A cowboy music enthusiast in his college days, he led a band while at Texas A&M, playing on the campus radio station.

In August, 1925, inspired by Vernon Dalhart's hit record, **The Prisoner's Song**, he recorded ten songs for Victor. His initial release – **When The Work's All Done This Fall** – sold nine hundred thousand copies. Further sessions (in 1926, 1927 and 1929) ensued, at which Sprague recorded mainly traditional cowboy material from the late nineteenth century.

A man of many talents – including insurance salesman, Army officer, coach, garage operator, etc – Sprague, who settled in Bryan, Texas, performed at various folk festivals during the 1960s and recorded for the German Folk Variety label in 1972. Following a short illness, he died at his home in 1978.

Recommended:
Carl T. Sprague (–/Bear Family)
Classic Cowboy Songs (–/Bear Family)

Joe Stampley

Born on June 6, 1943, in Springhill, Louisiana, Stampley was influenced by both country entertainers and rock'n'rollers like the Everly Brothers and Jerry Lee Lewis. His first forays into recordings were made in 1959 when he cut some rock'n'roll records for Imperial and Chess.

Becoming a member of the Uniques, a pop outfit that had hits in the late '60s with **Not Too Long Ago** and **All These Things**, Stampley started making a name for himself as a songwriter, signing with Gallico Music in Nashville. This led to a move towards country music and signing a contract with Dot Records in 1969.

He achieved a minor country hit via **Take Time To Know Her** (1971), finally making the Top 10 with **If You Touch Me (You've Got To Love Me)** in 1972 and achieving his first No.1 with **Soul Song** the following year. He achieved an impressive tally of major hits with such releases as **I'm Still Loving You** (1974), **Roll On Big Mama** (1975) – his first for Epic Records, **All These Things** (1976), **Everyday I Have To Cry Some** (1977), **Do You Ever Fool Around** (1978) and **Put Your Clothes Back On** (1979).

In 1979 Stampley teamed up with Moe Bandy, resulting in a No.1 hit with **Just Good Ole Boys** and the Top Vocal Duo award by the ACM (1979) and CMA (1980). Stampley consolidated his position in the '80s with such hits as **There's Another Woman** (1980), **Whiskey Chasin'** (1981), **Back Slidin'** (1982) and **Double Shot Of My Baby's Love** (1983), plus more duets with Moe Bandy, including the 1984 chart-topper **Where's The Dress**.

Recommended:
Soul Song (–/Ember)
Saturday Nite Dance (Epic/–)
Ten Songs About Her (Epic/–)
After Hours (Epic/–)
I'm Goin' Hurtin' (Epic/–)

Stanley Brothers

Responsible for some of the most beautiful harmony vocals ever to emerge from the bluegrass scene, guitarist and lead vocalist Carter Glen Stanley (born in McClure, Virginia, on August 27, 1925) and his brother, banjoist and vocalist Ralph Edmond Stanley (born in Stratton, Virginia, on February 25, 1927) formed an old-time band, the Stanley Brothers And The Clinch Mountain Boys, in 1946, and began broadcasting on radio station WCYB, Bristol, Virginia.

Recording for the small Rich-R-Tone label in 1948, they cut **Molly And Tenbrooks**. The band switched direction and played in the bluegrass style of Bill Monroe with Ralph Stanley utilizing the three-finger method of banjo playing popularized by Earl Scruggs.

In March 1949, the Stanleys signed for Columbia Records and began cutting a series of classic bluegrass sides (all vocals), retaining mandolin player and vocalist Pee Wee Lambert from their previous band and adding various fiddle and bass players at different sessions. George Shuffler replaced Lambert and became part of the Stanley Brothers' sound just prior to the band's last Columbia session in April 1952.

Throughout the '50s and '60s, the Stanleys recorded for such labels as Mercury, Starday and King, often cutting purely religious material. They also engaged on many tours, playing a prestigious date at London's Albert Hall as part of their European tour in March, 1966. But it was to be the Stanley Brothers' only British appearance. Carter Stanley died in a Bristol, Virginia hospital on December 1 that same year.

Since his brother's death, Ralph Stanley has kept the tradition of the Clinch Mountain Boys alive – though his music

Above: Bluegrass harmony vocal duo Carter (left) and Ralph Stanley.

has become increasingly traditional in character.

The music of the Stanleys has been reactivated by younger artists like Emmylou Harris, Dan Fogelberg and Chris Hillman, who have used many of the brothers' classics in their repertoires.

Recommended:
Recorded Live Volumes 1 & 2 (Rebel/–)
The Best Of (Starday/–)
Saturday Night And Sunday Morning (Freeland/–)

Ralph Stanley:
Ralph Stanley, A Man And His Music (Rebel/–)
Old Country Church (Rebel/–)

Kenny Starr

Loretta Lynn protégé Kenny Starr was born in Topeka, Kansas on September 21, 1953, his family later moving to Burlingame, Kansas, where Starr grew up.

At an early age he began visiting the local Veterans Of Foreign Wars hall, where he would unplug the jukebox and sing for nickles and dimes. By the age of nine he was leading his first band, the Rockin' Rebels. This was quickly superseded by another group, Kenny And The Imperials, which toured the area, earning Starr 10 to 15 dollars a night.

At 16 he became a country entertainer, initially leading a band called the Country Showman, later winning a talent contest held in Wichita after singing his version of Ray Price's **I Won't Mention It Again**.

Local promoter Hap Peebles saw Starr's performance on the show and asked if he would appear on a forthcoming Loretta Lynn and Conway Twitty concert, which Starr did, winning a standing ovation. After the concert, Loretta offered him a job in her own road show. She also helped him obtain a recording contract with MCA.

Left: Kenny Starr started his career by appearing with Loretta Lynn.

A singer-songwriter-guitarist, Starr had a No.1 country hit with his own **The Blind Man In The Bleachers** in January, 1976. He followed this with minor successes **Tonight I Face The Man Who Made It Happen** (1976), **Hold Tight** (1977) and **Slow Drivin'** (1978), before mysteriously fading from the limelight.

Recommended:
The Blind Man In The Bleachers (MCA/–)

The Statler Brothers

The Statlers vocal group originally consisted of Philip Balsley (born Augusta County, Virginia, August 8, 1939), Don Reid (born Staunton, Virginia, June 5, 1945), Harold Reid (born Augusta County, Virginia, August 21, 1939) and Lew DeWitt (born Roanoke County, Virginia, March 8, 1938). The Statlers – first Harold (bass), Lew (tenor) and Phil (baritone) – began singing together in 1955 at Lyndhurst Methodist Church in Staunton, Virginia. In 1960, Harold's younger brother Don joined the group – then known as the Kingsmen – and became front-man. The quartet passed an audition to become part of the Johnny Cash Show some three years later. At this point, they changed their name to the Statler Brothers after espying the name Statler on a box of tissues in a hotel room.

In 1965 they went further up the ladder after recording **Flowers On The Wall**, a song penned by DeWitt. This Columbia release became a Top 5 pop hit, also gaining a high place on the country charts. The group won two Grammy awards as a consequence.

Further hits followed – **Ruthless** and **You Can't Have Your Kate And Edith Too** (both 1967) proving among the most popular – but it was not until 1970 and a new recording contract with Mercury that the Statlers moved into top gear, immediately gaining a second cross-over hit with **Bed Of Roses**.

They became the undisputed kings of country vocal groups, winning a Grammy in 1972 for Best Vocal Performance for **Class Of '57** and scoring such major country hits as **I'll Go To My Grave Loving You** (1975), **Do You Know You Are My Sunshine**, **Who Am I To Say** (1978), **Charlotte's Web** (1980), **Don't Wait On Me** (1981) and **You'll Be Back (Every Night In My Dreams)** (1982). They were voted CMA Vocal Group Of The Year every year from 1972 to 1977, and then again in 1979 and 1980.

Due to ill-health, Lew DeWitt retired from the group in July 1982, though he did briefly pursue a solo career, releasing the album **On My Own** in 1985. He died from Crohn's disease in Waynesboro, Virginia on August 15, 1990. His place in the group was taken by Jimmy Fortune, who not only brought a high, clear tenor singing voice to the Statlers, but also a songwriting talent, penning several of their subsequent country hits. These continued with **Oh Baby Mine (I Get So Lonely)** and the chart-topping **Elizabeth** (1983), **Atlanta Blue** (1984), two more No.1s with **My Only Love** and **Too Much On My Heart**, plus a revival of **Hello Mary Lou** (all 1985), **Count On Me** (1986) and **Forever**

Left: The Statler Brothers have their own TV show on TNN.

(1987). While the hits started to become less frequent, the Statlers have maintained a very strong fan base, and, apart from 1983, have won the MCN fan-voted award as Top Vocal Group every year from 1971 through to 1992. Since 1991 they have hosted their own TV variety series on TNN, one of the most popular country shows of all time. The Statlers' name is associated with a certain meticulous attention to detail, an immediately recognizable sound and an almost unflagging compositional quality. Their most recent country hit came with **More Than A Name On A Wall**, a Top 10 entry in 1989, although they continue to produce excellent albums.

Recommended:
Today (Mercury/Mercury)
Sing Country Symphonies In E Major (Mercury/–)
The Originals (Mercury/–)
Pardners In Rhyme (Mercury/Mercury)
10th Anniversary (Mercury/–)
Music Memories And You (Mercury/–)
Four For The Show (Mercury/Mercury)
Radio Gospel Favorites (Mercury/–)
Words And Music (Mercury/–)

Oh Happy Day, Statler Brothers. Courtesy CBS Records.

Red Steagall

Russell 'Red' Steagall was born on December 22, 1937 in Gainesville, Texas. Polio struck when he was 15, leaving Steagall without the use of his left hand and arm. He used months of therapy and recuperation to master the guitar and mandolin, and later began playing in coffee houses during his stay at West Texas State University. His first job found him working for an oil company as a soil chemistry expert, with performing remaining a sideline until 1967 when **Here We Go Again**, a Steagall original penned with the aid of co-writer Don Lanier, provided Ray Charles with a chart record.

If You've Got The Time, I've Got The Song, Red Steagall. Courtesy Capitol Records.

Other Steagall songs subsequently found their way on to disc, and the Texan became involved on the song publishing side of the industry, eventually signing a recording contract with Dot in 1969.

He joined Capitol Records in 1972, coming up with such hit singles as **Party Dolls And Wine**, **Somewhere My Love** (both 1972) and **Someone Cares For You** (1974). He then rejoined ABC-Dot, finally making the Top 10 with **Lone Star Beer And Bob Wills Music** (1976). That success was short-lived, though he has recorded some fine albums, mainly with a western swing or cowboy feel.

In 1974, he helped Reba McEntire get started. He produced her first demo recordings in Nashville, which resulted in her being signed to Mercury Records. An in-demand rodeo performer, Red has become a real singing cowboy. He has his own ranch just outside Fort Worth, Texas, and plays a major role in keeping alive the music of the cowboy. In 1991 he was named 'Cowboy Poet of Texas' by the State Legislature, and is also responsible for organizing the annual Cowboy Gathering and Western Swing Festival in Fort Worth, Texas. Early 1993 saw Red gain a contract with Warner Brothers Records' spin-off label, Warner Western.

Recommended:
Red Steagall (MCA-Dot/–)
Cowboy Favorites (Delta/Silver Dollar)
Party Dolls And Wine (Capitol/–)
Lone Star Beer And Bob Wills Music (ABC-Dot/–)

Keith Stegall

A talented Nashville-based songwriter, Keith (born in 1955 in Wichita Falls, Texas, the son of Bob Stegall who played steel guitar for Johnny Horton) has seen his songs score on the country, pop, R&B, adult contemporary and jazz charts.

At the age of eight he made his stage debut on a country music show in Tyler, Texas, and before he reached his teens he had formed his own four-piece combo called the Pacesetters. Keith's family moved to Shreveport, Louisiana and he joined a folk group, the Cheerful Givers.

By this time he was busy writing songs while holding down a job as a night-club singer. Urged to make a move to Nashville, Stegall became a staff writer for CBS Songs (then called April/Blackwood) in 1978. He rapidly distinguished himself as a writer for the likes of Al Jarreau (**We're In This Love Together**), Leon Everette (**Hurricane**), Dr Hook (**Sexy Eyes**), Mickey Gilley (**Lonely Nights**) and the Commodores (**The Woman In My Life**).

After a brilliant start as a songwriter, Keith made a dismal recording artist, signing with Capitol in 1980 and making the lower rungs of the charts with **The Fool Who Fooled Around**. Never able to score higher than the mid-'50s with Capitol (and subsequently with its sister label EMI America), Keith signed with Epic in 1984 and made No.25 with **I Want To Go Somewhere**, followed by **Whatever Turns You On** (No.19), **California** (No.13) and cracked the Top 10 in 1985 with the self-penned **Pretty Lady**.

He has continued to concentrate on his writing, coming up with hits for such varied artists as Kenny Rogers, Juice Newton, Charley Pride, Johnny Mathis, Reba McEntire and Glen Campbell. Increasingly Stegall has turned to production, and has been responsible for all three of Alan Jackson's platinum albums.

Recommended:
Keith Stegall (Epic/–)

Ray Stevens

A talented singer-songwriter, arranger, producer and multi-instrumentalist, Stevens (born Ray Ragsdale in Clarkdale, Georgia, 1939) studied music at Georgia State University. He then moved to Nashville where he began recording such

Nashville, Ray Stevens. A self-produced album for Barnaby Records.

novelty hits as **Jeremiah Peabody's Poly Unsaturated Quick Dissolving Fast Acting Pleasant Tasting Green And Purple Pills** (1961), **Ahab The Arab** (1962), **Harry The Hairy Ape** (1963), **Mr Businessman** (1968) and **Gitarzan** (1969).

Since the end of the '60s Stevens, who has recorded for such labels as Judd, Mercury, Monument, Barnaby, Janus, Warner, RCA and MCA, has turned increasingly to country music, scoring with such discs as **Turn Your Radio On** (1971), **Nashville** (1973) and a semi-bluegrass version of **Misty** (1975), for which he was awarded a Grammy for Best Arrangement Accompanying A Vocalist.

From time to time he still produces pieces of sheer lunacy like **The Streak** (a pop No.1 in 1974), **Shriner's Convention** (1980), **Mississippi Squirrel Revival** (1984) and **It's Me Again, Margaret** (1985). In 1990 he opened his own theatre in Branson, Missouri, but, following the success of his 1992 Greatest Hits Comedy Video, which sold 1.6 million copies and topped the 'Billboard' video chart for months, Stevens was looking to expand his career in 1993 by working on a TV series and making a sideways step into movies.

Recommended:
Misty (Barnaby/Janus)
He Thinks He's Ray Stevens (MCA/MCA)
Beside Myself (MCA/MCA)

Above: Hard-drinking honky-tonker Gary Stewart lives out the lyrics to his songs in real life.

Gary Stewart

An exciting honky-tonk singer-songwriter whose vibrato-filled voice places him in a love or hate category, Stewart was born on May 28, 1945, in Letcher County, Kentucky, moving to Florida with his family when he was 12.

His recording career began in 1964 with **I Loved You Truly**. Stewart then went on the road, playing bass with the Amps, a rock outfit, eventually returning home to work for an aircraft firm.

In 1967 he met Bill Eldridge, an ex-rocker who had contacts in Nashville. Together they began writing songs, one becoming a minor hit for Stonewall Jackson. Others included: **Sweet Thang And Cisco**, a 1969 hit for Nat Stuckey; **When A Man Loves A Woman** and **She Goes Walking Through My Mind**, both 1970 Top 5 discs for Billy Walker; and other material for such artists as Kenny Price, Jack Greene, Johnny Paycheck, Cal Smith, Hank Snow and Warner Mack.

But, with his own recording contract with Kapp petering out, Stewart returned to Fort Pierce, Florida, leaving behind some demo sessions he had made of Motown material.

These demos were later heard by Roy Dea, a Mercury producer who was subsequently signed to RCA by Jerry Bradley. He immediately signed Stewart for his new label, his belief in the singer paying off when Stewart's **Drinkin' Thing**, a second-time-around release, became a Top 10 hit in 1974.

Since that time, Stewart has had other major chart discs with **Out Of Hand** (1975), **She's Actin' Single (I'm Drinkin' Doubles)** (1975), **In Some Room Above The Street** (1976), **Ten Years Of This** (1977) and **Whiskey Trip** (1978).

In 1982, Gary teamed up with singer-songwriter Dean Dillon, and the two wild honky-tonkers came up with some dynamic numbers, like **Brotherly Love** (1982) and **Smokin' In The Rockies** (1983), but failed to make their mark with the record-buying public.

A reputation for heavy drinking mixed with drug addiction saw Stewart's career fall apart. His wife left him and his son committed suicide as Stewart retreated to the Florida honky-tonk circuit, where he felt most at home. He recorded briefly for the local Red Ash Records in 1984, then resurfaced to national prominence at the end of the '80s, when he signed with the California-based Hightone Records. There was a handful of minor country hits, including **Brand New Whiskey** and **An Empty Glass** (1988), while the album **Brand New** made the charts in 1989.

Recommended:
Little Junior (RCA/–)
Brotherly Love – with Dean Dillon (RCA/–)
Battlefield (Hightone/–)
Brand New (Hightone/–)
Gary's Greatest (Hightone/–)

John Stewart

With over a dozen albums to his name as a solo performer, singer-songwriter John Stewart (born on September 5, 1939 in San Diego, California) has been a major influence on the American country and rock scene for more than 20 years.

He played rock'n'roll in his youth, but soon turned to folk music. A prolific writer, he performed a couple of his songs for the Kingston Trio backstage after one of their concerts. The Trio liked the songs **Green Grasses** and **Molly Dee**, which led to John being asked to form a similar folk trio by Roulette Records. the result was the Cumberland Three, who recorded three albums for the New York-based label between 1959 and 1961.

When founder member Dave Guard left the Kingston Trio at the end of 1961, John was asked to join them. This was after the days of **Tom Dooley**, but before the beat explosion, and the Trio were big news. However, despite appearing before mass audiences and playing an important part in the Trio, John never felt a real part of the Trio, or happy with the 'college fraternity' lifestyle they led. Indeed, for the whole six years he was with them, until the break-up in 1967, he was on a salary.

Cannons In The Rain, John Stewart. Courtesy RCA Records.

Judging by his subsequent music, the split must have been a welcome relief. He was far more in touch with other artists and once before nearly left the Trio to form a new group with John Phillips (of the Mamas And Papas) and Scott McKenzie. For a while he hung out with John Denver and, though they never went out as a duo, they recorded demos of **Leaving On A Jet Plane** and **Daydream Believer**. The latter song became a multi-million seller for the Monkees in 1967, Stewart surviving for years on the revenue from the song.

Stewart next found himself with Buffy Ford, and together they recorded **Signals Through The Glass**, an album based around the paintings of Andrew Wyeth. In the late '60s he worked on the Bobby Kennedy election campaign and utilized some of his experiences in the song **Omaha Rainbow**.

Throughout the '70s John recorded a series of albums as he moved through such labels as Capitol, Warners and RCA. **California Bloodlines**, recorded for Capitol in 1969 and produced by Nick Venet in Nashville, is counted as something of a milestone. Further albums featured the talents of such musicians and singers as James Taylor, Carole King, Doug Kershaw, Chris Darrow, James Burton, Glen D. Hardin, Buddy Emmons, Fred Carter Jr, Pete Drake and Charlie McCoy. But at the centre were John's songs.

He built up a sizeable cult following, especially in Britain, where a magazine, Omaha Rainbow, is named after one of his songs. Finally he achieved the commercial success he deserved when his second album for RSO Records, **Bombs Away**

Dream Babies, made the Top 10 album charts and his single **Gold** reached Top 5 and won John a gold disc in 1979.

This, however, turned out to be a short-lived success, and during the '80s John Stewart was once again relegated to cult hero status, a singer-songwriter who appeals to the more discerning music lovers. He penned Rosanne Cash's chart-topping **Runaway Train**, and she has worked closely with Stewart, even borrowing from his sound and style for some of her own recordings. He continues to please dedicated fans with regular album releases, which are now issued on small independent labels.

Recommended:
California Bloodlines (Capitol/Capitol)
The Phoenix Concerts (RCA/RCA)
Forgotten Songs Of Some Old Yesterday (RCA/RCA)
Bombs Away Dream Babies (RSO/RSO)
Blondes (Allegiance/Line)
Bullets In The Hourglass (Shanachie/–)
Cannons In The Rain/Wingless Angels (–/Bear Family)
Lonesome Picker Rides Again (Warner Bros/Line)

Redd Stewart

Henry Redd Stewart, born in Ashland City, Tennessee, on May 27, 1921, began his career by writing a song for a car dealer's commercial at the age of 14. He then formed and played in bands around the Louisville, Kentucky area until 1937, when Pee Wee King came to Louisville to play on radio station WHAS and signed Stewart as a musician. Eddy Arnold was the band's vocalist at the time, although Redd was Eddy's replacement when Arnold went solo. Then came Pearl Harbor, and Stewart was drafted for Army service in the South Pacific, during which period he wrote **A Soldier's Last Letter**, a major hit for Ernest Tubb.

After the war he rejoined King and began taking a serious interest in songwriting, teaming with King to write **Tennessee Waltz** (a hit for both King and Cowboy Copas but a 1950 six-million-seller for Patti Page), following this with **Slow Poke** (a 1951 gold disc for King), **You Belong To Me** (which, in the Joe Stafford version, topped the US charts for five weeks during 1952), and the reworked old fiddle tune, **Bonaparte's Retreat**.

Stewart's own career on disc has been less successful – despite stints with such labels as RCA, Starday, King and Hickory – and it seems that he will generally be remembered for his songwriting and his 30-year association with Pee Wee King.

Recommended:
The Best Of Pee Wee King And Redd Stewart (Starday/–)

Wynn Stewart

Born in Morrisville, Missouri on June 7, 1934, singer-songwriter Stewart received early singing experience in church, and at 13 appeared on KWTO, Springfield, Missouri. A year or so later his family moved to California and this was where he made his first recording at the age of 15.

During the mid-'50s, Stewart became signed to Capitol Records, later switching to Jackpot Records, a subsidiary of Challenge. His song **Above And Beyond** provided a big hit for Buck Owens in 1960, while some of Stewart's own Jackpot sides featured the voice of Jan Howard. For Challenge itself, Stewart provided such hits as **Wishful Thinking** (1959), **Big Big Day** (1961) and **Another Day, Another Dollar** (1962), around that period opening up his club in Las Vegas and appearing on his own TV show.

Some two and a half years later he sold the club and moved to California, signing once more for Capitol and promoting his discs by touring with a new band, the Tourists. Purveying his California style of honky-tonk and beer-stained ballads, Stewart made friends and influenced record buyers. The result was a flow of country chart entries that has included: **It's Such A Pretty World Today** (a No.1 in 1967), **'Cause I Have You** (1967), **Love's Gonna Happen To Me** (1967), **Something Pretty** (1968), **In Love** (1968), **World-Wide Travelin' Man** (1969) and **It's A Beautiful Day** (1970).

Moving through a succession of labels, he made a chart comeback with **After The**

Below: Redd Stewart, a songwriter whose Tennessee Waltz was a hit for Patti Page (1950) and others.

Above: Singer-songwriter Wynn Stewart was singing in his hometown church at five and was appearing regularly on radio at 13. He scored masses of hits in the '60s.

Storm, which reached the Top 10 in 1976. Subsequent releases failed to bring him the chart success he deserved, and Wynn continued recording for small labels up until shortly before his death at his home in Hendersonville, Tennessee, on July 17, 1985.

Recommended:
After The Storm (Playboy/–)
Baby It's Yours (Capitol/–)
Wishful Thinking (–/Bear Family)

Doug Stone

The man with one of the saddest voices in country music, Stone was born Doug Brooks in 1957, in Newnan, Georgia. He changed his name to avoid confusion with Garth Brooks, and, as he had just written a song **Heart Of Stone**, decided that Stone was a natural choice. Before arriving in Nashville in 1989, Doug had spent years playing the clubs of Georgia. A multi-instrumentalist who can play guitar, keyboards, fiddle and drums, he made his professional debut at 11, playing drums with the Country Rhythm Playboys. Prior to this, he had a moment of glory when he was seven, being invited by Loretta Lynn to get up onstage and play guitar with her. His mother used to sing around the house and encouraged young Doug, but, when he was 12, his parents separated, and he went to live with his father. After dropping out of high school, Doug had many jobs, mainly as an auto mechanic, but also worked as a carpenter, a lawn-mower mechanic and in a hamburger joint. He married in his early 20s, but family life and playing in dead-end bar bands was not easy, and later he divorced.

Doug built a little studio in his house, and also started his own dump truck business, as he chased a dream of being a singing star. In 1987 he lost everything, but his fortunes were revived when he was signed to an artist management contract by Phyllis Bennett. A demo reached producer Doug Johnson, and that was followed by a recording deal with Epic Records in Nashville in 1989.

His debut record, **I'd Be Better Off (In A Pine Box)**, a classic country weeper, rose to No.4 on the country charts in 1990. Further Top 10 success came with **Fourteen Minutes Old** and **These Lips Don't Know How To Say Goodbye**, then he hit the top with **In A Different Light** (1991). With his Stone Age Band, Doug took to the road. A hyperactive person, his fast-paced show with fancy dancing and pure showmanship was at odds with his slow, sad-tinged, country ballads which were dominating the country charts. He scored another Top 10 hit with **I Thought It Was You** (1991), then showed off his ability with an uptempo, honky-tonker in his second chart-topper, **A Jukebox With A Country Song** (1991). The hectic life on the road took its toll on Doug's health, and on the day of the 1992 ACM Awards Show, with him nominated for Top Male Artist and Top Song, he was undergoing emergency quadruple heart bypass surgery. Suitably recovered, Stone has continued in his hit-making way with a mixture of songs about romance carved in song, many with a witty twang about them. He scored Top 10 with **Made For Lovin' You** and **Warning Labels** (1992), and another No.1 with **Too Busy Being In Love** (1993). Stone's first three albums, all produced by Doug Johnson, have gone gold.

Recommended:
From The Heart (Epic/Epic)
The First Christmas (Epic/–)
I Thought It Was You (Epic/–)
Doug Stone (Epic/–)

Stoneman Family

One of the most famous groups in country music, the Stonemans revolved around Ernest V. 'Pop' Stoneman (born in Monorat, Carroll County, Virginia, on May 25, 1893), a carpenter who wrote to Okeh and Columbia seeking an audition in 1924.

A jew's-harp and harmonica player by the age of ten, and a banjoist and autoharp player in his teens, Pop was eventually heard by Okeh's Ralph Peer, who recorded some test sides in September, 1924, cutting a number of sides the following January. These included **The Sinking Of The Titanic**, one of the biggest-selling records of the '20s.

Between 1925 and 1929 Pop, sometimes with his wife or other members of his family, cut well over 200 titles for Okeh, Gennett, Paramount, Victor and other companies, also playing on dates with such acts as Riley Puckett and Uncle Dave Macon.

Having spent their royalties on cars and other luxuries, the Depression of 1929 hit the Stonemans hard. Only one recording date, featuring Pop and his son Eddie, emanated from this period, and Pop had to resume his former occupation as a carpenter in a Washington DC naval gun factory. Meanwhile, his wife Hattie struggled to bring up her family – which eventually numbered 13 children.

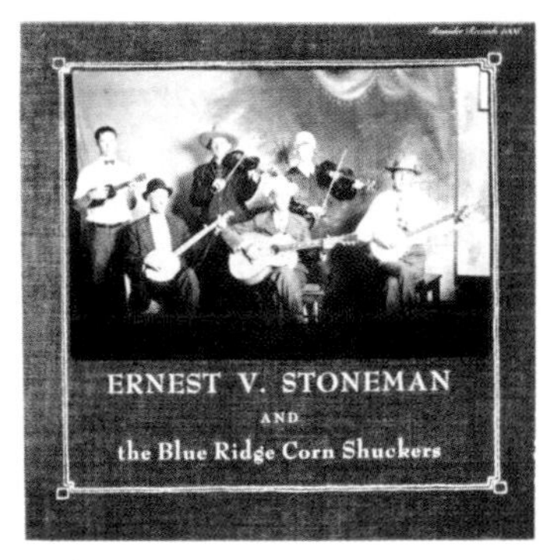

Ernest V. Stoneman And The Blue Ridge Corn Shuckers. Courtesy Rounder Records.

Several of the children became musicians and Pop formed a family band during the late '40s, playing in the Washington area and recording an album for Folkways in 1957 that helped spark off a whole new career.

Proving popular at the major folk festivals and on college dates, the Stonemans became an in-demand outfit, making their debut on the Grand Ole Opry and recording for Starday in 1962.

During the mid '60s the family moved to Nashville, appeared on the Jimmy Dean ABC-TV show, and were signed to appear in their own TV show, Those Stonemans, in 1966. A year later, they won the CMA award for the Best Vocal Group. The band then consisted of Pop (guitar, autoharp),

Above: George Strait is a huge star with a string of hits to his credit in the '80s and '90s.

Scotty (fiddle), Jim (bass), Van (guitar), Donna (mandolin) and Roni (banjo), the last-named becoming a star on the Hee Haw TV show.

However, a stomach ailment began to affect Pop and he died in Nashville on June 14, 1968, his last recording session having taken place that day.

Although the Stonemans were considered one of the finest semi-bluegrass bands, Pop's early record output with his Dixie Mountaineers featured nineteenth-century sentimental ballads, British traditional melodies, dance tunes, religious material and even a number of humorous sketches. He was reputed to have been the first musician to record with an autoharp.

Recommended:
In The Family (MGM/–)
The Stoneman Family (Folkways/–)
The Stonemans (MGM/–)
Stoneman's Country (MGM/–)
Tribute To Pop Stoneman (MGM/–)

Something Special, George Strait. Courtesy MCA Records.

George Strait

George Strait, born May 18, 1952, in Pearsall, Texas, emerged in the early '80s as one of the best exponents of unvarnished, clean-cut country music. When he first started recording in 1981, his authentic country sound, with twin fiddle breaks and strong steel guitar, seemed to breathe fresh air into the somewhat stale Nashville scene.

Born the second son of a junior high school teacher, George was raised on a ranch in Texas. After a short spell at college, George eloped with his high school sweetheart, Norma, and then joined the US Army.

While stationed in Hawaii, George started singing with a country band, using the songs of Merle Haggard, Bob Wills, George Jones and Hank Williams.

After his discharge in 1975, George returned to Texas and attended the Southwest Texas State University to complete his degree in agriculture. By this time he had been bitten by the music bug and, assembling his Ace In The Hole Band, was soon living a double life, attending classes by day and playing the clubs at night.

George and his band had built up a strong following on the southwest Texas honky-tonk circuit when, through the efforts of Erv Woolsey, a one-time MCA promotions man, he landed an MCA recording contract in early 1981. His first single, **Unwound**, reached the Top 10 in the country charts.

Strait spent more time at the top of the country singles charts than any other performer in the '80s, with more than two dozen records reaching No.1, including: **Fool Hearted Memory** (1982), **A Fire I Can't Put Out** (1983), **You Look So Good In Love**, **Does Fort Worth Ever Cross Your Mind** (1984), **The Chair** (1985), **It Ain't Cool To Be Crazy About You** (1986), **Ocean Front Property** (1987), **Famous Last Words Of A Fool** (1988), **Love Without End, Amen** (5 weeks, 1990) and **I've Come To Expect It From You** (5 weeks, 1990). He also amassed a huge number of gold and platinum albums, and was named CMA Male Vocalist Of The Year in 1985 and 1986, and Entertainer Of The Year in 1988 and 1990.

It was Strait's first album, **Strait Country**, that persuaded a rock-crazy kid named Garth Brooks to switch to country when he was in high school. From the beginning Strait has kept to his image of cowboy hat, straight-legged jeans and tailored western shirts, his sound and look becoming a much-duplicated symbol in late '80s country. There's never been any great themes in his many albums, just a 'Strait' country mix of those twin fiddles, steel guitar, Texas two-steps and sparsely produced ballads. He's never allowed the new crop of performers to affect his popularity, and the chart-topping singles have continued with **You Know Me Better Than That** (1991), **So Much Like My Dad** (1992) and **Heartland** (1993).

Strait reached a new plateau in his career when he took his first serious steps into the movies to star as country singer Rusty Wyatt Chandler in 'Pure Country', a 1992 film specially written for him. It became a major box office success and the soundtrack album, the first of his recordings to be produced by Tony Brown, became his biggest seller. He remains the biggest star of country music, with a level of consistency second-to-none. He grosses $10 million a year and broke Elvis' record for consecutive sell-out shows in Las Vegas.

Recommended:
Something Special (MCA/MCA)
Strait From The Heart (MCA/MCA)
Does Fort Worth Ever Cross Your Mind (MCA/MCA)
Holdin' My Own (MCA/MCA)
Chill Of An Early Fall (MCA/MCA)
Beyond The Blue Neon (MCA/MCA)
Livin' It Up (MCA/MCA)
Ocean Front Property (MCA/MCA)

Mel Street's Greatest Hits. Courtesy GRT Records.

Mel Street

One of the finest country singers to emerge in the '70s, Mel (born on October 21, 1933 in Grundy, West Virginia), never really achieved the success or recognition he so richly deserved.

He started out singing on local radio shows in the early '50s and, following his marriage, moved to Niagara Falls, New York. For several years he sang in a local night spot and eventually had enough money saved to return to West Virginia, where he opened his own automobile workshop.

With his four-piece band, Mel had his own television show in Bluefield, West Virginia called Country Showcase, at the same time working regularly in local clubs and honky-tonks. Eventually he gained a recording contract with the small Tandem Records, releasing his first single, **House Of Pride**, in 1970.

It was the other side of the record, Mel's self-penned **Borrowed Angel**, that gained most response from the public, eventually making the country Top 10 during 1972. During the next few years, Mel recorded for a variety of labels, achieving Top 20 hits with **Lovin' On Back Streets** (1972), **Walk Softly On The Bridges** (1973), **Forbidden Angel** (1974), **Smokey Mountain Memories** (1975), **I Met A Friend Of Yours Today** (1976) and **Close Enough For Lonesome** (1977).

An excellent song stylist who specialized in honky-tonk sagas and bar-room ditties, Mel was the first singer to record the songs of such writers as Eddie Rabbitt, Earl Thomas Conley and John Schweers, and was one of the best intrepreters of Bob McDill material, scoring with **Shady Rest**, **Barbara Don't Let Me Be The Last To Know** and filling his albums with McDill's songs.

Depressed due to a heavy workload and personal problems, Mel shot himself at his Hendersonville home on October 21, 1978 – his 45th birthday.

Recommended:
Smokey Mountain Memories (GRT/–)
Country Soul (Polydor/–)
Many Moods Of Mel Street (Sunbird/–)

Stringbean

Born in Annville, Kentucky, on June 17, 1915, Stringbean's real name was David Akeman, and he was the son of a fine banjo player. Stringbean made his own first banjo at the age of 12 and began playing professionally six years later in the Lexington area, eventually working with Cy Rogers' Lonesome Pine Fiddlers on radio station WLAP.

It was during this period that the performer became dubbed Stringbean and adopted a more comic direction with his act. During the late '30s, he worked with Charlie Monroe, then joined Bill Monroe on the Grand Ole Opry in July, 1942, staying with Monroe for three years.

Also known as the 'Kentucky Wonder', Stringbean was an outstanding banjo player in the style of Uncle Dave Macon and a long-time member of the Opry, but perhaps won even more fame through his appearances on the Hee Haw TV series. He died on November 10, 1973, he and his wife Estelle being brutally murdered on returning home from the Opry and discovering burglars in their house.

Recommended:
Salute To Uncle Dave Macon (Starday/–)
Me And My Old Crow (Nugget/–)

Marty Stuart

Proud to be a hillbilly singer, John Marty Stuart was born on September 30, 1958, in Philadelphia, Missouri. A multi-instrumentalist who is at home on guitar, bass, mandolin, fiddle and upright bass, he was raised on country and made his first professional appearance at 13 with Carl and Pearl Butler. A few months later Marty was touring with Lester Flatt And The Nashville Grass, making his debut on the Grand Ole Opry before he was 14. He stayed with Flatt for eight years, until the bluegrass legend died in 1979. Stuart opted then to stay in Nashville, where he built his reputation as a studio musician, before joining Johnny Cash's band in the early '80s. He married Cash's daughter, Cindy, but it was a short-lived and turbulent liaison and they were soon divorced.

Below: Marty Stuart keeps traditions while maintaining a contemporary edge in his music.

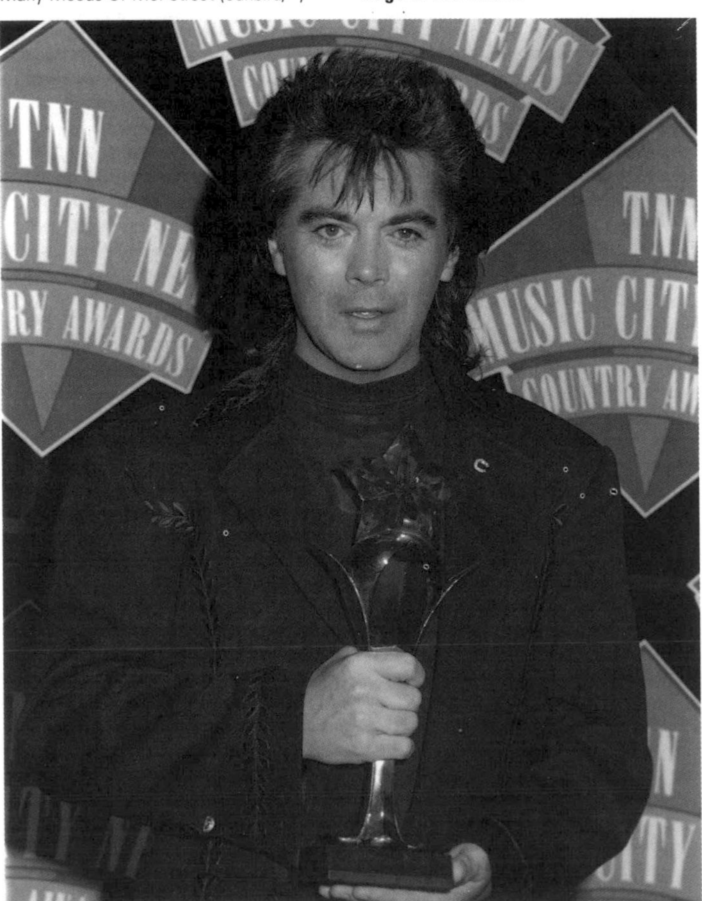

Stuart produced his first solo album, **Busy Bee Café**, in 1982. Released on the independent Sugar Hill label, the session band used attested to his reputation as picker, and included Doc Watson, Merle Watson and Johnny Cash on guitars, Jerry Douglas on dobro and Carl Jackson on banjo. Marty was on the road with Johnny Cash for the best part of six years, but he will managed to cram studio and concert work into his busy schedule, working for Bob Dylan, Billy Joel, Roger Miller, Willie Nelson, Emmylou Harris and others. He signed with Columbia Records in 1986, making the Top 20 with **Arlene**, but subsequent singles were only minor hits. His **Marty Stuart** album did poorly, and a follow-up, **Let There Be Country**, was never issued at the time. Deciding to get back to his roots, Marty hooked up with the Sullivans, a family gospel group with whom he had previously sung as a child. He played on and produced their **A Joyful Noise** album.

In 1989, he was signed by MCA Records, at this time building a high-profile image with his puffed-up, long, ebony hair – possibly the longest among country music performers. Initially his MCA recordings didn't take off, but with the help of videos that put across the visual side of Marty Stuart, he enjoyed Top 10 singles with **Hillbilly Rock**, **Little Things** (1990) and **Tempted** (1991). He made the headlines for his unique collection of country mementoes – he has a silver eagle tour bus (E.T.), which once belonged to Ernest Tubb, a trio of guitars that were originally owned by Lester Flatt, Hank Williams Sr and Clarence White, and sparkly rhinestone suits that were popular in the '50s and '60s. Far from trendy, Stuart describes himself as looking like "a cross between Roy Rogers, Porter Wagoner and Gene Autry", and his music as "rocking hillbilly music with a thump".

His career was given a big boost after he wrote **The Whiskey Ain't Workin'**, and sent it to Warner Brothers for Hank Williams Jr or Travis Tritt to record. The young Tritt decided to cut it, calling in Stuart to harmonize, and the result was a No.2 country smash in early 1992. Stuart and Tritt then got together for a highly successful 'No Hats Tour' that ran all the way through 1992. This helped Stuart's second MCA album, **Tempted**, to build up sales of more than 300,000, while his solo singles **Burn Me Down** and **Now That's Country** charted highly in 1992. A third MCA album, **This One's Gonna Hurt You**, found Stuart achieving a near flawless integration of southern rock, bluegrass, blues, honky-tonk, boogie and rockabilly. Travis dropped by to duet on the title song, another big-selling single. In the meantime, Marty Stuart is fulfilling his long-held ambition of taking country music, its legends and history, to a new younger audience. An accomplished writer and photographer, he has had his photos published in country music magazines and contributed several articles to music publications.

Recommended:
This One's Gonna Hurt You (MCA/–)
Hillbilly Rock (MCA/–)
Busy Bee Café (Sugar Hill/–)
Tempted (MCA/–)

Nat Stuckey

Perhaps an underrated performer – though he has been a consistent supplier of medium-sized hits – Nat Stuckey was born in Cass County, Texas, on December 17, 1937. Employed for some considerable time as a radio announcer, Stuckey also worked with a jazz group in 1957–58, becoming leader of a country band, the Corn Huskers, in 1958–59.

In 1966, Buck Owens recorded his fellow Texan's **Waitin' In The Welfare Line**. At this stage Stuckey, who had been working with the Louisiana Hayriders and recording for the Sims label, switched to Paula Records, scoring his own Top 10 hit with **Sweet Thang**.

Seven chart records later, in 1968, he label-hopped once more, this time signing for RCA and immediately scoring five major disc successes with **Plastic Saddle** (1968). **Joe And Mabel's 12th Street Bar And Grill**, **Cut Across Shorty**, **Sweet Thang And Cisco** and **Young Love** (all 1969), the last a duet with Connie Smith.

Stuckey never really enjoyed the quota of potent singles expected of him, only **She Wakes Every Morning With A Kiss** (1970) and **Take Time To Love Her** (1973) establishing his name in the upper regions of the charts.

During the mid-'70s, he became an MCA artist, scoring Top 20 hits with **Sun Comin' Up** (1976) and **The Days Of Sand And Shovels** (1978), and producing **Independence**, an album acclaimed by the critics. However, he still failed to achieve the kind of commercial success that his talent deserved. He recorded for a variety of small labels and even worked as a jingle singer. Nat Stuckey died of lung cancer on August 24, 1988.

Recommended:
Independence (MCA/MCA)
She Wakes Me With A Kiss Every Morning (RCA/RCA)

Joe Sun

Joe Sun is one of a talented breed of singer-songwriters who have unself-consciously absorbed influences across the spectrum from Hank Williams to Waylon Jennings.

In his music, he mixes soul, blues, honky-tonk, rock'n'roll, contemporary and traditional country into a sound that he describes as blues/country. Joe was born James Paulson on September 25, 1943, in Rochester, Minnesota. He arrived in Nashville in 1972 after spending time in college, the Air Force and in various jobs, which included a DJ stint at Radio WMAD in Madison, Wisconsin, and two years with a computer firm in Chicago.

While in Chicago he sang with a variety of semi-pro bands, working under the name Jack Daniels. Once in Nashville he gave himself five years to make it. For a

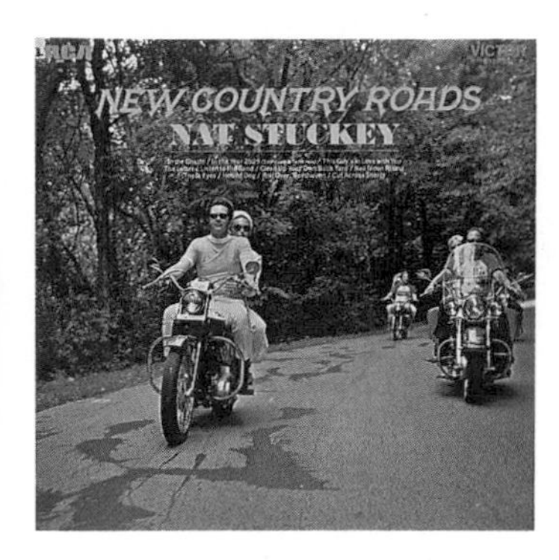

New Country Roads, Nat Stuckey. Courtesy RCA Records.

time he ran a small graphics business called The Sun Shop, then took up independent record promotions, which led to signing with Ovation towards the end of 1977.

His first single, **Old Flames (Can't Hold A Candle To You)**, came out in May 1978 and climbed steadily up the country charts, reaching the Top 20. Further hits followed with **I Came On Business For The King**, **I'd Rather Go On Hurtin'** and **Shotgun Rider** (1980). Then with his third album, **Livin' On Honky Tonk Time**, just released, Ovation closed down its record division.

Joe signed with Elektra in 1981, and though he has continued to make some fine records, gain rave reviews for his dynamic stage show and command a huge cult following in Britain, he has failed to achieve the commercial success he deserves.

Recommended:
The Sun Never Sets (–/Sonet)
Out Of Your Mind (Ovation/Ovation)
Hank Bogart Still Lives (–/Dixie Frog)

Billy Swan

A comparative unknown when his **I Can Help** single hit the charts in 1974, Swan turned out to have a long pedigree in southern music. Born on May 12, 1942, in Cape Giradeau, Missouri, he had written **Lover Please** at the age of 16. The song

Above: Sweethearts sisters Kristine and Janis started out in the '70s.

was recorded by his band of that time, Mirt Mirley And The Rhythm Steppers, but Clyde McPhatter made it a huge R&B hit.

Swan eventually tried his luck in Nashville and, taking odd jobs, he followed Kris Kristofferson as janitor at Columbia's studios. While working for Columbia Music, Swan became involved with Tony Joe White and produced that artist's first three, and most important, albums. He also backed Kris Kristofferson at the 1970 Isle of Wight Festival. Some time later he was to join Kinky Friedman's band for a while.

The 1974 album release of **I Can Help** revealed an artist with a liking for country, rock'n'roll and R&B. Swan made an

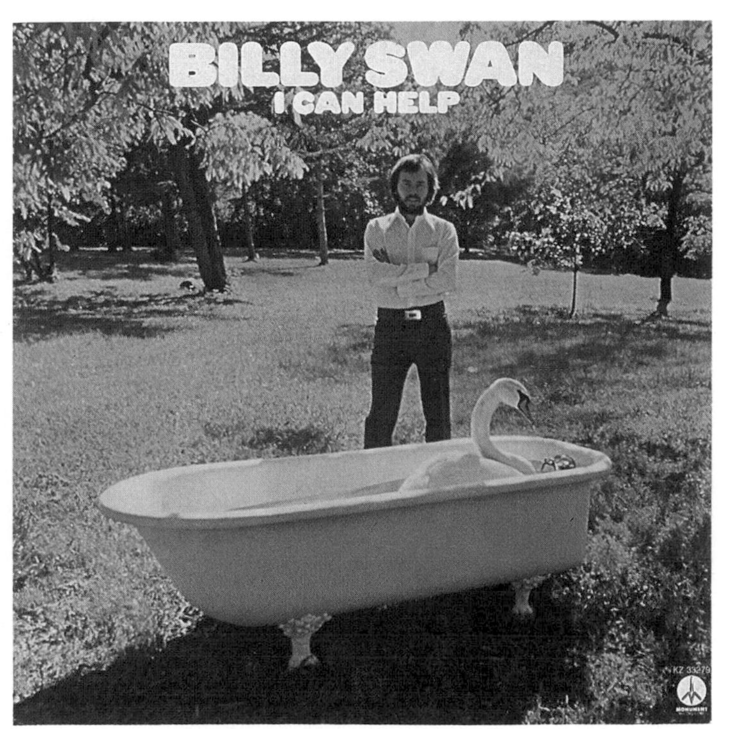

I Can Help, Billy Swan. Courtesy Monument Records.

impressive version of Presley's **Don't Be Cruel** and in 1975 he was successful with another pop single hit with **Everything's The Same**.

In the summer of 1978, Swan joined A&M Records, then in 1981 moved on to Epic Records, recording some fine country-rock sides and touring with such highly respected Nashville musicians as Kenny Buttrey and Charlie McCoy as backing musicians.

Billy was back in the musical headlines in the summer of 1986 when he tamed up with Randy Meisner, Jimmy Griffin and Robb Royer, plus other musicians, to form a new band, Black Tie, recording an album, **When The Night Falls**, for Bench Records. Swan now mostly tours with Kris Kristofferson.

Recommended:
I'm Into Lovin' You (Epic/–)
At His Best (Monument/–)
Rock'n'Roll Moon (Monument/Monument)
I Can Help (Monument/Monument)

Sweethearts Of The Rodeo

A modern duo that mixes strains of contemporary and traditional country into a rich musical sound of their very own, Sweethearts Of The Rodeo comprise sisters Janis Gill and Kristine Arnold, who grew up in southern California and started out harmonizing together in their early teens. They performed at shopping malls, pizza parlours and honky-tonks in the early '70s as the Oliver Sisters. In 1976, Emmylou Harris saw their act and invited the girls to sit in on one of her gigs, so they sang with the original Hot Band. Both girls married and put their musical ambitions to one side as they started to raise families. In 1983 Janis moved to Nashville with her husband, singer Vince Gill, and after a few months urged Kristine to move to Music City, convinced that this time the sisters could make an impact with their music. Kristine was the lead singer for the group, while Janis had been writing songs for a number of years.

Taking their name from the Byrds' country-rock album, Sweethearts Of The Rodeo entered the 1985 Wrangler Country Music Showdown, the world's largest talent contest, and were chosen as grand prize winners. This didn't net the expected recording contract, but a showcase gig at Nashville's Bluebird Café did the trick, and in 1986 they were signed to Columbia Records. They made their debut on the country charts with **Hey Doll Baby**, followed by such top 10 entries as **Since I Found You** (1986), **Midnight Girl/Sunset Town**, **Chains Of Gold** and **Gotta Get Away** (1987), as well as **Satisfy You**, **Blue To The Bone** (1988) and **I Feel Fine** (1989).

Their music, characterized by Kristine's distinct lead vocals and Janis' memorable harmony and guitar work, had a contemporary sound, but it still sounded hard country. Their recordings reverberated with swirling harmonies, staccato electric guitar and a driving rhythm section. Both are high-profile and dramatic-yet-playful onstage, with strong influences from the more colourful aspects of America's western heritage. Their stage costumes, which have played a big role in their image development, are designed and made by the girls themselves. When performing, Sweethearts Of The Rodeo present a kaleidoscope of colour in costumes which usually have a southwestern look. Due to their family commitments, Janis and Kristine are not able to tour as much as most country acts, and this has affected their record sales. They scored a handful of minor hits during 1990 and 1991, but, after cutting four high-quality albums, Sweethearts Of The Rodeo were dropped by Columbia in 1992. The following year they were signed to the independent Sugar Hill label, who are more interested in releasing quality music, than achieving mega-bucks with every release.

Recommended:
One Time, One Night (Columbia/CBS)
Sisters (Columbia/–)
Buffalo Zone (Columbia/CBS)

One Step Closer, Sylvia. Courtesy RCA Records.

Sylvia

One of the most attractive young ladies to have emerged on the Nashville scene in the early '80s, Sylvia was born Sylvia Kirby Allen, on December 9, 1956, in the small town of Kokomo, Indiana.

After graduating from high school in 1975, she headed for Nashville, armed with *a capella* demonstration tapes she had made. Sylvia took a part-time secretarial position with Tom Collins, at that time producer of Barbara Mandrell.

She soon persuaded Collins to use her on demo tapes of new songs being written by the songwriters under contract to the producer, and this led to her being used as a back-up vocalist on recording sessions by Ronnie Milsap, Barbara Mandrell and Dave And Sugar. Sylvia was finally offered an RCA contract in the summer of 1979, and her first release, **You Don't Miss A Thing**, made the country charts.

Further hits followed with **It Don't Hurt To Dream** and **Tumbleweed** (both 1980), then Sylvia hit the top of the charts with **Drifter** at the beginning of 1981, consolidating that success with **The Matador** (another No.1), **Heart On The Mend** and the attractive, pop-styled **Nobody**, a pop cross-over hit in 1982, which went on to win Sylvia a gold disc.

She maintained her stature as a pop-country star with such hits as **Like Nothing Ever Happened** (1982), **Snapshot** (1983), **I Never Quite Got Back (From Loving You)** (1984), **Fallin' In Love**, **Cry Just A Little Bit** (1985) and **I Love You By Heart** (a duet with Michael Johnson, 1986). Sylvia was not at ease with her pop-flavoured country. Her idol was Patsy Cline, and she wanted to move away from the disco-beat of **Nobody**. In 1986 she married and moved out of music. Then, as Sylvia Hutton, she returned to Nashville in 1992, writing with Craig Bickhardt and Verlon Thompson, and touring with a five-piece acoustic band.

Recommended:
Sweet Yesterday (RCA/RCA)
One Step Closer (RCA/–)
Drifter (RCA/–)

James Talley

A musical maverick, singer-songwriter James Talley, from Oklahoma, traded his carpentry skill for studio time to produce his breakthrough album, **Got No Bread, No Milk, No Money, But We Sure Got A Lot Of Love**, which he put out on his own label in 1975. Aided by some of Nashville's best session players, who donated their time, the response to the record was enthusiastic, with several major labels bidding for Talley's contract.

He signed with Capitol Records and during the next three years produced some timeless albums that pre-dated the 'New Traditionalist' movement of a dozen years later. Talley's music was steeped in rural traditions with threads of Jimmie Rodgers, Texas swing bands, country blues and Woody Guthrie all running through his work. However, his warm musical remembrances were considered revolutionary in the face of pop-country, and his recordings were shunned by Nashville, as if it was ashamed of its rich musical past. Undisturbed by this, Talley built up a cult following, very much in the mainstream of American troubadours. In many ways, his compositions marked a return to the sincerity of early country music, where attention was paid to the personal statement of the working man.

Talley carried this off with further conviction in his next album, **Tryin' Like The Devil**, in which every song related in one way or another to the plight of the worker and the enormous disparity between the rich and the poor. Not as mournful as it may sound, and like all of Talley's music, there was a warmth and humour as he utilized familiar musical forms – western swing, blues, classic rock'n'roll and country honky-tonk. He was to record two more albums for Capitol, but lack of airplay restricted sales and in 1978 the label dropped him. Talley has continued to write and make the occasional album, usually self-marketed, though a couple have surfaced on Bear Family Records in Germany. In 1992, he recorded his most unusual collection, **The Road To Torreon**. The result of a 20-year project with photographer Cavalliere Ketchum, the album centres upon the Hispanic culture of New Mexico.

Recommended:
Blackjack Choir/Ain't It Something (–/Bear Family)
Got No Bread/Tryin' Like The Devil (–/Bear Family)
American Originals (–/Bear Family)

Jimmie Tarlton

Though his name is now almost forgotten it was John James Rimbert Tarlton (born in Chesterfield County, South Carolina, 1892) who first recorded and arranged the old folk song **Birmingham Jail**.

The son of a sharecropper, Tarlton became proficient on banjo, guitar and harmonica while still a boy, his repertoire being drawn not only from the traditional material learned from his mother but also from the blues songs of the black workers. During his twenties, he began hoboing his way around the country, his route taking him to New York, Chicago and Texas, where he became an oil-field worker. After a spell in the cotton mills of Carolina and a trek through the Midwest with a medicine show, he opted for a full-time career in music. A 1926 partnership with Georgian guitarist Tom Darby proved eminently successful and resulted in a recording session for Columbia. In November, 1927, Darby and Tarlton recorded **Birmingham Jail** and **Columbus Stockade Blues**, the ensuing disc attaining impressive sales figures.

For the next three years, the duo continued to provide Columbia with discs, their contract finally terminating in 1930. And, though no recordings were made in 1931, dates with Victor (1932) and ARC (1933) followed. The partnership dissolved in 1933 when Darby returned to farming.

Tarlton, however, remained an active musician for many years, at one time working with Hank Williams in a medicine show. He was re-discovered by a new generation during the 1960s and began playing club dates and festivals, even cutting an album, **Steel Guitar Rag**, perhaps reminding everyone of his claim to be the first country steel guitar player. But it proved to be his final gesture: he died in 1973.

Recommended:
Darby And Tarlton (Old Timey/Bear Family)

Texas Tornados

A Tex-Mex quartet of veteran South Texas artists Freddy Fender (born June 4, 1937, San Benito, Texas), Flaco Jimenez (born March 11, 1939, San Antonio, Texas), Doug Sahm (born November 6, 1941, San Antonio, Texas) and Augie Meyers (born May 31, 1940, San Antonio, Texas), the Texas Tornados came together in December 1989, when they performed at a club in San Francisco as the Tex-Mex Revue. Such was their impact that by the following April they were recording their eponymous debut, having signed to Reprise Records. The foursome incorporated a wide variety of musical styles into a basic Tex-Mex structure that saw their first album garner a Grammy award in 1991. The follow-up, **Zone Of Our Own**, brought in a second nomination. An exciting live act, the Tornados attracted the same sort of unprecedented cross-cultural audience that Willie Nelson once bred, successfully blending Hispanics and Caucasians into a loyal group of fans.

All four members had enjoyed musical careers stretching back to the late '50s. Sahm and Meyers had been part of the Sir Douglas Quintet, while Fender had started out as a local R&B artist in Texas in 1957, and 20 years later had become a major country star. Jimenez, a legendary accordionist, is perhaps the best-known voice of Conjunto, a form of dance music that borrows from polkas and waltzes. Down through the years they had occasionally appeared at the same concerts and guested on each other's recordings. The core of the Tornados' success is their hypnotic blend of South

Back In The Swing Of Things, Hank Thompson. Courtesy MCA Records.

Texas' Mexican and Gringo musical cultures, such as blues, bar-room boogies, '50s rock'n'roll, doo-wop, swing, waltzes, polkas and Mexican folk.

The wide appeal of the group marked something of a comeback for the individual members, who all landed solo contracts with Reprise, producing solo albums of differing quality in an even wider cross-fusion of musical styles and sounds.

Recommended:
Texas Tornados (Warner Bros/Warner Bros)
Hanging On By A Thread (Reprise/–)
Zone Of Our Own (Reprise/–)
Flaco Jimenez – Partners (Reprise/Reprise)

B.J. Thomas

Born Billy Joe Thomas on August 27, 1942, in Hugo, Oklahoma, B.J. started out as a rocker, joining the Triumphs, a local band in Houston, Texas, at the age of 15. His first record with the group was titled **Lazy Man**, but it was with a Hank Williams song, **I'm So Lonesome I Could Cry** (a Scepter label release in 1966), that he obtained his first pop Top 10 hit.

Throughout the 1960s, Thomas continued logging pop Top 40 hits on Hickory and Scepter, the biggest of these being **Raindrops Keep Fallin' On My Head**, from the movie 'Butch Cassidy And The Sundance Kid', a US No. 1 in 1969, and a multi-award winner. From 1970 through to 1972, when he recorded **Rock And Roll Lullaby**, a single that featured the guitar of Duane Eddy, Thomas' name was a constant in the pop charts.

A switch from Scepter to Paramount signalled disaster for the Texan. His records failed to sell, and he was using everything from pills to cocaine. One of his lungs was pierced in a stabbing and by the mid-'70s he was bankrupt. Then came a turnabout. Billy Joe moved back into country and recorded Chips Moman and Larry Butler's **(Hey Won't You Play) Another Somebody Done Somebody Wrong Song**, a 1975 ABC Records release that became another pop No. 1. But B.J. was still on drugs when he cut the record and, in his autobiography 'Home Where I Belong', claims that he hardly remembers the session because he was using around 3,000 dollars worth of drugs each week during that period.

However, in January 1976, he became a born-again Christian and opted for a drug-free life. In 1977 he made a gospel album, also called **Home Where I Belong**, that saw him gaining a Grammy award. For a while, his name remained absent from the secular charts, but gradually his MCA releases began edging their way into the country Top 30 once more, via such singles as **Everybody Loves A Rain Song** (1978), **Some Love Songs Never Die** and **I Recall A Gypsy Woman** (1981). By 1983 he was back at the top. Signed to the Cleveland International label, he headed the country charts with **Whatever Happened To Old-Fashioned Love?** and **New Looks From An Old Lover**, following these with a Top 5 single in **She Meant Forever When She Said Goodbye**. And in 1984 he added to his tally with **The Whole World's In Love When You're Lonely** and **Rock And Roll Shoes**, a duet with Ray Charles.

Now a Grammy and Dove award-winner for his gospel releases, B.J. Thomas seems able to slot both sacred and secular songs into his repertoire and is equally happy playing both religious and country venues.

Recommended:
New Looks (Cleveland Int./Epic)
Home Where I Belong (Myrrh/Myrrh)
New Looks (Epic/Epic)
Midnight Minute (Reprise/–)

Above: Hank Thompson took western swing to the masses from the late '40s to the '70s.

Below: A guitar named Hank with its clean-shaven owner.

Hank Thompson

For 13 consecutive years (1953–1965), Thompson's Brazos Valley Boys won just about every western band poll and even today Thompson's influence pervades the country-rock scene.

Born Henry William Thompson in Waco, Texas, on September 3, 1925, he initially became a harmonica ace, winning many talent contests by his playing. Later he graduated to guitar, learning to play on a second-hand instrument costing only four dollars. During the early 1940s he began broadcasting on a local radio station and found a sponsor in a flour company. A few months later, in 1943, Thompson joined the Navy for a period of three years, upon discharge winning a spot on Waco station KWTX. He also formed a western swing band, the Brazos Valley Boys, and began recording for Globe Records in August 1946. The results of the session provided **Whoa Sailor**, a regional hit. This reached the ears of Tex Ritter, who then suggested to Capitol that they sign the Waco singer. In 1948, Thompson commenced a career with the label that was to last 18 years, scoring immediately with national hits in **Humpty Dumpty Heart** and **Today**, following these with **Green Light** and a remake of **Whoa Sailor** (1949). From then on came a perpetual stream of hits, the biggest being Thompson's version of a Carter-Warren song, **The Wild Side Of Life**, which became a million-seller in 1952. Though his last appearance on the US pop charts was with **She's A Whole Lot Like You**, back in mid-1960, Thompson continued to provide a non-stop flow of country chart winners for several years, these including **Oklahoma Hills**, **Hangover Tavern** (1961), **On Tap, In the Can Or In The Bottle, Smokey The Bear** (1968), **I've Come Awful Close** (1971), **Cab Driver** (1972) **The Older The Violin The Sweeter The Tune**, and **Who Left The Door To Heaven Open?** (1974). He quit Capitol for Warner Brothers in 1966, and then moved on to Dot in 1968.

Although record buyers have veered away from the western swing style that first brought Hank into prominence, he and his Brazos Valley Boys have continued to play an abundance of dates worldwide, also logging the occasional chart entry on such labels as ABC, MCA and Churchill. Hank has made a major contribution to country music, which was recognized when he was inducted into the Country Music Hall Of Fame in 1989.

Recommended:
The Best Of (Capitol/Capitol)
Sings The Gold Standards (Capitol/Capitol)
A Six Pack To Go (Capitol/Capitol)
Back In The Swing Of Things (Dot/ABC)
Capitol Collectors Series (Capitol/Capitol)
Songs For Rounders (Capitol/Stetson)

The Sue Thompson Story. Courtesy DJM Records.

Sue Thompson

Known as the lady with the itty-bitty voice, Sue Thompson always sounded like a teeny-bopper. Born Eva Sue McKee on July 19, 1926, in Nevada, Missouri, she grew up on a farm listening to country music or viewing western films. At seven she began playing guitar, After winning a San José talent contest during her high school days, she played a two-week engagement at a local theatre as a reward. Later she became a regular on Dude Martin's Hometown Hayride show, over San Francisco KGO-TV. And, after cutting sides with Martin's Round-Up Gang (she was married to Dude Martin for a time and later spent some years as the wife of Hank Penny), she signed to Mercury as a solo act.

Moving to LA, she appeared in cabaret and, in the late 1950s, on Red Foley's portion of the Opry. Following record dates with Columbia and Decca, Sue signed for Hickory Records in 1960 and had an initial hit with **Sad Movies**, a gold disc winner. Others followed including **Norman** (another million-seller), **James (Hold The Ladder Steady)**, **Paper Tiger**, **Have A Good Time** and **Angel, Angel**, most of her songs in pure pop vein, though she became increasingly country-oriented during the late 1960s. Her last hit of any size was **Never Naughty Rosie** in 1976.

Recommended:
The Sue Thompson Story (–/DJM)
Sweet Memories (–/Sundown)

Mel Tillis

Though Tillis has always had problems with his speech (having a life-long stutter), he has had little trouble putting words (and music) down on paper. His songwriting efforts include: **Detroit City**, **Honky Tonk Song**, **Ruby, Don't Take Your Love To Town**, **I'm Tired**, **One More Time**, **Crazy Wild Desire**, **A Thousand Miles Ago** and many other Top 10 entries.

Born in Tampa, Florida, on August 8, 1932, Tillis grew up in Tahokee, Florida. A drummer in the high school band, he later studied violin but opted out to become a footballer of some distinction. Next came a spell in the US Air Force, followed by a stint on the railroad. During this time Tillis developed his writing and performing ability and in 1957 headed for Nashville with three of his songs – all became hits for other singers. Tillis' own first hit disc came later in 1958 with **The Violet And The Rose**, a Columbia release. This was followed by **Finally** (1959), **Sawmill** (1959) and **Georgia Town Blues**, a duet with Bill Phillips (1960).

Following a switch to the Ric label and a Top 10 debut with **Wine** (1965), Mel became signed to Kapp (later Decca). He obtained Top 10 hits with: **Who's Julie?** (1968), **These Lonely Hands Of Mine** (1969), **She'll Be Hangin' Round Somewhere**, **Heart Over Mind** (both 1970). He then moved to MGM to continue his personal hit parade with: **Commercial Affection**, **Heaven Everyday** (1970), **Brand New Mister Me** (1971), **I Ain't Never** (1972), **Neon Rose**, **Sawmill** (1973), **Memory Maker** (1974) and **Woman In The Back Of My Mind** (1975), among others. To this can be added two hit duets with Sherry Bryce and **How Come Your Dog Don't Bite Nobody But Me?**, a diverting duet with Webb Pierce that hit the charts during 1963.

Heart Over Mind, Mel Tillis. Courtesy CBS Records.

Old Faithful, Mel Tillis. Courtesy MCA Records.

Adjudged CMA Entertainer Of The Year 1976, Mel moved to the MCA label and celebrated by logging a No. 1 single with **Good Woman Blues**, following this with other chart-toppers in **Heart Healer** (1977), **I Believe In You** (19789), **Coca Cola Cowboy** (1979) and, on the Elektra label, **Southern Rains** (1980). During 1981 he teamed with Nancy Sinatra for a duet album, **Mel and Nancy**. He switched back to MCA and went Top 10 with **In The Middle Of The Night** (1983) and **New Patches** (1984), also recording some sides with Glen Campbell.

One of the most prolific writers in country music, Mel is a versatile performer and one who has always been able to turn his speech impediment to good use, becoming, unbelievably, an in-demand guest on all of TV's top chat shows. Also a movie actor, he has appeared in several films, including 'WW And The Dixie Dance Kings' (1975) and 'Uphill All The Way' (1985).

Recommended:
New Patches (MCA/MCA)
I Believe In You (MCA/MCA)
The Best Of Mel Tillis And The Statesiders (Polydor/–)
American Originals (Columbia/–)

Pam Tillis

Born on July 24, 1957, in Plant City, Florida, Pam Tillis is the eldest of five children by country singer Mel Tillis and his wife Doris. She grew up in Nashville and made her first stage appearance at the age of eight, singing with Mel on the Grand Ole Opry. In high school and at the University of Tennessee in Knoxville, Pam was attracted to rock acts like the Eagles, Linda Ronstadt and Little Feat, rather than pure country music. During her teen years she began a short-lived career in country music, as a Stutterette, one of the back-up vocalists for Mel Tillis, but, being a rebel child she teamed up with a jazz pianist to form a fusion band and moved to California. Although her plans didn't work out, she began songwriting. She moved back to Nashville and signed a writer's contract with Tree Music in 1983. Occasionally Pam would play the Nashville clubs, and this resulted in her landing a record deal with Warner Brothers in 1984. Recording an album of pop songs, mainly self-written, Pam scored some country hits over a three-year period. However, it was as a songwriter and session singer that she was enjoying her biggest success.

Pam was in the studio working on a demo for **Someone Else's Trouble Now** (eventually a hit for Highway 101), when she was offered a recording contract with the newly opened Nashville office of Arista Records. The debut single, **Don't Tell Me What To Do**, raced up the charts in 1991, peaking at No. 4, and the album, **Put Yourself In My Place**, gained a gold disc. The album yielded more Top 10 singles with **One Of Those Things** and **Maybe It Was Memphis**. Rather surprisingly, the mandolin madrigal autobiography, **Melancholy Child**, was not released as a single. Overall, though, this was one of the best albums by a female singer in years. A package loaded with personality and the high gleam of experience, it served as an invigorating introduction to Pam. Proving that this was not to be a one-off experience, the next album, **Homeward Looking Angel**, was just as impressive, and Tillis enjoyed further chart success with **Shake The Sugar Tree** (1992) and **Let That Pony Run** (1993).

Girlish and garrulous, but with a gutsy poise unmatched in modern country music, and with her band the Mystic Biscuits, Pam Tillis has a knock-'em dead live show that is a powerhouse blend of rockabilly, country and blues, seasoned with a quick-witted dialogue she's obviously picked up from her father.

Recommended:
Put Yourself In My Place (Arista/–)
Homeward Looking Angel (Arista/–)

Floyd Tillman

One of country music's most successful songwriters, Floyd Tillman was born in Ryan, Oklahoma on December 8, 1914. He became singer, guitarist, mandolin and banjo player with the Mark Clark Orchestra and the Blue Ridge Playboys during the 1930s, signing for Decca in 1939 and cutting the self-written **It Makes No Difference Now**, a country classic.

During the late 1940s, Tillman, who became a Columbia recording artist in 1946, wrote such compositions as **I Love You So Much It Hurts** (1948), **Slipping Around** (1949) and **I'll Never Slip Around Again** (1949). The songs became even bigger hits when recorded by Jimmy Wakely, with big band vocalist Margaret Whiting duetting.

Though his name remained absent from the record charts during the 1950s, in 1960 Tillman scored with **It Just Tore Me Up**.

A honky-tonk hero and among the first to utilize electric guitar, Tillman was also responsible for such songs as **Each Night At Nine**, **I'll Keep On Lovin' You** and **Daisy Mae**. He was elected into the Nashville Songwriter's Hall Of Fame in 1970, and received an even greater honour when he was inducted into the Country Music Hall Of Fame in 1984.

Recommended:
Country Music Hall Of Fame (MCA/–)
Greatest Hits (Crazy Cajun/–)

Aaron Tippin

A former competitive bodybuilder, Aaron Tippin, born July 3, 1958, in Pensacola, Florida, is a real hillbilly singer with an ear-jarring South Carolina twang, which he developed while living in Traveler's Rest, near Greenville, South Carolina, where his pilot father moved his family in the mid-'60s. When he was ten, Aaron got his first guitar and later played banjo, but at the time was more intent on following in his father's footsteps and took to flying planes. He flew solo at 16 and by 19 had his multi-engine commercial licence, but when airlines started laying-off pilots in the late '70s, Tippin decided to try music full-time. By 1978 he was playing in local bluegrass bands the Dixie Ridge Runners and Tip And The Darby Hill Band. After completion of his aviation course, he found

Put Yourself In My Place, Pam Tillis. Courtesy Arista Records.

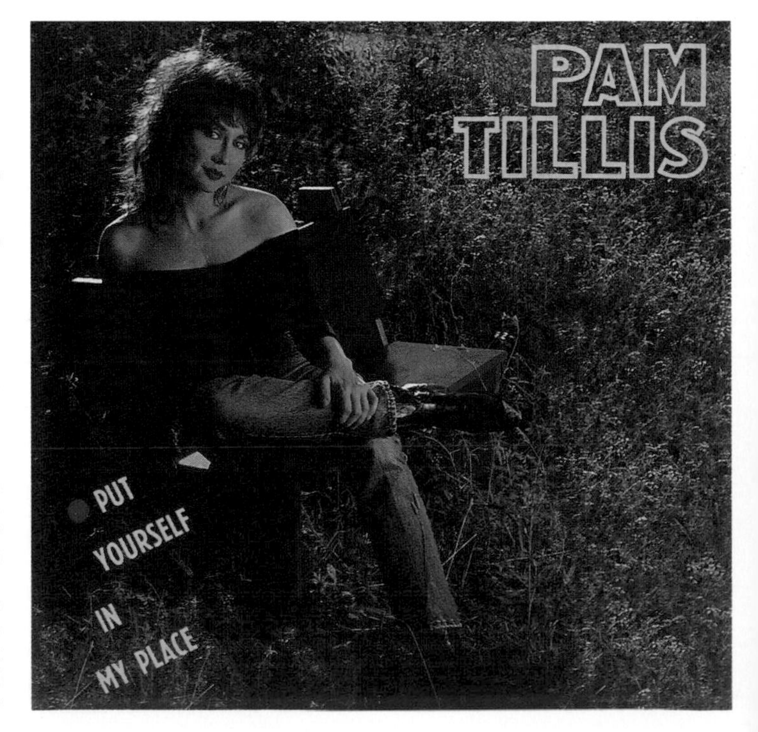

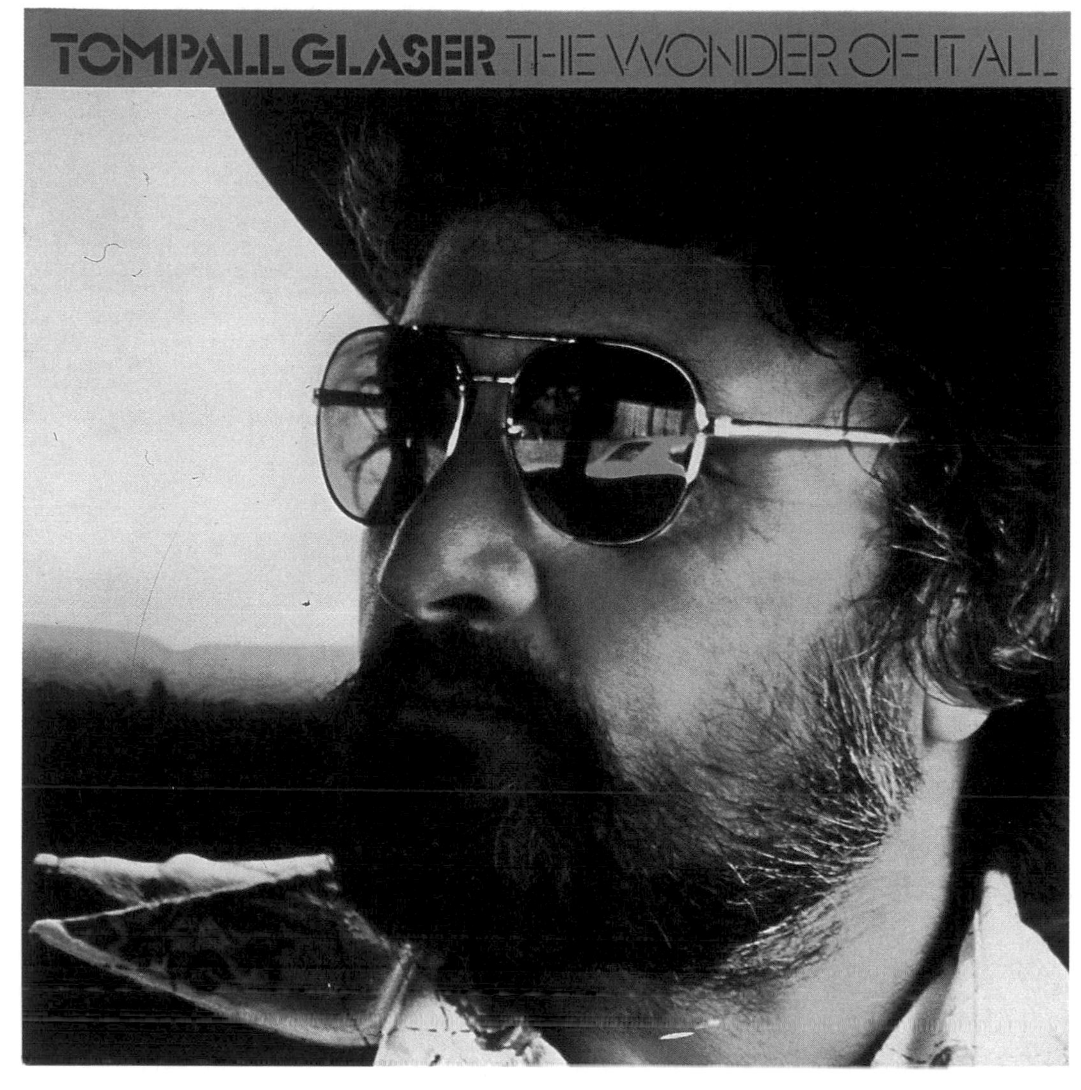

The Wonder Of It All, Tompall Glaser. Courtesy MCA Records.

work as a farm hand, truck driver and heavy equipment operator in an effort to support his family, while working the bars at weekends. His marriage disintegrated, so, in the mid-'80s, Aaron started regular trips to Nashville trying to break into music. He was working in an aluminium-rolling mill in Russellville, Kentucky by night, driving to Nashville to write songs by day. Signed to Acuff-Rose, he slowly started to get songs recorded by Charley Pride, Josh Logan and the Kingsmen.

In 1989, to save a few dollars, he recorded his own vocals for a new set of demos which were sent to RCA. They were impressed as much by the singer as the songs, and immediately signed Tippin to a recording contract. His first single, **You've Got To Stand For Something**, made the country Top 10 in 1990, and his debut album, carrying the same title, contained all his own songs and gained Tippin a gold album. Though this album received rave reviews, his follow-up singles were only minor hits. His vocal style, a raw blend of howls, yodels, bluesy slides and throaty crooning, didn't quite fit radio programming, but the fans decided that, regardless of radio, they were going to buy Tippin's records and watch his concerts. In 1992 he was back in the Top 10 with **There Ain't Nothing Wrong With The Radio**, a real uptempo number which was his first No.1, and **I Wouldn't Have It Any Other Way**. His second album, **Read Between The Lines**, also went gold.

You've Got To Stand For Something, Aaron Tippin. Courtesy RCA Records.

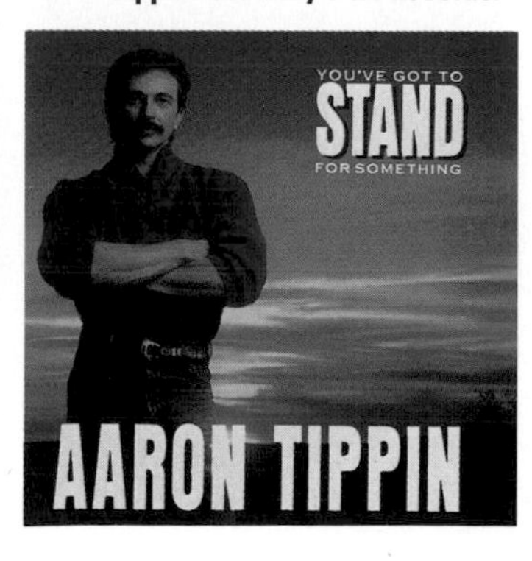

Recommended:

You've Got To Stand For Something (RCA/RCA)

Read Between The Lines (RCA/–)

Tompall And The Glaser Brothers

Born on a ranch in Spalding, Nebraska (Tompall Glaser on September 3, 1933, Charles 'Chuck' Glaser on February 3, 1936 and James Glaser on December 16, 1937), the brothers' early interest in music was generated by their parents, who were both country music devotees. During the early 1950s they formed a band and played local clubs and dance halls. As their reputation spread, they gained a 13-week series on KHAS, Hastings, Nebraska. Later, following a win on the Arthur Godfrey Talent Show, they travelled to Nashville in 1958 and were soon on tour with Marty Robbins. Signed by Decca, they recorded folk music (while performing country onstage) and were not entirely happy with their debut album, **This Land**.

But in 1962 they became members of the Grand Ole Opry and followed a country direction from then on. Also that year they toured with Johnny Cash, and the dates this package played at Las Vegas and Carnegie Hall provided the Glasers with an even wider audience.

By 1966 they had signed to MGM Records and begun a noteworthy recording career. Their delightful harmony vocals and acoustic guitar work were particularly suited to the more melodic country material and they kept abreast of the latest songwriting trends too. Their first hits came with songs like **Gone On The Other Hand** (1966) and **Through The Eyes Of Love** (1967), their first Top 10 being **Rings** (1971). Multi-award winners, they proved a popular live group and stopped the show at the 1970 Wembley Festival. But, after working seven days a week as a band and also trying to run a studio and an office, the brothers developed differences and split in 1972. Tompall cut a solo album called **Charlie** (1973) that indicated a change of direction, particularly in terms of lyrics. The title song spoke of a man's past and future lives and seemed prophetic.

Later albums saw Tompall in much the same mood, exploring themes that would not have been possible with the more melodic Glaser Brothers. He formed his Outlaw Band and appeared on the million-selling **The Outlaws**, along with Waylon Jennings and Willie Nelson, providing that album's grittiest track with a stomping version of Jimmie Rodgers' **T For Texas**.

Meanwhile, both Jim and Chuck were having some success in more mainstream country, all three Glaser brothers appearing in the country charts during 1974; Chuck with **Gypsy Queen**, Jim with **Fool Passin' Through** and **Forgettin' All About You**, and Tompall with **Texas Law Sez** and **Musical Chairs**.

Chuck, however, became increasingly involved in record production and management, but suffered a stroke that put him out of action for some time.

Although Tompall had a critically acclaimed solo career, there remained a bigger demand for the brothers as a threesome. So, in 1978, they got back together again, resuming their joint chart activities with a brace of mid-chart hits for Elektra in 1980. This time, it seemed, they were really going to achieve a breakthrough in record sales. Their 1981 version of Kris Kristofferson's **Lovin' Her Was Easier (Than Anything I'll Ever Do Again)** went to No.2 in the charts. There were other Top 20 records with **Just One Time** (1981) and **It'll Be Her** (1982). Then the Glasers split once again, Jim – who, with Jimmy Payne, had earlier penned the million-selling hit **Woman, Woman** for Gary Puckett – releasing a single on a new independent label, Noble Vision. Titled **When You're Not A Lady**, it not only went Top 20 in the charts, but, after a stay of 22 weeks, became the all-time longest-running debut release by a new record company in the history of those charts.

Charlie, Tompall Glaser. Courtesy MGM Records.

Proffering a soft, romantic line in country that contrasts with Tompall's method, Jim logged a country No. 1 with **You're Getting To Me Again** (1984) along with several other major chart climbers. In 1985, Noble Vision was absorbed into the MCA label, and, after scoring minor hits during the next two years, Jim's brief run of success came to an end.

Recommended:

Sing Great Hits From Two Decades (MGM/MGM)

The Award Winners (MGM/MGM)

Jim Glaser:

The Man In The Mirror (Noble Vision/–)

Tompall Glaser:

Charlie (MGM/MGM)

The Great Tompall And His Outlaw Band (MGM/MGM)

Tompall Glaser And His Outlaw Band (ABC/ABC)

The Outlaw (–/Bear Family)

The Rogue (–/Bear Family)

Diana Trask

An Australian who made it big in Nashville during the late 1960s and early 1970s, Diana Trask was born in Melbourne, Australia, on June 23, 1940. Winner of the top talent award at 16, she toured with a group before going to the US in 1959.

Then a pop vocalist, she later signed to Columbia Records, appeared on major TV shows and was even offered a film contract. However, she got married and returned to Australia. In the late 1960s, during a trip to the CMA convention, she became bitten by the country bug and stayed on in Nashville, having her first country hit with **Lock, Stock And Barrel**, a Dial release. She then joined Dot and cut an album, **Miss Country Soul**, and had Top 20 entries with **Say When**, **It's A Man's World**, **When I Get My Hands On You** (1973) and **Lean It All On Me** (1974).

She returned to Australia once more in 1975 and had her first hit there in 14 years with **Oh Boy**, a Festival release which went to No.2. Signed to Australian RCA in 1977 and Polydor in 1980, Diana was back in the US charts in '81 with **This Must be My Ship** and **Stirrin' Up Feelings**.

Recommended:
Diana's Country (Dot/–)
Miss Country Soul (Dot/–)

Merle Travis

Easily one of the most, if not *the* most, multi-talented men ever to enter the music business was Merle Travis, born in Rosewood, Mulenberg County, Kentucky, on November 29, 1917.

A singer and songwriter of major proportions and guitar stylist of monumental influence, he also proved adept as an actor, author and even cartoonist.

Merle learned the basics of his celebrated guitar style from Mose Rager, who, in turn, learned it from black railroad hand, fiddler and guitarist, Arnold Shultz. Merle adapted the finger style to a degree of complexity unknown in that era (it was to prove extremely influential to Chet Atkins and many others). His renown won him a job with a group called the Tennessee Tomcats before joining Clayton McMichen's Georgia Wildcats on WLW's Boone County Jamboree.

After a stint in the Marines, Travis relocated on the West Coast, perfecting his songwriting, appearing in minor roles in a host of westerns. He also signed with Capitol Records and had several of the biggest hits of the era: **Divorce Me C.O.D.** (1946), **So Round, So Firm, So Fully Packed** (1947) and several others which ranked on the charts, these including **Dark As A Dungeon** (1947) and **Sixteen Tons** (1947), a 1955 hit for Tennessee Ernie Ford.

Writer or co-writer of all his hits, he also co-wrote **No Vacancy** with Cliffie Stone and **Smoke! Smoke! Smoke!** with Tex Williams. He was equally adept at reworking folk tunes, and **John Henry**, **I Am A Pilgrim** and **Nine Pound Hammer** were all adapted by and integrated into the Travis style.

In the 1950s, Merle became a southern California fixture, appearing regularly on the Hometown Jamboree and Town Hall Party, making a striking appearance as a guitar-strumming sailor in the movie 'From Here To Eternity', where he introduced the song **Re-Enlistment Blues**. He moved to Nashville for a short while in the 1960s but later returned to California, using it as a base for frequent tours up to the time of his death in Tahlequah, Oklahoma, on October 20, 1983. Travis, whose last film appearance was in Clint Eastwood's 'Honky Tonk Man', was inducted into the Nashville Songwriters Hall Of Fame in 1970. Both Doc Watson and Chet Atkins named sons after him.

Above: Merle Travis parades a blues or two in 'From Here To Eternity'.

Recommended:
The Atkins-Travis Traveling Show – with Chet Atkins (RCA/RCA)
Walkin' The Strings (Capitol/Pathe Marconi)
Travis! (Capitol/Capitol)
Merle Travis And Joe Maphis (Capitol/Capitol)
Folk Songs Of The Hills (–/Bear Family)

Randy Travis

When Randy Travis hit Nashville with his **Storms Of Life** platinum album in 1986, country music was fumbling for a pop-flavoured identity as record sales had

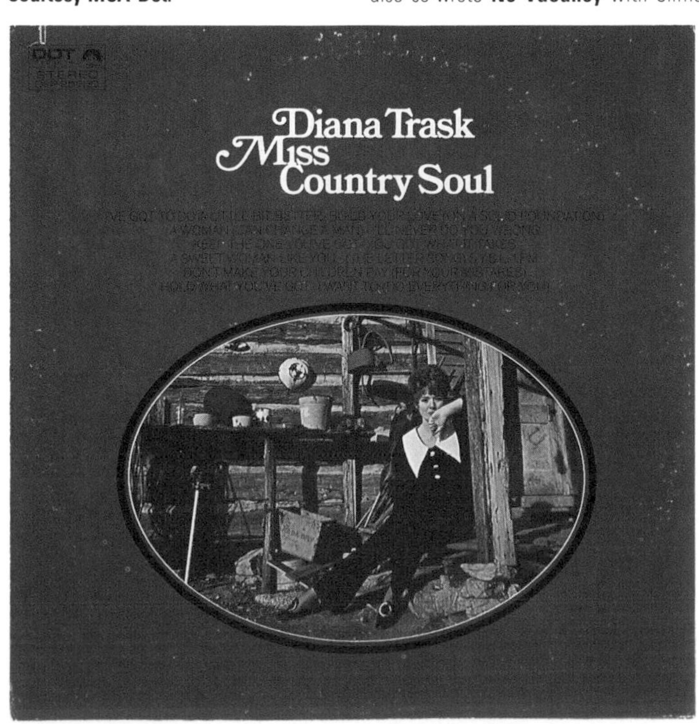

Miss Country Soul, Diana Trask. Courtesy MCA-Dot.

The Atkins-Travis Traveling Show. Courtesy RCA Records.

plummeted and the music had become very stale and cliché-ridden. Travis, sang straight-ahead country music with warmth and conviction, and, though he wasn't the best, he was young, good-looking, and in the right place at the right time.

Born Randy Bruce Traywick on May 4, 1959 in Marshville, North Carolina, Travis started singing and playing guitar when he was nine. Randy teamed up with his brothers Ricky and David, and, with their father arranging dates, they performed at local clubs. Frequently in trouble with the law, Travis appeared, while on probation, in a talent show at a Charlotte club owned by Lib Hatcher. Taking responsibility for the youngster as manager and guardian, Lib financed his first recordings as Randy Traywick, which were produced by Joe Stampley in Nashville and released on Paula Records. The single, **She's My Woman**, became a minor hit in 1979. Two years later Lib moved to Nashville to open a new club, The Nashville Palace, and develop Randy's career. Now working as Randy Ray, in 1982 he recorded his first album, **Randy Ray At The Nashville Palace**, while Hatcher took his demos around to every major label in Nashville. Eventually she persuaded Warner Brothers A&R executive Martha Sharp to see Randy performing. The label was looking for a new young artist to compete with George Strait and Ricky Skaggs and Martha signed him to a recording contract in 1985, suggesting a name change to Randy Travis.

The single **On The Other Hand**, released that summer, only did marginally better than his Paula release, but the next one, **1982**, made the Top 10 in 1986. The album **Storms Of Life** gained critical approval. **On The Other Hand** was singled out for special mention, so it was re-released and promptly climbed to the top of the charts. With his low-key, Lefty Frizzell-flavoured, pure traditional honky-tonk country vocals, Travis now dominated the charts with further No.1s, including: **Diggin' Up Bones** (1986), **Forever And Ever, Amen** (1987), **Too Gone Too Long** (1988), **Deeper Than The Holler** and **It's Just A Matter Of Time** (1989). With this success, Randy gained the CMA Horizon Award in 1986, and picking up both album and single award in 1987 for **Always And Forever** and **Forever And Ever, Amen**. His album sales sky-rocketed into platinum status, as Travis started attracting younger, female fans to country music.

Randy Travis was named CMA Male Vocalist in 1988, and also picked up his second Grammy. Initially not a great stage performer, he has gradually grown in confidence. Travis opened the floodgates for the 'New Traditionalists' who have dominated country music since the late '80s and given Nashville a new golden age. Randy has developed his songwriting, working closely with Don Schlitz and Alan Jackson, and maintained his success on the charts with singles **Hard Rock Bottom Of Your Heart** (1990), **Forever Together** (1991) and **If I Didn't Have You** (1992). Albums **Old 8 x 10**, **No Holding Back** and **High Lonesome** have each sold in excess of one million copies.

Recommended:
Old 8 × 10 (Warner Bros/Warner Bros)
Storms Of Life (Warner Bros/Warner Bros)
Heroes And Friends (Warner Bros/Warner Bros)
No Holding Back (Warner Bros/Warner Bros)
High Lonesome (Warner Bros/Warner Bros)

Above: In 1947 Ernest Tubb headed the first country show at Carnegie Hall.

Travis Tritt

Long-haired, blue-collar, biker-hero Travis Tritt was born on February 9, 1963 in Marietta, Georgia. With his rockin' country that celebrates the working-class South, his solid rock roots and influences that range from George Jones and Merle Haggard to the Allman Brothers and Lynyrd Skynyrd, Travis became one of country music's hottest acts of the early '90s. After graduation from high school in 1981, he went to work loading trucks, and within four years had worked his way up to a management position. By this time he was married, but, when the marriage didn't work out, he quit his job and began playing solo at various clubs. He worked Atlanta dinner clubs, rowdy honky-tonks and backroad greasy joints. Wherever he could, he worked his songs into his act, and it was through this that Tritt came to the attention of Danny Davenport, a local representative for Warner Brothers. Initially interested in some of Tritt's songs, Danny realized Tritt's potential as an entertainer when he saw him in front of an audience. Together at Davenport's home studio they began working on an album. When executives at Warners heard the tapes, they offered Travis Tritt a recording contract in early 1989.

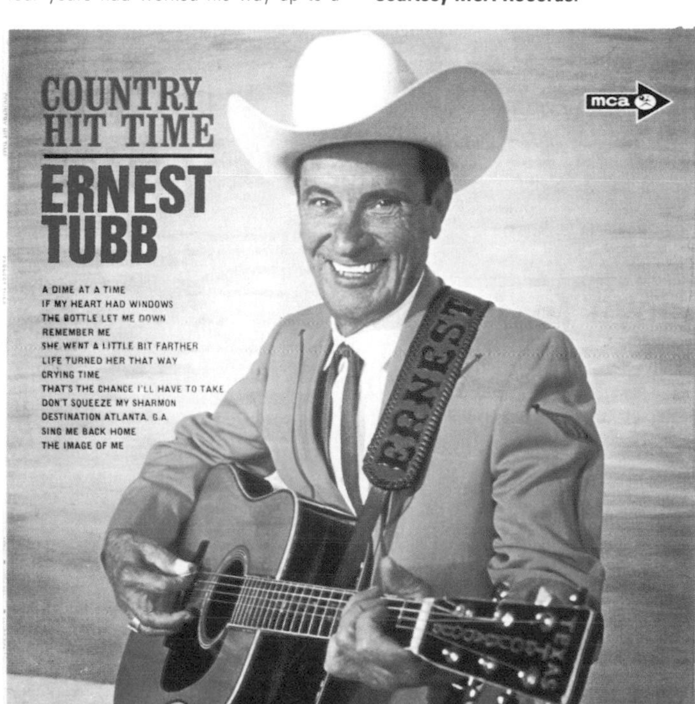

Country Hit Time, Ernest Tubb. Courtesy MCA Records.

His debut single, **Country Club**, went to No.9 on the country charts in late 1989. **Help Me Hold On** climbed to the top and Tritt enjoyed further big hits with **I'm Gonna Be Somebody** (1990), **Here's A Quarter (Call Someone Who Cares)** and **Anymore** (both 1991), as well as **Lord Have Mercy On The Working Man** and **Can I Trust You With My Heart** (both 1992). He picked up the CMA Horizon award in 1991, while **Country Club**, **It's All About To Change** and **T-R-O-U-B-L-E** have all gone platinum. In early 1992 he became the youngest member of the Grand Ole Opry. Following duet success with Marty Stuart on **The Whiskey Ain't Workin'**, the pair put together a 'No Hats Tour' that garnered so much critical acclaim that a national pay-per-view concert was filmed. **Bible Belt**, his rocking collaboration with Little Feat, was heard in Joe Pesci's 1992 film 'My Cousin Vinnie'. His white-hot version of the Elvis classic **Burnin' Love** was included in the soundtrack album for the motion picture 'Honeymoon In Vegas'. **Texas Flyer**, a song he originally recorded as **Dixie Flyer**, was re-recorded by him in honour of US Olympic athlete Lance Armstrong, and included on the **Barcelona Gold** album.

Recommended:
T-R-O-U-B-L-E (Warner Bros/–)
Country Club (Warner Bros/–)
It's All About To Change (Warner Bros/–)

Ernest Tubb

The sixth member to be elected to the Country Music Hall Of Fame and a regular member of the Opry from 1943 to the time of his death, Ernest Dale Tubb, the son of a Texas cotton farm overseer, was born in Crisp, Texas on February 9, 1914. Tubb's boyhood hero was the great Jimmie

Rodgers. Although he had dreams of emulating Rodgers and sang at various local get-togethers during his early teens, Tubb was almost 20 before he owned his first guitar. The year 1934 proved important to him, Tubb obtaining his initial dates on San Antonio KONO. During this period he married Lois Elaine Cook.

In 1935 Tubb's eldest son Justin was born and Ernest met Carrie Rodgers (Jimmie's widow). She and Ernest became good friends, Mrs Rodgers loaning him her husband's original guitar and also arranging an RCA recording session at which Tubb cut two sides: **The Passing Of Jimmie Rodgers** and **Jimmie Rodgers' Last Thoughts**. However, Tubb's luck was not always that good. His second son, Rodger Dale, was born in July 1938, but died after just a few weeks. Things began to look brighter after the birth of a daughter, Violet Elaine; Decca offered him a new record contract and he obtained a job on Fort Worth's KGKO.

It was at this stage that he became the Gold Chain Troubadour, earning 75 dollars a week promoting Universal's wares. It was a nickname which preceded his famous Texas Troubadour image. By 1941 he had also moved into movies, appearing in 'Fightin' Buckaroos'.

Next came his recording of **Walking The Floor Over You**, a self-penned composition. Released in autumn 1942, it became a million-seller, helping Tubb gain his first appearance on the Opry in December. He was to gain regular membership during 1943.

He continued logging successful discs and film appearances. Also, in 1947, he opened the first of his now famous record shops and commenced his Midnight Jamboree programme over WSM, advertising the shop and showcasing the talents of up and coming country artists.

Tubb married again in 1949, his new wife being Olene Adams, mother of Erlene, Olene, Ernest Jr, Larry and Karen Tubb. That year he appeared on hit records with the Andrews Sisters and Red Foley. He also achieved Top 10 placings with no less than five of his solo efforts, the biggest of these being **Slippin' Around** and **Blue Christmas**. From then through to 1969 he became the charts' Mr Consistency, thanks to such discs as **Goodnight Irene** (with Red Foley, 1950), **I Love You Because** (1950), **Missing In Action** (1952), **Two Glasses Joe** (1954), **Half A Mind** (1958), **Thanks A Lot** (1963), **Mr and Mrs Used-To-Be** (with Loretta Lynn, 1964) and **Another Story, Another Place** (1966). His only real absence from hit listing was between 1952 and 1954 when, following an exhausting Far East tour, Tubb suffered from an illness that kept him off the Opry.

An inveterate tourer, he and his Texas Troubadours played around 300 dates a year. An honest singer rather than a great one – emotionally he was a 10-point man, technically he came a lot further down the scale – when ET performed honky-tonk you could almost smell the booze. Much loved, when he set out to record his **Legend And Legacy** album for First Generation records in 1979, virtually everyone who was anyone in Nashville dropped by to see if they could help out. The album line-up eventually featured the names of Willie Nelson, Loretta Lynn, Vern Gosdin, Chet Atkins, Merle Haggard, Johnny Cash, Charlie Rich, Johnny Paycheck, Linda Hargrove, Marty Robbins, Conway Twitty, the Wilburn Brothers, Ferlin Husky, Waylon Jennings, Charlie Daniels, George Jones and many, many others. When he died, on September 6, 1984, the whole of Music City mourned the man writer Chet Flippo once accurately described as "honky-tonk music personified".

Recommended:
The Legend And The Legacy (First Generation/–)
The Ernest Tubb Story (MCA/MCA)
Honky Tonk Classics (Rounder/–)
The Country Hall Of Fame (–/MCA)
Let's Say Goodbye Like We Said Hello (–/Bear Family)
Live 1963 (Rhino/–)
The Yellow Rose Of Texas (–/Bear Family)

Justin Tubb

Eldest son of Ernest Tubb, singer-songwriter-guitarist Justin Tubb was born in San Antonio, Texas, on August 20, 1935. His father recorded one of his songs in 1952, and that year he and two of his cousins formed a group and began playing clubs in the Austin area, where Tubb was attending the University of Texas. But after just a year of college came the inevitable move to Nashville and a DJ job on a radio station in nearby Gallatin, Tennessee. Tubb not only spun discs, but also entertained his listeners with his own songs on air.

In 1953 he signed with Decca, the following year logging two hits, **Looking Back To See** and **Sure Fire Kisses**, both duets with Goldie Hill. Although Tubb became an Opry regular in 1955, his records sold only moderately well and he began to label hop, leaving Decca in 1959 and cutting sides for Challenge and Starday. Then, after a Top 10 Groove release in **Take A Letter Miss Gray** (1963), came a long association with RCA and some so-so chart visits with **Hurry, Mr Peters** (1965), **We've Gone Too Far Again** (1966) – both duets with Lorene Mann – and **But Wait There's More**, a solo item from 1967.

Once an inveterate tourer, Tubb has played in all but two states and has also appeared in several countries. During 1967 he took a show to the Far East, entertaining servicemen in Vietnam and other areas. Nowadays he tours less regularly but still appears on various country TV shows and is something of a fixture on the Opry. He has also enjoyed success as a writer, his most notable composition being **Lonesome 7-7203**, a No.1 for Hawkshaw Hawkins in 1963.

Recommended:
Justin Tubb, Star Of The Grand Ole Opry (Starday/–)
Justin Tubb (MCA–Dot/–)

Tanya Tucker

When she was nine years old, people at both MGM and RCA Records wanted to sign her. At 14 she had gained a Top 10 hit and a year later her face bedecked the cover of 'Rolling Stone'. Shortly after, she came up with the biggest country single in the land, also acquiring a reputation as a musical Lolita because of her penchant for songs equipped with provocative lyrics.

Born in Seminole, Texas on October 10, 1958, Tanya Denise Tucker, the daughter of a construction worker, spent her early years in Wilcox, Arizona, moving to Phoenix in 1967. There, Tanya and her father began attending as many country concerts as possible, visiting local fairs to hear Mel Tillis, Leroy Van Dyke, Ernest Tubb and others, Tanya often joining the stars onstage for an impromptu song.

Following a cameo role in the movie 'Jeremiah Johnson', Tanya, then 13, cut a demo tape that included her renditions of **For The Good Times**, **Put Your Hand In The Hand** and other songs. The results impressed Columbia's Billy Sherrill, who signed Tanya to the label and promptly produced her recording of Alex Harvey's **Delta Dawn**. The result was a 1972 Top 10 single, after which the Tucker–Sherrill partnership moved into further action to provide such chart-busters as **Love's The Answer**, **What's Your Mama's Name?**, **Blood Red And Going Down** (1973), **Would You Lay With Me (In A Field Of Stone)** and **The Man Who Turned My Mama On** (1974). **Would You Lay With Me**, one of the year's most controversial singles, also proved a hit of international proportions.

In 1976, following a million-dollar deal, Tanya signed for MCA, thus terminating her association with Sherill and creating some doubts as to her ability to survive without the guiding hand of the Columbia Svengali. But the doubts were quickly dispelled when **Lizzie And The Rainman**, **San Antonio Stroll** (1975), **You've Got Me To Hold On To** (1976), **Here's Some Love** (No.1, 1976), **It's A Cowboy Lovin' Night** (1977) and **Texas (When I Die)** (1978) all went Top 10.

In the late '70s, Tanya attempted to move further into the higher stakes of the rock field, but, even though she donned red tights for a highly publicized **TNT** album, things began falling apart a little. Equally publicized was her affair with Glen Campbell, with whom she recorded some duets before the twosome parted in 1981. Nevertheless, she had further solo Top 10 hits in 1980 with **Pecos Promenade** and **Can I See You Tonight**, then switched to Arista Records in 1982 for a disastrous association that saw her career plummet as a country star. To make matters worse, Tucker had also become addicted to alcohol and cocaine, and entered the Betty Ford clinic for treatment.

After a three-year absence from the charts, Tanya signed with Capitol Records and enjoyed the most successful period of her long career. She hit the charts with such No.1s as **Just Another Love** (1986), **I Won't Take Less Than Your Love** (with Paul Davis and Paul Overstreet, 1987), **If It Don't Come Easy**, **Strong Enough To Bend** (1988), **My Arms Stay Open All Night** (No.2, 1989), **Walking Shoes** (No.3, 1990), **Down To My Last Teardrop** (No.2, 1991) and **Two Sparrows In A Hurricane** (No.2, 1992). One of the most exciting female performers in country music, Tanya Tucker was named the CMA Female Vocalist Of The Year in 1991, but had to miss attending, as she was in hospital giving birth to her second child, Beau Grayson. A single mother, Tanya, or 'T' as she prefers to be known, carved her success in the late '80s by refusing to compromise. Her outstanding talent and distinctive vocal hiccup showed audiences that the little girl of **Delta Dawn** fame had blossomed into a forthright and beautiful woman.

Recommended:
Delta Dawn (Columbia/CBS)
Would You Lay With Me (Columbia/CBS)
Here's Some Love (MCA/MCA)
Strong Enough To Bend (Capitol/Capitol)
Can't Run From Yourself (Liberty/Liberty)
Tennessee Woman (Capitol/Capitol)
Love Me Like You Used To (Capitol/Capitol)
Lizzie And The Rainman (–/Cottage)

Above: Single-parent and country star Tanya Tucker successfully balances family and showbiz.

Strong Enough To Bend, Tanya Tucker. Courtesy Capitol Records.

Conway Twitty

Real name Harold Lloyd Jenkins, born in Friars Point, Mississippi, on September 1, 1933, Twitty learned guitar onboard a riverboat piloted by his country music-loving father. Almost signed by the Philadelphia baseball team, Twitty was drafted before the contract could be concluded and spent two years in the Army instead.

During the mid-1950s, he became a rock'n'roll singer, working on many radio stations and charting with a Mercury single **I Need Your Lovin'** (1957). Shortly after, he joined MGM Records and won a gold disc for **It's Only Make Believe**, one of 1958's biggest sellers. Extremely Presley-influenced at this point in his career, Twitty was hardly out of the pop charts between September 1958 and April 1961, also finding time to appear in three teen-angled movies – 'Sex Kittens Go To College', 'Platinum High School' and 'College Confidential'.

It was at this time that Twitty began writing country songs. His **Walk Me To The Door** was recorded by Ray Price in 1960. By June 1965, he himself was cutting country sides under Decca's Owen Bradley, at the same time settling down in Oklahoma City playing with a band known as the Lonely Blue Boys and (in June, 1966) commencing his own syndicated TV programme. During the late 1960s, Twitty moved to Nashville and began amassing an incredible number of hits – his solo chart-toppers alone including: **Next In Line** (1968), **I Love You More Today**, **To See My Angel Cry** (1969), **Hello Darlin'**, **15 Years Ago** (1970), **How Much More Can She Stand** (1971), **(Lost Her Love) On Our Last Date**, **I Can't Stop Loving You**, **She Needs Someone To Hold Her** (1972), **You've Never Been This Far Before** (1973), **There's A Honky Tonk Angel**, **I See The Want To In Your Eyes** (1974), **Linda On My Mind**, **Touch The Hand**, **This Time I've Hurt Her More Than She Loves Me** (1975), **After All The Good Is Gone**, **The Game That Daddies Play**, **I Can't Believe She Gives It All To Me** (1976), **Don't Take It Away**, **I May Never Get To Heaven**, **Happy Birthday Darlin'** (1979), **I'd Love To Lay You Down** (1980), **Rest Your Love On Me**, **Tight Fittin' Jeans** and **Red Neckin' Love Makin' Night** (1981), all Decca/MCA releases.

Additionally during this period, Conway, who became a vastly superior singer to the one known only to pop audiences, also fashioned an equally impressive number of hit duets with Loretta Lynn, hitting the No.1 spot with **After The Fire Is Gone**, **Lead Me On** (1971), **Louisiana Woman, Mississippi Man** (1973), **As Soon As I Hang Up The Phone** (1974) and **Feelin's** (1975). Voted Vocal Duo Of The Year by the CMA for four straight years in a row (1972–1975), Conway and Loretta also shared several business interests. An astute businessman, Twitty owned a music promotion company, a large slice of real estate and the Twitty City complex (a kind of theme park that included the homes of Conway, his four children and his mother), one of Nashville's major attractions since it opened in 1982.

Also in 1982 Conway quit MCA and moved to Elektra Records, immediately claiming three No.1s (with **The Clown**, **Slow Hand** and **The Rose**) in his first year with his new label. Switching to Warner Brothers in 1983, Twitty again obliged with three No.1s in a year during 1984, with **Somebody's Needin' Somebody**, **Ain't She Somethin' Else** and **I Don't Know About Love (The Moon Song)** – the last named featuring Conway's daughter, Joni Lee Twitty.

Twitty's success on the charts continued throughout the '80s, the most notable successes being **Don't Call Him A Cowboy** (1985) and **Desperado Love** (1986). He then re-joined MCA and continued with **I Want To Know You Before We Make Love** (1987), **I Wish I Was Still In Your Dreams** (1988), **She's Got A Single Thing In Mind** (1989), **Crazy In Love** (1990) and **I Couldn't See You Leavin'** (1991). In 1992 he felt his many business interests were getting in the way of his music. So, the tousle-haired tycoon put almost everything on the market. Twitty had just completed a new album, but, after an appearance at Branson, Missouri, was taken ill on his tour bus. Rushed to the Cox Medical Center in Springfield, Missouri, he had surgery to repair an abdominal aortic aneurysm, but he died on June 5, 1993.

Originally named after a famous silent film comedian, Twitty took his stage name from the towns of Conway (in Arkansas) and Twitty (in Texas). Made an honorary chief of the Choctaw nation in the early 1970s, he was also awarded the Indian name Hatako-Chtokchito-A-Yakni-Toloa – which translates into 'Great Man Of Country Music'. Apt for a singer who has, despite his pop heritage, never opted for cross-over appeal.

Recommended:
Classic Conway (MCA/MCA)
Songwriter (MCA/MCA)
Georgia Keeps Pulling On My Ring (MCA/MCA)
Conway (MCA/MCA)
Cross Winds (MCA/MCA)
Crazy In Love (MCA/–)
Even Now (MCA/–)
House On Old Lonesome Road (MCA/–)
Making Believe – with Loretta Lynn (MCA/–)
Borderline (MCA/–)

Above: The late Conway Twitty and his Twitty City, which celebrated its 10th birthday in 1992.

T. Texas Tyler

Tyler, real name David Luke Myrick, was born on June 20, 1916, near Mena, Arkansas. Educated in Philadelphia, he began his career at the age of 14, heading east and appearing on the Major Bowes Amateur Hour in New York during the 1930s. He became widely known as the 'Man With The Million Friends'.

Later came a further move to West Virginia, while in 1942 Tyler was in Louisiana, becoming a member of Shreveport KWKH's Hayride show. A period in the armed forces followed, Tyler settling down in the Hollywood area upon discharge and forming the T. Texas Western Dance Band, a popular unit.

It was during this period that Tyler wrote and recorded **Deck Of Cards**, a hit for Four Stars in 1948. A somewhat sentimental but ingenious monologue regarding a soldier who employed a deck of cards as his Bible, prayer book and almanac, Tyler's creation became a million-seller when recorded by Wink Martindale. The song also became a hit for Tex Ritter and British comedian Max Bygraves. Following this record, which won the 'Cashbox' award for the best country disc of 1948, Tyler came up with several more winners, the most potent of these being **Dad Gave The Dog Away** (1948), **Bumming Around** (1953), **Courting In The Rain** (1954) and his theme song, **Remember Me**.

During 1949, the Arkansas traveller appeared in 'Horseman Of The Sierras', a Columbia movie, and won a fair amount of acclaim from Range Round Up, his Los Angeles TV show. During the 1950s and 1960s he continued performing, both live and on TV. However, despite some worthwhile Starday releases, Tyler failed to place his name on the record charts during the later stages of his career. He died from natural causes on January 28, 1972, in Springfield, Missouri.

His Great Hits, T. Texas Tyler. Courtesy Hilltop Records.

Recommended:
T. Texas Tyler – His Great Hits (Hilltop/–)

Leroy Van Dyke

Van Dyke (born in Spring Fork, Missouri on October 4, 1929) was co-writer (with Buddy Black) and singer of **The Auctioneer**, a 1956 gold disc winner that incorporated a genuine high-speed auctioneering routine. He originally decided on a career in agriculture, obtaining a BS degree in that subject at the University of Missouri. After serving with Army intelligence during the Korean War, he became a livestock auctioneer and agricultural correspondent, utilizing his writing skills to pen songs. He sang **The Auctioneer** on a talent show and subsequently won a contract with Dot Records, his song providing his first release – ultimately a two and half million-seller.

A regular on the Red Foley TV Show, he later signed for Mercury, providing that label with **Walk On By**, yet another million-seller, in 1961. He followed this with **If A Woman Answers** and **Black Cloud**, both hits during the following year. After that his releases rarely charted impressively, only **Louisville** (1968) really making the grade.

Van Dyke, who made his film debut in 'What Am I Bid?' (1967), recorded for Warner Bros, Kapp, Decca and ABC-Dot after leaving Mercury in 1965, his last chart record of any size being **Texas Tea**, an ABC-Dot release in 1977.

Recommended:
Greatest Hits (MCA/–)
The Original Auctioneer (–/Bear Family)
The Auctioneer (–/Ace)

Townes Van Zandt

Legendary singer, songwriter and guitarist, Van Zandt, from Fort Worth, Texas, is very highly regarded for such classic songs as **Pancho And Lefty** (a hit duet for Willie Nelson and Merle Haggard) and **If I Needed You** (an Emmylou Harris and Don Williams duet). The son of a prominent oil family, he pursued a beatnik-like lifestyle in the early '60s and started performing in clubs in Houston.

In 1967 he was signed by the small Poppy Records label and recorded such albums as **Our Mother The Mountain**, which contained his own quirky folk-country songs. For a while he joined a trio called the Delta Mama Boys, but preferred to work solo. He then joined the Peace Corp, but returned to music in the late '70s. He has built up a cult following as a member of the thriving Texas singer-songwriter community. He cut an acclaimed **Live At The Old Quarter** album in 1977, but generally lived a reclusive life in a cabin in Tennessee. He was tempted back into music and recorded his first all-new album in Nashville with production by Jack Clement and Jim Rooney. Van Zandt has toured extensively, especially in Europe.

Recommended:
At My Window (Sugar Hill/Heartland)
High, Low And In Between (Tomato/Charly)
The Late Great (Tomato/Charly)
Live At The Old Quarter (–/Decal)

Randy Vanwarmer

Singer-songwriter Randy was born Randall Van Wormer on March 30, 1955, in Indian Hills, Colorado, but spent much of his life living in Cornwall, England. His father died in a car accident when Randy was ten, and his mother decided to get away from unhappy memories, uprooting the family and moving to Looe in Cornwall in 1967. It was at this time that young Randy started to play guitar and write songs, and in his teens he met up with Roger Moss, the pair working for several years as a duo. Endless trips to London with his song demos led to some songs being taken to Nashville, and in 1977 he signed with American independent Bearsville Records through Warner Bros, their London licensee. He began cutting an album, but just before the album was due for release, Bearsville pulled out of England, and Randy was left high-and-dry. He hopped on a plane and settled in Woodstock, New York, and signed directly to Bearsville in America. Several singles were released, and in 1979 Randy made a big breakthrough with the self-penned **Just When I Needed You Most**, which hit No.4 on the pop charts and crossed into the lower regions of the country listings. He recorded three albums during the next few years, but was unable to repeat that commercial success.

A move to Los Angeles saw Randy sign to a publishing company that had affiliations in Nashville, and in 1984 the Oak Ridge Boys took his **I Guess It Never Hurts To Hurt Sometimes** to the top of the country charts. The following year Randy moved to Nashville and had more writing success with **I Will Whisper Your Name** (Michael Johnson) and **Bridges And Walls** (Oak Ridge Boys). He also signed a recording contract with 16th Avenue Records, scoring a minor country hit with **I Will Hold You** (1988).

Recommended:
Every Now And Then (–/Etude)

Above: Porter Wagoner had a long association with Dolly Parton.

Porter Wagoner

Once a grocery store clerk, Wagoner (born in West Plains, Missouri, on August 12, 1930) used slow trading periods to pick guitar and sing. He was so impressive that he was engaged to promote the business over an early morning radio show.

His popularity on radio eventually led to a weekly series on KWTO, Springfield, in 1951. Wagoner later moved on to TV when KWTO became the home of Red Foley's Ozark Jubilee show. In August, 1952, he signed with RCA Records and, following several flops, had his first Top 5 hit with **A Satisfied Mind** three years later. Following two similarly successful singles in **Eat, Drink And Be Merry** (1955) and **What Would You Do (If Jesus Came To Your House)** (1956), the Missourian joined the Opry (1957). In 1960 he moved on to formulate his own TV show with singer Norma Jean (later replaced by Dolly Parton) and his band, the Wagonmasters.

Filmed in Nashville and initially syndicated to 18 stations, by the late '60s the programme was screened to over 100 outlets throughout the USA and Canada, establishing Wagoner's touring show as one of the most popular on the circuit.

Predominantly straight country in his own musical approach, although sometimes seemingly a catalyst for more startling innovations (Buck Trent first began playing electric banjo on the Wagoner programme while Porter had also been involved in some of Dolly Parton's more contemporary moves), he managed to gain a consistent foothold in the upper reaches of the charts throughout the years. He had Top 10 solo hits with: **Your Old Love Letters** (1961), **Misery Loves Company** (1962), **Cold Dark Waters** (1962), **I've Enjoyed As Much Of This As I Can Stand** (1962), **Sorrow On The Rocks** (1964), **Green, Green Grass Of Home** (1965), **Skid Row Joe** (1965), **The Cold Hard Facts Of Life** (1967), **Carroll County Accident** (1968) and **Big Wind** (1969). He also shared an impressive number of hit duets with Dolly Parton, Including **Burning The Midnight Oil** (1971), **Please Don't Stop Loving Me** (1974) and **Is Forever Longer Than Always?** (1976).

Wagoner's albums have included 'live' recordings made in 1964 and 1966; a bluegrass offering, cut in 1965; some 'downer' sessions, typified by such releases as **The Cold Hard Facts Of Life** and **Confessions Of A Broken Man**, both releases dealing with the seamier side of humanity; and a number of duet LPs with Skeeter Davis and Dolly Parton.

The successful partnership with Dolly Parton came to an end in 1974. Porter didn't want her to leave, but Dolly wanted to be free to develop her career. This parting marked Porter's rapid fall from the top and, with his records failing to make the Top 10, he finally left RCA in 1981. He recorded briefly for Warner/Viva during 1982 and 1983. Today, Porter is a very successful Nashville businessman, though

The Farmer, Porter Wagoner's 1973 tribute. Courtesy RCA Records.

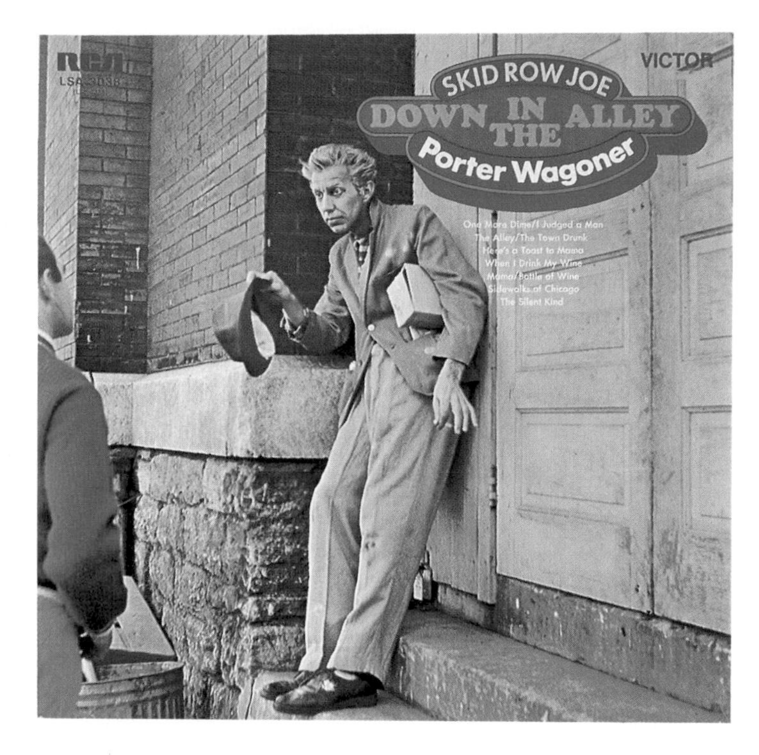

Down In The Alley, Porter Wagoner. Courtesy RCA Records.

he continues to be active in country music, both recording and performing.

Recommended:
Carroll County Accident (RCA/–)
Today (RCA/–)
Highway Heading South (RCA/RCA)
The Thin Man From The West Plains (–/Bear Family)

With Dolly Parton:
Porter And Dolly (RCA/RCA)
Two Of A Kind (RCA/RCA)

Jimmy Wakely

One of country music's major stars during the '40s and early '50s, James Clarence Wakely was born in a log cabin at Mineola, Arkansas, on February 16, 1914.

Raised and schooled in Oklahoma, where he took such jobs as a sharecropper, journalist and filling station manager, he became a professional musician during the mid-'30s, forming the Jimmy Wakely Trio with Johnny Bond and Scotty Harrell in 1937. The group appeared daily on Oklahoma City's WKY radio station. In 1940, Gene Autry guested on the show, liked the trio and signed them for his Melody Ranch CBS radio programme.

On Melody Ranch, Wakely quickly established himself as a star in his own right – eventually securing parts in over 50 movies (in 1948 he was nominated as the fourth most popular western film actor – only Roy Rogers, Gene Autry and Charles Starrett being rated higher).

After two years on the Autry show, he left to form his own band, employing such musicians as Cliffie Stone, Spade Cooley, Merle Travis and Wesley Tuttle. By 1949 he had become so popular that he beat both Frank Sinatra and Bing Crosby in the 'Billboard' pop vocalist poll, enjoying a huge hit with his version of Floyd Tillman's **Slippin' Around**. Recorded as a duet with pop vocalist Margaret Whiting, the disc soon became a million-seller for Capitol Records. Other hits with Margaret Whiting followed (including **I'll Never Slip Around Again**), the duo logging no less than seven Top 10 discs within two years.

Meanwhile, Wakely also did well in a solo capacity, such records as **I Love You So Much It Hurts** (1949), **I Wish I Had A Nickel** (1949), **My Heart Cries For You** (1950) and **Beautiful Brown Eyes** (1951) charting impressively. His 1948 hit, **One Has My Name, The Other Has My Heart**, in fact, started a whole cycle of 'cheatin' songs'.

But during the mid-'50s, Wakely's career seemed to run out of steam, and though he had a CBS networked radio show until 1958 and co-hosted a TV series with Tex Ritter in 1961, his record sales diminished, Wakely forming his own label, Shasta. However, in the mid-'70s, Wakely was still in showbiz, mainly playing to clubs in Los Angeles and Las Vegas, using an act that featured his children, Johnny and Linda Lee. Following a prolonged illness, Jimmy Wakely died on September 23, 1982, in Mission Hills, California.

Recommended:
Jimmy Wakely Country (Shasta/–)
Slippin' Around (Dot/–)
Big Country Songs (Vocalion/–)
Sante Fe Trail (–/Stetson)

Billy Walker

Once billed as the 'Travelling Texan – The Masked Singer Of Country Songs', William Marvin Walker was born in Ralls, Texas on January 14, 1929.

In 1944, at the age of 15, while Walker was attending Whiteface High School, New Mexico, he won an amateur talent show. The contest also gained him his own 15-minute Saturday radio show on KICA, Clovis, New Mexico, Walker hitchhiking 80 miles to play on the programme, then hitching his way home again.

Joining the Big D Jamboree in Dallas during 1949, he adopted his masked singer guise; the ploy worked, gaining the Texan a considerable following and a subsequent record contract from Columbia.

Other shows followed, Walker appearing on the Louisiana Hayride in the early '50s, the Ozark Jubilee between 1955 and 1960, and joining the Opry in 1960. His first hit disc came in 1954 with **Thank You For Calling**, but it was not until 1962 and the release of **Charlie's Shoes**, a nationwide No.1, that Walker began to dominate the charts.

The majority of his discs became Top 20 entries during the following decade, providing Columbia with such hits as **Willie The Weeper** (1962), **Circumstances** (1964), **Cross The Brazos At Waco** (1964) and **Matamoros** (1965), before signing with Monument and scoring with **A Million And One** (1966), **Bear With Me A Little Longer** (1966), **Anything Your Heart Desires** (1967), **Ramona** (1968) and **Thinking About You, Baby** (1969).

By 1970 Walker had joined MGM, gaining high chart placings with **When A Man Loves A Woman** (1970), **I'm Gonna Keep On Loving You** (1971) and **Sing A**

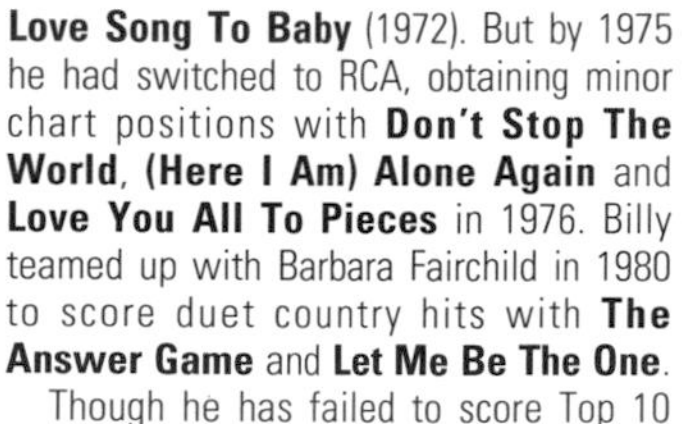

Love Song To Baby (1972). But by 1975 he had switched to RCA, obtaining minor chart positions with **Don't Stop The World**, **(Here I Am) Alone Again** and **Love You All To Pieces** in 1976. Billy teamed up with Barbara Fairchild in 1980 to score duet country hits with **The Answer Game** and **Let Me Be The One**.

Though he has failed to score Top 10 hits for many years, he has continued to record regularly for such minor labels as MRC, Scorpion, Caprice, Dimension and his own Tall Texan Records. Due to regular visits to Britain, he built up a whole new following in the 1980s, the likeable singer scoring with his Mexican-flavoured ballads that have played a major role in his long career. Walker has also made some film appearances, two of which were in 'Second Fiddle To A Steel Guitar' and 'Red River Round-Up'.

Recommended:
Alone Again (RCA/–)
Waking Up To Sunshine (Golden Memories/–)
Star Of The Grand Ole Opry (First Generation/–)
The Answer Game – with Barbara Fairchild (–/RCA)
Fine As Wine (MGM/–)
For My Friends (–/Bulldog)
Precious Memories (–/Word)

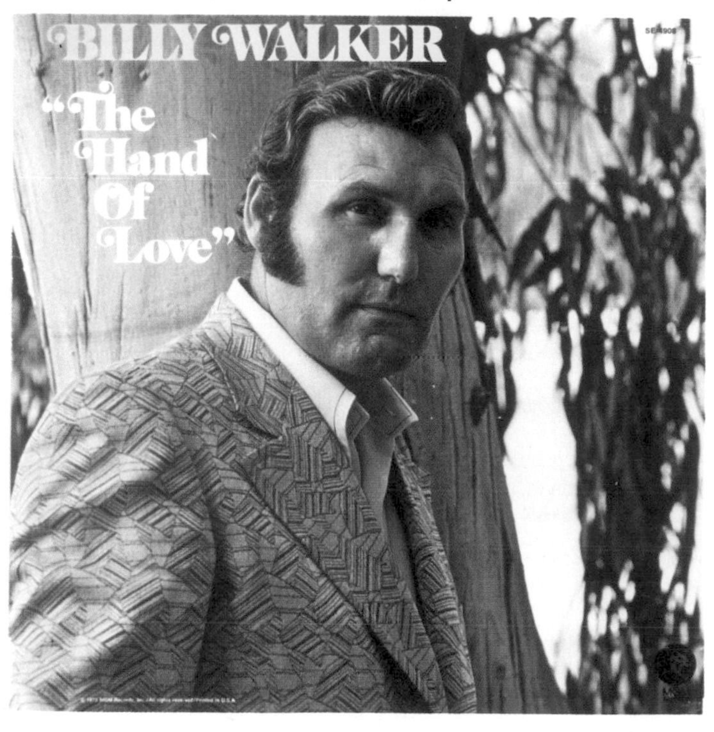

The Hand Of Love, Billy Walker. Courtesy MGM Records.

Charlie Walker

Born in Collins County, Texas, on November 2, 1926, Walker was a precocious singing and writing talent, becoming a good musician in his teens and joining Bill Boyd's Cowboy Ramblers in 1943. Later he was successful on radio, his announcing style sought after and getting him rated in 'Billboard's Top 10 Country Music Disc Jockey listing.

He signed with Columbia Records in the mid-'50s and in 1958 had his first big hit with **Pick Me Up On Your Way Down**. During the '60s and early '70s, he recorded for Columbia and Epic. Some of his hits were **Who'll Buy The Wine?** (1960), **Wild As A Wild Cat** (1965) and **Don't Squeeze My Sharmon** (1967). He also cut a series of honky-tonk titles that

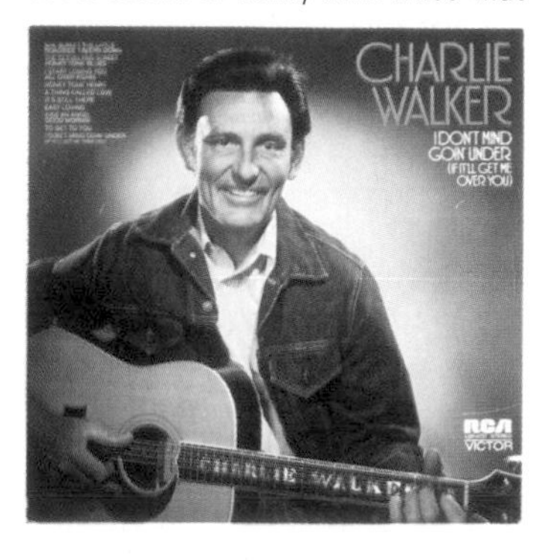

I Don't Mind Goin' Under, Charlie Walker. Courtesy RCA Records.

included **Close All The Honky Tonks** (1964), **Honky Tonk Season** (1969) and **Honky Tonk Women** (1970).

His announcing capabilities helped him gain many cabaret residencies, most notably at the Las Vegas Golden Nugget. A capable golfer, Walker has won respect as a knowledgeable golfing broadcaster.

In 1972 he became an RCA recording artist. His albums for the label included **Break Out The Bottle** and **I Don't Mind**

Going Under. A short spell with Capitol in 1974 proved fruitless, and since recording a couple of albums for Shelby Singleton's Plantation label in the late '70s, Charlie has concentrated on broadcasting.

Recommended:
Charlie Walker (MCA–Dot/–)

Jerry Jeff Walker

Originally a folkie operating out of New York, Jerry Jeff (real name Paul Crosby, born in Oneonta, New York, on March 16, 1942) became closely associated with the New Wave country movement emanating from Austin, Texas during the mid-'70s.

In 1966 he formed a rock group, Circus Maximus, with Austin songwriter Rob Runo, the band recording for Vanguard. However, Walker opted to become a solo act in 1968 and cut the self-penned **Mr Bojangles**, a memorable song regarding a street dancer he once met in a New Orleans jail, also providing Atco with an album of the same title. But although **Mr Bojangles** became a much-covered song and provided the Dirt Band with a Top 10 hit in 1970, Walker's career seemed to remain fairly stationary.

Signed to MCA in the early '70s, he mixed with fellow Texas singer-songwriters Guy Clark and Townes Van Zandt. With his own back-up unit, the Lost Gonzo Band, he recorded a series of good-timey, country albums that brought him a huge following across Texas.

Jerry Jeff split from the Lost Gonzo Band in 1977, but continued to record for MCA, later joining Elektra and forming the Bandito Band. In more recent years he has preferred to work as a solo performer, and has carved a new career as host of the popular TV series Austin City Limits.

Recommended:
Viva Terlingua (MCA/–)
Walker's Collectables (MCA/MCA)
It's A Good Night For Singing (MCA/MCA)
Too Old To Change (Elektra/–)
Hill Country Rain (Rykodisc/–)
Live At Gruene Hall (Rykodisc/–)
Navajo Rug (Rykodisc/–)

It's A Good Night For Singin', Jerry Jeff Walker's 1976 album. Courtesy MCA Records.

Jerry Wallace

Billed as 'Mr Smooth' – though he has been known to rock – Wallace is a one-time pop vocalist who swung into country music during the mid-'60s.

Born in Kansas City on December 15, 1933, singer-songwriter-guitarist Wallace was raised and educated in California. Following a brief term of service in the Navy, he made his first chart impact in 1958 when his recording of **How The Time Flies**, on Challenge, reached 11th place in the pop charts. The following year brought even more success when Wallace's version of **Primrose Lane**, a number later used as a theme for Henry Fonda's Smith Family TV series, became a million-seller.

After providing Challenge with 11 hit discs, Wallace signed for Mercury and cut more country-oriented material. **Life's Gone And Slipped Away** (1965) gained him his first country chart entry.

Since that time, he has cut sides for such labels as Liberty, Decca, MCA, MGM, 4-Star, Door Knob and BMA. His major country hits include: **The Morning After** (1971), **If You Leave Me Tonight, I'll Cry** (a No.1 in 1972), **Do You Know What It's Like To Be Lonesome?** (1973), **My Wife's House** (1974) and **I Miss You Already** (1977).

A performer on many top TV programmes, Wallace's voice has been heard on myriad commercials, while he has also turned up in such top-rated television shows as Hec Ramsey and Rod Serling's Night Gallery.

Recommended:
I Miss You Already (BMA/–)
The Golden Hits (4-Star/–)

Steve Wariner

Born in Kentucky on Christmas Day in 1954, Steve Noel Wariner grew up in a musical environment, with his father, a foundry worker, playing in a number of country bands in Indiana, where the family moved when Steve was young.

Before reaching his teens, Steve was playing bass in a little country band along with his father and an uncle. Although he was influenced by the music of George Jones and Merle Haggard, Steve's idol was guitar virtuoso Chet Atkins. When he finally signed a recording contract with RCA at the end of 1977, it was Chet who was to act as Steve's producer.

Prior to his move to Nashville and joining RCA, Steve had worked the road for three years with Dottie West and spent little more than two years as front-man for Bob Luman's band. An overlooked but very talented songwriter, Steve wrote his first RCA release, **I'm Already Taken**, which was later recorded by Conway Twitty.

Between April, 1978 and the end of 1982, Steve scored a series of minor country hits such as **So Sad (To Watch Good Love Go Bad)**, **Forget Me Not**, **The Easy Part's Over**, **Your Memory** and **By Now**, finally hitting the jackpot with **All Roads Lead To You**, a 1982 No.1. At this time Steve's recordings were

I Am Ready, Steve Wariner. Courtesy Arista Records.

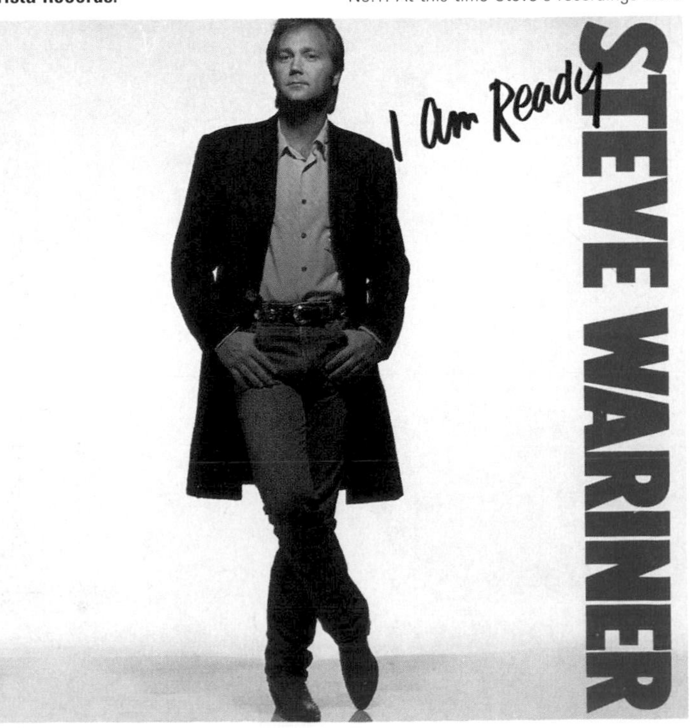

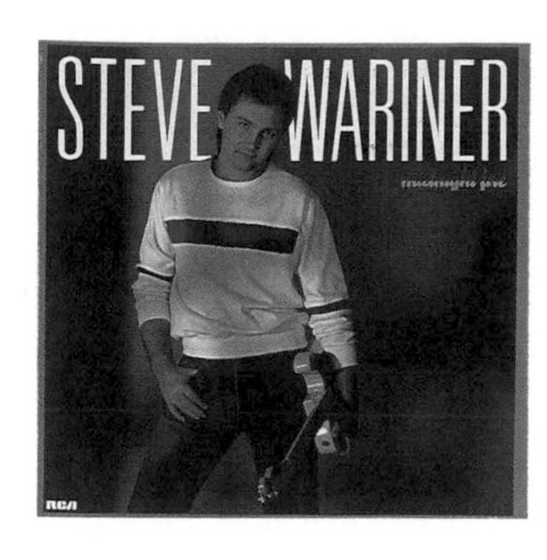

Midnight Fire, Steve Wariner. Courtesy RCA Records.

much like the Glen Campbell/Jimmy Webb pop-country classics of the '60s and were produced by Tom Collins.

A change of producer, with Norro Wilson and Tony Brown guiding Steve's recordings, led to a more country-styled approach and such Top 10 hits as **Don't Your Mem'ry Ever Sleep At Night**, **Midnight Fire** (both 1983) and **Lonely Women Make Good Lovers** (1984). When his contract with RCA came up for renewal at the end of 1984, he decided to make a move to MCA. For the first time he was allowed to play lead guitar on his own recordings, being turned loose on guitar solos, which proved to be nothing flashy but certainly well-executed and in keeping with the tone of the pieces.

The result has been such Top 10 hits as **Heart Trouble** and **What I Didn't Do** (both 1985), and No.1s with **Some Fools Never Learn** (1985), **You Can Dream Of Me, Life's Highway** (1986), **Smalltown Girl**, **The Weekend** and **Lynda** (all 1987). Wariner's wholesome good looks and pleasing personality made him a natural for such TV shows as Austin City Limits, Hee Haw, That Nashville Music and Country Music Comes Home. The hits continued with Top 10 winners **Baby I'm Yours**, **I Should Be With You** (1988) and two more No.1s with **Where Did I Go Wrong** and **I Got Dreams** (1989). In 1991 he teamed up with Mark O'Connor and the New Nashville Cats, vocalizing on the CMA award-winning **Restless**. He also moved over to Arista Records, with further Top 10 successes – **Leave Him Out Of This** (1991), **The Tips Of My Fingers** (1992) and **A Woman Loves** (1993).

Recommended:
Midnight Fire (RCA/–)
Life's Highway (MCA/–)
One Good Night Deserves Another (MCA/–)
I Am Ready (Arista/–)
Drive (Arista/–)
It's A Crazy World (MCA/MCA)
I Got Dreams (MCA/–)
I Should Be With You (MCA/–)

Doc Watson

Folk legend and heir to an old-time country tradition, guitarist-banjoist-singer Arthel (Doc) Watson was rediscovered in the boom folk years of the '60s and again in the '70s when the Nitty Gritty Dirt Band brought his music to the newly enthusiastic country-rock public.

Born in Deep Gap, North Carolina, on March 2, 1923, Doc was the son of a farmer who was prominent in the singing activities of the local Baptist church. Doc's grandparents lived with his immediately family and they taught the boy many traditional folk songs.

His first appearance was at the Boone, North Carolina, Fiddlers' Convention. He achieved fame locally but with country music 'smartening up' and with rock'n'roll finally hitting the scene, the mountain tradition was not foremost in the nation.

In 1960, some East Coast recording executives came to cut Clarence Ashley's String Band, and Ashley then appeared on a New York 'Friends Of Old Time Music' bill in 1961. Doc was invited along on the bill. As a consequence, he scored a solo gig at Gerde's Folk City in Greenwich Village, being rapturously received. In 1963, he consolidated his reputation considerably with an appearance on the Newport Folk Festival.

Doc relies heavily on traditional material. His voice, his guitar and banjo playing have a simplicity and intense profundity, almost making songs like **Tom Dooley** and **Shady Grove** his own.

During the '60s he recorded for Folkways and Vanguard Records. He has variously been heard on record with his mother, Mrs G. D. Watson, Jean Ritchie, his brother Arnold, Arnold's father-in-law Gaither Carlton and, most fruitfully, with son Merle, who sadly died when a tractor overturned on him at the Watson farm in Lenoir, North Carolina on October 23, 1985.

Although welded to an acoustic style which has been emulated widely by country-rock guitarists, Doc is no purist. His material runs the gamut of styles from bluegrass through western swing to a more commercial country-pop. During the 1970s he recorded for United Artists in Nashville with such session players as Joe Allen, Chuck Cochran, Johnny Gimble, Norman Blake and Jim Isbell, and even allowed his producer Jack Clement to surround him with strings on occasion.

A revered figure among old and young alike, drawing wild receptions quite out of keeping with his down-home musical style, Doc's concerts are virtually short courses in the history of American music, put across by using elements of field hollers, black blues, sacred music, mountain songs, gospel, bluegrass and even traces of jazz.

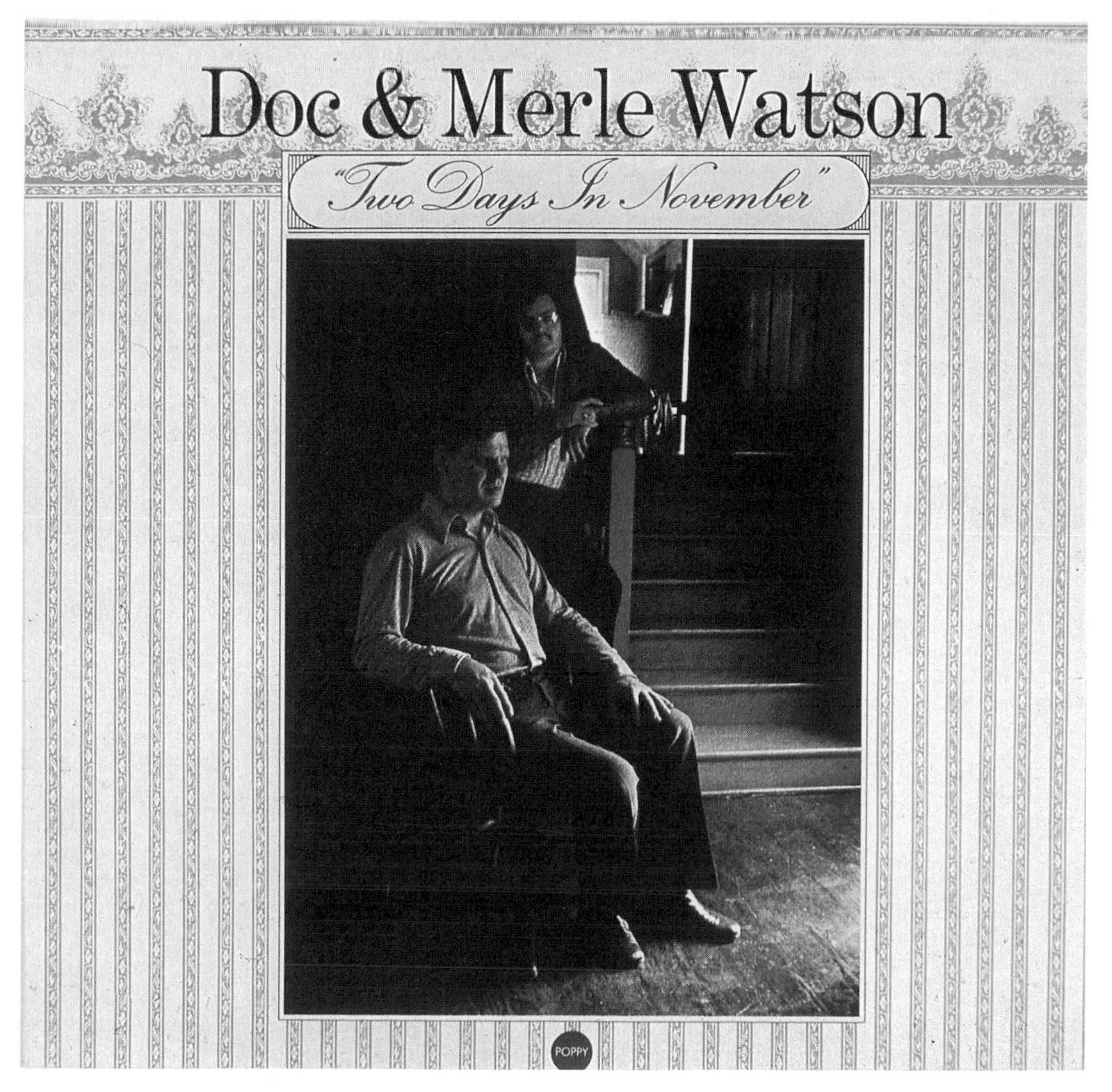

Two Days In November, Doc and Merle Watson. Courtesy UA.

Recommended:
Memories (UA/–)
Two Days In November (UA/UA)
Lonesome Road (UA/–)
The Watson Family Tradition (–/Topic)
In The Pines (–/Sundown)
Guitar Album (Flying Fish/–)
Ballads From Deep (Vanguard/Vanguard)
Songs For Little Pickers (Sugar Hill/–)

Gene Watson

A singer with an easy-flowing style and a penchant for tear-stained ballads, Watson (born on October 11, 1943 in Palestine, Texas) initially worked out of Houston, becoming a resident singer at the Dynasty Club and recording for various independent record labels, such as Resco and Wide World, during the early '70s.

Obtaining a regional hit with **Love In The Hot Afternoon** (previously recorded by Waylon Jennings, but never released), Gene signed with Capitol Records, who made the record into a Top 5 country chart success in 1975. One of those country singers, well-equipped vocally, who found it a long and difficult task to win any real recognition, Gene is not a man to be thrown off balance by the recurrence of his name on the charts. In a five-year association with Capitol he enjoyed more than a dozen Top 5 hits, including: **Paper Rosie** (1977), **Farewell Party** (1979), **Nothing Sure Looked Good On You** and **Bedroom Ballad** (both 1980).

Working both on the road and in the studio with his Farewell Party Band, Gene moved over to MCA Records in 1981. He continued to produce first-rate country recordings, choosing his material with meticulous care and singing in a kind of mellow style, but with enough distinctiveness to make it much more than easy-listening. As well as scoring Top 10 country hits with such songs as **Maybe I Should Have Been Listening** (1981), **This Dream's On Me** (1982), **Sometimes I Get Lucky And Forget** (1983), **Forever Again** (19894) and **Got No Reason Now For Going Home** (1985), he has endeared himself to British country fans.

Towards the end of 1985, Gene changed labels once again, moving to Epic Records and scoring a Top 10 hit with **Memories To Burn** (1985). Several minor hits and a change to Warner Brothers followed, seeing him back in the Top 10 with **Don't Waste It On The Blues** (1989). This success was short-lived, and one of country music's finest honky-tonk balladeers ended up without a major label deal. He was recording for the Canadian

Broadland label in 1992, still maintaining that quality of straight country music.

Recommended:
Beautiful Country (Capitol/–)
Old Loves Never Die (MCA/–)
Little By Little (MCA/–)
Memories To Burn (Epic/–)
Heartaches, Love And Stuff (MCA/–)
In Other Words (Broadland Canada/–)
At Last (Warner Bros/–)

Above: Gene Watson cut his first record at 18 years old.

Kevin Welch

One of Nashville's most respected songwriters, Kevin Welch (born on August 17, 1955, in Long Beach, California) established his reputation as a songwriter par excellence by penning hits for the Judds, Ricky Skaggs, Sweethearts Of The Rodeo and Don Williams. This Oklahoma-raised performer was restless in his youth, leaving home when he was 17 and dropping out of a college course to join a bluegrass band. By his mid-20s he'd already put a lot of miles on the road, that experience having paid off in a performance style and ear for powerful music that distinguish his work.

Moving to Nashville in the mid-'80s, Welch rapidly made his mark as a writer while still honing his skills as a performer. Working with his band, the Overtones, a unit comprised of crack studio musicians, he played the Nashville club circuit. He gained critical raves that landed him a recording contract with Warner/Reprise in 1988. One of country music's more inventive performers, Kevin's songs strive to recapture the romance and disillusionment of the road in lyrics full of stories about life-like characters. His eponymous debut album provided Welch with some minor country hits, including **Stay November** (1989), **Till I See You Again** (1990) and **True Love Never Dies** (1991). Touring with such diverse acts as the Oak Ridge Boys, Billy Bragg and Joe Ely also helped his reputation. During his 1991 European tour, his music was described as 'Western Beat'. It was a label that appealed precisely because of its subtle references. His music is western, not in a cowboy or swing sense, but because it is well-grounded in the roots of North America, and captures the heartbeat of American music. His second album was suitably titled **Western Beat**, and, as well as drawing critical praise from a wide cross-section of the media, has become a best-seller.

Because You Believed In Me, Gene Watson. Courtesy Capitol Records.

Recommended:
Western Beat (Reprise/–)
Kevin Welch (Reprise/–)

Freddy Weller

Born on September 9, 1947 in Atlanta, Georgia, Weller first achieved a fair degree of fame in the field of pop, both as a member of hit-parading rock group, Paul Revere And The Raiders, and as co-writer of many songs with Tommy Roe, including the million-sellers **Dizzy** (1968) and **Jam Up, Jelly Tight** (1969).

Once a bassist and guitarist with Joe South, Weller has also worked as a studio musician in Atlanta and toured as part of Billy Joe Royal's backup group. He became a country artist and signed to Columbia.

After achieving a Top 10 country hit with **Games People Play** in 1969, Weller enjoyed a successful patch through to 1971. **These Are Not My People**, **Promised Land**, **Indiana Lake** and **Another Night Of Love** all charted impressively during this period. In late 1974, Weller signed for Dot, scoring a couple of minor country hits.

He re-joined Columbia at the beginning of 1976, but was unable to make it back to the Top 10, only reaching the Top 30 with **Love Got In The Way** (1978) and **Fantasy Island** (1979), leading to him being dropped by the label in 1980.

Recommended:
Go For The Night (Columbia/–)
Roadmaster (Columbia/–)
The Promised Land (Columbia/–)
Back On The Street (–/Bulldog)

Kitty Wells

The acknowledged 'Queen Of Country Music', Kitty Wells (real name Muriel Deason) was born in Nashville, Tennessee, on August 30, 1918. As a child she sang gospel music at the neighbourhood church, at 14 learning to play guitar. Within a year she was playing at local dances, some time later obtaining her first radio dates.

While appearing on station WXIX's Dixie Early Birds show, she met Johnny Wright (Johnny And Jack), whom she married two years later (1938). By this time she had become a featured artist on the Johnny And Jack touring show, adopting the name Kitty Wells from a folk song called **Sweet Kitty Wells**.

With their backup unit the Tennessee Mountain Boys, Johnny And Jack and Kitty Wells toured widely during the late '30s and the war years of the '40s, their biggest breaks on radio coming in 1940 on WBIG, Greensboro, North Carolina, then later on WNOX Knoxville's Mid-Day Merry-Go-Round.

In 1947 came Johnny, Jack and Kitty's membership on Grand Ole Opry, after which they moved to Shreveport to become the stars of KWKH's new Louisiana Hayride. Five years later came an offer of a regular spot on the Opry, plus a record contract from Decca (she had previously recorded with RCA-Victor). The same year saw the release of **It Wasn't God Who Made Honky Tonk Angels**, an answer disc to Hank Thompson's **Wild Side Of Life**. This enabled Kitty to become the first female to have a No.1 country hit – although Patsy Montana's **I Wanna Be A Cowboy's Sweetheart** would have reached No.1 if charts had existed in 1935.

Since that time, Kitty Wells has amassed an amazing number of chart entries, including duets with Roy Drusky, Red Foley, Roy Acuff, Johnny Wright and Webb Pierce. The biggest of her solo successes were: **Paying For That Back Street Affair** (1953), **Making Believe** (1955), **Searching** (1956), **Jealousy** (1958), **Mommy For A Day** (1959), **Amigo's Guitar** (1959), **Left To Right** (1960), **Heartbreak U.S.A.** (1961), **Unloved, Unwanted** (1962), **Password** (1964) and **You Don't Hear** (1965).

Her awards are equally numerous and include 'Billboard's No.1 Country Music Female Artist Of The Year 1954–65, a 1974 Woman Of The Year award from the Nashville Association of Business and Professional Women, plus a citation for the Most Outstanding Tennessee Citizen in 1954.

Kitty, who has three children – Ruby, Carol Sue and Bobby – eventually

A Bouquet Of Country Hits, Kitty Wells. Courtesy MCA Records.

Above: Kitty Wells, the first woman to have a No.1 country hit.

terminated her long association with Decca during the mid-'70s, signing for the Macon-based Capricorn label and cutting an album, aptly named **Forever Young**. In 1976 she received the supreme accolade – being elected to the Country Music Hall Of Fame. Kitty still continues to tour all over America and records for her own Ruboca Records label, maintaining the style for which she is best known and refusing to be drawn into a modern pop-country sound.

Recommended:
Early Classics (Golden Country/–)
The Kitty Wells Story (MCA/MCA)
The Golden Years (Rounder/–)
Forever Young (Capricorn/–)
The Queen Of Country Music, 1949–1958 (–/Bear Family)
Country Music Hall Of Fame (MCA/–)

Special Delivery, Dottie West. Courtesy UA Records.

Dottie West

Known as the 'Country Sunshine' girl after writing and recording a song of that title for a Coke commercial, Dottie was born in McMinnville, Tennessee, on October 11, 1932, one of ten children. Farm raised, Dorothy Marie still had time to gain a college degree while helping to work the cotton and sugar cane fields.

She later incorporated these experiences into her songs, but one of Dottie's strengths has also been her ability to adapt pop stylings.

In the early 1950s, she studied music at Tennessee Tech and there met Bill West, her future husband. Bill was studying engineering but he played steel guitar and accompanied Dottie at college concerts. They later moved to Ohio where they appeared as a duo on local TV in the Cleveland area.

While visiting relatives in Nashville in 1959, they met some executives from Starday Records and were given a record contract. This resulted in local live appearances but little else.

Success eluded her until 1963 when Dottie, by this time signed to RCA, recorded **Let Me Off At The Corner**, a Top 30 disc. A year later came the big one – **Here Comes My Baby** – a West original that became covered by Perry Como, providing him with a pop hit and also earning Dottie a Grammy award.

Dottie then became an Opry regular (1964), arranged and worked with the Memphis and Kansas City Symphony Orchestra and provided RCA with such major solo hits as: **Would You Hold It Against Me?** (1966), **Paper Mansions** (1967), **Country Girl** (1968), **Forever Yours** (1970), **Country Sunshine** (1973) and **Last Time I Saw Him** (1974). She also recorded hit duets with Jim Reeves (**Love Is No Excuse** – 1964) and Don Gibson (**Rings Of Gold** and **There's A Story Goin' Round** – both 1969).

Between collecting numerous awards, making a few films, writing some 400 songs and commercials, and fitting in recording dates and several tours, she also found time to raise four children and marry again – her second husband was drummer Bryon Metcalf.

In 1978 Dottie teamed up with Kenny Rogers for a successful duet partnership which resulted in the Top 10 hits, **Everytime Two Fools Collide** (1978) and **All I Ever Need Is You** (1979). The pair were named CMA Vocal Duo in both 1978 and 1979. Dottie had signed with United Artists Records in 1976 and continued to score major country hits.

The red-haired beauty, who had helped launch the careers of Larry Gatlin and Steve Wariner, found herself something of a country sex queen in her late 40s, with full-colour centrespreads in several of the leading American magazines. Remaining a member of the Opry, Dottie notched up a few minor hits for the Permian label in 1984–85, then her career went into a sharp decline. She became caught up in a tragic spiral of disasters before a car wreck took her life on September 4, 1991. Hell-bent on destruction, Dottie was hooked on drugs and booze and reportedly owed over $1 million in back taxes, with the IRS taking many of her possessions, and for a time she lived in a parking lot on her tour bus.

In Session, Shelly West and David Frizzell. Courtesy WEA-Viva Records.

If It's All Right With You, Dottie West. Courtesy RCA Records.

Recommended:
High Times (Liberty/–)
Wild West (Liberty/–)
Special Delivery (UA/UA)
Carolina Cousins (RCA/RCA)
Everytime Two Fools Collide – with Kenny Rogers (UA/UA)

Shelly West

Daughter of Dottie West and her first husband, steel guitarist Bill West, Shelly (born on May 23, 1958 in Nashville, Tennessee) initially made an impression as the duet partner of David Frizzell on the 1981 chart-topper, **You're The Reason God Made Oklahoma**.

Shelly began performing in Dottie's shows in 1975, soon after her graduation from Nashville's Hillsboro High School. During her year and a half with Dottie, Shelly gradually worked her way up from harmony singing to her own solo spot.

Towards the end of 1977, she and Allen Frizzell (David and Lefty's younger brother who was employed as Dottie's front-man and guitarist) moved to California to pursue their own solo careers. They teamed up with David and worked honky-tonk club circuits.

Eventually David and Shelly began singing together and recorded a duet album featuring **You're The Reason God Made Oklahoma**, a song that Clint Eastwood insisted on using in his film, 'Any Which Way You Can'. Frizzell and West continued with such duet hits as: **Texas State Of Mind** (1981), **Another Honky Tonk Night On Broadway** (1982), **Cajun Invitation** (1983), **Another Dawn Breaking Over Georgia** (1984) and **Do Me Right** (1985).

These duets paved the way for Shelly's solo career, and she made her mark on the charts with **Jose Cuervo** and **Another Motel Memory** (both 1983), **Flight 309 To Nashville** (1984), **Now There's You** and **Don't Make Me Wait On The Moon** (both 1985). For a while Shelly was married to Allen Frizzell, but the couple separated and were divorced in 1985.

Recommended:
Don't Make Me Wait On The Moon (Warners-Viva/–)
West By West (Warners-Viva/–)
Carryin' On The Family Name – with David Frizzell (Warners-Viva/–)

Billy Edd Wheeler

Born in Whitesville, West Virginia, on December 9, 1932, Billy Edd is a college-educated country artist. He has a BA degree from Berea College, Kentucky, attended Yale Drama School and has been, variously, an editor, a music business executive, a Navy pilot and an instructor at Berea College.

The Kingston Trio had a Top 10 pop hit with his **Reverend Mr. Black** in 1963 and Wheeler himself enjoyed a rare hit single with **The Little Brown Shack Out Back**, which nearly (but not quite) topped the country charts in 1964, helping him earn an ASCAP writer's award.

Despite various changes of record company (he has been with such labels as Monitor, Kapp, UA and RCA), his only real influence upon the charts has been through songwriting. Johnny Cash and June Carter added to Wheeler's royalty cheque by recording his **Jackson**, a cross-over hit in 1967, and Kenny Rogers recorded the multi-million-selling pop-country smash **Coward Of The County** in 1979.

A collector of folk material – and author of a folk play – Billy Edd Wheeler was responsible for creating a special music room in the Mountain Hall Of Fame, Richwood, West Virginia.

Recommended:
Wild Mountain Flowers (Flying Fish/–)
Nashville Zodiac (UA/UA)
Love (RCA/–)

Clarence White

A rock musician with a bluegrass background, guitarist Clarence White was born in Lewiston, Maine on June 7, 1944. Raised in California, he played with the Country Boys at the age of ten, the group's other members being his brothers, Roland (16) and Eric (12).

A bluegrass unit, working at various barn dances and local functions in the Burbank area, the Country Boys materialized into the Kentucky Colonels in 1962. The line-up then was: Clarence (guitar), Roland (mandolin), Roger Bush (bass), Billy Ray Latham (banjo) and Leroy Mack (dobro).

Two albums were recorded before Clarence left in 1965 to become a session-man, appearing on disc with Ricky Nelson, the Everlys, the Byrds, Gene Clark, the Flying Burritos, Wynn Stewart and others.

After cutting a never-released solo album for the Bakersfield International label and working sporadically with Cajun Gib And Gene (Gib Guilbeau and Gene Parsons), White formed Nashville West, a short-lived country-rock unit that featured both Guilbeau and Parsons plus bassist Wayne Moore. But in September 1968, he became a regular member of the Byrds.

Returning to session work once more, White began fashioning a new solo album, also putting in some gigs with the re-formed Kentucky Colonels. However, the solo album was never completed; White was knocked down and killed by a drunken woman driver while loading equipment on to a van following a gig on July 14, 1973.

Recommended:
Clarence White And The Kentucky Colonels (Rounder/–)
Nashville West (Sierra Briar/Sundown)
Kentucky Colonels (–/UA)

Joy White

Red-headed Joy White is one of the new faces of country music who has brought years of experience to Nashville. Many country music authorities have predicted a bright future for this gutsy firecracker.

Joy was born in the small farming town of Turrell, Arkansas, but grew up in Mishawaka, Indiana. Her father was a guitar player and all of the family would get together to sing and play music. Joy sang in church as a child, harmonized on country songs at family picnics and by her teenage years was belting out rock'n'roll songs on the back of flatbed trucks. With her own bands, she toured extensively in the region around her hometown and also earned extra money performing jingles and commercials for local radio and TV.

Eventually Joy moved to Nashville in the late '80s, rapidly making an impact singing on demos and back-up vocals on recording sessions. In 1991 she landed a recording contract with Columbia Records, and those years of singing paid off with her debut album, **Between Midnight And Hindsight**. This dynamic set showed Joy to be a passionate vocalist, capable of delivering her own brand of hot-blooded country. The rockin' **Little Things** became a minor country hit at the end of 1992, in the first phase of establishing Joy White as one of the brightest talents of the '90s.

Recommended:
Between Midnight And Hindsight (Columbia/–)

The White Brothers, Clarence, Eric and Roland White. Courtesy Rounder Records.

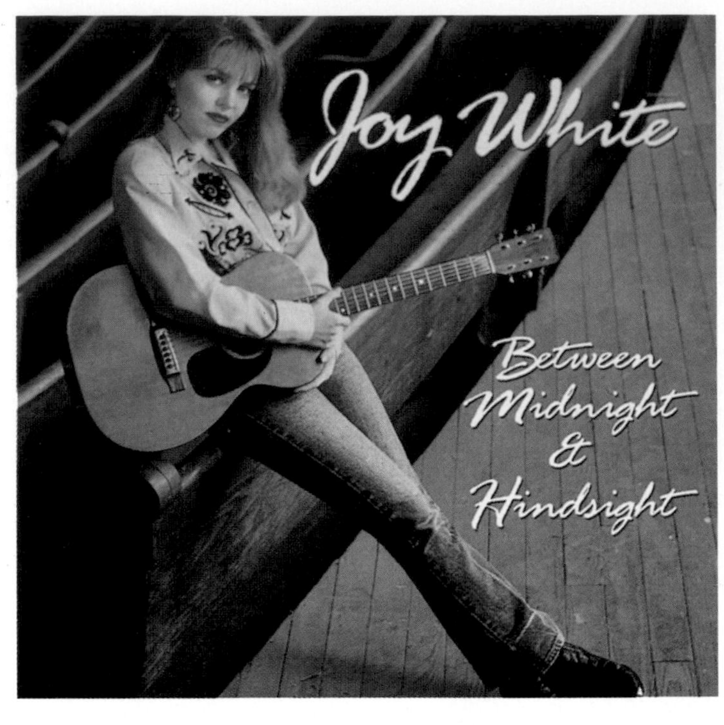

Between Midnight And Hindsight, Joy White. Courtesy Columbia Records.

The Whites

Buck White, piano, mandolin; Cheryl White Warren, acoustic bass; Sharon White Skaggs, acoustic guitar.

Although a distinctive and fascinating new group who emerged in the '80s, the Whites, in fact, go back a long way, with father Buck White having roots in western swing music and the distinction of playing piano on the original recording of Slim Willet's **Don't Let The Stars Get In Your Eyes** (1952).

Buck, who grew up in Texas, met his wife Pat in the early '50s, and music played a major role in their family life. At that time he played piano for various swing bands in and around the Wichita Falls area. In 1962 the family uprooted and moved to Arkansas where Buck became involved in bluegrass.

Working as Buck White And The Down Home Folks, the family group, consisting of Buck, Pat and their two daughters, Cheryl and Sharon, became popular on the bluegrass festival and other country circuits. In the early '70s, a move to Nashville was made and they recorded their first album, **Buck White And The Down Home Folks**, for County Records in April, 1972. It was to be another five years before they recorded a second album, **In Person**, by which time Cheryl and Sharon were able to work full-time with the group.

These were hard times for the family group, though, as bluegrass music was not exactly popular, and the two girls had to find various jobs to supplement their meagre showbusiness earnings. They were still recording regularly for Ridge Runner and Sugar Hill.

A short-lived recording contract with Capitol Records in 1982 came at the same time as they decided to change their name to the Whites. One single was released, **Send Me The Pillow You Dream On**, which made the charts, but Capitol decided not to release any further records.

Emmylou Harris and Ricky Skaggs eventually became involved in the group's career, with Emmylou inviting them to open her tour in support of her **Blue Kentucky Girl** album, and Ricky offering to produce an album for Warner-Curb. This resulted in the Whites being nominated in the CMA 1983 awards as Vocal Group Of The Year and achieving Top 10 country hits with **You Put The Blue In Me**, **Hangin' Around** (both 1983) and **I Wonder Who's Holding My Baby Tonight** (1984).

Curb Records amalgamated with MCA in 1984 and the Whites continued to score Top 10 hits on that label, with **Pins And Needles** (1984) and **If It Ain't Love (Let's Leave It Alone)** (1965). Members of the Grand Ole Opry since 1984, their style is based upon the distinctive harmonies of Cheryl and Sharon, the expert mandolin playing of Buck and the dobro work of Jerry Douglas, who has been playing with the Whites since the late '70s. In 1989 they moved on to the gospel-based Canaan Records.

Recommended:
Old Familiar Feeling (Warner-Curb/–)
Forever You (MCA-Curb/–)

Above: The Whites – the group consists of Buck White and his daughters Cheryl and Sharon. Buck is a fine blues pianist and expert mandolinist.

Whole New World (MCA-Curb/MCA)
Poor Folks Pleasure (–/Sundown)

Keith Whitley

Keith Whitley, one of the purest honky-tonk singers to make it to the top in country music in the '80s, was a sad victim of his own drinking excesses, which saw his career come to an end when he had finally made that long-awaited breakthrough to international acclaim and chart-topping records.

Though he was considered a 'New Traditionalist', Whitley, who was born on July 1, 1955, in Sandy Hook, Kentucky, had been around the scene for years before most of the newcomers had even picked up a guitar. Something of a child prodigy, he learned to play guitar at six and was on the radio with Buddy Starcher in Charleston, West Virginia, when eight. His real roots lay with honky-tonk music gleaned through his mother's record collection, but when young Keith started to play music, there were no honky-tonk musicians around, so he turned to bluegrass. In the late '60s, at age 13, together with his brother Dwight and fellow Kentuckian Ricky Skaggs, he formed the Lonesome Mountain Boys, a bluegrass group that rapidly built up a local reputation. They did a few shows with Ralph Stanley and the Clinch Mountain Boys, and in early 1971 Keith and Ricky joined the legendary band, recording a half dozen or so albums with Stanley, including a gospel collection called **Cry From The Cross**, which was named Bluegrass Album Of The Year in 1971.

Two years later Whitley formed his own band, the New Tradition, which enabled him to stretch the boundaries of bluegrass. As he branched out from bluegrass, he changed the band's name to the Country Store. In 1977 he was invited to join J.D. Crowe And The New South, a country/bluegrass fusion band. He often took lead vocals, both onstage and on recordings, leading to many recommending that he take his work to Nashville.

By this time he was married and also had a reputation for being a heavy drinker, but he heeded the advice, and signed with RCA Records in late 1982, but was still living in Kentucky. A year later he did move to Nashville, in order to actively pursue a solo career. He signed a writer's contract with Tree Music and worked as a writer and demo singer, waiting for his recording deal to take shape. Finally Whitley made his chart debut with **Turn Me To Love** (1984), followed in 1985 with a mini-LP, **A Hard Act To Follow**, which hardly set the world alight. He scored his first Top 20 hit with the lightweight **Miami, My Amy** at the end of 1985, followed by three Top 10 entries in **Ten Feet Away**, **Homecoming '63** (1986) and **Hard Livin'** (1987). In 1986, following a messy divorce, he married Lorrie Morgan, who was also struggling to establish herself as a country music performer, and, like her husband, was signed to RCA.

Working with various producers, Keith Whitley had recorded extensively, and though he had cut some fine material, including **Does Fort Worth Ever Cross Your Mind**, which was never released, his career seemed to be faltering. A new producer, Garth Fundis, was brought into the picture, and RCA gave Keith Whitley one last shot at making it. His third album, **Don't Close Your Eyes**, captured his hard honky-tonk country sound to perfection. Whether it was sad, tear-stained ballads or kickin' uptempo songs, Whitley was in top vocal form. The result was a gold album and chart-topping singles with **Don't Close Your Eyes** (Billboard's No.1 Country Hit, 1988), **When You Say Nothing At All** (1988) and **I'm No Stranger To The Rain** (1989).

By rights Whitley should have been on top of the world, but he was still drinking heavily. He lost his last battle with the bottle, on May 9, 1989, when an accidental overdose in his home at Brentwood took his life; the cruel irony was that he drank his life away at the very time his musical career seemed poised, after so many long years, for real stardom.

A week before his death, he had completed a new album, **I Wonder Do You Think Of Me**, which became a posthumous success with both the title song and **It Ain't Nothing'** topping the singles chart and **I'm Over You** making Top 5, while the album gained another gold disc. That album, plus the subsequent **Kentucky Bluebird**, a superb set put together by Garth Fundis in 1991 from previously unreleased tracks taken from original song demos or recordings that were not included on the singer's previous four albums, prove conclusively that country music was robbed of one of the finest pure-country song stylists, a singer who had never realized his vast unfulfilled potential.

The success of his recordings have continued long after his death with a **Greatest Hits** set in 1990 gaining a third gold disc, while a duet with Lorrie Morgan of **'Til A Tear Becomes A Rose** went Top 20 and was also named the CMA Vocal Event Of The Year in 1990. Another duet recording, **Brotherly Love**, recorded in 1987 with Earl Thomas Conley, but never issued, rose to No.2 on the charts in 1991. The question now remains as to how many more Keith Whitley recordings can RCA 'polish' up for release. There is little doubt that Keith Whitley is a singer whose music and memory will continue to endure for many years.

Below: Keith Whitley made a huge impact in the '80s, but died in 1989 due to an alcohol overdose.

Recommended:
Don't Close Your Eyes (RCA/RCA)
I Wonder Do You Think Of Me (RCA/RCA)
Kentucky Bluebird (RCA/RCA)

Ray Whitley

Despite a late start, Ray Whitley became a quite successful jack-of-all-trades in the era of the singing cowboy. Born on December 5, 1901, near Atlanta, Georgia,

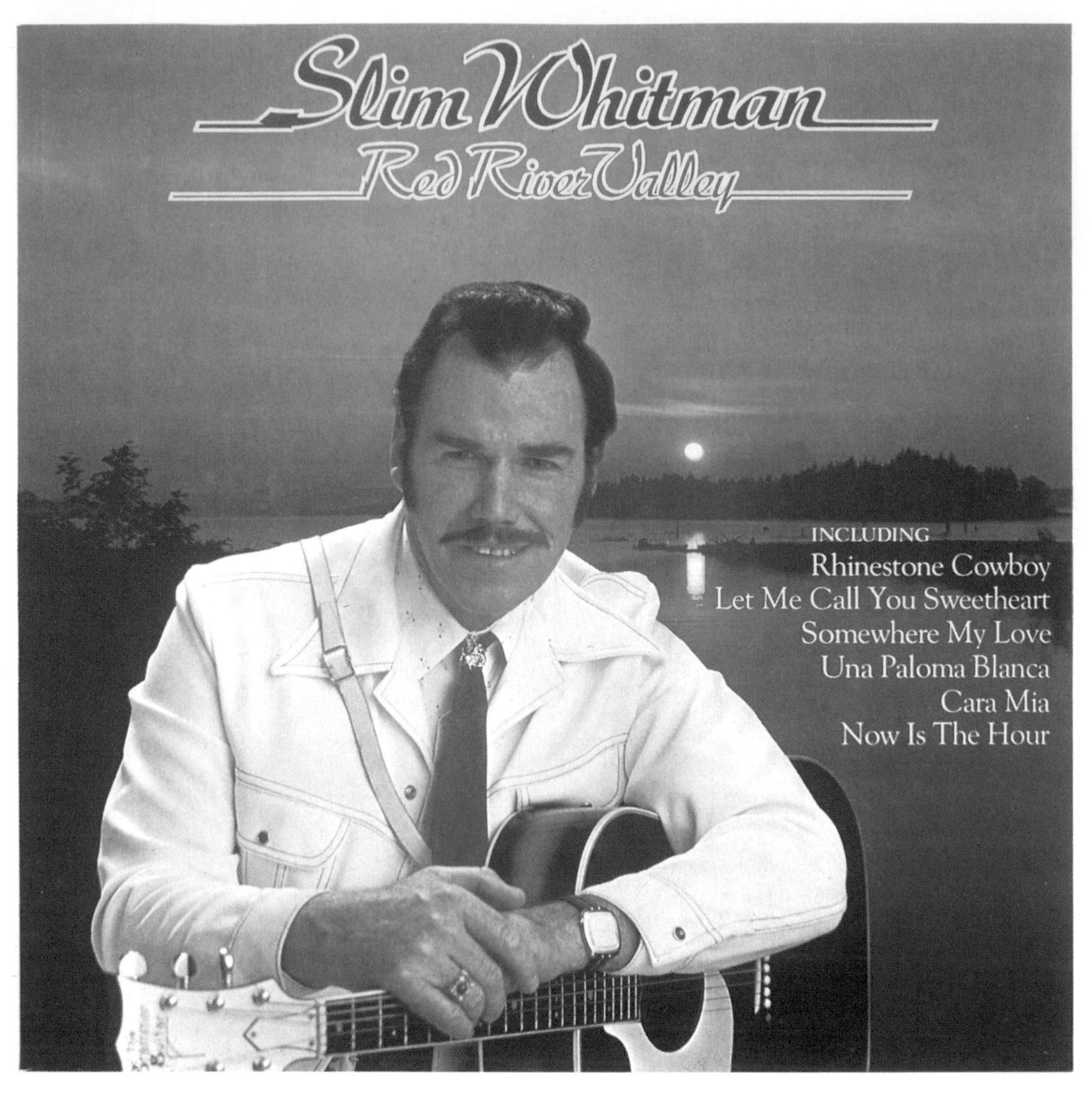

Red River Valley, Slim Whitman. Courtesy UA Records. Slim is a huge country figure in the US, where his albums are regularly advertised on television.

Whitley spent some time in the Navy and in Philadelphia and New York, where he pursued music as a hobby.

He auditioned for radio in New York City, and rapidly rose to co-host the WHN Barn Dance in the mid-1930s. Here he also recorded for the American Record Company complex, and also for Decca, his biggest hits being his theme song, **Blue Yodel Blues**, and **The Last Flight Of Wiley Post**.

One of the earliest singing cowboys to invade Hollywood, he appeared in films as early as 1936. He spent 1938–42 at RKO, where he made 18 musical shorts of his own and was the singing sidekick to George O'Brien and Tim Holt. His last role was as Watts, James Dean's manager in 'Giant'. Whitley was active musically during his film period, both as a cowboy singer and fronting a western swing band.

An active songwriter, he wrote or co-wrote with Fred Rose many of Gene Autry's big hits, including **Back In The Saddle Again**, **Lonely River**, **I Hang My Head And Cry** and **Ages And Ages Ago**.

In addition, Whitley also managed both the Sons Of The Pioneers and Jimmy Wakely for a time, and helped Gibson design and build their first J-200 guitar.

He was still turning up at western film festivals, singing and doing tricks with his bullwhip until shortly before his death in California on February 21, 1979.

Slim Whitman

Born Otis Dewey Whitman Jr in Tampa, Florida, on January 20, 1924, Whitman's early interest was in sport rather than music. He became a star pitcher with his Tampa high school team and hoped to make a career in baseball. However, on leaving school, he took a job in a meat-packing plant, where he met his wife.

Just prior to Pearl Harbor, Whitman became a shipyard fitter in Tampa, enlisting in the Navy during 1943. While in the Navy he learned guitar and entertained at shipboard events, upon return to civilian life splitting his time between baseball and entertaining when not working.

In 1946 he gained a contract with the Plant City Berries of the Orange Belt League – but his musical career also prospered via radio spots on Tampa WDAE and many local club bookings. It was at this point that Whitman opted for music as a full-time occupation and began establishing his reputation beyond the bounds of Tampa, in 1949 winning a record contract with RCA. After gaining some attention with **Casting My Lasso To The Sky**, he moved to Shreveport, becoming a regular on the Louisiana Hayride.

Country Style, Slim Whitman. Courtesy Music For Pleasure.

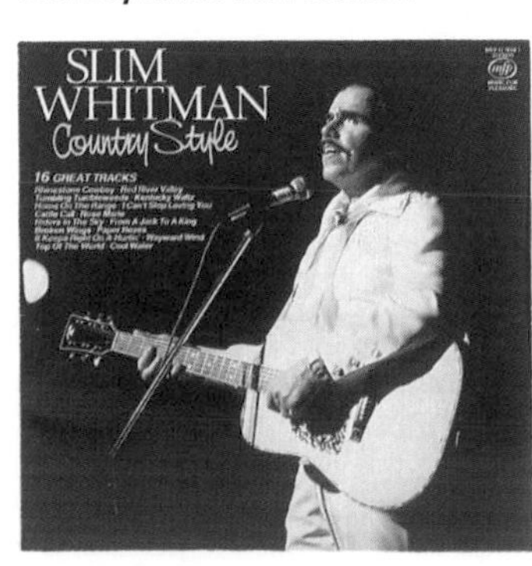

Later, in 1952, he signed for Imperial Records and staked an immediate claim to stardom with his semi-yodelled version of **Indian Love Call**. This was followed by **Rose Marie**, a massive seller. It provided Whitman with a gold disc and an audience ready to snap up such other offerings as **Secret Love** (1954), **Cattle Call** (1955), **More Than Yesterday** (1965), **Guess Who?** (1970) and **Something Beautiful** (1971).

Whitman's reliance on mainly sweet, romantic ballads, purveyed in a rich voice that switches easily to falsetto, had made him a worldwide favourite. But nowhere is he more popular than in Britain where he became the first country vocalist to perform at the London Palladium. His British tours are usually in the SRO bracket while such albums as **The Very Best Of Slim Whitman** and **Red River Valley** topped the UK charts in 1976 and 1977 respectively.

A special TV-advertised set of his most popular numbers became a big seller in America in 1983, introducing Slim to a whole new audience, while he has continued to record regularly for United Artists, Cleveland International and Epic.

Recommended:
15th Anniversary (Imperial/Liberty)
Yodeling (Imperial/Liberty)
Slim Whitman Collection (–/UA)
Songs I Love To Sing (Epic/Epic)
Angeline (Epic/Epic)
25th Anniversary Concert (–/UA)
Country Style (–/Music For Pleasure)
Rose Marie & Other Love Songs (–/Pickwick)
50 Original Tracks (–/Liberty)

Henry Whitter

One of the earliest of the country musicians to be recorded – it is claimed that his earliest recordings were pre-dated only by those of Eck Robertson – William Henry Whitter was born near Fries, Virginia, on April 6, 1892.

Working in a cotton mill to earn a living, he learned guitar, fiddle, piano, harmonica

A Portrait, Wilburn Brothers. Courtesy MCA Records.

Right: A consistent Top 10 hit-maker, Don Williams is known as the 'Gentle Giant Of Country Music'.

and organ, and began performing around the Fries area. In March 1923, he visited New York, gaining an audition with the General Phonograph Company and recording two numbers which were promptly shelved.

However, in the wake of Fiddlin' John Carson's success with his Okeh sides, Whitter was recalled to New York in December, 1923, there waxing nine numbers for Okeh release. The first of these, **The Wreck On The Southern Old '97**, backed with **Lonesome Road Blues**, was issued in January, 1924. Later that year, **Old '97** was to be recorded in the slightly different version by Vernon Dalhart.

Whitter continued to record as a soloist, an accompanist and as a bandleader, with Whitter's Virginia Breakdowners. He also formed a successful musical alliance with blind fiddler George Banman Grayson, recording with him several times between 1927 and 1929.

This partnership terminated when Grayson died in a road accident during the mid-'30s, and though Whitter continued in a solo role until the commencement of the 1940s, his health gradually deteriorated. He died from diabetes in North Carolina on November 10, 1941.

Wilburn Brothers

Once part of a family act that included their father, mother, elder brothers and a sister, Doyle (born in Thayer, Missouri, on July 7, 1930) and Teddy Wilburn (born in Thayer on November 30, 1931) began as hometown street corner singers. The Wilburn Family eventually toured the South and established a reputation that led to an Opry signing in 1941.

In the wake of the Korean War, Teddy and Doyle began working as a duo, touring with Webb Pierce and Faron Young. Obtaining a record contract with Decca, they scored a Top 10 disc in 1956 with **Go Away With Me**. This was the first of an impressive tally of hits that extended into the early '70s.

The Wilburns appealed to a wide audience, a fact reflected in their record sales of 1959 when three of their releases, **Which One Is To Blame?**, **Somebody Back In Town** and **A Woman's Intuition**, were major hits.

Founders of the Wil-Helm Talent Agency in conjunction with Smiley Wilson, the Wilburns found themselves representing many of Nashville's leading talents – including Loretta Lynn, whom they featured on their own highly popular TV show and for whom they obtained a recording contract with Decca.

Owners of Surefire Music, a publishing company, they have published the songs of Loretta Lynn, Johnny Russell and Patty Loveless, having no fewer than seven songs in the 'Coalminer's Daughter' film. They have enjoyed such Top 10 discs as **Trouble's Back In Town** (1962), **Tell Her No** (1963), **It's Another World** (1965) and **Hurt Her Once For Me** (1966). They continued to work the Opry throughout the '70s. Doyle Wilburn died from cancer on October 16, 1982, in Nashville. The family tradition has been maintained, Teddy still appearing on the Opry, occasionally accompanied by brothers Lester and Leslie.

Recommended:
A Portrait Of The Wilburn Brothers (MCA/–)
Sing Your Heart Out (Decca/–)
Country Gold (Decca/Stetson)

Doc Williams

Although he has never had a hit record, Doc Williams has been one of the most popular regional acts in country music. Along with his wife Chickie and their band the Border Riders, they still play hundreds of dates a year, filling houses in the Northeast and in the Canadian Maritimes, largely on the strength of their long-time association with WWVA and the Wheeling Jamboree.

Born Andrew J. Smik, of Bohemian descent, on June 26, 1914 Doc grew up in the musically rich area of eastern Pennsylvania. Except for brief stays at WREC in Memphis (1939) and WFMD in Frederick, Maryland (1945), he has remained in that area: Cleveland from 1934 to 1936, then Pittsburgh and finally Wheeling, from 1937.

In 1948 he married Jessie Wanda Crupe, who became known as Chickie, and they have had many regionally popular records on their own label, Wheeling: **Beyond The Sunset, Mary Of The Wild Moor**, **Silver Bells** and Doc's own song **Willie Roy, The Crippled Boy**.

A staunch traditionalist and a long-time spokesman for Wheeling, WWVA and the Wheeling Jamboree, Doc Williams has never achieved national success or huge record sales, but has been a great influence in the northeast and in Canada.

Recommended:
From Out Of The Beautiful Hills Of West Virginia (Wheeling/–)
Doc 'n' Chickie Together (Wheeling/–)
Wheeling Back To Wheeling (Wheeling/–)

Don Williams

Voted Country Artist Of The Decade by British fans in a 1980 poll, Don (born on May 27, 1939, near Plainview, Texas) has been called the 'Gentle Giant of Country Music', due to his laid-back personality and singing style. He first came to prominence with the pop-folk group, the Pozo Seco singers, in 1965. The group (comprised of Williams, Susan Taylor and Lofton Cline) had a major hit the following year with **Time**.

Between 1966 and 1967, the Pozos also had best-sellers with **I'll Be Gone**, **I Can Make It With You**, **Look What You've Done**, **I Believed It All** and **Louisiana Man**. But gradually their popularity waned and in 1971 Williams returned to Texas to join his father-in-law in his business.

A solo recording venture by former Pozo Susan Taylor had Don return to Nashville in a writing capacity, but he soon began singing once more. A solo album, **Don Williams Volume 1**, was released on Jack Clements' independent JMI label during 1973.

His debut solo single had been **Don't You Believe** (June, 1972) but it was his second, **The Shelter Of Your Eyes**, that gave him his first major bite at the county charts. Don finally made the country Top 10 with **We Should Be Together** and he topped the charts a few months later with **I Wouldn't Want To Live If You Didn't Love Me**.

He also hit the top spot with such songs as: **You're My Best Friend** and **Love Me Tonight** (both 1975); **'Til The Rivers All Run Dry** and **Say It Again** (both 1976); **Some Broken Hearts Never Mend** and **I'm Just A Country Boy** (both 1977); **Tulsa Time** (1978); **It Must Be Love** (1979); **I Believe In You** (a pop cross-over which hit No.24 on the pop charts in 1980); **Lord I Hope This Day Is Good** (1981); **If Hollywood Don't Need You** (1982); **Love Is On A Roll** (1983); **Stay Young** (1984) and **Walking A Broken Heart** (1985).

Harmony, Don Williams. Courtesy MCA Records.

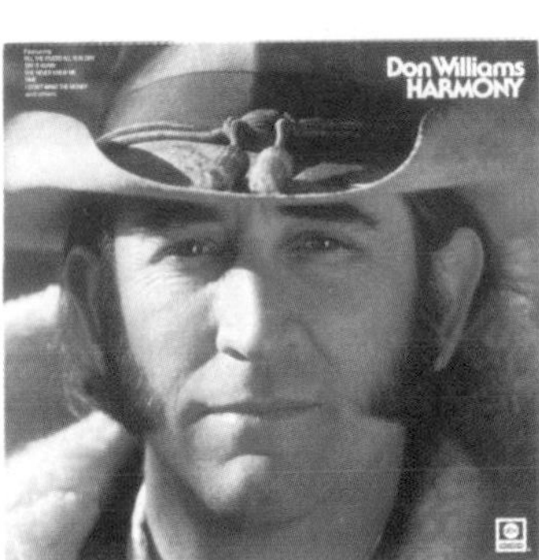

Surprisingly he made a British breakthrough with **I Recall A Gypsy Woman**, which made the British Top 10 in 1976, yet was never released as a single in America. He emerged as the British No.1 album-seller in 1978 (outselling every rock, country and pop act).

Don tries to keep his public appearances down to a minimum and spends much of his time writing songs, recording or tending his farm near Ashland City, Tennessee. He keeps his private life very much detached from his showbiz career, which resulted in him being dubbed the 'Reluctant Superstar'. In 1986 he made a label change, moving from MCA (formerly ABC-Dot) to Capitol, maintaining the familiar Don Williams' sound for such Top 10 entries as **Heartbeat In The Darkness** (No.1, 1986), **I'll Never Be In Love Again** (1987) and **Another Place, Another Time** (1988). He then moved on to RCA, scoring with **One Good Well** (1989), **Back In My Younger Days** (1990) and **Lord Have Mercy On A Country Boy** (1991).

Recommended:
The Best Of The Pozo Seco Singers (–/CBS Embassy)
Visions (ABC/ABC)
Portrait (MCA/MCA)
Listen To The Radio (MCA/MCA)
Café Carolina (MCA/MCA)
New Moves (Capitol/Capitol)
True Love (RCA/RCA)
It's Gotta Be Magic (–/Pickwick)
Traces (Capitol/Capitol)

Hank Williams Sr

One of the most charismatic figures in country music – his Opry performance of June 11, 1949, when his audience required him to reprise **Lovesick Blues** several times, is still considered as the Ryman's greatest moment – Hank was born Hiram King Williams in Georgia, Alabama, on September 17, 1923.

A member of the church choir at six, he was given a guitar by his mother a year later, receiving some tuition from Tee-Tot (Rufe Payne), an elderly black street musician. When barely a teenager he won $15 singing **WPA Blues** at a Montgomery amateur contest, then formed a band, the Drifting Cowboys, which played on station WSFA, Montgomery, for over a decade.

In 1946 Williams signed with Sterling Records, switching to the newly formed MGM label in 1947. Though virtually an alcoholic, he was booked as a regular on KWKH's Louisiana Hayride. After having scored with his recording of **Lovesick Blues**, he signed a contract with the Grand Ole Opry in 1949.

An early recording, **Move It On Over**, had already been a minor hit for Williams, but, after the runaway success of **Lovesick Blues** (a song waxed by yodeller Emmett Miller in 1925), he began cutting Top 10 singles with almost monotonous regularity.

With Fred Rose masterminding every Williams' recording session, arranging, playing, producing and often participating in the songwriting, such hits as **Wedding Belles**, **Mind Your Own Business**, **You're Gonna Change** and **My Bucket's Got A Hole In It** all charted during 1949. The following year provided: **I Just Don't Like This Kind Of Living'**, **Long Gone Lonesome Blues**, **Why Don't You Love Me?**, **Why Should We Try Anymore?** and **Moaning The Blues**.

These were followed by: **Cold, Cold Heart**, **Howlin' At The Moon**, **Hey Good Lookin'**, **Crazy Love**, **Baby We're Really In Love** (1951), **Honky Tonk Blues**, **Half As Much**, **Jambalaya**, **Settin' The Wood On Fire** and **I'll Never Get Out Of This World Alive** (1952). The last was ironically released just before his death (from a heart attack brought on by drinking) on New Year's Day, 1953.

He and his Drifting Cowboys had been booked to play a show in Canton, Ohio, and Williams hired a driver to chauffeur him through a snowstorm to the gig. He fell asleep along the way – but when the driver tried to rouse him at Oak Hill, Virginia, he was found to be dead. After his death, his records continued to sell in massive quantities. **Your Cheatin' Heart**, **Take These Chains From My Heart**, **I Won't Be Home No More** and **Weary Blue From Waitin'** all charted during the year that followed.

The last months of Williams' life – though financially rewarding – were ultra-tragic. A drug user in order to combat a spinal ailment caused by being thrown from a horse at the age of 17, he was fired from the Grand Ole Opry in August 1952 because of perpetual drunkenness. He was also divorced by his wife Audrey Sheppard – though he re-married to Billie Jean Jones soon after.

A difficult man to work with, being moody and uncommunicative, he was much respected and well loved by the country music fraternity. Over 20,000 people attended his funeral in Montgomery, at which Roy Acuff, Carl Smith, Red Foley and Ernest Tubb paid tribute in song.

His songs were well accepted in pop music as well – his compositions providing million-selling discs for Joni James (**Your Cheatin' Heart** – 1953), Tony Bennett (**Cold, Cold Heart** – 1951), Jo Stafford (**Jambalaya** – 1952). Williams' material has been recorded by rock bands, folk singers and black music acts.

Elected to the Country Music Hall Of Fame in 1961, his plaque reads: "The simple, beautiful melodies and straightforward plaintive stories in his lyrics of life as he knew it will never die."

His son, Hank Williams Jr. still carries on the tradition today and in 1964 provided the music to 'Your Cheatin' Heart', a Hollywood scripted film biography, in which George Hamilton portrayed Hank Sr.

Recommended:
Rare Takes And Radio Cuts (Polydor/–)
The Essential Hank Williams (–/MGM)
The Collector's Hank Williams (–/MGM)
Live At The Grand Ole Opry (MGM/MGM)
I Ain't Got Nothin' But Time (Polydor/ Polydor)
Country Store (–/Starblend)

Hank Williams Jr

Son of the late Hank Williams and his wife Audrey, Hank Jr was born in Shreveport, Louisiana, on May 26, 1949 – though he was taken to Nashville when only three months old and grew up there.

Just Me And My Guitar. Rare sides of Hank Williams. Courtesy CMF Records.

During his high school days he excelled in sports. When barely a teenager he toured with his mother's 'Caravan Of Stars' show, at the age of 15 having national hits via MGM releases **Long Gone Lonesome Blues** and **Endless Sleep**.

For many years he was forced to follow in his famous father's footsteps by doing endless versions of Hank Sr's songs. He fought against this in songs like **Standing In The Shadows (Of A Very Famous Man)**, which gave him a Top 5 country hit in 1966. Further major hits followed with: **It's All Over But The Crying** (1968), **Cajun Baby**, **I'd Rather Be Gone** (both 1969), **All For The Love Of Sunshine** (No.1 1970), **Ain't That A Shame** and **Eleven Roses** (both 1972), and **The Last Love Song** (1973).

In 1974 he moved from Nashville to Alabama, linked up with an old buddy named James R. Smith and began to forge a new Hank Williams Jr sound. He recorded the acclaimed **Hank Williams Jr And Friends** album, which featured country-rock musicians, including Charlie Daniels and Toy Caldwell.

On August 8, 1975 – about the time the album was being released – Hank Jr was involved in a climbing accident on a Montana mountain, suffering appalling head injuries. After two years of recuperation he re-emerged even stronger and made it back to the top with a series of gritty, best-selling country albums.

He changed labels in 1977, moving from MGM to Warner Bros. Although his singles failed to make much impression initially he

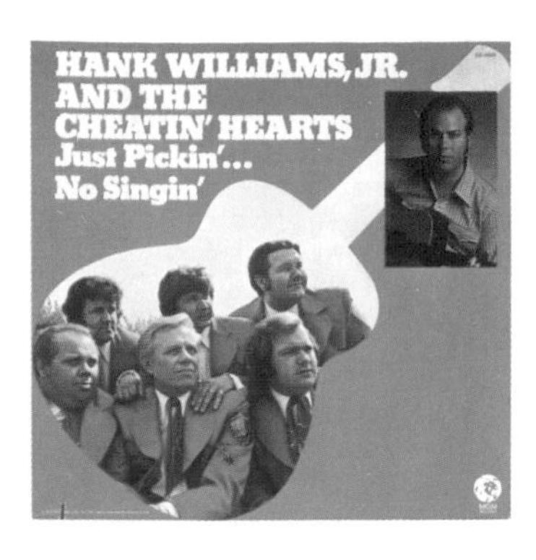

Just Pickin' ... No Singin', Hank Williams Jr And The Cheatin' Hearts.

finally made it back to the country Top 20 with **I Fought The Law** (1978) and scored Top 5 with **Family Tradition** and **Whiskey Bent And Hell Bound** (both 1979). He teamed up with Waylon Jennings on **The Conversation** (1979) and had such solo hits as **Women I've Never Had**, **Old Habits** (both 1980), **All My Rowdy Friends** (1981) and **A Country Boy Can Survive** (1982).

His brash, southern-styled country-rock, with macho lyrics, saw Hank Jr dominating the charts throughout the '80s with virtually every single making the Top 10. The most notable were **Leave Them Boys Alone** (with guest vocals by Waylon Jennings and Ernest Tubb, 1983), **Man Of Steel** (1984), **I'm For Love** (1985), **Mind Your Own Business** (1986), **Born To Boogie** (1987) and **If The South Woulda Won** (1988). This level of success, with many albums going gold or platinum and his shows being the wildest supported in country music, led to Hank Williams Jr finally gaining country music's top award, the CMA Entertainer Of The Year, in 1987 and 1988. Still unable to shake off the influence of his famous father, he duetted with Hank Sr on **There's A Tear In My Beer**, an obscure demo record from the early '50s that was discovered in 1988, and hit the Top 10 in 1989. However, the '90s found Hank Jr unable to score such major hits again, and in 1992 he signed with the re-activated Capricorn Records, though still working within the WEA family of labels.

Recommended:
Hank Williams Jr And Friends (MGM/MGM)
The New South (Warner Bros/–)
Habits Old And New (Elektra/–)
Man Of Steel (Warner Bros/–)
Are You Sure Hank Done It This Way (–/Warner Bros)
Five-O (Warner Bros/Warner Bros)
Maverick (Capricorn/–)
High Notes (Warner Bros/Warner Bros)
The Bocephus Box (Curb-Capricorn/–)

Lucinda Williams

Louisiana-born Lucinda Williams is a gutsy folk-country-blues performer, who is known more for her songs than her singing. Something of a folk troubadour, she has travelled extensively in America, playing music in the style to suit the region in which she was living at the time. Over the years this has given her music a rich, rootsy quality.

Lucinda's father is the well-known poet Miller Williams, and her mother was a pianist, though she never played professionally. During her childhood Lucinda mastered piano, zither, Hammond organ and guitar, and also started writing poetry and songs. By her teens she had become influenced by folk music, and later she absorbed folk-rock and the blues. Music became her way of making a living as she moved around from town to town to perform, living in such places as New Orleans, Los Angeles, Houston, Greenwich Village and Macon, Georgia.

Lucinda's first recordings were made for Smithsonian Folkways in the late '70s. They were mainly cover versions of country and blues classics with no definite sound of her own. Vocally she was a cross between Emmylou Harris and Bonnie Raitt. Already a prolific songwriter, from then on she was to record mainly her own songs, beginning with **Happy Woman Blues**, recorded in Houston, Texas, in 1980. A fine example of the country-blues style she was developing, this marked the first step towards wider acceptance. Later she moved to Los Angeles and in 1988 recorded a self-titled album for Rough Trade, which had her taking on a more contemporary edge. By now she was working with a band that included Gurf Morlix on various guitars, Donald Lindley on drums and Dr John Ciambotti on bass, a group of musicians who played with Lucinda for a number of years. This eponymous album contained the original versions of **Passionate Kisses**, which became a country hit for Mary-Chapin Carpenter, and **The Night's Too Long**, a Top 20 entry for Patty Loveless. A fourth album, **Sweet Old World**, for Chameleon Records, gained wider distribution through Elektra in 1992 and brought Lucinda much closer to country music. Her songs get right to the heart of people's emotions, heartaches and disasters, and her music is a fusion of roots that draws upon Williams' experiences of spending so many years on the road, honing and perfecting her musical craft.

Recommended:
Lucinda Williams (Rough Trade/Rough Trade)
Sweet Old World (Chameleon/Chameleon)

Tex Williams

Writer (with Merle Travis) and performer of **Smoke, Smoke, Smoke (That Cigarette)**, a 1947 hit that sold around two and a half million copies, Williams was predominantly a West Coast-based bandleader, who appeared in many films during the 1940s.

Born Sol Williams in Ramsey, Fayette County, Illinois, on August 23, 1917, he had his own one-man band and vocal show on radio WJBL, Decatur, Illinois, at the age of 13. He later toured throughout the States, Canada and Mexico with various western and hillbilly aggregations.

During the late 1930s he became Hollywood-based, there befriending Tex Ritter and working in films. After a long stay as lead vocalist and bass player with Spade Cooley, he formed his own band, the Western Caravan, in 1946, signing to the Capitol label.

Following the success of **Smoke, Smoke, Smoke**, the band's third release, Williams became a star and worked on scores of TV and radio shows, his band playing to capacity audiences at choice venues. Although his run of record successes seemed to peter out following the release of **Bluebird On Your Windowsill** (1949), he continued

Above: Lucinda Williams has provided hit songs for many country stars, including Patty Loveless.

recording for such labels as Decca and Liberty in the '50s.

A record deal with a Kentucky company, Boone, renewed Williams' acquaintance with the charts once more – albeit on a lower level. The singer scored good sales with **Too Many Tigers** (1965), **Bottom Of A Mountain** (1966), **Smoke, Smoke, Smoke** (1968) and several other titles. 1970 saw him signed to Monument, obtaining a Top 30 disc in 1971 with **The Night Miss Nancy Ann's Hotel For Single Girls Burned Down**. By the mid-'70s Williams had teamed up once more with Cliffie Stone, who had produced his Capitol recordings, and on Stone's Granite label came up with the album, **Those Lazy Hazy Days**. He died of lung cancer at his home in New Hall, California on October 11, 1985.

Recommended:
In Las Vegas (–/Sunset)
Oklahoma Stomp – with Spade Cooley (Club Of Spades/–)
Those Lazy Hazy Days (Granite/Pye)
Smoke, Smoke, Smoke (Capitol/Stetson)

Foy Willing

Born Foy Willingham in Bosque County, Texas in 1915, Foy aspired to a musical career while still in high school, appearing on radio as a solo singer and with a gospel quartet. He eventually found his way to New York City, where he appeared on radio for Crazy Water Crystals from 1933 to

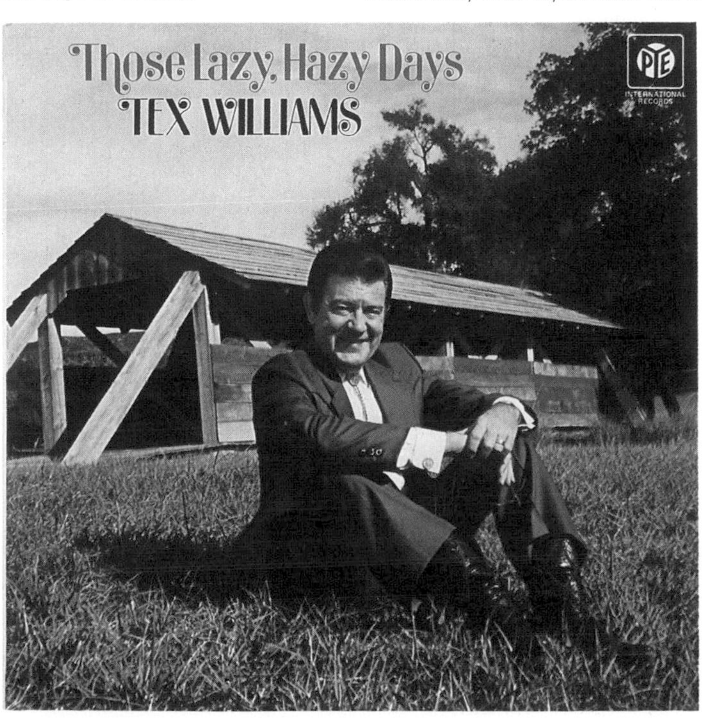

Those Lazy Hazy Days, Tex Williams. Courtesy PRT Records.

1935, when he returned to Texas to work in radio as an executive and announcer.

Willing moved to California in 1940 and founded the Riders Of The Purple Sage, originally composed of himself, Al Sloey and Jimmy Dean, although later members included Scotty Harrell, Fiddler Johnny Paul, accordionist/arranger Billy Leibert, accordionist Paul Sellers, guitarist Jerry Vaughn, clarinetist Neely Plumb and steel guitarist Freddy Traveres.

The group was formed in 1943 as cast members of the Hollywood Barn Dance, and throughout the rest of the 1940s they appeared on many radio shows including All Star Western Theater, the Andrews Sisters Show, the Roy Rogers Quaker Oats Show, and appeared in many Republic films with Monte Hale and Roy Rogers.

They recorded for Decca, Capitol, Columbia and Majestic Records. Their biggest hits were **No One To Cry To** and **Cool Water** on Majestic and **Texas Blues** and **Ghost Riders In The Sky** on Capitol. The Riders of The Purple Sage disbanded in 1952 when Willing left active performing, although there were a couple of quick albums on Roulette and Jubilee and a 1959 tour with Gene Autry.

Foy Willing was still recording, writing songs and appearing at western film festivals up until shortly before his death on June 24, 1978.

Willis Brothers

The Willis Brothers (Guy, born in Alex, Arkansas on July 15, 1915; Skeeter, born in Coalton, Oklahoma on December 20, 1917 and Vic, born in Schulter, Oklahoma on May 31, 1922) were originally known as the Oklahoma Wranglers. Their radio career commenced on KGEF, Shawnee, Oklahoma. In 1940, the trio moved on to become featured artists on the Brush Creek Follies Show on KMBC, Kansas City, Missouri.

With Guy as front-man and guitarist, Vic on accordion and piano, and Skeeter appearing in his role as the 'smilin' fiddler', the brothers appeared to be on the way to establishing a healthy reputation, but

Below: Foy Willing and the Riders Of The Purple Sage.

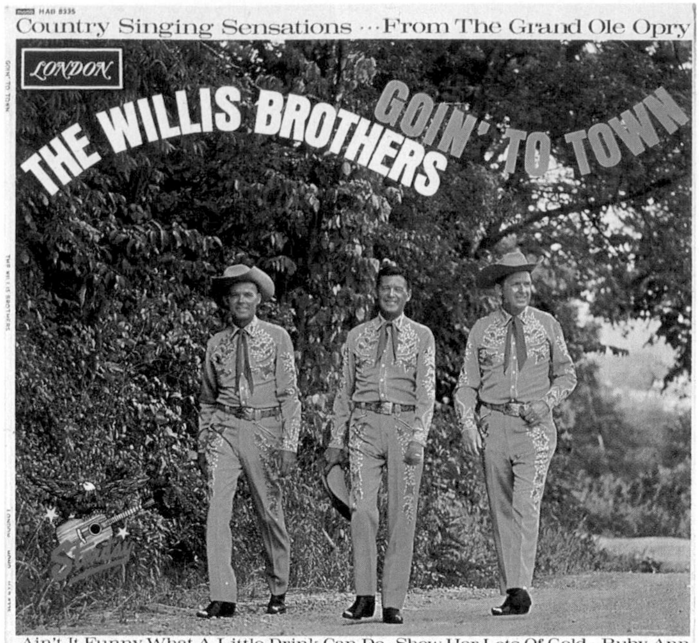

Goin' To Town, Willis Brothers. Courtesy London Records.

World War II intervened and the Willis Brothers joined the forces for four years.

They regrouped in 1946 and became Opry members until 1949, striking up an association with Eddy Arnold, on whose show they appeared for some eight years. Also during this period they became the first group to back Hank Williams on Sterling Records, and in a sense were the original Drifting Cowboys.

Rejoining the Opry in 1960, the group enjoyed some hit Starday singles throughout the '60s via such titles as **Give Me Forty Acres** (1964), **A Six Foot Two By Four** (1965), **Bob** (1967) and **Somebody Loves My Dog** (1967).

Often employed as session musicians, the Willis Brothers have recorded for several labels including RCA-Victor, Mercury, Sterling, Coral, Starday and MGM. They were the first featured act on the Jubilee USA shows in Springfield, Missouri. The brothers, who often impersonated other groups as part of their set, also had the distinction of being the initial country act to play a concert at Washington DC's Constitution Hall.

Following Skeeter's death from cancer in 1976 and Guy's retirement, younger brother Vic formed the Vic Willis Trio, using his accordion as lead instrument and performing modern country favourites in a close harmony style, which proved very popular at the Opry. Guy Willis died in Nashville on April 13, 1981.

Recommended:
Best Of The Willis Brothers (Starday/–)

Bob Wills

Leader of the finest western swing band ever to grace country music, and in fact the originator of the style, the late Bob Wills is acknowledged as having been one of the most influential performers in country music. Merle Haggard, Asleep At The Wheel, Alvin Crow and Red Steagall all borrowed from his repertoire, and Waylon Jennings payed tribute to the Texas fiddle man with the anthem, **Bob Wills Is Still The King**.

Born near Kosse, Limestone County, Texas on March 6, 1905, James Robert Wills was the first of ten children of a fiddle-playing father. In 1913 the Wills family moved to Memphis, Texas, where Jim Rob began playing fiddle at square dances, his initial instrument having been mandolin.

He lived on a West Texas family farm until 1929, when he went to Fort Worth, and became 'Bob' Wills after working in a medicine show that possessed one too many Jims. Forming a duo, the Wills Fiddle Band, with guitarist Herman Arnspiger during the summer of 1929, the fiddler and his band began playing for dances in the Fort Worth area.

During 1930 he added vocalist Milton Brown, the unit later metamorphosing into the Light Crust Doughboys. The Doughboys recorded for Victor in February, 1932, Brown leaving soon after, to be replaced by Tommy Duncan, a vocalist selected from over 70 other applicants.

Fired from the Doughboys in August 1933 – due to excessive drinking and his inability to get along with Doughboys' leader W. Lee O'Daniel – Wills took Duncan and his banjo-playing brother Johnnie Lee Wills with him, forming his own outfit, Bob Wills And His Playboys and gaining a regular spot on WACO in Waco. Although beset by legal problems, activated by Wills' ex-sponsors, the Burrus Mill and Elevator Company (the makers of Light Crust flour), the band struggled on, eventually making a base in Tulsa, where they became an institution on KVOO.

Now known as Bob Wills And His Texas Playboys, the band began recording for

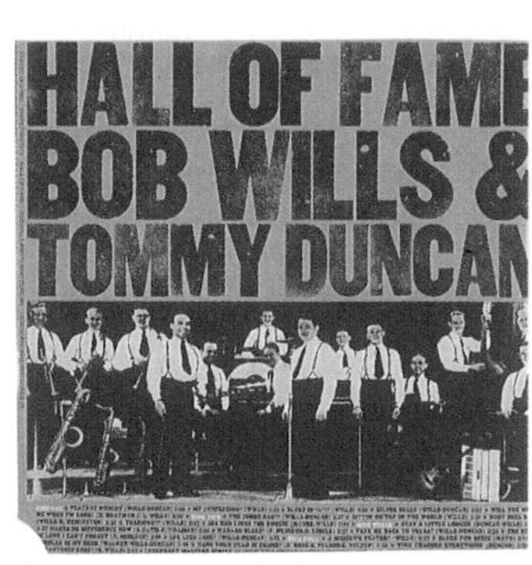

Hall Of Fame, Bob Wills and Tommy Duncan. Courtesy UA Records.

Brunswick, cutting some sides in Dallas during September, 1935. The record success, KVOO programmes and regular dances at Cain's Academy made their Tulsa years their most memorable.

A swing band with country overtones – the Playboys comprised 13 musicians by the mid-'30s and grew into an 18-piece during the 1940s – Wills' outfit played a miscellany of country ballads, blues and riffy jazz items with horns and fiddles vying for the front-line positions.

This sound proved tremendously popular and when, in April 1940, WIlls recorded his self-penned **San Antonio Rose** as a vehicle for Tommy Duncan's vocal artistry, the resulting disc became a million-seller.

Many more hits followed, but with the advent of Pearl Harbor, the band began to break up as its various members enlisted in the forces. Wills himself joined the Army in late 1942. Physically unfit for service life, he was discharged the following July, at which time he headed for California where he appeared on radio shows and made several films.

During the post-war period, with big bands generally fading, Wills was forced to use a smaller band, and began featuring fiddles, electric steel guitar and string instruments more prominently. He used Leon McAuliffe on steel much in the manner of jazz horn soloists. Again, the

Remembering The Greatest Hits Of Bob Wills. Courtesy MCA Records. Wills was the leader of the finest western swing band.

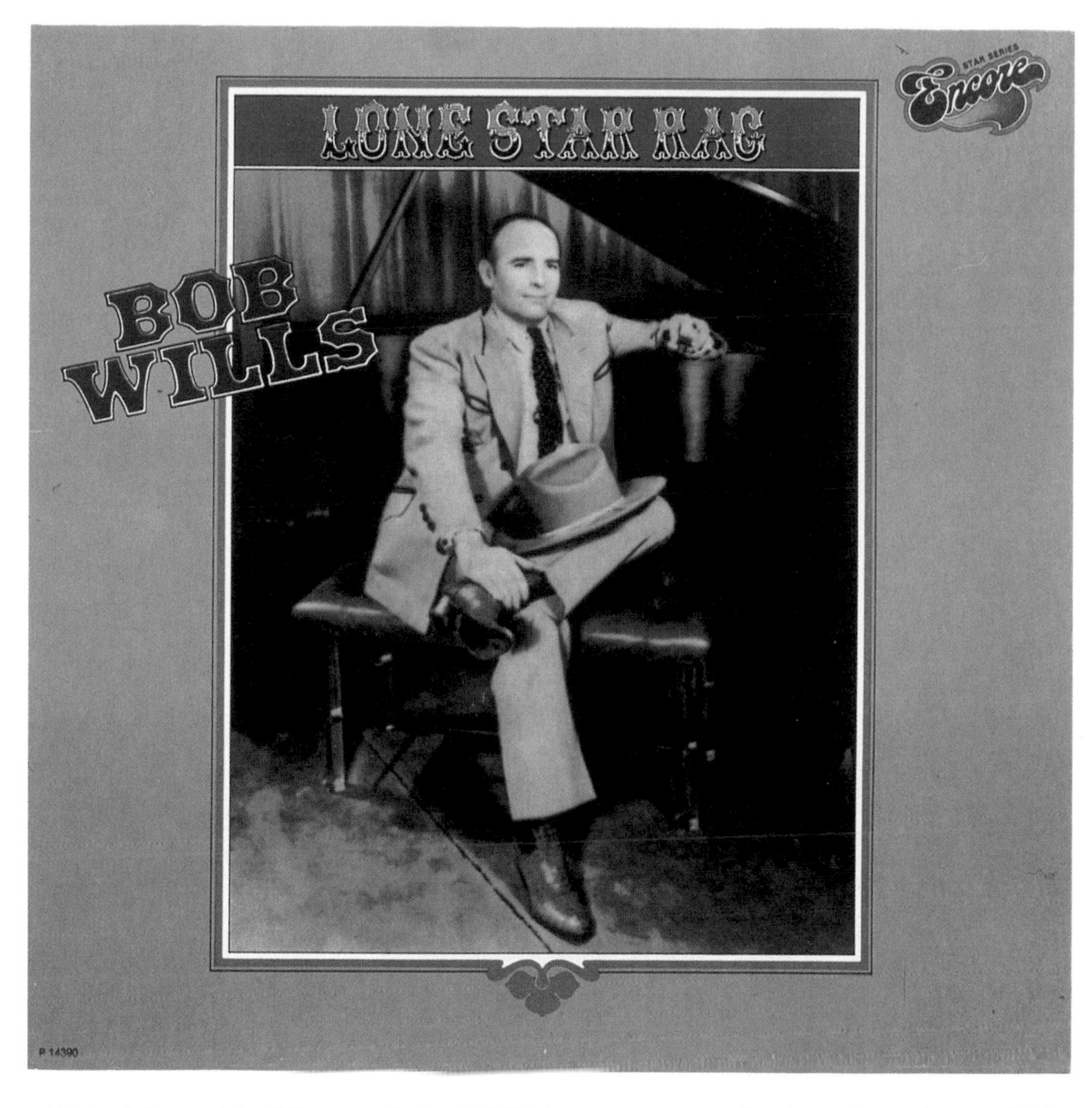

Lone Star Rag, Bob Wills. Courtesy CBS Records.

public loved the sound, but Wills was unsure where he was heading.

His health began to fail and during 1962 Wills suffered his first heart attack. Still he continued to tour with the Playboys, but in 1964 came another heart attack and he was forced to call a halt to his bandleading days.

Nevertheless, he still made a more limited number of appearances as a solo artist and continued making records, and he was honoured in the Country Music Hall Of Fame in October, 1968.

Also honoured by the State of Texas on May 30, 1969, Wills was paralysed by a stroke the very next day. But, although bedridden for many months, he fought back, and by 1972 began to appear at various functions, albeit in a wheelchair.

In December, 1973, he attended his last record date, many of the original Texas Playboys, plus Merle Haggard, taking part. In two days, 27 titles were cut for UA – but Wills was only present for a portion of the time, suffering a severe stroke after the first day. He never regained consciousness, although his actual death did not occur until 17 months later on May 13, 1975.

However, though Wills has gone his music lives on, the Playboys reforming several times since his death and many younger singers and bands recreating the Wills' sound both on record and in live shows.

Recommended:
Papa's Jumping (–/Bear Family)
Bob Wills (Columbia/CBS)
San Antonio Rose (Lariat/–)
Anthology (Columbia/CBS)
The Last Time (UA/–)
Leon McAuliffe Leads The Texas Playboys (Capitol/–)
Anthology, 1935–1973 (Rhino/–)

Johnnie Lee Wills

Although often cast in the shadow of his elder brother Bob, Johnnie Lee Wills actually carved out quite a long and successful career of his own. Born in east Texas in 1912, Johnnie Lee got his start playing tenor banjo with Bob as a member of the Light Crust Doughboys and left the band with him when Bob formed the Playboys.

Business got so good for Bob around 1940 that he formed a second band around Johnnie Lee, which grew to as many as 14 or 15 pieces. Called Johnnie Lee Wills And His Boys, they became extremely popular in the Tulsa area, and at one time or another contained many of the finest swing musicians of the era, including Leon Huff, Joe Holley and Jesse Ashlock.

They signed with Bullet Records in 1940 and had hits with **Rag Mop** and **Peter Cottontail**. They also recorded for Decca, Sims, RCA and a few smaller labels.

When western swing slumped in popularity in the 1950s, Johnnie Lee continued to run his famous Tulsa Stampede and opened a thriving western wear shop. He continued to appear at various western swing reunions until ill-health got the better of him in 1982, and he subsequently passed away on October 15, 1984.

Mac Wiseman

Malcolm B. Wiseman was born near Waynesboro, Virginia, on May 23, 1925. Country music in the area of the Shenandoah Valley in which Mac was raised was very much a folk art, and Mac learned music from people around him.

He has become known as a bluegrass artist, although his music also encompasses old-time, modern and even pop styles as well. Mac has floated in and out of the accepted Monroe–Earl Scruggs bluegrass style in his time and has indeed specialized in more traditional, sentimental material, such as **Jimmy Brown, The Newsboy** and **Letter Edged In Black**.

He attended the Shenandoah Conservatory of Music in Dayton, Virginia, and then joined the announcing staff of radio station WSVA in Harrisburg, Virginia, as newscaster and disc jockey. At this time he also wrote copy for station advertisements and worked nights with local country bands.

Possessing a warm, clear, tenor voice, Mac has been eagerly utilized by some of the top names in bluegrass, notably Bill Monroe and Flatt And Scruggs. Both acts included him live and on record. Mac began his career with another country music legend, Molly O'Day.

He has starred on Shreveport's Louisiana Hayride, Atlanta's WSB Barn Dance and Knoxville Tennessee Barn Dance, and has also guested on the Opry. He began recording on Dot Records in 1951 and his hits included: **'Tis Sweet To Be Remembered, Jimmy Brown, The Newsboy, Ballad Of Davy Crockett** and **Love Letters In The Sand**.

In 1957, Mac became Dot's Country A&R director and also ran the company's country music division for a few years. He recorded for Capitol Records during the early '60s, then moved back to Dot where he experimented with a string-filled album. Mac enjoyed a resurgence of popularity when he teamed up with Lester Flatt for some earthy bluegrass albums for RCA during the early '70s, and also built a large following in Britain with regular tours and record releases.

Apart from the bluegrass festivals he plays, Mac has been active behind the

Tulsa Swing, Johnny Lee Wills. Courtesy Rounder Records.

scenes in the industry. In recent years he has recorded for Churchill Records and CMH, maintaining a traditional bluegrass styling.

Recommended:
Early Dot Recordings Volume 1 (Country/–)
Golden Classics (Gusto/–)
Shenandoah Valley Memories (Canaan/Canaan)
Mac Wiseman Story (CMH/-)
Concert Favorites (RCA/RCA)
Lester'n'Mac – with Lester Flatt (RCA/RCA)
Grass Roots To Bluegrass (CMH/–)
Sings Gordon Lightfoot (CMH/–)

Del Wood

Probably the second (after Maybelle Carter) female country instrumentalist to achieve any real degree of fame, pianist Del Wood recorded a corny, ragtime version of **Down Yonder** (previously in a 1934 hit for Gid Tanner of the Skillet Lickers) on the Tennessee label in 1951 and came up with a million-seller.

Born Adelaide Hazelwood on February 22, 1920 in Nashville, Tennessee, Del Wood became an Opry member in 1951.

Above: Mac Wiseman, an outstanding bluegrass guitarist.

She remained with the show until shortly before her death on October 3, 1989, following a stroke on September 22. Perpetuator of several other best-selling discs in heavy-handed ragtime/honky-tonk style, Del has recorded for RCA, Mercury, Class, Decca and Lamb & Lion.

Recommended:
Ragtime Glory Special (Lamb & Lion/Lamb & Lion)
Tavern In The Town (Vocalion/–)

Sheb Wooley

A highly versatile performer, Wooley was voted CMA Comedian Of The Year in 1968 for his alter ego character Ben Colder. In 1964 he won a 'Cashbox' magazine award for "his outstanding contributions to country and popular music as a writer, recording artist and entertainer".

In the role of Pete Nolan, he co-starred in the TV series Rawhide. As Ben Colder he has scored with such recorded comedy hits as: **Don't Go Near The Eskimos** (1962), **Almost Persuaded No.2** (1966), **Harper Valley PTA (Later That Same Day)** (1968) and **15 Beers Ago** (1971).

Born in Erick, Oklahoma on April 10, 1921, Wooley spent his early years on his father's farm, becoming a competent horseman at the age of four and a rodeo rider during his teens. He formed his own band while still at school, later having his own network radio show for three years. In 1948, he was awarded his first major recording contract by MGM.

It was at this stage that Wooley moved to California and began working on 'Rocky Mountain', a film starring Errol Flynn. He has since been featured in more than 30 films. Wooley received considerable acclaim for his performance as the whiskey-drinking killer Ben Miller in 'High Noon'.

Co-star in 105 episodes of Rawhide, Wooley has appeared on countless TV shows. As a recording star under his own name (as opposed to releases using his Colder identity), he has enjoyed a six-week stay at the top the pop charts with his 1959 **Purple People Eater**, three years later toping the country charts with **That's My Pa**.

Recommended:
Blue Guitar (–/Bear Family)
Best Of Ben Colder (–/MGM)

Johnny Wright

Wright (born in Mt Juliet, Tennessee, on May 13, 1914) came from a musical family, his grandfather being a champion old-time fiddler and his father a five-string banjo player.

In 1933 he moved to nearby Nashville, there meeting and marrying Kitty Wells, also working with singer-guitarist Jack Anglin (born in Columbia, Tennessee on May 13, 1916) on radio station WSIX, Nashville and forming a duo, Johnny And Jack, with Anglin in 1938.

During the early '40s, Johnny And Jack toured with their band, the Tennessee Mountain Boys, playing on WBIG, Greensboro, North Carolina, WNOX, Knoxville and many other radio stations. By 1948 they joined the Grand Ole Opry, then left to become stars of Shreveport's Louisiana Hayride. Their popularity on the show led to an opportunity to rejoin Opry members in 1952, and Kitty, Johnny and Jack became regulars on the Opry for a period of 15 years.

Signed initially to Apollo, an R&B label (there cutting such sides as **Jolie Blon** and **Paper Boy**), Johnny And Jack switched to the more country-oriented RCA in the late '40s, scoring Top 20 hits with: **Poison Love** (1951), **Crying Heart Blues** (1951), **Oh Baby Mine (I Get So Lonely)** (1954), **Beware Of It** (1954), **Goodnight, Sweetheart, Goodnight** (1954), **Stop**

The World (1958), **Lonely Island Pearl** (1958) and **Sailor Man** (1959). Other hits included **Ashes Of Love** and **I Can't Tell My Heart That**.

Shortly after one last success with **Slow Poison**, a 1962 Decca release, Jack Anglin was killed (on March 8, 1963) in a car crash en route to a funeral service for Patsy Cline – at which point Wright formed a new roadshow and became a solo recording act, notching a chart No.1 in 1965 with **Hello Vietnam**.

His son Bobby proved a success during the late '60s. Wright formed the Kitty Wells–Johnny Wright Family Show in 1969, doing extensive tours, and he and Bobby recorded an album of Johnny And Jack material for Starday during 1977.

Recommended:
Here's Johnny And Jack (Vocalion/–)
All The Best Of Johnny And Jack (RCA/–)
Johnny And Jack And The Tennessee Mountain Boys (–/Bear Family)

Michelle Wright

Michelle Wright, a sensual, sexy lady, is one of only a handful of Canadian performers who have succeeded in making a big impact on the American country music scene. Born on July 1, 1961, in Merlin, Ontario, a small Canadian farming community, Michelle was bred on American music. She heard the rhythm and blues and Motown hits coming out of Detroit, just 45 minutes away, but especially she heard the strains of country music. Her mother was the singer in a country band called the Reflections for ten years, and her father played in a traditional country band with steel guitar, rhinestone suits and all the works. As a child she would often get up on the bandstand and sing, and by college days was working with a local band. This was followed by several years out on the road, working with pick-up bands, playing small clubs across Canada. In 1986, she won the CJBX London, Ontario 'Country Roads Talent Search'. That led to her signing to Canadian label Savannah Records and scoring several Canadian hits including **I Want To Count On You**, **New Fool At An Old Game** and **Rock Me Gently**.

She started gaining nominations from the Canadian Country Music Association in 1986, but it was not until 1989 that she really made a breakthrough, picking up her first Female Vocalist Of The Year award. It was a feat she was to repeat for the next three years. In 1992 she was also named Country Music Person Of The Year in recognition of her success in America. That success occurred because of the involvement of Nashville songwriters Rick Giles and Steve Bogard. The pair started providing Michelle with quality song material and also produced her recordings in Canada and Nashville. Her breakthrough in the States started to get under way when Michelle was signed to Arista Records in Nashville in early 1990, resulting in **New Kind Of Love** becoming her first US hit during that summer. Still based in Canada, Michelle scored a few minor hits during the next couple of years, then made a move to Nashville in order to develop her career further. The result was a Top 10 hit with **Take It Like A Man** (1992), and further success with **One Time Around** (1992), **He Would Be Sixteen** and **The Change** (1993).

A self-confessed one-time alcoholic, the husky-voiced songstress with the soulful edge possesses a strong ear for music that works best for her throaty voice, resulting in her second album, **Now And Then**, making a big impact on both the country and pop charts, gaining Michelle a gold disc in 1993.

Recommended:
Now And Then (Arista/Arista)
Do Right By Me (Savannah/Savannah)

Right: Canadian Michelle Wright gained a gold disc for her 1993 album Now And Then. The album made an impact on pop and country charts.

Left: Sheb Wooley played the killer Ben Miller in 'High Noon' and co-starred in the TV series Rawhide.

Tammy Wynette

One of the most successful female country singers of all time, Tammy Wynette was adjudged CMA Female Vocalist Of The Year for three consecutive years (1968–70), while her recording of **Stand By Your Man**, a No.1 in the US during 1968 and a major British hit in 1975, was the biggest selling single by a woman in the entire history of country music.

Tammy began life as Virginia Wynette Pugh, born near Tupelo, Mississippi on May 5, 1942. Her father died when she was but a few months old and her mother moved to Birmingham, Alabama, leaving her in the care of grandparents until the end of World War II. Brought up on a farm, Tammy learned to play the collection of instruments owned by her father, taking a lengthy series of music lessons with a view to a career in singing.

But, getting married at 17, she had little time for music during the next three years. Instead she became the mother of three children, her marriage breaking up before the third child was born. The baby, a girl, developed spinal meningitis, and Tammy had to supplement her earnings as a Birmingham beautician in order to pay off various bills incurred as a result of the child's ill health. She turned to music once more, becoming featured vocalist on station WBRC-TV's Country Boy Eddy Show during the mid-'60s, following this with some appearances on Porter Wagoner's syndicated TV programme.

Soon she began making the rounds of the Nashville-based record companies, in the meantime working as a club singer and a song-plugger in order to support her children. Following auditions for UA, Hickory and Kapp, Tammy was signed by Epic's Billy Sherrill and recorded **Apartment No.9**, a song written by Johnny Paycheck and Bobby Austin. Released in 1966, the disc proved a great success. The next release, **Your Good Girl's Gonna Go Bad** (1967), proved to be even stronger, becoming a Top 5 item. From then on it was plain sailing throughout the '60s as **I Don't Wanna Play House** (1967), **My Elusive Dreams** (with David Houston, 1967), **Take Me To Your World**, **D-I-V-O-R-C-E**, **Stand By Your Man** (all 1968), **Singing My Song** and **The Ways To Love A Man** (both 1969) qualified as chart-toppers.

Married five times, Tammy had several well publicized relationships with other partners, including Burt Reynolds and Rudy Gatlin of the Gatlin Brothers, but it was her marriage to singer George Jones that made most headlines. The pair announced their marriage on August 22, 1968 in order to silence gossips, but they were not in fact married until February 16, 1969. For several years they toured together and recorded several big-selling duets, but Jones' regular drinking bouts and Tammy's career aspirations but did not mix too well and they were divorced on March 13, 1975.

However, on disc, Tammy could do little wrong – **Run, Woman Run** (1970), **Good Lovin'** (1971) **'Til I Get It Right** (1973), **Woman To Woman** (1974), **'Til I Can**

Make It On My Own (1976), **Womanhood** (1978), **They Call It Making Love** (1979), **Another Chance** (1982), **Sometimes When We Touch** (a duet with Mark Gray, 1985) – being just some of her major hits during the 1970s and 1980s. Her **Greatest Hits** album (which remained in the charts for over 60 weeks) earned a platinum disc for sales in excess of one million.

Tammy married her longtime friend, and, for a time, record producer, George Richey, at her Florida home on July 6, 1978. In 1982 Tammy's career was captured in her biography 'Stand By Your Man' and made into a successful film.

Throughout the rest of the '80s Tammy seemed to be dogged by ill-health as her records began to fall off the charts. In 1986 she entered the Betty Ford Clinic for drug addiction, and also had several stomach operations. Even duet recordings with Ricky Skaggs, Randy Travis and Emmylou Harris failed to return her to the top. However, in 1992, when she recorded a strange duet with pop duo KLF and appeared in the group's video of **Justified And Ancient**, Tammy found herself high on the pop charts in both Britain and America again.

Below: Tammy Wynette recently moved into pop music with KLF.

Recommended:
Soft Touch (Epic/Epic)
Just Tammy (Epic/Epic)
No Charge (–/Embassy)
One Of A Kind (Epic/Epic)
Superb Country Sounds (–/Embassy)
Tears Of Fire (Epic/Epic)
Heart Over Mind (Epic/Epic)
Higher Ground (Epic/Epic)
Next To You (Epic/Epic)

With George Jones:
We're Gonna Hold On (Epic/Epic)
Let's Build A World Together (Epic/Epic)
Together Again (Epic/Epic)

Wynonna

The daughter of the Judds, country music's most popular duo of the '80s, Wynonna, who was born Christina Criminella on May 30, 1964 in Ashland, Kentucky, embarked on a highly successful solo career at the beginning of 1992. Signed to MCA/Curb, her debut album, simply titled, **Wynonna**, soared high on both the pop and country charts as it raced to a multi-platinum award.

Wynonna's eponymously titled debut solo album. Courtesy MCA Records.

A rebellious, but determined young lady, Wynonna had unwittingly been groomed for this success from her teenage years growing up in Kentucky, where her musical influences were very eclectic. They ranged from bluegrass, through mountain harmonies, big band music and Bonnie Raitt, whom she cites as one of the biggest influences on her own vocal style. She moved with her mother and younger sister Ashley to Franklin, Tennessee, near Nashville, at the age of 15. Her mother, Naomi Judd, encouraged her music, singing harmonies and trying to further her daughter's musical ambitions. After a high school talent contest came the now famous live audition in RCA's Joe Galante's office, when mother and daughter, backed only by Wynonna's guitar, impressed the Nashville veterans. Then there was the first show, playing to an audience of 10,000 people opening for the Statler Brothers in 1984. The Judds really took off during the next seven years. The hits started and kept coming – 23 of them – and the awards kept picking up. The Country Music Association's Award for Vocal Group in 1984 was the first of their seven CMA awards. They won the first of four Grammies in 1985. Album sales soared to over 10 million worldwide.

The Judds' career came to an end when Naomi announced plans to retire, due to chronic hepatitis. Musically, as well as emotionally, Wynonna had to face the world alone. It marked a complete new beginning with a new producer in Tony Brown, and a new studio band. Developing more of a white blues sound than the mountain country of the Judds, Wynonna immediately made an impact. Her first three singles, **She Is His Only Need**, **I Saw The Light** and **No One Else On Earth**, all became No.1s during 1992. A fourth, **My Strongest Weakness**, made the Top 5 in early 1993. Taking her music in different directions, Wynonna succeeded in keeping her own distinctive voice at the forefront of her new music, so that every song on her debut album, from the haunting gospel of **Live With Jesus** to the powerfully upbeat **I Saw The Light**, and the hard-edged **What It Takes** to the broken-hearted ballad **My Strongest Weakness**, sounds unmistakeably Wynonna.

She teamed up with Clint Black for the duet hit **A Bad Goodbye** in early 1993, then came a second album, **Tell Me Why**, which showcased the growing maturity, depth of feeling and emotion of Wynonna in a sparkling set of performances, highlighted yet again by the sheer strength of song material. The title song immediately became another Top 10 country hit, while the album passed the million sales mark within a few weeks of release.

Recommended:
Wynonna (MCA-Curb/Epic-Curb)
Tell Me Why (MCA-Curb/–)

Trisha Yearwood

Chic-looking Trisha Yearwood, who zoomed to stardom in 1991 with **She's In Love With The Boy**, is a country girl born on September 19, 1964, in Monticello, Georgia. The younger of two daughters of a banker father and school teacher mother, she grew up on a farm in the small central Georgia community, about an hour's drive from Atlanta, Athens and Macon. The music of Linda Ronstadt, Emmylou Harris and the Eagles were major influences during her teen years as she developed her own vocal style, making the rounds of area talent contests.

In 1985, after a stint at the University of Georgia, Trisha moved to Nashville to attend Belmont College, where she completed a music business course. Her first introduction to the music business was when she landed a job in the publicity department at MTM Records. Then came demo work for publishers. At this time she met a struggling Garth Brooks, who promised they'd work together if he ever made it big. True to his word, when Garth was recording his first album, Trisha was

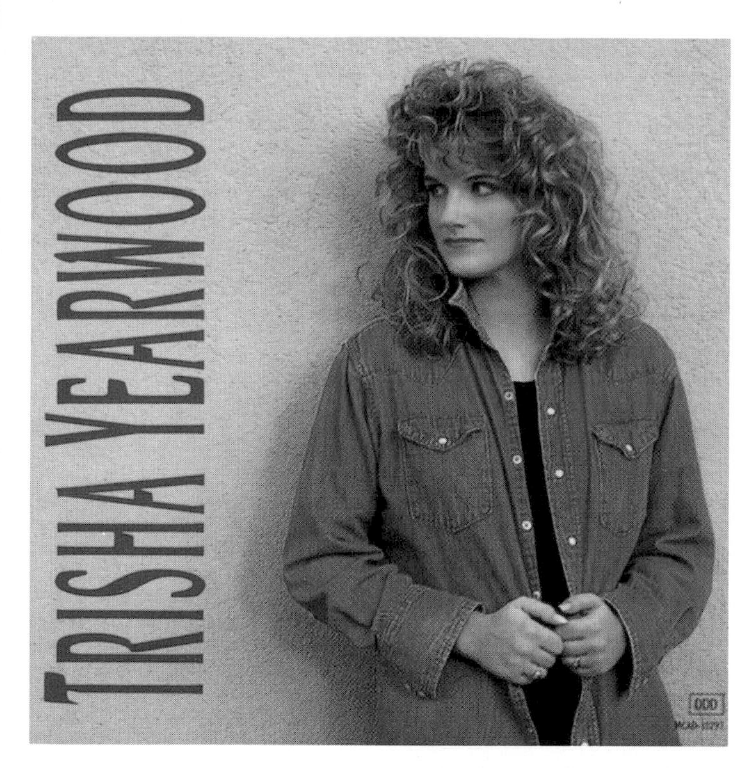

Trisha Yearwood's 1991 debut album. Courtesy MCA Records.

Below: Dwight Yoakam has recorded duets with various female partners.

called in to provide back-up vocals, and later Garth sang on her recordings, and also provided her with some songs.

With the help of Garth Fundis, a Nashville producer who had worked with Don Williams and Keith Whitley, Trisha played a showcase for Nashville's record industry and was signed to MCA Records in late 1990. The first single, **She's In Love With The Boy**, accompanied by a stunning video, raced to No.1 on the charts in 1991, followed by Top 10 entries with **Like We Never Had A Broken Heart** (1991), **That's What I Like About You**, **The Woman Before Me** and **The Wrong Side Of Memphis** (1992). The debut album, **Trisha Yearwood**, a stunning set of quality songs, had an extra edge that is absent from much of country music, giving her the added appeal to a younger audience. The end result is a powerful musical statement that saw the album make both the pop and country charts during 1991 as it eventually reached multi-platinum status. Sales of the album were obviously helped when Garth Brooks invited Trisha to be his opening act during his 1991 tour, but from then on this determined and talented young lady has made it to the top on her own terms. There have been further Top 10 entries with **Walkaway Joe** and **You Say You Will** (1993), while her second album, **Hearts In Armor**, featuring guest appearances by Garth Brooks, Don Henley, Vince Gill, Emmylou Harris and Raul Malo, has also reached platinum status.

Recommended:
Trisha (MCA/MCA)
Hearts In Armor (MCA/MCA)

Dwight Yoakam

A singer whose debut album for Reprise immediately climbed into the pop charts, Yoakam caught the attention of traditionalists and even punk rockers.

Born in Pikeville, Kentucky on October 23, 1956, Yoakam honed his style of country in bars and roadhouses along Route 23 in southern Ohio, then headed for L.A. because he felt Nashville music had become too slick; he claimed, "Those Buck Owens records in the late 1950s and 1960s were some of the hippest hillbilly stuff known to man – and Nashville has all but abandoned it."

He became an opener for roots-oriented rock acts like Los Lobos, the Blasters and Lone Justice, and released a six-track EP titled **Guitars, Cadillacs, Etc** on Oak Records during 1985. He then moved on to provide Reprise (practically a dead label after Sinatra and Neil Young moved off) with an attention-grabbing album.

Yoakam, who first sang what he calls 'hillbilly hymns' in a Pikeville church, found grudging acceptance in Nashville when singles such as **Honky Tonk Man** (1986), **Little Sister** (1987) and **Always Late With Your Kisses** (1988) all went Top 10 on the country charts. He teased Buck Owens out of premature retirement, and the pair enjoyed a chart-topping duet with **Streets Of Bakersfield** (1988), while Yoakam had his own solo hits with **I Sang Dixie** (No.1, 1989) and **I Got You** (1989). Albums **Hillbilly Deluxe** and **Buenas Noches From A Lonely Road** went platinum as Yoakam's uncompromising approach to country won over a whole legion of younger, rock-oriented fans.

The risk-taking nature of his country style can be perceived in his choice of duet partners, which includes Patty Loveless, k.d. lang, Maria McKee and the neo-folk-rock duo, the Indigo Girls. He has also cut a diverse range of songs and styles, from a stormy, swamp, rockin' **I Hear You Knockin'** to **Truckin'** on the Grateful Dead tribute album, **Deadicated**, to the Elvis signature tune **Suspicious Minds** on the 'Honeymoon In Vegas' soundtrack. Yet he has maintained country radio play, resulting in such country Top 10 entries as **You're The One** (1991) and **Ain't That Lonely Yet** (1993).

Recommended:
Guitars, Cadillacs, Etc (Reprise/Reprise)
Hillbilly Deluxe (Reprise/Reprise)
If There Was A Way (Reprise/Reprise)
Buenas Noches From A Lonely Room (Reprise/Reprise)
This Time (Reprise/Reprise)

Faron Young

Affectionately known as 'The Sheriff' Faron has been a stalwart of the country music industry for some three decades.

Born in Shreveport, Louisiana on February 25, 1932, he was raised on a farm

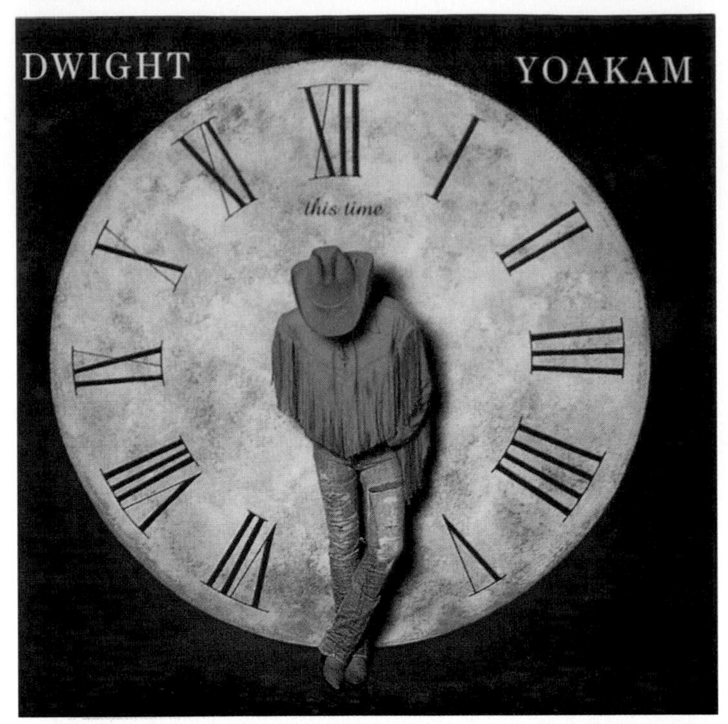

This Time, Dwight Yoakam. Courtesy Reprise Records.

outside the town and spent his boyhood days picking up guitar chords while out with the family's cows. He formed his first band at school.

After college he found that he had gained something of a reputation in Louisiana and was invited to join KWKH and subsequently the Louisiana Hayride itself. It was then that Webb Pierce employed him as a featured vocalist.

In 1951 he was signed by Capitol Records and had country hits with **Tattle Tale Eyes** and **Have I Waited Too Long?**. Joining the Opry in 1952, Young then spent from 1952 to 1954 in the US Army, touring widely to entertain the troops. His first major success came with the Ted Daffan song **I've Got Five Dollars And It's Saturday Night**, and with his bid for the teenage market in **Going Steady**.

The '50s saw him gaining many hits and massive popularity, obtaining a No.1 disc with **Sweet Dreams** and later with **Country Girl** (1959), following this in 1961 with another chart-topper in **Hello Walls**.

Throughout the '60s, the hits continued to flow. **Backtrack** (1961), **Three Days, The Comeback**, **Down By The River** (1962), **The Yellow Bandana**, **You'll Drive Me Back (Into Her Arms Again)** (1963), **Walk Tall** (1965), **Unmitigated Gall** (1966), **I Just Came To Get My Baby** (1968), **Wine Me Up** (1969) and **Keeping Up With The Joneses**, a 1964 duet with Margie Singleton, figured among his Top 10 successes for the Mercury label, to which Faron Young became contracted in 1961.

A versatile entertainer who presents an alive and amusing show with his own witty comments, Young retained his popularity over an extensive period, as such major '70s hits as **It's Four In The Morning**, **This Little Girl Of Mine** and **Just What I Had In Mind** clearly demonstrated.

He has appeared in a number of low-budget film productions, while his out-of-showbiz interests include a booking agency, a music publishing firm and magazine publishing – Young has been the owner of the monthly publication Music City News. He has even contributed to Nashville's rising skyline, owning the Faron Young Executive Building near Music Row.

Recommended:
The Man And His Music (Mercury/–)
Free And Easy (MCA/–)
The Capitol Years (–/Bear Family)
Sweethearts Or Strangers (Capitol/Stetson)

Below: Neil Young gained plaudits from critics with his 1985 Old Ways and moved close to country again with his 1992 offering, Harvest Moon.

Neil Young

A Canadian singer-songwriter, born on November 12, 1945 in Toronto, Neil worked as a folk singer in Canada in the early '60s before moving to California where he became involved in the West Coast pop movement.

He was a key member of country-rock band Buffalo Springfield, then became part of the soft-rock group Crosby, Stills, Nash And Young. Embarking on a solo career in the early '70s, he formed Crazy Horse, a multi-instrumental unit that mixed strands of country, blues, jazz and rock in a repertoire that proved highly successful in terms of the huge number of albums sold.

The eclectic Young surprised many when he travelled to Nashville at the beginning of 1985 to record a new album, **Old Ways**, which brought him widespread country acclaim. Teaming up with Willie Nelson, Waylon Jennings and several notable Music City musicians, he recorded **The Wayward Wind**, once a hit for Tex Ritter, plus a number of originals all in country mould, the result being a best-selling album. "Rock'n'roll has let me down," he claimed later. "It doesn't leave you a way to grow old gracefully."

He subsequently took to the road with a traditional country band, utilizing fiddles, banjo, steel and mandolin. It was a short-lived flirtation with country, and soon Young was flitting between musical styles, seemingly playing whatever kind of music he fancied from album to album. He moved once again closer to a country sound for his **Harvest Moon** album in 1992, which gained enthusiastic reviews and became his biggest-selling album in years.

Recommended:
Old Ways (Geffen/Geffen)
Harvest Moon (Warner Bros/Warner Bros)

Appendix

Lack of space prevents us from including a full entry on the following people, organizations and places, etc. Nevertheless, here is a brief run-down on the contributions they have made to country music.

A

WENDELL ADKINS – Born in Kentucky. Raised in Ohio, Wendell was a rock'n'roll teenage singer. Later he toured the Midwest with a band, moved to Florida, got noticed and ended up in Vegas. Following chunks of aid from Willie Nelson and David Allan Coe, he began an association with Gilley's club. He has recorded prolifically, but never made a major breakthrough.

BUDDY ALAN – Buck Owens' son, born in Tempe, Arizona, 1948, and a singer rarely out of the charts in the late '60s and early '70s. But when Dad stopped having hits, so too did 6'4" Buddy.

DANIELLE ALEXANDER – A singer-songwriter from Fort Worth, Texas, who enjoyed notable success in the late '80s, including a duet hit with Butch Baker. She started out singing jazz standards with her father in West Virginia, then sang in Texas bands, finally moving to Nashville, where she became a staff songwriter. Harold Shedd signed her to a Mercury contract in 1989, and since then Danielle has developed her own distinctive style.

SUSIE ALLANSON – A pop-styled country singer born in Minnesota in 1952, she performed in the musicals 'Hair' and 'Jesus Christ Superstar' before switching to country. Enjoyed several major country hits during the late '70s and early '80s, the biggest, **We Belong Together**, spending two weeks at No.2 in 1978.

TERRY ALLEN – Born in Wichita, Kansas in 1953, Allen became an art school teacher and singer. His first album (1975) was an incredible concept affair titled **Juarez**. Released by Fate Records of Chicago, the album became a cult record among country and rock fans alike. Later he cut an essential double album, **Lubbock (On Everything)**, which saw him working with the Joe Ely Band and adding to his reputation as a country songwriter of some piquancy.

AMAZING RHYTHM ACES – A Memphis-based country-rock band formed in the early 1970s, best known for **Third Rate Romance**, a Top 20 pop and country hit in 1975. They enjoyed further country hits with **Amazing Grace (Used To Be Her Favorite Song)** and **The End Is Not In Sight**, and released several critically acclaimed albums before disbanding in 1981.

AMERICAN RECORD COMPANY – A company that owned five labels of its own and, in addition, recorded material for Conqueror, the Sears-Roebuck label. Once owner of the finest country music catalogue in America, it eventually was taken over by Columbia.

LIZ ANDERSON – Famed country singer and songwriter, Liz, from Minnesota, was born in 1930 and at 16 married Casey Anderson. Parents of Lynn Anderson, they wrote country hits for Roy Drusky, Merle Haggard and others. Liz recorded extensively throughout the '60s and '70s and by the late '80s was hosting a country show on cable television in Nashville.

AREA CODE 615 – A supergroup comprising Nashville and Muscle Shoals sessionmen, Area Code 615 was originally formed by Mike Nesmith, who still owns unreleased tapes of Code's initial sessions. They later made two albums for Polydor, including **Stone Fox Chase**, which became the theme for BBC-TV's rock programme The Old Grey Whistle Test. Eventually they split up, and three members went on to form Barefoot Jerry.

JERRY ARHELGER – A Christian country music singer who has worked all over the world, although his base is in Wewahitchka, Florida. His CB/trucking song, **Breaker, Breaker, Sweet Jesus**, is reputed to have once logged over 250 call-in requests in one night on a Mississippi radio station.

Stacked Deck, Amazing Rhythm Aces. Courtesy ABC Records.

Bluer Than Midnight, Delia Bell. Courtesy County Records.

ARKIE THE ARKANSAS WOODCHOPPER – Longtime fixture on the National Barn Dance (1928–1970), his real name was Luther Ossenbrink. He was an all-rounder – a singer, guitarist, MC and square dance caller.

AUSTIN – A Texan city that, during the turn of the 1960s, became a centre for a new contemporary 'outlaw' form of country music. The changes evolved around Threadgill's Bar, where, around 1962, students from the university and musicians searching for an identity (such as Janis Joplin) gathered to play folk and country music. Later performers such as Michael Murphey, Steve Fromholtz and B. W. Stevenson came on like young, homespun idealists, using poetic imagery in their songs, while Kinky Friedman guyed country sounds and Commander Cody presented the whole of the southern music rainbow. The home of Willie Nelson, Austin – whose place in country music history is the subject of Jan Reid's fine book 'The Improbable Rise Of Redneck Rock' – still offers an alternative scene from that promulgated by Nashville.

B

DeFORD BAILEY – A black harmonica player, who opened the WSM Barn Dance on the night George D. Hay named the show 'The Grand Ole Opry'. Born in Carthage, Tennessee, in 1899, he remained on the Opry until 1941, but later drifted into obscurity. He operated a shoeshine stall in Nashville and made a brief television appearance on a blues show in the '60s. He died on July, 1982, and several Opry members were present at the funeral.

CARROLL BAKER – Born in Bridgewater, Nova Scotia, for many years she was Canada's 'Queen Of Country Music'. The winner of every award her country can offer, her record sales passed the million mark in 1983. Her main audience is, however, Canadian.

BAKERSFIELD – A Californian town touted as 'Nashville West' during the mid-1960s, when residents such as Buck Owens, Merle Haggard and Wynn Stewart achieved a fair degree in chart domination. The 'Bakersfield Sound' has been a major influence on West Coast country acts of the '80s and '90s.

EARL BALL – Born in Foxworth, Mississippi in 1941, Ball became a notable session pianist and producer who toured with the Johnny Cash Show for many years.

BANDANA – This Nashville-based band was formed in 1981 in the wake of Alabama's success. Signed to Warner Brothers they enjoyed several Top 20 country hits with their tight harmonies and smooth rock-based instrumental work. Following several personnel changes, the band split up in 1988.

AVA BARBER – A singer, born in Knoxville, Tennessee, 1954, who made it on the Lawrence Welk show in the mid-1970s. Signed to Ranwood Records, she had several hits, the biggest being a version of Gail Davis' **Bucket To The South** in 1978. In 1981 she signed with Oak Records.

GLENN BARBER – From Hollis, Oklahoma (born in 1935) but raised in Texas, Barber was a skilled carpenter who built his own recording studio. A multi-instrumentalist and singer-songwriter, he had hits for Sims and Starday in the mid-1960s before moving to Hickory and continuing his hit-making to the mid-1970s. By 1980 he had stopped having hit records and, in more recent times, has become a successful portrait painter.

BAREFOOT JERRY – Formed from the remnants of Area Code 615, Barefoot Jerry were headed by Wayne Moss who fashioned the band into an often inventive country-rock outfit that recorded variously for Capitol, Warner Bros and Monument, 1971–76.

RANDY BARLOW – A Detroit-born singer, Barlow worked as MC with Dick Clark's Caravan Of Stars in the mid-1960s. Then came a stint as a solo singer on the lounge circuit in California. The transition to country music came in the early 1970s, and by 1978 he was enjoying Top 10 success on Gene Autry's Republic Records, but within ten years he had faded from the country music scene.

SHANE BARMBY – A singer who was raised near Sacramento, California, Barmby, born in 1955, was named after the Alan Ladd movie 'Shane'. As a child he sang at rodeos and is a skilled trick ropester. In the early 1980s he landed bit parts in television series, but music was his first love. In 1987 he moved to Nashville, opened shows for Randy Travis, and signed with Mercury Records two years later.

J. J. BARRIE – A Canadian singer who became a genuine one-hit wonder in 1976 when his version of **No Charge** went to the top of the UK pop charts.

MOLLY BEE – Something of a child prodigy, Molly, who was born Molly Beachboard in Oklahoma City in 1939, appeared on Rex Allen's radio show in Tucson when only ten. In 1950 she became a regular on Cliffie Stone's Hometown Jamboree TV show. A successful actress and vocalist, she recorded for Capitol and MGM during the '60s and appeared in several musicals and TV variety shows. In the mid-1970s Molly recorded for Cliffie Stone's Granite label.

PHILOMENA BEGLEY – A superstar on the Irish country scene, Philomena was born in 1946 in County Tyrone, and her K-Tel album, **The Best Of Philomena Begley**, went platinum in her homeland. She has recorded duets with Ray Lynam, the pair frequently having appeared together onstage at Wembley. Philomena has remained a major star in Europe in a career that has spanned more than 30 years.

DELIA BELL – A bluegrass lady from Texas, who was raised in Hugo, Oklahoma, Delia possesses a voice that harks back to country's golden era. Loved by those who search the minor label catalogues, she came to the attention of Emmylou Harris in the early '80s and, unbelievably, got signed by Warner Brothers who released an album in 1983.

PINTO BENNETT – Bigger in Europe than the USA, Bennett, from Idaho, grew up on a ranch, listening to the music of Hank Williams and Lefty Frizzell. With his band, the Famous Motel Cowboys, he has been fusing honky-tonk with more contemporary sounds, mainly performing his own songs throughout Europe.

BEVERLY HILLBILLIES – This early California outfit had a radio show on KEJK, later KMPC, in 1928 and released a hit single the following year, **When The Bloom Is On The Sage/Red River Valley**. Zeke and Tom Manners, Hank Skillet, Elton Britt, Stuart Hamblen, Wesley Tuttle and Glen Rice were all members at various times.

BIG TOM – Born Tom McBride in Castleblaney, Ireland, he became a star in his native country where the song **Big Tom Is Still The King** topped the pop charts in 1980.

THE BINKLEY BROTHERS' DIXIE CLODHOPPERS – An early Grand Ole Opry string band that was severely under-recorded, they appeared on Nashville radio well into the mid-1930s, but only recorded during Victor's field trip to Nashville in 1928.

BILL BLACK COMBO – A rock'n'roll band formed in 1959 by bassist Bill Black, who was a regular sideman with Elvis Presley. They enjoyed pop success in the late '50s and the '60s. Black died of a brain tumor in 1965, but the band continued under the leadership of Bob Tucker, eventually edging towards country. By 1975 they cut their first country album, **Solid And Country**, with one track, **Boilin' Cabbage**, becoming the biggest instrumental of the year.

BLACKWOOD BROTHERS – The most famous gospel quartet in country music, formed in Mississippi in 1934. Originally comprising Roy, Doyle, James and Roy's son R.W., they regrouped after R.W. died in a 1954 plane crash, J. D. Sumner becoming the group's bass singer. All the original Blackwoods have died but the group continues on its highly successful way.

JACK BLANCHARD & MISTY MORGAN – A husband and wife team, both born in Buffalo, New York, they played mainly big band music prior to moving to Nashville in 1967, when Jack started producing records, songwriting and working as a newspaper cartoonist. He came up with the chart-topping **Tennessee Bird Walk** in 1969, which established the pair as a top-ranking act. Further hits came during the next six years, then the twosome faded from the scene.

BLASTERS – California-based rock/rockabilly band formed by Phil and Dave Alvin in 1979. They were brought up on Sun rockabilly and Chess R&B, and the band play an eclectic blend of roots music. Signed to Slash Records in 1981, their songs on albums **Non Fiction** and **Hard Line** dealt with the oppressed and under-privileged, combining downbeat lyrics with rock'n'roll excitement. Their **Long White Cadillac** is the story of Hank Williams' death.

BLUEGRASS – A development of traditional string band music formulated and popularized by Bill Monroe and his Blue Grass Boys. This music form was revitalized during the folk boom of the 1960s when a youthful element added touches of pop, the result being tagged 'New Grass'.

TONY BOOTH – From Tampa, Florida, a former member of Buck Owens' Buckaroos, Booth enjoyed consistent chart success throughout the '70s with his smooth honky-tonk style. His biggest hits came with **The Key's In The Mailbox** and **Lonesome 7-7203** in 1972.

JIMMY BOWEN – Once a hit-making rockabilly with Buddy Knox, Bowen became president of MCA Nashville in 1984 after a successful period as a Warner Brothers Nashville executive. By the early '90s he was heading up EMI's Nashville operations at Capitol-Liberty.

BOY HOWDY – A young four-man band formed in Los Angeles in 1991 comprising Jeffrey Steele (vocals, bass), Cary and Larry Park (guitars and vocals) and drummer Hugh Wright. Survivors of the tough Hollywood bar circuit, they play country music with a modernistic feel. Signed to Curb Records, their first album was **Welcome to Howdywood** in 1992.

KAREN BROOKS – A professional team roper and barrel racer on the rodeo circuit, Dallas-born Karen Brooks first established a reputation as a songwriter. She sang on the Austin, Texas scene in 1975, at the time married to singer-songwriter Gary P. Nunn. A few years later she moved to Nashville and signed with Warners, scoring several hits during the '80s. Later she joined up with Randy Sharp for a successful duet team.

CLARENCE 'GATEMOUTH' BROWN – A hugely talented black singer and multi-instrumentalist, Brown was born in Orange, Texas in 1924. Though often thought of as a bluesman, he plays great bluegrass and Cajun fiddle and has appeared at several international country festivals.

FRANK 'HYLO' BROWN – Guitarist and singer of impressive range who once worked with Flatt And Scruggs. His folksy style was well suited to bluegrass and his **Hylo Brown Meets The Lonesome Pine Fiddlers**, a Starday album reissued in 1985 by Gusto, is well worth hanging on to.

JANN BROWNE – California-based singer, born in Anderson, Indiana, in 1955 and raised in Shelbyville, Indiana. Her parents were professional square dancers and Jann was playing in local bands during her high school days. She moved to southern California in 1977, where she played the rough and rowdy honky-tonk circuit. From 1981 to 1983 she was the featured vocalist with Asleep At The Wheel. Later she worked sessions, including back-up vocals on Rosie Flores' LP. In 1988 she was signed to Curb Records, her sandpaper voice gaining her a hip, young country following.

ANITA BRYANT – Born in Barnsdall, Oklahoma in 1940, Anita was a former Miss Oklahoma (1958), who started out as a country singer. She had a million-seller with **Paper Roses** in 1960, the song that many years later was to prove a country hit for Marie Osmond. These days Anita is best known as a battler against gay rights.

WILMA BURGESS – Born in Orlando, Florida in 1939, Wilma attended Stetson University, Orlando, majoring in physical education, but opted for a music career. She moved to Nashville in 1960 and worked as a demo singer, eventually gaining a record deal with Decca. Specializing in dramatic ballads, Wilma enjoyed major chart success with **Baby** and **Misty Blue**.

T-BONE BURNETT – Born in Tokyo and raised in Fort Worth, Texas, Burnett entered the music business as a producer and sessionman. In 1980 he released his own album, **Truth Decay**, and has continued to record spasmodically since. He produced Elvis Costello's **King Of America**, and was half of the Coward Brothers with Costello.

DORSEY BURNETTE – A country rockabilly musician born in Memphis in 1932. Along with his younger bother Johnny, he helped form the Johnny Burnette Rock'n'Roll Trio in 1954. The two brothers wrote hit songs for Ricky Nelson, then embarked on solo careers. Dorsey scored some pop hits in 1960 and continued recording for several labels, producing minor country hits until he died of a heart attack in 1979.

TRACY BYRD – Born in Beaumont, Texas in 1967, but raised in nearby Vidor, Tracy is a Texas honky-tonk singer who performed in Texas clubs while studying at Lamar University, Beaumont. He replaced Mark Chesnutt at Cutter's night-club in Beaumont when the latter signed a Nashville record deal. Three years later, Tracy followed suit and in 1991 signed with MCA Records.

C

CACTUS BROTHERS – Combine an old-time seven-piece string band with rock'n'roll energy who can play everything from aged fiddle tunes to sweet soul classics, and there you have the Cactus Brothers, who played the Nashville clubs for several years prior to landing a Liberty Records contract in 1992. They appeared in the George Strait movie 'Pure Country', and have been described as "Nashville's answer to the Pogues". In their line-up is world-class dulcimer player David Schnaufer, who has several acclaimed solo albums to his credit.

CAJUN MUSIC – A product of Louisiana bayou area which merges jazz, blues, country and French folk music. Fiddle and accordion dominate with most vocals still in French.

J. J. CALE – A singer, songwriter and guitarist born in Oklahoma City in 1938, Cale's lazy country-blues, laid-back singing and understated guitar work have gained him a cult following and have also proved influential, most noticeably in much of Eric Clapton and Mark Knopfler's style. His songs have been well-covered in country music and his albums have a wonderful timeless country appeal.

STACY DEAN CAMPBELL – Born in New Mexico and raised in Texas and Oklahoma, Campbell has that modern-day James Dean glow about him that spills over into his music, which owes much to '50s rock'n'roll and rockabilly. He started out with a local Texas outfit, the Nickels, playing Southwest clubs before gaining Nashville attention in 1990. Signed to Columbia Records in 1991, Stacey Dean is on the threshold of a major career.

JUDY CANOVA – A sort of slicked-up Minnie Pearl of the 1940s, Judy Canova, born in Jacksonville, Florida in 1916, was a country comedienne successful on Broadway and in some two dozen movies. She also had a long-running radio show (1943–53) and recorded regularly into the 1950s.

CANYON – A Texas-based band featuring Steve 'Coop' Cooper on lead vocals, Canyon played the bar-room circuit for a number of years before landing a record deal with the Nashville-based 16th Avenue Records in 1987. With their smoothly executed country-rock sound, they scored a number of minor hits without breaking through to the Top 20.

WAYNE CARSON – A versatile songwriter, singer and guitarist, Wayne Carson Thompson, born in Denver in 1946, played in rock'n'roll and R&B bands during high school days, and for a time played lead guitar for Red Foley in the early '60s. He toured for two years with the All Star Hoot Nanny Road Show, then in 1967 he moved to Nashville, where he made an impact as a songwriter and session player. He has penned more than 100 country hits, his best-known song, **Always On My Mind**, was a CMA winner in 1982. Carson has recorded for Monument, Elektra and EMH without too much commercial success.

WILF CARTER – A pioneer of Canadian country music, singer-songwriter-guitarist Wilf Carter, a one-time Canadian cowboy, was born in Nova Scotia in 1904. He branched out from rodeo to radio in the early 1930s. He signed with RCA Victor at the same time and has sustained a recording career for more than 50 years, cutting sides for Bluebird, Decca, Starday and RCA, resulting in more than 40 albums. One of the many yodelling singers influenced by Jimmie Rodgers, he adopted the guise of 'Montana Slim' while working on a New York CBS radio show.

JOHNNY CARVER – A pop-styled country singer who started out in a gospel group with two aunts and an uncle, Carver was born in Jackson, Mississippi in 1940. The featured singer in the house band at the famed Palomino Club in California, he started scoring minor country hits in the late '60s, eventually making it into the Top 10 in the '70s. The warm-voiced Carver specialized in covers of pop hits such as **Tie A Yellow Ribbon** and **Afternoon Delight**.

PETE CASSELL – Cassell was a long-popular blind singer with a fine, smooth voice which anticipated the likes of Jim Reeves. Most popular in the '40s and early '50s, Pete recorded for Decca, Majestic and Mercury, but died all too young in 1953, a few years too early to cash in on the smooth Nashville Sound for which his voice was so well suited.

Hylo Brown Meets The Lonesome Pine Fiddlers. Courtesy Starday Records.

CONNIE CATO – Born in Carlinville, Illinois in 1955, Connie signed to Capitol at 17. This singer's big moment came in 1975 when her version of **Hurt** (one of her three hits that year) went Top 20. She was once known as 'Superskirt' and 'Superkitten' following 1974 hits of those titles.

BRYAN CHALKER – Excellent, deep-voiced British singer and guitarist who made several above-average albums, but became disenchanted with the UK club scene and eventually opted to become a country DJ on a Bristol radio station. For a while, Chalker also published one of Britain's leading country music magazines.

JEFF CHANCE – From El Campo, Texas, vocalist-guitarist-fiddler Jeff Barosh, along with his drummer brother Mick Barosh, formed Texas Pride. In 1985 the five-piece band changed its name to Chance and, signed to Mercury, enjoyed some notable country chart success. In 1988 Jeff Chance embarked on a solo career. Though his rugged good looks and dynamic stage show should have led to instant success, his breakthrough is still waiting to happen.

MARSHALL CHAPMAN – From Spartanburg, South Carolina, where she was born in 1949, Chapman moved to Nashville in 1973 and made an impact as a songwriter. Signed to Epic Records three years later, she came on tough and rocky as she performed in a slow, bluesy powerhouse way. Her best-known number, **Somewhere South Of Macon**, scraped into the country charts in 1977, but she was too different or ahead of her time, and Epic let her go.

THE CHIEFTAINS – Formed in the early '60s and one of the most influential and long-lived Irish folk bands, the Chieftains helped to spearhead a revival of Irish traditional music, and they have built up an international following. In 1992 they made a country music connection, appearing in concert in Nashville and recording **Another Country**, an album of country songs featuring guest stars Ricky Skaggs, Chet Atkins, Emmylou Harris, Don Williams, the Nitty Gritty Dirt Band and several other Nashville musicians.

LEW CHILDRE – A veteran of vaudeville. 'Doctor Lew', who was born in Opp, Alabama in 1901, was a Hawaiian guitar player and sophisticated rural comedian. A regular on the Opry from 1945 until shortly before his death in 1961, he worked with Stringbean between 1945–48 and was also an adept singer who sang old standards like **I'm Looking Over A Four Leaf Clover**.

CIMMARON – A sextet from Roanake, Virginia, Cimmaron were formed in 1979 by lead vocalist Bobby Smith. They play a rich blend of country-rock, honky-tonk and high-powered neo-traditional with superb country picking and bluegrass-flavoured vocal harmonies. Having released an album on their own labels, they signed with Alpine Records in 1992, immediately making the CMT and TNN video rotations with such numbers as **Can't You Just Stay Gone** and **Detroit Diesel**, which quickly generated its own country dance.

YODELLING SLIM CLARK – Born Raymond LeRoy Clark in Springfield, Massachusetts in 1917, Yodelling Slim worked for a long period as a woodman. Winner of the World Yodelling Championship in 1947, he began recording for such labels are Continental, Remington, Wheeling and Palomino, his albums including **Yodel Songs** (1960) and **Jimmie Rodgers Songs** (1965).

LEE CLAYTON – A product of the Nashville 'underground', Clayton, born in Russelville, Alabama in 1942, also played it tough and rough. Much of the time he was over-produced when all he really needed was his voice and his songs. He wrote **Ladies Love Outlaws**, a biggie for Waylon Jennings, and recorded several albums for Capitol during the '70s.

PAUL COHEN – A Decca talent scout, Cohen (born in 1908, died in 1970) was the first to recognize the potential of Nashville as a recording centre. He was elected to the Country Music Hall Of Fame in 1976.

Black And White Photograph, Corbin/ Hanner. Courtesy PolyGram Records.

B. J. COLE – Unquestionably the most successful sessionman in British country music, steelie Bryan Cole has played on hundreds of pop, rock and country records. He also formed his own label, Cow Pie.

COLLINS KIDS – Lorrie and Larry Collins from Oklahoma were a country-pop duo signed to Columbia Records in 1955, when they were still in their early teens. They recorded classic rockabilly singles and were regulars on Town Hall Party in Los Angeles. They split in 1964 and Larry became a successful country songwriter – **Delta Dawn** and **You're The Reason God Made Oklahoma**.

COLORADO – A five-piece band from Sutherland, Scotland. One of the most respected British bands, they have toured with many US stars, including Jean Shepard, Boxcar Willie, Melba Montgomery and Vernon Oxford. Voted Top British Country Group many times during the '80s, Colorado changed direction to encompass their Scottish roots music and have now become known as Caledonia.

COMPTON BROTHERS – Harry and Bill Compton, from St Louis, Missouri, won a 1965 talent contest that resulted in a Columbia recording contract. However, they later became associated with the Dot label, supplying minor hits during the late '60s and early '70s.

BILL CONLON – An Irish balladeer originally from Portaferry, County Down, but since 1987 based in North London. Bill Conlon has made a big impact throughout Europe with successful appearances in Germany, Switzerland and Holland, plus several albums which put him on par with some of Nashville's top country vocalists. He always surrounds himself with good musicians and his recordings have gained extensive BBC Radio 2 plays, with **I Don't Have Far To Fall** being chosen as Record Of The Week in 1991.

MERVYN CONN – The producer of the prestigious series of Wembley Country Festivals plus numerous European tours featuring just about every worthwhile name in country music. Often criticized, Conn is, nevertheless, the man who has done most to popularize country sounds in Britain.

CORBIN-HANNER BAND – Bob Corbin and Dave Hanner, from Ford City, Pennsylvania, met while at high school. They formed the Gravel Band and worked the Pittsburgh clubs. Later they moved to Nashville where they recorded for numerous labels from 1978, finally making an impact when they linked up with Mercury Records as Corbin-Hanner in 1990.

ELVIS COSTELLO – Born Declan McManus, Elvis changed his name and built up a strong following after an unsuccessful stint in a country-rock band. The Irish-born pop star fulfilled a life-long ambition when he appeared on the Grand Ole Opry in 1982 during an American tour. An ardent admirer of George Jones, Elvis took his band, the Attractions, to Nashville to record **Almost Blue**, an album of country standards produced by Billy Sherrill in 1980. He scored a British chart-topper with an update of George Jones' **A Good Year For The Roses**, and recorded a duet with Jones of **A Stranger In The House**.

COUNTRY MUSIC ASSOCIATION – The CMA is a trade association formed in 1958 by a cadre of businessmen, artists and DJs to further the cause of country music. The CMA Awards, which take place in Nashville each October, are the most prestigious in country music, the premier accolade being Entertainer Of The Year.

COUNTRY MUSIC FOUNDATION – A non-profit organization which operates Nashville's Country Music Hall Of Fame and Museum, plus the Country Music Foundation Library, etc. It sums up its goals as being "dedicated to the study and interpretation of country music's past through the display of artifacts and the collection and dissemination of data found on discs, tape, film and in printed material".

COUNTRY MUSIC HALL OF FAME – Country music's major shrine, based in Nashville, in which the greats of the past and present are honoured for their contributions to their chosen form of music. Two or three members are elected annually. Jimmie Rodgers, Bob Wills, Eddy Arnold, Roy Acuff, Kitty Wells, Minnie Pearl, Hank Williams, Uncle Dave Macon, Tex Ritter and Jimmie Davis are among those who have a commemorative plaque and portrait in the Hall, which is actually part of a larger museum.

COUNTRY-ROCK – A trend that grew out of late 1960s West Coast rock, particularly around LA. Gram Parsons can be credited as acting as a catalyst to many of the part-time mandolin players and steel guitarists who were scraping a living in rock-oriented California, while Bob Dylan also helped to encouraged the trend with his **Nashville Skyline** album, also leaking the fact that Hank Williams had always been one of his favourite singers.

BRENDAN CROKER – Yorkshire-born singer-songwriter, Brendan Croker started out in the late '60s playing country-blues in Yorkshire pubs with Steve Phillips and Mark Knopfler. A few years later he formed his own group, the Five O'Clock Shadows. Then in 1989 he reunited with Phillips and Knopfler in the Notting Hillbillies. A trip to Nashville in 1991 saw him recording with top Nashville pickers for the solo LP **The Great Indoors**. Wynonna Judd included his **What It Takes** on her first solo album.

HUGH CROSS – One of country music's earliest professional entertainers, Hugh Cross, born in eastern Tennessee in 1904, joined a medicine show at 16 and by the mid-1920s was a popular singer on radio and record. A guitarist, banjoist and songwriter, he joined the Cumberland Ridge Runners on the National Barn Dance from 1930–33, then he struck out on his own, appearing on several radio stations, eventually drifting into executive capacities.

J. C. CROWLEY – A songwriter, guitarist and singer from Galveston Bay, Texas, Crowley was a member of the pop group Player during the late '70s. He co-wrote their No.1 pop hit **Baby Come Back**. Moving towards country, he signed with RCA Nashville in 1988, making Top 20 on the country charts with **Paint The Town And Hang The Moon Tonight**.

MAC CURTIS – A hillbilly singer who became a leading rockabilly with the King label, Curtis (from Olney, Texas) still cuts fine country sides from time to time. During the mid-1970s he produced some of Ava Barber's sides for Ranwood.

Beneath The Texas Moon, J.C. Crowley. Courtesy RCA Records.

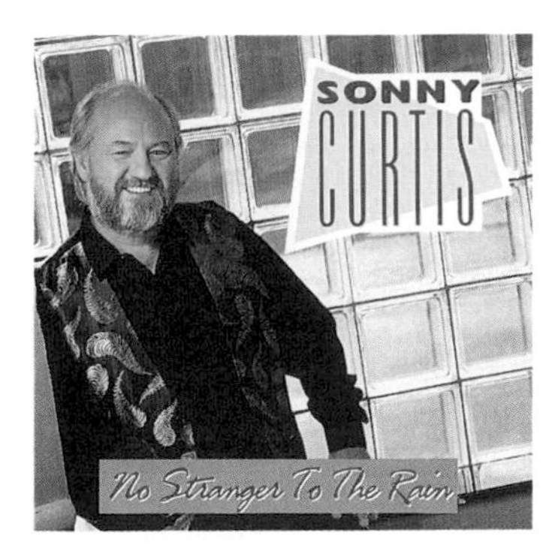

No Stranger To The Rain, Sonny Curtis. Courtesy Ritz Records.

SONNY CURTIS – Born in Meadow, Texas in 1937, singer, songwriter and guitarist Sonny Curtis will always be closely associated with Buddy Holly and the Crickets. He worked with Holly in 1956, then left to join Slim Whitman's group. A move to the West Coast in 1959 saw Curtis in the Crickets and penning pop/country hits for the Everly Brothers, Bobby Vee, Buddy Knox and others. Working both as a solo and occasional member of the Crickets, Curtis has recorded prolifically for various labels, making the country Top 20 with **Good Ol' Girls** in 1981. He has also penned such country hits as **I'm No Stranger To The Rain** (Keith Whitley) and **He Was On To Something** (Ricky Skaggs).

D

DICK DAMRON – A Canadian singer-songwriter, born in Bentley, Alberta, Damron is best known for his song, **Countryfied**, which became a US hit for George Hamilton IV in 1971.

DAVIS DANIEL – With his blonde hair, youthful features and light-hearted attitude, Davis Daniel seems like he could have walked straight off a California beach, surfboard in hand. In fact, he grew up in Illinois, lived for a while in Nebraska and Denver, before moving to Nashville in 1987. His real name is Daniel Davis, but he changed it to avoid confusion with Danny Davis and the Nashville Brass. He was driving a truck for Miller Beer Company when he landed a recording contract with Mercury in 1990. With his southern-flavoured vocal punch he made the Top 20 with **Crying Out Loud** in 1991.

JOHNNY DARRELL – Once a motel manager, Darrell (born in Cleburne County, Alabama, 1940) cut the original version of **Green, Green Grass Of Home** and also had the first hit renditions of **The Son Of Hickory Hollers Tramp** (1967) and **With Pen In Hand** (1968).

PAUL DAVIS – Born in Meridian, Mississippi in 1948, this singer-songwriter first made an impact as a contemporary country act with the self-penned **Ride 'Em Cowboy**, a pop hit on Bang Records in 1975. He enjoyed further pop success, then started to make an impression as a country writer. Based in Nashville he recorded chart-topping duets with Marie Osmond and Tanya Tucker and by the late '80s had established himself as an in-demand producer.

LAZY JIM DAY – Born in Creek, Kentucky in 1911, died in 1959, Day was one of the early stars of country radio. A singer, banjoist and guitarist, he originated the singing news routine and by the late 1930s was the leading comedian on the Opry.

EDDIE DEAN – A cowboy film star of the '40s, Edgar Dean Glossup was born in Posey, Texas in 1907 and began his musical career as a gospel singer in the early '30s. His older brother Jimmy appeared on the WLS National Barn Dance in Chicago in 1936 then they headed west to try their luck in films. Eddie joined Judy Canova's Radio Show and appeared in scores of films before gaining his own series of 20 films with PRC from 1946–48. He recorded prolifically for Decca, Mercury, Sage And Sand and Capitol, his biggest hits, **One Has My Name, The Other My Heart** (1948) and **I Dreamed Of A Hillbilly Heaven** (1955) are two genuine country classics which he co-wrote.

MARTIN DELRAY – A guitarist-singer from Texarkana, Arkansas, Martin Delray spent years working the clubs and honky-tonks before landing a recording contract with Atlantic Records in 1990. A hard country stylist with '50s rockabilly overtones, he tempted Johnny Cash along to the studios to add vocals to an update of his **Get Rhythm** in 1991.

IRIS DeMENT – A traditional-sounding singer-songwriter, Iris was born in Paragould, Arkansas in 1961. She grew up in a musical household in California. Her father played fiddle, mother sang and her elder sisters formed a group, the DeMent Sisters. In 1984 Iris was performing at folk clubs in Kansas City, later moving to Nashville where she sang on sessions for Emmylou Harris, Jann Brown and Nanci Griffith. Signed to Philo/Rounder Records, Jim Rooney produced her debut album, **Infamous Angel** (1992), which gained good reviews, but the traditional lilt of the music and Iris's hill-country vocals gained virtually no radio plays in America. British DJ Wally Whyton played a track on his BBC Radio Country Club programme and was inundated for requests for more. Warner Brothers in Nashville picked up her contract and re-promoted the album in 1993.

Infamous Angel, Iris DeMent. Courtesy Warner Bros Records.

JAMES DENNY – A one-time mail clerk, Denny was born in Buffalo Valley, Tennessee in 1911 and died in Nashville, 1963. He worked his way up the ladder to become talent director at Nashville's WSM radio. He also ran a booking agency business, at one point handling over 3,200 personal appearances throughout the world. He was elected to the Country Music Hall Of Fame in 1966.

KARL DENVER TRIO – This British country trio achieved pop group-styled stardom in 1961 when their Decca single, **Marcheta**, went Top 10 in the UK charts, to be followed by **Mexicali Rose** (1961), **Wimoweh** and **Never Goodbye** (both 1962). Two of the original threesome still work with the group.

SYDNEY DEVINE – Scotland's most successful country singer and a would-be Elvis, his **Doubly Devine** double-album on Philips went Top 20 in the UK during 1976, some 21 years after Devine first began his onstage career.

DIXIANA – From the heartlands of Greenville, South Carolina, Dixiana was formed in 1986. Lead singer Cindy Murphy was a longtime member of bluegrass outfit the Wooden Nickel Band, while brothers Mark and Phil Lister, as the Listers, hosted their own regional TV show in the mid-1970s. The other two members are Randall Griffith and Colonel Shuford. Signed to Epic Records in 1991, Dixiana have a hip image and traditional-flavoured contemporary country style, which ensures a bright future.

DOTTSY – For a while Dottsy Brodt (born in Seguin, Texas, 1953) swept all before her. A talent show winner at 12, by 14 she had her own weekly TV series, San Antonio. After quitting college in 1972 she appeared at Happy Shahan's Alamo Village in Bracketville, Texas, playing five shows a day. Later came tours with Johnny Rodriguez, and in 1974 she signed with RCA Records. Cute and country, petite and sweet, Dottsy seemed set to become a superstar. She had several major hits, the biggest being **(After Sweet Memories) Play Born To Lose Again** in 1977. Four years later she packed the music in to work with autistic and retarded children.

RUSTY DRAPER – Born Farrell Draper in Kirksville, Missouri, Rusty worked radio stations in Tulsa, Oklahoma, Des Moines, Iowa and Quincy, Illinois during his teens, then became singing MC at a club in San Francisco. A record deal with Mercury in 1951 resulted in his biggest hit, **Gambler's Guitar**, a pop and country Top 10 entry in 1953. He enjoyed further pop success in '50s, gaining several gold discs, but never made a major impact on the country charts, though he has recorded regularly in Nashville.

MARY DUFF – From Lobinstown, County Meath, Irish beauty Mary Duff has been one of Europe's top female country vocalists since linking up with Ritz Records in 1988, the following year winning the Euro Country Music Masters in Switzerland. Six years earlier she had started appearing with John Collier and New Dimension and Irish chart group Jukebox. Winning the Cavan Song Contest led to her signing with Ritz Records, tours with Daniel O'Donnell and her own bill-topping shows.

DUFFY BROTHERS – Ray and Leo Duffy, a British comedy duo from Peterborough, England, won the Marlboro Country Music competition and then went on tour with Marty Robbins and Tammy Wynette. Since then they have been voted Best British Country Duo for three years in a row, and also toured the word for two years as part of the Charley Pride road show.

JOHNNY DUNCAN – John Franklin Duncan from Oliver Springs, near Knoxville, Tennessee (where he was born in 1931), is better known in England, where he had a big pop hit with **Last Train To San Fernando** in 1957, and Australia, where he has lived since 1974. While in the US army he married an English woman and moved to England in 1955. Two years later he formed the Blue Grass Boys, in homage to Bill Monroe, and promoted as a skiffle band. He appeared on shows such as 6.5 Special and enjoyed several British pop hits in the late 1950s.

SLIM DUSTY – Australia's top-ranking country singer for four decades, Slim Dusty, born David Gordon Kirkpatrick in Dempsey, NSW, Australia in 1927, has sold more locally made records in his own land than any other artist. A champion of frontier ballads and music of the bush country, he has fought for survival since the beginning of his career. He rose to worldwide prominence with **A Pub With No Beer** (1958) and was made an MBE in 1970. He was still actively recording in the early 1990s.

BOB DYLAN – Born May 24, 1941, in Duluth, Minnesota, Dylan initially made the country connection through songwriting, with Johnny Cash and June Carter scoring his **It Ain't Me Babe** in 1964. He made two albums in Nashville – **John Wesley Harding** and **Nashville Skyline**, the latter providing his last hit single, **Lay Lady Lay**, in 1969. His music continues to touch on country, and many country performers have recorded his songs. In 1992 he co-wrote and duetted with Willie Nelson on their song about the plight of farmers, **Heartland**.

E

RAY EDENTON – A frequent winner of the NARAS Superpicker Band Rhythm Guitar award, Ray was born in Mineral, Virginia. He began playing professionally in 1946 and from 1952 to 1962 played on the Grand Ole Opry along with road shows for several stars. He is now a sought-after Nashville sideman.

DON EDWARDS – A pure cowboy singer, Edwards is America's modern cowboy troubadour. He performs authentic cowboy songs and writes modern classics full of the wit and wisdom of the West. He has been performing around the Southwest since the early 1960s and was signed to the new Warner Western label in 1992.

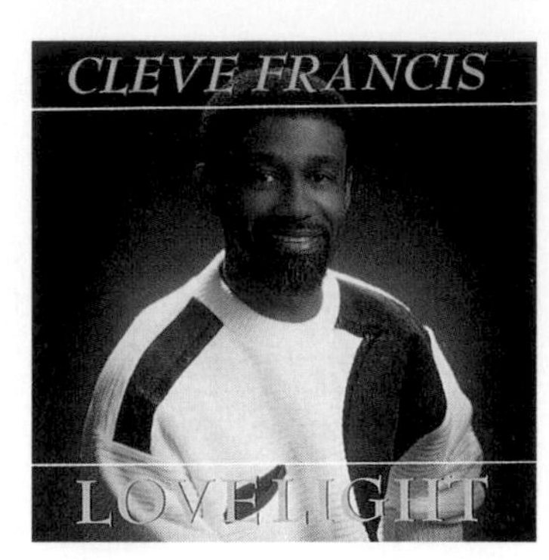

Lovelight, Cleve Francis. Courtesy Cottage Records.

JONATHAN EDWARDS – From Minnesota, where he was born in 1946, Edwards found success on the pop charts with **Sunshine**, a Top 20 entry in 1972. Labelled as part of the singer-songwriter genre, he always leaned towards country, having formed a bluegrass band, Sugar Creek, in 1965. He cut one of the first versions of **Honky Tonk Stardust Cowboy** in 1971. With his smooth tenor and delicate songs, he has record regularly for such labels as Atco, Atlantic and MCA, scoring some minor country hits in the late '80s.

RAMBLING JACK ELLIOTT – Born in Brooklyn, New York in 1931, Rambling Jack became known as an idealistic, bohemian folk singer, and a travelling troubadour. He flunked college to live in Greenwich Village, where he met Woody Guthrie, a friendship that proved fruitful in gaining him recognition. He played the Newport Festival in 1963, and continues to play colleges, coffee houses and folk festivals. He has been well received in Europe and enjoyed TV exposure and successful club bookings.

DARRYL AND DON ELLIS – This brother duo grew up in tiny Beaver Falls, Pennsylvania. Raised on a traditional country diet, they have had their own bands since Darryl was 15 and Don was 11. They made a move to Nashville in 1987, their airtight harmonies and insightful songwriting leading to a record deal with Epic four years later.

F

CHARLIE FEATHERS – An early Sun rockabilly artist, Feathers is still active in a family group in the Memphis area. Born in Hollow Springs, Mississippi, in 1932, he recorded for Flip, Sun, King, Kay, Memphis, Holiday Inn and Elektra Records, and is thought of as an artistic influence who somehow never found the right rockabilly record.

DICK FELLER – A singer-songwriter with a humorous cutting edge, Feller was born in Bronaugh, Missouri in 1943. A talented guitarist, he moved to Nashville in 1966, where he found work playing sessions or on the road with stars such as Mel Tillis and Warner Mack. He made an impact as a writer of hits for Johnny Cash and Jerry Reed, then came recordings with United Artists and Asylum in the '70s and such country hits as **The Credit Card Song**. A brilliant club performer, Dick Feller has built up a cult following in Europe.

DICK FORAN – Possessed of a fine voice – which sounded more at home on the Broadway stage than on the range – Foran was born in New Jersey in 1910, the son of a US Senator. He aspired to a career on the stage and later on film, but despite his singing cowboy films for Warner and Universal, he was also quite successful in high budget westerns and in other types of films. He retired at the start of the 1970s and lived in California up to the time of his death in 1979.

GERRY FORD – One of Scotland's most popular country performers, Gerry Ford was actually born in Athlone, County Westmeath, Eire in 1943. He moved to England in 1959, a few years later settling in Edinburgh where he joined the police force. A knowledgeable country music fanatic, he started singing seriously in the late '60s and has recorded prolifically since 1977, mainly in Nashville, where he is held in high esteem. Gerry is also a popular and successful DJ on BBC Radio Scotland.

LLOYD DAVID FOSTER – Texan singer-guitarist born in Wills Point in 1952, Foster played clubs on weekends, while driving a beer truck in Dallas by day. His first recordings were made for small Texas labels, then came stints with MCA and Columbia during the early 1980s, but that major breakthrough seems to have eluded him.

WALLY FOWLER – Cheerful, gladhanding Wally Fowler was born in Bartow County, Georgia in 1917. He first achieved success as a singer and songwriter, leading an Opry band called the Georgia Clodhoppers. He turned to gospel music, forming the Oak Ridge Quartet, forerunner of the Oak Ridge Boys, in the late '40s. In later years he turned his hand to gospel show promotion.

CLEVE FRANCIS – A practising cardiologist in Washington DC, hailing from Jennings, Louisiana, Cleve sang in a black gospel group during high school days. During medical college he sang in coffee and road houses, and recorded three independent label albums before making a breakthrough in 1992 when he was signed to Capitol/Liberty. Initially it was a video on CMT and TNN that brought Cleve Francis recognition and his first country hits.

J. L. FRANK – One of the great promoters of country music, Frank was born in Rossai, Alabama in 1900 and died in Detroit, 1952. Known as the 'Flo Ziegfeld of Country Music', he was instrumental in furthering the careers of Gene Autry, Roy Acuff, Ernest Tubb and many others. He was elected to the Country Music Hall of Fame in 1967.

RAYMOND FROGGATT – A distinctive British country singer-songwriter who started out as a pop writer, penning **Red Balloon**, a massive hit for the Dave Clark Five in 1968. A gritty, still slightly rock-oriented performer from Birmingham, he inspires fan adulation that is rare in British country music. He tours regularly, and performs his own, often excellent, material.

G

GEORGIA SATELLITES – An Atlanta-based rock band, the Georgia Satellites draw from a rich traditional American heritage with their guitar-heavy roots-type music. Influenced by a mixture of Dylan, Chuck Berry, George Jones and even ZZ Top, their music, which rocks with a fervour, made a commercial breakthrough in the late 1980s.

TERRI GIBBS – Born blind (in Augusta, Georgia, 1954), she was once a member of a group called Sound Dimension. A bluesy singer and pianist, Terri formed her own group in 1975 and gained a regular gig at Augusta's Steak and Ale House,

Where There's Smoke, Gibson/Miller Band. Courtesy Sony Music.

performing 50 songs a night. Signed to MCA in 1980, she came up with a Top 10 single **Somebody's Knockin'** later that year and won the CMA Horizon Award in 1981. For the next three years she maintained a steady flow of chart records without ever gaining the hit she undoubtedly deserved.

GIBSON-MILLER BAND – Five-man band led by Nashville songwriter Dave Gibson and Detroit rock guitarist Blue Miller, along with Bryan Grassmeyer, Steve Grossman and Mike Daly. They play what has been described as 'turbo-twang country'. Gibson is a noted Nashville songwriter with hits for Alabama, Tanya Tucker and Joe Diffie to his credit. Miller is a one-time guitar player with Bob Seger, while the other members have worked in bands with Vince Gill, Suzy Bogguss and Sweethearts Of The Rodeo. Signed to Epic Records in 1992, they have brought varied country and rock musical elements together for an exciting '90s country sound.

BRIAN GOLBEY – Sussex-born British singer, songwriter and multi-instrumentalist, Brian Golbey is one of Britain's finest country music entertainers. He first came to prominence in the mid-1960s at the Folk Voice conventions. He often teams up with banjo player Pete Stanley to play bluegrass, but is at his best with his own self-penned songs. He recorded many albums as a soloist, and in the mid-1970s was a member of folk-rock trio Cajun Moon for a short time.

TONY GOODACRE – A British perennial who started out on a Carrol Lewis talent show in 1957. Like many others, he has indulged in 'cover' albums of US hits (**Thanks To The Hanks**, **Roaming Round In Nashville**) but redeemed himself with **Written In Britain**, a release containing all British material by Terry McKenna and Pete Sayers.

GRAND OLE OPRY – The greatest show in country music, the Opry has been broadcast on Saturday nights over WSM since 1927 (though it formerly ran as the WSM Barn Dance from 1925). The show has survived many changes of location, first being housed in a WSM studio, then in a larger studio, then in the Hillboro Theater, before moving to the Dixie Tabernacle and, for a short spell, to the War Memorial Auditorium, before settling down for over 30 years at the Ryman Auditorium. It is today situated in the modern Grand Ole Opry house in the grounds of a huge amusement complex.

CLAUDE GRAY – A popular singer, guitarist and bandleader who was born in Henderson, Texas in 1932, he enjoyed a spate of best-selling singles in the early '60s when signed to Mercury Records. With his deep voice and impressive stature he had his biggest hit with **I'll Just Have Another Cup Of Coffee** in 1961, but was still charting as recently as 1982.

OTTO GRAY – Influential bandleader from Oklahoma, Gray organized the Oklahoma Cowboys, one of the very early professional country bands, in 1924, and the band lasted through to the mid-1940s. They presented a slick, very rehearsed show and achieved their greatest success in the Northeast, particularly over WGY in Schenectady, New York. They recorded for Gennett, Vocalion and Okeh.

GREAT PLAINS – This Minnesota/Oregon quartet, high on rock energy and cool with country twang, first got together as a band through playing with Michael Johnson in 1986. Jack Sundrud (lead singer, guitar), Russ Pahl (lead/steel guitar), Danny Dadmun Bixby (bass) and Michael Young (drums) were all impeccable musicians with credits on albums and tours by George Jones, Mary-Chapin Carpenter, etc. Signed to Columbia Records in 1991, they were just starting to get established when Pahl and Young left in May 1993, leaving the band's future in doubt.

RICKY LYNN GREGG – A former rock singer from Longview, Texas, long-haired Gregg built up a regional following in the Dallas/Fort Worth area. In 1983 he formed the Ricky Lynn Project and started touring. Three years later he joined rock band Head East as lead singer and wrote most of their 1989 album **Choice Of Weapons**.

Above: The Georgia Satellites mix up a popular brand of country-rock.

Then he moved towards country and formed a new outfit, Cherokee Thunder, in 1990, blending his rock roots with a solid country base. Jimmy Bowen signed him to Liberty Records in 1992.

CLINTON GREGORY – A vocalist and fiddle player from Martinsville in the backwoods of Virginia, Clinton Gregory is the son of a champion fiddle player and bootlegger. A former member of Suzy Bogguss' back-up band, he embarked on a fully fledged solo career signing with Step One Records in 1990 and scoring several country hits, the biggest being **(If It Weren't For Country Music) I'd Go Crazy**.

REX GRIFFIN – A popular singer, guitarist and songwriter best known for his composition **The Last Letter**, Rex (born in 1912) became popular over WSB in Atlanta and as host of the KRLD Texas Roundup. Probably the most fascinating aspect of his career was that he recorded **Lovesick Blues** for Decca, a record which went nowhere; nearly a decade later Hank Williams recorded Griffin's version identically, with tremendous success. Plagued with ill-health due to a lifelong drinking problem, Griffin died in 1959.

LEWIS GRIZZARD – Looking like a cross between Groucho Marx and Ernie Kovacs, Grizzard is a celebrated columnist, comedian, author and social commentator. With his southern-drenched humour he hits on such themes as southern living, religion and sex. Signed to Columbia Records he has been named country comedian by the CMA and is a popular concert performer.

ARLO GUTHRIE – Eldest child of Woody Guthrie, singer, guitarist and songwriter Arlo was born in Coney Island, New York in 1947. Though, broadly speaking, a folkie, he has always worked closely with country. His 1972 Top 20 pop hit was with Steve Goodman's **City Of New Orleans**, and the album **Last Of The Brooklyn Cowboys** (1973) featured country songs backed up by Buck Owens' Buckaroos.

H

THE HACKBERRY RAMBLERS – An early and influential Cajun band whose records, both in English and Cajun-French, helped win the music a wider audience in the '30s. In their prime they recorded mainly for Bluebird. They disbanded in 1939, but in the '60s leader Luderin Darbone reformed the group for appearances at folk festivals and weekend dances in local taverns.

THE HAGERS – Identical twins born in Chicago, Jim and John signed to Capitol in 1969 and notched minor hits through to 1971. Versatile and equipped with ready-to-please comedy routines, they made many appearances on Hee Haw, then headed for Hollywood.

MONTE HALE – Singing cowboy Monte Hale, born in San Angelo, Texas in 1921, starred in some 19 Republic westerns from 1945–51, making him one of the last singing cowboys in chronological terms. Although possessed of a strong, smooth voice, his records were not particularly successful. He toured as a singer with rodeos before bowing out of musical and acting careers while still a young man.

THERON HALE AND DAUGHTERS – Theron Hale (1883–1954) led one of the most interesting and popular of the early Opry bands from 1926 until the early 1930s. Unlike most of the raucous hoedown bands, their music was gentle and reminiscent of parlour music of the preceding century, highlighted by lovely twin fiddling.

THE HALEYS – Sisters Jo-Ann and Becky Haley from the West Yorkshire village of Harden, England, were still teenagers when they became professional country singers, initially as part of a trio, Applejack, in 1989, then as self-contained duo, the Haley Sisters, the following year. The British Country Music Association voted them Top British Duo in 1991 and 1992. Performing a wide variety of material, both girls share lead and harmony vocals. In 1993 they formed a back-up band and became known as the Haleys.

GEORGE HEGE HAMILTON V – Son of George Hamilton IV, Hege was born in Nashville in 1960 and has always been surrounded by country music. While attending the University of North Carolina he played in several college rock bands. Then he linked up with his father as guitarist and back-up vocalist in a package tour in 1983 that included Faron Young, Leroy Van Dyke and Dave Dudley, followed by many Opry appearances. In 1987 he landed a record deal with MTM Records. His album, **House Of Tears**, gained rave reviews as the lanky youngster took off on an epic promotional tour in a '65 Cadillac hearse. His folkabilly single, **She Says**, made the country charts in early 1988, then MTM closed down, leaving the talented singer-songwriter high and dry. He has since built up a cult following on the British circuit with country band Fever.

GUS HARDIN – A gutsy singer, Gus spent 11 years singing in Tulsa clubs. One-time mentor Leon Russell described her voice as "a cross between Otis Redding, Tammy Wynette and a truck driver". Signed to RCA in the early 1980s, she began logging a tally of Top 40 singles, gaining Top 10 records with **After The Last Goodbye** (1983) and **All Tangled Up In Love**, a duet with Earl Thomas Conley (1984).

JONI HARMS – Born and raised in Canby, Oregon, Joni Harms is a hometown beauty who won the Miss Northwest Rodeo title in 1979. Performing in clubs since a teenager, this talented singer-songwriter finally made the Nashville connection in 1988, when Jimmy Bowen signed her to the short-lived Universal Records. Her contract was switched to Capitol and she chalked up a sizeable hit with **I Need A Wife** in 1989.

ALEX HARVEY – Successful songwriter and quality soulful country vocalist, Alex Harvey was born in Brownsville, Tennessee, 1945. He obtained a degree in music at Murray State University, Kentucky, moved to Nashville in 1966 and established himself as a songwriter with such hits as **Delta Dawn**, **Reuben James** and **Tulsa Turnaround**. He gained a Capitol recording contract, but has had most success as a songwriter. His songs have been recorded by such stars as Helen Reddy, Tanya Tucker, Waylon Jennings and others.

HEE HAW – A syndicated TV show established in the summer of 1969. Full of cornporn humour supplied over the years by such funny men as Archie Campbell and Junior Samples plus musicians such as Buck Owens and Roy Clark, the show has continued to be highly popular, despite – or maybe because of – its total lack of anything that seems in the least sophisticated.

KELVIN HENDERSON – British, Bristol-born (1947) and -based singer, bandleader and radio broadcaster, Kelvin has been winning polls in the UK and various European countries for many years. He recorded several acclaimed albums, the best being **Black Magic Gun** in 1977, which perfectly captured the contemporary outlaw movement of the time. He had his own TV series, Country Comes West, and is an outspoken expert on country music.

TARI HENSLEY – Born Tari Dean Hodges in Independence, Missouri in 1953, her name is pronounced 'Terry'. An amateur singer as a teenager, she married bandleader Dan Hensley in 1972 and toured with his band for more than ten years. Eventually she landed a contract with Mercury Records in 1983 and scored several minor hits, but failed to make it into the Top 20.

GOLDIE HILL – Born in Karmes County, Texas in 1933, Goldie began her professional career during the early '50s, signing with Decca, appearing on Shreveport's Louisiana Hayride and having a Top 5 country hit with **Don't Let The Stars Get In Your Eyes** in 1953. She went into semi-retirement shortly after her marriage to Carl Smith in 1957, but is still remembered as one of the most popular female country singers of the '50s.

THE HILLSIDERS – A British band that began their long stay at the top in 1965. A Liverpool outfit, they have recorded a Chet Atkins-produced album with Bobby Bare and another with George Hamilton IV. They have had their own BBC-TV show, played a Royal Albert Hall date and even played a two-week engagement at the London Palladium.

STAN HITCHCOCK – A former DJ in Springfield, Missouri, Hitchcock, who was born in Pleasant Hope, Missouri in 1937, moved to Nashville in 1962. He landed a record deal with Epic in 1967 due to his successful TV series and made it into the country Top 20 with **Honey, I'm Home** in 1969. He continued recording well into the '70s, then switched to television backroom work and by the late '80s was programme director for Country Music Television.

BECKY HOBBS – Rebecca Hobbs was born in Bartlesville, Oklahoma in 1950. During high school she formed an all-girl band, the Four Faces Of Eve. She attended Tulsa University and became a member of another all-girl band, Surprise Package. A move to Baton Rouge in 1971 found her in a bar band, Swamp Fox. Two years later she moved to Los Angeles, writing songs and recording for MCA. She signed to Mercury in 1978 but didn't make Top 20 until she duetted with Moe Bandy on **Let's Get Over Them Together** in 1983. A vastly talented musician, singer and songwriter, Becky has never made the commercial impact she deserves. Her self-penned **Jones On The Jukebox** is a real classic country song.

ADOLPH HOFNER – A native Texan of German-Slavic descent, Adolph Hofner has had a long and fascinating career playing both western swing and ethnic dance music for Texas' large German-American community. He began his career in San Antonio in the '30s and continues, to this day, travelling five days a week within the Texas state line, sponsored by Pearl Beer.

DOYLE HOLLY – Born in Perkins, Oklahoma in 1936, he was a Kansas oilfield worker at 13 and joined the US army in 1953. Following discharge, he moved to Bakersfield, California, where he played in Johnny Burnette's band. He became a regular member of Buck Owens' Buckaroos from 1963–70. With his own band, the Vanishing Breed, he signed with Barnaby Records in the early '70s, registering several low-level hits.

BRUCE HORNSBY – Raised in Williamsburg, Virginia, singer, songwriter and pianist Bruce Hornsby leads the Range, a piano-based, jazz-influenced pop quintet which was formed in Los Angeles in 1982. Eclectic in their musical styles, they lean heavily towards country-rock and scored a country Top 40 entry with **Mandolin Rain** in 1987.

STEPHEN WAYNE HORTON – A singer-guitarist from Memphis, Horton is a throwback to '50s rock'n'roll. Playing the clubs in and around

15.25, The Hillsiders. Courtesy of the Hillsiders.

Above: Burl Ives, also an actor, had country hits from the '40s to the '60s.

Memphis finally led to a recording contract with Capitol in 1988. Though his first album was critically acclaimed, he has failed to impress record buyers with his turbo-charged country-rock.

JAMES HOUSE – A singer-songwriter from Sacramento, California, James began playing clubs as a single acoustic act at 18, eventually putting together the House Band to back up his country-rock repertoire. He became a staff writer with Unicity Publishing in Los Angeles and some of his country song demos reached Nashville. In 1988 he was signed to MCA. He has concentrated on his writing, penning hits for several major stars, while still touring and developing his own career.

RAY WYLIE HUBBARD – Born in Hugo, Oklahoma in 1946, Ray is a singer-songwriter whose **Up Against The Wall Redneck Mother** became the anthem of the Texas outlaw movement during the 1970s. His **Off The Wall** album, made for Willie Nelson's Lone Star label in 1978, is worthy of reasonable outlay – or even unreasonable outlay.

I

IDA RED – A British band based in Wales, Ida Red have successfully blended the best of traditional country into a multi-faceted contemporary styling. Bobbie Barnwell, a lady with several years experience as a singer and musician, plays guitar and accordion, her teenage daughter Sarah plays fiddle and harmonizes with her mother. Henry Nurdin is a master of every stringed instrument in sight, but is at his best as he dances about playing mandolin. The final member is Tim Smith, who plays mandolin, guitar, harmonica and adds vocals. The group have been building a healthy following since 1990.

FRANK IFIELD – A UK pop singer born in Coventry in 1937, but emigrated to Australia with his parents during World War II. He started working tent shows in 1950 and by 1955 had his own radio and TV shows and was recording straight country for the local Regal-Zonophone label. Within a few years he had become Australia's biggest recording star. He moved to England in 1959 and signed to Columbia (EMI), expanded his style to country-pop and within three years topped the UK pop charts with **I Remember You**. Utilizing his yodelling-falsetto developed in Australia, he enjoyed further pop success throughout the '60s, often updating country songs, and also broke into American pop and country charts. He toured regularly until a serious operation curtailed his singing activities in 1990.

THE IMPERIALS – A country gospel group that, during the mid-1960s, became back-up group for Elvis Presley. SInce 1975, they have been recording Christian music exclusively.

JERRY INMAN – Lead singer with the resident band at Hollywood's Palomino for several years, Jerry notched his first hit with the Chelsea label in 1974 and in the late '70s had a brace of mini-hits on Elektra before fading.

CHRIS ISAAK – A singer-songwriter born in Stockton, California in 1956, Isaak first made an impact as a member of rockabilly quartet Silvertone in the mid-1980s. As a solo performer he gained extensive radio plays for the haunting rockabilly anthem **Blue Hotel**, while an instrumental version of his **Wicked Game** was used in David Lynch's 'Wild At Heart' film. He has since provided incidental music for TV shows and films, while his parallel movie career has gathered strength with roles in 'Married To The Mob' and 'The Silence Of The Lambs'.

BURL IVES – Folksinger, actor, broadcaster and author, Burl Ives was born in Huntington Township, Illinois in 1909. He helped to keep folk music alive during the '40s with his radio broadcasts as the 'Wayfaring Stranger'. For a time a member of the Weavers, he has successfully balanced a career as actor ('East Of Eden', etc) and singer, registering country hits with **Riders In The Sky** (1949) **and Wild Side Of Life** (1952). He recorded in Nashville in the early '60s scoring his biggest hits with **A Little Bitty Tear** and **Call Me Mr In-Between**.

AUNT MOLLIE JACKSON – An early protest singer, Aunt Mollie was born Mary Magdalene Garland in Clay County, Kentucky in 1880. A member of a mining family, her mother died of starvation in 1886. Her brother, husband and son all died in pit accidents, and her father and another brother were blinded in the mines. She became a union organizer, singing at meetings and on picket lines, moving to New York in 1936 because she was blacklisted in Kentucky. Along with her sister, Sarah Ogan Gunning, she recorded a great wealth of material for the Library of Congress, though her only commercial disc was **Kentucky Miner's Wife**, a Columbia single. She died in 1960.

SHOT JACKSON – Owner of Sho-Bud guitar company, Shot was born in Wilmington, North Carolina. Previously a sideman with Johnny And Jack, Kitty Wells and Roy Acuff, Jackson remained with the latter until he and Acuff were injured in a near-fatal auto accident in 1965.

JANA JAE – A beautiful and talented fiddler, Jana won the National Lady's Fiddling Championship in 1973 and 1974. A part of the Buck Owens Show for a long period, she has since appeared on many top TV shows, including Hee Haw and has even played the Montreux Jazz Festival.

JASON AND THE SCORCHERS – A cowpunk outfit from Sheffield, Illinois and headed by Jason Ringenberg (vocals, guitar, harmonica). Nashville-based, they displayed a wild, breakneck-paced style of rock-oriented country on two EPs for the local Praxis label and in 1984 signed to EMI America, cutting an album **Lost And Found**. After moving closer to hard rock, the Scorchers split up in 1990 with Jason launching a solo career as 'Jason' and a hard-edged traditional country sound that has yet to find favour with country radio.

FRANK JENNINGS – A British singer, heavily influenced by Faron Young, his band, the Frank Jennings Syndicate, came into existence in 1970 and rose to be the best country unit in the UK, gaining a deal with EMI which did not last as long as it might have done.

JJ WHITE – Sisters Janice and Jayne White from northern California have been singing together most of their lives. Harmonizing was second nature as they sang in high school and performed at amateur talent nights. In 1990 they landed a recording contract with Curb Records and started charting country hit entries in the lower regions of the chart.

LOIS JOHNSON – From Knoxville, Tennessee, Lois Johnson worked on local radio from the age of 11 and was a regular on the WWVA Jamboree in Wheeling during her teen years. She joined the Hank Williams Jr Road Show in 1969, the pair recording several successful duets. Her biggest solo success came with **Loving You Will Never Grow Old**, a Top 10 entry in 1975.

ANTHONY ARMSTRONG JONES – Born Ronnie Jones in Ada, Oklahoma, 1950, something of a local child star, he was discovered by Conway Twitty in 1962 and worked shows with him for many years. His stage name came from the English photographer who married Princess Margaret. He has recorded for Chart, Epic and Air, making regular chart entries from 1969–86, the most notable being **Take A Letter Maria** in 1970.

BUELL KAZEE – Born in Burton Fork, Kentucky in 1900, Kazee was a college-educated, fully ordained minister of the church. He recorded during 1927–29 for Brunswick, singing and playing five-string banjo on such songs as **Hobo's Last Ride** and **Rock Island Line**. Author of a book, 'Faith In The Victory', he performed at many folk concerts and recorded some material for the Library of Congress. He died in 1976.

ROBERT EARL KEEN JR – A Texas singer-songwriter, born in Houston, 1956, Keen was at University with Lyle Lovett; the pair often performed together. Later he formed a bluegrass outfit, the Front Porch Boys, but has since become an acclaimed contemporary singer-songwriter. His song, **Sing One For Sister**, was picked up by Nanci Griffith. His own album, **West Textures**, about life in Texas, proved to be the ideal vehicle for his rough, gruff vocals.

TOBY KEITH – An emerging singer-songwriter from Oklahoma, this one-time cowboy signed with Mercury Records in 1992. Along with label-buddies John Brannen and Shania Twain, he was part of the 'Triple Play Tour', a whirlwind 1993 promotional tour to establish the acts. It worked for Toby Keith, as his self-penned single, **Should've Been A Cowboy**, soared to No.1 on the country charts. Definitely a major star of the future.

SANDY KELLY – Pretty Irish colleen from County Sligo, Sandy Kelly has been a major star in Ireland since the late '80s. A revival of Patsy Cline's **Crazy** topped the Irish charts and led to her own RTE TV series, which has been on air since 1990. Sandy has also made her mark in America, appearing at Fan Fair and in Branson, with Johnny Cash duetting with her on record and in concert. A past winner of the Country Euro-Masters, in 1993 Sandy embarked on her most ambitious project – 'Patsy Cline: A Musical Tribute', a show which was touring provincial theatres throughout Great Britain.

ANITA KERR – A singer born in Memphis, Tennessee in 1927 who got into the vocal group business early in life and later led the Anita Kerr Singers on records by Eddy Arnold, Jim Reeves, Chet Atkins, Skeeter Davis, Floyd Cramer and many other artists. By the 1970s she was based in Europe providing mainly MOR albums.

DON KING – A talented singer-songwriter-guitarist, Don was born in Omaha, Nebraska in 1954. He has never quite made the breakthrough to the big time. He moved to Nashville in 1974 and had a couple of Top 20 entries for Con Brio Records in 1977. He has since recorded for Epic, Bench Mark and 615, and written many hit songs for other major acts. He has his own thriving Don King Music Group publishing company in Nashville.

SID KING – Real name Sidney Erwin, Sid was born in Denton, Texas in 1936. He became leader of the Five Strings, a country outfit that started out as the Western Melodymakers. They edged into rockabilly, recorded for Starday , their repertoire including **Who Put The Turtle In Myrtle's Girdle**, and, in late 1954, they gained a Columbia contract, staying five years. Some of the Five Strings' radio shots can be heard on the Rollercoaster album, **Rockin' On The Radio.**

EDDIE KIRK – A one-time singer and guitarist with the Beverly Hillbillies, Kirk was an amateur flyweight boxer whose yodelling ability won him the National Yodelling Championship in 1935 and 1936. Born in Greeley, Colorado in 1919, he was a singer much in the smooth style of Eddy Arnold. Signed to Capitol Records, his biggest hits were **Candy Kisses** (a version of George Morgan's hit song) and **The Gods Were Angry With Me** in 1949.

Above: Sleepy La Beef found a rockabilly audience in the late '70s.

L

SLEEPY LA BEEF – The man mountain of rockabilly, a 6'6" singer-guitarist who once played the Swamp Monster in the movie 'The Exotic Ones'. The possessor of an amazing baritone voice, La Beef (from Smackover, Arkansas) has recorded for Starday, Columbia, Sun, Plantation, Rounder and other labels.

LaCOSTA – Elder sister of Tanya Tucker, LaCosta Tucker, born in Seminole, Texas in 1951, was working as a medical records technician in Toltrec, Arizona when Tanya hit with **Delta Dawn** in 1972. She joined her younger sister in Las Vegas and in 1974 LaCosta was signed to Capitol Records, scoring her biggest hit with **Get On My Love Train**, a Top 3 hit in 1974. More Top 20 hits came during the next few years, but by 1982, working as LaCosta Tucker, the hits stopped flowing and eventually she left the music business.

JOHN LAIR – A country music pioneer, Lair was born in Livingston, Kentucky in 1894 and died in 1985. He produced many country radio shows, formed the Cumberland Ridge Runners and, in 1937, together with the Duke Of Paducah, Red Foley and his brother Cotton Foley, bought and built the Renfro Valley Barn Dance.

CHARLIE LANDSBOROUGH – A British singer-songwriter born in Wrexham, Wales in 1941, but raised mainly in Birkenhead on Merseyside. He played in the Top Spots, a beat group in Liverpool in the early 1960s. Developing a soft, romantic style, he has provided hit songs for Foster And Allen and recorded occasionally since 1982. Signed to Ritz Records in 1992, Landsborough has his best opportunity to break through as both a songwriter and major British act.

CRISTY LANE – Born Eleanor Johnston in Peoria, Illinois in 1940, Cristy Lane has been one of the best-sold singers in country music, with her albums advertised on TV and in national magazines. Her husband, Lee Stoller, masterminded his wife's career, forming LS Records in 1972, with several major country hits following. In 1979 she signed with UA and scored a No.1 with **One Day At A Time** the next year Cristy's album of the same title is reputed to be the biggest-selling gospel album of all time.

RED LANE – Born in Bogalusa, Louisiana in 1939, this singer-songwriter and award-winning guitarist has worked with Merle Haggard's Strangers. He had a few hits of his own while recording for RCA in 1971–72.

NICOLETTE LARSON – A lady who initially found fame as back-up vocalist for Hoyt Axton and Linda Ronstadt, Larson was born in Helena, Montana in 1952 and raised in Kansas City. She moved to L.A. in 1974 and worked with the Nocturnes. Session work followed and a recording contract with Warner Brothers resulted in a Top 10 pop hit, **Lotta Love**, in 1979. A fan of country music, she moved to Nashville in 1984, signed to MCA, and started registering country hits. The most successful, **That's How You Know When Love's Right**,. featured Steve Wariner and made the Top 10 in 1986.

ALBERT LEE – Simply the finest country-rock guitarist ever to come out of Britain, he was born in Leominster, Herefordshire, 1943, but spent his teen years in London, where he played in various pop, rock and R&B bands. He appeared at London's Royal Albert Hall with Chet Atkins in 1969, played with the group Heads, Hands And Feet in the early '70s, and gained true recognition as a member of Emmylou Harris' Hot Band. Now living in America, he is a well-established Nashville session musician.

DICKEY LEE – A teenage Sun rockabilly who wrote **She Thinks I Still Care**, a country classic for George Jones, Dickey Lipscomb was born in Memphis, Tennessee in 1941. After recording for Sun Records in 1957–58, he went on to score pop hits for Smash in the early '60s with **Patches** being the biggest. A move to Nashville in 1970 saw him sign to RCA and enjoy an incredible run of country hits, including **Rocky**, a 1975 No.1. He had further hits for Mercury in the '80s, but has since devoted his time to his successful songwriting.

Left: Jason And The Scorchers split up in 1989 and Jason went solo.

ROBIN LEE – Born Robin Irwin in Nashville, it was only natural for this young lady to follow a country music career. At high school she sang with rock group the Practical Stylists, then she sang demos for music publishers. While still a teenager she signed with Evergreen Records, registering several low-level hits between 1983–86. She joined Atlantic Records in 1988 and broke into the Top 20 with the song **Black Velvet** in 1990.

ZELLA LEHR – A versatile entertainer, Zella Lehr, who was born in Burbank, California in 1951, worked in the family vaudeville act. the Crazy Lehrs, from the age of six. When the act split up she became a regular on TV's Hee Haw, from which she built a country music career. Playing club dates in Nashville led to a RCA recording contract and a Top 10 hit with Dolly Parton's **Two Doors Down** in 1978. She has registered further hits for Columbia and Compleat.

GORDON LIGHTFOOT – A Canadian folk singer born in Orilla, Ontario in 1938. Lightfoot was a major influence on the Nashville folk-country of the mid-1960s with his songs recorded by George Hamilton IV, Marty Robbins, Waylon Jennings, etc. He broke through as a recording star in the '70s with platinum albums and hit singles **If You Could Read My Mind**, **Sundown** and **Carefree Highway**.

THE LILLY BROTHERS – Mitchell 'Bea' B and Everett Lilly (born in Clear Creek, West Virginia in 1921 and 1924 respectively) are a bluegrass duo who began as the Lonesome Holler Boys on a Charleston radio show in 1939 but became residents in the Boston area, where they played for around 18 years. During 1973 they visited Japan, cutting several albums.

GEORGE LINDSAY – Comedian-character actor known to the world as 'Goober'. He was a regular on Hee Haw.

LaWANDA LINDSEY – Once a singer with her father's band, the Dixie Showboys, she signed to Nashville's Chart Records at 14 and has several hits in 1969–72. Later she moved on to Capitol, Mercury, etc., her biggest single to date being **Hello Out There** (1974).

LITTLE GINNY – An energetic British singer, born Ginnette Brown in Kingston-upon-Thames, England, she started out at 13 singing in country music clubs and in the Ivy Benson All-Girl Band. She had her own BBC-TV shows, then married Liverpool bass player Paul Kirkby and developed into a fine contemporary entertainer with her Room Service band. In 1986 she teamed up with Tammy Cline to form Two Hearts, a superb country duo with fast-moving stage show. She returned to a solo career in 1992.

HUBERT LONG – Born in 1923 and died in 1972, Long was elected to the Country Music Hall of Fame in 1979. Long began his career in a Texas dime store record department and later founded Nashville's first talent agency. He was the first person to serve as both president and chairman of the CMA.

LOUISIANA HAYRIDE – An influential show that originated on station KWKH, Shreveport, Louisiana, in 1948. The first programme featured Johnny And Jack, Kitty Wells, Bailes Brothers, etc. One of the first cast members to attain stardom was Hank Williams.

RAY LYNAM – Arguably Ireland's finest male country singer, he broke onto the scene in 1970 with a hard-country style developed from listening to Buck Owens and George Jones. Has recorded many duets with Philomena Begley. In 1980 he cut the **Music Man** album, which some hailed as "a milestone in Irish country music".

JUDY LYNN – A one-time teenage rodeo rider, national yodelling champion and beauty queen, Judy Lynn was born in Boise, Idaho, in 1936, the daughter of Joe Voiten, an ex-bandleader. She joined the Opry touring show in 1956 and began touring with her own eight-piece band four years later. Dressed in flamboyant western attire, she was a popular performer on the Nevada casino circuit for more than 20 years. Recorded for many labels from 1957–80, when she retired to become a church minister. She had only one major hit, with **Footsteps Of A Fool** in 1962.

M

DALE McBRIDE – A regional country star in Texas who never quite made the breakthrough to national stardom. Born in Bell County, Texas and raised in nearby Lampasas, McBride played guitar at 13 and was an original member of the Downbeats. For some time he worked in Jimmy Heap's Melody Masters. He recorded for Con Brio throughout the '70s, registering a dozen low-chart entries. His son, Terry McBride, is in McBride And The Ride.

MARY McCASLIN – A singer who grew up in California and became part of Linda Ronstadt's Stone Poneys before going solo and cutting a classic album for Barnaby. Her later albums of western-styled songs for Philo are exceptional.

DELBERT McCLINTON – Singer, songwriter, guitarist and harmonica player from Lubbock, Texas (born 1940), whose bluesy style has proved influential in country music. He played harp on Bruce Channel's 1962 pop hit **Hey! Baby**, and had songs covered by Waylon Jennings, Emmylou Harris, Vince Gill, etc. He duetted with Tanya Tucker on the 1993 country hit, **Tell Me About It**.

SKEETS McDONALD – A popular singer on the West Coast, McDonald is best known for his 1952 hit **Don't Let The Stars Get In Your Eyes**. Born in Greenaway, Arkansas in 1915, he began his career on local radio stations in Michigan before migrating to the West Coast after his World War II service. A longtime fixture on the Town Hall Party, he recorded for Capitol (1952–59) and Columbia (1959–67). He died of a heart attack in 1968.

PAKE McENTIRE – Dale Stanley McEntire, elder brother of Reba, was born in Chockie, Oklahoma in 1953. He sang at rodeos with Reba and another sister, Susie, as the Singing McEntires. Member of the Professional Cowboy Association since 1971, Pake continues to compete in roping events. Signed to RCA Records in 1985, he made Top 10 with **Savin' My Love For You** the following year.

WES McGHEE – Underrated in Britain where homegrown original country talent is ignored, this UK singer-songwriter, born in 1948, has made his mark in Texas, working with Ponty Bone, Butch Hancock and Kimmie Rhodes. He has recorded several invigorating albums on his own self-financed labels since the late '70s.

TIM McGRAW – Of Irish and Italian stock, the son of legendary baseball pitcher Tug McGraw, Tim was born in Delhi, Louisiana in 1969, but raised in Start, Louisiana. He moved to Nashville in 1989, played clubs and worked as a demo singer. He signed to Curb Records in 1990, and made his country chart debut with **Welcome To The Club** in 1992 and landed the opening spot in 1993's Honky Tonk Attitude Tour with Joe Diffie.

DON McLEAN – This talented singer-songwriter was born in New Rochelle, New York in 1945. He found overnight success in 1971 with **American Pie**, a pop classic that disguised his affection for country music. His 1973 album, **Playin' Favorites**, found him exploring his folk-country roots and since 1980 he has recorded in Nashville, making country Top 10 with **Cryin'** in 1981. A male version of Patsy Cline's **He's Got You** was a minor chart entry six years later.

TERRY McMILLAN – The man who has replaced Charlie McCoy as the most sought-after harmonica player. A NARAS Super Picker in 1975, his name appears on scores of records.

THE MAINES BROTHERS BAND – This country-rock family group from Texas have played a major role in the Lubbock music scene. The band began in the '50s with current members' father and uncle, James and Sonny Maines. The four brothers – Lloyd, Kenny, Steve and Donnie, got together in the late '70s with Richard Bowden, Gary Banks and Jerry Brownlow to record several albums for Texas Soul records. They linked up with Mercury in 1983, scoring their biggest hit with **Everybody Needs Love On Saturday Night** in 1985.

TIM MALCHAK – Originally a folksinger from Binghamton, New York, he worked in New York City and California before moving into country music in the early '80s. He teamed up with Dwight Rucker in 1980, and, as Malchak And Rucker, were the only country white/black duo to make the country charts. He went solo in 1986 and registered several Top 40 country entries, including **Colorado Moon**.

JAY DEE MANESS – Steel guitar supersessionman, born in Loma Linda, California in 1945, Maness has worked on hundreds of sides and with Buck Owens and Ray Stevens.

LINDA MARTELL – The first black female singer to appear on the Grand Ole Opry, she was initially a R&B singer born in Leesville, South Carolina, who included country material in her act. Shelby Singleton signed her to his Plantation Records and she registered three country hits, the biggest being **Color Him Father** in 1969.

WAYNE MASSEY – A singer-actor from Glendale, California, he played Johnny Drummond in TV's One Life To Live soap opera in the early '80s, which led to a recording contract with Polydor. He registered minor hits, but after he married Charly McClain and cut some duets, he made it into the Top 10 on Epic with **Just One Look In Your Eyes**.

MATHEWS, WRIGHT AND KING – A trio comprising Raymond Mathews, Woody Wright and Tony King. With their blend of three-part southern harmony with a bluegrass ring and a gospel spirit, they made an impact in 1992, gaining a Columbia record contract and several minor hits. Woody Wright was a former member of the Tennesseans, a Capitol act in 1979 and lead singer of Memphis, who charted in 1984.

THE MAVERICKS – A four-man, country-rock outfit from Miami, comprising lead singer Raul Malo, bassist Robert Reynolds, drummer Paul Deakin and guitarist David Lee Holt. Showcases in Nashville led to an MCA recording contract in 1992. Hard driving, hard country and hard nosed, their songs and their music exhibit great passion. Malo is a distinctive singer who is in great demand on Nashville sessions.

DONNA MEADE – Born in Chase City, Virginia in 1953, after building a regional following, Donna moved to Nashville and became a club singer at Buddy Killen's Bullpen Lounge. Signed to Mercury in 1987, she has registered several minor hits, but has yet to find that one song to take her into the big time.

TIM MENSY – A native of Virginia, this talented singer, songwriter and guitarist moved to Nashville in 1980 when he was 20 and joined Bandana. He spent several years on the fringes of mainstream country flirting with big time success. He wrote hit songs for Shenandoah, T. G. Sheppard, etc, played sessions and worked as a demo singer. Signed to Columbia Records in 1988, he had some minor success, but has made a bigger impact since joining Giant in 1992.

Dream Seekers, Matthews Wright And King. Courtesy Sony Music.

NED MILLER – A reluctant star, who refused to perform on stage, even when he had major hits, Miller made his mark initially as a songwriter. Born in Raines, Utah in 1925, he moved to California in the mid-1950s where he joined Fabor Records and wrote **Dark Moon** and **A Fallen Star**, two major country-pop hits. His original recording of **From A Jack To A King** in 1957 flopped, but when reissued in 1962 it became a massive pop-country seller in Britain and America. Further success followed with **Invisible Tears** and **Do What You Do Do Well**, but by the early '70s he had completely withdrawn from singing and songwriting.

BILLY MIZE – An ACM vice-president who was once host on Gene Autry's Melody Ranch show, Mize (born in Kansas City, Kansas in 1929) has always had TV connections, eventually heading his own production company. But he still tours with his Tennesseans and from 1966 through to 1974 charted consistently.

MOLLY AND THE HEYMAKERS – A quartet from Hayward, Wisconsin, comprising Molly Scheer, Andy Dee, Jeff Nelson and Joe Lindzius, they spent three years touring through Wisconsin, Minnesota and the usually rock-dominated Minneapolis, before gaining a Reprise record deal in 1989. With a blend of bluegrass, '60s California pop, rockabilly, honky-tonk, Cajun and western swing, they produce a lot of energy.

TINY MOORE – Another instrumental near-genius, born in Hamilton County, Texas in 1920, who plays both mandolin and fiddle, switching with ease from pure jazz to downhomey country. Once a Bob Wills Playboy, he later worked with Billy Jack Wills but these days plays many dates with Merle Haggard's Strangers.

THE MORRIS BROTHERS – Wiley and Zeke Morris comprised one of the fine duet acts which flooded country music in the mid-1930s. They became well-known for their smooth harmony singing and songs like **Salty Dog** and **Tragic Romance**. Their career lasted well into the '40s, although they retired to their native North Carolina and remained relatively inactive in later years. Wiley had a longer career playing mandolin and guitar for a number of bands, including Wade and J. E. Mainer, and with Charlie Monroe and his Kentucky Pardners.

TEX MORTON – One of Australia's greatest country artists, Morton was actually born in Nelson, New Zealand, in 1916. A fine singer and yodeller, he moved to Australia in 1932, making his first records for Regal Zonophone four years later and gaining instant popularity. In 1948 he settled in Canada, there becoming 'The Great Morton – The World's Greatest Hypnotist', also continuing with his singing career. In 1959 he returned to Australia, maintaining his reputation as a top-line entertainer until his death in Sydney in 1983.

JOHNNY AND JONIE MOSBY – Popular husband and wife duet team in the 1960s, Johnny from Arkansas and Jonie (born Janice Shields, 1940) from California had their own 'Country Music Time' on Los Angeles television and enjoyed several Top 20 hits for both Columbia and Capitol before fading from the scene to raise their family.

MOON MULLICAN – The originator of a highly personal two-finger piano style, Aubrey 'Moon' Mullican, born near Corrigan, Polk County, Texas, in 1909, influenced many later keyboard players, including Mickey Gilley and Jerry Lee Lewis. He gained his nickname 'Moon' because he slept by day and worked the clubs in Houston by night, then later appearing on radio and at clubs in the Louisiana-Texas area. By the mid-1940s Mullican had become a major solo attraction, his 1947 recording of **New Jole Blon** selling three million copies. He had another million-seller with **I'll Sail My Ship Alone** (1950) and joined the Grand Ole Opry. Later he toured as part of Governor Jimmie Davis' staff and band. Dogged by ill-health, the king of pumpin' piano died from a heart attack on January 1, 1967.

HEATHER MYLES – From Riverside, California, but raised in Texas, of Scottish-Canadian ancestry, this West Coast singer-songwriter

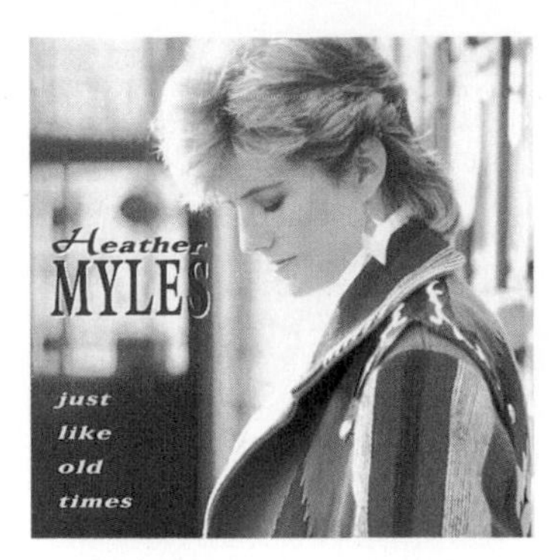

Just Like Old Times, Heather Myles. Courtesy HighTone Records.

started building a European following in 1992 after her debut album, **Just Like Old Times**, gained rave reviews. Re-interpreting the '60s Bakersfield Sound in her own distinctive way, Heather spent years playing the Texas honky-tonk circuit before relocating to California and forming the Cadillac Cowboys.

N

NASHVILLE – Known as Music City or the Country Capital, this Tennessee city started out as the Mecca of country music in 1925 when WSM began its Barn Dance. The first major artist to record in Nashville was Eddy Arnold in 1944, while Capitol was the first major label to commence a Nashville operation (1950). During the late '50s and '60s the famed 'Nashville Sound' took over and the producers, musicians and studios came up with a formula that not only saw such artists as Jim Reeves, Eddy Arnold, Chet Atkins and Floyd Cramer selling huge amounts of records to pop audiences, but also had scores of pop idols heading towards Tennessee in order to add a touch of saleable Nashville magic to their own wares. In 1974 came the opening of Opryland, a modern auditorium amid an extensive amusement park just a few miles outside the city.

NASHVILLE BLUEGRASS BAND – A five-man band that takes a bit of the twang out of bluegrass, they came to the fore in the late '80s, initially on Rounder, then later Sugar Hill Records. Mainstays of the group are Stuart Duncan, Pat Enright, Gene Libbea, Alan O'Bryant and Roland White, who produce modern bluegrass with hot picking.

BUCK AND TEX ANN NATION – To Buck (born Muskogee, Oklahoma, 1910) and Tex Ann (born Chanute, Kansas, 1916) goes the credit for starting the now popular country music parks, where Sunday afternoon crowds can picnic, relax and listen to country music. Active in the Northeast, Buck and Tex Ann opened their first park in 1934 and found their greatest success in Maine, of all the unlikely places.

NATIONAL BARN DANCE – One of the earliest and most influential of all radio barn dances for some years. Begun in 1924, it was later overshadowed by the Grand Ole Opry but survived until 1970. Early stars include Bradley Kincaid, Grace Wilson and Arkie The Woodchopper, these being followed by John Lair and the Cumberland Ridge Runners, Mac And Bob, Lulu Belle And Scotty and Red Foley.

TRACY NELSON – A bluesy country vocalist, born in Madison, Wisconsin in 1944, Tracy started out as a local folksinger, then joined R&B outfit the Fabulous Initiations. As founder member of Mother Earth in San Francisco in 1967, she became heavily country-oriented, and the band often recorded in Nashville. In 1973 she embarked on a solo career and duetted with Willie Nelson on the Grammy-nominated **After The Fire Is Gone**. Tracy has continued to record for MCA, Atlantic, Flying Fish and Adelphi.

OLIVIA NEWTON-JOHN – Lightweight English pop singer from Cambridge (born 1948), who grew up in Australia, but moved back to England in the late '60s. She enjoyed British and American pop success with country-flavoured songs in the early '70s, many crossing into the country charts. She won the CMA Female Vocalist Of The Year award in 1974, amid uproar from established country stars, but she soon turned her back on country and embarked upon a more lucrative film and pop-dance career.

EDDIE NOACK – A Texan honky-tonk singer and songwriter, Noack, who was born in Houston in 1930, never received due recognition during his lifetime. He recorded prolifically from 1949 up until shortly before his death in 1978, but gained more success as a writer, penning hits for Hank Snow, George Jones, Ernest Tubb, etc. He became an executive for the Nashville Songwriters' Association and has had several compilations of his early recordings released since his death.

NORMA JEAN – A highly regarded performer during the '60s, Norma Jean Beasler was born in Willston, Oklahoma in 1938 and raised in Oklahoma City, where she learned guitar and performed at square dances. By 1958 she had become a regular on Red Foley's Ozark Jubilee TV show – two years later she joined the Opry and also became a regular on Porter Wagoner's TV show. Soon her pure country voice was heard regularly in the charts, her biggest hit being **Let's Go All The Way** (1964). Replaced on the Wagoner show by Dolly Parton in 1967, Norma Jean's record sales dropped accordingly.

O

TIM O'BRIEN – A multi-talented bluegrass musician, singer and songwriter, from Wheeling, West Virginia, O'Brien started out in his early teens and throughout the '80s was lead vocalist with Hot Rize, with whom he also played mandolin, guitar and fiddle. Based in Colorado, the band recorded for Flying Fish and Sugar Hill before disbanding in 1990. Since then O'Brien has embarked on a solo career, with his own group, the O'Boys. He has also recorded duets with his older sister Mollie O'Brien and Kathy Mattea, who has turned some of his songs into country hits.

W. LEE O'DANIEL – Wilbert Lee O'Daniel, born in Malta, Ohio in 1890, became the first of many politicians to use the grass roots appeal of country music to propel him to high political office. An executive in the Burrus Mills Company, maker of Light Crust Flour, in 1930 he formed a band around a group of struggling musicians, the Light Crust Doughboys, to advertise the product over radio. Later he set up his own company and brand, Hillbilly Flour, and a new band, the Hillbilly Boys. In 1938 he waged a grass roots campaign for Governor of Texas, winning easily, and went on to serve a term in the US Senate.

JAMES O'GWYNN – Known as the 'Smiling Irishman Of Country Music', O'Gwynn was born in Winchester, Mississippi in 1928 and enjoyed considerable chart success in the late '50s and early '60s, making the Top 10 with **My Name Is Mud** in 1962 and appearing regularly on the Houston Jamboree, Louisiana Hayride and Grand Ole Opry.

Palomino Road's eponymous album. Courtesy Liberty Records.

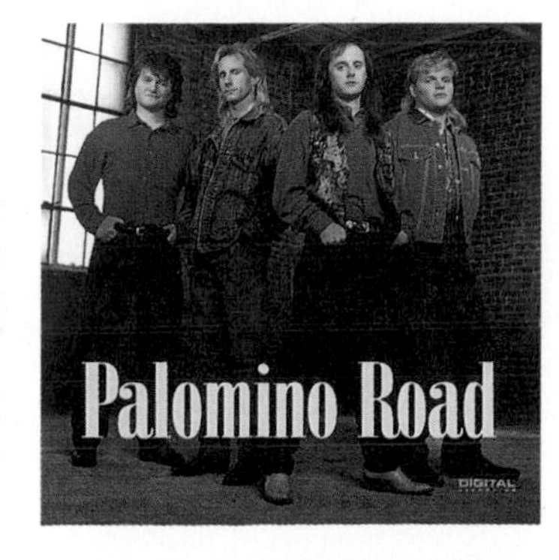

ORION – Masked mystery man who is the ultimate Elvis Presley clone, his real name is Jimmy Ellis from Orrville, Alabama. He was, naturally, recorded by Sun (revived by Shelby Singleton) from 1977 through to 1982, his albums being pressed as 'Collector's Edition Special Gold Vinyl'.

ROBERT ELLIS ORRALL – Singer, songwriter, pianist from Lynnfield, Massachusetts, Orrall first made an impact penning hits for Shenandoah, Carlene Carter, etc. Signed to RCA in 1992, he is building into one of the new major country acts of the future.

MARIE OSMOND – A member of the famous Osmond family, born Olive Marie Osmond in Ogden, Utah in 1959, she began appearing with her brothers at 14, and scored a major pop-country hit with **Paper Roses** at the same time. She co-starred with brother Donny in their own musical-variety TV series from 1976–80, and made a return to country in 1985 with Top 10 hits for Capitol. A tireless charity worker, Marie was bestowed with the prestigious Roy Acuff Community Service award.

JIM OWEN – Franklin, Tennessee singer-songwriter who is not only a Hank Williams soundalike but also an incredible lookalike. During the '80s, he logged a few hits for Sun.

TEX OWENS – Born in the Lone Star State in 1892, Owens was a popular star and co-host of several radio shows, but is best known and remembered for writing and singing his 1935 hit **Cattle Call**, on Decca. Owens died at his home in Baden, Texas in 1962. From a musical family, his daughter, Laura Lee, had a long career as Bob Wills' first girl singer, and in addition his sister was Texas Ruby, longtime Opry star.

P

TOM PACHECO – The 'Storyteller of the '90s', Tom Pacheco was born in Dartmouth, Massachusetts. His father, Tony, was a music store owner and guitar tutor, who had previously played in Europe with Django Reinhardt. Tom played folk music in Greenwich Village in the '60s and a few years later recorded for RCA and Columbia in a very modern country style, eventually moving on to play rock music. He travelled around playing music in Woodstock, Austin, Nashville and California, then settled in Ireland in 1987. Since then he has built a cult following as a singer-songwriter touring Europe and also working in and around Austin, Texas.

PATTI PAGE – Hardly a true country vocalist, Patti (born Clara Ann Fowler in Claremore, Oklahoma, 1927), had one of the biggest all-time country hits with **Tennessee Waltz**, in 1951. Really a pop singer, she switched to country at the start of the '70s and has since had a fair degree of success in the country charts, particularly during her stay with Shelby Singleton's Plantation label in the early '80s.

STU PAGE – A gritty British country singer born in Leeds, Yorkshire in 1954. He played with American bluegrass outfit the Warren Wilkeson Band in the mid-1970s, then formed Stu Page And Remuda in 1984, becoming one of Britain's most acclaimed modern country performers, mixing in fine original material with inspired covers of American country hits.

KEITH PALMER – A Missouri-born singer-songwriter, the adopted son of a fundamentalist, church-going farming couple, Palmer was raised in Corning, Arkansas. He moved to Nashville in 1989 and worked as a songwriter and demo singer. He penned **For My Broken Heart**, a No.1 for Reba McEntire, and he signed to Epic Records in 1991.

PALOMINO ROAD – High-energy country band of the '90s. Lead singer Ronnie Guilbeau, bassist David Frazier, guitarist J.T. Corenflo and drummer Chip Lewis had all been honing their skills in other bands, writing and playing sessions when they got together in early 1992. Following a showcase, they were signed by Jimmy Bowen to Liberty Records. Vocal blend harmonies, dynamic instrumental energy and rare songwriting skills make Palomino Road a band with a great future.

ANDY PARKER AND THE PLAINSMEN – Born near Mangum, Oklahoma in 1913, Andy Parker began his radio career at 16 in Elk City, Oklahoma. He assumed the role of singing cowboy on NBC's Death Valley Days from 1937–41. By 1946 he was based in L.A. and had formed a western harmony group, Andy Parker And The Plainsmen. They appeared in some eight films with Eddie Dean and signed with Capitol Records in 1947. The band had no particular big-selling records and Parker was forced to retire, suffering from a heart condition.

JIMMY PAYNE – A fine songwriter born in Arkansas in 1939, he played guitar with the Glaser Brothers' band and made his mark in Nashville as a writer of such hits as **Woman, Woman** (Gary Puckett & Union Gap, and the Glasers), **What Does It Take** (Skeeter Davis) and **My Eyes Can Only See As Far As You** (Charley Pride). He has recorded regularly since 1962, but has only achieved a few minor chart entries with **L.A. Angels** (1969), **Ramblin' Man** (1973) and **Turning My Love On** (1981).

PEARL RIVER – This Mississippi-based sextet have a bluesy, energy-laden vocal sound developed during a ten-year stint working small clubs and playing a mix of rock, R&B and country. Originally called the Toys, they took their new name from the river that runs through Mississippi. After being selected to appear at the Nashville Entertainments 'Music City Music Event' in 1992 they were signed to Liberty Records.

HERB PEDERSEN – Born in Berkeley, California in 1944, Pedersen is a singer-banjoist who has played with Flatt And Scruggs, the Dillards, Linda Ronstadt, Emmylou Harris, etc. He recorded solo albums for Epic and Sugar Hill, then joined the Desert Rose Band in 1986.

RALPH PEER – The most notable talent scout of the 1920s (born in Kansas City, Missouri in 1892; died in Hollywood, California in 1960). Peer discovered Jimmie Rodgers and the Carter Family, his 1923 sessions with Fiddlin' John Carson proving a landmark in country music history. In 1928 he formed Southern Music, a publishing company heavily involved in the publication of country songs.

PEGGY SUE – Younger sister of Loretta Lynn, born Peggy Sue Webb in Butcher's Hollow, Kentucky. For many years she toured with Loretta and co-wrote **Don't Come Home A'Drinkin'**. Signed to Decca Records in 1969, she had several minor hits. She later joined Door Knob Records and cut duets with her husband Sonny Wright, a one-time vocalist with Loretta's show.

RAY PENNINGTON – A talented singer, songwriter, record producer and executive, Pennington was born in Clay County, Kentucky in 1933. He made his TV debut in Cincinnati at the age of 16 and formed his own band in 1952, working local TV shows in such states as Kentucky, Indiana and Ohio. He moved to Nashville in 1964 and has recorded for Capitol, Monument and MRC, and has also been active as an executive at Step One Records.

BILL PHILLIPS – Longtime member of the Kitty Wells–Johnny Wright Roadshow, William Clarence Phillips was born in Canton, North Carolina in 1936. He moved to Nashville in 1957, making his mark as a songwriter, then signing with Columbia Records the following year. He appeared on the Grand Ole Opry, then moved over to Decca Records and began accruing chart entries, the most successful being **Put It Off Until Tomorrow** (1966) and **Little Boy Sad** (1969).

STU PHILLIPS – Longtime member of the Grand Ole Opry, Phillips was born in Montreal, Canada in 1933, but has lived in Nashville since 1964. He enjoyed several major hits on RCA during the late '60s, the most successful being **Juanita Jones** in 1967, the year he joined the Opry. He has remained a member ever since, though he has not recorded since 1970.

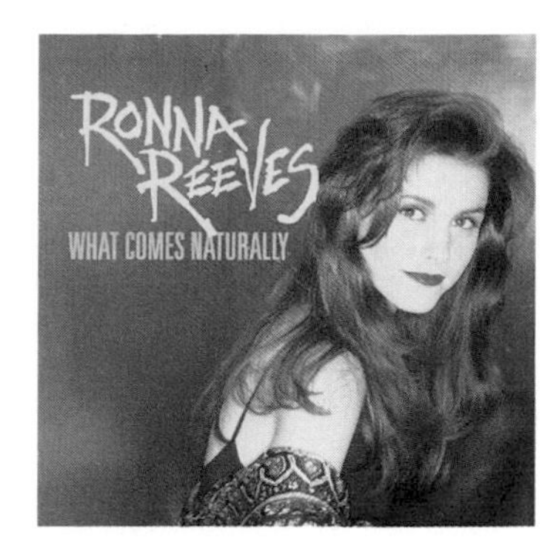

What Comes Naturally, Ronna Reeves. Courtesy PolyGram Records.

PIRATES OF MISSISSIPPI – This five-man outfit take the basic ingredients of country music and brand it with their own electric style. Named ACM's Top New Vocal Group in 1991, they played local gigs just for fun and unexpectedly landed a Capitol-Liberty recording contract, making it into the Top 20 with **Feed Jake**.

MARY K. PLACE – Born in Tulsa, Oklahoma, this actress portrayed a country singer called Loretta Haggers in the US TV series Mary Hartman, Mary Hartman (1976). Accepted as the real thing, she cut an album with Emmylou Harris and the Hot Band, and had a Top 5 hit with **Baby Boy** (1976).

POACHER – A British band who won the New Faces TV talent show in 1978. From Warrington, Lancashire, they became sponsored by the local vodka company and threatened an unbelievable breakthrough when, also in 1978, their version of **Darling**, on Republic, entered the US charts.

SANDY POSEY – Born in Jasper, Alabama in 1947, Sandy worked as a session singer in Nashville and Memphis before making an impact on the pop charts with **Single Girl** and **Born A Woman** in 1966. Country success came during the '70s when recording for Columbia, Monument and Warners. Nowadays Sandy works as a back up vocalist on the Nashville Network.

PRAIRIE OYSTER – This Canadian sextet play a cross between swamp rock, rockabilly and honky-tonk in a style described as 'Rock meets the twang of steel'. Having established themselves in Canada, where they were first formed in the mid-1970s, they regrouped in 1983 and made a breakthrough in America when they signed with RCA Records in Nashville in 1990 and started scoring country hits.

MALCOLM PRICE – A perennial on the UK country circuit, where he is warmly regarded. His albums for Decca, made in the early '60s with the Malcolm Price Trio, are ranked highly.

R

RED RECTOR – One of country music's top mandolin players, he was born in Marshall, North Carolina in 1929 and worked in the bands of Johnny And Jack, Bill Clifton, Charlie Monroe and Flatt And Scruggs. Throughout the '50s and '60s he played recording sessions, and by the '70s he took centre stage at festivals and was a recording star in his own right.

RONNA REEVES – A diminutive singer with sensually gravelly vocals from Big Springs, Texas, this lady put her first band together at 11. In the mid-1980s she opened shows for George Strait, then in 1991 landed a recording contract with Mercury. This led to touring with Billy Ray Cyrus as she has started to make an impact with her recordings.

KIMMIE RHODES – A multi-faceted Texas singer-songwriter, Kimmie started out in a gospel trio when she was six and by her late teens was playing the Austin clubs with the Jackalope Bros (Bobby Earl Smith and Joe Gracey). She toured with Al Dressen's Swing Revue in 1983–84, but now mainly works solo. Kimmie is something of an actress, painter, poet, dancer and writer and has appeared in films, videos and has sung and played on sessions in Austin. Her reflective **I Just Drove By** was picked up by Wynonna in 1993.

BOBBY G. RICE – A former rock'n'roller who turned to country, born in Boscobel, Wisconsin in 1944, Rice was part of a family band that had their own radio show on WRCO Richmond from 1957–64. He went solo and formed the Bobby Rice Band, scoring with country versions of pop oldies on Royal American in the early '70s. With his commercial, pop-slanted style, he had his biggest success with **You Lay So Easy On My Mind** (1972) on Metromedia, and has since placed hits for GRT, Republic, Sunbird and Charta.

TONY RICE – Acoustic guitarist and vocalist born in Danville, Virginia in 1951 into a musical family, he joined brothers Larry and Ronnie when he was nine as the Rice Brothers. Later came stints with J.D. Crowe's New South, Ricky Skaggs and Emmylou Harris. He has recorded regularly, both as a solo and with his own Tony Rice Unit for Rounder, Sugar Hill and Kaleidoscope.

PAUL RICHEY – Born in Promised Lane, Arkansas, Richey is a singer-songwriter, music publisher and Tammy Wynette's brother-in-law. As Wyley McPherson he broke into the charts with **Jedediah Jones** and **The Devil Inside** in 1982.

RIDERS IN THE SKY – A current-day cowboy trio who offer loving (if frequently tongue-in-cheek) renditions of Sons Of The Pioneers' songs plus material of a similar bent. Nashville-based, the group comprises Ranger Doug (guitar, vocals), Woody Paul Chrisman (fiddle, vocals) and Fred 'Too Slim' LaBour (bass, vocals). Ranger Doug in reality being Doug E. Green, the noted country music historian. Regulars on the Grand Ole Opry, Riders In The Sky also host 'Tumbleweed Theatre' a western movie programme on TNN, and have recorded for both Rounder and MCA.

BILLY LEE RILEY – Rockabilly singer and multi-instrumentalist from Pochantas, Arkansas, he recorded for Sun Records in mid-1950s, but never had any big hits. He was a powerful rock'n'roll singer and his influence has been immeasurable.

DENNIS ROBBINS – A gifted singer, songwriter and slide guitarist from Hazelwood, North Carolina, this one-time member of the Michigan pop group the Rockets made his initial impact in country in 1987 when he cut the original **Two Of A Kind (Workin' On A Full House)** on MCA Records. Two years later he was lead singer of Billy Hill, a band of Nashville sessionmen-songwriters. Robbins picked up his solo career again in 1991, signed to Giant Records, and has started to make a bigger impression.

HARGUS 'PIG' ROBBINS – CMA Instrumentalist Of The Year in 1976. A blind pianist from Spring City, Tennessee, he gained attention while playing Nashville clubs and became a top sessionman.

KENNY ROBERTS – A super yodeller known best in the North and Northeast, although he was actually born in Lenoir City, Tennessee in 1927. He has recorded for Decca, Coral, Dot, King and Starday, his biggest hits being the yodelling extravaganzas, **Chimebells** and **She Taught Me How To Yodel**.

ROCKABILLY – The first link between country and rock. Carl Perkins has demonstrated the link between the rhythms of certain Hank Williams' songs and early rock as part of his act, while chunks of pure rockabilly occur on the Ernie Ford boogies of the late '40s and early '50s.

FRED ROSE – Founder of the Acuff-Rose Music Publishing Co., Rose (born Evansville, Indiana in 1897, died in Nashville in 1954) was a one-time honky-tonk pianist who set up the publishing company in 1942. A fine songwriter – his credits include **Be Honest With Me**, **Blue Eyes Crying In The Rain**, **Take These Chains From My Heart**, **Tears On My Pillow**, **Settin' The Woods On Fire** and others – he often wrote in partnership with such people as Hank Williams, Ray Whitley and Hy Heath. In 1961 he posthumously became one of the first members of the Country Music Hall Of Fame, sharing the honour with Jimmie Rodgers and Hank Williams.

WESLEY ROSE – The son of Fred Rose, born in Chicago, Illinois in 1918, he has been responsible for expanding the whole horizon of Acuff-Rose's business and making it one of the most successful publishing companies in the world. Initially he moved into the world of record production, at the outset fashioning material for major labels but later forming his own Hickory Records. A music industry leader, Wesley Rose was one of the founder members of the CMA.

PETER ROWAN – Singer-songwriter and brilliant mandolinist, once a member of the Rowan Brothers but, in more recent times, the leader of his own band, switching from folk-rock through to bluegrass, pure country and Cajun. He often works with accordionist Flaco Jimenez.

LEON RUSSELL – Vocalist, songwriter, multi-instrumentalist, sessionman, producer and label chief, Russell was born in Lawton, Oklahoma in 1941. During the early '60s he played numerous sessions in California from Phil Spector through Bobby Vee, the Crickets to Frank Sinatra. He formed Shelter Records with British producer Denny Cordell in 1970, and recorded as Hank Wilson in a pseudo country-rock style. In 1976 he started his own Paradise Records, recorded duet hits with Willie Nelson, and also teamed up with New Grass Revival for shows and recordings.

TIM RYAN – Singer-guitarist from Montana, whose first taste of the big-time came at age 12 when he played lead guitar for Tex Williams. A move to Nashville in 1988 found him teaming up with veteran songwriter Alex Harvey. With his matinee idol looks, solid songwriting skills and an identifiable voice that can bend notes and slip into falsetto, he has registered several minor hits for Epic Records.

JOHN WESLEY RYLES – An in-demand Nashville-based singer, Ryles was born in Bastrop, Louisiana in 1950. He sang in the family gospel group the Ryles Singers, appearing on local radio while still a child. A regular on the Cowtown Hoedown Show in Fort Worth and the Big D Jamboree in Dallas, he then moved to Nashville in 1966. He has recorded prolifically since 1967 for Columbia (as John Wesley Ryles I), Plantation, ABC-Dot, MCA and Warners, scoring several Top 20 hits. He is also kept busy as a back-up singer on Nashville sessions.

S

JUNIOR SAMPLES – From Cumming, Georgia, Alvin Junior Samples (born in 1927) weighed nearly 300 lbs and was proud of being dubbed 'the world's biggest liar'. A country comedian who relied on storytelling, his single, **The World's Biggest Whopper**, became a mild hit in 1967 and opened the doors for a highly successful career. He was a regular on the Hee Haw Show until shortly before he died of a heart attack in 1983.

ART SATHERLEY – During the 1930s, Satherly (born in Bristol, England in 1889; died in 1986) helped provide ARC with one of the strongest country music catalogues in America, his signings including Gene Autry (1931). In 1938, when ARC became Columbia, Satherley was retained by the company and continued in an A&R role until his retirement in 1952, having added such acts as Lefty Frizzell, Marty Robbins, Little Jimmy Dickens, Bill Monroe and Carl Smith to the Columbia roster. Known as 'Uncle Art' he was elected to the Country Music Hall Of Fame in 1971 for his work as a record pioneer.

PETE SAYERS – Versatile British performer (born in Bath, Somerset, 1942) whose finest hour came in 1980 when he compered the Wembley Festival, gaining plaudits from even the most acid critics. A multi-instrumentalist of considerable ability and creator of such characters as the Phantom Of The Opry and the Lovely LaWanda, Sayers has headed his own BBC series, Pete Sayers Entertains.

Lisa Stewart's first album. Courtesy BNA Records.

DON SCHLITZ – The most successful Nashville songwriter of the '80s, Schlitz was born in Durham, North Carolina in 1952. He started pitching his songs in Nashville in the early '70s without too much success. Eventually he recorded **The Gambler** on the small Crazy Mammas label in 1978. Many covers came out with Kenny Rogers scoring a million-selling country-pop smash. Schlitz landed a short-lived Capitol recording contract, but has emerged as a highly inventive, commercial country writer.

JACK SCOTT – Rock'n'roll/country singer born Jack Scafone Jr in Windsor, Canada in 1936, but raised in Detroit. He formed a band, the Southern Drifters in early '50s and made his first recording for ABC-Paramount in 1957. Later Scott enjoyed notable pop success with country-flavoured rock-ballads. Though his recordings for Groove in 1963–64 veered closer to country, he has always been considered strictly a rock'n'roller.

TROY SEALS – One of Nashville's most consistent songwriters, Seals was born in Big Hill, Kentucky in 1938. He formed a band that played both rock and country in the late '50s and met and married pop singer Jo Ann Campbell, forming a duo. Later they moved to Nashville where he worked as a session guitarist and songwriter. He recorded in the '70s for Atlantic and Columbia, and, though his recordings were critically acclaimed, they flopped commercially. Writing with various partners, he has penned more than 200 Top 10 country hits.

DAWN SEARS – A petite lady with enormous vocal strength, she was a drummer in a band as a teenager and played the clubs in and around Minnesota. A move to Nashville in 1987 saw her become a regular on Ralph Emery's Early Morning TV show. This exposure resulted in a Warners' contract in 1992, but that breakthrough record has yet to surface.

SELDOM SCENE – A 'newgrass' supergroup formed by the Country Gentlemen's John Duffey in 1971, they played in the Washington D.C. area. Named the Seldom Scene because of their infrequent concert appearances, the group merged both traditional and current chart material, proving popular with the younger set. The original members along with Duffey were Mike Audridge (dobro), Ben Eldridge (banjo), John Sterling (guitars and vocals), plus former Country Gentlemen bassist Tom Gray.

KENNY SERRATT – Classic honky-tonk singer who missed out on the big-time. Born in Manila, Arkansas, he played at the Ramada in Hemet, California for 11 years, then moved to Montana from 1967 to 1972, where he worked as a lumberjack and rancher. Merle Haggard enticed him back into music and wrote and produced some of his initial hits in the early 1970s. He has recorded for MGM, Melodyland, Hitsville and MDJ, clocking up minor hits between 1972–81.

DOROTHY SHAY – Known as the Park Avenue Hillbilly, Dorothy's gimmick was to attire herself in exquisite gowns, then perform incongruous novelty hillbilly numbers. Very popular in the '40s, her biggest hit being **Feudin' And Fightin'** (1947). Born in 1923, she appeared in an Abbott and Costello movie 'Comin' Round The Mountain' in 1951 and in later life played a recurring role in the TV series The Waltons, but died in Santa Monica, California in 1978.

STEVE SHOLES – A&R manager for RCA's Country Music and R&B Division, Sholes (born in Washington DC in 1911; died in Nashville in 1968) was responsible for the label accumulating one of the most impressive country rosters in the world, signing Jim Reeves, Hank Snow, the Browns, Elvis Presley and Chet Atkins, the last eventually becoming Sholes' A&R assistant. He was elected to the Country Music Hall Of Fame in 1967.

RED SIMPSON – Singer-songwriter-impressionist, multi-instrumentalist and comedian who in the early '70s became associated with the Bakersfield crowd. Winner of Cashbox's New Male Vocalist award in 1972, he came up with an array of trucking songs and by 1979 could still be found charting with **The Flying Saucer Man And The Truck Driver**.

SHELBY SINGLETON – One of the shrewdest businessmen in Nashville, Singleton (born in Waskom, Texas in 1931) worked with Mercury Records and became vice-president, signing Jerry Lee Lewis and Charlie Rich along the way. In 1966 he resigned and formed his own production company, having his biggest success in 1968 when he produced Jeannie C. Riley's **Harper Valley PTA** for his own Plantation label. He has also set up other labels including SSS and Silver Fox, and in 1969 acquired Sun Records.

SIX SHOOTER – A teenage Nashville country band blessed with youthful good looks and barn-storming energy. Mainstays are Ronnie McDowell's son Ronnie Dean McDowell and his nephew Chris McDowell. They have toured with McDowell senior and appeared on the prime time sit-com Evening Shade. Their rock-rolled country is bringing in the younger generation as well as amazing the older country fans.

JIMMIE SKINNER – Born near Berea, Kentucky, his first taste of success came as a songwriter while working as a DJ in Knoxville, Tennessee. He signed as a performer with Mercury Records in the mid-1950s and had two Top 10 hits in 1957. Since then he recorded for Decca, Starday, King and Vetco and operated his successful Jimmie Skinner Music Center in Cincinnati for many years before moving to Nashville in the mid-1970s. He died in Hendersonville in 1979.

RUSSELL SMITH – Howard Russell Smith, from Lafayette, Tennessee, worked on local radio as a teenager. He formed the Amazing Rhythm Aces in the early '70s and was lead singer and main songwriter. He embarked on a solo career in Nashville in 1982, but though he has a distinctive soulful voice and recorded for Capitol and Epic, he has not made the impact expected. He has written dozens of country hits and helped form the quirky Run C&W group in 1992.

SISSY SPACEK – Award-winning movie actress (born in Quitman, Texas in 1950), who in her early days recorded for Roulette as Rainbo. Cast as Loretta Lynn in the film 'Coal Miner's Daughter' (1980), she did all the singing for the soundtrack and promptly got signed by Atlantic, cutting an above-average album **Hangin' Up My Heart**.

BUDDY SPICHER – Session fiddle-player whose career took off as a member of Area Code 615 in 1969. An in-demand picker ever since – he has even worked with Henry Mancini and the Pointer Sisters – he has recorded albums under his own name for Flying Fish and CMH Records.

BOBBY LEE SPRINGFIELD – Singer-songwriter born in Amarillo, Texas in 1953, who moved to Nashville as a teenager. He made his mark as a songwriter, penning hits for Marty Robbins, the Oak Ridge Boys, etc. As Bobby Springfield he signed to Kat Family Records, then adopted his full name when he joined Epic Records in 1986.

TERRY STAFFORD – A pop singer who returned to his country roots. Stafford was born in Hollis, Oklahoma and was raised in Amarillo, Texas. He worked in the Eugene Nelson Band in Texas as a teenager and moved to California in 1960 where he played with pop bands. Signed to Crusader Records, he had a massive pop hit with **Suspicion** in 1964. He began writing songs, with Buck Owens scoring with **Big In Vegas**. His own **Amarillo By Morning** was well-recorded in the '70s, before becoming a hit for George Strait in 1984. He recorded for Atlantic and Casino throughout the '70s.

KENNY STARR – A Loretta Lynn protégé, Starr was born in Topeka, Kansas in 1953. By the age of nine he was leading his own band, the Rockin' Rebels. Later he became a country entertainer with the Country Showmen, and was offered a job in Loretta Lynn's roadshow. A singer-songwriter-guitarist, he had a No.1 country hit with his own **The Blind Man In Bleachers** in

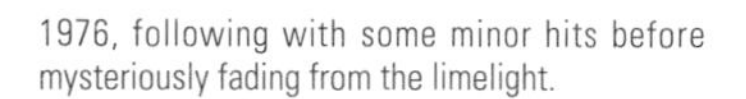

1976, following with some minor hits before mysteriously fading from the limelight.

STU STEVENS – British singer who, from time to time, looked likely to cross over and gain recognition from the pop fraternity, predominantly in 1979 when his version of Shel Silverstein and Even Steven's **Man From Outer Space** was picked up by MCA.

LISA STEWART – A Louisville, Mississippi native, this young lady, who was signed to BNA Records in 1991, has a wonderfully expressive vocal style with an alluring smoky quality that could see her emerge as one of the top female stars of the future.

CLIFFIE STONE – Born Clifford Gilpin Snyder in Burbank, California in 1917, Cliffie Stone has mainly been involved in the executive end of the music business. His father was a well-known banjo player-comedian known as Herman The Hermit, though Cliffie began his musical career as a bassist for big bands. He served as a DJ, MC and performer on several Los Angeles-area radio stations, and was bandleader and featured comedian on the Hollywood Barn Dance. In 1946 he linked up with the newly formed Capitol Records and stayed for over two decades, recording a half dozen albums and guiding the careers of Tennessee Ernie Ford and others. Owner of his own Central Songs publishing company, he co-wrote several hits, including **No Vacancy** and **Divorce Me C.O.D**. In the mid-1970s Cliffie Stone formed Granite Records in California. His son Curtis Stone was bassist with Highway 101.

J.D. SUMNER AND THE STAMPS – Legendary gospel group that grew out of the Stamps Quartet (formed in 1920). Headed by bass-voiced singer Sumner, they worked with Elvis Presley from 1971 up to the time of his death. Dave Rowland (of Dave And Sugar) and Richard Sterban (Oak Ridge Boys) both are former members of the Stamps Quartet.

JIMMY SWAGGART – Cousin of Jerry Lee Lewis and Mickey Gilley (born in Ferriday, Louisiana in 1935) who plays the same pumping piano but has aimed his music at the vast country-gospel audience, recording over 50 albums to date.

T

CARMOL TAYLOR – An underrated honky-tonk songwriter and singer, Taylor was born in Brilliant, Alabama in 1931 and worked at local shows and square dances from the age of 15. He tried his luck in Nashville in the mid-1960s and joined Al Gallico Music as a staff writer, penning hits for Charlie Walker, George Jones and David Wills. He recorded briefly for Elektra Records and others. He died of lung cancer in 1986.

CHIP TAYLOR – Brother of actor Jon Voight (born in New York in 1940), this singer-songwriter has penned hits for Waylon Jennings, Bobby Bare, Eddy Arnold, Jim Ed Brown and Floyd Cramer. Once a rockabilly singer with King, he did some excellent country-oriented albums with Warner Bros and Columbia.

TUT TAYLOR – A multi-instrumentalist (born in Milledgeville, Georgia in 1923), Robert 'Tut' Taylor is noted for his flat-picking dobro style. Also proficient on mandolin, fiddle, guitar, dulcimer, autoharp and banjo, he has provided back-up on scores of records and is renowned as a collector, builder and dealer in stringed instruments.

KAREN TAYLOR-GOOD – From El Paso, Texas, a singer-songwriter who worked her way up through the jingle jungle to work on sessions with George Jones, Dolly Parton, Conway Twitty and others, also working on the soundtracks of 'Best Little Whorehouse In Texas' and 'Smokey And The Bandit II'. In 1982 she and her manager formed Mesa Records, since when she has supplied a regular flow of mid-chart singles.

THE TENNEVA RAMBLERS – A relatively popular band of the late '20s and early '30s, originally known as the Jimmie Rodgers Entertainers, they were set to record for Ralph Peer in that historic week in August, 1927. At the last moment they defected from Rodgers and made up the new band name, which reflected the location of the session: Bristol, a city divided in half by the state line between Tennessee and Virginia. They were moderately successful in their recording efforts, but their decision to go it alone helped Rodgers' solo career.

AL TERRY – A Cajun-rocker born Alison Joseph Theriot in 1922, he hit it big when rockabilly came along with his **Good Deal Lucille**, which provided him with a tour with Red Foley and work with country package shows. However, despite the wide acceptance provided by the monster hit, Terry has remained pretty much a regional favourite.

UNCLE JIMMY THOMPSON – The first featured performer on the Saturday night barn dance show, which was to develop into the Grand Ole Opry, was Uncle Jimmy Thompson, born in Smith County, Tennessee in 1848. Primarily a farmer, he was frequent winner of a nationwide fiddle contest. Excited by the then new medium of radio, he applied – at the age of 78! – for a spot on WSM, and in 1925 his Saturday night show first came on the air. He stayed with the Opry (then still known as the WSM Barn Dance) until 1928, then toured a bit and recorded for both Columbia and Vocalion before passing away on February 17, 1931.

MARSHA THORNTON – A former singer with Country Music USA at Opryland, Marsha Thornton was born in Killen, Alabama in 1955. She left the Opryland show in 1988 and signed with MCA Records, scoring several minor chart entries, the best known being **A Bottle Of Wine And Patsy Cline** in 1990.

SONNY THROCKMORTON – A phenomenally successful country tunesmith, he was born James Fron Throckmorton in Carlsbad, New Mexico in 1941. He moved to Nashville in 1964 and worked as a staff writer with Tree Music, but it was to be ten years before he hit top gear, and has since penned more than 200 country hits. He recorded for Starcrest, Mercury and MCA in the late '70s, but has wisely concentrated on his writing – **Last Cheater's Waltz**, **Middle-Age Crazy** and **Can't You Hear That Whistle Blow** are just three of his classics.

JOHNNY TILLOTSON – A pop-country singer and songwriter, who is more country than many credit him. Born in Jacksonville, Florida in 1939, he was on local radio's Young Folks Revue from the age of nine. He appeared on the Toby Dowdy show in Jacksonville in his teens, leading to a recording contract with Cadence in 1958. He enjoyed several teen pop hits from 1959–65 but preferred country. Penned classic country ballads **It Keeps Right On A-Hurtin'** and **Out Of My Mind**, recorded in Nashville with album of country standards in 1962 that featured a young Charlie McCoy on harmonica. Continued to chart country on MGM, UA and Reward into the '80s.

Maybe The Moon Will Shine, Marsha Thornton. Courtesy MCA Records.

MITCHELL TOROK – Writer and singer of novelty country songs, Torok was born in Houston, Texas in 1929 and started playing guitar when he was 12. He first recorded in 1948 for small Texas labels, then in 1951 was signed to Abbott Records, producing the pop-country smash **Caribbean** in 1953 and writing **Mexican Joe**, a biggie for Jim Reeves.

BUCK TRENT – One of country's most proficient banjoists (born Charles Wilburn Trent in Spartanburg, South Carolina). He worked with Bill Carlisle in the late '50s and early '60s then moved on to become a member of Porter Wagoner's Wagonmasters. During 1973 Trent teamed up with Roy Clark and put out a number of albums for ABC. Some of these albums were solo and some all-banjo duets with Clark.

GRANT TURNER – Born in Abilene, Texas in 1912, Turner was the Dean of Opry announcers from 1945. He began his radio career at the age of 16 and joined Nashville's WSM in 1944. Elected to the Country Music Hall Of Fame in 1981, he died in Nashville on October 19, 1991.

WESLEY TUTTLE – Born in Lamar, Colorado, he became West Coast-based after several radio stints in the Midwest. He was signed to Capitol Records in 1945, one of his biggest selling records being **Crying In The Chapel**. He also recorded several duets with his wife Marilyn, later fading from the mainstream country scene when he began working as an evangelist.

U

DONNA ULISSE – Fashion-model beauty with a strong, sultry voice, Donna initially had plans for a modelling career, then opted for music. She was a regular weekend performer at a popular Virginia night-club before moving to Nashville in 1989. She sang back-up vocals on Nashville sessions, worked as a demo and commercial jingle singer, then landed an Atlantic Records contract in 1990.

UNCLE HENRY'S ORIGINAL KENTUCKY MOUNTAINEERS – A fine and popular old-time string band which, by making certain concessions to modernity, remained active well into the 1940s. 'Uncle Henry' Warren was born in Taylor County, Kentucky in 1903. The band began as early as 1928 and for the next 20 years played radio stations throughout the South. Uncle Henry's son, Jimmy Dale Warren, moved to the West Coast when the band split up and became lead singer with the Sons Of The Pioneers.

V

THE VAGABONDS – A smooth harmony trio who were with the Grand Ole Opry from 1931–38. They were unique at the time – they were all non-Southerners (all from the Midwest), and had acquired formal musical training. They were best known for their extremely popular **When It's Lamp Lighting Time In The Valley**, and they recorded a host of similar sentimental tunes for Bluebird and other smaller labels.

Below: Johnny Tillotson penned several classic country ballads.

RICK VINCENT – A true native of Bakersfield, Vincent is a throwback to the '60s and Buck Owens and Merle Haggard, yet with a real '90s energy to his music. This singer, songwriter and guitarist began playing in bands at 15 and worked clubs and colleges from California to the Carolinas, making several stops in Nashville to plug his songs. Eventually he moved to Nashville in 1989. After playing showcases and club dates, he gained a recording contract with Curb Records in 1991, making his debut on the charts with **Best Mistakes I Ever Made** (1992).

W

THE WAGONEERS – An Austin-based quartet formed in 1987 that rapidly gained an A&M recording contract for their blend of '50s-brand rockabilly and straight-ahead '60s country. Led by singer-songwriter Monte Warden, a veteran of the Austin music scene, they appeared in Austin City Limits and made a dent on the country charts with **I Wanna Know Her Again** in 1988.

HANK WANGFORD – Born Henry Hardman, he is also Dr Sam Hutt, an English gynaecologist. After befriending Gram Parsons, he formed a country band that became increasingly nutty, such monikers as Irma Cetas (the Vera Lynn of Vera Cruz), Brad Breath and Manley Footwear hiding the identities of various well-known sessioneers. A singer who has a love-hate relationship with country and its more 'sincere' aspects, he always seems just on the verge of making a breakthrough into commercial acceptance but, to date, has not quite made it.

B.B. WATSON – The first act signed to Nashville's new BNA label in 1991, B.B. (Bad Boy) Watson, whose real surname is Haskell, was raised in Shreveport, Louisiana. He has fronted his own Gulf Coast Cowboys Band in the Texas–Louisiana area since graduating from high school. His southern raspy voice with a gripping soulfulness is ideal for the powerful country themes he constructs in his songs. He has scored a major success with **Light At The End Of The Tunnel**.

Below: Jonny Young has one of the most distinctive voices in UK country.

JIM WEATHERLY – Singer-songwriter born in Pontotoc, Mississippi in 1943, Jim moved to Nashville in the late '60s. Though country acts cut his songs, it was soul singer Gladys Knight who put him on the map, scoring major pop hits with **Midnight Train To Georgia** and **The Best Thing That Ever Happened To Me**. Signed to Buddah Records, he recorded some great country-styled albums, making the country Top 10 with **I'll Still Love You**. He continues as a successful country tunesmith today, often co-writing with many of the new young hopefuls in Nashville.

DENNIS WEAVER – A character actor best known for his TV roles in Gunsmoke and McCloud, the easy-going Weaver acquired both a western image and a reasonable reputation as a country singer. Born in 1924, he hails from Joplin, Missouri. Following graduation from Oklahoma University he started on an acting career, appearing in several top films during the '50s. Recording his first country album for the Impress label in 1972, he subsequently signed for Ovation, recording in Nashville under the direction of Ray Pennington. Failing to come up with a hit single, Dennis Weaver returned to his first love, acting.

GORDIE WEST – Popular Canadian singer – though he was born in Skipton, England. A one-time power engineer, he worked with several country bands but eventually went solo, cutting his first album, **Alberta Bound**, during 1978.

SPEEDY WEST – One of the most recorded steelies in country music (born in Missouri in 1924). Once resident on Cliffie Stone's Hometown Jamboree, his 1955 Capitol album with Jimmy Bryant, **2 Guitars Country Style**, remains an indispensable instrumental item.

MICHAEL WHITE – The son of noted country tunesmith L.E. White (he wrote several of Conway Twitty's hits), Michael was born in Knoxville, Tennessee, but raised in Nashville. Following in his father's footsteps, he wrote his first songs at age 12, with two being recorded by Conway Twitty. He attended Lee College in Cleveland, Tennessee, pursuing a career in the Ministry, then opted for music. He sang lead in Gold Rush (a Top 40 covers band), then formed Fresh Horses (a southern rock outfit). Finally he returned to his musical roots, working as a demo singer and songwriter in Nashville. A record deal with Reprise in 1991 saw him placing his own versions of his songs on the charts.

WILD ROSE – All-girl, five-piece, Nashville-based band comprising well-established session singers/musicians Wanda Vick, Nancy Given-Prout, Kathy Mac, Pam Perry and Pamela Gadd. Following showcases and club dates around Nashville, they signed with Universal Records in 1989 and made the country Top 20 with **Breaking New Ground**.

LITTLE DAVID WILKINS – Born in Parsons, Tennessee. Another mail-order guitar player who started out with Sun Records in Memphis at 15. A chubby entertainer, perpetually on a diet, he has had hits for such labels as Plantation, MCA, Playboy and Epic.

CURLY WILLIAMS – Leader of the Georgia Peach Pickers, a popular Grand Ole Opry band in the '40s, he is best known for having written **Half As Much**, which, because it was popularized by Hank Williams, is frequently thought of as Hank's song. In the '50s he drifted out of performing.

LAWTON WILLIAMS – A former DJ in Detroit and Dearborn in the '40s, Williams (born in Troy, Tennessee in 1922) is best known as the writer of **Fraulein** and **Geisha Girl**. A talented singer, he had his own TV series in Fort Worth in the late '40s. Has recorded for several labels including Mercury and RCA, and was still touring regularly in the late '80s.

LEONA WILLIAMS – Born Leona Helton in Vienna, Missouri in 1943, she became part of the Helton family band at an early age. At 15 she married bass player Ron Williams, the duo becoming members of Loretta Lynn's back-up unit. In 1968 she signed with Hickory Records, this association lasting until 1974, when she joined MCA. A straight-down-the-line country singer, she joined the Merle Haggard roadshow as backing vocalist in 1975 and became not only Merle's duet partner, but his third wife (1978). It turned out to be a stormy marriage, and five years later they were divorced.

KELLY WILLIS – A young lady with a promising future, Kelly Willis was born in Virginia in 1969, but her music owes more to Texas dance music – roadhouse style. She joined her first band at 16, later moving to Austin, Texas, where Nanci Griffith put her in touch with MCA A&R chief Tony Brown, who immediately signed her up. Blending rockabilly abandon, hard-country emotion, Kelly uses her voice as another instrument, sliding and bending notes in classic country fashion.

DAVID WILLS – A classic honky-tonk singer who was born ten years ahead of his time in Pulaski, Tennessee in 1951. A prolific songwriter he was 'New Traditional' years before it was fashionable. Discovered by Charlie Rich, he gained an Epic recording contract in 1973, hitting Top 10 with **There's A Song On The Jukebox** and **From Barrooms To Bedrooms**. He has since recorded for UA, RCA and back to Epic, registering more than 20 minor hits.

NORRO WILSON – Norris Wilson, outstanding songwriter, singer, producer and music executive, was born in Scotsville, Kentucky in 1938. He enjoyed some regional success in the late '50s, then moved to Nashville in the early '60s, signing as a writer with Al Gallico Music, and recording for Mercury/Smash. He has cut records for several labels over the years, but it is as a writer of hits for Charlie Rich, Joe Stampley, Tammy Wynette, etc, that Wilson has built his reputation.

STEPHANIE WINSLOW – This accomplished fiddler and singer was born in Yankton, South Dakota in 1956. A child prodigy, she made her professional debut at age ten, and leans towards the showbiz/cabaret style of country. She has recorded for Warners, Primero and MCA throughout the '80s, making the Top 10 with **Say You Love Me**.

TOM WOPAT – Singer-actor who played Luke in the TV series Dukes Of Hazzard, Wopat (from Lodi, Wisconsin) recorded for EMI-America in Nashville during the late '80s, scoring several Top 20 country hits, including **The Rock And Roll Of Love**.

BOBBY WRIGHT – The son of Johnny Wright and Kitty Wells, Bobby was born in Charleston, West Virginia in 1942. He started making show business appearances while still a child, and gained the part of Willie in the TV series McHale's Navy, remaining with the show for a four-year run. A member of his parents' family show, he has recorded regularly since the late '60s, his most successful record, **Here I Go Again**, a Top 20 entry in 1971.

X

X-STATIONS – Powerful radio stations that operated just inside the Mexican border, cutting in on wavelengths used by US and Canadian stations. Many country singers including the Carter Family were helped on their way through border radio.

Y

SKEETS YANEY – One of country music's great regional stars, Skeets, who was a spectacular yodeller, was born Clyde Yaney in Mitchell, Indiana. A longtime star on KMOX in St Louis, he was one of the main members of the KMOX Barn Dance in the '30s and '40s. He died of cancer in 1978.

JONNY YOUNG – A British singer who possesses one of the sweetest, most distinctive voices in British country music. He started out in the early '60s in pop bands, forming the Jonny Young Four in 1967 with tight harmonies, and signed to RCA. The group was set for the big time, but a car accident, in which the bass player was killed, held up their career. They worked behind many American stars and recorded for several labels during the '70s. They re-formed as the Jonny Young Band and took on a more contemporary stance, still working the scene in the '90s.

Broken Heartland, Zaca Creek. Courtesy Giant Records.

Z

ZACA CREEK – West Coast quartet of the Foss Brothers – Gates (vocals), Scot (guitar), Jeff (keyboards) and James (bass). Their name comes from an underground stream in their hometown of Santa Ynez, California. Signed to Columbia in 1989, they scored some country hits, but have yet to make a big impact.